Annotated Teacher's Edition

HOLT

Chemistry

VISUALIZING MATTER

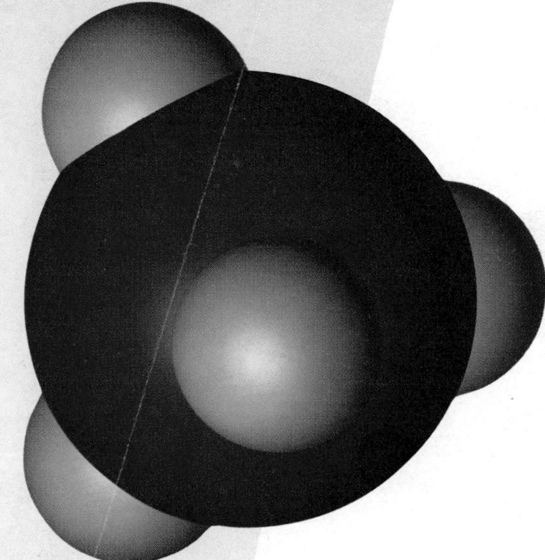

Salvatore Tocci

Claudia Viehland

HOLT, RINEHART AND WINSTON

Austin • New York • Orlando • Atlanta • San Francisco • Boston • Dallas • Toronto • London

Contributing Writers

Craig Gabler
Chemistry Teacher
Centralia High School
Centralia, WA

Jane Gallion
Author
Guildhouse Productions
Austin, TX

Dave Jaeger
Chemistry Teacher
Will C. Wood High School
Vacaville, CA

Thomas Lindberg
Research Chemist
The Upjohn Company
Kalamazoo, MI

Mark V. Lorson, Ph.D.
Chemistry Teacher
Jonathan Alder High School
Plain City, OH

R. Thomas Myers, Ph.D.
Emeritus Professor of Chemistry
Kent State University
Kent, OH

Suzanne Weisker
Science Teacher and Department Chair
Will C. Wood High School
Vacaville, CA

Jay A.Young, Ph.D.
Chemical Safety Consultant
Silver Spring, MD

For permission to reprint copyrighted material, grateful acknowledgment is made to the following sources:
Flinn Scientific, Inc., Batavia, IL, U.S.A.: From *Latex, The Preparation of a Rubber Ball*, Publication Number 430.00. Copyright ©1994 by Flinn Scientific Co., Inc. Reproduced for one-time use with permission from Flinn Scientific, Inc. All Rights Reserved. No part of this material may be reproduced or transmitted in any form or by any means, electronic or mechanical, including, but not limited to photocopy, recording, or any information storage and retrieval system, without permission in writing from Flinn Scientific, Inc.

Merck & Co., Inc., Rahway, NJ: From *The Merck Index: An Encyclopedia of Chemicals, Drugs and Biologicals,* Eleventh Edition, Susan Budavari, et al., Eds. Copyright ©1989 by Merck & Co., Inc.

Printed in the United States of America

ISBN 0-03-000193-5
45 032 99 989796

Acknowledgments

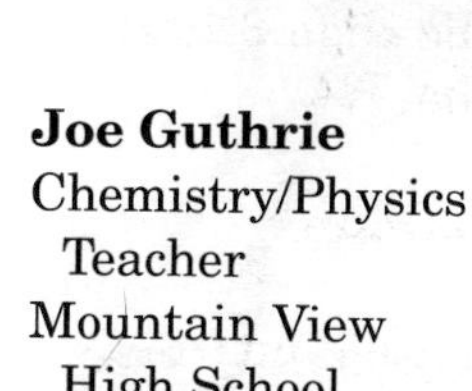

Editorial Director of Math and Science
Richard Monnard

Executive Editor
Ellen Standafer

Project Editors
Mark Grayson
Patty Coffey
Mary Ellen Teasdale
John Stokes

Editorial Staff
Laurie Baker
Jane Martin
Steve Oelenberger

Production
Beth Prevelige
Simira Davis
George Prevelige
Rose Degollado
Nancy Hargis

Design Development
Foca, Inc.
New York, NY

Lab Reviewers

Cheryl Epperson
Chemistry Teacher
Flour Bluff High School
Corpus Christi, TX

David L. Heiserman
Author
Sweethaven Publishing Services
Columbus, OH

Marilyn Lawson
Chemistry Teacher/ Science Department Chair
Gregory Portland High School
Portland, TX

A. Ruthe Tyson
Supervisor-Stanford Teacher Education Program
Stanford University
Stanford, CA

Kris Wynne-Jones
Photography Consultant
Chemistry Teacher
Poly Prep Country Day School
Brooklyn, NY

Text Reviewers

George Atkinson, Ph.D.
Associate Professor of Chemistry
University of Waterloo
Waterloo, Ontario, Canada

Judith A. Bazler, Ph.D.
Professor of Science Education/SMART Center Director
Lehigh University
Bethlehem, PA

Ted Branson
Chemistry Teacher
Hug High School
Reno, NV

Elizabeth Briggs
Chemistry Teacher
Ursuline Academy of Cincinnati
Cincinnati, OH

Patricia F. Buis, Ph.D.
Assistant Professor of Geology
University of Mississippi
Oxford, MS

G. Lynn Carlson, Ph.D.
Senior Lecturer in Chemistry
U. of Wisconsin-Parkside
Kenosha, WI

James A. Carroll, Ph. D.
Assistant Professor of Chemistry
University of Nebraska-Omaha
Omaha, NE

Tom Custer, Ph.D
Coordinator of Science
Anne Arundel County Public Schools
Annapolis, MD

William W. Duff
Chemistry Teacher
Baltimore School for the Arts
Baltimore, MD

Joanne Dunlap
Science Teacher
Concord High School
Concord, NH

David A. Ebert
Chemistry Teacher
Johnson Senior High School
Saint Paul, MN

David W. Eldridge, Ph.D.
Professor of Biology
Baylor University
Waco, TX

Neil Ellis
Chemistry/Physics Teacher
Raytown South High School
Raytown, MO

Jeffrey L. Engel
Chemistry Teacher
Madison County High School
Danielsville, GA

David J. Gamble
Chemistry Teacher
Robbinsdale Cooper High School
New Hope, MN

Joe Guthrie
Chemistry/Physics Teacher
Mountain View High School
Mountain View, CA

Robert Harriss
Professor of Environmental Chemistry
University of New Hampshire
Durham, NH

Smith L. Holt, Ph.D.
Professor of Chemistry
Oklahoma State University
Stillwater, OK

Robert Iverson
Chemistry Teacher
Irondale High School
New Brighton, MN

Anthony F. Kardis
Science Department Chair
Ladue Horton Watkins High School
St. Louis, MO

Shirley M. Kitchens
Chemistry/Physiology Teacher
Coronado High School
El Paso, TX

Samuel P. Kounaves, Ph.D
Associate Professor of Chemistry
Tufts University
Medford, MA

Alan Kousen
Science Consultant
Vermont Dept. of Education
Montpelier, VT

Janice Lane
Chemistry Teacher
St. Louis Park High School
Minneapolis, MN

Valerie Lang
Project Engineer
The Aerospace Corporation
Los Angeles, CA

Maureen Lemke
Chemistry Teacher
Navarro High School
Geronimo, TX

Doris I. Lewis, Ph.D.
Professor of Chemistry
Suffolk University
Boston, MA

Timothy Lincoln, Ph.D.
Associate Professor of Geology
Albion College
Albion, MI

Mark V. Lorson, Ph.D.
Chemistry Teacher
Jonathan Alder High School
Plain City, OH

Celia Marshak, Ph.D.
Chemistry Professor Emeritus,
College of Sciences
San Diego State University
San Diego, CA

Rachel McCaskill
Chemistry Teacher
Hillsborough High School
Tampa, FL

Michael A. McKinney, Ph.D.
Associate Professor of Chemistry
Marquette University
Milwaukee, WI

Audrey Miller, Ph.D.
Professor of Chemistry
University of Connecticut
Storrs, CT

Keith B. Oldham, Ph.D.
Professor of Chemistry
Trent University
Peterborough, Ontario, Canada

Lance Phillips, Ph.D.
Research Associate
Cornell University
Ithaca, NY

Terry M. Phillips, Ph.D.
Professor of Medicine/Immunochemistry
George Washington University Med. Ctr.
Washington, DC

Students apply the macroscopic observations and micromodels theme in developing a model for atomic structure that is consistent with experimental evidence. This study lays the developmental foundation for formula writing, nomenclature, chemical equations, and stoichiometry in Chapters 5 through 8.

J. J. Thomson's model of the atom

Niels Bohr's model of the atom

Students apply the classification and trends theme in examining the periodic table and the origins of the elements. Moles are first introduced here.

6 — Atomic number
C — Symbol
Carbon — Name
12.011 — Average atomic mass
$[He]2s^2 2p^2$ — Electron configuration

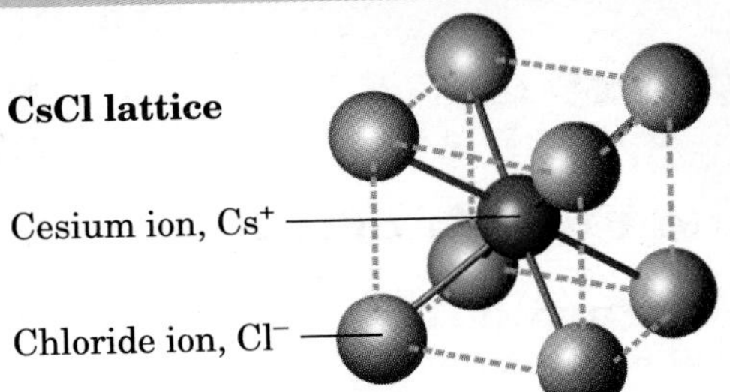

NaCl lattice
Sodium ion, Na^+
Chloride ion, Cl^-

CHAPTER 5

Students study ionic bonding, and then learn to write formulas, name ionic compounds, and apply the mole concept to solve percentage composition problems.

CHAPTER 6

Students study covalent bonding, and then learn to write formulas and name molecular compounds, which includes simple organic nomenclature. Students apply the mole concept to solve empirical formula problems.

Polyethylene $(C_2H_4)_n$ molecule

Students apply the conservation theme to balance equations and recognize reaction types.

Chemical reactions can be used to make products with useful properties that are different from those of the reactants.

$$C_6H_5OH + CO_2 + CH_3COOH \longrightarrow C_9H_8O_4 + H_2O$$

Phenol — Carbon dioxide — Acetic acid — Aspirin — Water

Students apply stoichiometric principles to solve problems using relevant examples.

With only one nut, you can only make a single nut-bolt pair, no matter how many bolts you have. The same is true for a chemical reaction in which one reactant is used up before the other.

Students apply the systems and interactions theme in exploring the effects of energy on chemical systems. Students examine the energy values of fuels including foods.

Calorimeter

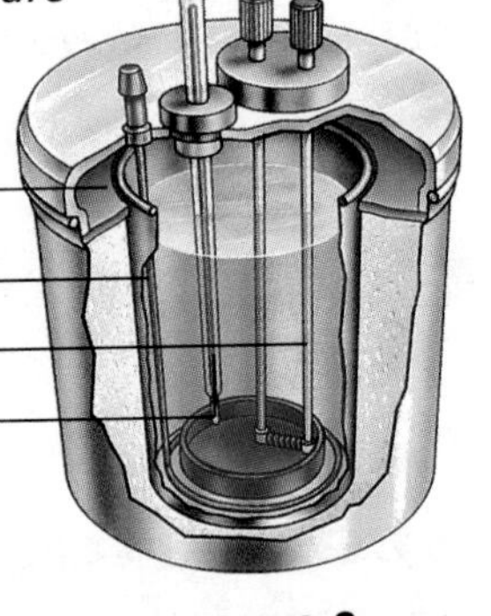

Students study KMT as a basis for the ideal gas model and make predictions using the gas laws. They apply the equilibrium and change theme in exploring condensation processes and properties of liquids.

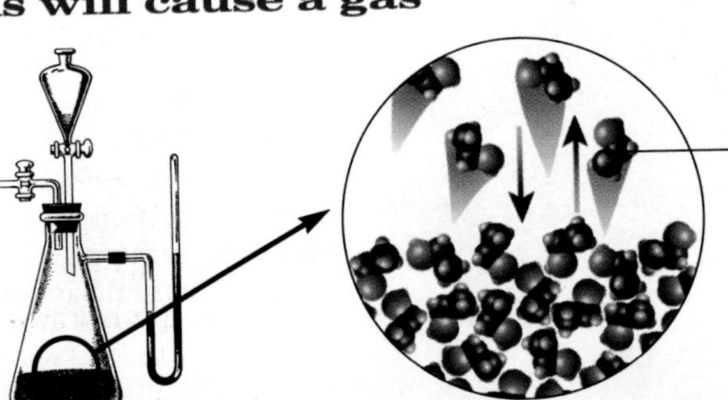

Vapor pressure of ethanol at equilibrium

CHAPTER 11

Students apply the macroscopic observations and micromodels theme to examine the properties of solutions and the applications of solubility principles.

Saltwater solution

Water molecule

Sodium ion, Na^+

Chloride ion, Cl^-

CHAPTER 12

Students apply the equilibrium and change theme to explore the properties of equilibrium systems and to study the behavior of systems in relation to Le Châtelier's principle.

Equilibrium system of NO_2 and N_2O_4

$$2NO_2(g) \rightleftharpoons N_2O_4(g)$$

0°C
very light brown

25°C
medium brown

100°C
dark brown

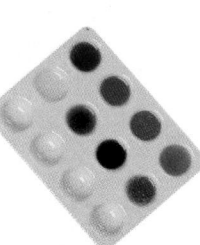

Possible collisions between two, four, five, and eight particles.

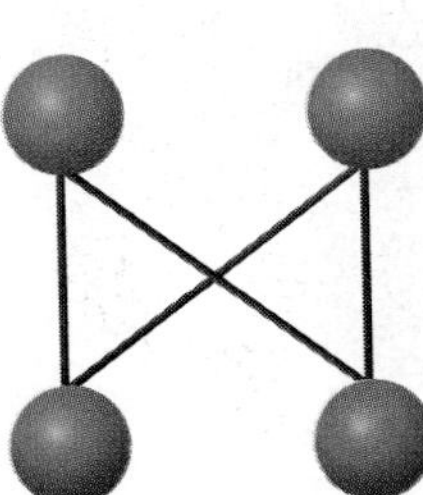
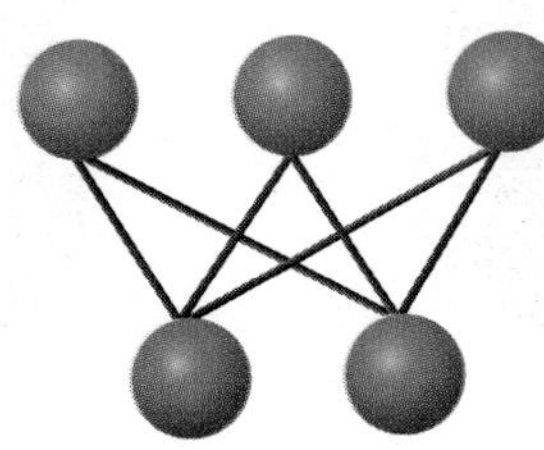
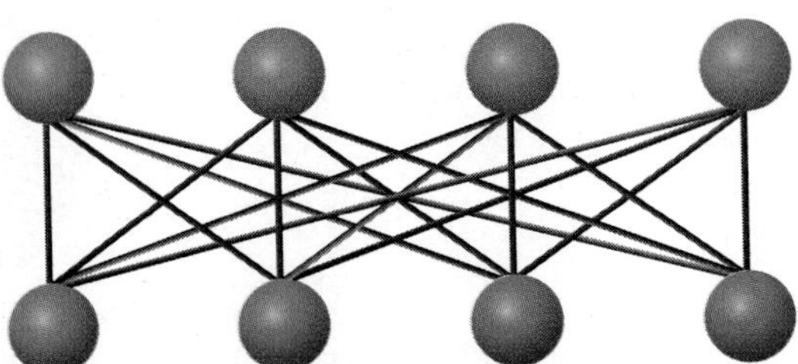

The activity series predicts electrons will flow from Al to Cu. A flow of electricity will light the bulb.

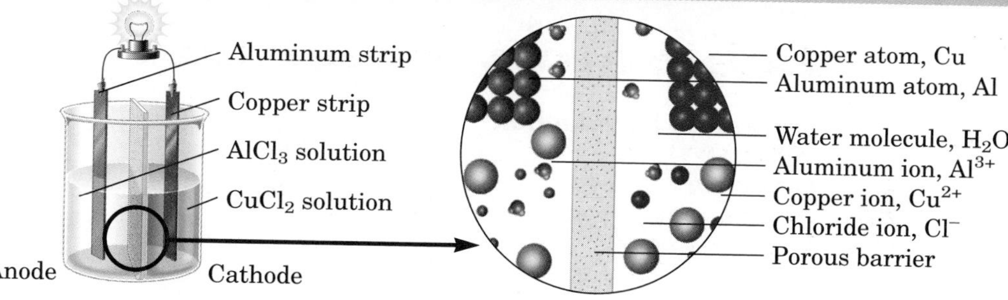

CHAPTER 15

Students apply the macroscopic observations and micromodels theme to explore the characteristics of electrochemical cells and redox reactions.

CHAPTER 16

Students apply the systems and interactions theme to explore the reactions of the nucleus.

Transmutation process of carbon-14 to nitrogen

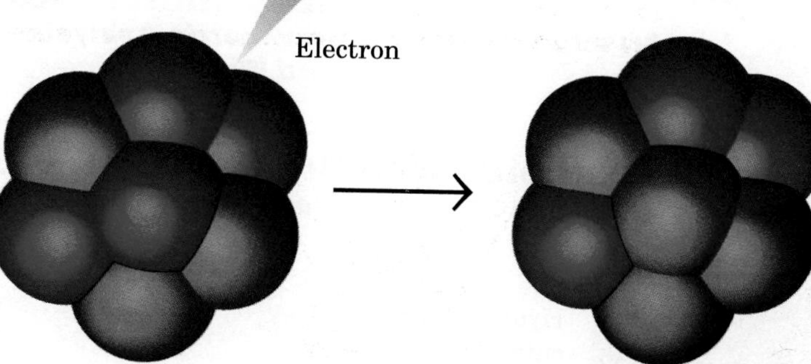

Student preparing for a lab

Teacher information regarding the laboratory program begins on page T643-1.

Hydrated cobalt chloride

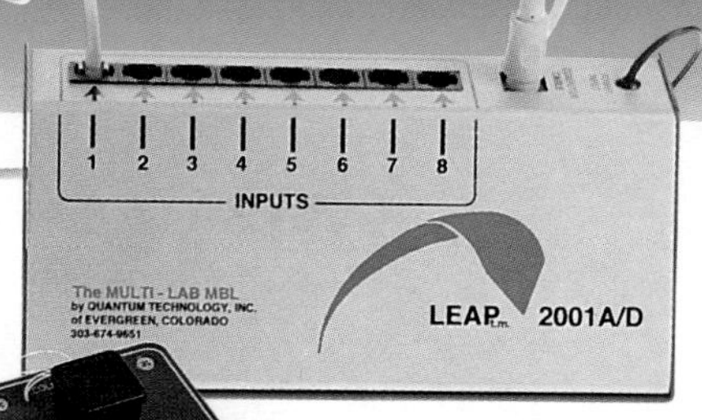

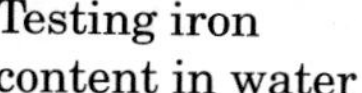
Testing iron content in water

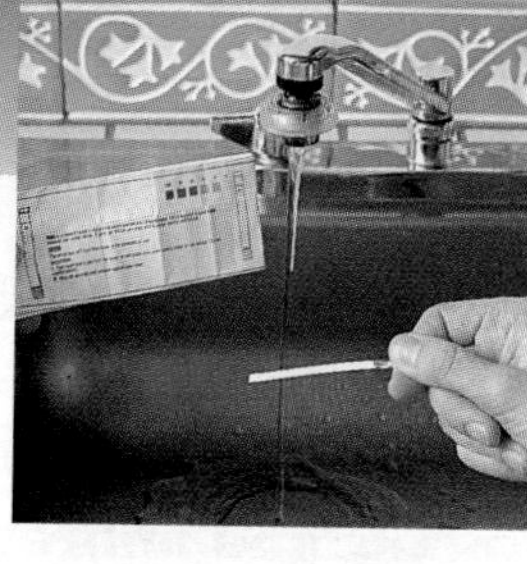

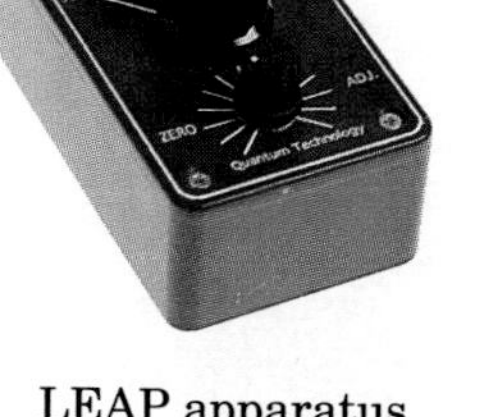
LEAP apparatus with colormeter

Heating water with a Bunsen burner

Various elements

Lead iodide precipitate

Calculator

Features

Consumer Focus

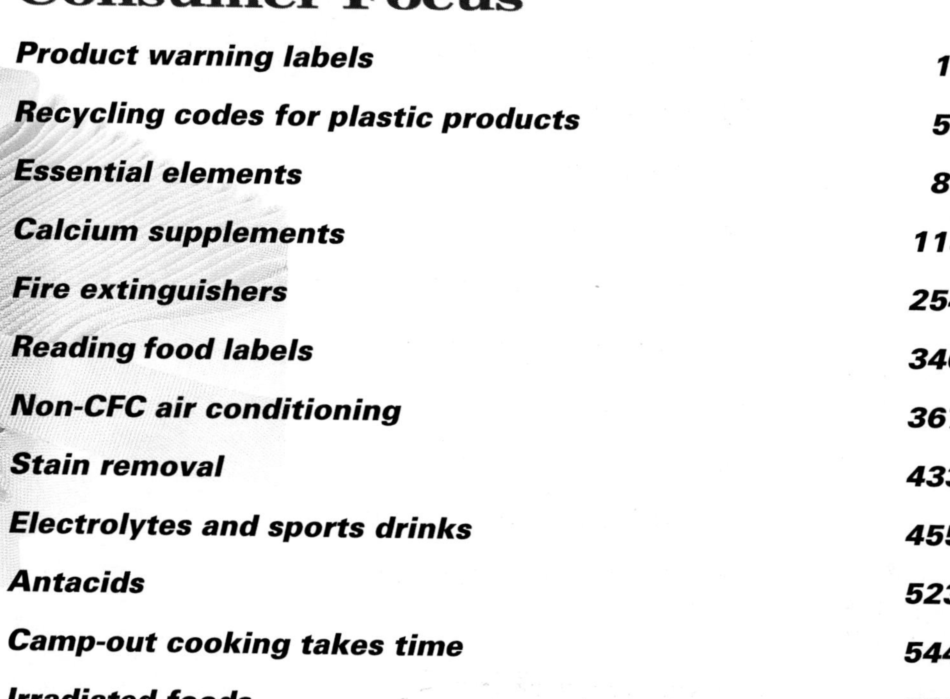

Fire hose used on Class A fires

Science, Technology, and Society

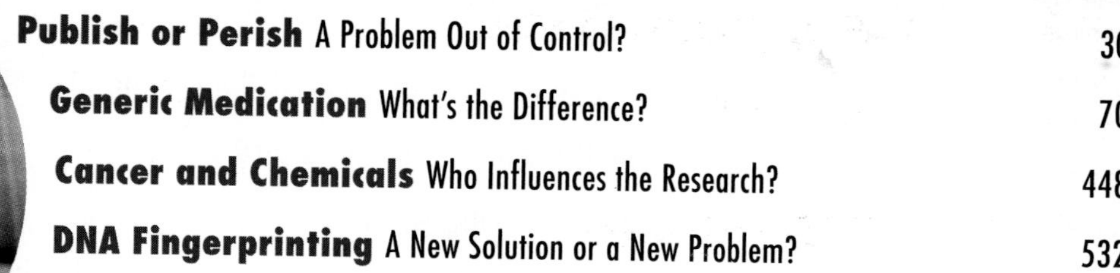

DNA fingerprint

HISTORICAL TIMELINES

Marie S. Curie

Albert Einstein and Niels Bohr

Dorothy Hodgkin

Yuan Tseh Lee

Holt Chemistry: Visualizing Matter

Chemistry has never been presented like this before

A totally new program, uniquely designed, to enable students to understand chemistry like never before.

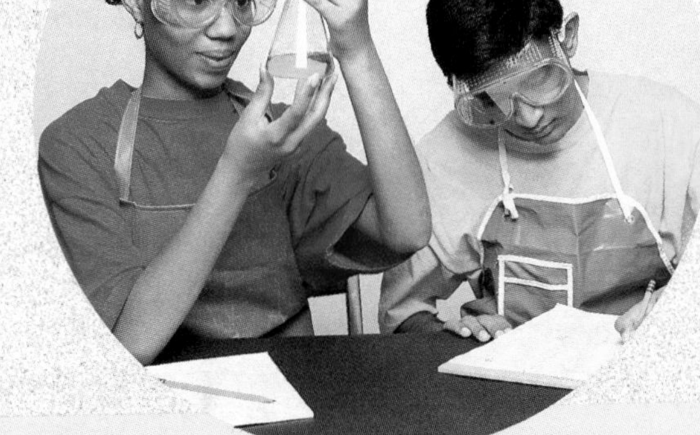

Using developmentally appropriate and relevant materials along with the latest teaching strategies, *Holt Chemistry: Visualizing Matter* gives students the solid grounding in basic chemical principles and skills that they need for college. Add to that foundation, a strategic problem solving strand, an innovative laboratory strand, and a holistic approach to evaluation and assessment, and you have the right combination for today's changing classroom.

Because it is new, *Holt Chemistry: Visualizing Matter* was developed to focus on a set of broad, developmentally appropriate instructional goals that combine mastery of chemistry content objectives, experiences in scientific inquiry, and connections to the real world and other curriculum areas.

Instructional goals

Students will attain:

- conceptual understanding of the section objectives
- scientific and chemical literacy
- problem-solving skills
- proficiency in the use of scientific methodology
- an understanding of the impact of science and technology on society
- proficiency in the use of technology
- proficiency in oral and written communication

Philosophical constructs

The Instructional Goals are based on the idea that students develop greater conceptual understanding when they are:

- active participants in the learning process
- given some responsibility for their own learning
- learning concepts that are meaningful in the context of their lives
- in an environment that encourages reflection and comparisons with you and their peers

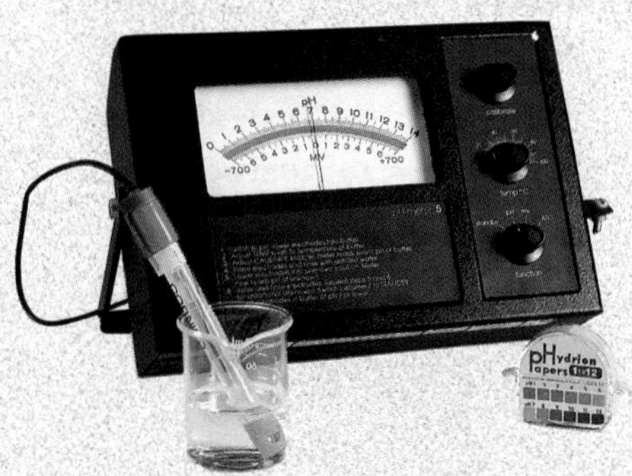

A streamlined concept load presented in the right sequence

Holt Chemistry: Visualizing Matter **was built from a very selective list of concepts to provide thorough instruction and to fit comfortably within the standard school year. With 16 chapters and the choice of over 70 new laboratory experiments, there is a wealth of material to do the job effectively.**

The course year at a glance

Chapter	Recommended Time Allotment*
1 The Science of Chemistry	5 days
2 Matter and Energy	13 days
3 Atomic Structure	10 days
4 Periodicity	7 days
5 Ionic Compounds	10 days
6 Covalent Compounds	13 days
7 Chemical Equations	7 days
8 Stoichiometry	15 days
9 Causes of Change	15 days
10 Gases and Condensation	15 days
11 Solutions	13 days
12 Chemical Equilibrium	13 days
13 Acids and Bases	15 days
14 Reaction Rates	10 days
15 Electrochemistry	10 days
16 Nuclear Chemistry	7 days
TOTAL	178 days

* including laboratory experiences, review, and assessment

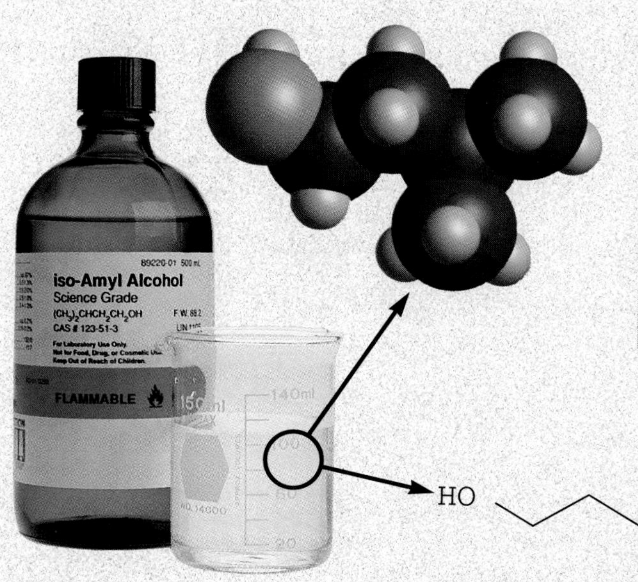

Isoamyl alcohol

Table of Contents

Key features

- **Atomic structure is covered early to lay the appropriate foundation for teaching formula writing, nomenclature, chemical equations, and stoichiometry**
- **Mole concept is presented early before chemical bonding.**
- **Formula writing and nomenclature are covered in two chapters to provide additional time to practice and refine this critical skill.**
- **Applications of the mole concept are found in every chapter after its introduction in Chapter 4. The mathematics concept load of every chapter has been carefully controlled to ensure success.**
- **Organic chemistry is integrated throughout the text. Organic examples bring relevance to every chapter.**
- **Energy is a cohesive theme throughout the text.**
- **You can teach all students need to know by covering just 15 of the 16 chapters.**

Scope and Sequence

Chapter	Energy	Mole concept
1 The Science of Chemistry	• conservation	
2 Matter and Energy	• forms of energy • energy effects on matter, relating mass and energy • exothermic vs. endothermic change	• the SI standard for amount of substance
3 Atomic Structure	• energy of the nucleus • energy levels in atoms • ground state vs. excited state • light energy emissions	
4 Periodicity	• ionization energy	• converting among grams, • moles, and numbers of particles • Avogadro's number
5 Ionic Compounds	• stability of ions • lattice energy	• percent composition • molar masses
6 Covalent Compounds	• stability of molecules • bond energy	• empirical and molecular formulas
7 Chemical Equations	• endothermic vs. exothermic reactions • energy as a reactant or product in an equation	• mole ratios
8 Stoichiometry		• mole ratios
9 Causes of Change	• heat vs. temperature • heat of fusion and vaporization • specific heat capacity • calorimetry • heats of reaction, formation, solution, combustion • enthalpy • reaction pathways • Boltzmann distributions • Hess's law • entropy and stability • Gibb's free energy	• heat of reaction and stoichiometry • entropy and stoichiometry • Gibb's free energy and stoichiometry
10 Gases and Condensation	• average kinetic energy of particles, and temperature • Boltzmann distributions	• ideal gas law • mole fraction • molar volume of gases
11 Solutions		• gas reaction stoichiometry • molarity • solution stoichiometry
12 Chemical Equilibrium	• heat of solution	• equilibrium concentrations in mol/L
13 Acids and Bases		• acid-base stoichiometry • molarity and pH
14 Reaction Rates	• reaction pathways • average KE of particles, and temperature	• concentration (M)
15 Electrochemistry	• electric energy from chemical energy • conservation	• mole ratios
16 Nuclear Chemistry	• stability of nuclei • mass defect and binding energy • strong force	

Organic and biochemistry	STS and applications
• structural representations of organic compounds • organic syntheses • properties and structures of pain relievers	• chemical industry • government regulation of chemicals • product warning labels • computer networks • ethical practices in research
	• recycling codes • generic medications
	• fireworks • dietary essential elements
	• calcium supplements • mercury poisoning
	• toxic waste disposal • drying agents
• alkanes • organic nomenclature • functional groups • polymers • aromatic compounds	• biodegradable substances
• respiration of glucose • photosynthesis • polymerization reactions	• fire extinguishers • corrosion protection • antacids
• artificial flavorings • combustion of octane	• fuel-air ratios for gasoline combustion • emission controls • air bags
• fuel values of carbohydrates, fats, and proteins	• caloric content of foods • reading food labels • determining basal metabolic rate
• chlorofluorocarbons, freons	• non-CFC air conditioning • greenhouse effect • Montreal Protocol
• triglyceride structure • saponification • detergents • fat-soluble and water-soluble vitamins • organochlorines	• PCB contamination • relative humidity • heat solution handwarmers • stain removal • emulsifying agents in foods • cancer and organochlorines
• blood pH • electrolytes in cells	• purifying drinking water • electrical safety • sports drinks • stain removal
• amino acids • fatty acids • nucleic acids • nitrogen bases • amines	• acid rain • antacids • DNA fingerprinting
• enzymes	• food preservation practices • camping stoves and reaction rate
	• bioluminescent reactions • composition and longevity of batteries
• biological effects of radiation	• irradiated foods • radioactive dating • radioactive imaging techniques

Chemistry has never been explained like this before

It starts with readable text that is interesting, engaging, and relevant.

The value of any text lies in its perceived usefulness to students. When the reading is interesting and the ideas are developed logically and clearly, students can depend on the text for help when you're not there. In *Holt Chemistry: Visualizing Matter*, each chapter opens with a high-interest story supported with a structure of well-organized lessons. Titles and headings highlight major ideas as the story of the chapter unfolds using many relevant contexts for introducing chemical concepts. All these techniques work together to make this text one that your students can depend on.

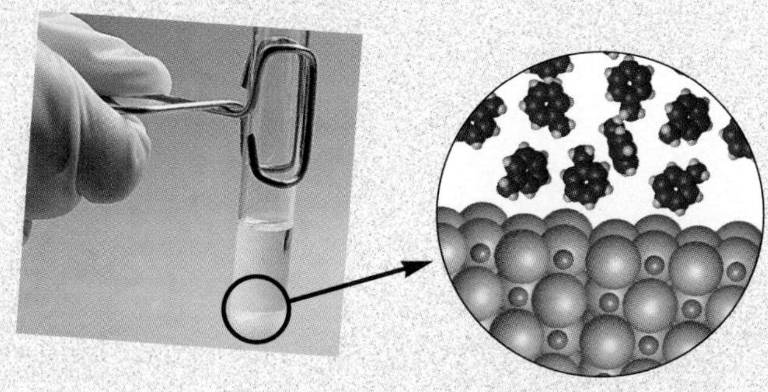

The Story sets up a set of questions to be explored as students build a conceptual framework while moving through the chapter.

CHAPTER 11

Solutions

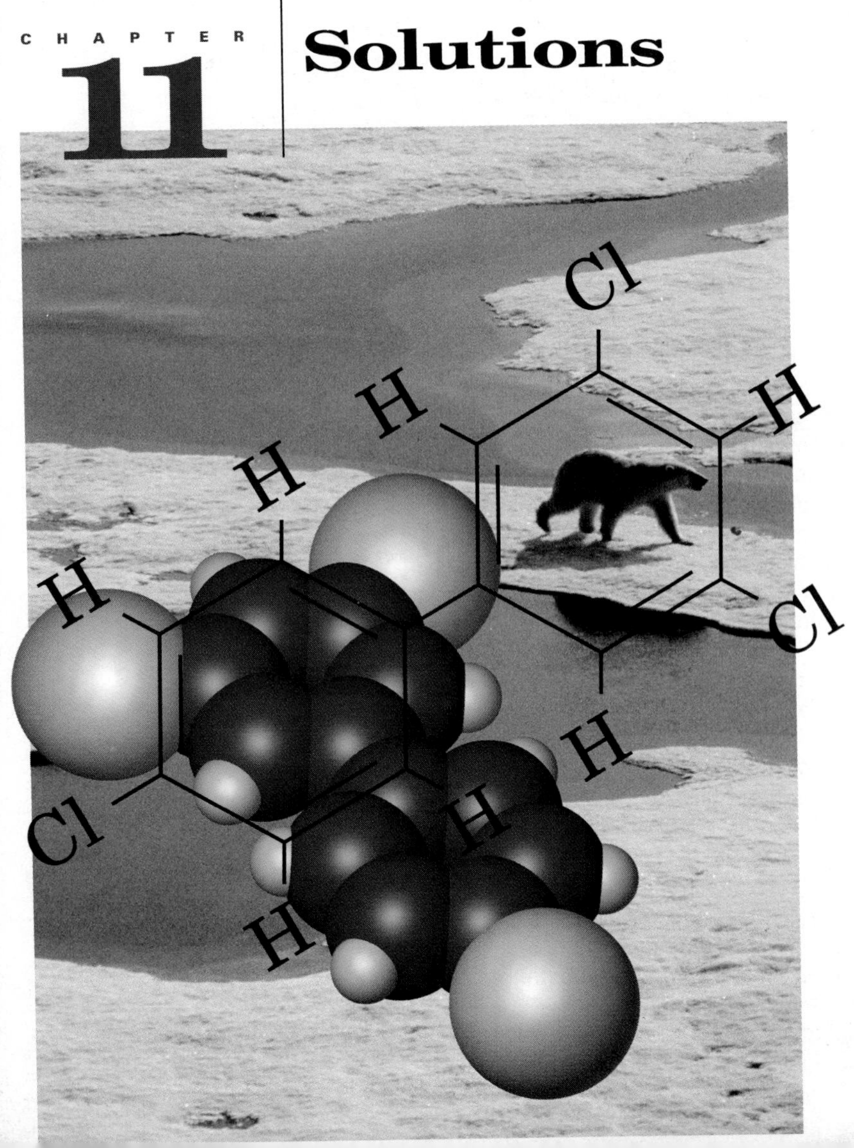

Solubility and the Arctic Bear Hunt

In northern Canada, a helicopter hovers over the snow. Dr. Malcolm Ramsay leans out of the doorway to keep sight. As Ramsay braces and aims his rifle, the bear lu the snow trying to escape. Ramsay fires, and the bear

Ramsay, who is an ecologist, signals the pilot to land in a clearing. He unloads his equipment, then shields his face from the helicopter-driven snowstorm as the pilot departs. By the time Ramsay reaches the bear, the tranquilizer-filled dart has done its job, and the animal lies quietly in the snow.

Ramsay begins a physical examination that will take hours. With a net, ropes, and a tripod, the bear is winched into the air to be weighed. Large syringes are filled with blood samples. A tooth is extracted; its growth rings will tell the animal's age. An electronic instrument records the body's resistance to electricity, which is an indication of the bear's muscle-to-fat ratio. A sample of fat tissue is taken. By the time the bear begins to stir, Ramsay has packed his gear and is waving to the returning helicopter. The polar bear will resume its migration to Hudson Bay, generally unharmed by the encounter.

The tissue samples are sent to a laboratory for analysis. The results Ramsay gets from the lab are worrisome. The bear's fat tissue contains polychlorinated biphenyls, better known as PCBs. This family of synthetic organic compounds has been used industrially in electrical and mechanical equipmen Originally scientists belie biologically inert, but in th ered that laboratory anim had higher than normal r and liver cancer.

Ramsay has found that of PCBs in the bear's fat t per million). So far, this c has caused the bear no ap entists are concerned abo because the way that PCB makes it possible for the c increase in the bears ever PCBs in the environment.

Ramsay, along with oth ing to answer some puzzl

- *How did the PCBs get ecosystem when there is in the bay area?*
- *Why did PCBs end up*

These questions can be that requires a knowledg mix and dissolve. In this explore the principles of then you will apply those some of the questions tha have about PCBs.

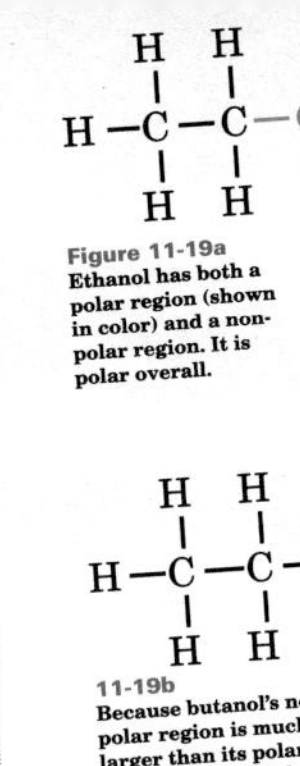

Figure 11-19a
Ethanol has both a polar region (shown in color) and a nonpolar region. It is polar overall.

There are many larger molecules that contain ... covalent bonds and other regions with nonpolar covalent bonds. Ethanol (ethyl alcohol), shown in Figure 11-19a, is a good example. The molecule contains one C—C bond, one C—O bond, one O—H bond, and five C—H bonds. Electronegativity calculations show that the C—C bond and the five C—H bonds are nonpolar, whereas the C—O and O—H are polar. The two polar bonds do not cancel each other, so the region around the oxygen is negative, and the region around the hydrogen next to the oxygen is positive. The remainder of the molecule is nonpolar.

Ethanol is found in beer, wine, and other alcoholic beverages. Because it mixes well with water, you might guess that the entire molecule is polar overall even though part of the molecule contains nonpolar bonds. The two polar bonds can create a dipole that can attract other polar molecules.

Experiments with a variety of alcohols suggest the following rule: *In determining solubility in water, a polar bond can have as much effect as a small nonpolar region.* This rule is very rough, but it is adequate for making many solubility predictions. Molecules with larger nonpolar regions, such as butanol in Figure 11-19b, are mostly nonpolar overall and dissolve only slightly in water. When it is important to know a chemical's precise solubility, you must either look it up in a chemical handbook or test it by adding small amounts to a solvent.

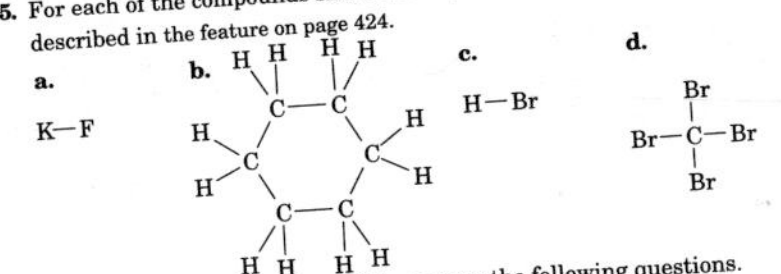

11-19b
Because butanol's nonpolar region is much larger than its polar region, it is nonpolar overall.

Section Review

15. For each of the compounds shown below, determine the particle type as described in the feature on page 424.

a. K—F
b. C_6H_{12} (ring structure)
c. H—Br
d. CBr_4

16. Use your answers from item **15** to answer the following questions. Would you expect compounds **a** and **b** to dissolve in each other? Would you expect compounds **b** and **d** to dissolve in each other? Would you expect compounds **c** and **d** to dissolve in each other? Estimate each compound's solubility in water.
17. Explain how water molecules can attract ions.
18. **Story Link**
Analyze the PCB molecule shown on page 402. Is the molecule polar or nonpolar?
19. **Story Link**
Why does the PCB molecule have a greater solubility in fats and oils? (Hint: fats and oils are mostly nonpolar substances.)

426 | Chapter 11

Story Links ►
in the Section Reviews, are questions and problems relating the topic of the story to the content of the section.

The Conclusion ►
provides closure to the questions posed in the story and opportunities for further exploration and research.

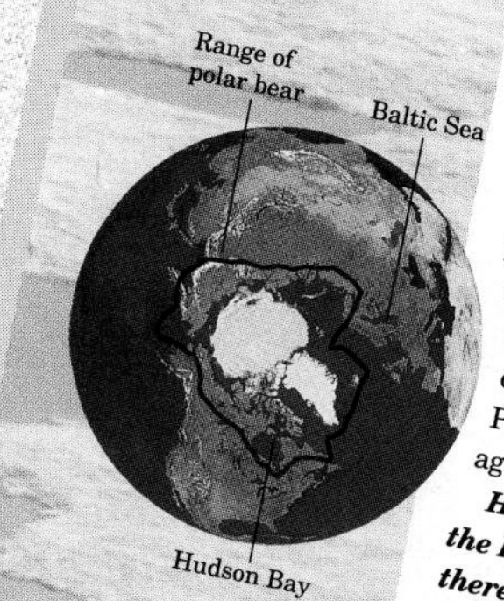

Conclusion: Solubility and the Arctic Bear Hunt

At the beginning of this chapter you followed Dr. Malcolm Ramsay as he examined a polar bear and took samples of its fat tissue to determine PCB content. This story posed some puzzling questions about the properties of PCBs. Look at those questions again.

How did the PCBs get into the Hudson Bay ecosystem when there is very little industry in the bay area? In past decades, the Baltic Sea was used as an ocean dump for PCBs and other wastes. It is likely that much of the discarded material remains in the mud at the bottom of the ocean. Because PCBs are slightly soluble in water, a small fraction of the PCBs dissolves and circulates in the ocean water.

Some PCBs are ingested by plankton, tiny marine animals. The plankton are eaten by larger plankton, which are eaten by fish. In addition to getting PCBs in their food, fish absorb some PCBs directly from the water. You may recall that oxygen is a nonpolar molecule and therefore has very low solubility in water. One liter of ocean water contains only about 3×10^{-4} mol of O_2, whereas the air we breathe contains 9×10^{-3} mol of O_2 per liter. This means that as fish process very large quantities of water through their gills, they extract PCBs as well as oxygen from the water.

Once PCB molecules are ingested by a fish, they diffuse out of the fish's watery blood, dissolve in the oily liver, and stay there. Over time, the concentration of PCBs in the liver increases. In effect, fish serve as vacuum cleaners, picking up low-concentration PCBs from the water and storing them to produce far higher concentrations in their bodies.

Why did the PCBs end up in the bear's fat? When a seal eats a fish, the oil-soluble PCBs end up in the seal's oily blubber in concen-trations that are even higher than those in fish. From there, it is just another step in the food chain to a polar bear's fat. The animals that are the most likely to have PCBs concentrated in their fatty tissues are polar bears. They are at the top of the food chain and live a long time.

Applying Concepts

1. What is the empirical formula of the PCB molecule on page 402?
2. Do you think this compound will float or sink in water? To estimate the density of the PCB molecule shown on page 402, find its molecular mass, and compare it with the molecular mass of biphenyl. Biphenyl, the related double-ring molecule with no chlorine atoms, has a density that is almost the same as water.

Research and Writing

Use the library to find out more about the following.

1. How can further buildup of PCBs in bears be prevented?
2. How can the harmful effects of PCBs already in bears be reduced?
3. What are the current procedures for PCB disposal?

440 | Chapter 11

Chemistry has never been developed like this before

Concepts are presented in ways that are effective for the learner.

For many students, the most effective presentations start with illustrations. To accommodate this learning style *Holt Chemistry: Visualizing Matter* incorporates a variety of techniques to blend words with pictures for those processes that are often difficult for beginning students to understand.

▼ ***How To* features** present difficult or complex processes using easy-to-follow illustrations.

How To

Carry out a titration

This procedure would be used to determine the unknown concentration of an acid using a standardized base solution. First, set up two clean burets as shown. Decide which of the burets will be used for the acid and the base. Rinse the acid buret three times with the acid to be used in the titration. Repeat this procedure in the base buret with the base solution to be used.

Fill the first to a point above the calibration mark with the acid of unknown concentration.

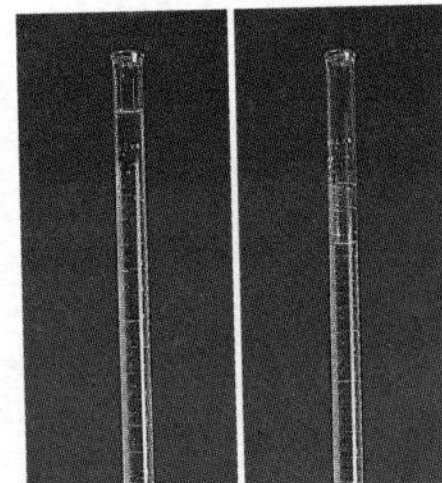

Release some acid from the buret to remove any air bubbles and to lower the volume to the calibrated portion of the buret.

Record the exact volume o acid in the buret to the ne 0.01 mL as your starting p

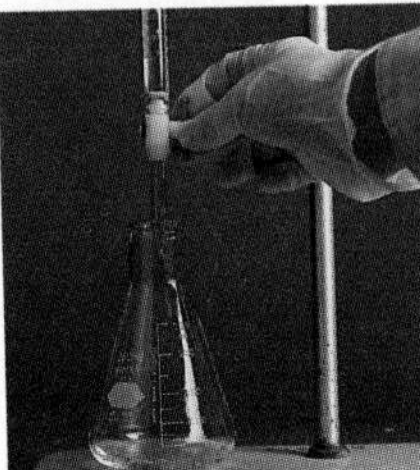

Release a predetermined volume of the acid (determined by your teacher or lab procedure) into a clean, dry Erlenmeyer flask.

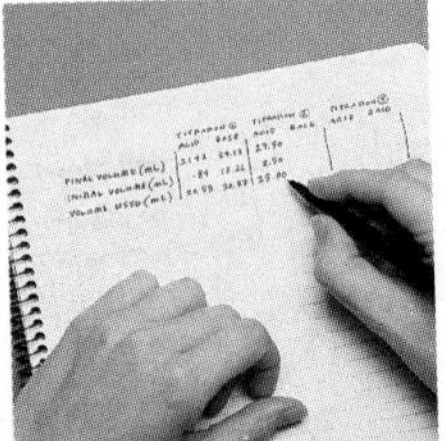

Re ac ne

512 | Chapter 13

How To

Write a balanced chemical equation

Chemical equations are sometimes more complex than the example you've just studied. But there is a step-by-step approach that can make balancing equations easier. These steps are outlined for the reaction in a safety match. This process is called balancing by inspection.

1. ***Write the word equation, showing all of the reactants and products.***

phosphorus + potassium chlorate ⟶

potassium chloride + phosphorus(V) oxide

2. ***Write the correct symbols and formulas for all of the reactants and products.***
In effect, change the word equation into a formula equation. When doing so, recall what you learned in Chapters 5 and 6 about formula writing. You may need to consult reference works for some formulas. Be sure to leave space to insert the coefficients.

$$?P_4 + ?KClO_3 \longrightarrow ?KCl + ?P_2O_5$$

3. ***Count the number of atoms of each element for both sides of the equation.***
List the kinds and numbers of atoms on each side of the equation.

Left side (reactants)	**Right side** (products)
P atoms: 4	P atoms: 2
K atoms: 1 ✓	K atoms: 1 ✓
Cl atoms: 1 ✓	Cl atoms: 1 ✓
O atoms: 3	O atoms: 5

4. ***Insert coefficients for atoms of one element at a time so that the law of conservation of mass is satisfied.***
For each element, the total number will be the least common multiple of the amounts on each side of the unbalanced equation. This number may be split ... ey contain the same element. Balancing all ... fficients requires a process of trial and error.

How To continued on the following page . . .

Chemical Equations | 241

How To

Draw Lewis structures

1. ***Determine the total number of valence electrons in the compound.***
This involves nothing more than adding the valence electrons of each atom. Consider the example of HCl. By now, you should realize that hydrogen has one valence electron, while chlorine has seven.
Total number of valence electrons: 1 + 7 = 8

2. ***Arrange the atoms' symbols to show how they are bonded and show valence electrons as dots.***
Be sure to distribute the paired dots so that the octet rule is followed, except in the case of hydrogen.

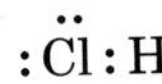

3. ***Compare the number of valence electrons used in the structure to the number available from step 1.***
There are two electrons in the covalent bond and six around the chlorine atom that are not shared.
Total number of electrons shown: 2 + 6 = 8

Flow charts ▶ provide a concrete scheme for classifying and making predictions. ▼

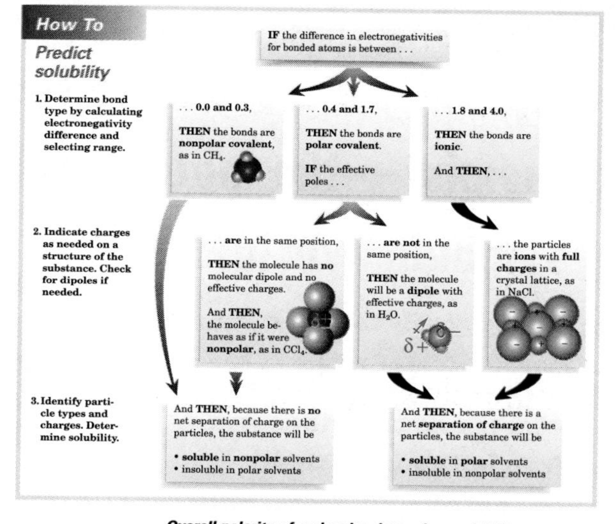

How To
Predict solubility
1. Determine bond type by calculating electronegativity difference and selecting range.
2. Indicate charges as needed on a structure of the substance. Check for dipoles if needed.
3. Identify particle types and charges. Determine solubility.
IF the difference in electronegativities for bonded atoms is between . . .
. . . 0.0 and 0.3, THEN the bonds are nonpolar covalent, as in CH₄.
. . . 0.4 and 1.7, THEN the bonds are polar covalent. IF the effective poles . . .
. . . 1.8 and 4.0, THEN the bonds are ionic. And THEN, . . .
. . . are in the same position, THEN the molecule has no molecular dipole and no effective charges. And THEN, the molecule behaves as if it were nonpolar, as in CCl₄.
. . . are not in the same position, THEN the molecule will be a dipole with effective charges, as in H₂O.
. . . the particles are ions with full charges in a crystal lattice, as in NaCl.
And THEN, because there is no net separation of charge on the particles, the substance will be
• soluble in nonpolar solvents
• insoluble in polar solvents
And THEN, because there is a net separation of charge on the particles, the substance will be
• soluble in polar solvents
• insoluble in nonpolar solvents

Overall polarity of molecules determines solubility

In this chapter you have focused on three types of substances: nonpolar molecules, molecular dipoles, and ionic crystals. Nonpolar molecules with either polar or nonpolar bonds carry no overall charge. When it comes to solubility, the full charges of ions and the partial charges of dipoles are grouped together because they attract each other. You can use the presence or absence of charges to predict whether one chemical will dissolve in another. Chemists sometimes summarize this rule by saying "like dissolves like."

- Many charged substances (ions and dipoles) will dissolve in other charged substances. (Some exceptions are described in **Table 11-1** on page 410.)
- A nonpolar substance will dissolve in another nonpolar substance.
- A polar or charged substance will not dissolve in a nonpolar substance.

These principles are summarized in the feature above, which can be used to help predict a substance's solubility.

424 | Chapter 11

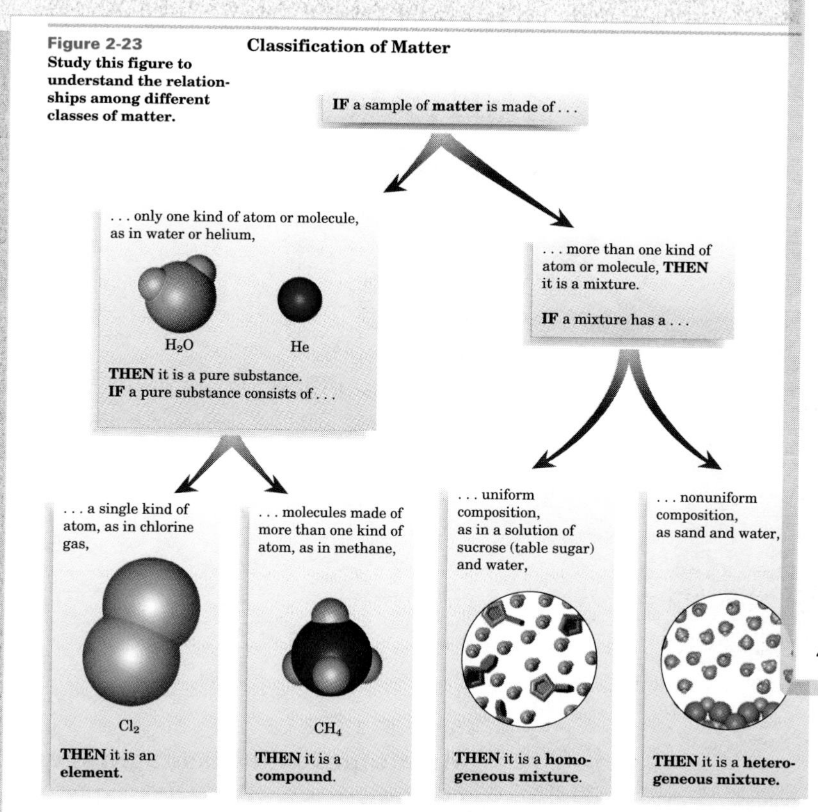

Figure 2-23
Study this figure to understand the relationships among different classes of matter.
Classification of Matter
IF a sample of matter is made of . . .
. . . only one kind of atom or molecule, as in water or helium,
H₂O
He
THEN it is a pure substance.
IF a pure substance consists of . . .
. . . more than one kind of atom or molecule, THEN it is a mixture.
IF a mixture has a . . .
. . . a single kind of atom, as in chlorine gas,
Cl₂
THEN it is an element.
. . . molecules made of more than one kind of atom, as in methane,
CH₄
THEN it is a compound.
. . . uniform composition, as in a solution of sucrose (table sugar) and water,
THEN it is a homogeneous mixture.
. . . nonuniform composition, as sand and water,
THEN it is a heterogeneous mixture.

Key problem solving ▶ approaches present a graphic summary of important quantitative concepts from the chapter. ▼

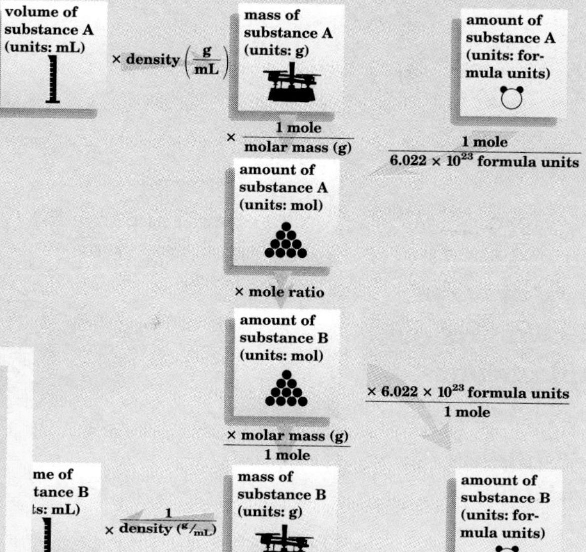

Key problem-solving approach: Stoichiometry calculations
When solving stoichiometry problems, select the variable given in your problem from the options near the top of this diagram. Then, follow the calculation pathway that leads to the proper units for the answer near the bottom of this diagram.
volume of substance A (units: mL)
× density (g/mL)
mass of substance A (units: g)
amount of substance A (units: formula units)
× 1 mole / molar mass (g)
1 mole / 6.022 × 10²³ formula units
amount of substance A (units: mol)
× mole ratio
amount of substance B (units: mol)
× 6.022 × 10²³ formula units / 1 mole
× molar mass (g) / 1 mole
volume of substance B (units: mL)
× 1 / density (g/mL)
mass of substance B (units: g)
amount of substance B (units: formula units)

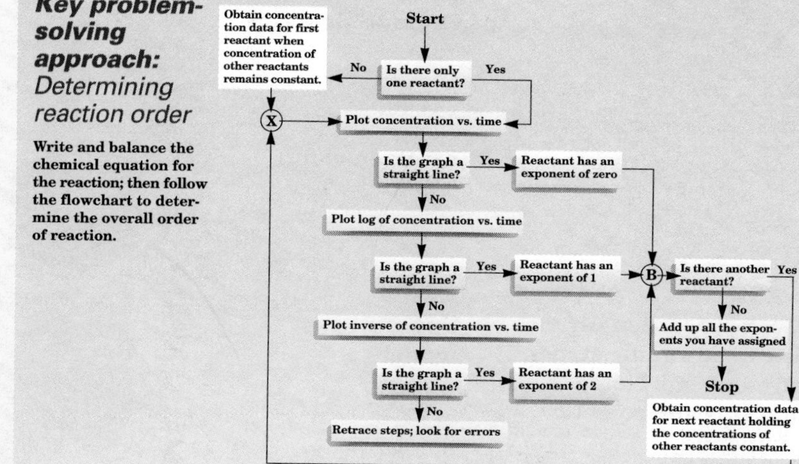

Key problem-solving approach: Determining reaction order
Write and balance the chemical equation for the reaction; then follow the flowchart to determine the overall order of reaction.
Start
Is there only one reactant?
No
Yes
Obtain concentration data for first reactant when concentration of other reactants remains constant.
Plot concentration vs. time
Is the graph a straight line?
Reactant has an exponent of zero
Plot log of concentration vs. time
Reactant has an exponent of 1
Plot inverse of concentration vs. time
Reactant has an exponent of 2
Retrace steps; look for errors
Is there another reactant?
Add up all the exponents you have assigned
Stop
Obtain concentration data for next reactant holding the concentrations of other reactants constant.

Chemistry texts have never looked like this before

Conceptual understanding requires practice in linking the worlds of the chemist—the macroscopic world, the microscopic world, and the symbolic world.

Chemistry texts have always had pictures and drawings, yet they often fell short in their instructional effectiveness. *Holt Chemistry: Visualizing Matter* has a unique illustration program that works closely with the text to make those ever important connections between macroscopic, microscopic, and symbolic representations.

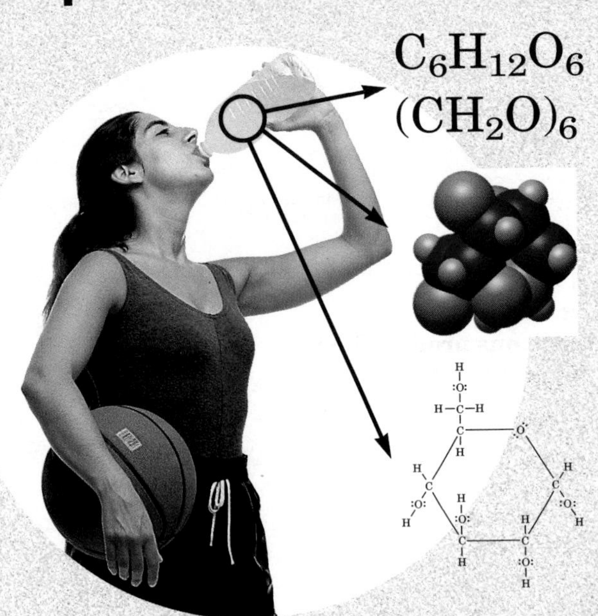

Symbolic representations

Students see the variety of ways that chemists represent phenomena using symbols—the language of chemistry.

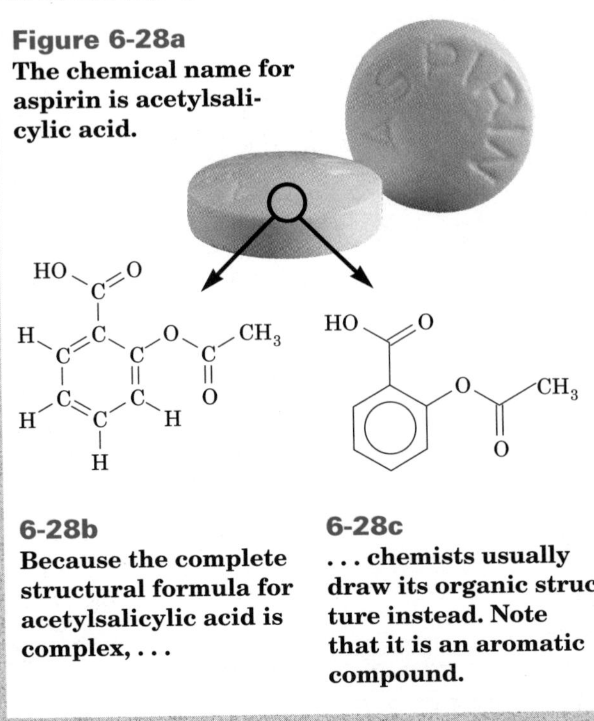

Figure 6-28a
The chemical name for aspirin is acetylsalicylic acid.

6-28b
Because the complete structural formula for acetylsalicylic acid is complex, . . .

6-28c
. . . chemists usually draw its organic structure instead. Note that it is an aromatic compound.

Table 6-12
Advantages and Disadvantages of Molecul

Type of representation	Example	Advanta
Chemical formula	C_6H_6	shows re
Lewis structure	H:C::C:H (Lewis structure of benzene)	shows ar ment of electrons molecule
Structural formula	H–C=C–H (structural formula of benzene)	shows a ment of and bon less com Lewis st
Organic structure	(hexagon)	shows a carbon c molecul than str
Electron-cloud model		shows sl shows p electron includin
Space-filling model		shows tl of molec space ta each mc than ele

Three-dimensional representations

All of the three-dimensional molecular representations throughout the text were drawn using the computer to provide the most accurate representations possible. Exact specifications and coordinates were used to produce images that are at the appropriate relative size and geometry.

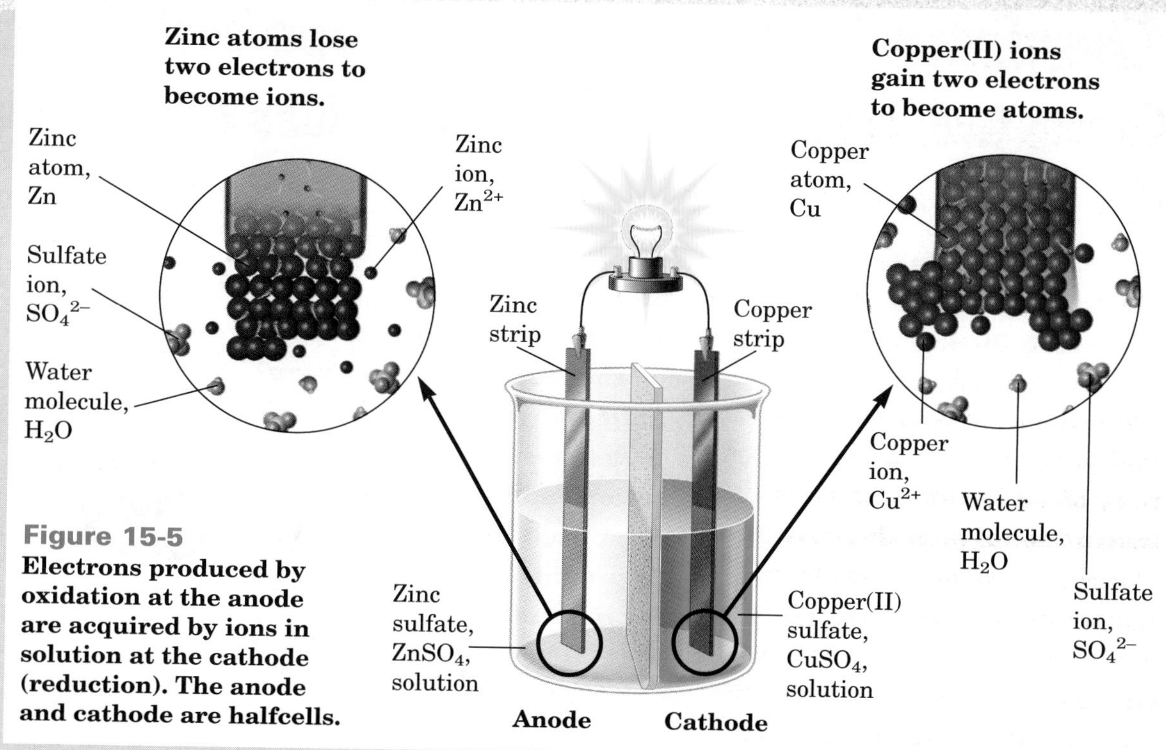

Figure 15-5
Electrons produced by oxidation at the anode are acquired by ions in solution at the cathode (reduction). The anode and cathode are halfcells.

Macro-micro connections

Illustrations start at the macro level and move to the micro level, linking what students see to the particle models that are the basis of understanding macroscopic phenomena.

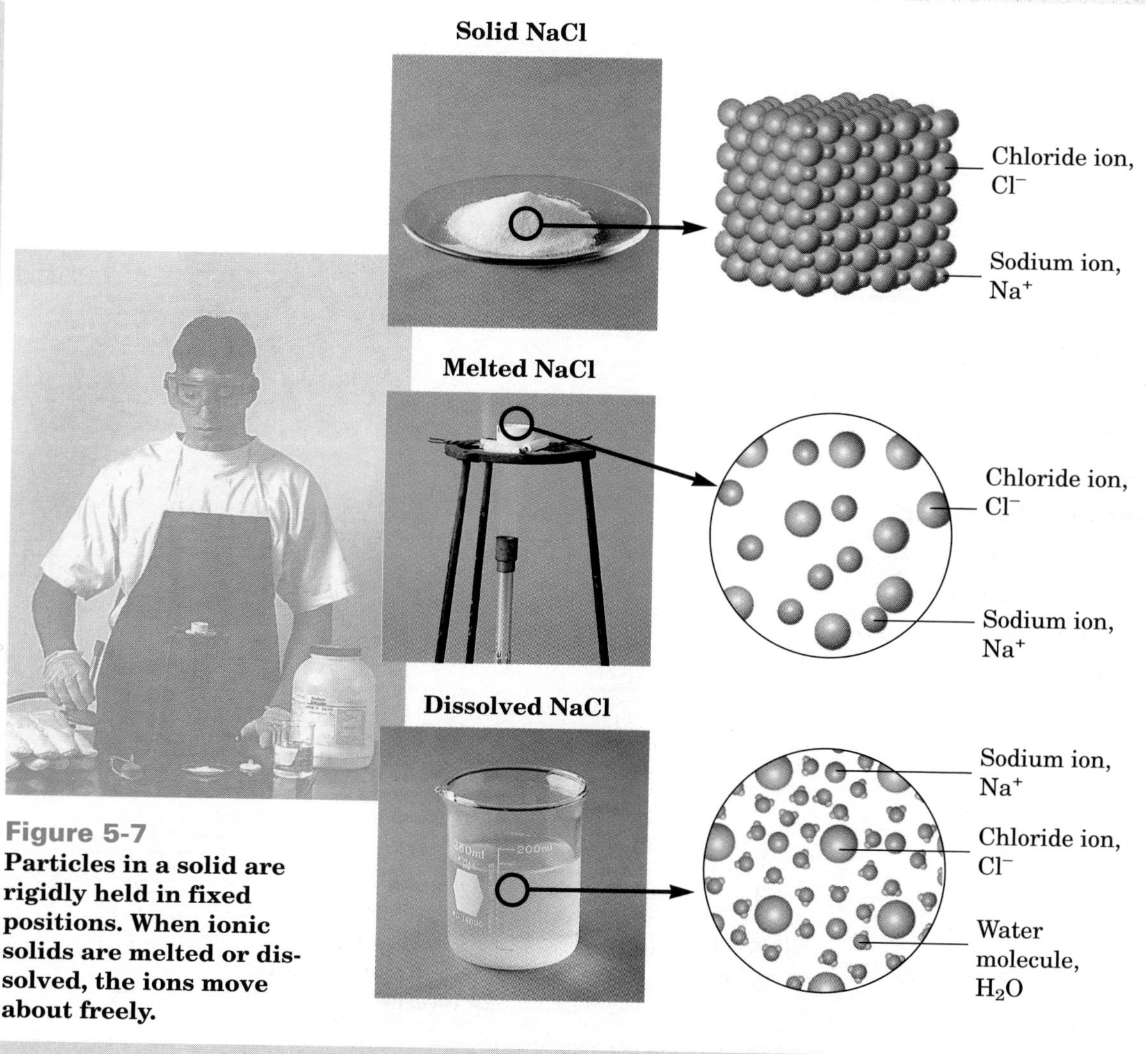

Figure 5-7
Particles in a solid are rigidly held in fixed positions. When ionic solids are melted or dissolved, the ions move about freely.

Chemistry labs have never been quite like this before

For the first time ever, the very best labs are right in the text.

Give your students the laboratory program that builds a framework for scientific inquiry and develops the communication skills that are critical to success in the real world. With *Holt Chemistry: Visualizing Matter*, students explore real science problems as the context for experimentation.

The laboratory program is uniquely designed to give students the tools needed to take on the responsibility of planning and implementing their own investigations. The Explorations and Investigations are organized to work in tandem to provide an effective model for inquiry.

More information regarding the structure and philosophy of the laboratory program can be found in the Laboratory Program Overview which begins on page T643-1.

▼ Explorations

provide the conceptual framework and laboratory techniques needed to conduct a full-scale inquiry for the problems posed in the Investigations. There are 19 Explorations in the Pupil's Edition.

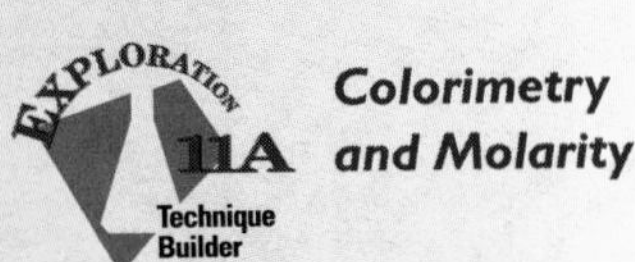

Colorimetry and Molarity

Objectives

Demonstrate **proficiency in preparing a solution and performing colorimetric easurements or observations.**

Relate **colorimetric measurements or observations to concentration.**

Determine **the molarity of a solution of unknown concentration.**

Situation

You are working in the quality control department of a pharmaceutical company. One of the company's products is a test solution for phenylketonuria. The test solution should contain iron(III) chloride, $FeCl_3$, at a concentration of 0.30 M. Lately there have been some problems in the production line in the factory. It is important that you analyze the solution to make sure it is of the proper concentration. If it is too dilute, the color change might not be noticeable enough. If it is too strong, the production process is wasting money by putting too much $FeCl_3$ in the solution.

Background

Doctors can use the $FeCl_3$ test solution to detect phenylketonuria in infants. People who have this disease are unable to break down the amino acid phenylalanine. If they eat foods containing too much phenylalanine, the toxic byproducts made when it is not completely broken down can make them sick. In the screening test, a few drops of the solution are sprinkled on a baby's wet diaper. If the solution turns a deep bluish green color, phenylpyruvic acid, a product of incomplete phenylketonuria metabolism, is present. Then a special diet with very small amounts of phenylalanine is prescribed for the child.

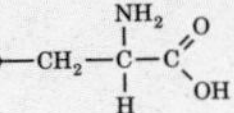

Phenylalanine

Phenyl-alanine

Problem

Like many solutions, $FeCl_3$ is colored, and in general, the more concentrated the solution, the darker its color will be. This is the basis for colorimetry, in which the color of a solution is used as a measure of its concentration. According to a relationship known as Beer's law, the amount of light of a specific wavelength that the solution absorbs, its absorbance, is proportional to its concentration. In some cases, there are slight deviations from Beer's law, so the graph of the relationship is a curve instead of a straight line. To determine the concentration of the sample pulled from the factory's production line, you must do the following.

- make several standard solutions of known concentration
- if special equipment such as a LEAP colorimeter or a spectrophotometer are available, make a graph of absorbance vs. concentration
- compare the solutions of known concentration to the unknown to determine the concentration of the unknown
- compare this extrapolation to the expected value, 0.30 M

736 | Exploration 11A

Safety

Always wear goggles and an apron to provide protection for your eyes and clothing. If you get a chemical in your eyes, immediately flush it out at the eyewash station while calling to your teacher. Know the locations of the emergency lab shower and eyewash and how to use them.

Do not touch any chemicals. If you get a chemical on your skin or clothing, wash it off at the sink while calling to your teacher. Make sure you carefully read the labels and follow the directions on all containers of chemicals that you use. Do not taste any chemicals or items used in the laboratory. Never return leftovers to their original containers; take only small amounts to avoid wasting supplies.

Always clean up the lab and all equipment after use, and dispose of substances according to proper disposal methods. Wash your hands thoroughly before you leave the lab after all lab work is finished.

Materials

- **Distilled water**
- **1.0 M HCl**
- **$FeCl_3 \cdot 6H_2O$ crystals**
- **10 mL graduated cylinder**
- **250 mL beaker**
- **250 mL volumetric flask**
- **Glass stirring rod**
- **Test tubes, 7**
- **Test tube rack**
- **Unknown solution**

Optional equipment

- **LEAP System with colormeter**
- **Spectrophotometer**
- **Cuvettes**
- **Lint-free wipes for cuvettes**

Preparation

1. ***Organizing Data***
   Prepare a data table in your lab notebook. It should have eight columns and five rows if you use the LEAP colormeter or a spectrophotometer, and four rows if this equipment will not be used. In the first row, label the boxes in the second through eighth columns *Test tube 0, Test tube 1, Test tube 2, Test tube 3, Test tube 4, Test tube 5,* and *Unknown*. In the first column, label the second box *mL 0.50 M* $FeCl_3$, the third box *mL* H_2O, and the fourth box *Estimates*. If you are using a colormeter or a spectrophotometer, label the fifth box in the first column *Measurements*. You will also need space for calculations.
2. Label seven test tubes *0, 1, 2, 3, 4, 5,* and *Unknown*. Label the beaker *Waste*.
3. If you will be using a spectrophotometer, turn it on now, because it must warm up for approximately 10 min.
4. Using a periodic table, determine the molar mass of $FeCl_3 \cdot 6H_2O$. Record it in your lab notebook.
5. Perform the necessary calculations to determine how many grams of $FeCl_3 \cdot 6H_2O$ would be needed to make 250 mL of a 0.50 M solution. Record this amount in your lab notebook.

Technique

Solution Preparation

6. Using the appropriate technique described in Chapter 11 and the amount calculated in step **5**, prepare 250 mL of a 0.50 M standard solution using $FeCl_3 \cdot 6H_2O$. Dilute the solution to 250 mL with 25 mL of 1.0 M HCl and more distilled water. Be sure to measure to the nearest 0.01 g. Record the amount used in your lab notebook.

Colorimetry and Molarity | 737

▼ Investigations involve students in designing their own experiments to meet the client's specifications. There are 23 Investigations in the Pupil's Edition.

Laboratory program philosophical constructs

A laboratory program is most relevant and effective when it emphasizes:

- developing laboratory techniques
- developing group interaction skills
- exploring meaningful questions and problems
- formulating experiments to seek answers to those questions and problems
- gathering, analyzing, and interpreting data
- communicating and critically reviewing results
- decision making based on economics
- using clearly stated standards to assess the lab experience

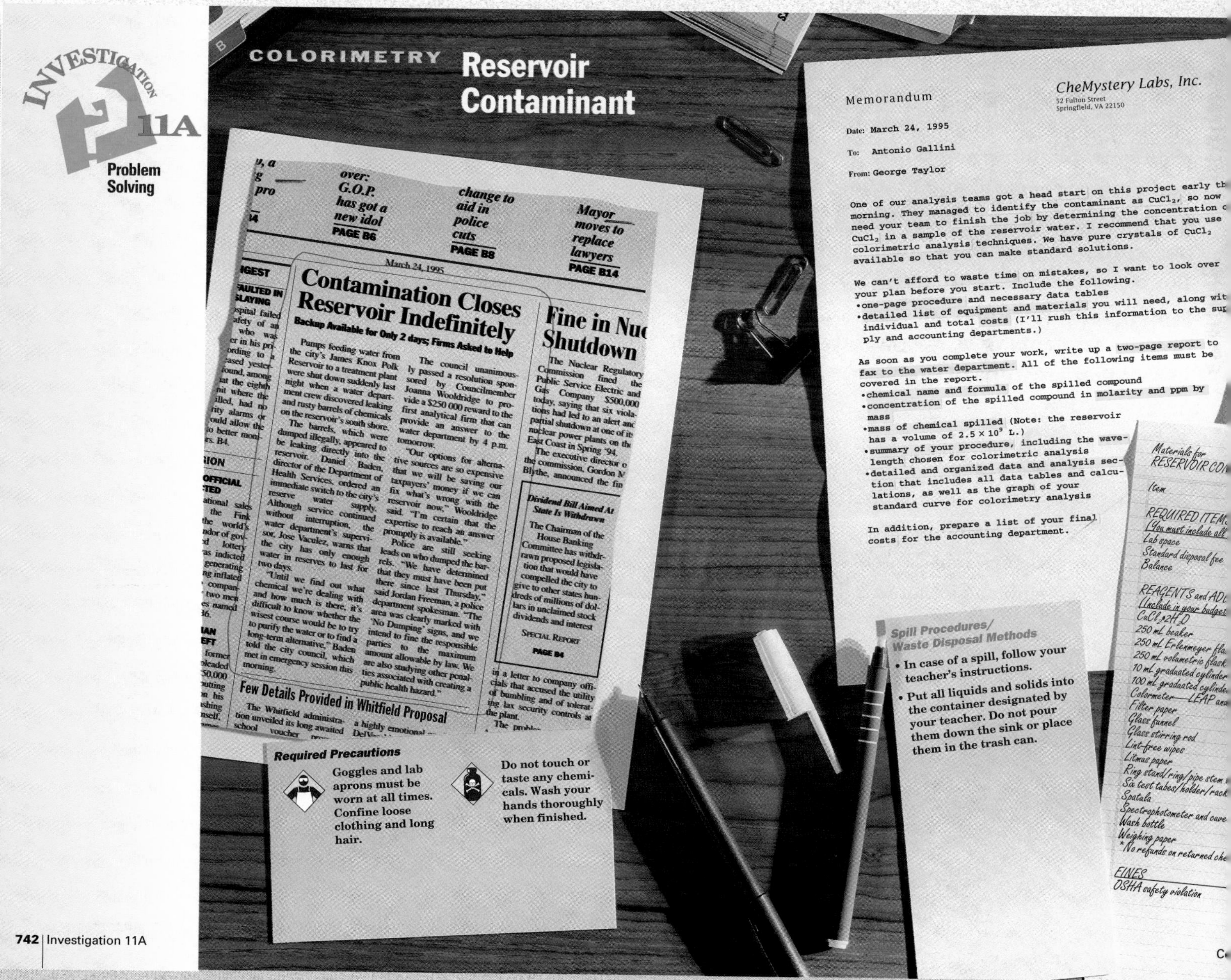

INVESTIGATION 11A

Problem Solving

COLORIMETRY

Reservoir Contaminant

over: G.O.P. has got a new idol PAGE B6

change to aid in police cuts PAGE B8

Mayor moves to replace lawyers PAGE B14

March 24, 1995

Contamination Closes Reservoir Indefinitely

Backup Available for Only 2 days; Firms Asked to Help

Pumps feeding water from the city's James Knox Polk Reservoir to a treatment plant were shut down suddenly last night when a water department crew discovered leaking and rusty barrels of chemicals on the reservoir's south shore.

The barrels, which were dumped illegally, appeared to be leaking directly into the reservoir. Daniel Baden, director of the Department of Health Services, ordered an immediate switch to the city's reserve water supply. Although service continued without interruption, the water department's supervisor, Jose Vaculez, warns that the city has only enough water in reserves to last for two days.

"Until we find out what chemical we're dealing with and how much is there, it's difficult to know whether the wisest course would be to try to purify the water or to find a long-term alternative," Baden told the city council, which met in emergency session this morning.

The council unanimously passed a resolution sponsored by Councilmember Joanna Wooldridge to provide a $250 000 reward to the first analytical firm that can provide an answer to the water department by 4 p.m. tomorrow.

"Our options for alternative sources are so expensive that we will be saving our taxpayers' money if we can fix what's wrong with the reservoir now," Wooldridge said. "I'm certain that the expertise to reach an answer promptly is available."

Police are still seeking leads on who dumped the barrels. "We have determined that they must have been put there since last Thursday," said Jordan Freeman, a police department spokesman. "The area was clearly marked with 'No Dumping' signs, and we intend to fine the responsible parties to the maximum amount allowable by law. We are also studying other penalties associated with creating a public health hazard."

Fine in Nuc[...] Shutdown

The Nuclear Regulatory Commission fined the Public Service Electric and Gas Company $500,000 today, saying that six violations had led to an alert and partial shutdown at one of its nuclear power plants on the East Coast in Spring '94.

The executive director of the commission, Gordon M[...] Blythe, announced the fin[...] in a letter to company officials that accused the utility of bumbling and of tolerating lax security controls at the plant.

Dividend Bill Aimed At State Is Withdrawn

The Chairman of the House Banking Committee has withdrawn proposed legislation that would have compelled the city to give to other states hundreds of millions of dollars in unclaimed stock dividends and interest

SPECIAL REPORT PAGE B4

Few Details Provided in Whitfield Proposal

The Whitfield administration unveiled its long awaited school voucher [...]

Required Precautions

Goggles and lab aprons must be worn at all times. Confine loose clothing and long hair.

Do not touch or taste any chemicals. Wash your hands thoroughly when finished.

Memorandum

CheMystery Labs, Inc.
52 Fulton Street
Springfield, VA 22150

Date: March 24, 1995

To: Antonio Gallini

From: George Taylor

One of our analysis teams got a head start on this project early th[is] morning. They managed to identify the contaminant as $CuCl_2$, so now [I] need your team to finish the job by determining the concentration o[f] $CuCl_2$ in a sample of the reservoir water. I recommend that you use colorimetric analysis techniques. We have pure crystals of $CuCl_2$ available so that you can make standard solutions.

We can't afford to waste time on mistakes, so I want to look over your plan before you start. Include the following.
- one-page procedure and necessary data tables
- detailed list of equipment and materials you will need, along wit[h] individual and total costs (I'll rush this information to the sup[...]ply and accounting departments.)

As soon as you complete your work, write up a two-page report to fax to the water department. All of the following items must be covered in the report.
- chemical name and formula of the spilled compound
- concentration of the spilled compound in molarity and ppm by mass
- mass of chemical spilled (Note: the reservoir has a volume of 2.5×10^9 L.)
- summary of your procedure, including the wavelength chosen for colorimetric analysis
- detailed and organized data and analysis section that includes all data tables and calculations, as well as the graph of your standard curve for colorimetry analysis

In addition, prepare a list of your final costs for the accounting department.

Spill Procedures/ Waste Disposal Methods

- In case of a spill, follow your teacher's instructions.
- Put all liquids and solids into the container designated by your teacher. Do not pour them down the sink or place them in the trash can.

Materials for RESERVOIR CO[...]

Item

REQUIRED ITEM[...]
(You must include all[...]
Lab space
Standard disposal fee
Balance

REAGENTS and AD[...]
(Include in your budget[...]
$CuCl_2 \cdot 2H_2O$
250 mL beaker
250 mL Erlenmeyer fla[...]
250 mL volametric flask
10 mL graduated cylinder
100 mL graduated cylinde[...]
Colormeter—LEAP an[...]
Filter paper
Glass funnel
Glass stirring rod
Lint-free wipes
Litmus paper
Ring stand/ring/pipe stem [...]
Six test tubes/holder/rack
Spatula
Spectrophotometer and cuve[...]
Wash bottle
Weighing paper
*No refunds on returned che[...]

FINES
OSHA safety violation

742 | Investigation 11A

Problem Solving has never been this meaningful before

Success starts with modeling the right techniques, strategies, and skills.

Conceptual understanding is truly attained when students can apply mathematical skills to solve chemical problems. We recognize that solving problems goes beyond the rote application of a memorized algorithm. *Holt Chemistry: Visualizing Matter* uses a three-step method that focuses on developing strategies and that provides the structural framework students need to develop logical reasoning skills. Using this model, your students will become skilled at analyzing problems, planning solutions, applying conceptual and mathematical models, and estimating and verifying their answers. Continuous exposure to good strategies and regular practice enables students to be successful in tackling any type of problem.

Solutions for the Sample Problems ▶ provide the in-depth explanations students need, and model the reasoning processes that can be used to solve similar problems.

Sample Problem 13H

Suppose that in the titration of 40. mL of vinegar, 20. mL of 0.50 M NaOH were needed to reach the equivalence point. What is the molarity of acetic acid in the vinegar? The molecular equation for the reaction is the following.

Sample Problem 11D

The structure of tetrachloroethylene is shown here. Use the electronegativity data in Table 11-3 and the feature on page 424 to determine if C_2Cl_4 is a molecular dipole. Describe its solubility.

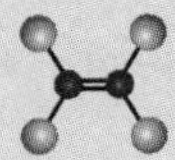

Sample Problem 8A
Mass-mass stoichiometry

Methyl salicylate, also known as "oil of wintergreen," is most often made in a synthesis reaction between methanol and salicylic acid. How many grams of salicylic acid are needed to produce 325 g of methyl salicylate, provided there is plenty of methanol available?

Salicylic acid + CH_3OH (Methanol) $\xrightarrow{HCl}$ Methyl salicylate + H_2O (Water)

1 List what you know

These facts were taken from the problem statement.

- **reactants:** salicylic acid, methanol
- **products:** methyl salicylate, water
- **mass of methyl salicylate produced:** 325 g
- **mass of salicylic acid needed:** ? g

These molar masses were calculated from the periodic table.

- **formula and molar mass of salicylic acid:** $C_7H_6O_3$, 138.13 g/mol
- **formula and molar mass of methyl salicylate:** $C_8H_8O_3$, 152.16 g/mol

2 Set up the problem

- To convert from the amount of product to the amount of reactant needed, first convert to moles and use the mole ratio to link the two amounts.

$$325\ \text{g}\ C_8H_8O_3 \times \overset{a}{\frac{1\ \text{mol}\ C_8H_8O_3}{152.16\ \text{g}\ C_8H_8O_3}} \times \overset{b}{\frac{1\ \text{mol}\ C_7H_6O_3}{1\ \text{mol}\ C_8H_8O_3}} \times \overset{c}{\frac{138.13\ \text{g}\ C_7H_6O_3}{1\ \text{mol}\ C_7H_6O_3}} = ?\ \text{g}\ C_7H_6O_3$$

Always remember to work stoichiometry problems by calculating with moles. Notice that all of the factors cancel to give you the units of the answer, g $C_7H_6O_3$.

a First, convert the given mass of the product into moles. The reciprocal of the molar mass (mol/g) of the product must be used in order to cancel the mass of $C_8H_8O_3$ and leave the amount of $C_8H_8O_3$ in moles.

b Next, to convert moles of product into moles of reactant, multiply by the mole ratio (1:1) from the balanced chemical equation.

c Last, convert the moles of reactant into mass using the molar mass of the reactant.

3 Estimate and calculate

- Before calculating, round off the numbers in the setup to make an estimate.

$$300 \times \frac{1}{150} \times \frac{1}{1} \times \frac{140}{1} \approx 280$$

- Use your calculator to work through the setup. Be sure to round to the correct number of significant figures, which is three.

$$325\ \text{g}\ C_8H_8O_3 \times \frac{1\ \text{mol}\ C_8H_8O_3}{152.16\ \text{g}\ C_8H_8O_3} \times \frac{1\ \text{mol}\ C_7H_6O_3}{1\ \text{mol}\ C_8H_8O_3} \times \frac{138.13\ \text{g}\ C_7H_6O_3}{1\ \text{mol}\ C_7H_6O_3} = ?\ \text{g}\ C_7H_6O_3$$

The answer is reasonably close to the estimate.

Calculator answer: 2.950331887×10^2

Answer to three significant figures: 295 g $C_7H_6O_3$

278 | Chapter 8

Special *How To* features ▶ present the structure that beginning students need for solving certain types of problems.

...refer to this page for ideas. The Sample Problems that follow can also provide helpful hints.

Remember that the *best way to convert from one substance to another is to relate the reaction stoichiometry **in moles***. If your data from the problem is in grams, convert it to moles first.

How To *Solve stoichiometry problems*

1. List what you know

Don't start using your calculator yet.

- Read the problem twice.
- Organize the information from the problem statement in a list or table.
- Identify what you are asked to find, and write down the units for the answer.
- For all substances you will be working with, write the formulas, and determine the molar masses. Check a periodic table if necessary.
- If there is a reaction, write an equation for it, making sure that it is balanced so that you'll have the correct mole ratios.
- List any conversion factors that you might need, such as molar masses, mole ratios, and unit conversions.

2. Set up the problem

Don't start using your calculator yet.

- First, analyze what needs to be done to get to the answer. See if there is any information not in the problem that you need for the answer.
- Identify which value given in the problem can be used as a starting point. Write it on the left side of a sheet of paper. On the right side of the paper, write an equals sign and then a question mark with the units of the answer. Fill in the conversion factors necessary to convert from what is given in the problem to what is sought in the answer.
- Most chemistry problems (and nearly all stoichiometry problems) require the amounts of substances to be in moles, so use the molar masses from step 1 to convert the amounts into moles, if necessary.
- If you need to change from amount of one substance to a different substance, use mole ratios derived from the balanced chemical equation. Remember that the mole ratio may not always be 1:1.
- Be sure to convert the data into appropriate units, such as *grams* instead of *kilograms*.

...factors, check to see how the units cancel each other. If they all cancel to give you the units you need for the answer, the setup is probably correct.

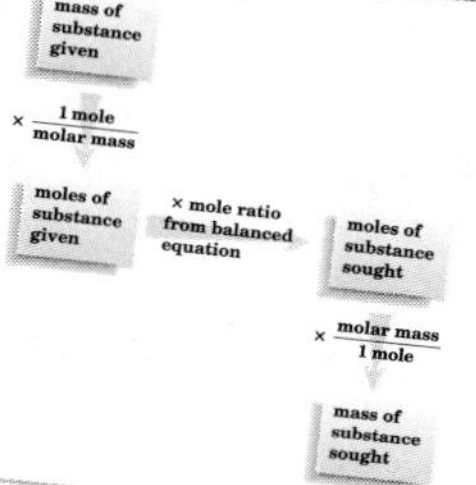

Figure 8-5 Most stoichiometry problems involve converting an amount into moles and using the mole ratio to calculate a corresponding amount of another substance in the reaction.

3. Estimate and calculate

Don't start using your calculator yet.

- First, estimate your answer. One way to do this is to round off the numbers in the problem setup and make a quick calculation. Another way is to compare conversion factors in the setup and decide whether the answer should be bigger or smaller than the beginning value.
- Then, begin your calculations by working through the problem setup you made in step 2. **In the examples in this book, numbers are not rounded off between steps in calculations.**
- When you have finished your calculations, *remember that you still don't have the correct answer*. You must round off and make sure that the answer has the correct number of significant figures.
- Always report the answer with the correct units, not just as a number.
- Compare your answer with the estimate. If they are not close, check all of your work.
- Make sure your answer is reasonable. For example, if you began with 5.3 mg of one reactant, and your answer is that it will make 725 g of a product, you know that you need to double-check all of your work.

Practice problems ▶ follow each Sample Problem. Answers are found in the back of the book.

Visual algorithims ▶ help reinforce text descriptions of routine problems.

Practice 8A

Answers for Concept Review items and Practice problems begin on page 841.

1. Tin(II) fluoride, also known as stannous fluoride, is added to some dental products to help prevent cavities. How many grams of tin(II) fluoride can be made from 55.0 g of hydrogen fluoride, HF, if there is plenty of tin?

$$Sn(s) + 2HF(aq) \longrightarrow SnF_2(aq) + H_2(g)$$

2. Fluorescein is a reddish powder used in some cosmetics, such as rouge and fingernail polish. It is made by combining phthalic anhydride with resorcinol. If a cosmetics company starts with 25.0 kg of resorcinol, how many kilograms of fluorescein can they produce, assuming that there is plenty of phthalic anhydride?

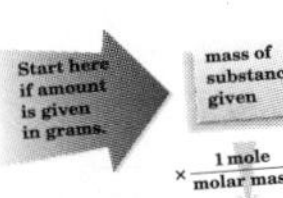
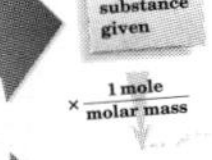
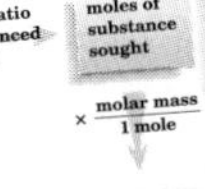
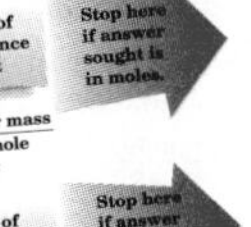

Two resorcinol + One phthalic anhydride —heat→ One fluorescein + 2H₂O Two water

Other stoichiometric calculations

Problems with amounts in moles

Some stoichiometry problems involve data or answers in mole amounts instead of mass amounts. These problems can be solved with an approach similar to that used to solve the problems having both data and answers in mass, but with a shortcut. There are fewer steps because one or both molar mass conversions are unnecessary, as shown in **Figure 8-6**, which has many of the same steps as **Figure 8-5**. If both the answer and the given data are in moles, the only conversion factor necessary to solve the problem is the mole ratio.

Solving Stoichiometry Problems With Moles or Grams

Start here if amount is given in grams. → mass of substance given → × 1 mole / molar mass → moles of substance given ← Start here if amount is given in moles. → × mole ratio from balanced equation → moles of substance sought → Stop here if answer sought is in moles. → × molar mass / 1 mole → mass of substance sought → Stop here if answer sought is in grams.

Figure 8-6 If a stoichiometry problem involves amounts in moles, solving the problem has fewer steps than problems with amounts in grams.

▼ Section Reviews include additional problems for practice.

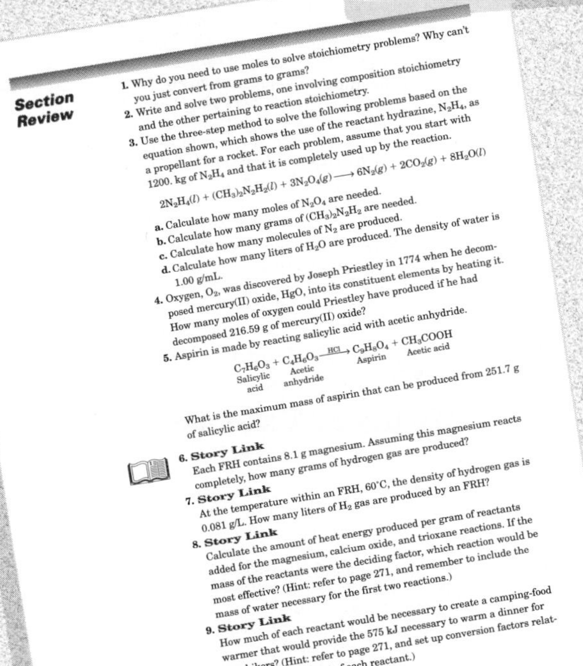

Section Review

1. Why do you need to use moles to solve stoichiometry problems? Why can't you just convert from grams to grams?
2. Write and solve two problems, one involving composition stoichiometry and the other pertaining to reaction stoichiometry.
3. Use the three-step method to solve the following problems based on the equation shown, which shows the use of the reactant hydrazine, N_2H_4, as a propellant for a rocket. For each problem, assume that you start with 1200. kg of N_2H_4 and that it is completely used up by the reaction.

$$2N_2H_4(l) + (CH_3)_2N_2H_2(l) + 3N_2O_4(g) \longrightarrow 6N_2(g) + 2CO_2(g) + 8H_2O(l)$$

a. Calculate how many moles of N_2O_4 are needed.
b. Calculate how many grams of $(CH_3)_2N_2H_2$ are needed.
c. Calculate how many molecules of N_2 are produced.
d. Calculate how many liters of H_2O are produced. The density of water is 1.00 g/mL.

4. Oxygen, O_2, was discovered by Joseph Priestley in 1774 when he decomposed mercury(II) oxide, HgO, into its constituent elements by heating it. How many moles of oxygen could Priestley have produced if he had decomposed 216.59 g of mercury(II) oxide?
5. Aspirin is made by reacting salicylic acid with acetic anhydride.

$$C_7H_6O_3 + C_4H_6O_3 \xrightarrow{HCl} C_9H_8O_4 + CH_3COOH$$

Salicylic acid, Acetic anhydride, Aspirin, Acetic acid

What is the maximum mass of aspirin that can be produced from 251.7 g of salicylic acid?

6. **Story Link** Each FRH contains 8.1 g magnesium. Assuming this magnesium reacts completely, how many grams of hydrogen gas are produced?
7. **Story Link** At the temperature within an FRH, 60°C, the density of hydrogen gas is 0.081 g/L. How many liters of H_2 gas are produced by an FRH?
8. **Story Link** Calculate the amount of heat energy produced per gram of reactants added for the magnesium, calcium oxide, and trioxane reactions. If the mass of the reactants were the deciding factor, which reaction would be most effective? (Hint: refer to page 271, and remember to include the mass of water necessary for the first two reactions.)
9. **Story Link** How much of each reactant would be necessary to create a camping-food warmer that would provide the 575 kJ necessary to warm a dinner for four hikers? (Hint: refer to page 271, and set up conversion factors relating energy and amount of each reactant.)

▼ Practice problems in the chapter review are correlated to a specific sample problem in the chapter where students can go for help. Answers to selected items are found in the back of the book.

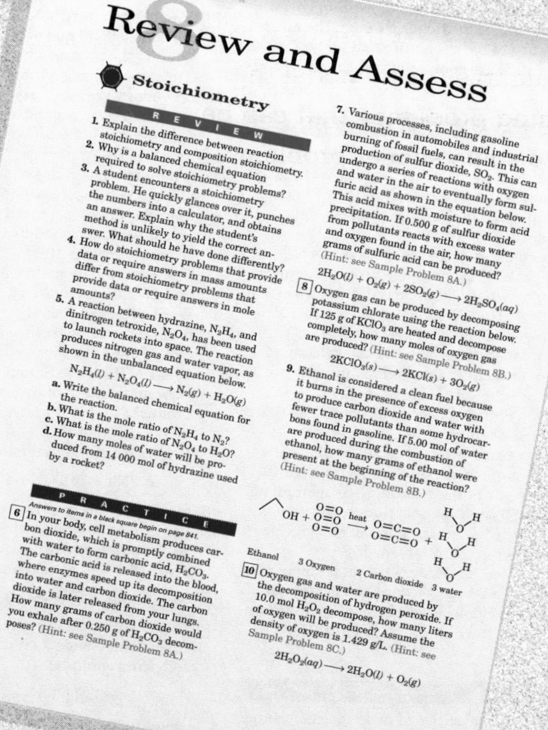

CHAPTER 8

Review and Assess

Stoichiometry

REVIEW

1. Explain the difference between reaction stoichiometry and composition stoichiometry.
2. Why is a balanced chemical equation required to solve stoichiometry problems?
3. A student encounters a stoichiometry problem. He quickly glances over it, punches the numbers into a calculator, and obtains an answer. Explain why the student's method is unlikely to yield the correct answer. What should he have done differently?
4. How do stoichiometry problems that provide data or require answers in mass amounts differ from stoichiometry problems that provide data or require answers in mole amounts?
5. A reaction between hydrazine, N_2H_4, and dinitrogen tetroxide, N_2O_4, has been used to launch rockets into space. The reaction produces nitrogen gas and water vapor, as shown in the unbalanced equation below.

$$N_2H_4(l) + N_2O_4(l) \longrightarrow N_2(g) + H_2O(g)$$

a. Write the balanced chemical equation for the reaction.
b. What is the mole ratio of N_2H_4 to N_2?
c. What is the mole ratio of N_2O_4 to H_2O?
d. How many moles of water will be produced from 14 000 mol of hydrazine used by a rocket?

PRACTICE

Answers to items in a black square begin on page 841.

6. In your body, cell metabolism produces carbon dioxide, which is promptly combined with water to form carbonic acid, H_2CO_3. The carbonic acid is released into the blood, where enzymes speed up its decomposition into water and carbon dioxide. The carbon dioxide is later released from your lungs. How many grams of carbon dioxide would you exhale after 0.250 g of H_2CO_3 decomposes? (Hint: see Sample Problem 8A.)
7. Various processes, including gasoline combustion in automobiles and industrial burning of fossil fuels, can result in the production of sulfur dioxide, SO_2. This can undergo a series of reactions with oxygen and water in the air to eventually form sulfuric acid as shown in the equation below. This acid mixes with moisture to form acid precipitation. If 0.500 g of sulfur dioxide from pollutants reacts with excess water and oxygen found in the air, how many grams of sulfuric acid can be produced? (Hint: see Sample Problem 8A.)

$$2H_2O(l) + O_2(g) + 2SO_2(g) \longrightarrow 2H_2SO_4(aq)$$

8. Oxygen gas can be produced by decomposing potassium chlorate using the reaction below. If 125 g of $KClO_3$ are heated and decompose completely, how many moles of oxygen gas are produced? (Hint: see Sample Problem 8B.)

$$2KClO_3(s) \longrightarrow 2KCl(s) + 3O_2(g)$$

9. Ethanol is considered a clean fuel because it burns in the presence of excess oxygen to produce carbon dioxide and water with fewer trace pollutants than some hydrocarbons found in gasoline. If 5.00 mol of water are produced during the combustion of ethanol, how many grams of ethanol were present at the beginning of the reaction? (Hint: see Sample Problem 8B.)

Ethanol + 3 Oxygen —heat→ 2 Carbon dioxide + 3 water

10. Oxygen gas and water are produced by the decomposition of hydrogen peroxide. If 10.0 mol H_2O_2 decompose, how many liters of oxygen will be produced? Assume the density of oxygen is 1.429 g/L. (Hint: see Sample Problem 8C.)

$$2H_2O_2(aq) \longrightarrow 2H_2O(l) + O_2(g)$$

Assessment has never been this comprehensive

A good assessment program measures what is important for students to learn.

Holt Chemistry: Visualizing Matter **includes a multifaceted assessment program that targets the instructional goals outlined on page T18. You can build a complete assessment program that fully measures students' achievements by selecting from the wide range of options shown.**

Pretest review materials

Chapter Review and Assess

In *Holt Chemistry: Visualizing Matter* you'll find a comprehensive set of review materials at the end of each chapter. Each review is divided into topics which include review items that focus on the literal, interpretive, and application levels of comprehension.

1 Review items focus on literal level comprehension.

2 Practice items provide additional work with the concepts presented in the Sample Problems. Each Practice item is correlated to a Sample Problem in the Chapter. All items with a boxed number are answered in the back of the student book.

3 Apply items focus on the application level of understanding.

Dalton's law and mole fractions

REVIEW

24. How are mole fractions used to find partial pressures of gas mixtures?
25. Gas is collected in the laboratory by bubbling it through water and trapping it in a container. How can you determine the pressure of the dry collected gas?
26. a. Why is it important to know the temperature of water when collecting a gas over it?
    b. How will the pressure of the dry gas change if the temperature of the water is increased?
27. Use the diffusion process to explain how a gaseous emission travels through the air.
28. Differentiate between diffusion and effusion.
29. What ratios are compared in Graham's law?

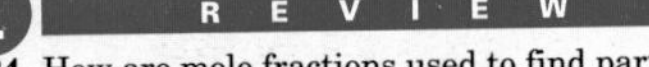

2 PRACTICE

30. Use **Table 10-5** to determine the partial pressure of oxygen collected over water if the temperature is 20.0°C and the total gas pressure is 98.0 kPa. (Hint: see Sample Problem 10D.)
31. The barometer at an indoor pool reads 105.00 kPa. If the temperature in the room is 30.0°C, what is the partial pressure of the "dry" air? (Hint: see Sample Problem 10D.)
32. A nitrogen molecule travels at 500. m/s a room temperature. What is the velocity of a helium molecule at the same temperature? (Hint: see Sample Problem 10E.)
33. A carbon dioxide molecule travels at 45.0 m/s at a certain temperature. What is the velocity of an oxygen molecule at the same temperature? (Hint: see Sample Problem 10E.)
34. Chlorine gas effuses through an opening at a rate 1.59 times slower than nitrogen gas. What is the molecular mass of chlorine gas? (Hint: see Sample Problem 10F.)
35. An unknown gas effuses through an opening at a rate 1.62 time slower than oxygen gas. What is the molecular mass of the unknown gas? (Hint: see Sample Problem 10F.)

3 APPLY

36. How do the velocities of neon molecules and krypton molecules compare when both gases are at the same temperature?

Ideal gas law, general gas law, and stoichiometry

1 REVIEW

37. What determines the value for the constant R?
38. Which gas variable is held constant when using the general gas law to analyze gas behavior?
39. How can a balanced chemical equation be used to determine the volume of a gas that will be produced from the reaction?

2 PRACTICE

40. Suppose you have a 500. mL container that contains 0.0500 mol of oxygen gas at 25°C. What is the pressure inside the container? (Hint: see Sample Problem 10G)
41. How many grams of oxygen gas in a 10.0 L container exert a pressure of 97.0 kPa at a temperature of 25.0°C? (Hint: see Sample Problem 10G.)
42. A helium balloon has a volume of 500. mL at STP. What will be its new volume if the temperature is increased to 325 K and its pressure is increased to 125 kPa? (Hint: see Sample Problem 10H.)
43. The air in a balloon has a volume of 3.00 L and exerts a pressure of 101. kPa at 300.0 K. What pressure does the air in the balloon exert if the temperature is increased to 400.0 K and the air is allowed to expand to 15.0 L? (Hint: see Sample Problem 10H.)
44. How many liters of hydrogen gas can be produced at 300.0 K and 104 kPa pressure if 20.0 g of sodium metal are reacted with water? (Hint: see Sample Problem 10I.)

$$2Na + 2H_2O \longrightarrow 2NaOH + H_2$$

45. How many liters of hydrogen gas can be produced at 290. K and 99.0 kPa if 17.0 g of potassium metal are reacted with water? (Hint: see Sample Problem 10I.)

$$2K + 2H_2O \longrightarrow 2KOH + H_2$$

46. Magnesium will burn in oxygen to form magnesium oxide as represented by the following equation.

$$2Mg + O_2 \longrightarrow 2MgO$$

What mass of magnesium will react with a 500.0 mL container of oxygen at 150°C and 70.0 kPa? (Hint: see Sample Problem 10J.)

47. One industrial method for producing nitric acid involves dissolving nitrogen dioxide in water.

$$3NO_2(g) + H_2O(l) \longrightarrow 2HNO_3(l) + NO(g)$$

What mass of water will react with a 1000.0 mL container of nitrogen dioxide at 25.0°C and 60.0 kPa? (Hint: see Sample Problem 10J.)

3 APPLY

48. An expandable weather balloon is filled with 90.0 L of hydrogen gas at ground level, where the pressure is 99.0 kPa and the temperature is 20.0°C. The balloon will burst if its volume exceeds 300.0 L. Can the balloon rise above the altitude of Mount Everest where the pressure drops to 32.1 kPa and the temperature decreases by 64.4°C.
49. Calculate the density of helium at standard temperature and pressure using the ideal gas law equation. Check you answer with a reference source of gas densities.
50. Suppose a certain automobile engine has a cylinder with a volume of 500.0 mL that is filled with air (21% oxygen) at a temperature of 55°C and a pressure of 101.0 kPa. What mass of octane must be injected to react with all of the oxygen in the cylinder?

$$2C_8H_{18}(l) + 25O_2(g) \longrightarrow 16CO_2(g) + 18H_2O(g)$$

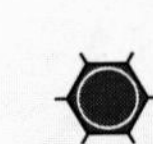

Condensation and liquids

1 REVIEW

51. Why do gases deviate from ideal gas behavior?
52. Use the following graph of the vapor pressure of water versus temperature to answer the following questions.
    a. At which point(s) does water boil at standard atmospheric pressure?
    b. At which point(s) is water only in the liquid phase?
    c. At which point(s) is water only in the vapor phase?
    d. At which point(s) is water in equilibrium with the liquid and vapor phases?

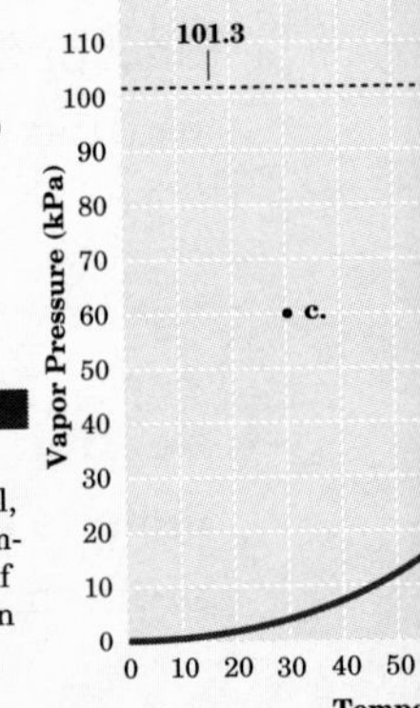

3 AP

53. Use the phase diagr mine what phase tra the following change
    a. At point a, pressu temperature rema
    b. At point b, tempe pressure is consta
    c. At point c, pressu perature is the sa

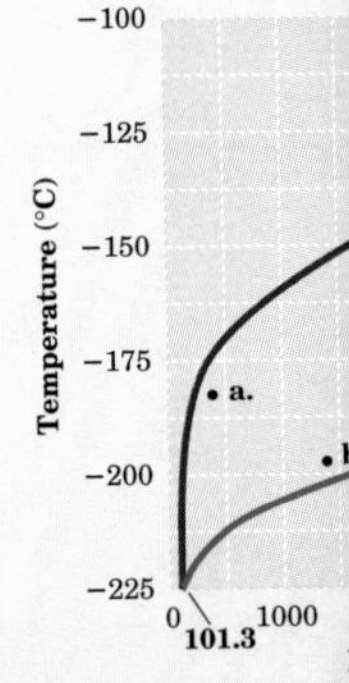

54. From the phase dia mine the physical s conditions.
    a. 2000 kPa and −1
    b. 3000 kPa and −1

Portfolio assessment

The nature and use of portfolios can vary greatly depending on teaching preferences and district requirements. In the *Holt Chemistry: Visualizing Matter* program, you are given a wide range of materials from which you and students can make selections to build a portfolio. Most of the portfolio suggestions involve writing projects therefore you'll find an analytic scale on the next page that can be used to assess student writing.

Building the portfolio

It is recommended that you establish the objectives of the portfolio at the beginning of the year and then make selections based on how well an assignment demonstrates the mastery of an objective. Portfolios should also include students' selections of their best work. Each student selection should be accompanied by a written, self-reflective rationale for why the selection was made.

Traditional assessment

Test Generator

Students learn best when they are told what they're expected to learn. Therefore, the Section Objectives for each chapter were used as the basis for the test items in the *Holt Chemistry: Visualizing Matter Test Generator*. This test item bank consists of more than 70 items per chapter from which to build chapter tests that focus on your areas of emphasis.

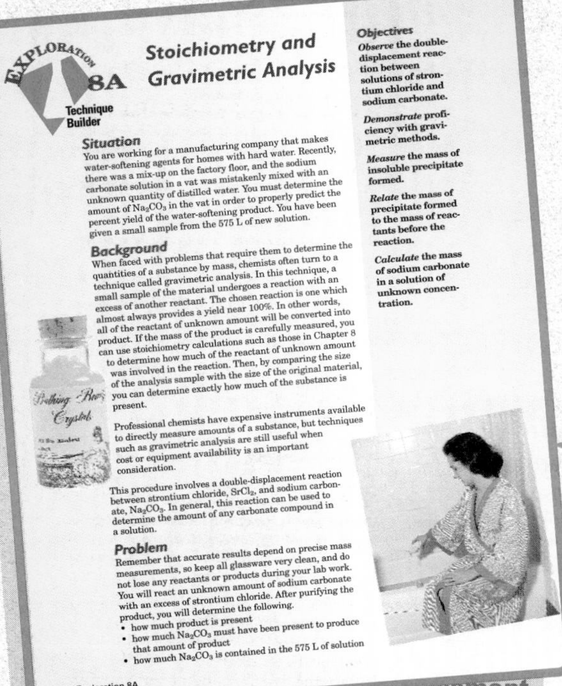

EXPLORATION 8A Technique Builder

Stoichiometry and Gravimetric Analysis

Situation

You are working for a manufacturing company that makes water-softening agents for homes with hard water. Recently, there was a mix-up on the factory floor, and the sodium carbonate solution in a vat was mistakenly mixed with an unknown quantity of distilled water. You must determine the amount of Na_2CO_3 in the vat in order to properly predict the percent yield of the water-softening product. You have been given a small sample from the 575 L of new solution.

Background

When faced with problems that require them to determine the quantities of a substance by mass, chemists often turn to a technique called gravimetric analysis. In this technique, a small sample of the material undergoes a reaction with an excess of another reactant. The chosen reaction is one which almost always provides a yield near 100%. In other words, all of the reactant of unknown amount will be converted into product. If the mass of the product is carefully measured, you can use stoichiometry calculations such as those in Chapter 8 to determine how much of the reactant of unknown amount was involved in the reaction. Then, by comparing the size of the analysis sample with the size of the original material, you can determine exactly how much of the substance is present.

Professional chemists have expensive instruments available to directly measure amounts of a substance, but techniques such as gravimetric analysis are still useful when cost or equipment availability is an important consideration.

This procedure involves a double-displacement reaction between strontium chloride, $SrCl_2$, and sodium carbonate, Na_2CO_3. In general, this reaction can be used to determine the amount of any carbonate compound in a solution.

Problem

Remember that accurate results depend on precise mass measurements, so keep all glassware very clean, and do not lose any reactants or products during your lab work. You will react an unknown amount of sodium carbonate with an excess of strontium chloride. After purifying the product, you will determine the following.

- how much product is present
- how much Na_2CO_3 must have been present to produce that amount of product
- how much Na_2CO_3 is contained in the 575 L of solution

Objectives

Observe the double-displacement reaction between solutions of strontium chloride and sodium carbonate.

Demonstrate proficiency with gravimetric methods.

Measure the mass of insoluble precipitate formed.

Relate the mass of precipitate formed to the mass of reactants before the reaction.

Calculate the mass of sodium carbonate in a solution of unknown concentration.

706 | Exploration 8A

The Assessment section, ► on the Teacher's Interleaf, highlights portfolio projects that are found in the Chapter Review and Assess section.

assessment

Performance assessment

1. Design an experiment to measure the percent yields for the reactions listed below. If your teacher approves your design, acquire the necessary materials, and carry out your plan to obtain percent yield data.

   a. $Zn(s) + 2HCl(aq) \longrightarrow ZnCl_2(aq) + H_2(g)$

   b. $2NaHCO_3(s) \xrightarrow{\Delta} Na_2CO_3(s) + H_2O(g) + CO_2(g)$

   c. $CaCl_2(aq) + Na_2CO_3(aq) \longrightarrow CaCO_3(s) + 2NaCl(aq)$

   d. $NaOH(aq) + HCl(aq) \longrightarrow NaCl(aq) + H_2O(l)$

   (Note: use only dilute NaOH and HCl, less concentrated than 1.0 mol/liter.)

2. Your teacher will give you an index card specifying a volume of a gas. Reactants to make the gas will also be listed. Describe exactly how you would make the gas from the reactants. Include a method of collecting the gas without allowing it to mix with air. Then specify how much of each reactant you need. Choose a limiting reactant and explain your choice. If your teacher approves your plan, obtain the necessary materials and make the gas. (Hint: look up the density of the gas in a chemical handbook.)

Portfolio projects

1. ***Research and communication***
   Research the composition of gasoline sold in your area. Contact a gasoline company to discover what formulations are used. Investigate whether the mixtures change by season or by geographic area. Find out if your area has any guidelines regarding gasoline additives that reduce air pollution. Present your findings to the class.
2. ***Cooperative activity***
   Investigate corporate, governmental, or private use of alternative fuel sources for vehicles. Hold a class debate to compare the costs and environmental effects of these alternative fuels.
3. ***Chemistry and you***
   Visit a car maintenance shop to find out how you can help reduce air pollution by increasing your car's efficiency. Make a checklist of tasks to perform regularly to help meet this goal.
4. ***Research and communication***
   Research the production of the following pollutants: methane, CH_4; mercury, Hg; lead, Pb; chlorine, Cl_2; and sulfur oxides, SO_x. Determine the stoichiometry involved in the production of each of these chemicals. Contact the EPA, and gather information on what you can do to help reduce pollution by these chemicals.

Stoichiometry | 309

Assessment Options

Traditional Assessment
Test Generator
Instructional Objectives Measured:
Content mastery
Select from items 1 to 75

Performance Assessment Options
(see pages T36 and T643-15 for scoring rubrics)
Students demonstrate mastery of objectives in a hands-on environment. You may want some of these materials in the Portfolio as well.

Investigation 8A and 8B (text)
Instructional Objectives Measured:
Content mastery, Use of scientific methodology, Problem solving skills, Proficiency in written communication

Performance Assessments, page 309
1. Designing an experiment
Instructional Objectives Measured:
Content mastery, Use of scientific methodology, Problem solving skills, Proficiency in written communication

2. Designing an experiment
Instructional Objectives Measured:
Content mastery, Use of scientific methodology, Problem solving skills, Proficiency in written communication

Portfolio Options
(see page T36 for scoring rubrics)
Students provide a written rationale for each selection made for the portfolio.

Concept Maps
Students use vocabulary from the Chapter Review to build a concept map.
Instructional Objectives Measured:
Content mastery

Formal Laboratory Reports
Exploration 8A and 8B (text)
Instructional Objectives Measured:
Content mastery, Use of scientific methodology, Problem solving skills, Proficiency in written communication

Portfolio Projects, page 309
Items 1 and 4. Research and communication
Instructional Objectives Measured:
Scientific and chemical literacy, Understanding the impact of science and technology on society, Proficiency in written communication

Item 2. Cooperative activity
Instructional Objectives Measured:
Scientific and chemical literacy, Understanding the impact of science and technology on society, Proficiency in oral communication

Item 3. Chemistry and you
Instructional Objectives Measured:
Scientific and chemical literacy, Understanding the impact of science and technology on society, Proficiency in written communication

Assessment: Measuring what students can do

Performance assessment

Developing the wide variety of skills associated with scientific inquiry has been difficult to measure adequately in the past. With the unique, new laboratory program for *Holt Chemistry: Visualizing Matter* you have ready-made performance assessments that provide a meaningful measure of students' skills.

Chapter Review and Assess performance tasks

Items in the Chapter Review and Assess section expand the options you have for assigning additional performance tasks. These items are less structured than the Investigations but require some of the same developmental steps. If you assign any of these items, it will be necessary to approve any plans (including safety measures) before students begin work. It is suggested that you use the criteria for the pre-lab Investigations provided on page T643-15.

Assessment Options

Traditional Assessment
Test Generator
Instructional Objectives Measured:
Content mastery
Select from items 1 to 75

Performance Assessment Options
(see pages T36 and T643-15 for scoring rubrics)
Students demonstrate mastery of objectives in a hands-on environment. You may want some of these materials in the Portfolio as well.

Investigation 8A and 8B (text)
Instructional Objectives Measured:
Content mastery, Use of scientific methodology, Problem solving skills, Proficiency in written communication

Performance Assessments, page 309
1. **Designing an experiment**
   Instructional Objectives Measured:
   Content mastery, Use of scientific methodology, Problem solving skills, Proficiency in written communication
2. **Designing an experiment**
   Instructional Objectives Measured:
   Content mastery, Use of scientific methodology, Problem solving skills, Proficiency in written communication

Portfolio Options
(see page T36 for scoring rubrics)
Students provide a written rationale for each selection made for the portfolio.

Concept Maps
Students use vocabulary from the Chapter Review to build a concept map.
Instructional Objectives Measured:
Content mastery

Formal Laboratory Reports
Exploration 8A and 8B (text)
Instructional Objectives Measured:
Content mastery, Use of scientific methodology, Problem solving skills, Proficiency in written communication

Portfolio Projects, page 309
Items 1 and 4. Research and communication
Instructional Objectives Measured:
Scientific and chemical literacy, Understanding the impact of science and technology on society, Proficiency in written communication

Item 2. Cooperative activity
Instructional Objectives Measured:
Scientific and chemical literacy, Understanding the impact of science and technology on society, Proficiency in oral communication

Item 3. Chemistry and you
Instructional Objectives Measured:
Scientific and chemical literacy, Understanding the impact of science and technology on society, Proficiency in written communication

Alternative assessment

Performance assessment

1. Design an experiment to measure the percent yields for the reactions listed below. If your teacher approves your design, acquire the necessary materials, and carry out your plan to obtain percent yield data.

   a. $Zn(s) + 2HCl(aq) \longrightarrow ZnCl_2(aq) + H_2(g)$

   b. $2NaHCO_3(s) \xrightarrow{\Delta} Na_2CO_3(s) + H_2O(g) + CO_2(g)$

   c. $CaCl_2(aq) + Na_2CO_3(aq) \longrightarrow CaCO_3(s) + 2NaCl(aq)$

   d. $NaOH(aq) + HCl(aq) \longrightarrow NaCl(aq) + H_2O(l)$

   (Note: use only dilute NaOH and HCl, less concentrated than 1.0 mol/liter.)

2. Your teacher will give you an index card specifying a volume of a gas. Reactants to make the gas will also be listed. Describe exactly how you would make the gas from the reactants. Include a method of collecting the gas without allowing it to mix with air. Then specify how much of each reactant you need. Choose a limiting reactant and explain your choice. If your teacher approves your plan, obtain the necessary materials and make the gas. (Hint: look up the density of the gas in a chemical handbook.)

Portfolio pr

1. ***Research***
   Research the
   in your area.
   to discover w
   Investigate
   by season or
   if your area
   gasoline add
   Present your
2. ***Cooperati***
   Investigate
   private use
   vehicles. Hol
   the costs and
   alternative f
3. ***Chemistry***
   Visit a car maintenance shop to find out how you can help reduce air pollution by increasing your car's efficiency. Make a checklist of tasks to perform regularly to help meet this goal.
4. ***Research and communication***
   Research the production of the following pollutants: methane, CH_4; mercury, Hg; lead, Pb; chlorine, Cl_2; and sulfur oxides, SO_x. Determine the stoichiometry involved in the production of each of these chemicals. Contact the EPA, and gather information on what you can do to help reduce pollution by these chemicals.

Investigations performance tasks

Each Investigation in the text and laboratory manual is a full scientific inquiry that puts the student in the role of a chemist to solve a real-world problem. You can be assured that each Investigation provides a comprehensive set of performance tasks. Scoring the Investigations is easy using the rubrics on page T643-15.

Laboratory skills assessed

The scoring rubrics are based on the following skills.

- **identifying problems**
- **designing experimental investigations to explore the problem**
- **identifying and controlling variables**
- **following instructions**
- **selecting the appropriate equipment and materials**
- **using correct techniques**
- **using appropriate safety measures**
- **recording accurate observations (both qualitative and quantitative)**
- **drawing valid conclusions**
- **evaluating methods**
- **presenting coherent results**

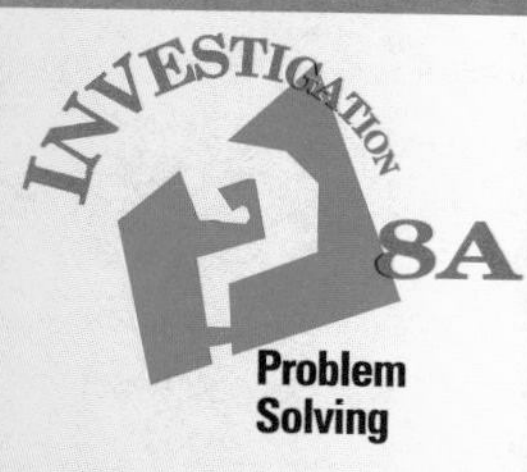

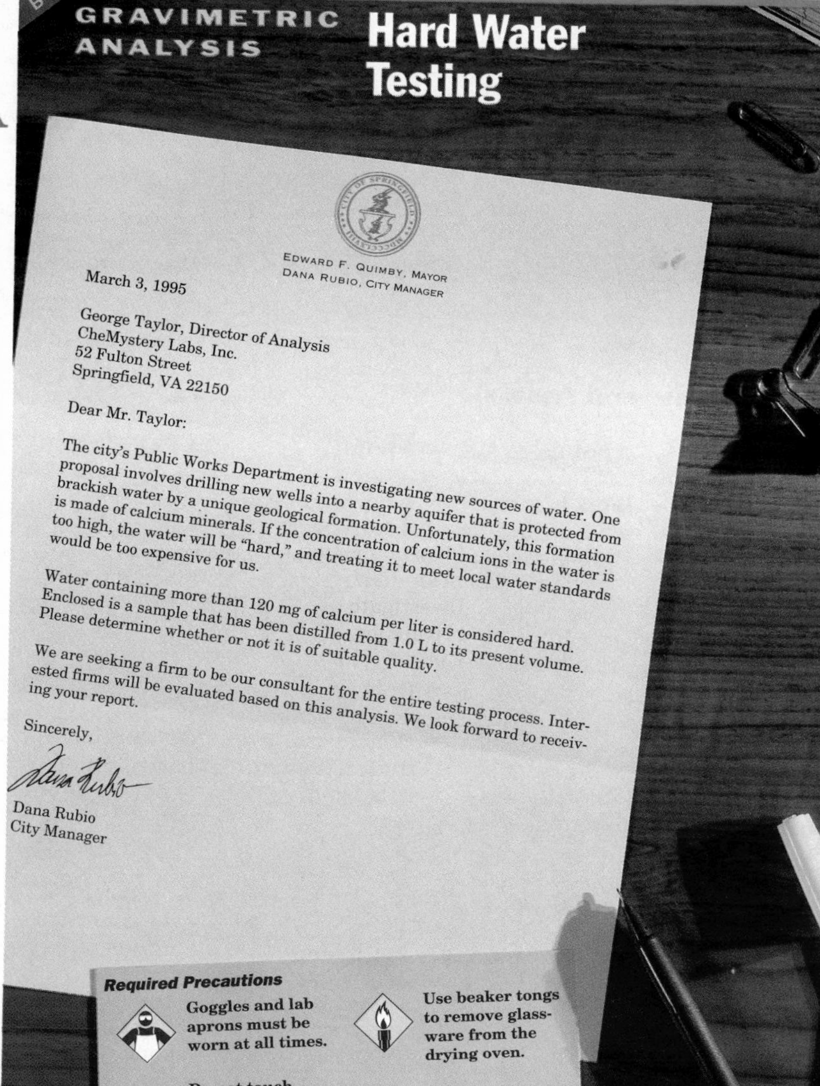

GRAVIMETRIC ANALYSIS

Hard Water Testing

EDWARD F. QUIMBY, MAYOR
DANA RUBIO, CITY MANAGER

March 3, 1995

George Taylor, Director of Analysis
CheMystery Labs, Inc.
52 Fulton Street
Springfield, VA 22150

Dear Mr. Taylor:

The city's Public Works Department is investigating new sources of water. One proposal involves drilling new wells into a nearby aquifer that is protected from brackish water by a unique geological formation. Unfortunately, this formation is made of calcium minerals. If the concentration of calcium ions in the water is too high, the water will be "hard," and treating it to meet local water standards would be too expensive for us.

Water containing more than 120 mg of calcium per liter is considered hard. Enclosed is a sample that has been distilled from 1.0 L to its present volume. Please determine whether or not it is of suitable quality.

We are seeking a firm to be our consultant for the entire testing process. Interested firms will be evaluated based on this analysis. We look forward to receiving your report.

Sincerely,

Dana Rubio
City Manager

Required Precautions

- Goggles and lab aprons must be worn at all times.
- Use beaker tongs to remove glassware from the drying oven.
- Do not touch or taste any chemicals. Wash your hands thoroughly when finished.

Memorandum

CheMystery Labs,
52 Fulton Street
Springfield, VA 22150

Date: March 4, 1995

To: Shane Thompson

From: George Taylor

We must do a very accurate and efficient job on this anal[...] this contract would be valuable for us in terms of both i[...] prestige. On the other hand, losing the contract to some [...] analysis firm would be awful!

We still don't have any capital expenditure funds for ela[...] equipment purchases, but we can solve this problem with [...] gravimetric analysis because calcium salts and carbonate [...] undergo double-displacement reactions to give insoluble c[...] carbonate as a precipitate.

Before you begin your work, I will need the following in[...] from you so that I can put together our bid.

- detailed one-page summary of your plan for the procedur[...] with all necessary data tables
- description of necessary calculations
- itemized list of equipment, with total costs (Our finan[...] planner tells me that even though we will bill the city [...] work, we can afford to spend only $200 000 on this pro[...]

After you complete the analysis, prepare a two-page report for Dana Rubio. Remember that this report will be seen by a variety of city officials, so be certain it projects the image we want to present. Make sure the following items are included.

- calculation of calcium concentration, in mg/L, for the aquifer water
- explanation of how you determined the amount of calcium in the sample, including measurements and calculations
- balanced chemical equation for the reaction
- explanations and estimations for any possible sources of error
- detailed invoice for services rendered and expenses incurred

Spill Procedures/ Waste Disposal Methods

- Solids must go in the trash can. Do not wash them down the sink.
- Liquids may be washed down the sink with an excess of water.
- Clean the area and all equipment after use.

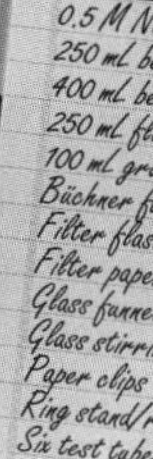

Assessment: Evaluating student's work

Explorations and investigations scoring rubrics

Because students are expected to present their laboratory plans and results based on specific guidelines for each experiment, a specific set of scoring rubrics for the laboratory program has been developed and can be found on page T643-15.

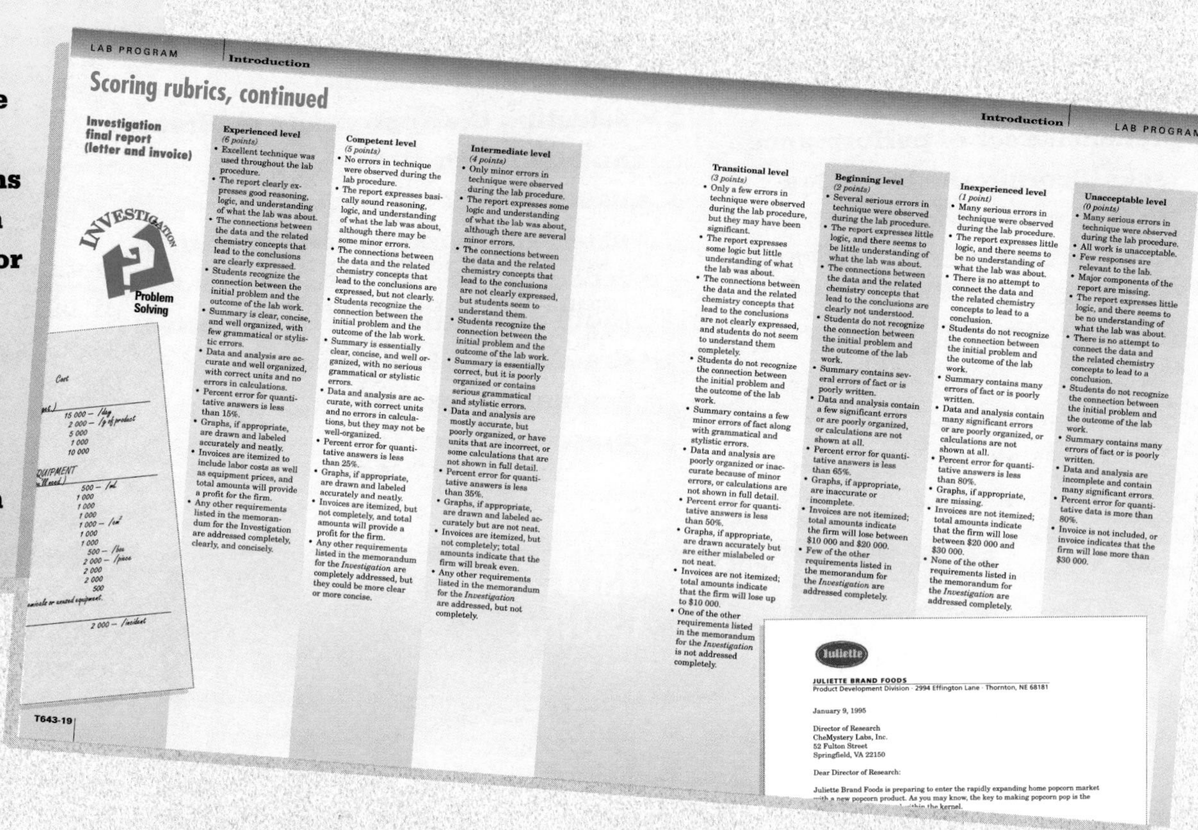

LAB PROGRAM | Introduction

Scoring rubrics, continued

Investigation final report (letter and invoice)

Experienced level
(6 points)
- Excellent technique was used throughout the lab procedure.
- The report clearly expresses good reasoning, logic, and understanding of what the lab was about.
- The connections between the data and the related chemistry concepts that lead to the conclusions are clearly expressed.
- Students recognize the connection between the initial problem and the outcome of the lab work.
- Summary is clear, concise, and well organized, with few grammatical or stylistic errors.
- Data and analysis are accurate and well organized, with correct units and no errors in calculations.
- Percent error for quantitative answers is less than 15%.
- Graphs, if appropriate, are drawn and labeled accurately and neatly.
- Invoices are itemized to include labor costs as well as equipment prices, and total amounts will provide a profit for the firm.
- Any other requirements listed in the memorandum for the Investigation are addressed completely, clearly, and concisely.

Competent level
(5 points)
- No errors in technique were observed during the lab procedure.
- The report expresses basically sound reasoning, logic, and understanding of what the lab was about, although there may be some minor errors.
- The connections between the data and the related chemistry concepts that lead to the conclusions are expressed, but not clearly.
- Students recognize the connection between the initial problem and the outcome of the lab work.
- Summary is essentially clear, concise, and well organized, with no serious grammatical or stylistic errors.
- Data and analysis are accurate, with correct units and no errors in calculations, but they may not be well-organized.
- Percent error for quantitative answers is less than 25%.
- Graphs, if appropriate, are drawn and labeled accurately and neatly.
- Invoices are itemized, but not completely, and total amounts will provide a profit for the firm.
- Any other requirements listed in the memorandum for the *Investigation* are completely addressed, but they could be more clear or more concise.

Intermediate level
(4 points)
- Only minor errors in technique were observed during the lab procedure.
- The report expresses some logic and understanding of what the lab was about, although there are several minor errors.
- The connections between the data and the related chemistry concepts that lead to the conclusions are not clearly expressed, but students seem to understand them.
- Students recognize the connection between the initial problem and the outcome of the lab work.
- Summary is essentially correct, but it is poorly organized or contains serious grammatical and stylistic errors.
- Data and analysis are mostly accurate, but poorly organized, or have units that are incorrect, or some calculations that are not shown in full detail.
- Percent error for quantitative answers is less than 35%.
- Graphs, if appropriate, are drawn and labeled accurately but are not neat.
- Invoices are itemized, but not completely; total amounts indicate that the firm will break even.
- Any other requirements listed in the memorandum for the *Investigation* are addressed, but not completely.

Transitional level
(3 points)
- Only a few errors in technique were observed during the lab procedure, but they may have been significant.
- The report expresses some logic but little understanding of what the lab was about.
- The connections between the data and the related chemistry concepts that lead to the conclusions are not clearly expressed, and students do not seem to understand them completely.
- Students do not recognize the connection between the initial problem and the outcome of the lab work.
- Summary contains a few minor errors of fact along with grammatical and stylistic errors.
- Data and analysis are poorly organized or inaccurate because of minor errors, or calculations are not shown in full detail.
- Percent error for quantitative answers is less than 50%.
- Graphs, if appropriate, are drawn accurately but are either mislabeled or not neat.
- Invoices are not itemized; total amounts indicate that the firm will lose up to $10 000.
- One of the other requirements listed in the memorandum for the *Investigation* is not addressed completely.

Beginning level
(2 points)
- Several serious errors in technique were observed during the lab procedure.
- The report expresses little logic, and there seems to be little understanding of what the lab was about.
- The connections between the data and the related chemistry concepts that lead to the conclusions are clearly not understood.
- Students do not recognize the connection between the initial problem and the outcome of the lab work.
- Summary contains several errors of fact or is poorly written.
- Data and analysis contain a few significant errors or are poorly organized, or calculations are not shown at all.
- Percent error for quantitative answers is less than 65%.
- Graphs, if appropriate, are inaccurate or incomplete.
- Invoices are not itemized; total amounts indicate the firm will lose between $10 000 and $20 000.
- Few of the other requirements listed in the memorandum for the *Investigation* are addressed completely.

Inexperienced level
(1 point)
- Many serious errors in technique were observed during the lab procedure.
- The report expresses little logic, and there seems to be no understanding of what the lab was about.
- There is no attempt to connect the data and the related chemistry concepts to lead to a conclusion.
- Students do not recognize the connection between the initial problem and the outcome of the lab work.
- Summary contains many errors of fact or is poorly written.
- Data and analysis contain many significant errors or are poorly organized, or calculations are not shown at all.
- Percent error for quantitative answers is less than 80%.
- Graphs, if appropriate, are missing.
- Invoices are not itemized; total amounts indicate that the firm will lose between $20 000 and $30 000.
- None of the other requirements listed in the memorandum for the *Investigation* are addressed completely.

Unacceptable level
(0 points)
- Many serious errors in technique were observed during the lab procedure.
- All work is unacceptable.
- Few responses are relevant to the lab.
- Major components of the report are missing.
- The report expresses little logic, and there seems to be no understanding of what the lab was about.
- There is no attempt to connect the data and the related chemistry concepts to lead to a conclusion.
- Students do not recognize the connection between the initial problem and the outcome of the lab work.
- Summary contains many errors of fact or is poorly written.
- Data and analysis are incomplete and contain many significant errors.
- Percent error for quantitative data is more than 80%.
- Invoice is not included, or invoice indicates that the firm will lose more than $30 000.

Juliette

JULIETTE BRAND FOODS
Product Development Division · 2994 Effington Lane · Thornton, NE 68181

January 9, 1995

Director of Research
CheMystery Labs, Inc.
52 Fulton Street
Springfield, VA 22150

Dear Director of Research:

Juliette Brand Foods is preparing to enter the rapidly expanding home popcorn market

T643-19

General performance tasks scoring rubric

(for items assigned from the Chapter Review and Assess)

6 Points	5 Points	4 Points	3 Points	2 Points	1 Point	0 Points
Work is superior, with no serious inaccuracies or inaccuracies are minor.	Work is effective, with no serious inaccuracies and few minor inaccuracies.	Work is satisfactory, with few serious inaccuracies and some minor ones, or with many minor inaccuracies.	Work is marginal, with some serious inaccuracies and some minor ones.	Work is unsatisfactory, with many serious inaccuracies.	Work is poor, with many serious inaccuracies.	No relevant responses were given.
Conceptual understanding is clearly demonstrated throughout the performance of the task.	Conceptual understanding is apparent throughout the student's work.	Conceptual understanding is somewhat apparent.	Conceptual understanding is not always apparent.	Conceptual understanding is lacking.	Conceptual understanding is not apparent.	
Reasoning is logical throughout.	Reasoning is generally logical.	Reasoning is somewhat weak from a logic perspective.	Reasoning is not always logical.	Reasoning is very weak.		
Communication is exemplary.	Communication is effective.	Communication is satisfactory.	Communication is marginal.	Communication is unsatisfactory.	Communication is very poor.	

Analytic scale for evaluating student writing

Assign a value of 1 to 4 for each of the stated criteria based on how well the project meets the criteria. Total the values and assign a grade based on a low score of 15 and a high score of 60.

4 Points
The paper clearly meets this standard.

3 Points
The paper indicates that the writer has made a serious effort and has been fairly successful.

2 Points
The paper indicates that the writer has made some effort, but with little success.

1 Point
The paper clearly does not achieve this standard.

Content

❍ The writing is likely to interest the intended audience.
❍ The writing achieves a clear purpose.
❍ The writing is unified and coherent.
❍ The writing reflects a thorough exploration of the subject and stays within the constraints of the project.
❍ The writing does not contain unrelated or distracting details.
❍ The writing effectively answers the question or problem stated for the project.

Organization

❍ The writing has a clear structure.
❍ The writing presents ideas and details in an effective order.
❍ The writing shows a clear connection between ideas.

Style

❍ The language suits the topic, audience, purpose, and occasion.
❍ The sentences are well worded, and not awkward.
❍ The paper does not reflect wordiness, clichés, unnecessary jargon, mixed metaphors, or other stylistic pitfalls.
❍ The writer's meaning is clear throughout the paper.

Grammar, Usage, and Mechanics

❍ The paper is relatively free of grammar and usage errors.
❍ The paper is relatively free of mechanical flaws and spelling errors.

Informal assessment direct observation checklist for groupwork

This checklist can be used for any Groupwork Strategies that you may assign and as a part of your evaluation of student's work in the lab. Assign a value from 1 to 4 to each of the stated criteria based on how well the group or individual group member meets the criteria. Total the values, and assign a grade.

4 Points
Strongly agree

3 Points
Agree

2 Points
Somewhat disagree

1 Point
Strongly disagree

❍ The group as a whole did a good job of staying on task.
❍ Each member of the group had a clearly stated role.
❍ Each member of the group contributed something useful to the group.
❍ Each member of the group treated the other members with respect.
❍ Each member of the group learned something from the assignment.

A Teacher's Annotated Edition with the help you need, where you need it

Planning guides, instructional strategies, and fully worked-out solutions to problems are at your fingertips.

Help for *planning* more effectively

Chapter interleaf pages are your organization guide to all the resources of the *Holt Chemistry: Visualizing Matter* program.

1 Chapter Overview
gives a quick synopsis of the chapter.

2 Concept Base
lists topics that should be covered prior to starting the chapter.

3 Looking Ahead
gives you a preview of related concepts in upcoming chapters.

4 Laboratory Equipment Needs
is your guide to finding the lists of materials and supplies you'll need for the demonstrations and lab experiences.

5 Planning Guide
gives you an overview of all the program resources categorized by text section.

6 Assessment Options
lists the many program resources that can be used for traditional, performance, and portfolio assessments. Each selection is designed to measure attainment of one or more of the instructional goals presented on page T18.

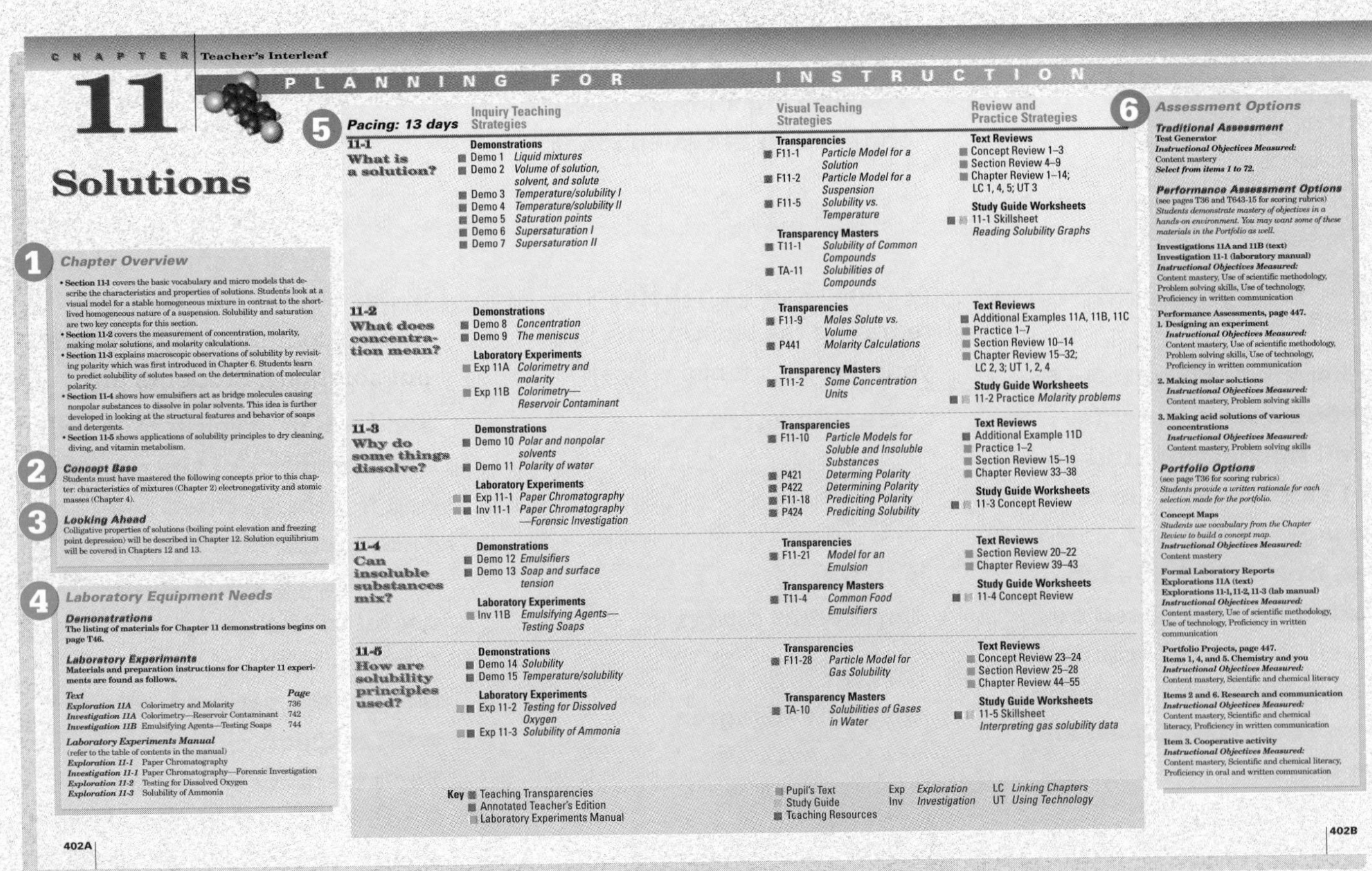

CHAPTER Teacher's Interleaf

11

Solutions

PLANNING FOR INSTRUCTION

1 Chapter Overview

- **Section 11-1** covers the basic vocabulary and micro models that describe the characteristics and properties of solutions. Students look at a visual model for a stable homogeneous mixture in contrast to the short-lived homogeneous nature of a suspension. Solubility and saturation are two key concepts for this section.
- **Section 11-2** covers the measurement of concentration, molarity, making molar solutions, and molarity calculations.
- **Section 11-3** explains macroscopic observations of solubility by revisiting polarity which was first introduced in Chapter 6. Students learn to predict solubility of solutes based on the determination of molecular polarity.
- **Section 11-4** shows how emulsifiers act as bridge molecules causing nonpolar substances to dissolve in polar solvents. This idea is further developed in looking at the structural features and behavior of soaps and detergents.
- **Section 11-5** shows applications of solubility principles to dry cleaning, diving, and vitamin metabolism.

2 Concept Base
Students must have mastered the following concepts prior to this chapter: characteristics of mixtures (Chapter 2) electronegativity and atomic masses (Chapter 4).

3 Looking Ahead
Colligative properties of solutions (boiling point elevation and freezing point depression) will be described in Chapter 12. Solution equilibrium will be covered in Chapters 12 and 13.

4 Laboratory Equipment Needs

Demonstrations
The listing of materials for Chapter 11 demonstrations begins on page T46.

Laboratory Experiments
Materials and preparation instructions for Chapter 11 experiments are found as follows.

Text		*Page*
Exploration 11A	Colorimetry and Molarity	736
Investigation 11A	Colorimetry—Reservoir Contaminant	742
Investigation 11B	Emulsifying Agents—Testing Soaps	744

Laboratory Experiments Manual
(refer to the table of contents in the manual)

Exploration 11-1	Paper Chromatography
Investigation 11-1	Paper Chromatography—Forensic Investigation
Exploration 11-2	Testing for Dissolved Oxygen
Exploration 11-3	Solubility of Ammonia

402A

5 Pacing: 13 days

Section	Inquiry Teaching Strategies	Visual Teaching Strategies	Review and Practice Strategies
11-1 What is a solution?	**Demonstrations** Demo 1 *Liquid mixtures* Demo 2 *Volume of solution, solvent, and solute* Demo 3 *Temperature/solubility I* Demo 4 *Temperature/solubility II* Demo 5 *Saturation points* Demo 6 *Supersaturation I* Demo 7 *Supersaturation II*	**Transparencies** F11-1 *Particle Model for a Solution* F11-2 *Particle Model for a Suspension* F11-5 *Solubility vs. Temperature* **Transparency Masters** T11-1 *Solubility of Common Compounds* TA-11 *Solubilities of Compounds*	**Text Reviews** Concept Review 1–3 Section Review 4–9 Chapter Review 1–14; LC 1, 4, 5; UT 3 **Study Guide Worksheets** 11-1 Skillsheet *Reading Solubility Graphs*
11-2 What does concentration mean?	**Demonstrations** Demo 8 *Concentration* Demo 9 *The meniscus* **Laboratory Experiments** Exp 11A *Colorimetry and molarity* Exp 11B *Colorimetry—Reservoir Contaminant*	**Transparencies** F11-9 *Moles Solute vs. Volume* P441 *Molarity Calculations* **Transparency Masters** T11-2 *Some Concentration Units*	**Text Reviews** Additional Examples 11A, 11B, 11C Practice 1–7 Section Review 10–14 Chapter Review 15–32; LC 2, 3; UT 1, 2, 4 **Study Guide Worksheets** 11-2 Practice *Molarity problems*
11-3 Why do some things dissolve?	**Demonstrations** Demo 10 *Polar and nonpolar solvents* Demo 11 *Polarity of water* **Laboratory Experiments** Exp 11-1 *Paper Chromatography* Inv 11-1 *Paper Chromatography —Forensic Investigation*	**Transparencies** F11-10 *Particle Models for Soluble and Insoluble Substances* P421 *Determing Polarity* P422 *Determining Polarity* F11-18 *Prediciting Polarity* P424 *Prediciting Solubility*	**Text Reviews** Additional Example 11D Practice 1–2 Section Review 15–19 Chapter Review 33–38 **Study Guide Worksheets** 11-3 Concept Review
11-4 Can insoluble substances mix?	**Demonstrations** Demo 12 *Emulsifiers* Demo 13 *Soap and surface tension* **Laboratory Experiments** Inv 11B *Emulsifying Agents—Testing Soaps*	**Transparencies** F11-21 *Model for an Emulsion* **Transparency Masters** T11-4 *Common Food Emulsifiers*	**Text Reviews** Section Review 20–22 Chapter Review 39–43 **Study Guide Worksheets** 11-4 Concept Review
11-5 How are solubility principles used?	**Demonstrations** Demo 14 *Solubility* Demo 15 *Temperature/solubility* **Laboratory Experiments** Exp 11-2 *Testing for Dissolved Oxygen* Exp 11-3 *Solubility of Ammonia*	**Transparencies** F11-28 *Particle Model for Gas Solubility* **Transparency Masters** TA-10 *Solubilities of Gases in Water*	**Text Reviews** Concept Review 23–24 Section Review 25–28 Chapter Review 44–55 **Study Guide Worksheets** 11-5 Skillsheet *Interpreting gas solubility data*

Key Teaching Transparencies; Annotated Teacher's Edition; Laboratory Experiments Manual; Pupil's Text; Study Guide; Teaching Resources; Exp *Exploration*; Inv *Investigation*; LC *Linking Chapters*; UT *Using Technology*

6 Assessment Options

Traditional Assessment
Test Generator
Instructional Objectives Measured:
Content mastery
Select from items 1 to 72.

Performance Assessment Options
(see pages T36 and T643-15 for scoring rubrics)
Students demonstrate mastery of objectives in a hands-on environment. You may want some of these materials in the Portfolio as well.

Investigations 11A and 11B (text)
Investigation 11-1 (laboratory manual)
Instructional Objectives Measured:
Content mastery, Use of scientific methodology, Problem solving skills, Use of technology, Proficiency in written communication

Performance Assessments, page 447.
1. **Designing an experiment**
   Instructional Objectives Measured:
   Content mastery, Use of scientific methodology, Problem solving skills, Use of technology, Proficiency in written communication
2. **Making molar solutions**
   Instructional Objectives Measured:
   Content mastery, Problem solving skills
3. **Making acid solutions of various concentrations**
   Instructional Objectives Measured:
   Content mastery, Problem solving skills

Portfolio Options
(see page T36 for scoring rubrics)
Students provide a written rationale for each selection made for the portfolio.

Concept Maps
Students use vocabulary from the Chapter Review to build a concept map.
Instructional Objectives Measured:
Content mastery

Formal Laboratory Reports
Explorations 11A (text)
Explorations 11-1, 11-2, 11-3 (lab manual)
Instructional Objectives Measured:
Content mastery, Use of scientific methodology, Use of technology, Proficiency in written communication

Portfolio Projects, page 447.
Items 1, 4, and 5. Chemistry and you
Instructional Objectives Measured:
Content mastery, Scientific and chemical literacy

Items 2 and 6. Research and communication
Instructional Objectives Measured:
Content mastery, Scientific and chemical literacy, Proficiency in written communication

Item 3. Cooperative activity
Instructional Objectives Measured:
Content mastery, Scientific and chemical literacy, Proficiency in oral and written communication

402B

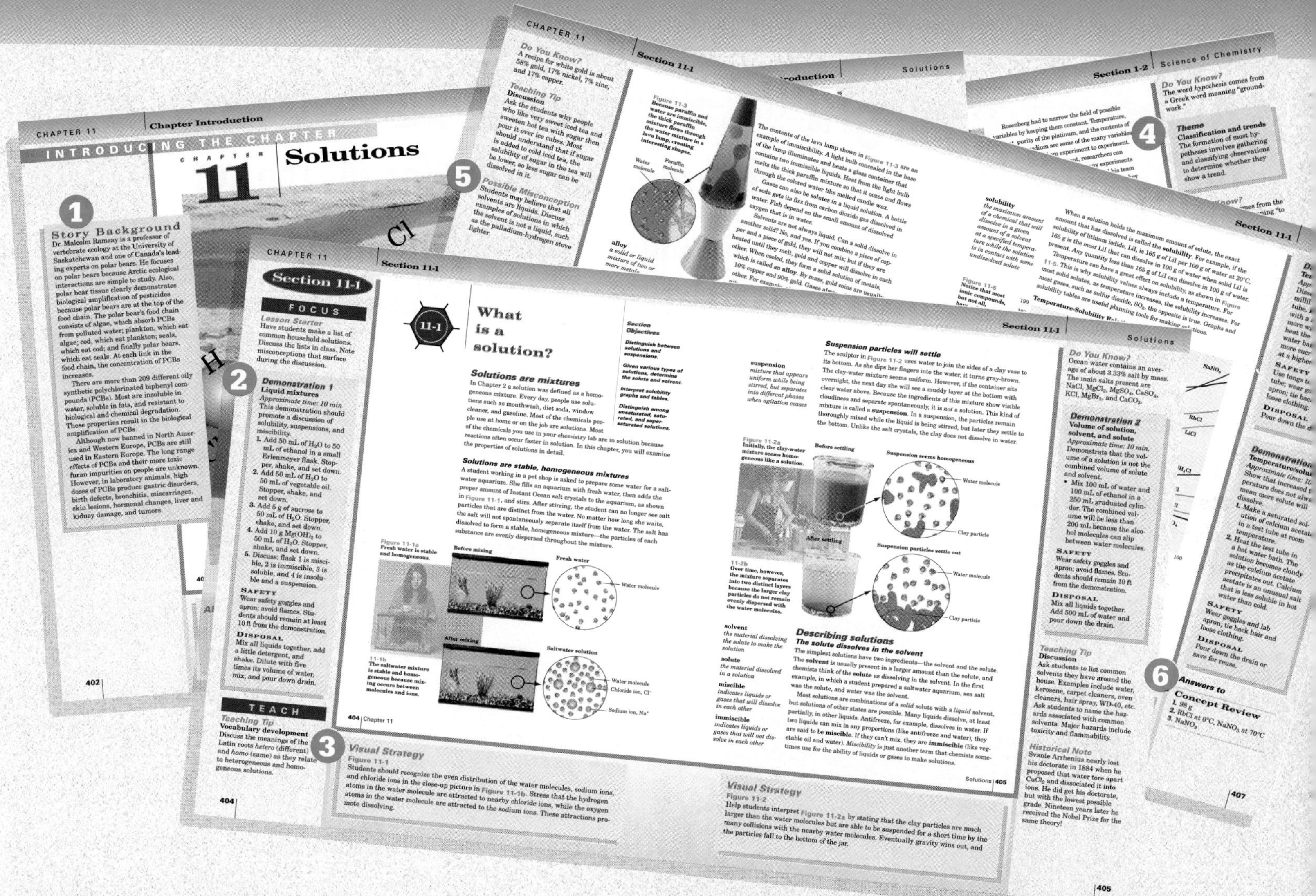

Help for *teaching* more effectively

Notes along the side and bottom margins provide concrete instructional strategies designed to increase student understanding and retention. Some of the more unique elements of the margins are as follows.

1 Story Background
gives you some additional information about the story theme of the chapter.

2 Demonstrations
provide the critical firsthand observations students often need to make sense of a concept. Thorough safety and disposal information is also provided.

3 Visual Strategies
provide lecture prompts for integrating the illustrations into your discussion of a concept.

4 Themes
provide lecture prompts for linking content to previously learned concepts.

5 Possible Misconceptions
alert you to work toward dispelling students' faulty concepts of an idea or topic.

6 Answers
for all problems and questions you assign students are found in this book. Your Teacher's Edition contains fully worked-out solutions. Students are given abbreviated answers to selected problems at the back of their text.

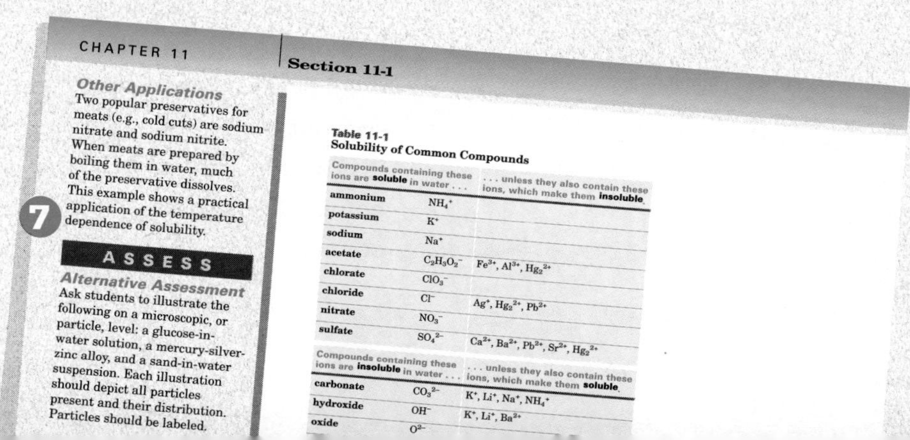

CHAPTER 11 | Section 11-1

Other Applications
Two popular preservatives for meats (e.g., cold cuts) are sodium nitrate and sodium nitrite. When meats are prepared by boiling them in water, much of the preservative dissolves. This example shows a practical application of the temperature dependence of solubility.

7

ASSESS

Alternative Assessment
Ask students to illustrate the following on a microscopic, or particle, level: a glucose-in-water solution, a mercury-silver-zinc alloy, and a sand-in-water suspension. Each illustration should depict all particles present and their distribution. Particles should be labeled.

Table 11-1
Solubility of Common Compounds

Compounds containing these ions are **soluble** in water . . .		. . . unless they also contain these ions, which make them **insoluble**.
ammonium	NH_4^+	
potassium	K^+	
sodium	Na^+	
acetate	$C_2H_3O_2^-$	Fe^{3+}, Al^{3+}, Hg_2^{2+}
chlorate	ClO_3^-	
chloride	Cl^-	Ag^+, Hg_2^{2+}, Pb^{2+}
nitrate	NO_3^-	
sulfate	SO_4^{2-}	Ca^{2+}, Ba^{2+}, Pb^{2+}, Sr^{2+}, Hg_2^{2+}
Compounds containing these ions are insoluble in water . . .		**. . . unless they also contain these ions, which make them soluble.**
carbonate	CO_3^{2-}	K^+, Li^+, Na^+, NH_4^+
hydroxide	OH^-	K^+, Li^+, Ba^{2+}
oxide	O^{2-}	

Help for *assessing* student progress more effectively

7 Alternative assessment
closes each text section with an idea that can be used to supplement the Section Reviews.

A chemistry program has never been this easy to use

The Teaching Resources with Organizer provides all the core resources you need in one convenient management system.

The Organizer

This portfolio provides a convenient place to store and carry all the *Holt Chemistry: Visualizing Matter* teaching resource materials for a chapter, including your own notes.

Study Guide

This student workbook is designed to provide the necessary review and practice students need to supplement the materials of the textbook in three critical areas—concept review, skill reinforcement, and problem-solving practice.

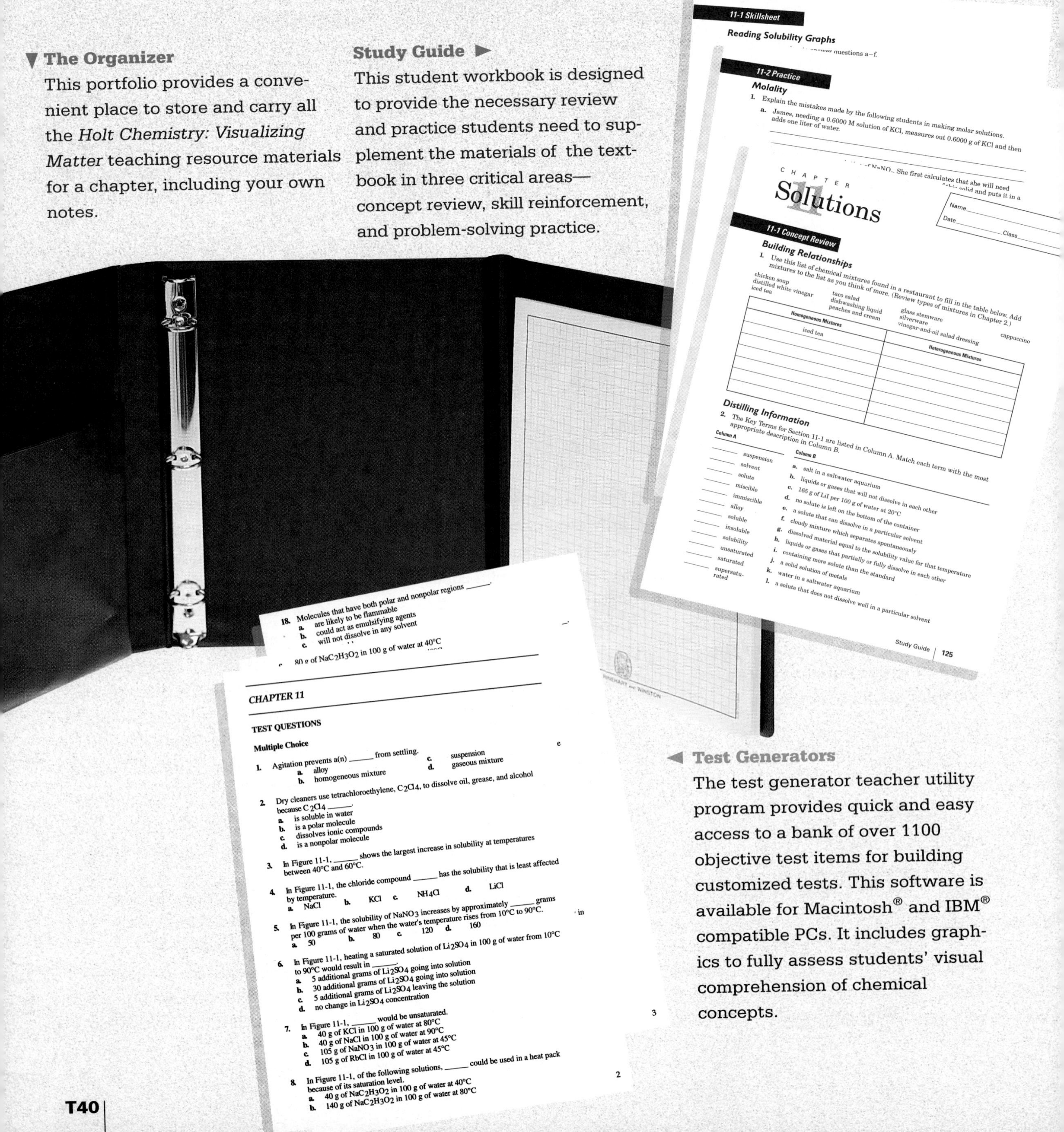

Test Generators

The test generator teacher utility program provides quick and easy access to a bank of over 1100 objective test items for building customized tests. This software is available for Macintosh® and IBM® compatible PCs. It includes graphics to fully assess students' visual comprehension of chemical concepts.

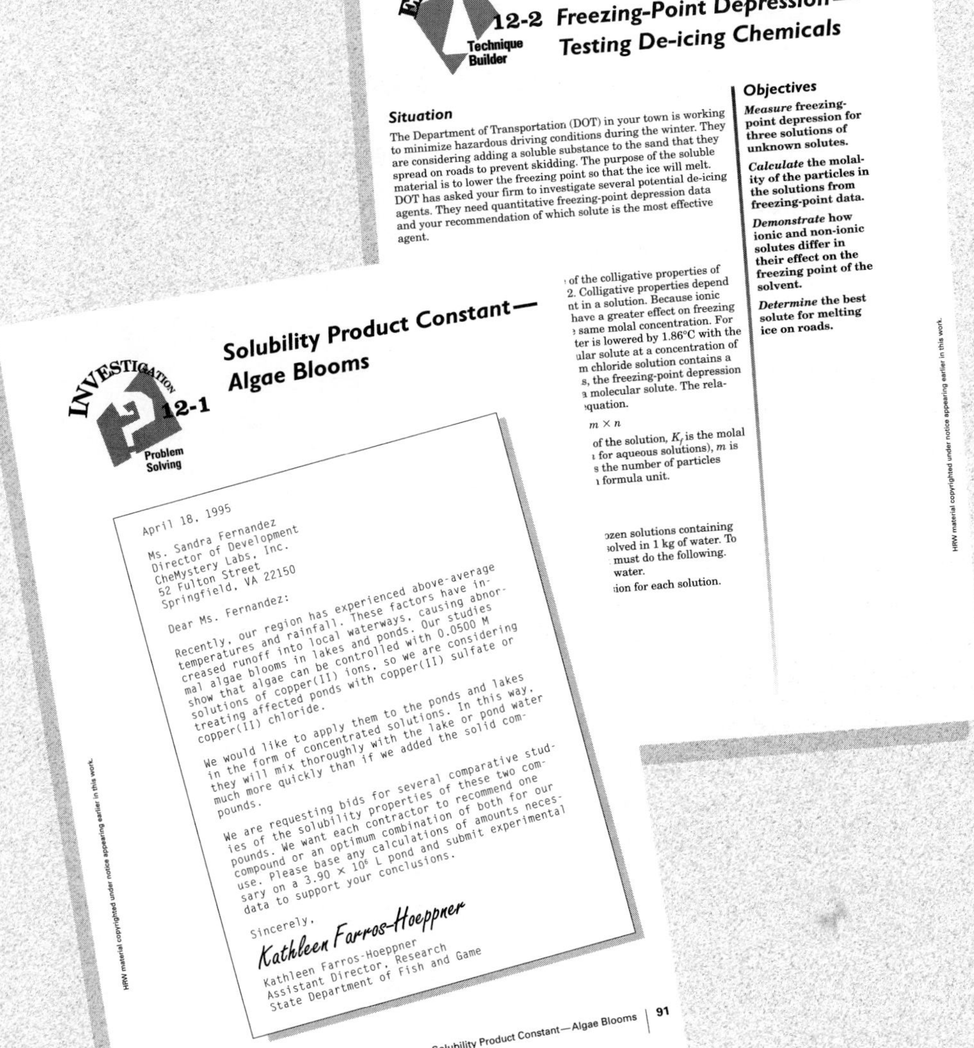

EXPLORATION 12-2 Technique Builder

Freezing-Point Depression—Testing De-icing Chemicals

Situation

The Department of Transportation (DOT) in your town is working to minimize hazardous driving conditions during the winter. They are considering adding a soluble substance to the sand that they spread on roads to prevent skidding. The purpose of the soluble material is to lower the freezing point so that the ice will melt. DOT has asked your firm to investigate several potential de-icing agents. They need quantitative freezing-point depression data and your recommendation of which solute is the most effective agent.

Objectives

Measure freezing-point depression for three solutions of unknown solutes.

Calculate the molality of the particles in the solutions from freezing-point data.

Demonstrate how ionic and non-ionic solutes differ in their effect on the freezing point of the solvent.

Determine the best solute for melting ice on roads.

INVESTIGATION 12-1 Problem Solving

Solubility Product Constant—Algae Blooms

April 18, 1995

Ms. Sandra Fernandez
Director of Development
CheMystery Labs, Inc.
52 Fulton Street
Springfield, VA 22150

Dear Ms. Fernandez:

Recently, our region has experienced above-average temperatures and rainfall. These factors have increased runoff into local waterways, causing abnormal algae blooms in lakes and ponds. Our studies show that algae can be controlled with 0.0500 M solutions of copper(II) ions, so we are considering treating affected ponds with copper(II) sulfate or copper(II) chloride.

We would like to apply them to the ponds and lakes in the form of concentrated solutions. In this way, they will mix thoroughly with the lake or pond water much more quickly than if we added the solid compounds.

We are requesting bids for several comparative studies of the solubility properties of these two compounds. We want each contractor to recommend one compound or an optimum combination of both for our use. Please base any calculations of amounts necessary on a 3.90×10^6 L pond and submit experimental data to support your conclusions.

Sincerely,

Kathleen Farros-Hoeppner

Kathleen Farros-Hoeppner
Assistant Director, Research
State Department of Fish and Game

Solubility Product Constant—Algae Blooms 91

◄ Laboratory Experiments

The supplemental student laboratory manual gives you 34 additional Explorations and Investigations with the same innovative design and structure that you find in the student text. These experiments are designed to supplement your existing laboratory program, and as such, the manual contains some experiments that require more sophisticated materials and procedures. As in the text, safety is emphasized, along with the recommendation that an Investigation should not be used without its accompanying Exploration. The laboratory manual is set up to be nonconsumable, so you can purchase reusable classroom sets or make multiple copies of the blackline masters as you need them.

The Laboratory Experiments Teacher's Edition provides the same detailed instructions and teacher support as those found in the Teacher's Edition of the text.

Teaching Transparencies ►

Concept development in the *Holt Chemistry: Visualizing Matter* program depends as heavily on the illustrations as it does on the text. To assist you in effectively using these important diagrams, the Teaching Transparencies package contains 100 color transparencies and over 80 black-and-white transparency masters of text illustrations.

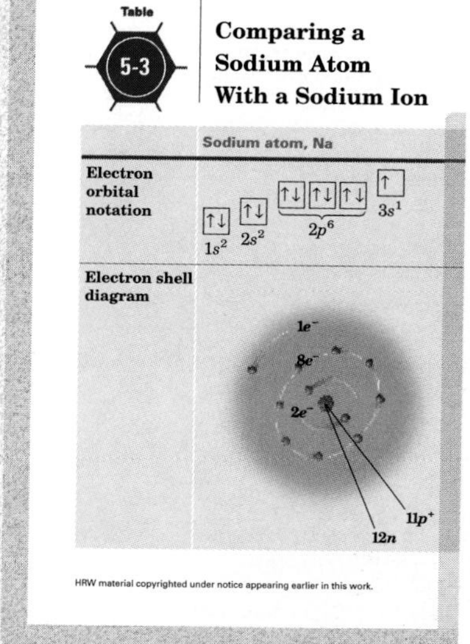

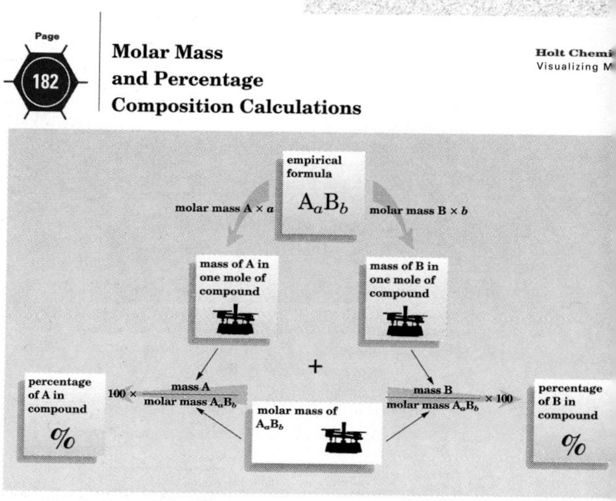

Using calculators and computers in the chemistry classroom

Calculators

It is assumed that students will be using calculators throughout the course. Though a scientific calculator is useful, most calculations require no more than a four-function calculator. It is assumed that students will have had adequate practice in using a calculator before they take chemistry. You may want to review calculations with numbers expressed in scientific notation because these may not be familiar to students.

Our philosophy is that it is more important to evaluate a student's ability to build and execute a rational plan for solving a problem than to simply evaluate the accuracy of a computation.

Programmable calculators provide the added advantage of requiring students to carefully plan the series of steps and operations needed to solve a problem. It is suggested that programming functions be used only after the student displays the appropriate conceptual knowledge base for the topic. Using them before this point will only increase a student's confusion and frustration. Programmable calculators may be used best in cooperative, group problem-solving sessions, a setting in which the programming task will not seem so formidable.

Step 1: $NO + NO \rightleftharpoons N_2O_2$

Step 2: $N_2O_2 + H_2 \rightleftharpoons N_2O + H_2O$

Step 3: $N_2O + H_2 \rightleftharpoons N_2 + H_2O$

When reactions are added, the K_{eq} of the overall reaction is equal to the product of the K_{eq} expressions for all the steps.

a. The K_{eq} expression for step 2 is as follows.

$$K_{eq} = \frac{[N_2O][H_2O]}{[N_2O_2][H_2]}$$

Explain the presence of water in the K_{eq} expression.

b. Write K_{eq} expressions for steps **1** and **3**, and determine the form of the K_{eq} expression for the overall reaction.

c. Write the balanced chemical equation for the overall reaction.

d. Why do concentration terms for the intermediates not appear in the K_{eq} expressions for the overall equation?

10. Theme: ***Systems and interactions***
Iodine monochloride, ICl, a substance used to determine the iodine value of fats and oils, is a gas above 92°C. ICl reacts with hydrogen gas to form iodine gas and hydrogen chloride gas.

$$ICl(g) + H_2(g) \longrightarrow I_2(g) + 2HCl(g)$$

The rate law for this reaction is: rate = $k[ICl][H_2]$. At 230°C k has a value of 0.163 L/mol•s. Raising the temperature to 240°C increases the value of k to 0.348 L/mol•s.

a. If the initial concentration for both ICl and H_2 is 0.50 M, compare the rates at 230°C and 240°C.

b. Compare the rates at 230°C for reactant concentrations of 0.50 M and 0.25 M.

c. Which has a greater effect on the reaction rate, a 10°C increase in temperature or a doubling of the concentration?

USING TECHNOLOGY

1. ***Graphics calculator***
Verify the decomposition of NO_2 is second order with respect to $[NO_2]$ by plotting the folowing reaction data.

Begin by clearing lists L1 through L4. Press [STAT] [4] [2nd] [1] [ENTER]. Repeat this key sequence three more times except press [2], [3], or [4] instead of [1]. Now enter the data you will plot. L_1 will contain time data, L_2 concentration data, L_3 log concentration data, and L_4 the inverse concentration data. Press [STAT] [1]. Press [0], because this is the first number in

Time (s)	$[NO_2]$
0	1.00×10^{-2}
60	0.683×10^{-2}
120	0.518×10^{-2}
180	0.418×10^{-2}
240	0.350×10^{-2}
300	0.301×10^{-2}
360	0.264×10^{-2}

[WINDOW] [▼] [0] [▼] [375] [▼] [10] [▼] [0] [▼] [0] [.] [0] [1] [▼] [0] [.] [0] [0] [0] [1]. To plot concentration vs. time press [2nd] [Y=] [1] [ENTER] [▼] [▶] [ENTER] [▼] [▶] [ENTER] [▼] [ENTER]. Reset the range limits and ***y***-axis scaling. Ymin = –3, Ymax = –2, and Yscl = 0.001. Plot log $[NO_2]$ vs time by pressing [Y=] [2nd] [2] [ENTER] [▼] [▶] [▶] [ENTER] [▼] [▶] [ENTER] [▼] [▶] [▶] [▶] [ENTER] [GRAPH]. Reset range limits and ***y***-axis scaling to plot $1/[NO_2]$ vs. time. Select plot 3 at the STAT PLOTS screen and set parameters. Press [GRAPH].

2. ***Graphics Calculator***
A reaction involving a single reactant, A, has the rate data shown to the right. Determine the order of the reaction by graphing the data. Use the instructions in item **1** as a guide.

Time (min)	[A]
5.0	0.100
5.63	0.090
6.16	0.080
7.00	0.070

3. ***Graphics Calculator***

Graphics calculators

With the changes in mathematics instruction, you will find over the next 10 years that many students will enter your chemistry class with a demonstrated proficiency in the use of the graphics calculator. Using *Holt Chemistry: Visualizing Matter* you can take advantage of this knowledge and skill by assigning the graphics calculator problems from the Chapter Review and Assess section. These problems provide an added dimension to some of the concepts presented in the chapter and further reinforce the links among data, graphs, and mathematical models.

The graphics calculator problems in this text are based on the functions of the Texas Instruments TI-82 graphics calculator. Instructions may need modification, depending on the type of calculator used.

◀ **Using Technology** sections in each Chapter review include items requiring the graphics calculator and computer software.

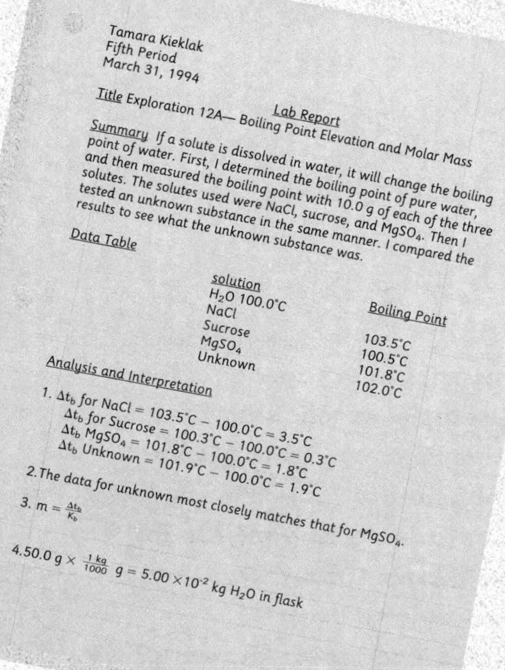

Tamara Kieklak
Fifth Period
March 31, 1994

Lab Report

Title Exploration 12A— Boiling Point Elevation and Molar Mass

Summary If a solute is dissolved in water, it will change the boiling point of water. First, I determined the boiling point of pure water, and then measured the boiling point with 10.0 g of each of the three solutes. The solutes used were NaCl, sucrose, and $MgSO_4$. Then I tested an unknown substance in the same manner. I compared the results to see what the unknown substance was.

Data Table

solution	Boiling Point
H_2O 100.0°C	
NaCl	103.5°C
Sucrose	100.5°C
$MgSO_4$	101.8°C
Unknown	102.0°C

Analysis and Interpretation

1. Δt_b for NaCl = 103.5°C − 100.0°C = 3.5°C
   Δt_b for Sucrose = 100.3°C − 100.0°C = 0.3°C
   Δt_b $MgSO_4$ = 101.8°C − 100.0°C = 1.8°C
   Δt_b Unknown = 101.9°C − 100.0°C = 1.9°C
2. The data for unknown most closely matches that for $MgSO_4$.
3. $m = \frac{\Delta t_b}{K_b}$
4. $50.0\ g \times \frac{1\ kg}{1000}\ g = 5.00 \times 10^{-2}$ kg H_2O in flask

Computer word processing programs

The unique nature of the *Holt Chemistry: Visualizing Matter* laboratory program provides a golden opportunity to integrate language-arts skills with science. We recommend that students use computers to prepare the various reports required for the labs. This practice gives students the opportunity to improve their organizational skills by setting up consistent formats for certain reports. In addition, students can exhibit their creative-writing talents in the reports required for the Investigations.

Computer spreadsheets and drawing programs

Commercial spreadsheet programs can be used in the same way as a programmable calculator. Spreadsheets can also be used for performing calculations to complete data tables for a laboratory experiment. Lab groups can be assigned the responsibility of setting up the programmed calculations for a particular experiment on a rotating basis. This strategy gets students more involved in the mechanics of the experiments and generally makes the results far more meaningful for the group that does the programming.

The Chapter Review and Assess sections throughout the text contain some items requiring the use of a computer drawing program. Though these items are optional, they provide students with opportunities to visually display their comprehension of a concept. If you use concept mapping as a study strategy, you may find that students enjoy preparing maps on the computer.

f(g) Scholar™ for *Holt Chemistry: Visualizing Matter*

f(g) Scholar™ is an integrated computer-based utility software program that combines the functions of a scientific calculator, spreadsheet, graphics calculator, and drawing package. It is appropriate to use *f(g)* Scholar™ for any problem in the text that requires a calculator, graphics calculator, or spreadsheet.

All the functions of *f(g)* Scholar™ can be imported into any popular word processing program. Therefore, this software is ideal for creating high-quality lab reports for the *Holt Chemistry: Visualizing Matter* Explorations and Investigations. *f(g)* Scholar™ can be used to create data tables or data can be entered into a preprogrammed spreadsheet for calculations. Data can be graphed and customized. Graphics can be accessed from the clip-art libraries of laboratory equipment or they can be drawn directly. All files can be exported to the word processing software to create the final report.

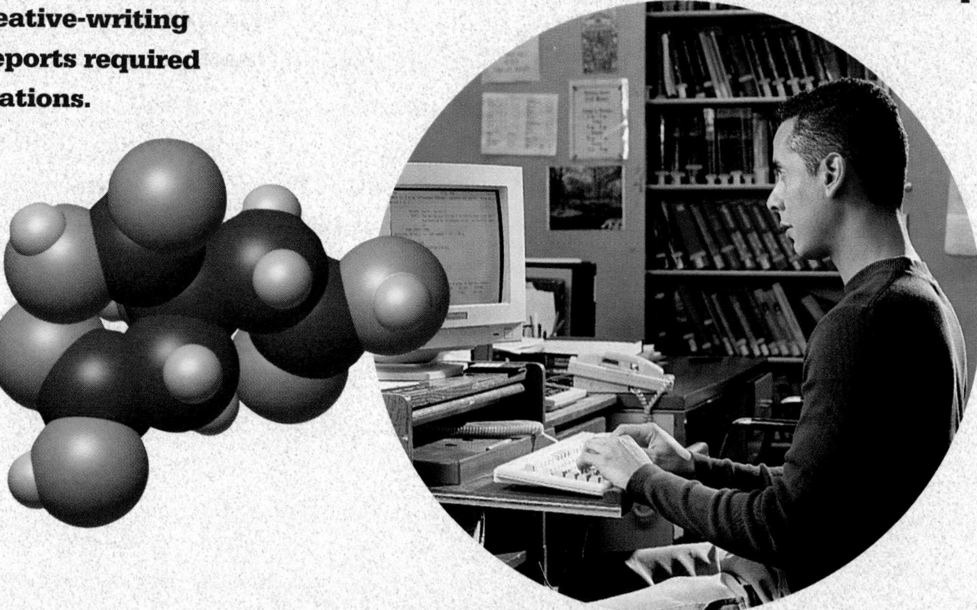

Reading and concept mapping in the chemistry classroom

Research has shown that the time spent helping students read technical material is critical to increasing comprehension and conceptual understanding. Two strategies are particularly important in the prereading phase of instruction: building on prior knowledge and establishing a purpose for reading.

Building on prior knowledge

The amount of prior knowledge students have about a topic directly influences their comprehension. *Help students identify the information they already have about a topic before they begin a chapter.* This activity assists them in relating new information to what they already know. The Tapping Prior Knowledge questions in the Teacher's Annotated Edition are designed to give you an idea of what students may already know about the topic to be covered.

The Possible Misconception sections in the Teacher's Annotated Edition alert you to common misconceptions of students so that you can direct your instruction at clarifying ideas and correcting faulty logic.

Establishing a purpose for reading

1. **Help students understand how information is organized in the textbook.**

   Teach students to preview a chapter by browsing through the chapter headings. The headings and illustrations give a clear picture of the progression of concept development. They also highlight important ideas in each subsection to help students build a basic structure that will be expanded as they read the chapter.

2. **Point out the basic features of *Holt Chemistry: Visualizing Matter.***

   Show students how to use the objectives, captions, and boldface vocabulary terms as guides to what is most important.

3. **Model effective study practices for students.**

   Use the ideas presented in Chapter 1 of the text. In addition to reviewing class notes, students can construct their own study guide questions from a reading preview, which can then be used to guide their reading of the chapter. Following this strategy, students will read with a purpose—a purpose of their own.

Concept mapping as an effective study strategy

Many secondary teachers focus so intently on teaching that they relieve their students of any responsibility for learning. Are your students responsible for their own learning? Do they know how to learn?

When faced with an overwhelming amount of new material, students resort to rote memorization as their learning approach. Concept mapping is a technique that helps students move beyond rote memorization.

Concept maps are suggested in the Portfolio Options sections of the Teacher's Annotated Edition for every chapter of the text. Students are given some instruction in concept mapping in Appendix C of the text.

Introducing concept mapping to the class

Have students read the Concept Mapping section in Appendix C. Let students know that you will expect them to learn and use this strategy. Be prepared for some initial resistance. If students understand that they will be graded on concept maps throughout the course, they will be more likely to learn the technique.

Making concept maps

Drawing concept maps involves the following steps.

1. Start with a list of concepts or ideas to be mapped. This list does not have to be complete, but it should be complete enough to allow you to choose the main idea of the map.
2. Look through the list to identify the concept words that relate directly to the main idea. Place these words below the main idea. Continue this procedure with all the words in your list and with any supporting ideas until all are placed in order of priority under the main idea.
3. Use lines to connect the concepts based on the relationships that link them.
4. Use connecting words to label the linking line so that the relationship between any two concepts is a clear and complete thought.
5. Look for all possibilities for cross-links in the map. Cross-links generally reflect a more thorough understanding of the relationships in the chapter.

Evaluating concept maps

When working with students to evaluate maps, emphasize that there is no single correct map for an idea. Maps should be evaluated according to the following general criteria.

1. **Are the concepts labeled appropriately?**
   Concepts should be represented succinctly—none should be longer than three words. The arrangement of concepts should be hierarchical from general to specific. Concepts should not be repeated on a map. A repeated concept should be represented with a cross-link.
2. **Are the linking words labeled appropriately?**
   Does the map show a clear distinction between concepts and links? Are the links between concepts meaningful? Do they correctly represent the relationships? Can complete ideas be traced through several links?
3. **Does the map show adequate branching?**
   Students' first maps generally look similar to outlines—linear in form and without adequate branching. Provide incentives for students to turn in highly branched maps, so that they will work hard to develop this skill.
4. **Does the map include cross-links?**
   The best maps show sufficient cross-links among concepts. Cross-links show that students know how multiple ideas are connected.

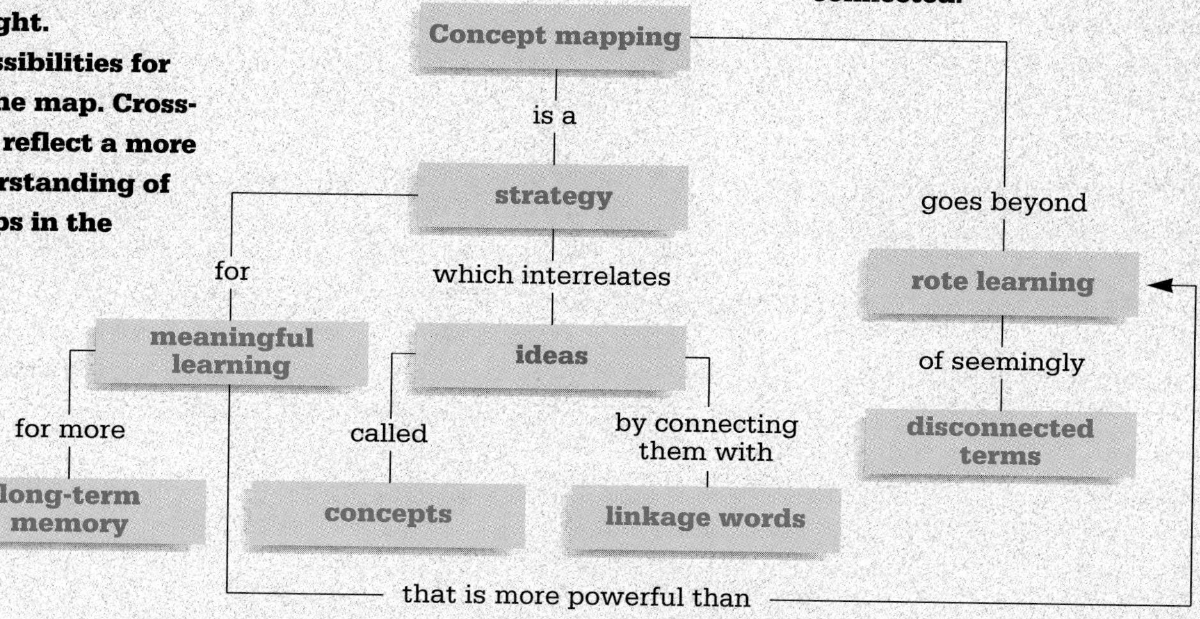

Demonstration Materials List

Chemicals

This list shows the chemicals needed for performing all of the laboratory demonstrations in the text. All of these items will be consumed during the course of the demonstrations.

For solutions, the list includes both descriptions of the concentrations and volumes required for the demonstration as well as the amount of pure concentrated reagent that should be ordered. Many chemical suppliers provide a variety of ready-made solutions. Ordering chemicals already diluted may be slightly more expensive, but it is much safer.

Item	Order quantity	Demo no.
Acetaldehyde	1 mL	15-4
Acetic acid, 1.0 M, 125 mL; 5% solution (vinegar), 530 mL	34 mL glacial acetic acid	1-1, 1-2, 2-4, 7-1, 8-3, 12-1, 13-6
$AgNO_3$, 0.1 M, 80 mL	12.5 g $AgNO_3$	5-3, 7-6, 12-3, 15-2
Alcohol, isopropyl	10 mL	4-5
Alum	210 g	5-GW
Aluminum metal, strip	0.5 × 3 cm	2-6, 15-3
Ammonia water, concentrated, 1 mL 1.0 M, 21 mL 10% solution (household)	52.5 mL conc. $NH_3 \cdot H_2O$ (NH_4OH) 500 mL	1-2, 10-1, 10-7, 12-8, 13-2, 13-6, 13-7
Arsenic	5 g	2-6, 3-1
Bleach, chlorine	10 mL	10-1
Brass metal	100 g	2-1
Brass metal, strip	0.5 × 3 cm	15-3
Bromcresol green	5 mL	13-9
$Ca(C_2H_3O_2)_2$ (sat'd solution), 30 mL	1.2 g $Ca(C_2H_3O_2)_2$	11-4
Calcium metal, **pea-sized pieces** turnings	 2 6 pieces	 5-1, 9-7 3-1, 4-3
CaO	0.005 g	13-5
Carbon, powdered	15 g	2-6, 3-1, 4-1
$CaSO_4 \cdot 2H_2O$ (gypsum)	5 g	4-2
Chalk	1 piece	13-4
Charcoal	1 piece	6-2
Copper wire (#12 or #14) or coat hanger	20 cm	2-6, 5-2, 5-3
Copper sheet	30 × 30 cm	2-5
Copper foil	10 × 15 cm	3-1, 4-1, 7-2, 7-6, 9-10, 15-2, 15-6

Item	Order quantity	Demo no.
$CuSO_4 \cdot 5H_2O$, 1.7 g 1.0 M, 15 mL 0.5 M, 100 mL	6.2 g 15 mL 100 mL	 2-9 7-1, 12-6, 15-1
Distilled water	as needed	throughout
Dry ice	100 g	4-5
Ethanol	175 mL	10-8, 11-1, 11-2, 11-4, 11-10
Filter paper	10 pieces	11-14, 12-3
Gold foil	5 × 5 cm	2-6, 3-1
H_2O_2, 3%	170 mL H_2O_2	8-2, 9-6, 14-8
H_2SO_4, conc., 125 mL 1.0 M, 100 mL 0.1 M, 200 mL	131.7 mL conc. H_2SO_4	7-1, 7-3, 9-9, 12-4, 14-7
HCl, conc. 3.0 M, 1 mL 1.0 M, 1.1 L 0.5 M, 50 mL 0.25 M, 50 mL 0.1 M, 620 mL	100 mL conc. HCl	throughout
Ice	2 L	9-9, 10-4, 11-15, 14-7
Iodine	3 g	2-6, 3-1, 4-1, 11-10
Iron metal filings strip	 45 g 1	 2-6, 3-1, 4-1, 9-1, 10-6 15-3
KH_2PO_4	6.8 g	13-9
KI, 1.0 M, 1 mL 0.1 M, 100 mL	1.3 g KI	14-8 10-1
KIO_3	1.8 g	14-7
$KMnO_4$, crystals, 0.25 g 0.05 M, 10 mL	0.33 g $KMnO_4$	10-8 15-4
KOH, 0.1 M, 295 mL	1.7 g KOH	13-9
Lead sheet	5 × 5 cm	2-6, 3-1

Item	Order quantity	Demo no.
LiCl	10 g	5-3
Litmus paper, red	5 pieces	1-2, 5-4
Litmus paper, blue	3 pieces	1-2
Lycopodium powder (or fine chalk dust)	1 g	4-4
Magnesium metal, powdered	0.5 g	9-7
Magnesium metal, strip	0.5 × 3 cm	2-6, 15-3
Magnesium ribbon	125 cm	3-1, 4-1, 4-3, 6-2, 7-5, 14-2, 14-3
Marble chips, $CaCO_3$	65 g	2-4, 7-4, 8-4, 13-4
Mercury	2 mL	3-1
Methyl orange	5 mL	13-9
$Mg(OH)_2$	10 g	11-1
MnO_2, powdered	0.05 g	9-6
Na_2CO_3	34 g	5-5, 8-1, 12-5
$Na_2S_2O_3 \cdot 5H_2O$, 35 g 1.0 M, 10 mL	37.5 g $Na_2S_2O_3 \cdot 5H_2O$	11-6, 11-10, 15-4
Na_2SO_4, 1.0 M, 250 mL	35.5 g Na_2SO_4	2-7
NaCl (rock salt)	2 g	4-2
NaCl (salt) saturated solution, 500 mL 0.5 M, 100 mL 0.2 M, 20 mL solid, 33 g	261.2 g NaCl	2-8, 6-1, 9-1, 12-1, 12-3, 12-6, 13-1, 15-1
$NaHCO_3$	76.5 g	1-1, 7-1, 8-3, 13-6, 14-6
NaOH pellets, 12.5 g 1.0 M, 1.5 L 0.25 M, 500 mL 0.1 M, 225 mL	78.4 g NaOH	throughout
Na_2SO_3	0.9 g	14-7
NH_4Cl	29 g	7-3, 11-5, 12-8
NH_4OH (see ammonia water)		
Nickel metal	10 g	3-1, 4-1
Oleic acid (0.5% in ethanol), 6 mL	0.1 g oleic acid	4-4
Onion's fusible metal	20 g	9-2
pH paper, full-range	1 roll	1-2, 13-3, 13-6
Phenolphthalein solution	20 mL	throughout

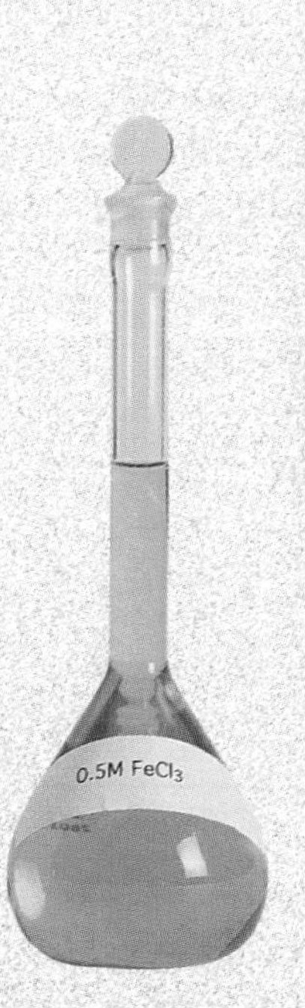

Item	Order quantity	Demo no.
Phosphorus, white	3 g	4-1
Phosphorus, red	2 g	2-6
Radiation source alpha $^{210}_{84}Po$, $^{228}_{90}Th$ beta $^{90}_{38}Sr$, $^{204}_{81}Tl$, $^{40}_{20}Ca$ gamma $^{22}_{11}Na$, $^{54}_{25}Mn$	 1 1 1	 4-5, 16-2 16-2 16-2
Silver foil	5 × 5 cm	2-6, 3-1
Sodium metal (pea size)	2 pieces	3-1, 3-2, 4-1
Starch	10 g	14-7
Steel wool	3 pieces	7-2, 15-3, 15-6
Sucrose (sugar), $C_{12}H_{22}O_{11}$	58 g	6-1, 11-1, 11-3, 11-8, 13-1
Sulfur (flowers)	30 g	3-1, 4-1, 7-2, 9-10
Tin foil	5 × 5 cm	3-1
Tin metal, strip	0.5 × 3 cm	2-6, 15-3
Tungsten metal	5 g	4-1
Vinegar (see acetic acid)		
Weighing paper	4 pieces	9-5, 14-4
Wooden splint	4	2-7, 10-6
Zinc metal, mossy, small pieces	25 g	7-1, 15-1
Zinc metal, sheet	11 × 10 cm	3-1, 4-1, 15-6
Zinc metal, strip	0.5 × 3 cm	2-6, 15-3
$Zn(NO_3)_2 \cdot 6H_2O$, 0.1 M, 30 mL	0.9 g $Zn(NO_3)_2 \cdot 6H_2O$	15-2

Equipment

This list shows the equipment and laboratory apparatus needed to perform all of the demonstrations in the text. All of these items will be reusable.

The absolute zero demonstrator and the specific heat demonstrator can be ordered through a scientific catalog, such as Sargent Welch and Fisher Scientific.

Item	Order quantity	Demo no.
Absolute zero demonstrator	1	10-4
Acetate, for overhead projector	1 sheet, 22 × 26 cm	9-6, 10-1
Aluminum, plate	1 × 10 × 10 cm	16-2
Animal fur	1 piece	11-11, 16-1
Balance, centigram	1	throughout
Balls, "Happy"	2	10-2
Barometer	1	14-2
Beaker, 50 mL	1	5-1, 5-5, 13-8
Beaker, 150 mL	3	6-1, 9-2, 12-8
Beaker, 250 mL	8	throughout
Beaker, 600 mL	3	throughout
Beaker, 1.5 L	1	3-2
Beaker, 2 L	2	10-3, 10-4
Bell jar	1	10-5. 10-9
Bottle, gas collection	4	7-4, 7-6, 11-13
Box lid, clear plastic 8 cm × 6 cm	1	10-1
Bunsen burner	1	throughout
Clamp, test-tube	3	6-1, 9-10, 10-4, 10-6, 12-5, 15-5
Clamp, buret	1	5-5, 7-2
Cobalt glass	50 × 50 cm	5-4
Conductivity tester, battery-powered	1	6-1, 12-1, 13-1
Crucible	1	5-3
Dish, shallow flat	1	12-7
Dropper	4	10-1, 12-3, 12-4, 13-9, 14-8
Electrolysis unit with DC power source	1	2-7, 12-4
Eudiometer tube, 50 mL	2	10-3, 14-2
Evaporating dish	1	2-8, 5-1, 5-4, 8-1
Flask, Erlenmeyer, 125 mL	4	7-3, 11-1

Item	Order quantity	Demo no.
Flask, Erlenmeyer, 250 mL	3	7-4, 10-9, 11-2, 12-5
Flask, volumetric 50 mL	1	11-9
Flask, volumetric 1 L	1	8-2
Forceps	1	11-10
Funnel	1	12-3
Geiger counter	1	16-2
Geissler tube with gas (e.g., H_2, He, N_2, Ne, Hg)	1 each	3-1, 3-3, 3-4, 4-1
Glass plate, flat	2, 5 × 5 cm each	1-4, 7-2, 7-4, 9-10
Glass stirring rod	4	9-2, 9-5, 11-1, 14-3
Glass tubing straight bent	 10 cm 4 × 4 cm	 2-9, 7-4 8-2
Graduated cylinder, 10 mL	1	14-2
Graduated cylinder, 50 mL	1	9-5
Graduated cylinder, 100 mL	3	9-8
Graduated cylinder, 250 mL	1	11-2
Graduated cylinder, 1 L	1	1-1, 10-8, 13-9
Graduated cylinder, 2 L	1	8-2
Hot plate	1	5-1, 5-4, 5-GW
LED (red or green)	1	15-6
Microscope	2	5-GW, 6-1
Mortar and pestle	2	1-2, 14-4
Nichrome wire	30 cm	6-1, 7-1
Ohmmeter, battery-operated	1	12-2
Overhead projector	1	throughout
Petri dish and cover	14	throughout
Power supply or Tesla coil	1	3-3, 3-4
Reaction plate, 96-well	1	10-1
Refrigerator freezer	1	12-5

Item	Order quantity	Demo no.
Ring	1	throughout
Ring stand	3	throughout
Rod, bakelite or ebonite	1	11-11, 16-1
Rod, glass	1	16-1
Rubber tubing	120 cm	2-9, 7-4, 12-7
Rubber tubing for faucet	1	11-11
Rubber tubing for gas burner	1	6-1
Scoopula	1	9-6
Spatula	1	10-8, 11-3
Specific heat demonstrator	1	9-3
Stirring rod	3	1-2, 9-8, 9-9, 11-3, 11-4, 12-6, 13-3
Stopper, cork, flask	1	5-2, 5-3, 12-2
Stopper, one-hole, test-tube	2	2-9, 8-2, 14-2
Stopper, test-tube	20	3-1, 11-1, 11-8, 11-10, 11-12

Item	Order quantity	Demo no.
Stopper, two-hole, #5	1	7-4
Striker	1	6-1
Test tube, Pyrex (20 × 150 mm)	20	throughout
Test tube, disposable	1	9-2
Test tube, large (Pyrex)	1	2-8, 11-6
Test-tube rack	1	6-1, 11-5
Thermometer, Celsius (nonmercury)	4	throughout
Thistle tube	1	7-4
Tongs, crucible	1	4-2, 6-2, 7-2, 7-5, 9-10
Tongs, test-tube	1	11-3, 11-4, 11-5
Trough	1	7-4, 8-2
Vacuum pump	1	10-5, 10-9
Voltmeter	1	15-3, 15-6
Watch glass	1	5-3, 5-4, 5-GW
Wire gauze	1	11-13
Wire gauze with ceramic center	2	throughout

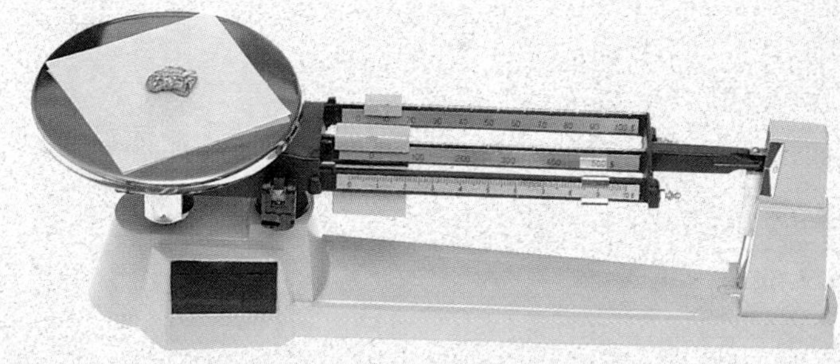

Miscellaneous materials

The demonstrations use many consumer products that will be consumed during the course of the demonstrations.

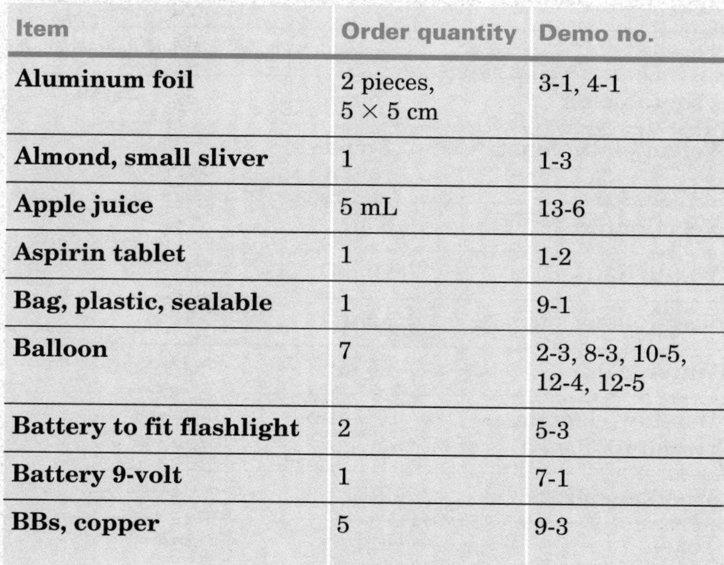

Item	Order quantity	Demo no.
Aluminum foil	2 pieces, 5 × 5 cm	3-1, 4-1
Almond, small sliver	1	1-3
Apple juice	5 mL	13-6
Aspirin tablet	1	1-2
Bag, plastic, sealable	1	9-1
Balloon	7	2-3, 8-3, 10-5, 12-4, 12-5
Battery to fit flashlight	2	5-3
Battery 9-volt	1	7-1
BBs, copper	5	9-3

Item	Order quantity	Demo no.
Bleach	1 mL	10-1
Board, plywood	1 × 10 × 10 cm	16-2
Bread	1 slice	10-5
Bulb to fit flashlight	1	5-3
Cabbage, red (1/4 head)	1	13-7
Can, coffee, 3 lb	1	6-2, 7-5
Candle	5	1-3, 7-4, 7-Tip, 9-3
Cardboard box with lid	1	16-3
Cardboard strip	2.5 × 18 cm	5-GW
Cheesecloth	1	5-GW

Item	Order quantity	Demo no.
Cup, paper (mini-size, e.g. ketchup serving)	2	9-4
Cup, plastic foam	6	1-4, 9-8, 9-9
Detergent	30 g	11-1
Detergent, liquid	50 mL	4-4
Dry cell, carbon-zinc	2	15-5
Dry cell, alkaline	2	15-5
Dry cell, 9-volt	1	12-4
Effervescent tablet (Alka-Seltzer)	4	14-1, 14-4
Fertilizer	2.5 g	5-4
Flashlight	3	15-5
Food coloring	60 mL	10-3, 12-7
Fruit (lemon, grapefruit, or apple)	1	15-6
Glass, tall drinking	1	2-2
Glass jar, large, with lid	1	4-5
Hand warmer (sports item)	1	11-7
Jelly bean	2	10-5
Lemon juice	5 mL	13-6
Light bulb with tungsten filament	1	3-4
Light bulb, fluorescent	1	3-4
Lighter fluid	5 mL	1-4
Luminescent light stick	1	14-5
Marker, permanent	2	1-4, 10-1, 11-4
Marker, water-soluble	1	11-14
Marshmallow	1	10-5
Match	5	2-7, 7-4, 10-6
Meter stick	1	10-2
Milk	5 mL	13-6
Milk of magnesia	5 mL	13-6
Mineral oil	30 mL	4-1
Mylar, heavy	1 sheet, 22 × 26 cm	10-1
Ni-cad rechargeable cell	2	15-5
Nylon hosiery	1	11-11, 16-1
Orange juice	5 mL	13-6
Paintbrush, small	1	13-8
Pan, flat	1	1-1

Item	Order quantity	Demo no.
Paper, blotter	1 piece, 12 × 12 cm	4-5
Paper, white	3 sheets, 22 × 26 cm	10-8, 11-13, 13-7, 13-8, 16-2
Paper clip	10	5-3, 15-6
Paper towels	as needed	throughout
Pennies	300	16-3
Plastic bottle, 2 L	3	8-3
Plastic wrap	2 sheets, 10 × 10 cm	10-8, 11-14
Potato	2	1-3, 15-3
Razor blade, single-edge	1	14-4
Ruler, metric	1	4-4
Sand	5 g	2-8
Sandpaper	4 pieces, 10 × 10 cm	2-5, 4-3, 6-2, 15-2
Shaving cream	25 mL	10-5
Soap, liquid	100 mL	1-1, 1-4, 11-12, 11-13
Soda	3 bottles	11-15
Spoon	1	2-2
Stopwatch (or clock with second hand)	3	14-1
Straw, long	2	2-2, 13-2
String	65 cm	5-GW, 11-14
Styrofoam	100 g	2-1
Styrofoam ball, small	2	16-1
Tape	1 roll	throughout
Thread	100 cm	14-2, 14-3, 16-1
Tray, cafeteria	1	4-4
Turnip	1	14-8
Vegetable oil	75 mL	11-1, 11-12
Velvet cloth, black	1 piece, 10 × 10 cm	4-5
Vermiculite	5 mL	9-1
Water, bottled	20 mL	13-3
Water, lake	5 mL	13-6
Water, rain	25 mL	13-3, 13-6
Window cleaner, ammonia-based	1 bottle	13-8
Wooden splints	2	10-6
Yeast	1 g	8-2

Optional materials and equipment

To help fit your situation and resources, the demonstrations contain alternative procedures for different types of equipment, as shown in this list.

Item	Quantity	Demo no.
Cloud chamber	1	4-5
KNO_3	10 g	5-3
Microscope, projecting	1	6-1
$NaNO_3$	10 g	5-3
Spectroscope	1	3-4
Water, softened	200 mL	12-2

Equipment and media suppliers

Aims Media
9710 DeSoto Avenue
Chatsworth, CA 91311

American Chemical Society
ACS Software
Distribution Office, Dept. 170
P.O. Box 57136
West End Station
Washington, DC 20037

Annenberg/CPB Project
P.O. Box 2345
South Burlington, VT 05407-2345

Central Scientific Company
11222 Melrose Avenue
Franklin Park, IL 60131-1364

Coronet Films and Video
108 Wilmot Road
Deerfield, IL 60015

Fisher Scientific Company
Educational Materials Division
4901 West LeMoyne Street
Chicago, IL 60651

Flinn Scientific, Inc.
P.O. Box 219
Batavia, IL 60510-0219

Frey Scientific Company
905 Hickory Lane
Mansfield, OH 44905

Journal of Chemical Education
JCE: Software
University of Wisconsin-Madison
Department of Chemistry
1101 University Avenue
Madison, WI 53706-1396

Lab Safety Supply, Inc.
401 S. Wright Road
P.O. Box 1368
Janesville, WI 53547-1368

Logal Software, Inc.
P.O. Box 1499
East Arlington, MA 02174-0022

Modern Talking Picture Service
5000 Park Street North
St. Petersburg, FL 33709

National Geographic Society
Educational Services Dept. 89
17th & M Streets, NW
Washington, DC 20036

National Science Teacher's Association
NSTA Special Publications
1742 Connecticut Avenue, NW
Washington, DC 20009

Optical Data Corporation
30 Technology Drive
Warren, NJ 07059

Quantum Technology
3015 Arena Drive
Evergreen, CO 80439

Sargent Welch
7300 N. Linden Avenue
Skokie, IL 60077

Science Kit and Boreal Labs
777 East Park Drive
Tonawanda, NY 14150

Total Science Safety System
JaKel, Inc.
585 Southfork Drive
Waukee, IA 50263

Ward's Natural Science Establishment, Inc.
5100 West Henrietta Road
P.O. Box 92912
Rochester, NY 14692

Videodiscovery, Inc
1700 Westlake Avenue, N
Seattle, WA 98109-3012

PLANNING FOR

1

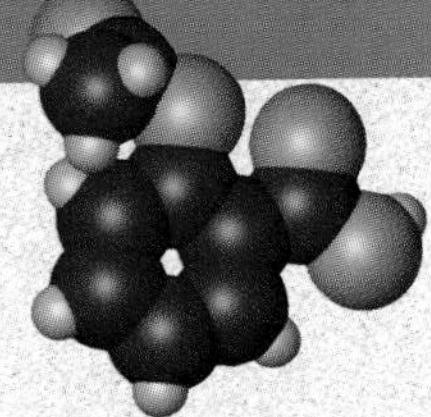

The Science of Chemistry

Chapter Overview

- **Section 1-1** uses the aspirin story to introduce the following: classification by properties, synthesis reactions, and the commercial development of a compound. Students encounter symbolic representations as they study the synthesis reactions for aspirin. They read about some of the things chemists do, and how their activities relate to the chemical industry. The role of government agencies in regulating chemicals is also presented.
- **Section 1-2** covers the activities of the scientific method using the example of the discovery of the anti cancer properties of cisplatin.
- **Section 1-3** introduces the broad themes that will be used to link ideas in this course, along with some useful study hints.

Concept Base

The history of aspirin is a useful example of the scientific method and provides opportunities for students to learn some chemistry as they examine the properties of a familiar pure substance. They also learn about organic reactions and the commercialization of a drug.

Looking Ahead

We assume students have seen chemical formulas and know that chemists use symbols to represent the elemental composition of compounds. Symbols, formulas, and nomenclature are covered in detail in Chapters 3, 5, and 6.

Laboratory Equipment Needs

Demonstrations

The listing of materials for Chapter 1 demonstrations begins on page T46.

Laboratory Experiments

Materials and preparation instructions for Chapter 1 are found as follows.

Text		*Page*
Exploration 1A	Laboratory Techniques	656
Investigation 1A	Conservation of Mass—Percentage of Water in Popcorn	660
Exploration 1B	Properties of Analgesics	662
Investigation 1B	Properties of Analgesics—Forensic Analysis	670

Laboratory Experiments Manual
(refer to the table of contents in the manual)
Exploration 1-1 Conservation of Mass

Pacing: 5 days

	Inquiry Teaching Strategies
1-1 What will I learn in chemistry?	**Demonstrations** ■ Demo 1 *Changes in matter* ■ Demo 2 *Litmus in acid and base* **Laboratory Experiments** ■ Exp 1A *Laboratory Techniques* ■ Inv 1A *Conservation of Mass—Percentage of Water in Popcorn* ■ ■ Exp 1-1 *Conservation of Mass*
1-2 How do scientists approach problems?	**Laboratory Experiments** ■ Exp 1B *Properties of Analgesics* ■ Inv 1B *Properties of Analgesics—Forensic Analysis*
1-3 How do I learn chemistry?	**Demonstrations** ■ Demo 3 *Experimentation*

Key ■ Teaching Transparencies
■ Annotated Teacher's Edition
■ Laboratory Experiments Manual

Visual Teaching Strategies

Transparency Masters
- F 1-4 *Aspirin Organic Synthesis*
- F 1-5 *Symbolic Representations of Benzene*
- T 1-2 *Top 25 Chemicals Produced in the United States*
- T1-3 *Top 25 Chemical Producers in the United States*
- P 651 *Safety Symbols*
- P 650 *Laboratory Reports*

Transparency Masters
- P 16 *Activities of the Scientific Method*
- F 1-12 *Structural Formulas for Acetaminophen and Ibuprofen*

Transparency Masters
- P 808 *Concept Mapping*
- P 809 *Concept Mapping*
- P 810 *Concept Mapping*

Review and Practice Strategies

Text Reviews
- Section Review 1–6
- Chapter Review 1–10

Study Guide Worksheets
- 1-1 Concept Review

Text Reviews
- Section Review 7–11
- Chapter Review 11–19

Study Guide Worksheets
- 1-2 Concept Review
- 1-2 Skillsheet *Using the Scientific Method*

Text Reviews
- Section Review 12–16
- Chapter Review 20–24

Study Guide Worksheets
- 1-3 Concept Review
- 1-3 Skillsheet *Study Skills*

Assessment Options

Traditional Assessment

Test Generator
Instructional Objectives Measured:
Content mastery
Select from items 1 to 60

Performance Assessment Options

(see pages T36 and T635-15 for scoring rubrics)
Students demonstrate mastery of objectives in a hands-on environment. You may want some of these materials in the Portfolio as well.

Investigation 1A and 1B (text)
Instructional Objectives Measured:
Content mastery, Use of scientific methodology, Problem solving skills, Use of technology, Proficiency in written communication

Performance Assessments, page 28

1. **Building study strategies**
   Instructional Objectives Measured:
   Content mastery, Problem solving skills
2. **Interpreting MSDS sheets**
   Instructional Objectives Measured:
   Content mastery, Scientific and chemical literacy, Proficiency in written communication
3. **Designing an experiment**
   Instructional Objectives Measured:
   Content mastery, Use of scientific methodology, Problem solving skills, Use of technology, Proficiency in written communication

Portfolio Options

(see page T36 for scoring rubrics)
Students provide a written rationale for each selection made for the portfolio.

Formal Laboratory Reports
Exploration 1A and 1B (text)
Exploration 1-1 (lab manual)
Instructional Objectives Measured:
Content mastery, Use of scientific methodology, Use of technology, Proficiency in written communication

Portfolio Projects, page 28
Items 1 and 3. Chemistry and you
Instructional Objectives Measured:
Content mastery, Scientific and chemical literacy, Understanding the impact of science and technology on society, Proficiency in oral and written communication

Item 2. Research and communication
Instructional Objectives Measured:
Scientific and chemical literacy, Understanding the impact of science and technology on society, Proficiency in oral and written communication

Science, Technology and Society, page 30
Instructional Objectives Measured:
Scientific and chemical literacy, Understanding the impact of science and technology on society, Problem solving skills, Proficiency in oral and written communication

■ Pupil's Text
■ Study Guide
■ Teaching Resources

Exp *Exploration*
Inv *Investigation*
LC *Linking Chapters*
UT *Using Technology*

INTRODUCING THE CHAPTER

CHAPTER 1

The Science of Chemistry

Story Background

Though salicylates could be obtained from natural plant sources (willow bark, the wintergreen plant, and the meadowsweet plant), they are expensive to isolate and produce from these sources. Fortunately, during the nineteenth century, German chemists had become expert in the development of synthetic forms of natural substances. German chemists were able to synthesize salicylic acid at one-tenth of the current market cost of producing the compound from natural sources. The point to emphasize in this chapter is that while natural products are held in high esteem, there are many instances where they are too costly for widespread consumption.

Some students may be aware of the current controversies surrounding taxol, the cancer chemotherapy treatment obtained from a yew tree that grows only in Northwest North America. As of this writing, taxol has not been produced synthetically.

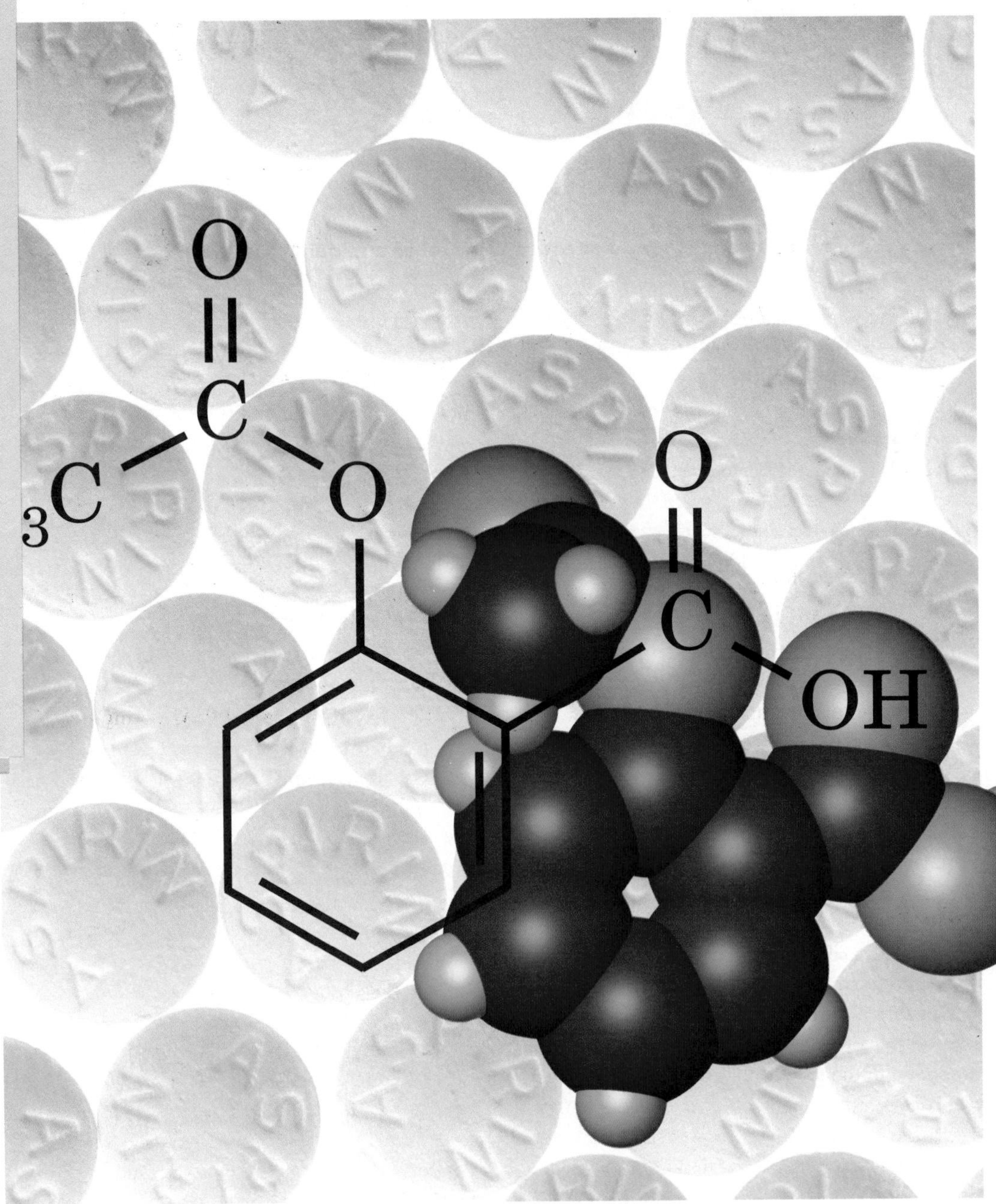

About the Illustration

The illustration shows one of the many structural representations of matter that students will see in this course. The planar structural model and the three-dimensional space-filling model for aspirin are shown.

A Wonder Drug

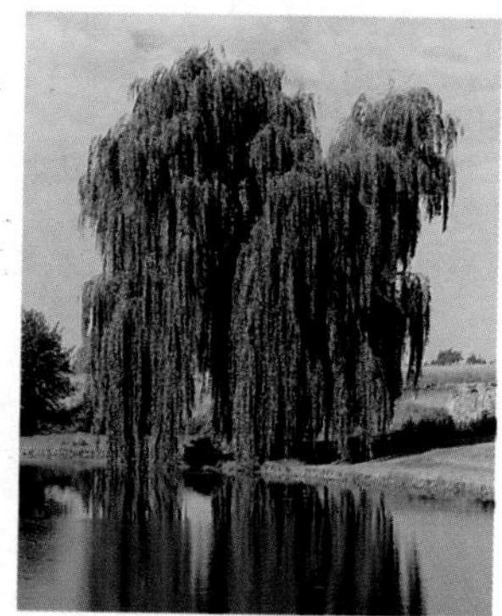

As far back as 2000 years ago, the Greeks and Romans relied upon a substance extracted from plants, particularly the bark of the willow tree, to relieve pain and treat fevers. This plant extract was used to treat various ailments throughout the world, even in North America where Native Americans used a bark extract to treat fever and pain.

In the 1760s, Edward Stone, an English clergyman and naturalist, wrote about how he used "twenty grains of powdered bark dissolved in water, administered every four hours" to treat 50 people suffering from acute, shiver-provoking illnesses. The results, he was happy to report, were excellent. Stone was unaware that his treatment would someday become a common prescription—two aspirin tablets every four hours. Stone was also unaware that his findings would set in motion a chain of events that would evolve into a billion-dollar industry.

Following up on Stone's report, German chemists in the early 1800s isolated a tiny amount of what they felt was the active ingredient in the bark extract and called it salicin, which is related to *Salix*, the genus name for willow. French chemists improved on the extraction process and were able to obtain about an ounce of salicin from three pounds of bark. Salicin could be converted to salicylic acid, the active form of the pain reliever.

Once its identity was known, a compound related to salicylic acid, sodium salicylate, was made available for public use. Unfortunately, sodium salicylate was found to be extremely irritating to the lining of the stomach. Those who took it often suffered stomach irritation or nausea and occasionally developed ulcers.

In the late 1800s, Felix Hoffmann, a German dye chemist, began looking for ways to reduce some of the unpleasant side effects of sodium salicylate. His interest in salicylate compounds came from wanting to help relieve the pain his father experienced from arthritis. His solution to the problem was to synthesize a different derivative of salicylic acid, acetylsalicylic acid.

Each year, Americans consume 30 billion tablets, equaling 16 000 tons of aspirin, at a cost of $2 billion. The story of how aspirin has become the most widely used drug in the world is also a story about chemistry.

You'll begin this course by studying some of the chemistry of aspirin. Along the way, you'll find the answers to some interesting questions.

- *What is aspirin?*
- *How does aspirin work?*
- *How is aspirin similar to other widely used pain relievers?*
- *How does buffered aspirin differ from regular aspirin?*
- *How does name-brand aspirin differ from generic brands?*

The science of chemistry has contributed much to our knowledge of the world and has enabled chemists to make tremendous changes in the way we live. These changes can lead to challenges that we all must face and, at times, to problems that we have yet to solve.

Tapping
Prior Knowledge

1. Have students write for five minutes about what they think chemistry is. Each student should submit at least one question that he or she hopes to have answered in this course. Keep the questions on hand for reference throughout the year.
2. Have students describe any methods that former teachers used that helped them learn the material being taught.

Section 1-1

FOCUS

Lesson Starter

Use the following demonstration to introduce students to some of the excitement of chemistry.

Demonstration 1

Changes in matter

Approximate time: 10 min

1. Put 3 drops of phenolphthalein dye in a 600 mL beaker before class. Add 200 mL of 0.25 M NaOH solution to the beaker containing the dye.
2. Now add 1 M HCl to the beaker until it turns clear again.
3. Add more NaOH to the beaker until it turns pink again. Continue adding acid and base until beaker is full.
4. Before class gently mix 20 mL of dishwashing liquid with 100 mL of vinegar. Place 50 g of baking soda, $NaHCO_3$, into a 1000 mL graduated cylinder, and place the cylinder in a larger container that will be able to catch the resulting bubbles. Add the vinegar-soap mixture, and watch the bubbles appear. Ask students what different properties they observed in each demonstration.

SAFETY

Wear safety goggles and a lab apron. Students should be at least 10 ft from the demonstration area.

DISPOSAL

Neutralize the acid or base leftover from each step, and pour them down the drain. Flush any residue with lots of water to dissolve it.

What will I learn in chemistry?

Section Objectives

Distinguish between the properties of acids and bases.

Classify the activities and fields of working chemists.

Classify compounds as organic or inorganic.

List three ways that the public is protected from chemical hazards.

Explain the difference between chemistry and technology.

Chemistry deals with the properties of chemicals

Although the development of aspirin as a commercial product began in a laboratory, not all chemistry is done in a lab. For example, the swimming pool maintenance person in Figure 1-1 runs tests to decide what **chemicals** must be added to the pool, and how much of each. A gardener runs tests to decide what chemicals will improve vegetable yields or treat plant diseases. An amateur photographer uses an acid stop bath to neutralize a base and stabilize the image on a negative. A baker experiments to determine the right mix of ingredients and temperature to make light and tasty bread. A consumer buying groceries for the week scans food labels for substances that can cause allergic reactions. Researchers throughout the world search for a drug that will be more effective than AZT in the treatment of HIV infections.

chemical
any substance formed by or used in a chemical reaction

It's not just the chemists who need to know something about chemistry. Because chemicals are an integral part of our everyday lives, everyone needs some background in chemistry.

Figure 1-1
Swimming pool maintenance work requires a knowledge of acidity and reactions to keep the pool in the correct chemical balance.

"I measure the pH to monitor the acid balance in the pool."

You use chemicals every day

Some people think chemicals cause the problems of pollution, cancer, and toxic waste and that they should be banned. But think for a moment about what that would mean—everything around you is a chemical. Imagine going to the supermarket to find fruits and vegetables grown without any chemical reactions. The produce section would be completely empty! In fact the entire supermarket would be empty because *all* foods contain chemical components. Artificial flavorings, colorings, and preservatives make food taste better, look better, and last longer before spoiling. Without chemicals you would have little to wear because most of your clothing is made of synthetic fibers. Even clothing made of natural fibers is the product of chemical reactions. Finally, consider your own body, which is a highly efficient chemical factory containing thousands of compounds. Aspirin, too, is a chemical.

Table 1-1
Some Properties of Acids and Bases

Water solutions of acids...	Water solution of bases...
taste sour, conduct electricity	taste bitter and feel slippery
turn blue litmus paper red	turn red litmus paper blue
have pH values less than 7	have pH values higher than 7
react with bases and certain metals to form salts	react with acids to form salts

Aspirin is classified by its properties

Chemistry focuses on the properties and reactions of matter like aspirin. Aspirin is a white crystalline compound with no odor and a slightly bitter taste. The chemical name for aspirin is acetylsalicylic acid. The term **acid** in the name signifies that it belongs to a group of compounds that have certain chemical properties, which are listed in **Table 1-1**. The degree to which a compound shows these properties is its acidity. Acidity can be expressed numerically as **pH**. In water, acids have pH values below 7. Another class of compounds called **bases** have pH values higher than 7 when in water. Bases have properties that are somewhat opposite of acids, as shown in **Table 1-1**.

The pH of aspirin in water is 2.7. Compare the acidity of aspirin to other familiar substances in **Figure 1-2**. Your stomach is an acidic environment because certain cells in the lining secrete hydrochloric acid, HCl. Hydrochloric acid solutions are more acidic than aspirin. Aspirin irritates the lining of the stomach. People with ulcers or other stomach disorders cannot use aspirin.

Bufferin is a trade name for buffered aspirin. Buffering keeps the pH of a solution somewhat constant even if some acid or base is added. So using buffered aspirin could reduce stomach irritation, but it does not eliminate the problem.

Aspirin can also be classified as an **organic compound** because it contains carbon. Hydrochloric acid is an example of an **inorganic compound** because it does not contain carbon.

acid
a class of compounds whose water solutions taste sour, turn blue litmus to red, and react with bases to form salts

pH
a quantitative expression of acidity with the standard for a neutral solution expressed as pH 7

base
a class of compounds that taste bitter, feel slippery in water solution, turn red litmus to blue, and react with acids to form salts

organic compound
any covalently bonded compound containing carbon (except carbonates and oxides)

inorganic compound
all compounds outside the organic family of compounds

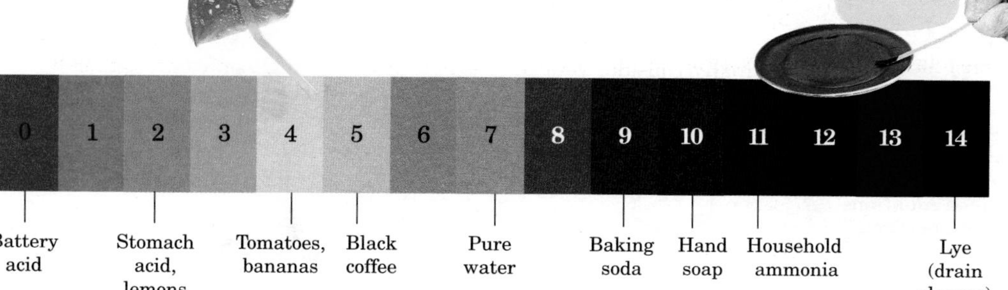

Figure 1-2
One way of measuring the pH of a substance is to use pH paper. The color of the paper at each pH is shown along with the pH measurements of some common materials.

TEACH

Possible Misconception

Many students think that chemicals differ from substances and materials in nature. Have students name their favorite foods. Show them that these are chemical in nature. Show them that cars, cosmetics, video games, sports equipment, and many other things they take for granted are chemicals.

Teaching Tip

Symbolic representations
Have students name all the acids they know. Write formulas for them on the board so that students can start to associate words with symbolic representations.

Content Rationale

Acid vs. base
Both pH and buffers are covered in more detail in Chapter 13, so it is not necessary to develop these concepts here.

Demonstration 2

Litmus in acid and base
Approximate time: 10 min

1. Place pieces of red and blue litmus paper in a beaker of vinegar.
2. Place pieces of red and blue litmus paper in a beaker of household ammonia.
3. Crush an aspirin tablet using a mortar and pestle. Add a small amount of water and the powdered aspirin to a watch glass and stir. Determine the pH with litmus paper and pH paper.

SAFETY
Wear safety goggles and a lab apron. Students should be at least 10 ft from the demonstration area.

DISPOSAL
Put the litmus paper and aspirin residue in the trash. Pour the liquid down the drain, and flush with water.

Visual Strategy

Figure 1-2
Have the students note that 7 is the middle of the scale. Many students have trouble remembering that 0 on the pH scale is the most acidic. Stronger acidity and a lower value on the pH scale seem contradictory to many students.

Do You Know?

Phenol is also used to make the common thermosetting plastic called Bakelite that is used to make electrical outlets and switches.

Content Rationale

Organic vs. inorganic compounds

Although high school chemistry courses traditionally focus on inorganic chemistry, we have elected to integrate organic chemistry where appropriate throughout the text. This strategy puts the focus of organic chemistry on concepts and reaction behavior rather than on nomenclature. There is no formal coverage of organic nomenclature in this course. However, the classification of organic compounds by functional group is presented in Chapter 6.

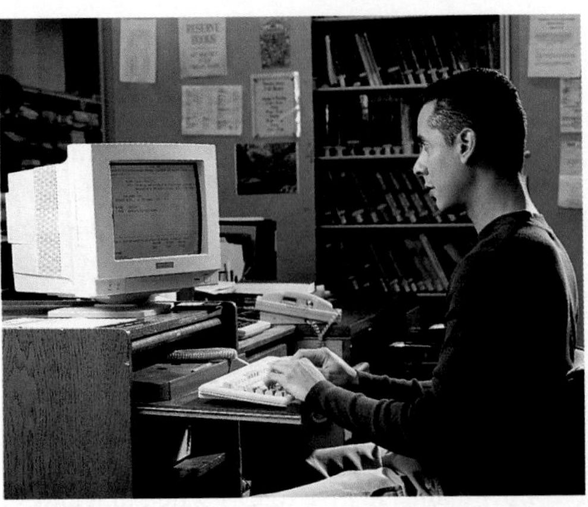

Figure 1-3
Searching the chemical literature is now easier and faster using on-line chemical databases.

Chemical reactions

Felix Hoffmann began working on an alternate salicylic acid compound to relieve his father's arthritic pain. At the time, sodium salicylate was being used and the standard dose was about 6–8 grams per day. The effects of this dosage were causing so much stomach discomfort that Hoffmann's father could no longer take it. Hoffmann searched for information about other known compounds that were related to sodium salicylate but less irritating. After testing the product on his father, he found that acetylsalicylic acid worked best. Hoffmann took his research results to the president of the company where he worked, the Bayer Company. As a result, the first commercial aspirin was marketed by the Bayer Company in 1899. The name *aspirin* was selected for acetylsalicylic acid using the letter *a* from *acetyl* and *spirin* from the *spiraea* plant—a natural source of salicylic acid.

Hoffmann's search of the chemical literature available on salicylate compounds was an important part of his work. A literature search is very often the first step of research chemists trying to create a new product or solve a problem. They often use Chemical Abstracts (CA), which is the most comprehensive on-line source of chemical information in the world. CA includes abstracted versions of articles from over 18 000 scientific journals published throughout the world.

synthesis
the process of building compounds from elementary substances through one or more chemical reactions

Aspirin is the product of chemical reactions

You know from previous science courses that compounds like water, H_2O, and sugar, $C_6H_{12}O_6$, are made from chemical reactions. **Figure 1-4** shows one **synthesis** process that is used to make aspirin.

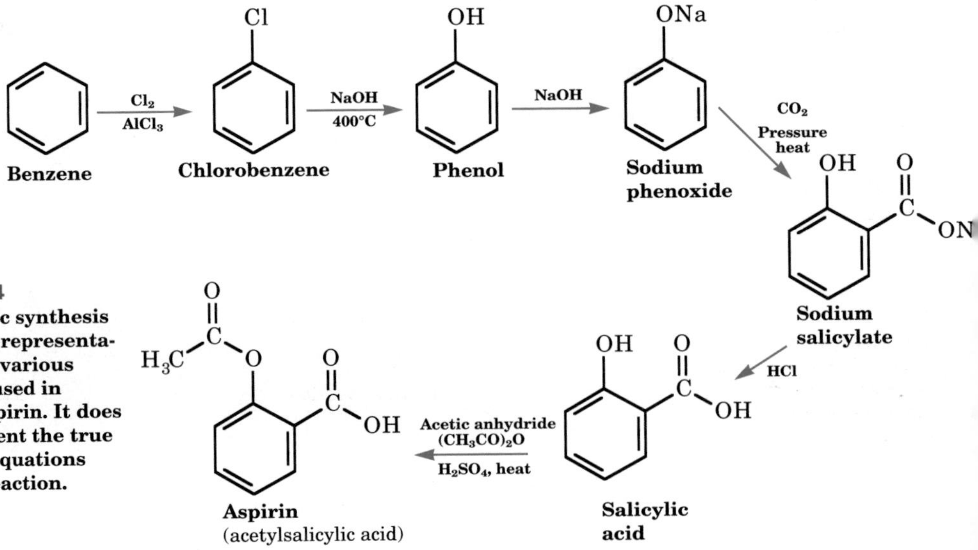

Benzene → (Cl_2, $AlCl_3$) → Chlorobenzene → (NaOH, 400°C) → Phenol → (NaOH) → Sodium phenoxide → (CO_2, Pressure, heat) → Sodium salicylate → (HCl) → Salicylic acid → (Acetic anhydride $(CH_3CO)_2O$, H_2SO_4, heat) → Aspirin (acetylsalicylic acid)

Figure 1-4
The organic synthesis shown is a representation of the various reactions used in making aspirin. It does not represent the true balanced equations for each reaction.

Visual Strategy

Figure 1-4

Point out how the organic synthesis shown differs from the way students have traditionally seen equations for reactions: (1) it takes multiple steps to get the final product; (2) the equations are not balanced; and (3) reagents used in the reactions are written above the arrow. Students should notice the structural similarities among the compounds and the prevelance of the benzene ring.

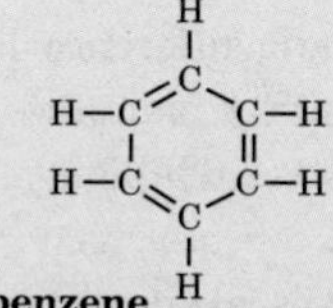

benzene

Figure 1-5a
Benzene is an organic compound with an unusual structure. The carbon atoms are arranged in a ring. Benzene and molecules that react similarly to benzene are called aromatic compounds.

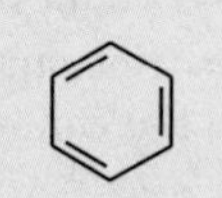

benzene

1-5b
This representation of benzene is an abbreviated structural formula showing carbon–carbon bonds. It is understood that one carbon atom is bonded to one hydrogen atom at each corner of the figure.

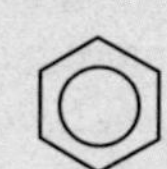

benzene

1-5c
This abbreviated structural formula more accurately depicts the bonding between carbon atoms in the ring. One carbon atom is bonded to one hydrogen atom at each corner of the figure.

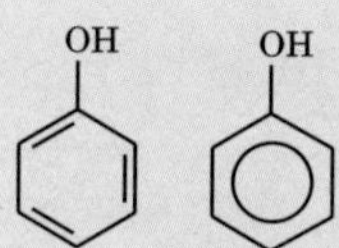

phenol is derived from benzene

1-5d
Structural representations of compounds derived from benzene are abbreviated in these forms to show the benzene ring and the atoms that substitute for hydrogen atoms.

reagent
a chemical used to convert one substance into another substance in a chemical reaction

Synthesizing aspirin involves a number of chemical reactions in which the product of one step is the reactant in the next step. Products of a reaction are always shown to the right of the arrow. Reactants are always shown to the left of the arrow. The information you see printed over each reaction arrow gives the conditions or the **reagents** used for the reaction to occur. The process starts with benzene, C_6H_6, which comes from coal tar and oil. It is a raw material for many familiar products like polystyrene and rubber. Chemists have several ways of representing benzene and compounds like it as shown in **Figure 1-5**.

Before a product is made available to the public, its synthesis is studied to determine the difficulties and costs of the process on a large scale. Because there is more than one way to make acetylsalicylic acid, as shown in **Figure 1-6**, alternative methods of production must be evaluated to determine which is most cost effective.

Willow branch

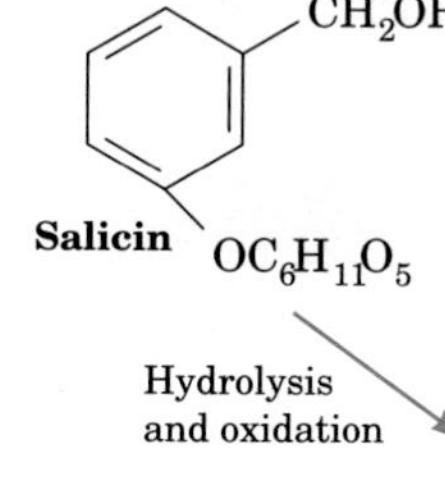

Salicin

OH O C CH_3 O

Methyl salicylate
(oil of wintergreen)

Hydrolysis and oxidation

Hydrolysis

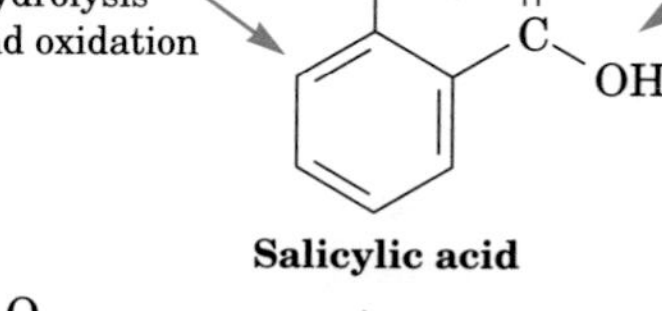

Salicylic acid

Addition of acetic anhydride

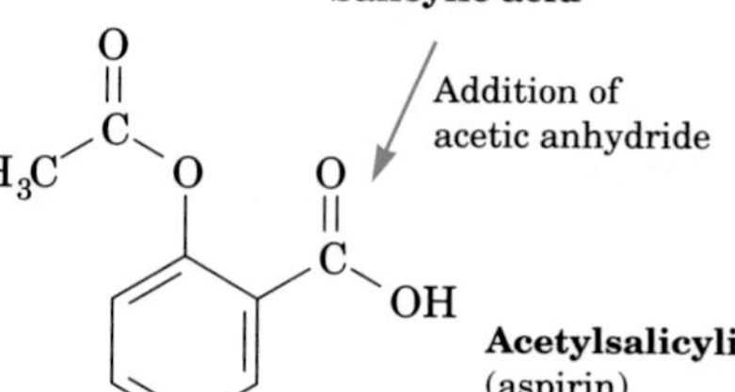

Acetylsalicylic acid
(aspirin)

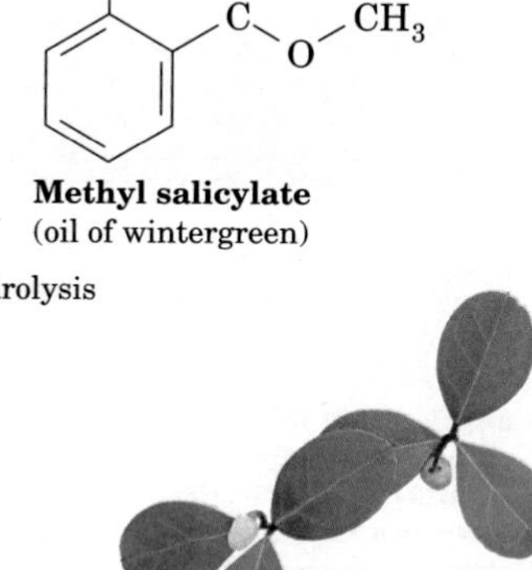

Wintergreen

Figure 1-6
Aspirin can also be synthesized starting with salicin in willow bark or methyl salicylate from the wintergreen plant.

Content Background

Oil of wintergreen is used as a flavoring agent in foods, beverages, pharmaceuticals, and perfumes. It is used as an ultraviolet-light absorber in sunscreens. However, its toxicity level by ingestion is about 30 mL for an adult and 10 mL for a child.

Possible Misconception

Many students will expect exactness in the representation of compounds and reactions. Using **Figure 1-5**, show them that chemists have several ways of representing the same substance, all of which are equally correct or valid. Symbols are the language of chemistry and, as in any language, there are often several words that can be used to describe an object.

Visual Strategy

Figure 1-6

Contrast these reactions with those in **Figure 1-4**. Even though the synthesis from benzene looks more complex, it is actually less expensive to produce aspirin by this reaction than by using natural raw materials. Ask students why they think the synthetic route is less expensive. (Natural raw materials may be in low supply and may be difficult to obtain, whereas raw materials for the synthesis are generally more abundant.)

Further Applications

Often in multistep reactions like the production of aspirin, which has seven steps, each step might be only 50% to 80% efficient. If you began with 100 kg of starting material and six steps were 80% efficient and one step was 50% efficient, how much aspirin would be produced?
$0.8^6 \times 0.5 \times 100$ kg = 13.1 kg
Chemical companies must also take the efficiency of reactions into account.

Most production processes begin with a piloting stage before full production begins. **Figure 1-7** shows that there is a dramatic difference between making a substance like nylon on a lab bench and in a manufacturing plant. Nylon production in the United States is about 2.6 billion pounds per year.

Chemists and chemical engineers who work on the large-scale development of a product want to make as much of it as possible, as cheaply as possible. Pilot programs for any new production process allow production chemists to monitor all aspects of the process to ensure that standards for the product can be met safely and economically.

When the product is a drug, the manufacturing process must adhere to the strict safety standards set by the U.S. Food and Drug Administration. Less than 1 out of 10 000 compounds synthesized by drug companies will ever make it to the consumer. Bringing a new drug to market generally takes about 10 years of research and testing before production can begin.

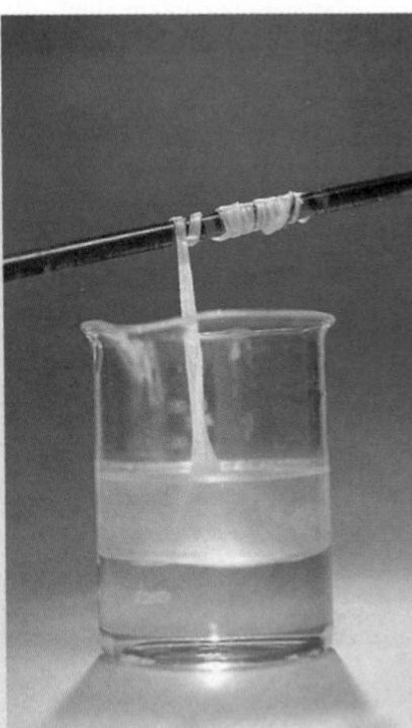

Figure 1-7a
Nylon was first produced on a small scale in 1935 by Dr. Wallace Carothers of Du Pont.

1-7b
The large-scale production of nylon results in a material that can be extruded into fibers for making nylon fabric. Nylon 66 and nylon 6 are the two most widely produced forms. Nylon 6 is prepared from another organic acid called aminocaproic acid.

Large-scale production considerations for chemicals

- What will the raw materials cost?
- Are the raw materials in adequate supply?
- What safety hazards are involved in the synthesis?
- Is it possible to carry out the reaction on a large scale? What kinds of equipment will be needed?
- How long will it take?
- What side reactions could take place? Do they need to be controlled?
- How pure will the product be? What steps must be taken to ensure a certain purity? What standards must be maintained for purity?
- Will the process involve the production and disposal of wastes or pollutants?
- What will it all cost?

Visual Strategy

Figure 1-7

Baking is a good example of the problems that can arise when quantities for a small batch are increased proportionately. Some recipes specifically state that doubling or tripling the quantities will give poor results.

Make students aware that chemists look first at the properties that a new chemical might possess for an industrial need. Then they determine ways to synthesize it.

Working in the field of chemistry involves the research, development, and production of new materials

The principles of chemistry are the foundation of the chemical industry, a multibillion-dollar business that employs chemists, chemical engineers, and chemical technicians to create new compounds and materials. Though you may think all chemists are research scientists, this aspect is only one of three major categories. **Figure 1-8** gives you some ideas about the variety of things chemists, engineers, and technicians do.

Research chemists work on the design and discovery of new materials. The drugs that will be used to treat or cure diseases like arthritis, diabetes, cancer, and AIDS are being designed and tested by research chemists. Work in this field requires a solid background in using computers. Today's chemists use computers extensively to design compounds, to simulate models of reactions to find out if they will work before going into the lab, and to communicate with other researchers concerning their progress in solving similar problems. To design experiments, researchers must be very familiar with today's sophisticated laboratory equipment.

Chemists, engineers, and technicians working in *development* are concerned with the task of developing full-scale production processes for new compounds and materials. They design the most inexpensive way of producing a compound or material that is also safe and environmentally sound.

Production chemists and technicians ensure that the new compound or material meets the predetermined standards set for its purity and other physical properties.

Jobs for chemists also exist in the areas of chemical sales and marketing, computer-software engineering, patent law, banking and finance, teaching, and technical writing.

Figure 1-8a
Research **focuses on the design, discovery, and preparation of new materials and products.**

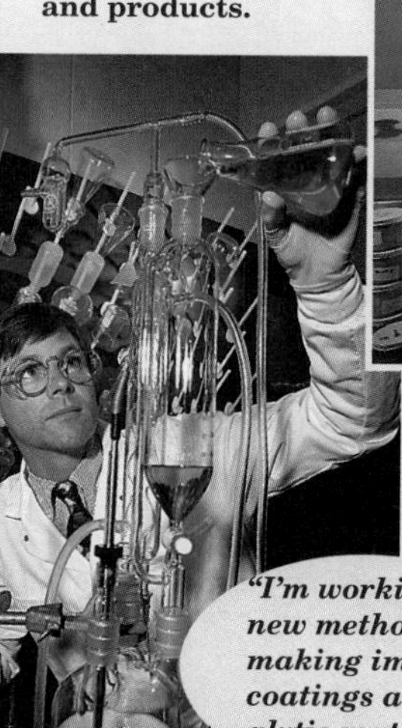

"I'm working on new methods for making improved coatings and catalytic materials."

"I take chemicals and make them into products that benefit us all."

1-8b
Development **focuses on the creation of methods for producing new products safely and economically.**

"I ensure that the equipment and processes we use deliver products that meet our customers' needs."

1-8c
Production **focuses on monitoring production processes to ensure that manufacturing and quality specifications are met.**

Content Background

Chemical engineers often find that reactions which are readily performed on a small scale in a laboratory do not function as well on a large scale. Part of a chemical engineer's job is to get these same reactions to work on a large scale by manipulating the reaction variables.

Do You Know?

A new, up-and-coming area of chemistry is polymer chemistry. As plastics and polymers become more important to industry, technology, and consumer products, more universities are specializing in polymer education and research. Many schools now have separate polymer departments.

Possible Misconception

Many students think that all scientists are researchers with graduate school backgrounds or doctorates. This section is designed to show them that chemists work in many fields outside the research laboratory. Many people in the chemical industry work as technicians who gain their skills from on-the-job training or by working through a 2-year degree program. To obtain more information on careers in chemistry write the American Chemical Society.
American Chemical Society
Education Division
1155 16th St., N.W.
Washington, D.C. 20036

Teaching Tip

What do chemists do?
Ask the class to raise a hand if they know someone who works in the field of chemistry. Try to arrange, throughout the year, for a few of these people to talk to the class about what they do. A better way to encourage student involvement is to have students interview the person and report to the class.

Do You Know?

The production of sulfuric acid in the United States was 80.31 billion lb in 1993, which is more than 325 lb of sulfuric acid produced per person. No wonder sulfuric acid is called the king of chemicals.

Teaching Tip

Symbolic vocabulary

Review the symbols for the elements using a periodic table (wall chart or page 111 of the text). Now ask students to interpret the formulas in **Table 1-2** by listing the elements in each formula on the board. Have them note the many combinations of carbon, hydrogen, and oxygen.

Cultural Connection

During World War I, there was a shortage of acetone, which was used to manufacture explosives. In England, Russian-born Jewish chemist Chaim Weizmann (1874–1952) devised a way for bacteria to manufacture the much needed acetone. Dr. Weizmann was also the first president of Israel.

Teaching Tip

The top 25 chemicals

Start a contest in which students keep a record in their notebooks of the number of times they encounter chemicals from the top 25 throughout the year. Award prizes for the most thorough records.

Table 1-2
Top 25 Chemicals in the U.S. (1993)

Rank	Name	Formula	Pounds produced (in billions)
1	sulfuric acid	H_2SO_4	80.31
2	nitrogen	N_2	65.29
3	oxygen	O_2	46.52
4	ethylene	C_2H_4	41.25
5	calcium oxide (lime)	CaO	36.80
6	ammonia	NH_3	34.50
7	sodium hydroxide	$NaOH$	25.71
8	chlorine	Cl_2	24.06
9	methyl *tert*-butyl ether	$C_5H_{12}O$	24.05
10	phosphoric acid	H_3PO_4	23.04
11	propylene	C_3H_6	22.40
12	sodium carbonate	Na_2CO_3	19.80
13	ethylene dichloride	$C_2H_4Cl_2$	17.95
14	nitric acid	HNO_3	17.07
15	ammonium nitrate	NH_4NO_3	16.79
16	urea	CN_2H_4O	15.66
17	vinyl chloride	C_2H_3Cl	13.75
18	benzene	C_6H_6	12.32
19	ethylbenzene	C_8H_{10}	11.76
20	carbon dioxide	CO_2	10.69
21	methanol	CH_3OH	10.54
22	styrene	C_8H_8	10.07
23	terephthalic acid	$C_8H_6O_4$	7.84
24	formaldehyde (37%)	CH_2O	7.61
25	xylene	C_8H_{10}	6.84

Source: *Chemical & Engineering News*, July 1994.

Chemicals production is a major industry

The chemical industry consists of those firms that supply chemical compounds that are raw materials for building other materials. **Table 1-2** is a list of the top 25 chemicals produced in the United States. Sulfuric acid, ranked first, is also the top inorganic chemical. It is the least expensive acid to produce and is therefore a desirable raw material for building other compounds. Its other uses are shown in **Figure 1-9**. Sulfuric acid is produced by burning sulfur and sulfur ores to produce sulfur trioxide, SO_3. Sulfur trioxide is converted to sulfuric acid by adding water.

$$SO_3 + H_2O \longrightarrow H_2SO_4$$

Nitrogen and oxygen, ranked second and third, are the two gases most prevalent in air. Both of these gases are obtained by liquifying air. Nitrogen and oxygen have different boiling points. As the liquid air mixture is warmed, the nitrogen boils off first, leaving the liquid oxygen behind.

Ethylene, ranked fourth, is the top organic chemical on the list. Most ethylene is used in making plastics. Ethylene is also a raw material for the production of ethylene dichloride (ranked 13th), vinyl chloride (ranked 17th), ethylbenzene (ranked 19th), and styrene (ranked 22nd).

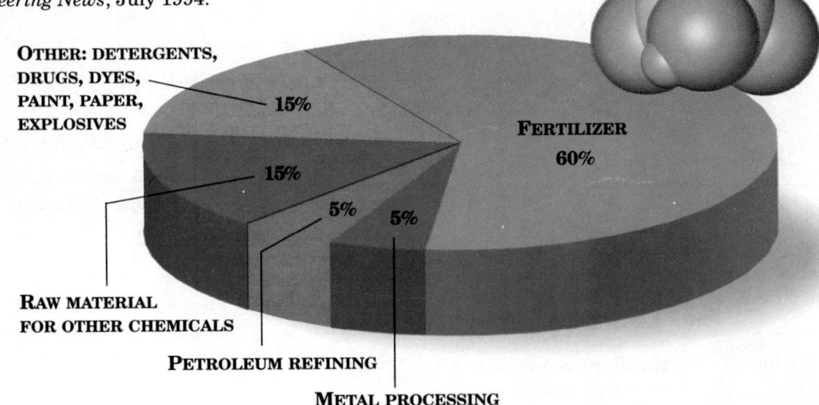

Figure 1-9
Most H_2SO_4 is used in fertilizers. When combined with salt, sulfuric acid can be used to make hydrochloric acid. It is a highly corrosive chemical. Outside of a lab about the only time you would encounter it is in the battery of a car.

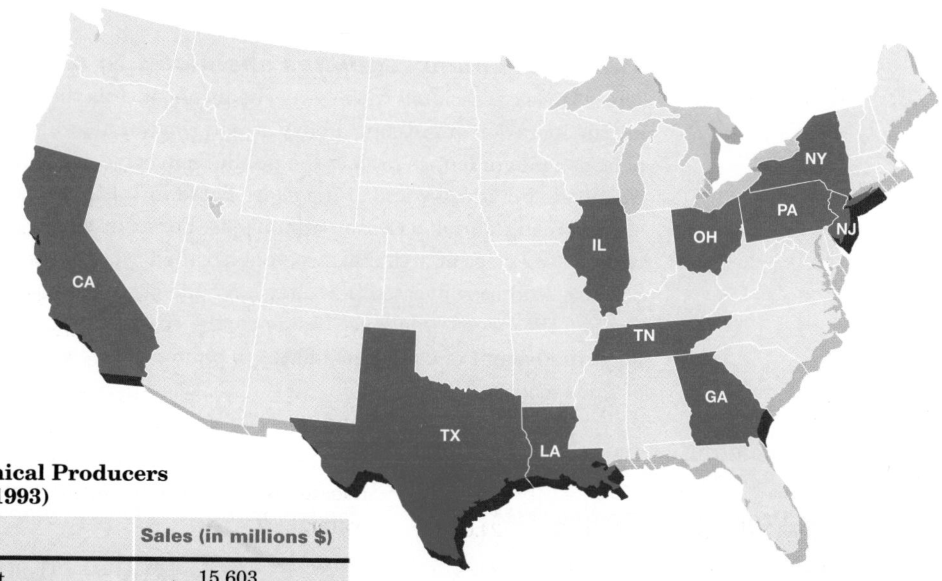

Figure 1-10
You can see that the top chemical producers are clustered by region. Texas and Louisiana, ranked first and second, occupy the region of the country where most petroleum refineries are found.

Table 1-3
Top 25 Chemical Producers in the U.S. (1993)

Rank	Name	Sales (in millions $)
1	Du Pont	15 603
2	Dow Chemical	12 524
3	Exxon	10 024
4	Hoechst Celanese	6 347
5	Monsanto	5 540
6	General Electric	5 042
7	Union Carbide	4 640
8	Occidental Petroleum	4 065
9	BASF	4 037
10	Eastman Chemical	3 903
11	Shell Oil	3 875
12	Amoco	3 773
13	ICI Americas	3 565
14	Mobil	3 533
15	Rohm and Haas	3 269
16	Arco Chemical	3 192
17	Miles	3 053
18	Air Products	2 906
19	W. R. Grace	2 895
20	AlliedSignal	2 791
21	Chevron	2 708
22	Ashland Oil	2 586
23	Ciba	2 519
24	Praxair	2 438
25	Phillips Petroleum	2 308

Source: *Chemical & Engineering News*, July 1994.

The chemical industry produces chemicals

Table 1-3 shows the top 25 chemical producers in the United States ranked by sales in millions of dollars. Some of the companies listed might be familiar to you, especially if you live in one of the top 10 chemical-producing states shown in **Figure 1-10**

Du Pont (ranked first) has held the top position since the companies were first ranked in 1969. Du Pont is classified as a diversified producer of chemicals, which means that it manufactures chemicals in more than one industrial division. Du Pont supplies industrial organic chemicals, industrial inorganic chemicals, and synthetic organic fibers like nylon and polyester. Dow Chemical (ranked second) is a producer of basic chemicals, and Exxon (ranked third) is a producer of petroleum chemicals. Hoechst Celanese (ranked fourth) is a major producer of synthetic organic fibers, plastics, and resins.

Groupwork Strategy

The top 25 companies

Divide the class into groups of three, and assign each group two of the chemical companies listed in **Table 1-3**. Using library resources, have each group determine the following.

1. Where are the companies located?
2. What types of chemicals do they produce?
3. How many divisions do they have?
4. How many people do they employ?
5. How has their business changed in the last 25 years?
6. What environmental hazards are caused by the chemicals they produce?

Teaching Tip

Enrichment: FDA ban

Have students research the ban on cyclamates. Who funded or did the research on cyclamates? How much material was fed to the rats and mice? What would happen if the same amount of sugar was fed to the animals? Does any country still use cyclamates?

Groupwork Strategy

Government agencies

Divide the class into groups of three, and assign each group one of the six government agencies listed in **Table 1-4**. Have the group find a mailing address, phone number, areas under each agency's jurisdiction, and the name of the local director. Have the group brainstorm instances that they feel would justify contacting each agency. What agencies have jurisdiction over the school they attend and why? Each group should also find three newspaper or magazine articles related to the agency they are studying.

The government regulates chemicals to reduce risk

There is risk associated with using chemicals, just as there is risk in having an operation or crossing a busy street. Government agencies whose policies and guidelines protect the public from some of the hazards of chemicals in industry and at home are listed in **Table 1-4**. Their policies are the result of risk assessment analysis. For example, in 1969 the widely used class of artificial sweeteners called cyclamates was banned because studies suggested that they have the potential to cause cancer. Though they were widely used in foods, the risk of cancer outweighed any potential benefit of using cyclamates to reduce the calorie content of some foods.

Table 1-4
Agencies Regulating Chemicals in the United States

Food and Drug Administration (FDA)

Purpose
Develops and enforces regulations regarding impurities and other hazards in foods, drugs, and cosmetics

Significant activities related to chemistry
- Approves new drugs for public release
- Sets labeling standards for drugs and foods (except meat and poultry)
- Sets regulations for the use of additives
- Inspects food manufacturers for sanitary conditions
- Validates product claims

Environmental Protection Agency (EPA)

Purpose
Works with state and local governments to enforce laws enacted by Congress regarding control of air and water pollution, solid wastes, pesticides, radiation, and toxic substances

Significant activities related to chemistry
- Sets limits for pollutants
- Sets restrictions on use of hazardous substances
- Sets tolerance levels for pesticides in foods; monitors those levels in humans, animals, and food plants

Occupational Safety and Health Administration (OSHA)

Purpose
Develops and enforces regulations regarding safe and healthy work environments

Significant activities related to chemistry
- Conducts site inspections to ensure safe conditions
- Imposes citations and penalties for noncompliance

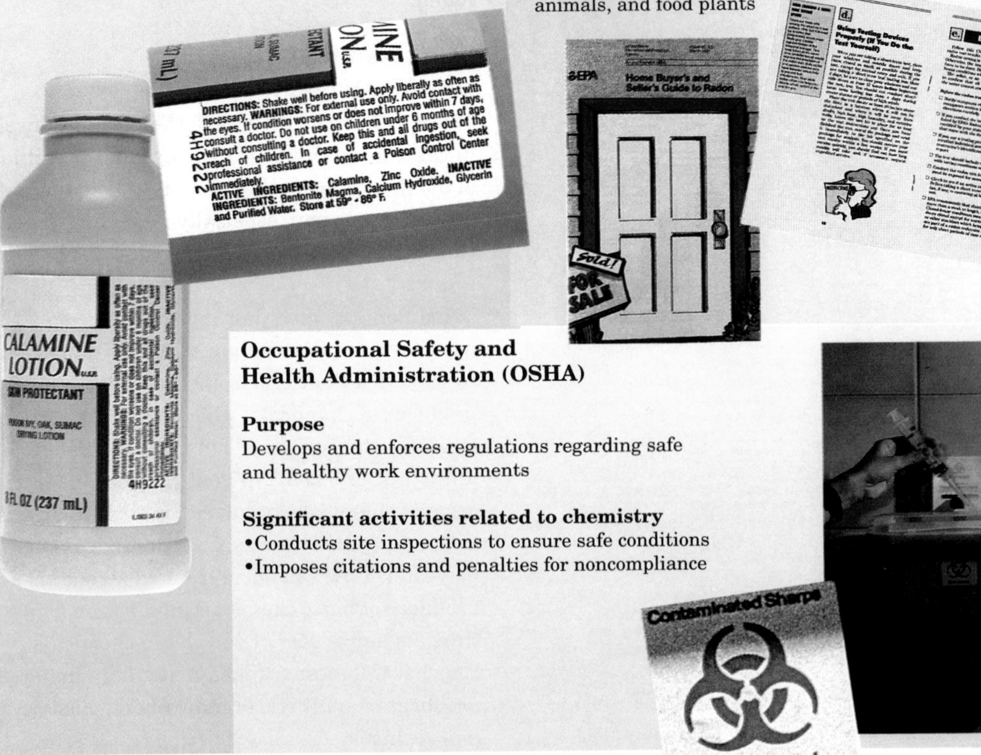

Agency for Toxic Substances and Disease Registry

Purpose
Works with state and federal agencies to collect and disseminate information regarding adverse health effects of hazardous substances in storage or of accidental release of hazardous substances through fire, explosions, or transportation accidents

Significant activities related to chemistry
- Maintains list of areas closed because of toxic contamination
- Assists in treatment of individuals exposed to hazardous substances

Consumer Product Safety Commission

Purpose
Evaluates safety of consumer products

Significant activities related to chemistry
- Enforces labeling of hazards on products
- Bans highly hazardous products

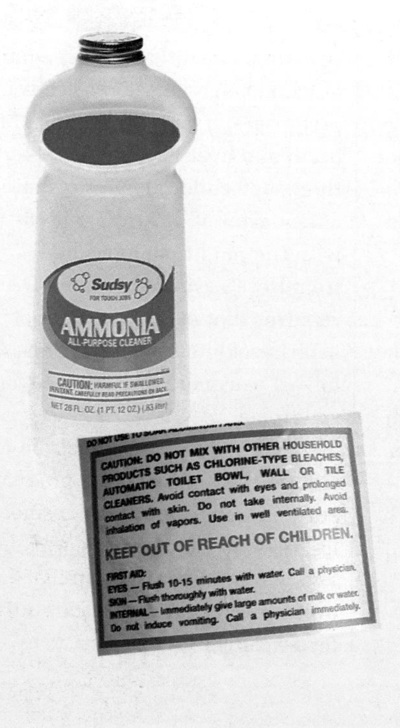

Department of Agriculture

Purpose
Works to maintain and improve agricultural productivity while protecting natural resources; ensures quality of food supply

Significant activities related to chemistry
- Analyzes foods for contaminants and pesticides
- Provides information on nutrient values in foods

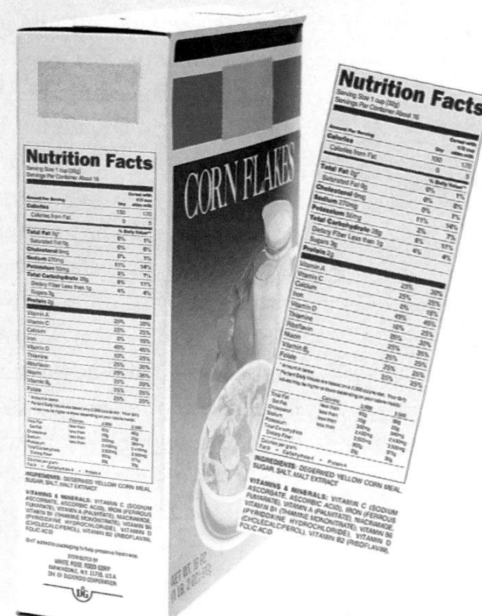

Teaching Tip

MSDS sheets

Obtain an MSDS from a chemical supplier if you don't have one. Review it with the students to point out where to find pertinent information that will affect them. Post the sheets in the room so that students can refer to them throughout the year.

Cultural Connection

One of the last alchemists was Johann Rudolf Glauber (1604-1668) of Germany. He discovered a method for producing hydrochloric acid using sulfuric acid and table salt. The by-product of the reaction, sodium sulfate, is still called Glauber's salt today. Glauber sold the sodium sulfate salt as an "elixir of life" although its best use is as a laxative.

Working with chemicals

To minimize your risk when working with chemicals in this course, you will be expected to pay strict attention to chemical labels, Material Safety Data Sheets (MSDS), and safety precautions. Chemists, like the one in **Figure 1-11**, know how important the MSDS is. You will be expected to consult these sheets frequently to understand the risks and hazards of using particular chemicals.

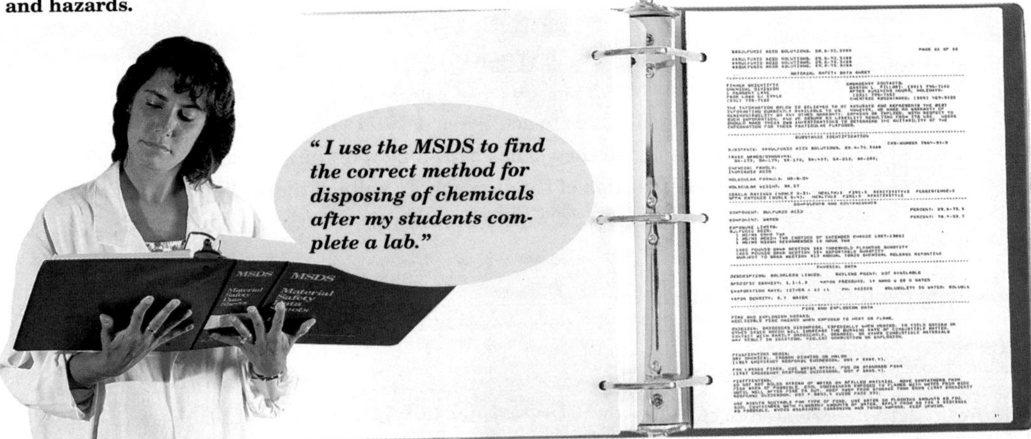
"I use the MSDS to find the correct method for disposing of chemicals after my students complete a lab."
MSDS
MSDS

Figure 1-11
People working with chemicals consult MSDS for precautions and hazards.

Consumer Focus

Product warning labels

Although aspirin is considered one of the safest over-the-counter drugs, there are still a few unpleasant side effects associated with its use. It may cause nausea or vomiting and may prolong bleeding. For this reason, aspirin should not be used by anyone who has just had a tonsillectomy or oral surgery, or by pregnant women or hemophiliacs. Some people, especially those with asthma, may be allergic to aspirin.

Some studies have also shown that children and adolescents with chickenpox or a severe case of flu may develop Reye's syndrome if they take aspirin. Symptoms of Reye's syndrome include sudden vomiting, violent headaches, and unusual behavior. As the condition progresses, it causes swelling of the brain and liver malfunction. More than 1000 children, approximately one-quarter of those diagnosed with the condition, have died from Reye's syndrome. The FDA requires that every container of aspirin sold in the United States have a warning label about Reye's syndrome.

You'll find warning labels on a wide variety of consumer products, such as foods, medicines, cosmetics, toys, cleaning agents, solvents, and paints. When you purchase a product, check the label carefully for warnings.

technology *the application of scientific knowledge for practical purposes*

Technology is using science to solve problems

Chemistry differs from **technology**, which is the use of scientific principles to solve a problem. Determining the structure of aspirin, an analytical process, is chemistry. Using aspirin to relieve pain and reduce fever is technology. Using the structure of aspirin to design other pain relievers is also technology.

Aspirin has a number of technological uses and has become the model for other pain relievers. This is an example of how chemists can take a product, modify it slightly, and come up with another substance with more useful properties and applications.

Because aspirin is a pure compound, there is no difference between the chemical action of name brands and generic brands. However, aspirin can vary from brand to brand in terms of dosage, buffering systems, and the time it takes to act. Some aspirin tablets have special coatings that do not dissolve in the acidic environment of the stomach, and therefore do not cause irritation of the stomach. In spite of these slight variations in brands, there is no difference between the relief you get from a generic brand and a name brand.

You can now buy aspirin-free pain relievers like ibuprofen and acetaminophen. Their structures are shown in **Figure 1-12**. Tylenol and Anacin-3 are brand names for products that both contain acetaminophen. This compound does not prolong bleeding and is less likely to cause stomach upset. Ibuprofen is the compound found in Advil and Nuprin. Like aspirin, ibuprofen may irritate the stomach and prolong bleeding. Stomach irritation can be reduced if products with ibuprofen are taken with food.

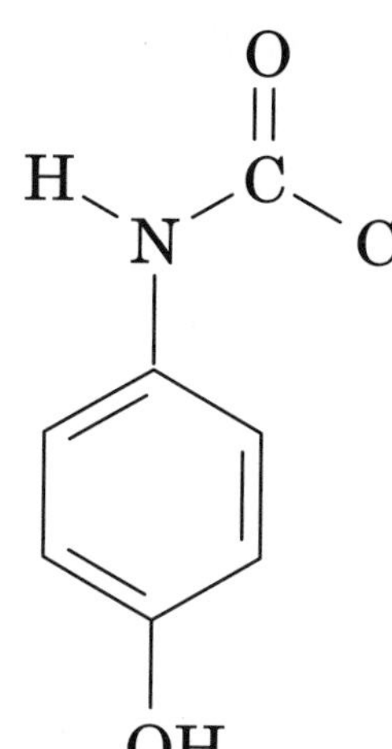

Acetaminophen

Figure 1-12a
Acetaminophen is a derivative of benzene which was shown in Figure 1-5. Notice that it differs from aspirin and ibuprofen in that it contains nitrogen.

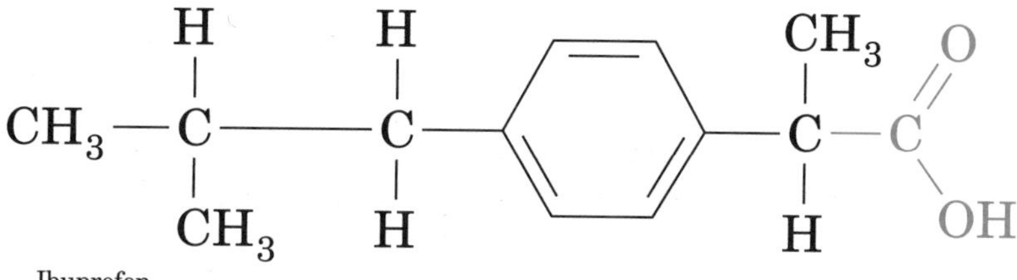

Ibuprofen

1-12b
Ibuprofen is also a derivative of benzene and, like aspirin, it is an acid. The carbon group shown in red on the right side of the molecule is the same group that gives aspirin its acidic properties.

Section Review

1. Classify all the compounds in **Table 1-2** as organic or inorganic.
2. Which compounds in **Table 1-2** are acids?
3. List four properties of H_2SO_4.
4. How does technology differ from chemistry?
5. List three ways that you can determine whether an unknown compound is an acid or base.

6. **Story Link**
What government agency regulates tamper-proof packaging for aspirin?

ASSESS

Alternative Assessment

Have students classify all of the substances listed in **Figure 1-4** as organic or inorganic.

Answers to Section Review

1. Organic: ethylene, methyl *tert*-butyl ether, propylene, ethylene dichloride, urea, vinyl chloride, benzene, ethylbenzene, methanol, styrene, terephthalic acid, formaldehyde, xylene
Inorganic: sulfuric acid, nitrogen, oxygen, calcium oxide, ammonia, sodium hydroxide, chlorine, phosphoric acid, sodium carbonate, nitric acid, ammonium nitrate, carbon dioxide (oxide of carbon)
2. Sulfuric acid, nitric acid, terephthalic acid
3. Possible responses are: tastes sour, conducts electricity in solution, has a pH of less than 7, reacts with bases and some metals to form salts, has the highest production of any chemical in the United States, and contains hydrogen, sulfur, and oxygen.
4. Technology is the application of chemical theory and principles. For example, the hardness of a diamond is due to its tetrahedral network structure. The study of that structure is chemistry. The use of the hardness property of diamonds for grinding is an example of technology.
5. Some possible responses are: test it with pH paper, test it with litmus paper, see if it reacts with an acid or base (look for pH changes). This is a good time to point out that tasting an unknown substance is *never* a good idea.
6. Consumer Product Safety Commission

Visual Strategy

Figure 1-12

Remind students that —CH_3 represents H–C(H)(H)–H. Show students how to count the number of bonds on a carbon atom in an organic structure. Then have them count the bonds on each carbon in both structures. What do they find?

Section 1-2

FOCUS

Lesson Starter

Observations of a candle
Place a candle in a holder on the front desk. Light the candle and have students discuss what causes the candle to burn. Gather enough information for students to build theories. Have them relate their observations to other things they have seen burning: wood, kerosene lanterns, etc. (When the candle wick is first lit, it burns; then wax melts, climbs the wick, nears the heat of the burning wick, and evaporates. The wax vapor and oxygen fuel the flame of the wick. The wick itself burns very little once the wax starts to burn.)

TEACH

Teaching Tip

Scientific method
Students have seen the scientific method presented many times and in many ways in other courses. The actual steps are not as important as the idea that there is an organized procedure and that experimental results are subjected to scrutiny until they are verified by other scientists.

How do scientists approach problems?

Section Objectives

Describe the processes in the scientific method.

Explain the purpose of controls in an experiment.

Describe the role of models in chemistry.

Distinguish between hypothesis, theory, and law.

Scientific knowledge is gained from experiments

When the early Romans and Greeks discovered how an extract prepared from willow bark could relieve a variety of ailments, they undoubtedly came upon this finding by trial and error. In their search, they must have tried many extracts prepared from a variety of plants. Most extracts were probably useless, and some may even have been lethal. In the end, their success was more the result of chance rather than the product of well-designed plans.

Although these Romans and Greeks placed great reliance on rational thought and logic, they rarely felt it necessary to test their findings or conclusions and were not inclined to experiment. Gradually, experiments became the crucial test for the acceptance of knowledge. Today, experiments are an integral part of research in all sciences, including chemistry. Science is distinguished from other fields of study in that it provides guidelines or methods for conducting research.

The way scientists carry out experiments or investigations is referred to as the *scientific method*. The scientific method is a logical approach to exploring a problem or question that has been raised through observation. In addition, this approach is designed to produce a solution or answer that can be tested, retested, and supported by experimentation. Although different representations are used to describe the scientific method, it consists of the fundamental activities outlined to the left. One distinguishing feature of science that separates it from other fields of study is the last step. The research findings of any study must be reproducible by other scientists for those findings to be valid.

The biggest difference in the way discoveries are really made is in how the question or problem is recognized. Sometimes questions arise from an accidental discovery. In turn, these questions may lead scientists to search for answers in one of several ways. As examples, let's look at how two important drugs were discovered.

Activities of the Scientific Method

Make observations and collect data that lead to a question.

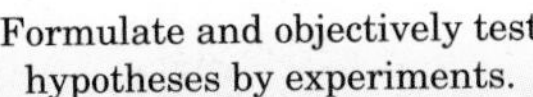

Formulate and objectively test hypotheses by experiments.

Interpret results and revise the hypothesis if necessary.

State conclusions in a form that can be evaluated by others.

Visual Strategy

Activities of the Scientific Method
Make students aware that this is a continuing process that cycles over and over until all new hypotheses have been exhausted or all the observations have been explained.

Scientific knowledge is sometimes gained by accident

In the 1920s, the British scientist Alexander Fleming was experimenting with bacteria that he was growing in special dishes. Before going on vacation, Fleming decided to leave the dishes as they were and clean them when he returned. When he returned from vacation, he noticed that some of the dishes had become contaminated with a blue mold. But what really caught Fleming's attention was that the bacteria had grown all over the dish except for the area around the mold.

Fleming hypothesized that the mold was secreting a substance that was lethal to the bacteria. He designed experiments to test his idea and found that he was correct. The substance secreted by the mold is what we now know as penicillin.

A similar accidental discovery was made some 40 years later. In this case, Dr. Barnett Rosenberg had been using platinum probes to study the effects of electric fields on living cells. He and his colleagues observed that bacteria did not reproduce near these probes. After conducting extensive experiments, they were able to show that a compound made from platinum formed near the electrodes. This compound, called cisplatin, is shown in Figure 1-13. Cisplatin was responsible for the changes in cell reproduction.

Rosenberg then reasoned that because cancer involves uncontrolled cell growth, perhaps cisplatin would be effective in treating cancer. The results of carefully planned experiments showed this reasoning to be correct. Cisplatin was approved by the Food and Drug Administration in 1979 for treatment of certain types of cancer.

The discovery of aspirin, penicillin, and cisplatin show that there is no *one* method that will guarantee an important discovery. However, there is a sequence of events that leads to the formation of a conclusion. That conclusion must be supported by data in order to be valid.

Figure 1-13a
Cisplatin is a compound made from platinum, chlorine, nitrogen, and hydrogen.

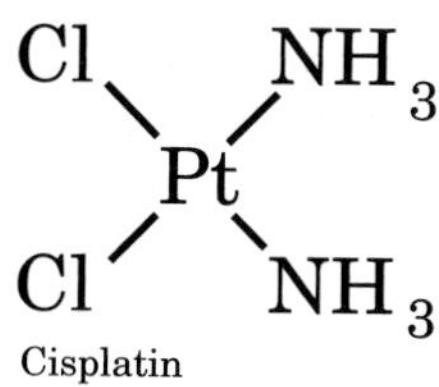

1-13b
The large cell in the center of the photograph is a cancer cell. Cisplatin affects the DNA of cancer cells to halt uncontrolled reproduction.

Historical Note

Alexander Fleming (1881–1955) investigated bacterial cultures and suspected the presence of an antibacterial effect, but he was never able to isolate the antibacterial substance.

Cultural Connection

Percy L. Julian (1898–1975) was an African American organic chemist who was the to first synthesize cortisone, testosterone, progesterone, and physostigmine.

Visual Strategy

Figure 1-13

Show students that this compound exists in two forms labeled *cis* and *trans*. Even though the composistion is the same, only the *cis* form affects cell growth; the *trans* form has no effect.

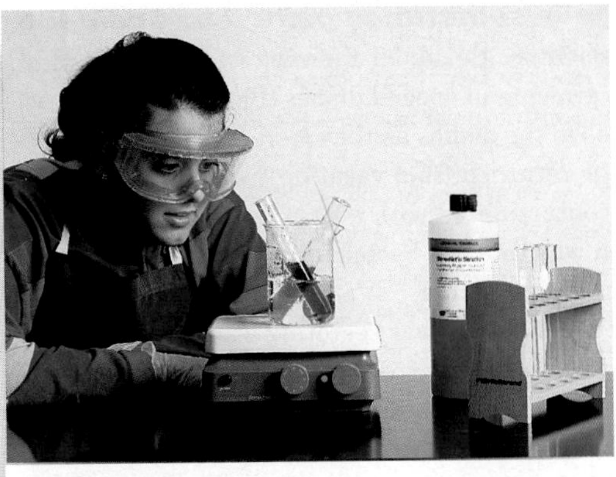

Figure 1-14
Carolyn is testing various food products claiming to be "sugar free" for the presence of sugar. If the test reagent turns from blue to red after reacting with the sample, sugar is present.

The scientific method involves making educated guesses

Recognizing and defining a problem or question stems from observation, as shown in **Figure 1-14**. The discovery of cisplatin and its usefulness in treating cancer resulted from a simple observation—bacteria did not reproduce near the platinum electrodes. That observation led to further experiments and tests.

Once observations have been made, they are analyzed. Scientists start by examining all the relevant information or data they have gathered. They look for related data to establish some relationship or conclusion. Scientists then try to come up with a reasonable explanation for what they have observed. Any explanation they propose must be testable. A reasonable explanation of some observation that can be tested is known as a **hypothesis**. Hypotheses are usually written using an *if–then* format that describes a cause-and-effect relationship. A testable hypothesis for Dr. Rosenberg's work with cisplatin is as follows: *If* cisplatin can slow or stop cell reproduction, *then* it could be effective in treating cancer, which is uncontrolled cell growth.

hypothesis
a proposition based on certain assumptions that can be evaluated scientifically

The scientific method involves experiments

An *experiment* is a process carried out under controlled conditions to test the validity of a hypothesis, as shown in **Figure 1-15**. Dr. Rosenberg's work with cisplatin initially involved the question of what was inhibiting the growth of the cells. The most important part of his experimentation was the control of *variables*. Before Rosenberg could come to the conclusion that cisplatin was indeed the reason that the growth of bacteria was affected, he had to measure the effects of all other possible causes. How did he know the effect was not caused by a simple temperature change in the culture medium? How did he know the culture had not been contaminated? How did he know there weren't changes in the amount of electric current? Because platinum is a generally unreactive (inert) element and the concentration of platinum compounds in the culture medium were very low, it initially seemed unlikely that platinum could be causing the cells to stop reproducing.

Figure 1-15
Monica and Desmond are experimenting to determine how to get the largest volume of popped corn from the fewest kernels. They hypothesize that *if* the volume of popped corn increases as the moisture in the kernels increases, *then* soaking the kernels in water should increase the popped volume of a fixed number of kernels.

18

Figure 1-16
Collaboration between scientists is common in the search for new information.

Rosenberg had to narrow the field of possible variables by keeping them constant. Temperature, current, purity of the platinum, and the contents of the culture medium are some of the many variables that could change from experiment to experiment. By keeping variables constant, researchers can isolate the key variable. It took many experiments over a two-year period for Rosenberg and his team to verify that the formation of cisplatin was the key variable that caused the cell growth to stop.

Rosenberg's initial work spawned further research of the anticancer properties of cisplatin. After years of testing by independent sources to verify the effectiveness of the drug, cisplatin was approved for use in cancer chemotherapy.

Data from experiments can lead to a theory

Any conclusion scientists make must come directly and solely from the data they obtain in their experiments. Many times, scientists discover that they are unable to arrive at a conclusion. In fact, they may discover that the data fail to support their hypothesis. In that case, they must reexamine the question and develop a new hypothesis to be tested.

theory
an explanation of an observation that is based on experimentation and reasoning

Any hypothesis that withstands repeated testing may become part of a theory. A **theory** is a broad generalization that is based on observation, experimentation, and reasoning. Because theories are not facts, but rather explanations, they can never be proven. A theory is considered successful if it explains *most* of what is observed and is modified as new information is discovered. For example, you will learn about atomic theory in Chapter 3. The idea that all matter is made of discrete particles called atoms that cannot be further divided was first proposed by Democritus about 400 B.C. Using experimentation in the early 1800s, John Dalton provided the data to support that theory. This theory has been modified as scientists have discovered a number of smaller particles that make up atoms, as illustrated in **Figure 1-17**.

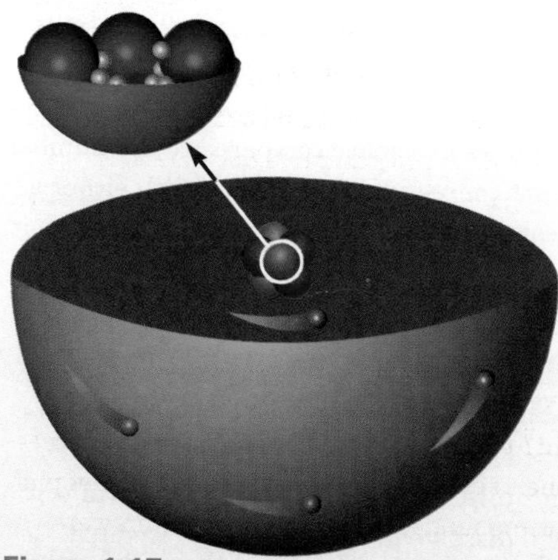

Figure 1-17
The discovery of quarks adds to the theory of atomic structure. A neutron is made of three quarks held together by small particles called gluons.

Top Quark, Last Piece in Puzzle Of Matter, Appears to Be in Place

By WILLIAM J. BROAD

The quest begun by philosophers in ancient Greece to understand the nature of matter may have ended in Batavia, Ill., with the discovery of evidence for the top quark, the last of 12 subatomic building blocks now believed to constitute all of the material world.

An international team of 439 scientists working at the Fermi National Accelerator Laboratory will announce the finding today, bringing nearly two decades of searching to a dramatic conclusion.

The Fermilab discovery, if confirmed, would be a major milestone for modern physics because it would complete the experimental proof of the gran... ...tical edific... known as...

in physics at Cornell University, said the finding was "a very big deal" that "makes the whole picture of subnuclear particles much more believable and better established."

"We've needed the top quark," he said. "It figures in all our calculations for further processes, and none of them would be right if it weren't there."

If the top quark could not be found, the Standard Model of theoretical physicists would collapse, touching off an intellectual crisis that would force scientists to rethink three decades of work in which governments around the globe had invested many billions of dollars.

All matter is made of atoms, but nearly a century ago physicists dis...red that atoms, long consid... ...he smallest uni...

Do You Know?
The word *hypothesis* comes from a Greek word meaning "groundwork."

Theme
Classification and trends
The formation of most hypotheses involves gathering and classifying observations to determine whether they show a trend.

Do You Know?
The word *theory* comes from the Greek word *theöria* meaning "to look at."

Teaching Tip
Theories
Ask students to name any theories they have already learned (theory of gravity, evolution, relativity, plate tectonics, etc.).

Groupwork Strategy
Scientific method
Divide the class into groups of three, and tell them they will have 10 minutes to work on the following problem. For many years heat was thought to be contained in a caloric fluid—things that give off a lot of heat when burned contain large amounts of caloric fluid, and things that give off small amounts of heat contain little caloric fluid. Caloric fluid was believed to move from hotter objects to colder objects. We still have the word *calorie* today as a remnant of the theory. Have each group propose a hypothesis and an experiment that might disprove the caloric theory and lead to a new theory. (Hint: friction gives off heat without burning.) The caloric theory was disproved by Count Rumford (1753–1814) née Benjamin Thompson.

Visual Strategy
Figure 1-16
Remind students that scientists keep written records of events and try to write down all observations even though they may seem unimportant at the time.

Teaching Tip

Models

Have students brainstorm other examples of models that they have encountered. Most students think models are objects or drawings. Show them that models can be explanations or even mathematical equations.

Stress that all models have some limitations. It is also important for students to realize that models can be smaller than the original object being studied. Examples include car models, models of chemical factories, models of the sun, and models of the universe.

ASSESS

Alternative Assessment

People often remark that hot water freezes faster than cold water. Have the students prepare a hypothesis and formulate some experiments that might explain this concept.

Answers to Section Review

7. to communicate their findings to the scientific community and to obtain the necessary verification of their work so that it is accepted by the scientific community
8. making observations and collecting data in order to formulate a question or hypothesis, test the hypothesis, interpret the results against the hypothesis, revise it if necessary, and communicate results
9. Models enable scientists to make predictions about systems, to build materials with desired properties, and to show concrete examples of abstract ideas.
10. Sample answer: If a higher temperature causes more sugar to dissolve in hot tea than in cold tea, then more sugar should dissolve in hot water, hot soda, or hot oil than in cold forms of these liquids.
11. He searched the literature for other salicylate compounds that were less acidic but showed effects similar to sodium salicylate.

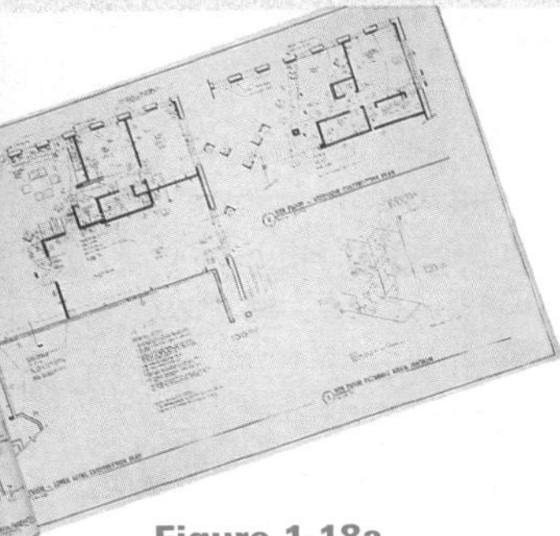

Figure 1-18a
Models take many forms. The blueprint is a two-dimensional model.

1-18b
The mouse is often used in research as a model for how humans react to drugs and treatments.

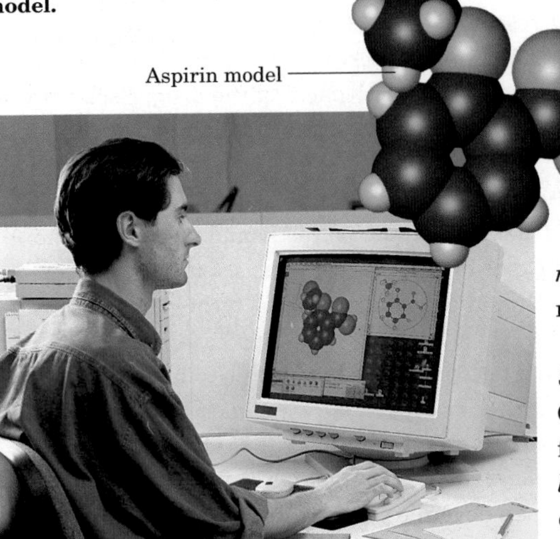

1-18c
Chemists today use computers to build molecules for specific uses. They also use simulation models to predict the results of reactions before entering the lab.

Models are like theories

Models play a major role in science. They can take many forms, from actual replicas to written descriptions. Figure 1-18 shows several types of models. Models can be used in various ways. The structure of acetylsalicylic acid serves as the model for the development of other chemical substances that could also work as painkillers. The wave-particle model describes the behavior of electrons as waves and as particles.

It is important to remember that models are not always exact. The model you see in Figure 1-18 represents a molecule. The atoms are not hard spheres, nor are they the sizes and colors shown. However, the model does show the geometrical arrangement of the atoms and their relative sizes.

When a model is an explanation, it explains *most* of the observations. Models, like theories, are refined as new information is discovered.

Scientific laws are based on facts

Certain facts in science always hold true. Such facts are labeled as scientific laws. A *scientific law* is a statement or mathematical expression of some consistency about the behavior of the natural world.

For example, motion is described by Newton's laws. Newton's first law states that if there is no net force acting on an object and if the object is at rest, it will stay at rest; and if the object is in motion, it will continue in its motion. This law explains the movement or lack of movement of all matter.

There are a limited number of laws in science compared to the number of theories and hypotheses. Do not confuse a scientific law with either a hypothesis or theory. A hypothesis *predicts* an event. A theory *explains* it. A law *describes* it.

Section Review

7. Give two reasons why chemists publish the results of their experiments.
8. What activities are part of the scientific method?
9. How do models help chemists acquire knowledge about the natural world?
10. You observe that more sugar will dissolve in hot tea than in iced tea. Formulate a hypothesis for this observation. Develop a plan to test your hypothesis.

11. **Story Link**
How did Felix Hoffmann build on other scientists' work with aspirin?

How do I learn chemistry?

Section Objectives

List one example of how you can use chemical knowledge.

List the characteristics of a system at equilibrium.

Describe the five themes that will be used in this text.

Make a study plan for this course.

Acquiring scientific knowledge

All of the advancements in communications technology have contributed to a rapid accumulation of scientific knowledge. Scientists depend heavily on the work of other scientists in their search for answers. Think back to Hoffmann's discovery of aspirin as a pain reliever. His original work was based on a literature search of the work done by other chemists on salicylate derivatives. Today that search would be much easier using an online database like Chemical Abstracts.

Rosenberg's work with cisplatin became the basis for further study by the National Cancer Institute and numerous other research facilities around the world working to find more effective cancer-fighting drugs. Today these scientists can easily talk to each other by modem using computer networks like Internet, the world's largest network. More than ever before, the ability to communicate information in an organized, understandable manner is essential to progress.

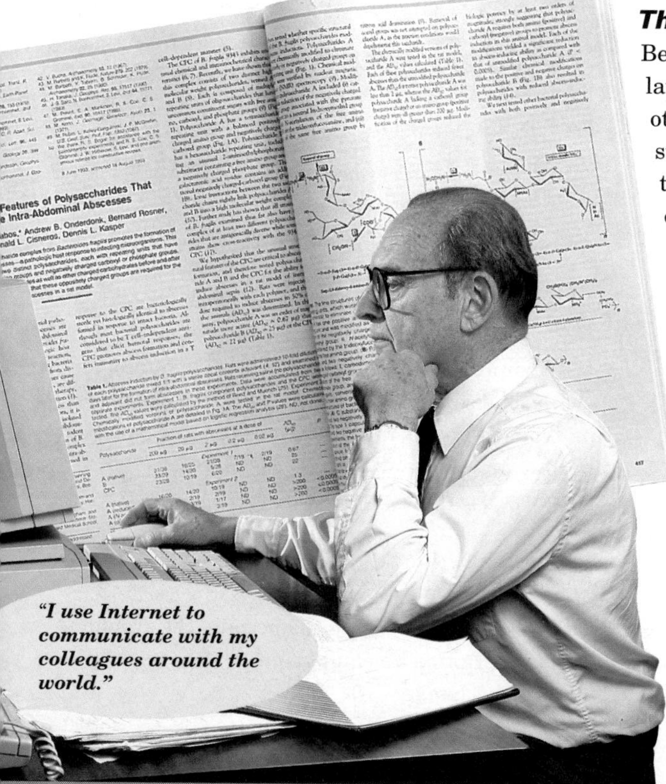

"I use Internet to communicate with my colleagues around the world."

Figure 1-19a
Scientists communicate their findings through published accounts in scientific journals.

1-19b
Scientists use computer networks to look up information and confer with other scientists.

Themes help unify your study

Because so much knowledge has accumulated over the years, chemistry, like any other science, has become a complex subject. However, that doesn't mean that chemistry is difficult to understand or impossible to learn. On the contrary, although understanding chemistry may be a challenge at times, it should never be an obstacle that you can't overcome, especially if you keep a few suggestions in mind. The wealth of information in any subject can be classified using a small number of fundamental ideas.

Section 1-3

FOCUS

Lesson Starter

Use the following demonstration to discuss how knowledge is acquired through trial-and-error testing.

Demonstration 3

Experimentation

Approximate time: 5 min

1. Show students a plastic foam cup, and make several marks on it using a permanent non-toxic marker. Ask students to propose ways to remove the marker ink. Show them that using soap and water does not work.
2. Show students a piece of glass, and make several marks on it with the same permanent non-toxic marker. Rub lighter fluid on the mark to remove it from the glass surface.
3. Rub some lighter fluid on the plastic foam cup. What happens? (It should dissolve.) Ask students what they learned and how they learned it?

SAFETY

Wear safety goggles, a lab apron, and plastic (not rubber) gloves. There should be no flames in the room at the time of the demonstration. Students should be at least 10 ft from the demonstration area. Limit the amount of lighter fluid to 25 mL. Use a 10 cm^2 piece of cloth to apply the lighter fluid to the cup.

DISPOSAL

Put the cup in the trash. Keep the cloth in the fume hood until the lighter fluid has evaporated. Return any remaining lighter fluid to the can for storage.

TEACH

Teaching Tip
Communication
Because reporting results is so fundamental to science, communication skills are as important to a scientist as scientific skills. Complete, thorough reporting of lab results and comprehensive organized note-taking are critical to students' success. All of the writing you receive from students should be graded for style and organization as well as content. Establish this criteria early so that students know what to expect.

Teaching Tip
Themes
Do not expect students to have a complete understanding of the themes at this stage in the course. The concepts represented by the themes are acquired as students encounter more and more examples of them. They will be highlighted throughout this teacher's guide for you to help students make the connections between ideas.

Teaching Tip
Equilibrium systems
Another example of equilibrium is a sealed soda bottle in which evaporation and condensation are constantly occurring. If the cap is removed, a change occurs in the system. It is also no longer a closed system.

system
all the parts that form a unified whole

equilibrium
a reversible reaction in which the rates of the forward and reverse reactions are equal

The themes in this text can help you organize what initially looks like a lot of unrelated information. Five themes have been selected to cover most of what you will learn this year. You will be given questions and problems throughout the text that will signal you to think about how a theme could be used to link some of the many things you have learned. Use these themes as a reference in your study throughout the year.

Macroscopic observations and micromodels
Chemistry focuses on the observable effects of the behavior of particles like atoms, molecules, and ions. The observable effects are in the macroscopic world and are acquired using the senses. Explaining those observations requires a model of what we believe to be occurring at the particle level. You will see numerous examples of this theme in the text illustrations. When you observe a substance, you'll see a model of its atoms, molecules, or ions, such as the one in **Figure 1-20**. When looking at a solid dissolving in water, you'll see models of the breakdown of the solid's ordered structure. Understanding events on the particle level enhances your ability to explain what you observe and to make predictions.

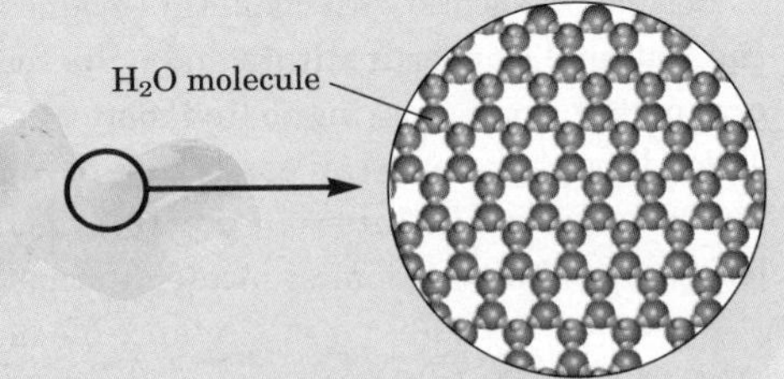

Figure 1-20
Ice consists of an open network of water molecules. Each water molecule is made from oxygen and hydrogen atoms.

Systems and interactions
A **system** is a collection of components that define something we choose to study. A system may be as large as the universe or as small as the contents of a test tube as in **Figure 1-21**. One goal of chemistry is to understand interactions within a system.

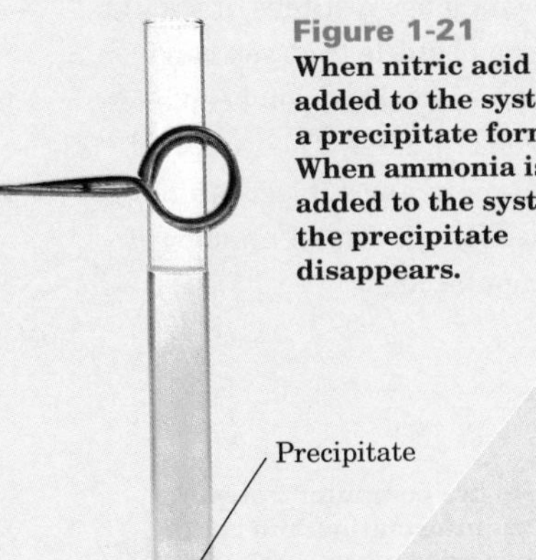

Figure 1-21
When nitric acid is added to the system, a precipitate forms. When ammonia is added to the system, the precipitate disappears.

Equilibrium and change
Equilibrium describes a state or condition of balance. In chemistry, equilibrium describes reversible processes occurring at the same rate. An example of an equilibrium system is shown in **Figure 1-22**, a glass of salt water. When you add salt to water until no more salt dissolves, you have what is called a solubility equilibrium system. Dissolved salt is in equilibrium with solid salt crystals at the bottom of the glass. The system appears to be in balance at the macroscopic level. However, the equilibrium model explains that dissolved salt is recrystallizing, while solid salt crystals are dissolving.

$$\textbf{dissolving} \rightleftharpoons \textbf{recrystallizing}$$

Because both processes occur at the same rate, the system is in equilibrium. Disrupting an equilibrium system usually produces noticeable change.

Visual Strategy
Figure 1-20
This illustration is respresentative of many others throughout the book that connect a micromodel to a macroscopic observation. Stress the releationship between macroscopic and microscopic as much as possible throughout the course.

Conservation

Although you will study numerous examples of change throughout this course, there are some fundamental properties of a system that are conserved, such as mass, energy, and charge. See **Figure 1-23**. Chemistry is more readily understandable and predictable because we know certain things cannot appear and disappear. For example, although mass is rearranged during a chemical reaction, it is conserved.

$$2H_2 + O_2 \rightarrow 2H_2O$$

Hydrogen Oxygen Water

Figure 1-23
Conservation of mass for the formation of water is symbolized by a balanced equation. The same number of each type of atom appears on both sides of the arrow.

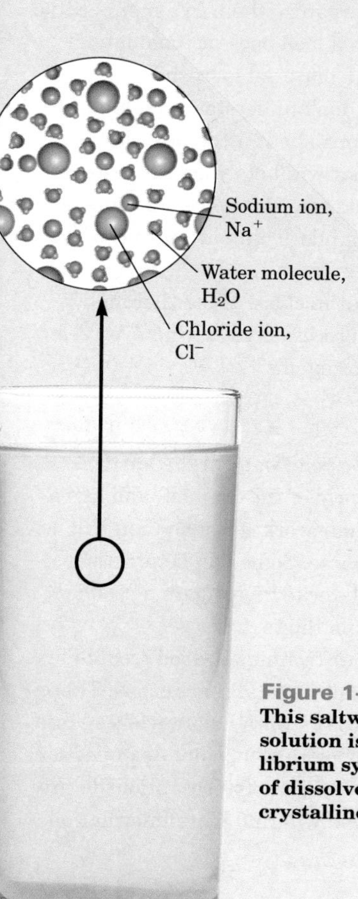

Figure 1-22
This saltwater solution is an equilibrium system of dissolved and crystalline salt.

Classification and trends

From your very first science course, you've been taught to classify. Developing categories based on similar characteristics or patterns of behavior will simplify your study and reduce the need to memorize. In this chapter, you learned that there are two large classes of compounds based on composition—organic and inorganic. You learned about two subclasses of compounds, acids and bases, each of which are characterized by a specific set of properties. If a compound is an acid, like those shown in **Table 1-5**, you can predict that when it is dissolved in water, it will have a pH of less than 7 and will turn blue litmus to red. You will get a lot of practice classifying, and your understanding of classification will be measured by your ability to make reasonable predictions.

Table 1-5
Common Acids and Their Formulas

Acid	Formula
sulfuric	H_2SO_4
nitric	HNO_3
hydrochloric	HCl
acetic	CH_3COOH
formic	$HCOOH$
phosphoric	H_3PO_4

Possible Misconception

Many students don't understand that the word *conserved* means "totally saved." Many think of conservation in terms of recycling, in which mass is not totally conserved.

Teaching Tip

Vocabulary

Remind students that they have used classification systems extensively in biology, especially in the taxonomic classification of living organisms.

Teaching Tips

Laboratory safety

Remind students that the chemistry laboratory can be a very dangerous place if proper safety precautions are not followed. Students should know what they are to do before they begin work. Plan periodic safety quizzes before students work in the lab. Assign a responsible student to be the safety chief whose job is to check all safety procedures before the lab begins.

Teacher expectations

If you aren't currently doing so, put your expectations for students in writing. Because a notebook is so critical to student success, make it mandatory. If students know that keeping a notebook has some effect on their grades, they will take it more seriously.

Concept maps

The technique of building concept maps, a powerful study aid, is covered in Appendix C. Assign concept maps frequently throughout the course so that students have plenty of opportunities to practice and perfect their technique.

Themes overlap

You probably recognized that these themes are not unrelated ideas but are interconnected. When describing an equilibrium situation, you look at the interactions of particles in a system and describe the equilibrium in terms of macroscopic observations and micromodels. Do not expect to fully understand all of these ideas at this point in the course. Thinking thematically takes time and requires practice. The Linking Chapters section in the Chapter Reviews will give you some practice.

Study tips

Your notebook as a resource

Good class notes are critical to your success in this course. The following are some guidelines for keeping up your notebook, which will save you time and frustration at exam time.

1. Listen during the lecture, and write down only those things that seem important. Don't try to write down everything.
2. Review your notes after the lecture, and organize them in a manner that allows you to make sense of them when you look back at them later. Revise any unclear explanations while the material is fresh in your mind. If there is still something that you don't understand, highlight the material to remind yourself to ask questions about it the next day in class. Use titles throughout your notes that will help you locate the information you need quickly. Some students rewrite their notes outside of class to keep the material fresh in their minds. Think of your notebook as a second textbook.
3. Keep notes on all the problems worked out in class. Make thorough notes for each step of the solution so the process is clear when you refer to these notes while working homework problems.

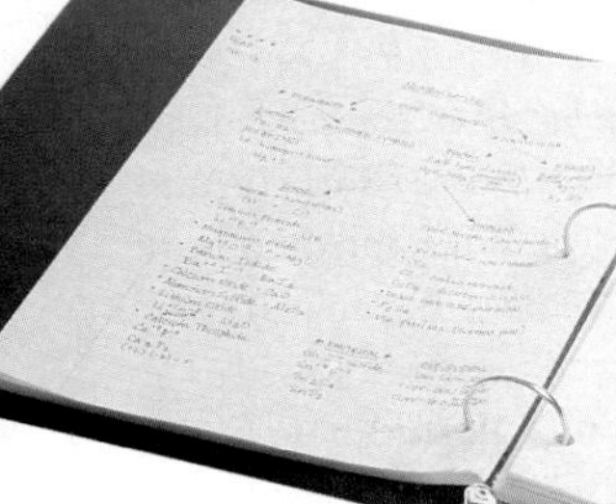

Figure 1-25
Review your class notes right away and keep your materials in good order.

Your work in the lab

Another important part of your study will be your work in the lab. It is there that you will be exposed to the real world of the chemist, and like a chemist, you will be expected to approach your work seriously. You will be assigned a variety of experiments in this course. Some will be practice in a particular technique that you will use later to investigate a problem. Others will reinforce the ideas you learn from the text.

Your success in the lab will depend on your technique. Good technique results in statistically valid data, which result in valid conclusions. Therefore, it is a good idea to be prepared for the day's work before walking into the lab. Know what you are supposed to do before you begin. Ask questions if you are unsure about a procedure. Observe all safety precautions to protect yourself and those working with you. You will find more information about your role in the lab on page 644.

How To
Succeed in learning chemistry

1. Learn the language.

Important terms and their definitions are listed in the margin of this book. Make sure you understand these definitions. Ask your teacher about them if they don't seem to make sense to you. In addition, writing formulas and using mathematical equations are also part of the language of chemistry. The Sample Problems will guide you through both of these skills, and the reviews will give you a lot of opportunity to practice.

2. Learn the material described by the Section Objectives.

Each section opens with a listing of what you are expected to learn. All questions and problems for that section relate to the objectives, so if you can work through the Reviews, you will know that you're ready for the test.

3. Use the illustrations.

The illustrations throughout the text will help you make the connections between what you see in the macroscopic world and micromodels. To practice thinking on the particle level, draw your own pictures to represent concepts or problems.

4. Take notes and review them frequently.

It is wise to review your notes as soon as possible after class. Write down questions on any material you don't understand so that you can ask them during the next class. Each chapter builds on the next, so don't fall behind! Remember that after the exam you can't forget what you've learned—you will use it in the next chapter. Use the Linking Chapters section in each Chapter Review to help reinforce necessary skills as you progress through the course.

DON'T CRAM FOR THAT EXAM!

5. Work all the problems you can for practice.

Always review the chapter's Sample Problems. If you have difficulty with a problem in the Chapter Review, the Hints refer you back to the Sample Problems. Answers to selected problems are in the back of the book so that you can see if you are on the right track. Never spend more than 15 minutes trying to solve a problem—get help! You won't pass the test if you can't work the problems.

6. Don't cram for an exam.

Section Review

12. What is a system? What is the system for the equilibrium example on page 22?

13. Bring in one example of a news story that you read recently in a newspaper or magazine that demonstrates why knowledge of chemistry is important.

14. Develop a study plan to use in this course.

15. What fundamental properties are conserved during chemical change?

16. List one example of something you did in the last month that involved chemistry.

Teaching Tip

Study techniques

Remind students that they should read the illustrations in the book and the captions that go with them to increase their understanding. It may also be necessary sometimes to read sections in the book more than once to fully comprehend the concepts.

It is far better to spend 30 minutes a day for 4 days studying for a test than to spend 2 hours the night before the test. Frequent review helps move information into long-term memory and allows time to link new information to that previously learned (constructivism).

Answers to Section Review

12. A system is all of the parts that form a unified whole; the system is the container of salt water.

13. Post interesting stories on the bulletin board.

14. Check students' plans for frequent reviews of notes.

15. mass and energy

16. Answers will vary.

ASSESS

Alternative Assessment

Have students explain the five major themes using examples.

Chemistry in Your Community

Consumer product price comparisons

Visit three drug stores in your area and do a pricing survey for aspirin, acetaminophen, and ibuprofen. Determine the cost per tablet for brand names at each store vs. the cost per tablet for generics. See if there are differences in dosage per tablet among the various brands. Prepare advertisements that might appear in a magazine or newspaper for the most economical brands in each pain-reliever category.

Answers to Applying Concepts

1. Organic: benzene, chlorobenzene, phenol, sodium phenoxide, sodium salicylate, salicylic acid, acetylsalicylic acid, acetic anhydride

   Inorganic: chlorine, aluminum chloride, sodium hydroxide, carbon dioxide (oxide of carbon), hydrogen chloride

Conclusion: A Wonder Drug

The story of aspirin is just one of many examples of how chemistry works. Let's go back to the questions posed on page 3 to build on the aspirin story.

What is aspirin? Aspirin is the common name for acetylsalicylic acid, $C_9H_8O_4$. It is a white crystalline compound with no odor and a slightly bitter taste. In moist air, acetylsalicylic acid will break down to form salicylic acid and acetic acid, which smells like vinegar.

How does aspirin work? In the 1970s, chemists working in London observed that many forms of tissue injury in the human body were followed by a release of a group of hormones known as prostaglandins. Prostaglandins are synthesized and released only when cells are injured, and are responsible for the redness, swelling, and fever that accompany an inflammation or infection. These compounds can also cause headache and pain throughout the body.

Chemists have suggested that aspirin somehow blocks the release of prostaglandins, but the details of how aspirin does this still remain a mystery.

How is aspirin similar to other widely used pain relievers? Aspirin has striking structural similarities to both acetaminophen and ibuprofen. All three of these pain relievers are organic compounds with a benzene ring as part of their structure.

How does buffered aspirin differ from regular aspirin? Because aspirin is an acid, it can increase the acidity level of the stomach. Buffers are mixtures that resist changes in pH when small amounts of acid or base are added. Buffers help maintain the stomach's overall acidity level even when some acid is added.

How does name-brand aspirin differ from generic brands? Because aspirin is a pure compound, there is no difference between the chemical action of name brands and generic brands in the body. Aspirin can vary from brand to brand in terms of dosage, buffering systems, and the time it takes to act.

Applying Concepts

1. Classify every compound listed in **Figure 1-4** as organic or inorganic.

Research and Writing

Use the library to find out more about the following.

1. Technology has created some challenges that we all must face and some problems that no one has yet solved. Choose a current event to highlight one such problem that chemists are currently trying to solve.
2. The Food and Drug Administration must approve any drug before it can be made available to the public. Find out how the FDA can determine that a drug is safe without testing it on humans.

CHAPTER 1

Highlights

Key terms

acid	**organic compounds**
base	**pH**
chemical	**reagent**
equilibrium	**synthesis**
hypothesis	**system**
inorganic compounds	**technology**
	theory

Key study strategy

Studying the textbook: The SQ3R method

Survey: Scan the headings throughout the chapter to learn what the chapter is about.

Question: Change each heading in the chapter to a question that you will answer while reading.

Read: Read slowly and with meaning to answer your question. Ask yourself what the author is expecting you to learn from each paragraph.

Recite: Say the answers to your questions out loud whenever possible. If you state your answers clearly, then you know the material and you've made some of it part of your long-term memory.

Review: This strategy also helps move information into your long-term memory. Using concept maps (described in Appendix C) is an active way to review material for long-term retention.

Key ideas

1-1 What will I learn in chemistry?

- Chemistry is the study of the structure of matter, its properties, and reactions.
- Chemists classify matter into categories based on common properties, such as organic compounds and inorganic compounds, or acids and bases.
- Chemists work in research, development, and production of new materials.
- Chemistry differs from technology, which is the application of scientific knowledge for practical purposes.
- Because chemicals pose risks to both individual health and the global environment, several government agencies issue policies and guidelines to protect the public from chemical hazards.

1-2 How do scientists approach problems?

- Scientists investigate a question or problem by reporting their observations in a way that other researchers can duplicate the conditions and verify the results.
- To explain scientific observations, scientists form a hypothesis, which is then tested by experiments with controlled variables. By changing only one variable at a time, scientists isolate the key variable.
- Theories and models are broad generalizations that explain most of the observations made of a particular phenomenon.
- While hypotheses and theories change with the discovery of new information, a limited number of principles hold true. These principles are scientific laws.

1-3 How do I learn chemistry?

- Five themes unify this text: macroscopic observations and micromodels, systems and interactions, equilibrium and change, conservation, and classification and trends.
- A successful study plan for this chemistry course includes learning the language of chemistry, mastering the section objectives, using the illustrations, working all practice problems, taking well-organized notes, and reviewing your notes promptly.

Closure Strategy

Have students collect all chemistry-related articles in one week from the local newspaper and magazines. Students should choose one story and write two paragraphs about why they found the article interesting and about its applications or effects on their lives.

1. Answers will vary, but may include cooking food, taking medications, applying cosmetics, wearing clothing fibers, or using hair products.
2. You could tell your friend that all substances are chemicals. Even "natural" materials are made of chemicals.
3. a. acid b. acid c. base d. base e. acid
4. a. inorganic b. inorganic c. organic d. organic e. inorganic f. organic g. organic h. organic
5. Usually chemists conduct a pilot program to study costs and possible hazards prior to any large-scale production of a chemical. When the pilot program is successful, they start production. After production has started, the safety standards set by government agencies must be followed.
6. Answers should include three of the following factors. A chemist must consider the availability and cost of raw materials, the method and equipment needed to manufacture the product on a large scale, the time required for production, the purification of the product, the hazards of production, the disposal of wastes, and the overall cost of production.
7. A product warning label describes possible hazards or adverse reactions that are associated with the use of the product. Like a product warning label, an MSDS contains hazard information, but it also includes important information on the safe handling of a substance.
8. a. Because both are classified as acids, they will have acidic properties, with a pH of less than 7.
   b. The reaction of two acids should result in an acidic product, pH < 7.

CHAPTER 1

Review and Assess

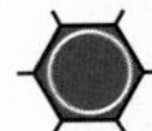

The field of chemistry

REVIEW

1. List four situations in which you used chemicals today.
2. Your friend mentions that she buys shirts made only of natural materials because she follows a chemical-free lifestyle. How would you respond?
3. The pH values for several products are listed below. Label each as acid, base, or neutral.
   a. grapes: pH = 4.0
   b. orange juice: pH = 3.7
   c. shampoo: pH = 10.0
   d. depilatory (hair remover): pH = 12.0
   e. bread: pH = 5.5
4. Classify each of the following compounds as organic or inorganic.
   a. HCl b. H_2SO_4 c. CH_3COOH
   d. $C_{12}H_{22}O_{11}$ e. H_2 f. CH_3Cl
   g. C_6H_6 h. $C_2F_2Cl_2$
5. Describe the stages involved in developing a new chemical product.
6. List three factors that chemists consider before proposing the full-scale production of a substance or product.
7. How does a product warning label differ from Material Safety Data Sheet (MSDS)?

APPLY

8. Aspartame is an artificial sweetener produced from the amino acids phenylalanine and aspartic acid.
   a. Predict the pH of phenylalanine and aspartic acid. Is it pH < 7, or pH = 7?
   b. Predict the pH of aspartame. Is it pH < 7, or pH = 7?

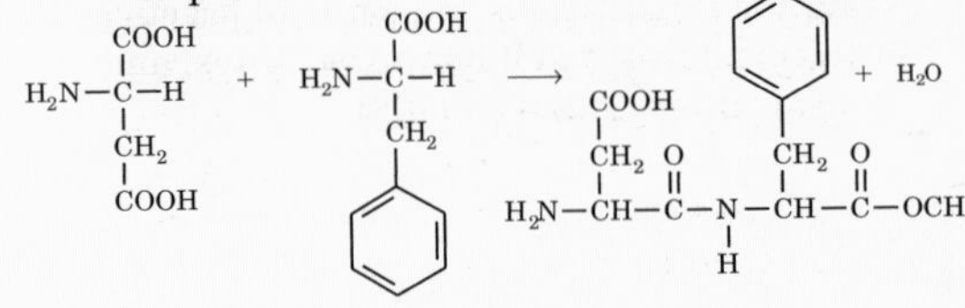

9. If new government regulations increased the cost of producing sulfuric acid, what industries would be most affected by the higher cost?
10. Classify the following examples as the result of science or technology.
    a. using a microwave oven
    b. determining the formula for a new polymer
    c. driving your car
    d. determining the mass of a white powder
    e. washing with soap
    f. dry cleaning a dress or suit

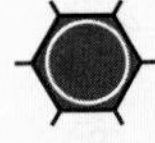

The scientific method

REVIEW

11. Identify the requirements of a good hypothesis.
12. How does the phrase "cause and effect" relate to the formation of a good hypothesis?
13. Classify these statements as observation, hypothesis, theory, or law.
    a. A system containing many particles will not go spontaneously from a disordered to an ordered state.
    b. The substance is silvery white, fairly hard, and is a good conductor of electricity.
    c. Bases taste bitter and feel slippery in water.
    d. If I pay attention in class, then I will succeed in this course.
14. What components are necessary for an experiment to be valid?
15. Explain the purpose of an experimental control.
16. Explain the relationship between models and theories.

APPLY

17. Explain the generalized statement: "No theory is written in stone."

18. a. The table below contains data from an experiment in which an air sample is subjected to different pressures. Propose a hypothesis that could be tested from this set of observations.
    b. What theories can be stated from the data in the table below?
    c. Are the data sufficient for the establishment of a scientific law. Why or why not?

The Results of Compressing an Air Sample

Volume $(cm)^3$	Pressure (kPa)	Volume × Pressure $(cm^3 \times kPa)$
100.0	33.3	3330
50.0	66.7	3340
25.0	133.2	3330
12.5	266.4	3330

19. You read in the paper that "experiments have shown that listening to music increases the efficiency of studying." Should you immediately start listening to music as you study? Explain.

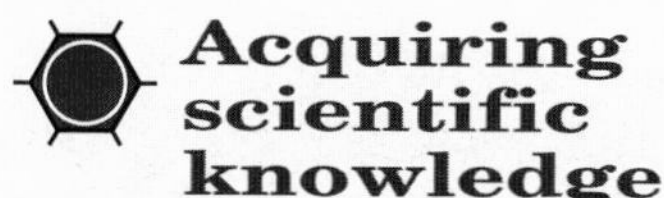

Acquiring scientific knowledge

REVIEW

20. Briefly describe the five themes used to organize the material in this textbook.
21. List at least five study skills that are required to successfully study chemistry.

APPLY

22. Name two social concerns that you have that can be better understood with the study of chemistry.
23. In music, the theme is a continually repeated melody that shows up throughout the piece. How do the themes in this book relate to a musical theme?
24. Why are themes important in scientific inquiry?

Alternative assessment

Performance assessment

1. Outline a study strategy for learning the important facts and understanding the concepts for this course. One effective strategy is to get a study partner. Another is to approach your study by assuming that you have to teach someone the material you are expected to learn. Discuss your plans with your teacher and use them throughout the course.
2. Select one of the chemicals in **Table 1-2** that is kept in the chemistry laboratory. Request a copy of the MSDS for that chemical from your teacher. Prepare your own product warning label from the MSDS. Compare your product warning label with the actual product warning label from the manufacturer.
3. Your teacher will provide you with an unknown chemical solution. Design an experiment that will classify the solution as an acid or a base.

Portfolio projects

1. ***Chemistry and you***
   Make a poster showing the types of product warning labels that are found on products in your home.
2. ***Research and communication***
   Clip a scientific article from a newspaper or magazine and paraphrase the article. Highlight and define chemical terms. Note the difference between the uses of chemistry and technology that appear in the article.
3. ***Chemistry and you***
   For one week, practice your observation skills by listing chemistry-related events around you. After your list is compiled, choose three events that are especially interesting or curious to you. Label three pocket portfolios, one for each event. As you progress through the chapters in this textbook, gather information that helps explain these events. Put pertinent notes, questions, figures, and charts in the folders. When you have enough information to explain each phenomenon, write a report and present it in class.

9. Because a very large quantity of sulfuric acid is used to manufacture fertilizer, the fertilizer-manufacturing industry and the agriculture industry that depends on fertilizer would probably be most affected.
10. a. technology b. science
    c. technology d. science
    e. technology f. technology
11. A good hypothesis explains an observation, is testable, and is used to make predictions.
12. Cause and effect is another way of stating a hypothesis and prediction. The hypothesis is an assumed cause of a phenomenon. The effect is another word for prediction.
13. a. law b. observation
    c. observation
    d. hypothesis
14. One variable must be tested at a time, and the others must be held constant. A control must be used for all variables being tested under all reaction conditions.
15. An experimental control provides a numerical reference or baseline for comparing the variable.
16. Models are theories, which help explain a concept.
17. Yes; theories are based on our current knowledge. They are subject to revision.
18. a. Possible hypotheses include the following. If the pressure increases to compress a sample of air, then the volume will decrease. If pressure and volume are inversely related, then their product will be constant.
    b. Pressure and volume are inversely proportional when temperature is held constant.
    c. No; more experimentation would be required to establish a scientific law.
19. No; before you accept the results of an experiment, you need to know more about how the experiment was conducted and how carefully the scientific method was followed. You also need to know if the results have been validated and accepted by other scientists.

20. The *systems and interactions* theme describes the cause-and-effect relationships that characterize the behavior of matter. The idea that the behavior of atomic and submicroscopic systems explains phenomenon at a macroscopic level is the focus of the *macroscopic observation and micromodels* theme. The *equilibrium and change* theme describes systems in a state of balance and the effects of stress on systems. *Conservation* defines the consistent patterns within a system. The *classification and trends* theme shows the systematic organization that categorizes matter and its behavior.

21. Answers will vary. They may include learning the language of chemistry, mastering the section objectives, using illustrations, taking and reviewing notes, and working all practice problems.

22. Answers will vary. Students might mention the role of chemistry (1) in determining and improving the quality of air and water; (2) in finding a cure for AIDS; and (3) in solving the problems of drug abuse.

23. A musical theme repeats throughout a song, just as the chemistry themes repeat throughout the chapters.

24. Themes link the many ideas and concepts of a field of study and provide a framework from which we can make predictions.

Performance assessment

1. Students' strategies will vary. However, the importance of taking good notes in class, reviewing those notes, and working through the assigned problems cannot be emphasized enough in reviewing strategies.

2. Answers will vary. Be sure the students' labels communicate hazards clearly.

3. Answers will vary. Be sure that all experimental designs include appropriate safety precautions. Students should use the laboratory safety section starting on page 651 as a guide. Have students use litmus and pH paper. Do not allow them to taste or touch the substance.

Portfolio projects

1. Grade posters based on the variety of products shown, organization, labeling, spelling, and overall artistic quality.

2. Grade the writing for its clarity and coherent organization, correct spelling and grammar usage, and correct usage of chemistry terminology and symbolism.

3. Collect the folders, and use them as a basis for assigning reports throughout the year by selecting one topic for each student from their folder.

Teaching Tip

Discussion

1. Start a discussion with students concerning all of the reasons why a researcher might alter data. Ask them to predict the potential damage that could be caused by such fraud.
2. Ask students why we have agencies like the FDA. Discuss the role of the FDA and other regulatory agencies in policing sound laboratory practices.
3. Have students go to the library and find more about the cold fusion incident cited in the story. What damage resulted from the false claims?

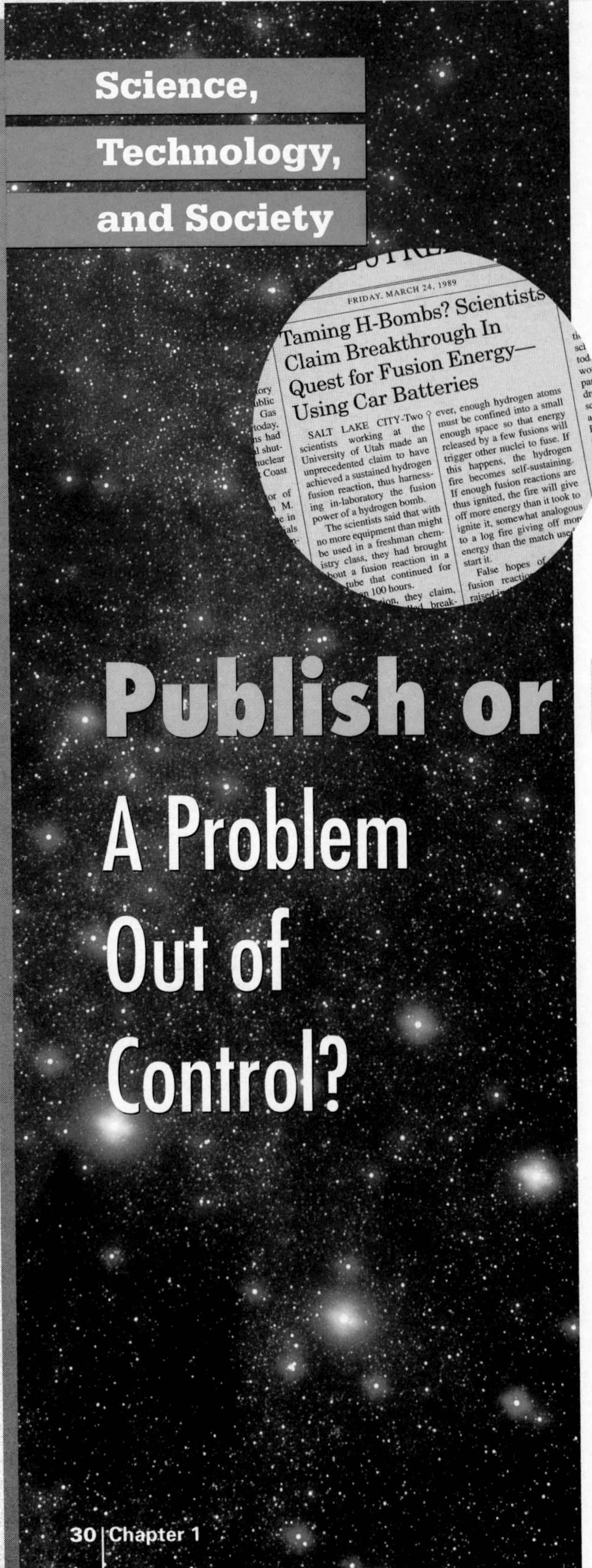

FRIDAY, MARCH 24, 1989

Taming H-Bombs? Scientists Claim Breakthrough In Quest for Fusion Energy—Using Car Batteries

SALT LAKE CITY-Two scientists working at the University of Utah made an unprecedented claim to have achieved a sustained hydrogen fusion reaction, thus harnessing in-laboratory the fusion power of a hydrogen bomb.

The scientists said that with no more equipment than might be used in a freshman chemistry class, they had brought about a fusion reaction in a tube that continued for ... 100 hours.

...ever, enough hydrogen atoms must be confined into a small enough space so that energy released by a few fusions will trigger other nuclei to fuse. If this happens, the hydrogen fire becomes self-sustaining. If enough fusion reactions are thus ignited, the fire will give off more energy than it took to ignite it, somewhat analogous to a log fire giving off more energy than the match used to start it.

False hopes of fusion reaction...

Science, Technology, and Society

On March 23, 1989, two chemists, B. Stanley Pons and Martin Fleischmann, claimed that they had produced a sustained thermonuclear fusion reaction, the same process that powers stars and exists briefly in the explosion of hydrogen bombs. This claim was astonishing in light of the fact that nuclear fusion had been pursued by physicists for over 40 years with expenditures of billions of dollars, all without success. Pons and Fleischmann had experimented for less than six years at a total cost of about $100 000. Furthermore, while conventional approaches required temperatures over 10 million kelvins for fusion to occur, Pons and Fleischmann produced fusion at room temperature with simple electrolysis equipment. Although the process they used was not clearly understood at the time, the expectation was that in a few years cold fusion would not only be harnessed to produce energy, but also change the way nuclear physics would be understood.

Publish or Perish

A Problem Out of Control?

Neither of these expectations came to pass. Nearly all attempts to repeat the experiment were unsuccessful. In the course of the next few months it was learned that Pons and Fleischmann had never performed any carefully controlled experiments in cold fusion and that the work they had done lacked reproducible data.

The cold fusion incident is a recent but hardly singular case of scientific dishonesty. For centuries improper practices have occurred in the sciences and have ranged from "trimming" actual data (to help them match the desired results) to the complete fabrication of findings.

Why would scientists mislead the public and the scientific community? Many influences can twist the work of reputable and normally honest scientists. One factor is called experimenter bias, which involves emphasizing data that fit an expectation, while overlooking, or sometimes deliberately suppressing data that contradict expectations. Experimenter bias is often unconscious and, as such, does not constitute deliberate deceit. For instance, Gregor Mendel, founder of the modern principles of heredity and genetics, may have classified some peas that were not clearly smooth or wrinkled or green or yellow into categories that gave the correct results. Thus, while Mendel's hypothesis was correct, his data were statistically "too good to be true."

Similarly, reputable scientists in the former Soviet Union and the United States believed in the 1960s that they had successfully formed a polymer (a chain of molecules linked together and the basic chemical form of plastics) out of ordinary water. It was finally determined that "polywater" was nothing more than water contaminated by silicon from the capillary tubes in which it was always observed.

Experimenter bias is embarrassing, but little more. Far more serious is the recent trend of professional self-preservation. As the number of scientists has increased in the last hundred years, so has the competition among them for available funding. The need to maintain a competitive edge has driven a number of scientists to alter data to yield more impressive results or, in some cases, to manufacture results completely. The excuse cited for this behavior is that scientists are overworked and pressed for time, so they draw conclusions of which they are absolutely certain with the intention of actually running the experiment later. All too often, the experimental values turn out to be nowhere near those that have been prematurely reported. This situation is worsened by researchers' fears that competitors may know of their work and may use this knowledge to publish findings first.

Many of these factors seem to have been at play in the cold fusion experiments.Pons, although hard working and dedicated, often worked so fast that he did not catch errors in his methods or analysis. His enthusiasm for new ideas, such as those suggested by Fleischman, may have actually encouraged experimental bias.

Concern over competitors was also involved. Pons and Fleischmann carried out their work at the University of Utah in Salt Lake City, while their principal rival, Steven Jones, was conducting similar experiments less than fifty miles away at Brigham Young University in Provo. The desire of administrators at each school to be the first to announce the ground-breaking work led to the publicizing of results before they had been sufficiently confirmed. Apparently they believed that there was too much at stake for them not to make the announcement.

What can be learned from the cold fusion episode? Obviously, more care must be taken in checking claims before they are published. Ironically, the speed with which cold fusion was debunked arose from the simplicity of the apparatus used. Scientists across the country had cold fusion reactors running within days of the announcement, and a flood of null results appeared before Pons' and Fleischmann's article was even published. If such efficiency can be employed in evaluating other claims that sound too good to be true, the effects of dishonesty and negligence may be reduced.

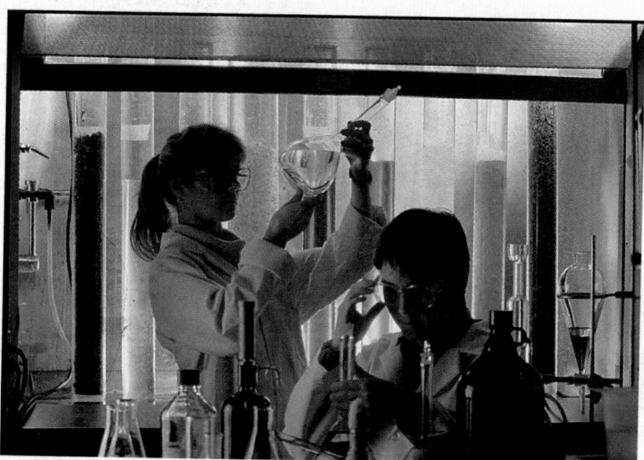

Researching the Issue

1. Martin Fleischmann has said, "If you really don't believe something deeply enough before you do an experiment, you will never get it to work." What are the strengths and weaknesses of this argument?
2. What methods could be used to reduce or eliminate experimenter bias? How could the "double blind" technique, which is often used in psychology, be modified for use in experimental chemistry?
3. Choose one of the following scientists and investigate claims that they misinterpreted or misreported research data. Consider whether the conclusions they reached were valid or invalid.
   - **a.** Isaac Newton
   - **b.** Sir Cyril Burt
   - **c.** Galileo Galilei
   - **d.** Robert Millikan
   - **e.** Alexander Gurwitch
   - **f.** René Blondlot
4. In some cases, test results for products (such as new drugs) have been falsified. What steps might be taken to safeguard against this practice?
5. The following books provide accounts of scientific dishonesty, its causes, and possible solutions.

   Kohn, Alexander. *False Prophets.* Oxford: Basil Blackwell Ltd., 1986.

   Broad, William & Wade, Nicholas. *Betrayers of the Truth.* New York: Simon and Schuster, 1982.

2

PLANNING FOR

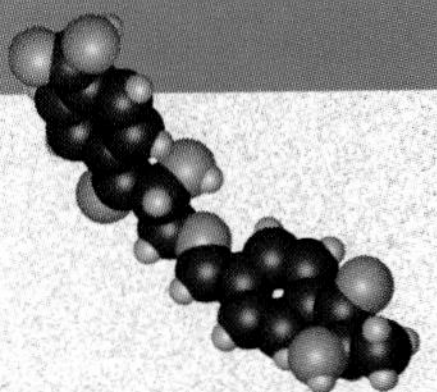

Matter and Energy

Chapter Overview

- **Section 2-1** introduces the concepts of matter and mass. Chemical and physical properties of matter, including physical state, are also explored.
- **Section 2-2** involves the definition of energy, forms of energy, calculation of kinetic energy, transformation of energy, and energy conservation. The energy changes that accompany chemical changes and changes of state are also discussed.
- **Section 2-3** is about the kinds and classes of matter including: atoms, molecules, elements, compounds, pure substances, allotropes, and heterogeneous and homogeneous mixtures. The distinction between intensive and extensive properties is also discussed.
- **Section 2-4** covers techniques for separating and identifying substances in mixtures. Density is introduced as an intensive property useful for identification. The graphical relationships among density, mass, and volume are also presented.
- **Section 2-5** teaches skills for making measurements, doing calculations, and reporting data, including: using SI units, following the rules for significant figures in measurements and calculations, and relating different temperature scales.

Concept Base

Students must bring the following skills to this chapter: solving simple algebraic equations, reading a metric ruler, and using a scientific calculator with the exponent key.

Looking Ahead

The structure of the atom will be discussed in Chapter 3, periodic properties of elements are discussed in Chapter 4, chemical bonding is discussed in Chapter 5.

Laboratory Equipment Needs

Demonstrations

The listing of materials for Chapter 2 demonstrations begins on page T46.

Laboratory Experiments

Materials and preparation instructions for Chapter 2 are found as follows.

Text		*Page*
Exploration 2A	Accuracy and Precision	672
Investigation 2A	Accuracy and Precision—Counterfeit Coins	676
Exploration 2B	Separation of Mixtures	678
Investigation 2B	Separation of Mixtures—Mining Contract	684

Laboratory Experiments Manual
(refer to the table of contents in the manual)

Exploration 2-1	Separation of Mixtures
Investigation 2-1	Separation of Mixtures—Tanker Truck Spill

Pacing: 13 days

Inquiry Teaching Strategies

2-1 What is matter?

Demonstrations
- ■ Demo 1 *Matter and mass*
- ■ Demo 2 *Index of refraction*
- ■ Demo 3 *Mass of gaseous matter*
- ■ Demo 4 *Evidence of chemical change I*
- ■ Demo 5 *Evidence of chemical change II*

2-2 How do matter and energy interact?

2-3 How is matter classified?

Demonstrations
- ■ Demo 6 *Elements*
- ■ Demo 7 *Separating water by electrolysis*

2-4 How are substances identified?

Demonstrations
- ■ Demo 8 *Separation of substances*
- ■ Demo 9 *Distillation*

Laboratory Experiments
- ■ ■ Exp 2-1 *Separation of Mixtures*
- ■ ■ Inv 2-1 *Separation of Mixtures—Tanker Truck Spill*

2-5 How should data be reported?

Laboratory Experiments
- ■ Exp 2A *Accuracy and Precision*
- ■ Inv 2A *Accuracy and Precision—Counterfeit Coins*
- ■ Exp 2B *Separation of Mixtures*
- ■ Inv 2B *Separation of Mixtures—Mining Contract*

Key
- ■ Teaching Transparencies
- ■ Annotated Teacher's Edition
- ■ Laboratory Experiments Manual

Visual Teaching Strategies

Review and Practice Strategies

Transparencies
- F 2-3 *Structural Features of Molecules*
- F 2-6 *Particle Models for Water*

Transparency Masters
- T 2-1 *Properties of Matter*

Text Reviews
- Section Review 1–5
- Chapter Review 1–9; LC 3; UT 2

Study Guide Worksheets
- 2-1 Concept Review
- 2-1 Skillsheet *Using Lab Equipment*

Transparencies
- F 2–17 *Evidence of Chemical Change*

Text Reviews
- Concept Review 6–8
- Section Review 9–11
- Chapter Review 10–21; LC 1

Study Guide Worksheets
- 2-2 Concept Review

Transparencies
- F 2-20 *Abundance of Elements in Earth's Crust*
- F 2-23 *Classifying Matter*

Text Reviews
- Concept Review 12–14
- Section Review 15–19
- Chapter Review 22–31; LC 5

Study Guide Worksheets
- 2-3 Concept Review

Transparency Masters
- T 2-4 *Recycling Codes*
- F 2-27 *Distillation Set Up*
- F 2-28 *Mass vs. Volume for Lead Samples*
- T A-9 *Density of Water*

Text Reviews
- Section Review 20–23
- Chapter Review 32–38; LC 4

Study Guide Worksheets
- 2-4 Concept Review
- 2-4 Skillsheet *Density*

Transparencies
- F 2-33 *Using Significant Figures*
- P 65 *SI Conversions*

Transparency Masters
- T A-1 *SI Measurement*
- T A-2 *Symbols and Abbreviations*
- P 805 *Determining Number of Significant Figures*
- P 806a *Making Calculations with Significant Figures*
- P 806b *Rounding*
- P 807 *Calculations with Numbers in Scientific Notation*

Text Reviews
- Concept Review 24–26
- Section Review 27–31
- Chapter Review 39–52; LC 2; UT 1

Study Guide Worksheets
- 2-5 Concept Review
- 2-5 Skillsheet *Accuracy and Precision*
- 2-5 Skillsheet *Making Useful Measurements*

■ Pupil's Text
■ Study Guide
■ Teaching Resources

Exp *Exploration*
Inv *Investigation*

LC *Linking Chapters*
UT *Using Technology*

Assessment Options

Traditional Assessment
Test Generator
Instructional Objectives Measured:
Content mastery
Select from items 1 to 60

Performance Assessment Options
(see pages T36 and T643-15 for scoring rubrics)
Students demonstrate mastery of objectives in a hands-on environment. You may want some of these materials in the Portfolio as well.
Investigation 2A and 2B (text)
Investigation 2-1 (laboratory manual)
Instructional Objectives Measured:
Content mastery, Use of scientific methodology, Problem solving skills, Use of technology, Proficiency in written communication

Performance Assessments, page 69

1. **Designing an experiment**
   Instructional Objectives Measured:
   Content mastery, Use of scientific methodology, Problem solving skills, Use of technology, Proficiency in written communication
2. **Designing an experiment**
   Instructional Objectives Measured:
   Content mastery, Use of scientific methodology, Problem solving skills, Use of technology, Proficiency in written communication
3. **Reference and research skills**
   Instructional Objectives Measured:
   Content mastery, Use of scientific methodology, Problem solving skills, Use of technology, Proficiency in written communication
4. **Describing substances by properties**
   Instructional Objectives Measured:
   Content mastery, Proficiency in written communication

Portfolio Options
(see page T36 for scoring rubrics)
Students provide a written rationale for each selection made for the portfolio.

Formal Laboratory Reports
Exploration 2A and 2B (text)
Exploration 2-1 (lab manual)
Instructional Objectives Measured:
Content mastery, Use of scientific methodology, Use of technology, Proficiency in written communication

Portfolio Projects, page 69
Items 1, 2, and 3. Chemistry and you
Instructional Objectives Measured:
Scientific and chemical literacy, Proficiency in written communication

Science, Technology and Society, page 70
Instructional Objectives Measured:
Scientific and chemical literacy, Understanding the impact of science and technology on society, Problem solving skills, Proficiency in oral and written communication

INTRODUCING THE CHAPTER

CHAPTER 2 Matter and Energy

Story Background

Plastics present the most severe recycling challenge of any large class of consumable materials. There are two major groups of plastics: the thermosets and the thermoplastics. Thermosets are rigid plastics that do not flow when they are heated. For this reason thermosets, which constitute about 13% of plastic production, cannot be recycled. A taillight for example, cannot be transformed into some other useful object of another shape. Thermoplastics, on the other hand, can sometimes be recycled because they consist of long-chain polymer molecules that slide past one another when heated. Thus, some thermosplastic products—such as soda bottles for example—can be transformed. Unfortunately however, many problems impede recycling of thermoplastics, as discussed in this chapter.

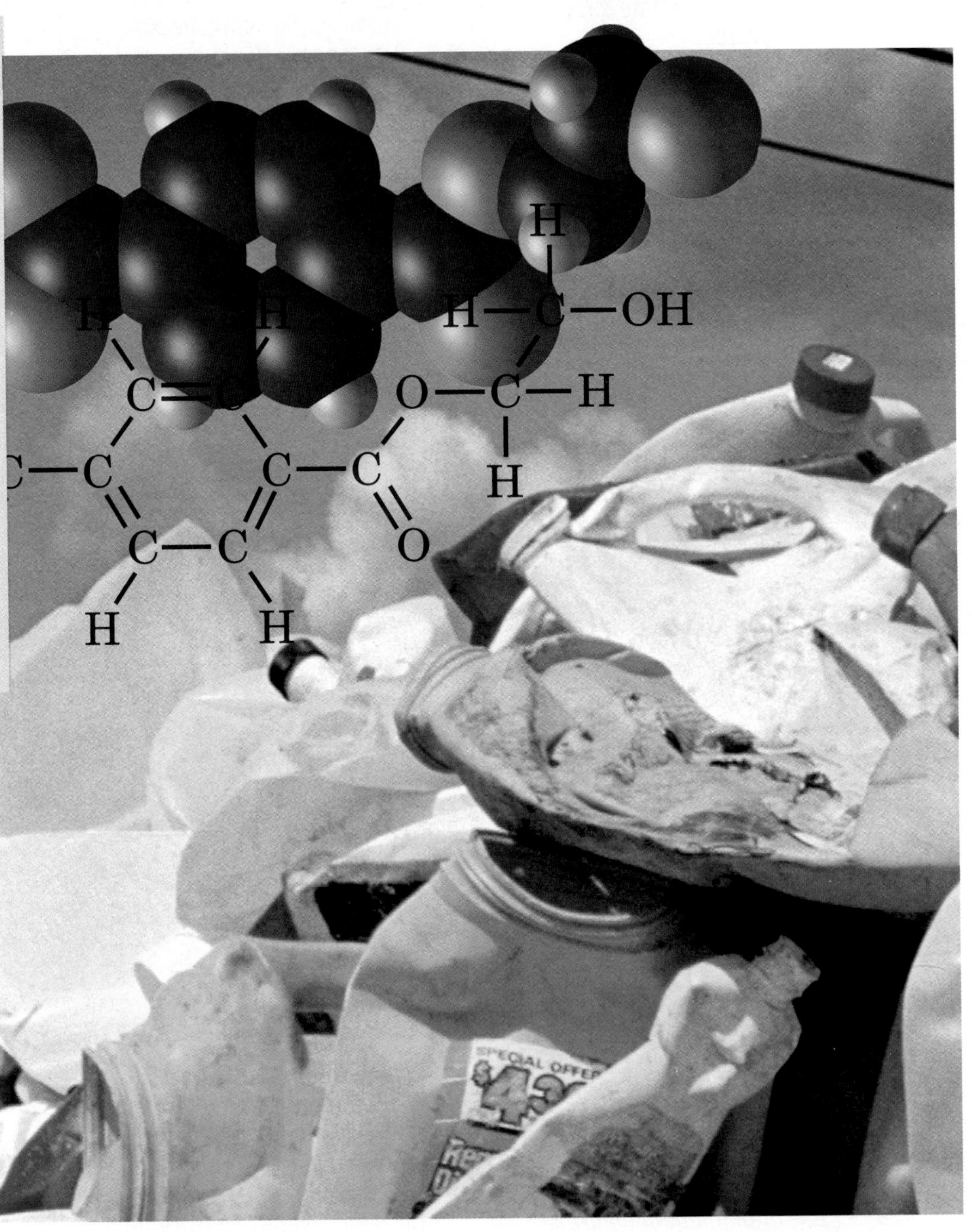

About the Illustration

The picture shows a polyethylene terephthalate, or PET, molecule on a background of plastic wastes sorted for recycling at a recycling center.

Recycling—Saving Matter and Energy

What happens to a plastic soda bottle that you put in the recycling bin? You could end up wearing it. New technology allows bottles made from PET (polyethylene terephthalate, a kind of plastic) to be converted to polyester. Polyester is a great material for many things, including T-shirts and fleecy jackets.

More likely, however, the plastic soda bottle you want to recycle will never be remade as a useful product. Odds are the bottle will end up in a landfill or incinerator. In the landfill, it will take up valuable space for, perhaps, hundreds of years. In the incinerator, the bottle will be burned, giving off exhaust gases that pollute the air.

On the other hand, when you toss an aluminum can into a recycling bin, the can will almost certainly be recycled. The average aluminum can deposited at a recycling center is melted down, reformed, and back on supermarket shelves in only six weeks.

Recycling aluminum saves energy and raw materials, reduces litter and pollution, saves landfill space, and is profitable too. Every business day, manufacturing companies make more than $2 million by recycling aluminum. When properly sorted by color, glass bottles are also easily and profitably recycled. Energy and raw materials are saved in the process.

What about paper? What happens to those newspapers and old homework pages that you dump in the recycling bin? The prospects for paper recycling are mixed. A lot of paper *is* recycled annually—with great energy and resource savings. For every ton of new paper made from 100% recycled wastepaper, 17 trees, 4100 kilowatt-hours of energy (enough to power the average home for six months), 7000 gallons of water, three cubic yards of landfill space, and taxpayers' dollars are saved. Yet more paper is collected than can actually be recycled. Excess paper piles up at some recycling collection sites with no place to go.

As you study this chapter, you will begin learning some chemistry that explains the successes and failures of current recycling efforts. Look especially for answers to the following questions.

- *How can materials be identified so they can be recycled or disposed of properly?*
- *How can mixtures be separated so that their components can be recycled?*
- *How is energy measured, and what do we mean when we say "recycling saves energy"?*
- *Why can some materials (such as aluminum) be recycled more successfully than others?*

As you will see, recycling is like all other processes of the physical world—it involves the interaction of matter and energy.

Tapping Prior Knowledge

- Have students do a "free-write" on the topics of matter and energy. Give students 5 min to write down everything they think they know about these topics, including questions and associations. Free-writes should not be graded, but used to assess students' knowledge and interests related to matter and energy.
- Determine students' ability to solve algebraic equations by asking them to solve the following for the variable x.
  - **a.** $v = xg$
  - **b.** $ab = xcd$
  - **c.** $s = t/x$
- Check students' facility with graphs by having them complete the Concept Review on graphing in Appendix B.

Section 2-1

FOCUS

Lesson Starter

Write the following words on the board.
peanut butter
water
fish
garbage
time
motion
the human brain
carbon dioxide gas
yourself
an idea
tree
energy

Ask students to sort the words into three categories.
1. *Matter*
2. *Not matter*
3. *Not sure*

Discuss responses. (Note that energy is *not* matter, though matter can be *measured* by its mass; mass can be converted to energy.) Finally, ask students why a clear definition of matter is important for studying chemistry.

Demonstration 1

Matter and mass

Approximate time: 15 min

This demonstration should motivate students to question their prior concepts of matter, mass, and volume.

1. Obtain several 100 g brass masses and several large pieces of Styrofoam which have been cut down to have a mass of 100 g.
2. Call students forward to compare each brass mass with a piece of Styrofoam. Have students hold one in each hand. Ask them how the two objects compare in volume. How do the objects compare in terms of matter? How do the objects compare in terms of mass?

What is matter?

Section Objectives

Describe matter by its properties.

Define mass, and state how to measure it.

Distinguish between chemical and physical properties, and give examples of each.

Draw models to represent solids, liquids, and gases on the particle level.

Describing matter

What do the stars, your brain, air, a peanut butter sandwich, and a plastic soda bottle have in common? All of these things are examples of matter. Matter, the "stuff" of which all things in the universe are composed, comes in a fantastic variety of forms.

matter
anything that has mass and volume

volume
the amount of space an object occupies

mass
the quantity of matter in an object

Matter has mass and volume

Matter is defined as anything that has mass and volume. What do we mean by *mass* and *volume*? **Volume** is simply the amount of space an object occupies. A grapefruit has more volume than a lemon, for example.

The **mass** of an object is a measure of how difficult it is to change the object's state of motion. You can compare masses of different objects by pushing or pulling them. Very massive objects are relatively hard to start or stop moving. Common units of mass are the gram (g), kilogram (kg), and milligram (mg). Mass also indicates the amount of matter in an object; the more massive an object is, the more matter it contains.

Note that mass is similar to weight, although strictly speaking they are not the same thing. The difference is that weight depends on gravity, while mass does not. Since gravity varies from place to place, weight also varies. But mass is constant because it is not determined by gravity.

Figure 2-1
The analytical balance is used to measure the mass of very small quantities of substances or to measure with extreme precision. As indicated by this analytical balance, the mass of the copper(I) chloride and flask is 73.3187 g.

A balance measures mass

Weight can be measured using a scale. To determine mass you must use a *balance*. In chemistry, you will generally be interested in mass rather than weight, so the balance is standard laboratory equipment. Two types of balances that you may use are the *analytical balance,* shown in **Figure 2-1**, and the *triple beam balance,* shown in **Figure 2-2** on the next page.

Although the term *weigh* is not literally equivalent to "determine the mass of," these expressions are often interchanged. In chemistry, when you hear the word *weigh,* you may actually need to determine mass. Check to be sure.

Visual Strategy

Figure 2-1 and Figure 2-2

Emphasize that the two balances shown are being used to determine the mass of the *same* quantity of matter. The analytical balance can provide a measurement to many more decimal places (usually to 0.0001 g, 100 times more precise than the triple-beam balance). As a result, the analytical balance can be used to measure much smaller samples of matter.

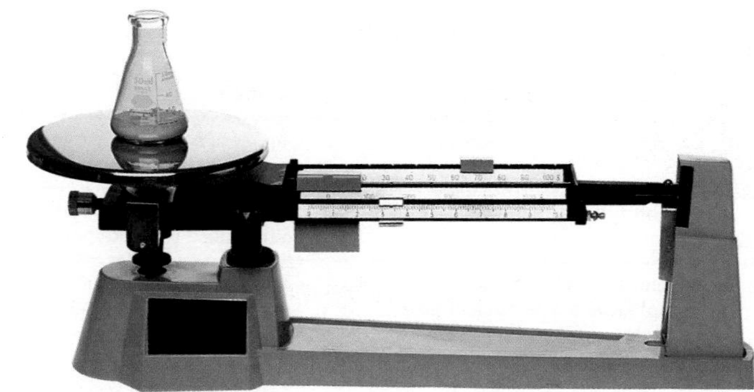

Figure 2-2
Mass can also be measured with a triple beam balance, though not as precisely as with an analytical balance. This triple beam balance shows that the mass of the flask and the copper(I) chloride it contains is 73.32 g.

atom
the basic unit of matter

molecule
a neutral group of atoms held together by chemical bonds

Atoms are basic units of matter

Imagine crushing some coffee beans in a grinder. The beans break into pieces. You keep grinding the beans until you get a fine powder. Each of the tiny grains contains billions and billions of submicroscopic **atoms**.

Matter in and around you is made of atoms. There are over 110 different kinds of atoms in the known universe. Yet, combined in various ways, atoms make up everything you can see, touch, smell, taste, or hear.

Atoms usually do not exist by themselves, but combine to form compounds. Some of these are clusters called **molecules**. A molecule may contain two atoms or two thousand, but in any case the molecule behaves as a unit. **Figure 2-3** shows three of the many kinds of inorganic and organic molecules in an orange. Most materials in your environment, like the orange, are made of many kinds of molecules, which are mixed together. An orange peel, for example, contains more than 100 different kinds of molecules.

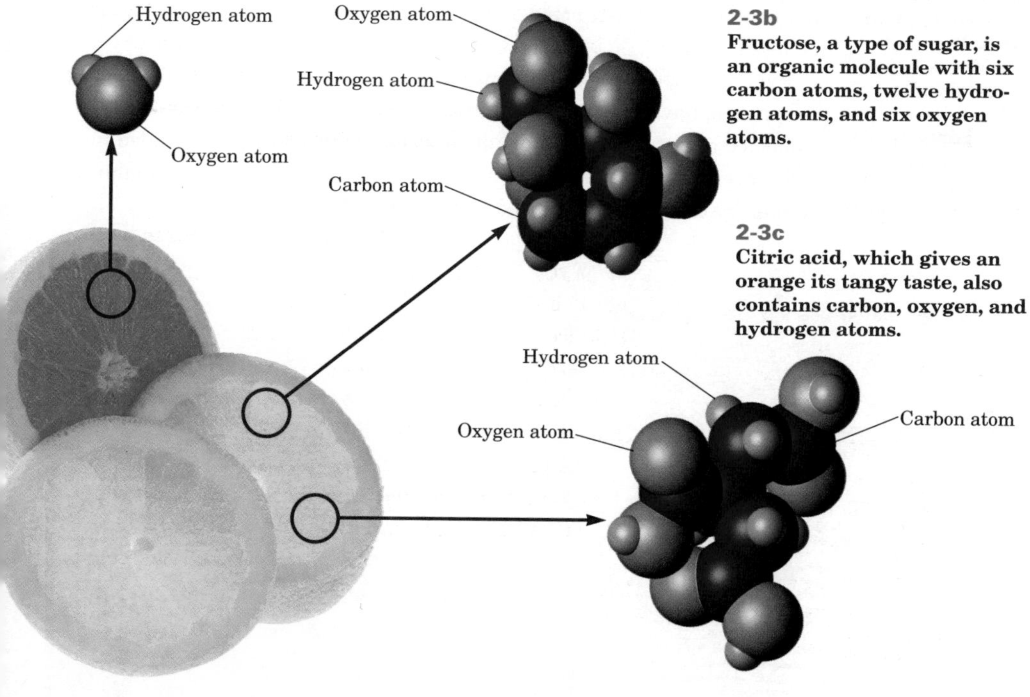

Figure 2-3a
One type of inorganic molecule found in an orange is water. It consists of two hydrogen atoms and one oxygen atom.

2-3b
Fructose, a type of sugar, is an organic molecule with six carbon atoms, twelve hydrogen atoms, and six oxygen atoms.

2-3c
Citric acid, which gives an orange its tangy taste, also contains carbon, oxygen, and hydrogen atoms.

3. Tell students to keep this demonstration in mind as they read Section 2-1.

SAFETY
No particular precautions required.

DISPOSAL
Save materials for reuse at a later time.

TEACH

Content Background

The text is careful to say that the astronaut's mass *approaches* weightlessness on the moon because true weightlessness could only occur if there were zero gravitational attraction between the astronaut and other massive bodies. True weightlessness could only occur if the astronaut were an infinite distance from other objects.

Teaching Tip

Vocabulary
To avoid imprecise and ambiguous terminology, use *determine the mass of* (e.g., "Determine the mass of 5 g of alum") rather than *weigh* ("Weigh 5 g of alum") when you are giving instructions for lab.

Do You Know?

The term "weigh station" is a misnomer. At such locations, fairly sensitive balances are used to determine masses, not weights, of cars and trucks.

Visual Strategy

Figure 2-3
Ask students what the individual colored spheres in the figure represent and what the clusters represent. Atoms are represented by different colors for clarity, and are not the colors shown in the models. Though the relative sizes of atoms are also represented by the space filling models, they are not the actual sizes. Ask students to name the atoms the two molecules have in common (oxygen, hydrogen).

Historical Note

Silver wires were used in early telegraph systems, but they were stolen so often that they were replaced with the less effective but less expensive copper wires.

Demonstration 2

Index of refraction

Approximate time: 5 min

Water has a high index of refraction compared to air so objects partially immersed in water appear to be bent.

1. Place a straw or spoon in a clear glass of water that is three-quarters full, and hold the spoon vertically so that there is little or no refraction.
2. Let the spoon fall to the side of the glass to show the classic "bent" spoon.

SAFETY

No particular precautions required.

DISPOSAL

None required.

Do You Know?

Gold is so malleable that it can be hammered into sheets about 50 atoms thick. At this thinness, it is transparent.

Table 2-1
Properties of Matter

Property	Description	Example
Electrical conductivity	ability to carry electricity	Copper is a good electrical conductor, so it is used in wiring.
Heat conductivity	ability to transfer energy as heat	Aluminum is a good heat conductor, so it is used to make pots and pans.
Density	mass-to-volume ratio of a substance; measure of how tightly matter is "packed"	Lead is a very dense material, so it is used to make sinkers for fishing line.
Melting point	temperature at which a solid changes state to become a liquid	Ice melts to liquid water at the melting point of water.
Boiling point	temperature at which a liquid boils and changes state to become a gas at a given pressure	Liquid water becomes water vapor at the boiling point of water.
Index of refraction	extent to which a given material bends light passing through it	The index of refraction of water tells you how much light slows and bends as it passes through water.
Malleability	ability to be hammered or beaten into thin sheets	Silver is quite malleable, so it is used to make jewelry.
Ductility	ability to be drawn into a thin wire	Tantalum is a ductile metal, so it is used to make fine dental tools.

Properties of matter

Matter can be described by its properties

Chemists describe different types of matter by listing their characteristics, or *properties*. Color, mass, volume, texture, transparency, flammability, and taste are a few of the many properties that are useful for describing matter. **Table 2-1** lists some other useful properties with which you may be less familiar.

Ever since the earliest humans sought flammable materials for making fire, people have been collecting information about properties of matter. Collected data from the history of chemical investigation is compiled in reference books that are easily available to you. A page from a popular chemistry reference book is shown in **Figure 2-4**.

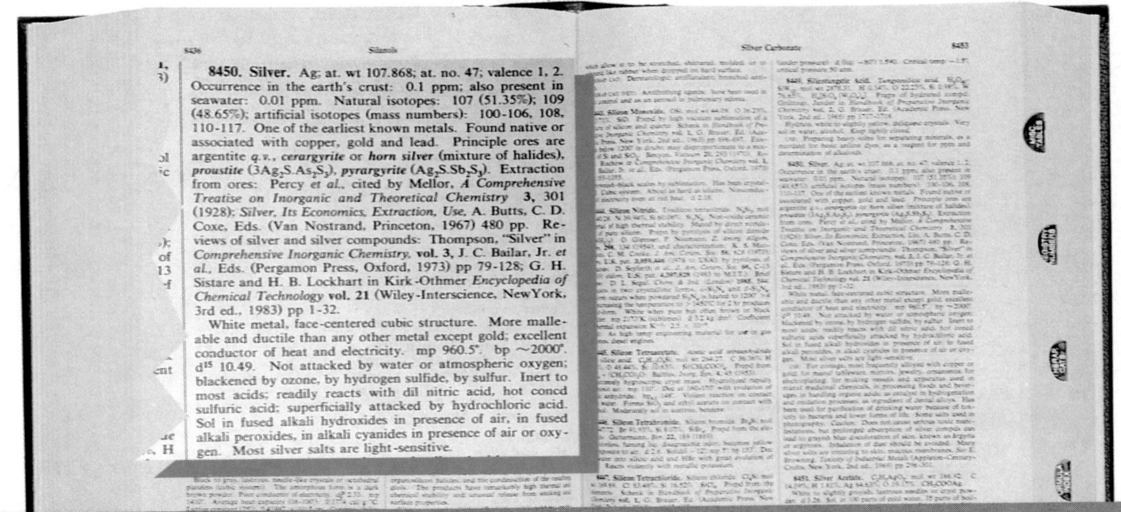

8450. Silver. Ag; at. wt 107.868; at. no. 47; valence 1, 2. Occurrence in the earth's crust: 0.1 ppm; also present in seawater: 0.01 ppm. Natural isotopes: 107 (51.35%); 109 (48.65%); artificial isotopes (mass numbers): 100-106, 108, 110-117. One of the earliest known metals. Found native or associated with copper, gold and lead. Principle ores are argentite *q.v.*, *cerargyrite* or ***horn silver*** (mixture of halides), *proustite* ($3Ag_2S.As_2S_3$), *pyrargyrite* ($Ag_2S.Sb_2S_3$). Extraction from ores: Percy *et al.*, cited by Mellor, *A Comprehensive Treatise on Inorganic and Theoretical Chemistry* **3**, 301 (1928); *Silver, Its Economics, Extraction, Use*, A. Butts, C. D. Coxe, Eds. (Van Nostrand, Princeton, 1967) 480 pp. Reviews of silver and silver compounds: Thompson, "Silver" in *Comprehensive Inorganic Chemistry*, **vol. 3**, J. C. Bailar, Jr. *et al.*, Eds. (Pergamon Press, Oxford, 1973) pp 79-128; G. H. Sistare and H. B. Lockhart in Kirk-Othmer *Encyclopedia of Chemical Technology* **vol. 21** (Wiley-Interscience, New York, 3rd ed., 1983) pp 1-32.

White metal, face-centered cubic structure. More malleable and ductile than any other metal except gold; excellent conductor of heat and electricity. mp 960.5°. bp ~2000°. d^{15} 10.49. Not attacked by water or atmospheric oxygen; blackened by ozone, by hydrogen sulfide, by sulfur. Inert to most acids; readily reacts with dil nitric acid, hot concd sulfuric acid; superficially attacked by hydrochloric acid. Sol in fused alkali hydroxides in presence of air, in fused alkali peroxides, in alkali cyanides in presence of air or oxygen. Most silver salts are light-sensitive.

Figure 2-4
Reference books catalog substances and describe them by their properties. This page from The Merck Index (Eleventh Edition, ©1989, Merck & Co., Inc.) gives a description of silver. How many properties in this list are familiar to you?

Visual Strategy

Figure 2-4

Ask students to read the description of silver shown. Ask them to identify all the properties listed in the passage. Put this list on the board. Make the point that chemists identify unknown substances by comparing the properties of an unknown with published data. Discuss the places one could go to obtain chemistry reference material and data.

Physical properties are useful for identifying things

Color, texture, shape, and mass can be observed or measured without changing the composition of matter. Properties such as these are called **physical properties**.

You rely on physical properties to identify things all the time. Using color, texture, and mass, it's easy to know which recycling bin a piece of white paper or a brown glass bottle belongs in, for example. Or, look at Figure 2-5. Even in a photograph, you can probably identify what material the objects shown are made from by observing some of their physical properties.

Figure 2-5
Can you identify what material these objects are made from? What properties did you use to identify them?

physical properties
properties that can be observed or measured without changing the composition of matter

state
the condition of being a gas, liquid, solid, plasma, or neutron star

State of matter is a physical property

An easily observed physical property is **state**. Solids, liquids, and gases are the three common states of matter. Figure 2-6 shows the solid, liquid, and gaseous states of water at the molecular level.

Solids have fixed volume and shape. These features result from the way solids are structured on a microscopic level. In the solid state, atoms and molecules are held tightly in a rigid structure but vibrate slightly about fixed positions.

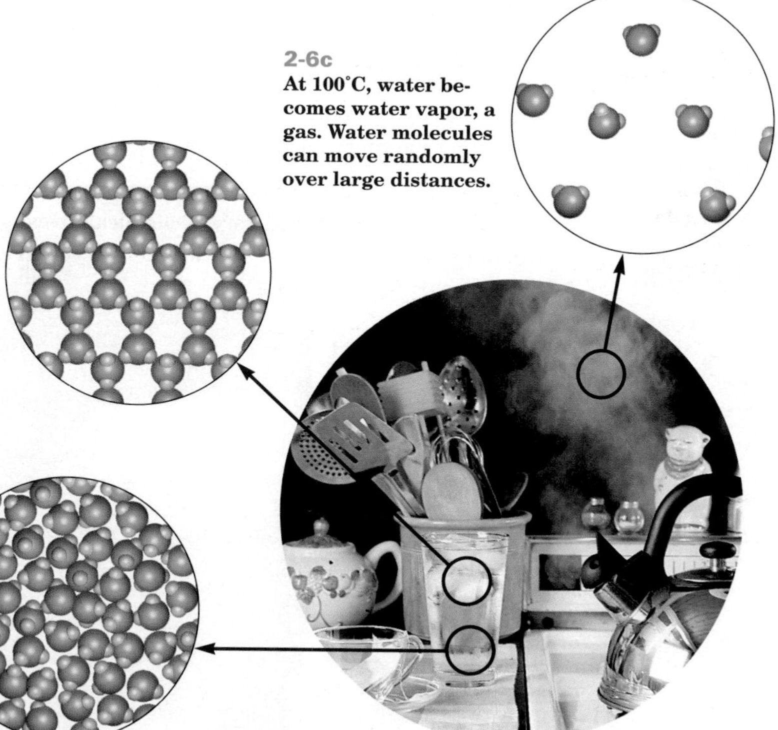

2-6c
At 100°C, water becomes water vapor, a gas. Water molecules can move randomly over large distances.

2-6a
Below 0°C, water molecules solidify to form ice. In the solid state, water molecules are held close together in a rigid structure.

Figure 2-6b
Between 0°C and 100°C, water exists mostly as a liquid. In the liquid state, water molecules are close together, but can move about freely.

Theme

Macroscopic observations and micromodels

The observable differences among the solid, liquid, and gaseous states owe much to their differing arrangements of atoms and molecules, as shown in Figure 2-6.

Demonstration 3

Mass of gaseous matter

Approximate time: 25 min

Disprove the misconception that air and gases have no mass.

1. Have a pair of students measure the mass of a deflated balloon and record the mass on the board.
2. Have one student blow up and tie the balloon.
3. Have students measure the mass of the inflated balloon, and record the mass on the board.
4. Discuss the discrepancy in mass. Where did the extra mass of the inflated balloon come from?

SAFETY
No particular precautions required.

DISPOSAL
None required.

Visual Strategy

Figure 2-5 and Figure 2-6

Have students identify the material in the photograph. Ask students to describe its distinguishing properties. List those properties on the board.

Ask students to compare and contrast the representations of the solid, liquid, and gas. Be sure that the relative spacing between molecules and the degree of orderliness are discussed.

Demonstration 4
Evidence of chemical change I

Approximate time: 10 min

This demonstration shows the reactivity of marble with acid.

1. Place a few small marble chips in a petri dish on an overhead projector.
2. Add enough vinegar to cover the chips.
3. Note the evidence of chemical change. What chemical property of marble is demonstrated?

SAFETY
Wear safety goggles and a lab apron. Students should stand back 10 ft.

DISPOSAL
Wearing safety goggles and a lab apron, decant the mixture, rinse the residue with tap water, decant the rinse water, and repeat the rinse. Pour the rinse water down the drain. Save the solid residue, wrap it in newspaper, and put it in the trash.

Demonstration 5
Evidence of chemical change II

Approximate time: 15 min

1. Sand and/or polish a large (30 cm × 30 cm) sheet of copper foil.
2. Hold the foil at a 45° angle over a Bunsen burner flame.
3. Have students note evidence of chemical change. What is the dark stain? What are the reactants in this reaction?

SAFETY
Wear safety goggles and a lab apron. Tie back loose hair and clothing.

DISPOSAL
Save the foil for reuse at a later time. Sweep up any flakes of copper oxide and put them in the trash.

Liquids have a fixed volume but variable shape. The particles in a liquid are not held together in the rigid manner characteristic of solids. Like ball bearings in oil, the particles of a liquid can slip and slide past one another. As a result, the liquid as a whole is able to flow. The distances between adjacent particles are constant on average, so the overall volume of the liquid is fixed.

Gases have no fixed shape or volume and therefore expand to fill any container they occupy. In the gaseous state, a few grams of helium would become evenly distributed in a small flask or throughout a large room. This behavior of gases results from the fact that their particles are not held to one another. Instead, they are free to move about.

At very high temperatures, matter can exist as a *plasma*, a fourth state of matter. In this high energy state, atoms are torn apart into smaller pieces. The sun, stars, much intergalactic matter, and the glowing interiors of fluorescent lights are in the plasma state. A fifth state of matter, that of the *neutron star*, has recently been discovered as well. Little is known about this state at the present time.

Chemical properties tell how substances interact

chemical properties *properties that can be observed only when substances interact with one another*

A material is not fully described by physical properties alone. Left out of the description is how it will behave when it is in contact with other materials. To describe this type of behavior, we refer to **chemical properties**, properties that are observable only when one substance interacts with another.

Reactivity with acid, for example, is a chemical property. Calcium hydroxide, the active ingredient in some antacid medications, such as the one shown in **Figure 2-7**, is very reactive with acids. It is used to neutralize excess hydrochloric acid that forms in your stomach. Reactivity with acid can sometimes have destructive effects, as shown in **Figure 2-8**.

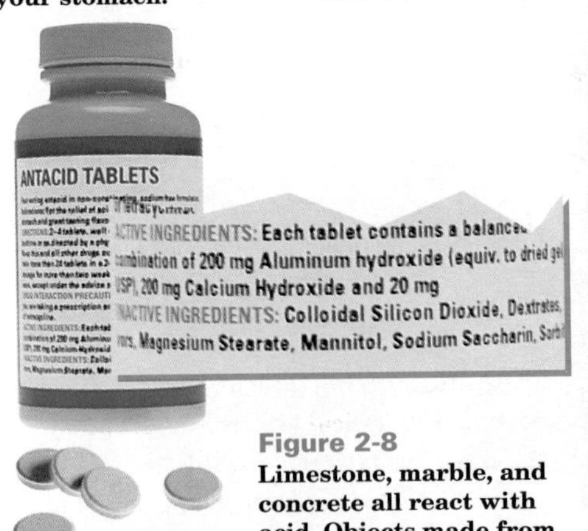

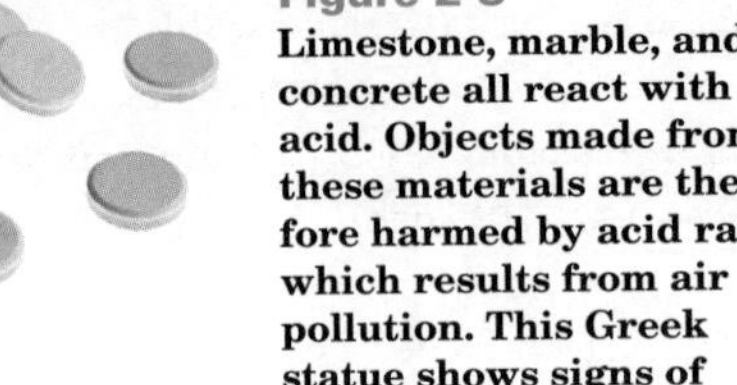

Figure 2-7
Some antacids contain calcium hydroxide as the active ingredient. Calcium hydroxide, a base, reacts with hydrochloric acid in your stomach.

Figure 2-8
Limestone, marble, and concrete all react with acid. Objects made from these materials are therefore harmed by acid rain, which results from air pollution. This Greek statue shows signs of damage by acid rain.

Visual Strategy
Figure 2-7

Stress that chemical reactions with observable chemical properties are constantly happening. Ask students to name other evidence of chemical reactions they recall witnessing in everyday life.

Figure 2-9a
The iron in this bicycle reacts with oxygen in the air to produce rust.

2-9b
Some kinds of fruit turn brown when they are exposed to the air because they contain compounds that react with oxygen.

Reactivity with oxygen is another chemical property. **Figure 2-9** shows how a bicycle and an apple react with oxygen.

Intensive properties result from the way matter is structured

Some properties of matter, such as mass, volume, and length, depend on the quantity of matter present. These are called *extensive* properties. But all chemical properties and many physical properties, including melting point, color, density, and texture, do not depend on the amount of matter present. These properties are called *intensive* properties.

For example, consider graphite, the material that makes up your pencil point. Pencil "leads" are made of graphite rather than lead. Graphite is a gray solid at room temperature and has a slippery texture regardless of whether it is in a pencil or a laboratory reagent bottle. Graphite has these intensive properties because of the way its atoms are arranged. **Figure 2-10** shows how the slippery texture of graphite results from its structure.

Figure 2-10
Graphite consists of sheets of carbon atoms stacked in layers. A pencil can glide across a piece of paper because the carbon layers that make up the point slide over one another easily.

Section Review

1. Draw a picture representing chlorine atoms in the gaseous state.
2. Tell whether the underlined property is a chemical or physical property. Iron is transformed into rust in the presence of air and water.
3. What properties of spinach distinguish it from ice cream?
4. Classify the following properties as physical or chemical and as extensive or intensive: volume, area, flammability, state, odor.

5. **Story Link**
What properties of aluminum distinguish it from glass?

Other Applications

Graphite is widely used in lubricants, foundry facings, shoe polish, brake linings, batteries, carbon motor brushes, and as a moderator in nuclear reactors.

ASSESS

Alternative Assessment

Have students prepare concept maps of the important ideas in the section. You may want to give students a head start by listing some of the terms you think they should use. Have students label the links among items on the concept map to explain how the items are related. Refer them to Appendix C if they need help.

Answers to Section Review

1. In the drawings, the atoms should be widely spaced and randomly oriented. Students may or may not know that chlorine is diatomic.
2. chemical property
3. taste, color, shape, odor, temperature, etc.
4. volume (physical, extensive); area (physical, extensive); flammability (chemical, intensive); state (physical, extensive); odor (chemical, intensive).
5. luster, electrical conductivity, color, shape, transparency, malleability, ductility, etc.

Visual Strategy

Figure 2-10
Use a deck of playing cards to model the sliding sheets that make up the structure of graphite. Tell students that the sliding sheets, as well as the high melting point, make graphite a useful lubricant for high-temperature mechanisms.

Section 2-2

FOCUS

Lesson Starter

Draw a roller coaster on the board for which the second peak is higher than the first peak. Ask students to comment on whether such a design would work. (It would not.) Indicate that their study of energy in this section will enable them to understand why such a roller coaster would not work. (The total energy of the roller coaster is the sum of the potential energy for each peak. The potential energy of the highest peak is greater than the initial energy at the lower peak.)

Teaching Tip

Energy transformations
Have the students brainstorm examples of transformations between kinetic and potential energy that they observe in everyday life. A roller coaster is a good example, as is a pendulum.

How do matter and energy interact?

Section Objectives

Distinguish among kinetic, potential, and other forms of energy.

Apply the conservation of energy and matter to systems.

Describe the energy changes that accompany changes of state.

Describe energy transfers in chemical reactions.

Much of chemistry is about creating new substances to improve food, fuel, health, and fashion. Chemists can put atoms and molecules together in different ways or take them apart. When atoms form molecules, and when molecules change into new ones or break down into atoms, energy is involved. Energy plays an essential role in chemistry. Energy is a broad concept, but a good basic description is that **energy** is the capacity to move or change matter.

energy
the capacity to do work

Forms of energy

Kinetic energy is energy of motion

kinetic energy
energy that moving objects possess by virtue of their motion

Kinetic energy is a type of energy that only moving objects have. A truck rolling down an alley sets objects in its way in motion or changes those objects as it smashes into them. The rolling truck has energy; it can move and change matter. Because the truck's energy arises from its motion, we call its energy *kinetic energy*.

The amount of kinetic energy, *KE*, in an object depends on its mass, m, and velocity, v. Kinetic energy is easy to calculate from the following equation.

$$\textbf{kinetic energy} = \frac{1}{2}\,\textbf{(mass)(velocity)}^2 \qquad \textbf{KE} = \frac{1}{2}\,\boldsymbol{mv^2}$$

Energy is measured in units of $kg{\cdot}m^2/s^2$, or *joules*, J. One joule is about equal to the energy you expend bringing a cheeseburger to your mouth. It is easy to apply the expression for kinetic energy to other situations as well. For example, a typical cheetah has a mass of 60 kg. Its peak speed, or velocity, is about 28 meters per second, m/s. Substituting these numbers into the expression below gives the kinetic energy of the cheetah in **Figure 2-11**

$$KE = \frac{1}{2}mv^2 = \frac{1}{2}(60\text{ kg})(28\text{ m/s})^2 = 2.4 \times 10^4\text{ J}$$

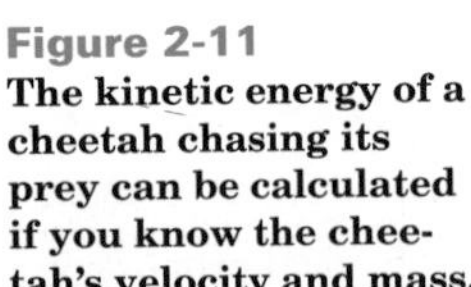

Figure 2-11
The kinetic energy of a cheetah chasing its prey can be calculated if you know the cheetah's velocity and mass.

The kinetic energy of a typical oxygen molecule in your classroom can also be easily calculated. At room temperature and standard pressure, an average oxygen molecule moves with a speed of about 400 m/s. Its mass is 5.3×10^{-26} kg. Substituting these numbers into the expression for kinetic energy gives us the kinetic energy of an average oxygen molecule.

$$KE = \frac{1}{2} mv^2 = \frac{1}{2} (5.3 \times 10^{-26} \text{ kg})(400 \text{ m/s})^2 = 4.2 \times 10^{-21} \text{ J}$$

The oxygen molecule's small mass gives it a small kinetic energy.

Potential energy is stored energy

potential energy
energy an object possesses because of its position

Potential energy is energy an object possesses because of its position. It is called *potential* energy because it has the potential, or ability, to make matter move and change once the energy is released.

An object's potential energy corresponds to the amount of force needed to keep it in position. For example, there is potential energy in a compressed spring. As shown in **Figure 2-12**, a force must be applied to the spring to hold it in place.

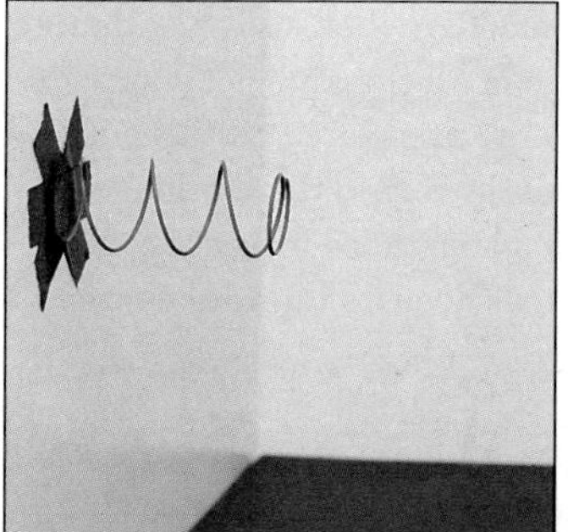

Figure 2-12a
The spring is at rest, neither stretched nor compressed.

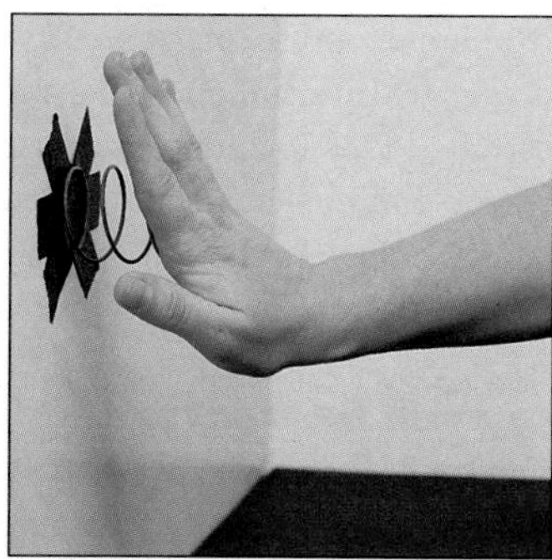

2-12b
The spring is compressed, which gives it potential energy.

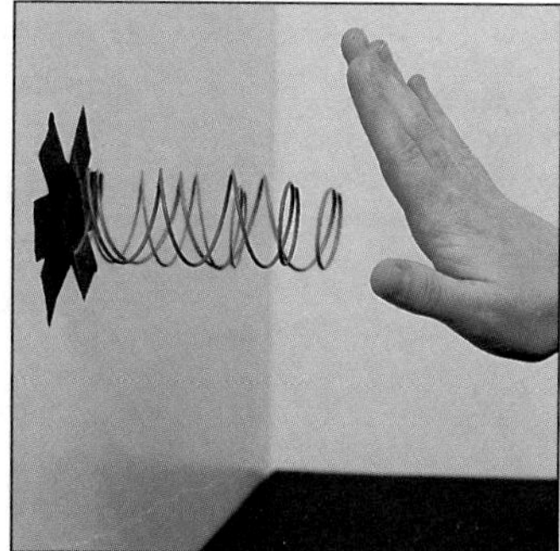

2-12c
Potential energy decreases as the force holding the spring in a compressed position is removed.

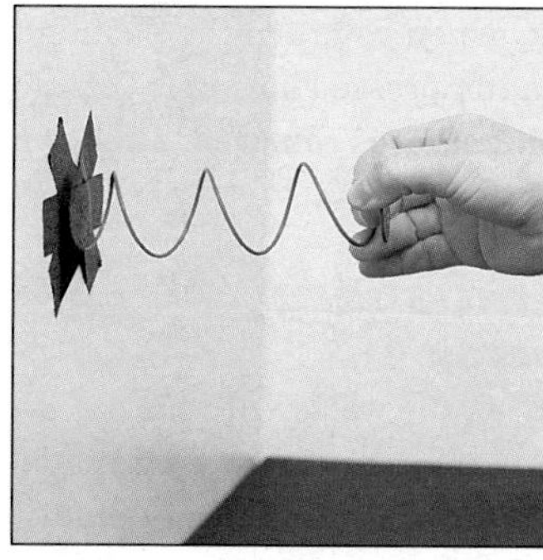

2-12d
The spring is stretched to give it potential energy.

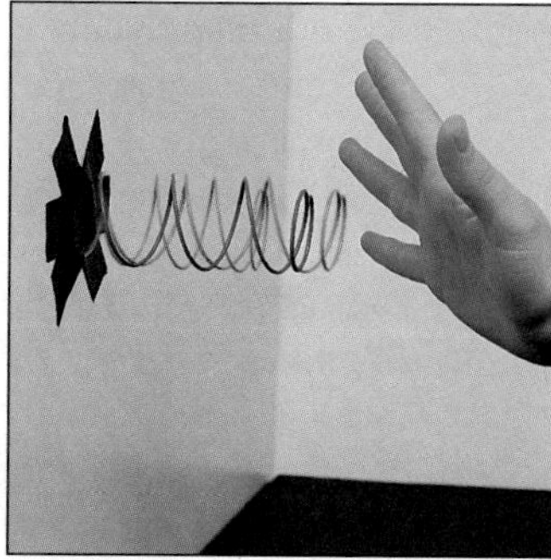

2-12e
Potential energy decreases as the force stretching the spring is removed, and the spring returns to its resting position.

Visual Strategy

Figure 2-12

Make the students aware that an oscillating spring is a good example of a device that employs kinetic-potential energy transformations. The oscillating spring is very similar to pendulum clocks and a playground swing. Some large, linear molecules oscillate much like these springs.

Do You Know?
Hoover Dam on the Colorado River is 221 m (725 ft) high and can supply 1.3×10^6 kW of energy. That's enough energy to light 1.3×10^7 100-W light bulbs. The dam was begun in 1930 and completed in 1936 using 3.4×10^6 m^3 (1.2×10^8 ft^3) of concrete.

Matter held in an elevated position is subject to the downward force of gravity, so forces are required to keep it in place. Therefore, elevated objects have potential energy. Which has more potential energy, a car parked on a small hill or the same car parked on a high mountaintop?

Potential energy also influences the behavior of atoms and molecules. Strong forces hold these tiny particles together, and as a result, clusters of such particles contain potential energy.

Kinetic and potential energy underlie other energy forms

There are many forms of energy, from wind to solar to nuclear to electric to sound energy. In the *macroscopic* world, the world of objects on a scale large enough to see, it makes sense to speak of these diverse forms of energy. But when you consider matter on the atomic level, you find that these diverse forms of energy are not truly distinct. They are just kinetic energy, potential energy, or combinations of these acting on small-mass particles. Kinetic and potential energy underlie electrical, solar, and the other seemingly different forms of energy that we encounter.

One of the important forms of energy to consider when studying matter is *internal energy*. Internal energy is the sum of all the kinds of energy inside a substance. This includes potential energy associated with the forces between atoms and molecules and kinetic energy due to the motions of atoms and molecules. As the temperature of a substance increases, the motion of its atoms and molecules increases, as does its internal energy.

Figure 2-13
This hydroelectric plant is at Hoover Dam, on the Colorado River in Arizona.

Energy can be transformed

The hydroelectric plant in Figure 2-13 shows how energy can be *transformed*, or changed from one form into another. Hydroelectric plants transform the potential energy of elevated bodies of water to the kinetic energy of rushing water and finally to electrical energy, which is delivered to homes or industry.

Energy is transformed continually all around you. The internal energy in a flashlight's battery is transformed to electrical energy and then to light energy when you turn the flashlight on, for example. Potential energy held in a compressed spring is transformed to kinetic energy when the compressing force is removed, as shown in Figure 2-12 on the preceding page.

Relating mass and energy

Mass is a form of energy

In 1905, Albert Einstein shook the world with his discovery of a third fundamental type of energy. The third kind of energy is a familiar quantity—mass! This idea is described in Einstein's famous equation.

$$E = mc^2$$

E represents the energy of an object, m is its mass, and c is the speed of light. Light travels a constant speed of 3×10^8 meters per second, m/s.

law of conservation of energy
the observed fact that in any chemical or physical process, energy is neither created nor destroyed

Besides being the amount of material in an object or a measure of the difficulty of changing the motion of an object, mass is also a form of energy. One way to look at this is to view the mass of an object as its "energy of being." Consider a basketball. If it is moving, it has kinetic energy. If it is elevated, it has potential energy. Even resting on the ground the basketball has energy, because there is internal energy in its atoms and molecules. But beyond all this, the ball has another kind of energy—mass energy. Its mass energy is not due to its motion or position, nor to the internal energy of its atoms and molecules, but to the very fact of its mass.

Mass, like other forms of energy, can be transformed. In transformations, a small amount of mass corresponds to a huge amount of other kinds of energy. In nuclear reactors, the masses of particles are converted to kinetic energy, which is then transformed to electric energy.

Figure 2-14
Energy is transferred as heat from the flame to the candle and the surrounding air.

Energy cannot be created or destroyed

In the previous examples, you found that energy can be transformed from one kind to another. Energy is also often transferred between objects. Energy transfer, as heat, happens when things that are at different temperatures are brought into contact, as in **Figure 2-14**.

When energy is transferred or changes form, none can be created or destroyed. The **law of conservation of energy** is the formal statement of this principle. The law of conservation of energy states that energy cannot be created or destroyed; it may be transformed or transferred from one object to another, but the total amount of energy in the universe never changes.

The law of conservation of energy can be applied to closed systems smaller than the entire universe as well. *Closed systems*, or well-defined groups of objects that are free to transfer energy only among one another, demonstrate energy conservation. Chemicals in a stoppered, insulated reaction flask approximate such a system. The total energy of the system does not change even though energy may be transformed or transferred among the atoms and molecules in the flask.

Energy transformation

6. A skier is in motion midway down a slope. Does the skier have kinetic energy, potential energy, or both kinetic and potential energy? Give the reason for your answer.
7. If an arrow is shot from a bow that has 40 J of potential energy, what will the kinetic energy of the arrow be?

8. **Story Link**
Certain trash incineration plants, called *high-technology resource recovery plants*, can use the heat given off by burning trash to produce electricity. This electricity can then be sold to nearby industry or housing developments. Describe the energy transformations that would be likely to occur in this process.

Answers for Concept Review items and Practice problems begin on page 841.

Answers to

Concept Review

6. The skier has both kinetic and potential energy. The potential energy arises from the skier's elevation. The kinetic energy comes from the skier's motion.
7. 40 J
8. Answers may vary but should include chemical energy (or internal energy) transformed to thermal energy and then to electrical energy.

Visual Strategy

Figure 2-14
The photograph shows that a range of temperatures can be found in the candle flame. Relate this to what students have learned about proper adjustment of the Bunsen burner to get the hottest flame.

Theme
Classification and trends
Almost every chemical reaction can be classified as exothermic or endothermic.

Teaching Tip
Endothermic vs. exothermic
Have students decide if the following are exothermic or endothermic.
a. charcoal burning (exo)
b. toilet bowl cleaner in water (exo)
c. baking a cake (endo)
d. using a gasoline engine (exo)

Do You Know?
Snow sublimes in the winter. Large piles of snow will appear to shrink without any signs of water running off. Iodine crystals will also sublimate rapidly when warmed to room temperature.

Teaching Tip
Conservation of energy
Discuss the following question with the students. If gasoline engines are only 23% efficient (they only use 23% of the available energy), where does the rest of the energy go? (The "missing" energy goes into heat, sound, and chemical reactions.)

Matter changes

Physical changes do not affect chemical composition

Figure 2-15
Glass bottles are crushed to tiny pieces, or cullet, in the recycling process. Crushing is a physical change because the chemical properties of the material are unaltered.

Before glass is recycled, it is crushed so that it will take up less space. This broken glass is called cullet, as shown in **Figure 2-15**. When it arrives at the factory, cullet is melted to the liquid state. The liquid glass is poured into molds in the shapes of new bottles and jars.

In the recycling process, glass undergoes a series of physical changes including melting and crushing. A **physical change** is a change that affects physical properties only. Chemical properties are unaffected. So, in recycling processes that involve only physical changes, recycled glass is chemically identical to glass made from raw materials.

physical change
a change that affects only physical properties

endothermic change
a physical or chemical change in which a system absorbs energy from its surroundings

exothermic change
a physical or chemical change in which energy is released by a system to its surroundings

Change of state is a physical change

For a gram of ice to change from the solid to the liquid state—to *melt*—333 J of energy are required. Why is this energy needed? To see why, picture what is happening on the molecular level. For a solid to melt, its atoms or molecules must begin to move vigorously enough to break partially free of their neighbors and break down the solid crystal structure. The atoms and molecules require energy input to increase their own kinetic energy. This energy input is most often supplied as heat.

Boiling, the rapid change of state from liquid to gas, consumes even more energy—2260 J per gram for water. In boiling, atoms or molecules in the liquid state need enough energy to break totally free of one another. Melting and boiling are called **endothermic** changes because they take in energy as heat from the surroundings. Melting and boiling both increase the internal energy of water molecules.

When a gas cools and becomes a liquid or a liquid cools and becomes a solid, it goes from a higher to a lower energy state. The energy given up by the gas or liquid is transferred to its surroundings as heat. Such changes, in which energy is given off rather than taken in by a system, are called **exothermic** changes. Exothermic changes occur when rain forms and liquid water freezes.

Figure 2-16
Change of state can be an exothermic or endothermic physical change, depending on the direction of energy flow between an object and its surroundings. When ice melts or liquid water vaporizes, energy is absorbed from the surroundings and endothermic change occurs.

Figure 2-17a
When vinegar and baking soda are combined, the solution *bubbles* and carbon dioxide forms.

2-17b
When sodium sulfide and cadmium nitrate are combined, cadmium sulfide, a yellow *precipitate* forms.

2-17c
When aluminum reacts with iron oxide in the clay pot, *heat and light* are produced. Melted material is collected in the skillet.

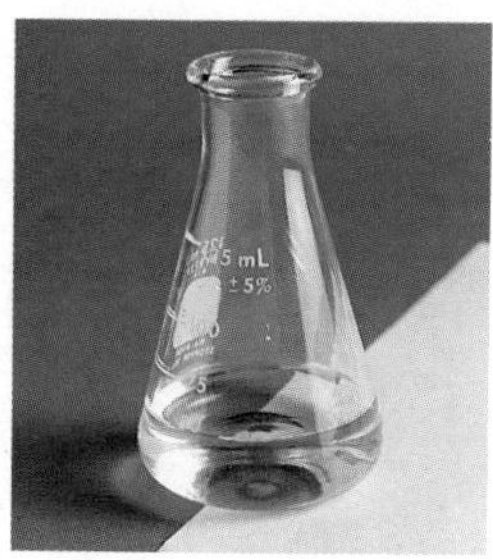

2-17d
When phenolphthalein is added to ammonia, a *color change* occurs.

chemical change
a change that produces one or more new substances

Chemical changes alter chemical composition

In a **chemical change**, one or more substances are changed into new ones. This change occurs at the atomic level, where atoms are rearranged without a loss or gain in the total number of atoms.

You learned in Chapter 1 that the substances undergoing chemical change are called reactants and those created in the reaction are called products. Figure 2-17 shows some of the signs that can indicate a chemical reaction has taken place.

All chemical changes are accompanied by transfers of energy. Like changes of state, chemical changes are either endothermic or exothermic, depending on whether they absorb or release energy. Reactions that produce heat are exothermic. Reactions that absorb heat are endothermic. In an endothermic reaction, the products have more energy than the reactants have. Which reaction shown in Figure 2-17 is obviously exothermic?

Section Review

9. Which has more internal energy: frozen or liquid orange juice? Explain the reason for your answer.

10. A few substances, such as solid carbon dioxide, or "dry ice," change directly from the solid to the gaseous state in a process called *sublimation*. Is sublimation an exothermic or endothermic physical change? Explain your answer.

11. Observe the photograph of the beakers. In the beaker on the right, an aluminum strip is immersed in nitric acid. In the beaker on the left, a copper strip is immersed in nitric acid. What evidence of a reaction do you see? Can you be sure no reaction is occurring in the beaker on the right?

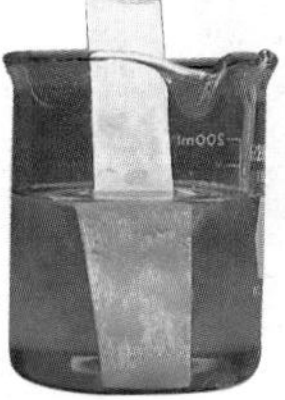

Possible Misconception

Students may think that any temperature change signals a chemical change. Physical changes are also classified as exothermic and endothermic.

ASSESS

Alternative Assessment

Take a trip to the library and have each student choose a method of energy production to collect information about. Ask each student to report on how the forms of energy and energy transformation come into play in their energy production method.

Answers to Section Review

9. Liquid orange juice has more internal energy due to the additional kinetic energy of its molecules.

10. Sublimation is an endothermic process because a substance must absorb energy from its surroundings to sublime.

11. There is no evidence of a reaction in the beaker on the right. However, the lack of visible evidence does not guarantee that a reaction is not taking place. The beaker on the left shows color changes, which are evidence that a chemical reaction is occurring.

Section 2-3

FOCUS

Lesson Starter

Have a collection of glass (brown, green, and clear) and plastic objects (milk jug, 2 L carbonated beverage bottle, plastic bag, cellophane, polyester, mylar, etc.) on display. Ask students to put the objects into at least four different groups. How are the items in each group alike or different? Discuss the groups from the aspect that no grouping idea can be wrong as long as the grouping scheme is consistently applied.

How is matter classified?

Section Objectives

Distinguish between pure substances and mixtures on the particle level.

Classify mixtures as heterogeneous or homogeneous.

Distinguish among atoms, elements, and compounds.

Explain how compounds can be distinguished from mixtures.

pure substance
matter composed of only one kind of atom or molecule

mixture
a collection of two or more pure substances physically mixed together

Recyclers classify matter—brown bottles in one collection bin, clear bottles in another. All materials must be classified and sorted before they can be deposited at a recycling center. Chemists have ways of classifying matter too. The categories they use are somewhat different from those you find at a recycling center, however.

Mixtures and pure substances

Matter can be classified as either a pure substance or a mixture

A **pure substance** is matter made of only one kind of atom or molecule. Carbon dioxide, hydrogen, and copper are all examples of pure substances. A **mixture**, on the other hand, is a collection of two or more pure substances physically mixed together that cannot be represented by a chemical formula.

The proportions of different substances in a mixture can vary. For instance, chicken soup is a mixture that may contain different relative amounts of celery, carrots, chicken, pepper, water, and other ingredients, depending on the recipe.

The properties of mixtures can vary because the proportions of the substances in them can vary. For example, gold is mixed with other metals in various proportions to obtain materials suitable for different purposes. Pure gold, which is also called *24 karat* gold, is too soft to keep its shape in jewelry, so it is mixed with other stronger metals to achieve necessary strength. The *alloy*, or solid mixture, of gold used in the finest jewelry is 18 karat gold; it contains 18 out of 24 parts, or 75% gold. The remaining six parts are usually copper, silver, or nickel. For even greater hardness and strength, 14 karat gold is used. As shown in **Figure 2-18**, 14 karat gold does not have quite as much of the brilliant gold color that is characteristic of pure gold or even 18 karat gold.

Figure 2-18a
The gold nugget contains a pure substance—gold. Pure gold, also called 24 karat gold, is too soft to be used for jewelry.

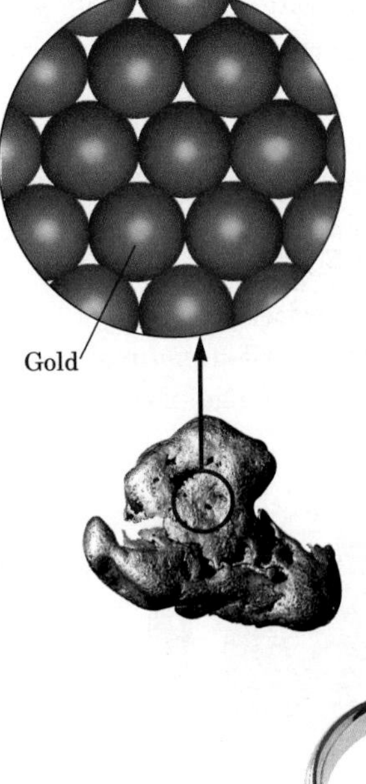

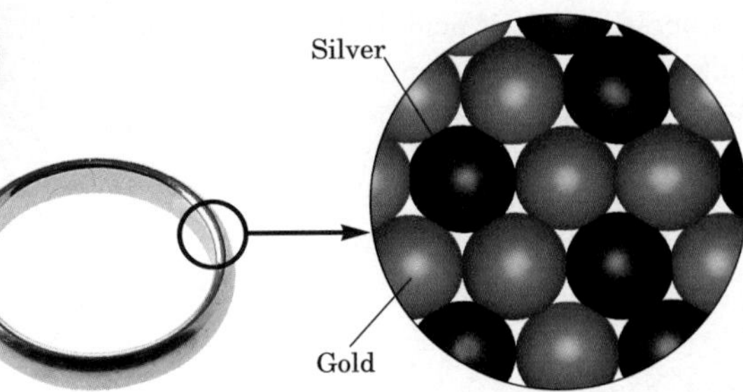

2-18b
A gold-silver alloy rated as 14 karat gold is 14/24 or 58.3% pure gold. This mixture is strong enough to be used for jewelry. Because 14 karat gold contains a smaller proportion of gold than pure gold, its gold color is less intense.

Visual Strategy

Figure 2-18

The microview of gold helps explain its malleability and ductility—two properties that make gold useful for jewelry. Also gold atoms reflect large amounts of light, making this metal highly lustrous and appealing.

A pure substance has specific chemical and physical properties. Any sample of pure water is clear, odorless, will produce bubbles of hydrogen gas if placed in contact with calcium, and has a boiling point of 100°C.* As you might imagine, one rarely finds perfectly pure substances in nature or in the laboratory. Therefore, "pure" is really a relative term. If a given substance is mixed with such a small amount of impurities that the impurities can be ignored, the substance is considered pure.

Mixtures can be further classified

Mixtures can be further classified as heterogeneous or homogeneous. A **homogeneous mixture** is one in which the substances are uniformly distributed. Gasoline, syrup, and saltwater are homogeneous mixtures. All regions of a homogeneous mixture are identical in their composition and their properties. Homogeneous mixtures that consist of substances mixed on the scale of individual particles are called *solutions*. Gasoline, tea, and salt water are all solutions, for example.

homogeneous mixture
a mixture containing substances that are uniformly distributed

heterogeneous mixture
a mixture containing substances that are not evenly distributed

phase
any part of a system that has uniform composition and properties

A **heterogeneous mixture** contains substances that are not evenly distributed. Some regions of a heterogeneous mixture have different properties from other regions. Heterogeneous mixtures can take many forms. Most of the things you see in **Figure 2-19** are heterogeneous mixtures.

Chemists sometimes also use the term *phase*. A **phase** is any part of a system that has uniform composition and properties. Mixtures contain more than one phase. Oil-and-vinegar salad dressing is a two-phase, heterogeneous mixture.

Many plastic products, such as plastic squeeze containers, are mixtures of molecules which present difficulties for recycling. If squeeze bottles are melted in the recycling process, the plastics run together and cannot be easily separated. The pure plastics in each layer are unrecoverable once they are mixed.

You may be surprised to learn that a plastic squeeze bottle is a heterogeneous mixture. If you examined the bottle microscopically you would see separate layers of different kinds of plastic. Without laboratory testing, there is no way to distinguish between homogeneous mixtures and heterogeneous mixtures that appear uniform to the naked eye.

*Water boils at 100°C at standard pressure. On a mountaintop, where the atmospheric pressure is lower than the standard pressure, water boils at a temperature lower than 100°C.

Figure 2-19a
Many of the things you see everyday, such as a mixed salad, oil-and-vinegar dressing, apple juice, and a plastic squeeze bottle are heterogeneous mixtures.

2-19b
A human skin cell is also an example of a heterogeneous mixture.

Theme
Classification and trends
Alloys are a subclass of two larger classes: metals and mixtures. Alloys therefore have the traits of mixtures (variable composition, properties of the substances composing them, etc.), as well as traits of metals (electric conductivity, etc.)

Possible Misconception
Students may confuse the definition of a pure substance with that of an element. Emphasize that both elements and compounds are pure substances. It may be helpful to draw pictures of a water molecule, hydrogen molecule, and neon atom on the board, indicating that these (and large collections of these) are all pure substances.

Teaching Tip
Recognizing pure substances
To help students grasp the definition of pure substances, ask them to brainstorm examples that you can list on the board. Discuss the list with students, circling the correct examples of pure substances. Tell students that they will learn the proper classifications of any incorrect items as they work through the chapter.

Teaching Tip
Recognizing homogeneous mixtures
Ask students to name some of their favorite foods and decide whether they are heterogeneous or homogeneous mixtures. If they are homogenous mixtures, are they also solutions?

Visual Strategy
Figure 2-19
Make the students aware that heterogeneous mixtures are not always easily discernible. Some heterogeneous mixtures appear homogeneous to the naked eye. Laboratory methods must be used to confirm apparent homogeneity.

Answers to

Concept Review

12. c, d, e, h, i, (accept f, g)
13. a. true
    b. false
    c. true
    d. true
14. a. Pictures should show only NO particles within the field of view. Spaces between particles should reflect those of the model for a gas.
    b. Pictures should show diatomic oxygen and diatomic nitrogen molecules within the field of view. Spaces between particles should reflect those of the model for a gas. The picture should show an even mixing of the gas molecules.

Concept Review

Homogeneous and heterogeneous mixtures

12. Consider the mixtures below. State which ones are definitely heterogeneous.

a. gelatin dessert	**b.** iced tea	**c.** concrete
d. topping on a pizza	**e.** toothpaste	**f.** creamy peanut butter
g. homogenized milk	**h.** a taco	**i.** chunky peanut butter

13. Mark the following statements true or false.
    a. All homogeneous mixtures appear to have a uniform composition.
    b. All mixtures that appear uniform are homogeneous.
    c. In a solution, no grains or particles are visible.
    d. Solutions are single-phase systems.
14. Draw picture-models representing the following.
    a. the pure substance nitric oxide (nitric oxide molecules contain one nitrogen atom and one oxygen atom)
    b. a homogeneous mixture of the two diatomic gaseous elements, oxygen and nitrogen.

Elements

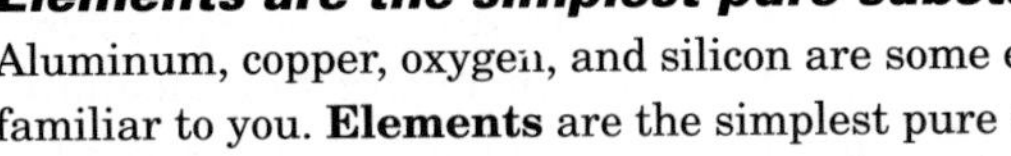

Elements are the simplest pure substances

elements
the 109 simplest substances from which more complex materials are made

Aluminum, copper, oxygen, and silicon are some elements that may be familiar to you. **Elements** are the simplest pure substances, because they contain only one kind of atom. For example, a piece of elemental silicon contains many billions of atoms, all of which are silicon atoms.

Every element has its own unique set of physical and chemical properties. Once it is purified, gold mined in Africa today is indistinguishable from gold panned in the California gold rush or gold discovered by the Aztecs of Mexico centuries ago. Pure gold, obtained from any source, is the same substance. It is an unreactive, soft metal that melts at 1064°C.*

Chem Fact
Gold is relatively uncommon in the Earth's crust. A million tons of earth contains, on average, only 10 pounds of pure gold. It is estimated that all the gold in the world would make a cube measuring 60 ft. on each edge.

A small number of elements make up most common substances

Although there are 109 different elements, only about a dozen compose the things we notice every day. By far, the most common element is hydrogen. More than 90% of the atoms in the known universe are hydrogen. The two elements oxygen and silicon make up more than 70% of the mass of the Earth's crust. Figure 2-20 shows the other elements whose masses contribute substantially to the Earth's crust. Living things are composed primarily of four elements: carbon, hydrogen, oxygen, and nitrogen. These four elements combine to create thousands of different molecules needed for life.

Figure 2-20
The pie chart below shows the abundances of various elements in the Earth's crust by mass.

OXYGEN 49.2%
SILICON 25.7%
ALUMINUM 8.1%
OTHER ELEMENTS 2.6%
MAGNESIUM 1.9%
POTASSIUM 2.4%
SODIUM 2.6%
CALCIUM 3.4%
IRON 4.7%

*The melting point is specified at a given pressure.

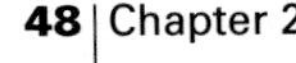

Demonstration 6
Elements
Approximate time: 10 min
Give students a concrete view of sample elements with this demonstration.

Place small quantities of the following elements in individual, covered petri dishes that are well sealed with tape and labeled: aluminum, magnesium, iron, iodine, gold leaf, silver foil, arsenic, red phosphorus, carbon, zinc, copper, tin, and lead. Some items may need to be polished.

SAFETY
Wear safety goggles and a lab apron while preparing the display. Students should be warned not to touch or open the petri dishes.

DISPOSAL
Save materials for reuse at a later time. Be sure to store the metals, phosphorus, and carbon with the combustibles in the storeroom. Store the iodine with the oxidizers.

Visual Strategy
Figure 2-20
Emphasize that while there are many elements on the periodic table, only a few are present on Earth in large quantities. As a comparison, you may want to list the approximate abundances of elements in sea water: chlorine, 1.9%; hydrogen, 10.8%; magnesium, 0.13%; sodium, 1.1%; and oxygen, 85.8%.

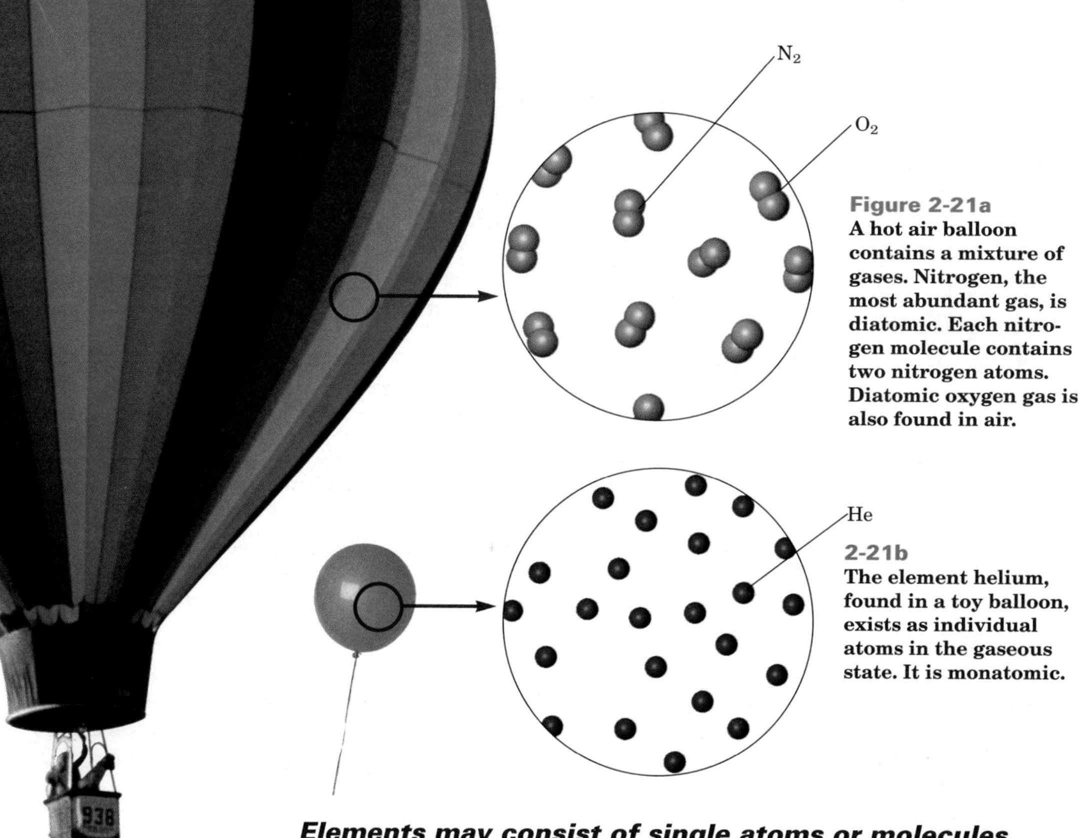

Figure 2-21a
A hot air balloon contains a mixture of gases. Nitrogen, the most abundant gas, is diatomic. Each nitrogen molecule contains two nitrogen atoms. Diatomic oxygen gas is also found in air.

2-21b
The element helium, found in a toy balloon, exists as individual atoms in the gaseous state. It is monatomic.

Historical Note

Prior to 1750, the 16 known elements were: antimony, arsenic, bismuth, carbon, cobalt, copper, gold, iron, lead, mercury, phosphorus, platinum, silver, sulfur, tin, and zinc.

Possible Misconception

Students may believe that elements are composed of a single atom. As counterexamples, emphasize the allotropes and diatomic gases in the text. These examples show that elements can indeed consist of more than just one individual atom.

Elements may consist of single atoms or molecules

An element may be made of individual atoms or molecules. For example, the helium gas in a toy balloon consists of individual atoms, as shown in **Figure 2-21**. If an element consists of molecules, those molecules contain just one type of atom. For example, the element nitrogen, found in air, exists in the molecular state. Each nitrogen gas molecule contains two nitrogen atoms joined together. For this reason, nitrogen is called a *diatomic* gas. Oxygen, another gas found in the air in appreciable amounts, is also diatomic.

Chem Fact
Triatomic oxygen, or ozone, can be formed by any high-voltage electrical discharge. This is why you can sometimes smell ozone around photocopy machines and very large electrical motors.

Some elements have allotropic forms

allotropes
different molecular forms of an element in the same physical state

A few elements, notably oxygen, phosphorus, sulfur, and carbon, are unusual because they exist as allotropes. **Allotropes** are different molecular forms of an element in the same physical state. One allotrope of oxygen is the diatomic oxygen gas you breathe. *Ozone* is another oxygen allotrope. Ozone molecules contain three atoms each.

The properties of allotropes can vary widely. Diatomic oxygen is a colorless, odorless gas essential to life. Ozone is a toxic, blue gas with a sharp odor. You can smell ozone after an intense thunderstorm. Lightning provides the energy to convert diatomic oxygen to ozone in the atmosphere.

Carbon has several interesting allotropes in the solid state. One allotrope is graphite, the gray solid with a slippery texture that you read about earlier in this chapter. Diamond is another carbon allotrope. The properties of graphite and diamond are, of course, very different.

Visual Strategy

Figure 2-21
Emphasize that both oxygen and nitrogen exist as diatomic molecules, whereas helium exists as individual atoms. Helium diffuses faster than nitrogen and oxygen through the semiporous walls of a rubber balloon because its particles are so small.

Other Applications

Among other uses, buckyballs and buckytubes are currently being studied as possible carrier molecules to deliver "packages" of medicines to different parts of the body.

Demonstration 7

Separating water by electrolysis

Approximate time: 15 min

1. Prior to class, prepare 200–250 mL of 1 M Na_2SO_4 solution.
2. Fill a standard electrolysis unit with the solution prepared in Step 1. Make more solution if needed. Apply a small DC current to the leads.
3. When the gas-collecting tubes are sufficiently full, turn off the current.
4. Make sure that all nearby flames are extinguished. Collect the O_2 gas in an inverted Pyrex test tube and quickly turn the tube upright . Place a glowing splint into the test tube. (It should glow brightly.)
5. Collect the H_2 in a similar fashion but leave it oriented with the mouth pointing downward. Bring a burning splint near the tube cup. (It should bark or burn quietly.)

SAFETY

Wear safety goggles, gloves, and a lab apron when preparing the solution and performing the demonstration. Have students stand back at least 10 ft.

DISPOSAL

Wearing safety goggles and a lab apron, pour the liquid down the drain.

Figure 2-22a The buckminsterfullerene molecule is commonly known as a buckyball. It is a carbon allotrope consisting of 60 carbon atoms arranged like a geodesic dome.

2-22b Buckytubes, discovered in 1991, resemble spiraling honeycombs with pentagons at the ends.

In the 1980s, a carbon allotrope consisting of molecules with 60 carbon atoms was discovered. This allotrope is shaped like the geodesic dome designed by the innovative American philosopher and engineer Buckminster Fuller. For this reason, the 60-carbon molecule shown in **Figure 2-22a** is called a "buckyball." More recently another allotropic form of carbon, the "buckytube," has been found. A buckytube is shown in **Figure 2-22b**. Buckyballs and buckytubes exhibit superconductivity (the ability to conduct electricity with no wasted energy).

Compounds

Compounds can be separated into elements

Pure substances that are composed of two or more different elements that are chemically combined are called **compounds**. Compounds are created when atoms of different elements join together in chemical reactions.

Carbon monoxide is an example of a molecular compound. Molecular compounds are made of molecules. Some compounds, such as table salt, are not made of separate molecules. Table salt, or sodium chloride, is an ionic compound. *Ionic compounds*, like molecular compounds, are made of at least two different elements that are chemically joined. However, an ionic compound is not made of discrete molecules. Instead its atoms are distributed in a continuous network. You will learn more about the differences between molecular and ionic compounds in Chapters 5 and 6.

Every compound has a unique set of properties

Because a compound is a pure substance, you know that it has a unique set of chemical and physical properties. You may be surprised to find out, however, that the properties of the compounds are often very different from those of the elements that compose them.

A set of elements can often form more than one compound. Carbon monoxide and carbon dioxide are examples of very different compounds made from carbon and oxygen. Carbon dioxide is used in some fire extinguishers to put out fires, while carbon monoxide burns when ignited. Carbon monoxide is extremely toxic, but carbon dioxide is the gas you exhale in normal respiration.

compounds
pure substances composed of two or more different elements

Compounds can be distinguished from mixtures

It is important to understand the difference between compounds and mixtures. There are three principal differences.

1. Mixtures are never made of only a single compound.
2. The properties of a mixture reflect the properties of the substances it contains, but the properties of a compound often bear no resemblance to the properties of the elements that compose it.
3. Compounds have a definite composition by mass of their combining elements. The iron-sulfur compound pyrite has a definite composition. The mass of pyrite is always 46.55% iron and 53.45% sulfur. In contrast, substances in a mixture can exist in any mass ratio. The sugar in a sugar-water mixture may make up 99% or 1% of the overall mass.

You can see how compounds and mixtures relate to other classes of matter in **Figure 2-23**.

Figure 2-23
Study this figure to understand the relationships among different classes of matter.

Classification of Matter

IF a sample of **matter** is made of . . .

. . . only one kind of atom or molecule, as in water or helium,

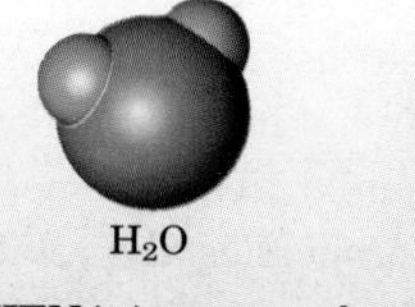

H_2O He

THEN it is a pure substance.
IF a pure substance consists of . . .

. . . more than one kind of atom or molecule, **THEN** it is a mixture.

IF a mixture has a . . .

. . . a single kind of atom, as in chlorine gas,

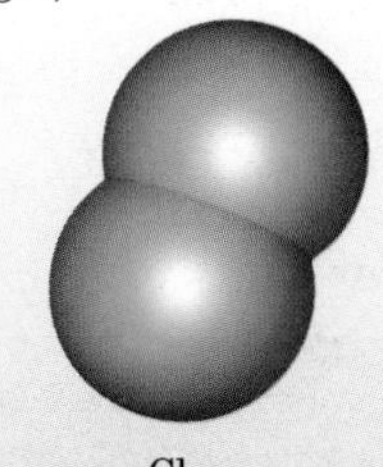

Cl_2

THEN it is an **element**.

. . . molecules made of more than one kind of atom, as in methane,

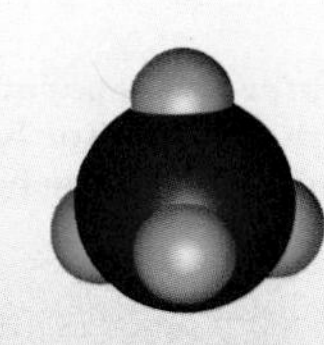

CH_4

THEN it is a **compound**.

. . . uniform composition, as in a solution of sucrose (table sugar) and water,

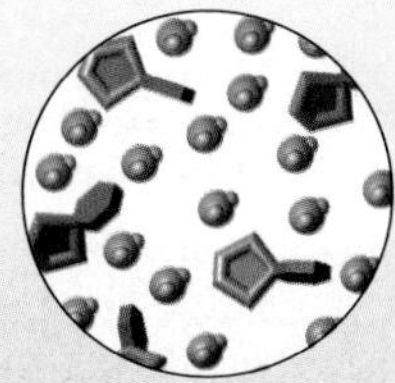

THEN it is a **homogeneous mixture**.

. . . nonuniform composition, as sand and water,

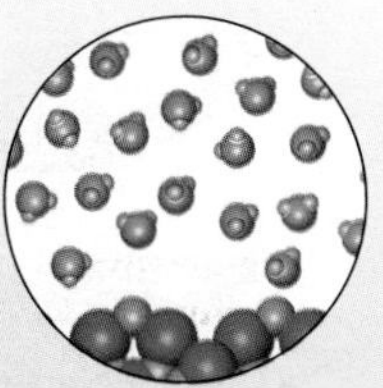

THEN it is a **heterogeneous mixture.**

Section Review

15. How many different compounds can a pure substance contain?
16. Bronze is a mixture of tin, copper, and zinc. To predict the properties of bronze, what information do you need about tin, copper, and zinc?
17. In what ways will a mixture of copper, oxygen, and sulfur be different from a compound of these three elements?

18. **Story Link**
    Why are the properties of aluminum always the same while the properties of a leaded-glass mixture may vary?
19. **Story Link**
    Why must aluminum and glass be sorted before they can be recycled?

ASSESS

Alternative Assessment

Have students work in pairs to develop quiz questions (with answers) for this section. You may want to give students parameters for developing good questions. Then give the class a quiz consisting of the best questions. Students whose questions are chosen are rewarded with easy questions!

Visual Strategy

Figure 2-23

Students often find it difficult to keep straight the definitions of *atom*, *molecule*, *element*, *compound*, *pure substance*, *heterogeneous mixture*, and *homogeneous mixture*. The organizer on this page can help show relationships among these classes of matter. The illustrations show the distinctions among types of matter on the particle level.

Answers to Section Review

15. none or one
16. The properties of tin, copper, and zinc must be known.
17. The properties of this mixture will reflect the properties of its component elements, but the properties of a compound made from these elements differ greatly from the elements themselves.
18. The properties of aluminum do not vary because aluminum is a pure substance. Leaded glass is a mixture so different samples of leaded glass may have different properties.
19. The containers must be sorted to avoid introducing impurities that could alter the properties of these materials.

Section 2-4

FOCUS

Lesson Starter

Draw a pie chart on the board showing that 3% of Earth's water is fresh, and 97% is salt water. Also state the following: "The consumption of fresh water by the human population increases each year. Yet, due to water pollution and other causes, the supply of fresh water has been steadily decreasing." Discuss the importance of separating mixtures in the context of desalinating ocean water to supply fresh water for human consumption.

Demonstration 8

Separation of substances

Approximate Time: 15 min

1. Place 4 g of NaCl and 5 g of sand into 20 mL of water in a large Pyrex test tube.
2. Ask students how the three substances could be separated. Tell them to think about the properties of the three substances. Ask them if filtering the sand will also separate the salt.
3. Ask for methods to separate the salt and water. An easy method is to place the solution in an evaporating dish (with the spout pointed in a safe direction) covered with a watch glass. Heat gently until dry. Salt crystals will form on the glass.

SAFETY
Wear safety goggles and a lab apron, and tie back loose clothing and long hair.

DISPOSAL
Save the sand for reuse at a later time. Dissolve the salt in water, and pour down the drain.

2-4

How are substances identified?

Section Objectives

List five methods of separating the components of a mixture.

List properties that can be used to identify a pure substance.

Calculate density and use density to identify pure substances.

Methods of separating mixtures

In order to recycle items in a pile of trash, you separate them by hand, using one or more physical properties to identify various substances. Or, if you need to separate steel cans from aluminum cans, you might use a magnet to identify and isolate them. (Steel is magnetic and aluminum is not.) There are a multitude of ways to separate mixtures. The method to use in a given situation depends on the properties and relative amounts of substances in a mixture.

Mixtures can be separated by physical means

The photographs in **Figure 2-24** and **Figure 2-25** on the next page show some of the many ways to separate a mixture using only physical means. When you *filter, evaporate, centrifuge,* or *decant* a mixture, you bring about only physical changes. The substances you obtain after separation are not chemically changed by the separation process.

Figure 2-24a
Coffee grounds and water are separated by filtration. Liquid and particles smaller than the filter holes pass through the filter. The *filtrate* is collected in the coffee pot.

2-24b
Magnetism can be used to separate magnetic components of a mixture, like iron and steel, from nonmagnetic components.

2-24c
In salt ponds such as this one, sea water evaporates. Sodium chloride, or salt, is left behind. This simple method of salt production is still widely used.

Visual Strategy

Figure 2-24

Stress with students that separation techniques are based on properties. The separation of coffee grounds and the salt in the salt pond is based on solubility. The metals are separated using the property of magnetism.

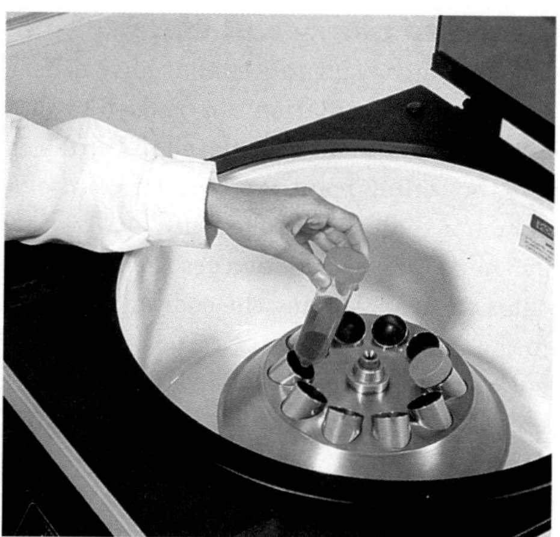

Figure 2-25a
A centrifuge is a tool used to separate matter of different densities. The centrifuge spins rapidly; the resulting centrifugal force pushes dense matter outward. Medical workers often separate the solid matter in blood from the liquid portion. Then tests can be performed on the solid cellular material.

2-25b
To decant a mixture, simply pour the liquids off and leave the solids behind. To do so, the solid contents must be well settled on the bottom of the container.

Chromatography is widely used in industry

Chromatography may be used to separate the components of a solution so that they can be identified. Chromatography works because different substances have different attractions to solvents and other media.

Paper chromatography can be used to separate black ink into the colored dyes it contains, as shown in **Figure 2-26**. How does this work? Water dissolves the ink. The ink-containing water travels up the paper through capillary action. Substances travel at different rates according to the attraction they have for the paper. Those most attracted to the paper travel the slowest. They appear in a relatively low position on the paper. Those that have the least affinity for the paper travel the fastest.

Research laboratories make great use of chromatography. The food industry uses gas chromatography to isolate and identify compounds that give foods their aromas. Some of these pleasant-smelling compounds can be processed for use in artificial flavors and fragrances.

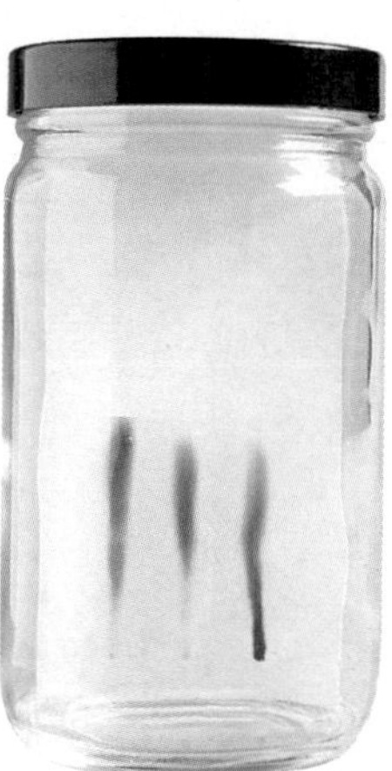

Figure 2-26
Chromatography can separate the dyes in different samples of black ink. First, ink marks are made on absorbant paper. Then, as shown in the middle photograph, solvent in the bottom of the jar rises through the paper, carrying the ink with it. The finished chromatogram, shown at right, displays different dyes found in each of the three black sample inks.

TEACH

Teaching Tip

Discussion

Discuss reasons for separating blood, as shown in the photograph in **Figure 2-25a**. People can donate whole blood or just plasma by having the blood centrifuged and the cells returned. The four blood types cannot all be interchanged, but plasma can be interchanged.

Cultural Connection

The earliest application of chromatography was performed by Russian botanist Michael Tswett in the early 1900s. He separated plant pigments, giving rise to the name "color writing."

Theme

Systems and interactions

In chromatography, there is an interaction between the solutes, solvent, and carrier medium. The intensity of these interactions depends on how fast the solute moves during the process.

Other Applications

The popcorn industry has increased its sales three-fold since the 1970s largely due to microwave popcorn. Food chemists were able to isolate the aromas in butter and popcorn that people like most. These flavorings are added to microwave popcorn to intensify its flavor and aroma. Who can resist the aroma of hot, buttery popcorn? Artificially flavored popcorn jelly beans are also popular.

Visual Strategy

Figure 2-25 and **Figure 2-26**

The separation technique most appropriate in a given case depends on the physical and chemical properties of the substances that are mixed.

Chromatography reveals the differing compositions of black inks made by different manufacturers. How could this technique help a forensic chemist link a particular pen to a handwritten note in a legal case?

Theme
Conservation
The saltwater mixture present before the distillation can be recombined after distillation to obtain the same mixture again.

Demonstration 9
Distillation
Approximate time: 20 min

1. Place 15 mL of a 1 M copper(II) sulfate solution into a Pyrex test tube fitted with a #5, 1-hole stopper with a single glass tube.
2. Run rubber tubing from the stopper to a Pyrex test tube that is three-quarters submerged in a 600 mL beaker of cold water.
3. Gently heat the copper(II) sulfate solution until distillation begins. Care must be taken to keep the copper(II) sulfate solution from boiling over.
4. When finished, remove the tube from the distillate to avoid drawing the cold water into the hot test tube.

SAFETY
Wear safety goggles and a lab apron, and tie back loose hair and clothing. Students should be a minimum of 10 ft. from the apparatus.

DISPOSAL
If the distillate is colorless, pour down the drain while wearing safety goggles and a lab apron. If the distillate is colored, add it to the liquid or solid residue in the larger test tube. Let water, if any, in the larger test tube evaporate and save the solid copper(II) sulfate for reuse at a later time.

Chem Fact
Crude oil is the greenish brown to black liquid pumped from underground that is later transported to oil refineries. There it is separated into simpler mixtures and substances, including kerosene and gasoline, which are used as fuel. Other ingredients obtained from crude oil serve as raw materials for plastics, certain drugs, and explosives.

Distillation can be used to purify salt water

Distillation is a method of separating substances that have different boiling points. **Figure 2-27** shows how distillation can be used to obtain pure water from a salt-water solution. Salt-water is placed in the distillation flask, where it is heated to the boiling point. The boiling point of sodium chloride, or salt, is much higher than the temperature at which the solution boils. Thus, the water in the salt-water mixture turns to steam, and rises upward in the distillation flask, leaving the sodium chloride behind. The steam flows through the condenser, a water-cooled glass tube. The steam condenses as it makes contact with the cooled surface of the condenser. The condensed steam is purified water in the liquid state. Pure water is collected in the receiving flask.

In a few locations throughout the world, distillation is used to obtain drinking water from seawater. However, the process requires a great deal of energy and is very expensive, so other ways of obtaining drinking water are preferred. Distillation is used on a wide scale in the petroleum industry to separate crude oil. The separated components of crude oil have many uses. Gasoline, heating oil, lubricants, and the materials from which plastics are made are all obtained from the separation of crude oil.

Figure 2-27a
During the first step of distillation, the salt water solution is heated to the boiling point of water.

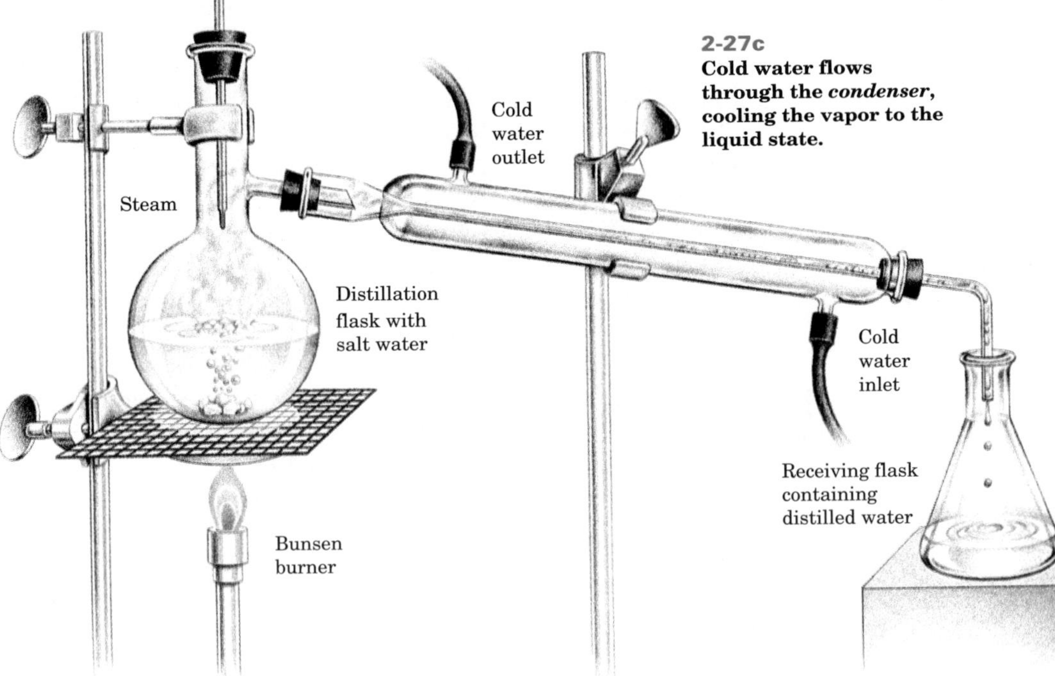

2-27c
Cold water flows through the *condenser*, cooling the vapor to the liquid state.

2-27b
The solution boils when it reaches its boiling point. Because the boiling point of water is lower from that of salt, pure water vapor rises and moves to the condenser while sodium chloride is left behind.

2-27d
Once the pure water vapor cools, it condenses back to the liquid state.

Visual Strategy
Figure 2-27
Discuss the purpose of the condenser which supplies a cool glass surface where condensation can occur. Also note that the substance with the lowest boiling point in the mixture boils off first, condenses first, and is collected first. In this case, the substance with the lowest boiling point is water.

Density is a constant ratio of mass to volume

A relationship between two quantities, such as mass and volume, can be represented on a graph. For example, consider a set of lead samples of different sizes. The mass and volume of each sample are listed in **Table 2-2**. The pairs of mass and volume values are represented as data points on the graph, as shown in **Figure 2-28**. Note that a straight line can be drawn through the data points. What does this mean?

Table 2-2
Mass and Volume Data for Samples of Lead

Sample number	Mass (g)	Volume (mL)
1	5.00	0.443
2	15.0	1.33
3	24.0	2.12
4	52.0	4.60
5	64.0	5.66
6	81.0	7.17
7	95.0	8.41
8	101	8.94
9	142	12.6
10	153	13.5

The rising straight line from left to right indicates that mass increases at a constant rate as volume increases. As the volume of lead doubles, its mass doubles; as the volume triples, mass also triples, etc. In other words, mass is *directly proportional* to volume for lead.

The *slope* of a graph of directly proportional quantities, or the degree to which the graph slants, equals the ratio of the quantity plotted on the vertical axis (y) divided by the corresponding quantity plotted on the horizontal axis (x). This is mathematically stated as follows.

$$\textbf{slope} = \frac{y}{x}$$

Note that the slope of the graph of directly proportional quantities is constant. The slope of the mass-versus-volume graph for lead has a constant value of 11.3. Pick any data point on the graph, and divide the mass value by the corresponding volume value, and you will get this result.

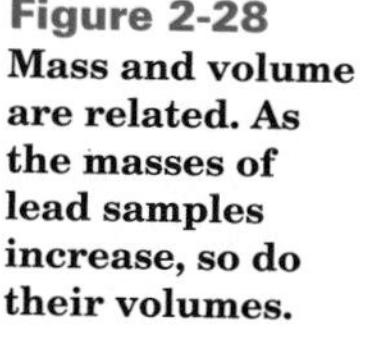

Mass vs. Volume for Lead Samples

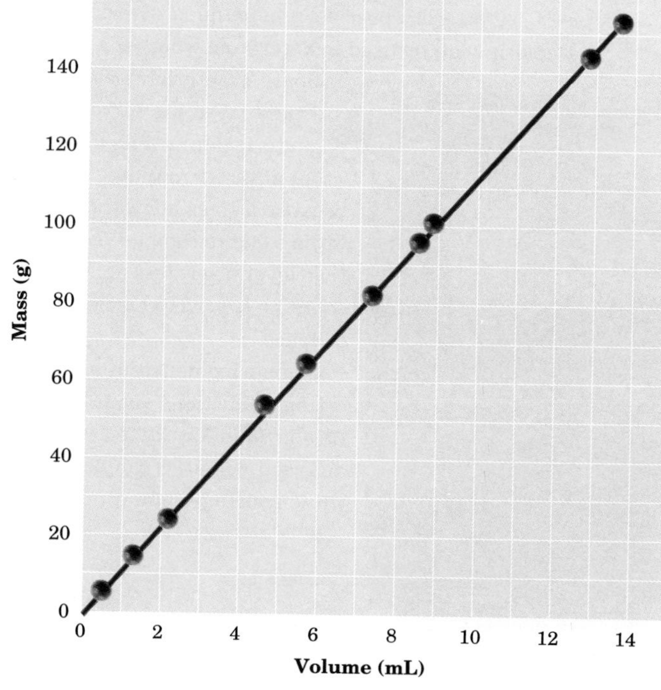

Figure 2-28
Mass and volume are related. As the masses of lead samples increase, so do their volumes.

Teaching Tip

Reinforce the concept that density is an intensive property. Pick up an eraser and say "This eraser has a mass of about 50 g and a volume of 200 mL. What is its density?" Write "50 g/200 mL = 0.25 g/mL" on the board. "Imagine we cut the eraser in half. What is its density?" After some answers are offered, show the calculation with the new values for mass and volume that prove that the density is unchanged.

Visual Strategy

Figure 2-28

Help students understand the significance of linear relationships and direct proportions. Explain that a constant slope shows that two quantities are related through a simple proportionality, and the slope of the line is equal to the constant of proportionality. Sometimes the slope represents a physical quantity, as in the case of density.

Teaching Tip

Estimating volumes of irregularly shaped objects
Write a list of volume formulas for simple geometric shapes on the blackboard. Distribute some irregularly shaped objects. Show students how to estimate the volumes of irregularly sized objects by matching them to the regular geometric form that is most similar in shape. Have students make the necessary measurements and calculate the volume of each object. Discuss the error factor involved this estimation process.

Table 2-3
Densities of Various Substances

Substance	Density (g/mL) at 25°C
aluminum	2.70
carbon dioxide gas (1 atm)	0.00197
copper	8.94
ethyl alcohol	0.789
gold	19.3
hydrogen gas (1 atm)	0.00089
iron	7.86
osmium	22.6
sodium chloride	2.164
sucrose (table sugar)	1.587
water	0.997

The constant slope of a mass-versus-volume graph has a special name—*density*. Density is the ratio of mass to volume, as expressed below.

$$\text{density} = \frac{\textit{mass}}{\textit{volume}} \qquad D = \frac{m}{V}$$

Density is usually expressed in units of grams per milliliter, g/mL. Thus, the density of lead is 11.3 g/mL. As noted earlier in this chapter, density can be interpreted conceptually as well as mathematically. In conceptual terms, density is a measure of how tightly matter is packed. **Table 2-3** shows the density of some interesting substances. Osmium, a blue-white metal, is the densest substance on Earth. It is so dense that a piece of osmium the size of a football would be too heavy for you to lift.

Density can be used to identify substances

Because the density of a substance is the same for all samples of it, density can be used to identify substances. For example, suppose you find a bracelet on the street that appears to be silver. You wonder if it could be pure. To find out, you take the bracelet into the lab and measure its mass with a balance—210 g. Next you measure its volume—20.0 mL. The density of the bracelet is calculated as follows.

$$D = \frac{m}{V} = \frac{210 \text{ g}}{20.0 \text{ mL}} = 10.5 \text{ g/mL}$$

Next, you look up the density of silver in a reference table and find that it is 10.5 g/mL. The bracelet you found is silver!

Densities can be used to determine whether a given material will sink or float. A material floats if its density is lower than that of the surrounding medium. The material sinks if its density is greater than that of its surrounding fluid. Ice floats on water because ice has a lower density than water. Check **Figure 2-6** to compare ice and water on the particle level. Ice is less dense than water because it contains fewer H_2O molecules in a given amount of space.

Comparing densities allows you to make interesting predictions about floating and sinking. As shown in **Figure 2-29**, two or more liquids of different densities can be layered if you are careful while pouring them.

Figure 2-29
Substances with lower densities float in or on top of substances with higher densities.

Visual Strategy

Figure 2-29
Point out that the mixture shown does not separate by states of matter, but rather by density. For example, solid wood floats easily on liquid corn oil. Ask students to recall what happens when a cork is put into water.

Consumer Focus

Recycling codes for plastic products

More than half of the states in the U.S. have enacted laws that require plastic products to be labeled with numerical codes that identify the type of plastic. These codes are shown in **Table 2-4**. Used plastic products can be sorted by these codes and properly recycled or processed. Knowing what the numerical codes mean will give you an idea of how successfully a given plastic product can be recycled. This may affect your decision to buy, or not to buy, particular items.

Table 2-4
Recycling Codes

Recycling code	Type of plastic	Physical properties	Example	Uses
1	polyethylene terephthalate (PET)	usually glossy, tough, rigid; drips when burned; sinks in water	soda bottles	fiberfill for jackets, sleeping bags, carpet
2	high density polyethylene (HDPE)	rough surface, semi-rigid; drips and gives off white smoke when heated	milk containers	furniture, toys, trash cans
3	polyvinyl chloride	very glossy, semi-rigid; does not burn; sinks in water	plastic bags	usually not recycled
4	low density polyethylene (LDPE)	low gloss, flexible; drips and gives off white smoke when burned; floats on water	clear bottles for cooking oils, peanut butter jars, shampoo, liquor	not recycled
5	polypropylene	low gloss, semi-rigid; drips and gives off white smoke when burned	heat-proof containers, insulated clothing, marine rope	not recycled
6	polystyrene (P/S, PS)	glossy, brittle to semi-rigid; does not drip when burned; gives off black smoke when burned; sinks in water unless in the form of foam	fast-food containers	recycled for permanent outdoor furniture such as park benches

Section Review

20. Can iron oxide (rust) be separated into oxygen and iron by chromatography? Why or why not?
21. Ocean water can be made suitable for drinking by boiling the water, and then condensing the vapor back to the liquid state. What is the name of this process?
22. Graph the data in the table to the right to determine the mass of 6.0 mL of iron. Also determine the density of iron by whatever method you choose.

Mass (g)	Volume (mL)
15.6	2.0
27.3	3.5
31.2	4.0
41.3	5.3

23. **Story Link**
You have an unlabeled container that looks like it is made of aluminum. How can you be sure whether it belongs in the aluminum recycling bin?

Teaching Tip
Types of plastics
Have an example of each type of plastic for display. Polypropylene is a common plastic rope material. Polyvinyl chloride, also called PVC, is used in new construction for drain pipes.

Teaching Tip
Recycling
Have students collect data for one week on the number of metal cans, glass bottles, and plastic containers the are thrown in the trash at home. Have them classify their data by metal type, glass color, and plastic type (recycling number). Ask students for three ways to increase the amount they recycle or three ways to decrease the amount of trash they produce at home.

ASSESS

Alternative Assessment
Propose this story to the class. A local plastic recycler typically grinds all of the plastic into small grains and places them into six different containers. When the plastic is ready to be recycled, the boss realizes that the six containers were not labeled and that all the plastic pellets appear somewhat similar. How can the plastics in each container be identified? (Answers: by density, or a combination of other intrinsic properties such as color, melting point, solubility, etc.)

Answers to Section Review
20. No; rust is a compound, not a mixture.
21. distillation

Answers are continued on page 69A.

Section 2-5

FOCUS

Lesson Starter

Before class, prepare several measuring sticks (one for each group) of various lengths from old meter sticks or paint stirrers. If. possible, paint the sticks so that all markings are covered, or cover the markings with masking tape. Divide the class into groups of three. Pick a common object that can be easily measured by each group (such as the height of the chalkboard). Ask each group to measure the height of the object as accurately as possible with the measuring stick. Compare the results. Who has the correct answer? Why are standard units needed?

Theme

Classification and trends

Measurement is based on seven base units. From these units, most other units can be derived.

How should data be reported?

Section Objectives

List the basic SI units and the quantities they describe.

Use metric prefixes to convert metric units.

Distinguish between accuracy and precision in measurement.

Use significant figures in measurements and calculations.

Standard units

Properties are described most exactly in numerical terms. You can describe a person's height, mass, or age in words: tall or short, heavy or light, old or young. But such descriptions are easily misinterpreted because they are inexact and subjective. From a child's perspective, you may be considered "old." A grandparent, however, would probably describe you as quite young.

Numbers make descriptions more exact. You may have used the expression, "On a scale of one to ten, it's about a four," or "On a scale of one to ten, that's a minus two!" These descriptions are still vague because it is unclear what the numbers refer to. This is why standard units are useful.

Table 2-5
SI Base Units

Quantity	Unit	Symbol
Length	meter	m
Mass	kilogram	kg
Time	second	s
Electric current	ampere	A
Thermodynamic temperature	kelvin	K
Amount of substance	mole	mol
Luminous intensity	candela	cd

SI units are used in science

In 1960, the scientific community adopted a subset of the metric system to use as the standard scientific system of measurement units. This is the "Systeme Internationale" (SI). It features seven base units. Combinations of the base units can be used to describe nearly all physical measurements. Although only five are used extensively in chemistry, all seven base units are shown in **Table 2-5**.

Base units and prefixes establish appropriate scale

Any SI unit can be modified with prefixes to match the scale of the object being measured. Meters may be suitable to express a person's height. Millimeters (0.001 m) are more appropriate for measuring the diameter of a living cell. Atoms and molecules are generally measured in picometers or nanometers. The nuclear decay of an atom happens in picoseconds; the blink of an eye takes milliseconds. A human lifetime (assuming 75 years) is about 2.4 gigaseconds. Metric prefixes are given in **Table 2-6**.

Table 2-6
Some SI Prefixes

Prefix	Symbol	Meaning
giga	G	billion
mega	M	million
kilo	k	thousand
deci	d	tenth
centi	c	hundredth
milli	m	thousandth
micro	μ	millionth
nano	n	billionth
pico	p	trillionth

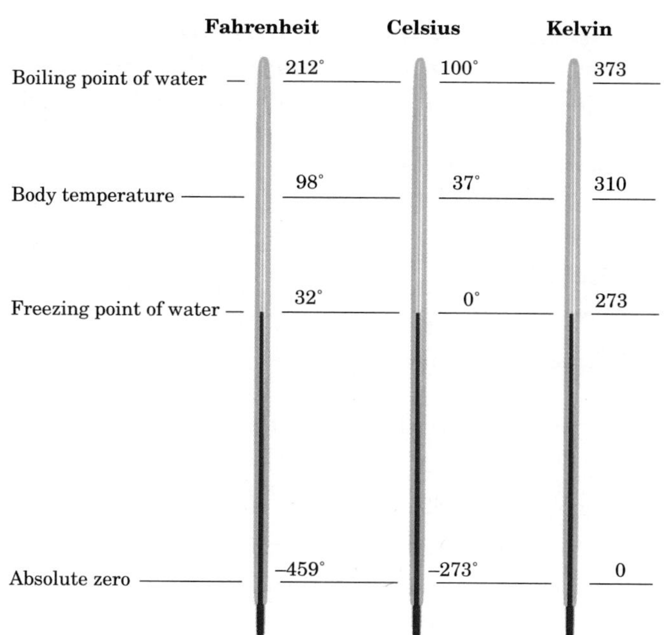

Figure 2-30
Three different scales are used to measure temperature. The Fahrenheit scale is used to measure temperature on a daily basis in the United States. The Celsius scale is the metric scale used to measure temperature. The Kelvin scale is used in scientific work.

Kelvins are used to measure temperature

You are probably familiar with the Fahrenheit and Celsius temperature scales. Notice that the SI unit of temperature is the **kelvin** (abbreviated K, with no degree symbol). Kelvins are the same size as Celsius degrees, but the zero on each temperature scale does not correspond to the same actual temperature.

The Celsius scale designates zero degrees as the freezing point of water. Absolute zero, or 0 K, is the coldest temperature theoretically possible. **Figure 2-30** compares the Fahrenheit, Celsius, and Kelvin scales. It is simple to convert a temperature from Celsius degrees to kelvins.

$$\mathbf{°C + 273.16 = K, \text{ or } K - 273.16 = °C}$$

You will study the basis for the absolute temperature scale in Chapter 10.

Derived units can be obtained for any quantity

Because the seven SI base units cannot measure every observable property, *derived units* are created by multiplying or dividing the seven base units in various ways. The derived unit for area, for example, is the square meter, m^2. It is obtained by multiplying a distance unit by a distance unit, as in the case of a rectangular surface.

$$\textbf{area units: } m \times m = m^2$$

Volume is expressed in the derived SI unit cubic meters, m^3. It is obtained by multiplying length, width, and height.

$$\textbf{volume units: } m \times m \times m = m^3$$

In the traditional metric system, the liter, L, is the volume unit. The liter and milliliter, mL, are widely used in scientific work even though the cubic meter is the SI unit. In the health-related fields, the cubic centimeter, cm^3, is often used instead of the milliliter.

Teaching Tip

Spatial relationships

Build a cubic meter from eight meter sticks. Use four student volunteers to show the size of a cubic meter. Students will be surprised at its size. Tell them that a cubic meter of water has a mass of 1000 kg.

Visual Strategy

Figure 2-30

Have students note that the three temperature scales are just different ways of expressing the same average translational kinetic energy of atoms or molecules.

Do You Know?

The meniscus effect of water in a glass container is due to water's attraction for the glass. When water is placed in a plastic graduated cylinder, the climbing effect is greatly reduced. When mercury is placed in a glass or plastic container, the meniscus effect is reversed due to greater attractions of mercury atoms to each other than to the walls of the container.

Reliability of measurements

Measurements must be made with care

When you measure the mass of an object or its volume or length, you obtain a number with units. How accurate is this number? How close to the true value will the measurement be? No experimentally obtained value is exact because all measurements are subject to errors—instrument errors, method errors, and human errors.

If a measuring device is not calibrated (or scaled) correctly, instrument errors will be introduced. For instance, a scale may be printed incorrectly on a graduated cylinder. Or a balance may stick and not read small masses correctly. Also, measurements must be made using a standard, or agreed-on, method. Reading the volume of the liquid in a graduated cylinder must be done the same way every time, as shown in **Figure 2-31**. Can you see why?

Human error can occur by chance or through bias. People inevitably make mistakes, even when care is taken. Human errors can also be caused by sheer carelessness. Experimenters usually have some idea about what answer they would like to see; this introduces bias. Recognizing one's own biases is an important part of research. But because it is impossible to detect all of one's own biases, experimental work can never be completely free of them.

Figure 2-31
The position of the meniscus indicates this volume should be read as 50.0 mL of solution.

accuracy
the extent to which a measurement approaches the true value of a quantity

precision
the degree of exactness or refinement of a measurement

Accuracy differs from precision

There are two things to consider when making and reporting a measurement: accuracy and precision. The **accuracy** of a measurement is the extent to which it approaches the true value and is free from error. The **precision** of a measurement is how exactly or sharply stated the measurement is. A measurement made on a very fine scale is more precise than a coarse measurement made with a blunt instrument. An analytical balance measures mass with greater precision than a triple beam balance, as **Figure 2-2** shows. **Figure 2-32** illustrates the difference between accuracy and precision.

Figure 2-32
Precision relates to exactness. Accuracy relates to how close a measurement is to the true value.

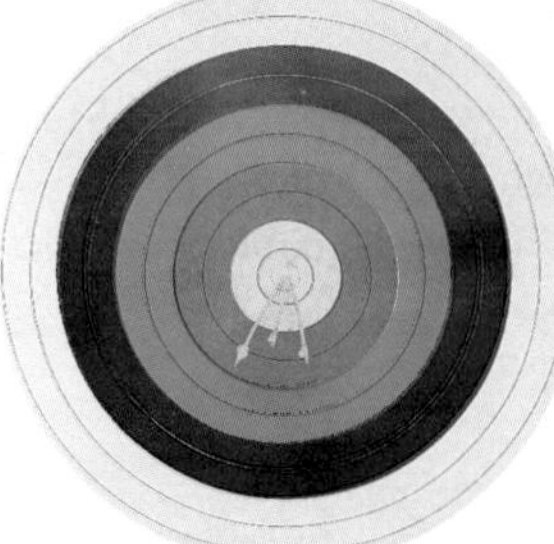

Good accuracy, good precision

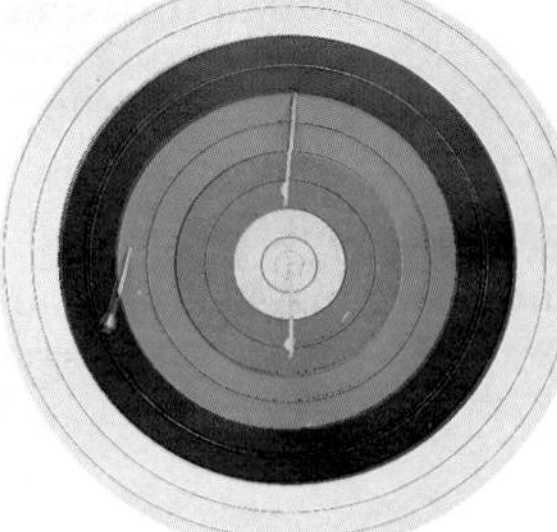

One accurate result, poor precision overall

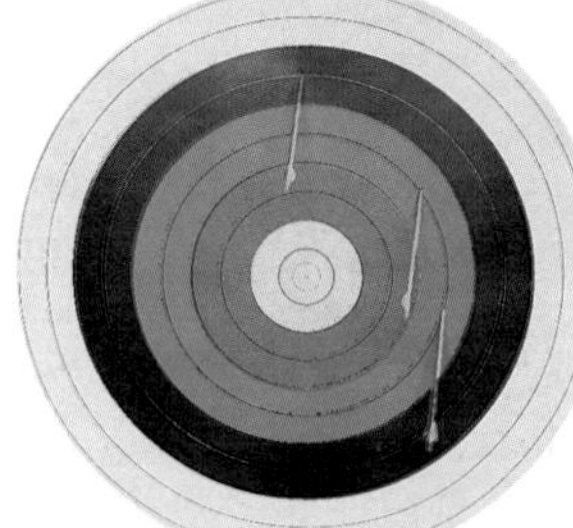

Poor precision, poor accuracy

Visual Strategy

Figure 2-31 and **Figure 2-32**

Make sure students realize they should always read from the bottom of the meniscus.

Accuracy and precision are most often illustrated by using arrows on a target. On the target on the left, the arrows are accurate and precise. On the target in the middle, two arrows are far from the center, so they are inaccurate. On the target on the right, the arrows are neither precise nor accurate.

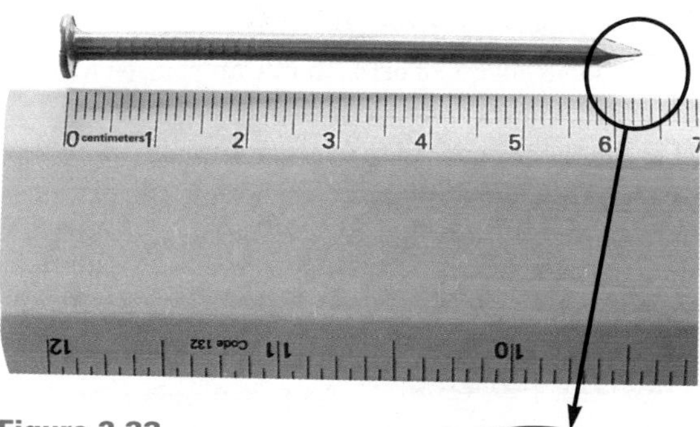

Figure 2-33
By comparing the length of the nail to the scale on the ruler, you can see that the nail is more than 6.3 cm but less than 6.4 cm long. To report this measurement in significant figures, report *6.3*, which you have actually measured, plus the next digit, which you estimate. The final measurement, correctly reported to three significant figures, is 6.35 cm.

Measurements must be reported to the correct number of significant figures

If a measurement is reported with either too many or too few digits, it is not possible to tell how precise the measurement really is. To indicate precision, significant figures are used. Significant figures are those digits in a measurement that have actually been measured by comparison with a scale, plus one estimated digit. For example, if you were measuring temperature with a thermometer marked in intervals of 1°C, you would report the temperature to the nearest tenth of a degree—25.1°C, for example. The procedure for expressing a measurement in significant figures is shown in **Figure 2-33**. Note that the number of significant figures you obtain is limited by how finely divided your measuring scale is.

As the fourth rule in **Table 2-7** indicates, you cannot tell whether zeros to the right of a nonzero number but to the left of a decimal are significant just by looking at a measurement. But there are two ways to write such numbers so that you can tell whether such zeros are significant.

1. Put a decimal point after zeros that are significant.
2. Use scientific notation.

For example, if all the zeros in the measurement 1100 m are significant, you can write the measurement as either 1100. m or in scientific notation, as 1.100×10^3 m. (For more information on how to use scientific notation, please see **Appendix B.**)

Table 2-7
Rules for Determining Whether Zeros Are Significant

Rule	Examples
1. Zeros appearing between nonzero digits are significant.	a. 40.7 L has three significant figures. b. 87 009 mi. has five significant figures.
2. Zeros that appear in front of nonzero digits are not significant.	a. 0.095 897 m has five significant figures. b. 0.0009 kg has one significant figure.
3. Zeros at the end of a number and to the right of a decimal are significant.	a. 85.00 g has four significant figures. b. 9.000 000 000 has 10 significant figures.
4. Zeros at the end of a number but to the left of a decimal may or may not be significant. If such a zero has been measured or is the first estimated digit, it is significant. On the other hand, if the zero has not been measured or estimated but is just a placeholder, it is *not* significant.	a. 2000 m may contain from one to four significant figures, depending on how many zeros are placeholders.

Teaching Tip

Length measurements
Pass out meter sticks or centimeter rulers (or photocopies of centimeter rulers) when discussing uncertainty in measurement. Students can form a concept of the length of a meter and also view the uncertainty area between millimeter marks on the measuring device.

Additional Examples

Students must be able to properly determine the number of significant digits in a number before they can successfully use significant digits in addition or multiplication problems. Have the students determine the number of significant digits for the following.
345 (3)
0.93 (2)
340 (2)
0.097 (2)
39.5 (3)
0.09305 (4)
38.05 (4)
0.00760 (3)
139.0 (4)
1.008 (4)
10001 (5)
15.96 (4)
10100 (3)
1.95×10^{-8} (3)
89050 (4)
3.9×10^3 (2)

Visual Strategy

Figure 2-33
Help students estimate the distance between the certainty lines. Remind them that although no one can be truly "certain" of the correct value, care must be taken so the value can be typically estimated as a 1 in 10 value.

Teaching Tip

Significant digits

Refer students to Appendix B on significant digits for additional explanation and practice.

Significant figures are used in calculations

Each measured quantity has some degree of error in it. Can you see why? When measurements are added, subtracted, multiplied, or divided, small errors may either cancel each other or add up to a larger error. To avoid the possibility of adding extra error, significant figures are used in calculations.

The procedure for obtaining significant figures in addition and subtraction is different from the procedure used for multiplication and division. **Table 2-8** shows the procedure for each. See Appendix B for more examples of calculations with significant figures and for information on rounding numbers.

Table 2-8
Rules for Using Significant Figures in Calculations

Operation	Rule	Example
Addition or subtraction	The answer can have no more digits to the right of the decimal point than there are in the measurement with the smallest number of digits to the right of the decimal point.	3.95 + 2.879 + 213.6 = 220.429 $\xrightarrow{\text{round off}}$ 220.4
Multiplication or division	The answer can have no more significant figures than there are in the measurement with the smallest number of significant figures.	12.257 × 1.162 = 14.2426 $\xrightarrow{\text{round off}}$ 14.24

Now can you put together all the rules for using significant figures to perform a laboratory task such as determining the density of, say, a sugar cube? This is how you might go about the task: Measure the mass of the sugar cube with a triple beam balance. Following the procedure for correct measuring summarized in **Figure 2-33**, you determine that the mass of the cube is 2.08 g. Now determine the volume of the sugar cube. To do this, you can measure the length of one of the cube's edges. Your ruler shows the length is 1.1 cm. The volume of a cube can be found from the following expression.

$$\text{Volume} = \text{length (cm)} \times \text{length (cm)} \times \text{length (cm)}$$

If you apply the rules for multiplying with significant figures listed in **Table 2-8**, you find that the volume of a sugar cube, correctly expressed, is 1.3 cm^3, or 1.3 mL. Now use your measurements, the expression for density, and the rules in **Table 2-8** to determine the density of a sugar cube and report the result in the correct number of significant figures.

$$D = m/V = 2.08 \text{ g}/1.3 \text{ mL} = 1.6 \text{ g/mL}$$

The exercises below will give you more practice with significant figures.

Calculating with significant figures

24. Express the following calculations in the proper number of significant figures.

a. 129/29.2 **b.** 30.8/45.0 **c.** 0.098/45.4 **d.** 3.45/0.78
e. 1.551 × 3.260 × 4.9001 **f.** 3.02 × 500. × 0.0023

25. A 102 kg boulder is falling at a velocity of 20. m/s. How much kinetic energy does the boulder possess?

26. A block of rock salt has a mass of 10.7 g. Each edge of the block has a length of 5.00 cm. What is the density of the rock salt?

Answers to

Concept Review

24. a. 4.42 **b.** 0.684
c. 2.2×10^{-3} **d.** 4.4
e. 24.78 **f.** 3.5

25. $\frac{1}{2}(102 \text{ kg})(20. \text{ m/s})^2 = 2.0 \times 10^4 \text{ kg·m}^2/\text{s}^2$

26. $V = (5 \text{ cm})^3 = 125 \text{ cm}^3$
$D = 10.7 \text{ g}/125 \text{ cm}^3 = 8.56 \times 10^{-2} \text{ g/cm}^3$

$$\begin{array}{r} 6.25 \\ \times\ 4.75 \\ \hline 3125 \\ 43750 \\ 250000 \\ \hline 29.6875 \end{array}$$

Answer: 29.7

Figure 2-34
A calculator does not round an answer to the correct number of significant figures. In this case, the answer should have only three significant figures.

Calculators do not increase the precision of calculated values

A calculator often exaggerates the precision of your calculations by reporting too many digits. The calculator in **Figure 2-34** was used to calculate the area of a room 6.25 m long and 4.75 m wide. Because the length and width have three significant figures, the area must also have three significant figures. The correctly rounded product is 29.7 m^2. Compare this answer with the number that the calculator reports. The answer provided by the calculator is misleading. It implies that we know the size of the room to a greater degree of precision than we actually do.

Section Review

27. a. Convert 302°C to kelvins.
b. Convert 185 K to degrees Celsius.

28. Suggest appropriate SI units and prefixes for measuring the following objects.
a. the length of a textbook
b. the volume of a bathtub (about 30 gallons of water)
c. the mass of an eyelash
d. the volume of an aluminum can

29. Speed is calculated from length and time according to the following equation. What derived SI unit is used to describe speed?
speed = distance / time

30. A quarter (25¢) has a mass of about 5.65 g. Express this mass in milligrams, nanograms, and kilograms.

31. Round each of these measurements to three significant figures.
a. 24.590 **b.** 24.353 **c.** 3.002 **d.** 956.789 **e.** 67.963

ASSESS

Alternative Assessment

Have students find the problems associated with the following story. (Read the story several times and write the data on the board.) Three students are asked to find the volume of a typical classroom using standard meter sticks. Someone suggests that they split up the work so that one student measures the height, another measures the width, and a third measures the length. The first student records a height of 2.943 m. The second student records a width of 5 m. The third student records a length of 8.9635972 m. The volume of the room is found to be 131.89933 m^3. Discuss these measurements with respect to accuracy, precision, and significant digits.

Answers to Section Review

27. a. 302°C + 273 = 575 K
b. 185 K − 273 = −88°C

28. a. cm **b.** m^3 (accept also L or kL)
c. μg **d.** m^3 (accept also cm^3 or mL)

29. meters per second, m/s

30. 5.65 g/(1 × 10^{-3} g/mg) = 5.65 × 10^3 mg
5.65 g/(1 × 10^{-9} g/ng) = 5.65 × 10^9 ng
5.65 g/(1 × 10^3 g/kg) = 5.65 × 10^{-3} kg

31. a. 24.6 **b.** 24.4
c. 3.00 **d.** 957
e. 68.0

Chemistry in Your Community

To relate the recycling issue to the local community, use one or more of the following exercises.

1. Ask students to locate the nearest recycling center. To do this, they can call the city hall or the department of sanitation or public works, or they can check the phone book. Have students obtain answers to the following questions.
   a. What is the address of the nearest recycling center? What are its hours of operation?
   b. What materials does the center accept?
   c. How should materials be sorted or packaged?
   d. How well should the materials be cleaned?
   e. How much does the center pay for each type of recyclable item?
2. Have students design a recycling plan they can use at home. (Where should recycling bins be placed for convenience? What materials must be sorted? How will materials be transported to the recycling center?)
3. If there is no recycling program at school, ask students to design one that will work with local resources.

Answers to Applying Concepts

1. Polyester can be made of recycled PET materials.
2. Because mass and energy are conserved, matter never really disappears.

Conclusion: Recycling—Saving Matter and Energy

Throughout this chapter, you have been learning about matter and energy and applying your knowledge to recycling. In the introduction, you were asked to keep several questions in mind. Look at these questions again.

How can materials be identified so they can be disposed of properly? As you have seen, materials can be identified by measuring density and other physical and chemical properties and comparing the results to published data. Also, some plastics can be identified by recycling codes.

How can mixtures be separated so that their components can be recycled? Distillation, filtration, evaporation, and magnetic separation are often useful. For most of your solid household recyclables, however, separation by hand is still the most practical method. Better yet, set up bins for different types of recyclables so that you do not have to sort waste materials that have been mixed together.

How is energy measured, and what do we mean when we say "recycling saves energy"? In scientific work, energy is usually measured in joules or in the SI base units $kg{\cdot}m^2/s^2$. To "save" energy is to transfer and transform it wisely, for we know from the law of conservation of energy, that the total amount of energy in the universe can never change.

Why can some materials (such as aluminum) be recycled more successfully than others? You've considered many scientific answers to this question. One reason aluminum and glass can be easily recycled is that these materials can be melted and remolded without altering their chemical compositions. But plastic products are often mixtures of various plastics that mingle together when heated and cannot then be purified.

Recycling involves economics as well as science. The economic reality is that without consumer demand for recycled products, there will be no supply of them because manufacturers cannot stay in business. For recycling to work, there must be reliable consumer demand. How can you help make this happen?

Applying Concepts

1. Why might wearing polyester clothing be a way to support recycling?
2. "Nothing is ever really thrown away." Explain, in scientific terms, what this statement means.

Research and Writing

Use community resources to research the following questions.

1. New plastic soda bottles cannot now be made from used ones. Why not? What research efforts are underway to meet this challenge?
2. Why is some recycled paper stockpiled at recycling collection centers? How will the United States government's recent pledge to buy more recycled paper impact the paper-recycling industry?

CHAPTER 2

Highlights

Key terms

accuracy	law of conservation of energy
allotropes	mass
atom	matter
chemical change	mixture
chemical properties	molecule
compounds	phase
elements	physical change
endothermic change	physical properties
energy	potential energy
exothermic change	precision
heterogeneous mixture	pure substance
homogeneous mixture	state
kinetic energy	volume

Key ideas

2-1 What is matter?

- Matter is anything that has mass and volume.
- Chemical properties can be observed only when one substance interacts with another. Physical properties can be observed without changing the chemical composition of a substance.
- The relative positions and motions of atoms or molecules determine physical state.

2-2 How do matter and energy interact?

- Kinetic energy is the energy of motion, and potential energy is stored energy. These give rise to other forms of energy.
- Energy can change form but cannot be created or destroyed.
- Changes in which a system absorbs energy are endothermic changes. Changes in which a system releases energy are exothermic changes.

2-3 How is matter classified?

- Pure substances contain only one type of atom or molecule.
- Uniformly distributed mixtures are homogeneous; uneven mixtures are heterogeneous.
- Compounds, unlike mixtures, are pure substances with a definite composition.

2-4 How are substances identified?

- Components of a mixture can be separated and identified by comparing their properties to a standard.

2-5 How should data be reported?

- There are seven SI base units. Other SI measurements are derived from these base units.
- Accuracy expresses how close a measured value is to the true value, and precision expresses the exactness of a measurement.

Key problem-solving approach: *SI conversions*

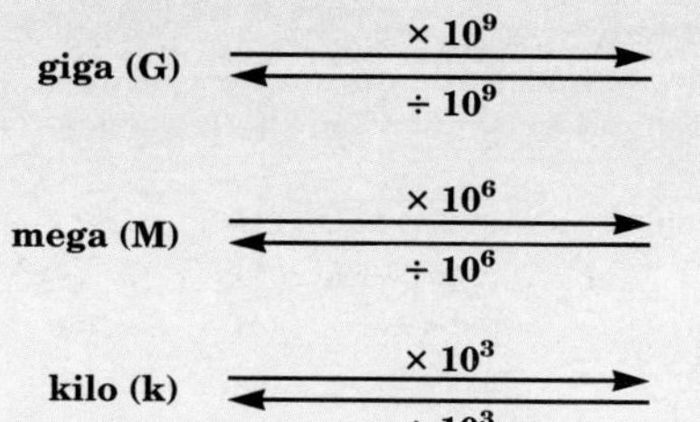

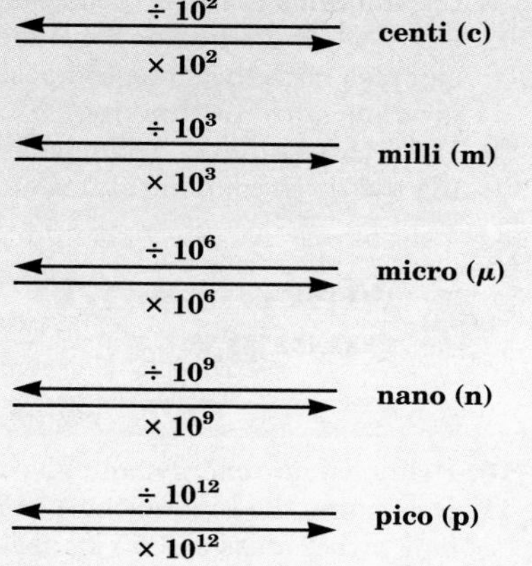

*Kilogram, the base unit for mass, does not appear in this list because it has a different set of conversion values (1 kg = 1000 g).

Closure Strategy

Select three familiar materials such as copper, plastic, and aspirin. Have students use **Figure 2-23** to classify the materials. Have students write their reasons for selecting one branch of the flow chart over another in working through the scheme for each example.

1.a. Matter is anything that has mass and volume.
b. Mass is the amount of matter in an object and is independent of gravity. Weight depends on gravity, which varies throughout the universe.
c. The atom is the fundamental building block of matter.

2.a. Answers will vary, but should include only those properties that can be measured or observed without changing the composition of a material, such as color, physical state, density, texture, luster, malleability, conductivity, etc.
b. A chemical property describes how substances interact. One possible example is the brown color of an apple exposed to oxygen that is the result of oxidation.

3. Intensive properties (such as texture) do not depend upon the amount of a substance present, whereas extensive properties (such as volume) do.

4. The five states of matter are: solid—fixed volume and shape; liquid—fixed volume, variable shape; gas—no fixed shape or volume; plasma—high energy state; and neutron star—properties unknown.

5. A scale measures weight, not mass and the scale reading would be affected by the gravity of the moon which differs from that of Earth.

6. Answers will vary. Physical properties include (but are not limited to) the taste, smell, color, and texture of the item. Chemical properties could be observable if the object is able to interact with another substance.

CHAPTER 2

Review and Assess

The nature of matter

REVIEW

1. a. What is matter?
   b. How do weight and mass differ?
   c. What is the building block of matter?
2. a. Name three physical properties.
   b. Define *chemical property*, and give an example.
3. Differentiate between intensive and extensive properties. Give an example of each.
4. Describe the five states of matter.

APPLY

5. An astronaut plans to bring home a massive moon-rock collection. Mission control on Earth wants to know the mass of the collection in advance. Why can't the astronaut use a scale to obtain the data requested by mission control?
6. Pick an object you can see right now. List three physical properties of the object that you can observe. Can you also observe a chemical property of the object? Explain.
7. Compare the physical and chemical properties of salt and sugar. What properties do they share? Which properties could you use to distinguish between salt and sugar?
8. A student checks the volume, melting point, and shape of two unlabeled samples of matter and finds that the measurements are identical. From this he concludes that the samples have the same composition. What is wrong with his thinking?
9. Draw three atomic-level models representing the three common states of matter.

Matter and energy

REVIEW

10. Define *kinetic energy*, and give an example.
11. Determine the kinetic energy of a bowling ball with a mass of 4.55 kg as it rolls down an alley at 3.20 m/s.
12. Define *potential energy*, and give an example of an object that possesses it.
13. Give an example of energy transformation that you witnessed today.
14. What is the relationship between mass and energy?
15. State the law of conservation of energy.
16. Differentiate between an exothermic and an endothermic process.
17. Label the following processes as endothermic or exothermic changes.
   a. evaporating water
   b. condensing water vapor
   c. cooling bath water
   d. evaporating perspiration

APPLY

18. Show that an object with a mass of 8.00 kg can have the same kinetic energy as an object with a mass of 32.0 kg.
19. Explain how a skateboard resting on the ground contains energy. If the skateboard were moving, would it have more energy than it did at rest? Explain.
20. Water evaporates from a puddle on a hot, sunny day faster than on a cold, cloudy day. Explain this phenomenon in terms of interactions between matter and energy.
21. Label each of the following as a physical or a chemical change.
   a. Coffee beans are processed in a grinder.
   b. Carbonated soda pop loses its fizz.
   c. A short-order cook fries an egg.
   d. Water expands when frozen, decreasing its density.
   e. Heated calcium carbonate decomposes and releases carbon dioxide gas.
   f. Rain droplets high in the atmosphere form hail.
   g. Hydrogen peroxide decomposes into water and oxygen gas in the presence of light.

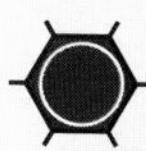

The classification of matter

REVIEW

22. Indicate whether the substances below are compounds or elements.
    a. SO_2 b. S_8 c. C_{60} d. CH_4
23. Differentiate between pure substances and mixtures. Give examples of each.
24. Compare and contrast the properties of homogeneous and heterogeneous mixtures.
25. Tell which of the following items are heterogeneous mixtures. Tell which items could be homogeneous mixtures.
    a. copper penny b. bean burrito
    c. paint d. diamond
    e. plastic wrap f. sea water
    g. blood h. tea
26. Give an example of each type of solution.
    a. liquid solvent/ liquid solute
    b. liquid solvent/solid solute
    c. solid solvent/solid solute
27. Name three elements that exist as molecules.
28. How are atoms and molecules related to one another?

APPLY

29. Choose a mixture. Tell how its properties compare with the properties of the pure substances that compose it.
30. Identify each phase in the following mixtures.
    a. chunky peanut butter b. soda pop
    c. a sponge
31. Diamond and graphite are both allotropic forms of carbon. Why do their physical properties differ?

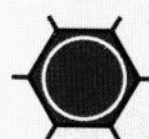

Separation and identification

REVIEW

32. Describe four different methods of separating mixtures. Give an example of a mixture that could be separated using each of the separation techniques you described.
33. a How would you describe density to a friend unfamiliar with the concept?
    b. State the mathematical expression for density.
    c. Is density an intensive or extensive property?
34. Describe how different substances can be distinguished by density.
35. Calculate the volume of 2.00 g of a substance with a density of 4.54 g/mL.

APPLY

36. For each pair below, indicate the substance with the greater density. Explain your answer.
    a. rubber stopper and cork
    b. ice cube and water
    c. rubber stopper and automobile tire
37. Substances A and B are colorless, odorless liquids that are nonconductors and flammable. The density of substance A is 0.97 g/mL; the density of substance B is 0.89 g/mL. Are A and B the same substance? Explain.
38. A forgetful student leaves an uncapped watercolor marker on an open notebook. Upon returning, she discovers the leaking marker has produced a rainbow of colors on the top page.
    a. Is the ink a pure substance or a mixture? How do you know?
    b. What separation technique is involved?

Reporting data

REVIEW

39. a. Why are measurements often more useful than word descriptions?
    b. There are two parts to every measurement. One of these is the numerical part. What is the other necessary part of every measurement?
40. Distinguish between *precision* and *accuracy*.
41. a. What is an SI base unit?
    b. Name the five most common base units used in chemistry and the quantity that each represents.
    c. Explain what derived units are. Give an example of one.
42. What is the purpose of significant figures?
43. a. Contrast human and instrument error.
    b. Describe a situation where each type of error could have disastrous results.
44. a. If you add a series of numbers, how many significant figures can the sum have?
    b. If you multiply a series of numbers, how many significant figures can the product have?

7. Both salt and sugar are white, granular solids at room temperature. Properties that could be used to tell these substances apart include: taste, density, reactivity with various chemicals, and crystal structure which can be observed using a magnifying glass.
8. Different substances often share some physical properties. To determine more conclusively whether the substances are the same, he should measure density and check a wider range of physical properties. An extensive property, such as volume, is of little use in determining the identity of a substance.
9. The models should have the following characteristics: the solid should show an orderly arrangement of many particles in close proximity; the liquid model should show a similar number of particles randomly arranged in close proximity; and the gas model should show few particles, randomly spaced, at some distance from each other.
10. Kinetic energy is the energy of motion. All examples should convey movement, such as an avalanche, falling rain, a moving train, or a spinning dancer.
11. $\frac{1}{2}(4.55\ \text{kg})(3.20\ \text{m/s}^2) = 23.3\ \text{J}$
12. Potential energy is energy due to position. Student examples should reveal an understanding of energy held in readiness, such as a stretched rubber band.
13. Answers will vary, but should show an understanding of potential energy becoming kinetic, or vice versa.
14. The relationship between mass and energy is stated in Einstein's equation: $E = mc^2$. The equation reveals that energy is equivalent to mass times the square of the speed of light.

15. Energy cannot be created or destroyed, but can change form.

16. Endothermic processes absorb energy from the surroundings, and exothermic processes release energy into the surroundings.

17. a. endothermic
b. exothermic
c. exothermic
d. endothermic

18. An 8.00 kg object moving with a velocity of 4.00 m/s has the same kinetic energy as a 32.0 kg object moving with a velocity of 2.00 m/s.

$\frac{1}{2}(8.00 \text{ kg})(4.00 \text{ m/s})^2 = 64.0 \text{ J}$

$\frac{1}{2}(32.0 \text{ kg})(2.00 \text{ m/s})^2 = 64.0 \text{ J}$

19. There is internal energy in the skateboard because the atoms and molecules composing the board have kinetic and potential energy. If it is moving, the skateboard has additional kinetic energy.

20. The sun's energy increases the motion of the water molecules until they have the energy to break free of one another. On warm days, more of the sun's energy reaches Earth.

21. a. physical
b. physical
c. chemical
d. physical
e. chemical
f. physical
g. chemical

22. a. compound
b. element
c. element
d. compound

23. A pure substance is composed of only one type of particle (atom, ion pair, or molecule), such as lead. A mixture, such as salt and water, is a physical combination of two or more substances.

APPLY

45. Perform the following calculations, and the express answers in the correct number of significant figures.
a. $12.4 \times 7.943 + 0.0064$
b. $(246.83 / 26.3) - 1.349$
c. $0.1273 - 0.000008$

46. Express normal body temperature (37.0°C) in terms of the Kelvin scale.

47. Give the appropriate prefix and base unit for each of the indicated measurements.
a. thickness of a dime
b. length of a highway
c. volume of a thimble
d. mass of a staple
e. diameter of a hamburger

48. What derived units are appropriate for expressing the following?
a. kinetic energy
b. density
c. potential energy
d. speed
e. pressure
f. the rate of water flow

49. Why can a measured number never be exact?

50. Why is the Kelvin temperature scale more convenient than the Celsius scale when investigating low-temperature superconducting materials?

51. Is it possible for a number to be too small or too large to be expressed adequately in the metric system? Explain your answer.

52. a. The inch was originally defined as the distance between the knuckles and thumb of King Edgar (A.D. 959–975). Discuss the practical limitations of this early unit of measurement.
b. Invent a length unit of your own that is practical and accurate. Discuss how your unit avoids the problems created by King Edgar's early version of the inch.

Linking chapters

1. ***Scientific laws***
The law of conservation of energy is a scientific law rather than a scientific theory. What does this tell you about how conservation of energy can be applied?

2. ***Succeeding in chemistry***
Review the guidelines for success stated on page 25. Write a summary of how you have applied each step in your study of Chapter 2. Be specific in your answers; for example, describe what you learned by looking at particular illustrations.

3. ***Chemical reactions***
Describe the organic synthesis of aspirin (which you studied in Chapter 1) in terms of the following vocabulary terms from Chapter 2: atom, molecule, compound, element, pure substance, mixture, and chemical change.

4. ***Scientific method***
State a hypothesis that you could test using a separation technique such as distillation, chromatography, or filtration. Also explain how the separation technique would help you prove or disprove the hypothesis.

5. **Theme: *Systems and interactions***
Use the diagram below to answer the following questions.
a. Is the change from **a** to **b** a physical or chemical change?
b. Is the mixture in **b** homogeneous or heterogeneous?
c. What method would you use to separate mixture **b** into its pure substances?

a.

b.

USING TECHNOLOGY

1. ***Graphics calculator***
Density can be graphically represented as a straight line with a positive slope when the x value denotes volume and the y value denotes mass. Set the x-range of your calculator at 0–60 and the y-range at 0–160. Select the y= key, and enter the function $y = 2.70x$.
   a. Press GRAPH and use the TRACE mode to estimate the volume of aluminum (density of 2.70 g/mL) for the following masses: 15.0 g, 75.0 g, and 125 g.
   b. Calculate the true volumes of these masses. Note the accuracy of your calculations.
   c. Enlarge the range by a factor of 4. Estimate the volumes using the TRACE mode. Repeat this step after reducing the range by a factor of 4. What happens to your calculator's precision when you change the range?
2. ***Computer art***
Use a computer art program to model each of the concepts listed below.
   a. Illustrate the particles of a substance changing state from solid to liquid to gas.
   b. Represent a chemical reaction in which one atom and one molecule react to produce two molecules.
   c. Draw models of homogeneous and heterogeneous mixtures on the particle level.

Alternative assessment

Performance assessment

1. Using what you know about separation techniques, construct an experiment to separate a mixture of water, salt, sand, and oil. Outline your procedure and show it to your teacher. When the teacher approves the outline, obtain the mixture and test the procedure.
2. Your teacher will make several unknown samples of pure metals available to you. Describe to your teacher exactly how you will gather information about the densities of the metals. When your procedure is approved, test the metals, and use chemistry reference books to identify them.
3. Using the *Handbook of Chemistry and Physics,* compile a list of the physical and chemical properties of several substances available in your lab. Use a computer flowchart program to outline the procedures you could follow to identify the substances.
4. Describe as completely as possible all of the physical properties of a simple household object. Highlight any subjective parts of your description. Compare language to numerical data as a basis for useful scientific description.

Portfolio projects

1. ***Chemistry and you***
Describe in a journal the types of matter you come in contact with daily. Separate these into large categories, such as pure substances and mixtures, and then break them down into smaller groups, such as elements, compounds, solutions, and alloys. When done, total the entries in each category. With a chart or graph, relate the relationship between your totals and the relative abundance of these substances on Earth.
2. ***Chemistry and you***
   a. Using outside resources, trace the history of antimatter research from the hypothesis of its existence through experimental proofs to plans for the space station construction of Astromag (Particle Astrophysics Magnet Facility).
   b. Collect information on possible uses of the energy produced by collisions of matter with antimatter.
3. ***Chemistry and you***
We often don't notice expressions of commonplace measurements. What units do you use each day without thinking about them? Keep a record of all units that you use in a one day. Comment on the difficulties you might encounter if standard units did not exist.

24. Homogeneous and heterogeneous mixtures can both vary in composition. However, homogeneous mixtures are uniform in composition, and heterogeneous materials vary in composition throughout the material.

25. **a.** could be homogeneous
b. heterogeneous
c. heterogeneous (paint settles in a can)
d. could be homogeneous
e. could be homogeneous
f. heterogeneous
g. heterogeneous
h. could be homogeneous

26. Answers will vary. Some possible responses are:
a. gasoline and oil in a lawnmower
b. salt water
c. 14-karat gold jewelry composed of gold and silver

27. Answers will vary. Possible answers include oxygen, O_2; sulfur, $S_{.}$, nitrogen, N_2; chlorine, $Cl_{.}$, and fluorine, F_2.

28. Atoms are the basic units of matter. Molecules are clusters of atoms.

29. Answers will vary. One possible choice is salt water. Salt contributes to the taste and conductivity of the mixture. The remaining properties (a colorless liquid and ability to flow) are contributed by water.

30. **a.** solid and liquid
b. liquid and gas
c. solid and gas

31. Their different crystalline structures give rise to different physical properties. Physical properties depend on the way the atoms are arranged.

Answers from page 57

22.

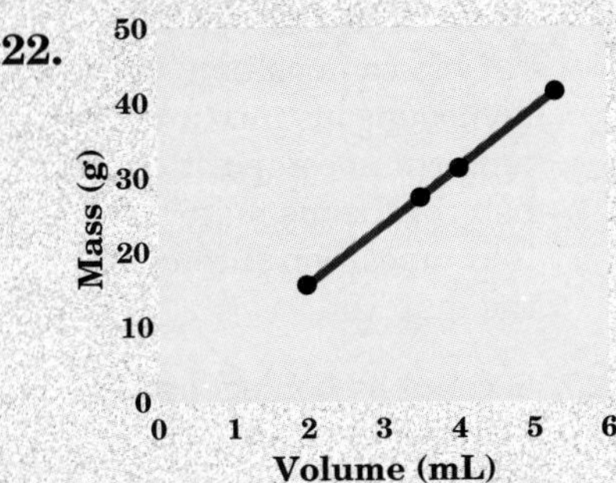

$m = 47$ g; $D = 7.8$ g/mL

23. Determine the density of the metal and see if it matches the density for aluminum.

Answers from page 59

32. Distillation is a process used to separate substances according to their boiling points. Chromatography separates substances according to their solubility and affinity to the stationary phase (paper). Filtration is a process by which substances are separated according to particle size. Centrifuging is a process by which substances are separated by density. Possible applications include crude oil (distillation); leaf pigments in biology lab (chromatography); sifting gravel through a screen (filtration), and separating plasma from cellular material in blood (centrifuging).

33. **a.** Answers will vary, but should reflect the fact that density is the ratio between the amount of substance present (mass) and the volume that it occupies.

b. $\text{density} = \frac{\text{mass}}{\text{volume}}$

c. Density is an intensive property.

34. Each known pure substance has a characteristic density. Therefore, even if substances appear similar, their densities will differ.

35. $\frac{2.00 \text{ g}}{4.54 \text{ g/ mL}} = 0.441 \text{ mL}$

36. a. Because the rubber stopper sinks in water and the cork floats, the rubber stopper must be more dense.
b. Because an ice cube floats in water, the water must be more dense.
c. Because both objects are made of the same substance, vulcanized rubber, the density is the same.

37. No; the different densities show that the substances are not the same.

38. **a.** Because the ink separated into its components, it is a mixture.
b. This scenario illustrates the chromatographic separation of ink.

39. **a.** Words are inexact and subjective. Numbers have exact meanings.
b. The other necessary part of every measurement is the unit.

40. Precision reveals how exactly or sharply stated a measurement is. Accuracy refers to how closely the measurement approaches the true value for the quantity.

41. **a.** An SI base unit is one of the seven units by which all measurements can be expressed in SI.
b. The most common base units and the quantities they describe are: kilogram—mass, meter—length, liter—volume, second—time, and kelvin—temperature.
c. A derived unit is a combination of base units that corresponds to a physical quantity. For example, the derived unit for density is g/mL.

42. Significant figures denote the precision of a measurement.

43. **a.** Human error is error due to a mistake or inaccuracy caused by an individual. Human errors can be visual errors or common mistakes in judgment. Instrument error is generated within the instrument itself and is due to faulty equipment, an improper standard, and other causes.
b. Both types of error can cause faulty conclusions. Answers will vary on the situations.

44. **a.** In addition, the answer can have no more digits to the right of the decimal than the figure with the least number of significant figures.
b. In multiplication, the answer is rounded off to the number of digits in the number with the least number of significant figures.

45. **a.** 98.5
b. 8.04
c. 0.1273

46. 37.0°C + 273.16 = 310.2 K

47. **a.** micro(meters)
b. kilo(meters)
c. milli(liters)
d. milli(grams)
e. centi(meters)

48. **a.** $\frac{\text{kg}\cdot\text{m}^2}{\text{s}^2}$

b. $\frac{\text{g}}{\text{mL}}$

c. $\frac{\text{kg}\cdot\text{m}^2}{\text{s}^2}$

d. m/s
e. lb/ft^2
f. L/s

49. All measurements estimate the last figure. No scales, meter sticks, or other measuring instruments can have an infinite number of divisions, which is what would be needed to determine exact numbers.

50. Very low temperatures are expressed as positive numbers on the kelvin scale.

51. No; if technology increases to a point where smaller or larger prefixes become necessary, new prefixes could be added to those already in use.

52. **a.** Not every thumb was the same length as Edgar's; therefore, the "inch" was not exactly the same every where. Also, if the King had rheumatoid arthritis, the length of his thumb could change.
b. Answers will vary. Units should relate to a standard that can be consistently duplicated.

Linking chapters

1. Because conservation of energy is a scientific law, it is consistently applicable to any situation involving energy.

2. Answers will vary. Look for items consistent with the list on page 25 of the text. You might have students share with others techniques that work well.

3. Answers will vary. Check that all the listed vocabulary words have been used and that the process for making aspirin is correct.

4. Answers will vary. Evaluate for understanding of the chosen separation technique and for understanding of the term *hypothesis*.

5. **a.** The change is physical.
   b. The mixture is homogeneous.
   c. Distillation would separate the mixture.

USING TECHNOLOGY

1. **a.** $y = 15.5$, $x = 5.74$; $y = 75.8$, $x = 28.1$; $y = 125.8$, $x = 46.6$
   b. 5.56 g, 27.8 g, 46.3 g
   c. Precision is reduced when the range is increased.

2. Answers will vary, but check for the following. The highly ordered solid becomes very random as a gas. There should be a lot of distance between the gas particles. Note whether the homogeneous mixture shows a uniform distribution of particles. The heterogeneous mixture should show a random distribution of particles.

Performance assessment

1. The water and salt can be separated by evaporation. The water and oil are separated by decantation. The sand is separated by filtration.
2. Metals chosen should not react with water. Students can determine the densities using the water displacement method.
3. Students' flow charts should include testing solubility, density, and reactivity with certain compounds such as acids.
4. Answers will vary, but should include subjective descriptions such as color, approximate size, texture, shape, and feel. Quantitative descriptions provide more exact data.

Teaching Tip

Discussion

The use of generic drugs is an interesting study in consumer behavior and economics. Because generic brands of consumer goods were once perceived to be of low quality, consumers were far less eager to save money when it came to medications. This is a good opportunity to reinforce the principles regarding pure substances from Chapter 2. For example, make students aware that any product labeled as aspirin must be acetylsalicylic acid. The action of that compound as an analgesic is the same regardless of who manufactures it. Advertised differences for brands of aspirin have to do with dosage, additives to reduce side effects, and additives to reduce other symptoms for which aspirin is ineffective.

1. Ask the class for a show of hands if their families request generic drugs when the option is available from the pharmacist. Ask students to share any information they might have concerning the price difference between generic and brand name prescription medication.
2. Ask students to voice their opinions about any concerns they might have regarding the use of generic drugs. Use these ideas to start a discussion about the safety of generics and the role of the FDA in maintaining product quality.
3. Ask the class if their opinions about generic drugs have changed since the discussion of the feature.

Science, Technology, and Society

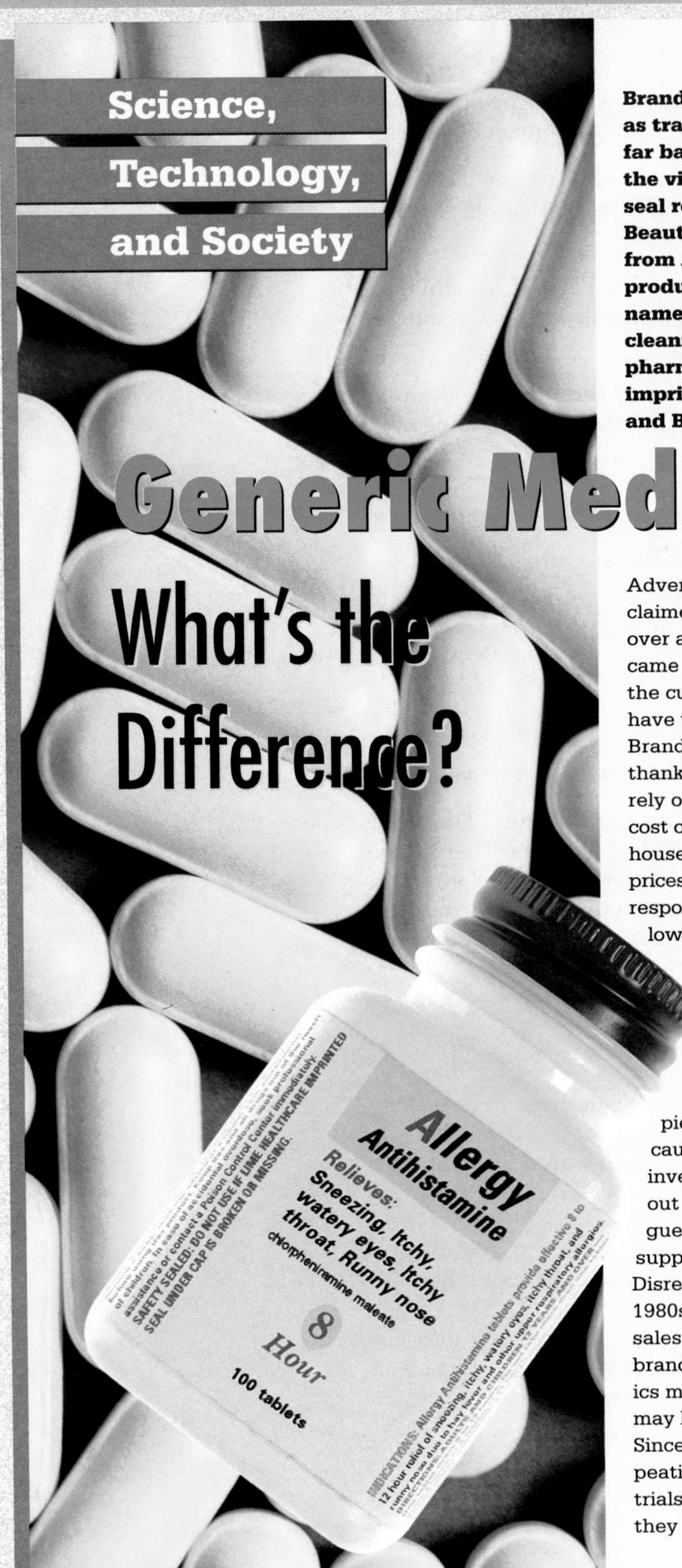

Generic Medication

What's the Difference?

Brand names on manufactured products are as traditional as the products themselves. As far back as the Egyptian dynasties, wine from the vineyards of the Pharaoh Ahknaton bore a seal reading "Nefer Nefer Nefer," or "Thrice Beautiful" to identify the best wine to emerge from Ahknaton's vineyards. Today, many products are readily identifiable by their names and labels. Food items, cosmetics, cleaning products, and over-the-counter pharmaceuticals bear manufacturer's imprints like General Mills, Lever Brothers, and Bayer to identify them to the customer.

Advertising as an industry was founded on the claimed superiority of one manufacturer's product over another. To the consumer, a brand name came to represent reliability. No matter where the customer purchased it, the product would have the same purity, dosage, and contents. Brand name items were readily identifiable, thanks to advertising, and consumers came to rely on this promise of consistent quality. But the cost of advertising these products to make them household words also brought about increased prices. In the 1970s and 80s, large grocery chains, responding to recession and consumer demand for lower prices, began to package their own brands. "House brand" tomato sauce, dishwashing liquid, and other staples were quickly followed by a new product line, commodities without any brand names at all.

The pharmaceutical industry long occupied a unique place in American business. Because the drug companies held patents on their inventions, they were able to raise prices without fear of competition. The drug companies argued that this level of profit was necessary to support their research and development efforts. Disregarding this argument, the FDA in the mid-1980s moved to promote competition in drug sales by allowing the sale of generic clones of brand-name medications. Provided these generics meet the bioequivalency requirement, they may be dispensed in place of brand-name drugs. Since generic clones are produced without repeating the expensive, time-consuming clinical trials needed to certify the brand-name drug, they are cheaper to produce and to buy.

Drug manufacture, unlike the production of other generic items, is supervised by the FDA. Generic drugs must be rigorously tested and approved before sale, and must be chemically identical to the name brand. The FDA's supervisory role was particularly important in the early days of generic drug manufacture, as some companies cut corners to rush their products onto the market. The scandal resulting when a few generic drug companies resorted to bribery and substitution to gain FDA approval for their products caused even more stringent procedures to be put in place. These regulations ensured that generics, were exactly equivalent, but much less costly. The FDA currently approves about 200 generic drugs annually, and a staff of nearly 50 chemistry reviewers tests and reviews all such drugs before they are approved for sale.

A drug's bioequivalence rating guarantees that there is no real difference in the chemical composition of the drug, nor in its action in the body. For example, you know that aspirin has the formula $C_9H_8O_4$. No matter what brand of aspirin you buy, the chemical composition of the drug is still $C_9H_8O_4$. In actual practice, a brand-name version of thioridazine, called Mellaril, produced by Sandoz, Inc., was analyzed with its generic equivalent. The variations found between the brand-name and generic drug were as many as you would find between different lots of thioridazine, according to a regulatory letter from the FDA to Sandoz. Because all drugs are inherently variable in individual patients, a statistical variation of plus or minus 20 percent is standard, and is the basis for bioequivalency. The average statistical variation for generics is a mere 3.5 percent, giving generic drugs only a minute difference from the original, and consumers an average 52.8 percent break in the purchase price of generics over brand-name prescriptions.

Today, generic drugs account for 60 to 85 percent of the market for new prescriptions in the United States. More and more, providers are offering consumers a choice of pharmaceuticals when they fill their prescriptions. Continuing education courses are giving pharmacists information and techniques for educating consumers on the use of generic drugs, and overcoming their fears about bioequivalency. Mindful of rising health care costs, employer benefit specialists in companies encourage their members to choose generics over prescription medications to save money, an advantage to both the company and its employees.

As expiration dates on patents for dozens of brand-name medications grow closer, generic drug manufacturers hasten to develop generic equivalents and some, such as Merck and Co., have formed subsidiaries to supply the generics market with bioequivalent drugs of their own manufacture. Merck's West Point Pharmaceutical Division produces the generic equivalent of Dolibid, Merck's brand-name anti-inflammatory on which the patent expired in 1989.

Drug Topics magazine reported in March, 1994, that there are at least 170 vendors of generic drugs in the United States, offering about 2000 economically priced products. Six such firms have formed a consortium to investigate generic drug manufacturing opportunities in Russia, where there has been a critical shortage of drugs and vaccines. The consortium, MIR Pharmaceuticals, Inc., is conducting a feasibility study of cost and problems associated with setting up an FDA-standard drug manufacturing facility in Russia. There are some $60 billion in drug patents due to expire within the next five years, which will make these substances fair game for manufacture in generic forms.

Generics are here to stay. Reputable companies produce them, doctors prescribe them by checking a box on the prescription blank. Grateful consumers use them to affordably maintain their health.

Researching the Issue

1. How long does a new medication remain under patent?
2. When a generic equivalent comes on the market, how does the FDA ensure that it is identical to a name brand medication?
3. How are generics manufactured outside the United States regulated for purity?
4. How long does it typically take to bring generic substitutes to market after the patent on a drug expires?
5. State your views as to who should make the decision to prescribe a generic medication—the patient, the doctor, the pharmacist?

3

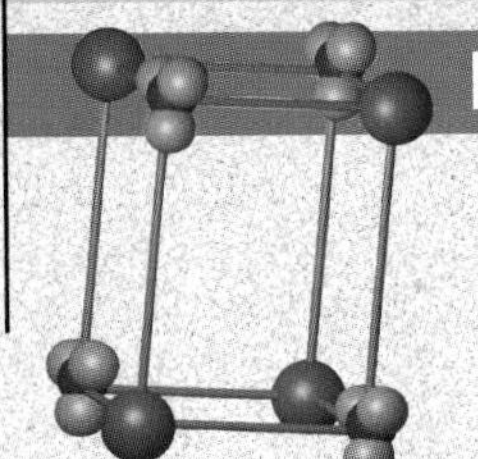

PLANNING FOR

Atomic Structure

Chapter Overview

- **Section 3-1** introduces the periodic table as the organization scheme for the known elements. Students learn that elements in the same group or region will have some similar properties.
- **Section 3-2** starts from the macroscopic world of elements to the sub microscopic world of atoms. The development of the modern atomic model is presented along with the current quantum model.
- **Section 3-3** describes how electrons are arranged in atoms and shows how to write and interpret quantum numbers, electron configurations, and orbital diagrams.

Concept Base

Students must have mastered the following concepts prior to this chapter: the differences between elements and atoms (Chapter 2); operations with exponential numbers (Chapter 2 and Appendix B).

Looking Ahead

Periodic trends are examined in depth in Chapter 4. Both Chapters 3 and 4 provide the necessary foundation students need for chemical bonding in Chapters 5 and 6.

Laboratory Equipment Needs

Demonstrations

The listing of materials for Chapter 3 demonstrations begins on page T46.

Laboratory Experiments

Materials and preparation instructions for Chapter 3 are found as follows.

Text		*Page*
Exploration 3A	Flame Tests	686
Investigation 3A	Flame Tests—Identifying Materials	690

Pacing: 10 days	Inquiry Teaching Strategies
3-1 How are the elements organized?	**Demonstrations** ■ Demo 1 *Elements of the periodic table* ■ Demo 2 *Reactivity of sodium in water*
3-2 What is the basic structure of an atom?	**Demonstrations** ■ Demo 3 *Electron flow* ■ Demo 4 *Emission spectra* **Laboratory Experiments** ■ Exp 3A *Flame Tests* ■ Inv 3A *Flame Tests—Identifying Minerals*
3-3 How do the structures of atoms differ?	

Key ■ Teaching Transparencies ■ Annotated Teacher's Edition ■ Laboratory Experiments Manual

Visual Teaching Strategies

Transparencies
- T 3-2 *Regions of the Periodic Table*

Transparency Masters
- T 3-3 *Essential Elements*

Review and Practice Strategies

Text Reviews
- Concept Review 1–2
- Section Review 3–7
- Chapter Review 1–8; LC 3

Study Guide Worksheets
- 3-1 Concept Review
- 3-1 Skillsheet *Identifying and Using Regions in the Periodic Table*

Transparencies
- F 3-10 *Law of Conservation of Mass*
- F 3-11 *Law of Multiple Proportions*
- F 3-16 *Gold Foil Experiment*
- F 3-22 *Electromagnetic Spectrum*
- F 3-25 *Probability Plots for Hydrogen*

Transparency Masters
- F 3-23 *Wave Comparison*

Text Reviews
- Section Review 8–14
- Chapter Review 9–23, 29–34; LC 1–2; UT 1

Study Guide Worksheets
- 3-2 Concept Review

Transparencies
- F 3-30 *Shapes of* s, p, *and* d *Orbitals*
- F 3-31 *Orbitals of the Carbon Atom*
- F 3-32 *Electron Configuration Sequence Model*
- F 3-33 *Relative Energies of Orbitals*

Transparency Masters
- T 3-6 *Number of Electrons Accommodated in Electron Energy Levels and Sublevels*
- P 101 *Writing Electron Configurations*

Text Reviews
- Section Review 15–19
- Chapter Review 24–28, 35–44; UT 2

Study Guide Worksheets
- 3-3 Concept Review
- 3-3 Skillsheet *Atomic Numbers, Mass Numbers, and Quantum Numbers*
- 3-3 Practice *Writing Electron Configurations*

Pupil's Text
Study Guide
Teaching Resources

Exp *Exploration*
Inv *Investigation*
LC *Linking Chapters*
UT *Using Technology*

Assessment Options

Traditional Assessment

Test Generator
Instructional Objectives Measured:
Content mastery
Select from items 1 to 75

Performance Assessment Options

(see page T36 and T643-15 for scoring rubrics)
Students demonstrate mastery of objectives in a hands-on environment. You may want some of these materials in the Portfolio as well.

Investigation 3A (text)
Instructional Objectives Measured:
Content mastery, Use of scientific methodology, Problem solving skills, Use of technology, Proficiency in written communication

Performance Assessments, page 105

1. **Designing an experiment**
   Instructional Objectives Measured:
   Use of scientific methodology, Problem solving skills, Use of technology
2. **Reference and research skills**
   Instructional Objectives Measured:
   Content mastery, Use of scientific methodology, Proficiency in written communication

Portfolio Options

(see page T36 for scoring rubrics)
Students provide a written rationale for each selection made for the portfolio.

Concept Maps
Students use vocabulary from the Chapter Review to build a concept map.
Instructional Objectives Measured:
Content mastery

Formal Laboratory Reports
Exploration 3A (text)
Instructional Objectives Measured:
Content mastery, Use of scientific methodology, Use of technology, Proficiency in written communication

Portfolio Projects, page 105
Items 1 and 4. Research and communication
Instructional Objectives Measured:
Scientific and chemical literacy, Proficiency in written communication

Item 2. Cooperative activity
Content mastery, Problem solving skills

Item 3. Chemistry and you
Instructional Objectives Measured:
Scientific and chemical literacy, Proficiency in written communication

INTRODUCING THE CHAPTER

CHAPTER 3 Atomic Structure

Story Background

Pyrotechnics, the "science of fire", includes the research and development of many similar devices, ranging from safety matches to hazard flares to fireworks. Fireworks, like other pyrotechnic devices, utilize oxidation-reduction reactions between an oxygen source (oxidizer) and a fuel (reducing agent). The oxygen source and fuel are typically solid chemicals that must be mixed together. When ignited, the chemicals melt and vaporize so that they can react. Rapid energy release occurs as the reaction takes place.

The brilliant flashes of light produced by fireworks consist of electromagnetic radiation with wavelengths between 380 nm and 780 nm. Fireworks produce light through atomic emission and molecular emission. Also, fireworks may produce light—most notably bright white flashes—through incandescence. For example, white light is produced when a reactive metal, such as magnesium, is used as a fuel. Solid metal oxide particles are created when the metal fuel is oxidized. The metal oxide particles glow white-hot in the intense heat generated in the reaction.

The history of fireworks manufacturing is filled with tragic incidents. In 1983, the Grucci fireworks manufacturing plant in Long Island (one of the nation's major manufacturers) was destroyed in an explosion. Students should be reminded that fireworks must be handled with extreme caution. Every year, thousands of Americans are injured or killed in accidents involving fireworks.

About the Illustration

The ionic compound represented in the illustration is strontium carbonate, $SrCO_3$, also called *strontianite*. Strontium carbonate emits red light. Strontium chloride and strontium hydroxide are two other strontium compounds used in fireworks that emit red light.

Excited Atoms and the Fourth of July

The national anthem blasts through loudspeakers, and the show is about to begin. It's July 4th, and hundreds have come to watch the fireworks display. Crack! A rocket flares and red stars burst forth. Kaboom!

A dazzling white chrysanthemum-shaped shower of light fills the sky. The show continues, but wait, something is missing. You've seen red and white fireworks, but what about blue?

You may see a burst of blue, but it will not be as brilliant as the red, white, orange, green, gold, pink, and yellow displays. Today's pyrotechnic technology has not yet produced a bright, eye-popping blue. The search for one keeps fireworks manufacturers scrambling.

Fireworks, originally invented in China, have used a similar basic design for hundreds of years. The design consists of a shell that fits inside a cannon-like mortar; black powder beneath the shell that is a mixture of 15 parts charcoal, 10 parts sulfur, and 75 parts potassium nitrate; and a fuse that feeds into the powder. When the fuse is lit, the powder ignites and explodes, producing gas that expands behind the trapped shell, thrusting it high into the sky. When it reaches a safe altitude, a time-delayed fuse lights within the shell, setting off chemical reactions between an oxygen source and a fuel within the shell. These reactions release tremendous energy and create the spectacular effects you witness.

Some recent advances would astound pyrotechnicians of the past, however. For example, fireworks can be ignited indoors at rock concerts and other entertainment events. Some fireworks turn off and on in midair with intermittent color. Fuses can be lit by remote with computer-timed switches, improving fireworks safety. But despite all these advances, there are still no really good blue fireworks. Why not?

Bill Page, the resident chemist at Astro Pyrotechnics fireworks manufacturing firm, answers the question. The colors of fireworks, Page explains, are generated by atoms or molecules that are present in the gaseous state when the chemical contents of shells react. Different elements, either in the elemental state or combined in compounds, produce different colors. Compounds of strontium produce red fireworks, aluminum compounds produce bright white light, barium compounds produce green light, and sodium atoms produce yellow light. A copper compound, copper chloride, is the best producer of blue light yet identified. But copper chloride is unstable at high temperatures.

A brilliant blue display may be achieved someday by Bill Page or some other pyrotechnic chemist—maybe by you. To get started, consider the following questions.

- *What is light, and how do various colors of light differ?*
- *What is going on at the level of atoms and molecules when fireworks produce colored light?*
- *How does the instability of copper chloride at high temperatures interfere with its ability to emit blue light?*

The answers to these questions, as you will see, involve atomic structure.

Tapping Prior Knowledge

1. How do atoms of different elements differ? How are they the same?
2. Have students draw pictures of how they think atoms should look. Repeat this activity at the end of the Chapter. Have students describe how their models have changed as a result of studying the Chapter.

Section 3-1

FOCUS

Lesson Starter

Put a transparency of the periodic table on an overhead projector or refer students to a periodic table posted in your classroom. Ask students to share what they've learned previously about the periodic table. List their responses on the board to start the discussion of the table.

Demonstration 1

Elements of the periodic table

Approximate time: 15 min

1. Place small samples of the elements listed below in stoppered test tubes. Label each test tube with the name of each element.
2. Use the following substances as the samples: sodium under dry mineral oil, magnesium ribbon (polished), calcium turnings, iron, nickel, copper foil, silver foil, gold foil, zinc, mercury, aluminum, carbon, tin, lead, arsenic, sulfur, iodine, neon in a Geissler tube, helium in a Geissler tube, and hydrogen in a Geissler tube.

Students should be familiar with most of these substances and should notice a variety of different properties. If a wide desk is available, the samples can be set up as they would appear on the periodic table.

SAFETY

Do not open containers of elements and do not allow students to handle containers. To avoid breaking the fragile Geissler tube, enclose it in a transparent shield.

DISPOSAL

Save for reuse at a later date.

How are the elements organized?

Section Objectives

Describe the organization of the modern periodic table.

Use the periodic table to obtain information about the properties of elements.

Explain how the names and symbols of elements are derived.

Identify common metals, nonmetals, metalloids, and noble gases.

Properties of elements relate to atomic structures

As early as 400 B.C., Greek philosophers proposed that matter is made of atoms, extremely small particles that cannot be broken down into anything smaller. The word *atom* actually comes from the Greek word for "indivisible." Today, the atom is still recognized as a fundamental unit of matter. But scientists know that the atom is actually divisible and contains dozens of subatomic particles. Three of these particles are of major importance in chemistry: protons, neutrons, and electrons.

All atoms have similar structures. Protons and neutrons cluster together to form a central core, or *nucleus.* Electrons inhabit the space surrounding the nucleus. Beyond this basic similarity, however, there is variation among the structures of different elements. One variation is the number of protons. Each of the 109 chemical elements has a characteristic number of protons. For example, silicon atoms, such as those shown in **Figure 3-1**, contain 14 protons. Neon atoms contain 10 protons each, while oxygen atoms have 8 protons. As you will discover later in this chapter, electrons are arranged differently in atoms of different elements.

The structures of different elements vary, but not in a wild or random fashion. The structures differ in gradual and regular ways. For example, a hydrogen atom contains one proton, helium has two protons, lithium has three protons, and so on, up to element 109. The number of protons in atoms increases steadily in increments of one proton.

The properties of an element depend on its structure. Thus, you might infer that the behavior of elements also varies gradually from one element to the next. This is indeed the case. The orderly variation in the structure and properties of atoms is represented in the *periodic table of the elements*, a complete chart of all the elements in the known universe.

Figure 3-1a
Individual silicon atoms are visible in this image, provided by a scanning tunneling microscope. Silicon atoms are semiconductors—their electrical conductivity is between that of the metals and nonmetals.

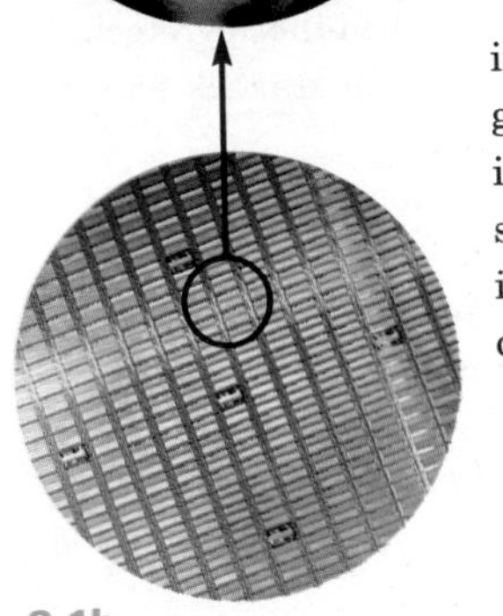

3-1b
This wafer of pure silicon will be used to fabricate integrated circuits.

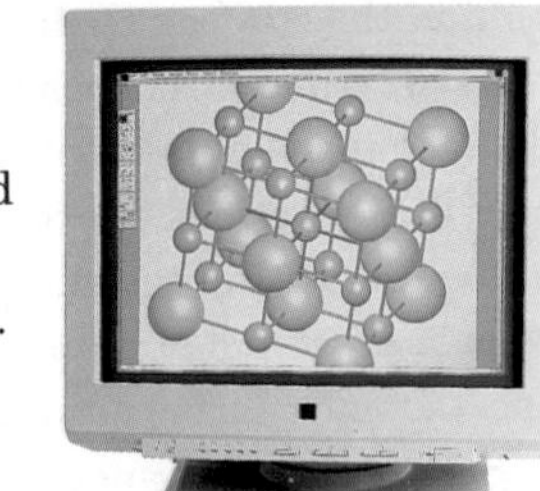

3-1c
Computers are built from integrated circuits or "computer chips." Integrated circuits take advantage of the semiconductivity of silicon—a property that reflects the structure of silicon atoms.

Visual Strategy

Figure 3-1

Make the students aware that every element has unique properties which determine how that element is used.

The scanning tunneling microscope provides images of atoms and bonds on a solid surface. Scientists now use the microscope in constructing semiconductor circuits. The image shown is like a contour map of the surface atoms.

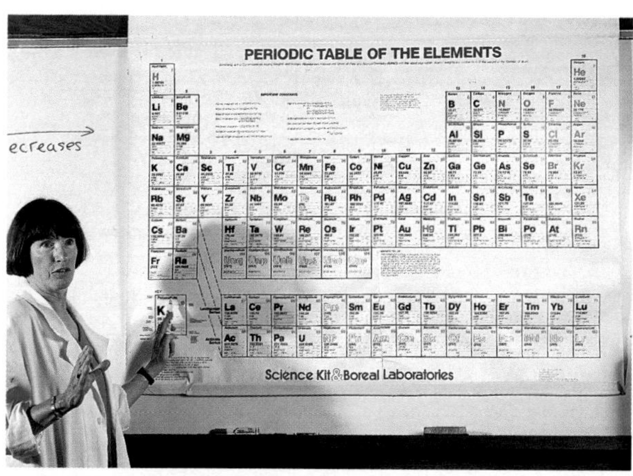

Figure 3-2
The periodic table is an organized display of all the chemical elements.

The periodic table shows all the elements

The periodic table is shown in **Figure 3-2**. It contains a complete listing of all known elements and the chemical symbols that represent them. The symbol for chlorine is Cl, for example. Hydrogen has the symbol H and silicon has the symbol Si.

The symbols for the elements consist of either one, two, or three letters, as shown in **Table 3-1** on the next page. In the case of a single letter, the symbol is always a capital letter. In the case of symbols with two and three letters, only the first letter is capitalized. Elements represented by symbols containing three letters have not been officially named. These three-letter symbols will be changed when the International Union of Pure and Applied Chemistry (IUPAC) gives the elements official names.

It is easy to see that the symbols for chlorine, hydrogen, and silicon are logical. But you may wonder how the names and symbols of some elements could possibly relate to each other. For example, tungsten, which is used to make light bulb filaments, has the symbol W. The key to the apparent mismatch is that tungsten was discovered in Germany and named from the German word *Wolfram*.

Other elements are named in honor of a person or after the place where the element was discovered. Curium (Cm), for example, is a radioactive element named in honor of Marie and Pierre Curie, the discoverers of radioactivity. Californium (Cf) was discovered in California. Can you identify where francium and berkelium were discovered?

The periodic table is organized by properties

group
a series of elements that form a column in the periodic table

The periodic table is arranged so that elements with similar properties fall within the same **group**. Group 1, the column of elements at the left edge of the periodic table, contains the following elements: lithium, sodium, potassium, rubidium, cesium, and francium. All of the Group 1 elements have low densities, low melting points, and good electrical conductivity. As **Figure 3-3** shows, Group 1 elements are shiny until they react with air and are soft enough to be cut with a knife. They also react violently with cold water.

Figure 3-3
Sodium is soft enough to be cut with a knife. It is shiny until it reacts with oxygen in air to form a dull surface.

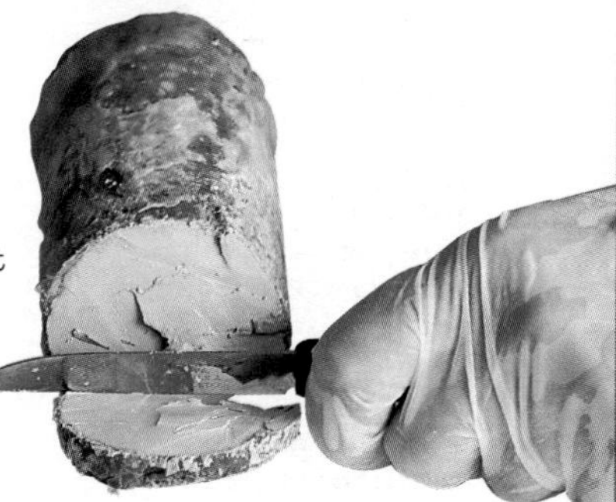

TEACH

Teaching Tip
Poster assignment
Each student is assigned an element and is told to make a poster about its properties and structure. The posters can then be displayed throughout the room. Students can be called on throughout the year to be the resource person for their element when it is discussed in class.

Teaching Tip
Compounds vs. atoms
Students learned in Chapter 2 that the properties of compounds depend on the arrangement of their atoms. In this chapter, students learn that the properties of atoms are related to the arrangement of their subatomic particles. Check that students recognize the important distinction between these two parallel ideas.

Theme
Classification and trends
The periodic table is one of the best examples in chemistry of how trends are used for classification. Elements on the periodic table are arranged according to family trends in properties and according to increasing atomic number.

Visual Strategy
Figure 3-2
Have students note that every element with a two-letter symbol begins with a capital letter and is followed by a lower case letter. If only uppercase letters were used, then a symbol like CO could either be cobalt or carbon monoxide.

Do You Know?

The International Union of Pure and Applied Chemistry (IUPAC) is a voluntary, nonprofit association of national organizations representing chemists of 45 countries. It was formed in 1919 to help produce uniform practices in the field of chemistry.

Historical Note

Many elements were known and named by the Greeks and Romans. Therefore, names of these elements don't seem to fit their symbols. Some of these elements include: aurum–gold; argentum–silver; hydrargyrum–mercury; plumbum–lead; stannum—tin; cuprum–copper; and ferrum–iron.

Do You Know?

It is the IUPAC who ultimately decides the names of elements.

Teaching Tip

Denoting atomic masses

Observant students will notice that parentheses appear around the atomic masses of certain elements in the 2-page periodic table in Chapter 4, but parentheses do not appear around the atomic masses of elements in **Table 3-1**. The parentheses denote atomic masses of the most stable or common isotopes of certain unstable elements, which are not covered until Chapter 4.

Historical Note

The periodic table was proposed by Russian chemist Dmitri Mendeleev in 1869. Mendeleev correctly used his table to predict the properties of elements not yet discovered. So great was this contribution that after his death, Mendeleev's coffin was draped with a cloth periodic table.

Table 3-1
Element Names and Symbols

Element	Symbol	Atomic number	Atomic mass
actinium	Ac	89	227.0278
aluminum	Al	13	26.981539
americium	Am	95	243.0614
antimony	Sb	51	121.757
argon	Ar	18	39.948
arsenic	As	33	74.92159
astatine	At	85	209.9871
barium	Ba	56	137.327
berkelium	Bk	97	247.0703
beryllium	Be	4	9.012182
bismuth	Bi	83	208.98037
boron	B	5	10.811
bromine	Br	35	79.904
cadmium	Cd	48	112.411
calcium	Ca	20	40.078
californium	Cf	98	251.0796
carbon	C	6	12.011
cerium	Ce	58	140.115
cesium	Cs	55	132.90543
chlorine	Cl	17	35.4527
chromium	Cr	24	51.9861
cobalt	Co	27	58.93320
copper	Cu	29	63.546
curium	Cm	96	247.0703
dysprosium	Dy	66	162.50
einsteinium	Es	99	252.083
erbium	Er	68	167.26
europium	Eu	63	151.965
fermium	Fm	100	257.0951
fluorine	F	9	18.9984032
francium	Fr	87	223.0917
gadolinium	Gd	64	157.25
gallium	Ga	31	69.723
germanium	Ge	32	72.61
gold	Au	79	196.96654
hafnium	Hf	72	178.49
helium	He	2	4.002602
holmium	Ho	67	164.93032
hydrogen	H	1	1.00794
indium	In	49	114.818
iodine	I	53	126.90447
iridium	Ir	77	192.22
iron	Fe	26	55.847
krypton	Kr	36	83.80
lanthanum	La	57	138.9055
lawrencium	Lr	103	262.11
lead	Pb	82	207.2
lithium	Li	3	6.941
lutetium	Lu	71	174.967
magnesium	Mg	12	24.3050
manganese	Mn	25	54.93805
mendelevium	Md	101	258.10
mercury	Hg	80	200.58
molybdenum	Mo	42	95.94
neodymium	Nd	60	144.24
neon	Ne	10	20.1797
neptunium	Np	93	237.0482
nickel	Ni	28	58.6934
niobium	Nb	41	92.90638
nitrogen	N	7	14.00674
nobelium	No	102	259.1009
osmium	Os	76	190.23
oxygen	O	8	15.9994
palladium	Pd	46	106.42
phosphorus	P	15	30.973762
platinum	Pt	78	195.08
plutonium	Pu	94	244.0642
polonium	Po	84	208.9824
potassium	K	19	39.0983
praseodymium	Pr	59	140.90765
promethium	Pm	61	144.9127
protactinium	Pa	91	231.03588
radium	Ra	88	226.0254
radon	Rn	86	222.0176
rhenium	Re	75	186.207
rhodium	Rh	45	102.90550
rubidium	Rb	37	85.4678
ruthenium	Ru	44	101.07
samarium	Sm	62	150.36
scandium	Sc	21	44.9655910
selenium	Se	34	78.96
silicon	Si	14	28.0855
silver	Ag	47	107.8682
sodium	Na	11	22.989768
strontium	Sr	38	87.62
sulfur	S	16	32.066
tantalum	Ta	73	180.9479
technetium	Tc	43	97.9072
tellurium	Te	52	127.60
terbium	Tb	65	158.92534
thallium	Tl	81	204.3833
thorium	Th	90	232.0381
thulium	Tm	69	168.93421
tin	Sn	50	118.710
titanium	Ti	22	47.88
tungsten	W	74	183.84
unnilquadium	Unq	104	261.11
unnilpentium	Unp	105	262.114
unnilhexium	Unh	106	263.118
unnilseptium	Uns	107	262.12
unniloctium	Uno	108	265
unnilennium	Une	109	266
uranium	U	92	238.0289
vanadium	V	23	50.9415
xenon	Xe	54	131.29
ytterbium	Yb	70	173.04
yttrium	Y	39	88.90585
zinc	Zn	30	65.39
zirconium	Zr	40	91.224

period
a series of elements that form a horizontal row in the periodic table

The rows of the periodic table are called **periods**. Elements close to each other in the same period are more similar than those farther apart in the same period. For example, the properties of the elements of Group 1 vary significantly from those of Group 17, but are somewhat similar to those observed in the Group 2 elements. Consider the elements potassium, calcium, and bromine. Potassium and calcium, as members of Group 1 and Group 2 respectively, share properties not possessed by bromine, a member of Group 17.

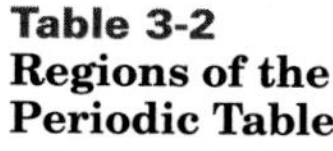

Using chemical names and symbols

1. Write the chemical symbols for the following elements.
   a. lithium **b.** barium **c.** chromium
   d. phosphorus **e.** plutonium **f.** sulfur
2. State the name of each of the following elements, which are represented by their symbols.
   a. H **b.** Cl **c.** Fe **d.** Cf

Regions of the periodic table

The periodic table contains regions of similar elements

The periodic table is like a map of the United States. It features individual elements (states), groups (the Pacific coastal or Rocky Mountain states), and periods (the Corn Belt or Sun Belt states). Just as the United States can be divided into regions (the northeast or the southwest), so can the periodic table. **Table 3-2** shows the four regions of the periodic table—metals, nonmetals, metalloids, and noble gases.

Table 3-2
Regions of the Periodic Table

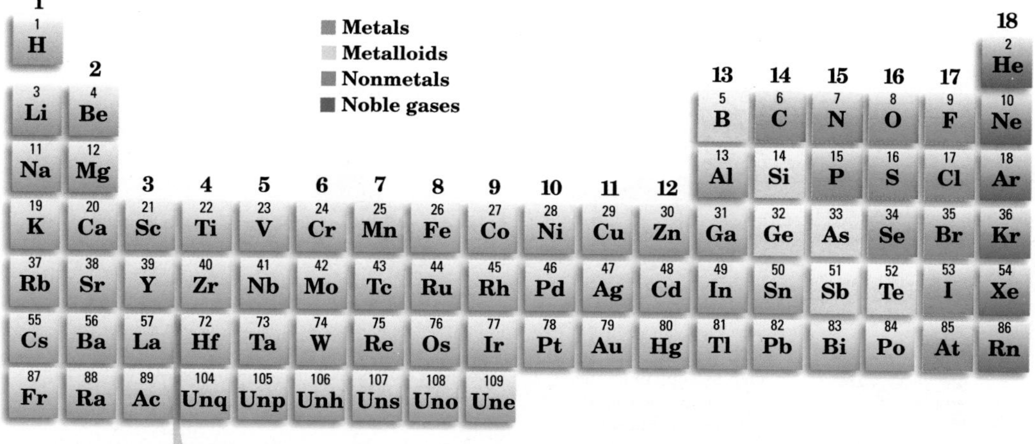

Demonstration 2
Reactivity of sodium in water
Approximate time: 5 min

1. Fill a 250 mL beaker half full with tap water.
2. Take a small piece of sodium (about 1/2 the size of a pea), drop it quickly into the water and cover with a 1.5 L beaker. **Stand back.** Dim the lights for a greater effect.

SAFETY
Wear safety goggles and a lab apron. Students should stand back at least ten feet.

DISPOSAL
Neutralize the residue with 1 M acid and pour down the drain

Answers to Concept Review

1. a. Li
b. Ba
c. Cr
d. P
e. Pu
f. S
2. a. hydrogen
b. chlorine
c. iron
d. californium

Teaching Tip
Group numbers
Point out to the students that the Group numbers start at the left and are counted towards the right, for a total of 18 groups, neglecting the lanthanides and actinides.

Visual Strategy
Table 3-2
Take time to review the regions of the table. Make students aware that elements situated on the "borders" of a region may have properties characteristic of both regions. The elements denoted as metalloids can vary depending on how the characteristics of the region are defined.

Groupwork Strategy

Divide the class into groups of three and pass out copies of old magazines with pictures of people. Ask each student in the group to cut out 10 pictures. Have each group classify the 30 pictures into at least 6 different categories. Discuss with the students: On what basis were the categories chosen? What was done with pictures that could fit more than one category? Is it possible to define a category of individuals for which there are no pictures?

26
Fe

Figure 3-4
Iron is the most abundant, widely used, and least expensive metal. The earliest iron artifacts date to 3000 B.C.

Properties and Uses of Selected Metals

Silver

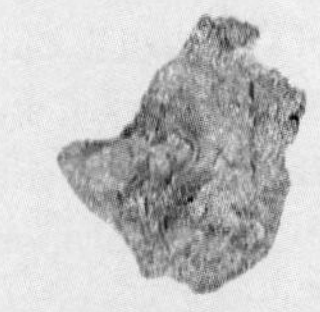

Silver, like other metals, has excellent reflective properties (luster). It is therefore used to plate mirrors. Silver is also ductile and malleable so it can be hammered, molded, or drawn into shape for jewelry.

Mercury

Mercury is the only metal that is a liquid at room temperature. Because mercury conducts heat and expands and contracts evenly with temperature change, it has long been used in thermometers.

Titanium

Titanium is strong, has low mass, and resists corrosion. It is therefore used in high performance jet engines and missiles. Also, because it does not react with flesh or bone, titanium is used to make surgical pins.

Metals form the largest region of the table

metal
any of a class of elements that generally are solid at room temperature, have a grayish color and shiny surface, and conduct electricity

The largest region of the periodic table contains elements classified as **metals**. Note that this region is shaded blue in **Table 3-2**. Metals are excellent conductors of heat and electricity. They also share other distinguishing properties that make them useful in technology, as shown in **Figure 3-4**. Metals are also generally lustrous, ductile, and malleable.

Nonmetals are the second largest region of the table

nonmetal
any chemical element that is neither a metal, metalloid, or a noble gas

If you recall the properties of metals, you can probably guess what properties nonmetals have just from their name. **Nonmetal** elements have varying properties but are generally poor conductors of heat and electricity. Most are gases or are brittle solids at room temperature. One nonmetal, iodine, is shown in **Figure 3-5**. Where are the nonmetals located in the periodic table, **Table 3-2**?

53
I

Figure 3-5
Iodine is a nonmetal. Tincture of iodine, a mixture of alcohol and iodine, is a familiar antiseptic used to treat cuts and scrapes.

Properties and Uses of Selected Nonmetals

Selenium

Selenium is used in light-sensitive devices such as copy machines because its electrical conductivity increases in response to light shining on it. Its various forms range from a metallic gray solid to one with a red glassy appearance.

Phosphorus

Phosphorus is known for phosphorescence—the most common form glows in the dark. Other forms are used in fireworks and matches. Most phosphorus is used to produce fertilizer. Phosphorus is very toxic and must be handled carefully.

Carbon

Carbon atoms join with other atoms to form strong structural units. Carbon atoms combine with one another to create diamonds, which are so hard that they can be used in cutting tools. Carbon combines quickly with oxygen at high temperatures.

Visual Strategy

Figure 3-4 and **Figure 3-5**

Have students compare and contrast the observable characteristics of the metals and nonmetals shown. They should notice that selenium looks somewhat metallic. Have them note Se's position on the periodic table.

Content Background

The elements included in the metalloid region of the periodic table depend on how the term *metalloid* is defined.

Historical Note

The first inert gas found to react was xenon which was reacted with fluorine in 1962 (although the possibility of reacting was originally suggested as early as 1932 by Linus Pauling). In a short period of time other xenon compounds as well as radon and krypton compounds were formed.

Figure 3-6
Boron, a metalloid, conducts electricity only at high temperatures. The mineral borax, which contains boron, is widely used in laundry and cleaning products.

Properties and Uses of Selected Metalloids

Tellurium

Tellurium has some properties of metals and some of nonmetals. Pure tellurium has a metallic luster yet, like a nonmetal, it is easily ground into a powder. It is a relatively rare element found in the Earth's crust.

Germanium

Germanium is a brittle, crystalline solid that retains its luster in air. Germanium is also an important semiconductor. It is slightly more conductive than silicon so circuits fabricated with germanium operate faster than silicon.

Arsenic

Arsenic is a gray solid that tarnishes in air. When oxidized it has a garlic odor. Because it is very toxic, it has been used as a pesticide. Arsenic is also used in electronics fabrication because of its metalloid properties.

Metalloids have properties of both metals and nonmetals

By now you realize that metals appear toward the left in the periodic table and nonmetals appear toward the right. Sandwiched in between these two regions are the **metalloids**. Locate the metalloids in the periodic table, **Table 3-2**. As the name suggests, a metalloid is an element that has some properties of metals and others of nonmetals. Several metalloids, most importantly silicon and germanium, are used in semiconducting devices, because their moderate electrical conductivity is just right for this delicate operation. Boron, shown in **Figure 3-6**, is another useful metalloid.

metalloid
an element having properties of metals as well as nonmetals

noble gas
an element that exists in the gaseous state at normal temperatures and is nonreactive with other elements

Noble gases are on the far right of the periodic table

Several elements are extremely unreactive or inert—they rarely form compounds with other elements. Such elements compose a group known as the **noble gases**. Krypton, used to make the lighting shown in **Figure 3-7**, is a noble gas. Find the noble gas region of the periodic table, **Table 3-2**. Which of the elements found in this group do you recognize?

Figure 3-7
Krypton is used in lighting products. Some krypton is used as an inert atmosphere for light bulbs. Mixed with argon, it is used in fluorescent lamps.

Properties and Uses of Selected Noble Gases

Neon

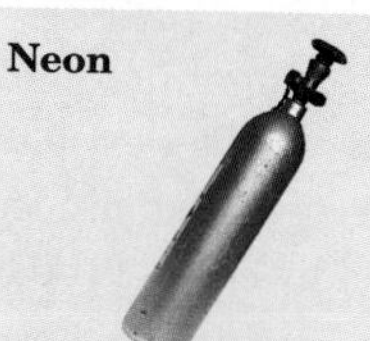

Neon emits brilliant red light when stimulated by electric current so it is useful for advertising signs. Neon is a monatomic gas. It is the fourth most abundant element in the universe.

Helium

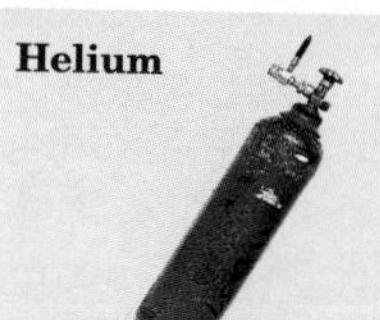

Helium is a light, nonreactive gas so it is used in both weather balloons and toy balloons. It is plentiful, inexpensive, and harmless. Helium has the lowest melting point of any element.

Radon

Radon is an odorless, colorless, tasteless radioactive gas. It is emitted from certain rocks underground, and can accumulate in homes and basements. Overexposure to radon can cause cancer.

Visual Strategy

Figure 3-6

Students are generally surprised by the metallic appearance of metalloids. Emphasize that while metals are good conductors of electricity, metalloids are semiconductors.

Cultural Connection

The Japanese use gold foil as a decorative addition to foods. The trace element gold is found mainly in joint areas of the body.

Goiters are rare among peoples living near oceans who eat seafood that contains iodine. Inland-dwelling peoples are historically much more apt to have goiters due to a lack of iodine in their diet.

ASSESS

Alternative Assessment

Have pairs of students make flash cards with names of elements on one side and the abbreviation of the element and its region of the periodic table on the other side of the card. Students can practice using the cards for ten minutes. Then the flash cards are put away and have each student state the name, symbol, and region of 15 elements they studied with their flash cards.

Answers to Section Review

3. The properties of each element are unique because the structure of each element is unique.
4. Zinc and mercury are most likely to have similar properties because they are in the same chemical family. (Accept also copper and zinc, which are from neighboring groups.)
5. Bromine, nickel, silver, tin, einsteinium, carbon, and krypton
6. Metals: cesium
   Nonmetals: carbon, selenium
   Metalloids: boron, antimony
   Noble gases: radon
7. It would probably be a metalloid as it has properties of both metals and nonmetals.

Consumer Focus

Essential elements

Of the more than 100 elements in the universe, 33 are essential to your health. These elements are highlighted in red in **Table 3-3**. Carbon, hydrogen, oxygen, and nitrogen are present in the largest amounts. Together, they make up 96% of your body mass. These elements are the major components of proteins, carbohydrates, lipids, and nucleic acids. Such compounds provide the structural materials your body needs for energy and for growth and repair.

Calcium, phosphorus, potassium, chlorine, sodium, and magnesium are the major minerals. They have a variety of important functions. For example, calcium is needed for your blood to clot when you bleed, and for your muscles to contract when you move. Other minerals help build hormones.

Some elements are called *trace elements* because only very small amounts—100 mg or less per day—are required. Trace elements include iron, manganese, copper, iodine, zinc, cobalt, chromium, nickel, silicon, tin, and vanadium.

To be sure you are getting enough of the major minerals and trace elements, you can take a multivitamin/mineral supplement. You should check with your doctor before taking such a supplement and take only the recommended dosage. These substances can be very toxic if you overdose. A spoonful of arsenic oxide, for example, is lethal, as you might know from mystery novels.

A healthy diet is safer than a supplement. If you eat a diet rich in animal protein, legumes, nuts, fruits, and vegetables, you are probably getting all the elements you need. Leafy, green vegetables are one especially good source of minerals and trace elements. The standard advice "eat your spinach! " is based on good science as well as common sense.

Table 3-3
Elements Needed for Health

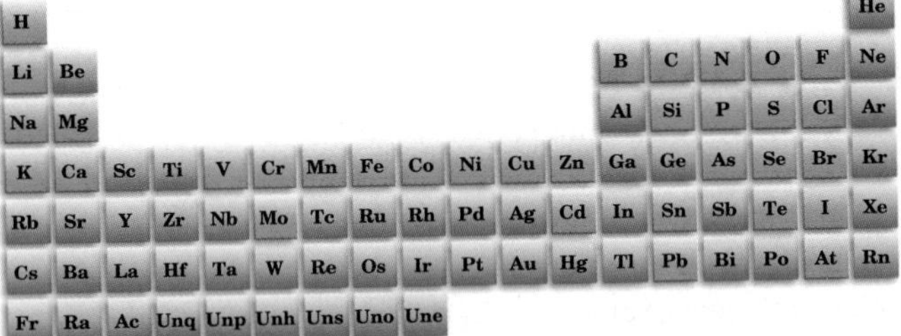

H																	He
Li	Be											B	C	N	O	F	Ne
Na	Mg											Al	Si	P	S	Cl	Ar
K	Ca	Sc	Ti	V	Cr	Mn	Fe	Co	Ni	Cu	Zn	Ga	Ge	As	Se	Br	Kr
Rb	Sr	Y	Zr	Nb	Mo	Tc	Ru	Rh	Pd	Ag	Cd	In	Sn	Sb	Te	I	Xe
Cs	Ba	La	Hf	Ta	W	Re	Os	Ir	Pt	Au	Hg	Tl	Pb	Bi	Po	At	Rn
Fr	Ra	Ac	Unq	Unp	Unh	Uns	Uno	Une									

Section Review

3. Why are the properties of each element unique?
4. Which of the following elements would most likely have similar properties: zinc, copper, lead, and mercury? Explain the reason for your answer.
5. Name the elements represented by each of the following symbols: Br, Ni, Ag, Sn, Es, C, and Kr.
6. Identify each of the following elements as either a metal, nonmetal, metalloid, or noble gas: radon, carbon, selenium, cesium, antimony, and boron.
7. Assume that a new element was recently discovered. This element has a shiny luster and conducts electricity only moderately. In which of the four major categories would you place this element? Defend your choice.

What is the basic structure of an atom?

Section Objectives

Infer the existence of atoms from the laws of definite composition, conservation of mass, and multiple proportions.

List the five basic principles of Dalton's atomic theory.

Describe models of the atom.

Compare and contrast the properties of electrons, protons, and neutrons.

Explain the particle-wave nature of electrons.

Describe the quantum model of the atom.

Building the atomic model

Figure 3-8b shows a nuclear weapon exploding. Because the "atom bomb" and atomic energy have had such a huge impact on modern life, the present era has been described as the "atomic age." In such an age, many people take the existence of atoms for granted. But no one has ever directly seen an atom. Scientists infer their existence from an enormous amount of evidence.

To infer is to draw a conclusion based on observations that you make or evidence that you gather. For example, you may have had a cup of coffee or tea this morning. After one sip, you may have realized you had too much sugar in it. Even if you had the most powerful microscope, you could search and never find a trace of sugar. But you know it's there. So you sometimes accept things, or infer their existence, without ever actually seeing them, as long as you have some evidence for their existence. In the case of the coffee or tea, you know that sugar is in it because it tastes sweet. Similarly, scientists accept atoms without actually seeing them. But scientists, too, require evidence to support their inferences. There is plenty of experimental evidence for atoms, as you will see.

Figure 3-8a
This internationally recognized symbol warns of the presence of radioactivity. Radioactivity is produced by unstable atoms.

3-8b
The tremendous energy of the atomic nucleus is exploited by the atomic bomb. The first bomb was built in Los Alamos, New Mexico, as part of the "Manhattan Project."

3-8c
Nuclear power plants tear atoms apart, thereby releasing the atom's enormous energy for use in homes and industry.

Section 3-2

FOCUS

Lesson Starter

Present students with a shoebox that has some type of noisy, rattling object inside. Wrap the box with black construction paper. Pass the box around the room, and ask students to describe what they think is inside the box. Ask for suggestions as to how to determine what is inside the box without opening it. Relate the box to the structure of the atom. How is the atom studied when it is not visible?

TEACH

Teaching Tip

Inferences

Perform the following exercise to describe an inference. Ask the students how they would prove to you that there is air in the room. Ask for specific evidence.

Possible Misconception

Many people mistakenly believe that the result of a meltdown in a nuclear power plant would be a nuclear explosion, similar to that caused by an atomic bomb. This is not the case. The hazard presented by a meltdown is the release of dangerous amounts of radioactive material, not a nuclear explosion.

Visual Strategy

Figure 3-8

Make students aware that the fissioning of elements such as uranium is constantly occurring in the environment—it does not occur just in nuclear bombs and reactors. The lack of uranium and other transuranium elements in the environment is due to continual fissioning of these elements into other elements.

Figure 3-9a
Table sugar (sucrose) is composed of 42.1% carbon, 51.4% oxygen, and 6.5% hydrogen. Table sugar contains exact proportions of these elements . . .

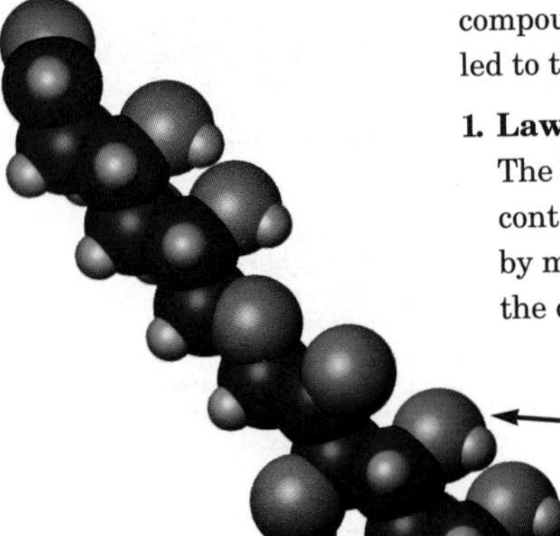

Evidence supporting the atomic theory

Recall that the idea that matter is made from atoms can be traced to Greek philosophers as far back as 400 B.C. But the ancient Greeks were better theorists than experimenters. Experimental results supporting the existence of atoms did not appear until more than 2000 years later in eighteenth-century Europe. There, early chemistry investigators—the first true chemists—noticed certain characteristics shared by all chemical compounds. Their observations of compounds and chemical reactions led to three laws that describe how compounds are formed.

1. Law of definite composition

The law of definite composition states that a compound contains the same elements in exactly the same proportions by mass regardless of the size of the sample or source of the compound. **Figure 3-9** illustrates this law.

3-9b
. . . regardless of the size of the sample or source of it.

2. Law of conservation of mass

The law of conservation of mass states that when two or more elements react to produce a compound, the total mass of the compound is the same as the sum of masses of the individual elements. The law of conservation of mass is illustrated in **Figure 3-10**.

Figure 3-10
The total mass of a system remains the same whether elements are combined, separated, or rearranged.

Combination of atoms:

S + O_2 ⟶ SO_2

Sulfur atom, Oxygen molecule, Sulfur dioxide molecule

32 mass units + 32 mass units = 64 mass units

Separation of atoms:

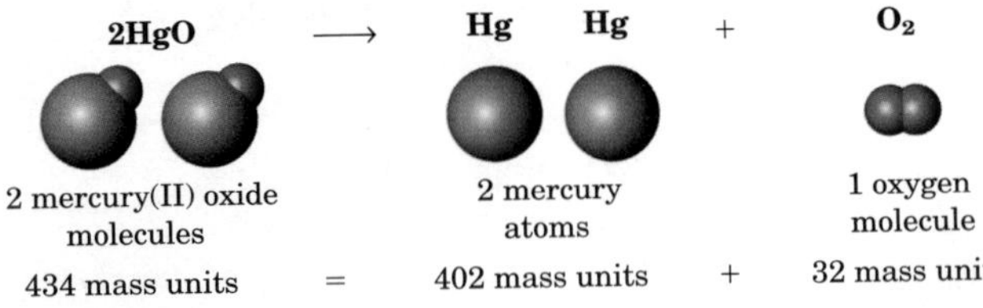

Rearrangement of atoms:

H_2CO_3 ⟶ H_2O + CO_2

1 carbonic acid molecule, 1 water molecule, 1 carbon dioxide molecule

62 mass units = 18 mass units + 44 mass units

Visual Strategy

Figure 3-10

Students will use the law of conservation of mass to verify their answers to stoichiometric problems in Chapter 8. Familiarize students with this law by carefully working through the numbers in the illustration.

Teaching Tip

Updating Dalton's theory

Ask students how Dalton's ideas have been revised given what we now know about them. For example, in statement 2 of the text: Isotopes of elements do have slightly different properties. In statement 5 of the text: Atoms can be subdivided, created, and destroyed in nuclear processes.

3. Law of multiple proportions

The law of multiple proportions applies to different compounds made from the same elements. It states that the mass ratio for one of the elements that combines with a fixed mass of the other element can be expressed in small whole numbers. **Figure 3-11** illustrates the law of multiple proportions.

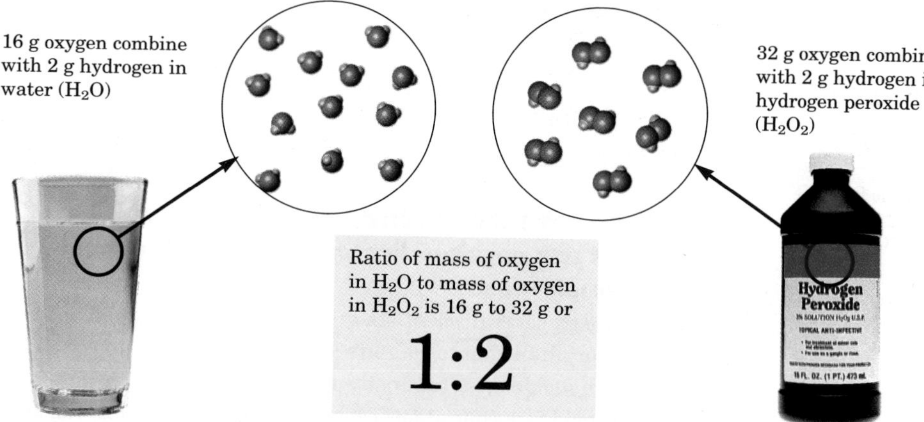
16 g oxygen combine with 2 g hydrogen in water (H_2O)
32 g oxygen combine with 2 g hydrogen in hydrogen peroxide (H_2O_2)
Ratio of mass of oxygen in H_2O to mass of oxygen in H_2O_2 is 16 g to 32 g or
1:2
Hydrogen Peroxide

Figure 3–11
Water and hydrogen peroxide are both formed from the elements hydrogen and oxygen. The ratio of the masses of oxygen in these two compounds is 1:2.

Dalton's atomic theory includes five principles

Taken together, these three laws represent much of the quantitative data obtained by chemists in the 1700s. But as useful as these laws were, no one realized their true potential until the English chemist John Dalton showed that, collectively, the laws demonstrate the existence of atoms.

Dalton argued that the three experimental laws could not be explained without assuming that all compounds are made from tiny particles such as atoms. This reasoning led to the development of the modern theory of the atom. Dalton's early atomic theory, published in the early 1800s, includes five basic principles.

1. All matter is made of indivisible and indestructible atoms.
2. Atoms of a given element are identical in their physical and chemical properties.
3. Atoms of different elements have different physical and chemical properties.
4. Atoms of different elements combine in simple, whole-number ratios to form chemical compounds.
5. Atoms cannot be subdivided, created, or destroyed when they are combined, separated, or rearranged in chemical reactions.

Dalton's proclamation attracted much interest from other chemists, who conducted experiments to test the new theory. While some exceptions to Dalton's atomic theory were eventually discovered, the theory itself has never been discarded, only modified and expanded as the world of the atom was explored.

Visual Strategy

Figure 3-11

Emphasize that it is the ratio of oxygen in H_2O to oxygen in H_2O_2 that is of interest. The micromodels show that the amount of hydrogen in H_2O and in H_2O_2 is the same but that the amount of oxygen varies in a small, whole-number proportion.

Demonstration 3
Electron flow
Approximate time: *2 min*
This demonstration shows movement of electrons in a gas-filled tube.

1. Connect a Geissler tube containing a common gas to a Geissler tube power supply or Tesla coil.
2. Mention that as the electrons move through the tube, they excite gas atoms which give off light. You may wish to save a more complete explanation of the light emission for later.

SAFETY
Wear safety goggles and a lab apron. Students should stand back at least ten feet. Be aware of the risk of potentially lethal electric shock. All electrical connections should be covered with insulating tape and there should be no bare wires. To avoid breaking of the fragile Geissler tube, enclose it in a transparent shield.

DISPOSAL
Save for reuse at a later date.

Teaching Tip
Vocabulary development
Remind the students that they have heard the word *cathode*. A computer screen is called a CRT or cathode-ray tube.

Historical Note
English physicist William Crookes used a tube of this sort that contained a metal strip between the cathode and anode. The placement of the metal strip caused a shadow to be produced at the anode end of the tube and therefore showed that the particles moved from negative to positive, which was contrary to the views held about electricity all the way back to Benjamin Franklin's time.

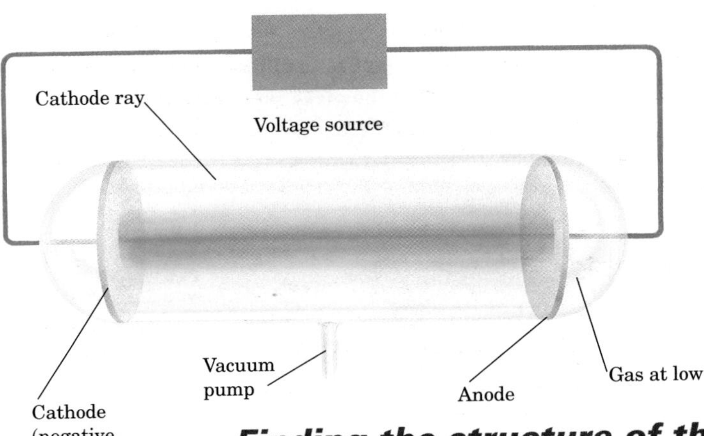

Figure 3-12
Electric current flows from the cathode to the anode in a cathode-ray tube when the electrodes are connected to a voltage source.

Finding the structure of the atom
Electrons are negatively charged particles that have a small mass

Experiments by several scientists in the middle of the nineteenth century led to a major alteration to Dalton's atomic theory. The atom was found not to be indivisible after all. Instead, it consists of smaller particles.

The first evidence that atoms consist of smaller particles was obtained by researchers whose main interest was electricity rather than the structure of the atom. These researchers studied the flow of electric current in glass tubes such as the one shown in **Figure 3-12**. Notice the metal disks called *electrodes* that are placed at each end of the tube. When the electrodes were connected to a source of voltage and most of the gas was removed from the tube, current flowed in the tube, and the remaining gas began to glow. A glowing beam always originated at the negative electrode, or **cathode**, and traveled toward the positive electrode, or **anode**. For this reason, the glowing beams were named *cathode rays*, and the apparatus in which they were observed became known as a *cathode-ray tube*. Today, cathode-ray tubes have various practical applications, including television sets, computer monitors, and radar screens, as shown in **Figure 3-13**.

cathode
a negative electrode through which current flows

anode
a positive electrode through which current flows

Researchers observed that a small paddle wheel placed in the paths of cathode rays rolled from the cathode toward the anode. This suggested that the rays must be composed of small, individual particles that could push the paddle wheel down the tube.

Figure 3-13
Cathode rays "paint" the pictures on television screens and computer CRT (cathode-ray tube) screens.

Visual Strategy
Figure 3-12 and Figure 3-13
Make students aware that a power source is needed to supply the electrons that flow from the cathode. Because the particles would be stopped by gas molecules in air on their way to the anode, the tube is evacuated. Make students aware that by controlling the direction of electron flow with magnetic and electric fields, different parts of a CRT screen can be made to light up.

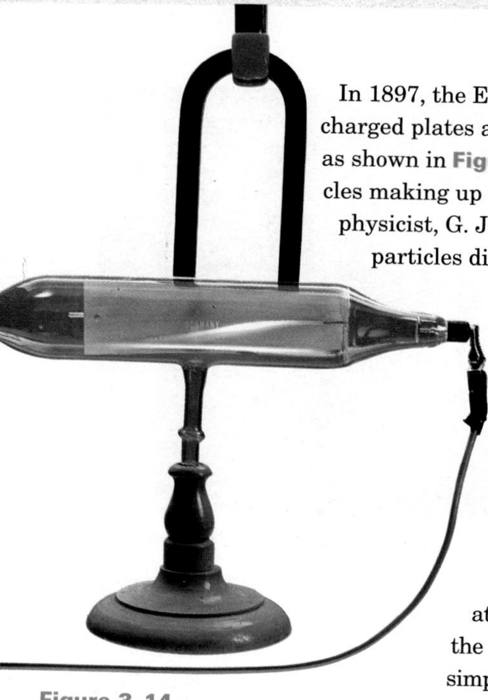

In 1897, the English physicist J. J. Thomson discovered that electrically charged plates and magnets deflected the straight paths of cathode rays, as shown in **Figure 3-14**. The direction of deflection shows that the particles making up cathode rays must be negatively charged. The English physicist, G. Johnstone Stoney named the small, negatively charged particles discovered in the cathode ray tube experiments *electrons*.

Experiments were later performed to determine the charge and mass of the electron. The electron was discovered to have a mass nearly 2000 times smaller than the mass of the smallest atom, hydrogen. The lightness of electrons implied that atoms must contain other, heavier matter to account for most of their mass.

There was additional evidence that atoms are made up of other, yet-to-be-identified kinds of matter. Atoms were known to be electrically neutral. This meant that an atom must contain some positively charged matter to balance the negative charges of its electrons. Thomson developed a simple early model of the atom based on all this information. His model of the atom, which is shown in **Figure 3-15**, was named the "plum pudding model" because it resembled plum pudding, a British dessert that consists of a ball of sweet bread with pieces of fruit embedded in it. Thomson envisioned the atom as a ball of positive charge with negatively charged electrons embedded inside.

Figure 3–14
Note the deflection of the cathode ray. Magnets near the cathode-ray tube cause the beam to be deflected as shown. The direction of the deflected beam with respect to the north and south poles of the magnets indicates that the particles making up the beam have a negative charge.

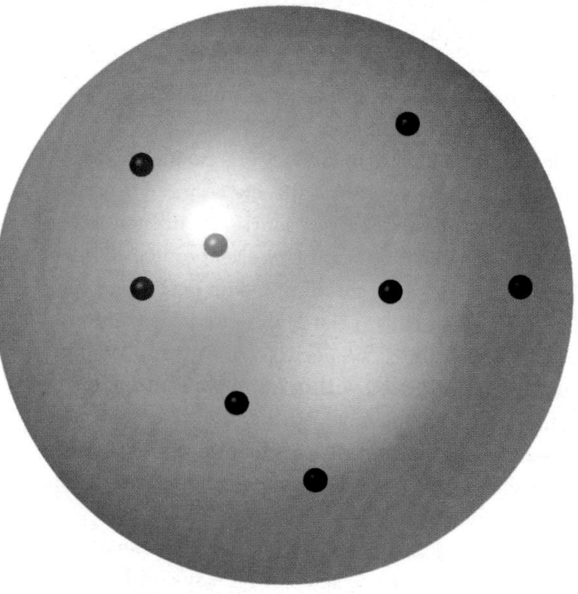

Figure 3–15a
J. J. Thomson's model of the atom featured negatively charged electrons embedded in a ball of positive charge.

3-15b
Thomson's model of the atom is often called the plum pudding model of the atom. Can you see the resemblance?

Possible Misconception

Many students will believe that electrons flow from the positive pole to the negative pole. Remind students that electrons are negative and are attracted to the positive pole of the battery. A further confusing aspect is the definition of current flow which is from positive to negative.

Visual Strategy

Figure 3-15

Make students aware that although this theory may seem rather odd or humorous, it was the best explanation possible for events documented at the time in chemistry and physics. Humorous names are still used in science, such as the names of the elementary particles "charm" and "strangeness".

Teaching Tip

Analogy

Have students imagine that 50 billiard balls are hung from strings at various heights throughout the front half of the room. A student is then blindfolded and given 5 tennis balls to throw towards the front of the room. Ask students what would happen to the tennis balls? Though some would go through unscathed, some would be deflected to the side, and in some cases would hit head on and the tennis ball would be deflected backwards.

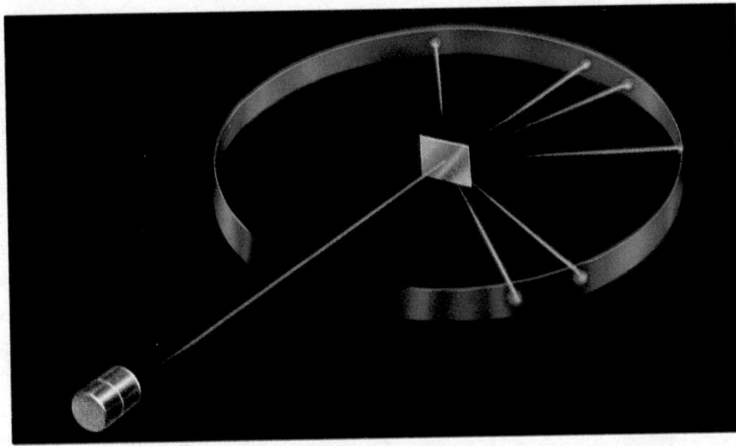

Figure 3–16
In Rutherford's famous experiment, small positively charged particles were targeted at a thin foil of gold atoms. Note how the gold atoms scatter these small particles.

Each atom has a positively charged inner core

A student of J. J. Thomson, Ernest Rutherford, assembled a research team to perform an experiment that ultimately disproved the plum pudding model of the atom. This now-famous "gold-foil experiment" is represented in Figure 3-16. Rutherford's team directed a beam of tiny positively charged particles, called *alpha particles*, at a very thin gold foil sheet. The team measured the angles at which the alpha particles were deflected from their straight-line paths as they emerged from the foil. This procedure was much like determining the shape of an object hidden under a table by rolling marbles at it, as shown in Figure 3-17.

nucleus
the central region of an atom made up of protons and neutrons

The researchers discovered that most of the alpha particles they shot at the foil passed straight through it. Rutherford reasoned that these alpha particles went straight through because they were traveling through empty regions of the foil. The researchers also observed that some particles did not pass straight through the foil but were slightly deflected. But what really surprised the scientists were the alpha particles that were actually scattered back.

Rutherford reasoned that the deflections resulted from electrical repulsion between the positively charged alpha particles and the positively charged matter contained in the atom. However, if the positive charge were spread out within the atom as the plum pudding model described, the backward scattering of the alpha particles would not have been possible. The positive charge and the mass must be concentrated within atoms to scatter the alpha particles backward. Rutherford had discovered the positively charged core of the atom. The tiny central region of the atom was named the **nucleus**, from the Latin word meaning "little nut." Electrons, Rutherford supposed, traveled in the space surrounding the nucleus in a way similar to the motion of the planets around the sun.

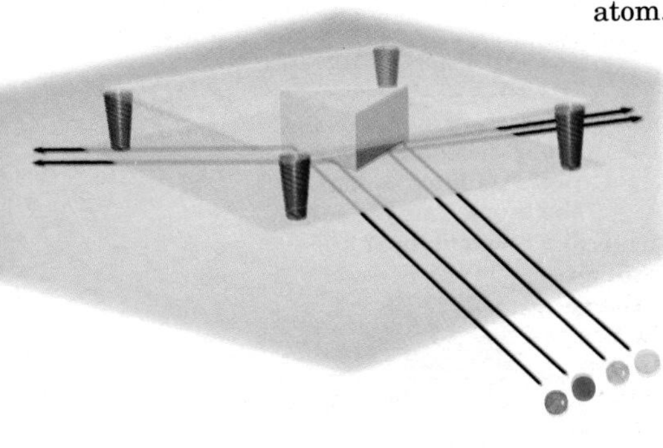

Figure 3–17
Marbles are rolled at a hidden object. The shape of the object can be determined from the directions of the scattered marbles.

Visual Strategy

Figure 3-16 and Figure 3-17

Make students aware that over 99% of the alpha particles traveled through the gold foil. The appearance of only a few deflections provided the observations on which Rutherford built his theory.

Figure 3-18
If the nucleus of an atom were the size of a marble, then the whole atom would be the size of a football stadium.

The nucleus, then, is the dense central portion of the atom that contains all of its positive charge and nearly all of its mass but occupies only a small fraction of its volume. By measuring the fraction of alpha particles that were deflected and their angles of deflection, the radius of the nucleus was calculated to be at least 10 000 times smaller than the radius of the whole atom. To put this relationship in perspective, examine **Figure 3-18**.

Electrons occupy energy levels within an atom

energy level or principal energy level
a specific energy or group of energies that may be possessed by an electron in an atom

Classical physics predicted that, because all accelerating charged particles radiate energy, the circulating electrons in Rutherford's planetary model must also radiate energy. As electrons radiated, their energy would be lost, and they would not be able to stay in orbit. The electrons would spiral toward the nucleus and collapse into it. The atom would be destroyed in a billionth of a second. Yet most atoms were known to remain stable for thousands of years. Clearly, something was wrong with Rutherford's model of the atom.

In 1913, a young Danish physicist named Niels Bohr came up with a new model that addressed the deficiencies of Rutherford's model of the atom. Bohr proposed that electrons in an atom can reside only in certain **energy levels**. The rungs of a ladder are similar to the energy levels within an atom, as shown in **Figure 3-19**. A person can move up or down the ladder only by standing on its rungs; it is impossible to stand between them. Similarly, the Bohr model of an atom postulated that an electron can reside only in certain energy levels within an atom, as shown in **Figure 3-20** on the next page.

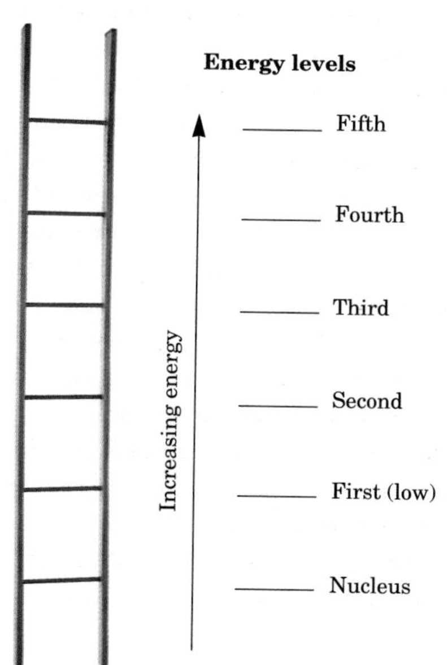

Figure 3-19
The energy levels of electrons within an atom are like the rungs of a ladder. Just as you cannot stand between the rungs of a ladder, electrons can reside only in certain energy levels in an atom.

Visual Strategy
Figure 3-18
This is a good opportunity to drive home the point that the atom is mostly empty space. Students should also realize that as atomic size increases, the amount of empty space in the atom also increases.

Possible Misconception
Most students will have the Bohr atom as their concept of the atom. Remind them that the modern model of the atom will continue to undergo revision as more subatomic particles are discovered.

Demonstration 4
Emission spectra
Approximate time: 5 min

1. Obtain as many Geissler tubes containing gases as possible to show the students the variations in excited states. (Examples include H_2, He, N_2, Ne, Hg, etc.)
2. Excite the tubes with a Geissler tube power supply or Tesla coil to show that the electrons can be excited to higher energy levels.
3. Take an incandescent light bulb and attach it to the power supply. (Nothing happens.) Ask the students how this bulb works. (The tungsten filament is thermally excited—this is the reason this bulb is only 30% efficient.)
4. Take a fluorescent tube and excite it with the power supply. Why does it glow? (Electrons are more easily excited.) Why is it cheaper to operate than incandescent bulbs?
5. If available, have the students view the spectra using spectroscopes.

SAFETY
Wear safety goggles and a lab apron. Students should stand back at least 10 ft. Be aware of the potentially lethal risk of electric shock: all electrical connections should be covered with insulating tape and there should be no bare wires. To avoid breaking the fragile Geissler tubes, enclose each in a transparent shield.

DISPOSAL
Save the equipment for reuse at a later date.

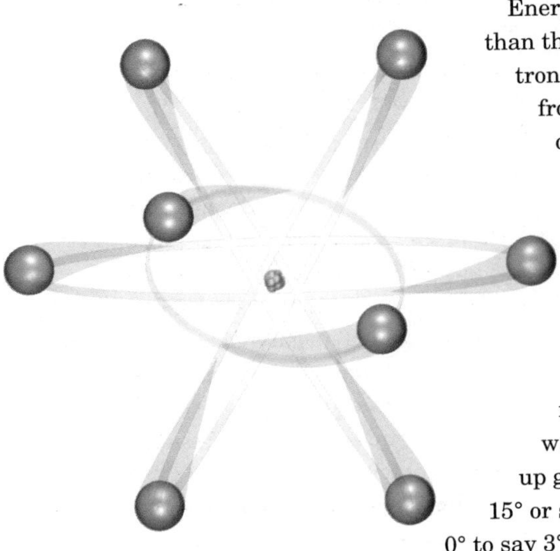

Energy levels close to the nucleus have lower energy than those farther from the center of the atom. An electron must gain the right amount of energy to move from a lower to a higher energy level. The amount of energy an electron needs to make such a jump between energy levels is called a *quantum*. Thus, the Bohr model shows that the energy of an electron in an atom is *quantized*. It cannot have any arbitrary value of energy; only particular values are possible.

The quantization of energy in an atom is unlike anything you can experience directly on the macroscopic level. If a child's swing behaved this way when you pushed it, the swing would not move up gradually from a vertical position to an angle of 15° or so. A quantized swing would jump abruptly from 0° to say 3°, to 9°, and so on and would never be found at any angles in between!

Figure 3-20
The model of the atom proposed by Niels Bohr included quantized energy levels for the electrons. Although the Bohr model has been revised in part, the quantum concept of the atom is still considered valid today.

Neutrons add mass to the nucleus

The positively charged nuclear particles that repelled the alpha particles in Rutherford's experiment were found to be quite heavy. The mass of one such particle is almost 2000 times that of an electron. Scientists call these heavy particles *protons*.

The mass of the proton presented a dilemma because the masses of all atoms besides hydrogen were known to be larger than the total mass of their protons and electrons. Clearly there must be yet another particle that contributed to the mass of an atom. The search for a third subatomic particle was soon underway. This particle, however, proved to be more difficult to detect than either electrons or protons because it was electrically neutral. Some 30 years after the discovery of electrons, Irene Joliot-Curie (daughter of the famous Pierre and Marie Curie) discovered that when beryllium was bombarded with alpha particles, a beam with high penetrating power was formed. A British scientist, James Chadwick, found that this penetrating beam was made of particles that had approximately the mass of protons. Also, the beam was not deflected by electric or magnetic fields. Chadwick deduced that the beam was composed of neutrons—neutral particles that have mass equal to that of protons.

With the discovery of the neutron, the picture of an atom was thought to be complete. An atom consists of a nucleus containing protons with positive charges and neutrons with no electrical charge. Protons and neutrons together make up nearly all the mass of an atom. Most of the volume of an atom consists of empty space in which electrons, with negative charges and insignificant mass, are located. Is this model of the atom still accepted as truly complete?

Visual Strategy

Figure 3-20

The concept that most students have for an atom is that of the Bohr model shown. Emphasize that this model has a historical significance. However, it could not be used to explain observed spectra in multi-electron atoms. Further study showed that the idea of a well defined orbit for an electron was flawed.

Figure 3-21a
The bright-line emission spectrum is unique to each element, just like a fingerprint is unique to each person. The bright-line emission spectrum for calcium . . .

3-21b
. . . is easily distinguished from the bright-line emission spectrum for mercury.

The modern view of the atom

Electrons can be described as particles or waves

Electrons were first recognized as particles because they were seen to push a paddle wheel down a cathode-ray tube. Other experiments on the light emitted by energized, gaseous atoms also indicate that electrons are particles.

When all electrons are in the lowest possible energy levels, an atom is said to be in its **ground state**. When the atom absorbs energy so that its electrons are boosted to higher energy levels, the atom is said to be in an **excited state**. Experiments on gaseous atoms showed that they absorb energy from an electric discharge to reach excited states. Later, the gaseous atoms return to the ground state by emitting the energy they had absorbed, now in the form of light.

ground state
the lowest energy state of a quantized system

excited state
the condition of an atom in a state higher than the ground state

The light emitted by an element when its electrons return to a lower energy state can be viewed as a *bright-line emission spectrum,* shown in **Figure 3-21**. The bright-line emission spectrum is obtained by passing the light that an element emits through a prism to separate the various colors it contains. Each colored band represents the light energy released by an electron when it returns from a high energy state to a lower one. The energy of each band of colored light is equal to the difference between the original and final energy levels of the electron.

Sunlight, when passed through a prism, produces the *visible spectrum*—all the colors of light that can be perceived by the human eye, as shown in **Figure 3-22**.

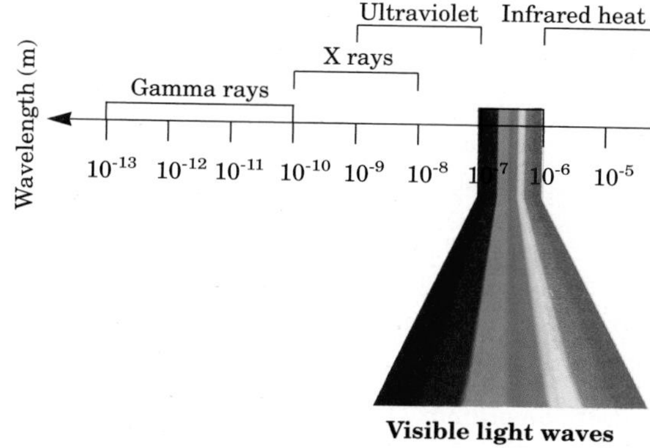

Figure 3-22
The electromagnetic spectrum is composed of electromagnetic radiation over a broad range of wavelengths. The visible spectrum is only a small portion of the electromagnetic spectrum. Wavelengths of visible light are in the range of 10^{-7} m.

Visual Strategy

Figure 3-21

Make students aware that these spectra are characteristic and unique to each element. Unknown samples can be identified by producing emission spectra of the sample and matching the spectral lines to known spectra to determine the elements present.

Possible Misconception

Students don't generally understand the dual nature of light. Therefore students should be instructed that light is neither a wave or particle but a different entity that has some properties of both.

Teaching Tip

Analogy

Electron microscopes provide greater resolution because the size of the reflected electron is much smaller than the size of reflected visible light. An analogy would be trying to look at the structure of a system of bowling pins by throwing a bowling ball at them. Because the bowling ball knocks the pins over, the ball is deflected slightly. If marbles were thrown at the bowling pins, most marbles would be greatly deflected and the structure of the object could be determined by noting the pattern of the deflections.

electromagnetic spectrum
the total range of electromagnetic radiation, ranging from the longest radio waves to the shortest gamma waves

The visible spectrum is a small portion of the larger **electromagnetic spectrum**. The electromagnetic spectrum consists of the various classes of electromagnetic waves: microwaves, radio waves, X rays, gamma rays, infrared radiation, and ultraviolet radiation, in addition to visible light. All electromagnetic waves are essentially the same—they are electromagnetic vibrations traveling through space in the form of waves. However, the *wavelength* and *frequency* of different kinds of electromagnetic waves differ, as **Figure 3-23** shows. The wavelength of a wave is the distance between two identical points on the wave. Frequency is the number of wavelengths that pass a certain point in a given period of time—usually a second. Wavelength and frequency are inversely related.

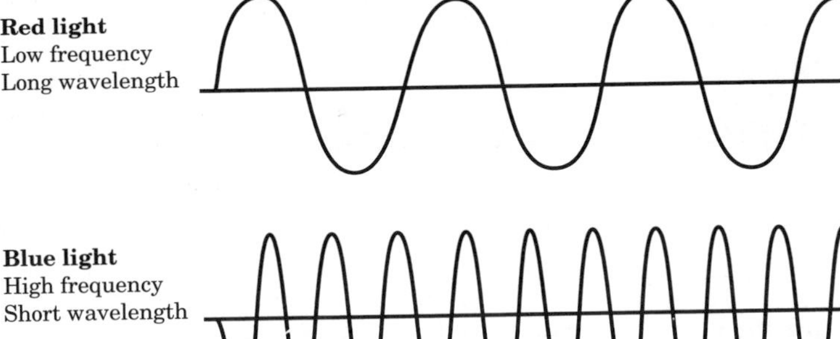

Figure 3-23
The frequency and wavelength of a wave are inversely related. As wavelength increases, frequency decreases. The higher the energy of a wave, the shorter its wavelength and the higher its frequency.

All electromagnetic waves are created by accelerating charged particles. The radio waves that transmit the signals that a radio picks up are produced by the motions of electrons up and down along an antenna wire. The visible spectrum is produced by the "jumping" and "falling" of electrons in atoms. In "falling" from a state of higher energy to a lower state, the electron moves closer to the nucleus. Electrons, then, have properties of particles, as evidenced by their behavior in cathode-ray tubes and their ability to "jump" and "fall" among high- and low-energy orbitals.

Electrons also have properties of waves. However, electrons have tiny wavelengths, on the order of 10^{-10} m. Thus, electron waves are much smaller than visible light waves. Because of this, the wave behavior of electrons can be observed only on the scale of very small objects—such as atoms. The electron microscope provides evidence of the wave behavior of electrons. The image in **Figure 3-24** was taken with an electron microscope, which uses electron waves to explore details of matter too small to be seen with an ordinary light microscope.

Figure 3-24
This electron micrograph shows the germination of two grains of pollen at a magnification of 1100×.

Because electrons have dual properties of both particles *and* waves, a new model for atomic structure had to be developed. This model was produced by quantum theory.

Visual Strategy

Figure 3-23

Be sure students understand the meaning of the terms *wavelength*, *frequency*, and *amplitude*. Make a transparency of this figure from the blackline master in the *Teaching Transparencies* package. Mark 1 wavelength on each drawing. Be sure students understand that frequency involves time and that 2 waves can have the same wavelength but different amplitudes.

quantum theory
the field of physics based on the idea that energy is quantized and that this has significant effects on the atomic level

orbital
a region of an atom in which there is a high probability of finding electrons

Quantum theory provides a modern picture of the atom

The description of the atom that is built on the wave properties of electrons is the **quantum theory**. The quantum theory uses the complex mathematical equations of *quantum mechanics* that describe waves. The quantum mechanical model of the atom is an essentially mathematical one, and it cannot be represented by anything that exists in the macroscopic world.

Like the Bohr model, the quantum mechanical model predicts quantized energy levels for electrons. Unlike the Bohr model, however, the quantum mechanical model that scientists use today does not describe the exact path an electron takes around the nucleus. It is concerned with the *probability*, or likelihood, of finding an electron in a certain position.

When all of the probable locations of an electron are plotted as points on a graph, they indicate a region where the electron can be found, as **Figure 3-25** shows. Such regions are called **orbitals**. Each orbital can be occupied by up to two electrons. Note that the boundaries of an orbital are fuzzy because the probability of finding an electron changes gradually throughout the region. Regions with the densest concentrations of points are areas where the probability of finding electrons is highest. Due to their fuzzy boundaries and gradual shading, orbitals are sometimes called "electron clouds." Just as the blade of a fan could be anywhere within its blurred path, the electron could be anywhere within an electron cloud. The quantum mechanical model of the atom is therefore often called the "electron cloud model."

Figure 3-25a
This probability plot shows the hydrogen atom in its ground state. The electron is most likely to be found in regions where the dots are closest together.

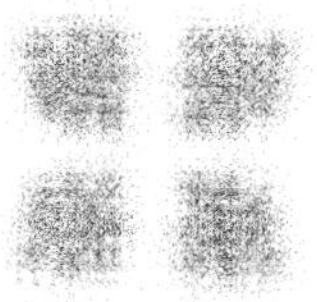

3-25b
This probability plot shows hydrogen in an excited state. The region where the electron is most likely found is no longer spherical.

It is conventional to draw a surface around the model of an electron cloud to designate where an electron can be found 90% of the time, as shown in **Figure 3-26**. With this helpful technique, it is possible to simplify the representation of electron orbitals while maintaining a basically accurate picture of the atom.

The position and velocity of an electron can be measured, but the precision of these measurements is limited. The quantum theory shows that regardless of how sophisticated technology becomes, it will never be possible to measure both the electron's position and its velocity simultaneously. This fact is known as the *Heisenberg uncertainty principle*: you can never know exactly where an electron is if you know exactly how fast it is moving. Conversely, you can never know exactly how fast an electron is moving if you know exactly where it is. The quantum theory shows that there is a limit to what we can know about events inside an atom.

Figure 3-26
The exact location of an electron cannot be determined. Instead, electron orbitals indicate the areas where electrons have the highest probability of being located. Though orbitals have fuzzy boundaries, it is conventional to draw closed surfaces over them that show where the electron has a 90% probability of being found. The cloud model of hydrogen in its ground state is most often represented as a closed sphere, as shown here.

Content Background

The position and velocity of an electron cannot be determined simultaneously due to the fact that any attempt to measure the electron will cause the electron to alter its path.

Teaching Tip

Heisenberg's principle
A bumper sticker available from the American Chemical Society states "Heisenberg may have slept here!" Discuss the meaning of this statement with students. The Heisenberg uncertainty principle indicates that one could never be sure whether Heisenberg slept there or not.

Visual Strategy

Figure 3-26
Emphasize the 3-dimensional nature of the atom, and that electrons occupy a spherical volume. This helps move students away from the Bohr model and orbits.

ASSESS

Alternative Assessment

Have students build a time line for the major events in the development of our modern theory of the atom. Start with events in the text. Award extra credit to those students who go outside the text to build a more complete chronology.

Answers to Section Review

8. Possible responses might include that rain can be inferred from raindrops collected on surfaces such as sidewalks. Such evidence is indirect, like the evidence for atoms.
9. The law of definite composition, the law of conservation of mass, and the law of multiple proportions
10. The first, third, and fourth of Dalton's principles listed on page 83 still apply to the structure of the atom.
11. Models are useful in simplifying complex phenomena and in providing the basis for further experimentation.
12. The major differences relate to charge and mass. Neutrons are electrically neutral, protons are positive, and electrons are negative. The mass of the proton and neutron are comparable, but the mass of the electron is about 2000 times smaller.
13. The particle nature of electrons is supported by their behavior in cathode-ray tubes; their wave nature is supported by the development of electron microscopes.
14. The quantum theory features orbitals of various shapes compared to Bohr's circular orbits, and the quantum view holds that the location of electrons cannot be known precisely.

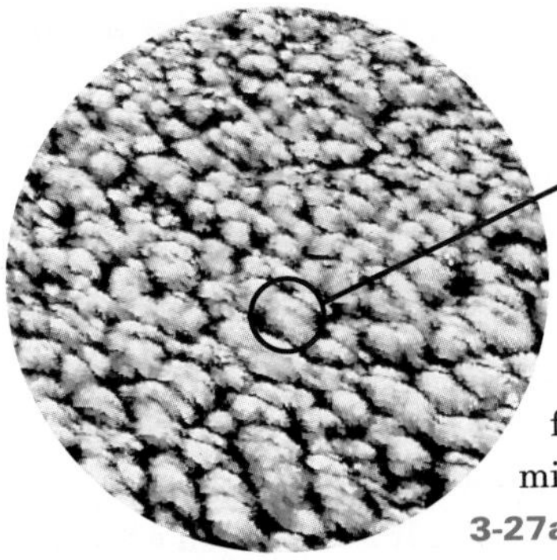

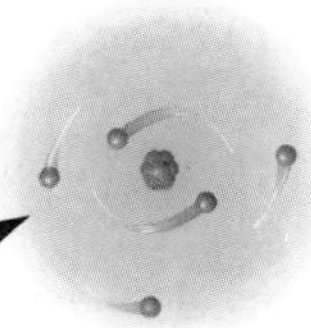

3-27b
The modern atomic model describes the atom as a central cluster of neutrons and protons surrounded by electrons that travel in orbitals.

Figure 3-27a
The scanning tunneling microscope allows scientists to "see" the surface of boron atoms.

The atom is a scientific model

John Dalton showed that the existence of atoms can be inferred from experimental evidence. Now, with the scanning tunneling microscope, there is even better evidence that atoms exist, as **Figure 3-27a** shows.

However, no one really knows what the internal structure of an atom looks like. Because scientists cannot look directly at an atom, the best they can do is design clever experiments to probe the atom's inner design. From these experiments, the scientists build a model to account for the experimental findings, as shown in **Figure 3-27b**.

The model of the atom has been revised many times. Dalton's indivisible spheres were replaced with the plum pudding model, which was replaced by Rutherford's model of the atom, the Bohr model, and finally the electron cloud model. Each model was built on experimental data. Once the model was created, it served as a springboard for further experiments designed to test it. When the experimental tests revealed flaws, the model was revised.

Models play a key role in science. As you have seen in the case of the atom, models provide guidelines for research that leads to new knowledge. Models also help simplify and organize phenomena that would otherwise be too complex to understand.

Current experimental research in atomic physics and chemistry may show that flaws exist in the model of atomic structure that scientists accept today. Nonetheless, the current model, like its predecessors, has greatly advanced our understanding of the structure of atoms and our ability to predict how they will behave.

Section Review

8. How might you infer that it has rained without actually having seen the rain? How is this like the belief that atoms exist?
9. What laws first enabled scientists to infer the existence of atoms?
10. Which of Dalton's five principles still apply to the structure of an atom?
11. How are models useful in understanding atomic structure?
12. Describe the major differences between electrons, protons, and neutrons.
13. What evidence supports the particle nature of electrons? the wave nature of electrons?
14. How does the quantum theory's depiction of atomic structure differ from the Bohr model?

How do the structures of atoms differ?

Section Objectives

Decipher an element's atomic number and mass number in terms of its atomic structure.

Decipher the information provided by the four quantum numbers with respect to the location of electrons in atoms.

Define the Pauli exclusion principle in terms of the arrangement of electrons.

Use the periodic table to write the electron configurations and orbital diagrams for various atoms.

Atomic number and mass number

If all atoms consist of electrons, protons, and neutrons, then how are the atoms of one element different from those of another element? As you know, atoms differ not in the kinds of subatomic particles they contain, but in how many of each kind there are.

Each element has an atomic number

The number of protons in the nucleus of an atom is known as its **atomic number**. Hydrogen, which has atomic number 1, contains one proton in its nucleus. Oxygen has atomic number 8 and contains eight protons. The periodic table lists the atomic number of each element. **Figure 3-28** shows the hydrogen and oxygen blocks of the periodic table. Notice that the atomic number of each element is given.

If you know an element's atomic number, you also know how many electrons are contained in an electrically neutral atom of that element. Such an atom must contain the same number of electrons as protons. Thus, nitrogen, with atomic number 7, has seven protons and seven electrons.

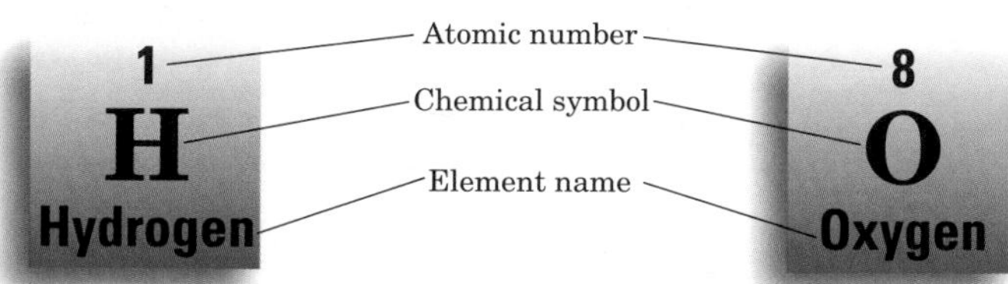

Figure 3-28
The periodic table lists each element's name, symbol, and atomic number. Atomic mass and electron configuration—which you will learn about in this chapter—are also given on most periodic tables.

Isotopes of the same element have different mass numbers

The number of protons that an element contains is fixed, so all atoms of the same element have the same atomic number. However, the number of neutrons can vary. Atoms of the same element that have the same number of protons but different numbers of neutrons are called **isotopes**.

Consider the simplest element, hydrogen. Three isotopes of hydrogen exist. They are named *protium, deuterium,* and *tritium*. Each hydrogen isotope contains one proton in its nucleus. But protium contains no neutrons, while deuterium has one neutron, and tritium has two. Despite their differences, isotopes have very similar chemical properties. Can you see why this is true?

Chem Fact
Although scientists in Nazi Germany possessed the theoretical know-how to produce an atomic bomb, their progress was slowed by several key practical deficiencies. One was their inability to produce the required amounts of water with hydrogen atoms that have the heavy deuterium nucleus. This *heavy water* was needed to control the rate of nuclear reactions. American scientists, on the other hand, did find ways to produce large amounts of heavy water.

atomic number
the number of protons in the nucleus of an atom

isotope
one of two or more atoms having the same number of protons but different numbers of neutrons

Section 3-3

FOCUS

Lesson Starter
Pass out sheets of graph paper and have the students make graphs of mass number versus atomic number for the first 36 elements. The atomic number should be on the x-axis and the mass number should be on the y-axis. Have the students connect the points. Then, on the same graph, have the students draw a straight line passing through the origin that has a slope equal to one. Use points (0, 0) and (36, 36). Ask students what the straight line represents and how the number of neutrons compares to the number of protons and electrons for heavier elements. Ask them to predict how the graph would look for the first 72 elements.

TEACH

Possible Misconception
Make it clear to students that the identity of the atom is determined by the number of protons, not the number of electrons or neutrons. The number of electrons and neutrons can vary, and the element will still be the same element; but if the number of protons changes, then it becomes a different element.

Content Background

The isotopes of hydrogen have specific names because of their great interest to chemists. These are the only isotopes in which major mass differences are found. In particular, deuterium has a mass twice as large as protium. By substituting deuterium atoms for protium atoms on molecules, chemical reactions can be followed by tracking the deuterium during the reaction. This tracking can often be accomplished by using mass spectroscopy and infrared spectroscopy which are sensitive to the differences in mass.

Teaching Tip

Determining mass number

For many elements, students can easily determine the mass number of the most common isotope by rounding the atomic mass to the nearest whole number. Make the students aware that atoms cannot have partial neutrons.

Additional Examples

Determine the number of protons, neutrons, and electrons for the most common isotope of each of the following atoms: N, F, Kr, and Au.

N has 7*p*, 7*e*, 7*n*
F has 9*p*, 9*e*, 10*n*
Kr has 36p, 36*e*, 48*n*
Au has 79*p*, 79*e*, 118*n*

mass number
the total number of protons and neutrons in the nucleus of an atom

In addition to having an atomic number, each element also has a mass number. The **mass number** is the total number of protons and neutrons in the atomic nucleus.

An isotope is represented by its chemical symbol with two additional numbers written to the left of it. The mass number is written as a superscript, and the atomic number is written as a subscript.

A second way to represent isotopes is to use both the element's name and mass number. Deuterium, for example, is described as hydrogen-2 and tritium as hydrogen-3. Two isotopes of oxygen are oxygen-16 and oxygen-18. **Table 3-4** shows how to represent isotopes.

Table 3-4
Isotopes of Hydrogen

Name	Number of neutrons	Symbol	Representation
protium	0	${}^{1}_{1}H$	hydrogen-1
deuterium	1	${}^{2}_{1}H$	hydrogen-2
tritium	2	${}^{3}_{1}H$	hydrogen-3

The number of neutrons in an element is easy to find

If you know the atomic number of an element, you know how many protons and electrons it contains. If you also know the mass number, you can determine how many neutrons the element contains.

For example, consider the element sodium. It has the atomic number 11 and a mass number of 23. From its atomic number, you can quickly determine that a sodium atom has 11 protons and 11 electrons. To determine the number of neutrons, simply subtract the atomic number from the mass number. Sodium has 23 − 11, or 12, neutrons.

Labeling electrons in atoms

Quantum numbers are used to differentiate among electrons

In quantum theory, each electron in an atom is assigned a set of four quantum numbers. Three of these numbers, like coordinates on a map, give the location of the electron. The fourth quantum number describes the orientation of an electron in an orbital.

The first quantum number, called the *principal quantum number*, is represented by the letter n. This number describes the energy level that the electron occupies. The principal quantum number is assigned a positive integer starting with 1 (n = 1, 2, 3, 4, . . .). Generally speaking, the larger the value of n, the farther away from the nucleus and the higher the energy of the electron. For example, an electron for which $n = 2$ occupies a higher energy level than an electron for which $n = 1$. **Figure 3-29** shows various energy levels of the hydrogen atom and their relative spacing with respect to the nucleus. These are not distances.

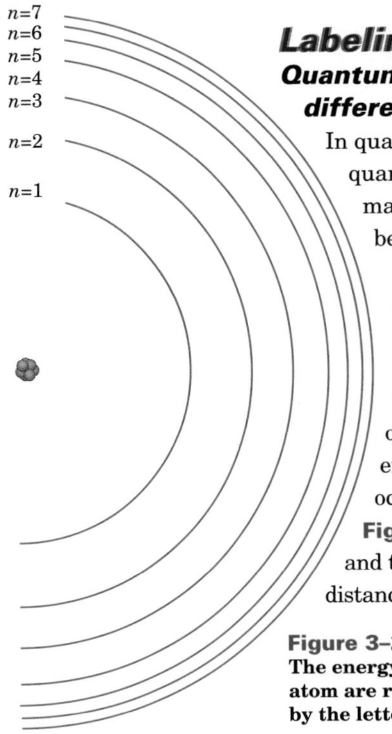

n=7
n=6
n=5
n=4
n=3
n=2
n=1

Figure 3–29
The energy levels of an atom are represented by the letter *n*.

Visual Strategy

Figure 3-29

Though energy levels are represented by circles in the figure, emphasize that the model is showing relative position and not orbits.

Table 3-5
Letter Designations for Values of ℓ

ℓ	Letter
0	*s*
1	*p*
2	*d*
3	*f*

Quantum numbers are also used to describe the shapes of atomic orbitals. A second quantum number is designated by the letter ℓ. It provides a code for the shapes of orbitals.

Values of ℓ are often expressed as letters rather than numbers. **Table 3-5** shows number values of ℓ along with their letter designations. The lowest energy orbital in each energy level is designated as *s*. It has a spherical shape as shown in **Figure 3-30**. There is only one *s* orbital allowed in each energy level. There are three orbitals designated *p* in each energy level. Each *p* orbital is shaped like a dumbbell. The *p* orbitals have greater energy than *s* orbitals within a particular energy level. Orbitals designated *d* and *f* have more complex shapes and still higher energy. Notice the shapes of *d* orbitals, shown in **Figure 3-30**.

As you can see below, orbitals (other than *s* orbitals) can be oriented in a number of different ways. A *p* orbital can exist in one of three possible orientations, for example. A third quantum number, the magnetic quantum number, is designated m_ℓ. The value of m_ℓ tells you the electron's position by designating the spatial orientation of the orbital that the electron occupies.

Electrons that have the same values of *n* and ℓ but not necessarily m_ℓ are said to be in the same **sublevel** of the atom. The 2*p* sublevel contains a maximum of three orbitals, 2*px*, 2*py*, 2*pz*. These orbitals are oriented along the *x*, *y*, and *z* axes respectively.

sublevel
one orbital or a group of orbitals within an energy level which have the same value of ℓ

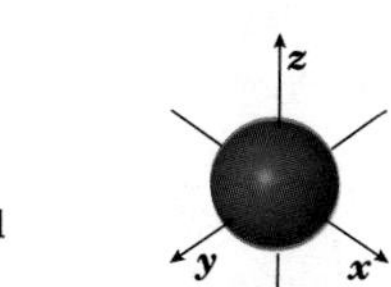

Figure 3-30a
The *s* orbital is shaped like a sphere.

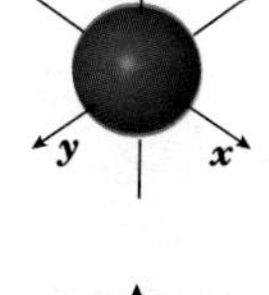

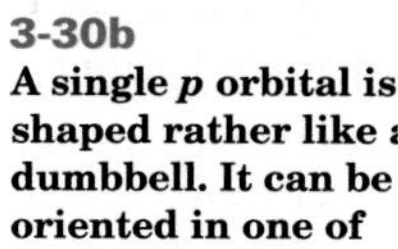

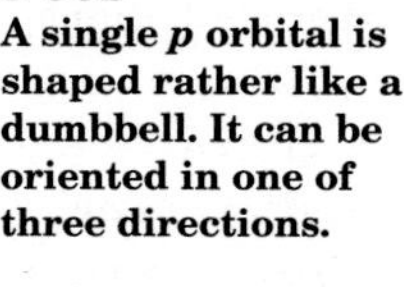

3-30b
A single *p* orbital is shaped rather like a dumbbell. It can be oriented in one of three directions.

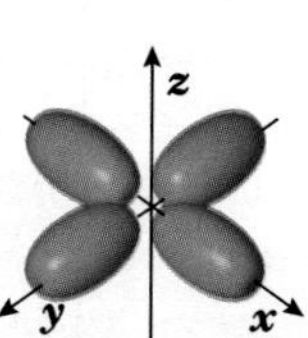

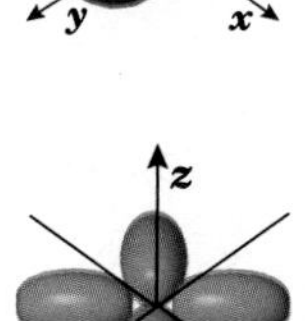

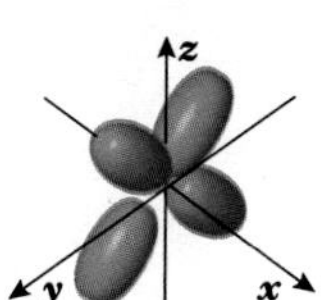

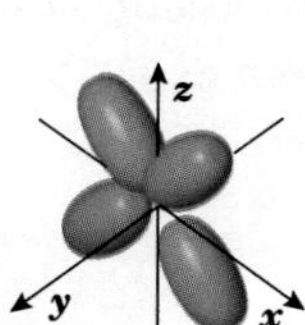

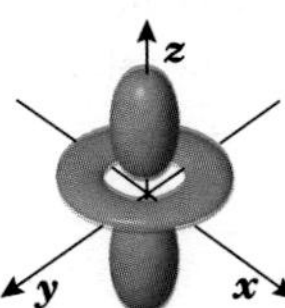

3-30c
A single *d* orbital can assume any of the forms represented here. Notice that four of the five possible *d* orbital shapes are the same, but they are oriented differently in space.

Teaching Tip
Analogy
The four quantum numbers can be thought of as the specific address of the electron in which four pieces of information are needed to specify where the electron is located—for example the town, street, block, and house number.

Possible Misconception
Many students cannot see the three-dimensional nature of the *xyz* axes as they are drawn on the board. Use three meter sticks held in the center with one hand to show how the *z*-axis is coming out of the board.
This is also a good time to show that the p_x, p_y, and p_z orbitals are different because they have different locations in space.

Visual Strategy
Figure 3-30
Make students aware that the *p* orbital can exist on any of the three axes, and as the *p* fills up, it will exist on all three axes. Remind the students that these shapes are based on plots of the probable locations of electrons in atoms.

Teaching Tip

Analogy

Students are sure to ask what happens if an electron enters an orbital that already contains an electron spinning in the same direction. The answer is that one of the electrons will change its spin direction. Remind students that spin is relative to your location. If you have a glass table and spin a basketball on it, the view from above the ball would be opposite of the view from below the ball. Therefore, merely rotating the electron 180° will reverse the spin.

Theme

Classification and trends

Because the periodic table is organized along the lines of energy sublevels, it is possible to predict the orbital configuration of an element from the periodic table.

Table 3-6
Number of Electrons Accommodated in Electron Energy Levels and Sublevels

Principal energy level	Sublevels available	Number of orbitals in sublevel $(2\ell + 1)$	Number of electrons possible in sublevel $[2(2\ell + 1)]$	Total electrons possible for energy level $(2n^2)$
1	*s*	1	2	2
2	*s*	1	2	8
	p	3	6	
3	*s*	1	2	18
	p	3	6	
	d	5	10	
4	*s*	1	2	32
	p	3	6	
	d	5	10	
	f	7	14	
5	*s*	1	2	50
	p	3	6	
	d	5	10	
	f	7	14	
	*g**	9	18	
6	*s*	1	2	72
	p	3	6	
	d	5	10	
	*f**	7	14	
	*g**	9	18	
	*h**	11	22	

*These orbitals are not used in the ground state of any known element.

To describe the motion of an electron, there is a fourth quantum number, m_s, called the spin quantum number. This number labels the orientation of the electron. Electrons in an orbital spin in opposite directions. These directions are arbitrarily designated as + 1/2 and –1/2.

No two electrons have an identical set of four quantum numbers. This statement, first made by the German chemist Wolfgang Pauli, is appropriately referred to as the Pauli exclusion principle. The Pauli exclusion principle ensures that no more than two electrons can be found within a particular orbital. There is also a maximum number of electrons allowed in each sublevel and principal energy level. **Table 3-6** summarizes the possible sublevels and orbitals that can be found in each principal energy level. Also note that the maximum numbers of electrons in orbitals, sublevels, and principal energy levels can be calculated from expressions involving quantum numbers.

So far you have learned that the way to designate a single electron in an atom is to use its quantum numbers. All elements except hydrogen contain more than one electron, as illustrated in **Figure 3-31**. To represent all of the electrons in an atom, you can use the Pauli exclusion principle along with quantum numbers to write an overall electron configuration for the element.

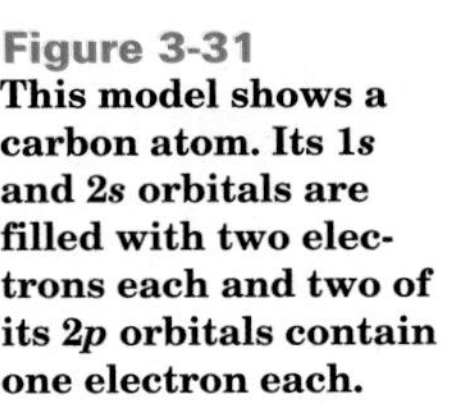

Figure 3-31
This model shows a carbon atom. Its 1*s* and 2*s* orbitals are filled with two electrons each and two of its 2*p* orbitals contain one electron each.

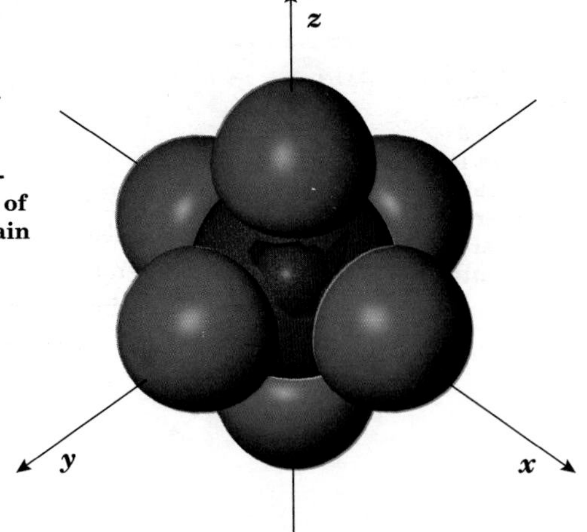

Visual Strategy

Figure 3-31

Be sure students understand that an orbital exists, as a physical entity, only when it is occupied by an electron.

Orbital diagrams and electron configurations are models for electron arrangements

Orbital diagrams are used to show how electrons are distributed within sublevels and to show the direction of spin. In an orbital diagram, each orbital is represented by a box, and each electron is represented by an arrow. The direction of spin is represented by the direction of the arrow. The orbital diagram for hydrogen, which contains one electron, would be represented as follows.

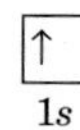

An electron configuration is an abbreviated form of the orbital diagram. The electron configuration for hydrogen would be $1s^1$. The coefficient 1 indicates that the sole electron occupies the first energy level, and the superscript 1 indicates that one electron occupies the *s* orbital within that first energy level.

How do you write the orbital diagram and electron configuration for boron? The atomic number of boron is 5. It has 5 electrons; this means there are electrons in the first and second energy levels. There are two electrons in the first ($n = 1$) principal energy level, both in an *s* orbital. There are three electrons in the second principal energy level, two in an *s* orbital and one in a *p* orbital. The electron configuration for boron can be written $1s^2 2s^2 2p^1$. The orbital diagram can be written as follows.

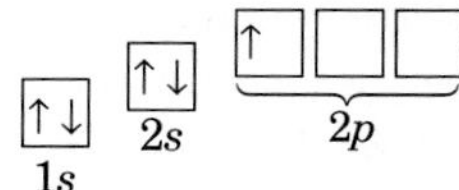

Neon, with 10 electrons, could be represented with the electron configuration $1s^2 2s^2 2p^6$ or with the following orbital diagram.

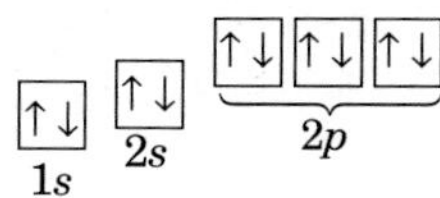

Hund's rule
the most stable arrangement of electrons is that with the maximum number of unpaired electrons, all with the same spin direction

Electrons are arranged in accordance with **Hund's rule**, which states that orbitals of equal energy are each occupied by one electron before any pairing occurs by adding a second electron. In this way, repulsion between electrons in a single orbital is minimized. All electrons in singly occupied orbitals must have the same spin. When two electrons occupy an orbital they have opposite spins.

You can write electron configurations yourself as long as you keep in mind the maximum number of electrons allowed in each orbital and the order in which orbitals are filled. For an atom in the ground state, electrons fill the orbitals beginning with the lowest energy orbital before filling orbitals with successively higher energies.

Teaching Tip

Symbolic representations

Give students lots of practice in drawing orbital diagrams to represent electron configurations. Once they master these pictorial representations, they will be more successful in writing and interpreting the shorthand forms of electron configurations.

Teaching Tip

Analogy

Students are often confused why the $4s^2$ orbital comes before the $3d^{10}$ orbital. An analogy can be used to explain this. Let principal quantum numbers be apartment floors with angular quantum numbers being the style of apartment. The *s* is a simple apartment, the *p* is a plush apartment, and the *d* is a deluxe apartment. A deluxe apartment is fancier than a plush apartment; therefore, it costs more energy dollars to live in a *d* apartment than a *p* or *s* apartment. It also costs more energy dollars to live on the higher floors of the building. What occurs on the fourth floor is that although the 4*s* apartment is on a higher floor, it is a very simple apartment. It costs more energy dollars to live in the fancier 3*d* apartment. Because we are concerned with saving energy dollars, the 4*s* apartment will be leased before the more lavish 3*d* apartments.

Additional Examples

Have students determine the electron configuration for ${}_{9}F$, ${}_{17}Cl$, ${}_{53}I$, and ${}_{35}Br$. They should note that all of the electron configurations terminate in the p^5.

F, $1s^22s^22p^5$

Cl, $1s^22s^22p^63s^23p^5$

I, $1s^22s^22p^63s^23p^64s^23d^{10}4p^65s^2 4d^{10}5p^5$

Br, $1s^22s^22p^63s^23p^64s^23d^{10}4p^5$

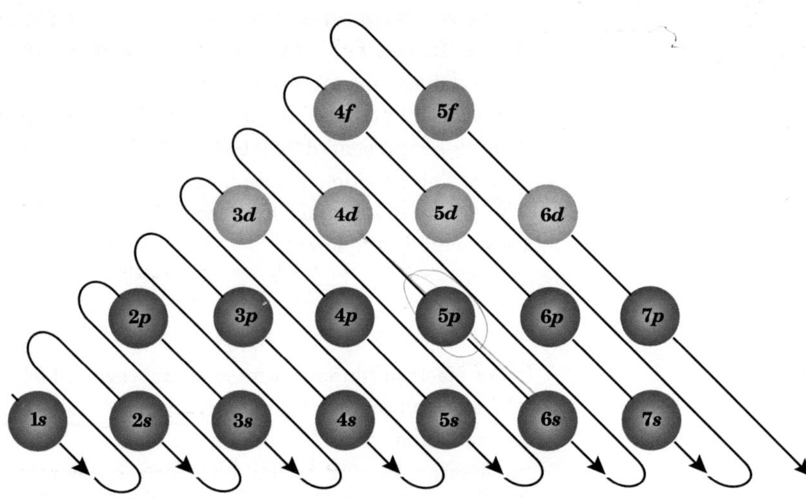

Figure 3-32 **Electrons fill orbitals in the order indicated by the arrow. Note that there is some overlap of energy levels. The 3*d* orbitals actually have higher energy than the 4*s* orbital and the 7*s* orbital has lower energy than the 6*d* orbital, for example.**

Figure 3-32 shows a sequence model that can be used to write electron configurations. Notice that principal energy levels do not always fill in order. You can see in **Figure 3-33**, which shows the energy distribution of orbitals, that there are areas where principal energy levels overlap. Notice that even though the 3*d* orbital is in a lower principal energy level, it is higher in energy than the 4*s*. Therefore a 4*s* orbital generally fills before a 3*d* orbital.

The electron configuration for gallium, $1s^22s^22p^63s^23p^64s^23d^{10}4p^1$, shows why you need to understand **Figure 3-32** and **Figure 3-33** to write its orbital diagram.

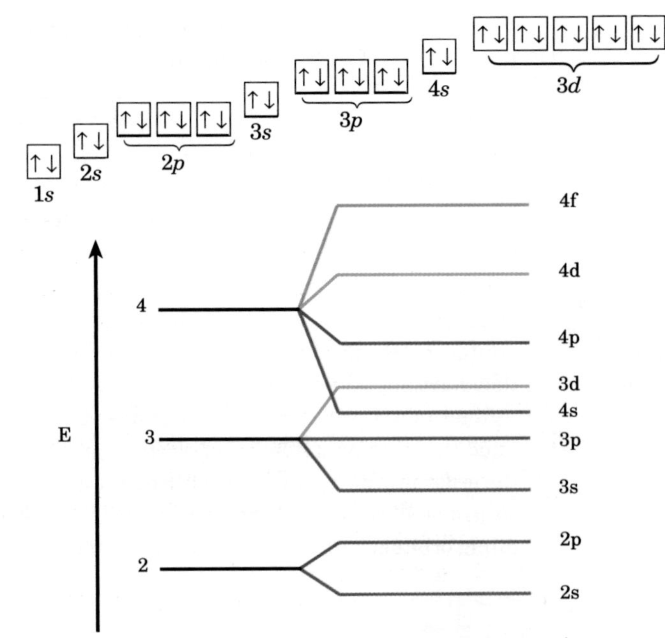

Figure 3-33 **Energy generally increases as *n*, the principal energy level, increases. It is possible, however for the lowest sublevel of *n* = 4 to be below the highest sublevel of *n* = 3. This occurs in potassium and calcium atoms, where successive electrons enter the 4*s* rather than the 3*d* sublevel.**

Visual Strategy

Figure 3-32

The diagram representing order of orbital filling is a very useful generalization though there are a few exceptions, including copper and chromium. In both these cases, only one electron occupies the 4*s* orbital before the 3*d* level begins to fill. The ground state electron configuration of chromium is $1s^22s^22p^63s^23p^63d^54s^1$. The ground state electron configuration for copper is $1s^22s^22p^63s^23p^63d^{10}4s^1$.

How To
Write an electron configuration

1. *Locate the element whose electron configuration you wish to write in the periodic table.*
For example, locate oxygen. Use the atomic number to determine the number of electrons in one atom. Oxygen has eight electrons.

2. *Fill orbitals in the proper order with electrons.*
Keep in mind which sublevels and orbitals are found in each energy level. This is described in **Table 3-6**. Keep filling orbitals using the order shown in **Figure 3-32** until you have placed all the electrons of the element whose configuration you are writing. The electron configuration for oxygen is as follows.

$$1s^22s^22p^4$$

3. *Check that the total number of electrons in the electron configuration equals the atomic number.*
For oxygen, the total number of electrons shown in the electron configuration is 2 + 2 + 4, or 8. This is equal to the atomic number of oxygen.

Electron configurations can be written in terms of noble gases

The electron configuration for every element can be found on the periodic table on pages 110–111. Notice that the configurations are abbreviated to save space even further. For example, the electron configuration for sodium, Na, could be written as $1s^22s^22p^63s^1$. On the periodic table, you can see the configuration for sodium is represented as follows.

$$[Ne]3s^1$$

The sodium block is the first one in the table that shows filling of the 3*s* orbital. The preceding noble gas, neon, ends the row in the previous period by completely filling the 2*p* orbital. You can use the symbol for neon to represent all the filled orbitals through 2*p*. When you express an element's electron configuration in terms of a noble gas, you are using the same abbreviated notation shown on the periodic table.

Section Review

15. Determine the number of electrons, protons, and neutrons in each of the following: $^{235}_{92}U$, $^{106}_{46}Pd$, and $^{133}_{55}Cs$.
16. What do the quantum numbers n, ℓ, m_ℓ, and m_s represent?
17. Write the electron configurations for phosphorus and nickel. Then draw the orbital diagrams for these elements.
18. Write the complete electron configurations for magnesium, sulfur, and potassium. Then write their electron configurations using the symbols for the noble gases.
19. What element is represented by $[Ne]3s^23p^6$? What does it have in common with Ne?

Groupwork Strategy
Divide students into groups of three, and have them determine the electron configuration and electron orbital notation for the following atoms: Sc, K, P, B, and Ce.
Sc, $1s^22s^22p^63s^23p^64s^23d^1$
K, $1s^22s^22p^63s^23p^64s^1$
P, $1s^22s^22p^63s^23p^3$
B, $1s^22s^22p^1$
Ce, $1s^22s^22p^63s^23p^64s^23d^{10}4p^6$ $5s^24d^{10}5p^66s^24f^2$

Additional Examples
Write the abbreviated noble gas electron configuration for following atoms: Ca and Sn.
Ca is $[Ar]3s^2$; 18 + 2 = 20
Sn is $[Kr]5s^24d^{10}5p^2$; 36 + 2 + 10 + 2 = 50

ASSESS

Alternative Assessment
Give students electron configurations for a few relatively simple atoms such as lithium, boron, and oxygen. Students should draw a model of each atom, showing orbitals present, their shapes, and their relative orientations.

Answers to Section Review
15. $^{235}_{92}U$ has 92 protons and electrons and 143 neutrons; $^{106}_{46}Pd$ has 46 protons and electrons and 60 neutrons; $^{133}_{55}Cs$ has 55 protons and electrons and 78 neutrons.
16. n represents the energy level or distance from the nucleus; ℓ represents the shape of an orbital; m_l represents the orientation of an orbital; m_s represents electron spin.
17. *See page 105A.*
18. Magnesium: $1s^22s^22p^63s^2$ or $[Ne]3s^2$
Sulfur: $1s^22s^22p^63s^23p^4$ or $[Ne]3s^23p^4$
Potassium: $1s^22s^22p^63s^23p^64s^1$ or $[Ar]4s^1$
19. Argon; both neon and argon are noble gases

Chemistry In Your Community

Your students can investigate the field of pyrotechnics by performing the following activities.

1. Find out about the local laws and regulations governing the use of fireworks.
2. Determine whether fireworks manufacturers or other pyrotechnical firms exist in your area. If so, invite a scientist or technician from the firm to visit your class as a guest lecturer.
3. Obtain fireworks sold legally in your area. Bring them to class and allow students to read the labels. Based on the list of ingredients or advertised effects, have students hypothesize about what chemical elements the fireworks contain. Are the warning labels sufficient to protect people from the potential safety hazards of using these materials? Does the Consumer Products Safety Commission regulate fireworks?

Answers to Applying Concepts

1. yellow requires sodium compounds; green requires barium compounds; and red requires strontium compounds
2. They emit different amounts of energy when the boosted electrons fall back to the ground state. This energy corresponds to light of specific wavelengths.

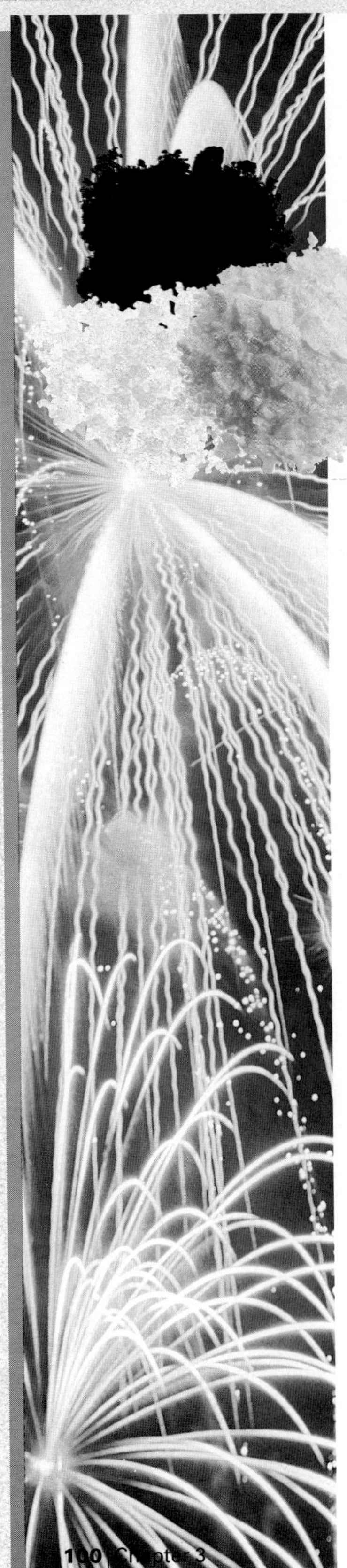

Conclusion: Excited Atoms and the Fourth of July

What is light, and how do various colors of light differ? Visible light consists of electromagnetic radiation having wavelengths between 3.8×10^{-7} m and 7.8×10^{-7} m. The longest visible waves appear red, shorter waves appear blue, and the shortest visible waves are violet.

What is going on at the level of atoms and molecules when fireworks produce colored light? The colors are generated either by atoms or molecules present in the gas state when a firework shell explodes. In the former case, *atomic emission* occurs: atoms are excited so that their electrons are bumped from the ground state to a higher energy state. The electrons quickly return to the ground state, shedding the energy they had absorbed as light of a specific wavelength. Sodium is a potent atomic light emitter. Sodium atoms heated to more than 1800°C give off yellow light with a wavelength of 5.89×10^{-7} m. The process is so efficient that it tends to overwhelm other atomic or molecular light emissions in a fireworks explosion. Yellow fireworks explosions are the easiest to produce and, consequently, are very popular.

As with atomic emission, *molecular emission* involves a transition from a ground state to an excited one. Gaseous molecules radiate light when they are heated to a temperature high enough to reach an excited state; the electrons in their atoms must absorb energy to receive an energy boost. Then electrons fall to a lower energy state, and light of a particular wavelength is emitted.

How does the instability of copper chloride at high temperatures interfere with its ability to emit blue light? Copper chloride is unstable—it decomposes into copper and chlorine atoms—at a temperature only slightly higher than that at which it emits light. Therefore, if the temperatures generated by the exploding shell and reacting chemicals are not precisely controlled, copper chloride crystals absorb so much heat energy that they break apart and never emit blue light.

Applying Concepts

1. Fireplace logs that burn various colors are available in many supermarkets. How might a manufacturer treat a log so that it burns with a bright yellow flame? a green flame? a red flame?
2. Why do different elements produce light of different colors?

Research and Writing

Use the library to find out more about the following.

1. By what process do hand-held sparklers emit light? How does this process differ from atomic and molecular emission?
2. Describe three applications of pyrotechnic technology, other than fireworks.

CHAPTER 3

Highlights

Key terms

anode
atomic number
cathode
electromagnetic spectrum
energy level
excited state
ground state
group
Hund's rule
isotope
mass number
metal
metalloid
noble gas
nonmetal
nucleus
orbital
period
quantum theory
sublevel

Key ideas

3-1 How are the elements organized?

- The periodic table organizes all known elements according to atomic structure so that elements with similar properties fall within the same group.
- The periodic table can be divided into four regions of elements with similar characteristics: metals, metalloids, nonmetals, and noble gases.

3-2 What is the basic structure of an atom?

- Three laws support the existence of the atom: the law of definite composition, conservation of mass, and multiple proportions.
- Atoms consist of particles with positive and negative charges. The positive charge is confined to a central nucleus that accounts for most of the atom's mass but has a radius 10000 times smaller than that of the whole atom.
- The nucleus also contains neutrons, particles with no charge.
- Electrons, particles with a negative charge and a mass 2000 times smaller than the mass of hydrogen, travel about the nucleus. Discrete amounts of energy are released or absorbed when electrons move from one energy level to another.
- Quantum theory describes the wave nature of electrons.

3-3 How do the structures of atoms differ?

- The atomic number is the number of protons contained in each atom of that element. The mass number is the total number of protons and neutrons in an atom.
- A unique set of four quantum numbers describes each electron in an atom. The Pauli exclusion principle states that no two electrons can have the same four quantum numbers.

Key problem-solving approach: *Determining electron configurations*

Determine the number of electrons in the atom. Then, use the diagram on the right to determine the order for filling. Finally, fill the orbitals using the energy level diagram at the bottom as a guide.

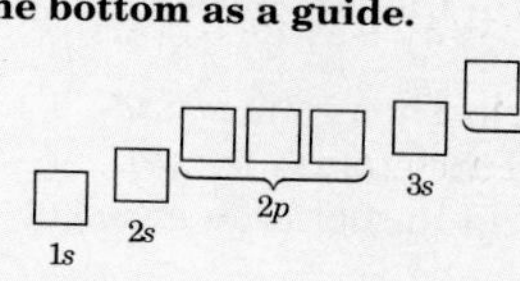

Number of Electrons

Determine the Order of Filling Orbitals

Closure Strategy

Present students with a written quiz.

1. Using a periodic table: name 1 metal, 1 nonmetal, 1 metalloid, and 1 noble gas; name 3 members of the same group; and 3 members of the same period.
2. List 5 major findings or events dealing with atomic theory that occurred between the time of Dalton's and Heisenberg's findings.
3. Determine the electron configuration for the last electron of the following elements: S, Pt, Sr, K, and Al.

1. The table is organized to follow the periodic law: the properties of the elements are a function of increasing atomic number.

2. a. The four regions include metals, nonmetals, metalloids, noble gases.
   b. Metals are conductors of heat and electricity. They are generally shiny and gray. Nonmetals are poor conductors of heat and electricity. Metalloids have conductivity properties between that of metals and nonmetals, hence they are described as semiconductors. Noble gases have low reactivity.

3. Many elements have symbols that are derived from the name of the element in another language.

4. aluminum, Al, metal; carbon, C, nonmetal; fluorine, F, nonmetal; iron, Fe, metal; lead, Pb, metal; silver, Ag, metal

5. a. It would be in the metalloid region.
   b. The element is germanium because it and silicon are the only metalloids in Group 14.

6. Carbon is a nonmetal, and lead is a metal.

7. a. metalloid
   b. nonmetal
   c. noble gas
   d. metal

8. Tin is a metal that will conduct heat from the drink and transfer it to the surroundings. The ceramic cup has nonmetallic properties and acts as a better insulator to keep heat within the cup.

9. The law of definite composition explains this.

10. Because ibuprofen is a compound, it will have the same composition no matter where it is manufactured.

CHAPTER 3

Review and Assess

Classifying the elements

REVIEW

1. What is the basis for the organization of the periodic table?
2. a. What are the four regions of the periodic table?
   b. Describe the general characteristics of each region.
3. Propose a reason why sodium, Na; copper, Cu; and lead, Pb; have symbols that do not relate directly to their names.
4. Complete the following table.

Element	Symbol	Region of the table
aluminum		
	C	
fluorine		
iron		
	Pb	
	Ag	

APPLY

5. One substance used in dental alloys is grayish white, shiny, and a poor conductor of electricity.
   a. In what region of the periodic table would this element be found?
   b. If this element has chemical properties similar to those of silicon, what is its name?
6. Carbon is the chemical basis for all life on the Earth. Lead is a poison that can cause muscle deterioration and brain damage. Although these two elements are in the same group, some of their properties are very different. Propose a reason for this.
7. Identify the region of the periodic table that would include an element suitable for the following uses.
   a. An element that slightly conducts electricity and is used to make the diodes that allow you to run a battery-powered radio from a wall socket
   b. A highly reactive gas that helps to purify drinking water
   c. A highly unreactive gas enclosed in lightbulbs to prevent the filament from evaporating
   d. An electrical conductor that connects the electrodes in a hearing aid to its battery
8. Explain why a cold soft drink would warm up to room temperature much faster in a tin cup than in a ceramic cup, made mostly of SiO_2.

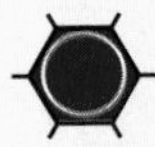

Atoms in compounds

REVIEW

9. Identify the law that explains why the water in a raindrop falling on Phoenix, Arizona, and the water flowing through the Nile Delta in Egypt both contain two hydrogen atoms for every oxygen atom.

APPLY

10. Ibuprofen, $C_{13}H_{18}O_2$, that you take for headaches is manufactured in Michigan and contains 75.69% carbon, 8.80% hydrogen, and 15.51% oxygen. If you are vacationing in Europe, how do you know that the ibuprofen you buy at a store in Munich, Germany, is the same as the chemical you buy at home?
11. Relate the law of definite composition to the law of multiple proportions.

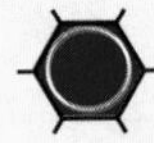

Models of the atom

REVIEW

12. a. What flaws exist in Dalton's model of the atom?
    b. What flaws exist in Thomson's plum pudding model of the atom?
    c. What flaws exist in Rutherford's model of the atom?

d. What flaws exist in Bohr's model of the atom?
e. Do any flaws exist in the modern quantum model of the atom? Explain.

13. For each letter in the diagram below describe what happens to the alpha particles. Explain what can be inferred about the structure of the atom from each instance.

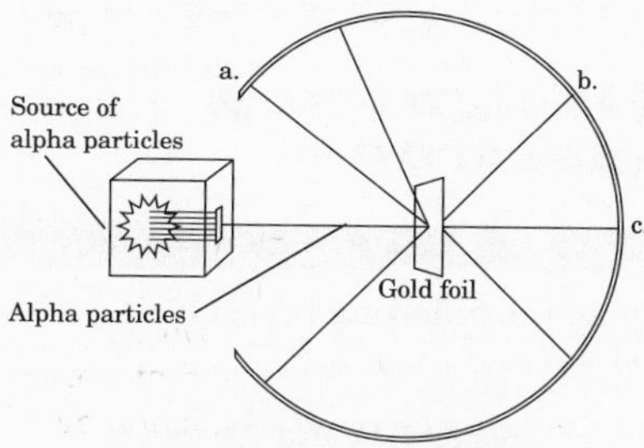

14. a. Which of Dalton's principles was contradicted by the work of J. J. Thomson?
b. Which of Dalton's principles was contradicted by the bombs that were dropped on Hiroshima and Nagasaki, Japan?
c. Which of Dalton's principles is contradicted by a doctor using radioactive isotopes to trace chemicals in the body?
d. Do any of Dalton's principles still hold completely true today? If so, which ones hold true?

15. Identify the models illustrated below.

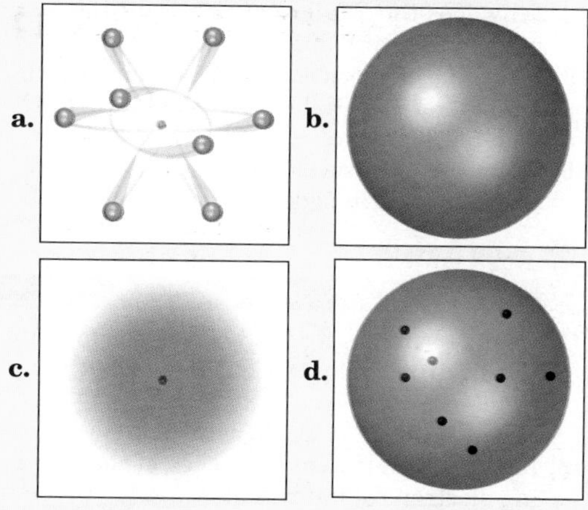

16. Which of the following did Rutherford's experiments suggest?
a. Atoms contain negatively charged electrons.
b. Electrons travel around the nucleus.
c. Atoms contain a positively charged nucleus.
d. Alpha particles exist.

17. What is the relationship between bright-line emission spectra and Bohr's atomic model?

18. Is it possible to verify the current model of the atom by direct observation? Explain.

APPLY

19. What atomic model explains why light from excited gases gives off a bright-line spectrum?

20. If a lightning bolt strikes your power lines while your color television is running, your screen may become magnetized, causing the color to become unbalanced. This problem can be fixed with a magnetic instrument known as a degausser.
a. Refer to the diagram below to explain how you know that the ray creating the picture on screen is composed of charged particles.
b. How is this related to the development of the plum-pudding atomic model?

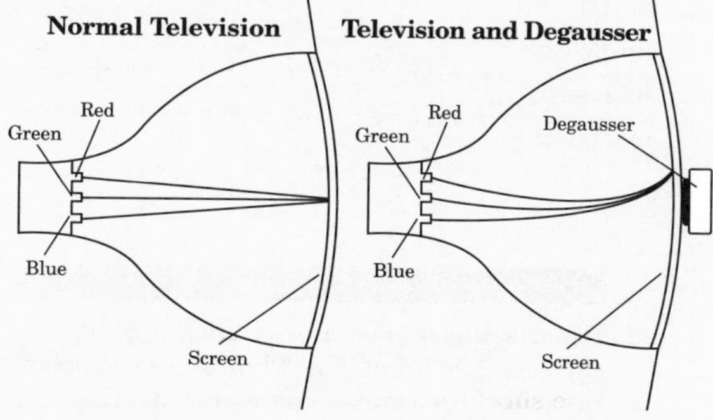

21. Regarding his famous gold foil experiment, Rutherford said, "It is about as incredible as if you had fired a fifteen-inch shell at a piece of tissue paper, and it came back and hit you." Why was Rutherford so amazed by the results of his experiment?

22. How were the atomic models developed when no one had seen the atom?

23. How are the frames that run through a movie projector similar to the quantum concept?

11. Each different compound conforming to the law of multiple proportions has a definite composition.

12. a. The atom is not solid, and it is composed of subatomic particles.
b. There actually is a positive nucleus, and the electrons move.
c. Some force must keep the electrons from colliding with the nucleus.
d. Electrons do not orbit the nucleus in fixed paths.
e. The modern quantum model is a theory and flaws could exist in the model.

13. a. The alpha particle struck a positive nucleus and bounced back.
b. The alpha particle passed close to a positive nucleus and was deflected slightly.
c. The alpha particle passed through the empty space of the atom.

14. a. Thomson contradicted Dalton's idea that atoms cannot be subdivided.
b. The development of bombs shows that the atom is destructible and divisible.
c. Radioactivity shows that atoms of a given element are not all identical in their physical and chemical properties.
d. Yes. Atoms of different elements have different physical and chemical properties, and atoms of different elements combine in simple, whole-number ratios to form chemical compounds.

15. a. Bohr
b. Dalton
c. quantum
d. Thomson

16. c

17. Each bright line represents a quanta of energy released when electrons move to a lower level in Bohr's model.

18. There is currently no proof because we are unable to see the actual interior of an atom.

19. Only the Bohr model and the quantum mechanical model explain the bright-line spectrum because the other models do not consider the quantum concept.

20. a. Beams of charged particles will be deflected by a magnet.
b. J. J. Thomson's experiments showed that a beam of electrons will bend in the presence of a magnet.

21. The gold foil was so thin, Rutherford thought that the high powered beam of alpha particles would penetrate the foil completely. However, the gold foil not only deflected the particles, but bounced some of them back completely.

22. Models of the atom were developed by observing the behavior of atoms at the macroscopic level, then developing explanations for these observations. This lead to making inferences about the structure.

23. A quantum is a specific amount of energy. The movie appears continuous because the individual frames are of a specific size and they run through the projector at a specific rate. The movie could be considered quantized because the frames are all the exact same size.

24. Quantum mechanics allows us to describe electrons by their wave and particle characteristics.

25. Electrons are located by quantum numbers.

Parts of the atom

REVIEW

24. Why is it necessary to use quantum mechanics to describe electrons?
25. How is the location of an electron within an atom described?
26. Write nuclear symbols for the isotopes of uranium with the following numbers of neutrons.
a. 142 neutrons
b. 143 neutrons
c. 146 neutrons
27. Three isotopes of oxygen are listed below. Identify how many protons, neutrons, and electrons are in each atom.
a. $^{16}_{8}O$ b. $^{17}_{8}O$ c. $^{18}_{8}O$
28. Complete the table below.

Isotope	Number of protons	Number of electrons	Number of neutrons	Atomic number
carbon–12				
carbon–14				
tin–119				
tin–120				
lithium–7				
sodium–23				

APPLY

29. Use the diagram below to indicate what each component of J. J. Thomson's cathode-ray tube illustrates about electrons. Explain your answer.

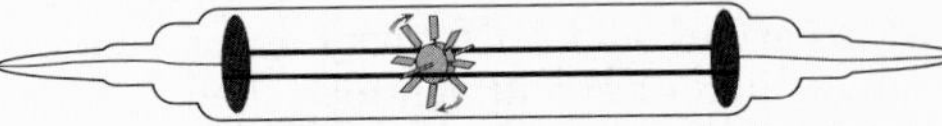

30. How is an atom like a city?
31. How does Heisenberg's uncertainty principle make it impossible to know completely the exact structure of an atom?
32. a. Differentiate between an orbit and an orbital.
b. How do these words relate to the correct description of electrons in an atom?
33. What would happen to poisonous chlorine gas if the following alterations were made?
a. A proton is added to each atom.
b. An electron is added to each atom.
c. A neutron is added to each atom.

34. In the diagram on the right, indicate which subatomic particles would be found in areas **a** and **b**.

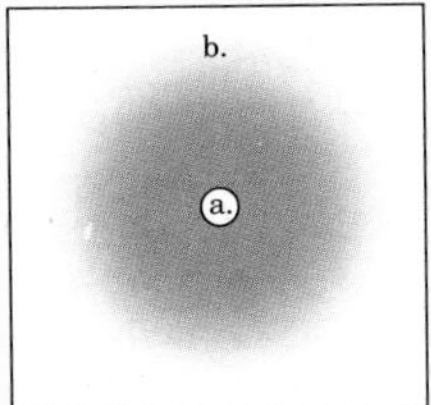

Designating electrons

REVIEW

35. Complete the following table about quantum numbers.

Type	Symbol	Values	Description
principal			
orbital shape			
magnetic			
spin			

36. What do the electron configurations of the noble gas elements have in common?
37. The element sulfur has an electron configuration of $1s^22s^22p^63s^23p^4$.
a. What does the superscript *6* refer to?
b. What does the letter *s* refer to?
c. What does the coefficient *3* refer to?
38. Why does the Pauli exclusion principle include the word *exclusion*?
39. Write the electron configuration for calcium, a nutrient essential to healthy bone growth and development.
40. Write the electron configuration for copper which is used in pennies.

APPLY

41. Use the symbols for the noble gases to write the electron configurations for the following elements.
a. Zr b. U c. Rn
42. Identify the element with each of the following electron configurations.
a. $1s^22s^22p^2$
b. $1s^22s^22p^63s^23p^1$
c. $1s^22s^22p^63s^23p^64s^23d^{10}4p^65s^14d^{10}$
43. Why are noble gases used to abbreviate electron configurations?
44. How are quantum numbers like an address?

Linking chapters

1. **Theme: *Conservation***
A Japanese company recently developed a signboard that uses no wiring and no power source. The characters are illuminated by absorbing surrounding light and then radiating different colors. Use the concept of a quantum to explain how this sign can emit light with no power source.

2. **Theme: *Micromodels***
Explain the different macroscopic observations that led to the following.
a. the discovery of the nucleus
b. the quantum concept
c. the discovery of electrons

3. **Theme: *Classification and trends***
Scientists are continually attempting to create new elements. If more elements were created past number 109, would they all be metals? Explain.

USING TECHNOLOGY

1. ***Computer art***
Use a computer art program to illustrate the orbitals in electron configurations using the diagram on page 98 as a model. Draw and write the configurations for C, Na, and Ti.

2. ***Computer program***
Write a spreadsheet program that will sum up the exponents for any electron configuration you write.

Alternative assessment

Performance assessment

1. Your teacher will provide you with ionic substances containing the following ions: Na^+, Li^+, Be^{2+}, Mg^{2+}, K^+, Ca^{2+}, and Sr^{2+}. Develop criteria to judge the identity of the different ions. Show your criteria and safety precautions to your teacher for approval.
2. Use the safety appendix to write the necessary precautions for performing flame tests and for working with open flames. Your teacher will give you an unknown ion to identify.
3. Organize the characteristics of metals, nonmetals, metalloids, and noble gases on a poster. Your teacher will provide you with several unknown elements. Classify them by the characteristics noted on your poster.

Portfolio projects

1. ***Research and communication***
Research several elements that have a symbol that is inconsistent with their English name. Some examples include silver (Ag), gold (Au), and mercury (Hg). Compare the reasoning for these names to the reasoning for symbols and names that coincide.
2. ***Cooperative activity***
Build your own full-scale model of a particular isotope of an atom. Be sure to use a wide open area for this project. Look up the radius of the nucleus and atom, and build the atom with relative sizes. Discover how far away another atom would be if the atoms existed in the solid state.
3. ***Chemistry and you***
Heisenberg's uncertainty principle applies to atomic-sized systems. However, it reminds us that uncertainty exists elsewhere. Write a description of instances in your life when major uncertainty existed. Could uncertainty have been avoided in any of these situations?
4. ***Research and communication***
"Neon" signs actually use substances other than neon, which provides a bright red light. Research the different colors used for these signs. Determine the chemicals that are utilized. Design your own sign, and identify which gases you would use to achieve the desired color scheme.

26. a. $^{234}_{92}U$
b. $^{235}_{92}U$
c. $^{238}_{92}U$

27. a. 8 protons, 8 electrons, and 8 neutrons
b. 8 protons, 8 electrons, and 9 neutrons
c. 8 protons, 8 electrons, and 10 neutrons

28. 6, 6, 6, 6
6, 6, 8, 6
50, 50, 69, 50
50, 50, 70, 50
3, 3, 4, 3
11, 11, 12, 11

29. The paddle wheel shows that electrons have a particle nature because they move an object when they hit it.

30. Answers will vary. Possible answer: Most of the population of a city is located near the center. As one travels away from the city the population density is reduced. This is similar to the mass of an atom being concentrated in the center, while the probability of finding an electron decreases as distance from the nucleus increases.

31. Determining the position of an electron would require hitting it with a heavier particle. The impact of the heavier particle would alter the speed and position of the electron and would not then give an accurate reading of position or velocity. Heisenberg tells us that there is always some degree of uncertainty in our measurements regarding the position and velocity of the electron.

Continued from page 99.

17. Phosphorus: $1s^22s^22p^63s^23p^3$

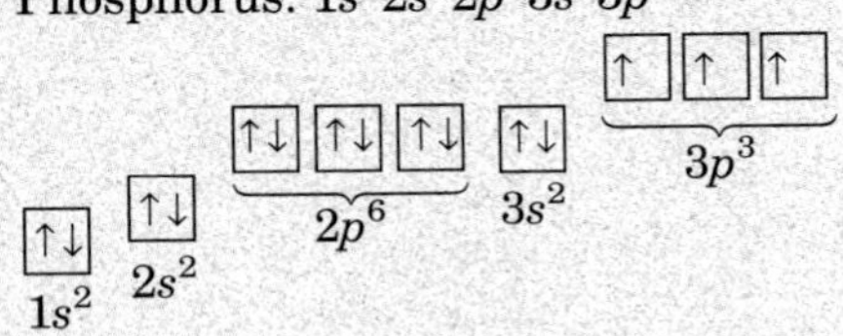

Nickel: $1s^22s^22p^63s^23p^63d^84s^2$

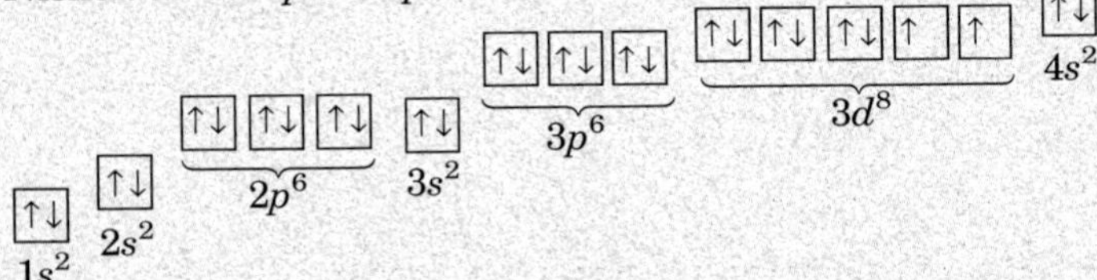

Continued from page 105.

32. a. An orbit is an exact path around something. An orbital is the high probability volume for a particular electron, not an exact path.
b. *Orbit* implies a knowledge of velocity and location, so it is an incorrect to be used in describing the movement of electrons. *Orbital* describes the probability of location and velocity, and most accurately reflects our current model for the atom.

33. a. Chlorine would become argon gas.
b. Chlorine would form chloride ions.
c. Chlorine would form a different isotope of chlorine.

34. a. protons and neutrons
b. electrons

35. a. n; positive integers beginning with 1; energy level
b. ℓ; *s, p, d,* or *f*; type of orbital
c. m_ℓ; spatial orientation; designating spatial orientation
d. m_s; $+1/2$ or $-1/2$; direction of electron spin

36. Each noble gas electron configuration ends with a filled *p* orbital.

37. a. the number of electrons in a sub level
b. the type of orbital
c. the energy level

38. Electrons are excluded from an orbital that already contains an identical electron.

39. $1s^22s^22p^63s^23p^64s^2$

40. $1s^22s^22p^63s^23p^64s^13d^{10}$

41. a. [Kr] $5s^24d^2$
b. [Rn] $7s^26d^15f^3$
c. [Rn]

42. a. carbon
b. aluminum
c. silver

43. The noble gases have filled energy levels and fall at the end of a completed period.

44. The quantum numbers of an electron describe the location of an electron like an address describes the location of a person's residence.

Linking chapters

1. The light shining on the sign excites the electrons in the material. When these electrons return to the ground state, they release the quantum of energy that gives the sign light.

2. a. Alpha particles bounced off gold foil.
b. Bright line spectra were observed.
c. Cathode rays were observed.

3. No. Most would be metals, but some metalloids, nonmetals, and noble gases could appear depending on the atomic number of the new element.

USING TECHNOLOGY

1. Answers will vary. Be sure the students accurately show the *p* and *d* orbitals.

2. Answers will vary. Be sure students' programs use correct configurations.

Performance assessment

1. Answers will vary. Plans should describe the flame test procedure, using Exploration 3A on page 686 as a model. Plans should show adherence to safe laboratory procedures.

2. Answers will vary. Check for the appropriate listings of characteristics for each category. Note the criteria the student uses in classifying each unknown.

Portfolio projects

1. Answers will vary. The names are often foreign and use a characteristic properly to describe the element.

2. Answers will vary. Be sure that the scale for relative size of the parts of the atom is accurate.

3. Answers will vary. Some may include what locker they get, whether it will rain, the amount of food in a serving, or fries in an order. Students should conclude that although some uncertainty can be eliminated, much cannot.

4. Answers and designs will vary. Sodium vapor produces yellow, mercury vapor produces blue, and hydrogen produces pink.

PLANNING FOR

1
H
Hydrogen
1.00794
$1s^1$

4 Periodicity

Chapter Overview

- **Section 4-1** covers the development of the periodic table and the arrangement of elements with similar properties as groups within the periodic table. Each family of the periodic table is profiled.
- **Section 4-2** examines periodic trends relating to atomic size, ionization energy, electron affinity and electronegativity.
- **Section 4-3** discusses the origins of the chemical elements resulting from stellar nucleosynthesis, and the creation of artificial elements in particle accelerators.
- **Section 4-4** explains the concepts of average atomic mass and the mole. The fundamental conversions among moles, molar mass, and numbers of atoms are presented in the context of the three step method used throughout the book for solving problems.

Concept Base

Students must have mastered the following concepts prior to this chapter: using conversion factors, expressing results to the correct number of significant figures (Chapter 2); drawing and interpreting electron configurations (Chapter 3); naming and locating the particles of the atom (Chapter 3).

Looking Ahead

Chemical bonding is introduced in Chapters 5 and 6.

Laboratory Equipment Needs

Demonstrations

The listing of materials for Chapter 4 demonstrations begins on page T46.

Pacing: 7 days	Inquiry Teaching Strategies
4-1 What makes a family of elements?	**Demonstrations** ■ Demo 1 *Families of the periodic table* ■ Demo 2 *Orange-yellow light of sodium* ■ Demo 3 *Reactivity of magnesium and calcium*
4-2 What trends are found in the periodic table?	**Demonstrations** ■ Demo 4 *Size of an oleic acid molecule*
4-3 How are elements created?	**Demonstrations** ■ Demo 5 *Cloud chamber*
4-4 Can atoms be counted or measured?	

Key ■ Teaching Transparencies
■ Annotated Teacher's Edition
■ Laboratory Experiments Manual

INSTRUCTION

Visual Teaching Strategies / Review and Practice Strategies

Visual Teaching Strategies	Review and Practice Strategies
Transparency Masters ■ T 4-1 *Properties of Some Elements Predicted by Mendeleev* ■ T 4-2 *Periodic Table* ■ T A-6 *The Elements* ■ T A-7 *Properties of Common Elements*	**Text Reviews** ■ Section Review 1–5 ■ Chapter Review 1–17; LC 1, 3, 4 **Study Guide Worksheets** ■ ■ 4-1 Concept Review ■ ■ 4-1 Skillsheet *Exploring Electron Configuration and Periodicity*
Transparencies ■ T 4-5 *Atomic Radii* ■ F 4-23 *Stellar Nucleosynthesis: Formation of Carbon-12* **Transparency Masters** ■ F 4-15 *Trends in Atomic Radii* ■ F 4-17 *Trends in Ionization Energies* ■ T 4-7 *Summary of Periodic Trends*	**Text Reviews** ■ Section Review 6–10 ■ Chapter Review 18–28 **Study Guide Worksheets** ■ ■ 4-2 Concept Review ■ ■ 4-2 Skillsheet *Periodic Trends*
	Text Reviews ■ Section Review 11–15 ■ Chapter Review 29–36; LC 2 **Study Guide Worksheets** ■ ■ 4-3 Concept Review ■ ■ 4-3 Skillsheet *Nuclear Reactions*
Transparencies ■ P 142 *Solving Mole Problems*	**Text Reviews** ■ Additional Examples 4A, 4B, 4C, 4D, 4E, 4F ■ Practice 1–12 ■ Section Review 16–23 ■ Chapter Review 37–59; UT 1–3 **Study Guide Worksheets** ■ ■ 4-4 Skillsheet *Moles and Atomic Mass* ■ ■ 4-4 Practice *Average Atomic Mass*; *Conversions: Moles and the Number of Atoms*; *Conversions: Moles and Mass*

■ Pupil's Text
■ Study Guide
■ Teaching Resources
Exp *Exploration*
Inv *Investigation*
LC *Linking Chapters*
UT *Using Technology*

Assessment Options

Traditional Assessment

Test Generator
Instructional Objectives Measured:
Content mastery
Select from items 1 to 75

Performance Assessment Options

(see page T36 and T643-15 for scoring rubrics)
Students demonstrate mastery of objectives in a hands-on environment. You may want some of these materials in the Portfolio as well.

Performance Assessments, page 147

1. **Concept mastery**
   Instructional Objectives Measured:
   Content mastery
2. **Designing a flow scheme**
   Instructional Objectives Measured:
   Content mastery, Problem solving skills

Portfolio Options

(see page T36 for scoring rubrics)
Students provide a written rationale for each selection made for the portfolio.

Concept Maps
Students use vocabulary from the Chapter Review to build a concept map.
Instructional Objectives Measured:
Content mastery

Portfolio Projects, page 147
Items 1 and 4. Research and communication
Instructional Objectives Measured:
Content mastery, Scientific and chemical literacy, Problem solving skills, Proficiency in oral and written communication

Item 2. Cooperative activity
Instructional Objectives Measured:
Content mastery, Scientific and chemical literacy, Problem solving skills, Proficiency in oral and written communication

Item 3. Chemistry and you
Instructional Objectives Measured:
Scientific and chemical literacy, Problem solving skills, Proficiency in oral and written communication

INTRODUCING THE CHAPTER

CHAPTER 4

Periodicity

Story Background

This is a good time to discuss the high toxicity of mercury, and mercury poisoning. Mercury poisoning can occur not only with contact from the metal itself, but also by breathing mercury vapors, which was the case for Decker in the story. Contact with both organic and inorganic mercury compounds can also result in mercury poisoning. Methyl mercury compounds caused a large outbreak of poisonings in Japan in the 1950s among people who ate fish and shellfish containing high levels of mercury compounds. This type of poisoning has also been a concern for people who consume fish from the Great Lakes region of the United States.

Absorbed mercury is concentrated in the kidneys and ultimately results in kidney dysfunction. Death occurs from the accumulation of toxic wastes in the blood. Chronic mercury poisoning is characterized by tremors, numbness, inflammation in the mouth, loss of appetite, and mental depression.

While mercury can appear to be fun to play with, it is a deadly poison and should not be touched or left uncovered. Mercury spills must be cleaned up quickly and completely to prevent the release of vapors.

About the Illustration

Alloys and amalgams are mixtures of metals produced by heating the parts of the mixture to high temperatures. A similar process is used, in reverse, to separate the components of an alloy or amalgam, which is the subject of the story.

Periodic Trends and a Medical Mystery

George Decker seemed to be in perfect health. He was 35 years old, did not smoke, and had no history of serious illness. But one day after cooking something on his stove, Decker began to cough, wheeze, and gasp for air. His girlfriend rushed him to the hospital for treatment.

Tests revealed that Decker had pneumonia in both lungs. Pneumonia is an inflammation of lung tissue. Normally the lungs are as light and airy as cotton candy. Tiny air sacs called alveoli provide the lungs with oxygen-rich air. Pneumonia, however, causes the air sacs to become filled with white blood cells and cellular debris, making breathing difficult. Most pneumonias are caused by organisms such as bacteria, fungi, or viruses. But tissue samples taken from Decker's lungs showed no evidence of microorganisms. What caused Decker to develop pneumonia was a mystery.

Decker's condition became worse. The air sacs in his lungs continued to fill up with solid materials. As a result, his lungs thickened so much that they could not hold enough oxygen to sustain life. Eventually, Decker died.

A pathologist was called in to perform an autopsy. The autopsy revealed that Decker's lungs, as expected, were heavy and thick. But still no trace of pneumonia-causing microorganisms was found. Baffled, the pathologist scrutinized Decker's hospital records. What caught his attention was the fact that Decker had just finished cooking when he began having breathing problems. What Decker had been cooking—in fact what he had cooked many times before—turned out to be what caused the pneumonia.

By talking to Decker's employer, the pathologist discovered that Decker collected old gold and silver dental fillings. Both gold and silver dental fillings are not pure but are actually mixtures of metals. Gold fillings consist primarily of gold mixed with palladium. Silver dental fillings are composed of silver and mercury, with added copper or other metals. Decker would take the fillings home, place them in a pot, and cook them on his stove to extract the precious metals. The pure gold and silver could be sold for a sizable profit. Decker had been employed in a metal shop where his job was to pour molten aluminum into casts to form ingots. His experience at work provided him with the know-how to extract the gold and silver from dental fillings. But Decker's ignorance about certain physical and chemical properties of the metals he was working with proved fatal.

In this chapter, you will explore the elements, including the metals used in dental fillings, that form the periodic table. Then you will be able to answer the following questions.

- *What properties of elements can be determined from the periodic table?*
- *How do trends in the periodic table explain George Decker's death?*

Periodic trends are not only interesting and important to your study of chemistry, but as you will see, they can even be a matter of life and death!

Tapping ***Prior Knowledge***

Find out how much of the content covered in this chapter is familiar to students by having them describe what they know about an element by looking at its block on the periodic table. List the responses on the board, and have students identify any noticeable trends.

Section 4-1

FOCUS

Lesson Starter

Display examples of as many different kinds of fruit as possible. (Or search magazines for pictures of as many kinds of fruit as possible.) Divide the class into groups of three and have them classify the fruit at least three different ways. They need to write down each scheme and the justification for the choices. Pool the class results to show the different groupings and to determine whether items can fit in more than one family.

Demonstration 1

Families of the periodic table

Approximate time: 10 min

Obtain elements from each of the *s*- and *p*- block families and the transition elements and place them in well-sealed test tubes for display in the front of the room. Use the following elements: sodium (very small piece covered with mineral oil containing no water); magnesium ribbon (polished with sandpaper); aluminum; carbon; phosphorus; sulfur; iodine (before covering with the lid, cover the bottom with plastic wrap to help slow sublimation); Geissler tubes with helium, neon, or argon; iron filings; tungsten; nickel; copper foil (polished with sandpaper); and zinc sheet (polished with sandpaper).

SAFETY

Wear safety goggles, gloves, and a lab apron while handling chemicals. Do not allow students to handle the test tubes or Geissler tubes.

DISPOSAL

Save materials for reuse at a later time.

What makes a family of elements?

Section Objectives

Describe how the modern periodic table is organized.

State the periodic law.

Explain why elements in the same family of the periodic table have similar properties.

Describe characteristic properties of the alkali metals, alkaline-earth metals, transition metals, actinides, lanthanides, halogens, and noble gases.

Relate the properties of various elements to their electron configurations.

Families of elements

How could you tell if two people you have never seen before are members of the same biological family? Chances are you would look for physical features that they have in common. Or the way that two people behave might provide a clue as to whether they are related. Both people might have similar facial expressions or share certain striking mannerisms. Thus, both physical and behavioral traits can indicate whether people are related. The same is true for elements. Elements that are members of the same family of the periodic table share structural and behavioral characteristics.

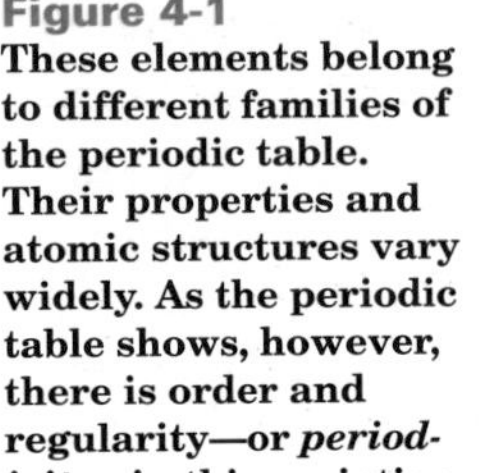

Figure 4-1
These elements belong to different families of the periodic table. Their properties and atomic structures vary widely. As the periodic table shows, however, there is order and regularity—or *periodicity*—in this variation.

Dmitri Mendeleev invented the periodic table

As **Figure 4-1** shows, the properties of elements vary widely. But the elements do not differ from one another in a random fashion. Their properties vary in an orderly way. Because of this, they can be grouped as families of similar elements whose properties vary only slightly from one to the next. Though other chemists had proposed groupings for some elements, it was not known before 1871 that this kind of regularity—or *periodicity*—among the elements was so widespread.

Dmitri Mendeleev, a Russian chemistry professor, discovered periodicity one evening while he sat writing a chemistry book for his students. To organize his thoughts about the elements, Mendeleev wrote the name of each element on a separate card along with its properties. He shuffled the cards about on his desk looking for patterns among them. To his delight, he discovered that when the elements were put in rows such that atomic mass increased from left to right, columns of elements with similar properties could be built. **Figure 4-2** shows the organization of the elements as originally conceived by Mendeleev. Notice the gaps in the table for elements with masses of 45, 68, 70, and 180.

Mendeleev's Organization of the Known Elements

			Ti = 50	Zr = 90	? = 180
			V = 51	Nb = 94	Ta = 182
			Cr = 52	Mo = 96	W = 186
			Mn = 55	Rh = 104,4	Pt = 197,4
			Fe = 56	Ru = 104,4	Ir = 198
			Ni = Co = 59	Pd = 106,6	Os = 199
H = 1			Cu = 63,4	Ag = 108	Hg = 200
	Be = 9,4	Mg = 24	Zn = 65,2	Cd = 112	
	B = 11	Al = 27,4	? = 68	Ur = 116	Au = 197?
	C = 12	Si = 28	? = 70	Sn = 118	
	N = 14	P = 31	As = 75	Sb = 122	Bi = 210?
	O = 16	S = 32	Se = 79,4	Te = 128?	
	F = 19	Cl = 35,5	Br = 80	J = 127	
Li = 7	Na = 23	K = 39	Rb = 85,4	Cs = 133	Tl = 204
		Ca = 40	Sr = 87,6	Ba = 137	Pb = 207
		? = 45	Ce = 92		
		?Er = 56	La = 94		
		?Yt = 60	Di = 95		
		?In = 75,6	Th = 118?		

Figure 4-2
Mendeleev's table grouped elements with similar properties into columns called "groups" or "families." Elements are arranged in order of increasing atomic mass.

Visual Strategy

Figure 4-2

The illustration shows a representation of Mendeleev's original organization for the elements. Ask students what the gaps in the table represent (shown by question marks). Part of Mendeleev's fame came from his ability to predict the properties of these missing elements.

Table 4-1
Some Elements Predicted by Mendeleev

Predicted elements	Element and year discovered	Properties	Predicted properties	Observed properties
Ekaaluminum	gallium 1875	density of metal	6.0 g/mL	5.96 g/mL
		melting point	low	30°C
		oxide formula	Ea_2O_3	Ga_2O_3
Ekaboron	scandium 1877	density of metal	3.5 g/mL	3.86 g/mL
		oxide formula	Eb_2O_3	Sc_2O_3
		solubility of oxide	dissolves in acid	dissolves in acid
Ekasilicon	germanium 1886	melting point	high	900°C
		density of metal	5.5 g/mL	5.47 g/mL
		color of metal	dark gray	grayish white
		oxide formula	EsO_2	GeO_2
		density of oxide	4.7 g/mL	4.70 g/mL
		chloride formula	$EsCl_4$	$GeCl_4$

Mendeleev postulated that the gaps represented missing elements. He predicted that the gaps would be filled by elements not yet discovered. He also predicted the properties of these missing elements. **Table 4-1** shows three "missing" elements, along with some of their properties, that were discovered just as Mendeleev predicted.

The modern periodic table is based on the periodic law

In addition to gaps, Mendeleev's table contained some irregularities. In a few cases, putting elements in order of increasing mass placed elements in columns where they did not seem to fit. Their properties were not similar to those of the other elements in the same family. Mendeleev assumed these irregularities were due to errors in atomic mass measurement. Yet these mass measurements were subsequently shown to be quite accurate.

Forty years later, the young English scientist Henry Moseley removed the irregularities in Mendeleev's table. Moseley discovered that each element has a unique nuclear charge and, therefore, a different atomic number. When Moseley arranged the elements in order of increasing atomic number, rather than according to atomic mass as Mendeleev had done, the irregularities disappeared. Hence, in today's version of the periodic table, elements are placed in order of increasing atomic number.

The most common form of the periodic table used by scientists today is shown in **Table 4-2**, on the next two pages. It is based on the **periodic law**, which states that the physical and chemical properties of the elements are periodic functions of their atomic numbers.

periodic law
properties of elements tend to change with increasing atomic number in a periodic way

TEACH

Teaching Tip

Discussion
Ask the students to name as many similar properties of families from the periodic table as possible. Keep track of the characteristics on the board. A good starting point is to mention the antiseptic characteristics of the halogen family: Cl, Br, F, and I.

Historical Note

German chemist Julius Lothar Meyer (1830–1895) arranged the order of the elements by their atomic volume; and produced findings similar to those of Mendeleev. Unfortunately, Meyer published his findings in 1870, one year after Mendeleev published his.

Cultural Connection

Eka- comes from the Sanskrit word for *"one,"* meaning the next one in a family of the periodic table. For example, francium was called *ekacesium*.

Theme

Classification and trends
Mendeleev based his predictions about the missing elements on the trends of the elements that were present in his period table.

Visual Strategy

Table 4-1
Make the students aware of the close approximations between the predicted element and the actual element. Ask students to propose why these elements had not yet been discovered.

Teaching Tip

Discussion

Make the student aware that not all periodic tables have the information arranged in the block the same way. Some tables have the atomic number in a corner or under the element's name.

Table 4-2

Periodic Table of the Elements

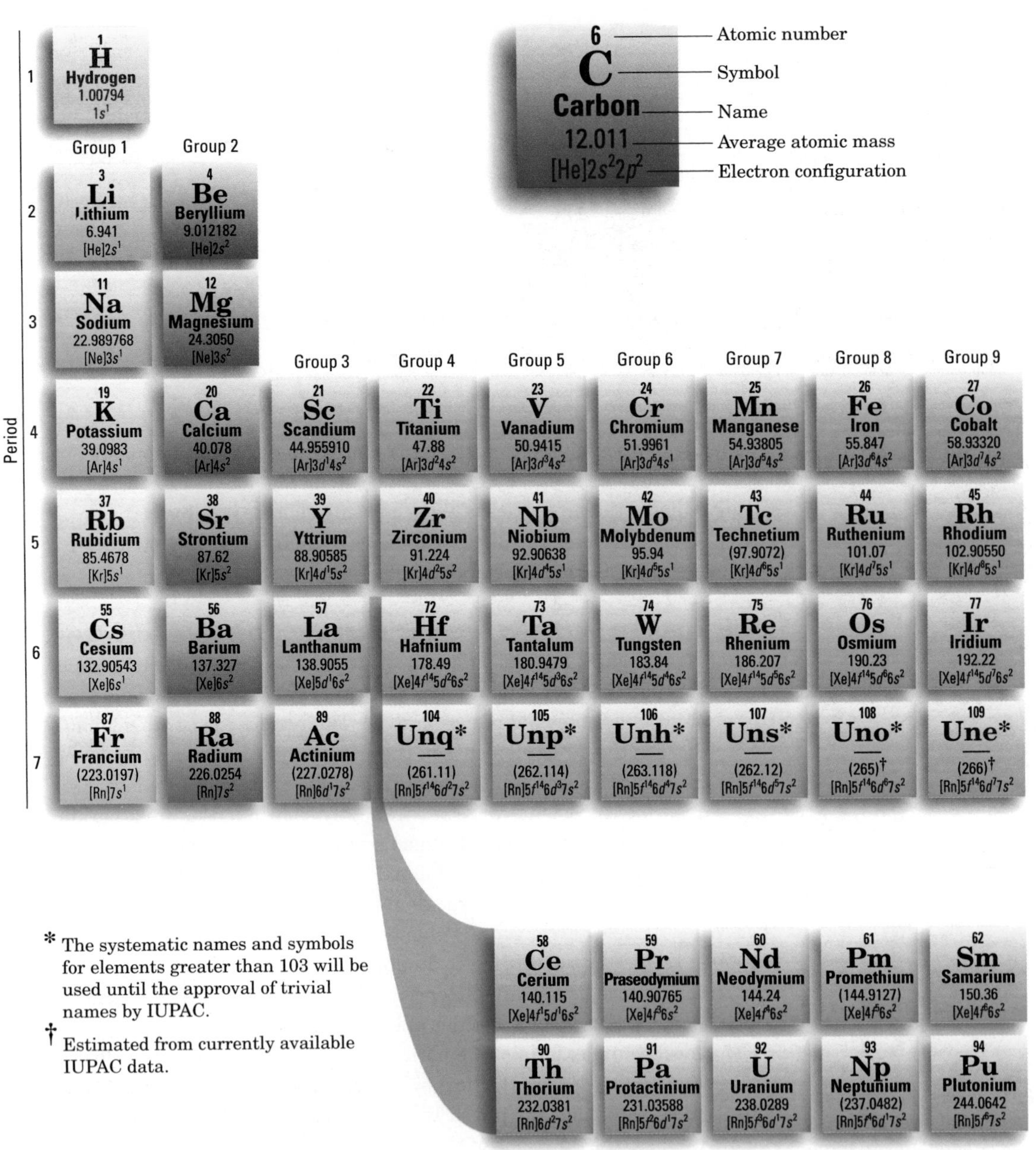

6 — Atomic number
C — Symbol
Carbon — Name
12.011 — Average atomic mass
[He]$2s^2 2p^2$ — Electron configuration

Period	Group 1	Group 2	Group 3	Group 4	Group 5	Group 6	Group 7	Group 8	Group 9
1	1 **H** Hydrogen 1.00794 $1s^1$								
2	3 **Li** Lithium 6.941 [He]$2s^1$	4 **Be** Beryllium 9.012182 [He]$2s^2$							
3	11 **Na** Sodium 22.989768 [Ne]$3s^1$	12 **Mg** Magnesium 24.3050 [Ne]$3s^2$							
4	19 **K** Potassium 39.0983 [Ar]$4s^1$	20 **Ca** Calcium 40.078 [Ar]$4s^2$	21 **Sc** Scandium 44.955910 [Ar]$3d^1 4s^2$	22 **Ti** Titanium 47.88 [Ar]$3d^2 4s^2$	23 **V** Vanadium 50.9415 [Ar]$3d^3 4s^2$	24 **Cr** Chromium 51.9961 [Ar]$3d^5 4s^1$	25 **Mn** Manganese 54.93805 [Ar]$3d^5 4s^2$	26 **Fe** Iron 55.847 [Ar]$3d^6 4s^2$	27 **Co** Cobalt 58.93320 [Ar]$3d^7 4s^2$
5	37 **Rb** Rubidium 85.4678 [Kr]$5s^1$	38 **Sr** Strontium 87.62 [Kr]$5s^2$	39 **Y** Yttrium 88.90585 [Kr]$4d^1 5s^2$	40 **Zr** Zirconium 91.224 [Kr]$4d^2 5s^2$	41 **Nb** Niobium 92.90638 [Kr]$4d^4 5s^1$	42 **Mo** Molybdenum 95.94 [Kr]$4d^5 5s^1$	43 **Tc** Technetium (97.9072) [Kr]$4d^6 5s^1$	44 **Ru** Ruthenium 101.07 [Kr]$4d^7 5s^1$	45 **Rh** Rhodium 102.90550 [Kr]$4d^8 5s^1$
6	55 **Cs** Cesium 132.90543 [Xe]$6s^1$	56 **Ba** Barium 137.327 [Xe]$6s^2$	57 **La** Lanthanum 138.9055 [Xe]$5d^1 6s^2$	72 **Hf** Hafnium 178.49 [Xe]$4f^{14} 5d^2 6s^2$	73 **Ta** Tantalum 180.9479 [Xe]$4f^{14} 5d^3 6s^2$	74 **W** Tungsten 183.84 [Xe]$4f^{14} 5d^4 6s^2$	75 **Re** Rhenium 186.207 [Xe]$4f^{14} 5d^5 6s^2$	76 **Os** Osmium 190.23 [Xe]$4f^{14} 5d^6 6s^2$	77 **Ir** Iridium 192.22 [Xe]$4f^{14} 5d^7 6s^2$
7	87 **Fr** Francium (223.0197) [Rn]$7s^1$	88 **Ra** Radium 226.0254 [Rn]$7s^2$	89 **Ac** Actinium (227.0278) [Rn]$6d^1 7s^2$	104 **Unq*** — (261.11) [Rn]$5f^{14} 6d^2 7s^2$	105 **Unp*** — (262.114) [Rn]$5f^{14} 6d^3 7s^2$	106 **Unh*** — (263.118) [Rn]$5f^{14} 6d^4 7s^2$	107 **Uns*** — (262.12) [Rn]$5f^{14} 6d^5 7s^2$	108 **Uno*** — (265)† [Rn]$5f^{14} 6d^6 7s^2$	109 **Une*** — (266)† [Rn]$5f^{14} 6d^7 7s^2$

58 **Ce** Cerium 140.115 [Xe]$4f^1 5d^1 6s^2$	59 **Pr** Praseodymium 140.90765 [Xe]$4f^3 6s^2$	60 **Nd** Neodymium 144.24 [Xe]$4f^4 6s^2$	61 **Pm** Promethium (144.9127) [Xe]$4f^5 6s^2$	62 **Sm** Samarium 150.36 [Xe]$4f^6 6s^2$
90 **Th** Thorium 232.0381 [Rn]$6d^2 7s^2$	91 **Pa** Protactinium 231.03588 [Rn]$5f^2 6d^1 7s^2$	92 **U** Uranium 238.0289 [Rn]$5f^3 6d^1 7s^2$	93 **Np** Neptunium (237.0482) [Rn]$5f^4 6d^1 7s^2$	94 **Pu** Plutonium 244.0642 [Rn]$5f^6 7s^2$

* The systematic names and symbols for elements greater than 103 will be used until the approval of trivial names by IUPAC.

† Estimated from currently available IUPAC data.

Atomic masses listed in this table reflect the precision of current measurements. However, atomic mass measurements throughout the text have been rounded to two places to the right of the decimal.

Visual Strategy

Table 4-2

Make students aware of the information available from this table, including atomic number, symbol, name, atomic mass, family, group, electron configuration, and whether the element is a metal, nonmetal, or metalloid. Also point out the positions of the actinide and lanthanide series and that these elements are found at the base of the table to keep it from being too wide.

Metals
- Alkali metals
- Alkaline-earth metals
- Transition metals
- Other metals

Metalloids
- Metalloids

Nonmetals
- Halogens
- Other nonmetals

Noble gases
- Noble gases

Group 10	Group 11	Group 12	Group 13	Group 14	Group 15	Group 16	Group 17	Group 18
								2 **He** Helium 4.002602 $1s^2$
			5 **B** Boron 10.811 $[He]2s^22p^1$	6 **C** Carbon 12.011 $[He]2s^22p^2$	7 **N** Nitrogen 14.00674 $[He]2s^22p^3$	8 **O** Oxygen 15.9994 $[He]2s^22p^4$	9 **F** Fluorine 18.9984032 $[He]2s^22p^5$	10 **Ne** Neon 20.1797 $[He]2s^22p^6$
			13 **Al** Aluminum 26.981539 $[Ne]3s^23p^1$	14 **Si** Silicon 28.0855 $[Ne]3s^23p^2$	15 **P** Phosphorus 30.973762 $[Ne]3s^23p^3$	16 **S** Sulfur 32.066 $[Ne]3s^23p^4$	17 **Cl** Chlorine 35.4527 $[Ne]3s^23p^5$	18 **Ar** Argon 39.948 $[Ne]3s^23p^6$
28 **Ni** Nickel 58.6934 $[Ar]3d^84s^2$	29 **Cu** Copper 63.546 $[Ar]3d^{10}4s^1$	30 **Zn** Zinc 65.39 $[Ar]3d^{10}4s^2$	31 **Ga** Gallium 69.723 $[Ar]3d^{10}4s^24p^1$	32 **Ge** Germanium 72.61 $[Ar]3d^{10}4s^24p^2$	33 **As** Arsenic 74.92159 $[Ar]3d^{10}4s^24p^3$	34 **Se** Selenium 78.96 $[Ar]3d^{10}4s^24p^4$	35 **Br** Bromine 79.904 $[Ar]3d^{10}4s^24p^5$	36 **Kr** Krypton 83.80 $[Ar]3d^{10}4s^24p^6$
46 **Pd** Palladium 106.42 $[Kr]4d^{10}5s^0$	47 **Ag** Silver 107.8682 $[Kr]4d^{10}5s^1$	48 **Cd** Cadmium 112.411 $[Kr]4d^{10}5s^2$	49 **In** Indium 114.818 $[Kr]4d^{10}5s^25p^1$	50 **Sn** Tin 118.710 $[Kr]4d^{10}5s^25p^2$	51 **Sb** Antimony 121.757 $[Kr]4d^{10}5s^25p^3$	52 **Te** Tellurium 127.60 $[Kr]4d^{10}5s^25p^4$	53 **I** Iodine 126.90447 $[Kr]4d^{10}5s^25p^5$	54 **Xe** Xenon 131.29 $[Kr]4d^{10}5s^25p^6$
78 **Pt** Platinum 195.08 $[Xe]4f^{14}5d^96s^1$	79 **Au** Gold 196.96654 $[Xe]4f^{14}5d^{10}6s^1$	80 **Hg** Mercury 200.59 $[Xe]4f^{14}5d^{10}6s^2$	81 **Tl** Thallium 204.3833 $[Xe]4f^{14}5d^{10}6s^26p^1$	82 **Pb** Lead 207.2 $[Xe]4f^{14}5d^{10}6s^26p^2$	83 **Bi** Bismuth 208.98037 $[Xe]4f^{14}5d^{10}6s^26p^3$	84 **Po** Polonium (208.9824) $[Xe]4f^{14}5d^{10}6s^26p^4$	85 **At** Astatine (209.9871) $[Xe]4f^{14}5d^{10}6s^26p^5$	86 **Rn** Radon (222.0176) $[Xe]4f^{14}5d^{10}6s^26p^6$

63 **Eu** Europium 151.965 $[Xe]4f^76s^2$	64 **Gd** Gadolinium 157.25 $[Xe]4f^75d^16s^2$	65 **Tb** Terbium 158.92534 $[Xe]4f^96s^2$	66 **Dy** Dysprosium 162.50 $[Xe]4f^{10}6s^2$	67 **Ho** Holmium 164.93032 $[Xe]4f^{11}6s^2$	68 **Er** Erbium 167.26 $[Xe]4f^{12}6s^2$	69 **Tm** Thulium 168.93421 $[Xe]4f^{13}6s^2$	70 **Yb** Ytterbium 173.04 $[Xe]4f^{14}6s^2$	71 **Lu** Lutetium 174.967 $[Xe]4f^{14}5d^16s^2$
95 **Am** Americium (243.0614) $[Rn]5f^77s^2$	96 **Cm** Curium (247.0703) $[Rn]5f^76d^17s^2$	97 **Bk** Berkelium (247.0703) $[Rn]5f^97s^2$	98 **Cf** Californium (251.0796) $[Rn]5f^{10}7s^2$	99 **Es** Einsteinium (252.083) $[Rn]5f^{11}7s^2$	100 **Fm** Fermium (257.0951) $[Rn]5f^{12}7s^2$	101 **Md** Mendelevium (258.10) $[Rn]5f^{13}7s^2$	102 **No** Nobelium (259.1009) $[Rn]5f^{14}7s^2$	103 **Lr** Lawrencium (262.11) $[Rn]5f^{14}6d^17s^2$

Mass numbers in parentheses are those of the most stable or most common isotope.

111

Historical Note

Argon was the first noble gas to be discovered in 1894. It makes up about 1% of the atmosphere. Radon was the last to be discovered in 1900. English physicist William Ramsay (1852–1916) was awarded the Nobel Prize in chemistry in 1904 for his work with the noble gases.

The periodic table can be used to determine the electron configuration of each element

The periodic table contains a wealth of information about each element. Examine the key to **Table 4-2** for the information it gives about carbon: atomic number, symbol, name, atomic mass, and electron configuration. Notice that a shorthand form is used to show electron configuration.

Perhaps the most important thing to notice about the periodic table is the similarity of electron configurations of elements in most families. Look at the periodic table to see this for yourself. Recall from Chapter 3 that the electron configuration of an element determines its chemical behavior. Do you see why elements of the same family have similar properties? Their chemical properties are similar because their electron configurations are similar, as the periodic table shows.

Chem Fact

Radon is the heaviest noble gas. Its isotopes are radioactive, making radon the only radioactive gas at normal temperatures and pressures.

Family characteristics

Group 18 elements are the noble gases

You were introduced to the Group 18 elements in Chapter 3. You learned that these elements, listed in **Figure 4-3a**, have a special name: *noble gases*. The noble gases were called the *inert gases* because they were thought to be unreactive. No stable compounds of three noble gases—helium, neon, and argon—have ever been prepared. The other noble gases—xenon, krypton, and radon—have very low reactivity because in recent years a few compounds have been made from them.

Noble gas atoms are characterized by electron configurations featuring full *s* and *p* orbitals in the highest principal energy level. From the low reactivity we observe in noble gases, we can infer that the noble gas electron configuration is very *stable*, or resistant to change. Indeed, we observe that when atoms of other groups lose or gain electrons to form compounds, they achieve an electron configuration like that of the noble gases. The stability of the element in the compound increases compared to the stability of the element alone.

Despite the low reactivity of the noble gases, they have many uses. Neon and argon, for example, are used in advertising signs. And helium, which is not flammable because it is unreactive, is the preferred lighter-than-air gas used in airships and weather balloons.

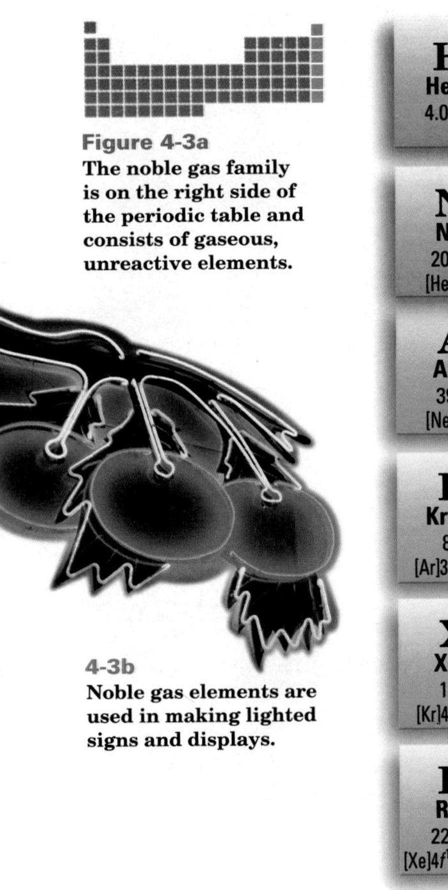

Figure 4-3a
The noble gas family is on the right side of the periodic table and consists of gaseous, unreactive elements.

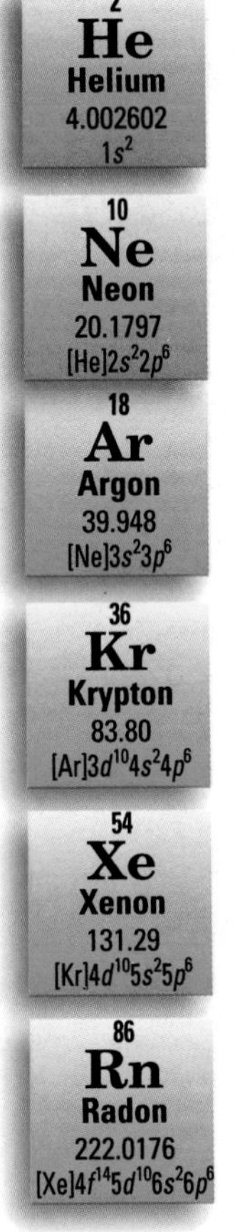

4-3b
Noble gas elements are used in making lighted signs and displays.

Visual Strategy

Figure 4-3

Neon emits red light when its electrons are excited, such as with the application of a high voltage. Ordinarily, it is a colorless, odorless, and tasteless gas. True neon signs are red. Other colors are obtained by using other gases.

alkali metals
highly reactive metallic elements which form alkaline solutions in water, burn in air, and belong to Group 1 of the periodic table

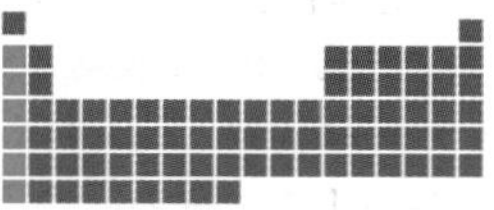

Figure 4-4a
The alkali metals are located on the left edge of the periodic table.

3
Li
Lithium
6.941
[He]$2s^1$

11
Na
Sodium
22.989768
[Ne]$3s^1$

19
K
Potassium
39.0983
[Ar]$4s^1$

37
Rb
Rubidium
85.4678
[Kr]$5s^1$

55
Cs
Cesium
132.90543
[Xe]$6s^1$

87
Fr
Francium
(223.0197)
[Rn]$7s^1$

4-4b
Potassium is stored in oil to prevent it from reacting with moisture and oxygen in air.

Group 1 is also known as the alkali metals

Group 1 elements, listed in **Figure 4–4a**, are soft, highly reactive metals. They are so soft that they can be cut with a knife. Once cut, their shiny surfaces dull quickly as they react with oxygen in air. In the laboratory, alkali metals are usually stored under oil or kerosene to protect them from moisture and oxygen in air. As you can see from **Figure 4-4b**, when added to cold water, an alkali metal reacts vigorously.

In ancient times, people discovered that ashes mixed with water produce a slippery solution that can remove grease. In the Middle Ages such solutions were named *alkaline*, after the Arab word for ashes, *al-qali*. We now know that the ashes that produce alkaline solutions contain compounds of the Group 1 elements. The Group 1 elements are called the **alkali metals** because they produce alkaline solutions and because they have metallic properties.

The reactivity of alkali metals results from their characteristic electron configurations, which feature a single electron in the highest energy level. When an alkali metal atom loses this single electron, the next-lowest energy level becomes the outer shell. This energy level has the same electron configuration as the noble gases. Thus, by losing a single electron, an alkali-metal atom achieves the stable, nonreactive electron configuration of the noble gases.

The alkali metals share many physical properties, as **Table 4-3** shows. Additionally, due to their metallic nature, solid alkali metals are good conductors of electricity.

The single electron can easily be bumped off an alkali metal atom when electrical energy is applied. In the gaseous state at high voltages, the motion of dislodged electrons makes up electric current flow. Sodium vapor, for example, conducts electricity. When an electric current passes through sodium vapor, sodium atoms emit energy in the form of orange-yellow light. Light of this color penetrates fog very well. Therefore, sodium vapor is used in street lamps and in fog lights on cars.

Table 4-3
Physical Properties of Alkali Metals

Element	Melting point (°C)	Boiling point (°C)	Density (g/cm³)	Atomic radius (pm)
lithium	179	1336	0.53	152
sodium	98	883	0.97	186
potassium	64	758	0.86	227
rubidium	39	700	1.53	248
cesium	28	670	1.90	265
francium	27	677	unknown	unknown

Historical Note

Potassium was first produced by English chemist Humphry Davy on October 6, 1807, using electrolysis of molten potassium carbonate. One week later he produced sodium by electrolysis using sodium carbonate.

Do You Know?

Sodium vapor lights are inexpensive to operate because it takes very little electrical energy to dislodge the electrons of the sodium atom. Unfortunately, most people don't find the orange-yellow light very appealing so sodium vapor lights are used mainly where light but not its color is important, such as in public parking lots.

Demonstration 2

Orange-yellow light of sodium

Approximate time: 10 min

The light is seen better in a darkened room.

1. Use a pair of crucible tongs to hold a piece of calcium sulfate (gypsum or dry wall board) in the burner flame. Note the low color production.
2. Use a pair of crucible tongs to hold a piece of rock salt in the burner flame.
3. Have the students note the bright orange-yellow flame. Ask the students why sodium was chosen for lighting. (easy to excite and gives off a lot of light)

SAFETY
Wear safety goggles and a lab apron. Students should be at least 10 ft from the demonstration area as the rock salt may splatter when heated.

DISPOSAL
Put cooled specimens in the trash.

Historical Note

Davy also separated Group 2 metals from their oxides. In 1808 he separated magnesium from magnesia, and calcium from lime. The word *calcium* is from the Latin word for lime.

Demonstration 3

Reactivity of magnesium and calcium

Approximate time: 10 min

1. Place two petri dishes 1/2 full of water on an overhead projector.
2. Add two drops of phenolphthalein indicator to each petri dish. Explain to students that phenolphthalein is like litmus, but it can change from clear to red. Red is the color that denotes the presence of a base.
3. Place a 2 cm piece of magnesium ribbon polished with sandpaper in a petri dish. Place a small piece of fresh calcium metal (the size of a BB) in another petri dish. Be sure the indicator is in contact with the metals in both dishes. The dish with calcium should turn pink while the dish with magnesium should turn pink very slowly, if at all.

SAFETY

Wear safety goggles and a lab apron. Students must stand back at least 10 ft.

DISPOSAL

Combine the liquids and suspended solids, neutralize the mixture with 1 M acid and pour down the drain.

The lightest alkali metal is lithium. Although it has only one electron in its outer energy level just as the other alkali metals do, it is less reactive. The reason for this has to do with the small size of lithium atoms. Because it is a small atom, the outer electron and the nucleus are relatively close to one another. Consequently the electron is held tightly and is not readily removed. This makes lithium just right for certain applications, including an experimental electric car. The car gets its energy from a lithium-based battery. Lithium compounds are used in medications for psychological depression, though the biochemical processes that account for its success are not yet fully understood.

alkaline-earth metals
reactive, metallic elements which belong to Group 2 of the periodic table

Group 2 is also known as the alkaline-earth metals

Group 2 elements are known as the **alkaline-earth metals**. They are all harder, denser, stronger, and have higher melting points than the alkali metals. Check **Figure 4-5a** to see what elements belong to Group 2. How are their electron configurations similar to each other? How do their electron configurations compare with those of Group 1 elements?

The alkaline-earth metals are all reactive. However, they are not as reactive as the Group 1 metals. Why is this so? Group 2 metals are less reactive because they have to lose two electrons to achieve a noble gas electron configuration, while the alkali metals have to lose only one electron.

Consider the alkaline-earth metal magnesium. If the surface of an object made from magnesium is exposed to air, it reacts with the oxygen in air to form the compound magnesium oxide. The magnesium oxide serves as a protective layer that prevents the remaining magnesium metal from corroding. Also, magnesium is lighter than other structural metals, yet still very strong. For all of these reasons, magnesium has a wide variety of practical applications, from the building of aircraft and missiles to the manufacture of ladders and tools.

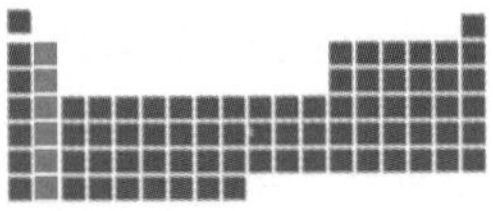

4 **Be** Beryllium 9.012182 [He]$2s^2$

12 **Mg** Magnesium 24.3050 [Ne]$3s^2$

20 **Ca** Calcium 40.078 [Ar]$4s^2$

38 **Sr** Strontium 87.62 [Kr]$5s^2$

56 **Ba** Barium 137.327 [Xe]$6s^2$

88 **Ra** Radium (226.0254) [Rn]$7s^2$

Figure 4-5a
The alkaline-earth metals are the second column of elements from the left edge of the periodic table.

4-5b
Emeralds contain the Group 2 element beryllium.

Figure 4-6
The Taj Mahal in India was commissioned in 1632 by emperor Shāh Jahān as a tomb for his wife. The central structure and domes are made of pure white marble, which contains calcium.

Visual Strategy

Figure 4-5

Point out that impurities in jewels often give them their colors. Several colors of diamonds can be found (pink, yellow, blue) due to minute traces of impurities.

The best known alkaline-earth metal is calcium. Calcium-containing compounds, such as those in limestone and marble, are common in the Earth's crust. Marble was used in the construction of the Taj Mahal, shown in **Figure 4-6**, on the previous page, because it is hard and durable.

Consumer Focus

Calcium supplements

Calcium plays an important role in bone formation. Too little calcium in the diet, especially during periods of bone growth, may lead to soft bones that break easily. Milk is a well-known source of calcium. Fortunately, most children in the United States get enough calcium by drinking milk and eating a balanced diet to avoid noticeably weak bones.

However, many Americans suffer the disabling effects of weak bones later in life. More than 25 million people have osteoporosis, a disease that primarily affects people over 50, in which the bones lose mass, and break easily. Osteoporosis-related hip fractures are an especially serious problem for many elderly Americans. Osteoporosis also carries an expense to society of over $10 billion a year.

Medical investigators agree that the best way to prevent osteoporosis in later life is for children and young adults to consume more calcium. Additional calcium consumed during the childhood and teenage years leads to development of more massive bones. People who start off with stronger, more massive bones can withstand the loss of bone mass that inevitably occurs with aging better than those who start off with weaker bones. The need for extra calcium is greatest between ages 9 and 18 when young people lay down 37 percent of their adult bone mass.

To obtain the calcium they need, many people take calcium supplements in tablet or liquid form. However, investigations have shown that some calcium supplements are potentially dangerous because they contain lead, a poisonous metal that can destroy brain cells and stunt growth.

If you think you may need more calcium, check the level in **Table 4-4** then monitor your diet. Eat plenty of calcium-rich foods, such as dairy products and broccoli. If you want to take a supplement, choose a product that has been approved by the Food and Drug Administration. And remember: Never exceed the recommended dosage without proper medical supervision.

Table 4-4
Optimal Daily Intake of Calcium*

Group		Optimal Daily Intake (in milligrams of calcium)
Children		800
Teenagers		1200 to 1500
Men	25–50	800
	51–65	1000
	over 65	1500
Women	25–50	1000
	51–65	1500
	pregnant and nursing	1900

*Optimal Daily Intake values are the amounts of calcium needed to build and maintain bone mass in order to prevent disease.

Other Applications

Calcium sulfate is commonly called plaster of Paris or gypsum and is widely used in the building industry to make wallboard. Magnesium sulfate is commonly called Epsom salts and is used for fireproofing, textiles, mineral waters, ceramics, paper, and is sold in the drugstore as a laxative and soaking aid.

Cultural Connections

The Taj Mahal was built as a mausoleum outside Agra, India. It was started in 1632 and completed in 1643. A mausoleum of black marble for the emperor was planned but never built.

Groupwork Strategy

Chemical families

Divide the class into groups of three (or four if necessary) and assign each group one chemical family. Select from Groups 1, 2, 13, 14, 15, 16, 17, or 18. Using the textbook, the CRC handbook, and other reference books, have each group determine six properties for their chemical family. Each group should write their findings on the board to compile a list of properties of all families of the periodic table.

Do You Know?

Vitamin D aids in the deposition of calcium in bones. This is one reason vitamin D is added to milk.

Theme

Equilibrium and change

The equilibrium of the body's vitamin and mineral levels must be maintained, or the body will begin to alter its standard operating procedures.

Teaching Tip

Galvanized metal

Ask the students to name some items they know to be galvanized with zinc. Ask about the purpose of the zinc. Tell them that it oxidizes before iron. In doing so, zinc protects iron from attack by oxygen. Ask students what uses of iron nails would require them to be galvanized.

Cultural Connection

Iron played a very important role in the history of civilization. Bronze weapons were no match for the iron weapons in the Iron Age. The Dorians, a Greek tribe that was equipped with iron weapons overcame the Mycenaean Greeks and penetrated all the way to Canaan. These Greeks became the well-known Philistines of Biblical times.

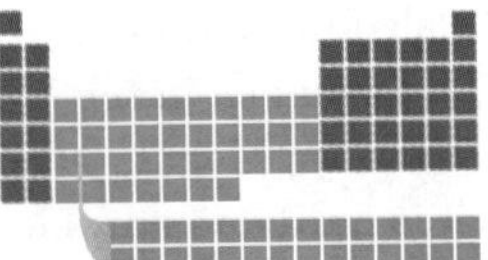

Figure 4-7a
The transition elements are found in the middle of the periodic table.

21 Sc Scandium 44.955910 $[Ar]3d^14s^2$	22 Ti Titanium 47.88 $[Ar]3d^24s^2$	23 V Vanadium 50.9415 $[Ar]3d^34s^2$	24 Cr Chromium 51.9961 $[Ar]3d^54s^1$	25 Mn Manganese 54.93805 $[Ar]3d^54s^2$	26 Fe Iron 55.847 $[Ar]3d^64s^2$	27 Co Cobalt 58.93320 $[Ar]3d^74s^2$
39 Y Yttrium 88.90585 $[Kr]4d^15s^2$	40 Zr Zirconium 91.224 $[Kr]4d^25s^2$	41 Nb Niobium 92.90638 $[Kr]4d^45s^1$	42 Mo Molybdenum 95.94 $[Kr]4d^55s^1$	43 Tc Technetium (97.9072) $[Kr]4d^65s^1$	44 Ru Ruthenium 101.07 $[Kr]4d^75s^1$	45 Rh Rhodium 102.90550 $[Kr]4d^85s^1$
57 La Lanthanum 138.9055 $[Xe]5d^16s^2$	72 Hf Hafnium 178.49 $[Xe]4f^{14}5d^26s^2$	73 Ta Tantalum 180.9479 $[Xe]4f^{14}5d^36s^2$	74 W Tungsten 183.84 $[Xe]4f^{14}5d^46s^2$	75 Re Rhenium 186.207 $[Xe]4f^{14}5d^56s^2$	76 Os Osmium 190.23 $[Xe]4f^{14}5d^66s^2$	77 Ir Iridium 192.22 $[Xe]4f^{14}5d^76s^2$
89 Ac Actinium (227.0278) $[Rn]6d^17s^2$	104 Unq — (261.11) $[Rn]5f^{14}6d^27s^2$	105 Unp — (262.114) $[Rn]5f^{14}6d^37s^2$	106 Unh — (263.118) $[Rn]5f^{14}6d^47s^2$	107 Uns — (262.12) $[Rn]5f^{14}6d^57s^2$	108 Uno — (265) $[Rn]5f^{14}6d^67s^2$	109 Une — (266) $[Rn]5f^{14}6d^77s^2$

58 Ce Cerium 140.115 $[Xe]4f^15d^16s^2$	59 Pr Praseodymium 140.90765 $[Xe]4f^36s^2$	60 Nd Neodymium 144.24 $[Xe]4f^46s^2$	61 Pm Promethium (144.9127) $[Xe]4f^56s^2$	62 Sm Samarium 150.36 $[Xe]4f^66s^2$
90 Th Thorium 232.0381 $[Rn]6d^27s^2$	91 Pa Protactinium 231.03588 $[Rn]5f^26d^17s^2$	92 U Uranium 238.0289 $[Rn]5f^36d^17s^2$	93 Np Neptunium (237.0482) $[Rn]5f^46d^17s^2$	94 Pu Plutonium (238.0496) $[Rn]5f^67s^2$

4-7b
Which one of these iron buckets was galvanized?

transition elements
metallic elements that have varying properties and belong to Groups 3 through 12 of the periodic table

Groups 3 through 12 contain the transition elements

Elements in Groups 3 through 12 are called the **transition elements**. Like the elements belonging to Groups 1 and 2, the transition elements are all metals. However, the transition elements are not as reactive as elements belonging to Groups 1 and 2. With the exception of mercury, the transition metals are harder, denser, and have higher melting points than the Group 1 and 2 metals.

Check **Figure 4-7a** to see that there are many irregularities in the electron configurations of the transition elements. Unlike Group 1 and 2 elements, the transition elements do not all have the same number of electrons in their highest occupied energy level. Many of them have two electrons in the *s* sublevel of their highest occupied energy level, but some have only one, and palladium has none. Because the transition elements are filling the *d* orbitals with electrons, they are sometimes called *d*-block elements.

While looking at the transition elements, you probably recognized a number of them, including iron and zinc. Zinc is one of industry's most useful metals. Large amounts of it are used to *galvanize* iron products, or coat them with a protective layer of zinc. The galvanized bucket shown in **Figure 4-7b** will not rust. Zinc is more reactive than iron, so zinc is oxidized (reacts to form a compound containing oxygen) before iron.

Visual Strategy

Figure 4-7

The transition elements fill *s* and *p* orbitals with electrons at the same time, which leads to the variety of oxidation states that are characteristic of transition metals. The series of 10 elements exist as a transition from the highly reactive metals to the less reactive metals. Make the students aware that other metals can also form protective coatings out of oxides.

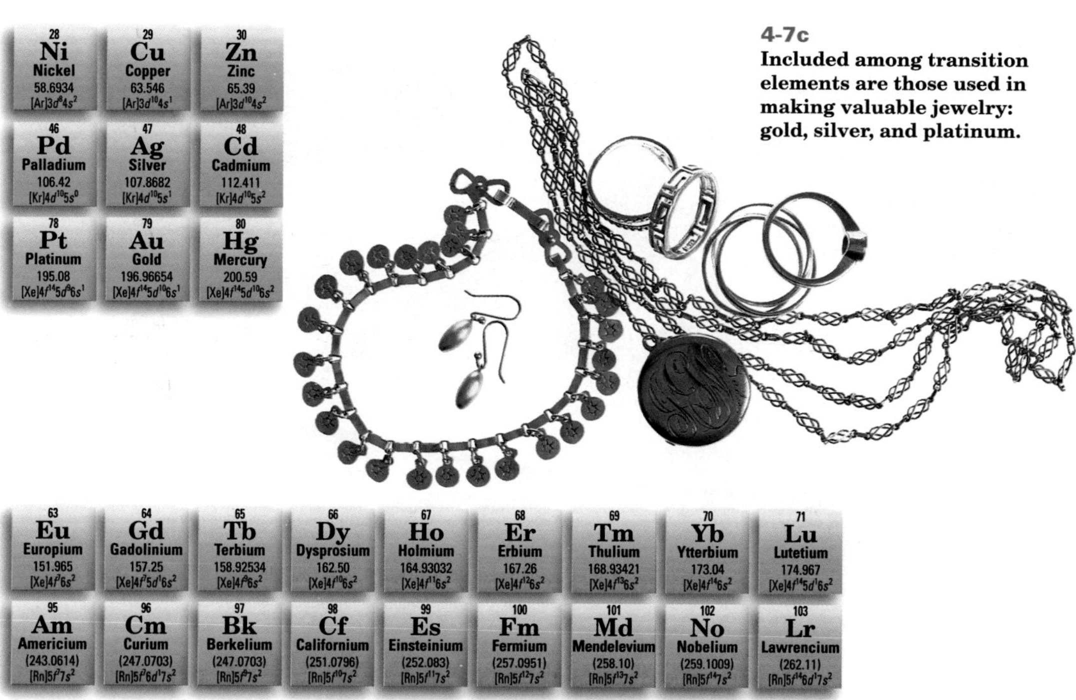

4-7c
Included among transition elements are those used in making valuable jewelry: gold, silver, and platinum.

Do You Know?
All members of the lanthanide series have an oxidation state of 3+ and very similar chemical properties.

Historical Note
All transuranium elements (elements with atomic number greater than 92) have been discovered since 1940. Most were discovered under the leadership of American physicist Glenn Theodore Seaborg.

As you learned in Chapter 3, parts of two periods of transition elements are placed toward the bottom of the periodic table to keep the table conveniently narrow. These rows of elements are referred to as the lanthanides and actinides. The **lanthanides** include elements 58 through 71. The name of element 57 will give you a clue as to how the lanthanides got their name. Check the periodic table to identify the two groups of elements on either side of the lanthanides. Examine the periodic table again to verify the similarities of their electron configurations.

lanthanides
*shiny, metallic elements with atomic numbers 58 through 71 that fill the 4*f *orbitals*

The lanthanides are shiny, reactive metals. Some have practical uses. For example, compounds of some lanthanide metals are used to make the phosphor dots in television tubes. When bombarded by a beam of electrons, the phosphors emit light of various colors. These colors combine to create the brilliant images you see on a color television screen.

actinides
*metallic elements with atomic numbers 90 through 103 that fill the 5*f *orbitals*

Elements 90 through 103 constitute the **actinides**. The name of element 89 tells you how the actinides got their name. In the unique case of the actinides, nuclear structure is of more practical importance than electron configuration. All of the actinides have an unstable arrangement of protons and neutrons in the nucleus. All actinides have radioactive forms. Perhaps the best-known actinide is uranium. Its nuclear disintegration provides the energy for power plants, submarines, and aircraft carriers.

Visual Strategy
Figure 4-7c
Make students aware that jewelry is made from alloys of gold, silver, or platinum due to the softness of the pure metals. Sterling silver is an alloy containing 7.5% copper. Gold and platinum will not corrode in the air but sterling silver will corrode slowly (silver will corrode fairly quickly by itself).

Groupwork Strategy

Tin and food packaging

Divide the class into groups of three for a brainstorming session that will last 10 minutes. Remind students that there are no wrong ideas and all ideas are to be accepted and discussed. Present this supposition: The world's supply of tin to make tin cans has been found to contain a radioactive isotope that may cause mutations in 1 out of every 1 000 000 people.

1. How should this problem be addressed?
2. What other packaging could be developed that is durable, yet biodegradable?

Have each group quickly report their ideas to the class.

Do You Know?

It is dangerous to buy dented or bent cans containing food because it is possible for the tin layer to be scratched or to flake off. If this happens, the can could rust and open. Also, the inside of most food cans is coated with varnish or plastic. When the can is dented or bent, this lining can break. The acids in the food can then attack the metal in the can, making the food unsuitable for consumption.

Theme

Systems and interactions

The interactions of the elements in families containing metalloids are not very predictable because possible combinations could include metals, metalloids, and nonmetals all present at once.

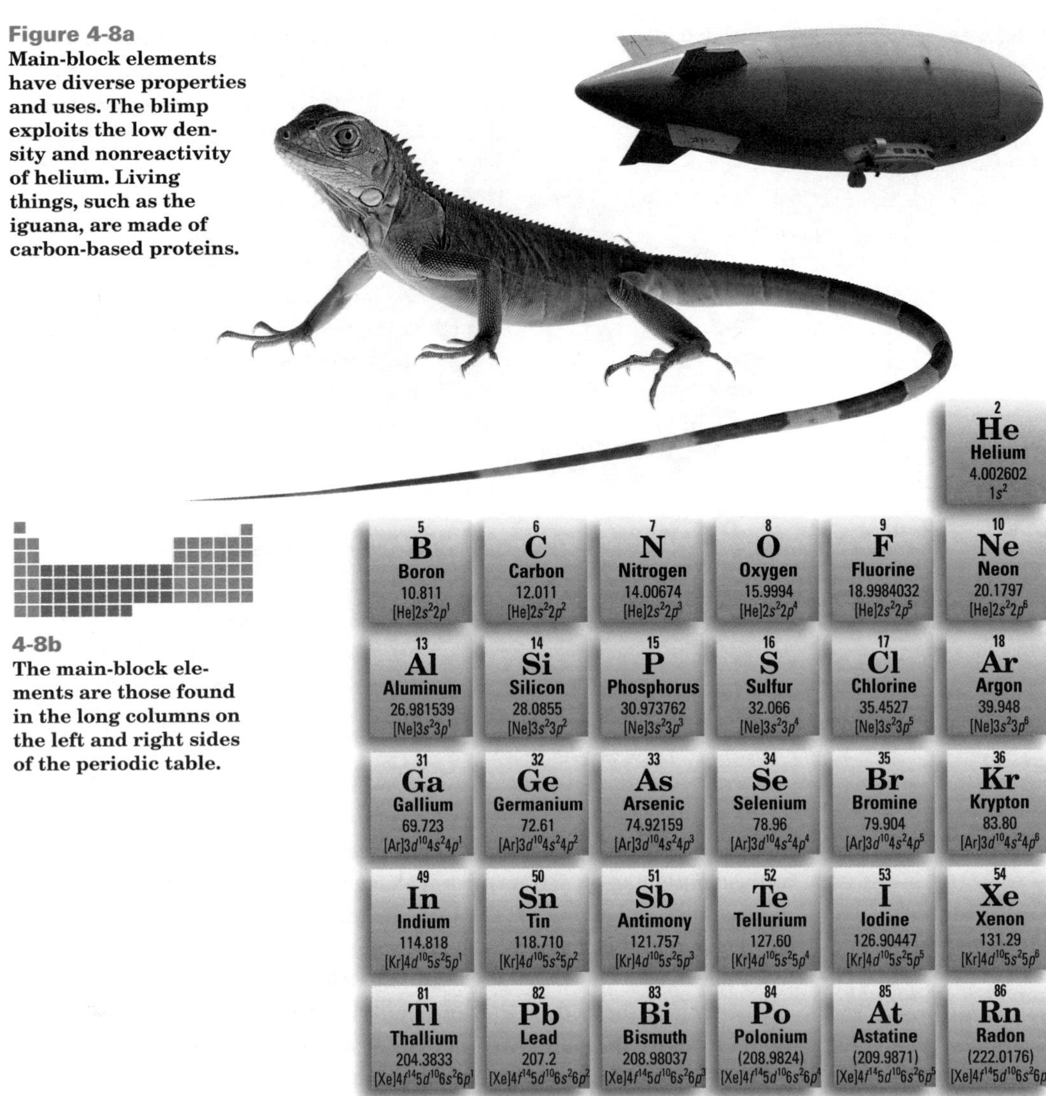

Figure 4-8a Main-block elements have diverse properties and uses. The blimp exploits the low density and nonreactivity of helium. Living things, such as the iguana, are made of carbon-based proteins.

4-8b The main-block elements are those found in the long columns on the left and right sides of the periodic table.

Main-block elements include groups 13 through 18

Along with elements in Groups 1 and 2, those in Groups 13 through 18 are called **main-block elements**. The main-block elements shown in **Figure 4-8b** are also called the *representative elements* because they represent a wide range of chemical and physical properties.

main-block elements
elements that represent the entire range of chemical properties and belong to Groups 1, 2, and 13 through 18 in the periodic table

Within Groups 13 to 18, properties vary systematically. For example, these elements include metals, metalloids, nonmetals, and noble gases. You are probably familiar with many of the elements found in Groups 13 to 18, including carbon, nitrogen, oxygen, aluminum, tin, and lead. Two of the elements, silicon in Group 14 and oxygen in Group 16, account for four of every five atoms found near the surface of the Earth.

Visual Strategy

Figure 4-8

Remind students that Groups 13 through 16 are known by the top member of each group, *i.e.*, the boron family, carbon family, nitrogen family, and oxygen family.

halogens
elements that combine with most metals to form salts and that belong to Group 17 of the periodic table

Two families of elements within Groups 13 through 18 have special names. The Group 18 elements, as you know, are also called the *noble gases*. The elements of Group 17 are known as the **halogens**. The halogens combine easily with metals, especially the alkali metals, to form compounds known as *salts*. *Halogen* is derived from Latin and means *salt-former*. Common table salt is made of a halogen, chlorine, in chemical combination with an alkali metal, sodium.

Figure 4-9a
Hydrogen sits apart from the other elements in the periodic table.

The halogens are the most reactive nonmetal elements. Their electron configurations show why. The highest energy level in a halogen atom is just one electron short of being identical to the noble gas configuration. When halogens react chemically, they gain the extra electron they need to obtain the full stable electron arrangement of the noble gases.

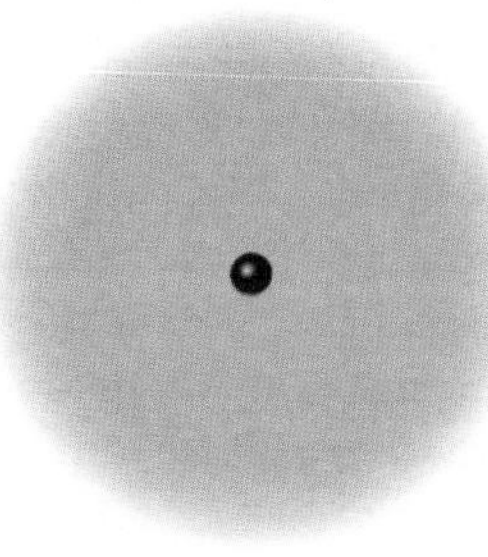

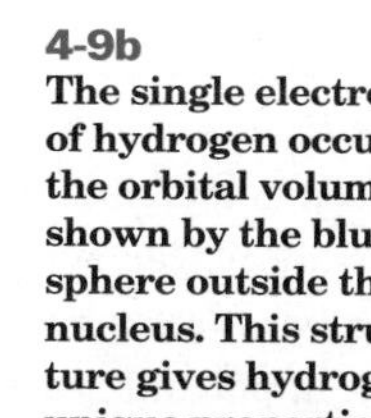

4-9b
The single electron of hydrogen occupies the orbital volume shown by the blue sphere outside the nucleus. This structure gives hydrogen unique properties and puts this element in a family of its own.

One element forms its own chemical family

The element that "sits" by itself in the periodic table is the most common element in the universe—hydrogen. Hydrogen is usually considered to be a chemical family by itself because it behaves unlike any other element.

The cloud model for hydrogen shown in **Figure 4-9** has one electron. Hydrogen reacts very rapidly with most other elements, including oxygen, as you can see in **Figure 4-10**. Because it reacts so easily, hydrogen in its free state is rare, while compounds containing hydrogen are very common. Water is the most abundant hydrogen compound. Combined with carbon and oxygen, hydrogen is present in fats, proteins, and carbohydrates. The main industrial use of hydrogen is to combine it with nitrogen to make ammonia. Large quantities of ammonia, in turn, are used in the production of fertilizers.

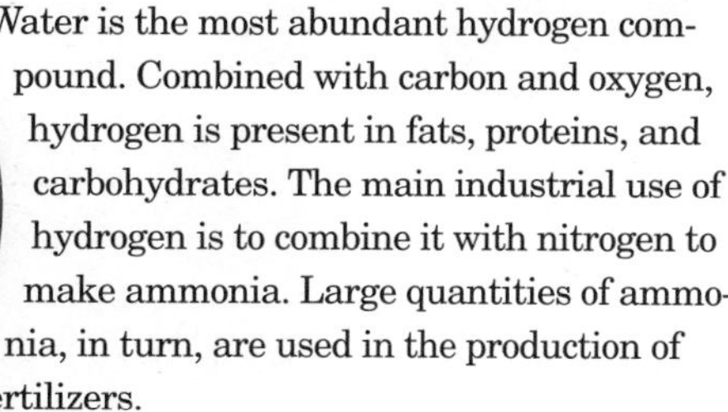

Figure 4-10
When hydrogen reacts with oxygen in the space shuttle, enough energy is released to lift the shuttle into orbit.

Section Review

1. State the periodic law.
2. Why do the elements belonging to a particular group exhibit similar chemical behavior?
3. How do the alkali metals differ from the alkaline-earth metals?
4. Explain what the transition metals have in common with respect to their electron configurations.
5. What features do all of the halogens have in common?

Teaching Tip

Vocabulary development
Many students will recognize the term *halogen* from halogen light bulbs on cars and flashlights. Others may recognize the term *halite* which is derived from halogen. Halite is the name given to sodium chloride found in nature.

ASSESS

Alternative Assessment

Divide the class into research groups of three and assign each group a transition family from Groups 3 to 12. Have them determine the uses for the elements, where they are found, how they are mined, and any environmental hazards they pose. Have them attempt to determine the cost of the element. Have them present group reports lasting not more than 5 minutes per group. Have only two or three group presentations per day to keep interest. Select presenters by drawing names from a box so that everybody is ready on the first presentation day.

Answers to Section Review

1. Properties of elements are a periodic function of atomic number.
2. They exhibit similar behavior because they have the same number of electrons in their outer energy levels.
3. The alkaline-earth metals are less reactive than the alkali metals.
4. The transition elements contain electrons in their outermost *s* sublevel and nearby *d* sublevel.
5. The halogens are all reactive and combine with metals to form salts. Also, the halogens are one electron short of achieving noble gas electron configurations.

Visual Strategy

Figure 4-10
Tell students that the two smaller outer fuel tanks on the shuttle contain solid rocket fuel. The large tank contains liquid hydrogen and liquid oxygen.

Section 4-2

FOCUS

Lesson Starter
Begin with a discussion of the term *trends*. Ask students to define what a trend is and to describe some trends they observe, such as trends in fashion, behavior, color, and food. Discuss how trends are used to classify and how this idea applies to trends they will observe based on the arrangement of elements in the periodic table.

TEACH

Demonstration 4
Size of an oleic acid molecule
Approximate time: 30 min
This demonstration is based on the ability of an oleic acid solution to spread out, forming a film that is one molecule thick.

1. Clean a cafeteria tray with soap and water. Fill the tray to a depth of with 1 cm with water.
2. Gently spread lycopodium powder or fine chalk dust over the surface of the water and disperse it by blowing (too much dust will impede the oleic acid).
3. Add one drop of 0.5% oleic acid solution (an oleic acid and alcohol solution) to the tray. When it stops spreading, measure the two largest diameters of the film. Record the measurement in centimeters.
4. The thickness of the film can be calculated by dividing the volume of oleic acid in the drop by the area of the film. Have a student determine the number of drops of oleic acid that equal 5 mL.

What trends are found in the periodic table?

Section Objectives

Describe the trends seen in the periodic table with respect to atomic radius, ionization energy, electron affinity, and electronegativity.

Relate trends of the periodic table to the atomic structures of the elements.

Periodic trends

The people shown in **Figure 4-11** are not arranged in any particular manner. How would you sort them so that some trend, or steady change, would be apparent in the group? Perhaps you could arrange them according to height so that a trend from taller to shorter people would be apparent. Or, you could arrange them by age.

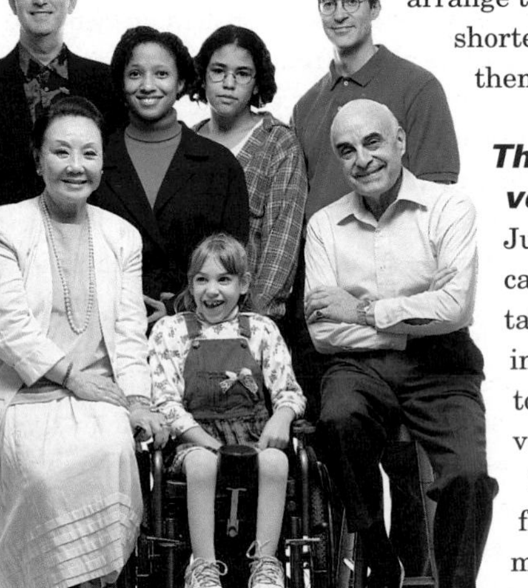

Figure 4-11
How could you arrange these people to reveal a trend in their ages?

The periodic table contains vertical and horizontal trends

Just as various trends can be seen among people, so they can be seen among elements. As you know, in the periodic table, elements are placed side by side in order of increasing atomic number. This arrangement makes it possible to view certain important trends among them. There are vertical trends within groups and there are horizontal trends which you can spot as you look across the table from left to right. A vertical trend among the alkali metals is shown in **Figure 4-12**. Knowing what the trends are among the elements enables you to make predictions about their chemical behavior.

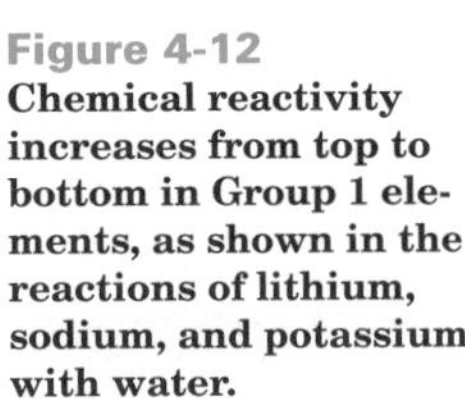

Figure 4-12
Chemical reactivity increases from top to bottom in Group 1 elements, as shown in the reactions of lithium, sodium, and potassium with water.

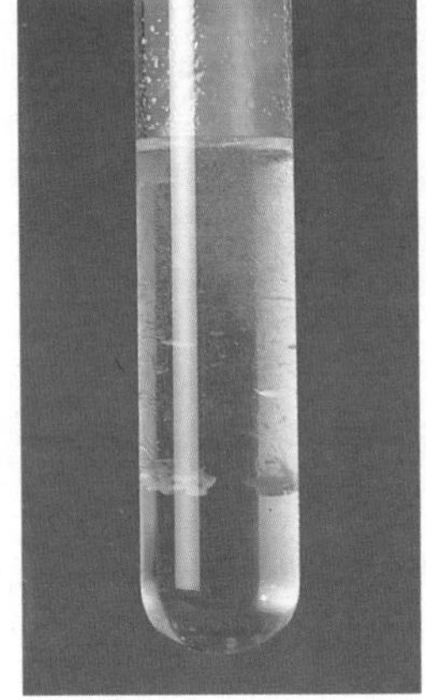

Lithium

Sodium

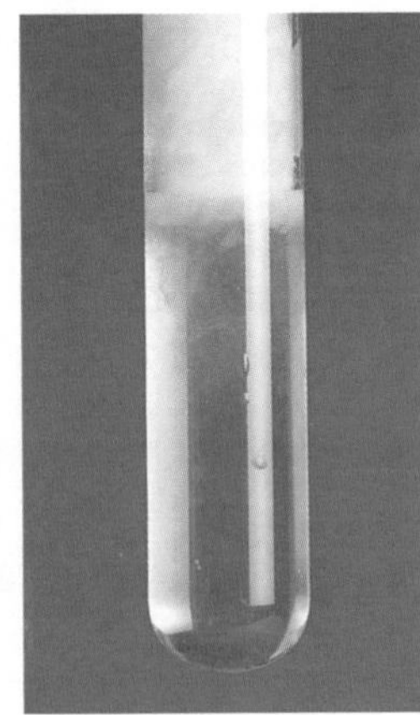

Potassium

Atomic radius increases within a family

atomic radius
one-half of the distance from center to center of two like atoms

Recall that the nucleus of an atom occupies only a small fraction of its volume. Electrons outside the nucleus determine the atom's boundaries. These electrons travel in regions that are pictured as "clouds." Imagine how difficult it would be to measure the size of a real cloud. The same is true of measuring the size of an atom. Its outer boundaries are fuzzy and variable—just like those of a cloud.

Chemists calculate **atomic radius** in several ways. The radii of some atoms are determined by measuring the distance between centers of like atoms that are joined together in a diatomic molecule. Other measurements are taken from bond lengths of atoms in compounds. Data in tables can vary because there is no firm agreement on these measurements. Use **Figure 4-13** to calculate atomic radii. Note that atoms are represented as simple spheres when atomic size is calculated this way. This is an oversimplification, but a useful one. In many instances, atoms are best represented as spheres for the sake of simplicity.

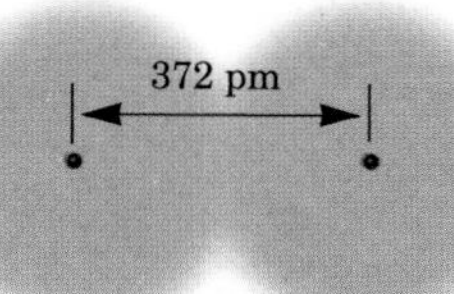

Sodium atoms

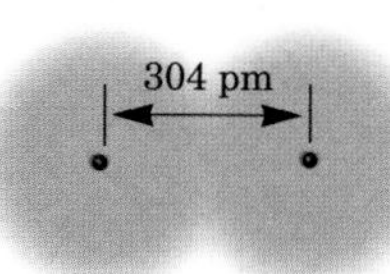

Lithium atoms

Figure 4-13
The distances between centers of like atoms joined as molecules are shown. Use this data to calculate the atomic radius for both a lithium atom and a sodium atom. Use units of picometers, pm.

Do atomic radii show a periodic trend? **Table 4-5** shows the atomic radii of main-block atoms. Notice that, generally, *atomic radius increases as you progress down through the elements in each group.* If you check the electron configurations of elements in the periodic table, you'll see that as you move down any group, another principal energy level is added. It is easy to see that as principal energy levels are added, the atomic radius gets bigger because electrons are added to energy levels farther away from the nucleus.

Table 4-5
Comparing Atomic Radii*

1	2	13	14	15	16	17	18
1 H 37							2 He 50
3 Li 152	4 Be 112	5 B 85	6 C 77	7 N 70	8 O 73	9 F 72	10 Ne 71
11 Na 186	12 Mg 160	13 Al 143	14 Si 118	15 P 110	16 S 103	17 Cl 100	18 Ar 98
19 K 227	20 Ca 197	31 Ga 135	32 Ge 122	33 As 120	34 Se 119	35 Br 114	36 Kr 112
37 Rb 248	38 Sr 215	49 In 167	50 Sn 140	51 Sb 140	52 Te 142	53 I 133	54 Xe 131
55 Cs 265	56 Ba 222	81 Tl 170	82 Pb 146	83 Bi 150	84 Po 168	85 At (140)	86 Rn (141)

*Radius in picometers

Use this information, to determine the volume of one drop by dividing 5 mL by the number of drops. This volume is then divided by 200, which is the dilution factor of the oleic acid, to provide the volume of oleic acid in the drop. The area of the film is determined by averaging the total diameter measurements and using the formula for the area of a circle, πr^2. Divide the volume of the drop by the area of the circle to get the thickness of the film.

SAFETY
Wear safety goggles and a lab apron. Students must stand back at least 10 ft. Ensure that there are no flames in the room. Do not use more than 10 mL of the oleic acid solution.

DISPOSAL
Pour the water with the film on it down the drain. Save the oleic acid solution for reuse at a later time, or dilute it with 100 mL of water, and pour it down the drain. Clean the tray with soap and water.

Historical Note
French physicist, Jean Baptiste Perrin (1870-1942) first produced a close estimate for the size of an atom in 1908 based on an equation suggested by Albert Einstein in 1905.

Theme
Systems and interactions
The interaction between the electrons of nearby atoms keep the atoms a minimum distance apart.

Visual Strategy
Figure 4-13 and Table 4-5
Emphasize that the measurement shown in the figure is the distance between two nuclei and must be divided by 2 to determine the approximate radius of the atom. Emphasize that there is no firm agreement on exact values for atomic radii due to the various methods used to make the measurements and to the fact that the atom has no fixed boundary.

shielding effect
the reduction of the attractive force between a nucleus and its outer electrons due to the blocking effect of inner electrons

There is another reason why atomic size increases within a family. Descending within a group, the number of occupied orbitals between the nucleus and the outermost energy level increases. The added inner electrons reduce the attraction between the outer electrons and the nucleus. This **shielding effect** allows the outer electrons to be farther away from the nucleus, such that the size of the atom increases.

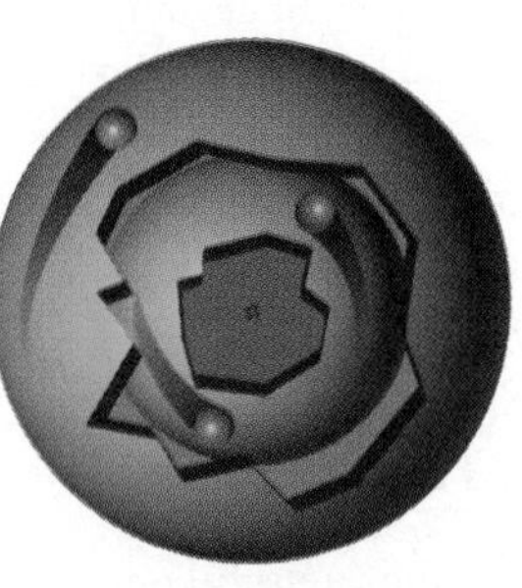

Figure 4-14
Energy levels are shown as hollow spheres and electrons are shown as moving spheres in this representation of the lithium atom. Electrons in intermediary energy levels interfere with the attraction between the nucleus and outer energy level electrons.

Atomic size decreases from left to right across a period

As **Table 4-5** shows, the atomic radii of elements follow another periodic trend. *Atomic radii generally decrease as you move across a period from left to right*. Why is this so?

In crossing a period from left to right, each atom gains one more proton and one more electron. No principal energy levels are added, so electrons enter the same energy level. As the number of protons increases across a period, the positive charge of the nucleus increases. As a result, the nucleus exerts a greater pull on all of the electrons in the atom. Hence, as electrons are added within a period, they are pulled closer to the nucleus, and atomic size decreases. When atomic radii are plotted against atomic number, as in **Figure 4-15**, this periodic trend becomes clear.

Notice from the graph in **Figure 4-15** that this trend is less pronounced in periods where there are many electrons between the nucleus and outermost energy level. As you move from left to right in a period, one proton is added to the nucleus and one electron is added to the outer energy level. As you proceed from left to right in the period, the effective positive charge increases gradually. Therefore, the electrons are pulled closer to the nucleus. As the electrons are pulled closer to the nucleus, they get closer to each other and repulsions occur. Finally, a stage is reached where the electrons won't come closer, and the size of the atom tends to level off.

Figure 4-15
Because elements in Period 2 have fewer electrons between their nuclei and outermost energy levels compared to those in Period 5, the shielding effect is less. Thus the decrease in atomic size across Period 2 is more pronounced than across Period 5.

Atomic Radii of Sample Elements

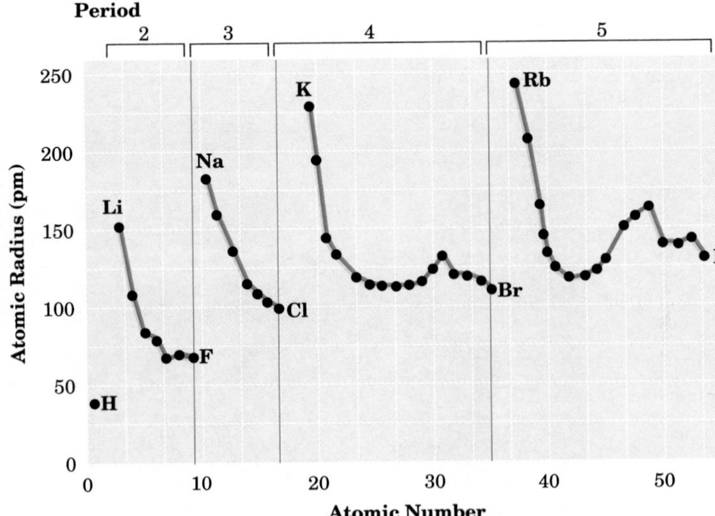

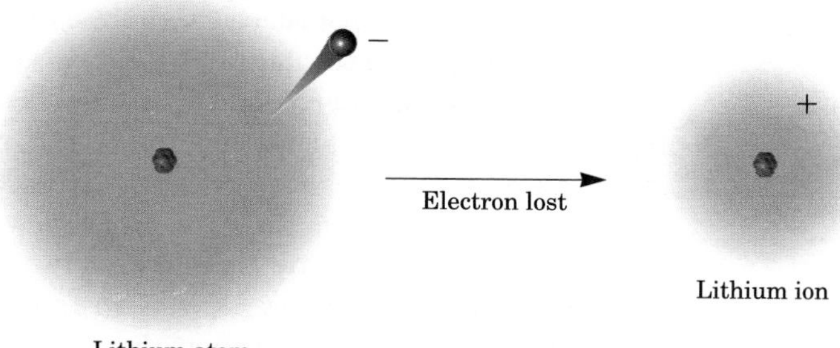

Figure 4-16
When enough energy is absorbed by a lithium atom, the atom loses an electron to become a positive ion.

Ionization energy follows a periodic trend

Recall that atoms are normally electrically neutral. However, an atom may lose or gain electrons to become an electrically charged **ion**. Imagine you can reach into an atom, hold the nucleus with one hand, and remove the outermost electron, creating an ion. The energy you would use to remove the electron is the **ionization energy** of the atom. For any element (A), the process of removing an electron, shown in **Figure 4-16** can be represented as follows.

ion
an atom or group of atoms that has gained or lost one or more electrons to acquire a net electric charge

ionization energy
the amount of energy needed to remove an electron from a specific atom or ion in its ground state in the gas phase

$$\mathbf{A + energy \longrightarrow A^+ + e^-}$$

In the process shown above, a neutral atom absorbs energy equal to its ionization energy. As a result, the neutral atom acquires a positive charge and becomes a positive ion. The ionization energies of elements display periodic trends. **Figure 4-17** shows what happens to ionization energy as you move across a period. **Figure 4-18** on the next page summarizes the trend: *Ionization energy generally decreases as you move down a group of elements but increases across a period.*

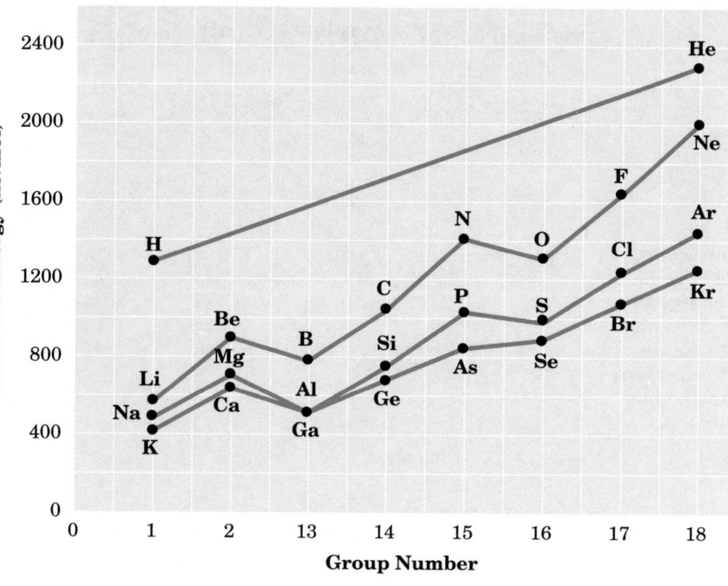

Figure 4-17
Ionization energies of main-block elements belonging to the first four periods are shown. The periodic trend in ionization energy is the opposite of that for atomic size. This makes sense if you think about changes in electron configuration that occur in groups and periods.

Possible Misconception

Many students will have trouble remembering that ionization energy concerns the loss of electrons only—especially when ions can have positive and negative charges. They also have trouble understanding that the loss of an electron is represented by the following equation.

$$A + \textit{energy} \longrightarrow A^+ + e^-$$

Because the electron is shown as being added to the right side of the equation, students can become confused. They also have problems remembering that when electrons are removed, atoms become more positive. Most students intuitively feel atoms should become more negative (like on a number line) as they lose electrons instead of becoming more positive. List the number of protons and electrons for some atoms and their accompanying ions. Show how the charges add to give a neutral or charged particle.

Visual Strategy

Figure 4-17

Make students aware that each trend line on the graph describes a different period from the periodic table.

Teaching Tip

Students will remember the meaning of electron affinity much more easily if they understand the definition of affinity. Explain that most students have an affinity (desire or liking) for ice cream or money.

Possible Misconception

Some students have difficulty understanding that energy given off is denoted as a negative value, whereas energy absorbed is denoted by a positive value.

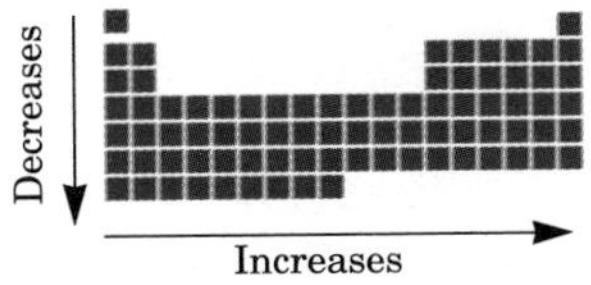

Figure 4-18
Ionization energy generally decreases from top to bottom and increases from left to right across the periodic table.

If you recall the major change in electron configuration as you proceed through a group, you should see why the ionization energy decreases. The electrons that are removed from atoms of each succeeding element in a group are in higher energy levels. Therefore, they are farther from the nucleus and more easily removed. In addition, the shielding effect has an impact. As you move down a group of elements, more and more electrons lie between the nucleus and the electrons in the highest occupied energy level. These additional electrons help to shield the outermost electrons from the attractive forces of the nucleus.

But why does the ionization energy generally increase as you move through a period from left to right? Recall that as you move through a period, the nuclei of succeeding elements exert more pull on their electrons. This stronger pull means that more energy is required to remove an electron. This is similar to what happens when someone tries to pull away something you are holding. The harder you hold onto it, the more difficult it is to remove.

Electron affinity decreases within a family and increases within a period

electron affinity *the energy change that accompanies the addition of an electron to an atom in the gas phase*

The ability of an atom to attract and hold an extra electron is its **electron affinity**. Electron affinity is measured as the energy change that occurs when an electron is added to an atom as shown in **Figure 4-19**. For an atom of any element (A), the attraction of an electron can be represented as in the following equation. Note that when an atom gains an extra electron, it acquires a negative charge and becomes a *negative ion*.

$$\mathbf{A + e^- \longrightarrow A^- + energy}$$

Electron affinity can have either a positive or negative numerical value. A negative value indicates that an atom releases energy when it gains an electron. A positive electron affinity means that energy must be added to the atom for the electron to be added. Thus, the more negative its electron affinity is, the more easily an atom can take in an extra electron.

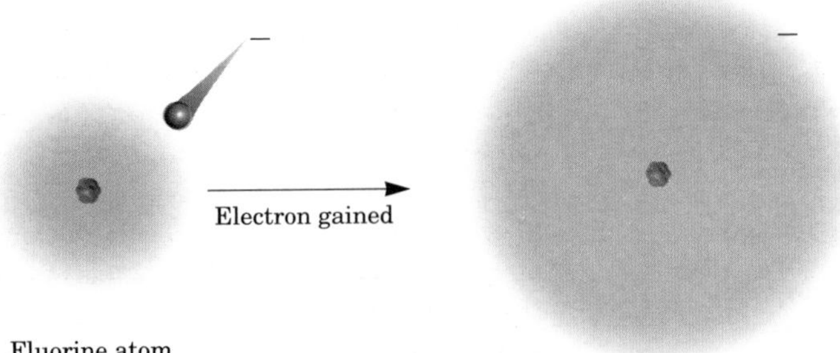

Figure 4-19
A fluorine atom gains an electron to become a negatively charged ion. Fluorine gives off energy in the process. Therefore, the electron affinity of fluorine has a negative value.

Table 4-6
Electron Affinity Values (in kJ/mol)

1	2		13	14	15	16	17
1 H −73							
3 Li −60	4 Be 0		5 B −27	6 C −154	7 N −7	8 O −141	9 F −329
11 Na −53	12 Mg 0		13 Al −42	14 Si −134	15 P −72	16 S −200	17 Cl −349
19 K −48	20 Ca 0		31 Ga −29	32 Ge −120	33 As −77	34 Se −195	35 Br −325
37 Rb −47	38 Sr 0		49 In −29	50 Sn −107	51 Sb −101	52 Te −190	53 I −295
55 Cs −45	56 Ba 0		81 Tl −19	82 Pb −35	83 Bi −91	84 Po −183	85 At −270

By studying **Table 4-6**, you can see that *electron affinity values generally become more negative as you move from left to right across a period.* This trend can be explained by changes in nuclear charge, atomic radius, and shielding effect. From left to right across a period, nuclear charge increases, atomic radius decreases, and the shielding effect remains constant, so the attractive force that the nucleus can exert on another electron increases.

From top to bottom within a group, electron affinity tends to become less negative. Again, nuclear charge, atomic radius, and the shielding effect account for the observed trend. As you go down a group the effective nuclear charge increases. This effect is more than offset by the larger size. The result is decreased attraction for an added electron which results in a decreased electron affinity.

Electronegativity decreases within a family and increases within a period

electronegativity *the tendency for an atom to attract electrons to itself when it is combined with another atom*

The **electronegativity** of an atom is the tendency of an atom to attract electrons to itself when it is chemically combined with another element. Electronegativity is expressed in terms of a relative scale with arbitrarily selected standard units. Fluorine, the most electronegative element, is assigned an electronegativity value of 4.0. Electronegativity values for the other elements are calculated in relation to this value. **Figure 4-20**, on the next page, shows the general electronegativity trend of the periodic table. The actual values can be found on **Table 11-3**, page 418. The noble gases are not included in these figures because they do not form a significant number of chemical compounds.

Content Background

Electronegativity can also be related to the size of the atom. The smaller the atom in a family or series, the greater its electronegativity. A small atom has a stronger force of attraction for its own electrons, and for other atoms' electrons.

Historical Note

Linus Pauling (1901–1994) developed the quantitative scale for electronegativity using bond-strength.

ASSESS

Alternative Assessment

Give each student a copy of the periodic table. Without referring to class notes or the text, students should draw arrows indicating periodic trends in ionization energy, electronegativity, shielding effect and atomic radius.

Answers to Section Review

6. It is difficult to measure the size of an atom because it does not have a firm boundary.
7. Metals are less electronegative than nonmetals. Metals have lower ionization energies than nonmetals.
8. Except for atomic radius, each of these properties show decreasing values down a group of elements and increasing values from left to right across a period. Atomic radius increases moving down a group and decreases moving left to right across a given period.
9. The atom that loses an electron is positively charged and smaller. The atom that gains an electron is negatively charged and larger.
10. a. fluorine, carbon, iodine, lithium, potassium, rubidium
    b. calcium, germanium, bromine, nitrogen, oxygen, fluorine
    c. cesium, beryllium, krypton, helium

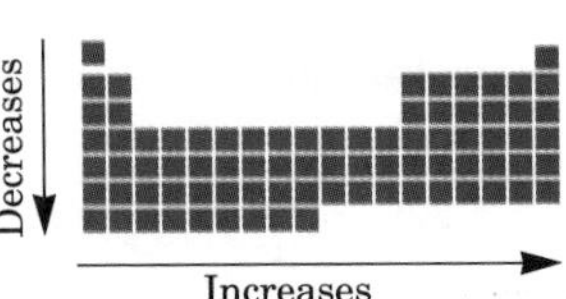

Figure 4-20
Electronegativity values tend to decrease down a group and increase across a period.

As **Figure 4-20** shows, electronegativity follows a periodic trend: *electronegativity values generally decrease going down a group and increase going across a period.* The least electronegative element is cesium, found toward the lower left corner of the table. The most electronegative element is fluorine, located in the upper right corner of the periodic table (excluding the unreactive noble gases).

Table 4-7 summarizes the periodic trends in atomic radius, ionization energy, shielding effect, and electronegativity. You will see the importance of these trends in the chapters that follow as you study how elements react to form compounds.

Table 4-7
Summary of Periodic Trends

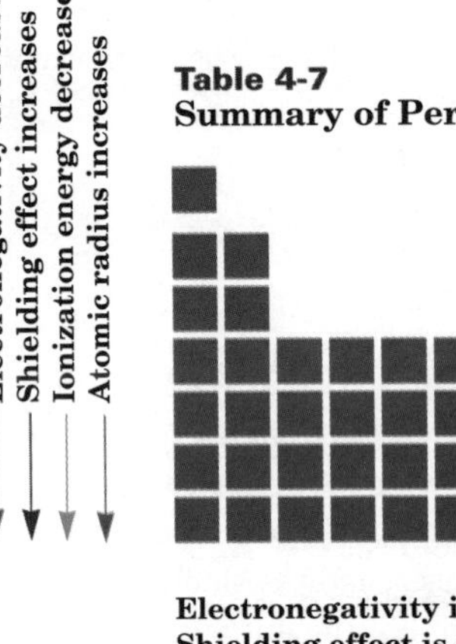

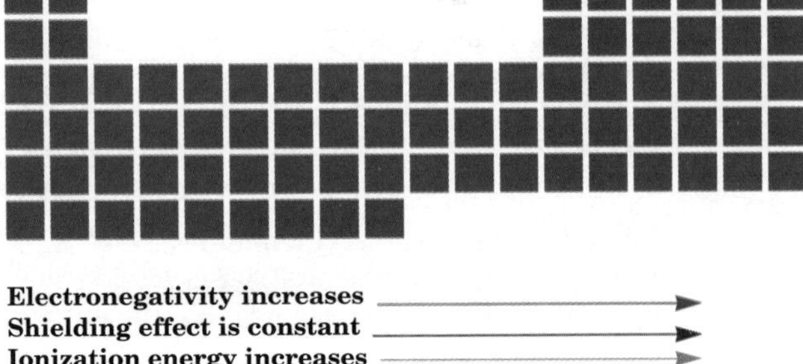

Section Review

6. Why is it difficult to measure the size of an atom?
7. How do metals and nonmetals generally compare with respect to their ionization energies and electronegativities?
8. What trends are evident in atomic radius, ionization energy, and electronegativity as you proceed down a group of elements? How do each of these trends progress as you move across a period?
9. When an atom loses an electron to become an ion, what happens to its electric charge? To its size? When an atom gains an electron to become an ion, what happens to its charge? To its size?
10. a. Arrange the following atoms to show a trend of increasing atomic radius: potassium, carbon, rubidium, iodine, fluorine, and lithium.
    b. Arrange these elements in terms of increasing electronegativity values: fluorine, nitrogen, calcium, germanium, oxygen, and bromine.
    c. Arrange these elements in order of increasing ionization energy: beryllium, helium, krypton, and cesium.

Visual Strategy

Figure 4-20

The table of electronegativities is found in Chapter 11 on page 418. Refer students to this table to confirm the trend shown in this figure.

How are elements created?

Section Objectives

Distinguish between naturally occurring and synthetic elements.

Describe how the naturally occurring elements are formed.

Explain the term **nuclear reaction.**

Explain how scientists use particle accelerators to create synthetic elements.

The origins of naturally occurring elements

You have learned much about the uses of the elements. You have also learned how to read the periodic table to obtain information about the properties and structures of elements. But two fundamental questions remain: Where do the elements come from? How were they created?

Natural and synthetic elements are created in different ways

Of the over 110 elements currently known, only elements up to number 92, uranium, occur naturally.* The remaining elements in the periodic table are all *synthetic*. These elements are made by teams of research scientists working in vast laboratories around the world. One such laboratory, the Stanford Linear Accelerator, is 3 km long and is shown in Figure 4-21.

The origins of both natural and synthetic elements are equally fascinating. While the creation of synthetic elements requires technology and human ingenuity on the grandest scale, the naturally occurring elements were forged in violent unions of matter and energy in deep space. With the exception of hydrogen and trace amounts of other light elements, all of the natural elements found on Earth were manufactured in the interiors of stars that exploded long before our solar system came into being. Hydrogen was created as an immediate consequence of the Big Bang, the initial explosion that is believed to have launched the universe.

* Four elements with atomic numbers less than 92 are not truly naturally occurring. These elements are: Fr, Pm, Tc, and At. Technetium (Tc) is produced in the laboratory and does not exist naturally. The other three elements exist naturally in minute amounts, but samples are obtained from artificial sources.

Figure 4-21
The Stanford Linear Accelerator Center is a national facility for research in subatomic physics. It is located at Stanford University in California and includes several particle accelerators. Particle accelerators are the largest machines on Earth. Synthetic elements are created in particle accelerators during energetic collisions of small particles.

Section 4-3

FOCUS

Lesson Starter

Tell students that the alchemists wanted to change metals to gold. Tell students that it is now possible (but not economically feasible) to turn metals into gold. Remind them that 400 years ago, no one knew about the strucure of an atom or the existence of protons, electrons, and neutrons.

TEACH

Do You Know?

The age of the sun has been estimated to be 9.5 billion years old with approximately 6 billion years left in its life span. The age of a star can be estimated by its color and by its percentage composition of helium.

Historical Note

Even a hundred years ago it was known that the sun was not "burning." It had been estimated that even though the sun weighs 6×10^{30} kg, it would take only approximately 400 years for it to burn out and historical records in Europe alone went back further than that. At this point, it was suggested that the sun must be undergoing some other process than burning.

Content Background

The first fusion reactions were produced in the 1950s with the production of the first fusion bomb also known as the hydrogen bomb or H-bomb. Unfortunately, they cannot be called controlled fusion reactions. Even today, fusion reactions are difficult to maintain for more than a few millionths of a second because of the extremely high temperatures necessary (>20 million °C) and the production of a magnetic bottle that holds the reaction in place.

Cultural Connection

Many of the chemists and physicists who worked on the Allied fission project of the late 1930s and early 1940s were German Jews who sought refuge in Britain and the United States to escape persecution. After the war, many of these people were interested in peaceful uses for nuclear energy, not destructive ones.

Elements are created through nuclear fusion

Stars form when clouds of dust and hydrogen gas condense under the influence of gravity. As stellar material condenses, pressure builds and temperatures within a young star reach millions of degrees. The spectra of stars show that they are composed chiefly of hydrogen and helium. The star nearest the Earth, the sun, is a great ball of hydrogen and helium gas that glows because of its high temperature in the same way that a burning coal glows when it is "red hot."

If the stars' high temperatures and resulting glow were simply the result of the gravitational contraction of stellar material, the stars would shine for only about 30 million years. Yet, rocks have been found on Earth indicating that the solar system is at least 4 billion years old. So, from what source do the stars derive the energy they need to shine for such a long time?

nuclear reaction *a reaction that involves a change in the nucleus of an atom, as opposed to a chemical reaction which involves changes to the arrangement of electrons that surround the nucleus*

The principal source of stellar energy is *nuclear fusion*. Nuclear fusion occurs when the nuclei of two or more atoms join together, or *fuse*, to form the nucleus of a larger atom. The basic fusion process in most stars typically follows this pattern: four hydrogen nuclei combine to produce one helium nucleus, represented in Figure 4-22. Nuclear fusion is one kind of **nuclear reaction**. In a nuclear reaction, an atomic nucleus gains or loses protons, thereby becoming a different element.

The mass of the helium nucleus formed in the nuclear fusion process, however, is slightly less than the mass of the four hydrogen nuclei that went into it. The small amount of "missing" mass is converted to energy according to Einstein's famous equation: $E = mc^2$. Hence the mass of combining nuclei, which is converted to energy in the nuclear fusion process, supplies the enormous energy that stars use to shine.

Nuclear fusion is not only the principal source of energy for stars, but also the process by which elements heavier than hydrogen are created. Most of the helium in the universe, as noted above, was produced inside stars as hydrogen nuclei undergo nuclear fusion. The sun alone converts about 400 million tons (3×10^{14} g) of hydrogen into helium every second.

Figure 4-22a **Nuclear fusion in the sun has provided the Earth with energy for billions of years. Fusion reactions also create elements heavier than hydrogen.**

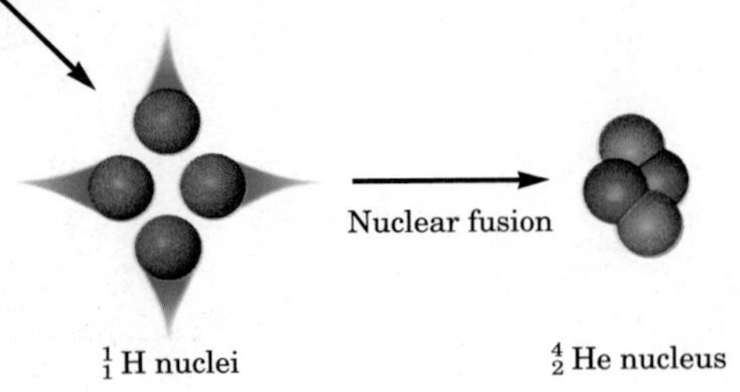

4-22b **The single helium nucleus has less mass than the four hydrogen nuclei from which it is formed. The small amount of mass "lost" during fusion is converted to energy.**

Visual Strategy

Figure 4-22

Emphasize that the alpha particle produced in fusion consists of two protons and two neutrons, not the original four protons.

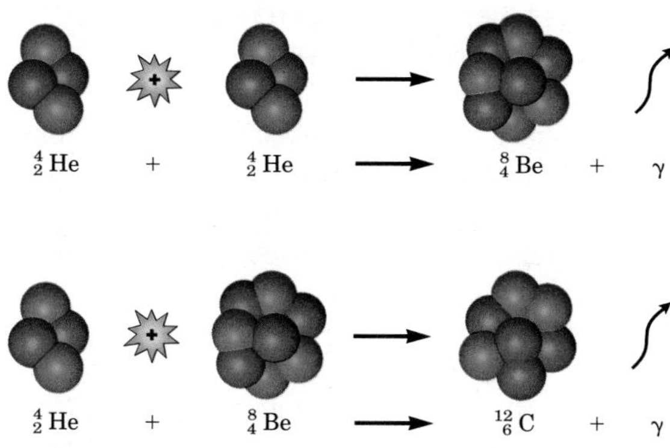

Figure 4-23
Carbon-12 is formed in stars through these nuclear reactions.

Fusion reactions other than the fusion of four hydrogen nuclei to create a helium nucleus also occur. Depending on the temperature of a star, its mass, and the stage of its development, many other fusion reactions may take place. For example, the fusion reactions illustrated in **Figure 4-23** can occur at temperatures above 10^8 K. This process is important only in stars hotter than the Earth's sun. Note that in the first part of the process, two helium nuclei fuse to produce an isotope of beryllium (as well as electromagnetic energy in the form of gamma rays, which are symbolized by the lowercase Greek letter gamma, γ). In the second step, the newly created beryllium isotope fuses with a helium nucleus to produce a carbon-12 isotope.

When a star uses up all of the elements that fuel its nuclear fusion, the star is no longer stable, and it dies in a last, great explosion. The elements forged within the star are flung into space. When planets eventually condense from this material, they take up the rich array of elements found in the stellar debris. Elements heavier than Fe are produced by supernovas.

The relative abundances of the elements found in the universe have been calculated from meteorites that have fallen from space and from other sources. The results are shown graphically in **Figure 4-24**. This graph closely matches the relative abundance of elements found in stars. The close match is strong evidence that the elements distributed throughout the universe are the remnants of stars that have long since exploded.

Figure 4-24
Hydrogen is the most abundant element in the universe. Which is the least abundant natural element?

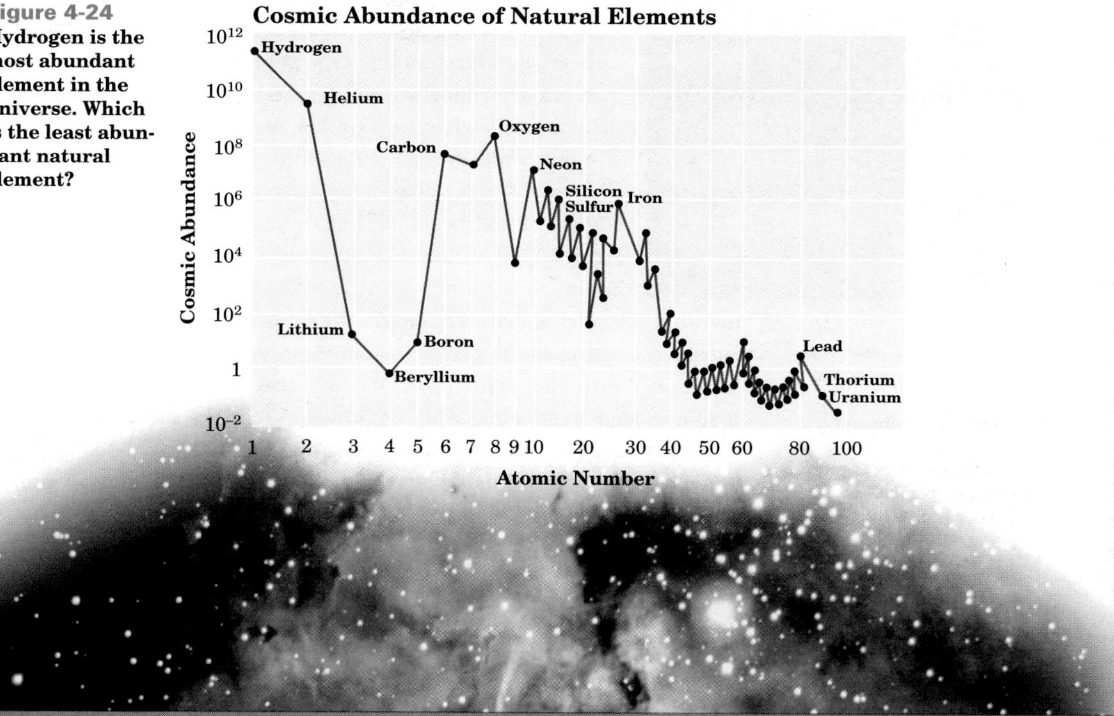

Visual Strategy

Figure 4-24

Have students compare the values on this graph to the abundances of elements in the Earth's crust, **Figure 4-20** on page 48. What similarities and differences do they notice?

Demonstration 5

Cloud chamber

Approximate time: 15 min

1. To make a simplified Wilson cloud chamber, get a large glass jar with lid, a piece of black velvet, a piece of blotter paper, some alcohol, and a flat piece of dry ice.
2. Cut the velvet to just fit inside the lid. Cut the blotter paper, and glue it to the bottom of the jar (or obtain a commercially produced cloud chamber).
3. Add rubbing alcohol to the blotter paper drop-by-drop until it is saturated but not dripping.
4. Place the jar lid on the dry ice. Place a packaged nuclear source (obtained from a science supply house) in the lid, and screw on the jar. (The jar should be sitting upside down on the dry ice.)
5. The jar should become saturated with alcohol vapor, and the emitted nuclear particles should leave white streaks.

SAFETY

Wear safety goggles and a lab apron. Students must stand back at least 10 ft. Ensure that there are no flames in the room. Have no more than 100 mL of alcohol in the room. Do not touch the radioactive source. Use cloth or leather gloves when handling the dry ice. To break the dry ice, wrap it in toweling, place it on a firm surface, and strike it with a hammer.

DISPOSAL

Return leftover alcohol to the stock bottle. Allow the dry ice to sublime in a restricted location not accessible to students.

The origins of synthetic elements

The first artificial isotope was created in 1919

Figure 4-25a
Numerous synthetic elements were originally created by scientists working at Lawrence Berkeley Laboratory on the University of California campus. Some of these Berkeley scientists are shown here.

In Chapter 3, you read about how scientists working in the early 1900s shot extremely small, positively charged particles at a thin sheet of gold foil and discovered the atomic nucleus. These small particles, called *alpha particles*, are fast-moving helium nuclei. Alpha particles are represented by the symbol for helium: ${}^{4}_{2}\text{He}$.

In 1919, alpha particles were used in another important experiment on atomic nuclei. This time, alpha particles were targeted at nitrogen atoms. The alpha particles reacted with the nitrogen nuclei. As a result, an oxygen isotope was created. This experiment was the first instance in which one element was transformed into another in the laboratory. This transformation of nitrogen is represented by the following equation.

$$ {}^{14}_{7}\text{N} + {}^{4}_{2}\text{He} \longrightarrow {}^{17}_{8}\text{O} + {}^{1}_{1}\text{H} $$

Besides oxygen, what other product was made by bombarding nitrogen with alpha particles?

Elements heavier than uranium are synthetic

Today, scientists change one element into another by bombarding nuclei with various small particles, such as protons, neutrons, alpha particles, and beta particles (which are fast-moving electrons). The bombarding particles, known collectively as "nuclear bullets," undergo nuclear reactions with the nuclei at which they are aimed. Many isotopes of the naturally occurring elements—as well as numerous *synthetic* elements—have been made this way. Indeed, all of the synthetic elements with atomic numbers greater than 92 (uranium) but less than 101 (mendelevium) were created through this type of process.

4-25b
These tracks show the paths taken by colliding particles in the bubble chamber of a particle accelerator. Such bubble chamber tracks are used to detect particles created in collisions.

In order for "nuclear bullets" and nuclei to undergo a nuclear reaction when they collide, rather than just bounce off one another, they must be moving very fast and have a lot of energy. Accelerating particles to the required speeds is a major technological feat that is accomplished by particle accelerators, such as the Stanford Linear Accelerator shown in **Figure 4-21** on page 127, or the cyclotron located at Lawrence Berkeley Laboratory on the University of California campus, which is the source of the photograph in **Figure 4-25a**.

Elements with atomic numbers 101 and greater are also synthetic, but they have been created by a different process. To make these elements, accelerators hurl entire nuclei at one another. For example, nobelium, atomic number 102, is created by crashing together carbon and curium nuclei. This reaction is represented by the following equation.

$$ {}^{12}_{6}\text{C} + {}^{244}_{96}\text{Cm} \longrightarrow {}^{254}_{102}\text{No} + 2{}^{1}_{0}n $$

Visual Strategy

Figure 4-25

Make the students aware that physicists and chemists study particle-chamber tracks like this to find telltale signs of small particles. As with other parts of atomic theory, we cannot view the particles themselves, but we can tell where the particles have been and how they behaved and interacted while moving.

Unq
Credit for the discovery of this radioactive synthetic metal will eventually be assigned to Russian scientists at the Joint Institute for Nuclear Research at Dubna, or to scientists at the University of California at Berkeley, depending on which team can provide the best evidence of having created this element.

Une
On August 29, 1982, element 109 was made and identified by scientists at the Heavy Ion Research Laboratory, in Darmstadt, West Gemany.

Mendelevium
Synthesized in 1955 by G. T. Seaborg, A. Ghiorso, B. Harvey, G. R. Choppin, and S. G. Thompson at the University of California, Berkeley; named in honor of the inventor of the periodic table.

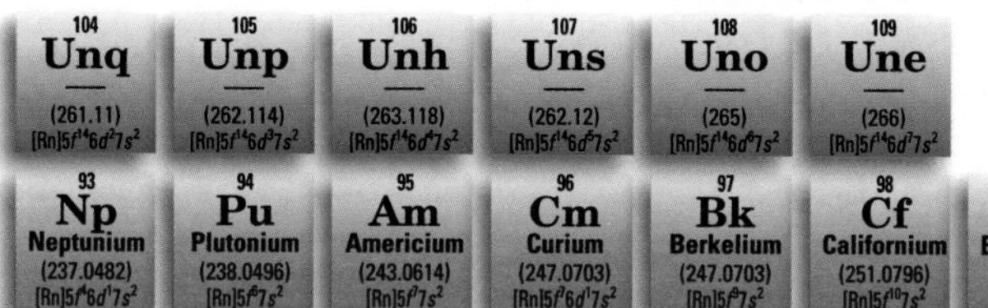

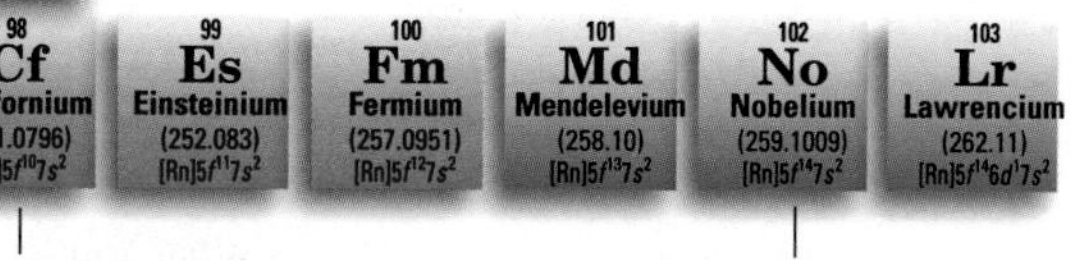

Curium
Synthesized in 1944 by G. T. Seaborg, R. A. James, and A. Ghiorso at the University of California, Berkeley; named in honor of Marie and Pierre Curie.

Californium
Synthesized in 1950 by G. T. Seaborg, S. G. Thompson, A. Ghiorso, and K. Street, Jr. at the University of California, Berkeley; named in honor of the state of California.

Nobelium
Synthesized in 1958 by A. Ghiorso, G. T. Seaborg, T. Sikkeland, and J. R. Walton; named in honor of Alfred Nobel, discoverer of dynamite and founder of the Nobel Prize.

Figure 4-26
All the highlighted elements are synthetic. Those shown in red were created by colliding moving particles with stationary targets. The elements shown in blue were created by colliding nuclei.

Though other discoveries have been reported, the discovery of element 109 has been thoroughly verified and accepted. Element 109 is extremely unstable. Only three atoms of element 109 have ever been produced, and they existed for only a short time, only 0.0034 second, just long enough to be spotted and identified.

Today, scientists are attempting to create some "superheavy" elements that may be more stable than element 109. Why should these superheavy elements be stable? To understand why, you need to know that protons and neutrons are arranged in alternating shells within an atomic nucleus. Each shell can contain no more than a certain maximum number of particles. When its shells are filled, a nucleus is stable. If a nucleus contains unfilled shells, it is unstable and breaks apart spontaneously.

If element 114 were created, calculations show that the shells in its nucleus would be filled. Therefore, element 114 should be stable. Look at a periodic table to see that there is an empty spot where element 114 could fit, under lead in Group 14. But can element 114 exist? No one knows for sure. More experiments are needed to find out.

Section Review

11. Define the term *naturally occurring element*. Where are the naturally occurring elements found in the periodic table?
12. How are the naturally occurring elements created?
13. What evidence is there that the elements are made in stars?
14. What is a synthetic element?
15. Describe the two types of processes scientists use to create synthetic elements.

ASSESS

Alternative Assessment

Ask each student to write a brief science fiction short story describing a jouney into space to observe the formation of various elements in stars. Students should describe the particles and processes they would observe in as much detail as possible.

Answers to Section Review

11. Naturally ocurring elements are those made through natural processes, as opposed to artificial elements that are produced by scientists in laboratories. The naturally occurring elements are found through uranium, atomic number 92.
12. The naturally ocurring elements are created through stellar fission and fusion.
13. The abundances of elements found in stars closely match the abundances of elements found in the universe at large.
14. A synthetic element is one that is produced in the laboratory.
15. Scientists can accelerate subatomic particles at stationary targets or cause two or more nuclei to collide to form a larger nucleus.

Section 4-4

FOCUS

Lesson Starter

Divide the class into groups of three and provide each group with identical packs of 5 strips of paper of different lengths. Have the groups use the strips in brainstorming some type of measuring system. Students will be forced to invent a standard and use it to measure their paper strips. Let each group share its strategy with the class.

TEACH

Teaching Tip

Analogy

An analogy can be presented using fruit baskets containing apples, grapefruits, and peaches. Begin by setting up these parameters. A basket filled with apples would have a mass of 4 kg. If it were filled with grapefruits, it would have a mass of 5 kg. If it were filled with peaches, it would have a mass of 4.75 kg. Notice that the fruits are all about the same size, similar to the size relationships for isotopes. Ask students to estimate the mass of the fruit basket if it contained 50% apples and 50% grapefruits.

$(0.50 \times 4 \text{ kg}) + (0.50 \times 5 \text{ kg}) = 4.5 \text{ kg}$

Most students will be able to intuitively guess the answer is 4.5 kg. Then ask the students to change the fruit to 40% apples, 15% grapefruits, and 45% peaches and determine the mass of the basket.

$(0.40 \times 4 \text{ kg}) + (0.15 \times 5 \text{ kg}) + (0.45 \times 4.75 \text{ kg}) = 4.49 \text{ kg}$

Relate this analogy to isotopes to show students that the average atomic mass of an element must be greater than smallest isotopic mass and less than the greatest isotopic mass.

Can atoms be counted or measured?

Section Objectives

Explain the relationship between atomic mass and atomic mass units.

Use a periodic table to determine the average atomic mass for an element.

Use the mole as a counting unit for large numbers of atoms.

Solve problems with conversions between moles, Avogadro's number, and molar mass.

Calculate the mass of a single atom.

Finding mass measurements in the periodic table

Mendeleev initially arranged the periodic table according to atomic masses. Although Moseley changed this so that the table is now arranged according to atomic number, the periodic table is still your first source for information on the masses of atoms. You learned in Chapter 3 that the mass of an atom can be expressed as its mass number. But a mass number indicates the total number of protons and neutrons present, not their actual mass. Usually, mass is expressed in mass units.

Atomic mass is expressed in atomic mass units

Measured in grams, the masses of atoms are extremely small. A carbon-12 atom, for example, has a mass of only 1.99×10^{-23} g. Such extremely small numbers can be a nuisance in calculations. Therefore, instead of using actual atomic masses, chemists find it more convenient to work with relative atomic masses. In determining relative atomic masses, one atom is arbitrarily chosen as the standard and assigned a value. The masses of all the other atoms are then expressed in relation to this standard value. Relative scales are easy to establish and use, as you can see from examining **Figure 4-27**.

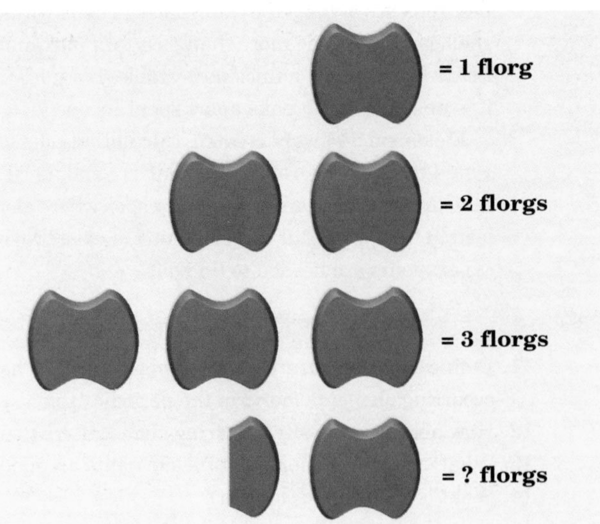

Figure 4-27
In establishing a relative scale, a standard is selected and everything is then compared to this standard. What is the standard used in this illustration?

atomic mass unit
one-twelfth the mass of the carbon-12 isotope

In establishing the relative scale for atomic masses, scientists agreed upon the carbon-12 atom as the standard. A single carbon-12 atom, which has an actual mass of 1.99×10^{-23} g, was arbitrarily assigned the value of 12 atomic mass units. Based on this, one **atomic mass unit** is exactly 1/12 of the mass of a carbon-12 atom. So, in terms of the actual mass values of atoms, one atomic mass unit equals 1.66×10^{-24} g.

The atomic mass unit is represented by the symbol *amu*. The atomic mass of a carbon-12 atom is therefore 12 amu. Hydrogen-1, which has 1/12 of the mass of carbon-12, has an atomic mass of 1 amu. The atomic masses of all the other elements are expressed in the same way—relative to the atomic mass arbitrarily assigned to carbon-12. The mass of an atom expressed in atomic mass units (amu) is called the **atomic mass.**

atomic mass
the mass of an atom in atomic mass units

The periodic table lists average atomic mass

If you check the atomic mass given for either carbon or hydrogen on a periodic table, you'll notice something odd. The value listed for carbon is 12.01, not just 12; that for hydrogen is 1.00794, not simply 1. In fact, most of the atomic masses listed in the periodic table are not whole numbers. Most atomic masses in the periodic table are not whole numbers because the mass listed for any element is the weighted average of the masses of all its naturally occurring isotopes.

Weighted averages, like relative scales, are not complicated. Your teacher uses a weighted average when calculating the average for a set of grades for a class. The same procedure is used for determining the average atomic mass of an element. Just as a weighted grade average depends on two factors (students and grades), the average atomic mass depends on two factors—in this case, the mass and relative abundance of each isotope. Consider the example of copper. Its average atomic mass is computed below.

Sample Problem 4A
Calculating average atomic mass

Copper has two naturally occurring isotopes: copper-63 and copper-65. The relative abundance of copper-63 is 69.17%; the atomic mass of copper-63 is 62.94 amu. The relative abundance of copper-65 is 30.83%; its atomic mass is 64.93 amu. Determine the average atomic mass for copper.

Calculate

- **Use a weighted average such that the atomic mass of each isotope is multiplied by its relative abundance. The sum of these products is divided by 100.**

$$\frac{(69.17 \times 62.94\text{ amu}) + (30.83 \times 64.93\text{ amu})}{100} = \text{Average atomic mass of copper} = 63.55\text{ amu}$$

Practice 4A

Answers for Concept Review items and Practice problems begin on page 841.

1. **Calculate the average atomic mass for silicon if 92.21% of its atoms have a mass of 27.98 amu, 4.70% have a mass of 28.98 amu, and 3.09% have a mass of 29.97 amu.**
2. **Calculate the average atomic mass of oxygen. Oxygen has three naturally occurring isotopes: oxygen-16 with a mass of 15.99 amu; oxygen-17 with a mass of 17.00 amu; and oxygen-18 with mass of 18.00 amu. The relative abundances are 99.76%, 0.038%, and 0.20%, respectively.**

Additional Examples for Sample Problem 4A

Determine the average atomic mass for both carbon and uranium given the following information.

a. The abundance of ^{12}C is 98.90% with a mass of 12.000 amu. The abundance of ^{13}C is 1.1% with a mass of 13.003335 amu.

b. The abundance of ^{235}U is 0.720% with a mass of 235.043924 amu. The abundance of ^{238}U is 99.28% with a mass of 238.050784 amu.

Solutions are at the bottom of the page.

Teaching Tip

Enrichment

To show how large a mole is, help students to determine the number of moles of water drops in the Pacific Ocean. This will require them to find the area of the Pacific Ocean and its average depth in a resource book. They then need to find the number of water drops that equal 1 mL.

Sample data
Surface area—8×10^7 mi^2
Average depth (estimated)—14 000 ft
Drops/mL—20–22 drops
Answer: 1.945×10^{25} drops = 32 moles of water drops

Answers to Practice 4A

1. 28.09 amu
2. 15.99 amu

*Full solutions for items **1** and **2** are found on page 147A.*

Solutions for Additional Examples 4A

a. (0.9890)(12.000) + (0.011)(13.003335) = 12.011037 = 12.01
b. (0.00720)(235.043924) + (0.99280)(238.050784) = 238.029134 = 238.03

Table 4-8
Counting Units

Units	Example
1 dozen eggs	
1 ream of paper	
1 mole	6.022×10^{23} particles

The mole

A mole is a huge number

All calculations involving atoms so far have dealt with terms that are defined in relation to infinitesimally small particles: atomic number relates to the number of protons; mass number relates to the number of protons and neutrons; and the atomic mass unit standard relates to an isotope of carbon. Rather than always working on the atomic scale, chemists often find it useful to work with a unit that represents a large collection of atoms. Such a unit serves as a bridge between the invisible world of atoms and the macroscopic world of materials and objects. This unit is called a *mole* and is abbreviated *mol*. How big is a mole?

A **mole** is a collection of $6.022\ 137 \times 10^{23}$ particles, as shown in **Table 4-8**. The mole is usually rounded to 6.022×10^{23}. You may wonder why 6.022×10^{23} particles was chosen as the counting unit for atoms or molecules. What is so special about this number? The answer is that there is exactly 1 mol of atoms in the atomic mass of an element when that mass is expressed in grams. For example, the atomic mass of carbon-12 is 12.00 amu. At the same time, 1 mol, or 6.022×10^{23}, carbon-12 atoms has a mass of 12.00 g.

mole
the fundamental SI unit used to measure the amount of a substance

In recognition of the value and importance of this number, scientists named it in honor of Amedeo Avogadro (1776–1856), an Italian scientist whose ideas were crucial to the early development of chemistry, as explained in **Figure 4-28**. **Avogadro's number** is the number of particles, $6.022\ 137 \times 10^{23}$, in exactly 1 mol of a pure substance.

Avogadro's number
6.022×10^{23}, the number of particles in a mole

To appreciate how large Avogadro's number is, imagine that all 5 billion people on Earth were to do nothing but count the atoms in 1 mol of an element. If each person counted at the rate of one atom per second, it would take about 4 million years to count all of the atoms in that 1 mol!

Figure 4-28
Avogadro's number was determined by a German physicist nine years after Avogadro's death. Count Amedeo Avogadro was a lawyer whose interests turned to mathematics and physics. He later became a professor of these subjects. Avogadro was the first to propose the concept of molecules. Unfortunately, the significance of Avogadro's ideas was not recognized until after his death.

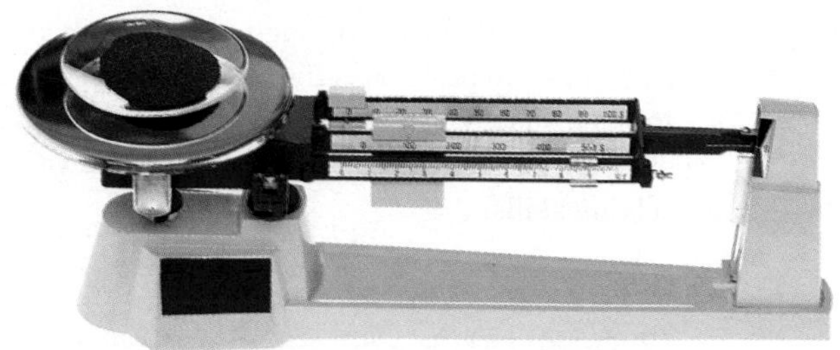

Figure 4-29
One mole of iron has a mass of 55.85 g and contains 6.022×10^{23} iron atoms. The mass shown represents a mole of Fe and the mass of the watch glass.

Moles can be converted to number of atoms and vice versa

Avogadro's number is useful as a factor in converting from a given number of moles to the equivalent number of atoms. The conversion works just like when you change a quantity like 20 dozen eggs to the equivalent number of individual eggs. If 1 dozen is 12, then 20 dozen is 20×12. The conversion of 5 mol to the equivalent number of atoms works the same way. If 1 mol is equivalent to 6.022×10^{23} atoms, then 5 mol would contain $5(6.022 \times 10^{23})$ atoms.

If you were given a number of atoms and needed to find the number of moles, you would again use Avogadro's number. The reverse calculation works the same way as when you have 350 objects and you want to express this value in dozens. If 1 dozen is 12, then you divide the total number of objects by 12 to determine the number in dozens.

$$\frac{350}{12} = 29.2 \text{ dozen}$$

To convert the number of atoms to moles you divide the number of atoms by Avogadro's number. For example, if you have 1×10^{10} atoms, you would divide by 6.022×10^{23} to get the number of moles. **Figure 4–30** shows another way to look at these conversions. The relationship of 1 mol to 6.022×10^{23} atoms can be written as two equivalent unit factors.

$$\frac{6.022 \times 10^{23} \text{ atoms}}{1 \text{ mol}} \qquad \frac{1 \text{ mol}}{6.022 \times 10^{23} \text{ atoms}}$$

Factors like the ones in **Figure 4–30** are helpful because you can keep track of units as a way of checking your work. You generally multiply the factor that has the correct units for the answer in the numerator by the given quantity. You can see how this process works in the feature and Sample Problems that follow.

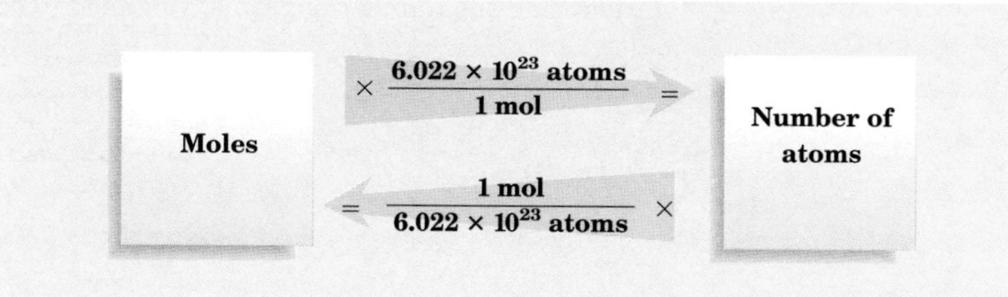

Figure 4-30
By using Avogadro's number, 6.022×10^{23}, you can determine moles from number of atoms and number of atoms from moles.

How To

Use the sample problems in this text

The Sample Problems throughout this book are set up to help you learn to solve problems effectively. The process for solving most Sample Problems is divided into three steps. Each step emphasizes a specific set of activities for you to follow that will help you develop your problem-solving skills.

1. List what you know

Don't start using your calculator yet. The biggest mistake that beginning chemistry students make is taking numbers from a problem statement and using the calculator before they know what the problem means.

- Read the problem twice.
- Organize the information given in the problem statement. What is given? What are you asked to find?
- List any conversion factors you might need such as Avogadro's number or molar masses from the periodic table.

2. Set up the problem

Don't start using your calculator yet.

- First, analyze what needs to be done to get the answer. Identify the value given in the problem that is your starting point. Write the units needed for your answer. Then determine the relationships needed to get from the given value to the answer.
- Write down your setup with all the conversion factors. Check to see how the units cancel each other. If they all cancel to give you the units needed for your answer, the setup is probably correct.

3. Calculate and verify

- Make an estimate of your answer by rounding the numbers in the setup and making a quick calculation. Another way of making an estimate is to look at the conversion factors in your setup and decide whether your answer should be larger or smaller than the beginning value.
- Work through your setup from Step 2. In the examples in this book that require multiple calculations, the numbers are not rounded between steps.
- When you finish your calculations, round off the answer to the correct number of significant figures.
- Verify your answer to make sure it is *reasonable*. For example, if you make a conversion of grams to moles and the number of moles you get is larger than the number of grams, you need to double-check your work.

Sample Problem 4B
Converting moles to number of atoms

Determine how many atoms are present in 2.5 moles of silicon.

1 List what you know
- **moles of Si** = 2.5
- **number of atoms Si** = ?
- **Avogadro's number** = 6.022×10^{23} atoms in one mole

2 Set up the problem
- **Use Figure 4-30, to determine which factor will take you from moles to the number of atoms.**

$$\frac{6.022 \times 10^{23} \text{ atoms}}{1 \text{ mol Si}}$$

- **Multiply the number of moles by this factor.**

$$2.5 \text{ mol Si} \times \frac{6.022 \times 10^{23} \text{ atoms Si}}{1 \text{ mol Si}} = ?$$

3 Calculate and verify
- **Solve and cancel like units in the numerator and denominator.**

$$2.5 \cancel{\text{mol Si}} \times \frac{6.022 \times 10^{23} \text{ atoms Si}}{1 \cancel{\text{mol Si}}} = 1.5 \times 10^{24} \text{ atoms Si}$$

- **The answer has the correct units and is more than Avogadro's number of atoms, which makes sense because you started with more than one mole.**

Sample Problem 4C
Converting number of atoms to moles

Convert 3.01×10^{23} atoms of silicon to moles of silicon.

1 List what you know
- **number of atoms Si** = 3.01×10^{23}
- **moles of Si** = ?
- **Avogadro's number** = 6.022×10^{23} atoms in one mole

2 Set up the problem
- **Use Figure 4-30 to determine which factor will take you from the number of atoms to moles.**

$$\frac{1 \text{ mol}}{6.022 \times 10^{23} \text{ atoms}}$$

- **Multiply the number of atoms by this factor.**

$$3.01 \times 10^{23} \text{ atoms Si} \times \frac{1 \text{ mol Si}}{6.022 \times 10^{23} \text{ atoms Si}} = ?$$

3 Calculate and verify
- **Solve and cancel like units in the numerator and denominator.**

$$3.01 \times 10^{23} \cancel{\text{atoms Si}} \times \frac{1 \text{ mol Si}}{6.022 \times 10^{23} \cancel{\text{atoms Si}}} = 0.500 \text{ mol Si}$$

- **The answer has the correct units and is less than 1 mol, which makes sense because you started with less than Avogadro's number of atoms.**

Practice 4B

1. **How many atoms are present in 3.7 moles of sodium?**
2. **How many atoms are present in 155 moles of arsenic?**

4C

3. **How many moles of xenon are equivalent to 5.66×10^{26} atoms?**
4. **How many moles of silver are equivalent to 2.888×10^{15} atoms?**

Additional Examples for Sample Problem 4B

Determine the number of atoms in the following.

a. 3.5 mol Na
b. 0.78 mol of Mg
c. 22 mol of Al

Solutions are at the bottom of the page.

Additional Examples for Sample Problem 4C

Convert the following to moles.

a. 4.78×10^{22} atoms of Ag
b. 6.85×10^{23} atoms of Hg
c. 1.23×10^{26} atoms of He

Solutions are at the bottom of the page.

Answers to Practice 4B and 4C

1. $3.7 \text{ mol Na} \times \frac{6.022 \times 10^{23} \text{ atoms Na}}{1 \text{ mol Na}} =$ 22.28×10^{23} atoms Na = 2.2×10^{24} atoms Na
2. $155 \text{ mol As} \times \frac{6.022 \times 10^{23} \text{ atoms As}}{1 \text{ mol As}} =$ $933 \times 10^{23} = 9.33 \times 10^{25}$ atoms As
3. $5.66 \times 10^{26} \text{ atoms Xe} \times \frac{1 \text{ mol}}{6.022 \times 10^{23} \text{ atoms Xe}} =$ $0.9399 \times 10^{3} \text{ mol} = 9.40 \times 10^{2}$ mol Xe
4. $2.888 \times 10^{15} \text{ atoms Ag} \times \frac{1 \text{ mol}}{6.022 \times 10^{23} \text{ atoms Ag}} =$ $0.4796 \times 10^{-8} \text{ mol} = 4.796 \times 10^{-9}$ mol Ag

Solutions for Additional Examples 4B

a. 2.1×10^{24} atoms Na **b.** 4.7×10^{23} atoms Mg **c.** 1.3×10^{25} atoms Al

Solutions for Additional Examples 4C

a. 0.0794 mol Ag **b.** 1.14 mol Hg **c.** 204 mol He

Full solutions for Additional Examples 4B and 4C are found on page 147A.

Additional Examples for Sample Problem 4D

Convert the following mole amounts to masses in grams.

a. 3.8 mol of F
b. 8.95 mol of Ba
c. 0.655 mol of Fe

Solutions are at the bottom of the page.

Moles can be converted to mass and vice versa

molar mass
the mass in grams of one mole of a given substance

The relationship between moles and mass is appropriately referred to as the *molar mass*. The **molar mass** is the mass in grams of 1 mol of a substance. One mole of carbon-12 atoms, for example, has a molar mass of 12.000 g. Because a random sample of an element includes isotopes, the molar mass for an element is the same as its average atomic mass. For example, one molar mass of carbon is 12.01 g. The periodic table provides the molar mass value for each element.

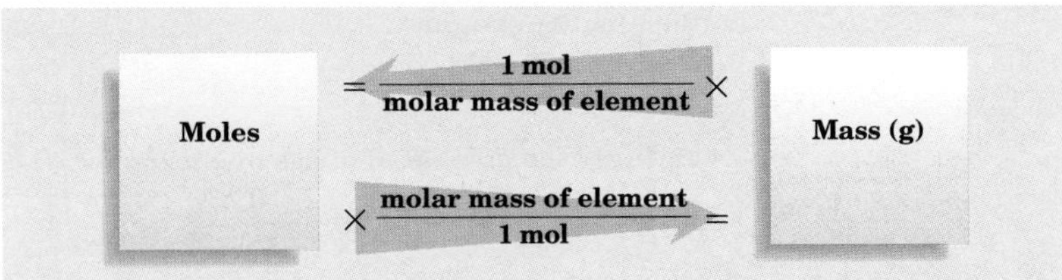

Figure 4-31
By using the molar mass of an element, you can convert between moles and the mass in grams of an element.

Like Avogadro's number, molar mass can be used as a conversion factor in chemical calculations. **Figure 4-31** shows the relationship between moles and the mass in grams of an element. Consider a problem where you must determine the mass in grams of 3.50 mol of the element copper. First, you must check the periodic table to see that the average atomic mass for copper is 63.55 amu. This means that the molar mass of copper is 63.55 g. (Note: we have rounded to two decimal places to the right of the decimal point.) Next, you set up the problem, using the molar mass as a conversion factor, as shown below.

Sample Problem 4D
Converting moles to mass

Determine the mass in grams of 3.50 moles of the element copper.

1 List what you know

- **moles of Cu** = 3.50 mol
- **mass of Cu** = ?
- **molar mass of copper** = 63.55 g

2 Set up the problem

- **Use Figure 4-31 to determine which factor will take you from the number of moles to the number of grams.**

$$\frac{63.55 \text{ g Cu}}{1 \text{ mol Cu}}$$

- **Multiply the number of moles by this factor .**

$$3.50 \text{ mol Cu} \times \frac{63.55 \text{ g Cu}}{1 \text{ mol Cu}} = ?$$

3 Calculate and verify

- **Solve and cancel like units in the numerator and denominator.**

$$3.50 \cancel{\text{mol Cu}} \times \frac{63.55 \text{ g Cu}}{1 \cancel{\text{mol Cu}}} = 222 \text{ g Cu}$$

- **The answer has the units specified by the problem. An answer of more than 63.55 g (the molar mass) makes sense because you started with more than 1 mol.**

Solutions for Additional Examples 4D

a. $3.8 \text{ mol F} \times \frac{19.00 \text{ g F}}{1 \text{ mol F}} = 72 \text{ g F}$

b. $8.95 \text{ mol Ba} \times \frac{137.33 \text{ g Ba}}{1 \text{ mol Ba}} = 1230 \text{ g Ba}$

c. $0.655 \text{ mol Fe} \times \frac{55.85 \text{ g Fe}}{1 \text{ mol Fe}} = 36.6 \text{ g Fe}$

Sample Problem 4E
Converting mass to moles

Determine the number of moles represented by 237 g of copper atoms.

1 List what you know

- mass of Cu = 237 g
- moles of Cu = ?
- molar mass of copper = 63.55 g

2 Set up the problem

- Use Figure 4-30 to determine which factor will take you from the number of grams to moles.

$$\frac{1 \text{ mol Cu}}{63.55 \text{ g Cu}}$$

- Multiply the number of grams by this factor.

$$237 \text{ g Cu} \times \frac{1 \text{ mol Cu}}{63.55 \text{ g Cu}} = ?$$

3 Calculate and verify

- Solve and cancel like units in the numerator and denominator.

$$237 \cancel{\text{g Cu}} \times \frac{1 \text{ mol Cu}}{63.55 \cancel{\text{g Cu}}} = 3.73 \text{ mol Cu}$$

- The answer has the units specified by the problem. An answer of more than 1 mol makes sense because you started with more than the molar mass of copper.

Practice 4D

1. Find the mass in grams of 8.6 moles of bromine atoms.
2. Find the mass in grams of 7.55 moles of silicon atoms.

4E

3. How many moles are in 38 g of carbon atoms?
4. How many moles are in 2 g of hydrogen atoms?

The average mass of atoms can be calculated from molar mass

Now that you know how to convert between moles, Avogadro's number, and molar mass, you can calculate the mass of a single atom of any element, as shown by the model in Figure 4-32. For example, the mass in grams of a single silicon atom can be calculated as shown on the next page.

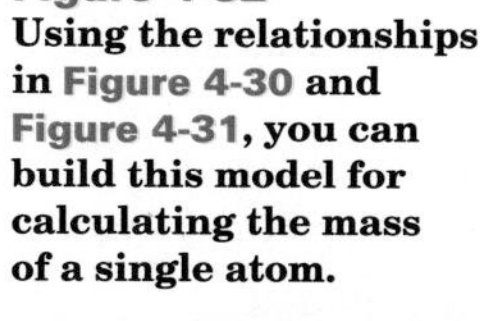

Figure 4-32 Using the relationships in Figure 4-30 and Figure 4-31, you can build this model for calculating the mass of a single atom.

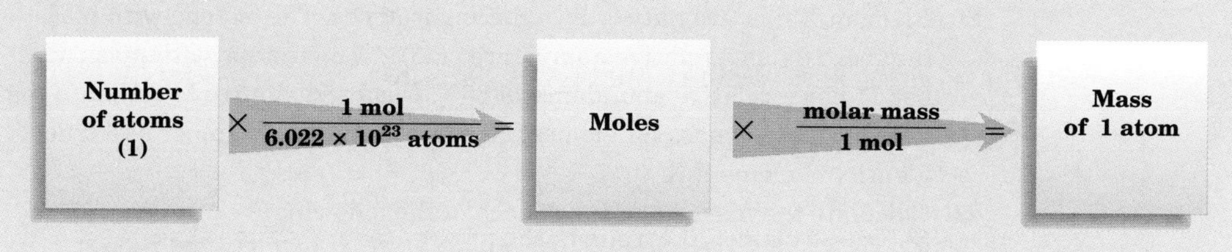

Additional Examples for Sample Problem 4E

Convert the following masses to moles.

a. 238 g Mn
b. 5.4 g Ti
c. 114.3 g Ne

Solutions are at the bottom of the page.

Answers to Practice 4D and 4E

1. $8.6 \text{ mol Br} \times \frac{79.90 \text{ g Br}}{1 \text{ mol Br}} = 690 \text{ g Br}$

2. $7.55 \text{ mol Si} \times \frac{28.09 \text{ g Si}}{1 \text{ mol Si}} = 212 \text{ g Si}$

3. $38 \text{ g C} \times \frac{1 \text{ mol}}{12.01 \text{ g C}} = 3.2 \text{ mol C}$

4. $2 \text{ g H} \times \frac{1 \text{ mol}}{1.01 \text{ g H}} = 2 \text{ mol H}$

Solutions for Additional Examples 4E

a. $238 \text{ g Mn} \times \frac{1 \text{ mol Mn}}{54.94 \text{ g Mn}} = 4.33 \text{ mol Mn}$

b. $5.4 \text{ g Ti} \times \frac{1 \text{ mol Ti}}{47.88 \text{ g Ti}} = 0.11 \text{ mol Ti}$

c. $114.3 \text{ g Ne} \times \frac{1 \text{ mol Ne}}{20.18 \text{ g Ne}} = 5.664 \text{ mol Ne}$

Additional Examples for Sample Problem 4F

Determine the mass of 1 atom of the following.

a. gold
b. boron
c. zinc

Solutions are at the bottom of the page.

Answers to Practice 4F

1. 1.68×10^{-24} g/atom H
2. 2.523×10^{-22} g/atom Eu

*Full solutions for items **1** and **2** are found on page 147A.*

ASSESS

Alternative Assessment

Play chemistry bingo. Number pieces of paper 1 through 36, and place them in a can. Have students draw a number. That number is the atomic number of the element they are to research. Each student should write a descriptive paragraph about the element that contains hints that can be used to identify the element. These paragraphs should be due the next day. Students are then given a blank card containing the number of rows and columns that is close to the number of students in class. For example, if the class has 28 students, the card could have 5 columns with 5 rows giving 25 boxes. Students then label each box using symbols for the elements, in any order they wish. Next, the student-made clues are read to the class and the students determine the identity of the element. The rules are like bingo in which the completion of a vertical, horizontal, or diagonal row wins the game.

Answers to Section Review

16. It is based on the standard of the carbon-12 isotope.

Answers are continued on page 147A.

Sample Problem 4F
Finding the mass of an atom

Find the mass of a single silicon atom.

1 List what you know

- **molar mass of silicon** = 28.09 g
- **Avogadro's number** = 6.022×10^{23} atoms in one mole
- **mass of one silicon atom** = ?

2 Set up the problem

- **You know the number of atoms in 28.09 g. You can divide the mass by the number of atoms to get the mass per atom. Or you can express the relationships above as factors that will give g/atom as an answer.**

$$\frac{28.09 \text{ g Si}}{1 \text{ mol Si}} \qquad \frac{1 \text{ mol Si}}{6.022 \times 10^{23} \text{ atoms Si}}$$

3 Calculate and verify

- **Solve and cancel like units in the numerator and denominator.**

$$\frac{28.09 \text{ g Si}}{\cancel{1 \text{ mol Si}}} \times \frac{\cancel{1 \text{ mol Si}}}{6.022 \times 10^{23} \text{ atoms Si}} = 4.665 \times 10^{-23} \text{ g/atom Si}$$

- **The answer has the units specified by the problem. A very small value for mass makes sense because you are calculating the mass of a single atom.**

Practice 4F

1. **Find the average mass of hydrogen atoms in grams.**
2. **Find the average mass of europium atoms in grams.**

More than 99% of the mass of a silicon atom is due to the protons and neutrons in its nucleus. The electrons that occupy most of the volume of the atom contribute very little to its mass. However, their number and arrangement determine most of the chemical behavior of atoms, which you will learn more about in the following chapter.

Section Review

16. Why was the relative scale of atomic masses established?

17. Distinguish among atomic mass, atomic mass unit, and average atomic mass.

18. What do isotopes of the same element have in common? How are they different?

19. Write the isotopic symbols for argon-36, argon-38, and argon-40.

20. Element X has two naturally occurring isotopes. The isotope with mass number 10 has a relative abundance of 20%. The isotope with mass number 11 has a relative abundance of 80%. Use these figures to estimate the average atomic mass for element X. State the atomic number and true identity of element X.

21. Calculate the mass in grams of each of the following.
a. 1.38 mol N **b.** 6.022×10^{23} atoms of Ag **c.** 2.57×10^{8} mol S

22. Calculate the number of atoms present in each of the following.
a. 2 mol Fe **b.** 40.1 g Ca **c.** 4.5 mol boron-11

23. Calculate the average mass in grams of Pt atoms.

Solutions for Additional Examples 4F

a. (196.97 g Au/1 mol Au)(1 mol Au/ 6.022×10^{23} atoms) = 3.271×10^{-22} g Au
b. (10.81 g B/1 mol B) (1 mol B/ 6.022×10^{23} atoms) = 1.795×10^{-23} g B
c. (65.39 g Zn/1 mol Zn) (1 mol Zn/ 6.022×10^{23} atoms) = 1.086×10^{-22} g Zn

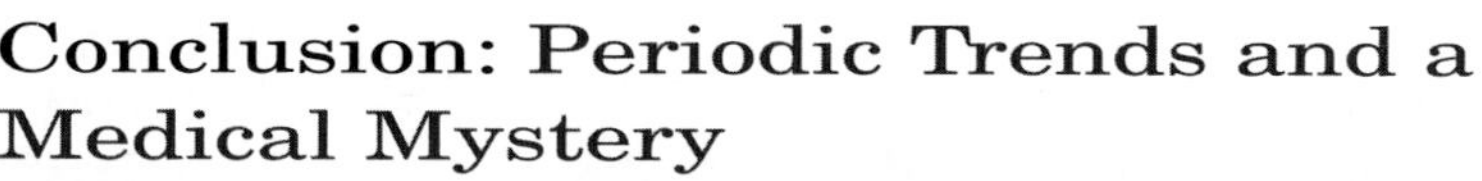

Conclusion: Periodic Trends and a Medical Mystery

Now that you have completed the chapter, reconsider the introductory questions and the mysterious death of George Decker.

What properties of the elements can be determined from the periodic table? Atomic radius, ionization energy, electron affinity, and electronegativity values are some trends you have studied. Also, melting point and boiling point vary in a periodic manner. Melting points generally increase as you move across from Group 3 to Group 6. Then, beginning with Group 7, melting points start to decrease. At Group 12, melting points drop dramatically and then continue to decrease across the table.

How do trends in the periodic table explain George Decker's death? Gold fillings contain gold and palladium. As the periodic trend in melting points would predict, gold melts at a lower temperature than palladium. To separate these metals, Decker heated them to the melting point of gold and then removed the pure liquid gold.

George Decker applied a similar process to silver fillings, with deadly results. Silver fillings contain silver, copper, and mercury. To separate silver from copper, this mixture must be heated to 962° C, the melting point of silver. This temperature is not only higher than the melting point of mercury, it is also higher than the *boiling point* of mercury. Thus, mercury evaporates before copper and silver can be separated. And mercury vapors are deadly.

Gaseous mercury atoms, once they are inhaled, enter red blood cells and are then transported throughout the body. Mercury is fairly difficult to oxidize, as its position in the periodic table shows. But once in body cells, enzymes supply the energy needed to oxidize mercury. Mercury ions create chaos by reacting with numerous substances in the body. This causes tissue damage and inflammation that result in pneumonia.

Applying Concepts

1. Assume that Decker had inhaled a total of 0.375 mol of mercury. How many atoms of mercury is this?
2. How many years would it take to count all the mercury atoms that Decker inhaled if all 250 million people in the United States were to count them at a rate of one atom per second?

Research and Writing

Use the library to find out more about the following.

1. You learned that the metals used in dental fillings are all transition metals. Report on other practical uses these transition metals have.
2. In the process of discovering elements, some early chemists actually tasted them or inhaled their vapors. Profile one such scientist, and describe the consequences of his or her dangerous laboratory practices.

Chemistry in Your Community

Have a few students interview their local dentists to find out what substances are used to fill teeth. What is the average mass of silver or gold amalgam used in a dental filling? How many moles of silver or gold is this? Have students check the prices of each metal in the newspaper to determine the monetary value of the amount of silver or gold used in a filling. How much are the metals in the fillings worth? From this, have students determine the cost of an individual gold or silver atom.

Have students do some research concerning the toxicity of amalgams and the growing concerns about their use.

Answers to Applying Concepts

1. $0.375 \text{ mol} \times \frac{6.022 \times 10^{23} \text{ atoms}}{1 \text{ mol Hg}} =$ 2.26×10^{23} atoms
2. 2.26×10^{23} atoms/ 2.5×10^{8} people = 9.0×10^{14} atoms/person

   (9.0×10^{14} atoms counted at 1 per s) = 9.0×10^{14} s

   (9.0×10^{14} s)(1 min/60 s) (1 h/60 min)(1 day/24 h) (1 yr/365 day) = 2.8×10^{7} yr

Closure Stategy

Ask students to prepare a concept map for the chapter. Concept maps should include most vocabulary terms along with other integral terms or concepts.

CHAPTER 4

Highlights

Key terms

actinides	**ion**
alkali metals	**ionization energy**
alkaline-earth metals	**lanthanides**
atomic mass	**main-block elements**
atomic mass unit	**molar mass**
atomic radius	**mole**
Avogadro's number	**nuclear reaction**
electron affinity	**periodic law**
electronegativity	**shielding effect**
halogens	**transition elements**

Key ideas

What makes a family of elements?

- According to the periodic law, properties of elements are periodic functions of their atomic numbers.
- In the periodic table, elements are ordered left to right by increasing atomic number. Elements with similar properties are grouped vertically in families.
- Elements of the same family have similar characteristic properties because they have similar electron configurations. For example, noble gases are unreactive because they have full *s* and *p* orbitals in the highest energy level.

What trends are found in the periodic table?

- The atomic radius of the elements decreases as you move left to right across a period and increases down a periodic table group.
- Ionization energy, electron affinity, and electronegativity generally increase as you move left to right across a period and decrease as you move down a group.

How are elements created?

- Naturally occurring elements (atomic numbers 1–92) were formed in the interior of stars. Synthetic elements (atomic numbers above 93) are made by research scientists.
- All elements larger than helium are formed by nuclear reactions in which nuclei gain or lose protons to become different elements.
- Synthetic elements are made in particle accelerators, which launch particles at speeds fast enough to generate nuclear reactions.

Can atoms be counted or measured?

- An atomic mass unit is 1/12 of the mass of a carbon-12 atom. The atomic mass of an atom is the mass of that atom expressed in atomic mass units.
- Chemists represent large collections of atoms using moles. One mole is equal to 6.022×10^{23} particles.
- The mass in grams of one mole of a substance is called the molar mass.

Key problem-solving approach: Using Avogadro's number

When solving molar conversion problems, begin at the box with the appropriate units. Follow the calculation pathway to the desired units.

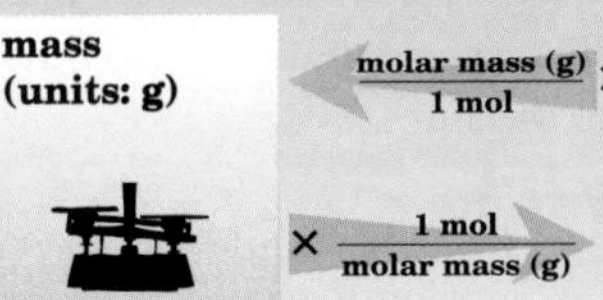

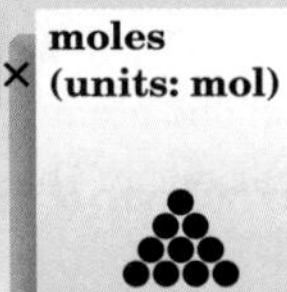

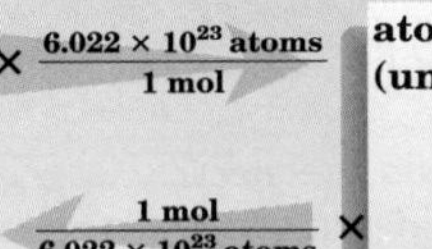

CHAPTER 4

Review and Assess

Organization of the periodic table

REVIEW

1. How do chemists use the periodic law to classify elements?
2. Yttrium, which follows strontium, has an atomic number one greater than strontium. Barium is 18 atomic numbers after strontium, but falls directly beneath it in the periodic table. Does strontium share more properties with yttrium or barium? Explain your answer.
3. a. What determines the vertical arrangement of the periodic table?
   b. What determines the horizontal arrangement of the periodic table?
4. All halogens are highly reactive. What causes this similarity among the halogens?
5. a. What property do the noble gases share?
   b. How does this property relate to the electron configuration of the noble gases?
6. Why is beryllium, a highly reactive metal, placed in Group 2?
7. Use the periodic table to describe the properties of the following elements.
   a. bromine, Br b. barium, Ba
   c. xenon, Xe d. tungsten, W
   e. rubidium, Rb f. neptunium, Np
   g. promethium, Pm
8. Argon differs from both chlorine and potassium by one proton each. Compare the reactivity and electron configurations of these three elements.
9. a. How do the electron configurations of the transition metals differ from the electron configurations of the metals in Groups 1 and 2?
   b. How do the electron configurations of the actinide and lanthanide series differ from the electron configurations of the other transition metals?
10. a. What groups make up the main-block elements?
    b. Why are the main-block elements also called the representative elements?
11. a. Why is hydrogen in a family by itself?
    b. Some periodic tables place hydrogen above the alkali metals, and some place it above the halogens. Explain the reasoning behind both of these placements.

APPLY

12. Compare the modern periodic table to Mendeleev's periodic table in **Figure 4-2**.
    a. List the differences between Mendeleev's periodic table and the modern table.
    b. Identify the discrepancies in Mendeleev's table that were rectified in Moseley's table.
13. While at an amusement park, you inhale helium from a balloon to make your voice squeaky. A friend says this practice is dangerous because the helium will react with your blood and produce toxic compounds. Is your friend correct? Explain.
14. a. What is happening to the sodium atom shown in the diagram below?
    b. How will the electron configuration of the atom change when the atom becomes an ion?
    c. Would a potassium atom behave in a similar way? Explain.

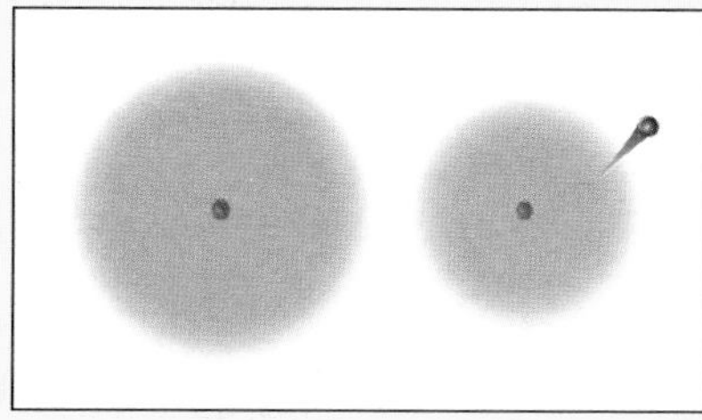

15. Why would you never expect a Ca^{+} or Na^{2+} ion to exist?
16. Calcium ions, Ca^{2+}, play an important role in muscle relaxation. Potassium ion, K^{+}, is also found in your body. Why is there no danger of K^{+} and Ca^{2+} reacting with each other?

1. According to the periodic law, the physical and chemical properties of elements are periodic functions of their atomic numbers. Chemists use this periodicity to group elements with chemical similarities into families.

2. Barium and strontium are in the same group and have similar electron configurations. Therefore, the properties of Ba and Sr are more similar than those of Y and Sr.

3. a. Elements are arranged vertically according to reactivity.
   b. Elements are arranged horizontally in order of increasing atomic mass.

4. All the halogens have seven valence electrons.

5. a. Noble gases have a low reactivity.
   b. Noble gases have a low reactivity because their atoms have filled *s* and *p* orbitals.

6. Beryllium falls into Group 2 because it has two electrons in its outer *s* orbital and has characteristics similar to other alkaline-earth metals.

7. a. Bromine is a nonmetal that reacts with metals to form salts.
   b. Barium is reactive metal that will lose two electrons in forming compounds.
   c. Xenon is a noble gas and therefore it has a low reactivity.
   d. Tungsten is not as reactive as metals in Groups 1 and 2. It is harder, denser, and has a higher melting point than metals in Groups 1 and 2.
   e. Rubidium is a soft, highly reactive metal. It reacts with oxygen in the air.
   f. Neptunium is synthetic and radioactive.
   g. Promethium is a shiny, reactive metal.

8. Argon has a very stable electron configuration with filled *s* and *p* orbitals. Thus, it has a low reactivity. Chlorine is one electron short of having filled *s* and *p* orbitals. It is therefore very reactive to achieve a stable electron configuration. Potassium has one more electron than the stable *s* and *p* filled orbital. It is also very reactive.

9. a. Unlike Group 1 and 2 elements, transition elements do not all have the same number of electrons in their outermost orbitals.
   b. The *f* orbitals are filling for the lanthanides and actinides, while the *d* orbitals are filling for the other transition metals.

10. a. Groups 1, 2, and 13 through 18 make up the main-block elements.
    b. They are called representative elements because the properties of these elements vary systematically within these sections of the table.

11. a. Hydrogen is in a family by itself because its properties somewhat resemble both Group 1 and Group 17 elements.
    b. Like the alkali metals, hydrogen has one electron in the s orbital. Also, like the halogens, hydrogen is only one electron from the configuration of a noble gas.

12. a. Mendeleev organized the elements by atomic weight while the modern periodic table is organized by atomic number.
    b. Elements such as potassium, nickel, iodine, and protactinium are in a different order than in Mendeleev's table, because they have atomic weights that are less than the preceding element.

13. Helium, in Group 18, has a very low reactivity.

17. You read a science fiction story about an alien race of silicon-based life-forms. Use information from the periodic table to hypothesize why the author chose silicon over other elements. (Hint: life on Earth is carbon based.)

Periodic trends

REVIEW

18. a. Why don't scientists define atomic radius as the radius of a single electron cloud?
    b. What periodic trends occur for atomic radius?
19. a. Use an analogy of a football team's offensive line protecting the quarterback to explain the shielding effect of electrons.
    b. How does the shielding effect alter atomic size?
20. How does the periodic trend in atomic radius relate to the addition of electrons?
21. Define ionization energy, electron affinity, and electronegativity.
22. a. What periodic trends exist for ionization energy?
    b. How does this trend relate to different energy levels?
23. What happens to electron affinity values as you move left to right across a period?

APPLY

24. a. Examine the graph below. Explain the patterns you see among the transition metals and among the nonmetals.

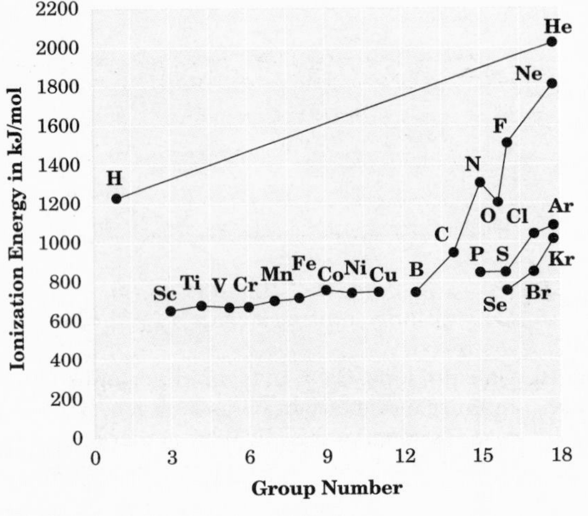

    b. Which element on this graph would be the best conductor of electricity?
    c. What possible reasons might exist for not using certain elements as conductors?
25. Identify which trends in the diagrams below describe atomic radius, ionization energy, or electronegativity.

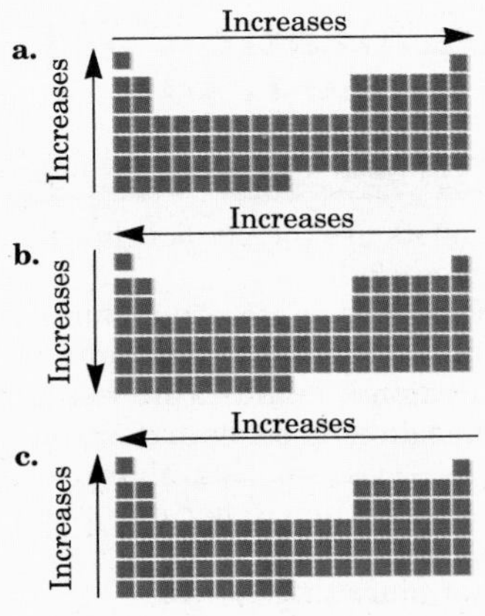

26. a. The number of nonmetals in each group increases as you move across the periodic table from Group 13 to Group 18. Using the concept of atomic size, explain this observation.
    b. Will there ever be a metallic noble gas? Explain.
27. How do the trends in atomic radius relate to the following?
    a. ionization energy
    b. electronegativity
28. Name three periodic trends you encounter in your life.

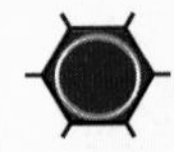

Creating the elements

REVIEW

29. When two elements are involved in a nuclear reaction, a different element is created. How does this happen?
30. How does the nuclear fusion process appear to create energy? Is energy really created?
31. What two significant features characterize the elements with an atomic number of 93 or greater?
32. Cite two reasons why hydrogen is involved in the most common nuclear fusion reaction.

APPLY

33. Compare the two charts below.
   a. Why are the most abundant elements in Earth's crust not the most abundant in the universe?
   b. Explain how the elements in stars formed the elements in Earth's crust.

Elemental Abundance in Universe **Elemental Abundance on Earth**

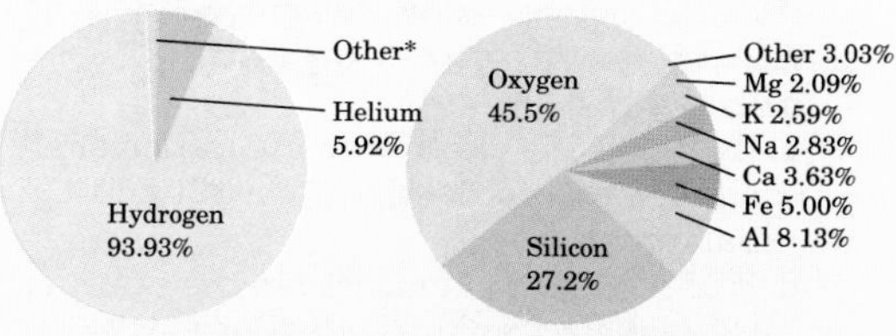

*Oxygen 0.075%, Carbon 0.047%, Nitrogen 0.0094%, Neon 0.0087%, Magnesium 0.0042%, Silicon 0.0030%, Other 0.0027%.

34. Irene Joliot Curie created the first artificial radioactive isotope, phosphorus-30, in 1934 when she bombarded aluminum-27, a shiny metal with conductive properties, with helium nuclei. The resulting product was a nonmetal with completely different properties. What caused the change in properties?

35. In 1987, Russian scientist Yury Organessian claimed to have synthesized element 110 using two different reactions. Predict the element that he used to bombard the starting elements in each reaction.
   a. starting with uranium, U, nuclei
   b. starting with thorium, Th, nuclei

36. Years ago, many people dreamed of transforming lead, an abundant metal, into gold, a rare and highly prized metal. Could gold be made from lead using the nuclear processes described in this chapter?

Atomic mass units

REVIEW

37. a. How is the carbon-12 atom used to define atomic mass units?
   b. How are atomic mass units and atomic mass related?

38. Why is the atomic mass of nitrogen 14.0067 and not 14.?

39. What needs to be determined before the atomic mass of an element can be verified?

PRACTICE

Answers to items in a black square begin on page 841

40 The lithium found in a hearing aid battery has two naturally occurring isotopes. Lithium-6 has a mass of 6.015121 amu and an abundance of 7.42%. Lithium-7 has a mass of 7.016003 amu and an abundance of 92.58%. Calculate the atomic mass of lithium. (Hint: see Sample Problem 4A.)

41. Silver found in jewelry has two naturally occurring isotopes. Silver-107 has a mass of 106.905092 amu and an abundance of 51.35%. Silver-109 has a mass of 108.904757 amu and an abundance of 48.65%. Calculate the atomic mass of silver. (Hint: see Sample Problem 4A.)

APPLY

42. Create your own relative scale that is similar to the atomic mass unit.

43. a. Several elements in the periodic table have atomic masses very close to a whole number. What is the likely cause?
   b. What could be the cause of atomic masses that are not close to an integer?

The mole

REVIEW

44. a. How are moles related to atomic mass?
   b. How is Avogadro's number related to moles?

45. How would you determine the number of atoms in 2 mol of gallium?

46. How is molar mass related to average atomic mass?

PRACTICE

47 How many atoms of oxygen enter your lungs when you inhale 5.00×10^{-2} mol of oxygen atoms? (Hint: see Sample Problem 4B.)

48. How many atoms are in the 6.75×10^{-2} mol of mercury within the bulb of a thermometer? (Hint: see Sample Problem 4B.)

49 How many moles of gold are in 1.00 L of sea water if there are 1.50×10^{17} atoms in the sample? (Hint: see Sample Problem 4C.)

50. How many moles are in a copper penny containing 1.80×10^{21} atoms? (Hint: see Sample Problem 4C.)

14. a. The sodium atom is losing an electron in its outermost orbital.
   b. The sodium ion configuration will be the same as neon, a noble gas.
   c. A potassium atom would behave in a similar way because potassium, like sodium, is a Group 1 element.

15. A calcium atom needs to lose two electrons to form an ion with a noble gas configuration. A sodium atom needs to lose only one electron to achieve noble gas configuration. Ca^{+} has the unstable configuration of an alkali metal. Na^{2+} has the unstable configuration of a halogen.

16. Potassium and calcium ions are both positively charged and will therefore repel each other rather than attract.

17. Silicon is the element most chemically similar to carbon because it falls directly beneath carbon in the periodic table. If the chemical properties of carbon are characteristic of life, then a carbon-like element such as silicon could be the basis of another life form.

18. a. Atomic radius is 1/2 the distance between the nuclei of two atoms of the same element. Because electron clouds do not have a definite edge, the radii of the clouds are best estimated using 1/2 the distance be-tween bonded atoms. The distance between nuclei can be measured more precisely.
   b. Atomic radii generally increase down a group and decrease from left to right across a period.

19. a. The shielding effect is the screening of outer electrons of an atom by inner electrons. The defensive line of a football team screens the outer players from the quarterback like the electrons screen the outer electrons from the attractive force of the nucleus.
    b. Because of the shielding effect, outer electrons are farther from the nucleus, and therefore, the size of the atom increases as electrons are added.

20. As electrons are added within the same energy level, the outer electrons are pulled closer to the nucleus by the increased nuclear charge, causing a decrease in atomic radius.

21. Ionization energy is the amount of energy needed to remove an electron from an atom. Electron affinity is the energy released when an electron is added to an atom. Electronegativity is the attraction of an atom to the electrons in a chemical bond. Ionization energy deals with losing an electron, electron affinity deals with gaining an electron, and electronegativity deals with attracting electrons within a bond.

22. a. Ionization energy decreases down a group and increases across a period.
    b. As atoms in a group get larger, and each successive energy level is farther from the influence of the nucleus. Therefore, it takes less energy to remove electrons from the larger atoms in a group than from the smaller atoms.

23. Electron affinity generally becomes more negative as you move left to right across a period.

51. A neon sign contains 4.50×10^{-2} mol of Ne gas. How many grams of gas are in that sign? (Hint: see Sample Problem 4D.)
52. How many grams are in the 2.00 mol phosphorus used to coat your television screen? (Hint: see Sample Problem 4D.)
53. A cup of hot chocolate contains 35.0 mg sodium. How many moles of sodium are in the chocolate? (Hint: see Sample Problem 4E.)
54. One cup of whole milk contains 290. mg calcium. How many moles of calcium are in the drink? (Hint: see Sample Problem 4E.)
55. Element 106, has a molar mass of 263.12 g/mol. What is the average mass of a single atom of element 106? (Hint: see Sample Problem 4F.)
56. Element 107, is produced by bombarding bismuth with chromium. What is the average mass of the chromium atom used to produce element 107? (Hint: see Sample Problem 4F.)

APPLY

57. Relate the mole to two other counting units.
58. The mass of one mole of toy poodles is 9.00×10^{27} g. What is the mass of a single toy poodle?
59. In 1986, the national debt was 2.00×10^{12} dollars. How many moles of dollars is this?

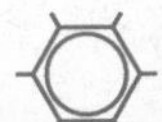 Linking chapters 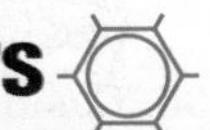

1. **Theme: *Classification and trends***
   Explain the nature of the observations that Dmitri Mendeleev used to first construct his periodic table.
2. ***Allotropes***
   How do the allotropes of an element differ from the isotopes of an element?
3. ***Electron configuration and quantum numbers***
   a. How does the electron configuration of all alkali metals differ from that of the halogens?
   b. What quantum numbers are the same for the last electron in all alkaline-earth metals?
   c. What quantum numbers are the same for the last electron in all noble gases?
4. **Theme: *Classification and trends***
   a. What trends were first used to classify the elements?
   b. What trends were discovered after the elements were classified in the periodic table?

USING TECHNOLOGY

1. ***Graphics calculator***
   Calculations from moles to grams can be done on the graphics calculator using a simple program. Press PRGM ▶ ▶ ENTER to open a new program from the program menu. Name the program "MOLES TO G." Then enter the following program. Prompt M: "MX" ⟶ Y1 (Keystrokes: PRGM; ▶ ▼ ENTER; M; ENTER; " M X "; STO; Y-VARS; ENTER; ENTER; QUIT.) To run the program, press PRGM; ENTER; ENTER and enter the molar mass of your element. Press GRAPH and use the TRACE function to read the appropriate y-value that corresponds to the x-value of moles.
   a. How many grams of Fe are in 210. mol?
   b. How many grams of S are in 13.0 mol?
   c. How many grams of P are in 1.25 mol?

Alternative assessment

Performance assessment

1. Your teacher will give you a notecard that identifies the electronegativity, ionization energy, and electron affinity of an element. Identify the element by analyzing these periodic trends.
2. You are given a sample of an unknown element. Design a set of procedures that you could use to determine if the unknown is a metal, nonmetal, or metalloid. Organize your procedures in the form of a flow chart like the one you saw in Chapter 2 for classifying matter.

Portfolio projects

1. ***Research and communication***
   Research some methods chemists might initially have used to arrive at Avogadro's number. Then compare these methods with modern methods. What other element besides carbon has been used as a basis for the mole? Study the reasons for basing all measurements on carbon.
2. ***Cooperative activity***
   Construct your own periodic table or obtain one of the posters available that shows related objects in a periodic arrangement like vegetables or fruits. Describe the organization of the table and the trends it exemplifies. Use this table to make predictions about your subject matter.
3. ***Chemistry and you***
   Many minerals in food are elements. They are essential to your health. Examine the product labels of foods that you eat. Determine which elements are represented in your food and what their function is in the body. Make a poster of foods that are good sources of the minerals you need.
4. ***Research and communication***
   Research the use of nuclear power. Analyze the most common method of utilizing nuclear fuel. Research the waste products and the efficiency of the energy production. Have a class debate concerning the use of nuclear power. Be sure to address the concerns of waste disposal, safety, and availability of the fuel source.

2. ***Graphics calculator***
   Write a similar program to convert grams to moles. Open a new program, and name it "G TO MOLES." Enter the following program. Prompt M: "(F(1,M))X": ⟶ Y1 (Keystrokes: PRGM; ▶; ▼; ENTER; M; ENTER; "; (; 1; ÷; M;); X; "; STO; Y-VARS; ENTER; ENTER; QUIT.) To run the program, press PRGM; ENTER; ENTER; and enter the molar mass of your element. Press GRAPH and use the TRACE function to read the appropriate y-value that corresponds to the x-value of mass.
   a. How many moles of carbon are in 25.0 g?
   b. How many moles of lead are in 135 g?
   c. How many moles of gold are in 55.0 g?
3. ***Computer spreadsheet***
   Create a computer spreadsheet that will calculate the number of atoms, the number of moles, and the mass of a sample when the given quantity is atoms, moles, or grams.

24. a. The transition metals on the graph have similar low ionization energies. Nonmetal ionization energies increase moving left to right across a period.
b. scandium
c. Some elements may be too rare, too reactive, or too soft or brittle to be drawn into wires. Elements that do not have metallic properties would not have the electron arrangements needed to be a good conductor.

25. a. describes ionization energy and electronegativity
b. describes atomic radius
c. describes nothing

26. a. Atoms that have tightly held electrons have more nonmetallic character. As you travel across the periodic table, the atomic size decreases because nuclear charge is increasing. These factors cause electrons to be more tightly held. Thus, the elements on the right side of the table should show increasing nonmetallic character.
b. If a noble gas were sufficiently large that electrons could be removed with minimum energy, the noble gas could theoretically act as a metal. However, this is not a reality with the known elements.

27. a. As atomic radius increases, ionization energy decreases.
b. As atomic radius increases, electronegativity tends to decrease.

28. Answers will vary. Some may include the seasons, the regularity of the school year, or a daily schedule.

29. In a nuclear reaction, the nucleus of an atom changes by gaining or losing protons to become a different element.

Answers from page 133

Answers to Practice 4A

1. $\frac{(92.21 \times 27.98 \text{ amu}) + (4.70 \times 28.98 \text{ amu}) + (3.09 \times 29.97 \text{ amu})}{100} = 28.09 \text{ amu}$

2. $\frac{(99.76 \times 15.99 \text{ amu}) + (0.038 \times 17.00 \text{ amu}) + (0.20 \times 10.00 \text{ amu})}{100} = 15.99\text{amu}$

(Note: Due to rounding of isotope masses, this value varies slightly from that stated in the periodic table.)

Answers from page 137

Solutions for Additional Examples 4B

a. $3.5 \text{ mol Na} \times \frac{6.022 \times 10^{23} \text{ atoms}}{1 \text{ mol Na}} = 2.1 \times 10^{24} \text{ atoms Na}$

b. $0.78 \text{ mol Mg} \times \frac{6.022 \times 10^{23} \text{ atoms}}{1 \text{ mol Mg}} = 4.7 \times 10^{23} \text{ atoms Mg}$

c. $22 \text{ mol Al} \times \frac{6.022 \times 10^{23} \text{ atoms}}{1 \text{ mol Al}} = 1.3 \times 10^{25} \text{ atoms Al}$

Solutions for Additional Example 4C

a. $4.78 \times 10^{22} \text{ atoms of Ag} \times \frac{1 \text{ mol Ag}}{6.022 \times 10^{23} \text{ atoms}} = 0.0794 \text{ mol Ag}$

b. $6.85 \times 10^{23} \text{ atoms of Hg} \times \frac{1 \text{ mol Hg}}{6.022 \times 10^{23} \text{ atoms}} = 1.14 \text{ mol Hg}$

c. $1.23 \times 10^{26} \text{ atoms of He} \times \frac{1 \text{ mol He}}{6.022 \times 10^{23} \text{ atoms}} = 204 \text{ mol He}$

Answers from page 140

Answers to Practice 4F

1. $\frac{1.01 \text{ g H}}{1 \text{ mol H}} \times \frac{1 \text{ mol H}}{6.022 \times 10^{23} \text{ atoms H}} =$
$0.1677 \times 10^{-23} \text{ g/atom H} = 1.68 \times 10^{-24} \text{ g/atom H}$

2. $\frac{151.96 \text{ g Eu}}{1 \text{ mol Eu}} \times \frac{1 \text{ mol Eu}}{6.022 \times 10^{23} \text{ atoms Eu}} =$
$25.234 \times 10^{-23} \text{ g/atom Eu} = 2.523 \times 10^{-22} \text{ g/atom Eu}$

Continued from page 140

Answers to Section Review

17. Atomic mass is the mass of an atom expressed in g or amu; an atomic mass unit is 1/12 the mass of a carbon-12 atom. The average atomic mass of an element is the weighted average of all the isotopes of an element.
18. Isotopes of the same element have the same numbers of protons and electrons but differ in the numberof neutrons.
19. $^{36}_{18}Ar$, $^{38}_{18}Ar$, $^{40}_{18}Ar$
20. 10.8 amu, boron, atomic number 5
21. **a.** 19.3 g **b.** 107.87 g **c.** 8.24×10^9 g
22. **a.** 1×10^{24} **b.** 6.02×10^{23} **c.** 2.7×10^{24}
23. 3.24×10^{-22} g

Continued from page 1..

30. The "missing mass" that is a result of the fusion process is converted to energy.

31. All the elements with atomic numbers 93 or greater are synthetic and radioactive.

32. Hydrogen is involved in the most common fusion process because it reacts rapidly with most elements and it is the most abundant element in the universe.

33. **a.** Most matter in space composes stars, which are far too hot to support the silicon and oxygen atoms found on Earth.
b. Successive nuclear reactions create larger atoms from hydrogen and helium.

34. The number of protons in the nucleus changed. This changed the element into a different element.

35. **a.** argon
b. calcium

36. No. Lead has a higher atomic number than gold. Protons would have to be removed from the nucleus instead of added, as described in the text.

37. **a.** One amu is 1/12 the mass of a carbon-12 atom.
b. The mass of an atom expressed in amu is the atomic mass.

38. The atomic mass of nitrogen is 14.006 because it is a weighted average of the masses of nitrogen isotopes.

39. The different isotopes of an element and the abundance of each must be known.

40. $\frac{(7.42 \times 6.015121 \text{ amu}) + (92.58 \times 7.016003)}{100} = 6.94 \text{ amu}$

41. $\frac{(51.35 \times 106.905092) + (48.65 \times 108.904757)}{100} = 107.9 \text{ amu}$

42. Answers will vary. Answers should show an understanding of creating a standard base unit that is easily converted. For example, a full gas tank could be standardized to represent a specific number of gallons. All gas tank volumes could be expressed in terms of the standard.

43. **a.** The abundance of one isotope greatly outweighs others.
b. Several different isotopes could exist in relatively equal abundances.

44. **a.** The atomic mass of an element in grams contains 1 mol of atoms.
b. Avogadro's number is the number of particles in 1 mol of a substance.

45. Multiply Avogadro's number by two.

46. The molar mass of an element is the same as its average atomic mass.

47. $5.00 \times 10^{-2} \text{ mol O} \times \frac{6.022 \times 10^{23} \text{ O atoms}}{1 \text{ mol O}} = 3.01 \times 10^{22} \text{ O atoms}$

48. $6.75 \times 10^{-2} \text{ mol Hg} \times \frac{6.022 \times 10^{23} \text{ Hg atoms}}{1 \text{ mol Hg}} = 4.06 \times 10^{22} \text{ Hg atoms}$

49. 1.50×10^{17} atoms Au $\times \dfrac{1 \text{ mol Au}}{6.022 \times 10^{23} \text{ atoms Au}} = 2.49 \times 10^{-7}$ mol Au

50. 1.80×10^{21} atoms Cu $\times \dfrac{1 \text{ mol Cu}}{6.022 \times 10^{23} \text{ atoms Cu}} = 2.99 \times 10^{-3}$ mol Cu

51. 4.50×10^{-2} mol Ne $\times \dfrac{20.18 \text{ g Ne}}{1 \text{ mol Ne}} = 0.908$ g Ne

52. 2.00 mol P $\times \dfrac{30.97 \text{ g P}}{1 \text{ mol P}} = 61.9$ g P

53. 35.0 mg Na $\times \dfrac{1 \text{ g}}{1000 \text{ mg}} \times \dfrac{1 \text{ mol Na}}{22.99 \text{ g Na}} = 1.52 \times 10^{-3}$ mol Na

54. 290. mg Ca $\times \dfrac{1 \text{ g}}{1000 \text{ mg}} \times \dfrac{1 \text{ mol Ca}}{40.08 \text{ g Ca}} = 7.24 \times 10^{-3}$ mol Ca

55. $\dfrac{263.12 \text{ g } 106}{1 \text{ mol } 106} \times \dfrac{1 \text{ mol } 106}{6.022 \times 10^{23} \text{ atoms}} = 4.369 \times 10^{-22}$ g/atom 106

56. $\dfrac{52.00 \text{ g Cr}}{1 \text{ mol Cr}} \times \dfrac{1 \text{ mol Cr}}{6.022 \times 10^{23} \text{ atoms}} = 8.635 \times 10^{-23}$ g/atom Cr

57. Answers will vary. Examples may include a dozen eggs, a ream of paper, or a bushel of corn all units representing a larger number of individual units.

58. $\dfrac{9.00 \times 10^{27} \text{ g poodle}}{1 \text{ mol poodle}} \times \dfrac{1 \text{ mol poodles}}{6.022 \times 10^{23} \text{ poodles}} = 1.49 \times 10^{4}$ g/poodle

59. 2.00×10^{12} dollars $\times \dfrac{1 \text{ mol}}{6.022 \times 10^{23} \text{ dollars}} = 3.32 \times 10^{-12}$ mol

Linking chapters

1. Mendeleev organized the elements based on observations he made of properties that recurred within the elements he observed.

2. Allotropes of an element are different physical forms of that element. Isotopes deal with individual atoms of an element that have different numbers of neutrons and therefore different masses.

3. a. All alkali metals have one *s* electron in the outermost energy level, and all halogens have two s electrons and five *p* electrons in the outermost energy level.
   b. The last electron in all alkaline-earth metals has an ℓ number of *s* and the same m_ℓ number because there is only one orbital in s sublevel.
   c. The last electron in all noble gases has an ℓ number of *p*.

4. a. reactivity and weight
   b. atomic size, ionization energy, electron affinity, electron arrangement (configuration), and electronegativity

USING TECHNOLOGY

1. a. 1.17×10^{4} g Fe
   b. 417 g S
   c. 38.7 g P

2. a. 2.08 mol C
   b. 0.652 mol Pb
   c. 0.279 mol Au

3. Answers will vary. Be sure that the spreadsheet is programmed correctly to give accurate answers.

Performance assessment

1. Check students' work by having them write down the handbook data that they used to identify their unknowns.

2. Answers will vary. Be sure the reasoning for identifying the chemical follows a logical pattern.

Portfolio projects

1. Check students' work for good source material. Many students will discover that oxygen was used once as a basis for atomic mass units.

2. Answers will vary. Students' explanations of their tables should reflect a periodic arrangement and discernable trends. Some subjects that students might use to construct a periodic table include baseball cards, comic books, cars, or types of clothing.

3. Answers will vary. Students' work should demonstrate a comprehensive listing of the minerals found in foods and their functions. Additional credit can be given to those students who determine dietary recommendations for minerals.

4. Answers will vary. Be sure students analyze both fusion and fission processes. Make sure students consider both the safety and efficiency of nuclear power. Students should recognize that at this point in time there are no permanent disposal sites for nuclear waste.

5

PLANNING FOR

Ionic Compounds

Chapter Overview

- **Section 5-1** describes stability as the driving force behind ion formation.
- **Section 5-2** describes the electron transfer process that forms ions. Students look at visual models for an explanation of the properties of salts and the packing patterns that occur in their unit cells. A scaled representation of the Born-Haber cycle sums up the importance of the lattice energy in salt formation.
- **Section 5-3** identifies common monatomic ions and describes how their binary compounds are named.
- **Section 5-4** identifies common polyatomic ions and describes how their compounds are named. Students follow separate flowcharts to name an ionic compound when the formula is known or to write the formula when the name is known.
- **Section 5-5** uses familiar salts to present percent composition and water of hydration.

Concept Base

Students must have mastered the following concepts prior to this chapter: potential energy, heat, endothermic and exothermic change (Chapter 2); atomic structure and determining electron configuration (Chapter 3); atomic masses (Chapter 4).

Looking Ahead

Bonding through shared electron pairs will be described in Chapter 6. Factors that affect stability in physical and chemical changes will be described in Chapter 9. Conductivity of salt solutions in voltaic cells and the electron transfer process will be discussed further in Chapter 15.

Laboratory Equipment Needs

Demonstrations

The listing of materials for Chapter 5 demonstrations begins on page T46.

Laboratory Experiments

Materials and preparation instructions for Chapter 5 are found as follows.

Text		*Page*
Exploration 5A	Percentage Composition of Hydrates	692
Investigation 5A	Hydrates—Gypsum and Plaster of Paris	696

Pacing: 10 days	**Inquiry Teaching Strategies**
5-1 Why do atoms form bonds?	**Demonstrations** ■ Demo 1 *The synthesis of two salts illustrates relative stability* ■ Demo 2 *Stability accompanies lowered potential energy*
5-2 What holds a salt together?	**Demonstrations** ■ Demo 3 *Ions but not ionic compounds conduct electricity*
5-3 How do you name salts?	
5-4 What is a polyatomic ion?	
5-5 How are salts used and measured?	**Demonstrations** ■ Demo 4 *Analysis of fertilizer* ■ Demo 5 *Heating a hydrate* **Laboratory Experiments** ■ Exp 5A *Percentage Composition of Hydrates* ■ Inv 5A *Hydrates—Gypsum and Plaster of Paris*

Key ■ Teaching Transparencies
■ Annotated Teacher's Edition
■ Laboratory Experiments Manual

Visual Teaching Strategies / Review and Practice Strategies

Visual Teaching Strategies	Review and Practice Strategies
Transparency Masters ■ T 5-1 *Orbital Notation for Three Noble Gases* ■ T 5-2 *Electron Orbital Notation for Argon and Potassium*	**Text Reviews** ■ Section Review 1–6 ■ Chapter Review 1–14; LC 1–2; UT 1 **Study Guide Worksheets** ■ ■ 5-1 Concept Review ■ ■ 5-1 Skillsheet *Using the Octet Rule* *Using a Bond Energy Table*
Transparencies ■ T 5-3 *Comparing Sodium Atom with Sodium Ion* ■ F 5-5 *Electron Orbital Notation for Chlorine* ■ F 5-6 *Sodium Chloride* ■ F 5-7 *Particle Models for NaCl* ■ F 5-8 *NaCl and CsCl Lattices* ■ F 5-9 *Unit Cells* ■ F 5-10 *Energy Changes*	**Text Reviews** ■ Section Review 7–17 ■ Chapter Review 15–24 **Study Guide Worksheets** ■ ■ 5-2 Concept Review ■ ■ 5-2 Skillsheet *Modeling Atoms, Ions, and Ionic Compounds*
Transparency Masters ■ T 5-6 *Common Monatomic Ions* ■ F 5-12 *Variable Oxidation States for Transition Metal Ions*	**Text Reviews** ■ Concept Review 18–21 ■ Practice 1 ■ Section Review 22–28 ■ Chapter Review 25–27, 31, 35–39 **Study Guide Worksheets** ■ ■ 5-3 Practice *Writing Ionic Formulas*
Transparencies ■ P 173 *Naming Ionic Compounds* ■ P 174 *Writing Formulas for Ionic Compounds* **Transparency Masters** ■ T 5-8 *Some Polyatomic Ions* ■ T A-5 *Common Ions*	**Text Reviews** ■ Practice 1 ■ Section Review 29–31 ■ Chapter Review 28–30, 32–34 **Study Guide Worksheets** ■ ■ 5-4 Concept Review ■ ■ 5-4 Practice *Formulas with Polyatomic Ions*
Transparencies ■ P 182 *Percentage Composition Calculations*	**Text Reviews** ■ Practice 1–7 ■ Section Review 32–34 ■ Chapter Review 40–61 **Study Guide Worksheets** ■ ■ 5-5 Practice *Calculating with Molar Mass*

■ Pupil's Text
■ Study Guide
■ Teaching Resources

Exp *Exploration*
Inv *Investigation*

LC *Linking Chapters*
UT *Using Technology*

Assessment Options

Traditional Assessment

Test Generator
Instructional Objectives Measured:
Content mastery
Select from items 1 to 75

Performance Assessment Options

(see pages T36 and T643-15 for scoring rubrics)
Students demonstrate mastery of objectives in a hands-on environment. You may want some of these materials in the Portfolio as well.

Investigation 5A (text)
Instructional Objectives Measured:
Content mastery, Use of scientific methodology, Problem solving skills, Proficiency in written communication

Performance Assessments, page 187

1. Designing an experiment
Instructional Objectives Measured:
Content mastery, Problem solving skills, Use of scientific methodology, Proficiency in written communication

2. Devising classification criteria
Instructional Objectives Measured:
Content mastery, Problem solving skills, Use of scientific methodology, Proficiency in written communication

Portfolio Options

(see page T36 for scoring rubrics)
Students provide a written rationale for each selection made for the portfolio.

Concept Maps
Students use vocabulary from the Chapter Review to build a concept map.
Instructional Objectives Measured:
Content mastery

Formal Laboratory Reports
Exploration 5A (text)
Instructional Objectives Measured:
Content mastery, Use of scientific methodology, Proficiency in written communication

Portfolio Projects, page 187
Items 1 and 4. Research and communication
Instructional Objectives Measured:
Content mastery, Scientific and chemical literacy, Use of technology, Proficiency in oral and written communication

Item 2. Chemistry and you
Instructional Objectives Measured:
Content mastery, Scientific and chemical literacy, Proficiency in written communication

Item 3. Cooperative activity
Instructional Objectives Measured:
Content mastery, Scientific and chemical literacy, Proficiency in oral and written communication

INTRODUCING THE CHAPTER

CHAPTER 5 Ionic Compounds

Story Background

Salt that is left behind when shallow seas evaporate becomes buried by layers of sediment. As these sediment layers accumulate, the temperature and pressure on the salt deposit increases until the salt becomes fluid. In time, enough pressure is exerted on the salt that it flows upward through the less dense sediment. Upward movement of the salt continues as more sediment accumulates, so the salt mass remains at or near the surface of the surrounding sediment unless the source of the salt is exhausted during growth. In this case, the salt growth stops, and the dome becomes buried.

Salt domes have a roughly circular or elliptical cross section. Their horizontal dimensions are of the same magnitude as their vertical dimensions. A Gulf Coast dome may begin as far as 1.5 km below ground level and be approximately 1.6–10 km wide and 4.5–9.0 km deep. The first 600 m of these domes may be cap rock, which consists of limestone overlying anhydrite. Cavities made in the domes by solution-mining techniques have a storage capacity of about 4.8×10^6 m^3. These cavities have been used to store liquified petroleum gas, liquified propane, and crude oil. Because of their impermeability, these cavities have also been considered for the disposal of hazardous waste.

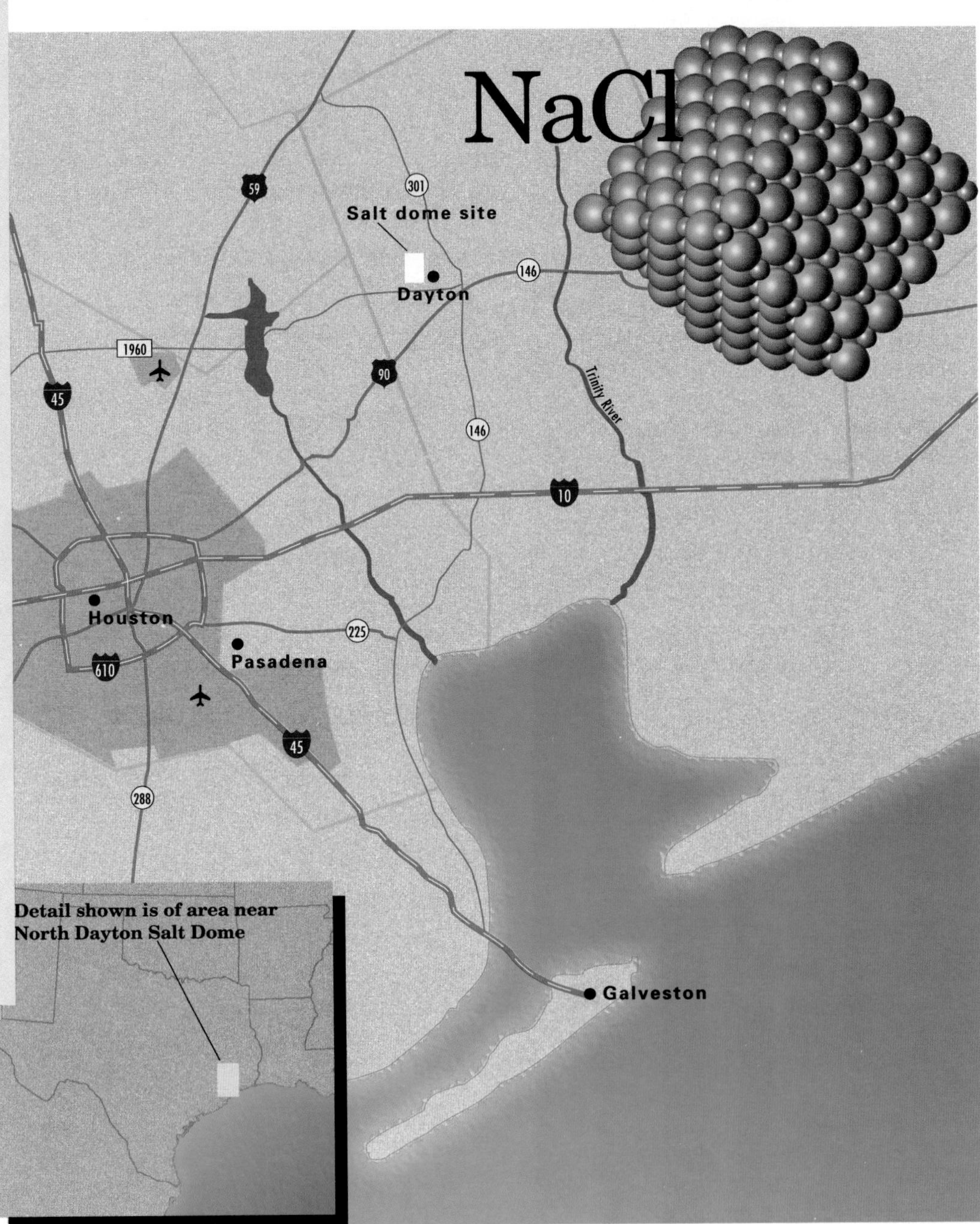

About the Illustration

Houston is one of the five most populated cities in the United States. Nevertheless, the underground salt domes in nearby Dayton were proposed as toxic-waste burial sites in the early 1990s.

Salt Domes and Toxic Wastes

In the early 1990s, a scientific showdown took place in Texas. Scientists and engineers with a disposal firm proposed a plan to store toxic wastes hundreds of meters underground, while other scientists argued that the plan was too risky. How could nearby residents and the state government know who to believe?

Many communities are facing a disposal crisis. Toxic compounds containing lead, mercury, and arsenic are used in making many products, including car batteries, fluorescent light bulbs, and thermometers. When these products are discarded, they can't be put into ordinary landfills because the toxins slowly seep into the ground water, poisoning the water supply for plants, animals, and people. If they are burned, metals are released into the atmosphere, poisoning the air. Finding a safe way to dispose of these toxic wastes has seemed impossible.

In 1990, Hunter Industrial Facilities, Inc. applied for a permit to bury toxic wastes near Houston, Texas, in a new way that they claimed was safe. Hunter's scientists and engineers proposed drilling a hole to gouge out narrow, deep caverns 250 m beneath the surface. Toxic wastes would be combined with ash and cement to form solid pellets, which would be dumped down the shaft. Then, the shaft opening would be sealed with reinforced concrete.

Because of their immense size, each cavern could hold the contents of up to 4 million barrels of toxic wastes. But the claims for safety were based neither on the size nor depth of the caverns, but rather on what the walls of these caverns would be made of—salt, NaCl.

Similar salt domes have served for years as storage sites for oil and gas. In the United States, potentially explosive natural gases, such as propane, are currently stored in more than 1000 caverns in 26 states. A nuclear device that was one-third the size of the atomic bomb dropped on Hiroshima was detonated in a dome in Mississippi. No measurable radiation has leaked into nearby drinking water supplies, even after 20 years.

While some scientists believed that the unique properties of salt will provide a safe storage area for toxic wastes, other scientists opposed the idea. They did not believe that the salt domes would adequately confine the wastes. Many of their arguments against the project near Houston were also based on the chemical properties of salt.

To understand how scientists can be on both sides of this controversy, you need to understand how each group interpreted the properties of salts when answering the following questions.

- *Would the salt react and combine dangerously with the hazardous wastes, and leak into the areas surrounding the domes?*
- *Could outside influences cause a break in the salt cavern walls?*

Tapping Prior Knowledge

Prepare students for the concepts they will need for this chapter by discussing the following topics.

Discussion topics

1. Some reactions give off heat, while other reactions take on heat from their surroundings. What role do you think heat might play in compound formation?
2. Atoms react to form compounds. Could the reverse situation occur? Explain by describing the sort of forces you feel hold compounds together.
3. Why might atoms need to combine with each other? How could they go about combining?

Explore everyday experiences that are relevant to this chapter with the following questions.

Experience-based items

1. List some objects with a crystalline shape.
2. Name as many household products as possible by their chemical names.
3. If water is a poor conductor, why must swimmers leave a pool during a thunderstorm?

Section 5-1

FOCUS

Lesson Starter

Begin the lesson by laying one of three rectangular blocks flat on a tabletop. Stand one block upright. Stand one block on a corner, and let it fall to the tabletop. Knock over the upright block. Discuss stability in terms of balance and block positions. Now push one block just over the edge of the tabletop so that it becomes unbalanced and falls to the floor. Stability is relative. Stable products from one reaction can become reactants in a separate reaction.

Demonstration 1

The synthesis of two salts illustrates relative stability

Approximate time: 25 min

1. Pour 25 mL of H_2O into a small beaker.
2. Use forceps to add a pea-sized piece of Ca metal to the water.
3. After the reaction is complete, pour about 10 mL of the beaker's contents into an evaporating dish. Set the dish on a hot plate. Evaporate the water, leaving $Ca(OH)_2$.
4. Add 2 drops of phenolphthalein to the solution in the beaker.
5. Add 0.1 M HCl dropwise until the solution turns colorless. Add 3 more drops of HCl. Pour 10 mL of this liquid into an evaporating dish. Place the dish on a hot plate. Evaporate the liquid, leaving $CaCl_2$.

Why do atoms form bonds?

Section Objectives

Explain why the atoms of noble gases have chemical stability.

Recognize that most isolated atoms achieve stability by forming chemical bonds.

Relate chemical stability, energy, bond formation, and the octet rule.

Compare the bond energies of various chemical compounds, including salts.

Stability

In considering the question of why atoms form bonds, the best place to start may be to examine what is special about those atoms that generally do not form bonds. In Chapter 4, you learned that the noble gases, the members of Group 18, have one thing in common—chemically, they are very unreactive. They exist as individual atoms, and they rarely form chemical bonds.

Chem Fact
Some potassium compounds are used as salt substitutes for low-sodium diets. However, overuse of such compounds can lead to muscular weakness and heart attacks.

Most other elements behave very differently from the noble gases. Potassium, for example, is a very reactive metal. Potassium reacts violently with a variety of substances, including the oxygen present in air. Potassium also reacts violently with water, as shown in **Figure 5-1**. One product of this reaction is a compound dissolved in the beaker which is more chemically stable than the potassium metal; unlike the original potassium metal, it does not react with water or air. The forces of attraction that held the atoms in the water molecules and the atoms in the potassium metal together are broken, and the atoms of each are attracted in new ways to form new substances that are chemically more stable. These forces of attraction are called *chemical bonds.*

Figure 5-1
Potassium metal is chemically unstable. When dropped in water, bonds between the potassium atoms and within the water molecules are broken, and new bonds between atoms are formed to make a new, more stable substance.

Filled energy levels make noble gas atoms stable

To understand why noble gas atoms do not form bonds under ordinary circumstances, examine **Table 5-1**. What do you notice about the number of electrons in the outermost energy levels of each noble gas? You may recall from Chapter 3 that the *s* and *p* orbitals of the outermost occupied energy level (except when it is the first one) are filled when they contain eight electrons.

Table 5-1
Electron Orbital Notations for Three Noble Gases

Noble gas	Helium, He	Neon, Ne	Argon, Ar
Third energy level			[↑↓] 3s [↑↓][↑↓][↑↓] 3p
Second energy level		[↑↓] 2s [↑↓][↑↓][↑↓] 2p	[↑↓] 2s [↑↓][↑↓][↑↓] 2p
First energy level	[↑↓] 1s	[↑↓] 1s	[↑↓] 1s

Helium, with only two electrons, contains the maximum number that can be found in its only energy level. (Recall from Chapter 3 that the first energy level holds only two electrons.) The other noble gases also have eight electrons filling the *s* and *p* orbitals in their outermost occupied energy levels. They do not have room for additional electrons in their outermost *s* and *p* orbitals. Eight electrons in outer levels help make atoms stable.

Table 5-2
Electron Orbital Notations for Argon and Potassium

Element	Argon, Ar	Potassium, K
Fourth energy level		[↑] 4s
Third energy level	[↑↓] 3s [↑↓][↑↓][↑↓] 3p	[↑↓] 3s [↑↓][↑↓][↑↓] 3p
Second energy level	[↑↓] 2s [↑↓][↑↓][↑↓] 2p	[↑↓] 2s [↑↓][↑↓][↑↓] 2p
First energy level	[↑↓] 1s	[↑↓] 1s

The octet rule predicts reactivity of atoms

The presence of eight electrons in the outermost energy level is known as a *stable octet*. An extremely useful rule in chemistry is based on stable octets. The most likely reactions are those that yield products whose atoms have stable octets. Once an atom has a stable octet, it does not form bonds as easily, and it is said to be *less reactive*.

On the other hand, an atom that does not have eight electrons filling the *s* and *p* orbitals in the outermost level is said to be *more reactive* because it will form chemical bonds with other atoms to achieve a stable octet.

Now look at the electron orbital notation for potassium in **Table 5–2**. Notice that it differs from that of argon by a single electron. It is this single electron that makes potassium so reactive. If a potassium atom loses this electron, the electron orbital notation of the resulting particle is the same as that of argon. It is also more stable, like argon. The reactivity shown by potassium in losing an electron to achieve a noble gas notation follows an observed principle called the **octet rule**.

octet rule
main-block elements form bonds by rearranging electrons so that each atom has a stable octet in its outermost energy level

Other numbers and arrangements of electrons can also contribute some stability to atoms, especially in the case of the transition metals and other atoms with unfilled *d* orbitals. However, the octet rule is the best indicator of stability for the main-block elements.

SAFETY
Wear safety goggles and a lab apron. Dilute spills with water, if necessary. Then, wearing gloves, soak up the mess with cloth or paper toweling; rinse the toweling, neutralize the rinsings with 1 M NaOH or HCl, and pour down the drain. Do not use baking soda for acid spills.

DISPOSAL
Combine the $Ca(OH)_2$, $CaCl_2$, and all liquids; dissolve or suspend solids. Neutralize with 1 M HCl or NaOH as appropriate, and pour down the drain.

TEACH

Historical Note

Helium, first identified in 1868 through spectroscopic analysis of the sun, was not discovered on Earth until 1891 because of its lack of reactivity.

Teaching Tip

The Hindenberg and chemical stability

The hydrogen-filled Hindenberg airship exploded over Lakehurst, New Jersey in 1937. Hydrogen atoms are unstable, or reactive, because they do not have a noble gas configuration. Helium, a stable, nonreactive noble gas, is now used in airships instead of hydrogen.

Historical Note

Gilbert N. Lewis proposed the octet rule in 1936.

Do You Know?

Potassium is stored under water-free mineral oil because it reacts with water vapor in the air, but white phosphorous is stored under water because it reacts with oxygen but not the water vapor in the air.

Demonstration 2

Stability accompanies lowered potential energy

Approximate time: 15 min

Making a system's center of gravity closer to the ground can be likened to lowering the potential energy in a chemical system.

1. Insert pointwise, a 20 cm piece of solid copper wire (#12 or #14) or coat hanger wire into a cork stopper.
2. Have several students try to balance the system point first on one finger for 20 s without dropping it.
3. After they have tried, bend the system into a C shape.
4. Walk around the room with the system hanging on your finger while you begin a discussion about stability.

SAFETY

Wear safety goggles for step 1 while sticking the wire into the cork.

DISPOSAL

Put into trash or save for next time.

Potential energy and stability

When potassium is dropped in water, it changes from a metal that is chemically unstable into part of a compound that is more stable. But more is going on than just the rearrangement of electrons and atoms. You can also tell from **Figure 5-1** that this chemical change is very exothermic.

Although atoms gain stability by forming a chemical bond, keep in mind that they do lose something—energy. When two atoms form a chemical bond, energy is released. As you can see in **Figure 5-2**, when the individual potassium atoms lose electrons and attain a stable octet in the reaction with water, they release some energy in the form of heat and light, and a lower potential energy level is attained.

The change in energy in this reaction is important because it is *easily quantifiable and measurable*. Energy is the ability to do work. But changes in stability are harder to measure and quantify. Chemical stability can be considered the inability to cause a chemical change—the opposite of energy. Thus, a measurement of a change in energy can be a useful indicator for a change in stability.

You may know that natural changes that result in a decrease of potential energy are favored. For example, water tends to flow downhill and never uphill. The potassium compound in the beaker in **Figure 5-1** will not revert to potassium metal and water without an input of energy. Another way to examine these examples is to say that changes that maximize stability are favored. Thus, bond formation for most individual atoms, except the noble gases, is a process that is favored. By rearranging electrons so that atoms acquire eight electrons in the outermost energy level, these atoms become more stable.

Atoms Release Energy and Gain Stability in Bonding

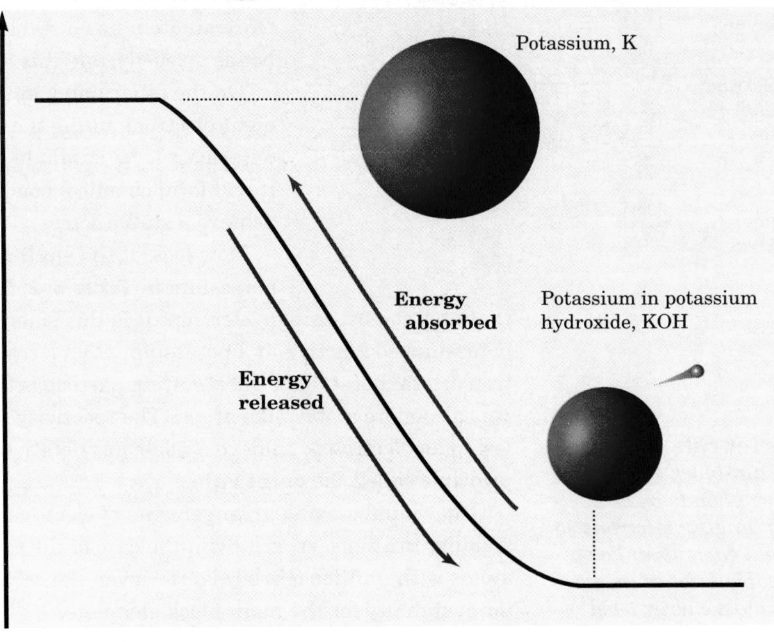

Figure 5-2
A potassium atom is reactive because it has an electron beyond an octet. When added to water, potassium forms bonds to achieve an octet. In the new substance, potassium hydroxide, the potassium is more stable and has a lower potential energy.

Breaking bonds requires energy

It takes energy to overcome the forces of attraction in a bond and separate the atoms. The energy required to break a chemical bond is known as **bond energy** and is measured in kilojoules per mole (kJ/mol).

bond energy
the energy required to change the atoms in one mole of bonds from a bonded state to an unbonded state

Atoms that are held by a strong bond are very stable and have a low energy state. Therefore, large amounts of energy must be supplied to bring the compound to a high enough energy state to break the bond.

Conversely, atoms held by a weak bond (low bond energy) are not as stable and have a higher energy state. Relatively little energy is required to break apart these atoms.

Electron transfers create charged particles

In many chemical reactions, an atom may give up electrons from its outermost energy level. Other atoms will accept these additional electrons to fill their outermost levels.

Recall that atoms are electrically neutral because they have the same number of protons and electrons. If an atom gives up or accepts electrons, it will have an imbalance of electric charges.

As a result of donating and accepting electrons, particles with positive and negative charges are formed. These positive and negative charges will bring the particles together to form one type of bond. *Salts*, such as those described in the beginning of the chapter, contain this type of bond.

Section Review

1. Explain why atoms of the noble gases do not easily form bonds, unlike most other atoms.
2. State the octet rule. Explain its relationship to bond formation.
3. What happens to the energy and stability of an atom when it forms a chemical bond?
4. How could each of the following atoms achieve a stable octet?
   a. calcium, Ca **b.** oxygen, O **c.** iodine, I **d.** xenon, Xe
5. Identify the particle below that has lost an electron. What charge does this particle now have? Which particle has gained an electron? What will the charge be on this particle?

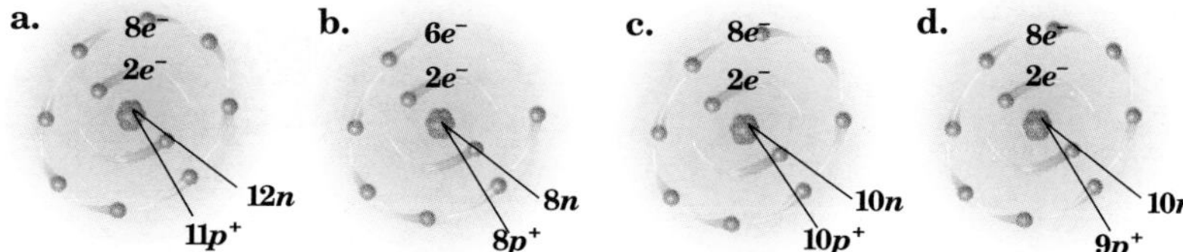

6. **Story Link**
   Why are the high bond energies of salts used as an argument by some scientists to support the plan to bury toxic wastes in salt domes?

ASSESS

Alternative Assessment

Draw an energy diagram that represents the relative stability of magnesium and its compounds in the following series of reaction equations. (Hint: use **Figure 5-2** as a guide.)

$$5Mg + O_2 + N_2 \xrightarrow[\Delta]{air} Mg_3N_2 + 2MgO$$

$$Mg_3N_2 + 6H_2O \longrightarrow 3Mg(OH)_2 + 2NH_3$$

$$Mg(OH)_2 \xrightarrow[\Delta]{air} MgO + H_2O$$

Answers to Section Review

1. Noble gases, except He, already have an octet of outershell electrons.
2. Representative elements lose or gain electrons to achieve noble gas configuration in most of their compounds.
3. The potential energy decreases, and the stability increases.
4. **a.** lose $2e^-$
   b. gain $2e^-$
   c. gain $1e^-$
   d. no change
5. particle **a**, 1+; particle **d**, 1−
6. A high bond energy means high thermal stability, so the salt will not likely react with toxic wastes.

Section 5-2

FOCUS

Lesson Starter

To begin this lesson, show that ionic compounds are held by attraction between charges. Blow up a small rubber balloon and charge it statically by rubbing it on your own hair or on animal fur. Take the balloon near someone who has long hair (preferably a blonde as blond hair is lighter in weight) and show a small attraction for the hair. Recharge the balloon, stick it on the wall, and let it hang there. Forces of attraction between positive and negative charges hold the balloon in place. Discuss how this is like ionic bonding in crystals.

TEACH

Chapter Connections

Forming salts

Salts are also formed during acid-base neutralization reactions, which are discussed in Chapter 13. Because of this, there are thousands of compounds that fall under the heading of salt.

Do You Know?

The Waste Isolation Pilot Plant (WIPP), located in a salt deposit 26 mi. from Carlsbad, New Mexico, is a facility that the U.S. Department of Energy built for the safe disposal and retrieval of transuranic (TRU) waste . The salt beds are said to be well suited for TRU waste disposal because:

1. they lie 2150 ft beneath the desert surface and are in an area of little earthquake activity.
2. they are not found near running groundwater.
3. they are easily mined.
4. they are self-sealing around voids, so the waste stored there will be encapsulated naturally over time.

Tests are being conducted in labs around the country to determine the viability of WIPP for its designated purpose.

What holds a salt together?

Section Objectives

Describe how sodium and chlorine atoms achieve stable octets by forming a chemical bond with one another.

Distinguish among ions, cations, and anions.

Explain how an ionic bond is formed.

Describe the properties of ionic compounds, including the crystal lattice structure of salts.

Summarize the energy changes that occur when an ionic compound is formed.

Electron transfers

Chem Fact

Sodium chloride makes up more than 60% of the particles dissolved in human blood plasma.

Sodium is a reactive Group 1 metal, much like the potassium metal discussed earlier. Chlorine, a Group 17 nonmetal, is a poisonous gas composed of pairs of chlorine atoms bonded together. When the two are placed together, a violent exothermic reaction occurs, leaving behind a white residue, as shown in **Figure 5-3**. This compound, formed from two dangerous elements, is something you probably eat every day—table salt. Chemists call it *sodium chloride* instead of salt because the word *salt* could be used to describe any one of thousands of different compounds. Yet all salts have something in common. They are formed from atoms in the same way and therefore share certain properties.

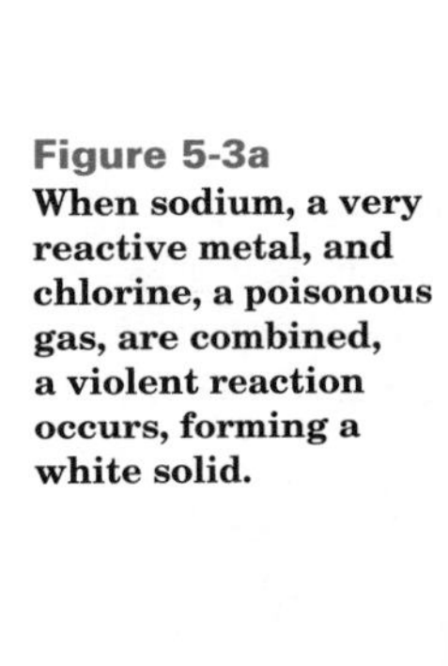

Figure 5-3a When sodium, a very reactive metal, and chlorine, a poisonous gas, are combined, a violent reaction occurs, forming a white solid.

5-3b The solid crystals that form are sodium chloride, the salt used to enhance the flavor of foods.

5-3c Using a technique called X-ray diffraction, chemists have collected data that can be used to build a model of a sodium chloride crystal.

Visual Strategy

Figure 5-3

Compare the potential energies of **Figure 5-3a** and **Figure 5-3b**. Comment on how you know the reaction is exothermic. Ask students to speculate about how they would quantify the heat given off and how this heat might relate to bond energy. Students should begin to realize energy changes are quantifiable so that the Born-Haber cycle for NaCl on page 161 becomes a logical closure for ions and lattice formation.

Figure 5-4
This scanning electron micrograph of sodium chloride has been magnified 840×. (Note: the colors seen are an artifact of the electron micrograph process.)

All salts are made of charged particles formed by donating electrons. Sodium chloride is an excellent example of how such a bond forms and how the nature of this bond determines the properties of a salt. Sodium chloride consists of cube-shaped crystals that are hard and brittle. **Figure 5-4** shows a photograph of sodium chloride taken with a high-powered microscope. Sodium chloride must be heated to 801°C before it melts and to 1413°C before it boils. Like many other salts, sodium chloride dissolves in water, and when dissolved or molten, it conducts electricity. To understand why salts have such similar properties, recall what you know about energy, stability, the octet rule, and the electron orbital notations of sodium and chlorine.

Sodium atoms lose electrons to achieve stable octets

Look at the electron orbital notations shown in **Table 5-3**. How could the sodium atom most easily achieve a stable octet? You may suggest that the sodium atom could take on seven additional electrons or lose the one electron in its outermost level. Removing one electron requires plenty of energy, but still much less energy than gaining seven. By removing this one electron, the sodium atom would then have a different arrangement of electrons. In effect, the former highest energy level would be stripped away if this electron were removed.

After losing an electron, sodium's electron configuration is $1s^2 2s^2 2p^6$. There's another element that has this same configuration in its natural state without having to lose an electron. Check a periodic table or table of electron configurations to identify this element. In what group is this element found?

Table 5-3
Comparing a Sodium Atom With a Sodium Ion

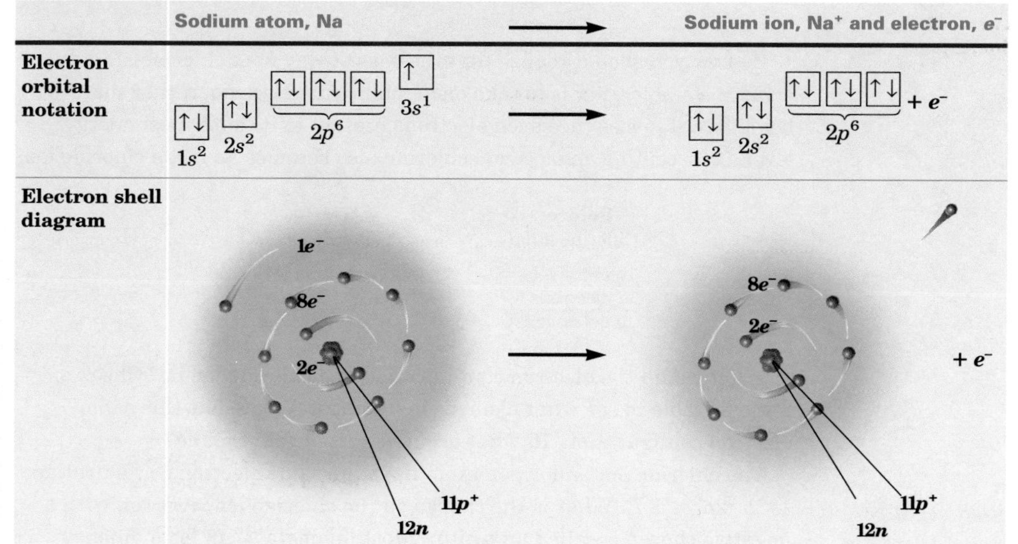

	Sodium atom, Na	→	Sodium ion, Na^+ and electron, e^-
Electron orbital notation	[↑↓] $1s^2$ [↑↓] $2s^2$ [↑↓][↑↓][↑↓] $2p^6$ [↑] $3s^1$	→	[↑↓] $1s^2$ [↑↓] $2s^2$ [↑↓][↑↓][↑↓] $2p^6$ + e^-
Electron shell diagram		→	+ e^-

Possible Misconception

The Na^+ ion has *not* become neon, it only has the stable electron structure of neon. It still has the same number of protons and neutrons as Na.

Do You Know?

The typical barrel of petroleum contains 42 gal, which could lead to 168 million gal of toxic waste. This would be equivalent to a space the size of a football field flooded about 142.5 m (467.4 ft) high.

Teaching Tip

Ion formation and electron configuration

Write the electron configuration of the noble gases on the board as column headings. Divide the class into groups of 6. Call out an atom by name. Have the groups work out the electronic configuration and identify the noble gas with which it is isoelectronic. One group's runner then goes to the board and writes the atomic symbol under the proper noble gas. Let the other groups identify the charge this atom has as an ion.

Visual Strategy

Table 5-3

Using **Table 5-3** as a guide, have the students draw diagrams for ^{20}Ca and ^{13}Al and determine the ion charge for each.

Theme

Classification and trends
Metals lose electrons to become stable like the preceding noble gas, while nonmetals tend to gain electrons to become stable like the following noble gas.

Teaching Tip

Modeling atoms, ions, and ionic compounds
Model atoms, ions, and ionic compounds on an overhead projector using circles cut from transparency film. Circles 6 mm in diameter represent outer-shell electrons. Mark each circle with a horizontal line so that electrons are recognized as negatively charged. Circles 10 cm in diameter represent the core of each atom modeled. Using different core colors will reinforce atom identity. Mark the center of these circles with the number of p^+ and n.

Possible Misconception

In an effort to simplify, students may believe that elements form only one ion. Warn them that nonmetals and most transition elements have more than one ionic form.

Students are sometimes confused by the sign on an ion's charge. Remind students that the addition of an electron gives a negative charge because the electron itself is negatively charged.

Recall that atoms of the noble gases, with the exception of helium, have stable octets in their natural state. A sodium atom, however, achieves a stable octet by having an electron stripped away, thus revealing its next lower energy level with its eight electrons.

Before	**After**
Sodium atom	Sodium ion
11 protons (+)	11 protons (+)
11 electrons (–)	10 electrons (–)
0 net charge	1+ net charge

In losing an electron, sodium loses a negative charge. As a result, sodium is no longer neutral; the number of protons does not equal the number of electrons. The particle with a 1+ charge produced when the Na atom loses an electron is called an ion. Any ion with a positive charge is called a **cation**. Metals tend to form cations.

cation
ion with a positive charge

Chlorine atoms gain electrons to achieve stable octets

Although chlorine is found in nature as Cl_2, the bond can be broken by light, heat, or electrical energy, yielding two individual Cl atoms. How would you describe the electron configuration of the chlorine atom shown in **Figure 5-5**? How would this chlorine atom achieve a stable octet?

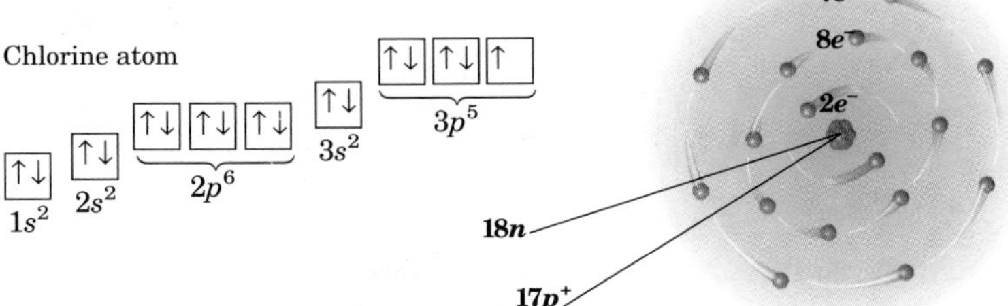

Figure 5-5
Look at the electron orbital notation and electron shell diagram of a chlorine atom. How many additional electrons does it need to achieve a stable octet?

By now you should realize that the easiest way for a chlorine atom to achieve a stable octet is to take on an additional electron, rather than being forced to lose the seven electrons present in its outermost energy level. By accepting an electron, chlorine also becomes an ion, a chloride ion.

Before	**After**
Chlorine atom	Chloride ion
17 protons (+)	17 protons (+)
17 electrons (–)	18 electrons (–)
0 net charge	1– net charge

What would the electron configuration of a chloride ion be? Check a periodic table to see what element in its natural state has this same electron configuration. In what group is this element found?

The chloride ion and argon atom have the same electron configuration, $1s^22s^22p^63s^23p^6$. What is the charge on the chloride ion? Any ion with a negative charge is called an **anion**. Most nonmetals can form anions.

anion
ion with a negative charge

Oppositely charged ions attract and bond to each other

The force of attraction between the 1+ charge on a sodium ion and the 1– charge on a chloride ion is called an **ionic bond**. *All salts are held together by ionic bonds.*

ionic bond
bond formed by the attraction of oppositely charged ions

The structures of salts show that the attractions are between more than just a single cation and a single anion. This attraction is so great that many sodium and chloride ions are pulled together in a tightly packed structure. The tight packing of the ions causes pieces of salt to have a distinctive crystal shape, such as the one shown in **Figure 5-6**.

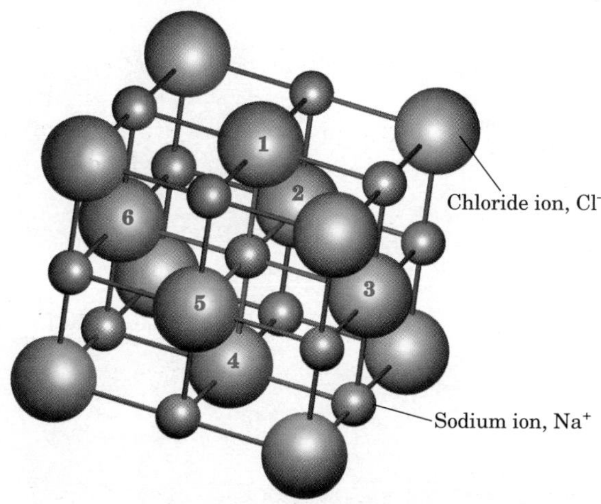

Figure 5-6
In a sodium chloride crystal, each sodium ion is surrounded by six chloride ions. Similarly, each chloride ion is surrounded by six sodium ions.

How many chloride anions surround each sodium cation? How many sodium cations, in turn, surround each chloride anion? When these ions come together, the oppositely charged ions are closer together than similarly charged ones. As a result, the attraction between oppositely charged ions is much greater than the repulsion between ions with the same charge. In addition, because the forces extend even further into the crystal, the force of attraction for ions in a crystal is 1.76 times as strong as for a single Na^+—Cl^- pair.

In NaCl and other compounds with ionic bonds, the amount of charge from cations equals the amount from anions, so that the compound itself is electrically neutral. Any chemical compound that has such a balance of oppositely charged ions is called an **ionic compound**.

ionic compound
chemical compound composed of cations and anions combined so that the total positive and negative charges are equal

Explaining the properties of salts

Salts are made of ions

Now that you have studied a specific salt, sodium chloride, and the types of bonds that hold it together, you can use what you already know to predict and understand the properties of other **salts**.

salt
an ionic compound containing cations other than H^+ and anions other than OH^- or O^{2-}

To conduct an electric current, a substance must have charged particles that can move about freely. Recall that the particles in a solid have some vibrational motion, but remain in relatively fixed locations. Thus, ionic solids such as salts are not good electric conductors because the ions cannot move very much, except to vibrate.

Teaching Tip

Binary ionic compound formation and properties of atoms and ions

Point out that the simplest way for atoms to react is for an atom of fairly low ionization potential, such as Na, to meet an atom of fairly high electron affinity, such as Cl. Then one or more electrons are transferred, and ions are formed.

Point out that when atoms become ions, their properties are drastically altered. The potassium in Section 5-1 went from a silver solid to a colorless ion in solution. To form the white compound, NaCl, sodium went from a silverish gray soft solid to a colorless ion, and chlorine went from a yellowish green gas to a colorless ion.

Historical Note

The independent existence of ions was proposed by Arrhenius. In 1916 Walther Kossel proposed that atoms lose or gain electrons to achieve the structure of an inert gas and that the compounds formed consist of ions.

Visual Strategy

Figure 5-6

Every Cl^- is surrounded by 6 Na^+ and every Na^+ by six Cl^-. Oppositely charged ions are located above, below, right, left, front, and back of the ion being considered, as illustrated by the numbered Cl^- ions.

Demonstration 3

Ions but not ionic compounds conduct electricity

Approximate time: 15 min

1. Construct the following cork-and-clip assembly for use as a conductivity apparatus. Insert two steel paper clips through the cork so that they do not touch each other. Spinning the paper clips should help drill them through the cork. Attach wires to the clips, bulb, and battery as shown below.

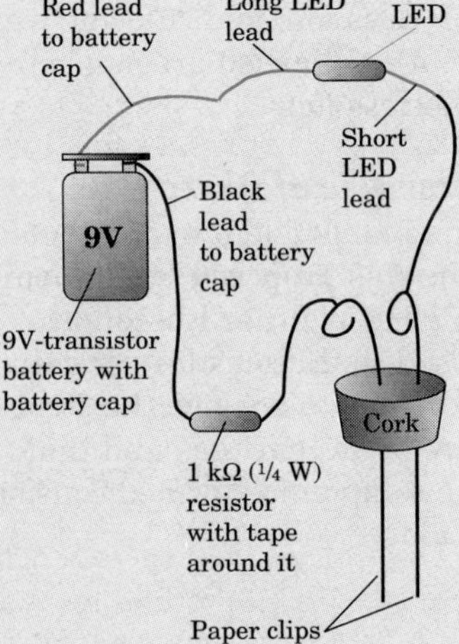

2. Fill a small, uncracked, porcelain crucible about half full with lithium chloride, LiCl.
3. Insert the leads of a conductivity tester. Be sure the leads do not touch each other. The unlit bulb indicates that the solid salt does not conduct electricity.
4. Remove the tester.
5. Heat the crucible until the LiCl is melted. Then lower the leads to the conductivity apparatus into the LiCl. The bulb lights up.
6. Fill a 250 mL beaker with 100 mL of distilled water. Place the leads of the conductivity tester into the beaker. The bulb does not glow.
7. Add 10 g of LiCl to the beaker and retest. The bulb should glow.

However, if the ions could move about, ionic compounds would become good conductors. If a crystal is melted, the ions that make up the crystal can move past each other. No longer held in rigid positions by ionic bonds, these ions can move about freely, as shown in **Figure 5–7**, and thus conduct electricity. Similarly, when a crystal dissolves, its ions are no longer held tightly in a crystal and can therefore conduct electricity.

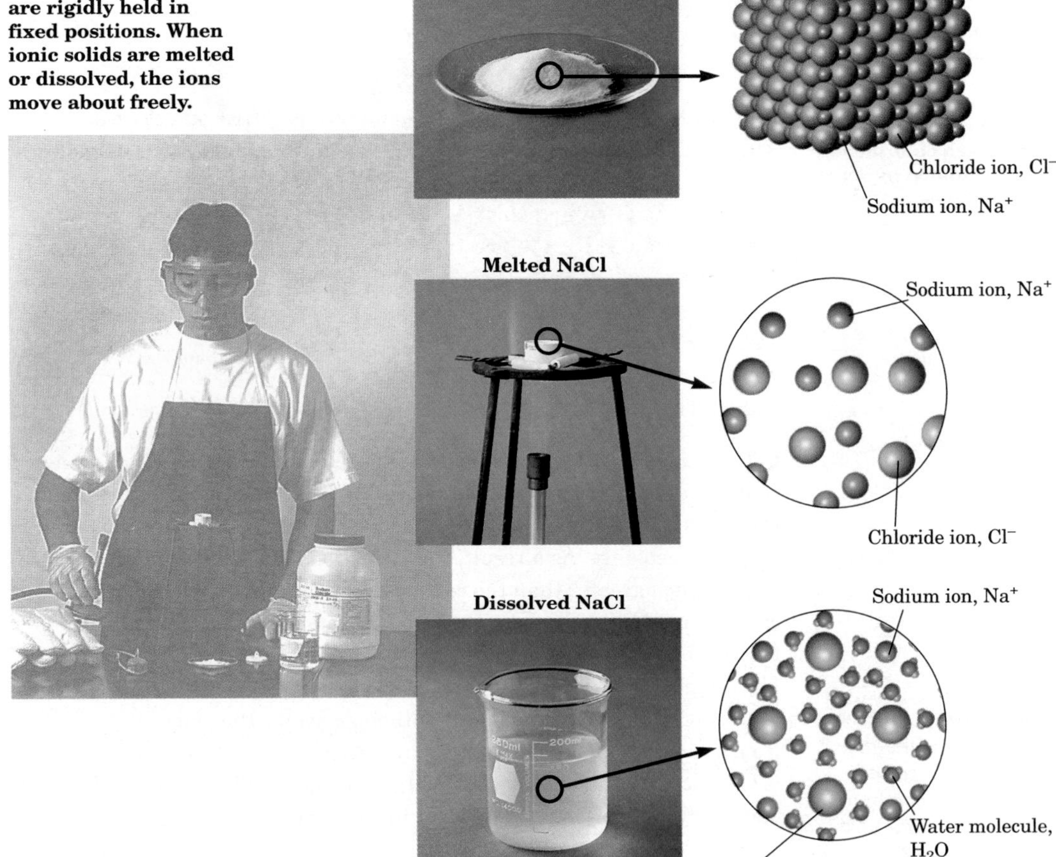

Figure 5-7
Particles in a solid are rigidly held in fixed positions. When ionic solids are melted or dissolved, the ions move about freely.

Salts have ordered packing arrangements

crystal lattice *three-dimensional arrangement of atoms or ions in a crystal*

The ions in a salt form repeating patterns, with each ion held rigidly in place by strong attractive forces that bond it to several oppositely charged ions. No matter which ions are present, the order in which they are packed is repeated throughout the salt. The pattern found in the arrangement of ionic compounds is called a **crystal lattice**.

Visual Strategy

Figure 5-7

You may wish to emphasize this sequence of pictures with the results from Demonstration 3. You may also wish to compare ion movement in crystals with ion movement in metals shown in **Figure 5-11**.

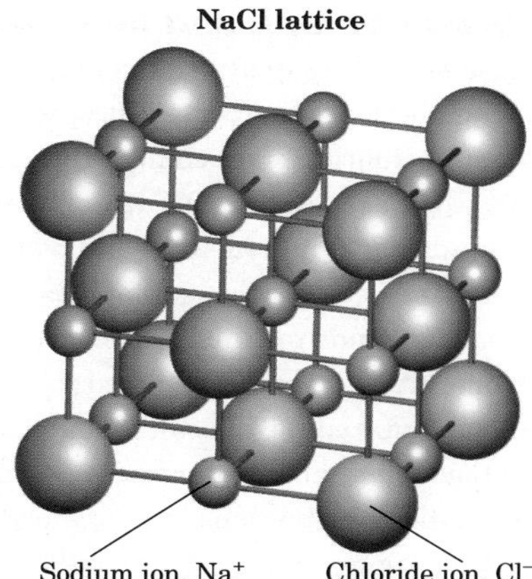

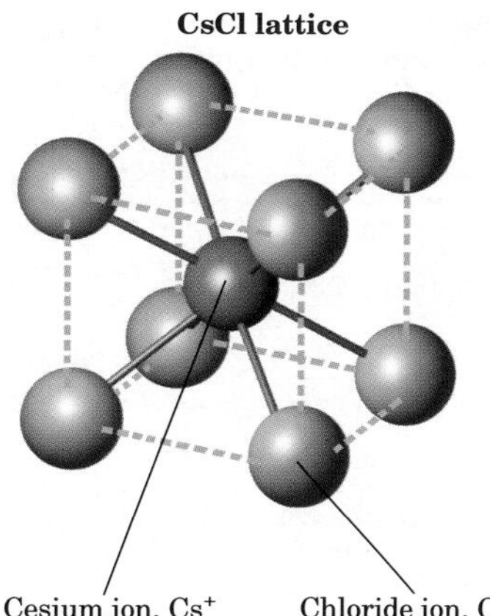

Figure 5-8
The crystal lattice structure of sodium chloride differs from that of cesium chloride.

Not all salts, however, have the same crystal lattice structure as sodium chloride. In cesium chloride, CsCl, each cation is surrounded by *eight* anions instead of *six* anions, as shown in **Figure 5–8**. Despite this difference, both sodium chloride and cesium chloride crystals have a structural similarity. The crystals of both salts are made of simple repeating ionic units that yield shapes which are cubic.

unit cell
the simplest portion of a crystal lattice that portrays the three-dimensional structure of the entire lattice

The simplest repeating unit of a crystal is known as a **unit cell**. Both sodium chloride and cesium chloride crystals are classified as part of a cubic crystal system. However, the unit cells making up the cubic crystals of sodium chloride differ from those in cesium chloride. These two salt crystals have different numbers of nearest neighbors because their ions have different sizes. Cesium ions are larger than sodium ions. The larger cesium ions have room for more chloride ions around them, thus causing a cesium chloride unit cell to have a slightly different spatial arrangement.

Examine **Figure 5–9**, which shows three different ways that atoms can be arranged in the unit cells to form a cubic crystal system. For sodium chloride, if you look at either the chloride ions alone or the sodium ions alone, each type forms a face-centered cubic unit cell. For cesium chloride, the ions of cesium form a simple cubic arrangement intertwined with a simple cubic arrangement of chloride ions.

Figure 5-9
The unit cells of crystals with a cubic shape can have one of three possible arrangements of particles.

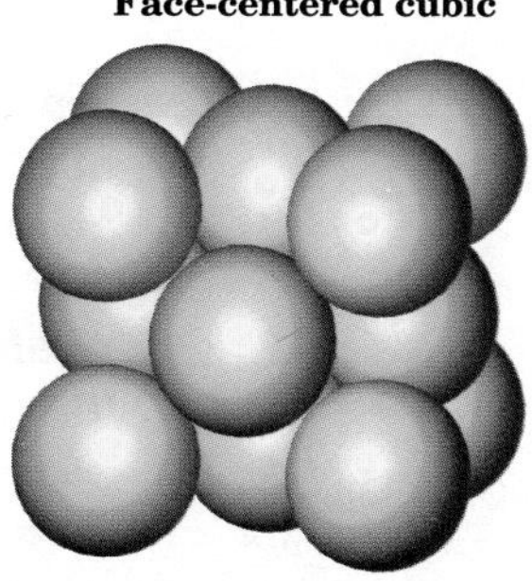

Variation: Recover the solid salt by covering the solution beaker with a watch glass and heating strongly until evaporation is complete. Retest the solid salt to show nonconductivity. Other salts that can be used are $NaNO_3$, KNO_3, and $AgNO_3$.

SAFETY
Wear safety goggles and a lab apron. Use a battery-powered conductivity tester only. Dilute spills with water, if necessary. Then, wearing gloves, soak up the mess with toweling. Rinse the toweling, and pour the rinsings down the drain.

DISPOSAL
Put tape over the battery's terminals and save for the next time, or dispose of according to supplier's or manufacturer's directions. The chemicals should be dissolved in water and poured down the drain.

Further Applications

One industrial method for making NaOH is to dissolve NaCl in H_2O and apply energy in the form of electricity. The initial reaction produces the elements, but Na immediately combines with H_2O to produce NaOH and H_2. The Cl_2 and H_2 are removed, and the dilute lye solution is concentrated into NaOH pellets.

Groupwork Strategy

Students can get a rough idea of the cubic shape of NaCl by looking at table salt crystals through a microscope. Ask students why the corners are rounded and not precise. (They are roughed up during mining and shipping.)

Teaching Tip

Melting points for ionic compounds

Other examples are NaCl, 801°C; NaBr, 750°C; NaI, 662°C; K_2CO_3, 891°C; $CaCl_2$, 772°C; Na_2CO_3, 851°C; and $CaSO_4$, 1450°C.

Do You Know?

Many ionic compounds decompose before boiling, but their high melting points are still a sign of their strong ionic bonding. $MgSO_4$ decomposes at 1124°C; KCl melts at 772°C and sublimes at 1500°C; NH_4Cl sublimes at 350°C; and $CaCO_3$ decomposes at 825°C.

Theme

Classification and trends

The higher the melting point, the stronger the crystal lattice, and the more thermally stable the salt is.

Possible Misconception

Students need to know that many exothermic reactions require energy to begin the reaction. To begin breaking the bonds in gasoline in an engine, we must apply a spark. The new bonds are more stable, so the reaction is exothermic, and the excess energy can be used for power.

Table 5-4
Comparing Melting and Boiling Points of Compounds

Compound	Melting point (°C)	Boiling point (°C)
calcium iodide	784	1100
carbon tetrachloride	–23	77
hydrogen fluoride	–83	19.5
hydrogen sulfide	–85.5	–61
iodine monochloride	27	97
magnesium fluoride	1261	2239
methane	–182	–164

Salts do not melt and boil easily

Because of the strong attraction between their ions, all ionic compounds share certain properties. The boiling point for sodium chloride is much higher than that for water—1413°C compared with 100°C. Similarly, most other ionic compounds have high melting and boiling points, as you can see from **Table 5-4**. Because the ions in such a compound form strong bonds to a number of different ions, a considerable amount of energy is required to break them apart, as would be required for the salts to melt or boil. This is especially true for small ions formed by atoms. However, in some compounds, the ions are very large, and the forces of attraction are weaker, resulting in lower boiling points.

Salts are hard and brittle

Like most other salts, table salt is fairly hard and brittle. Both properties are shared by ionic compounds with small ions because of the bonding between ions.

No matter which unit cell a crystal has, ions are arranged in a repeating pattern, forming layers. As long as the layers stay in a fixed position relative to one another, the ionic compound will be hard because it will take a lot of energy to break all of the bonds between the ions.

However, if a force moves one layer slightly, ions with the same charge will be next to each other. What would happen as a consequence of such a realignment of ions? When a force is large enough to reposition the ions, the repulsive forces between two layers cause them to break apart.

Energy and ionic bonding

Energy can be released or absorbed when ions form

Removing electrons from atoms requires an input of energy. Recall from Chapter 4 that this energy is known as ionization energy. The ionization energy to remove one electron from each atom in a mole of sodium atoms is 495.8 kJ/mol. On the other hand, adding electrons to atoms releases energy. This energy release results from the affinity certain atoms have for electrons. A mole of chlorine atoms, for example, releases 348.6 kJ/mol when an electron is added to the outermost energy level of each atom.

Forming ions is only one part of bonding

If adding an electron to a chlorine atom cannot supply enough energy to remove an electron from a sodium atom, why does an ionic bond form?

Forming an ionic bond actually involves several steps, which are shown in **Figure 5-10** on the next page. The chief driving force for the reaction is the last step in the formation of an ionic compound, in which the ions come together to form the crystal lattice. Keep in mind that the starting materials are sodium atoms and chlorine gas, which is in the form of pairs of bonded chlorine atoms. The final product consists of sodium chloride salt crystals.

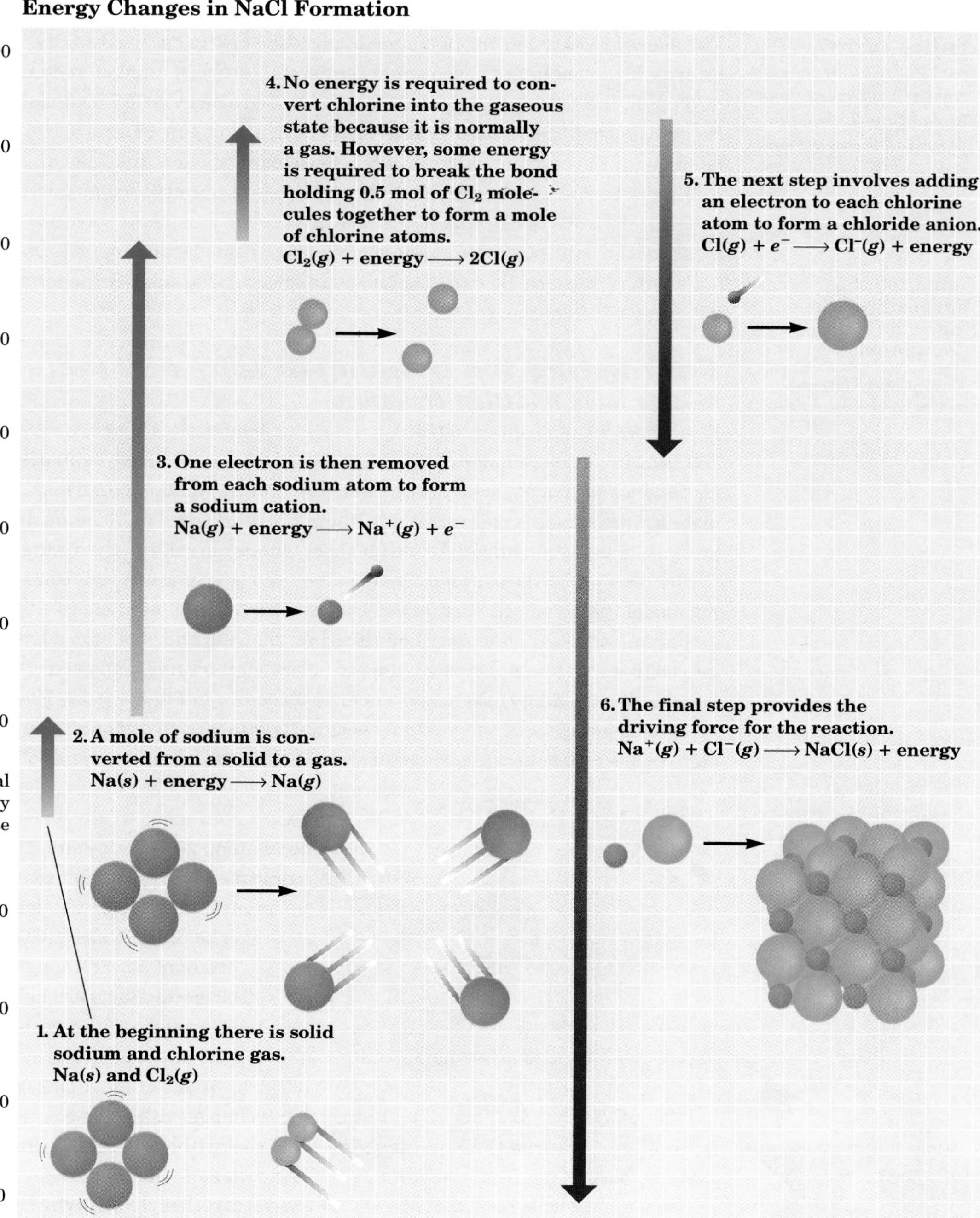

Energy Changes in NaCl Formation
800
700
600
500
400
300
200
100
initial energy state
-100
-200
-300
-400
4. No energy is required to convert chlorine into the gaseous state because it is normally a gas. However, some energy is required to break the bond holding 0.5 mol of Cl_2 molecules together to form a mole of chlorine atoms.
$Cl_2(g)$ + energy ⟶ $2Cl(g)$
5. The next step involves adding an electron to each chlorine atom to form a chloride anion.
$Cl(g) + e^- \longrightarrow Cl^-(g)$ + energy
3. One electron is then removed from each sodium atom to form a sodium cation.
$Na(g)$ + energy ⟶ $Na^+(g) + e^-$
6. The final step provides the driving force for the reaction.
$Na^+(g) + Cl^-(g) \longrightarrow NaCl(s)$ + energy
2. A mole of sodium is converted from a solid to a gas.
$Na(s)$ + energy ⟶ $Na(g)$
1. At the beginning there is solid sodium and chlorine gas.
$Na(s)$ and $Cl_2(g)$

Figure 5-10
Of the steps involved in making sodium chloride, steps 2–4 are endothermic, and the last two are exothermic. Combining the steps and the energy changes provides the net reaction, which is exothermic. In other words, the final energy state of NaCl is lower than the initial energy state of the reactants.

Chapter Connections

The energy changes for NaCl lattice formation involve electron affinity, ionization potential, and endothermic and exothermic reactions, which were encountered in Chapters 2 and 4.

Do You Know?

The series of reactions diagrammed in **Figure 5-10** are known collectively as the Born-Haber cycle.

Visual Strategy

Figure 5-10

Summarize the formation of NaCl by going through the steps in the diagram. Compare your result with the value in the table on page 162. Although energy needs to be added for the reaction to start, the overall reaction is exothermic because the lattice energy is a large negative quantity. Ask students how much energy would be given off if 0.5 mol of NaCl were formed. (206 kJ)

Chapter Connections
Bond strength and electronegativity
Remind students that the bond strength increases as the electronegativity difference increases, so K—F is a stronger bond than K—Cl.

ASSESS

Alternative Assessment

1. Arrange the lattice energies listed in **Table 5-5** according to increasing potential energy.
2. Explain why a face-centered cubic crystal takes up more room than a body-centered cubic crystal.
3. Graph a Born-Haber cycle for $MgCl_2$.

lattice energy
energy released when a crystal containing one mole of an ionic compound is formed from gaseous ions

Forming bonds in a crystal lattice releases energy

The energy released in the last step is called the lattice energy. **Lattice energy** is the energy released when the crystal lattice of an ionic solid is formed. In the case of sodium chloride, the lattice energy is 787.5 kJ/mol, far greater than the input of energy in steps 2, 3, and 4 in **Figure 5-10**. However, if the crystal lattice did not release energy when it formed, there would not be enough energy solely from step 5 for sodium atoms to give up their electrons to chlorine atoms. Lattice energy provides a way to measure the bond strength in ionic compounds. Some lattice energies of ionic compounds are given in **Table 5-5**. In general, the smaller the ion, or the greater the charge, the higher the lattice energy.

Table 5-5 Lattice Energies for Some Ionic Compounds

Compound	kJ/mol
LiCl	861.3
LiBr	817.9
LiI	759
NaCl	787.5
NaBr	751.4
NaI	700
CaF_2	2634.7
MgO	3760
MgS	3160
CaO	3385
CaS	2775

Contrasting the bonding in salts and metals

Metals have a lattice structure

Even though salts contain metal ions, they have properties that are very different from those of metal atoms bonded to each other. In general, metals conduct electricity, even in the solid state. Metals tend to be softer, more *ductile* (capable of stretching without breaking), and more *malleable* (capable of being pounded into a thin sheet without breaking) than salts.

The reasons for these properties of metals can be explained by a bonding model that is specific to metals. The atoms in a metal are held together in a crystal by the forces of attraction that the nuclei of atoms have for electrons. In a metal crystal, the atoms are all the same, and each metal atom has many other atoms nearby, as shown in **Figure 5-11**. However, the bonding electrons are not specifically attached to each individual atom in the lattice because the orbitals in the outermost energy levels of adjacent atoms overlap. The bonding electrons are not attached to any one metal atom, as in ionic salts, but are free to move throughout the crystal. This is why metals can conduct electricity.

Similarly, even though the metal atoms are held in somewhat fixed positions in the crystal, the positions of the atoms can change within the electron pool without breaking the attractions that hold the metal together. This is why metals can be bent or stretched without shattering.

Figure 5-11
The ability of metals to be hammered into shape depends on the softness that results from the bonding among the metal atoms.

Section Review

7. Which of the particles shown below are cations? Give reasons for your answers.

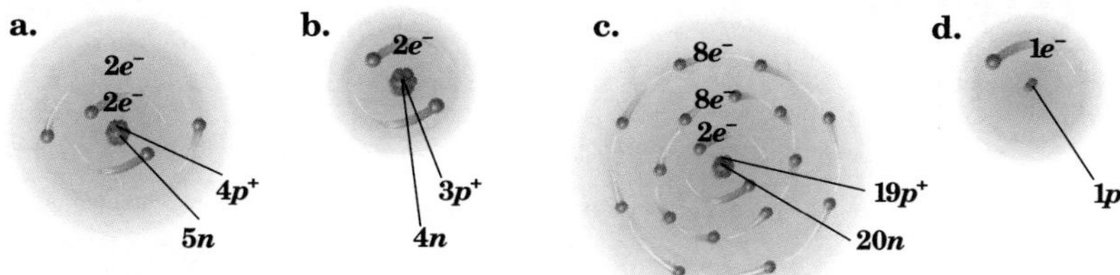

8. Which of the particles shown below are anions? Give reasons for your answers.

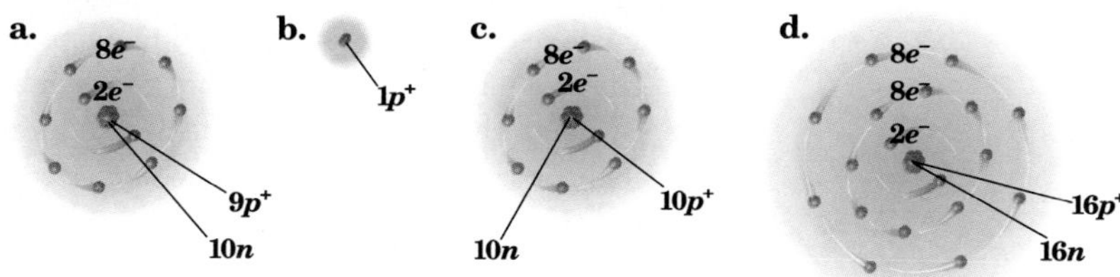

9. Give the charges for ions formed from elements of the following groups.
   a. Group 1 **b.** Group 2 **c.** Group 16 **d.** Group 17
10. Draw the electron shell notations for the ions of the following elements: magnesium, oxygen, and bromine. Indicate whether each is a cation or anion.
11. Suggest a reason why the structure of a unit cell of LiF differs from that of a unit cell of LiCl.
12. **Table 5-4** on page 160 lists several compounds along with their boiling points. Which ones would you classify as ionic compounds?
13. With the knowledge that ionic compounds are both hard and brittle, which of the compounds shown below would you classify as ionic? Explain your answers.

a. 

b. 

c. 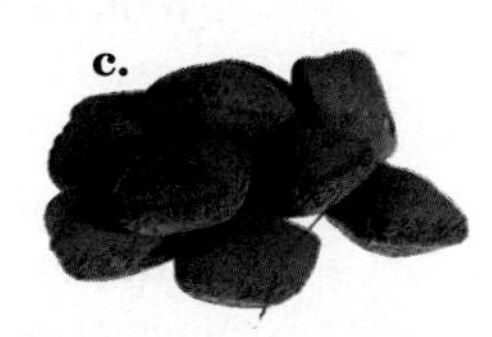

d. 

14. What role does lattice energy play in forming an ionic compound?
15. Name two properties of metals, and explain each in terms of the bonding model for metals.

16. **Story Link**
    Explain how the hardness of salts can be used to argue in favor of the salt-dome toxic-waste plan.
17. **Story Link**
    Explain how the brittleness of salts can be used to argue against the salt-dome toxic-waste plan.

Answers to Section Review

7. **b** and **c** are cations; the number of e^- is less than the number of p^+.
8. **a** and **d** are anions; the number of e^- is greater than the number of p^+.
9. **a.** 1+, **b.** 2+, **c.** 2–, **d.** 1–
10.

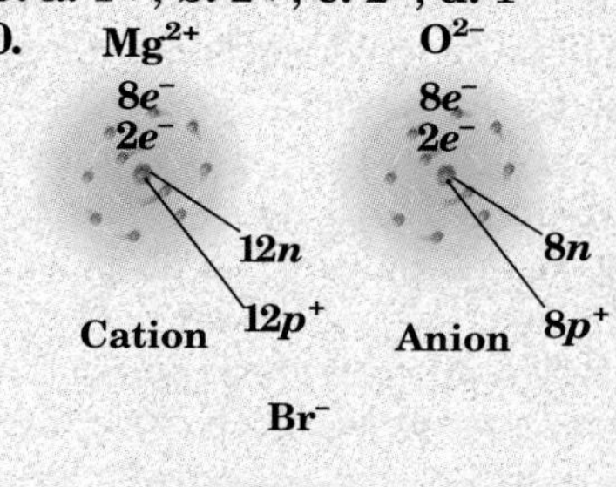

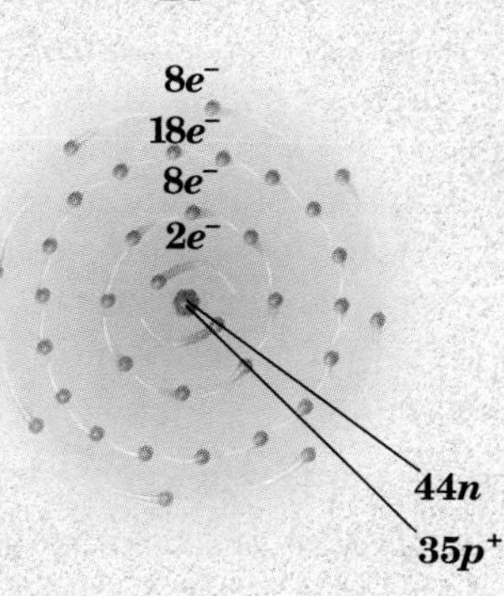

Anion

11. Size of ionic radii; Cl^- is much larger than F^-.
12. calcium iodide and magnesium fluoride
13. **a.** ionic
    b. not ionic; soft and fluid
    c. not ionic; brittle but soft
    d. not ionic; hard but not brittle
14. The lattice energy has such a large negative value that the overall reaction is exothermic.
15. **Properties:** conduct electricity, soft, ductile, malleable. **Reasons:** Bonding electrons are not attached to any one metal atom but are free to move. Metal atoms are not in fixed positions, so they can shift positions within the electron pool.
16. The salts would not be easily penetrable.
17. Toxic waste could penetrate the crystalline structure of the salt where breaks or abrasions occur, and work its way through the dome into the surrounding ground and water tables.

Section 5-3

FOCUS

Lesson Starter

Chemical nomenclature for ionic compounds

This board exercise points out the importance of order in chemical names. Write these on the board: Washington George Carver, Einstein Albert, and Madame Marie Curie. Have the students identify the people. Discuss the importance of name order and correct spelling of the name for recognition of the person. Extend the discussion to include ionic compounds. Cations and anions can be thought of as first and last names, respectively, for ionic compounds. The order as well as the spelling is important for recognition of the compound.

TEACH

Teaching Tip

Vocabulary development

Students can be helped to remember that cations are positive by drawing a *plussy* cat.

This helps also them to say the word cation properly.

Students can be helped to remember anions are negative by drawing an *antion* with a long body.

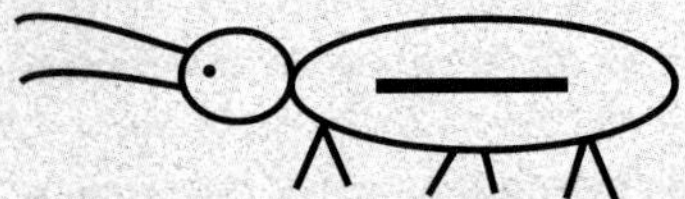

Cultural Connection

Chemical nomenclature is identical worldwide. This makes understanding new ideas and compounds from other countries much easier.

How do you name salts?

Section Objectives

Use the rules for writing and naming monatomic ions.

Apply the rules for writing and naming binary ionic compounds.

Recognize and explain why certain formulas are invalid.

Naming ions

A salt may be any one of thousands of chemical compounds. So how do chemists describe which salt they're discussing or working with in the laboratory?

Chemists use a system to write and to name the ions and the salts they form. This system is based on the chemical symbols listed in the periodic table. By using this system, chemists have a uniform standard to represent chemical compounds. The system makes communication easier, saves time, and simplifies calculations.

monatomic ion
cation or anion formed from a single atom

Monatomic ions are made from a single atom

Many elements found in the main groups on the periodic table form ions with noble gas notations by either losing or gaining electrons. In fact, the periodic table can be helpful in determining how these elements form ions. For example, Group 1 metals lose one electron to produce cations with a 1+ charge. Group 2 metals lose two electrons to produce cations with a 2+ charge.

On the other side of the periodic table, Group 16 nonmetals form ionic compounds by gaining two electrons to produce anions with a 2– charge. Finally, Group 17 nonmetals gain one electron to form anions with a 1– charge. Group 18 elements usually don't form ions.

Any ion formed when an atom gains or loses electrons is known as a **monatomic ion**. To write the chemical symbol for a monatomic ion you must indicate both the symbol for the element and its charge. **Table 5-6** lists the chemical symbols and charges for several monatomic ions. Notice how the symbol is written for an ion with a charge greater than either 1+ or 1–. For example, the symbol for the magnesium ion is written as Mg^{2+}, not Mg^{+2} or Mg^{++}.

The system for naming chemical substances, whether ions or compounds, is known as nomenclature. The nomenclature for monatomic ions is straightforward. A monatomic cation is named using the element's name followed by *ion*. Na^+ ion is *sodium ion* and Mg^{2+} ion is *magnesium ion*.

A monatomic anion is named by dropping the ending of the element's name and adding the suffix *-ide*. Cl^- ion is *chloride ion*, S^{2-} ion is *sulfide ion*, and N^{3-} ion is *nitride ion*.

Table 5-6
Common Monatomic Ions

Positive charge	Ion name
1+	cesium ion, Cs^+ lithium ion, Li^+ potassium ion, K^+ rubidium ion, Rb^+ sodium ion, Na^+
2+	barium ion, Ba^{2+} calcium ion, Ca^{2+} magnesium ion, Mg^{2+} strontium ion, Sr^{2+}
3+	aluminum ion, Al^{3+}

Negative charge	Ion name
1–	bromide ion, Br^- chloride ion, Cl^- fluoride ion, F^- iodide ion, I^-
2–	oxide ion, O^{2-} sulfide ion, S^{2-}
3–	nitride ion, N^{3-}

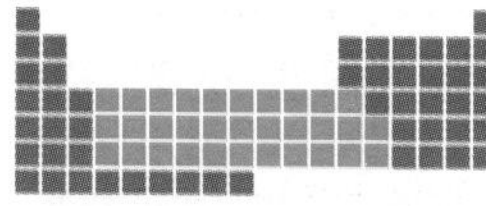

The octet rule cannot always predict charges

Determining how an element forms an ion is not always as easy as checking a periodic table and applying the octet rule. Some ions, especially those of transition metals, can reach stability without noble gas configurations. In addition, some elements can form ions in more than one way. For example, iron may lose either two or three electrons, and copper may lose either one or two electrons. In such cases, you will have to refer to a table that lists ions having more than one charge. **Figure 5-12** lists most of the elements that have more than one common charge or that are difficult to predict.

Ti^{2+} Ti^{3+} Ti^{4+}	V^{2+} V^{3+} V^{4+} V^{5+}	Cr^{2+} Cr^{3+} Cr^{6+}	Mn^{2+} Mn^{3+} Mn^{4+} Mn^{7+}	Fe^{2+} Fe^{3+}	Co^{2+} Co^{3+}	Ni^{2+} Ni^{3+}	Cu^{+} Cu^{2+}	Zn^{2+}	Ga^{3+}	
Zr^{4+}	Nb^{3+} Nb^{5+}	Mo^{6+}	Tc^{4+} Tc^{6+} Tc^{7+}	Ru^{3+}	Rh^{3+}	Pd^{2+} Pd^{4+}	Ag^{+}	Cd^{2+}	In^{3+}	Sn^{2+} Sn^{4+}
Hf^{4+}	Ta^{5+}	W^{6+}	Re^{4+} Re^{6+} Re^{7+}	Os^{3+} Os^{4+}	Ir^{3+} Ir^{4+}	Pt^{2+} Pt^{4+}	Au^{+} Au^{3+}	Hg_2^{2+} Hg^{2+}	Tl^{+} Tl^{3+}	Pb^{2+} Pb^{4+}

Figure 5-12
Many metals can form ions with more than one charge.

Nomenclature for ions that can have other charges

How do chemists name ions with more than one common charge? Different names are needed, and to solve this problem, chemists use a system involving Roman numerals to indicate charge. Copper can form compounds containing either Cu^{+} ions or Cu^{2+} ions. The former is written as copper(I) ion and is read as *copper one ion*. The latter is written as copper(II) ion and is read as *copper two ion*. Roman numerals are not used for atoms that always form ions with the same charge. For example, the Mg^{2+} is never referred to as *magnesium(II) ion*, because magnesium only forms ions with that charge.

Symbols for monatomic ions

18. How would you write the symbol for the ion formed when an oxygen atom gains two additional electrons?

19. How would you write the chemical symbol for the ion formed when a phosphorus atom gains three electrons?

20. How would you write and name the ions for tin?

21. For each group below, list only those ion charges that all members of the group can form.

a. Group 5 **b.** Group 7 **c.** Group 9 **d.** Group 11 **e.** Group 12

Answers for Concept Review items and Practice problems begin on page 841.

Teaching Tip

Macroscopic observation of crystal structure

Display crystals of various size and shape such as NaCl and quartz. Attractive crystals can be purchased at science museums and through chemical catalogs.

Groupwork Strategy

Practicing nomenclature

Have students make flashcards for each ion in **Table 5-6** and **Table 5-8**. The name of the ion should be on one side of the card, while the symbol and charge for the ion are on the reverse side of the card. Have students practice associating the name of the ion with its symbol and charge by quizzing each other with their cards. Then hand each group a slip of paper on which you have written the name (or formula) for 3–5 ionic compounds. Have the group identify the formula (or name) for each compound. If you use compounds found in the home, you may wish to ask students how these compounds are used or where they are found. Some examples are $Mg(OH)_2$ (milk of magnesia), NaCl (table salt), NH_4OH (ammonia water), $HC_2H_3O_2$ (vinegar), and $NaHCO_3$ (baking soda).

Teaching Tip

Multiple ion formation

Transition metals can have multiple charges because the electrons in the next-to-outermost shell are loosely bound and very close in energy to the valence shell electrons.

Possible Misconception

Sometimes students believe that the Roman numeral in the name of a compound represents the subscript for that cation in the compound's formula. Emphasize that Roman numerals represent the charge on the cation and that Arabic numerals are used for subscripts.

Answers to

Concept Review

18. O^{2-}

19. P^{3-}

20. Sn^{2+}, tin(II) ion, and Sn^{4+}, tin(IV) ion

21. a. 5+, **b.** 4+,7+ **c.** 3+, **d.** 1+, **e.** 2+ (from **Figure 5-12**)

Teaching Tip
More examples of ionic bonding
Metal oxides (rust and tarnish), lime, chalk, rubies, sapphires, and Ag_2S (silver tarnish) are examples of ionic bonding.

Teaching Tip
Naming ionic compounds
Determine names for the following formulas: $MgBr_2$, Al_2S_3, MnO_2, VCl_3, TiO_2, W_2O_3, CrO_3, $NiCl_2$, and Hg_2Cl_2.
Solutions: magnesium bromide; aluminum sulfide; manganese(IV) oxide; vanadium(III) chloride; titanium(IV) oxide; tungsten(VI) oxide; chromium(III) oxide; nickel(II) chloride; and mercury(I) chloride.

Naming ionic compounds

Binary compounds are made from two elements

binary compound *compound composed of two elements*

Recall that table salt is made of ions of two elements—sodium and chlorine. Ionic compounds consisting of two elements are known as **binary compounds**. Naming binary compounds is easy—you just combine the names of the two ions. The cation is written first and named first. For example, the ionic compound composed of a cesium ion (Cs^+) and a chloride ion (Cl^-) is called cesium chloride. A barium ion (Ba^{2+}) and an oxide ion (O^{2-}) compose barium oxide.

Writing chemical formulas for binary ionic compounds can be simple or somewhat more involved, depending on which two ions are combined. Recall that an ionic compound, although composed of charged ions, is neutral because the total charge of the cations equals the total charge of the anions. Consequently, in sodium chloride, Na^+ ions and Cl^- ions are present in a 1:1 ratio. For every sodium ion with a single positive charge, there is a chloride ion with a single negative charge. Thus, the positive and negative charges cancel each other. Similarly, in barium oxide, Ba^{2+} ions and O^{2-} ions are present in a 1:1 ratio. The formulas are written as NaCl and BaO to show that in each compound the cation and anion are present in a 1:1 ratio. Notice that in writing the chemical formula, the cation is always written first. The absence of subscript numbers alongside the chemical symbols implies that there is one of each ion in the formula, a 1:1 ratio. Also notice that the charges are not usually written as a part of the formula.

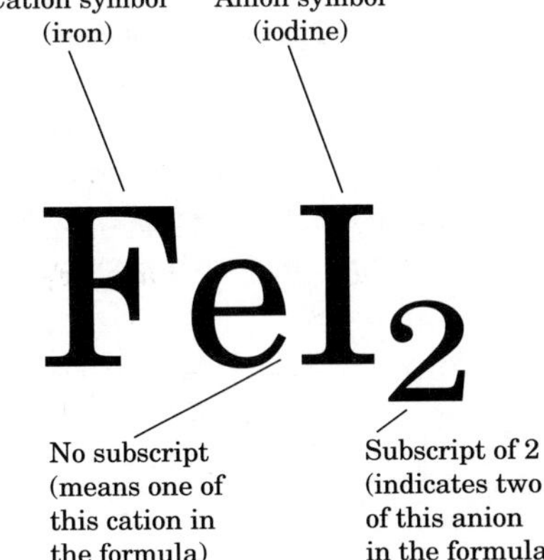

Figure 5-13
Iron(II) iodide is a binary compound with a 1:2 ratio of iron to iodide ions.

Ionic compounds have balanced charges

Remember that in an ionic compound, the charges balance to produce a neutral compound. If you are given the ions in an ionic compound and their charges, you can determine which subscripts will be needed to write the formula. The key is to determine the total number of positive and negative charges necessary for each ion in the formula. This number is the least common multiple of the charges of the individual ions.

For example, if an ionic compound contains manganese(IV) ions, Mn^{4+}, and oxide ions, O^{2-}, the least common multiple of the charges would be four. Because the total number of charges necessary for each ion in the formula is four, it will require one manganese(IV) cation for four positive charges, and two oxide anions for four negative charges. The correct formula for the compound is MnO_2.

How would you write the formula for the binary compound made from iron(II) ions, Fe^{2+}, and iodide ions, I^-? Obviously, you need twice as many iodide ions, each with a 1– charge, to balance the iron(II) ions, each with a 2+ charge. Therefore, the ratio of Fe^{2+} ions to I^- ions must be 1:2. The formula must show this ratio. As shown in **Figure 5-13**, this is done by using a subscript to the right of the symbol for iodine, indicating twice as many I^- ions as Fe^{2+} ions: FeI_2.

Sample Problem 5A
Writing ionic formulas

What is the formula for aluminum oxide?

1 List
- **Symbol for aluminum ion from Table 5-6:** Al^{3+}
- **Symbol for oxide ion from Table 5-6:** O^{2-}

2 Set up
- **Write the symbols for the ions, side by side, with the cation first.**

$$Al^{3+}O^{2-}$$

3 Solve
- **Find the least common multiple of the ions' charges.**
The least common multiple of three and two is six—to get a neutral compound, you would need a total of six positive charges and six negative charges.
- **To get six positive charges:** you need two Al^{3+} ions because $2 \times 3+ = 6+$.
- **To get six negative charges:** you need three O^{2-} ions because $3 \times 2- = 6-$. Therefore, the ratio of Al^{3+} to O^{2-} should be 2:3.
- **The formula would then be written as follows.**

$$Al_2O_3$$

Practice 5A
Answers for Concept Review items and Practice problems begin on page 841.

1. **Determine the formulas for the ionic compounds shown below.**

a. 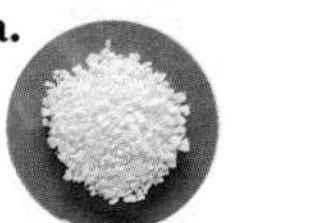
Calcium chloride

b. 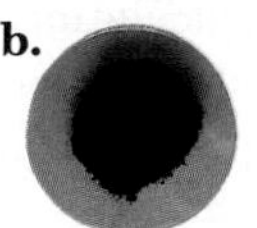
Iron(II) oxide

c. 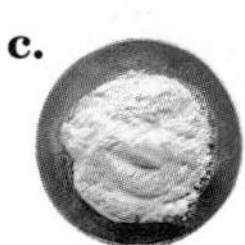
Magnesium oxide

Chemical formulas must reflect actual composition of compounds
The formula for sodium chloride is always written as NaCl and never as $NaCl_2$, and the formula for calcium fluoride is always shown as CaF_2 and never as Ca_2F_4. The reason in each case is different.

In the case of sodium chloride, why does the crystal show a 1:1 ratio of Na and Cl and not a 1:2 or 1:3 ratio? After all, the more chloride ions that bind to sodium ions, the greater the lattice energy. In fact, the lattice energy that would be produced by $NaCl_2$ could be as much as three times that produced by NaCl, if sodium could form an Na^{2+} ion.

But Na^{2+} ions and $NaCl_2$ do not form naturally. To understand why, recall that sodium loses a single 3*s* electron to attain a stable octet. To remove an additional 2*p* electron and create an Na^{2+} ion would dismantle the stable octet. You already know that a decrease in stability results from an input of energy. But if the energy required to remove an additional electron is less than that released when the lattice is formed, it could still happen.

Additional Example for Sample Problem 5A
Determine the formulas for the following compound names: aluminum nitride, strontium phosphide, calcium chloride, and rubidium iodide.

Answers to Practice 5A
1. **a.** $CaCl_2$
b. FeO
c. MgO

Teaching Tip
Al_2O_3 and aluminum siding
Ask the students if they have aluminum siding on their house. Ask what happens when you brush against it. A light colored film of Al_2O_3 usually comes off the siding and leaves a mark that remains different from the rest of the siding for years.

Solutions for Additional Example 5A
AlN, Sr_3P, $CaCl_2$, and RbI.

Content Background

Terminology

The term *formula unit* is used for ionic compounds, and the term *molecular formula* (Chapter 6) is used for covalent compounds. The different terminology arises from the different types of bonding. A molecular formula gives the symbols for the combined elements and subscripts that tell *how many* of each atom are present. Formula units, by contrast, are only the *ratios* in which the ions are combined. Although formula units are often identical to the empirical formula, molecular formulas generally are not.

Table 5-7
Ionization Energies for Sodium, Magnesium, and Aluminum

Element	Ionization Energy (kJ/mol) First	Second	Third	Fourth
sodium	496	4562	6910	9543
magnesium	738	1451	7733	10 543
aluminum	578	1817	2745	11 577

In Chapter 4, ionization energy was described as the amount of energy required to remove the outermost electron from a mole of neutral atoms. Chemists have also measured and recorded how much energy it takes to remove additional electrons from the ion formed. These other ionization energies, such as the second, third, and fourth ionization energies, are shown for sodium, magnesium, and aluminum in **Table 5-7**.

As you can see, the amount of energy required to remove a $2p$ electron from a sodium ion (the second ionization energy) is almost 10 times greater than that required to remove the sole $3s$ electron. The amount of energy released if Na^{2+} ions formed a crystal lattice with Cl^- ions does not come close to that needed to remove the $2p$ electron. There is simply not enough energy to make $NaCl_2$. Similarly, Mg^{3+} and Al^{4+} require too much energy to occur naturally. *Chemical formulas should always describe compounds that can really exist.*

Chem Fact
Calcium fluoride is one of the compounds that can be added to drinking water to help fight tooth decay.

Ionic formulas show the smallest ratio of ions

In the case of CaF_2, however, there is enough energy to make Ca_2F_4. Why isn't it correct to write Ca_2F_4 as the formula for calcium fluoride? After all, it's just as stable as CaF_2. Ca_2F_4 is a neutral compound, as can be verified by calculating the total charge for the calcium ions ($2 \times 2+ = 4+$) and for the fluoride ions ($4 \times 1- = 4-$). To understand why writing the formula of calcium fluoride as Ca_2F_4 is incorrect, consider the crystal shown in **Figure 5-14**.

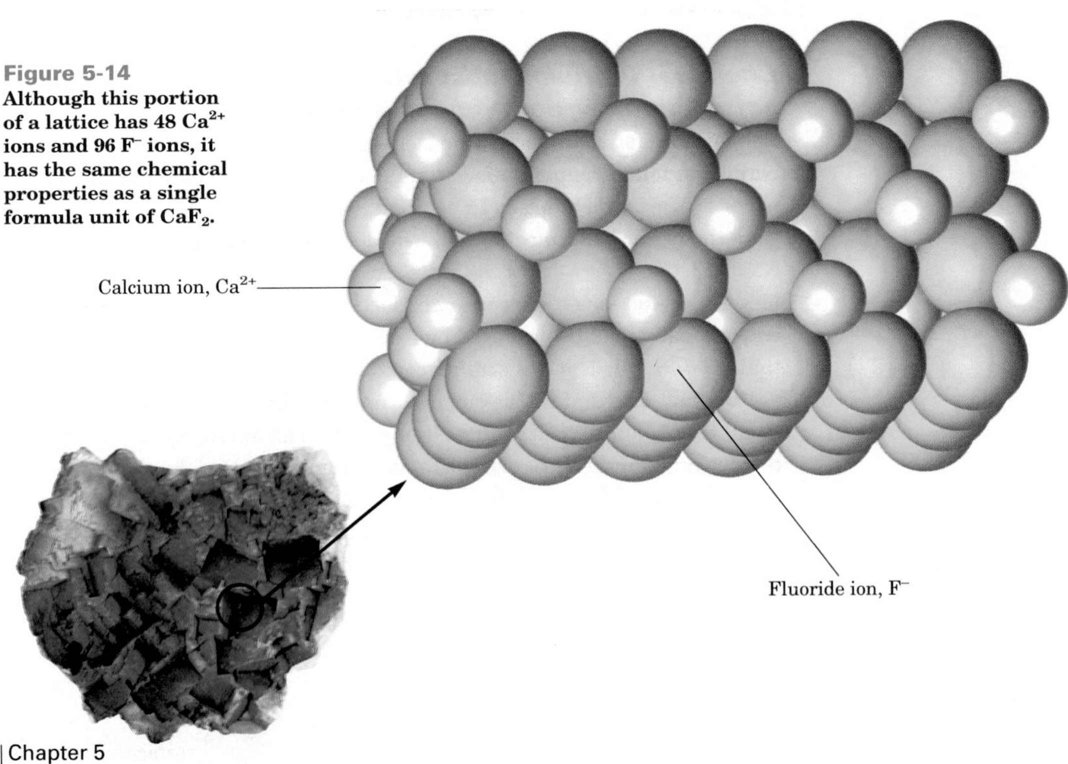

Figure 5-14
Although this portion of a lattice has 48 Ca^{2+} ions and 96 F^- ions, it has the same chemical properties as a single formula unit of CaF_2.

As you may recall, ionic compounds are composed of ions arranged in an orderly, repeating pattern. The model of a calcium fluoride crystal, shown in **Figure 5-14**, could be described as $Ca_{48}F_{96}$ because there are 48 Ca^{2+} ions and 96 F^- ions shown in the model. However, if the crystal were cut into two pieces (each representing $Ca_{24}F_{48}$), each smaller crystal would have the same physical and chemical properties as the larger crystal.

empirical formula
simplest whole-number ratio of atoms that matches the relative ratio found in a chemical compound

Because either of those formulas describes a compound with similar chemical properties (as do others such as $Ca_{12}F_{24}$, Ca_3F_6, and even CaF_2), you should *express all ionic formulas as the lowest whole-number ratio of the atoms that form the compound*, rather than using such large numbers. In this way, everyone will use the same formula for the same compound no matter how much of it is present. This "simplest" formula is also known as the **empirical formula**.

formula unit
simplest collection of atoms from which a compound's formula can be established

In the case of calcium fluoride, the formula must be written as CaF_2 because the lowest whole-number ratio is 1:2. CaF_2 is the simplest formula for calcium fluoride. CaF_2 also represents the **formula unit** for calcium fluoride. One formula unit of sodium chloride is one sodium ion plus one chloride ion (NaCl).

Section Review

22. Many chemical compounds have common names. Give the chemical names for each of the following.
a. lime, CaO **b.** alumina, Al_2O_3 **c.** calcite, $CaCO_3$
d. chalcocite, Cu_2S **e.** magnesia, MgO

23. What are the names of the following monatomic ions?
a. Cr^{3+} **b.** Co^{2+} **c.** H^+ **d.** Pb^{4+}

24. Write formulas for the ionic compounds formed from the following.
a. magnesium and iodine **b.** rubidium and sulfur
c. Fe^{3+} and O^{2-} **d.** Pb^{2+} and O^{2-}

25. Name the following pairs of compounds.
a. $SnCl_2$, $SnCl_4$ **b.** MnO, MnO_2 **c.** FeO, Fe_2O_3

26. Give the chemical names for each of the following ionic compounds.
a. KCl **b.** CaI_2 **c.** Li_2O **d.** Cu_2O

27. Rewrite the following to represent the correct formula unit for each compound.
a. CaBr **b.** Ag_2I_2 **c.** NaO_2 **d.** Al_4Br_{12}

28. Element Z can form ions with either a 1+ or 2+ charge. Element Y can form ions with either a 2– or 3– charge. Write all of the possible formulas for the binary compounds that can be formed from these two elements.

ASSESS

Alternative Assessment

Name the binary compounds listed below. Then identify the cation and anion present and give the total number of ions present in the formula.
AlF_3, K_2O, Mg_3N_2, BaS

Answers to Section Review

22. a. calcium oxide
b. aluminum oxide
c. calcium carbonate
d. copper(I) sulfide
e. magnesium oxide

23. a. chromium(III) ion
b. cobalt(II) ion
c. hydrogen ion
d. lead(IV) ion

24. a. MgI_2
b. Rb_2S
c. Fe_2O_3
d. PbO

25. a. tin(II) chloride and tin(IV) chloride
b. manganese(II) oxide and manganese(IV) oxide
c. iron(II) oxide and iron(III) oxide

26. a. potassium chloride
b. calcium iodide
c. lithium oxide
d. copper(I) oxide

27. a. $CaBr_2$
b. AgI
c. Na_2O
d. $AlBr_3$

28. Z_2Y, Z_3Y, ZY, Z_3Y_2

Section 5-4

FOCUS

Lesson Starter

On the board write the names of the following compounds but not the formulas.

Salts with monatomic anions
calcium carbide, CaC_2
calcium sulfide, CaS
aluminum phosphide, AlP
magnesium nitride, Mg_3N_2

Salts with polyatomic anions
calcium carbonate, $CaCO_3$
calcium sulfate, $CaSO_4$
aluminum phosphate, $AlPO_4$
magnesium nitrate, $MgNO_3$

Have students compare the two lists. Then write the formulas next to each compound name. Compare the formulas. Formulas in both groups consist of cations and anions. The cations are the same in both groups. Therefore, the oxygen atoms and the "original" anion must be acting together as a single unit.

What is a polyatomic ion?

Section Objectives

Distinguish between monatomic and polyatomic ions.

Identify polyatomic ions in formulas and in compound names.

Use the proper rules for naming and writing formulas for ionic compounds containing polyatomic ions.

Polyatomic ions

The student shown in **Figure 5-15** works at a garden center. While stocking the fertilizer display, he notices that fertilizer contains potassium compounds, nitrogen compounds, and phosphorus compounds. From his study of chemistry, he recognizes these to be ionic compounds.

Most of the potassium is in the form of K_2CO_3. The formula for the compound that supplies nitrogen is NH_4NO_3. The phosphorus is contained in $Ca(H_2PO_4)_2$. These ionic compounds have formulas that are much more complicated than the binary compounds you saw in the previous section. However, with a few additional rules, they can be almost as easy to decipher and understand.

Each of these salts contains a special type of ion. Unlike monatomic ions, these ions are not composed of a single element. Rather, they are **polyatomic ions** that are formed from two or more elements. A polyatomic ion consists of two or more atoms that act as a single ion. Because they act as single ions, they are given special names.

polyatomic ion
ion made of two or more atoms bonded together that function as a single ion

Figure 5-15
Fertilizer contains compounds such as K_2CO_3, NH_4NO_3, and $Ca(H_2PO_4)_2$. All of these are made of polyatomic ions.

Visual Strategy

Figure 5-15

It is possible to buy fertilizers to do specific jobs. For example, the first number on the label is the percentage of nitrogen, which is necessary for leaf growth and greening. Lawn fertilizer that is used mainly for greening the grass has a high percentage of nitrogen, such as 30-0-0.

Polyatomic ions can be cations or anions

Compounds made of polyatomic ions also contain cations and anions. The only difference is that a polyatomic ion is made of two or more atoms. To write or name formulas with polyatomic ions, you must still begin by determining which parts of a compound are cations and which are anions.

K_2CO_3 contains potassium, a Group 1 metal that almost always forms cations with a 1+ charge. Since ionic compounds are neutral, the remainder of the formula must balance the two positive charges from the two potassium ions in the K_2 part of the formula. Therefore, CO_3 must be a polyatomic anion with a charge of 2–, as shown in **Figure 5-16**.

Figure 5-16a This potassium carbonate formula unit . . .

$$K_2CO_3 \rightarrow K^+ \; K^+ \qquad K_2CO_3 \rightarrow CO_3^{2-}$$

5-16b . . . is made of two potassium ions, each with a charge of 1+, . . .

5-16c . . . and a single carbonate ion with a charge of 2–.

Compounds with polyatomic ions must be neutral

Fertilizer also contains ammonium nitrate, NH_4NO_3, which has two polyatomic ions—ammonium and nitrate. Look at **Table 5-8** and you will see that the ammonium ion is written as NH_4^+. The charge on the ion is 1+. The nitrate ion has the formula NO_3^-, and the charge on this ion is 1–.

To balance the anion and cation charges in the formula for ammonium nitrate, you will need one of each ion. In this way, the single positive charge of the ammonium cation is balanced by the single negative charge of the nitrate anion, resulting in a neutral compound. Thus, the formula for the compound should be written as NH_4NO_3, indicating that the ions are present in a 1:1 ratio.

For polyatomic ions, it is important to understand that although the charges are written to the right of the formula, the charge applies to the ion as a whole and not to any individual element. For example, the 1– charge on NO_3^- applies to the whole ion, and not just to the oxygen atoms.

Table 5-8
Some Polyatomic Ions

Negative charge	Ion name and formula	
1–	acetate ion, $C_2H_3O_2^-$ bromate ion, BrO_3^- chlorate ion, ClO_3^- chlorite ion, ClO_2^- cyanide ion, CN^- dihydrogen phosphate, $H_2PO_4^-$ hydrogen carbonate ion (bicarbonate ion), HCO_3^-	hydrogen sulfate ion, HSO_4^- hydroxide ion, OH^- hypochlorite, ClO^- nitrate ion, NO_3^- nitrite, NO_2^- perchlorate, ClO_4^- permanganate, MnO_4^-
2–	carbonate ion, CO_3^{2-} chromate ion, CrO_4^{2-} dichromate ion, $Cr_2O_7^{2-}$ hydrogen phosphate, HPO_4^{2-}	oxalate ion, $C_2O_4^{2-}$ peroxide ion, O_2^{2-} sulfate ion, SO_4^{2-} sulfite ion, SO_3^{2-}
3–	arsenate ion, AsO_4^{3-}	phosphate ion, PO_4^{3-}
Positive charge	**Ion name and formula**	
1+	ammonium ion, NH_4^+	
2+	dimercury(I) ion, Hg_2^{2+}	

TEACH

Groupwork Strategy

Polyatomic ions act as a single unit in compound formation

Fill a sheet of paper with transfers of the following patterns.

A

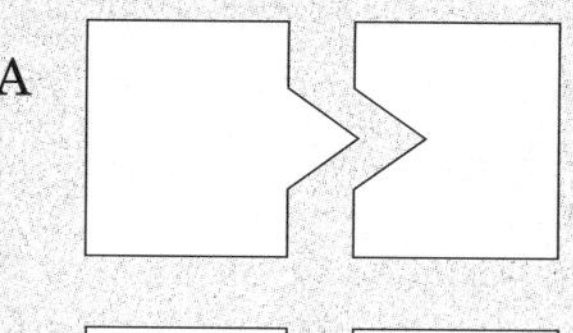

B

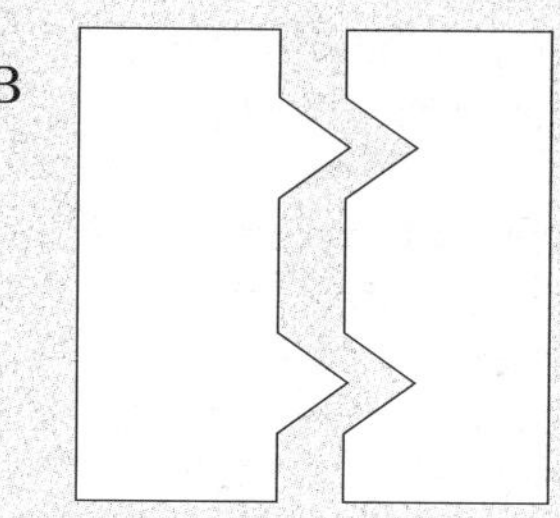

C

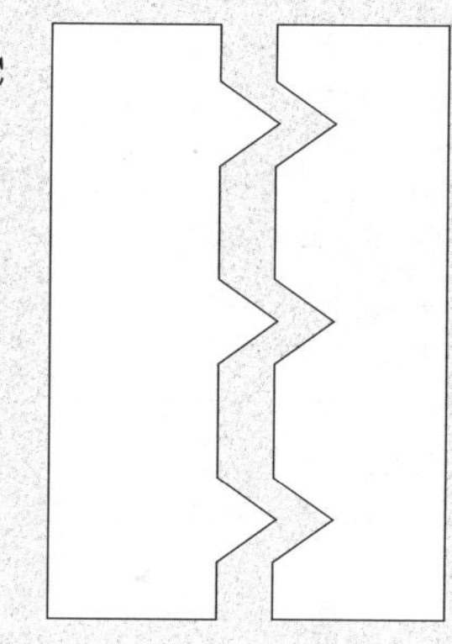

Give each group a copy of the paper. Have students label the blocks with the names or formulas of ions in **Table 5-8**. The blocks with tabs are anions. Each tab is a 1– charge. The blocks with notches are cations. Each notch is a 1+ charge. Use the patterns to form assigned compounds such as $K_2Cr_2O_7$, $AlAsO_4$, $BaSO_4$, $Ca(OH)_2$, HCN, H_3PO_4, and $(NH_4)_2CO_3$.

Teaching Tip

Nomenclature

To illustrate that the order of elements in polyatomic ions is important, pronounce a name backwards: Mary ⟶ Yram (e•ram). Ions, like names, have a conventional way of writing and pronouncing that name.

Teaching Tip

Subscripts and suffixes

Remind students that subscripts on atoms in the formulas for polyatomic ions can not be changed. You may wish to emphasize the meaning of the parentheses by determining the number of atoms in several compounds containing polyatomic ions. Some examples are $Mg(HCO_3)_2$, $Sr(HSO_4)_2$, $(NH_4)_2SO_4$, and $Na_2HAl(PO_4)_2$. (Answers: 11, 13, 15, 14).

Although the *-ate* ending on anions indicates the presence of oxygen in addition to another element, it has little to do with the charge or the formula. You may wish to list several anions that end in *-ate* and point out the dissimilarity between charge, number of oxygen atoms, and the ending.

Do You Know?

Polyatomic ions that end in *-ate* were derived from acids whose names end in *-ic*. Polyatomic ions that end in *-ite* were derived from acids whose names end in *-ous*.

Parentheses group polyatomic ions

The third compound found in this fertilizer seems complicated: $Ca(H_2PO_4)_2$. The number of subscripts may seem intimidating, but it can be deciphered if you remember that polyatomic ions act as a unit. Therefore, *to show more than one polyatomic ion, parentheses are used*. The subscripts attached to the parentheses refer to everything within the parentheses.

This compound contains a calcium ion, Ca^{2+}. From **Table 5-8**, you see that $H_2PO_4^-$ is a dihydrogen phosphate ion with a 1– charge. In the complete formula, the subscript *2* at the end of the parentheses means that each formula unit contains two dihydrogen phosphate ions (everything inside the parentheses), as shown in **Figure 5-17**. In this way, the charges balance to give a neutral compound. Even though these polyatomic ions are large and complex, they are arranged in a crystal lattice just like ions in binary ionic compounds.

Figure 5-17a
This calcium dihydrogen phosphate formula unit . . .

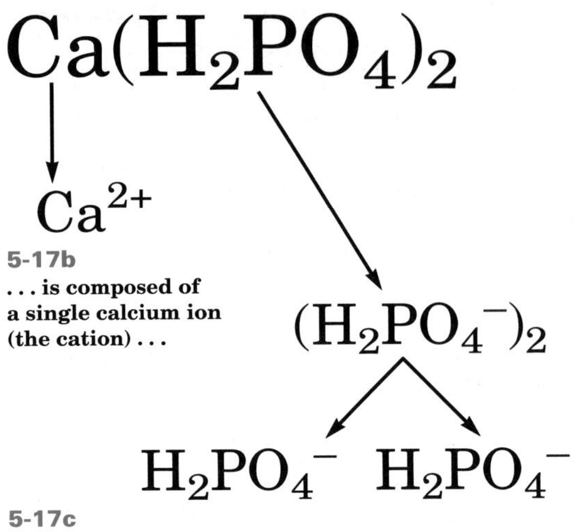

$$Ca(H_2PO_4)_2$$

$$Ca^{2+}$$

5-17b
. . . is composed of a single calcium ion (the cation) . . .

$$(H_2PO_4^-)_2$$

$$H_2PO_4^- \quad H_2PO_4^-$$

5-17c
. . . and two dihydrogen phosphate ions (the anions).

Compounds with polyatomic ions

A set of rules explains the naming of polyatomic ions

Ionic compounds containing polyatomic ions are named in the same manner as binary ionic compounds. The cation name is listed first, followed by the anion name.

Look again at **Table 5-8**. Do you notice something unusual about some of the polyatomic ions listed in this table? In some cases, two different polyatomic ions are formed by the same two elements. For example, chlorine and oxygen atoms can combine to form either ClO_3^- ions, ClO_2^- ions, or ClO^- ions . Sulfur can combine with oxygen to form either SO_3^{2-} ions or SO_4^{2-} ions.

You may have noticed that one of the two atoms in every case was oxygen. In naming these compounds, the ions with larger numbers of oxygen atoms are named with the *-ate* ending, and the ions with smaller numbers of oxygen atoms are named with the *-ite* ending. Notice that the ending does not tell you how many oxygen atoms are present in the polyatomic ion.

The rules for writing formulas for ionic compounds involving polyatomic ions are similar to those for binary compounds. When a polyatomic ion has a subscript, the use of parentheses is necessary to make it clear that the subscript refers to the whole ion, as with $Ca(H_2PO_4)_2$. Parentheses are not needed if only a single polyatomic ion is used in the formula, as with K_2CO_3 and NH_4NO_3. On the other hand the formula for ammonium sulfate, $(NH_4)_2SO_4$, requires the use of parentheses. Note that parentheses can be used with polyatomic cations or anions.

Chem Fact
Calcium sulfate, $CaSO_4$, is an ingredient in plaster and wallboard; calcium sulfite, $CaSO_3$, is used as a preservative in some fruit juices.

Sample Problem 5B

Writing formulas with polyatomic ions

Write the formula for ammonium carbonate.

1 *List*

- **Symbol for ammonium ion:** $NH_4^{\ +}$
- **Symbol for carbonate ion:** $CO_3^{\ 2-}$

2 *Set up and solve*

- **Write the symbols for the ions, side by side, with the cation first.**

$$NH_4^{\ +}\ CO_3^{\ 2-}$$

- **To get a neutral ionic compound, you need two ammonium ions, each with a 1+ charge, for every carbonate ion with a 2– charge.** To show this 2:1 ratio, you would write the formula as follows.

$$(NH_4)_2CO_3$$

Practice 5B

1. Write formulas for the following.
   a. aluminum sulfate b. magnesium hydroxide c. copper(II) acetate
   d. hydrogen peroxide e. iron(III) sulfide f. lead(II) phosphate

Rules for naming and writing ionic formulas

At this point you may feel that naming and writing the formulas for ionic compounds are complicated processes. However, the same basic steps apply, whether the ionic compound is polyatomic or not. By working your way through the steps you will be led in the right direction for naming a particular ionic compound.

How To

Name an ionic compound

Identify the cation and anion, using **Table 5-6**, **Figure 5-12**, and **Table 5-8**.

IF the cation . . .

. . . is a metal that **can have more than one charge,** such as copper, which can form either Cu^+ or Cu^{2+} ions, . . .

THEN determine the cation charge in this formula, using **Figure 5-12** as a guide.

THEN write the **cation name first, using Roman numerals** for the charge,

and **THEN** write the **anion name last**.

$Cu(NO_3)_2$ is copper(II) nitrate.

. . . is either a **polyatomic ion**, such as ammonium, NH_4^+, or a metal that can have **only one charge**, such as sodium, Na^+, . . .

. . . **THEN** write the **cation name first**,

and **THEN** write the **anion name last**.

NH_4Cl is ammonium chloride.
NaOH is sodium hydroxide.

Additional Example for Sample Problem 5B

Write formulas for the following four compounds that contain polyatomic ions: calcium hydroxide, potassium sulfate, copper(I) carbonate, and barium nitrate.

Solution for Additional Example 5B
$Ca(OH)_2$, K_2SO_4, Cu_2CO_3, and $Ba(NO_3)_2$

Answers to Practice 5B

1. **a.** $Al_2(SO_4)_3$
   b. $Mg(OH)_2$
   c. $Cu(C_2H_3O_2)_2$
   d. H_2O_2
   e. Fe_2S_3
   f. $Pb_3(PO_4)_2$

Do You Know?

The most-produced compound in the United States is sulfuric acid, much of which is used to make nitric acid, which is used in turn to make fertilizers.

Visual Strategy

How to Name an Ionic Compound

Guide students through the flowchart using the following examples.
Names: tin(II) oxide, iron(III) chloride, zinc sulfate, strontium hydrogen carbonate, ammonium dichromate, calcium phosphate, and lead(IV) sulfide.
Formulas: SnO, $FeCl_3$, $ZnSO_4$, $Sr(HCO_3)_2$, $(NH_4)_2Cr_2O_7$, $Ca_3(PO_4)_2$, PbS_2.

ASSESS

Alternative Assessment

Prefixes and suffixes tell the reader something about the substance being discussed. Study this list of ions whose names end in *ate*: NO_3^-, SO_4^{2-}, PO_3^{3-}, CrO_4^{2-}, ClO_3^-. Name the ions, and comment on their similarities and differences. Use your comments to name Na_2SiO_3.

Similarly, writing a formula for an ionic compound involves a series of steps that remain the same for all ionic compounds. By working your way through the steps, you will be led through the rules governing the writing of the formula for a particular ionic compound.

How To
Write a formula for an ionic compound

Identify the cation and anion names in the compound name.

Write the symbols for the cation and anion side by side.

IF the cation name . . .

. . . is followed by Roman numerals, as in lead(IV) chloride, **THEN** assign that amount of charge to the cation, such as Pb^{4+}, and determine the anion charge, based on **Table 5-6** or **Table 5-8**.

. . . **is not followed by Roman numerals, THEN** determine the ions' charges, based on **Table 5-6** or **Table 5-8**.

And **THEN** find the **least common multiple** of the ions' charges. For example, in aluminum oxide, the least common multiple of the charges for Al^{3+} and O^{2-} would be 6.

IF polyatomic ions . . .

. . . **are not present**, use **subscripts** to indicate how many of each ion would be necessary to have the amount of charge designated by the least common multiple.

Aluminum oxide is Al_2O_3.
$2 \times 3+ = 6+$
$3 \times 2- = 6-$
Lead(IV) chloride is $PbCl_4$.
$1 \times 4+ = 4+$
$4 \times 1- = 4-$

. . . **are present**, use **subscripts and parentheses** to indicate how many of each ion would be necessary to have the amount of charge designated by the least common multiple.

Iron(III) acetate is $Fe(C_2H_3O_2)_3$.
$1 \times 3+ = 3+$
$3 \times 1- = 3-$
Magnesium hydroxide is $Mg(OH)_2$.
$1 \times 2+ = 2+$
$2 \times 1- = 2-$

Section Review

29. What is the main difference between monatomic and polyatomic ions?

30. Name the following compounds.
a. $NaNO_3$ **b.** $Ca(NO_2)_2$ **c.** $Fe(OH)_3$ **d.** $AlAsO_4$ **e.** $(NH_4)_2SO_4$

31. Write the formulas for the following compounds.
a. iron(II) chlorate **b.** sodium bicarbonate
c. mercury(I) acetate **d.** ammonium hydroxide
e. copper(II) phosphate **f.** aluminum oxalate

Answers to Section Review

29. Monatomic ions are composed of only one type of atom. Polyatomic ions are composed of two or more atoms bonded together and acting as a single unit.

30. **a.** sodium nitrate
b. calcium nitrite
c. iron(III) hydroxide
d. aluminum arsenate
e. ammonium sulfate

31. **a.** $Fe(ClO_3)_2$
b. $NaHCO_3$
c. $Hg_2(C_2H_3O_2)_2$
d. NH_4OH
e. $Cu_3(PO_4)_2$
f. $Al_2(C_2O_4)_3$

Visual Strategy
How to Write a Formula for an Ionic Compound

Guide students through the flowchart using the following examples: calcium sulfate, ammonium carbonate, copper(II) nitrate, strontium chloride, sodium sulfide, calcium nitride, and manganese(IV) oxide.

Answers: $CaSO_4$, $(NH_4)_2CO_3$, $Cu(NO_3)_2$, $SrCl_2$, Na_2S, Ca_3N_2, and MnO_2.

How are salts used and measured?

Section Objectives

Use chemical formulas to determine the molar mass of an ionic compound.

Determine the percentage composition of an element.

Interpret the chemical formula for a hydrated crystal.

Comparing masses

Salts have many uses. You already know that sodium chloride is used in food and is used to melt ice and snow on roads. You may not realize that sodium chloride, like many other salts, is also a valuable starting material for many other useful substances, as shown in **Table 5-9**.

If you are using sodium chloride to make chlorine gas, you need to know how many sodium and chloride ions you have because the number of chloride ions will determine how much chlorine gas can be made. As you have read, salts are made of ions arranged in a crystal lattice, a three-dimensional pattern. In the case of NaCl, each Na^+ ion is surrounded by six Cl^- ions, and each Cl^- ion is surrounded by six Na^+ ions. But if you have a 55 g sample of sodium chloride, how can you count the number of NaCl units you have if you can't tell which Na^+ ion goes with which Cl^- ion?

Although chemists rarely deal with individual NaCl pairs, calculations are treated as if they did because the 1:1 ratio of Na^+ ions to Cl^- ions will be true for all samples of NaCl. Similarly, because the ratio of ions in a formula unit for any ionic compound will be the same for all samples of that compound, most calculations with salts and other ionic compounds are done as if the compound were actually a pile of individual formula units, rather than an interconnected crystal lattice.

Table 5-9
Some Uses of NaCl in Chemical Synthesis

Reactants	Products	Uses
NaCl, H_2O	NaOH	chemical manufacturing, paper production, water treatment, petroleum refining, soap manufacturing
	Cl_2 gas	chemical manufacturing, water treatment, paper bleach, bleach manufacturing
NaCl, $CaCO_3$	Na_2CO_3	glass manufacturing, soap, detergent, water softener, paper production, photography
	$CaCl_2$	drying agent, glue manufacturing, ice and snow melter, concrete
NaCl, H_2SO_4	Na_2SO_4	chemical reagent, dyeing and printing
	HCl gas	chemical reagent, steel manufacturing, metal cleaners, chemical manufacturing, food processing, oil recovery

Section 5-5

FOCUS

Lesson Starter

Begin the lesson with the following demonstration.

Demonstration 4
Qualitative analysis of fertilizer
Approximate time: 25 min

1. **N from NH_4^+ ion:** Add about 2.0 g of fertilizer to an evaporating dish. Warm gently on a hot plate. Then add about 15.0 mL of 1 M NaOH solution. Place a moist piece of red litmus paper on the underside of a watch glass and cover the evaporating dish. If the litmus paper turns blue, NH_3 is present. If the paper does not change color, heat gently. False positives will be produced if the NaOH spatters onto the litmus paper.
2. **K^+ ion:** Dissolve about 0.5 g of fertilizer in 10–15 mL of 1 M HCl. Test this solution in a flame. View the flame test through purple cobalt glass. If K^+ ions are present, they will be seen as a short-lived purple flame.

SAFETY
Wear safety goggles and a lab apron. Dilute spills with water, if necessary. Then, while wearing gloves, soak up the mess with cloth or paper toweling. Rinse the toweling, neutralize the rinsings with 1 M HCl or NaOH, and pour down the drain. Do not use baking soda on acid spills.

DISPOSAL
Step 1: Warm gently in the hood to drive off residual NH_3. Neutralize with 1 M HCl and pour down the drain. Step 2: Neutralize with 1 M NaOH and pour down the drain.

TEACH

Teaching Tip

Familiar salts and their uses

Write the words *Salt* and *Use* side by side as column heads on the chalkboard or overhead. Begin the list with zinc oxide, ZnO, an astringent, and Epsom salts, $MgSO_4{\cdot}12H_2O$, a water softener often used to soak tired feet. Ask students to complete the list.

Do You Know?

Barium nitrate is used in pyrotechnics (fireworks) to produce green light.

Teaching Tip

Compound composition and molar mass

It is a good idea to practice counting atoms in a molecule before attempting to determine molar mass. Write the following formulas on the board, and ask students how many of each atom are present: $CaSO_3$, $K_2Cr_2O_7$, $Al(NO_3)_3$, $Al_2(SO_3)_3$, and $CuSO_4{\cdot}5H_2O$.

Additional Example for Sample Problem 5C

Calculate the molar mass for the following compounds.

a. $Al_2(SO_4)_3$
b. $Ca(OH)_2$
c. Fe_2O_3
d. $CuSO_4{\cdot}5H_2O$

Solutions are at the bottom of the page.

Answers to Practice 5C

1. **a.** 122.55 g/mol
   b. 234.06 g/mol
   c. 132.17 g/mol
   d. $NaHCO_3$; 84.01 g/mol
   e. $K_2Cr_2O_7$; 294.20 g/mol
   f. $Mg(ClO_4)_2$; 223.21 g/mol

*Full solutions for items **a–f** are on page 187A.*

A pure substance has a specific molar mass

molar mass *sum of the molar masses of all atoms represented by 1 mol of formula units*

A formula such as $CaCl_2$ can represent either one Ca^{2+} ion and two Cl^- ions or a mole of Ca^{2+} ions and two moles of Cl^- ions. The **molar mass,** expressed in grams, of a compound can be calculated from the molar masses of the elements. To determine the molar mass of an ionic compound, you must add the individual molar masses of all the atoms shown in the formula. Although ions may have slightly different masses than their corresponding atoms, such differences are small enough that they can be disregarded. Molar masses of compounds with polyatomic ions are calculated the same way.

Sample Problem 5C Calculating molar mass

The subscript 2 means that in every mole of compound there are 2 moles of NO_3^- ions, each containing 1 mole of N and 3 moles of O atoms.

What is the molar mass of barium nitrate, $Ba(NO_3)_2$?

1 List

- **Interpret the formula.**

Every mole of the compound contains 1 mol of Ba, 2 mol of N, and 6 mol of O.

- **Check the periodic table for the molar mass of each element.**

Ba molar mass: 137.33 g/mol
N molar mass: 14.01 g/mol
O molar mass: 16.00 g/mol

2 Set up

- $\text{1 mol Ba} \times \frac{\text{137.33 g Ba}}{\text{1 mol Ba}} = \text{mass of 1 mol Ba}$

- $\text{2 mol N} \times \frac{\text{14.01 g N}}{\text{1 mol N}} = \text{mass of 2 mol N}$

- $\text{6 mol O} \times \frac{\text{16.00 g O}}{\text{1 mol O}} = \text{mass of 6 mol O}$

3 Estimate and calculate

- **Estimate by adding the individual molar masses rounded off to the nearest whole number.**

137:	137
+ (2 × 14):	+28
+ (6 × 16):	+96
	261

- **Add the actual masses calculated using the setups from Step 2.**

Mass of 1 mol Ba	=	137.33 g
Mass of 2 mol N	=	28.02 g
Mass of 6 mol O	=	96.00 g
Molar mass of $Ba(NO_3)_2$	=	261.35 g/mol

Practice 5C

1. **Calculate the molar mass of the following compounds.**
   a. $KClO_3$ **b.** $Ca(H_2PO_4)_2$ **c.** $(NH_4)_2SO_4$
   d. sodium hydrogen carbonate **e.** potassium dichromate
   f. magnesium perchlorate

Solutions for Additional Example 5C

a. 342.17 g/mol
b. 74.10 g/mol
c. 159.70 g/mol
d. 249.72 g/mol

*Full solutions for items **a–d** are found on page 187A.*

Describing mass relationships within a compound

If you were in the business of making chlorine, you would want to know exactly what mass of chlorine you could make from a given mass of sodium chloride. If you needed a lot of salt to produce a very small amount of chlorine, your business wouldn't be very profitable. One way to determine this would be to look at the *percentage composition* of NaCl.

To determine percentage composition, remember that the molar mass is 100% of the mass represented by the formula. The percentage of each element in the compound is simply the molar mass of the element divided by the molar mass of the whole compound and multiplied by 100.

$$\frac{\textbf{mass of element in compound}}{\textbf{total mass of compound}} \times 100 = \textbf{percentage by mass of element}$$

For example, sodium chloride is 39.34% sodium and 60.66% chlorine.

Sample Problem 5D
Determining percentage composition from a formula

Copper(I) sulfide is found in nature as the mineral chalcocite, a copper ore. What is the percent composition of pure chalcocite?

1 *List*

- **Compound name:** copper(I) sulfide
- **From the periodic table, you can determine that the formula is Cu_2S.**

Cu molar mass: 63.55 g/mol

S molar mass: 32.07 g/mol

2 *Set up*

- **Find the molar mass for Cu_2S.**

Mass of 2 mol Cu:	127.10 g
Mass of 1 mol S:	+ 32.07 g
Molar mass of Cu_2S:	159.17 g/mol

- **Use the molar masses of Cu and S to set up calculations of percentage composition for each element in the compound.**

$$\text{percentage Cu} = \frac{\text{mass of 2 mol Cu}}{\text{molar mass of } Cu_2S} \times 100 = \frac{127.10 \text{ g Cu}}{159.17 \text{ g } Cu_2S} \times 100$$

$$\text{percentage S} = \frac{\text{mass of 1 mol S}}{\text{molar mass of } Cu_2S} \times 100 = \frac{32.07 \text{ g S}}{159.17 \text{ g } Cu_2S} \times 100$$

3 *Calculate and verify*

- **Calculate the percentage composition for each element by using the setups from Step 2 and rounding to the correct number of significant figures.**

79.8517308% = 79.85% Cu

20.14826915% = 20.15% S

- **To check your work, be sure the sum of the percentages is very close to 100%.**

Practice 5D

1. **Chalcopyrite has the formula $CuFeS_2$. What is the percentage composition of this compound?**
2. **Will you get more copper from the same mass of pure chalcopyrite or pure chalcocite? Explain your answer.**

Chalcopyrite

Teaching Tip

Percentage composition

Before introducing percentage composition, have the class solve the following experience-based problems. What is the percentage of girls in the class? What is the percentage of boys wearing glasses in the class? What is the percentage of athletic shoes to total shoes?

A nonchemical example, such as the following, may help students visualize percentage composition involving several elements. Question: A man buys 200 kg of fruit for the zoo in the following amounts: 29% bananas, 35% oranges, 19% grapes, 15% watermelons, and 2% kiwi fruit. How many kilograms of each did he buy? (58 kg, 70 kg, 38 kg, 30 kg, and 4 kg, respectively)

Do You Know?

Vast deposits of chalcocite are found in Montana, Arizona, Utah, Nevada, Alaska, Chile, Mexico, and Europe.

Additional Examples for Sample Problem 5D

a. Calculate the percentage composition by mass for the following: $Mg(OH)_2$, NH_4NO_3, and $Mg(ClO_3)_2$.

b. If an ore is 18% bauxite (Al_2O_3), how many kilograms of aluminum will be recovered per metric ton (1000 kg) of ore?

Solutions are at the bottom of the page.

Answers to Practice 5D

1. 34.6% Cu, 30.4% Fe, 34.9% S
2. pure chalcocite because the percentage of copper is greater

Solutions for Additional Example 5D

a. 41.68% Mg, 54.86% O, 3.46% H; 35.00% N, 5.05% H, 59.96% O; 12.71% Mg, 37.08% Cl, 50.20% O

b. 18% of 1000 kg = 180 kg. Bauxite is 52.92% Al; 52.92% of 180 kg is 95.26 kg of aluminum.

*Full solutions for item **a** is found on page 187A.*

Demonstration 5

Water and mass are lost when a hydrate is heated

Approximate time: 20 min

1. Place a Pyrex test tube in a small beaker.
2. Add about 3 g of $Na_2CO_3{\bullet}10H_2O$ to the test tube.
3. Set the beaker and test tube on a balance and record the total mass on the board.
4. Remove the Pyrex test tube from the beaker, and attach it to a ring stand with a buret clamp. It should be mounted horizontally with the mouth pointed slightly downward.
5. Heat strongly for 3 min. Note the water vapor coming out of the test tube. Make sure most of the moisture is gone from the mouth of the tube by heating gently, if necessary.
6. Cool for a few minutes.
7. Remeasure the mass of the beaker. Record this new mass and compare it to the mass before heating.

SAFETY
Wear safety goggles and a lab apron.

DISPOSAL
Dissolve in water, and pour down the drain.

Answers to Practice 5E

1. Mn_2O_3
2. $Cd(OH)_2$

*Full solutions for items **1** and **2** are found on page 187A.*

Formulas can be determined from composition data

When a new chemical substance is discovered, its discoverers often do not know its formula or its composition. Analytical chemists use chemical reactions to break down the compound into its elements and analyze the mass of each element present. From this information they can calculate the chemical formula of the compound. As with most other chemistry calculations, the key is to convert the percentages by mass into amounts by *moles*. Then, compare the mole amounts to find the simplest whole-number ratio. The best way to do this is to divide each amount in moles by the smallest of the mole amounts. This will give a coefficient of one for the atoms present in the smallest amount. After this step, additional multiplication may be necessary to achieve whole numbers of atoms. These numbers will be the coefficients in the formula.

Sample Problem 5E
Determining formula from composition data

As part of a science fair project, Antonio is analyzing the contents of fresh alkaline batteries. He has determined that one ingredient is a black powdery compound, of 63% manganese and 37% oxygen. What is the compound's formula?

1 List

- **percentage Mn:** 63%
- **percentage O:** 37%

2 Set up

- **Find the moles of manganese and oxygen present.**

In percentage problems, it is often convenient to assume you are working with a 100. g sample of the substance.

$$63 \text{ g Mn} \times \frac{1 \text{ mol Mn}}{54.94 \text{ g Mn}} = 1.146705497 \text{ mol Mn} = 1.1 \text{ mol Mn}$$

$$37 \text{ g O} \times \frac{1 \text{ mol O}}{16.00 \text{ g O}} = 2.3125 \text{ mol O} = 2.3 \text{ mol O}$$

3 Calculate

- **To determine the simplest ratio of moles in the compound, select the smallest number of moles, and divide other numbers of moles by it.**

$$\frac{1.1 \text{ mol Mn}}{1.1} = 1.0 \text{ mol Mn}$$

$$\frac{2.3 \text{ mol Mn}}{1.1} = 2.1 \text{ mol O}$$

1.0 mol Mn: 2.1 mol O is nearly the same as 1 mol Mn: 2 mol O.

- **Write the formula using the smallest whole number ratio of oxygen to manganese.**

$$MnO_2$$

Practice 5E

1. **While analyzing a dead alkaline battery, Antonio finds a compound of 70% manganese and 30% oxygen. What is its formula?**
2. **Find the formula for an ingredient of rechargeable batteries that has the following percentage composition: 21.9% O, 1.4% H, and 76.7% Cd.**

Percentage composition tells how much water is in a hydrate

Some salts have the ability to bind water molecules within their lattice structure. These compounds are known as hydrated crystals. The anhydrous salts are used as drying agents because they can absorb so much water as they form hydrated crystals. Others change color when hydrated and can serve as moisture indicators. As shown in **Figure 5-18**, the desiccant packed with a camera lens or most electronic equipment contains silica gel, a hydrate of SiO_2. In fact, many pure anhydrous salts will absorb moisture from the air to form hydrates.

Figure 5-18
When you unpack a new camera lens, CD player, or television, the box usually contains one or more packets of desiccant. The packet absorbs moisture in the air so that the equipment will not be damaged.

The copper sulfate shown in **Figure 5-19** is an example of a salt that can form a hydrate. In copper sulfate pentahydrate, for every formula unit of copper sulfate, five molecules of water are trapped, as seen in the formula $CuSO_4{\cdot}5H_2O$. Notice that when writing the formula for a hydrate, a number is placed in front of the formula for water. This number indicates how many units of water are present for every formula unit of the crystal.

Another hydrate is sodium carbonate decahydrate, $Na_2CO_3{\cdot}10H_2O$. In problems with hydrates, it is important to remember that 25.0 g of Na_2CO_3 contains more sodium carbonate formula units than 25.0 g of $Na_2CO_3{\cdot}10H_2O$. In the anhydrous salt, all 25.0 g are Na_2CO_3, but in the hydrate, the $\cdot 10H_2O$ in the formula tells you that water makes up some of the 25.0 g, with the remainder being Na_2CO_3.

Figure 5-19a
Anhydrous copper sulfate, $CuSO_4$, is a white powder, . . .

5-19b
. . .but when water is added, a blue hydrate, $CuSO_4{\cdot}5H_2O$, is formed.

Figure 5-19c
Anhydrous cobalt chloride has a lavender or blue color, . . .

5-19d
. . . but when water is added, the red or pink hydrate, $CoCl_2{\cdot}6H_2O$, is formed.

Groupwork Strategy

Growing alum crystals

To a 600 mL beaker, add 475 mL of water. Stir in 114 g (4 oz) of alum. On a hot plate, heat gently while stirring. Add more alum until no more dissolves (about 86 g). Cool the solution. With pot holder, pour about 5 mL of the solution into a watch glass. Add about 10 g of alum to the solution remaining in the beaker. Cover the beaker with cheesecloth. Set it in a cool, dry place, and let it stand. Allow the liquid in the watch glass to evaporate, and select the largest crystal from those that have begun to grow. Tie a string around the crystal, and suspend it from a piece of cardboard so that it is in the main stock solution. The alum crystals should begin to grow within 24 h and may continue to grow for a few weeks. View some of the more sturdy crystals under a microscope. The formula for household alum is $NaAl(SO_4)_2{\cdot}12H_2O$.

SAFETY

Participating students should wear safety goggles and a lab apron. Wear gloves when crystals are handled.

DISPOSAL

Save the crystals for display at the school, or put them into the trash. Solutions may be poured down the drain. If students wish to take home ONE crystal (and *only one* crystal), it should be put in a sealed container labeled to identify the contents, and with the admonition "Do not open, return to (teacher's name) at (school name) before (date one month or less from current date)."

Do You Know?

Household alum is a hydrated double salt with the chemical formula $NaAl(SO_4)_2{\cdot}12H_2O$. In the early 1920s, alum was added to single-acting baking powder to make a double-acting baking powder. The addition of alum lowered the manufacturing cost and increased the the rising action of the powder, which made the product attractive to both consumers and manufacturers.

Additional Examples for Sample Problem 5F

a. What percentage of $CuSO_4{\cdot}5H_2O$ is H_2O?

b. $MnCl_2$ is what percentage of $MnCl_2{\cdot}2H_2O$?

c. How much water could 100 g of anhydrous $MnCl_2$ absorb if the hydrated form is $MnCl_2{\cdot}2H_2O$?

Solutions are at the bottom of the page.

Answers to Practice 5F

1. $\frac{86.3 \text{ g } Na_2CO_3}{37.03\%} = 233 \text{ g hydrate}$

2. $\frac{33.5 \text{ g } H_2O}{18.02 \text{ g/mol}} \times \frac{1 \text{ mol } CaCl_2}{6 \text{ mol } H_2} \times \frac{110.97 \text{ g } CaCl_2}{1 \text{ mol}} = 34.4 \text{ g } CaCl_2$

ASSESS

Alternative Assessment

When anhydrous $CaCl_2$ is fully hydrated, it is removed from the desiccator, heated in an oven, cooled, and returned to the desiccator for reuse.

a. What is the purpose for heating in the oven?

b. If the molar mass of the hydrate is 147.02 g, what is the formula for the hydrate?

c. What percentage of the hydrate is water?

Answers to Section Review

32. a. 158.18 g; $Ca(C_2H_3O_2)_2$
b. 357.46 g; $Fe_3(PO_4)_2$
c. 277.33 g
d. 105.98 g

33. a. 40.50% Zn, 19.86% S, 39.64% O
b. 27.93% Fe, 24.06% S, 48.01% O
c. 25.45% Cu, 12.84% S, 57.66% O, 4.04% H

34. anhydrous $CaCl_2$ because water is a greater percentage of this crystal's mass

*Full solution to item **33a** is found on page 187A.*

Sample Problem 5F
Percentage composition of hydrates

What percentage of hydrated sodium carbonate, $Na_2CO_3{\cdot}10H_2O$, is Na_2CO_3?

1 *List*

- **Determine the molar mass of Na_2CO_3.**

$$2 \text{ mol Na} \times \frac{22.99 \text{ g Na}}{1 \text{ mol Na}} = 45.98 \text{ g}$$

$$1 \text{ mol C} \times \frac{12.01 \text{ g C}}{1 \text{ mol C}} = 12.01 \text{ g}$$

$$3 \text{ mol O} \times \frac{16.00 \text{ g O}}{1 \text{ mol O}} = 48.00 \text{ g}$$

Molar mass of Na_2CO_3 = 105.99 g/mol

- **Determine the mass of 10 mol H_2O.**

$$20 \text{ mol H} \times \frac{1.01 \text{ g H}}{1 \text{ mol H}} = 20.2 \text{ g}$$

$$10 \text{ mol O} \times \frac{16.00 \text{ g O}}{1 \text{ mol O}} = 160.0 \text{ g}$$

Mass of 10 mol H_2O = 180.2 g

- **Determine the molar mass of $Na_2CO_3{\cdot}10H_2O$.**

105.99 g for 1 mol Na_2CO_3
\+ 180.2 g for 10 mol H_2O
286.2 g/mol for $Na_2CO_3{\cdot}10H_2O$

2 *Set up*

- **Set up the percentage.**

$$\text{percentage } Na_2CO_3 = \frac{\text{molar mass of } Na_2CO_3}{\text{molar mass of } Na_2CO_3{\cdot}10H_2O} \times 100$$

3 *Estimate and calculate*

- **Calculate the answer.**

$$\text{percentage } Na_2CO_3 = \frac{105.99}{286.2} \times 100 = 37.03354298\% = 37.03\% \ Na_2CO_3$$

Practice 5F

1. **How many grams of hydrated sodium carbonate would be needed to supply the same number of formula units of Na_2CO_3 as 86.3 g of anhydrous sodium carbonate? (Hint: use the percentage calculated in Step 3 above.)**

2. **Anhydrous $CaCl_2$ is used as a drying agent because it forms hydrates such as $CaCl_2{\cdot}6H_2O$. How many grams of $CaCl_2$ would it take to absorb 33.5 g of water if all of it were converted to $CaCl_2{\cdot}6H_2O$?**

Section Review

32. Find the molar mass of each of the following.
a. calcium acetate **b.** iron(II) phosphate
c. $Al(ClO_3)_3$ **d.** Na_2CO_3

33. Find the percentage composition of each of the following.
a. zinc sulfate **b.** $Fe_2(SO_4)_3$ **c.** $CuSO_4{\cdot}5H_2O$

34. Gram for gram, which will absorb more water: anhydrous $CaCl_2$ forming $CaCl_2{\cdot}6H_2O$ or anhydrous $Co_3(PO_4)_2$ forming $Co_3(PO_4)_2{\cdot}8H_2O$? Explain your answer.

Solutions to Additional Examples 5F

a. $\frac{5 \times (18.02 \text{ g/mol } H_2O)}{249.72 \text{ g/mol hydrate}} \times 100 = 36.08\%$

b. $\frac{125.84 \text{ g/mol } MnCl_2}{161.88 \text{ g/mol hydrate}} \times 100 = 77.73\ \%$

c. $100 \text{ g } MnCl_2 \times \frac{36.04 \text{ g } H_2O}{125.84 \text{ g } MnCl_2} = 28.64 \text{ g}$

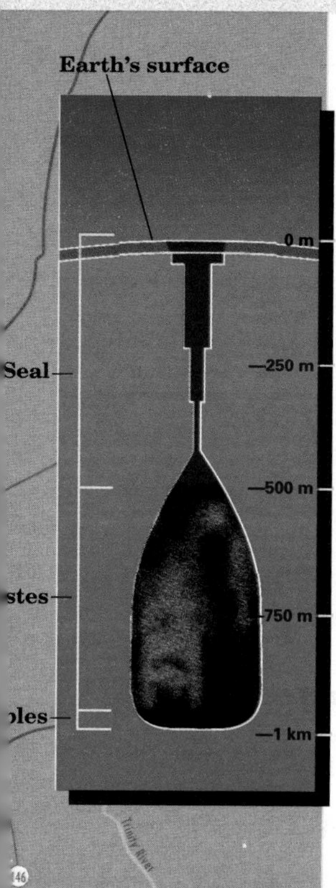

Conclusion: Salt Domes and Toxic Wastes

Remember the properties of salts as you examine the arguments for and against burying toxic wastes in salt domes.

Would the salt react and combine dangerously with the hazardous wastes and leak into the areas surrounding the domes? Those proposing the plan said there would be no dangerous reaction because the wastes would be sealed in airless caverns. To be sure that no air would leak in, nitrogen gas was to have been pumped in to check for leaks. Without oxygen, the wastes should be less reactive. The atoms in a salt have already achieved stable octets and should be nonreactive.

In addition, those in favor of the plan said that the crystal lattice would form an impenetrable barrier that would prevent the toxic wastes from escaping.

However, opponents of the project pointed out that even stable, nonreactive atoms or compounds can combine under certain circumstances. The salt in some domes may have impurities that are difficult to detect and that may weaken the crystal lattice. Also, wastes may react differently from the pure compounds stored in other salt domes.

Could outside influences cause a break in the salt cavern walls? The many salt domes currently in use have not leaked yet. Although salts are brittle under normal conditions, salt domes are under such high pressure that the salt can "flow" when it deforms, instead of cracking.

Storage caverns are usually built away from the edges and top of a salt dome, so most of the salt would have to dissolve in water before a leak would occur. Right now, this dissolving is an extremely slow geologic process because there is water only near the Earth's surface.

However, in 1993 the Texas Natural Resource Conservation Commission ruled against granting the permit. Hunter appealed the decision to the Texas Attorney General but lost the appeal as well.

Applying Concepts

1. A typical salt dome is about 620 m in diameter and 2.0×10^3 m deep. If the density of the salt, NaCl, is 2.2 g/cm^3, how many grams of NaCl are there? How many moles of the salt is this?
2. If 1.0 L of water is needed to dissolve 350 g of NaCl, how many liters of water would it take to dissolve the entire dome?

Research and Writing

Use the library to find out the following.

1. What other ideas are being suggested for the storage of hazardous wastes? Who should pay for taking care of the wastes?
2. Research the number and average size of salt domes in the United States and the amount of toxic wastes produced annually. If the domes were used to store wastes, how long would it take to fill them?

Chemistry in Your Community

Every household gets rid of waste products daily as it discards containers and materials from the products used. Assign groups of students to investigate each of the following areas of waste disposal and hold a panel discussion at a local club meeting, PTA meeting, or teachers' in-service meeting.

1. What is hazardous waste, and how does it differ from nonhazardous waste?
2. What type of precautions are taken in your community to ensure safe disposal of hazardous wastes?
3. What action is being taken to lessen the amount of material needing disposal?
4. How do local gas stations and oil-change shops dispose of their used products?
5. Contact a local hospital to determine if any special precautions are needed and used to dispose of hospital waste.

Answers to Applying Concepts

1. Assume the dome is cylindrical. Answers may vary because of rounding.

$$V_{cyl} = \pi r^2 h$$

$$= (3.14) \times \left(\frac{620 \text{ m}}{2}\right)^2 \times (2.0 \times 10^3 \text{ m}) = 6.0 \times 10^8 \text{ m}^3$$

$$\text{g of NaCl} = V_{cyl} \times D$$

$$= 6.0 \times 10^8 \text{ m}^3 \times \frac{2.2 \times 10^6 \text{ g}}{1 \text{ m}^3} = 1.3 \times 10^{15} \text{ g}$$

$$\text{mol of NaCl} = \text{g/molar mass} = \frac{1.3 \times 10^{15} \text{ g}}{58.44 \text{ g/mol}} = 2.2 \times 10^{13} \text{ mol}$$

2. From item 1, there are 1.3×10^{15} g of NaCl in this dome.

$$V_{water} = (1.3 \times 10^{15} \text{ g NaCl}) \times \frac{1.0 \text{ L } H_2O}{350 \text{ g NaCl}} = 3.7 \times 10^{12} \text{ L}$$

Closure Strategy

Magnesium oxide and sodium fluoride have the same crystal structure as NaCl. Show that the cations in these two compounds are isoelectronic, as are the anions. Sketch the unit cell for each compound, and explain the observation that MgO is nearly twice as hard as NaF and has a boiling point nearly three times as large as NaF.

CHAPTER 5

Highlights

Key terms

anion	ionic compounds
binary compounds	lattice energy
bond energy	molar mass
cation	monatomic ion
crystal lattice	octet rule
empirical formula	polyatomic ion
formula unit	salt
ionic bond	unit cell

Key ideas

Why do atoms form bonds?

- Atoms without filled outermost energy levels will form chemical bonds to achieve a stable octet.
- An input of energy is needed to break chemical bonds between atoms. Energy is released when bonds form between atoms.

What holds a salt together?

- The opposite charges of anions and cations attract, forming a tightly packed substance of bonded ions called a crystal lattice, which determines the properties of ionic compounds.
- A salt will form if the energy released to create anions and form the crystal lattice is greater than the energy absorbed to create cations.

How do you name salts?

- Ionic compounds are named by joining the cation and anion names.
- The subscripts in the formula for an ionic compound indicate the lowest electrically neutral whole-number ratio of cations to anions.

What is a polyatomic ion?

- Polyatomic ions are two or more atoms bonded together and functioning as a single unit.
- Parentheses are used to group polyatomic ions in a chemical formula with a subscript.

How are salts used and measured?

- Molar mass can be calculated from the chemical formula of a compound.
- The percentage composition gives the mass percentages of the elements within a compound.

Key problem-solving approach: *Molar mass and percent composition*

To determine molar mass or percent composition, start with the formula and follow the appropriate calculation pathway.

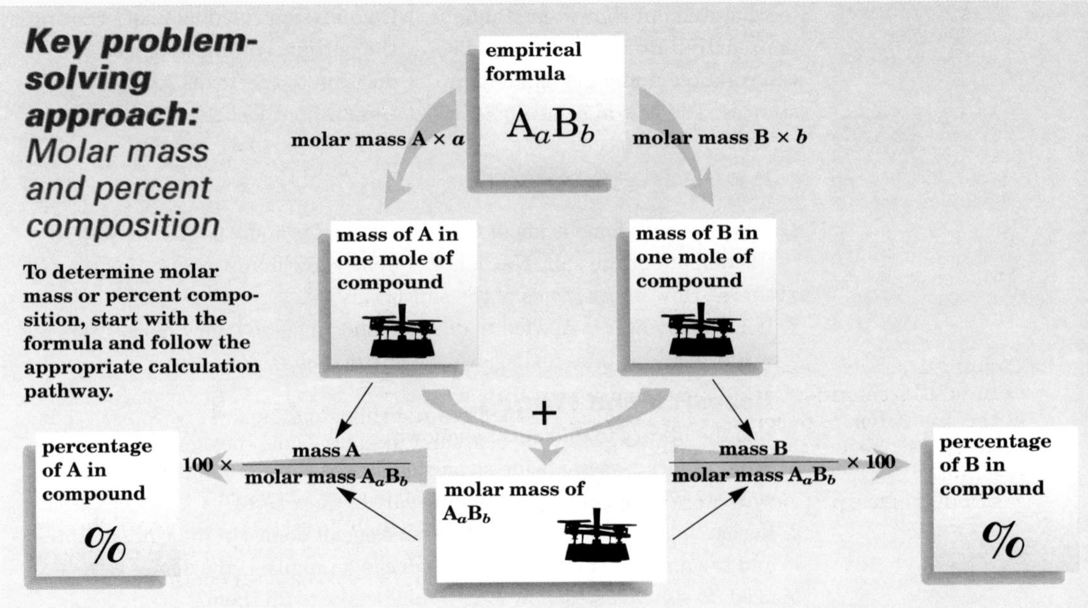

CHAPTER 5

Review and Assess

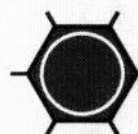

Bond formation

REVIEW

1. Predict the reactivity of atoms of the following elements. Explain your predictions. (Hint: write the electron configuration for each atom.)
   a. lithium, Li **b.** helium, He **c.** nitrogen, N
2. When a certain substance is dropped into water, no reaction is apparent. How would you describe the stability and energy of the atoms making up this substance?
3. How might atoms of the following elements achieve stability?
   a. Magnesium, Mg (atomic number 12)
   b. Bromine, Br (atomic number 35)
   c. Oxygen, O (atomic number 8)
   d. Iodine, I (atomic number 53)
   e. Aluminum, Al (atomic number 13)
   f. Nitrogen, N (atomic number 7)
4. What happens to the energy level and stability of two bonded atoms when they are separated and become individual atoms?
5. Propose a reason why magnesium forms Mg^{2+} ions and not Mg^{6-} ions.
6. Complete the table below.

Atom	Ion	Noble gas configuration
S		
Be		
I		
Rb		
O		
Sr		
F		

7. Cadmium and chlorine can combine to form cadmium(II) chloride, $CdCl_2$, which is used in the production of television picture tubes. Identify the effect of the following on this reaction.
   a. stability of the products compared to the reactants
   b. energy

APPLY

8. Which has greater potential energy, a noble gas or a metal? Explain your answer.
9. With the exception of technetium, all elements with an atomic number less than 93 are naturally occurring. Do you expect to be able to find all of these elements as pure substances? Explain.
10. Your lab partner believes that all ions are stable. How would you explain what is wrong with his belief?
11. Why are most metals found in nature as ores that need refining and not as pure metals?
12. A classmate insists that sodium gains a positive charge when it becomes an ion because it gains a proton. Explain this student's error.
13. Which diagram below illustrates the electron shell diagram for a potassium ion found in the nerve cells of your body? (Hint: potassium's atomic number is 19.)

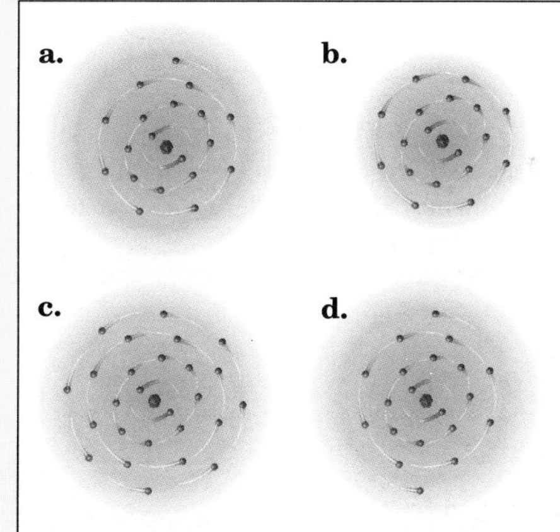

14. Free radicals are atoms with electrons that are not in pairs; thus, they do not have an octet. Some people believe that these free radicals are responsible for cancer, the effects of aging, and the depletion of the ozone layer. What would cause these free radicals to be so environmentally dangerous?

1. **a.** Li is highly reactive due to its single isolated electron in the outermost energy level.
   b. He is nonreactive due to its stable noble gas configuration.
   c. Although not as reactive as Li, N is reactive; N needs to gain 3 electrons to achieve a stable octet.

2. Because there was no reaction, the atoms apparently have relatively high stability. Thus, they must be at a somewhat low energy level.

3. **a.** Mg could lose two electrons to form a stable octet.
   b. Br could gain one electron to form a stable octet.
   c. O could gain two electrons to form a stable octet.
   d. Iodine could gain one electron to form a stable octet.
   e. Aluminum could lose three electrons to form a stable octet.
   f. Nitrogen could gain three electrons to form a stable octet.

4. The energy level is increased, and stability is lowered.

5. Losing two electrons takes less energy than gaining six electrons.

6. S, S^{2-}, Ar; Be, Be^{2+}, He; I, I^-, Xe; Rb, Rb^+, Kr; O, O^{2-}, Ne; Sr, Sr^{2+}, Kr; F, F^-, Ne.

7. **a.** Stability increases.
   b. Energy decreases.

8. A metal contains greater potential energy because metals can release energy as they react. Noble gases do not react and therefore do not release energy.

9. No. Many elements are highly reactive and form compounds readily. Therefore, the compounds rather than the elements occur naturally.

10. Ions that do not produce a noble gas electron configuration are not stable.

11. Most metals are reactive, so they occur naturally only as compounds within ores.

12. Ions are produced by gaining and losing electrons. The gain or loss of a proton would change the element.

13. b

14. To achieve stability, these free radicals will react with other chemicals to complete an octet.

15. a. $SrCl_2$ has a 1:2 ratio because two Cl^- ions are needed to balance the one Sr^{2+} ion.
b. RbCl has a 1:1 ratio because one Cl^- ion is needed to balance the one Rb^+ ion.
c. $AlCl_3$ has a 1:3 ratio.
d. BaO has a 1:1 ratio.
e. Al_2O_3 has a 2:3 ratio.
f. AlN has a 1:1 ratio.

16. Because each ion is strongly attracted to several surrounding ions of opposite charge, it takes a tremendous amount of energy to break all of the attractions.

17. To conduct electricity, ionic substances must be in solution or in the liquid state. It would be hard to contain the solution in a circuit board, and the temperature would have to be very high to maintain a liquid state.

18. No. The internal structure of salt will always be the same.

19. Li^+ ion: $1s^2$
I^- ion: $1s^2\ 2s^2\ 2p^6\ 3s^2 3p^6\ 4s^2\ 3d^{10}\ 4p^6\ 5s^2\ 4d^{10}\ 5p^6$

20. Ionic: KI, $AlPO_4$, NaBr
Non-ionic: Fe, CH_4, $C_6H_{12}O_6$

21. a. Liquid, will conduct electricity.
b. Solid, will not conduct electricity.
c. Solution, will conduct electricity.

22. No. Ionic substances will always form consistent unit cells.

Structure and properties of ionic compounds

REVIEW

15. Determine the ratio of cations to anions for the following compounds.
    a. strontium chloride, an ingredient in fireworks
    b. rubidium chloride, an ingredient in gasoline
    c. aluminum chloride, an ingredient in wood preservatives
    d. barium oxide, a substance used in making detergents for lubricating oils
    e. aluminum oxide, an ingredient in some dental cements
    f. aluminum nitride, a substance used in making semiconductors
16. Why do ionic compounds ordinarily have such high melting and boiling points?
17. Under the right conditions, ionic substances can conduct electricity very well. Describe these conditions, and explain why an ionic substance would be a poor choice as a conductor for a computer circuit board.
18. Does a cube of salt have a different unit cell than powdered salt? Explain your answer.

APPLY

19. The electron configurations for a lithium atom is $1s^22s^1$. The configuration for an iodine atom is $1s^22s^22p^63s^23p^64s^23d^{10}4p^65s^24d^{10}5p^5$. Write the electron configurations for the ions that form lithium iodide, a substance used in photography.
20. Use the table below to identify the chemicals as ionic or non-ionic.
21. Label the drawings below as representing solid, liquid, or solution. State whether the substance pictured will or will not conduct electricity.

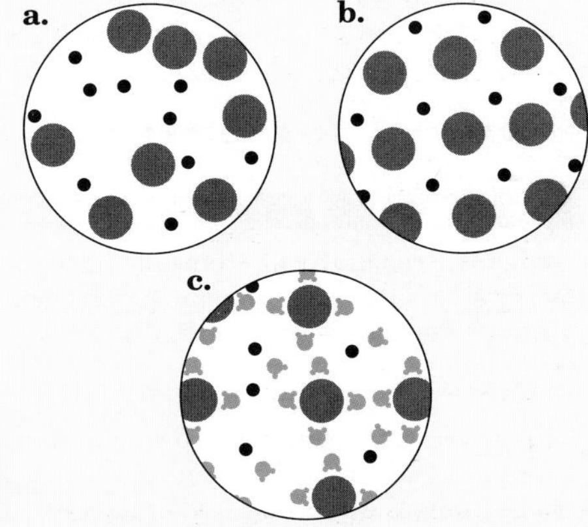

22. Sodium chloride can be prepared by reacting sodium metal with chlorine gas. Another way to prepare it is to combine sodium hydroxide and hydrochloric acid and allow any water present to evaporate. Will the method of preparation affect the unit cell and crystalline structure of NaCl?
23. Although metals and salts have similar lattice structures, metals make good materials for electrical wiring. Why aren't salts used instead?
24. The equations below show the energy changes that are necessary for calcium bromide, $CaBr_2$, to be formed from the elements calcium and bromine. Is formation of $CaBr_2$ an exothermic or endothermic process? Explain.

$$Ca(s) + 8.5\text{ kJ} \longrightarrow Ca(g)$$
$$Ca(g) + 1735\text{ kJ} \longrightarrow Ca^{2+}(g) + 2e^-$$
$$Br_2(g) + 193\text{ kJ} \longrightarrow 2Br(g)$$
$$2Br(g) + 2e^- \longrightarrow 2Br^-(g) + 650\text{ kJ}$$
$$Ca^{2+}(g) + 2Br^-(g) \longrightarrow CaBr_2(s) + 2176\text{ kJ}$$

Substance	State at room temperature	Conducts electricity at room temperature	Melting point (°C)	Conducts electricity as a liquid
KI	solid	no	680	yes
Fe	solid	yes	1535	yes
$AlPO_4$	solid	no	1460	yes
CH_4	gas	no	−182.6	no
NaBr	solid	no	755	yes
$C_6H_{12}O_6$	solid	no	83	no

Formulas and nomenclature

REVIEW

25. Complete the table below, and then use it to answer the following questions.

Element	Ion	Name of ion
barium	Ba^{2+}	
chlorine		chloride ion
chromium	Cr^{3+}	
fluorine	F^-	
manganese		manganese(II) ion
oxygen		oxide ion

a. Give the formula for an ingredient of electrical batteries containing manganese and chlorine.
b. Give the formula for a compound containing chromium and fluorine used to treat silk.
c. Give the formula for a compound containing barium and oxygen used to manufacture lubricating oil detergents.

26. Why are there no rules for naming Group 18 ions?

27. Refer to **Table 5-7**. Explain the large break that appears between different ionization energies for different elements.

28. Why is the strontium nitrate found in roadside emergency flares represented as $Sr(NO_3)_2$ rather than SrN_2O_6?

29. What is the difference between the chlorite ion and the chlorate ion? (Hint: refer to **Table 5-8**)

30. Give the name, formula, and charge for the ions found in the following compounds.
a. $NaClO$ **b.** $(NH_4)_2CO_3$
c. K_2HPO_4 **d.** $CuCN$
e. FeC_2O_4 **f.** $MnC_2H_3O_2$
g. Hg_2SO_4

PRACTICE

Answers to items in a black square begin on 841

31 Give formulas for the following compounds.
a. aluminum fluoride, used in ceramics
b. magnesium oxide, an antacid
c. calcium sulfide, used in luminous paints
d. strontium bromide, an anticonvulsant
(Hint: see Sample Problem 5A.)

32. Give formulas for the following compounds.
a. cadmium(II) bromide, used in process engraving
b. palladium(II) chloride, used in some photographic toning solutions
c. vanadium(V) oxide, an ingredient in yellow glass
d. cobalt(II) sulfide, used as a catalyst
(Hint: see Sample Problem 5A.)

33 Give formulas for the following compounds.
a. potassium hydrogen phosphate, an ingredient in nondairy creamers
b. strontium nitrate, an ingredient in red safety flares
c. lithium sulfate, an antidepressant
d. magnesium dihydrogen phosphate, used to make wood fireproof.
(Hint: see Sample Problem 5B.)

34. Give formulas for the following compounds.
a. ammonium acetate, a meat preservative
b. mercury(I) nitrate, used to blacken brass
c. titanium(III) sulfate, used as a stain remover
d. chromium(III) phosphate, a green pigment
(Hint: see Sample Problem 5B.)

APPLY

35. Determine the subscripts that are most likely in the formulas for ionic substances of the following elements.
a. an alkali metal and a halogen
b. an alkaline-earth metal and a halogen
c. an alkali metal and a member of Group 16
d. an alkaline-earth metal and a member of Group 16

36. Use **Figure 5-12** to provide all possible formulas for the ionic compounds listed.
a. iron chloride **b.** copper oxide
c. tin fluoride

37. Explain what is wrong with each of the following chemical formulas.
a. $RbCl_2$ **b.** $Ge_{12}S_{24}$ **c.** $NaCs$ **d.** $NaNe$

38. Explain the error contained in each of these descriptions of barium sulfide.
a. Ba_2S **b.** BaS_2 **c.** Ba_2S_2

39. How many atoms of each element are contained in a single formula unit of iron(III) formate, $Fe(CHO_2)_3{\cdot}H_2O$, a compound used as a preservative in fodder?

23. Salts are too brittle and do not conduct electricity unless melted or dissolved.

24. Exothermic; overall the exothermic steps release more energy than is absorbed by the endothermic steps.

25. barium ion, Cl^-, chromium(III) ion, fluoride ion, Mn^{2+}, O^{2-}
a. $MnCl_2$
b. CrF_3
c. BaO

26. Group 18 contains the noble gases, which ordinarily do not form ions.

27. The large break occurs after enough electrons have been lost so that the atom has an octet. Removing another electron would destroy the stable octet.

28. The atoms in the nitrate ion act as a unit and are, therefore, grouped as a unit.

29. The chlorate ion contains one more oxygen atom than the chlorite ion.

30. **a.** Na^+, sodium ion
ClO^-, hypochlorite ion
b. NH_4^+, ammonium ion
CO_3^{2-}, carbonate ion
c. K^+, potassium ion
HPO_4^{2-}, hydrogen phosphate ion
d. Cu^{2+}, copper(II) ion
CN^-, cyanide ion
e. Fe^{2+}, iron(II) ion
$C_2O_4^{2-}$oxalate ion
f. Mn^+, manganese(I) ion
$C_2H_3O_2^-$, acetate ion
g. Hg_2^{2+}, mercury(I) ion
SO_4^{2-}, sulfate ion

31. **a.** aluminum ion, Al^{3+}; fluoride ion, F^-
The least common multiple is 3.
Al^{3+}: $3+ \times 1 = 3+$;
F^-: $1- \times 3 = 3-$
AlF_3
b. MgO
c. CaS
d. $SrBr_2$

32. a. cadmium, Cd^{2+}; bromide, Br^-
least common multiple is 2
Cd^{2+}: $2+ \times 1 = 2+$;
Br^- : $1- \times 2 = 2-$
$CdBr_2$
b. $PdCl_2$
c. V_2O_5
d. CoS

33. a. potassium: K^+; hydrogen phosphate: HPO_4^{2-}
least common multiple is 2
K^+: $1+ \times 2 = 2+$;
HPO_4^{2-}: $2- \times 1 = 2-$
K_2HPO_4
b. $Sr(NO_3)_2$
c. Li_2SO_4
d. $Mg(H_2PO_4)_2$

34. a. ammonium, NH_4^+; acetate, $C_2H_3O_2^-$
least common multiple is 1
NH_4^+: $1+ \times 1 = 1+$;
$C_2H_3O_2^-$: $1- \times 1 = 1-$
$NH_4C_2H_3O_2$
b. $Hg_2(NO_3)_2$
c. $Ti_2(SO_4)_3$
d. $CrPO_4$

35. a. alkali metal, 1; halogen, 1
b. alkaline-earth metal, 1; halogen, 2
c. alkali metal, 2; Group 16, 1
d. alkaline-earth metal, 1; Group 16, 1

36. a. $FeCl_2$ and $FeCl_3$
b. CuO and Cu_2O
c. $SnBr_2$ and $SnBr_4$

37. a. The ratio of rubidium to chloride ions should be 1:1, as in RbCl.
b. The subscripts should be in the smallest whole-number ratio, as in GeS_2.
c. Two positive ions alone cannot form an ionic substance.
d. The noble gas neon does not form ions or compounds.

38. a. Sulfur forms an S^{2-} ion not an S^{4-} ion. Thus, barium sulfide has only one barium ion and one sulfide ion in the formula unit. (Alternate argument: Barium forms a Ba^{2+} ion not a Ba^{1+} ion.)

Molar mass and percentage composition

REVIEW

40. When purifying metals from mined ores, why is it important to know the percentage composition of the ore?
41. Potassium chlorate, $KClO_3$, decomposes to give oxygen, O_2, and potassium chloride, KCl. How many grams of oxygen can be produced by the decomposition of 150. g of $KClO_3$? The percentage composition for $KClO_3$ is 28.93% Cl, 31.91% K, and 39.17% O.
42. What is the percentage composition of ammonium nitrate, NH_4NO_3, a common fertilizer?
43. How would you go about determining the formula for a compound containing only barium, sulfur, and oxygen?
44. Which of the following molar masses should be used in problems with hydrated copper sulfate, $CuSO_4$•$5H_2O$?
a. 159.62 g/mol b. 177.64 g/mol
c. 249.72 g/mol d. 339.82 g/mol

PRACTICE

45 What is the molar mass of barium carbonate? (Hint: see Sample Problem 5C.)
46. While working at a toothpaste factory, you are instructed to add 200. mol of SnF_2, an ingredient to prevent tooth decay, to a batch of product. If you know the molar mass of SnF_2, you can determine the number of grams needed. What is the molar mass of SnF_2? (Hint: see Sample Problem 5C.)
47 Iron pyrite, FeS_2, is a a shiny golden crystal often called fool's gold. What is the molar mass of iron pyrite? (Hint: see Sample Problem 5C.)
48. Tin(IV) oxide, SnO_2, is the ingredient that gives fingernail polish its characteristic luster. What is the percentage composition of SnO_2? (Hint: see Sample Problem 5D.)
49 Some antacids use compounds of calcium, a mineral that is often lacking in the diet. What is the percentage composition of calcium carbonate, a common antacid ingredient? (Hint: see Sample Problem 5D.)
50. A superconductor has the formula $YBa_2Cu_3O_7$. What is the percentage composition of the substance? (Hint: see Sample Problem 5D.)
51 During winter vacation, you work at a ski resort covering icy sidewalks with a substance containing 26.2% N, 7.5% H, and 66.3% Cl. What is the formula for this compound? (Hint: see Sample Problem 5E.)
52. Magnetite is an iron ore with natural magnetic properties. It contains 72.4% Fe and 27.6% O. What is the formula for magnetite? (Hint: see Sample Problem 5E.)
53 Phosphorus forms two oxides. One has 56.34% P and 43.66% O. The other has 43.64% P and 56.36% O. What are the empirical formulas for these compounds? (Hint: see Sample Problem 5E.)
54. What percentage of hydrated cobalt(II) chloride, $CoCl_2$•$6H_2O$, which is used as a humidity indicator, is $CoCl_2$? (Hint: see Sample Problem 5F.)
55 What percentage of blue-violet hydrated chromium(III) sulfate, $Cr_2(SO_4)_3$•$18H_2O$, a paint pigment, is $Cr_2(SO_4)_3$? (Hint: see Sample Problem 5F.)
56. Soil in several regions has a reddish tinge due to the presence of iron in the form of limonite, Fe_2O_3•$3/2H_2O$. What percentage of limonite is water? (Hint: see Sample Problem 5F.)

APPLY

57. Which iron ore has more pure iron per kilogram of ore, Fe_2O_3 or Fe_3O_4?
58. Which substance, anhydrous cobalt chloride, $CoCl_2$, or anhydrous magnesium chloride, $MgCl_2$, will absorb the most water per gram if both form a hexahydrate with six moles of water per mole of salt?
59. What percentage of ammonium carbonate, $(NH_4)_2CO_3$, an ingredient in smelling salts, is the ammonium ion, NH_4?
60. Which should yield a higher percentage of pure aluminum per gram, aluminum phosphate or aluminum chloride?
61. Aluminum chlorohydrate is an ingredient used in some antiperspirants. In its anhydrous form, it has 30.93% Al, 45.86% O, 20.32% Cl, and 2.89% H.
a. What is the empirical formula for aluminum chlorohydrate ?
b. Aluminum chlorohydrate forms a hydrated compound with 2.00 mol H_2O / 1.00 mol compound. What is the percentage composition of the hydrate?

Linking chapters

1. ***Electron configurations***
Scientists have been able to create fluoride compounds with the noble gases krypton and xenon. However, fluorine will not form compounds with helium or neon.
   a. Use the different electron configurations of the noble gases to explain why some can form compounds.
   b. Could fluorine form a compound with radon? Explain.
2. ***Periodic trends***
Use **Table 5-7** to predict trends of first, second, third, and fourth ionization energies for the following series of elements.
   a. K, Ca, Ga
   b. Li, Na, K
   c. Cs, Ba, Tl

USING TECHNOLOGY

1. ***Graphics calculator***
Ionic substances have high melting points. The periodic trends of melting points can be used to predict the properties of different substances. Press WINDOW and set Xmin = 0, Xmax = 10, Xscl = 1, Ymin = 0, Ymax = 1000, and Yscl = 100. Press STAT 4 on your calculator and clear lists L_1 and L_2. Then press STAT 1 and enter the values 1, 2, 3, and 4 in L_1. Enter the following melting points in L_2: 712°C for $MgCl_2$, 772°C for $CaCl_2$, 868°C for $SrCl_2$, and 963°C for $BaCl_2$. Press STAT ▶ 3 for two-variable statistics. Set Xlist = L_1, Ylist = L_2 and Freq = 1. Press STAT ▶ 5 ENTER Y= VARS 5 ▶ ▶ 7 to calculate a line for the data. Then press ZOOM 9 to display the graph. Press TRACE and use the arrow keys to answer the following questions.
   a. What trend is displayed?
   b. $BeCl_2$ has a melting point of 405°C. Is this likely to be an ionic substance like the others? Explain.
   c. Predict the melting point of the ionic substance $RaCl_2$.

Alternative assessment

Performance assessment

1. Your instructor will give you a notecard with one of the following formulas on it: $NaC_2H_3O_2•3H_2O$, $MgCl_2•6H_2O$, $LiC_2H_3O_2•2H_2O$, and $MgSO_4•7H_2O$. Design an experiment to determine the percentage of water by mass in the hydrated salt described by the formula. Be sure to explain what steps you will take to ensure that the salt is completely dry. If your teacher approves your design, obtain the salt. What percentage of water does it contain?
2. Devise a set of criteria that will allow you to classify the following substances as ionic or non-ionic: $CaCO_3$, Cu, H_2O, NaBr, and C (graphite). Show your criteria to your instructor. If it is approved, obtain some of the listed substances. Are these substances ionic or non-ionic?

Portfolio projects

1. ***Research and communication***
Create your own model of an ionic bond in the form of an analogy, picture, mechanical model, or computer program. Present your model to the class, and explain the ways in which your model does or does not match experimental observations.
2. ***Chemistry and you***
Ions play an important physiological role in your body. Select one such ion, and write a report detailing its function. Be sure to include recent medical information.
3. ***Cooperative activity***
Keep the ingredients labels from all of the food products you eat in one day. Make a list of all of the salts contained in each one. Compile a master list for the whole class, identifying which salts were eaten by the most people. Research the properties and uses of the salts that were most frequently eaten, and as a class, create an information poster describing the functions of these compounds.
4. ***Research and communication***
Many people follow low-sodium diets. However, they still desire a flavor enhancer like common table salt. Research the different types of salt substitutes and the physiological effects of each. Determine which is the safest salt substitute, and organize your information into a report.

b. Barium forms a Ba^{2+} ion not a Ba^{4+} ion. Thus, barium sulfide does not have two sulfide ions in the formula unit. (Alternative argument: Sulfur forms an S^{2-} ion not an S^- ion.)
c. Ba_2S_2 is not the empirical formula because the subscripts are not in the smallest whole-number ratio.

39. 1 Fe atom, 3 C atoms, 7 H atoms, 8 O atoms.

40. The percentage composition can be used to predict how much metal will be produced from the ore.

41. $150.\ g\ KClO_3 \times 0.3917 = 58.8\ g\ O_2$

42. percentage N:

$$\frac{28.02\ g\ N}{80.06\ g\ NH_4NO_3} \times 100 = 35.01\%$$

percentage H:

$$\frac{4.04\ g\ H}{80.06\ g\ NH_4NO_3} \times 100 = 5.05\%$$

percentage O:

$$\frac{48.00\ g\ O}{80.06\ g\ NH_4NO_3} \times 100 = 59.96\%$$

43. First, determine the percentage of each element contained in a sample. Then, calculate the number of moles of each. The numbers from the smallest whole-number ratio of amounts can be used as subscripts in the formula.

44. c

45. $1\ mol\ Ba \times \frac{137.33\ g\ Ba}{1\ mol\ Ba} = 137.33\ g\ Ba$

$$1\ mol\ C \times \frac{12.01\ g\ C}{1\ mol\ C} = 12.01\ g\ C$$

$$3\ mol\ O \times \frac{16.00\ g\ O}{1\ mol\ O} = 48.00\ g\ O$$

molar mass of $BaCO_3$:
137.33 g + 12.01 g + 48.00 g
= 197.34 g/mol

46. $1\ mol\ Sn \times \frac{118.71\ g\ Sn}{1\ mol\ Sn} = 118.71$

$$2\ mol\ F \times \frac{19.00\ g\ F}{1\ mol\ F} = 38.00\ g\ F$$

molar mass of SnF_2: 118.71 g + 38.00 g

= 156.71 g/mol

Answers from page 176

Answers for Additional Examples 5C

a. molar mass of $Al_2(SO_4)_3 = 2\left(\frac{26.98\text{ g Al}}{1\text{ mol Al}}\right) + 3\left(\frac{32.07\text{ g S}}{1\text{ mol S}}\right) + 12\left(\frac{16.00\text{ g O}}{1\text{ mol O}}\right)$
$= \frac{342.17\text{g}}{1\text{ mol}}$

b. molar mass of $Ca(OH)_2 = 1\left(\frac{40.08\text{ g Ca}}{1\text{ mol Ca}}\right) + 2\left(\frac{16.00\text{ g O}}{1\text{ mol O}}\right) + 2\left(\frac{1.01\text{ g H}}{1\text{ mol H}}\right)$
$= \frac{74.10\text{g}}{1\text{ mol}}$

c. molar mass of $Fe_2O_3 = 2\left(\frac{55.85\text{ g Fe}}{1\text{ mol Fe}}\right) + 3\left(\frac{16.00\text{ g O}}{1\text{ mol}}\right) = \frac{159.70\text{g}}{1\text{ mol}}$

d. molar mass of $CuSO_4{\cdot}5H_2O =$

$1\left(\frac{63.55\text{ g Cu}}{1\text{ mol Cu}}\right) + 1\left(\frac{32.07\text{ g S}}{1\text{ mol S}}\right) + 9\left(\frac{16.00\text{ g O}}{1\text{ mol O}}\right) + 10\left(\frac{1.01\text{ g H}}{1\text{ mol H}}\right) = \frac{249.72\text{g}}{1\text{ mol}}$

Answers to Practice 5C

a. $1\left(\frac{39.10\text{ g K}}{1\text{ mol K}}\right) + 1\left(\frac{35.45\text{ g Cl}}{1\text{ mol Cl}}\right) + 3\left(\frac{16.00\text{ g O}}{1\text{ mol O}}\right) = \frac{122.55\text{ g}}{1\text{ mol KClO}_3}$

b. $1\left(\frac{40.08\text{ g Ca}}{1\text{ mol Ca}}\right) + 4\left(\frac{1.01\text{ g H}}{1\text{ mol H}}\right) + 2\left(\frac{30.97\text{ g P}}{1\text{ mol P}}\right) + 8\left(\frac{16.00\text{ g O}}{1\text{ mol O}}\right)$
$= \frac{234.06\text{ g}}{1\text{ mol Ca(H}_2\text{PO}_4)_2}$

c. $2\left(\frac{14.01\text{ g Na}}{1\text{ mol Na}}\right) + 8\left(\frac{1.01\text{ g H}}{1\text{ mol H}}\right) + 1\left(\frac{32.07\text{ g S}}{1\text{ mol S}}\right) + 4\left(\frac{16.00\text{ g O}}{1\text{ mol O}}\right)$
$= \frac{132.17\text{ g}}{1\text{ mol (NH}_4)_2\text{SO}_4}$

d. $1\left(\frac{22.99\text{ g Na}}{1\text{ mol Na}}\right) + 1\left(\frac{1.01\text{ g H}}{1\text{ mol H}}\right) + 1\left(\frac{12.01\text{ g C}}{1\text{ mol C}}\right) + 3\left(\frac{16.00\text{ g O}}{1\text{ mol O}}\right)$
$= \frac{84.01\text{ g}}{1\text{ mol NaHCO}_3}$

e. $2\left(\frac{39.10\text{ g K}}{1\text{ mol K}}\right) + 2\left(\frac{52.00\text{ g Cr}}{1\text{ mol Cr}}\right) + 7\left(\frac{16.00\text{ g O}}{1\text{ mol O}}\right) = \frac{294.20\text{ g}}{1\text{ mol K}_2\text{Cr}_2\text{O}_7}$

f. $1\left(\frac{24.31\text{ g Mg}}{1\text{ mol Mg}}\right) + 2\left(\frac{35.45\text{ g Cl}}{1\text{ mol Cl}}\right) + 8\left(\frac{16.00\text{ g O}}{1\text{ mol O}}\right) = \frac{223.21\text{ g}}{1\text{ mol Mg(ClO}_4)_2}$

Answers from page 177

Answers for Additional Examples 5D

a. molar mass of $Mg(OH)_2$ = 58.33 g

$\frac{24.31\text{ g Mg}}{58.33\text{ g Mg(OH)}_2} \times 100 = 41.68\%$

$\frac{32.00\text{ g O}}{58.33\text{ g Mg(OH)}_2} \times 100 = 54.86\%$

$\frac{2.02\text{ g H}}{58.33\text{ g Mg(OH)}_2} \times 100 = 3.46\%$

molar mass of NH_4NO_3 = 80.06 g

$\frac{28.02\text{ g N}}{80.06\text{ g NH}_4\text{NO}_3} \times 100 = 35.00\%$

$\frac{4.04\text{ g H}}{80.06\text{ g NH}_4\text{NO}_3} \times 100 = 5.05\%$

$\frac{48.00\text{ g O}}{80.06\text{ g NH}_4\text{NO}_3} \times 100 = 59.96\%$

molar mass of $Mg(ClO_3)_2$ = 191.21 g

$\frac{24.31\text{ g Mg}}{191.21\text{ g Mg(ClO}_3)_2} \times 100 = 12.71\%$

$\frac{70.90\text{ g Cl}}{191.21\text{ g Mg(ClO}_3)_2} \times 100 = 37.08\%$

$\frac{96.00\text{ g O}}{191.21\text{ g Mg(ClO}_3)_2} \times 100 = 50.21\%$

Answers from page 178

1. $\frac{70\text{ g Mn}}{54.94\text{ g/mol}} = 1.27\text{ mol};\ \frac{1.27\text{ mol}}{1.27\text{ mol}} = 1$

$\frac{30\text{ g O}}{16.00\text{ g/mol}} = 1.88\text{ mol};\ \frac{1.88\text{ mol}}{1.27\text{ mol}} = 1.5$

whole-number ratio 2:3

Mn_2O_3

2. $\frac{76.7\text{ g Cd}}{112.41\text{ g/mol}} = 0.682\text{ mol};\ \frac{0.682\text{ mol}}{0.682\text{ mol}} = 1$

$\frac{21.9\text{ g O}}{16.00\text{ g/mol}} = 1.37\text{ mol};\ \frac{1.37\text{ mol}}{0.682\text{ mol}} = 2$

$\frac{1.4\text{ g H}}{1.01\text{ g/mol}} = 1.39\text{ mol};\ \frac{1.39\text{ mol}}{0.682\text{ mol}} = 2$

whole-number ratio 1:2:2

$Cd(OH)_2$

Answers from page 180

33. a. $\frac{65.39\text{ g/mol Zn}}{161.46\text{ g/mol ZnSO}_4} \times 100 = 40.50\%\text{ Zn}$

$\frac{32.07\text{ g/mol S}}{161.46\text{ g/mol ZnSO}_4} \times 100 = 19.86\%\text{ S}$

$\frac{4(16.00\text{ g/mol O})}{161.46\text{ g/mol ZnSO}_4} \times 100 = 39.64\%\text{ O}$

Continued from page 187

47. mass of Fe: 1 mol $\times \frac{55.85\text{ g Fe}}{1\text{ mol}} = 55.85\text{ g}$

mass of S: 2 mol $\times \frac{32.07\text{ g S}}{1\text{ mol}} = 64.14\text{ g}$

molar mass FeS_2: 55.85 g + 64.14 g = 119.99 g

48. molar mass SnO_2: 118.71 g + 32.00 g = 150.71 g/mol

percentage Sn: $\frac{118.71\text{ g Sn}}{150.71\text{ g SnO}_2} \times 100 = 78.77\%\text{ Sn}$

percentage O: $\frac{32.00\text{ g O}}{150.71\text{ g SnO}_2} \times 100 = 21.23\%\text{ O}$

49. molar mass $CaCO_3$: 40.08 g + 12.01 g + 48.00 g = 100.09 g/mol

percentage Ca: $\frac{40.08\text{ g Ca}}{100.09\text{ g CaCO}_3} \times 100 = 40.04\%\text{ Ca}$

percentage C: $\frac{12.01\text{ g C}}{100.09\text{ g CaCO}_3} \times 100 = 12.00\%\text{ C}$

percentage O: $\frac{48.00\text{ g O}}{100.09\text{ g CaCO}_3} \times 100 = 47.96\%\text{ O}$

50. molar mass $YBa_2Cu_3O_7$: 88.91 g + 274.66 g + 190.65 g + 112.00 g = 666.22 g

percentage Y: $\frac{88.91\text{ g Y}}{666.22\text{ g YBa}_2\text{Cu}_3\text{O}_7} \times 100 = 13.35\%\text{ Y}$

percentage Ba: $\frac{274.66 \text{ g Ba}}{666.22 \text{ g } YBa_2Cu_3O_7} \times 100 = 41.23\%$ Ba

percentage Cu: $\frac{190.65 \text{ g Cu}}{666.22 \text{ g } YBa_2Cu_3O_7} \times 100 = 28.62\%$ Cu

percentage O: $\frac{112.00 \text{ g O}}{666.22 \text{ g } YBa_2Cu_3O_7} \times 100 = 16.81\%$ O

51. $26.2 \text{ g N} \times \frac{1 \text{ mol N}}{14.01 \text{ g N}} = 1.87 \text{ mol N}$

$7.50 \text{ g H} \times \frac{1 \text{ mol H}}{1.01 \text{ g H}} = 7.43 \text{ mol H}$

$66.3 \text{ g Cl} \times \frac{1 \text{ mol Cl}}{35.45 \text{ g Cl}} = 1.87 \text{ mol Cl}$

$\frac{7.43}{1.87} = 3.97$

$\frac{1.87}{1.87} = 1.00$

1.00 mol N: 3.97 mol H: 1.00 mol Cl

NH_4Cl

52. $72.4 \text{ g Fe} \times \frac{1 \text{ mol Fe}}{55.85 \text{ g Fe}} = 1.30 \text{ mol Fe}$

$27.6 \text{ g O} \times \frac{1 \text{ mol O}}{16.00 \text{ g O}} = 1.73 \text{ mol O}$

$\frac{1.73}{1.30} = 1.33$

$\frac{1.30}{1.30} = 1.00$

1.00 mol Fe: 1.33 mol O

3.00 mol Fe: 3.99 mol O

Fe_3O_4

53. first compound

mol P: $56.33 \text{ g P} \times \frac{1 \text{ mol P}}{30.97 \text{ g P}} = 1.819 \text{ mol P}$

mol O: $43.66 \text{ g O} \times \frac{1 \text{ mol O}}{16.00 \text{ g O}} = 2.729 \text{ mol O}$

$\frac{1.819}{1.819} = 1.000$

$\frac{2.729}{1.819} = 1.500$

1.00 mol P: 1.50 mol O

2.00 mol P: 3.00 mol O

P_2O_3

second compound

mol P: $43.66 \text{ g P} \times \frac{1 \text{ mol P}}{30.97 \text{ g P}} = 1.410 \text{ mol P}$

mol O: $56.34 \text{ g O} \times \frac{1 \text{ mol O}}{16.00 \text{ g O}} = 3.521 \text{ mol O}$

$\frac{1.410}{1.410} = 1.000$

$\frac{3.521}{1.410} = 2.497$

1.00 mol P: 2.50 mol O

2.00 mol P: 5.00 mol O

P_2O_5

54. $1 \text{ mol Co} \times \frac{58.93 \text{ g Co}}{1 \text{ mol Co}} = 58.93 \text{ g Co}$

$2 \text{ mol Cl} \times \frac{35.45 \text{ g Cl}}{1 \text{ mol Cl}} = 70.90 \text{ g Cl}$

$CoCl_2$ molar mass: 58.93 g + 70.90 g = 129.83 g/mol

$12 \text{ mol H} \times \frac{1.01 \text{ g H}}{1 \text{ mol H}} = 12.12 \text{ g H}$

$6 \text{ mol O} \times \frac{16.00 \text{ g O}}{1 \text{ mol O}} = 96.00 \text{ g O}$

$6H_2O$ molar mass: 12.12 g + 96.00 g = 108.12 g/mol

$CoCl_2{\cdot}6H_2O$ molar mass: 129.83 g + 108.12 g = 237.95 g/mol

$\frac{129.83 \text{ g/mol}}{237.95 \text{ g/mol}} \times 100 = 54.56\%$ $CoCl_2$

55. $2 \text{ mol Cr} \times \frac{52.00 \text{ g Cr}}{1 \text{ mol Cr}} = 104.00 \text{ g Cr}$

$3 \text{ mol S} \times \frac{32.07 \text{ g S}}{1 \text{ mol S}} = 96.21 \text{ g S}$

$12 \text{ mol O} \times \frac{16.00 \text{ g O}}{1 \text{ mol O}} = 192.00 \text{ g O}$

$Cr_2(SO_4)_3$ molar mass: 104.00 g + 96.21 g + 192.00 g = 392.21 g/mol

$36 \text{ mol H} \times \frac{1.01 \text{ g H}}{1 \text{ mol H}} = 36.36 \text{ g H}$

$18 \text{ mol O} \times \frac{16.00 \text{ g O}}{1 \text{ mol O}} = 288.00 \text{ g O}$

18 H_2O molar mass: 36.36 g + 288.00 g = 324.36 g/mol

$Cr_2(SO_4)_3{\cdot}18H_2O$ molar mass: 392.21 g + 324.36 g = 716.57 g/mol

$\frac{392.21 \text{ g/mol}}{716.57 \text{ g/mol}} \times 100 = 54.73\%$ $Cr_2(SO_4)_3$

56. Fe_2O_3 molar mass: 2(55.85 g/mol) + 3(16.00 g/mol) = 159.70 g/mol

3/2 H_2O molar mass: (3/2)(2)(1.01) g/mol) + (3/2)(1)(16.00 g/mol) = 27.03 g/mol

$Fe_2O_3{\cdot}3/2H_2O$ molar mass: 159.70 g + 27.03 g = 186.73 g/mol

percentage due to H_2O: $\frac{27.03 \text{ g/mol}}{186.73 \text{ g/mol}} \times 100 = 14.48\%$

57. Fe_2O_3 molar mass: 2(55.85 g/mol) + 3(16.00 g/mol) = 159.70 g/mol

Fe_2O_3 yields 111.70 g/mol Fe

$\frac{111.70 \text{ g/mol}}{159.70 \text{ g/mol}} \times 100 = 69.94\%$ Fe

Fe_3O_4 molar mass: 3(55.85 g/mol) + 4(16.00 g/mol) = 231.55 g/mol

Fe_3O_4 yields 167.55 g/mol Fe

$\frac{167.55 \text{ g/mol}}{231.55 \text{ g/mol}} \times 100 = 72.36\%$ Fe

Fe_3O_4 yields more iron per kilogram of ore than does Fe_2O_3

58. Because both substances will absorb equal amounts of water, the less massive chemical will absorb a greater weight percentage of water. $MgCl_2$ is only 95.2 g/mol compared to 129.83 g/mol for $CoCl_2$. $MgCl_2$ will absorb the most water per gram.

59. NH_4^+ ion molar mass: 14.01 g N + 4(1.01 g) H = 18.05 g/mol

smelling salts molar mass:
2(14.01 g N) + 8(1.01 g H) + 12.01 g C + 3(16.00 g O) = 96.11 g/mol

$$2 \times \frac{18.05 \text{ g/mol}}{96.11 \text{ g/mol}} \times 100 = 37.56\% \ NH_4^+$$

60. aluminum phosphate, $AlPO_4$ molar mass:
26.98 g/mol + 30.97 g/mol + 4(16.00 g/mol) = 121.95 g/mol

percentage aluminum: $\frac{26.98 \text{ g/mol}}{121.95 \text{ g/mol}} \times 100 = 22.12\%$ Al

aluminum chloride, $AlCl_3$ molar mass:
26.98 g/mol + 3(35.45 g/mol) = 133.33 g/mol

percentage aluminum: $\frac{26.98 \text{ g/mol}}{133.33 \text{ g/mol}} \times 100 = 20.24\%$ Al

$AlPO_4$ yields a higher aluminum percentage.

61. a. $30.93 \text{ g Al} \times \frac{1 \text{ mol Al}}{26.98 \text{ g Al}} = 1.146 \text{ mol Al}$

$45.86 \text{ g O} \times \frac{1 \text{ mol O}}{16.00 \text{ g O}} = 2.866 \text{ mol O}$

$20.32 \text{ g Cl} \times \frac{1 \text{ mol Cl}}{35.45 \text{ g Cl}} = 0.5732 \text{ mol Cl}$

$2.89 \text{ g H} \times \frac{1 \text{ mol H}}{1.01 \text{ g H}} = 2.86 \text{ mol H}$

$\frac{1.146}{0.5732} = 1.999$

$\frac{2.866}{0.5732} = 5.000$

$\frac{0.5732}{0.5732} = 1.000$

$\frac{2.86}{0.5732} = 4.99$

2.00 mol Al: 5.00 mol O: 1.00 mol Cl: 5.00 mol H

$Al_2O_5ClH_5$

b. molar mass: 2(26.98) g Al+ 5(16.00) g O + 35.45 g Cl+ 5(1.01) g H = 174.46 g/mol

percentage Al: $\frac{53.96 \text{ g Al}}{174.46 \text{ g } Al_2O_5ClH_5} \times 100 = 30.93\%$ Al

percentage O: $\frac{80.00 \text{ g O}}{174.46 \text{ g } Al_2O_5ClH_5} \times 100 = 45.86\%$ O

percentage Cl: $\frac{35.45 \text{ g Cl}}{174.46 \text{ g } Al_2O_5ClH_5} \times 100 = 20.32\%$ Cl

percentage H: $\frac{5.05 \text{ g H}}{174.46 \text{ g } Al_2O_5ClH_5} \times 100 = 2.89\%$ H

Linking chapters

1. a. The outermost electrons in krypton and xenon are farther from the nucleus than those of neon and helium. Because of this distance, the electrons of krypton and xenon are not attached to the core of the atom as strongly as are those of neon and helium. Therefore, an element with a very high electron affinity can remove the outermost electrons.

b. Because radon is even larger than xenon, its attraction for the outermost electrons is weaker. Thus, radon is likely to form a compound with fluorine.

2. a. Ionization energy will increase from first to fourth for all three atoms. The greatest increase for K occurs between first and second ionization energies. Ca's greatest increase occurs between second and third, and Ga has its largest increase between third and fourth.

b. Ionization energy will increase from the first to the fourth for all three atoms. Each atom will have a marked increase between first and second ionization energies. All four ionization energies will increase from Li to K.

c. Ionization energy increases from first to fourth for all three atoms. Cs's greatest increase occurs between first and second ionization energies. Ba will have its largest increase between second and third. Tl will have its largest increase between third and fourth.

USING TECHNOLOGY

1. a. The melting point rises as a larger alkaline-earth metal bonds with chlorine.

b. No. Its melting point is lower than predicted by the trend of the other ionic chloride compounds.

c. 1041°C

Performance assessment

1. Answers will vary. The procedure should include measuring the mass of the salt before and after removing the water and reheating until a constant mass is attained.
2. Answers will vary. Be sure the testing criteria is safe and appropriate. When possible, students should use simpler observations such as hardness and crystalline structure, instead of melting and boiling points.

Portfolio projects

1. Answers will vary. Be sure the model relates to at least one aspect of the structure or properties of ionic compounds.
2. Answers will vary. Some ions chosen may be sodium, potassium, calcium, or hydrogen carbonate.
3. Answers will vary. Be sure that students recognize some common names for salts.
4. Answers will vary. Be sure that students address the problems that coincide with some substitutes; for example, Chinese restaurant syndrome which is caused by monosodium glutamate.

PLANNING FOR

6 Covalent Compounds

Chapter Overview

- **Section 6-1** compares and contrasts ionic and covalent bonding. Intermolecular forces are introduced and ordered by relative strength.
- **Section 6-2** shows how to draw Lewis dot structures for compounds with and without resonance, and for polyatomic ions. The concept of oxidation number is introduced. How to name a molecular compound is presented.
- **Section 6-3** distinguishes between empirical, molecular, and structural formulas.
- **Section 6-4** covers valence shell electron pair repulsion theory. Students learn to identify the molecular geometry of molecules that have one, two, three, or four substituents bound to the central atom.
- **Section 6-5** discusses the versatile bonding of the carbon atom, introduces organic structures and functional groups, and extends formula writing.

Concept Base

Students must have mastered the following concepts prior to this chapter: atomic structure and determining electron configurations (Chapter 3); atomic masses and electronegativity (Chapter 4); ionic bonding and percent composition (Chapter 5).

Looking Ahead

Interrelationships between polarity and molecular shape will be discussed in Chapter 11. Using oxidation numbers to identify and balance oxidation-reduction reactions will be covered in Chapter 15.

Laboratory Equipment Needs

Demonstrations

The listing of materials for Chapter 6 demonstrations begins on page T46.

Laboratory Experiments

Materials and preparation instructions for Chapter 6 are found as follows.

Text		*Page*
Exploration 6A	Polymers and Toy Balls	698
Investigation 6A	Polymers—Toy Trampoline	702

Laboratory Experiments Manual
(refer to the table of contents in the manual)

Investigation 6-1	Covalent and Ionic Bonding—Ceramics Fixative
Exploration 6-2	Viscosity of Liquids
Investigation 6-2	Viscosity—New Lubricants

Pacing: 13 days

Section	Inquiry Teaching Strategies
6-1 Why do atoms share electrons?	**Demonstrations** ■ Demo 1 *Comparing types of bonds* ■ Demo 2 *Bond formation* **Laboratory Experiments** ■ ■ Inv 6-1 *Covalent and Ionic Bonding—Ceramics Fixative* ■ ■ Exp 6-2 *Viscosity of Liquids* ■ ■ Inv 6-2 *Viscosity—New Lubricants*
6-2 How are molecules specified?	
6-3 How are formulas represented?	
6-4 How can you tell the shape of a molecule?	
6-5 How do bonds and properties relate?	**Laboratory Experiments** ■ Exp 6A *Polymers and Toy Balls* ■ Inv 6A *Polymers—Toy Trampoline*

Key ■ Teaching Transparencies
■ Annotated Teacher's Edition
■ Laboratory Experiments Manual

Visual Teaching Strategies		Review and Practice Strategies
Transparencies F 6-2 *Comparing Crystals* F 6-3 *Forces Between Particles* F 6-5 *Bond Length and Stability* T 6-4 *Electronegativity* F 6-6 *Predicting Bond Character*		**Text Reviews** Section Review 1–6 Chapter Review 1–11; LC 2; UT 1 **Study Guide Worksheets** 6-1 Concept Review 6-1 Skillsheet *Using Tables*
Transparencies F 6-17 *Symbolic Representations of Glucose*		**Text Reviews** Additional Examples 6A, 6B Practice 1–6 Concept Review 7–9 Section Review 10–15 Chapter Review 12–23 **Study Guide Worksheets** 6-2 Concept Review 6-2 Practice *Lewis structures*
Transparency Masters T 6-7 *Comparing Empirical and Molecular Formulas*		**Text Reviews** Additional Examples 6C, 6D Practice 1–5 Section Review 16–19 Chapter Review 24–36; LC 1 **Study Guide Worksheets** 6-3 Concept Review 6-3 Practice
Transparencies F 6-24 *Symbolic Representations of Methane: The Tetrahedron*		**Text Reviews** Section Review 20–24 Chapter Review 37–41 **Study Guide Worksheets** 6-4 Concept Review 6-4 Practice
Transparencies F 6-26 *Diamond and Graphite* F 6-27 *Resonance Structures* **Transparency Masters** T 6-9 *Properties of Diamond and Graphite* T 6-10 *Names of Alkanes* T 6-11 *Functional Groups* F 6-28 *Structural Formulas for Aspirin*		**Text Reviews** Section Review 25–28 Chapter Review 42–52 **Study Guide Worksheets** 6-5 Concept Review 6-5 Practice *Deciphering Organic Structures*

Pupil's Text — Study Guide — Teaching Resources
Exp *Exploration* — Inv *Investigation*
LC *Linking Chapters* — UT *Using Technology*

Assessment Options

Traditional Assessment

Test Generator
Instructional Objectives Measured:
Content mastery
Select from items 1 to 75

Performance Assessment Options

(see pages T36 and T643-15 for scoring rubrics)
Students demonstrate mastery of objectives in a hands-on environment. You may want some of these materials in the Portfolio as well.

Investigation 6A (text)
Investigations 6-1 and 6-2 (laboratory manual)
Instructional Objectives Measured:
Content mastery, Use of scientific methodology, Problem solving skills, Use of technology, Proficiency in written communication

Performance Assessments, page 231

1. **Designing an experiment**
   Instructional Objectives Measured:
   Content mastery, Problem solving skills, Use of scientific methodology, Scientific and chemical literacy, Proficiency in written communication
2. **Designing an experiment**
   Instructional Objectives Measured:
   Content mastery, Problem solving skills, Use of scientific methodology, Proficiency in written communication

Portfolio Options

(see page T36 for scoring rubrics)
Students provide a written rationale for each selection made for the portfolio.

Concept Maps
Students use vocabulary from the Chapter Review to build a concept map.
Instructional Objectives Measured:
Content mastery

Formal Laboratory Reports
Exploration 6A (text)
Exploration 6-2 (lab manual)
Instructional Objectives Measured:
Content mastery, Use of scientific methodology, Use of technology, Proficiency in written communication

Portfolio Projects, page 231
Items 1 and 5. Chemistry and you
Instructional Objectives Measured:
Content mastery, Scientific and chemical literacy, Proficiency in oral and written communication

Items 2 and 4. Research and communication
Instructional Objectives Measured:
Content mastery, Scientific and chemical literacy, Proficiency in oral and written communication

Item 3. Cooperative activity
Instructional Objectives Measured:
Content mastery, Use of scientific methodology, Problem solving skills, Proficiency in oral and written communication

INTRODUCING THE CHAPTER

CHAPTER 6 Covalent Compounds

Story Background

Low density polyethylene, LDPE, is formed when ethylene polymerizes in the presence of a catalyst at 200°C and 2000 atm pressure. The catalyst initiates the reaction by removing a pi electron from ethylene's double bond, creating a free radical (**a**). Ethylene units are trapped by the free radical end of the molecule and the lone electron is transferred down the chain (**b**); in this way, the chain grows.

a.

$$\begin{array}{c} \text{H}\quad\text{H} \\ |\quad\ \ | \\ -\text{C}-\text{C}\cdot \\ |\quad\ \ | \\ \text{H}\quad\text{H} \end{array}$$

b.

$$\begin{array}{c} \text{H}\quad\text{H} \\ |\quad\ \ | \\ -\text{C}-\text{C}\cdot \\ |\quad\ \ | \\ \text{H}\quad\text{H} \end{array} + \begin{array}{c} \text{H}\qquad\ \text{H} \\ \text{C}=\text{C} \\ \text{H}\qquad\ \text{H} \end{array} \rightarrow \begin{array}{c} \quad\ \text{H}\quad\text{H}\quad\text{H}\quad\text{H} \\ \quad\ |\quad\ \ |\quad\ \ |\quad\ \ | \\ \text{H}-\text{C}-\text{C}-\text{C}-\text{C}\cdot \\ \quad\ |\quad\ \ |\quad\ \ |\quad\ \ | \\ \quad\ \text{H}\quad\text{H}\quad\text{H}\quad\text{H} \end{array}$$

This rationale accounts for the linear structure for polyethylene first described in 1936. However, investigation revealed side chains on LDPE, about 15–30 per 1000 carbon atoms. Side chains are believed to form when the radical end of the chain turns back on itself and "bites off" a hydrogen on a neighboring C atom. If the polymerization occurs in the presence of a catalyst without heating or pressurizing the gas, the occurence of side chains decreases to about 1 per 1000 C atoms. Side chains affect how closely the main polymer chains can pack together; hence, they affect the polymer's crystallinity and density. Polyethylene formed under low pressure is called high density polyethylene or HDPE. HDPE has the higher degree of crystallinity, so it is harder, tougher, more tear resistant, and less likely to be reheated for molding than LDPE.

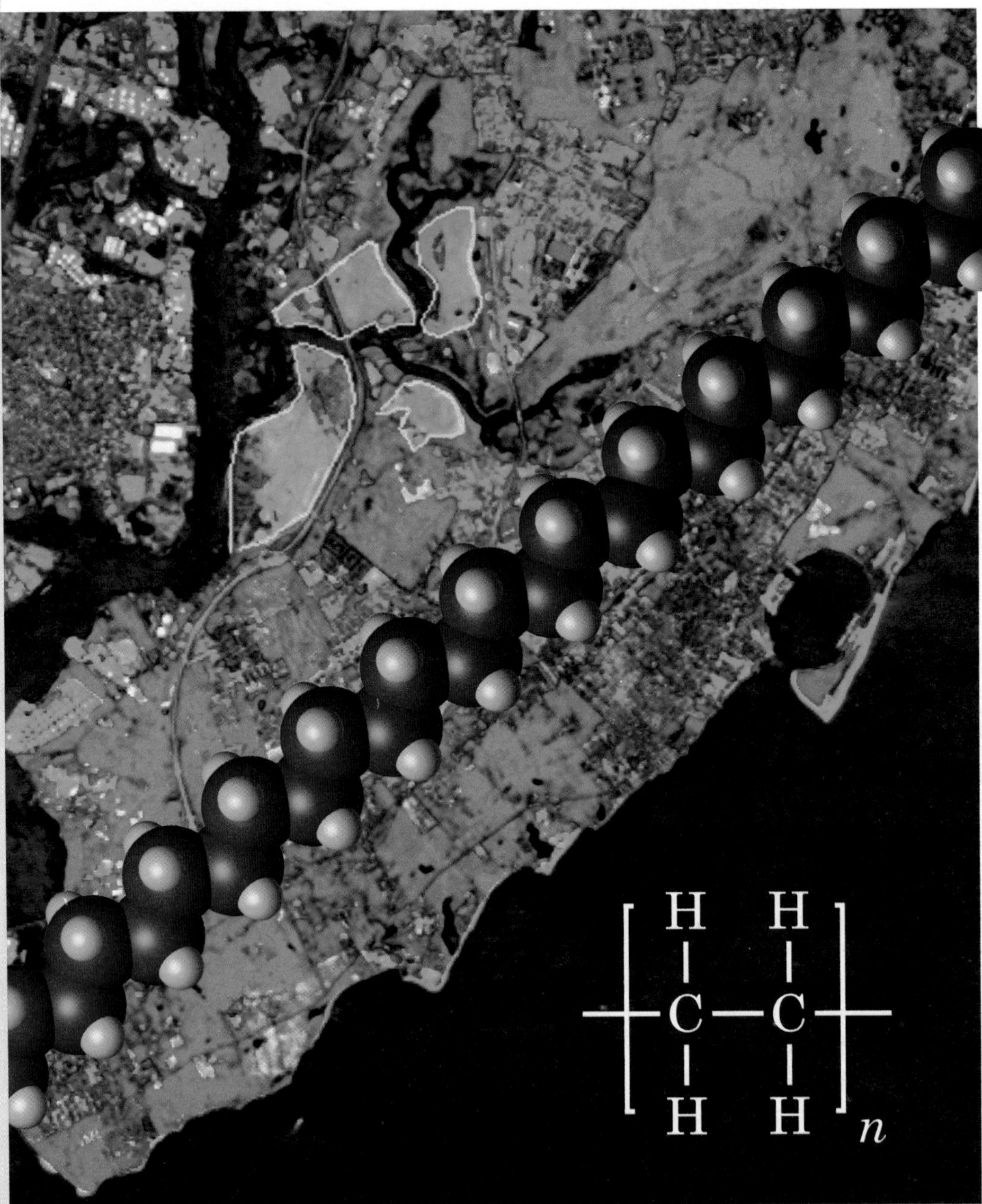

About the Illustration

Viewed from the air, the Fresh Kills landfill, like thousands of other landfills in the United States, contains slowly degrading polyethylene products.

Molecular Bonds and Trash Bags

What is the largest structure ever made by humans? It's not the Great Wall of China, which extends for a total of 6300 km and can even be seen by astronauts at an altitude of 1000 km. It is actually the Fresh Kills landfill in Staten Island, New York, shown from a satellite, at left.

Although a landfill is not a structure in the same way as the Great Wall, this site does represent the largest collection of materials ever accumulated for a single purpose by humans. Begun in 1948 atop a swamp, the Fresh Kills landfill receives nearly 13 000 metric tons of garbage every day. As a result, it covers nearly 20 km^2 and holds 6.5×10^7 m^3 of refuse.

Eventually, it will tower over 130 m at its tallest point. If this much trash were stacked on a thousand football fields, it would pile up as high as a 21-story building.

Landfills such as Fresh Kills are so big because Americans produce a lot of garbage. Some studies have found that on average each American family of four produces 9 kg of garbage per day, or 3.5 metric tons per year. Even if the waste isn't hazardous, there are basically only three ways of dealing with all this garbage—burn it, recycle it, or bury it.

Burying garbage has been the most commonly used method since humans first started producing trash. But today, places to bury garbage are quickly filling up. Since 1978, some 14 000 landfills in the United States have been filled up and shut down. The opening of new landfills is not keeping pace with the closing of old ones. As populations increase, some communities face garbage crises, with nowhere to put their refuse.

Many people say that minimizing the amount of garbage produced is the best solution; that way, there would be less to bury. Moreover, the garbage that does get buried should be biodegradable so that microorganisms can break it down quickly once it's in a landfill. Biodegradable products would also take up less space as they decompose in landfills.

Trash bags have become a target in the search for biodegradable materials because Americans use a lot of them. For example, just raking all of the leaves from a single 23 m maple tree can fill 10 large plastic bags. But the plastic trash bags are made from polyethylene, a material that does not break down easily.

Chemically, this type of plastic consists of carbon and hydrogen atoms bonded together to form long chains, often as long as 7000 carbon-hydrogen units. These carbon-hydrogen units are very stable—they are hard to break.

Scientists are seeking ways to change polyethylene so that it will break down more easily in landfills, yet still hold up to daily use.

But before looking at how scientists are trying to accomplish this task, you need to understand how compounds such as plastics are built from individual atoms. Questions you will need to explore include the following.

- *What is the composition of the material currently being used in trash bags?*
- *How do bags labeled as biodegradable differ?*

Tapping Prior Knowledge

Prepare students for the concepts they will need for this chapter with the following discussions and activities.

1. Ask students how the bonding between atoms in oil or plastic wrap compares with the bonding in salts. Extend this inquiry with the following two questions. How could bonding account for the differences in physical characteristics? How could these two compounds be named if chemical nomenclature is used to describe the bonding between atoms?
2. Discuss the importance of perspective and depth to a draftsman. What techniques do artists use to convey depth and perspective in their drawings? Set out simple objects such as a cube or pyramid. Ask students to sketch these shapes. Ask them to draw a template that they can cut out and glue or tape together to form the simple shape.
3. Relating the complete structural formula for an organic molecule with its abbreviated structural formula may take some practice. You can stimulate student interest and confidence by displaying several pictures or models of dinosaur, fish, or other animal skeletons. Ask students to identify these skeletons. Then point out that they were able to do this because they mentally supplied the filler for the skeleton. In much the same way, they will supply C atoms and H atoms in abbreviated organic structures.

Section 6-1

FOCUS

Lesson Starter

Tell a story about two entrepreneurs, each of whom wishes to start a restaurant. One wants to serve Chinese food and the other wants to serve Italian food. One of the two people owns a suitable, large building, while the other person owns an abundance of cooking equipment. How can these two people improve their situation? Relate this to the covalent bond between hydrogen and chlorine.

Demonstration 1

Comparing types of bonds

Approximate time: 10 min

1. Fill a Pyrex test tube one-fourth full of table sugar (sucrose) and another test tube one-fourth full of sodium chloride.
2. Ask the students for a variety of ways to tell the two compounds apart.
3. Using either a projecting microscope or several stereo or standard microscopes, have the students view the two compounds.
4. Pour a little of each compound into separate 150 mL beakers that are half full of water. Check with a conductivity tester.
5. Heat a nichrome wire, dip it into the sugar crystals, and place it in the flame. Reheat the wire, dip it into the salt, and place it in a flame.
6. Heat both test tubes in a flame. Why did one compound melt?

Why do atoms share electrons?

Section Objectives

Compare and contrast the properties of substances with covalent and ionic bonds.

Identify the forces acting on two covalently bonded atoms.

Explain the changes that occur in stability and energy as a covalent bond forms.

Use electronegativity values to determine the nature of a chemical bond.

Distinguish between intermolecular bonds and covalent and ionic bonds.

Comparing types of bonds

Hydrogen and oxygen are stable separately, but when combined, all it takes is a spark to create a violent explosion, as shown in **Figure 6-1**. Afterward, only tiny droplets of moisture are left. By applying some of what you learned in Chapter 5, you can explain much of what is happening in such an explosion.

You know that breaking bonds requires energy. The initial spark provides the energy to break the bonds that already exist in oxygen gas and hydrogen gas. But after that small input of energy, there is a huge release of energy, and a new product is formed—water. The release of energy suggests two things: first, new bonds are being formed as atoms are rearranged; and second, the product made by rearranging atoms, water, is lower in energy and more stable than the reactants, hydrogen and oxygen.

So far, this seems like the same situation as with ionic compounds described in Chapter 5, but water has very different properties than salts, as can be seen in **Table 6-1**. These different properties are shared by many other substances as well.

Figure 6-1a
Hydrogen is a colorless gas that is relatively stable on its own, . . .

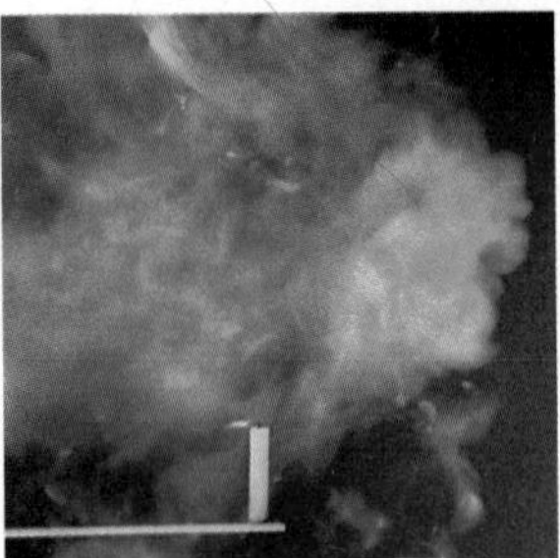

6-1b
. . . but in the presence of oxygen, a spark or small flame can cause an explosion, with the production of water.

Table 6-1
Properties of Salt, Water, Iodine, and Hydrogen

Substance	Physical state (at room temperature)	Melting point	Boiling point	Electrical conductivity
NaCl	solid	high (801°C)	high (1413°C)	high (when melted or dissolved)
H_2O	liquid	low (0°C)	low (100°C)	very low (for pure water)
I_2	solid	low (113.5°C)	low (184.3°C)	very low
H_2	gas	very low (–259.3°C)	very low (–252.8°C)	very low

Safety
Wear safety goggles and lab apron. Tie back hair and loose clothing. Have students stand *at least* 10 ft from the demonstration area. Use only a battery-powered conductivity tester. Do not use one that runs on 110/115 V A/C power even if it is stepped down by a transformer.

Disposal
Put solids into trash. Pour liquids down the drain.

Molecular compounds have covalent bonds

Based on the differences in **Table 6-1**, it seems that the bonds in water must be different from those in salt. Lower melting and boiling points for water suggest that the attractions among particles are weaker than those in salt. The fact that pure liquid water does not conduct much current suggests that there are few ions present. To explain the properties of compounds like water, iodine, and hydrogen (all of which have some similar properties), you need a new bonding model.

covalent bond
bond formed when atoms share pairs of electrons

The properties of water can be explained by considering a bond made through electron sharing. In a **covalent bond**, atoms do not lose or gain electrons. Instead, they share pairs of electrons to achieve stability, often by filling their outer energy levels so that they have stable octets.

Can the covalent model explain why the properties of these compounds with covalent bonds differ so much from those with ionic bonds? Why are the melting and boiling points of molecular compounds so much lower than those of salts?

molecular compound
substance consisting of atoms that are covalently bonded

The force that holds together two atoms in a covalent bond can be as strong as the force between an individual cation and anion. However, in an iodine molecule, the two iodine atoms are covalently bonded only to each other and not to any other iodine atoms. Compare the structures of sodium chloride and iodine in **Figure 6-2**. Notice that even though iodine and salt are both solids, iodine exists as individual molecules, whereas NaCl exists as an extended crystal lattice. Each ion in the salt lattice is bonded to six adjacent ions. Because covalent compounds are often in the form of individual molecules, they are called **molecular compounds.**

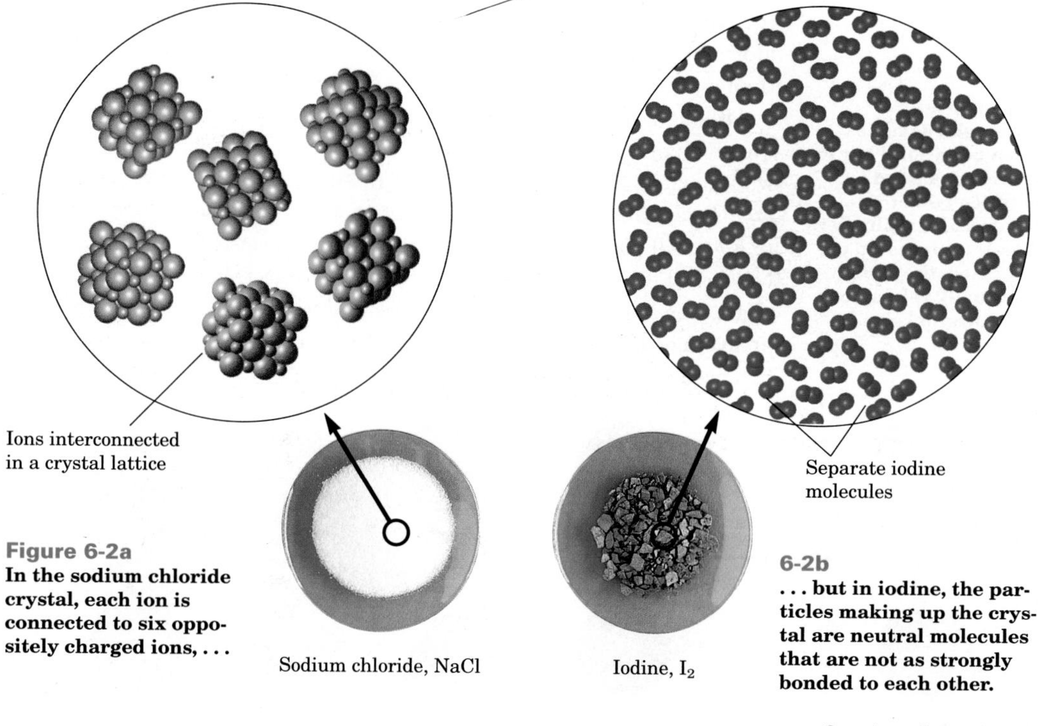

Figure 6-2a
In the sodium chloride crystal, each ion is connected to six oppositely charged ions, . . .

6-2b
. . . but in iodine, the particles making up the crystal are neutral molecules that are not as strongly bonded to each other.

Visual Strategy
Figure 6-2
Notice that the iodine atoms are bonded in discrete groups of two, but the sodium and chloride ions are attracted to each other in such a way that each ion is surrounded by six ions of unlike charge. This arrangement extends until no further attractions can be made.

TEACH

Demonstration 2

Bond formation

Approximate time: 10 min

1. Clean a 10 cm piece of Mg ribbon with sandpaper.
2. Darken the lights, and remind the students not to look directly at the Mg while it is burning.
3. Using crucible tongs to hold the strip, light the Mg ribbon, and hold it inside a coffee can. The light should still be quite visible as the ionic MgO bond forms.
4. Using tongs, place a piece of charcoal into the burner flame, and attempt to get the charcoal glowing. The formation of covalent bonds in CO_2 is much less energetic than the formation of ionic bonds in MgO.

SAFETY

Wear safety goggles and a lab apron. Have students stand back at least 10 ft. Cover any sharp edges on the rim of the coffee can with tape.

DISPOSAL

Immerse the charcoal in water to be sure it is extinguished, and put it and the cooled MgO into the trash.

Teaching Tip

Electron interactions

You may wish to point out that the electrons in an H_2 molecule are not static. They are moving close to the speed of light, and because the electrons are charged particles sharing the same covering, or vail, that encompasses the two H nuclei, the movement of one electron affects the movement of the other electron.

Forces of electric attraction make a covalent bond

Why are atoms attracted and held together in a covalent bond? Consider all of the forces in the simplest covalent bond, that between two hydrogen atoms. An attractive force occurs between the electron on the first hydrogen atom and the proton on the second hydrogen atom. However, a repulsive force would also occur—the two electrons would repel each other, as would the two protons. At first, you might think that these attractive and repulsive forces would cancel each other, causing the two individual atoms to remain separate. But experiments have shown that most hydrogen exists as two atoms bonded together.

Figure 6-3
When two hydrogen atoms come into contact, the particles within one atom attract and repel particles within the other atom. When these forces balance, a bond forms.

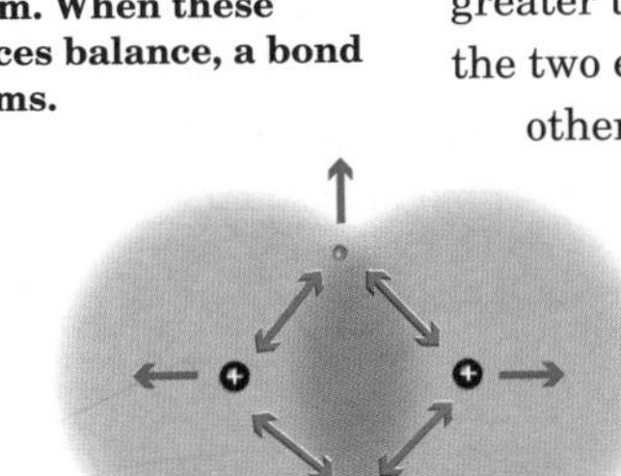

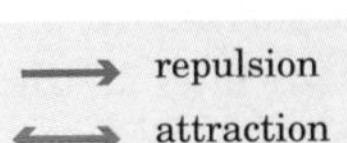

As you can see in **Figure 6-3**, the distance between two protons is greater than the distance between a proton and either electron. Similarly, the two electrons are each closer to the two protons than they are to each other. Consequently, there is an attraction between an electron of one atom and the proton of the second atom and another attraction between the first atom's proton and the second atom's electron. Taken together, the attractions are much greater than the repulsions between the two protons and between the two electrons. This net attraction holds the two hydrogen atoms together and is the basis of the covalent bond that forms between them.

Another reason that attractive forces between atoms can overcome the repulsive forces is that another factor besides charge becomes important. You learned in Chapter 3 that the two electrons sharing an orbital must have different values for the spin quantum number. For convenience, they are said to have opposite "spins," even though the term does not mean that they spin in opposite directions. A pair of electrons shared between atoms in a stable covalent bond have opposite spins and occupy less space than a pair of electrons in an orbital on only one atom.

Although covalent bonds are often drawn or modeled as rigid sticks connecting the atoms, a more useful model shows the bond to be somewhat flexible, like a spring, as shown in **Figure 6-4**. If the two hydrogen atoms start to move apart, the attractive forces between the electrons and protons will pull them back. If the atoms are too close, the repulsive forces between the two protons and between the two electrons will push them apart. The atoms actually vibrate back and forth, but they vibrate around an average distance, at which the attractive and repulsive forces are balanced.

Figure 6-4
Although chemists often use a solid bar to represent a bond between two atoms, bonds are actually flexible. If compressed or stretched, they will eventually return to their original size.

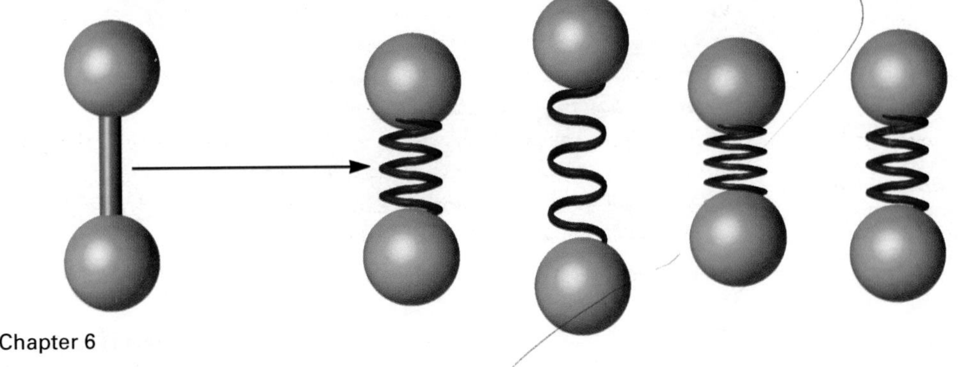

Visual Strategy

Figure 6-4

Atoms that are bonded vibrate so that the bond between them is stretched and compressed, much like two Ping Pong balls attached to the ends of a Slinky or strong spring. As the temperature of the molecule's surroundings rises, the energy of the oscillations increases, as does the distance that the atoms move when oscillating. Finally, the bond will break.

Figure 6-5
A bond forms when atoms are a certain distance from each other. At this distance, the atoms are in a low energy state. If they are closer together or farther apart, they will be in an unstable situation.

Bond Length Represents a Position of Stability

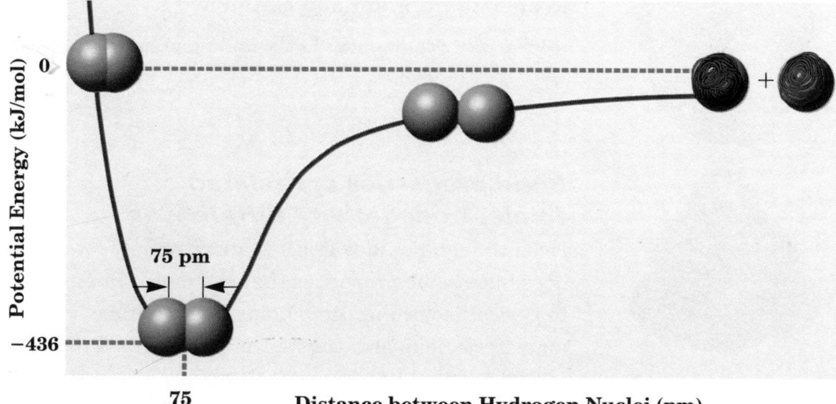

Bond length and energy are inversely related

Another way to consider this model is to examine the potential energy levels for two hydrogen atoms at different distances, as shown in **Figure 6-5**. At what point on the graph do the two atoms have the most stability? This point, representing the lowest potential energy level on the graph, also represents the average distance between the two atoms when they are covalently bonded. Atoms that are farther apart than this distance tend to get closer because of the attractive forces, and atoms that are closer together than this distance tend to get farther apart because the repulsive forces grow stronger.

bond length
average distance between the nuclei of two bonded atoms

The two bonded atoms vibrate back and forth, but as long as their energy remains near this minimum potential energy level, they are covalently bonded to each other. The average distance that separates them is known as their **bond length.**

Because hydrogen atoms are more stable when bonded than when isolated, energy is released as the bond between them is formed. **Table 6-2** lists the bond lengths and bond energies for various atoms connected by covalent bonds. Keep in mind that bond lengths are never really fixed values because the atoms vibrate. Bond lengths can also vary depending on what other bonds are present in a molecule. The bond lengths listed in **Table 6-2** represent average values. Bond energy is the energy required to break a chemical bond to produce individual atoms, each keeping its own electrons.

Bond energy is the best indicator of the strength of the force of attraction in a bond. In general, the closer two atoms are, the greater the bond energy that is required to separate them, but there are exceptions to this rule. Large atoms usually have longer bond lengths and lower bond energies than smaller atoms. As a result, it is often easier to break a bond between two large atoms than between two small atoms.

Refer again to the graph shown in **Figure 6-5**. Trace the line on the graph that represents the separation of the two covalently bonded hydrogen atoms. Notice that 436 kJ of energy are needed to make a mole of hydrogen molecules so unstable that their bonds break.

Table 6-2
Average Bond Lengths and Energies

Bond	Length (pm)	Energy (kJ/mol)
H—H	75	436
H—C	109	418
H—Cl	127	432
H—Br	142	366
C—O	143	341
C—C	154	332
H—I	161	298
C—Cl	177	326
C—Br	194	276
Cl—Cl	199	243
Br—Br	229	193
I—I	266	151

Teaching Tip
Mental modeling of a covalent bond
Have students imagine that they have two strongly magnetic tennis balls. The balls cannot easily be shoved into each other or pulled apart. They are currently at their optimum low-energy state. When the balls are moved, they return to the same low-energy position.

Theme
Classification and trends
As bond length increases, the energy of the bond decreases. Thus, atoms that can bond closer together have a much greater bond strength.

Teaching Tip
Bond length vs. bond energy
List separately all of the entries in **Table 6-2** in which C is present. Point out that as the bond length increases, the bond energy decreases.

Visual Strategy
Figure 6-5
Trace the events on the energy curve, pointing out that nature likes things at an energy minimum. In the plot, this energy minimum occurs when the atoms are spaced 75 pm apart. Past this point the repulsion of like charges pushes the atoms apart. Follow the bonding process in terms of bond length, and compare with the bond oscillations in **Figure 6-4**.

Theme
Classification and trends
Bonding between elements from opposite ends of the periodic table is ionic, while bonding between elements from the same end of the periodic table is covalent.

Teaching Tip
Periodicity and bond polarity
Electronegativity increases as you go up and toward the right in the periodic table. Use this as a rule of thumb to predict bond character from the periodic table.

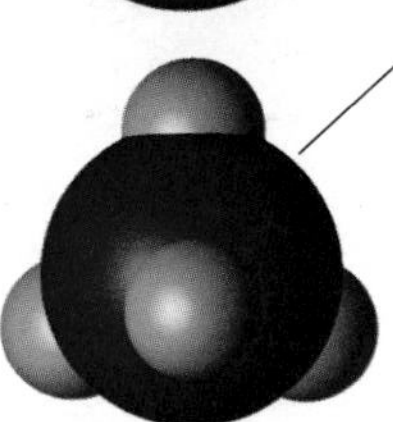

Table 6-3
Properties of Water and Methane

Substance	Melting point (°C)	Boiling point (°C)	Reaction to electric field
H_2O	0.0	100.0	aligned
CH_4	−182.6	−161.4	not aligned

Bond properties are related to electronegativity differences

Even though liquid water has an extremely low electrical conductance, it does have some properties that are more similar to ionic compounds than to typical molecular compounds like methane, CH_4. Water will respond to an electric field, as a substance with charged particles will. Unlike molecular compounds of similar molar mass, such as methane, which has very low boiling and melting points, water remains a liquid for a much larger temperature range. See **Table 6-3**.

Once more, these properties can be explained by expanding the model used for bonding. Few chemical compounds are either totally *molecular* or totally *ionic*. These terms actually refer to the extremes at either end of a continuous spectrum of bonding. The bonds in many compounds, like water, have some features of both types of bonds. For example, even though electrons are shared between the hydrogen and oxygen atoms in a water molecule, they are not shared evenly. In each bond, the oxygen atom attracts the electrons much more than the hydrogen atom does. Because the electrons are unequally shared, the molecule behaves as if each oxygen atom has a partial negative charge, and each hydrogen atom has a partial positive charge.

To determine whether this uneven sharing will be very small or so large that the substance is made of fully charged ions, you can compare the atom's relative ability to pull electrons in a bond toward itself. You learned in Chapter 4 that electronegativity is a periodic property. The electronegativity table, **Table 6-4**, is an attempt to summarize observations and measurements of how strongly certain atoms attract electrons while in a covalent bond.

Table 6-4
Electronegativity Values

0–0.9 | 1.0–1.9 | 2.0–2.9 | > 2.9

1 H 2.1																	
3 Li 1.0	4 Be 1.5											5 B 2.0	6 C 2.5	7 N 3.0	8 O 3.5	9 F 4.0	
11 Na 0.9	12 Mg 1.2											13 Al 1.5	14 Si 1.8	15 P 2.1	16 S 2.5	17 Cl 3.0	
19 K 0.8	20 Ca 1.0	21 Sc 1.3	22 Ti 1.5	23 V 1.6	24 Cr 1.6	25 Mn 1.5	26 Fe 1.8	27 Co 1.8	28 Ni 1.8	29 Cu 1.9	30 Zn 1.6	31 Ga 1.6	32 Ge 1.8	33 As 2.0	34 Se 2.4	35 Br 2.8	
37 Rb 0.8	38 Sr 1.0	39 Y 1.2	40 Zr 1.4	41 Nb 1.6	42 Mo 1.8	43 Tc 1.9	44 Ru 2.2	45 Rh 2.2	46 Pd 2.2	47 Ag 1.9	48 Cd 1.7	49 In 1.7	50 Sn 1.8	51 Sb 1.9	52 Te 2.1	53 I 2.5	
55 Cs 0.7	56 Ba 0.9	57 La 1.1	72 Hf 1.3	73 Ta 1.5	74 W 1.7	75 Re 1.9	76 Os 2.2	77 Ir 2.2	78 Pt 2.2	79 Au 2.4	80 Hg 1.9	81 Tl 1.8	82 Pb 1.8	83 Bi 1.9	84 Po 2.0	85 At 2.2	
87 Fr 0.7	88 Ra 0.9	89 Ac 1.1															

Atoms with large electronegativity values, such as fluorine, chlorine, and oxygen, tend to attract electrons in a bond more strongly than atoms with low electronegativity values, such as sodium, magnesium, and lithium. The greater the difference in electronegativities between two atoms, the more ionic character the bond they form will have.

Consider cesium fluoride, CsF, as an example. The electronegativity value for Cs is 0.70; that for F is 4.00. The difference between these two electronegativity values is 3.30. **Figure 6-6** indicates that this bond has properties that are much more like those of ionic bonds, rather than those of covalent bonds. When the difference in electronegativities between two atoms in a bond is much greater than 2.1, the bond is classified as mostly ionic.

Next, consider the example of the bond formed between silicon and oxygen. Because oxygen has an electronegativity value of 3.5 and silicon's value is 1.8, the difference is 1.7. This difference places it near the threshold of being classified as an ionic bond. Use **Figure 6-6** to describe the nature of the bond between silicon and oxygen.

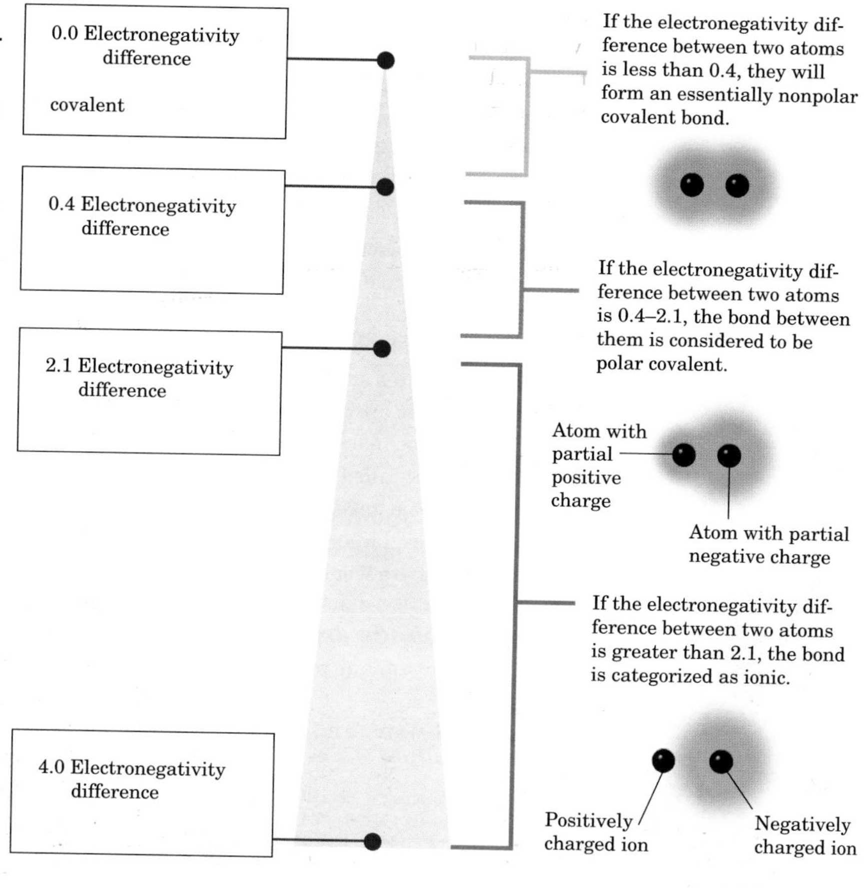
0.0 Electronegativity difference
covalent
0.4 Electronegativity difference
2.1 Electronegativity difference
4.0 Electronegativity difference
If the electronegativity difference between two atoms is less than 0.4, they will form an essentially nonpolar covalent bond.
If the electronegativity difference between two atoms is 0.4–2.1, the bond between them is considered to be polar covalent.
Atom with partial positive charge
Atom with partial negative charge
If the electronegativity difference between two atoms is greater than 2.1, the bond is categorized as ionic.
Positively charged ion
Negatively charged ion

Figure 6-6
Differences in electronegativity can be used as a rough measure to predict the properties of the bond. In general, the greater the electronegativity differences, the more ionic properties the bond will have.

Visual Strategy

Figure 6-6

Determine the bond character for the following molecules: Rb—Br (2.0), Na—Cl (2.1), F—Cl (1.0), Ca—S (1.5), Mg—O (2.3), Cl—Cl (0.0), the H—O bond in water (1.4), and the C—H bond in methane (0.4).

Teaching Tip

Percent ionic character

Very few bonds are 100% covalent. All other bonds have both ionic and covalent characteristics. The greater the percent ionicity of the bond, the more pronounced the ionic characteristics will be.

The following formula is used to calculate the percent ionic character of bonds.

$$P = 18(\chi a - \chi b)^{1.4}$$

χ_a and χ_b are the electronegativies of the bonding atoms. Electronegativities are listed in **Table 11-3**.

For the O—H bond:

$$P = 18(3.5 - 2.1)^{1.4}$$
$$= 18(1.4)^{1.4} = 18(1.6) = 29\%.$$

Chapter Connections

Polarity of bonds vs. polarity of molecules

Polarity of bonds depends on electronegativity differences between the bonded atoms. Polarity of molecules depends on spatial symmetry. Therefore, you may wish to caution students that although the bonds within a compound are polar, the compound itself may be nonpolar. Rules for determining polarity of molecules are discussed in Chapter 14.

Can you think of an example in which the difference in electronegativities between two atoms would be exactly zero? Obviously, when the two atoms that bond are of the same element, no difference in electronegativity exists. This means that both hydrogen atoms in a hydrogen molecule have an equal attraction for the electrons they share. When a covalent bond forms between two atoms with equally shared electrons, the bond is said to be a **nonpolar covalent bond.**

nonpolar covalent bond
covalent bond in which the bonding electrons are shared equally between the two bonding atoms

Covalent bonds with uneven electron sharing are polar

When atoms of different elements bond, the sharing can never be truly equal. As an example, consider carbon and oxygen. The oxygen atom, with an electronegativity of 3.5, will have a greater affinity for the electrons than will the carbon atom, with an electronegativity of 2.5. One of several reasons why is that the oxygen atom has more protons than the carbon atom, but they have an equal number of electron energy levels.

Because the attractive force for bonding electrons is not equal, the electrons will not be shared equally. When a covalent bond forms with unequally shared electrons, the bond is said to be a **polar covalent bond.** The uneven sharing causes the more electronegative atom to have a partial negative charge. The other atom will have a partial positive charge.

polar covalent bond
covalent bond in which the bonding electrons are more strongly attracted by one of the bonding atoms

Intermolecular forces

Weak attractions also form between molecules

All atoms and molecules attract each other. But the **intermolecular forces** between molecules or atoms are usually not as strong as covalent or ionic bonds because they *do not* involve transferring or sharing electrons. One good measure of the strength of intermolecular forces in a substance is the boiling point. The more strongly the particles are attracted, the higher the boiling point will be. For *nonpolar* substances, the attraction is a weak one, roughly in direct proportion to the number of electrons. The forces between nonpolar substances are called *London forces*.

intermolecular forces
attraction resulting from forces between molecules

If the molecules are *polar,* then there is an additional force because the positive end of one molecule attracts the negative end of a nearby molecule, and so on. This is called the *dipole force.* The more polar the molecule, the stronger the dipole force. The farther apart molecules are, the weaker the dipole force. Thus, large polar molecules which cannot get as close to each other tend to have a weaker force of attraction than smaller ones.

Hydrogen bonds are strong intermolecular forces

Water has many unique properties that are not shared by similar substances such as hydrogen sulfide, H_2S, as shown in **Table 6-5**. The much higher melting and boiling points shown for water indicate that water molecules have a particularly strong polar-polar attraction for each other, much stronger than the attractions between H_2S molecules.

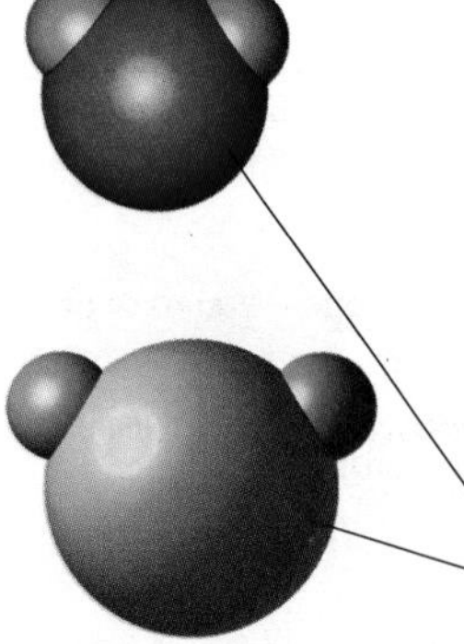

Table 6-5
Properties of Water and Hydrogen Sulfide

Substance	Melting point (°C)	Boiling point (°C)
H_2O	0.0	100.0
H_2S	–85.5	–60.3

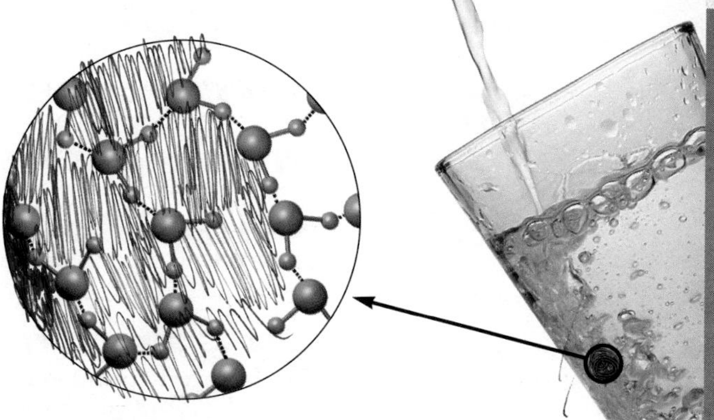

Figure 6-7
Many of water's unique properties are a result of the hydrogen bonds that form between different water molecules.

hydrogen bond
attraction occurring when a hydrogen atom bonded to a strongly electronegative atom is also attracted to another electronegative atom, often of a different molecule

These especially strong intermolecular forces are called **hydrogen bonds**. Hydrogen bonds can form when hydrogen, the smallest atom, is bonded within a molecule to fluorine, oxygen, or nitrogen (the smallest and most electronegative atoms). Because of the small size of the hydrogen atoms, the partial positive charge on the hydrogen atom of one molecule is able to come very close to the partial negative charge of a different electronegative atom, as shown in **Figure 6-7**. As with other bonds, the closer the opposite charges are to each other, the stronger the attraction between them is.

The name *hydrogen bond* is confusing because, like other intermolecular forces, these types of attraction are considerably weaker than chemical bonds between atoms. Hydrogen bonding is an important factor in many of the chemical reactions that make life possible. The DNA strands in your cells' genes are able to match up time and time again because of hydrogen bonds. Hydrogen bonding also enables enzymes to attract and latch onto the chemicals they change to make energy for life.

Section Review

1. Describe the attractive and repulsive forces that occur as two atoms are brought closer together.
2. What happens to the stability and energy of two atoms as they form a covalent bond?
3. List three differences between molecular and ionic compounds, and explain how they relate to the differences in bond types.
4. Use **Figure 6-6** and electronegativity values from **Table 6-4** to classify the bonds between the following.
   a. Cs and Br **b.** H and S **c.** Ca and O **d.** Si and Cl
5. Compare the behavior of bonding electrons in a nonpolar covalent bond with that of bonding electrons in a polar covalent bond.

6. **Story Link**
   Recall that the chemical compound in plastic—polyethylene—consists of many units with carbon-hydrogen bonds. Use **Figure 6-6** and electronegativity values from **Table 6-4** to determine the type of these bonds.

Visual Strategy
Figure 6-7
Hydrogen bonds are not true bonds; they are electrostatic attractions between the polar ends of molecules; so, hydrogen bonding can take place in all directions. The force of attraction decreases as the distance between an H atom and its neighboring O atoms increases. These attractions loosely but effectively tie water molecules together so that the boiling and melting points are higher than they would be in the absence of intermolecular attractions.

ASSESS

Alternative Assessment

1. Fluorine can be substituted for H in many hydrocarbon compounds because of its small size. The compounds are then less reactive. Account for this decrease in reactivity in terms of bond strength.
2. Have the students write for five minutes about the similarities of hydrogen bonding between molecules and taking a little sister or brother along on a dinner date to a restaurant where 15 other couples are doing the same.

Answers to Section Review

1. Refer to **Figure 6-5**. Nuclei are positively charged, and electrons are negatively charged. Like charges repel; unlike charges attract. When two atoms are far apart, the attraction between unlike charges predominates. At close distances, the atoms are pushed apart because like charges are too close to coexist. At an ideal distance, bonding occurs.
2. Stability increases; potential energy decreases.
3. Because covalent compounds are bonded in discrete units, their boiling points, melting points, and conductivity are significantly lower than those of ionic compounds which are bonded in long arrays.
4. **a.** ionic
   b. polar covalent
   c. ionic
   d. polar covalent
5. Electrons in nonpolar covalent bonds spend equal time about the two nuclei because the nuclei attract the electrons equally. Electrons in polar covalent bonds spend more time around the nucleus that has the greater electron affinity.
6. polar covalent; $2.5 - 2.1 = 0.4$

Section 6-2

FOCUS

Lesson Starter

Begin the lesson with this story of sharing to gain stability. A set of twins graduated from high school and decided to live on their own. Each twin had a monthly income of $1000 and wanted a two-bedroom apartment. However, their separate monthly living costs would be $2000. To fullfill their dream while maintaining a stable lifestyle, the twins would have to share a two bedroom apartment. Relate this story to two hydrogen atoms, each having one electron but needing two electrons for stability.

TEACH

Teaching Tip

Identifying valence electrons

Locate the element in the periodic table. Count the *s* and *p* electrons from left to right along the element's row. Stop when you arrive at the assigned element.

For students who have trouble identifying valence electrons from electron configurations, have them underline the largest Arabic numeral as often as it occurs and then add the superscripts together. The electron configuration for Cl has two underlined terms: $1s^22s^22p^6\underline{3s^2}\underline{3p^5}$. The number of valence electrons is 2+ 5, or 7. Calcium's electron configuration has 1 underlined term: $1s^22s^22p^63s^23p^6\underline{4s^2}$. There are two valence electrons.

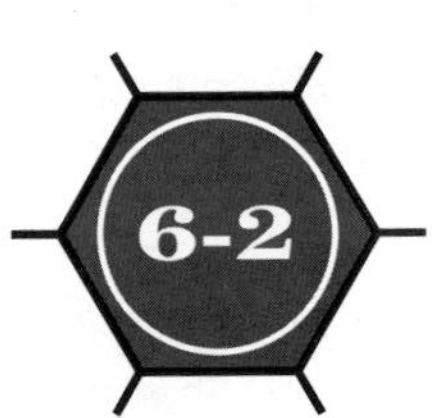

How are molecules specified?

Section Objectives

Draw Lewis structures to show the arrangement of valence electrons among the atoms in a molecular compound.

Recognize exceptions to the octet rule and the limitations of Lewis structures.

Determine the oxidation number for each atom in a molecular compound.

Use both prefixes and Roman numerals to name molecular compounds.

Lewis structures

valence electron *electron present in the outermost energy level of an atom*

Electronegativity differences provide only limited information about the nature of a bond between two atoms. To develop a model that explains how some atoms combine, you can use dots to represent the **valence electrons** that are involved in bond formation.

Each hydrogen atom, with an electron configuration of $1s^1$, has only one valence electron. This valence electron is shown in **Figure 6-8**. The lone valence electron can also be represented by a single dot.

Figure 6-8
A hydrogen atom and its single valence electron can be represented by the letter *H* and a dot.

$$\mathrm{H}\cdot$$

Using a pair of dots to represent the pair of valence electrons shared in a bond, a hydrogen molecule would be drawn as follows.

$$\mathrm{H}:\mathrm{H}$$

Note that the two dots are placed between the two hydrogen atoms to indicate that they are shared by both atoms. Hydrogen does not share more than two electrons.

Next, consider a chlorine atom, with the electron configuration $1s^22s^22p^63s^23p^5$, as shown in **Figure 6-9**. Chlorine has 17 electrons, but 2 electrons are in the first energy level, and 8 are in the second energy level. The 7 electrons in the third and outermost energy level are the valence electrons. Thus, a structure for chlorine can be drawn as follows.

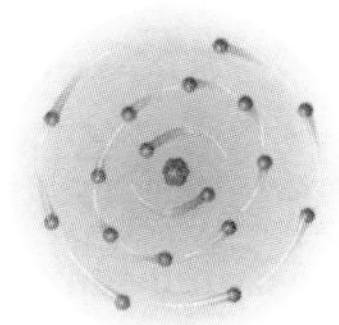

Figure 6-9
A chlorine atom has 17 electrons, 7 of which are valence electrons. The chlorine nucleus and the inner 10 electrons can be represented by the symbol Cl. The 7 valence electrons can be represented by 7 dots.

$$:\ddot{\underset{\cdot\cdot}{\mathrm{Cl}}}\cdot$$

If there are two chlorine atoms, each would need just one additional electron to satisfy the octet rule. Neither can take an electron from the other, so they share. These atoms can achieve stability by sharing a pair of electrons to form a covalent bond as shown.

$$:\ddot{\underset{\cdot\cdot}{\mathrm{Cl}}}:\ddot{\underset{\cdot\cdot}{\mathrm{Cl}}}:$$

Identify the two dots that represent the pair of electrons that are shared. Notice that each chlorine atom also has electrons that are not part of the bond; they are called **unshared pairs.** How many unshared pairs does each chlorine atom have?

unshared pair *pair of electrons that is not involved in covalent bonding, but instead belongs exclusively to one atom*

The use of these pairs of dots to indicate shared and unshared pairs makes it easier to check whether an atom has a stable octet. If the symbol is surrounded by an unshared or shared pair on the left, top, right, and bottom, the atom represented has a stable octet.

When drawing electron-dot diagrams, the pair of dots representing a shared pair or covalent bond can also be shown by a long dash, representing a **single bond**.

single bond
sharing of one pair of electrons between two atoms

$$H:H \qquad H—H$$

$$:\ddot{\underset{\cdot\cdot}{Cl}}:\ddot{\underset{\cdot\cdot}{Cl}}: \qquad :\ddot{\underset{\cdot\cdot}{Cl}}—\ddot{\underset{\cdot\cdot}{Cl}}:$$

Lewis structure
diagram showing the arrangement of valence electrons among the atoms in a molecule

A **Lewis structure** represents a chemical formula: the nuclei and inner-shell electrons are represented by the element's atomic symbol, and covalent bonds are represented by pair of dots or by dashes. Unshared pairs are represented by pairs of dots adjacent to only one atomic symbol. Lewis structures, which can be drawn for either ionic or molecular compounds, can help you understand how atoms engage in bonds.

How To
Draw Lewis structures

1. Determine the total number of valence electrons in the compound.

This involves nothing more than adding the valence electrons of each atom. Consider the example of HCl. By now, you should realize that hydrogen has one valence electron, while chlorine has seven.

Total number of valence electrons: $1 + 7 = 8$

2. Arrange the atoms' symbols to show how they are bonded and show valence electrons as dots.

Be sure to distribute the paired dots so that the octet rule is followed, except in the case of hydrogen.

$$:\ddot{\underset{\cdot\cdot}{Cl}}:H$$

3. Compare the number of valence electrons used in the structure to the number available from step 1.

There are two electrons in the covalent bond and six around the chlorine atom that are not shared.

Total number of electrons shown: $2 + 6 = 8$

4. Change to a single dash each pair of dots that represents two shared electrons.

$$:\ddot{\underset{\cdot\cdot}{Cl}}—H$$

5. Be sure that all atoms, with the exception of hydrogen, follow the octet rule.

The hydrogen atom has two valence electrons, and the chlorine atom is surrounded by eight valence electrons. The octet rule is satisfied.

Historical Note

Gilbert N. Lewis (1875–1946) and Irving Langmuir (1881–1957) independently developed the idea of two chlorine atoms sharing an electron to stabilize each other's orbits.

Teaching Tip

Shared electrons

Electrons in covalent bonds are covered by the same vail. That is, nuclei of both atoms contributing to the bond attract the shared electrons unequally. By sharing an electron, the atom has given up sole ownership of that electron.

Additional Examples for Sample Problem 6A

Determine the Lewis structure for:

a. F_2
b. H_2O
c. SCl_2
d. AsF_3
e. SiH_4

Solutions are at the bottom of the page.

Lewis structures can involve many atoms

When drawing a Lewis structure for a molecule with more than two atoms, you must first decide how to arrange the atoms. Keep in mind the following guidelines.

- Hydrogen and halogen atoms usually bond to only one other atom in a molecule and are usually on the outside or end of a molecule.
- The atom with the smallest electronegativity is often the central atom.
- When a molecule contains more atoms of one element than the others, these atoms often surround the central atom.

Sample Problem 6A Lewis structures

Draw the Lewis structure for iodomethane, CH_3I.

1 Calculate the total number of valence electrons

1 C atom with 4 electrons = 1 × 4 = 4
3 H atoms with 1 electron = 3 × 1 = 3
1 I atom with 7 electrons = 1 × 7 = 7
14 valence electrons

2 Arrange the atoms

- **Follow the guidelines to determine the central atom.** It is most likely that carbon is in the middle of the structure.

```
  ..
 :I:
  ..
H:C:H
  ..
  H
```

3 Compare the number of electrons used with the number of valence electrons available

- **All of the 14 available valence electrons were used.**

7 × 2 = 14 electrons

4 Change dots to dashes where appropriate

```
   ..
  :I:
   |
 H—C—H
   |
   H
```

5 Check to see that the octet rule has been followed

Practice 6A

Answers for Concept Review items and Practice problems begin on page 841.

1. **Draw the Lewis structures for iodine monochloride, ICl, and hydrogen bromide, HBr.**
2. **Try drawing the Lewis structures for dichloromethane, CH_2Cl_2, and methanol, CH_3OH.**
3. **Draw a Lewis structure for cyclohexane, C_6H_{12}. (Hint: the carbon atoms form a six-membered ring.)**

Two atoms can share more than one electron pair

Because carbon has four valence electrons in its outermost energy level, it can form bonds with up to four atoms to satisfy the octet rule. In the molecule of ethane, C_2H_6, shown in **Figure 6-10**, each carbon atom has bonded with another carbon atom and three hydrogen atoms. Each hydrogen atom is bonded to a carbon atom and shares one pair of electrons with the carbon atom.

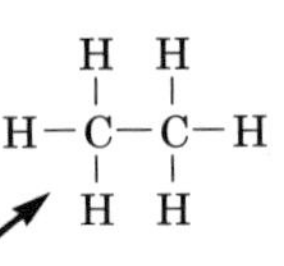

Figure 6-10
Astronomers believe that liquid ethane, C_2H_6, covers the surface of Titan, Saturn's largest moon.

Answers to Practice 6A

1. :I:Cl: H:Br: (with lone pairs shown as dots)

2. H:C:Cl: (with H above and :Cl: below C)   H:C:O:H (with H above and H below C)

3.
```
        H  H
         \/
  H      C      H
   \   /   \   /
H—C         C—H
    |         |
H—C         C—H
   /   \   /   \
  H      C      H
         /\
        H  H
```

Solutions for Additional Examples 6A

a. :F—F: (each F with three lone pairs) **b.** H—O—H (bent, O with two lone pairs) **c.** :Cl—S—Cl: (bent, S with two lone pairs, each Cl with three lone pairs) **d.** :F—As—F: with :F: below As (As with one lone pair) **e.** H—Si—H with H above and H below Si

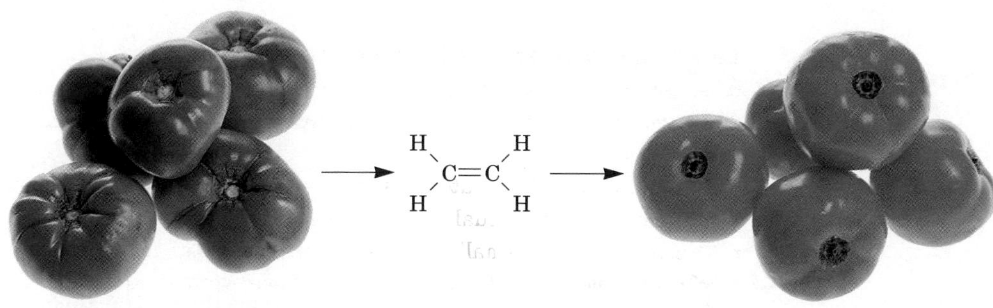

Figure 6-11a
Ethene, C_2H_4, also called ethylene, is a hormone found in most plants. Tomatoes release ethylene as they ripen.

6-11b
In turn, this ethylene causes other tomatoes to ripen more quickly. The produce industry uses ethylene as a ripening agent because often fruit must be picked before it is ripe for shipping purposes.

double bond
covalent bond formed by the sharing of two pairs of electrons between two atoms

triple bond
covalent bond formed by the sharing of three pairs of electrons between two atoms

In addition to carbon, other elements, including, nitrogen, oxygen, and occasionally sulfur, can share more than one pair of electrons with other atoms to satisfy the octet rule. For example, if two carbon atoms share two pairs of electrons (four electrons total) instead of just one pair, a double covalent bond or **double bond** can form. The molecule of ethene shown in **Figure 6-11** consists of two carbon atoms that have formed a double bond and four hydrogen atoms.

Two carbon atoms are also able to share three pairs of electrons. The molecule with two carbon atoms that have formed a **triple bond** and two hydrogen atoms is called ethyne, shown in **Figure 6-12**. Although the names for ethane, ethene, and ethyne may not seem very different, the substances have very different Lewis structures, characteristics, and uses. Nitrogen also can form triple bonds, as in nitrogen gas, N_2, and hydrogen cyanide, HCN.

If no arrangement of single bonds provides an appropriate Lewis structure, it could be that the molecule contains multiple bonds. Sample Problem 6B shows how to deal with such molecules.

Figure 6-12a
Spelunkers exploring caves often use carbide lamps. In these lamps, calcium carbide, CaC_2, reacts with water to form ethyne, C_2H_2.

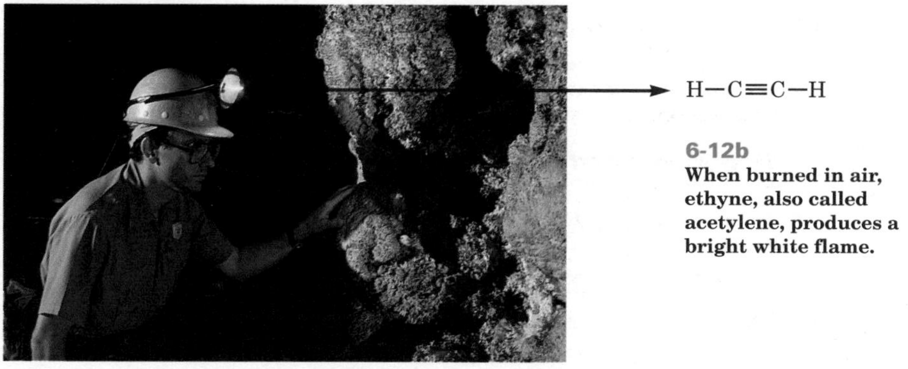

6-12b
When burned in air, ethyne, also called acetylene, produces a bright white flame.

Further Applications

Calcium carbide, CaC_2, is formed when coal is heated in the presence of lime, CaO, at elevated temperatures. Dripping water onto CaC_2 forms acetylene, C_2H_2, and $Ca(OH)_2$. A screw valve regulates the drip rate of the water and, therefore, the amount of acetylene produced.

$$CaC_2 + H_2O \longrightarrow Ca(OH)_2 + C_2H_2$$

Historical Note

Chloroprene, invented by Julius A. Nieuwland in the late 1920s, was the first synthetic rubber. Du Pont developed the rubber under the name neoprene.

Do You Know?

Derivatives of acetylene include acetaldehyde, acetic acid, acetone, water-base paints, vinyl fabric and floor coverings, drycleaning solvents, and aerosol insecticide sprays.

Further Applications

An oxygen-acetylene torch is useful for welding and cutting metal because acetylene burns with an exceptionally hot flame (3300°C or 6000°F).

Visual Strategy

Figure 6-10, Figure 6-11, and **Figure 6-12**

Removing two H^+ ions from the ethane molecule forces the carbon atoms to form a second bond between them; the result is ethene. Removing another two H^+ ions forms the triple bond of ethyne. The remaining H atoms still try to stay as far away from one another as possible. Formation of multiple bonds increases the rigidity of the molecule. Can a fourth bond form between the carbons? (no)

Teaching Tip
Multiple bonds in Lewis structures
The presence of multiple bonds can be quickly identified by drawing Lewis dot structures in the following manner.

1. Give every atom (except H) an octet.
2. Count the total number of electrons, T.
3. Add up the number of valence electrons, V.
4. Subtract V from T.
5. If your answer is 0, no multiple bonds are present.
6. For each 2-electron difference, one pi bond is formed.

Additional Examples for Sample Problem 6B

Draw Lewis structures for:

a. C_3H_6O
b. CO
c. O_2
d. N_2F_2
e. $B_3N_3H_6$

Solutions are at the bottom of the page.

Sample Problem 6B Lewis structures with multiple bonds

Draw the Lewis structure for formaldehyde, CH_2O.

1 *Calculate the total number of valence electrons*

2 H atoms with 1 electron = 2
1 C atom with 4 electrons = 4
1 O atom with 6 electrons = 6
12 valence electrons

2 *Arrange the atoms*

- **Follow the guidelines given earlier to determine the central atom.**

It is most likely that carbon is in the middle of the structure. Place the other atoms around carbon. It seems like it would be easy to add shared and unshared pairs as shown, but it is important to check this approach.

H:C̤̈:Ö̤:
H

3 *Compare the number of electrons used to the number of valence electrons available*

- **electrons used:** 14
- **valence electrons available:** 12

Obviously, something is wrong with this Lewis structure. Two electrons must be eliminated without violating the octet rule. *Whenever too many electrons have been used, try connecting atoms with a double bond or other multiple bond.*

In this case, by removing two unshared electrons from both the carbon and oxygen atoms and by forming a double covalent bond between carbon and oxygen, only 12 electrons are used. Now the number used equals the number available.

H:C̤::O̤:
H

- **electrons used:** 12
- **valence electrons available:** 12

4 *Change dots to dashes where appropriate*

- **Carbon forms two single bonds and one double bond, for a total of four covalent bonds.**

H—C=O̤:
|
H

5 *Check to see that the octet rule is not violated*

Remember that the double bond represents four electrons shared by carbon and oxygen. As a result, the octet rule has been followed.

Practice 6B

1. **Draw the Lewis structures for the following molecules.**
   a. **carbon dioxide, CO_2** b. **tetrachloroethene, C_2Cl_4**
   c. **nitrogen, N_2** d. **hydrogen cyanide, HCN**
2. **Benzene, C_6H_6, contains a ring of six carbon atoms. Draw one possible Lewis structure for benzene.**
3. **Benzoic acid, C_6H_5COOH, contains a benzene ring, but one of the hydrogen atoms has been replaced with a —COOH group. Draw a Lewis structure for benzoic acid.**

Answers to Practice 6B

1. a. :Ö=C=Ö:
   b. :C̤̈l and :C̤̈l on one C, C=C, :C̤̈l: and :C̤̈l: on the other C
   c. :N≡N:
   d. H:C≡N:
2. Ring of six C atoms with alternating double bonds, each C bonded to one H
3. Benzene ring (alternating double bonds) with H on five carbons and —C(=Ö)—Ö—H on the sixth

Solutions for Additional Examples 6B

a. 26 e^- are needed to make all single bonds, but only 24 valence e^- are available, so one double bond must be formed.

H—C(H)(H)—C(=Ö)—C(H)(H)—H H—C(H)(H)—C(H)(H)—C(=O)—H H₂C=C(H)—C(H)(H)—Ö—H H₂C=C(H)—C(H)(:O:—H)—H

Answers are continued on page 231A.

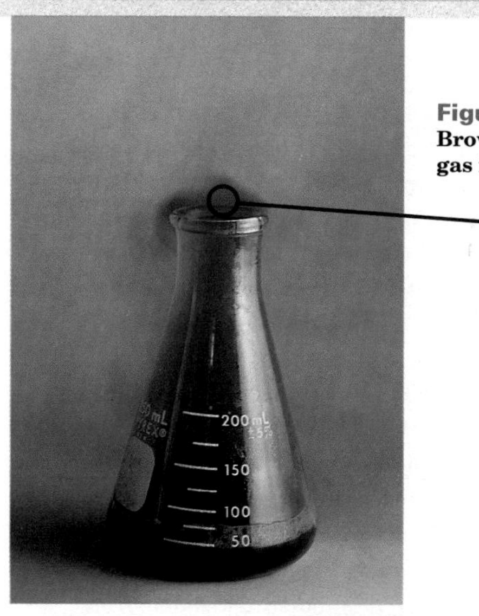

Figure 6-13a
Brown nitrogen dioxide gas is extremely poisonous.

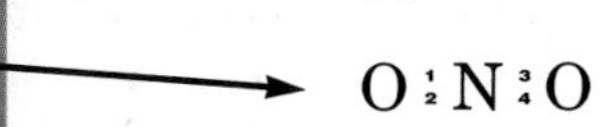

O:N:O

6-13b
Although it may initially seem easy to start drawing a Lewis structure for nitrogen dioxide, . . .

$\ddot{\underset{\cdot\cdot}{O}}=\underset{\cdot}{N}-\ddot{\underset{\cdot\cdot}{O}}: \longleftrightarrow :\ddot{\underset{\cdot\cdot}{O}}-\underset{\cdot}{N}=\ddot{\underset{\cdot\cdot}{O}}$

6-13c
. . . no structure obeys the octet rule. Nitrogen dioxide is considered a resonance hybrid of the structures shown.

Sometimes no single Lewis structure is correct

At some point you may come across a molecule like NO_2, as shown in **Figure 6-13a**, which has 17 valence electrons. As shown in **Figure 6-13b**, you can arrange the 3 atoms and 4 of the electrons with no trouble. However, no matter how you try, the 13 remaining electrons cannot be placed so that each atom has 8 valence electrons, even if you use double bonds. Instead, you get two equivalent structures.

In each of the structures shown in **Figure 6-13c**, a single electron is left unpaired on the nitrogen atom, which does not have a stable octet. Since there were an odd number of electrons to begin with, this situation is unavoidable. From Chapter 3 you know that, ordinarily, atoms have paired electrons. Atoms with unpaired electrons are somewhat unstable, although they can exist. Similarly, situations without an octet are unstable, but they too are possible. *Before such a rule-breaking structure is considered accurate, either an octet must be impossible, or experimental measurements must indicate that a non-octet structure is more likely.*

Chem Fact
Nitrogen dioxide, NO_2, is a compound that gives smog its brownish color. The presence of an unpaired electron is one factor contributing to its reactivity.

Note also that the double bond is on the left in one of the structures in **Figure 6-13c** and on the right in the other. The fact that the two structures are equally likely indicates that neither structure alone is correct. This is not uncommon. Even some other molecules that obey the octet rule can have several possible Lewis structures, each of which is equally likely.

Such molecules are evidence that Lewis structures are merely models that cannot accurately portray all cases. When more than one Lewis structure for a molecule can be drawn, the molecule is said to be a *resonance hybrid.* Originally, chemists believed that the molecule resonated between the different structures, like a plucked guitar string vibrating back and forth. Now, however, chemists treat the molecule as if it had a structure that was the *average* of these structures. Measurements show that in NO_2, the N—O bonds are identical, with bond lengths between those of single and double bonds. The fact that the NO_2 molecule has an odd electron is evidence of an important rule: molecules that exhibit resonance are more stable than one would predict from each of the contributing Lewis structures.

Teaching Tip
Resonance structures
Resonance is found in structures that require a double bond to the central atom but the placing of that double bond involves a choice between two atoms of the same element.

Teaching Tip
Determining bond order
The bond order (BO) in a covalent molecule can be found by adding up the number of bonds to the central atom and dividing the result by the number of atoms bonded to the central atom. For example, ozone has 3 bonds and 2 bonding atoms; the BO is 1.5. CO_2 has 4 bonds and 2 bonding atoms; the BO is 2. N_2 has 3 bonds and 1 bonding atom; the BO is 3. CH_4 has 4 bonds and 4 bonding atoms; the BO is 1.

Visual Strategy
Figure 6-13
To emphasize that the real structure is somewhere between the two extremes shown, calculate the bond order by dividing the number of bonds to N, 3, by the number of bonding atoms, 2. The result, 1.5, is neither 1 nor 2 as shown in the resonance drawing. The presence of the odd electron on N makes this compound very reactive and classifies it as a free radical.

Teaching Tip

Additional Examples

Determine the Lewis structures for the following polyatomic ions.

a. ClO_4^-
b. SO_3^{2-}
c. NO_3^-
d. N_3^-
e. $C_2O_4^{2-}$

*Full solutions for items **a–e** are found on page 231A*

Answers to

Concept Review

7. 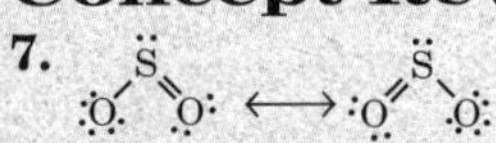

8. 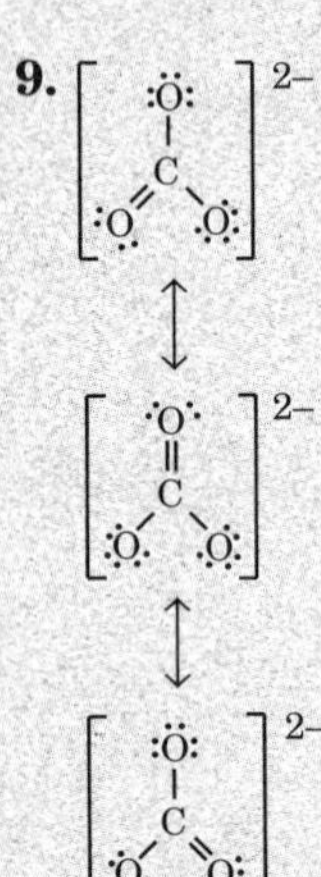

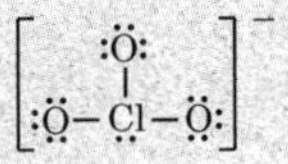

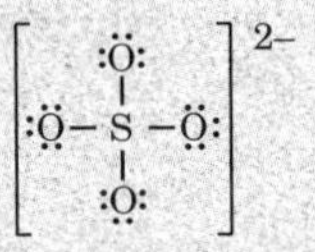

9.

Figure 6-14a
An ammonium cation is formed when a hydrogen ion is combined with an ammonia molecule. Smelling salts often contain an ionic compound composed of an ammonium cation and a carbonate anion.

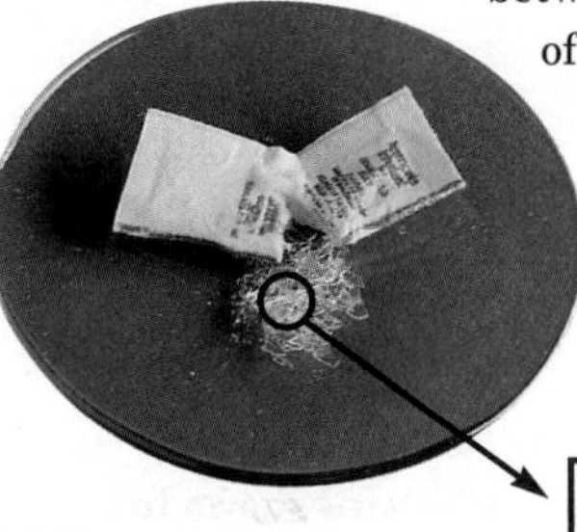

6-14b
The brackets around the Lewis structure of a polyatomic ion, such as ammonium, indicate that the ion has a charge, in this case a positive one.

Lewis structures can be drawn for polyatomic ions

The polyatomic ions you studied in Chapter 5 are held together by covalent bonds just like those in a molecular compound. The only difference in drawing a Lewis structure for a polyatomic ion is that you must alter the number of electrons to reflect the total charge of the polyatomic ion.

The ammonium cation, which is contained in the smelling salts shown in **Figure 6-14**, is prepared from an ammonia molecule and a hydrogen ion, which is only a proton. In this case, a covalent bond is formed between the hydrogen ion and the unshared pair on the nitrogen atom of the ammonia molecule.

Unlike other bonds, in which each atom provides one electron to a pair that is later shared, the nitrogen atom provides both of the electrons that form one of the four covalent bonds in NH_4^+. This special type of bond is called a *coordinate covalent bond*. Once the coordinate covalent bond is formed, it is indistinguishable from the other covalent bonds in the ammonium cation even though the nitrogen atom contributed both of the shared electrons.

In Lewis structures for polyatomic ions, brackets are put around the polyatomic ion to indicate that the charge is for the polyatomic ion as a whole, not for a specific atom.

If you check the periodic table, you'll see that nitrogen normally has five valence electrons, while each hydrogen usually has only one valence electron to contribute in forming a bond. This equals nine electrons. But since the charge on the polyatomic cation is 1+, it must be missing one of these nine electrons.

For *cations,* the positive charge is due to a smaller number of electrons, so you must *subtract* the charge from the count of electrons. For *anions,* the negative charge is due to additional electrons, so you must *add* the charge to the number of electrons.

Chem Fact
In 1993, more than 16 billion kg of ammonia were produced in the United States, mostly for use in fertilizers. Ammonia ranks sixth among the most-produced chemicals.

1 N atom with 5 valence electrons:	$1 \times 5 =$	5 electrons
4 H atoms with 1 valence electron each:	$4 \times 1 =$	+ 4 electrons
total electrons (if a neutral substance):		9 electrons
charge of 1+ means 1 less electron:		−1 electron
total electrons for the polyatomic cation:		8 electrons

The calculated number, eight electrons, matches the eight electrons shown in the Lewis structure for the ammonium ion in **Figure 6-14b**.

Concept Review

Resonance and polyatomic ions

7. Try drawing two resonance hybrid structures for SO_2.
8. Draw the Lewis structures for ClO_3^- and SO_4^{2-}.
9. Draw two of the resonance structures for CO_3^{2-}.

Answers for Concept Review items and Practice problems begin on page 841.

Visual Strategy

Figure 6-14

Dissect the bonding in this structure to help students see the source for the charge on this ion: 3 covalent bonds between N and 3 H atoms, and 1 coordinate covalent bond between N and H^+ ion. One of the hydrogens brought only a proton with it and not an electron, making the overall charge positive due to the shortage of one electron.

Naming molecular compounds

Although there are several steps to writing the Lewis structures for molecular compounds, naming a compound is relatively straightforward, especially for compounds made of only two elements. Such compounds can be named in one of two ways, both of which are similar to the methods used to name the salts that were described in Chapter 5, as shown in **Figure 6-15**.

Figure 6-15
Chemists had to invent a system for naming compounds that symbolically portrayed both the similarities and the differences in compounds such as these.

P_2O_3
diphosphorus **tri**oxide
phosphorus(III) oxide

P_2O_5
diphosphorus **pent**oxide
phosphorus(V) oxide

Table 6-6
Prefixes for Naming

Prefix	Number of atoms
mono-	1
di-	2
tri-	3
tetra-	4
penta-	5
hexa-	6
hepta-	7
octa-	8
nona-	9

One naming system uses prefixes, roots, and suffixes

Chemists often name molecular compounds using the system of prefixes shown in **Table 6-6**. Prefixes and suffixes are usually attached to root words of the elements in a compound. For example, the two oxides of carbon, CO and CO_2, are named carbon monoxide and carbon dioxide, respectively. The first element named is usually the one with the lowest electronegativity value. The first word in the compound name is the first element's name. If the molecule contains only one atom of the first element given in the formula, the prefix *mono-* is omitted in naming that element. For example, no prefix is used with carbon in either compound. The root *oxy-* is given the *-ide* ending, just as for anions in an ionic compound. A prefix is used to tell you how many oxygen atoms each compound has.

The Stock system of naming uses oxidation numbers

Another way to name molecular compounds resembles the Stock system. Using the method from Chapter 5, Cu^+ ions form copper(I) chloride, CuCl, while Cu^{2+} ions form copper(II) chloride, $CuCl_2$.

Covalently bonded atoms *do not* give and take electrons but rather *share* them, so these atoms do not obtain full positive or negative charges like ions do. However, because the sharing is often unequal, the atom with the greatest attraction for electrons will pull the shared electrons toward itself, giving it a fractional negative charge. The atom that has less of an attraction for the electrons will have a slight positive charge. However, now we act as if the entire charge has been transferred. This charge is known as the **oxidation number** of the atom and can be used in naming the molecular compound.

oxidation number
apparent charge assigned to an atom based on the assumption of complete transfer of electrons

Oxidation numbers can also be used as tools for analyzing reduction-oxidation reactions, a special type of chemical reaction that will be studied further in Chapter 15. For now, learning a little about how to assign oxidation numbers will help you name and identify molecular compounds.

Teaching Tip
Naming molecular compounds
Practice naming molecular compounds with the following additional examples: CCl_4, SO_2, SO_3, ClF_3, PCl_3, AsF_5, and SiO_2.
Answers: carbon tetrachloride, sulfur dioxide, sulfur trioxide, chlorine trifluoride, phosphorus trichloride, arsenic pentafluoride, and silicon dioxide

Theme
Macro observations and micromodeling
Chemists adopt naming systems that attempt to describe the actual relationships between atoms.

Chapter Connections

Rules for assigning oxidation numbers (oxidation states) are on pages 588–589 of Chapter 15.

Teaching Tip

Determining oxidation numbers

Oxidation numbers can be tracked with a tally row above the individual elements.

Determine the oxidation states for the elements in the following compounds: H_2SO_3, $Al_2(SO_3)_3$, and N_2O_5.

H = +1, S = +4, O = –2;
Al = +3, S = +4, O = –2; and
N = +5, O = –2

ASSESS

Alternative Assessment

1. These five substances exist: ClF_3, SF_6, XeF_2, PCl_2F_3, SiF_6^{2-}. These five substances do not exist: CF_6^{2-}, NCl_2F_3, NeF_2, OF_6, FBr_3. From the Lewis dot structure for each compound and the electron configuration for each central atom, propose a reason why some elements can violate the octet rule and others cannot. (Below period 2, elements have empty *d* orbitals available for bonding.)
2. Form groups of 5–7 students each. Provide each student with four 3 in paper circles. Each circle represents an electron, and each person represents an atom. Have each group use their appendages and paper electrons to act out the bonding in CH_4, CO_2, C_2H_2, C_2H_4.

Answers to Section Review

10. a. 2
 b. 6
 c. 7

Answers are continued on page 231A.

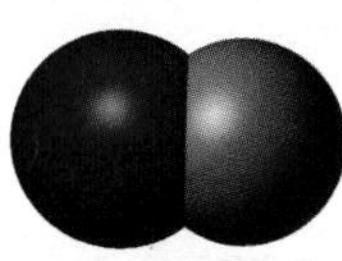

Figure 6-16a
To balance the –2 oxidation number of the oxygen atom in carbon monoxide, the carbon atom must have an oxidation number of +2.

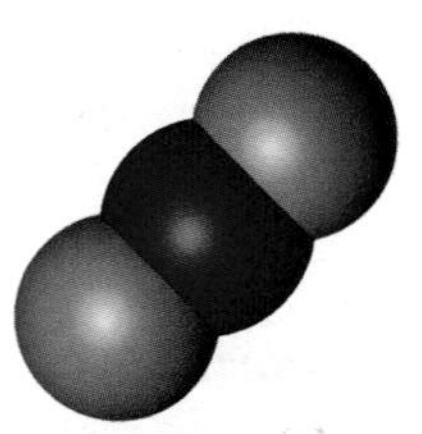

6-16b
In CO_2, the oxidation number of oxygen is also –2. Because there are two oxygen atoms in each molecule, this carbon atom must have an oxidation number of +4.

Determining oxidation numbers from a formula

When most atoms bond with oxygen, the oxygen atom attracts the shared pair of electrons more strongly than the other atoms do. These two electrons, along with their negative charges, spend most of the time near the oxygen atom. *As a result, oxygen is usually assigned an oxidation number of −2.* Similarly, hydrogen atoms do not have a high electronegativity, and the electron they donate in bonding usually spends most of the time around the other atom in the bond. *Thus, hydrogen is usually assigned an oxidation number of +1.* Other atoms tend to have oxidation numbers similar to the charges they have when in ionic compounds because that is often the number needed to complete an octet. The element with the higher electronegativity will have a negative oxidation number. For example, in NH_3, N is −3 and H is +1.

Before using Roman numerals to name a molecular compound, you need to determine the oxidation number for each atom. This can be tricky because some elements can have more than one oxidation number, depending on the other atoms in the compound. *The sum of the oxidation numbers in a neutral compound must equal zero. For a polyatomic ion, the sum must be equal to the charge on the ion.* Using this knowledge and the rules of thumb for the oxidation numbers of hydrogen, oxygen, and the halogens, you can predict the oxidation numbers of most other atoms in a compound or polyatomic ion.

As an example, consider CO and CO_2, the molecules shown in **Figure 6-16**. Using Roman numerals to represent the oxidation numbers, CO could be named carbon(II) oxide, and CO_2 could be named carbon(IV) oxide. But like many common covalent compounds, CO and CO_2 are usually named using the prefix method—carbon monoxide and carbon dioxide.

Section Review

10. Write the number of valence electrons for atoms with the following electron configurations.
 a. $1s^22s^2$ **b.** $1s^22s^22p^63s^23p^4$ **c.** $1s^22s^22p^63s^23p^5$
11. Draw Lewis structures for the following.
 a. bromine, Br_2 **b.** ozone, O_3
 c. sulfur trioxide, SO_3 **d.** hydronium ion, H_3O^+
12. Using the system of prefixes, name the following molecular compounds.
 a. SiO_2 **b.** SO_2 **c.** CF_4 **d.** N_2O_3
13. Give the oxidation numbers of each atom in the following compounds.
 a. PBr_5 **b.** PBr_3 **c.** H_2SO_4 **d.** SO_3 **e.** CCl_4
14. Write formulas for the following compounds.
 a. dinitrogen tetroxide **b.** phosphorus trichloride
 c. dinitrogen pentoxide

15. **Story Link**
 Try drawing the Lewis structure of the $C_{12}H_{26}$ molecule, which is like a fragment of polyethylene. The carbon atoms form the "backbone" of this compound. Visualize what 7000 carbon-hydrogen units might look like. This may help you understand why plastic does not break down very easily.

How are formulas represented?

Section Objectives

Given composition data, determine the empirical and molecular formulas of a molecular compound.

Given the molecular formula, determine the structural formula for a molecular compound.

Empirical formulas

How many different ways could you describe what happened at a basketball game? You could tell someone the final score, give a detailed scoring history for the game, or even videotape it. Any one of these ways would give someone an idea of what the game was like. Each way, though, would provide the person with different information.

How many different ways could you describe the glucose in the sports drink that the basketball player in **Figure 6-17** is drinking? You may not think there are as many ways to describe a simple molecular compound as there are to describe a basketball game, but chemists do have a number of ways to represent a molecule. Moreover, each way provides different information about the molecule. You have seen two possible ways to represent a molecule—Lewis structures and chemical formulas.

Figure 6-17a **The formula for the glucose molecule found in this sports drink . . .**

$C_6H_{12}O_6$

$(CH_2O)_6$

6-17b **. . . can be shown a number of different ways.**

In fact, chemists can write the formula for the same molecular compound in various ways. The simplest formula, or *empirical formula,* consists of the symbols for the elements combined with subscripts showing the smallest whole-number ratio of the atoms. For example, CH_2O is the empirical formula for glucose, $C_6H_{12}O_6$. Formulas for ionic compounds are really empirical formulas showing the smallest whole-number ratio of ions.

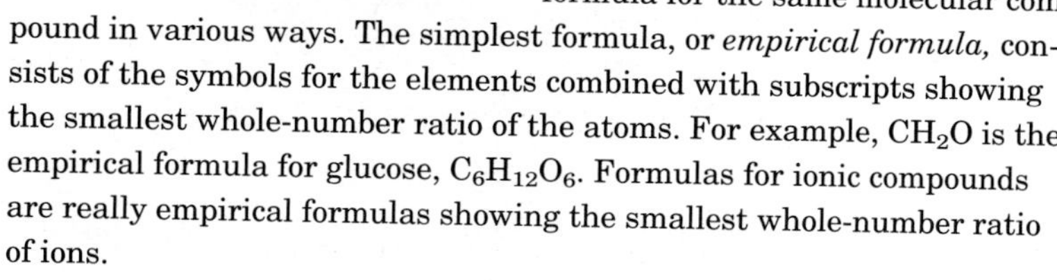

Empirical formulas are determined experimentally

Chemists sometimes have to determine the empirical formula for a compound recently discovered in nature. Similarly, after chemists synthesize a new compound, they often analyze it to obtain its empirical formula and check their work. They do this by analyzing composition data to determine what elements are present and their percentage by mass, just as was discussed in Chapter 5. By translating these mass percentages into mole ratios, chemists can determine the empirical formula for the compound, as shown in Sample Problem 6C on the next page.

Section 6-3

FOCUS

Lesson Starter

Begin this lesson by writing the following fictitious recipe for vegetarian stew on the board.

84 potatoes, 63 carrots, 15 onions, 3 heads of garlic, 27 turnips, 42 pieces of celery, 9 cans of green beans, 6 cans of tomato puree, 6 diced peppers, 9 cans of corn, and 6 cans of lima beans. Serves 81.

Ask students to determine the simplest whole-number ratio for the ingredients, the number of people this simplest recipe serves, and the ingredient ratio that would serve 324 people.

TEACH

Teaching Tip

Rules for determining empirical formulas from percentage composition

Many students like to have specific rules when solving applied math problems. The following rules are for empirical formula determination.

1. Assume a 100 g sample, so that "%" becomes "g".
2. Divide g by atomic mass.
3. Divide each result (mole) by the smallest result present (mole ratio).
4. Look for whole-number ratios.

Additional Examples for Sample Problem 6C

a. Determine the empirical formula for a compound that is 56.6% K, 8.7% C, and 34.7% O.

b. An oxide is 69.9% Fe and 30.1% O. What is its empirical formula?

Solutions are at the bottom of the page.

Answers to Practice 6C

1. $K_2Cr_2O_7$; moles 0.6793 K, 0.6810 Cr, 2.377 O; mole ratio 1:1:3.5; whole-number ratio 2:2:7
2. P_2O_5; moles 0.1431 P, 0.3573 O; mole ratio 1:2.5; whole-number ratio 2:5
3. Na_2SO_4; moles 1.409 Na, 0.7063 S, 2.812 O; mole ratio 2:1:4

Sample Problem 6C Calculating empirical formulas from percentage composition

Chemical analysis of 2-propanol, also known as isopropanol or rubbing alcohol, indicates that it is 60.0% C, 13.4% H, and 26.6% O. What is its empirical formula?

1 List what you know

- **60.0% C means:** 60.0 g C in every 100. g of compound
- **13.4% H means:** 13.4 g H in every 100. g of compound
- **26.6% O means:** 26.6 g O in every 100. g of compound
- **molar mass for C:** 12.00 g/1 mol C
- **molar mass for H:** 1.01 g/1 mol H
- **molar mass for O:** 16.00 g/1 mol O

2 Set up conversion from mass to moles

- **To convert mass to moles, multiply by the reciprocal of the molar mass to cancel *g*, leaving *mol* in the numerator. Remember to follow significant figure rules.**

$$60.0\ \text{g C} \times \frac{1\ \text{mol C}}{12.00\ \text{g C}} = 5.00\ \text{mol C}$$

$$13.4\ \text{g H} \times \frac{1\ \text{mol H}}{1.01\ \text{g H}} = 13.26732673\ \text{mol H} = 13.3\ \text{mol H}$$

$$26.6\ \text{g O} \times \frac{1\ \text{mol O}}{16.00\ \text{g O}} = 1.6625\ \text{mol O} = 1.66\ \text{mol O}$$

3 Compare amounts of each element in moles

At this point, the formula could be written as $C_5H_{13.3}O_{1.66}$, but empirical formulas have whole numbers for subscripts.

- **The empirical formula is the simplest whole-number mole ratio of the atoms. Divide the moles of each element by the moles of the element with the smallest amount to find a whole-number ratio.**

1.66 is the smallest value.

$$\frac{1.66\ \text{mol O}}{1.66} = 1.00\ \text{mol of O}$$

$$\frac{5.00\ \text{mol C}}{1.66} = 3.012048193\ \text{mol C} = 3.01\ \text{mol of C.}$$

$$\frac{13.3\ \text{mol H}}{1.66} = 8.012048193\ \text{mol H} = 8.01\ \text{mol of H.}$$

For every 1.00 mol of O, there are 3.01 mol of C and 8.01 mol of H.

The formula contains atoms in this ratio: 1 O : 3 C : 8 H.

Often, the numbers do not always divide out perfectly because of slight errors of measurement or rounding, but they are usually close enough for a reasonable guess.

- **The empirical formula for isopropanol must be C_3H_8O.**

4 Verify your work

- **Use the empirical formula and molar masses to calculate the percentage composition of a single mole of isopropanol. Does it match the values from the problem?**

Practice 6C

1. **What is the empirical formula of a compound that contains 26.56% potassium, 35.41% chromium, and 38.03% oxygen?**
2. **Determine the empirical formula for a compound that contains 5.717 g of O and 4.433 g of P.**
3. **What would be the empirical formula for a compound that contains 32.38% sodium, 22.65% sulfur, and 44.99% oxygen?**

Solutions for Additional Examples 6C

a. K_2CO_3

K: $\frac{56.6\ \text{g}}{39.10\ \text{(g/mol)}} = 1.45\ \text{mol}$; $\frac{1.45\ \text{mol}}{0.724\ \text{mol}} = 2.00$

C: $\frac{8.7\ \text{g}}{12.01\ \text{(g/mol)}} = 0.724\ \text{mol}$; $\frac{0.724\ \text{mol}}{0.724\ \text{mol}} = 1.00$

O: $\frac{34.7\ \text{g}}{16.00\ \text{(g/mol)}} = 2.17\ \text{mol}$; $\frac{2.17\ \text{mol}}{0.724\ \text{mol}} = 3.00$

b. Fe_2O_3

Fe: $\frac{69.9\ \text{g}}{55.85\ \text{(g/mol)}} = 1.25\ \text{mol}$; $\frac{1.25\ \text{mol}}{1.25\ \text{mol}} = 1.00$

O: $\frac{30.1\ \text{g}}{16.00\ \text{(g/mol)}} = 1.88\ \text{mol}$; $\frac{1.88\ \text{mol}}{1.25\ \text{mol}} = 1.50.$

This is not a whole number ratio, but if doubled, the ratio will be 2:3.

Molecular formulas

molecular formula
gives type and actual number of atoms in a chemical compound

Empirical formulas can be compared to the score of a basketball game. They just give you an idea of what is in the compound by indicating the simplest whole-number ratio of atoms. But they do not tell you exactly how many atoms of each element are present in a molecule of the compound. For that information, you would need a **molecular formula.** A molecular formula can be compared to a detailed scoring history of the game. Just as a play-by-play will give you a better idea of which players scored the most baskets, the molecular formula will give you a better idea of the molecule's properties based on exactly how many of each atom there will be in a molecule.

With some compounds, the empirical formula is the same as the molecular formula. Such is the case with water, H_2O. But for most compounds, the two formulas are not the same. Consider the three molecular compounds in **Table 6-7**. The first compound is toxic and causes cancer. The second gives vinegar its sour taste. The third is a type of sugar found in sports drinks and other foods. All three compounds have the same empirical formula, CH_2O, but because each has a different molecular formula, they each have very different properties.

Chem Fact
Formaldehyde, CH_2O, is used as a disinfectant, for hardening photographic gels and plates, to improve dyeing on fabrics, and to make polymers.

Molecular formulas are determined from molar mass

Notice that in **Table 6-7** the molecular formulas are multiples of the empirical formulas.

$$x\ \textbf{(empirical formula)} = \textbf{molecular formula}$$

Note that x can be any whole number. For formaldehyde, x is 1; for acetic acid, x is 2; and for glucose, x is 6. A similar relationship exists between the molar masses of the empirical formula and that of the compound. If you know both the empirical formula mass and the molecular formula mass, you can determine the number by which you must multiply the empirical formula to get the molecular formula for a compound.

$$\frac{\textbf{molecular formula mass}}{\textbf{empirical formula mass}} = x$$

Table 6-7
Comparing Empirical and Molecular Formulas

Compound	Empirical formula	Molecular formula	Molar mass (g)	Representation
formaldehyde	CH_2O	CH_2O (1 times empirical formula)	30.03	
acetic acid	CH_2O	$C_2H_4O_2$ (2 times empirical formula)	60.06	
glucose	CH_2O	$C_6H_{12}O_6$ (6 times empirical formula)	180.18	

Additional Examples for Sample Problem 6C

a. The empirical formula for mercury(I) chloride is found to be HgCl. The molar mass is found to be 472.08 g/mol. What is the molecular formula?

b. The empirical formula for trichloroisocyanuric acid, the active ingredient in dry household bleaches, is OCNCl. The molar mass is 232.42 g/mol. What is the molecular formula?

Solutions are at the bottom of the page.

Answers to Practice 6D

1. $\frac{78.11 \text{ g/mol}}{13.02 \text{ g/mol}} = 5.999$

$6(C_1H_1) = C_6H_6$

2. Moles from percentages: 6.373 mol C 12.01 mol H, 0.7081 mol O

Empirical formula: $C_9H_{17}O_1$

$\frac{282.45 \text{ g/mol}}{141.26 \text{ g/mol}} = 1.9995$

$2(C_9H_{17}O_1) = C_{18}H_{34}O_2$

Sample Problem 6D Determining molecular formula

The empirical formula of a compound containing phosphorus and oxygen was found to be P_2O_5. Experiments show that the molar mass of the compound is 283.89 g/mol. What is the molecular formula of the compound?

1 List what you know

- **empirical formula:** P_2O_5
- **molar mass of unknown compound** = 283.89 g/mol
- **1 mol P** = 30.97 g P
- **1 mol O** = 16.00 g O
- **Determine the formula mass for the empirical formula, P_2O_5. Be sure to follow significant figure rules.**

2 mol P = 2 × 30.97 g P	=	61.94 g
5 mol O = 5 × 16.00 g O	=	80.00 g
empirical formula mass	=	141.94 g/mol

2 Set up the problem

- **Use the equation from page 209 to solve for x, the factor relating the empirical and molecular formulas.**

$$x = \frac{\text{molecular formula mass}}{\text{empirical formula mass}} = \frac{283.89 \text{ g/mol}}{141.94 \text{ g/mol}} = 2.0001 \approx 2$$

- **Once x has been calculated, multiply the empirical formula by it to get the molecular formula.**

$x(\text{empirical formula}) = 2(P_2O_5) = P_4O_{10}$

3 Verify your work

- **Calculate the formula mass for the molecular formula, and compare it to the experimental formula mass given in the problem.**

Practice 6D

1. **Determine the molecular formula of a compound having an empirical formula of CH and a molar mass of 78.11 g/mol.**
2. **A compound has the following composition: 76.54% C, 12.13% H, 11.33% O. If its molar mass is 282.45 g/mol, what is its molecular formula?**

polymer
large molecule made of many repeated small subunits, each of which is a small molecule or group of atoms

Polymers are large molecules made of repeating units

To understand the structure of the polyethylene in trash bags, take a closer look at the name of the material. The name *polyethylene* can be broken down into two parts: *poly*, which means "many," and *ethylene,* which is the name for a molecule with the molecular formula C_2H_4. Thus, polyethylene means "many C_2H_4," an accurate description of the molecule is shown in **Figure 6-18**. Polyethylene is a typical **polymer,** a substance composed of many repeating groups of atoms. The units that link to form a polymer are called *monomers*. For example, C_2H_4 is the monomer for polyethylene.

Many of the molecules that make life possible, such as DNA, proteins, and starch, are also polymers made of different types of monomers. Polymers with many useful properties can also be made in a lab or factory, using a variety of different monomers. The one thing all these natural and artificial polymers have in common is that they are all held together with covalent bonds. These bonds allow the polymer molecules to resist being broken down.

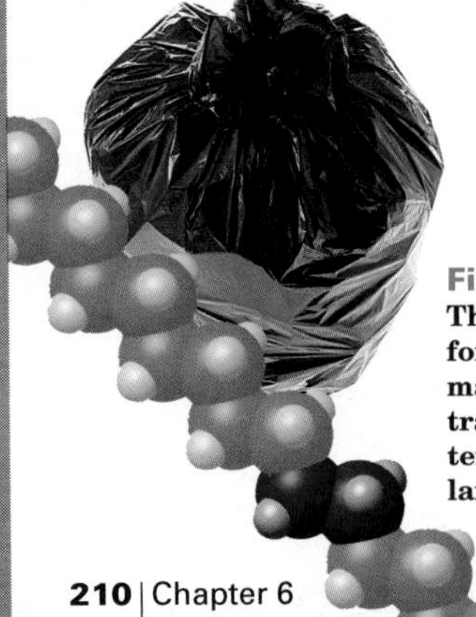

Figure 6-18
The molecular formula for polyethylene, the material used to make trash bags, can be written as $(C_2H_4)_n$, with n as large as 3500.

Solutions for Additional Example 6C

a. mass of HgCl is 236.04 g/mol.

$$\frac{\text{molar mass}}{\text{empirical mass}} = \frac{472.08 \text{ (g/mol)}}{236.04 \text{ (g/mol)}} = 2.0000$$

$2(HgCl) = Hg_2Cl_2$

b. mass of OCNCl is 77.46 g/mol.

$$\frac{\text{molar mass}}{\text{empirical mass}} = \frac{232.42 \text{ (g/mol)}}{77.47 \text{ (g/mol)}} = 3.00$$

$3(OCNCl) = O_3C_3N_3Cl_3$

Structural formulas

structural formula *indicates the spatial arrangement of atoms and bonds within a molecule*

Molecular formulas tell you what types of atoms are present and the number of each that are in the compound. But to know how the atoms are connected together, you need a **structural formula.** A structural formula can be compared to a videotape of a basketball game. Just as the videotape will give you a thorough idea of where each player was during the game, a structural formula will give you much more detailed information about the positions of atoms within a molecule.

Properties depend on atoms and arrangement

Compare the empirical, molecular, and structural formulas of the two molecules shown in **Table 6-8**. Both have the same kinds and number of atoms. The only difference is the way they are arranged. Yet this small structural difference makes a huge difference in their chemical properties. Diethyl ether was once used as an anesthetic, while 1-butanol can be used to dissolve varnish. Structural formulas are similar to Lewis structures, but unshared pairs of electrons are not shown. A dash represents a single bond shared between two atoms.

Table 6-8
Comparing 1-Butanol and Diethyl Ether

Compound	Empirical formula	Molecular formula	Structural formula	Melting point (°C)	Boiling point (°C)	Density (g/mL)
1-butanol	$C_4H_{10}O$	$C_4H_{10}O$	H H H H H–C–C–C–C–OH H H H H	–90	117	0.810
diethyl ether	$C_4H_{10}O$	$C_4H_{10}O$	H H H H H–C–C–O–C–C–H H H H H	–116.3	34.6	0.713

Section Review

16. Determine the empirical formula for each of the following.
    a. a compound containing 63.50% Ag, 8.25% N, and 28.25% O
    b. a compound found to contain 111.16 g of Fe and 63.84 g of S
17. Determine the molecular formula for each of the following.
    a. a compound with a molar mass of 86.17 g/mol that contains 83.62% carbon and 16.38% hydrogen
    b. a compound with a molar mass of 92.01 g/mol that contains 0.608 g of nitrogen and 1.388 g of oxygen
18. Draw a structural formula for each of the following molecules.
    a. C_2H_2 b. CCl_4 c. C_2H_6O (2 possible structures)

19. **Story Link**
    The molecular formula for the polyethylene used in trash bags can be as much as 3500 times its empirical formula. What is the molar mass for such a polyethylene molecule?

Further Applications

Low density polyethylene (0.915 g/cm^3) is used for packaging film, coating paper and wire, drum liners, toys, trash bags, chewing-gum base, and squeeze bottles. High density polyethylene (0.95 g/cm^3) is used for blow-molded products, injection-molded products, piping, fibers, and containers for gasoline and oil. Low molecular weight polyethylene (mass from 2000 g to 5000 g) is used as a mold-release agent for rubber and plastics, paper coatings, liquid polishes, and textile finishing agents.

ASSESS

Alternative Assessment

Have students investigate the addition polymerization process and report their findings in two parts: the general characteristics of the process and an application of the process to polyethylene.

Answers t… Sectio…view

1… moles 0.5887 Ag, 0.5890 N, 1.766 O; mole ratio 1:1:3; $AgNO_3$
b. moles 1.991 Fe, 1.991 S; mole ratio 1:1; FeS

17. a. moles 6.968 C, 16.21 H
empirical formula C_3H_7

$$\frac{86.17}{43.07} = 2\,;\ 2(C_3H_7) = C_6H_{14}$$

b. moles 0.04340 N, 0.08675 O
empirical formula NO_2

$$\frac{92.01}{46.01} = 2;\ 2(NO_2) = N_2O_4$$

18. a.
```
H       H
  \   /
   C=C
  /   \
H       H
```
b.
```
       :Cl:
        |
  :Cl — C — Cl:
        |
       :Cl:
```
c.
```
    H   H
    |   |
H — C — C — Ö — H
    |   |
    H   H

    H       H
    |       |
H — C — Ö — C — H
    |       |
    H       H
```

19. empirical formula, C_2H_4
(3500)(28.06 g/mol) = 98 210g/mol

Visual Strategy

Figure 6-18

The zigzagging of the molecules is an attempt to represent the tetrahedral shape of the carbon bonding system.

Section 6-4

FOCUS

Lesson Starter

Begin this lesson with a group puzzle exercise. Before class draw twenty 1.5 in × 1.5 in boxes on a sheet of paper. Draw a circle in each box. Write the letter C inside six of the boxes and add the four bonds as shown. Write the letter H inside the remaining 14 boxes and add a single bond as shown.

Divide students into groups of three. Provide each group with a copy of the paper and a pair of scissors. Ask them to develop as many different molecules as possible in 5 min using all of the atoms and all of the bonds. Compare the findings. Ask the students how they could do this exercise three dimensionally. Would anything change? How would it be different?

TEACH

Teaching Tip

Visualizing molecular shapes Because many students have difficulty visualizing three-dimensional objects, you may want to display several familar objects and have the students sketch them. Alternatively, you might have them construct a three-dimensional model of a two-dimensional design.

You may also wish to reinforce three-dimensional shapes of molecules by modeling SO_2, CO_2, and H_2 with a teacher's demonstration model kit.

How can you tell the shape of a molecule?

Section Objectives

Relate the chemical properties of a molecule to its shape.

Explain the basis of the VSEPR theory.

Predict the shape of a molecule from its Lewis structure.

Chem Fact
The sense of smell is related to a sense organ that distinguishes the shapes of molecules.

Molecular shapes

Although you have learned a lot about the structures of molecules, you have not really studied their three-dimensional orientation or molecular geometry. Shape is an important factor in determining the chemical properties of a molecule. Hemoglobin, a polymer contained in your red blood cells, transports oxygen to all parts of your body. If a genetic mutation causes the shape of hemoglobin to change, the result could be sickle cell anemia, as shown in **Figure 6-19**. This condition is often so serious that children who inherit the disorder from both parents seldom live past the age of two.

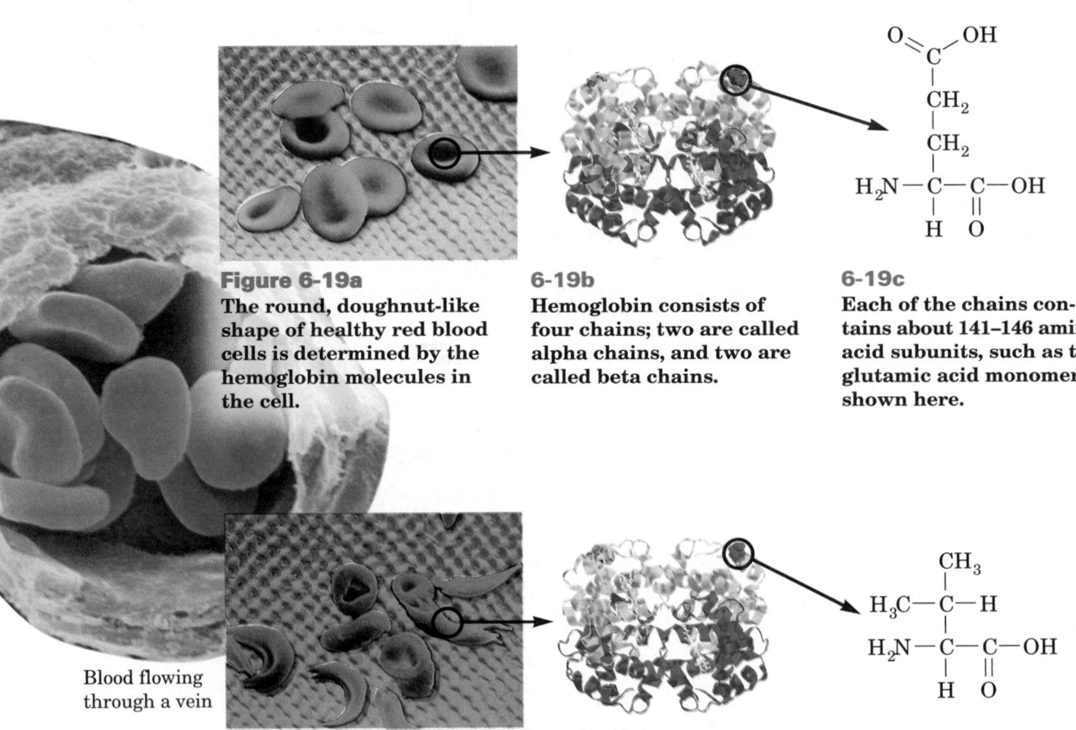

Figure 6-19a The round, doughnut-like shape of healthy red blood cells is determined by the hemoglobin molecules in the cell.

6-19b Hemoglobin consists of four chains; two are called alpha chains, and two are called beta chains.

6-19c Each of the chains contains about 141–146 amino acid subunits, such as the glutamic acid monomer shown here.

6-19d Because of their shape, sickle cells clog small blood vessels so that not as much oxygen reaches body cells.

6-19e The sickled cell shape is the result of the slight changes in the shape of hemoglobin molecules found in red blood cells.

6-19f These changes in the shape of hemoglobin are caused by a genetic mutation, in which another amino acid, valine, is substituted for one of the glutamic acid monomers in each of the beta chains.

Visual Strategy

Figure 6-19

Point out differences between the two amino acids shown to emphasize that even small changes in molecular structure can drastically alter the shape and function of a molecule. In this case the mutated cell cannot travel through blood vessels as easily as the normal red blood cell, nor can it bind the same amount of O_2.

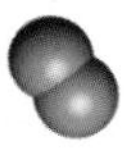

Figure 6-20 Molecules with only two atoms, such as H_2, can only have one shape.

Figure 6-21 Even though carbon dioxide and sulfur dioxide have the same number of atoms, they have different shapes.

Carbon dioxide

Sulfur dioxide

Shapes cannot be predicted from molecular formulas

Molecules with relatively simple molecular formulas also have simple shapes. In molecules with only two atoms, such as the H_2 molecule shown in **Figure 6-20**, only one shape is possible. But for molecules of more than two atoms, molecular shapes become more complicated. In such cases, you need to know more than their molecular formulas.

There is usually no obvious relationship between the molecular formula of a compound and its shape. Consider the two molecules carbon dioxide, CO_2, and sulfur dioxide, SO_2. Both contain three atoms, two of which are oxygen atoms. Yet they have different shapes, as shown in **Figure 6-21**. With such similarities in their atomic makeup, why is CO_2 linear, while SO_2 is bent? The answer lies in the arrangement of the valence electrons, especially unshared pairs.

After performing many tests that were designed to detect the shape of different molecules, scientists built a theory that summarized their findings. **VSEPR theory** states that electrostatic repulsion between the valence-level electron pairs causes these pairs to be oriented as far apart as possible. This positioning of pairs represents the most stable arrangement because it will take an input of energy to overcome the repulsive forces and push the electron pairs closer together. Although useful predictions about shape can be made with this theory, there is evidence that this theory does not provide a good description of electron cloud behavior.

VSEPR (Valence Shell Electron Pair Repulsion) theory *system for predicting molecular shape based on the idea that pairs of electrons orient themselves as far apart as possible*

How To

Determine molecular shape from Lewis structures

1. Draw the Lewis structure for the molecule.

2. Count the number of electron clouds surrounding the central atom.

- Each single bond counts as an electron cloud.
- Each multiple bond generally counts as a single electron cloud.
- Each nonbonding electron pair must be considered an electron cloud. These aid in determining the geometry, but only bonded atoms are included in the shape.

3. Apply the appropriate geometry based on the number of electron clouds.

- **Two electron clouds: linear**
  Consider the case of CO_2, which has the following Lewis structure.

$$\ddot{\underset{\cdot\cdot}{O}}=C=\ddot{\underset{\cdot\cdot}{O}}$$

Each double bond is counted as a single electron cloud, even though it contains four electrons. The orientation that will keep the two electron clouds on either side of the carbon atom as far apart as possible is the straight-line arrangement, shown in **Figure 6-22**. For molecules with only two electron clouds surrounding the central atom, this is the likeliest geometry, with a bond angle of 180°.

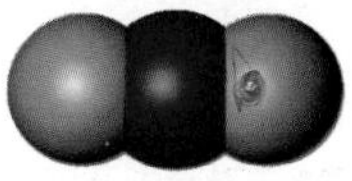

Figure 6-22 According to VSEPR theory, a molecule with two electron clouds around a central atom is most likely to be arranged in a straight line, as shown in this model of carbon dioxide.

How To continued on following page . . .

Content Background

The ability of VSEPR theory to accurately model the bonding in a molecule decreases as the size or complexity of the bonded substance increases.

Teaching Tip

Molecular vs. electron pair geometry

Geometrical determination depends on how you view the molecule. If you are the central atom looking out toward your appendages, your view never changes because a nonbonding electron pair is seen as an appendage, as is a bonding electron pair. However, if you are viewing the molecule from the outside looking in, only bonding pairs are seen, so your perspective changes as the number of bonding pairs changes. The electron pair geometry for CH_4, NH_3, H_2O, and BrO^- is tetrahedral, but their molecular geometries decrease in symmetry. These symmetries are, respectively, tetrahedral, trigonal pyramidal, bent or V-shaped, and linear. Construct a chart for ClO_4^-, ClO_3^-, ClO_2^-, and ClO^- with the following column heads.

Lewis dot structure
No. of bonding pairs
No. of nonbonding pairs
Molecular geometry
Electron pair geometry

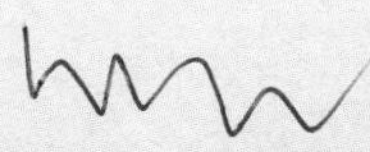

Teaching Tip

Bonding orbitals and geometric shape

Point out that the geometry is determined by the sigma framework only; pi bonds are not counted because they come about as a result of the sigma framework.

Content Background

Atomic orbitals are described mathematically by wave functions. These wave functions can be added together or subtracted from one another to describe mixed or hybrid orbitals. The characteristics of the hybrid orbitals reflect the type of orbitals from which the hybrids are made. The *s*, *p*, and *d* orbitals can be hybridized. The number of orbitals blended together equals the number of bondable pairs on the central atom. So linear geometry uses *sp* hybrid orbitals. Trigonal planar geometry uses sp^2 hybrid orbitals, and tetrahedral geometry uses sp^3 hybrid orbitals.

The hybridization for each C atom in polyethylene is sp^3. Hybrid orbital combinations sp^3d and sp^3d^2 produce trigonal bipyramidal and octahedral geometries, respectively.

How To continued from the previous page . . .

- **Three electron clouds: trigonal planar**

  Now consider the Lewis structure for sulfur trioxide, SO_3.

  :Ö:
  |
  Ö=S—Ö:

  SO_3 has three electron clouds surrounding the central atom. The most likely arrangement for this molecule is shown in **Figure 6-23**. This shape is called a "trigonal planar" arrangement of electron pairs.

Figure 6-23
When a molecule consists of a central atom surrounded by three other atoms, as in SO_3, a trigonal planar arrangement is most likely.

- **Four electron clouds: tetrahedral**

  The situation for molecules with four electron clouds, like methane, CH_4, is somewhat trickier because instead of having a planar shape that can be drawn on a paper like the others, it has a three-dimensional shape called a *tetrahedron*. Several different views of a tetrahedron are shown in **Figure 6-24**. The shape can be described as a tripod with a fourth leg sticking straight up. Such an arrangement of electron pairs will yield bond angles of 109.5°.

4. Adjust bond angles to account for unbonded pairs

The SO_2 molecule shown in **Figure 6-21** has one unbonded pair and two bonds to oxygen atoms. As a result, it has three electron clouds and falls into the "trigonal planar" category, with angles between the electron pairs of about 120°. However, an unshared pair of electrons occupies more space than a shared pair. As a result of their greater size, the force of repulsion between unshared pairs is slightly greater than the repulsive force between shared pairs. Consequently, the electrons shared between each O atom and the central S atom are pushed together, causing the bond angle for SO_2 to be 119.5°, instead of 120°. The shape is based only on the atoms, so it is described as *bent*. A similar effect is observed for double and triple bonds, because their electron clouds take up more space than the electron clouds for a single-bond pair.

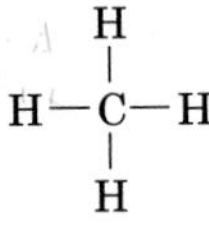

Figure 6-24a
A tetrahedral shape can't be shown accurately with a planar structural formula.

6-24b
The true shape can only be shown by a three-dimensional model, or . . .

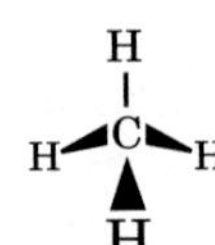

6-24c
. . . in other ways that emphasize its three-dimensionality, such as with wedge-shaped bonds that seem to recede or protrude from the plane of the page.

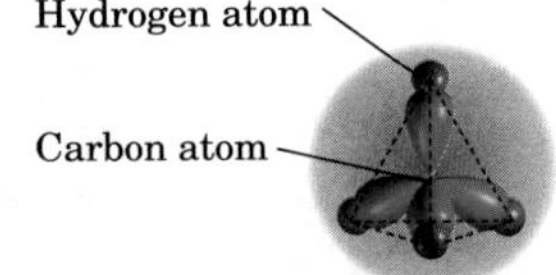

6-24d
Another way to portray the tetrahedral shape shows the location of the orbitals of the atoms within the molecule.

Visual Strategy

Figure 6-24

Point out that the similarities in structure as you move from **a** to **d**. Why would more than one representation be needed? The tetrahedral shape of CH_4 results from the four H atoms achieving maximum separation from one another.

Although this presentation of VSEPR theory can be very useful in predicting the shape of compounds made from main-block elements, it does not explain the rules for molecules that have more than an octet of valence electrons. In addition, VSEPR theory does not always work for some complex molecules. However, this simplified discussion should explain the shapes of most of the molecules in this chemistry text.

Sample Problem 6E
Predicting molecular shapes

Determine the shape and approximate bond angles for the following compounds: NH_3 and H_2O.

1 Draw Lewis structures

H:N:H   H:O:
 H   H

2 Count the electron clouds

- **NH_3 has three bonds to H atoms and one unbonded pair:** four electron clouds
- **H_2O has two bonds to H atoms and two unbonded pairs:** four electron clouds

3 Apply the proper VSEPR geometry

- **Both atoms will have a tetrahedral arrangement of electron clouds, giving bond angles of about 109.5°.**

4 Account for unbonded pairs

- **For ammonia, consider the tetrahedron that is similar to a tripod with a fourth leg sticking straight up. If the unbonded electron pair is assigned the position of the leg sticking up, the shape that is left is pyramidal, or a pyramid with three triangular sides and a triangular base.**

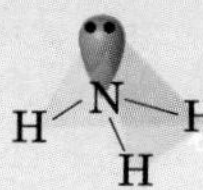

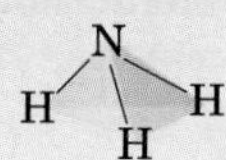

- **For water, the shape will be bent. This will be true regardless of which two legs of the tetrahedron the two unbonded pairs are assigned to.**

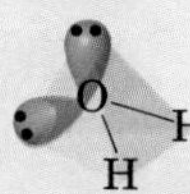

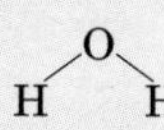

- **Due to the effects of the unbonded pairs, ammonia and water both have bond angles slightly less than 109.5°. (Ammonia has bond angles of 107.3°, and water has an angle of 104.5°.)**

Section Review

20. How might changing the shape of a molecule affect its properties?
21. Predict the molecular shape of Br_2 and HBr.
22. What is the relationship between the VSEPR theory and molecular shape?
23. Predict the molecular shapes of SCl_2, PI_3, and NCl_3.
24. What is the angle for the C—C—C bond in C_3H_8? Using this information, explain why organic structural formulas often show chains of carbon atoms as zigzag lines instead of straight lines.

Additional Examples for Sample Problem 6E

Determine the molecular shape for the following compounds: BF_3, $CHCl_3$, NNO, Br_2, $AlCl_3$, HCl, PH_3, H_3O^+, GaI_3, CF_3Cl, and CO.

Solutions are at the bottom of the page.

Teaching Tip

Electron pairs and geometry
Nonbonded pairs of electrons on the central atom lower the symmetry of the molecule, so both NH_3 and H_2O have tetrahedral electron pair geometry but their molecular geometries are pyramidal and bent, respectively.

ASSESS

Alternative Assessment

1. Have students determine the geometries of 3–5 polyatomic ions.
2. Have pairs of students build a structure that has four corners using six toothpicks that touch each other in just one point. Use clay to hold toothpicks together. (forms a tetrahedron)

Answers to Section Review

20. In the example of hemoglobin, the mutant cell binds and delivers O_2 to the cells less effectively than normal red blood cells. The sickle shape cannot move through the tubular blood vessels as easily as the round cells.
21. Br_2 is linear. HBr is linear.
22. VSEPR theory predicts the most stable molecular arrangement because it assumes maximum separation for bonded ions.

Answers are continued on page 231B.

Solutions for Additional Examples 6E
linear: Br_2, HCl, CO, NNO
trigonal planar: BF_3, $AlCl_3$, GaI_3
trigonal pyramidal: PH_3, H_3O^+
tetrahedral: $CHCl_3$, CF_3Cl

Section 6-5

FOCUS

Lesson Starter

Have students work in groups of three for 5 min to determine the makeup, properties, and some uses for the compounds in Figure 6-25 and Figure 6-26. Graphite's structure makes it excellent for lubricants, pencil lead, dry inks, paints, and motor brushes. It can also be made into fibers and used for fishing poles, tennis rackets, golf clubs, baseball bats, heating pads, and jet-engine components. Diamond's tough and durable structure is used in surgical knives, windows in space probes, oil-well drill bits, grinding steel, glass and metal cutting, abrasive wheels, and jewelry. Uses for cellulose include wood fibers, paper, ethanol production, insulation, fuel, plastics, lacquers, and nitrocellulose explosives. Polyethylene is found in wire coating, chewing-gum base, squeeze bottles, film, packaging, plastic wrap, piping, and polishes.

How do bonds and properties relate?

Section Objectives

Describe the properties and structure of a covalent network.

Relate the physical properties of organic compounds to their chemical structure.

Classify organic molecules by their functional groups.

Decipher abbreviated structural formulas for organic compounds by identifying functional groups.

List the advantages and disadvantages of various structural models.

Carbon and bonding

Remember that carbon generally forms four bonds and has the Lewis structure shown in Figure 6-25. When it forms single bonds, the compounds will reflect a tetrahedral geometry. Everything you see in Figure 6-26 contains substances with carbon atoms bonded to one another. Carbon often forms long chain compounds, giving rise to a wide variety of compounds with different properties. The differences in properties are related to the ways that carbon atoms are bonded.

$\cdot\dot{\underset{\cdot}{C}}\cdot$

Figure 6-25
Because a carbon atom has four valence electrons, it can form bonds with up to four other atoms.

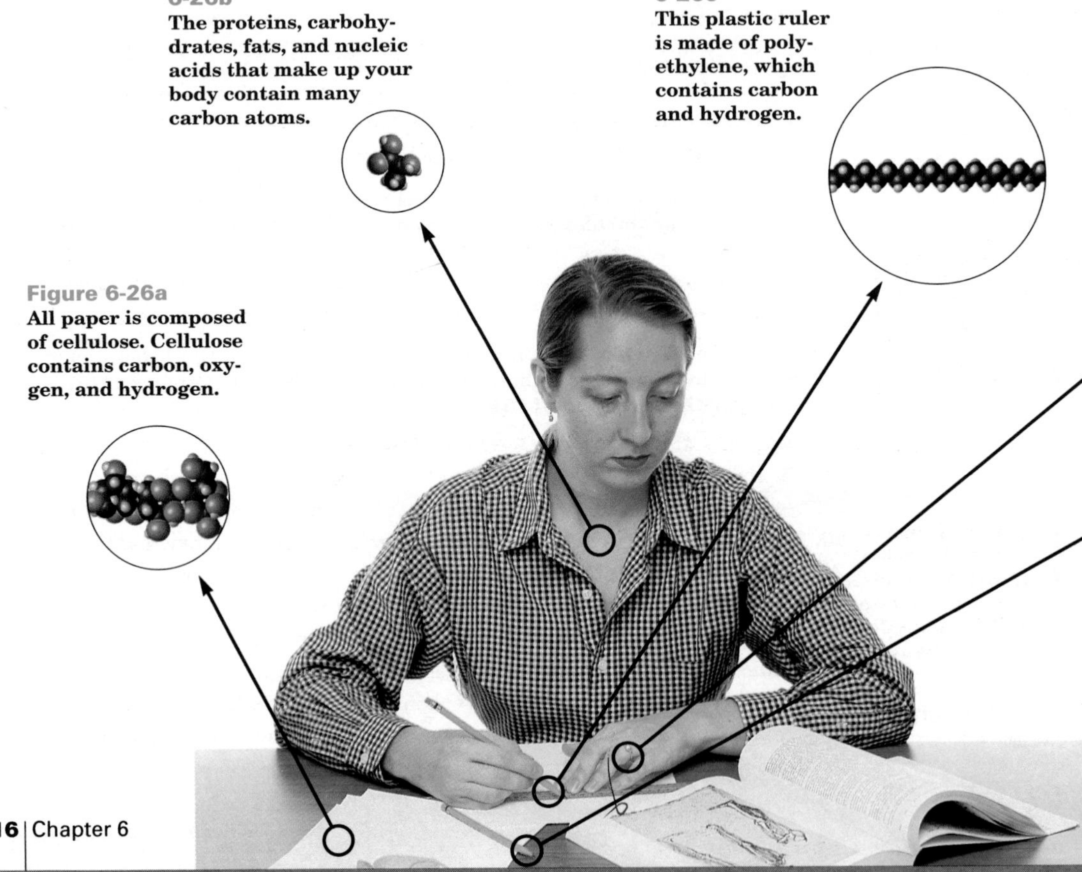

6-26b
The proteins, carbohydrates, fats, and nucleic acids that make up your body contain many carbon atoms.

6-26c
This plastic ruler is made of polyethylene, which contains carbon and hydrogen.

Figure 6-26a
All paper is composed of cellulose. Cellulose contains carbon, oxygen, and hydrogen.

216 | Chapter 6

Carbon can form covalent networks

Some of the properties of diamond and graphite are shown in **Table 6-9**. You might expect each substance to have properties similar to molecular compounds, because the carbon-carbon bonds in each are nonpolar covalent. But graphite and diamond each have some properties of molecular substances and some properties that are unusual for molecular compounds. Although graphite is soft like most molecular substances, it conducts electricity and has a high melting point. Diamond is hard, like an ionic crystal, but it is not as brittle.

Table 6-9
Properties of Diamond and Graphite

Property	Diamond	Graphite
Mohs' hardness scale rating	10 (top of scale)	0.5–1
Density	3.51 g/mL	2.23 g/mL
Melting point	(changes to graphite)	(sublimes at 3652°C)
Conductivity	low	high

Unlike compounds consisting of individual molecules, graphite and diamond are examples of network solids, in which all atoms are covalently linked to others, somewhat like the crystal lattice of ionic compounds. Because every atom is bonded to several other atoms, melting requires enough energy to break all of these bonds.

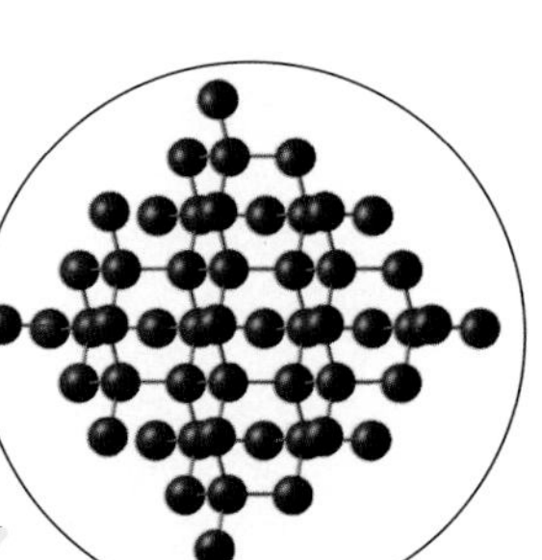

6-26d
Diamond's hardness is due to the many strong covalent bonds in its network structure.

Figure 6-26d shows the bonding between the carbon atoms in diamond. Each carbon atom in diamond is bonded to four other carbon atoms in three dimensions, at tetrahedral angles. This tetrahedral arrangement creates a network of covalently bonded carbon atoms that gives diamond its great strength and hardness. This arrangement also makes diamond denser than graphite.

Notice that in graphite, as shown in **Figure 6-26e**, the atoms are arranged in layers. Each carbon atom is covalently bonded to three other carbon atoms in a layer forming a strongly bonded covalent network of carbon atoms. London forces exist between layers, which can slide over each other. The ability of the layers to slide over one another not only accounts for graphite's tendency to crumble but also makes graphite a good lubricant, especially at temperatures that are too high for lubricating oils and petroleum jellies. Some of the bonding electrons on each carbon atom are very mobile, allowing graphite to conduct electricity, just as metals do.

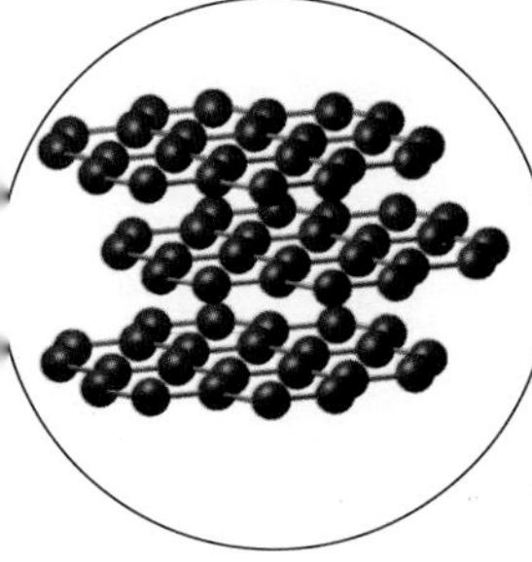

6-26e
The soft and crumbly properties of graphite stem from the relative weakness of the London forces between different layers of carbon atoms.

Organic chemistry

Organic compounds are important to life

The molecules found in all living things—proteins, carbohydrates, nucleic acids, and lipids—all contain "backbones" that include carbon atoms linked together. In addition, other atoms can be bonded to these carbon backbones, forming a special category of compounds known as **organic compounds**.

Organic compounds, with few exceptions, are molecular compounds that contain carbon and, usually, hydrogen. Other atoms, such as oxygen, nitrogen, sulfur, and phosphorus, are also frequently found in organic compounds. More than 6 000 000 organic compounds have been identified and named.

organic compound
compound containing carbon

TEACH

Further Applications

Graphite is mixed with varying amounts of clay to make pencil lead. The more clay that is present in the lead, the higher the number of the pencil. Although found in nature, much of the graphite used for pencils and industry is made in electric furnaces.

Historical Note

In 1772, Antoine Lavoisier heated a diamond in a closed vessel until it disappeared. The only product formed was CO_2. He concluded that diamond is pure carbon.

Do You Know?

Although extremely hard, diamonds have a relatively low density of 3.50 g/cm^3.

Historical Note

Friedrich August Kekulé von Stradonitz defined organic chemistry in 1861 as the chemistry of carbon compounds.

Theme

Classification and trends
Organic molecules are classified according to the characteristic bonds and atoms attached to carbon.

Visual Strategy

Figure 6-26
The layered structure of graphite and the hardness of diamond result from differences in the bonding between C atoms. In graphite, each C atom is bonded to three other C atoms in a trigonal planar arrangement, so there is no cross-linking between layers. In diamond, every C atom is bonded to four other C atoms in a tetrahedral arrangement.

Further Applications

The octane rating of gasoline is determined by running mixtures of octane and heptane in a standardized engine and measuring the power output. For example, a mixture of 90% octane and 10% heptane would have an octane rating of 90. An octane rating of 100 would be pure octane. Gasoline is then run in the same engine, and the power output is compared to that of octane and heptane. Gasoline that has very little octane present but still has a high "octane rating" has been developed.

Teaching Tip

Because several of the gases on this list are familar to students, you may wish to discuss their uses and properties.

Theme

Classification and trends

The alkane series has the general formula C_nH_{2n+2}, where n is equal to the number of C atoms.

Hydrocarbons are the simplest organic compounds

Hydrocarbons are composed of only hydrogen and carbon. Gasoline is a mixture of hydrocarbons, including octane. Hydrocarbons are grouped into categories based on the bonding between carbon atoms. The simplest hydrocarbons are those whose carbon atoms are connected in a long chain by single covalent bonds. These hydrocarbons are called *alkanes,* and they are a large part of all petroleum products. The suffix *-ane* is a part of their name and indicates that only single bonds are present. The names and formulas of the first 10 straight-chain alkanes are shown in **Table 6-10**. The names of most other organic compounds are based on these names.

Table 6-10
Names of Alkanes

Number of carbon atoms	Name of alkane	Molecular formula	Structural formula	Melting point (°C)	Boiling point (°C)
1	methane	CH_4	H H—C—H H	−182	−161
2	ethane	C_2H_6	H H H—C—C—H H H	−172	−88
3	propane	C_3H_8	H H H H—C—C—C—H H H H	−187.7	−42.1
4	butane	C_4H_{10}	H H H H H—C—C—C—C—H H H H H	−138.4	−0.5
5	pentane	C_5H_{12}	H H H H H H—C—C—C—C—C—H H H H H H	−129.7	36.1
6	hexane	C_6H_{14}	H H H H H H H—C—C—C—C—C—C—H H H H H H H	−95	69
7	heptane	C_7H_{16}	H H H H H H H H—C—C—C—C—C—C—C—H H H H H H H H	−90.6	98.4
8	octane	C_8H_{18}	H H H H H H H H H—C—C—C—C—C—C—C—C—H H H H H H H H H	−56.8	125.7
9	nonane	C_9H_{20}	H H H H H H H H H H—C—C—C—C—C—C—C—C—C—H H H H H H H H H H	−51	150.8
10	decane	$C_{10}H_{22}$	H H H H H H H H H H H—C—C—C—C—C—C—C—C—C—C—H H H H H H H H H H H	−29.7	174.1

Visual Strategy

Table 6-10

Compare the melting or boiling points with the length of the carbon chain and identify the trend that occurs. Point out that the number of carbons in the chain is identified by its latin prefix.: *dec* for 10, *non* for 9, *oct* for 8, *hept* for 7, *hex* for 6, *pent* for 5, *but* for 4, *prop* for 3, *eth* for 2, and *meth* for 1.

Functional groups determine properties of organic molecules

Most organic compounds, especially those used by living things, consist of carbon atoms that are covalently bonded to other types of atoms, especially oxygen and nitrogen atoms. Combinations of carbon, hydrogen, oxygen, and nitrogen atoms can also form groups known as **functional groups.** Many properties of organic chemicals depend on functional groups, so organic compounds are classified by the functional groups they contain. **Table 6-11** gives an overview of some common functional groups.

functional group
group of atoms that determines an organic molecule's chemical properties

The compounds in each class have similar chemical properties. For example, almost all alcohols can lose a water molecule, H_2O, to form an *alkene,* a compound containing a double bond.

The way a compound is named provides clues about its functional groups and its structure. For example, heptanol is a derivative of heptane, a 7-carbon alkane. The *-ol* ending tells you that this compound is an alcohol and contains the —OH functional group that is characteristic of all alcohols. Because of this functional group, heptanol has physical and chemical properties that are more similar to other alcohols than to the heptane from which it is derived.

Table 6-11
Some Common Organic Functional Groups

Name	Chemical symbol	Naming suffix or prefix	Example	Properties and uses
Alcohol	—C—OH	*-ol*	$H_3C—C(OH)(H)—CH_3$ isopropanol	some properties similar to water, able to form hydrogen bonds, useful precursor for many compounds
Ether	—C—O—C—	*ether*	$H_3C—CH_2—O—CH_2—CH_3$ diethyl ether	volatile solvents used in anesthetics, no hydrogen bonding
Aldehyde	—C(=O)—H	*-al*	H—C(=O)—H methanal or formaldehyde	reactive compounds formed by oxidizing alcohols, used to preserve tissues and in polymers, no hydrogen bonding
Organic acid	—C(=O)—OH	*-oic acid* or *-ate* for ionized form	$H_3C—C(=O)—OH$ ethanoic acid or acetic acid	usually weakly acidic compounds, formed by oxidizing aldehydes, combined with glycerol in fats, strong hydrogen bonding
Ketone	—C(=O)—	*-one*	$H_3C—C(=O)—CH_3$ acetone	polar solvents used in paints and textile processing, no hydrogen bonding
Amine	NH_2 on —C—; —C—NH—C—; —C—N(—C—)—C—	*-amine, amino-, -ine,* or *azo-*	$H_2N—CH_3$ (H—N(H)—C(H)(H)—H) methylamine	slightly basic compounds with some similarities to ammonia, often have unpleasant, fishy smell, weak hydrogen bonding

Teaching Tip

Familiar uses of organic compounds

A familiar use for each example listed in **Table 6-11** can be given to the students so they realize they have already seen or used these organic compounds.

alcohol: wood alcohol for cleaning oil-based paints, or rubbing alcohol (isopropanol)
ether: diethyl ether is one of the ingredients in starting fluid for gasoline engines
aldehyde: formaldehyde is used to make formalin, which is used to embalm animals
organic acid: acetic acid is diluted to 5% to make vinegar
ketone: acetone is used as a paint thinner and in some nail polish removers
amine: amino acids used to build proteins in the body

Groupwork Strategy

Drawing structural formulas

In groups of three, have students determine possible structures for C_2H_6O and C_3H_8O.

C_2H_6O

```
   H  H               H        H
   |  |               |        |
H—C—C—O—H          H—C—O—C—H
   |  |               |        |
   H  H               H        H
```

[—OH group can be attached to either C atom]

C_3H_8O

```
                              H
                              |
   H  H  H                 H  O  H
   |  |  |                 |  |  |
H—C—C—C—O—H           H—C—C—C—H
   |  |  |                 |  |  |
   H  H  H                 H  H  H

   H     H  H
   |     |  |
H—C—O—C—C—H
   |     |  |
   H     H  H
```

[O atom can be to the right or left of the central atom]

Visual Strategy

Table 6-11

Combine the structural formulas listed in **Table 6-10** with the functional groups in **Table 6-11** to reinforce the identification of functional groups and the naming of straight-chain compounds. Some examples of structures and their names appear on page 231B.

Historical Note

Benzene was first synthesized during the 1850s by Pierre Eugène Marcelin Berthelot, but Friedrich August Kekulé is credited with correctly determining the resonance bonding structure of benzene, C_6H_6.

Do You Know?

Benzene, a recognized carcinogen, is readily absorbed through the skin.

To name an organic chemical, you simply take the prefix from the alkane name for the hydrocarbon with the same number of carbons. Then you add a suffix that describes the compound's functional group.

Some organic molecules contain more than one type of functional group. Amino acids, which are the building blocks of proteins in living things, are a good example. One part of the molecule has an amine group, and another part has an organic acid group. Amino acids have the characteristics of both amines and organic acids. One end is weakly basic, and the other end is weakly acidic.

Organic structures

Carbon atoms can also form ring structures

Carbon atoms that covalently bond to one another can also form ring structures. These rings are like pieces of the carbon network in graphite. Benzene, C_6H_6, is one of the most important organic compounds consisting of a ring of carbon atoms. Notice that the double bonds in this structure can be validly drawn in either of the ways shown in **Figure 6-27a**. Benzene is an example of a resonance hybrid. It has these resonance structures because of the arrangement of the orbitals in the benzene molecule. The electrons involved in double bonds are in orbitals that extend above and below the plane of the molecule. These orbitals are so close, they actually overlap.

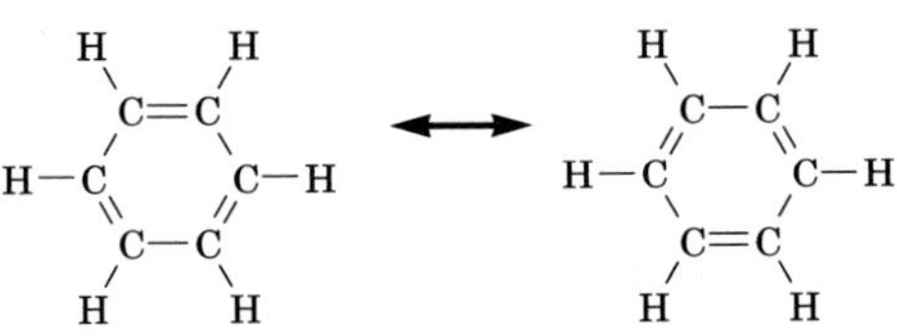

Figure 6-27a
Benzene has two possible resonance structures.

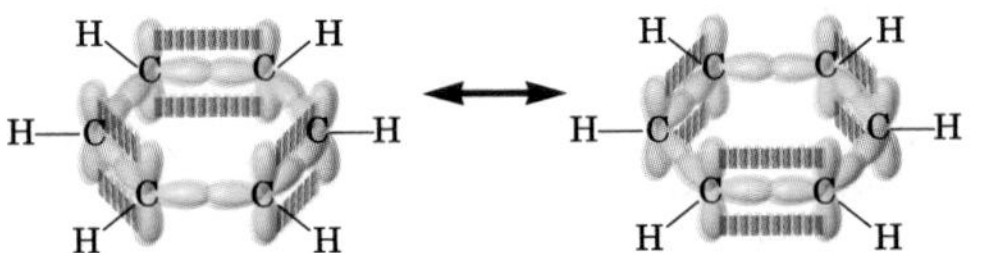

6-27b
Because the electrons involved in the double bonds are contained in orbitals that all touch each other, each resonance structure is equally likely.

Figure 6-28a
The chemical name for aspirin is acetylsalicylic acid.

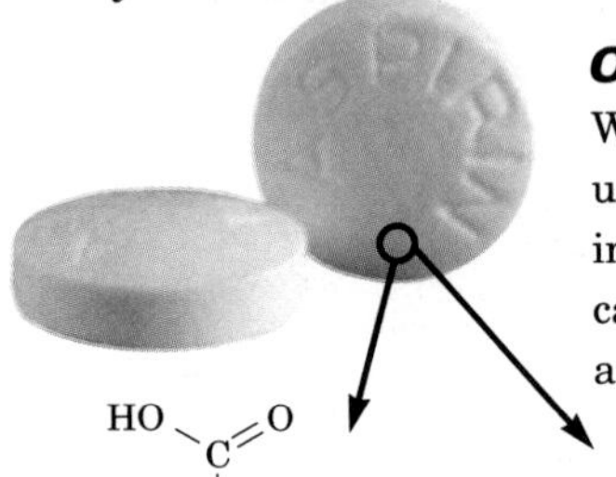

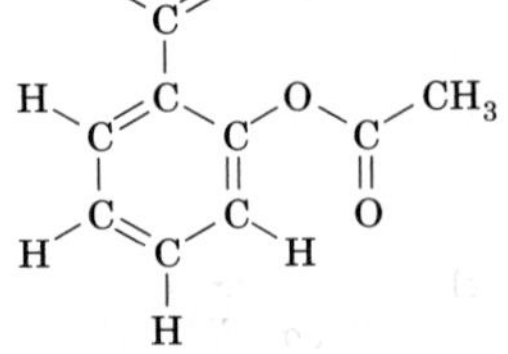

Organic structures are often abbreviated

When chemists draw structural formulas for organic molecules, they often use a short-hand notation that leaves out the carbon and hydrogen atoms in the main part of the molecule and shows only carbon-carbon bonds. You can figure out how many hydrogen atoms and carbon-hydrogen bonds there are by remembering that carbon tends to form four bonds. Atoms other than hydrogen are always shown. The zigzag pattern of the carbon chain indicates where two tetrahedrally oriented C—C bonds are joined at a carbon atom. Although organic structures may not seem useful for smaller molecules, they are very helpful when it comes to more complicated molecules such as the one shown in **Figure 6-28**. Because some atoms aren't shown, the functional groups are easier to recognize.

6-28b
Because the complete structural formula for acetylsalicylic acid is complex, . . .

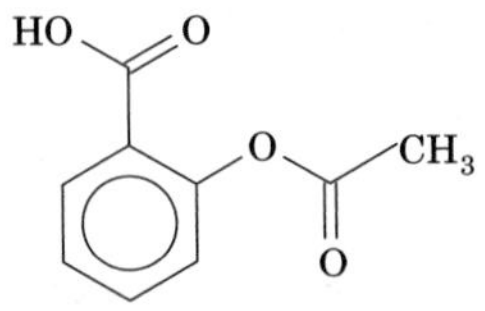

6-28c
. . . chemists usually draw its organic structure instead. Note that it is an aromatic compound.

Visual Strategy

Figure 6-28c

This is a shorthand notation for drawing molecules. Where the lines bend, a C atom is found. Hydrogen atoms are not shown, so the viewer must mentally supply any H atoms needed to account for all four bonds to each C atom. This system requires practice. Practice drawing structures for the alkanes in **Table 6-10**. Practice interpreting organic structures with the seven structures on page 231C.

Molecules can be portrayed in many ways

So far, you have seen a variety of different ways to portray atoms, ions, and molecules, from Lewis structures to three-dimensional views of molecules. Each structure has aspects that accurately describe some aspects of the bonded atoms, but like all models, each has its shortcomings, some of which are described in **Table 6-12**. One disadvantage shared by all of these models is that they attempt to portray the three-dimensional structure of atoms on a flat page. Another disadvantage is that these models cannot show motion. All molecules are constantly moving. Not only do molecules move from one place to another, but bonds within each molecule are constantly stretching, compressing, bending, and twisting. Today, molecular modeling of some large molecules is done on computers so that chemists can look at a molecule from a variety of angles and in motion.

Table 6-12
Advantages and Disadvantages of Molecular Models

Type of representation	Example	Advantages	Disadvantages
Chemical formula	C_6H_6	shows relative numbers of each kind of atom in a molecule or formula unit	does not show bonds, atom sizes, or actual shape
Lewis structure	H H C::C H:C C:H C:C H H	shows arrangement of valence electrons in a molecule	does not show actual shape of molecule or atom sizes; electron position is not definite as is implied; larger molecules can be too complicated to show
Structural formula	H H C=C H–C C–H C–C H H	shows arrangement of all atoms and bonds in a molecule; less complicated than Lewis structures	does not necessarily show actual shape of molecule or atom sizes; ignores unbonded pairs; larger molecules can be too complicated to show
Organic structure		shows arrangements of carbon chains in organic molecules; less complicated than structural formulas	does not necessarily show actual shape of molecule or atom sizes; does not show all atoms and bonds
Electron-cloud model		shows shape of molecule; shows probability clouds of electrons in each molecule, including nonbonding pairs	difficult to label individual atoms; larger molecules can be too complicated to show; bonds are not clearly indicated
Space-filling model		shows three-dimensional shape of molecule; shows most of the space taken by electrons in each molecule; less complicated than electron-cloud models	uses false colors to differentiate among elements; bonds are not clearly indicated; parts of large molecules may be hidden

Groupwork Strategy

Divide the class into groups of three. Have them determine the electron dot structure, structural formula, and organic structure for propane, C_3H_8, and diethyl ether, $CH_3CH_2OCH_2CH_3$.

Additional Examples for Sample Problem 6F

Have students determine the molecular formula of:

a. caffeine
b. histamine
c. epinephrine

Solutions are at the bottom of the page.

ASSESS

Alternative Assessment

Investigate the source for the optical properties of gemstones. How do the structure(s) for other gemstones compare to the structure for diamond? Write a one page synopsis of your findings.

Answers to Section Review

25. Each C atom in graphite is connected to three other C atoms, resulting in trigonal planar geometry. The planar sheets are not connected and can glide over one another. Each C atom in diamond is bonded to four other C atoms, resulting in tetrahedral geometry and a rigid, three-dimensional array.

26. All three structures are made from atoms of C, H, and O. They differ in their bonding arrangements. An alcohol has an —OH group; an aldehyde has a carbonyl group, —C(=O)—, and an organic acid has a carboxyl group, —C(=O)OH.

27. Organic structures are condensed structural formulas in which C—C bonds are drawn as line segments, and C—H bonds are implied.

28. a. amino acid; groups are carboxylic acid, amine, amide; molar mass is 144.1 g; line formula is $C_5H_8N_2O_3$

b. aldehyde; groups are alkene, six-membered ring, aldehyde; molar mass is 132.08 g; line formula is C_9H_8O

Sample Problem 6F
Deciphering organic structures

Unlike most other protein monomers, tryptophan cannot be manufactured by the human body, so it must be available in the diet. For this reason, tryptophan is called an essential amino acid. What is the molecular formula of tryptophan?

1 ***Place carbon atoms at all unmarked bond angles***

2 ***Place hydrogen atoms in C—H bonds so that each carbon has four bonds***

3 ***To determine the molecular formula, count the number of each type of atom***

$C_{11}H_{12}N_2O_2$

Section Review

25. Explain why graphite is very soft, while diamond is extremely hard, even though both are made entirely of carbon atoms in covalent networks.

26. Compare and contrast the structures of an alcohol, an aldehyde, and an organic acid, all of which contain oxygen and hydrogen in their functional groups.

27. Explain how organic structures of molecules can make functional groups easier to identify.

28. Classify these organic molecules by identifying their functional groups. Then, calculate the molar mass of each.

a. asparagine **b.** cinnamaldehyde

Solutions for Additional Examples 6F

a. $C_5H_4N_4O_2$

b.

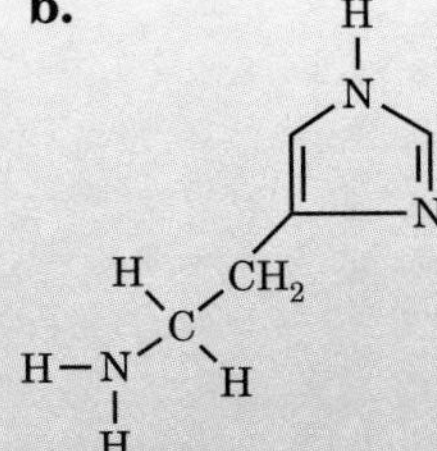

$C_5H_9N_3$

c. $HOCHCH_2NHCH_3$

$C_9H_{13}NO_3$

Conclusion: Molecular Bonds and Trash Bags

MISCELLANEOUS 20%

PAPER 50%

PLASTIC 10%

From your study of molecular bonds, you can now examine some of the problems in making trash bags that biodegrade.

What is the composition of the material currently being used in trash bags? Trash bags are made of polyethylene, a polymer that can consist of more than 3500 subunits arranged in long chains. Polyethylene is a hydrocarbon, meaning it is composed of atoms of hydrogen and carbon only.

Because most of the chemicals that living things use for food are smaller organic chemicals or have many functional groups, most of their enzymes are not useful in breaking down large hydrocarbon chains like polyethylene.

Even if bacteria in the landfill start to decompose polyethylene, they must start at either end of the long molecule, slowly breaking off only one of the 7000 carbon atoms in the chain at a time. As a result, it takes a very long time for an entire polyethylene molecule to be broken down.

How do bags labeled as biodegradable differ? Chemists have recently made it possible for bacteria to attack polyethylene at more places than just the edges of the bag. To accomplish this, they make the bags out of a mixture of cornstarch and polyethylene.

Bacteria consider these cornstarch molecules to be a "tasty treat." Most living things have developed enzymes that allow them to use starch and other compounds with many —OH functional groups as energy sources.

When the bacteria consume these cornstarch molecules, a "hole" in the bag is left behind. This, in turn, leaves behind smaller pieces of polyethylene, which are easier to break down.

Applying Concepts

1. How many points of attack would there be if 25 starch molecules were equally spaced among 3500 units of ethylene monomers?
2. When microorganisms break down polyethylene, they usually change it into several shorter molecules known as fatty acids. Fatty acids are organic acids containing long hydrocarbon chains. Draw the structural formula for a fatty acid with the formula $C_9H_{19}COOH$.

Research and Writing

Use the library to find out more about the following.

1. Research the properties of cornstarch. Give a reason why biodegradable trash bags are not made entirely of cornstarch instead of polyethylene.
2. Check to see what compounds, other than cornstarch, are being considered as possible candidates to make trash bags biodegradable. What tests are used to measure these against bags made of regular polyethylene and polyethylene with cornstarch?

Chemistry In Your Community

Oxygen forms covalent bonds with the hydrocarbon fuels burned to power cars and heat homes. Most of the compounds formed from the combustion process are released directly into the air. Relate this possible pollution problem to your local community by performing one or more of the following tasks.

1. Locate the refinery nearest your home. Determine how much petroleum is refined there, and compare this amount with the amount of petroleum refined worldwide.
2. Determine how the cost of consumer products made from refined petroleum has varied with the cost of crude oil over the past three to five years.
3. Investigate the refining process for crude oil. Determine how fuel oil and propane are produced and transported to the consumer. How much butane is used in an average year to operate the heating system of a mobile home?
4. If gasoline storage tanks are located in your area, find out the average capacity of the tanks and how many tanks can be grouped in one storage area.

Answers to Applying Concepts

1. 50; for each starch molecule "eaten" by an enzyme, two points on the polyethylene chain are exposed, as shown in the figure.

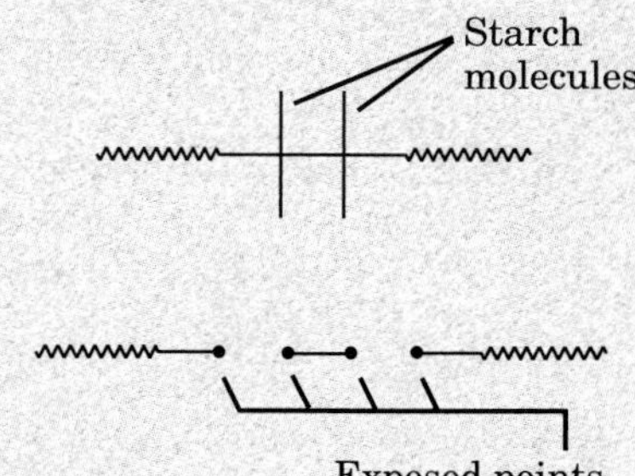

2.

```
  H  H  H  H  H  H  H  H   O
  |  |  |  |  |  |  |  |  //
H—C——C——C——C——C——C——C——C——C—O—H
  |  |  |  |  |  |  |  |
  H  H  H  H  H  H  H  H
```

Content Background

The growth of carbon chemistry

Although carbon is one of the earliest elements to be discovered, and substances containing carbon have been recognized from the earliest times, the actual recognition of chemical behavior on the part of carbon is a fairly recent development. Scheele's discoveries of carbon compounds were the first systematic additions to those compounds long recognized, if not truly understood, such as acetic acid or vinegar. In the 115 years between Scheele's first discovery and the first edition of Beilsteins' *Handbook of Organic Chemistry* in 1881, the number of known organic compounds increased to 15 000. In the next 115 years, it increased to about 20 times that amount!

Teaching Tip

Discussion

Have students brainstorm as to some of the possible applications of fullerenes. Although no industrial production or use has emerged yet for these unusual compounds, the potential is unquestionable. (Possible suggestions include placing a radioisotope within a "buckeyball" as a type of radioactive tracer or using large "buckeytubes" as microscopic capillary tubes.)

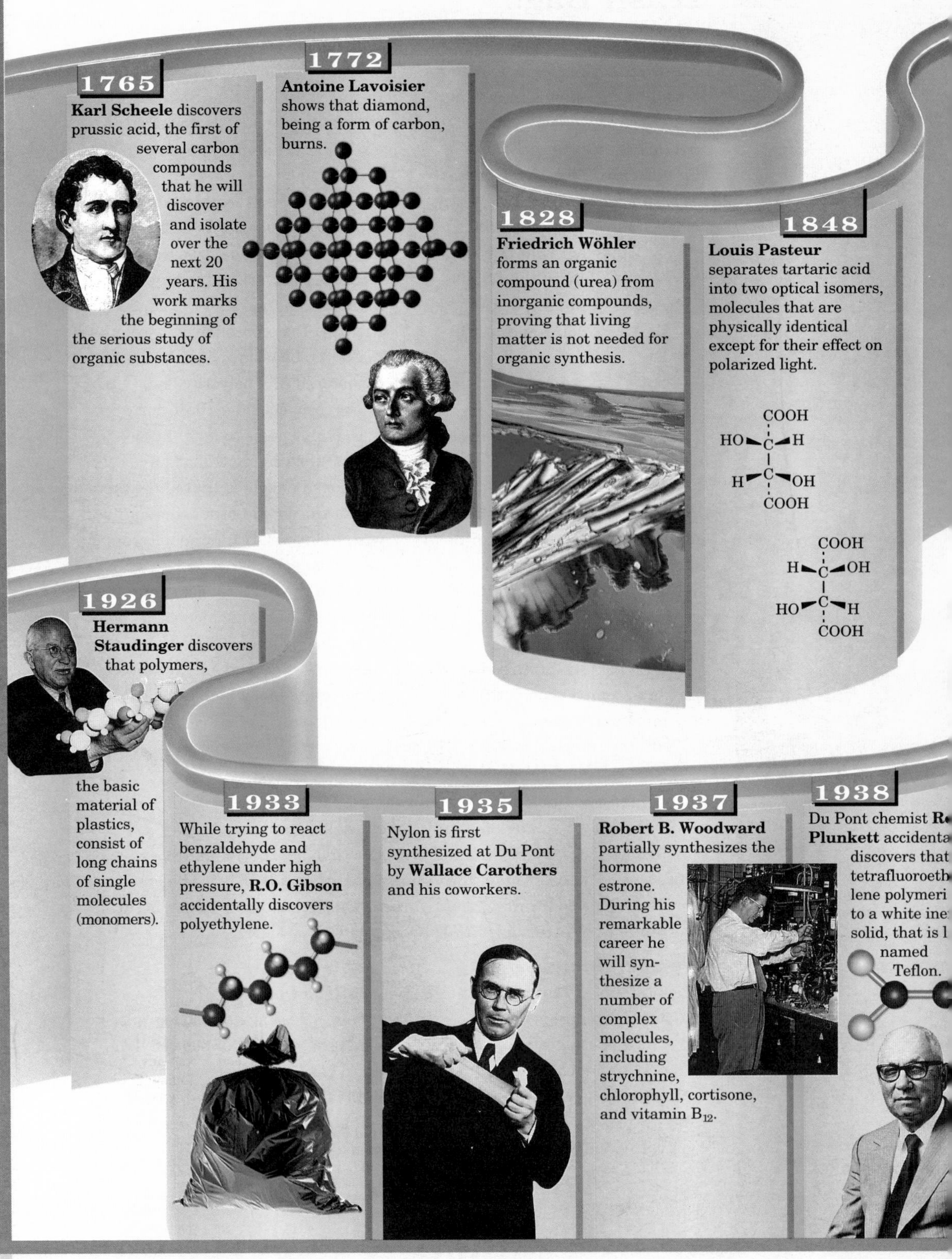

Breakthroughs in Carbon Chemistry
1765
Karl Scheele discovers prussic acid, the first of several carbon compounds that he will discover and isolate over the next 20 years. His work marks the beginning of the serious study of organic substances.
1772
Antoine Lavoisier shows that diamond, being a form of carbon, burns.
1828
Friedrich Wöhler forms an organic compound (urea) from inorganic compounds, proving that living matter is not needed for organic synthesis.
1848
Louis Pasteur separates tartaric acid into two optical isomers, molecules that are physically identical except for their effect on polarized light.
COOH
HO
C
H
H
C
OH
COOH
COOH
H
C
OH
HO
C
H
COOH
1926
Hermann Staudinger discovers that polymers, the basic material of plastics, consist of long chains of single molecules (monomers).
1933
While trying to react benzaldehyde and ethylene under high pressure, R.O. Gibson accidentally discovers polyethylene.
1935
Nylon is first synthesized at Du Pont by Wallace Carothers and his coworkers.
1937
Robert B. Woodward partially synthesizes the hormone estrone. During his remarkable career he will synthesize a number of complex molecules, including strychnine, chlorophyll, cortisone, and vitamin B12.
1938
Du Pont chemist R
Plunkett accidenta
discovers that
tetrafluoroeth
lene polymeri
to a white ine
solid, that is l
named
Teflon.

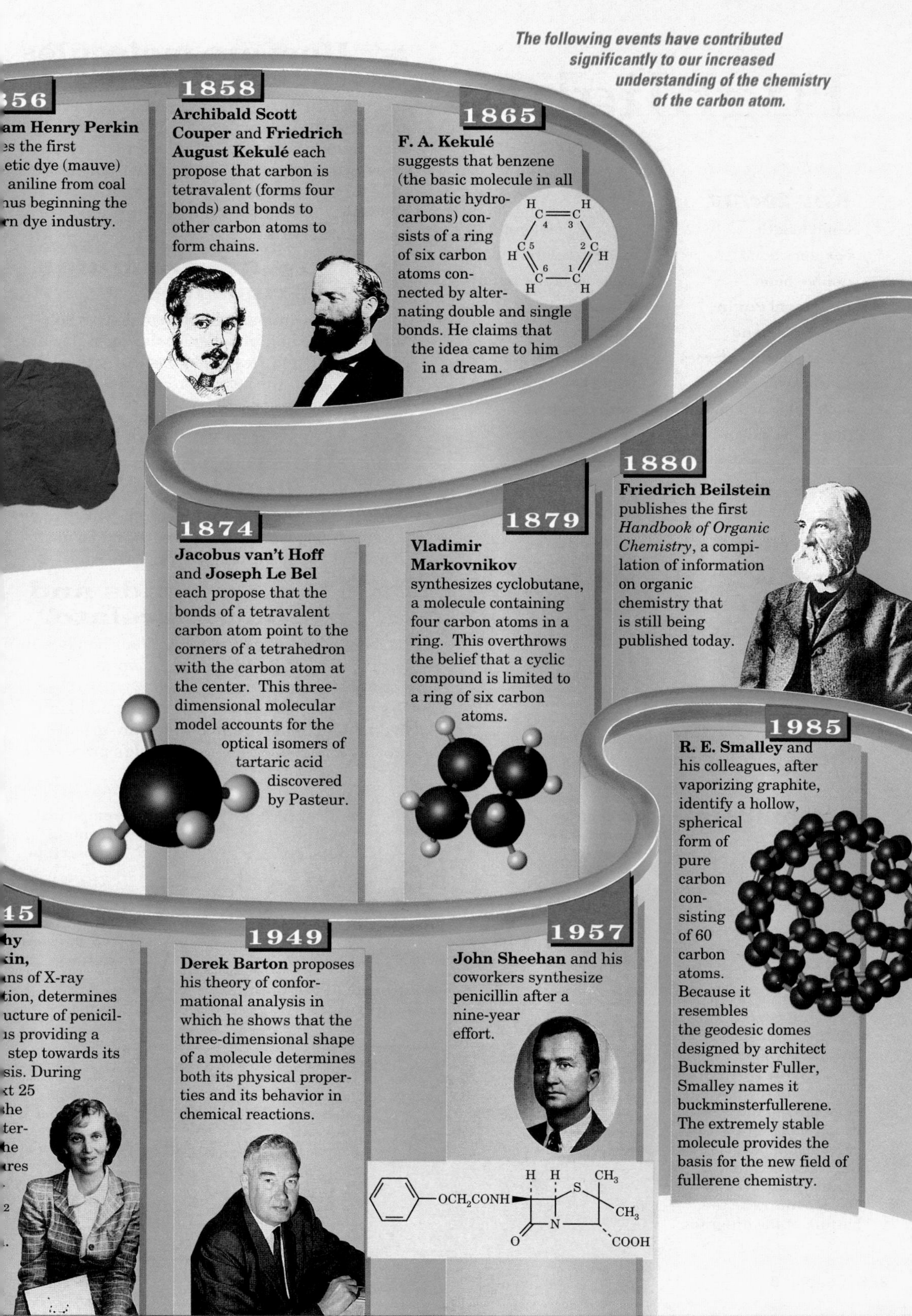

The following events have contributed significantly to our increased understanding of the chemistry of the carbon atom.
1858
Archibald Scott Couper and Friedrich August Kekulé each propose that carbon is tetravalent (forms four bonds) and bonds to other carbon atoms to form chains.
1865
F. A. Kekulé suggests that benzene (the basic molecule in all aromatic hydrocarbons) consists of a ring of six carbon atoms connected by alternating double and single bonds. He claims that the idea came to him in a dream.
1874
Jacobus van't Hoff and Joseph Le Bel each propose that the bonds of a tetravalent carbon atom point to the corners of a tetrahedron with the carbon atom at the center. This three-dimensional molecular model accounts for the optical isomers of tartaric acid discovered by Pasteur.
1879
Vladimir Markovnikov synthesizes cyclobutane, a molecule containing four carbon atoms in a ring. This overthrows the belief that a cyclic compound is limited to a ring of six carbon atoms.
1880
Friedrich Beilstein publishes the first Handbook of Organic Chemistry, a compilation of information on organic chemistry that is still being published today.
1949
Derek Barton proposes his theory of conformational analysis in which he shows that the three-dimensional shape of a molecule determines both its physical properties and its behavior in chemical reactions.
1957
John Sheehan and his coworkers synthesize penicillin after a nine-year effort.
1985
R. E. Smalley and his colleagues, after vaporizing graphite, identify a hollow, spherical form of pure carbon consisting of 60 carbon atoms. Because it resembles the geodesic domes designed by architect Buckminster Fuller, Smalley names it buckminsterfullerene. The extremely stable molecule provides the basis for the new field of fullerene chemistry.

Content Background

Chemistry and chance

Although far from being a haphazard area of chemistry, carbon chemistry has had more than its share of happy accidents. To list but a few that pertain to the timeline: Perkin's discovery of mauve resulted from a frustrated attempt to synthesize quinine, a drug used to alleviate fever; Kekulé's dream of snakes forming a ring provided him with the idea for benzene (he claimed to have had a similar vision when he suggested tetravalent carbon's formation of molecular chains); and Smalley's discovery of buckminsterfullerene. A similar instance not mentioned in the timeline is when Christian Schönbein, who is said to have used a cotton apron to wipe up nitric acid spill, discovered nitrocellulose, or guncotton, when the dried apron exploded.

Teaching Tip

Discussion

Emphasize that the expansiveness of carbon chemistry, as well as the number of accidental successes in the field, is the result of carbon's four unpaired electrons and its subsequent ability to form a wide variety of covalent bonds.

Closure Strategy

Propose these questions to the class. If carbon dioxide has a bond angle of 180°, and sulfur trioxide has a bond angle of 120°, why does methane not have a bond angle of 90°? How does VSEPR theory account for these bond angles?

CHAPTER 6

Highlights

Key terms

bond length	organic compound
covalent bond	oxidation number
double bond	polar covalent bond
functional group	polymer
hydrogen bond	single bond
intermolecular forces	structural formula
Lewis structure	triple bond
molecular compound	unshared pair
molecular formula	valence electrons
nonpolar covalent bond	VSEPR theory

Key ideas

Why do atoms share electrons?

- Covalent bonds form as atoms share pairs of electrons.
- The higher the electronegativity difference, the more ionic the bond character.
- Intermolecular forces, such as hydrogen bonds, dipole forces, and London forces, link molecules to each other.

How are molecules specified?

- A Lewis structure models the valence electron arrangement in a molecule. In most molecules, each atom is surrounded by an octet of electrons. Some molecules have more than one valid Lewis structure.
- Molecules are named using either prefixes or Roman numerals.

How are formulas represented?

- Empirical formulas show the smallest whole-number ratio of atoms. Molecular formulas are multiples of empirical formulas. Structural formulas show the spatial arrangement of atoms in molecules.

How can you tell the shape of a molecule?

- VSEPR theory states that electron pairs are as far apart as possible. This theory can be used to predict the molecular shape.

How do bonds and properties relate?

- Carbon, with its ability to form four bonds, can form molecules and covalent networks in a variety of shapes.
- Organic compounds can be named and identified by functional groups, which are groups of atoms responsible for the molecule's properties.

Key problem-solving approach: *Calculations with percentage composition*

When solving problems involving percentage composition, empirical formula, or mass, select the variable that is given from the options near the top of this diagram. Then follow the calculation pathway to the required answer quantity shown among the options near the bottom of the diagram.

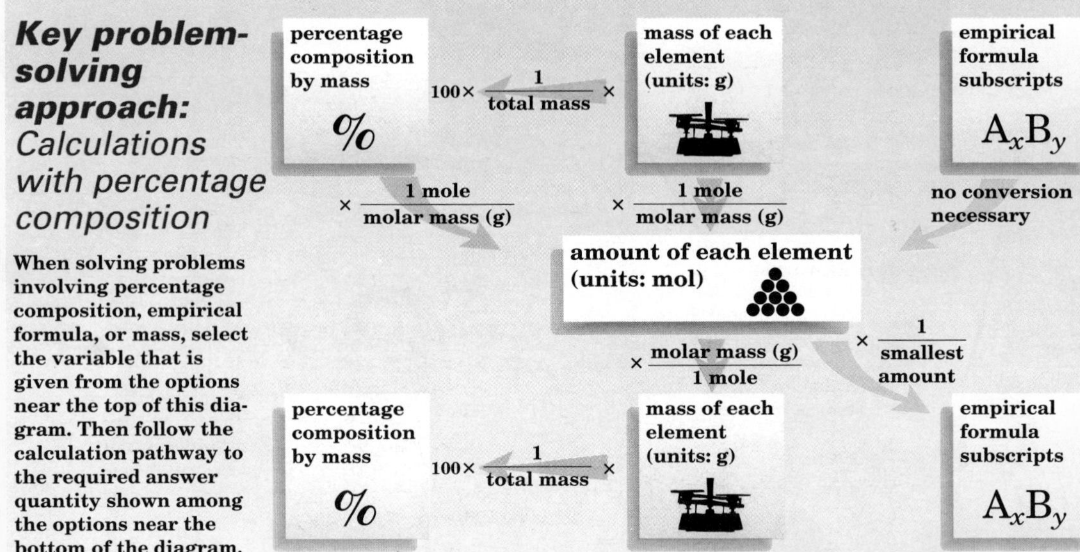

CHAPTER 6

Review and Assess

Chemical bonds

REVIEW

1. a. How does a covalent bond differ from an ionic bond?
   b. How do molecular compounds differ from ionic compounds?
2. a. Draw and label the forces that affect atoms in a covalent bond.
   b. Explain why bonding electrons are able to come relatively close to each other.
   c. Why is a spring a better model than a stick for a covalent bond?
3. a. Which is likely to have the higher bond energy: an O—C bond in HO—CH_3 with a bond length of 141 pm or an O—C bond in HO—C_6H_5 with a bond length of 136 pm?
   b. Which is likely to have the longer bond length: an H—O bond in H—OH with a bond energy of 498 kJ/mol or an H—O bond in H—OC_2H_5 with a bond energy of 436 kJ/mol?
4. Why are bond lengths given in average, not exact, values?
5. a. Explain why the melting and boiling points of molecular compounds are usually lower than those of ionic compounds.
   b. Which will have the lower boiling point, HF or HCl? (Hint: HF is more polar than HCl, and is a smaller molecule.)
6. a. Why are intermolecular forces weaker than covalent or ionic bonds?
   b. How does a hydrogen bond develop between two molecules?
7. What type of bonds do line segments *ab*, *cd*, and *de* represent in the figure below?

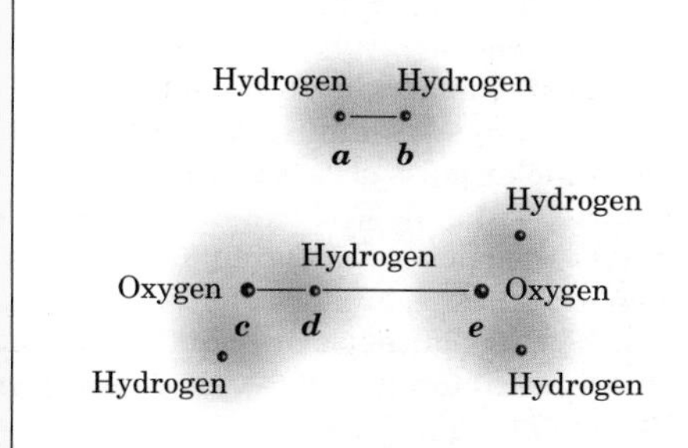

APPLY

8. Which graph below represents the inverse relationship between bond length and bond energy?

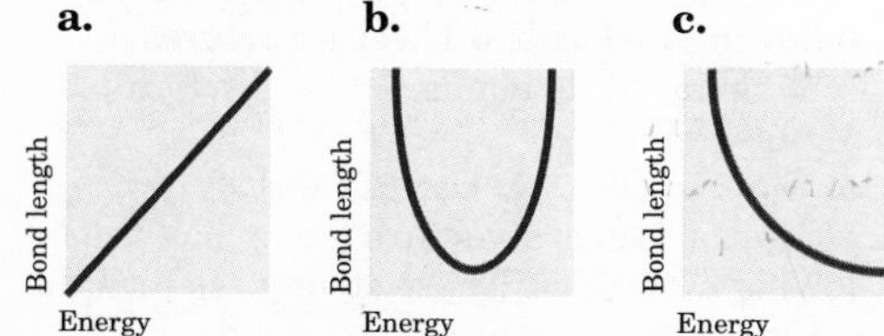

9. Use **Table 6-4** and electronegativity differences to determine whether the bonds between the following pairs of elements are ionic, polar covalent, or nonpolar covalent.
   a. Na—F b. K—Cl c. N—O
   d. H—I e. Al—O f. S—O
   g. Cl—Br
10. When a positively charged plastic comb is brought close to a stream of water, the water is deflected by the comb. Using the electronegativity difference between hydrogen and oxygen, explain why this phenomenon occurs.
11. Arrange the following pairs from strongest to weakest attraction.
    a. polar molecule and polar molecule
    b. nonpolar molecule and nonpolar molecule
    c. polar molecule and ion
    d. ion and ion

Lewis structures

REVIEW

12. Determine the number of valence electrons in Mg, K, and Al. Draw the electron-dot diagram for each.
13. Using atoms *A* and *B*, draw a Lewis structure for a diatomic molecule containing each of the following types of bonds. (Hint: assume that each atom has an octet, and remember to show unbonded pairs.)
    a. single bond b. double bond
    c. triple bond
14. What term is used when molecules have more than one valid Lewis structure?

1. a. An ionic bond is due to attraction between positive and negative ions, while a covalent bond results from atoms sharing electrons.
   b. Molecular compounds have covalent bonds and often contain individual molecules. Ionic compounds have ionic bonds and contain ions bonded within a crystal lattice structure.

2. a. Student drawings should show the force of attraction between electrons of one atom and the protons of another atom and the force of repulsion between like charges within each atom.
   b. The force of attraction between electrons and nuclei is stronger than the force of repulsion between electrons. Repulsive forces are reduced when two electrons that are close together have opposite spins.
   c. The spring better represents the bond's flexibility. Like a spring, a bond can stretch or compress, but will resist forces tending to shorten or lengthen it.

3. a. The O—C bond in HO—C_6H_5 has a greater bond energy.
   b. The O—H bond in H—OC_2H_5 has a greater bond length.

4. Because the atoms vibrate, stretch, and pull along the bond, the bond length is not fixed.

5. a. Covalently bonded molecules have only weak intermolecular forces while ionic compounds exist as crystal lattices with each ion bonded to several others. The lattice requires more energy to break apart.

b. Hydrogen bonds can form with fluorine, giving the compound HF stronger intermolecular forces and thus a higher boiling point than HCl.

6. a. because intermolecular forces do not involve transferring or sharing electrons

b. When a hydrogen atom bonds to a very electronegative element, the partial positive charge on the hydrogen atom can be attracted to the partial negative charge on another molecule.

7. Bond *ab* is a nonpolar covalent bond, *cd* is a polar covalent bond, and *de* is an intermolecular bond or, more specifically, a hydrogen bond.

8. Graph C represents the bond length–bond energy relationship.

9. a. 4.0 − 0.9 = 3.3, ionic
b. 3.0 − 0.8 = 2.2, ionic
c. 3.5 − 3.0 = 0.5, polar covalent
d. 2.5 − 2.1 = 0.4, polar covalent
e. 3.5 − 1.5 = 2.0, polar
f. 3.5 − 2.5 = 1.0, polar covalent
g. 3.0 − 2.8 = 0.2, nonpolar covalent

10. The water molecule's polar covalent bond results in the molecule having a partial positive end and a partial negative end. The negative end is attracted to the positively charged comb.

11. The sequence of attractions from strongest to weakest are d, c, a, and b.

12. Mg has two valence electrons. K has one valence electron, and Al has three valence electrons.
Mg: K· Al:

13. a. :A—B:
b. A=B
c. :A≡B:

PRACTICE

Answers to items in a black square begin on page 841

15 Draw the Lewis structure for each of the following molecules. (Hint: see Sample Problem 6A.)
a. NF_3 **b.** $GeCl_4$ **c.** CH_3OH

16. Draw the Lewis structure for each of the following molecules. (Hint: see Sample Problem 6A.)
a. PCl_3 **b.** CCl_2F_2 **c.** CH_3NH_2

17 Draw the Lewis structure for each of the following molecules. These structures include multiple bonds. (Hint: see Sample Problem 6B.)
a. O_2 **b.** CS_2 **c.** C_2H_5COOH

18. Draw the Lewis structure for each of the following molecules. These structures include multiple bonds. (Hint: see Sample Problem 6B.)
a. N_2 **b.** CH_3COCH_3 **c.** CH_3CHO

APPLY

19. Draw the Lewis structure for each of the molecules below.
a. a refrigerant with one C atom and four F atoms
b. a photographic chemical with three H atoms, one N atom, and one O atom
c. a natural-gas ingredient with two C atoms and six H atoms

20. Draw the Lewis structure for each of the following polyatomic ions.
a. OH^- **b.** H_3CCOO^- **c.** BrO_3^-

21. Write a possible Lewis structure for SF_6, which does not follow the octet rule.

22. NO_3^- has several resonance structures. Draw two.

23. Explain what is wrong with the following structures, then correct each one.

a. H—H—S:

b.

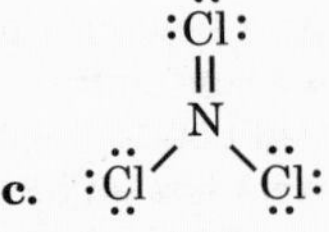

c.

Names and formulas of molecular compounds

REVIEW

24. Give the oxidation number for each atom in the following compounds.
a. H_3PO_4 **b.** CO_2 **c.** As_2O_5 **d.** HNO_3

25. Using the system of prefixes, name the following molecular compounds.
a. SF_4 **b.** SF_6 **c.** P_2O_5 **d.** XeF_4

26. Write formulas for the following molecular compounds.
a. bromine(III) fluoride
b. carbon(IV) chloride
c. carbon(IV) oxide
d. nitrogen(V) oxide
e. phosphorus pentachloride

27. What is the empirical formula for each of the following compounds?
a. $C_{12}H_{24}O_6$ **b.** Hg_2I_2 **c.** C_6H_6

28. a. How is a structural formula different from a molecular formula?
b. How is a molecular formula related to an empirical formula?

29. a. Name three objects that a classmate is wearing or using that are made of polymers.
b. Why is the formula for polyethylene sometimes written $(C_2H_4)_n$?

30. Which formula—empirical, molecular, or structural—is most useful in determining the chemical properties of a molecule? Explain your answer.

PRACTICE

31 Chemical analysis of citric acid shows that it contains 37.51% C, 4.20% H, and 58.29% O. What is its empirical formula? (Hint: see Sample Problem 6C.)

32. Chemical analysis of tetraethyl lead, an additive once used in gasoline, shows that it contains 29.71% C, 6.22% H, and 64.07% Pb. What is its empirical formula? (Hint: see Sample Problem 6C.)

33 What is the molecular formula of the molecule that has an empirical formula of CH_2O and a molar mass of 120.12 g/mol? (Hint: see Sample Problem 6D.)

34. What is the molecular formula of *para*-dichlorobenzene, an ingredient in moth balls that has a molar mass of 147.00 g/mol and an empirical formula of C_3ClH_2? (Hint: see Sample Problem 6D.)

APPLY

35. In each of the following, element X forms a molecular compound with the formula shown. What is a likely oxidation number for X in each compound?
a. XF **b.** Na_2XO_4 **c.** XCl_2 **d.** H_2XO_3

36. A 175.0 g sample of a flavor enhancer, monosodium glutamate (MSG), contains 56.15 g C, 9.43 g H, 74.81 g O, 13.11 g N, and 21.49 g Na. What is its empirical formula?

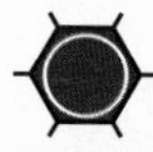

Molecular shapes

REVIEW

37. Explain why VSEPR theory predicts molecular shapes with the largest bond angles possible.

38. Name the molecular shapes shown below.

a. **b.** **c.**

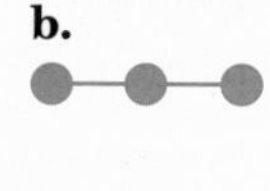

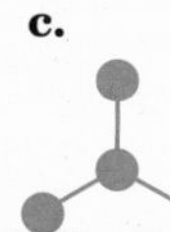

PRACTICE

39 Draw the shape and approximate bond angles for the following molecules. (Hint: see Sample Problem 6E.)
a. CCl_4 **b.** $BeCl_2$ **c.** PH_3

40. Draw the shape and approximate bond angles for the following polyatomic ions. (Hint: see Sample Problem 6E.)
a. NO_2^- **b.** NO_3^- **c.** NH_4^+

APPLY

41. Use the following Lewis structures to help you describe the molecular shape of **a**, **b**, and **c**. (The Lewis structure does not necessarily represent the geometry of the molecule.)

a. H—C≡C—H

b. :B̤̈r—P̤̈—B̤̈r:
        |
      :B̤̈r:

c. :C̤̈l—Ö̤—C̤̈l:

Organic compounds

REVIEW

42. a. Why does carbon form more compounds than chlorine does?
b. Why is diamond hard and strong?
c. Why is graphite slippery?

43. Use **Table 6-10** to name the following alkanes.
a. C_7H_{16} **b.** C_4H_{10} **c.** $C_{10}H_{22}$ **d.** C_3H_8

44. Use **Table 6-11** to identify the type of functional group from the name for each of the following organic compounds.
a. propanol **b.** acetic acid **c.** propanal **d.** hexanone **e.** butanamine

45. Classify the functional groups of the organic compounds shown below.

a. C—C—C—C(=O)—OH

b.

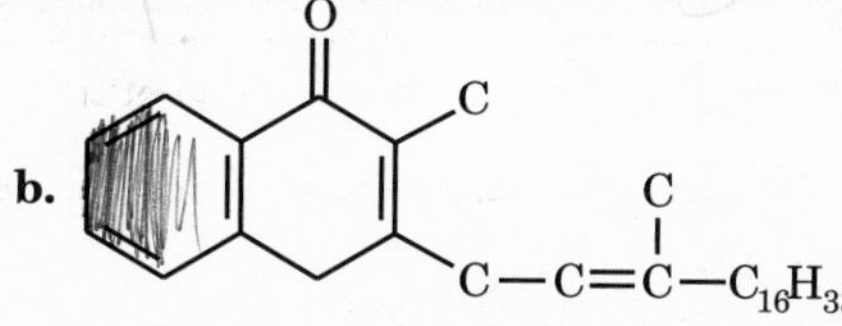

c.

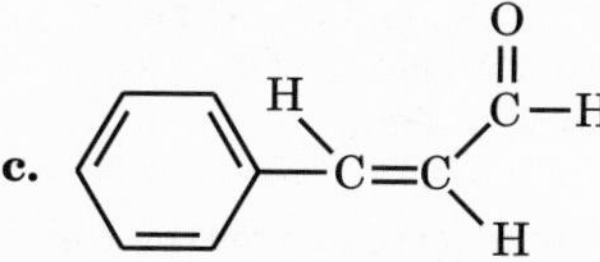

d. $(CH_3)_3N$ drawn as H—C(H)(H)—N bonded to two CH₃ groups

e.

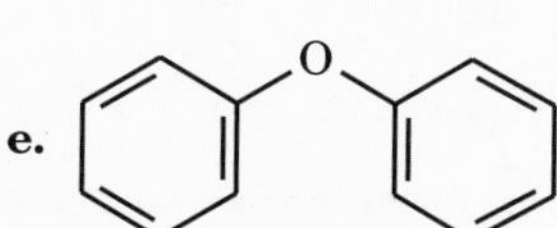

46. Compare the advantages of structural formulas with those of chemical formulas.

47. a. Which alkanes are gases at room temperature, 25°C? (Hint: use **Table 6-10**.)
b. Which do you predict will boil first: $C_{12}H_{26}$ or $C_{15}H_{32}$?

14. A Lewis structure is merely a model that has limitations. As a result, for some molecules, the actual structure is a blend of two or more possible Lewis structures.

15. a. :F̤:N̤:F̤: with :F̤: below N
b. :C̤l: above, :C̤l:G̤e:C̤l:, :C̤l: below
c. H above, H:C̤:H, :Ö: below, H below

16. a. :C̤l:P̤:C̤l: with :C̤l: below P
b. :C̤l: above, :C̤l:C̤:F̤:, :F̤: below
c. H above, H:C̤:N̤:H, H H below

17. a. Ö=Ö
b. S̤=C=S̤
c. H—C(H)(H)—C(H)(H)—C(=Ö)—Ö—H

18. a. :N≡N:
b. H—C(H)(H)—C(=Ö)—C(H)(H)—H
c. H—C(H)(H)—C(=Ö)—H

19. a. :F̤: above, :F̤:C̤:F̤:, :F̤: below
b. H—N̤(H)—Ö—H
c. H—C(H)(H)—C(H)(H)—H

20. a. [:Ö:H]⁻

b. 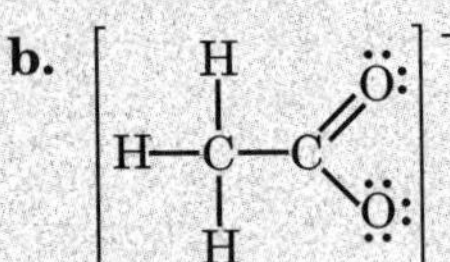

c. [:Ö:Br:Ö: / :Ö:]⁻

21. 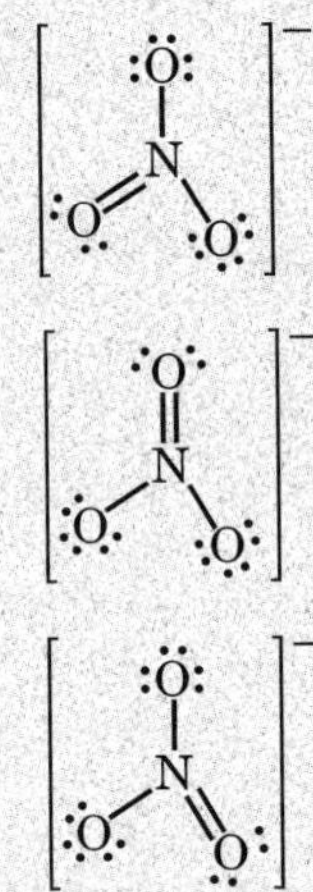

22. Accept any two of the following.

[:Ö–N(=O:)–Ö:]⁻ (three resonance structures of NO_3^-)

23. a. Accept any of the following: hydrogen is never a central atom; hydrogen does not form more than one bond; wrong number of valence electrons.

:S̈:H
H

b. octet rule not followed for carbon atom

H–C(=O)–Ö–H

c. wrong number of valence electrons on Cl

:C̈l–N̈–C̈l:
|
:C̈l:

24. a. H = +1, P = +5, O = –2
b. C = +4, O = –2
c. As = +5, O = –2
d. H = +1, N = +5, O = –2

PRACTICE

48 The organic structure for proline, an amino acid, is shown below. Draw its structural formula. (Hint: see Sample Problem 6F.)

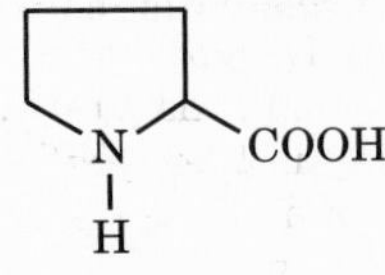

49. The organic structure for vitamin A is shown below. Draw its structural formula. (Hint: see Sample Problem 6F.)

H_3C CH_3 … OH

APPLY

50. Name the following organic compounds.
a. (OH)
b. (NH_2)
c. (O, OH)
d. Write molecular formulas for the compounds shown in **a**, **b**, and **c**.

51. Name the following organic compounds.
a. (O)
b. (O, H)
c. (OH)
d. Write molecular formulas for the compounds shown in **a**, **b**, and **c**.

52. Draw structures for the following.
a. butanoic acid, found in butter
b. nonanol, used in making artificial lemon oil
c. 2-pentanone, a solvent (Hint: the number tells you which carbon the functional group is attached to.)
d. Write molecular formulas for the compounds in **a**, **b**, and **c**.

Linking chapters

1. ***Molar mass***
Find the molar mass for each of the following polyatomic ions.
a. $[SO_4]^{2-}$ (:Ö–S–Ö: with :O: above and below)
b. $[NO_3]^{1-}$ (:Ö–N(=O:)–Ö:)
c. $[ClO_4]^{1-}$ (:Ö–Cl–Ö: with :Ö: above and :O: below)
d. [:C≡N:]$^{1-}$

2. **Theme: *Classification and trends***
The table below shows different bond energies for single, double, and triple bonds.
a. Which is the shortest type of bond?
b. Predict how the N—O bond would compare to the N═O bond in terms of bond energy and bond length.

	Carbon and carbon	Oxygen and oxygen	Carbon and oxygen	Carbon and nitrogen
Single	347 kJ/mol	142 kJ/mol	360 kJ/mol	305 kJ/mol
Double	611 kJ/mol	498 kJ/mol	728 kJ/mol	615 kJ/mol
Triple	837 kJ/mol	N/A	N/A	891 kJ/mol

USING TECHNOLOGY

1. ***Graphics calculator***
The relationship between bond length and bond energy can be shown graphically. Select the WINDOW key on your calculator, and use the following range: Xmin = 50, Xmax = 300, Xscl = 50, Ymin = 100, Ymax = 500, Yscl = 50. Enter the statistics from **Table 6-2** using the STAT mode:
a. Press STAT 1 and enter the data from **Table 6-2** in lists L_1 and L_2.
b. Press STAT ▶ 5 ENTER to compose a best-fit line.
c. Press Y= VARS 5 ▶ ▶ 7 to copy the equation, and press GRAPH to draw.
What general relationship exists between bond length and bond energy?

Alternative assessment

Performance assessment

1. Devise a set of experiments to study how well biodegradable plastics break down. If your teacher approves your plan, conduct a class experiment to test the procedure on products labeled “biodegradable.”
2. Your teacher will make available unlabeled samples of the organic chemicals listed in the table below. Develop an experiment to identify each of the unknown chemicals using the properties listed in **Table 6-11** as a guide. If your teacher approves your plan, identify the unknown substances.

Chemical	Structure	Functional group
Benzoic acid	(benzene ring–C(=O)–OH)	organic acid
Ethyl alcohol	(–OH)	alcohol
Hexane diamine	H_2N–(chain)–NH_2	amine

d. Press TRACE ▶ to move the cursor along the best-fit line. Predict the bond lengths of the H—S bond (368 kJ/mol), C—N bond (305 kJ/mol), and N—O bond (230 kJ/mol).

e. Press STATPLOT ENTER and set Plot 1 “On” for a scatterplot with “Xlist”= L_1 and “Ylist” = L_2. Then press GRAPH to compare your original data points to the best-fit line. Press TRACE and use the arrow keys to determine which two data points are farthest from the line. Describe what limitations there are on the general relationship between bond length and bond energy.

2. ***Computer presentation***
Using presentation software, create a tutorial that explains how to draw Lewis structures and provides plenty of practice.

Portfolio projects

1. ***Chemistry and you***
Inventory the food you consume in a single day. Compare the content labels from those foods, and then list the most commonly used chemicals in them. With the aid of your teacher and some reference books, try to classify the organic chemicals by their functional groups.
2. ***Research and communication***
Many everyday materials could be reused or recycled. However, recyclable products are continually deposited into landfills like the one at Fresh Kills in the story at the beginning of chapter. Research the waste-management methods of your community. Write local waste-management officials to find out what has been done to reduce the amount of landfill space needed for the community’s refuse. With your class, organize a plan to publicize the recycling programs available.
3. ***Cooperative activity***
As a class or small group, research preservatives for various foods. Examine their chemical structure. Determine a way to test for organic functional groups of possibly hazardous preservatives.
4. ***Research and communication***
Covalently bonded solids such as silicon, an element used in computer components, are harder than pure metals. Research theories that explain the hardness of covalently bonded solids and their usefulness in the computer industry. Present your findings to the class.
5. ***Chemistry and you***
Because margarine spreads are now used more often than margarine sticks and contain a smaller percentage of fat than margarine sticks (80% fat) or butter, baking recipes are being changed to specify butter instead of margarine. Research the percentage of fat in margarine spreads available to you. Test the effect of using spreads with varying percentages of fat in a recipe. What is the minimum percentage of fat that will still provide adequate results? Try several recipes. Are some recipes more sensitive to ingredient changes than others?

25. a. sulfur tetrafluoride
b. sulfur hexafluoride
c. diphosphorus pentoxide
d. xenon tetrafluoride

26. a. BrF_3
b. CCl_4
c. CO_2
d. N_2O_5
e. PCl_5

27. a. C_2H_4O
b. HgI
c. CH

28. a. Both give the number and type of atoms in a molecule, but only the structural formula shows how the atoms are arranged.
b. The molecular formula is a multiple of the empirical formula.

29. a. Answers will vary. Students may name articles of clothing, shoes, or school supplies made of synthetics, plastics, or other polymers.
b. The number of ethylene groups in a polyethylene molecule can vary.

30. The structural formula is the most useful because it shows the arrangement and bonding of the atoms.

31. $37.51\text{ g C} \times \frac{1\text{ mol C}}{12.01\text{ g C}} = 3.123\text{ mol C}$

$\frac{3.123\text{ mol C}}{3.123} = 1.000\text{ mol C}$

$4.20\text{ g H} \times \frac{1\text{ mol H}}{1.01\text{ g H}} = 4.16\text{ mol H}$

$\frac{4.16\text{ mol H}}{3.123} = 1.33\text{ mol H}$

$58.29\text{ g O} \times \frac{1\text{ mol O}}{16.00\text{ g O}} = 3.643\text{ mol O}$

$\frac{3.643\text{ mol O}}{3.123} = 1.167\text{ mol O}$

1.000 mol C: 1.33 mol H: 1.167 mol O
whole-number ratio: 6 mol C: 8 mol H: 7 mol O

$C_6H_8O_7$

Answers from page 202

Solutions for Additional Examples 6B

b. :C≡O:

c. $\ddot{O}=\ddot{O}$

d. :F̤—N̈=N̈—F̤:

e. 30 e^- are available; three double bonds in a ring structure;. also known as inorganic benzene

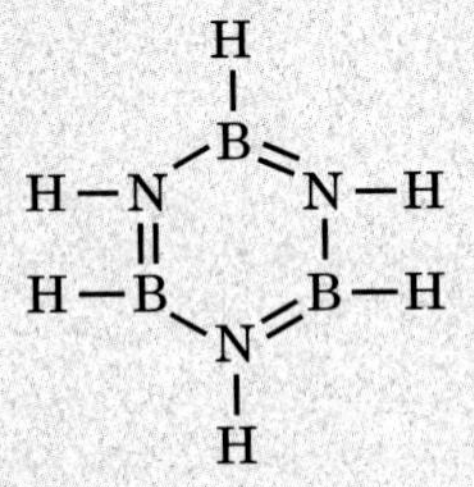

Answers from page 204

Teaching Tip

Additional Examples

a. 32 e^-; no resonance

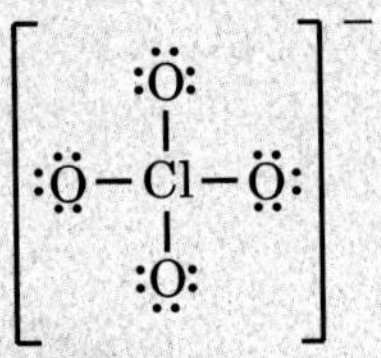

b. 216 e^-; resonance

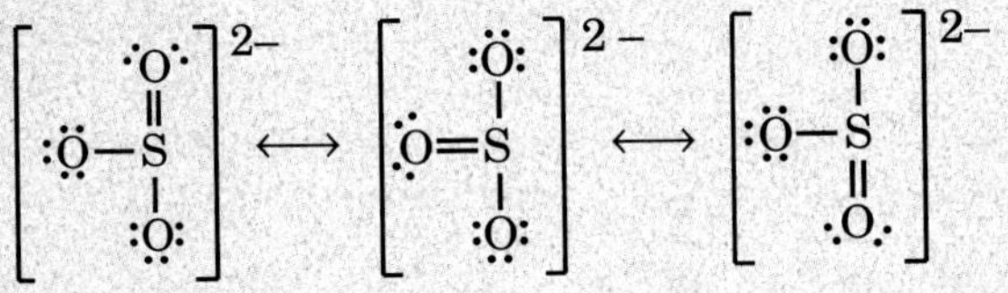

c. 24 e^-; resonance

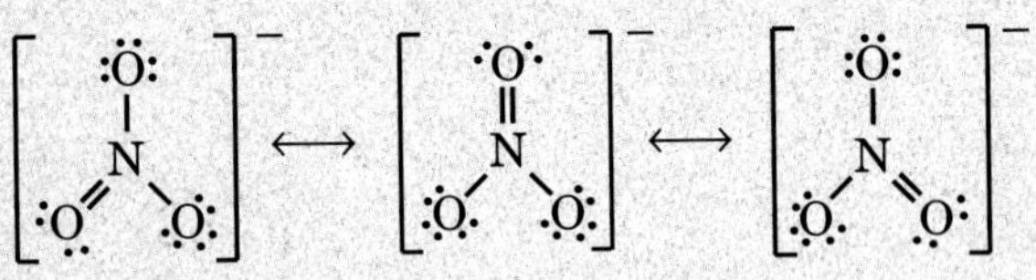

d. 16 e^-; resonance

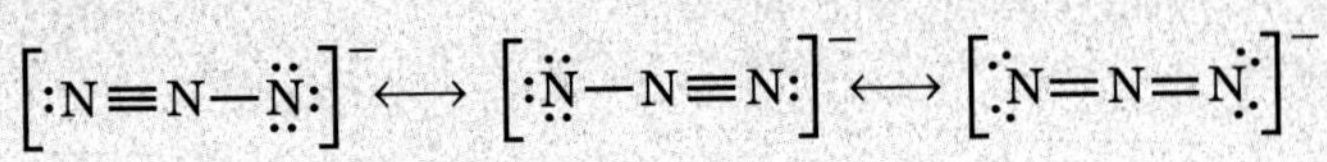

[:N≡N—N̤̈:]$^-$ ⟷ [:N̤̈—N≡N:]$^-$ ⟷ [N̤=N=N̤]$^-$

e. 34 e^-; resonance

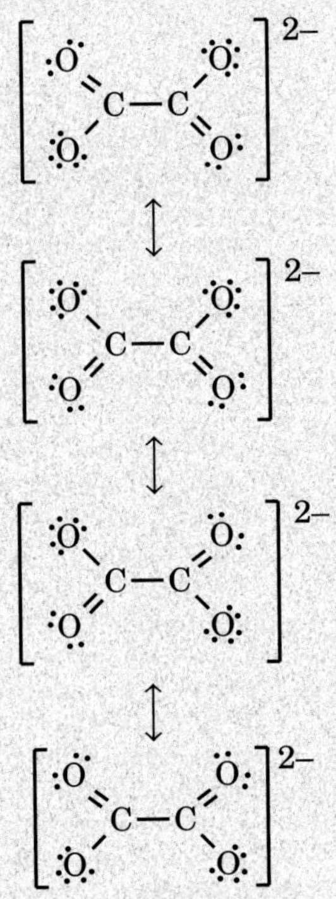

Answers from page 206

Answers to Section Review

11. a. :B̤̈r—B̤̈r:

b.

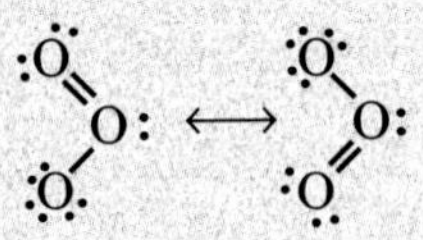

c.

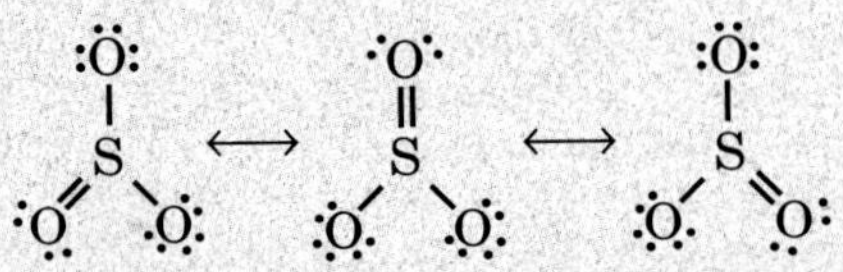

d.

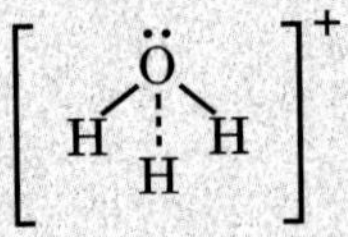

12 a. silicon dioxide
b. sulfur dioxide
c. carbon tetrafluoride
d. dinitrogen trioxide

13. a. P = +5, Br = –1
b. P = +3, Br = –1
c. H = +1, S = +6, O = –2
d. S = +6, O = –2
e. C = +4, Cl = –1

14. a. N_2O_4
b. PCl_3
c. N_2O_5

15. H—C—C—C—C—C—C—C—C—C—C—C—C—H (each C bonded to an H above and below)

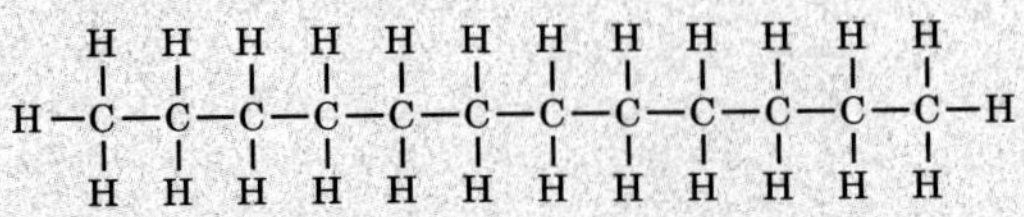

Answers from page 215

Answers to Section Review

23. SCl_2: bent, 2 lone pairs on S
PI_3: pyramidal, 1 lone pair on P
NCl_3: pyramidal, 1 lone pair on N

24. 109.5°; each C atom is singly bonded to four other atoms, resulting in a tetrahedral shape. Linear geometry requires an angle of 180°.

Use with page 219.

```
    H       H  H
    |       |  |
 H—C—O—C—C—H
    |       |  |
    H       H  H
```

methyl ethyl ether

```
    H     O
    |    //
 H—C—C—H
    |
    H
```

ethanal

```
    H  H  H  H  H    O
    |  |  |  |  |   //
 H—C—C—C—C—C—C—OH
    |  |  |  |  |
    H  H  H  H  H
```

hexanoic acid

```
    H  H  H  H       H  H  H  H
    |  |  |  |       |  |  |  |
 H—C—C—C—C—O—C—C—C—C—H
    |  |  |  |       |  |  |  |
    H  H  H  H       H  H  H  H
```

dibutyl ether

```
    H  H  H  H  H  H  H    O
    |  |  |  |  |  |  |   //
 H—C—C—C—C—C—C—C—C
    |  |  |  |  |  |  |   \
    H  H  H  H  H  H  H    OH
```

octanoic acid

```
    H  H  H  H  H  H    O
    |  |  |  |  |  |   //
 H—C—C—C—C—C—C—C
    |  |  |  |  |  |   \
    H  H  H  H  H  H    OH
```

heptanoic acid

```
    H  H  O  H
    |  |  ||  |
 H—C—C—C—C—H
    |  |     |
    H  H     H
```

butanone

```
    H  H  H  H  H  H
    |  |  |  |  |  |
 H—C—C—C—C—C—C—OH
    |  |  |  |  |  |
    H  H  H  H  H  H
```

1-hexanol

```
    H  H  H  H  OH H
    |  |  |  |  |  |
 H—C—C—C—C—C—C—H
    |  |  |  |  |  |
    H  H  H  H  H  H
```

2-hexanol

```
    H  H  H  H  H    O
    |  |  |  |  |   //
 H—C—C—C—C—C—C—H
    |  |  |  |  |
    H  H  H  H  H
```

hexanal

```
    H  H     H  H
    |  |  ..  |  |
 H—C—C—N—C—C—H
    |  |  |  |  |
    H  H  H  H  H
```

diethyl amine

```
    H
    |
 H—C—OH
    |
    H
```

methanol

```
    H  H  H  O  H
    |  |  |  ||  |
 H—C—C—C—C—C—H
    |  |  |     |
    H  H  H     H
```

2-pentanone

```
    H  H  H  H    O
    |  |  |  |   //
 H—C—C—C—C—C—OH
    |  |  |  |
    H  H  N  H
          ..
         /  \
        H    H
```

3-amino-pentanoic acid

Use with page 220.

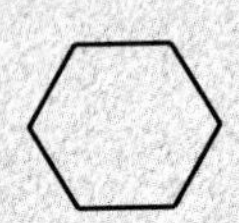

cyclohexane

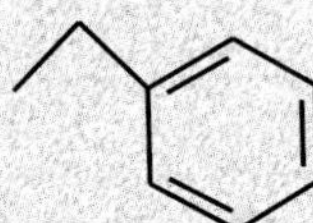

ethyl benzene

O OH NH$_2$

phenylalanine or α-amino-β-phenyl propionic acid

O OH

benzoic acid

O H

butanal

O

methyl ethyl ether

Continued from page 231

32. $64.07\ \text{g Pb} \times \frac{1\ \text{mol Pb}}{207.2\ \text{g Pb}} = 0.3092\ \text{mol Pb}$

$\frac{0.3092\ \text{mol Pb}}{0.3092} = 1.000\ \text{mol Pb}$

$29.71\ \text{g C} \times \frac{1\ \text{mol C}}{12.01\ \text{g C}} = 2.474\ \text{mol C}$

$\frac{2.474\ \text{mol C}}{0.3092} = 8.001\ \text{mol C}$

$6.22\ \text{g H} \times \frac{1\ \text{mol H}}{1.01\ \text{g H}} = 6.16\ \text{mol H}$

$\frac{6.16\ \text{mol H}}{0.3092} = 19.9\ \text{mol H}$

8 mol C: 20 mol H: 1 mol Pb

$C_8H_{20}Pb$

33. The molar mass for the empirical formula is 30.03 g/mol.

$\frac{120.12\ \text{g/mol}}{30.03\ \text{g/mol}} = 4$

The molecular formula is $4 \times CH_2O$, or $C_4H_8O_4$.

34. The molar mass for the empirical formula is 73.50 g/mol.

$\frac{147.00\ \text{g/mol}}{73.50\ \text{g/mol}} = 2.000$

The molecular formula is $2 \times C_3ClH_2$, or $C_6Cl_2H_4$.

35. a. 1+ **b.** 6+ **c.** 2+ **d.** 4+

36. $\frac{13.11\ \text{g N}}{1} \times \frac{1\ \text{mol N}}{14.01\ \text{g N}} = 0.9358\ \text{mol N}$

$\frac{0.9358\ \text{mol N}}{0.9348} = 1.00\ \text{mol N}$

$\frac{21.49\ \text{g Na}}{1} \times \frac{1\ \text{mol Na}}{22.99\ \text{g Na}} = 0.9348\ \text{mol Na}$

$\frac{0.9348\ \text{mol Na}}{0.9348} = 1.00\ \text{mol Na}$

$\frac{56.15\ \text{g C}}{1} \times \frac{1\ \text{mol C}}{12.01\ \text{g C}} = 4.675\ \text{mol C}$

$\frac{4.675\ \text{mol C}}{0.9348} = 5.00\ \text{mol C}$

$\frac{9.43\ \text{g H}}{1} \times \frac{1\ \text{mol H}}{1.01\ \text{g H}} = 9.33\ \text{mol H}$

$\frac{9.33\ \text{mol H}}{0.9348} = 9.98\ \text{mol H}$

$\frac{74.81\ \text{g O}}{1} \times \frac{1\ \text{mol O}}{16.00\ \text{g O}} = 4.676\ \text{mol O}$

$\frac{4.676\ \text{mol O}}{0.9348} = 5.00\ \text{mol O}$

$C_5H_{10}O_5NNa$

37. The larger the bond angle for the bonds within a molecule, the farther apart the electron pairs will be, and the weaker the forces of repulsion among them.

38. a. tetrahedral **b.** linear **c.** trigonal planar

39. a.

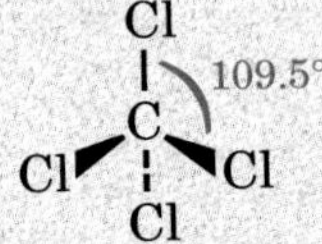

b. 180° Cl—Be—Cl

c. P H H H 109.5°

40. a.

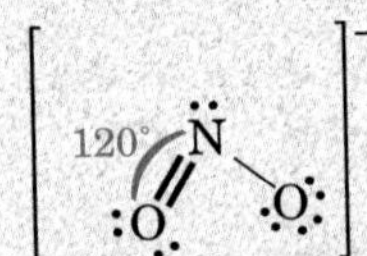

b.

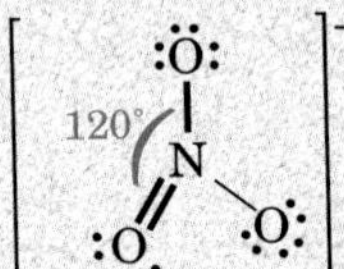

c.

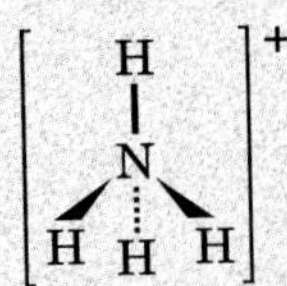

41. Molecule A is linear, molecule B is pyramidal, like a four-sided pyramid with triangular sides, and molecule C is bent.

42. a. Because it can form four bonds, carbon can easily form long-chain compounds and rings; it can bond with itself and many other elements.
b. A tetrahedral arrangement makes a strong three-dimensional network of covalently bonded carbon atoms that are attracted to each other.
c. The trigonal planar network of covalently bonded carbon atoms exist in layers. Only weak intermolecular bonds hold the layers together, so they can slide against each other.

43. a. heptane
b. butane
c. decane
d. propane

44. a. alcohol
b. organic acid
c. aldehyde
d. ketone
e. amine

45. a. organic acid
b. ketone; double bond
c. aldehyde; double bond
d. amine
e. ether

46. Structural formulas show the arrangement of all atoms and bonds in a molecule, while chemical formulas indicate only the number and kinds of atoms.

47. a. Methane, ethane, propane, and butane are gases at room temperature.
b. $C_{12}H_{26}$ will boil first.

48.

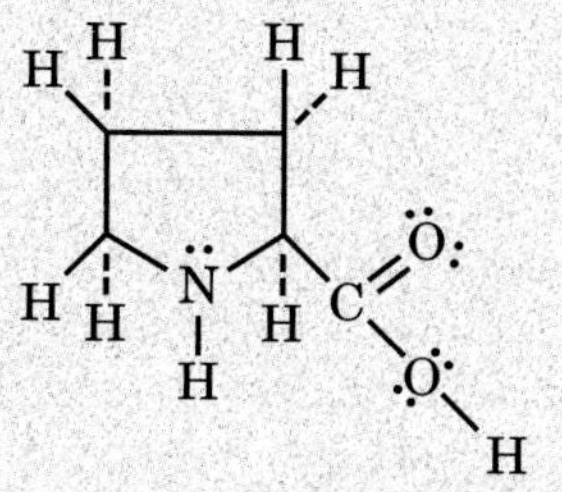

49.

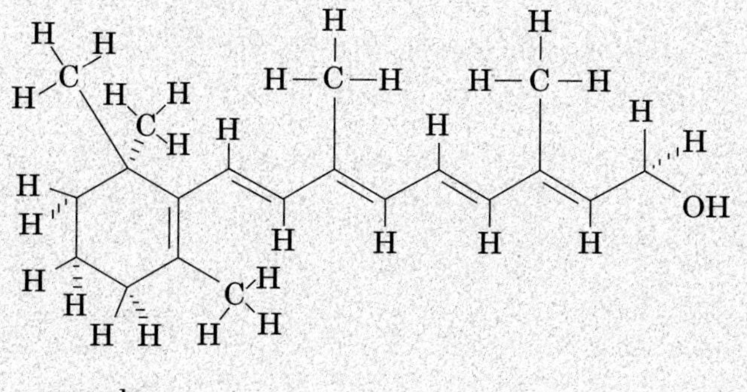

50. a. nonanol
b. butanamine
c. hexanoic acid
d. $C_9H_{20}O$, $C_4H_{11}N$, $C_6H_{12}O_2$

51. a. butanone
b. butanal
c. butanol
d. C_4H_8O, C_4H_8O, and $C_4H_{10}O$

52. a.

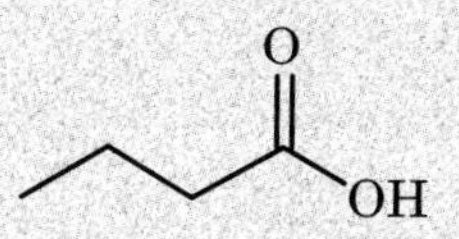

b.

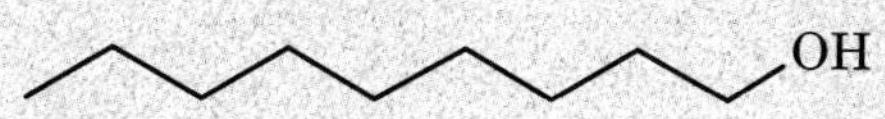

c.

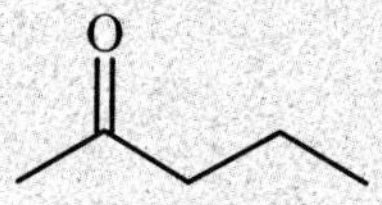

d. $C_4H_8O_2$, $C_9H_{20}O$, and $C_5H_{10}O$

Linking chapters

1. a. S: 1×32.07 g/mol = 32.07 g/mol
O: 4×16.00 g/mol = 64.00g/mol
96.07 g/mol
b. N: 1×14.01 g/mol = 14.01 g/mol
O: 3×16.00 g/mol = 48.00 g/mol
62.01 g/mol
c. Cl: 1×35.45 g/mol = 35.45 g/mol
O: 4×16.00 g/mol = 64.00 g/mol
99.45 g/mol
d. C: 1×12.01 g/mol = 12.01 g/mol
N: 1×14.01 g/mol = 14.01 g/mol
26.02 g/mol

2. a. Triple bonds are shortest.
b. The N═O bond is shorter and requires more energy to break than the N—O bond.

USING TECHNOLOGY

1. c. An inverse relationship exists between bond length and bond energy.
d. H—S, 135 pm; C—N, 175 pm;, and N—O, 217 pm.
e. The H—H and H—Cl bonds are the furthest from the line. They are also the shortest and highest in energy of the bonds listed in **Table 6-2**. Thus the inverse relationship is not as easy to predict for short, high-energy bonds.

2. Answers will vary. Be sure Lewis structures are correct.

Performance assessment

1. Verify experimental procedure is safe and will differentiate biodegradability.

2. Verify procedures are safe and will differentiate benzoic acid, ethyl alcohol, diethyl ether, and aniline.

Portfolio projects

1. Answers will vary. Verify the classification into functional groups of chemicals commonly used in food.

2. Answers will vary. Verify that a cost-benefit analysis is done to justify recycling.

3. Answers will vary. Check that tests will identify only the preservative.

4. Answers will vary. Verify that the theories explain hardness only of substances with covalent bonds.

5. Answers will vary. Fat content of margarine spreads ranges from 0% to 70%. In general, the lower the fat content, the greater the effect on recipes, but some recipes are more sensitive than others.

7 Chemical Equations

PLANNING FOR

Chapter Overview

- **Section 7-1** establishes the definitions of exothermic and endothermic reactions, examines the different types of reactions that can occur, and emphasizes the conservation principles regarding matter and energy. The activation energy of a reaction and reaction reversibility are also given specific attention.
- **Section 7-2** covers the principles and techniques involved in balancing chemical equations.
- **Section 7-3** examines the concepts of enthalpy change in a reaction, reaction completion, and the different phases of reactants and products.
- **Section 7-4** examines the applications of chemical reactions and the various conditions under which reactions may or may not occur (attention is given in particular to the fire extinguishing abilities of carbon dioxide). The five basic types of reactions (combustion, decomposition, synthesis, single displacement, and double displacement) are explored in detail.

Concept Base

Students must have mastered the following concepts prior to this chapter: covalent bonding (Chapter 6); ionic bonding (Chapter 5); diatomic molecules and polyatomic molecules (Chapter 6); the law of conservation of matter (Chapter 2); the number of atoms and molecules in a mole (Avogadro's number) (Chapter 4); and the basic properties of valence with regard to the families of the periodic table (Chapter 4).

Looking Ahead

The use of the balanced equation in determining stoichiometric relationships will be discussed in Chapter 8. Enthalpy and the causes of reactions will be further examined in Chapter 9.

Laboratory Equipment Needs

Demonstrations

The listing of materials for Chapter 7 demonstrations begins on page T46.

Laboratory Experiments

Materials and preparation instructions for Chapter 7 are found as follows.

Text		*Page*
Investigation 7A	Single Displacement Reactions—Industrial Waste Recycling	704

Laboratory Experiments Manual

(refer to the table of contents in the manual)

Exploration 7-1	Chemical Reactions and Solid Fuel

Pacing: 7 days	Inquiry Teaching Strategies
7-1 What is a chemical reaction?	**Demonstrations** ■ Demo 1 *Chemical reactions* ■ Demo 2 *Production of copper (II) sulfide*
7-2 How are reactions written?	
7-3 What information is in an equation?	**Demonstrations** ■ Demo 3 *Exothermic and endothermic reactions*
7-4 How can reactions be used?	**Demonstrations** ■ Demo 4 *Single displacement generation of CO_2* ■ Demo 5 *Reaction of CO_2 with Mg* ■ Demo 6 *Displacement of silver with copper* **Laboratory Experiments** ■ Inv 7A *Single Displacement Reactions—Industrial Waste Recycling* ■ ■ Exp 7-1 *Chemical Reactions and Solid Fuel*

Key ■ Teaching Transparencies
■ Annotated Teacher's Edition
■ Laboratory Experiments Manual

Visual Teaching Strategies	Review and Practice Strategies
	Text Reviews ■ Section Review 1–7 ■ Chapter Review 1–5; LC 1 **Study Guide Worksheets** ■ ■ 7-1 Concept Review
	Text Reviews ■ Additional Example 7A ■ Practice 1–2 ■ Section Review 8–13 ■ Chapter Review 6–16; LC 2 **Study Guide Worksheets** ■ ■ 7-2 Concept Review ■ ■ 7-2 Skillsheet *Conservation of Mass* ■ ■ 7-2 Practice *Balancing Chemical Equations*
Transparency Masters ■ T 7-3 *Symbols Used in Chemical Equations*	**Text Reviews** ■ Section Review 14–18 ■ Chapter Review 17–22; UT 1 **Study Guide Worksheets** ■ ■ 7-3 Concept Review
Transparencies ■ P 265 *Determining Reaction Types* **Transparency Masters** ■ T 7-8 *Activity Series*	**Text Reviews** ■ Concept Review 19–21 ■ Additional Examples 7B, 7C ■ Practice 1–6 ■ Section Review 22–26 ■ Chapter Review 23–38; LC 3 **Study Guide Worksheets** ■ ■ 7-4 Practice *Equations of Basic Reaction Types*

■ Pupil's Text
■ Study Guide
■ Teaching Resources

Exp *Exploration*
Inv *Investigation*

LC *Linking Chapters*
UT *Using Technology*

Assessment Options

Traditional Assessment

Test Generator
Instructional Objectives Measured:
Content mastery
Select from items 1 to 75

Performance Assessment Options

(see page T36 and T643-15 for scoring rubrics)
Students demonstrate mastery of objectives in a hands-on environment. You may want some of these materials in the Portfolio as well.

Investigation 7A (text)
Instructional Objectives Measured:
Content mastery, Use of scientific methodology, Problem solving skills, Proficiency in written communication

Performance Assessments, page 269
1. Designing an experiment
Instructional Objectives Measured:
Content mastery, Scientific and chemical literacy, Problem solving skills, Use of scientific methodology, Proficiency in written communication

Portfolio Options

(see page T36 for scoring rubrics)
Students provide a written rationale for each selection made for the portfolio.

Concept Maps
Students use vocabulary from the Chapter Review to build a concept map.
Instructional Objectives Measured:
Content mastery

Formal Laboratory Reports
Exploration 7-1 (lab manual)
Instructional Objectives Measured:
Content mastery, Scientific and chemical literacy, Problem solving skills, Use of scientific methodology, Use of technology, Proficiency in written communication

Portfolio Projects, page 269
Items 1 and 2. Chemistry and you
Instructional Objectives Measured:
Content mastery, Scientific and chemical literacy, Proficiency in written communication

Items 3 and 4. Research and communication
Instructional Objectives Measured:
Content mastery, Scientific and chemical literacy, Proficiency in oral and written communication

INTRODUCING THE CHAPTER

CHAPTER 7 Chemical Equations

Story Background

The task of restoring the Statue of Liberty was part of a project initiated in 1981 (the hundredth anniversary of the statue's construction) by the National Park Service. The project was completed in 1986.

Many chemical factors involving both the statue's deterioration and endurance had to be considered. For instance, while formation of iron(III) oxide led to the rusting of the iron ribs, the formation of the patina on the copper skin protected it from any further degradation. This protection is the result of a balanced mixture of the compounds brochantite ($CuSO_4{\cdot}3Cu(OH)_2$) and antlerite ($CuSO_4{\cdot}2Cu(OH)_2$). Yet changes in the proportions of these two compounds suggest that the patina, and thus the few millimeters of copper skin beneath it, might be endangered. The statue's north side has a higher concentration of antlerite, which is more susceptible to dust erosion. Debate continues on whether this degradation is caused by or completely unrelated to acid rain.

The Statue of Liberty's restoration exemplifies both the multiple applications of chemical reactions in the practical art of restoring public structures and the serious effects that both chemical and physical aspects of the environment have on the permanence of these structures.

About the Illustration

This 300-ton aluminum scaffolding surrounded the Statue of Liberty during the last two years of its restoration. Of the three Statues of Liberty built by Frederic Bartholdi, the one in New York is the largest. The other two are in Paris, France.

Reactivity and the Statue of Liberty

Several people climbed inside the Statue of Liberty even though it was closed to visitors that day in 1985. One man pointed an unusual tool at the inside of the statue. White powder shot out of the tool, scouring the interior surface. What was he doing? Why?

The man was not a vandal bent on destroying the statue. He was part of a team working with the National Park Service on an $85-million project to restore the 100-year-old statue and museum.

The tool was a combination blaster/vacuum cleaner that used sodium hydrogen carbonate, $NaHCO_3$, to remove a layer of tar from the inside of the statue. The tar was supposed to have kept salt water from leaking into the statue, where it could make contact between the statue's copper outer shell and its iron internal support framework. But over time the tar deteriorated, and the statue leaked.

A team of workers led by architect John Robbins, architectural conservator Fran Gale, and metallurgist Norman Nielson had searched for something rough enough to remove the tar but gentle enough to avoid damaging the copper skin, which is only slightly thicker than a penny. Before deciding on $NaHCO_3$, the team tried ground walnut shells, corncobs, glass beads, rice, and even sugar. Scouring all 300 sheets of copper in the statue required 40 tons of $NaHCO_3$.

When the statue was built in 1886, the French sculptor Frederic Bartholdi chose copper because of its softness and malleability, the same properties that make it difficult to clean. After being hammered into shape, the sheets were riveted together and hung on an iron structure, an internal tower surrounded by an armature framework of 1800 iron bars. The structure was designed by the French engineer Alexandre-Gustave Eiffel, who also designed the famous tower in Paris that bears his name. Iron was the strongest and most flexible material available at a reasonable cost.

Copper and iron, like other metals, can react to form ionic compounds. The green coating, or *patina*, on the outside of the statue resulted when the copper reacted to form several green copper compounds, including CuO, $CuSO_4 \cdot 3Cu(OH)_2$, and $CuSO_4 \cdot 2Cu(OH)_2$. These compounds do not dissolve well in water, and they therefore protect the remaining copper from reacting with substances in precipitation and the air.

The iron supports had also reacted to form a series of hydrates, such as $Fe_2O_3 \cdot 3H_2O$, which is often referred to as iron(III) hydroxide, because of its empirical formula, $Fe(OH)_3$. This compound is also commonly known as "rust." It is a weak, flaky compound that crumbles easily. Some of the iron bars had rusted to only one-half of their original thickness.

- *Why did the iron support beams corrode so much, while the thin copper skin did not?*
- *How could future corrosion be prevented?*

In this chapter, you will study reactions and how they can be controlled, information that will enable you to answer these questions.

Tapping Prior Knowledge

Use the following questions to start students thinking about and discussing the concepts to be covered.

Experience-based questions

1. Why is it a common practice to paint iron railings to prevent rusting?
2. Why do most plastics burn?
3. Why is water not recommended for extinguishing oil fires?

Section 7-1

FOCUS

Lesson Starter
Begin the lesson by having students list three familiar processes that are chemical reactions. Discuss the lists in class.

Demonstration 1

Chemical reactions

Approximate time: 5 min

1. Label three petri dishes *1*, *2*, and *3*, and place them at the corners of an overhead projector.
2. Half-fill dish *1* with vinegar. Add 1 g of baking soda, $NaHCO_3$. This produces CO_2 gas.
3. Half-fill dish *2* with 0.5 M $CuSO_4$. Add 5 g of mossy zinc. The zinc replaces the copper in solution.
4. Half-fill dish *3* with 1 M H_2SO_4. Attach one end of two 10 cm pieces of nichrome wire to each pole of a 9 V battery. Place the free ends of the wires in the acid to start electrolysis. Hydrogen and oxygen evolve at the wires.

SAFETY
Safety goggles and a lab apron must be worn. Students must be 10 ft or more from demonstration. In case of an acid spill, dilute the spill with water and, wearing gloves, soak up the spill with cloth or paper towels. Rinse the towels, neutralize the rinsings with 1 M NaOH, and flush down the drain.

DISPOSAL
Dish *1*—Pour down the drain and flush with water. Dish *2*—Add 1 M NaOH to precipitate copper, and filter. Neutralize filtrate with 1 M HCl, and pour down the drain. Put copper in trash.

What is a chemical reaction?

Section Objectives

Explain what a chemical reaction is.

Distinguish between evidence that suggests a chemical reaction may have occurred and conclusive proof of a chemical change.

Use a word equation to describe a chemical reaction.

Explain why some reactions are endothermic and others are exothermic.

Relate conservation of mass to the rearrangement of atoms in a chemical equation.

Explain why reactants must be brought into contact for a reaction to occur.

Chemical change

You can't avoid chemical reactions. You see them taking place all the time—in rusting iron, in leaves that change color in the fall, in milk that turns sour, and in exhaust fumes emitted from cars. Chemical reactions are happening not only all around you, but also inside of you. Digestion and respiration each involve chemical reactions.

A *chemical reaction* is simply a chemical change—the process by which one or more substances are changed into one or more different substances. In any chemical reaction, the original substances are known as *reactants,* while the resulting substances are called *products.* One chemical reaction that you probably see quite often is shown in **Figure 7-1**.

Evidence of a chemical change

To understand what is involved in a chemical reaction, consider **Figure 7-1**. When a safety match is struck against a matchbook cover, several observations can be made. Heat and light are released in the form of flames; the substance on the match head is enveloped in the flames and changes color; and a puff of unpleasant-smelling smoke rises from the match. In addition, a hissing noise can be heard.

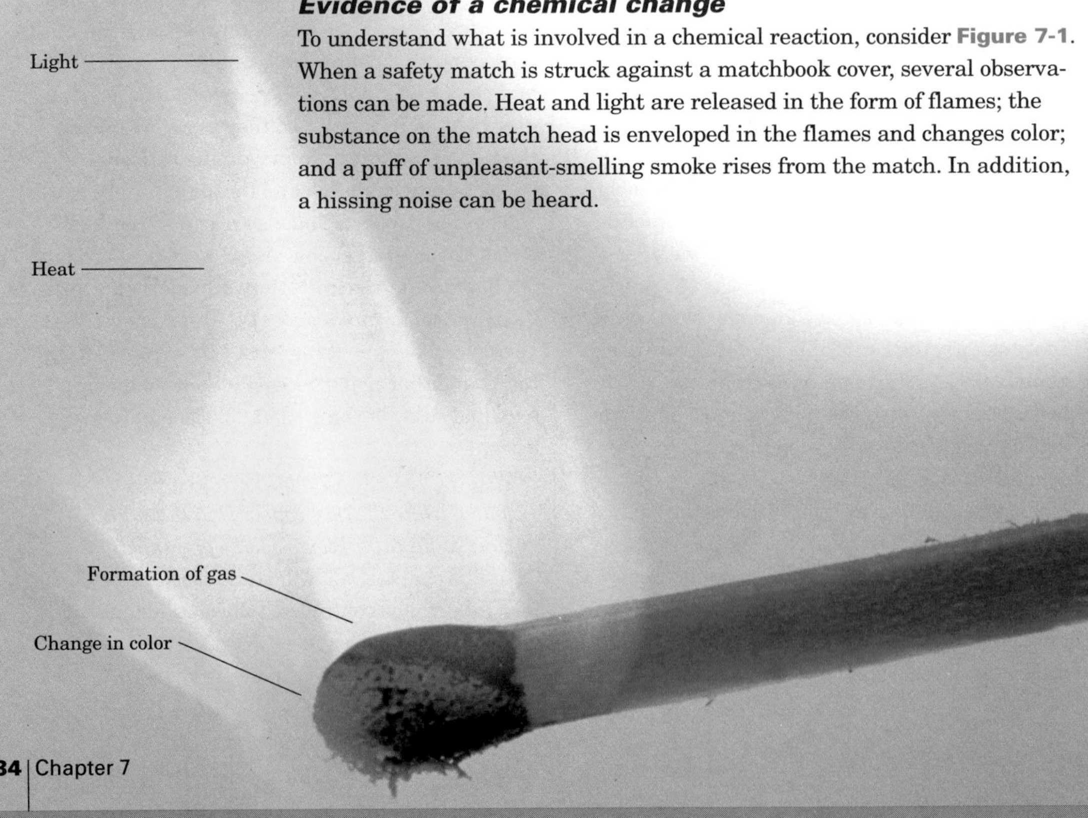

Figure 7-1
Chemical changes occur as a match burns.

Visual Strategy

Figure 7-1
Make students aware that lighting a match is a fairly violent reaction that leads in turn to all of the chemical processes observed. Does the match burn hotter when initially lighted or when it is half-way burnt? (When initially lighted, because at that point its activation energy is reached.)

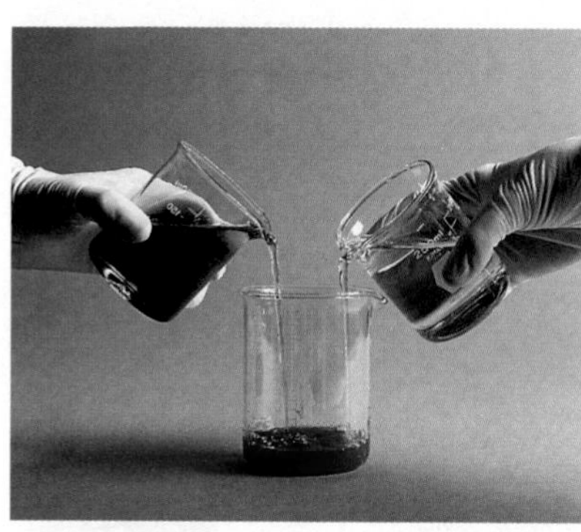

Figure 7-2
The formation of a solid precipitate when two solutions are mixed is an indication of a chemical change.

How do you know what has happened to the match? It appears to have undergone a chemical change because the products left at the end appear very different from the reactants present at the beginning. The same is true of the solutions shown mixing in **Figure 7-2**. In general, the changes that are noticeable in **Figure 7-1** and **Figure 7-2** are evidence that a chemical reaction *may have occurred*.

But this kind of evidence only suggests that a chemical change may have occurred. It cannot be considered absolute proof that a chemical change has taken place. For proof, you need to analyze the products to verify that their *physical and chemical properties* are different from those of the reactants.

For example, a change in heat energy would certainly be involved in the melting of ice. However, analysis of the water that forms would reveal that it has the same chemical properties as the ice that melted. Consequently, you can be certain that no chemical reaction takes place in the melting of ice, even though there is a noticeable change in heat.

Chemical reactions release or absorb energy

Why are heat, light, and noise all considered evidence that a chemical change may have occurred? Each indicates an energy change. Most reactions either release or absorb energy. Consider the reaction discussed at the beginning of Chapter 6, in which hydrogen and oxygen react to produce water in an exothermic reaction. This exothermic reaction can be described by the following word equation, which names the reactants and the products.

$$\underbrace{\text{Hydrogen} + \text{Oxygen}}_{\text{Reactants}} \longrightarrow \underbrace{\text{Water} + \text{Energy}}_{\text{Products}}$$

Recall from Chapter 2 that according to the law of conservation of energy, energy cannot be created or destroyed. So where does the energy that this reaction produces come from?

In Chapter 6, the energy changes in the chemical reaction were linked to breaking and forming bonds. Energy, in the form of a spark, is needed to start the process of breaking the covalent bonds in the hydrogen and oxygen molecules. When new bonds form among the atoms to produce the water molecules, energy is released in the form of heat and light.

But in the case of exothermic reactions such as the synthesis of water, the bonds holding the products together are stronger than those in the reactants. Recall from your study of chemical bonding in Chapters 5 and 6 that the stronger a bond is, the more energy that is released when the bond is formed. In addition, the atoms in the bond are more stable once the bond is formed. In this case, more energy is released in bond formation than was absorbed to break bonds, and the overall reaction is exothermic. The great majority of chemical reactions in nature are exothermic.

Dish *3*—Neutralize the acid with 1 M NaOH, and pour down the drain.

TEACH

Demonstration 2
Production of copper(II) sulfide

Approximate time: 10 min

1. Darken the room so that the reaction will be seen easily.
2. Cut a piece of copper foil 1 cm × 10 cm.
3. Fill a Pyrex test tube 2 cm full with sulfur flowers.
4. Clamp the test tube vertically onto a ring stand, and heat vigorously.
5. Using crucible tongs, insert the copper strip into the test tube, and wait for the sulfur vapors to reach the copper. The bright light indicates the formation of copper(II) sulfide.
6. Remove the copper strip, and let it cool for inspection. Note how the copper has been replaced by copper(II) sulfide.

SAFETY

Safety goggles and a lab apron must both be worn. Students must be 10 ft or more from demonstration. This demonstration must be performed in a properly operating fume hood. If the sulfur vapors catch on fire in the test tube, cover the tube with a glass plate until the fire is out.

DISPOSAL

Clean the copper with steel wool, and save for reuse. Put CuS, steel fragments from cleaning, and the ruined test tube in the trash.

Theme
Conservation
Chemicals in a reaction are changed into different substances with phases (gas, liquid, solid) that may be different, but the number of atoms is the same as in the original reactants. Molecular bond energy is also conserved as part of the overall system energy.

Teaching Tip
Mnemonic device
A convenient device to help students remember the processes involved with endothermic and exothermic reactions is that heat goes **in**to an **en**dothermic reaction and **ex**its an **ex**othermic one.

Possible Misconception
Some students may believe that because certain reactions require heat to initiate them, they are automatically endothermic. Stress the fact that a reaction's classification is based on the *total* energy in a reaction. If more heat is put into the reaction than is released when the reaction is over, it is endothermic. If more heat is obtained than was initially added, the reaction is exothermic. Point out that many exothermic reactions require heat to start them at room temperature.

Teaching Tip
Discussion
Light a candle, and let students determine whether the reaction of the burning candle is exothermic or endothermic. Try to stimulate the discussion with arguments like "Because energy had to be put into the reaction to start it, it must be endothermic," or (touching the candle's base) "The candle itself doesn't seem warmer, so the reaction must not be exothermic." This can help determine if the class conceptually understands this to be an exothermic reaction. Safety glasses and a lab apron are required for anyone handling the candle.

Energy and the Water Reaction

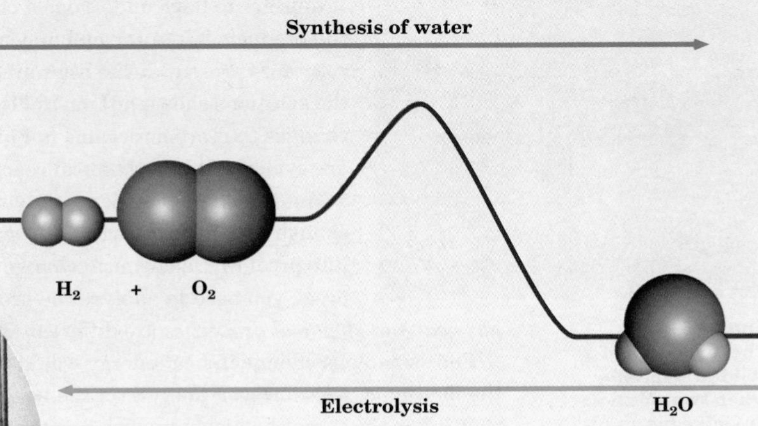

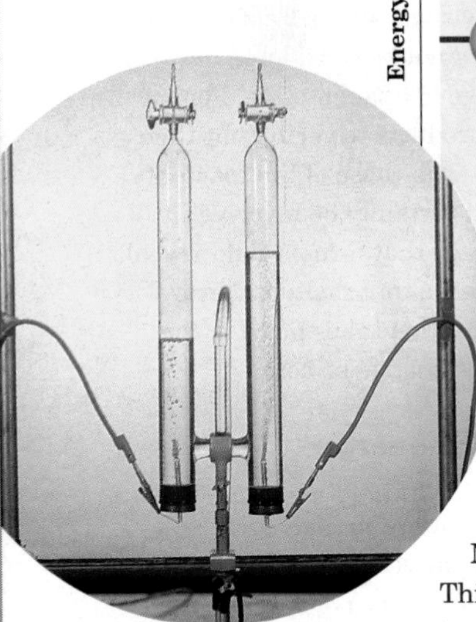

Figure 7-3
Breaking down water in electrolysis requires an input of energy that is the same size as the output of energy from the synthesis reaction of water.

But under the right conditions, endothermic reactions can occur. One necessary condition is that an adequate amount of energy must be available as indicated by the following word equation.

$$\text{water} + \text{energy} \longrightarrow \text{hydrogen} + \text{oxygen}$$

Notice that this reaction is the opposite of the exothermic reaction. This reaction is known as *electrolysis* because electrical energy is added to water to break it down. The same amount of energy that was released when the water molecules were formed must be absorbed, as shown in **Figure 7-3**. Then, as hydrogen and oxygen molecules form again, the same amount of energy that was absorbed to break the bonds in these molecules will be released. The net result will be that the same amount of energy released in making the water will be absorbed in breaking it down again.

Chem Fact
Because water is a very poor conductor of electricity, a small amount of an ionic substance must be added to the water for electrolysis to occur.

Atoms are rearranged in a chemical change
Recall from Chapter 2 the law of conservation of mass—mass is never created or destroyed. But where does the new substance with new properties formed in a chemical reaction come from? Where does the old substance go?

In ordinary chemical changes, atoms do not change into other atoms; nor do they appear or disappear. *The bonding patterns among the atoms are merely rearranged.* Consider the word equation for the match reaction.

$$\text{potassium chlorate} + \text{phosphorus} \longrightarrow \text{potassium chloride} + \text{phosphorus(V) oxide} + \text{energy}$$

In the match-head reaction, bonds are broken, atoms are rearranged, and then new bonds are formed. Both sides of the equation contain potassium, chlorine, phosphorus, and oxygen atoms (oxygen is in potassium chlorate and phosphorus(V) oxide). There are no types of atoms in the products that are not found in the reactants.

Visual Strategy
Figure 7-3
Draw attention to the difference between the plateaus at each end of the reaction. A small spark produces enough energy for the H_2 and O_2 molecules to reach the top of the activation-energy "peak" and form a water molecule. The heat released from water formation causes more H_2 and O_2 to form more water and heat. When water is decomposed, more electrical energy is consumed than heat is produced.

Particles must collide for a chemical reaction to occur

Chem Fact
In "strike anywhere" matches, all the ingredients are on the match head. Some commonly used ingredients are P_4S_3, $KClO_3$, and $K_2Cr_2O_7$.

Consider what happens when a safety match is lit. One reactant is on the match head. The other is on the matchbook's striking surface. The reaction will not begin unless the two are brought together by striking the match head across the striking surface, as shown in **Figure 7-4**.

When the reactants are brought together, their particles collide. If these collisions happen with enough energy, the existing bonds in the reactants can be broken. In effect, such collisions leave behind unbonded atoms that can form new bonds and make products. If the collision is too gentle, the molecules will simply bounce off one another unchanged. The reason that the match must be slid across the striking surface is that the heat from friction is a source of energy to get the reaction started.

Some unwanted reactions can be prevented by knowing what happens when reactants are brought together. Matches are kept inside a matchbook or box to isolate them from the striking surface on the outside. The chances of inadvertently lighting a match are thereby greatly reduced.

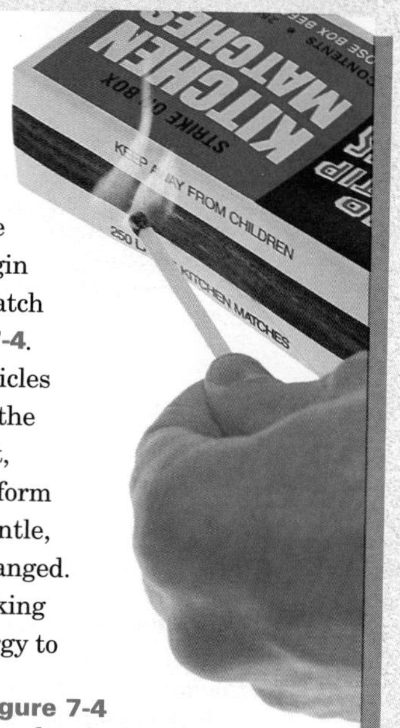

Figure 7-4
The chemical reactants for igniting safety matches are potassium chlorate, $KClO_3$, on the match head and phosphorus, P_4, on the striking surface.

Section Review

1. What is a chemical reaction?
2. What is the only way to prove that a chemical reaction has occurred?
3. In photosynthesis, plants make oxygen and sugars from carbon dioxide, water, and the energy from sunlight. Write a word equation for this process.
4. Decide whether the reactions that occur in each of the following situations are exothermic or endothermic. Explain your choice.
   a. the "burning" of gasoline in a car's engine
   b. the creation of oxygen and sugars from carbon dioxide and water by plants through photosynthesis
   c. fireworks exploding in the sky
5. When an orange substance is heated, it reacts by giving off nitrogen gas, green crystals of chromium(III) oxide, and water vapor. What atoms were contained in the original substance?
6. Explain which would be good places to store potassium.
   a. in the bottom of a sealed jar of mineral oil
   b. in the air near an open window
   c. in a glass jar that contains only oxygen

7. **Story Link**
   In the description of the Statue of Liberty, what evidence was given that chemical reactions had occurred?

Other Applications

The idea of separated reactants is also used in the preparation of commercial baking powder, which uses alum or cream of tartar as the dry acid, baking soda, and a drying agent such as cornstarch. Water causes the acid and soda to react strongly with each other, so the cornstarch is added to absorb water molecules from the air.

ASSESS

Alternative Assessment

Have students develop a concept map relating energy release and absorption, exothermic and endothermic reactions, electronegativity, electron affinity, and covalent and ionic bonding.

Answers to Section Review

1. A chemical reaction is the process by which one or more substances are changed into one or more different substances.
2. To prove that a chemical reaction has occurred, one must verify that the products have chemical properties different from those of the reactants.
3. carbon dioxide + water + energy ⟶ oxygen + sugars
4. **a.** exothermic
   b. endothermic
   c. exothermic
5. chromium, oxygen, nitrogen, and hydrogen
6. **a.** Suitable for storage; the hydrocarbon molecules in oil isolate potassium from water and oxygen.
   b. Unsuitable for storage; water vapor in air reacts with potassium to form potassium hydroxide, KOH, and hydrogen gas.
   c. Unsuitable for storage; potassium reacts with oxygen to form K_2O.
7. Iron became iron oxide trihydrate, $Fe_2O_3{\cdot}3H_2O$. The copper skin reacted to form green copper oxides and copper-sulfate–copper-hydroxide complexes.

Section 7-2

FOCUS

Lesson Starter

Ask students to think of a solution to the following problem. A group of children wish to sell banana splits to their neighbors. Each banana split requires 1 bowl, 1 spoon, 1/2 banana, 3 scoops of ice cream, 3 tablespoons each of chocolate, strawberry, and butterscotch sauces, 1 cup of whipped cream, and 1 maraschino cherry. Have each student write a word equation that describes the synthesis of one banana split. (1 bowl + 1 spoon + 1/2 banana + 3 scoops of ice cream + 3 tablespoons of chocolate sauce + 3 tablespoons of strawberry sauce + 3 tablespoons of butterscotch sauce + 1 cup of whipped cream + 1 maraschino cherry = 1 banana split) If a gallon of ice cream holds 24 scoops, and all of it is used, how much of each ingredient is needed? (8 times as much as is needed for one banana split.) How many banana splits can be made? (8) How many whole bananas are required? (4)

Possible Misconception

It may be necessary to remind the class that many elements are found in nature as diatomic molecules and not as free atoms. Common elements found as diatomic molecules include: H_2, N_2, O_2, F_2, Cl_2, Br_2, and I_2.

TEACH

Teaching Tip

Chemical subscripts

Stress to students that subscripts can never be altered when trying to balance an equation. Doing so will change the structure, formula, and molar mass of the molecule.

How are reactions written?

Section Objectives

Translate word equations into formula equations.

Relate conservation of mass to a balanced equation.

Distinguish between coefficients in a chemical equation and subscripts in a chemical formula.

Properly balance formula equations.

Accurate chemical equations

Reactions can be described with word equations, but it is more convenient to use the chemical symbols and formulas for elements and compounds. In Chapters 5 and 6 you learned that a correctly written formula matches the actual composition of the bonded atoms. Similarly, a correctly written **chemical equation** describes exactly which and how many atoms are rearranged during the course of a reaction.

chemical equation *symbols that describe a chemical reaction and indicate identities and relative amounts of reactants and products*

Atoms and mass are conserved in chemical reactions

The word equation for the water reaction is shown in **Table 7-1**. Beneath the word equation is a formula equation for the same reaction. The molecular models are also shown.

Table 7-1
Different Ways to Write the Water Reaction

Representation	Reactants	Yield (or "produce")	Product
Word equation	hydrogen + oxygen	⟶	water
Formula equation	H_2 + O_2	⟶	H_2O
Molecular models	+	⟶	

Imagine the reaction of the hydrogen molecules and oxygen molecules, portrayed by the models in **Table 7-1**. First, the reactants must collide so that the bonds holding them together can be broken. Only then can the individual atoms in the reactants form new bonds to make water molecules.

Look again at the models for the formula equation in **Table 7-1**. How many of each type of atom will there be when all of the bonds in these reactants are broken? How many are there in the product, once the new bonds are formed?

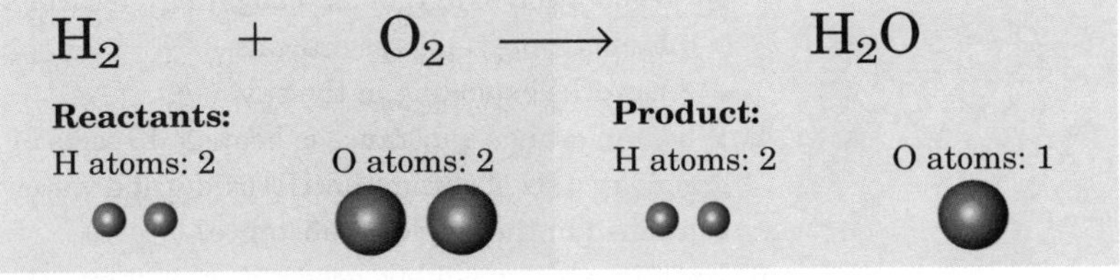

There are *two* oxygen atoms in the reactants but only *one* in the product. One of the oxygen atoms has apparently disappeared, in violation of the law of conservation of mass, as shown in **Figure 7-5** on the next page. *But this is not what has actually happened.* Measuring the masses of the product and the reactants shows that they remain the same. Somehow the formula equation must be adjusted so that it does not appear as if an oxygen atom simply disappeared. The law of conservation of mass must be taken into consideration.

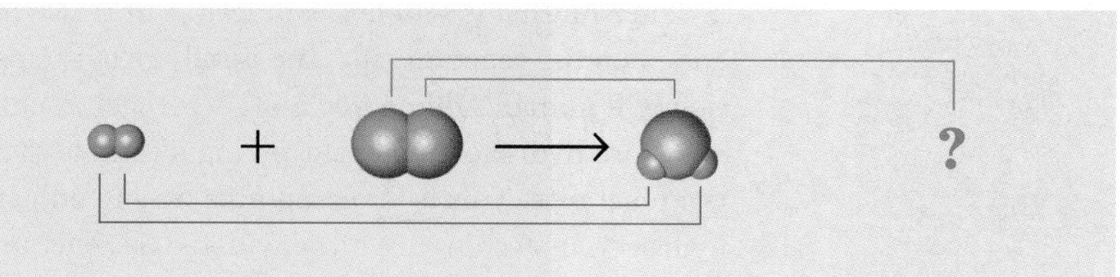

Figure 7-5
One hydrogen molecule and one oxygen molecule can form one water molecule, but there is an oxygen atom in the reactants that is not accounted for in the products.

Coefficients indicate amounts of reactants and products

When trying to write a correct equation, DO NOT change the products or reactants either by adding new ones or by changing the subscripts and identities of the formulas in the equation.

The difficulty with the equation can be solved by adjusting the numbers of particles to get equal numbers of each atom on either side of the arrow. To balance both of the oxygen atoms, you would need *two* water molecules, each containing one oxygen atom, as shown in **Figure 7-6**.

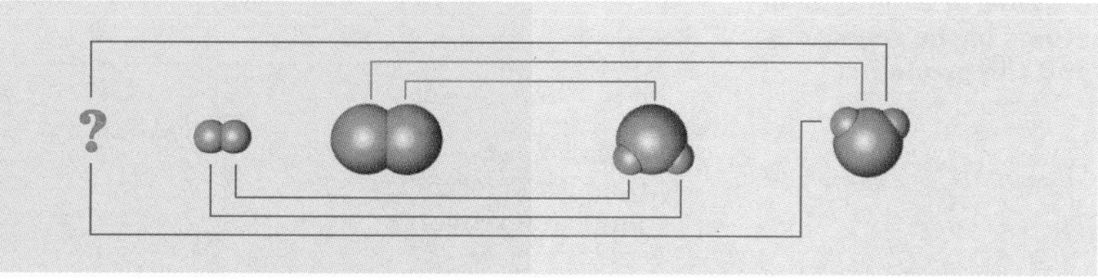

Figure 7-6
The extra oxygen atom can be accounted for by showing *two* water molecules as products, but now there are more hydrogen atoms in the product than in the reactants.

This can be done by writing the number *2* in front of the formula for the water molecule.

$$H_2 + O_2 \longrightarrow 2H_2O$$

coefficient
numeral used in a chemical equation to indicate relative amounts of reactants or products

Numerals placed in front of a formula in this way are called **coefficients**. Just as a subscript indicates the number of atoms within a molecule or formula unit, a coefficient indicates the relative number of molecules, formula units, or atoms required for a complete reaction. Because a coefficient refers to an entire unit, such as molecules or formula units, *you should never insert a coefficient between the elements in a chemical formula.* Is the addition of a coefficient to the product, water, enough to solve the conservation of mass problem in this equation?

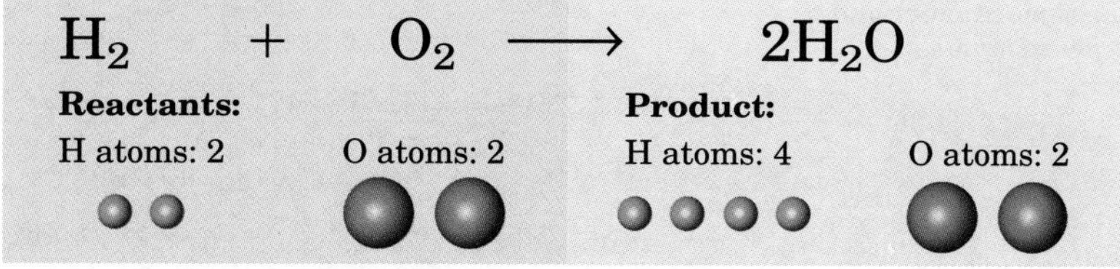

Teaching Tip

Application

Students have used coefficients in math class for several years and understand that $8 \times Y$ equals Y multiplied by 8. Relate this to an example like 3NaOH, in which NaOH is multiplied by 3. Emphasize the fact that each subscript is multiplied by the coefficient to determine the total number of atoms present. For example, $8Ca(NO_3)_2$ contains 8 calcium atoms, 16 nitrogen atoms, and 48 oxygen atoms. It is helpful to remind students that $2H_2O$ could be written $2H_2O_1$; the number of hydrogen atoms is 2×2, or 4, and the number of oxygen atoms is 2×1, or 2. Obviously, one oxygen atom is present because O is in the formula. When learning to balance equations, it is sometimes advantageous, although improper, to include subscripts of 1.

Visual Strategy

Figure 7-5 and Figure 7-6

Stress that the large number of atoms present makes it possible to determine that two more hydrogen atoms are needed (and available) to combine with the remaining oxygen atom. Also stress that both atoms of a diatomic hydrogen molecule are not likely to join with the same oxygen atom, as is pictured, because of the large number of oxygen atoms present even in a small sample.

Theme
Conservation
Balancing equations is a conservation task in which coefficients are manipulated until the numbers of all elements on both sides of the equation are equal.

The equation is still not balanced. But if two more hydrogen atoms are added on the reactant side, the numbers of atoms on both sides of the equation will match. Adding *two* more hydrogen atoms is the same as adding *one* more hydrogen molecule, giving a total of *two* hydrogen molecules in the reactants. This is an acceptable way to adjust the equation because hydrogen molecules are already listed as reactants. *No new reactants or products are being added*. Only the relative amounts have changed.

Now the law of conservation of mass is no longer being violated, as shown in **Figure 7-7**. No atoms are left unaccounted for, and all of the reactants and products are a part of the reaction.

Notice that if no number is written as a coefficient on a final balanced equation, the number is understood to be 1. In the hydrogen-oxygen equation, for example, one oxygen molecule is reacting.

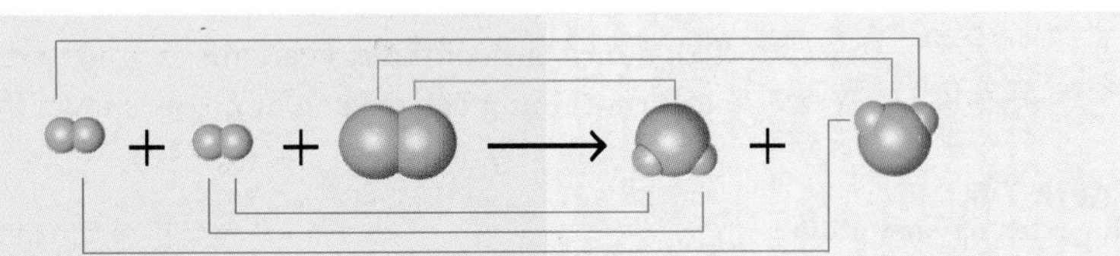

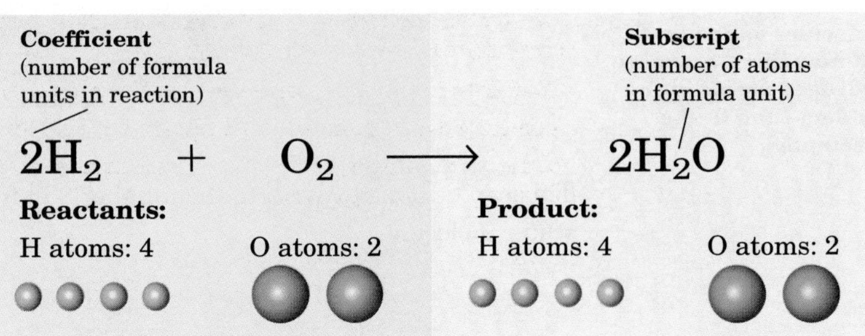

Figure 7-7
The properly balanced equation has the same number of each type of atoms in the reactants and the products.

At first, you may think something has been lost because three molecules of reactants produce only two molecules of product. But compare this to the building of a car, as shown in **Figure 7-8**. Many parts, including four wheels, one engine, and one frame, are required to make only one car, but that doesn't mean something is lost in the product. Before and after the car is made, the number and types of specific *parts* are equal, but the *overall number* of separate objects need not be the same.

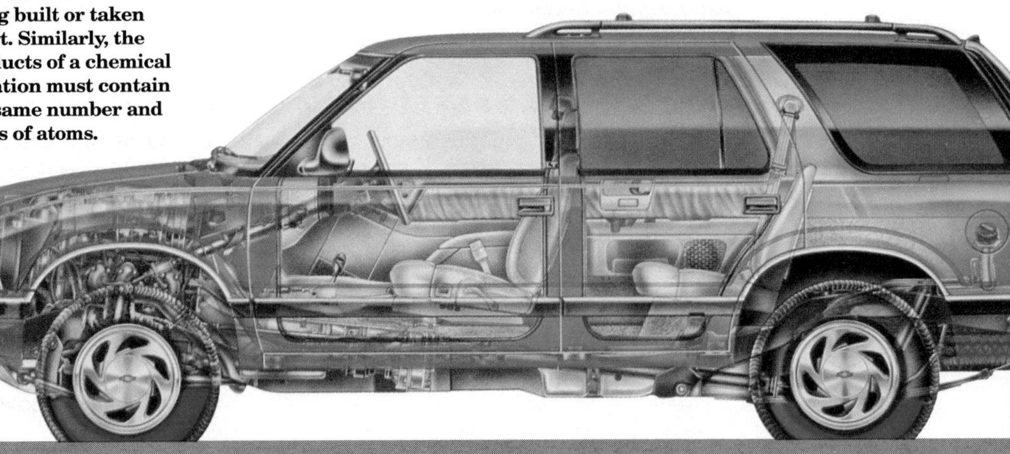

Figure 7-8
A car contains the same total number and types of parts whether it is being built or taken apart. Similarly, the products of a chemical equation must contain the same number and types of atoms.

Visual Strategy
Figure 7-8
Stress the advantage of using whole numbers for coefficients, as they reflect the fundamental nature of the component materials. For example, in the case of the automobile, while the ratio of tires to engines may be expressed as either 4:1 or 1:1/4, only the former will result in the components for a complete auto.

For this reason, conservation of mass is tested by checking the amounts and types of *atoms* (not molecules) that are present. This process of inserting coefficients to satisfy the law of conservation of mass is referred to as "balancing an equation."

Balancing a formula equation turns it into a *chemical equation*. A chemical equation uses symbols and formulas to show the reactants and products and indicates the relative amounts by using coefficients.

How To

Write a balanced chemical equation

Chemical equations are sometimes more complex than the example you've just studied. But there is a step-by-step approach that can make balancing equations easier. These steps are outlined for the reaction in a safety match. This process is called balancing by inspection.

1. *Write the word equation, showing all of the reactants and products.*

phosphorus + potassium chlorate ⟶

potassium chloride + phosphorus(V) oxide

2. *Write the correct symbols and formulas for all of the reactants and products.*

In effect, change the word equation into a formula equation. When doing so, recall what you learned in Chapters 5 and 6 about formula writing. You may need to consult reference works for some formulas. Be sure to leave space to insert the coefficients.

$$?P_4 + ?KClO_3 \longrightarrow ?KCl + ?P_2O_5$$

3. *Count the number of atoms of each element for both sides of the equation.*

List the kinds and numbers of atoms on each side of the equation.

Left side (reactants)	**Right side** (products)
P atoms: 4	P atoms: 2
K atoms: 1 ✓	K atoms: 1 ✓
Cl atoms: 1 ✓	Cl atoms: 1 ✓
O atoms: 3	O atoms: 5

4. *Insert coefficients for atoms of one element at a time so that the law of conservation of mass is satisfied.*

For each element, the total number will be the least common multiple of the amounts on each side of the unbalanced equation. This number may be split among different reactants if they contain the same element. Balancing all of the reactant and product coefficients requires a process of trial and error.

How To continued on the following page . . .

Teaching Tip

Application

Stress to students that equations with improper formulas cannot be balanced correctly (that is, the equation, though balanced, is nonsensical). As an example, show the balanced equation for the synthesis of sodium chloride.

$$2Na + Cl_2 \longrightarrow 2NaCl$$

What would happen if the formula for salt were incorrectly written as Na_2Cl?

$$\_Na + Cl_2 \longrightarrow \_Na_2Cl?$$

$$4Na + Cl_2 \longrightarrow 2Na_2Cl$$

The error in the subscript of the product causes the coefficient for Na to be too large, and the ratio of sodium to chlorine is incorrect. To see this clearly, count the total valence charges for each of the components (1+ for Na, 1− for Cl). Are they balanced on each side of the equation, as required? (No; on each side the charge is 2+ rather than 0.)

Content Background

Balancing polyatomic ions
Help students to recognize that polyatomic ions with subscripts often appear in parentheses on both sides of an equation.

$$Al_2(SO_4)_3 + Ba(NO_3)_2 \longrightarrow BaSO_4 + Al(NO_3)_3$$

can be thought of as

$$Al_2X_3 + BaY_2 \longrightarrow BaX + AlY_3.$$

The balanced equation is written as follows.

$$Al_2(SO_4)_3 + 3Ba(NO_3)_2 \longrightarrow 3BaSO_4 + 2Al(NO_3)_3$$

Show students that there are three SO_4^{2-} groups and six NO_3^- groups on each side of the equation.

Groupwork Strategy

Pair off students, and have one student in each pair balance the odd-numbered equations while the other student balances the even-numbered equations. Have team members check each other's work.

1. $HCl + Cu \longrightarrow CuCl_2 + H_2$
   $2HCl + Cu \longrightarrow CuCl_2 + H_2$
2. $NaClO_3 \longrightarrow NaCl + O_2$
   $2NaClO_3 \longrightarrow 2NaCl + 3O_2$
3. $Mg(OH)_2 \longrightarrow MgO + H_2O$
   $Mg(OH)_2 \longrightarrow MgO + H_2O$
4. $Pb(NO_3)_2 + NaCl \longrightarrow PbCl_2 + NaNO_3$
   $Pb(NO_3)_2 + 2NaCl \longrightarrow PbCl_2 + 2NaNO_3$
5. $Cu + AgNO_3 \longrightarrow Ag + Cu(NO_3)_2$
   $Cu + 2AgNO_3 \longrightarrow 2Ag + Cu(NO_3)_2$
6. $HCl + CaCO_3 \longrightarrow H_2O + CO_2 + CaCl_2$
   $2HCl + CaCO_3 \longrightarrow H_2O + CO_2 + CaCl_2$
7. $Fe + H_2O \longrightarrow Fe_3O_4 + H_2$
   $3Fe + 4H_2O \longrightarrow Fe_3O_4 + 4H_2$
8. $Al_2(SO_4)_3 + Ca(OH)_2 \longrightarrow Al(OH)_3 + CaSO_4$
   $Al_2(SO_4)_3 + 3Ca(OH)_2 \longrightarrow 2Al(OH)_3 + 3CaSO_4$
9. $Na + F_2 \longrightarrow NaF$
   $2Na + F_2 \longrightarrow 2NaF$
10. $F_2 + AlBr_3 \longrightarrow Br_2 + AlF_3$
    $3F_2 + 2AlBr_3 \longrightarrow 3Br_2 + 2AlF_3$

How To continued from the previous page . . .

Table 7-2
Tips for Balancing Equations

Step	Rule
1	First, balance the types of atoms that appear in only one reactant and only one product.
2	Balance the remaining types of atoms one at a time.
3	Balance H atoms and O atoms after most of the other elements have been balanced.
4	If the same polyatomic ions appear on both sides of the equation, treat them as if they were single units, like monatomic ions.

The tips shown in **Table 7-2** may help organize your thinking. In this reaction, all atoms appear in only one reactant and one product. The only valid suggestion is to leave the O atoms until the end. Starting with phosphorus, there are four atoms in the reactants and two in the products. Adding a coefficient of 2 to the products will help balance the equation.

$$?P_4 + ?KClO_3 \longrightarrow ?KCl + 2P_2O_5$$

Left side (reactants)		**Right side** (products)	
P atoms:	4 ✓	P atoms:	4 ✓
K atoms:	1 ✓	K atoms:	1 ✓
Cl atoms:	1 ✓	Cl atoms:	1 ✓
O atoms:	3	O atoms:	10

5. ***Repeat steps 3 and 4 until the law of conservation of mass holds for all of the elements in the equation.***

It now appears that only the oxygen atoms are out of balance. Use the coefficient 10 for the reactant side, and multiply the coefficient already on the product side by 3, so that the least common multiple, 30, is the number of oxygen atoms on each side of the equation. *Do not be concerned if adding coefficients causes an imbalance in other atoms that were previously balanced.*

$$?P_4 + 10KClO_3 \longrightarrow ?KCl + 6P_2O_5$$

Left side (reactants)		**Right side** (products)	
P atoms:	4	P atoms:	12
K atoms:	10	K atoms:	1
Cl atoms:	10	Cl atoms:	1
O atoms:	30 ✓	O atoms:	30 ✓

The coefficient for the KCl product will be 10 since that will balance the potassium and chloride ions in the reactant. This will balance most of the reactants and products.

$?P_4 + 10KClO_3 \rightarrow 10KCl + 6P_2O_5$

Left side (reactants)	**Right side** (products)
P atoms: 4	P atoms: 12
K atoms: 10 ✓	K atoms: 10 ✓
Cl atoms: 10 ✓	Cl atoms: 10 ✓
O atoms: 30 ✓	O atoms: 30 ✓

The last coefficient to be readjusted is for phosphorus. *Note that it may take several adjustments before the right combination of coefficients is found to balance the whole equation.*

$3P_4 + 10KClO_3 \rightarrow 10KCl + 6P_2O_5$

Left side (reactants)	**Right side** (products)
P atoms: 12 ✓	P atoms: 12 ✓
K atoms: 10 ✓	K atoms: 10 ✓
Cl atoms: 10 ✓	Cl atoms: 10 ✓
O atoms: 30 ✓	O atoms: 30 ✓

6. ***Make sure that there are equal numbers of atoms of each element on both sides of the equation.***
Yes, the law of conservation of mass has been satisfied. If it hasn't, try working through these steps again.

Sample Problem 7A
Balancing chemical equations

Cellular respiration is the process that your body uses to get energy from the food you eat. In cellular respiration, sugars such as glucose, $C_6H_{12}O_6$, react with oxygen. The net result is an increase of energy and the production of carbon dioxide and water. Write the balanced chemical equation for cellular respiration.

1 Write the word equation

glucose + oxygen ⟶ carbon dioxide + water

2 Write the formula equation

$?C_6H_{12}O_6 + ?O_2 \longrightarrow ?CO_2 + ?H_2O$

3 Count the number of atoms for each element

Left side (reactants)	**Right side** (products)
C atoms: 6	C atoms: 1
H atoms: 12	H atoms: 2
O atoms: 8	O atoms: 3

4 Insert coefficients for atoms of one element at a time

Begin with carbon, because it appears in only one reactant and one product.

$C_6H_{12}O_6 + ?O_2 \longrightarrow 6CO_2 + ?H_2O$

Remember when no coefficient is written, as for $C_6H_{12}O_6$, one formula unit is involved.

Left side (reactants)	**Right side** (products)
C atoms: 6 ✓	C atoms: 6 ✓
H atoms: 12	H atoms: 2
O atoms: 8	O atoms: 13

Sample Problem 7A continued on the following page . . .

Additional Examples for Sample Problem 7A

For the given word equations, write the formula equations, and show the steps for balancing the chemical equations.

a. propane + oxygen ⟶ carbon dioxide + water

b. chromium(III) sulfate + potassium phosphate ⟶ potassium sulfate + chromium(III) phosphate

Solutions are at the bottom of the page.

Other Applications

The optimum oxygen to acetylene ratio (5:2) of an oxy-acetylene torch (used for cutting and welding metals) can be determined by looking at the balanced equation.

$2C_2H_2 + 5O_2 \longrightarrow 4CO_2 + 2H_2O$

ASSESS

Alternative Assessment

Have each student write a short essay explaining how to balance the following equation.

$\_Cr(NO_3)_3 + \_CuSO_4 \longrightarrow \_Cr_2(SO_4)_3 + \_Cu(NO_3)_2$

Solutions for Additional Examples 7A

a. $C_3H_8 + 5O_2 \longrightarrow 3CO_2 + 4H_2O$

b. $Cr_2(SO_4)_3 + 2K_3PO_4 \longrightarrow 3K_2SO_4 + 2CrPO_4$

Sample Problem 7A continued from the previous page . . .

5 Repeat steps 3 and 4 until the law of conservation of mass holds for the equation

Work with hydrogen next, because it appears in only one reactant and one product.

$$C_6H_{12}O_6 + ?O_2 \longrightarrow 6CO_2 + 6H_2O$$

Left side (reactants)	**Right side** (products)
C atoms: 6 ✓	C atoms: 6 ✓
H atoms: 12 ✓	H atoms: 12 ✓
O atoms: 8	O atoms: 18

Work with oxygen last, because it is in more than one of the reactants and more than one of the products.

$$C_6H_{12}O_6 + 6O_2 \longrightarrow 6CO_2 + 6H_2O$$

Left side (reactants)	**Right side** (products)
C atoms: 6 ✓	C atoms: 6 ✓
H atoms: 12 ✓	H atoms: 12 ✓
O atoms: 18 ✓	O atoms: 18 ✓

6 Make sure that there are equal numbers of atoms of each element on both sides of the equation

Practice 7A

Answers for Concept Review items and Practice problems begin on page 841.

1. **Balance the following equations.**
   a. $CaSi_2 + SbCl_3 \longrightarrow Si + Sb + CaCl_2$ **b.** $C_2H_2 + O_2 \longrightarrow CO_2 + H_2O$
   c. $Al + CH_3OH \longrightarrow (CH_3O)_3Al + H_2$
2. **When calcium is added to water, calcium hydroxide and hydrogen gas are formed. When calcium hydroxide is heated, water and calcium oxide are the products. Write balanced chemical equations that describe this series of changes.**

Section Review

8. Explain the differences among a word equation, formula equation, and chemical equation.
9. Use diagrams of particles to explain why four atoms of silver can produce only two formula units of silver sulfide, even with an excess of sulfur atoms.
10. Indicate which of the numbers in the following equation are coefficients and which are subscripts. Which can be adjusted to balance an equation? Why?

$$3H_2SO_4 + 2Al \longrightarrow Al_2(SO_4)_3 + 3H_2$$

11. Rewrite the following word equations as formula equations.
    a. Hydrogen reacts with chlorine to produce hydrogen chloride gas.
    b. Aluminum and iron(III) oxide react to produce aluminum oxide and iron.
    c. Potassium chlorate decomposes to yield potassium chloride and oxygen.
    d. Calcium hydroxide and hydrochloric acid react to produce calcium chloride and water.
12. Balance the equations you wrote for item **11**.
13. **Story Link**
    Write balanced equations for the following word reactions.
    a. Copper and oxygen make copper(II) oxide.
    b. Iron and oxygen make iron(III) oxide.

Answers to Practice 7A

1. **a.** $3CaSi_2 + 2SbCl_3 \longrightarrow 6Si + 2Sb + 3CaCl_2$
   b. $2C_2H_2 + 5O_2 \longrightarrow 4CO_2 + 2H_2O$
   c. $2Al + 6CH_3OH \longrightarrow 2Al(CH_3O)_3 + 3H_2$
2. $Ca + 2H_2O \longrightarrow Ca(OH)_2 + H_2$
   $Ca(OH)_2 \xrightarrow{\Delta} CaO + H_2O$

Answers to Section Review

8. A word equation lists only the reaction components. A formula equation gives the actual chemical structure of the components. A chemical equation describes exactly which and how many atoms are rearranged in a reaction.
9. Ag^+ Ag^+ Ag^+ Ag^+ + S^{2-} S^{2-} S^{2-} S^{2-} S^{2-} S^{2-} = Ag—S—Ag Ag—S—Ag S^{2-} S^{2-} S^{2-} S^{2-}
10. 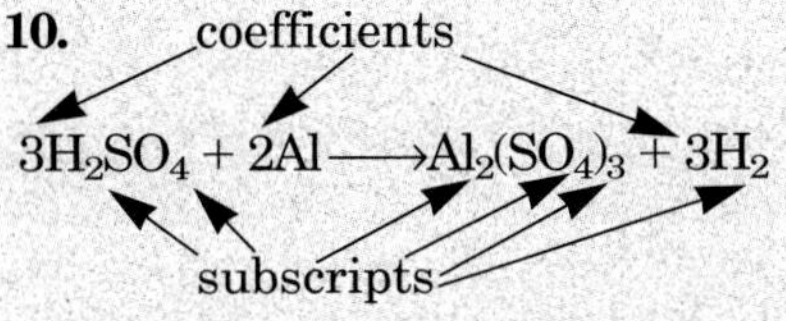

    Coefficients can be adjusted to balance chemical equations; changing subscripts changes the formula represented.
11. **a.** $H_2 + Cl_2 \longrightarrow HCl$
    b. $Al + Fe_2O_3 \longrightarrow Al_2O_3 + Fe$
    c. $KClO_4 \longrightarrow KCl + O_2$
    d. $Ca(OH)_2 + HCl \longrightarrow CaCl_2 + H_2O$
12. **a.** $H_2 + Cl_2 \longrightarrow 2HCl$
    b. $2Al + Fe_2O_3 \longrightarrow Al_2O_3 + 2Fe$
    c. $KClO_4 \longrightarrow KCl + 2O_2$
    d. $Ca(OH)_2 + 2HCl \longrightarrow CaCl_2 + 2H_2O$
13. **a.** $2Cu + O_2 \longrightarrow 2CuO$
    b. $4Fe + 3O_2 \longrightarrow 2Fe_2O_3$

7-3 What information is in an equation?

Section Objectives

Use information from a chemical equation to describe the energy change involved in a reaction.

List states of matter for reactants and products.

Interpret a chemical equation in terms of the relative number of molecules involved and the moles of reactants and products.

Derive mole ratios from a balanced chemical equation.

Use mole ratios and ΔH values to calculate energy changes in reactions.

Equations as instructions

Equations are like chemical recipes in that they describe a process that makes something new out of the ingredients. You've already seen that equations can tell you about the identities of products and reactants. Therefore, it is important to be sure the formulas are written correctly so that these substances are correctly identified. But equations contain much more information.

The ingredients label on the cookie package shown in **Figure 7-9** identifies the ingredients in cookies, but it is not a recipe. Using only this list would make it very difficult to actually make cookies. The list doesn't tell you that the cookies need to bake in the oven, nor does it tell you at what temperature the oven must be set. The order in which the ingredients should be mixed is not specified. Perhaps most important, the amounts of each ingredient are missing. Cookies containing more salt than flour would not be very tasty.

A recipe contains information about the amounts and forms of the ingredients, the order in which they are to be added, all of the instructions needed to make the cookies, and an indication of how many cookies can be made using the amounts given. Like a recipe, many chemical equations contain more than just the identities of the reactants and products. When read properly, the equations contain clues about whether the reaction they describe will occur, how to make the reaction occur, and how much of the reactants and products will be involved.

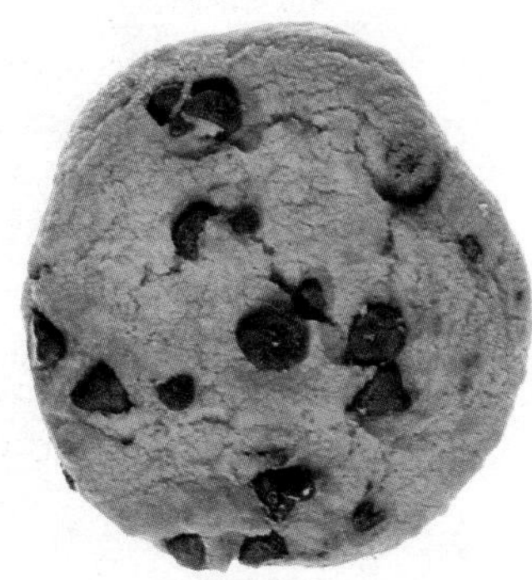

INGREDIENTS: ENRICHED WHEAT FLOUR (CONTAINS NIACIN, REDUCED IRON, THAMINE MONONITRATE [VITAMIN B_1], RIBOFLAVIN [VITAMIN B_2], SWEET CHOCOLATE DROPS (SUGAR, CHOCOLATE, COCOA BUTTER, DEXTROSE, AND SOY LECITHIN–AN EMULSIFIER), VEGETABLE SHORTENING (PARTIALLY HYDROGENATED SOYBEAN AND/OR COTTONSEED OILS), SUGAR, BROWN SUGAR, HIGH FRUCTOSE CORN SYRUP, SALT, BAKING SODA, WHEY, NATURAL AND ARTIFICIAL FLAVOR AND COCOA PROCESSED WITH ALKALI).

Figure 7-9
To make the cookies, you need a recipe with instructions, not just a list of ingredients. Like a recipe, a chemical equation can contain instructions about how to make reactions happen.

Section 7-3

FOCUS

Lesson Starter

Divide the class into small groups, and explain that they are going to start a small business selling a baked item that someone in their group knows how to make. The group is to create a flowchart that will show step by step how to arrive at the finished product by combining all of the ingredients in the proper order. Give the class 10 min to finish the project and to name their company. Ask one or two groups to share their flowchart with the class. Check to see if the information to make their item is complete. This flowchart can be related to chemical equations, which provide the information and steps needed to complete reactions. Emphasize that cooking is really just "chemistry in the kitchen."

Visual Strategy

Figure 7-9

Draw attention to the fact that the ingredients found in the cookie interact to produce the structure, flavor, and texture of chocolate chip cookies. The cookie is the product of a reaction involving the reactants flour, butter, sugar, brown sugar (sugar with molasses), salt, baking soda, vanilla, eggs, and chocolate.

TEACH

Demonstration 3

Exothermic and endothermic reactions

Approximate time: 5 min

This demonstration shows the release and absorption of heat when crystals are dissolved.

1. Fill two 125 mL Erlenmeyer flasks with 50 mL of water.
2. Invite two students to measure the temperature of the water in a flask and record the temperatures on the board.
3. Add 2 g of NaOH pellets to one flask and 10 g of NH_4Cl crystals to the other flask. Let the crystals dissolve for 2 min, making sure not to use the thermometers as stirring rods. Have the students measure the new temperatures and record them on the board.
4. Discuss what happened in each flask. Ask why the changes took place. How could the dissolution rate of NaOH be increased? (Cool it.) How could the dissolution rate of NH_4Cl be increased? (Warm it.)

SAFETY

Lab aprons and safety goggles should be worn by the participating students and teacher. The rest of the class must be 10 ft from the demonstration. In case of an alkali spill, dilute the spill with water and, while wearing gloves, soak up the spill with cloth or paper towels. Rinse the towels, neutralize the rinsings with 1 M HCl, and flush down the drain.

Figure 7-10a
The energy released by the synthesis reaction for two moles of water is as much energy used in 7.5 minutes of walking up stairs, . . .

7-10b
. . . 14.5 minutes of basketball, . . .

Energy changes in equations

You have seen that energy can be produced in exothermic reactions and absorbed in endothermic reactions.

Exothermic: $2H_2 + O_2 \longrightarrow 2H_2O + 572\ kJ$
Endothermic: $2H_2O + 572\ kJ \longrightarrow 2H_2 + O_2$

The most common reactions in chemistry are exothermic. Endothermic ones can occur, but they are less likely or require special conditions. Notice that the special condition required in the reaction of water to form hydrogen and oxygen is an input of 572 kJ of energy, usually in the form of electricity. This is the same amount of energy in the activities shown in Figure 7-10.

Another way that energy changes are indicated is in terms of ΔH, the amount of heat energy released or absorbed when the reaction takes place at normal atmospheric pressure. Positive ΔH values indicate an endothermic reaction, in which energy is absorbed. Negative ΔH values indicate an exothermic reaction, in which energy is released. Such values are often designated as amounts of energy per mole of a product or reactant. Thus, the ΔH values are half of the heat energy values noted in the first set of equations, which were for two moles of water. In cases in which energy amounts are measured per mole of a substance, fractional coefficients may be used. Note that the coefficient refers to *one-half of a mole* of oxygen, *not* one-half of a molecule.

7-10c
. . . or 16.5 minutes of swimming.

Exothermic: $H_2 + \frac{1}{2}O_2 \longrightarrow H_2O$

$$\Delta H = -286\ kJ/mol\ H_2O$$

Endothermic: $H_2O \longrightarrow H_2 + \frac{1}{2}O_2$

$$\Delta H = +286\ kJ/mol\ H_2O$$

Visual Strategy

Figure 7-10

Point out that the amount of time required to expend a certain amount of energy is inversely proportional to the strenuousness of the task. The physical definition of power describes how much energy is converted over an amount of time (usually joules per second). Multiplying power by time gives the amount of energy used during that time.

One way to remember the meaning of positive and negative ΔH values is to consider them as a part of a recipe: *the* ΔH *value contains part of the instructions on how to get the reaction to happen.* If ΔH is a positive number, energy must be added to the reactants and absorbed as the reaction takes place. If ΔH is a negative number, energy will be lost by the reactants, and this energy will be released.

In the example of the synthesis of water, because the ΔH has a negative value, 286 kJ of energy are lost by the reactants and released as heat when the reaction occurs. In the reaction that breaks down water into hydrogen and oxygen, the positive ΔH value indicates that the 286 kJ of energy must be added to the reactants for the reaction to take place.

Chem Fact
Many combustion reactions take place with reactants in the gaseous state. Even though gasoline is a liquid, its vapors are what burns in combustion.

Reaction conditions in equations

Having all of the substances listed in an equation is not always enough to be certain the reaction will occur. Often, reactions require special conditions. It could be that the reactant substances must be gases, or that the reaction requires high pressure. Without such conditions, some reactions won't occur at all. Others will occur, but will take place so slowly that they are not useful. In addition to the chemical formulas for the reactants and products, chemical equations often contain symbols that indicate the states of matter for the reactants and products. Other notations in an equation, usually over the reaction arrow, can indicate special conditions that are required for the reaction to happen efficiently. **Table 7-3** lists the notations or symbols that are commonly used in chemical equations. Using these notations, the balanced equation for the safety match reaction discussed earlier in the chapter can be written as shown below.

$$3P_4(s) + 10KClO_3(s) \xrightarrow{\text{heat (friction)}} 10KCl(s) + 6P_2O_5(s)$$

Table 7-3
Symbols Used in Chemical Equations

Reactants and Products: Symbol	Meaning
(s) or (cr)	solid or crystal
(l)	liquid
(g)	gas
(aq)	in aqueous solution (dissolved in water)
$\downarrow$	solid precipitate product forms
$\uparrow$	gaseous product forms

Reaction Conditions: Symbol	Meaning
$\longrightarrow$	"produces" or "yields," indicating result of reaction
$\rightleftarrows$	reaction in which products can re-form into reactants; final result is a mixture of products and reactants
$\xrightarrow{\Delta}$ or $\xrightarrow{\text{heat}}$	reactants are heated
$\xrightarrow{1.0 \times 10^8 \text{kPa}}$	pressure at which reaction is carried out
$\xrightarrow{0°\text{C}}$	temperature at which reaction is carried out
$\xrightarrow{\text{Pd}}$	chemical formula of a catalyst added to speed up a reaction
$\xrightarrow[e^-]{}$	electrolysis

DISPOSAL
In the fume hood combine the NaOH and NH_4Cl solutions and add 150 mL of water. Dissolve 6 g of NaOH pellets in the solution. Heat and boil the solution gently for 5 min. Neutralize with 1 M H_2SO_4 and pour down the drain.

Content Background
Enthalpy change for gases
The definition of change in enthalpy of a gas is the sum of the change in energy and the pressure times the change in the gas volume.

$$\Delta H = \Delta E + P\Delta V$$

Content Background
Completion of reactions
Reactions tend to go to completion if one of the products is effectively removed from the system. This can be accomplished when a gas or precipitate is produced. One reaction that tends to go to completion is the decomposition of sugar. Completion is also reached if one of the products is only slightly ionized, as in the neutralization reaction of HCl and NaOH.

$$H_3O^+ + Cl^- + Na^+ + OH^- \longrightarrow Na^+ + Cl^- + 2H_2O$$

Theme
Systems and interactions
Reactants in the gaseous state tend to react faster than those in either liquid or solid states.

Teaching Tip
Conservation
Stress to students that conservation laws apply to the reactants and products of chemical reactions in terms of grams and atoms, but not in terms of moles or molecules. This can be made clear by pointing out that the numbers in the *Moles* and *Molecules* rows in **Table 7-4** do not add up, but the numbers in the *Mass* and *Total mass* rows, as well as the total number of atoms in the *Equation* row, do.

Figure 7-11a
The cellular respiration reaction that breaks down glucose is used as an energy source by animals, such as this dog . . .

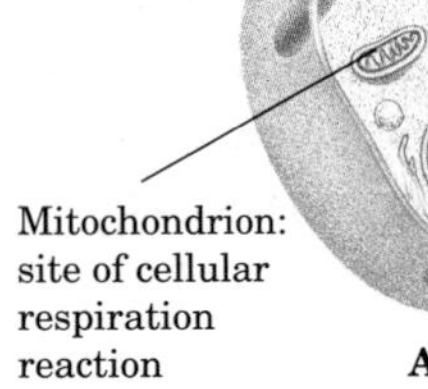

Mitochondrion: site of cellular respiration reaction

Animal cell

Quantitative relationships

Equations reveal the numbers of formula units involved

Take another look at the equation for cellular respiration introduced in Sample Problem 7A.

$$C_6H_{12}O_6(aq) + 6O_2(g) \longrightarrow 6CO_2(g) + 6H_2O(l) + \text{energy}$$

This reaction is used by living things to release stored energy. **Figure 7-11** shows where this reaction takes place at the cellular level in both plants and animals. The coefficients used to balance the equation serve not only to satisfy the law of conservation of mass, but also to *predict* the amount of each reactant that is required to produce a certain amount of product. The coefficients in a chemical equation tell you the relative numbers of formula units of reactants and products in the reaction. For example, the equation for cellular respiration tells you that *one* molecule of glucose, $C_6H_{12}O_6$, reacts with *six* molecules of oxygen, O_2, to produce *six* molecules of carbon dioxide, CO_2, and *six* molecules of water, H_2O. Another way to read the equation is in terms of *moles*. This approach will give you all of the information you see in **Table 7-4**.

As you can see from **Table 7-4**, mass is conserved in terms of grams. **Figure 7-12** on the next page demonstrates how atoms are conserved. Remember that the numbers of moles and molecules do not necessarily remain the same on both sides of the equations, because the molecules are being created and destroyed as the atoms are rearranged. In this case, one big molecule, $C_6H_{12}O_6$, is being used to create many smaller molecules, $6CO_2$ and $6H_2O$.

Table 7-4
Information From a Balanced Chemical Equation

Equation:	$C_6H_{12}O_6$	$+ 6O_2$	$\longrightarrow$	$6CO_2$	$+ 6H_2O$
Moles	1	+ 6	$\longrightarrow$	6	+ 6
Molecules	$(6.02 \times 10^{23}) \times 1$	$+ (6.02 \times 10^{23}) \times 6$	$\longrightarrow$	$(6.02 \times 10^{23}) \times 6$	$+ (6.02 \times 10^{23}) \times 6$
Mass (grams)	180.18×1	$+ 32.00 \times 6$	$\longrightarrow$	44.01×6	$+ 18.02 \times 6$
Total mass (grams)	180.18	+ 192.00	$\longrightarrow$	264.06	+ 108.12

7-11b
. . . and plants, such as this golden pothos vine.

Mitochondrion: site of cellular respiration reaction

Plant cell

$$C_6H_{12}O_6 + 6O_2 \longrightarrow 6H_2O + 6CO_2$$

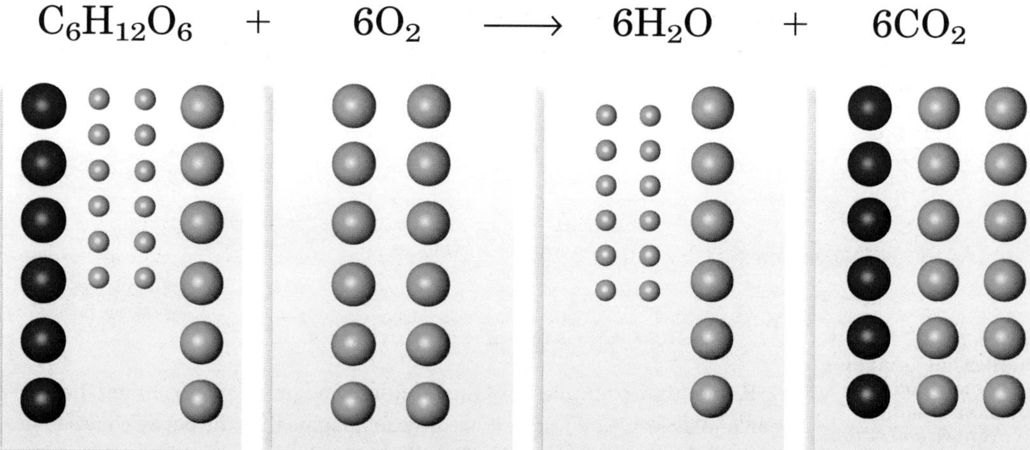

Figure 7-12
Atoms are neither created nor destroyed in the cellular respiration reaction. They are only rearranged to form new products.

Balanced equations show proportions

What would happen if you had *two* molecules of glucose instead of *one*? What would be required to ensure that the reaction went to completion? To answer this question, consider the models of the reactants and products shown in **Figure 7-12**. One way to determine the outcome would be to run the reaction one molecule at a time. The first glucose molecule would require six O_2 molecules and would produce six CO_2 molecules and six H_2O molecules. As the second molecule reacted, it would require an additional six O_2 molecules and would produce six more CO_2 molecules and six more H_2O molecules. In short, the reaction of two glucose molecules would require and produce 12 molecules of each of the other reactants and products, as shown in **Table 7-5**. For three glucose molecules, another six of each of the other molecules will be required for a complete reaction giving a total of 18 molecules of each type. A similar calculation can be made for four glucose molecules, which require 24 molecules of each type. Look closely at the proportions of reactants and products. You should be able to detect a pattern in the table. How many molecules of each reactant and product will be necessary for five glucose molecules?

Table 7-5
Amounts for Additional Molecules of Glucose

	Amount Consumed		Amount Produced	
	$C_6H_{12}O_6$	O_2	H_2O	CO_2
First glucose molecule	1 molecule	6 molecules	6 molecules	6 molecules
Two glucose molecules	2 molecules	12 molecules	12 molecules	12 molecules
Three glucose molecules	3 molecules	18 molecules	18 molecules	18 molecules
Four glucose molecules	4 molecules	24 molecules	24 molecules	24 molecules

Teaching Tip

Mole ratios

Remind students of the electrolysis of water that they learned about in Chapter 6. If they followed the reaction for some time, they ought to have noticed that the production of hydrogen exceeded that of oxygen by about 2 to 1. By looking at the balanced equation they can see that the ratio should be 2:1.

$$2H_2(g) + O_2(g) \longrightarrow 2H_2O(g)$$

(Remember that in a ratio of 2 volume units to 1, each volume unit of gas contains the same number of molecules—one of Avogadro's findings.)

Visual Strategy

Figure 7-12

Students should be aware that if the amount of O_2 available for the reaction is insufficient, the reaction will not always proceed as planned. In this case, not all of the sugar would undergo combustion. In the combustion reaction for graphite, however, insufficient oxygen will lead to the production of CO instead of CO_2.

Possible Misconception

When converting from molecules to moles, students should recall that a mole is simply 6.022×10^{23} particles. Therefore, the ratio of particles for each component of the reaction can always be considered the same as a mole ratio.

ASSESS

Alternative Assessment

Have students draw a concept map that shows all of the relationships in a balanced chemical equation.

Ratio for one banana split is 1:1:1:3:1

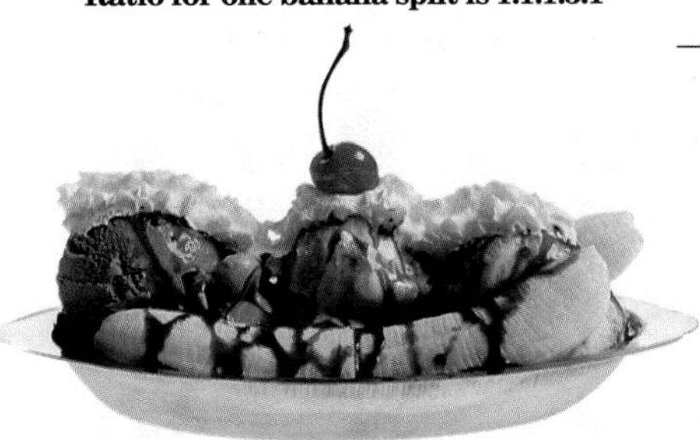

1 cherry
1 squirt of whipped cream
1 spoon of hot fudge
3 scoops of ice cream
1 banana

4 cherries
4 squirts of whipped cream
4 spoons of hot fudge
12 scoops of ice cream
4 bananas

Ratio for four banana splits is 4:4:4:12:4

4:4:4:12:4 ratio is the same as 1:1:1:3:1 ratio

Figure 7-13
Whether one is making one banana split or four banana splits, the proportions of the ingredients remain the same.

Rather than counting all of the reactants and products or running the reaction one molecule at a time, it is easier to imagine that the entire equation has been multiplied by two. If you wanted to know how much of the reactants and the products would be required and produced by 64 glucose molecules, you would multiply the entire equation by 64. This situation is similar to the situation shown in **Figure 7-13**.

$$C_6H_{12}O_6 + 6O_2 \longrightarrow 6CO_2 + 6H_2O \text{ (equation)}$$

$$2C_6H_{12}O_6 + 12O_2 \longrightarrow 12CO_2 + 12H_2O \text{ (equation} \times 2)$$

$$64C_6H_{12}O_6 + 384O_2 \longrightarrow 384CO_2 + 384H_2O \text{ (equation} \times 64)$$

Relative amounts in equations can be expressed in moles

You have already seen how to interpret the cellular respiration reaction in terms of molecules. In addition, the equation for the cellular respiration reaction also indicates that 1 mol of $C_6H_{12}O_6$ reacts with 6 mol of O_2 to produce 6 mol of CO_2 and 6 mol of H_2O. This can be expressed as a ratio of reactants and products.

$$1 \text{ mol } C_6H_{12}O_6 : 6 \text{ mol } O_2 : 6 \text{ mol } CO_2 : 6 \text{ mol } H_2O$$

Thus, for any pair of molecules or formula units, a *mole ratio* can be determined by comparing the coefficients from the balanced chemical equation. For example, the mole ratio of glucose to carbon dioxide would be 1:6 because glucose's coefficient is 1 and carbon dioxide's coefficient is 6.

$$1 \text{ mol } C_6H_{12}O_6 : 6 \text{ mol } CO_2$$

What other mole ratios can you establish for this reaction? Be sure to express mole ratios in the lowest whole-number ratios.

Mole ratios can be multiplied just like equations. They can also be used like conversion factors to compare amounts of substances. In this way, you can determine exactly how much of a reactant is needed for the reaction or how much of a product can be expected. For example, 3 mol of $C_6H_{12}O_6$ would create 18 mol of CO_2.

$$3 \text{ mol } \cancel{C_6H_{12}O_6} \times \frac{6 \text{ mol } CO_2}{1 \text{ mol } \cancel{C_6H_{12}O_6}} = 18 \text{ mol } CO_2$$

Visual Strategy

Figure 7-13

Remind students of the banana-split problem in the Lesson Starter of Section 7-2. Mention that the banana splits pictured above are slightly different from (are an allotrope of) those previously described. Have students write the proportions of the nine ingredients in the banana splits of Section 7-2 (2:2:1:6:2:6:6:6:2). Stress that this recipe produces two banana splits, each using half of a banana.

Energy is usually expressed in amounts per mole

The reaction shown below is used by living things to release stored energy. But how much energy can it release? The ΔH value for this reaction is given along with the balanced equation.

$$C_6H_{12}O_6(aq) + 6O_2(g) \longrightarrow 6CO_2(g) + 6H_2O(l)$$

$$\Delta H = -2870 \text{ kJ/mol glucose}$$

Because this ΔH value is expressed in terms of *kJ/mol of glucose,* it can be treated as part of a mole ratio. For every mole of glucose, 2870 kJ of heat energy will be released.

$$2870 \text{ kJ} : 1 \text{ mol } C_6H_{12}O_6 : 6 \text{ mol } O_2 : 6 \text{ mol } CO_2 : 6 \text{ mol } H_2O$$

Thus, you can calculate how much energy is provided by a reaction that produces 24 mol of H_2O or any other amount of any product or reactant.

$$24 \cancel{\text{mol } H_2O} \times \frac{2870 \text{ kJ}}{6 \cancel{\text{mol } H_2O}} = 11\,480 \text{ kJ}$$

Section Review

14. Which of the following equations represent exothermic reactions?

a. $H_2(g) + Cl_2(g) \longrightarrow 2HCl(g) + 185 \text{ kJ/mol } H_2$

b. $Al_2O_3(s) + 1675.7 \text{ kJ} \longrightarrow 2Al(s) + \frac{3}{2}O_2(g)$

c. $C(s) + H_2O(g) \longrightarrow CO(g) + H_2(g)$
$\Delta H = +131.3 \text{ kJ/mol}$

d. $SnCl_2(s) + Cl_2(g) \longrightarrow SnCl_4(s)$
$\Delta H = -186.2 \text{ kJ/mol } SnCl_4$

15. Explain the states of the reactants and products and the other conditions that are described by each of the following equations.

a. $2KClO_3(s) + 156 \text{ kJ} \xrightarrow{\text{heat}} 2KCl(s) + 3O_2(g)$

b. $CaCO_3(s) \xrightarrow{\Delta} CaO(s) + CO_2(g)$
$\Delta H = +178.1 \text{ kJ/mol}$

c. $NH_4NO_3(s) \xrightarrow{200°C} N_2O(g) + 2H_2O(g)$
$\Delta H = -100. \text{ kJ/mol}$

d. $N_2(g) + 3H_2(g) \xrightarrow[10^7 \text{ Pa, } 500°C]{\text{Ru, C catalyst}} 2NH_3(g)$
$\Delta H = -92.0 \text{ kJ/mol } N_2$

16. Identify all possible mole ratios in each of the reactions in item **15**.

17. Using the mole ratios in item **16,** determine how many moles of each product could be formed given the following reactant amounts.

a. reaction **a** with 16.0 mol $KClO_3$

b. reaction **b** with 6.00 mol $CaCO_3$

c. reaction **c** with 7.00 mol NH_4NO_3

d. reaction **d** with 9.00 mol H_2

18. For each of the reactions in item **15** and amounts in item **17**, calculate how many kilojoules of energy would be involved. Also indicate whether the change would be exothermic or endothermic.

Answers to Section Review

14. a, d

15. a. For every 2 mol of solid potassium chlorate, 156 kJ of heat are added. Oxygen gas evolves from the reactant, leaving solid potassium chloride.

b. Heat (178.1 kJ for every mole of reactant) is added to solid calcium carbonate. Gaseous carbon dioxide evolves, leaving solid calcium oxide.

c. Solid ammonium nitrate, when heated to 200°C, dissociates into gaseous dinitrogen oxide (nitrous oxide) and water vapor. 100 kJ of heat are released for each mole of reactant.

d. Nitrogen and hydrogen gasses react in the presence of ruthenium and carbon catalysts at a temperature of 500°C and a pressure of 10^7 Pa to produce ammonia gas and 92 kJ of heat per mole of nitrogen.

16. a. 2 mol $KClO_3$: 156 kJ : 2 mol KCl : 3 mol O_2

b. 1 mol $CaCO_3$: 178.1 kJ : 1 mol CaO : 1 mol CO_2

c. 1 mol NH_4NO_3 : 1 mol N_2O : 2 mol H_2O : 100. kJ

d. 1 mol N_2 : 3 mol H_2 : 2 mol NH_3 : 92.0 kJ

17. a. 16.0 mol of $KClO_3$ yield 16.0 mol of KCl and 24.0 mol of O_2.

b. 6.00 mol of $CaCO_3$ yield 6.00 mol each of CaO and CO_2.

c. 7.00 mol of NH_4NO_3 yield 7.00 mol of N_2O and 14.00 mol of H_2O.

d. 9.00 mol of H_2 react with 3.00 mol of N_2 to produce 6.00 mol of NH_3.

18. a. +1250 kJ, endothermic

b. +1070 kJ, endothermic

c. −700. kJ, exothermic

d. −276 kJ, exothermic

Section 7-4

FOCUS

Lesson Starter

Have students write two equations for producing CO_2 from $CaCO_3$. One reaction combines HCl with $CaCO_3$.

$$CaCO_3 + 2HCl \longrightarrow CaCl_2 + CO_2 + H_2O$$

The other reaction heats $CaCO_3$ to a temperature of 900°C.

$$CaCO_3 \longrightarrow CaO + CO_2$$

Discuss some of the uses of the different products.

Demonstration 4

Single displacement generation of CO_2

Approximate time: 15 min

1. Set up an apparatus for gas generation by inserting a thistle tube through the rubber stopper until it is 1 cm from the bottom of the flask.
2. Insert a fire-polished glass tube through the stopper, and attach it to the rubber tubing.
3. Fill the trough two-thirds full of water, and place the tubing and three inverted, water-filled collection bottles in the trough.
4. Place 50 g of marble chips ($CaCO_3$) into the flask. Add enough distilled water to cover the bottom of the thistle tube.
5. Add 1 M HCl a little at a time into the thistle tube while collecting three bottles of CO_2 by water displacement.
6. Discard the first bottle and cover the second and third bottles with glass plates. Store bottles upright. **Save one bottle for use in Demonstration 5.**

How can reactions be used?

Section Objectives

Categorize reactions as belonging to one (or more) of five basic types of chemical reactions.

Write chemical equations representing each type of chemical reaction.

Use an activity series to predict whether a given reaction will occur and what its products will be.

Write total and net ionic equations for double displacement reactions.

Putting reactions to work

You already know that chemical reactions involve rearrangements of atoms to produce substances with new properties. Sometimes a reaction is desirable, especially if the new substances have special uses because of their properties. For example, aspirin can be made in a multi-step process from acetic acid, carbon dioxide, and phenol, a highly toxic compound. This synthesis is summarized in **Figure 7-14**.

$$C_6H_5OH + CO_2 + CH_3COOH \longrightarrow C_9H_8O_4 + H_2O$$

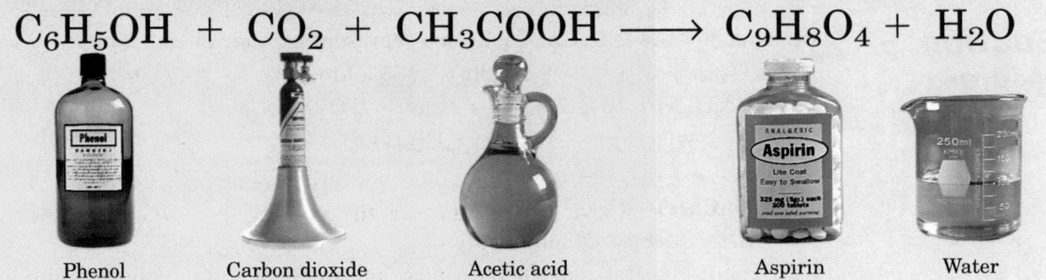

Figure 7-14 **Chemical reactions can be used to make products that have useful properties that are different from those of the reactants.**

Another reason for wanting a reaction to proceed is to use the energy it releases. The reaction between methane and oxygen produces the energy for a Bunsen burner's flame to boil water or melt glass.

Conversely, some reactions are undesirable, and the goal is to prevent them from occurring. Sometimes the same reactants can lead to several different reactions with different products, only one of which is useful.

The key to understanding and controlling all reactions lies in grouping together reactions that follow similar patterns, just as elements with similar properties were classified together.

combustion *exothermic reaction that usually involves oxygen to form the oxides of the elements in a reactant*

Oxygen combines with other elements in combustion

Most reactions that are used for energy are **combustion** reactions, which usually involve the production of oxide compounds. The heat from the phosphorus and potassium chlorate safety match reaction from Section 7-1 is great enough to ignite the sulfur on the match head, which undergoes combustion to produce sulfur dioxide. Combustion reactions are exothermic, releasing a large amount of heat and light.

$$S_8(s) + 8O_2(g) \longrightarrow 8SO_2(g) + \text{energy}$$

Oxide of reactant

Figure 7-15
The complete combustion of any hydrocarbon, such as methane, yields carbon dioxide and water.

Another example of a combustion reaction is the reaction between methane and oxygen in a Bunsen burner, as shown in **Figure 7-15**. Note that the carbon from methane makes carbon dioxide, and the hydrogen makes water (hydrogen oxide). This production of water occurs when any hydrocarbon is burned.

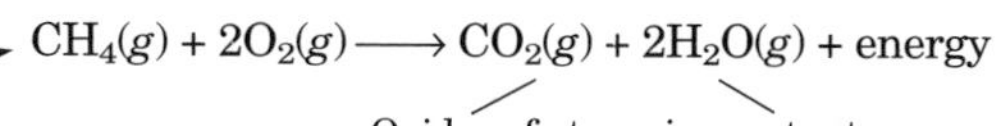

$$CH_4(g) + 2O_2(g) \longrightarrow CO_2(g) + 2H_2O(g) + \text{energy}$$

Oxides of atoms in reactants

When a match is lit and the phosphorus and sulfur reactions get the cardboard or wood core of the match lit, the cellulose in the matchstick begins to burn. Cellulose is a polymer chain of glucose units. The following equation represents the combustion of 20 glucose units, but cellulose chains can be thousands of units long.

$$(C_6H_{10}O_5)_{20} + 120O_2 \longrightarrow 120CO_2 + 100H_2O + \text{energy}$$

Oxides of atoms in reactants

Although the combustion reaction indicates that the only products are water and carbon dioxide, ashes usually remain after a match burns. The equation is not often followed exactly. Combustion is incomplete, and some carbon that has not reacted with oxygen remains. Also, some carbon monoxide, CO, is produced in incomplete combustion. For wood or cardboard to burn so that only carbon dioxide and water are produced, the temperature must be high. *Remember that not all reactions are complete and special conditions are often required to maximize the reaction.* For combustion reactions, the higher the temperature, the more likely it is to be complete.

In all of the combustion examples in this section, binary oxides are formed. **Table 7-6** lists common oxides of nonmetals, some of which are formed in combustion reactions. The more complete the reaction, the more likely it is that the oxides near the bottom of the table will be formed.

Chem Fact
If gas furnaces or car engines aren't well ventilated, CO can form from incomplete combustion. The symptoms of CO poisoning are headaches, nausea, and difficulty sensing and thinking.

Table 7-6
Oxides of Some Nonmetals

Carbon	Nitrogen	Phosphorus	Sulfur
carbon(II) oxide (carbon monoxide), CO	nitrogen(I) oxide, N_2O	phosphorus(III) oxide, P_2O_3 (or P_4O_6)	sulfur(IV) oxide, SO_2
carbon(IV) oxide (carbon dioxide), CO_2	nitrogen(II) oxide, NO	phosphorus(V) oxide, P_2O_5 (or P_4O_{10})	sulfur(VI) oxide, SO_3
	nitrogen(III) oxide, N_2O_3		
	nitrogen(IV) oxide, NO_2 (or N_2O_4)		
	nitrogen(V) oxide, N_2O_5		

Concept Review

Predicting the products of combustion

19. Many stoves and water heaters get energy by burning natural gas, a mixture that contains propane, C_3H_8. Write a balanced equation for the combustion of propane to form CO_2 and H_2O.

Answers for Concept Review items and Practice problems begin on page 841.

7. Light two candles. Cover one candle with one of the two bottles of CO_2. Cover the other candle with an air-filled bottle. Note that the candle in the CO_2 bottle is extinguished sooner.

SAFETY
Safety goggles and a lab apron must be worn. Students must be 10 ft or more from demonstration. In case of an acid spill, dilute the spill with water and, wearing gloves, soak up the spill with cloth or paper towels. Rinse the towels, neutralize the rinsings with 1 M NaOH, and flush down the drain.

DISPOSAL
See Demonstration 5, page 254.

Content Background
Product uses
Calcium chloride, $CaCl_2$, the solid product from the reaction shown in Demonstration 4, is used as a dehydrating agent and as a component in antifreeze. Calcium oxide, CaO, or quicklime, the solid product from the decomposition of calcium carbonate, has long been used in making plaster. In post-imperial Rome, many of the marble monuments were torn down, and the marble was heated to produce quicklime.

Other Applications
Incomplete combustion
If air instead of pure oxygen is used in the combustion process, nitrogen-oxide compounds, many of which are poisonous, are produced. Some nitrogen-oxide compounds are also light sensitive and turn brown when exposed to sunlight, causing the brown smog that appears in some cities.

Visual Strategy
Figure 7-15
Stress that the hot, luminous flame is a visual clue to the presence of a combustion reaction. Also emphasize that in the example depicted, the natural gas used in a burner is a 98% hydrocarbon mixture containing 85% methane, 9% ethane, 3% propane, and 1% butane. The remaining 2% is a mixture of nitrogen and helium.

Answers to
Concept Review

19. $C_3H_8 + 5O_2 \xrightarrow{\Delta} 3CO_2 + 4H_2O$

Demonstration 5
Reaction of CO_2 with Mg
Approximate time: 5 min

This demonstration shows why a CO_2 fire extinguisher should not be used on burning magnesium. The reaction that results

$$2Mg + CO_2 \longrightarrow 2MgO + C,$$

is demonstrated. ***Use the CO_2 collected in Demonstration 4.***

1. Place the remaining bottle of CO_2 upright in a coffee-can shield.
2. Light a 5 cm piece of magnesium ribbon while holding it with crucible tongs, and insert it into the gas bottle. Note the black carbon and white magnesium oxide products in the bottle.

SAFETY
Safety goggles with ultraviolet filters and a lab apron must be worn. Students must be 10 ft or more from demonstration. Remind students (and yourself) not to look directly at the burning magnesium. It gives off ultraviolet light that can damage the eyes.

DISPOSAL
Neutralize any residual acid in the gas-generating apparatus with 1 M NaOH, and pour the liquid down the drain. Collect the un-reacted $CaCO_3$ and $CaCl_2$ from the flask, and any carbon and MgO in the collection bottle, and throw it all in the trash.

Consumer Focus

Fire extinguishers

The key to most firefighting strategies is to control the combustion reaction by separating the reactants, fuel and oxygen, so that the molecules can no longer collide and react. Different types of fuels require different types of firefighting methods. Most fire extinguishers display codes indicating which types of fires they can put out.

Water, used on Class A fires involving solid fuels such as wood, cools the fuel so that it does not react as readily. The steam that is produced helps to displace the oxygen in the air around the fire. A Class B fire, in which the reactant is a liquid or gas, is put out best by a fire extinguisher containing CO_2 gas. Because CO_2 is more dense than O_2, it forms a layer underneath the O_2, cutting off the oxygen supply for the combustion reaction.

Class C fires involving a "live" electric circuit can also be extinguished by CO_2. Liquid water cannot be used, or there will be a danger of electric shock. Some Class C fire extinguishers contain a dry chemical which smothers the fire by interrupting the chain reaction that is occurring. For example, a competing reaction may take place with the contents of the fire extinguisher and the intermediates of the reaction. Class C fire extinguishers usually contain compounds such as ammonium dihydrogen phosphate, $(NH_4)H_2PO_4$, or sodium hydrogen carbonate, $NaHCO_3$.

Finally, Class D fires involve burning metals. These fires cannot be extinguished with CO_2 or water because they may react with some hot metals. For these fires, nonreactive dry powders are used to cover the metal and keep it separate from the oxygen in the air. One kind of powder contains finely ground sodium chloride crystals mixed with a special polymer that allows the crystals to adhere to any surface, even vertical ones. Another is finely ground graphite powder, which not only cuts off the oxygen supply but also absorbs heat.

Most fire extinguishers can be used with more than one type of fire. Check the fire extinguishers in your home and school for these codes. Be sure you know which kinds of fires they can put out.

Visual Strategy
Fire Extinguishers

Stress that fire extinguishers are to be used only on the types of fires specified on the extinguishers' display codes. In the case of chemical fires, conditions of the fire need to be taken into account. For example, a carbon dioxide fire extinguisher would not successfully extinguish burning magnesium because magnesium burns at a temperature high enough to decompose carbon dioxide.

Synthesis and decomposition reactions

Compounds are made in synthesis reactions

So far in this book, you have examined several synthesis reactions, in which a new compound is made from several reactants. For example, in Chapter 5, the reaction in which sodium metal and chlorine gas form sodium chloride crystals was discussed in great detail.

In this chapter, the reaction of sulfur in a match head with oxygen from the air to form a new compound, sulfur dioxide, was referred to as a combustion reaction. But this reaction involves atoms of two elements forming molecules of a new compound. Therefore, the sulfur-oxygen reaction can be described as either a synthesis reaction or a combustion reaction. *Some reactions can fit more than one category.*

The compound formed in a synthesis reaction can be made from more than one element as reactants, or it can be made from smaller molecules. As in the photosynthesis reaction, *the key feature of a synthesis reaction is that one of the compounds produced is larger or more complex than any of the reactant substances.*

$$6CO_2(g) + 6H_2O(l) \longrightarrow C_6H_{12}O_6(aq) + 6O_2(g)$$

More complex compound

Chem Fact
The production of one component of acid rain involves a series of synthesis reactions, starting with sulfur impurities in coal and other fuels, that form sulfuric acid.

$2SO_2(g) + O_2(g) + 2H_2O(l) \longrightarrow 2H_2SO_4(aq)$

Sample Problem 7B
Predicting products of a synthesis reaction

Write a balanced chemical equation for the synthesis reaction between lithium metal and liquid bromine.

1 List what you know

- **reactants:** $Li(s) + Br_2(l)$
- **reaction type:** synthesis
- **product(s):** ?

2 Identify the products

- **Because it is a synthesis reaction, a compound containing the same atoms as the reactants will be among the products. Use information about compounds and formulas to determine the likeliest product.**

From Chapter 4, you know that Li is a metal and Br is a nonmetal.

From Chapters 5 and 6, you know that this combination is likely to form a salt with ionic bonds. When ionic bonds are formed, Li tends to form Li^+ ions, and Br_2 tends to form Br^- ions. The salt produced is probably LiBr.

3 Write and balance the equation

- **Write a formula equation, and then balance it.**

$?Li(s) + ?Br_2(l) \longrightarrow ?LiBr(s)$

$2Li(s) + Br_2(l) \longrightarrow 2LiBr(s)$

- **Check to see that the question has been answered properly.**

Practice 7B

1. **Write a balanced equation for the synthesis reaction between carbon, C, and sulfur, S_8.**
2. **Write a balanced equation for the synthesis reaction that produces iron(III) oxide.**

Solution for Additional Example 7B

$2Zn(s) + O_2(g) \longrightarrow 2ZnO(s)$

Theme
Classification and trends
Reactions can be classified into five basic categories: synthesis, decomposition, single displacement, double displacement, and combustion.

Teaching Tip
Analogy
Many reactions can be described as "chemicals going to a chemical dance." Most simple synthesis reactions are the combination of two reactants into one product (similar to two people meeting at the dance and leaving as a couple). A decomposition reaction is one in which the couple's relationship decomposes at the dance (most students have seen this one by now). A single-displacement reaction can only occur if a more active person, say a cation, comes along and displaces another cation who is "dancing" with an anion. A double-displacement reaction occurs when two couples change partners (again, cations always combine with anions).

Additional Example for Sample Problem 7B
Write a balanced equation for the synthesis reaction between zinc and oxygen. The element with the lower electronegativity tends to be the cation (in this instance, zinc). Zinc has a 2+ charge, and oxygen has a 2– charge.

Solution is at the bottom of the page.

Answers to Practice 7B

1. $4C + S_8 \longrightarrow 4CS_2$
2. $4Fe + 3O_2 \longrightarrow 2Fe_2O_3$

Do You Know?

Many compounds have been discovered by accident during the synthesis of other compounds, including many dyes that were in great demand throughout Europe in the 1800s. Unfortunately, the production of synthetic dyes put many large tropical plantations and laborers out of work because the dyes could be made synthetically at a cost lower than the cost of growing plants and extracting the dyes from them.

Theme

Macroscopic observations and micromodeling

The polymer production of the cellulose molecule allows the plant to continue growing.

Historical Note

In 1869 John Wesley Hyatt (1837–1920), in an attempt to find a substitute for the fragile ivory in billiard balls, discovered the first synthetic polymer. Called celluloid, it is a material that can be molded into shape. In 1884, Louis M. H. Bernigaud pushed the same substance through tiny holes in a nozzle, making small threads that he called rayon.

Groupwork Strategy

Have students form groups of three or four and use library sources to prepare short group reports on the synthetic poly-mers mentioned in **Figure 7-17**. Suggest that they obtain samples of their products for the others to view. Questions to be addressed should include the following. How is the product used? How is it made? How much is made? Is it an important polymer in the world today? Did it lead to the development of other polymers?

polymerization *chemical reaction in which many simple molecules combine in chains to form a very large molecule*

Polymerization reactions are a form of synthesis

Plants use the glucose produced in photosynthesis both as an energy source and as material to help build their structure. Plants and wood are made mostly of cellulose. To make cellulose, glucose molecules produced in photosynthesis are bonded to form long chains. The same reaction occurs over and over again, bonding one glucose molecule to another and gradually building up long, strong fibers of cellulose composed of thousands of glucose units, as shown in **Figure 7-16**.

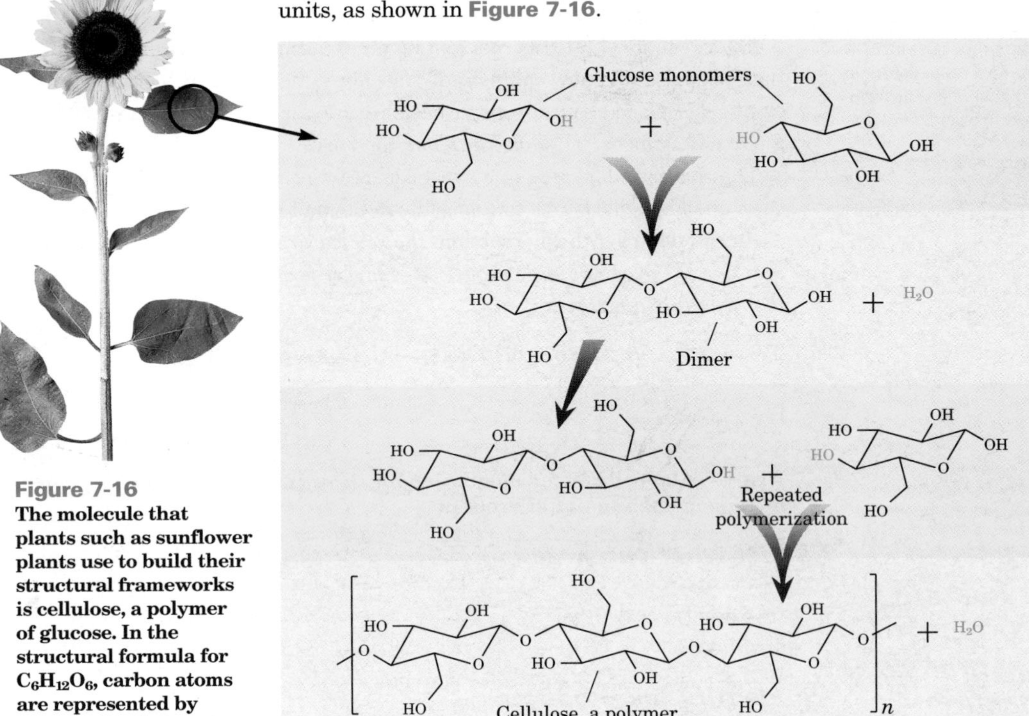

Figure 7-16
The molecule that plants such as sunflower plants use to build their structural frameworks is cellulose, a polymer of glucose. In the structural formula for $C_6H_{12}O_6$, carbon atoms are represented by every unlabeled vertex.

This type of reaction is called **polymerization**. In such reactions, many similar repeating units, known as *monomers*, are linked to form very large molecules, called *polymers*. In the cellulose-building reaction, the glucose molecules are the monomers, and the cellulose chains are the polymers.

In living things, polymerization reactions are responsible for building many molecules of life, especially protein and DNA. Chemists can also synthesize many polymers. The polyethylene used to make trash bags is a polymer. Other artificially made polymers include the examples shown in **Figure 7-17**.

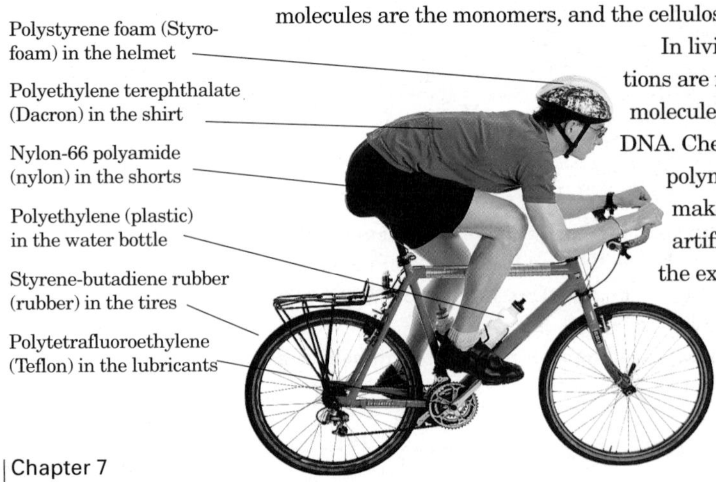

Polystyrene foam (Styrofoam) in the helmet

Polyethylene terephthalate (Dacron) in the shirt

Nylon-66 polyamide (nylon) in the shorts

Polyethylene (plastic) in the water bottle

Styrene-butadiene rubber (rubber) in the tires

Polytetrafluoroethylene (Teflon) in the lubricants

Figure 7-17
Natural and artificial polymers are used in many consumer products.

Visual Strategy

Figure 7-16

Stress that the formation of a polymer is like adding links onto a chain. There is no end to the length of the chain that can be produced, and there can be different uses for different lengths of a chain.

In most polymers, like polyethylene and cellulose, the monomers are all identical. In other cases, such as proteins, different monomers may be combined. Although the amino acid monomers that make up proteins appear to be very different, each one has an amino functional group and an organic acid functional group, so the monomers all link in the same way, forming a "backbone" of carbon, nitrogen, and oxygen atoms. A polymer with three amino acids is called a *tripeptide*.

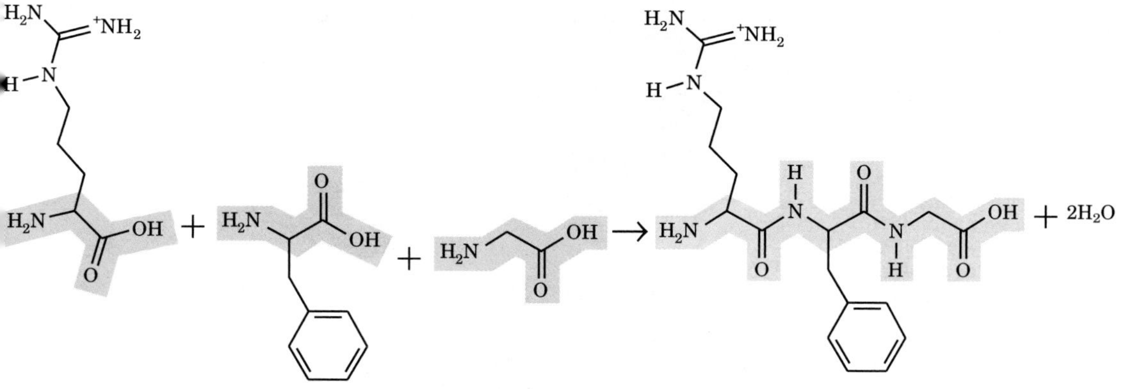

Arginine + Phenylalanine + Glycine ⟶ Tripeptide + Water

Compounds are broken down during decomposition

When $NaHCO_3$ is heated, it is broken down, and its atoms are rearranged to form three products. Such reactions, in which a compound is broken into different parts, are known as **decomposition** reactions.

decomposition
chemical reaction in which a single compound is broken down to produce two or more simpler substances

$$2NaHCO_3(s) \longrightarrow Na_2CO_3(s) + H_2O(g) + CO_2(g)$$

Reactant compound broken into several products

Metal carbonates decompose to yield CO_2. Some metal oxides decompose to yield O_2. Other common products of decompositions include H_2 and N_2.

The electrolysis of water is another decomposition reaction. Like most decomposition reactions, this reaction is endothermic. The energy required for this reaction may be supplied by a battery. Remember that the hydrogen and oxygen are produced in a 2:1 ratio, which you can predict by looking at the coefficients of the products in the balanced equation.

$$2H_2O(l) \longrightarrow 2H_2(g) + O_2(g)$$

More than one product, each less complex

A decomposition reaction can yield products that are elements, as in the case of the electrolysis of water, or products that are less complex substances than the reactants, as in the $NaHCO_3$ reaction.

The key feature of a decomposition reaction is that a reactant compound is broken into two or more less complex formula units.

Additional Examples

a. Write a balanced equation for the decomposition of hydrogen peroxide.

$$2H_2O_2 \longrightarrow 2H_2O + O_2$$

b. Predict the decomposition of mercury(II) oxide.

$$2HgO \longrightarrow 2Hg + O_2$$

Demonstration 6
Displacement of silver with copper

Approximate time: 5 min

This demonstration provides a visually vivid example of a single-displacement reaction.

1. Place a petri dish half-full of 0.1 M silver nitrate, $AgNO_3$, on the overhead projector.
2. Cut copper foil into six 1 cm^2 pieces.
3. Add the copper foil to the silver nitrate solution. (The solution should become blue as the copper replaces the silver, which will form clumps at the edges of the copper foil.)

SAFETY

Safety goggles and a lab apron must be worn. Students must be 10 ft or more from demonstration. Remember that silver nitrate will stain the skin if left in contact with it for even a short time.

DISPOSAL

Let the mixture evaporate completely and put it in a labeled bottle or flask, stating the contents, source (demonstration number and chapter), and the phrase "for future use." **Do not dispose of in a landfill or in any other way.**

Figure 7-18
Aluminum undergoes a displacement reaction with solutions of copper compounds to form copper metal and a solution of an aluminum compound. Some of the copper metal adheres to the aluminum foil, and the rest falls to the bottom of the beaker.

Displacement reactions

Elements trade places in single-displacement reactions

When hydrogen gas is needed in the lab, the reaction of zinc and hydrogen chloride is used.

$$Zn(s) + 2HCl(aq) \longrightarrow ZnCl_2(aq) + H_2(g)$$

single-displacement *chemical reaction in which one element replaces another element in a compound*

One way to think of this reaction is as a **single-displacement** reaction. Before the reaction begins, atoms of zinc are in the elemental form, and atoms of hydrogen are combined in a compound with chlorine. After the reaction, the atoms of hydrogen are in the elemental form, and atoms of zinc are combined in a compound with chlorine. In effect, the zinc atoms and hydrogen atoms switch places.

Single displacements can also occur with elements other than hydrogen. For example, if a piece of aluminum foil is scraped and then added to a solution of copper(II) chloride, reddish copper metal starts to form on the aluminum foil, and some falls to the bottom of the container. Aluminum is displacing copper in the compound, leaving behind copper metal, as shown in **Figure 7-18**.

$$2Al(s) + 3CuCl_2(aq) \longrightarrow 2AlCl_3(aq) + 3Cu(s)$$

Chem Fact
In electrical wiring in buildings, it is dangerous to join copper and aluminum wiring because of the metals' tendency to undergo this type of reaction.

Recall the example of potassium and water from Chapter 5. When the two are combined, there is a violent reaction, a gas is formed, and a solution remains. The equation for the reaction shows that potassium is displacing some of the hydrogen atoms in water, producing hydrogen gas and an ionic compound, potassium hydroxide, that dissolves in water. The heat released by the reaction is great enough to ignite the hydrogen so that it undergoes combustion with oxygen in the air.

$$2K(s) + 2HOH(l) \longrightarrow 2KOH(aq) + H_2(g) \quad \text{Single-displacement reaction}$$

$$2H_2(g) + O_2(g) \longrightarrow 2H_2O(g) \quad \text{Combustion reaction}$$

Visual Strategy
Figure 7-18

Make students aware that the more active aluminum is displacing the copper to produce free copper and combined aluminum. The number of copper atoms differs because the oxidation state of Cu is 2+ and the oxidation state of Al is 3+.

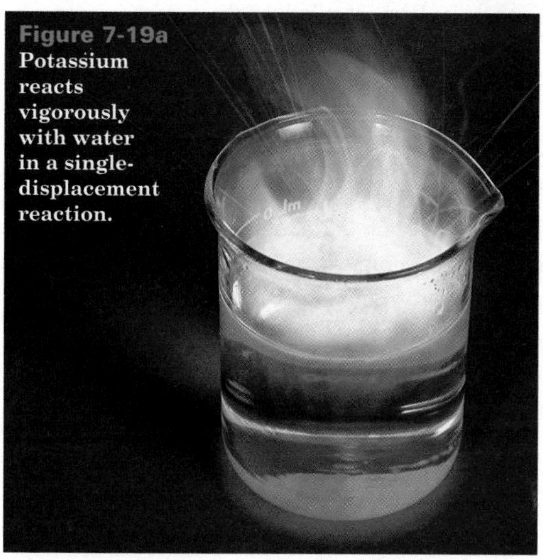

Figure 7-19a Potassium reacts vigorously with water in a single-displacement reaction.

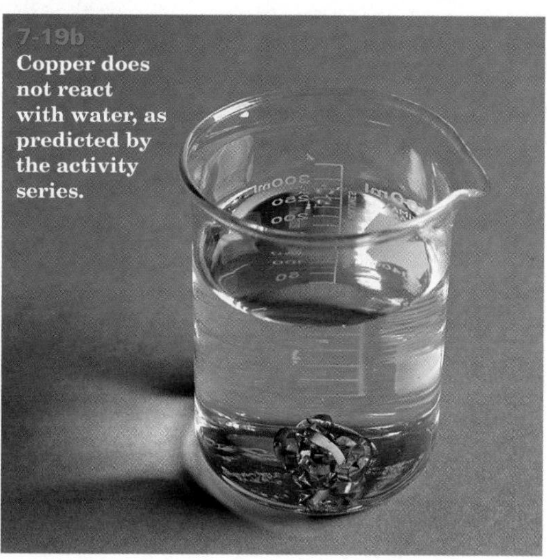

7-19b Copper does not react with water, as predicted by the activity series.

activity series *arrangement of elements in the order of their tendency to react with water and acids*

Table 7-7
Activity Series

Element	Reactivity
Li Rb K Ba Ca Na	react with cold H_2O and acids, replacing hydrogen
Mg Al Mn Zn Cr Fe	react with acids or steam, but usually not liquid water, to replace hydrogen
Ni Sn Pb	all react with acids, but not water, to replace hydrogen
H_2 Cu Hg	all react with oxygen to form oxides
Ag Pt Au	mostly unreactive

Halogen	Reactivity
F_2 Cl_2 Br_2 I_2	listed from most reactive to least reactive

An activity series summarizes and predicts reactions

Recall what happened when the piece of potassium was dropped in water. A rapid single displacement occurred as potassium replaced hydrogen in a compound. **Figure 7-19** shows what happens when a piece of copper, another metal, is dropped into water—nothing. No bubbles of hydrogen gas are formed. The single-displacement reaction does not occur.

$$Cu(s) + 2H_2O(l) \longrightarrow \text{no reaction}$$

Being able to write a chemical equation, therefore, does not necessarily mean that the reaction will actually take place.

Years of scientific experiments involving single-displacement reactions have been summarized in the **activity series**. The activity series is a list of elements organized according to their tendency to react (their "activity"). The more readily an element reacts with other substances, the greater its activity.

The more active elements tend to be more stable in a compound than in elemental form. If a more active element, such as potassium, and a compound containing a less active element, such as H_2O (which contains hydrogen), are brought together, the more active element will replace the less active element in the compound. In the activity series in **Table 7-7**, the elements are arranged in order of activity, with the most active element at the top. In general, any element can displace those below it from compounds but not those above it. There are a few exceptions to these rules, but this basic pattern is the most important thing to remember.

For example, potassium is near the top of the list, far above hydrogen. It can replace hydrogen in most compounds, including water. This confirms our observation of a violent reaction when potassium is added to water. But copper is farther down on the list, below hydrogen. It cannot replace hydrogen in compounds. This is why we observed no reaction when copper was added to water.

Theme
Equilibrium and change
Elements are displaced only by more active elements. The farther apart elements are on the activity series, the more violent the reaction will be when they are mixed because one will more easily displace the other.

Possible Misconception
Students need to be made aware that the more active element (the one higher on the list) needs to be a free element in order to replace a combined element in a molecule. Many times students will not recognize a "no reaction" situation because they forget that the free element must be the more reactive element for a reaction to occur. Remind the class that cations always replace cations and anions always replace anions.

Teaching Tip
Application
Make students aware that the halogen activity-series list is actually a separate list, and not a continuation of the list above it. All of the metals displace cations, while all of the halogens displace anions. Examples should be given to emphasize this fact.

$$2K + 2HCl \longrightarrow 2KCl + H_2$$

$$F_2 + 2HCl \longrightarrow 2HF + Cl_2$$

Teaching Tip
Rule of thumb
It is often easier for students to visualize the reactivity of water if it is written HOH instead of H_2O. When written as HOH, it is easier to see that a H atom is available to be replaced by more active metals, such as K.

$$2K + 2HOH \longrightarrow 2KOH + H_2$$

Visual Strategy
Figure 7-19b
Stress that the copper will be slowly oxidized by dissolved oxygen in the water but not by the oxygen atoms in the water molecule itself. The reaction could be written as follows.

$$Cu(s) + 2H_2O(l) \longrightarrow \text{no reaction}$$

Other Applications

Large bars of magnesium are often hung on ocean-going ships to help slow down the formation of Fe_3O_4. Magnesium is a more active metal than iron, and MgO will form more readily than Fe_3O_4.

Other Applications

One of the most impressive single-displacement reactions is the thermite reaction, which is used to weld large pieces of steel together.

$$Al + Fe_3O_4 \longrightarrow Fe + Al_2O_3$$

Although this is a highly exothermic reaction, a very high temperature is needed to start the reaction, which results in molten iron.

Answers to

Concept Review

20. a. $Ba(s) + 2H_2O(l) \longrightarrow Ba(OH)_2(s) + H_2(g)$

b. $2Rb(s) + 2H_2O(l) \longrightarrow 2RbOH(s) + H_2(g)$

c. $Zn(s) + 2H_2O(l) \longrightarrow Zn(OH)_2(s) + H_2(g)$

21. a. $4Au(s) + 3O_2(g) \longrightarrow 2Au_2O_3(s)$

b. $2Mn(s) + O_2(g) \longrightarrow 2MnO(s)$
(Other reactions can produce Mn_3O_4, MnO_2, Mn_2O_7, Mn_2O_3, and MnO_3.)

c. $2Pb(s) + O_2(g) \longrightarrow 2PbO(s)$
(Other reactions can produce PbO_2, Pb_3O_4, Pb_2O_3, and Pb_2O.)

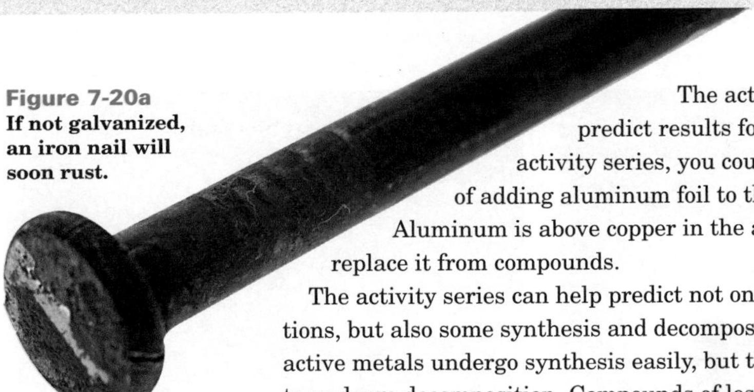

Figure 7-20a
If not galvanized, an iron nail will soon rust.

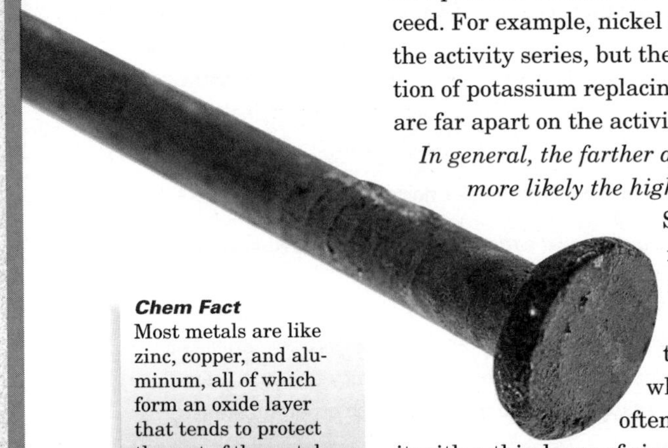

7-20b
After a nail is galvanized, the zinc layer corrodes preferentially because it is more active than iron. As a result, zinc(II) hydroxide, $Zn(OH)_2$ is formed.

The activity series enables you to predict results for many reactions. With the activity series, you could have predicted the result of adding aluminum foil to the copper chloride solution. Aluminum is above copper in the activity series and can replace it from compounds.

The activity series can help predict not only single-displacement reactions, but also some synthesis and decomposition reactions. In general, active metals undergo synthesis easily, but their compounds do not tend to undergo decomposition. Compounds of less active metals undergo decomposition easily, but are more difficult to synthesize. For example, calcium metal reacts so quickly with the oxygen in air that it forms an oxide tarnish while you watch. But copper wire can remain exposed to air for many years before forming copper oxide.

Controlling reactions with the activity series

Along with predicting whether a reaction will occur, the activity series also provides indications of how easily and quickly the reaction will proceed. For example, nickel can replace tin, which is immediately below it in the activity series, but the reaction is not as violent and fast as the reaction of potassium replacing hydrogen in water. Potassium and hydrogen are far apart on the activity series.

In general, the farther apart two elements are on the activity series, the more likely the higher one will replace the lower one in compounds. Similarly, synthesis reactions with potassium metal occur much faster than those with nickel or tin.

Knowledge of the activity series can be used to prevent *corrosion,* the synthesis reaction in which metals form oxides and hydroxides. Iron is often galvanized to protect it from rusting by coating it with a thin layer of zinc, as shown in **Figure 7-20**. The iron will react only after most of the zinc has reacted. A similar approach is used to prevent corrosion in underground pipes, ship hulls, and cars.

Chem Fact
Most metals are like zinc, copper, and aluminum, all of which form an oxide layer that tends to protect the rest of the metal. Iron is the only commonly used metal that is susceptible to continued rusting in the presence of air and water.

Concept Review

Using the activity series to predict reactions

20. If a single-displacement reaction would occur between $H_2O(l)$ and the following metals, write a balanced chemical equation for the reaction.
a. Ba **b.** Rb **c.** Zn

21. If a synthesis reaction would occur between $O_2(g)$ and the following metals, write a balanced chemical equation for the reaction.
a. Au **b.** Mn **c.** Pb

Visual Strategy

Figure 7-20
Make students aware that this process is called galvanization and is commonly used as an anticorrosion measure for nails and garbage cans.

Sample Problem 7C
Using the activity series

Write a balanced chemical equation for the reaction that will occur if the following substances are mixed together.

Cu(*s*)

$Zn(NO_3)_2$(*aq*)

$Cu(NO_3)_2$(*aq*)

Zn(*s*)

1 Check the activity series for the more reactive metal

- **The metal higher on the activity series in Table 7-7 is most likely to displace the other.**

Zinc is higher than copper. This means that zinc metal will form a zinc compound as a product, displacing copper in a compound. So the zinc metal and copper nitrate will be the reactants.

$$Cu(NO_3)_2(aq) + Zn(s) \longrightarrow Cu(s) + Zn(NO_3)_2(aq)$$

2 Check other combinations of reactants for reactions

Cu will not react with $Cu(NO_3)_2$, nor will Zn react with $Zn(NO_3)_2$. Because Cu is lower than Zn on the activity series, Cu will not react to replace Zn in $Zn(NO_3)_2$. No other combination of reactants will result in a reaction.

Practice 7C

Using Table 7-7, arrange the following reactions in order of decreasing tendency to occur. Indicate which would not occur at all. For those reactions that would occur, predict the products that would be formed.

1. **$Zn(s) + CuCl_2(aq) \longrightarrow$**
2. **$Mg(s) + Pb(NO_3)_2(aq) \longrightarrow$**
3. **$Ni(s) + Al_2(SO_4)_3(aq) \longrightarrow$**
4. **$Cu(s) + AgNO_3(aq) \longrightarrow$**

double-displacement
chemical reaction in which two elements in different compounds exchange places

In double-displacement reactions, atoms are exchanged between compounds

In **Figure 7-21**, two colorless solutions are being mixed. Where they mix, a bright yellow cloud is formed. This is a **double-displacement** reaction. The colorless solution in the graduated cylinder contains potassium iodide. The solution in the beaker contains lead(II) nitrate. There are actually two products formed: the solid, yellow ionic compound lead(II) iodide, PbI_2, and potassium nitrate, KNO_3, which dissolves easily in water and can't be seen in **Figure 7-21**.

$$2KI(aq) + Pb(NO_3)_2(aq) \longrightarrow PbI_2(s) + 2KNO_3(aq)$$

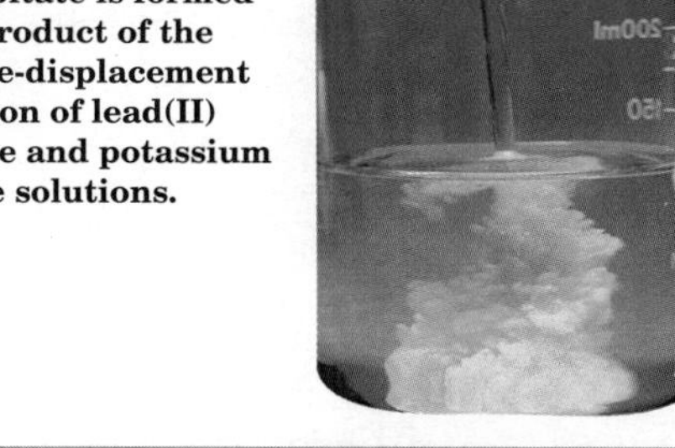

Figure 7-21
The yellow cloud of lead(II) iodide precipitate is formed as a product of the double-displacement reaction of lead(II) nitrate and potassium iodide solutions.

From this equation, it appears that the atoms within the reactants have been exchanged to form the products. The main reason the reaction occurs so dramatically is that one of the products is a solid that does not dissolve in water. Because the compound does not dissolve in water, there is little chance for the ions in PbI_2 to collide with K^+ and NO_3^- ions and re-form KI and $Pb(NO_3)_2$.

Additional Examples for Sample Problem 7C

a. Which of the following could replace Ag in cold water: Al, Au, Rb, Cr, K, Ni, Li, Sn? Which could replace Ag in the presence of steam?
b. What would be the reaction between K(*s*) and AgCl(*aq*)?
c. What would be the reaction between steam and Fe?
d. What reaction will occur when Fe is mixed with Al_2O_3 and heated?

Solutions are at the bottom of the page.

Answers to Practice 7C

Reactions are listed from most to least likely.

2. $Mg(s) + Pb(NO_3)_2(aq) \longrightarrow Pb(s) + Mg(NO_3)_2(aq)$
1. $Zn(s) + CuCl_2(aq) \longrightarrow Cu(s) + ZnCl_2(aq)$
4. $Cu(s) + 2AgNO_3(aq) \longrightarrow 2Ag(s) + Cu(NO_3)_2(aq)$
3. $Ni(s) + Al_2(SO_4)_3(aq) \longrightarrow$ no reaction

Solutions for Additional Examples 7C

a. Rb, K, Li; Al, Rb, Cr, K, Li
b. $K(s) + AgCl(aq) \longrightarrow Ag(s) + KCl(aq)$
c. $2Fe + 6HOH \longrightarrow 2Fe(OH)_3 + 3H_2$
d. Fe is lower in the activity series and, therefore, is less active than Al. There will be no reaction.

Teaching Tip
Vocabulary
Acid-base reactions fall under the category of "neutralization reactions" because the two components neutralize each other and produce a salt and water.

ASSESS

Alternative Assessment
Gold is mined in many third-world countries by using mercury to remove the gold. Divide the class into groups of three, and have each group use library resources to determine the processes used to refine gold, and current uses of gold. Have each group decide how the mercury used in this method of mining gold affects the environment, and if the need for gold justifies the use of mercury.

In Chapter 5, you learned that when ionic compounds such as KI and $Pb(NO_3)_2$ dissolve, the solutions actually contain mixtures of ions.

$$KI(aq) = K^+(aq) + I^-(aq)$$

$$Pb(NO_3)_2(aq) = Pb^{2+}(aq) + 2NO_3^-(aq)$$

Thus, the real reaction is better indicated by individually listing all of the ions in the reactants and products in a total ionic equation.

$$2K^+(aq) + 2I^-(aq) + Pb^{2+}(aq) + 2NO_3^-(aq) \longrightarrow PbI_2(s) + 2K^+(aq) + 2NO_3^-(aq)$$

When these four ions are mixed, two of them, Pb^{2+} and I^-, bond to form a solid. The other two ions, K^+ and NO_3^-, are on both sides of the total ionic equation. Because they remain unchanged in the reaction, they are called *spectator ions*. Total ionic equations can be long and difficult to decipher, so double-displacement reactions are often written using *net ionic equations*, which indicate only the ions involved in the reaction. Spectator ions, such as K^+ and NO_3^-, are omitted.

$$2I^-(aq) + Pb^{2+}(aq) \longrightarrow PbI_2(s)$$

The precipitate, lead(II) iodide, PbI_2, is responsible for the yellow cloud produced when the solutions mix.

Double-displacement reactions always involve the formation either of a molecular compound such as water or the creation of a gas or precipitate.

You have already learned a little about acids and bases and their properties. You may even know that they are "opposites" that can *neutralize* each other. If equally acidic and basic solutions are combined, the resulting solution will have properties of neither acids nor bases. But why have the properties changed?

Consider the reaction that occurs when a person takes an antacid product to settle an upset stomach caused by the secretion of too much acid by the cells lining the stomach.

$$Mg(OH)_2(aq) + 2HCl(aq) \longrightarrow MgCl_2(aq) + 2H_2O(l)$$

Antacids contain bases, such as magnesium hydroxide, to neutralize the hydrochloric acid produced by the stomach, as shown in **Figure 7-22**.

Figure 7-22
After taking an antacid for indigestion, double displacement reactions provide the relief you get. The stomach acid (modeled here by an acid solution in a beaker) reacts with antacid to make a salt and water, substances with neutral pH.

Visual Strategy
Figure 7-22
Ask students if $MgCl_2$ is poisonous? (No.) Although many compounds could be used to neutralize stomach acid, such as lye, NaOH, why should these compounds never be used? (They either damage tissues as they pass through the alimentary canal, or they react to form toxic products.)

But remember that if you combine $Mg(OH)_2$ and HCl in a solution, you are really combining solutions containing four different ions. Not only that, but the hydroxide anion and hydrogen cations react to form water. All that remains when an acid has completely reacted with a base in a double-displacement reaction is water and a salt.

Recall from Chapter 5 that a salt is a compound made when atoms donate and accept electrons to become ions and bond together. A salt can also be defined as a product of a double-displacement reaction between an acid and a base. In fact, there is no detectable difference between a solution made from the double-displacement neutralization reaction and one made by dissolving magnesium chloride in water.

The net ionic equation for the neutralization reaction includes only the hydroxide anion and hydrogen cation because they are changed into a water molecule by the reaction.

$$H^+(aq) + OH^-(aq) \longrightarrow H_2O(l)$$

Section Review

22. Balance the following equations, and identify which of the five types of reactions each represents.
   a. $Cl_2(g) + KBr(aq) \longrightarrow KCl(aq) + Br_2(l)$
   b. $CaO(s) + H_2O(l) \longrightarrow Ca(OH)_2(aq)$
   c. $AgNO_3(aq) + K_2SO_4(aq) \longrightarrow Ag_2SO_4(s) + KNO_3(aq)$
   d. $NH_3(g) + O_2(g) \xrightarrow{Pt} NO(g) + H_2O(g)$
   e. $H_2SO_4(aq) + KOH(aq) \longrightarrow K_2SO_4(aq) + H_2O(l)$
23. Write balanced chemical equations for the following reactions.
   a. synthesis of sulfuric acid, H_2SO_4, from water and sulfur trioxide
   b. combustion of butane, C_4H_{10}
   c. decomposition of potassium chlorate to form potassium chloride and oxygen
   d. single-displacement reaction for zinc and copper(II) sulfate
   e. double-displacement reaction for silver nitrate and sodium chloride
24. Explain how to use an activity series to predict chemical behavior.
25. Predict whether the following reactions would occur. For each that would, complete and balance the equation.
   a. $Ni(s) + H_2O(l) \longrightarrow$
   b. $Br_2(l) + KI(aq) \longrightarrow$
   c. $Mg(s) + Cu(NO_3)_2(aq) \longrightarrow$
   d. $FeO(s) \longrightarrow$
   e. $Ba(s) + O_2(g) \longrightarrow$
26. Write both the total ionic and net ionic equations for each of the following.
   a. Potassium chloride and silver nitrate solutions react to yield the ionic compound potassium nitrate, which dissolves, and silver chloride, which is insoluble.
   b. When copper(II) chloride reacts with ammonium phosphate in aqueous solutions, soluble ammonium chloride forms, and copper(II) phosphate precipitates out of the solution.

Answers to Section Review

22. a. $Cl_2(g) + 2KBr(aq) \longrightarrow 2KCl(aq) + Br_2(l)$
   single displacement
   b. $CaO(s) + H_2O(l) \longrightarrow Ca(OH)_2(aq)$
   synthesis
   c. $2AgNO_3(aq) + K_2SO_4(aq) \longrightarrow Ag_2SO_4(s) + 2KNO_3(aq)$
   double displacement
   d. $4NH_3(g) + 5O_2(g) \xrightarrow{Pt} 4NO(g) + 6H_2O(g)$
   combustion
   e. $H_2SO_4(aq) + 2KOH(aq) \longrightarrow K_2SO_4(aq) + 2H_2O(l)$
   double displacement
23. a. $SO_3(g) + H_2O(g) \longrightarrow H_2SO_4(g)$
   b. $2C_4H_{10}(g) + 13O_2(g) \longrightarrow 8CO_2(g) + 10H_2O(g)$
   c. $KClO_4(s) \longrightarrow KCl(s) + 2O_2(g)$
   d. $Zn(s) + CuSO_4(aq) \longrightarrow ZnSO_4(aq) + Cu(s)$
   e. $AgNO_3(aq) + NaCl(aq) \longrightarrow AgCl(s) + NaNO_3(aq)$
24. In single-displacement reactions, if the activity of the free element is greater than that of the element in the compound, the reaction will take place.
25. a. No reaction
   b. $Br_2(l) + 2KI(aq) \longrightarrow 2KBr(aq) + I_2(aq)$
   c. $Mg(s) + Cu(NO_3)_2(aq) \longrightarrow Mg(NO_3)_2(aq) + Cu(s)$
   d. No reaction
   e. $2Ba(s) + O_2(g) \longrightarrow 2BaO(s)$

Answers are continued on page 269A.

Chemistry in Your Community

Displacement reactions are slowly destroying marble statues, monuments, and metal bridges. Have students determine where any old statues, monuments, or bridges are located in your area, and have them gather enough information to perform the following tasks.

1. Learn what materials were used in building or sculpting the structures. Find out when they were built.
2. Use the information found in item 1, along with library research, to suggest causes of deterioration.
3. Write equations for the chemical reactions that are taking place during deterioration.
4. Perform additional research to discover what measures are being taken to retard these reactions, and what steps are used to restore and repair the decaying or broken items.

Theme

Systems and interactions

The total system of the Statue of Liberty needs to be taken into account when repair or replacement of components is necessary. Producing the required changes can also cause other interactions in the system, such as those corrected by the nitric acid bath.

Answers to Applying Concepts

1. As in galvanization, the zinc's higher activity with respect to iron will cause the zinc to oxidize before the iron does. Thus, the iron will be preserved as long as enough zinc is available to react.
2. $HOOC(CH_2)_nCOHN(CH_2)_nNH_2$

Conclusion: Reactivity and the Statue of Liberty

At the beginning of this chapter, you read about problems associated with the restoration of the Statue of Liberty.

Why did the iron support beams weaken so much, while the copper skin did not? When the tar coating wore away, the salt water acted as an electrical conductor between the copper and iron. Because copper is below iron in the activity series, iron will corrode preferentially, just like zinc on a galvanized nail. The copper will not begin to react until nearly all of the iron reacts.

How could future corrosion be prevented? During restoration, the iron bars had to be replaced a few at a time so that the statue was not left unsupported. Two alloys, or mixtures of metals, were used to replace the rusted iron. One of them, ferallium, is a very strong steel-aluminum alloy that resists corrosion. It was used to connect the framework to the statue's structural tower. The original iron framework was designed with many twists and turns, so 316L stainless steel, which is more flexible than ferallium but also is corrosion-resistant, was used as a replacement.

The bars of the structure were also painted with sealants to prevent corrosion. First, an undercoat containing potassium silicate and zinc dust was applied. On top of that, an epoxy polyamide polymer was used. The sealants had to be water based, or fumes could build up to toxic or explosive levels within the statue. They also had to stick to each other and to the bars. Finally, the top coat needed to be resistant to graffiti.

To keep the copper shell from touching the new frame, a layer of solid polytetrafluoroethylene (Teflon) was inserted between them.

Finally, the cracks between the copper sheets needed to be sealed to keep out the sea spray. A silicone sealant used in construction was chosen because it is long lasting, withstands changes in temperature, and doesn't corrode or discolor the statue's copper skin.

Applying Concepts

1. Using the activity series, explain why zinc dust is an important part of the undercoat for the structure.
2. The top coat of paint on the steel framework is composed of the monomers shown in the equation below. Write a formula for the product. (Hint: only the ends of each molecule will react.)

$$HOOC(CH_2)_nCOOH + H_2N(CH_2)_nNH_2 \longrightarrow ? + H_2O$$

Research and Writing

Use the library to find out more about the following.

1. Report on a chemical or a process used in a specific restoration project.
2. Some chemistry technology can be used on artworks or documents. Report on how such techniques are used to detect forgeries.

CHAPTER 7

Highlights

Key terms

activity series	decomposition
chemical equation	double-displacement
coefficient	polymerization
combustion	single-displacement

Key ideas

What is a chemical reaction?

- A chemical reaction involves rearrangements of atoms from the reactants into new substances, the products.
- The only way to prove that a change was a chemical change is to demonstrate that the new substances produced have different properties from those of the reactants.
- An exothermic reaction releases energy to the environment because the bonds holding the products together are stronger than the bonds holding the reactants together.

How are reactions written?

- A formula equation uses atomic symbols to describe the rearrangement of atoms of the reactants to form the products in a chemical reaction.
- The mass, number, and types of atoms remain the same on both sides of a balanced equation.
- Coefficients indicating amounts of reactants and products can be adjusted to balance a chemical equation. Subscripts within a chemical formula cannot be changed.

What information is in an equation?

- Notations in a chemical equation can indicate special conditions required for a reaction to occur, as well as the states of matter for the substances in the reaction.
- Chemical equations describe energy changes in a reaction. ΔH indicates the amount of energy released or absorbed as heat for each mole of product or reactant.
- The coefficients in a balanced equation indicate the relative amounts required for each substance. They can be interpreted in terms of formula units or as a mole ratio.

How can reactions be used?

- Five basic types of chemical reactions include combustion, synthesis, decomposition, single-displacement, and double-displacement.
- Polymerization reactions are synthesis reactions that build large molecules from many repeated units.
- An activity series predicts single-displacement reactions. Any element on the list will displace those below it in compounds.
- Double-displacement reactions can be represented by net ionic equations indicating only the ions that are changed by the reaction.

Key problem-solving approach: *Reaction types*

To classify a chemical reaction, compare the reaction to the description of each of the five basic reaction types. The proper description points to the reaction type and an example of that reaction.

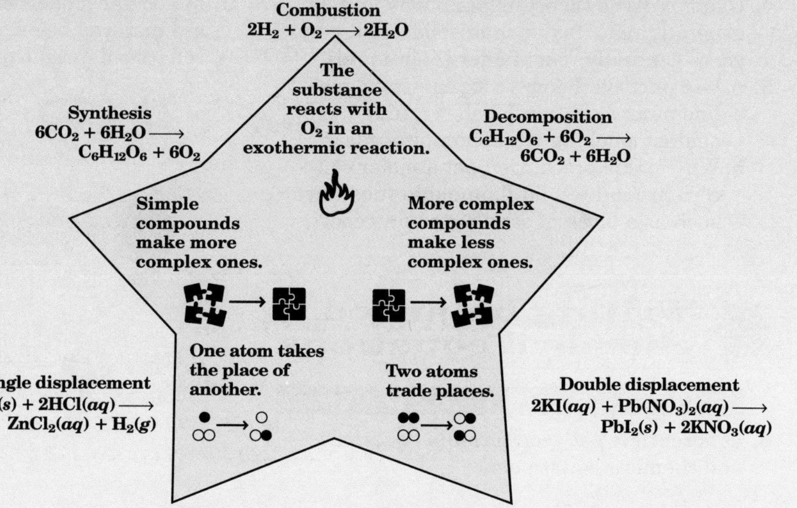

Closure Strategy

1. If 15 mol of hydrogen gas are provided for the reaction $H_2(g) + Cl_2(g) \longrightarrow 2HCl(g)$, how many moles of chlorine are required for a complete reaction, and how much hydrogen chloride is produced? How much of each would be needed or produced if 5.25 mol of hydrogen were used? (15 mol of Cl_2 are needed; 30 mol of HCl are produced. 5.25 mol H_2 and 5.25 mol Cl_2 will yield 10.5 mol of HCl.)
2. Have students complete and balance each of the following reactions
   - **a.** Synthesis, $H_2O + SO_2$
     $H_2O + SO_2 \longrightarrow H_2SO_3$
   - **b.** Decomposition, $Mg(OH)_2$
     $Mg(OH)_2 \longrightarrow MgO + H_2O$
   - **c.** Single displacement, $K + Fe(OH)_3$
     $3K + Fe(OH)_3 \longrightarrow 3KOH + Fe$
   - **d.** Double displacement, $Na_2S + ZnCl_2$
     $Na_2S + ZnCl_2 \longrightarrow 2NaCl + ZnS$
   - **e.** Combustion, $C_6H_6 + O_2$
     $2C_6H_6 + 15O_2 \longrightarrow 12CO_2 + 6H_2O$

1. a. calcium oxide + water ⟶ calcium hydroxide + heat
   b. The bonds in the products must be stronger because more energy was released in forming new bonds than was absorbed in breaking old ones.
   c. Evidence of a chemical reaction includes the change in properties of the substances in the reaction, and the rearrangement of atoms in the molecules.
2. Reactions in nature usually form stable bonds, which have low energies. When reactants change into more stable products, they give up energy, and the reaction is exothermic.
3. a. Reactants are not destroyed, and products are not created. The atoms are merely rearranged. The energy results from the formation of products that have more stable bonds than those of the reactants.
   b. The correct statement should read, "The atoms of the reactants rearranged to form the products. Energy was released in the process."
4. Because the molecules in a liquid move around more than those in a solid, more atoms can collide with each other, and thus react. Gases have the most movement and have the greatest probability of collision and reaction.
5. a. The liquid gasoline evaporates into the gaseous state. When these vapors mix with air, the molecules of gasoline and oxygen are in close proximity and require very little energy to react exothermically.
   b. Solids do not release as many particles into the air. Thus, an ignitable gaseous mixture is not formed.

CHAPTER 7

Review and Assess

The nature of chemical reactions

REVIEW

1. Calcium oxide, CaO, is an ingredient in cement mixes. When water is added, CaO reacts to form calcium hydroxide, $Ca(OH)_2$, which can be used as an egg preservative. The properties of CaO are different from those of $Ca(OH)_2$.
   a. Write a word equation for this exothermic chemical reaction.
   b. Are the bonds stronger in the reactants or the products? Explain your answer.
   c. What evidence is there that this is a chemical reaction?
2. Use the concepts of bond energy and stability to explain why most naturally occurring chemical reactions are exothermic.
3. A student writes the following statement in a lab report: The atoms of the reactants are destroyed, and the atoms of the products are created; energy is also created.
   a. Explain the scientific inaccuracies in the student's statement.
   b. How could the student correct the inaccurate statement?

APPLY

4. Using particle theory, explain why liquids generally react faster than solids, and why gases generally react faster than liquids.
5. a. Use particle theory to explain why gasoline pumps contain labels warning against smoking while pumping gas.
   b. Why is this precaution not a major concern around a solid flammable substance such as a block of candle wax or wood?

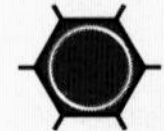

Writing balanced chemical equations

REVIEW

6. Differentiate between formula equations and chemical equations.
7. Why must coefficients, and not subscripts, be changed to balance a chemical equation?
8. The white paste that lifeguards rub on their noses to prevent sunburn contains zinc oxide, $ZnO(s)$, as an active ingredient. Zinc oxide is produced by burning zinc sulfide.

$$2ZnS(s) + 3O_2(g) \longrightarrow 2ZnO(s) + 2SO_2(g)$$

   a. What is the coefficient for sulfur dioxide?
   b. What is the subscript for oxygen gas?
   c. How many atoms of oxygen react?
   d. How many atoms of oxygen appear in the sulfur dioxide molecules?
9. Balance the following equations.
   a. $CaH_2(s) + H_2O(l) \longrightarrow Ca(OH)_2(aq) + H_2(g)$
   b. $CH_3CH_2C{\equiv}CH(g) + Br_2(l) \longrightarrow CH_3CH_2CBr_2CHBr_2(l)$
   c. $Pb(NO_3)_2(aq) + NaOH(aq) \longrightarrow Pb(OH)_2(s) + NaNO_3(aq)$
10. Translate the following word equations into balanced formula equations.
    a. silver nitrate + potassium iodide ⟶ silver iodide + potassium nitrate
    b. nitrogen dioxide + water ⟶ nitric acid + nitrogen monoxide
    c. silicon tetrachloride + water ⟶ silicon dioxide + hydrochloric acid
11. Molecular models of chemical reactions are pictured below. Correct the drawings to reflect balanced equations.

a.
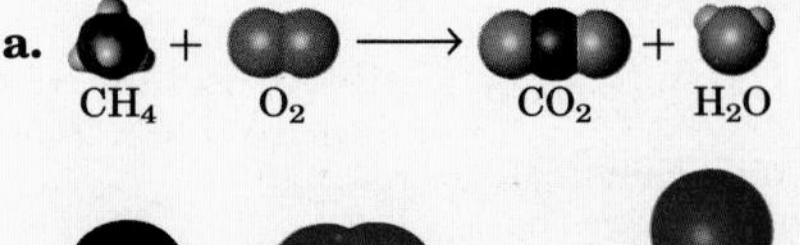

b.
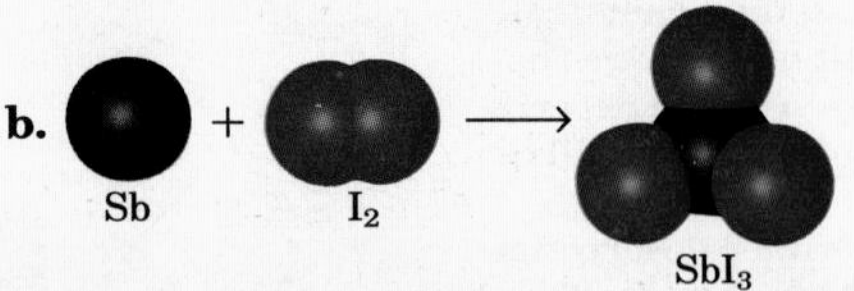

c. + ⟶
H_2 N_2 NH_3

PRACTICE

Answers to items in a black square begin on page 841.

12 Carbon tetrachloride is used as an intermediate chemical in the manufacture of other chemicals. It is prepared by reacting chlorine gas with methane. Hydrogen chloride is also formed in this reaction. Write the balanced chemical equation for the production of carbon tetrachloride. (Hint: see Sample Problem 7A.)

13. Sodium hydroxide is produced commercially by the electrolysis of aqueous sodium chloride. Hydrogen and chlorine gases are also produced. Write the balanced chemical equation for the production of sodium hydroxide. (Hint: see Sample Problem 7A.)

APPLY

14. Iron chloride, $FeCl_3$, is a chemical used in photography. It can be produced by reacting iron and chlorine. Identify the chemical equation that is balanced, and explain what is wrong with the two that are not.
   a. $Fe(s) + Cl_3(g) \longrightarrow FeCl_3(s)$
   b. $2Fe(s) + 3Cl_2(g) \longrightarrow 2FeCl_3(s)$
   c. $Fe(s) + 3Cl_2(g) \longrightarrow Fe2Cl_3(s)$

15. What product is missing in the following equation?

$$MgO + 2HCl \longrightarrow MgCl_2 + ?$$

16. How many moles of HCl can be made from 6.15 mol of H_2 and an excess of Cl_2?

Decoding chemical equations

REVIEW

17. What is the significance of the positive or negative sign before the ΔH value in a chemical reaction?

18. Aluminum sulfate, $Al_2(SO_4)_3$, is used in fireproofing fabrics and the manufacture of antiperspirants. It can be formed from a reaction with H_2SO_4.

$$Al_2O_3(s) + 3H_2SO_4(aq) \longrightarrow Al_2(SO_4)_3(aq) + 3H_2O(l)$$

   a. How many moles of $Al_2(SO_4)_3$ would be produced if 6 mol of H_2SO_4 reacted with an excess of Al_2O_3?
   b. How many moles of Al_2O_3 are required to make 2 mol of H_2O?
   c. If 588.0 mol of Al_2O_3 react with excess H_2SO_4, how many moles of each of the products would be produced?

APPLY

19. Sucrose, $C_{12}H_{22}O_{11}$, is the sugar used to sweeten many foods. Inside the body it is broken down to produce H_2O and CO_2.

$$C_{12}H_{22}O_{11} + 12O_2 \longrightarrow 12CO_2 + 11H_2O \quad \Delta H = -5.65 \times 10^3 \text{ kJ}$$

   a. List all of the ratios that can be derived from this equation.
   b. Three teaspoons (0.0695 mol) of table sugar are mixed in a glass of iced tea. How much energy will the person drinking this iced tea gain from the sugar?

20. Copper(II) nitrate, $Cu(NO_3)_2$, is used to give a dark finish to items made of copper metal, making them appear antique. Copper(II) nitrate can be produced by reacting copper metal with nitric acid. Complete the following table which analyzes this reaction.

$$3Cu(s) + 8HNO_3(aq) \longrightarrow 3Cu(NO_3)_2(aq) + 4H_2O(l) + 2NO(g)$$

Equation	Cu	HNO_3	$Cu(NO_3)_2$	H_2O	NO
Moles					
Formula units					
Molar mass (g/mol)					
Total mass (g)					
State of matter					

21. 1,3-butadiene gas is an intermediate in the production of synthetic rubber, which is widely used in many products, from electrical insulation to car tires. This gas can be obtained by putting butane gas under high pressure and heat. Refer to the diagram below to answer the following questions.
   a. Is this reaction exothermic or endothermic?
   b. If 236.5 kJ of heat are absorbed, what would be ΔH for the reaction?
   c. Write the full equation for the reaction, along with the ΔH value.

Heat energy + [Butane] $\xrightarrow{\text{Catalyst}}$ [1,3-butadiene] + H—H, H—H [Hydrogen gas]

22. Why do you breathe hard during exercise?

6. A formula equation shows the reactants and the products in a reaction. A chemical equation indicates the amount of each reactant and product involved in that reaction.

7. Coefficients balance the equation by changing the number of each type of formula unit on each side of the equation so that the numbers and types of atoms are the same for all of the reactants and all of the products. Changing subscripts would change the composition of each formula unit.

8. a. 2
   b. 2
   c. 6
   d. 4

9. a. $CaH_2(s) + 2H_2O(l) \longrightarrow Ca(OH)_2(aq) + 2H_2(g)$
   b. $CH_3CH_2C{\equiv}CH(g) + 2Br_2(l) \longrightarrow CH_3CH_2CBr_2CHBr_2(l)$
   c. $Pb(NO_3)_2(aq) + 2NaOH(aq) \longrightarrow Pb(OH)_2(s) + 2NaNO_3(aq)$

10. a. $AgNO_3 + KI \longrightarrow AgI + KNO_3$
   b. $3NO_2 + H_2O \longrightarrow 2HNO_3 + NO$
   c. $SiCl_4 + 2H_2O \longrightarrow SiO_2 + 4HCl$

11. Check student drawings to ensure all atoms are accounted for.
   a. $CH_4 + 2O_2 \longrightarrow CO_2 + 2H_2O$
   b. $2Sb + 3I_2 \longrightarrow 2SbI_3$
   c. $3H_2 + N_2 \longrightarrow 2NH_3$

12. formula equation:
$CH_4(g) + Cl_2(g) \longrightarrow CCl_4(l) + HCl(g)$
balanced equation:
$CH_4(g) + 4Cl_2(g) \longrightarrow CCl_4(l) + 4HCl(g)$

13. formula equation:
$NaCl(aq) + H_2O(l) \xrightarrow[e]{} NaOH(aq) + Cl_2(g) + H_2(g)$
balanced equation:
$2NaCl(aq) + 2H_2O(l) \xrightarrow[e]{} 2NaOH(aq) + Cl_2(g) + H_2(g)$

14. a. This is improperly balanced because the subscript for chlorine has been changed. There is no such element (at least at *STP*) as Cl_3.
b. This is the properly balanced equation.
c. This is improperly balanced because a coefficient has been added within the formula unit $FeCl_3$ instead of in front of it.

15. The missing product is water, H_2O.

16. $H_2(g) + Cl_2(g) \longrightarrow 2HCl(g)$
Twice as many moles of HCl can be made, which is 12.3 mol.

17. The ΔH sign indicates whether heat is taken in or released by a system in a reaction.

18. a. 2 mol $Al_2(SO_4)_3$
b. $\frac{2}{3}$ mol Al_2O_3
c. 588.0 mol $Al_2(SO_4)_3$ and 1764 mol H_2O

19. a. 1 mol $C_{12}H_{22}O_{11}$: 12 mol O_2:12 mol CO_2: 11 mol H_2O:5.65×10^3 kJ

b. $0.0695 \text{ mol } C_{12}H_{22}O_{11} \times \frac{5.65 \times 10^3 \text{ kJ}}{1 \text{ mol } C_{12}H_{22}O_{11}} = 393 \text{ kJ}$

20. Moles 3, 8, 3, 4, 2
Formula units
1.807×10^{24}, 4.818×10^{24}, 1.807×10^{24}, 2.409×10^{24}, 1.204×10^{24}
Molar mass
63.55 g/mol, 63.02 g/mol, 187.57 g/mol, 18.02 g/mol, 30.01 g/mol
Total mass
190.65 g, 504.16 g, 562.71 g, 72.08 g, 60.02 g
State of matter
solid, in solution, in solution, liquid, gas

21. a. endothermic
b. $\Delta H = +236.5$ kJ
c. $C_4H_{10}(g) \xrightarrow{\text{catalyst}/\Delta} C_4H_6(g) + 2H_2(g)$
$\Delta H = +236.5$ kJ

22. You need more oxygen to react with glucose in the energy-producing respiration reaction.

Classifying chemical reactions

REVIEW

23. Explain the difference between single-displacement and double-displacement reactions.
24. Use **Table 7-7** to predict which metal would be the best to plate components of a boat to prevent corrosion.
a. Rb b. Cr c. Ba
25. Use **Table 7-7** to predict which metal would be the best choice as a container for concentrated acid.
a. Sn b. Mn c. Pt
26. How do total and net ionic equations differ?
27. Which ions in a total ionic equation are called spectator ions? Why?
28. The saline solution used to soak contact lenses is primarily NaCl dissolved in water. Which of the following ways to designate the solution is incorrect?
a. $NaCl(aq)$ b. $NaCl(s)$
c. $Na^+(aq) + Cl^-(aq)$
29. What are some of the characteristics of each of these five common chemical reactions?
a. combustion
b. synthesis
c. decomposition
d. single-displacement
e. double-displacement

PRACTICE

30. Write the balanced chemical equation for the synthesis of the antacid magnesium oxide from magnesium metal and oxygen gas. (Hint: see Sample Problem 7B.)
31. Write the balanced chemical equation for the synthesis of the pollutant nitrogen dioxide from nitrogen monoxide and oxygen gas. (Hint: see Sample Problem 7B.)
32. Write a balanced chemical equation for the reaction that will occur if the following ingredients are mixed together. (Hint: see Sample Problem 7C.)

$Pb(s)$, $PbCl_2(aq)$, $Zn(s)$, and $ZnCl_2(aq)$

33. Write a balanced chemical equation for the reaction that will occur if the following ingredients are mixed together. (Hint: see Sample Problem 7C.)

$Rb(s)$, $RbNO_3(aq)$, $Zn(s)$, and $Zn(NO_3)_2(aq)$

APPLY

34. What other reaction type(s) can be used to categorize combustion reactions? Explain.
35. Terephthalic acid and ethylene glycol, shown below, can combine to synthesize a single monomer of the polyethylene terephthalate polymer used in film production. During the reaction, a molecule of water is formed. Write the balanced formula equation for this reaction.

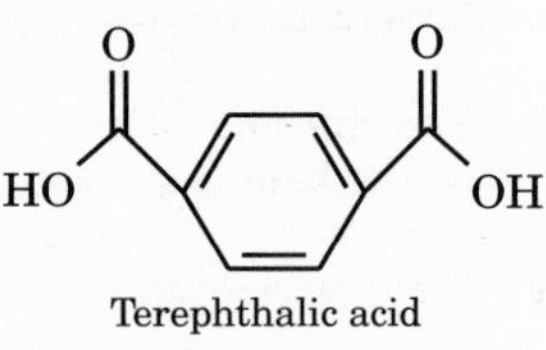

Terephthalic acid

HO OH
Ethylene glycol

36. Use **Table 7-7** to predict whether the following reactions are possible.
a. $Ni(s) + MgSO_4(aq) \longrightarrow NiSO_4(aq) + Mg(s)$
b. $6Li(s) + Al_2(SO_4)_3(aq) \longrightarrow 3Li_2SO_4(aq) + 2Al(s)$
c. $Ba(s) + 2H_2O(l) \longrightarrow Ba(OH)_2(aq) + H_2(g)$
37. a. Write the total and net ionic equation for the reaction in which the antacid $Al(OH)_3$ neutralizes stomach acid.
b. Identify the spectator ions in this reaction.
c. What would be the advantages of using $Al(OH)_3$ as an antacid rather than $NaHCO_3$, which undergoes the following reaction with stomach acid?

$NaHCO_3(aq) + HCl(aq) \longrightarrow NaCl(aq) + H_2O(aq) + CO_2(g)$

38. When heated, tungsten metal usually forms an oxide compound in a synthesis reaction. Light-bulb filaments are made of tungsten. Use the information in the diagram to explain why the light-bulb filament does not form an oxide compound.

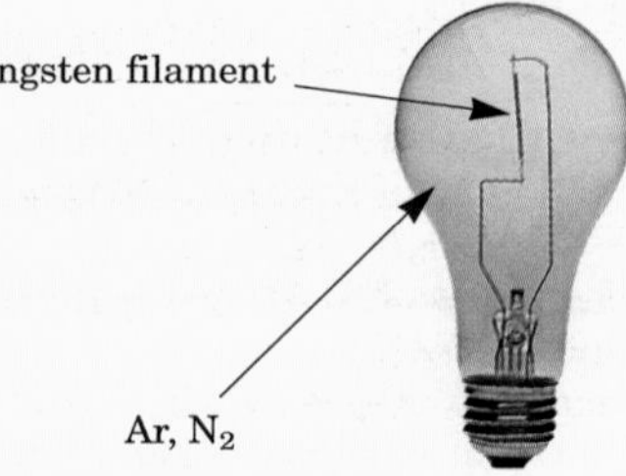

Linking chapters

1. ***Conservation of energy***
When wood is burned, energy is released in the forms of heat and light. Explain why this change does not violate the law of conservation of energy.

$$[C_6H_{10}O_5]_n + 6nO_2 \longrightarrow 5nH_2O + 6nCO_2$$

2. ***Structural formulas***
Neoprene is a polymer used to make shoes and gloves.
a. Identify the monomer in this portion of a neoprene molecule by drawing its Lewis structure.
b. What is the empirical formula of neoprene?

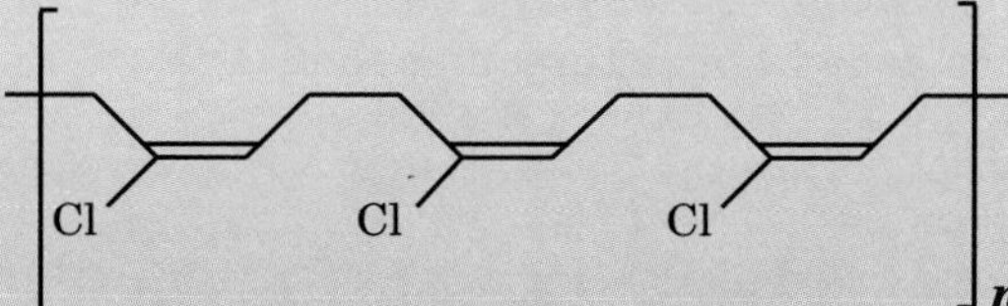

3. **Theme: *Classifications and trends***
Although cesium is not included in the activity series in **Table 7-7**, where would you expect it to appear based on its position in the periodic table?

USING TECHNOLOGY

1. ***Graphics calculator***
In a given chemical equation, the ΔH value can be expressed as part of a mole ratio. This ratio can be graphically represented as the slope of a straight line. The graph can be used to predict how much energy is transferred when a given amount of substance reacts. The combustion of 1.00 mol of methane, CH_4, has a ΔH of −890. kJ. Using the WINDOW function, set the x-range from 0 to 20 with a scale of 2 and the y-range from −20 000 to 0 with a scale of 2000. Select the Y= key and enter the function Y_1 = (−) 890 X, T, θ, which is $y = -890x$. (Hint: use the negative key, not the subtraction key.) Press GRAPH and use the TRACE mode to estimate the kilojoules released when the following amounts of methane are burned.
a. 5.00 mol **b.** 8.00 mol **c.** 15.00 mol

Alternative assessment

Performance assessment

Design an experiment to test different antacids on the market. Include $NaHCO_3$, $Mg(OH)_2$, $CaCO_3$, and $Al(OH)_3$ in your data. Discover which one neutralizes the most acid and what byproducts are formed. Show your experiment to your teacher for approval. If your experiment is approved, obtain the necessary chemicals from your teacher and test your procedure.

Portfolio projects

1. ***Chemistry and you***
For one day, record situations that show evidence of a chemical change. Identify the reactants and the products, and determine whether there is proof of a chemical reaction. Classify each of the chemical reactions according to the five common reaction types in the chapter.
2. ***Chemistry and you***
Research safety tips for dealing with fires. Create a poster or brochure about fire safety that explains both these tips and their basis in science.
3. ***Research and communication***
Much of the energy used in the United States comes from the combustion of hydrocarbon fuels. Other sources of energy are available, however, including the sun, wind, and water. As a group, research either hydrocarbon fuels or alternative fuels. Analyze the reactions involved in using and producing the fuel. For example, solar power produces no byproducts; however, the production of solar panels should be considered. Have the class debate which power source is the safest, the most economical, the most efficient, and the most convenient.
4. ***Research and communication***
Many products are labeled biodegradable. Choose several biodegradable items on the market, and research the decomposition reactions involved. Be sure to take into account any special conditions that must occur for the substance to biodegrade. Present your information to the class to help inform the students about what products are best for the environment.

23. A single-displacement reaction involves one element taking the place of another element within a compound. A double-displacement reaction involves two ions from different substances trading places.

24. **b.** Cr

25. **c.** Pt

26. Total ionic equations account for all of the ions contributed by the components. Net ionic equations show only those ions that form a new compound in the reaction.

27. Spectator ions are the ions that are present on both the product and reactant sides of the equation, but are unchanged by the reaction. They are therefore considered “spectators” to the reaction.

28. **b.** NaCl(s)

29. Possible answers include:
a. A chemical reacts exothermically with oxygen.
b. The products are more complex than the reactants.
c. The reactants are more complex than the products.
d. An element in a compound is replaced by another element.
e. Two ions in separate compounds trade places.

30. $2Mg(s) + O_2(g) \longrightarrow 2MgO(s)$

31. $2NO(g) + O_2(g) \longrightarrow 2NO_2(g)$

32. Zinc is higher than lead on the activity series, so zinc will displace lead from compounds.
$Zn(s) + PbCl_2(aq) \longrightarrow ZnCl_2(aq) + Pb(s)$

33. Rubidium is higher than zinc on the activity series, so rubidium will displace zinc from compounds.
$2Rb(s) + Zn(NO_3)_2(aq) \longrightarrow 2RbNO_3(aq) + Zn(s)$

Answer from page 263

26. a. $K^+(aq) + Cl^-(aq) + Ag^+(aq) + NO_3^-(aq) \longrightarrow K^+(aq) + NO_3^-(aq) + AgCl(s)$
(total ionic equation)
$Ag^+(aq) + Cl^-(aq) \longrightarrow AgCl(s)$
(net ionic equation)

b. $3Cu^{2+}(aq) + 6Cl^-(aq) + 6NH_4^+(aq) + 2PO_4^{3-}(aq) \longrightarrow 6NH_4^+(aq) + 6Cl^-(aq) + Cu_3(PO_4)_2(s)$
(total ionic equation)
$3Cu^{2+}(aq) + 2PO_4^{3-}(aq) \longrightarrow Cu_3(PO_4)_2(s)$
(net ionic equation)

Continued from page 269

34. When oxygen combines with a substance to form a more complex molecule, the reaction is a synthesis reaction. When oxygen helps break larger chemicals into less complex ones, the reaction is a decomposition reaction.

35. $C_8H_6O_4(s) + C_2H_6O_2(l) \longrightarrow C_{10}H_{10}O_5(s) + H_2O(l)$

36. a. The reaction is not possible because nickel is less active than magnesium and will not displace Mg in compounds.
b. The reaction is possible because lithium is more active than aluminum and will displace Al in compounds.
c. The reaction is possible because barium is more active than hydrogen and will displace H in water.

37. a. total: $Al^{3+}(aq) + 3OH^-(aq) + 3H^+(aq) + 3Cl^-(aq) \longrightarrow Al^{3+}(aq) + 3Cl^-(aq) + 3H_2O(l)$
net: $3H^+(aq) + 3OH^-(aq) \longrightarrow 3H_2O(l)$
b. Al^{3+} and Cl^-
c. $Al(OH)_3$ produces water and a salt. $NaHCO_3$ produces water, a salt, and a gas, which could cause discomfort.

38. The light-bulb filament is sealed off from the oxygen in the air and, therefore, cannot combine with it to form an oxide compound. The gases in the bulb do not undergo a reaction with tungsten, even at high temperature.

Linking chapters

1. During the reaction the reactants are converted into more stable products. This process involves the breaking of chemical bonds and the formation of new bonds that require less energy. The surplus energy is released as heat and light.

2. a.

```
 [  H          H  ]
 [  :          :  ]
—[  C          C  ]—
 [  :  \      / :  ]
 [  H   C == C  H  ]
 [    :Cl:   H     ]
```

b. C_4H_5Cl

3. Cesium would be in the highest reactivity group of the activity series, where the other alkali metals are listed also.

USING TECHNOLOGY

1. **a.** -4.45×10^3 kJ
b. -7.12×10^3 kJ
c. -1.34×10^4 kJ

Performance assessment

Answers will vary. Be sure that students' testing schemes are logical and appropriate, and that proper laboratory procedures are followed.

Portfolio projects

1. Answers will vary. Some examples may be fire or car engines combusting gasoline.
2. Answers will vary. Evaluate according to the scientific accuracy and the feasibility of the safety measure.
3. Answers will vary. Be sure that each group researches the environmental effects of producing as well as consuming the energy.
4. Answers will vary. Be sure that students consider reaction conditions such as light, moisture, and heat.

8

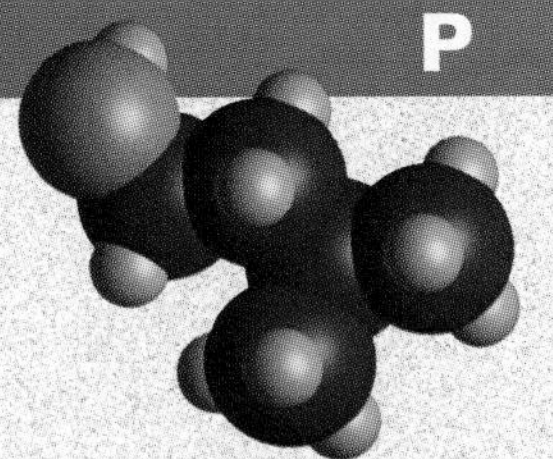

Stoichiometry

Chapter Overview

- **Section 8-1** uses realistic problems to introduce students to the stoichiometric relationships chemists use to determine quantities needed or produced in reactions. The text presents an in-depth explanation of the thinking process used to solve these kinds of problems. The three step method, first introduced in Chapter 4 with moles, is further reviewed and reinforced to focus students' attention on the development of problem-solving strategies. The visual algorithms in this section provide additional reinforcement using a different learning modality.
- **Section 8-2** extends the isoamyl acetate situation presented in Section 8-1 as students explore the concept of limiting reactants. This concept is further developed when students calculate percent yields for reactions.
- **Section 8-3** provides further reinforcement of the chemical concepts covered in previous lessons, as students explore the real world stoichiometric problems related to auto air bags.

Concept Base

Students must have mastered the following concepts prior to this chapter: calculating molar masses (Chapter 4); writing formulas (Chapters 5 and 6); writing and balancing chemical equations (Chapter 7).

Looking Ahead

Students will use the concepts from Chapter 8 in most every chapter that follows it. Therefore, it is critically important that students master stoichiometric calculations. Enthalpy, entropy, and free energy will be introduced in Chapter 9. Students will explore equilibrium, rates, and yields in Chapters 12 and 14.

Laboratory Equipment Needs

Demonstrations

The listing of materials for Chapter 8 demonstrations begins on page T46.

Laboratory Experiments

Materials and preparation instructions for Chapter 8 are found as follows.

Text		*Page*
Exploration 8A	Stoichiometry and Gravimetric Analysis	706
Investigation 8A	Gravimetric Analysis—Hard Water Testing	710
Exploration 8B	Stoichiometry of Reactions	712
Investigation 8B	Stoichiometry—Viral Disease Alert	716

Pacing: 15 days	**Inquiry Teaching Strategies**
8-1 How much can a reaction produce?	**Demonstrations** ■ Demo 1 *Stoichiometry of a sodium carbonate reaction* ■ Demo 2 *Stoichiometric production of oxygen* **Laboratory Experiments** ■ Exp 8A *Stoichiometry and Gravimetric Analysis* ■ Inv 8A *Gravimetric Analysis—Hard Water Testing* ■ Exp 8B *Stoichiometry of Reactions* ■ Inv 8B *Stoichiometry—Viral Disease Alert*
8-2 How much does a reaction really produce?	**Demonstrations** ■ Demo 3 *Limiting reactants*
8-3 How can stoichiometry be used?	**Demonstrations** ■ Demo 4 *Decomposition of marble chips*

Key ■ Teaching Transparencies ■ Annotated Teacher's Edition ■ Laboratory Experiments Manual

Visual Teaching Strategies

Transparencies
- F 8-5 *Solving Mass-Mass Stoichiometry Problems*
- F 8-6 *Solving Stoichiometry Problems With Moles or Grams*
- F 8-7 *Solving Various Types of Stoichiometry Problems*

Transparencies
- F 8-15 *Fuel-Air Ratios in an Engine*
- P 304 *Stoichiometry Calculations*

Review and Practice Strategies

Text Reviews
- Additional Examples 8A, 8B, 8C
- Practice 1–6
- Section Review 1–9
- Chapter Review 1–13; LC 1; UT 1

Study Guide Worksheets
- 8-1 Practice
  Mass-mass Stoichiometry
  Mole-mass Stoichiometry
  Stoichiometry Calculations
  Using Density

Text Reviews
- Additional Examples 8D, 8E, 8F
- Practice 1–4
- Section Review 10–16
- Chapter Review 14–28; LC 2; UT 2

Study Guide Worksheets
- 8-2 Practice
  Finding the Limiting Reactant
  Percent Yield

Text Reviews
- Additional Examples 8G, 8H, 8I
- Practice 1–7
- Section Review 17–22
- Chapter Review 29–39

Study Guide Worksheets
- 8-3 Practice
  Practical Uses of Stoichiometry

Pupil's Text
Study Guide
Teaching Resources

Exp *Exploration*
Inv *Investigation*

LC *Linking Chapters*
UT *Using Technology*

Assessment Options

Traditional Assessment

Test Generator
Instructional Objectives Measured:
Content mastery
Select from items 1 to 75

Performance Assessment Options

(see pages T36 and T643-15 for scoring rubrics)
Students demonstrate mastery of objectives in a hands-on environment. You may want some of these materials in the Portfolio as well.

Investigation 8A and 8B (text)
Instructional Objectives Measured:
Content mastery, Use of scientific methodology, Problem solving skills, Proficiency in written communication

Performance Assessments, page 309

1. **Designing an experiment**
   Instructional Objectives Measured:
   Content mastery, Use of scientific methodology, Problem solving skills, Proficiency in written communication
2. **Designing an experiment**
   Instructional Objectives Measured:
   Content mastery, Use of scientific methodology, Problem solving skills, Proficiency in written communication

Portfolio Options

(see page T36 for scoring rubrics)
Students provide a written rationale for each selection made for the portfolio.

Concept Maps
Students use vocabulary from the Chapter Review to build a concept map.
Instructional Objectives Measured:
Content mastery

Formal Laboratory Reports
Exploration 8A and 8B (text)
Instructional Objectives Measured:
Content mastery, Use of scientific methodology, Problem solving skills, Proficiency in written communication

Portfolio Projects, page 309
Items 1 and 4. Research and communication
Instructional Objectives Measured:
Scientific and chemical literacy, Understanding the impact of science and technology on society, Proficiency in written communication

Item 2. Cooperative activity
Instructional Objectives Measured:
Scientific and chemical literacy, Understanding the impact of science and technology on society, Proficiency in oral communication

Item 3. Chemistry and you
Instructional Objectives Measured:
Scientific and chemical literacy, Understanding the impact of science and technology on society, Proficiency in written communication

INTRODUCING THE CHAPTER

CHAPTER 8 Stoichiometry

Story Background

The FRH was used by soldiers in the 1990–1991 Desert Storm conflict. The FRH was developed jointly by the U.S. Army and Zesto Therm, a company based in Cincinnati, Ohio.

To achieve the high exothermic reaction needed to warm food, the magnesium undergoes a supercorrosion process shown by the equation in the photograph.

The NaCl used in the FRH enhances the corrosion of magnesium, which is naturally protected by an oxide coating. The Cl^- ion in NaCl reacts with $Mg(OH)_2$ to form MgOHCl. The magnesium oxide coating is soluble in the MgOHCl, so it dissolves. Once the coating is eaten away, water can continue to react with the magnesium.

The role of iron in the FRH is not completely understood, but scientists think that it provides a site for the exchange of ions and electrons between magnesium and water.

About the Illustration

Soldiers in the field are shown using an FRH, "Flameless Ration Heater." The supercorrosion of magnesium, which is the reaction shown, is highly exothermic and provides the energy to heat the food.

Mass Relationships and Feeding the Army

At dinnertime, the soldiers were miles from the nearest field kitchen. They each pulled out a meal pouch, inserted it into a small bag, and added water from a canteen to the bag. The contents grew warm, and within 15 minutes, they were eating hot meals.

The dinner is one of the U.S. Army's "Meal, Ready-to-Eat," or MRE, rations. Each MRE is a complete main dish within a pouch made of aluminum foil and plastic. In 1985, researchers at the Army's Natick Research, Development, and Engineering Center in Massachusetts invented a new way to heat MREs using a special "Flameless Ration Heater," or FRH.

Each FRH is a plastic sleeve containing a paperboard-covered pad with holes in it. Within the pad are metal particles embedded in a polymeric matrix. The metal contains a 90% magnesium and 10% iron alloy, a homogeneous mixture of the two elements. The entire assembly has a mass of 20.0 g, of which about 8.1 g are magnesium. A little NaCl is also included.

Most people think of water as being used to cool things, but in this case, the water reacts with magnesium in an exothermic single-displacement reaction. The energy released by this reaction in the FRH warms the meal to 60°C.

$$Mg(s) + 2H_2O(l) \longrightarrow Mg(OH)_2(s) + H_2(g) + 353\ kJ$$

According to the activity series, this reaction ordinarily doesn't occur. But when Fe, a less active metal, is in contact with it, Mg corrodes preferentially, just like the zinc coating on galvanized iron. The cardboard pad keeps the water and magnesium in contact. Earlier models of the FRH used more than 60 mL of water, but the latest version requires only 45 mL.

The necessity of heating meals with equipment that is lightweight and easy to use is one shared by the Army and people involved in outdoor activities such as camping and backpacking.

A heater similar to the FRH is also available for campers. It uses the synthesis reaction that combines calcium oxide and water.

$$CaO(s) + H_2O(l) \longrightarrow Ca(OH)_2(s) + 64.4\ kJ$$

Before the FRH, the Army issued bars of trioxane, $C_3H_6O_3$, to warm meals. The bars, which came in 15 g and 30 g sizes, were burned. Complete combustion of a 30 g bar releases about 500 kJ of energy.

MREs could not be heated directly by burning trioxane. Instead, the burning trioxane heated a metal cup of water containing the MRE pouch.

- *Why was the magnesium-water reaction of the FRH chosen over the other methods?*
- *Why is there a difference between the amount of water that is added and the amount of water that the reaction requires?*

To answer these questions, you need to understand how the amounts and masses of substances in a reaction compare. In this chapter, you will learn to predict the amount of reactant needed for a reaction, and you will apply these principles to determine the most effective way to heat an MRE.

Tapping **Prior Knowledge**

Ask the following questions to lead into the idea of proportions and stoichiometric relationships.

1. What do the coefficients of an equation represent?
2. If a recipe for cookies is doubled to make a larger batch, how does that change affect the number of cookies that can be made?
3. If a cookie recipe calls for 2 eggs and you have only 1 egg, what could you do to adjust the recipe for 1 egg? How would the adjustment affect the number of cookies you could make?

Section 8-1

FOCUS

Lesson Starter

Use the demonstration that follows to trace mass relationships in a simple reaction. Tell students that you are going to turn Na_2CO_3 into NaCl using the following reaction.

$$Na_2CO_3(s) + 2HCl(aq) \longrightarrow 2NaCl(aq) + H_2O(l) + CO_2(g)$$

Demonstration 1

Stoichiometry of a sodium carbonate reaction

Approximate time: 20 min

1. Determine the mass of a Pyrex evaporating dish to the thousandth of a gram, and record the mass on the board.
2. Add 1 g of anhydrous Na_2CO_3, and record the mass of the dish and salt.
3. Add 1 M HCl dropwise until the bubbling stops. Then add a little more HCl to make sure the reaction has gone to completion.
4. Place the evaporating dish on a piece of wire gauze on a ring stand and in a fume hood. Boil the liquid to dryness (heat gently at first).
5. Record the mass of the dish and dried NaCl.

Some students will be surprised that the mass of the NaCl is greater than the Na_2CO_3. Ask students how this is possible and what happens to the excess HCl.

Save the results for Section 8-2, page 289, which covers theoretical and actual yields.

SAFETY

Wear safety goggles and a lab apron. Students should stand back 10 ft.

DISPOSAL

Wrap the NaCl in newspaper, and put it in the trash.

How much can a reaction produce?

Section Objectives

Distinguish between composition stoichiometry and reaction stoichiometry.

Apply a three-step method to solve stoichiometry problems.

Use mole ratios and molar masses to create conversion factors for solving stoichiometry problems.

Stoichiometry

stoichiometry *mass and quantity relationships among reactants and products in a chemical reaction*

In Chapter 7, you learned that reactions can be used to make new products, to break down reactants, or to provide a source of energy. But the predictions you made in Chapter 7 relate only to the *identities* and *relative amounts* of the products and reactants. It is also very useful to predict exactly *how much mass* of a substance will be involved in a reaction. Such predictions are a part of chemistry known as **stoichiometry**.

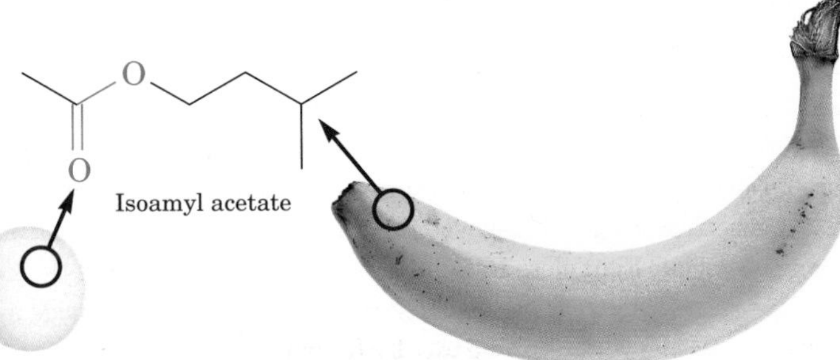

Figure 8-1
Isoamyl acetate is the compound primarily responsible for a banana's fruity flavor. It is used in a variety of artificially flavored products.

Making chemicals by the kilogram

The makers of artificial flavors for items such as banana ice cream or banana-flavored bubble gum need to use stoichiometry when they make the banana flavoring. The substance providing most of the taste and smell of bananas is actually a single compound known as 3-methylbutyl acetate, or isoamyl acetate. The structure of this compound is shown in **Figure 8-1**. Isoamyl acetate is a member of a group of organic compounds known as esters. Esters are among the organic chemicals responsible for the aromas and flavors of fruits. The ester functional group is indicated in red on the structure in **Figure 8-1**.

Because there are so many compounds in a banana, it is difficult to isolate and purify natural flavoring. It is actually more cost-efficient to synthesize isoamyl acetate in the lab and use it as an artificial banana flavoring. The equation for the synthesis reaction is shown below. The flavor chemicals shown in **Figure 8-2** on the following page are also often synthesized as artificial flavors in the lab.

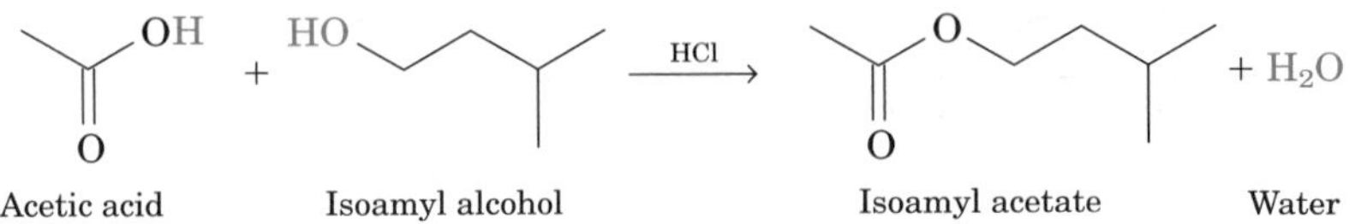

Figure 8-2
Many flavors and aromas are due to organic chemicals such as esters, aldehydes, alcohols, and ketones.

Consider the following question: ***If a flavoring manufacturer has 45.0 kg of isoamyl alcohol and enough acetic acid to react with all of it, what is the maximum number of kilograms of banana flavoring that can be made?*** Questions that deal with amounts in reactions are examples of *reaction stoichiometry.* Using what you have read so far in this chapter and what you learned in Chapters 2, 5, 6, and 7, you *already know everything necessary* to find the answer to this question. Reaction stoichiometry is simply a new way to apply skills such as writing chemical formulas, calculating formula masses, and converting from mass to moles.

Organize what you already know

When solving a problem like this one, a good way to start is to make a table or list of all of the information given in the problem, as well as what answer you are trying to find. *Remember to include units for all quantities.* This will help you later when you are trying to decide how to solve the problem. The sections of **Table 8-1** highlighted in pink include this starting material. The red question mark in the table indicates the answer that is sought. Problems like this one are often known as *mass-mass problems* because the data given is a mass amount and the answer sought is also a mass amount.

Start by figuring out more than just what the problem tells you. In Chapter 6, you learned how to get the molecular formulas for the reactants and products from the organic structures shown on the previous page for isoamyl alcohol and acetic acid.

In Chapter 5, you learned how to use the periodic table to calculate the molar masses of these reactants and products. Such calculations are called *composition stoichiometry* because they describe the mathematical relationships among the elements that make up a substance. The results are shown in the rest of **Table 8-1**. *For almost all chemistry problems of this type, you will need the molar masses of the substances involved.*

Table 8-1
Some Data for the Banana-Flavoring Problem

Reactants	Amount	Formula	Molar mass
isoamyl alcohol	45.0 kg	$C_5H_{11}OH$	88.17 g/mol
acetic acid	excess	CH_3COOH	60.06 g/mol
Products	**Amount**	**Formula**	**Molar mass**
isoamyl acetate	? kg	$CH_3COOC_5H_{11}$	130.21 g/mol
water	not given	H_2O	18.02 g/mol

TEACH

Theme
Conservation
Stoichiometric relationships are based on conservation.

Do You Know?
Stoichiometry is from the Greek *stoicheion,* which means "to measure the elements."

Cultural Connection
Many cultures have developed stoichiometric techniques of making metal alloys. The ancient Egyptians used a copper-tin alloy of bronze by 3000 B.C. The Hittites of Asia Minor developed steel around 1500 B.C. The Japanese and Chinese made fine steel. These metals were used for making weapons, which enabled these cultures to rule their regions.

Teaching Tip
Problem-solving techniques
Expert problem solvers list all of the variables for a problem in an organized fashion. Make your students expert problem solvers by insisting that they organize the data from the problem. This is the purpose of the *List* phase in the sample problems throughout this book.

Additional Examples
Determine the formula masses of $NaHCO_3$ and $(NH_4)_2SO_4$.

$NaHCO_3$
Na 1 × 22.99 = 22.99
H 1 × 1.01 = 1.01
C 1 × 12.01 = 12.01
O 3 × 16.00 = 48.00
84.01

$(NH_4)_2SO_4$
N 2 × 14.01 = 28.02
H 8 × 1.01 = 8.08
S 1 × 32.07 = 32.07
O 4 × 16.00 = 64.00
132.17

Visual Strategy
Figure 8-1
Be sure students are aware that a carbon atom exists at each corner and at the ends of the lines in the structural formula. These carbon atoms are bonded to the appropriate number of hydrogen atoms. Have students rewrite the structural formula showing all of the carbon and hydrogen atoms.

Additional Examples

Balance the following equations.

a. $HCl(aq) + Ba(s) \longrightarrow BaCl_2(aq) + H_2(g)$

$2HCl(aq) + Ba(s) \longrightarrow BaCl_2(aq) + H_2(g)$

b. $K_2SO_4(aq) + Ca(NO_3)_2(aq) \longrightarrow KNO_3(aq) + CaSO_4(s)$

$K_2SO_4(aq) + Ca(NO_3)_2(aq) \longrightarrow 2KNO_3(aq) + CaSO_4(s)$

Theme

Conservation

Balanced equations ensure the conservation of atoms, which is necessary to keep molar ratios in the correct proportions.

Teaching Tip

Unit factors

Remind students that when they see 1000 g/1 kg, the whole fraction is equal to 1. Also, the quantity that they are generally asked to find is on top while the unit they are converting from is on the bottom. Finally, remind the students that while units cancel, the integers do not.

The last piece of information that you need is the balanced chemical equation for the reaction. In Chapter 7, you learned that a balanced chemical equation can provide you with the mole ratios for the substances in the reaction so that you can compare relative amounts. In the balanced equation for isoamyl acetate synthesis, no coefficients are written. This means that all of the coefficients are equal to 1, so all of the mole ratios are 1:1.

$$C_5H_{11}OH(l) + CH_3COOH(l) \xrightarrow{HCl} CH_3COOC_5H_{11}(l) + H_2O(l)$$

One mole of isoamyl alcohol and one mole of acetic acid react to make one mole of isoamyl acetate and one mole of water. *For almost all chemistry problems, you need a balanced chemical equation so that you can find the mole ratios.*

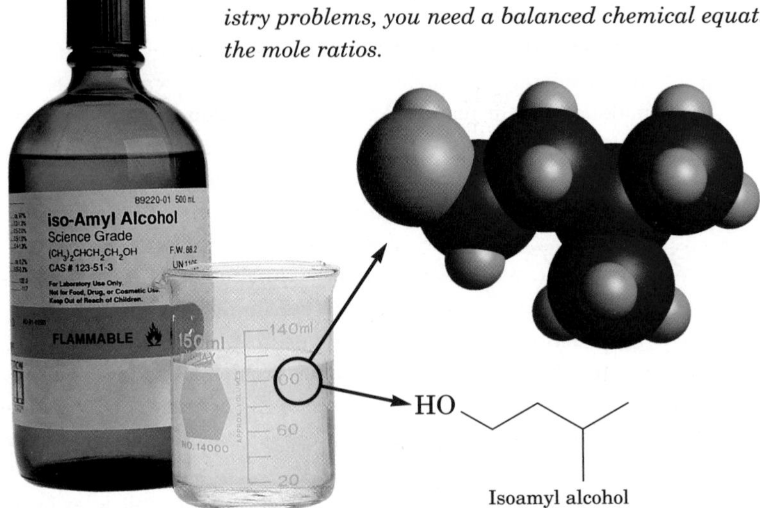

Figure 8-3
Isoamyl alcohol, one of the reactants in the banana-flavoring synthesis reaction, has many uses. It is used as a solvent and in synthesizing artificial silk, lacquers, and other materials. Recall that the —OH functional group is characteristic of an alcohol.

Use conversion factors on data from the problem

1. Check the units given

Right now, all of your data is in grams except the amount actually given in the problem, 45.0 kg. Having data with units that do not match is a clue that you need to convert the units—in this case, from kilograms to grams. Recall from Chapter 2 that such conversions can be achieved by multiplying by a conversion factor that connects the units—for example, 1000 g equals 1 kg. But how do you know which factor to use?

$$\frac{1000\ g}{1\ kg} \text{ or } \frac{1\ kg}{1000\ g}\ ?$$

The correct factor is the one that has the units you already have, *kg*, on the bottom and the units you want to have, *g*, on the top—the factor on the left. Remember that when the value from the problem is multiplied by this conversion factor, the unwanted units, *kg*, cancel and leave the desired units, *g*.

$$45.0\ \cancel{kg} \times \frac{1000\ g}{1\ \cancel{kg}} = 4.50 \times 10^4\ g \text{ isoamyl alcohol reactant}$$

Visual Strategy

Figure 8-3

Isoamyl alcohol can be represented using a structural formula or a three-dimensional space-filling model. Help students interpret both representations by having them note the presence of carbon and hydrogen atoms on the space-filling model and matching the atoms to their positions on the structural formula. Make students aware that the —OH group is characteristic of all alcohols.

2. Determine the mole ratio

You want to know the amount of isoamyl acetate product from the given amount of reactant. In Chapter 7, you learned that the mole ratio provides information about the relative amounts of the reactants and products shown in **Figure 8-3** on the previous page and **Figure 8-4**. From the balanced chemical equation for the reaction, you know that for every mole of isoamyl alcohol used in the reaction, a mole of isoamyl acetate will be produced. But right now you know the amount of isoamyl alcohol only as a mass in *grams*. To compare amounts of reactant and product, you must first convert the amount of the reactant to *moles*. Use the molar mass as the conversion factor to make this change. *Grams* belong on the bottom of the conversion factor because you want to cancel it out, leaving the desired units, *mol*.

$$\frac{\cancel{88.17\ \text{g}}}{\cancel{1\ \text{mol isoamyl alcohol}}} \quad \text{or} \quad \frac{1\ \text{mol isoamyl alcohol}}{88.17\ \text{g}}$$

$$4.50 \times 10^4\ \cancel{\text{g}} \times \frac{1\ \text{mol}}{88.17\ \cancel{\text{g}}} = 510.\ \text{mol isoamyl alcohol}$$

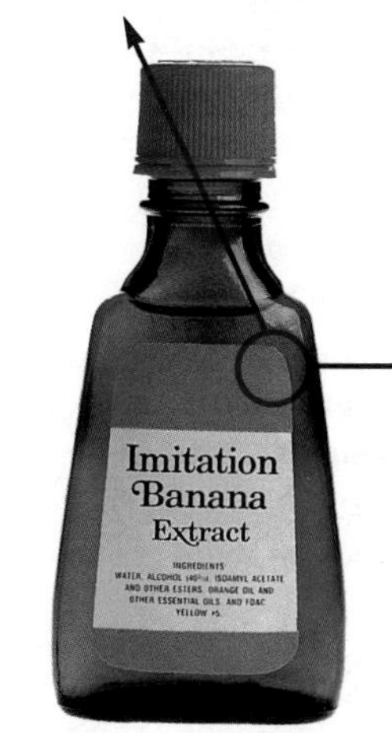

Figure 8-4
Besides being used as a flavoring agent, isoamyl acetate is also used as an ingredient in air fresheners, as a solvent, and as an ingredient in the manufacture of artificial leather, artificial pearls, and waterproof varnishes.

3. Use the mole ratio to calculate moles

Now you can use the mole ratio to calculate the amount of product that can be made from a reactant. In this problem, the calculation is not very difficult because it involves a 1:1 mole ratio. However, it can be tricky with more difficult reactions. Be certain that the mole ratio is written like a conversion factor that cancels the given units and leaves the units of the substance of interest.

$$510.\ \cancel{\text{mol isoamyl alcohol}} \times \frac{1\ \text{mol isoamyl acetate}}{\cancel{1\ \text{mol isoamyl alcohol}}} = 510.\ \text{mol isoamyl acetate}$$

4. Use molar mass to calculate grams

Finally, you know the amount of isoamyl acetate produced in this reaction. But check the problem and your data list on page 273. You were asked for the number of *kilograms* of isoamyl acetate, not the number of *moles*. Once again, some conversions are needed. First, convert the moles of isoamyl acetate to grams using the molar mass, as in Chapter 5. Then, convert from grams to kilograms using the correct conversion factor, as in Chapter 2.

$$510.\ \cancel{\text{mol}} \times \frac{130.21\ \cancel{\text{g}}}{1\ \cancel{\text{mol}}} \times \frac{1\ \text{kg}}{1000\ \cancel{\text{g}}} = 66.4\ \text{kg isoamyl acetate product}$$

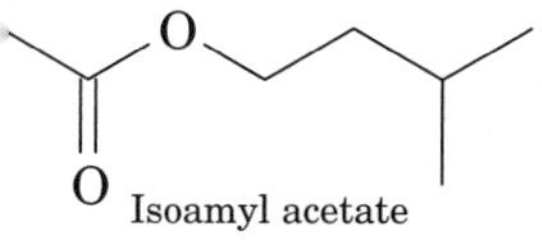

Isoamyl acetate

Imitation Banana Extract

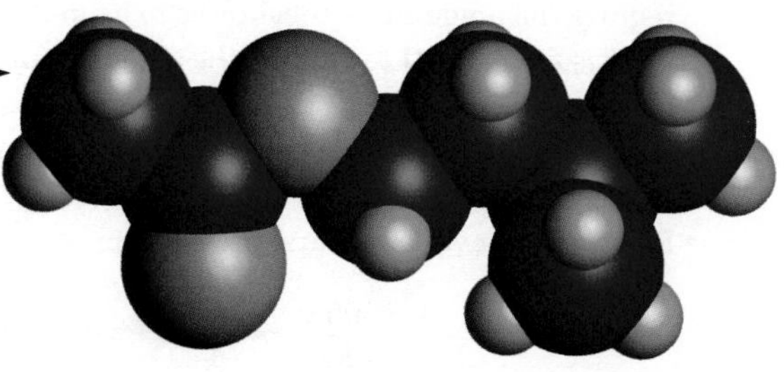

Teaching Tip

Problem solving

Stress that when doing stoichiometry problems, moles are always involved as intermediaries. The mole ratio is the most critical factor in setting up a correct problem solution. It is also the point that presents the most confusion to students. To ensure understanding, keep referring back to the recipe idea from the beginning of the chapter so that students think in proportional terms. The Additional Examples that follow focus primarily on mole ratios.

Additional Examples

Use these three balanced equations to determine how much of the highlighted compound is formed.

a. Given 4 moles of HCl.
$2HCl + Ba \longrightarrow \mathbf{BaCl_2} + H_2$

$$4\ \text{mol HCl} \times \frac{1\ \text{mol BaCl}_2}{2\ \text{mol HCl}} = 2\ \text{mol BaCl}_2$$

b. Given 25 moles of K_2SO_4.
$\mathbf{2KNO_3} + CaSO_4 \longrightarrow K_2SO_4 + Ca(NO_3)_2$

$$25\ \text{mol K}_2\text{SO}_4 \times \frac{2\ \text{mol KNO}_3}{1\ \text{mol K}_2\text{SO}_4} = 50\ \text{mol KNO}_3$$

c. Given 7 moles of $MgCO_3$.
$2AlCl_3 + 3MgCO_3 \longrightarrow \mathbf{Al_2(CO_3)_3} + 3MgCl_2$

$$7\ \text{mol MgCO}_3 \times \frac{1\ \text{mol Al}_2(\text{CO}_3)_3}{3\ \text{mol MgCO}_3} = 2.33\ \text{mol Al}_2(\text{CO}_3)_3$$

Teaching Tip

Problem solving

Research shows that students are more successful at solving problems when they have adopted a general heuristic. As an expert problem solver, do not underestimate the difficulties students have in understanding a problem statement. Focus your initial instruction on analyzing problem statements to model how experts approach problems.

Remind students that it is far more important to correctly determine the setup for solving a problem than to actually work out the calculations to get an answer. It might be helpful to assign some problems for which students are graded on the setups only. In this way, students will get the idea that the answer itself is not the most important thing. Several of the problem-solving worksheets in the ***Holt Chemistry: Visualizing Matter Study Guide*** focus on the stepwise development of problem setups. You may also want to use this strategy when you construct test items.

It is sometimes helpful to "force" students to set up their problems in a very specific style so that they learn to do the problems correctly by following the style or organization that experts use. This will allow the students to better help one another in peer teaching situations. Later, when they fully understand the concepts, they can begin to use their own shortcuts.

It will probably be necessary to help students decide whether an answer is reasonable or not. Many students have done very little estimating and often will think that 1.2 and 1.5 are not close results.

Theme

Classification and trends

Most stoichiometry problems involve moles as intermediary steps.

Three-step method

In solving stoichiometry problems, it is important to think about what the problem means and not just to multiply and divide the given values with your calculator. In Chapter 4 you learned about the three-step method used in solving the Sample Problems. Because the ability to solve stoichiometry problems is crucial to your success in chemistry, the method is repeated here for you to review before working with the Sample Problems that follow. As you work through your homework problems, it may be useful to refer to this page for ideas. The Sample Problems that follow can also provide helpful hints.

Remember that the *best way to convert from one substance to another is to relate the reaction stoichiometry* ***in moles***. If your data from the problem is in grams, convert it to moles first.

How To

Solve stoichiometry problems

1. List what you know

Don't start using your calculator yet.

- Read the problem twice.
- Organize the information from the problem statement in a list or table.
- Identify what you are asked to find, and write down the units for the answer.
- For all substances you will be working with, write the formulas, and determine the molar masses. Check a periodic table if necessary.
- If there is a reaction, write an equation for it, making sure that it is balanced so that you'll have the correct mole ratios.
- List any conversion factors that you might need, such as molar masses, mole ratios, and unit conversions.

2. Set up the problem

Don't start using your calculator yet.

- First, analyze what needs to be done to get to the answer. See if there is any information not in the problem that you need for the answer.
- Identify which value given in the problem can be used as a starting point. Write it on the left side of a sheet of paper. On the right side of the paper, write an equals sign and then a question mark with the units of the answer. Fill in the conversion factors necessary to convert from what is given in the problem to what is sought in the answer.
- Most chemistry problems (and nearly all stoichiometry problems) require the amounts of substances to be in moles, so use the molar masses from step **1** to convert the amounts into moles, if necessary.
- If you need to change from amount of one substance to a different substance, use mole ratios derived from the balanced chemical equation. Remember that the mole ratio may not always be 1:1.
- Be sure to convert the data into appropriate units, such as *grams* instead of *kilograms*.

- A plan that works for solving most reaction stoichiometry problems is summarized in **Figure 8-5**.
- When you have finished writing down your plan with all of the conversion factors, check to see how the units cancel each other. If they all cancel to give you the units you need for the answer, the setup is probably correct.

Solving Mass-Mass Stoichiometry Problems

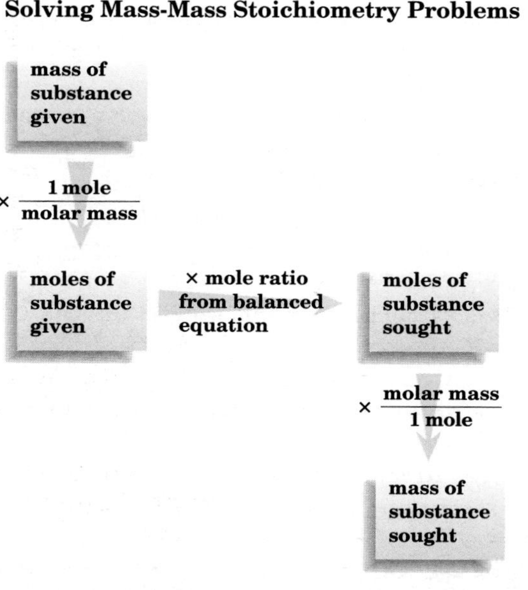

Figure 8-5
Most stoichiometry problems involve converting an amount into moles and using the mole ratio to calculate a corresponding amount of another substance in the reaction.

3. Estimate and calculate

Don't start using your calculator yet.

- First, estimate your answer. One way to do this is to round off the numbers in the problem setup and make a quick calculation. Another way is to compare conversion factors in the setup and decide whether the answer should be bigger or smaller than the beginning value.

Then, begin your calculations by working through the problem setup you made in step **2**. **In the examples in this book, numbers are not rounded off between steps in calculations.**

- When you have finished your calculations, *remember that you still don't have the correct answer*. You must round off and make sure that the answer has the correct number of significant figures.
- Always report the answer with the correct units, not just as a number.
- Compare your answer with the estimate. If they are not close, check all of your work.
- Make sure your answer is reasonable. For example, if you began with 5.3 mg of one reactant, and your answer is that it will make 725 g of a product, you know that you need to double-check all of your work.

Teaching Tip
Calculators

Students generally overemphasize the value of the calculator in becoming successful problem solvers. Stress that the calculator is strictly a computational device. It cannot work out a problem on its own. The thought process used to build a problem setup is the key to problem-solving success. That is why the generalized model in this text does not prompt for the calculator until Step 3, which is after the problem is set up.

Visual Strategy
Figure 8-5

Make students aware that the mole ratio is the basis for the stoichiometric calculations and is found directly in the middle of the flowchart.

Additional Examples for Sample Problem 8A

a. Limestone ($CaCO_3$) will react with most acids to form a calcium salt, water, and carbon dioxide. Determine the amount of water produced if 65.2 g of calcium carbonate are allowed to react with excess phosphoric acid according to the following reaction.

$$CaCO_3 + H_3PO_4 \longrightarrow Ca_3(PO_4)_2 + H_2O + CO_2$$

b. Oxygen can be produced by the decomposition reaction of potassium chlorate. How much oxygen is produced when 49.89 g of potassium chlorate are decomposed?

$$KClO_3 \longrightarrow KCl + O_2$$

Solutions are at the bottom of the page.

Possible Misconception

Many students use their calculators inefficiently with numbers that are multiplied in the denominator of fractions. Many students will work out problems like

$$\frac{62 \times 70}{15 \times 35}$$

by dividing the product of the numerator by the product of the denominator. Instead they can be told that if the number is in the numerator, it is multiplied, and if it is in the denominator, it is divided. The problem then becomes straight forward when using a calculator by pressing the following keys:

$62 \times 70 \div 15 \div 35 =$

Sample Problem 8A
Mass-mass stoichiometry

Methyl salicylate, also known as "oil of wintergreen," is most often made in a synthesis reaction between methanol and salicylic acid. How many grams of salicylic acid are needed to produce 325 g of methyl salicylate, provided there is plenty of methanol available?

$$\text{Salicylic acid} + \underset{\text{Methanol}}{CH_3OH} \xrightarrow{HCl} \text{Methyl salicylate} + \underset{\text{Water}}{H_2O}$$

❶ List what you know

These facts were taken from the problem statement.

- **reactants:** salicylic acid, methanol
- **products:** methyl salicylate, water
- **mass of methyl salicylate produced:** 325 g
- **mass of salicylic acid needed:** ? g

These molar masses were calculated from the periodic table.

- **formula and molar mass of salicylic acid:** $C_7H_6O_3$, 138.13 g/mol
- **formula and molar mass of methyl salicylate:** $C_8H_8O_3$, 152.16 g/mol

❷ Set up the problem

- **To convert from the amount of product to the amount of reactant needed, first convert to moles and use the mole ratio to link the two amounts.**

$$325 \text{ g } C_8H_8O_3 \times \overset{a}{\frac{1 \text{ mol } C_8H_8O_3}{152.16 \text{ g } C_8H_8O_3}} \times \overset{b}{\frac{1 \text{ mol } C_7H_6O_3}{1 \text{ mol } C_8H_8O_3}} \times \overset{c}{\frac{138.13 \text{ g } C_7H_6O_3}{1 \text{ mol } C_7H_6O_3}} = ? \text{ g } C_7H_6O_3$$

ⓐ First, convert the given mass of the product into moles. The reciprocal of the molar mass (mol/g) of the product must be used in order to cancel the mass of $C_8H_8O_3$ and leave the amount of $C_8H_8O_3$ in moles.

ⓑ Next, to convert moles of product into moles of reactant, multiply by the mole ratio (1:1) from the balanced chemical equation.

ⓒ Last, convert the moles of reactant into mass using the molar mass of the reactant.

Always remember to work stoichiometry problems by calculating with moles. Notice that all of the factors cancel to give you the units of the answer, g $C_7H_6O_3$.

❸ Estimate and calculate

- **Before calculating, round off the numbers in the setup to make an estimate.**

$$300 \times \frac{1}{150} \times \frac{1}{1} \times \frac{140}{1} \approx 280$$

- **Use your calculator to work through the setup. Be sure to round to the correct number of significant figures, which is three.**

$$325 \text{ g } C_8H_8O_3 \times \frac{1 \text{ mol } C_8H_8O_3}{152.16 \text{ g } C_8H_8O_3} \times \frac{1 \text{ mol } C_7H_6O_3}{1 \text{ mol } C_8H_8O_3} \times \frac{138.13 \text{ g } C_7H_6O_3}{1 \text{ mol } C_7H_6O_3} = ? \text{ g } C_7H_6O_3$$

Calculator answer: 2.950331887×10^2

Answer to three significant figures: 295 g $C_7H_6O_3$

The answer is reasonably close to the estimate.

Solutions for Additional Example 8A

a. $3CaCO_3 + 2H_3PO_4 \longrightarrow Ca_3(PO_4)_2 + 3H_2O + 3CO_2$

$$65.2 \text{ g } CaCO_3 \times \frac{1 \text{ mol } CaCO_3}{100.09 \text{ g } CaCO_3} \times \frac{3 \text{ mol } H_2O}{3 \text{ mol } CaCO_3} \times \frac{18.02 \text{ g } H_2O}{1 \text{ mol } H_2O} = 11.7 \text{ g } H_2O$$

b. $2KClO_3 \longrightarrow 2KCl + 3O_2$

$$49.89 \text{ g } KClO_3 \times \frac{1 \text{ mol } KClO_3}{122.55 \text{ g } KClO_3} \times \frac{3 \text{ mol } O_2}{2 \text{ mol } KClO_3} \times \frac{18.02 \text{ g } O_2}{1 \text{ mol } O_2} = 11.00 \text{ g } O_2$$

Practice 8A

Answers for Concept Review items and Practice problems begin *on page 841.*

1. Tin(II) fluoride, also known as stannous fluoride, is added to some dental products to help prevent cavities. How many grams of tin(II) fluoride can be made from 55.0 g of hydrogen fluoride, HF, if there is plenty of tin?

$$Sn(s) + 2HF(aq) \longrightarrow SnF_2(aq) + H_2(g)$$

2. Fluorescein is a reddish powder used as a coloring agent in some cosmetics, such as shampoo. It is made by combining phthalic anhydride with resorcinol. If a cosmetics company starts with 25.00 kg of resorcinol, how many kilograms of fluorescein can they produce, assuming that there is plenty of phthalic anhydride?

2 (resorcinol, HO, OH) + (phthalic anhydride, O, O, O) $\xrightarrow{\text{heat}}$ (fluorescein, HO, O, OH, O, C=O) + $2H_2O$

Two resorcinol | One phthalic anhydride | One fluorescein | Two water

Other stoichiometric calculations

Problems with amounts in moles

Some stoichiometry problems involve data or answers in mole amounts instead of mass amounts. These problems can be solved with an approach similar to that used to solve the problems having both data and answers in mass, but with a shortcut. There are fewer steps because one or both molar mass conversions are unnecessary, as shown in **Figure 8-6**, which has many of the same steps as **Figure 8-5**. If both the answer and the given data are in moles, the only conversion factor necessary to solve the problem is the mole ratio.

Figure 8-6
If a stoichiometry problem involves amounts in moles, solving the problem has fewer steps than problems with amounts in grams.

Solving Stoichiometry Problems With Moles or Grams

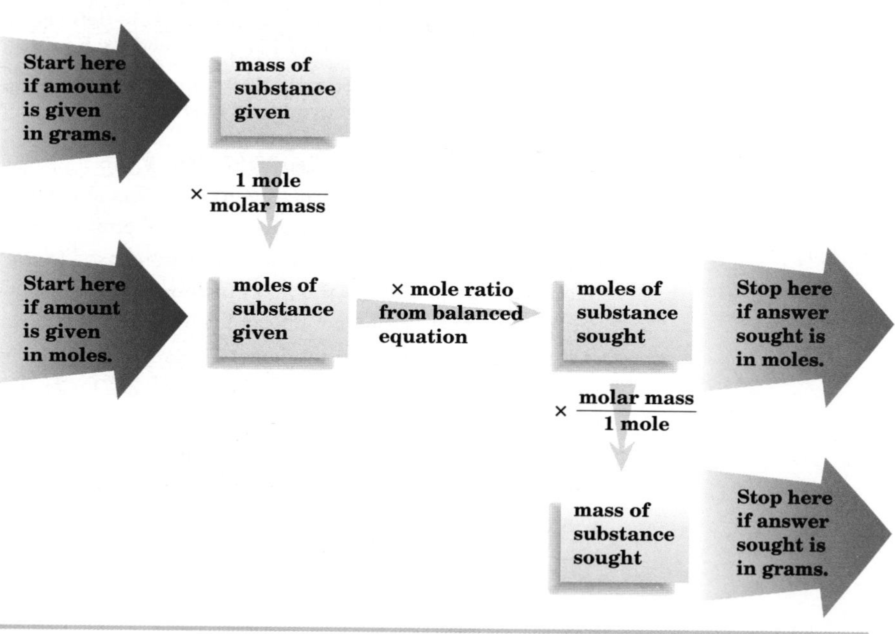

Visual Strategy

Figure 8-6

Go through the flowchart with students to be sure they understand its organization and how to apply it to different problems. Visual algorithms can be very helpful to many students in helping them recognize the overall process of working through a multistep stoichiometry problem.

Answers to Practice 8A

1. $55.0\ g\ HF \times \frac{1\ mol}{20.1\ g\ HF} \times \frac{1\ mol\ SnF_2}{2\ mol\ HF} \times \frac{156.7\ g\ SnF_2}{1\ mol\ SnF_2} = 215\ g\ SnF_2$

2. $25.00\ g\ C_6H_4(OH)_2 \times \frac{1\ mol\ C_6H_4(OH)_2}{110.1\ g\ C_6H_4(OH)_2} \times \frac{1\ mol\ C_{20}H_{12}O_5}{2\ mol\ C_6H_4(OH)_2} \times \frac{332.3\ g\ C_{20}H_{12}O_5}{1\ mol\ C_{20}H_{12}O_5} =$

$37.72\ kg\ C_{20}H_{12}O_5$

Teaching Tip

Problem solving

The sum of the masses on the reactant side must be equal to the sum of the masses on the product side. This is an excellent verification strategy that students can use to determine if the problem was worked out correctly. Errors typically come from incorrect balancing, incorrect formula masses, and multiplication errors.

Groupwork Strategy

Pair students and present them with these three stoichiometry problems. Have each group determine the amounts of all unknown products and reactants. Then have them compare the sum of the masses on the reactant side with the sum of the masses on the product side.

a. $Mg(ClO_3)_2 \longrightarrow$ **$MgCl_2$** (49.7 g) $+ O_2$

b. $NaI +$ **Cl_2** (52.6 g) $\longrightarrow NaCl + I_2$

c. $K_2SO_4 +$ **$Ca(NO_3)_2$** (288.6 g) $\longrightarrow CaSO_4 + KNO_3$

Answers for Groupwork Strategy items are found on page 309A.

Additional Examples for Sample Problem 8B

a. Propane, C_3H_8, burns in oxygen to produce carbon dioxide and water. What mass of propane could be combusted by 8.35 mol of oxygen?

$$C_3H_8 + O_2 \longrightarrow H_2O + CO_2$$

b. Limestone, $CaCO_3$, can be decomposed with heat to form lime, CaO, and carbon dioxide. How many moles of lime would be formed from the decomposition of 20.1 kg of limestone?

$$CaCO_3 \longrightarrow CaO + CO_2$$

Solutions are at the bottom of the page.

Answers to Practice 8B

1. $1.03 \times 10^{-2}\ \text{mol } O_2 \times \frac{6\ \text{mol } CO_2}{6\ \text{mol } O_2} \times \frac{44.01\ \text{g } CO_2}{1\ \text{mol } CO_2} = 0.453\ \text{g } CO_2$

2. $1.03 \times 10^{-2}\ \text{mol } O_2 \times \frac{6\ \text{mol } H_2O}{6\ \text{mol } O_2} \times \frac{18.02\ \text{g } H_2O}{1\ \text{mol } H_2O} = 0.186\ \text{g } H_2O$

Sample Problem 8B
Mole-mass stoichiometry

The human body needs at least 1.03×10^{-2} mol O_2 every minute. If all of this oxygen is used for the cellular respiration reaction that breaks down glucose, how many grams of glucose does the human body consume each minute?

$$C_6H_{12}O_6(s) + 6O_2(g) \longrightarrow 6CO_2(g) + 6H_2O(l)$$

1 List

- **amount of O_2 used each minute:** 1.03×10^{-2} mol O_2
- **molar mass of $C_6H_{12}O_6$:** 180.18 g/mol
- **mole ratio:** 1 mol $C_6H_{12}O_6$:6 mol O_2
- **mass of glucose needed each minute:** ? g

Because oxygen is already given in a mole amount, the molar mass of O_2 is unnecessary.

2 Set up

- **First, analyze what needs to be done to get the answer. You are given moles of oxygen, so you can convert to moles of glucose using the mole ratio. Then, you can use the molar mass to determine the mass in grams of the glucose.**

$$1.03 \times 10^{-2}\ \text{mol } O_2 \times \frac{1\ \text{mol } C_6H_{12}O_6}{6\ \text{mol } O_2} \times \frac{180.18\ \text{g } C_6H_{12}O_6}{1\ \text{mol } C_6H_{12}O_6} = ?\ \text{g } C_6H_{12}O_6$$

3 Estimate and calculate

- **Before calculating, round off the numbers in the setup to make an estimate.**

$$0.01 \times \frac{1}{6} \times \frac{180}{1} \approx 0.3$$

Then use your calculator to work through the setup. Be sure to round to the correct number of significant figures, which is three.

The answer is reasonably close to the estimate.

Answer to three significant figures:

$$1.03 \times 10^{-2}\ \text{mol } O_2 \times \frac{1\ \text{mol } C_6H_{12}O_6}{6\ \text{mol } O_2} \times \frac{180.18\ \text{g } C_6H_{12}O_6}{1\ \text{mol } C_6H_{12}O_6} = 0.309309\ \text{g } C_6H_{12}O_6$$

Practice 8B

1. **Use the information in Sample Problem 8B to determine how many grams of carbon dioxide would be produced each minute.**
2. **Use the information in Sample Problem 8B to determine how many grams of water would be produced each minute.**

Using density with stoichiometry

Remember that the key to solving any reaction stoichiometry problem is to always calculate in moles. Once the number of moles is determined, conversion factors such as molar mass can be used to convert to the mass in grams. Similarly, once the mass is known, the density of a substance can be used to convert from mass to volume.

Recall from Chapter 2 that density is defined as the mass of a substance per unit volume, expressed mathematically as $D = m/V$. Once again, the key is to use the density value to set up a conversion factor that will cancel the units in the measurement you have and leave the units of the measurement for the answer, as shown in Sample Problem 8C. Once more, you already know everything you need to solve such problems from your study of concepts earlier in the book.

Solutions for Additional Example 8B

a. $8.35\ \text{mol } O_2 \times \frac{1\ \text{mol } C_3H_8}{5\ \text{mol } O_2} \times \frac{44.11\ \text{g } C_3H_8}{1\ \text{mol } C_3H_8} = 73.7\ \text{g } C_3H_8$

b. $20\,100\ \text{g } CaCO_3 \times \frac{1\ \text{mol } CaCO_3}{100.09\ \text{g } CaCO_3} \times \frac{1\ \text{mol CaO}}{1\ \text{mol } CaCO_3} = 201\ \text{mol CaO}$

Sample Problem 8C
Stoichiometry calculations with density

In the space shuttles, the CO_2 that the crew exhales is removed from the air by a reaction within canisters of lithium hydroxide. On average, each astronaut exhales about 20.0 mol of CO_2 daily. What volume of water will be produced when this amount of CO_2 reacts with an excess of LiOH? (Hint: the density of water is about 1.00 g/mL.)

$$CO_2(g) + 2LiOH(s) \longrightarrow Li_2CO_3(aq) + H_2O(l)$$

1 List

- **amount of CO_2:** 20.0 mol CO_2
- **density of H_2O:** 1.00 g/mL
- **molar mass of H_2O:** 18.02 g/mol
- **mole ratio:** 1 mol CO_2:1 mol H_2O
- **volume of H_2O produced:** ? mL

2 Set up

- **First, analyze what needs to be done to get the answer. You are given moles of CO_2, so you can convert to moles of water using the mole ratio. Then, use the molar mass to find out the mass of water.**

$$20.0 \text{ mol } CO_2 \times \frac{1 \text{ mol } H_2O}{1 \text{ mol } CO_2} \times \frac{18.02 \text{ g } H_2O}{1 \text{ mol } H_2O} \times \frac{\quad}{\quad} = ?\text{ mL } H_2O$$

- **Next, convert from mass of H_2O to volume of H_2O using the density as the conversion factor. The correct conversion factor will have *g H_2O* in the denominator and *mL H_2O* in the numerator.**

$$20.0 \text{ mol } CO_2 \times \frac{1 \text{ mol } H_2O}{1 \text{ mol } CO_2} \times \frac{18.02 \text{ g } H_2O}{1 \text{ mol } H_2O} \times \frac{1 \text{ mL } H_2O}{1.00 \text{ g } H_2O} = ?\text{ mL } H_2O$$

3 Estimate and calculate

- **Before calculating, round off the numbers in the setup to make an estimate.**

$$20 \times \frac{1}{1} \times \frac{18}{1} \times \frac{1}{1} \approx 360$$

- **Then, use your calculator to work through the setup. Be sure to round to the correct number of significant figures, which is three.**

Answer to three significant figures: 360. mL H_2O

Practice 8C

1. **The reaction that causes cake batter to rise involves the production of CO_2 from $NaHCO_3$. How many liters of CO_2 gas will be created when 15.0 g $NaHCO_3$ are heated? (Note: at baking temperature, the density of CO_2 is about 1.10 g/L.)**

$$2NaHCO_3(s) \longrightarrow H_2O(g) + Na_2CO_3(s) + CO_2(g)$$

2. **A common ingredient used in some sunscreens is *p*-aminobenzoic acid, or PABA, which can absorb some of the ultraviolet radiation of the sun. It is made from *p*-nitrobenzoic acid, which is a solid with a density of 1.58 g/mL. The reaction actually has several intermediate steps but can be summarized as shown below. What is the maximum mass of PABA that can be made from 500. mL of *p*-nitrobenzoic acid crystals?**

p-nitrobenzoic acid (benzene ring with COOH and NO_2) $+ 3H_2 \xrightarrow{Pt}$ *p*-aminobenzoic acid (benzene ring with COOH and NH_2) $+ 2H_2O$

Additional Examples for Sample Problem 8C

Solve the following problems for the unknown variable.

a. density = 5.92 g/mL
volume = 52.7 mL
mass = ?

b. density = ?
volume = 85.44 mL
mass = 133.2 g

c. density = 8.14 g/mL
volume = ?
mass = 8765 g

d. When pentane burns in oxygen, it produces carbon dioxide and water. If 85.5 g of pentane, C_5H_{12}, are burned, what volume of carbon dioxide is produced?(Assume the CO_2 is allowed to cool to room temperature.) Density of carbon dioxide is approximately 1.997 g/L.

$$C_5H_{12} + O_2 \longrightarrow CO_2 + H_2O$$

e. Magnesium burns in oxygen to produce magnesium oxide. How much magnesium will burn in the presence of 189 mL of oxygen? The density of oxygen is 1.429 g/L.

$$Mg + O_2 \longrightarrow MgO$$

Solutions are at the bottom of the page.

Answers to Practice 8C

1. $15.0 \text{ g } NaHCO_3 \times \frac{1 \text{ mol } NaHCO_3}{84.01 \text{ g } NaHCO_3} \times \frac{1 \text{ mol } CO_2}{2 \text{ mol } NaHCO_3} \times \frac{44.01 \text{ g } CO_2}{1 \text{ mol } CO_2} = 3.93 \text{ g } CO_2$

$\frac{3.93 \text{ g}}{1.10 \text{ g/L}} = 3.57 \text{ L } CO_2$

2. 648 g $NH_2C_6H_4CO_2H$

*Full solution for item **2** is found on page 309A.*

Solutions for Additional Examples 8C

a. 312 g **b.** 1.559 g/mL **c.** 1080 mL

d. $85.5 \text{ g } C_5H_{12} \times \frac{1 \text{ mol } C_5H_{12}}{72.17 \text{ g } C_5H_{12}} \times \frac{5 \text{ mol } CO_2}{1 \text{ mol } C_5H_{12}} \times \frac{44.01 \text{ g } CO_2}{1 \text{ mol } CO_2} \times \frac{1 \text{ L } CO_2}{1.997 \text{ g } CO_2} = 131 \text{ L } CO_2$

e. $0.189 \text{ L } O_2 \times \frac{1.429 \text{ g } O_2}{1 \text{ L } O_2} \times \frac{1 \text{ mol } O_2}{32.00 \text{ g } O_2} \times \frac{2 \text{ mol Mg}}{1 \text{ mol } O_2} \times \frac{24.30 \text{ g Mg}}{1 \text{ mol Mg}} = 0.410 \text{ g Mg}$

Demonstration 2

Stoichiometric production of oxygen

Approximate time: 10 min

1. Fit a 1 L volumetric flask with a one-hole stopper that contains a 4 × 4 cm glass bend.
2. Run rubber tubing from this bend to an inverted 2 L graduated cylinder that is filled with water and placed in a container that is half filled with water. The container should be large enough to hold the water expelled from the cylinder. Support the inverted cylinder with a large ring and ringstand.
3. Determine the mass of the volumetric flask and add approximately 100 mL of 3% H_2O_2. Add 1 g of yeast to the H_2O_2, and stopper the flask.
4. Swirl the flask until all bubbling has ceased, indicating the completion of the reaction. Measure the amount of oxygen gas collected (it should be about 1000 mL).

The density of oxygen at 20°C is about 1.331 g/L and the density of 3% H_2O_2 is approximately 1 g/mL. The volume of oxygen can be calculated as follows.

$$3 \text{ g } H_2O_2 \times \frac{1 \text{ mol } H_2O_2}{34.02 \text{ g } H_2O_2} \times$$

$$\frac{1 \text{ mol } O_2}{2 \text{ mol } H_2O_2} \times \frac{32.00 \text{ g } O_2}{1 \text{ mol } O_2} \times$$

$$\frac{1 \text{ L } O_2}{1.331 \text{ g } O_2} = 1.060 \text{ L } O_2 = 1.0 \text{ L } O_2$$

SAFETY

Wear safety goggles and a lab apron. Students should remain back at least 10 ft.

DISPOSAL

Pour everything down the drain.

[Suggested by Philip A. Clift in "Sampling Stoichiometry, The Decomposition of Hydrogen Peroxide," *The Science Teacher*, Vol. 59, pp. 23–25.]

Calculating the number of atoms or formula units

Just as molar mass, density, and mole ratios can be used as conversion factors in problems, Avogadro's number, 6.022×10^{23}, can be used to calculate the number of atoms or formula units participating in a reaction. Again, the two points to remember are the same as with other stoichiometry problems.

- Be certain that you work with moles.
- Make sure that you set up the conversion factors so that the quantity sought is in the numerator and the quantity given is in the denominator.

Figure 8-7 summarizes the approaches that can be taken in stoichiometric calculations. Notice that several steps are the same as in **Figure 8-5** and **Figure 8-6**.

Solving Many Types of Stoichiometry Problems

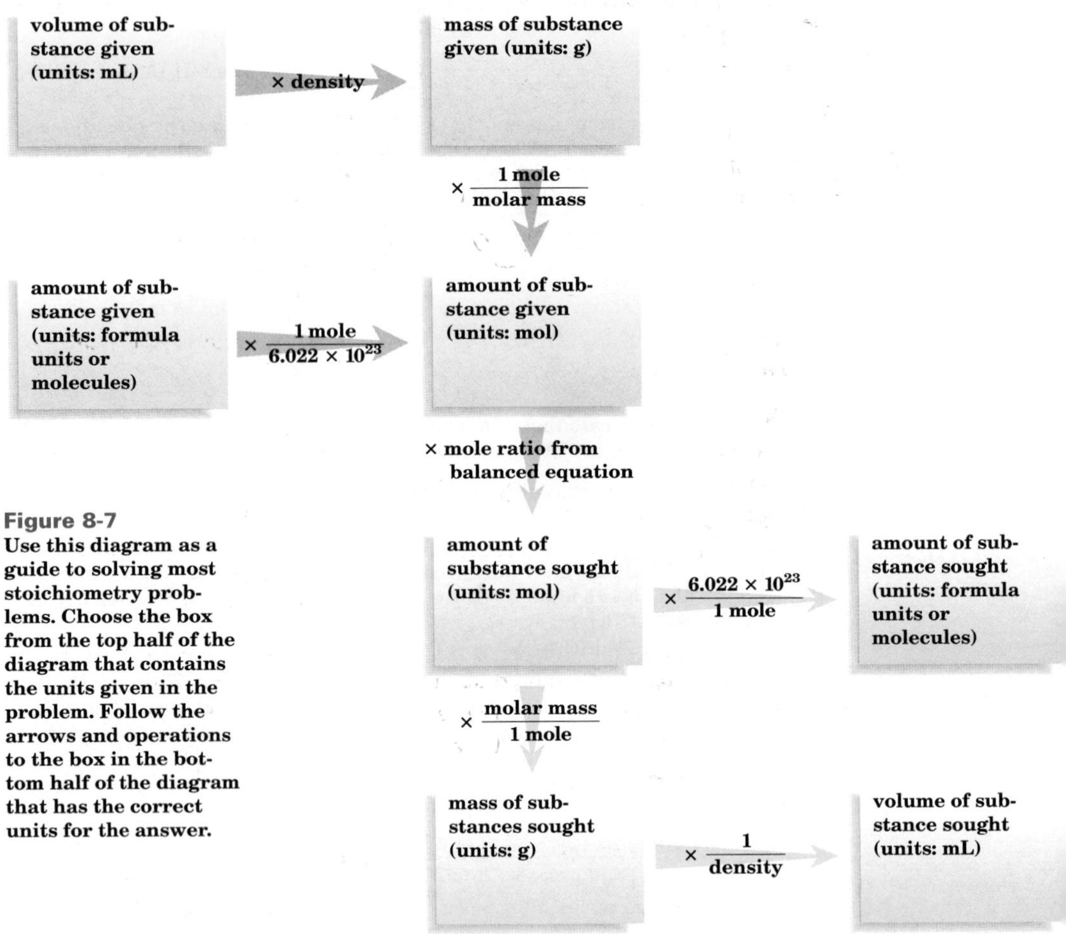

Figure 8-7
Use this diagram as a guide to solving most stoichiometry problems. Choose the box from the top half of the diagram that contains the units given in the problem. Follow the arrows and operations to the box in the bottom half of the diagram that has the correct units for the answer.

Section Review

1. Why do you need to use moles to solve stoichiometry problems? Why can't you just convert from grams to grams?
2. Write and solve two problems, one involving composition stoichiometry and the other pertaining to reaction stoichiometry.
3. Use the three-step method to solve the following problems based on the equation shown, which shows the use of the reactant hydrazine, N_2H_4, as a propellant for a rocket. For each problem, assume that you start with 1200. kg of N_2H_4 and that it is completely used up by the reaction.

$$2N_2H_4(l) + (CH_3)_2N_2H_2(l) + 3N_2O_4(g) \longrightarrow 6N_2(g) + 2CO_2(g) + 8H_2O(l)$$

   a. Calculate how many moles of N_2O_4 are needed.
   b. Calculate how many grams of $(CH_3)_2N_2H_2$ are needed.
   c. Calculate how many molecules of N_2 are produced.
   d. Calculate how many liters of H_2O are produced. The density of water is 1.00 g/mL.
4. Oxygen, O_2, was discovered by Joseph Priestley in 1774 when he decomposed mercury(II) oxide, HgO, into its constituent elements by heating it. How many moles of oxygen could Priestley have produced if he had decomposed 216.59 g of mercury(II) oxide?
5. Aspirin is made by reacting salicylic acid with acetic anhydride.

$$\underset{\text{Salicylic acid}}{C_7H_6O_3} + \underset{\text{Acetic anhydride}}{C_4H_6O_3} \xrightarrow{HCl} \underset{\text{Aspirin}}{C_9H_8O_4} + \underset{\text{Acetic acid}}{CH_3COOH}$$

   What is the maximum mass of aspirin that can be produced from 251.7 g of salicylic acid?

6. **Story Link**
   Each FRH contains 8.1 g magnesium. Assuming this magnesium reacts completely, how many grams of hydrogen gas are produced?
7. **Story Link**
   At the temperature within an FRH, 60°C, the density of hydrogen gas is 0.081 g/L. How many liters of H_2 gas are produced by an FRH?
8. **Story Link**
   Calculate the amount of heat energy produced per gram of reactants added for the magnesium, calcium oxide, and trioxane reactions. If the mass of the reactants were the deciding factor, which reaction would be most effective? (Hint: refer to page 271, and remember to include the mass of water necessary for the first two reactions.)
9. **Story Link**
   How much of each reactant would be necessary to create a camping-food warmer that would provide the 575 kJ necessary to warm a dinner for four hikers? (Hint: refer to page 271, and set up conversion factors relating energy and amount of each reactant.)

ASSESS

Alternative Assessment

Ask students to determine the amount of sodium to add to 50. g of fluorine to make sodium fluoride, which is added to drinking water to help stop tooth decay. They are to proceed according to the methods used in class. No calculators are to be used, and the answer should be estimated. Provide these values as molar masses: Na = 23 g/mol, F = 19 g/mol.

$$2Na + F_2 \longrightarrow 2NaF$$

$$50.\ g\ F_2 \times \frac{1\ mol\ F_2}{38\ g\ F_2} \times \frac{2\ mol\ Na}{1\ mol\ F_2} \times \frac{23\ g\ Na}{1\ mol\ Na} = 61\ g\ Na$$

Answers to Section Review

1. Moles provide a way of equating numbers of particles. One mole of any substance has the same number of particles as one mole of another substance. Equal numbers of grams for two substances will not have equivalent numbers of particles.
2. Answers will vary. Check that students know the difference between a mole calculation for an isolated substance and one which uses the relationships resulting from a reaction.

Answers are continued on page 309A.

Section 8-2

FOCUS

Lesson Starter

Use the following demonstration to introduce the concept of limiting reagents. Various amounts of CO_2 gas will be produced, showing the need for correct proportions.

Demonstration 3

Limiting reactants

Approximate time: 15 min

Note: Run this demonstration before class to determine if the quantities specified will work with the size of balloons you use. Change amounts accordingly.

1. Obtain three 2 L plastic carbonated soda bottles. Remove the labels for viewing, and number the bottles *1*, *2*, and *3*.
2. Stretch three 9–12 in. balloons by inflating them. Before class, measure out the following quantities of materials: 10 g of $NaHCO_3$, 4 g of $NaHCO_3$, 7.5 g of $NaHCO_3$, 50 mL of vinegar, 100 mL of vinegar, 150 mL of vinegar.
3. Fill the soda bottles with 50 mL, 150 mL, and 100 mL vinegar, respectively.
4. Fill the balloons with 10 g, 4 g, and 7.5 g of $NaHCO_3$, respectively. Twist the balloons to seal in the $NaHCO_3$; then attach each balloon to the top of a bottle. Bottle *1* has 50 mL of vinegar and 10 g of $NaHCO_3$. Bottle *2* has 150 mL of vinegar and 4 g of $NaHCO_3$. Bottle *3* has 100 mL of vinegar and 7.5 g $NaHCO_3$.

How much does a reaction really produce?

Section Objectives

Distinguish between a limiting reactant and an excess reactant.

Identify the limiting reactant in a problem, and calculate the theoretical yield.

Distinguish between theoretical yield and actual yield.

Given the actual yield and the quantity of the limiting reactant, calculate the percent yield.

Use percent yield to calculate actual yield.

Leftover reactants

You now know how to use mole ratios and other conversion factors to figure out how much of a product should be produced by a chemical reaction. So far, such reactions and calculations have described an *ideal situation*. Some assumptions were made: there was always enough of the other reactants so that none would be left over, and all reactions went to completion, producing only the products indicated in the equation. There were no competing reactions.

These assumptions are useful when trying to learn how reactions and equations work and how to make predictions from them. But to treat real reactions accurately, you must take additional factors into account, such as the amounts of all reactants and the completeness of the reaction.

Figure 8-8
With only one nut, you can only make a single nut-bolt pair, no matter how many bolts you have. The same is true for a chemical reaction in which one reactant is used up before the other.

Reactants combine in specific whole-number ratios

Have you ever tried to assemble a bicycle only to discover that you didn't have enough nuts for the bolts that were supplied, as shown in **Figure 8-8**? For every bolt, you needed a nut. In other words, the shortage of nuts limited what you could do. Or have you ever been in a car that ran out of gas? The car needed both gas and oxygen from the air to run, so once the gas was gone, it wouldn't run, no matter how much oxygen was present.

A similar limitation occurs in all chemical reactions. When a chemical reaction occurs, the reactants are not always present in amounts equal to their mole ratios. For example, reconsider the reaction used to make the banana-flavored ester, isoamyl acetate.

$$\text{Acetic acid} + \text{Isoamyl alcohol} \xrightarrow{HCl} \text{Isoamyl acetate} + H_2O \text{ (Water)}$$

According to this equation, for every mole of acetic acid, a mole of isoamyl alcohol is needed. What would happen if there were 20 mol of acetic acid and only 1 mol of isoamyl alcohol? How much isoamyl acetate could be made?

excess reactant *reactant that will not be used up in a reaction that goes to completion*

limiting reactant *reactant that is consumed first in a reaction that goes to completion*

One mole of acetic acid will react with one mole of isoamyl alcohol. After that, the isoamyl alcohol reactant would be used up, and none would be available to react with the other 19 moles of acetic acid. Obviously, the reaction would stop after forming only one mole of each product even if there were 20 moles of one reactant at the start. An excess of this reactant would be left over, as shown in **Figure 8-9**.

An **excess reactant** is the reactant that is not completely used up in a chemical reaction. In the preceding paragraph, acetic acid was the excess reactant. There will be some of this reactant left over when the reaction is complete. On the other hand, isoamyl alcohol is known as the limiting reactant. A **limiting reactant** is a reactant that is used up first and thus limits the amount of other reactants that can participate in a chemical reaction to make products. (Sometimes, chemists refer to it as the *limiting reagent*.)

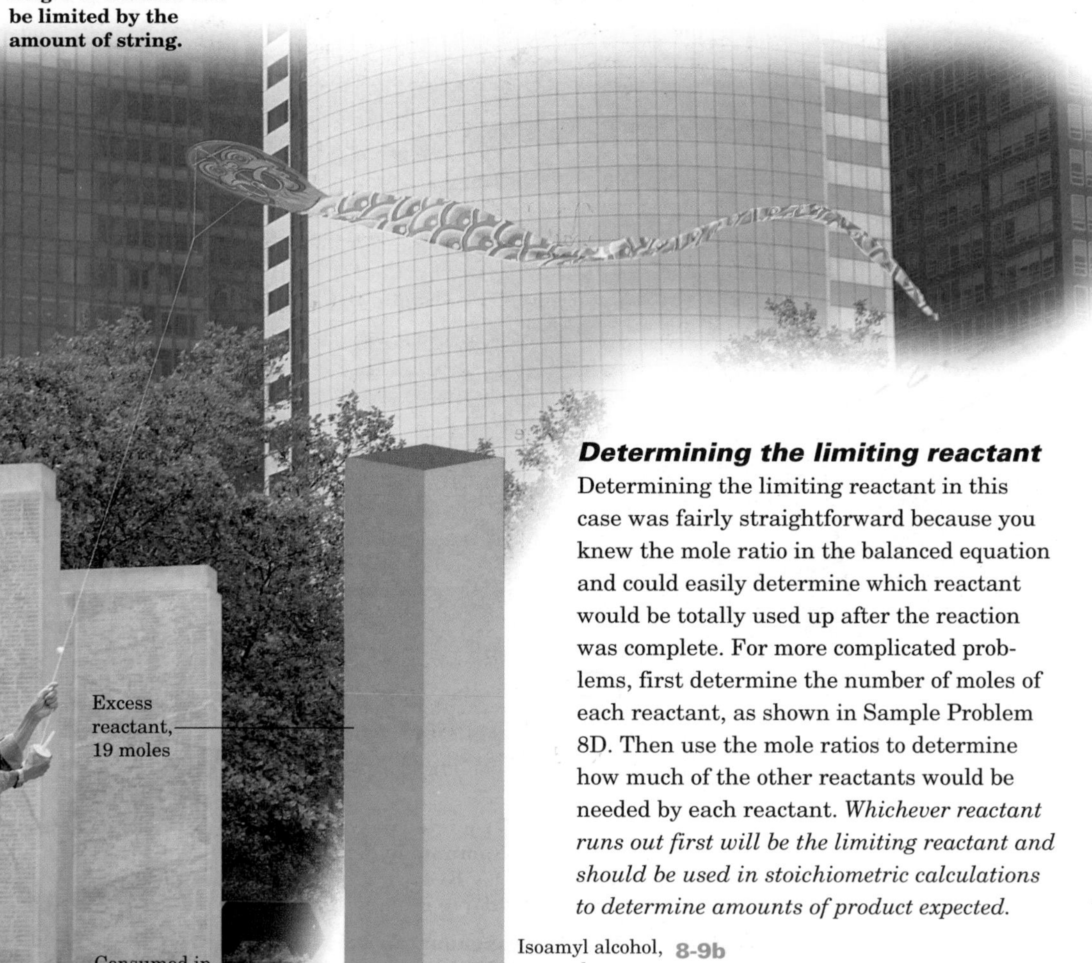

Figure 8-9a
Flying a kite can be an analogy for a reaction with limiting reactants. No matter how strongly the wind blows, the height of the kite will be limited by the amount of string.

8-9b
Similarly, in a chemical reaction, once one of the reactants is used up, the reaction will stop, no matter how much of the other reactant remains.

Determining the limiting reactant

Determining the limiting reactant in this case was fairly straightforward because you knew the mole ratio in the balanced equation and could easily determine which reactant would be totally used up after the reaction was complete. For more complicated problems, first determine the number of moles of each reactant, as shown in Sample Problem 8D. Then use the mole ratios to determine how much of the other reactants would be needed by each reactant. *Whichever reactant runs out first will be the limiting reactant and should be used in stoichiometric calculations to determine amounts of product expected.*

Visual Strategy

Figure 8-8

Make students aware that, like a kite's string, the limiting reagent controls the reaction.

5. Secure the balloons to the bottle with tape or a wire twist tie. Untwist the balloons, and let the $NaHCO_3$ fall into the vinegar. Remain back as the balloons fill in case one breaks. Bottles 1 and 2 should produce about half of the volume of bottle 3. Bottle 1 has limited vinegar, while bottle 2 has limited $NaHCO_3$. Bottle 3 should have the correct stoichiometric amount to produce 2 L of CO_2 at STP.

SAFETY
Wear safety goggles and a lab apron. Students should remain back at least 10 ft.

DISPOSAL
Pour the solutions down the drain. Throw the balloons and bottles into the trash.

TEACH

Groupwork Strategy

Analogy

Present students with the idea that they are going into the business of making pizzas. List the ingredients for one pizza on the board.

400 g flour
50 mL of water
10 g yeast
120 g of sauce
250 g of cheese
5 g of oregano
5 g of basil

Divide the class into groups of four, and present them with the following information on a handout. Their business has purchased 500 pizza pans, 60 kg of sauce, 100 kg of cheese, 2.5 kg of basil, 2.5 kg of oregano, 5 kg of yeast, 25 L of water, and 200 kg of flour.

Give students 10 minutes to work out the following problems. Which of the ingredients is the limiting agent? (cheese) How many pizzas can be made? (400) If each pizza sells for \$3.50, how much money would the firm gross? (\$1400) If you were able to get more cheese, what is the maximum number of pizzas that could be made? (500)

Theme
Conservation
Mass is always conserved during chemical reactions. Excess reagents remain unused.

Possible Misconception
Excess reagents
Make students aware that the extra vinegar in Bottle 2 of Demonstration 3 is still part of the reaction system. Students often mistakenly believe that when one of the reactants is used up, the other disappears too.

Additional Examples for Sample Problem 8D

a. Hydrogen sulfide gas can be formed by the action of HCl on FeS, forming $FeCl_2$ as a product. How many molecules of hydrogen sulfide are formed if 130.5 g of FeS are mixed with excess HCl?

$$FeS + HCl \longrightarrow FeCl_2 + H_2S$$

b. Iron(III) oxide can be reduced with carbon monoxide to form molten iron and carbon dioxide. If you had 180.0 kg of iron(III) oxide and 110.0 kg of carbon monoxide, how many kilograms of iron would be produced?

$$Fe_2O_3 + CO \longrightarrow Fe + CO_2$$

Solutions are at the bottom of the page.

Answers to Practice 8D

1. There is not enough $Al(OH)_3$.
2. N_2 is the limiting reactant.

Full solutions for items ***1*** *and* ***2*** *are found on page 309B.*

Sample Problem 8D
Determining the limiting reactant

Carbon monoxide can be combined with hydrogen to produce methanol, CH_3OH. Methanol is used as an industrial solvent, as a reactant in synthesis, and as a clean-burning fuel for some racing cars. If you had 152.5 kg CO and 24.50 kg H_2, how many kilograms of CH_3OH would be produced?

❶ List

- **balanced chemical equation for reaction:** $CO(g) + 2H_2(g) \longrightarrow CH_3OH(l)$
- **amount of CO:** 152.5 kg CO
- **amount of H_2:** 24.50 kg H_2
- **molar mass of CO:** 28.01 g/mol
- **molar mass of H_2:** 2.02 g/mol
- **molar mass of CH_3OH:** 32.05 g/mol
- **amount of CH_3OH produced:** ? kg CH_3OH

❷ Set up

- **Before you can do any calculations to convert reactants to products, you need to determine which reactant is the limiting reactant. First, figure out how many moles of each reactant are present.**

$$152.5 \text{ kg CO} \times \frac{1000 \text{ g}}{1 \text{ kg}} \times \frac{1 \text{ mol CO}}{28.01 \text{ g CO}} = 5.444 \times 10^3 \text{ mol CO present}$$

$$24.50 \text{ kg } H_2 \times \frac{1000 \text{ g}}{1 \text{ kg}} \times \frac{1 \text{ mol } H_2}{2.02 \text{ g } H_2} = 1.213 \times 10^4 \text{ mol } H_2 \text{ present}$$

- **Then, for one of these reactant amounts, figure out how much of the first reactant would be needed to use up all of the second reactant. Compare this with the amount actually present.**

$$1.213 \times 10^4 \text{ mol } H_2 \times \frac{1 \text{ mol CO}}{2 \text{ mol } H_2} = 6.065 \times 10^3 \text{ mol CO needed}$$

Because the amount of CO needed to use up the H_2 is larger than the amount of CO present, the CO will be the limiting reactant.

It does not matter which of the two reactant amounts you choose. If you started with CO first, you would find that 1.089×10^4 mol of H_2 were needed. Because the amount of H_2 needed was less than the amount present, CO will be the limiting reactant.

- **Use the limiting reactant to set up the stoichiometric calculation.**

$$5.444 \times 10^3 \text{ mol CO} \times \frac{1 \text{ mol } CH_3OH}{1 \text{ mol CO}} \times \frac{32.05 \text{ g } CH_3OH}{1 \text{ mol } CH_3OH} \times \frac{1 \text{ kg}}{1000 \text{ g}} = ? \text{ kg } CH_3OH$$

❸ Estimate and calculate

- **Before calculating, round off the numbers in the setup to make an estimate.**

$$5500 \times \frac{1}{1} \times \frac{30}{1} \times \frac{1}{1000} \approx 165$$

- **Use your calculator to work through the setup. Be sure to round to the correct number of significant figures, which is four.**

Answer to four significant figures: 174.5 kg CH_3OH

Practice 8D

1. **The hydrochloric acid, HCl, secreted in your stomach can be neutralized by taking an antacid like aluminum hydroxide, $Al(OH)_3$. This is a double-replacement reaction. If 34.0 g HCl are secreted and 12.0 g $Al(OH)_3$ are taken, is there enough $Al(OH)_3$ to react with all of the HCl?**
2. **Ammonia, NH_3, is used throughout the world as a fertilizer. To manufacture ammonia, nitrogen, N_2, is combined with hydrogen, H_2, in a synthesis reaction. If 92.7 kg N_2 and 265.8 kg H_2 are used, which is the limiting reactant?**

Solutions for Additional Example 8D

a. 8.940×10^{23} molecules H_2S

b. 125.9 kg Fe

Full solutions for items ***a*** *and* ***b*** *are found on pages 309A.*

Cost is a factor in selecting the limiting reactant

When any chemical reaction is actually carried out in industry, the least abundant reactant, which is often more expensive, is the limiting reactant. The most abundant reactant, which is usually cheaper, is the excess reactant. In this way the expensive reactant will be completely used up and not wasted, while some of the cheaper reactant will be left over.

In addition to being a way to maintain cost-efficiency, this principle can be used to control which reactions happen. One example is the production of the cider and cider vinegar shown in **Figure 8-10** from apple juice. At first, the original apple juice is kept where there is no oxygen. When no oxygen is present, the microorganisms in the apple juice cannot use the cellular respiration reaction as a source of energy. Instead, they use an alternative pathway in which the glucose is fermented, or broken down, into molecules of ethanol. The resulting solution is hard cider.

Figure 8-10 **Makers of cider vinegar use special tanks and equipment to pump air through hard cider so that oxygen, O_2, is the excess reactant.**

In the next step in the production of apple cider vinegar, the presence of plenty of oxygen is used as a means to control another reaction. Once the ethanol in hard cider is exposed to air, it slowly reacts with atmospheric oxygen to produce acetic acid, resulting in cider vinegar. Because the oxygen in the air costs nothing and is abundant, the makers of cider vinegar pump air through hard cider as they make it into vinegar. In this way, oxygen is not the limiting reactant. The ethanol is the limiting reactant, so it will be entirely consumed in the reaction.

$$CH_3CH_2OH + O_2 \longrightarrow CH_3COOH + H_2O$$

Ethanol Oxygen Acetic acid Water

Similarly, the manufacturers of banana flavoring discussed at the beginning of the chapter tend to use acetic acid as the excess reactant because it costs much less than isoamyl alcohol. For example, acetic acid costs about $11.00 for 500 mL (525 g). But 500 mL (406 g) of isoamyl alcohol costs about $16.00. When compared mole for mole, isoamyl alcohol is almost three times as expensive as acetic acid. With an excess of acetic acid, none of the more expensive isoamyl alcohol is wasted by being leftover.

Incomplete reactions

Measuring what a chemical reaction actually produces

Although equations tell you what *should* happen, they can't always tell you what *will* happen. For example, a reaction will stop once the limiting reactant has been used up, regardless of how much of the other reactants is present.

Sometimes reactions don't match equations in other ways. The amount expected from stoichiometry calculations is called the **theoretical yield**. But, in some cases, the **actual yield**, the amount of product actually created, is not what was expected—it can be much less than the theoretical yield. But how can this be? For an example, reconsider the banana-flavoring problem.

theoretical yield *calculated maximum amount of product possible from a given amount of reactant*

actual yield *measured amount of product actually produced from a given amount of reactant*

Do You Know?

Cider vinegar is made into white (clear) vinegar by filtering and distillation. The brown color of cider vinegar comes from the oxidized apple pieces distributed throughout the vinegar.

Teaching Tip

Pricing for acetic acid and isoamyl alcohol

Acetic acid comes as a 99.8% solution with a density of 1.05 g/mL. The cost per mole is as follows.

$$\frac{\$11.00}{500 \text{ mL}} \times \frac{1 \text{ mL}}{1.05 \text{ g}} \times \frac{60.06 \text{ g}}{1 \text{ mol}} = \$1.26/\text{mol}$$

Isoamyl alcohol, $C_5H_{12}O$, comes as a 100% solution with a density of 0.813 g/mL. The cost per mole is as follows.

$$\frac{\$16.00}{500 \text{ mL}} \times \frac{1 \text{ mL}}{0.813 \text{ g}} \times \frac{88.17 \text{ g}}{1 \text{ mol}} = \$3.47/\text{mol}$$

Possible Misconception

Students often confuse the terms *theoretical findings* with *actual findings*. Actual findings are actual measurements. Theoretical findings are calculations and predictions.

Theme
Equilibrium and change
Reactions will often reach an equilibrium state, which results in a lower yield of product. Small equilibrium constants favor the reactants, resulting in lower yields.

Theme
Systems and interactions
A chemical reaction system is not as simple as the balanced equation implies. Many competing side reactions interfere with the production of the desired product.

Table 8-2
Predictions and Measurements for Isoamyl Acetate Synthesis

Reactants	Formula	Mass	Amount present	Amount needed for reaction
isoamyl alcohol	$C_5H_{11}OH$	500. g	5.67 mol (limiting reactant)	5.67 mol
acetic acid	CH_3COOH	1.25×10^3 g	20.8 mol	5.67 mol
Products	**Formula**	**Amount expected**	**Theoretical yield (mass expected)**	**Actual yield (mass produced)**
isoamyl acetate	$CH_3COOC_5H_{11}$	5.67 mol	738 g	454 g
water	H_2O	5.67 mol	102 g	62.8 g

When a worker at the flavoring factory mixes 500. g of isoamyl alcohol and 1.25×10^3 g of acetic acid, the stoichiometry calculations that account for isoamyl alcohol being a limiting reactant give the results summarized in the second half of **Table 8-2**. But when the actual yield is measured, it is much less than was expected. What went wrong? Of course, the first things to check are the calculations and measurements. But even after checking them, the amounts come out the same. In science, calculations and theories are useful only as ways to predict actual measurements and observations.

In this case, the calculations were not wrong, but they were based on an assumption that was not supported by the observations. Stoichiometric calculations like the ones in **Table 8-2** assume that a reaction will use up all of the limiting reactant to make the products indicated in the equation.

But in cases such as the one shown in **Table 8-2**, the reaction does not completely use up the reactants to make only the products in the equation. If it had, there would be more water and isoamyl acetate present at the end. What happened here? In this case the problem is the equation for the reaction. Although the balanced chemical equation is correct by itself, it describes only part of what is going on. Reactions other than the synthesis of isoamyl acetate are also occurring and producing other products. In fact, the main reaction in competition with the synthesis reaction is a decomposition reaction that is the reverse of the synthesis reaction. In other words, the products are re-forming the reactants, as shown below. The net effect is that at the end of the reaction, instead of having only products, the result is a mixture of products and reactants.

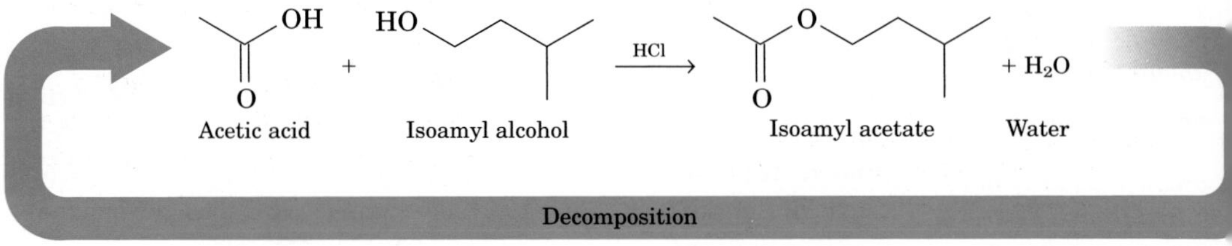

Acetic acid + Isoamyl alcohol $\xrightarrow{HCl}$ Isoamyl acetate + H_2O (Water)

Decomposition

Percent yield is a way to describe reaction efficiency

If the reaction creates a mixture of products and reactants, how can any worthwhile predictions be made about the results of this reaction? Suppose a worker in the flavoring factory kept track of the results of several attempts to make isoamyl acetate. The results are shown in **Table 8-3**.

As you can see, the results are reasonably consistent—the actual yield and the theoretical yield occurred in roughly the same proportion each time. A good way to express this is with **percent yield**, which describes how much of the theoretical yield is actually produced.

percent yield
ratio of actual yield to theoretical yield, multiplied by 100

$$\textbf{percent yield} = \frac{\textbf{actual yield}}{\textbf{theoretical yield}} \times 100$$

The percent yield figures for the values shown in **Table 8-3** are 61.5%, 62.5%, 61.2%, 61.2%, and 62.1%. These values are all reasonably close, but why aren't they equal? Along with the possibility that competing side reactions will lower the actual yield and the percent yield, some of the product may be lost, especially if it is a gas or has to be purified from a mixture. As a result, the percent yield figures are usually averaged over several different trials. For this set of values, the average is 61.7%.

Table 8-3
Data From Several Trials of Isoamyl Acetate Synthesis

Mass of isoamyl alcohol used (g)	Theoretical yield of isoamyl acetate (g)	Actual yield of isoamyl acetate (g)
500.	738	454
500.	738	461
500.	738	452
500.	738	452
500.	738	458

The use of these different ways to describe yields is similar to the use of statistics to describe how frequently a player scores in sports. For example, if you try 20 layups during a basketball game, your theoretical yield is 40 points. That is the maximum number of points you could get from the 20 layups. However, if you make only 10 of those shots, your actual yield is 20 points. Your shooting percentage, or percent yield, is 50%. The batting average shown on the back of baseball cards like the one in **Figure 8-11** is similar to percent yield because it is a ratio of the number of hits to the number of times at bat.

Chemists cannot predict percent yields by looking at a balanced equation or by performing stoichiometric calculations. Precise percent yields must be measured experimentally. However, after closely observing many different reactions, experienced chemists can recognize patterns and reaction types that allow them to make pretty good estimates of percent yield.

Figure 8-11
A baseball player's batting average can be considered similar to a percent yield, because it is a ratio of the number of hits achieved to the number of hits possible.

Teaching Tip
Percent yield
Use the results from the Lesson Starter for Section 8-1 to determine the percent yield for the reaction of sodium carbonate with hydrochloric acid to produce sodium chloride, water, and carbon dioxide.

$$Na_2CO_3 + 2HCl \longrightarrow 2NaCl + H_2O + CO_2$$

If 2 g of sodium carbonate are used, the theoretical yield is calculated as follows.

$$2.000 \text{ g } Na_2CO_3 \times \frac{1 \text{ mol } Na_2CO_3}{106.01 \text{ g } Na_2CO_3} \times \frac{2 \text{ mol NaCl}}{1 \text{ mol } Na_2CO_3} \times \frac{58.45 \text{ g NaCl}}{1 \text{ mol NaCl}} = 2.205 \text{ g NaCl}$$

The actual yield is around 2.100 g. This leads to a percent yield as follows.

$$\% \text{ yield} = \frac{2.100}{2.205} \times 100 = 95.2\%$$

Theme
Systems and interactions
Percent yield is an indicator of the effects of side reactions on the main reactions.

Additional Examples for Sample Problem 8E

a. Nitrogen and hydrogen at high temperatures and pressures are converted into ammonia using the Haber process. The process was developed by Fritz Haber (1868–1934) and allowed Germany to continue to manufacture explosives after their nitrate supplies were cut off.

$$H_2 + N_2 \longrightarrow NH_3$$

When 400. kg of H_2 are added to an excess of N_2, 1040 kg of NH_3 are produced. What is the percent yield of the reaction?

b. A standard laboratory preparation of iodine is the following reaction.

$$NaI + MnO_2 + H_2SO_4 \longrightarrow Na_2SO_4 + MnSO_4 + H_2O + I_2$$

What is the percent yield of I_2 if the actual yield of I_2 was 39.78 g when the amount of NaI used was 62.55 g?

Solutions are at the bottom of the page.

Answers to Practice 8E

1. $$550.\text{ g } C_6H_5CH_3 \times \frac{1 \text{ mol } C_6H_5CH_3}{92.15 \text{ g } C_6H_5CH_3} \times \frac{1 \text{ mol } C_6H_4CH_3NO_2}{1 \text{ mol } C_6H_5CH_3} \times \frac{137.15 \text{ g } C_6H_4CH_3NO_2}{1 \text{ mol } C_6H_4CH_3NO_2} = 819 \text{ g } C_6H_4CH_3NO_2$$

$$\frac{305 \text{ g}}{819 \text{ g}} \times 100 = 37.2\%$$

2. Because the formula mass and mole ratios are the same as item 1, the theoretical yield for o-nitrotoluene is also 819 g.

$$\frac{468 \text{ g}}{819 \text{ g}} \times 100 = 57.1\%$$

Sample Problem 8E
Calculating percent yield

A student is synthesizing aspirin by adding 200. g of salicylic acid to an excess of acetic anhydride. Calculate the percent yield if 96.0 g of aspirin are produced. (Hint: before you can determine the percent yield, you must first calculate the theoretical yield of aspirin based on the amount of the limiting reactant supplied.)

Salicylic acid + Acetic anhydride $\xrightarrow{HCl}$ Aspirin, acetylsalicylic acid + Acetic acid

❶ List

- **amount of limiting reactant:** 200. g $C_7H_6O_3$
- **molar mass of reactants:** 138.13 g/mol $C_7H_6O_3$; 102.10 g/mol $C_4H_6O_3$
- **molar mass of products:** 180.17 g/mol $C_9H_8O_4$; 60.06 g/mol $C_2H_4O_2$
- **mole ratio:** 1 mol $C_7H_6O_3$:1 mol $C_9H_8O_4$
- **actual yield:** 96.0 g $C_9H_8O_4$
- **theoretical yield:** ? g $C_9H_8O_4$
- **percent yield:** ? %

❷ Set up

- **Before you can calculate percent yield, you must calculate theoretical yield by using the molar masses and mole ratios as conversion factors.**

$$200.\text{ g } C_7H_6O_3 \times \frac{1 \text{ mol } C_7H_6O_3}{138.13 \text{ g } C_7H_6O_3} \times \frac{1 \text{ mol } C_9H_8O_4}{1 \text{ mol } C_7H_6O_3} \times \frac{180.17 \text{ g } C_9H_8O_4}{1 \text{ mol } C_9H_8O_4} = ?\text{ g } C_9H_8O_4$$

Calculator answer: 260.8701947 g $C_9H_8O_4$

Properly rounded answer: 261 g $C_9H_8O_4$

- **Calculate percent yield from the actual and theoretical yields.**

$$\frac{96.0 \text{ g actual}}{261 \text{ g theoretical}} \times 100 = ?\% \text{ yield}$$

❸ Estimate and calculate

- **Check the initial calculation of theoretical yield, and estimate the percent yield by rounding off the numbers.**

$$200 \times \frac{1}{140} \times \frac{1}{1} \times \frac{180}{1} \approx$$

$$\frac{10}{7} \times 180 \approx 270$$

$$\frac{90}{270} \times 100 \approx 33\%$$

- **When calculating, be sure to use the correct number of significant figures.**

Calculator answer: 36.7816092

Properly rounded answer: 36.8% yield

Solutions for Additional Example 8E

a. $$400\,000 \text{ g } H_2 \times \frac{1 \text{ mol } H_2}{2.02 \text{ g } H_2} \times \frac{2 \text{ mol } NH_3}{3 \text{ mol } H_2} \times \frac{17.04 \text{ g}}{1 \text{ mol } NH_3} = 2\,250\,000 \text{ g } NH_3$$

$$\% \text{ yield} = \frac{1\,040\,000 \text{ g}}{2\,250\,000 \text{ g}} \times 100 = 46.2\%$$

b. $$62.55 \text{ g } I_2 \times \frac{1 \text{ mol } I_2}{253.80 \text{ g } I_2} \times \frac{2 \text{ mol NaI}}{1 \text{ mol } I_2} \times \frac{149.90 \text{ g NaI}}{1 \text{ mol NaI}} = 73.88 \text{ g NaI}$$

$$\% \text{ yield} = \frac{39.78 \text{ g}}{73.88 \text{ g}} \times 100 = 53.84\%$$

Practice 8E

1. **One step in making *para*-aminobenzoic acid, PABA, an ingredient in some suntan lotions, involves replacing one of the hydrogen atoms in a toluene molecule with an —NO_2 group, directly opposite the —CH_3 group. Calculate the percent yield if 550. g of toluene added to an excess of nitric acid provides 305 g of the nitrotoluene product.**

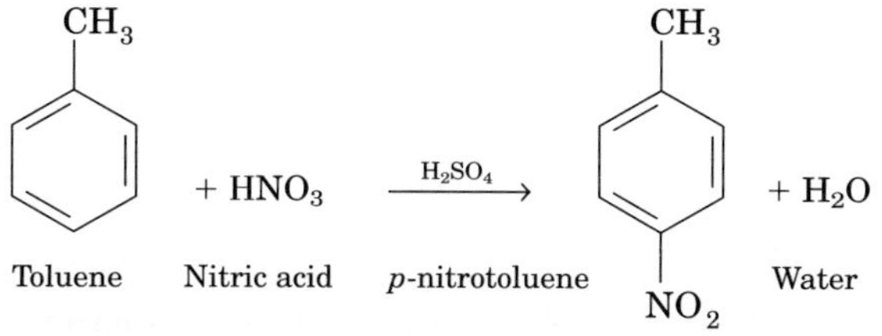

CH_3 CH_3

$+ HNO_3 \xrightarrow{H_2SO_4}$ $+ H_2O$

Toluene Nitric acid *p*-nitrotoluene Water

NO_2

2. **One reason for the low yield in the reaction shown above is a competing side reaction that produces another nitrotoluene product, but one which cannot be used to make PABA. After 550. g of toluene reactant were used with an excess of nitric acid, 468 g of this other product remained. Calculate its percent yield.**

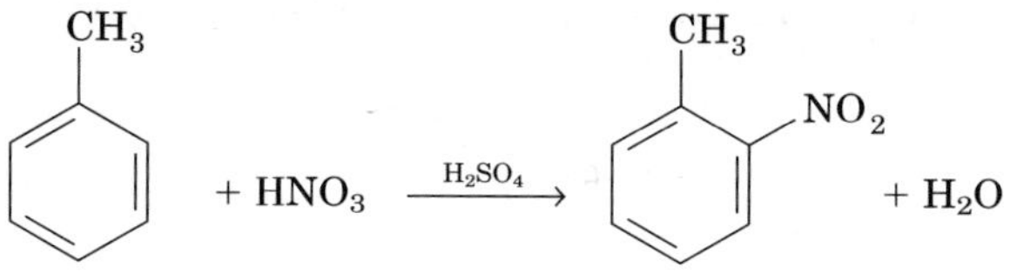

CH_3 CH_3

NO_2

$+ HNO_3 \xrightarrow{H_2SO_4}$ $+ H_2O$

Toluene Nitric acid *o*-nitrotoluene Water

Percent yield figures can be used to predict actual yield

Once the percent yield has been measured, chemists can use it to evaluate the efficiency of a chemical reaction. Just as a basketball player would like to have the highest possible shooting percentage, a chemist would like to have the maximum percent yield. If the percent yield is too low, the chemist might be able to make changes in reaction conditions to increase the actual yield. Such changes will be explored in later chapters.

Perhaps more important, once the percent yield has been measured for the reaction, it can be used as a conversion factor in stoichiometry to make predictions about any reaction, *even those that do not go to completion.* For example, the chemical equation for the synthesis of isoamyl acetate can be written to include the percent yield, as shown below. With this information, the actual yield of the reaction can be calculated.

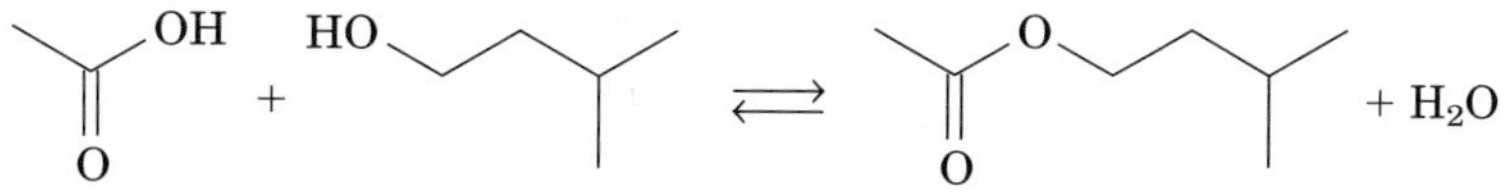

OH HO O

$+ \rightleftharpoons$ $+ H_2O$

O O

Acetic acid Isoamyl alcohol Isoamyl acetate: 61.7% Water

After calculating the theoretical yield of the reaction in the usual way, you can use the percent yield as a conversion factor. The easiest way to do this is to write it as a number of grams of actual yield per 100 g of theoretical yield.

$$61.7\% \text{ yield} = \frac{61.7 \text{ g actual yield}}{100 \text{ g theoretical yield}} \times 100$$

Additional Example for Sample Problem 8F

Huge quantities of sulfur dioxide are produced from zinc sulfide in the following reaction.

$$ZnS + O_2 \longrightarrow ZnO + SO_2$$

If the typical yield is 86.78%, how much SO_2 should be expected if 4897 g of ZnS are used?

Solution is at the bottom of the page.

For example, 150. g of isoamyl alcohol should provide a theoretical yield of 221 g of isoamyl acetate. But you know from the equation with the percent yield to expect only about 61.7% of that isoamyl acetate to actually be formed.

$$221 \text{ g theoretical yield} \times \frac{61.7 \text{ g actual yield}}{100 \text{ g theoretical yield}} = 136 \text{ g actual yield}$$

Sample Problem 8F *Using percent yield*

A more efficient way to prepare the molecule that was used to produce PABA for suntan lotions involves a slightly different starting material, known as isopropylbenzene. This reaction usually has a 91.2% yield. How many grams of the product, *para*-nitro-isopropylbenzene, can you expect if 775 g of isopropylbenzene react with an excess of nitric acid?

Isopropylbenzene + HNO_3 (Nitric acid) $\xrightarrow{H_2SO_4}$ p-nitro-isopropylbenzene (NO_2) + H_2O (Water)

1 List

- **This is a case in which a table for organizing data is useful.**

Substance	Formula	Molar mass	Mass present	Mole ratio
isopropylbenzene	C_9H_{12}	120.21 g/mol	775 g	1
para-nitro-isopropylbenzene	$C_9H_{11}NO_2$	165.21 g/mol	? g	1

Percent yield: 91.2%

2 Set up

- **First, set up the calculation for the theoretical yield.**

$$775 \text{ g } C_9H_{12} \times \frac{1 \text{ mol } C_9H_{12}}{120.21 \text{ g } C_9H_{12}} \times \frac{1 \text{ mol } C_9H_{11}NO_2}{1 \text{ mol } C_9H_{12}} \times \frac{165.21 \text{ g } C_9H_{11}NO_2}{1 \text{ mol } C_9H_{11}NO_2} \times \frac{\quad}{\quad} = ? \text{ g } C_9H_{11}NO_2$$

- **Then, use the percent yield as a conversion factor to calculate actual yield.**

$$775 \text{ g } C_9H_{12} \times \frac{1 \text{ mol } C_9H_{12}}{120.21 \text{ g } C_9H_{12}} \times \frac{1 \text{ mol } C_9H_{11}NO_2}{1 \text{ mol } C_9H_{12}} \times \frac{165.21 \text{ g } C_9H_{11}NO_2}{1 \text{ mol } C_9H_{11}NO_2} \times \frac{91.2 \text{ g actual}}{100 \text{ g theoretical}} = ? \text{ g } C_9H_{11}NO_2$$

3 Estimate and calculate

- **Round the numbers to make an estimate.**

$$800 \times \frac{1}{120} \times \frac{1}{1} \times \frac{165}{1} \times (0.9) \approx 990$$

- **Calculate, and round to the correct number of significant figures.**

Calculator answer: 971.3869728

Correctly rounded answer: 971 g $C_9H_{11}NO_2$

Solutions for Additional Example 8F

$$4897 \text{ g ZnS} \times \frac{1 \text{ mol ZnS}}{97.46 \text{ g ZnS}} \times \frac{1 \text{ mol } SO_2}{1 \text{ mol ZnS}} \times \frac{64.07 \text{ g } SO_2}{1 \text{ mol } SO_2} = 3219 \text{ g } SO_2$$

$$\% \text{ yield} = \frac{\text{actual}}{\text{theoretical}} \times 100$$

$$86.78\% = \frac{\text{actual}}{3219 \text{ g}} \times 100 = 2793 \text{ g } SO_2$$

Section Review

10. Titanium(IV) oxide, TiO_2, is used as a pigment in paints and as a whitening and coating agent for paper. It can be made by reacting O_2 with $TiCl_4$.

$$TiCl_4(s) + O_2(g) \longrightarrow TiO_2(s) + 2Cl_2(g)$$

a. If 4.5 mol of $TiCl_4$ react with 3.5 mol of O_2, identify both the limiting and excess reactants.

b. How many moles of excess reactant will remain if the reaction goes to completion?

c. How many moles of each product should be formed if the reaction goes to completion?

11. How much nitric acid, HNO_3, is produced when NO_2 gas is bubbled under pressure through 100. g of H_2O? (Hint: assume the reaction goes to completion.)

$$3NO_2(g) + H_2O(l) \longrightarrow 2HNO_3(aq) + NO(g)$$

12. When phosphorus burns in the presence of oxygen, P_4O_{10} is produced. In turn, P_4O_{10} reacts with water to produce phosphoric acid, which is one of the compounds found in acid precipitation.

$$P_4O_{10}(g) + H_2O(l) \longrightarrow H_3PO_4(aq)$$

a. Write a balanced chemical equation for this reaction.

b. When 100. g of P_4O_{10} are reacted with 200. g of H_2O, what is the theoretical yield of phosphoric acid?

c. If the actual yield is 126.24 g of H_3PO_4, what is the percent yield for this reaction?

13. If 50.0 g of benzaldehyde react with an excess of acetaldehyde to make cinnamaldehyde and the percent yield is 84.5%, how many grams of cinnamaldehyde will be produced?

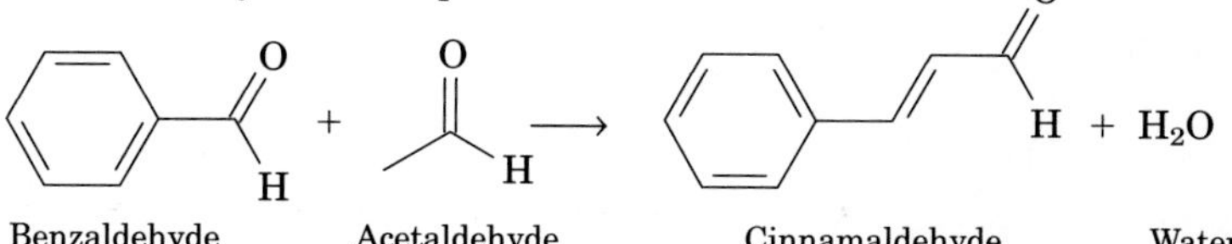

14. Suggest two reasons why theoretical yields are rarely achieved in chemical reactions.

15. **Story Link**
When using an FRH, soldiers add 45 mL of water. How much water is theoretically necessary for the reaction of the 8.1 g of magnesium in the FRH? (Hint: the density of water is 1.0 g/mL.)

16. **Story Link**
Using your answer to item **15**, state whether water or magnesium is the limiting reactant.

ASSESS

Alternative Assessment

Divide the class into groups of two, and present them with the following situation. They have a business that manufactures calcium chloride. Calcium chloride is produced according to the following reaction.

$$CaO + HCl \longrightarrow CaCl_2 + H_2O$$

a. How much $CaCl_2$ can be produced if there are 1467 kg of lime, CaO, and 2500 kg of HCl available for the reaction?

$$1\,467\,000 \text{ g CaO} \times \frac{1 \text{ mol CaO}}{56.08 \text{ g CaO}} =$$

26 160 mol CaO

$$2\,500\,000 \text{ g HCl} \times \frac{1 \text{ mol HCl}}{36.46 \text{ g HCl}} =$$

68 570 mol HCl

$$26\,160 \text{ mol CaO} \times \frac{2 \text{ mol HCl}}{1 \text{ mol CaO}} =$$

52 320 mol HCl

68 570 mol > 52 320 mol

HCl is in excess,f so CaO is the limiting agent.

$$26\,160 \text{ mol CaO} \times \frac{1 \text{ mol } CaCl_2}{1 \text{ mol CaO}} \times \frac{110.98 \text{ g } CaCl_2}{1 \text{ mol } CaCl_2} = 2\,903\,000 \text{ g} =$$

2903 kg $CaCl_2$

b. If the yield of calcium chloride is usually 79.78% and calcium chloride sells for $1.35 per kilogram, how much will the company make on the sales?

$$2903 \text{ kg } CaCl_2 \times \frac{\$1.35}{\text{kg } CaCl_2} = \$3919.05$$

Answers to Section Review

Answers are found on page 309B.

Section 8-3

FOCUS

Lesson Starter

Find out how many students' families have cars with airbags. Ask if anyone in the class has been in a car when an airbag was triggered. Ask students to share their experiences. Ask them where they think the gas that inflates the bag comes from. Refer back to the responses as you cover the stoichiometry of the air bag in this section.

TEACH

Teaching Tip

Uses of stoichiometry

For an integrated economics and chemistry lesson, see "Silver Science" by William J. Sumrall, *The Science Teacher*, Vol. 58, pp. 36–39. This article presents four stoichiometric problems and their solutions using silver prices quoted from the newspaper.

Cultural Connection

Alchemists were prevalent in the early 1300s, as was the fear that the alchemists could turn metals into gold and ruin the economy of the region. Therefore, Pope John XXII declared a ban on alchemy.

How can stoichiometry be used?

Section Objectives

Relate volume calculations in stoichiometry to the inflation of automobile safety air bags.

Use the concept of limiting reactants to explain why changing fuel-air ratios affects engine performance.

Use percent yield to compare the efficiency of pollution-control mechanisms in cars.

Stoichiometry and cars

So far in your study of stoichiometry, you have examined a number of chemical reactions with practical applications—from banana flavoring to cosmetics to aspirin. But stoichiometry's practical importance goes beyond factories and laboratories. When you drive a car, you are depending on stoichiometry to keep you safe and make the car work efficiently with the smallest possible impact on the environment.

Air-bag design depends on stoichiometric precision

Air bags are designed to protect occupants in a car from injuries during a high-speed front-end collision, as shown in **Figure 8-12**. When inflated, they gently slow down the occupants of a car so that they do not strike the steering wheel, windshield, or instrument panel as hard as they would without the air bag. Stoichiometry and the principles of reactions studied in Chapter 7 are used by air-bag designers to make certain that air bags do not under-inflate or over-inflate. Bags that under-inflate do not provide enough protection for the occupants, and bags that over-inflate can cause injury or may even rupture, making them useless. To adequately protect occupants, air bags must fully inflate within one-tenth of a second after impact. The systems that make an air bag work this quickly are shown in **Figure 8-13** on the next page. A front-end collision transfers energy to a crash sensor that signals the firing of the ignitor, which is similar to a small blasting cap. The ignitor provides heat energy to start a reaction in a mixture called the *gas generant*, which forms a gaseous product. The ignitor also raises the temperature and pressure within the reaction chamber, a metal vessel, so that the reaction occurs at a rate fast enough to fill the bag before the occupant strikes it. This reaction chamber releases the gas into the bag while a high-efficiency filter keeps the reactants and the solid products away from the occupant.

Figure 8-12
When used in combination with seat belts, air bags can lessen the severity of injuries in the event of a front-end collision.

Visual Strategy

Figure 8-12

Make students aware that cars are made to crumple in order to absorb some force of a crash, reducing the amount of force that the human body absorbs. Fortunately, people suffer fewer injuries in crashes today, but unfortunately, the cars sustain more damage.

For most current systems, the gas generant is a solid mixture of sodium azide, NaN_3, plus an oxidizer. The gas that inflates the bag is almost entirely nitrogen gas, N_2, which is produced in the following decomposition reaction.

$$2NaN_3(s) \longrightarrow 2Na(s) + 3N_2(g)$$

However, this reaction alone cannot inflate the bag fast enough, and the sodium metal produced is a dangerously reactive substance. Oxidizers such as ferric oxide, Fe_2O_3, are included in the gas generant so that they can immediately react with the sodium metal in a single-displacement reaction. This exothermic reaction also raises the temperature more than a hundred degrees so that the gas fills the bag faster.

$$6Na(s) + Fe_2O_3(s) \longrightarrow 3Na_2O(s) + 2Fe(s)$$

But even sodium oxide is unsafe because it is an extremely basic substance. Eventually, it reacts with carbon dioxide, CO_2, and moisture from the air to form sodium hydrogen carbonate, or baking soda.

$$Na_2O(s) + 2CO_2(g) + H_2O(g) \longrightarrow 2NaHCO_3(s)$$

The volume of gas needed to fill an air bag of a certain volume depends on the amount of gas available and the density of the gas. Gas density, in turn, depends on temperature. To calculate the amount of gas generant necessary, air-bag designers must know the stoichiometry of the reactions and account for energy changes in the reaction, which may change the temperature, and thus the density, of the gas.

Figure 8-13
When a crash occurs, a switch in the sensor closes, and electricity from the battery or the backup power supply flows to the inflator behind the steering wheel. The ignitor within the inflator heats up rapidly, igniting the gas generant to start the sodium azide decomposition reaction. The nitrogen gas that is made inflates the air bag.

Do You Know?
The inflation of an airbag takes about 0.04 s.

Theme
Systems and interactions
The airbag was designed to be part of the entire system of safety features of the car and to interact with the other features to afford the greatest safety possible.

Do You Know?
About 56 L of nitrogen are produced during the explosive decomposition of the sodium azide.

It takes an acceleration of about $-98\ m/s^2$ to set off the airbag.

Visual Strategy
Figure 8-13
Make students aware that the explosive force of the pressurized gas breaks open the air-bag container and fills the bag. The release of a gas during a chemical reaction also causes the pressure that propels cannon balls and bullets from cannons and guns.

Additional Example for Sample Problem 8G

Marble, $CaCO_3$, reacts with hydrochloric acid and is changed to calcium chloride, water, and carbon dioxide. If the density of carbon dioxide is 1.997 g/L, what volume of carbon dioxide is produced when a 4.000 g of marble are reacted with HCl?

$$CaCO_3 + HCl \longrightarrow CaCl_2 + H_2O + CO_2$$

Solution is at the bottom of the page.

Answers to Practice 8G

1. $2NaN_3 \longrightarrow 2Na + 3N_2$

$6Na + Fe_2O_3 \longrightarrow 3Na_2O + 2Fe$

$92.2\text{ g }NaN_3 \times \frac{1\text{ mol }NaN_3}{65.02\text{ g }NaN_3} \times \frac{2\text{ mol Na}}{2\text{ mol }NaN_3} \times \frac{22.99\text{ g Na}}{1\text{ mol Na}} = 32.6\text{ g Na}$

$32.6\text{ g Na} \times \frac{1\text{ mol Na}}{22.99\text{ g Na}} \times \frac{1\text{ mol }Fe_2O_3}{6\text{ mol Na}} \times \frac{159.7\text{ g }Fe_2O_3}{1\text{ mol }Fe_2O_3} =$

$37.7\text{ g }Fe_2O_3$

2. $6Na + Fe_2O_3 \longrightarrow 3Na_2O + 2Fe$

$Na_2O + 2CO_2 + H_2O \longrightarrow 2NaHCO_3$

$37.7\text{ g }Fe_2O_3 \times \frac{1\text{ mol }Fe_2O_3}{159.7\text{ g }Fe_2O_3} \times \frac{3\text{ mol }Na_2O}{1\text{ mol }Fe_2O_3} \times \frac{61.98\text{ g }Na_2O}{1\text{ mol }Na_2O} =$

$43.9\text{ g }Na_2O$

$43.9\text{ g }Na_2O \times \frac{1\text{ mol }Na_2O}{61.98\text{ g }Na_2O} \times \frac{2\text{ mol }NaHCO_3}{1\text{ mol }Na_2O} \times \frac{84.01\text{ g }NaHCO_3}{1\text{ mol }NaHCO_3} =$

$119\text{ g }NaHCO_3$

3. $\frac{119\text{ g }NaHCO_3}{2.20\text{ g/mL}} = 54.1\text{ mL }NaHCO_3$

4. 179. g $NaHCO_3$, 128 g CH_3COOH

5. 38.4 g H_2O, 175 g CH_3COONa

*Full solutions for items **4** and **5** are found on page 309B.*

Sample Problem 8G
Stoichiometry and density: air bags

Assume that 65.1 L of N_2 gas are needed to inflate an air bag to the proper size. How many grams of NaN_3 must be included in the gas generant to generate this amount of N_2? (Hint: the density of N_2 gas at this temperature is about 0.916 g/L.)

1 List

- **balanced chemical equation:** $2NaN_3(s) \longrightarrow 2Na(s) + 3N_2(g)$
- **volume of N_2:** 65.1 L N_2
- **density of N_2:** 0.916 g/L
- **molar mass of NaN_3:** 65.02 g/mol
- **molar mass of N_2:** 28.02 g/mol
- **mole ratio:** 2 mol NaN_3: 3 mol N_2
- **mass of reactant:** ? g NaN_3

2 Set up

- **First, the product amount must be converted from a volume to a mass using density and then from a mass to amount in moles using molar mass. Next, the mole ratio is used to determine moles of reactant needed. Then, the mass of reactant is calculated using molar mass as a conversion factor.**

$$65.1\text{ L }N_2 \times \frac{0.916\text{ g }N_2}{1\text{ L }N_2} \times \frac{1\text{ mol }N_2}{28.02\text{ g }N_2} \times \frac{2\text{ mol }NaN_3}{3\text{ mol }N_2} \times \frac{65.02\text{ g }NaN_3}{1\text{ mol }NaN_3} = ?\text{ g }NaN_3$$

3 Estimate and calculate

- **Round off to make an estimate.**

$$66 \times \frac{1}{1} \times \frac{1}{30} \times \frac{2}{3} \times \frac{66}{1} \approx$$

$$\frac{22}{10} \times \frac{44}{1} \approx 96.8$$

- **Calculate, and round to the correct number of significant digits.**

Calculator answer: 92.2495035 g

Properly rounded answer: 92.2 g NaN_3

Practice 8G

1. **How much Fe_2O_3 must be added to the gas generant for this amount of NaN_3?**
2. **Calculate the mass of sodium hydrogen carbonate produced in the air-bag reaction.**
3. **The density of $NaHCO_3$ is 2.20 g/mL. What volume of $NaHCO_3$ is produced by the reaction?**
4. **How many grams of sodium hydrogen carbonate and acetic acid would be needed to inflate the same air bag with CO_2 gas? CO_2 gas has a density 1.57 times that of N_2 gas at this temperature. (Hint: in other words, 1.57 times as many grams of CO_2 will be required to inflate the same air bag.)**

$$NaHCO_3(s) + CH_3COOH(aq) \longrightarrow CH_3COONa(aq) + H_2O(l) + CO_2(g)$$

Sodium hydrogen carbonate | Acetic acid | Sodium acetate | Water | Carbon dioxide

5. **How many grams of water and sodium acetate would be produced by the reaction of $NaHCO_3$ and CH_3COOH to fill an air bag with CO_2?**

Solution for Additional Example 8G

$$4.000\text{ g }CaCO_3 \times \frac{1\text{ mol }CaCO_3}{100.09\text{ g }CaCO_3} \times \frac{1\text{ mol }CO_2}{1\text{ mol }CaCO_3} \times \frac{44.01\text{ g }CO_2}{1\text{ mol }CO_2} = 1.759\text{ g }CO_2$$

$$1.759\text{ g }CO_2 \times \frac{1\text{ L }CO_2}{1.997\text{ g }CO_2} = 0.881\text{ L }CO_2$$

Engine efficiency depends on reactant proportions

Even if you never have to depend on the stoichiometric calculations that are planned by air-bag designers, any time you drive a car, you are using stoichiometry to control how fast the car moves.

You may already know that a car's engine "burns" gasoline to make the engine run and the car move. The faster the engine runs, the more quickly you will use up the gasoline. But how does the liquid in the gas tank make the engine run? Why does pressing on the accelerator make it run faster? What happens when an engine is "flooded" by having too much gas in it before it is started?

The answers to all these questions require a knowledge of stoichiometry and of the combustion reaction in which gasoline burns. Because the main purpose for the combustion of gasoline is to provide energy, it will be included as one of the products in the following word equation.

$$\text{gasoline} + \text{air} \longrightarrow \text{carbon dioxide} + \text{water} + \text{energy}$$

Actually, this word equation does not identify the reactants very well. Gasoline is not a single chemical but a mixture of many different hydrocarbons, each containing between 5 and 12 carbon atoms.

On average, the gasoline used for automobiles can be treated as if it were pure isooctane (2,2,4-trimethylpentane), whose structure is shown in **Figure 8-14**. This molecule has a molar mass that is about the same as the weighted average of the molecules within the gasoline.

In addition, the only component of air that the gasoline uses in combustion is oxygen, which is only 20.9% of air by volume. So, a satisfactory way to portray the combustion reaction would be to indicate the actual proportions of the reactants and products in the following balanced chemical equation.

$$2C_8H_{18}(g) + 25O_2(g) \longrightarrow 16CO_2(g) + 18H_2O(g) + 10\,900 \text{ kJ}$$

The two reactants must be mixed in a mole ratio that is close to the one shown in the balanced chemical equation for efficient combustion. If too much excess reactant is leftover, the engine will not perform well and may even stop firing.

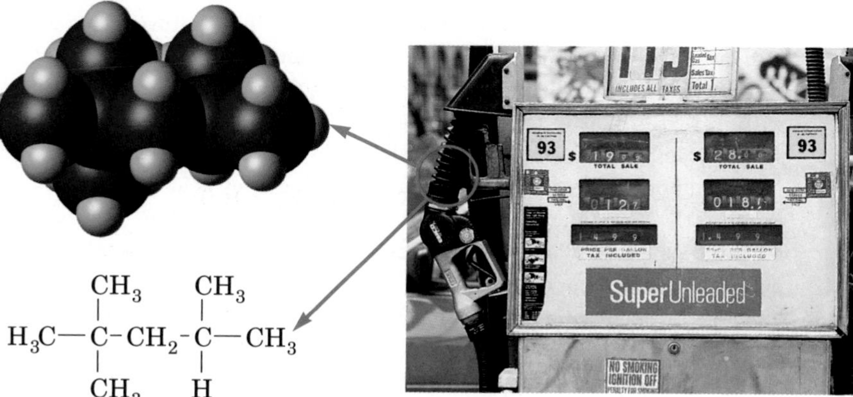

Figure 8-14
Isooctane, one of the components found in gasoline, has some properties as a pure substance that are similar to those of the mixture, gasoline.

Demonstration 4

Decomposition of marble chips

Approximate time: 10 min (accompanies Additional Example for Sample Problem 8G)

1. Place a petri dish on an overhead projector, and fill it half full with 1 M HCl.
2. Have a student determine the mass of 5–10 marble chips to the nearest thousandth of a gram.
3. Add the marble chips to the petri dish. Ask what gas is being given off. How do you know when the reaction is finished? Leave this reaction running while you work out the example on page 296.

SAFETY
Wear safety goggles and a lab apron. Students should remain back 10 ft.

DISPOSAL
Decant the liquid, and wash the remaining chips, if any. Combine the washings with the decanted liquid. Neutralize the 1 M NaOH, and pour this mixture down the drain. The leftover chips can be put in the trash.

Content Background

The formation of water from this reaction is the main reason why the tailpipes of cars rust. If a car is not driven a sufficient distance (more than 10 miles) to completely heat up the tailpipe assembly, then the water will condense on the walls of the pipes in the exhaust system and the muffler.

Theme

Conservation

Although energy is always conserved, some energy is lost by a subsystem as it escapes into the larger system. In the case of an engine as a subsystem, energy is always escaping into the engine's surroundings.

Content Background

Cars with fuel injection use the engine's computer to control the oxygen-fuel ratio.

Engine running at normal speeds

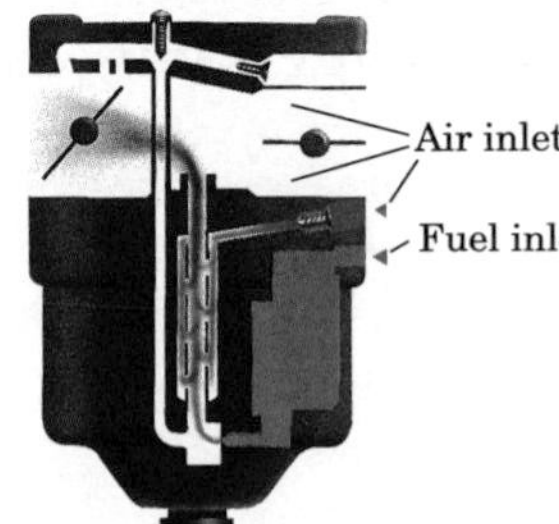

Engine starting

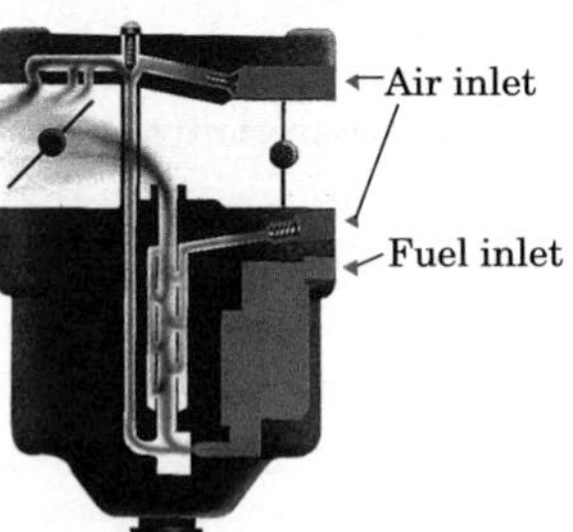

Engine idling

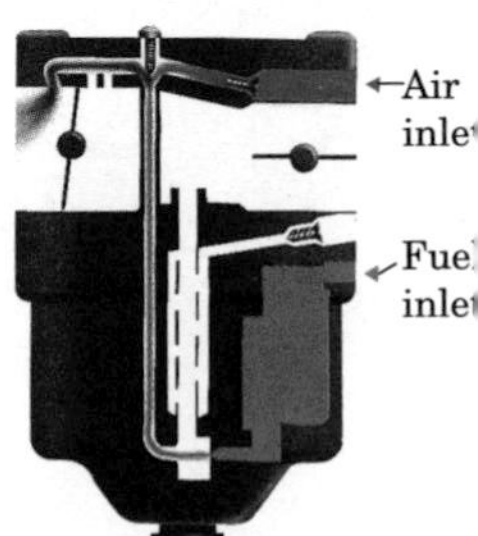

1:16 fuel-air ratio by mass
(1:13.2 isooctane-oxygen mole ratio)

Figure 8-15a
Under ordinary running conditions, an engine's fuel-air ratio is maintained at 1:16 by the carburetor, instead of the 1:12.5 stoichiometric mole ratio, so that the mixture is slightly lean, meaning that gasoline is the limiting reactant. Oxygen is in excess.

1:2 fuel-air ratio by mass
(1:1.7 isooctane-oxygen mole ratio)

8-15b
When an engine is starting, the mixture is very rich, a 1:2 ratio of fuel to air. Oxygen is the limiting reactant, and gasoline is in excess.

1:9 fuel-air ratio by mass
(1:7.4 isooctane-oxygen mole rat

8-15c
When an engine is idling, th reaction mixture is kept rich at a 1:9 fuel-air ratio, meanin that oxygen is the limiting reactant. In this way, there will always be plenty of fuel for the combustion reaction.

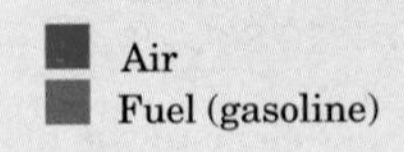

Either a carburetor, as shown in **Figure 8-15**, or a fuel-injection system, depending on the car's year and model, is responsible for feeding gasoline and air to the engine in variable but precise mixtures.

As an engine runs under ordinary conditions, the fuel-air ratio is kept relatively close to the 1:12.5 mole ratio for the balanced equation. For maximum fuel economy, gasoline is the limiting reactant by a slight amount. In such a "lean" mixture, it is more likely that all of the gasoline molecules will react to provide energy and none will be leftover. This will also minimize pollution due to incomplete combustion.

Other conditions, such as starting or idling, require "rich" combinations with a greater relative amount of fuel so that there is enough fuel to support combustion. The excess gasoline also cools the engine.

As you push down on the gas pedal when an engine is idling, it opens the throttle valve, increasing the flow of both air and gasoline. If either reactant is restricted too much, the engine might stall. For example, if you pump the gas pedal too much before starting, the mixture will contain almost entirely gasoline and may not burn because there will not be enough oxygen. On the other hand, if there is too much oxygen and not enough gasoline, the engine will stall just as if the car were out of gas.

Visual Strategy

Figure 8-15

Make students aware that liquid gasoline is not normally placed in the cylinders to explode. In order for the gasoline to explode, it must be in the vapor state. The carburetor mixes oxygen with the volatile gasoline to make a vapor mixture for the cylinder explosion.

Sample Problem 8H
Stoichiometry calculations: air–fuel ratio

How many liters of air must react with 1.000 L of isooctane in order for combustion to occur completely? At 20°C, the density of isooctane is 0.6916 g/mL, and the density of oxygen is 1.331 g/L. (Hint: remember to use the percentage of oxygen in air.)

1 List

- **A table will help organize the data.**

Reactant	Formula	Molar mass	Density	Volume	Mole ratio
isooctane	C_8H_{18}	114.26 g/mol	0.6916 g/mL	1.000 L	2
oxygen	O_2	32.00 g/mol	1.331 g/L	? L	25

Percentage oxygen in air: 20.9% oxygen

Volume of air needed: ? L

2 Set up

- **First, the volume of isooctane must be converted to mass and then to moles. Then, the mole ratio is used to convert to moles of oxygen. The moles of oxygen must be converted to mass and then to volume of oxygen. Then, the percentage of oxygen in air is used to convert to volume of air.**

This problem looks complicated, but if you break down the process into smaller pieces, it is not difficult to understand.

$$1.000\ \text{L}\ C_8H_{18} \times \frac{1000\ \text{mL}}{1\ \text{L}} \times \frac{0.6916\ \text{g}\ C_8H_{18}}{1\ \text{mL}\ C_8H_{18}} \times \frac{1\ \text{mol}\ C_8H_{18}}{114.26\ \text{g}\ C_8H_{18}} \times$$

$$\frac{25\ \text{mol}\ O_2}{2\ \text{mol}\ C_8H_{18}} \times \frac{32.00\ \text{g}\ O_2}{1\ \text{mol}\ O_2} \times \frac{1\ \text{L}\ O_2}{1.331\ \text{g}\ O_2} \times \frac{100\ \text{L air}}{20.9\ \text{L}\ O_2} = ?\ \text{L air}$$

3 Estimate and calculate

- **Round off to make an estimate.**

$$1 \times 1000 \times 0.7 \times \frac{1}{115} \times \frac{25}{2} \times 32 \times \frac{1}{1.33} \times 5 \approx$$

$$700 \times \frac{5}{46} \times 32 \times \frac{3}{4} \times 5 \approx 700 \times \frac{25}{46} \times 24 \approx$$

$$350 \times 25 = 8750$$

- **Calculate, and round off correctly.**

Rounded answer: 8.704×10^3 L air

Practice 8H

1. **Indicate whether the following fuel-air ratios are too rich in fuel. Then indicate what volume of the limiting reactant would need to be added to bring them to the proper stoichiometric ratio.**
   a. **3.000 L of isooctane: 35 000. L of air**
   b. **24.75 L of isooctane: 2.00×10^5 L of air**
   c. **57.3 mL of isooctane: 400. L of air**
2. **Rather than isooctane alone, a better model for the complex mixture in gasoline would be a 9:1 molar mixture of isooctane and *n*-heptane, C_7H_{14}.**
   a. **Write balanced chemical equations for the combustion of isooctane and *n*-heptane.**
   b. **If the density of the isooctane-heptane mixture is 0.6908 g/mL, how many liters of air are needed for the complete combustion of 1.000 L of the mixture?**

Additional Example for Sample Problem 8H

Propane torches are used by plumbers to solder copper tubing. How much air is required to react with 2.00 L of propane? (Density of oxygen is 1.331 g/L at 20°C; density of propane is 1.83 g/L at 20°C; and air is 20.9% oxygen.)

$$C_3H_8 + O_2 \longrightarrow CO_2 + H_2O$$

Solution is at the bottom of the page.

Answers to Practice 8H

1.a. $3.000\ \text{L}\ C_8H_{18} \times \frac{1000\ \text{mL}}{1\ \text{L}} \times \frac{0.6916\ \text{g}\ C_8H_{18}}{1\ \text{mL}\ C_8H_{18}} \times \frac{1\ \text{mol}\ C_8H_{18}}{114.26\ \text{g}\ C_8H_{18}} \times \frac{25\ \text{mol}\ O_2}{2\ \text{mol}\ C_8H_{18}} \times \frac{32.00\ \text{g}\ O_2}{1\ \text{mol}\ O_2} \times \frac{1\ \text{L}\ O_2}{1.331\ \text{g}\ O_2} \times \frac{100\ \text{L air}}{20.9\ \text{L}\ O_2} = 26\ 110\ \text{L of air}$

no, 35 000 L > 26 110 L

b. $24.75\ \text{L}\ C_8H_{18} \times \frac{1000\ \text{mL}}{1\ \text{L}} \times \frac{0.6916\ \text{g}\ C_8H_{18}}{1\ \text{mL}\ C_8H_{18}} \times \frac{1\ \text{mol}\ C_8H_{18}}{114.26\ \text{g}\ C_8H_{18}} \times \frac{25\ \text{mol}\ O_2}{2\ \text{mol}\ C_8H_{18}} \times \frac{32.00\ \text{g}\ O_2}{1\ \text{mol}\ O_2} \times \frac{1\ \text{L}\ O_2}{1.331\ \text{g}\ O_2} \times \frac{100\ \text{L air}}{20.9\ \text{L}\ O_2} = 215\ 400\ \text{L of air}$

yes, 2.00×10^5 L < 2.154×10^5 L

c. $57.3\ \text{mL}\ C_8H_{18} \times \frac{0.6916\ \text{g}\ C_8H_{18}}{1\ \text{mL}\ C_8H_{18}} \times \frac{1\ \text{mol}\ C_8H_{18}}{114.26\ \text{g}\ C_8H_{18}} \times \frac{25\ \text{mol}\ O_2}{2\ \text{mol}\ C_8H_{18}} \times \frac{32.00\ \text{g}\ O_2}{1\ \text{mol}\ O_2} \times \frac{1\ \text{L}\ O_2}{1.331\ \text{g}\ O_2} \times \frac{100\ \text{L air}}{20.9\ \text{L}\ O_2} = 499\ \text{L of air}$

yes, 400. L < 499 L

Answers are continued on page 309B.

Solution for Additional Example 8H

$$2.00\ \text{L}\ C_3H_8 \times \frac{1.83\ \text{g}\ C_3H_8}{1\ \text{L}\ C_3H_8} \times \frac{1\ \text{mol}\ C_3H_8}{44.11\ \text{g}\ C_3H_8} \times \frac{5\ \text{mol}\ O_2}{1\ \text{mol}\ C_3H_8} \times \frac{32.00\ \text{g}\ O_2}{1\ \text{mol}\ O_2} \times \frac{1\ \text{L}\ O_2}{1.331\ \text{g}\ O_2} \times \frac{100\ \text{L air}}{20.9\ \text{L}\ O_2} = 47.7\ \text{L of air}$$

Theme

Systems and interactions
Production of pollutants from using air rather than oxygen as reactant is part of the inefficiency of combustion engines.

Table 8-4
Clean Air Act Targets for 1996 Air Pollution*

Pollutant	Cars	Light trucks	Motorcycles
hydrocarbons	0.25 g/km	0.50 g/km	5.0 g/km
carbon monoxide	2.1 g/km	2.1–3.1 g/km, depending on truck size	12 g/km
oxides of nitrogen (NO_2)	0.25 g/km	0.25–0.68 g/km, depending on truck size	not regulated

*Note: excludes standards for diesel-powered vehicles. EPA standards for cars and light trucks are actually measured in grams per mile at 75°F.

Car designers use stoichiometry to control pollution

Automobiles are the primary source of air pollution in many parts of the world. To reduce the amount of photochemical smog and other pollution caused by automobile exhaust, Congress passed the Clean Air Act in 1968. This act was amended in 1990 to set new, more restrictive emission control standards for automobiles driven in the United States. **Table 8-4** lists the latest standards for exhaust pollution, which are issued by the U.S. Environmental Protection Agency.

But where do all of these pollutants come from if the combustion of gasoline produces only carbon dioxide, water, and energy? Although the equation for the combustion of isooctane shows most of what happens when gasoline burns, it does not show the whole story. For example, if the fuel-air mixture is improperly balanced and there is not enough oxygen, as when a car is started, some carbon monoxide will be produced instead of carbon dioxide. In cold weather, cars need even more fuel to start than usual, and more carbon monoxide is formed.

If combustion is incomplete, or if gasoline fumes leak out of the tank, hydrocarbons can be released into the atmosphere. In addition, in the high-pressure and high-temperature environment of the engine, nitrogen oxides are formed from the nitrogen and oxygen in air.

One of the Clean Air Act standards limits the amount of nitrogen oxides, especially NO_2, that a car can emit. Such compounds can combine with water in the atmosphere to produce acids that are a part of acid rain. In addition, these compounds can react with unburned hydrocarbons to produce irritating chemicals. Because these chemicals are produced in reactions that are catalyzed by the energy from the sun's ultraviolet light, they form what is referred to as photochemical smog. Photochemical smog can make your eyes burn, and it can be harmful for people with respiratory or heart problems. As you can see in **Figure 8-16**, photochemical smog is easy to detect around many cities. However, it is caused by very small amounts of pollutants. The typical percentage of pollutants in smoggy air is about 0.003%.

Figure 8-16
Smog is a problem for many cities in the United States. It not only is unpleasant to look at, but also can make breathing difficult for many people.

300 Chapter 8

Visual Strategy

Figure 8-16
Some cities like Los Angeles often suffer from thermal inversions that trap the photochemical smog above the city. When this happens, the smog continues to build up until it is eventually blown eastward into the mountains. Thermal inversions occur when a layer of warm air forms over a layer of cooler air.

Formation of photochemical smog begins when NO_2 molecules absorb light. The NO_2 decomposes as it absorbs this energy and produces oxygen atoms and nitrogen(II) oxide molecules. In turn, the oxygen atoms produced by the NO_2 decomposition react with oxygen molecules in the air to produce ozone.

$$NO_2(g) \xrightarrow{\text{ultraviolet light}} NO(g) + O(g)$$

$$O_2(g) + O(g) \longrightarrow O_3(g)$$

You may already know that ozone in the upper atmosphere serves as a protective shield against the sun's ultraviolet rays. But closer to the Earth, ozone is a very reactive molecule that can crack rubber, corrode metals, and damage living tissues. In addition, ozone can undergo a complex series of reactions with any hydrocarbons that are not completely burned by a car's engine. The products of these reactions also contribute to photochemical smog.

Automobile manufacturers use stoichiometry to predict when adjustments will be necessary to keep exhaust emissions within legal limits. Moreover, car manufacturers must be sure that all of these stoichiometric concerns are met without raising costs to the point where the manufacturers begin to lose their share of the consumer market.

Because the units in **Table 8-4** are *grams per kilometer*, auto manufacturers must first take into account how much fuel the vehicle will burn in order to move a certain distance. Automobiles with better gas mileage will use less fuel per kilometer. If the reaction progresses in the same way, according to stoichiometry, more fuel-efficient cars should also have slightly lower emissions per kilometer.

Most cars have catalytic converters that treat the exhaust before it is released to the air. The platinum, palladium, or rhodium found in these converters, as shown in **Figure 8-17**, assists in the decomposition of NO_2 into N_2 and O_2, harmless gases already found in the air. Catalytic converters also decrease emissions of CO and hydrocarbons. But catalytic converters perform at their best in warm weather and when the ratio of air and fuel in the engine is very close to the proper stoichiometric ratio. Newer model cars include on-board computers and oxygen sensors to make sure that the proper ratio is maintained so that the engine and the catalytic converter work at top efficiency.

Figure 8-17
The catalytic converters used in automobiles effectively decrease nitrogen oxides, carbon monoxide, and hydrocarbons in exhaust, unless the converter is exposed to leaded gasoline or conditions of extreme heat.

Catalytic converter containing ceramic pellets coated with platinum, palladium, or rhodium catalysts

Content Background

A catalytic converter is placed before the muffler in the exhaust system so that it heats up sufficiently to cause the catalytic reduction of nitrogen oxides. The catalytic converter can become so hot that cars with catalytic converters should not be parked over leaves, paper, or other flammable objects.

It is fairly easy for the catalytic material to become fouled from substances such as lead compounds, which greatly lower the effectiveness of the catalyst. Lead compounds in gasoline slow the explosion rate and coat everything in the cylinder and exhaust line. Leaded gasoline also fouls spark plugs faster, which requires more frequent tune-ups.

Theme

Equilibrium and change
Catalytic converters alter equilibrium by lowering the energy of activation needed to decompose larger nitrogen compounds.

Visual Strategy

Figure 8-17
Photochemical smog occurs when light-sensitive nitrogen oxide compounds (resulting from the use of air instead of pure oxygen in engines) turn brown, imparting a dirty look to the air. Catalytic converters lower the amount of photochemical smog produced, although it is still a problem.

Additional Examples for Sample Problem 8I

Race cars often burn ethanol, C_2H_5OH, for added performance.

a. If the car holds 100. L of ethanol and all of the carbon in it forms carbon dioxide, what volume of carbon dioxide is added to the air?

b. If all of the carbon dioxide then mixes with water to form carbonic acid, what mass of carbonic acid is formed? (Density of ethanol is 0.816 g/mL and the density of CO_2 is 1.997 g/L)

$$C_2H_5OH + O_2 \longrightarrow CO_2 + H_2O$$

$$H_2O + CO_2 \longrightarrow H_2CO_3$$

Solutions are at the bottom of the page.

ASSESS

Alternative Assessment

1. Have students develop eight good questions *with answers* regarding the concepts or applications of stoichiometry to chemistry, the environment, or everyday life.
2. Have students determine four ways that stoichiometry can be used to calculate quantities at home, work, or school or in the environment.

Answers to Section Review

17. 15.8 L N_2

18. 1851 g O_2

19. 36.6 g O_2

20. 54.9 g O_2

21. 145. g CO

22. From **Table 8-4**, the CO level is 2.1 g/km, which exceeds the 145 g/km from item **21**.

*Full solutions for items **17–21** are found on page 309C.*

The contents of gasoline can also be adjusted to decrease pollution. Volatile hydrocarbon compounds like benzene easily evaporate into the air. Some newer blends of gasoline use fewer volatile organic compounds. The addition of ethanol, C_2H_5OH, to fuel can help fuel burn with lower CO levels. Many cities in the United States are required to use these blends of gasoline as part of the Clean Air Act in an effort to improve the air quality.

Sample Problem 8I
Calculating yields: pollution

How much ozone could be produced from 3.50 g of NO_2 contained in a car's exhaust? (Hint: first calculate how much O is produced from NO_2. Then, use this value to calculate how much O_3 is produced.)

1 List

- **amount of reactant:** 3.50 g NO_2
- **molar mass for NO_2:** 46.01 g/mol
- **molar mass for O_3:** 48.00 g/mol
- **amount of products:** ? g O_3
- **balanced equations:** $NO_2(g) \longrightarrow NO(g) + O(g)$
  $O_2(g) + O(g) \longrightarrow O_3(g)$

2 Set up

- **First, convert the mass of NO_2 into moles so that the mole ratio can be used to calculate moles of O. Then, use the mole ratio for the second equation to calculate moles and then grams of O_3.**

$$3.50\ \cancel{g\ NO_2} \times \frac{1\ \cancel{mol\ NO_2}}{46.01\ \cancel{g\ NO_2}} \times \frac{1\ \cancel{mol\ O}}{1\ \cancel{mol\ NO_2}} \times \frac{1\ \cancel{mol\ O_3}}{1\ \cancel{mol\ O}} \times \frac{48.00\ g\ O_3}{1\ \cancel{mol\ O_3}} = ?\ g\ O_3$$

3 Estimate and calculate

- **Round off to make an estimate.**

$$3.5 \times \frac{1}{46} \times \frac{1}{1} \times \frac{1}{1} \times \frac{48}{1} \approx 3.5$$

- **Calculate the answer, and round to the correct number of significant figures.**

Calculator answer: 3.651~~380135~~

Correct answer: 3.65 g O_3

Section Review

17. Calculate how many liters of N_2 gas would be produced if 22.4 g of NaN_3 were placed inside an air-bag ignitor.
18. Determine how many grams of O_2 must be provided by a car's carburetor or fuel-injection system so that 528.7 g of C_8H_{18} would be completely combusted.
19. Calculate how many grams of O_2 could react with 18.3 g of O produced from the NO_2 emitted in a car's exhaust.
20. How many grams of O_3 would be produced by the O_2 and O in item **19**?
21. Assume that 74.0 g of isooctane must be combusted to drive a car for 1.0 km. What is the theoretical yield of carbon monoxide, if it is assumed that all of the carbon atoms in the isooctane form CO?
22. Check the data listed in **Table 8-4**. Using your answer from item **21**, what is the maximum percent yield of CO permitted by law to be released in this car's exhaust?

Solutions for Additional Example 8I

a. $100\ L\ C_2H_5OH \times \frac{816\ g\ C_2H_5OH}{1\ L\ C_2H_5OH} \times \frac{1\ mol\ C_2H_5OH}{46.08\ g\ C_2H_5OH} \times \frac{2\ mol\ CO_2}{1\ mol\ C_2H_5OH} \times \frac{44.01\ g\ CO_2}{1\ mol\ CO_2} \times \frac{1\ L\ CO_2}{1.997\ g\ CO_2} =$ 78 100 L of CO_2

b. $78\ 100\ L\ CO_2 \times \frac{1.997\ g\ CO_2}{1\ L\ CO_2} \times \frac{1\ mol\ CO_2}{44.01\ g\ CO_2} \times \frac{1\ mol\ H_2CO_3}{1\ mol\ CO_2} \times \frac{62.03\ g\ H_2CO_3}{1\ mol\ H_2CO_3} = 220\ 000\ g\ of\ H_2CO_3$

FRH

Conclusion: Mass Relationships and Feeding the Army

At the beginning of this chapter, you learned about how the Army invented a new way to make sure that soldiers get hot meals.

Why was the magnesium-water reaction of the FRH chosen over the other methods? Soldiers have a great deal of equipment to carry. The less mass devoted to warming equipment, the better. Soldiers also operate under hectic conditions. The simpler the method is to use, the better.

From the story, you know that the reaction of 1 mol of Mg produces more than five times as much energy as the reaction of 1 mol of CaO. When equal masses are compared, the advantage is even greater.

But mole for mole and gram for gram, trioxane releases far more energy than does the Mg reaction. Why was this method discarded?

Trioxane must be used outside, it must be lit with matches or a lighter, and it takes longer to heat a meal, which makes it difficult to use in dangerous situations. On the other hand, an FRH can be used safely in vehicles, tents, ships, planes, and buildings.

Why is there a difference between the amount of water that is added and the amount of water the reaction requires? To warm a meal in the FRH, 45 mL of water are added, even though the balanced chemical equation requires only 12 mL.

Water is an excess reactant. Thus, the more expensive magnesium will be the limiting reactant that is completely consumed in the reaction.

Another reason for the extra water is to provide efficient heat transfer over a larger area. Also, the paperboard and the porous pad soak up some of the water, so it is not available for the reaction.

The designers of this system balanced many competing concerns. If too little water is added, the reaction will not go to completion, some magnesium will be wasted, and not as much energy will be generated. On the other hand, excess water absorbs heat that could be warming the meal.

Applying Concepts

1. Before the development of the FRH, soldiers heated their meals using the combustion reaction of trioxane, shown at right. Write the balanced chemical equation for this reaction.

Research and Writing

Use the library to find out more about the following.

1. In many wars, food poisoning contributes substantially to the number of casualties. Compare the percentages for several wars. Give a reason for any pattern you find.
2. Check the listings for portable stoves in a catalog of camping goods. Determine what fuels they use, and write balanced chemical equations for the reactions.

Chemistry in Your Community

1. Visit a local appliance store to determine the efficiency and energy output of various types of kitchen stoves. Determine the cost per energy output and determine the "best buy."
2. Have students work in groups making posters that summarize a survey they make of new model cars to compare air-bag options. The following are some suggested catagories.
   - **a.** no air bags
   - **b.** driver's side only
   - **c.** driver's and passenger's sides
   - **d.** side-impact air bags
   - **e.** other

   Have each group design an advertisement for the car that they feel provides the best overall air-bag options and the best air-bag options for the price.

Answers to Applying Concepts

1. $C_3H_6O_3 + 3O_2 \longrightarrow 3CO_2 + 3H_2O$

Closure Strategy

Formulating test review questions

Divide the class into groups of two to four students. Each group develops three stoichiometry problems that are used to review for the Chapter test. Students can be awarded points for the most challenging and creative problems. Have the groups trade problems and work them out on their own. Review as many of them as you can in class after students have finished working.

CHAPTER 8

Highlights

Key terms

actual yield	percent yield
excess reactant	stoichiometry
limiting reactant	theoretical yield

Key ideas

8-1 How much can a reaction produce?

- Reaction stoichiometry compares the mass and quantity of substances in a chemical equation. Composition stoichiometry describes the relationship among the elements within a substance.
- Stoichiometry problems can be solved with conversion factors created from mole ratios, molar masses, density, and Avogadro's number.

8-2 How much does a reaction really produce?

- Once the limiting reactant has been used up, no more product can be formed, no matter how much of the other reactant(s) remains.
- The theoretical yield is the calculated maximum amount of product possible from a given amount of reactants.
- The actual yield is the amount of product experimentally measured after the reaction of a given amount of reactants.
- Percent yield, 100 times the ratio of actual yield to theoretical yield, describes reaction efficiency.

8-3 How can stoichiometry be used?

- Stoichiometry can be used by automobile designers to maximize a car's safety and performance, while minimizing its environmental impact.

Key problem-solving approach: *Stoichiometry calculations*

When solving stoichiometry problems, select the variable given in your problem from the top options near the top of this diagram. Then, follow the calculation pathway that leads to the proper units for the answer near the bottom of this diagram.

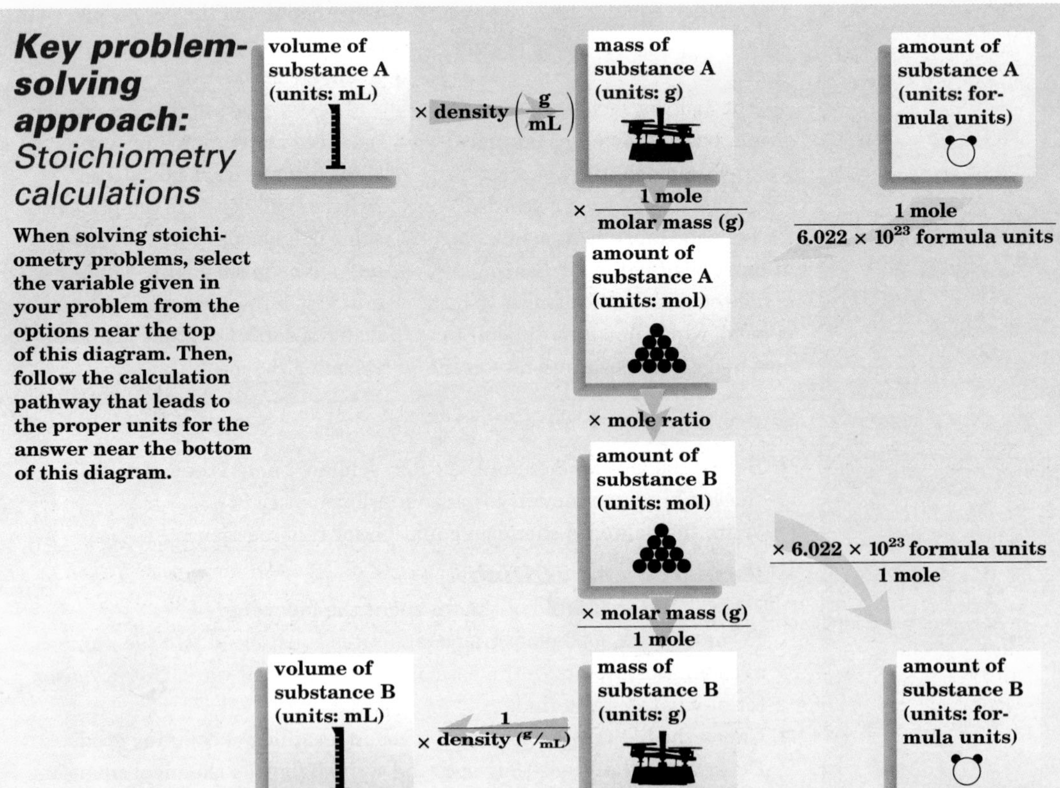

CHAPTER 8

Review and Assess

Stoichiometry

REVIEW

1. Explain the difference between reaction stoichiometry and composition stoichiometry.
2. Why is a balanced chemical equation required to solve stoichiometry problems?
3. A student encounters a stoichiometry problem. He quickly glances over it, punches the numbers into a calculator, and obtains an answer. Explain why the student's method is unlikely to yield the correct answer. What should he have done differently?
4. How do stoichiometry problems that provide data or require answers in mass amounts differ from stoichiometry problems that provide data or require answers in mole amounts?
5. A reaction between hydrazine, N_2H_4, and dinitrogen tetroxide, N_2O_4, has been used to launch rockets into space. The reaction produces nitrogen gas and water vapor, as shown in the unbalanced equation below.

$$N_2H_4(l) + N_2O_4(l) \longrightarrow N_2(g) + H_2O(g)$$

a. Write the balanced chemical equation for the reaction.
b. What is the mole ratio of N_2H_4 to N_2?
c. What is the mole ratio of N_2O_4 to H_2O?
d. How many moles of water will be produced from 14 000 mol of hydrazine used by a rocket?

PRACTICE

Answers to items in a black square begin on page 841.

6. In your body, cell metabolism produces carbon dioxide, which is promptly combined with water to form carbonic acid, H_2CO_3. The carbonic acid is released into the blood, where enzymes speed up its decomposition into water and carbon dioxide. The carbon dioxide is later released from your lungs. How many grams of carbon dioxide would you exhale after 0.250 g of H_2CO_3 decomposes? (Hint: see Sample Problem 8A.)
7. Various processes, including gasoline combustion in automobiles and industrial burning of fossil fuels, can result in the production of sulfur dioxide, SO_2. This can undergo a series of reactions with oxygen and water in the air to eventually form sulfuric acid as shown in the equation below. This acid mixes with moisture to form acid precipitation. If 0.500 g of sulfur dioxide from pollutants reacts with excess water and oxygen found in the air, how many grams of sulfuric acid can be produced? (Hint: see Sample Problem 8A.)

$$2H_2O(l) + O_2(g) + 2SO_2(g) \longrightarrow 2H_2SO_4(aq)$$

8. Oxygen gas can be produced by decomposing potassium chlorate using the reaction below. If 125 g of $KClO_3$ are heated and decompose completely, how many moles of oxygen gas are produced? (Hint: see Sample Problem 8B.)

$$2KClO_3(s) \longrightarrow 2KCl(s) + 3O_2(g)$$

9. Ethanol is considered a clean fuel because it burns in the presence of excess oxygen to produce carbon dioxide and water with fewer trace pollutants than some hydrocarbons found in gasoline. If 5.00 mol of water are produced during the combustion of ethanol, how many grams of ethanol were present at the beginning of the reaction? (Hint: see Sample Problem 8B.)

OH + O=O, O=O, O=O —heat→ O=C=O, O=C=O + H–O–H, H–O–H, H–O–H

Ethanol | 3 Oxygen | 2 Carbon dioxide | 3 water

10. Oxygen gas and water are produced by the decomposition of hydrogen peroxide. If 10.0 mol H_2O_2 decompose, how many liters of oxygen will be produced? Assume the density of oxygen is 1.429 g/L. (Hint: see Sample Problem 8C.)

$$2H_2O_2(aq) \longrightarrow 2H_2O(l) + O_2(g)$$

1. Composition stoichiometry deals with only one substance, while reaction stoichiometry deals with relationships resulting from a reaction. It includes determining the amount of another product or reactant from a given amount of product or reactant.
2. The coefficients in a balanced chemical equation give the mole ratios needed for stoichiometry calculations.
3. First, the student did not read the problem carefully. Second, he used his calculator too soon. He should have read the problem several times and then listed what information was given and what information he was asked to find. Based on this data, he should have developed a plan for how to set up the problem using conversion factors and mole ratios that combine to give the units for the answer. Then the calculator could be used for the final computation.
4. Stoichiometry problems with data in grams must include extra steps to convert mass to moles, and then moles to mass in order to get an answer.
5. a. $2N_2H_4 + N_2O_4 \longrightarrow 3N_2 + 4H_2O$
   b. 2 mol N_2H_4:3 mol N_2
   c. 1 mol N_2O_4:4 mol H_2O
   d. Because each mole of hydrazine produces twice as many moles of water, 28 000 mol of water are produced.
6. $H_2CO_3 \longrightarrow H_2O + CO_2$

$$0.250\ \text{g}\ H_2CO_3 \times \frac{1\ \text{mol}\ H_2CO_3}{62.03\ \text{g}\ H_2CO_3} \times \frac{1\ \text{mol}\ CO_2}{1\ \text{mol}\ H_2CO_3} \times \frac{44.01\ \text{g}\ CO_2}{1\ \text{mol}\ CO_2} = 0.177\ \text{g}\ CO_2$$

7. $$0.500\ \text{g}\ SO_2 \times \frac{1\ \text{mol}\ SO_2}{64.07\ \text{g}\ SO_2} \times \frac{2\ \text{mol}\ H_2SO_4}{2\ \text{mol}\ SO_2} \times \frac{98.09\ \text{g}\ H_2SO_4}{1\ \text{mol}\ H_2SO_4} = 0.765\ \text{g}\ H_2SO_4$$

8. $125 \text{ g KClO}_3 \times \frac{1 \text{ mol KClO}_3}{122.55 \text{ g KClO}_3} \times \frac{3 \text{ mol O}_2}{2 \text{ mol KClO}_3} = 1.53 \text{ mol O}_2$

9. $C_2H_5OH + 3O_2 \longrightarrow 2CO_2 + 3H_2O$

$5.00 \text{ mol H}_2\text{O} \times \frac{1 \text{ mol C}_2\text{H}_5\text{OH}}{3 \text{ mol H}_2\text{O}} \times \frac{46.08 \text{ g C}_2\text{H}_5\text{OH}}{1 \text{ mol C}_2\text{H}_5\text{OH}} = 76.8 \text{ g C}_2\text{H}_5\text{OH}$

10. $10.0 \text{ mol H}_2\text{O}_2 \times \frac{1 \text{ mol O}_2}{2 \text{ mol H}_2\text{O}_2} \times \frac{32.00 \text{ g O}_2}{1 \text{ mol O}_2} \times \frac{1 \text{ L O}_2}{1.429 \text{ g O}_2} = 112 \text{ L O}_2$

11. $25.0 \text{ mol NH}_3 \times \frac{4 \text{ mol NO}}{4 \text{ mol NH}_3} \times \frac{30.01 \text{ g NO}}{1 \text{ mol NO}} \times \frac{1 \text{ L NO}}{1.340 \text{ g NO}} = 560. \text{ L NO}$

12. No. Zinc and hydrogen must be compared by moles, not mass.

13. a. The conversion factor is wrong because it compares reactants and products with mass, not moles. The final answer has a multiplication error.

b. The conversion factor is wrong because it does not cancel out grams. The reciprocal should be used.

c. Nothing is wrong; the conversion factors are correct.

14. A limiting reactant is used up before the excess reactant and thus limits a reaction.

15. The only reactions without limiting reactants are those that have precise proportions of different reactant amounts that match the mole ratios.

16. a. 1.5 mol $ZnCO_3$

b. Because more moles of $ZnCO_3$ are required for the reaction, $ZnCO_3$ is the limiting reactant, and $C_6H_8O_7$ is the excess reactant.

c. 60.0 mol $ZnCO_3$ and 40.0 mol $C_6H_8O_7$

d. $ZnCO_3$ is the limiting reactant because 6.0 mol of $ZnCO_3$ will react with only 4.0 mol of $C_6H_8O_7$, leaving the rest as excess.

11. One of the intermediate steps in the production of nitric acid is the reaction between ammonia and oxygen. If 25.0 mol of ammonia gas react with excess oxygen, how many liters of NO will be produced? Assume the density of NO is 1.340 g/L. (Hint: see Sample Problem 8C.)

$$4NH_3(g) + 5O_2(g) \longrightarrow 4NO(g) + 6H_2O(g)$$

APPLY

12. Hydrogen gas can be produced by adding an active metal, such as zinc, to hydrochloric acid. A student reports that for every kilogram of zinc reacted, 1.0 kg of hydrogen gas is evolved. Is this correct? Explain why or why not.

$$Zn(s) + 2HCl(aq) \longrightarrow ZnCl_2(aq) + H_2(g)$$

13. Explain what, if anything, is wrong with the setups shown below, which are for problems relating to the production of ammonium hydrogen phosphate, a common fertilizer.

$$H_3PO_4(aq) + 2NH_3(g) \longrightarrow (NH_4)_2HPO_4(aq)$$

a. $10.00 \text{ g NH}_3 \times \frac{1 \text{ g (NH}_4)_2\text{HPO}_4}{2 \text{ g NH}_3} = 20.00 \text{ g (NH}_4)_2\text{HPO}_4$

b. $10.00 \text{ g NH}_3 \times \frac{17.04 \text{ g NH}_3}{1 \text{ mol NH}_3} = 170.4 \text{ mol NH}_3$

c. $10.00 \text{ L NH}_3 \times \frac{0.761 \text{ g NH}_3}{1 \text{ L NH}_3} \times \frac{1 \text{ mol NH}_3}{17.04 \text{ g NH}_3} = 0.447 \text{ mol NH}_3$

Limiting reactants

REVIEW

14. Differentiate a limiting reactant from an excess reactant.

15. Do all reactions have a limiting reactant? Explain.

16. Answer the following questions about the production of zinc citrate, $Zn_3(C_6H_5O_7)_2$, an ingredient in toothpaste.

$$\underset{\text{zinc(II) carbonate}}{3ZnCO_3(s)} + \underset{\text{citric acid}}{2C_6H_8O_7(aq)} \longrightarrow \underset{\text{zinc(II) citrate}}{Zn_3(C_6H_5O_7)_2(aq)} + \underset{\text{water}}{3H_2O(l)} + \underset{\text{carbon dioxide}}{3CO_2(g)}$$

a. How many moles of $ZnCO_3$ are needed to react with 1 mol $C_6H_8O_7$?

b. If there is 1 mol $ZnCO_3$ and 1 mol $C_6H_8O_7$, which is the limiting reactant?

c. How many moles of $ZnCO_3$ and $C_6H_8O_7$ are required to produce 20.0 mol $Zn_3(C_6H_5O_7)_2$?

d. If there are 6.0 mol $ZnCO_3$ and 10.0 mol $C_6H_8O_7$, which is the limiting reactant?

PRACTICE

17 When copper metal is added to silver nitrate solution, silver metal and copper(II) nitrate are produced. If 100. g of copper metal are added to a solution containing 100. g of silver nitrate, how many grams of silver metal will be produced? (Hint: see Sample Problem 8D.)

18. A fruit-scented air freshener can be made by reacting butanoic acid with methanol to produce methyl butanoate and water. How many grams of methyl butanoate can be produced if 50.0 g of butanoic acid react with 40.0 g of methanol? (Hint: see Sample Problem 8D.)

Butanoic acid + $H_3C{-}OH$ (Methanol) $\longrightarrow$ Methyl butanoate + Water

APPLY

19. Identify the limiting reactant and the excess reactant in the following situations.

a. firewood burning in a campfire

b. stomach acid breaking down food

c. sulfur compounds from the air tarnishing silver

d. NO_2 reacting with water vapor to produce acid rain

20. A perfume manufacturer needs to produce large supplies of nitrobenzene for the creation of different fragrances. The nitrobenzene can be made by reacting benzene with nitric acid in the presence of a catalyst. If benzene costs \$1.75/mol and nitric acid costs \$0.57/mol, which should the perfume manufacturer choose to be the limiting reactant?

$$C_6H_6 + HNO_3 \longrightarrow C_6H_5NO_2 + H_2O$$

Percent yield

REVIEW

21. a. Differentiate theoretical yield from actual yield.
b. How is actual yield determined?
c. How is theoretical yield determined?

22. Why do many chemical reactions produce less than the amount of product predicted by stoichiometry?

PRACTICE

23 Magnesium is obtained from sea water. $Ca(OH)_2$ is added to sea water, precipitating $Mg(OH)_2$. This is filtered out, and reacted with HCl to produce $MgCl_2$. This is dried, fused, and electrolyzed, producing Mg and Cl_2 as shown in the equation below. If 185.0 g of magnesium are recovered from 1000. g of magnesium chloride, what is the percent yield for this reaction? (Hint: see Sample Problem 8E.)

$$MgCl_2(l) \longrightarrow Mg(s) + Cl_2(g)$$

24. The combustion of methane produces carbon dioxide and water. Assume that 2.00 mol CH_4 are burned in the presence of excess oxygen. What is the percent yield if the reaction produces 80.0 g CO_2? (Hint: see Sample Problem 8E.)

$$CH_4(g) + 2O_2(g) \longrightarrow CO_2(g) + 2H_2O(g)$$

25 Coal gasification is a process that converts coal into methane gas. If this reaction has a percent yield of 85.0%, how much methane can be obtained from 1250 g of carbon? (Hint: see Sample Problem 8F.)

$$2C(s) + 2H_2O(l) \longrightarrow CH_4(g) + CO_2(g)$$

26. If the percent yield for the coal gasification process in item 25 is increased to 95.0%, how much methane can be obtained from 2750 g of carbon? (Hint: see Sample Problem 8F.)

APPLY

27. A sandpaper company uses silicon carbide, SiC, to make its product. Reacting silicon dioxide with graphite yielded 30.0 kg of SiC. The theoretical yield is 998 mol. What is the percent yield?

28. a. Can actual yield ever exceed theoretical yield? Explain.
b. In the lab, you run an experiment that appears to have a percent yield of 115%. Propose possible reasons for this result.

Practical uses of stoichiometry

REVIEW

29. Use stoichiometry to explain the following problems that a lawn mower may have.
a. A lawn mower fails to start because the engine floods.
b. A lawn mower stalls after starting cold and idling.

30. Use stoichiometry to explain why a 4.00 kg firework would produce more light than a 2.00 kg firework containing the same proportion of reactants.

PRACTICE

31 Phosphate baking powder is a mixture of starch, sodium hydrogen carbonate, and calcium dihydrogen phosphate. When mixed with water, phosphate baking powder releases carbon dioxide gas, causing a dough or batter to bubble and rise.

$$2NaHCO_3(aq) + Ca(H_2PO_4)_2(aq) \longrightarrow 2Na^+(aq) + Ca^{2+}(aq) + 2HPO_4^{2-}(aq) + 2CO_2(g) + 2H_2O(l)$$

If 0.750 L of CO_2 is needed for a cake and each kilogram of baking powder contains 168 g of $NaHCO_3$, how many grams of baking powder must be used to generate this amount of CO_2? The density of CO_2 at baking temperature is about 1.25 g/L. (Hint: see Sample Problem 8G.)

17. $Cu(s) + 2AgNO_3(aq) \longrightarrow 2Ag(s) + Cu(NO_3)_2(aq)$

$100.\ g\ Cu \times \frac{1\ mol\ Cu}{63.55\ g\ Cu} =$

1.57 mol Cu present

$100.\ g\ AgNO_3 \times \frac{1\ mol\ AgNO_3}{169.88\ g\ AgNO_3} =$

0.589 mol $AgNO_3$ present

$1.57\ mol\ Cu \times \frac{2\ mol\ AgNO_3}{1\ mol\ Cu} =$

3.14 mol $AgNO_3$ needed

Because there is less $AgNO_3$ than is needed to react with the Cu, $AgNO_3$ is the limiting reactant.

$0.589\ mol\ AgNO_3 \times \frac{2\ mol\ Ag}{2\ mol\ AgNO_3} \times \frac{107.87\ g\ Ag}{1\ mol\ Ag} = 63.5\ g\ Ag$

18. $C_4H_8O_2 + CH_3OH \longrightarrow C_5H_{10}O_2 + H_2O$

$50.0\ g\ C_4H_8O_2 \times \frac{1\ mol\ C_4H_8O_2}{88.12\ g\ C_4H_8O_2} =$

0.567 mol $C_4H_8O_2$ present

$40.0\ g\ CH_3OH \times \frac{1\ mol\ CH_3OH}{32.05\ g\ CH_3OH} =$

1.25 mol CH_3OH present

$0.567\ mol\ C_4H_8O_2 \times \frac{1\ mol\ CH_3OH}{1\ mol\ C_4H_8O_2} =$

0.567 mol CH_3OH needed

Because more CH_3OH is present than is needed, it is in excess, and $C_4H_8O_2$ is the limiting reactant.

$0.567\ mol\ C_4H_8O_2 \times \frac{1\ mol\ C_5H_{10}O_2}{1\ mol\ C_4H_8O_2} \times \frac{102.15\ g\ C_5H_{10}O_2}{1\ mol\ C_5H_{10}O_2} = 57.9\ g\ C_5H_{10}O_2$

19. a. The wood is limiting, and oxygen in the air is in excess.
b. The food is limiting, and the acid is in excess.
c. The silver is limiting, and the sulfur compounds are in excess.
d. The NO_2 is limiting, and the water is in excess.

20. Equal mole amounts of the reactants are consumed in the reaction. Benzene is the limiting reactant because it is more expensive per mole.

21. a. Theoretical yield is the maximum calculated amount of product that could be produced, while actual yield is the measured amount actually produced.

b. Precise actual yields are mass measurements from a balance.

c. Theoretical yield is calculated from the balanced chemical equation for the reaction.

22. If the reverse reaction occurs, reactants are re-formed from the products. Other side reactions may also occur, forming other products.

23. $1000.\ \text{g MgCl}_2 \times \frac{1\ \text{mol MgCl}_2}{95.20\ \text{g MgCl}_2} \times \frac{1\ \text{mol Mg}}{1\ \text{mol MgCl}_2} \times \frac{24.30\ \text{g Mg}}{1\ \text{mol Mg}} = 255.3\ \text{g Mg}$

$\frac{185.0\ \text{g Mg actual}}{255.3\ \text{g Mg predicted}} \times 100 = 72.46\%\ \text{yield}$

24. $2.00\ \text{mol CH}_4 \times \frac{1\ \text{mol CO}_2}{1\ \text{mol CH}_4} \times \frac{44.01\ \text{g CO}_2}{1\ \text{mol CO}_2} = 88.0\ \text{g CO}_2$

$\frac{80.0\ \text{g CO}_2\ \text{actual}}{88.0\ \text{g CO}_2\ \text{predicted}} \times 100 = 90.9\%\ \text{yield}$

25. percent yield: 85.0%

$1250\ \text{g C} \times \frac{1\ \text{mol C}}{12.01\ \text{g C}} \times \frac{1\ \text{mol CH}_4}{2\ \text{mol C}} \times \frac{16.04\ \text{g CH}_4}{1\ \text{mol CH}_4} \times \frac{85.0\ \text{g CH}_4\ \text{actual}}{100.\ \text{g CH}_4\ \text{theoretical}} = 710.\ \text{g CH}_4$

26. percent yield: 95.0%

$2750\ \text{g C} \times \frac{1\ \text{mol C}}{12.01\ \text{g C}} \times \frac{1\ \text{mol CH}_4}{2\ \text{mol C}} \times \frac{16.04\ \text{g CH}_4}{1\ \text{mol CH}_4} \times \frac{95.0\ \text{g CH}_4\ \text{actual}}{100.\ \text{g CH}_4\ \text{theoretical}} = 1740\ \text{g CH}_4$

27. $998\ \text{mol SiC} \times \frac{40.10\ \text{g SiC}}{1\ \text{mol SiC}} \times \frac{1\ \text{kg}}{1000\ \text{g}} = 40.0\ \text{kg SiC}$

$\frac{30.0\ \text{kg actual}}{40.0\ \text{kg theoretical}} \times 100 = 75.0\%\ \text{yield}$

32. The addition of yeast can make bread rise because the yeast produces CO_2 from glucose, $C_6H_{12}O_6$, according to the equation below. Assume that 0.50 L of carbon dioxide is required for a loaf of bread. How many grams of $C_6H_{12}O_6$ must be broken down by yeast to produce this amount of CO_2? The density of CO_2 at baking temperature is about 1.25 g/L. (Hint: see Sample Problem 8G, and balance the equation.)

$$C_6H_{12}O_6(s) \longrightarrow C_2H_5OH(l) + CO_2(g)$$

33 Plaster of Paris, $CaSO_4 \cdot 1/2\,H_2O$, has many uses, including castings and dental cement. It can be obtained by heating gypsum, $CaSO_4 \cdot 2H_2O$. How many liters of water vapor evolve when 5.00 L of gypsum are heated at 110°C to produce plaster of Paris? At 110°C the density of $CaSO_4 \cdot 2H_2O$ is 2.32 g/mL, and the density of water vapor is 0.581 g/L. (Hint: see Sample Problem 8H.)

$$2CaSO_4 \cdot 2H_2O(s) \longrightarrow 2CaSO_4 \cdot \tfrac{1}{2}H_2O(s) + 3H_2O(g)$$

34. Builders and dentists must store plaster of Paris, $CaSO_4 \cdot 1/2\,H_2O$, in airtight containers to prevent it from absorbing water vapor and changing back into gypsum, $CaSO_4 \cdot 2H_2O$. If 7.50 kg of plaster of Paris absorbed excess water vapor, what volume of gypsum would form? The density of $CaSO_4 \cdot 2H_2O$ is 2.32 g/mL. (Hint: see Sample Problem 8H.)

$$2CaSO_4 \cdot \tfrac{1}{2}H_2O(s) + 3H_2O(g) \longrightarrow 2CaSO_4 \cdot 2H_2O(s)$$

35 A common additive to motor oil is the substance *p-tert*-butylphenol. It is produced by two reactions, starting with the reactants chlorobenzene and sodium hydroxide. How many moles of *p-tert*-butylphenol could be produced if 500. g of NaOH are available? (Hint: see Sample Problem 8I.)

Step 1: chlorobenzene (Cl on benzene ring) + NaOH ⟶ phenol (OH on benzene ring) + NaCl

Step 2: phenol (OH on benzene ring) + $H_3C{-}C(CH_3)(OH){-}CH_3$ ⟶ *p-tert*-butylphenol (OH and $C(CH_3)_3$ para on benzene ring) + H_2O

36. Gold can be recovered from sea water by reacting the water with an active metal such as zinc, which is refined from zinc oxide. The zinc displaces the gold in the water. What mass of gold can be recovered if 2.00 g of ZnO and an excess of sea water are available? (Hint: see Sample Problem 8I.)

$$2ZnO(s) + C(s) \longrightarrow 2Zn(s) + CO_2(g)$$

$$2Au^{3+}(aq) + 3Zn(s) \longrightarrow 3Zn^{2+}(aq) + 2Au(s)$$

APPLY

37. Explain the stoichiometry involved in blowing air on the base of a dwindling campfire to keep the coals burning.

38. Why would it be unreasonable for an amendment to the Clean Air Act to call for 0% pollution emissions from cars with combustion engines?

39. While working in a lab to synthesize acetaminophen, a common pain reliever, you discover that the amount of reactants planned for the reaction was based on the assumption that the entire theoretical yield would be produced. What should you do to be certain that as much acetaminophen as was expected is actually made? Explain.

Linking chapters

1. *Recognizing reaction types*
Determine the conversion factors needed for the following problems. Be sure to write a balanced equation first.

a. How many grams of oxygen gas are evolved from the decomposition of a known amount of water in moles?

b. How many moles of hydrochloric acid are needed to completely react with a known mass of zinc in a single-displacement reaction?

c. How many moles of calcium carbonate are produced in a double-displacement reaction between a known mass of calcium nitrate and potassium carbonate?

2. Theme: *Equilibrium and change*
In some synthesis reactions, part of the product decomposes into reactants. How is the reversal similar to the dissolving and recrystallizing of sugar in iced tea that illustrated equilibrium and change in Chapter 1?

USING TECHNOLOGY

1. ***Graphics calculator***
The ratios used in stoichiometry can be expressed as the slope of a straight line graph. For example, the complete combustion of ethanol is represented below.

$$C_2H_5OH(g) + 3O_2(g) \longrightarrow 2CO_2(g) + 3H_2O(g)$$

$\Delta H = -1368$ kJ

For every mole of ethanol, 1368 kJ of energy are released. The mole ratio is 1368 kJ:1 mol C_2H_5OH. By graphing a function with this slope, you can predict how much energy can be obtained from any amount of ethanol. Select the **WINDOW** key on your calculator, and set Xmin to 0, Xmax to 20, and Xscl to 2. Set Ymin to 0, Ymax to 30 000, and Yscl to 3000. Select the **Y=** key and enter the function $Y_1 = (1368 \div 1)$ **X,T,Θ**. This represents the equation $y = 1368x$, with y equal to energy and x equal to number of moles ethanol. Press **GRAPH** and use the **TRACE** mode to estimate the energy released during the combustion of the following amounts of ethanol.
a. 5.5 mol **b.** 18 mol **c.** 0.75 mol

2. ***Computer spreadsheet***
Percent yield will allow you to predict what the actual yield of an experiment will be. Create a spreadsheet that will calculate the actual yield from any percent yield and theoretical yield.

Alternative assessment

Performance assessment

1. Design an experiment to measure the percent yields for the reactions listed below. If your teacher approves your design, acquire the necessary materials, and carry out your plan to obtain percent yield data.

a. $Zn(s) + 2HCl(aq) \longrightarrow ZnCl_2(aq) + H_2(g)$

b. $2NaHCO_3(s) \xrightarrow{\Delta} Na_2CO_3(s) + H_2O(g) + CO_2(g)$

c. $CaCl_2(aq) + Na_2CO_3(aq) \longrightarrow CaCO_3(s) + 2NaCl(aq)$

d. $NaOH(aq) + HCl(aq) \longrightarrow NaCl(aq) + H_2O(l)$

(Note: use only dilute NaOH and HCl, less concentrated than 1.0 mol/liter.)

2. Your teacher will give you an index card specifying a volume of a gas. Reactants to make the gas will also be listed. Describe exactly how you would make the gas from the reactants. Include a method of collecting the gas without allowing it to mix with air. Then specify how much of each reactant you need. Choose a limiting reactant and explain your choice. If your teacher approves your plan, obtain the necessary materials and make the gas. (Hint: look up the density of the gas in a chemical handbook.)

Portfolio projects

1. ***Research and communication***
Research the composition of gasoline sold in your area. Contact a gasoline company to discover what formulations are used. Investigate whether the mixtures change by season or by geographic area. Find out if your area has any guidelines regarding gasoline additives that reduce air pollution. Present your findings to the class.

2. ***Cooperative activity***
Investigate corporate, governmental, or private use of alternative fuel sources for vehicles. Hold a class debate to compare the costs and environmental effects of these alternative fuels.

3. ***Chemistry and you***
Visit a car maintenance shop to find out how you can help reduce air pollution by increasing your car's efficiency. Make a checklist of tasks to perform regularly to help meet this goal.

4. ***Research and communication***
Research the production of the following pollutants: methane, CH_4; mercury, Hg; lead, Pb; chlorine, Cl_2; and sulfur oxides, SO_x. Determine the stoichiometry involved in the production of each of these chemicals. Contact the EPA, and gather information on what you can do to help reduce pollution by these chemicals.

28. a. No. The theoretical yield is the maximum amount that can be obtained. Thus, actual yield will always be less.
b. Answers will vary. For example, there is a possibility of human or instrument error. Also, the product may be contaminated, thus appearing to have more mass.

29. a. The ratio of gasoline to air is too high. The combustion reaction will not take place.
b. The ratio of air to gasoline is too high. The combustion reaction will not take place.

30. Twice as much material will produce twice as much energy in the form of light.

31. $0.750\text{ L } CO_2 \times \frac{1.25\text{ g } CO_2}{1\text{ L } CO_2} \times \frac{1\text{ mol } CO_2}{44.01\text{ g } CO_2} \times \frac{2\text{ mol } NaHCO_3}{2\text{ mol } CO_2} \times \frac{84.01\text{ g } NaHCO_3}{1\text{ mol } NaHCO_3} \times \frac{1000\text{ g baking powder}}{168\text{ g } NaHCO_3} =$ 10.7 g baking powder

32. $C_6H_{12}O_6(s) \longrightarrow 2C_2H_5OH(l) + 2CO_2(g)$

$0.50\text{ L } CO_2 \times \frac{1.25\text{ g } CO_2}{1\text{ L } CO_2} \times \frac{1\text{ mol } CO_2}{44.01\text{ g } CO_2} \times \frac{1\text{ mol } C_6H_{12}O_6}{2\text{ mol } CO_2} \times \frac{180.18\text{ g } C_6H_{12}O_6}{1\text{ mol } C_6H_{12}O_6} = 1.3\text{ g } C_6H_{12}O_6$

33. $5.00\text{ L } CaSO_4{\cdot}2H_2O \times \frac{1000\text{ mL}}{1\text{ L}} \times \frac{2.32\text{ g } CaSO_4{\cdot}2H_2O}{1\text{ mL } CaSO_4{\cdot}2H_2O} \times \frac{1\text{ mol } CaSO_4{\cdot}2H_2O}{172.19\text{ g } CaSO_4{\cdot}2H_2O} \times \frac{3\text{ mol } H_2O}{2\text{ mol } CaSO_4{\cdot}2H_2O} \times \frac{18.02\text{ g } H_2O}{1\text{ mol } H_2O} \times \frac{1\text{ L } H_2O}{0.581\text{ g } H_2O} =$ 3130 L H_2O

34. $7.50\text{ kg } CaSO_4{\cdot}\frac{1}{2}H_2O \times \frac{1000\text{ g}}{1\text{ kg}} \times \frac{1\text{ mol } CaSO_4{\cdot}\frac{1}{2}H_2O}{145.16\text{ g } CaSO_4{\cdot}\frac{1}{2}H_2O} \times \frac{2\text{ mol } CaSO_4{\cdot}2H_2O}{2\text{ mol } CaSO_4{\cdot}\frac{1}{2}H_2O} \times \frac{172.19\text{ g } CaSO_4{\cdot}2H_2O}{1\text{ mol } CaSO_4{\cdot}2H_2O} \times \frac{1\text{ mL } CaSO_4{\cdot}2H_2O}{2.32\text{ g } CaSO_4{\cdot}2H_2O} \times \frac{1\text{ L}}{1000\text{ mL}} =$ 3.83 L $CaSO_4{\cdot}4H_2O$

Answers from page 279

a. $49.7\ g\ MgCl_2 \times \frac{1\ mol\ MgCl_2}{95.20\ g\ MgCl_2} \times \frac{1\ mol\ Mg(ClO_3)_2}{1\ mol\ MgCl_2} \times$

$\frac{191.20\ g\ Mg(ClO_3)_2}{1\ mol\ Mg(ClO_3)_2} = 99.8\ g\ Mg(ClO_3)_2$

$49.7\ g\ MgCl_2 \times \frac{1\ mol\ MgCl_2}{95.20\ g\ MgCl_2} \times \frac{3\ mol\ O_2}{1\ mol\ MgCl_2} \times$

$\frac{32.00\ g\ of\ O_2}{1\ mol\ O_2} = 50.1\ g\ of\ O_2$

b. $52.6\ g\ Cl_2 \times \frac{1\ mol\ Cl_2}{70.90\ g\ Cl_2} \times \frac{2\ mol\ NaI}{1\ mol\ Cl_2} \times \frac{149.90\ g\ NaI}{1\ mol\ NaI} = 222\ g\ NaI$

$52.6\ g\ Cl_2 \times \frac{1\ mol\ Cl_2}{70.90\ g\ Cl_2} \times \frac{2\ mol\ NaCl}{1\ mol\ Cl_2} \times \frac{58.45\ g\ NaCl}{1\ mol\ NaCl} = 86.7\ g\ NaCl$

$52.6\ g\ Cl_2 \times \frac{1\ mol\ Cl_2}{70.90\ g\ Cl_2} \times \frac{1\ mol\ I_2}{1\ mol\ Cl_2} \times \frac{253.80\ g\ I_2}{1\ mol\ I_2} = 188\ g\ of\ I_2$

c. $288.6\ g\ Ca(NO_3)_2 \times \frac{1\ mol\ Ca(NO_3)_2}{164.10\ g\ Ca(NO_3)_2} \times \frac{1\ mol\ K_2SO_4}{1\ mol\ Ca(NO_3)_2} \times$

$\frac{174.27\ g\ K_2SO_4}{1\ mol\ K_2SO_4} = 306.5\ g\ K_2SO_4$

$288.6\ g\ Ca(NO_3)_2 \times \frac{1\ mol\ Ca(NO_3)_2}{164.10\ g\ Ca(NO_3)_2} \times \frac{1\ mol\ CaSO_4}{1\ mol\ Ca(NO_3)_2} \times$

$\frac{136.15\ g\ CaSO_4}{1\ mol\ CaSO_4} = 239.4\ g\ CaSO_4$

$288.6\ g\ Ca(NO_3)_2 \times \frac{1\ mol\ Ca(NO_3)_2}{164.10\ g\ Ca(NO_3)_2} \times \frac{2\ mol\ KNO_3}{1\ mol\ Ca(NO_3)_2} \times$

$\frac{101.11\ g\ KNO_3}{1\ mol\ KNO_3} = 355.6\ g\ KNO_3$

Answers from page 281

2. $500.\ mL \times 1.58\ g/mL = 790.\ g\ C_6H_4(NO_2)COOH$

$790\ g\ C_6H_4(NO_2)COOH \times \frac{1\ mol\ C_6H_4(NO_2)COOH}{167.13\ g\ C_6H_4(NO_2)COOH} \times$

$\frac{2\ mol\ NH_2C_6H_4CO_2H}{2\ mol\ C_6H_4(NO_2)COOH} \times \frac{137.15\ g\ NH_2C_6H_4CO_2H}{1\ mol\ NH_2C_6H_4CO_2H} = 648\ g\ NH_2C_6H_4CO_2H$

Answers from page 283

3. Check for adherence to the three-step method. Do not expect students to solve all problems using a single setup. Novices are generally more successful when they break a complex problem into multiple steps. Setups and answers only are given here.

a. $1200.\ kg = 1.200 \times 10^6\ g\ N_2H_4$

$1.200 \times 10^6\ g\ N_2H_4 \times \frac{1\ mol\ N_2H_4}{32.06\ g\ N_2H_4} \times \frac{3\ mol\ N_2O_4}{2\ mol\ N_2H_4} \times \frac{92.02\ g\ N_2O_4}{1\ mol\ N_2O_4} =$

$5.166 \times 10^6\ g\ N_2O_4$

b. $1.200 \times 10^6\ g\ N_2H_4 \times \frac{1\ mol\ N_2H_4}{32.06\ g\ N_2H_4} \times \frac{1\ mol\ (CH_3)_2N_2H_2}{2\ mol\ N_2H_4} \times$

$\frac{60.12\ g\ (CH_3)_2N_2H_2}{1\ mol\ (CH_3)_2N_2H_2} = 1.13 \times 10^6\ g\ (CH_3)_2N_2H_2$

c. $1.200 \times 10^6\ g\ N_2H_4 \times \frac{1\ mol\ N_2H_4}{32.06\ g\ N_2H_4} \times \frac{6\ mol\ N_2}{2\ mol\ N_2H_4} \times$

$\frac{6.022 \times 10^{23}\ molecules}{1\ mol\ N_2} = 6.762 \times 10^{28}\ molecules\ N_2$

d. $1.200 \times 10^6\ g\ N_2H_4 \times \frac{1\ mol\ N_2H_4}{32.06\ g\ N_2H_4} \times \frac{8\ mol\ H_2O}{2\ mol\ N_2H_4} \times \frac{18.02\ g\ H_2O}{1\ mol\ H_2O} =$

$2.698 \times 10^6\ g\ H_2O$

$\frac{2.698 \times 10^6\ g\ H_2O}{1.00\ g/mL} = 2.698 \times 10^6\ mL\ H_2O$

$2.698 \times 10^6\ mL\ H_2O \times \frac{1\ L}{1000\ mL} = 2.698 \times 10^3\ L\ H_2O$

4. $2HgO \longrightarrow 2Hg + O_2$

$216.59\ g\ HgO \times \frac{1\ mol\ HgO}{216.59\ g\ HgO} \times \frac{1\ mol\ O_2}{2\ mol\ HgO} = 0.500\ 00\ mol\ O_2$

5. $251.7\ g\ C_7H_6O_3 \times \frac{1\ mol\ C_7H_6O_3}{138.13\ g\ C_7H_6O_3} \times \frac{1\ mol\ C_9H_8O_4}{1\ mol\ C_7H_6O_3} \times \frac{180.17\ g\ C_9H_8O_4}{1\ mol\ C_9H_8O_4} =$

$328.31\ g\ C_9H_8O_4$

6. $Mg + 2H_2O \longrightarrow Mg(OH)_2 + H_2$

$8.1\ g\ Mg \times \frac{1\ mol\ Mg}{24.30\ g\ Mg} \times \frac{1\ mol\ H_2}{1\ mol\ Mg} \times \frac{2.02\ g\ H_2}{1\ mol\ H_2} = 0.67\ g\ H_2$

7. $\frac{0.67\ g\ H_2}{0.081\ g/L} = 8.3\ L\ H_2$

8. 353 kJ per 1 mol Mg + 2 mol H_2O = 353 kJ per 60.34 g reactants

$\frac{353\ kJ}{60.34\ g\ reactants} = 5.85\ kJ/g$

64.4 kJ per 1 mol CaO + 1 mol H_2O = 64.4 kJ per 74.10 g reactants

$\frac{64.4\ kJ}{74.10\ g\ reactants} = 0.870\ kJ/g\ reactants$

30 g $C_3H_6O_3$ = 500 kJ

$\frac{500\ kJ}{30\ g\ C_3H_6O_3} = 17\ kJ/g$

The trioxane reaction provides more energy per gram.

9. $\frac{575\ kJ}{64.4\ kJ/mol\ reactants} = 8.93\ mol\ reactants$

$8.93\ mol\ CaO \times \frac{56.08\ g\ CaO}{1\ mol\ CaO} = 501\ g\ CaO$

$8.93\ mol\ H_2O \times \frac{18.02\ g\ H_2O}{1\ mol\ H_2O} = 161\ g\ H_2O$

Answers from page 286

Solutions for Additional Examples 8D

a. $130.5\ g\ FeS \times \frac{1\ mol\ FeS}{87.92\ g\ FeS} \times \frac{1\ mol\ H_2S}{1\ mol\ FeS} \times \frac{6.023 \times 10^{23}\ molecules\ H_2S}{1\ mol\ H_2S} =$

$8.940 \times 10^{23}\ molecules\ H_2S$

b. $180\ 000\ g\ Fe_2O_3 \times \frac{1\ mol\ Fe_2O_3}{159.70\ g\ Fe_2O_3} = 1127\ mol\ Fe_2O_3$

$110\ 000\ g\ CO \times \frac{1\ mol\ CO}{28.01\ g\ CO} = 3927\ mol\ CO$

$1127\ mol\ Fe_2O_3 \times \frac{3\ mol\ CO}{1\ mol\ Fe_2O_3} = 3381\ mol\ CO$

CO is in excess, so Fe_2O_3 is the limiting agent.

$1127\ mol\ Fe_2O_3 \times \frac{2\ mol\ Fe}{1\ mol\ Fe_2O_3} \times \frac{55.85\ g\ Fe}{1\ mol\ Fe} \times \frac{1\ kg}{1000\ g} = 125.9\ kg\ Fe$

Answers to Practice 8D

1. $Al(OH)_3 + 3HCl \longrightarrow AlCl_3 + 3H_2O$

$$34.0 \text{ g HCl} \times \frac{1 \text{ mol HCl}}{36.46 \text{ g HCl}} = 0.933 \text{ mol HCl}$$

$$12.0 \text{ g Al(OH)}_3 \times \frac{1 \text{ mol Al(OH)}_3}{78.01 \text{ g Al(OH)}_3} = 0.154 \text{ mol Al(OH)}_3$$

$$0.933 \text{ mol HCl} \times \frac{1 \text{ mol Al(OH)}_3}{3 \text{ mol HCl}} = 0.311 \text{ mol Al(OH)}_3 \text{ needed}$$

0.154 mol $Al(OH)_3$ is less than 0.311 mol $Al(OH)_3$; there is not enough $Al(OH)_3$

2. $N_2 + 3H_2 \longrightarrow 2NH_3$

$$92.7 \text{ kg } N_2 \times \frac{1000 \text{ g}}{1 \text{ kg}} = 9.27 \times 10^4 \text{ g } N_2$$

$$9.27 \times 10^4 \text{ g } N_2 \times \frac{1 \text{ mol } N_2}{28.02 \text{ g } N_2} = 3.31 \times 10^3 \text{ mol } N_2$$

$$265.8 \text{ kg } H_2 \times \frac{1000 \text{ g}}{1 \text{ kg}} = 2.66 \times 10^5 \text{ g } H_2$$

$$2.66 \times 10^5 \text{ g } H_2 \times \frac{1 \text{ mol } H_2}{2.02 \text{ g } H_2} = 1.32 \times 10^5 \text{ mol } H_2$$

$$3.31 \times 10^3 \text{ mol } N_2 \times \frac{3 \text{ mol } H_2}{1 \text{ mol } N_2} = 9.93 \times 10^3 \text{ mol } H_2$$

9.93×10^3 mol H_2 is less than 1.32×10^5 mol H_2; therefore N_2 is the limiting reactant.

Answers from page 293

10. a. $$4.5 \text{ mol TiCl}_4 \times \frac{1 \text{ mol } O_2}{1 \text{ mol TiCl}_4} = 4.5 \text{ mol } O_2$$

$TiCl_4$ is in excess; only 3.5 mol $TiCl_4$ will be used. O_2 is the limiting reactant.

b. 4.5 mol $TiCl_4$ − 3.5 mol $TiCl_4$ = 1.0 mol excess $TiCl_4$

c. 3.5 mol TiO_2 formed and 2×3.5 mol, or 7.0 mol, Cl_2 formed

11. $$100. \text{ g } H_2O \times \frac{1 \text{ mol } H_2O}{18.02 \text{ g } H_2O} \times \frac{2 \text{ mol HNO}_3}{1 \text{ mol } H_2O} \times \frac{63.02 \text{ g HNO}_3}{1 \text{ mol HNO}_3} =$$

700 g HNO_3

12. a. $P_4O_{10}(g) + 6H_2O(l) \longrightarrow 4H_3PO_4(aq)$

b. $$100. \text{ g } P_4O_{10} \times \frac{1 \text{ mol } P_4O_{10}}{283.88 \text{ g } P_4O_{10}} \times \frac{6 \text{ mol } H_2O}{1 \text{ mol } P_4O_{10}} = 2.11 \text{ mol } H_2O$$

$$200. \text{ g } H_2O \times \frac{1 \text{ mol } H_2O}{18.02 \text{ g } H_2O} = 11.1 \text{ mol } H_2O$$

P_4O_{10} is the limiting reactant.

$$100. \text{ g } P_4O_{10} \times \frac{1 \text{ mol } P_4O_{10}}{283.88 \text{ g } P_4O_{10}} \times \frac{4 \text{ mol } H_3PO_4}{1 \text{ mol } P_4O_{10}} \times \frac{98.00 \text{ g } H_3PO_4}{1 \text{ mol } H_3PO_4} =$$

138 g H_3PO_4

c. $$\frac{126.24 \text{ g}}{138 \text{ g}} \times 100 = 91.5\%$$

13. $$50.0 \text{ g} C_6H_5CHO \times \frac{1 \text{ mol } C_6H_5CHO}{106.13 \text{ g } C_6H_5CHO} \times \frac{1 \text{ mol } C_9H_8O}{1 \text{ mol } C_6H_5CHO} \times$$

$$\frac{132.17 \text{ g } C_9H_8O}{1 \text{ mol } C_9H_8O} = 62.3 \text{ g } C_9H_8O \text{ theoretical yield}$$

$$\frac{x \text{ g } C_9H_8O}{62.3 \text{ g } C_9H_8O} \times 100 = 84.5\%$$

x = 52.6 g C_9H_8O actual yield

14. Reactions reach an equilibrium state, or side reactions occur that use up reactants and reduce the actual yield for a reaction.

15. $$8.1 \text{ g Mg} \times \frac{1 \text{ mol Mg}}{24.30 \text{ g Mg}} \times \frac{2 \text{ mol } H_2O}{1 \text{ mol Mg}} \times \frac{18.02 \text{ g } H_2O}{1 \text{ mol } H_2O} =$$

12. g H_2O, which equals 12. mL H_2O

16. Mg is the limiting reactant.

Answers from page 296

4. D of CO_2 = D of $N_2 \times 1.57$

0.916 g/L × 1.57 = 1.44 g/L

$$65.1 \text{ L } CO_2 \times \frac{1.44 \text{ g } CO_2}{1 \text{ L } CO_2} \times \frac{1 \text{ mol } CO_2}{44.01 \text{ g } CO_2} \times \frac{1 \text{ mol NaHCO}_3}{1 \text{ mol } CO_2} =$$

2.13 mol $NaHCO_3$

mol $NaHCO_3$ = mol CH_3COOH

$$2.13 \text{ mol NaHCO}_3 \times \frac{84.01 \text{ g NaHCO}_3}{1 \text{ mol NaHCO}_3} = 179. \text{ g NaHCO}_3$$

$$2.13 \text{ mol } CH_3COOH \times \frac{60.06 \text{ g } CH_3COOH}{1 \text{ mol } CH_3COOH} = 128. \text{ g } CH_3COOH$$

5. All mole ratios are 1 to 1.

2.13 mol $NaHCO_3$ require 2.13 mol H_2O and 2.13 mol CH_3COONa.

$$2.13 \text{ mol } H_2O \times \frac{18.02 \text{ g } H_2O}{1 \text{ mol } H_2O} = 38.4 \text{ g } H_2O$$

$$2.13 \text{ mol } CH_3COONa \times \frac{82.04 \text{ g } CH_3COONa}{1 \text{ mol } CH_3COONa} = 175. \text{ g } CH_3COONa$$

Answers from page 299

2. a. Note that this is a mixture and that two equations are needed.

$2C_8H_{18}(g) + 25O_2(g) \longrightarrow 16CO_2(g) + 18H_2O(g)$

$2C_7H_{14}(g) + 21O_2(g) \longrightarrow 14CO_2(g) + 14H_2O(g)$

b. $$1.000 \text{ L} \times \frac{1000 \text{ mL}}{1 \text{ L}} \times \frac{0.6908 \text{ g}}{1 \text{ mL}} = 690.8 \text{ g mixture}$$

690.8 g represents 9 parts C_8H_{18} and 1 part C_7H_{14}

690.8 g = 621.7 g C_8H_{18} and 69.1 g C_7H_{14}

$$621.7 \text{ g } C_8H_{18} \times \frac{1 \text{ mol } C_8H_{18}}{114.26 \text{ g } C_8H_{18}} \times \frac{25 \text{ mol } O_2}{2 \text{ mol } C_8H_{18}} \times \frac{32.00 \text{ g } O_2}{1 \text{ mol } O_2} \times$$

$$\frac{1 \text{ L } O_2}{1.331 \text{ g } O_2} \times \frac{100 \text{ L air}}{20.9 \text{ L } O_2} = 7824 \text{ L of air}$$

$$69.1 \text{ g } C_7H_{14} \times \frac{1 \text{ mol } C_7H_{14}}{98.21 \text{ g } C_7H_{14}} \times \frac{21 \text{ mol } O_2}{2 \text{ mol } C_7H_{14}} \times \frac{32.00 \text{ g } O_2}{1 \text{ mol } O_2} \times$$

$$\frac{1 \text{ L } O_2}{1.331 \text{ g } O_2} \times \frac{100 \text{ L air}}{20.9 \text{ L } O_2} = 850. \text{ L of air}$$

total air = 7824 L + 850.0 L = 8674 L

Answers from page 302

17. $2NaN_3(s) + 2Na(s) \longrightarrow 3N_2(g)$

$$22.4 \text{ g } NaN_3 \times \frac{1 \text{ mol } NaN_3}{65.02 \text{ g } NaN_3} \times \frac{3 \text{ mol } N_2}{2 \text{ mol } NaN_3} \times \frac{28.02 \text{ g } N_2}{1 \text{ mol } N_2} = 14.5 \text{ g } N_2$$

$$\frac{14.5 \text{ g } N_2}{0.916 \text{ g/L}} = 15.8 \text{ L } N_2$$

18. $2C_8H_{18}(g) + 25O_2(g) \longrightarrow 16CO_2(g) + 18H_2O(g)$

$$528.7 \text{ g } C_8H_{18} \times \frac{1 \text{ mol } C_8H_{18}}{114.26 \text{ g } C_8H_{18}} \times \frac{25 \text{ mol } O_2}{2 \text{ mol } C_8H_{18}} \times \frac{32.00 \text{ g } O_2}{1 \text{ mol } O_2} \times$$

$$\frac{1 \text{ L } O_2}{1.331 \text{ g } O_2} \times \frac{100 \text{ L air}}{20.9 \text{ L } O_2} = 1851 \text{ g } O_2$$

19. $O_2(g) + O(g) \longrightarrow O_3(g)$

$$18.3 \text{ g O} \times \frac{1 \text{ mol O}}{16.00 \text{ g O}} \times \frac{1 \text{ mol } O_2}{1 \text{ mol O}} \times \frac{32.00 \text{ g } O_2}{1 \text{ mol } O_2} = 36.6 \text{ g } O_2$$

20. $18.3 \text{ g O} \times \frac{1 \text{ mol O}}{16.00 \text{ g O}} \times \frac{1 \text{ mol } O_3}{1 \text{ mol O}} \times \frac{48.00 \text{ g } O_3}{1 \text{ mol } O_3} = 54.9 \text{ g } O_3$

21. $2C_8H_{18}(g) + 17O_2(g) \longrightarrow 16CO(g) + 18H_2O(g)$

$$74.0 \text{ g } C_8H_{18} \times \frac{1 \text{ mol } C_8H_{18}}{114.26 \text{ g } C_8H_{18}} \times \frac{16 \text{ mol CO}}{2 \text{ mol } C_8H_{18}} \times \frac{28.01 \text{ g CO}}{1 \text{ mol CO}} = 145. \text{ g CO}$$

22. From **Table 8-4**, the CO level is 2.1 g/km, which exceeds the 145 g/km from item **21**.

Continued from page 309

35. $500. \text{ g NaOH} \times \frac{1 \text{ mol NaOH}}{40.00 \text{ g NaOH}} \times \frac{1 \text{ mol phenol}}{1 \text{ mol NaOH}} \times$

$$\frac{1 \text{ mol } p\text{–}tert\text{butylphenol}}{1 \text{ mol phenol}} = 12.5 \text{ mol } p\text{–}tert\text{butylphenol}$$

36. $2.00 \text{ g ZnO} \times \frac{1 \text{ mol ZnO}}{81.39 \text{ g ZnO}} \times \frac{2 \text{ mol Zn}}{2 \text{ mol ZnO}} \times \frac{2 \text{ mol Au}}{3 \text{ mol Zn}} \times \frac{196.97 \text{ g Au}}{1 \text{ mol Au}} =$

3.23 g Au

37. Near the center of the fire, oxygen is the limiting reactant. The added oxygen from your breath reacts with the wood to produce more heat.

38. To call for 0% pollution would require 100% yield for the combustion reaction of gasoline. This does not take into account actual yield or the stoichiometry of nitrogen from the air reacting in the hot engine.

39. You should increase the amount of reactants. Otherwise, less than the expected amount of acetaminophen will be produced. The reactant amounts shoud be multiplied by $\frac{100}{\text{percent yield}}$.

Linking chapters

1. a. $2H_2O(l) \longrightarrow 2H_2(g) + O_2(g)$

$$\text{mol } H_2O \times \frac{1 \text{ mol } O_2}{2 \text{ mol } H_2O} \times \frac{32.00 \text{ g } O_2}{1 \text{ mol } O_2} = \text{g } O_2$$

b. $Zn(s) + 2HCl(aq) \longrightarrow ZnCl_2(aq) + H_2(g)$

$$\text{g Zn} \times \frac{1 \text{ mol Zn}}{65.39 \text{ g Zn}} \times \frac{2 \text{ mol HCl}}{1 \text{ mol Zn}} = \text{mol HCl}$$

c. $Ca(NO_3)_2(aq) + K_2CO_3(aq) \longrightarrow CaCO_3(s) + 2KNO_3(aq)$

$$\text{g } Ca(NO_3)_2 \times \frac{1 \text{ mol } Ca(NO_3)_2}{102.09 \text{ g } Ca(NO_3)_2} \times \frac{1 \text{ mol } CaCO_3}{1 \text{ mol } Ca(NO_3)_2} = \text{mol } CaCO_3$$

2. Both the synthesis reaction and the dissolution of sugar-involve reversible processes. At equilibrium, the forward reaction occurs at the same rate as the reverse reaction, so no net change is visible.

USING TECHNOLOGY

1. a. about 7.5×10^3 kJ
b. about 2.5×10^4 kJ
c. about 1.0×10^3 kJ

2. Answers will vary. Spreadsheets should have adjustable percent yield values and theoretical yield values.

Performance assessment

1. Answers will vary. They must include theoretical yield calculations and safe laboratory procedures. Be sure that students use small amounts of reagents and diluted acid solutions to minimize hazards. Note: all products should be neutralized to a pH of 7 and washed down the drain with a 20-fold excess of water.

2. Answers will vary. Be sure that the students specify a proper means of collecting the gas. Also, experiments must include safe laboratory procedures. Recommended choices are O_2 from 3% H_2O_2 and KI catalyst or CO_2 from $NaHCO_3$ and CH_3COOH.

Portfolio projects

1. Answers will vary according to gasoline distributors in the area. Most companies alter composition based on the weather.

2. Answers will vary. Be sure to have students study the production of these fuels as well as the output.

3. Answers will vary. Some suggestions might be to use premium gasoline, regularly check the engine, and frequently change oil and air filters.

4. Answers will vary. Be sure that the stoichiometry accounts for the pollutant.

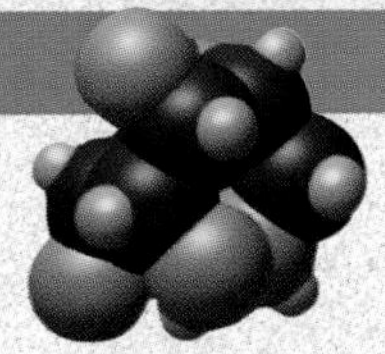

PLANNING FOR

Causes of Change

Chapter Overview

- **Section 9-1** discusses the role of heat in phase changes, specific heat capacity, and calorimetry, and distinguishes between heat and temperature.
- **Section 9-2** presents heat as a driving force behind change, and introduces Hess's law as a means for predicting how endothermic or exothermic a reaction will be.
- **Section 9-3** shows how the degree of randomness within a system affects change.
- **Section 9-4** describes the interrelationship of enthalpy and entropy in producing change, and introduces free energy as a measure of reaction spontaneity.
- **Section 9-5** shows how to determine caloric intake, how to estimate your energy expenditures for a specified caloric intake, and how to read a food label.

Concept Base

Students must have mastered the following concepts prior to this chapter: heat, temperature, kinetic energy, exothermic, and endothermic change (Chapter 2); mole (Chapter 3); stoichiometry (Chapter 8).

Looking Ahead

How activation energy affects reaction rate, and how a catalyst affects the ΔH of a reaction will be discussed in Chapter 14. Production of electric energy, a form of free energy, from chemical reactions will be discussed in Chapter 15.

Laboratory Equipment Needs

Demonstrations

The listing of materials for Chapter 9 demonstrations begins on page T46.

Laboratory Experiments

Materials and preparation instructions for Chapter 9 are found as follows.

Text		*Page*
Exploration 9A	Calorimetry and Hess's Law	718
Investigation 9A	Calorimetry—Biological Incubator	724
Exploration 9B	Energy Requirements and Food Testing	726
Investigation 9B	Food Testing—Baby-Food Label Fraud	730

Laboratory Experiments Manual
(refer to the table of contents in the manual)

Exploration 9-1	Specific Heat Capacity
Investigation 9-1	Specific Heat Capacity—Terrorist Investigation
Exploration 9-2	Constructing a Heating/Cooling Curve
Investigation 9-2	Constructing a Cooling Curve—Melting Oils
Exploration 9-3	Heat of Fusion
Exploration 9-4	Heat of Solution

Pacing: 15 days

Inquiry Teaching Strategies

9-1 How does energy affect change?

Demonstrations
- ■ Demo 1 *A hand warmer*
- ■ Demo 2 *Low melting metals*
- ■ Demo 3 *Specific heats of metals*
- ■ Demo 4 *Boiling water*

Laboratory Experiments
- ■ Exp 9A *Calorimetry, Hess's Law*
- ■ Inv 9A *Calorimetry—Biological Incubator*
- ■ ■ Exp 9-1 *Specific Heat Capacity*
- ■ ■ Inv 9-1 *Specific Heat Capacity—Terrorist Investigation*
- ■ ■ Exp 9-2 *Constructing Heat-*
- ■ ■ Inv 9-2 *ing/Cooling Curves*

9-2 How does enthalpy drive changes?

Demonstrations
- ■ Demo 5 *Spontaneous changes*
- ■ Demo 6 *Energy of activation*
- ■ Demo 7 *Activation energy*
- ■ Demo 8 *Hess's law*

Laboratory Experiments
- ■ ■ Exp 9-3 *Heat of Fusion*
- ■ ■ Exp 9-4 *Heat of Solution*

9-3 How does entropy drive changes?

Demonstrations
- ■ Demo 9 *Entropy change for ice*

9-4 How do enthalpy and entropy interact?

Demonstrations
- ■ Demo 10 *Production of copper sulfide*

9-5 What are your energy needs?

Laboratory Experiments
- ■ Exp 9B *Energy Requirements and Food Testing*
- ■ Inv 9B *Food Testing—Baby-Food Label Fraud*

Key
- ■ Teaching Transparencies
- ■ Annotated Teacher's Edition
- ■ Laboratory Experiments Manual

INSTRUCTION

Visual Teaching Strategies	Review and Practice Strategies
Transparencies ■ F 9-5 *Calculating Energy Changes for a Heating Curve* **Transparency Masters** ■ F 9-2 *Temperature vs. Time Data from Ice Cube Experiment* ■ T 9-6 *Transfers of Heat for Different Types of Changes*	**Text Reviews** ■ Additional Examples 9A, 9B, 9C ■ Practice 1–6 ■ Section Review 1–8 ■ Chapter Review 1–18; LC 1 **Study Guide Worksheets** ■ ■ 9-1 Concept Review ■ ■ 9-1 Practice *Changes in Heat Energy*
Transparency Masters ■ F 9-10 *Enthalpy Changes Over the Course of an Exothermic Reaction* ■ F 9-12 *Reaction Pathway for Potassium-Water Reaction* ■ F 9-13 *Distribution of Kinetic Energy Among Particles*	**Text Reviews** ■ Concept Review 9–10 ■ Practice 1–2 ■ Section Review 11–18 ■ Chapter Review 19–26; LC 2, 3 **Study Guide Worksheets** ■ ■ 9-2 Skillsheet *Reaction Pathways Activation Energy* ■ ■ 9-2 Practice
Transparencies ■ F 9-16 *Models for Low Entropy and High Entropy Systems*	**Text Reviews** ■ Practice 1–2 ■ Section Review 19–22 ■ Chapter Review 27–35 **Study Guide Worksheets** ■ ■ 9-3 Practice
Transparencies ■ F 9-20 *Models for Water* **Transparency Masters** ■ T 9-8 *Relating Enthalpy and Entropy to Spontaneity* ■ T A-12 *Thermodynamic Properties*	**Text Reviews** ■ Practice 1–4 ■ Section Review 23–27 ■ Chapter Review 36–45; LC 4; UT 1 **Study Guide Worksheets** ■ ■ 9-4 Practice *Free Energy*
Transparencies ■ P 346 *Nutritional Labeling* ■ P 348 *Energy Calculations* **Transparency Masters** ■ T 9-9 *Energy Expended (Cal/day)* ■ T 9-10 *Estimating Energy for Activity Level*	**Text Reviews** ■ Section Review 28–31 ■ Chapter Review 46–51 **Study Guide Worksheets** ■ ■ 9-5 Skillsheet *Calculating Daily Energy Needs*

■ Pupil's Text　　Exp *Exploration*　　LC *Linking Chapters*
■ Study Guide　　Inv *Investigation*　　UT *Using Technology*
■ Teaching Resources

Assessment Options

Traditional Assessment

Test Generator
Instructional Objectives Measured:
Content mastery
Select from items 1 to 75

Performance Assessment Options

(see pages T36 and T643-15 for scoring rubrics)
Students demonstrate mastery of objectives in a hands-on environment. You may want some of these materials in the Portfolio as well.

Investigations 9A and 9B (text)
Investigations 9-1 and 9-2 (laboratory manual)
Instructional Objectives Measured:
Content mastery, Use of scientific methodology, Problem solving skills, Use of technology, Proficiency in written communication

Performance Assessments, page 353
1. **Designing an experiment**
   Instructional Objectives Measured:
   Content mastery, Problem solving skills, Use of scientific methodology
2. **Developing an experimental procedure**
   Instructional Objectives Measured:
   Content mastery, Problem solving skills, Use of scientific methodology

Portfolio Options

(see page T36 for scoring rubrics)
Students provide a written rationale for each selection made for the portfolio.

Concept Maps
Students use vocabulary from the Chapter Review to build a concept map.
Instructional Objectives Measured:
Content mastery

Formal Laboratory Reports
Explorations 9A and 9B (text)
Explorations 9-1, 9-2, 9-3, 9-4 (lab manual)
Instructional Objectives Measured:
Content mastery, Use of scientific methodology, Use of technology, Proficiency in written communication

Portfolio Projects, page 353
Items 1, 4, and 5. Chemistry and you
Instructional Objectives Measured:
Content mastery, Scientific and chemical literacy, Proficiency in oral and written communication

Items 2 and 3. Research and communication
Instructional Objectives Measured:
Content mastery, Scientific and chemical literacy, Proficiency in oral and written communication

INTRODUCING THE CHAPTER

CHAPTER 9 Causes of Change

Story Background

Heat can be used as the propelling agent in heat engines which include automobile engines and steam turbines. The maximum theoretical efficiency of a heat engine is given by the following equation.

$$\mathbf{Eff} = \frac{T_2 - T_1}{T_2}$$

In rough terms, T_2 is the cylinder temperature, and T_1 is the exhaust temperature. T_2 is limited by the mechanical strength of metals at high temperatures, and T_1 cannot be lower than ambient temperature. Consequently, the best efficiency of a steam power plant is 40%, while an auto's best efficiency is about 25%. Electric motors, however, are 100% efficient in theory and 90% in practice. This means that electric engines should be almost four times as efficient as combustion engines. However, the source of the electricity used to power the car affects the overall efficiency of the system, which can be as low as 25%, or no better than the present automobile engines. Another possibility for improving engine efficiency is the use of sunlight to produce hydrogen that can be used in fuel cells for propulsion. Fuel cells have an efficiency of 60% or better and are nonpolluting.

About the Illustration

The Mazda Miata shown here runs on hydrogen fuel. Its modified engine design looms in the background.

Energy and Low-Emission Cars

Would you like a car like this one? The car looks like many other sports cars of the same model available from dealers around the country. It can go from 0 to 60 mph in 10.5 s. But it's different—very different. And as much as you might love to drive or own such a car, unfortunately you can't—at least not yet.

This car is different not because of its shape or its stereo, but because of what is under the hood. It has a 1300 cm^3, 120 horsepower, two-rotor engine that doesn't use gasoline—it is powered instead by hydrogen.

Scientists have long been searching for alternative sources of energy to replace petroleum products that produce toxic emissions and that will eventually be exhausted. As an alternative for gasoline, scientists and engineers have experimented with solar-powered and battery-powered vehicles.

However, with current technology, solar-powered vehicles can be unreliable because they can't run unless the sun has been shining. Battery-powered vehicles are expected to be more expensive than gasoline-powered cars because the batteries are made of expensive and heavy materials. In addition, pollution can be created when the electricity needed to recharge the batteries is generated.

As a fuel, hydrogen's big advantage is that it produces very low toxic emissions. When gasoline is used, several undesirable products, including hydrocarbons and oxides of nitrogen and carbon, are released into the atmosphere. These compounds cause acid precipitation, contribute to photochemical smog, and promote the "greenhouse effect." When pure hydrogen is burned, however, no hydrocarbons or carbon oxides are emitted. The levels of nitrogen oxides are also diminished.

For the moment, only a few of these cars will be produced as prototypes. Mass production is still years or even decades away.

With all these advantages, why aren't hydrogen cars available now? One challenge is the expense of producing and distributing the huge supplies of hydrogen that will be needed. Another is the high cost of producing the car.

Even if a cost-effective method of producing hydrogen were available, the problem of safe storage would remain. Hydrogen is hard to store because it is a very reactive gas. In the presence of a spark, it combines easily and rapidly with oxygen, resulting in an explosive reaction that releases a large quantity of energy.

In this chapter, you will study the driving forces that cause reactions to occur so that you can answer the following questions.

- *What drives the hydrogen combustion reaction?*
- *How can this reaction be controlled so that it does not happen accidentally?*

Tapping Prior Knowledge

Prepare students for the concepts they will need for this chapter with the following discussion topics.

1. Water heated on a stove to make spaghetti quits boiling almost immediately when removed from the stove, but begins boiling again when returned to the stove. How can you explain this observation?
2. Account for why steam burns are so much more severe than hot water burns.
3. Compare the path you take to school with the path home. Are they always the same? Do the ending and starting points differ or remain the same?
4. Have a student walk from his or her desk to the board and back several times. Have students compare the forward and reverse paths. Place a step stool in the student's path and have him or her walk to the board and back again. Was this trip harder or easier than the first trip? Which trip took more time and why?
5. Have students give their impressions of the phrase "counting Calories."
6. Ask students to write down the foods that make up their favorite meal. Make a list on the board of the most commonly eaten foods. Then classify them as primarily fat, protein, or carbohydrate. Why would you feel more full after eating a large baked potato than after eating a thin slice of roast beef?

Section 9-1

FOCUS

Lesson Starter

Begin the lesson by showing that heat for the hand warmer in **Demonstration 1** is generated when Fe is oxidized according to the following equation.

$$4Fe + 3O_2 \longrightarrow 2Fe_2O_3 + heat$$

NaCl provides electrolytes to catalyze the reaction. The vermiculite, a hydrated silicate of Al, Mg, and Fe, expands when heated. This expanded form is often used for its insulating properties, as in this demonstration.

Demonstration 1

An inexpensive hand warmer

Approximate time: 5 min

1. Place 25 g of iron powder in a small plastic sealable bag, add 1 g of sodium chloride, seal the bag, and shake to mix.
2. Add 1 tablespoon of vermiculite, reseal, and shake again.
3. Activate the hand warmer by adding 5 mL of water, sealing, and squeezing or shaking the bag thoroughly to mix the contents. Heat will be produced in about a minute.
4. Pass the bag around the class.

SAFETY

Wear safety goggles and a lab apron.

DISPOSAL

Put in trash.

[Summerlin, Borgford, and Ealy. *Chemical Demonstrations,* Vol. 2, ACS,1988.]

How does energy affect change?

Section Objectives

Distinguish between heat and temperature.

Explain how a calorimeter measures changes in energy.

Explain what is meant by the caloric content of foods.

Calculate the amount of heat energy released or absorbed in a chemical reaction or physical change.

List examples of different heats of reaction.

Use heats of reaction in stoichiometric calculations.

Heat in physical changes

If an ice cube is left out on a table at room temperature, no one is surprised that it melts and changes into water. No matter how long the water sits on the table, it will not become ice again as long as it is held at room temperature. If an ice cube and glass of water are left in a freezer at –10°C, the ice cube will not melt, and the water in the glass will freeze.

In other words, the physical change of melting and freezing can be considered a reversible change, one that does not happen in only one direction. The direction of the change of state can be reversed merely by adjusting the temperature. The change from water to ice (and back again) can be summarized by the following equation.

$$H_2O(l) \rightleftharpoons H_2O(s)$$

Pure water has a melting/freezing point of 0°C. Below 0°C, freezing will occur, and the equation written above runs from left to right. Eventually, only solid will remain. Above 0°C, melting will occur, and the equation runs from right to left so that only liquid remains.

Melting and boiling occur at specific temperatures

To understand why changing the temperature will affect the state of matter of water, you need to know the difference between heat energy and temperature. Consider the experiment shown in Figure 9-1. Two beakers are placed on hot plates. One contains a single ice cube; the other contains eight ice cubes. Times and temperatures are measured as each beaker is heated, with the hot plates set on medium. This way, the hot plates deliver the same amount of energy to each beaker each second.

Figure 9-1a
In this experiment, one beaker containing eight ice cubes . . .

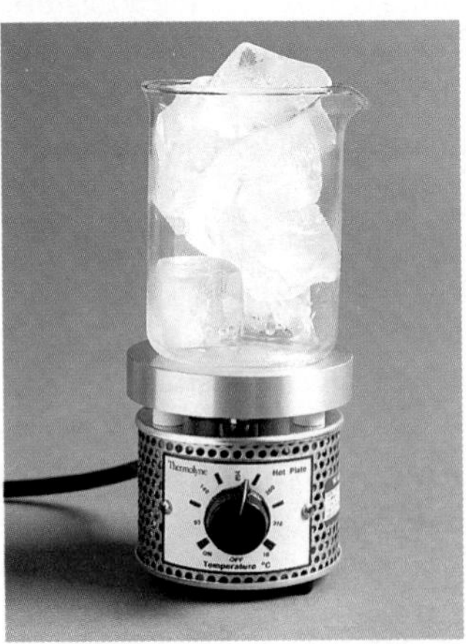

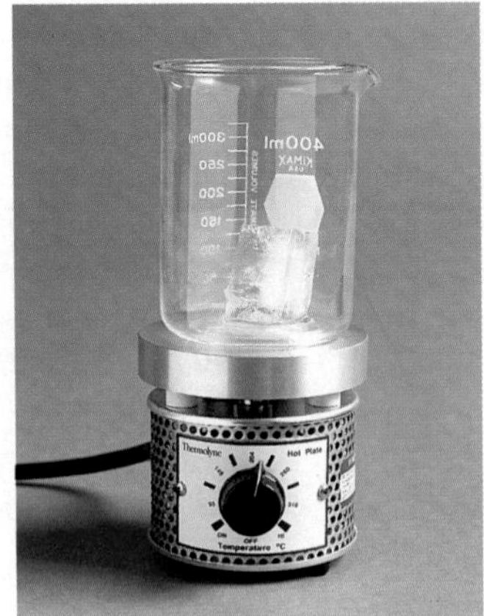

9-1b
. . . and another containing only one ice cube were both heated.

Table 9-1
Data from Ice Cube Experiment

Beaker with one ice cube		Beaker with eight ice cubes	
Temperature	Time (s)	Temperature	Time (s)
0°C (start)	0	0°C (start)	0
0°C (ice disappears)	25	0°C (ice disappears)	190
25°C	33	25°C	250
50°C	41	50°C	310
75°C	49	75°C	370
100°C (water begins boiling)	57	100°C (water begins boiling)	429
100°C (water disappears)	226	100°C (water disappears)	1701

In each case, the solid ice cubes melt, making liquid water. Eventually, the liquid boils, forming gaseous steam until there is nothing left in the beakers. The data are summarized in **Table 9-1** and represented by the graph shown in **Figure 9-2**.

Because the hot plates deliver heat at a constant rate, similar time intervals represent similar amounts of heat transferred by the hot plates. The quantity of heat transferred in the first 30 s is the same as the heat transferred between 200 s and 230 s.

First, compare the differences between the data for one ice cube and for eight ice cubes. From the graph, you can tell that the temperature at which melting and boiling occur are the same, regardless of the amount of ice used. But the wide difference in time indicates that different amounts of heat were needed. Each sample of ice melted at 0°C, but only 25 s of heat were required to completely melt the lone ice cube. A much longer time, 190 s, was required to melt eight ice cubes completely. The amount of heat that the hot plate transferred during 57 s was enough to melt the water in one ice cube and bring it to boiling, but not enough to melt the eight ice cubes entirely. In fact, the eight ice cubes required about eight times as much heat to be transferred before they boiled. Yet boiling occurred at the same temperature, 100°C, for both samples of ice.

Temperature vs. Time Data from Ice Cube Experiment

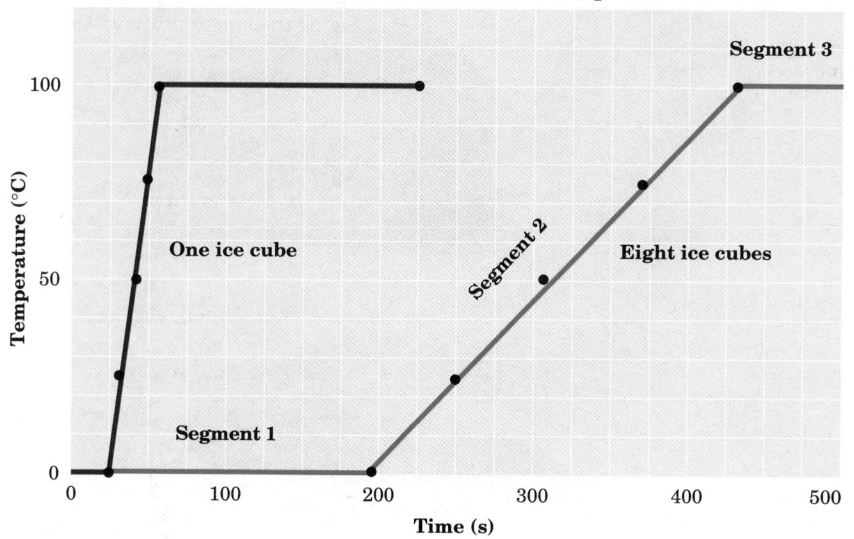

Figure 9-2
When ice is heated, its temperature does not change until all of it has melted. Then the temperature rises steadily until it reaches the boiling point, rising no further until all of the water has boiled away.

TEACH

Demonstration 2

Low melting metals: Onion's fusible metal

Approximate time: 10 min

1. Obtain a piece of Onion's fusible metal (available from Flinn Scientific) that has been allowed to cool in a flat position. Pass it around the room, and allow students to attempt to bend it.
2. Place the metal into a 150 mL beaker of boiling water. The metal should liquefy, and you can push it around with a glass stirring rod to show that it is liquefied.
3. Pour most of the water into a waste beaker. Pour the metal *slowly* into cold water.
4. Pass the solidified metal around to gloved students for inspection. Metallic crystals may be seen on the surface if cooling was sufficiently slow.

SAFETY
Observe all precautions described on the label and in the MSDS for Onion's fusible metal. Wear plastic or rubber gloves. Students should also wear gloves when they handle the metal.

DISPOSAL
Save the Onion's fusible metal for next time.

Visual Strategy

Figure 9-2

Start at the bottom of the graph and go up. Point out the change in state that occurs at the flat portions of the graph. Stress that these flat segments mean phase changes are accompanied by constant temperature. Stress also that the greater the amount of mass present, the longer the system stays at the temperature of the phase change. This is shown for 0°C but not for 100°C.

Next examine the shape of the plots. Both plots have the same basic shape: flat, then slanted upward, and then flat again. Because the plot for the eight ice cubes is larger, we will examine its shape in further detail. What happens in the three different parts of the plot?

In the first segment, heat was transferred for 190 s, but the temperature of the sample did not change at all—it remained at 0°C until there was no ice present. The heat that was transferred during the next 190 s raised the temperature of the sample, now entirely liquid water, by almost 80°C. The upward slope of the second segment shows that, as heat was transferred, the temperature of the water increased. At 429 s the temperature reached 100°C, and boiling began, as indicated in the third segment of the graph. Once more, the temperature held steady during boiling, despite the large amounts of heat being added.

Modeling melting, heating, and boiling

But why is the temperature constant during melting or boiling? You know from Chapter 2 that ice is a solid, consisting of a rigid arrangement of water molecules in a crystal. The molecules vibrate but cannot move from one place to another. Liquid water consists of water molecules that are still attracted to one another but not rigidly held together. They are able to move from one position to another as well as vibrate.

For ice to melt, energy must be added to cause the vibrations to increase enough so that the molecules can break loose from their positions in the ice crystal. Because temperature is a measure of the average kinetic energy of the particles in a substance, temperature will increase only after the forces holding the ice crystals together are overcome. When the particles are free to move, the heat energy that is added is converted to energy of motion for the particles. No matter how much heat energy is transferred, the temperature will not rise above 0°C until all of the solid melts.

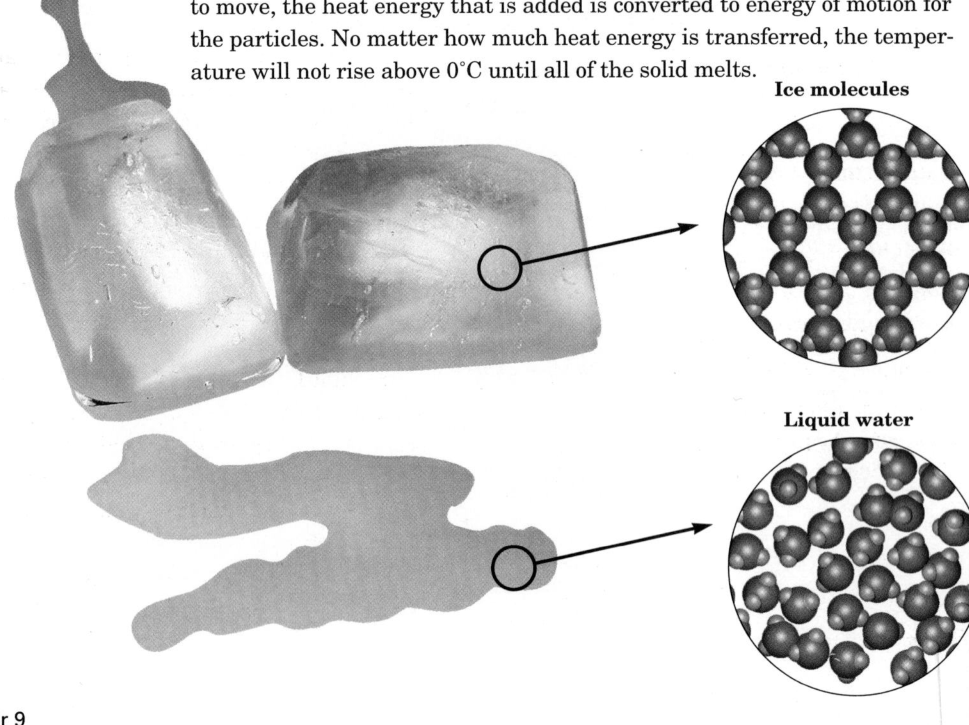

Figure 9-3
As ice absorbs energy from the surroundings, the vibrations of the molecules in the lattice increase. When the force of the vibrations is greater than the force of attraction between molecules in the lattice, the lattice breaks down to form liquid water, a process you see as melting.

Visual Strategy

Figure 9-3

The molecules in the liquid are farther apart and have greater movement than the molecules in the solid because the absorbed heat overcomes the intramolecular forces. You may wish to refer back to hydrogen bonding in **Figure 6-7** on page 197.

Chem Fact
If you put pressure on ice, its melting point is decreased. It is one of very few substances which behave in this fashion.

Once all of the ice was melted, the heat energy transferred by the hot plate caused an increase in the temperature of the sample. During this process, the temperature climbed steadily until it reached the boiling point. As you learned in Chapter 2, gases consist of individual molecules with only weak attractions for each other. For boiling to occur, the molecules needed more energy, which was transferred as heat by the hot plate, to overcome the intermolecular forces holding the liquid water molecules together. The high-energy water molecules were moving so rapidly that they moved right out of the pan and into the air as a gas. Because heat energy was used to overcome the intermolecular forces, none was available to raise the temperature past 100°C.

temperature
a measure of the average kinetic energy of random motion of the particles in a sample of matter

heat
the total of kinetic energy of random motion of molecules, atoms, or ions in a substance

specific heat capacity
the amount of heat energy required to increase the temperature of one gram of a substance by one degree Celsius

Heat and temperature are different

The temperatures at melting and boiling were the same for one ice cube or eight ice cubes. ***Temperature*** *is a measure of the average kinetic energy of random motion of the particles in a substance.*

The beaker with eight ice cubes required more heat to reach the melting and boiling temperatures because it had more particles in it. ***Heat*** *is a measure of the total amount of energy transferred from an object of high temperature to one of low temperature.* The energy was transferred from the hot plate to the water molecules in the beaker. A change in temperature, which means a change in average kinetic energy values, is usually accompanied by a transfer of energy in the form of heat. However, the more particles there are, the greater the amount of heat that must be transferred to raise the average kinetic energy, the temperature, of all the particles.

Specific heat capacity relates temperature changes to heat changes

The **specific heat capacity** of a substance is the amount of heat energy required to raise the temperature of one gram of that substance by one degree Celsius. Its symbol is c_p. The p in the symbol stands for "at constant pressure." The specific heat capacity can vary if pressure is not kept constant. It can also vary at different temperatures, but for most purposes, it is reasonable to assume the specific heat capacity remains constant.

Specific heat capacity is a physical property of matter. Each pure substance has a characteristic specific heat capacity. Scientists have measured and recorded specific heat capacities for many substances, including those shown in **Table 9-2** Specific heat capacities must be measured; they cannot be determined from a chemical formula.

Table 9-2
Specific Heat of Selected Substances

Substance	Specific heat capacity, c_p (J/g•°C at 25°C unless noted)
water, $H_2O(l)$	4.180
ethanol, $C_2H_5OH(l)$	2.438
methane, $CH_4(g)$	2.200
isooctane, $C_8H_{18}(l)$	2.093
ice, $H_2O(s)$ at -5°C	2.077
steam, $H_2O(g)$ at 100°C	2.042
aluminum, Al(s)	0.897
table salt, NaCl(s)	0.865
graphite, C(s)	0.714
iron, Fe(s)	0.449
silver, Ag(s)	0.235
mercury, Hg(l)	0.139
tungsten, W(s)	0.132

Teaching Tip
Direction of heat flow
Heat flows from a greater source to lesser source, i.e., from hot to cold, in an effort to reach thermal equilibrium.

Historical Note
Joseph Black, a Scottish chemist and physician, was the first to differentiate between heat and temperature.

Teaching Tip
Heat content vs. temperature
Ask students to imagine a bathtub filled with warm H_2O at a temperature of 35°C. Dip a cup of H_2O from the bathtub. What will be the temperature of the H_2O in the cup? (35°C) Which amount of H_2O, the cup or bathtub, can melt the greatest amount of ice? (the bathtub) Emphasize that although two things have the same temperature, they do not necessarily have the same heat content.

Demonstration 3
Specific heats of metals
Approximate time: *5 min*

1. Obtain a specific heat demonstration device, which consists of a central hub from which five identical metal rods project. The metals are typically brass, copper, steel, nickel, and aluminum.
2. Before class, drip some candle wax into the indentation at the end of each rod and affix a copper BB on top of the wax.
3. Place the center of the hub directly over a burner flame and keep track of the order in which the BBs drop.

SAFETY
Wear safety goggles and a lab apron.

DISPOSAL
Save the metals for next time. The wax can be put in the trash.

Groupwork Strategy

Practice calculating ΔH

In groups of two, have students determine the amount of energy needed to increase the temperature of 100 g of each substance in **Table 9-2** by 20°C.

Additional Example for Sample Problem 9A

Assuming the density of water is 1.0 g/mL and is constant over the temperature range, answer the following questions.

a. How much heat is lost by 4.0 L of water that is cooled from 87°C to 21°C?

b. If 980 kJ of energy are added to 6.2 L of water at 18°C, what is the final temperature of the water?

Solutions are at the bottom of the page.

Answers to Practice 9A

1. $\Delta H = m \times c_p \times \Delta t$

   $\Delta H = (200 \text{ g}) \times (4.180 \text{ J/g}{\cdot}°\text{C}) \times (37°\text{C} - 65°\text{C}) = -23\,408 \text{ J} = -23.4 \text{ kJ}$

2. $\Delta H = m \times c_p \times \Delta t$

   $\Delta H = (425 \text{ g}) \times (0.897 \text{ J/g}{\cdot}°\text{C}) \times (200°\text{C} - 25°\text{C}) = 66\,714 \text{ J} = 66.7 \text{ kJ}$

Figure 9-4
Metals typically have low heat capacities. When a sterling silver spoon is left in a cup of hot tea, it rapidly reaches the temperature of the tea.

Substances with low specific heat capacities require less energy to feel hot than those with high specific heat capacities. For example, a piece of silver changes temperature faster than a piece of iron with the same mass. For this reason, a sterling silver spoon such as the one shown in **Figure 9-4** will warm up more quickly that a stainless steel one containing mostly iron.

If the temperature change and the mass of the substance undergoing the change are given, the specific heat capacity can be used to calculate the change in heat. The symbol for "change in" is the Greek letter delta, Δ. Thus, the change in temperature can be symbolized as Δt.

change in heat = change in temperature × mass × specific heat capacity

$$\textbf{change in heat} = \Delta t \times m \times c_p$$

Sample Problem 9A
Specific heat capacity

Imagine that you're working outdoors on a hot, humid day. If you drink four glasses of ice water at 0°C, how much heat energy is transferred as this water is brought to body temperature? Assume that each glass contains 250. g of water and that your body temperature is 37°C.

1 List

- **mass of 4 glasses of water:** $m = 4 \times 250.\text{ g} = 1.00 \times 10^3 \text{ g } H_2O$
- **change in water temperature:** $\Delta t = 37°\text{C} - 0°\text{C} = 37°\text{C}$
- **specific heat capacity of water:** $c_p = 4.180 \text{ J/g}{\cdot}°\text{C}$
- **quantity of heat energy transferred:** ? J

2 Set up

- **Use the specific heat capacity equation to calculate the amount of heat energy transferred.**

  $\Delta t \times m \times c_p$ = change in heat energy

  $(37°\text{C})(1.00 \times 10^3 \text{ g})(4.180 \text{ J/g}{\cdot}°\text{C}) = ?$ J

3 Estimate and calculate

- **Round off to make an estimate.**

  $40 \times 1000 \times 4 \approx 1.6 \times 10^5$ J

- **Calculate, and round off correctly.**

  Calculator answer: 1.5466×10^5

  Rounded answer: 1.5×10^5 J = +150 kJ

Practice 9A

Answers for Concept Review items and Practice problems begin on page 841.

1. **How much heat energy is released to your body when a cup of hot tea containing 200. g of water is cooled from 65°C to body temperature, 37°C?**
2. **How much heat energy is needed to raise the temperature of a 425 g aluminum baking sheet from room temperature, 25°C, to a baking temperature of 200°C?**

Solutions for Additional Example 9A

a. $\Delta H = m \times c_p \times \Delta t$

$\Delta H = (4000 \text{ g})(4.180 \text{ J/g}{\cdot}°\text{C})(-66°\text{C}) = -1104 \text{ kJ}$

b. $\Delta H = m \times c_p \times \Delta t$

$$\frac{\Delta H}{m \times c_p} = t_{final} - t_{init} \qquad t_{final} = \frac{980\,000 \text{ J}}{(6200 \text{ g})(4.180 \text{ J/g}{\cdot}°\text{C})} + 18°\text{C} = 55.8°\text{C}$$

Amounts of heat transferred can be measured

Melting and boiling also involve energy transfer, just like changing the temperature of the liquid water did. The amounts of energy required for melting, heating, and boiling various substances have been measured, as shown in **Table 9-3**.

Table 9-3
Molar Heat Data for Some Substances

Name	Symbol	Description	Examples: Mercury, Hg	Examples: Ethanol, C_2H_5OH	Examples: Water, H_2O
Heat of fusion	ΔH_{fus}	Energy needed to melt one mole	+2.29 kJ/mol mp = −38.8°C	+5.02 kJ/mol mp = −114.1°C	+6.00 kJ/mol mp = 0.0°C
Heat change for 1.0°C		Energy needed to raise temperature of one mole by 1.0°C	2.80×10^{-2} kJ/mol	1.12×10^{-1} kJ/mol	7.53×10^{-2} kJ/mol
Heat of vaporization	ΔH_{vap}	Energy needed to boil one mole	+59.1 kJ/mol bp = 357°C	+38.6 kJ/mol bp = 78.3°C	+40.6 kJ/mol bp = 100.0°C

Using this table, you can predict how much heat energy would be required to repeat the ice cube experiment for exactly one mole of water molecules. First, enough energy must be added to melt the ice, changing it into liquid water. This amount is the heat of fusion, 6.00 kJ/mol. Then, the liquid water must be heated from 0.°C to 100.°C. From **Table 9-3**, the amount of energy, or heat change, required for 100°C is 100 times 7.53×10^{-2} kJ/mol, or 7.53 kJ/mol. Last, the liquid water at 100.°C must be converted to steam at 100.°C, requiring the heat of vaporization, 40.6 kJ/mol. Adding these values (6.00 kJ/mol + 7.53 kJ/mol + 40.6 kJ/mol) gives you a prediction that the total amount of heat energy transferred during the experiment is 54.1 kJ/mol of H_2O. **Figure 9-5** shows how to calculate the amounts of heat energy needed for each step in a heating curve.

Calculating Energy Changes for Various Parts of a Heating Curve

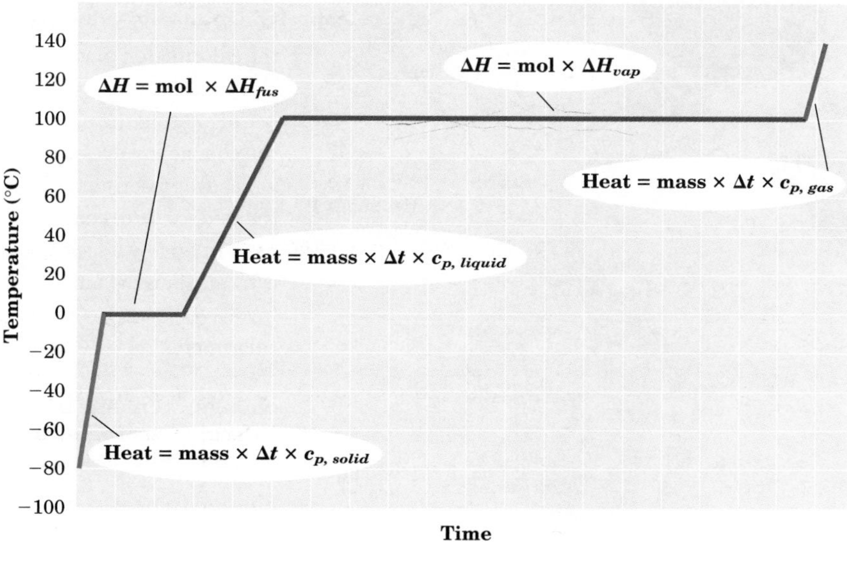
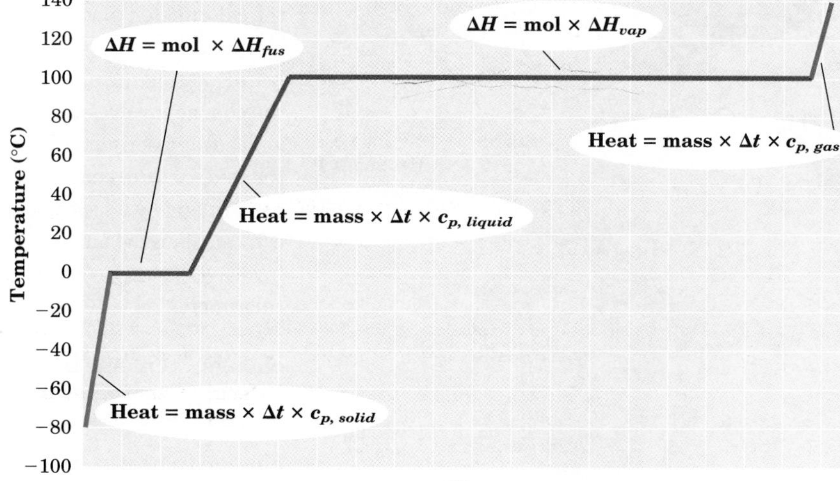

Figure 9-5
Differing amounts of heat energy are required in the different parts of the process going from heating ice to heating steam.

Demonstration 4

Boiling water in a paper cup

Approximate time: 10 min

1. Place a small paper cup (like those used for ketchup at fast food restaurants) on a wire gauze above a bunsen burner and watch it burn.
2. Fill a second paper cup 3/4 full with water, and place it on the wire gauze above the burner. The water will boil, keeping the temperature of the cup below the combustion temperature of paper (233°C or 451°F). The cup will not burn until the water is completely evaporated.
3. Students may wonder if this works with Styrofoam cups; it does not. Try this demo in a hood because of the fumes from the heated Styrofoam. Do not use much water because most of it will run out.

SAFETY
Wear safety goggles and a lab apron. Students stand 10 ft back from the demonstration area.

DISPOSAL
Pour water down the drain. Put cups in the trash.

Visual Strategy

Figure 9-5

Assemble the individual equations on the graph into the full equation for calculating the heat lost when a solid becomes a gas. Following the heating curve, calculate ΔH values for 100 g H_2O heated from –20°C to 120°C and then cooled from 120°C to –20°C. Point out that ΔH_{vap} and ΔH for condensation differ in sign but not quantity. The same is true for ΔH_{fus} and ΔH for melting, and for the two ΔH_{rxn}.

Content Background

Combustion of organic compounds and bomb calorimetry

Combustion is a general term applied to any chemical change in which light and heat are produced; however, it usually refers to burning in the presence of oxygen. Combustion reactions involving hydrocarbons are highly exothermic. The heat energy released per mole of hydrocarbon burned is the heat of combustion, ΔH_{comb}, for that hydrocarbon. ΔH_{comb} values are assigned, by convention, a positive value, so they are equal in magnitude but opposite in sign to ΔH^0 values for combustion reactions.

Combustion reactions are the basis for the use of hydrocarbons as fuels. Large amounts of gases are produced along with large amounts of heat when hydrocarbons burn. The rapid formation of these gases at high temperature and pressure drives the pistons or turbine blades in an internal combustion engine. If excess oxygen is present, the products are CO_2 and H_2O. If insufficient oxygen is present, other products such as carbon monoxide or elemental carbon will form because combustion is incomplete. Many modern cars control the air intake electronically to optimize the fuel to oxygen ratio.

A bomb calorimeter can be used to measure the heat that evolves when a substance burns at constant volume. The mass of the substance to be burned is determined before the substance is placed inside the bomb, a steel vessel capable of withstanding high temperatures and pressures. The bomb is sealed, pressurized with oxygen to ensure complete combustion, and placed inside a metal bucket filled with water, the mass of which is known.

Heat in chemical changes

Energy can be interconverted into other forms

Isooctane combustion can be used as a model for the burning of gasoline. The equation for this reaction includes an energy change of 10 990 kJ.

$$2C_8H_{18}(l) + 25O_2(g) \longrightarrow 16CO_2(g) + 18H_2O(l) + 10\,990\text{ kJ}$$

But how was this amount determined? Although chemical handbooks list energy changes for many such chemical reactions and physical changes, these energy values were first determined by measuring them experimentally.

Changes in heat reflect the movement of energy from one system to the surroundings. Energy released during the combustion of isooctane is in the form of heat and moves from the fuel and oxygen to the products and their surroundings, including the engine and the air. Some of this energy can be transferred into other forms, such as the kinetic energy that moves the pistons which turn the car's axle. Similarly, heat energy moves from the water in an ice tray to the air inside a freezer as it freezes. *Such interconversions are possible whether the energy was generated by a physical change or a chemical change.*

Measuring changes in heat energy

If the transfer of energy between a system and its surroundings occurs only in the form of heat, the changes in energy can be easily measured. A device to measure heat energy is known as a *calorimeter*. **Figure 9-6** shows the basic parts of one type of calorimeter. This calorimeter can be used to measure the specific heat capacity of materials, by measuring the water's temperature change as a warm object is placed in the calorimeter.

Another type of calorimeter, a *bomb calorimeter*, can be used to measure changes in heat energy during combustion reactions. A substance to be oxidized is placed in a "bomb," a chamber in the center of the calorimeter, along with excess oxygen. An electrical heating element initiates the reaction.

Calculating the amount of heat energy for the change involves the same procedure whichever calorimeter is used. By multiplying the temperature change by the mass of water present and by its specific heat capacity, you can calculate the amount of heat energy involved in the change occurring within the calorimeter.

$$\textbf{change in heat} = \textbf{change in temperature} \times \textbf{mass } \mathbf{H_2O} \times \textbf{specific heat capacity of } \mathbf{H_2O}$$

$$\textbf{change in heat} = \Delta t_{H_2O} \times m_{H_2O} \times c_{p,H_2O}$$

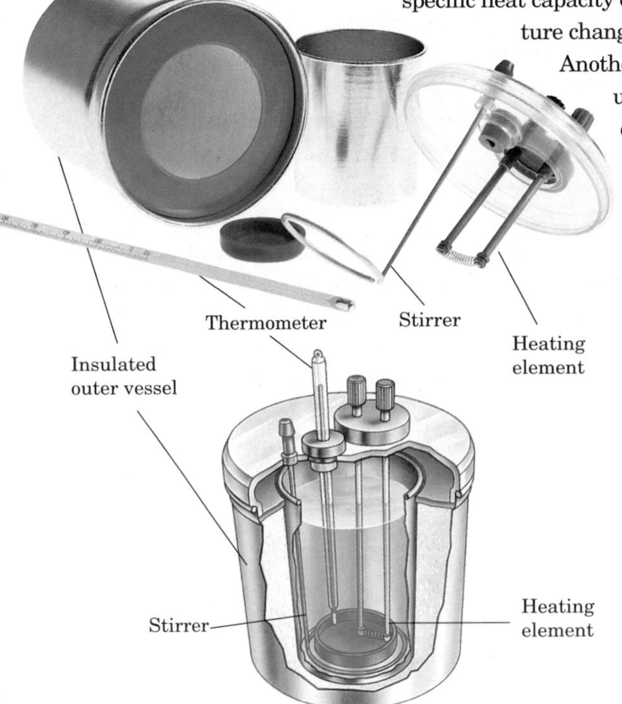

Figure 9-6
The heat energy released by a change taking place in the calorimeter can be measured by determining how much warmer it makes the surrounding water.

Table 9-4
Energy and Nutrient Breakdown of Some Foods*

Food and measure	Calories	Grams carbohydrate	Grams protein	Grams fat
Apple, 1 medium	81	21.1	0.3	0.5
Cheeseburger, 4.1 oz.	310	31.2	15.0	13.8
Cola soft drink, 6 oz.	72	19.0	0	0
French fries, 3.4 oz.	320	36.3	4.4	17.1
Potato, baked, 7.1 oz.	220	51.0	4.7	0.2
Potato chips, 1 oz.	150	15.0	1.0	10.0
Pretzels, 1 oz.	110	22.0	2.0	2.0
Yogurt with fruit, 8 oz.	240	43.0	9.0	3.0
Yogurt, low-fat frozen, 3 oz.	100	22.0	4.0	0.8

*These figures have been rounded and represent average values.

You get energy from carbohydrates, proteins, and fats

The substances your body needs in order to grow and maintain life come from the nutrients in food. There are six classes of nutrients in food—carbohydrates, proteins, lipids, water, vitamins, and minerals. Of these, carbohydrates, proteins, and fats are major sources of energy for the body.

The energy content of food is calculated from data collected by burning the dry food sample in a bomb calorimeter. Your body does not burn food as efficiently as the calorimeter, so the data from calorimetry must be adjusted slightly. **Table 9-4** gives the energy content of some foods expressed in Calories.

Table 9-5
Energy Content of Carbohydrates, Proteins, and Fats

Nutrient	Cal/g	kJ/g
Carbohydrate	4	17
Protein	4	17
Fat	9	38

Up to this point you have been using energy measurements expressed in kilojoules. So what does it mean that one medium, plain baked potato supplies about 220 Calories? Most reference tables for foods give information in dietary Calories (which are kilocalories) rather than kilojoules. One dietary Calorie equals the following.

$$1 \text{ Cal}^* = 4.184 \text{ kJ}$$

Using this conversion, the energy of the baked potato is equivalent to 920 kJ. Using this same conversion, a cheeseburger at 310 Calories supplies 1297 kJ of energy. **Table 9-5** gives the caloric equivalent of 1 g of the three energy nutrients. These are standard general conversion factors that can be used to estimate the energy content of a typical diet, but not of a specific food. Notice that fats provide more energy per gram than carbohydrates and proteins. Fats are the energy storage compounds in the body.

*The large Calorie is written with a capital C and is equivalent to 1000 cal or 1 kcal.

The water bucket is sealed and sits inside a sealed, insulated chamber so that all of the heat evolved by the burned substance is absorbed by the water and the calorimeter. The reaction is initiated by a spark from a small, electronically heated wire in the calorimeter. A stirrer keeps the water moving so that the heat is distributed evenly and the temperature is uniform. The temperature is measured when it becomes constant. The heat evolved during the reaction is equal to the heat absorbed by the water plus the heat absorbed by the calorimeter. The heat capacity for the calorimeter, sometimes called the calorimeter constant, is determined experimentally by adding a known amount of heat and measuring the rise in temperature of the calorimeter and the water it contains. When computing the energy content of the substance burned, the energy of the spark used to initiate the reaction must be subtracted out.

When the substance burned in a bomb calorimeter is a food, the heat of combustion is referred to as the energy content or the caloric value of that food. Any food can be burned in a calorimeter, but some foods must be dried first.

Measuring the heat given off when the food is burned provides a direct measure of the energy stored in the food's chemical bonds, so the process is called direct calorimetry. Alternatively, the amount of energy released during burning can be determined indirectly by measuring the amount of oxygen consumed during the reaction. This process is indirect calorimetry.

Teaching Tip

Vocabulary

You may wish to draw attention to the footnote concerning the capitalization of the *C* in Calorie.

Additional Example for Sample Problem 9B

What is the heat of combustion of sucrose, $C_{12}H_{22}O_{11}$, if 1.500 g of sucrose raise the temperature of 3.00 kg of water in a bomb calorimeter from 18.00°C to 19.97°C?

Solutions are at the bottom of the page.

Answers to Practice 9B

1. Heat gained by H_2O is equal to the heat lost by oatmeal.

   $\Delta H = m \times c_p \times \Delta t$

   $\Delta H = (2.5 \times 10^3 \text{ g}) \times (4.180 \text{ J/g·°C}) \times (27.2°\text{C} - 25.0°\text{C}) = 22\,990 \text{ J} = 23 \text{ kJ}$

2. $\Delta H = m \times c_p \times \Delta t$

   $18.5 \times 10^3 \text{ J} = (2.50 \times 10^3 \text{ g}) \times (4.180 \text{ J/g·°C}) \times (X°\text{C} - 25.0°\text{C})$

   $X°\text{C} = \frac{18.5 \times 10^3}{(2.50 \times 10^3)(4.180)} + 25.0 =$

   26.8°C

Sample Problem 9B
Calorimetry calculations with food

A nutritional chemist burns one pulverized peanut with a mass of 0.887 g in a bomb calorimeter. The calorimeter contains 2.50 kg of water, and its temperature increases from 25.0°C to 27.0°C as the peanuts burn. What is the energy content of the peanuts?

1 List

- **mass of water:** 2.5 kg = 2.50×10^3 g
- **change in temperature:** 27.0°C − 25.0°C = 2.0°C
- **specific heat capacity of water:** 4.180 J/g•°C
- **heat energy transferred:** ? kJ

2 Set up

- **Use the specific heat capacity equation to calculate the amount of heat energy transferred.**

  $\Delta t \times m \times c_p$ = change in heat energy

  $$(2.0\not{°\text{C}})(2.50 \times 10^3 \not{\text{g}})(4.180 \not{\text{J}}/\not{\text{g}}\text{•}\not{°\text{C}})\left(\frac{1 \text{ kJ}}{1000 \not{\text{J}}}\right) = ? \text{ kJ}$$

3 Estimate and calculate

- **Round off to make an estimate.**

  $$5000 \times 4 \times \frac{1}{1000} \approx 20$$

- **Calculate and round off correctly.**

  Calculator answer: 20.9

  Rounded answer: +21 kJ

Practice 9B

1. **What is the energy content of a 1.28 g sample of oatmeal that raises the temperature of 2.50 kg of water within a calorimeter from 25.0°C to 27.2°C?**
2. **Predict the final temperature of 2.50 kg of water within a calorimeter if the water is at 25.0°C before a 1.8 g piece of dried peach with an energy content of 18.5 kJ is burned.**

heat of reaction
the amount of heat energy absorbed or released during a chemical change

Stoichiometry and heat calculations

Certain physical and chemical changes are frequently used as energy sources, and chemists have recorded the amounts of heat energy involved in these changes in reference works such as chemical handbooks. This way, the change in heat energy can be determined without measuring it yourself. These listings are sometimes referred to as tables of *heats* of whatever process is being monitored. For example, a term used to describe the energy associated with a chemical reaction is **heat of reaction**. Heat energies involved in specific types of reactions can also have special names, as shown in **Table 9-6**.

Table 9-6
Transfers of Heat for Different Types of Changes

Heat	Symbol for a mole	Description
Heat of reaction	ΔH_r or ΔH_{rxn}	heat energy absorbed or released during a reaction
Heat of formation	ΔH_f^0	heat energy absorbed or released during synthesis of one mole of a compound from its elements at 298 K and 1 atm of pressure*
Heat of solution	ΔH_{sol}	heat energy absorbed or released when a substance dissolves in a solvent
Heat of combustion	ΔH_{comb}	heat energy released when a substance reacts with oxygen to form CO_2 and H_2O

*Because elements are not synthesized from themselves, their ΔH_f^0 is defined to be zero.

Solution for Additional Example 9B

For 1.500 g of sucrose: $\Delta H = m \times c_p \times \Delta t$

$\Delta H = (3000 \text{ g})(4.180 \text{ J/g·°C})(1.970°\text{C}) = 24\,700 \text{ J}$

The molar mass of sucrose is 342.34 g/mol, so for 1.00 mol of sucrose:

$$\Delta H = \frac{24.700 \text{ kJ}}{1.500 \text{ g}} \times \frac{342.34 \text{ g sucrose}}{1 \text{ mol}} = 5637 \text{ kJ/mol}$$

Heat refers to the total flow of energy during a chemical or physical change. The greater the quantity of the substance undergoing a change, the more heat will be transferred. For example, the heat of combustion for methane is 891 kJ/mol.

$$CH_4(g) + 2O_2(g) \longrightarrow CO_2(g) + 2H_2O(l)\ \Delta H_{comb} = -891\text{ kJ}$$

For 10.0 mol of methane, 8910 kJ of heat (10 times the heat of combustion) will be produced.

$$10CH_4(g) + 20O_2(g) \longrightarrow 10CO_2(g) + 20H_2O(l)\ \Delta H_{comb} = -8910\text{ kJ}$$

For 100. mol of methane, 89 100 kJ of heat will be produced.

$$100CH_4(g) + 200O_2(g) \longrightarrow 100CO_2(g) + 200H_2O(l)\ \Delta H_{comb} = -89\ 100\text{ kJ}$$

Because the amount of heat energy transferred is expressed in terms of kJ/mol of one of the products or reactants, this amount can be used in stoichiometry problems as if it were another mole ratio, as shown in Sample Problem 9C.

Sample Problem 9C *Stoichiometry and heat transfers*

What are the minimum masses of oxygen and gasoline that must be delivered by the carburetor or fuel-injection system to a car's engine in order for it to generate the 5000. kJ of energy needed to pass a slow-moving truck? (For this problem, assume that the gasoline is entirely isooctane, as shown in the equation.)

$$2C_8H_{18}(l) + 25O_2(g) \longrightarrow 16CO_2(g) + 18H_2O(l)\ \Delta H_{comb} = -1.099 \times 10^4\text{ kJ}$$

1 List

- **change in heat:** $\Delta H = -1.099 \times 10^4$ kJ
- **total energy needed:** 5000. kJ
- **molar masses of reactants:** 114.26 g C_8H_{18}/mol
  32.00 g O_2/mol
- **mass of C_8H_{18}:** ? g C_8H_{18}
- **mass of O_2:** ? g O_2

2 Set up

- **First, determine the proportionality factor for the reaction. The reaction shown in the problem produces more energy than you need. Because energy is a product, it can be treated as a limiting factor in the reaction.**
  $1.099 \times 10^4\text{ kJ} > 5000.\text{ kJ}$
- **Therefore, determine the mole amounts of each reactant that would be required to provide that much energy, starting with isooctane.**
  $$5000.\text{ kJ} \times \frac{2\text{ mol } C_8H_{18}}{1.099 \times 10^4\text{ kJ}} = 0.9099\text{ mol } C_8H_{18}$$
- **Perform a similar calculation to determine the necessary amount of oxygen.**
  $$5000.\text{ kJ} \times \frac{25\text{ mol } O_2}{1.099 \times 10^4\text{ kJ}} = 11.37\text{ mol } O_2$$

Sample Problem 9C continued on following page...

Additional Examples for Sample Problem 9C

a. A 5000 W generator fueled by propane, C_3H_8, produces 5.000 kJ/s. How many grams of O_2 and C_3H_8 are needed to operate this generator for 1 h?

$$C_3H_8 + 5O_2 \longrightarrow 4H_2O + 3CO_2 + 2.200 \times 10^3\text{ kJ}$$

b. A butane lighter is used to heat 500 g of water from 18.0°C to 55.0°C. How much butane and oxygen are needed to heat the water?

$$2C_4H_{10} + 13O_2 \longrightarrow 8CO_2 + 10H_2O + 5713\text{ kJ}$$

Solutions are at the bottom of the page.

Solutions for Additional Example 9C

a. $\frac{18\ 000\text{ kJ}}{2.200 \times 10^3\text{ kJ}} = 8.182$; The energy needed is 18 000 kJ/h. 8.182(1 mol)(44.11 g/mol) = 360.9 g C_3H_8

8.182(5 mol)(32.00 g/mol) = 1309 g O_2.

b. $\frac{77\text{ kJ}}{5713\text{ kJ}} = 0.0135$; The energy needed is (500 g H_2O)(4.180 J/g·°C)(37.0°C) = 77 kJ.

1.35×10^{-2} (2 mol)(58.14 g/mol) = 1.57 g C_4H_{10}; 1.35×10^{-2} (13 mol)(32.00 g/mol) = 5.62 g O_2

Answers to Practice 9C

1. mass of C_8H_{18} = 420 g
   mass of O_2 = 1500 g
2. 1750 kJ

Full solutions for items 1–2 are found on page 353A.

ASSESS

Alternative Assessment

Construct a heating curve for isooctane or mercury. Use your curve to calculate the heat of reaction when the temperature increases from –25°C to 75°C. (Hint: refer to **Figure 9-5**.)

Answers to Section Review

1. Temperature is a measure of the average kinetic energy of the particles' random motion. Heat is a measure of the total amount of heat transferred from an object of high temperature to one of low temperature.
2. enthalpy of fusion, enthalpy of reaction, enthalpy of vaporization
3. $(1.00 \times 10^3 \text{ kg}) \times (4.180 \text{ J/g•°C}) \times (2.0°\text{C}) = 8.36 \text{ kJ}$
4. $\Delta H = 4.5 \times 10^6$ kJ
5. $2C_4H_{10} + 13O_2 \longrightarrow 8CO_2 + 10H_2O$

   $\frac{10 \text{ g butane}}{58.14 \text{ g/mol}} \times \frac{2.878 \times 10^3 \text{ kJ}}{1 \text{ mol}} = 4.95 \text{ kJ}$
6. a. $\frac{17.1 \text{ g fat}}{38 \text{ kJ/g}} = 650 \text{ kJ}$

   b. $\frac{10 \text{ g fat}}{38 \text{ kJ/g}} = 380 \text{ kJ}$

   c. 2 g fat produces 76 kJ

   d. 13.8 g fat produces 524.4 kJ
7. frozen yogurt:

   $(8 \text{ oz}) \times \frac{0.8 \text{ g fat}}{3 \text{ oz}} \times \frac{38 \text{ Cal}}{1 \text{ g fat}} = 80.94 \text{ Cal}$

   yogurt with fruit: 114 Cal
8. 428 g H_2

*Full solutions for items **4** and **8** are found on page 353A.*

Sample Problem 9C continued from previous page...

- **Now calculate the masses of C_8H_{18} and O_2 using the molar masses.**

  0.9099 mol C_8H_{18} × 114.26 g/mol = **?** g C_8H_{18}

  11.37 mol O_2 × 32.00 g/mol = **?** g O_2

3 Estimate and calculate

- **Round off to make an estimate.**

  1 × 114 ≈ 114 g C_8H_{18}

  11 × 32 ≈ 352 g O_2

- **Calculate, and round off correctly.**

  Calculator answer: 1.03965174×10^2 g C_8H_{18}
  3.6384×10^2 g O_2

  Rounded answer: 104.0 g C_8H_{18}
  363.8 g O_2

Practice 9C

1. **Most automobile engines are only about 25% efficient, meaning that only 25% of the energy released by the combustion reaction is used for movement. Using this efficiency, what masses of oxygen and isooctane would need to be delivered to the engine for the car's kinetic energy to increase by 5000. kJ?**
2. **Chemical reactions such as respiration help provide energy for movement and keep your body warm. On average, each person in the United States consumes 110. g of sugar daily. If all of this sugar is assumed to be glucose, how much heat energy from glucose is transferred to one person through respiration each day?**

$$C_6H_{12}O_6(s) + 6O_2(g) \longrightarrow 6CO_2(g) + 6H_2O(l) \ \Delta H_{rxn} = -2870 \text{ kJ}$$

Section Review

1. What is the difference between heat and temperature?
2. Identify each of the following symbols: ΔH_{fus}, ΔH_{rxn}, and ΔH_{vap}.
3. How much energy is released by a reaction that raises the temperature of 1.00 kg of water in a calorimeter from 25.0°C to 27.0°C?
4. A swimming pool measures 6.0 m × 12.0 m and has a uniform depth of 3.0 m. The pool is full of water at a temperature of 20.°C. How much heat energy must be released by the pool's heater to raise the water temperature to 25°C?
5. Gaseous butane, C_4H_{10}, is burned in cigarette lighters. Write the balanced chemical equation for the combustion of butane. Butane's heat of combustion is 2.878×10^3 kJ/mol. How many kilojoules of heat would be provided by the combustion of 10.0 g of butane?
6. Use **Table 9-4** and **Table 9-5** to calculate the number of kilojoules provided by the fat in one serving of each of the following foods.
   a. french fries **b.** potato chips **c.** pretzels **d.** cheeseburger
7. Use **Table 9-4** to calculate the calories from fat in an 8 oz. serving of frozen yogurt versus the same size serving of yogurt with fruit.

8. **Story Link**
   Is more energy released when 428 g of H_2 or 428 g of isooctane, C_8H_{18}, react with an excess of oxygen?

$$2H_2(g) + O_2(g) \longrightarrow 2H_2O(l) \ \Delta H_{comb} = -571.6 \text{ kJ}$$

$$2C_8H_{18}(l) + 25O_2(g) \longrightarrow 16CO_2(g) + 18H_2O(l) \ \Delta H_{comb} = -10\,990 \text{ kJ}$$

How does enthalpy drive changes?

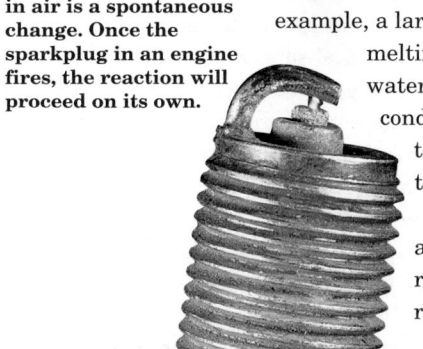

Section Objectives

Explain the relation among heat of reaction, enthalpy change, and stability.

Interpret an energy diagram of the reaction pathway for an exothermic reaction.

Explain the relationship between activation energy and chemical changes.

Use the kinetic molecular theory to explain why some exothermic reactions do not require an input of energy.

Use Hess's law to solve problems involving physical and chemical changes.

Driving forces

spontaneous change *a change that will occur because of the nature of the system, once it is initiated*

The melting of ice on a tabletop is an example of a **spontaneous change**. Spontaneous changes occur primarily in one direction. Once ice melts, it won't refreeze unless the temperature changes. In other words, ice melting is not reversible at room temperature.

There are several important ideas to keep in mind about spontaneous changes. First, spontaneity does not mean that a reaction occurs quickly. For example, even though ice melts spontaneously at 1°C, the change usually takes place very slowly. Another related point is that some spontaneous reactions occur only when a small amount of energy is added to the system. The combustion of a gasoline-air mixture is a spontaneous change. Even though an outside influence, a spark, as shown in **Figure 9-7**, must be provided to get the reaction going, once the combustion has occurred, the reaction is not reversible under ordinary conditions. No one has ever observed carbon dioxide and water vapor in the air spontaneously re-forming oxygen and isooctane or other components of gasoline.

Chem Fact
The change of diamond to graphite is a spontaneous reaction under normal conditions, but it is extremely slow.

However, some reactions can be reversed if conditions change. For example, a large drop in temperature will induce freezing instead of melting. In reality, many chemical changes are like the ice-to-water physical change: they are reversible, depending upon the conditions. Some reactions that are spontaneous and appear to be complete under normal conditions can be reversed if the conditions are changed to an extreme.

Figure 9-7

Under most conditions, the burning of gasoline in air is a spontaneous change. Once the sparkplug in an engine fires, the reaction will proceed on its own.

So far, you have explored some details of how reactions and other changes release or absorb heat. But why do some reactions happen, and others don't? Are there ways to induce reactions that wouldn't take place otherwise?

The answers to these questions rely upon a model that will describe the driving forces, or underlying tendencies, that cause or resist change.

Heat transfers are an important part of the model of driving forces. But to understand how to use heat flow and temperature to predict and control reactions, a more complete model is necessary.

Section 9-2

FOCUS

Lesson Starter
Begin the lesson by observing the rise in temperature when NaOH dissolves in water in the following demonstration. Ask students why the NaOH does not recrystallize when the temperature of the solution returns to room temperature.

Demonstration 5
Spontaneous changes
Approximate time: 5 min

1. Fill a 25 mm × 125 mm Pyrex test tube with 25 mL tap water and record the temperature.
2. Measure the mass of 1 g of NaOH pellets on a sheet of weighing paper.
3. Add the NaOH to the water, stir gently, and determine what changes occur in the NaOH and the temperature of the solution.

SAFETY
Wear safety goggles and a lab apron. Use a nonmercury thermometer. Follow first aid instructions on the label and in the MSDS if NaOH gets in the eyes or on skin or clothing.

DISPOSAL
Neutralize the solution with 1 M HCl, and pour it down the drain.

TEACH

Do You Know?

The word *enthalpy* comes from the Greek *enthalpein* meaning "to warm in."

Teaching Tip

You may wish to point out that **Figure 9-8a** is an exothermic reaction and associate it with the energy diagrams in **Figure 9-10** or **Figure 9-12**.

Exothermic Reaction $C_{12}H_{22}O_{11}(s) \xrightarrow{H_2SO_4} 12C(s) + 11H_2O(l)$

Before **After**

Figure 9-8a
The total enthalpy of any substance can never be calculated or measured, but the *change* in enthalpy can be measured. This reaction is exothermic, and energy is released. The products must have less enthalpy than the reactants have.

Endothermic Reaction $Ba(OH)_2{\cdot}8H_2O(s) + 2NH_4NO_3(s) \longrightarrow 2NH_3(aq) + 10H_2O(l) + Ba(NO_3)_2(aq)$

Before **After**

9-8b
This reaction absorbs so much heat that the condensation on the flask freezes it to the block of wood. The products have more enthalpy than the reactants have.

Enthalpy is the total of a substance's energy

enthalpy
total energy content of a system

In an exothermic change, such as the combustion of isooctane, the heat of reaction is negative, indicating heat energy is released. This implies that somehow the reactants have more energy "stored" within them than the products do. The name given to the total energy that substances contain at constant pressure is **enthalpy**, represented by the symbol H. Any information about enthalpies is gained only by comparing systems before and after a change, as shown in **Figure 9-8**. Thus, the actual total enthalpy of a substance, whether product or reactant, can never be measured or calculated directly.

Enthalpy change is equal in magnitude to heat of reaction at constant pressure

enthalpy change
heat energy released or absorbed when a physical or chemical change occurs at constant pressure

For most physical and chemical changes, the enthalpy of the system before the change is different from that after the change. This **enthalpy change** is represented by the symbol ΔH. In fact, at constant pressure, the enthalpy change for a chemical change is the same amount as the heat of reaction. Thus, ΔH can be read as "enthalpy change," which is a comparison of the total enthalpies of the products and the reactants.

$$\Delta H = H_{products} - H_{reactants}$$

Chemical equations sometimes include the value of ΔH. Exothermic reactions are assigned negative ΔH values, and endothermic reactions have positive ΔH values.

Exothermic

$$C_6H_{12}O_6(s) + 6O_2(g) \longrightarrow 6CO_2(g) + 6H_2O(l) \qquad \Delta H = -2870 \text{ kJ}$$

Endothermic

$$6CO_2(g) + 6H_2O(l) \longrightarrow C_6H_{12}O_6(s) + 6O_2(g) \qquad \Delta H = +2870 \text{ kJ}$$

Compare the enthalpies of a sample of water before and after condensation, as shown in **Figure 9-9**. As water vapor, the water molecules are widely separated. For the sample as a whole, this corresponds to a high enthalpy state. After condensing into a liquid, the molecules are much closer to each other, which corresponds to a lower enthalpy state. The difference in enthalpy between these two states equals the amount of heat change during the process, the heat of vaporization. Thus, the condensation of water vapor releases heat. The enthalpy decreases.

$$H_2O(g) \longrightarrow H_2O(l) \qquad \Delta H = -40.6 \text{ kJ/mol}$$

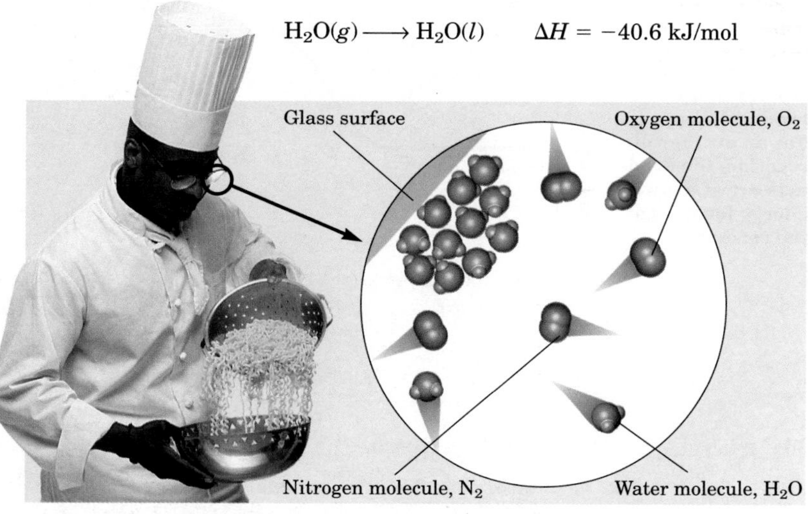

Figure 9-9
Condensation is an exothermic process. Gaseous water molecules come together on the cool surface of the man's glasses, forming a liquid, which results in a lower enthalpy state.

Decreasing enthalpy drives some spontaneous changes

A reaction that liberates energy is more likely to occur than one that does not. This is one of the underlying *driving forces* for chemical and physical changes. In general, changes that involve a decrease in enthalpy are favored. The more energy released by an exothermic reaction, the greater the decrease in enthalpy, and the more likely it is that the reaction will be spontaneous.

Because energy is released in all exothermic reactions, the products of an exothermic reaction have a lower energy state than that of the reactants and are more stable from an energy perspective. This was true of the potassium and water reaction described in Chapter 5 and of the combustion of hydrogen described in Chapter 6.

Possible Misconception

If students are confused about why exothermic reactions have negative ΔH values, remind them that the use of a negative or positive sign shows only the *direction* of heat flow. Heat flows *out* into the surroundings as products are formed, leaving them with a lower heat content than the reactants.

Do You Know?

When the dew forms, tremendous amounts of energy are added to the air, so the temperature of the air varies only slightly from the temperature at which the dew forms. Because of this, weather forecasters are able to determine the low temperature for the day from the dew point.

Demonstration 6

Energy of activation and the decomposition of H_2O_2

The catalyst, MnO_2, lowers the energy of activation for the decomposition of H_2O_2.

Approximate time: *5 min*

1. Write the following equation on the overhead.

$$H_2O_2 \longrightarrow H_2O + O_2$$

2. Fill two petri dishes half full with 3% hydrogen peroxide (drugstore strength) and place them under the equation.
3. Sprinkle no more than 0.05 g of powdered manganese dioxide, MnO_2, into one of the dishes. The evolution of oxygen will now be perceptible.

SAFETY
Wear safety goggles and a lab apron. Use only clean MnO_2.

DISPOSAL
Decant the liquid, and pour it down the drain. Put the leftover MnO_2 into the trash.

Reaction pathways

Most reactions need energy to get started

The combustion of isooctane in gasoline is a strongly exothermic reaction. If such a change in enthalpy is a driving force for chemical reactions, why doesn't isooctane react instantly when stored in a can with air? Why is a spark necessary to ignite the gasoline and oxygen?

From the discussion in Chapter 6 of the combustion of H_2 and O_2 to make H_2O, you already know the answer. If new bonds are to form to make products, the reactants must be brought together. Because of the electrons surrounding atoms, the reactants tend to repel each other, requiring energy to overcome these forces. This initial input of energy is called **activation energy**. The symbol for activation energy is E_a or E_{act}.

activation energy *the minimum amount of energy that must be supplied to a system to start a chemical change*

One way to explore this idea is to examine a reaction pathway, which shows the relationship between energy and the progress of the reaction. The reaction pathway shown in **Figure 9-10** is the general pattern for all exothermic reactions.

Exothermic Reaction Pathway

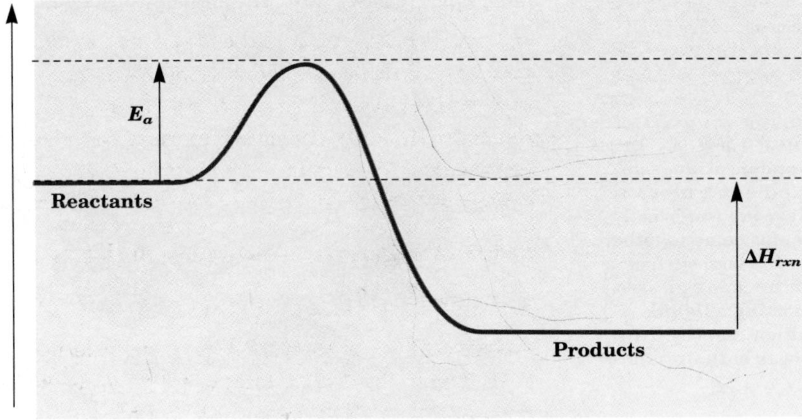

Figure 9-10 A reaction pathway shows how the relative energy levels of the reactants and products compare. For an exothermic reaction, the products are at a lower energy level than the reactants.

The spark provides initial activation energy needed to bring the reactants together. The transfer of energy from the spark to the reaction system causes an increase in the energy level, as shown by the peak on the reaction pathway. Energy is released as the products separate. The released energy is able to serve as activation energy for other reactant molecules. The net change in bond energy between the reactants and the products is the heat of reaction, or the enthalpy change.

Concept Review

Using reaction pathways

Use **Figure 9-10** to answer the following questions.

9. Explain whether ΔH_{rxn} would change if E_a were increased, but the reactants and the products remained at the same enthalpy levels.
10. Explain why the reverse reaction is less likely than forward reaction. (Hint: what is E_a for the reverse reaction?)

Answers for Concept Review items and Practice problems begin on page 841.

Answers to Concept Review

9. No; the amount of heat in the initial and final states has not changed, so ΔH_{rxn} has not changed.
10. E_a for the reverse reaction is greater than that for the forward reaction, so more heat is required for the reverse reaction. The reactants are at a higher enthalpy than the products, so they are less thermodynamically stable.

Visual Strategy

Figure 9-10

Almost all reactions require an input of energy to start them. The upward movement of the curve represents the energy received by the system. The downward movement of the curve represents energy that the system releases. When the output is greater than the input, the reaction is exothermic.

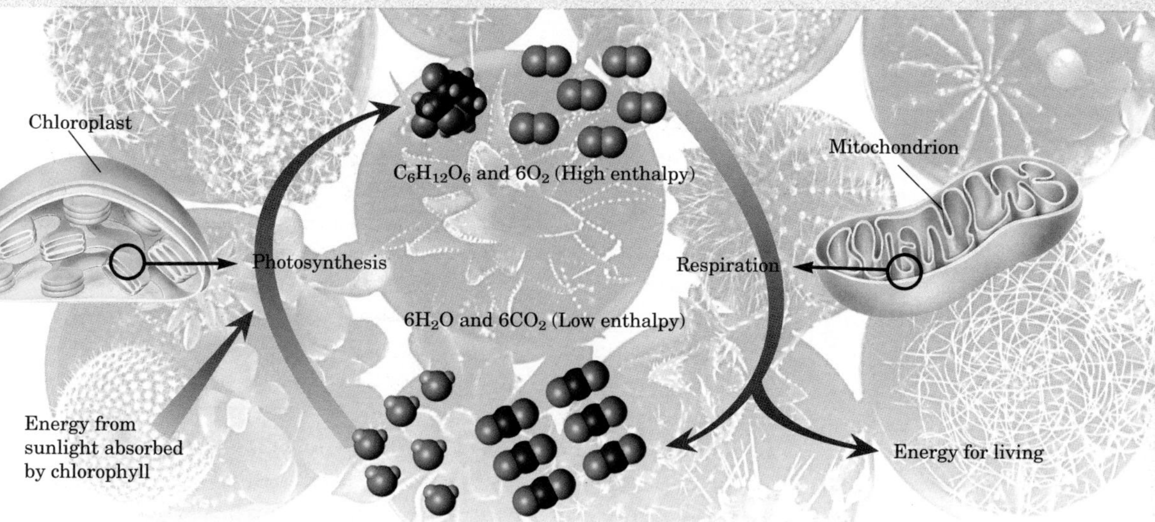

Figure 9-11 **Photosynthesis is an endothermic reaction taking place in a chloroplast. Respiration is an exothermic reaction taking place in a mitochondrion. Both require activation energy.**

Endothermic changes require a source of energy

In general, exothermic changes are favored over endothermic changes, but many endothermic changes can take place. For example, plants use the endothermic photosynthesis reaction to make sugar. This sugar, when decomposed through cellular respiration, provides the energy that plants and other living things need to live, as shown in **Figure 9-11**. You may remember that the equations for these two reactions are the reverse of one another. *The change in enthalpy of the reverse reaction will be numerically equal to that of the forward reaction, but the sign will be changed.*

Photosynthesis: $6CO_2(g) + 6H_2O(l) \longrightarrow C_6H_{12}O_6(s) + 6O_2(g) \quad \Delta H = +2870 \text{ kJ}$

Respiration: $C_6H_{12}O_6(s) + 6O_2(g) \longrightarrow 6CO_2(g) + 6H_2O(l) \quad \Delta H = -2870 \text{ kJ}$

How is it that photosynthesis, an endothermic reaction, can occur, even though the driving force of enthalpy favors the opposite reaction? You know that plants require sunlight to survive. This sunlight is the source of the energy for the photosynthesis reaction. Endothermic reactions occur if there is a sufficient source of energy available.

In plants, chlorophyll molecules absorb energy from sunlight and transform it into chemical energy that can be used to overcome several activation energy barriers and to drive the complex series of reactions that constitute photosynthesis.

Energy is unevenly distributed among particles

If reactions need activation energy to get started, how is it that some spontaneous reactions, like the reaction of potassium and water to form potassium hydroxide, proceed without added energy, whereas some, like the isooctane combustion reaction, do require added energy?

$$2K(s) + 2H_2O(l) \longrightarrow 2KOH(aq) + H_2(g) \qquad \Delta H = -393 \text{ kJ}$$

Actually, the only difference between the potassium-water reaction and other reaction pathways is that the activation energy for the potassium reaction is lower, as shown in **Figure 9-12** on the next page.

Teaching Tip
Effective collisions
When the reactants reach the top of the energy hill, they can fall backward as easily as they can go forward. Energy requirements for an effective collision are discussed in Chapters 10 and 14.

Teaching Tip
Diagrams for endothermic reactions
An endothermic reaction can be modeled by tracing the curve in **Figure 9-10** backward. Compare the input and output of energy. Point out how the heat of the reverse reaction differs from that of the exothermic forward reaction.

Further Applications
Cold packs feel cold because they absorb heat energy from the skin to drive their reaction(s). If a good energy source is not available, the reaction in the cold pack will proceed very slowly.

Historical Note
German chemist Richard Willstätter was the first person to determine the structure of chlorophyll.

Demonstration 7

Activation energy

The activation energy for the reaction between Ca and H_2O is so low that the reaction occurs at room temperature, but Mg requires boiling H_2O because the activation energy for this reaction is much higher.

Approximate time: 5 min

1. Fill two 250 mL beakers half full of warm water.
2. Place 0.5 g of powdered or granular Mg into the beaker, and cover with wire gauze.
3. Using tongs, place a piece of calcium half the size of a pea into the water, cover with wire gauze, and stand back until the reaction is finished. Why did one reaction "work" while the other one did not?

SAFETY

Wear safety goggles and a lab apron. Have students stand 10 ft back from demonstration area.

DISPOSAL

Pour the water off the Mg into the drain. Dry the Mg and save for next time. After all the Ca has reacted, neutralize the solution with 1 M HCl, and pour it down the drain.

Do You Know?

A curve of this shape is also known as a Gaussian distribution after German mathematician Karl Friedrich Gauss (1777–1855).

Chapter Connections

Velocity and kinetic energy distributions for molecules are discussed further in Chapters 10 and 14.

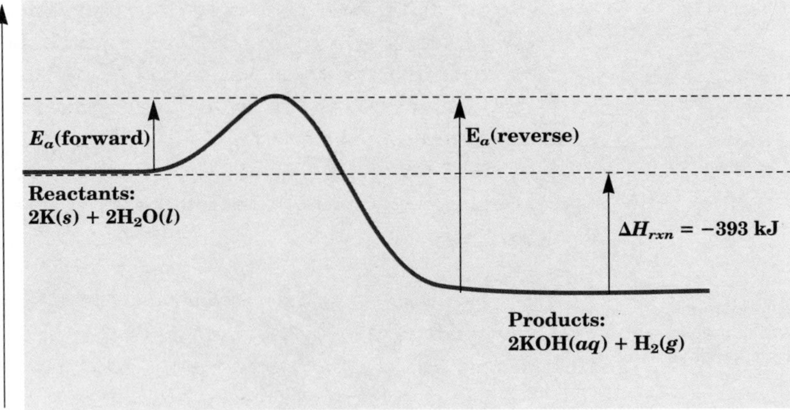

Figure 9-12 The reaction pathway for the potassium-water reaction requires little activation energy, so it is easy to get started once the reactants are in contact.

From the kinetic molecular theory, you already know that particles, such as molecules, atoms, and ions, are constantly in motion and possess different amounts of energy, depending on how rapidly they are moving. This, in turn, is related to their temperature. The higher the temperature of the substance, the faster, on average, the particles are moving.

Even though a sample of matter, such as a chunk of potassium, has a characteristic average energy that depends on its temperature, not all of its particles have that same energy. Most have an energy close to that of the average, but some have more, and some have less. That is why the graph of kinetic energy versus number of particles has the shape similar to a normal distribution curve: a peak in the middle with a "shoulder" on each side. This situation is summed up in **Figure 9-13**. In Chapter 14, you will see exactly how changing the temperature can change the shape of this curve.

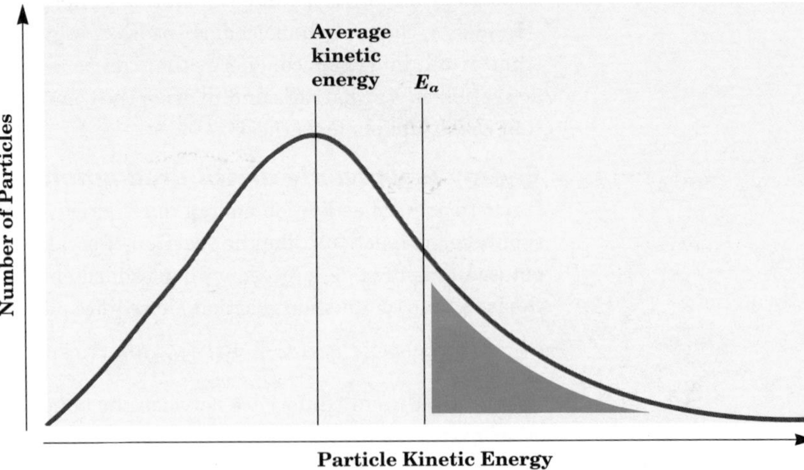

Figure 9-13 In any sample of particles, some will have kinetic energies higher than the necessary activation energy for a reaction. These particles will react immediately.

Visual Strategy

Figure 9-12

Compare this energy diagram with the one in **Figure 9-10**. Point out that the difference between the input and output energies is the heat of reaction. Refer back to the photograph of this reaction in **Figure 5-1**, and point out that the products are also more thermodynamically stable than the reactants.

According to the law of conservation of energy, particles can convert their energy of motion into other forms of energy, including the activation energy necessary to make a reaction happen. If the activation energy has the value indicated by the line marked E_a on the horizontal axis of the graph, the molecules represented by the shaded zone already have enough energy to react. This explains why heat, light, or a spark is not needed for the reaction to occur—some of the particles have enough activation energy to begin the reaction without requiring any additional input of energy.

If the reaction is exothermic, as the first particles react, more energy is released as the products are formed. The potassium and water particles that didn't have enough energy to react absorb this energy in the form of heat, which increases their motion and kinetic energy. Some of these particles now have enough energy to react and release still more energy. This process continues until nearly all of the particles have reacted.

Calculating enthalpy changes

Combining enthalpy changes using Hess's law

Hess's law
the total enthalpy change for a chemical or physical change is the same whether it takes place in one or several steps

Any two processes that start with the reactants in the same condition and finish with the products in the same condition will involve the same enthalpy change. This statement is the basis of **Hess's law,** which states that the overall enthalpy change in a reaction is equal to the sum of the enthalpy changes for the individual steps in the process. If the heats of reaction for each of these smaller reactions are known, then they can be combined to give the heat change for the overall reaction.

How can you predict the enthalpy change for the decomposition of 5.00 mol of ice to produce hydrogen gas and oxygen gas from the resulting water?

$$2H_2O(s) \longrightarrow 2H_2(g) + O_2(g)$$

So far in this chapter, several different values for enthalpy changes associated with water have been discussed; they are summarized in **Table 9-7**. According to the principle behind Hess's law, if the products and reactants are the same for two reactions, the overall enthalpy change will be the same. In effect, it doesn't matter whether the ice decomposes into hydrogen and oxygen gas directly or melts first. Either way, you will be starting with ice and ending with hydrogen and oxygen, so the net enthalpy change will be the same. The enthalpy changes in **Table 9-7** can be considered to be individual steps that are each a part of a greater overall change. Enthalpy changes for steps can be added together to describe a net enthalpy change. This addition can occur whether the enthalpy changes are for physical changes or for chemical changes.

Table 9-7
Some Enthalpy Changes Involving H_2O

Reaction	Enthalpy change (kJ)
$2H_2(g) + O_2(g) \longrightarrow 2H_2O(l)$	$\Delta H_{rxn} = -571.6$
$2H_2O(l) \longrightarrow 2H_2(g) + O_2(g)$	$\Delta H_{rxn} = +571.6$
$H_2O(l) \longrightarrow H_2O(g)$	$\Delta H_{vap} = +40.6$ (at 100°C)
$H_2O(s) \longrightarrow H_2O(l)$	$\Delta H_{fus} = +6.00$ (at 0°C)

Demonstration 8

Hess's law and the heat of neutralization for NaOH

Approximate time: 10 min

1. Write the following three reactions on the board:
   $NaOH + H_2O \longrightarrow Na^+ + OH^- + H_2O + \Delta H_1$
   $Na^+ + OH^- + H_3O^+ + Cl^- \longrightarrow 2H_2O + Na^+ + Cl^- + \Delta H_2$
   $NaOH + H_3O^+ + Cl^- \longrightarrow 2H_2O + Na^+ + Cl^- + \Delta H_3$
   These equations obey Hess's law, so if equivalent moles are used, the sum of the temperature changes for equations 1 and 2 should equal the temperature change for equation 3.
2. **Reaction 1**: Place 50 mL of distilled water into a Styrofoam cup, and record the temperature. Add 1 g of NaOH, stir until dissolved, record the temperature, and determine the change.
3. **Reaction 2**: Pour 25 mL of 1 M HCl into a Styrofoam cup, and record the temperature. Add 25 mL of 1 M NaOH, stir, record the temperature, and determine the change.
4. **Reaction 3**: Pour 50 mL of 1 M HCl into a Styrofoam cup, and record the temperature. Add 1 g of NaOH pellets. Stir until dissolved, record the temperature, and determine the change.
5. Compare $\Delta T_1 + \Delta T_2$ with ΔT_3.

SAFETY
Wear safety goggles and a lab apron. Use a nonmercury thermometer. Follow first aid instructions on the label and in the MSDS if acid or base gets in the eyes or on skin or clothing.

DISPOSAL
Rinse out the Styrofoam cups, and put them into the trash. Combine all liquids, neutralize them with 1 M HCl or NaOH as appropriate, and pour down the drain.

Historical Note

The paths of chemistry and physics crossed when Swiss-Russian chemist Germain Henri Hess (1802–1850) announced in 1840 that the path of a reaction is independent of the total energy (Hess's law). Hess is often referred to as the father of thermochemistry.

The last equation in **Table 9-7** is the only one that begins with ice, like the net reaction on the previous page. If it is multiplied by two, it will match the amount of ice in the net reaction. The second equation in **Table 9-7** is the only one containing the products, hydrogen and oxygen gas, that are also the products of the net reaction. If you add these equations and cancel substances that appear in the same form and same amount on both sides of the equations, you get the following result.

$$2H_2O(s) \longrightarrow 2H_2O(l) \qquad \Delta H_{fus} = +12.00 \text{ kJ}$$
$$2H_2O(l) \longrightarrow 2H_2(g) + O_2(g) \qquad \Delta H_{rxn} = +571.6 \text{ kJ}$$

$$2H_2O(s) + \cancel{2H_2O(l)} \longrightarrow \cancel{2H_2O(l)} + 2H_2(g) + O_2(g) \quad \Delta H_{total} = +583.6 \text{ kJ}$$

This equation matches the net reaction. Thus, decomposing ice involves the same net energy as first melting it and then decomposing the liquid water. But the question asked how much enthalpy was released when *5.00 mol H_2O,* not *2.00 mol H_2O,* reacted. A simple stoichiometry calculation provides the answer.

$$\frac{583.6 \text{ kJ}}{2.00 \cancel{\text{mol}} \text{ H}_2\text{O}} \times 5.00 \cancel{\text{mol}} \text{ H}_2\text{O} = 1460 \text{ kJ}$$

Another way of interpreting Hess's law states that the enthalpy change of a reaction is equal to the enthalpy of formation of the products minus the enthalpy of formation of the reactants, when stoichiometric relationships are taken into account.

$$\Delta H_{rxn} = \text{sum of } \Delta H^o_{f,\ products} - \text{sum of } \Delta H^o_{f,\ reactants}$$

This approach treats a reaction as if all of the reactants were decomposed into elements. Then, those elements proceed to form the bonds needed to make the products. The superscript zero indicates standard conditions.

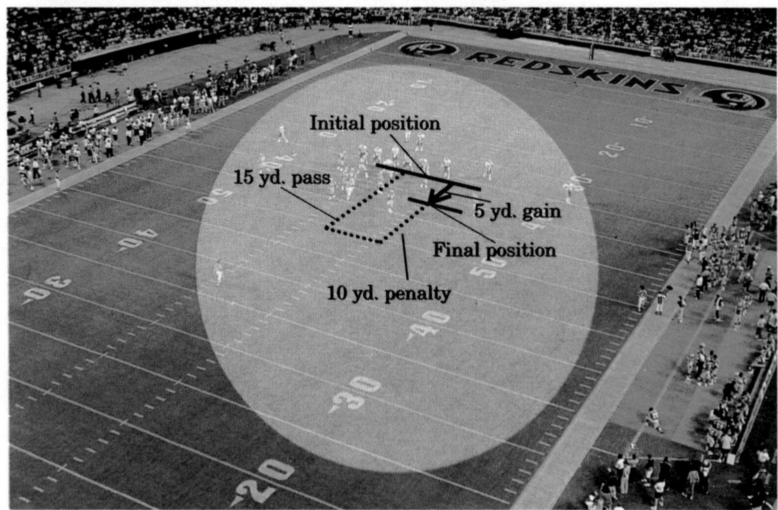

Figure 9-14
In football, as in Hess's law, only initial and final conditions matter. If a football team gains 15 yards on a pass play but has a 10-yard penalty, the team's field position is no different than if the team had gained only 5 yards during the play.

Sample Problem 9D
Heats of formation and Hess's law

The Romans used CaO as a mortar for building. CaO was mixed with water to produce $Ca(OH)_2$, which reacted slowly with CO_2 in air to form limestone, $CaCO_3$. This reaction is represented as follows.

$Ca(OH)_2(s) + CO_2(g) \longrightarrow H_2O(g) + CaCO_3(s)$

Determine the ΔH value for this reaction using the data given below.

Substance	ΔH_f^o (kJ/mol)	Substance	ΔH_f^o (kJ/mol)
$Ca(OH)_2$ (s)	−986.1	$H_2O(g)$	−241.8 kJ
$CO_2(g)$	−393.5	$CaCO_3(s)$	−1206.9 kJ

1 List

- **Reactants:**

1 mol $Ca(OH)_2(s)$: $\Delta H_f^o = -986.1$ kJ
1 mol $CO_2(g)$: $\Delta H_f^o = -393.5$ kJ

- **Products:**

1 mol $H_2O(g)$: $\Delta H_f^o = -241.8$ kJ
1 mol $CaCO_3(s)$: $\Delta H_f^o = -1206.9$ kJ

- **Heat of reaction:** $\Delta H_{rxn} = ?$ kJ

2 Set up

- **According to Hess's law, the sum of the ΔH_f^o values for the products minus the sum of the ΔH_f^o values of the reactants will give the net ΔH value for the reaction.**

[sum of ΔH_f^o values for products] − [sum of ΔH_f^o values for reactants] = ? kJ

$[(-241.8 \text{ kJ}) + (-1206.9 \text{ kJ})] - [(-986.1 \text{ kJ}) + (-393.5 \text{ kJ})] = ?$ kJ

3 Estimate and calculate

- **Round off to make an estimate.**

$[(-240) + (-1200)] - [(-990) + (-390)] \approx$

$[-1440] - [-1380] \approx -60$

- **Calculate, and round off correctly.**

$\Delta H_{rxn} = -69.1$ kJ/mol

Practice 9D

1. **Under extreme pressure, graphite can be converted to diamond. Because both are forms of pure carbon, each undergoes combustion to make carbon dioxide. Given the following heats of reaction for the combustion reactions, calculate the enthalpy change for converting graphite to diamond.**

$C(graphite) + O_2(g) \longrightarrow CO_2(g)$ $\quad \Delta H_{rxn} = -393.5$ kJ/mol C

$C(diamond) + O_2(g) \longrightarrow CO_2(g)$ $\quad \Delta H_{rxn} = -395.4$ kJ/mol C

2. **Calculate the heat of reaction for changing 1.00 mol of silica, SiO_2, that has been extracted from sand into the pure silicon that is needed to make computer chips. Use the following data.**

$SiO_2(s) + 2C(s) \longrightarrow$ impure $Si(s) + 2CO(g)$ $\quad \Delta H_{rxn} = +689.9$ kJ

impure $Si(s) + 2Cl_2(g) \longrightarrow$ pure $SiCl_4(g)$ $\quad \Delta H_{rxn} = -657.0$ kJ

pure $SiCl_4(g) + 2Mg(s) \longrightarrow 2MgCl_2(s)$ + pure $Si(s)$ $\quad \Delta H_{rxn} = -625.6$ kJ

Additional Examples for Sample Problem 9D

ΔH data is in Appendix A, **Table A-11**.

a. How much energy is required to turn 10 mol of steam into elemental hydrogen and oxygen?

$2H_2O(g) \longrightarrow 2H_2(g) + O_2(g)$

b. What is the heat of reaction for the decomposition of potassium chlorate into potassium chloride and oxygen? ΔH_f for $KClO_3(s)$ is equal to –390.83 kJ/mol.

$2KClO_3(s) \longrightarrow 2KCl(s) + 3O_2(g)$

c. What is the heat of reaction for the reduction of iron(III) oxide with carbon monoxide?

$Fe_2O_3(s) + 3CO(g) \longrightarrow 2Fe(s) + 3CO_2(g)$

Solutions are at the bottom of the page.

Answers to Practice 9D

1. $C_{(graphite)} \longrightarrow C_{(diamond)}$

$\Delta H_{rxn} = (-393.5 \text{ kJ/mol}) + (+395.4 \text{ kJ/mol}) = 1.9 \text{ kJ/mol}$

2. $SiO_2(s) + 2C(s) + 2Cl_2(g) + 2Mg(s) \longrightarrow 2CO(g) + 2MgCl_2(s)$ + pure $Si(s)$

$\Delta H_{rxn} = (+689.9 \text{ kJ}) + (-657.0 \text{ kJ}) + (-625.6 \text{ kJ}) = -592.7 \text{ kJ}$

Solutions for Additional Examples 9D

a. Use the equations in **Table 9-7** on page 329.
$\Delta H_{rxn} = +571.6 \text{ kJ} + 2(-40.6) = +490.4$ kJ for 2 moles of steam.
Ten moles of steam, $5(+490.4 \text{ kJ}) = 2452$ kJ

b. ΔH_{rxn} = sum of ΔH_f^o of products − sum of ΔH_f^o of reactants
$\Delta H_{rxn} = [2(-435.87) + 3(0)] - [2(-390.83)] = -90.08$ kJ/mol

c. ΔH_{rxn} = sum of ΔH_f^o of products − sum of ΔH_f^o of reactants
$\Delta H_{rxn} = (-1180.53) - (-824.2 + -331.59) = -24.74$ kJ/mol

ASSESS

Alternative Assessment

Have students write one or two paragraphs that describe the role of bond energy in determining the value of ΔH for a specific reaction and compare the heat content of the products from an exothermic reaction with the heat content of products from an endothermic reaction.

Answers to Section Review

11. a. NO_2; less heat is absorbed when the elements combine to form N_2O_4 than when they form NO_2. This places NO_2 higher on the enthalpy axis.
    b. $\Delta H_{rxn} = (+9.1) - 2(33.1) = -57$ kJ/mol

12.

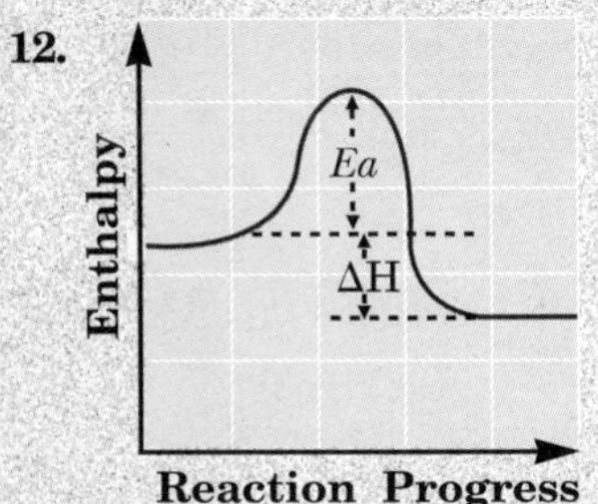

13. a. friction created when the match is struck
    b. energy of the UV light waves
    c. kinetic energy of dissolved particles

14. $\Delta H_{rxn} = \Delta H_f$ of products $- \Delta H_f$ of reactants $= (-393.5) - (-110.5) = -283.0$ kJ/mol

15. greater than

16. -325.1 kJ; -186.2 kJ; -511.3 kJ

17. $2H_2O(s)$ + heat $\longrightarrow 2H_2(g) + O_2(g)$

0 | +12 kJ | $2H_2O(s)$
100 | $2H_2O(l)$
200 | +583.6 kJ
300
400 | +571.6 kJ
500
600 | $2H_2(g)$, $O_2(g)$

18. ΔH_{rxn} is not affected.

Section Review

11. Nitrogen dioxide, $NO_2(g)$, has a heat of formation of 33.1 kJ/mol. Dinitrogen tetroxide, $N_2O_4(g)$, has a heat of formation of 9.1 kJ/mol. Which compound is likely to have the highest enthalpy? What is the enthalpy change for synthesizing N_2O_4 from NO_2?
12. Draw and label an energy pathway for a typical exothermic reaction.
13. Describe the sources of activation energy in each of the following reactions.
    a. A match is struck against a match cover and bursts into flame.
    b. When exposed to ultraviolet light, hydrogen and chlorine combine to form hydrogen chloride.
    c. Vinegar and baking soda react instantly to make carbon dioxide and sodium acetate solution.
14. Use Hess's law to calculate the heat of combustion of carbon monoxide to form carbon dioxide, given that the heat of formation for carbon monoxide is -110.5 kJ/mol and that of carbon dioxide is -393.5 kJ/mol. (Hint: remember that the heat of formation represents the enthalpy needed to synthesize a substance from its elements at 298 K and 1 atm. Under those conditions, oxygen is already in the form of oxygen molecules, so no synthesis reaction is necessary. Thus, its heat of formation is defined to be 0.)
15. Is the activation energy greater than, equal to, or less than the change in enthalpy of an endothermic reaction? (Hint: examine **Figure 9-10**, the energy diagram for an *exothermic* process.)
16. The diagram below represents the interpretation of Hess's law for the following reaction.

$$Sn(s) + 2Cl_2(g) \longrightarrow SnCl_4(l)$$

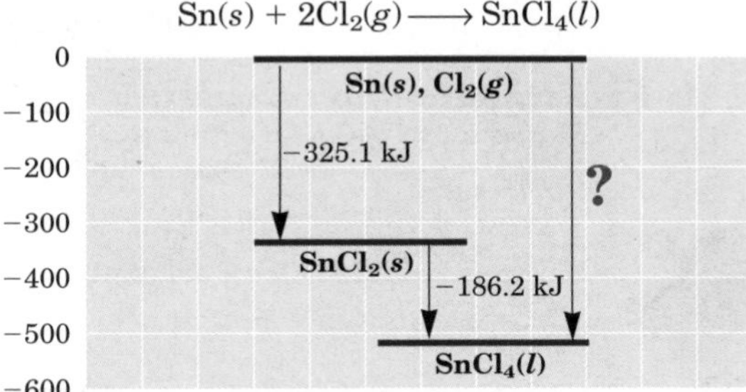

Use the diagram to determine the ΔH for each step and the net reaction

$$Sn(s) + Cl_2(g) \longrightarrow SnCl_2(l) \qquad \Delta H = ?$$
$$SnCl_2(s) + Cl_2(g) \longrightarrow SnCl_4(l) \qquad \Delta H = ?$$
$$Sn(s) + 2Cl_2(g) \longrightarrow SnCl_4(l) \qquad \Delta H = ?$$

17. Use the data in **Table 9-7** to construct a diagram like that in item **16** to graphically show the decomposition of ice to form gaseous hydrogen and oxygen. Your diagram should include the enthalpy change for each step along with the total enthalpy change for the net reaction.
18. The activation energy needed for a particular reaction can be reduced by using a catalyst. How would a reduction of E_a affect ΔH_{rxn}?

How does entropy drive changes?

Section Objectives

Define entropy, and provide examples of high- and low-entropy situations.

Explain the relation between a change in entropy and stability.

Use standard entropy values in calculations.

Entropy and stability

You learned in Chapter 2 that solids have a very specific and rigid arrangement that is less flexible than the arrangement of liquids. Thus, an ice cube left on a countertop spontaneously changes from an ordered substance (a solid) to a less-ordered substance (liquid). The tendency in nature is to more disorder. The measure of the degree of disorder is called **entropy**. For a system that can exist in two states, the one with higher entropy or disorder tends to be more stable. Usually, it is far easier to go from the low-entropy state to the high-entropy state. For example, it is easy to imagine throwing an intact jigsaw puzzle on the floor and having the pieces get mixed up. However, one rarely can throw a jumbled puzzle on the floor and have all the pieces bounce into place. Examples of high and low entropy states are shown in **Figure 9-15**.

entropy
a measure of the randomness or disorder of a system

Low entropy

High entropy

Figure 9-15
By comparing the photographs you can see changes from low entropy to high entropy. Melted ice, the broken pitcher, and the jumbled jigsaw puzzle all represent an increase in the disorder of the system.

Section 9-3

FOCUS

Lesson Starter

In the following demonstration the heat of solution for H_2SO_4 in water causes the temperature increase in the first cup, but the temperature goes down in the second cup because the heat is being used to increase the entropy of the system.

Demonstration 9

Entropy change for ice

Approximate time: 5 min

1. Before class, make 9 M H_2SO_4 by *slowly* adding 100 mL of concentrated H_2SO_4 to 100 mL of H_2O while stirring. Cool to room temperature before beginning demonstration.
2. Place 100 g of water at 0°C into a Styrofoam cup.
3. Record the temperature.
4. Add 100 mL of 9 M H_2SO_4 to the cup and stir.
5. After 1 min, record the temperature again.
6. Repeat steps **2–5** for100 g of ice at 0°C.

SAFETY

When diluting H_2SO_4, use a hood that is known to operate properly. Wear goggles, lab apron, gloves, and face shield. Be sure that an operating safety shower or eyewash station are nearby and that another person is present to assist or call for help. For the demonstration wear safety goggles, face shield, and lab apron. Use a non-mercury thermometer. Have students stand 15 ft away from the demonstration area. Follow first aid instructions on the labels and in the MSDS if acid or base gets in the eyes or on skin or clothing.

DISPOSAL

Neutralize the solutions with 1 M NaOH, and pour them down the drain.

TEACH

Do You Know?

The word *entropy* is derived from the Greek *tropia*, meaning "a turning toward."

Historical Note

German physicist Rudolf Julius Emanuel Clausius (1822–1888) coined the term *entropy* for the ratio of a system's heat content to its absolute temperature.

Chem Fact
According to the second law of thermodynamics, the total entropy of the universe in always increasing.

Entropy is given the symbol S. All physical and chemical changes involve a change in entropy, or ΔS. Although entropy is often described in terms of randomness or disorder, a better way to understand it is in terms of the number of ways a substance can be arranged. For example, there is only one way to assemble the pieces of a jigsaw puzzle correctly to form a completed picture. But there are a great number of ways to arrange the pieces when they are just thrown in a box.

Another example of a physical change with an increase in entropy is the sugar-water system shown in **Figure 9-16**. Immediately after sugar is added, the system is initially very ordered. All the sugar molecules are in one region at the bottom of the pitcher. All the water molecules are in a different region, the rest of the pitcher. Eventually, the two types of molecules spontaneously form a more disordered system, in which sugar and water molecules are homogeneously mixed.

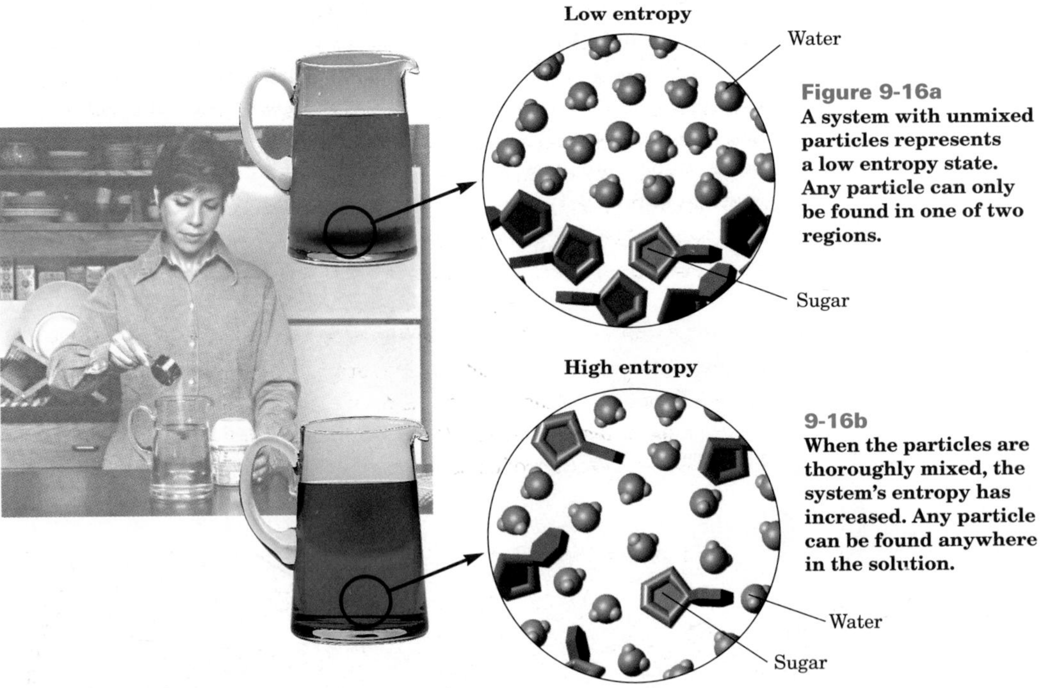

Figure 9-16a
A system with unmixed particles represents a low entropy state. Any particle can only be found in one of two regions.

9-16b
When the particles are thoroughly mixed, the system's entropy has increased. Any particle can be found anywhere in the solution.

Entropy's effects increase with temperature

If equal amounts of sugar are added to a glass of cold water and a glass of hot water, the sugar will dissolve more rapidly in the hot water. Why? The liquid water molecules possess kinetic energy, and as they move randomly, they collide with the sugar molecules and mix with them. The ongoing collisions of water and sugar molecules help sugar dissolve. The hot water has more kinetic energy and faster moving molecules than the cold water has. There are more collisions between water and sugar molecules in hot water than in the cold water.

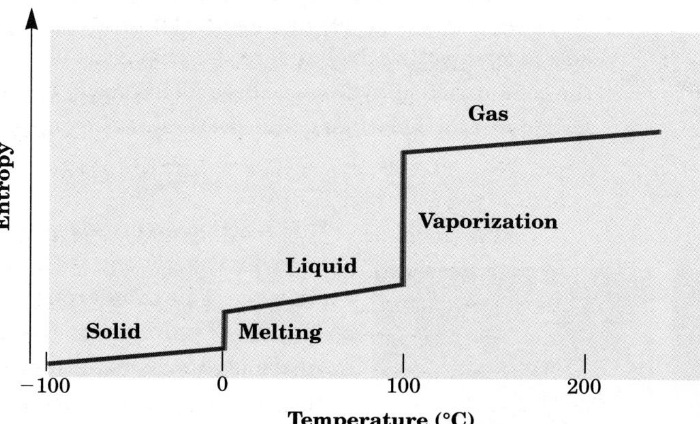

Figure 9-17
Notice that the entropy of a solid, liquid, or gas changes very little when heated compared to the high increase in entropy when ice melts and water vaporizes.

Entropy changes for chemical changes can be predicted

From what you know about entropy, you can make some educated guesses about how entropy will affect chemical reactions. The key is to remember that high entropy situations, which are randomized and disordered or which have more possibilities for arrangement, are generally preferred or are more stable, as long as other factors are kept equal.

Solids are very low in entropy, and liquids are higher in entropy, as shown by the graph in **Figure 9-17**. Gases are much higher in entropy than liquids because gas particles are farther apart. Chemical changes that produce gases are favored when entropy is the main consideration. Not only are the particles less ordered in a gas, but when there is more than one gaseous product, they will mix homogeneously. If liquids are formed, they may or may not mix, depending upon whether they can dissolve. When only solids are formed, they cannot mix easily. When a reaction results in more particles, there are more ways they can be arranged. These considerations are illustrated in **Figure 9-18**.

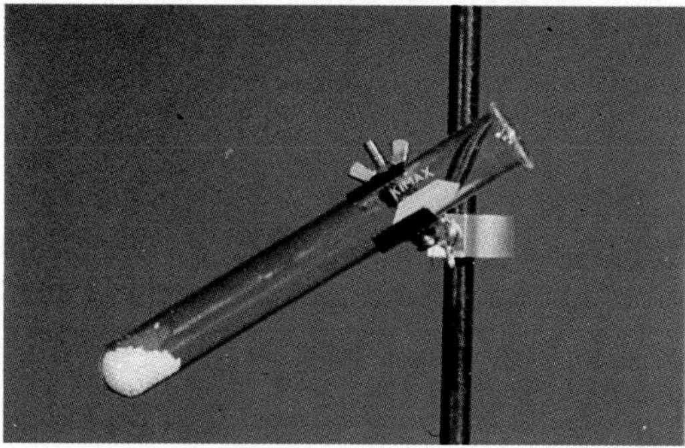

Figure 9-18a
The decomposition of ammonium nitrate, $NH_4NO_3(s)$, involves a change from a low-entropy system with one solid reactant . . .

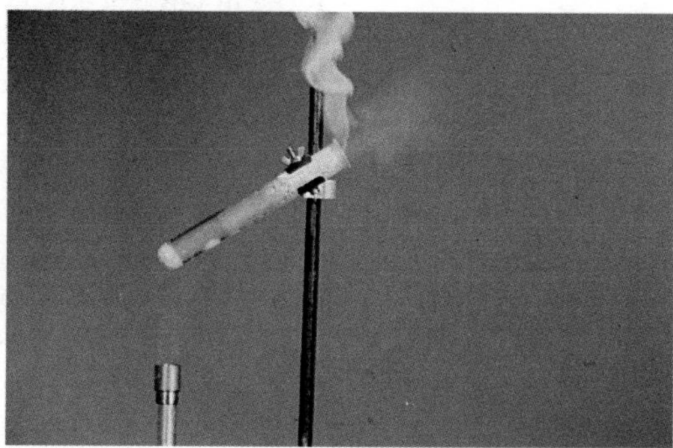

9-18b
. . . to a high-entropy system with two gaseous products and one liquid product.

Further Applications

Jell-O brand gelatin is made by dissolving the heavily sugared gelatin in hot water and then cooling it.

Cultural Connection

As early as 940 B.C., the Chinese produced salt by piping natural gas from where it seeped from the ground to the shoreline where they burned it to evaporate brine.

Visual Strategy

Figure 9-17

As the temperature increases, the ice-to-water reaction becomes more entropy driven and occurs more easily. Likewise, the water-to-steam reaction occurs more easily as the entropy effect is increased. Point out the vertical portions of this graph and stress that entropy changes occur *during* a phase change. Compare with **Figure 9-5** on page 317 to show phase changes are primarily entropy driven.

Additional Examples for Sample Problem 9E

Entropy values are in Appendix A, **Table A-11**.

a. What is the change in entropy for the decomposition of potassium chlorate? ΔS for $KClO_3(s)$ is equal to 143.7 J/mol•K.

$$2KClO_3(s) \longrightarrow 2KCl(s) + 3O_2(g)$$

b. What is the change in entropy for the reduction of iron(III) oxide by carbon monoxide?

$$Fe_2O_3(s) + 3CO(g) \longrightarrow 2Fe(s) + 3CO_2(g)$$

Solutions are at the bottom of the page.

Content Background

Absolute Zero

The misconception that absolutely all motion ceases at absolute zero is a common one. However, the non-thermal zero-point vibrations remain, due to quantum effects. They are unavailable to do work or release energy because they represent the vibrations and configurations of the particles in their lowest possible ground states. For this fundamental energy to be released, there would have to be an even lower ground state available for the particles to move to.

Entropy's effects can be overcome

The effects of a decrease in entropy can be overcome if enough enthalpy decrease, the other driving force for change, is involved. For example, in the combustion of hydrogen, three molecules in the gaseous state react to form only two molecules, thus decreasing the entropy.

$$2H_2(g) + O_2(g) \longrightarrow 2H_2O(g)$$

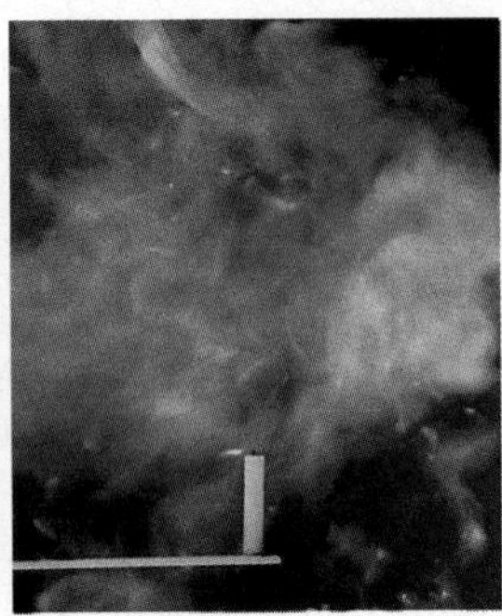

Figure 9-19
Although the combustion of hydrogen creates a product with lower entropy than the reactants, the reaction occurs because the heat energy released by the reaction causes a larger increase in the entropy of the surroundings: the air and balloon.

This reaction occurs, despite the decrease in entropy for the system, because the reaction is exothermic. This exothermic reaction releases heat, as shown in **Figure 9-19**. The heat released by an exothermic change increases the particle motion in the surroundings, resulting in an increase in entropy in the system and surroundings combined that is greater than the decrease in entropy in the system alone.

The same concept explains how refrigerators and air conditioners work. They make a specific system cooler and lower in entropy, but only by heating the surroundings and increasing the entropy there.

Entropy can be quantified

Unlike enthalpy, absolute values of entropy can be measured. Entropy involves thermal motion. Absolute zero, or 0 K (−273°C), is the temperature at which all motion in solids ceases (except for zero-point vibrations, whose energy cannot be utilized for anything) and the entropy is considered to be zero. Changes in entropy can be measured in comparison with this zero baseline, and chemistry handbooks contain tables of entropy values for elements and compounds.

Entropy has units of J/K. Like enthalpy, entropy can be used in stoichiometric calculations or Hess's law changes if the entropy values are given in units of J/K•mol. The key is to remember that, like enthalpy, the net entropy change is equal to the sum of the entropies of the products minus the sum of the entropies of the reactants when stoichiometric relationships are taken into account.

$$\Delta S = S_{products} - S_{reactants}$$

For example, at 100.°C, the entropy of liquid water is 87 J/K•mol, and that of steam is 196 J/K•mol. The entropy change, ΔS, is equal to the difference between the two. The positive value of ΔS indicates that entropy increases as water evaporates or boils.

$$H_2O(l) \longrightarrow H_2O(g)$$

$$\Delta S = S_{products} - S_{reactants}$$

$$\Delta S = 196 \text{ J/K•mol} - 87 \text{ J/K•mol} = +109 \text{ J/K•mol}$$

Solutions for Additional Examples 9E

a. ΔS = sum of ΔS^o of products − sum of ΔS^o of reactants
$\Delta S = [2(82.6) + 3(204.8)] - [2(143.7)] = +492.2$ J/K

b. ΔS = sum of ΔS^o of products − sum of ΔS^o of reactants
$\Delta S = [2(27.1) + 3(213.4)] - [89.9 + 3(197.7)] = +11.4$ J/K

Sample Problem 9E
Entropy calculations

Calcium carbonate, $CaCO_3$, decomposes into calcium oxide and carbon dioxide. What is the entropy change for this reaction?

$CaCO_3(s) \longrightarrow CO_2(g) + CaO(s)$

Substance	S (J/K•mol)
$CaCO_3(s)$	+92.9
$CO_2(g)$	+213.8
$CaO(s)$	+38.2

1 List
- **The necessary pieces of data are already shown in the table.**

2 Set up
- **The sum of the S values for the products minus the sum of the S values of the reactants will give the net ΔS for the reaction.**

[sum of S values for products] − [sum of S values for reactants] = ? J/K•mol

[(38.2 J/K•mol) + (213.8 J/K•mol)] − [92.9 J/K•mol] = ? J/K•mol

3 Estimate and calculate
- **Round off to make an estimate.**

[40 + 210] − [90] ≈ [250] − [90] ≈ 160

- **Calculate, and round off correctly.**

ΔS = +159.1 J/K•mol

Because this reaction involves a single solid reactant producing two products, one of which is gaseous, the answer should be a large positive number, reflecting an increase in entropy.

Practice 9E

1. **What is the entropy change for converting graphite to diamond? Is this likely to be a spontaneous change?**

   graphite: S = +5.7 J/K•mol
   diamond: S = +2.4 J/K•mol

2. **Calculate the entropy change for the single replacement reaction that occurs at 25°C between copper chloride crystals and iron metal. (Hint: remember to adjust the entropy values to account for the mole ratios in the balanced equation.)**

   $3CuCl_2(s) + 2Fe(s) \longrightarrow 2FeCl_3(s) + 3Cu(s)$

Substance	S (J/K•mol)
$CuCl_2(s)$	+108.1
$Fe(s)$	+27.3
$FeCl_3(s)$	+142.3
$Cu(s)$	+33.2

Section Review

19. Give two examples of a change in entropy. Indicate whether the change is an increase or decrease in entropy of the system.
20. What kind of change in entropy promotes a spontaneous physical or chemical change?
21. Describe what happens to the entropy of ice as it melts to form water, the water is heated, and the water boils to form water vapor.
22. The entropy of 2-propanol (also known as isopropyl alcohol or rubbing alcohol) is 309.2 J/K•mol in the gaseous state (at 25°C), and the entropy change for evaporation at this temperature is 128.1 J/K•mol. What is the entropy for liquid 2-propanol?

Answers to Practice 9E

1. ΔS = (+2.4) − (+5.7) = −3.3 J/K; no
2. ΔS = [2(+142.3)+3(+33.2)]−[3(+108.1)+2(+27.3)]J/K•mol = +5.3 J/K

ASSESS

Alternative Assessment

You have seen that enthalpy and entropy are involved when ice changes to steam. Develop a visual aid that shows the combined effect of enthalpy and entropy over a range of −20°C to 120°C.

Answers to Section Review

19. Answers will vary. Steam condensing to a liquid is accompanied by a decrease in entropy. Burning graphite to produce CO_2 gas is accompanied by an increase in entropy.
20. A positive change, or increase, in entropy promotes spontaneous reactions.
21. The entropy increases during the melting and boiling processes because they are phase changes. The entropy changes very little while the water is being heated.
22. The entropies of the gaseous and liquid states of 2-propanal differ by ΔS_{vap}. Because 2-propanol is going from the gas phase to the liquid phase, ΔS_{vap} is subtracted from +309.2 J/K•mol to get +181.1 J/K•mol.

Section 9-4

FOCUS

Lesson Starter

Begin the lesson with a discussion that directs students toward developing the Gibbs free energy relationship. Write the following equation on the board.

$2H_2O \longrightarrow 2H_2 + O_2$

$\Delta H = +572$ kJ/mol

$\Delta S = +327$ J/mol

Place an H-tube water electrolysis unit on the demonstration table. Explain what it is used for and that 483 kJ/mol of free energy must be put in for the unit to operate. Tell the students the following facts.

1. Enthalpy and entropy are often competing energy factors.
2. The magnitude of entropy increases or decreases as the temperature increases or decreases.
3. If the total energy change is negative, the reaction is spontaneous.

Have students speculate on how this reaction can be completed without using an electrolysis unit. Guide the discussion with **Table 9-8** and the following equation.

$X = \Delta H - T\Delta S$

The temperature needed to outweigh the enthalpy effect is approximately 1482°C.

SAFETY

If this apparatus is operated, use only a 9 V "transistor radio" battery. Do NOT use any other source of current. Wear safety goggles and lab apron. Students should stand 10 ft back from demonstration area.

Demonstration 10

Production of copper sulfide from the elements

Approximate time: 5–10 min

A darkened room will help the students see the bright, red glow of the exothermic reaction between Cu and S in which entropy increases.

1. Cut a 1 cm × 10 cm piece of copper foil.

How do entropy and enthalpy affect change?

Section Objectives

Summarize the conditions that favor a spontaneous chemical reaction.

Define free energy.

Calculate the change in free energy to predict whether a reaction will occur spontaneously.

Describe how chemical reactions can be controlled.

Entropy and enthalpy

So far, you've learned about two driving forces that can affect chemical and physical changes—entropy and enthalpy. Most changes involve both enthalpy and entropy. It is the interplay of these two driving forces that determines whether or not a physical or chemical change will actually happen. Changes that are spontaneous are often referred to as *thermodynamically favored.* It is important to remember that even spontaneous changes can take a long time to occur. What the terms *spontaneous* and *thermodynamically favored* mean is that the opposite change will not occur unless reaction conditions change.

Predicting the combined effects of entropy and enthalpy

When entropy and enthalpy oppose each other, the dominant one determines the outcome. But how do you know which will predominate? One way to find out is to make some observations.

Think of what happens when ice melts. The water molecules go from a solid, in which they are held in a rigid pattern within a crystal, to a liquid, in which they are free to move. This is an endothermic process that produces disorder because the increased freedom of movement in the liquid state means that there are more ways to arrange the molecules than there are in a solid, as shown in **Figure 9-20**.

The change in entropy encourages melting, but the change in enthalpy discourages it. The observation that this change happens spontaneously only if the temperature is above 0°C signifies that the entropy change is dominant over the enthalpy change at higher temperatures.

Solid ice

Melting

Figure 9-20a Melting is endothermic, but involves an increase in entropy. Melting is favored at higher temperatures.

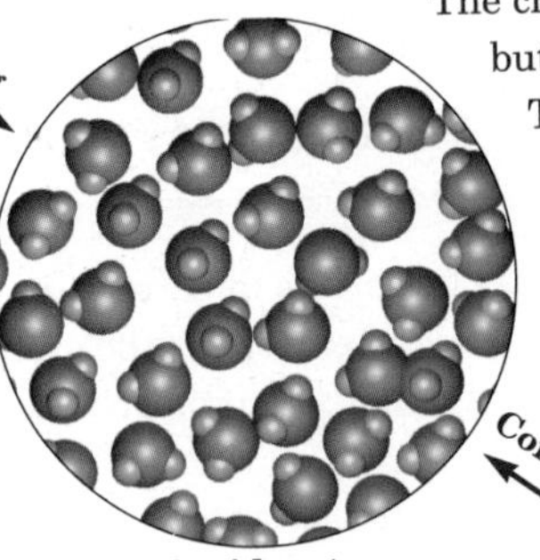

Condensation

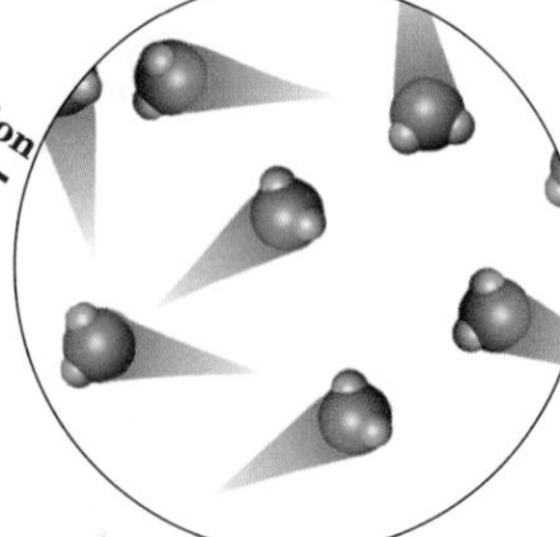

9-20b Condensation is exothermic, but involves a decrease in entropy. Condensation is favored at lower temperatures.

Visual Strategy

Figure 9-20

Translate the written descriptions for these two processes into mathematical symbols, and compare them to the entries in **Table 9-8**.

Table 9-8
Relating Enthalpy and Entropy to Spontaneity

ΔH	ΔS	Spontaneity	Example
– value Exothermic	+ value (disordering)	Always spontaneous	$2K(s) + 2H_2O(l) \longrightarrow 2KOH(aq) + H_2(g)$
– value Exothermic	– value (ordering)	Spontaneous at *lower* temperatures	$H_2O(g) \longrightarrow H_2O(l)$
+ value Endothermic	+ value (disordering)	Spontaneous at *higher* temperatures	$H_2O(s) \longrightarrow H_2O(l)$
+ value Endothermic	– value (ordering)	Never spontaneous	$16CO_2(g) + 18H_2O(l) \longrightarrow 2C_8H_{18}(l) + 25O_2(g)$

Now think of what happens when steam condenses. At the beginning, the system has high entropy. The particles in a gas have a great freedom of movement, so there are many ways in which they can be arranged. But when the system becomes liquid, the particles have less freedom of movement because they are closer together. As a result, the system has less entropy. The change in entropy discourages condensation, but the exothermic change in enthalpy favors it. This change is observed to happen spontaneously at normal atmospheric pressure, but only if the temperature is below 100°C. This observation indicates that enthalpy is the stronger driving force at lower temperatures. **Table 9-8** summarizes the thermodynamic spontaneity of possible combinations of enthalpy and entropy changes.

Chem Fact
The discussion assumes that enthalpy and entropy remain constant with temperature. This is not technically true, but unless there is a change of state, the effects of temperature on values of ΔH and ΔS are small enough that they can be disregarded.

Free energy

Free energy is a combination of entropy and enthalpy

Depending on the temperature, either entropy or enthalpy can be the cause of a change. How can you tell which will make a change happen?

A quantity that relates enthalpy and entropy in a way that indicates which predominates is **free energy**. Think of free energy as the quantity of energy that is available or stored to do useful work or to cause a change. The free energy of a system is represented by the letter G, in honor of the American chemist J. Willard Gibbs, who developed the equation to predict whether a reaction will occur spontaneously. This equation is used to calculate the change in free energy, represented by ΔG. If the value of ΔG is negative, the change is spontaneous. If it is positive, the change will not occur spontaneously. If ΔG is zero, this is an indication that neither the forward nor the reverse reaction is thermodynamically favored; the final result will be a state of dynamic equilibrium, with a mixture of products and reactants.

free energy
a quantity of energy related to the capacity of a system to do work, which can be used to predict spontaneity

$$\Delta G = \Delta H - (T\Delta S)$$

ΔG: positive value means change is not spontaneous

ΔG: negative value means change is spontaneous

Recall that ΔH represents the enthalpy change, T represents the temperature expressed in kelvins, and ΔS denotes the entropy change.

2. Fill a Pyrex test tube 2 cm deep with sulfur flowers. Clamp the test tube vertically to a ring stand. Heat strongly.
3. Using crucible tongs, hold the Cu strip over the test tube so that it is hit by the sulfur vapors. Point out that no reaction has taken place.
4. Insert the Cu strip into the hot, viscous sulfur vapors inside the test tube. The bright red glow shows the formation of CuS.
5. Remove the Cu strip and let it cool for inspection.

SAFETY
Wear safety goggles and a lab apron. Students stand 10 ft away from the demonstration area. If the sulfur vapors catch fire in the test tube, cover the tube with a piece of glass plate to smother.

DISPOSAL
Put the damaged test tube and used Cu foil in the trash.

TEACH

Teaching Tip

Predicting spontaneity
ΔH for the CuS reaction in the above demonstration is –271 kJ/mol. Calculate the entropy change for this reaction, and use **Table 9-8** to predict the outcome of the reaction before running the demonstration. ΔS values are 33.15 J/mol for Cu, 66.5 J/mol for CuS, and 167.821 J/mol for $S(g)$.

Do You Know?

Also termed *Gibbs free energy* in many texts, ΔG was once written as ΔF.

Possible Misconception

ΔH cannot be used to predict spontaneity even though most exothermic reactions are spontaneous.

Additional Examples for Sample Problem 9F

a. What is the free energy change for the decomposition of potassium chlorate if the reaction is carried out at 950°C?

$2KClO_3(s) \longrightarrow 2KCl(s) + 3O_2(g)$

$\Delta H = -90.1$ kJ/mol
$\Delta S = +492.2$ J/mol•K

b. What is the minimum temperature necessary for the reduction of iron(III) oxide with carbon monoxide to occur spontaneously?

$Fe_2O_3(s) + 3CO(g) \longrightarrow 2Fe(s) + 3CO_2(g)$

$\Delta H = +24.7$ kJ/mol
$\Delta S = +11.4$ J/mol•K

Solutions are at the bottom of the page.

Answers to Practice 9F

1. $\frac{(\Delta H - \Delta G)}{\Delta S} = T$

$T = (135.5 + 0.1) \div (+0.1488) = +911.3\ K = +638.3°C$

2. $\Delta G = \Delta H - T\Delta S$

$\Delta G = (-393.5) - (298) \times (+0.003) = -394$ kJ/mol
It is spontaneous.

3. raise the temperature

4. $\Delta G = \Delta H - T\Delta S$

$\Delta G = -851.5$ kJ/mol $- (448\ K \cdot 0.0385$ kJ/mol•K)

$\Delta G = -869$ kJ/mol

Remember that *when making free energy calculations, the temperature must be in kelvins.* Instructions for converting temperatures were given in Chapter 2. Also remember that ΔS values are given in J/K•mol or J/K and ΔH values are in kJ/mol or kJ. Be certain to convert ΔS values from J/K into kJ/K.

Sample Problem 9F
Calculating free energy changes

Graphite and steam can react to produce water gas, a fuel which is a mixture of several gases, but mostly of carbon monoxide and hydrogen gas. When water gas is made industrially, the reaction is carried out at temperatures near 900°C. Are the products or reactants favored at this temperature? (Hint: assume that ΔH and ΔS are constant at all temperatures.)

$H_2O(g) + C(graphite) \longrightarrow CO(g) + H_2(g)$ $\quad \Delta H = +135.5$ kJ $\quad \Delta S = +148.8$ J/K

1 List

- **temperature:** 900°C + 273 = 1173 K
- **enthalpy change:** +135.5 kJ
- **entropy change expressed in kJ/K:** $148.8\ J/K \times \frac{1\ kJ}{1000\ J} = 0.1488\ kJ/K$
- **free energy change:** ? kJ

2 Set up

- **To determine the answer, you must calculate the value for the free energy change for the reaction at 900°C and check whether or not it is negative. If it is negative, the reaction will be spontaneous, and the products will be favored.**

$\Delta G = \Delta H - T\Delta S = ?$ kJ

$\Delta G = 135.5\ kJ - \left(1173\ K \times \frac{0.1488\ kJ}{K}\right) = ?$ kJ

3 Estimate and calculate

- **Round off to make an estimate.**
  $135 - (1200 \times 0.150) \approx 135 - 180 \approx -45$
- **Calculate, and round off correctly.**
  Calculator answer: $135.5 - 174.5424 = -39.0424$
  Correctly rounded answer: -39.04 kJ
  Because the free energy change is negative, the reaction is spontaneous, and the products are favored.

Practice 9F

1. **Holding the temperature of the reactants in the water-gas reaction at 900°C is expensive. What is the lowest temperature that the manufacturers could use and still have the reaction be spontaneous? (Hint: assume that ΔS and ΔH do not change and $\Delta G = -0.1$ kJ.)**
2. **Calculate ΔG for the following reaction at 25°C. Is the reaction spontaneous?**
   $C(s) + O_2(g) \longrightarrow CO_2(g)$ $\quad \Delta H^o = -393.5$ kJ/mol $\quad \Delta S^o = +3.0$ J/K•mol
3. **A sample of ammonium chloride, NH_4Cl, is at 25°C. What must be done to make the decomposition reaction spontaneous?**
   $NH_4Cl(s) \longrightarrow NH_3(g) + HCl(g)$ $\quad \Delta H^o = +176$ kJ/mol $\quad \Delta S^o = +285$ J/K•mol
4. **The thermite reaction used in some welding applications has the following enthalpy and entropy changes at 25°C. Assuming ΔS and ΔH are constant, calculate ΔG at 175°C.**
   $Fe_2O_3(s) + 2Al(s) \longrightarrow 2Fe(s) + Al_2O_3(s)$ $\quad \Delta H^o = -851.5$ kJ/mol, $\Delta S^o = +38.5$ J/K•mol

Solutions to Additional Examples 9F

a. $\Delta G = \Delta H - T\Delta S$

$\Delta G = -90.1\ kJ/mol - (1223\ K)(0.4922\ kJ/mol\cdot K) = -692\ kJ/mol$

b. $\frac{(\Delta H - \Delta G)}{\Delta S} = T$

$T = \frac{[(+24.7\ kJ/mol) - (0\ kJ/mol)]}{0.0114\ kJ/mol\cdot K} = -2167\ K$

Comparing free energy changes

Values for free energy changes are similar to entropy and enthalpy in that they are usually tabulated as an amount *per mole.* As a result, these values can often be used in stoichiometric calculations. Additionally, if a reaction can be broken into two smaller reactions, its total free energy change will be equal to the sum of the component free energy changes.

For example, at 25°C, ΔG for the combustion of gaseous hydrogen to make water vapor is −228.6 kJ/mol. For the condensation of water vapor at 25°C to form liquid water, ΔG is −8.5 kJ/mol. The ΔG value involved in the change when making liquid water from gaseous hydrogen and oxygen is the sum of these two values, −237.1 kJ/mol.

$$H_2(g) + \tfrac{1}{2}O_2(g) \longrightarrow H_2O(g) \qquad \Delta G = -228.6 \text{ kJ/mol}$$
$$H_2O(g) \longrightarrow H_2O(l) \qquad \Delta G = -8.5 \text{ kJ/mol}$$
$$H_2(g) + \tfrac{1}{2}O_2(g) \longrightarrow H_2O(l) \qquad \Delta G = -237.1 \text{ kJ/mol}$$

Controlling reactions

How to influence what happens in a reaction

Any factor that affects the change in free energy of a reaction will influence how the reaction or change proceeds. One of the factors in the free energy equation is temperature. The reasons why temperature can have this effect, described earlier in this section, are clear if you examine the equation closely.

As shown in **Figure 9-21**, at low temperatures, the $T\Delta S$ part of the equation is smaller and will have less of an influence on the value of the free energy change. At high temperatures, the opposite is true. In cases where the effects of enthalpy and entropy changes tend to force the reaction in different directions, the temperature can be the factor that determines how a reaction will occur.

Figure 9-21
At lower temperatures, particles are moving slowly, and free energy is determined mostly by the enthalpy change. At higher temperatures, particles are moving faster, and the contribution of the entropy change becomes a more important factor than the enthalpy change.

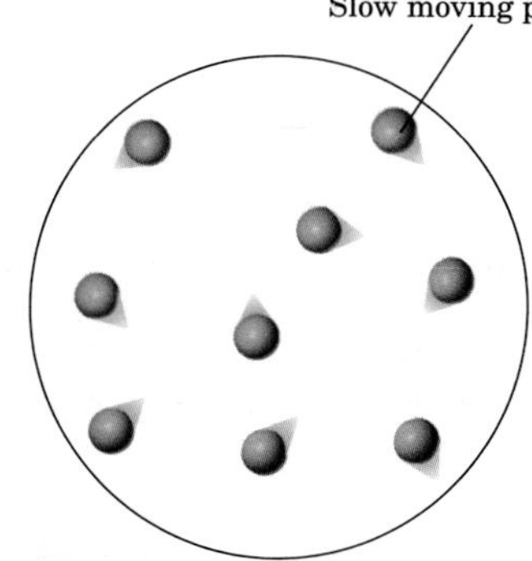

Low temperatures—small T value means less randomness

$\Delta G = \Delta H - T\Delta S$

ΔH plays larger role

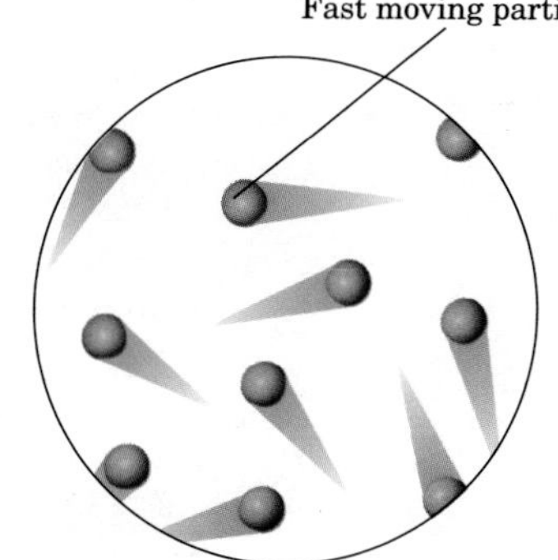

High temperatures—large T value means more randomness

$\Delta G = \Delta H - T\Delta S$

$T\Delta S$ plays larger role

Teaching Tip

Remind students that if the coefficient in the balanced chemical equation is other than 1, the ΔG values must be divided by that coefficient to get per mole quantities.

Theme

Macro observations and micromodeling

The spontaneity of a reaction at a given temperature can be explained by changes in both the enthalpy and entropy of the system.

ASSESS

Alternative Assessment

1. Develop a concept map for classroom display regarding the topics of heat, temperature, specific heat, enthalpy, entropy, stoichiometry, free energy. Also include a flowchart describing problem-solving techniques for entropy, enthalpy, and free energy problems.
2. Break the class into groups of three. Have the groups prepare a list of advantages and disadvantages of using gasoline or diesel fuel in an internal combustion engine, or of using solar or steam energy to heat a home.

Just as water will freeze below 0°C and ice will melt above 0°C, many chemical reactions occur in one direction at low temperatures and in the other direction at high temperatures. For example, consider the reaction of an oxygen atom, O, combining with a molecule of oxygen, O_2, to produce ozone, O_3, as shown in **Figure 9-22**. When present in the lower atmosphere, ozone is a pollutant that can cause smog. However, in the stratosphere, ozone provides a blanket that absorbs the sun's ultraviolet radiation.

$$O_2(g) + O(g) \longrightarrow O_3(g) \qquad \Delta H = -106.5 \text{ kJ} \qquad \Delta S = -127.4 \text{ J/K}$$

At room temperature, 298 K, the effects of the enthalpy change are dominant, and the exothermic synthesis reaction is spontaneous, as indicated by the negative value for the free energy change of the reaction.

$$\Delta G = \Delta H - T\Delta S$$

$$\Delta G = -106.5 \text{ kJ} - (298 \text{ K})(-0.1274 \text{ kJ/K})$$

$$\Delta G = -106.5 \text{ kJ} - (-38.0 \text{ kJ}) = -68.5 \text{ kJ}$$

But at higher temperatures such as 1000. K, the reverse is true. The effects of the entropy change, which favor the decomposition of one molecule into an atom and a molecule, are dominant, and the reverse reaction is spontaneous. Then the value of the free energy change is positive.

$$\Delta G = \Delta H - T\Delta S$$

$$\Delta G = -106.5 \text{ kJ} - (1000. \text{ K})(-0.1274 \text{ kJ/K})$$

$$\Delta G = -106.5 \text{ kJ} - (-127.4 \text{ kJ}) = +20.9 \text{ kJ}$$

It is even possible to calculate the temperature at which neither reaction will be thermodynamically favored. At this temperature, ΔG will have the value of zero. For this to be true, ΔH and $T\Delta S$ must be equal so that you have zero left after subtracting. You already know from the two previous calculations that this temperature must be somewhere between 298 K and 1000. K. (Again, assume ΔH and ΔS are constant to get an approximate answer.)

$$\Delta G = \Delta H - T\Delta S = 0 \text{ kJ}$$

$$T\Delta S = \Delta H \text{ if } \Delta G = 0$$

$$T = \frac{\Delta H}{\Delta S} \quad \frac{-106.5 \text{ kJ}}{-0.1274 \text{ kJ}} = 836 \text{ K}$$

$$= 563°\text{C}$$

Figure 9-22a
Ozone can form as a result of electrical sparks such as lightning. The lightning provides the energy necessary to form O atoms from molecules. Ozone synthesis is thermodynamically favored over decomposition near room temperature. Ozone near the surface of the Earth is harmful.

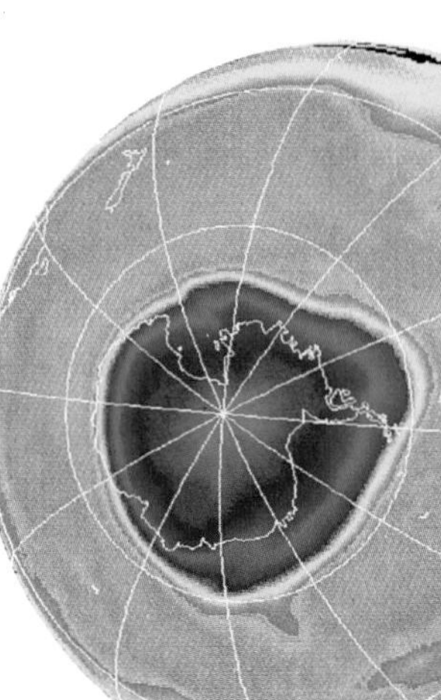

9-22b
Ozone in the stratosphere shields Earth from ultraviolet radiation. At these altitudes, ozone reacts with pollutants such as chlorine from chlorofluorocarbons in a different reaction that favors decomposition.

Temperature is not the only reaction condition that can affect the spontaneity of a reaction. Anything that changes the values of the enthalpy and entropy changes will have an effect upon the value of the free energy change. Earlier, it was mentioned that enthalpy change is equal to the heat of reaction, provided that the pressure is held constant. Pressure changes will have an effect on the enthalpy change, which will in turn affect the free energy change of the reaction. If this effect is big enough, it can affect the spontaneity. Other factors, such as the nature of the reactants, can also affect the energy and spontaneity of a reaction.

Section Review

23. Using **Table 9-8**, predict what conditions, if any, are needed to make the following changes spontaneous.

a. exothermic reaction with an increase in entropy

b. enthalpy and entropy are both increased

c. enthalpy and entropy are both decreased

24. Define *free energy,* and explain how a change in free energy is calculated.

25. From the following values, compute the ΔG value for each reaction, and predict whether each reaction will occur spontaneously.

Reaction	ΔH (kJ)	Temperature	ΔS (J/K)
1	+125	293 K	+35
2	−85.2	127 K	+125
3	−275	500°C	+450

26. How might you induce an endothermic reaction that does not usually occur spontaneously?

27. Story Link

For use in cars, hydrogen gas can be prepared in several ways, such as by the decomposition of water or hydrogen chloride.

$$2H_2O(l) \longrightarrow 2H_2(g) + O_2(g)$$
$$2HCl(g) \longrightarrow H_2(g) + Cl_2(g)$$

Use the following data to determine whether these reactions can occur spontaneously. Assume that ΔH and ΔS are constant.

	ΔH_f^o (kJ/mol)	S^o (J/K•mol)
$H_2O(l)$	−285.8	+70.0
$H_2(g)$	0	+130.7
$O_2(g)$	0	+205.1
$HCl(g)$	−92	+187
$Cl_2(g)$	0	+223.1

Answers to Section Review

23. a. no changes

b. high temperature

c. low temperature

24. a. Free energy is the quantity of energy that is available or stored to do work or to cause change. Free energy is calculated from the following equation.
$\Delta G = \Delta H - T\Delta S$

25. a. ΔG = 115 kJ; nonspontaneous

b. ΔG = −101 kJ; spontaneous

c. ΔG = −623 kJ; spontaneous

26. add heat so that $T\Delta S$ overcomes ΔH

27. a. nonspontaneous;
$\Delta H = 0 - 2(-285.8) = +571.6$ kJ

$\Delta S = [2(130.7) + 205.1] - 2(70.0) = +326.5$ J

$\Delta G = (+571.6) - (298)(+0.3265) = +474.3$ kJ

b. nonspontaneous;
$\Delta H = 0 - 2(-92) = +184$ kJ

$\Delta S = [2(130.7) + 223.1] - 2(187) = +110.5$ J

$\Delta G = (+184) - (298)(+0.1105) = +151.1$ kJ

Section 9-5

FOCUS

Lesson Starter

Begin the lesson by having students determine the average number of normal breaths taken in a minute. Determine a class average, and then fill a 1 L graduated cylinder with water. Have three students separately determine the amount of gas exhaled using water displacement. **(Be sure to use a clean piece of tubing.)** Determine the average value. If 20% of air is O_2, and only about 25% of the O_2 present is used in each breath, calculate the liters of O_2 used in one day. Use the following equation to determine the amount of glucose burned by the body with this daily amount of oxygen.

$$C_6H_{12}O_6 + 6O_2 \longrightarrow 6CO_2 + 6H_2O$$

SAFETY

All students wear safety goggles and a lab apron.

What are your energy needs?

Section Objectives

Describe the factors that affect daily energy requirements.

Estimate daily energy expenditure.

Estimate daily caloric intake.

Interpret food labeling for nutrient content.

Evaluate your diet against recommended guidelines.

Calculating your energy needs

As a living organism, your body requires energy to carry out the processes that keep you alive. Every movement you make involves the contraction of muscles and requires energy. Even when you are asleep, your body uses energy to keep you breathing and your heart beating. Obviously, the more you move, the more energy you need. In this section, you'll find out how much energy it takes to keep you going. Using this information and what you learned about the energy content of foods in Section 9-1, you can determine whether your energy requirements balance your energy intake.

basal metabolic rate
one's resting energy expenditure measured in the morning, at least 12 hours after the last meal

Estimating your metabolic rate

Your daily energy needs are determined by three factors—basal metabolic rate, activity level, and the thermic effect of food. By calculating the energy equivalents of these three factors, you can determine your daily energy needs.

Basal metabolic rate (BMR) includes the energy needed for all life sustaining reactions except the digestion and absorption of food. BMR is not constant but varies with factors like age, gender, stress level, and general health. To measure basal metabolic rate accurately, measurements must be taken before any activity or intake of food. The measurements collected under these controlled conditions give the **resting metabolic rate** (RMR). A resting metabolic rate is a basal metabolic rate measured after five to six hours of no food or exercise. BMRs and RMRs usually differ by less than 3%. Estimated RMRs for the general population are given in **Table 9-9**.

resting metabolic rate
energy expended by a person at rest in a thermally neutral environment

Table 9-9
Energy Expended at Resting Metabolic Rate (Cal/day)

Body mass	Male age 10–18	18–30	30–60	>60	Body mass	Female age 10–18	18–30	30–60	>60
50 kg (110 lb.)	1526	1444	1459	1162	**50 kg (110 lb.)**	1356	1231	1264	1121
57 kg (125 lb.)	1648	1551	1540	1256	**57 kg (125 lb.)**	1441	1334	1325	1195
64 kg (140 lb.)	1771	1658	1621	1351	**64 kg (140 lb.)**	1527	1437	1386	1268
70 kg (155 lb.)	1876	1750	1691	1423	**70 kg (155 lb.)**	1600	1525	1438	1331
77 kg (170 lb.)	1998	1857	1772	1526	**77 kg (170 lb.)**	1685	1628	1499	1404
84 kg (185 lb.)	2121	1964	1853	1621	**84 kg (185 lb.)**	1771	1731	1560	1478
91 kg (200 lb.)	2243	2071	1935	1716	**91 kg (200 lb.)**	1856	1833	1621	1552

Table 9-10
Estimating Energy for Activity Level (in Relation to Resting Metabolic Rate)

Activity level	Activity factor
Resting (sleeping, reclining)	RMR × 1.0
Very light (seated and standing activities, painting, driving, lab work, typing, sewing, ironing, cooking, playing cards, playing a musical instrument)	RMR × 1.5
Light (walking on a level surface, garage work, carpentry, house cleaning, child care, golf, sailing, table tennis)	RMR × 2.5
Moderate (walking 3.5–4 mph, weeding and hoeing, carrying a load, cycling, skiing, tennis, dancing)	RMR × 5.0
Heavy (walking uphill with a load, chopping wood, manual digging, basketball, climbing, football, soccer)	RMR × 7.0

Source: Adapted from *Recommended Dietary Allowances*, 10th Edition

Figure 9-23
You can find calorie information based on serving sizes in the nutritional facts panel on food packages.

Thermic effect of food is estimated as a percentage of caloric intake

You can use **Table 9-10** to estimate the effects of activity on daily energy requirements. The table provides a realistic estimate as long as you don't exaggerate your activity level.

The thermic effect of food is the energy required for digestive processes. *Ten percent of the total caloric intake is an accepted figure for estimating the thermic effect of food.* Accurate measurement of this effect depends on the percentages of carbohydrates, fats, and proteins consumed. Fat is stored more efficiently than carbohydrate and protein.

Determining caloric intake

You can find data just about anywhere on the caloric equivalents of most of the foods you eat. Fortunately, most of the packaged food you get from the grocery store provides a lot of nutritional information. The cans in **Figure 9-23** each have labels in the style now mandated by the Food and Drug Administration (FDA).

Keep a food diary for a day. Use Calorie data on each of the foods you eat to determine your caloric intake for that day. The food diary and labels are also useful in doing a dietary analysis of how your caloric consumption is subdivided into carbohydrates, proteins, and fats.

How To
Estimate your energy expenditure

1. Use **Table 9-9** to estimate your RMR for 24 hours. Convert this value to Cal/h.
2. Use **Table 9-10** to calculate your activity level in Calories. Divide the day into the number of hours for your various levels of activity. For each level of activity, multiply the time by the factor from **Table 9-10** and by the RMR in Cal/h calculated in step **1**.
3. Determine the thermic effect of food by estimating total caloric intake (use food value tables) and multiplying by 0.10.
4. Add the results of steps **2** and **3** to get your total energy expenditure.

TEACH

Groupwork Strategy
Assign lunch menus from the foods listed in **Table 9-4** to groups of 4. Have students use **Table 9-5**, **Table 9-9**, and **Table 9-10** to determine how long it would take them to burn all of the Calories in their assigned lunch.

Teaching Tip
Caloric intake
Discuss the importance of caloric balance in maintaining weight and the nutritional effects of a caloric imbalance. You may also wish to invite a local nutritionist to present a lecture on this topic.

Do You Know?
When so few calories are consumed that adipose tissue disappears, cachexia results. Cachexia is caused by malnutrition, illness, and certain hormone imbalances. When cachexia results from an emotional or psychological disorder, it is called anorexia nervosa, and, contrary to popular opinion, it affects men as well as women.

ASSESS

Alternative Assessment

Estimate your energy expenditure for 3–5 days. Present your findings in the form of a report.

Answers to Section Review

28. basal metabolic rate, activity level, thermic effect of food

29. Assuming that a bag contains 1 oz of potato chips, your excess caloric intake is 150 Cal.

$$\frac{1 \text{ day}}{150 \text{ Cal}} \times \frac{3500 \text{ Cal}}{1 \text{lb}} = 23.3 \text{ days}$$

30. You lose 0.106 lb/day, so you need 9.43 days to lose 1 lb.

31. a. protein = 131.6 Cal;
carbohydrate = 496.8 Cal;
fat = 416.7 Cal;
total Cal = 1045 Cal

b. accounts for about half of the total calories

c. $\frac{1\text{h}}{370 \text{ Cal}} \times 10.45 \text{ Cal} = 2.82 \text{ h}$

Consumer Focus

Reading food labels

The FDA's standardization of nutritional labeling has made it possible to compare the nutritional compositions of similar foods, and it is also easier to determine how a particular food affects your daily caloric intake and how it matches sound dietary guidelines.

Nutrition Facts
1. Serving Size 4 crackers (28g)
Servings Per Container About 16
2. Amount Per Serving
Calories 120 Calories from Fat 25
3. % Daily Value*

4. **Total Fat** 3g	5%
Saturated Fat 0.5g	3%
Polyunsaturated Fat 0g	
Monounsaturated Fat 1g	
Cholesterol 0mg	0%
Sodium 180mg	8%
Total Carbohydrate 22g	7%
Dietary Fiber 1g	4%
Sugars 7g	
Protein 2g	

5. Vitamin A 0% • Vitamin C 0%
Calcium 2% • Iron 6%
6. * Percent Daily Values are based on a 2,000 calorie diet. Your daily values may be higher or lower depending on your calorie needs:

	Calories:	2,000	2,500
Total Fat	Less than	65g	80g
Sat Fat	Less than	20g	25g
Cholesterol	Less than	300mg	300mg
Sodium	Less than	2400mg	2400mg
Total Carbohydrate		300g	375g
Dietary Fiber		25g	30g

7. Calories per gram:
Fat 9 • Carbohydrate 4 • Protein 4

1. Serving size—The amounts listed now more closely approximate the amounts people eat. Serving sizes have been standardized for the same types of foods to allow for easy comparison.
2. Calories from fat—The fat Calories are based on grams of fat multiplied by the Calories per gram for fat (9 Cal/g).
3. Percent daily values—This section provides a breakdown of the nutrient content with respect to a 2000 Cal diet. Thus, 8 g of fat represent 12% of the recommended daily fat intake for someone on a 2000 Cal diet.
4. Total fat—A sum and breakdown of the fat content by degree of saturation are given.
5. Vitamin and mineral content—This breakdown allows you to compare foods as sources of vitamins and minerals. It also helps you in determining whether your diet provides enough of these nutrients. A value of 10% means the food is a "good source" of that nutrient. Values of 20% and higher mean that the food is "high" in that nutrient.
6. Dietary recommendations based on percent daily values—This information has been standardized to reflect caloric consumption at realistic levels. These values can be used to calculate recommendations at other caloric levels.
7. Calories per gram—The energy equivalent of these nutrients is constant and independent of the type of food.

Section Review

28. Name three factors that determine your daily energy needs.

29. You gain 1 lb. for every 3500 Cal you consume over what you expend. Assume your energy intake and expenditures are in balance and then you decide to add a bag of potato chips to your lunch every day. How long will it take you to gain 1 lb.? (Hint: use **Table 9-4**.)

30. Assume your energy intake and expenditures are in balance and then you decide to walk for one hour each day. If an hour of walking requires 370 Cal, how long will it take you to lose 1 lb.?

31. A typical fast-food lunch consisting of a cheeseburger, french fries, and a large soft drink has the following nutritional breakdown.

Protein	32.9 g
Carbohydrate	124.2 g
Fat	46.3 g

a. Calculate the total number of Calories in this lunch.

b. How does this nutrient content compare to the daily recommendations for someone on a 2000 Cal diet?

c. How long would you have to walk to burn the Calories in this lunch?

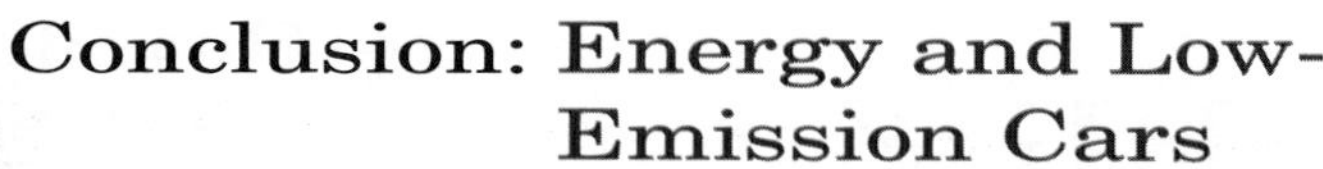

Conclusion: Energy and Low-Emission Cars

At the beginning of this chapter, you explored hydrogen-powered cars.

What drives the hydrogen combustion reaction? All chemical reactions that occur are driven by one or both of two forces: an increase in entropy or a decrease in enthalpy. The exothermic combustion reaction is driven by a large decrease in enthalpy even though entropy decreases because three molecules (2 H_2 and 1 O_2) combine to form only two molecules (2 H_2O).

How can this reaction be controlled so that it does not happen accidentally? It would be difficult to maintain the extreme conditions necessary for the reaction to be non-spontaneous. Instead, the hydrogen is absorbed within a compound, a metal hydride containing hydrogen, manganese, and titanium, until it is needed. Hydrogen fits within the relatively widely-spaced metal atoms, where it is more stable.

The absorption of hydrogen is strongly exothermic. As the vehicle is refueled, water is circulated around the storage tanks to absorb the heat.

The reverse process, the release of hydrogen, is endothermic. To provide the energy necessary, coolant water that has absorbed heat from the engine is circulated around the storage tank.

One "tankful" of the metal hydride fuel provides only about half of the mileage provided by a full tank of gas. Because of the complicated fuel system, a hydrogen-powered car costs at least one-and-a-half times as much as its gasoline-powered counterpart.

The only place where humans depend on hydrogen for energy is in space. The energy supplied by the combustion of hydrogen is used by spacecraft such as the space shuttle.

Despite the remaining obstacles, engineers at several car companies believe that use of hydrogen-powered cars will be common someday.

Applying Concepts

1. Which of the following is the most efficient fuel, in terms of energy output per unit mass of fuel?

Fuel	ΔH_{comb} (kJ/mol) at 298 K
methane	+891
methanol	+726
propane	+2219
1-propanol	+2021

Research and Writing

Use the library to find out more about the following.

1. Some car manufacturers are experimenting with compressed natural gas (CNG) as an alternative fuel. What advantages and disadvantages does CNG offer compared to gasoline and hydrogen gas?

Chemistry in Your Community

Automobiles are easily recognizable as energy users. The periodic stop at the gas station is one indicator of the amount of energy expended by our cars. Conduct a survey of 10 different models of cars and trucks.

1. Plot graphs of the fuel mileage versus the mass of car, the engine displacement (L or in^3), the list price, the size of gas tank, and the size of the tire. Study the information found in the graphs, and identify the trends you observe.
2. Which models can run on 87 octane, and which need a higher octane? Use the current gasoline prices for the octane rating of the car being studied to determine the fuel cost for 15 000 mi.

Answers to Applying Concepts

1. methane

Closure Strategy

Differentiate between ΔG, ΔH, and ΔS, and identify how each is affected when a substance such as water changes phases.

CHAPTER 9

Highlights

Key terms

activation energy	**heat of reaction**
basal metabolic rate	**Hess's law**
enthalpy	**resting metabolic rate**
enthalpy change	**specific heat capacity**
entropy	**spontaneous change**
free energy	**temperature**
heat	

Key Ideas

How does energy affect change?

- Heat and temperature are not the same thing. Heat is the measure of total energy transferred from one system to another; temperature is the measure of average kinetic energy in a sample of matter.
- The specific heat capacity for a substance describes how much heat energy is required to cause a temperature change.
- Heat of reaction—the energy released or absorbed during a chemical reaction or a physical change—is measured with a calorimeter.
- The amount of heat involved in a physical or chemical change can be expressed as a ratio in stoichiometric calculations.

How does enthalpy drive changes?

- When heat is released during a change, enthalpy is decreased. Decreasing enthalpy drives most chemical reactions.
- All reactions require activation energy, which can be provided by adding energy to the system or by converting the kinetic energy a substance already has.
- Total changes in enthalpy can be calculated by adding together the enthalpy changes for each part of a reaction.

How does entropy drive changes?

- Entropy, a measure of the disorder of a system, increases with increasing temperature.
- A higher entropy level in a reaction results in a more stable product.

How do entropy and enthalpy affect change?

- Entropy and enthalpy values can be mathematically combined with temperature to describe the free energy of a system and predict whether the reaction is spontaneous.

What are your energy needs?

- Resting metabolic rate (RMR) is the amount of energy needed to keep the body running at its most basic level.
- The thermic effect of food is the energy needed for the chemical reaction of digestion.

Key problem-solving approach: Calculations involving energy

When solving energy calculations, begin at the top by writing the balanced equation. Then, follow the arrows to calculate the desired quantity.

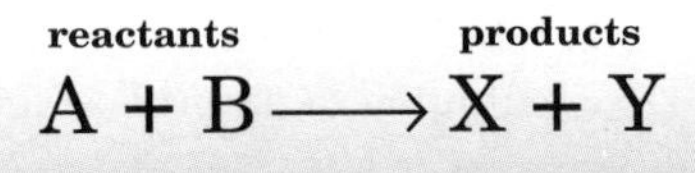

Enthalpy

$\Delta H_{rxn} = (\Delta H_{fX} + \Delta H_{fY}) - (\Delta H_{fA} + \Delta H_{fB})$

$+\Delta H$ = endothermic

$-\Delta H$ = exothermic

Entropy

$\Delta S_{rxn} = (S_X + S_Y) - (S_A + S_B)$

$+\Delta S$ = disordering

$-\Delta S$ = ordering

$\times \frac{1\ kJ}{1000\ J}$

Free Energy

$\Delta G = \Delta H_{rxn} - T\Delta S_{rxn}$

$+\Delta G$ = nonspontaneous

$-\Delta G$ = spontaneous

CHAPTER 9

Review and Assess

Heat and energy

REVIEW

1. What is the difference between heat and temperature?
2. How is heat of solution related to heat of reaction?
3. Indicate whether the change occurring in the following situations is due to specific heat capacity, heat of fusion, or heat of vaporization.
   a. a pot of boiling coffee
   b. ice cream melting on a hot day
   c. a burner on the stove heating up
4. Which illustration below shows the energy flow during an exothermic reaction, and which shows the energy flow during an endothermic reaction?

   a. $CH_4 + O_2 \longrightarrow CO_2 + H_2O$

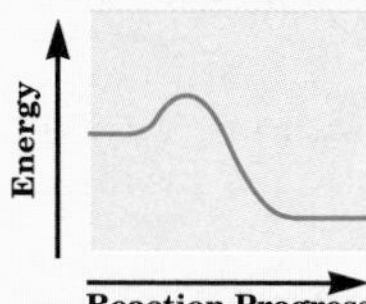

   b. $H_2O + CO_2 \longrightarrow C_6H_{12}O_6 + O_2$

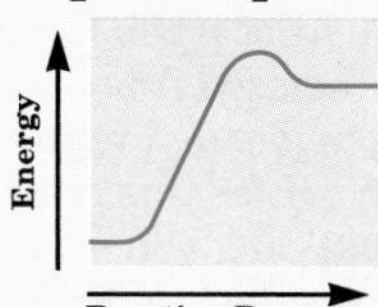

5. A man with a mass of 80 kg requires 2500 Cal/ day to keep his body working efficiently. What is his required daily energy in kJ?
6. Complete the following table.

Food and measure	kJ	Calories
carrot, 1, trimmed	50.2	
jelly doughnut, 1		220.
whole milk, 1 cup	656.9	
rice cake, 1		60.0

PRACTICE

Answers to items in a black square begin on page 841.

7. If you fill a bathtub with 200. kg of water at 44°C, how much heat energy is lost as the water cools to a temperature of 21°C? (Hint: see Sample Problem 9A.)
8. On a cold winter day with a temperature of 4°C, you pick up a penny from the ground and put it in your pocket. If the penny has a mass of 1.85 g, how much heat energy must be transferred to the coin to warm it to your body temperature, 37°C? The specific heat of copper is 0.385 J/g•°C. (Hint: see Sample Problem 9A.)
9. A nutritional chemist burns a saltine cracker in a calorimeter containing 2.50 kg of water. The temperature increases from 25.0°C to 29.8°C. What is the energy content of the cracker in Calories and in kilojoules? (Hint: see Sample Problem 9B.)
10. Predict the final temperature of 2.50 kg of water in a calorimeter if the water is at 25.00°C before 0.5 oz. of noodles containing 54.0 Cal are burned. (Hint: see Sample Problem 9B.)
11. When suffering from a fever, your body temperature rises from 37°C to 40°C, using 787 kJ of energy in the process. Assume that your body burns only glucose to raise your temperature. How many grams of glucose are consumed?

$$C_6H_{12}O_6(s) + 6O_2(g) \longrightarrow 6H_2O(l) + 6CO_2(g)$$

$\Delta H = -2870$ kJ/mol
(Hint: see Sample Problem 9C.)

12. A mountain climber, wanting a drink of water, must melt the snow from the mountain with a propane burner. How many grams of propane would the mountain climber have to use to generate the 33.5 kJ needed to melt the snow for 100 mL of water if her burner heats with 50% efficiency?

$$C_3H_8(g) + 5O_2(g) \longrightarrow 4H_2O(l) + 3CO_2(g)$$

$\Delta H = -2220$ kJ
(Hint: see Sample Problem 9C.)

1. Heat is a measure of total energy transferred; temperature is a measure of average kinetic energy.
2. The heat of reaction is the amount of energy transferred during the course of a reaction. Heat of solution is the amount of energy transferred when a substance dissolves in a solvent.
3. a. heat of vaporization
   b. heat of fusion
   c. specific heat capacity
4. a. exothermic
   b. endothermic
5. 10 460 kJ/day
6. carrot, 12 Cal;
   jelly doughnut, 920. kJ;
   whole milk, 157.0 Cal;
   rice cake, 251 kJ
7. $(-23°C)(2.00 \times 10^5 \text{ g}) \times (4.180 \text{ J/g}\cdot°C) = -1.9 \times 10^7 \text{ J}$
8. $(33°C)(1.85 \text{ g})(0.385 \text{ J/g}\cdot°C) = +23.5 \text{ J}$
9. $4.8°C \times \frac{4.180 \text{ J}}{1 \text{ g}\cdot°C} \times \frac{2.50 \times 10^3 \text{ g}}{1} \times$

   $= 50. \text{ kJ}$

   $50. \text{ kJ} \times \frac{1 \text{ Cal}}{4.184 \text{ kJ}} = 12 \text{ Cal}$
10. $\Delta t = 54.0 \text{ Cal} \times \frac{1000 \text{ cal}}{1 \text{ Cal}} \times \frac{1 \text{ g}\cdot°C}{1 \text{ cal}} \times \frac{1}{2.50 \times 10^3 \text{ g}} = 21.6°C$

    $t_f = 25.00°C + 21.6°C = 46.6°C$
11. energy change ratio:
    2870 kJ:1 mol $C_6H_{12}O_6$
    molar mass: 180.18 g/mol

    $787 \text{ kJ} \times \frac{1 \text{ mol } C_6H_{12}O_6}{2870 \text{ kJ}} \times \frac{180.18 \text{ g } C_6H_{12}O_6}{1 \text{ mol } C_6H_{12}O_6} = 49.4 \text{ g } C_6H_{12}O_6$
12. energy change ratios:
    2220 kJ:1 mol C_3H_8
    molar mass: 44.11 g/mol
    efficiency: 50%

    $33.5 \text{ kJ} \times \frac{1 \text{ mol } C_3H_8}{2220 \text{ kJ} \times 50\%} \times \frac{44.11 \text{ g } C_3H_8}{1 \text{ mol } C_3H_8} = 1.33 \text{ g } C_3H_8$

13. The water inside the peanut's calorimeter will rise to a higher temperature than the water in the celery's calorimeter.

14. a. $6.00\ kJ + 100(7.53 \times 10^{-2}\ kJ) + 40.6\ kJ = 54.1\ kJ$
b. $395.8(2.80 \times 10^{-2}\ kJ) + 59.1\ kJ = 70.2\ kJ$
c. $5.00\ mol \times 38.6\ kJ/mol = 193\ kJ$

15. a. specific heat capacity of water
b. heat of fusion of steel components, specific heat capacity of solid and molten steel

16. *See page 353A.*

17. The concrete surrounding the pool has a lower specific heat capacity than the water in the pool.

18. Answers will vary. Be certain students properly identify system and surroundings. Various exothermic reactions are used in furnaces (systems) to heat the rest of a house (surroundings). When a home is airconditioned, an endothermic change in the air conditioner (system) cools the air (surroundings). Then, the process is reversed outside the house so that the reaction is exothermic, releasing heat to the outdoors (surroundings).

19. Enthalpy is the total energy content of a system. This is unmeasurable. Enthalpy change is a measurable amount of heat released or absorbed when a physical or chemical change occurs.

20. The more enthalpy decreases, the more likely the reaction is spontaneous.

21. Some exothermic reactions appear not to require added energy because they gain activation energy from the kinetic energy of some particles.

The full solution for item ***16*** *is found on page 353A.*

APPLY

13. An ounce of peanuts has more Calories than an ounce of celery. If equal masses of peanuts and dried celery were burned in separate calorimeters, how would the temperature of the water inside one calorimeter compare to the temperature of the water in the other?

14. Refer to **Table 9-3** to answer the following questions.
a. How much energy is required to change 1 mol of ice at 0°C to steam at 100°C?
b. How much energy is required to change 1 mol of liquid mercury at –38.8°C to mercury vapor at 357°C?
c. How much energy is required to boil 5.00 mol of ethanol at 78.3°C?

15. Name the type(s) of heat data needed by the following people to complete their jobs.
a. a pool operator who needs to regulate water temperature of a heated swimming pool
b. a steel mill worker who needs to regulate the heat of the melting furnace

16. Refer to **Table 9-3** and draw and label a temperature vs. heat curve for 1 mol Hg being heated from a solid at –38.8°C to a gas at 357°C.

17. The concrete that surrounds a swimming pool takes less time to heat up than the water in the pool on a hot summer day. What can you infer about the specific heat capacity values for concrete and water?

18. During an exothermic change, energy is lost from the reaction system and gained by the surroundings as heat. The opposite happens in an endothermic change. Explain how the use of both processes controls the temperature of your home. (Hint: for each process, carefully consider what the system is and what the surroundings are.)

Enthalpy changes

REVIEW

19. What is the difference between enthalpy and an enthalpy change?

20. What is the relationship between enthalpy and spontaneous reactions?

21. Why do some exothermic reactions appear to require no added activation energy?

22. Use the following terms to label the energy diagram for the production of phosgene, $COCl_2$, a chemical used in the production of organic supplies: activation energy, ΔH, reactants, products.

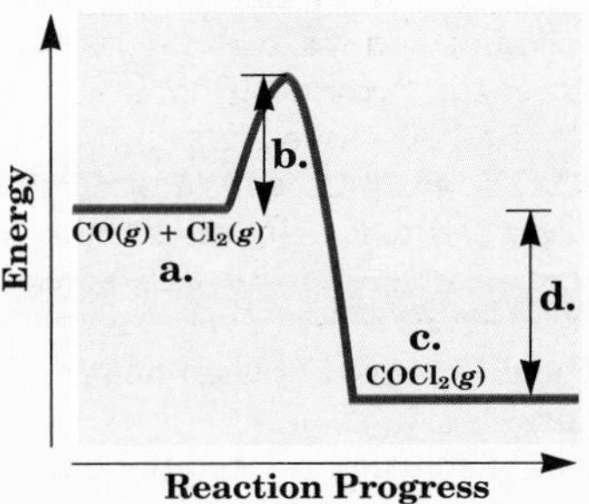

PRACTICE

23 In the engine of your car, nitrogen and oxygen can combine to form nitrogen oxides, chemicals that contribute to pollution. Below is a reaction that forms nitrogen dioxide from previously formed nitrogen monoxide. Determine the ΔH value for this reaction using the heats of formation given. (Hint: see Sample Problem 9D.)

$$2NO(g) + O_2(g) \longrightarrow 2NO_2(g)$$

Substance	ΔH^o_f (kJ/mol)
$NO(g)$	+90.2
$O_2(g)$	0
$NO_2(g)$	+33.2

24. Pentane, the smallest hydrocarbon found in gasoline, can be synthesized from 2-pentanol and hydrogen gas in a multistep process. Determine the ΔH for this reaction using the heats of formation given for the equation below. (Hint: see Sample Problem 9D.)

$$C_5H_{11}OH(l) + H_2(g) \longrightarrow C_5H_{12}(l) + H_2O(l)$$

Substance	ΔH^o_f (kJ/mol)
$C_5H_{11}OH(l)$	–365.2
$H_2(g)$	0
$C_5H_{12}(l)$	–173.5
$H_2O(l)$	–285.8

A P P L Y

25. Some fire extinguishers work by evolving an unreactive gas, like CO_2, that smothers the flame. The contents also work as a cooling agent. Using your knowledge of activation energy, explain the importance of including a cooling agent in fire extinguishers.

26. The gasification of coal is a method of producing methane. The entire reaction has four separate steps, with heats of reaction of + 188 kJ, –54 kJ, –221 kJ, and –75 kJ, respectively. What is the ΔH of this entire reaction for the gasification of coal?

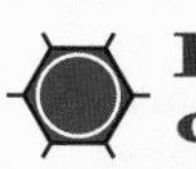

Entropy changes

R E V I E W

27. Give three examples of each of the following conditions.
 a. high entropy **b.** low entropy

28. Which provides a greater number of ways to arrange particles: a high-entropy or low-entropy system?

29. Why can absolute values of entropy be calculated while absolute values of enthalpy cannot be?

30. What is the difference between entropy units and enthalpy units?

P R A C T I C E

31. Solid iron will slowly oxidize into iron oxide, Fe_2O_3, if it is left untreated in the open air. What is the entropy change for this reaction? (Hint: see Sample Problem 9E.)

$$4Fe(s) + 3O_2(g) \longrightarrow 2Fe_2O_3(s)$$

Substance	S (J/K•mol)
$Fe(s)$	27.3
$O_2(g)$	205.2
$Fe_2O_3(s)$	87.4

32. Ammonium nitrate is a common fertilizer that can become a powerful explosive when it undergoes rapid decomposition. What is the entropy change for this reaction? (Hint: see Sample Problem 9E.)

$$2NH_4NO_3(s) \longrightarrow 2N_2(g) + 4H_2O(g) + O_2(g)$$

Substance	S (J/K•mol)
$NH_4NO_3(s)$	151.1
$N_2(g)$	191.6
$H_2O(g)$	188.8
$O_2(g)$	205.2

A P P L Y

33. Oil of citronella, a volatile liquid, is often burned to repel insects. For each molecule of oil of citronella, several molecules of carbon dioxide and water are produced. Identify whether entropy increases or decreases in this reaction, and list three pieces of evidence for this change in entropy.

34. Predict whether ΔS of the changes in the following examples is positive or negative.
 a. A chemist heats a chemical in order to determine its melting point.
 b. A petroleum distillery fractionally distills crude oil by boiling off different volatile hydrocarbons.
 c. The condensation of water vapor on a humid day causes a glass to appear to sweat.

35. Which graph shown below depicts the mathematical relationship between S and temperature for a system which is not undergoing a chemical reaction or a phase change?

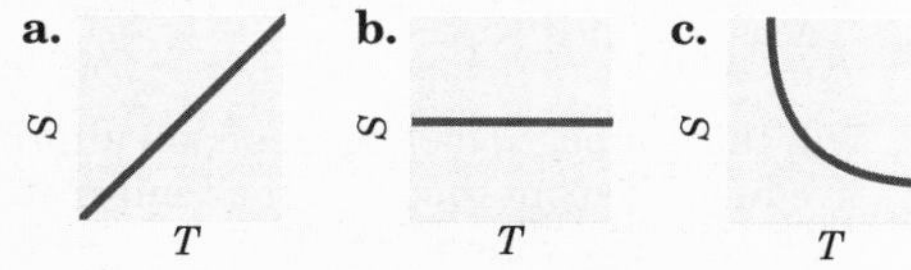

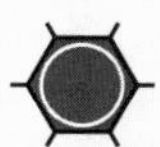

Free energy

R E V I E W

36. What types of enthalpy and entropy changes always give rise to a spontaneous change?

37. What free-energy condition must occur for a reaction to be spontaneous?

38. Name three reaction conditions that can control a reaction by affecting its spontaneity.

39. Is a solution of a substance representative of a high-entropy system or a low-entropy system compared to the pure substance and pure solvent?

22. **a.** reactants
 b. activation energy
 c. products
 d. ΔH

23. reactants:
 1 mol NO: $\Delta H_f = +90.2$ kJ
 1 mol O_2: $\Delta H_f = 0$ kJ

 product:
 1 mol NO_2: $\Delta H_f = +33.2$ kJ

 $\Delta H_{rxn} = [2(+33.2 \text{ kJ})] - [2(+90.2 \text{ kJ}) + 0 \text{ kJ}] = -114$ kJ

24. reactants:
 1 mol $C_5H_{11}OH$: $\Delta H_f = -365.2$ kJ
 1 mol H_2: $\Delta H_f = 0$ kJ

 products:
 1 mol C_5H_{12}: $\Delta H_f = -173.5$ kJ
 1 mol H_2O: $\Delta H_f = -285.8$ kJ

 $\Delta H_{rxn} = [(-173.5 \text{ kJ}) + (-285.8 \text{ kJ})] - [(-365.2 \text{ kJ}) + 0 \text{ kJ}] = -94.1$ kJ

25. Acting as a cooling agent, fire extinguishers withdraw heat from the reactants, thus lowering their kinetic energy so that they cannot attain the activation energy necessary to react.

26. –162 kJ

27. **a.** Answers will vary. Be certain students' answers describe a disordered system, such as a gas.
 b. Answers will vary. Be certain students' answers describe an ordered system, such as a solid crystal lattice.

28. high entropy

29. Entropy is defined as zero at absolute zero. We can compare all other entropy values to this standard. We do not have a standard for zero enthalpy.

30. Entropy units are J/K•mol, and enthalpy units are kJ/mol.

31. $\Delta S = [2(87.4 \text{ J/K})] - [4(27.3 \text{ J/K}) + 3(205.2 \text{ J/K})] = -550$ J/K

32. $\Delta S = [2(191.6 \text{ J/K}) + 4(188.8 \text{ J/K}) + (205.2 \text{ J/K})] - [2(151.1 \text{ J/K})] = 1041$ J/K

33. This change involves an increase in entropy. Three pieces of evidence include the number of product molecules is greater than the number of reactant molecules, the high temperatures of combustion, and a liquid becoming a gas.

34. a. positive
b. positive
c. negative

35. a

36. If the change produces both a decrease in enthalpy and an increase in entropy, it is always spontaneous.

37. The free energy must be a negative value.

38. Temperature, pressure, and the nature of the reactants alter the free energy of a reaction. By controlling these variables, you can control the spontaneity of the reaction.

39. high entropy

40. a. Entropy units must be converted into kJ/K•mol, or enthalpy must be converted into J/mol.
b. correct
c. Temperature must be converted into K.
d. Enthalpy and entropy are in the wrong parts of the equation.

41. Because entropy is multiplied by temperature, high temperatures mean more randomness within the system, so entropy will have a greater influence on the free energy of a reaction at high temperatures than enthalpy.

42. temperature:
100.°C + 273 = 373 K

enthalpy change: +140.5 kJ

entropy change:
–38.9 J/K = –0.0389 kJ/K

ΔG = 140.5 kJ – 373 K (–0.0389 kJ/K) = +155 kJ

The reaction is not spontaneous.

40. What, if anything, is wrong with the following setups for calculating free energy for this reaction?

$$2Na(s) + Cl_2(g) \longrightarrow 2NaCl(s)$$

$\Delta H = -822.4$ kJ/mol
$\Delta S = -181.5$ J/K•mol

a. $\Delta G = -822.4$ kJ/mol – (298 K) (–181.5 J/K•mol)
b. $\Delta G = -822.4$ kJ/mol – (323 K) (–181.5 J/K•mol) (1 kJ/1000 J)
c. $\Delta G = -822.4$ kJ/mol – (15°C) (–181.5 J/K•mol) (1 kJ/1000 J)
d. $\Delta G = -181.5$ J/K•mol – (273 K) (–822.4 kJ/mol) (1000 J/1 kJ)

41. At high temperatures does enthalpy or entropy usually have a greater effect on a reaction's free energy? Explain your answer.

PRACTICE

42. Titanium chloride, $TiCl_4$, is an intermediate chemical formed in the production of titanium metal, a substance used extensively in aerospace technology. Will the reaction for the production of $TiCl_4$ be spontaneous if it is carried out at 100.0°C? Show your work. (Hint: see Sample Problem 9F.)

$$TiO_2(s) + 2Cl_2(g) \longrightarrow TiCl_4(l) + O_2(g)$$

$\Delta H = +140.5$ kJ
$\Delta S = -38.9$ J/K

43. Will the combustion of benzene, an alternative fuel, be spontaneous if the temperature is 25°C? Show your work. (Hint: see Sample Problem 9F.)

$$2C_6H_6(l) + 15O_2(g) \longrightarrow 12CO_2(g) + 6H_2O(l)$$

$\Delta H = -6535$ kJ
$\Delta S = -439.1$ J/K

APPLY

44. Draw a graph with ΔH on the x-axis and ΔS on the y-axis. In each quadrant of the graph, indicate what conditions, if any, are necessary to make a reaction spontaneous for each combination of positive and negative ΔH and ΔS values.

45. Gold can be dissolved using nitrosyl chloride, NOCl. The entropy change for the spontaneous reaction to make NOCl at a low temperature is negative. Is the enthalpy positive or negative?

Energy and your body

REVIEW

46. What does the human body's basal metabolic rate represent?

47. What happens to your energy requirements as your activity level increases?

48. How does the thermic effect of food differ from BMR?

APPLY

49. Support the following statements, using what you know about metabolism and the thermic effect of food.
a. Men have a higher BMR than women.
b. A person living in Alaska has a higher BMR than a person living in Florida.

50. Use **Table 9-9** and **Table 9-10** to answer the following questions.
a. A 16-year-old, 50 kg girl rides her bicycle for 2 h. How many Cal/h does she expend?
b. About how many Calories do you expend in 1 h of cleaning your room?
c. A 24-year-old, 84 kg basketball player plays for 30 min. How many Cal/h does he expend?

51. How long could a 30-year-old 75 kg cyclist ride using the energy from a meal with 50.1 g protein, 145.3 g carbohydrate, and 29.7 g fat?

Linking chapters

1. ***Bond energy***
Energy diagrams show the relative bond energies of reactants and products. How does the concept of enthalpy relate to bond energy?

2. ***Chemical reactions***
For a chemical reaction to take place, particles must collide with each other. Gasoline-air mixtures don't react at cool temperatures, but do react at warmer temperatures. What other requirement is there for reaction collisions?

3. **Theme: *Systems and interactions***
Inside a calorimeter, two systems interact. What are these two systems, and how do they interact?

4. **Theme: *Equilibrium and change***
Oxygen forms ozone, and at the same time, the ozone decomposes to form oxygen. At 836°C the reaction of oxygen gas to form ozone has a ΔG value of zero. Which of these reactions, if any, is absent?

USING TECHNOLOGY

1. ***Graphics calculator***
Free energy can be predicted by examining a graph of the free energy equation. This function can be graphed on a calculator. Press PRGM ▶ ▶ ENTER to open a new program from the program menu on your graphics calculator, and give it a name. Then enter the following program.
:Prompt H,S
:"H–(X/1000)S" → Y_1.
Remember to use the "ALPHA" key for entering the letters. (Keystrokes: PRGM; ▶; 2; H,S; ENTER; "H–(X÷1000)S"; STO; 2nd; Y-VARS; ENTER; ENTER). The temperature at which ΔG will be zero is the value of x at the y-intercept. Run the program by pressing 2nd; QUIT; PRGM; select the program by pressing ENTER; ENTER. Enter the desired ΔH and ΔS values, press GRAPH, and then TRACE until $y = 0$. Determine the temperature at which the following reactions will have free energy values of zero. (Hint: see owner's manual to reset the WINDOW function to show the correct parts of the graph.)
a. $C_7H_8(l) \longrightarrow C_7H_8(g)$
$\Delta H = +\ 38.0$ kJ/mol
$\Delta S = +\ 99.7$ J/K•mol
b. $PCl_3(g) + Cl_2(g) \longrightarrow PCl_5(g)$
$\Delta H = -87.9$ kJ/mol
$\Delta S = -170.25$ J/K•mol
c. $H_2(g) \longrightarrow 2H(g)$
$\Delta H = +435.9$ kJ/mol
$\Delta S = +\ 98.74$ J/K•mol

Alternative assessment

Performance assessment

1. Design an experiment to measure the heat capacities of zinc, copper, and iron. If your teacher approves your design, obtain the materials needed. When finished, compare your experimental values to values from a chemical handbook or reference source.
2. Develop a procedure to measure the ΔH of a reaction. If your teacher approves, test your procedure by measuring the ΔH value of the following reaction. Determine the accuracy of your method by comparing your ΔH to the accepted ΔH value.

$$NaC_2H_3O_2(s) \longrightarrow Na^+(aq) + C_2H_3O_2^-(aq)$$

Portfolio projects

1. ***Chemistry and you***
During the course of a week, keep a journal of the different reactions you encounter. Estimate the sign and order of magnitude for ΔH, ΔS, and ΔG values for the reactions.
2. ***Research and communication***
Conduct research on different mechanisms that lower the entropy of a system. Decide which are most effective. Present your findings to the class.
3. ***Research and communication***
The chapter story covers several alternative fuels for cars. Conduct research on the public and private uses of these fuels in both automobiles and other combustion engines. Determine which ones can be used and produced with the least environmental impact. Present your findings to the class.
4. ***Chemistry and you***
Evaluate your own diet based on information found in **Table 9-4**, **Table 9-5**, **Table 9-9** and **Table 9-10**. You may also want to research such diet issues as starchy carbohydrates versus sugary carbohydrates, complete versus incomplete proteins, and saturated versus unsaturated fats. Determine a realistic weekly plan of diet and exercise that will help you balance energy needs and food consumption. Try this plan and report to the class how you feel after one week.
5. ***Chemistry and you***
Evaluate the labels of serveral different types of pet food. Explain which one you believe to be most nutritious and give your reasons.

43. temperature:
25.°C + 273 = 298 K

enthalpy change: –6535 kJ

entropy change:
–439.1 J/K = –0.4391 kJ/K

$\Delta G = -6535$ kJ – 298 K (–0.4391 kJ/K) = –6404 kJ

The reaction is spontaneous.

44.

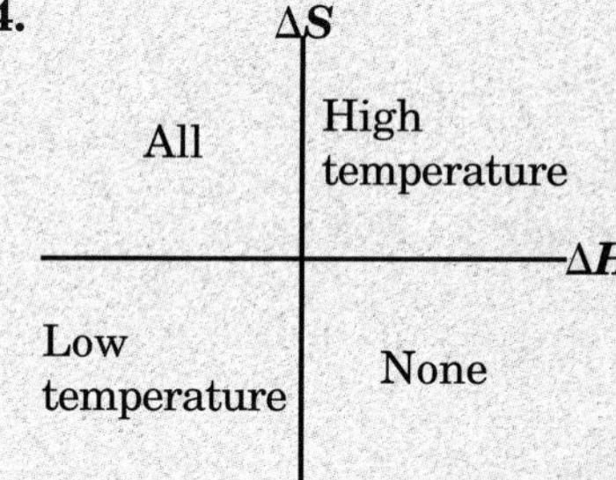

45. negative

46. The basal metabolic rate in the human body represents the energy consumed by all life-sustaining reactions, excluding digestion and absorption of food.

47. As your activity level increases, your energy requirements increase.

48. The thermic effect of food is the energy needed for digestive and catabolic reactions to break down the food, while the BMR is the energy needed for all life-sustaining reactions except digestion.

49. a. On average, men are more muscular, and muscles use more energy than fat.
b. In a colder climate, more energy is expended to keep the body warm.

50. a. $\frac{1356 \text{ Cal}}{1 \text{ day}} \times \frac{1 \text{ day}}{24 \text{ h}} \times 5 \times 2 \text{ h} =$ 565 Cal

b. Answers should range from 141 Cal to 237 Cal; answers should be calculated using the following equation.
$\frac{\text{RMR}}{24} \times 2.5 = \text{Cal/h}$

c. $\frac{1964 \text{ Cal}}{1 \text{ day}} \times \frac{1 \text{ day}}{24 \text{ h}} \times 0.5 \text{ h} \times 7 =$ 286 Cal

Answers from page 322

Answers to Practice 9C

1. $5000.\text{ kJ kinetic energy} \times \frac{100\text{ kJ released}}{25\text{ kJ kinetic energy}}$

$\times \frac{2\text{ mol } C_8H_{18}}{1.099 \times 10^4\text{ kJ released}} \times \frac{114.26\text{ g } C_8H_{18}}{1\text{ mol } C_8H_{18}} = 420\text{ g } C_8H_{18}$

$5000.\text{ kJ kinetic energy} \times \frac{100\text{ kJ released}}{25\text{ kJ kinetic energy}}$

$\times \frac{25\text{ mol } O_2}{1.099 \times 10^4\text{ kJ released}} \times \frac{32.00\text{ g } O_2}{1\text{ mol } O_2} = 1500\text{ g } O_2$

2. $\frac{110\text{ g } C_6H_{12}O_6}{180.12\text{ g/mol}} \times \frac{2870\text{ kJ}}{1\text{ mol}} = 1750\text{ kJ}$

Answers to Section Review

4. $V_{pool} = (12)(6)(3)\text{ m}^3$

$m_{water} = (216\text{ m}^3)\left(\frac{10^2\text{ cm}}{1\text{ m}}\right)^3 \frac{1\text{ g}}{1\text{ cm}^3} = 216 \times 10^6\text{ g}$

$\Delta H = (216 \times 10^6\text{ g})(4.180\text{ J/g}\cdot{}^\circ\text{C})(5.0^\circ\text{C}) = 4.5 \times 10^9\text{ kJ}$

8. $428\text{ g } H_2 \times \frac{1\text{ mol } H_2}{2.02\text{ g } H_2} \times \frac{571.6\text{ kJ}}{2\text{ mol}} = 60\,556\text{ kJ}$

$428\text{ g } C_8H_{18} \times \frac{1\text{ mol } C_8H_{18}}{114.18\text{ g } C_8H_{18}} \times \frac{10\,990\text{ kJ}}{2\text{ mol}} = 20\,598\text{ kJ}$

Answers from page 350

16.

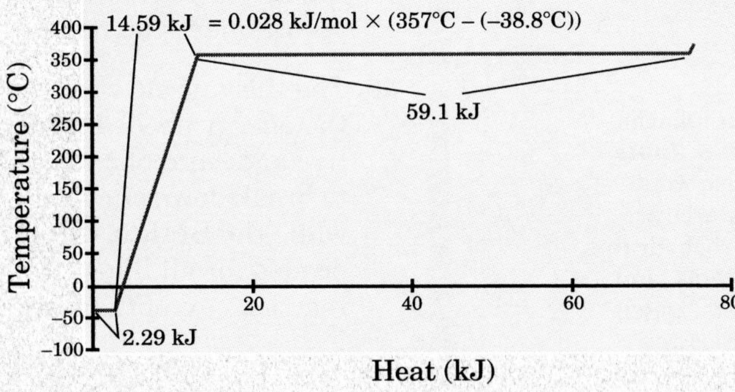

Answers from page 353

51. Calories from carbohydrates:

$145.3\text{ g} \times \frac{4\text{ Cal}}{1\text{ g}} = 581.2\text{ Cal}$

Calories from protein:

$50.1\text{ g} \times \frac{4\text{ Cal}}{1\text{ g}} = 200.4\text{ Cal}$

Calories from fat:

$29.7\text{ g} \times \frac{9\text{ Cal}}{1\text{ g}} = 267.3\text{ Cal}$

Total Calories available: 1048.9

$1048.9\text{ Cal} \times \frac{1\text{ day}}{1964\text{ Cal}} \times \frac{24\text{ h}}{1\text{ day}} \times \frac{1}{7} = 1.83\text{ h}$

Linking chapters

1. If more energy is released when the bonds of products form than is absorbed to break the bonds of the reactants, the enthalpy of reaction is negative.
2. If the particles collide with enough kinetic energy to overcome the activation energy, the reaction will occur spontaneously.
3. The two systems are the reaction in the bomb and the temperature change of the surrounding water. As one loses heat, the other one gains it.
4. Both of these reactions actually happen. However, they happen at the same rate, so the system appears to have no reaction.

USING TECHNOLOGY

1. a. ≈ 383 K
b. ≈ 511 K
c. ≈ 4425 K

Performance assessment

1. Answers will vary. Be sure your students use a safe procedure and available materials. A calorimeter can easily be made from a foam cup.
2. Answers will vary. Be sure students follow proper safety procedures, including proper personal protective equipment and chemical safety.

Portfolio projects

1. Answers will vary. They may include food digestion, gas combustion in cars, or sweating during exercise.
2. Answers will vary. Research could include air conditioners, freezers, or dehumidifiers.
3. Answers will vary. Be sure that the students focus their research not only on the use of the fuels but also on the means to produce them. For example, battery-powered cars provide a clean source of fuel, but the production of the batteries and their disposal can have harmful effects on the environment.
4. Answers may vary, but after one week of a balanced plan students should report feeling more energetic.
5. Answers will vary. Accept all reasonable justifications.

10

PLANNING FOR

Gases and Condensation

Chapter Overview

- **Section 10-1** describes the basic properties of gases, kinetic molecular theory, and the four descriptive variables for gases from a microscopic and macroscopic perspective. The roles of gases in global atmospheric chemistry are given special attention.
- **Section 10-2** covers the gas laws of Charles and Boyle, and Dalton's law of partial pressures. The basis of the absolute temperature scale is introduced in establishing Charles's law. The connection between mole fraction of a gas and molar volume is examined. Gas mobility in the forms of diffusion and effusion is also explored.
- **Section 10-3** presents the ideal gas law, the ideal gas law constant, and the combined gas law. The role of stoichiometry in determining quantities of reactant and product gases is examined.
- **Section 10-4** focuses on condensation of gases and their deviation from the ideal gas model at high pressures and low temperatures. The dependence of boiling and freezing points of liquids on their vapor pressure and the atmospheric pressure is explored, as are liquid volatility and phase diagrams. Surface tension of liquids is also discussed.

Concept Base

Students must have mastered the following concepts prior to this chapter: the law of conservation of matter, kinetic energy, direct and indirect graphical relationships (Chapter 2); the number of atoms and molecules in a mole (Avogadro's number); moles (Chapter 4); balancing chemical equations (Chapter 7); stoichiometry (Chapter 8).

Looking Ahead

The behavior of liquids and solutions will be further examined in Chapters 11 and 12.

Laboratory Equipment Needs

Demonstrations

The listing of materials for Chapter 10 demonstrations begins on page T46.

Laboratory Experiments

Materials and preparation instructions for Chapter 10 are found as follows.

Text		*Page*
Investigation 10A	Gas Diffusion—Industrial Spill	732
Investigation 10B	Gas Stoichiometry—Fire Suppression System	734

Laboratory Experiments Manual
(refer to the table of contents in the manual)

Exploration 10-1	Gas Pressure-Volume Relationship
Investigation 10-1	Gas Temperature-Volume Relationship—Balloon Flight
Exploration 10-2	Masses of Equal Volumes of Gases

Pacing: 15 days — Inquiry Teaching Strategies

10-1 What are characteristics of gases?

Demonstrations
- Demo 1 *Gaseous diffusion*
- Demo 2 *Elastic and inelastic collisions*
- Demo 3 *The barometer*

Laboratory Experiments
- Inv 10A *Gas Diffusion—Industrial Spill*

10-2 What behaviors are described by the gas laws?

Demonstrations
- Demo 4 *Determing absolute zero*
- Demo 5 *Volume changes with a vacuum pump*
- Demo 6 *Amount of oxygen in air*
- Demo 7 *Diffusion of HCl and NH_3*
- Demo 8 *Diffusion of $KMnO_4$ in water*

Laboratory Experiments
- Exp 10-1 *Gas Pressure-Volume Relationship*
- Inv 10-1 *Gas Temperature-Volume Relationship—Balloon Flight*
- Exp 10-2 *Masses of Equal Volumes of Gases*

10-3 How do the gas laws fit together?

Laboratory Experiments
- Inv 10B *Gas Stoichiometry—Fire Suppression System*

10-4 What conditions will cause a gas to condense?

Demonstrations
- Demo 9 *Boiling water in a vacuum*

Key
- Teaching Transparencies
- Annotated Teacher's Edition
- Laboratory Experiments Manual

Visual Teaching Strategies / Review and Practice Strategies

Visual Teaching Strategies	Review and Practice Strategies
Transparencies ■ F 10-1 *Comparing Relative Volumes* **Transparency Masters** ■ T 10-1 *Composition of Dry Air* ■ F 10-8 *Comparing Molecular Speeds of Different Gases* ■ F 10-9 *Comparing Molecular Speeds of a Gas* ■ F 10-15 *Volume vs. Quantity* ■ T A-8 *Densities of Gases*	**Text Reviews** ■ Section Review 1–11 ■ Chapter Review 1–11; LC 3, 7; UT 2 **Study Guide Worksheets** ■ ■ 10-1 Concept Review
Transparencies ■ F 10-22 *Gas Mixtures* ■ F 10-23 *Particle Model for a Gas Collected over Water* **Transparency Masters** ■ F 10-17 *Volume vs. Temperature* ■ F 10-18 *Charles's Law* ■ F 10-20 *Pressure vs. Volume* ■ F 10-21 *Pressure vs. Reciprocal of Volume* ■ T 10-6 *Vapor Pressure of Water* ■ T A-3 *Vapor Pressures of Water at Selected Temperatures*	**Text Reviews** ■ Additional Examples 10A, 10B, 10C, 10D, 10E, 10F ■ Practice 1–13 ■ Section Review 12–22 ■ Chapter Review 12–36; LC 5; UT 1 **Study Guide Worksheets** ■ ■ 10-2 Concept Review ■ ■ 10-2 Practice *Charles's Law* *Boyle's Law* *Dalton's Law of Partial Pressure* *Graham's Law*
	Text Reviews ■ Additional Examples 10G, 10H, 10I, 10J ■ Practice 1–12 ■ Section Review 23–33 ■ Chapter Review 37–50; LC 1, 2, 4, 6; UT 3 **Study Guide Worksheets** ■ ■ 10-3 Practice *Ideal Gas Law and Gas Stoichiometry*
Transparencies ■ F 10-27 *Measuring the Vapor Pressure of a Liquid* ■ P 395 *Gas Law Calculations* **Transparency Masters** ■ F 10-28 *Vapor Pressures*	**Text Reviews** ■ Section Review 34–39 ■ Chapter Review 51–54 **Study Guide Worksheets** ■ ■ 10-4 Skillsheet *Vapor Pressure* *Phase Diagrams*

■ Pupil's Text ■ Study Guide ■ Teaching Resources

Exp *Exploration*, Inv *Investigation*, LC *Linking Chapters*, UT *Using Technology*

Assessment Options

Traditional Assessment

Test Generator
Instructional Objectives Measured:
Content mastery
Select from items 1 to 75

Performance Assessment Options

(see page T36 and T643-15 for scoring rubrics)
Students demonstrate mastery of objectives in a hands-on environment. You may want some of these materials in the Portfolio as well.

Investigation 10A and 10B (text)
Investigation 10-1 (laboratory manual)
Instructional Objectives Measured:
Content mastery, Use of scientific methodology, Problem solving skills, Use of technology, Proficiency in written communication

Performance Assessments, page 401

1. **Designing an experiment**
   Instructional Objectives Measured:
   Content mastery, Use of scientific methodology, Problem solving skills, Use of technology, Proficiency in written communication
2. **Matching microviews**
   Instructional Objectives Measured:
   Content mastery
3. **Matching graphs**
   Instructional Objectives Measured:
   Content mastery

Portfolio Options

(see page T36 for scoring rubrics)
Students provide a written rationale for each selection made for the portfolio.

Concept Maps
Students use vocabulary from the Chapter Review to build a concept map.
Instructional Objectives Measured:
Content mastery

Formal Laboratory Reports
Exploration 10-1 and 10-2 (lab manual)
Instructional Objectives Measured:
Content mastery, Use of scientific methodology, Use of technology, Proficiency in written communication

Portfolio Projects, page 401
Items 1 and 4. Reseach and communication
Instructional Objectives Measured:
Content mastery, Understanding the impact of science and technology on society, Proficiency in written and oral communication

Items 2, 3, and 5. Chemistry and you
Instructional Objectives Measured:
Content mastery, Scientific and chemical literacy, Proficiency in written and oral communication

INTRODUCING THE CHAPTER

CHAPTER 10 Gases and Condensation

Story Background

Balloons have been used to explore the upper atmosphere since 1931, when Auguste Piccard developed an airtight, pressurized cabin that allowed crews to ascend to stratospheric levels. Piccard realized that a balloon expands in volume at higher altitudes and so used a much larger balloon than had been employed in previous ascents. The vast volume could then be partially filled with enough gas to provide lift for the cabin and still have enough room to expand once external atmospheric pressure had dropped.

Currently, two types of balloons are used for stratospheric research: superpressure and zero-pressure balloons. Superpressure balloons are inflated fully and then sealed so that their volume and pressure are kept constant. The balloon ascends until the density of the air outside the the balloon equals that inside, at which point the balloon no longer rises. Zero-pressure balloons, by contrast, have variable volumes and pressures and so can float at any altitude during their flight. They are the largest balloons and are able to carry the heaviest payloads.

Besides being able to place instruments in the part of the atmosphere to be studied, balloons are a relatively inexpensive way of testing satellite equipment prior to launch. Because the equipment is retrievable, adjustments can be made until the instrument is ready to be placed in orbit.

About the Illustration

This zero-pressure balloon for probing the upper atmosphere is made of a 0.02 mm thick polyethylene film and has a volume of 1.1 million m^3. The balloon reaches full inflation at an altitude of about 37 km.

Ascent Into the Ozone

In the still air, the sun begins to rise above the horizon. The busy team of scientists, awake for hours, monitors atmospheric conditions to be sure they are right. Today looks like a great day for the research team—today, there will be a launch.

As the sun rises, a growing silver form appears in the distance. A strange hissing sound fills the air. The shape, now recognizable as a huge balloon, takes form as it inflates with helium.

The ground crew continues to monitor atmospheric conditions as other scientists turn their attention to the expanding balloon. When the top of the partially inflated balloon extends about 250 m into the air, it is ready for launch. Though the balloon is only a small fraction of the size it will be when completely inflated, it still is capable of lifting a payload of telescopes, cameras, and recording equipment with a total mass of over 700 kg. With the data provided by these instruments, scientists can derive a clearer picture of what is happening to the upper atmosphere.

When the lead lines are removed, the balloon will ascend to about 38 500 m above Earth's surface. This mission, like many others before it, involves collecting data on the ever-changing ozone layer in the atmosphere. After collecting data for several days, the payload instruments will be retrieved by scientists when the payload falls back to Earth. The data collected will become part of the monthly log scientists use to make predictions about the ozone layer.

With the advent of satellites you might expect the use of balloons to be obsolete. Though satellites are used to record atmospheric composition data and meteorological conditions, balloons offer several advantages over satellites. Balloons are far less expensive to build and to launch. The cost of launching a satellite includes the price of the satellite itself and the rocket used to propel it into orbit. Another advantage of using balloons to measure the ozone layer is that the balloon can collect data from within the layer, while a satellite can merely provide data from an aerial perspective above the layer.

The principles used in launching balloons were discovered nearly 150 years ago and are the subject of this chapter. In studying these ideas, you will find that there are simple answers to seemingly complicated questions like the following.

- *Why is the balloon only partially inflated at launch?*
- *How do scientists know how much to inflate the balloon?*
- *Why are scientists monitoring the ozone layer?*
- *What gases are thought to be responsible for changes in the ozone layer?*
- *What's being done to protect the ozone layer?*

Tapping ***Prior Knowledge***

Use the following questions to start students thinking about and discussing the concepts to be covered.

1. Name three common applications of expanded and compressed gases. Are the gases in a gaseous state at all times?
2. What do a bathyscaphe (a chamber used for exploring the sea at extreme depths), a submarine, a jet airplane, and a space shuttle have in common (besides being vehicles)? (Hint: each has to accommodate the same type of passenger.)
3. Does water boil at the same temperature at all altitudes? If not, how does the boiling temperature vary with altitude?

Section 10-1

FOCUS

Lesson Starter

Begin the lesson by comparing the actual molar volume of an ideal gas with that of 1 mol of water. Show the class two boxes, one with a volume of 22.4 L (approximately 28.5 cm × 28.5 cm × 28.5 cm) and the other with a volume of 0.018 L (about 2.6 cm × 2.6 cm × 2.6 cm).

Demonstration 1

Gaseous diffusion

Approximate time: 15 min

This demonstration shows the rapid movement of chlorine and ammonia gases.

1. Prepare the cover of a 96-well, flat-bottomed, plastic tissue-culture plate (or the lid to a clear plastic, 8 cm × 6 cm box) by drilling two small holes in opposite corners of the plate.
2. Using the plate size as a guide, draw a grid of 1 cm^2 squares on a heavy sheet of clear Mylar.
3. Prepare a 0.1 M solution of KI/phenolphthalein by adding 2 mL of 1% phenolphthalein solution to 100 mL of 0.1 M potassium iodide.
4. Place 1 drop of 0.1 M KI/phenolphthalein in each square of the grid, leaving two 2 cm × 2 cm areas at opposite corners unfilled.
5. Cover the grid with the plate so that the holes lie over the unfilled squares and the drops of KI/phenolphthalein are exposed only to the air under the plate.

What are characteristics of gases?

Section Objectives

Describe the general properties of gases.

Describe the role of Earth's atmosphere in trapping radiant energy.

Identify the causes of ozone depletion over Antarctica.

Explain the assumptions of kinetic molecular theory.

Explain the energy distribution diagram for a gas at different temperatures.

Describe the four variables that define a gaseous system.

Describe how a barometer is used to measure pressure.

Properties of gases

Gases are the least complex state of matter. In elementary school you learned that gases can be compressed and that they expand to fill their containers. This means that the volume of a gas is variable. Recall that both solids and liquids have essentially fixed volumes and cannot be easily compressed because there is very little space between particles, as shown in the models in **Figure 10-1**. Therefore, gases typically have very low densities compared with liquids and solids. In fact, at standard conditions 1 mol of oxygen gas (32.0 g) has a volume of roughly 22 400 mL compared with 1 mol of water (18.0 g) with a volume of 18.0 mL or 1 mol of aluminum metal (27.0 g) with a volume of 10.0 mL.

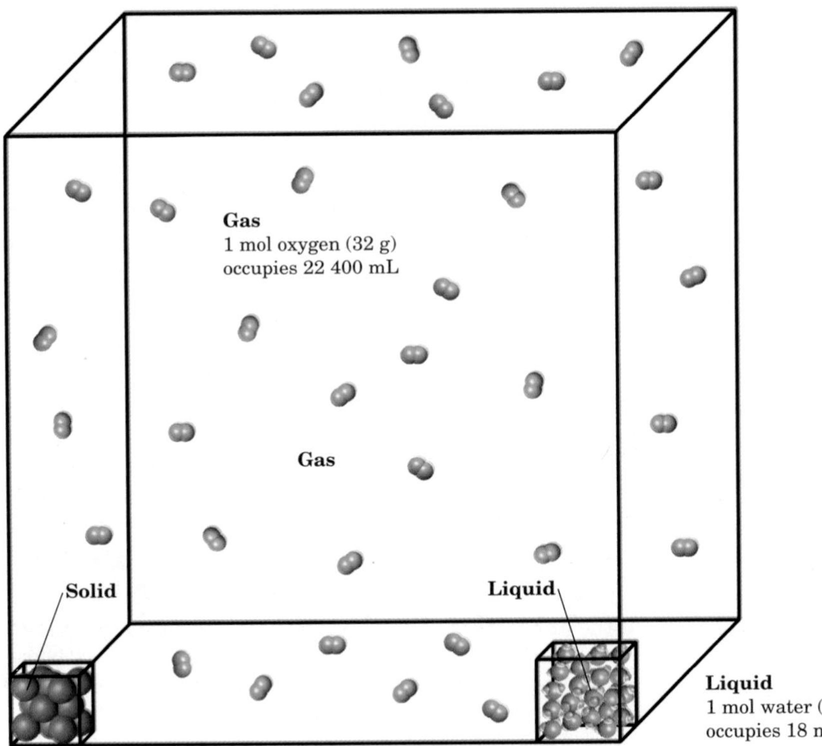

Figure 10-1
Though all three substances are similar in mass, the volume occupied by the gas is substantially greater than that of the solid or liquid. Gases are easily compressed because there is so much space between particles compared to a solid or liquid.

Visual Strategy

Figure 10-1

Emphasize that the volume of 1 mol of either liquids or solids depends on how tightly the particles are packed. Gases, many of whose behavior is close to ideal, always have a molar volume of 22.4 L. Stress that the particles of a gas barely interact, while the particles in a liquid or solid are close enough to each other to be affected by attractive forces.

Table 10-1
Composition of Dry Air at Sea Level

Gas	Formula	Percentage by volume
nitrogen	N_2	78.084
oxygen	O_2	20.948
argon	Ar	0.934
carbon dioxide	CO_2	0.033
neon	Ne	0.001 82
helium	He	0.000 52
methane*	CH_4	0.000 2
krypton	Kr	0.000 1
hydrogen	H_2	0.000 05
carbon monoxide*	CO	0.000 01
xenon	Xe	0.000 008
ozone*	O_3	0.000 002
ammonia*	NH_3	0.000 001
nitrogen dioxide*	NO_2	0.000 000 1
sulfur dioxide*	SO_2	0.000 000 02

* Denotes components with variable concentrations.

The atmosphere is a sea of gases

The molecular composition of the atmosphere provides the perfect conditions for life to exist on Earth. For example, the oxygen we breathe is generally abundant in supply. The atmosphere provides protection from the sun, and it helps keep Earth at a comfortable temperature. **Table 10-1** is a listing of the gases found in our atmosphere.

How does this layer of gases keep the Earth warm? More than half of the sun's radiant energy that is directed toward Earth is absorbed by the surface. **Figure 10-2** shows a model for what happens to this energy as it passes through the atmosphere. Some of the radiant energy never reaches Earth's surface because it is deflected back into space by clouds and particles in the atmosphere. Some of the radiant energy that does penetrate the atmosphere will be reflected into space after striking materials on the ground such as snow and concrete. Earth radiates most of this absorbed energy back into the atmosphere as infrared radiation, which accomplishes a very important task—it warms our atmosphere.

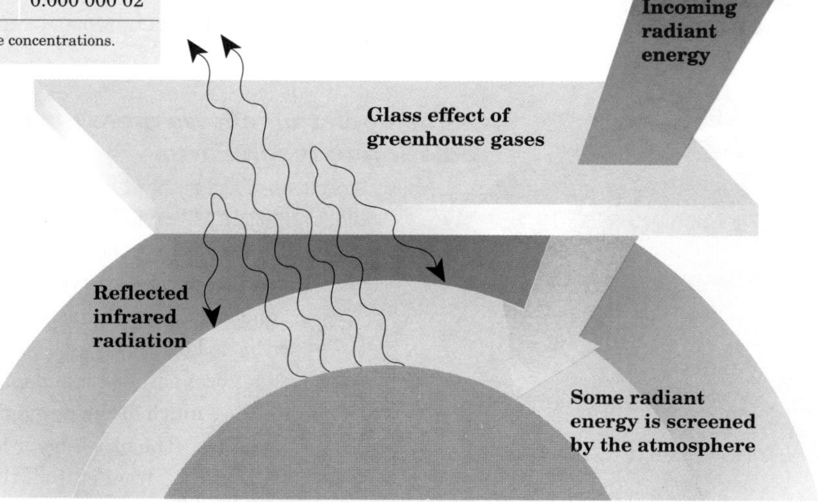

Figure 10-2
Greenhouse gases in the atmosphere can trap radiant energy by reflecting it back to Earth.

Greenhouse gases trap radiant energy

The absorption of infrared energy and the trapping of radiant energy by molecules in the atmosphere is a mechanism known as the **greenhouse effect**. This effect is caused largely by the presence of carbon dioxide. However, levels of water, methane, and ozone also contribute to trapping radiant energy much like the glass on a greenhouse. They hold in the heat by trapping radiant energy that passes through the atmosphere. As the levels of greenhouse gases in the atmosphere increase, the temperature of the Earth could increase as well.

greenhouse effect
an increase in the warming effects of infrared radiation absorption brought about by an increase in levels of carbon dioxide and other greenhouse gases in the atmosphere

6. Add 3 drops of chlorine bleach to one hole, followed by 2 drops of 3 M HCl. Add 3 drops of concentrated ammonia, NH_3, to the other hole. Cover both of the holes with small pieces of tape.
7. The strong acid will release chlorine from the bleach. The chlorine will then displace the iodine in KI. This will turn the indicator solution a violet-brown color. The strongly basic ammonia will cause the phenolphthalein to turn red.
8. Note that the ammonia, whose molecular mass is 17.04 g/mol, diffuses more rapidly than chlorine gas, whose molecular mass is 70.90 g/mol.

SAFETY
Because of the concentrations of HCl and NH_3 used in this demonstration, a face shield, gloves, and a lab apron must be worn. Students must be 10 ft or more from demonstration.

DISPOSAL
Rinse the plate and Mylar grid, making sure to combine all liquids into one vessel. Neutralize the solution with 1 M HCl or NaOH as appropriate, and pour down the drain.

[Source: Epp, Lyons, and Brooks, *J. of Chem. Ed.*, Vol.66, p. 436.]

TEACH

Teaching Tip

Rule of thumb
Since gas particles are approximately 10 times farther apart than liquid or solid particles, the volume of a substance in the gas phase is roughly 1000 ($10 \times 10 \times 10$) times greater than the volume of the same amount of the substance in the liquid phase.

Possible Misconception

The matter of global warming is complicated by the number of contributing factors. Periodic variations in solar activity and volcanic activity are two factors that, along with the greenhouse effect, determine how much the Earth's atmosphere heats or cools within a given year.

Content Background

Sulfate aerosols and climate

One of the ironies of burning carbon fuels is that sulfur impurities are released into the atmosphere, producing SO_2, one gas responsible for acid rain. Because it reflects sunlight, SO_2 helps to cool the Earth's atmosphere and partially counteracts the greenhouse warming caused by CO_2. This, however, is not a solution to the problem posed by the greenhouse effect because it affects only regions of high industrialization, while the greenhouse effect occurs throughout the atmosphere. This results in a higher overall temperature in the Southern Hemisphere than in the Northern Hemisphere, which could result in more violent storms or droughts. SO_2 is also a contributor to the depletion of ozone, O_3.

mutagen
any substance or agent that causes a noticeable increase in the frequency of mutations

The 1980s produced a decade-long warming trend, ending with 1990 being the warmest year on record. The possibility that this trend will continue has raised concerns about global warming—the idea that the average global temperature could increase by as much as 3°C during the next 100 years.

An increase of 3°C over 100 years may not seem to be cause for too much concern. However, some computer models predict that an average global temperature increase from 17°C to 20°C could trigger massive droughts in inland regions and cause catastrophic floods in coastal areas as glaciers and polar caps melt. The weather could also become unpredictable.

Carbon dioxide levels are rising in the atmosphere

What is causing this warming effect? During the past century, human activities have been largely responsible for nearly a 15 percent increase in the CO_2 level of the atmosphere. One of these activities is burning rainforest lands, which not only increases CO_2 from the combustion process, but reduces the amount of plant life on Earth that can consume CO_2 for photosynthesis.

However, it is the burning of fossil fuels, not rain forests, that releases the most CO_2 into the atmosphere. Many everyday activities involve the combustion of fossil fuels. Driving a car, cooking on an outdoor grill or gas stove, using a power lawn mower, and using a gas clothes dryer all involve combustion processes that release CO_2. Most electricity is generated by burning fossil fuels.

Ozone depletion allows more ultraviolet radiation to reach Earth

About 9 percent of the radiant energy emitted by the sun is ultraviolet radiation. Ultraviolet radiation is classified as a **mutagen**, because the ionizing radiation can cause a mutation or various types of skin cancers, including one form that is almost always fatal.

Fortunately, Earth is shielded by a layer of ozone in the stratosphere that screens out much of the damaging ultraviolet radiation. The ozone layer is extremely thin and fragile. If all the ozone molecules were placed on the Earth's surface, they would form a layer only 3 mm thick, which is about the thickness of the cover on this book! **Figure 10-3** shows a model for the parts of the atmosphere so that you can see the relative location of the ozone layer.

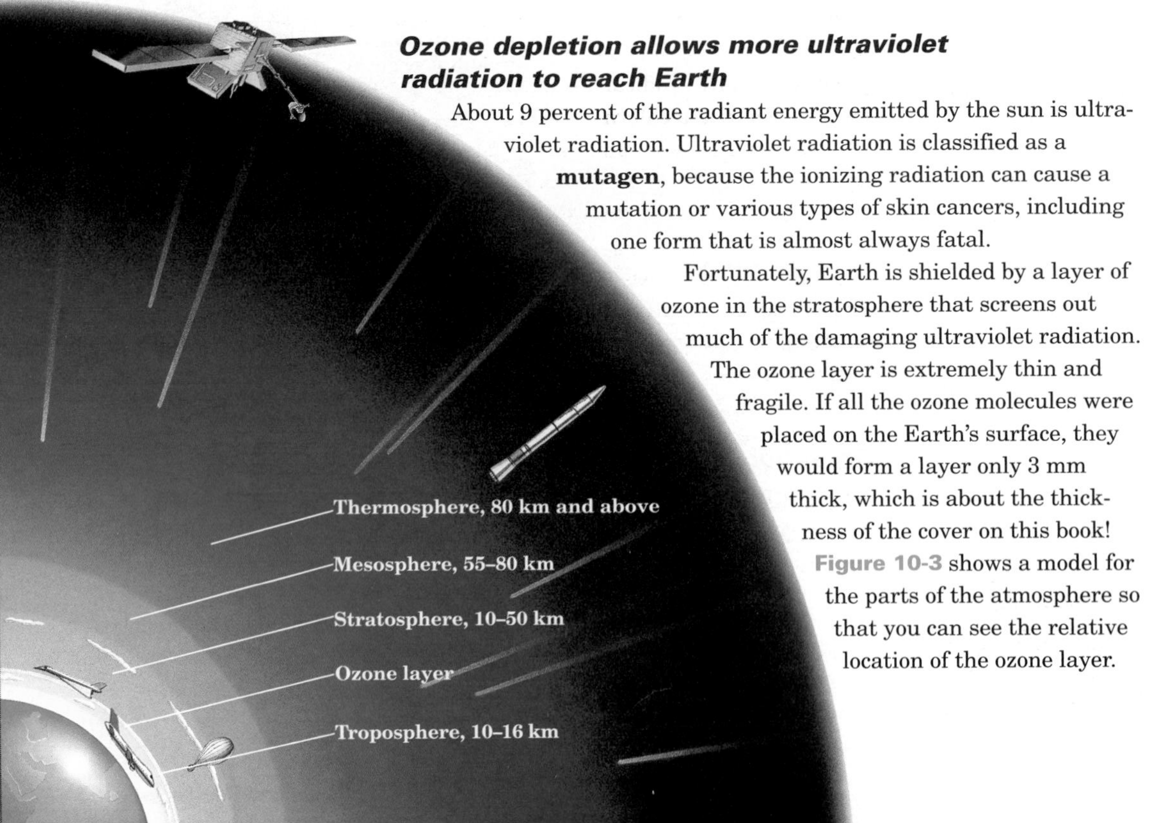

Figure 10-3
The atmosphere is divided into five regions. The troposphere is that region closest to Earth where all our weather patterns form. The ozone layer is part of the stratosphere. Satellites can be found in the thermosphere. Rockets also ascend into the thermosphere. Balloons can be found in the stratosphere. The SST travels in the stratosphere at about 15 km. Jet aircraft travel in the troposphere at about 12 km from Earth.

Visual Strategy

Figure 10-3

Stress that the ozone layer's formation is the result of ultraviolet radiation interacting with molecular oxygen to form O_3. Within the stratosphere the density of oxygen is high enough for ozone to form, with the highest concentration occuring around 25 km above the Earth's surface.

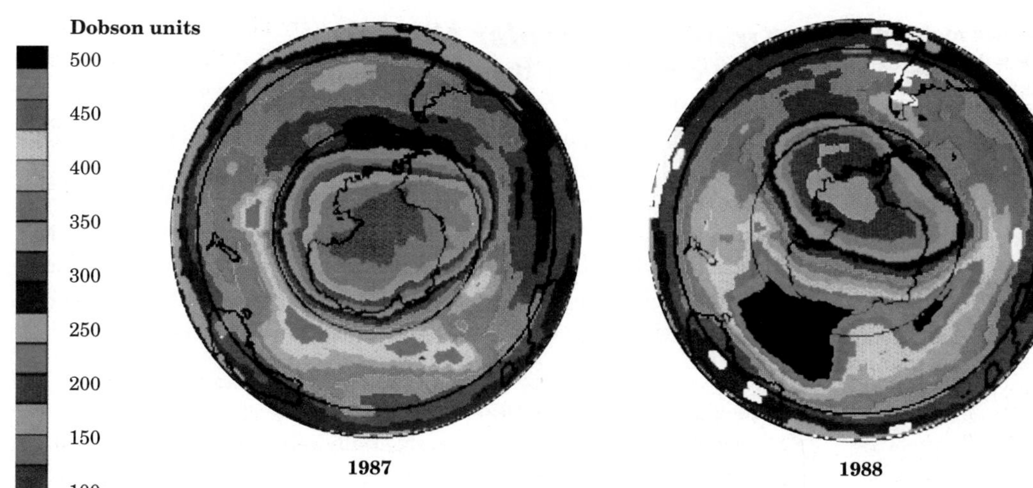

Figure 10-4
These satellite photos show the difference in ozone levels over the Antarctic for 1987 and 1988. Low levels of ozone are represented by shades of pink, as shown by the color scale on the left.

A hole in the ozone occurs normally over the Antarctic on a yearly cycle as shown in **Figure 10-4**. Nearly all the ozone in the lower stratosphere disappears for about six weeks in September and October each year. In the years between 1977 and 1984, scientists found ozone levels decreased by more than 40 percent each spring. Further study showed that organic chlorine compounds called **chlorofluorocarbons** (CFCs), like the ones shown in **Figure 10-5**, are causing the ozone to break down. CFCs are gases used primarily as coolants in air-conditioning systems and refrigerators, and as propellants for aerosol sprays. When released, these gases drift up to the stratosphere where energy from ultraviolet light is able to break the bond holding a chlorine atom to a CFC molecule. The Cl atom produced is a free radical atom that reacts with an ozone molecule to form oxygen and a chlorine–oxygen free radical. *Free radicals* are highly reactive particles with an unpaired electron, as shown in the following equation.

chlorofluorocarbons
a family of organic compounds in which the hydrogen atoms have been replaced by fluorine and chlorine

$$\underset{\text{Free radical}}{\mathrm{Cl}\cdot} + \mathrm{O_3}(g) \longrightarrow \underset{\text{Free radical}}{:\ddot{\underset{..}{\mathrm{Cl}}}:\ddot{\underset{..}{\mathrm{O}}}\cdot} + \mathrm{O_2}(g)$$

As a result of this reaction, the ozone hole gets larger each fall. In 1993, the region of the ozone hole spread to over 23 million square kilometers, almost the size of the entire North American continent.

To halt the destruction of the ozone layer, 74 nations signed the Montreal Protocol that calls for ending the production of all CFCs by 1996. However, CFCs now in use will continue to accumulate in the stratosphere until the year 2005. Scientists estimate that it will be the year 2060 before the chlorine concentration returns to the level it was in the 1970s when the ozone hole was first discovered.

```
      Cl                   Cl
      |                    |
Cl—C—Cl             F—C—Cl
      |                    |
      F                    F
```

Trichlorofluoromethane
Freon 11

Dichlorodifluoromethane
Freon 12

Figure 10-5
Freon 11 and Freon 12 were commonly used chlorofluorocarbon compounds. Freon 11 is a liquid below 23.7°C. Freon 12 is a gas at room temperature, 25°C.

Content Background
Ozone replenishment
Once chlorofluorocarbons, sulfur dioxide, and nitrogen oxides (yet another contributor to ozone depletion and a common gas in aircraft and automobile exhaust) are removed from the ozone layer, the high-energy ultraviolet radiation from the sun will begin replenishing the ozone layer.

Possible Misconception

The purpose of classifying gases as ideal is to make their properties more understandable. Nevertheless, students are often confused by the concept of an ideal gas because they expect real substances to have the characteristics of an ideal gas. Inform the class that because an ideal gas is a theoretical gas, it is possible to have particles with zero volume, perfectly elastic collisions, and no interactive forces.

Content Background

Ideal vs. real gases

Of all real gases, hydrogen exhibits behavior that is the most nearly ideal. However, it too condenses when enough pressure is applied.

Possible Misconception

For most real gases, the interactive forces between particles is attractive. However, if a gas consists of ionized particles, as in the case of a plasma, repulsion can also become important. Plasmas are often treated as a fourth state of matter.

Theme

Conservation

The energy of collisions is conserved. In an elastic collision, energy is conserved mechanically (the speed of the particle is the same before and after the collision). In an inelastic collision, part of the energy is converted into heat.

Do You Know?

In England, rubber bands are called elastic bands. However, when rubber bands are stretched, they become warm, signifying that they are undergoing an inelastic process.

Content Background

Macroscopic elastic collision

An example of a nearly elastic collision is that of a steel ball dropped on a large, flat piece of steel.

kinetic molecular theory
the theory that explains the behavior of gases at the molecular level

ideal gas
a model that effectively describes the behavior of real gases at conditions close to standard temperature and pressure; a gas for which the product of the pressure and volume is proportional to the absolute temperature

Kinetic molecular theory

Kinetic molecular theory is based on assumptions for an ideal gas

You know that the purpose of models in science is to make predictions. A model should explain the facts we know and correctly predict events that have not yet occurred. The **kinetic molecular theory** (KMT) was developed in the mid-1800s and is still extremely useful as a model for predicting gas behavior. This theory is based on assumptions about a theoretical gas often referred to as the **ideal gas**. The ideal gas is a model in itself and is described by the following statements, which are also the basis of KMT.

1. Ideal gas particles are so small that the volume of the individual particles if they were at rest is essentially zero when compared with the total volume of the gas.
2. Ideal gas particles are in constant, rapid, random motion, moving in straight lines in all directions until they collide with other particles, as shown in the model in **Figure 10-6**.
3. There are no attractive or repulsive forces between particles, and collisions between particles are elastic.
4. The average kinetic energy of the particles is directly proportional to the absolute temperature (measured in kelvins).

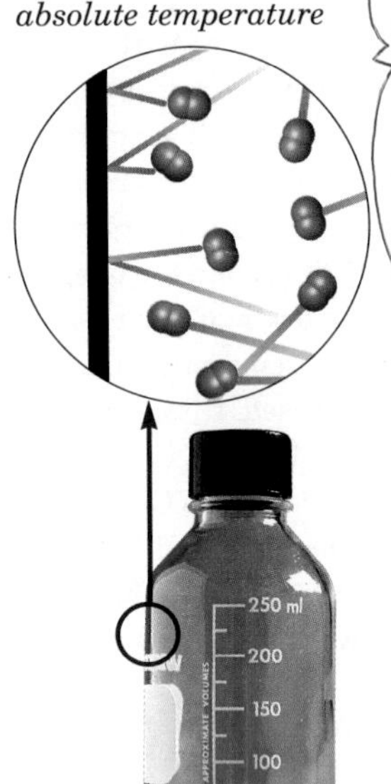

Figure 10-6
Gas particles travel in straight lines until they collide with each other or the walls of their container.

When we say that the volume of the particles at rest must be almost zero, we don't mean that the volume is actually zero. It is just so small that it's negligible compared with the volume that the moving gas particles occupy. When you realize that a few drops of perfume can evaporate to fill a whole room, the "zero volume" part of this assumption makes sense. However, if gas particles really had no volume, the liquid or solid formed when they condensed would also have no volume! This assumption is not perfectly true.

Moving gas particles have kinetic energy. The average velocity of oxygen molecules is calculated to be about 400 m/s at 298 K. This calculation is based on the kinetic energy equation, which relates kinetic energy to one-half of the mass multiplied by the square of the velocity.

$$KE = \frac{1}{2}mv^2$$

Collisions between ideal particles are described as "elastic" because we assume that total energy is conserved; no energy is emitted when two gas particles collide, and none is lost (on average) when particles collide with the walls. If the collisions were not considered elastic, the colliding particles would continuously lose energy, slow down, and eventually stop, but this doesn't happen. Collisions between ordinary objects are not elastic. For example, a ball loses some of its kinetic and potential energy on each bounce. Not even "Super Balls" bounce with totally elastic collisions, as shown in **Figure 10-7**.

Figure 10-7
Although a Super Ball bounces much higher than a tennis ball or basketball, even the Super Ball does not undergo elastic collisions. However, we assume collisions of ideal gas particles are elastic.

The assumption that gas particles do not attract each other is not realistic. Most particles have some degree of electrical attraction for other particles. However, the strength of that attraction depends on the mass and distance between the particles. Most common gases are composed of small molecules that are very far apart compared with those of a solid or liquid. The attraction between gas particles is so weak that at normal temperatures and pressures the particles are moving too quickly to be affected. If the molecules are pushed close together and slowed down enough, the attractive forces between particles begin to affect the particles' behavior, and condensation occurs. All gases eventually condense into liquids with a measurable volume when subjected to high enough pressure and low enough temperature. However, the ideal gas defined by our model never condenses, no matter how much we compress or cool it.

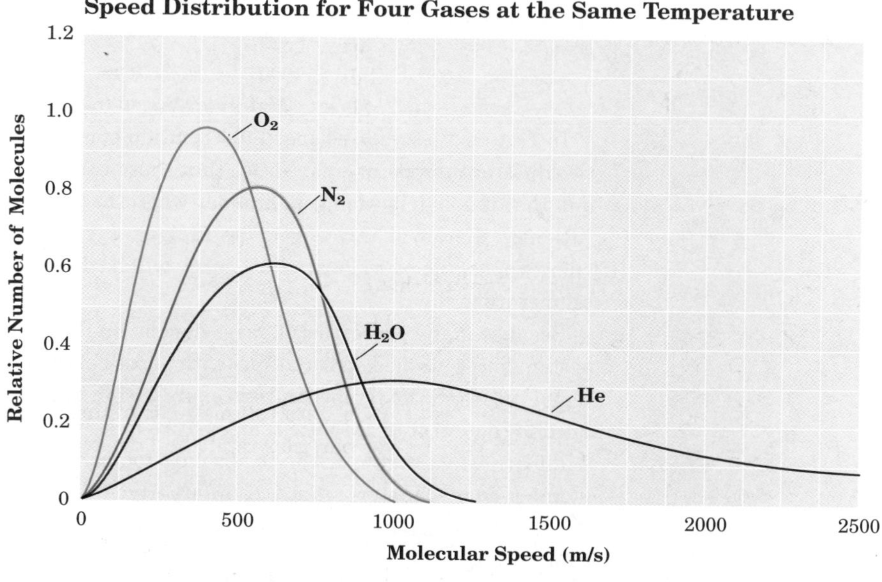

Figure 10-8 At the same temperature all four gases have the same average kinetic energy. Each gas has a distribution of molecular speeds. The shape of each curve varies because the molecular mass of each gas differs. Light gases (He) have more molecules moving at high speeds than the heavier gases (O_2 and N_2).

Four variables describe a gas

The temperature of a gas determines the average kinetic energy of the particles

An important aspect of the kinetic theory of gases is that the average kinetic energy of the particles depends on the temperature of the gas. Experiments in the late 1800s led to the discovery that not all gas particles of the same substance travel at the same speed. **Figure 10-8** shows how speed is distributed in four different samples of gas molecules all at the same temperature. The average kinetic energy of all four gases is the same, however the curves differ in shape. You should notice that the steepest curve is for O_2, the gas with the largest molecular mass. The most flattened curve is He, the gas with the lowest molecular mass. The data points used in making each curve were calculated from an equation called the Maxwell-Boltzmann energy distribution. Notice that the curve for each gas is not symmetrical.

Demonstration 2

Elastic and inelastic collisions

Approximate time: 5 min

1. Obtain a pair of "Happy Balls" from a scientific supply house. (One ball is made of a very flexible rubber like that used in "Super Balls." The other ball is made of a very inelastic rubber used for cushioning collisions and for shoes.)
2. Place the "Super Ball" in your left pocket and the inelastic ball in your right pocket.
3. Remove and bounce the "Super Ball" on the tabletop several times while initiating a discussion on elastic collisions. Return the "Super Ball" to your left pocket.
4. Ask for two volunteers to help determine the ball's elasticity by measuring how high it bounces. Ask one student to hold the meter stick, while the other drops the ball.
5. Remove the inelastic ball from your right pocket, hand it to the student, and stand back. Ask students what has happened to the ball? Discuss the types of collisions for each ball. Have students describe some uses for each kind of rubber.

SAFETY
Safety goggles are recommended for all participants. The rest of the class must stand back at least 10 ft.

DISPOSAL
None.

Historical Note
Physicists James Clerk Maxwell (1831–1879) and Ludwig Boltzmann (1844–1906) used the idea that gases consist of a large number of randomly moving particles (the kinetic molecular theory) to derive Boyle's law.

Visual Strategy

Figure 10-8
Stress that while the individual particles have an average speed equal to or greater than 400 m/s (about 900 mi/h), they are also experiencing approximately 10^{10} collisions per second. The constant change in the direction of each particle causes the gas to spread outward at a slower speed than the average molecular speed.

Teaching Tip

Analogy

Remind students that temperature is proportional to the average kinetic energy of the particles. Some particles travel slowly at a given instant, while others travel quickly. To clarify this concept, have students imagine that they are wearing rubber-balloon suits that allow them to run into and bounce off the walls and each other. If half of the class started running at 5 m/s and the other half started running at 10 m/s, and if they collided continuously for several minutes, what would happen to everyone's speed with each collision? (It would change.) At the end of several minutes, would half of the class still be running at 5 m/s and the other half at 10 m/s? (No; the speeds should be randomly dispersed.)

Content Background

Units of volume

Strictly speaking, 1 L does not equal 1000 cm^3. Specifically, 1 L = 1000 mL, 1 mL = 1.000 027 cm^3, and 1 m^3 = 1 000 000 cm^3 = 999 973 mL = 999.973 L.

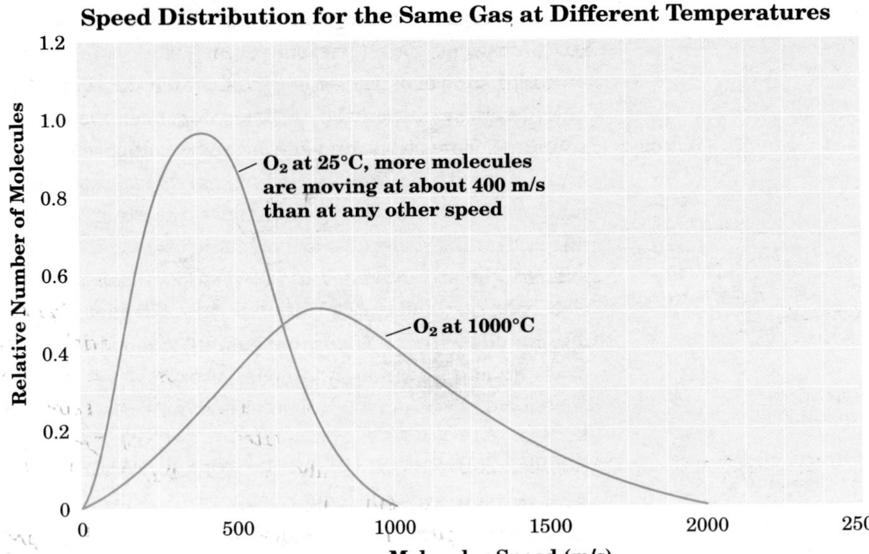

Figure 10-9 Increasing the temperature of a sample of gas shifts the energy distribution in the direction of greater average kinetic energy.

In **Figure 10-9** you see speed-distribution curves for the same gas at two different temperatures. Notice that these curves are not symmetrical and that the distribution gets broader when the temperature is increased. Although there will always be a few high-energy molecules and a few low-energy ones, the average kinetic energy will be directly proportional to the temperature.

Because many properties of gases depend on the temperature of the system being studied, calculations with gases should include a specified temperature. If the temperature is expressed in degrees Celsius, you will need to convert it to absolute temperature in kelvins. In Section 10-2 you will learn why the conversion to absolute temperature is necessary. Remember from Chapter 2 that 0 K differs from 0°C by 273 (it's actually 273.16). *For calculations in this text we will round the conversion value to 273.*

$$\mathbf{K = {^\circ C} + 273.}$$
$$\mathbf{T = t + 273.}$$

The symbol T is used to represent absolute temperature; the symbol t represents Celsius temperature.

The volume of a gas is derived from linear measurements

Volume is not one of the SI base units but is derived from linear measurements using the following equation.

$$\textbf{volume } (V) = \textbf{length} \times \textbf{width} \times \textbf{height}$$

Gas volumes can be expressed in traditional metric units such as liters (L) and milliliters (mL) or in SI units such as cubic meters (m^3) or cubic centimeters (cm^3). Because the volumes of most containers used to hold gases are not easily determined, gas volume measurements are often made by the volume displacement of water.

Visual Strategy

Figure 10-9

Make sure students understand that molecular speed is inversely related to molar mass. The greater the mass, the slower the speed.

pressure
the force exerted per unit area

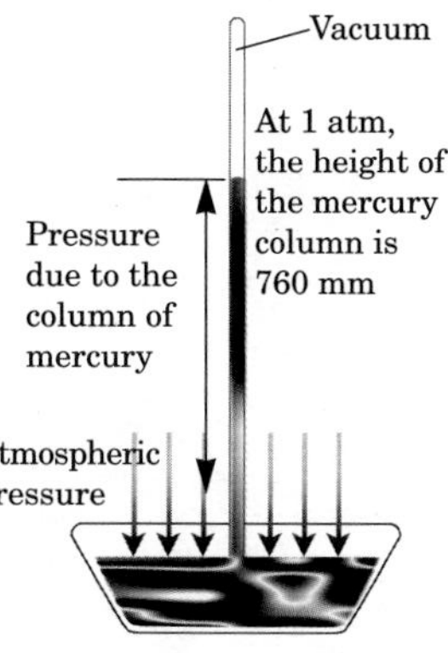

Figure 10-10
Although water could be used in a barometer, the atmosphere would support a column of water about 10.3 m high, which is too high to be practical. Mercury has a density 13.7 times greater than water, so the mercury barometer is less than 1 m high.

The pressure of a gas is the force of its particles exerted over an area

You can't see wind, but you can feel its force as it pushes against you. Air pushes against the walls of a tire and rushes out if you open the valve to release it. **Pressure**, P, is defined as a force exerted over a specified area.

$$\text{pressure} = \frac{\text{force}}{\text{unit area}}$$

In relation to gases, pressure is a measure of the total force exerted by the moving particles of a gas as they collide with the walls of the container. In **Figure 10-6** you saw a model for particles colliding with the walls of a container. Each collision exerts a force. The sum of all the forces over the surface area of the container is the pressure exerted by the gas. Because each collision contributes to the total pressure, the pressure is proportional to the number of collisions. As the number of collisions increases, the pressure increases.

Why does pumping air into a car tire increase the pressure in the tire? Adding more particles increases the number of collisions on the inside surface of the tire, which increases the pressure in the tire.

The atmosphere exerts a pressure on Earth

The atmosphere that surrounds Earth is a sea of air—a mixture of gases that exerts a force on the surface of the Earth. This force is measured by its effect on the height of a mercury column in a tube, as in **Figure 10-10**. The level of mercury in the column rises or falls depending on the force exerted by gas molecules in air on the mercury surface. This force can be seen on a small scale with the model in **Figure 10-11**. The column of air exerts a force over a certain area of the Earth, creating atmospheric pressure. One standard atmosphere is defined as the pressure of the atmosphere that will support a column of mercury that is 760 mm high.

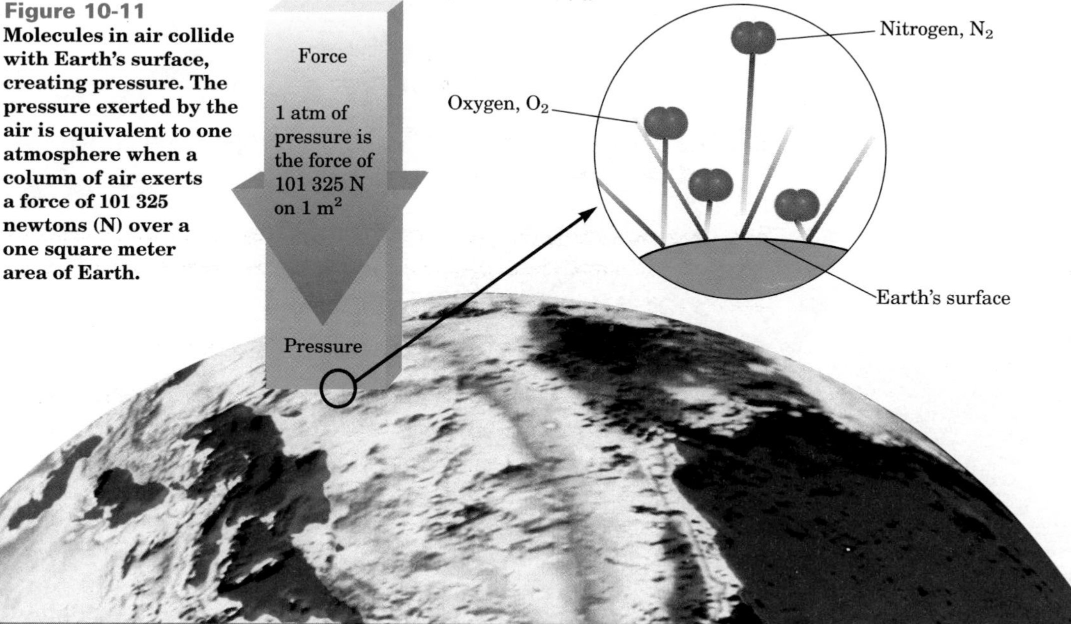

Figure 10-11
Molecules in air collide with Earth's surface, creating pressure. The pressure exerted by the air is equivalent to one atmosphere when a column of air exerts a force of 101 325 newtons (N) over a one square meter area of Earth.

Visual Strategy

Figure 10-10

Emphasize that the so-called vacuum found above the water or mercury in an eudiometer tube is not a total vacuum, but only a partial vacuum. The water or mercury begins to vaporize immediately and thus partially fills the space.

Demonstration 3

The barometer

Approximate time: 5 min

1. Half fill a 50 mL eudiometer tube with tap water colored with food dye.
2. Cover the eudiometer tube mouth with a forefinger, invert the tube, and place it upside down in a clear, 1 L graduated cylinder filled with similarly colored water. Once the tube is submerged, uncover the tube mouth.
3. Adjust the position of the eudiometer until the water levels inside and outside the tube are equal. Have a student determine the volume of the gas within the tube by noting the number on the tube at the boundary between the liquid and gas. Have the student record the value on the board.
4. Let the eudiometer rest on the bottom of the cylinder, and record the new gas volume.
5. Lift the eudiometer until the 50 mL mark is level with the fluid level in the cylinder, and record the new gas volume. Ask students why these three readings are different. (When pushed down, the water in the cylinder applies pressure to the gas in the eudiometer. When lifted, the outside air pressure cannot support the raised column of water, so the gas volume increases.)

Safety

Lab aprons and safety goggles should be worn by the participating student and teacher. The rest of the class must be 10 ft or more from the demonstration.

Disposal

Pour water down the drain.

Theme

Macroscopic observations and micromodeling

The changes in the level of the liquid of a barometer reflect changes in the gas pressure, which is a measure of the energy of the gas particles.

Cultural Connection

The barometer was invented by the Italian physicist and mathematician Evangelista Torricelli (1608–1647) in 1643. The pressure unit *torr* was named in his honor.

Historical Note

Many early scientists believed that the liquid was suspended in eudiometer tubes by invisible ropes. Robert Boyle showed this to be untrue by inserting a long wire into a eudiometer tube and wiggling it back and forth. Since the wire did not catch on the rope, the rope did not exist. It is for simple and straightforward experiments such as this that Boyle is considered the "father of modern chemistry."

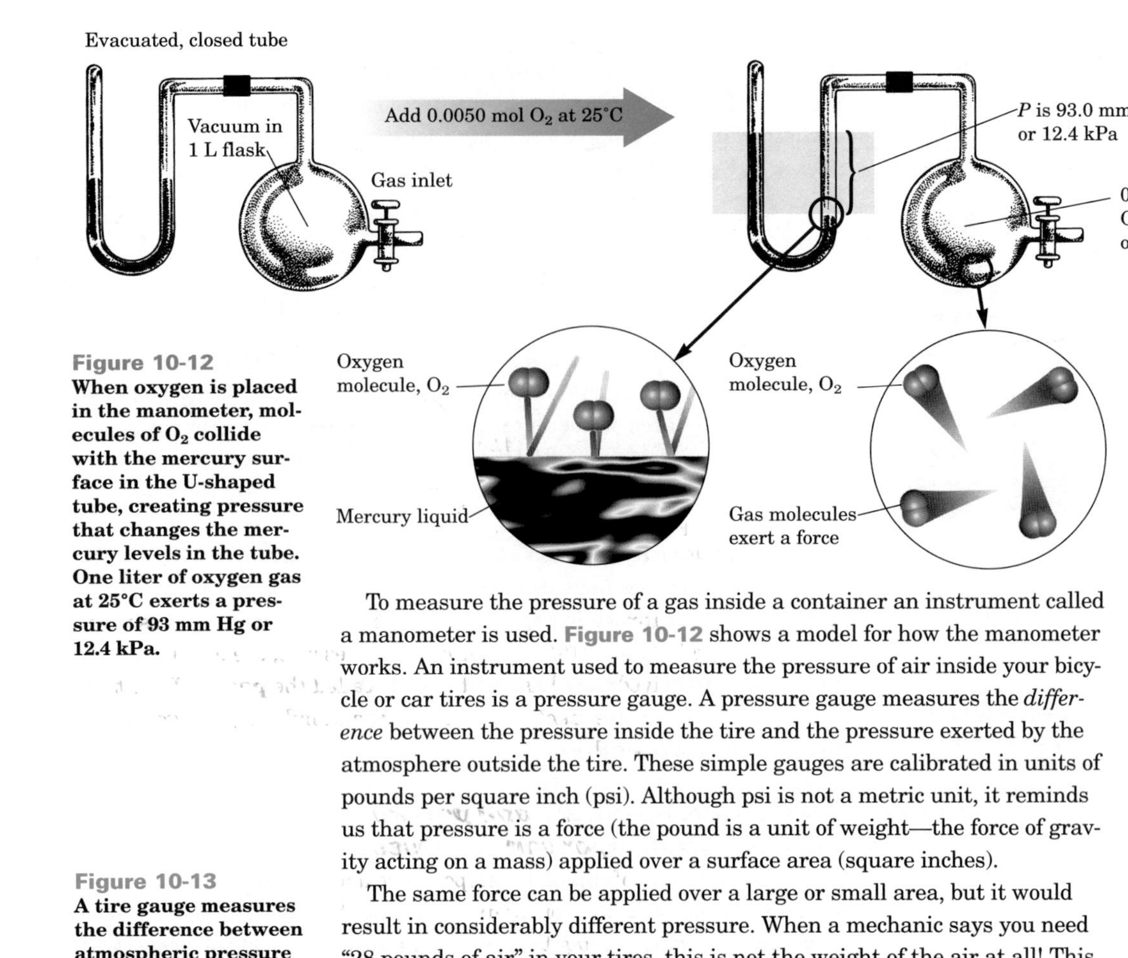

Figure 10-12
When oxygen is placed in the manometer, molecules of O_2 collide with the mercury surface in the U-shaped tube, creating pressure that changes the mercury levels in the tube. One liter of oxygen gas at 25°C exerts a pressure of 93 mm Hg or 12.4 kPa.

To measure the pressure of a gas inside a container an instrument called a manometer is used. Figure 10-12 shows a model for how the manometer works. An instrument used to measure the pressure of air inside your bicycle or car tires is a pressure gauge. A pressure gauge measures the *difference* between the pressure inside the tire and the pressure exerted by the atmosphere outside the tire. These simple gauges are calibrated in units of pounds per square inch (psi). Although psi is not a metric unit, it reminds us that pressure is a force (the pound is a unit of weight—the force of gravity acting on a mass) applied over a surface area (square inches).

The same force can be applied over a large or small area, but it would result in considerably different pressure. When a mechanic says you need "28 pounds of air" in your tires, this is not the weight of the air at all! This means that the pressure inside your tire must be 28 psi above atmospheric pressure. Atmospheric pressure is approximately 14.7 psi at sea level. Figure 10-13 shows that a tire gauge reads zero when the pressures inside and outside the tire are equal.

Figure 10-13
A tire gauge measures the difference between atmospheric pressure and the internal pressure of air in the tire.

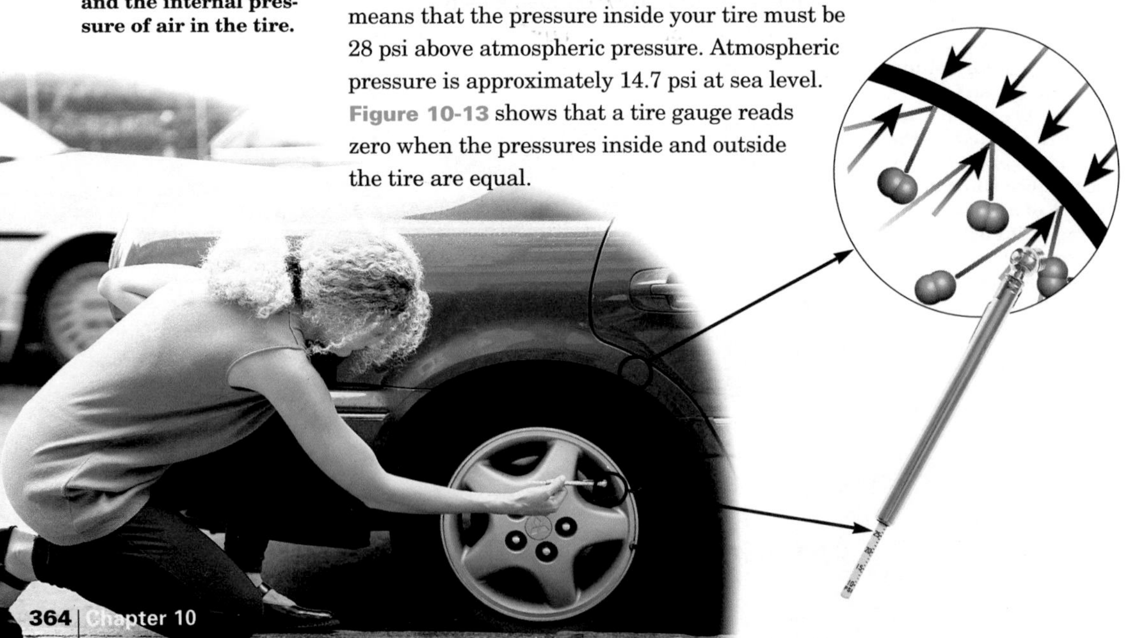

Visual Strategy

Figure 10-12

Stress that a manometer measures *excess* pressure. A perfect vacuum is never obtainable in practice, so there are always a few particles on the evacuated side of the mercury column. Likewise, when the flask is evacuated, there are a few particles on that side so that the forces on each side of the mercury column balance each other. As gas is added to the flask, the mercury is pushed toward the evacuated end.

Table 10-2
Pressure Units

Unit	Abbreviation	Definition	Equivalence to one atmosphere
atmospheres	atm	1 atm is the pressure that supports a column of mercury 760 mm high.	
bars	bar	1 bar is a force of 100 000 N/m^2.	1.013 bar
millimeters of mercury	mm Hg	1 mm Hg is the pressure that supports a column of mercury 1 mm high at 0°C.	760 mm Hg
pascals	Pa	1 Pa is the force of 1 N/m^2.	1.01325×10^5 Pa or 101.325 kPa
pounds per square inch	psi	1 psi is the force of 1 lb/in^2.	14.7 psi
torr	torr	1 torr is the pressure that supports a column of mercury 1 mm high at 0°C.	760 torr

pascal
a unit of pressure equal to the force of 1 N/m^2

In addition to atmospheres, a variety of units are used to express pressure as shown in **Table 10-2**. The SI unit for pressure, the one we will use most often, is derived from base units and is called the **pascal**, Pa. Atmospheres will be used frequently as well—bars and millibars are generally used in meteorology only.

Standard temperature and pressure are used for comparison purposes

standard temperature and pressure (*STP*)
standard conditions for a gas of 0°C and 1.0000 atm

To study the effects of changing temperature and pressure on a gas, it is useful to have a standard for comparison. Scientists have specified a set of standard conditions, giving them the name **standard temperature and pressure**, or ***STP***. *STP* represents a pressure and temperature that are fairly easy to reproduce in any laboratory throughout the world.

$$STP = 0°\text{C and } 1.0000 \text{ atm}$$

Because there are other units used to describe pressure, as shown in **Table 10-2**, *STP* is equivalent to the following.

$$STP = 273.16 \text{ K and } 101.325 \text{ kPa}$$
$$STP = 273.16 \text{ K and } 1.0000 \text{ atm}$$
$$STP = 273.16 \text{ K and } 760.0 \text{ mm Hg}$$
$$STP = 273.16 \text{ K and } 760.0 \text{ torr}$$

Millimeters of mercury, mm Hg, was once generally accepted as the unit of pressure. However, it seemed odd to use a length measurement (mm) to express a derived quantity, force per unit area. A unit called torr gradually replaced mm Hg for this reason, though you will still see mm Hg used. One millimeter of Hg is equivalent to one torr. When a problem specifies *STP* conditions, select the pressure value with units that match the other pressure units given in the problem.

Content Background
Derived SI units of pressure
Because the *pascal* is a fairly small unit, the *kilopascal*, kPa, will be the SI pressure unit used most often in this textbook.

Teaching Tip
Mnemonic device
Most students are able to guess the conditions for standard temperature and pressure if given the chance. Remind the class that a standard must be reproduced easily anywhere in the world. At this point most will guess a pressure of 1 atm and temperatures of either 0°C or 100°C. 100°C is not used because the boiling point of water is very pressure dependent.

Historical Note

Amadeo Avogadro (1776–1856) published his hypothesis in 1811, but it remained unrecognized for almost 50 years because it was not published in German, the major language among chemists from around 1800 until after World War II.

Content Background

Particle number and gas behavior

Gaseous systems containing equal numbers of moles behave the same in terms of pressure, volume, and temperature fluctuations.

Possible Misconception

Emphasize that equal moles of gases contain equal numbers of particles and have the same volume. This is true only for gases.

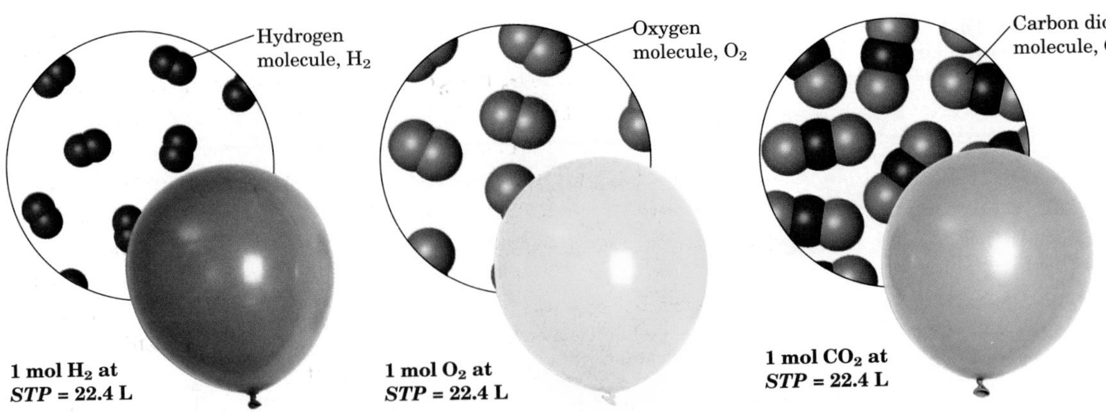

Figure 10-14
Equal numbers of moles of gas at *STP* have equal volumes.

Gases with equal volumes under the same conditions have equal numbers of particles

In the early 1800s, Amedeo Avogadro was one of several scientists studying the behavior of gases. Some important fundamental relationships in chemistry were determined from this study. Avogadro's experiments led him to hypothesize that *equal volumes of gases under the same conditions have equal numbers of molecules*, which is represented by the model in **Figure 10-14**. Though this sounds simple, it is a very important hypothesis, and it applies only to gases. **Avogadro's principle** allows you to say that two balloons of the same size at the same temperature and pressure contain the same number of gas particles. This statement will hold true for any two gases, no matter what their identity!

Avogadro's principle
equal volumes of different gases under the same conditions have the same number of molecules

From Avogadro's hypothesis, we reason that the volume of a gas at a given temperature and pressure is directly proportional to the quantity of the gas in moles. This means that as the number of moles increases, the volume increases. **Figure 10-15** shows a graph of this proportionality.

molar volume
the volume of one mole of a substance at STP

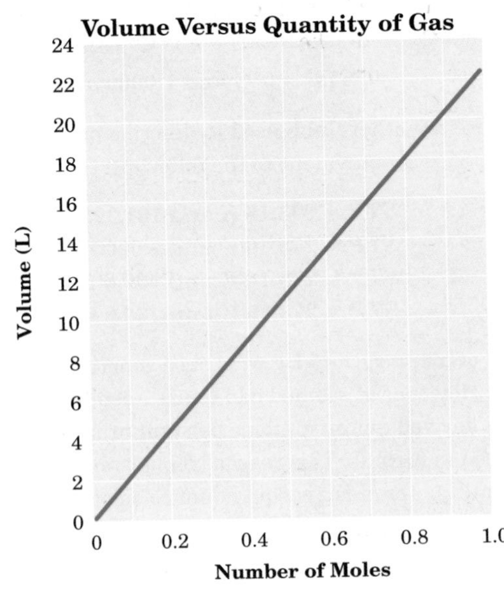

Figure 10-15
The graph shows there is a direct relationship between the volume and quantity of gas. Whenever the quantity of gas is increased, the volume will increase.

Remember that the *amount* of gas is measured in moles, not grams.

Because gases are often compared at *STP*, the volume of one mole of gas at *STP* is called the **molar volume**. At *STP*, the molar volume of any gas is approximately 22.4 L. Notice that the slope of the graph in **Figure 10-15** is 22.4. Because the graph shows a direct relationship and the graph line passes through the origin, the following mathematical relationship holds.

$$\frac{\text{volume}}{\text{number of moles}} = k \text{ (constant)}$$

Visual Strategy

Figure 10-15

Make sure students are aware that the graph depicts the behavior of an ideal gas. Real gases will have slightly different values.

Consumer Focus

Non-CFC air conditioning

In keeping with the limits set for CFC's by the Montreal Protocol, car manufacturers have developed new air-conditioning systems that use a hydrofluorocarbon, 1,1,1,2 tetrafluoroethane, called R-134a, as the refrigerant instead of the traditional R-12 (dichlorodifluoromethane).

R-134a has low toxicity and appears to have little effect on the ozone layer. However, the chemical differences between R-12 and R-134a result in some major problems for consumers.

The major disadvantage is that the properties of R-134a make it such that it *cannot* be used in an air-conditioning system that was built for R-12 because R-12 and R-134a are chemically incompatible. In addition R-134a systems operate at slightly higher pressures and are slightly less efficient than R-12 systems. Therefore, R-134a requires a complete redesign of the existing air-conditioning system including all new compressors, evaporators, and condensers that are engineered to operate at higher pressures with a less efficient coolant.

Recharging and flushing an existing R-12 air-conditioning system is expensive because R-12 is now more costly and limited in supply.

Servicing an R-134a system won't exactly be inexpensive, because service equipment also had to be re-engineered. Because non-CFC air conditioning is a new area, consumers should have work performed at a reputable service station by knowledgeable mechanics.

Section Review

1. Why are atmospheric scientists concerned about burning fossil fuels?
2. Give two physical properties that clearly distinguish gases from liquids or solids.
3. Determine the density in g/mL of a 0.70 g sample of oxygen gas that occupies 5.0 L. Compare the density of oxygen gas to the density of water, assuming both substances are at *STP*.
4. What part of a CFC reacts with ozone?
5. List three ways that an ideal gas differs from a real gas.
6. If a tire gauge reads 20.0 psi, what is the actual pressure in the tire expressed in psi?
7. Describe the relationship between molecular energy and temperature.
8. What does it mean to say that the barometric reading in the room is 756 mm Hg?
9. What are the four variables discussed in this section that can be used to describe a gas quantitatively?

10. **Story Link**
    How high above the Earth will the helium balloon have to ascend before it reaches the ozone level in the stratosphere?
11. **Story Link**
    The density of helium at *STP* is 0.1875 g/L. If the balloon filled with helium at *STP* has a volume of 1.5×10^9 L, what mass of helium is in the balloon?

ASSESS

Alternative Assessment

Have students write a half-page explanation of the kinetic molecular theory and the need for standard temperature and pressure.

Answers to Section Review

1. Burning fossil fuels produces carbon dioxide, which absorbs and re-emits infrared radiation reflected from the Earth, causing the atmosphere to warm (the greenhouse effect).
2. Gases have a very low density compared to liquids or solids. Unlike solids and liquids, equal volumes of gases under identical conditions contain equal numbers of particles. Also, gases are highly expandable and *physically* behave alike in response to temperature and pressure changes.
3. $D_{O_2} = 1.4 \times 10^{-4}$ g/mL
   $D_{H_2O} = 1$ g/mL (7143 times greater)
4. Chlorine reacts with ozone to form molecular oxygen and a chlorine-oxygen free radical.
5. Ideal gases have negligiable volumes; undergo elastic collisions; and do not attract each other.
6. For P_{atm} of 14.7 psi, the pressure of air in the tire is 20.0 + 14.7 = 34.7 psi.
7. The average kinetic energy of gas particles is directly proportional to absolute temperature.
8. The height of the mercury column in a barometer is the measure of the atmospheric pressure that can support the column. To support 760 mm of mercury, or 760 mm Hg, the pressure at 0°C must be 1 atm. At 756 mm Hg, the pressure is 0.995 atm.

Answers are continued on page 401A.

Section 10-2

FOCUS

Lesson Starter

Start the lesson by reminding the class that decreasing the temperature decreases the average energy of the particles. Ask what happens if $T = 0$ K? (Particle energy would become zero.) What happens to the gas pressure? (No energy means no motion and therefore no force or pressure.)

Demonstration 4

Determining absolute zero

Approximate time: 20 min

1. Obtain an absolute zero demonstrator (it looks like a toilet bowl float attached to a pressure gauge with a handle and gas valve).
2. Place a 2 L beaker half full of water on a ring stand. Open the demonstrator valve, immerse the copper ball into the water, and support the demonstrator on the ring stand with a clamp. Heat the water until it boils for 10 min.
3. Have a student record the pressure reading from the gauge and the water temperature with a thermometer.
4. Close the valve, remove the demonstrator from the boiling water, insert it into a 2 L beaker of ice water, and open the valve. Have the students note the immediate drop in pressure.
5. Clamp the demonstrator in place for 5 min. Have the same student record the gas pressure and temperature of the ice water.

What behaviors are described by the gas laws?

Section Objectives

Relate the ideal gas model to observed behavior of real gases.

Use basic gas relationships to solve mathematical problems.

Gas laws

You will now study a series of mathematical relationships that relate the four variables described in the last section to kinetic molecular theory. Each of the four variables is represented by a symbol.

P = pressure exerted by the particles
V = volume occupied by the particles
T = temperature in kelvins of the particles
n = number of moles of the particles

These symbols define the mathematical expressions known as the gas laws. Using the gas laws, you can make predictions about how the volume of a fixed amount of gas will respond to changes in pressure and temperature.

Charles's law provides a basis for absolute temperature

You already know that heating a gas will make it expand. The particles move faster when they are heated and push with more force against the walls of their container. When a balloon is inflated at room temperature, the walls expand until the pressures inside and outside the balloon are about equal. We can use the balloon shown in **Figure 10-16** to determine how much the volume of a fixed amount of gas changes for each degree that the temperature changes.

Figure 10-16a The balloon is inflated at 25°C.

10-16b At –5°C, the amount of gas in the balloon has not changed, but the volume it occupies has decreased.

Visual Strategy

Figure 10-16

Point out to students that this illustration is a great example of a direct graphical relationship. As the temperature goes up, the balloon's volume goes up; when the temperature goes down, the volume goes down. Because the ability of the gas in the balloon to rise above air is related to its density, the balloon literally goes up or down with temperature.

Table 10-3
Volume–Temperature Data for a Gas at Constant Pressure

Temperature (°C)	Volume (mL)
273	1094
100	747
10	568
1	545
0	545
–1	546
–73	403
–173	199
–223	100

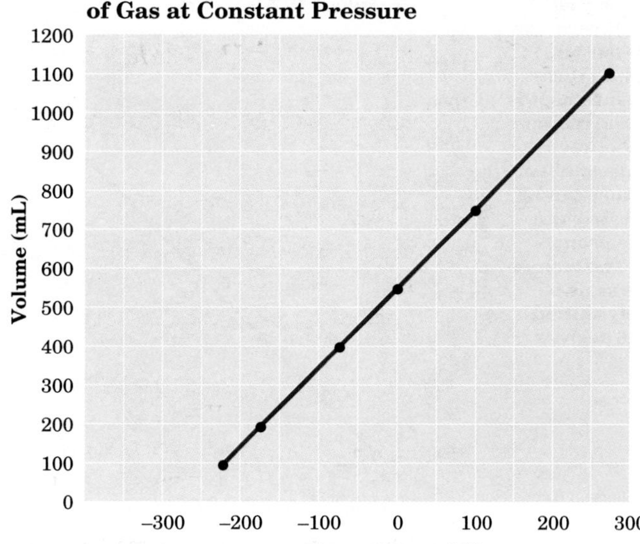

Figure 10-17
The graph of the data from Table 10-3 shows volume increases with temperature.

Table 10-3 shows some data for an experiment in which a fixed amount of gas in an expandable container is warmed and cooled. The volume of the gas changes in response to the changes in temperature. The pressure exerted by the gas is kept constant at 101.3 kPa so that you can study the effects of temperature on volume.

The data indicate that as temperature decreases, the volume of the gas decreases. When the temperature drops from 0°C to –223°C, the volume drops from 545 mL to 100 mL. And, as temperature increases, the volume of the gas increases. A graph of the data in **Table 10-3** is shown in **Figure 10-17**. Notice that while the points are such that you can draw a best-fit line for the data, the relationship is not a direct proportion because the best-fit line does not pass through the origin. If you extend the line so that it intercepts the x-axis, the temperature at that point is –273°C. A direct proportion would mean that volume varies directly as temperature if there is some constant k such that

$$V = kT$$

If you examine the data in **Table 10-3**, you see that you cannot find a consistent value for k such that the equation holds true for every set of data points. If we change the temperature scale such that the x-intercept at –273 now becomes zero, the data can be replotted to produce a graph that represents a direct proportion. **Figure 10-18** on the next page shows that graph with the x-axis relabeled. The temperature scale has been shifted such that –273°C is now equal to 0; the graph now represents a direct proportion. The new temperature scale differs from the Celsius scale by a constant of 273 and is called the **absolute temperature scale**. Remember that the actual difference between the absolute and Celsius scales is 273.16.

absolute temperature scale
a temperature measurement made relative to absolute zero—the lowest possible temperature

6. On a graph in which the x- and y-axes are labeled *Temperature* and *Pressure*, respectively, plot the two data points, and extrapolate the line to 0 K. The result can also be determined by using the slope equation, $y = mx + b$, where $m = (y_2 - y_1)/(x_2 - x_1)$.

SAFETY
Lab aprons and safety goggles should be worn by the participating student and teacher. The rest of the class must be at least 10 ft away from the demonstration.

DISPOSAL
When the boiling water has cooled, pour it and the ice water down the drain.

TEACH

Teaching Tip

Enrichment
A safe and instructive project for students to try at home is to have them inflate a balloon, place it in the refrigerator for 30 min, and then note its volume. Suggest that they do the same thing using the freezer. This can be even more helpful if it is done the night before the discussion. (Extra credit points can be given for doing the experiment, and parents or siblings can be involved by signing a note saying they witnessed the activity.)

Content Background
Extraction to absolute zero

Explain to students that the data can be extrapolated to zero only because it is theorized that the data will continue to decrease linearly as both quantities approach zero.

Possible Misconception

Many students will question why there is a bottom to the temperature scale, but no top. Remind them that if all of the energy is removed, then the system will be at its "coldest," and the absolute lowest temperature will be reached. However, it is always possible to add more energy to raise the absolute temperature.

Teaching Tip
Negative volumes

Discuss with students the problem of trying to arrive at a negative volume. What would it look like? Is it possible? Reinforce the idea that although numbers can be negative, many concepts using numbers cannot be.

Possible Misconception
Thermodynamic temperature

Stress that when a value for temperature is required for a thermodynamics problem, it must be expressed in kelvins. To make this clear, have students use Charles's law ($V/T = constant$) to determine, without converting the temperature to kelvins, the volume of a gas at $T = 0°C$ and $T = -10°C$.

Historical Note

The relationship $P_1/T_1 = P_2/T_2$, in which the volume is kept constant, is often called Gay-Lussac's law. Joseph Louis Gay-Lussac was greatly interested in the study of gases and was also a pioneer in ballooning. He rose as high as 7000 m (4.35 mi) in a balloon.

Teaching Tip
Problem solving

Students may have problems when the unknown variable is in the denominator. The easiest solution for this is to flip the fractions so that the denominators become the numerators, and vice versa.

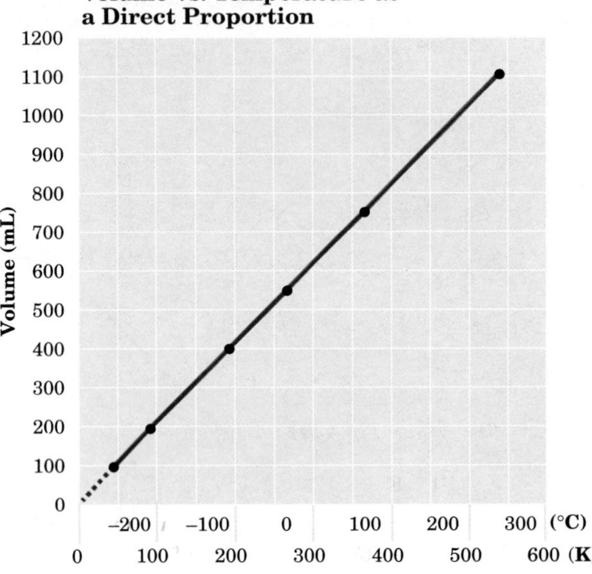

Figure 10-18
By changing the temperature scale on the graph so that the x intercept is at 0, the relationship between volume and temperature can be expressed as a direct proportion. The temperature scale is merely shifted 273.16 degrees.

Table 10-4
Volume–Temperature Data for a Gas at a Constant Pressure

Volume (mL)	Temperature (°C)	Temperature (K)	V/T
1094	273	546	2.00
748	100	373	2.01
568	10	283	2.01
545	1	274	1.99
545	0	273	2.00
546	–1	272	2.01
403	–73	200	2.02
199	–173	100	1.99
100	–223	50	2.00

Absolute zero in kelvins is the lowest possible temperature. Why? Theoretically, any temperature lower than 0 K (–273.16°C) would represent a negative volume. In **Table 10-4** the original data has been changed to reflect the new temperature scale and the mathematical relationship for a direct proportion, which is expressed mathematically as follows.

$$\frac{\textbf{volume }(V)}{\textbf{temperature }(T)} = \textbf{constant }(k)$$

$$\frac{V}{T} = k$$

If volume divided by temperature is a constant, we can equate this ratio for gases under two different sets of conditions. The initial set of conditions is labeled V_1 and T_1, and another set of conditions is labeled V_2 and T_2.

$$\frac{V_1}{T_1} = k \text{ and } \frac{V_2}{T_2} = k$$

The initial and final conditions can be set equal to each other.

$$\frac{V_1}{T_1} = \frac{V_2}{T_2}, \text{ where } P \text{ is constant}$$

Charles's law
the volume of a gas at constant pressure is directly proportional to the absolute temperature

This relationship is often called **Charles's law**, though it is the work of the French chemists Jacques Charles and Joseph Gay-Lussac that resulted from their interest in ballooning.

Visual Strategy
Figure 10-18

Stress that a major advantage of the Celsius and absolute temperature scales is that the size of a degree is identical for both systems. This allows for easy conversion from one to the other.

Sample Problem 10A
Solving volume–temperature problems

A sample of gas occupies 24 m^3 at 100. K. What volume would the gas occupy at 400. K?

1 List

- $V_1 = 24\ m^3$
- $T_1 = 100.\ K$
- $T_2 = 400.\ K$
- $V_2 = ?$
- **Temperature increases, which means the volume should increase also.**

2 Set up

- **Use Charles's law and solve for V_2.**

$$\frac{V_1}{T_1} = \frac{V_2}{T_2} \text{ therefore, } \frac{24\ m^3}{100.\ K} = \frac{V_2}{400.\ K}$$

3 Estimate and calculate

$$V_2 = 24\ m^3 \times \frac{400.\ K}{100.\ K}$$

- **Note that the ratio of temperatures is greater than 1, which means the final volume will be greater, as predicted.**

$V_2 = 96\ m^3$

Practice 10A

Answers for Concept Review items and Practice problems begin on page 841.

1. **Gas in a balloon occupies 2.5 L at 300. K (about room temperature). At what temperature will the balloon expand to 7.5 L?**
2. **The balloon from item 1 is dipped into liquid nitrogen that is at a temperature of 80. K. What volume will the gas in the balloon occupy at this temperature?**

Boyle's law describes a pressure-volume relationship

We can use a syringe to study the effect of pressure on the volume of a gas. A fixed amount of gas is trapped in the syringe. If you push on the plunger, as in **Figure 10-19**, you are compressing the trapped gas particles into a smaller space. In the smaller space, the particles collide with the walls more often, and more collisions result in a higher pressure.

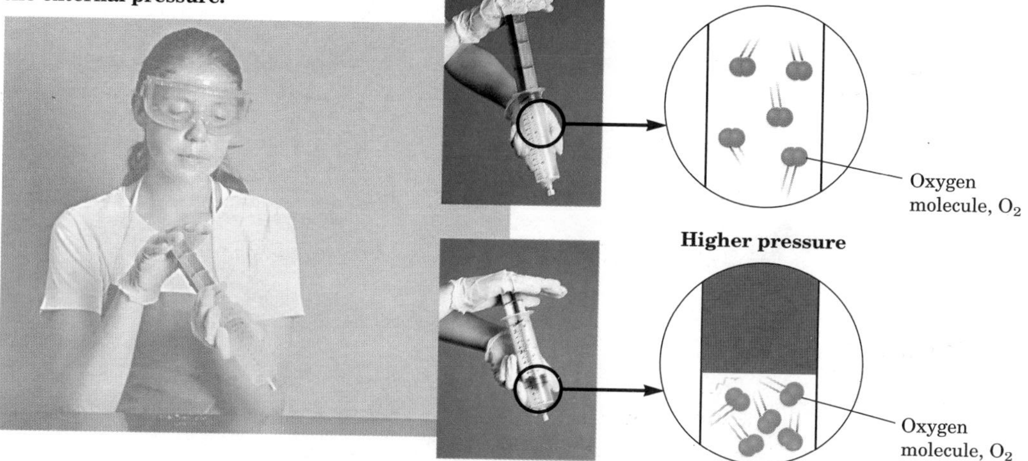

Figure 10-19
To study the effects of pressure on volume, a fixed amount of gas can be placed in a syringe. Increasing the pressure on the syringe compresses the gas to the point where its internal pressure equals the external pressure.

Solutions for Additional Examples 10A

a. $\frac{V_1}{T_1} = \frac{V_2}{T_2} = \frac{852\ mL}{189\ K} = \frac{V_2}{293\ K}$, or $V_2 = \frac{(852\ mL)(293\ K)}{189\ K} = 1.32 \times 10^3\ mL$

b. $T_1 = 23°C = (23 + 273) = 296\ K$

$\frac{V_1}{T_1} = \frac{V_2}{T_2} = \frac{234.9\ L}{296\ K} = \frac{200.0\ L}{T_2}$, or $T_2 = \frac{(200.0\ L)(296\ K)}{234.9\ L} = 252\ K$, or $-21°C$

Additional Examples for Sample Problem 10A

a. A sample of gas has a volume of 852 mL at 189 K. What volume does the gas occupy at 293 K?

b. A sample of gas has a volume of 234.9 L at a temperature of 23°C. What temperature is necessary to have a volume of 200.0 L?

Solutions are at the bottom of the page.

Answers to Practice 10A

1. $\frac{V_1}{T_1} = \frac{V_2}{T_2} = \frac{2.5\ L}{300.\ K} = \frac{7.5\ L}{T_2}$, or

$T_2 = \frac{(7.5\ L)(300.\ K)}{2.5\ L} = 9.0 \times 10^2\ K$

2. $\frac{V_1}{T_1} = \frac{V_2}{T_2} = \frac{2.5\ L}{300.\ K} = \frac{V_2}{80.\ K}$, or

$V_2 = \frac{(2.5\ L)(80.\ K)}{300.\ K} = 0.67\ L$

Teaching Tip

Discussion

Remind the class (or perform the demonstration again) that as the eudiometer was pushed to the bottom of the graduated cylinder, the pressure exerted on the gas in the tube increased, and its volume decreased. When the eudiometer was lifted, the gas pressure decreased, and the volume increased. This is an example of Boyle's law and has probably been performed (albeit not too rigorously) by everyone, either as they have washed a glass or played with cups in the tub as a child.

Historical Note

Otto von Guericke (1602–1686) of Magdeburg, Prussia, invented a pump to remove air from containers. In 1654, he demonstrated that when two hollow copper hemispheres (called Magdeburg hemispheres) were fitted together with a flat, greased flange and were evacuated, they could not be separated, even by two teams of horses. This showed the great force exerted by the atmosphere.

Demonstration 5

Volume changes with a vacuum pump

Approximate time: 15 min

1. Gather the following objects to demonstrate how they will be affected by lowering the pressure in a vacuum pump: a balloon, shaving cream, a few marshmallows, some jelly beans, and a slice of bread.
2. Discuss with students what should happen to each object before placing it in the pump. The balloon is a good object to start with, followed by the marshmallows. (Both of these objects will swell, but the bread will do nothing. The shaving cream will expand almost to the top of the bell jar, and if the pressure is released slowly, it often goes completely back into the beaker.)

SAFETY

Because of the possibility of the bell jar shattering under low pressures, all students must wear safety goggles. Teacher must wear safety goggles and a lab apron. Students must be 10 ft or more from demonstration.

DISPOSAL

Throw all tested materials in the trash.

Cultural Connection

In Cameroon, Africa, carbon dioxide from underground magma rises and accumulates on the floor of Lake Nyos. Periodically, the pressure of the carbon dioxide gas overcomes the pressure of the deep water, and the carbon dioxide rises to the surface, where the expanded gas drifts along the ground, silently suffocating all oxygen-dependent life in its path. Scientists are searching for methods to release the gas before the pressure becomes too great.

Table 10-5 Pressure–Volume Data

Pressure (kPa)	Volume (mL)	*PV*
100	500	50 000
150	333	49 950
200	250	50 000
250	200	50 000
300	166	49 800
350	143	50 500
400	125	50 000
450	110	49 500

If you push with a fixed pressure, the volume inside will decrease until the pressure inside the syringe equals the total pressure (applied and atmospheric) outside it. When the pressures are equal, the plunger won't move any farther. You can measure the volume changes as you increase the pressure on the gas. To note how these two conditions affect each other, you need to keep the temperature of the system constant.

Table 10-5 shows pressure-volume data for an experiment of this type working with a larger volume of gas. Looking at these data, you see that an increase in pressure results in a decrease in volume. You should also notice that the product of each pair of pressure and volume measurements stays essentially constant, within the precision of the data.

Boyle's law
the volume of a gas at constant temperature is inversely proportional to the pressure

Because the product of pressure times volume is a constant, it means that *PV* under one set of conditions equals *PV* for another set of conditions as long as the temperature and the amount of gas remain constant. This relationship was identified by Robert Boyle and is known as **Boyle's law**. Boyle's law is represented by the following equation.

$$PV = k$$

Graphing the data in **Figure 10-20** gives a hyperbola, which is representative of an inverse proportion. As one variable increases, the other decreases proportionately. A treatment similar to Charles' law is:

$$P_1V_1 = P_2V_2 \text{, where } T \text{ is constant.}$$

Figure 10-21 on the next page shows a graph of the same data with pressure on the x-axis and the reciprocal of volume ($1/V$) on the y-axis. Note that we now have a straight-line graph passing through the origin, which means that P and $1/V$ are directly proportional.

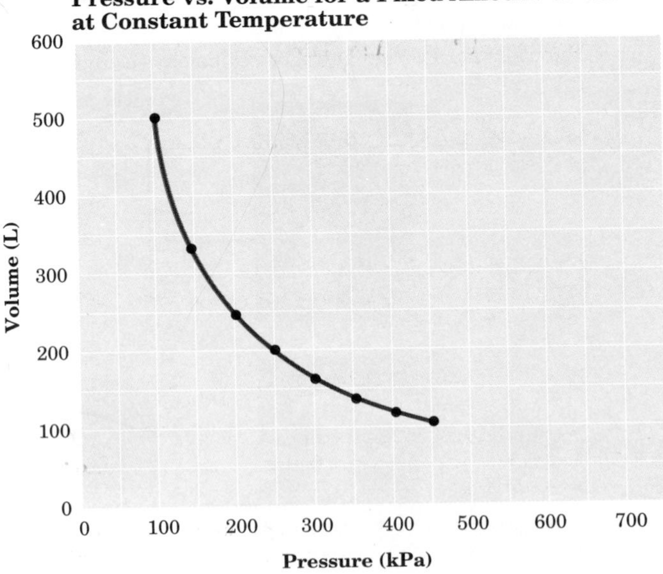

Figure 10-20 The pressure versus volume graph represents an inverse relationship—as pressure increases, volume decreases.

Visual Strategy

Figure 10-20 and Figure 10-21

Make students aware that these two graphs provide the same information. Point out that **Figure 10-20** provides a good visual example of an inverse, or indirect, relationship. When one variable goes up, the other goes down. A plot of one variable versus the inverse of the other variable will give a direct, linear relationship (**Figure 10-21**), which can be used to easily determine values from a graph.

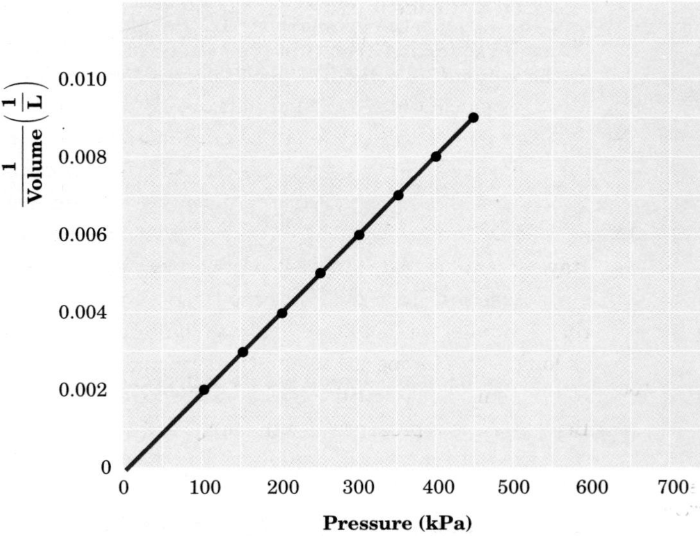

Figure 10-21 When pressure data is graphed with the reciprocal of volume (1/V), the graph shows that pressure and the reciprocal of volume are directly proportional.

Sample Problem 10B
Solving pressure–volume problems

The gas in a balloon has a volume of 4 L at 100 kPa. The balloon is released into the atmosphere, and the gas in it expands to a volume of 8 L. What is the pressure on the balloon at the new volume?

1 List

- $V_1 = 4$ L
- **The volume expands, and V_2 is 8 L.**
- $P_1 = 100$ kPa
- $P_2 = ?$
- **Because the gas expands, the pressure decreases.**

2 Set up

- **Use Boyle's law to determine P_2.**

$$P_1V_1 = P_2V_2, \text{ and } P_2 = \frac{P_1V_1}{V_2}$$

- **Multiply pressure by the ratio of volumes that will give a pressure decrease.**

$$P_2 = 100 \text{ kPa} \times \frac{4 \not{L}}{8 \not{L}}$$

3 Estimate and calculate

- **The volume ratio is 1:2, so we would expect the final pressure to be half the original pressure.**

$P_2 = 50$ kPa, which is consistent with the estimate.

Practice 10B

1. **The gas in a 10.0 L container exerts a pressure of 100. kPa. What pressure is needed to compress the gas to 2.0 L while keeping the temperature constant?**
2. **If the pressure of a 2.5 m^3 sample of a gas is 1.5 atm, what volume will the gas occupy if the pressure is changed to 7.5 atm?**

Content Background
Derivation of Boyle's law
Some students will appreciate seeing this simple derivation. Since $P_1V_1 = k$, a constant value, and $P_2V_2 = k$ for the same gas sample, then $P_1V_1 = P_2V_2$.

Other Applications
Divers have dissolved gases in their blood just as people on the surface do. However, problems develop when the diver is in deep water and surfaces too quickly. As the diver rises, the water pressure drops rapidly, and the bubbles of dissolved gas expand. These gases can get trapped in the body's joints and can cause permanent damage or death. The condition, called the *bends*, is named for the curling or bending of the diver in response to the pain in the joints.

Additional Examples for Sample Problem 10B

a. A syringe has a sample of gas with a volume of 3.95 mL at a pressure of 760 torr. What is the pressure of the gas in the syringe if the volume is changed to 2.53 mL?

b. A partially inflated weather balloon has a volume of 1.50×10^3 L at 0.988 atm of pressure. What is the volume of the balloon when released to a height where the pressure is 0.441 atm?

Solutions are at the bottom of the page.

Answers to Practice 10B

1. $P_1V_1 = P_2V_2$
   (100 kPa)(10.0 L) = P_2(2.0 L)
   $P_2 = 5.0 \times 10^2$ kPa
2. $P_1V_1 = P_2V_2$
   (1.5 atm)(2.5 m^3) = (7.5 atm)V_2
   $V_2 = 0.50$ m^3

Solutions for Additional Examples 10B

a. $P_1V_1 = P_2V_2$; (760 torr)(3.95 mL) = P_2(2.53 mL)

$$P_2 = \frac{(760 \text{ torr})(3.95 \text{ mL})}{2.53 \text{ mL}} = 1.19 \times 10^3 \text{ torr}$$

b. $P_1V_1 = P_2V_2$; (0.988 atm)(1.50×10^3 L) = (0.441 atm)V_2

$$V_2 = \frac{(0.988 \text{ atm})(1.50 \times 10^3 \text{ L})}{0.441 \text{ atm}} = 3.36 \times 10^3 \text{ L}$$

Additional Example for Sample Problem 10C

The volume of an air bubble released by the diver at a depth of 115 m is about 108 mL. If the bubble stays together until it reaches the surface, what will the bubble's volume be?

Solution is at the bottom of the page.

Answers to Practice 10C

1. $P_1V_1 = P_2V_2$

$(1.0 \text{ atm})(2.0 \text{ L}) = (0.27 \text{ atm})V_2$

$V_2 = \frac{(1.0 \text{ atm})(2.0 \text{ L})}{0.27 \text{ atm}} = 7.4 \text{ L}$

2. $P_2 = (10.0 \text{ m})\left(\frac{100 \text{ kPa}}{10.2 \text{ m}}\right) +$

$101.3 \text{ kPa} = 199.3 \text{ kPa}$

$P_1V_1 = P_2V_2$

$(101.3 \text{ kPa})(4 \text{ L}) = (199.3 \text{ kPa})V_2$

$V_2 = \frac{(101.3 \text{ kPa})(4 \text{ L})}{199.3 \text{ kPa}} = 2 \text{ L}$

The balloon compresses to a 2 L volume.

3. $P_1V_1 = P_2V_2$

$P_2 = \frac{(22.5 \text{ kPa})(155 \text{ cm}^3)}{90.0 \text{ cm}^3} = 38.8 \text{ kPa}$

4. $P_1V_1 = P_2V_2$

$V_2 = \frac{(0.500 \text{ atm})(300. \text{ mL})}{0.750 \text{ atm}} =$

200.0 mL

5. $P_1V_1 = P_2V_2$

$V_2 = \frac{(1.0 \text{ atm})(5.0 \text{ L})}{0.5 \text{ atm}} = 10. \text{ L}$

Sample Problem 10C
Boyle's law for divers

Divers know that the pressure exerted by the water increases with depth. The pressure increases about 100 kPa every 10.2 m, so that at 10.2 m below the surface, the pressure is 201 kPa; at 20.4 m, the pressure is 301 kPa, and so forth. Given that the volume of a balloon is 4.0 L at *STP*, what is the volume 50. m below the water's surface at the same temperature?

1 List

- $P_1 = 101.3 \text{ kPa}$
- **Pressure increases by 100 kPa every 10.2 m of water depth.**
- $V_1 = 4.0 \text{ L}$
- ***T*** is constant
- **distance** = 50. m
- $P_2 = ?$
- $V_2 = ?$
- **Increased pressure should result in decreased volume.**

2 Set up

- **The pressure increases steadily as you descend. If the pressure increases 100 kPa for 10.2 m, you can calculate the pressure increase per meter of depth.**

$\frac{100 \text{ kPa}}{10.2 \text{ m}} = 9.8 \text{ kPa/m}$, which is rounded to 10 kPa/m

- **At 50. m, the total pressure, P_2, is equal to the pressure at the surface, which is 101.3 kPa, plus the pressure from the increase in depth for 50 m.**

$P_2 = 101.3 \text{ kPa} + (50. \text{ m} \times 10 \frac{\text{kPa}}{\text{m}}) = 600 \text{ kPa}$

- **Use Boyle's law to determine V_2.**

$(101.3 \text{ kPa})(4.0 \text{ L}) = (600 \text{ kPa})(V_2)$

$V_2 = 4.0 \text{ L} \times \frac{101.3 \text{ kPa}}{600 \text{ kPa}}$

3 Estimate and verify

- **V_2 should decrease as predicted because the pressure ratio is less than 1.**

$4.0 \text{ L} \times \frac{101.3 \text{ kPa}}{600 \text{ kPa}} = 0.67532 \text{ L}$, calculator answer

$V_2 = 0.70 \text{ L}$, rounded answer

Practice 10C

1. **A 2.0 L balloon at a pressure of 1.0 atm on Earth's surface ascends 10 km into the atmosphere, where the pressure is 0.27 atm. What is the volume of the balloon at that altitude (assuming the temperature stays the same)?**
2. **What happens to the volume of a 4 L balloon inside a sinking boat that descends to the bottom of a lake 10.0 m below the surface?**
3. **A flask containing 155 cm^3 of hydrogen was collected at a pressure of 22.5 kPa. Under what pressure would the gas have a volume of 90.0 cm^3 at constant temperature?**
4. **If the pressure exerted on a 300. mL sample of hydrogen gas at constant temperature is increased from 0.500 atm to 0.750 atm, what will be the final volume of the sample?**
5. **A helium balloon has a volume of 5.0 L at a pressure of 1.0 atm. The balloon is released and reaches an altitude of 6.5 km at a pressure of 0.50 atm. If the gas temperature remains the same, what is the volume of the balloon?**

Solution for Additional Example 10C

$P_1 = (115 \text{ m})(\frac{100 \text{ kPa}}{10.2 \text{ m}}) + 101.3 \text{ kPa} = 1.23 \times 10^3 \text{ kPa}$

$P_1V_1 = P_2V_2 = (1.23 \times 10^3 \text{ kPa})(108 \text{ mL}) = (101.3 \text{ kPa})V_2$

$V_2 = 1.31 \times 10^3 \text{ mL}$

Dalton's law applies to mixtures of gases

If you put 1 mol of oxygen and 1 mol of nitrogen into a flask, the total pressure in the flask is the sum of the pressures exerted by each gas. Each gas exerts pressure just as if the other one were not present, as shown by the model in **Figure 10-22**. Though this idea might seem odd, it makes sense when you think of the gas mixture as the total number of particles. A mole of each gas will have 6.022×10^{23} particles. If you put both gases in one container, there are twice the number of particles. The 2 mol (total) of gas exert twice as much pressure as either one did alone, assuming a fixed volume and temperature. Therefore, in a mixture of gases, the total pressure is the sum of the partial pressures, which describes a relationship known as **Dalton's law of partial pressures**. **Partial pressures** are the pressures due to each gas in the mixture. For a mixture of gases labeled *a*, *b*, and *c*, the partial pressure equation would be written as follows.

Dalton's law of partial pressures *the total pressure in a gas mixture is the sum of the partial pressures of the individual components*

partial pressures *the pressure of an individual gas in a gas mixture that contributes to the total pressure of the mixture*

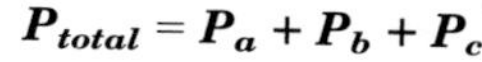

$$P_{total} = P_a + P_b + P_c$$

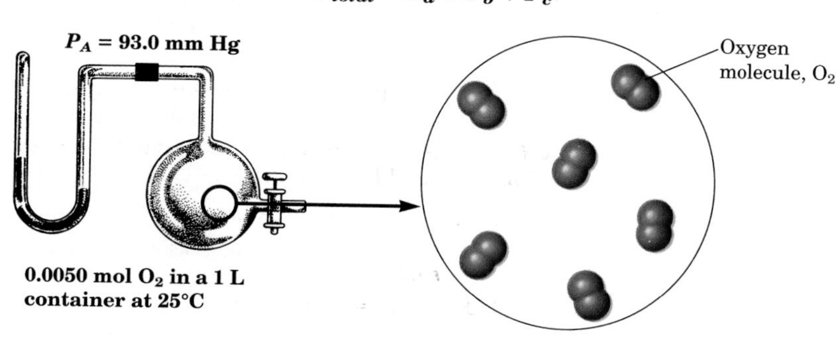

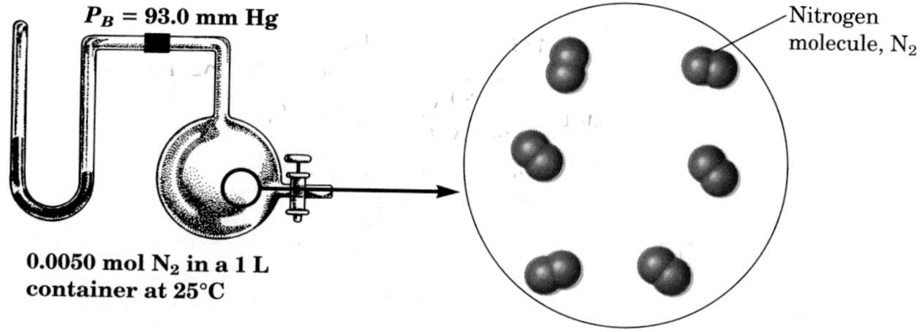

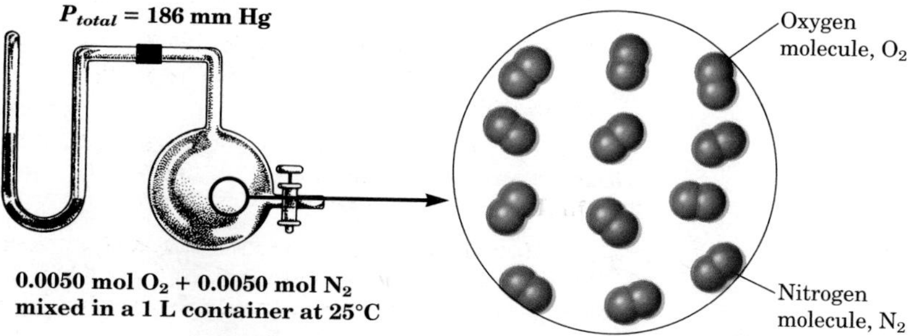

Figure 10-22
If samples of oxygen and nitrogen gas in separate containers are mixed, the pressure exerted by the gas mixture is the sum of the pressures exerted by each gas individually. At the same temperature the energy of both gases is the same. Therefore, they exert the same pressure.

Teaching Tip
Discussion
An excellent example of partial pressures is the total pressure of the atmosphere, which consists of the partial pressures of its component gases, mainly nitrogen, oxygen, argon, carbon dioxide, and water vapor.

Visual Strategy
Figure 10-22
Make students aware that equal moles of gas at the same temperature and volume will have equal pressure, and therefore the total pressure is simply the sum of the component pressures.

Possible Misconception

Stress that gases collected over water are considered "wet" due to the addition of water vapor. "Dry" gases sold in canisters have had the water vapor removed. One method of removing the water vapor is to pass the gas through an anhydrous compound, like anhydrous calcium chloride, that absorbs the water.

Demonstration 6

Amount of oxygen in air

Approximate time: 5 min for setup

1. Wet the inside of a test tube, and fill it 1/10 full with iron filings or fine steel wool that has been rinsed free of residual oil. Gently shake the tube to disperse the filings onto the wet tube walls.
2. Place a finger over the test tube mouth, invert the tube, and gently push it into a 250 mL beaker of water so that the mouth is 1 cm below the surface. Clamp the tube in place.
3. As the iron filings oxidize, oxygen will be removed from the air. The water level will rise about 1/5 of the tube length, an amount equal to the volume occupied by the oxygen. The effect will reach completion by the next day.
4. On the following day, test to see if there is any oxygen left in the tube. Place your finger over the tube mouth, remove the tube from the water, and, keeping the tube inverted, insert a glowing splint. The splint should go out immediately.

SAFETY

Safety goggles and a lab apron must be worn. Students must be 10 ft or more from the demonstration.

DISPOSAL

Throw rusty iron into the trash.

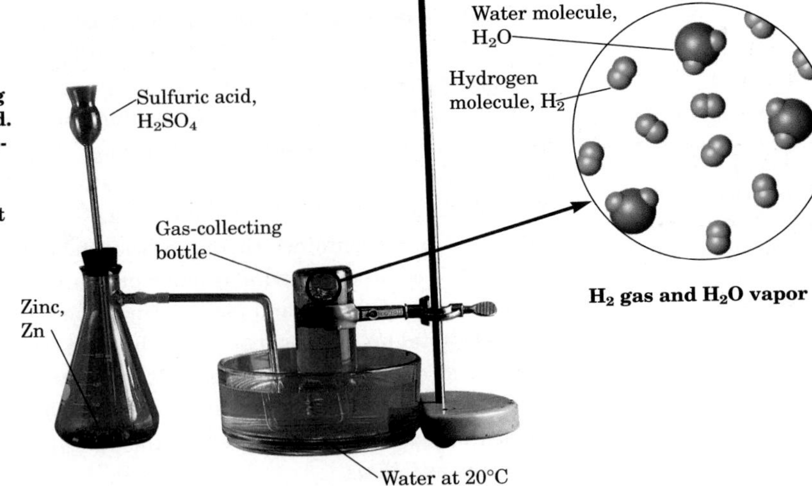

Figure 10-23 **Hydrogen can be collected by water displacement by reacting zinc with sulfuric acid. The hydrogen gas produced displaces the water in the gas collecting bottle where it is mixed with water vapor.**

Dalton's law is most often used to calculate the pressure of a gas collected over water. **Figure 10-23** shows such a setup for collecting hydrogen. The collection flask contains hydrogen gas and some water vapor that adds to the total pressure. Therefore, to find the pressure of the hydrogen alone, you need to subtract the pressure due to the water vapor, which can be found in **Table 10-6**. For example, the system above is at 20°C. The total pressure of the gases in the collecting bottle can be adjusted so that it is the same as the air pressure in the room. If the pressure reading for the room is 98.2 kPa, then the pressure exerted by the hydrogen gas is calculated as follows.

98.2 kPa	total pressure
− 2.3 kPa	pressure of H_2O at 20°C
95.9 kPa	pressure of H_2 at 20°C

Table 10-6
Vapor Pressure of Water at Various Temperatures

Temperature (°C)	Pressure H_2O (kPa)	Temperature (°C)	Pressure H_2O (kPa)
0	0.61	55	15.75
5	0.87	60	19.93
10	1.23	65	25.02
15	1.71	70	31.18
20	2.34	75	38.56
25	3.17	80	47.37
30	4.25	85	57.82
35	5.63	90	70.12
40	7.38	95	84.53
45	9.59	100	101.32
50	12.34	105	120.79

Visual Strategy

Figure 10-23

Make sure students understand that this method of gas collection does not work well for gases that are soluble in water. Water-soluble gases can be collected by mercury displacement.

Sample Problem 10D
Dalton's law of partial pressure for gas collected over water

Hydrogen gas is collected over water at a total pressure of 95.0 kPa. The volume of hydrogen collected is 28 mL at 25°C. What is the partial pressure of hydrogen gas?

1 List

- $P_{total} = 95.0$ kPa
- **The flask contains both hydrogen and water vapor.**
- **From Table 10-6, $P_{water\ vapor}$ at 25°C is 3.17 kPa**
- $V = 28$ mL

2 Set up

$P_{total} = P_{hydrogen} + P_{water\ vapor}$

95.0 kPa $= P_{hydrogen} +$ 3.17 kPa

3 Estimate and verify

- **The partial pressure of hydrogen alone must be less than the total pressure.**

$P_{hydrogen} = 91.8$ kPa

Practice 10D

1. **A gas is collected by water displacement at 50°C and a barometric pressure of 95.00 kPa. What is the pressure exerted by the dry gas?**
2. **Oxygen gas is collected by water displacement from the reaction of Na_2O_2 and water. The O_2 displaces 318 mL of water at 23°C and 1.0000 atm. What is the pressure of the dry O_2?**

Mole fractions are used in determining partial pressures

mole fraction
the number of moles for an individual substance compared with the total number of moles in the mixture expressed as a ratio

The partial pressure of any gas in a mixture can be calculated using the **mole fraction** of that gas in the mixture. A mole fraction can be represented mathematically as follows.

$$\textbf{mole fraction of gas A} = \frac{\textbf{moles of gas A}}{\textbf{total number of moles of gas}}$$

For example, if gas A contributes one-fifth of the total number of particles, we say that the mole fraction of gas A is 0.2. The partial pressure of gas A is one-fifth, or 0.2, times the total pressure.

The tissues and cells in your lungs cannot absorb oxygen efficiently unless this gas exerts a pressure above 14 kPa. At sea level, the pressure due to oxygen in the atmosphere is typically about 20 kPa. At high altitudes oxygen is still about 20% of the total number of moles, so the mole fraction of O_2 at any altitude is about the same. What changes at high altitudes is the partial pressure of O_2. The partial pressure of oxygen at high altitudes is lower than at sea level. The partial pressure of O_2 is 0.2 times the total pressure. If the partial pressure falls well below 14 kPa, oxygen is not absorbed as efficiently. This is why novice climbers often experience "altitude sickness" from insufficient oxygen. Lightheadedness and shortness of breath are two common symptoms of this illness. People who live at high altitudes become acclimated by producing more red blood cells to carry oxygen. Pilots in fighter jets, which are not pressurized as much as passenger jets, must breathe supplemental oxygen to ensure that the partial pressure of oxygen is about 20 kPa.

Solution for Additional Example 10D

$P_{total} = P_{H_2O} + P_{CO_2}$
P_{H_2O} at 50°C = 12.34 kPa
175.93 kPa = 12.34 kPa + P_{CO_2}
P_{CO_2} = (175.93 kPa − 12.34 kPa) = 163.59 kPa

Content Background

Vapor pressure of water
Make students aware that the vapor pressure of water boiling at 100°C is equal to 1 atm.

Possible Misconception

Stress that whether the enclosed system is a 2 L bottle of water or an indoor swimming pool, as long as the system is closed and the temperature remains unchanged, the partial pressure due to the water vapor is the same.

Additional Example for Sample Problem 10D

The pressure of carbon dioxide above a water solution in a closed bottle is 175.93 kPa when the temperature is 50°C. What is the pressure due to the carbon dioxide alone?

Solution is at the bottom of the page.

Answers to Practice 10D

1. $P_{total} = P_{water} + P_{dry\ gas}$
   P_{water} at 50°C = 12.34 kPa
   95.00 kPa = 12.34 kPa + $P_{dry\ gas}$
   $P_{dry\ gas}$ = (95.00 kPa − 12.34 kPa)
   = 82.66 kPa
2. $P_{total} = P_{H_2O} + P_{O_2}$
   P_{H_2O} at 23°C = 2.81 kPa
   P_{total} = 1.0000 atm = 101.32 kPa
   P_{O_2} = (101.32 kPa − 2.81 kPa)
   = 98.51 kPa

Teaching Tip

Analogy
Use an example of buying fruit at the grocery store to teach the concept of mole fraction. For example, Sam buys 0.3 mol of bananas, 0.8 mol of grapes, 0.2 mol of cherries, and 0.7 mol of apples at the store. What (mole) fraction of the purchase consisted of cherries? (The total number of moles is 0.3 + 0.8 + 0.2 + 0.7, or 2.0. The mole fraction of cherries is 0.2/2.0, or 0.1.)

Demonstration 7

Diffusion of HCl and NH_3

Approximate time: 2 min

1. Place a bottle of 1 M HCl next to a bottle of household ammonia water, $NH_3(aq)$. **Do not dilute concentrated ammonia for this demonstration.**
2. Remove the lids of both bottles, and watch for the formation of a white cloud of NH_4Cl.
3. If needed, blow across the HCl bottle toward the ammonia, taking care not to breath in the fumes of either gas.

SAFETY
Safety goggles and a lab apron must be worn. Students must be 10 ft or more from demonstration. Demonstration must be performed in a properly operating fume hood.

DISPOSAL
Cap the bottles, and return them to storage for future use. If disposal is necessary, combine all liquids, neutralize with 1 M HCl or NaOH as appropriate, and pour down the drain.

Theme

Equilibrium and change
Gases will continue to diffuse until the system has reached equilibrium.

Demonstration 8

Diffusion of $KMnO_4$ in water

Approximate times:
setup—3 min
total time—1 week

1. Fill a clear, 1 L graduated cylinder with water to within 4 mm of the top.
2. Let the water stand until its surface is still.

Gases diffuse to fill their containers

We notice the expansion of some gases through our sense of smell. When you open a bottle of household ammonia, it doesn't take long for the odor of ammonia gas, NH_3, to fill the room. If you spill a little paint thinner or take hot cookies out of the oven, within a moment or two the odor is recognizable throughout the room. Gaseous molecules of the compounds responsible for the smell are traveling at high speeds in all directions and are mixing quickly with the air in a process called **diffusion**. Mixtures of gases do not separate but stay uniformly mixed, as shown by the models in **Figure 10-24**. The most familiar example of a gaseous mixture is air. All the particles in the mixture are in motion. Experiments show that light molecules move faster than heavy ones. However, because they move randomly, gases stay mixed even if they have different molecular masses. Otherwise, the atmosphere would be in layers: carbon dioxide, oxygen, nitrogen, argon, water vapor, and helium.

diffusion
the process by which particles disperse from regions of higher concentration to regions of lower concentration

The diffusion model explains how odors travel across a room. The rapidly moving particles are constantly colliding with other particles, which causes their movement to be an erratic zigzag kind of pattern, as you saw in the particle model in **Figure 10-6** on page 360.

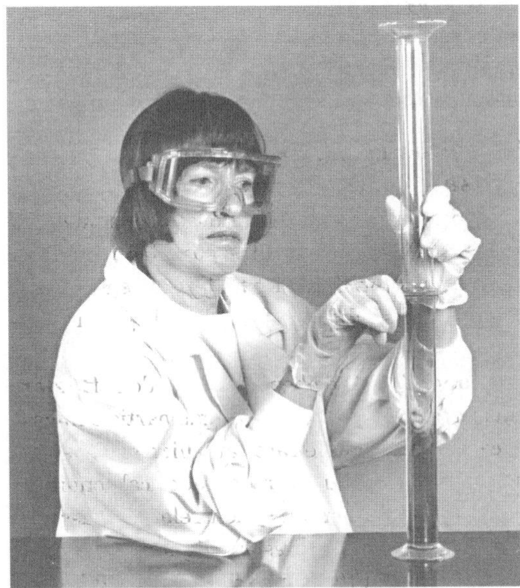

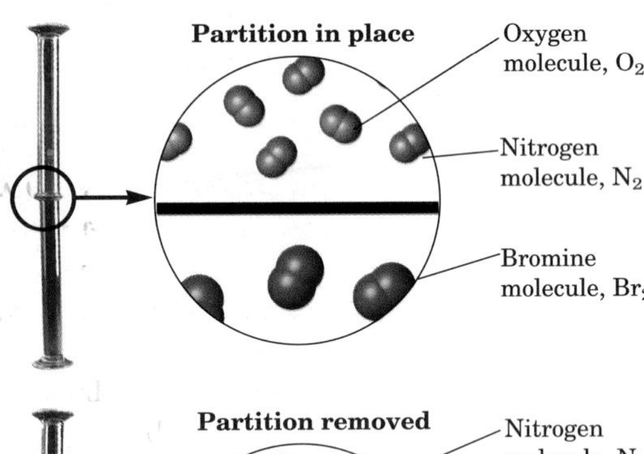

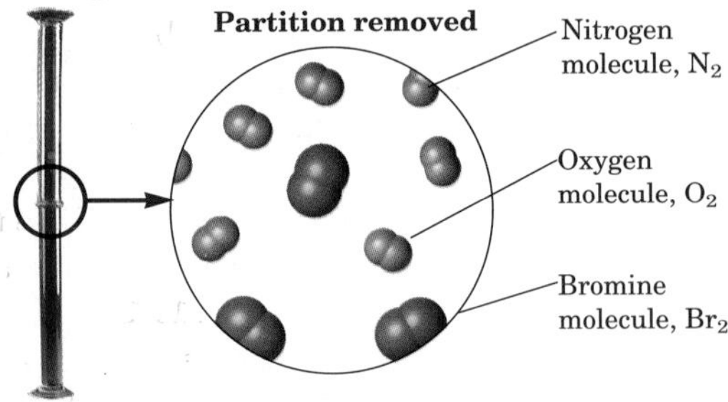

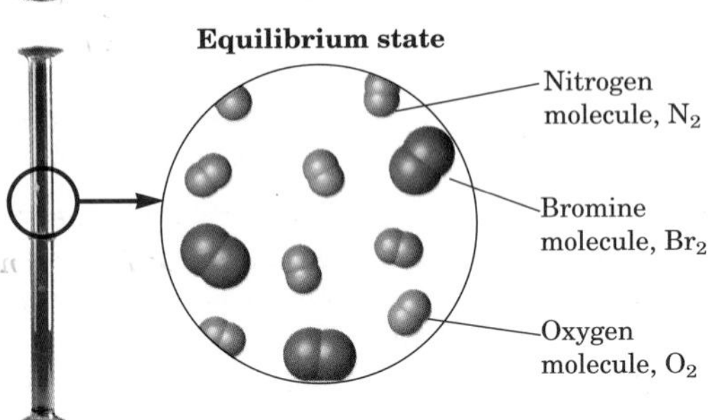

Figure 10-24
Bromine vapor diffuses up the glass cylinder until it hits the partition. The volume above the partition is filled with air. However, once the partition is removed, the vapor continues to move up the tube. Its particles collide with those in air until it uniformly fills the space.

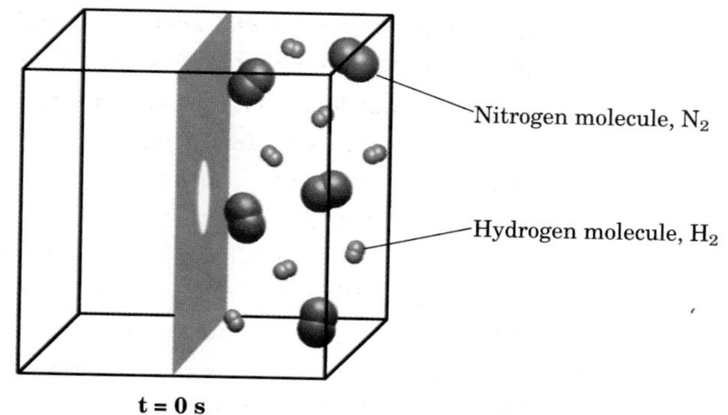

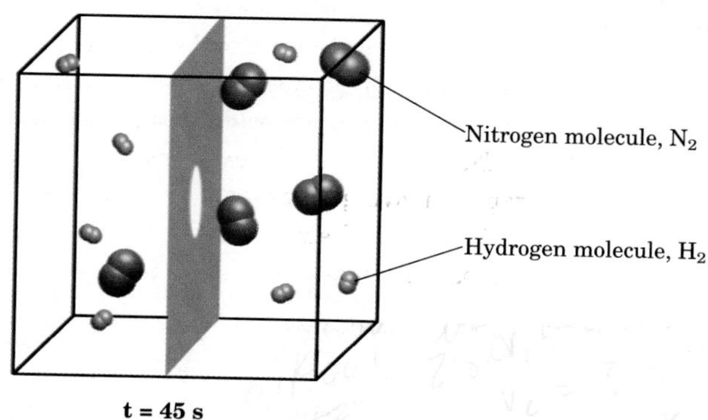

Figure 10-25
At the same temperature hydrogen molecules travel faster than nitrogen molecules because they are lighter. As a result, if a hole is made in the barrier, more hydrogen molecules will pass through the hole in a given time than will nitrogen molecules. Hydrogen molecules effuse at a faster rate (particles per second) than nitrogen molecules.

Graham's law applies to the rate at which gases diffuse and effuse

If we compared the rates at which different gases diffuse, we would find that the speed is related to the molecular mass of the gas. Gases with light particles diffuse quickly. So we would predict that at the same temperature, helium (4.00 g/mol) would diffuse faster than krypton (83.80 g/mol).

effusion
the motion of a gas through an opening into an evacuated chamber

The same relationship also applies to the phenomenon called **effusion**, which is shown by the model in **Figure 10-25**. Once a gas particle finds a hole in the apparatus, it escapes. The relative rates at which two gases labeled A and B effuse from the same container under identical conditions of temperature and total pressure depend only on their relative masses. In Section 10-1, you learned that heavy molecules and light molecules have the same average kinetic energy at the same temperature. If kinetic energy is represented by $1/2\ mv^2$, then we can equate the kinetic energy of gas A and gas B by the following relationship.

Graham's law of effusion
the rates of effusion for two gases are inversely proportional to the square roots of their molar masses at the same temperature and pressure

$$\frac{1}{2}m_A v_A{}^2 = \frac{1}{2}m_B v_B{}^2$$

Rearranging this equation gives a mathematical relationship referred to as **Graham's law of effusion.**

$$\frac{v_A}{v_B} = \sqrt{\frac{m_B}{m_A}}$$

3. Prepare a strip of paper 10 cm long and 4–6 mm wide. Fold the strip so that it resembles a *W*.
4. Arrange the strip over the top of the cylinder so that the center fold dips under the water's surface and the "wings" drape over the cylinder's sides.
5. Tape the strip in place to prevent it from moving.
6. With a spatula place **one** small crystal of $KMnO_4$ in the center fold of the paper, under the surface of the water.
7. Seal the top of the cylinder with plastic wrap or aluminum foil.
8. Note the dispersion of the purple $KMnO_4$ out and downward as it dissolves in the water. The motion of the water molecules (Brownian motion) acts to change the direction of the $KMnO_4$ particles, causing diffusion.

SAFETY
Safety goggles and a lab apron must be worn. Students must be 10 ft or more from the demonstration.

DISPOSAL
Toss the paper strip, tape, and foil or plastic into the trash. Add 1 mL of ethyl or denatured alcohol to the purple liquid, and stir gently. When the liquid turns brown, pour it down the drain.

Historical Note
Scottish chemist Thomas Graham (1805–1869) was the first to study the mixing of two gaseous substances.

Teaching Tip
Problem solving
Mention that it is permissible to square both sides of the equation to get rid of the square root if doing so eases calculations. $(v_a/v_b)^2 = m_b/m_a$

Visual Strategy
Figure 10-25
Make students aware that diffusion and effusion will continue until the concentrations of both gases are equal throughout the system. Also point out that if either of the molecules were monatomic, they would spread out more rapidly (specifically, by a factor of $\sqrt{2}$, or 1.41).

Additional Example for Sample Problem 10E

An ammonia molecule, NH_3, travels at about 658 m/s at room temperature. How fast would a molecule of hydrogen sulfide, H_2S, travel under the same conditions?

Solution is at the bottom of the page.

Additional Example for Sample Problem 10F

A carbon dioxide molecule travels at about 409 m/s at room temperature. What is the molar mass of a molecule that travels at 322 m/s under the same conditions?

Solution is at the bottom of the page.

Answers to Practice 10F

1. $m_1 = 20.18$ g/mol Ne

$m_2 = 58.14$ g/mol C_4H_{10}

$$\frac{v_2}{v_1} = \sqrt{\frac{m_1}{m_2}} = \sqrt{\frac{20.18 \text{ g/mol}}{58.14 \text{ g/mol}}} =$$

$$\sqrt{0.347} = 0.589 = \frac{v_2}{400 \text{ m/s}}$$

$v_2 = (0.589)(400 \text{ m/s}) = 236$ m/s

2. $m_1 = 34.09$ g/mol H_2S

$m_2 = 152.16$ g/mol $C_8H_8O_3$

$m_3 = 106.13$ g/mol C_7H_6O

$v_1 = 450$ m/s

$$\frac{v_2}{v_1} = \sqrt{\frac{m_1}{m_2}} = \sqrt{\frac{34.09 \text{ g/mol}}{152.16 \text{ g/mol}}} =$$

$$\sqrt{0.224} = 0.473 = \frac{v_2}{450 \text{ m/s}}$$

$v_2 = (0.473)(450 \text{ m/s}) = 213$ m/s

$$\frac{v_3}{v_1} = \sqrt{\frac{m_1}{m_3}} = \sqrt{\frac{34.09 \text{ g/mol}}{106.13 \text{ g/mol}}} =$$

$$\sqrt{0.321} = 0.567 = \frac{v_3}{450 \text{ m/s}}$$

$v_3 = (0.567)(450 \text{ m/s}) = 255$ m/s

Hydrogen sulfide will be smelled first, followed by benzaldehyde, and then methyl salicylate.

Sample Problem 10E
Comparing molecular speeds

An oxygen molecule travels at about 480 m/s at room temperature. How fast would a molecule of sulfur trioxide, SO_3, travel at the same temperature?

1 List

- $m_{O_2} = 32.00$ g/mol
- $m_{SO_3} = 80.07$ g/mol
- $v_{O_2} = 480$ m/s
- $v_{SO_3} = ?$

2 Set up

- **Use Graham's law.**

$$\frac{v_{SO_3}}{480 \text{ m/s}} = \sqrt{\frac{32.00 \text{ g/mol}}{80.07 \text{ g/mol}}}$$

3 Estimate and calculate

- **SO_3 is heavier, so its velocity should be slower than that of oxygen.**

$$v_{SO_3} = 480 \text{ m/s}\sqrt{\frac{32.00 \text{ g/mol}}{80.07 \text{ g/mol}}} = 3.0 \times 10^2 \text{ m/s}$$

Sample Problem 10F
Determining molecular masses

An unknown gas effuses through an opening at a rate 3.53 times slower than nitrogen gas. What is the molecular mass of the unknown gas?

1 List

- **N_2 moves 3.53 times faster than the unknown gas.**
- **If v represents the velocity of the unknown gas speed, then the velocity of $N_2 = 3.53v$**
- $m_{N_2} = 28.02$ g/mol
- **mass of the unknown, $m_x = ?$**

2 Set up

- **Use the equation for Graham's law.**

$$\frac{v_{N_2}}{v_x} = \sqrt{\frac{m_x}{m_{N_2}}} \qquad \frac{3.53v}{v} = \sqrt{\frac{m_x}{28.02 \text{ g/mol}}}$$

- **Square both sides of the equation to simplify the expression.**

$$\frac{(3.53v)^2}{v^2} = \frac{m_x}{28.02 \text{ g/mol}} \qquad 3.53^2 = \frac{m_x}{28.02 \text{ g/mol}}$$

$$m_x = (28.02 \text{ g/mol})(3.53^2)$$

3 Estimate and calculate

- **The molecular mass of the unknown gas is greater than that of N_2 because its velocity is less than that of nitrogen.**

$$m_x = (28.02 \text{ g/mol})(3.53^2) = 3.49 \times 10^2 \text{ g/mol}$$

Practice 10F

1. **Given that neon gas travels at 400 m/s at a given temperature, calculate the rate of diffusion for butane, C_4H_{10}, at the same temperature.**
2. **Hydrogen sulfide, H_2S, has a very strong rotten egg odor. H_2S particles travel at about 450 m/s. Methyl salicylate, $C_8H_8O_3$, has a wintergreen odor. Benzaldehyde, C_7H_6O, has an almond odor. If vapors for these three substances were released at the same time from across a room, which would you smell first, and why?**

Solution for Additional Example 10E
v of H_2S = 465 m/s

Solution for Additional Example 10F
m of unknown molecule = 71.0 g/mol

Full solutions for Additional Examples 10E and 10F are found on page 401A.

Section Review

12. A container contains 20.0 g of oxygen at 40.0 kPa. Twenty grams of neon are added to the container.
    - a. Calculate the mole fraction of each gas.
    - b. Calculate the partial pressure of neon and the total pressure of the gas mixture.
    - c. Explain why the pressure exerted by oxygen does not change when neon is added to the container.
13. A certain mass of gas occupies 2.0 L at 60 kPa. What pressure allows the gas to expand to 6.0 L?
14. A 3.5 L sample of a gas has a temperature of 227°C and a pressure of 160 kPa.
    - a. What is the new volume if the gas is held at the same pressure but cooled to 27°C?
    - b. What is the new pressure if the gas is held at the same volume but heated to 500°C?
    - c. What is the new volume if the gas is adjusted to *STP*?
15. What temperature will increase the pressure of a gas in a 450 mL container from 60 kPa at 300 K to 320 kPa?
16. A gas occupies 4.7 m^3 at 250 Pa.
    - a. What is the gas volume at 4800 Pa?
    - b. What is the gas volume in liters at 1.0 atm?
17. Think about what happens to a gas bubble that enters the bloodstream of a diver at 50 m below the surface. What happens to the volume of the bubble as the diver ascends to the surface and the pressure is reduced?
18. Using the ideas of the kinetic molecular theory, which of the following gases is best for filling automobile tires: CO_2, air, or He? Explain your answer.
19. Why do your ears sometimes pop when you drive down a steep incline or descend in an airplane?
20. Inexpensive balloons filled with air deflate within a few days because the pores in the walls of the balloons are larger than most of the molecules that are found in air. Explain why helium-filled balloons collapse more quickly than those filled with air.

21. **Story Link**
    At 40 km above the Earth, the pressure is about 2.87×10^2 Pa. A helium balloon launched from Earth's surface, at 1.000 atm of pressure, has a volume of 30.0 m^3. What would be the volume of the balloon at 40 km? You can assume that the temperature of the He stays constant.
22. **Story Link**
    Half of the gas in a 20.0 L tank of helium at 35°C and 1050 kPa is released into a weather balloon while keeping the temperature constant and the balloon pressure at 100. kPa. What is the volume of the weather balloon?

ASSESS

Alternative Assessment

Have students list 10 ways in which gas laws can be applied. (examples: refrigeration, submarine ballast, gas liquification, weather balloons, etc.)

Answers to Section Review

12. a. mole fraction of O_2 = 0.387
    mole fraction of Ne = 0.613
    b. P_{gas} = mole fraction $\times P_{tot}$
    P_{tot} = (40.0 kPa)/(0.387)
    = 103 kPa
    P_{neon} = 63 kPa
    c. Pressure depends on the number of gas particles. The number of oxygen molecules remains unchanged, so the partial pressure is unchanged.
13. P = 20 kPa
14. a. T_1 = 227°C = 500 K
    T_2 = 27°C = 300 K
    V_2 = 2.1 L
    b. T_2 = 500°C = 773 K
    P_2 = 247 kPa
    c. V_2 = 3.0 L
15. $T = 1.6 \times 10^3$ K
16. a. $V = 0.24$ m^3
    b. V = 12 L
17. The bubble expands, occupying a greater volume within the bloodstream.
18. CO_2 is best. The smaller the mass of a gas particle, the greater its average velocity at a given temperature, and the quicker it effuses through flaws and leaks. Helium effuses from a tire the fastest of the three gases; carbon dioxide effuses the slowest.
19. Even a small altitude shift changes the pressure outside the eardrum. As you descend, the air pressure increases until air rushes into the inner ear and equalizes pressures on both sides of the eardrum. The rush of air causes the "pop."
20. Helium, being monatomic and smaller than diatomic nitrogen (air), passes through the porous walls of the balloon more easily .
21. $V = 1.06 \times 10^4$ m^3
22. V = 105 L

Section 10-3

FOCUS

Lesson Starter

Tell students the following story. As you start to fire up the grill you discover that your 20 L tank of propane is nearly empty, so you decide to have it filled. The problem is that at this time of night the only propane dealer open is Pretty Shady Propane. When you arrive there you realize that you must be on your chemistry guard at this establishment, so you demand that a pressure gauge be used to show when the propane tank is full. When the tank is brought out and you see that the gauge reads full, you pay the attendant. However, when you lift the tank you notice that it is extremely light and very hot. How did Pretty Shady Propane swindle you? (By heating the gas, its pressure was increased, causing the gauge to show the same pressure as for a full tank at room temperature. Tell your students that this situation is imaginary and that propane tanks should never be heated!)

TEACH

Theme

Classification and trends

Boyle's, Charles's, and Gay-Lussac's laws can be classified under the ideal gas law.

Teaching Tip

Problem solving

Point out that the ideal gas law can be substituted for Boyle's, Gay-Lussac's, or Charles's laws by omitting the variable that remains constant in those equations. Many students will appreciate this because they no longer have to remember four different equations.

How do the gas laws fit together?

Section Objectives

Use the ideal gas law and general gas law to solve gas problems.

Apply your knowledge of reaction stoichiometry to solve gas stoichiometry problems.

Combining the gas laws

You have thus far studied a series of relationships that can be used to predict the behavior of a gas under changing conditions. Each relationship includes at least one of the four variables we use to measure changes in a gas—pressure, P; volume, V; temperature, T; and number of moles, n. Mathematically, it should be possible to combine all the gas laws you studied in the last section to build an equation in which pressure, volume, temperature, and the number of moles can vary.

The ideal gas law relates all four gas variables

ideal gas law
the equation of state for an ideal gas in which the product of the pressure and volume is proportional to the product of the absolute temperature and the amount of gas expressed in moles

The connection among P, V, n, and T exists in the mathematical combination of Boyle's law, Charles's law, and Avogadro's law. This gas law, called the **ideal gas law,** is most often used when the amount of the gas may change. It is represented mathematically by the following equation.

$$PV = nRT$$

R is a proportionality constant. The value you use for R in calculations depends on the units given for pressure and volume in a problem. Because we use kilopascals and liters most of the time in this text, the value of R that you will use most often is the following.

$$R = \frac{8.314 \text{ L•kPa}}{\text{mol•K}}$$

R can also be expressed as

$$R = \frac{0.0821 \text{ L•atm}}{\text{mol•K}}$$

Sample Problem 10 G
Ideal gas law

A sample of carbon dioxide with a mass of 0.250 g was placed in a 350 mL container at 400. K. What is the pressure exerted by the gas?

1 List

- **There is no change of conditions for the gas, so this is a straightforward "equation of state" problem.**
- ***P*** = ?
- ***V*** = 350 mL
- ***T*** = 400. K
- ***n*** = to be calculated from 0.250 g, and the formula mass, 44.01 g/mol, of CO_2
- ***R*** = 8.314 L•kPa/mol•K

Sample Problem 10G continued on page 383

Sample Problem 10G continued from page 382

2 Set up

- **The mass in grams must be converted into an amount in moles.**

$$\text{mol } CO_2 = \frac{0.250 \text{ g } CO_2}{44.01 \text{ g/mol } CO_2} = 5.68 \times 10^{-3} \text{ mol } CO_2$$

- **The volume unit must be compatible with the value of R used, therefore the volume must be converted from milliliters to liters.**

1 mL = 0.001 L

350 mL = 0.35 L

- **Use the ideal gas law equation.**

$PV = nRT$

- **Substitute the values given in the problem into the equation.**

$(P)(0.35 \text{ L}) = (5.68 \times 10^{-3} \text{ mol})(8.314 \text{ L•kPa/mol•K})(400. \text{ K})$

3 Calculate

- **Solve for pressure by dividing each side of the equation by the volume to simplify the equation.**

$$P = \frac{(5.68 \times 10^{-3} \text{ mol})(8.314 \text{ L•kPa/mol•K})(400. \text{ K})}{(0.35 \text{ L})}$$

$P = 54$ kPa

Practice 10G

1. **A 500 g block of dry ice (solid CO_2) vaporizes to a gas at room temperature. Calculate the volume of gas produced at 25°C and 975 kPa.**
2. **Calculate the volume of 1 mol of CO_2 at *STP*.**
3. **Average lung capacity for humans is about 4.0 L. At 37°C (body temperature) and 110 kPa, how many moles of oxygen gas could your lungs hold?**
4. **If 40 g of methane, CH_4, is confined to 2500 mL at 200°C, what pressure does it exert?**
5. **At what temperature will 7.0 mol of helium gas exert a pressure of 1.2 atm in a 25.0 m^3 tank?**

The general gas law is derived from the ideal gas law

The ratio of PV/nT equals the ideal gas constant R. Thus, for the same gas under two sets of conditions it is possible to write the following expressions.

$$R = \frac{P_1V_1}{n_1T_1} \text{ and } R = \frac{P_2V_2}{n_2T_2}$$

Because both equations equal R, then both expressions can be set equal to each other. And if the amount of gas, n, does not change, n can be cancelled from both equations, and the expression is simplified to get the following.

$$\frac{P_1V_1}{T_1} = \frac{P_2V_2}{T_2}$$

This relationship is often referred to as the *general gas law* or *combined gas law*. It is often used in calculations for determining the volume of a gas at *STP*.

Additional Examples for Sample Problem 10G

a. How many moles of gas are in a balloon that has a volume of 15.9 L at a pressure of 149 kPa and temperature of 28°C?

b. What mass of ammonia, NH_3, is required to fill a 14.88 L bottle to a pressure of 199 kPa at 25°C?

Solutions are at the bottom of the page.

Answers to Practice 10G

1. $n = \dfrac{500 \text{ g } CO_2}{44.01 \text{ g/mol } CO_2} = 11.4 \text{ mol } CO_2$

$T = 25°\text{C} = 298 \text{ K}$

$V = nRT/P$

$$= \frac{(11.4 \text{ mol})\left(8.314 \frac{\text{kPa•L}}{\text{mol•K}}\right)(298 \text{ K})}{975 \text{ kPa}} =$$

29.0 L of CO_2

2. $T = 0°\text{C} = 273 \text{ K}$

$P = 101.3 \text{ kPa}$

$V = nRT/P$

$$= \frac{(1.00 \text{ mol})\left(8.314 \frac{\text{kPa•L}}{\text{mol•K}}\right)(273 \text{ K})}{101.3 \text{ kPa}} =$$

22.4 L

3. $T = 37°\text{C} = 310 \text{ K}$

$n = PV/RT$

$$= \frac{(110 \text{ kPa})(4.0 \text{ L})}{\left(8.314 \frac{\text{kPa•L}}{\text{mol•K}}\right)(310 \text{ K})} = 0.17 \text{ mol}$$

4. $n = \dfrac{40 \text{ g } CH_4}{16.05 \text{ g/mol } CH_4} = 2.5 \text{ mol } CH_4$

$T = 200°\text{C} = 473 \text{ K}$

$V = 2500 \text{ mL} = 2.500 \text{ L}$

$P = nRT/V$

$$= \frac{(2.5 \text{ mol})\left(8.314 \frac{\text{kPa•L}}{\text{mol•K}}\right)(473 \text{ K})}{2.500 \text{ L}} =$$

3.9×10^3 kPa

5. $V = 25.0 \text{ m}^3 = 25.0 \times 10^3 \text{ L}$

$T = PV/nR$

$$= \frac{(1.2 \text{ atm})(25.0 \times 10^3 \text{ L})}{(7.0 \text{ mol})\left(0.0821 \frac{\text{atm•L}}{\text{mol•K}}\right)} =$$

5.2×10^4 K

Solutions for Additional Examples 10G

a. $T = 28°\text{C} = 301 \text{ K};\ n = PV/RT = \dfrac{(149 \text{ kPa})(15.9 \text{ L})}{\left(8.314 \frac{\text{kPa•L}}{\text{mol•K}}\right)(301 \text{ K})} = 0.947 \text{ mol}$

b. $T = 25°\text{C} = 298 \text{ K};\ n = PV/RT = \dfrac{(199 \text{ kPa})(14.88 \text{ L})}{\left(8.314 \frac{\text{kPa•L}}{\text{mol•K}}\right)(298 \text{ K})} = 1.20 \text{ mol}$

mass of NH_3 = number of moles of NH_3 × molar mass of NH_3 = (1.20 mol)(17.04 g/mol) = 20.4 g

Additional Examples for Sample Problem 10H

a. A gas has a volume of 18.5 L at 85.5 kPa and 296 K. What is the volume of the gas at *STP*?

b. A balloon full of helium gas has a volume of 132 L at 99.7 kPa and 30°C. What temperature is required for the balloon to have a volume of 176 L at a pressure of 77.6 kPa?

c. A bread bag is inflated to a volume of 3.89 L at 111 kPa and 23°C. If the volume drops to 3.05 L at a temperature of 4°C, what is the new pressure?

Solutions are at the bottom of the page.

Answers to Practice 10H

1. $\frac{P_1V_1}{T_1} = \frac{P_2V_2}{T_2} =$

$\frac{(100.\text{ kPa})(2.0\text{ m}^3)}{100.\text{ K}} = \frac{(200.\text{ kPa})V_2}{400.\text{ K}}$

$V_2 = 4.0\text{ m}^3$

2. $T_1 = 25°\text{C} = 298\text{ K}$

$T_2 = 225°\text{C} = 498\text{ K}$

$\frac{P_1V_1}{T_1} = \frac{P_2V_2}{T_2} = \frac{(900\text{ kPa})(8.00\text{ L})}{298\text{ K}}$

$= \frac{P_2(2.00\text{ L})}{498\text{ K}}$

$P_2 = 6.02 \times 10^3\text{ kPa}$

3. $\frac{P_1V_1}{T_1} = \frac{P_2V_2}{T_2} =$

$\frac{(500.\text{ Pa})(850.\text{ mL})}{500.\text{ K}} = \frac{(200.\text{ Pa})(700.\text{ mL})}{T_2}$

$T_2 = 165\text{ K}$

4. $\frac{P_1V_1}{T_1} = \frac{P_2V_2}{T_2} = \frac{(500\text{ kPa})(45\text{ m}^3)}{750\text{ K}} =$

$\frac{(101.3\text{ kPa})V_2}{273\text{ K}}$

$V_2 = 81\text{ m}^3$

Sample Problem 10H
Using the combined gas law

A helium balloon with a volume of 410. mL is cooled from 27°C to –27°C. The pressure on the gas is reduced from 110. kPa to 25. kPa. What is the volume of the gas at the lower temperature and pressure?

❶ List

- $P_1 = 110.\text{ kPa}$ $\quad P_2 = 25.\text{ kPa}$
- $V_1 = 410.\text{ mL}$ $\quad V_2 = ?$
- $T_1 = 27°\text{C} = 300\text{ K}$ $\quad T_2 = -27°\text{C} = 246\text{ K}$
- **The temperature decreases, so the volume should decrease; at the same time, the pressure decreases, so the gas expands. The effects are opposites but do not necessarily cancel each other completely.**

❷ Set up

- **Because the amount of gas does not change, use the combined gas law equation.**

$$\frac{P_1V_1}{T_1} = \frac{P_2V_2}{T_2}$$

$$\frac{(110.\text{ kPa})(410.\text{ mL})}{(300\text{ K})} = \frac{(25.\text{ kPa})(V_2)}{(246\text{ K})}$$

❸ Estimate and verify

- **The pressure ratio is larger than the temperature ratio, so the pressure reduction will have a greater effect on the volume than the increase in temperature. The new volume should be larger than 400 mL.**

$$V_2 = 410.\text{ mL} \times \frac{246\text{ K}}{300\text{ K}} \times \frac{110.\text{ kPa}}{25.\text{ kPa}}$$

$$V_2 = 1.5 \times 10^3\text{ mL}$$

Practice 10H

1. A gas occupies 2.0 m³ at 100. K, exerting a pressure of 100. kPa. What volume would the gas occupy at 400. K if the pressure is increased to 200. kPa?
2. An 8.00 L sample of neon gas at 25°C exerts a pressure of 900 kPa. If the gas is compressed to 2.00 L and the temperature is raised to 225.°C, what will the new pressure be?
3. A sample of methane that initially occupies 850. mL at 500. Pa and 500. K is compressed to a volume of 700. mL. To what temperature will the gas need to be cooled to lower the pressure of the gas to 200. Pa?
4. A sample of carbon dioxide occupies 45 m³ at 750 K and 500 kPa. What is the volume of this gas at *STP*?

Gas stoichiometry

Gas volumes can be determined from mole ratios in balanced equations

Avogadro's work showed us that the mole ratio of two gases at the same temperature and pressure is the same as the volume ratio. This relationship greatly simplifies the calculation of the volume of product or reactant in a chemical reaction involving gaseous reactants and products. For example, how many moles of hydrogen would react with 4 mol of nitrogen according to the following equation?

$$3H_2(g) + N_2(g) \longrightarrow 2NH_3(g)$$

Solutions for Additional Examples 10H

a. $\frac{P_1V_1}{T_1} = \frac{P_2V_2}{T_2} = \frac{(85.5\text{ kPa})(18.5\text{ L})}{296\text{ K}} = 5.34\text{ kPa•L/K} = \frac{(101.3\text{ kPa})V_2}{273\text{ K}}$, or $V_2 = \frac{(5.34\text{ kPa•L/K})(273\text{ K})}{101.3\text{ kPa}} = 14.4\text{ L}$

b. $T_2 = 314\text{ K} = 41°\text{C}$

c. $P_2 = 132\text{ kPa}$

*Full solutions for items **b** and **c** are found on page 401A.*

The mole ratio of hydrogen to nitrogen from the chemical reaction is 3 to 1, so the proportion is 3 mol H_2 to 1 mol N_2.

$$\frac{3 \text{ mol } H_2}{1 \text{ mol } N_2} = \frac{x \text{ mol } H_2}{4 \text{ mol } N_2} \qquad x = 12 \text{ mol } H_2$$

This ratio is also equivalent to the volume ratio. If the question had been, What volume of hydrogen would react with 4 liters of nitrogen? the approach would be the same. Because temperature and pressure are not specified, it is assumed that they are constant and that the mole ratio will be the same as the volume ratio.

$$\frac{3 \text{ mol } H_2}{1 \text{ mol } N_2} = \frac{x \text{ L } H_2}{4 \text{ L } N_2} \qquad x = 12 \text{ L } H_2$$

Sample Problem 10I
The ideal gas equation and stoichiometry

How many liters of hydrogen gas will be produced at 280. K and 96.0 kPa if 40.0 g of sodium react with excess hydrochloric acid according to the following equation?

$$2Na(s) + 2HCl(aq) \longrightarrow 2NaCl(aq) + H_2(g)$$

1 List

- V = ?
- P = 96.0 kPa
- n = to be calculated from mole ratio by converting 40.0 g Na to moles
- **molar mass of Na** = 22.99 g/mol
- R = 8.314 L•kPa/mol•K
- T = 280. K

2 Set up

- **First, use the mole ratio in the chemical equation to determine the number of moles of hydrogen that can be produced. Second, use the ideal gas law to calculate the volume of the hydrogen under the temperature and pressure conditions given. The mole ratio between sodium and hydrogen is 2 to 1.**

$$\frac{40.0 \text{ g Na}}{22.99 \text{ g/mol}} = 1.74 \text{ mol Na}$$

$$1.74 \text{ mol Na} \times \frac{1 \text{ mol } H_2}{2 \text{ mol Na}} = 0.870 \text{ mol } H_2$$

- **Then use the ideal gas equation.**

$$(96.0 \text{ kPa})(V) = (0.870 \text{ mol } H_2)(8.314 \text{ L·kPa/mol·K})(280. \text{ K})$$

3 Estimate and verify

$$V = \frac{(0.870 \text{ mol } H_2)(8.314 \text{ L·kPa/mol·K})(280. \text{ K})}{(96.0 \text{ kPa})}$$

$$V = 21.1 \text{ L of } H_2$$

- **The answer of about 20 L is reasonable because it is close to the volume of 1 mol at *STP*.**

Theme
Systems and interactions

The total system of a gas can be described by pressure, volume, temperature, and number of gas particles (moles).

Content Background
Gay-Lussac's law of combining volumes of gases

This law states that at constant temperature and pressure, gaseous reactants and products can be expressed in small, whole-number ratios.

Additional Example for Sample Problem 10I

In the combustion reaction of 149 g of propane with excess oxygen, what volume of carbon dioxide is produced at *STP*?

$$C_3H_8 + 5O_2 \longrightarrow 3CO_2 + 4H_2O$$

Solution is at the bottom of the page.

Solution for Additional Example 10I

The mole ratio of CO_2 to C_3H_8 is 3:1.

$$n = \frac{149 \text{ g } C_3H_8}{44.11 \text{ g/mol } C_3H_8} \times \frac{3 \text{ mol } CO_2}{1 \text{ mol } C_3H_8} = 10.1 \text{ mol } CO_2$$

$$V = nRT/P = \frac{(10.1\text{mol})\left(8.314\frac{\text{kPa·L}}{\text{mol·K}}\right)(273 \text{ K})}{101.3 \text{ kPa}} = 226 \text{ L of } CO_2$$

Additional Example for Sample Problem 10J

A student wishes to prepare oxygen by using the thermal decomposition of potassium chlorate, $KClO_3$. Knowing that the gas will have a temperature of 700°C and a pressure of 98.6 kPa, how much potassium chlorate will be necessary to produce 125 mL of oxygen?

$$2KClO_3 \longrightarrow 2KCl + 3O_2$$

Solution is at the bottom of the page.

Answers to Practice 10I and 10J

1. Assume that the pressure and temperature for each gas is the same.

$$\frac{V_1}{V_2} = \frac{n_1}{n_2} = \frac{1 \text{ mol } H_2}{2 \text{ mol HCl}} = \frac{3.0 \text{ L}}{V_2}$$

$V_2 = 2 \times 3.0 \text{ L} = 6.0 \text{ L}$

2. a. $\frac{V_1}{V_2} = \frac{n_1}{n_2} = \frac{5 \text{ mol } O_2}{4 \text{ mol } NH_3} = \frac{V_1}{15 \text{ L}}$

$V_1 = (5/4)(15 \text{ L}) = 19 \text{ L } O_2$

b. $\frac{n_1}{n_2} = \frac{6 \text{ mol } H_2O}{4 \text{ mol } NH_3} = \frac{n_1}{3.5 \text{ mol } NH_3}$

$n_1 = 5.25 \text{ mol } H_2O$

c. $n_{NH_3} = PV/RT =$

$$\frac{(100 \text{ kPa})(15 \text{ L})}{\left(8.314 \frac{\text{kPa}\cdot\text{L}}{\text{mol}\cdot\text{K}}\right)(350 \text{ K})} = 0.52 \text{ mol}$$

n_{H_2O} produced in reaction =

$$0.52 \text{ mol } NH_3 \times \frac{6 \text{ mol } H_2O}{4 \text{ mol } NH_3} =$$

0.78 mol

mass of H_2O produced =

(0.78 mol)(18.02 g/mol) = 14 g

Answers are continued on page 401A.

Sample Problem 10J
The ideal gas equation and stoichiometry

Magnesium metal will "burn" in carbon dioxide to produce elemental carbon and magnesium oxide.

$$2Mg(s) + CO_2(g) \longrightarrow 2MgO(s) + C(s)$$

What mass of magnesium will react with a 250 mL container of carbon dioxide gas at 77°C and 65 kPa?

1 List

- P = 65 kPa
- V = 250 mL = 0.25 L
- n = ?
- R = 8.314 L•kPa/mol•K
- t = 77°C, therefore, T = 350 K
- **molar mass of Mg** = 24.30 g/mol

2 Set up

- **Like Sample Problem 10I, this is a two-part problem. Start by determining the amount of CO_2 in the container using the ideal gas equation. Then you must use the mole ratio and formula mass of magnesium to calculate the mass of magnesium needed. The amount of CO_2 is calculated from the ideal gas equation.**

$PV = nRT$

$(65 \text{ kPa})(0.25 \text{ L}) = (n)(8.314 \text{ L}\cdot\text{kPa/mol}\cdot\text{K})(350 \text{ K})$

$$n = \frac{(65 \text{ kPa})(0.25 \text{ L})}{(8.314 \text{ L}\cdot\text{kPa/mol}\cdot\text{K})(350 \text{ K})}$$

$n = 5.6 \times 10^{-3} \text{ mol } CO_2$

3 Estimate and verify

$$5.6 \times 10^{-3} \text{ mol } CO_2 \times \frac{2 \text{ mol Mg}}{1 \text{ mol } CO_2} \times \frac{24.30 \text{ g Mg}}{1 \text{ mol Mg}} = 0.27 \text{ g Mg}$$

Practice 10I

1. **The formation of hydrogen chloride occurs in the following reaction.**

$$H_2(g) + Cl_2(g) \longrightarrow 2HCl(g)$$

If 3.0 L of H_2 are used, how many liters of HCl can be produced? What do you have to assume to answer this question?

2. **Ammonia and oxygen react to form nitrogen monoxide and water, as shown by the following equation.**

$$4NH_3(g) + 5O_2(g) \longrightarrow 4NO(g) + 6H_2O(l)$$

Assume that the temperature and pressure stay constant at 350. K and 100. kPa.
a. What volume of O_2 is needed to burn 15 L of ammonia?
b. How many moles of H_2O are produced when 3.50 mol of NH_3 burn?
c. How many grams of H_2O are produced when 15 L of NH_3 burn?

10J

3. **Solid LiOH has been used in spacecraft to remove exhaled CO_2 from the environment as shown by the following equation.**

$$2LiOH(s) + CO_2(g) \longrightarrow Li_2CO_3(s) + H_2O(l)$$

How many grams of LiOH must be used to absorb the carbon dioxide that exerts a partial pressure of 5.0 kPa at 15°C in a space laboratory that is 4.0 m × 2.5 m × 8.0 m?

Solution for Additional Example 10J

mass of $KClO_3$ required = 0.124g

A full solution is found on page 401A.

Section Review

23. A sample of carbon dioxide has a mass of 35.0 g and occupies 2.5 L at 400 K. What pressure does the gas exert?
24. What volume is occupied by 0.45 g of nitrogen measured at 100 kPa and 25°C?
25. A sample of nitrogen gas exerts 2.5 atm at 250 K.
    a. What is the density of the gas, expressed in g/L?
    b. What is the density of the gas, in g/L, if the pressure is increased to 5.0 atm?
26. How many moles of sulfur dioxide, SO_2, are contained in a 4.0 L container at 450 K and 5 kPa?
27. During a chemical reaction, nitrogen gas was collected by displacement of water at 17°C. The gas displaced 75.0 mL of water, and the atmospheric pressure was 97.25 kPa.
    a. What was the pressure of the nitrogen alone?
    b. How many moles of nitrogen were collected?
    c. What mass of nitrogen was collected?
28. A 10.0 L tank of helium gas is filled to a pressure of 400 psi. Assuming no loss of gas, how many balloons, each containing 4.0 L of helium at 100 kPa, could you fill?
29. Solid iron(III) hydroxide decomposes to produce iron(III) oxide and water vapor. If 0.75 L of water vapor are produced at 227°C and 1.0 atm,
    a. how many grams of iron(III) hydroxide were used?
    b. how many grams of iron(III) oxide are produced?
30. Assume that 13.5 g of Al(*s*) react with HCl(*aq*) according to the following balanced equation at *STP*.

$$2Al(s) + 6HCl(aq) \longrightarrow 2AlCl_3(aq) + 3H_2(g)$$

    a. How many moles of Al react?
    b. How many moles of H_2 are produced?
    c. How many liters of H_2 are produced at *STP*?
31. Air is 20.9% oxygen by volume.
    a. How many liters of air are needed to complete the combustion of 25.0 L of octane vapor, $C_8H_{18}(g)$?
    b. What volume of each product is produced?
32. Methanol, CH_3OH, is made by causing carbon monoxide and hydrogen gases to react at high temperature and pressure. If 450. mL of CO and 825 mL of H_2 react,
    a. which reactant is in excess?
    b. how much of that reactant remains when the reaction is complete?
    c. what volume of $CH_3OH(g)$ is produced?
33. A 4.0 L tank of helium at the carnival supply store is intended to fill 100 balloons to a volume of 3.0 L each at 25°C and 745 torr. These values take into account a 4% loss of gas when the balloons are being filled. What must the original pressure inside the tank be if it is filled at 0°C?

ASSESS

Alternative Assessment

Divide the class into groups of three, and have them use the concepts of ideal gas behavior to explain the operation of a refrigerator in which the gas, freon, is repeatedly cooled and heated. Library sources may be required.

Answers to Section Review

23. $P = 1.1 \times 10^3$ kPa

24. $V = 0.40$ L

25. a. $D = 3.4$ g/L
b. $D = 6.8$ g/L

26. $n = 5 \times 10^{-3}$ mol

27. a. $P_{H_2O} = 1.94$ kPa
$P_{N_2} = (97.25\text{ kPa} - 1.94\text{ kPa}) = 95.31$ kPa
b. $n_{N_2} = 2.96 \times 10^{-3}$ mol
c. mass of nitrogen = 8.29×10^{-2} g

28. $P = 400\text{ psi} \times \frac{101.3\text{ kPa}}{14.7\text{ psi}} =$ 2.76×10^3 kPa

$P_1V_1 = (2.76 \times 10^3\text{ kPa})(10.0\text{ L}) =$ $2.76 \times 10^4\text{ kPa·L} = P_2V_2 =$ $(100\text{ kPa})(V_2)$

$V_2 = 276$ L total.

The tank cannot be completely emptied, so 10.0 L must remain. The total volume available is 266 L, enough for 66 balloons.

29. a. $2Fe(OH)_3 \longrightarrow Fe_2O_3 + 3H_2O$

Under conditions given, 1.8×10^{-2} mol of H_2O is produced, requiring 1.2×10^{-2} mol of $Fe(OH)_3$, or 1.3 g $Fe(OH)_3$.
b. 0.96 g Fe_2O_3

30. a. 0.500 mol Al reacts
b. 0.750 mol H_2 is produced
c. $V_{STP} = 16.8$ L H_2

Answers are continued on page 401A.

Section 10-4

FOCUS

Lesson Starter

Discuss with the class what assumptions of the kinetic molecular theory do not match what they know about real gases. (Particles do have volume, collisions are inelastic, and there are attractive forces between the particles.) In light of these differences, why is the ideal gas model used at all? (It is easy to apply and allows us to make predictions that under certain conditions, such as low pressure, closely match reality.)

TEACH

Theme

Systems and interactions
Real gases exhibit many interactions between particles, causing deviations from the "ideal" model.

What conditions will cause a gas to condense?

Section Objectives

List the conditions under which gases deviate from ideal behavior.

Relate attractive forces to boiling point, volatility, and surface tension.

Interpret a phase diagram, and describe the significance of the triple point.

Relate volatility and vapor pressure.

Relate vapor pressure and temperature.

Describe sublimation.

Define boiling point in terms of vapor pressure.

Forces of attraction

All the mathematical equations you have used to make predictions of the conditions for gases are based on the kinetic molecular theory. The predictions we make using these mathematical models come relatively close to reality under most conditions. However, gases deviate from ideal behavior when the molecules are close. For example, 1 mol of CO_2 at 100 atm (10 132 kPa) and 50°C has a predicted volume of about 265 mL. However, if this gas is subjected to those conditions in the laboratory, the actual volume is measured at about 130 mL. Why is the actual volume so much smaller than predicted? Let's go back to the assumptions of the kinetic molecular theory of gases presented on page 360. There we identified two assumptions that were not exactly true for real gases.

1. The volume of the gas particles is negligible.
2. There are no attractive forces between particles.

The CO_2 described above is under high pressure and is compressed into a small volume. Similar deviations between predicted and actual gas volumes occur at low temperatures. At low temperatures, gas particles slow down. At high pressures, gas volumes are reduced, the gases still differ from the ideal because of particle size, as shown in **Figure 10-26**. In both cases the conditions are such that particle size and the attractive forces between particles significantly affect particle volume.

If the temperature is low enough and the pressure is high enough, the attractive forces between particles will be so strong that the gas will condense to form a liquid or solid.

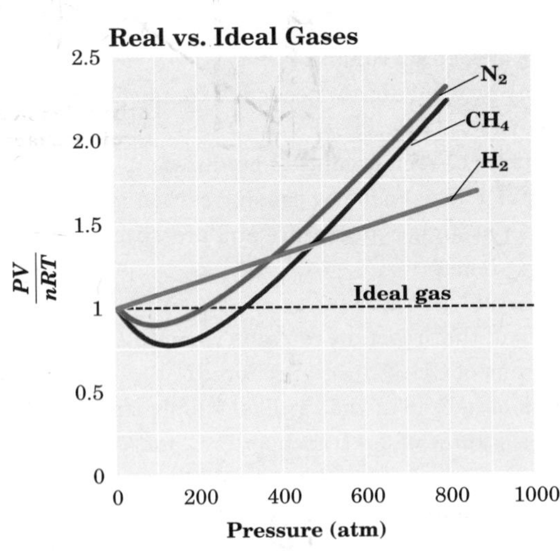

Figure 10-26 The ratio of ***PV/nRT*** for any ideal gas is one, which is represented by the dashed line. Using actual measurements for real gases, you see that they deviate from the ideal.

Vapor pressure increases with temperature

A common gas that condenses easily is water vapor. You can see water vapor condense on any cold surface, such as a window. For example, your mirror fogs after a hot shower, and your glasses fog when you come in from the cold.

Even though water exists as a liquid at room temperature and standard pressure, you know that in an open container, water evaporates over a period of time until it finally "disappears." The water vapor formed during evaporation is like any other gas in that it exerts pressure and expands and contracts with temperature. You may have noticed, however, that in a closed system, water does not appear to evaporate.

In a closed container, water will evaporate until the air in the container is said to be *saturated* with water vapor. When the air is saturated with vapor, the vapor condenses to a liquid as fast as the liquid evaporates, and the two processes of evaporation and condensation continue at equal rates.

$$\text{evaporation} \rightleftharpoons \text{condensation}$$

equilibrium vapor pressure
a measure of the tendency of the particles of a liquid substance to enter the gas phase at a given temperature

This equation represents an equilibrium system. The pressure due to the water vapor reaches a maximum value (at a given temperature), called the **equilibrium vapor pressure.** The equilibrium vapor pressure at any given temperature is an actual measurement taken from an apparatus like the one shown in **Figure 10-27**.

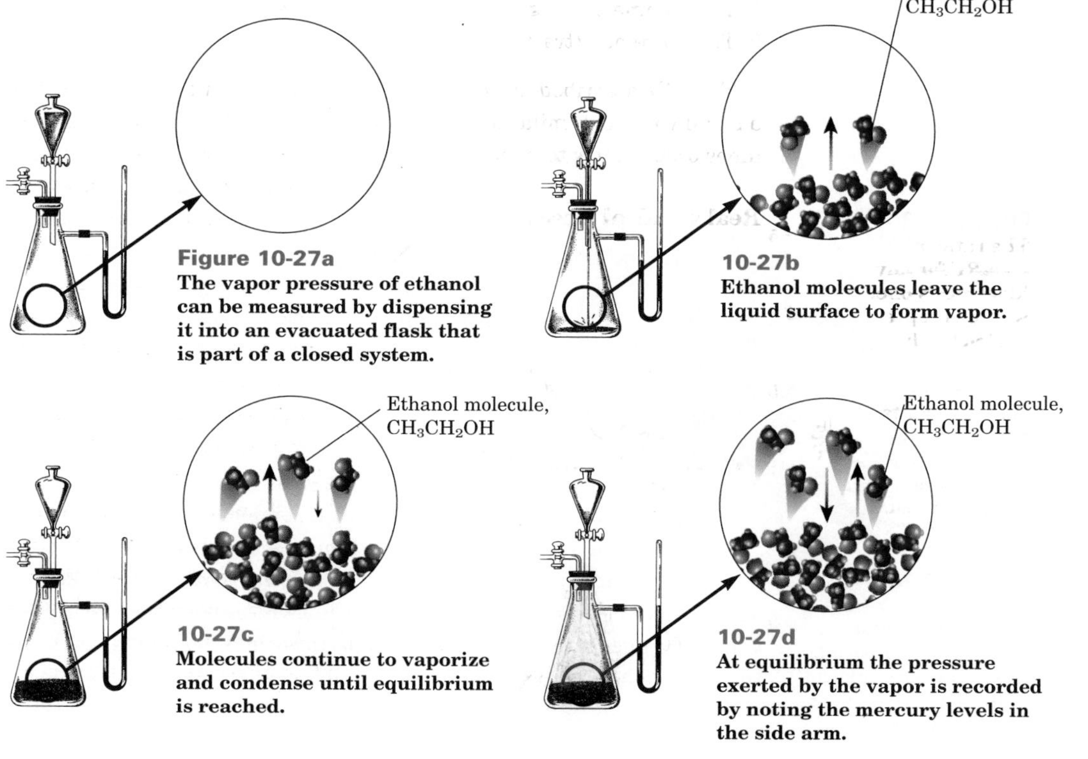

Figure 10-27a The vapor pressure of ethanol can be measured by dispensing it into an evacuated flask that is part of a closed system.

10-27b Ethanol molecules leave the liquid surface to form vapor.

10-27c Molecules continue to vaporize and condense until equilibrium is reached.

10-27d At equilibrium the pressure exerted by the vapor is recorded by noting the mercury levels in the side arm.

Visual Strategy

Figure 10-27

Make students aware that the U-shaped bend in the glass is the manometer discussed earlier in the chapter. The vacuum in the flask removes any gases or vapors that are initially present so that they do not contribute to the pressure (recall Dalton's law of partial pressure).

Demonstration 9

Boiling water in a vacuum

Approximate time: 10 min

1. Halfway fill a 250 mL Erlenmeyer flask with warm water, and place a nonmercury thermometer in it.
2. Place the flask on top of the vacuum-pump platform, position it so that the thermometer is visible, and cover with a bell jar. You may wish to put some paper towels on the bottom of the platform in case some water bumps out during boiling.
3. Ask a student to come forward and read the thermometer.
4. Run the vacuum pump until the water boils.
5. Ask the student to read the thermometer again. Why does the water continue to cool? (Evaporation is a cooling process.)
6. Shut down the vacuum pump. Before you remove the flask ask whether or not it is hot. Remind the students that the concepts of "boiling" and "heat" do not necessarily go together.

SAFETY
The teacher must wear a lab apron, and everyone must wear safety goggles in case the bell jar shatters under the vacuum. Students must be at least 10 ft away from the demonstration.

DISPOSAL
Pour water down the sink.

Teaching Tip

Discussion

Make students aware that water boils below 100°C in the mountains because the air pressure is lower, which means that the water does not have to get as hot to boil. This is also why food takes longer to cook at high altitudes.

Other Applications
The inverse of the previous demonstration is the pressure cooker. The increase in pressure, as shown in **Figure 10-28**, causes an increase in boiling point. Once the water in the cooker boils, its high temperature causes the food to cook faster.

Teaching Tip
Discussion
The following is a description of what occurs in a closed soda bottle. The liquid is constantly evaporating, while the gas is constantly condensing. When the lid is removed, the system tries to reach equilibrium with the conditions in the room, and more and more liquid evaporates in order to reach equilibrium.

Content Background
Volatile substances
Volatile liquids or solids require only small amounts of energy to enter the gaseous phase.

As the temperature of the water increases, its maximum vapor pressure increases. **Figure 10-28** shows how vapor pressure increases with temperature. When the vapor pressure reaches the atmospheric pressure, the liquid begins to boil. The temperature at which boiling occurs is called the boiling point.

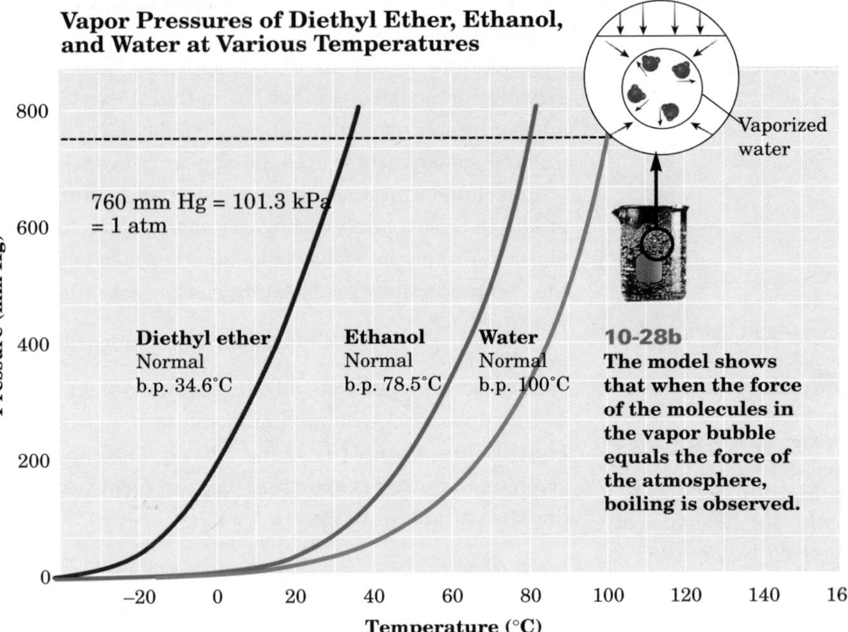

Figure 10-28a
A liquid boils when its vapor pressure equals the pressure of the atmosphere. If the pressure in a room is about 1 atm (760 mm Hg) each liquid boils at its normal boiling point.

Vapor Pressures of Diethyl Ether, Ethanol, and Water at Various Temperatures

10-28b
The model shows that when the force of the molecules in the vapor bubble equals the force of the atmosphere, boiling is observed.

Volatile substances have high vapor pressures

volatile
a term used to describe a substance that is readily vaporized at low temperature

Like density, vapor pressure is a characteristic property of a substance. Substances that have high vapor pressures at low temperatures are described as **volatile**. You can generally recognize volatile substances by odor because they vaporize easily. For example, gasoline, turpentine, moth balls, acetone (the substance in nail-polish remover), and ethanol are volatile substances.

Perfumes include volatile substances to carry the scent. Perfumes can contain up to 90% alcohol, which evaporates very quickly. The essence of most perfumes is a blend of several ingredients with different volatilities to make the scent last over time. Of these ingredients, citrus oils, herbs, and flower extracts tend to have the lowest molecular masses and are the most volatile, so you smell them first. Flower oils from roses, lilies, violets, and jasmine have higher molecular masses and give the perfume a longer lasting scent.

As you might expect, substances that vaporize easily have weaker attractive forces between particles than substances with low vapor pressures. Stronger attractive forces keep particles in a condensed state, resulting in lower vapor pressures.

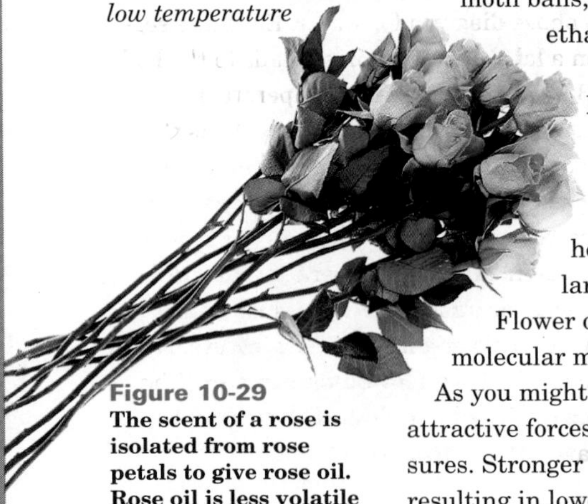

Figure 10-29
The scent of a rose is isolated from rose petals to give rose oil. Rose oil is less volatile than some other scents and thus lasts longer.

Visual Strategy
Figure 10-28
Emphasize that boiling can occur anywhere along the line between liquid and vapor. If the pressure is below the line or the temperature is far enough to the right of the line, then the liquid will boil instantaneously.

Figure 10-30 The phase diagram for water can be used to predict the physical state of water at different pressures and temperatures. Note that this diagram is not drawn to scale.

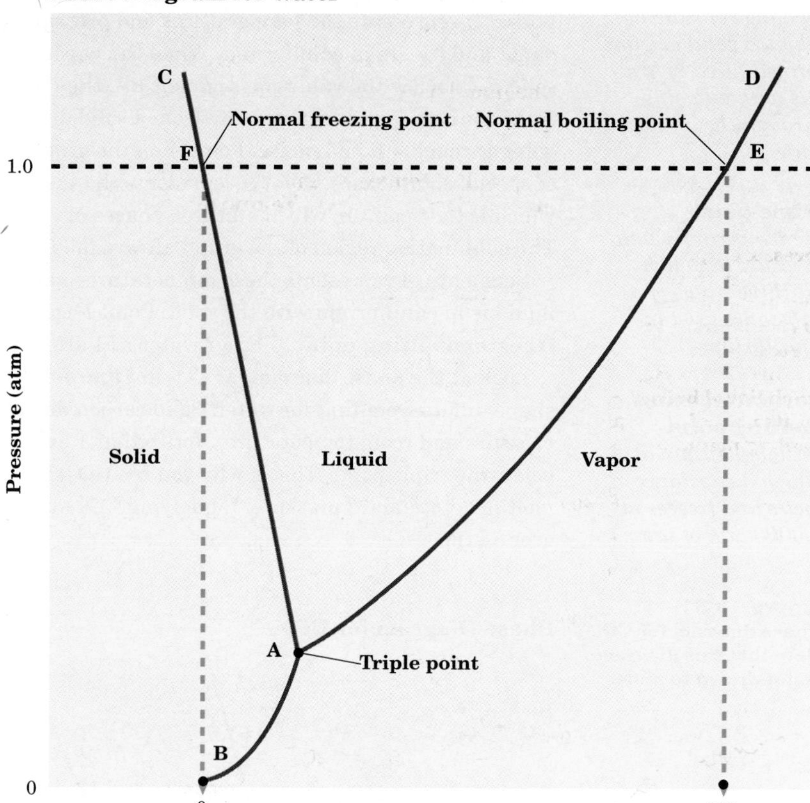

Content Background

Phase diagrams

Phase diagrams can be used as models to predict the properties and phases of substances. They can be determined for most chemical substances.

Phase diagrams relate temperature and pressure to physical state

phase diagram *a graphic representation of the relationships between the physical state of a substance and its pressure and temperature*

normal boiling point *the temperature at which a substance boils at 1.0000 atm of pressure*

A **phase diagram** is a graph that shows the pressures and temperatures at which different phases of a substance are at equilibrium with each other. Look at Figure 10-30, the phase diagram for water. Each line segment on the graph has been given a label. Point *E* corresponds to the boiling point of water at 1 atm (101.3 kPa). At the boiling temperature, the liquid and vapor are in equilibrium with each other. Segment *AD* is the vapor pressure–temperature curve for the liquid phase. Any point along segment *AD* represents the change of state from a liquid to a vapor and the boiling point of water at the corresponding pressure. You can now see why boiling point is more correctly defined as the temperature at which the vapor pressure equals the external or atmospheric pressure. If the pressure is greater than 1 atm, water boils at a higher temperature. If the pressure is less than 1 atm, water boils at a lower temperature. The **normal boiling point** (which you see listed in reference tables) represents the boiling point of a substance at 1 atm, or 101.3 kPa.

Visual Strategy

Figure 10-30

Stress that the phase diagram shown consists of just two variables: pressure and temperature. Explain that many phase diagrams include volume as a third variable and that they are depicted either as a series of three two-dimensional diagrams, each showing the phase changes for any two of the variables, or as a three-dimensional representation, which is called a *PVT* diagram.

Other Applications

Other good examples of materials that sublimate include iodine crystals, mothballs, and moth flakes. Moth flakes are used in urinals to mask odors. They do not dissolve because they are very nonpolar. Instead they slowly sublimate and freshen the air.

Possible Misconception

Many students will have trouble understanding that all three states of matter can exist at the same time at the triple point. Suggest that critical points be compared to the apex of a pyramid, where a step in any direction places one on a different wall of the structure. At the triple point, a shift in any direction will result in a different physical state of the substance.

sublimation
a change of state in which a solid is transformed directly to a gas without going through the liquid state

triple point
the temperature and pressure at which all three states of a substance exist in equilibrium

normal freezing/ melting point
the temperature at which a substance melts and freezes at 1.0000 atm of pressure

Segment *BA* is the vapor pressure–temperature curve for the solid phase. It represents the temperatures and pressures at which water vapor and ice are in equilibrium. When the vapor pressure above the solid falls below the values on segment *BA*, then ice will sublime. **Sublimation** is a change of state from a solid directly to a gas without going through a liquid phase. Point *A* on the graph denotes coordinates of special significance. Point *A* represents the **triple point** of water—which is that point in which all three phases of water exist in equilibrium. The sublimation region of the graph always falls below the triple point.

Segment *CA* represents those temperatures and pressures at which the liquid is in equilibrium with the solid. Point *F* corresponds to the **normal freezing/melting point**, 0°C, for water at 1 atm (101.3 kPa).

Look at the phase diagram for CO_2 in **Figure 10-31**. Note how the shape differs from that for water. Solid carbon dioxide sublimes at normal pressure and room temperature. Notice that 1 atm on the *y*-axis falls well below the triple point. This is why you see CO_2 subliming rather than melting at standard pressure. Liquefying CO_2 would require a pressure of over 5 atm.

Figure 10-31
Phase diagram for CO_2. Note that this diagram is not drawn to scale.

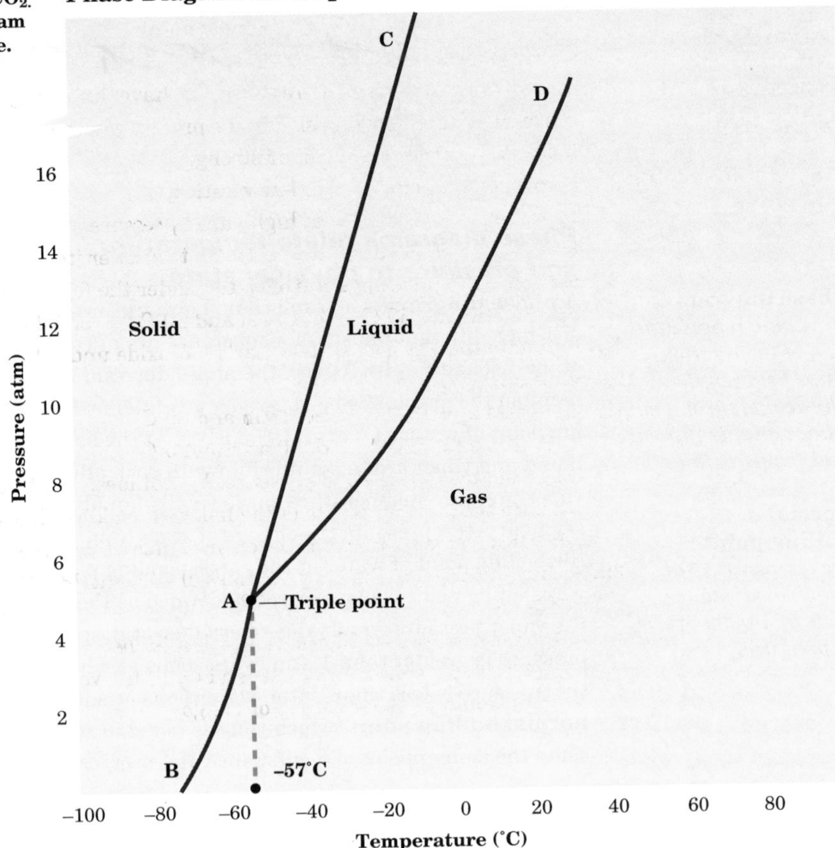

surface tension
the measure of a liquid's tendency to decrease its surface area to a minimum

Attractive forces in liquids result in surface tension

The forces between particles in liquids and solids are stronger than those in gases. The forces among particles on the surface are especially interesting. In **Figure 10-32** you see an example of the strength of these forces shown by the **surface tension** of water. The model shows that particles within the liquid are affected by forces of attraction from all sides. In contrast, particles on the surface are pulled only by particles in the same plane and below. Because the forces among particles at the surface are distributed unevenly, the surface can support objects that exert a small force over a large area.

Figure 10-32a
The weight of the needle is supported by the surface tension of . . .

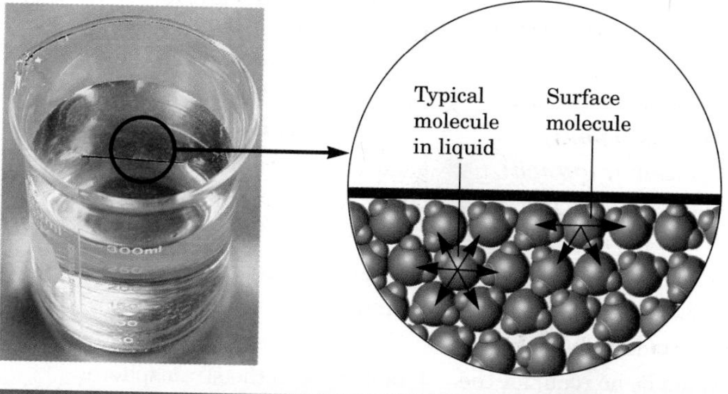

10-32b
. . .water molecules at the surface which are pulled inward and closer together by the unequal distribution of forces.

Section Review

34. Under what conditions do real gases behave differently than ideal gases?

35. For each of the following situations predict whether the attractive forces between particles are weak or strong.

a. low boiling point **b.** low volatility
c. low surface tension **d.** high vapor pressure

Use **Figure 10-30** and **Figure 10-31** to answer items **36** and **37**.

36. Estimate the physical state of water under the following conditions.

a. 0.75 atm and 50°C **b.** 1 atm and 80°C **c.** 0.25 atm and 100°C

37. Estimate the physical state of carbon dioxide under the following conditions.

a. 10 atm and 50°C **b.** 1 atm and 80°C
c. 10 atm and –80°C **d.** 10 atm and 0°C

38. When heated at standard pressure, I_2 sublimes. What does this mean? Look up the melting points of the halogens in a handbook. Explain the differences in melting point in terms of attractive forces.

39. A chunk of dry ice, CO_2, with a mass of 400 g at 78.5°C, sublimes to a gas at 1.0 atm.

a. If the pressure is held constant, what is the volume of CO_2 at 22°C?
b. If the dry ice sublimed in a trash bag with a volume of 25 L at 22°C, what would be the pressure of the gas?

Content Background

Surface tension

Mercury's surface tension is so great that it will form small spheres that roll easily and are very difficult to pick up.

ASSESS

Alternative Assessment

Have students develop a flowchart showing how to solve gas-related problems when given different starting information. Be sure that the charts include pressure, volume, temperature, number of moles and molecules, and stoichiometry.

Answers to Section Review

34. When particles are close to each other, as when a gas is at high pressure or low temperature, their behavior deviates from that of an ideal gas.

35. a. weak attraction (easily evaporated)
b. strong attraction (easily condensed)
c. weak attraction (easily "spread out" in liquid phase)
d. weak attraction (easily evaporated)

36. a. liquid
b. liquid
c. vapor

37. a. gas
b. gas
c. solid
d. gas

38. Iodine goes directly from a solid state to a gaseous one. Melting points increase as atomic radii increase. In the solid state the larger halogen molecules experience a stronger attractive force than do the smaller molecules and thus require higher temperatures for melting.

39. a. n for 400 g CO_2 = 9.09 mol

$T = 22°C = 295\ K$

$PV = nRT = 220\ atm{\cdot}L$

$V = 220\ L$

b. $P_1V_1 = P_2V_2$

$$P_2 = \frac{(1.0\ atm)(220\ L)}{25\ L} = 8.8\ atm$$

Visual Strategy

Figure 10-32

Emphasize that the slightly unbalanced forces found among the molecules near and at the surface allow them to support the needle's weight.

Chemistry in Your Community

Visit a neighborhood automotive service station, dealership, or repair shop, and observe the precautions and techniques necessary for the removal and refilling of R-12, R-134a (the new replacement refrigerant), and other refrigeration products for automobile air-conditioning systems. Find out where the company sends the chemicals for recycling or disposal. Inquire about the amount of paperwork required by the federal and state governments. Find out what effects the phasing-out of chlorofluorocarbons (CFCs) has had on business. Outline the differences between the R-12 refrigeration systems and the new refrigeration systems using R-134a.

Answers to Applying Concepts

1. molecular mass of CCl_2F_2 = 120.91 g/mol

   molecular mass of Cl_2 = 70.90 g/mol

   mole fraction of Cl_2 = 70.90/120.91 = 0.5864
   mass of Cl_2 molecules = 0.5864 × 365 g = 214 g

2. $n_{CCl_2F_2} = \frac{365\text{ g}}{120.91\text{ g/mol}} = 3.02\text{ mol}$,

   number of CCl_2F_2 molecules =

   $$\left(\frac{6.022 \times 10^{23}\text{ molecules}}{1\text{ mol}}\right)(3.02\text{ mol})$$

   $= 1.82 \times 10^{24}$ CCl_2F_2 molecules, or 3.64×10^{24} Cl atoms available for reaction with ozone

3. number of O_3 molecules destroyed by 365 g of CCl_2F_2 = $(10\,000)(3.64 \times 10^{24})$ = 3.64×10^{28}

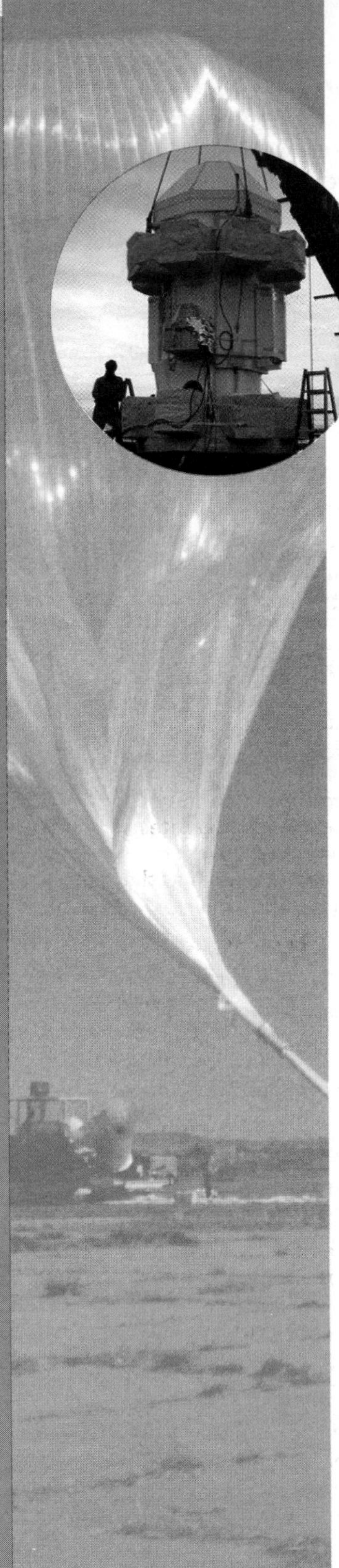

Conclusion: Ascent Into the Ozone

In addition to collecting data on the ozone layer, balloons have been used to chart the Milky Way galaxy, collect dust from comets and meteorites, and photograph the sun. New balloons are being designed to lift a 100 cm telescope into the atmosphere to explore the center of our galaxy. These new balloons have volumes of over 1 million m^3 and are able to lift payloads of over 3500 kg.

Why is the balloon only partially inflated at launch? The volume of the balloon increases as it ascends into the atmosphere and the pressure on the balloon decreases. If the balloon were completely inflated at the launch site, there would be no room for the expanding helium as the balloon ascended, and the pressure of the gas inside the balloon would cause it to burst.

How do scientists know how much to inflate the balloon? They calculate the mass and volume of gas needed at launch based on the volume they want the balloon to have at its cruising altitude.

Why are scientists monitoring the ozone layer? Destruction of the ozone layer increases the percentage of UV rays that strike Earth's surface, which can result in an increased incidence of mutations and skin cancers.

What gases are thought to be responsible for changes in the ozone layer? Chlorofluorocarbons, or CFCs, break down in the stratosphere to form chlorine that reacts with ozone to form oxygen gas. One chlorine atom is capable of reacting with as many as 10 000 ozone molecules. Halons, which are organic compounds containing bromine and nitrogen oxides, have also been shown to react with ozone molecules in the stratosphere.

What's being done to protect the ozone layer? The Montreal Protocol stipulates the following.

- 100% phaseout of CFCs, CCl_4, and CH_3CCl_3 by 1996
- Production freeze of CFCs by 1996; 100% phaseout by 2030
- Halons phased out in 1994
- Production freeze of methyl chloride for 1995.

Applying Concepts

1. The refrigerant released when fixing a typical automobile-air-conditioning system contains CCl_2F_2. If 365 g of CCl_2F_2 are released into the atmosphere from one air-conditioning unit, calculate the mass of chlorine atoms made available to react with ozone.
2. Calculate the number of chlorine atoms released in that 365 g sample.
3. If one chlorine atom can destroy up to 10 000 ozone molecules, calculate the number of ozone molecules destroyed by the chlorine atoms from item **2**.

Research and Writing

Use the library to find out more about the following.

1. Hydrochlorofluorocarbons, HCFCs, are being considered as alternatives to CFCs. HCFCs will react in the troposphere before reaching the stratosphere and the ozone layer. Find out why these compounds will also be phased out by the Montreal Protocol.

CHAPTER 10 Highlights

Key terms

absolute temperature scale	Graham's law of effusion	normal freezing point
Avogadro's principle	greenhouse effect	partial pressure
Boyle's law	ideal gas	pascal
Charles's law	ideal gas law	phase diagram
chlorofluorocarbons	kinetic molecular theory	pressure
Dalton's law of partial pressures	molar volume	STP
diffusion	mutagen	sublimation
effusion	normal boiling point	surface tension
equilibrium vapor pressure		triple point
		volatile

Key problem-solving approach: *Using gas laws*

The ideal gas law can be used to solve gas law problems involving a single gas under changing conditions. Match the situation described by the problem with the arrow that represents those conditions. Follow the path to the correct relationship.

Boyle's law

$PV=k$

Charles's law

$\frac{V}{T}=k$

***P* and *V* change**
***n*, *R*, *T* are constant**

Ideal gas law

$PV=nRT$

***T* and *V* change**
***P*, *n*, *R* are constant**

***P*, *V*, and *T* change**
***n* and *R* are constant**

General gas law

$\frac{PV}{T}=k$

Key ideas

What are characteristics of gases?

- Gases have variable volumes and very low densities.
- When released into the atmosphere, some gases lead to environmental problems such as the greenhouse effect and ozone depletion.
- The kinetic molecular theory is a model of gas behavior based on assumptions about a theoretical gas known as an ideal gas.
- The pressure, temperature, volume, and number of moles of a gas are four variables that define a gaseous system.

What behaviors are described by the gas laws?

- Charles's law explains the direct relationship between the volume and temperature of a gas.
- Boyle's law explains the indirect relationship between the volume and pressure of a gas.
- Dalton's law relates the partial pressures of the individual gases in a gas mixture to the total pressure of the mixture.
- Graham's law states that the effusion rates for two gases are inversely proportional to the square roots of their molar masses at the same temperature and pressure.

How do the gas laws fit together?

- The ideal gas law relates pressure, volume, temperature, and moles of a gas.
- The general gas law relates pressure, volume, moles, and temperature of a gas sample under two sets of conditions.

What conditions will cause a gas to condense?

- The behavior of real gases deviates from ideal at very high pressures and very low temperatures.
- The attractive forces between particles in a liquid affect the boiling point, volatility, and surface tension of that liquid.
- A phase diagram, a graph relating the physical state of a substance to its pressure and temperature, shows the triple point, and the conditions that result in sublimation.
- A substance with a high volatility has a high vapor pressure at low temperatures.

Closure Strategy

1. A gas is heated in a closed container. What will happen to the pressure? (It increases.)
2. Why do aerosols coming out of the can feel cold when they hit your skin? (The drop in pressure lowers the temperature.)
3. Have the class explain this situation (it may have happened to them). A student comes home from school and is thirsty for a soda, but it isn't cold and there are no ice cubes. She places the soda in the freezer unit and sets the timer for 45 minutes. After 45 minutes she removes the can from the freezer, shakes it, and determines that the soda is still liquid. Upon opening the can, she finds the soda is frozen. What happened? (The increase in pressure in the can kept the soda from freezing at its normal freezing point of 0°C. When the pressure was lowered suddenly, the temperature fell below the freezing point and the liquid instantly froze.)

1. a. The kinetic molecular theory assumes that the volume of an ideal gas particle is zero.
   b. The motion of a gas particle is constant, rapid, and along a straight line. The direction changes when the gas particle collides with other particles. On the average, the particles in a gas move in all directions.
2. Complete combustion of a fossil fuel yields carbon dioxide and water vapor. Both of these gases are known to absorb and re-emit infrared radiation in the atmosphere, causing atmospheric warming (the greenhouse effect).
3. a. The graph would become broader and lower, and the peak would lie more to the right.
   b. The average molecular speed is about 400 m/s.
   c. No. Graph C is symmetrical and the peak is not broad enough.
4. a. Four variables are temperature, pressure, volume, and amount of gas.
   b. Other variables that affect gases are also affected by temperature.
5. a. The level of mercury in the barometer rises or falls as the atmospheric pressure increases or decreases, respectively.
   b. Mercury is 13.7 times as dense as water, so a mercury column less than a meter high responds to atmospheric pressure as effectively as a column of water that is nearly 14 m in height. This permits a mercury barometer to be of a practical size.
6. All of the gases in **Table 10-1** have a molar volume equal to 22.4 L at *STP.*
7. Air conditioning, refrigeration, and aerosol spray technologies have been changed so that substances other than CFCs are used.

CHAPTER 10

Review and Assess

Describing a gas

REVIEW

1. a. What assumption does the kinetic molecular theory make about the volume of an ideal gas particle?
   b. Describe the motion of an ideal gas particle.
2. How does the combustion of a fossil fuel contribute to the greenhouse effect?
3. Use **Figure 10-9** and the graphs below to answer the following questions.
   a. What would happen to the shape of **Graph A** if the temperature of the gas increased from 300 K to 1000 K?
   b. Estimate the average molecular speed for the gas in **Graph B.**
   c. Is **Graph C** an accurate representation of molecular speed? Explain

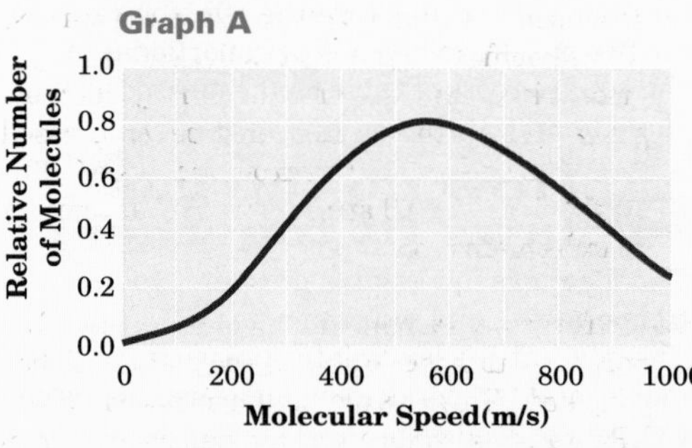

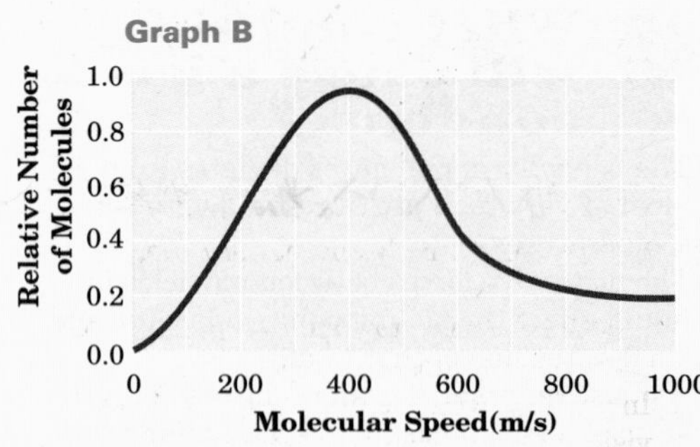

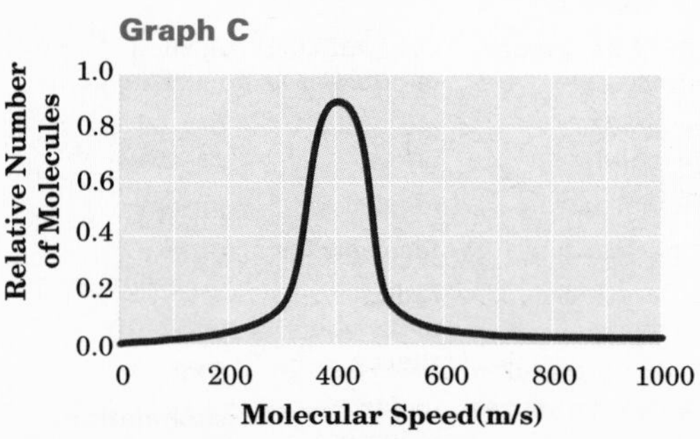

4. a. List the four variables used to describe a gas.
   b. Why must a specific temperature be stated when working with gas data?
5. a. Briefly describe how a barometer works.
   b. Why is mercury used instead of water to measure air pressure?
6. How do the molar volumes of the atmospheric gases in **Table 10-1** compare at STP?

APPLY

7. What technologies have been affected by the Montreal Protocol, which was signed by 74 nations?
8. A ball is dropped from a distance of 2.0 meters above the floor.
   a. If the collision is elastic, how high will the ball bounce?
   b. If the collision is inelastic, will the ball bounce higher or lower than 2.0 meters?
9. Gas companies often store their fuel supplies in liquid form in large storage tanks. Liquid nitrogen is used to keep the temperature low enough for the fuel to remain condensed in liquid form. Although continuous cooling is expensive, storing a condensed fuel as a liquid is still more economical than storing it as a gas. Give one reason why storing a liquid is more economical than storing a gas.

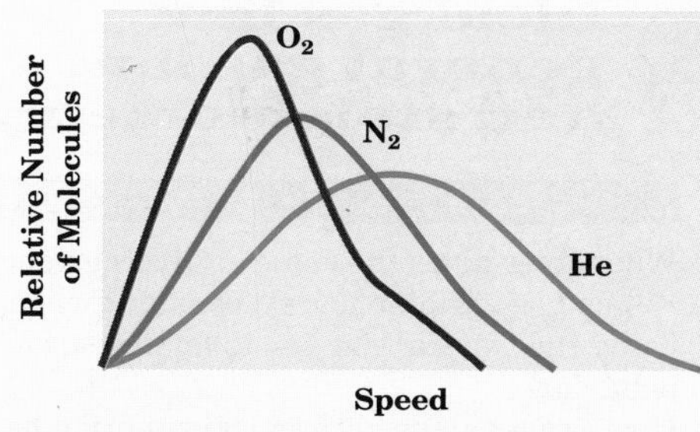

10. Use the graph above to answer the following questions.
   a. If all three gases are at the same temperature, which one has the highest molar mass and which one has the lowest molar mass?
   b. If the three graphs represent one gas at three different temperatures, which represents the lowest temperature and which curve represents the highest temperature?

11. Below is a diagram of two instruments, a mercury barometer and a water barometer. Which instrument is measuring a higher pressure? Explain.

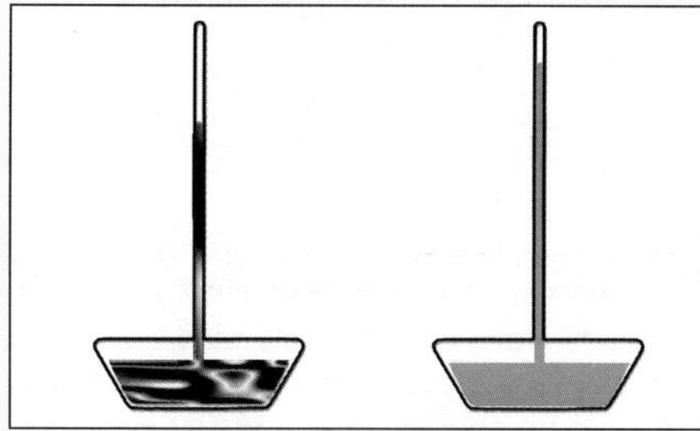

Charles's law and Boyle's law

REVIEW

12. a. Describe the relationship between the volume and temperature of the gas in **Table 10-4**.
   b. List reasons why the values in the V/T column of **Table 10-4** are not exactly the same.
   c. How do the values in the *Temperature (°C)* column compare with those in the *Temperature (K)* column of **Table 10-4**?
   d. Which gas law is verified by the data in **Table 10-4**?

13. What type of relationship between pressure and volume does Boyle's law express?

PRACTICE

Answers to items in black squares begin on page 841.

14. A child receives a balloon filled with 2.30 L of helium from a vendor at an amusement park. The temperature outside is 311 K. What will be the volume of the balloon when the child brings it home to an air-conditioned house at 295 K? (Hint: see Sample Problem 10A.)

15. A 1.5 L pocket of air with a temperature of 295 K rises in the air. What will be the volume of the air pocket if the temperature is decreased to 275 K and the pressure is not changed? (Hint: see Sample Problem 10A.)

16. A small 2.00 L fire extinguisher has an internal pressure of 506.6 kPa at 25°C. What volume of methyl bromide, the fire extinguisher's main ingredient, is needed to fill an empty fire extinguisher at standard pressure if the temperature remains constant? (Hint: see Sample Problem 10B.)

17. A child brings an inflatable ball on a small plane. Before take-off, the 2.00 L ball has a pressure of 101.3 kPa. The pilot flies the plane at an altitude where the air pressure is 75.0 kPa. What is the volume of the ball at this altitude if the temperature in the plane does not change? (Hint: see Sample Problem 10B.)

18. At a deep sea station 200. m below the surface of the Pacific Ocean, workers live in a highly pressurized environment. How many liters of gas at STP on the surface must be compressed to fill the underwater environment with 2.00×10^7 L of gas at 20.0 atm? Assume that temperature remains constant. (Hint: see Sample Problem 10C for information.)

19. A child carries a ball with a volume of 4.00 L from the surface of a pool to the bottom of the deep end. The pressure on the surface is 100. kPa. To what volume will the ball shrink when the pressure at the bottom of the pool is 135. kPa? Assume that temperature is constant. (Hint: see Sample Problem 10C.)

APPLY

20. Use the kinetic molecular theory to explain Charles's law.

21. Use Charles's law to explain the danger of throwing an aerosol can into a fire.

22. In a science fiction story, the main character visits a planet where the temperature is less than absolute zero. Use Charles's law to explain why this temperature is unrealistic.

23. Use Boyle's law to explain why "bubble wrap" pops when you squeeze it.

8. a. The ball loses no energy in an elastic collision, so it returns to the height from which it was dropped, or 2.0 m.
   b. In an inelastic collision some energy is converted into heat, so the ball does not move as fast after impact as it did before. It therefore bounces to a level lower than 2.0 m.

9. Liquid natural gas has about 600 times less volume than natural gas. The economic savings in storage space compensate for the added cost and risk of using liquid natural gas.

10. a. O_2 has the highest molar mass. He has the lowest molar mass.
   b. The curve at the far left is the gas at its lowest temperature. The curve at the far right is the gas at its highest temperature.

11. The mercury barometer (on the left) is measuring a higher pressure. Assuming that each barometer is 1 m tall, the column of water would need to be more than 10 m high to measure a gas pressure equal to that measured by the mercury barometer .

12. a. **Table 10-4** shows that the volume of a gas divided by the temperature yields a constant value.
   b. Answers will vary, but could include:
      1) Slight human errors in measurement.
      2) Attraction between gas particles varies from one gas to another, and varies for a single gas with change in temperature.
      3) The precision of the instruments may vary with changing temperature.
   c. Each value in the Temperature (K) column is 273 degrees greater than the corresponding value in the Temperature (°C) column.
   d. Charles's law

13. Boyle's law expresses the indirect relationship that exists between the volume and the pressure of a gas at a constant temperature.

14. $V_2 = V_1 \times \frac{T_2}{T_1} = 2.30\ \text{L} \times \frac{295\ \text{K}}{311\ \text{K}} =$ 2.18 L

15. $V_2 = V_1 \times \frac{T_2}{T_1} = 1.5\ \text{L} \times \frac{275\ \text{K}}{295\ \text{K}} = 1.4\ \text{L}$

16. $V_2 = \frac{P_1 V_1}{P_2} = \frac{(506.6\ \text{kPa})(2.00\ \text{L})}{101.3\ \text{kPa}} =$ 10.0 L

17. $V_2 = \frac{P_1 V_1}{P_2} = \frac{(101.3\ \text{kPa})(2.00\ \text{L})}{75.0\ \text{kPa}} =$ 2.70 L

18. $V_2 = \frac{P_1 V_1}{P_2} =$ $\frac{(20.0\ \text{atm})(2.00 \times 10^7\ \text{L})}{1.00\ \text{atm}} =$ 4.00×10^8 L

19. $V_2 = \frac{P_1 V_1}{P_2} = \frac{(100\ \text{kPa})(4.00\ \text{L})}{135\ \text{kPa}} =$ 2.96 L

20. According to the kinetic molecular theory, the kinetic energy of particles increases as the temperature increases. The particles also spread out farther because of their increased average energy and speed. The two properties, temperature and volume, increase directly, so that their ratio is constant.

21. The volume of gas inside the can will expand as the temperature increases, causing the can to burst or explode.

22. Charles's Law states the following relationship.
V/T = constant
If the temperature is a negative number, the corresponding volume of the gas would also be negative. Negative volumes are not attainable and are thus physically meaningless.

23. Pressing a bubble in the packing reduces the volume of air inside the bubble and increases the pressure. The pressure eventually becomes so large that the plastic bubble is ruptured and "pops".

Dalton's law and mole fractions

REVIEW

24. How are mole fractions used to find partial pressures of gas mixtures?
25. Gas is collected in the laboratory by bubbling it through water and trapping it in a container. How can you determine the pressure of the dry collected gas?
26. a. Why is it important to know the temperature of water when collecting a gas over it?
    b. How will the pressure of the dry gas change if the temperature of the water is increased?
27. Use the diffusion process to explain how a gaseous emission travels through the air.
28. Differentiate between diffusion and effusion.
29. What ratios are compared in Graham's law?

PRACTICE

30. Use **Table 10-6** to determine the partial pressure of oxygen collected over water if the temperature is 20.0°C and the total gas pressure is 98.0 kPa. (Hint: see Sample Problem 10D.)
31. The barometer at an indoor pool reads 105.00 kPa. If the temperature in the room is 30.0°C, what is the partial pressure of the "dry" air? (Hint: see Sample Problem 10D.)
32. A nitrogen molecule travels at 500. m/s at room temperature. What is the velocity of a helium molecule at the same temperature? (Hint: see Sample Problem 10E.)
33. A carbon dioxide molecule travels at 45.0 m/s at a certain temperature. What is the velocity of an oxygen molecule at the same temperature? (Hint: see Sample Problem 10E.)
34. Chlorine gas effuses through an opening at a rate 1.59 times slower than nitrogen gas. What is the molecular mass of chlorine gas? (Hint: see Sample Problem 10F.)
35. An unknown gas effuses through an opening at a rate 1.62 time slower than oxygen gas. What is the molecular mass of the unknown gas? (Hint: see Sample Problem 10F.)

APPLY

36. How do the velocities of neon molecules and krypton molecules compare when both gases are at the same temperature?

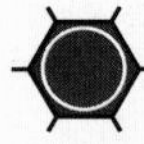

Ideal gas law, general gas law, and stoichiometry

REVIEW

37. What determines the value for the constant R?
38. Which gas variable is held constant when using the general gas law to analyze gas behavior?
39. How can a balanced chemical equation be used to determine the volume of a gas that will be produced from the reaction?

PRACTICE

40. Suppose you have a 500. mL container that contains 0.0500 mol of oxygen gas at 25°C. What is the pressure inside the container? (Hint: see Sample Problem 10G)
41. How many grams of oxygen gas in a 10.0 L container exert a pressure of 97.0 kPa at a temperature of 25.0°C? (Hint: see Sample Problem 10G.)
42. A helium balloon has a volume of 500. mL at STP. What will be its new volume if the temperature is increased to 325 K and its pressure is increased to 125 kPa? (Hint: see Sample Problem 10H.)
43. The air in a balloon has a volume of 3.00 L and exerts a pressure of 101. kPa at 300.0 K. What pressure does the air in the balloon exert if the temperature is increased to 400.0 K and the air is allowed to expand to 15.0 L? (Hint: see Sample Problem 10H.)
44. How many liters of hydrogen gas can be produced at 300.0 K and 104 kPa pressure if 20.0 g of sodium metal are reacted with water? (Hint: see Sample Problem 10I.)
$$2Na + 2H_2O \longrightarrow 2NaOH + H_2$$
45. How many liters of hydrogen gas can be produced at 290. K and 99.0 kPa if 17.0 g of potassium metal are reacted with water? (Hint: see Sample Problem 10I.)
$$2K + 2H_2O \longrightarrow 2KOH + H_2$$
46. Magnesium will burn in oxygen to form magnesium oxide as represented by the following equation.
$$2Mg + O_2 \longrightarrow 2MgO$$
What mass of magnesium will react with a 500.0 mL container of oxygen at 150°C and 70.0 kPa? (Hint: see Sample Problem 10J.)

47. One industrial method for producing nitric acid involves dissolving nitrogen dioxide in water.

$$3NO_2(g) + H_2O(l) \longrightarrow 2HNO_3(l) + NO(g)$$

What mass of water will react with a 1000.0 mL container of nitrogen dioxide at 25.0°C and 60.0 kPa? (Hint: see Sample Problem 10J.)

APPLY

48. A plastic weather balloon is filled with 90.0 L of hydrogen gas at ground level, where the pressure is 99.0 kPa and the temperature is 20.0°C. The balloon will burst if its volume exceeds 300.0 L. Can the balloon rise above the altitude of Mount Everest where the pressure drops to 32.1 kPa and the temperature decreases by 64.4°C?

49. Calculate the density of helium at standard temperature and pressure using the ideal gas law equation. Check your answer with a reference source of gas densities.

50. Suppose a certain automobile engine has a cylinder with a volume of 500.0 mL that is filled with air (21% oxygen) at a temperature of 55°C and a pressure of 101.0 kPa. What mass of octane must be injected to react with all of the oxygen in the cylinder?

$$2C_8H_{18}(l) + 25O_2(g) \longrightarrow 16CO_2(g) + 18H_2O(g)$$

Condensation and liquids

REVIEW

51. Why do gases deviate from ideal gas behavior?

52. Use the following graph of the vapor pressure of water versus temperature to answer the following questions.

a. At which point(s) does water boil at standard atmospheric pressure?

b. At which point(s) is water only in the liquid phase?

c. At which point(s) is water only in the vapor phase?

d. At which point(s) is liquid water in equilibrium with water vapor?

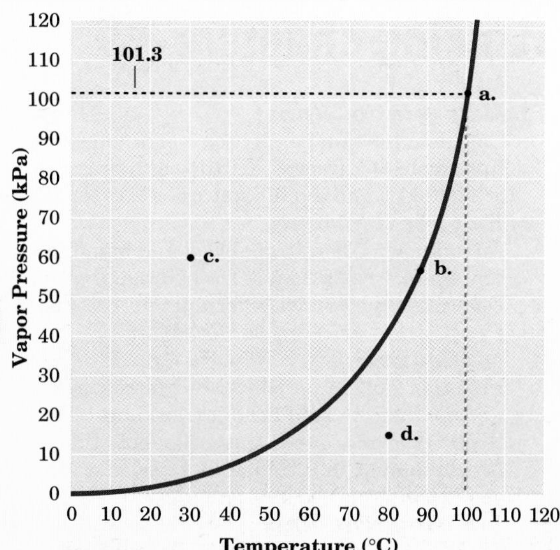

APPLY

53. Use the phase diagram for oxygen to determine what phase transition would occur if the following changes occurred.

a. At point **a**, pressure is decreased, and temperature remains constant.

b. At point **b**, temperature is decreased, and pressure is constant.

c. At point **c**, pressure is increased, and temperature is the same.

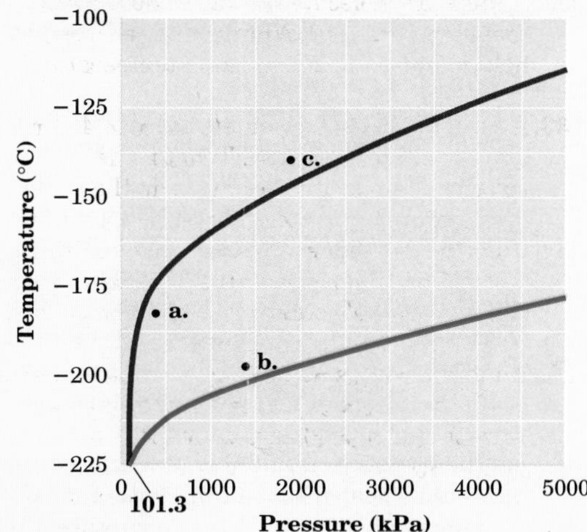

54. From the phase diagram for oxygen, determine the physical state under the following conditions.

a. 2000 kPa and −180°C

b. 3000 kPa and −120°C

24. Mole fractions represent the number of moles, and thus particles, of a gas as a part of the total number of moles, or particles, of gas in a mixture. The total pressure of the gas is the result of collisions between the gas particles and the container, so it follows that the more there is of a given gas, the larger its contribution to the total pressure. The measure of that contribution is the partial pressure, which can be described by the following equation.

P_{gas} = (mole fraction of gas) × P_{total}

The total pressure is the sum of the partial pressures of the component gases.

25. Measure the total gas pressure, which is the same as the atmospheric pressure, at which the gas is collected. Measure the temperature of the displaced water, and use **Table 10-6** to determine the water vapor pressure. Subtract the partial pressure due to water vapor molecules from the total pressure to find the partial pressure of the dry gas.

26. a. Water vapor pressure, like the pressure of all gases, varies with temperature.

b. At high temperatures (> 60°C) the water vapor pressure increases significantly, causing the dry gas partial pressure to decrease.

27. Rapidly moving gas particles collide with other gas particles and objects, so that they move through the air in a zigzag pattern.

28. Diffusion describes how gas particles move through another gas or liquid. Effusion is the movement of a gas through an opening.

29. The ratio of the molar masses and the ratio of the square of the average velocities are compared.

30. Partial pressure of water at 20.0°C = 2.34 kPa, so 98.0 kPa − 2.34 kPa = 95.7 kPa

31. Partial pressure of water at 30.0°C = 4.25 kPa, so 105.00 kPa − 4.25 kPa = 100.75 kPa

32. $\frac{v_{He}}{v_{N_2}} = \sqrt{\frac{m_{N_2}}{m_{He}}}$

$\frac{v_{He}}{500\text{ m/s}} = \sqrt{\frac{28.02\text{ g/mol}}{4.00\text{ g/mol}}}$

$v_{He} = 1.32 \times 10^3$ m/s

33. $\frac{v_{O_2}}{v_{CO_2}} = \sqrt{\frac{m_{CO_2}}{m_{O_2}}}$

$\frac{v_{O_2}}{45.0\text{ m/s}} = \sqrt{\frac{44.01\text{ g/mol}}{32.00\text{ g/mol}}}$

$v_{O_2} = 52.8$ m/s

34. $\frac{v_{Cl_2}}{v_{N_2}} = \sqrt{\frac{m_{N2}}{m_{Cl_2}}}$

$\frac{v_{Cl_2}}{1.59(v_{Cl_2})} = \sqrt{\frac{28.02\text{ g/mol}}{m_{Cl_2}}}$

$m_{Cl_2} = 70.8$ g/mol

35. $\frac{v_{O_2}}{v_x} = \sqrt{\frac{m_x}{m_{O_2}}}$

$\frac{1.62(v_x)}{v_x} = \sqrt{\frac{m_x}{32.00\text{ g/mol}}}$

$m_x = 84.0$ g/mol

36. Using Graham's Law

$\frac{v_{Ne}}{v_{Kr}} = \sqrt{\frac{m_{Kr}}{m_{Ne}}} = \sqrt{\frac{83.8\text{ g/mol}}{20.2\text{ g/mol}}} = 2.04$

Neon atoms have a velocity that is 2.04 times greater than the velocity of krypton atoms at the same temperature.

37. The value of R is dependent upon the units used to express the volume (V), number of moles (n), the temperature (T), and the pressure (P) of a gas. By using 1 mol of a gas at STP, a measurement of V yields a value for R.

38. The number of moles of a gas is held constant when using the general gas law.

39. The molar ratio of one gas to another as determined by the coefficients in a balanced equation equals the ratio of the volumes of the gases.

40. $PV = nRT$

$T = 25°\text{C} = 298\text{ K}$

$P = \frac{nRT}{V}$

$= \frac{(0.0500\text{ mol})\left(8.314\ \frac{\text{L•kPa}}{\text{mol•K}}\right)(298\text{ K})}{0.500\text{ L}}$

$= 248$ kPa

Linking chapters

1. ***Mole relationships***
Calculate the number of atoms of argon gas that fill a 1.0 L fluorescent tube at a pressure of 100.0 kPa and a temperature of 298 K.

2. ***Empirical formula***
An unknown gas was analyzed and found to contain 82% carbon and 18% hydrogen. Determine the empirical formula for the gas.

3. ***Properties related to electron configuration***
In Period 2 of the periodic table, nitrogen, oxygen, fluorine, and neon are gases at STP. Neon is the only one of these elements that is not diatomic in the gaseous state. How do you explain this observation?

4. ***Limiting reactants***
Carbon monoxide is only partially oxidized; further oxidation of carbon monoxide produces carbon dioxide.

$$2CO(g) + O_2(g) \longrightarrow 2CO_2(g)$$

Suppose that 500.0 mL of CO at 101 kPa and 15°C is reacted with 500.0 mL of O_2 at the same pressure and temperature. Which gas would be the limiting reactant?

5. **Theme: *Macroscopic observations and micromodels***
As a car travels down the road, the air pressure inside the tires may increase. Prepare a model that illustrates what happens to the tires as the car moves.

6. ***Stoichiometry***
Calculate the volume of gas released at STP from the reaction in a car airbag, when 92.0 g of NaN_3 decomposes according to the following reaction.

$$2NaN_3(s) \longrightarrow 2Na(s) + 3N_2(g)$$

7. ***Micromodels***
Draw a micromodel for each of the following.
a. balloon filled with helium
b. balloon filled with air
c. balloon filled with oxygen

USING TECHNOLOGY

1. ***Graphics calculator***
The different gas laws can be compared at STP conditions and graphed on your calculator. Graph Gay-Lassac's law using $P_2 = 1$ atm and $T_2 = 273$ K as constants. Press the Y= key and enter the function $y = x/273$. Press the WINDOW key and set the WINDOW to Xmin = 0, Xmax = 500, Xscl = 50, Ymin = 0, Ymax = 2.5, Yscl = 1. Press the GRAPH key to display the function. Use the TRACE mode to determine the effect of increasing temperature (x value) on the pressure (y value) of an enclosed gas.
a. Would a sealed scuba tank with a maximum stress point of 5.0 atm rupture after warming to 315 K?
b. Will a glass bulb with a stress point of 1.5 atm shatter if it is heated to 150°C?

2. ***Computer art***
Using a computer art program, sketch how gas particles in a closed system may act when the pressure, temperature, volume, or the number of gas particles is changed. Use the kinetic molecular model as a guide.

3. ***Computer spreadsheet***
Create a spreadsheet that will run calculations using the general gas law equation or ideal gas law equation.

Alternative assessment

Performance assessment

1. Design an experiment to measure the molar mass of the gas found inside a disposable butane lighter. Measure the mass and volume of the gas at a known temperature and pressure. Use the ideal gas law equation $PV = nRT$ to find molar mass. Compare your answer with the known molar mass of butane. Try different brands of lighters to compare results. Other reactions that produce a gas can also be used.
2. Match the microviews below with the appropriate description. Microviews can be matched more than once.
   a. model for neon gas
   b. model for chlorine gas
   c. model for gas mixture
   d. model for gas collected over water
   e. model for a pure gas
   f. model for a volatile liquid

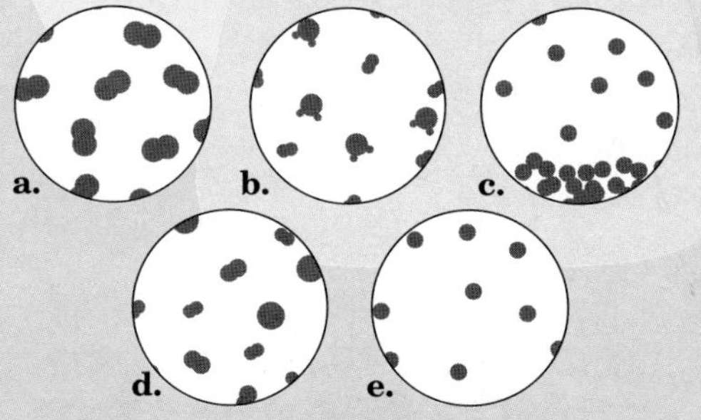

3. Match the graphs below with the appropriate description. Graphs can be matched more than once.
   a. a graph showing a directly proportional relationship
   b. a graph with a slope = 0
   c. a graph showing an inversely proportional relationship
   d. a graph with a constant slope

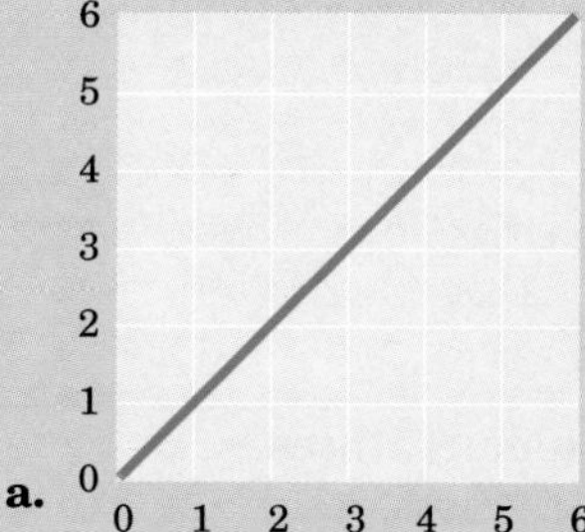

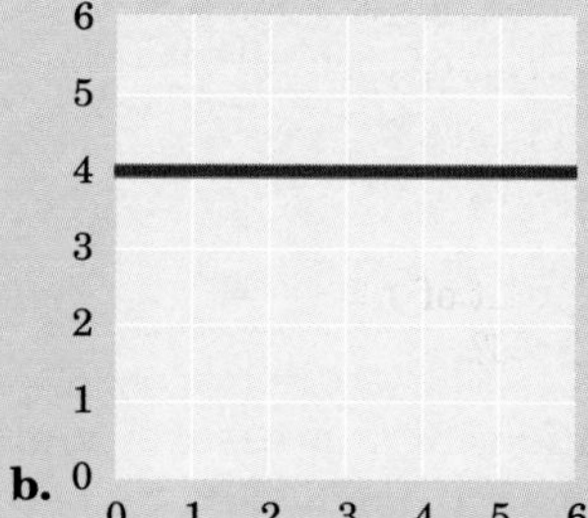

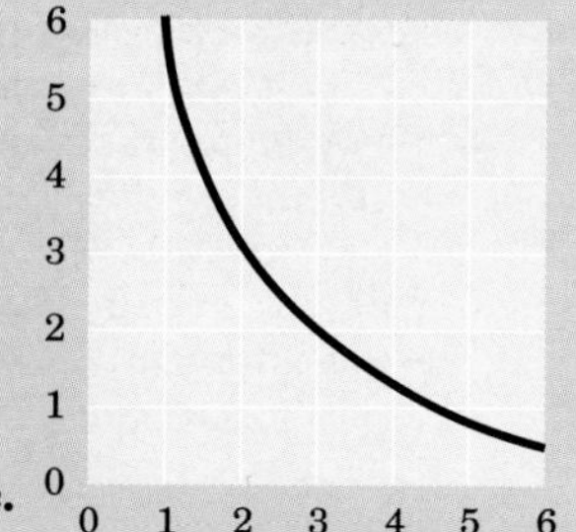

Portfolio projects

1. ***Research and communication***
   For more than 300 years, gases have been an important part of the evolution of chemistry and physics. Research the important contributions of the following scientists.
   a. Robert Boyle (1627–1691)
   b. Jacques Charles (1746–1823)
   c. Joseph Gay-Lussac (1778–1850)
   d. Stanislao Cannizzaro (1826–1910)
   e. Amedeo Avogadro (1776–1856)
   f. James Clerk Maxwell (1831–1879)
2. ***Chemistry and you***
   Locate a hot air balloon group to discuss with them how the gas laws are used to fly their balloons. The group may be willing to give a demonstration. Report your experiences to your class.
3. ***Chemistry and you***
   Industrial gases are shipped in canisters of various colors that are specific to the gas. Find out what color code is used for each gas.
4. ***Research and communication***
   Some commercial gases are produced by the liquification and fractionation of air. Find out how these processes are used to separate a mixture into its component gases.
5. ***Chemistry and you***
   Commercial dry cake mixes will generally have two sets of preparation instructions. Go to the grocery store and look at some boxed, dry cake mixes. What is the purpose for having two sets of instructions? How do they differ and why do they differ?

41. $T = 25.0°C = 298.0\ K$

$$n = \frac{PV}{RT} = \frac{(97.0\ kPa)(10.0\ L)}{\left(8.314\ \frac{L\cdot kPa}{mol\cdot K}\right)(298.0\ K)} = 0.392\ mol\ O_2$$

$$m_{O_2} = 0.392\ mol \times \frac{32.00\ g}{1\ mol} = 12.5\ g$$

42. $$\frac{P_1V_1}{T_1} = \frac{P_2V_2}{T_2}$$

$$\frac{(101.3\ kPa)(500.\ mL)}{273\ K} = V_2 \times \frac{125\ kPa}{325\ K}$$

$V_2 = 482\ mL$

43. $$\frac{P_1V_1}{T_1} = \frac{P_2V_2}{T_2} = \frac{(101\ kPa)(3.00\ L)}{300.0\ K} = P_2 \times \frac{15.0\ L}{400.0\ K}$$

$P_2 = 26.9\ kPa$

44. number of moles of Na =

$$\frac{20.0\ g\ Na}{22.99\ g/mol} = 0.870\ mol$$

$$0.870\ mol\ Na \times \frac{1\ mol\ H_2}{2\ mol\ Na} = 0.435\ mol\ H_2$$

$$V = \frac{nRT}{P} = \frac{(0.435\ mol\ H_2)\left(8.314\ \frac{L\cdot kPa}{mol\cdot K}\right)(300.0\ K)}{(104\ kPa)}$$

$= 10.4\ L\ H_2$

45. number of moles of K =

$$\frac{17.0\ g\ K}{39.1\ g/mol} = 0.435\ mol$$

$$0.435\ mol\ K \times \frac{1\ mol\ H_2}{2\ mol\ K} = 0.218\ mol\ H_2$$

$$V = \frac{nRT}{P} = \frac{(0.218\ mol\ H_2)\left(8.314\ \frac{L\cdot kPa}{mol\cdot K}\right)(290.\ K)}{(99.0\ kPa)}$$

$= 5.31\ L\ H_2$

46. $T = 150°C = 423\ K$

$$n = \frac{PV}{RT} = \frac{(70.0\ kPa)(0.500\ L)}{\left(8.314\ \frac{L\cdot kPa}{mol\cdot K}\right)(423\ K)}$$

$= 9.95 \times 10^{-3}\ mol\ O_2$

$$9.95 \times 10^{-3}\ mol\ O_2 \times \frac{2\ mol\ Mg}{1\ mol\ O_2} \times \frac{24.30\ g\ Mg}{1\ mol\ Mg} = 0.484\ g\ Mg$$

Answers from page 367

Answers to Section Review

9. volume, pressure, temperature, and quantity (in moles of gas)

10. At least 10 km, with maximum concentration at about 25 km.

11. mass of He = 2.8×10^8 g

Answers from page 380

Solution for Additional Example 10E

$m_1 = 17.04$ g/mol NH_3, $m_2 = 34.09$ g/mol H_2S

$$\frac{v_2}{v_1} = \sqrt{\frac{m_1}{m_2}} = \sqrt{\frac{17.04 \text{ g/mol}}{34.09 \text{ g/mol}}} = \sqrt{0.4998} = 0.707 = \frac{v_2}{658 \text{ m/s}}$$

$v_2 = (0.707)(658 \text{ m/s}) = 465 \text{ m/s}$

Solution for Additional Example 10F

$v_1 = 409$ m/s for CO_2, $v_2 = 322$ m/s for unknown molecule

$$\frac{m_2}{m_1} = \left(\frac{v_1}{v_2}\right)^2 = \left(\frac{409 \text{ m/s}}{322 \text{ m/s}}\right)^2 = (1.27)^2 = 1.61 = \frac{m_2}{44.01 \text{ g/mol}}$$

$m_2 = (1.61)(44.01 \text{ g/mol}) = 71.0 \text{ g/mol}$

Answers from page 384

Solutions for Additional Examples 10H

b. $T_1 = 30°C = 303$ K

$$\frac{P_1V_1}{T_1} = \frac{P_2V_2}{T_2} = \frac{(99.7 \text{ kPa})(132 \text{ L})}{303 \text{ K}} = 43.4 \text{ kPa•L/K} = \frac{(77.6 \text{ kPa})(176 \text{ L})}{T_2},$$

$$\text{or } T_2 = \frac{(77.6 \text{ kPa})(176 \text{ L})}{(43.4 \text{ kPa•L/K})} = 314 \text{ K} = 41°C$$

c. $T_1 = 23°C = 296$ K; $T_2 = 4°C = 277$ K

$$\frac{P_1V_1}{T_1} = \frac{P_2V_2}{T_2} = \frac{(111 \text{ kPa})(3.89 \text{ L})}{296 \text{ K}} = 1.46 \text{ kPa•L/K} = \frac{P_2(3.05 \text{ L})}{277 \text{ K}},$$

$$\text{or } P_2 = \frac{(1.46 \text{ kPa•L/K})(277 \text{ K})}{3.05 \text{ L}} = 132 \text{ kPa}$$

Continued from page 386

Answers to Practice 10I and 10J

3. $V_{CO_2} = 4.0 \text{ m} \times 2.5 \text{ m} \times 8.0 \text{ m} = 80 \text{ m}^3 = 80 \times 10^3 \text{ L}$

$T = 15°C = 288$ K

$$n_{CO_2} = PV/RT = \frac{(5.0 \text{ kPa})(8.0 \times 10^4 \text{ L})}{\left(8.314 \frac{\text{kPa•L}}{\text{mol•K}}\right)(288 \text{ K})} = 1.7 \times 10^2 \text{ mol}$$

moles of LiOH required

$$n_{LiOH} = 1.7 \times 10^2 \text{ mol } CO_2 \times \frac{2 \text{ mol LiOH}}{1 \text{ mol } CO_2} = 3.4 \times 10^2 \text{ mol LiOH}$$

mass of LiOH required = $(3.4 \times 10^2 \text{ mol})(23.95 \text{ g/mol}) = 8.1 \times 10^3$ g

Solution for Additional Example 10J

The mole ratio of $KClO_3$ to O_2 is 2:3.

number of moles of $O_2 = n = \frac{PV}{RT}$

$T = 700°C = 973$ K

$$n = \frac{(98.6 \text{ kPa})(0.125 \text{ L})}{\left(8.314 \frac{\text{kPa•L}}{\text{mol•K}}\right)(973 \text{ K})} = 1.52 \times 10^{-3} \text{ mol } O_2$$

$$n_{KClO_3} = 1.52 \times 10^{-3} \text{ mol } O_2 \times \frac{2 \text{ mol } KClO_3}{3 \text{ mol } O_2} = 1.01 \times 10^{-3} \text{ mol } KClO_3$$

mass of $KClO_3$ required = $(1.01 \times 10^{-3} \text{ mol } KClO_3)(122.55 \text{ g/mol } KClO_3) = 0.124$ g

Answers from page 387

Answers to Section Review

31. a. $2C_8H_{18} + 25O_2 \longrightarrow 16CO_2 + 18H_2O$

V of O_2 needed for reaction =

$$25 \text{ L } C_8H_{18} \times \frac{25 \text{ mol } O_2}{2 \text{ mol } C_8H_{18}} = 3.1 \times 10^2 \text{ L} = 20.9\% \text{ of } V_{air}$$

$V_{air} = 1.5 \times 10^3$ L

b. $V_{CO_2} = 200$ L

$V_{H_2O} = 225$ L

32. a. $CO + 2H_2 \longrightarrow CH_3OH$

Only 413 L of CO are needed, so CO is in excess.

b. 37 mL of CO remain.

c. $V = 413$ mL of CH_3OH

33. $$P_2 = \frac{P_1V_1T_2}{V_2T_1} = \frac{(745 \text{ torr})(5.0 \times 10^2 \text{ L})(273 \text{ K})}{\left(\frac{96}{100} \times 4.0 \text{ L}\right)(298 \text{ K})} = 8.9 \times 10^4 \text{ torr}$$

Continued from page 401

47. $T = 25.0°C = 298.0$ K

$$n = \frac{PV}{RT} = \frac{(60.0 \text{ kPa})(1.000 \text{ L})}{\left(8.314 \frac{\text{L•kPa}}{\text{mol•K}}\right)(298.0 \text{ K})} = 2.42 \times 10^{-2} \text{ mol } NO_2$$

$$2.42 \times 10^{-2} \text{ mol } NO_2 \times \frac{1 \text{ mol } H_2O}{3 \text{ mol } NO_2} \times \frac{18.02 \text{ g } H_2O}{1 \text{ mol } H_2O} = 0.145 \text{ g } H_2O$$

48. $T_1 = 20.0°C = 293.0$ K

$T_2 = 293.0 \text{ K} - 64.4°C = 228.6$ K

$$\frac{P_1V_1}{T_1} = \frac{P_2V_2}{T_2} = \frac{(99.0 \text{ kPa})(90.0 \text{ L})}{293.0 \text{ K}} = V_2 \times \frac{32.1 \text{ kPa}}{228.6 \text{ K}}$$

$V_2 = 217$ L

Since the volume is only 217 L at the altitude of Mt. Everest, the balloon will rise above this altitude before bursting.

49. $$D_{He} = \frac{m_{He}}{V} = \frac{P \times \text{molar mass He}}{RT} = \frac{(101.3 \text{ kPa})\left(\frac{4.002 \text{ g}}{1 \text{ mol}}\right)}{\left(8.314 \frac{\text{L•kPa}}{\text{mol•K}}\right)(273.2 \text{ K})} = 0.1785 \text{ g/L}$$

D_{He} listed in *CRC Handbook of Chemistry and Physics* = 0.1785 g/L

50. $T = 55°C = 328$ K

$$n = \frac{PV}{RT} = \frac{(101.0 \text{ kPa})(0.500 \text{ L})}{\left(8.314 \frac{\text{L•kPa}}{\text{mol•K}}\right)(328 \text{ K})} = 1.85 \times 10^{-2} \text{ mol air}$$

$$1.85 \times 10^{-2} \text{ mol air} \times \frac{0.21 \text{ mol } O_2}{1 \text{ mol air}} \times \frac{2 \text{ mol } C_8H_{18}}{25 \text{ mol } O_2} \times \frac{114.26 \text{ g } C_8H_{18}}{1 \text{ mol } C_8H_{18}} =$$

3.6×10^{-2} g of C_8H_{18}

51. Gases under high pressures and at low temperatures do not behave as ideal gases because attractive forces between particles affect the volume.

52. a. point **a**
b. point **c**
c. point **d**
d. point **a** and point **b**

53. a. changes from a liquid to a gas
b. changes from a liquid to a solid
c. changes from a gas to a liquid

54. a. liquid
b. gas

Linking chapters

1. $n = \frac{PV}{RT} = \frac{(100.0 \text{ kPa})(1.0 \text{ L})}{\left(8.314 \frac{\text{L•kPa}}{\text{mol•K}}\right)(298 \text{ K})} = 0.04 \text{ mol argon}$

number of atoms of Ar = 0.04 mol argon × $\frac{6.022 \times 10^{23} \text{ atoms}}{1 \text{ mol}}$ =

2.4×10^{22} atoms

2. Assuming that the unknown sample has a mass of 100 g, then

$82\% \text{ C} = 82 \text{ g C} = 82 \text{ g} \times \frac{1 \text{ mol C}}{12.01 \text{ g}} = 6.8 \text{ mol C}$

$18\% \text{ H} = 18 \text{ g H} = 18 \text{ g} \times \frac{1 \text{ mol H}}{1.01 \text{ g}} = 18 \text{ mol H}$

The mole ratio of hydrogen to carbon is $\frac{18}{6.8} = 2.65 \approx 2\frac{2}{3} = \frac{8}{3}$

so there are 8 H atoms for every 3 C atoms and the formula is therefore C_3H_8.

This formula also makes sense with regard to the possible bond configurations.

3. Neon is a noble gas with a stable octet of valence electrons. The other elements bond covalently and form diatomic molecules in order to establish a stable octet.

4. At the same temperature and pressure, the amount of the two gases in moles can be determined from their volumes. Because there are equal volumes of CO and O_2 (500.0 mL), there are equal numbers of moles of CO and O_2. Two moles of carbon monoxide are needed for every mole of oxygen, so CO is the limiting reactant. 250 mL of O_2 will remain unreacted.

5. Friction between the tire and the road increases the average kinetic energy of the gas particles in the tire, which in turn causes the particles to collide more frequently with the inner wall of the tire. This increase in collision number on the microscopic level appears macroscopically as an increase in pressure.

6. number of moles of $NaN_3 = \frac{92.0 \text{ g}}{65.02 \text{ g/mol}} = 1.41 \text{ mol}$

$1.41 \text{ mol NaN}_3 \times \frac{3 \text{ mol N}_2}{2 \text{ mol NaN}_3} = 2.12 \text{ mol N}_2$

$V = \frac{nRT}{P} = \frac{(2.12 \text{ mol N}_2)\left(8.314 \frac{\text{L•kPa}}{\text{mol•K}}\right)(273 \text{ K})}{(101.3 \text{ kPa})} = 47.5 \text{ L N}_2$

7. a. 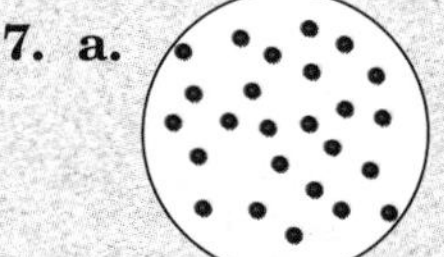 b. 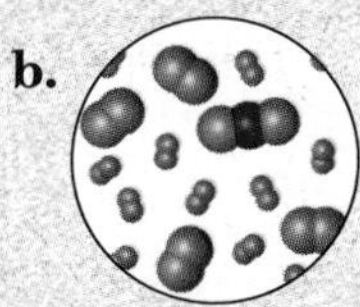 c. 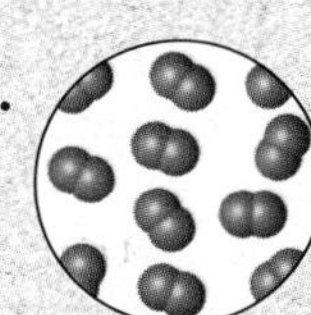

USING TECHNOLOGY

1. a. No. The pressure is only 1.15 atm.
   b. Yes, it will shatter. The pressure at $T = 150°\text{C} = 423 \text{ K}$ is 1.55 atm.

2. The model should show gas particles moving in straight lines before and after collisions, with direction changed randomly by collisions. The particles should have a wide range of speeds initially, but after several collisions most of the particles should have a single, average speed. The more a particle's speed departs from the average speed, the smaller the number of particles that should have that speed. The particles should move faster with increased temperature and pressure, and decreased volume. An increase in the number of particles should increase the frequency with which collisions occur.

3. The spreadsheet should have entries available for each of the four variables in the ideal gas law equation, and for the six variables in the general gas law equation.

Performance assessment

1. Assess the design in terms of the method used for collecting the butane gas and measuring its volume, and in terms of the method for determining the total mass of butane in the lighter. Note whether the ambient pressure and temperature at which the experiment is performed is taken into account.

2. **a.** e **b.** a **c.** b, d **d.** b **e.** a, e **f.** c

3. **a.** a **b.** b **c.** c **d.** a, b

Portfolio projects

1. a. Boyle's work on the relation between the volumes and pressures of gases should be included.
   b. Charles's work on the relation between the volumes and temperatures of gases should be included. His work on the first gas balloons is also worth noting.
   c. Gay-Lussac's work on combining volumes and the discovery that volume ratios in gaseous reactions are whole numbers should be included. The connection between Gay-Lussac's and Dalton's models and Dalton's rejection of Gay-Lussac's work is also worth noting.
   d. Cannizzaro's revival of Avogadro's work and his distinction between atoms and molecules should be included.
   e. Avogadro's hypothesis, in which he stated that equal volumes of gases under identical conditions contain the same number of particles, should be included. The proposal that molecules of elemental gases would solve Dalton's problem of "half" atoms and the development of Avogadro's theoretical work from Gay-Lussac's experiments should also be noted.
   f. Maxwell's development of the kinetic molecular theory, explanation of velocity distributions of gases, and understanding of the dependence of particle energy on temperature should be included.

2. The report should include factors such as the temperature of the hot air in the balloon, the average time it takes to heat the air to that temperature, the balloon's volume, and the weight the balloon is capable of lifting.

3. Answers will vary depending on the system used by the manufacturer. For example, though green is generally used for oxygen, this is not absolutely the case throughout the industry.

4. A discussion of the use of increased pressure and lowered temperature to liquefy air should follow arguments similar to those in Section 10-4. Features of fractionation, such as the recondensation of gas particles on cool surfaces, loss of energy of more massive particles upon recondensing, and separation of massive components from mixtures of lighter substances should be included.

5. The discussion should describe the differences between high-altitude directions and standard directions. Answers will vary depending on the brand of mix chosen, but common features are the addition to the mix of flour and extra water when used at altitudes over 3500 ft. The discussion should emphasize lower evaporation temperatures at high elevations and the effect that adding water and flour has in countering the loss of water that results from "low-temperature" evaporation.

11

PLANNING FOR

Solutions

Chapter Overview

- **Section 11-1** covers the basic vocabulary and micro models that describe the characteristics and properties of solutions. Students look at a visual model for a stable homogeneous mixture in contrast to the short-lived homogeneous nature of a suspension. Solubility and saturation are two key concepts for this section.
- **Section 11-2** covers the measurement of concentration, molarity, making molar solutions, and molarity calculations.
- **Section 11-3** explains macroscopic observations of solubility by revisiting polarity which was first introduced in Chapter 6. Students learn to predict solubility of solutes based on the determination of molecular polarity.
- **Section 11-4** shows how emulsifiers act as bridge molecules causing nonpolar substances to dissolve in polar solvents. This idea is further developed in looking at the structural features and behavior of soaps and detergents.
- **Section 11-5** shows applications of solubility principles to dry cleaning, diving, and vitamin metabolism.

Concept Base

Students must have mastered the following concepts prior to this chapter: characteristics of mixtures (Chapter 2) electronegativity and atomic masses (Chapter 4).

Looking Ahead

Colligative properties of solutions (boiling point elevation and freezing point depression) will be described in Chapter 12. Solution equilibrium will be covered in Chapters 12 and 13.

Laboratory Equipment Needs

Demonstrations

The listing of materials for Chapter 11 demonstrations begins on page T46.

Laboratory Experiments

Materials and preparation instructions for Chapter 11 experiments are found as follows.

Text		*Page*
Exploration 11A	Colorimetry and Molarity	736
Investigation 11A	Colorimetry—Reservoir Contaminant	742
Investigation 11B	Emulsifying Agents—Testing Soaps	744

Laboratory Experiments Manual
(refer to the table of contents in the manual)

Exploration 11-1	Paper Chromatography
Investigation 11-1	Paper Chromatography—Forensic Investigation
Exploration 11-2	Testing for Dissolved Oxygen
Exploration 11-3	Solubility of Ammonia

Pacing: 13 days — Inquiry Teaching Strategies

11-1 What is a solution?

Demonstrations
- Demo 1 *Liquid mixtures*
- Demo 2 *Volume of solution, solvent, and solute*
- Demo 3 *Temperature/solubility I*
- Demo 4 *Temperature/solubility II*
- Demo 5 *Saturation points*
- Demo 6 *Supersaturation I*
- Demo 7 *Supersaturation II*

11-2 What does concentration mean?

Demonstrations
- Demo 8 *Concentration*
- Demo 9 *The meniscus*

Laboratory Experiments
- Exp 11A *Colorimetry and molarity*
- Exp 11B *Colorimetry—Reservoir Contaminant*

11-3 Why do some things dissolve?

Demonstrations
- Demo 10 *Polar and nonpolar solvents*
- Demo 11 *Polarity of water*

Laboratory Experiments
- Exp 11-1 *Paper Chromatography*
- Inv 11-1 *Paper Chromatography—Forensic Investigation*

11-4 Can insoluble substances mix?

Demonstrations
- Demo 12 *Emulsifiers*
- Demo 13 *Soap and surface tension*

Laboratory Experiments
- Inv 11B *Emulsifying Agents—Testing Soaps*

11-5 How are solubility principles used?

Demonstrations
- Demo 14 *Solubility*
- Demo 15 *Temperature/solubility*

Laboratory Experiments
- Exp 11-2 *Testing for Dissolved Oxygen*
- Exp 11-3 *Solubility of Ammonia*

Key
- Teaching Transparencies
- Annotated Teacher's Edition
- Laboratory Experiments Manual

Visual Teaching Strategies

Review and Practice Strategies

Transparencies
- F11-1 *Particle Model for a Solution*
- F11-2 *Particle Model for a Suspension*
- F11-5 *Solubility vs. Temperature*

Transparency Masters
- T11-1 *Solubility of Common Compounds*
- TA-11 *Solubilities of Compounds*

Text Reviews
- Concept Review 1–3
- Section Review 4–9
- Chapter Review 1–14; LC 1, 4, 5; UT 3

Study Guide Worksheets
- 11-1 Skillsheet *Reading Solubility Graphs*

Transparencies
- F11-9 *Moles Solute vs. Volume*
- P441 *Molarity Calculations*

Transparency Masters
- T11-2 *Some Concentration Units*

Text Reviews
- Additional Examples 11A, 11B, 11C
- Practice 1–7
- Section Review 10–14
- Chapter Review 15–32; LC 2, 3; UT 1, 2, 4

Study Guide Worksheets
- 11-2 Practice *Molarity problems*

Transparencies
- F11-10 *Particle Models for Soluble and Insoluble Substances*
- P421 *Determing Polarity*
- P422 *Determining Polarity*
- F11-18 *Prediciting Polarity*
- P424 *Prediciting Solubility*

Text Reviews
- Additional Example 11D
- Practice 1–2
- Section Review 15–19
- Chapter Review 33–38

Study Guide Worksheets
- 11-3 Concept Review

Transparencies
- F11-21 *Model for an Emulsion*

Transparency Masters
- T11-4 *Common Food Emulsifiers*

Text Reviews
- Section Review 20–22
- Chapter Review 39–43

Study Guide Worksheets
- 11-4 Concept Review

Transparencies
- F11-28 *Particle Model for Gas Solubility*

Transparency Masters
- TA-10 *Solubilities of Gases in Water*

Text Reviews
- Concept Review 23–24
- Section Review 25–28
- Chapter Review 44–55

Study Guide Worksheets
- 11-5 Skillsheet *Interpreting gas solubility data*

Pupil's Text	Exp *Exploration*	LC *Linking Chapters*
Study Guide	Inv *Investigation*	UT *Using Technology*
Teaching Resources		

Assessment Options

Traditional Assessment

Test Generator
Instructional Objectives Measured:
Content mastery
Select from items 1 to 72.

Performance Assessment Options

(see pages T36 and T643-15 for scoring rubrics)
Students demonstrate mastery of objectives in a hands-on environment. You may want some of these materials in the Portfolio as well.

Investigations 11A and 11B (text)
Investigation 11-1 (laboratory manual)
Instructional Objectives Measured:
Content mastery, Use of scientific methodology, Problem solving skills, Use of technology, Proficiency in written communication

Performance Assessments, page 447.

1. **Designing an experiment**
   Instructional Objectives Measured:
   Content mastery, Use of scientific methodology, Problem solving skills, Use of technology, Proficiency in written communication
2. **Making molar solutions**
   Instructional Objectives Measured:
   Content mastery, Problem solving skills
3. **Making acid solutions of various concentrations**
   Instructional Objectives Measured:
   Content mastery, Problem solving skills

Portfolio Options

(see page T36 for scoring rubrics)
Students provide a written rationale for each selection made for the portfolio.

Concept Maps
Students use vocabulary from the Chapter Review to build a concept map.
Instructional Objectives Measured:
Content mastery

Formal Laboratory Reports
Explorations 11A (text)
Explorations 11-1, 11-2, 11-3 (lab manual)
Instructional Objectives Measured:
Content mastery, Use of scientific methodology, Use of technology, Proficiency in written communication

Portfolio Projects, page 447.
Items 1, 4, and 5. Chemistry and you
Instructional Objectives Measured:
Content mastery, Scientific and chemical literacy

Items 2 and 6. Research and communication
Instructional Objectives Measured:
Content mastery, Scientific and chemical literacy, Proficiency in written communication

Item 3. Cooperative activity
Instructional Objectives Measured:
Content mastery, Scientific and chemical literacy, Proficiency in oral and written communication

INTRODUCING THE CHAPTER

CHAPTER 11 Solutions

Story Background

Dr. Malcolm Ramsay is a professor of vertebrate ecology at the University of Saskatchewan and one of Canada's leading experts on polar bears. He focuses on polar bears because Arctic ecological interactions are simple to study. Also, polar bear tissue clearly demonstrates biological amplification of pesticides because polar bears are at the top of the food chain. The polar bear's food chain consists of algae, which absorb PCBs from polluted water; plankton, which eat algae; cod, which eat plankton; seals, which eat cod; and finally polar bears, which eat seals. At each link in the food chain, the concentration of PCBs increases.

There are more than 209 different oily synthetic polychlorinated biphenyl compounds (PCBs). Most are insoluble in water, soluble in fats, and resistant to biological and chemical degradation. These properties result in the biological amplification of PCBs.

Although now banned in North America and Western Europe, PCBs are still used in Eastern Europe. The long range effects of PCBs and their more toxic furan impurities on people are unknown. However, in laboratory animals, high doses of PCBs produce gastric disorders, birth defects, bronchitis, miscarriages, skin lesions, hormonal changes, liver and kidney damage, and tumors.

About the Illustration

Polar bears live along the frozen shores and in the icy waters of the Arctic Ocean. This polar bear is lumbering across Canada's Ellesmere Island, which is just west of Greenland. The chemical structure shown above is a PCB molecule.

Solubility and the Arctic Bear Hunt

In northern Canada, a helicopter hovers over the snow. Dr. Malcolm Ramsay leans out of the doorway to keep a polar bear in sight. As Ramsay braces and aims his rifle, the bear lumbers through the snow trying to escape. Ramsay fires, and the bear runs faster.

Ramsay, who is an ecologist, signals the pilot to land in a clearing. He unloads his equipment, then shields his face from the helicopter-driven snowstorm as the pilot departs. By the time Ramsay reaches the bear, the tranquilizer-filled dart has done its job, and the animal lies quietly in the snow.

Ramsay begins a physical examination that will take hours. With a net, ropes, and a tripod, the bear is winched into the air to be weighed. Large syringes are filled with blood samples. A tooth is extracted; its growth rings will tell the animal's age. An electronic instrument records the body's resistance to electricity, which is an indication of the bear's muscle-to-fat ratio. A sample of fat tissue is taken. By the time the bear begins to stir, Ramsay has packed his gear and is waving to the returning helicopter. The polar bear will resume its migration to Hudson Bay, generally unharmed by the encounter.

The tissue samples are sent to a laboratory for analysis. The results Ramsay gets from the lab are worrisome. The bear's fat tissue contains polychlorinated biphenyls, better known as PCBs. This family of synthetic organic compounds has been used industrially in electrical and mechanical equipment for decades. Originally scientists believed that PCBs were biologically inert, but in the 1970s they discovered that laboratory animals exposed to PCBs had higher than normal rates of birth defects and liver cancer.

Ramsay has found that the concentration of PCBs in the bear's fat tissue is 8 ppm (parts per million). So far, this concentration of PCBs has caused the bear no apparent harm. But scientists are concerned about the polar bears because the way that PCBs dissolve in fat makes it possible for the concentration to increase in the bears even if the amount of PCBs in the environment does not change.

Ramsay, along with other scientists, is working to answer some puzzling questions.

- *How did the PCBs get into the Hudson Bay ecosystem when there is very little industry in the bay area?*
- *Why did PCBs end up in the bear's fat?*

These questions can be answered by research that requires a knowledge of how oily chemicals mix and dissolve. In this chapter, you will explore the principles of chemical solubility, and then you will apply those principles to answer some of the questions that environmentalists have about PCBs.

Tapping
Prior Knowledge

Remind students of concepts they will need for this chapter by giving the following review quiz.

Review quiz

1. Name at least two ways that mixtures differ from compounds.
2. Why is salt mixed with water classified as a solution?

Remind students of everyday experiences relevant to this chapter with the following questions.

Experience-based questions

1. Why do you use soap when you wash your hands?
2. What do you observe when oil and water are mixed together? Why does this happen?
3. Ask students to share any experiences they have had with removing stains from clothing. What solvents did they use and ask them to speculate as to why they were or were not successful in removing the stain.

Section 11-1

FOCUS

Lesson Starter

Have students make a list of common household solutions. Discuss the lists in class. Note misconceptions that surface during the discussion.

Demonstration 1

Liquid mixtures

Approximate time: 10 min

This demonstration should promote a discussion of solubility, suspensions, and miscibility.

1. Add 50 mL of H_2O to 50 mL of ethanol in a small Erlenmeyer flask. Stopper, shake, and set down.
2. Add 50 mL of H_2O to 50 mL of vegetable oil. Stopper, shake, and set down.
3. Add 5 g of sucrose to 50 mL of H_2O. Stopper, shake, and set down.
4. Add 10 g $Mg(OH)_2$ to 50 mL of H_2O. Stopper, shake, and set down.
5. Discuss: flask 1 is miscible, 2 is immiscible, 3 is soluble, and 4 is insoluble and a suspension.

SAFETY

Wear safety goggles and apron; avoid flames. Students should remain at least 10 ft from the demonstration.

DISPOSAL

Mix all liquids together, add a little detergent, and shake. Dilute with five times its volume of water, mix, and pour down drain.

TEACH

Teaching Tip

Vocabulary development

Discuss the meanings of the Latin roots *hetero* (different) and *homo* (same) as they relate to heterogeneous and homogeneous solutions.

What is a solution?

Section Objectives

Distinguish between solutions and suspensions.

Given various types of solutions, determine the solute and solvent.

Interpret solubility graphs and tables.

Distinguish among unsaturated, saturated, and supersaturated solutions.

Solutions are mixtures

In Chapter 2 a solution was defined as a homogeneous mixture. Every day, people use solutions such as mouthwash, diet soda, window cleaner, and gasoline. Most of the chemicals people use at home or on the job are solutions. Most of the chemicals you use in your chemistry lab are in solution because reactions often occur faster in solution. In this chapter, you will examine the properties of solutions in detail.

Solutions are stable, homogeneous mixtures

A student working in a pet shop is asked to prepare some water for a saltwater aquarium. She fills an aquarium with fresh water, then adds the proper amount of Instant Ocean salt crystals to the aquarium, as shown in **Figure 11-1**, and stirs. After stirring, the student can no longer see salt particles that are distinct from the water. No matter how long she waits, the salt will not spontaneously separate itself from the water. The salt has dissolved to form a stable, homogeneous mixture—the particles of each substance are evenly dispersed throughout the mixture.

Figure 11-1a
Fresh water is stable and homogeneous.

Before mixing

Fresh water

Water molecule

After mixing

Saltwater solution

Water molecule

Chloride ion, Cl^-

Sodium ion, Na^+

11-1b
The saltwater mixture is stable and homogeneous because mixing occurs between molecules and ions.

Visual Strategy

Figure 11-1

Students should recognize the even distribution of the water molecules, sodium ions, and chloride ions in the close-up picture in **Figure 11-1b**. Stress that the hydrogen atoms in the water molecule are attracted to nearby chloride ions, while the oxygen atoms in the water molecule are attracted to the sodium ions. These attractions promote dissolving.

Suspension particles will settle

The sculptor in **Figure 11-2** uses water to join the sides of a clay vase to its bottom. As she dips her fingers into the water, it turns gray-brown. The clay-water mixture seems uniform. However, if the container sits overnight, the next day she will see a muddy layer at the bottom with clear water above. Because the ingredients of this mixture show visible cloudiness and separate spontaneously, it is *not* a solution. This kind of mixture is called a **suspension**. In a suspension, the particles remain thoroughly mixed while the liquid is being stirred, but later they settle to the bottom. Unlike the salt crystals, the clay does not dissolve in water.

suspension
mixture that appears uniform while being stirred, but separates into different phases when agitation ceases

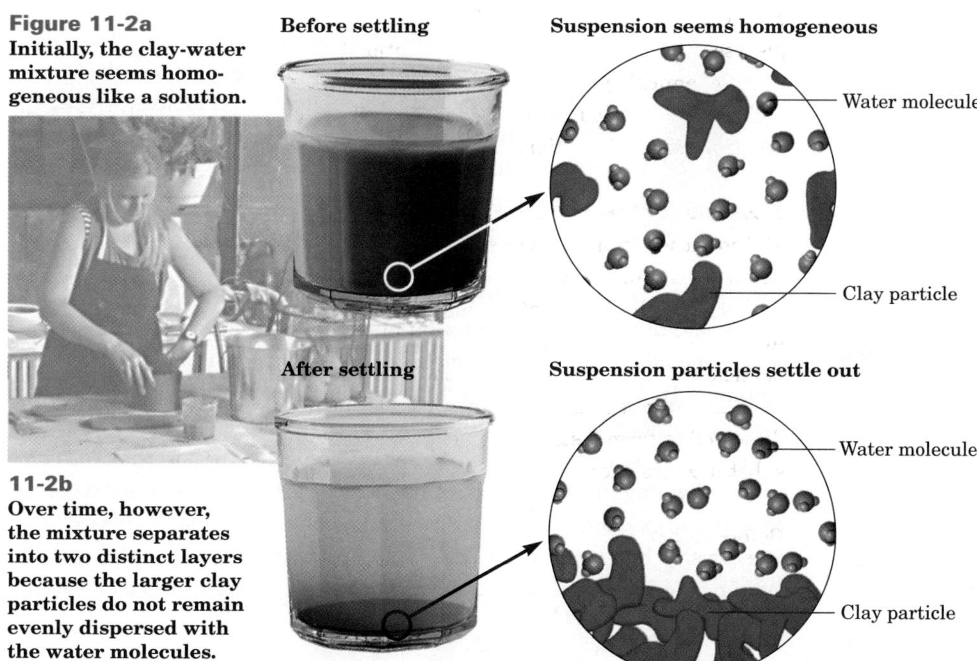

Figure 11-2a
Initially, the clay-water mixture seems homogeneous like a solution.

11-2b
Over time, however, the mixture separates into two distinct layers because the larger clay particles do not remain evenly dispersed with the water molecules.

Describing solutions

The solute dissolves in the solvent

The simplest solutions have two ingredients—the solvent and the solute. The **solvent** is usually present in a larger amount than the solute, and chemists think of the **solute** as dissolving in the solvent. In the first example, in which a student prepared a saltwater aquarium, sea salt was the solute, and water was the solvent.

Most solutions are combinations of a *solid* solute with a *liquid* solvent, but solutions of other states are possible. Many liquids dissolve, at least partially, in other liquids. Antifreeze, for example, dissolves in water. If two liquids can mix in any proportions (like antifreeze and water), they are said to be **miscible**. If they can't mix, they are **immiscible** (like vegetable oil and water). *Miscibility* is just another term that chemists sometimes use for the ability of liquids or gases to make solutions.

solvent
the material dissolving the solute to make the solution

solute
the material dissolved in a solution

miscible
indicates liquids or gases that will dissolve in each other

immiscible
indicates liquids or gases that will not dissolve in each other

Do You Know?

Ocean water contains an average of about 3.33% salt by mass. The main salts present are NaCl, $MgCl_2$, $MgSO_4$, $CaSO_4$, KCl, $MgBr_2$, and $CaCO_3$.

Demonstration 2

Volume of solution, solvent, and solute

Approximate time: 10 min.

Demonstrate that the volume of a solution is not the combined volume of solute and solvent.

- Mix 100 mL of water and 100 mL of ethanol in a 250 mL graduated cylinder. The combined volume will be less than 200 mL because the alcohol molecules can slip between water molecules.

SAFETY
Wear safety goggles and apron; avoid flames. Students should remain 10 ft from the demonstration.

DISPOSAL
Mix all liquids together. Add 500 mL of water and pour down the drain.

Teaching Tip

Discussion
Ask students to list common solvents they have around the house. Examples include water, kerosene, carpet cleaners, oven cleaners, hair spray, WD-40, etc. Ask students to name the hazards associated with common solvents. Major hazards include toxicity and flammability.

Historical Note

Svante Arrhenius nearly lost his doctorate in 1884 when he proposed that water tore apart $CuCl_2$ and dissociated it into ions. He did get his doctorate, but with the lowest possible grade. Nineteen years later he received the Nobel Prize for the same theory!

Visual Strategy

Figure 11-2
Help students interpret **Figure 11-2a** by stating that the clay particles are much larger than the water molecules but are able to be suspended for a short time by the many collisions with the nearby water molecules. Eventually gravity wins out, and the particles fall to the bottom of the jar.

Do You Know?
A recipe for white gold is about 58% gold, 17% nickel, 7% zinc, and 17% copper.

Teaching Tip
Discussion
Ask the students why people who like very sweet iced tea sweeten hot tea with sugar then pour it over ice cubes. Most should understand that if sugar is added to cold iced tea, the solubility of sugar in the tea will be lower, so less sugar can be dissolved in it.

Possible Misconception
Students may believe that all solvents are liquids. Discuss examples of solutions in which the solvent is not a liquid, such as the palladium-hydrogen stove lighter.

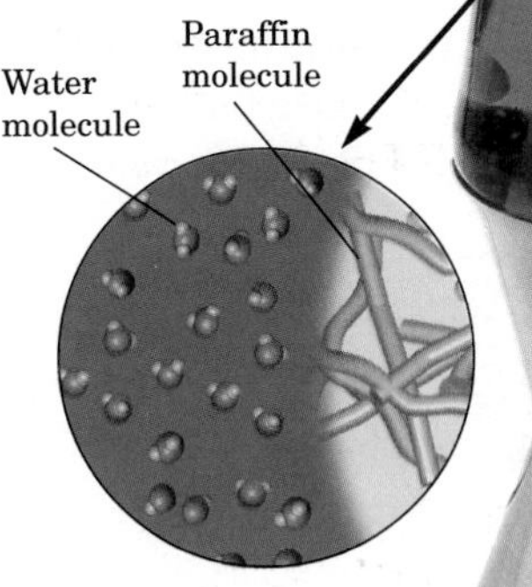

Figure 11-3 **Because paraffin and water are immiscible, the thick paraffin mixture flows through the water mixture in a lava lamp, creating interesting shapes.**

The contents of the lava lamp shown in **Figure 11-3** are an example of immiscibility. A light bulb concealed in the base of the lamp illuminates and heats a glass container that contains two immiscible liquids. Heat from the light bulb melts the thick paraffin mixture so that it oozes and flows through the colored water like melted candle wax.

Gases can also be solutes in a liquid solution. A bottle of soda gets its fizz from carbon dioxide gas dissolved in water. Fish depend on the small amount of dissolved oxygen that is in water.

Solvents are not always liquid. Can a solid dissolve in another solid? No, and yes. If you combine a piece of copper and a piece of gold, they will not mix; but if they are heated until they melt, gold and copper will dissolve in each other. When cooled, they form a solid mixture of metals, which is called an **alloy**. By mass, gold coins are usually 10% copper and 90% gold. Gases always dissolve in each other. For example, air is a solution of gases including nitrogen, oxygen, argon, and carbon dioxide.

alloy
a solid or liquid mixture of two or more metals

Solubility is the maximum that can dissolve

Salt is said to be *soluble* in water because the solid salt seems to disappear if enough water is added; no solid is left on the bottom of the container. By looking at **Figure 11-4**, you can see that a solute may be **soluble**, partly soluble, or **insoluble** in a particular solvent.

soluble
can be dissolved in a particular solvent

insoluble
does not dissolve appreciably in a particular solvent

The terms *soluble*, *partly soluble*, or *insoluble* are vague—there are no precise dividing lines among the three categories. When more precision is needed, the exact amount of a solute can be measured and expressed numerically. For example, at 20°C (about room temperature), 36.0 g of sodium chloride is the most that will normally dissolve in 100 g of water.

Figure 11-4 **The copper(II) nitrate in the test tube on the left is soluble in water. The copper(II) acetate in the center test tube is partly soluble in water. The copper(II) hydroxide in the test tube on the right is insoluble in water. (The black ring visible in the right test tube is a small amount of copper(II) oxide, formed as copper(II) hydroxide decomposes upon heating.)**

Visual Strategy
Figure 11-4
Point out to students that they can judge the relative solubility of the compounds in **Figure 11-4** in terms of either the amount of solid left undissolved or the darkness of the solution's color.

solubility
the maximum amount of a chemical that will dissolve in a given amount of a solvent at a specified temperature while the solution is in contact with some undissolved solute

When a solution holds the maximum amount of solute, the exact amount that has dissolved is called the **solubility**. For example, if the solubility of lithium iodide, LiI, is 165 g of LiI per 100 g of water at 20°C, 165 g is the *most* LiI that can dissolve in 100 g of water when solid LiI is present. *Any* quantity less than 165 g of LiI can dissolve in 100 g of water.

Temperature can have a great effect on solubility, as shown in **Figure 11-5**. This is why solubility values always include a temperature. For most solid solutes, as temperature increases, the solubility increases. For most gases, such as sulfur dioxide, SO_2, the opposite is true. Graphs and solubility tables are useful planning tools for making solutions.

Temperature-Solubility Relationships*

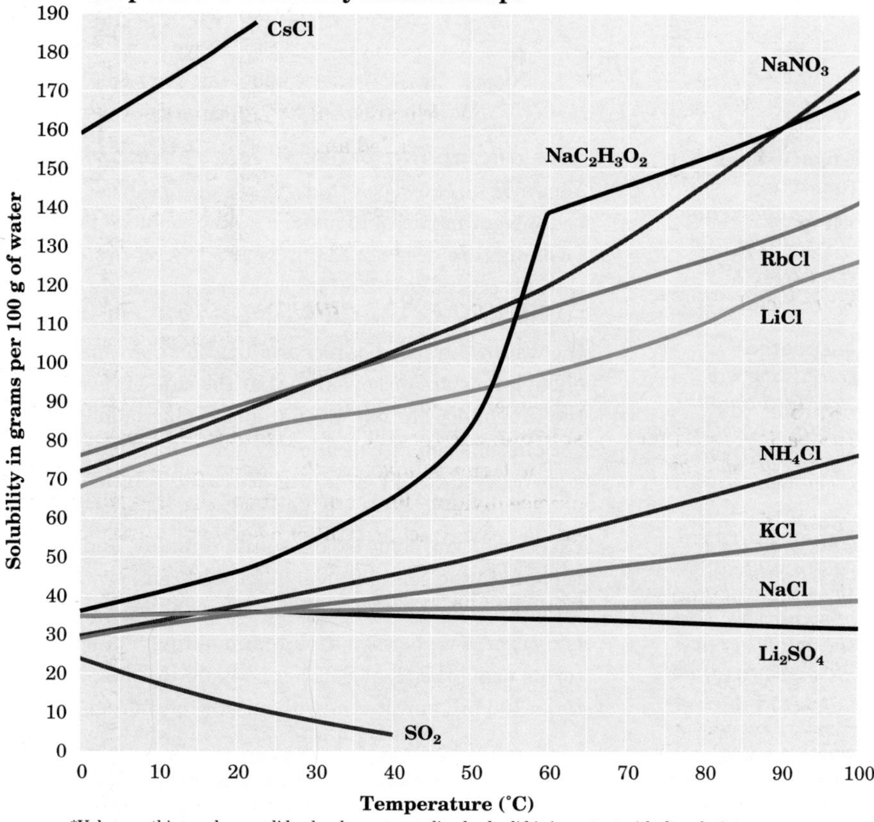

Figure 11-5
Notice that most ionic compounds, but not all, have solubilities in water that increase as temperature increases. The opposite is true for gases such as sulfur dioxide, SO_2.

*Values on this graph are valid only when some undissolved solid is in contact with the solution. For SO_2, values are valid when pressure due to SO_2 and water vapor is 101.325 kPa.

Concept Review

Using solubility graphs

Use **Figure 11-5** to answer the following questions.

1. What is the solubility of lithium chloride at 60°C?
2. Does rubidium chloride or sodium nitrate have a greater solubility at 0°C? at 70°C?
3. Which solute's solubility in water is most temperature dependent: sodium nitrate, potassium chloride, or sodium chloride?

Answers for Concept Review items and Practice problems begin on page 841.

Demonstration 3

Temperature/solubility I

Approximate time: 10 min

Dissolve sucrose in a few milliliters of water in a test tube. Keep adding sucrose with a spatula until no more will dissolve. Then heat the test tube in a hot water bath, and show that more sucrose can be added at a higher temperature.

SAFETY
Use tongs to hold the test tube; wear goggles and lab apron; tie back hair and loose clothing.

DISPOSAL
Pour down the drain.

Demonstration 4

Temperature/solubility II

Approximate time: 10 min

Show that increased temperature does not always mean more solute will dissolve.

1. Make a saturated solution of calcium acetate in a test tube at room temperature.
2. Heat the test tube in a hot water bath. The solution becomes cloudy as the calcium acetate precipitates out. Calcium acetate is an unusual salt that is less soluble in hot water than cold.

SAFETY
Wear goggles and lab apron; tie back hair and loose clothing.

DISPOSAL
Pour down the drain or save for reuse.

Visual Strategy

Figure 11-5

To fully utilize **Figure 11-5**, check that students understand that not all solubilities increase with increased temperature (e.g. $LiSO_4$). Also, students should be able to read information from the graph at any given temperature.

Answers to Concept Review

1. 98 g
2. RbCl at 0°C, $NaNO_3$ at 70°C
3. $NaNO_3$

Demonstration 5

Saturation points

Approximate time: 40 min

This demonstration can be set up at the beginning of class and allowed to run during class.

1. To four separate Pyrex test tubes, add 10 mL of distilled H_2O and one of the following: 3 g, 4 g, 5 g, and 6 g of NH_4Cl. Mark each test tube with the amount of ammonium chloride it contains.
2. Place the test tubes in a water bath using a 600 mL beaker on a ring stand. The bath water should be just above the water level in the test tubes.
3. Heat the beaker using a burner or hot plate until the solutes in the four test tubes have dissolved.
4. Remove the heat source and place a nonmercury thermometer in each test tube. Give the thermometers a chance to warm up, and then remove the test tubes from the beaker (with thermometers intact) and place them side by side in a test tube rack.
5. When the first student notices crystallization, read the temperature of the crystallizing solution. You may wish to graph the solubility curve at the end of class.

SAFETY

Wear goggles and lab apron; tie back hair and loose clothing. Students should remain 10 ft from the demonstration.

DISPOSAL

Combine all liquid and solids into an Erlenmeyer flask. Evaporate the water and save the ammonium chloride for future use.

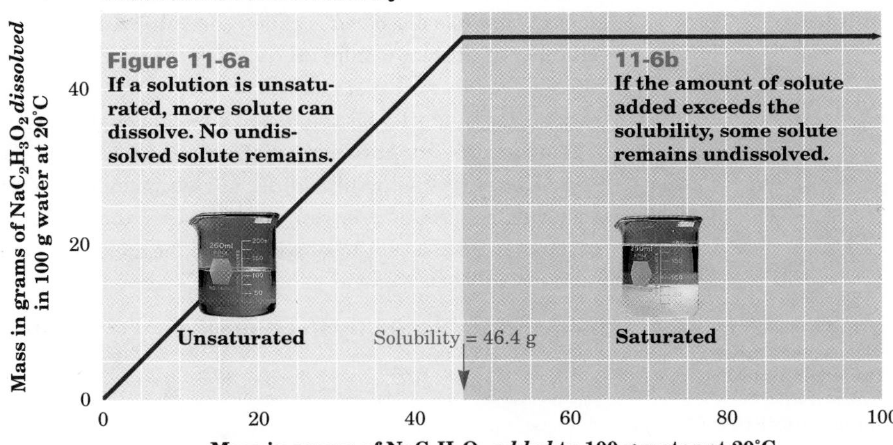

Figure 11-6a If a solution is unsaturated, more solute can dissolve. No undissolved solute remains.

11-6b If the amount of solute added exceeds the solubility, some solute remains undissolved.

unsaturated *containing less than the standard amount of solute specified by the solubility at a given temperature*

saturated *containing the standard amount of solute specified by the solubility at a given temperature*

supersaturated *containing more than the standard amount of solute specified by the solubility at a given temperature*

The dissolving process is a reversible reaction

You have already seen that when salt dissolves in water, its ions mix with the solvent particles to make a stable, homogeneous mixture. But this uniform mixture does not happen instantly. At first, the ions on the edge of the salt crystal break off into water and quickly dissolve. Some of these ions bump back into the crystal and recrystallize. As more ions dissolve in the water, the speed of recrystallization increases. At the solubility value, enough ions are dissolved so that the rate of recrystallization is equal to the rate of dissolution. Even if more salt is added, no more will dissolve.

Solutions can be classified by how they relate to the solubility value, as shown in **Figure 11-6**. In **unsaturated** solutions, the amount of solute is so much less than the solubility that the dissolving is complete. In a **saturated** solution, some excess solute remains, and the amount that dissolves is equal to the solubility value for that temperature. **Supersaturated** solutions are able to contain more than the usual solubility, as long as there is no excess undissolved solute remaining. **Figure 11-7** explains these terms using the Heat Solution, a hand warmer. The Heat Solution contains 100 mL of water and about 60 g of sodium acetate, $NaC_2H_3O_2$.

Figure 11-7a At 100°C, 60 g of $NaC_2H_3O_2$ will dissolve completely in the 100 mL of water contained in a Heat Solution pack.

11-7b When the solution is cooled to 20°C, $NaC_2H_3O_2$ does not recrystallize . . .

11-7c . . . unless the solution is disturbed. Clicking the disk in the center of the pack triggers rapid exothermic recrystallization.

11-7d The heat pack can be re-used if it is heated above the saturation point again.

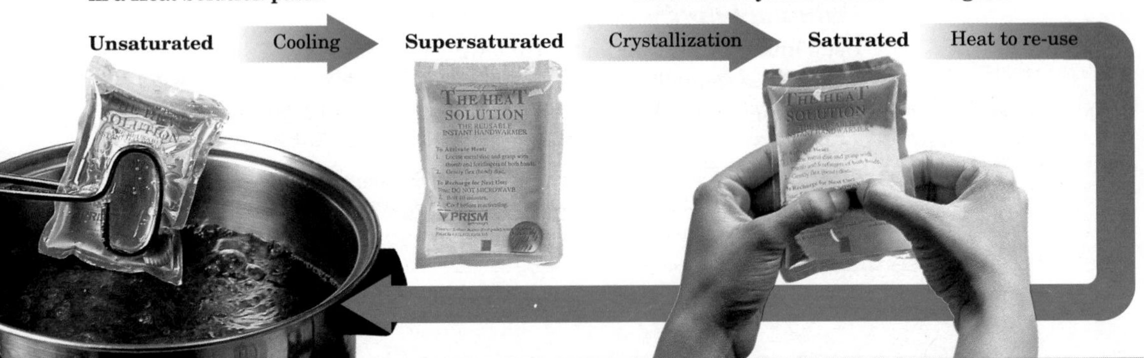

Visual Strategy

Figure 11-6

Help students to understand the two halves of **Figure 11-6**. In the left part, as solute is added, it all dissolves. Once the solubility is reached, no more can dissolve. The mass of dissolved solute will remain the same, as shown by the flat line on the graph. Any excess solute will remain undissolved.

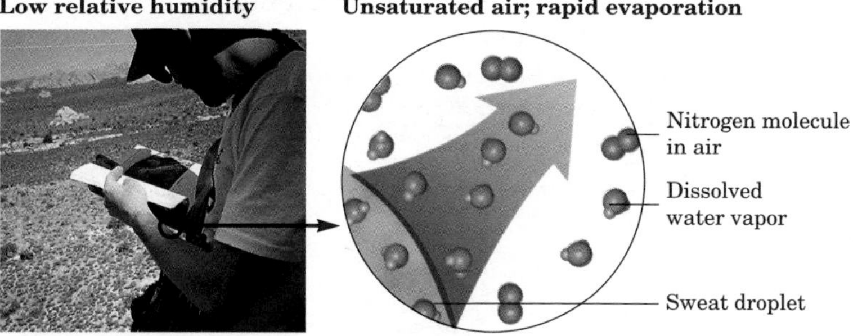

Figure 11-8a
On a 35°C day in New Mexico with a low relative humidity (37%), perspiration will evaporate quickly, cooling the body.

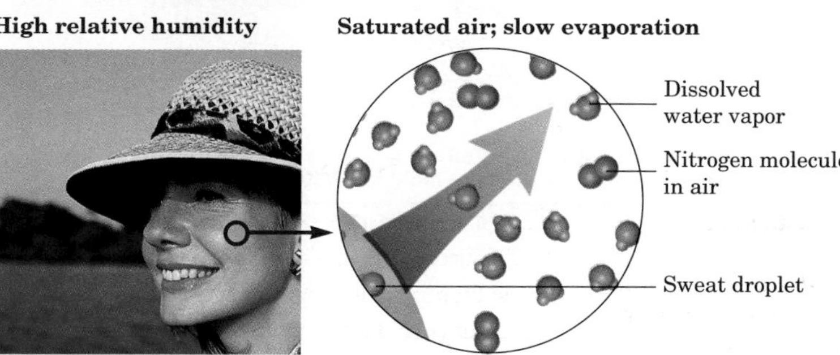

11-8b
On a 35°C day in Louisiana with a high relative humidity (92%), perspiration evaporates more slowly because moisture condenses from the air almost as fast as it evaporates from the body.

An example of saturation is the relative humidity figure given in the weather report. At any given temperature the air can hold a certain amount of water in the form of gaseous vapor. As long as the relative humidity is less than 100%, the air is unsaturated because it holds less than the maximum amount of water vapor. **Figure 11-8** shows why the air feels "sticky" when it is very hot, and the relative humidity is high. When the relative humidity reaches 100%, the air is saturated and can hold no more water vapor in solution. If more water is present, it will come out of solution as fog, rain, dew, frost, or other precipitation.

Actual practice shows that solutions contain more dissolved solute than the usual solubility limits. This can be achieved by carefully cooling a nearly saturated solution. Such supersaturated solutions are not breaking the rules. Solubility was defined as the maximum that can dissolve *when the solution is in contact with undissolved solute*. The supersaturated solution no longer holds any undissolved solute. Supersaturated solutions are not very stable. If you disturb the solution, you will see rapid crystallization of solids or bubbling of gases.

Some substances don't dissolve

It is easy to get the *misimpression* that all substances dissolve in water. You know that copper, zinc, and iron do not dissolve in water and are classified as insoluble. Some metal salts dissolve slightly but, for practical purposes, are also classified as insoluble in water. **Table 11-1** on the next page summarizes the solubility of many compounds and can be very useful. For example, if an experiment requires silver ions, it would be a waste of time to try to use silver chloride, but silver nitrate would serve your purpose very well.

Chem Fact
Many tests for analyzing dissolved metal ions make use of solubility principles. A sample is mixed with reagents that form an insoluble solid precipitate with some metal ions.

Demonstration 6
Supersaturation I
Approximate time: 30 min
Students will enjoy seeing a supersaturated solution "crash out." Fill a large demonstration-size test tube two-thirds full with sodium thiosulfate pentahydrate. Heat until all of the solid appears to melt as it dissolves in its own water molecules. Let it cool to room temperature. Then drop one small seed crystal of $Na_2S_2O_3 \cdot 5H_2O$ into the solution. The tube will become warm as the crystals "crash out."

SAFETY
Wear goggles and lab apron; tie back hair and loose clothing. Students should remain 10 ft from the demonstration.

DISPOSAL
Pour down the drain or save for reuse.

Demonstration 7
Supersaturation II
Approximate time: 10 min
Hand warmers can be used to demonstrate supersaturation. They can be purchased at ski shops and camping stores.

SAFETY
Read the label and follow all safety precautions.

DISPOSAL
Follow disposal directions on label.
Note: If the label does not provide safety and disposal information, ask the manufacturer or use a different brand that is labeled.

Visual Strategy
Figure 11-8
Help students identify the differences in the two parts of **Figure 11-8**. In the top half, there is very little water vapor dissolved in the air, so evaporation is rapid. In the bottom half, there is a lot of water vapor dissolved in the air, so evaporation is slow. Point out that the amounts of water vapor in the figure have been exaggerated to make the differences apparent.

Other Applications

Two popular preservatives for meats (e.g., cold cuts) are sodium nitrate and sodium nitrite. When meats are prepared by boiling them in water, much of the preservative dissolves. This example shows a practical application of the temperature dependence of solubility.

ASSESS

Alternative Assessment

Ask students to illustrate the following on a microscopic, or particle, level: a glucose-in-water solution, a mercury-silver-zinc alloy, and a sand-in-water suspension. Each illustration should depict all particles present and their distribution. Particles should be labeled.

Answers to Section Review

4. a. suspension
   b. suspension
   c. solution
5. a. solute—carbon dioxide; solvent—water
   b. solute—sugar; solvent—water
6. a. supersaturated
   b. saturated
   c. unsaturated
7. a. yes
   b. no
   c. yes
   d. no
8. yes; yes; yes; silver chloride (AgCl)
9. (most soluble) CsCl, LiCl, NH_4Cl, NaCl, $PbCl_2$ (least soluble)

Table 11-1
Solubility of Common Compounds

Compounds containing these ions are **soluble** in water . . .		. . . unless they also contain these ions, which make them **insoluble**.
ammonium	NH_4^+	
potassium	K^+	
sodium	Na^+	
acetate	$C_2H_3O_2^-$	Fe^{3+}, Al^{3+}, Hg_2^{2+}
chlorate	ClO_3^-	
chloride	Cl^-	Ag^+, Hg_2^{2+}, Pb^{2+}
nitrate	NO_3^-	
sulfate	SO_4^{2-}	Ca^{2+}, Ba^{2+}, Pb^{2+}, Sr^{2+}, Hg_2^{2+}
Compounds containing these ions are insoluble in water . . .		**. . . unless they also contain these ions, which make them soluble.**
carbonate	CO_3^{2-}	K^+, Li^+, Na^+, NH_4^+
hydroxide	OH^-	K^+, Li^+, Ba^{2+}, Na^+
oxide	O^{2-}	
phosphate	PO_4^{3-}	K^+, Na^+, NH_4^+
silicate	SiO_3^{2-}	K^+, Na^+
sulfide	S^{2-}	K^+, Na^+, NH_4^+
sulfite	SO_3^{2-}	K^+, Na^+, NH_4^+

Section Review

4. Identify the following as solutions or suspensions.
   a. muddy river water **b.** orange juice
   c. chlorinated water in a swimming pool
5. Name the solute(s) and solvent in these solutions.
   a. carbonated water **b.** sugar water
6. Categorize the following solutions as saturated, unsaturated, or supersaturated. (Hint: refer to **Figure 11-5**.)
   a. 38 g of NaCl stirred into 100 g of water at 99°C, then cooled to 25°C
   b. 50 g of NaCl added to 100 g of water, with some left undissolved on the bottom
   c. 10 g of NaCl stirred into 100 g of water
7. Use **Table 11-1** to determine whether the following compounds are soluble in water.
   a. copper(II) sulfate **b.** copper(II) sulfite
   c. iron(III) chloride **d.** aluminum acetate
8. If you mixed silver nitrate with water, would it dissolve? If you mixed potassium chloride with water, would it dissolve? If you mixed the contents of these two combinations, would anything settle to the bottom?
9. Using **Table 11-1** and **Figure 11-5**, place the following compounds in order from most soluble to least soluble: LiCl, NH_4Cl, $PbCl_2$, CsCl, NaCl.

What does concentration mean?

Section Objectives

Relate various ways of expressing concentration data.

Recognize that units of concentration represent a ratio.

Use units of molarity in stoichiometric calculations.

Describe the procedure for making solutions of different molarities.

Concentration as a ratio

concentration *ratio of solute to solvent or solution*

To describe saturation precisely, a quantitative measure is needed—units that measure **concentration**. The important feature of all concentration measures is the ratio expressed. There are many ratios that express how much of some substance is present. Certain expressions are favored in medicine, others in pollution control, and others in biological research, as described in **Table 11-2**. Because chemists are concerned with the interaction of substances, they must know not only how much is present, but also how the amount compares, molecule for molecule, with another chemical.

Table 11-2
Concentration Units

Name	Abbr.	Units	Uses
grams/100. g	g/100. g	$\frac{\text{g solute}}{\text{100. g solvent}}$	solubility descriptions, medical products
mass percent or "weight percent"	%	$\frac{\text{g solute}}{\text{100. g solution}}$	biological research
parts per million	ppm	$\frac{\text{g solute}}{\text{1 000 000. g solution}}$ *	small concentrations
parts per billion	ppb	$\frac{\text{g solute}}{\text{1 000 000 000. g solution}}$ *	very small concentrations, as in pollutants or contaminants
parts per trillion	ppt	$\frac{\text{g solute}}{\text{1 000 000 000 000. g solution}}$ *	extremely small concentrations, as in isotopes used as tracers in medicine
molarity	M	$\frac{\text{mol solute}}{\text{L solution}}$	laboratory chemistry, where the solute may undergo a chemical change according to a mole ratio
molality	m	$\frac{\text{mol solute}}{\text{kg solvent}}$	calculation of special properties such as boiling-point elevation and freezing-point depression

*volume for gases

Molarity

molarity *concentration unit, expressed as moles of solute per liter of solution*

The mole is the unit used to count atoms and molecules and to relate the amounts of reactants and products in chemical reactions. The mole is also the basis of the concentration measurement called **molarity**.

$$\textbf{molarity} = \frac{\textbf{moles solute}}{\textbf{liters of solution}} \qquad \mathbf{M} = \frac{\textbf{mol}}{\textbf{L}}$$

Even though molarity is just *one* method of measuring concentration, it can be represented by *four* labels: *molarity*, *molar*, *M*, and *mol/L*. These labels all indicate the concentration of a solution as a ratio of moles of solute per liter of solution.

Section 11-2

FOCUS

Lesson Starter

Display several solutions of potassium permanganate with various concentrations. Ask students to guess which solution is most concentrated. Most will intuitively know that the most concentrated solution has the deepest color. Ask students to formalize their intuitive concept of concentration first by developing an equation to express it and second by postulating possible concentration units.

Demonstration 8

Concentration

Approximate time: 20 min

1. Place 1 g, 2 g, and 3 g of sucrose into three separate test tubes.
2. Add 10 mL of water to each tube. Stopper each tube.
3. Have a student shake each tube until the sugar dissolves.
4. Now discuss and compare the concentrations of the solutions by using the words *dilute* and *concentrated*. For example, compare tubes 1 and 2 by asking which is more concentrated. Then compare tubes 2 and 3 by asking which is more dilute. Point out that the same tube can thus be considered both dilute and concentrated. Have students suggest methods to eliminate the inconsistencies.

SAFETY
The teacher and student demonstrators should wear goggles and lab aprons.

DISPOSAL
Pour down the drain.

TEACH

Teaching Tip

Discussion

Ask students what it means to say coffee or tea is too strong. How could the statement be more precise?

Demonstration 9

The meniscus

Approximate time: 15 min

Pass around a volumetric flask. Review the procedure for filling the flask so that the bottom of the meniscus touches the inscribed volume mark. Discuss the fact that aqueous solutions form a meniscus because the polar attraction of water for glass causes the solution to climb. Ask the students why this effect is much more prevalent in glass containers than plastic ones.

SAFETY

The teacher and students should wear safety goggles.

DISPOSAL

No disposal required.

Possible Misconception

Molarity

Many students have a difficult time rationalizing that the molarity of a solution is constant regardless of its volume. Make 1 L of a 1 M solution. Pour samples of various volumes into beakers. Ask which beaker has the highest molarity. Refer students to **Figure 11-9** and discuss the constant slope of the graph.

Figure 11-9
If several samples of different volumes are taken from the same solution, the moles of solute and liters of solution are in direct proportion. The ratio (molarity) is a constant, as you can see by selecting any point along the line.

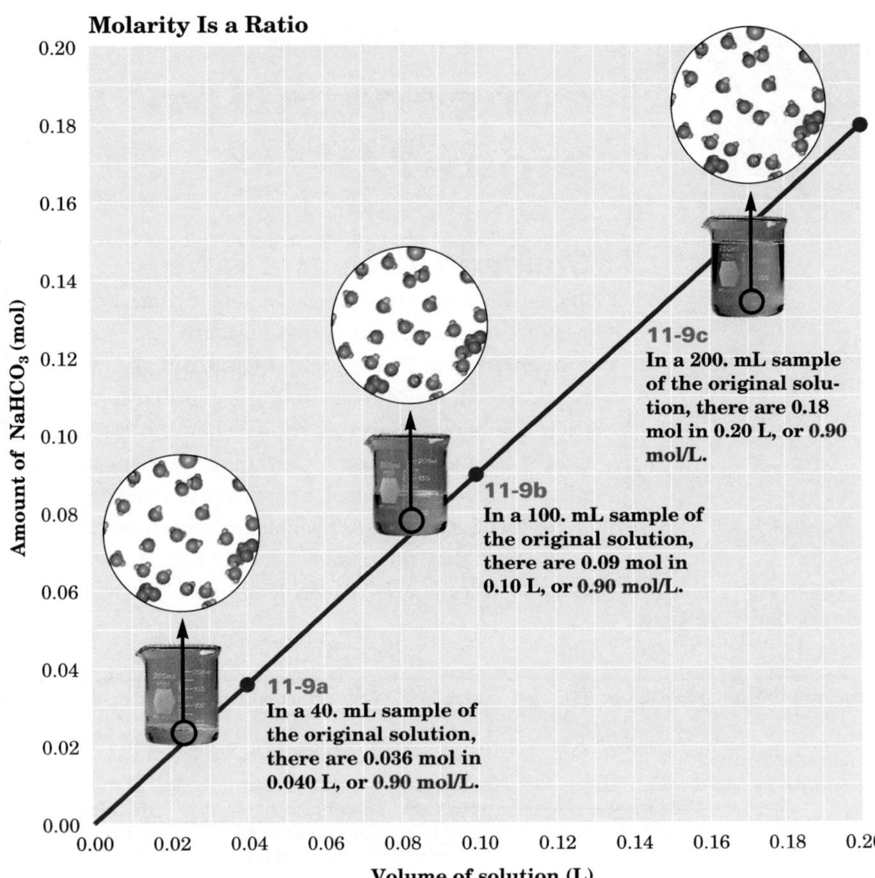

Molarity Is a Ratio
Amount of $NaHCO_3$ (mol)
0.00
0.02
0.04
0.06
0.08
0.10
0.12
0.14
0.16
0.18
0.20
Volume of solution (L)
0.00 0.02 0.04 0.06 0.08 0.10 0.12 0.14 0.16 0.18 0.20
11-9a
In a 40. mL sample of the original solution, there are 0.036 mol in 0.040 L, or 0.90 mol/L.
11-9b
In a 100. mL sample of the original solution, there are 0.09 mol in 0.10 L, or 0.90 mol/L.
11-9c
In a 200. mL sample of the original solution, there are 0.18 mol in 0.20 L, or 0.90 mol/L.

Determining molarity

It is easy to calculate molarity when the volume of solution is exactly 1.0 L. Of course, most situations involve more or less than 1.0 L of solution. Keep in mind that molarity is a *ratio*, the moles of solute *divided by* the liters of solution. Once you make a 0.9 M solution, the ratio of amount of solute to volume of solution is *constant*, regardless of how much of the solution you sample. This idea is shown in **Figure 11-9**.

Before calculating the molarity of a solution, be sure the amount of solute and the volume of solution are expressed in the proper units.

$$\frac{\text{mol solute}}{\text{L solution}}$$

To prepare a solution of a given concentration, if you just mix a liter of solvent and the proper amount of solute in moles, the final solution will have a volume slightly greater or smaller than 1.00 L depending on the attractive forces among the solute and solvent particles. The definition of molarity is based on the number of moles in the final volume of the solution. To make a solution that contains exactly 1.00 L when finished, it is necessary to follow the very specific procedure shown on the next page.

Visual Strategy

Figure 11-9

Point out to students that although the volumes of solution and amounts of solute change in each beaker, as indicated by the line on the graph, the proportion of the two stays the same for a solution of the same concentration, as shown by the microviews of each solution.

How To

Prepare a 0.5000 molar solution

Start by calculating the mass of $CuSO_4 \cdot 5H_2O$ needed. To make this solution, each liter will require 0.5000 mol of $CuSO_4 \cdot 5H_2O$. To convert this amount to a mass, multiply by the molar mass of $CuSO_4 \cdot 5H_2O$.

Add some solvent to the $CuSO_4 \cdot 5H_2O$ to dissolve it, then pour it into a 1.000 L volumetric flask.

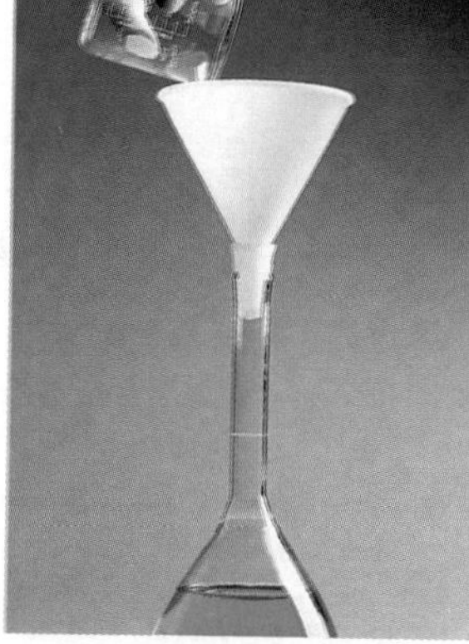

Rinse the weighing beaker with more solvent, and pour it into the flask until the solution nears the neck of the flask.

Put the stopper in the flask, and swirl it thoroughly.

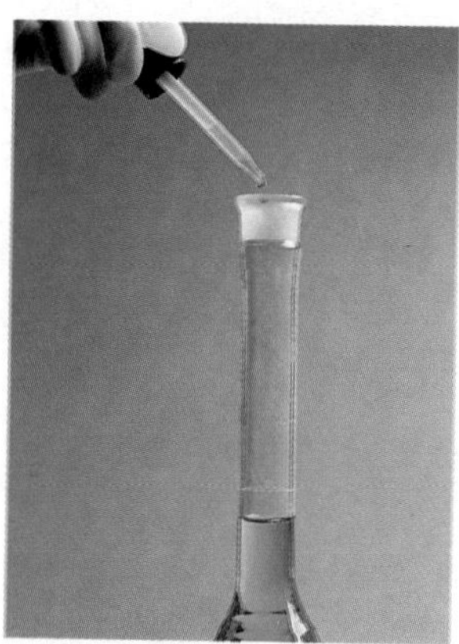

Carefully fill exactly to the 1.000 L mark with more solvent.

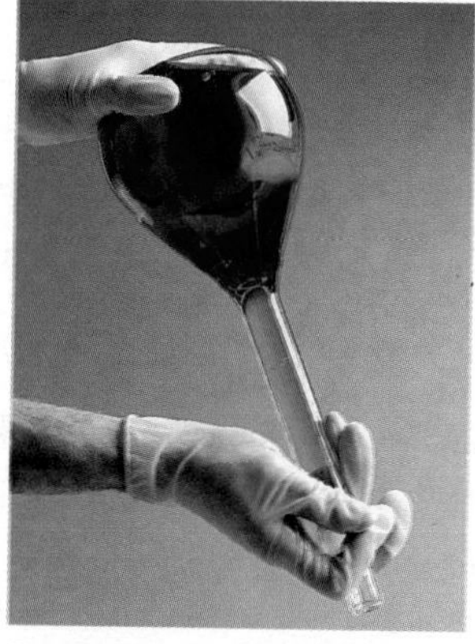

Restopper, and invert at least 10 times to ensure complete mixing.

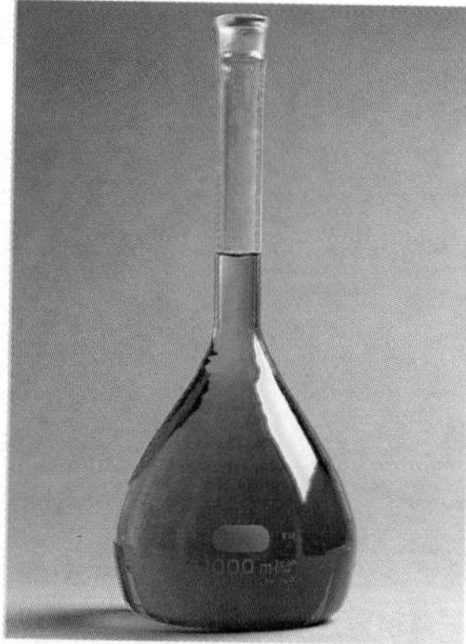

The resulting solution has 0.5000 mol of solute dissolved in 1.000 L of solution—a 0.5000 M concentration.

Teaching Tip

Discussion

Write a balanced equation on the chalkboard, such as the following.

$$2KCl + MgS \longrightarrow K_2S + MgCl_2$$

Stress that the coefficients represent molar relationships.

Visual Strategy

How to Prepare a 0.500 Molar Solution

Point out the number of steps that promote precision. First, calculations are made to be certain the correct amounts will be used. Then, the weighing beaker is rinsed with solvent to be sure all of the solute is poured into the flask. To be certain the solute is fully dissolved, the flask is swirled thoroughly before it is completely full. Only then can a precise determination of solution volume be made.

Additional Example for Sample Problem 11A

Find the molarity of a $Mg(NO_3)_2$ solution in which 25.0 g of $Mg(NO_3)_2$ are dissolved in 1350 mL of water.

Solutions are at the bottom of the page.

Sample Problem 11A Calculating molarity

What is the molarity of a potassium chloride solution that has a volume of 400. mL and contains 85.0 g of KCl?

1 List

- **solution volume:** 400. mL
- **solute mass:** 85.0 g KCl
- **solute molar mass:** $\frac{74.55 \text{ g KCl}}{1 \text{ mol KCl}}$
- **solution concentration:** ? M KCl = $\frac{? \text{ mol KCl}}{\text{L}}$

2 Set up

- **Because the units of molarity are mol/L, the mass must be converted to an amount in moles and the volume of solution must be converted to liters.**

$$\overset{a}{\frac{85.0 \text{ g KCl}}{400 \text{ mL}}} \times \overset{b}{\frac{1000 \text{ mL}}{1 \text{ L}}} \times \overset{c}{\frac{1 \text{ mol KCl}}{74.55 \text{ g KCl}}} = \frac{? \text{ mol KCl}}{\text{L}}$$

a Molarity is moles of solute per volume of solution. Place the solute quantity over the solution amount, even though they are not in the correct units yet.

b Multiply by a conversion factor that will cancel the *mL* to give units of *L* in the denominator.

c Use the conversion factor from the molar mass that will cancel the mass of KCl to give the amount in moles.

3 Estimate and calculate

- **Before calculating, round off the numbers in the setup, and solve the math in your head.**

$$\frac{85}{400} \times \frac{1000}{1} \times \frac{1}{75} \approx \frac{17}{6} \approx \text{almost } 3$$

The answer is close to the estimate.

- **Use your calculator to find the actual answer, rounding to the correct number of significant figures, which is three.**

$$\frac{2.850\cancel{435949} \text{ mol KCl}}{\text{L}} = 2.85 \text{ M}$$

Additional Example for Sample Problem 11B

Find the amount of $AgNO_3$ needed to make 375 mL of a 0.750 M solution.

Solutions are at the bottom of the page.

Sample Problem 11B Variations with molarity: mass of solute

Sodium thiosulfate, $Na_2S_2O_3$, is used in developing photographic film. It is known to photographers as hypo or "fixer." How many grams of $Na_2S_2O_3$ are needed to make 100.0 mL of a 0.250 M solution?

1 List

- **solution volume:** 100.0 mL
- **solution concentration:** 0.250 M $Na_2S_2O_3$
- **solute mass:** ? g $Na_2S_2O_3$
- **solute molar mass:** $\frac{158.12 \text{ g } Na_2S_2O_3}{1 \text{ mol } Na_2S_2O_3}$

2 Set up

- **First, analyze what needs to be done to get the answer. You are trying to find out the mass in grams, and you are given a volume. First convert the volume to amount in moles using molarity.**

$$\frac{100.0 \text{ mL}}{1} \times \frac{1 \text{ L}}{1000 \text{ mL}} \times \frac{0.250 \text{ mol } Na_2S_2O_3}{1 \text{ L}} = 0.0250 \text{ mol } Na_2S_2O_3$$

Solution for Additional Example 11A

$$\frac{25.0 \text{ g } Mg(NO_3)_2}{1350 \text{ mL}} \times \frac{1000 \text{ mL}}{1 \text{ L}} \times \frac{1 \text{ mol } Mg(NO_3)_2}{148.32 \text{ g } Mg(NO_3)_2} = 0.125 \text{ M}$$

Solution for Additional Example 11B

$$\frac{375 \text{ mL}}{1} \times \frac{1 \text{ L}}{1000 \text{ mL}} \times \frac{0.750 \text{ mol } AgNO_3}{1 \text{ L}} \times \frac{169.88 \text{ g } AgNO_3}{1 \text{ mol } AgNO_3} = 47.8 \text{ g } AgNO_3$$

Note that this step is just like the mole-mass conversions from Chapter 3.

Then the number of moles can be converted to mass using the molar mass.

$$0.0250 \text{ mol } Na_2S_2O_3 \times \frac{158.12 \text{ g } Na_2S_2O_3}{1 \text{ mol } Na_2S_2O_3} = ? \text{ g } Na_2S_2O_3$$

- **Check the two steps to be sure that all units cancel except the proper unit for the answer, *g $Na_2S_2O_3$*.**

3 Estimate and calculate

- **Use your calculator to find the precise answer. Round off to the correct number of significant figures.**

$$0.0250 \text{ mol } Na_2S_2O_3 \times \frac{158.12 \text{ g } Na_2S_2O_3}{1 \text{ mol } Na_2S_2O_3} = 3.953 = 3.95 \text{ g } Na_2S_2O_3$$

- **Check the reasonableness of the answer.**

The number $\frac{0.250 \text{ mol } Na_2S_2O_3}{1 \text{ L}}$ is the same as $\frac{\frac{1}{4} \text{ mol } Na_2S_2O_3}{1 \text{ L}}$.

A quarter of the molecular weight (160 g) would be about 40 g. Because the solution is one-tenth of a liter, we only need one-tenth of that amount, about 4.0 g.

Sample Problem 11C
Variations with molarity: solution stoichiometry

How many mL of a 0.500 M solution of copper(II) sulfate, $CuSO_4$, are needed to react with an excess of aluminum, Al, to provide 11.0 g of copper?

1 List

- **solution concentration of reactant:** 0.500 M $CuSO_4$
- **mass of product:** 11.0 g Cu
- **solution volume:** ? mL $CuSO_4$ solution
- **balanced equation for reaction:**
  $3CuSO_4(aq) + 2Al(s) \rightarrow 3Cu(s) + Al_2(SO_4)_3(aq)$
- **molar mass of $CuSO_4$:** 159.62 g/mol
- **molar mass of Cu:** 63.55 g/mol
- **mole ratio:** 3 mol $CuSO_4$:3 mol Cu

Because Al is in excess, it is not a factor in solving the problem.

2 Set up

- **You are trying to find the solution volume from the mass of a substance in a reaction. The amount of Cu produced is given in grams. By converting it to moles, you can determine the number of moles of $CuSO_4$ needed because the mole ratio of Cu to $CuSO_4$ is 3 to 3.**

$$\frac{11.0 \text{ g Cu}}{1} \times \frac{1 \text{ mol Cu}}{63.55 \text{ g Cu}} \times \frac{3 \text{ mol } CuSO_4}{3 \text{ mol Cu}} \times \frac{\quad}{\quad} \times \frac{\quad}{\quad} = ? \text{ mL } CuSO_4 \text{ solution}$$

- **Convert the amount in moles to a volume, using the reciprocal of molarity.**

$$\frac{1 \text{ L solution}}{0.500 \text{ mol } CuSO_4}$$

- **Then, convert *L of solution* to *mL of solution*.**

$$\frac{11.0 \text{ g Cu}}{1} \times \frac{1 \text{ mol Cu}}{63.55 \text{ g Cu}} \times \frac{3 \text{ mol } CuSO_4}{3 \text{ mol Cu}} \times \frac{1 \text{ L solution}}{0.500 \text{ mol } CuSO_4} \times \frac{1000 \text{ mL solution}}{1 \text{ L solution}}$$

$$= ? \text{ mL } CuSO_4 \text{ solution}$$

Check that all units cancel, except the answer unit, mL solution.

Sample Problem 11C continued on next page . . .

Additional Examples for Sample Problem 11C

a. Determine the volume of 0.250 M $NaHCO_3$ needed to react with 5.00 g of H_2SO_4 and produce Na_2SO_4, CO_2, and H_2O.

b. Suppose that 48.0 mL of 0.395 M Na_2SO_4 are added to excess $Ca(NO_3)_2$ to produce $NaNO_3$ and insoluble $CaSO_4$. Find the mass of $CaSO_4$ produced.

Solutions are at the bottom of the page.

Solutions for Additional Examples 11C

a. $2NaHCO_3 + H_2SO_4 \longrightarrow Na_2SO_4 + 2H_2O + 2CO_2$

$$\frac{5.00 \text{ g } H_2SO_4}{1} \times \frac{1 \text{ mol } H_2SO_4}{98.09 \text{ g } H_2SO_4} \times \frac{2 \text{ mol } NaHCO_3}{1 \text{ mol } H_2SO_4} \times \frac{1 \text{ L}}{0.250 \text{ M } NaHCO_3} \times \frac{1000 \text{ mL}}{1 \text{ L}} = 408 \text{ mL } NaHCO_3$$

b. $Ca(NO_3)_2 + Na_2SO_4 \longrightarrow 2NaNO_3 + CaSO_4$

$$\frac{48.0 \text{ mL } Na_2SO_4}{1} \times \frac{1 \text{ L}}{1000 \text{ mL}} \times \frac{0.395 \text{ mol } Na_2SO_4}{1 \text{ L}} \times \frac{1 \text{ mol } CaSO_4}{1 \text{ mol } Na_2SO_4} \times \frac{136.15 \text{ g } CaSO_4}{1 \text{ mol } CaSO_4} = 2.58 \text{ g } CaSO_4$$

Answers to Practice Problems

1. $\frac{125 \text{ g NaOCl}}{2.00 \text{ L}} \times \frac{1 \text{ mol NaOCl}}{74.44 \text{ g NaOCl}} =$ 0.840 M

2. $15.3 \text{ g NaHCO}_3 \times \frac{1 \text{ mol NaHCO}_3}{84.01 \text{ g NaHCO}_3} \times \frac{1 \text{ L}}{0.375 \text{ mol NaHCO}_3} \times \frac{1000 \text{ mL}}{1 \text{ L}} =$ 486 mL

3. $256 \text{ g CH}_3\text{COOH} \times \frac{1 \text{ mol CH}_3\text{COOH}}{60.06 \text{ g CH}_3\text{COOH}} \times \frac{1 \text{ L}}{2.50 \text{ mol CH}_3\text{COOH}} = 1.70 \text{ L}$

4. $75.0 \text{ mL} \times \frac{1 \text{ L}}{1000 \text{ mL}} \times \frac{0.250 \text{ mol CuSO4•5H}_2\text{O}}{1 \text{ L}} \times \frac{249.72 \text{ g CuSO}_4\text{•5H}_2\text{O}}{1 \text{ mol CuSO}_4\text{•5H}_2\text{O}} =$ 4.68 g $CuSO_4•5H_2O$

5. $3.50 \text{ L} \times \frac{1.15 \text{ mol C}_{12}\text{H}_{22}\text{O}_{11}}{1 \text{ L}} \times \frac{342.35 \text{ g C}_{12}\text{H}_{22}\text{O}_{11}}{\text{mol C}_{12}\text{H}_{22}\text{O}_{11}} \times \frac{1 \text{ kg}}{1000 \text{ g}} =$ 1.38 kg $C_{12}H_{22}O_{11}$

6. $2H_3PO_4 + Ca(OH)_2 \longrightarrow Ca(H_2PO_4)_2 + 2H_2O$

$7.50 \text{ L H}_3\text{PO}_4 \times \frac{5.00 \text{ mol H}_3\text{PO}_4}{1 \text{ L H}_3\text{PO}_4} \times \frac{1 \text{ mol Ca(H}_2\text{PO}_4)_2}{2 \text{ mol H}_3\text{PO}_4} \times \frac{234.06 \text{ g Ca(H}_2\text{PO}_4)_2}{1 \text{ mol Ca(H}_2\text{PO}_4)_2} =$ 4390 g $Ca(H_2PO_4)_2$

$7.50 \text{ L H}_3\text{PO}_4 \times \frac{5.00 \text{ mol H}_3\text{PO}_4}{1 \text{ L H}_3\text{PO}_4} \times \frac{2 \text{ mol H}_2\text{O}}{2 \text{ mol H}_3\text{PO}_4} \times \frac{18.02 \text{ g H}_2\text{O}}{1 \text{ mol H}_2\text{O}} =$ 676 g H_2O

7. $Pb^{2+} + 2NaOH \longrightarrow Pb(OH)_2 + 2Na^+$

$\frac{0.285 \text{ g Pb(OH)}_2}{12.5 \text{ L}} \times \frac{1 \text{ mol Pb(OH)}_2}{241.2 \text{ g Pb(OH)}_2} \times \frac{1 \text{ mol Pb}^{2+} \text{ ions}}{1 \text{ mol Pb(OH)}_2} = 9.45 \times 10^{-5} \text{ M}$

ASSESS

Alternative Assessment

Have students make a concept map of the following vocabulary terms: *suspension, solvent, solute, miscible, immiscible, soluble, insoluble, solubility, unsaturated, saturated, supersaturated, concentration, molarity.*

Sample Problem 11C continued from previous page . . .

3 Estimate and calculate

- **Round off, and estimate.** $\frac{10}{1} \times \frac{1}{60} \times \frac{3}{3} \times \frac{1}{0.5} \times \frac{1000}{1} \approx 300$
- **Use the calculator, and record the proper significant figures. Because the starting value, 11.0 g Cu, has only three significant figures, the answer should also have three.**

$$\frac{11.0 \text{ g Cu}}{1} \times \frac{1 \text{ mol Cu}}{63.55 \text{ g Cu}} \times \frac{3 \text{ mol CuSO}_4}{3 \text{ mol Cu}} \times \frac{1 \text{ L solution}}{0.500 \text{ mol CuSO}_4} \times \frac{1000 \text{ mL solution}}{1 \text{ L solution}}$$

$= 346.1847107 \text{ mL} = 346 \text{ mL CuSO}_4$ solution

Practice 11A

Answers for Concept Review items and Practice problems begin on page 841.

1. What is the molarity of a sodium hypochlorite bleach that contains 125 g NaOCl in 2.00 L of solution?

2. How many milliliters of a 0.375 M solution of sodium hydrogen carbonate, $NaHCO_3$, are needed to provide 15.3 g of $NaHCO_3$?

3. How many liters of a 2.50 M solution of acetic acid, CH_3COOH, are needed to provide 256 g of acetic acid?

11B

4. How many grams of copper sulfate pentahydrate, $CuSO_4•5H_2O$, will be needed to make 75.0 mL of 0.250 M solution?

5. How many kilograms of table sugar (sucrose: $C_{12}H_{22}O_{11}$) will be needed to make 3.50 L of a 1.15 M solution?

11C

6. The calcium phosphate used in fertilizers can be made according to the following unbalanced equation.

$$H_3PO_4(aq) + Ca(OH)_2(s) \longrightarrow Ca(H_2PO_4)_2(aq) + H_2O(l)$$

How many grams of each product would there be if 7.50 L of 5.00 M phosphoric acid reacted with an excess of calcium hydroxide?

7. The workers at a hazardous waste collection site have collected 12.5 L of solution from an automobile battery that contains a trace of lead(II) ions. They add an excess of sodium hydroxide to make insoluble lead hydroxide, which will not leach into ground water. If the completed reaction provides 0.285 g of lead(II) hydroxide, what was the initial molar concentration of the lead(II) ions in the solution?

Section Review

10. What is the molar concentration of these solutions?
    a. 0.0750 moles of $NaHCO_3$ in a volume of 115 mL of solution
    b. 0.750 moles of $NaC_2H_3O_2$ in a volume of 115 mL of solution

11. a. If a solution has 3.0 mol of solute in 2.0 L of solution, what is its molar concentration?
    b. How many moles would there be in 350 mL of solution?

12. Methanol can react according to the following equation.

$$CH_3OH(l) + HBr(aq) \longrightarrow CH_3Br(g) + H_2O(l)$$

If there are exactly 2.25 mol of CH_3OH, how many liters of 0.500 M HBr will be needed?

13. Describe in your own words exactly how you would prepare 1.00 L of a 0.85 molar solution of formic acid, HCOOH.

14. **Story Link**
On page 403, you read that Dr. Ramsay's tests on bear fat revealed a PCB concentration of 8 ppm. How many milligrams of PCB would there be in 35 g of this bear fat?

Answers to Section Review

10. a. 0.652 M
    b. 6.52 M
11. a. 1.5 M
    b. 0.52 mol
12. 4.5 L HBr
13. Mix 39.1 g of formic acid in less than 1.0 L of water in a 1.0 L volumetric flask, then add enough water to make exactly 1.0 L.
14. 0.3 mg

Why do some things dissolve?

Section Objectives

Determine polarity of solutes using electronegativity and molecular geometry.

Describe the solvation process.

Explain the solubility of ionic solutes and polar solutes in water.

Explain the insolubility of nonpolar solutes in water.

Testing solubility

Some substances dissolve, while others do not. The easiest way to determine whether something will dissolve is to add it to a solvent and watch what happens. **Figure 11-10** shows a series of situations with one solute and two solvents. The solute is lithium chloride, LiCl, a white solid. One solvent is toluene, $C_6H_5CH_3$, a clear liquid with a slight petroleum odor; the other is the most popular solvent in chemistry—water, H_2O. Lithium chloride and water are the only pair that dissolve.

Chem Fact
Toluene, once common in school labs, is now considered a dangerous chemical. In 1991, the EPA urged industry to cut toluene emissions in half by 1995.

Will it dissolve? Using models to explain why

The model for explaining solubility is based on the very simple idea of attractions between particles. Lithium chloride is composed of ions with large electrical charges, which attract each other strongly. Water is composed of polar molecules with a slightly positive end and a slightly negative end.

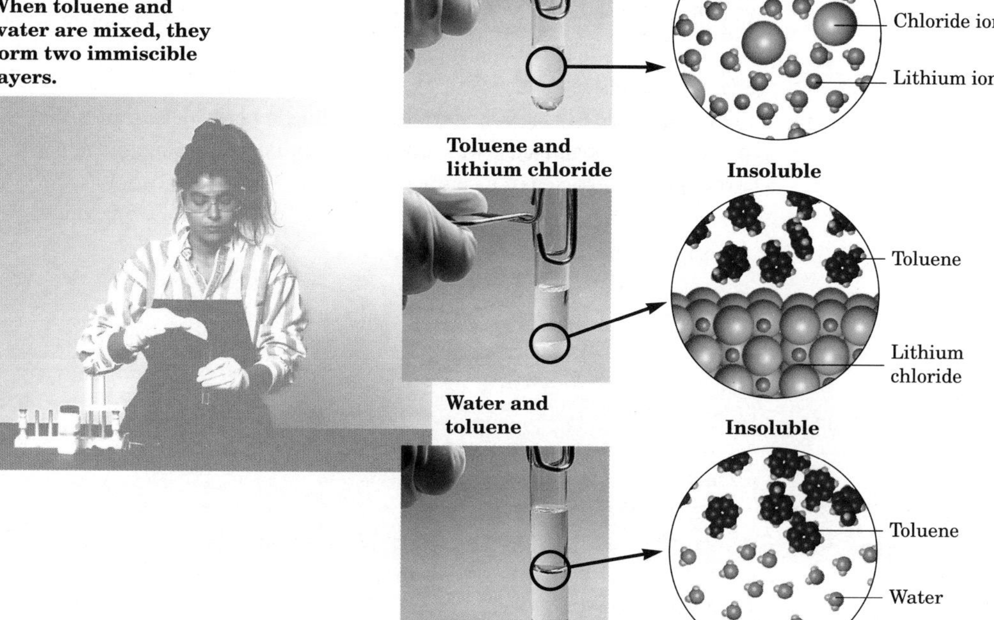

Figure 11-10
When water and lithium chloride, LiCl, are mixed, LiCl dissolves. When toluene and LiCl are mixed, LiCl does not dissolve. When toluene and water are mixed, they form two immiscible layers.

Section 11-3

FOCUS

Lesson Starter
Write the expression "like dissolves like" on the chalk board. Ask students, in pairs, to discuss this statement and conjecture its meaning. Have one student from each pair report their ideas to the larger class.

Demonstration 10
Polar and nonpolar solvents
Approximate time: 15 min
This demonstration should provoke discussion about solvent-solute interactions.

1. Place 10 mL of H_2O and 10 mL of ethanol into separate Pyrex test tubes. Using forceps, add no more than 2 small crystals of I_2 to each tube.
2. Stopper and shake each tube. The nonpolar iodine should impart a purple hue to the less polar ethanol but not to the very polar water.

Variation
Do this demonstration in Petri dishes on the platform of an overhead projector. View coloration on the screen. Slight stirring may be necessary.

SAFETY
Wear goggles and lab apron. Students should remain 10 ft from the demonstration.

DISPOSAL
Slowly add 1 M sodium thiosulfate solution to iodine solution while stirring until mixture is colorless. Then pour down the drain.

Visual Strategy
Figure 11-10
Focus students' attention on the polar attraction between the hydrogen in water and the chloride ion, and between the oxygen in water and the lithium ion. Stress the lack of attraction between nonpolar toluene and polar lithium chloride. Also stress the lack of attraction and interaction between water and toluene.

TEACH

Teaching Tip

Rule of thumb

A quick rule of thumb regarding the polarity of chemical bonds is that if combining elements come from opposite ends of the periodic table, the bonding is ionic. If the elements are from the same side of the periodic table, bonds are covalent.

Content Background

Alcohol dissolves equally well in water and in turpentine. An alcohol has polar bonds at its hydroxide end and nonpolar bonds within its hydrocarbon end, so it dissolves equally well in polar and nonpolar solvents.

Possible Misconception

Some students may conceive of the solution process as a chemical reaction and even view solutions as compounds. Emphasize that solutions are mixtures and that they are not formed through chemical reactions but through a process which is essentially very thorough mixing on the particle level.

Nevertheless, the attraction between ions such as lithium and chloride and water molecules is strong enough that the water molecules can pull the ions away from the solid crystal. This attraction causes lithium chloride to dissolve in water.

What happens when lithium chloride and toluene are mixed? Toluene molecules carry essentially no electrical charge, so they are not strongly attracted to the ions in the crystal. But ions within the crystal are attracted strongly to each other. In the absence of any similar attraction, the ions remain embedded in the solid crystal.

In the third example you saw the combination of the two solvents, toluene and water. There is very little attraction between the molecules of toluene and water, but this factor alone is not why the liquids do not dissolve. The real obstacle is the attraction that each water molecule has for other water molecules. Imagine a single water molecule about to penetrate the layer of toluene. Ahead of it are toluene molecules for which the water molecule has very little attraction or repulsion. Behind the water molecule, however, are water molecules with partial electrical charges that attract each other. These other molecules pull the water molecule back into the water layer.

Predicting attractions

To predict whether two substances will dissolve, you must first predict attractions by determining whether the particles contain electrical charges. This process begins with investigating the types of chemical bonds. In Chapter 6 you learned that there are several fundamental types of bonds, including ionic, polar covalent, and nonpolar covalent. We will examine these to see how each affects solubility. Because metallic substances usually dissolve only in other metals, you will not consider solubility and metallic bonds. You may also recall from Chapter 6 that electronegativity is the key to predicting bond type. The electronegativities of the elements are given in **Table 11-3**.

The difference in electronegativities of the elements determines the type of chemical bond they will form, and it indicates the magnitude of charge on the particles in a compound.

Table 11-3
Electronegativity Values

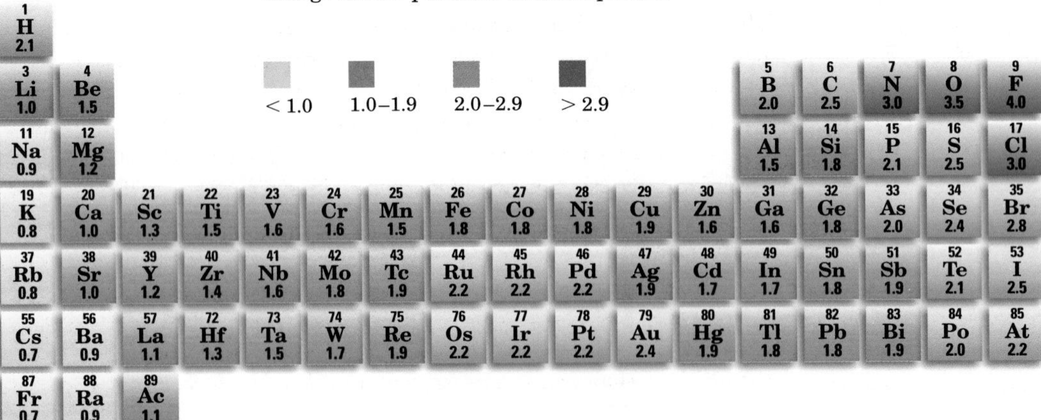

1 H 2.1																
3 Li 1.0	4 Be 1.5											5 B 2.0	6 C 2.5	7 N 3.0	8 O 3.5	9 F 4.0
11 Na 0.9	12 Mg 1.2											13 Al 1.5	14 Si 1.8	15 P 2.1	16 S 2.5	17 Cl 3.0
19 K 0.8	20 Ca 1.0	21 Sc 1.3	22 Ti 1.5	23 V 1.6	24 Cr 1.6	25 Mn 1.5	26 Fe 1.8	27 Co 1.8	28 Ni 1.8	29 Cu 1.9	30 Zn 1.6	31 Ga 1.6	32 Ge 1.8	33 As 2.0	34 Se 2.4	35 Br 2.8
37 Rb 0.8	38 Sr 1.0	39 Y 1.2	40 Zr 1.4	41 Nb 1.6	42 Mo 1.8	43 Tc 1.9	44 Ru 2.2	45 Rh 2.2	46 Pd 2.2	47 Ag 1.9	48 Cd 1.7	49 In 1.7	50 Sn 1.8	51 Sb 1.9	52 Te 2.1	53 I 2.5
55 Cs 0.7	56 Ba 0.9	57 La 1.1	72 Hf 1.3	73 Ta 1.5	74 W 1.7	75 Re 1.9	76 Os 2.2	77 Ir 2.2	78 Pt 2.2	79 Au 2.4	80 Hg 1.9	81 Tl 1.8	82 Pb 1.8	83 Bi 1.9	84 Po 2.0	85 At 2.2
87 Fr 0.7	88 Ra 0.9	89 Ac 1.1														

Visual Strategy

Table 11-3

Remind students of the periodic trend in electronegativity—the most electronegative elements are in the upper right corner of the periodic table (neglecting the noble gases), while the least electronegative are in the lower left corner.

Content Background

Ionic charge underlies many properties of solutions that are formed from ionic solids. The greater the ionic charge, the stronger the attraction an ion has for other charged particles, the greater the lattice energy of the resulting ionic solid, and the greater the heat of solution.

Ionic bonds have strong attractions

Two elements will form an ionic bond when the difference in the electronegativities is 2.1 or more. This type of bond does not result in molecules but in ionic crystals composed of millions of rows of alternating positive and negative ions.

The particles that make up an ionic crystal have full electrical charges. In many cases the charges are 1+ and 1–, but some are stronger, such as the 2+ and 3+ charges on the Mg^{2+} ion and Al^{3+} ion. All these ions have very strong attractions for other charged ions and molecules with partial charges, and this attraction will determine how they dissolve.

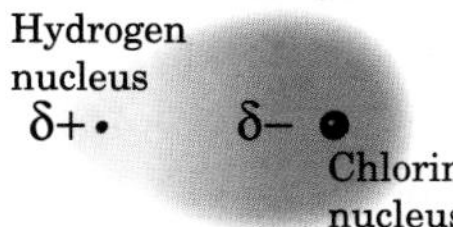

Figure 11-11
This model for a polar covalent bond shows an electron cloud that is more dense near the chlorine nucleus, resulting in partial charges.

Nonpolar covalent bonds have no charges

Two atoms will form a nonpolar covalent bond when the difference in the electronegativities is in the range of 0 through 0.4. Because the atoms in a nonpolar covalent bond share the electrons equally, neither is more positive or negative than the other. Nonpolar covalent bonds, as the name suggests, carry no electrical charge. In the absence of other attractions, nonpolar molecules will dissolve in each other.

Polar covalent bonds have weak charges

Two atoms will form a polar covalent bond when the difference in the electronegativities is in the range of 0.4 through 2.1. When chlorine, which has an electronegativity of 3.0, bonds with hydrogen, which has an electronegativity of 2.1, the atoms share electrons unequally. Because the shared electrons spend more time near the nucleus of the chlorine atom than the hydrogen atom, the chlorine atom carries a slight negative charge, and the hydrogen atom carries a slight positive charge. These partial charges are represented by δ– and δ+ in **Figure 11-11**. They are not as strong as the full negative and positive charges on ions, but they are strong enough to attract other particles with charges, such as other polar molecules and ions. Thus, you would expect molecules held together by polar covalent bonds to dissolve ionic compounds and each other.

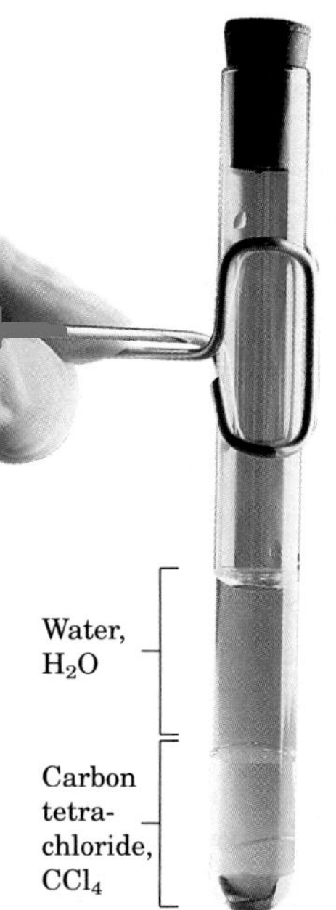

Figure 11-12
Even though carbon tetrachloride, CCl_4, and water, H_2O, both have polar covalent bonds, they do not dissolve in each other.

Testing predictions in the lab

The model of solubility based on attractions and differences in bond type has led to several generalizations about solubility. So far, you expect that polar covalent molecules and ions will dissolve together because they contain charges that can attract. You saw an example of this when the LiCl dissolved in water in **Figure 11-10**. Similarly, you could predict that ions and polar covalent molecules will not dissolve in molecules such as toluene that contain nonpolar covalent bonds and no charges. The experiments in **Figure 11-10** support this prediction.

But are these rules enough to predict all cases? What would happen if you mixed water with carbon tetrachloride, CCl_4? Both contain polar covalent bonds, so you predict they will dissolve. But if you try this in the lab, the results may surprise you: they *don't* dissolve, as shown in **Figure 11-12**.

Visual Strategy

Figure 11-11

Point out the surface formed where the water and carbon tetrachloride meet. This is almost as clearly visible as the water-air interface. Ask students which liquid is more dense. (The CCl_4 must be more dense, because it is at the bottom of the tube.)

Teaching Tip

Enrichment

You may wish to include the effects of lone pairs in your presentation on molecular shape and polarity. If so, explain that completely symmetrical molecules are nonpolar. As symmetry decreases, the polarity of molecules increases. Symmetry is decreased by the presence of lone pairs on the central atom of a molecule and by the attachment of unlike atoms to the central atom.

Examples: CCl_4 is nonpolar, but $CHCl_3$ is polar. NH_4^+ is nonpolar, but NH_3 is polar. SO_3 is nonpolar, but SO_2 is polar.

Any scientist's first instinct is to double-check and make sure that the chemicals are not contaminated and that the procedure was carried out properly. Even when double-checked repeatedly, CCl_4 and H_2O do not dissolve. The prediction was incorrect.

In this case the theory is not totally wrong, but it is incomplete. It turns out that the type of bond is not the only factor that determines solubility. The shape of the molecule is also important.

Molecular shape affects polarity

In a molecule with only two atoms, such as HCl, the atoms can be arranged in only one way—a straight line. However, if a molecule has three atoms, more than one arrangement is possible, as you may recall from your study of molecular shapes in Chapter 6.

A molecule of hydrogen sulfide, H_2S, is arranged in a V shape (bent), whereas a molecule of carbon dioxide, CO_2, has a straight-line (linear) configuration. Both of these molecules, shown in **Figure 11-13**, contain polar covalent bonds. You might expect that both of them would act like particles containing charges and dissolve easily in water. But nearly three times more H_2S dissolves in water than CO_2! Having polar covalent bonds is not enough to ensure solubility. *The molecule must also be polar overall.*

Figure 11-13a
Molecules that contain three atoms can be arranged in two different ways. H_2S exists in a V-shaped (bent) arrangement, . . .

δ+
Hydrogen nucleus
δ+
Hydrogen nucleus
δ–
Sulfur nucleus

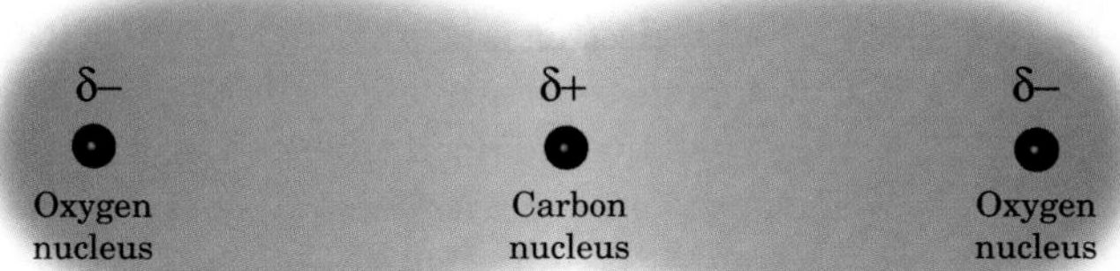

11-13b
. . . but CO_2 exists in a straight-line (linear) arrangement.

Determining effective poles

Examine the models of the CO_2 and H_2S molecules. Each molecule appears to have three different partial charges. But a nearby molecule or ion does not distinguish among all of the individual charges within another small molecule. It is affected by only an overall partial positive charge at one part of the molecule and an overall partial negative charge elsewhere. These average overall charges are called "effective poles."

H_2S and CO_2 have different solubilities resulting from the locations of their effective poles. To determine the location of the effective poles in a molecule, draw its structure with the correct shape, and then connect all of the partial positive charges. The center of the line or shape they form will be the effective positive pole. Do the same for the negative charges to determine the effective negative pole.

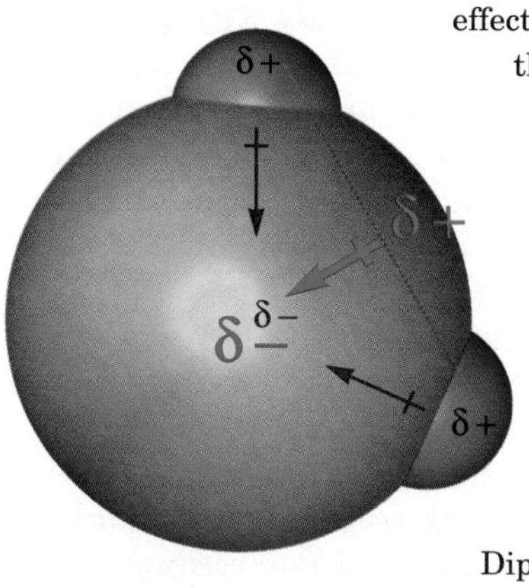

Try this first with the molecule of H_2S shown in **Figure 11-14**. When you connect the two partial positive charges, you get a straight line. The effective positive pole will be at the midpoint of the line connecting the hydrogen atoms. In **Figure 11-14**, and in the molecules on the next few pages, *partial charges are shown in black, and effective poles are shown in blue*. Although the effective positive pole may seem to be a "phantom" on one edge of the sulfur atom, the molecule will behave as though it had a single partial positive charge centered there.

All of the partial negative charges are on the center of the sulfur atom, so the effective negative pole is also there. Such a molecule is called a molecular *dipole*, because its two effective poles are on opposite sides of the molecule. Dipoles are indicated by an arrow pointing from positive to negative.

Dipoles can also be drawn for individual bonds. The molecular dipole is the sum of the individual bond dipoles. In **Figure 11-14** and in the molecules on the next few pages, *bond dipoles are shown in black and molecular dipoles are shown in blue*.

If this polar H_2S molecule is approached by an anion or the negatively charged end of another molecular dipole, the positive portion of H_2S will be attracted, and the negative portion of H_2S will be repelled. These forces help a particle containing negative charges to dissolve in H_2S molecules.

Figure 11-14
Other negative charges are attracted to an H_2S molecule as if it had an effective positive pole midway between its two hydrogen atoms.

δ Partial charge
δ Effective pole
Bond dipole
Molecular dipole

Effective poles can cancel each other

Now consider the locations of effective poles on the CO_2 molecule shown in **Figure 11-15**. The midpoint of the line connecting the two partial negative charges is at the same place as the partial positive charge. In this molecule, the negative and positive effective poles are at the same point.

If a CO_2 molecule is approached by a negative charge, the negative charge will be attracted by the positive effective pole located at the center of the carbon atom. However, it will also be repelled by the negative effective pole, which is also located at the center of the carbon atom. These forces will cancel each other completely. The net effect will be as if the molecule had no effective poles at all.

Examine the bond dipoles in **Figure 11-15**. Because they point in opposite directions, they cancel each other. There is no molecular dipole. In other words, *molecules whose poles are at the same location behave the same way as molecules with nonpolar covalent bonds.* This explains why CO_2 is much less soluble in H_2O than H_2S is. Even though all three molecules have the same type of bonds, CO_2 molecules are *nonpolar* overall, and H_2O and H_2S molecules are dipoles.

Figure 11-15
Other charges are not attracted or repelled by a CO_2 molecule because the effective poles are at the same place and cancel each other.

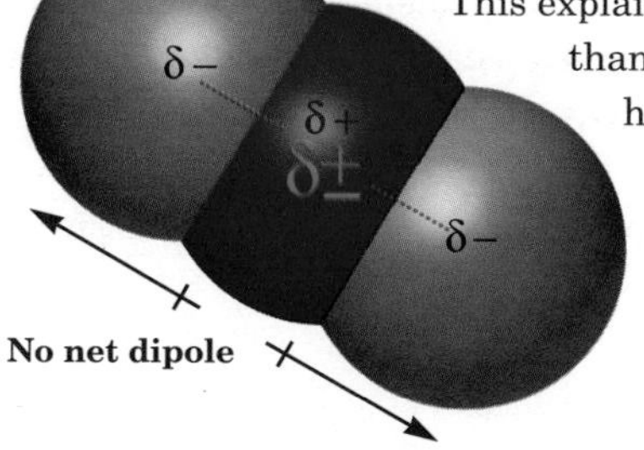

No net dipole

Teaching Tip

Enrichment

If you discuss the effects of lone pairs on polarity (see *Teaching Tip* page 420) you can extend the discussion with the following flowchart, which also incorporates Lewis dot structures.

Determining Polarity

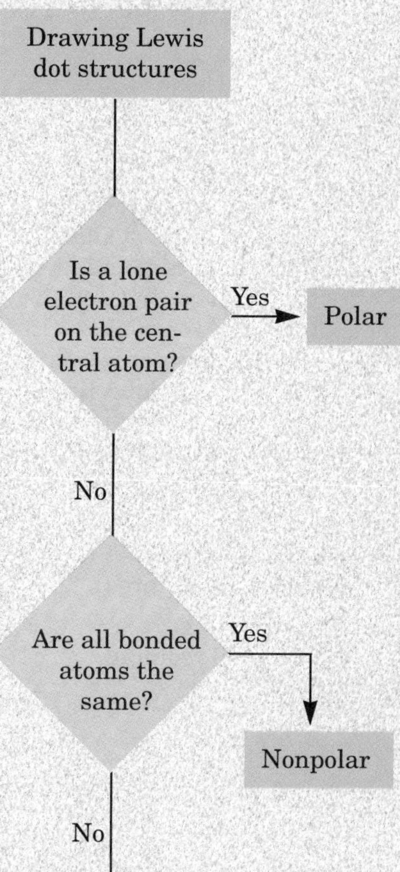

Visual Strategy

Figure 11-14 and **Figure 11-15**

Another way to relate partial charges and molecular dipoles is to determine all of the bond dipoles and then add them using vector addition techniques, placing the arrows head to tail. If there is a net displacement from the starting point, there is the molecular dipole.

Demonstration 11
Polarity of water
Approximate time: 10 min
This demonstration shows that water has a polar structure.

1. Obtain a Bakelite static electricity rod and animal fur or a glass static electricity rod and an old pair of nylon hosiery.
2. Adjust an easily viewed faucet to have as slow a flow as possible but still have a steady stream. It will help to install a tubing connector to the faucet if one is available.
3. Charge the rod, and bring it near (but not touching) the point where the water leaves the faucet. A deflection of the water toward the rod can be seen about a foot below the rod.

SAFETY
No specific safety precautions required.

DISPOSAL
No disposal required.

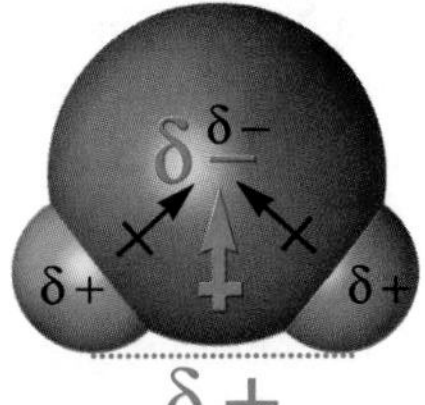

Figure 11-16
Electronegativity differences indicate that the bonds in water are polar covalent with partial charges on each end.

Reexamining water and carbon tetrachloride

Are the differences in H_2O and CCl_4 solubility also due to differences in overall polarity? To find out, first determine the polarity of the water molecule. Make a diagram with the correct bent molecular shape, as shown in **Figure 11-16**. All of the partial negative charges are on the oxygen atom, *so the effective charge will be at the same point.* "Average" the two positive charges. The effective positive charge is located in the space between the two hydrogen atoms. Because this space is not at the same location as the effective negative pole, water is a dipole.

Next, examine the molecule of carbon tetrachloride, CCl_4. We know from Chapter 6 that the shape of CCl_4 is a tetrahedron, with four bond angles of 109.5°. But this shape is three dimensional and is difficult to draw accurately on paper. *When determining polarity of molecules with four bond angles of 109.5° such as the CCl_4 molecule, the essential point, that CCl_4 is a symmetrical molecule, is more easily conveyed by representing the molecule as a square*, as in **Figure 11-17**. Because CCl_4 contains four identical bonds, you must average the locations of the four negative charges to find the location of the effective negative pole. As before, connect the negative partial charges, forming a square. The effective negative pole is at the center of the square, on the central carbon atom.

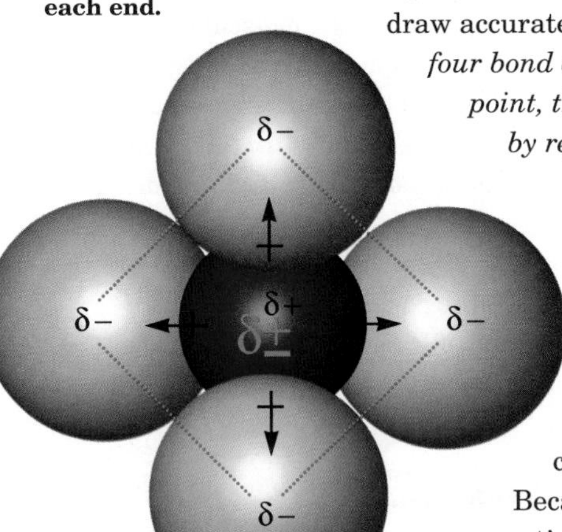

Figure 11-17
The difference in electronegativity indicates polar covalent bonds in CCl_4. Although CCl_4 has a tetrahedral shape in three dimensions, the lack of polarity in CCl_4 is easier to visualize using a two-dimensional square.

The partial positive charges are also all on the central carbon atom, so the effective positive pole is there as well. Because the molecule is symmetrical, the effective positive and negative poles are at the same location and cancel each other. As a result, the molecule cannot attract another charged particle or repel it. This makes CCl_4 a nonpolar molecule overall, with very little attraction for charges.

Now that you know the true nature of carbon tetrachloride, you can make a prediction that actually matches what happens. Water and carbon tetrachloride should not dissolve, because H_2O is a dipole. Even though carbon tetrachloride has polar bonds, the molecule as a whole is nonpolar. In other words, CCl_4 behaves more like the toluene molecules discussed at the beginning of this section than like water. The features of polar covalent bonds are given in **Figure 11-18** on the next page.

Completing the model for solutions

Now that the behavior of CCl_4 has been explained, you can add what you've learned about polar covalent bonds and dipoles to the model of solutions. This improved model can be used to predict solubilities for many substances.

Visual Strategy
Figure 11-17
The CCl_4 tetrahedron has been simplified to the two dimensions. You can use models to more accurately portray its shape. First, "average" the charges for the two pairs of Cl atoms. The "average" will be the midpoint of an imaginary line connecting the atoms. Then average those "average" charges. The effective negative and positive charges are in the center, on the carbon atom.

Figure 11-18
Molecules with polar covalent bonds can be either polar or nonpolar overall, depending on the location of their effective poles.

Predicting Polarity

IF the difference in electronegativity for bonded atoms is between . . .

. . . **0.4 and 2.1**,

THEN the bonds are **polar covalent**, as in the following.

$\overset{\delta+}{C}—\overset{\delta-}{Cl}$ and $\overset{\delta+}{H}—\overset{\delta-}{O}$

IF the effective poles . . .

. . . **are** in the same position,

THEN the molecule has **no** molecular dipole and no effective charges.

And **THEN**, the molecule is **nonpolar**, as in CCl_4.

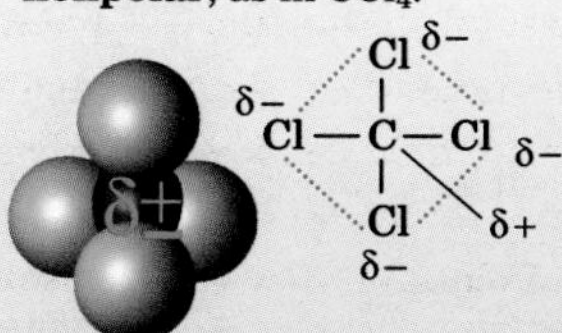

. . . are **not** in the same position,

THEN the molecule will be a **dipole** with effective charges, as in H_2O.

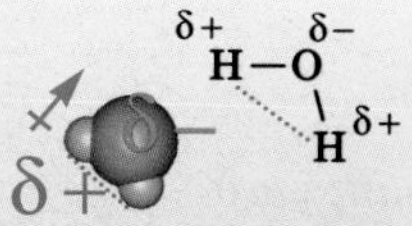

Predicting dipoles

Use these steps for a molecule with polar covalent bonds.

1. For each polar covalent bond, place a δ− on the atom with higher electronegativity and a δ+ on the atom with lower electronegativity.
2. Place a δ+ at the position of the effective positive pole by averaging the locations of the individual positive poles.
   - The average location of two poles is the midpoint of the line connecting the partial positive charges.
   - The average location of three poles is the center of the triangle formed by connecting the partial positive charges.
   - The average location of four or more poles is the center of the shape formed by connecting the partial positive charges.
3. Place a δ− at the position of the effective negative pole by averaging the locations of the individual negative charges using the method in step 2.
4. If the effective positive pole is in a different location than the effective negative pole, the molecule is a dipole that can attract charges. Otherwise, the molecule is nonpolar.

Teaching Tip

Discussion

Many students are familiar with ammonia cleaning products. Have them determine whether NH_3 is polar or nonpolar. Explain that ammonia is polar by discussing the structure of ammonia and by pointing out that it dissolves well in water.

Teaching Tip

Discussion

Continue the discussion of the phrase mentioned in the Lesson Starter, "like dissolves like," in light of what students have learned in the chapter. Examine sets of molecules to see how alike they are and how this relates to solubility.

Example: The general formula for an alcohol is R—OH, in which R stands for an organic alkyl group. Compare this to the formula for water, H—OH. Discuss how solubilities of alcohols are affected by the hydrocarbon tail (R group). The longer the R group and the less polar the molecule, the less like water the alcohol is; and the alcohol's solubility (compared to water) decreases also.

Do You Know?

Although the C—H *bond* has an electronegativity difference of 0.4, it is considered nonpolar covalent, as are hydrocarbon molecules such as methane.

How To

Predict solubility

IF the difference in electronegativities for bonded atoms is between . . .

1. Determine bond type by calculating electronegativity difference and selecting range.

. . . **0.0 and 0.4,**

THEN the bonds are **nonpolar covalent,** as in CH_4.

. . . **0.4 and 2.1,**

THEN the bonds are **polar covalent.**

IF the effective poles . . .

. . .**2.1 and 4.0,**

THEN the bonds are **ionic.**

And **THEN,** . . .

2. Indicate charges as needed on a structure of the substance. Check for dipoles if needed.

. . . **are** in the same position,

THEN the molecule has **no** molecular dipole and no effective charges.

And **THEN,** the molecule behaves as if it were **nonpolar,** as in CCl_4.

. . . **are not** in the same position,

THEN the molecule will be a **dipole** with effective charges, as in H_2O.

δ− δ+

. . . the particles are **ions** with **full charges** in a crystal lattice, as in NaCl.

3. Identify particle types and charges. Determine solubility.

And **THEN,** because there is **no** net separation of charge on the particles, the substance will be

- **soluble** in **nonpolar** solvents
- insoluble in polar solvents

And **THEN,** because there is a net **separation of charge** on the particles, the substance will be

- **soluble** in **polar** solvents
- insoluble in nonpolar solvents

Overall polarity of molecules determines solubility

In this chapter you have focused on three types of substances: nonpolar molecules, molecular dipoles, and ionic crystals. Nonpolar molecules with either polar or nonpolar bonds carry no overall charge. When it comes to solubility, the full charges of ions and the partial charges of dipoles are grouped together because they attract each other. You can use the presence or absence of charges to predict whether one chemical will dissolve in another. Chemists sometimes summarize this rule by saying "like dissolves like."

- Many charged substances (ions and dipoles) will dissolve in other charged substances. (Some exceptions are described in **Table 11-1** on page 410.)
- A nonpolar substance will dissolve in another nonpolar substance.
- A polar or charged substance will not dissolve in a nonpolar substance. These principles are summarized in the feature above, which can be used to help predict a substance's solubility.

Visual Strategy

How to Predict Solubility

Ask students to predict the solubility of the following substances by using the graphic organizer: NaF, CO_2, CO, O_2.

Answers: NaF and CO are soluble in substances with charged particles; CO_2 and O_2 are soluble in substances with uncharged particles.

Sample Problem 11D *Analyzing molecules for polarity*

The structure of tetrachloroethylene is shown here. Use the electronegativity data in Table 11-3 and the feature on page 424 to determine if C_2Cl_4 is a molecular dipole. Describe its solubility.

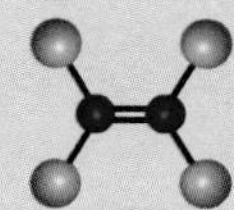

For C_2Cl_4, the difference in electronegativities for bonded atoms is between . . .

1. Determine bond type by calculating electronegativity difference and selecting range.

There are two types of bonds in this molecule: the C=C double bond and the C—Cl single bond.

. . . **0.0 and 0.4** for the C=C bond (2.5 − 2.5 = 0.0).

So **THEN** the C=C bonds are **nonpolar covalent.**

. . . **0.4 and 2.1** for the C—Cl bonds (3.0 − 2.5 = 0.5).
So **THEN** the C—Cl bonds are **polar covalent**.

For C_2Cl_4, the effective poles **are** in the same position.

2. Indicate charges as needed on a structure of the substance.

*Using the rules for effective poles, the effective negative pole is at the center of the rectangular shape formed by the chlorine atoms. The effective positive pole is halfway between the two carbon atoms. Both poles are at the same place, so there is **no molecular dipole**.*

So **THEN**, C_2Cl_4 has **no** molecular dipole and no effective charges.

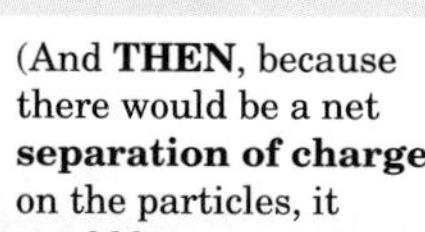

And **THEN**, C_2Cl_4 behaves as if it were **nonpolar** covalent.

(**IF** they **were not** in the same position,

THEN the molecule would be a **dipole**, with effective charges.)

3. Identify particle types and charges to determine solubility.

The molecule as a whole is nonpolar, even though it contains polar covalent bonds. It will behave as a nonpolar substance.

And **THEN**, because there is **no** net separation of charge on the particles, C_2Cl_4 will be

- **soluble** in **nonpolar** solvents
- insoluble in polar solvents

(And **THEN**, because there would be a net **separation of charge** on the particles, it would be

- **soluble** in **polar** solvents
- insoluble in nonpolar solvents)

Practice 11D

1. **The structure of boron trichloride, BCl_3, is shown at right. Use Table 11-3 and the feature on page 424 to determine if BCl_3 is polar.**
2. **Sulfur dioxide, SO_2, is represented at right. Use Table 11-3 and the feature on page 424 to determine whether it will dissolve in water.**

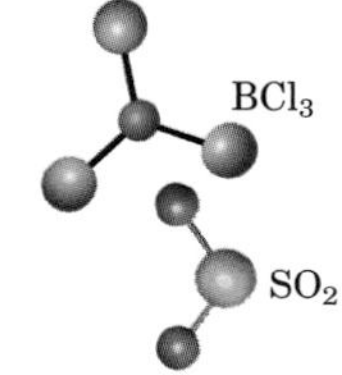

Additional Examples for Sample Problem 11D

Analyze the following molecules for polarity.
a. acetylene, C_2H_2
b. sulfur trioxide, SO_3

Solutions are at the bottom of the page.

Answers to Practice 11D

1. B—Cl bonds are polar covalent because their electronegativity difference is 1.0. The bonds are symmetrically arranged about B, so the molecule is nonpolar.
2. SO_2 is a polar covalent molecule, so it should dissolve in water.

Solution for Additional Examples 11D

a. Calculate the electronegativity difference for C≡C: 2.5 – 2.5 = 0. This fits the 0.0 – 0.4 range, so the bond is nonpolar covalent. Now repeat the process for the C—H bond: 2.5 – 2.1 = 0.4. This fits both the 0.0–0.4 range and the 0.4–2.1 range, so the bond is nonpolar covalent. Therefore, acetylene contains no charged particles and is nonpolar.

b. Sulfur trioxide has a trigonal planar structure. For the S—O bonds, the electronegativity difference is 3.5 – 2.5 = 1.0. This fits the 0.4–2.1 range, so the bond is polar covalent. Because the molecule is symmetrical, it is not a dipole.

Other Applications

To remove condensed water from fuel tanks during cold winter weather, a drying agent is added. Methanol, CH_3OH, also known as wood alcohol, is usually the drying agent used. The polar end of the methanol molecule dissolves in water, and the nonpolar end dissolves in gasoline, thus effectively removing the water from the tanks by carrying it through the fuel system.

ASSESS

Alternative Assessment

Have students work in groups of four. Have each student develop a molarity problem modeled after the Sample Problems, including the answer to the problem. Have students trade problems and review solutions in their groups. For accountability, pick one problem from each group, and assign it to all members of the group as a "real" quiz once the groups have finished developing, solving, and discussing the problems among themselves.

Answers to Section Review

15. a. ions
b. nonpolar molecule
c. dipole
d. nonpolar molecule

16. a. no
b. yes
c. KF—soluble in water; C_6H_{12}—not soluble in water; HBr—soluble in water; CBr_4—not soluble in water

17. Because of their shape, water molecules are dipoles that can attract charged ions.

18. nonpolar

19. Like fats and oils, PCBs are nonpolar; PCBs dissolve in fats and oils and vice versa.

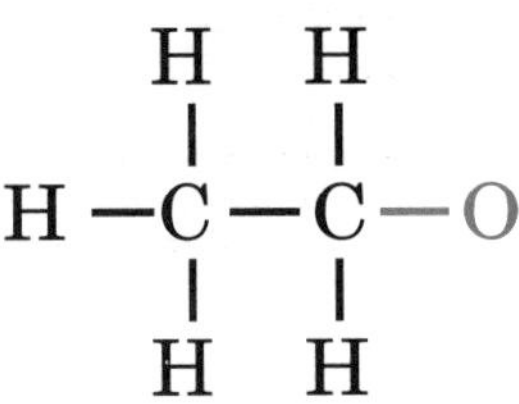

Figure 11-19a
Ethanol has both a polar region (shown in color) and a nonpolar region. It is polar overall.

Molecules with two types of bonds

There are many larger molecules that contain some regions with polar covalent bonds and other regions with nonpolar covalent bonds. Ethanol (ethyl alcohol), shown in **Figure 11-19a**, is a good example. The molecule contains one C—C bond, one C—O bond, one O—H bond, and five C—H bonds. Electronegativity calculations show that the C—C bond and the five C—H bonds are nonpolar, whereas the C—O and O—H are polar. The two polar bonds do not cancel each other, so the region around the oxygen is negative, and the region around the hydrogen next to the oxygen is positive. The remainder of the molecule is nonpolar.

Ethanol is found in beer, wine, and other alcoholic beverages. Because it mixes well with water, you might guess that the entire molecule is polar overall even though part of the molecule contains nonpolar bonds. The two polar bonds can create a dipole that can attract other polar molecules.

H H H H
H—C—C—C—C—O—H
H H H H

11-19b
Because butanol's nonpolar region is much larger than its polar region, it is nonpolar overall.

Experiments with a variety of alcohols suggest the following rule: *In determining solubility in water, a polar bond can have as much effect as a small nonpolar region.* This rule is very rough, but it is adequate for making many solubility predictions. Molecules with larger nonpolar regions, such as butanol in **Figure 11-19b**, are mostly nonpolar overall and dissolve only slightly in water. When it is important to know a chemical's precise solubility, you must either look it up in a chemical handbook or test it by adding small amounts to a solvent.

Section Review

15. For each of the compounds shown below, determine the particle type as described in the feature on page 424.

a. KF (*s*)

b. (cyclohexane ring: six C atoms, each bonded to two H atoms)

c. H—Br

d. Br—C(Br)(Br)—Br

16. Use your answers from item **15** to answer the following questions. Would you expect compounds **a** and **b** to dissolve in each other? Would you expect compounds **b** and **d** to dissolve in each other? Estimate each compound's solubility in water.

17. Explain how water molecules can attract ions.

18. Story Link
Analyze the PCB molecule shown on page 402. Is the molecule polar or nonpolar?

19. Story Link
Why does the PCB molecule have a greater solubility in fats and oils? (Hint: fats and oils are mostly nonpolar substances.)

Can insoluble substances mix?

Section Objectives

Describe the role of emulsions in mixing nonpolar solutes with water.

List the chemical characteristics of an emulsifying agent.

Relate the action of soaps and detergents to the properties of an emulsifying agent.

Overcoming insolubility

You have probably heard the saying, "Oil and water don't mix." It's true. You can demonstrate it the next time you prepare dinner by making oil and vinegar salad dressing.

In a small bottle or carafe, add olive oil to a depth of about 3 cm, then add about 15 mL (3 teaspoons) vinegar, which is a solution of acetic acid and water. You will see the vinegar sink through the oil and collect at the bottom, as in **Figure 11-20**. This observation indicates two things.

1. Vinegar sinks because it is more dense than olive oil.
2. The two liquids are immiscible (insoluble).

emulsion
colloidal-sized droplets (about 100 nm wide) of one liquid suspended in another liquid

Continue adding vinegar until you have about half as much vinegar as oil. If you cap the carafe and shake well, you will see a turbulent mixture of oil and vinegar. What you *can't* see are millions of microscopic droplets of vinegar surrounded by oil. This mixture is called an **emulsion**. The droplets formed in an emulsion are colloidal particles. They are smaller than the particles in a suspension, but larger than the particles in a solution.

This salad-dressing emulsion is a temporary condition. As soon as you stop shaking the container, the droplets of oil begin to rise, the vinegar sinks, and the layers of each re-form. Clearly, the olive oil and vinegar do not dissolve.

To complete your salad dressing, you may wish to add small amounts of salt, chopped onion, mustard powder, pepper, or other spices. Shake the carafe to form an emulsion just before pouring the dressing. This will ensure that you get all of the flavors: vinegar, olive oil, and the spices that dissolve in each solvent.

Figure 11-20
Oil and vinegar will mix if shaken vigorously. However, the particles of oil and vinegar do not make a homogeneous solution.

Section 11-4

FOCUS

Lesson Starter
Have students answer the question "Can insoluble substances mix?" by writing a short paragraph. Discuss the responses as a class.

TEACH

Demonstration 12
Emulsifiers
Approximate time: 10 min
Demonstrate the formation of an emulsion.

1. Place 25 mL of H_2O and 25 mL of vegetable oil in a 250 mL Erlenmeyer flask.
2. Stopper, and shake vigorously. Note that the two layers separate again after a few minutes.
3. Add a few squirts of liquid soap, and shake again. An emulsion should form instead of layers.

SAFETY
Wear safety goggles and lab apron.

DISPOSAL
Pour down the drain.

Other Applications
Other common emulsions are floor and glass waxes, drugs, paints, shortenings, textile and leather dressings, milk, and butter.

Visual Strategy
Figure 11-20 and **Figure 11-3**
Refer back to **Figure 11-3** now that students understand the role of polarity in solubility. Point out that the immiscibility in the lava lamp is due to a polar-nonpolar (H_2O-paraffin) interaction. This is similar to the immiscibility displayed in **Figure 11-20**. The paraffin movement is due to density changes as the paraffin heats and cools.

Do You Know?

Lecithin is also used in margarine, mayonnaise, chocolate, candy, baked goods, animal feeds, paints, leaded gasoline, printing inks, soaps, cosmetics, as a mold release for plastics, blending agent in oils and resins, for rubber processing, and in textile fiber lubricants.

Teaching Tip

Emulsifiers in food

Have students make a list of all the processed foods containing emulsifiers or emulsifying agents that they find in the home pantry or refrigerator. Make a composite list on the board and categorize the functions of the emulsifiers.

Emulsions: making oil and water mix

The next time you are at the grocery store, look at the commercial oil and vinegar salad dressings. You will see two types. The traditional type has distinct layers. The newer type, shown in **Figure 11-21**, has no layers because the oil and vinegar have been mixed in such a way that they stay mixed. The secret of the new style is the addition of a chemical that keeps the oil and vinegar in an emulsion.

Lecithin, which is found in all living organisms, is a family of several similar molecules that stabilize emulsions. Lecithin is obtained commercially by extracting it from soybeans. As you can see in **Figure 11-21**, lecithin molecules have three major branches. The two longer branches are nonpolar, but the shorter, third branch is polar.

Chem Fact
The proteins in eggs are used as emulsifiers in most baking recipes.

When lecithin is added to pure water or pure oil, it dissolves in oil but not in water. Its real value, however, lies in its ability to attract and dissolve in both oil and water when these solvents are present together. Each molecule of lecithin dissolves partly in the oil and partly in the water, linking two immiscible liquids. Lecithin stabilizes emulsions, preventing the layers from re-forming. Lecithin is an **emulsifying agent**.

emulsifying agent *stabilizes an emulsion that would otherwise separate into different phases*

It is often said that emulsifying agents make oil and water mix. It is more accurate to say that when oil and water are vigorously agitated, they mix to form an emulsion and that the emulsifying agent keeps the oil and water mixed. Some other emulsifying agents found in food are shown in **Table 11-4** on the next page.

Figure 11-21
An oil and vinegar emulsion can be stabilized by adding an emulsifying agent, such as lecithin. The emulsifying agent acts as a bridge between the oil and the watery vinegar to make them mix.

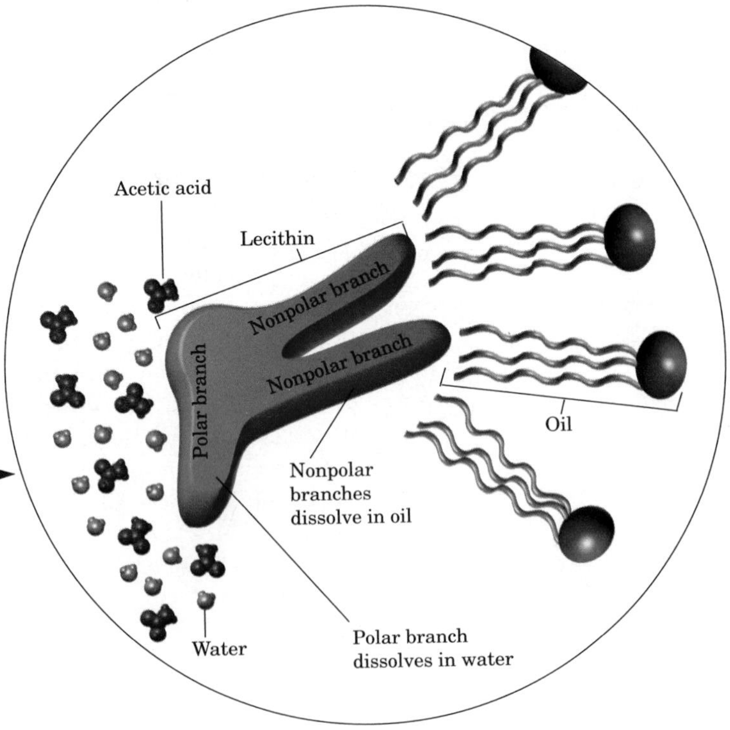

Visual Strategy

Figures 11-21

Once students understand that different parts of the lecithin molecule have different solubilities, have students identify on **Figures 11-21** the forces that will oppose the creation of oil and water layers.

Table 11-4
Some Common Food Emulsifiers*

Emulsifying agent types	Other names used in labeling	Uses
Arabic gum	acacia, Australian gum, gum senegal	beverages, candy, chewing gum
Carrageenan (dried seaweed plant)	chondrus, genugel, Irish gum, Irish moss gelose, pellugel, viscarin	artificially sweetened jellies and jams, cheese products, chocolate products, dessert gels, ice cream
Glycerides		baked goods, frozen desserts, fudge, ice cream, lard, margarine, peanut butter, whipped topping
Polysorbates	capmul, glycosperse, liposorb, monitan, monooleate or monostearate, polyoxyethylene sorbitan, polysorban 60 or 80	baked goods, barbecue sauce, chewing gum, cottage cheese, cream fillings, dried gelatin mix, frozen desserts, ice cream, icings, margarine, pickles, salad dressings, shortenings, vitamin supplements
Sodium stearoyl lactylate		baked good mixes, dehydrated potatoes, imitation cheese, snack dips, toppings, waffles
Sorbitan c	arlacel, crill, durtan, emsorb, sorbon, sorbitan monostearate, sorgen, span	cakes, fillings, icings, whipped topping
Tragacanth gum		baked goods, citrus beverages, condiments, fruit fillings, oils, relishes, salad dressings
Xanthan gum		baked goods, batter or breaded mixes, beverages, canned chili, desserts, jams, jellies, milk products, pizza topping mixes, salad dressings, stews (canned or frozen)

*Note: many of these emulsifiers also serve several other purposes in foods.

Soap is an emulsifying agent

Have you ever been dirty from working outside? If you washed your hands with water alone, you were probably only partly successful, as most of the dirt remained. To get really clean, plain water is not enough. You have to use soap and water, for reasons relating to emulsions.

Perspiration is a mixture of water and oils. Water cools your body by evaporation; oils keep your skin soft. However, over a period of time, the oils accumulate and coat your skin with an oily layer in which flakes of old skin, dirt, and bacteria become embedded.

Oil and water are immiscible, so the water may never remove the oil and dirt on your skin. To remove all of the dirt, you must first emulsify the oil by scrubbing, then stabilize it with an emulsifying agent like soap. Only then can the soap-and-oily-dirt emulsion be rinsed away from your skin by more water.

Demonstration 13
Soap and surface tension

Approximate time: 15 min

Show how soap affects the surface tension of water.

1. Fill a gas collection bottle half full with water. Cover the top with a small square of notebook paper. While holding the paper in place, invert the bottle over a sink. The paper should stay in place due to the air pressure pushing up on the paper. (A common misconception is that a partial vacuum holds up the paper.)
2. Repeat using a wire gauze with no asbestos center, making sure that it is very flat. The screen, too, should hold back the water due to the high surface tension of water.
3. Remove the screen, and add a few squirts of liquid soap or detergent to the water.
4. Shake well, replace the screen, and invert. The screen should not stay in place because the interactions between water molecules have been lessened by the soap.

SAFETY
Wear safety goggles.

DISPOSAL
Pour down the drain.

Multicultural Perspective

Soap was used in ancient Sumeria (the region that is now Iran and Iraq) in 2500 B.C. Later, the Greek physician Galen made soap from the reaction of fat with an alkali. He hailed the ability of soap to clean the body and to medicate.

Historical Note

Cincinnati, Ohio, was locally known as Porkopolis in the 1830s and 1840s because of its large slaughterhouses for the hog industry. By 1850, Cincinnati had become the nation's chief pork-packing center. This led to an abundance of animal fats. The abundance of raw material along with a good locale on the Ohio River made Cincinnati the soap-manufacturing center of the United States, and the home of such companies as Proctor and Gamble.

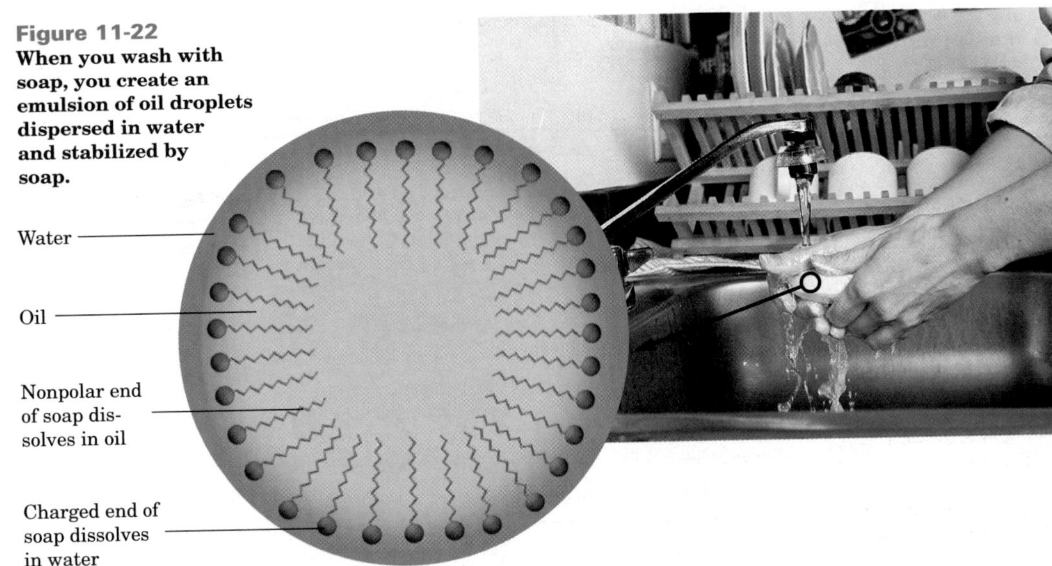

Figure 11-22
When you wash with soap, you create an emulsion of oil droplets dispersed in water and stabilized by soap.

When you wash with soap, as shown in **Figure 11-22**, one end of the soap molecule dissolves in the oil while the other end dissolves in the surrounding water. The oil droplet stays suspended in the water and can be easily washed away.

Soap was apparently discovered thousands of years ago, and soap making is one of the oldest chemical industries. The two critical ingredients are lye, NaOH, or potash, KOH, and oils or fat. The North American settlers shown in **Figure 11-23** made their own potash by making a hole in the bottom of a barrel, filling the barrel with wood ashes, pouring water into the barrel, and then collecting the mild solution of potash that dripped from the bottom. Fat from slaughtered cows or pigs was combined with the potash, and then heated and stirred in a large kettle. The mixture was poured into a wooden frame, where it hardened into soap.

Figure 11-23
The steps in soap making are shown here, from making potash in the barrel at the right, boiling it with fat in the kettle in the middle, and cutting the soap into blocks at the left.

Although the settlers didn't know it, the chemical reaction that took place broke down the fat molecules into smaller molecules with charged and nonpolar ends, as shown in **Figure 11-24** on the next page. Such molecules make an effective emulsifying agent. Today, most soaps are made from vegetable oils instead of animal fat. One company was so pleased with the soap they made from palm and olive oils they named it Palmolive.

Visual Strategy

Figure 11-22

Be sure students understand that the oil-soap-water diagram in **Figure 11-22** is an extension of the idea shown in **Figure 11-21**. Instead of only one emulsifier molecule, there are many, but the interactions are the same.

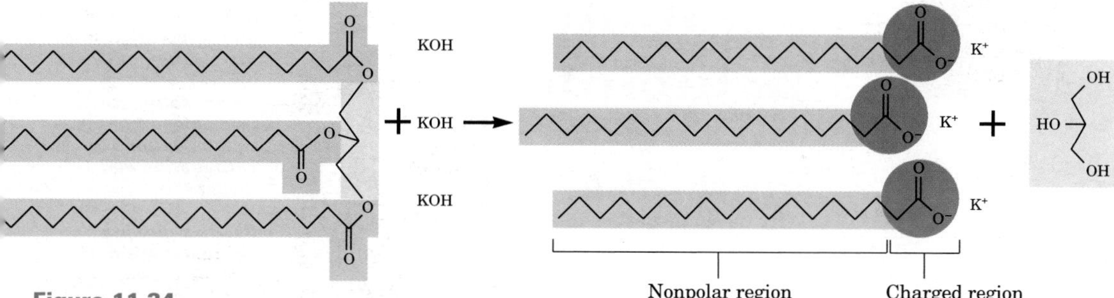

Figure 11-24
To make soap, an animal fat molecule reacts with potassium hydroxide (potash) to produce three soap molecules and a glycerine molecule.

"Hard" water destroys soap's emulsifying abilities

Even though soap has been used for centuries, it is not the ideal cleaning agent. If you get your water from an underground source, you may have "hard" water, which contains high concentrations of calcium and magnesium ions. In hard water, Ca^{2+} and Mg^{2+} cations are strongly attracted to the anionic end of soap. When cations attach to the anionic soap, they form an insoluble compound. You may have seen this scum in the form of bathtub rings. If you use soap and hard water in the washing machine, the scum clings to fabric and makes the clothes look dingy.

Detergents emulsify better in hard water

In the 1930s, chemists developed soap substitutes that work in hard water without forming insoluble compounds. They added other chemicals that enhance the basic cleaning power of the substitutes, and they named the mixture "detergent." Today, almost all laundry products are detergents.

To replace soap, chemists turned to a class of synthetic compounds that have long molecules with one end that is charged and soluble in water and another end that is nonpolar and insoluble.

surfactant
a compound that stabilizes an emulsion by acting at the surface between two immiscible substances

Like soap, these compounds, called surface active agents, or **surfactants,** act at the surface that separates oil and water. In a detergent, a surfactant plays the same role as soap in stabilizing emulsions of oil in water. **Figure 11-25** shows the structure of one surfactant.

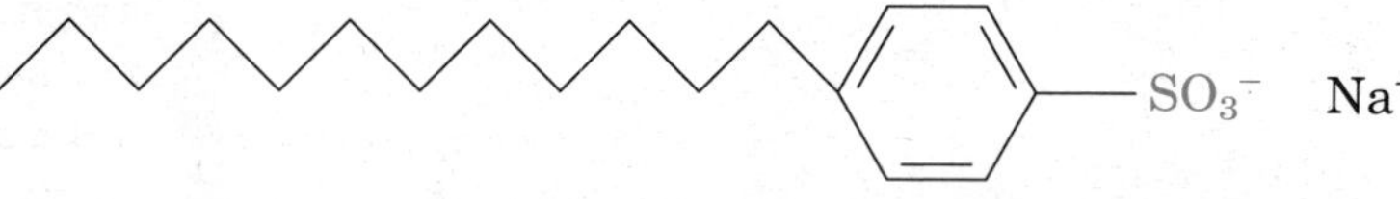

Figure 11-25
Sodium dodecylbenzene sulfonate is an anionic surfactant. This ionic compound does not make an insoluble scum with calcium or magnesium ions.

Section Review

20. How does an emulsion differ from a homogeneous mixture?
21. Explain why a substance containing only ionic bonds would not be a good emulsifier for mixtures of water and oil.
22. Contrast the behavior of soap and detergent in pure and "hard" water.

Visual Strategy

Figure 11-25

Ask students which part of sodium dodecylbenzene sulfonate in **Figure 11-25** dissolves in water. Students should identify the polar SO_3 group. If students suggest the benzene group, point out that the rounded shape used for polar regions in **Figure 11-22** and **Figure 11-24** are merely conventions. Shape alone does not determine polarity.

ASSESS

Alternative Assessment

Assess students' understanding of this section by having them write a summary of it. Summaries should include all vocabulary terms and major practical applications, such as soaps and salad dressings.

Answers to Section Review

20. Unlike a homogeneous mixture, an emulsion is a mixture of insoluble substances. Some emulsions separate into individual phases.
21. An ionic substance dissolves in water but not in oil, so it could not stabilize an emulsion.
22. Soaps and detergents behave similarly in "soft" water. Detergents work well in "hard" water, but the Ca^{2+} ions and Mg^{2+} ions of "hard" water are attracted to the polar region of soap and form an insoluble "scum."

Section 11-5

FOCUS

Lesson Starter

Start class with a review quiz to remind students of chapter topics that are extended in this section. Ask the following questions.

1. Why does oil-based paint dissolve in paint thinner but not in water?
2. Give an example of a polar molecule. Show its structure, and explain how its structure accounts for its polarity.
3. Repeat question 2 for a nonpolar molecule.

Demonstration 14

Solubility

Approximate time: 45 min

Show that the solubility of ink varies with different solvents.

1. Obtain four pieces of paper chromatography tape, each 15 cm long. (If this is not available, use strips cut from filter paper.) Cut each piece to form a pointed end.
2. Tape a piece of string across each of two 600 mL beakers. Fold and hang each strip on the string so that the tip of the paper is just above the bottom of the beaker.
3. Fill one beaker with water to a depth of 1 cm and the other beaker to a depth of 1 cm with ethanol.
4. With a water soluble black marker, place a small dot near the pointed end of two of the strips. When dry, reapply two more times. Repeat for the other two strips, but use a permanent black marker.

How are solubility principles used?

Section Objectives

Relate solvent type and solubility to the dry-cleaning process.

Relate solubility principles of gases to applications in carbonation and diving.

Relate solubility to the metabolism of vitamins A and C in the body.

Describe the conditions under which vitamin toxicity could occur.

Dry cleaning

Why do some clothes need to be dry cleaned, while others do not? Washing with water and detergents cleans most clothes just fine. But there are three situations in which dry cleaning may be necessary: if your clothes have a stubborn stain, such as ink or rust; if you have spilled something greasy on your clothes; and if the label recommends it.

Certain fabrics, especially silk and wool, do not respond well to water. They may shrink, take on stubborn wrinkles, or lose their shape. To avoid these effects, they must be cleaned without water.

Cleaners use solubility principles to remove stains

Dry cleaners are experts at removing stubborn stains. Some dry cleaners will ask you which food was spilled on a garment. Knowing the composition of the stain helps them decide how to treat it. Removing a stain, such as oil or grease, that doesn't dissolve in water involves two steps, as shown in **Figure 11-26**. First, the stain is treated with a substance that loosens the stain when the area is brushed. The stain is removed when the garment is washed in a mechanical dry cleaner.

If a garment has a water-soluble stain, it is first treated with a stain remover that is specific for that stain. The stain is then flushed away with a steam gun. The garment is allowed to dry, and then it is placed in the dry-cleaning machine to remove any other stains that are insoluble in water. *Dry cleaning does involve liquids.* The process uses a nonpolar solvent, tetrachloroethylene (perchloroethylene), instead of water.

Figure 11-26a
These pants have a spaghetti sauce stain on them, so this man is loosening the stain by brushing liquid solvent into the spot.

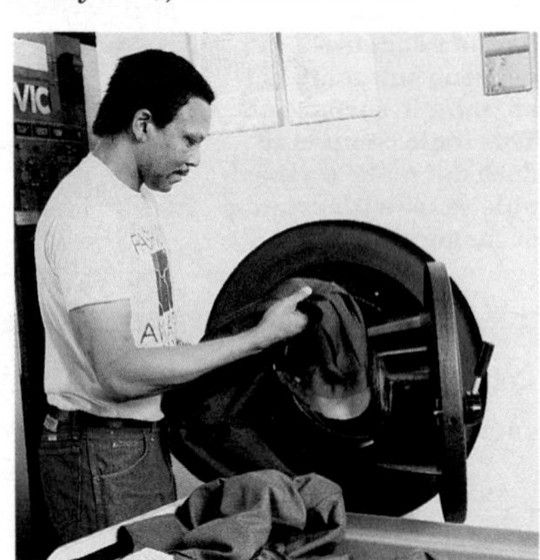

11-26b
The pants are then loaded into a dry-cleaning machine and are washed with a nonpolar solvent.

Chem Fact
To meet the OSHA standard and other federal guidelines, dry cleaners must operate airtight machines so that no C_2Cl_4 escapes.

Dry cleaning is said to have started in 1825, when a kerosene lamp was knocked over onto a greasy tablecloth at the house of a dye-factory owner, Jean-Baptiste Jolly. When the cloth dried, the area of the kerosene spill was noticeably cleaner. In the early decades of dry cleaning, some firms dipped the clothes first in kerosene and then in gasoline, both excellent degreasers. But they are also dangerous to use because they are very flammable.

A good dry-cleaning solvent must be nonpolar so that it can dissolve oil and grease, which trap the water-insoluble particles in the cloth fibers. Tetrachloroethylene, C_2Cl_4, shown in **Figure 11-27**, is favored because it is a versatile solvent of oil, grease, and alcohols; it is not flammable; and it is volatile, so it can be recycled. When the solvent is pumped from the dry-cleaning machine, it contains a certain amount of dissolved oils from clothes. It can be reused up to 200 times, but if it is reused excessively it stops taking oil out and starts putting it into the clothes. The solvent can be recycled by distillation: when heated, C_2Cl_4 vaporizes, leaving the oils behind; when cooled, the C_2Cl_4 vapors condense to form the clean solvent.

Tetrachloroethylene is a suspected carcinogen, and the United States Occupational Safety and Health Administration has said that dry-cleaning workers should not inhale air containing more than 200 ppm of the solvent.

Figure 11-27
The most commonly used dry-cleaning solvent is tetrachloroethylene, C_2Cl_4.

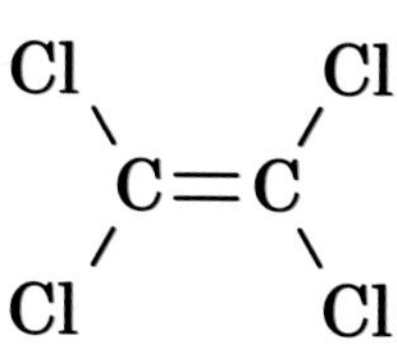

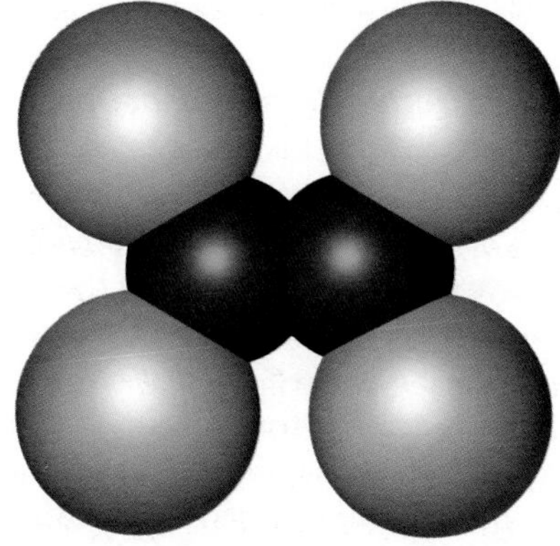

Consumer Focus

Stain removal

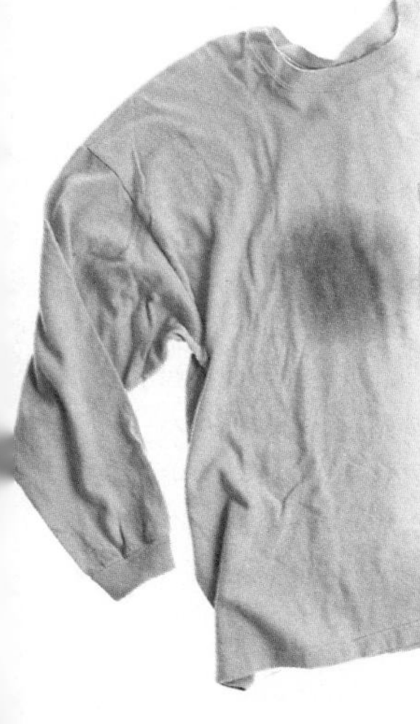

When you remove stains, you can see solubility principles in action. Beverage stains can be very stubborn because they contain so many different compounds. To remove stains of coffee, soft drinks, or wine, experts suggest the following series of solvents.

Prepare a 1.0 L solution of warm water and about 5 mL of hand dishwashing detergent, made acidic with about 20 mL of vinegar. Soak and occasionally agitate the fabric in this solution for about 30 min. Rinse with water. Soak in ethanol or isopropanol (sold as rubbing alcohol) for 15 min. Rinse with water. Soak for 30 min. in a solution of about 20 mL of enzyme-containing laundry detergent mixed with about a quart of warm water. If any stain remains, mix a solution of one part chlorine bleach (by volume) to about four parts of water. Test the fabric to be sure the bleach won't change the color, and then soak the fabric in the bleach solution for 2 min. Rinse with water.

5. Place one strip dotted with each marker in the beakers, being sure the dots do not go below the surface of the liquid. Cover with plastic wrap and let the solvent climb up the paper until it almost reaches the string.
6. Finally, remove the strips, and let them dry. While they are drying, compare the movement rates, degrees of separation, and order of colors. If this demonstration is begun at the beginning of the period, it should be finished before the end.

SAFETY
Wear safety goggles and lab apron; avoid flames. Use markers labeled "nontoxic."

DISPOSAL
Dispose of paper strips in trash, but let the alcohol-soaked strip dry first. Mix the two liquids, then dilute to five times the original volume with water, stir, and pour down the drain.

Groupwork Strategy
Have students work in groups of three or four. Ask each student in the group to use the stain removal procedure described in the Consumer Focus to remove a different fabric stain as a homework assignment. For example, one student in the group might try a grease stain, another student could try an ink stain, and a third could try a coffee stain. (Note: enzyme-containing laundry detergent or scrubbing powder is available in most supermarkets.) Have students bring their fabric samples to class. In class, have group members compare their results. Each group should compile a report explaining any observed variations in stain removal for the various stains.

TEACH

Other Applications

a. Heated fish tanks need aerators, while unheated goldfish and guppy bowls do not because O_2 is far less soluble in heated water.

b. Fast-food restaurants often use hot water to make ice cubes that are clear. The clarity comes from the fact that hot water will contain less gas than cold water. The hot water also constantly defrosts the refrigeration unit.

Demonstration 15

Temperature/solubility

Approximate time: 10 min

To demonstrate that temperature affects the solubility of a gas in a liquid, place a bottle of soda in an ice bath, another bottle in a warm-water bath (maximum temperature 35°C), and a third bottle in a room-temperature water bath. Open the bottles, and watch the gas escape.

SAFETY

Students should remain 10 ft from the demonstration. Do not allow anyone to drink soda.

DISPOSAL

Pour down the drain.

Teaching Tip

Discussion

Stress that the solubility of gases in liquids is the opposite of the rule for the solubility of solids in liquids. Gases are less soluble as the temperature of the solution increases.

Gas solubility

If you look at an unopened soda bottle, the liquid inside has no bubbles. But what happens once you open an ice-cold soda bottle? First, there is a hissing sound of gas escaping. Then you can see bubbles rising in the liquid. When you first taste it, there is plenty of "fizz" and it is tart. Later, after it is warm, it tastes "flat."

Why does this happen? Where did the bubbles come from? What can you do to keep soda from going flat? If you apply what you already know about gases and solutions, you can answer these questions and others. You may already know that the bubbles in soda are due to carbon dioxide gas, CO_2. Because the bubbles are not visible in the unopened soda bottle, shown in **Figure 11-28a**, you know the bottle's contents, including the CO_2, are a solution that is stable and homogeneous. Part of the CO_2 reacts with water to form carbonic acid, H_2CO_3, which gives the soda a tart taste.

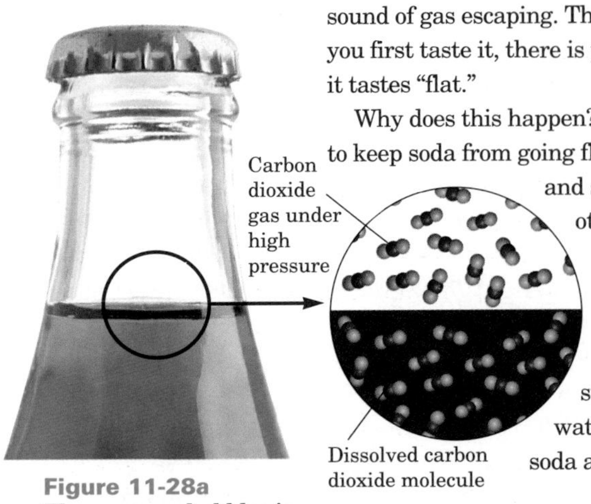

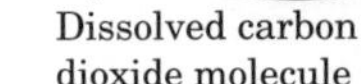

Figure 11-28a
There are no bubbles in an unopened bottle of soda because carbon dioxide is dissolved in the liquid.

11-28b
When the bottle cap is removed, the pressure inside the bottle decreases rapidly. Carbon dioxide escapes due to the lower solubility of CO_2.

Carbonation depends on pressure

From Chapter 10, you know that if gas escapes from a container, such as a soda bottle, the pressure is lowered. After this decrease in pressure, the bubbles of CO_2 are seen coming out of the solution.

Chemists have figured out the details of the relationship between pressure and gas solubility. Henry's law sums up the basic idea: *the amount of gas dissolved in a solvent is proportional to the partial pressure of that gas over the solvent.* In an unopened soda bottle, there is a higher partial pressure of CO_2 in the neck of the bottle, above the liquid, than there is in the air outside of the bottle. When the bottle is opened, the CO_2 escapes into the air, and the partial pressure of CO_2 in the bottle's neck soon equals that of the CO_2 in the air. As a result of this decrease in the partial pressure of CO_2, the solubility of CO_2 in the soda is reduced, the solution becomes supersaturated, and the excess CO_2 bubbles out of solution, as shown in **Figure 11-28b**.

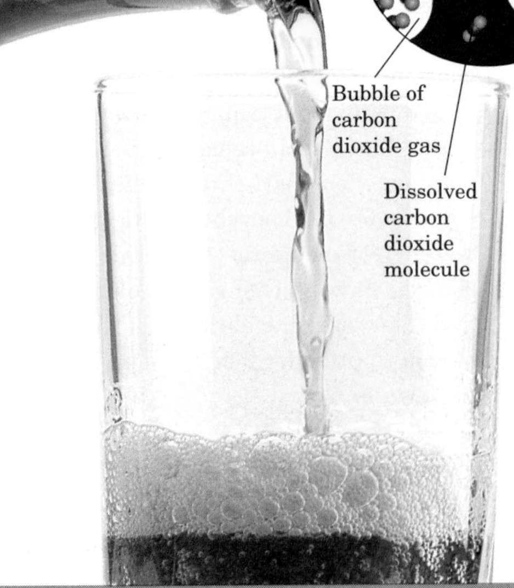

Gas solubility decreases with increased temperature

After the soda bottle is open and the soda gets warmer, it has fewer bubbles and tastes flat. Similarly, if you compare the taste of a newly opened warm soda with that of a newly opened cold soda, the warm soda will taste flat.

Warm soda is flat because there is less CO_2 dissolved in it. Chemists have found that other dissolved gases follow a similar pattern: *solubility decreases with increasing temperature*. Recall from **Figure 11-5** on page 407 that this pattern is the *opposite* of what you know about most solid solutes, which dissolve better in hot solvents.

Visual Strategy

Figure 11-28

Students should be aware that before the bottle shown in the figure is opened, the system is at equilibrium with CO_2 gas constantly dissolving into and out of solution. When opened, the system is no longer at equilibrium, and the solution releases CO_2, trying to re-establish a new equilibrium rate.

Gas solubility and diving

The principles of gas solubility are used in a variety of situations other than soda bottles. Knowledge of these principles is a life-or-death matter for the underwater diver shown in **Figure 11-29**.

When divers are underwater, they are subject to more pressure than at the surface. As a result, the pressure of the air that they breathe must be increased to force the air out of the air tank and into their lungs.

The diver's blood is a water-based mixture. Gases such as oxygen are carried by hemoglobin molecules within the blood cells in this solution. But some oxygen and other gases, such as nitrogen, which makes up most of air, can dissolve in the blood just like carbon dioxide in soda.

If a diver breathing pressurized air rises to the surface too quickly, decompression sickness, called "the bends," can result, causing intense pain and possibly even death.

The sudden decrease from high air pressure to normal air pressure is similar to the decrease that takes place when you open a soda bottle. Gases, such as nitrogen, that dissolve in blood, body fluids, and nonpolar fat bubble out of solution. Such bubbles in the blood can block blood flow in smaller blood vessels, press on tissues, and affect nerve impulses.

Figure 11-29
Underwater divers breathing pressurized air must return to the surface gradually to prevent "the bends."

Most deep-sea divers prevent "the bends" by breathing a mixture that contains helium instead of compressed air, which contains nitrogen. Because helium is less soluble than nitrogen, less of it dissolves in the body tissues. Then, if they have to return to the surface rapidly, there will be less gas bubbling out of solution.

Concept Review

Solubility of gases and carbonation

23. Explain why the carbon dioxide in the air is not enough to recarbonate an opened bottle of soda.
24. What can you do to keep soda from getting flat? Explain how your suggestions work.

Vitamins and solubility

Vitamins act to control chemical reactions within the body. Some vitamins promote the occurrence of specific chemical reactions that are a part of metabolism. Other vitamins prevent harmful reactions that could cause damage to the body. Without vitamins, the beneficial reactions do not occur, and deficiency diseases develop.

The body needs only small quantities of vitamins, about 0.4 g per kilogram of food that you eat. Essential vitamins are not made in the body; therefore, you must get them from the food you eat or by taking vitamin supplements. Not getting enough vitamins can make you sick, but so can getting too much. The reasons relate to solubility.

Historical Note

The manufacture of PCBs was ceased in 1976 in the United States, but companies were allowed to continue to use all existing stock.

Teaching Tip

Why boiled water tastes flat
Ask students to conjecture why boiled water tastes flat, and why lake fish sometimes die in large numbers on hot days.

Answers: Boiled water tastes flat because the dissolved air has been expelled. Fish die because there is less oxygen in lakes on hot days.

Answers to Concept Review

23. Because CO_2 is only moderately soluble in water at normal air pressure, higher CO_2 pressures are necessary for more CO_2 to dissolve in the water.
24. Keep the soda covered tightly, and keep it cold. Less CO_2 remains in solution as the temperature increases, and if the bottle is uncapped more CO_2 escapes.

Content Background

Though excess vitamin C is usually excreted by the body, doses of more than a few grams of vitamin C often produce diarrhea.

Teaching Tip

Discussion

Discuss methods for cooking vegetables that minimize the loss of vitamin C and vitamin B complex, which are water soluble. One way is to cook vegetables in as little water as possible to minimize the quantity of vitamins that leach into the cooking medium. Steaming is preferable to boiling. Another technique is to pan-fry vegetables as quickly as possible.

Multicultural Perspective

The traditional Chinese diet is rich in vitamins and minerals because vegetables are a main staple. Also, the water-soluble vitamins are largely retained during cooking because Chinese cooks use the wok, in which vegetables cook quickly at high heat.

Figure 11-30
Lemons, oranges, grapefruits, and limes are rich sources of vitamin C.

Vitamin C dissolves in water

People on poor diets have suffered from scurvy at least since the time of the Crusades, nearly 1000 years ago. Soldiers and sailors were especially susceptible to scurvy, which causes fatigue and bleeding of the gums and prevents wounds from healing. Some ships' captains claimed that lemon juice would prevent the disease.

The first controlled study of the effects of diet on scurvy was in 1747 when James Lind, a doctor in the British Royal Navy, compared the reactions of sailors suffering from scurvy when given cider, garlic, mustard, vinegar, ocean water, oranges, or lemons. The sailors who received citrus fruits recovered. We now know that citrus fruits, such as those in **Figure 11-30**, are rich in vitamin C. The structure of this important molecule is shown in **Figure 11-31**.

Researchers have found that vitamin C has several vital functions, but none is more important than its role in the synthesis of collagen. Collagen is the protein that makes up tendons, gives strength to teeth and bones, and, most importantly, knits endothelial cells together to make blood vessels.

To synthesize collagen where it is needed, vitamin C must be distributed throughout the body. Vitamin C is easily transported by the blood because it is water soluble. Important information about vitamin C and other water-soluble vitamins is shown in **Table 11-5** on the next page.

Healthy people have 5 mg or more of vitamin C per liter of blood plasma. Scurvy symptoms appear when this concentration falls to 1.5 mg/L. If an adult consumes the recommended dietary amount, 60 mg per day, the concentration in the blood plasma reaches about 8 mg/L.

But some people prefer to take a lot of vitamin C. Is it possible to take too much? Probably not. When intake increases, the vitamin C level rises to about 15 mg/L of plasma, but then the kidneys begin excreting the vitamin faster. Most people's blood concentration will not exceed 20 mg/L even if they consume 10 or 20 times the recommended amount. Your kidneys are so good at regulating water-soluble vitamins that overdosing on vitamin C is very unlikely.

Chem Fact
1 cup canned orange juice: 86 mg vitamin C

1 baked potato with skin: 26 mg vitamin C

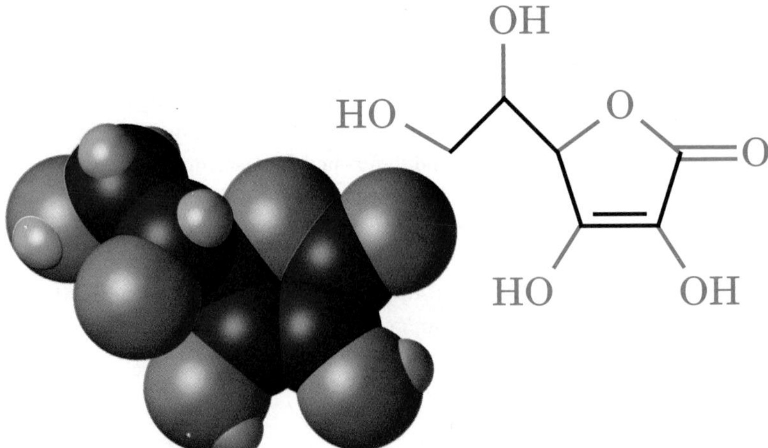

Figure 11-31
Vitamin C's most common form is ascorbic acid, shown here. Because ascorbic acid contains almost as many polar bonds as nonpolar bonds, it is very soluble in water but insoluble in fat and oil.

Visual Strategy

Figure 11-31 and Figure 11-32

Have students draw complete structural formulas for vitamins A and C based on the organic structures shown in **Figure 11-31** and **Figure 11-32**. Then have students identify which atoms are hidden in the three-dimensional models shown.

Table 11-5
Some Water-Soluble Vitamins

Vitamin	Chemical formula	Molecular mass (g/mol)	Reference daily intake (RDI)*
Folic acid	$C_{19}H_{19}N_7O_6$	441.40	180 μg
Vitamin B_1 (thiamine mononitrate)	$C_{12}H_{17}N_5O_4S$	327.36	1.2 mg (1200 μg)
Vitamin B_2 (riboflavin)	$C_{17}H_{20}N_4O_6$	376.36	1.4 mg (1400 μg)
Niacin	$C_6H_5NO_2$	123.11	1.6 mg (1600 μg)
Vitamin B_6 (pyridoxine hydrochloride)	$C_8H_{12}ClNO_3$	205.64	1.4 mg (1400 μg)
Vitamin B_{12} (cyanocobalamin)	$C_{63}H_{88}CoN_{14}O_{14}P$	1355.38	2.0 μg
Vitamin C (ascorbic acid)	$C_6H_8O_6$	176.12	60 mg (60 000 μg)

*The RDI is the average intake recommended for individuals by the U. S. Food and Drug Administration.

Other Applications
Nobel laureate Linus Pauling was a well-known proponent of megadoses of vitamin C as a cure for the common cold. He stated that his personal intake was 10 g per day. The recommended daily intake is 60 mg!

Vitamin A dissolves in oils and fats

Chem Fact
1 spear of broccoli: 233 μg vitamin A

1 whole, scraped, raw carrot: 2025 μg vitamin A

For decades, children have been told, "Eat your carrots so you'll have good eyes." There is some truth to the saying because carrots, and some other vegetables, contain carotenes that your body easily converts into vitamin A, which is shown in Figure 11-32. If you don't get enough vitamin A, you may lose your ability to see at night—a disorder called night blindness.

Vitamin A affects vision because your body uses it to make one of the vital light-sensitive molecules in the retina. Vitamin A is also needed for normal growth of skin, respiratory tract, cornea, and other epithelial tissues.

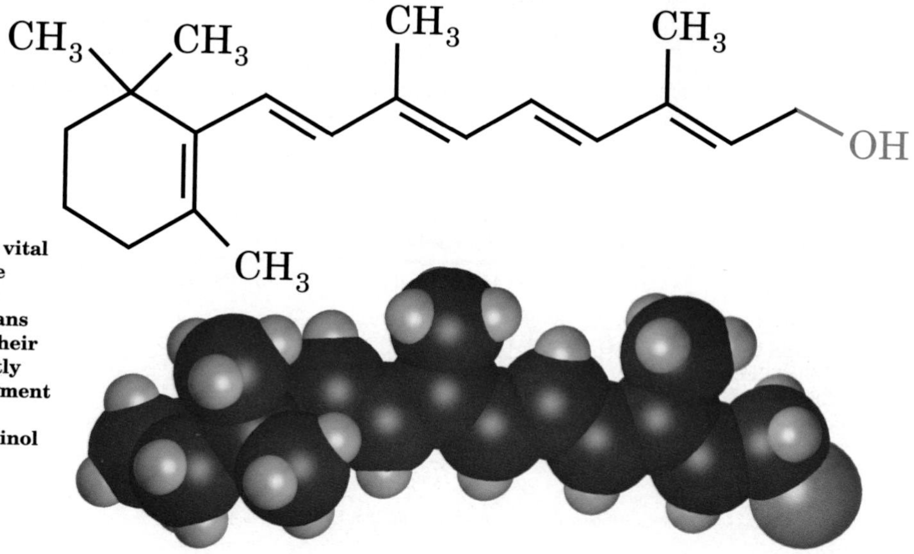

Figure 11-32
Vitamin A is also known as retinol because it plays a vital role in helping the retina of your eye detect light. Humans get about half of their vitamin A indirectly from the plant pigment carotene, which consists of two retinol molecules linked together.

Possible Misconception

Many students may believe that "natural" vitamins are chemically different than "synthetic vitamins." Though obtained from different sources, natural and synthetic vitamins are identical. The body cannot tell the difference between "ascorbic acid" (synthetic vitamin C) and "rose hips" (natural vitamin C) because the two have an identical chemical make-up.

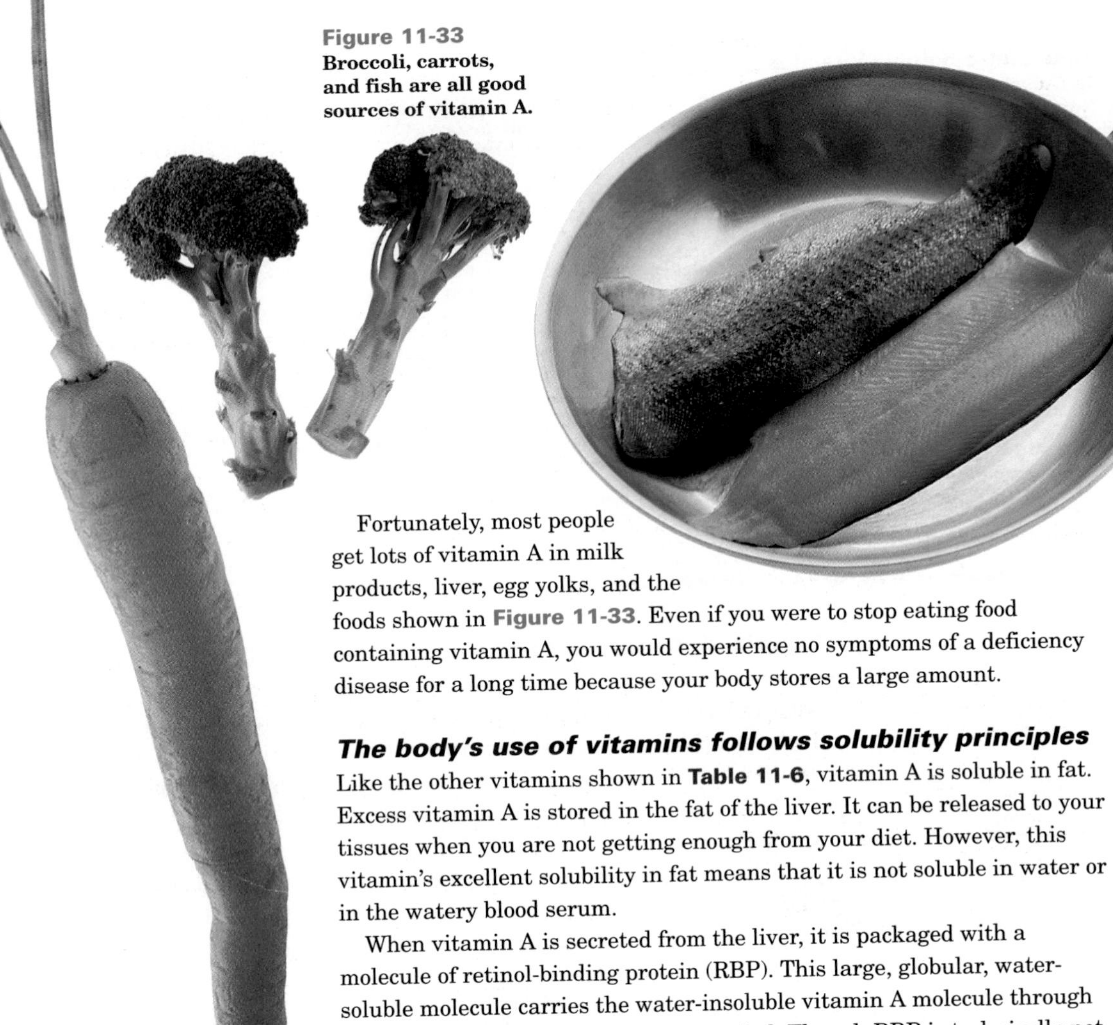

Figure 11-33
Broccoli, carrots, and fish are all good sources of vitamin A.

Fortunately, most people get lots of vitamin A in milk products, liver, egg yolks, and the foods shown in **Figure 11-33**. Even if you were to stop eating food containing vitamin A, you would experience no symptoms of a deficiency disease for a long time because your body stores a large amount.

The body's use of vitamins follows solubility principles

Like the other vitamins shown in **Table 11-6**, vitamin A is soluble in fat. Excess vitamin A is stored in the fat of the liver. It can be released to your tissues when you are not getting enough from your diet. However, this vitamin's excellent solubility in fat means that it is not soluble in water or in the watery blood serum.

When vitamin A is secreted from the liver, it is packaged with a molecule of retinol-binding protein (RBP). This large, globular, water-soluble molecule carries the water-insoluble vitamin A molecule through the blood to the tissues where it is needed. Though RBP is technically not an emulsifying agent, it serves a similar function: it makes a nonpolar substance mix with water.

Table 11-6
Some Fat-Soluble Vitamins

Vitamin	Chemical formula	Molecular mass (g/mol)	Reference daily intake (RDI)*	Toxicity level
Vitamin A (retinol)	$C_{20}H_{30}O$	286.44	875.0 µg	> 15 000 µg/day
Vitamin D (cholecalciferol and related compounds)	$C_{27}H_{44}O$	384.62	6.5 µg	> 5000 µg/day
Vitamin E (α-tocopherol)	$C_{29}H_{50}O_2$	430.69	9.0 mg (9000 µg)	> 600 mg/day (> 600 000 µg/day)

*The RDI is the average intake recommended for individuals by the U. S. Food and Drug Administration.

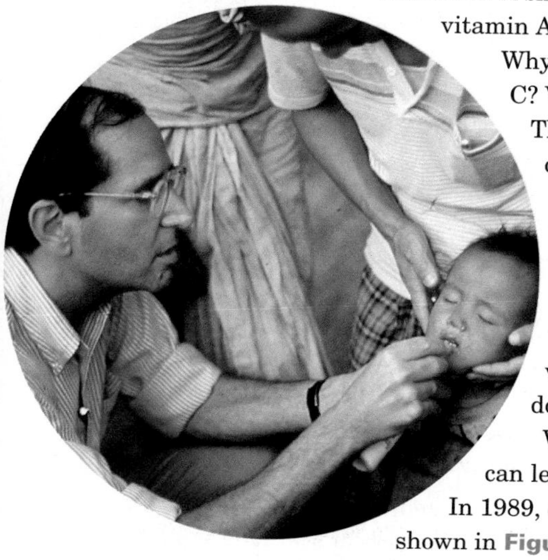

Figure 11-34
In Indonesia, Alfred Sommer examines a child for signs of vitamin A deficiency. Because vitamin A is fat soluble, it can be stored in the liver for long periods. Sommer believes that two vitamin A capsules per year, at a cost of one cent each, could prevent most cases of xerophthalmia.

Normally, because vitamin A is present in food only in small amounts, it is not possible to get too much. But sometimes people become enthusiastic about improving their diet and start taking large quantities of vitamin supplements or tablets. Large doses of vitamin A, taken for many weeks, can overload the storage capacity of the liver and raise the concentration of the vitamin throughout the body to toxic levels. Toxicity symptoms include dry, rough skin, cracked lips, hair loss, birth defects, headaches, and muscle and joint pain.

A bizarre case of extreme vitamin A poisoning was reported by Arctic explorers who killed and ate a polar bear. Members of the expedition who ate the bear's liver became very ill, and three of them lost patches of skin. Much later, scientists discovered that polar bear liver is extremely rich in vitamin A. A single quarter-pound (113 g) serving provides enough vitamin A to meet an adult's dietary requirement for 2.5 years!

Why is it possible to overdose on vitamin A but not vitamin C? Vitamin A is fat soluble; vitamin C is water soluble. The kidneys remove chemicals that are toxic or too concentrated from the blood. You would expect, therefore, that the kidneys would remove excesses of both vitamins A and C. But while the kidneys do a meticulous job of regulating small water-soluble molecules, they are far less effective in processing large molecules, including many proteins. This means that vitamin A is only gradually metabolized, or broken down by chemical reactions within the cells.

When vitamin A is completely absent from the diet, it can lead to blindness and hasten death by other ailments. In 1989, experts such as research physician Alfred Sommer, shown in **Figure 11-34**, estimated that 500 000 children in underdeveloped nations are blinded annually by xerophthalmia, a disintegration of the cornea due to the lack of vitamin A.

Section Review

25. Which compound would be the most effective dry-cleaning agent? Explain your answer.

a. isopropanol, $CH_3CH(OH){-}CH_3$ **b.** acetic acid, $CH_3C(=O){-}OH$

c. cyclohexane, C_6H_{12}

26. What would you do to remove the dissolved gases in a sample of water?

27. Many fruit juices contain vitamin C. Why don't any contain substantial amounts of vitamin A?

28. Vitamin E is a long, chainlike molecule with several hydrocarbon rings on one end. Its formula is $C_{29}H_{50}O_2$. Would you expect levels of vitamin E to be easily controlled by the kidneys? Why?

29. Story Link
Bears, like people, are mammals. Given what you've learned about the kidneys, explain why PCBs aren't flushed from the bears' bodies.

ASSESS

Alternative Assessment

Ask each student to prepare a report profiling a practical application of solubility principles that they utilize in their daily lives. Students' reports should include a microscopic as well as macroscopic description, all relevant vocabulary, and illustrations, if appropriate. For best results, have students provide an example that has not been discussed in detail in the text.

Answers to Section Review

25. Cyclohexane would be the best dry-cleaning agent because it is nonpolar and able to dissolve oily dirt and grease.

26. Heat the water; solubility of gases decreases with an increase in temperature.

27. Juices are mostly water, and vitamin A does not dissolve in water.

28. Kidneys are good at regulating water-soluble vitamins, but vitamin E is fat soluble.

29. PCBs are soluble in fats and oils, so they are not washed out by the kidneys.

Chemistry in Your Community

The PCBs discussed in this chapter are, of course, just one class of chemical compounds that contribute to water pollution. To have students investigate water pollution at the local level, and see how this relates to solubility, do the following activities.

1. Pose the following questions to the class:
   - What are the industrial sources of water pollution in the local area?
   - What types of chemicals must be removed in the treatment of waste water?
   - How is this accomplished?

   Students can brainstorm ways to find the information as part of the activity. You may want to supply phone books, reference books (such as *Embracing the Earth* by Mark T. Harris, The Noble Press, 1991), or other available resources. Students can write letters or make phone calls to appropriate agencies to find answers to the questions as a homework assignment.
2. Have students find out what techniques are used locally to treat water-soluble pollutants (most ionic substances). Compare these to techniques used to treat non-water-soluble pollutants, such as oil. (This information can be obtained through local water regulatory agencies.)
3. Ask students to determine whether local water supplies are contaminated with substances that bioaccumulate. If so, check local agencies to learn how wildlife is monitored for toxicity.
4. If possible, schedule a visit to water and waste-water treatment plants.

Conclusion: Solubility and the Arctic Bear Hunt

At the beginning of this chapter you followed Dr. Malcolm Ramsay as he examined a polar bear and took samples of its fat tissue to determine PCB content. This story posed some puzzling questions about the properties of PCBs. Look at those questions again.

How did the PCBs get into the Hudson Bay ecosystem when there is very little industry in the bay area? In past decades, the Baltic Sea was used as an ocean dump for PCBs and other wastes. It is likely that much of the discarded material remains in the mud at the bottom of the ocean. Because PCBs are slightly soluble in water, a small fraction of the PCBs dissolves and circulates in the ocean water.

Some PCBs are ingested by plankton, tiny marine animals. The plankton are eaten by larger plankton, which are eaten by fish. In addition to getting PCBs in their food, fish absorb some PCBs directly from the water. You may recall that oxygen is a nonpolar molecule and therefore has very low solubility in water. One liter of ocean water contains only about 3×10^{-4} mol of O_2, whereas the air we breathe contains 9×10^{-3} mol of O_2 per liter. This means that as fish process very large quantities of water through their gills, they extract PCBs as well as oxygen from the water.

Once PCB molecules are ingested by a fish, they diffuse out of the fish's watery blood, dissolve in the oily liver, and stay there. Over time, the concentration of PCBs in the liver increases. In effect, fish serve as vacuum cleaners, picking up low-concentration PCBs from the water and storing them to produce far higher concentrations in their bodies.

Why did the PCBs end up in the bear's fat? When a seal eats a fish, the oil-soluble PCBs end up in the seal's oily blubber in concentrations that are even higher than those in fish. From there, it is just another step in the food chain to a polar bear's fat. The animals that are the most likely to have PCBs concentrated in their fatty tissues are polar bears. They are at the top of the food chain and live a long time.

Applying Concepts

1. What is the empirical formula of the PCB molecule on page 402?
2. Do you think this compound will float or sink in water? To estimate the density of the PCB molecule shown on page 402, find its molecular mass, and compare it with the molecular mass of biphenyl. Biphenyl, the related double-ring molecule with no chlorine atoms, has a density that is almost the same as water.

Research and Writing

Use the library to find out more about the following.

1. How can further buildup of PCBs in bears be prevented?
2. How can the harmful effects of PCBs already in bears be reduced?
3. What are the current procedures for PCB disposal?

Answers to Applying Concepts

1. The empirical formula is $C_{12}H_7Cl_3$.
2. It will sink because the PCB molar mass is greater than the molar mass of biphenyl, and thus the PCB will be more dense than water. The molar mass of the PCB molecule shown is 257.54 g/mol; the molar mass of biphenyl ($C_{12}H_{10}$) is 154.22 g/mol.

CHAPTER 11

Highlights

Key terms

alloy
concentration
emulsifying agent
emulsion
immiscible
insoluble
miscible
molarity
saturated
solubility
soluble
solute
solvent
supersaturated
surfactant
suspension
unsaturated

Key problem-solving approach: Calculations with molarity

When solving molarity problems, select the variable given in your problem from the options near the top of this diagram. Then, follow the calculation pathway that leads to the proper units for the answer near the bottom of this diagram.

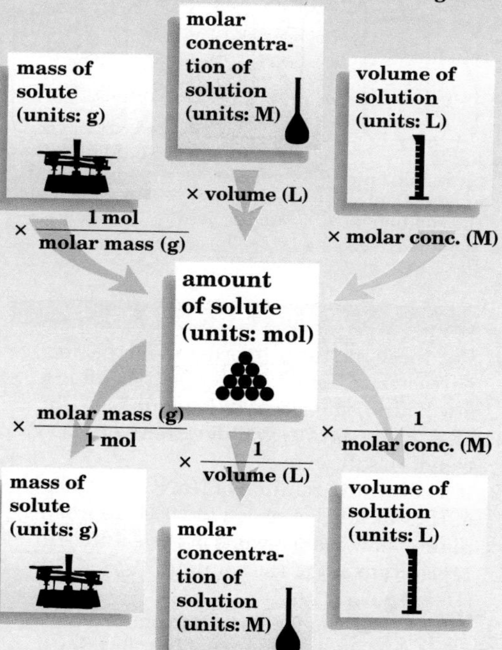

Key ideas

11-1 What is a solution?

- Solutions stay mixed; suspensions will settle.
- A solute is dissolved in a solvent to make a solution.
- Solubility is the amount of a solute that will ordinarily dissolve in a solvent at a certain temperature.
- A solution can be described qualitatively by whether it contains more solute than the solubility (supersaturated), as much solute as the solubility (saturated), or less solute than the solubility (unsaturated).

11-2 What does concentration mean?

- A common concentration measure in chemistry is molarity, which is the moles of solute per liter of solution.

11-3 Why do some things dissolve?

- Polar solvents dissolve ionic and polar solutes by attracting them into the solvent. Nonpolar substances will not dissolve in water.
- The polarity of a molecular substance can be predicted by determining the polarity of the individual bonds and the effective poles of the particle.

11-4 Can insoluble substances mix?

- An emulsion is a mixture of two insoluble substances. Ordinarily, emulsions will form two phases on standing.
- Emulsifying agents stabilize emulsions by dissolving partly in a polar droplet and partly in a nonpolar droplet. This linking prevents the droplets from recombining to form two phases.

11-5 How are solubility principles used?

- Dry cleaning uses nonpolar solvents to dissolve solutes such as oily dirt and greasy stains without harming delicate fabrics.
- Gases are more soluble at increased pressures and lower temperatures.
- Vitamins can be water soluble (polar) or fat soluble (nonpolar). Water-soluble vitamins are regulated by the kidneys. Fat-soluble vitamins link with special proteins to travel in the blood.
- Fat-soluble vitamins are stored in the liver.

Closure Strategy

Formulating quiz questions
Have each student develop two questions (with answers) from each section in the chapter for a quiz. Compile a quiz from the best questions. Students whose questions appear on the quiz are rewarded with easy questions!

1. **a.** A solution is a stable and homogeneous mixture.
   b. A solute is a substance dissolved in a solution. A solvent is a substance that dissolves another to make a solution.
   c. Answers can vary. Check that the examples are stable and homogenous.

2. **a.** A suspension is a heterogeneous mixture with particles that may settle out on standing.
   b. Answers can vary. Check that only unstable mixtures are given.

3. **a.** It is the amount of a substance that can dissolve in a given amount of a solvent at a specified temperature, with undissolved solute present.
   b. Temperature (and pressure for gases) must be specified for solubility.

4. **a.** about 32 g
   b. about 123 g
   c. about 35 g

5. about 68°C

6. **a.** insoluble
   b. insoluble
   c. insoluble
   d. soluble
   e. Lead(II) acetate should be used because it will dissolve.

7. **a.** unsaturated
   b. unsaturated
   c. saturated
   d. saturated
   e. pure solvent
   f. pure solvent

8. **a.** unsaturated
   b. saturated
   c. unsaturated

9. A supersaturated solution has more solute dissolved in it than the solubility would allow at the specified temperature given.

10. **a.** supersaturated
    b. saturated
    c. unsaturated

CHAPTER 11

Review and Assess

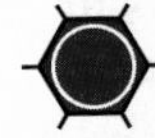

Properties of solutions

REVIEW

1. **a.** What is a solution?
   b. Identify and define the two components of a solution.
   c. Give two examples of common solutions.
2. **a.** How does a suspension differ from a solution?
   b. Give two examples of common suspensions.
3. **a.** What is meant by the solubility of a substance?
   b. What condition(s) must be specified in citing solubility levels?
4. Use **Figure 11-5** to determine the solubility of each of the following, in grams per 100 g of H_2O.
   a. KCl at 10°C **b.** $NaNO_3$ at 60°C
   c. Li_2SO_4 at 30°C
5. Use **Figure 11-5** to determine the temperature at which the solubility of NH_4Cl is 60 g $NH_4Cl/100\ g\ H_2O$.
6. Use **Table 11-1** to determine which of the following lead compounds are insoluble.
   a. lead(II) chloride **b.** lead(II) carbonate
   c. lead(II) sulfide **d.** lead(II) acetate
   e. If you needed to make a solution of Pb^{2+}, which of the above solutes would you use? Why?
7. The drawings below represent macroscopic and microscopic views of a solution being made by adding increasing amounts of solute. Match the microscopic and macroscopic views and label them *pure solvent*, *unsaturated*, or *saturated*.

a.
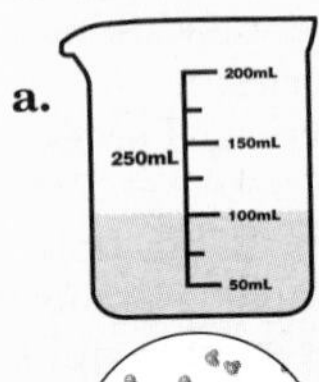

b.
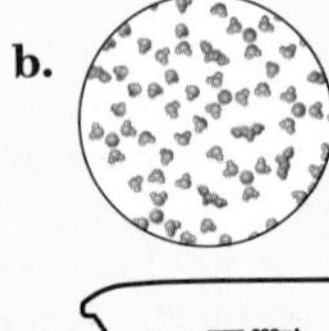

c.
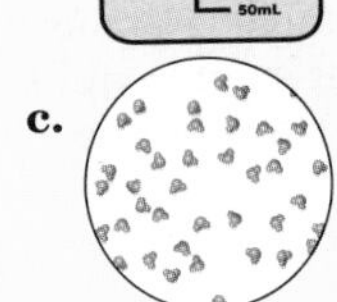

d.
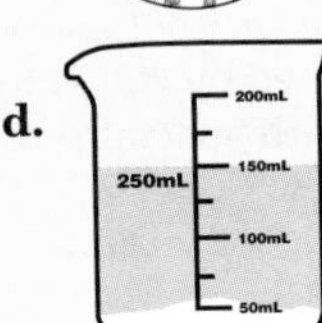

e.
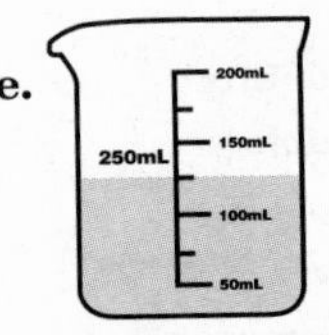

f.
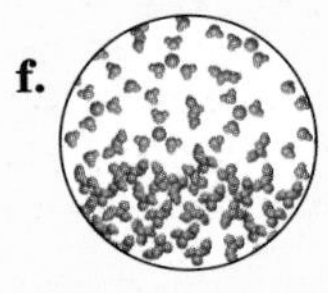

8. Use **Figure 11-5** to decide whether each description represents a saturated or unsaturated solution.
   a. 40 g of KCl added to 100 g of H_2O at 50°C
   b. 120 g of NH_4Cl added to 100 g of H_2O at 80°C
   c. 100 g of RbCl added to 100 g of H_2O at 55°C
9. What is a supersaturated solution?
10. In each of the diagrams shown below, the same amount of sodium acetate has been added to different solutions of sodium acetate. Indicate whether each of the original solutions was unsaturated, saturated, or supersaturated.

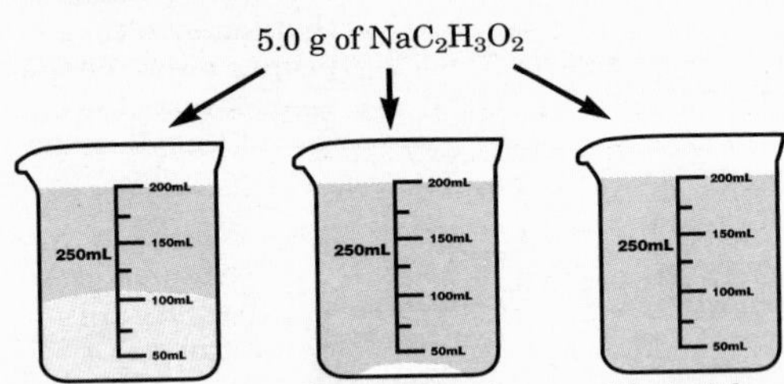

a. Rapid recrystallization of more than the additional amount
b. Additional amount remains undissolved
c. Most of the additional amount dissolves

APPLY

11. Put these mixtures in order of increasing particle size: muddy water (settles after a few hours), sugar water, ketchup (settles after a few days), sand in water (settles rapidly), salt water.
12. If a saturated solution of $NaNO_3$ in 100 g of H_2O at 60°C is cooled to 10°C, approximately how many grams of the solute will precipitate out of the solution? (Use **Figure 11-5**.)
13. Why would you feel more uncomfortable on a hot, humid day than on a hot, dry day?

14. Plot a solubility graph for $AgNO_3$ with the data below. Plot grams of solute (in increments of 50) per 100 g of H_2O on the vertical axis and the temperature in degrees Celsius on the horizontal axis.
   a. Estimate the solubility of $AgNO_3$ at 30°C, 55°C, and 75°C.
   b. At what temperature would the solubility of $AgNO_3$ be 275 g per 100 g of H_2O?
   c. If 98.5 g of $AgNO_3$ were added to 100 g of H_2O at 10°C, would the resulting solution be saturated or unsaturated?

Grams solute per 100 g H_2O	Temperature (°C)
122	0.
216	20.
311	40.
440	60.
585	80.
733	100.

Concentration and molarity

REVIEW

15. Describe in detail how you would make 250. mL of a 0.500 M solution of NaCl with equipment found in your chemistry lab.
16. a. How many moles and grams of NaOH are in 1.00 L of a 2.50 M NaOH solution?
   b. To make 1.00 L of 2.50 M NaOH, will you need more than 1.00 L of H_2O, less than 1.00 L of H_2O, or exactly 1.00 L of H_2O? Explain your answer.
17. You are determining the concentration of a solution of salt water by evaporating different samples and measuring the mass of salt that remains. Your data are shown below. Calculate the average molar concentration of the original solution.

Sample volume (mL)	Mass of NaCl (g)
25	2.9
50.	5.6
75	8.5
100.	11.3

18. The actual molarity of the solution in item 17 is determined to be 2.00 M. Calculate the percent error in the data.

PRACTICE

Answers to items in a black square begin on page 841.

19. Determine the molarity of each of the following solutions. (Hint: see Sample Problem 11A.)
   a. 3.5 mol $NaNO_3$ in 1.0 L of solution
   b. 5.0 mol KOH in 2.0 L of solution
   c. 0.20 mol Na_2S in 0.25 L of solution
20. Determine the molarity of each of the following solutions. (Hint: see Sample Problem 11A.)
   a. 0.25 mol $FeCl_3$ in 2.0 L of solution
   b. 0.015 mol $KMnO_4$ in 350 mL of solution
   c. 3.5×10^{-4} mol $NaC_2H_3O_2$ in 25 mL of solution
21. How many moles of each solute would be required to prepare each of the following solutions? (Hint: see Sample Problem 11B.)
   a. 1.0 L of a 4.0 M $AgNO_3$ solution
   b. 2.50 L of a 0.500 M HCl solution
   c. 400. mL of a 0.250 M HNO_3 solution
22. How many moles of each solute would be required to prepare each of the following solutions? (Hint: see Sample Problem 11B.)
   a. 30.0 mL of a 0.0100 M H_3PO_4 solution
   b. 5.0 mL of an 18 M H_2SO_4 solution
   c. 250. mL of a 1.00×10^{-4} M CH_3OH solution
23. Determine the molarity of each of the following solutions. (Hint: see Sample Problem 11A.)
   a. 20.0 g of NaOH in enough H_2O to make 2.00 L of solution
   b. 14.0 g of NH_4Br in enough H_2O to make 150. mL of solution
   c. 65.0 g of $CuCl_2$ in enough H_2O to make 300. mL of solution
24. Determine the molarity of each of the following solutions. (Hint: see Sample Problem 11A.)
   a. 3.50 g CCl_4 in enough C_6H_{12} to make 500. mL of solution
   b. 0.150 g $CuSO_4{\cdot}5H_2O$ in enough H_2O to make 25.0 mL of solution
   c. 3.00×10^{-4} g $AlCl_3$ in enough H_2O to make 2.00 mL of solution
25. How many grams of each solute would be required to make each of the following solutions? (Hint: see Sample Problem 11B.)
   a. 1.00 L of a 1.50 M NaCl solution
   b. 2.25 L of a 3.50 M NaOH solution
   c. 750.0 mL of a 2.50 M K_2SO_4 solution
26. How many grams of each solute would be required to make each of the following solutions? (Hint: see Sample Problem 11B.)
   a. 1.50 mL of a 0.300 M $Ca(NO_3)_2$ solution
   b. 25 mL of a 1.5 M LiCl solution
   c. 3.00×10^{-4} L of a 2.50×10^{-2} M $C_6H_{12}O_6$ solution

11. in order of increasing particle size: salt water, sugar water, ketchup, muddy water, sand in water
12. about 40 g of $NaNO_3$.
13. On a humid day, the already saturated air does not allow quick evaporation of perspiration to cool the body.
14. *See page 447A.*
15. Answers can vary, but should include the following steps.
   1. Mass of solute should be carefully measured in a beaker.
   2. Solute should be dissolved in a small amount of solvent and poured into a volumetric flask.
   3. The measuring beaker should be rinsed into the volumetric flask. More solvent should be added, and the flask should be shaken until the solute dissolves. Then, the volume of the flask should be brought to exactly 250 mL.
16. a. 2.50 moles and 100. g
   b. Less than 1.00 L of water will be needed because the volume of the solution will expand as solute is added.
17. 1.9 M
18. 5%
19. a. 3.5 M $NaNO_3$
   b. 2.5 M KOH
   c. 0.80 M Na_2S
20. a. 0.12 M $FeCl_3$
   b. 0.043 M $KMnO_4$
   c. 0.014 M $NaC_2H_3O_2$
21. a. 4.0 mol $AgNO_3$
   b. 1.25 mol HCl
   c. 0.100 mol HNO_3
22. a. 3.00×10^{-4} mol H_3PO_4
   b. 0.090 mol H_2SO_4
   c. 2.50×10^{-5} mol CH_3OH
23. a. 0.250 M NaOH
   b. 0.953 M NH_4Br
   c. 1.61 M $CuCl_2$

*Full solutions for items **14** and **18–23** are found on page 447A.*

24. a. 0.0455 M CCl_4
 b. 0.0240 M $CuSO_4$
 c. 1.13×10^{-3} M $AlCl_3$

25. a. 87.7 g NaCl
 b. 315 g NaOH
 c. 327 g K_2SO_4

26. a. 0.0738 g $Ca(NO_3)_2$
 b. 1.6 g LiCl
 c. 1.35×10^{-3} g $C_6H_{12}O_6$

27. 698 g $Ca_3(PO_4)_2$ and 243 g H_2O

28. 86.7 g CdS

29. Take 20 mL of 0.50 M NaOH and dilute to 100 mL to make 0.1 M NaOH solution.

30. 0.025 L of oil

31. 2.20×10^{-2} g present, 1.25×10^{-2} g allowed

32. 0.540 g new, 45.0 g old

33. a. Substances with similar polarity will dissolve in each other.
 b. The shape of the molecule can have an effect on its overall polarity.

34. Water molecules are dipoles with separation of charges, so they can attract and dissolve either dipoles or ions.

35. *See page 447B.*

36. C_4H_{10} (2.5 – 2.1 = 0.4) will dissolve in CS_2 (2.5 – 2.5 = 0) because both are nonpolar substances. The small C—H dipoles cancel each other.

37. $CsCl_2$ (3.0 – 0.7 = 2.3) is ionic and will not dissolve in nonpolar C_6H_{12} (2.5 – 2.1 = 0.4). The small C—H dipoles in C_6H_{12} cancel each other.

38. a. Because the polar region (4 bonds) is nearly as large as the nonpolar region (5 bonds), it will be soluble in water.
 b. Because the polar region (3 bonds) is so much smaller than the nonpolar region (40 bonds), it will not be soluble in water.

Full solutions for items ***24–32*** *and* ***35*** *begin on page 447A.*

27. What mass of each product results if 750. mL of 6.00 M H_3PO_4 react with an excess of $Ca(OH)_2$ according to the following equation? (Hint: see Sample Problem 11C.)

$$2H_3PO_4(aq) + 3Ca(OH)_2(aq) \longrightarrow Ca_3(PO_4)_2(s) + 6H_2O(l)$$

28. A yellow pigment in some artists' oil paints is cadmium sulfide. How much cadmium sulfide would be made if 350. mL of 2.00 M $(NH_4)_2S$ reacted with 400. mL of 1.50 M $Cd(NO_3)_2$ according to the following equation? (Hint: see Sample Problem 11C.)

$$(NH_4)_2S(aq) + Cd(NO_3)_2(aq) \longrightarrow 2NH_4NO_3(aq) + CdS(s)$$

A P P L Y

29. You have 500 mL of a 0.5 M NaOH solution. For an experiment, you need 0.1 M NaOH. Describe how you would make 100 mL of a 0.1 M NaOH solution from the 0.5 M NaOH.
30. The owner's manual for an outboard motor gives the proportion for mixing gasoline and oil as 100 to 1. You have 2.5 L of gasoline. How much oil do you need to mix with this much gasoline?
31. A shipment of shark meat was destroyed after it was found to contain 1.76 ppm methyl mercury, $[HgCH_3]^+$, which is higher than the legal limit of 1.00 ppm.
 a. If a shark has a mass of 12.5 kg, what mass of methyl mercury is present in the shark?
 b. What is the maximum number of grams of methyl mercury the shark meat could contain and still be safe?
32. Heavy exposure to lead or lead salts is toxic. To prevent lead poisoning, the current standard for lead in paint is 600. ppm, though some older paints had concentrations of 5.00×10^4 ppm. On average, a bucket of paint has a mass of 900. g.
 a. Calculate the maximum mass of lead in a bucket of new paint.
 b. Calculate the mass of lead in a bucket of old paint, based on the concentration given.

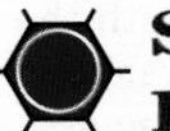

Solubility and polarity

R E V I E W

33. a. Explain what is meant by the chemists' expression "like dissolves like."
 b. Explain why another factor must be taken into account when polar covalent bonds are present.
34. Explain why water is a good solvent for both polar molecules and ionic crystals.
35. Copy the table below onto another piece of paper. Then, using the procedure outlined in the feature on page 424, fill in the blank spaces in this table.

Substance	Electronegativity difference	Bond type	Structural formula	Particle type	Sol- bilit wat
NaBr					ye
N_2				nonpolar molecule	
SO_2			O–S=O		
$CaCl_2$		ionic			
C_2H_6	2.5 – 2.1 = 0.4				

P R A C T I C E

36. Butane, C_4H_{10}, is added to carbon disulfide, CS_2. Will the butane dissolve? (Hint: see Sample Problem 11D.)
37. Cesium chloride, CsCl, is added to cyclohexane, C_6H_{12}. Will the cesium chloride dissolve? (Hint: see Sample Problem 11D.)

A P P L Y

38. For the compounds shown below, determine the number and ratio of polar and nonpolar bonds, then estimate each compound's solubility in H_2O.

 a. Ethylene glycol (H, H, H, H, OH, OH)

 b. Myristic acid (OH, O)

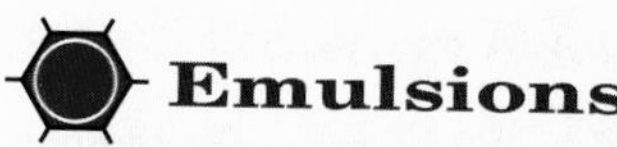

Emulsions

REVIEW

39. Explain why it is inaccurate to say that emulsifying agents actually make insoluble chemicals dissolve.

40. Give two examples of emulsifying agents, and explain how each performs its function.

41. Explain why detergents are better emulsifying agents than soaps, particularly in hard water.

APPLY

42. Butter contains about 80% oily butterfat and 16% water. During the process of making butter, it is churned, or stirred very quickly, for a long period of time. Explain why the churning step is necessary.

43. Which of these substances would make the best emulsifying agent? Explain your answer.

a. $CaCO_3{\cdot}6H_2O$

b. $(CH_3)_3C{-}O{-}C_2H_5$

c. CH_3 (on benzene ring)

d. $\left[(CH_3)_2{-}\underset{\displaystyle C_2H_5}{\underset{|}{N}}{-}(CH_2)_{15}CH_3\right]^+ Br^-$

Applications of solubility

REVIEW

44. What are the chemical characteristics of a good dry-cleaning solvent? How do they differ from those of an emulsifying agent?

45. Using the data below, plot a solubility graph for the following gases with grams of solute per 100 g of H_2O on the vertical axis and temperature in degrees Celsius on the horizontal axis. Then answer the questions at the top of the next column.

Gas	20°C	30°C	40°C	50°C	60°C
CO_2	0.169	0.126	0.097	0.076	0.058
H_2S	0.38	0.30	0.24	0.19	0.15
Cl_2	0.73	0.57	0.46	0.39	0.33

a. Estimate the solubility of each gas at 45°C.

b. At what temperature would the solubility of Cl_2 be 0.50 g per 100 g of H_2O?

c. If a solution contains 0.100 g of CO_2 in 100 g of H_2O at 35°C, is it unsaturated, saturated, or supersaturated?

d. Assuming that other gases follow the pattern of these three gases, should you heat or cool water if you are trying to dissolve a gas in it?

46. What do vitamins do in the body?

47. Name one source of vitamin A and one of vitamin C.

48. Explain the role of solubility in accounting for the fact that you are more likely to overdose on vitamin A than on vitamin C.

49. How does vitamin deficiency differ from vitamin toxicity?

PRACTICE

50 Ascorbic acid, vitamin C, has the formula $C_6H_8O_6$. What is the molarity of a plasma solution of ascorbic acid if the concentration is 5.0 mg/L? (Hint: see Sample Problem 11A.)

51. A cup (250 mL) of whole milk contains about 0.4 mg of riboflavin, $C_{17}H_{20}N_4O_6$. What is the molarity of riboflavin in the milk? (Hint: see Sample Problem 11A.)

APPLY

52. Some laundry detergents have enzyme additives. What role could these enzymes play in cleaning?

53. Crayon companies recommend treating wax stains on clothes by spraying them with WD-40 lubricant, applying dishwashing liquid, and then washing them. Explain why.

54. To produce clear ice cubes instead of cloudy ones, should hot or cold water be used? Explain your answer.

55. Some industrial plants use water from nearby rivers as a coolant. When that water is returned to the river, it is a few degrees warmer. Explain how this practice could cause a fish kill. (Hint: fish use the oxygen dissolved in water.)

39. It is inaccurate to say that an emulsifying agent makes insoluble chemicals dissolve because the liquids remain immiscible. The emulsifier keeps them mixed by partially dissolving in adjacent drops of each immiscible liquid.

40. Answers can vary. Check that they indicate that the emulsifying agent can dissolve in and link both polar and nonpolar substances.

41. Although soaps and detergents behave similarly in "soft" water, "hard" water contains Ca^{2+} and Mg^{2+} ions, which form insoluble compounds with soaps but not detergents.

42. Stirring emulsifies the suspension of fat and water and keeps it from separating until the oily fat hardens and the resulting butter emulsion is stabilized.

43. Choice d is the best emulsifier because it has both charged and nonpolar ends.

44. The most important characteristic of a dry-cleaning solvent is that it be nonpolar. An emulsifying agent needs both polar and nonpolar regions.

45. *See page 447B.*

46. Vitamins act to control chemical reactions within the body.

47. Answers can vary. Fish, carrots, and broccoli are some sources of vitamin A. Lemons, oranges, grapefruits, and limes are some sources of vitamin C.

48. Vitamin C is water soluble, and the excess is removed by the kidneys; vitamin A is fat soluble and accumulates in the body, especially the liver.

*Full solutions for item **45** are found on page 447B.*

49. Vitamin deficiency is an illness due to a lack of a vitamin, while vitamin toxicity is due to an excess of vitamins.

50. 2.8×10^{-5} M $C_6H_8O_6$

51. 4×10^{-6} M $C_{17}H_{20}N_4O_6$

52. Enzymes break down and dissolve the oily dirt and stains in fabrics.

53. The WD-40 lubricant is a nonpolar solvent that dissolves the wax, and the dishwashing liquid is an emulsifying agent for the lubricant-wax mixture so the stain washes away with water.

54. Cold water can have more dissolved gases than hot water. The hot water would produce clear ice cubes since it would not have gases present in the water at freezing.

55. Warmer water returned to a river would have less oxygen dissolved in it. This lack of oxygen could kill the fish.

*Full solutions for items **50** and **51** are found on page 447B.*

Linking chapters

1. **a.** no precipitate: both products are aqueous
   b. $CaCl_2(aq) + H_2CO_3(aq) \longrightarrow CaCO_3(s) + 2HCl(aq)$
   precipitate: $CaCO_3(s)$
   c. $3Na_2S(aq) + 2FeBr_3(aq) \longrightarrow 6NaBr(aq) + Fe_2S_3(s)$
   precipitate: $Fe_2S_3(s)$

2. 0.0467 M

3. **a.** 300. mL HCl
   b. 4.46 mL HCl
   c. 25.6 mL HCl

4. **a.** KCl, KNO_3, $AgNO_3$, NaCl, $NaNO_3$
   b. HCl, KOH, NaOH
   c. Solutes with endothermic heats of solution, listed in choice a, would dissolve better in hot solvents because they absorb heat while dissolving.

5. **a.** The heat of solution for RbCl will be more endothermic than that for KCl.

Linking chapters

1. ***Double-replacement reactions***
   Use **Table 11-1** to predict whether any precipitate will form when the following solutions are mixed. If a precipitate does form, what would you expect it to be?
   a. KOH and $NaNO_3$ **b.** $CaCl_2$ and H_2CO_3
   c. Na_2S and $FeBr_3$
2. ***Molarity of a gas***
   Air is a solution of nitrogen, oxygen, and other gases. A 5.00 L sample of air contains 20.9% oxygen by volume. The density of oxygen is 1.429 g/L. What is the molarity of oxygen in the air sample?
3. ***Gas stoichiometry***
   When small amounts of hydrogen gas are required in a laboratory, chemists frequently use a reaction similar to the one shown.
   $$2HCl(aq) + Zn(s) \longrightarrow ZnCl_2(aq) + H_2(g)$$
   If 0.250 M HCl is used along with an excess of zinc, how many milliliters of solution will be required to generate the following amounts of hydrogen gas?
   a. 0.0375 mol H_2
   b. 12.5 mL of H_2 at 2.50 atm and 310 K
   c. 101 mL of H_2 at 0.750 atm and 15°C
4. ***Endothermic and exothermic processes***
   The table to the right shows the heats of solution when the solutes are dissolved in H_2O. Use it to answer the following.
   a. Which solutes have endothermic heats of solution?
   b. Which solutes would increase the temperature of the solvent?
   c. Which solutes would dissolve better in hot solvents? Cite reasons for your answer.

Solute	Heat of solution (kJ/mol)
hydrogen chloride, HCl	–74.84
potassium chloride, KCl	+17.22
potassium hydroxide, KOH	–57.61
potassium nitrate, KNO_3	+34.89
silver nitrate, $AgNO_3$	+22.59
sodium chloride, NaCl	+3.88
sodium hydroxide, NaOH	–44.51
sodium nitrate, $NaNO_3$	+20.50

5. ***Classification and trends***
   Use the table for item **4** to answer these questions.
   a. The heat of solution for rubidium chloride, RbCl, isn't given in the table. Predict how its heat of solution compares to potassium chloride, KCl. (Hint: examine the chart closely, looking for patterns, and also examine a periodic table.)
   b. Arrange the following compounds in order of their heats of solution, beginning with the most exothermic: NaF, LiF, RbF, KF.

USING TECHNOLOGY

1. ***Graphics calculator***
   Molarity can be graphically represented as a straight line with a positive slope. A 0.500 M solution of NaOH can be graphed using the function $y = 0.5x$, where x equals the volume of the solution in liters. Select the [y =] key of the graphics calculator, and enter the function.
   a. Press [GRAPH] and use the [TRACE] mode to estimate the number of moles of NaOH in 300. mL of solution, 750. mL of solution, and 75.0 mL of solution.
   b. Calculate the mass of NaOH in each solution sample using the mole estimates you found in item **a**.
2. ***Graphics calculator***
   Use the graphics calculator to graph the mole amounts and volumes in a 1.00 M solution of $CuCl_2$. Use the graph to estimate how you would prepare 1.00 M solutions of $CuCl_2$ with the following volumes.
   a. 350. mL **b.** 150. mL
   c. 3.00 L **d.** 750. mL
3. ***Computer art***
   Using a computer art program, sketch the particles involved in the process of dissolving a solid in water. Include views showing the initially pure water, the addition of the solid, the solid beginning to dissolve, and the final state, in which the solid is completely dissolved to form an unsaturated solution.

Alternative assessment

Performance assessment

1. Design an experiment to identify an unknown substance that is CsCl, RbCl, LiCl, NH_4Cl, KCl, or NaCl. (Hint: examine the solubility graph, **Figure 11-5**.) If your instructor approves your design, get a sample from the instructor, and carry out your plan to identify it.
2. Your instructor will give you an index card describing a solution required in a lab. Describe exactly how you would make the solution. If your instructor approves your plan, obtain what you need from the instructor, and make the solution.
3. The more concentrated an aqueous solution of ethylene glycol is, the more dense it will be. Mechanics make use of this fact when they check the antifreeze in a car's radiator. Using the data for ethylene glycol solutions shown below, graph density vs. concentration. Then, determine the concentration of a solution with a density of 1.030 g/mL. What density would you predict for a 2.50 M solution?

Concentration (M)	Density (g/mL)
0.65	1.003
1.30	1.008
3.30	1.024
7.50	1.057

4. ***Computer spreadsheet***
   Many reagent chemicals used in the lab are sold in the form of concentrated aqueous solutions, as shown in the table below. Different volumes are diluted to 1.00 L to make less concentrated solutions. Create a spreadsheet that will calculate the volume of concentrated reagent needed to make 1.00 L solutions of any molar concentration that you enter.

Reagent	Concentration (M)
H_2SO_4	18
HCl	12.1
HNO_3	16
H_3PO_4	14.8
CH_3COOH	17.4
NH_3	15

Portfolio projects

1. ***Chemistry and you***
   For one week, record instances in which you used a mixture of some type. Identify whether each is a solution, suspension, emulsion, or a heterogeneous mixture. If possible, identify the different parts of each mixture.
2. ***Research and communication***
   Emergency response teams working with oil spills use chemical and physical properties of oil and water along with solubility principles to prevent spills from spreading and to clean them up. Research the techniques used, and explain why they work. Present your findings to the class.
3. ***Cooperative activity***
   Find out how waste motor oil is collected in your community and where it goes after it is collected. Hold a class debate on whether the community's procedure is safe for the environment. If you find that the procedure is safe, work with your classmates to produce a publicity campaign to promote the use of oil recycling facilities. If you find that it is not safe, create a pamphlet to make the public aware of why recycling oil is important, and how other communities have succeeded.
4. ***Chemistry and you***
   Keep a seven-day diary of what you eat. Consult a table of food values to determine what your vitamin intake is and whether it matches or exceeds the RDIs for the vitamins listed in **Table 11-5**.
5. ***Chemistry and you***
   Read the labels of three different over-the-counter vitamin supplements sold at your local pharmacy. Copy the amount of vitamin A supplied in one tablet. What dosage of each supplement could cause vitamin A toxicity? (Hint: to convert IUs to µg, multiply by 0.025.)
6. ***Research and communication***
   There are many arguments for and against the use of vitamin supplements. Find out what they are. What is the role of the FDA in the regulation of vitamin supplements? Prepare the arguments for or against increased regulation of vitamin supplements, and present them in a class debate.

b. Starting with the most exothermic: LiF, NaF, KF, RbF

*Full solutions for Linking chapters items **2** and **3** are found on page 447B.*

USING TECHNOLOGY

1. **a.** 0.150 mol NaOH; 0.375 mol NaOH; 0.0375 mol NaOH
   b. 6.00 g NaOH; 15.0 g NaOH; 1.50 g NaOH
2. **a.** Dissolve 47.1 g $CuCl_2$ to make 350. mL.
   b. Dissolve 20.2 g $CuCl_2$ to make 150. mL
   c. Dissolve 403 g $CuCl_2$ to make 3.00 L.
   d. Dissolve 101 g $CuCl_2$ to make 750. mL.
3. Students' depictions should clearly show particle-level interactions and distinguish between solvent and solute particles.
4. Answers will vary.

Alternative assessment

For guidelines and information relating to the implementation and evaluation of Performance assessment and Portfolio projects, please turn to page T36 at the front of this book.

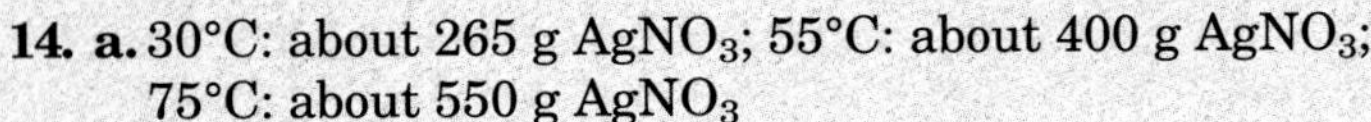

14. a. 30°C: about 265 g $AgNO_3$; 55°C: about 400 g $AgNO_3$; 75°C: about 550 g $AgNO_3$

b. about 32°C

c. unsaturated

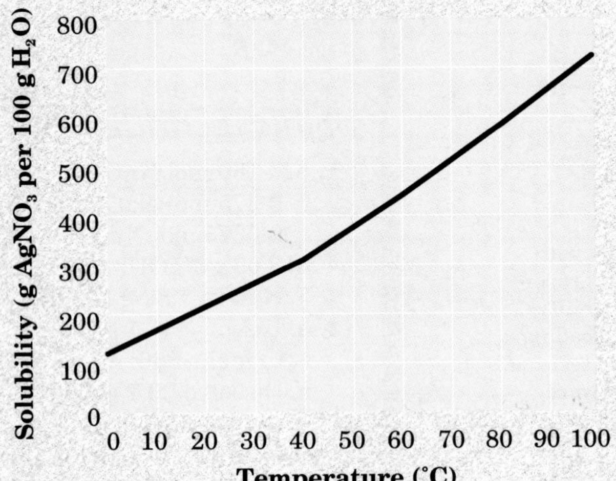

18. $\frac{2.00 - 1.9}{2.00} \times 100 = 5\%$

19. a. $\frac{3.5 \text{ mol } NaNO_3}{1.0 \text{ L}} = 3.5 \text{ M } NaNO_3$

b. $\frac{5.0 \text{ mol KOH}}{2.0 \text{ L}} = 2.5 \text{ M KOH}$

c. $\frac{2.0 \text{ mol } Na_2S}{0.25 \text{ L}} = 8.0 \text{ M } Na_2S$

20. a. $\frac{0.25 \text{ mol } FeCl_3}{2.0 \text{ L}} = 0.12 \text{ M } FeCl_3$

b. $\frac{0.015 \text{ mol } KMnO_4}{350 \text{ mL}} \times \frac{1000 \text{ mL}}{1 \text{ L}} = 0.043 \text{ M } KMnO_4$

c. $\frac{3.5 \times 10^{-4} \text{ mol } NaC_2H_3O_2}{25 \text{ mL}} \times \frac{1000 \text{ mL}}{1 \text{ L}} = 0.014 \text{ M } NaC_2H_3O_2$

21. a. $1.0 \text{ L} \times \frac{4.0 \text{ mol } AgNO_3}{1 \text{ L}} = 4.0 \text{ mol } AgNO_3$

b. $2.50 \text{ L} \times \frac{0.500 \text{ mol HCl}}{1 \text{ L}} = 1.25 \text{ mol HCl}$

c. $400. \text{ mL} \times \frac{1 \text{ L}}{1000 \text{ mL}} \times \frac{0.250 \text{ mol } HNO_3}{1 \text{ L}} = 0.100 \text{ mol } HNO_3$

22. a. $30.0 \text{ mL} \times \frac{1 \text{ L}}{1000 \text{ mL}} \times \frac{0.0100 \text{ mol } H_3PO_4}{1 \text{ L}} = 3.00 \times 10^{-4} \text{ mol } H_3PO_4$

b. $5.0 \text{ mL} \times \frac{1 \text{ L}}{1000 \text{ mL}} \times \frac{18 \text{ mol } H_2SO_4}{1 \text{ L}} = 0.090 \text{ mol } H_2SO_4$

c. $250. \text{ mL } 3 \frac{1 \text{ L}}{1000 \text{ mL}} \times \frac{1.00 \times 10^{-4} \text{ mol } CH_3OH}{1 \text{ L}} = 2.50 \times 10^{-5} \text{ mol } CH_3OH$

23. a. $\frac{20.0 \text{ g NaOH}}{2.00 \text{ L}} \times \frac{1 \text{ mol}}{40.00 \text{ g NaOH}} = 0.250 \text{ M NaOH}$

b. $\frac{14.0 \text{ g } NH_4Br}{150. \text{ mL}} \times \frac{1000 \text{ mL}}{1 \text{ L}} \times \frac{1 \text{ mol}}{97.95 \text{ g } NH_4Br} = 0.953 \text{ M } NH_4Br$

c. $\frac{65.0 \text{ g } CuCl_2}{300. \text{ mL}} \times \frac{1000 \text{ mL}}{1 \text{ L}} \times \frac{1 \text{ mol}}{134.45 \text{ g } CuCl_2} = 1.61 \text{ M } CuCl_2$

24. a. $\frac{3.50 \text{ g } CCl_4}{500. \text{ mL}} \times \frac{1000 \text{ mL}}{1 \text{ L}} \times \frac{1 \text{ mol}}{153.81 \text{ g } CCl_4} = 0.0455 \text{ M } CCl_4$

b. $\frac{0.150 \text{ g } CuSO_4 \cdot 5H_2O}{25.0 \text{ mL}} \times \frac{1000 \text{ mL}}{1 \text{ L}} \times \frac{1 \text{ mol}}{249.72 \text{ g } CuSO_4 \cdot 5H_2O} =$

$0.0240 \text{ M } CuSO_4$

c. $\frac{3.00 \times 10^{-4} \text{ g } AlCl_3}{2.00 \text{ mL}} \times \frac{1000 \text{ mL}}{1 \text{ L}} \times \frac{1 \text{ mol}}{133.33 \text{ g } AlCl_3} = 1.13 \times 10^{-3} \text{ M } AlCl_3$

25. a. $1.00 \text{ L} \times \frac{1.50 \text{ mol NaCl}}{1 \text{ L}} \times \frac{58.44 \text{ g NaCl}}{1 \text{ mol}} = 87.7 \text{ g NaCl}$

b. $2.25 \text{ L} \times \frac{3.50 \text{ mol NaOH}}{1 \text{ L}} \times \frac{40.00 \text{ g NaOH}}{1 \text{ mol}} = 315 \text{ g NaOH}$

c. $750.0 \text{ mL} \times \frac{1 \text{ L}}{1000 \text{ mL}} \times \frac{2.50 \text{ mol } K_2SO_4}{1 \text{ L}} \times \frac{174.27 \text{ g } K_2SO_4}{1 \text{ mol}} =$

$327 \text{ g } K_2SO_4$

26. a. $1.50 \text{ mL} \times \frac{1 \text{ L}}{1000 \text{ mL}} \times \frac{0.300 \text{ mol } Ca(NO_3)_2}{1 \text{ L}} \times \frac{164.10 \text{ g } Ca(NO_3)_2}{1 \text{ mol}} =$

$0.0738 \text{ g } Ca(NO_3)_2$

b. $25 \text{ mL} \times \frac{1 \text{ L}}{1000 \text{ mL}} \times \frac{1.5 \text{ mol LiCl}}{1 \text{ L}} \times \frac{42.39 \text{ g LiCl}}{\text{mol}} = 1.6 \text{ g LiCl}$

c. $3.00 \times 10^{-4} \text{ L} \times \frac{2.50 \times 10^{-2} \text{ mol } C_6H_{12}O_6}{1 \text{ L}} \times \frac{180.18 \text{ g } C_6H_{12}O_6}{\text{mol}} =$

$1.35 \times 10^{-3} \text{ g } C_6H_{12}O_6$

27. $750. \text{ mL} \times \frac{1 \text{ L}}{1000 \text{ mL}} \times \frac{6.00 \text{ mol } H_3PO_4}{1 \text{ L}} \times \frac{1 \text{ mol } Ca_3(PO_4)_2}{2 \text{ mol } H_3PO_4} \times$

$\frac{310.18 \text{ g } Ca_3(PO_4)_2}{1 \text{ mol } Ca_3(PO_4)_2} = 698 \text{ g } Ca_3(PO_4)_2$

$750. \text{ mL} \times \frac{1 \text{ L}}{1000 \text{ mL}} \times \frac{6.00 \text{ mol } H_3PO_4}{1 \text{ L}} \times \frac{6 \text{ mol } H_2O}{2 \text{ mol } H_3PO_4} \times \frac{18.02 \text{ g } H_2O}{1 \text{ mol } H_2O} =$

$243 \text{ g } H_2O$

28. $350. \text{ mL} \times \frac{1 \text{ L}}{1000 \text{ mL}} \times \frac{2.00 \text{ mol } (NH_4)_2S}{1 \text{ L}} = 0.700 \text{ mol } (NH_4)_2S$

$400. \text{ mL} \times \frac{1 \text{ L}}{1000 \text{ mL}} \times \frac{1.50 \text{ mol } Cd(NO_3)_2}{1 \text{ L}} =$

$0.600 \text{ mol } Cd(NO_3)_2$ [limiting reactant]

$0.600 \text{ mol } Cd(NO_3)_2 \times \frac{1 \text{ mol CdS}}{1 \text{ mol } Cd(NO_3)_2} \times \frac{144.48 \text{ g CdS}}{1 \text{ mol CdS}} = 86.7 \text{ g CdS}$

29. Take 20 mL of 0.5 M NaOH and dilute to 100 mL to make 0.1 M NaOH solution.

$100 \text{ mL} \times \frac{1 \text{ L}}{1000 \text{ mL}} \times \frac{0.1 \text{ mol NaOH}}{1 \text{ L}} \times \frac{1 \text{ L}}{0.50 \text{ mol NaOH}} \times \frac{1000 \text{ mL}}{1 \text{ L}} =$

20 mL of 0.5 M NaOH diluted in 100 mL of solution

30. $2.5 \text{ L gas} \times \frac{1 \text{ L oil}}{100 \text{ L gas}} = 0.025 \text{ L of oil}$

31. $1.76 \text{ ppm} = \frac{1.76}{1 \times 10^6}$

$12.5 \text{ kg} \times \frac{1 \times 10^3 \text{ g}}{1 \text{ kg}} \times \frac{1.76}{1 \times 10^6} = 0.0220 \text{ g methyl mercury}$

$12.5 \text{ kg} \times \frac{1 \times 10^3 \text{ g}}{1 \text{ kg}} \times \frac{1.00}{1 \times 10^6} = 0.0125 \text{ g methyl mercury}$

32. new paint lead mass: $900. \text{ g} \times \frac{600.}{1 \times 10^6} = 0.540 \text{ g}$

old paint lead mass: $900. \text{ g} \times \frac{5.00 \times 10^4}{1 \times 10^6} = 45.0 \text{ g}$

35.

Substance	Electro-negativity difference	Bond type	Structural formula	Particle type	Solubility in water
NaBr	1.9	ionic	N.A.	ion	yes
$\mathbf{N_2}$	0	nonpolar covalent	N≡N	nonpolar molecule	no
$\mathbf{SO_2}$	1.0	polar covalent	O–S=O	dipoles	yes
$\mathbf{CaCl_2}$	2.0	ionic	N.A.	ion	yes
$\mathbf{C_2H_6}$	2.5 – 2.1 = 0.4	nonpolar covalent	$H_3C–CH_3$	nonpolar	no

45. **a.** Solubilities at 45°C: 0.42 g Cl_2/100 g H_2O, 0.21 g H_2S/100 g H_2O, and 0.086 g CO_2/100 g H_2O.
b. about 36°C
c. unsaturated
d. cool

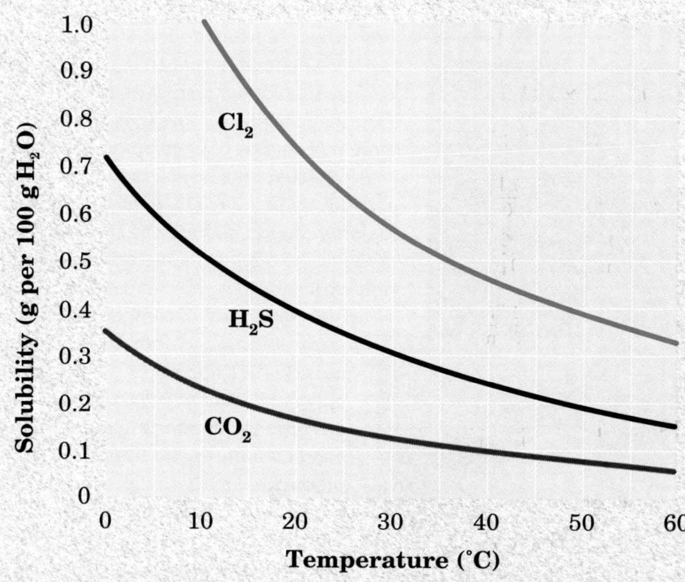

50. $\frac{5.0 \text{ mg } C_6H_8O_6}{1 \text{ L}} \times \frac{1 \text{ g}}{1000 \text{ mg}} \times \frac{1 \text{ mol } C_6H_8O_6}{176.14 \text{ g } C_6H_8O_6} = 2.8 \times 10^{-5} \text{ M } C_6H_8O_6$

51. $\frac{0.4 \text{ mg } C_{17}H_{20}N_4O_6}{250 \text{ mL}} \times \frac{1 \text{ g}}{1000 \text{ mg}} \times \frac{1000 \text{ mL}}{1 \text{ L}} \times \frac{1 \text{ mol } C_{17}H_{20}N_4O_6}{376.41 \text{ g } C_{17}H_{20}N_4O_6} =$
$4 \times 10^{-6} \text{ M } C_{17}H_{20}N_4O_6$

Linking chapters

2. $5.00 \text{ L} \times \frac{20.9}{100} \times \frac{1.429 \text{ g } O_2}{1 \text{ L}} \times \frac{1 \text{ mol}}{32.00 \text{ g } O_2} = 0.0467 \text{ M}$

3. **a.** $0.0375 \text{ mol } H_2 \times \frac{2 \text{ mol HCl}}{1 \text{ mol } H_2} \times \frac{1 \text{ L HCl}}{0.250 \text{ mol HCl}} \times \frac{1000 \text{ mL}}{1 \text{ L}} =$

300. mL HCl

b. $n = \frac{PV}{RT}$

$n = \frac{(12.5 \times 10^{-3} \text{ L } H_2)(2.50 \text{ atm})}{(0.0821 \text{ L•atm/mol•K})(310 \text{ K})} = 1.23 \times 10^{-3} \text{ mol } H_2$

$1.23 \times 10^{-3} \text{ mol } H_2 \times \frac{2 \text{ mol HCl}}{1 \text{ mol } H_2} \times \frac{1 \text{ L HCl}}{0.250 \text{ mol HCl}} \times \frac{1000 \text{ mL}}{1 \text{ L}} =$

9.84 mL HCl

c. $n = \frac{PV}{RT}$

$n = \frac{(0.750 \text{ atm})(101 \times 10^{-3} \text{ L } H_2)}{(0.0821 \text{ L•atm/mol•K})(288 \text{ K})}$

$n = 3.20 \times 10^{-3} \text{ mol } H_2 \times \frac{2 \text{ mol HCl}}{1 \text{ mol } H_2} \times \frac{1 \text{ L HCl}}{0.250 \text{ mol HCl}} \times \frac{1000 \text{ mL}}{1 \text{ L}} =$

25.6 mL HCl

Content Background

The two objectives of this feature are to make students aware of potential conflicts of interest that affect scientific research and to give students experience in analyzing potentially biased writing. Although the process of science is structured to reduce bias and prejudice, the nature of the social environment in which science takes place precludes a perfect system of checks and balances.

Teaching Tip

Analyzing biased writing
The writer appears to be biased against chemical companies. Have students pick out statements that appear to be biased or lacking factual evidence.

Teaching Tip

Debating the issue
Divide the class into two teams to debate the issue of whether personnel from chemical companies should have any decision-making authority regarding the funding of independent research on the possible hazards of the chemicals produced by their firms.

Science, Technology, and Society

About 182 000 women in the United States will be found to have breast cancer this year. Already 1.8 million women have been diagnosed with the disease and perhaps another 1 million have it but do not know it yet. Standard medical treatments have had little impact on the long-range outcome of the disease: 25 percent of women with breast cancer die within 5 years of their diagnosis. Forty percent die within 10 years. The death rate from metastatic breast cancer—disease that has spread to other parts of the body—has remained unchanged for more than 40 years.

Cancer and Chemicals

Who Influences Research?

However, what has changed is a woman's chance of developing breast cancer at some time in her life. In 1940, that chance was 1 in 20. The National Cancer Institute estimates that today, by the time a woman is 85 years old, her chance is 1 in 8. Scientists have identified several factors that appear to increase a woman's chance of developing breast cancer.

Altogether known risk factors like heredity account for just 20–30% of all breast cancers. The majority of women with the disease have none of these. As a result, some researchers now think that exposure to some chemicals may be responsible. One group of compounds is highly suspected—organochlorines. Organochlorines include DDT and other pesticides; PCBs, and polyvinyl chloride (PVC). About 11 000 different organochlorines are manufactured and used by industries. Others are formed as by-products of industrial processes like bleaching paper pulp, disinfecting wastewater, and burning garbage containing plastics and other chlorinated materials. Organochlorines are highly toxic, slow to degrade, and they accumulate in fatty breast tissue. Studies have found 177 different organochlorines in samples of fat, blood, semen, and mother's milk in residents of the United States and Canada.

A 1993 study by Mary S. Wolff and coworkers at Mt. Sinai Hospital in New York found that women who get breast cancer had blood levels of DDE, a residue of DDT, that were 35 percent higher than those in women without the disease. Women with 19 parts per billion (ppb) DDE had four times the cancer risk of women with 2 ppb DDE. Although DDT use in the United States was banned in

1972, trace amounts remain in food and soil. Also, the pesticide is still manufactured here and sold to other countries with less strict environmental laws. DDT sprayed on crops in these countries has been found to be carried huge distances in air and water.

Before 1976, the breast cancer death rate among Israeli women under age 44 was unusually high. At the same time, high concentrations of three pesticides—DDT, BHC, and lindane—were found in women's breast milk and in Israeli milk and dairy products. In 1976, use of the three pesticides was banned. By 1978, DDT levels in breast milk dropped 43%, lindane dropped 90%, and BHC dropped 98%. Less than a decade later, the breast cancer rate for Israeli women under age 44 fell 30%. In the time period studied, Israel was the only one of 28 countries in which the death rate from breast cancer actually declined.

Federal law requires the National Cancer Institute to expand its research program on preventing cancers caused by exposures to carcinogens. But at a congressional budget hearing in October of 1993, NCI Director Samuel Broder testified that the NCI spent only 1% of its almost $2 million budget on environmental cancer studies, and he gave no indications that the NCI planned to shift its priorities. NCI research will continue to emphasize the treatment and cure of cancer. Of the part of the budget given to researching causes of the disease, the two areas to receive the most emphasis will remain the links to diet and smoking.

Many health specialists have criticized this approach. They want additional funds devoted to research into identifying carcinogens and studying the link between exposure and the incidence of cancer. Some critics have speculated that the ties between the chemical companies and the organizations devoted to cancer research are too close. These critics point to the interlocking relationships among cancer researchers, policy makers, and industry representatives. They question appointments such as that in the 1980s of Armand Hammer, then president of Occidental Petroleum, to a position as chair of the President's Cancer Advisory Board.

The fear often expressed by critics is that lawmakers may be unduly influenced by lobbyists for the chemical industry. Critics also worry that chemical companies have a vested interest in protecting their own industry. Will chemical companies be willing to support research that may identify their products as probable causes of cancer?

Industry spokespeople, on the other hand, claim a role in helping to identify and solve the problem of carcinogens. They argue that the funds donated by the chemical companies for research are essential to finding both the causes and treatment of cancer. The chemical companies view their involvement as stemming from a natural, communal interest—not as a conflict of interest. From this viewpoint, everyone involved is focused on the problem but working from a different perspective.

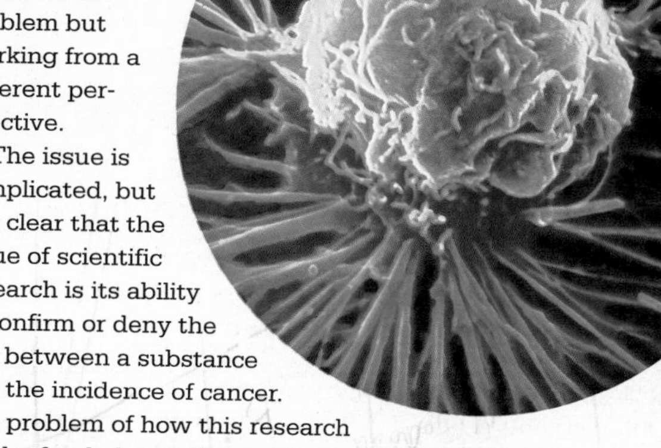

The issue is complicated, but it is clear that the value of scientific research is its ability to confirm or deny the link between a substance and the incidence of cancer. The problem of how this research will be funded and directed is the issue. To be effective, a researcher needs a degree of independence, and critics of the current policy worry that researchers will be reluctant to "bite the hand that feeds them." As long as chemical companies are directly involved, many health specialists worry that the objectivity of the research may be compromised.

Researching the Issue

1. Go to the library and find at least 5 books and/or magazine articles on the subject of how cancer research should be funded or directed. Write a statement that summarizes the debate, and list evidence from your reading that should be considered in making a judgment.
2. Find the structural formula for the hormone estrogen. In what ways are organochlorines structurally similar to estrogen?
3. If you were a researcher, what guidelines would you develop to protect the integrity of your research? Write a position statement to be given to potential contributors indicating how they could and could not interact with your research.

PLANNING FOR

12 Chemical Equilibrium

Chapter Overview

- **Section 12-1** examines the properties of electrolytes and nonelectrolytes in solution and their affects on electrical conductance. Models for ion formation and mobility in solution are explored for ionic and molecular compounds to explain conductance. Colligative properties of solutions are discussed, with emphasis on boiling point elevation and freezing point depression.
- **Section 12-2** distinguishes between reversible and completion reactions. Dynamic and static equilibrium and rate constants are discussed. System equilibria and responses to stresses (Le Châtelier's principle) are examined. This study is extended to complex ions.
- **Section 12-3** examines calculations involving equilibrium constants. Calculations and applications of solubility-product constants are also examined. Use of the common ion effect to shift the equilibrium of a reaction system is also discussed.

Concept Base

Students must have mastered the following concepts prior to this chapter: covalent bonding (Chapter 6); ionic bonding (Chapter 5); reactions (Chapter 7); molar relationships (Chapter 8); enthalpy and entropy (Chapter 9); solubility, solution concentration, and polarity (Chapter 11).

Looking Ahead

Electrolytic behavior is explored more thoroughly in Chapter 13 (acid-base reactions) and in Chapter 15 (redox reactions). Rates of reaction are treated in Chapter 14.

Laboratory Equipment Needs

Demonstrations

The listing of materials for Chapter 12 demonstrations begins on page T46.

Laboratory Experiments

Materials and preparation instructions for Chapter 12 are found as follows.

Text		*Page*
Exploration 12A	Boiling-Point Elevation and Molar Mass	746
Investigation 12A	Boiling-Point Elevation—Contaminated Sugar	750
Exploration 12B	Equilibrium Expressions	752
Investigation 12B	Equilibrium Expressions—Iron Content of Tea	758

Laboratory Experiments Manual
(refer to the table of contents in the manual)

Exploration 12-1	Solubility-Product Constant
Investigation 12-1	Solubility-Product Constant—Algae Blooms
Exploration 12-2	Freezing-Point Depression—Testing De-icing Chemicals
Investigation 12-2	Freezing-Point Depression—Making Ice Cream

Pacing: 13 days

	Inquiry Teaching Strategies
12-1 What happens in an aqueous solution?	**Demonstrations** ■ Demo 1 *Conductance of solutions* ■ Demo 2 *Conductance of tap and distilled water* **Laboratory Experiments** ■ Exp 12A *Boiling-Point Elevation and Molar Mass* ■ Inv 12A *Boiling-Point Elevation—Contaminated Sugar* ■ ■ Exp 12-2 *Freezing-Point Depression—Testing De-icing Chemicals* ■ ■ Inv 12-2 *Freezing-Point Depression—Making Ice Cream*
12-2 What is an equilibrium system?	**Demonstrations** ■ Demo 3 *Completion reactions* ■ Demo 4 *Reversible reactions* ■ Demo 5 *Temperature effects on gas production* ■ Demo 6 *Complex ion equilibria*
12-3 How is equilibrium measured?	**Demonstrations** ■ Demo 7 *Kinetics simulation* ■ Demo 8 *Common ion effect* **Laboratory Experiments** ■ Exp 12B *Equilibrium Expressions* ■ Inv 12B *Equilibrium Expressions—Iron Content of Tea* ■ ■ Exp 12-1 *Solubility Product Constant* ■ ■ Inv 12-1 *Solubility Product Constant—Algae Blooms*

Key ■ Teaching Transparencies
■ Annotated Teacher's Edition
■ Laboratory Experiments Manual

Visual Teaching Strategies

Transparencies
- F 12-2 *Particle Models for Electrolytes and Nonelectrolytes in Solution*
- F 12-6 *Particle Models for Strong and Weak Electrolytes in Solution*

Review and Practice Strategies

Text Reviews
- Concept Review 1–3
- Section Review 4–10
- Chapter Review 1–10; LC 4

Study Guide Worksheets
- 12-1 Concept Review
- 12-1 Skillsheet
  Ions in Chemical Equations
  Colligative Properties

Transparencies
- F 12-10 *Particle Model for the Formation of a Precipitate*

Text Reviews
- Section Review 11–14
- Chapter Review 11–21; LC 2, 5

Study Guide Worksheets
- 12-2 Concept Review
- 12-2 Skillsheet
  Complex Ions
  Interpreting Le Châtelier's Principle

Transparencies
- F 12-23 *Particle Model for the Common Ion Effect*

Transparency Masters
- P 480 *Equilibrium Calculations*

Text Reviews
- Concept Review 15–16
- Additional Examples 12A, 12B, 12C
- Practice 1–7
- Section Review 17–21
- Chapter Review 22–38; LC 1, 3; UT 1, 2

Study Guide Worksheets
- 12-3 Concept Review
- 12-3 Skillsheet
  Writing K_{eq} Equations
  Writing K_{sp} Equations
  Interpreting Constants
- 12-3 Practice
  Calculating K_{eq} from Concentrations
  Calculating Concentrations from K_{eq}
  Calculating Concentrations from K_{sp}

Pupil's Text | Study Guide | Teaching Resources
Exp *Exploration* | Inv *Investigation*
LC *Linking Chapters* | UT *Using Technology*

Assessment Options

Traditional Assessment

Test Generator
Instructional Objectives Measured:
Content mastery
Select from items 1 to 75

Performance Assessment Options

(see page T36 and T643-15 for scoring rubrics)
Students demonstrate mastery of objectives in a hands-on environment. You may want some of these materials in the Portfolio as well.

Investigation 12A and 12B (text)
Investigation 12-1 and 12-2 (laboratory manual)
Instructional Objectives Measured:
Scientific and chemical literacy, Problem solving skills, Use of scientific methodology, Understanding the impact of science and technology on society, Use of technology, Proficiency in written communication

Performance Assessments, page 485

1. **Designing an experiment**
   Instructional Objectives Measured:
   Problem solving skills, Use of scientific methodology, Proficiency in written communication
2. **Designing an experiment**
   Instructional Objectives Measured:
   Problem solving skills, Use of scientific methodology, Proficiency in written communication
3. **Designing an experiment**
   Instructional Objectives Measured:
   Problem solving skills, Use of scientific methodology, Proficiency in written communication

Portfolio Options

(see page T36 for scoring rubrics)
Students provide a written rationale for each selection made for the portfolio.

Formal Laboratory Reports
Exploration 12A and 12B (text)
Exploration 12-1 and 12-2 (lab manual)
Instructional Objectives Measured:
Scientific and chemical literacy, Problem solving skills, Use of scientific methodology, Understanding the impact of science and technology on society, Use of technology, Proficiency in written communication

Portfolio Projects, page 485
Items 1, 2, and 5. Research and communication
Instructional Objectives Measured:
Scientific and chemical literacy, Understanding the impact of science and technology on society, Proficiency in oral and written communication

Item 3. Cooperative activity
Instructional Objectives Measured:
Content mastery, Proficiency in oral communication

Item 4. Chemistry and you
Instructional Objectives Measured:
Content mastery, Scientific and chemical literacy, Understanding the impact of science and technology on society, Proficiency in written communication

INTRODUCING THE CHAPTER

CHAPTER 12 Chemical Equilibrium

Story Background

The story shows that drowning is not the only danger associated with sailing on seas and oceans. Dehydration poses a threat peculiar to oceans, seas, and inland "seas" (such as the Great Salt Lake), all of which contain high concentrations of dissolved salts.

The salt in sea water is the result of hydrogen chloride gas rising from the Earth's interior during the first 1.5 billion years of its existence. When the temperature of the surface fell below 100°C, many gases in the atmosphere (such as H_2O and HCl) condensed. The dissociation of HCl resulted in a high concentration of chloride ions (≈19 g/kg of sea water). The chloride reacted with minerals that contained cations, such as sodium, so that various salts formed.

For many species of marine (saltwater) fish, sea water poses the same threat that it does to humans. Even if these fish did not drink the water, they would find the concentration of salt in their bodies shifting toward equilibrium with the concentration of salt in sea water and away from the concentration required for them to survive. This happens because the skins of fish are semipermeable and allow water to enter or leave the fish until the salt concentrations on both sides of the skin are equal. As a result of this equilibration process, the fish lose water, just as humans lost at sea without fresh water do. Fortunately for the fish, their kidneys provide a special mechanism for expelling excess salt that returns their internal salt concentrations to their proper levels.

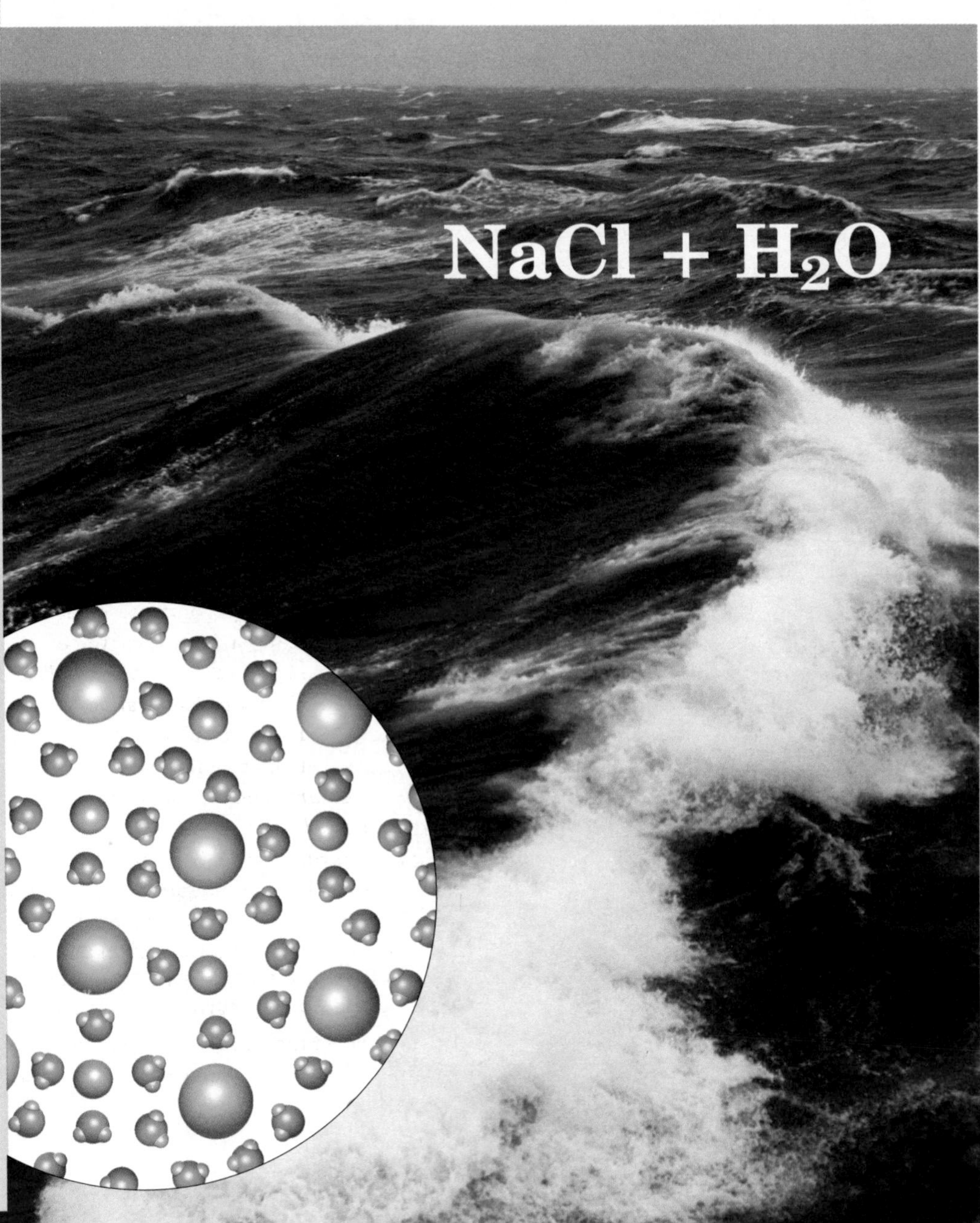

About the Illustration

The waves shown are breaking on the surface of the turbulent North Atlantic Ocean. The second largest of the Earth's oceans, the Atlantic is noted for its violent storms, which have posed a threat to sea travelers for centuries.

Sea Water and Equilibrium

Boating accidents at sea leave people with water all around them but none to drink. Sea water harms the body's equilibrium system, so a person must drink desalinated water to keep from becoming very ill.

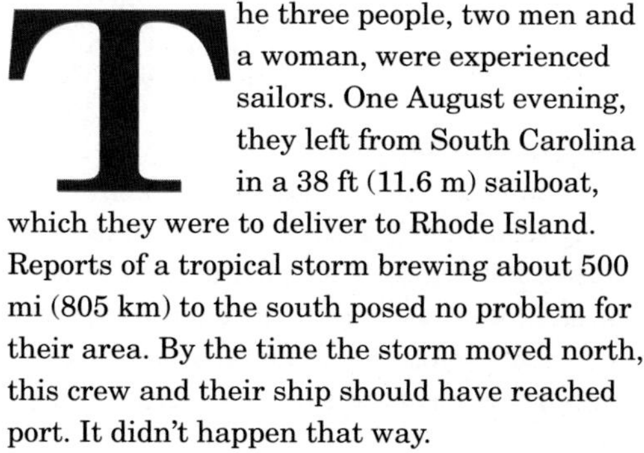

The three people, two men and a woman, were experienced sailors. One August evening, they left from South Carolina in a 38 ft (11.6 m) sailboat, which they were to deliver to Rhode Island. Reports of a tropical storm brewing about 500 mi (805 km) to the south posed no problem for their area. By the time the storm moved north, this crew and their ship should have reached port. It didn't happen that way.

The seas were calm the first night, but by the next day the winds had picked up. The storm had developed into a hurricane and had overtaken the vessel, which was more than 100 mi (160.9 km) from shore. Disaster struck the boat and its crew as winds exceeded 100 mph (44.7 m/s). Waves picked up the vessel and slammed it down on its side. Water poured into the boat. Within seconds, the three crew members filled a canvas bag with food, water, and supplies; then they scrambled into a life raft.

Waves crashed down on them and flipped the raft several times. The supplies were lost, and the raft threatened to sink. The three people quickly drank the only water they had so that they could use the container to bail out the sea water and try to keep the raft afloat. The following day, the winds died down and the waves dropped to 20 ft (6.1 m).

The three people took stock of their situation. They had some flares and two paddles, but no radio, food, or water. On the sixth day at sea, the thirsty crew began to fantasize; the men talked about eating watermelons, and the woman dreamed of chocolate shakes. On the 11th day at sea, all three were severely dehydrated and near death. They knew, however, that even though they were surrounded by water, they couldn't drink it. If they did, it would definitely make their medical condition even worse.

You will learn in this chapter why drinking sea water would increase dehydration. Questions you should consider while you study this chapter include the following.

- *How would drinking sea water affect the aqueous equilibrium in the body?*
- *Can sea water be made safe to drink?*
- *How do you know when water is safe to drink?*

Unfortunately, the stranded sailors had no means of making the sea water drinkable, and because it never rained, they could not collect any water to drink. Suffering from severe dehydration on their 11th day at sea, the crew was finally rescued by a Coast Guard helicopter near noon.

Tapping
Prior Knowledge

Experience-based questions

1. Why isn't distilled water a good conductor of electricity?
2. When sugar is placed in water it dissolves. Explain this dissolving process.
3. Why is salt thrown on icy steps and roads in the winter?
4. How can a reaction be made to go to completion?

Section 12-1

FOCUS

Lesson Starter

Prepare for the demonstration by writing the formula for water, H_2O, acetic acid, CH_3COOH, hydrochloric acid, HCl, and sodium chloride, NaCl, on the board. With the class determine the type of bonding for each compound by using electronegativity values. Ask what types of compounds conduct electricity.

Demonstration 1

Conductance of solutions

Approximate time: 10 min

This demonstration should clarify the conditions that produce conductance.

1. Halfway fill four labeled beakers with 1 M acetic acid, 1 M hydrochloric acid, 0.5 M sodium chloride solution, and distilled water, respectively.
2. Take out the battery-operated conductivity tester, and examine each solution to show whether the pre-demonstration answers are correct.
3. Ask what happened to the covalently bonded compounds that caused them to conduct electricity. (The compounds dissociated to form ions.)

SAFETY

Safety goggles and a lab apron must be worn. Students must be 10 ft or more from demonstration. In case of an acid spill, dilute the spill with water and, while wearing gloves, soak up the spill with cloth or paper towels. Rinse the towels, neutralize the rinsings with 1 M NaOH, and flush down the drain.

What happens in an aqueous solution?

Section Objectives

Describe how conductance can be measured.

Distinguish among the properties of nonelectrolytes, weak electrolytes, and strong electrolytes.

Describe the theory of ionization.

Compare and contrast the dissociation of ionic and molecular compounds in aqueous solutions.

List two colligative properties, and explain how each can be demonstrated.

Conductance

As you know from what you have studied in previous chapters, when a substance dissolves in water, the result is an aqueous solution that is a homogeneous mixture. Is there a way to further classify solutions in terms of solutes? Think back to Chapters 5 and 6, where you learned about ionic and molecular compounds. One way to further classify solutions is by whether or not they conduct electricity.

Some solutions conduct electricity

The measurement of a solution's ability to conduct electricity is known as **conductance**. You have probably seen an apparatus similar to that shown in **Figure 12-1** to measure conductance. The apparatus uses the solution being tested as a "continuance" of the wires carrying the electric current to complete the circuit. When a solution contains charged particles that can continue "carrying" the electricity, the bulb will light. This type of solution contains an **electrolyte**.

conductance
the measurement of a solution's ability to conduct electrical energy

electrolyte
any substance that, when it is dissolved in a solution, will conduct an electric current by means of movement of ions

Figure 12-1
The space between the electrodes in the conductivity apparatus prevents the conductance of electricity. The presence of ions in a solution being tested provides the path for conductance.

Visual Strategy

Figure 12-1

Emphasize that the solution does not produce electricity; it merely conducts electricity.

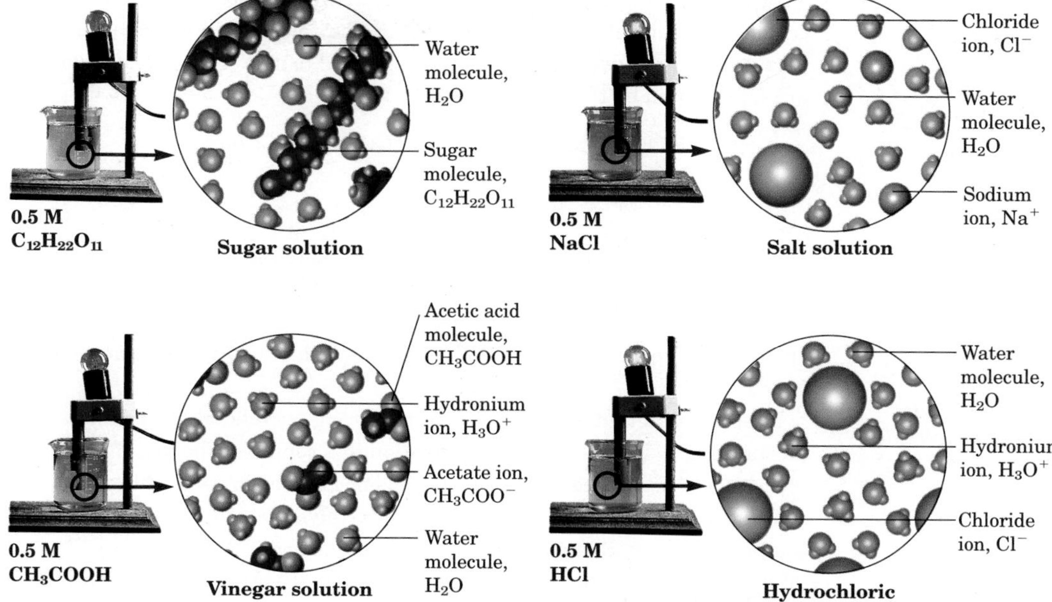

Figure 12-2
For solutions at the same concentration, the brightness of the bulb is an indication of the strength of the electrolyte.

Vinegar, which is a dilute acetic acid solution, is considered a **weak electrolyte** because it displays low conductance, lighting the bulb dimly. Salt, or sodium chloride, is considered a **strong electrolyte** because it exhibits high conductance in aqueous solution, lighting the bulb brightly. When a solution does not contain charged particles to continue "carrying" the electricity, the bulb will not light up. The solute in this type of solution is called a **nonelectrolyte**. An example of a nonelectrolyte is sugar. When dissolved in distilled water, sugar produces no measurable conductance in the apparatus with the light bulb. How do you think tap water would compare to the solutions shown in **Figure 12-2** in terms of conducting electricity?

weak electrolyte
a compound that experiences only a small degree of dissociation in an aqueous solution

strong electrolyte
a substance that is completely or largely dissociated in an aqueous solution

nonelectrolyte
a substance that, when dissolved in an aqueous solution, will not conduct an electric current

A simple way to compare the strength of electrolytes is to use the conductivity apparatus to measure the conductance of various solutions at the same concentration. By comparing the brightness of the bulb under these conditions, you can determine which substances are weak or strong electrolytes. The strength of an electrolyte depends on its tendency to form ions in a solution at a specific concentration. Weak electrolytes form solutions with few ions. Acetic acid is a weak electrolyte. A water solution of acetic acid contains acetic acid molecules, water molecules, acetate ions, and hydronium ions, as shown in **Figure 12-2**. Because the solution contains both molecules and ions, it can be represented by the following equation.

$$CH_3COOH(aq) + H_2O(l) \rightleftharpoons CH_3COO^-(aq) + H_3O^+(aq)$$

The double arrow shows that both reactants and products exist in the solution and that the conversion to ions is not complete. In fact, about 99% of the acetic acid remains as molecules.

DISPOSAL
Combine all liquids, neutralize with 1 M NaOH, and pour down the drain.

TEACH

Content Background

Measurement of conductance
Ohmmeters powered by dry cells can also be used as "conductance" meters, although the solutions may affect the probes. Placing the probes through a cork will help to keep the probes spaced evenly for all experiments. The inverse of resistance is conductivity. The unit for conductivity, humorously enough, is the mho—the reverse of ohm.

Teaching Tip

Discussion
Remind students that weak and strong are only relative terms and not specific indicators of an electrolyte's conductance.

Possible Misconception

A nonelectrolyte does indeed conduct a small amount of electricity, but it is so small an amount that it is not easily measured.

Visual Strategy

Figure 12-2
Make students aware that solution strength (concentration) is independent of whether the solution is a strong or weak electrolyte.

Demonstration 2

Conductance of tap and distilled water

Approximate time: 5 min

This demonstration shows the presence of electrolytes in tap water.

1. Fill one beaker with tap water and another with distilled water.
2. Using a dry cell-powered ohmmeter with the probes fitted through a cork, measure each solution's resistance.
3. If softened water is available, place it in a third beaker and measure its resistance.

SAFETY

Safety goggles, a lab apron, and rubber gloves must be worn. Students must be 10 ft or more from demonstration.

DISPOSAL

Pour water down the sink. Save cork for reuse.

Possible Misconception

Stress that an electrolyte's strength depends on how easily it dissociates, not on its concentration in solution. Also note that electrolyte strength is independent of molecular complexity. Compounds with polyatomic ions can either be weak or strong electrolytes.

Other Applications

New hairdryers come with permanently attached "ground fault interrupter circuit breakers," which lessen the chance of electrocution. If a change occurs between the ground and neutral lines, the breaker cuts the circuit within a few thousandths of a second, hopefully before electrocution occurs.

Both HCl and NaCl are strong electrolytes. Remember that HCl is covalent or molecular. NaCl is ionic. When they dissolve in water, all the particles of those two compounds are converted to ions. These conversions are represented by the following equations.

$$HCl(aq) + H_2O(l) \longrightarrow H_3O^+(aq) + Cl^-(aq)$$

$$NaCl(s) \xrightarrow{H_2O} Na^+(aq) + Cl^-(aq)$$

Note that the reaction arrow is written pointing to the products side of the equation to represent a complete reaction for the formation of ions. Strong electrolytes form more ions in the solution and display strong conductance. HCl actually reacts with water to form an ion called the **hydronium ion**, H_3O^+.

hydronium ion
a hydrogen ion covalently bonded to a water molecule, written as H_3O^+

Keep in mind that the use of the term "strong" as related to electrolytes has nothing to do with their concentration in a solution. The use of the term "weak" as related to electrolytes has nothing to do with dilute solutions. For example, HCl is a strong electrolyte in solution, whether the solution is very concentrated (10 M) or very dilute (0.000 001 M). A very dilute solution, however, has low conductance. On the other hand, CH_3COOH is a weak electrolyte in solution, no matter how concentrated or dilute the solution.

Tap water conducts electricity

The instructions with a hair dryer or electric shaver warn you about the danger of dropping it into a tub or sink of water. When using extension cords outdoors, you are cautioned to keep them away from water. Obviously, electricity and water are a hazardous combination that, at times, can prove fatal. **Figure 12-3** provides a reminder of this danger.

Have you ever wondered why water is something you should avoid when you're dealing with electricity? Actually, it's not the water that's the problem. It's what's in the water. You have already learned that water is a good solvent. Unlike distilled water, which does not conduct enough electricity to light the bulb in the conductivity apparatus, tap water contains various dissolved salts and minerals. The "harder" the water, the more salts and minerals it contains. Although these dissolved salts and minerals generally do not give tap water high conductance, remember that it is still a conductor of electricity. That's why you have to be careful when using electricity near water.

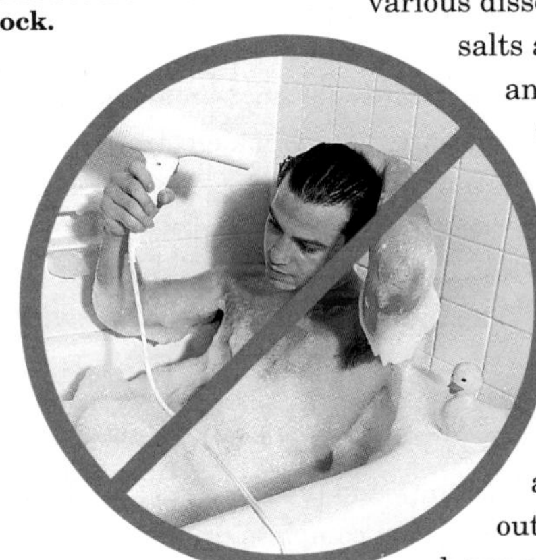

Figure 12-3
An electrical appliance is dangerous when it comes in contact with water because it can cause severe electric shock.

Although it doesn't display high conductance, tap water can still conduct enough electric current to injure or kill someone. As you enjoy the pool, lake, or beach when a thunderstorm is in the area, be sure to keep in mind that chlorinated water, ground water, and salt water are very good conductors of electricity. Make sure you get out of the water during a storm; don't think that lightning can harm you only if it is extremely close.

Consumer Focus

Electrolytes and sports drinks

Whenever you work hard or participate in sports activities, you sweat and need to drink water to replace the lost fluids. Sweat is more than just water. It has a salty taste. As you saw with the conductivity tests, salt is an electrolyte. If you take salt tablets when you sweat a lot, you also need to remember to drink lots of water.

You may have seen commercials showing professional athletes drinking sports drinks, which combines electrolytes in a flavored drink to solve the problem of replenishing fluids lost while sweating. Look at labels when you shop for something to quench your thirst after exercising vigorously. Check the label for electrolytes such as potassium ions (K^+), sodium ions (Na^+), and calcium ions (Ca^{2+}).

The fluid in your body's tissues is an aqueous electrolytic solution. The electrolytes serve important roles in various physiological functions, including nerve impulse conduction and muscle contraction.

At rest, a nerve cell uses "ion pumps" to maintain a high concentration of Na^+ ions outside the cell and K^+ ions inside. These ions do not diffuse through the membrane when the nerve cell is at rest. However, when the cell is stimulated, ions pass through the cell membrane, moving Na^+ ions inside and K^+ ions outside the nerve cell. When the cell returns to its resting state, Na^+ ions are again "pumped" out and the K^+ ions are "pumped" in, readying the cell for the next stimulus.

Muscle contractions also depend on the movement of ions. When muscle cells are stimulated, Ca^{2+} ions go into the cells and cause protein fibers to slide together, resulting in a muscle contraction. When Ca^{2+} ions move back outside the cells, the muscle relaxes.

Obviously, your nerves and muscles need electrolytes. When you sweat, you also lose electrolytes (that salty taste). If you replenish your lost fluids *and electrolytes* with a sports drink, be careful. Too much, like too little, can be dangerous. In fact, some people have to watch their intake of certain electrolytes, especially Na^+ and Ca^{2+}. Some problems include high blood pressure and kidney problems.

Consult a doctor for your particular needs, and check labels accordingly.

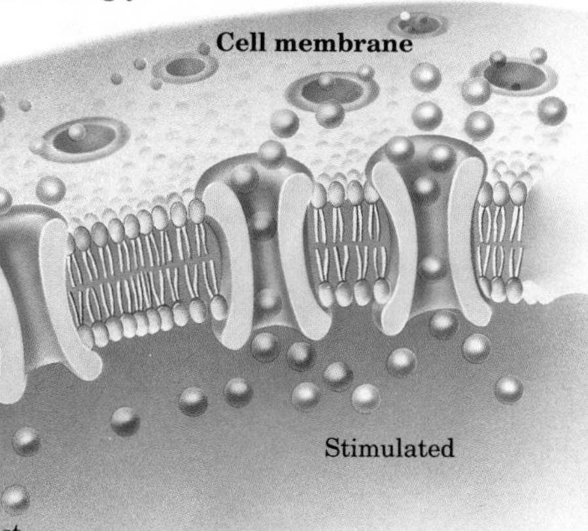

Other Applications

It has been noted recently by the medical community that most backyard athletes do not lose enough salt during the course of exercise to warrant the use of electrolyte-replenishing drinks. The biggest need of the body is fluid (water) replenishment.

Visual Strategy

Electrolytes in cells

Stress that the pumping of the Na^+ and K^+ ions in the cells is very dynamic and that the body is constantly striving to reestablish equilibrium conditions.

Historical Note

The theory of ionization, proposed in 1884 by Svante Arrhenius (1859–1927) as his doctoral dissertation, met with considerable resistance. Even in 1903, when Arrhenius was nominated to share the Nobel Prizes for both physics and chemistry, physicists blocked him from sharing the prize for physics.

Teaching Tip

Analogy

Dissociation may be compared to a group of green students (solute) and blue students (solvent, in this case water). The green students are clustered in a group (crystal). A group of blue students approach and are able to attract and hold onto individual green students more effectively than green students can hold onto each other. Several of these blue students will remain with the now *hydrated* green student as he or she drifts through the solution. New green students are then exposed to the attractions of new blue students.

Teaching Tip

New application of previous information

Remind students of Demonstration 3 in Chapter 7, in which sodium hydroxide and potassium permanganate crystals were dissolved in water and the change in temperature of the solutions was noted. Ask what the recipient or source of heat in those reactions was. (chemical bonds undergoing dissociation) Ask why a larger change occurred as the samples were dissolved more thoroughly. (As solvent contact increased, so did the opportunity for dissociation.)

theory of ionization *the explanation of the process by which electrolytes break apart in solution in the form of freely moving ions that can conduct an electrical current*

Ions in solution

All electrolytes form ions in water

The ability of solutions to conduct an electric current is explained by the **theory of ionization**, which states that electrolytes in solution break apart to form ions. These ions, or charged particles, move freely in the solution and, consequently, conduct an electric current.

Chemists recognize that electrolytes can be either ionic or molecular compounds. You know that ions are released when salt, an ionic compound, dissolves. Compounds such as hydrochloric acid are molecular electrolytes. You observed that both the NaCl and HCl solutions conduct electricity in the conductivity apparatus, so you know that solutions of both ionic and molecular compounds can be electrolytes.

Figure 12-4a
When a soluble salt dissolves, it dissociates.

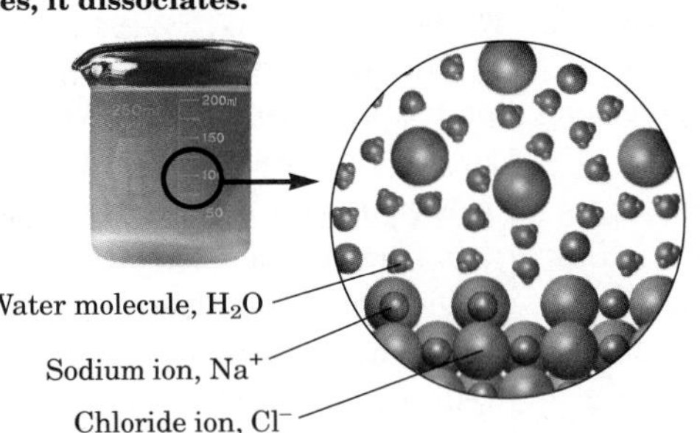

12-4b
As ions leave the crystal surface, they are surrounded by water molecules, forming hydrated ions.

hydration *the process by which water molecules surround each ion as it moves into solution*

dissociation *a process using energy to separate a compound into ions in water*

Ionic compounds dissociate in water

Figure 12-4 illustrates what happens when sodium chloride, or table salt, dissolves in water. At the surface of a salt crystal, the positive ends of water molecules attract Cl^- ions, while the negative ends attract Na^+ ions. Attractive forces between water and the ions are just about as strong as forces holding the ions in the crystal, so ions are drawn away from the crystal surface and into solution. In this case there is a fairly large increase in entropy. The free energy change is negative. The process by which water molecules surround each ion as it moves into solution is known as **hydration**, and the ions are said to be *hydrated*.

As ions become hydrated and diffuse into solution, other ions in the crystal become exposed to water. These ions will also move away from the crystal surface, and, eventually, the entire crystal dissolves.

The separation of ions that occurs when an ionic compound dissolves in water is called **dissociation**. The dissociation process for NaCl can be shown by the following chemical equation.

$$NaCl(s) \longrightarrow Na^+(aq) + Cl^-(aq)$$

Table 12-1
Enthalpy of Solution

Substance	Enthalpy of solution (kJ/mol)
$AgNO_3$	+22.77
$CuSO_4$	–67.81
KCl	+17.58
$KClO_3$	+42.03
KNO_3	+35.7
KOH	–54.59
LiCl	–35.0
Li_2CO_3	–12.8
NaCl	+4.27
NaOH	–41.6
NH_4NO_3	+25.5

Energy is involved in the dissociation process

The dissociation of an ionic compound requires the addition of energy to separate ions at the crystal surface and to separate water molecules. Energy is liberated when the ions are hydrated because hydration is a process in which opposite charges are drawn together.

As you can see from **Table 12-1**, dissociation can be either an endothermic or exothermic process. The heat of solution for a solution with water as the solvent depends on the solute ions involved because the energy needed to separate water molecules is always the same.

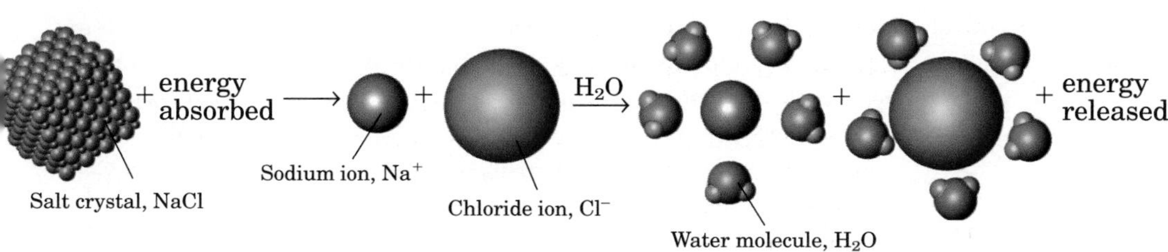

Figure 12-5
Individual ions are separated from the solid lattice by energy before they are hydrated by water molecules.

Figure 12-5 shows that energy is required to separate ions from the lattice. Energy is released when hydrated ions are released from the lattice. The heat of solution represents the net energy change for these two processes. In all cases where the enthalpy change is positive, an increase in entropy causes the process to be spontaneous. But how can the entropy of solution be negative? Small ions form a tightly arranged group with the molecules such that there is actually an increase in order when the ions dissolve.

Table 12-2
Heat of Hydration

Ion charge	Radius (picometers)	Heat of hydration (kJ/mol)	Trend
Li^+	65	523	hydration energy decreases as the radius increases
Na^+	96	418	
Cl^-	181	361	
Ca^{2+}	168	293	hydration energy increases as the charge increases (and the radius decreases)
Mg^{2+}	65	1940	
Al^{3+}	55	4690	

As you can see in **Table 12-2**, more energy is usually released in the hydration of ions that are smaller or higher in charge. Notice that although the Mg^{2+} ion is about the same size as the Li^+ ion, the Mg^{2+} ion, with its higher charge, releases almost four times as much energy when hydrated as the Li^+ ion does.

Molecular compounds may also dissociate in water, involving energy in the process

Most molecular electrolytes are polar covalent molecules. Unlike ionic compounds, polar covalent electrolytes *react* with water. Consider what happens to acetic acid, a polar covalent molecule, when it is mixed with water, as shown in **Figure 12-6**. Though you might expect the more concentrated solution to be a better conductor, that is not the case. Why?

Figure 12-6a
The conductivity of a solution is determined by the degree of dissociation rather than by the solution's concentration.

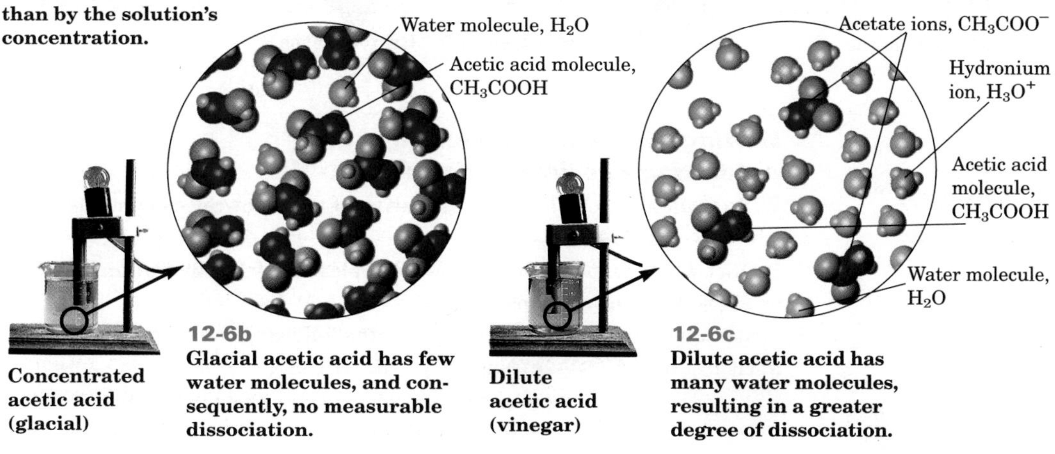

Concentrated acetic acid (glacial)

12-6b
Glacial acetic acid has few water molecules, and consequently, no measurable dissociation.

Dilute acetic acid (vinegar)

12-6c
Dilute acetic acid has many water molecules, resulting in a greater degree of dissociation.

Teaching Tip
Rule of thumb
Make students aware that when compounds with the same cation dissociate, the process may be either exothermic or endothermic, as is the case with KOH and KNO_3. In general, anions are better predictors of whether dissociation is endothermic or exothermic.

Theme
Conservation
The energy released or absorbed by the dissociation process equals the amount of energy originally needed to extract the molecule from aqueous solution. Energy is conserved in the dissociation of both ionic and molecular compounds.

Content Background
Charge, bonding, and energy
Ions with more charges must break more bonds in their respective ion pairs when dissociated. Because more bonds are broken, more energy is released.

Possible Misconception
Molecular compounds can be classified as either electrolytes or nonelectrolytes even though their covalent bonding encourages nonelectrolytic behavior.

Do You Know?
Vinegar is a 5% acetic acid solution.

Visual Strategy
Figure 12-5
Stress that each ion is surrounded by the hydrating water molecules and that this interaction between the solute and water molecules permits hydration to occur. Make students aware of the orientation of the water molecules around each ion. The positive hydrogen atoms are attracted to the anion, while the negative oxygen atoms are attracted to the cation.

Teaching Tip

The hydronium ion

Remind students that hydrogen has only one electron and one proton, and that if it donates the electron the remaining proton (H^+) migrates to a nearby water molecule and covalently bonds with the electron pair of the highly electronegative oxygen atom. The result is a water molecule with an extra proton, which gives this new hydronium ion a positive charge.

Possible Misconception

Make students aware that the double arrow means that the reaction is continuously proceeding in both directions.

Concentrated (glacial) acetic acid (17 M) contains relatively few water molecules in the solution, as shown in **Figure 12-6b**. These few water molecules cannot bring about any measurable dissociation of acetic acid molecules. However, in the dilute solution of acetic acid (0.1 M), many more water molecules are present, as shown in **Figure 12-6c**. The negative ends of these water molecules exert enough force to break a covalent bond in the acetic acid molecule and form the hydronium ion, H_3O^+. Acids, as a group, form hydronium ions in water solutions.

As you can see in **Figure 12-7**, there are two arrows in the equation for the formation of the hydronium ion, indicating a *reversible reaction*. The arrows in both directions indicate that the reaction can go both ways. You will learn more about this in the next section.

When molecular compounds form charged particles in solution, the process is also referred to as dissociation. The dissociation of ionic and molecular compounds is different. When an ionic compound dissolves, the ions, which are already present, are released and become dissociated from each other. When a molecular compound dissolves in water, ions that were not originally present in the undissolved compound form from the reaction of the compound with water.

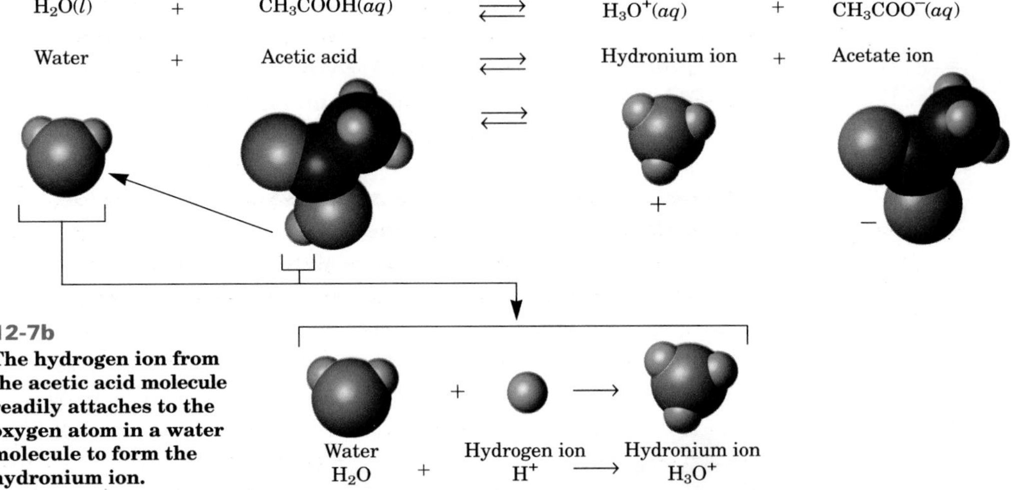

Figure 12-7a
A hydronium ion is formed when a hydrogen ion from an acid covalently bonds to an oxygen atom in a water molecule.

12-7b
The hydrogen ion from the acetic acid molecule readily attaches to the oxygen atom in a water molecule to form the hydronium ion.

Writing dissociation equations and predicting energy requirements

1. Write equations to show the dissociation of the following compounds.
   a. $CaCl_2$ **b.** $Al_2(SO_4)_3$
2. Predict whether more energy would be released by the hydration of $MgCl_2$ or $AlCl_3$. Explain your answer.
3. Write equations for the dissociation of the following compounds.
   a. HCl **b.** H_2CO_3 **c.** Na_2CrO_4 **d.** NH_4NO_3

Answers for Concept Review items and Practice problems begin on page 841.

Answers to

Concept Review

1. **a.** $CaCl_2(s) \xrightleftharpoons{H_2O} Ca^{2+}(aq) + 2Cl^-(aq)$

   b. $Al_2(SO_4)_3(s) \xrightleftharpoons{H_2O} 2Al^{3+}(aq) + 3SO_4^{2-}(aq)$

2. Because more energy is usually released in the hydration of ions that are smaller in size or higher in charge, $AlCl_3$, with a cation charge of 3+, should release more energy during hydration than $MgCl_2$, with a cation charge of 2+.

3. **a.** $HCl(aq) + H_2O(l) \rightleftharpoons H_3O^+(aq) + Cl^-(aq)$

   b. $H_2CO_3(aq) + H_2O(l) \rightleftharpoons H_3O^+(aq) + HCO_3^-(aq)$

   c. $Na_2CrO_4(s) \xrightleftharpoons{H_2O} 2Na^+(aq) + CrO_4^{2-}(aq)$

   d. $NH_4NO_3(aq) + H_2O(l) \rightleftharpoons NH_3(aq) + H_3O^+(aq) + NO_3^-(aq)$

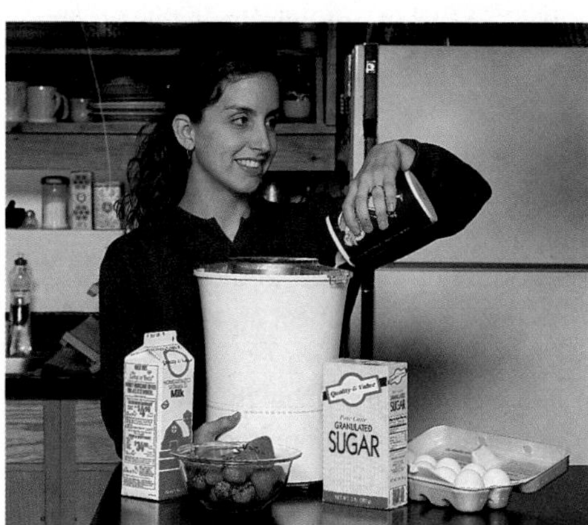

Figure 12-8
An ice cream maker uses an ice-salt-water mixture to provide a temperature that is colder than ice and water alone.

Colligative properties
Solutes can change some properties of a solvent

When you cook pasta or vegetables in boiling water, recipes generally call for the addition of salt to the water. While this is done simply to flavor the food being cooked, the addition of salt has a slight effect on the boiling process. The boiling point of water is 100°C at normal air pressure of 1 atm, or 101.3 kPa. Once water boils vigorously, its temperature stays relatively constant at 100°C. The temperature does not continue to rise as boiling continues. Adding salt to the water elevates its boiling point. Water with a little salt in it boils at a slightly higher temperature than pure water. Food will cook somewhat faster in salted water because the temperature is slightly higher than in plain water.

A similar effect is noted with the freezing point of water. However, in the case of freezing, the effect is in the opposite direction. Water freezes at 0°C at normal air pressure of 1 atm, or 101.3 kPa. No matter how much ice you have, its temperature stays relatively constant until it melts. When you are making homemade ice cream, as in **Figure 12-8**, salt is mixed with the ice that surrounds the ice cream tub to reduce the temperature at which water freezes. The temperature of the ice-salt mixture drops below 0°C, which is better for making ice cream.

These two properties of solutions, the boiling-point elevation and the freezing-point depression, are known as *colligative properties*. Any property which is determined by the number of particles in solution is known as a **colligative property**. Colligative properties depend only on the concentrations of solute particles rather than the identity of the actual solute. Colligative properties, which include osmotic pressure, freezing-point depression, and boiling-point elevation, change with the amount of solute added to the solvent. Let's look at a model now to explain why these changes occur.

colligative property
a physical property that is dependent on the number of particles present rather than on the size, mass, or characteristics of those particles

Recall from Chapter 10, and **Figure 10-28**, that the boiling point of a liquid is the temperature at which the vapor pressure of the liquid equals the atmospheric pressure. Adding salt, or any nonvolatile solute, lowers the vapor pressure such that more heat energy must be added to raise the vapor pressure of the liquid phase to that of the atmospheric pressure, thereby elevating the boiling point of the solvent.

Recall that the freezing point of a substance is the temperature at which the vapor pressures of the liquid and solid are equal. Refer to the phase diagram for water, **Figure 10-30**. Because salt lowers the vapor pressure of water, the vapor pressures of water and ice can only be equal at a temperature lower than that for pure water. Therefore, salt is said to depress the freezing point of water.

Possible Misconception
Many students are led to believe that adding salt to water greatly increases the boiling point. To raise the boiling point just 2°C they must add 234 g (over 0.5 lb!) to 2 L of water. Adding a pinch of salt to water just adds flavoring.

Teaching Tip
Discussion
Ask students why salt is placed on the icy roads to hasten melting. (It lowers the freezing point of the salt-ice solution, thus melting the ice.) Why are some melting agents made of $CaCl_2$ instead of NaCl, which is cheaper? ($CaCl_2$ depresses the freezing point 50% more than NaCl.) Why isn't salt used below −10°C? (This temperature is below that at which the maximum concentration of the solution is reached, and therefore freezing will occur anyway.)

Theme
Systems and interactions
Adding additional molecules to solutions causes interactions that interfere with the vaporization of the compound. This causes the system to require more energy to vaporize.

Teaching Tip
Rule of thumb
The larger the number of dissociated particles available, the greater the colligative effects on boiling and freezing points.

ASSESS

Alternative Assessment
Have students write short definitions for and give examples of these terms: electrolyte, nonelectrolyte, dissociation, hydration, colligative properties, and freezing-point depression.

Answers to Section Review

4. 1 mol of salt (Sodium chloride produces two particles in solution; table sugar, sucrose, produces only one.)
5. The lamp in the apparatus remains dark. An insoluble salt does not dissociate in solution.
6. Concentrated acetic acid contains few water molecules with which to react and form ions, so it has low conductance. Diluting the acid promotes ion formation and hence conductance.
7. Ionically bonded compounds do not react with the polar solvent but are transferred through the solution by the solvent molecules. Covalently bonded molecular compounds react with water to form hydronium ions.
8. Water molecules remove the ionically bonded solute molecules individually from the surface of their crystal structure and then cluster around the solute molecules by orienting their negative ends toward the cations and their positive ends toward the anions (the process of hydration). In covalently bonded molecular compounds, water molecules break the covalent bond between hydrogen and the rest of the molecule, leaving a proton and an anion. Water then reacts with the proton to form a hydronium ion.
9. The four most abundant ions in sea water are Cl^-, Na^+, SO_4^{2-}, and Mg^{2+}. While most elements may be found in sea water, few exist in concentrations above 10 $\mu g/L$, which is too little for feasible recovery. Heavy elements tend to precipitate and accumulate on the ocean floor. Some of these elements, such as manganese, have been recovered.
10. If lightning strikes the ocean surface, the electricity will be conducted through the sea water and anything in contact with it.

Colligative properties depend on solute particle concentration

Any solute, including nonelectrolytes such as sugar, will affect the colligative properties of a solution. The effect depends on the number of solute particles that dissolve in a given amount of solvent. The more solute particles in solution, the greater the lowering of the freezing point or the raising of the boiling point for the solution. For example, adding 100 formula units of NaCl to water will have about twice the effect on a colligative property as adding 100 molecules of $C_{12}H_{22}O_{11}$ to water.

Examine **Figure 12-9** to discover the reason for this difference. The following equations show why the number of solute particles that form in solution depends on the chemical nature of the solute.

$$C_{12}H_{22}O_{11} \xrightarrow{H_2O} C_{12}H_{22}O_{11} \text{ (one particle)}$$

$$NaCl \xrightarrow{H_2O} Na^+ + Cl^- \text{ (two particles)}$$

$$CaCl_2 \xrightarrow{H_2O} Ca^{2+} + 2Cl^- \text{ (three particles)}$$

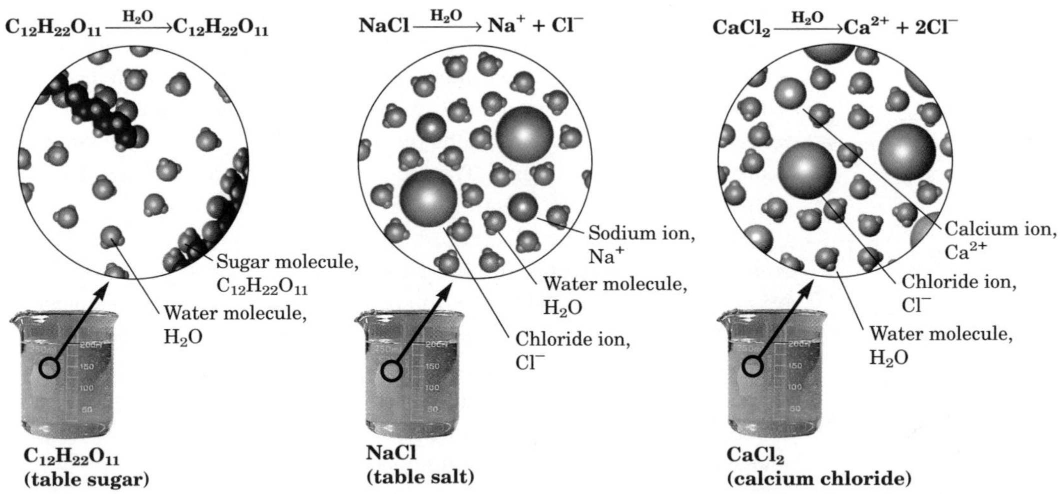

Figure 12-9 Colligative properties depend on the number of particles formed by the solute in the solution. Comparison of each substance shown illustrates this concept.

Section Review

4. Would you get a higher boiling point by adding one mole of table sugar or one mole of table salt to equal masses of water?
5. Describe what would happen if an insoluble salt were added to the water in the conductance apparatus shown in **Figure 12-1**.
6. Explain the conductance of concentrated and dilute acetic acid solutions.
7. State the theory of ionization for both ionic and molecular compounds.
8. Describe the process of dissolving ionic and molecular compounds in water.

9. **Story Link**
Name some substances dissolved in sea water. Which might be worth reclaiming? What might lie undissolved on the ocean floor?
10. **Story Link**
Explain why the three people stranded for 11 days on a raft had to be concerned about electricity from lightning during a thunderstorm.

Visual Strategy

Figure 12-9

Emphasize the number of particles available after the dissociation of each type of solute.

What is an equilibrium system?

Section Objectives

Distinguish between reactions that go to completion and reversible reactions.

Determine which reaction is favored when a stress is applied to an equilibrium system using LeChâtelier's principle.

Describe what happens when complex ions are present in aqueous equilibrium.

Completion and reversible reactions

When you look at the components of a chemical reaction, you are looking at how compounds and elements recombine to form products. **Figure 12-10** illustrates an example of this process. In studying reactions, you may have observed changes such as the formation of a precipitate or the evolution of a gas. These types of reactions are described as going to completion. Two examples are shown in the following equations.

$$NaHCO_3(aq) + HC_2H_3O_2(aq) \longrightarrow NaC_2H_3O_2(aq) + CO_2(g) + H_2O(l)$$
Formation of a gas

$$NaCl(aq) + AgNO_3(aq) \longrightarrow AgCl(s) + NaNO_3(aq)$$
Formation of a precipitate

Figure 12-10
Some ions in solution can form a precipitate in a reaction that goes to completion.

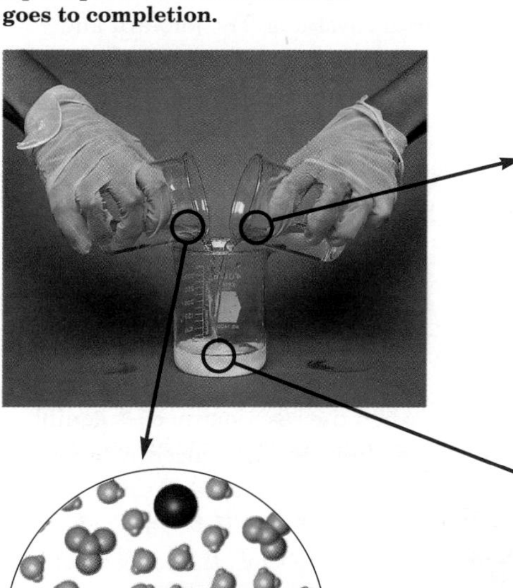

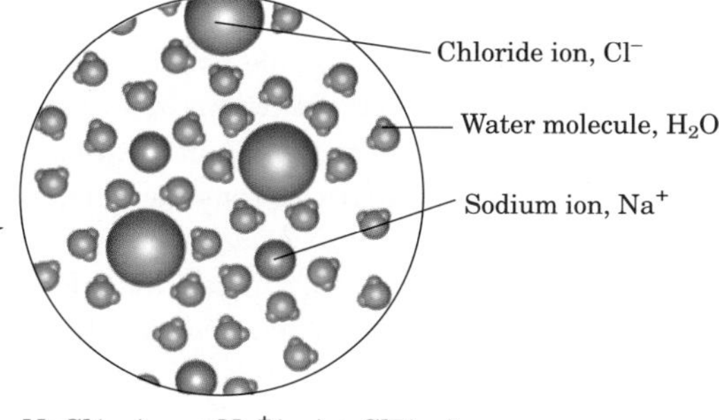

$NaCl(aq) \longrightarrow Na^+(aq) + Cl^-(aq)$
Sodium chloride is a soluble salt that dissociates in water to form ions.

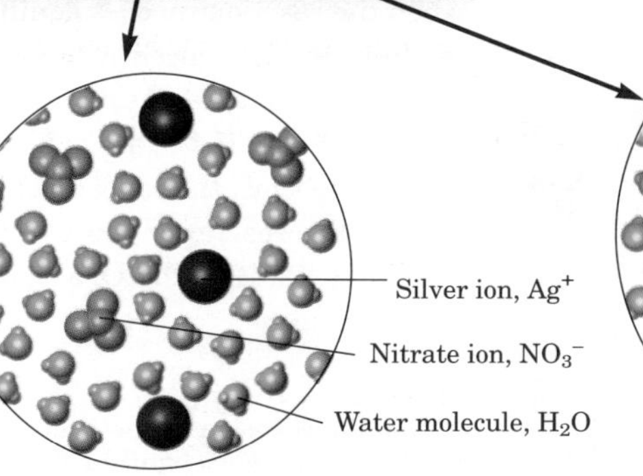

$AgNO_3(aq) \longrightarrow Ag^+(aq) + NO_3^-(aq)$
Silver nitrate is a soluble salt that dissociates in water to form ions.

Water molecule, H_2O
Nitrate ion, NO_3^-
Silver ion, Ag^+
Chloride ion, Cl^-
Sodium ion, Na^+
Silver chloride, AgCl

$Ag^+(aq) + Cl^-(aq) \longrightarrow AgCl(s)$
Silver chloride is an insoluble salt that forms when Ag^+ ions can react with Cl^- ions.

Visual Strategy

Figure 12-10

Make students aware that Na^+ and NO_3^- ions are still dissociated throughout the solution and that in a reaction like this they are called *spectator ions* because they are not involved directly in the reaction.

Section 12-2

FOCUS

Lesson Starter

Start the lesson by asking students to list two everyday processes or chemical reactions that are reversible and can thus be returned to their original state. Have them then list two irreversible processes.

Demonstration 3

Completion reactions

Approximate time: 5 min

1. Place 15 mL of 0.1 M $AgNO_3$ solution in a test tube.
2. Add 15 mL of 0.2 M NaCl solution. A white AgCl precipitate that is photoreactive forms.
3. Filter the solution into another test tube.
4. Add 5 mL more of 0.2 M NaCl solution to show that the reaction is complete and that no more $AgNO_3$ is present.
5. Add 1 drop of $AgNO_3$ to show what happens when Ag^+ is present.

SAFETY

Safety goggles and a lab apron must be worn. Students must be 10 ft or more from demonstration.

DISPOSAL

Add a few drops of NaCl solution until all of the silver ions have precipitated. Decant the liquid from the AgCl crystals, and evaporate the liquid. Combine the remaining $AgNO_3$ and NaCl crystals with the AgCl crystals and put them in a bottle labeled *For future use* and listing contents, demonstration number and chapter. **Do not dispose of the crystals in a landfill.**

Teaching Tip

Stress that the reaction in Demonstration 3 is a completion reaction by asking students what would happen if more AgCl were added to the solution. Adding more reactants yields more of the products, but adding more products does not produce the reactants.

Demonstration 4

Reversible reactions

Approximate time: 5 min

The demonstration shows the reversbility of the following reaction (Ind is phenolphthalein).

$H_2In\,(aq) + 2OH^-(aq) \rightleftharpoons In^{2-}(aq) + 2H_2O(l)$

1. Place 2 drops phenolphthalein solution in a test tube half-filled with distilled water.
2. Add a few drops of 0.1 M NaOH solution to the test tube until the solution turns a light pink.
3. Add 2 drops of NaOH (reactant) to the solution. Note that it becomes red.
4. Add 2 drops of phenolphthalein (reactant) to the test tube. The red of the solution should become darker.
5. Pour the contents of the tube into a 500 mL beaker and add water (product) until the solution is again a light pink or colorless.

SAFETY

Safety goggles and a lab apron must be worn. Students must be 10 ft or more from demonstration. In case of a base spill, dilute the spill with water and, while wearing gloves, soak up the spill with cloth or paper towels. Rinse the towels, neutralize the rinsings with 1 M HCl, and flush down the drain.

DISPOSAL

Neutralize the solution with 0.1 M HCl and pour down the drain.

Reversible reactions reach equilibrium

Some reactions do not run to completion. Instead, the products that are formed in a chemical reaction may re-form the original reactants. If the reactants and products are kept in a closed system where nothing can escape, the reactions may go in either direction. The following are examples of reversible reactions.

$$H_2(g) + I_2(g) \rightleftharpoons 2HI(g)$$

$$2NO_2(g) \rightleftharpoons N_2O_4(g)$$

Notice that the arrows go in both directions, indicating a reversible reaction. Reversible reactions explain why some cold and hot packs can be reused. By reversing the reactions, the reactants are re-formed, and the reactions can occur again, either absorbing or releasing heat energy.

Another example of a reversible reaction is the recharging of a battery. The interior of the automobile battery has lead plates, lead(IV) oxide plates, and a sulfuric acid solution, H_2SO_4. When the battery is used, the forward reaction occurs, releasing energy in the form of electricity. The reverse reaction occurs when electricity from an outside source is fed into the circuit. This reversible reaction is represented by the following equation.

$$Pb(s) + PbO_2(s) + 2H_2SO_4(aq) \rightleftharpoons 2PbSO_4(s) + 2H_2O(l) + \text{energy}$$

Reversible reactions may be equilibrium systems. The forward and reverse reactions occur at equal rates, and there are no overall changes as long as the conditions remain the same. At equilibrium, the total amount of particles remains constant.

Because our model of a system at equilibrium has constant concentrations, we can use this information to determine the completeness of the reaction and to make predictions. Consider the reaction that represents the dissociation of acetic acid.

$$CH_3COOH(aq) + H_2O(l) \rightleftharpoons CH_3COO^-(aq) + H_3O^+(aq)$$

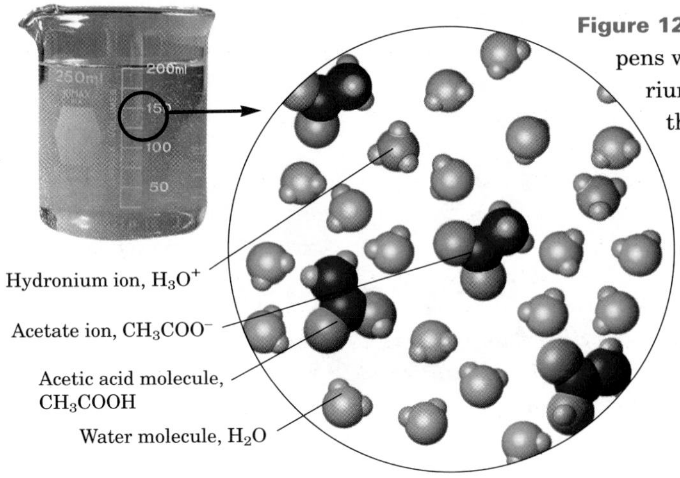

Figure 12-11 shows a model of what happens when the reaction reaches equilibrium. Initially, the concentrations of the two products, represented by CH_3COO^- and H_3O^+, are zero. At the same time, the concentrations of the two reactants, represented by CH_3COOH and H_2O, are at a maximum.

Figure 12-11a
This beaker contains an aqueous solution of acetic acid in water.

12-11b
When the reaction of acetic acid and water reaches equilibrium, the system contains CH_3COOH, H_2O, H_3O^+, and CH_3COO^-.

Figure 12-12
The rates of the forward and reverse reactions reach equilibrium after a period of time.

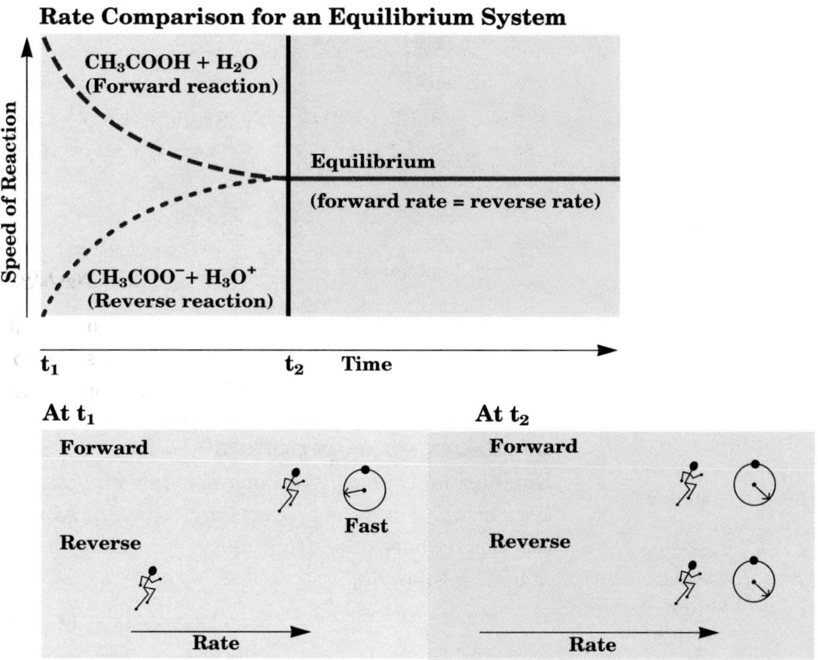

At t_1 the rates are depicted by the forward reaction going fast and the reverse reaction starting from zero.

At t_2 the rates are equal. They remain equal at all times greater than t_2.

Over time, as shown in **Figure 12-12**, the rate of the forward reaction decreases as the reactants are used up. Meanwhile, the rate of the reverse reaction increases as the products are formed. When the two reaction rates become equal, equilibrium of the system is established. At equilibrium, the individual concentrations of the reactants and the products undergo no further changes as long as conditions remain the same. However, chemical activity continues to occur even when equilibrium is reached. Although the macroscopic properties no longer change (the system appears unchanged), the forward and reverse reactions are still occurring. In other words, the reaction does not stop at equilibrium, but maintains a constant concentration or ratio of reactants to products. The reactants continue forming products, and, at the same rate, the reverse reaction continues, with the products forming reactants. That's why any chemical equilibrium system is really a dynamic process, not a static one.

To understand the difference between a dynamic and static process, imagine that you and a friend are each holding a baseball. With each of you just holding your own baseball, the situation can be described as static. But if you throw the balls back and forth, action occurs; the situation is described as dynamic. If you continue to throw the baseballs, your situation continues to be dynamic.

Visual Strategy

Figure 12-12

Stress that although the reaction rates are equal, this does not mean that equal amounts (concentrations) of reactants for both the forward and reverse reaction are present. At equilibrium the number of dissociation reactions in a given amount of time equals the number of recombination reactions in the same amount of time.

TEACH

Theme

Equilibrium and change

The release of a gas or the formation of a precipitate removes ions from the reaction so that equilibrium can no longer be reestablished and the reaction continues to completion.

Content Background

Reversible reactions vs. completion reactions

Another good example of a *reversible* reaction is that of carbonic acid decomposing into carbon dioxide and water and then re-forming into carbonic acid.

$$H_2CO_3(aq) \rightleftharpoons H_2O(l) + CO_2(aq)$$

This reaction takes place in a sealed soda bottle. When the bottle is opened and the CO_2 is allowed to escape, the reaction goes to *completion*, and the soda goes flat.

$$H_2CO_3(aq) \longrightarrow H_2O(l) + CO_2(g)$$

Teaching Tip

Analogy

Help the students understand the concept of a *dynamic equilibrium* by considering a washroom at a major sporting event. Assume there is a constant movement of people into and out of the washroom. The system is at equilibrium because the number of people in the washroom stays constant and the number of people in the stadium stays constant. It is also dynamic because different people are going in and out at different times. The system is changing but stable.

Teaching Tip
Definition
Reversible reactions are characteristic of equilibrium systems.

Demonstration 5
Temperature effects on gas production
This demonstration should help students understand how a system's equilibrium depends on temperature.
Perform steps 1–4 a day before the demonstration.
Approximate time: 15 min
1. Inflate and stretch three balloons several times.
2. Stretch each balloon over the mouth of a 250 mL Erlenmeyer flask to be sure they fit.
3. Fill each balloon with 10 g of sodium carbonate, Na_2CO_3.
4. Place 50 mL of 1 M HCl in each flask. Store one flask in the freezer at 0°C, one in the refrigerator at 15°C, and one at 30°C or room temperature. Place a thermometer in each flask so that the temperatures can be monitored.

Perform steps 5–7 as the actual demonstration.
Approximate time: 10 min
5. Twist each balloon so that it can be placed onto a flask without any sodium carbonate falling into the flask. Place the balloons over the mouths of the three flasks.
6. Clamp the three flasks to separate ring stands, making sure that no sodium carbonate falls out of the balloons.
7. Ask for three student volunteers. Have the students release the sodium carbonate into the flasks at the same time. **All volunteers must then stand back.** The balloons will inflate at different rates.

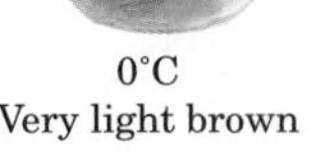

Figure 12-13
Different temperatures can cause an equilibrium system to shift and reach a new equilibrium. More NO_2 is present when the brown color is more dominant, and more N_2O_4, which is colorless, makes the solution a lighter color.

System equilibria

Equilibria can favor the formation of reactants or products

Le Châtelier's principle
a principle stating that if a system at equilibrium is disturbed by applying stress, the system will adjust in such a way as to counter the stress

Equilibrium is established in a reversible reaction when the amounts of the reactants and the products are constant. An example of a reversible reaction on page 462 involved oxides of nitrogen as represented by the following equation.

$$\underset{\text{Brown}}{2NO_2(g)} \rightleftharpoons \underset{\text{Colorless}}{N_2O_4(g)}$$

Figure 12-13 shows this equilibrium system at three different temperatures. At 25°C, the system at equilibrium is a medium color of brown, but when cooled to 0°C, the color of the solution in the flask is lighter. We infer that some of the brown NO_2 is no longer there and that more of the N_2O_4 has formed. At 100°C, you can see that the contents of the flask are darker. We infer that more of the brown NO_2 has formed at the higher temperature.

Because our model involves a reversible reaction, we can use it to explain the color changes that were observed. The color lightens at 0°C because there is more N_2O_4 than at 25°C. Therefore, the reduction in temperature must cause the reaction to shift from its original equilibrium state to favor the forward reaction. But why does this shift occur?

Figure 12-14
Henri Le Châtelier (1850–1936) was particularly interested in the relationship of science to industry and in how to obtain the maximum yield from any chemical reaction. His research provided us with the principle that has greatly increased the efficiency of many industrial processes.

Stresses alter equilibrium

When a system is at equilibrium, it will stay that way until something changes this condition. In the reaction between NO_2 and N_2O_4, the condition that changed is temperature. An important principle, developed by the French chemist Henri Louis Le Châtelier, shown in **Figure 12-14**, enables us to predict how changes in a system at equilibrium will affect the status of the system. **Le Châtelier's principle** states that when a system at equilibrium is disturbed by applying stress, it attains a new equilibrium position to accommodate the change and relieve the stress. Examples of stress that can be used to change a system at equilibrium include changes in concentration, temperature, or pressure.

Visual Strategy
Figure 12-13
Point out that the equilibrium of this system can also be affected by the concentrations of the components. If one or the other of the two gases could be introduced into the bottle without losing any of the gas already contained, the equilibrium position would be shifted toward the gas with the lower concentration.

Temperature affects equilibrium systems

Temperature changes can cause equilibrium shifts. Consider the reaction between NO_2 and N_2O_4 as the temperature is decreased from 25°C to 0°C. The exothermic process at 25°C is shown by the following equation.

$$2NO_2(g) \xrightleftharpoons{\text{at } 25^\circ C} N_2O_4(g) + 58.8 \text{ kJ} \qquad 2NO_2 \xrightleftharpoons{\text{at } 25^\circ C} N_2O_4$$

Le Châtelier's principle states that a system will react to relieve any stress placed on it. In the previous reaction, energy is one of the products. Therefore, by lowering the temperature, one of the products (energy) is being removed. As the temperature continues to drop, the system reacts in such a way as to relieve the stress. To accomplish this, more NO_2 is converted to N_2O_4.

When the rate of the forward reaction is greater than the rate of the reverse reaction, the forward arrow can be shown as longer than the reverse arrow. As the gas becomes lighter, the equation is written as follows.

Brown Colorless

$$2NO_2(g) \xrightleftharpoons{\text{at } 0^\circ C} N_2O_4(g) \qquad 2NO_2 \xrightleftharpoons{\text{at } 0^\circ C} N_2O_4$$

Shift to the right

The system reaches a new equilibrium state at the new temperature, with constant macroscopic properties. As long as this temperature is maintained, the gas in the flask will remain a light brown. The rates of the forward and reverse reactions are again equal.

On the other hand, by increasing the temperature to 100°C, more heat energy is added to the system. It reacts in such a way as to relieve the stress, and absorbs some of the heat. The equilibrium shifts left and forms more brown NO_2.

When the reverse reaction is favored, that arrow can be shown as longer. As the gas becomes darker brown, the equation is written as follows.

Brown Colorless

$$2NO_2(g) \xrightleftharpoons{\text{at } 100^\circ C} N_2O_4(g) \qquad 2NO_2 \xrightleftharpoons{\text{at } 100^\circ C} N_2O_4$$

Shift to the left

The reaction reaches a new equilibrium state for the system at the new temperature, with constant macroscopic properties. As long as this temperature is maintained, the gas in the flask will remain a dark brown.

Pressure changes may alter gaseous equilibrium systems

According to the equation at the top of this page, 2 mol of NO_2 react to produce 1 mol of N_2O_4. When this equilibrium mixture is subjected to an increase in pressure, it relieves the stress by favoring the reaction that produces fewer gas molecules. Fewer gas molecules will exert less pressure. Observations show that an increase in pressure will cause the system to shift to the right forming more N_2O_4 molecules. Conversely, a decrease in pressure will favor the reaction that produces the most gas molecules, so the system shifts to the left, forming more NO_2 molecules.

SAFETY

Safety goggles and lab aprons must be worn by all participants in the demonstration. Students must be 10 ft or more from demonstration. In case of an acid spill, dilute the spill with water and, while wearing gloves, soak up the spill with cloth or paper towels. Rinse the towels, neutralize the rinsings with 1 M NaOH, and flush down the drain.

DISPOSAL

Combine all liquids. If solution is alkaline, pour it down the drain. If solution is still acidic, neutralize it with 1 M NaOH and then pour it down the drain.

[Source: Summerlin, Borgford, and Ealy. *Chemical Demonstrations: A Sourcebook for Teachers*, Vol. 2, ACS, 1988.]

Teaching Tip

Analogy

How temperature affects the equilibrium rate can be understood with the following illustration. Imagine two rooms connected by two doorways. Each door swings in only one direction, so that the exit for one room is the entrance for the other. The rooms are crowded with people, and the flow into each room is dictated by the size of the doorways. If the rooms are warmed, one of the doorways becomes much bigger while the other becomes smaller. The flow of people into the first room exceeds the flow out until a new equilibrium is eventually established. Similarly, when the rooms are cooled, the doors reverse size, so that now the flow of people out of the first room exceeds the incoming flow, and the crowd moves toward the second room until a new equilibrium is established. The rate at which the doors change size is a property of the particular reaction.

Historical Note

Henri Louis Le Châtelier (1850–1936) was a French chemist who was translating a work by American physicist Josiah Willard Gibbs (1839–1903) when he proposed his principle in 1888. As it turned out, Gibbs' theories of thermodynamics explained Le Châtelier's principle quite well.

Demonstration 6

Complex ion equilibria

Approximate time: 5 min

This demonstration shows the shift in the equilibrium for a complex ion solution.

1. Fill a test tube halfway with a 1 M $CuSO_4$ solution. The blue color is characteristic of the $[Cu(H_2O)_4]^{2+}$ ion.
2. Add NaCl crystals and stir the solution. Continue to add salt until the solution turns a deep green. The new color indicates the replacement of the water ligands with chloride ions to form $[CuCl_4]^{2-}$ ions.
3. Add more water to shift the equilibrium again to the blue color.

$$[Cu(H_2O)_4]^{2+} + 4Cl^- \rightleftharpoons [CuCl_4]^{2-} + 4H_2O$$

SAFETY

Safety goggles and a lab apron must be worn. Students must be 10 ft or more from demonstration.

DISPOSAL

Add an excess of 1 M NaOH to precipitate the copper as $Cu(OH)_2$. Filter and wash the precipitate with water. Neutralize the combined filtrate and washings with 1 M HCl, and pour down the drain. Wrap the wet $Cu(OH)_2$ crystals in newspaper and place in the trash.

Figure 12-15
The effects of a reversible reaction can be observed in certain sunglasses. When more light energy is added, the forward reaction is favored, forming more silver atoms (the darker color).

Practical use of Le Châtelier's principle

Have you ever worn sunglasses like the pair shown in **Figure 12-15**? If you have, then you know that in bright light they get darker. The lenses of these glasses contain the ionic compound AgCl. In the presence of light energy, AgCl decomposes, as shown by the following equation.

$$AgCl(s) + \text{energy} \rightleftharpoons Ag^\circ + Cl^\circ$$

Notice that this reaction is a decomposition and not a dissociation. The difference is that neutral atoms are formed in an AgCl decomposition. Ions are formed from an AgCl dissociation. In the reverse reaction, the Ag° and the Cl° atoms react to form AgCl(*s*). When more energy is added, a stress is placed on the system. To relieve this stress, the forward reaction is momentarily favored, shifting the equilibrium to the right, as shown by the following equation.

complex ions
an ion having a structure in which a central atom or ion is bonded by coordinate bonds to other ions or molecules (called ligands)

$$AgCl(s) + \text{energy} \xrightarrow[\leftarrow]{\longrightarrow} Ag^\circ + Cl^\circ$$

As the system adjusts to reach a new equilibrium, more Ag° atoms are formed. The Ag° atoms make the lenses darker and reduce the amount of sunlight that can penetrate them.

With less light energy, the equilibrium shifts to the left as indicated by the following equation.

$$AgCl(s) + \text{energy} \xrightarrow[\longleftarrow]{\rightarrow} Ag^\circ + Cl^\circ$$

When this happens, fewer Ag° atoms are present, and the lenses lighten and allow more light to pass through them.

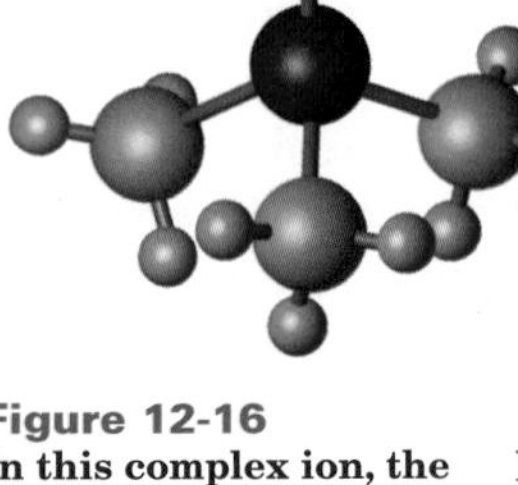

Figure 12-16
In this complex ion, the ammonia molecules are the ligands bonded to the central zinc ion.

Complex ion equilibria

Changes in concentration alter equilibrium systems

Complex ions are composed of a central metal ion combined with a specific number of other ions or polar molecules. For example, $[Zn(NH_3)_4]^{2+}$, shown in **Figure 12-16**, is a complex ion. The formation of a complex ion generally involves a reversible reaction that reaches equilibrium. Complex ions are interesting because many are colored, and their reactions can be tracked by color changes. In this section, you will see how changing the concentration of a reactant or product in a complex-ion equilibrium system also follows Le Châtelier's principle.

Visual Strategy

Figure 12-16

Emphasize that most complex ions are able to change their structure because of the complex bonding. Point out that the reactivity and properties of a complex ion can be explained in part by its structure.

Table 12-3
Examples of Complex Ions

Complex ion	Color
$[Co(NH_3)_5Cl]^{2+}$	violet
$[Co(NH_3)_5H_2O]^{3+}$	red
$[Co(NH_3)_6]^{3+}$	yellow-orange
$[Co(CN)_6]^{3-}$	pale yellow
$[Ni(NH_3)_6]^{2+}$	blue-violet
$[Cu(NH_3)_4]^{2+}$	blue-purple
$[Cu(H_2O)_4]^{2+}$	green-blue
$[Fe(CN)_6]^{4-}$	yellow
$[Fe(CN)_6]^{3-}$	red
$[Fe(SCN)(H_2O)_5]^{2+}$	deep red

In our example, zinc is the central metal ion that is bonded to four ammonia molecules. The molecules (or ions) that are bonded to the central metal atom are often referred to as **ligands**. Some common *ligands* are ions such as Cl^-, F^-, and CN^-, or polar molecules such as NH_3 and H_2O. More examples of complex ions are listed in **Table 12-3**.

Figure 12-17 shows a solution containing a complex ion of nickel. As you can see, this particular complex ion consists of a Ni^{2+} ion with six water molecules and produces the green color in the solution. When an excess of a concentrated ammonia solution is added to this green solution, a blue-violet color forms. This blue-violet color is characteristic of another complex ion of nickel. Ammonia molecules have replaced the water molecules that surround the nickel ion, as shown by the following equation.

$$\underset{\text{Green}}{[Ni(H_2O)_6]^{2+}(aq)} + \underset{\text{Colorless}}{6NH_3(aq)} \rightleftharpoons \underset{\text{Blue-violet}}{[Ni(NH_3)_6]^{2+}(aq)} + \underset{\text{Colorless}}{6H_2O(l)}$$

ligand
a functional group, atom, or molecule that is attached to the central atom of a complex ion

Each metal complex ion has a distinct color that makes it easy to identify which ion is present in a higher concentration in the solution. This reaction for complex ions is reversible and eventually reaches equilibrium. However, the equilibrium can be shifted in either direction, depending on which stress is applied. For example, if more NH_3 is added to the aqueous solution, the stress forces the equilibrium to shift to the right. Thus the color of the system will change to blue-violet. How can you get the system to change to green?

Some other aqueous equilibria reactions involving complex ions are shown below.

$$\underset{\text{Colorless}}{[Zn(H_2O)_4]^{2+}} + H_2O \rightleftharpoons \underset{\text{Colorless}}{[Zn(H_2O)_3(OH)]^+} + H_3O^+$$

$$\underset{\text{Light blue}}{[Cu(H_2O)_4]^{2+}} + 4NH_3 \rightleftharpoons \underset{\text{Deep blue-purple}}{[Cu(NH_3)_4]^{2+}} + 4H_2O$$

$$\underset{\text{Green}}{[Cr(H_2O)_3Cl_3]} + 3H_2O \rightleftharpoons \underset{\text{Violet}}{[Cr(H_2O)_6]^{3+}} + 3Cl^-$$

$$\underset{\text{Pink}}{[Co(H_2O)_6]^{2+}} + 4Cl^- \rightleftharpoons \underset{\text{Deep blue}}{[CoCl_4]^{2-}} + 6H_2O$$

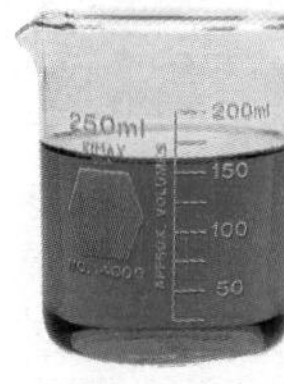

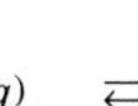

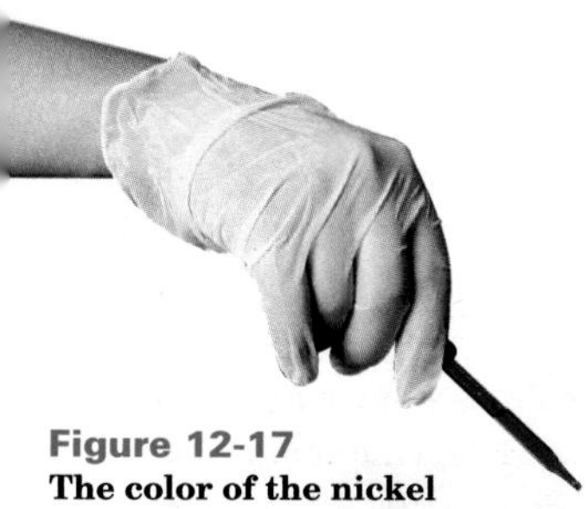

Figure 12-17
The color of the nickel complex makes it easy to see which complex ion is favored in the solution. This reaction can be reversed by adding more water to the nickel ammonia complex to return it to the green solution.

$$6NH_3(aq) + [Ni(H_2O)_6]^{2+}(aq) \rightleftharpoons [Ni(NH_3)_6]^{2+}(aq) + 6H_2O(l)$$

Historical Note

German-Swiss chemist Alfred Werner developed the theory of coordinated compounds (complex ions) in 1891. According to Werner, the idea came to him in his sleep, waking him with a start at 2 A.M.

ASSESS

Alternative Assessment

Ask students to suggest how Haber used Le Châtelier's principle to obtain ammonia from elemental nitrogen and hydrogen. All three components are gases.

$$N_2 + 3H_2 \rightleftharpoons 2NH_3$$

(High pressure accompanied by high temperature favors the side of the reaction that produces the smallest volume of gas. Because the molar volume of a gas is constant, the side of the equation with the fewest moles of gas will be favored. There are 4 mol of reactants (nitrogen, hydrogen) but only 2 mol of ammonia, so ammonia production is favored, and equilibrium shifts to the right. Removal of the ammonia will shift the reaction further to the right.)

Visual Strategy

Figure 12-17

Make students aware of the change in the structural formula of $[Ni(H_2O)_6]^{2+}$ to $[Ni(NH_3)_6]^{2+}$. Stress that the color changes that occur in complex ion interactions when concentrations change help to illustrate the establishment of new states of equilibrium.

Answers to Section Review

11. Gaseous carbon dioxide escapes from the reaction, so equilibrium (essential for reaction reversibility) is never reached.
12. System equation:

$$Na^+(aq) + Cl^-(aq) \underset{}{\overset{H_2O}{\rightleftharpoons}} NaCl(s) + \text{heat.}$$

As salt recrystallizes, the equilibrium position moves toward the right.

13. a. Equilibrium shifts to the right, producing more $[Cu(NH_3)_4]^{2+}$.
    b. Equilibrium shifts to the right, producing $[CoCl_4]^{2-}$.
    c. AgCl is insoluble, so adding $AgNO_3$ to the solution removes Cl^- by forming AgCl. Equilibrium moves to the right, producing more $[Cr(H_2O)_6]^{3+}$ and Cl^-.
    d. Equilibrium shifts to the right, producing more H_3O^+ and $[Zn(H_2O)_3(OH)]^+$.
    e. Equilibrium shifts to the right as acidic H_3O^+ is neutralized, water is formed, and $[Zn(H_2O)_3(OH)]^+$ is produced.
14. Any process that shifts the reaction in item **12** far to the right would produce a complete reaction. Examples of this are distillation or cooling the solution until the salt recrystallizes.

Figure 12-18
The color changes on the treated cloth for a simple weather indicator are shown for both dry and humid conditions.

Practical uses of complex ions: weather prediction and stain removal

Have you ever seen a weather indicator that appears blue when the weather is supposed to stay dry and turns pink when it is supposed to rain? The simple little item involves some interesting chemistry. The color changes are shown in **Figure 12-18**. The weather indicator has a piece of fabric or paper that has been soaked in a solution of cobalt(II) chloride, $CoCl_2$. When the moisture in the air is low, the cobalt(II) chloride forms a tetrahedral complex, $[CoCl_4]^{2-}$, that is blue. When the humidity is high, the $[CoCl_4]^{2-}$ reacts with water and forms a complex ion that has six water molecules attached to cobalt, $[Co(H_2O)_6]^{2+}$, and that ion appears pink. The color change in the weather indicator can be explained by Le Châtelier's principle and the following chemical equation.

$$\underset{\text{Blue}}{[CoCl_4]^{2-}} + 6H_2O \rightleftharpoons \underset{\text{Pink}}{[Co(H_2O)_6]^{2+}} + 4Cl^-$$

Rust stains are difficult to remove from fabrics in regular washing but can be bleached out with an oxalic acid solution. Iron oxide combines with oxalic acid to form a complex ion, $[Fe(C_2O_4)_3]^{3-}$, as shown by the following equation.

$$\underset{\text{Rust}}{Fe_2O_3(s)} + \underset{\text{Oxalic acid}}{6H_2C_2O_4(aq)} + 3H_2O(l) \rightleftharpoons \underset{\text{Complex ion}}{2[Fe(C_2O_4)_3]^{3-}(aq)} + 6H_3O^+(aq)$$

The complex ion is soluble and can be washed off the fabric with detergent. Oxalic acid works in a similar manner to remove some iron-based ink stains on fabrics, which would not ordinarily wash out with detergents alone. Lemon juice, containing citric acid, can usually produce the same results as oxalic acid.

Section Review

11. Explain why the following reaction runs to completion rather than being reversible.

$$H_2CO_3(aq) \rightarrow H_2O(l) + CO_2(g)$$

12. When a hot, saturated solution of salt is allowed to cool, any excess salt could recrystallize as the solution cools. Explain what is happening in terms of the solution equilibrium if the salt does recrystallize.
13. Using the equations for complex ions (on page 467), make a hypothesis about how each system would change under the following conditions.
    a. Adding NH_3 to the copper complex ion system.
    b. Removing $[CoCl_4]^{2-}$ from the cobalt complex ion system.
    c. Adding $AgNO_3$ to the chromium complex ion system. (Hint: check your solubility table.)
    d. Adding water to the zinc complex ion system.
    e. Adding a base to the zinc complex ion system.

14. **Story Link**
    What would be one way to separate water from salt that would be an example of a reaction that goes to completion? Explain the process.

Visual Strategy

Figure 12-18

The cobalt chloride humidity indicator is an excellent example of how some complex ion behavior is dependent on water concentration. Many complex ion reactions require water as a ligand and therefore depend on water concentration (e.g., humidity) to determine the position of equilibrium, which in turn determines the ion's chemical structure and color.

How is equilibrium measured?

Section Objectives

Write K_{eq} expressions for any reaction in equilibrium.

Use K_{eq} values to calculate concentrations of reactants or products.

Write K_{sp} expressions for the solubility of slightly soluble salts.

Calculate K_{sp} values given the concentrations of ions in aqueous equilibrium.

The equilibrium constant K_{eq}

Look again at the reaction of acetic acid and water.

$$CH_3COOH(aq) + H_2O(l) \rightleftharpoons CH_3COO^-(aq) + H_3O^+(aq)$$

At equilibrium, the reactants continue to form the products, and the products are reacting to re-form the reactants at the same rate. So, even though reactions continue to occur at equilibrium, the rates are equal and effectively cancel each other out, resulting in no net chemical change. Consequently, the equilibrium concentrations of reactants and products remain constant. At equilibrium, the forward rate ($rate_f$) equals the reverse rate ($rate_r$), with the equilibrium point based on a specific temperature and pressure.

Doug never thought joining the circus would be this hard!

K_{eq} quantitatively describes equilibrium

When working with equilibrium systems, we make calculations involving the concentrations of the products and the reactants. We express the concentration of a substance by placing it in brackets. For example, to specify the concentration of OH^- ions in water as 1.0×10^{-7} M, you write it as $[OH^-] = 1.0 \times 10^{-7}$ M. When solving equilibrium problems, the formula or numerical value for a compound or ion is enclosed in brackets to represent a concentration expressed in mol/L or M.

The following is the reaction for acetic acid and water.

$$CH_3COOH(aq) + H_2O(l) \rightleftharpoons CH_3COO^-(aq) + H_3O^+(aq)$$

Chem Fact

Over two-thirds of the acetic acid manufactured is used in the production of either vinyl acetate or cellulose acetate.

The rate of the forward reaction ($rate_f$) can be expressed using the product of the constant, k_f, and the concentrations of the reactants.

$$rate_f = k_f\,[CH_3COOH][H_2O]$$

The term k_f is a constant for the forward reaction and depends on the size and kind of ion or molecule involved in the reaction. The concentrations of the reactants are enclosed in brackets.

Similarly, the rate of the reverse reaction ($rate_r$) can be expressed by the following equation.

$$rate_r = k_r\,[CH_3COO^-]\,[H_3O^+]$$

The term k_r is a constant for the reverse reaction and depends on the size and kind of ion or molecule involved in the reaction. The concentrations of the products are enclosed in brackets.

Section 12-3

FOCUS

Lesson Starter

Begin the lesson by mentioning the old (and not very safe) practice of siphoning gasoline from a fairly full tank to an empty gas can.

Demonstration 7

Kinetics simulation

Approximate time: 15 min

1. Fill a shallow container half full with water to which food coloring has been added, and submerge a 60 cm piece of rubber tubing in it.
2. Place 450 mL of colored water (using the same food coloring as before) into a 600 mL beaker. Set the beaker on a raised, stable support about 5 cm high.
3. Place a second 600 mL beaker next to and 4 cm below the first.
4. Place a finger over each end of the tubing, put one end into the beaker with colored water, and remove that finger. Place the other end into the empty beaker, and let the water flow.
5. Discuss with students when and where the flow will stop. (By raising either beaker, one can see the system's reversibility.)
6. Remove water from the lower beaker to show that water will flow into it again.

SAFETY

Safety goggles must be worn. Students must be 10 ft or more from the demonstration.

DISPOSAL

Pour water down the sink.

[Source: Hansen and Krause. *J. Chem. Ed.*, Vol. 61, p. 804, 1984.]

TEACH

Content Background

Rate constants

Make students aware that the rate constants, k_f and k_r, are characteristic of the conditions and reaction in question. Every reaction will have a different rate constant.

Possible Misconception

Stress to students that even though capital and lowercase forms of the letter k are used, the rate and equilibrium constants represent different concepts and should not be confused with each other. They are related *only* because they both depend on the concentrations of products or reactants, so the equilibrium constant, K_{eq}, equals the ratio of the forward-reaction rate constant to the reverse-reaction rate constant, or k_f / k_r.

Teaching Tip

Reinforcement

Remind students that equilibrium does not mean that the amounts of reactants and products are equal. It means that the rates of conversion (the number of reactions per unit of time) are equal. Only a K_{eq} of 1 would mean that the products of the concentrations of reactants are equal to the products of the concentrations of products.

Historical Note

Norwegian chemists Cato Maximilian Guldberg (1836–1902) and Peter Waage (1833–1900) published the idea of equilibrium constants in 1863. Unfortunately, their work was published in Norwegian, and it went unnoticed until 1879, when it was translated into German.

Because the two rates are equal at equilibrium, you can show the two equations as equal.

$$\mathbf{rate_f = rate_r}$$

We can substitute the expressions in the equation as follows.

$$k_f\,[CH_3COOH][H_2O] = k_r\,[CH_3COO^-][H_3O^+]$$

At this point, we can simplify all of these ideas into a workable equation. If you divide both sides by the same expression, $k_r\,[CH_3COOH][H_2O]$, you get an expression that can then be simplified to the following equation.

$$\frac{k_f}{k_r} = \frac{[CH_3COO^-][H_3O^+]}{[CH_3COOH][H_2O]}$$

You can then rewrite this equation as the final relationship for the equilibrium expression because the ratio k_f/k_r is also a constant.

$$K_{eq} = \frac{[CH_3COO^-][H_3O^+]}{[CH_3COOH][H_2O]}$$

equilibrium constant, or K_{eq} *for a reversible reaction, a number expressing the relationship between the mathematical product of the molar concentrations of the products divided by the mathematical product of the molar concentrations of the reactants, each raised to the power of its coefficient in the balanced equation*

K_{eq} is known as the **equilibrium constant**. The equilibrium constant is the ratio (at equilibrium) of the mathematical product of the concentrations of the products to the mathematical product of the concentrations of the reactants.

Each concentration is raised to the power of its coefficient in the equation

A chemical equation can be written in the following general form.

$$\mathbf{aA + bB \rightleftharpoons cC + dD}$$

The equilibrium constant for this equation is written as follows.

$$K_{eq} = \frac{[C]^c[D]^d}{[A]^a[B]^b}$$

Note that the products, represented as C and D, appear in the numerator of the expression, while the reactants, represented as A and B, appear in the denominator. Also notice that each concentration is raised to the power equal to the coefficient of the substance in the equation. Each coefficient in a chemical equation becomes a superscript in the equilibrium constant expression for that equation. When a reaction occurs in dilute aqueous solution, the concentration of water is omitted. The $[H_2O]$ is almost constant and does not need to be included in the K_{eq} calculation.

Determining K_{eq} for substances at equilibrium

In chemistry, we can express the equilibrium constant for any chemical reaction using the general equation. For example, the K_{eq} expression for the formation of ammonia can be determined from its equation as follows.

$$N_2(g) + 3H_2(g) \rightleftharpoons 2NH_3(g)$$

$$K_{eq} = \frac{[NH_3]^2}{[N_2][H_2]^3}$$

Table 12-4
Equilibrium Constants

Equation	K_{eq} value	Temperature
$N_2O_4(g) \rightleftharpoons 2NO_2(g)$	5.6×10^{-3}	0°C
	5.9×10^{-3}	25°C
	8.3×10^{-1}	55°C
$CO_2(g) + H_2(g) \rightleftharpoons CO(g) + H_2O(g)$	2.0	1120°C
	4.40	1727°C
$N_2(g) + O_2(g) \rightleftharpoons 2NO(g)$	4.5×10^{-31}	25°C
	6.7×10^{-10}	627°C
	1.7×10^{-3}	2027°C
$CH_3COOH + H_2O \rightleftharpoons H_3O^+ + CH_3COO^-$	1.8×10^{-5}	25°C
$2HI(g) \rightleftharpoons H_2(g) + I_2(g)$	1.5×10^{-2}	350°C
	1.8×10^{-2}	425°C
	2.2×10^{-2}	1123°C

The actual value for a particular K_{eq} is found experimentally. A chemist analyzes the equilibrium mixture and calculates the concentrations of all substances involved in the reaction. When determining the K_{eq} value for a particular reaction, the chemist must also specify the temperature under which the concentrations are measured. The K_{eq} value will change if the temperature changes. **Table 12-4** gives some values for equilibrium constants. You can see that the same reactions have different values at different temperatures.

Writing equilibrium expressions

15. You can write an equilibrium expression for any reaction that is balanced by using the general equation for the equilibrium constant. Write equilibrium expressions for the following reactions.
    a. $HOCl(aq) \rightleftharpoons H^+(aq) + ClO^-(aq)$
    b. $[Cu(H_2O)_4]^{2+}(aq) + 4NH_3(aq) \rightleftharpoons [Cu(NH_3)_4]^{2+}(aq) + 4H_2O(l)$
16. Write equilibrium expressions for all reactions listed in **Table 12-4**.

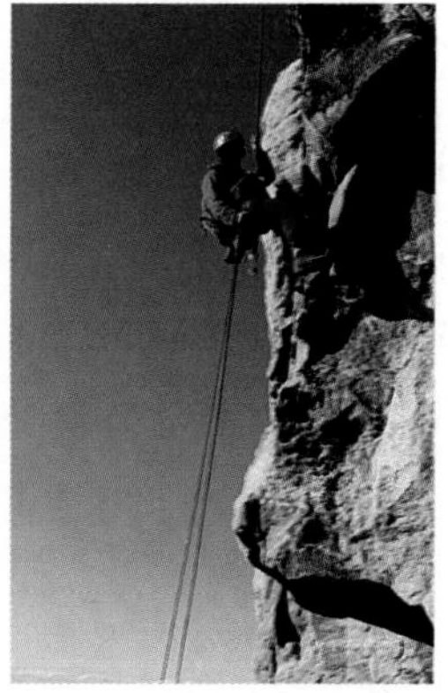

Figure 12-19
An example of equilibrium is a descent down the face of a cliff, involving a double-rope system secured above the person in a method called "rappelling."

Using the value of K_{eq} to determine which direction is favored in a reaction

Rappelling, shown in **Figure 12-19**, is an example of opposing forces maintaining an equilibrium. If the two rope forces are equal, there is no movement for the climber. This situation is similar to having a K_{eq} value of one in a chemical equation. For a K_{eq} value of one, the product of the concentrations of the products (the numerator) and the product of the concentrations of the reactants (the denominator) have the same value. (Remember, these equilibrium concentrations are raised to the correct powers.)

What does a K_{eq} value that is greater than one tell you about which reaction is favored at equilibrium? In reactions that have K_{eq} values larger than one, the forward reaction is favored. For reactions that have K_{eq} values smaller than one, the reverse reaction is favored. Consequently, by knowing the value of K_{eq} for a reaction, you know which reaction is favored. Using known values of K_{eq}, we can make calculations and predictions about equilibrium systems without having to do lab work or make actual measurements.

Possible Misconception

Emphasize that if the reaction was written in the reverse order, the new K_{eq} would be the inverse of the old K_{eq}.

Teaching Tip

Application
Point out that K_{eq} values are measured by doing a systematic study of the reaction system. K_{eq} can be used to predict the favored direction of a reaction.

Answers to
Concept Review

15. a. $K_{eq} = \frac{[ClO^-][H^+]}{[HClO]}$

b. $K_{eq} = \frac{[[Cu(NH_3)_4]^{2+}]}{[[Cu(H_2O)_4]^{2+}]\,[NH_3]^4}$

16. a. $K_{eq} = \frac{[NO_2]^2}{[N_2O_4]}$

b. $K_{eq} = \frac{[CO][H_2O]}{[CO_2][H_2]}$

c. $K_{eq} = \frac{[NO]^2}{[N_2][O_2]}$

d. $K_{eq} = \frac{[CH_3COO^-][H_3O^+]}{[CH_3COOH]}$

e. $K_{eq} = \frac{[H_2][I_2]}{[HI]^2}$

Visual Strategy

Figure 12-19
In rappelling, the equilibrium is maintained by balancing the downward force of gravity with the upward pull from the climber's arm muscles.

Content Background
Biochemical equilibrium
This process can be explained by realizing that the equilibrium expression is a constant proportion: the product of the product concentrations divided by the product of the reactant concentrations, or $K_{eq} = [p]/[r]$. If something is added to increase $[r]$, this change causes a stress to the equilibrium because the value for K_{eq} is now smaller than it should be. Some of the reactants will change into products, raising $[p]$ and lowering $[r]$, which will raise the value of K_{eq} back to its original value. Thus, by adding reactant or removing product, the formation of more product is favored.

Theme
Equilibrium and change
The body is constantly trying to establish equilibrium conditions in the many reactions that occur within it.

Content Background
pH and body chemistry
The kidneys are very sensitive to the pH of the blood and will begin removing and neutralizing the H_3O^+. The neutralized acid then fills the bladder. This accounts for a person's need to urinate immediately after a CO_2-producing activity such as vigorous crying or strenuous exercise.

Carbon dioxide and carbonic acid are part of a biochemical equilibrium system

Respiration is an essential function of living organisms. Carbon dioxide given off by your cells during respiration diffuses into your blood, where some of it combines with water to form the compound carbonic acid, H_2CO_3.

$$CO_2(g) + H_2O(l) \rightleftharpoons H_2CO_3(aq)$$

In turn, H_2CO_3 reacts with the water that makes up most of the plasma of your blood. The equation for this reaction is written as follows.

$$H_2CO_3(aq) + H_2O(l) \rightleftharpoons H_3O^+(aq) + HCO_3^-(aq)$$

This process is illustrated in **Figure 12-20**, showing how your breathing and blood plasma pH level are part of a biochemical equilibrium system.

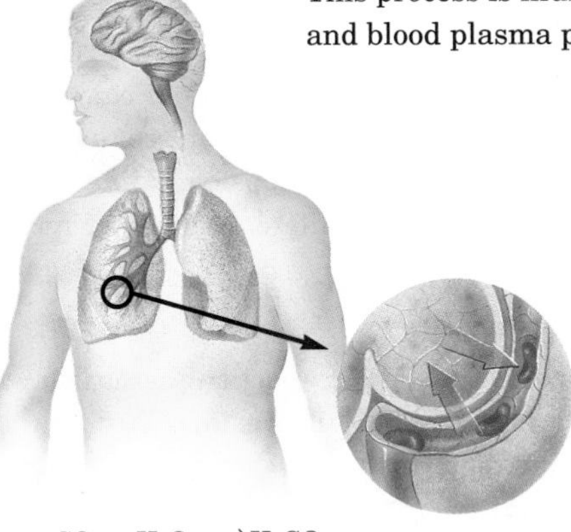

$CO_2 + H_2O \rightleftharpoons H_2CO_3$
$H_2CO_3 + H_2O \rightleftharpoons H_3O^+ + HCO_3^-$

Figure 12-20a
In solution, carbonic acid exists in equilibrium with H_2O and CO_2. However, this equilibrium can be altered, depending on the nature of the stress.

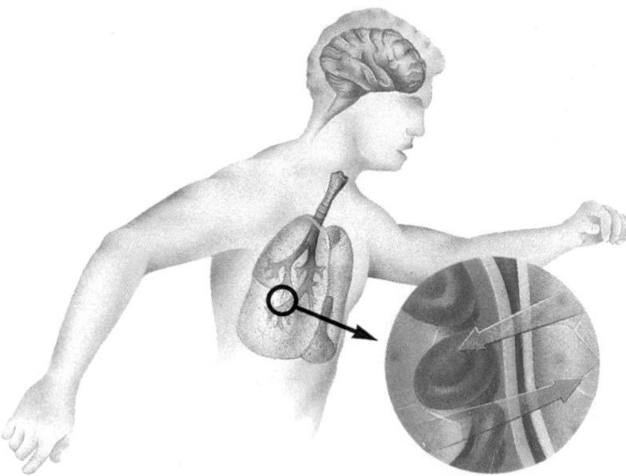

$CO_2 + H_2O \rightleftharpoons H_2CO_3$
$H_2CO_3 + H_2O \rightleftharpoons H_3O^+ + HCO_3^-$

12-20b
If you exercise vigorously, more CO_2 is produced in your body. An increase in the concentration of CO_2 will place stress on the first equation, forcing the equilibrium to shift to the right so that the formation of H_2CO_3 will be favored. As the concentration of H_2CO_3 increases, stress is placed upon the second equation, forcing the equilibrium to shift to the right so that more H_3O^+ and HCO_3^- are produced.

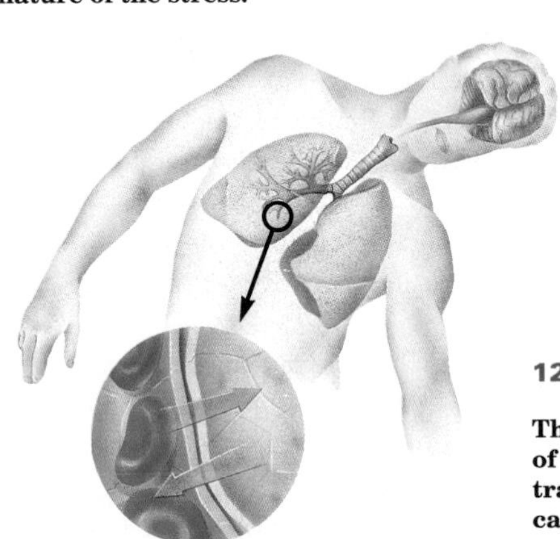

$CO_2 + H_2O \rightleftharpoons H_2CO_3$
$H_2CO_3 + H_2O \rightleftharpoons H_3O^+ + HCO_3^-$

12-20c

The net result of an increase in your rate of respiration is an increase in the concentration of both the H_3O^+ and HCO_3^- ions carried in your blood. The pH level of your blood plasma decreases due to the increased hydronium ion concentration. Your brain sends a message to exhale CO_2 more rapidly in order to return the system to its original state.

Visual Strategy
Figure 12-20
Stress that these reactions may be written as the following overall reaction.

$$CO_2 + 2H_2O \rightleftharpoons H_3O^+ + HCO_3^-$$

Your body is sensitive to even small changes in the pH level of your blood plasma. When a drop in the pH level is detected, your brain responds by sending a signal to increase your rate of breathing. The faster you breathe, the more CO_2 you exhale. By exhaling more CO_2, the initial stress of an increased CO_2 concentration is reduced, and the original equilibrium will be reestablished.

Suppose, however, that someone is hyperventilating. As a result of breathing very rapidly, the person exhales CO_2 at a very rapid rate, shifting both equilibria reactions to the left. As a result, the concentration of both H_3O^+ and HCO_3^- ions drops, and the pH level of the blood plasma rises. Your body responds by constricting the cerebral blood vessels, which reduces the flow of blood to your brain and causes dizziness. When normal breathing is restored, as depicted in **Figure 12-21**, the pH level returns to normal, and there is no longer a feeling of dizziness.

Figure 12-21
Hyperventilation can cause dizziness. A person with this condition should sit down and attempt to restore normal breathing by lowering the head and breathing slowly into cupped hands or a small paper bag.

Sample Problem 12A
Calculating K_{eq} for the carbonic acid system at equilibrium

At a temperature of 25°C, the following concentrations of the reactants and products for the reaction involving carbonic acid and water are present: $[H_2CO_3] = 3.3 \times 10^{-2}$ M, $[H_3O^+] = 1.1 \times 10^{-5}$ M, and $[HCO_3^-] = 7.1 \times 10^{-1}$ M. The $[H_2O]$ is not considered because it is constant. What is the K_{eq} value for the following reaction at equilibrium in a dilute aqueous solution?

$$H_2CO_3(aq) + H_2O(l) \rightleftharpoons H_3O^+(aq) + HCO_3^-(aq)$$

1 List

- $[H_2CO_3] = 3.3 \times 10^{-2}$ M
- $[H_3O^+] = 1.1 \times 10^{-5}$ M
- $[HCO_3^-] = 7.1 \times 10^{-1}$ M
- $K_{eq} = ?$

2 Set up

- **Write the K_{eq} expression for this reaction.**

$$K_{eq} = \frac{[H_3O^+][HCO_3^-]}{[H_2CO_3]}$$

- **Substitute the given values.**

$$K_{eq} = \frac{[1.1 \times 10^{-5}][7.1 \times 10^{-1}]}{[3.3 \times 10^{-2}]}$$

3 Estimate and calculate

- **A rough estimate of the answer is approximately 2.5×10^{-4}.**

$$K_{eq} = \frac{[1.1 \times 10^{-5}][7.1 \times 10^{-1}]}{[3.3 \times 10^{-2}]} = 2.4 \times 10^{-4}$$

- **The K_{eq} value for this reaction is 2.4×10^{-4}, which is consistent with the rough estimate.**

Content Rationale

The equilibrium problems in this section give concentration values at equilibrium. In most equilibrium problems, the initial reactant and product concentrations are given, so that with the value for K_{eq} the equilibrium concentrations can be determined. If the change between initial and equilibrium concentrations is defined as x, reactant concentrations decrease and product concentrations rise by an amount nx, where n is the coefficient number for that particular reactant or product. For example, for the reaction

$$2HI \rightleftharpoons H_2 + I_2$$

the equilibrium expression is

$$K_{eq} = \frac{([H_2]_0 + x)([I_2]_0 + x)}{([HI]_0 - 2x)^2}$$

where $[HI]_0$, $[H_2]_0$, and $[I_2]_0$, are the initial concentrations of reactant and products, respectively. Often the initial product concentrations equal zero. For $K_{eq} < 10^{-6}$, x is so small that the reactant concentrations are effectively unchanged. (See Sample Problem 13F for an example in which x is significant.) Item 1 of Using Technology, Chapter 12 Review, provides an example for solving problems of this type.

Additional Example for Sample Problem 12A

The Haber process enabled Germany to manufacture ammonia for explosives during World War I. At a pressure of 1000 atm and temperature of 723 K, the reaction for the Haber process is as follows.

$$N_2 + 3H_2 \rightleftharpoons 2NH_3$$

If the concentration of nitrogen is 2.95 M, the concentration of the hydrogen is 7.68 M, and the concentration of the ammonia is 5.78 M, what is the K_{eq} value for this reaction?

Solution is at the bottom of the page.

Solution for Additional Example 12A

$$K_{eq} = \frac{[NH_3]^2}{[N_2][H_2]^3} = \frac{[5.78]^2}{[2.95][7.68]^3} = 2.50 \times 10^{-2}$$

Additional Examples for Sample Problem 12B

a. The K_{eq} for the reaction

$$N_2O_4 \rightleftarrows 2NO_2$$

at room temperature is 1.70×10^2. If the concentration of N_2O_4 is 2.30×10^{-1} M, what is the concentration of NO_2?

b. At 2027°C the K_{eq} for the reaction

$$N_2 + O_2 \rightleftarrows 2NO$$

is 1.70×10^{-3}. If the concentration of NO is 0.523 M, what is the concentration of O_2?

Solutions are at the bottom of the page.

Answers to Practice 12A and 12B

1. $K_{eq} = \frac{[CO_2][H_2]}{[CO][H_2O]} = \frac{[0.012][0.012]}{[0.010][0.020]} =$ 0.720

2. $K_{eq} = \frac{[NO_2]^2}{[N_2O_4]} = \frac{[1.2 \times 10^{-1}]^2}{[4.0 \times 10^{-2}]} =$ 3.6×10^{-1}

3. $K_{eq} = \frac{[CH_3OH]}{[CO][H_2]^2}$

$290 = \frac{[CH_3OH]}{[0.025][0.080]^2}$

$[CH_3OH] = 0.046$ mol/L

4. $K_{eq} = \frac{[NO]^2}{[N_2][O_2]}$

1.7×10^6

$\frac{[NO]^2}{[1.8 \times 10^{-3}][4.2 \times 10^{-4}]} =$

$\frac{[NO]^2}{7.56 \times 10^{-7}}$

or

$[NO] = \sqrt{(1.7 \times 10^6)(7.56 \times 10^{-7})} =$ 1.1 mol/L

Sample Problem 12B
Calculating concentrations using the equilibrium constant

Ammonia gas is extremely soluble in water. An aqueous solution of ammonia is known as ammonium hydroxide.

$$NH_3(g) + H_2O(l) \rightleftarrows NH_4^+(aq) + OH^-(aq)$$

K_{eq} for this reaction equals 1.8×10^{-5} at a temperature of 298 K. If the equilibrium concentration of NH_3 is 6.82×10^{-3} M, calculate the concentration of ammonium ion at equilibrium. The concentration of H_2O is constant and need not be included in the K_{eq} calculations.

1 List

- $K_{eq} = 1.8 \times 10^{-5}$
- $[NH_3] = 6.82 \times 10^{-3}$ M
- $[NH_4^+] = [OH^-] = ?$

2 Set up

- **Write the K_{eq} expression for this reaction.**

$$K_{eq} = \frac{[NH_4^+][OH^-]}{[NH_3]}$$

- **Substitute the known values in this equation.**

$$1.8 \times 10^{-5} = \frac{[NH_4^+][OH^-]}{[6.82 \times 10^{-3}]}$$

- **Then solve for $[NH_4^+][OH^-]$.**

$[NH_4^+][OH^-] = (1.8 \times 10^{-5})(6.82 \times 10^{-3})$

3 Estimate and calculate

- **Rounding the numbers for estimation, you get 14×10^{-8}.**
$[NH_4^+][OH^-] = (1.8 \times 10^{-5})(6.82 \times 10^{-3}) = 1.2 \times 10^{-7}$ M
- **The concentration of each ion, $[NH_4^+]$ and $[OH^-]$, is found by taking the square root of 1.2×10^{-7} M. (The concentrations of the two ions are equal.)**
$[NH_4^+] = \sqrt{1.2 \times 10^{-7}}$ M
- **The concentration of ammonium ion at equilibrium is 3.5×10^{-4} M.**

Practice 12A

Answers for Concept Review items and Practice problems begin on page 841.

1. **For the following reaction, equilibrium is established at a certain temperature when the following concentrations are present: [CO] = 0.010 mol/L, $[H_2O]$ = 0.020 mol/L, $[CO_2]$ = 0.012 mol/L, and $[H_2]$ = 0.012 mol/L. Calculate the K_{eq} value for this reaction.**

$$CO(g) + H_2O(g) \rightleftarrows CO_2(g) + H_2(g)$$

2. **For the system involving N_2O_4 and NO_2 at equilibrium at a temperature of 100°C, the concentration of N_2O_4 is 4.0×10^{-2} mol/L, and the concentration of NO_2 is 1.2×10^{-1}mol/L. What is the K_{eq} value for the reaction to form NO_2?**

12B

3. **Methanol can be prepared by the reaction of H_2 and CO at high temperatures, according to the following equation.**

$$CO(g) + 2H_2(g) \rightleftarrows CH_3OH(g)$$

What is the concentration of $CH_3OH(g)$ if $[H_2]$= 0.080 mol/L and [CO]= 0.025 mol/L at 700 K, and K_{eq} = 290?

4. **Using a K_{eq} value of 1.7×10^6 at 2027°C for the following reaction, what is the equilibrium concentration of nitrogen monoxide when the concentrations of nitrogen and oxygen at equilibrium are 1.8×10^{-3} mol/L and 4.2×10^{-4} mol/L, respectively?**

$$N_2(g) + O_2(g) \rightleftarrows 2NO(g)$$

Solutions for Additional Examples 12B

a. $K_{eq} = \frac{[NO_2]^2}{[N_2O_4]} = \frac{[NO_2]^2}{[2.30 \times 10^{-1}]} = 1.70 \times 10^2$; $[NO_2] = \sqrt{(1.70 \times 10^2)(2.30 \times 10^{-1})} = 6.25$ mol/L

b. In this problem the concentrations of N_2 and O_2 are equal, so $[O_2][N_2]$ may be written as $[O_2]^2$.

$K_{eq} = \frac{[NO]^2}{[N_2][O_2]} = \frac{[NO]^2}{[O_2]^2} = \frac{[0.523]^2}{[O_2]^2} = 1.70 \times 10^{-3}$; $[O_2]^2 = \frac{(0.523)^2}{1.70 \times 10^{-3}}$; $[O_2] = [N_2] = 12.7$ mol/L

The solubility-product constant K_{sp}

K_{sp} is an equilibrium constant for slightly soluble ionic substances

Recall from Chapter 11 that solubility refers to the maximum amount of a solute that will dissolve per unit volume of a solvent under various conditions. In Chapter 11, solubility was expressed in units of grams of solute per 100 g of water. As you will see in this section, solubility can also be expressed in terms of equilibrium expressions.

solubility-product constant, or K_{sp} *the equilibrium constant for a solid in equilibrium with its ions in a saturated solution; used for substances that are described as "insoluble" because they are only very slightly soluble*

In Chapter 11, you learned that salts differ in degree of solubility in water; some are highly soluble, some are only slightly soluble, and others are almost insoluble. Actually, most "insoluble" salts will dissolve to some extent in water and are more accurately described as being *very slightly soluble.* In such cases, an extremely small quantity of the salt saturates the solution, usually less than 0.1 g of salt per 100 g of water.

An example of a slightly soluble salt is barium sulfate, $BaSO_4$. The equation for this salt in aqueous equilibrium is the following.

$$BaSO_4(s) \rightleftharpoons Ba^{2+}(aq) + SO_4^{2-}(aq)$$

The K_{eq} expression is written as shown by the following equation.

$$K_{eq} = \frac{[Ba^{2+}][SO_4^{2-}]}{[BaSO_4]}$$

However, the concentration of $BaSO_4$ in the solid state is a constant. Consequently, this equation can be simplified and written as follows.

$$K_{eq} = [Ba^{2+}][SO_4^{2-}]$$

This is the solubility of a pure solid ionic compound, so the equilibrium constant is called the **solubility-product constant**, or K_{sp}. The expression now reads as follows.

$$K_{sp} = [Ba^{2+}][SO_4^{2-}]$$

Table 12-5 Solubility-Product Constants at 25°C

Salt	K_{sp}
Ag_2CO_3	8.45×10^{-12}
Ag_2CrO_4	1.12×10^{-12}
Ag_2S	1.09×10^{-49}
AgBr	5.35×10^{-13}
AgCl	1.77×10^{-10}
AgI	8.51×10^{-17}
$Al(OH)_3$	1.98×10^{-31}
$BaSO_4$	1.07×10^{-10}
$Ca_3(PO_4)_2$	2.07×10^{-33}
$CaSO_4$	7.10×10^{-5}
CuS	1.27×10^{-36}
$Fe(OH)_3$	2.64×10^{-39}
FeS	1.59×10^{-19}
$MgCO_3$	6.82×10^{-6}
$MnCO_3$	2.24×10^{-11}
$Ni_3(PO_4)_2$	4.73×10^{-32}
PbS	9.04×10^{-29}
$PbSO_4$	1.82×10^{-8}
$SrSO_4$	3.44×10^{-7}
$ZnCO_3$	1.19×10^{-10}
ZnS	2.93×10^{-25}

Each ion concentration is raised to the power of its coefficient in the equation

The equation for a slightly soluble ionic substance in a saturated solution can be written in the following general form.

$$A_aB_b(s) \rightleftharpoons aA^+ + bB^-$$

The general equation for the solubility-product constant, K_{sp}, is as follows.

$$K_{sp} = [A^+]^a[B^-]^b$$

Look at the K_{sp} values in **Table 12-5**. The K_{sp} value for a slightly soluble salt reveals the relative solubilities of its ions. Compounds with larger K_{sp} values are more soluble than those with smaller K_{sp} values.

Theme

Classification and trends

The classification of certain salts as insoluble or slightly soluble permits the amount of ions in solution to be predicted from the solubility-product constants.

Other Applications

Note that because of low K_{sp} values, chemicals precipitate out of solution in massive deposits as the oceans dry up. This has occurred throughout geologic history, such as when limestone, $CaCO_3$, and gypsum, $CaSO_4$, were deposited in thick layers as the seas slowly evaporated.

475

Additional Example for Sample Problem 12C

At 25°C the K_{sp} for the dissociation of Ag_2CO_3 is 8.45×10^{-12}. What is the ion concentration of Ag^+ if the CO_3^{2-} concentration is 3.40×10^{-5} M?

Solution is at the bottom of the page.

Keep in mind that the K_{sp} expression has limited applications. The K_{sp} equation can be very useful for calculating the concentrations of ions formed from slightly soluble salts. However, K_{sp} cannot be applied successfully to salts that are more soluble. These simple equations do not hold when concentrations of ions are high, as is the case of highly soluble salts.

Sample Problem 12C Calculating concentrations using K_{sp}, the solubility-product constant

Consider the dissociation of the salt CaF_2.

$CaF_2(s) \rightleftarrows Ca^{2+}(aq) + 2F^-(aq)$

At a temperature of 298 K, the concentration of Ca^{2+} ions is 2.2×10^{-4} M, and K_{sp} for CaF_2 is 1.46×10^{-10}. Calculate the concentration of F^- in the solution.

1 List

- $K_{sp} = 1.46 \times 10^{-10}$
- $Ca^{2+} = [2.2 \times 10^{-4}]$
- $F^- = ?$

2 Set up

- **Write the K_{sp} expression for the reaction.**
  $K_{sp} = [Ca^{2+}][F^-]^2$
- **Substitute the known values into this equation.**
  $1.46 \times 10^{-10} = [2.2 \times 10^{-4}][F^-]^2$
- **Rearrange to solve for $[F^-]^2$.**
  $[F^-]^2 = \dfrac{1.46 \times 10^{-10}}{2.2 \times 10^{-4}}$
- **Solve for $[F^-]$ by determining the square root of the expression.**
  $[F^-]^2 = 0.664 \times 10^{-6}$ (determine the square root)

Refer to the instruction book for your calculator to find out how to determine the square root of a number with an exponent.

3 Estimate and calculate

- **To approximate the square root, change the value to 67×10^{-8}, which has a square root of roughly 8×10^{-4}.**
  $[F^-]^2 = 0.664 \times 10^{-6}$ (determine the square root)

  $[F^-] = 8.1486195 \times 10^{-4}$, which is rounded to 8.1×10^{-4}

  $[F^-] = 8.1 \times 10^{-4}$
- **The F^- ion concentration is 8.1×10^{-4} M.**

Practice 12C

1. **What is the K_{sp} value for $Co_3(PO_4)_2$ if the concentrations at equilibrium and standard temperature are determined to be 1.60×10^{-6} M for Co^{2+} ions and 2.24×10^{-9} M for PO_4^{3-} ions?**
2. **At 298 K, the K_{sp} value for barium carbonate is 2.58×10^{-9}. What is the concentration of barium ions in a saturated solution of this salt? The concentration of CO_3^{2-} ions is 7.12×10^{-5} M, and the dissociation equation is as follows.**
   $BaCO_3(s) \rightleftarrows Ba^{2+}(aq) + CO_3^{2-}(aq)$
3. **The K_{sp} value for silver carbonate is 8.14×10^{-12} at 298 K. The concentration of carbonate ions is 2.85×10^{-6} M; what is the concentration of silver ions?**

Answers to Practice 12C

1. $K_{sp} = [Co^{2+}]^3[PO_4^{3-}]^2 = [1.60 \times 10^{-6}]^3[2.24 \times 10^{-9}]^2 = (4.10 \times 10^{-18})(5.02 \times 10^{-18}) = 2.06 \times 10^{-35}$
2. $K_{sp} = [Ba^{2+}][CO_3^{2-}]$
   $2.58 \times 10^{-9} = [Ba^{2+}][7.12 \times 10^{-5}]$
   or
   $[Ba^{2+}] = \dfrac{2.58 \times 10^{-9}}{7.12 \times 10^{-5}} = 3.62 \times 10^{-5}$ M
3. $Ag_2CO_3 \rightleftarrows 2Ag^+ + CO_3^{2-}$
   $K_{sp} = [Ag^+]^2[CO_3^{2-}]$
   $8.14 \times 10^{-12} = [Ag^+]^2[2.85 \times 10^{-6}]$
   or
   $[Ag^+] = \dfrac{\sqrt{8.14 \times 10^{-12}}}{\sqrt{2.85 \times 10^{-6}}} = 1.69 \times 10^{-3}$ M

Solution for Additional Example 12C

$Ag_2CO_3 \rightleftarrows 2Ag^+ + CO_3^{2-}$;
$K_{sp} = [Ag^+]^2[CO_3^{2-}] = 8.45 \times 10^{-12} = [Ag^+]^2[3.40 \times 10^{-5}]$

or

$[Ag^+] = \dfrac{\sqrt{8.45 \times 10^{-12}}}{\sqrt{3.40 \times 10^{-5}}} = 4.99 \times 10^{-4}$ M

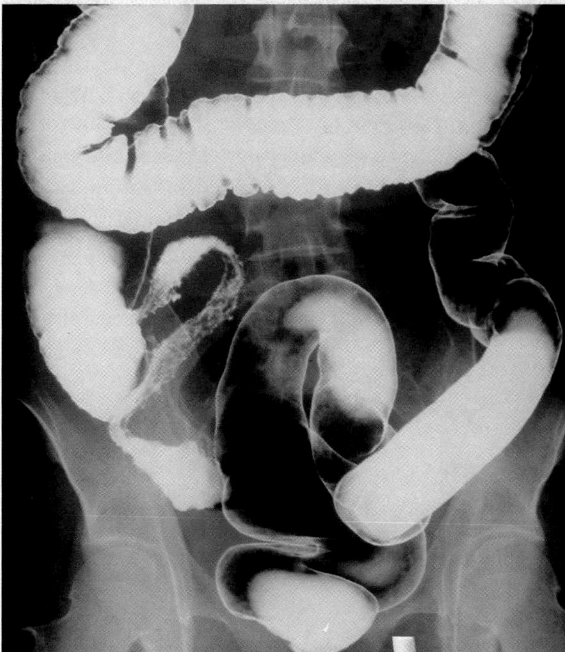

Figure 12-22a
The light areas on an X ray of the digestive tract show the insoluble barium sulfate.

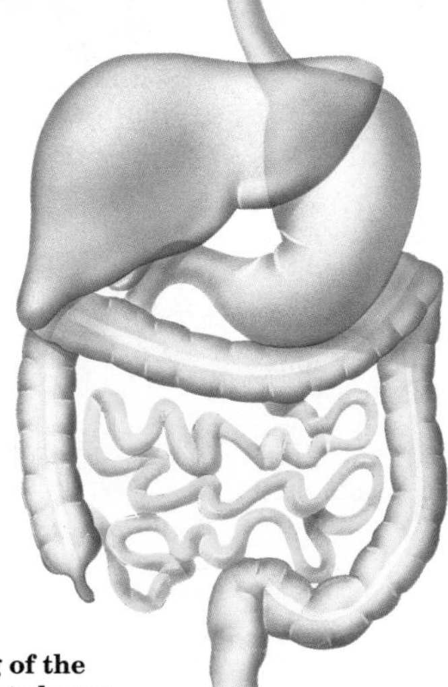

12-22b
This drawing of the digestive tract shows the same area that is shown in the X ray.

Common ion effect

Practical uses of equilibrium constants

Barium sulfate, $BaSO_4$, is an insoluble salt, which means that very little of it will actually dissolve in solution. Doctors use $BaSO_4$ as an X-ray contrast medium. The patient ingests solid $BaSO_4$ powder that is suspended in water and will appear as light areas on the X-ray film. By studying these light areas, like those shown in **Figure 12-22**, doctors can diagnose problems in the patient's digestive tract. Doctors must make sure that almost no Ba^{2+} ions are present to dissolve in a person's body fluids because Ba^{2+} ions are poisonous. Knowing the solubility-product constant (K_{sp}) for $BaSO_4$, calculations are made to determine the amount of $BaSO_4$ needed for the procedure. To ensure that the concentration of Ba^{2+} ions will not exceed the safety level, a soluble salt such as Na_2SO_4, which also contributes SO_4^{2-} ions, is added to increase the SO_4^{2-} ion concentration and reduce the Ba^{2+} ion concentration.

Na_2SO_4 dissociates as shown by the following equation.

$$Na_2SO_4(s) \longrightarrow 2Na^+(aq) + SO_4^{2-}(aq)$$

Increasing the concentration of Na_2SO_4 results in an increase in the concentration of SO_4^{2-} ions in the solution. These SO_4^{2-} ions affect the chemical reaction involving the dissociation of $BaSO_4$. Examine the following equation.

$$BaSO_4(s) \rightleftharpoons Ba^{2+}(aq) + SO_4^{2-}(aq)$$

common ion
an ion that comes from two or more substances making up a chemical solution

An increase in the SO_4^{2-} ions shifts the equilibrium to the left, and this shift causes Ba^{2+} ions to combine with SO_4^{2-} ions, forming solid $BaSO_4$. By having $BaSO_4$ precipitate out of solution, there is a lower concentration of Ba^{2+} ions in the solution. Adding the Na_2SO_4 has the effect of reducing the concentration of Ba^{2+} ions in the solution. Because SO_4^{2-} ion is present in both $BaSO_4$ and Na_2SO_4, sulfate is the **common ion** present in both substances dissolved in the solution.

Demonstration 8

Common ion effect

Approximate time: 5 min

This demonstration shows how an excess of ammonium ions, NH_4^+, reduces the concentration of hydroxide ions, OH^-.

1. Place a petri dish on an overhead projector, and fill it halfway with 1 M NH_4OH.
2. Add 2 drops of phenolphthalein. The solution should be red.
3. Stir a little solid NH_4Cl into the solution; the red color will disappear. The NH_4^+ ions suppress the ionization of the aqueous ammonia.

SAFETY

Safety goggles, lab apron, and rubber gloves must be worn (NH_4Cl is highly toxic). Students must be 10 ft or more from the demonstration.

DISPOSAL

Add an excess of 1 M NaOH to convert the NH_4Cl into $NH_3(aq)$. In an operating fume hood, warm the solution to expel the NH_3. Neutralize the liquid with 1 M HCl and pour it down the drain.

[Source: Fillinger. *J. Chem. Ed.*, Vol. 8, pp. 1852–1855, 1931.]

Possible Misconception

Some students will have trouble grasping the concept of the common ion effect. First, remind them that because the solubility of $BaSO_4$ is very low, it is not involved in the equation for the solubility-product constant. It follows that the product of the two ion concentrations is equal to the constant, K_{sp}. A constant value cannot change, so if one of the concentrations rises, the other must drop. By means of this principle, the common ion effect can be used to increase or decrease the concentration of a chosen ion.

Visual Strategy

Figure 12-22

Point out that, in addition to the concerns for the patient's safety, it is essential to keep the barium in the form of the "insoluble" salt because neither of the ions will function as an effective X-ray contrast medium. The purpose of using this compound requires that it remains undissociated.

ASSESS

Alternative Assessment

Have students write a short summary of the concepts in this section (equilibrium equations, solubility-product equations, common ion effect). Be sure to have them include vocabulary terms and practical applications of the concepts studied.

Answers to Section Review

17. If all but one component's concentration is known, knowing K_{eq} for the given conditions allows one to calculate the unknown concentration. K_{eq} provides qualitative information about whether the reaction under the given conditions favors product or reactant formation (a large K_{eq} value favors products; a small K_{eq} value favors reactants).

18. a. $K_{eq} = \frac{[Ag^+][Cl^-]}{[AgCl]}$

b. $K_{eq} = \frac{[Pb^{2+}][NO_3^-]^2}{[Pb(NO_3)_2]}$

c. $K_{eq} = \frac{[[Ni(H_2O)_6]^{2+}][SO_4^{2-}]}{[NiSO_4]}$

19. The concentration of the solid, undissolved salt does not change, so that any change in the equilibrium of the solution occurs only with changes in ion concentrations.

20. $K_{sp} = [Al^{3+}][OH^-]^3$

21. $K_{sp} = [A^{3+}]^2[B^{2-}]^3 = [2.7 \times 10^{-3}]^2[1.1 \times 10^{-6}]^3 = 9.7 \times 10^{-24}$

common ion effect *a process in which an ionic compound becomes less soluble upon the addition of one of its ions by adding another compound*

The displacement of an equilibrium caused by the presence of more than one source for a reactant or product ion is called the **common ion effect**. Remember from Le Châtelier's principle that changing the concentration of ions can have an effect on the equilibrium of the system. In **Figure 12-23,** Na_2SO_4 is added to the saturated solution of $BaSO_4$. The increase in the concentration of SO_4^{2-} ions causes the equilibrium to shift to the left so that less $BaSO_4$ dissolves in the presence of the common ion, thereby causing a decrease in the Ba^{2+} ion concentration in the solution.

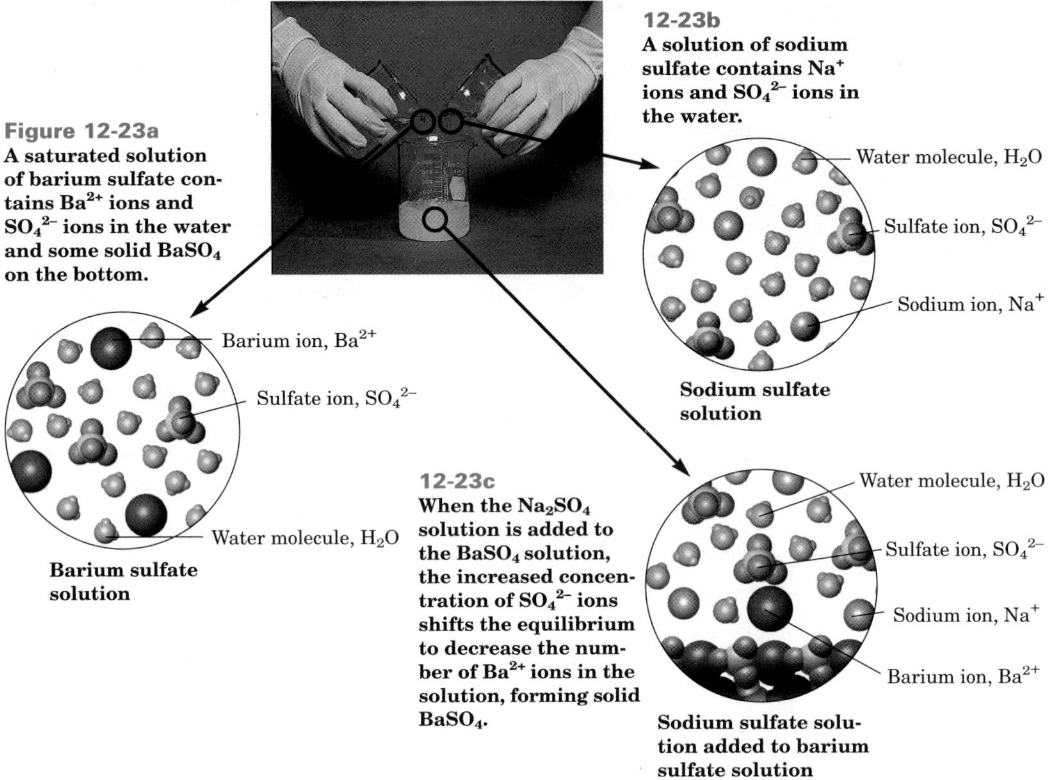

Figure 12-23a A saturated solution of barium sulfate contains Ba^{2+} ions and SO_4^{2-} ions in the water and some solid $BaSO_4$ on the bottom.

12-23b A solution of sodium sulfate contains Na^+ ions and SO_4^{2-} ions in the water.

12-23c When the Na_2SO_4 solution is added to the $BaSO_4$ solution, the increased concentration of SO_4^{2-} ions shifts the equilibrium to decrease the number of Ba^{2+} ions in the solution, forming solid $BaSO_4$.

Section Review

17. What information does the K_{eq} value for a reaction provide?

18. Write the K_{eq} expression for each of the following reactions.

a. silver chloride in an aqueous solution

b. the dissociation of lead(II) nitrate in aqueous solution

c. $NiSO_4(aq) + 6H_2O(l) \rightleftharpoons [Ni(H_2O)_6]^{2+}(aq) + SO_4^{2-}(aq)$

19. Why does the concentration of undissolved salt not appear in the K_{sp} expression?

20. Write the solubility-product expression for the slightly soluble salt aluminum hydroxide, $Al(OH)_3$.

21. Calculate K_{sp} for the following hypothetical reaction.

$$A_2B_3(s) \rightleftharpoons 2A^{3+}(aq) + 3B^{2-}(aq)$$

The equilibrium concentration of A^{3+} is 2.7×10^{-3} M, and the concentration of B^{2-} is 1.1×10^{-6} M.

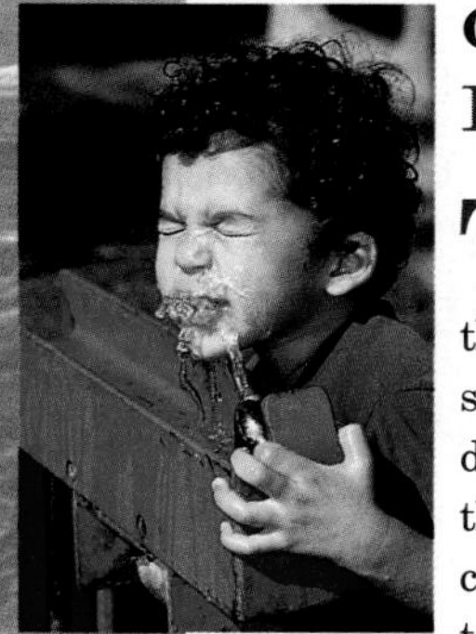

Conclusion: Sea Water and Equilibrium

The chapter began with a story about three people who were stranded without drinkable water. Even though your body contains between 55% to 65% fluid by weight, losing just 2% makes you feel extremely thirsty, and death is inevitable with a 20% loss. The people rescued had lost nearly 20% of their body fluids.

The principles learned in this chapter can be used to answer the questions presented in the story.

How would drinking sea water affect the aqueous equilibrium in the body? Body fluids contain ions. Normally, water flows across cell membranes into areas with a higher concentration of dissolved electrolytes, including Na^+, Cl^-, Ca^{2+}, Mg^{2+}, and K^+ ions. Drinking sea water, with its high concentration of some of these ions, causes an electrolytic imbalance. As a result, water would move out of cells. Brain cells would shrink, causing seizures, coma, and then death. Hospital care is required to restore normal electrolyte balance.

Can sea water be made safe to drink? Many substances dissolved in sea water are electrolytes. The methods used to remove these substances include ion exchange and electrodialysis membranes. The latter method makes use of salt's electrolytic properties to collect ions along a membrane in an electric field. The ions are separated from sea water, making it salt-free and safe to drink.

How do you know when water is safe to drink? You can have water samples tested at city water treatment plants. Many substances come into contact with water before it goes to a processing plant. Various processes remove such substances before it becomes available to the public. Chemists approve community water supplies.

Applying Concepts

1. Sea water contains a high concentration of Na^+ and Cl^- ions formed from the dissociation of NaCl. A reference for K_{sp} values does not list NaCl. What does that tell you about NaCl?
2. When people suffer from dehydration, their hearts must beat harder to maintain circulation. Why?

Research and Writing

Use the library to do the following.

1. Prepare a report to show how aqueous equilibrium processes are involved in the formation of stalagmites and stalactites in caves.
2. Sea water contains many salts and minerals. Research what is being done to extract minerals from the oceans.
3. Some people stranded at sea have survived for more than 100 days. Report on what methods they used to obtain drinkable water.

Answers to Applying Concepts

1. The degree of dissociation for NaCl is so great that the reaction strongly favors products (the ions). K_{sp} is so large that it is indeterminable.
2. As the amounts of water in the body decrease during dehydration, the amount of solid salts increases. This leads to a "thickening" of the blood, requiring more force to move it through the circulatory system.

Closure Strategy

1. Have students determine (either individually or in small groups) five reasons why the flow of chemical solutions into major lakes and oceans needs to be closely monitored and controlled. Solubility, solution equilibrium, and chemical toxicity need to be taken into consideration. (Many salts contain metals that are hazardous even in relatively small concentrations, and as ions in solution they are more easily introduced into the bodies of humans and animals. High concentrations may lead to ion precipitation and accumulation on the ocean floor, thus contaminating sea life (e.g., shellfish). Oceans are so large that they can acquire vast amounts of pollutants before toxic effects become noticeable. Oceans and lakes are large and therefore hard to clean up. Adding certain chemicals may upset equilibria that already exist in the water, thus affecting plant and animal life.)

2. *Chelating agents* are ligands with several bonding sites that allow them to form complex ions with metal ions in solution. Have students use library resources to research the chemistry of chelation, and have them write a three-page report on their findings. The role of metal chelate compounds in biochemistry (e.g., chlorophyll), the use of chelating agents in the treatment of metal poisoning and environmental pollution, and the effectiveness of different agents with respect to different metals should all be included in the reports.

CHAPTER 12 Highlights

Key terms

- colligative property
- common ion
- common ion effect
- complex ions
- conductance
- dissociation
- electrolyte
- equilibrium constant
- hydration
- hydronium ion
- Le Châtelier's principle
- ligand
- nonelectrolyte
- solubility-product constant
- strong electrolyte
- theory of ionization
- weak electrolyte

Key ideas

What happens in an aqueous solution?

- Substances that can conduct electricity when they dissolve in a solution are called electrolytes.
- Electrolytes form ions in a solution. Strong electrolytes undergo large dissociation; weak electrolytes undergo small dissociation.
- Electrolytes are either ionic or molecular compounds that dissociate in water.
- Both electrolytes and nonelectrolytes can affect colligative properties, such as boiling point elevation and freezing point depression, which depend on the number of particles in solution.

What is an equilibrium system?

- Reactions go to completion when products include such things as the formation of a precipitate or the evolution of a gas.
- Reactions that are reversible can re-form the reactants. When the forward and reverse reactions occur at equal rates, the reaction is at equilibrium.
- A system stays at equilibrium until conditions change. Le Châtelier's principle explains that stress, including temperature, pressure, and change in concentration, can shift equilibrium.

How is equilibrium measured?

- The equilibrium constant K_{eq} is a number that expresses the relationship between the concentrations of substances in a reversible reaction at a given temperature.
- The solubility-product constant K_{sp} is the equilibrium constant for a solid in equilibrium with its ions in a saturated solution of a slightly soluble salt.
- Adding a common ion to a solution can shift the equilibrium to reduce the concentration of a substance in solution.

Key problem-solving approach: *Equations using constants*

For an equilibrium problem, analyze the type of equilibrium presented. Follow the arrows in the diagram to determine the proper equilibrium expression to use.

Sample Equation

$$aA + bB \rightleftharpoons cC + dD$$

Equilibrium problem

Slightly soluble salt in water → **Solubility product constant (K_{sp})** → **Problem will use** $K_{sp} = [C]^c[D]^d$

Other reversible reactions → **Equilibrium constant (K_{eq})** → **Problem will use** $K_{eq} = \frac{[C]^c[D]^d}{[A]^a[B]^b}$

CHAPTER 12

Review and Assess

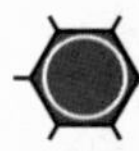

Properties of aqueous solutions

REVIEW

1. a. What property identifies all electrolytes?
   b. How could you safely determine whether an unknown aqueous solution contains an electrolyte?
2. Rate the solutions below from most conductive to least conductive.
   a. 1.0 M KOH solution
   b. 10. M ammonia solution
   c. 5.0 M glucose solution
3. a. How does a strong electrolyte differ from a weak electrolyte?
   b. Distinguish between the use of the terms *strong* and *weak* with the terms *dilute* and *concentrated* in describing solutions containing electrolytes.
   c. Which of the following solutes would show conductivity properties when dissolved in water: NaCl, KCl, $C_{12}H_{22}O_{11}$, $Mg(NO_3)_2$, CCl_4, and $Al_2(SO_4)_3$?
4. a. Describe the hydration process on the particle level for an ionic compound in water.
   b. Describe the dissociation process on the particle level for a molecular compound in water.
5. What determines whether a dissociation is endothermic or exothermic?

APPLY

6. When you touch an appliance that has a short circuit, an electric current travels through your body, giving you a shock. What feature of the human body allows this to happen?
7. Sports medicine utilizes heats of solution in hot and cold packs. Many of these products work by dissolving an ionic substance in water. Study **Table 12-1**, and suggest the best substances to use for hot packs and for cold packs. Explain the reasons for your choices.
8. Suggest a method for making water boil at 100°C in high-altitude areas.
9. In northern Utah the temperature falls well below freezing in the winter. Would you expect the Great Salt Lake to be safe for ice skating? Explain your answer.
10. Rank the following substances from greatest to least ability to change the colligative properties of water. Explain your reasons.
    a. NaOH b. $C_6H_{12}O_6$ c. $AlBr_3$

Solutions in equilibrium

REVIEW

11. a. Compare the rate of the forward reaction to the rate of the reverse reaction for a solution as it approaches equilibrium.
    b. Compare the rate of the forward reaction to the rate of the reverse reaction for a solution in equilibrium.
12. a. How does Le Châtelier's principle relate to reversible reactions?
    b. Does Le Châtelier's principle affect reactions that go to completion?
13. Use **Figure 12-12** to answer the following questions.
    a. What is happening to the rate of formation of H_3O^+ before the system reaches equilibrium?
    b. When is the rate of the forward reaction the greatest?
    c. How would the addition of sodium acetate, $NaCH_3COO$, affect the equilibrium described in the figure?
14. Draw two diagrams that depict the difference between the microscopic and macroscopic events that occur when a reaction is at equilibrium.
15. Identify the ligands in the following complex ions.
    a. $[AlF_6]^{3-}$ (Al bonded to six F) b. $[Ni(CN)_4]^{2-}$ (Ni bonded to four CN)

1. a. The ability to conduct electricity is a property shared by all electrolytes.
   b. An apparatus similar to the one in **Figure 12-1** could be used. If the bulb lights when the wires are submerged in the unknown solution, the solution contains an electrolyte.
2. From the most to least:
   a. 1.0 M KOH (strong electrolyte),
   b. 10. M ammonia (weak electrolyte), and
   c. 5.0 M glucose (nonelectrolyte).
3. a. Strong electrolytes dissociate completely in solution. Weak electrolytes partially dissociate.
   b. Strong and weak refer to the degree of dissociation of an electrolyte and are independent of concentration. Dilute and concentrated refer to the concentration of the electrolyte in solution. Thus it is possible to have a concentrated solution of a weak electrolyte.
   c. NaCl, KCl, $Mg(NO_3)_2$, $Al_2(SO_4)_3$
4. a. Answers will vary. The central focus of the hydration model should be water molecules surrounding ions, with the oxygen atoms oriented toward cations and the hydrogen atoms oriented toward anions.
   b. Answers will vary. The central focus of the dissociation model will be the rearrangement of atoms and the formation of hydronium ions.
5. The size and charge of the ions, which determine the ease with which they are dissociated, and the energy involved in hydration, determine whether the dissociation is exothermic or endothermic.

6. Because the human body contains electrolytes, it can conduct electrical currents.

7. Answers will vary. Substances with high positive heats of solution should be suggested for cold packs. Substances with high negative heats of solution should be suggested for hot packs. (Other suggestions, such as avoiding the use of hazardous substances like NaOH, should also be considered.)

8. Add a solute to raise the boiling point, a colligative property.

9. No. Because the Great Salt Lake is a salt solution, its freezing point has been reduced. Depending on the number of salt particles present, this solution may not freeze completely, even at extremely low temperatures, and would not be safe for skating.

10. From the greatest to the least:
c. $AlBr_3$ (four ions),
a. NaOH (two ions), and
b. $C_6H_{12}O_6$ (one molecule).

11. a. Before equilibrium, the forward reaction rate is greater than the reverse reaction rate. As the system approaches equilibrium, the forward rate decreases and the reverse rate increases.
b. At equilibrium, the forward reaction rate equals the reverse reaction rate.

12. a. Le Châtelier's principle states that if stress is applied to a system, the system reacts to accommodate the change and reaches a new equilibrium. Stress can be caused by changes in temperature or concentration of reactants and products.
b. No. If a reaction reaches completion, it is not at equilibrium. Le Châtelier's principle applies only to reactions that reach equilibrium.

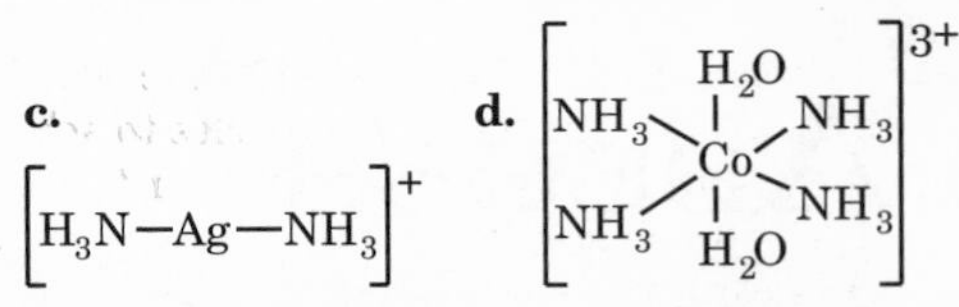

APPLY

16. Identify the stress on each reaction below.
   a. the color of a shirt changing according to the heat surrounding it
   b. the chemical in a pool test kit turning red in response to the chloride ions in the water
   c. a mood ring changing color when you get angry

17. A home economics teacher instructs the class to add several teaspoons of sugar to their homemade lemonade and then stir until the sugar stops dissolving. Is this instruction chemically correct? Explain.

18. Ethyl acetate, $CH_3COOC_2H_5$, is used to make artificial silk and is prepared by reacting acetic acid, CH_3COOH, and ethanol, C_2H_5OH. Water is also produced in this reaction.
   a. Write the balanced equation for this reaction.
   b. If the production of ethyl acetate is not sufficient, what would increase the yield?
   c. Why is it a good idea to remove ethyl acetate as it forms?

19. Identify which of the following reactions will go to completion. Give a reason for your answer.
   a. dripping water depositing limestone in caves
   b. CO_2 bubbles forming as hydrochloric acid reacts with limestone
   c. saline solution dissolving all of the mineral deposits on a contact lens

20. Carbonated beverages contain carbonic acid in solution. The end result of the dissociation of this acid is represented by the following equation.

$$H_2CO_3(aq) \rightleftharpoons H_2O(l) + CO_2(g)$$

What changes would cause a shift to the right?

21. Use Le Châtelier's principle and the reaction below to answer the following questions.

$$2SO_2(g) + O_2(g) \rightleftharpoons 2SO_3(g)$$

   a. SO_2 reacts with oxygen to form SO_3. How can you minimize the amount of SO_2 that re-forms?
   b. How can you produce oxygen from this reaction?
   c. How can you ensure that the amount of SO_3 formed remains unchanged?

Quantitative equilibrium

REVIEW

22. Although the K_{eq} value is known as the equilibrium constant, **Table 12-4** shows that one substance can have several different K_{eq} values. Why is it still called a constant?

23. Predict the concentration(s) of the reactants and the products in the decomposition reaction of ammonium bisulfide when K_{eq} is greater than 1, less than 1, and equal to 1.

$$NH_4HS(s) \rightleftharpoons NH_3(g) + H_2S(g)$$

24. Use **Table 12-4** to determine whether the following equations favor the forward or the reverse reaction at the given temperature.
   a. $2HI(g) \rightleftharpoons H_2(g) + I_2(g)$ at 350°C
   b. $N_2(g) + O_2(g) \rightleftharpoons 2NO(g)$ at 25°C
   c. $CH_3COOH(l) \rightleftharpoons H^+(aq) + CH_3COO^-(aq)$ at 25°C
   d. $N_2O_4(g) \rightleftharpoons 2NO_2(g)$ at 0°C
   e. $N_2O_4(g) \rightleftharpoons 2NO_2(g)$ at 55°C

25. Why can the term *insoluble* be a misleading description for most salts that do not appear to dissolve in water?

26. Write equilibrium constant expressions for the following reactions.
   a. $2NO_2(g) \rightleftharpoons N_2O_4(g)$
   b. $[Cu(NH_3)_4]^{2+} \rightleftharpoons Cu^{2+}(aq) + 4NH_3(aq)$
   c. $AgCl(s) \rightleftharpoons Ag^+(aq) + Cl^-(aq)$

27. Relate Le Châtelier's principle to the common ion effect.

PRACTICE

Answers to items in a black square begin on page 841.

28. Vinegar, a solution of acetic acid, CH_3COOH, and water, is used in varying concentrations for different household tasks. If the concentration of the acetic acid solution at equilibrium is 3.00 M, the H_3O^+ concentration at equilibrium is 7.22×10^{-3} M, and the CH_3COO^- concentration is 7.22×10^{-3} M, what is the K_{eq} value for acetic acid? (Hint: see Sample Problem 12A.)

$$CH_3COOH(aq) + H_2O(l) \rightleftharpoons H_3O^+(aq) + CH_3COO^-(aq)$$

29. Aniline, $C_6H_5NH_2$, is a weak base. If the concentration of aniline is 6.00 M at equilibrium, and the concentrations of both $C_6H_5NH_3^+$ and OH^- are 5.08×10^{-5} M, what is the K_{eq} value for aniline? (Hint: see Sample Problem 12A.)

$$C_6H_5NH_2(l) + H_2O(l) \rightleftharpoons C_6H_5NH_3^+(aq) + OH^-(aq)$$

30 Benzoic acid, C_6H_5COOH, is a slightly soluble acid used in food preservation. K_{eq} for benzoic acid is 6.30×10^{-5} at 25°C. For a solution of benzoic acid that has an equilibrium concentration of 2.00 M, what are the concentrations of H_3O^+ and $C_6H_5COO^-$? (Hint: see Sample Problem 12B.)

O OH (benzene ring) $+ H_2O \rightleftharpoons$ O O^- (benzene ring) $+ H_3O^+$

31. Phenol, C_6H_5OH, is used as a disinfectant in many household cleaning solutions. If the equilibrium concentration of phenol is 1.61×10^{-3} M, what are the concentrations of H_3O^+ and $C_6H_5O^-$? K_{eq} for phenol is 1.60×10^{-10} at 25°C. (Hint: see Sample Problem 12B.)

32 Aluminum hydroxide, $Al(OH)_3$, is used in the process of waterproofing fabrics. If the Al^{3+} concentration is 2.63×10^{-9} M, what is the concentration of OH^-? The K_{sp} value for $Al(OH)_3$ is 1.30×10^{-33} at 288 K. (Hint: see Sample Problem 12C.)

33. Silver sulfide, Ag_2S, is an ingredient used to make ceramics. If the Ag_2S in your dog's ceramic water dish dissolves in the water, giving a Ag^+ concentration of 4.93×10^{-17} M, what concentration of S^{2-} will be in your dog's water? The K_{sp} value for Ag_2S is 6.00×10^{-50} at 288 K. (Hint: see Sample Problem 12C.)

APPLY

34. What is wrong with the following expression for K_{eq}?

$$NH_3(g) + O_2(g) \rightleftharpoons NO(g) + H_2O(g)$$

$$K_{eq} = \frac{[NH_3][O_2]}{[NO][H_2O]}$$

35. Lactic acid, $HC_3H_5O_3$, a by-product of cell metabolism, causes your muscles to ache shortly after vigorous exercise. Write an equilibrium equation for the dissociation of lactic acid in muscle tissue.

O OH (carbonyl) / HO—C—H / CH_3 $+ H_2O \rightleftharpoons$ O O^- (carbonyl) / HO—C—H / CH_3 $+ H_3O^+$

36. The figure below shows the results of adding different chemicals to distilled water.
   a. Which substance(s) is (are) completely soluble in water?
   b. Is it correct to say that AgCl is completely insoluble? Explain.

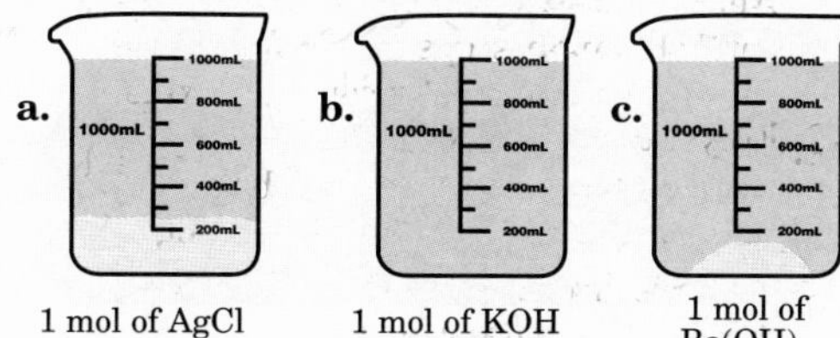

1 mol of AgCl | 1 mol of KOH | 1 mol of $Ba(OH)_2$

37. Pb^{2+} in drinking water can result in nerve damage. Trace amounts of Pb^{2+} in a water supply system can be detected by increasing the concentration of Cl^- ions to form $PbCl_2$ as a precipitate. Assuming that all of the chemicals listed in the table below are equally available at low cost, which would be most useful in testing for Pb^{2+} in your water supply? Explain.

Name	Formula	K_{sp}
silver chloride	AgCl	1.77×10^{-10}
lead sulfate	PbS	9.04×10^{-29}
sodium chloride	NaCl	high solubility
lead sulfate	$PbSO_4$	1.82×10^{-8}

38. a. Are any of the K_{sp} values listed in the table above greater than 1?
   b. Explain this observation.

13. a. It is increasing.
   b. The rate is greatest at the beginning of the reaction.
   c. Equilibrium would change to form more CH_3COOH and H_2O.

14. Drawings should show the microscopic level as a dynamic process with both reactions occurring. The macroscopic level should be static, showing no change. (The total amount of particles remains constant; there is a constant ratio of reactants to products.)

15. a. F^-
   b. CN^-
   c. NH_3
   d. H_2O and NH_3

16. a. heat (temperature)
   b. chloride ions (concentration)
   c. body heat (temperature)

17. No. The sugar continues to dissolve and crystallize because it is in equilibrium.

18. a. $CH_3COOH + C_2H_5OH \rightleftharpoons CH_3COOC_2H_5 + H_2O$
   b. Add more of the reactants or remove either of the products.
   c. Removal of the ethyl acetate would place a stress on the system. This stress would be relieved by increasing the rate for the forward reaction. This results in increased production of ethyl acetate.

19. **a**, **b**, and **c** (assuming that all three systems are open)

20. A shift to the right could be caused by increasing temperature or changing concentration (removal of either product, or addition of more carbonic acid).

21. a. Add more O_2, and remove SO_3 as it forms.
   b. Add excess SO_3.
   c. Remove SO_3 from the reaction.

22. K_{eq} is a constant for a substance at a given temperature and pressure.

23. For the reaction shown, if K_{eq} is greater than 1, the concentrations of the products are greater than the concentration of the reactant. When K_{eq} is less than 1, the concentration of the reactant is greater than the concentrations of the products. When K_{eq} is equal to 1, the product of the concentrations of the products is equal to the concentration of the reactant.

24. a. reverse
b. reverse
c. reverse
d. forward
e. reverse

25. Most salts are at least very slightly soluble. Although they appear not to dissolve, a minute portion does dissolve until equilibrium is reached.

26. a. $K_{eq} = \frac{[N_2O_4]}{[NO_2]^2}$

b. $K_{eq} = \frac{[Cu^{2+}][NH_3]^4}{[[Cu(NH_3)_4]^{2+}]}$

c. $K_{eq} = \frac{[Ag^+][Cl^-]}{[AgCl]}$

27. When added to a solution, a common ion becomes more concentrated, so that a stress is placed on the system. The equilibrium shifts to relieve the stress by reducing the concentration of the common ion through the formation of a precipitate.

28. $K_{eq} = \frac{(7.22 \times 10^{-3})(7.22 \times 10^{-3})}{(3.00)} =$

1.74×10^{-5}

29. $K_{eq} = \frac{(5.08 \times 10^{-5})(5.08 \times 10^{-5})}{(6.00)} =$

4.30×10^{-10}

30. $6.30 \times 10^{-5} = \frac{[H_3O^+][C_7H_5O_2^-]}{2.00}$

$[H_3O^+][C_7H_5O_2^-] = 1.26 \times 10^{-4}$

$\sqrt{1.26 \times 10^{-4}} = [H_3O^+] = [C_7H_5O_2^-]$

$\sqrt{1.26 \times 10^{-4}} = 1.12 \times 10^{-2}$ M

Linking chapters

1. ***Saturated solutions***
   a. Relate the behavior of sparingly soluble salts to the concept of saturation.
   b. How does Le Châtelier's principle explain the behavior of a saturated solution that forms more solid solute precipitate when a common ion is added?
2. ***Percent yield***
   Using Le Châtelier's principle, explain how to increase the percent yield of a reversible chemical reaction.
3. ***Free energy***
   a. What does the equilibrium constant have in common with free energy?
   b. What is the ΔG of a reaction at equilibrium?
4. ***Bond energy***
   Heat of solution occurs when an ionic substance dissolves in water. Explain the energy transfers involved in the heat of dissociation and heat of hydration that result in the heat of solution.
5. **Theme: *Equilibrium and change***
   a. If a process is in equilibrium, does this mean there are no changes?
   b. If a reversible reaction favors the formation of the products, are any reactants being re-formed?

USING TECHNOLOGY

1. ***Graphics calculator***
   Equations using the equilibrium constant can be analyzed using a quadratic equation and a graphics calculator. The amount of product formed can be predicted from the beginning amount of reactant. For example, predict how many H_3O^+ and CHO_2^- ions will be formed from 3.00 M formic acid.

   $$HCHO_2 + H_2O \rightleftharpoons H_3O^+ + CHO_2^-$$

   $$K_{eq} = 1.8 \times 10^{-4}$$

   First, write out the equilibrium equation with $[HCHO_2]$ as $3.00 - x$ (the original concentration less the dissociated ions), $[H_3O^+]$ as x, and $[CHO_2^-]$ as x. Then, solve the equation for zero. You should get the quadratic equation $x^2 + (1.8 \times 10^{-4})x - 5.4 \times 10^{-4} = 0$. Press the **Y=** key and enter the quadratic equation. Then press **GRAPH** and use the **TRACE** key to find the value of the x-intercept. This will equal the concentration of H_3O^+ and the concentration of CHO_2^-. $[HCHO_2]$ will be 3.00 minus the x-intercept.

   a. What concentrations of H_3O^+ and CHO_2^- will result from 3.00 M $HCHO_2$?
   b. What concentrations of H_3O^+ and CHO_2^- will result from 5.00 M $HCHO_2$?
   c. What concentrations of H_3O^+ and ClO_2^- will result from the dissociation of 3.00 M $HClO_2$? (Hint: K_{eq} for $HClO_2$ is 1.2×10^{-2}.)
2. ***Data analysis***
   Data analysis equipment can measure the concentrations of specific ions in solution. The H_3O^+ concentration of a solution can be analyzed using a pH meter. First, calculate the theoretical $[H_3O^+]$ for the reaction of each of the following acids with water using the K_{eq}. Analyze your data by organizing a table ranking the acids from highest $[H_3O^+]$ to the lowest $[H_3O^+]$.

Acid	K_{eq}
3.00 M HCN	$K_{eq} = 4.0 \times 10^{-10}$
3.00 M HF	$K_{eq} = 6.7 \times 10^{-4}$
3.00 M HOCl	$K_{eq} = 2.95 \times 10^{-8}$
3.00 M $HC_2H_2ClO_2$	$K_{eq} = 1.35 \times 10^{-3}$
3.00 M HNO_2	$K_{eq} = 5.13 \times 10^{-4}$

Alternative assessment

Performance Assessment

1. Your instructor will give you an index card with a specific equilibrium reaction on it. Describe how you would alter the reaction to produce either more of the products or more of the reactants. Show your method to your instructor. If your method is approved, obtain the necessary materials from your instructor, and perform the experiment.
2. Devise an experiment to test the conductance properties of certain substances that are soluble in water. Show your experimental procedure to your instructor for approval. When your procedure is approved, test the conductance properties of the following substances.
   a. glucose, $C_6H_{12}O_6$
   b. potassium permanganate, $KMnO_4$
   c. aspartame, $C_{14}H_{18}N_2O_5$
   d. calcium saccharin, $CaC_{14}H_8N_2O_6S_2$
3. Study the colligative properties of different substances. Develop a procedure to test the effect of both ionic and molecular substances in solution on boiling point and freezing point. Show your procedure to your instructor for approval. When your procedure is approved, obtain the following materials to test your experiment and complete the table below. CAUTION: *Ethanol and acetone are flammable. Do not use around an open flame.*

Solvent	Solute	Concentration	Property	Effect
Water	NaCl		freezing point	
Water	sucrose		freezing point	
Ethanol	NaCl		boiling point	
Ethanol	sucrose		boiling point	
Acetone	NaCl		boiling point	
Acetone	sucrose		boiling point	

Portfolio projects

1. ***Research and communication*** Analyze your daily intake of foods. Identify the substances as ionic or molecular. Study the physiological effects of each substance, and indicate which are the most helpful to your body. Use this information to choose foods rich in the vitamins and solutes that you need. Make a poster of your findings.
2. ***Research and communication*** Colligative properties are often used in the production of automobile supplies. The most familiar example is antifreeze. Research the development of antifreeze and the different substances used to obtain its properties. Study the weather in your area to determine the best type of antifreeze and the most appropriate concentration to use. Make a chart to present your results.
3. ***Cooperative activity*** As a group, develop a unique method of demonstrating the difference between static and dynamic equilibrium. Show how Le Châtelier's principle will work for one and not the other. Perform your demonstration for the rest of your class. If possible, perform your demonstration for a different science class that has not yet been introduced to the concept of dynamic equilibrium.
4. ***Chemistry and you*** Study the many uses of complex ions. Analyze products you use that contain complex ions. Write an essay that explains how your life might be different without complex ions.
5. ***Research and communication*** Many different pollutants are in the form of ions dissolved in water. Industries must remove these ions from the water to avoid exposing the environment to harmful chemicals. Research the use of solubility principles and equilibrium to clean up the chemicals listed below. Write a report about the effect that these chemicals have on our environment.
   a. Pb^{2+}
   b. Hg_2^{2+} and Hg^{2+}
   c. Cl^-
   d. SO_4^{2-}
   e. PO_4^{3-}

31. $1.60 \times 10^{-10} = \frac{[H_3O^+][C_6H_5O^-]}{1.61 \times 10^{-3}}$

$[H_3O^+][C_6H_5O^-] = 2.58 \times 10^{-13}$

$\sqrt{2.58 \times 10^{-13}} = [H_3O^+] = [C_6H_5O^-]$

$\sqrt{2.58 \times 10^{-13}} = 5.08 \times 10^{-7}$ M

32. $K_{sp} = [Al^{3+}][OH^-]^3$

$1.30 \times 10^{-33} = 2.63 \times 10^{-9} \times [OH^-]^3$

$[OH^-] = 7.91 \times 10^{-9}$ M

33. $K_{sp} = [Ag^+]^2[S^{2-}]$

$6.00 \times 10^{-50} = (4.93 \times 10^{-17})^2\,[S^{2-}]$

$[S^{2-}] = 2.47 \times 10^{-17}$ M

34. The equation is not balanced, so the exponents for the various concentrations in the equilibrium constant expression are incorrect. Also, the concentrations for the products must be in the numerator, rather than in the denominator as shown. Similarly, the concentrations for the reactants must be in the denominator instead of the numerator. The proper equation is $4NH_3(g) + 5O_2(g) \rightleftharpoons 4NO(g) + 6H_2O(g)$, and the equilibrium constant is expressed as follows.

$$Keq = \frac{[NO]^4[H_2O]^6}{[NH_3]^4[O_2]^5}$$

35. $HC_3H_5O_3 + H_2O \rightleftharpoons H_3O^+ + C_3H_5O_3^-$

$$Keq = \frac{[H_3O^+][C_3H_5O_3^-]}{[HC_3H_5O_3]}$$

36. **a.** Only KOH is completely soluble in water.
b. No, because some AgCl does dissolve before equilibrium is established.

37. NaCl would work best because it will ionize completely. The Cl^- ions will cause most of the lead to precipitate out of solution as $PbCl_2$, and no other toxic ions will be introduced in the process.

38. a. No.

b. A K_{sp} value greater than 1 means that the salt is very soluble. The values in the short table listed in item **37** are taken from **Table 12-5**, which is a list of slightly soluble substances. There is no K_{sp} value given for sodium chloride beccause it dissolves completely.

Linking Chapters

1. a. "Insoluble" salts dissolve very slightly before equilibrium between the dissolution and precipitation processes is reached. Because of this these salts reach their saturation points very quickly.

b. A saturated solution is in equilibrium between solid solute and ions. The addition of the common ion favors the reaction that forms more solid solute. For example, more NaCl can be made to crystallize out of solution with the addition of Cl^- or Na^+ ions.

2. Any reaction will produce only a portion of its theoretical yield. This actual yield can be used to calculate percent yield. In a reversible reaction, the amount of product can be increased by causing stress on the left, or reactant, side of the equation, thus increasing the percent yield. This is most easily achieved by continually removing the desired product from the reaction.

3. a. Both the equilibrium constant and free energy can be used to predict the tendency for a reaction to occur.

b. $\Delta G = 0$

4. When a substance dissolves, energy is required to separate the ions from the solid. This energy, the heat of dissociation, is used to break the ionic bond. Energy in the form of heat of hydration is part of the process of water molecules surrounding an ion. The net energy change for these two processes represents the heat of solution, which can be either positive or negative.

5. a. No, there are still changes because the forward and reverse reactions are both occurring.

b. Yes, they are constantly re-formed, but at a smaller rate than that at which the products are formed.

USING TECHNOLOGY

1. a. 2.3×10^{-2} M H_3O^+, 2.3×10^{-2} M CHO_2^-, and 2.98 M $HCHO_2$

b. 3.0×10^{-2} M H_3O^+, 3.0×10^{-2} M CHO_2^-, and 4.97 M $HCHO_2$

c. 0.18 M H_3O^+, 0.18 M ClO_2^-, and 2.82 M $HClO_2$

2. Numerical answers will vary according to equipment available. The ranking of acids from most H_3O^+ produced to least H_3O^+ produced is $HC_2H_2ClO_2$, HF, HNO_2, HOCl, and HCN. The larger the K_{eq} value, the more H_3O^+ produced.

Performance assessment

1. Answers will vary. Be sure students follow safe laboratory procedures.

2. Answers will vary. Potassium permanganate and calcium saccharin should have electrolytic properties.

3. Answers will vary. Boiling points should be raised, and freezing points should be depressed.
CAUTION: *Ethanol and acetone are flammable. Do not use around an open flame.*

Portfolio projects

1. Answers will vary according to the students' diets.

2. Answers will vary according to geography and climate. Students should include ethylene glycol as the current antifreeze. Older antifreezes include kerosene, honey, and salt.

3. Answers will vary. Demonstrations for dynamic equilibrium should successfully show how stress on a system causes the rate of either the forward or reverse reaction to increase and the rate of the opposing reaction to decrease until the system reestablishes equilibrium in accordance to Le Châtelier's principle. The demonstration should also emphasize that this process takes place on the microscopic level, while static equilibrium should be shown to be what dynamic equilibrium appears to be on the macroscopic level.

4. Answers will vary. Complex ions are used in coloring materials, meteorology, chelating compounds, foods and beverages, and stain removal products.

5. Answers will vary according to the industries in the area.

PLANNING FOR

13 Acids and Bases

Chapter Overview

- **Section 13-1** reviews the operational definitions of acids and bases presented in Chapter 1 and relates these properties to those of electrolytes, which were studied in Chapter 12. Students then contrast the model proposed by Arrhenius with that of Brønsted and Lowry. The subject of anhydrides is introduced in the context of pollutants and solving the problem of acid rain. The amphiprotic nature of water is used to introduce the concept of a conjugate acid–base pair.
- **Section 13-2** extends the coverage of equilibrium from Chapter 12 as students explore the behavior of weak acid and weak base equilibrium systems. The concept of pH, introduced in Chapter 1, is now presented from an quantitative perspective.
- **Section 13-3** covers the neutralization process and its application to titrations. The equilibrium behavior of indicators is also introduced.
- **Section 13-4** differs from the traditional presentation of acids and bases to give students some exposure to organic acids and bases. Buffers are introduced from a biochemical perspective. To extend the coverage of Chapter 9, some nutritional chemistry is also introduced.

Concept Base

Students must have mastered the following concepts prior to this chapter: functional groups of organic compounds (Chapter 6); stoichiometric relationships (Chapter 8); and equilibrium systems (Chapter 12).

Looking Ahead

Students will explore rates in Chapter 14, redox in Chapter 15, and nuclear reactions in Chapter 16.

Laboratory Equipment Needs

Demonstrations

The listing of materials for Chapter 13 demonstrations begins on page T46.

Laboratory Experiments

Materials and preparation instructions for Chapter 13 are found as follows.

Text		*Page*
Exploration 13A	Acid-Base Titration of an Eggshell	760
Investigation 13A	Acid-Base Titration—Industrial Spill	764
Exploration 13B	Buffering Capacity	766
Investigation 13B	Buffering Capacity—Viral Growth Medium	772

Laboratory Experiments Manual
(refer to the table of contents in the manual)

Investigation 13-1	Measuring pH—Home Test Kit
Investigation 13-2	Measuring pH—Acid Precipitation Testing
Investigation 13-3	Acid-Base Titration—Vinegar Tampering Investigation

Pacing: 15 days — Inquiry Teaching Strategies

13-1 What are acids and bases?

Demonstrations
- ■ Demo 1 *Ionization of acidic solutions*
- ■ Demo 2 *Production of an acid from an acid anhydride*
- ■ Demo 3 *Acid rain*
- ■ Demo 4 *Effects of acid rain on calcium carbonate*
- ■ Demo 5 *Producing a base from a base anhydride*

13-2 How are weak acids and bases compared?

Demonstrations
- ■ Demo 6 *pH*

Laboratory Experiments
- ■■ Inv 13-1 *Measuring pH—Home Test Kit*
- ■■ Inv 13-2 *Measuring pH—Acid Precipitation Testing*

13-3 What is a titration?

Demonstrations
- ■ Demo 7 *Red cabbage indicator*
- ■ Demo 8 *Indicators*

Laboratory Experiments
- ■ Exp 13A *Acid-Base Titration of an Eggshell*
- ■ Inv 13A *Acid-Base Titration—Industrial Spill*
- ■■ Inv 13-3 *Acid-Base Titration—Vinegar Tampering Investigation*

13-4 What are the biochemical roles of some acids and bases?

Demonstrations
- ■ Demo 9 *Buffers in solution*

Laboratory Experiments
- ■ Exp 13B *Buffering Capacity*
- ■ Inv 13B *Buffering Capacity—Viral Growth Medium*

Key
- ■ Teaching Transparencies
- ■ Annotated Teacher's Edition
- ■ Laboratory Experiments Manual

Visual Teaching Strategies	Review and Practice Strategies
	Text Reviews ■ Concept Review 1–5 ■ Section Review 6–11 ■ Chapter Review 1–10; LC 1, 5, 6 **Study Guide Worksheets** ■ ■ 13-1 Concept Review ■ ■ 13-1 Skillsheet *Using Tables*
Transparencies ■ F 13-11 *pH Scale Relationships* ■ P 525 *Equilibrium and pH Calculations*	**Text Reviews** ■ Additional Examples 13A, 13B, 13C, 13D, 13E, 13F, 13G ■ Practice 1–8 ■ Section Review 12–19 ■ Chapter Review 11–30; UT 2 **Study Guide Worksheets** ■ ■ 13-2 Concept Review ■ ■ 13-2 Practice *Calculating pH*
Transparencies ■ F 13-15 *Titration Curve for a Strong Acid and Strong Base* ■ F 13-16 *Titration Curve for a Weak Acid and Strong Base* ■ F 13-17 *Molecular and Ion Forms of Phenolphthalein*	**Text Reviews** ■ Additional Example 13H ■ Practice 1–4 ■ Section Review 20–24 ■ Chapter Review 31–49; LC 2–4, 7, 8 **Study Guide Worksheets** ■ ■ 13-3 Concept Review ■ ■ 13-3 Practice *Calculations with Titration Data*
Transparencies ■ F 13-19 *Essential Amino Acids* ■ F 13-22 *Structural Components of DNA and RNA* ■ F 13-25 *Fatty Acid Composition of Dietary Fats* **Transparency Masters** ■ F 13-21 *Zwitterion Form of Alanine at High and Low pH* ■ F 13-24 *Comparing Structures of Saturated and Unsaturated Fatty Acids*	**Text Reviews** ■ Concept Review 25 ■ Section Review 26–33 ■ Chapter Review 50–57; UT 1 **Study Guide Worksheets** ■ ■ 13-4 Concept Review ■ ■ 13-4 Skillsheet *Working with Amino Acid Structures*

■ Pupil's Text   Exp *Exploration*   LC *Linking Chapters*
■ Study Guide   Inv *Investigation*   UT *Using Technology*
■ Teaching Resources

Assessment Options

Traditional Assessment

Test Generator
Instructional Objectives Measured:
Content mastery
Select from items 1 to 75

Performance Assessment Options

(see pages T36 and T634-35 for scoring rubrics)
Students demonstrate mastery of objectives in a hands-on environment. You may want some of these materials in the Portfolio as well.

Investigation 13A and 13B (text)
Investigation 13-1, 13-2, and 13-3 (laboratory manual)
Instructional Objectives Measured:
Content mastery, Use of scientific methodology, Problem solving skills, Use of technology, Proficiency in written communication

Performance Assessments, page 531

1. **Designing an experiment**
   Instructional Objectives Measured:
   Use of scientific methodology, Problem solving skills, Scientific and chemical literacy, Use of technology, Proficiency in written communication
2. **Testing pH**
   Instructional Objectives Measured:
   Use of scientific methodology
3. **Designing an experiment**
   Instructional Objectives Measured:
   Use of scientific methodology, Scientific and chemical literacy

Portfolio Options

(see page T36 for scoring rubrics)
Formal Laboratory Reports
Exploration 13A and 13B (text)
Instructional Objectives Measured:
Content mastery, Use of scientific methodology, Use of technology, Proficiency in written communication

Portfolio Projects, page 531
Items 1 and 3. Chemistry and you
Instructional Objectives Measured:
Content mastery, Scientific and chemical literacy, Use of scientific methodology, Use of technology, Understanding the impact of science and technology on society, Proficiency in written communication

Item 2. Research and communication
Instructional Objectives Measured:
Scientific and chemical literacy, Proficiency in written communication

Item 4. Cooperative activity
Instructional Objectives Measured:
Scientific and chemical literacy, Understanding the impact of science and technology on society, Proficiency in oral communication

Science, Technology, and Society, page 532
Instructional Objectives Measured:
Scientific and chemical literacy, Understanding the impact of science and technology on society, Problem solving skills, Proficiency in oral and written communication

INTRODUCING THE CHAPTER

CHAPTER 13 Acids and Bases

Story Background

The practice of embalming has religous and cultural roots. The most skilled embalmers were the ancient Egyptians. The most elaborate procedures were reserved for royalty because the ancient Egyptians believed in life after death.

All vital organs were removed, treated with alcohol, and placed in separate jars or vessels. The body cavities were filled with myrrh and other aromatic resins and perfumes. All incisions were closed, and the body was placed in KNO_3 for 70 days. The body was then washed, wrapped in cotton cloth, and dipped in a gummy substance. The mummy was then placed in its tomb.

If students have read *Jurassic Park,* ask them what they learned from this novel about the structure and properties of DNA. Talk about why it is not possible to create an organism from ancient DNA.

You may also want to touch on the subject of the Science, Technology, and Society feature found at the end of this chapter. The STS describes another application of DNA technology that is currently in the news: DNA fingerprinting and its use in criminal cases.

About the Illustration

Other naturally mummified remains have been found buried in the high, dry mountains of South America.

Ancient Acids

With a sterile scalpel, Svante Pääbo scraped tissue from an ancient mummy. This was the end of his long quest for permission to test an idea that first came to him when he was a student in Sweden in the early 1980s. Now he finally had access to 23 of the best-preserved mummies in the state museums in Berlin, Germany.

As a student, Pääbo had wondered whether the techniques that were being used to study the genetic material of living organisms could be applied to dead organisms. He also wanted to know if it was possible to extract DNA from dead tissue like the skins of human mummies preserved in museums throughout the world. But getting tissue samples from archaeological specimens isn't easy. Most museum curators refused to give Pääbo permission, knowing that he would have to destroy a part of the specimen in order to study its genetic material.

Deoxyribonucleic acid, DNA, has been called the master molecule of life. It is the primary component of your chromosomes, and it serves as the carrier of the hereditary information stored in your genes. For 50 years, scientists have been studying the DNA of living organisms, trying to understand how it controls the activities of the cell. Only recently have Svante Pääbo and other scientists begun to study DNA from dead and even extinct organisms. Such DNA is known as ancient DNA. Ancient DNA is recovered from ancient bones, organisms frozen in the Arctic tundra, or creatures trapped in amber.

An insect preserved in amber for 125 million years was the basis for a popular book and movie in which dinosaurs were brought back from extinction. But it's the ancient DNA of our human ancestors that most intrigues scientists like Pääbo. Scientists of this new breed are known as molecular archaeologists. The mummies of people who lived in Egypt more than 4400 years ago provided molecular archaeologists with their first look at ancient human DNA. A human brain that survived intact for nearly 7000 years in a Florida sinkhole was a source of even more ancient DNA.

By studying DNA from sources like these, scientists hope to shed new light on the pathways of human development at the molecular level. The work is painstaking and often frustrating. Precious samples of rare DNA can easily be contaminated by something as simple as a skin cell sloughed off a researcher's body. Also, because there is so little DNA to work with, it is not always possible to repeat experiments in order to establish confidence in the results. Nevertheless, molecular archaeologists are excited about successes they have had with ancient plant and animal DNA and are optimistic that their new field will add another dimension to the study of human origins.

In this chapter, you will explore some of the chemical characteristics of a group of compounds that includes DNA—acids. Recall that DNA's full name is deoxyribonucleic acid. From what you learn in this chapter, you will be able to answer the following questions.

- *What makes DNA an acid?*
- *Is DNA a strong or weak acid?*
- *Why have some ancient DNA fragments been preserved better than others?*

To answer this last question, you will need to take a look at the second group of compounds that are examined in this chapter—bases.

Tapping **Prior Knowledge**

1. What causes acid rain?
2. Describe the properties of acids and bases. How do you determine whether a substance is an acid or base?
3. How does an organic acid differ in structure from an inorganic acid?
4. In the following reaction, how many grams of $CaCl_2$ will be produced from 18 g of HCl?

$$HCl + Ca(OH)_2 \longrightarrow CaCl_2 + H_2O \quad (\approx 27.4\ g)$$

Section 13-1

FOCUS

Lesson Starter

Have students bring in samples of various fruits, juices, and beverages to test their acidity. Use Hydrion paper to determine the pH of each of the samples. Make a chart showing the results of the tests. Students are generally surprised by the level of acidity in some foods. Remind students never to taste any substance brought into the laboratory.

TEACH

Teaching Tip

Properties of acids and bases
Ask students to recall as many properties of acids as possible. List them on the board. Ask students to recall all the acids they know. Write them on the board.

Theme

Classification and trends
The class of compounds called acids is defined by characteristic properties.

Demonstration 1

Ionization of acidic solutions

Approximate time: 10 min

1. Fill three 250 mL beakers with distilled water.
2. Write the formulas $C_{12}H_{22}O_{11}$, HCl, and NaCl on the board. Ask students to describe the type of bonding in each.
3. Add 2 g of sugar to beaker 1, 10 mL of 1 M HCl to beaker 2, and 2 g of salt to beaker 3.
4. Test each solution with the battery-powered conductivity tester. Ask students why HCl is a conductor even though it is covalent. Have them recall what they learned about molecular electrolytes in Chapter 12.

13-1

What are acids and bases?

Section Objectives

Describe the characteristic properties of aqueous acids and bases.

Distinguish between the Arrhenius and Brønsted definitions of acids and bases.

Describe how some compounds can function as both a Brønsted acid and a Brønsted base.

Explain the difference between a strong acid and a weak acid and between a strong base and a weak base.

Identify conjugate acid-base pairs.

Relate the reactions of anhydrides to acid rain.

Acids and bases

Have you ever had difficulty swallowing an aspirin tablet? If you have, then you know how unpleasant aspirin tastes. You learned in Chapter 1 that aspirin is acetylsalicylic acid. When this acid dissolves in saliva, it tastes sour. To mask the sour taste, manufacturers of children's aspirin add both natural and artificial sweeteners. The foods and beverages shown in Figure 13-1 also contain acids and you know that some of these taste sour too. A sour taste is a characteristic property of all acids in aqueous solution.

Figure 13-1
Fruits contain citric acid, lactic acid, and ascorbic acid. Carbonated cola drinks contain benzoic acid and sorbic acid. Teas contain tannic acid.

Acids have distinctive properties

In addition to a sour taste, most acids share several other properties. For example, many aqueous acids react with carbonate in rocks to produce carbon dioxide gas as they dissolve the rock. Acids react with some metals to produce hydrogen gas, as you can see in Figure 13-2. Because the metal reacts with the acid to replace hydrogen, the acid must contain hydrogen. In fact, many acids contain hydrogen.

Figure 13-2
Sulfuric acid reacts with zinc to produce hydrogen gas.

Figure 13-3 shows that aqueous acid solutions conduct electricity. You learned in Chapter 12 that conductivity depends on the presence of ions in solution. Both acids in Figure 13-3 form ions in solution, but one acid forms more ions than the other. The degree to which an acid dissociates determines whether it is strong or weak.

Figure 13-3
Strong acids, like hydrochloric acid on the left, dissociate completely and are therefore better conductors, as shown by the brightly lit bulb on the left. Weak acids, like acetic acid on the right, partially dissociate, so the bulb on the right is dimly lit.

Visual Strategy

Figure 13-3
Remind students that the brighter the bulb, the greater the degree of dissociation. Refer back to the micromodels for HCl and CH_3COOH ionization in Chapter 12.

Table 13-1
Some Strong and Weak Acids

Strong acids	Weak acids
chloric acid, $HClO_3$	acetic acid, CH_3COOH
hydrobromic acid, HBr	boric acid, H_3BO_3
hydrochloric acid, HCl	hydrocyanic acid, HCN
hydroiodic acid, HI	hydrofluoric acid, HF
nitric acid, HNO_3	hydrosulfuric acid, H_2S
perchloric acid, $HClO_4$	hypochlorous acid, HClO
periodic acid, HIO_4	nitrous acid, HNO_2
permanganic acid, $HMnO_4$	oxalic acid, $H_2C_2O_4$
sulfuric acid, H_2SO_4	phosphoric acid, H_3PO_4
tetrafluoroboric acid, HBF_4	sulfurous acid, H_2SO_3

Any acid that dissociates completely in aqueous solution is considered a *strong acid*. Hydrochloric acid is a common strong acid. Because HCl is less dangerous than other strong acids, it is used for tough cleaning jobs. *Weak acids* are dissociated only partially in aqueous solutions. Because relatively few ions form, weak acids exhibit poor conductivity. **Table 13-1** lists the names of some strong and weak acids.

Bases have distinctive properties

Recall from Chapter 1 that you learned about some of the characteristics of bases. They taste bitter and feel slippery. They cause litmus to change from red to blue and they show pH values between 7 and 14. Bases also react with acids to form salts. Household ammonia, used for cleaning, is a base. If you have ever used household ammonia, you know it feels slippery. **Figure 13-4** shows some common household products that consist of aqueous solutions of bases.

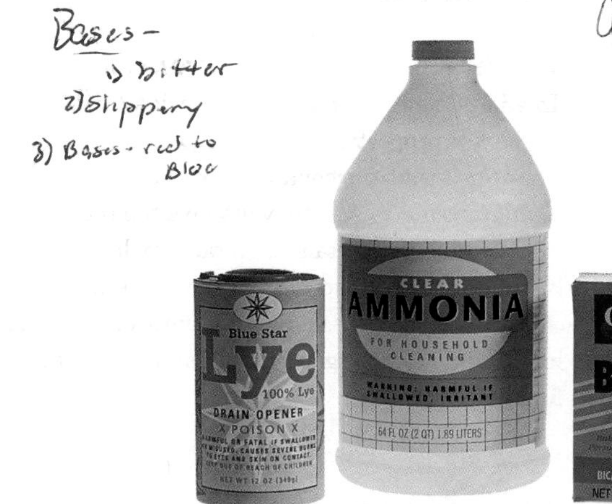

Figure 13-4
Some household cleaning products contain bases and are able to dissolve grease and oil.

Another important property of acids and bases is their ability to neutralize each other. This means that each loses its characteristic properties (such as sour taste or slippery feeling) as a result of neutralization. You will learn more about neutralization later in this chapter.

Table 13-2
Some Strong and Weak Bases

Strong bases	Weak bases
barium hydroxide, $Ba(OH)_2$	ammonia, NH_3
calcium hydroxide, $Ca(OH)_2$	aniline, $C_6H_5NH_2$
potassium hydroxide, KOH	potassium carbonate, K_2CO_3
sodium hydroxide, NaOH	sodium carbonate, Na_2CO_3
trisodium phosphate, Na_3PO_4	trimethylamine, $(CH_3)_3N$

The degree to which a base forms ions in aqueous solutions determines whether it is a strong or weak base. Any base that dissociates completely is a *strong base*. On the other hand, a *weak base* dissociates only partially in aqueous solution. Because relatively few ions are present, a weak base is not a good conductor of electricity. **Table 13-2** lists the names of some strong and weak bases.

SAFETY
Wear safety goggles and a lab apron. Students should remain back at least 10 ft.

DISPOSAL
Neutralize the acid with 1 M NaOH. Pour all the solutions down the drain.

Theme
Macroscopic observations and micromodels
The dissociation of acids to produce ions is shown on a macroscopic level by the completion of the current to light the bulb of the conductivity tester. The theory of ionization explains this observation.

Do You Know?
The common name for hydrochloric acid when sold in hardware stores is muriatic acid. This name is considered obsolete according to current chemical nomenclature.

Theme
Classification and trends
Acids and bases are classified as either strong or weak.

Do You Know?
Strong bases actually begin to dissolve the surface of skin tissue, making the skin feel slippery.

Sodium hydroxide is used to clean drain pipes because it is very effective at dissolving fats, oils, and hair. It is also good at dissolving soap scum, which will clog the pipe.

Theme
Classification and trends
The class of compounds called bases is defined by characteristic properties.

Historical Note

Svante August Arrhenius (1859–1927) presented his theory of ionic dissociation as his Ph.D. thesis in 1884. His theory was met with considerable resistance, however he won the Nobel Prize for chemistry in 1903.

Teaching Tip

Hydrogen ions

Remind students that H^+ is another way to represent a proton, which is a hydrogen atom that has lost its only electron.

Teaching Tip

Dot structures

Draw the electron dot structure for the NO_3^- ion to show that it indeed accepts an electron from a nearby hydrogen atom but does not accept the proton.

Theme

Conservation

During dissociation, charge and atoms are conserved.

Teaching Tip

Dot structures

Draw the electron dot structure for both K and OH to show why K is willing to give up an electron to become K^+ and OH is willing to accept an electron to become OH^-.

$$K^+[:\ddot{O}:H]^-$$

Content Rationale

The introduction of non-hydroxide bases on this page sets the stage for the introduction of Brønsted-Lowry acid-base theory on page 495.

Theme

Systems and interactions

When acids and bases dissociate, they interact to form water molecules.

Figure 13-5 **Svante August Arrhenius, 1859–1927, received the Nobel Prize in 1903 for his work on the theory of electrolytic dissociation.**

Arrhenius definitions of acids and bases

Acids in water produce H^+; bases in water produce OH^-

In the 1880s, the Swedish chemist Svante Arrhenius, shown in **Figure 13-5**, introduced the theory of ionization, which you learned in Chapter 12. This theory states that in aqueous solutions both molecular and ionic compounds can break apart to form ions. Arrhenius found that his theory of ionization explained the behavior of acids and bases.

Arrhenius reasoned that the characteristic properties of acids are the properties of hydrogen ions in solution. Thus, an Arrhenius acid is defined as any compound that contains hydrogen and dissociates in aqueous solution to form H^+ ions. Nitric acid dissociates completely, as shown by the following equation.

$$HNO_3(aq) \longrightarrow H^+(aq) + NO_3^-(aq)$$

Arrhenius also reasoned that the characteristic properties of bases are the properties of OH^- ions in solution. An Arrhenius base is any compound that contains OH^- ions and dissociates to produce OH^- ions in aqueous solution. For example, as potassium hydroxide dissolves in water, it dissociates as shown by the following equation.

$$KOH(s) \xrightarrow{H_2O} K^+(aq) + OH^-(aq)$$

hydroxide ion
the OH^- anion

The OH^- ion is known as the **hydroxide ion**. You should not confuse the hydroxide ion, OH^-, with the hydroxyl group, —OH. You learned in Chapter 6 that the hydroxyl group is a functional group that forms a covalent bond to carbon in the organic compounds known as alcohols.

If you look in **Table 13-2**, you see several compounds that do not include the hydroxide ion in the formula. Ammonia is probably the most common weak base. Though its formula does not include the hydroxide ion, ammonia reacts with water to form hydroxide ions, as shown by the following equation.

$$NH_3(aq) + H_2O(l) \longrightarrow NH_4^+(aq) + OH^-(aq)$$

Bottles of ammonia water are frequently labeled as ammonium hydroxide.

The class of organic bases known as *amines* are related to ammonia in structure. Trimethylamine, $(CH_3)_3N$, reacts similarly to ammonia when added to water, as shown by the following equation.

$$H_3C-\underset{\displaystyle CH_3}{\underset{|}{N}}-CH_3 + H_2O \longrightarrow \left[H_3C-\underset{\displaystyle CH_3}{\underset{|}{\overset{\displaystyle H}{\overset{|}{N}}}}-CH_3\right]^+ + OH^-$$

In the Arrhenius theory, neutralization is the reaction of $H^+(aq)$ and $OH^-(aq)$ to form water, as shown by the following equation.

$$H^+(aq) + OH^-(aq) \longrightarrow H_2O(l)$$

Chem Fact

Nitric acid is ranked 14th among top chemicals produced in the United States. Most nitric acid is produced from ammonia, which is ranked sixth in production.

Acid and base anhydrides

Some oxides react with water to produce acids

You've heard the saying "What goes up must come down." This expression certainly applies to gases emitted on Earth that rise into the atmosphere. Recall from Chapter 10 that gases are produced by the burning of fossil fuels. Some 80% of all the energy used in the United States is obtained by burning fossil fuels. Coal and oil burned by power plants and gasoline burned by cars release sulfur oxides. These sulfur oxide compounds react with moisture in the atmosphere to produce acids.

$$SO_2(g) + H_2O(l) \longrightarrow \underset{\text{Sulfurous acid}}{H_2SO_3(aq)}$$

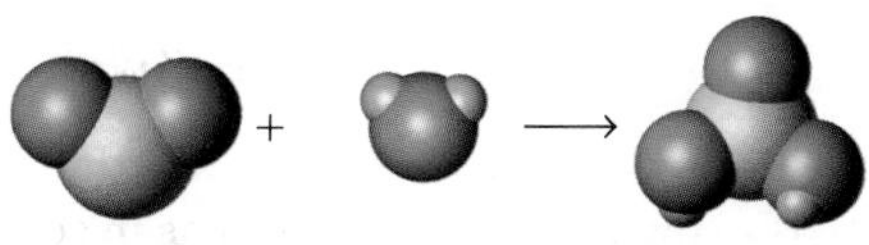

$$SO_3(g) + H_2O(l) \longrightarrow \underset{\text{Sulfuric acid}}{H_2SO_4(aq)}$$

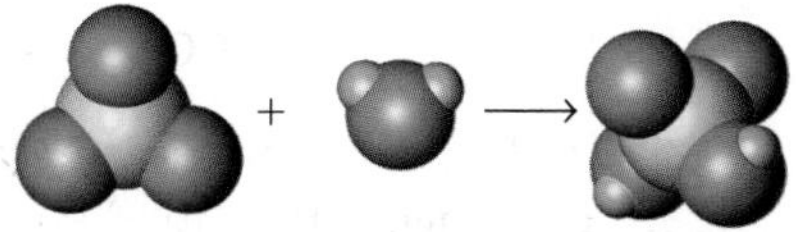

acid anhydride *an oxide that forms an acid when reacted with water*

Substances such as SO_2 and SO_3 belong to a class of compounds called acid anhydrides. An **acid anhydride** is an oxide that reacts with water to form an acid. When fossil fuels are burned, the combustion reactions produce some acid anhydrides and other oxides that are released into the atmosphere, where they form acids that fall back to Earth in the form of acid precipitation. Acid anhydrides are usually oxides of nonmetals in the upper right portion of the periodic table.

An oxide of nitrogen also plays a part in increasing the acidity of rainfall. Nitrogen dioxide is also released into the atmosphere as a product of the burning of fossil fuels. It reacts with water in the atmosphere to produce nitric acid, as shown by the following equation.

$$3NO_2(g) + H_2O(l) \longrightarrow NO(g) + \underset{\text{Nitric acid}}{2HNO_3(aq)}$$

Even though the equation shows an oxide forming an acid, NO_2 is not an acid anhydride. You'll notice that when the sulfur oxides react with water, they form acids exclusively. In the reaction of nitrogen dioxide and water, we get an acid along with nitrogen oxide. The oxides of nitrogen that are acid anhydrides are N_2O_5 and N_2O_3. They react with water as shown by the following equations.

$$N_2O_5(s) + H_2O(l) \longrightarrow 2HNO_3(aq)$$
$$N_2O_3(g) + H_2O(l) \longrightarrow 2HNO_2(aq)$$

Theme

Classification and trends

Acid and base anhydrides are characterized as nonmetal and metal oxides, respectively.

Teaching Tip

Anhydrides

Students should notice that most acid anhydrides are nonmetal oxides, such as carbon dioxide, sulfur dioxide, sulfur trioxide, nitrogen dioxide, etc.

Demonstration 2

Production of an acid from an acid anhydride

Approximate time: 5 min

1. Add about 100 mL of distilled water to a 250 mL beaker. Add one drop of phenolphthalein indicator.
2. Add 5 drops of 1 M $NH_3(aq)$ solution.
3. Use a clean, long straw or piece of glass tubing, and blow through the solution until the solution becomes clear again.

$$CO_2 + H_2O \longrightarrow H_2CO_3$$

SAFETY

Wear safety goggles and a lab apron. Students should remain back at least 10 ft.

DISPOSAL

Pour the solution down the drain.

Demonstration 3

Acid rain

Approximate time: 5 min

Have some students bring in samples of rainwater, bottled water, and tap water to test pH.

SAFETY

Do not allow students to collect water samples from other sources because they could contain pathogens.

DISPOSAL

Pour all solutions down the drain. Throw the Hydrion paper into the trash.

Demonstration 4

Effects of acid rain on calcium carbonate

Approximate time: 5 min

1. Fill a petri dish half full with 1 M HCl, and place it on an overhead projector.
2. Add several small marble chips. Note the evolution of CO_2 gas.
3. Add a piece of chalk to show its rapid decomposition.

Discuss ways that marble objects are protected from exposure to acid rain.

SAFETY

Wear safety goggles and a lab apron. Students should remain back at least 10 ft.

DISPOSAL

Rinse the marble and chalk, then throw them into the trash. Combine the rinse water with the NaOH solution. Neutralize with 1 M acid, and pour the solution down the drain.

Pollutants cause acid rain

Normally, the pH of rainwater is about 5.5. However, the acids formed in the atmosphere can lower this pH to 3. The most acidic rainfall in the United States occurred in Wheeling, West Virginia, where the pH was measured at 1.5. Acid rain is such a large-scale environmental problem because it is difficult to control. The areas producing the pollutants that cause acid rain usually don't experience its effects. Pollutants that are released from factories in highly industrialized areas may travel hundreds of miles, carried by the wind, before they fall back to Earth. In the United States, the heavy burning of fossil fuels in the industrialized Midwest produced very acidic rainfall in the Northeast and Canada. **Figure 13-6** illustrates how acids formed in industrial areas can have an effect many miles away.

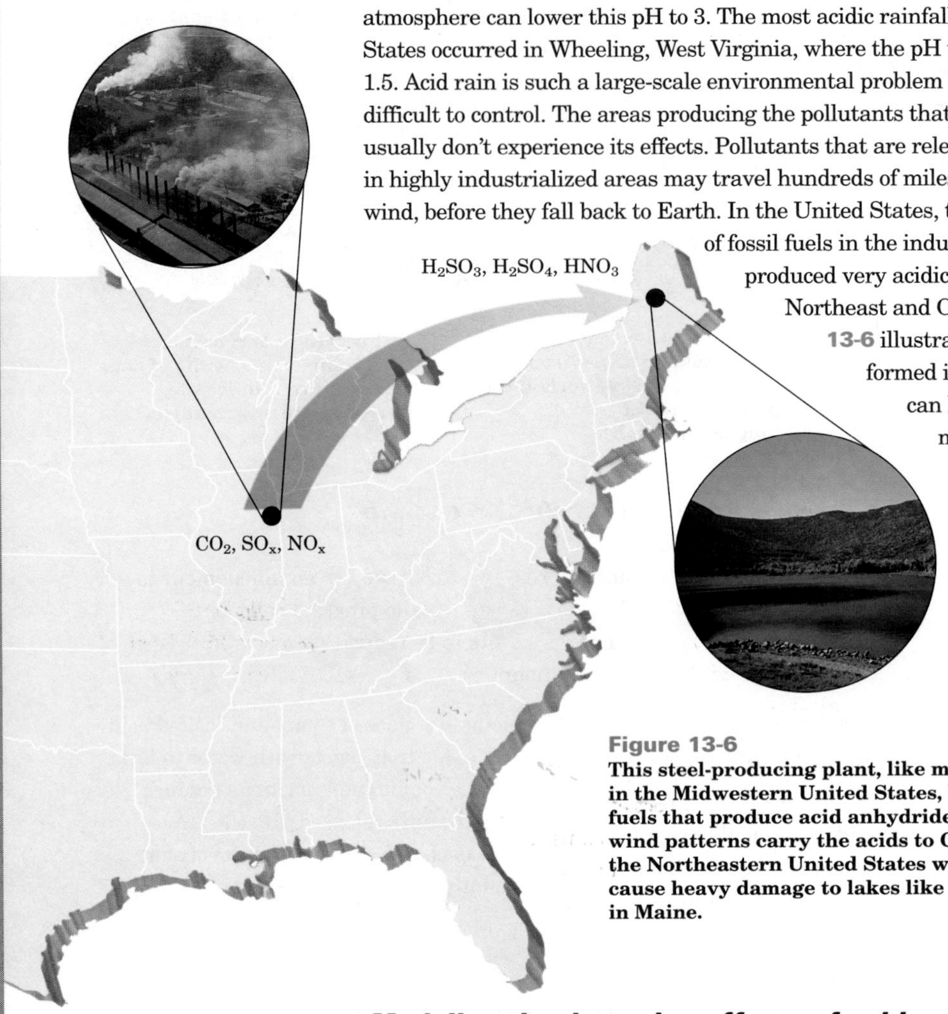

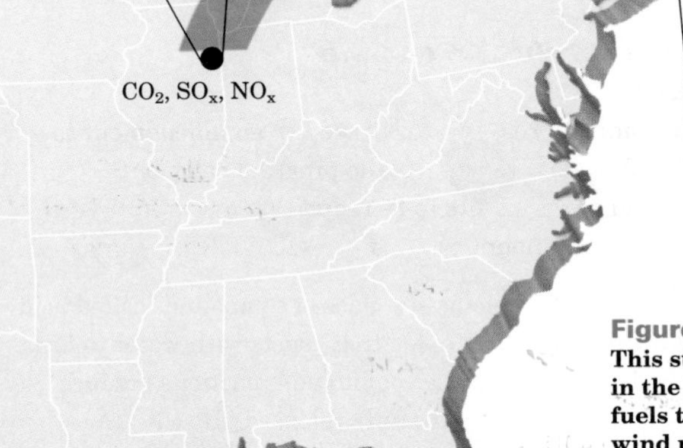

Figure 13-6
This steel-producing plant, like many others in the Midwestern United States, burns fossil fuels that produce acid anhydrides. Prevailing wind patterns carry the acids to Canada and the Northeastern United States where they cause heavy damage to lakes like this one in Maine.

Figure 13-7
The calcium carbonate in seashells is decomposed by the action of an acid.

Modeling the damaging effects of acid precipitation

What happens when a piece of chalk or a seashell is placed in an acid? Chalk and seashells are made of calcium carbonate. Recall that one property of acids is their reaction with carbonates. You can see in **Figure 13-7** how calcium carbonate reacts with the hydronium ion in aqueous acids. The equation for this reaction is shown as follows.

$$CaCO_3(s) + 2H_3O^+(aq) \longrightarrow Ca^{2+}(aq) + CO_2(g) + 3H_2O(l)$$

Calcium carbonate in marble buildings and statues reacts slowly with acidic rainwater. Notice in **Figure 13-8a** on the next page how acids affect marble. The impact of acid precipitation is widespread. In addition to damaging marble, it is causing fish to die in major lakes and rivers. Acid rain also causes food crops to grow more slowly and forests to be thinned out, as shown in **Figure 13-8c**.

Visual Strategy

Figure 13-6

Make students aware that the pollution problems of an area can be caused by pollution formed hundreds of miles away.

Figure 13-8a
Marble structures such as this statue in Brooklyn, New York, are slowly decomposing from acid precipitation.

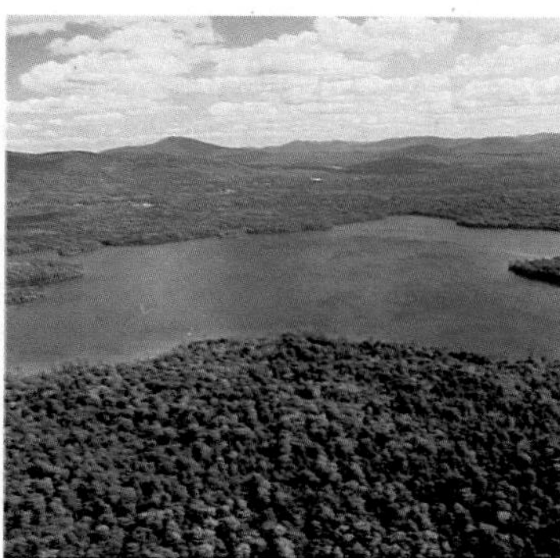

13-8b
Acid rain has reduced the pH of this lake in the Adirondack Mountains of New York to the point where the lake can no longer support life.

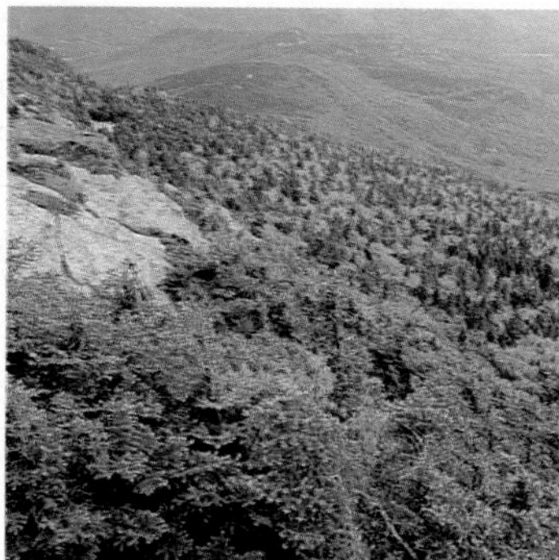

13-8c
This area in Vermont is beginning to show the effects of acid rain. Notice the reddish areas on many of the evergreens.

Recent measurements show a decline in acid precipitation

In 1990 the Acid Rain Control Program was passed as an amendment to the Clean Air Act of 1970. By the year 2000, the program calls for SO_2 emissions by industry and power plants to be reduced to about 15 million tons a year, less than half the amount emitted in 1980. Improvements in the air quality have already been observed.

Solving the problem with base anhydrides

To counteract the effect of acid precipitation on aquatic life, calcium oxide, commonly known as lime, is sometimes added to ponds and lakes. Calcium oxide, CaO, belongs to a class of compounds called base anhydrides. A **base anhydride** is an oxide that reacts with water to form a base. When calcium oxide is added to water, the base calcium hydroxide, $Ca(OH)_2$, is produced.

base anhydride
an oxide that forms a base when reacted with water

$$CaO(s) + H_2O(l) \longrightarrow Ca(OH)_2(s)$$

Calcium hydroxide is also called "slaked" lime. It is a strong base, but it is only slightly soluble in water. In aqueous solution, the slightly soluble $Ca(OH)_2$ dissociates to form Ca^{2+} and OH^- ions.

$$Ca(OH)_2(s) \longrightarrow Ca^{2+}(aq) + 2OH^-(aq)$$

Because this solution contains OH^- ions, it can neutralize acidic solutions.

How can you predict whether an oxide is a basic anhydride or an acidic anhydride? In general, oxides of metals in groups 1 and 2 produce bases when they react with water. For example, magnesium oxide, MgO, produces magnesium hydroxide, $Mg(OH)_2$, in aqueous solution. Oxides of nonmetals often produce acids in water solution. Carbon dioxide reacts with water to form carbonic acid, H_2CO_3.

Chem Fact
Lime ranked fifth in 1993 among top chemicals produced in the United States.

Groupwork Strategy

Divide the class into groups of three, and give them 5 min to brainstorm solutions to the following questions.

1. How can samples of the same rainfall have different pH values?
2. Scientists have usually found that the rain at the beginning of a storm is more acidic than rain collected near the end. Why?
3. Why do lakes have a slightly basic pH ?

Teaching Tip

Base anhydrides
Have students notice that most base anhydrides are metal oxides, such as calcium oxide, magnesium oxide, sodium oxide, etc.

Demonstration 5

Producing a base from a base anhydride
Approximate time: 2 min

1. Fill a petri dish half full of water, and place it on an overhead projector. Add one drop of phenolphthalein.
2. Sprinkle a small pinch, about 0.005 g, of CaO into the solution. The solution will turn pink from the production of calcium hydroxide, $Ca(OH)_2$.

SAFETY
Wear safety goggles and a lab apron. Students should remain back at least 10 ft.

DISPOSAL
Add 1 M HCl slowly until all the solid has dissolved and the solution is clear and colorless. Pour the solution down the drain.

Theme
Classification and trends
Acids are classified as monoprotic, diprotic, or triprotic.

Teaching Tip
Nomenclature
Binary acids are easy to name and recognize. The name has the prefix *hydro-*, the root name of the element, and the suffix *-ic*, as in HCl, *hydro*chlor*ic* acid, and H_2Te, *hydro*tellur*ic* acid.

Do You Know?
Ternary acids that contain oxygen are also known as oxyacids. Oxyacids that were commonly derived from minerals in the ground, like sulfuric and phosphoric acids, are also known as mineral acids.

Theme
Equilibrium and change
An equilibrium is reached in the reaction of weak acids with water.

Answers to
Concept Review
1. $Na_2O + H_2O \rightarrow 2NaOH$
   $Cl_2O_7 + H_2O \rightarrow 2HClO_4$
2. chloric acid, nitric acid, perchloric acid, periodic acid, permanganic acid, tetrafluoroboric acid, acetic acid, hydrocyanic acid, hypochlorous acid, nitrous acid
3. CsO, base
   TeO, acid
   I_2O_5, acid
   SrO, base

Types of acids
Monoprotic and polyprotic acids differ in relative number of ionizable hydrogen ions

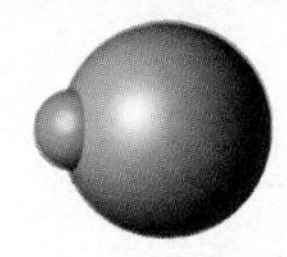

Figure 13-9a Hydrochloric acid is classified as a monoprotic, binary acid.

Compare the moles of acid with the moles of base for the reactions of HCl, $H_2C_2O_4$, and H_3PO_4 shown below.

$$HCl(aq) + KOH(aq) \longrightarrow KCl(aq) + H_2O(l)$$

$$H_2C_2O_4(aq) + 2KOH(aq) \longrightarrow K_2C_2O_4(aq) + 2H_2O(l)$$

$$H_3PO_4(aq) + 3KOH(aq) \longrightarrow K_3PO_4(aq) + 3H_2O(l)$$

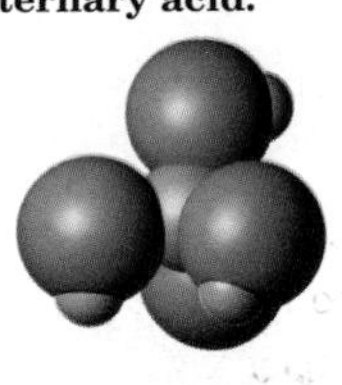

Figure 13-9b Phosphoric acid is classified as a polyprotic, ternary acid.

H_3PO_4 is a weak acid and HCl is a strong acid (shown in **Figure 13-9**). However, one mole of H_3PO_4 will react with three times more base than one mole of HCl does. See the mole ratios in the above equations. H_3PO_4 has three hydrogen ions per molecule to react with the OH^- ions of a base. Acids that can supply more than one hydrogen ion per molecule are called *polyprotic* acids. Acids that can supply only one hydrogen ion per molecule, like hydrochloric acid, are *monoprotic* acids. Notice that it is common to write acidic hydrogens first in the formula for inorganic acids. Can you explain why H_2SO_4 and $H_2C_2O_4$ are *diprotic* acids? Why is acetic acid, $HC_2H_3O_2$, a monoprotic acid?

Binary and ternary acids differ in the number of elements per molecule

Acids can also be categorized by numbers of constituent elements. For example, HCl is a binary acid. A *binary* acid is an acid composed of only two elements, one of which is hydrogen. Some other binary acids are HBr, HI, HF, H_2S, H_2Se, and H_2Te. Acids that contain hydrogen and two other elements, like H_2SO_4 and H_3PO_4, are known as *ternary* acids.

Figure 13-9c The carboxyl group of acetic acid is highlighted in red. Acetic acid is a monoprotic, ternary, carboxylic acid.

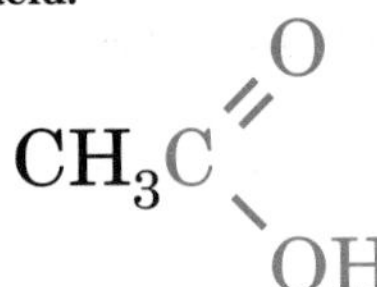

Carboxylic acids all contain the carboxyl group

In Chapter 1 you encountered a number of acids that were organic compounds. Organic acids contain the carboxyl group and for that reason organic acids are often called carboxylic acids. Acetic acid is the most familiar carboxylic acid. It is shown in **Figure 13-9c**.

Anhydrides and types of acids
1. Write equations for the formation of NaOH and $HClO_4$ from their anhydrides, Na_2O and Cl_2O_7.
2. Which of the acids listed in **Table 13-1** are monoprotic, ternary acids?
3. Use the periodic table to predict whether the following anhydrides will form acids or bases in water: Cs_2O, TeO_3, I_2O_5, and SrO.

Answers for Concept Review items and Practice problems begin on page 841.

Visual Strategy
Figure 13-9
Make students aware that binary acids like H_2S, can also be diprotic. Have students note once again the structure of the carboxyl group as characteristic of organic acids.

Stress that the number of ionizable hydrogen atoms for an acid will be important when studying neutralization reactions later in the chapter.

Brønsted acids and bases

Acids donate H^+ ions; bases accept H^+ ions

Chemists came to realize that H^+ ions are protons, and protons cannot exist in aqueous solution. Also, the fact that an ammonia solution is a base is explained by describing ammonia as ammonium hydroxide. However, anhydrous ammonia can neutralize HCl.

$$NH_3(g) + HCl(g) \longrightarrow NH_4Cl(s)$$

Therefore, NH_3 must be a base. These observations led to a change in the definition of acids and bases. In the 1920s, the Danish chemist Johannes Brønsted and the English chemist Thomas Lowry independently suggested a new definition. A *Brønsted acid* is a hydrogen ion donor, and a *Brønsted base* is a hydrogen ion acceptor.

To understand the definition of a Brønsted acid, consider the reaction of acetic acid with water.

$$\underset{\text{Acetic acid}}{CH_3COOH(aq)} + \underset{\text{Water}}{H_2O(l)} \rightleftharpoons \underset{\text{Acetate ion}}{CH_3COO^-(aq)} + \underset{\text{Hydronium ion}}{H_3O^+(aq)}$$

Acetic acid, CH_3COOH, is an organic acid. The equation shows that a hydrogen ion, H^+, is transferred from the –COOH group on acetic acid to water. You can visualize this transfer more clearly with the help of electron-dot diagrams.

$$\text{H:}\overset{\text{H}}{\underset{\text{H}}{\ddot{\text{C}}}}\text{:C}\genfrac{}{}{0pt}{}{\cdot\ddot{\text{O}}\cdot}{\ddot{\text{O}}\text{:}\boxed{\text{H}}} + \text{H:}\ddot{\text{O}}\text{:}\underset{\text{H}}{} \rightleftharpoons \left[\text{H:}\overset{\text{H}}{\underset{\text{H}}{\ddot{\text{C}}}}\text{:C}\genfrac{}{}{0pt}{}{\cdot\ddot{\text{O}}\cdot}{\ddot{\text{O}}\text{:}}\right]^- + \left[\text{H:}\overset{\text{H}}{\underset{\text{H}}{\ddot{\text{O}}}}\text{:}\right]^+$$

Brønsted acid Brønsted base

Because CH_3COOH donates a hydrogen ion, it acts as a Brønsted acid. Notice that H_2O accepts a hydrogen ion and is therefore the Brønsted base in this reaction.

Another example of a Brønsted base can be seen in this equation for the reaction between ammonia and water.

$$NH_3(aq) + H_2O(l) \rightleftharpoons NH_4^+(aq) + OH^-(aq)$$

In this reaction a hydrogen ion, H^+, is transferred from H_2O to NH_3. Again, you can picture this transfer with the help of electron-dot diagrams.

$$\text{H:}\underset{\text{H}}{\ddot{\text{N}}}\text{:H} + \text{:}\overset{\boxed{\text{H}}}{\ddot{\text{O}}}\text{:H} \rightleftharpoons \left[\text{H:}\overset{\text{H}}{\underset{\text{H}}{\ddot{\text{N}}}}\text{:H}\right]^+ + \text{:}\ddot{\text{O}}\text{:H}^-$$

Brønsted base Brønsted acid

Because NH_3 accepts the hydrogen ion, it functions as a Brønsted base. Notice that in this reaction H_2O donates a hydrogen ion and acts as a Brønsted acid.

Teaching Tip

Amphiprotic water

Stress that one of water's greatest properties is the ability to behave as both a weak acid and a weak base. It should be no surprise that the pH of water is found in the middle of the pH scale.

Be sure students understand the reactions on this page. For additional examples, draw dot structures for the carbonic acid reactions on page 496.

Theme

Equilibrium and change

An equilibrium is reached in the reaction of weak bases with water.

Do You Know?
Amphoteric is a synonym of amphiprotic.

Teaching Tip
Conjugate acid-base pairs
Emphasize that the stronger an acid, the weaker its conjugate base. Likewise, the stronger a base, the weaker its conjugate acid.

Answers to Concept Review

4. a. base, CH_3NH_2; acid, H_2O
   b. HSO_4^-, acid; base, H_2O

5. a. base
   b. acid

Amphiprotic compounds

Some compounds can behave both as Brønsted acids and Brønsted bases

If you define acids and bases according to the Arrhenius definitions, you can be sure that an acid will always behave as an acid and a base will always behave as a base. This is true because, in Arrhenius's definition, acids contain ionizable hydrogen, and bases produce OH^- in solution. The Brønsted definitions expand the class of substances that are called bases. The OH^- ion is only one of many bases according to the new definition. This leads to some interesting consequences. Take another look at the two electron-dot representations on page 495. In the first reaction H_2O is a Brønsted base, while in the second reaction H_2O is a Brønsted acid. Any substance that can act as either an acid or a base according to the Brønsted definition is described as **amphiprotic**.

amphiprotic
having the property of behaving as an acid and base

Identifying Brønsted acids and bases

4. In the following equations, identify which reactant is the Brønsted acid and which reactant is the Brønsted base.
   a. $CH_3NH_2(aq) + H_2O(l) \rightleftharpoons CH_3NH_3^+(aq) + OH^-(aq)$
   b. $HSO_4^-(aq) + H_2O(l) \rightleftharpoons H_3O^+(aq) + SO_4^{2-}(aq)$
5. In the following reactions, identify whether H_2O behaves as a Brønsted acid or a Brønsted base.
   a. $HClO_4(aq) + H_2O(l) \longrightarrow H_3O^+(aq) + ClO_4^-(aq)$
   b. $H_2O(l) + SO_3^{2-}(aq) \rightleftharpoons HSO_3^-(aq) + OH^-(aq)$

Conjugate acid-base pairs

Reaction of a Brønsted acid and base produces the conjugate base and acid

In Chapter 12 you learned about the equilibrium system between H_2CO_3 and H_2O in the plasma of your blood.

$$H_2CO_3(aq) + H_2O(l) \rightleftharpoons H_3O^+(aq) + HCO_3^-(aq)$$

Compare that equilibrium system with the reactions of the carbonate ion and the hydrogen carbonate ion with an acid.

$$CO_3^{2-}(aq) + H_3O^+(aq) \rightleftharpoons HCO_3^-(aq) + H_2O(l)$$

$$HCO_3^-(aq) + H_3O^+(aq) \rightleftharpoons H_2CO_3(aq) + H_2O(l)$$

Notice that the second reaction is the reverse of the equilibrium system in your blood. In that system, H_2CO_3 donates a hydrogen ion, so it is a Brønsted acid. Water accepts the hydrogen ion, so it is a Brønsted base. But take another look at the equations. The product formed after H_2CO_3 donates a hydrogen ion is the hydrogen carbonate ion, HCO_3^-.

Notice that in the second reaction the hydrogen carbonate ion acts as a Brønsted base, accepting a hydrogen ion from H_3O^+. The hydrogen carbonate ion, HCO_3^-, is called the conjugate base of the acid H_2CO_3, and

H_2CO_3 is called the conjugate acid of the base, HCO_3^-. A **conjugate base** is the ion or molecule that is formed when a Brønsted acid has given up a hydrogen ion. A **conjugate acid** is the ion or molecule that is formed when a Brønsted base accepts a hydrogen ion.

conjugate base
the particle formed when an acid has donated a H^+ ion

conjugate acid
the particle formed when a base has accepted a H^+ ion

Now identify the product formed when H_2O, acting as a base, accepts a hydrogen ion from H_2CO_3. The H_3O^+ ion that is formed is the conjugate acid of the base H_2O. The hydronium ion, H_3O^+, now has a hydrogen ion to donate and is a Brønsted acid in the reverse reaction. The relationship between an acid and its conjugate base and between a base and its conjugate acid can be seen by labeling each reactant and product as shown in the following examples.

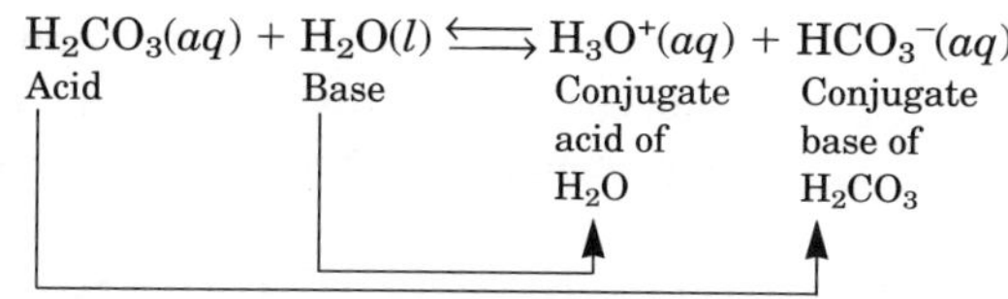

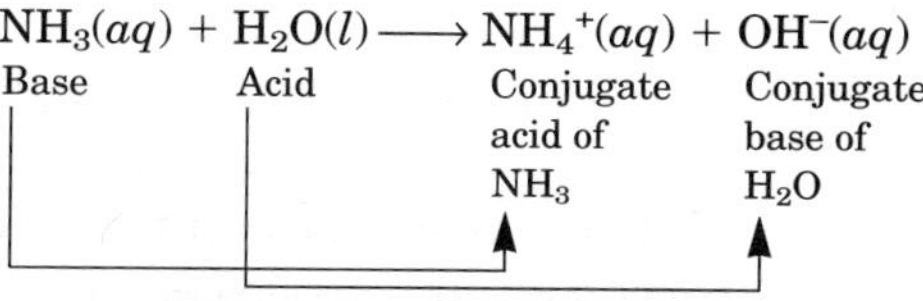

Theme

Equilibrium and change
An equilibrium is established between an conjugate acid-base pair.

ASSESS

Alternative Assessment

Have students list four properties of acids, four properties of bases, define an Arrhenius acid and base, and define a Brønsted acid and base.

Section Review

6. Explain how the Arrhenius definitions of acids and bases differ from the Brønsted definitions.
7. What is an amphiprotic substance?
8. CH_3COOH is a monoprotic acid, yet the formula shows four hydrogen atoms. Why is acetic acid classified as a monoprotic acid?
9. In the following reactions, label the conjugate acid-base pairs.
   a. $H_3PO_4(aq) + NO_2^-(aq) \rightleftharpoons HNO_2(aq) + H_2PO_4^-(aq)$
   b. $CN^-(aq) + HCO_3^-(aq) \rightleftharpoons HCN(aq) + CO_3^{2-}(aq)$
   c. $HCN(aq) + SO_3^{2-}(aq) \rightleftharpoons HSO_3^-(aq) + CN^-(aq)$
   d. $H_2O(l) + HF(aq) \rightleftharpoons F^-(aq) + H_3O^+(aq)$
10. Predict whether an aqueous solution of Na_2O is acidic or basic.

11. **Story Link**
   DNA contains four different organic bases. The structural formulas for two of them are shown below. Identify the sites in each compound where hydrogen ions could be accepted.

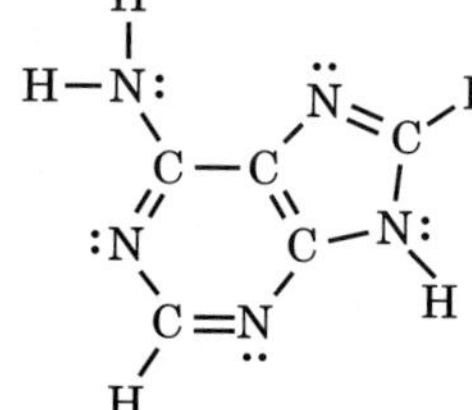

Answers to Section Review

6. Arrhenius definitions describe all acids as having H^+ and bases as having OH^-. Brønsted defines acids as proton donors and bases as proton acceptors; therefore bases do not have to include OH^- to exhibit the characteristics of a base.
7. An amphiprotic substance can act as an acid or base.
8. Acetic acid donates only one hydrogen ion per molecule in acid-base reactions.
9. a. acid, H_3PO_4; base, $H_2PO_4^-$ base, NO_2^-; acid, HNO_2
   b. base, CN^-; acid, HCN acid, HCO_3^-; base, CO_3^{2-}
   c. acid, HCN; base, CN^- base, SO_3^{2-}; acid, HSO_3^-
   d. base, H_2O; acid, H_3O^+ acid, HF; base, F^-
10. $Na_2O + H_2O \rightarrow 2NaOH$ Na_2O will be basic in water.
11. Hydrogen ions could be accepted at all sites on N or O atoms that have nonbonding pairs. The first compound has six possible sites. The second compound has five possible sites.

Section 13-2

FOCUS

Lesson Starter

Refer students back to the macromodels for strong acids and weak acids from Chapter 12 on pages 453 and 458 to review the equilibrium reactions of weak acids with water. You want students to realize that both molecules and ions coexist at equilibrium. The extent to which a forward or reverse reaction is favored is described by the equilibrium constant.

TEACH

Demonstration 6

pH

Approximate time: 10 min

Prepare small samples of the following for pH testing with Hydrion paper: 1 M NaOH in a dropper bottle, household ammonia, milk of magnesia, lake water, baking soda, distilled water, milk, rainwater, apple juice, orange juice, vinegar, lemon juice, and 1 M HCl in a dropper bottle. Have students note the range of pH. Dry compounds can be tested by wetting the Hydrion paper first.

SAFETY

Wear safety goggles and a lab apron. Students should remain back at least 10 ft.

DISPOSAL

Combine all liquids, and neutralize with 1 M acid or 1 M base as appropriate. Pour the neutralized solution down the drain.

Teaching Tip

Reviewing K_{eq}

Remind students that they worked with equilibrium constants in chapter 12.

$$K_{eq} = \frac{[C]^c[D]^d}{[A]^a[B]^b}$$

The dissociation constant, K_a, is set up the same way.

How are weak acids and bases compared?

Section Objectives

Explain how dissociation constants are a measure of the relative strength of weak acids and bases.

Explain and use the pH scale to calculate the concentrations of H_3O^+ and OH^- given the pH of a solution.

Use the pH and molarity of a weak acid solution to calculate K_a.

Dissociation constants

acid-dissociation constant
a quantity derived from the ratio of the concentrations of the products and reactants at equilibrium for a weak acid equilibrium system

The equilibrium constant for the reaction of an aqueous weak acid with water is known as the **acid-dissociation constant**, or K_a. Consider the equilibrium established by acetic acid in vinegar. The equation for the equilibrium system is written below.

$$CH_3COOH(aq) + H_2O(l) \rightleftharpoons CH_3COO^-(aq) + H_3O^+(aq)$$

The equilibrium expression for this reaction is written as follows.

$$K_a = \frac{[CH_3COO^-][H_3O^+]}{[CH_3COOH]}$$

The concentration of H_2O does not appear in the equilibrium expression. The essentially constant concentration of water has already been incorporated into the numerical value of K_a.

K_a and K_b values can be used to compare acid and base strength

At 25°C, the acid-dissociation constant, K_a, for acetic acid has been experimentally determined to be 1.76×10^{-5}. The K_a value for acetic acid can be compared with K_a values for other weak acids as a measure of their relative strengths. **Table 13-3** lists some acid-dissociation constants.

Table 13-3
Selected Acid-Dissociation Constants

Acid name	Acid formula	Reaction temperature (°C)	K_a
acetic	CH_3COOH	25	1.75×10^{-5}
arsenic	H_3AsO_4	18	5.62×10^{-3}
boric	H_3BO_3	20	7.3×10^{-10}
citric	$H_3C_6H_5O_7$	20	7.10×10^{-4}
formic	$HCOOH$	20	1.77×10^{-4}
hydrogen peroxide	H_2O_2	25	2.4×10^{-12}
nitrous	HNO_2	12.5	4.6×10^{-4}
selenous	H_2SeO_3	25	3.5×10^{-2}
sulfurous	H_2SO_3	18	1.54×10^{-2}

Table 13-4
Selected Base Dissociation Constants

Base name	Base formula	Reaction temperature (°C)	K_b
ammonia	NH_3	25	1.77×10^{-5}
aniline	$C_6H_5NH_2$	25	4.27×10^{-10}
asparagine	$C_4H_4O_3(NH_2)_2$	20	1.63×10^{-12}
dimethylamine	$(CH_3)_2NH$	25	5.4×10^{-4}
pyridine	C_5H_5N	25	1.77×10^{-9}
silver hydroxide	$AgOH$	25	1.0×10^{-2}
trimethylamine	$(CH_3)_3N$	25	6.45×10^{-5}

Similarly, equilibrium constants can be determined for weak bases in aqueous solution. These are called **base-dissociation constants**, or K_b. Some are shown in **Table 13-4**.

Look at the equation for the equilibrium system in a bottle of aqueous ammonia used for cleaning. The base-dissociation constant expression would be written as follows.

$$NH_3(aq) + H_2O(l) \rightleftharpoons NH_4^+(aq) + OH^-(aq)$$

$$K_b = \frac{[NH_4^+][OH^-]}{[NH_3]}$$

base-dissociation constant
a quantity derived from the ratio of the concentrations of the products and reactants at equilibrium for a weak base equilibrium system

The equation for the base-dissociation constant for an organic base follows the same pattern. The dissociation equation of trimethylamine and the equation for its K_b are written as follows.

$$(CH_3)_3N(aq) + H_2O(l) \rightleftharpoons (CH_3)_3NH^+(aq) + OH^-(aq)$$

$$K_b = \frac{[(CH_3)_3NH^+][OH^-]}{[(CH_3)_3N]}$$

pH and [H_3O^+]

The concentration of H_3O^+ in solution can be measured

pH
the negative logarithm of the hydronium ion concentration of an aqueous solution; used to express acidity

The pH is a number that is derived from the concentration of hydronium ions in solution. Defined in mathematical terms, **pH** is the negative logarithm (to the base 10) of the H_3O^+ concentration.

$$pH = -log[H_3O^+]$$

For example, if the H_3O^+ concentration in a solution equals 1.0×10^{-12} M, then the pH of the solution is calculated as follows.

$$pH = -log[1.0 \times 10^{-12}]$$

Using a calculator to determine the log of 1.0×10^{-12}, you get –12. Therefore the pH is as follows.

$$pH = -log[1.0 \times 10^{-12}] = -(-12) = 12$$

The neutral point on the pH scale, pH 7, is neither acid or base. However, any solution with a pH of 7 has equal concentrations of H_3O^+ ion and OH^- ion. Let's look at pure water, which has a pH of 7, to see how this is possible.

Theme

Classification and trends
Weak acids and bases can be further classified by their K_a and K_b values.
Make students aware that the closer the value of K_a or K_b is to zero, the weaker the acid or base is.

Possible Misconception

When it is said that a small K_a means that the reverse reaction is favored, this means that the formation of the acid molecule is preferred.

Theme

Macroscopic observations and micromodels
The dissociation of water is the model that explains the presence of ions detected in pure water.

Possible Misconception

When students see the equation for the dissociation of water, they often think that all the water is dissociating. Remind them that by definition there are only 1×10^{-7} mol of H_3O^+ ions per liter of water. In 1 L (approximately 1000 g) of water, there are 1000 g ÷ 18g/mol, or 55.5 mol, of water that have not dissociated.

Possible Misconception

It is useful to show students some examples of logarithms. For example, show that the logarithm of 100 = 2, 1000 = 3, 10 000 = 4, and 100 000 = 5. Then show them that the logarithm of 200 = 2.301, which is the same as 2 × 100, so the log of 2 plus the log of 100 = 0.301 + 2 = 2.301. Likewise, the logarithm of 2000 = 3.301, 20 000 = 4.301, and 200 000 = 5.301.

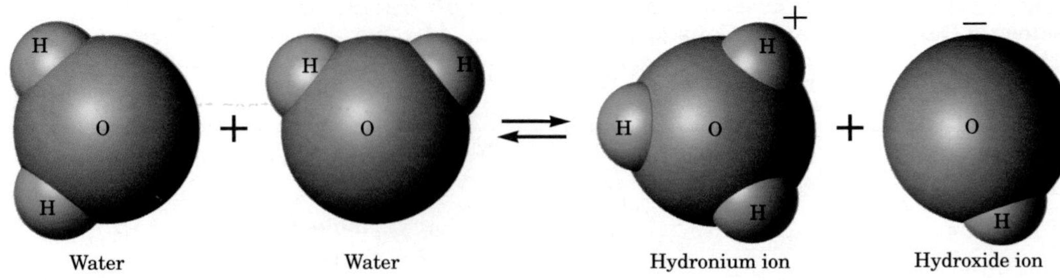

Figure 13-10
The reaction of two molecules produces a hydronium ion and a hydroxide ion. Even pure water at pH 7 contains a small concentration of each of these ions.

Water itself can dissociate

In Chapter 11, you learned that water acts as a solvent in dissolving various solutes. In Chapter 12, you learned that water can bring about the dissociation of both ionic and molecular compounds in solution but pure water does not conduct electricity. Yet there are some ions in pure water. The ions are present in such small quantities, however, that they cannot be detected by a simple conductivity apparatus. You learned that water is an amphiprotic compound. It can act as an acid or base, both of which are quite weak. Water has a very slight tendency to react with itself, which can be described as a Brønsted acid-base reaction. One of the reacting water molecules is a weak acid, the other is a weak base. In **Figure 13-10** you can see how two water molecules interact to produce ions. A hydrogen ion is transferred from one water molecule to another water molecule, forming a hydronium ion and a hydroxide ion. These ions can conduct an electric current, but because they are present in such small quantities, a sensitive measuring instrument known as a microammeter is needed to detect their presence.

Measurements taken with sensitive instruments show that pure water contains 1.0×10^{-7} mol of H_3O^+ ions per liter of solution. This concentration is consistent with a pH measurement for pure water of 7, because when pH is calculated from the H_3O^+ ion concentration you get the following.

$$pH = -log[1.0 \times 10^{-7}] = -(-7.0) = 7.0$$

The calculation shows that the pH of pure water matches the actual measurement of the pH.

If you look again at **Figure 13-10**, you can see that for every H_3O^+ ion formed when water dissociates, one OH^- ion is also formed. Thus pure water contains 1.0×10^{-7} mol of OH^- ions per liter along with 1.0×10^{-7} mol of H_3O^+ ions per liter. The H_3O^+ ion is a Brønsted acid and OH^- ion is a Brønsted base. Pure water contains equal concentrations of the acid and base and, therefore, water is said to be neutral. So if the pH of any solution at 25°C equals 7, then the solution is neutral, and that solution has a H_3O^+ ion concentration that is equal to 1.0×10^{-7} M. That same solution also has a OH^- ion concentration equal to 1.0×10^{-7} M.

Visual Strategy

Figure 13-10

Make students aware that not every water molecule undergoes dissociation; only about 1 in 10 000 000 do.

As $[H_3O^+]$ increases, the pH decreases

When an acid is added to pure water, the concentration of H_3O^+ ions will increase to a value greater than 1.0×10^{-7} M. If an acid added to pure water causes $[H_3O^+]$ to increase to 1.0×10^{-5} M, then the pH will equal 5. Notice that the H_3O^+ ion concentration increased by a factor of 100, from 1.0×10^{-7} M to 1.0×10^{-5} M and the pH decreased by two units, from 7 to 5. Each one-unit change in the pH represents a ten-fold change in the H_3O^+ ion concentration.

As $[H_3O^+]$ decreases, the pH increases

The addition of a base to pure water causes the concentration of H_3O^+ ions to drop below 1.0×10^{-7} M, and the pH rises. For example, if the addition of NaOH causes $[H_3O^+]$ to decrease from 1.0×10^{-7} to 1.0×10^{-9}, the pH increases from 7 to 9.

Each pH value represents a relationship between $[H_3O^+]$ and $[OH^-]$

The pH of a solution does not tell you whether the solution contains a strong or a weak acid. It tells you the concentration of hydronium ions from which you can calculate the concentration of OH^- ions in the solution as well. You can determine the pH of an aqueous solution in several ways. Examine **Figure 13-11** and compare the pH values that were obtained by testing various solutions with a pH meter. Each pH value represents a specific hydronium ion concentration. Even the most acidic solutions contain some hydroxide ions too. You can see in **Figure 13-11** how an increasing hydronium ion concentration results in a decreasing hydroxide ion concentration. At every pH the following relationship holds.

$$[H_3O^+][OH^-] = 10^{-14}$$

Figure 13-11
A pH meter measures pH to a greater degree of precision than pH paper. The pH value of various common substances are shown along with their pH paper indicator color. The relationship between $[H_3O^+]$ and $[OH^-]$ is shown for each pH value. Notice as $[H_3O^+]$ increases, $[OH^-]$ decreases and vice versa.

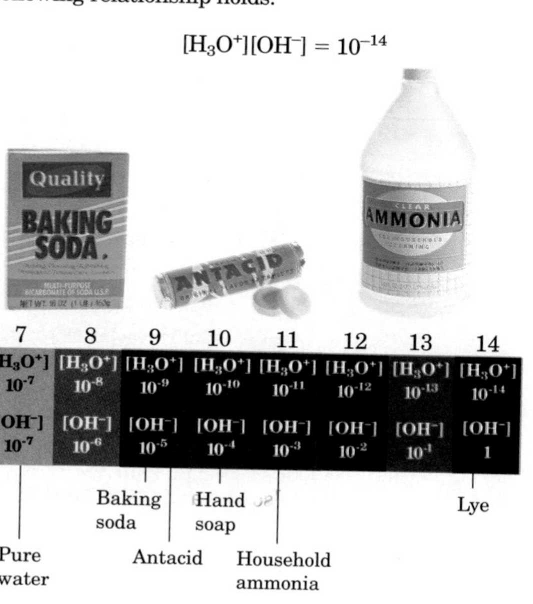

pH	0	1	2	3	4	5	6	7	8	9	10	11	12	13	14
$[H_3O^+]$	1	10^{-1}	10^{-2}	10^{-3}	10^{-4}	10^{-5}	10^{-6}	10^{-7}	10^{-8}	10^{-9}	10^{-10}	10^{-11}	10^{-12}	10^{-13}	10^{-14}
$[OH^-]$	10^{-14}	10^{-13}	10^{-12}	10^{-11}	10^{-10}	10^{-9}	10^{-8}	10^{-7}	10^{-6}	10^{-5}	10^{-4}	10^{-3}	10^{-2}	10^{-1}	1

Theme
Equilibrium and change
The product of the H_3O^+ and OH^- concentrations in aqueous solutions is always equal to 1×10^{-14}.

Teaching Tip
Logarithmic scale
Stress that the pH scale is a logarithmic scale, so an increase in hydroxide concentration by a factor of 10 means a pH increase of 1. Likewise, an increase in hydronium ion concentration by a factor of 10 means a pH decrease of 1.

Theme
Systems and interactions
The addition of a concentrated acid or base to a system will cause a large change in the acidity or basicity of the system.

Do You Know?
Note that the pH of stomach acid is about 2. Therefore it is no wonder that indigestion causes a burning sensation. The throat and tongue sustain acid burns because they are not made of the same type of tissue found in the stomach. Bulimics incur throat and tongue tissue damage. They also develop dental problems because stomach acid slowly eats away at the calcium in the teeth.

Heartburn is the result of stomach acid escaping through the sphincter muscle of the upper stomach and causing a burning sensation in the lower esophagus. Because this area is behind the heart, people have associated this burning effect with the heart instead of with the stomach and esophagus.

Visual Strategy
Figure 13-11
Be sure students understand the relationship between H_3O^+ and OH^-. They will use this mathematical relationship to calculate the pH of a base in Sample Problem 13E.

Teaching Tip

Calculators and problem solving

Most students using regular scientific calculators will have to enter the information in the reverse order of the pH = $-\log[H_3O^+]$ equation. For instance, to find the pH of a substance with a H_3O^+ concentration equal to 1×10^{-5} M, the students must enter the information as the following: [1] [EE] [–5] [log] [±]. The answer 5 will be on the screen.

Additional Examples for Sample Problem 13B

a. What is the pH of a 0.00050 M solution of HBr, a strong acid?

b. What is the pH of a solution if $[H_3O^+] = 2.45 \times 10^{-3}$ M?

Solutions are at the bottom of the page.

Sample Problem 13A
Calculating pH

Refer to the instruction booklet for your calculator to find out how to determine the log of an exponential number.

What is the pH of a 0.0001 M solution of HCl, a strong acid?

❶ List

- **HCl is completely dissociated.**
- **$[H_3O^+]$ of a 0.0001 M HCl solution** = 0.0001 M or 1.0×10^{-4} M
- **pH** = ?

❷ Set up

- **Write the equation for pH.**
  pH = $-log[1 \times 10^{-4}]$
- **Use your calculator to determine the log of 1.0×10^{-4}.**
  The log of 10^{-4} is −4.

❸ Calculate

pH = −(−4) = 4.0

Sample Problem 13B
Calculating pH

What is the pH of a solution if $[H_3O^+] = 3.4 \times 10^{-5}$ M?

❶ List

- **$[H_3O^+]$** = 3.4×10^{-5} M
- **pH** = ?

❷ Set up

- **Write the equation for pH.**
  pH = $-log[3.4 \times 10^{-5}]$

❸ Calculate and verify

- **Use your calculator to determine the log of 3.4×10^{-5}.**
  pH = $-log[3.4 \times 10^{-5}]$ = – (–4.4685), which is rounded to – (– 4.5) = 4.5

The pH of a 1.0×10^{-5} M acid solution is 5. Thus the pH of a solution containing a greater concentration of H_3O^+ ions would be more acidic and have a pH less than 5. Therefore, a pH of 4.5 is reasonable.

Sample Problem 13C
Calculating $[H_3O^+]$

Refer to the instruction booklet for your calculator to find out how to determine the antilog of a number.

The pH of a solution is measured with a pH meter and determined to be 9.00. What is the H_3O^+ ion concentration?

❶ List

- **pH** = 9.00
- **$[H_3O^+]$** = ?

❷ Set up

$$9.00 = -log[H_3O^+]$$
$$-9.00 = log[H_3O^+]$$
$$antilog(-9.00) = [H_3O^+]$$

❸ Calculate

- **Use your calculator to determine the antilog of −9.00.**
  $[H_3O^+] = 1.00 \times 10^{-9}$ M

Solutions for Additional Examples 13B

a. pH = $-\log[5.0 \times 10^{-4}] = 3.3$

b. pH = $-\log[2.45 \times 10^{-3}] = 2.61$

Sample Problem 13D
Calculating $[H_3O^+]$

Refer to the instruction booklet for your calculator to determine the antilog of a number.

The pH of a solution is measured with a pH meter and determined to be 7.52. What is the H_3O^+ ion concentration?

1 List

- **pH** = 7.52
- **$[H_3O^+]$** = ?

2 Set up

$7.52 = -\log[H_3O^+]$

$-7.52 = \log[H_3O^+]$

$antilog(-7.52) = [H_3O^+]$

3 Calculate

- **Use your calculator to find the antilog of –7.52.**
  $[H_3O^+] = 3.01995 \times 10^{-8}$ M which is rounded to 3.02×10^{-8} M

Sample Problem 13E
Calculating the pH of a basic solution

What is the pH of a 0.025 M NaOH solution?

1 List

- **Concentration of NaOH:** 0.025 M
- pH: ?

2 Set up

- **This problem differs from the previous ones because you are starting with a base and the concentration represents the concentration of OH^- ions in the solution. Start by determining $[H_3O^+]$ of the solution.**
- **For any solution**
  $[H_3O^+][OH^-] = 10^{-14}$
- **In this problem $[OH^-]$ = 0.025 M.**
- **Therefore,**
  $$[H_3O^+] = \frac{10^{-14}}{[0.025]} = \frac{10^{-14}}{2.5 \times 10^{-2}} = 0.40 \times 10^{-12} = 4.0 \times 10^{-13}$$
- **Now calculate pH. Start with the following equation.**
  $pH = -\log[4.0 \times 10^{-13}]$
- **Use your calculator to determine the log of 4.0×10^{-13}.**
  The log of 4.0×10^{-13} is –12.3979, which is rounded to –12.4.

3 Calculate

$pH = -(-12.4) = 12.4$

Practice

Answers for Concept Review items and Practice problems begin on page 841.

13A 1. **Determine the pH of the following solutions of strong acids.**
a. 0.00001 M HNO_3 **b.** 0.01 M HNO_3

13B 2. **Determine the pH of the following solutions.**
a. 2.50×10^{-6} M HNO_3 **b.** 8.750×10^{-4} M HCl

13C 3. **Determine $[H_3O^+]$ of the following solutions.**
a. drain cleaner with a pH of 13.0 **b.** ammonia with a pH of 11.0

13D 4. **Determine $[H_3O^+]$ of the following solutions.**
a. human blood with a pH of 7.4 **b.** apple juice with a pH of 3.5

13E 5. **Determine the pH of the following base solutions.**
a. 0.0005 M KOH **b.** 2.80×10^{-4} M NaOH

Additional Examples for Sample Problem 13D

a. What is the $[H_3O^+]$ of a solution with a pH of 5.85?

b. The pH of a solution was found to be 11.3. What is the $[H_3O^+]$?

Solutions are at the bottom of the page.

Answers to Practice 13A–13E

1. a. 5 b. 2
2. a. 5.60 b. 3.058
3. a. 1×10^{-13} b. 1.00×10^{-11}
4. a. 4.0×10^{-8} b. 3.16×10^{-4}
5. a. 10 b. 10.5

Solutions for Additional Examples 13D

a. $antilog(-5.85) = [H_3O^+] = 0.00000141$ M $= 1.41 \times 10^{-6}$ M

b. $antilog(-11.3) = [H_3O^+] = 5.01 \times 10^{-12}$ M

Do You Know?

The purpose of walking after a strenuous activity is to allow the blood to remove excess lactic acid from the muscle cells. Excess lactic acid in muscle cells can cause cramping.

Additional Examples for Sample Problem 13F

a. When the concentration of selenic acid is 0.05 M, the pH of the acid is 1.72. Calculate the K_a for selenic acid.

$$H_2SeO_4 + H_2O \rightleftharpoons H_3O^+ + HSeO_4^-$$

b. When the concentration of nitrous acid is 0.08 M, the pH of the acid is 2.4. Calculate the K_a for nitrous acid.

$$HNO_2 + H_2O \rightleftharpoons H_3O^+ + NO_2^-$$

Solutions are at the bottom of the page.

Theme

Systems and interactions
Because weak acids do not totally dissociate, their interactions and effects on pH can be determined using K_a.

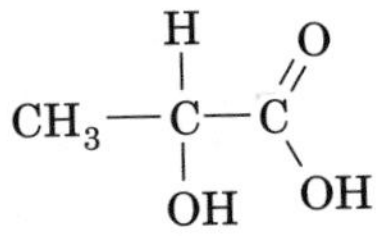

$$CH_3-\underset{OH}{\overset{H}{C}}-C(=O)OH$$

Figure 13-12
Lactic acid is a carboxylic acid, as shown by the carboxyl group in its structural formula.

Calculating dissociation constants

Using pH to calculate K_a

Have you ever gone to your refrigerator for milk only to find that it had turned sour? The sour taste comes from lactic acid, a monoprotic organic acid that is present in sour milk. Lactic acid reacts with water as shown by the following equation.

$$\underset{\text{Lactic acid}}{CH_3CHOHCOOH(aq)} + H_2O(l) \rightleftharpoons H_3O^+(aq) + \underset{\text{Lactate ion}}{CH_3CHOHCOO^-(aq)}$$

Sample Problem 13F
Calculating the dissociation constant for an acid

Assume that enough lactic acid is dissolved in sour milk to give a solution concentration of 0.100 M lactic acid. A pH meter shows that the pH of the sour milk is 2.43. Calculate K_a for the lactic acid equilibrium system.

1 List

- **The lactic acid concentration is initially 0.100 M. However at equilibrium the lactic acid concentration will be less than 0.100 M. The amount that it is reduced depends on the amount of lactic acid that dissociates.**
- $[CH_3CHOHCOOH] < 0.100$ M
- pH = 2.43
- $[H_3O^+] = [CH_3CHOHCOO^-]$
- $K_a = \dfrac{[H_3O^+][CH_3CHOHCOO^-]}{[CH_3CHOHCOOH]} = ?$

2 Set up

- **To solve for K_a, you need to know the equilibrium concentration of each substance in the K_a expression. From the pH, you can calculate the value of $[H_3O^+]$, which is also the value of $[CH_3CHOHCOO^-]$.**

$$2.43 = -\log[H_3O^+]$$
$$-2.43 = \log[H_3O^+]$$
$$\text{antilog}(-2.43) = [H_3O^+]$$

- **Use your calculator to determine the antilog of –2.43.**
$[H_3O^+] = 3.71535 \times 10^{-3}$ M, which is rounded to 3.72×10^{-3} M
- **According to the equation, H_3O^+ and $CH_3CHOHCOO^-$ are present in a 1:1 molar ratio. Thus,**
$[CH_3CHOHCOO^-] = 3.72 \times 10^{-3}$ M
- **The lactic acid concentration at equilibrium equals the initial concentration of lactic acid minus the amount that dissociates. According to the equation, there is a 1:1 ratio between the amount of H_3O^+ formed and the amount of lactic acid that dissociated. Thus if 3.72×10^{-3} M of H_3O^+ are formed in this reaction, then 3.72×10^{-3} M of lactic acid must have dissociated. Thus $[CH_3CHOHCOOH]$ present at equilibrium is found as follows.**
0.100 M – 0.00372 M = 0.09628 M = 9.6×10^{-2} M
- **You now have all the concentrations to substitute in the equilibrium expression for K_a.**

3 Calculate

$$K_a = \frac{[H_3O^+][CH_3CHOHCOO^-]}{[CH_3CHOHCOOH]} = \frac{(3.72 \times 10^{-3})(3.72 \times 10^{-3})}{(9.6 \times 10^{-2})} = 1.4 \times 10^{-4}$$

Solutions for Additional Examples 13F

a. $K_a = \dfrac{[H_3O^+][HSeO_4^-]}{[H_2SeO_4]}$; $1.72 = -\log[H_3O^+]$; $[H_3O^+] = 0.019$ M $= [HSeO_4^-]$; $K_a = \dfrac{[0.019][0.019]}{[0.05 - 0.019]} = 1.16 \times 10^{-2}$

b. $K_a = \dfrac{[H_3O^+][NO_2^-]}{[HNO_2]}$; $2.4 = -\log[H_3O^+]$; $K_a = \dfrac{[0.0040][0.0040]}{[0.08 - 0.0040]} = 2.1 \times 10^{-4}$

Practice 13F

1. Calculate K_a for lactic acid if a 0.2800 M solution has a pH of 2.21.
2. Most berries have a tart taste because they contain benzoic acid. Calculate the concentration of benzoic acid molecules in berries with a pH of 4.96. The concentration of benzoate ions is 0.0140 M, and K_a for benzoic acid is 6.3×10^{-5}. The equation for the equilibrium reaction is written as follows.

$$C_6H_5COOH(aq) + H_2O(l) \rightleftharpoons C_6H_5COO^-(aq) + H_3O^+(aq)$$

Dissociation constant of water

Calculating the $[OH^-]$ or $[H_3O^+]$ of an aqueous solution

You learned that pure water dissociates only slightly, as shown by the following equation.

$$H_2O(l) + H_2O(l) \rightleftharpoons H_3O^+(aq) + OH^-(aq)$$

K_w is the symbol used to represent the equilibrium constant of this reaction.

$$K_w = [H_3O^+][OH^-]$$

In pure water, $[H_3O^+]$ and $[OH^-]$ equal 1.00×10^{-7} M. By substituting these values into the equilibrium expression, you can determine K_w for water.

$$K_w = [1.00 \times 10^{-7}][1.00 \times 10^{-7}] = 1.00 \times 10^{-14}$$

dissociation constant for water *the ion product constant for water that is equal to 1.00×10^{-14}*

The **dissociation constant for water,** K_w, at 25°C is 1.00×10^{-14}. You have already used K_w to calculate the concentration of H_3O^+ ions in base solutions to determine their pH. This same idea is used to calculate the concentration of OH^- ions if you know $[H_3O^+]$.

Sample Problem 13G *Calculating $[OH^-]$*

The pH of sea water is 8.3. What is $[OH^-]$ of sea water?

1 List

- pH = 8.3
- $[H_3O^+][OH^-] = 1.00 \times 10^{-14}$
- $[OH^-] = ?$

2 Set up

$$8.3 = -log[H_3O^+]$$
$$-8.3 = log[H_3O^+]$$
$$antilog(-8.3) = [H_3O^+]$$

3 Calculate

- **Use your calculator to determine the antilog of −8.3.**
  $[H_3O^+] = 5.01187 \times 10^{-9}$ M, which is rounded to 5.0×10^{-9} M
- **Substitute all values into the expression for K_w.**
  $1.00 \times 10^{-14} = [5.0 \times 10^{-9}][OH^-]$

$$[OH^-] = \frac{1.00 \times 10^{-14}}{5.0 \times 10^{-9}} = 2.0 \times 10^{-6} \text{ M}$$

Practice 13G

1. Calculate $[OH^-]$ for the following solutions.

   a. 0.001 M HNO_3
   b. milk with a pH of 6.7
   c. 1.5×10^{-4} M KOH
   d. soap with a pH of 10.8
   e. strawberries with a pH of 3.5
   f. lemon juice with a pH of 2.3

Teaching Tip

pH scale

Show students how the range for the pH scale is derived by calculating pH from $[H_3O^+]$ = 1 M for the acid and, in a separate problem, by calculating pH from $[OH^-]$ = 1 M for the base.

Theme

Conservation

The product of the H_3O^+ and OH^- concentrations in aqueous solutions is always equal to 1×10^{-14}.

The OH^- concentration can be determined from the H_3O^+ concentration because the product of the concentrations is a constant.

Additional Example for Sample Problem 13G

The pH of milk of magnesia is 10.5. What is the $[OH^-]$?

Solution is at the bottom of the page.

Answers to Practice 13F

1. 1.39×10^{-4}
2. 2.4×10^{-3} M

Practice 13G

1. a. 1×10^{-11}
   b. 5.0×10^{-8}
   c. 1.5×10^{-4}
   d. 6.31×10^{-4}
   e. 3.2×10^{-11}
   f. 2.0×10^{-12}

Solutions for Additional Example 13G

$antilog(-10.5) = [H_3O^+]$; $[H_3O^+] = 3.16 \times 10^{-11}$ M

$[H_3O^+][OH^-] = 1 \times 10^{-14}$; $[3.16 \times 10^{-11}][OH^-] = 1 \times 10^{-14}$

$$[OH^-] = \frac{1 \times 10^{-14}}{[3.16 \times 10^{-11}]}$$

$[OH^-]$ = 0.000 316 M

Multicultural Perspective

Scottish chemist Thomas Graham (1805–1869) studied various forms of phosphoric acid in 1833 and showed that it was possible to replace more than one of the hydrogen ions present in the molecule.

ASSESS

Alternative Assessment

Have students write for 3 min on the following.

1. Why is HCl considered a strong acid, and why is CH_3COOH considered a weak acid?
2. What does numerical value for K_a represent?

Answers to Section Review

12. Strong acids and bases are completely dissociated, so there is no equilibrium between the ionized and molecular forms of the strong acid or base. Dissociation constants describe the ratio of product and reactant concentrations at equilibrium.

13. pH is a measure of acidity that is calculated as the negative logarithm of the hydronium ion concentration.

14. Water is capable of donating and accepting protons.

15. The pH 11 solution has a larger $[OH^-]$. It is 100 times larger. The pH 8 solution has a larger $[H_3O^+]$. It is 100 times larger.

16. $[OH^-] = 1.3 \times 10^{-11}$ M

17. $K_a = 3.43 \times 10^{-5}$

18. acid, H_3PO_4; base $H_2PO_4^-$
base, H_2O; acid, H_3O^+
acid, $H_2PO_4^-$; base, HPO_4^{2-}
base, H_2O; acid, H_3O^+
acid, HPO_4^{2-}; base PO_4^{3-}
base, H_2O; acid, H_3O^+

19. H_3PO_4 is the strongest acid, PO_4^{3-} is the strongest base.

*Full solutions for items **15–17** are found on page 531A.*

Polyprotic acids

pH and K_a values of polyprotic acid solutions

You learned that polyprotic acids can donate more than one hydrogen ion. For example, carbonic acid can donate two hydrogen ions, as shown by the following equations.

$$H_2CO_3(aq) + H_2O(l) \rightleftharpoons H_3O^+(aq) + HCO_3^-(aq)$$

$$HCO_3^-(aq) + H_2O(l) \rightleftharpoons H_3O^+(aq) + CO_3^{2-}(aq)$$

Both reactions reach equilibrium so both reactions have acid dissociation constants. The K_a value for the first reaction is 4.3×10^{-7}, while the K_a value for the second reaction is 5.6×10^{-11}. Carbonic acid is a very weak acid, as indicated by the small K_a values for both reactions. Donating a second hydrogen ion is much more difficult than donating the first one as shown by the smaller K_a for the second reaction. For simple polyprotic acids, each successive loss of a hydrogen ion is 10^4 to 10^6 times more difficult than the previous loss. This means that the first dissociation of a polyprotic acid produces up to a million times more H_3O^+ ions than the second dissociation. Thus, the pH of a polyprotic acid solution depends primarily on the dissociation of the first hydrogen ion.

Section Review

12. Explain why dissociation constants apply only to weak acids and bases and not to strong acids and bases.
13. Define the pH of a solution.
14. How does the dissociation of water demonstrate the amphiprotic nature of its molecules?
15. One aqueous base has a pH of 8. Another has a pH of 11. Which solution has the larger concentration of OH^- ions? How many times larger? Which solution has the larger concentration of H_3O^+ ions? How many times larger?
16. An aqueous solution is found to have a pH of 3.1. Determine both $[H_3O^+]$ and $[OH^-]$ of this solution.
17. A 0.0160 M solution of a monoprotic acid has a pH value of 3.14. Calculate K_a for this acid.
18. Phosphoric acid is a triprotic acid that can go through three successive dissociation reactions.

$$H_3PO_4(aq) + H_2O(l) \leftrightarrows H_3O^+(aq) + H_2PO_4^-(aq) \quad K_a = 7.1 \times 10^{-3}$$

$$H_2PO_4^-(aq) + H_2O(l) \leftrightarrows H_3O^+(aq) + HPO_4^{2-}(aq) \quad K_a = 6.2 \times 10^{-8}$$

$$HPO_4^{2-}(aq) + H_2O(l) \leftrightarrows H_3O^+(aq) + PO_4^{3-}(aq) \quad K_a = 4.5 \times 10^{-13}$$

Identify the conjugate acid-base pairs for each of the successive dissociation equations for phosphoric acid.

19. Which of the pairs in item **18** contains the strongest acid? Which contains the strongest base?

Section 13-3

FOCUS

Lesson Starter

Ask students for their ideas about the meanings of the word *neutralization* (in common usage, to neutralize the enemy; in chemistry, to neutralize a base). Work with students to refine their ideas in order to describe the characteristics and effects of the process shown in **Figure 13-13**.

What is a titration?

Section Objectives

Write an equation for a neutralization reaction.

Describe how an acid-base titration is performed.

Determine the molarity of an acid or base solution using titration data.

Use Le Châtelier's principle to explain why indicators change color.

Distinguish between equivalence point and end point.

Neutralization

The acid shown in **Figure 13-13a** can dissolve metals; the base can unclog a drain. But if you mix these two solutions, you get an aqueous salt solution that cannot dissolve metals or unclog drains. To understand why this happens, look at the molecular equation for the reaction.

$$HCl(aq) + NaOH(aq) \longrightarrow NaCl(aq) + H_2O(l)$$

How would you classify this reaction? Notice that the products are a salt (NaCl) and water. This reaction is known as a neutralization reaction according to the Arrhenius concept of acids and bases. A **neutralization reaction** in aqueous solution is a reaction of an acid and a hydroxide base to produce a salt and water. Look at the ionic equations for this reaction.

neutralization reaction
a reaction between an acid and a hydroxide base in which H^+ and OH^- react to form H_2O

Ionic equation

$$H_3O^+(aq) + Cl^-(aq) + Na^+(aq) + OH^-(aq) \longrightarrow Na^+(aq) + Cl^-(aq) + 2H_2O(l)$$

Net ionic equation

$$H_3O^+(aq) + OH^-(aq) \longrightarrow 2H_2O(l)$$

Figure 13-13a
The pH paper turns red because the solution in this beaker is acidic.

13-13b
The pH paper turns blue because the solution in this beaker is basic.

13-13c
The pH paper turns green because the solution in this beaker is neutral.

Visual Strategy

Figure 13-13

Students should recognize that the ion that characterizes the acid and the ion that characterizes the base react to form the neutral compound and water.

TEACH

Possible Misconception

Make students aware that there are many indicators available and that the colors of indicators and their end points vary.

Multicultural Perspective

German chemist Jeremias Benjamin Richter (1762–1807) was the first to quantitatively study neutralization reactions. In 1792, he published his work, which stated that a fixed weight of one chemical reacted with a fixed weight of another chemical.

Demonstration 7

Red cabbage indicator

Approximate time: 15 min

1. Shred 1/4 of a head of red cabbage and boil it in 500 mL of water for 5 min or until the liquid turns dark blue.
2. Let the solution cool and pour it into a beaker. Put the beaker on an overhead projector.
3. Add a few drops of vinegar.
4. Add some household ammonia solution to the beaker.
5. Add more vinegar to the beaker until the solution changes color.
6. You can soak some paper strips in the cabbage indicator to make your own indicator paper.

SAFETY
Wear safety goggles and a lab apron. Students should remain back at least 10 ft.

DISPOSAL
Pour the solutions down the drain.

Theme

Classification and trends
Indicators are used to classify solutions as acids or bases.

salt
a compound with an ionic lattice that is formed from an acid when H^+ is replaced by a metal ion or cation

You learned that a salt is defined as an ionic compound having a crystal lattice structure. You can now see that a **salt** can also be defined as a compound consisting of the anion of an aqueous acid and the cation of an aqueous hydroxide base. If you were to evaporate the solution formed from the neutralization reaction, you would obtain solid NaCl. Neutralization reactions are one way of preparing salts. What salt can be made by mixing H_2SO_4 and KOH?

Titration is used to find the concentration of an acid or base solution

titration
an analytical procedure used to determine the concentration of a sample by reacting it with a standard solution

titration standard
a solution of precisely known concentration; also called a titrant

indicator
(as applied to acid-base chemistry) a substance that reversibly changes color depending on the pH

transition interval
the pH range over which an indicator exhibits different colors for its acidic and alkaline forms

In **Figure 13-13**, the concentrations of both the HCl and NaOH solutions were known when the solutions were mixed. But sometimes the concentration of either the acid or the base is not known. In that case, you can carry out a titration to determine the unknown concentration. A **titration** is a procedure in which a solution of known concentration is used to determine the concentration of a second unknown solution.

The solution of known concentration is the **titration standard** on which the analysis is based. For example, you could use a standard solution of NaOH to titrate an unknown concentration of HCl. The feature on page 512 shows how to carry out the titration. Notice that an indicator is added to the acid solution being titrated. Any substance in solution that changes its color as it reacts with either an acid or a base is called an **indicator**. Many dyes, as well as natural substances like the juice of red cabbage, can act as indicators as shown in **Figure 13-14**. Red cabbage juice turns red with vinegar but blue-green with baking soda. Some indicators produce a variety of colors depending on the pH of the solution.

Selecting the proper indicator is important because each indicator changes its color over a particular range of pH values. You can see this in **Table 13-5**. The pH range over which an indicator changes color is called its **transition interval.** The figure also shows that indicators are classified into three types based on the types of titrations and their transition intervals. What would be an appropriate indicator to use for titrating NaOH with HCl—a strong base and strong acid? Bromthymol blue would be a good choice. Notice in **Table 13-5** that bromthymol blue is yellow in an acidic solution, blue in a basic solution, and green in transition.

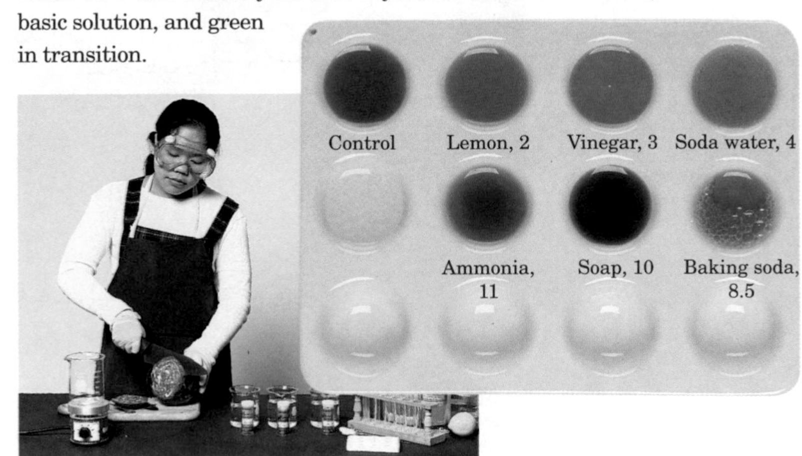

Figure 13-14
The juice from red cabbage leaves can be used as an indicator to measure the pH of household items. A set of color standards is made using the indicator in solutions of known pH from 1 through 14. When the household items are tested, they can be compared to the colors of the known standards to determine the pH of each.

Visual Strategy

Figure 13-14

Make students aware that the cabbage juice is changing in response to the amount of H_3O^+ ions in solution.

Table 13-5
Color Ranges of Selected Indicators Used in Titrations

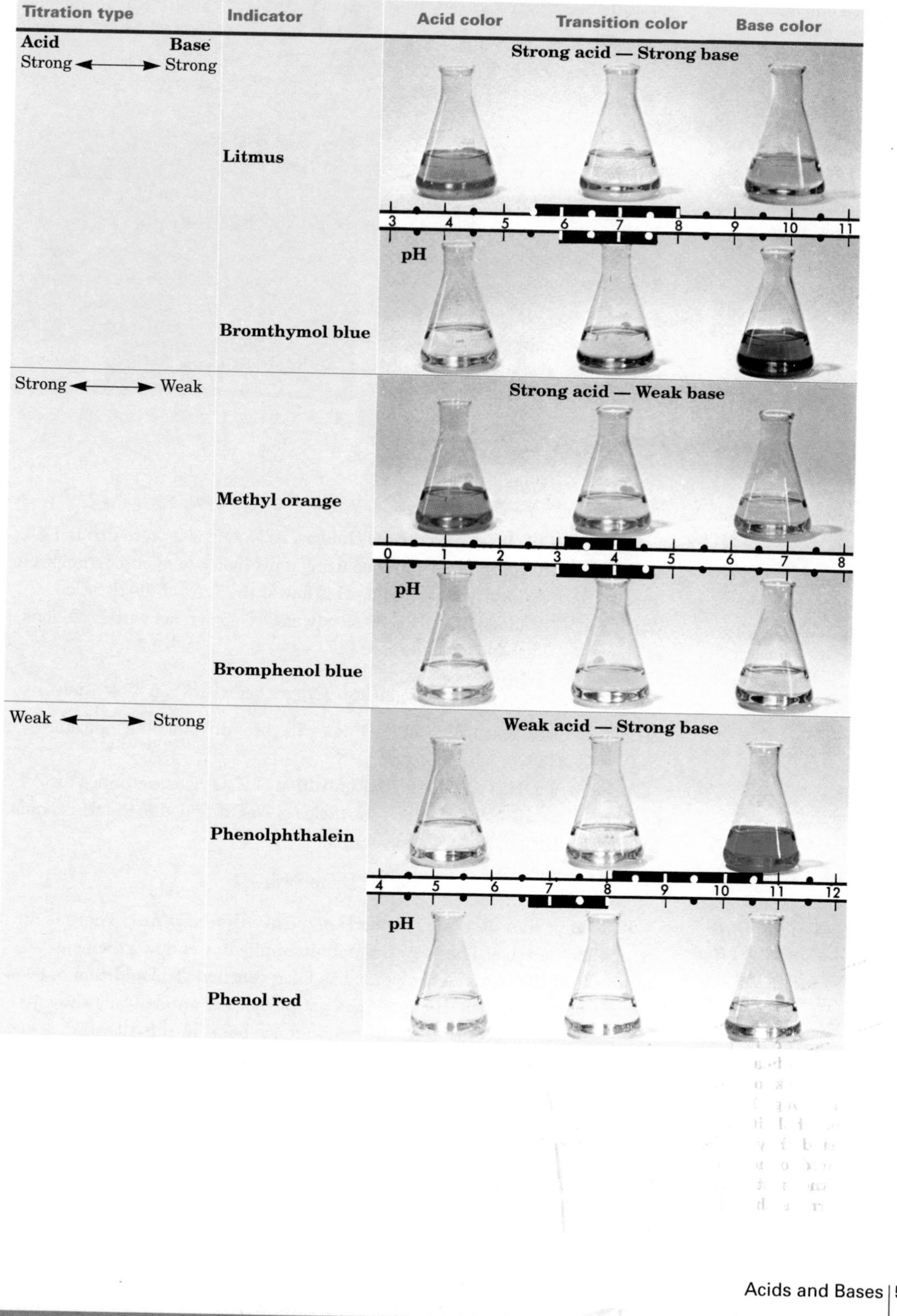

Titration type	Indicator	Acid color	Transition color	Base color
Acid Strong ↔ Base Strong	Litmus	Strong acid — Strong base		
	Bromthymol blue			
Strong ↔ Weak	Methyl orange	Strong acid — Weak base		
	Bromphenol blue			
Weak ↔ Strong	Phenolphthalein	Weak acid — Strong base		
	Phenol red			

Do You Know?

The active ingredient in many laxatives is phenolphthalein. Remind students to be careful with phenolphthalein indicator because the phenolphthalein can be absorbed through the skin.

Theme

Systems and interactions
An indicator changes color in response to its interaction with an acid or base.

Visual Strategy

Table 13-5

Make sure students realize that the pH ranges shown underneath the flasks are shifted for the three different solutions.

Teaching Tip

Perform the titration of HCl with NaOH depicted by **Figure 13-15**, so that students can see how the data points in the graph are derived.

Theme

Systems and interactions

The titration is based on the interaction of equal numbers of acid and base molecules.

Demonstration 8

Indicators

Approximate time: 5 min

1. The day before class, mix a few drops of phenolphthalein indicator with 20 mL of distilled water.
2. Make a sign with a saying of your choice on a large piece of white paper using the indicator-water mixture and allow the paper to dry.
3. Hang the paper on the wall or chalkboard and have a spray bottle of ammonia-based window cleaner on the desktop.
4. Pick up the cleaner and spray the sign. The letters should appear in red. Ask students why this happened. Other indicators can be used to make other colors for the sign.

SAFETY

Wear safety goggles and a lab apron. Students should remain back at least 10 ft. Spray the cleaner away from the students.

DISPOSAL

Throw the sign into the trash.

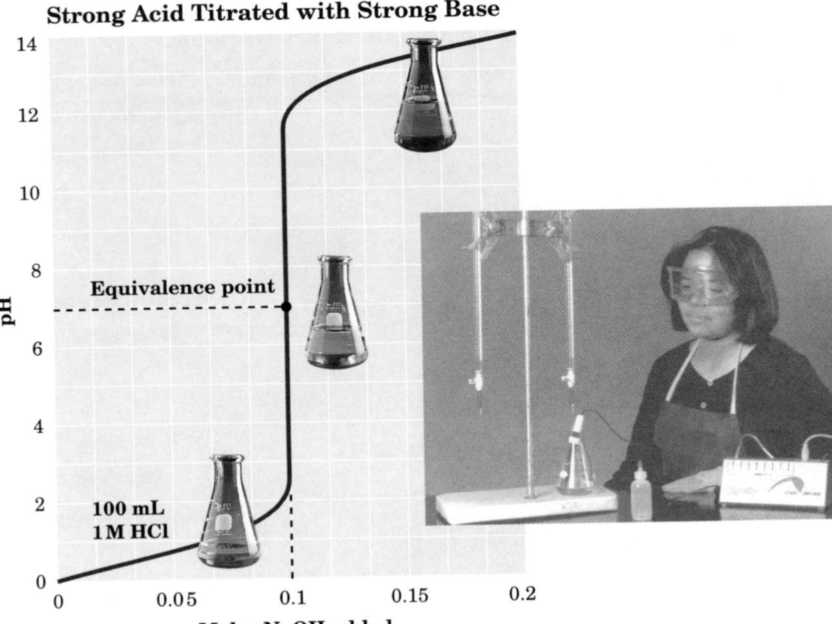

Figure 13-15 The titration curve for a strong acid and strong base reflects the set of pH data points collected using a pH probe. The rapid rise in pH at 0.1 moles NaOH added occurs with the addition of just a few drops of base. The curve has been smoothed to represent the best fit of the data.

Figure 13-15 shows how the pH changes as NaOH is used to titrate HCl. The data for this graph was obtained using a pH meter to monitor changes in the acid flask as base is added. The pH is low at the start of the titration. When NaOH is added, the pH rises slowly as OH^- ions react with H_3O^+ ions, as shown by the following equation.

$$OH^-(aq) + H_3O^+(aq) \longrightarrow 2H_2O(l)$$

As long as there is an excess of H_3O^+ ions, the bromthymol blue indicator stays yellow.

As more NaOH is added, the concentration of H_3O^+ ions continues to decrease until the point at which the moles of NaOH added to the flask equal the moles of HCl in the original solution.

$$\text{mol HCl} = \text{mol NaOH}$$

equivalence point *the point in a titration process where the moles of standard are stoichiometrically equivalent to the moles of substance titrated*

This point in a titration is called the **equivalence point**. There is no excess acid or base. The bromthymol blue indicator is now green signalling that the equivalence point has been reached. The addition of even a small amount of NaOH causes the pH to rise dramatically, as shown in **Figure 13-15**. Notice that the equivalence point in this titration is reached when the pH is 7. The equivalence point for a titration of a strong acid with a strong base will be at pH 7. If more NaOH is added and the equivalence point is passed, the pH continues to rise very sharply for a bit and then more gradually as additional NaOH makes the solution increasingly basic.

Visual Strategy

Figure 13-15

Make students aware that 100 mL of 1 M HCl contains 0.1 mol HCl and thus requires 0.1 mol of NaOH for complete neutralization.

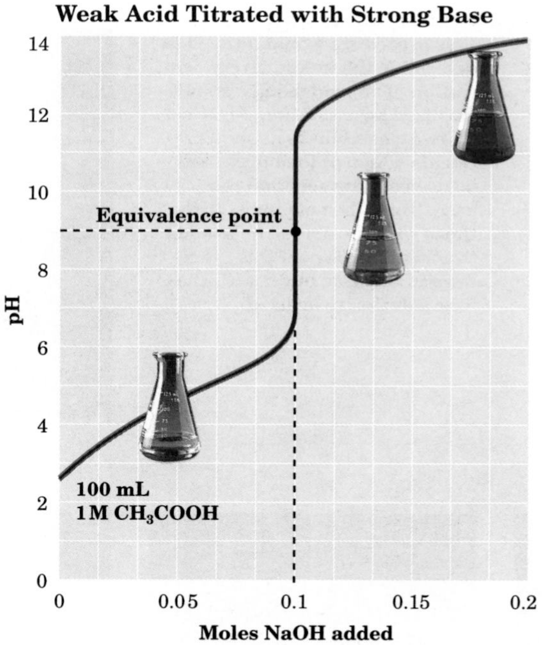

Figure 13-16
The titration curve for a weak acid and strong base shows that the titration begins at a higher pH than that for the curve in Figure 13-15. The equivalence point occurs when the solution is basic at pH 8.7.

Titrating a weak acid with a strong base

The titration you just examined represents only one of three possible types of acid-base titrations, as shown in the following diagram.

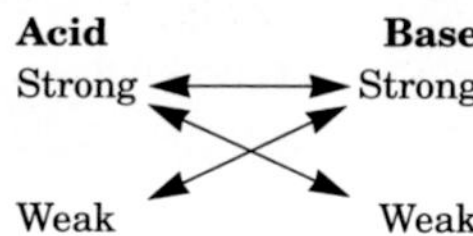

Suppose you want to determine the molarity of acetic acid, CH_3COOH, in vinegar by titrating it with a standard solution of NaOH. In this case, you would be titrating a weak acid with a strong base. To choose an appropriate indicator, you would check **Table 13-5**. Notice that phenolphthalein is an indicator that can be used in the titration of a weak acid by a strong base because the color range of the phenolphthalein falls within the pH range for the equivalence point.

Compare **Figure 13-16** with **Figure 13-15** to see how this titration differs from one between a strong acid and a strong base. The pH of the weak acid solution is higher than the strong acid solution at the same concentration. The addition of NaOH causes only a gradual change in the pH as H_3O^+ ions react with the added OH^- ions. The equivalence point is reached when equal molar amounts of CH_3COOH and NaOH have combined. However, notice in **Figure 13-16** that the equivalence point in this titration occurs at a pH of 8.7. Why is it not at 7? Continued addition of NaOH past the equivalence point causes an immediate, rapid rise in pH and then a gradual increase as the solution becomes more basic, just as with a strong acid–strong base.

Theme
Systems and interactions
The interaction of a strong base with a weak acid causes a shift in the position of the equivalence point related to pH.

Teaching Tip
Weak acid–strong base titration
Remind students that the conjugate of the weak acid is the strong base. CH_3COOH titrated with NaOH will produce H_2O and the strong base CH_3COONa helping to make the overall solution more basic and accounting for the increased pH of the equivalence point.

Content Rationale
Normality
Normality is not covered in this text, because it is an obsolete unit and no longer used in most college texts. It is far more useful to cover reactions of diprotic and triprotic acids as simple stoichiometric problems, rather than increasing the complexity of a problem by introducing another concentration unit.

Visual Strategy
Figure 13-16
Have students notice that when using a weak acid and strong base, the equivalence point is shifted toward a higher pH. This is why different indicators are necessary to cover a range of end points.

Teaching Tip

Titrations

Make students aware that if too much indicator is used, the indicator will affect the equivalence point of the reaction because it is an acid.

Discuss with students why at least three repetitions are needed when doing a titration, whether the results agree or not.

How To

Carry out a titration

This procedure would be used to determine the unknown concentration of an acid using a standardized base solution. First, set up two clean burets as shown. Decide which of the burets will be used for the acid and the base. Rinse the acid buret three times with the acid to be used in the titration. Repeat this procedure in the base buret with the base solution to be used.

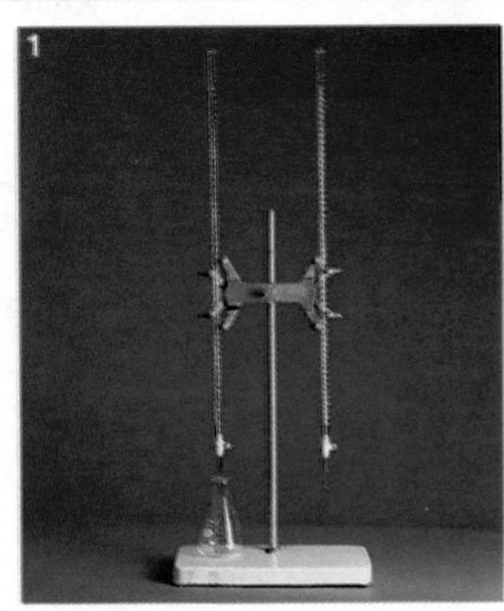

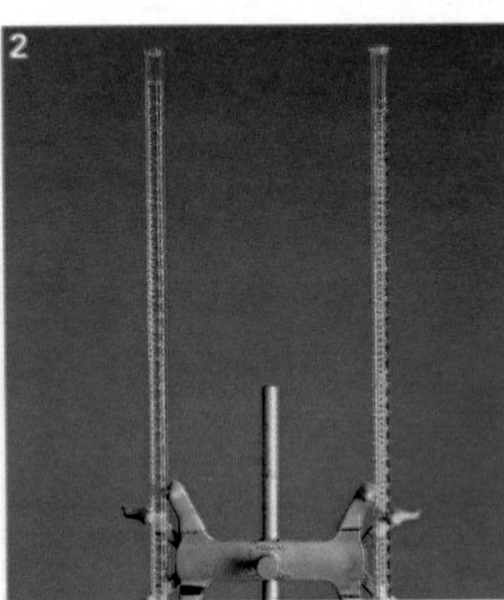

Fill the first to a point above the calibration mark with the acid of unknown concentration.

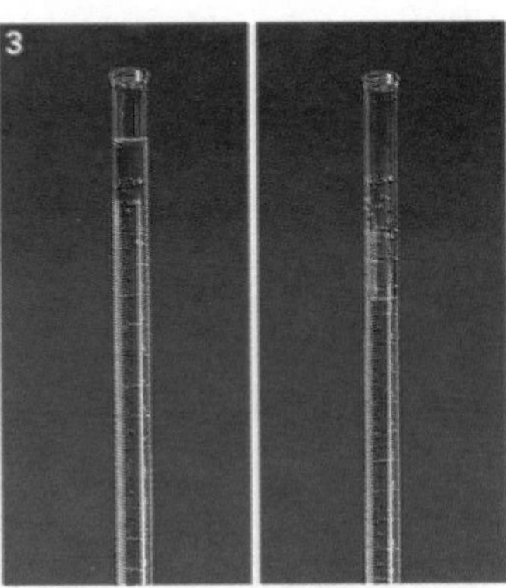

Release some acid from the buret to remove any air bubbles and to lower the volume to the calibrated portion of the buret.

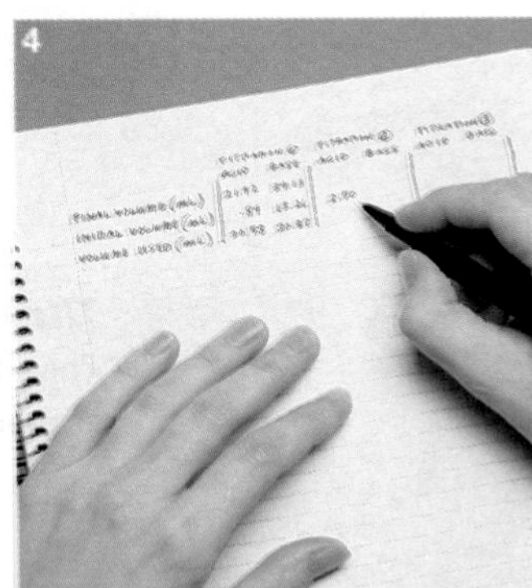

Record the exact volume of the acid in the buret to the nearest 0.01 mL as your starting point.

Release a predetermined volume of the acid (determined by your teacher or lab procedure) into a clean, dry Erlenmeyer flask.

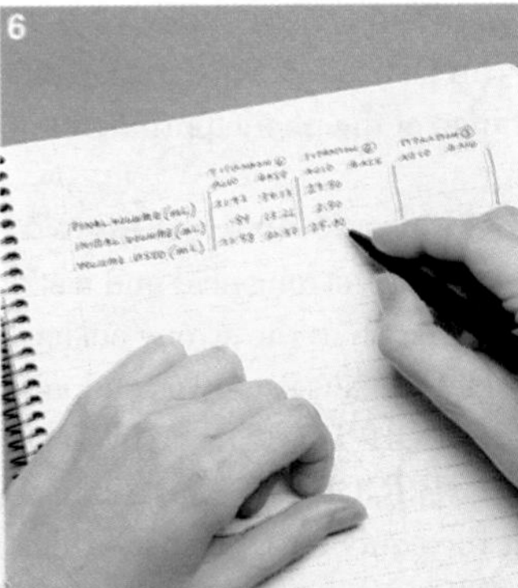

Record the exact volume of the acid released into the flask to the nearest 0.01 mL.

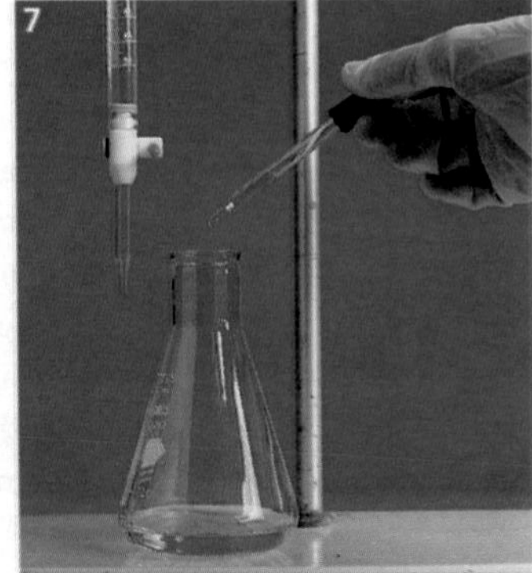

Add three drops of the appropriate indicator (in this case phenolphthalein) to the flask.

Visual Strategy

How to Carry Out a Titration

Help students understand that the buret can be read to the hundredth of a milliliter by breaking the space between tenths of a milliliter into 10 equal spaces.

Fill the other buret with the standardized base solution to a point above the calibration mark. The concentration of the standardized base is known to a certain degree of precision because it was previously titrated with an exact mass of a solid acid.

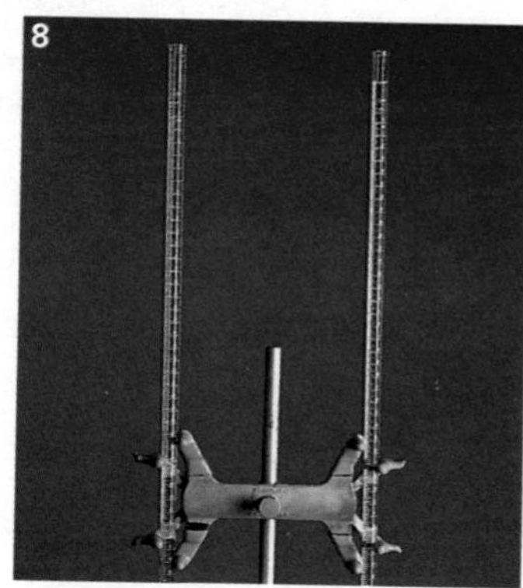

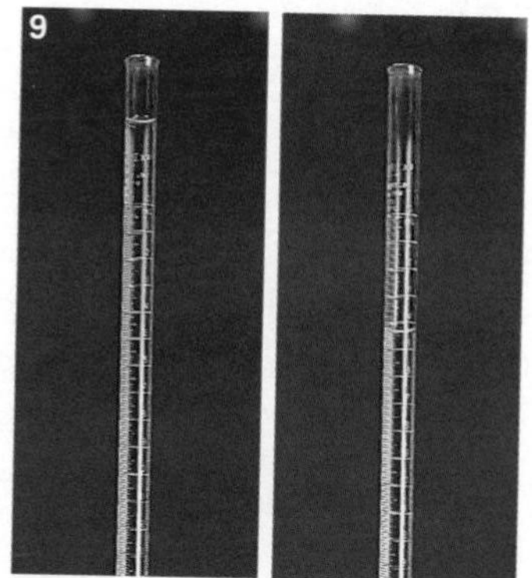

Release some base from the buret to remove any air bubbles and to lower the volume to the calibrated portion of the buret.

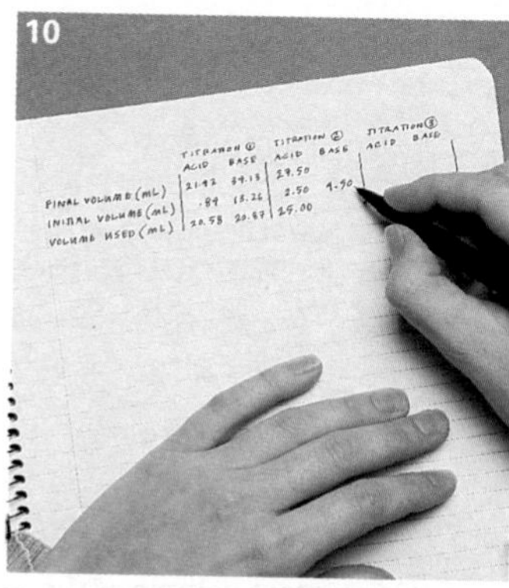

Record the exact volume of the base to the nearest 0.01 mL as your starting point.

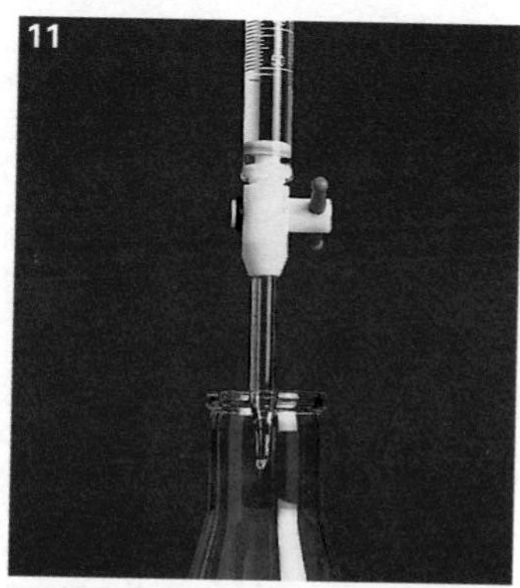

Place the Erlenmeyer flask under the base buret as shown. Notice that the tip of the buret extends into the mouth of the flask.

Slowly release base from the buret into the flask while *constantly* swirling the contents of the flask. The pink color of the indicator should fade with swirling.

The titration is nearing the endpoint when the pink color stays for longer periods of time. At this point, add base drop by drop.

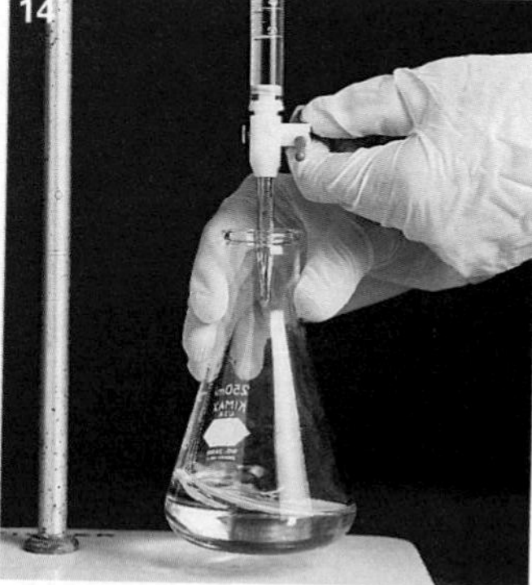

The equivalence point is reached when a very light pink color remains after 30 seconds of swirling.

Record the volume to the nearest 0.01 mL. Repeat the titration until you have three results that agree within 0.05 mL.

513

Additional Examples for Sample Problem 13H

a. 20.0 mL of 0.100 M HCl are titrated with 19.5 mL of an NaOH solution. What is the molarity of the NaOH solution?

$HCl + NaOH \longrightarrow NaCl + H_2O$

b. 50.5 mL of 0.230 M HCl are needed to titrate 39. mL of KOH. What is the molarity of the KOH solution?

$HCl + KOH \longrightarrow KCl + H_2O$

c. 35.5 mL of 0.345 M H_2SO_4 are used to titrate 42.7 mL of KOH. What is the molarity of the KOH?

$H_2SO_4 + 2KOH \longrightarrow 2H_2O + K_2SO_4$

Solutions are at the bottom of the page.

Answers to Practice

1. $HCl + NaOH \longrightarrow NaCl + H_2O$

$50\ mL\ HCl \times \frac{1\ L\ HCl}{1000\ mL} \times \frac{0.40\ mol\ HCl}{1\ L} = 0.02\ mol\ HCl$

The mole ratio is 1:1.

$\frac{0.02\ mol\ NaOH}{35.\ mL\ NaOH} \times \frac{1000\ mL}{1\ L} =$

0.57 M NaOH

2. 37 mL
3. 0.157 M CH_3COOH
4. 0.0422 M HCl

*Full solutions for items **2–4** are found on page 531A.*

Sample Problem 13H
Calculating concentration from titration data

Even though moles of acid and base are not specifically asked for in the problem statement, this problem is similar to other stoichiometry problems. You need moles.

Suppose that in the titration of 40. mL of vinegar, 20. mL of 0.50 M NaOH were needed to reach the equivalence point. What is the molarity of acetic acid in the vinegar? The molecular equation for the reaction is the following.

$$\underset{\text{Sodium hydroxide}}{NaOH(aq)} + \underset{\text{Acetic acid}}{CH_3COOH(aq)} \rightleftharpoons \underset{\text{Sodium acetate}}{NaCH_3COO(aq)} + \underset{\text{Water}}{H_2O(l)}$$

1 List

- **volume of NaOH** = 20. mL
- **concentration of NaOH** = 0.50 M
- **volume of vinegar** = 40. mL
- **mole ratio** = 1 mol NaOH: 1 mol CH_3COOH
- **moles of NaOH** = ?
- **moles of CH_3COOH** = ?
- **concentration of CH_3COOH** = ?

2 Set up

- **Both the volume and molarity of the base are known. From these values you can calculate the moles of NaOH used to titrate the acid.**

$$20.\ mL\ NaOH \times \frac{1\ L\ NaOH}{1000\ mL\ NaOH} \times \frac{0.50\ mol\ NaOH}{1\ L\ NaOH} = 0.010\ mol\ NaOH$$

- **The balanced chemical equation shows that 1 mol of NaOH is needed for every 1 mol of CH_3COOH. Therefore, the amount of CH_3COOH in the vinegar is also 0.010 mol.**

3 Calculate and verify

- **Use the volume of the vinegar to calculate its molarity. There are 0.010 mol CH_3COOH in 40. mL of vinegar.**

$$\frac{0.010\ mol\ CH_3COOH}{40.\ mL\ CH_3COOH} \times \frac{1000\ mL\ CH_3COOH}{1\ L\ CH_3COOH} = 0.25\ M\ CH_3COOH$$

The mole ratio for this reaction is 1:1. It makes sense that the CH_3COOH concentration would be half of the NaOH concentration, because there is twice as much vinegar used in the titration.

Practice 13H

1. **In the titration of 35. mL of liquid drain cleaner containing NaOH, 50. mL of 0.40 M HCl must be added to reach the equivalence point. What is the molarity of the base in the cleaner?**
2. **Calculate how many milliliters of 0.25 M $Ba(OH)_2$ must be added to titrate 46. mL of 0.40 M $HClO_4$. (Hint: write the equation for the reaction and note the mole ratios.)**
3. **A 15.5 mL sample of 0.215 M KOH was titrated with an acetic acid solution. It took 21.2 mL of the acid to reach the equivalence point. What is the molarity of the acetic acid solution?**
4. **A 20.0 mL sample of an HCl solution is titrated with 27.4 mL of a standard solution of $Ba(OH)_2$. The concentration of the standard is 0.0154 M. What is the molarity of the HCl? The equation for the reaction is the following.**

$$2HCl(aq) + Ba(OH)_2(aq) \rightleftharpoons BaCl_2(aq) + 2H_2O(l)$$

Solutions for Additional Examples 13H

a. 0.103 M NaOH

b. 0.292 M KOH

c. 0.571 M KOH

*Full solutions for items **a–c** are found on page 531A.*

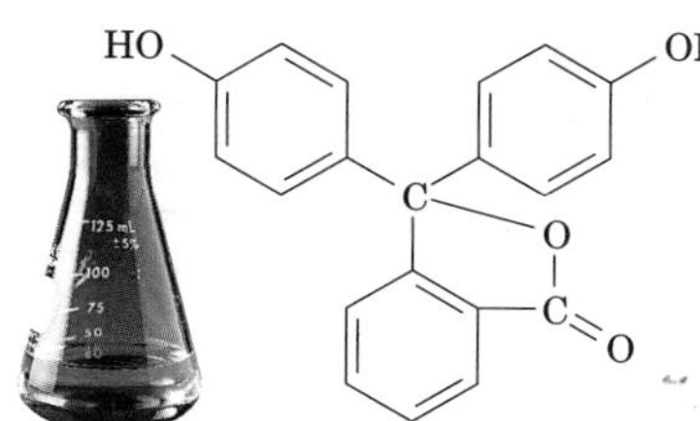

Le Châtelier's principle explains why indicators change color

What causes the color changes shown in **Table 13-5**? Indicators used in titrations are either weak acids or weak bases. Phenolphthalein is a weak acid that can be represented by the simplified formula HIn. Like all weak acids, phenolphthalein ionizes to a slight extent in aqueous solution and forms an equilibrium system. The addition of acid or base to the equilibrium system places a stress on the system. Le Châtelier's principle can be used to explain the color change for phenolphthalein.

$$\underset{\text{Clear}}{HIn(aq)} + H_2O(l) \rightleftharpoons H_3O^+(aq) + \underset{\text{Magenta}}{In^-(aq)}$$

HIn donates a hydrogen ion to the base H_2O. The In^- ion is the conjugate base of HIn. Undissociated HIn is clear in aqueous solution, while the In^- ion is magenta or deep pink in color.

In an acidic solution, H_3O^+ ions are present in excess. These react with the Brønsted base, In^-. As a result, the formation of HIn is favored, and the solution is clear. As a base is added in a titration in which phenolphthalein is the indicator, OH^- ions react with H_3O^+ ions to produce water molecules.

Figure 13-17
The molecular form of phenolphthalein is in equilibrium with the ionized form, which is magenta in color. Addition of acid favors the reverse reaction, and the colorless molecular form predominates.

$$OH^-(aq) + H_3O^+(aq) \rightleftharpoons 2H_2O(l)$$

As the H_3O^+ ions are used to form water, the equilibrium shifts to the right because H_3O^+ is being removed.

$$HIn(aq) + H_2O(l) \rightleftharpoons H_3O^+(aq) + In^-(aq)$$

As this shift occurs, the magenta color of In^- ions appears. The indicator has reached its end point. The **end point** of an indicator is the pH at which the indicator changes color. To be useful in a titration, the end point of an indicator must be close to the equivalence point of the titration. Refer again to **Table 13-5**, and recall that bromthymol blue was used as the indicator in the titration of a strong acid with a strong base. Notice that the point at which bromthymol blue changes from yellow to blue is close to the pH of the equivalence point of a strong-acid–strong-base titration.

end point
a point in a titration indicated by the color change of an indicator

Section Review

20. Write balanced equations for the neutralization of the following.
a. HNO_3 and $Ca(OH)_2$ **b.** KOH and H_2SO_4

21. Refer to **Table 13-5** to identify an indicator for titrating a strong acid with a strong base and another for titrating a weak base with a strong acid.

22. If 20 mL of 0.010 M aqueous HCl are required to titrate 30 mL of an aqueous solution of NaOH, what is the molarity of the NaOH solution?

23. Explain how a color change of an indicator illustrates Le Châtelier's principle.

24. Explain the difference between end point and equivalence point. Why is it important that both occur at approximately the same pH in a titration?

Theme
Macroscopic observations and micromodels
The color change of an indicator is explained by the dynamic equilibrium between conjugate acid-base pairs.

ASSESS

Alternative Assessment
Have students describe how stoichiometry can be used to determine the amount of base present when the amount of acid is known. Why is a pH meter more useful than an indicator for determining the equivalence point of a titration?

Answers to Section Review

20. a. $2HNO_3 + Ca(OH)_2 \longrightarrow Ca(NO_3)_2 + 2H_2O$
b. $2KOH + H_2SO_4 \longrightarrow K_2SO_4 + 2H_2O$

21. strong acid and strong base: litmus, bromthymol blue; weak base and strong acid: methyl orange, bromphenol blue

22. $HCl + NaOH \longrightarrow NaCl + H_2O$

$$20\ \text{mL HCl} \times \frac{1\ \text{L}}{1000\ \text{mL}} \times \frac{0.010\ \text{mol HCl}}{1\ \text{L}} = 0.0002\ \text{mol HCl}$$

The mole ratio is 1:1.

$$\frac{0.0002\ \text{mol NaOH}}{30\ \text{mL}} \times \frac{1000\ \text{mL}}{1\ \text{L}} = 7 \times 10^{-3}\ \text{M NaOH}$$

23. An indicator, in solution, exists in molecular and ionized forms. The addition of an acid or base shifts the equilibrium to favor one form, resulting in the appearance of the color of the predominant form.

24. The end point describes the pH range in which an indicator changes color. The equivalence point is the point in a titration at which the moles of acid are stoichiometrically equal to the moles of base.

Section 13-4

FOCUS

Lesson Starter

Use the following demonstration to show students the properties of buffers.

Demonstration 9

Buffers in solution

Approximate time: 10 min

1. Make a buffer solution at pH 7 by dissolving 6.8 g of KH_2PO_4 in water. Add 295 mL of 0.1 M KOH and dilute the solution to 1 L.
2. Place 50 mL of distilled water and 50 mL of the buffer solution into two separate 250 mL beakers and place them on an overhead projector.
3. Add 2–3 drops of phenolphthalein indicator to both beakers. Using a dropper, add 0.1 M NaOH drop by drop to both solutions. The water solution should change color after a few drops, but the buffer solution will require over 7 mL of the base before a color change occurs.
4. Repeat the demonstration using methyl orange or bromcresol green as the indicator and add 0.1 M HCl drop by drop. Again a few drops of HCl are needed in the water solution, but several mL are needed to change the indicator's color in the buffer solution.

SAFETY

Wear safety goggles and a lab apron. Students should remain back at least 10 ft.

DISPOSAL

Combine all solutions, and neutralize the mixture with 1 M HCl or 1 M NaOH as appropriate. Pour the neutralized solution down the drain.

Source: D. Humphreys, *Demonstrating Chemistry*, McMaster University; Hamilton, Ontario, 1983.

13-4

What are the biochemical roles of some acids and bases?

Section Objectives

Explain how buffers keep pH relatively constant in the body.

Describe the behavior of amino acids in response to changes in pH.

List the structural components of nucleic acids.

Distinguish between saturated and unsaturated fatty acids.

Describe two functions of fatty acids in the body.

List sources of organic bases in the body.

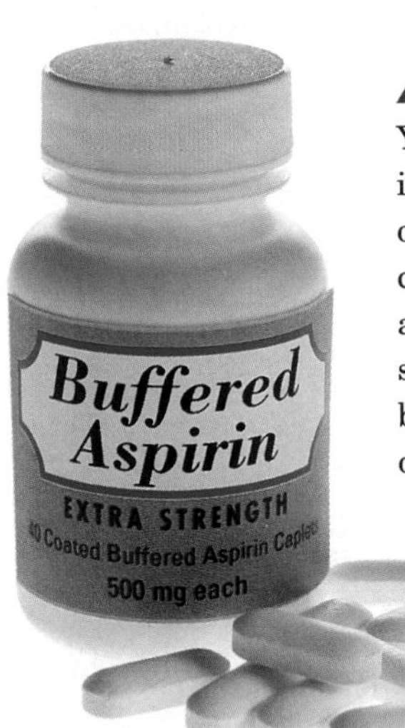

Figure 13-18
Buffered products can maintain a relatively constant pH with small additions of acids or bases.

Acids and bases in the body

Your body is a sea of acids and bases. The proteins in your hair, nails, cell membranes, and other parts of your body consist of amino acids. Enzymes that catalyze reactions in your body are composed of amino acids. Hydrochloric acid is secreted in your stomach to aid in the digestion of foods. Organic bases are major components of DNA and products of the digestion of proteins.

Enzymes that catalyze the many reactions in your body can function only within a limited pH range. Therefore, maintaining pH within the body is critical to life. You learned in other chapters that foods or drugs, such as aspirin, can change acidity levels in the body. Buffered aspirin, shown in **Figure 13-18**, was developed to keep the pH of the stomach at a more constant level and to prevent damage to the stomach lining. Your body contains a number of buffering systems like that in buffered aspirin to keep your internal pH where it should be. Without buffers, the thousands of reactions that take place in your body each day would not occur quickly enough to keep you alive.

Some solutions can maintain their pH when small quantities of acid or base are added

buffer
a system that is able to withstand small additions of acid or base without a significant change in pH; the system is composed of a conjugate acid-base pair

A **buffer** is a system consisting of a conjugate acid-base pair that is capable of resisting changes in pH when small amounts of acid or base are added. A buffer solution contains similar concentrations of a weak acid and one of its soluble salts, or a weak base and one of its soluble salts. For example, a buffer can be made from acetic acid and one of its soluble salts, such as sodium acetate. Acetic acid will react with water to form the equilibrium system shown by the following equation.

$$CH_3COOH(aq) + H_2O(l) \rightleftharpoons CH_3COO^-(aq) + H_3O^+(aq)$$

The soluble sodium acetate salt will dissociate completely in water, as shown by the following equation.

$$NaCH_3COO(s) \xrightarrow{H_2O} CH_3COO^-(aq) + Na^+(aq)$$

When you put the acid solution and salt together in the same system, both reactions take place. There are two things you should notice about this system.

Buffer - resists sudden change in pH
weak acid + salt, weak acid { CH3COOH / NaCH3COO-
CH3COOH + HOH → CH3COO + H3O
CH3COO + HOH → CH3COOH + OH

1. The system contains high concentrations of both the molecular and ion forms of the acid.
2. The acetate ion is a product of both reactions.

Let's look at how the acetic-acid–acetate-ion buffer system is able to absorb small amounts of acid or base without a significant change in pH. If OH^- ions (a base) are added, they will react with the molecular acid, as shown by the following equation.

$$CH_3COOH(aq) + OH^-(aq) \rightleftharpoons CH_3COO^-(aq) + H_2O(l)$$

If H_3O^+ ions (an acid) are added, they will react with the acetate ion in solution, as shown by the following equation.

$$CH_3COO^-(aq) + H_3O^+(aq) \rightleftharpoons CH_3COOH(aq) + H_2O(l)$$

A buffer is limited in the amount of acid or base that it can accept without changing pH. The pH range over which a buffer can respond effectively to the additions of an acid or base is called the *buffer capacity*. The buffer capacity of the acetic-acid–acetate-ion buffer system is in the pH range between 4.3 and 5.3.

Your body contains several buffer systems to handle the result of many acid–base reactions that occur as part of your metabolism. The pH of the blood rarely rises above 7.45 or drops below 7.35. One component of your blood that maintains this pH range is the buffer system consisting of a weak acid, carbonic acid, and its conjugate base, hydrogen carbonate ions, HCO_3^-. The hydrogen carbonate ions react with H_3O^+ ions from the acid, as shown by the following equilibrium equation.

$$HCO_3^-(aq) + H_3O^+(aq) \rightleftharpoons H_2CO_3(aq) + H_2O(l)$$

According to Le Châtelier's principle, the addition of an acid places stress on the equilibrium system. There is a net shift to the right as HCO_3^- neutralizes the added acid. If a base is added to the system, it reacts with carbonic acid molecules to form hydrogen carbonate ions, HCO_3^-.

$$H_2CO_3(aq) + OH^-(aq) \rightleftharpoons HCO_3^-(aq) + H_2O(l)$$

Excess OH^- ions are neutralized by forming water and the pH of your blood remains close to 7.4.

The phosphate buffer system in the cytoplasm of your cells works in much the same way.

$$H_2PO_4^-(aq) + H_2O(l) \rightleftharpoons H_3O^+(aq) + HPO_4^{2-}(aq)$$

The phosphate buffer system is effective in the pH range from 6.4 to 7.4.

Chem Fact
When blood pH falls below 7.4, a condition called acidosis occurs. A person with acidosis can faint and go into a coma if the condition persists.

Concept Review

Buffer action

25. The phosphate buffer system consists of the dihydrogen phosphate ion, $H_2PO_4^-$, and the hydrogen phosphate ion, HPO_4^{2-}. Describe how this system reacts to maintain a constant pH when OH^- ions are added.

Theme
Systems and interactions
The interactions of the chemicals of the body are kept under control by enzymes.

Teaching Tip
Make students aware that the K_w for water is lower than the K_a for acetic acid (1.76×10^{-5}), so the formation of water is preferred over the formation of acetic acid.

Theme
Macroscopic observations and micromodels
The ability of a buffer to maintain pH with the addition of some acid or base is explained by the equilibrium model for weak acids.

Teaching Tip
Make students aware that the addition of H_3O^+ causes the formation of CH_3COOH and H_2O due to the low K_a and K_w of these compounds.

Do You Know?
Both the kidneys and the lungs help to offset the changing pH of the blood.

Answers to

Concept Review

25. $H_2PO_4^- + OH^- \longrightarrow HPO_4^{2-} + H_2O$

An added OH^- ion accepts a proton from $H_2PO_4^-$ to form H_2O.

Do You Know?

Of the more than 80 amino acids found in nature, only 20 serve as building blocks for proteins.

amino acid
a carboxylic acid that is the structural subunit of a protein

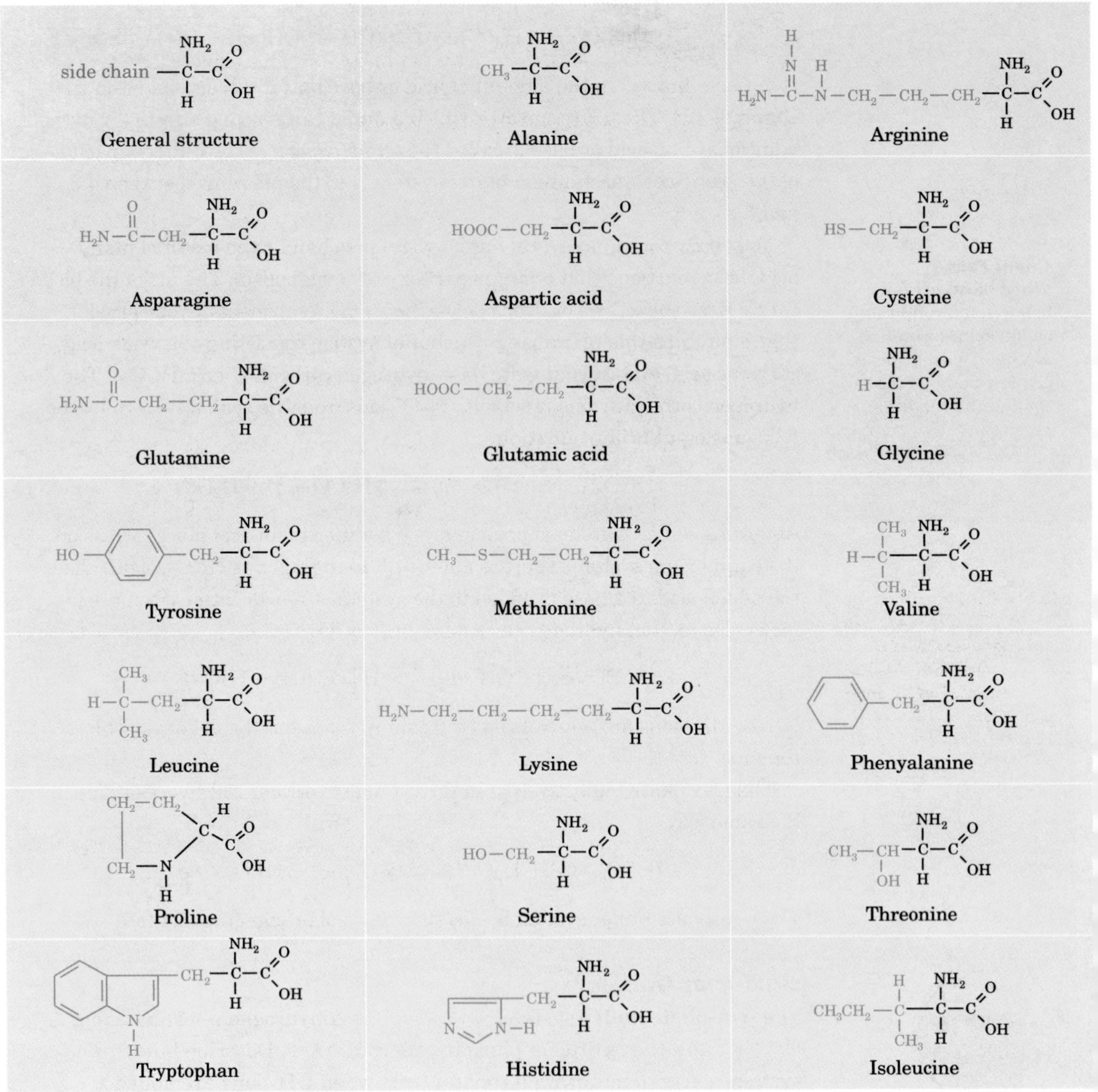

Figure 13-19
All amino acids fit the same general formula. The 20 essential amino acids are shown with all *R*-groups shown in pink.

Amino acids are weak organic acids that are the subunits of proteins

Amino acids are organic acids made mostly of carbon, hydrogen, oxygen, and nitrogen. All proteins in the body are built from a collection of 20 different amino acids arranged in different combinations. Amino acids have the general structure shown in the lavender box of **Figure 13-19**. Note that all amino acids have an amine group, —NH_2 or —NH, and the carboxyl group, —COOH, but differ in the nature of the *R*-group. You can see these groups by looking at the structural formulas of the 20 amino acids also shown in **Figure 13-19**.

General structure	Alanine	Arginine
Asparagine	Aspartic acid	Cysteine
Glutamine	Glutamic acid	Glycine
Tyrosine	Methionine	Valine
Leucine	Lysine	Phenyalanine
Proline	Serine	Threonine
Tryptophan	Histidine	Isoleucine

Visual Strategy

Figure 13-19

Students should notice the structural similarities and differences between the acid structures shown. Ask students to identify the *R*-group in each acid.

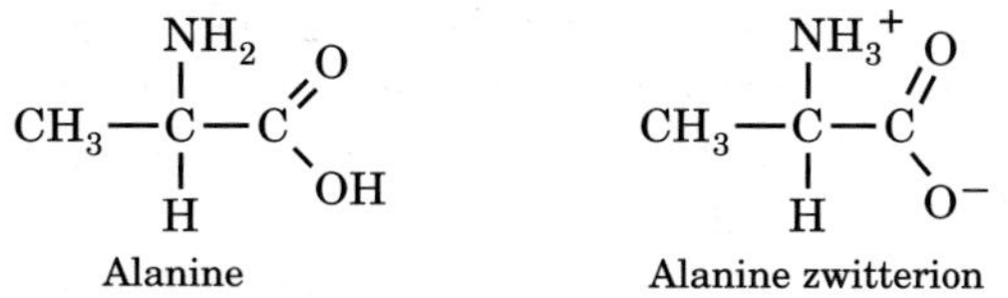

Teaching Tip
Emphasize that the zwitterion carries both a positive and a negative charge, which makes it dipolar.

Figure 13-20
The zwitterion forms as a result of a hydrogen ion transfer from the carboxyl group to the amine group.

$$CH_3-C(NH_2)(H)-C(=O)OH \qquad CH_3-C(NH_3^+)(H)-C(=O)O^-$$

Alanine — Alanine zwitterion

The acid properties of amino acids come from the carboxyl group, the functional group found in all organic acids. Their behavior is somewhat different than that of other acids of similar molecular mass and appears to be more like that of a salt than that of a neutral molecular compound. For example, even the smallest amino acids, glycine and alanine, have melting points close to 300°C. The solubilities of amino acids vary greatly depending on the *R*-groups. However, glycine is very soluble in water, but not in nonpolar solvents. To describe this behavior, chemists suggest that solid amino acids exist as dipolar ions that are formed from the transfer of a hydrogen ion from the carboxyl group to the amine group, as shown in **Figure 13-20**. This form of an amino acid is called a **zwitterion** and functions as a very polar molecule even though it has no net charge. The positive charge is centered on the nitrogen of $—NH_3$. The negative charge is centered between the oxygen atoms.

zwitterion
the reactive dipolar form of an amino acid

Zwitterions respond to changes in pH. In the presence of acids, the $—COO^-$ group accepts hydrogen ions. In the presence of bases, the $—NH_3^+$ group donates hydrogen ions. Because a zwitterion can act as an acid or base, as shown in **Figure 13-21**, it is amphiprotic. Amino acids donate hydrogen ions from the carboxyl group, but accept hydrogen ions from the amine group. As a result, amino acids demonstrate some slight buffering effects in the body. K_a values for amino acids range from 1.5×10^{-2} for histidine to 1.5×10^{-3} for tryptophan.

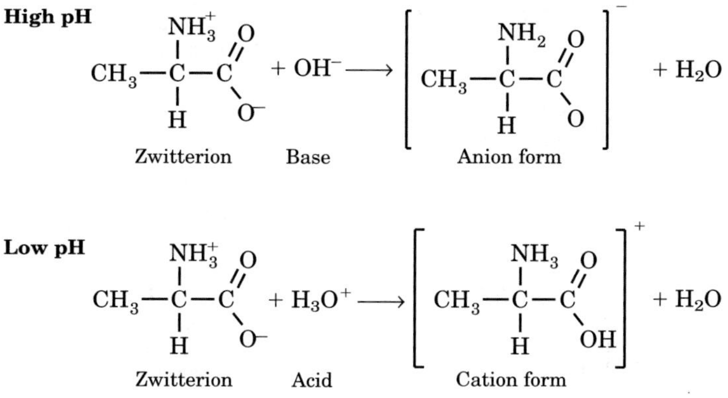

Figure 13-21
The zwitterion form of alanine is amphiprotic. In the presence of a base, it will shift to its anion form. In the presence of an acid, it will shift to its cation form.

High pH

$$CH_3-C(NH_3^+)(H)-C(=O)O^- + OH^- \longrightarrow [CH_3-C(NH_2)(H)-C(=O)O]^- + H_2O$$

Zwitterion — Base — Anion form

Low pH

$$CH_3-C(NH_3^+)(H)-C(=O)O^- + H_3O^+ \longrightarrow [CH_3-C(NH_3)(H)-C(=O)OH]^+ + H_2O$$

Zwitterion — Acid — Cation form

Visual Strategy
Figure 13-20
Have students notice the positive and negative charges on the zwitterion.

nucleic acid
a biological polymer consisting of phosphoric acid, a 5-carbon sugar, and four nitrogen bases

Nucleic acids are made from acid and base components

You should know from your biology class that the two most important **nucleic acids** in the body are DNA and RNA. The actual names of both of these compounds indicate that they are acids—deoxyribonucleic acid and ribonucleic acid. Both are polymers consisting of the same structural subunits: phosphoric acid, a five-carbon sugar, and four organic nitrogen bases (amines), as shown in **Figure 13-22**. DNA and RNA differ from each other in the composition of their sugars and in one nitrogen base. DNA contains deoxyribose. RNA contains ribose. Three of the nitrogen bases are common to both DNA and RNA, but only DNA has thymine, and only RNA has uracil.

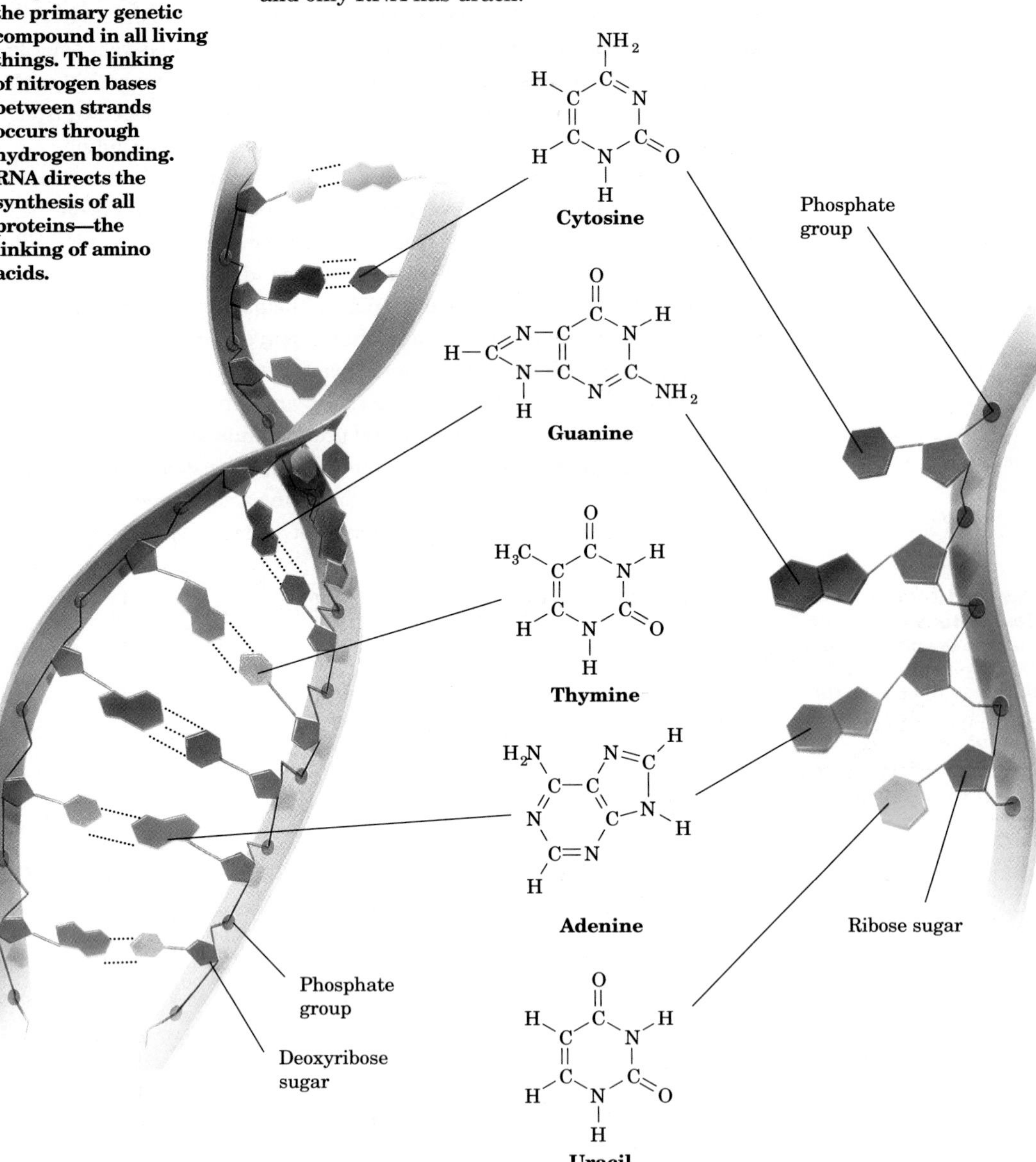

Figure 13-22
DNA and RNA have similar molecular components. DNA is the primary genetic compound in all living things. The linking of nitrogen bases between strands occurs through hydrogen bonding. RNA directs the synthesis of all proteins—the linking of amino acids.

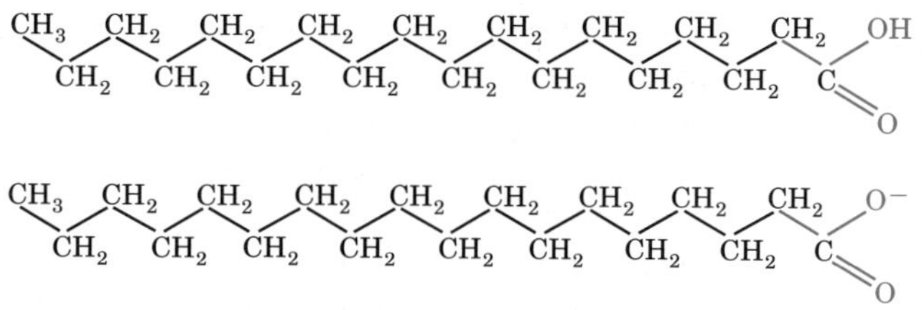

```
CH3  CH2  CH2  CH2  CH2  CH2  CH2  CH2  CH2   OH
   CH2  CH2  CH2  CH2  CH2  CH2  CH2  CH2  C
                                             O

CH3  CH2  CH2  CH2  CH2  CH2  CH2  CH2  CH2   O−
   CH2  CH2  CH2  CH2  CH2  CH2  CH2  CH2  C
                                             O
```

Figure 13-23
Fatty acids are long chain hydrocarbons that also contain the carboxyl group. As Brønsted acids, they donate hydrogen ions, resulting in long-chain fragments with a charged end.

Fatty acids are the subunits of lipids

fatty acid
the long chain carboxylic acid subunit of a fat and oil

One of the structural components of animal fats and vegetable oils are compounds called **fatty acids.** Fatty acids consist of long chains of hydrocarbons with a carboxyl group, —COOH, at one end, as shown in **Figure 13-23**. The length of the chain determines the physical properties of the fatty acid. Short-chain fatty acids and unsaturated long-chain fatty acids are generally found in oils which are liquids (from four to seven carbon atoms long). Long-chain fatty acids (more than 12 carbon atoms long) are found in fats, which are generally solids at room temperature. Fatty acids are not very soluble in water even though the carboxyl group is very polar. Solubility of fatty acids decreases as the length of the fatty-acid chain increases.

saturated fatty acid
the long chain carboxylic acid subunit of a fat and oil with no double bonds

The fats and oils in your diet contain varying amounts of two classes of fatty acids. These classes are saturated and unsaturated fatty acids, and they differ from each other by the relative numbers of hydrogen atoms bonded to the carbon chain. The structure of a **saturated fatty acid** is shown in **Figure 13-24**.

unsaturated fatty acid
a fatty acid with one or more carbon-carbon double bonds

Compare its structure to an **unsaturated fatty acid** of similar length. A *polyunsaturated* fatty acid has more than one carbon-carbon double bond. The hydrogenation of oils involves addition of hydrogen atoms to the double bonds of an unsaturated fatty acid. This process causes the oil to be more solid. Margarines and shortenings are made from hydrogenated oils. Hydrogenated fats last longer before spoiling than unsaturated fats.

Figure 13-24
Saturated fatty acids have all carbons linked with single bonds. Unsaturated fatty acids will have one or more carbon–carbon double bonds in the chain.

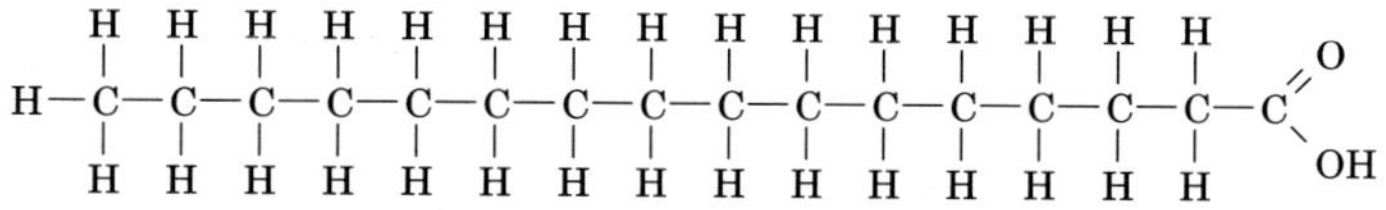

```
  H  H  H  H  H  H  H  H  H  H  H  H  H  H  H      O
  |  |  |  |  |  |  |  |  |  |  |  |  |  |  |    //
H-C--C--C--C--C--C--C--C--C--C--C--C--C--C--C--C
  |  |  |  |  |  |  |  |  |  |  |  |  |  |  |    \
  H  H  H  H  H  H  H  H  H  H  H  H  H  H  H      OH
```

Palmitic acid, saturated

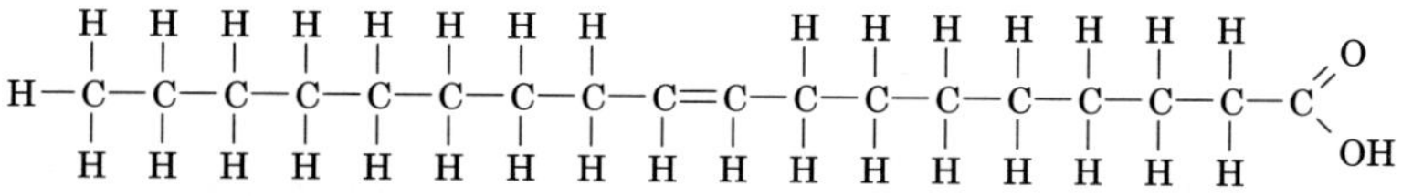

```
  H  H  H  H  H  H  H  H        H  H  H  H  H  H  H      O
  |  |  |  |  |  |  |  |        |  |  |  |  |  |  |    //
H-C--C--C--C--C--C--C--C--C=C--C--C--C--C--C--C--C--C
  |  |  |  |  |  |  |  |  |  |  |  |  |  |  |  |  |    \
  H  H  H  H  H  H  H  H  H  H  H  H  H  H  H  H  H      OH
```

Oleic acid, unsaturated

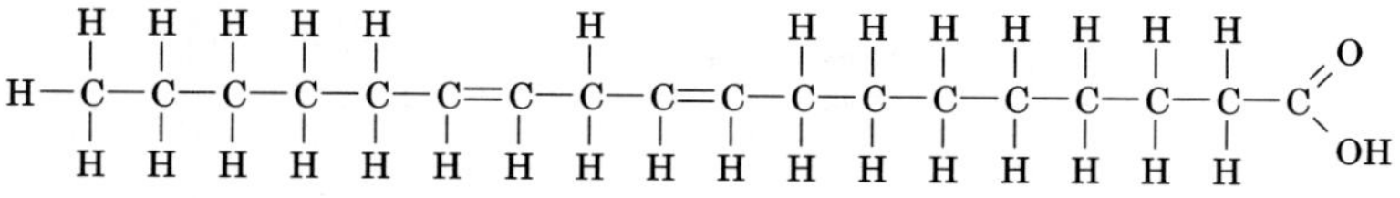

```
  H  H  H  H  H        H        H  H  H  H  H  H  H      O
  |  |  |  |  |        |        |  |  |  |  |  |  |    //
H-C--C--C--C--C--C=C--C--C--C=C--C--C--C--C--C--C--C--C
  |  |  |  |  |  |  |  |  |  |  |  |  |  |  |  |  |    \
  H  H  H  H  H  H  H  H  H  H  H  H  H  H  H  H  H      OH
```

Linoleic acid, polyunsaturated

Teaching Tip

Show students that saturated fatty acids have the general formula C_nH_{2n+1} COOH.

Do You Know?

Nickel is the catalyst used in the hydrogenation of shortenings like Crisco™.

Theme

Classification and trends
Fatty acids are classified as saturated or unsaturated.

Theme

Systems and interactions
Because fatty acids are polar on one end and nonpolar on the other, they are able to interact with both polar and nonpolar compounds.

Teaching Tip

Emphasize that the structure of the lipid bilayer results in the semipermeability of the cell membrane.

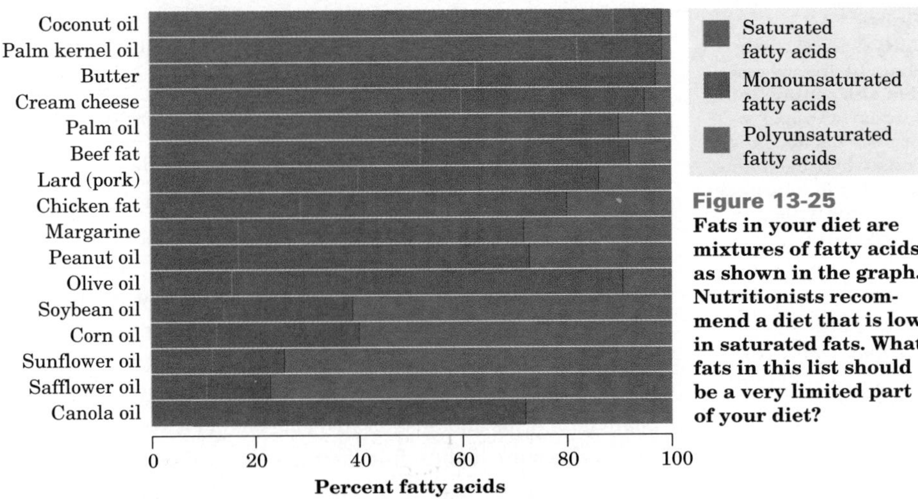

Figure 13-25
Fats in your diet are mixtures of fatty acids, as shown in the graph. Nutritionists recommend a diet that is low in saturated fats. What fats in this list should be a very limited part of your diet?

Most fatty acids in foods are in the form of triglycerides. You looked at a triglyceride structure when you studied soap-making in Chapter 11. Foods can also contain *monoglycerides* and *diglycerides*. When you see the fat content on a food label, the figure generally represents a mix of triglycerides. **Figure 13-25** shows the saturated fat composition of some familiar fats and oils found in a typical diet.

Fatty acids are also a component of the lipid bilayer that makes up the membrane of a cell. The fatty acid is the nonpolar end of a larger compound, as depicted by the model in **Figure 13-26**.

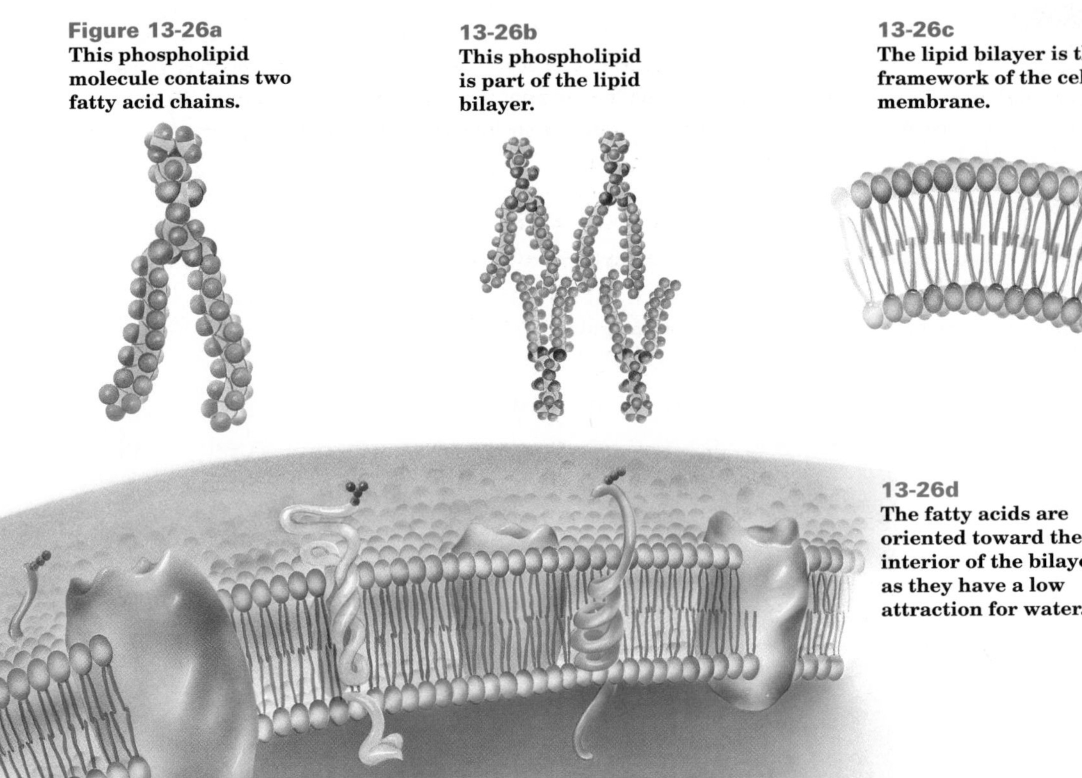

Figure 13-26a
This phospholipid molecule contains two fatty acid chains.

13-26b
This phospholipid is part of the lipid bilayer.

13-26c
The lipid bilayer is the framework of the cell membrane.

13-26d
The fatty acids are oriented toward the interior of the bilayer as they have a low attraction for water.

Consumer Focus

Antacids

You learned in Chapter 7 that antacids are used to neutralize excess stomach acid.

Although antacids contain a variety of ingredients, they all contain a base to counteract the stomach acid. The base is either sodium hydrogen carbonate, $NaHCO_3$, calcium carbonate, $CaCO_3$, aluminum hydroxide, $Al(OH)_3$, or magnesium hydroxide, $Mg(OH)_2$.

In any antacid, it is the carbonate or hydroxide ion that provides the alkalinity needed to offset excess stomach acid. However, the metal ion that is used in the antacid is also important. Antacids containing $NaHCO_3$ are the most potent and can seriously disrupt the acid-base balance in your blood. Consequently, some antacid manufacturers have substituted $CaCO_3$ for $NaHCO_3$. However, if taken in large amounts, calcium can promote kidney stones. Too much aluminum ingested from antacid products containing $Al(OH)_3$ can interfere with the body's absorption of important chemicals, including the phosphorus needed for healthy bones. Concern has also been raised about a possible connection between ingesting aluminum and Alzheimer's disease. Excess magnesium from antacids with $Mg(OH)_2$ may pose problems for people with kidney trouble.

You should know the active ingredient in any antacid product before you purchase it, and you should never use any antacid for a prolonged period of time. It's best to avoid the need for an antacid in the first place. You can minimize the production of excess stomach acid by eating a healthy diet, avoiding stress, and limiting your consumption of coffee, fatty foods, and chocolate.

Section Review

26. Describe how a lactic-acid–lactate-ion buffer reacts with both H_3O^+ and OH^- ions. (Hint: see the formula for lactic acid on page 504.)

27. Draw the structural formula for aspartic acid. Label the *R* group, the amine group, and the carboxyl group on your structure.

28. Draw the structural formula for the zwitterion form of aspartic acid and how it reacts in the presence of OH^- ions.

29. Label each of the following fatty acids as saturated or unsaturated.

a. $CH_3CH_2CH_2CH_2CH_2CH_2CH_2CH_2CH_2CH_2CH_2COOH$

b. $CH_3CH_2CH_2CH_2CH_2CH_2CH{=}CHCH_2CH_2CH_2CH_2CH_2CH_2CH_2COOH$

c.

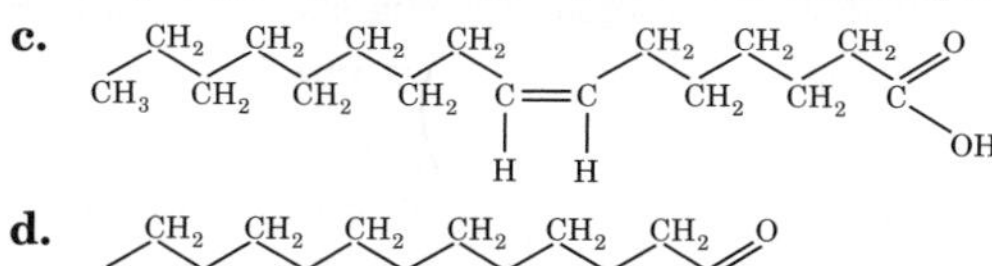

d. $CH_3CH_2CH_2CH_2CH_2CH_2CH_2CH_2CH_2CH_2CH_2CH_2COOH$ (drawn as zigzag chain)

30. Draw the ionized form of the fatty acid from item **29a**.

31. What is it that is saturated in a saturated fatty acid?

32. What functional group is characteristic of organic bases?

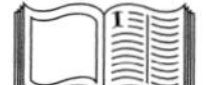

33. Story Link

What component of DNA makes it an acid?

ASSESS

Alternative Assessment

Have students explain the importance and role of acids in the body.

Answers to Section Review

26. $CH_3CHOHCOO^- + H_3O^+ \longrightarrow CH_3CHOHCOOH + H_2O$

$CH_3CHOHCOOH + OH^- \longrightarrow CH_3CHOHCOO^- + H_2O$

27. $HOOC{-}CH_2{-}CH(NH_2){-}COOH$ — labels: Amine group (NH_2); Carboxyl group

28. $HOOC{-}CH_2{-}CH(NH_3^+){-}COO^- + OH^- \longrightarrow [HOOC{-}CH_2{-}CH(NH_2){-}COO]^-$

29. **a.** saturated
b. unsaturated
c. unsaturated
d. saturated

30. $CH_3CH_2CH_2CH_2CH_2CH_2CH_2CH_2CH_2CH_2CH_2COO^-$

31. Carbon atoms are completely saturated with hydrogen atoms at all available bond sites. No double or triple carbon-carbon bonds exist that could be broken to add more hydrogen atoms to the structure.

32. amine group

33. phosphoric acid

Chemistry in Your Community

Have students visit a local grocery store or drugstore and determine the active ingredients in six different antacid products. Have them develop a method to determine the amount of acid that each antacid product is able to neutralize (assuming the recommended adult dosage). After you have reviewed and approved their method, proceed to determine the amount of acid neutralized per dose of each antacid. Findings can be reported in the form of a poster.

Answers to Applying Concepts

1\.

Thymine

Adenine

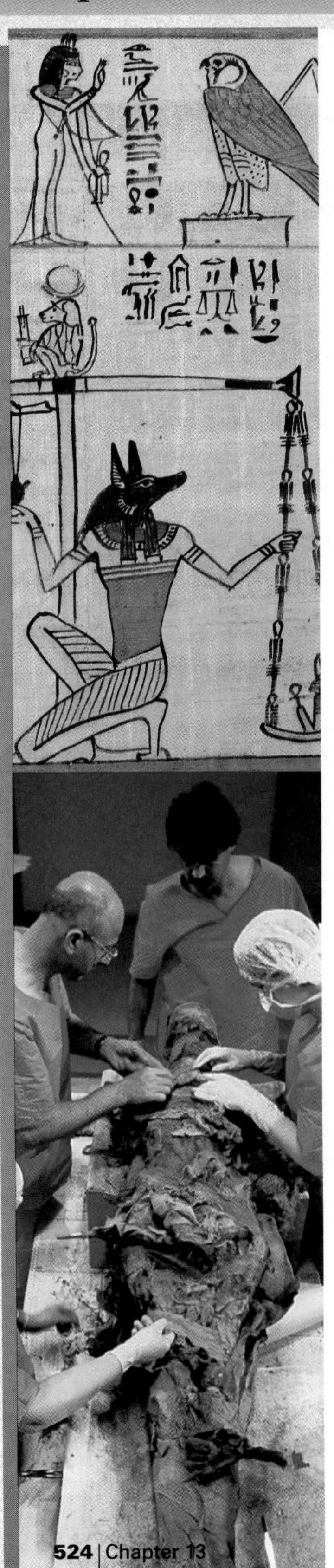

Conclusion: Ancient Acids

You began your study of acids and bases by reading about a new field of science that has recently emerged—molecular archaeology. Scientists in this field are interested in the study of just one particular acid—DNA.

What makes DNA an acid? DNA consists of two giant strands of bases linked together in a helical formation. A chain of alternating ribose sugar molecules and phosphoric acid molecules, H_3PO_4, forms the backbone that holds the bases in place. Under ordinary pH levels in biological systems, DNA exists in the conjugate base form. The phosphoric acid molecules, having donated their hydrogen ions, form anions called phosphate groups. Thus, DNA contains thousands of sites where H^+ ion transfer can take place. Therefore DNA fits the definition of a Brønsted acid.

Is DNA a strong or weak acid? The K_a value for DNA is approximately 1×10^{-1} or 0.1. DNA is a weak acid, but it is stronger than phosphoric acid itself.

Why have some ancient DNA fragments been preserved better than others? Ancient DNA is retrieved in very small fragments from organisms that have been subjected to the forces of nature for thousands or even millions of years. The preservation of the organism is the result of the action of a compound known as tannic acid, commonly found in peat bogs. Unfortunately, tannic acid destroys DNA.

But even in a peat bog a body may be found that contains some ancient DNA if there are limestone rocks in the area. Limestone is a mineral containing calcium carbonate, $CaCO_3$. $CaCO_3$ neutralizes acids and tends to preserve the DNA fragments in the soft tissues from destruction by tannic acid.

Applying Concepts

1. Bases such as those shown on page 520 hold the two strands of DNA together by means of hydrogen bonds. Make a drawing that shows how these bonds are formed between thymine and adenine.

Research and Writing

1. Why do molecular archaeologists like Svante Pääbo think the idea of bringing extinct species back into existence is more a matter of science fiction than scientific research?
2. Molecular archaeologists are attempting to test the proposal that all living humans are descended from one woman, nicknamed Eve, who lived in Africa some 200 000 years ago. To do so, they are analyzing the DNA found in specialized structures called mitochondria that are found in the cytoplasm of cells. Report on the latest findings in this research.

CHAPTER 13

Highlights

Key terms

- acid anhydride
- acid-dissociation constant
- amino acid
- amphiprotic
- base anhydride
- base-dissociation constant
- buffer
- conjugate acid
- conjugate base
- dissociation constant for water
- end point
- equivalence point
- fatty acid
- hydroxide ion
- indicator
- neutralization reaction
- nucleic acid
- pH
- salt
- saturated fatty acid
- titration
- titration standard
- transition interval
- unsaturated fatty acid
- zwitterion

Key ideas

What are acids and bases?

- Strong acids and bases dissociate completely in aqueous solution; weak ones dissociate only partially.
- A Brønsted acid is a hydrogen ion donor, and a Brønsted base is a hydrogen ion acceptor. Some compounds can function as both.
- A conjugate acid forms when a Brønsted base accepts a hydrogen ion. A conjugate base forms when a Brønsted acid releases a hydrogen ion.
- An acid anhydride reacts with water to form an acid. A base anhydride reacts with water to form a base.

How are weak acids and bases compared?

- Dissociation constants measure the relative strength of weak acids and bases. The smaller the numerical value of this equilibrium constant, the weaker the acid or base.
- The pH measures the hydronium ion concentration in a solution.
- Water, an amphiprotic substance, dissociates to form hydronium ions and hydroxide ions.

What is a titration?

- When combined, acids and bases neutralize each other to produce water and a salt.
- The concentration of a solution may be determined through a titration with a solution of known concentration.
- The equivalence point of a titration can be determined by the end point of an indicator.
- The end point of an indicator occurs when the indicator changes color. The equivalence point of a titration occurs when stoichiometric amounts of acid and base are present.

What are the biochemical roles of some acids and bases?

- Amino acids, nucleic acids, and fatty acids are biological molecules that function within the pH range of living systems.
- Buffers maintain pH in living systems.

Key problem-solving approach: *Concentrations and pH conversions*

When converting between concentrations and pH values, begin at the appropriate box, and follow calculation commands to the desired value.

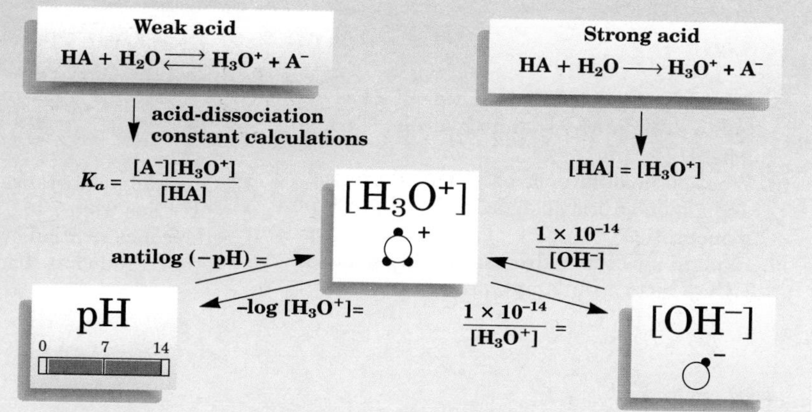

Closure Strategy

Students should work in small groups to make a concept map using as many of the terms from the following list as possible.

Term list
acid anhydride
amphiprotic
base anhydride
conjugate acid
conjugate base
endpoint
equivalence point
hydroxide ion
indicator
neutralization reaction
pH
salt
titration

Photocopy the best maps and distribute them to students to use when studying for the chapter test.

1. An acid tastes sour, reacts with metals to produce hydrogen gas, and conducts electricity in water solution. A base tastes bitter, feels slippery, and conducts electricity in water solution.

2. *See page 531A.*

3. N_2O_5 is an acid anhydride.

4. *See page 531A.*

5. NaH is not a base by the Arrhenius definition because it does not contain the –OH group.

6. Monoprotic acids contain one hydrogen atom per molecule of acid. Diprotic acids contain two hydrogen atoms per acid molecule. H_2CO_3 is diprotic.

7. HCO_3^- acts as an amphiprotic ion because it can be a proton donor or proton acceptor in reactions. Therefore, it reacts as both an acid or base.

8. $HCN + H_2O \longrightarrow H_3O^+ + CN^-$

acid + base $\longrightarrow$ conjugate acid + conjugate base

9. NaOH dissolves oils and neutralizes acids.

10. For the reverse reaction, the acetate ion is a proton acceptor, thereby reacting as a base.

11. The value represents the ratio of dissociated ions to the molecular form of a weak acid or base. The smaller a K_a or K_b value is, the weaker the acid or base.

*Full solutions for items **2** and **4** are found on page 531A.*

CHAPTER 13

Review and Assess

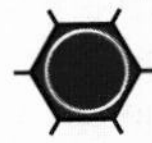

Comparing acids and bases

REVIEW

1. Compare the properties of an acid to those of a base.
2. Arrhenius was the first scientist to define the properties of acids and bases. Fill in the table below using Arrhenius's definitions.

	Ions released into solution by dissociation of acid or base	Acid or base
HCl		
$Ba(OH)_2$		

3. What is the acid anhydride in the following equation?

$$N_2O_5 + H_2O \longrightarrow 2HNO_3$$

4. Complete the table below by indicating the acid or base formed by the hydration of the oxides given.

Oxide	Substance formed when hydrated	Acid or base?
SO_3	H_2SO_4	acid
BaO		
P_2O_5		
SeO_3		
K_2O		
Li_2O		

5. How does Arrhenius's definition of a base fail to explain why sodium hydride, NaH, is a base?
6. What is the difference between a diprotic and a monoprotic acid? Is H_2CO_3 diprotic or monoprotic?
7. Explain how the hydrogen carbonate ion, HCO_3^-, acts as an amphiprotic ion.

APPLY

8. Write an equation for the reaction between hydrocyanic acid, HCN, and water. Label the acid, base, conjugate acid, and conjugate base.
9. Explain why sodium hydroxide, NaOH, is used in oil refining to remove acids.
10. Explain how the acetate ion formed as a product in the forward reaction can act as a base in the reverse reaction.

$$CH_3COOH(aq) + H_2O(l) \rightleftharpoons CH_3COO^-(aq) + H_3O^+(aq)$$

Strengths of weak acids and weak bases

REVIEW

11. What does the numerical value for K_a and K_b represent?
12. Using the table below, answer the following questions.
    a. Which is the strongest acid? Why?
    b. Which acid dissociates the least? Why?

Acid	Formula	Value of K_a
arsenic	H_3AsO_4	5.62×10^{-3}
acetic	CH_3COOH	1.76×10^{-5}
benzoic	C_6H_5COOH	6.5×10^{-5}
cyanic	HCNO	7.4×10^{-4}
phosphoric	H_3PO_4	7.5×10^{-3}
hydrofluoric	HF	6.8×10^{-4}
carbonic	H_2CO_3	4.3×10^{-7}

13. You add a substance to pure water, and the pH rises from 7 to 9. What has happened to the concentration of H_3O^+ ions? Is the substance added an acid or a base?

14. Classify the acids by placing a check in the columns that apply.

Acid	Formula	Triprotic	Diprotic	Monoprotic
arsenic	H_3AsO_4	✔		
acetic	CH_3COOH			
sulfurous	H_2SO_3			
phosphoric	H_3PO_4			
hydrofluoric	HF			

Acid	Binary	Ternary	Strong	Weak
arsenic		✔		✔
acetic				
sulfurous				
phosphoric				
hydrofluoric				

15. Explain the relationship between the dissociation of water and the pH of a neutral solution.
16. What is the relationship between the forward and reverse reactions and the strength of an acid or a base?
17. Why is the first dissociation constant for a polyprotic acid used to calculate the pH value for an aqueous solution of that acid?

A P P L Y

18. A solution of HCl has a pH of 1. You add NaOH to the solution, and the pH rises to 1.18. What has happened? What does this have to do with the relationship of acids to bases along the pH scale? Can you estimate what amounts of HCl and NaOH might have been used?
19. Use two arbitrary pH values and the mathematical method for determining pH to show how the concentration of H_3O^+ increases as pH decreases.
20. Write an equilibrium expression for the dissociation of benzoic acid C_6H_5COOH.

P R A C T I C E

Answers to items in a black square begin on page 841.

21 What is the pH of a 0.00256 M HCl solution? (Hint: see Sample Problem 13B.)
22. What is the pH of a solution with $[H_3O^+]$ of 6.35×10^{-10} M? (Hint: see Sample Problem 13B.)
23 What is the pH of a seawater solution that has a hydronium ion concentration of 1.35×10^{-8} M? (Hint: see Sample Problem 13B.)
24. What is the pH of lemon juice if $[H_3O^+] = 0.00363$ M? (Hint: see Sample Problem 13B.)
25 The average pH of precipitation in an area near a power plant is 3.45. What is $[H_3O^+]$? (Hint: see Sample Problem 13D.)
26. The pH of a glass of soda water is 4.3. What is $[H_3O^+]$? (Hint: see Sample Problem 13D.)
27 What is the pH of a 2.5×10^{-5} M KOH solution? (Hint: see Sample Problem 13E.)
28. What is the pH of an ammonia solution that has a hydroxide ion concentration of 10^{-3} M? (Hint: see Sample Problem 13E.)
29 A solution of 5% acetic acid has a concentration of 0.83 M and K_a of 1.8×10^{-5}. What is the pH of the solution? (Hint: see Sample Problem 13F.)
30. What is the concentration of all of the components of a benzoic acid solution if K_a is 6.5×10^{-5}, pH is 2.96, and C_6H_5COOH is 0.020 M? (Hint: see Sample Problem 13F.)

Titrations

R E V I E W

31. Write the complete molecular and net ionic equations for the neutralization of sulfuric acid with potassium hydroxide.
32. Describe the role of an indicator in the titration of an unknown acid against a standardized base.
33. Why is the transition interval of an indicator critically important when performing different titrations?
34. How is the color change of an indicator related to pH?
35. Refer to the table below.
    a. Which indicator would be the best choice for a titration with an end point at a pH of 4.0?
    b. Which indicators would work best for a titration of a weak base with a strong acid?

Indicator	Acid color	Base color	pH range of color
thymol blue	red	yellow	1.2–2.8
bromphenol blue	yellow	blue	3.0–4.6
bromcresol green	yellow	blue	2.0–5.6
bromthymol blue	yellow	blue	6.0–7.6
phenol red	yellow	red	6.6–8.0
alizarin yellow	yellow	red	10.1–12.0

12. a. The strongest acid is phosphoric. It has the largest numerical value for K_a, therefore it dissociates most in aqueous solution.
    b. Carbonic acid dissociates least because it has the smallest value for K_a.

13. The concentration of H_3O^+ has decreased by 100 times. To reduce the H_3O^+ concentration, a base must have been added.

14. *See page 531A.*

15. Pure water dissociates into equal numbers of hydronium and hydroxide ions; it is therefore neutral. Any neutral solution will have a pH value equal to that of pure water, 7.

16. The smaller the K_a or K_b is, the less forward the reaction is favored. As the relative strength of an acid or base decreases, the reverse reaction is more favored over the forward reaction.

17. The values for second dissociation constants are very small compared to the value for K_a for the first dissociation. The amount of H^+ contributed from the second dissociation is negligible compared to the amount from the primary dissociation.

18. The NaOH has neutralized some of the HCl, making the solution slightly less acidic. You can estimate that only a small amount of NaOH was used because HCl normally has a pH of 1. The addition of NaOH only raised the pH slightly.

19. pH values = 3 and 5

$pH = -\log [H_3O^+]$
$pH = -\log [H_3O^+]$

$3 = -\log [H_3O^+]$
$5 = -\log [H_3O^+]$

$[H_3O^+] = 0.001$ M
$[H_3O^+] = 0.00001$ M

The concentration of hydronium ion when the pH is 3 is 100 times stronger than when the pH is 5.

20. $C_6H_5COOH + H_2O \rightleftharpoons H_3O^+ + C_6H_5COO^-$

21. 2.59

22. 9.20

23. 7.87

24. 2.44

25. antilog (–3.45) = 3.55×10^{-4}

26. antilog (–4.3) = 5.01×10^{-5}

27. $\frac{1.00 \times 10^{-14}}{2.5 \times 10^{-5}} = 4.0 \times 10^{-10}$ M H_3O^+

pH = 9.4

28. $\frac{1.00 \times 10^{-14}}{1 \times 10^{-3}} = 1 \times 10^{-11}$ M H_3O^+

pH = 11

29. $[H_3O^+] = x$; $[CH_3COO^-] = x$
$[CH_3COOH] = 0.83\ M - x$
(x is negligible in this problem)

$1.8 \times 10^{-5} = \frac{x^2}{[0.83]}$; $x = 3.9 \times 10^{-3}$

$pH = -\log[3.9 \times 10^{-3}]$

pH = 2.4

30. $C_6H_5COOH + H_2O \longrightarrow C_6H_5COO^- + H_3O^+$

K_a is not needed in this problem. It is given as a distractor.

pH = 2.96 = $[H_3O^+]$ of 1.1×10^{-3}

The mole ratio is 1 to 1.

$[C_6H_5COO^-] = 1.1 \times 10^{-3}$

$[C_6H_5COOH] = 0.020\ M - 1.1 \times 10^{-3} = 1.9 \times 10^{-2}$

31. $H_2SO_4(aq) + 2KOH(aq) \longrightarrow 2H_2O(l) + 2K_2SO_4(aq)$

$H_3O^+(aq) + OH^-(aq) \longrightarrow H_2O(l)$

32. The indicator changes color within a specific pH range. If the endpoint of the indicator is in the same pH range as the equivalence point for the titration, it will signal the end of the titration.

33. The transition range should be over a narrow pH range and the color change should be dramatic to estimate the equivalence point accurately.

34. Indicators are weak acids and bases whose equilibrium reactions are sensitive to changes in pH.

36. Why is a different indicator used for a strong acid/strong base titration than for a weak acid/strong base titration?

37. Why does an indicator need to be a weak acid or base?

38. How is the transition interval of an indicator related to the end point of a titration?

39. A 50 mL solution of CH_3COOH is titrated with a 1.000 M NaOH solution. Describe what species (molecules and ions) are present in the titration flask at the following points in the titration.
a. pH 6 **b.** pH 9 **c.** pH 11

40. Using **Figure 13-16** as a guide, sketch a titration curve for 50 mL of 1 M solution of the weak base, NH_3, with 1.000 M HCl.

41. For the following titrations, select the best indicator from these choices: bromphenol blue, bromthymol blue, phenol red.
a. formic acid, HCOOH, with NaOH
b. perchloric acid with LiOH
c. sulfuric acid with potassium hydroxide
d. ethylamine, $C_2H_5NH_2$, with hydrochloric acid

P R A C T I C E

42 Formic acid is used in the production of leather. Residue from leather production is tested to determine the concentration of formic acid in excess. A 50.0 mL sample of formic acid, HCOOH, is neutralized by 35.0 mL of 0.250 M NaOH. What is the molarity of the formic acid solution? (Hint: see Sample problem 13H.)

43. You wish to determine the molarity of a lactic acid solution. A 150. mL sample of lactic acid, $CH_3CHOHCOOH$, is titrated with 125 mL of 0.75 M NaOH. What is the molarity of the acid sample? (Hint: see Sample Problem 13H.)

44 If 35.40 mL of 1.000 M HCl are neutralized by 67.3 mL of NaOH, what is the molarity of the NaOH solution? (Hint: see Sample Problem 13H.)

45. If 50.00 mL of 1.000 M H_2SO_4 are neutralized by 35.4 mL of KOH, what is the molarity of the KOH solution? (Hint: see Sample Problem 13H.)

46 If 18.5 mL of a 0.350 M H_2SO_4 solution neutralizes 12.5 mL of aqueous LiOH, how many grams of LiOH were used to make the LiOH solution?

47. A student is preparing a standardized NaOH solution for use in a titration. About 2 g of NaOH are dissolved in 500 mL of distilled water. This solution is titrated against an oxalic acid dihydrate solution, $H_2C_2O_4{\cdot}2H_2O$, which is made by dissolving 1.125 g of the solid acid in 40 mL of distilled water. The titration requires 174 mL of NaOH.
a. Write the equation for the neutralization reaction.
b. Calculate the molarity of the NaOH standard. (Hint: oxalic acid is a diprotic acid.)

A P P L Y

48. In the titration of a weak acid with a strong base, the pH of the equivalence point is not 7, but in the basic range. Use the concept of conjugate acids and bases to explain why the pH is in the basic range.

49. A student passes an end point in a titration. Is it possible to add an additional measured amount of the unknown and continue the titration? Explain how this process might work. How would the answer for the calculation of the molarity of the unknown differ from the answer the student would get if the titration had been run successfully?

Biochemical roles of acids and bases

R E V I E W

50. Explain how weak acids in the body can help maintain the buffering capacity of a living system.

51. How would you test to see if a system is buffered?

52. The pH of your cells needs to be between 6.4 and 7.4 for the cells to function properly. How can this pH be kept constant even with DNA and RNA, two nucleic acids, inside the cell?

53. What structural feature of small fatty acids allows them to be more soluble in water than larger fatty acids?

A P P L Y

54. Enzymes and proteins are long chains of amino acids.
a. How do the structures and functions of proteins and enzymes change with pH?
b. Describe why an enzyme's activity changes with changes in pH.

55. Identify the fatty acids in the following fats as saturated, unsaturated, or polyunsaturated.

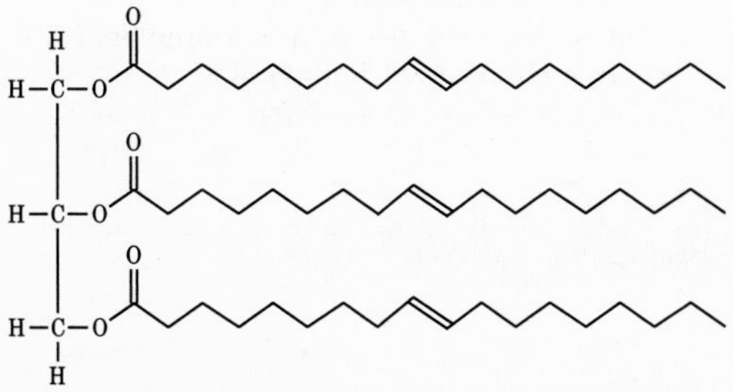

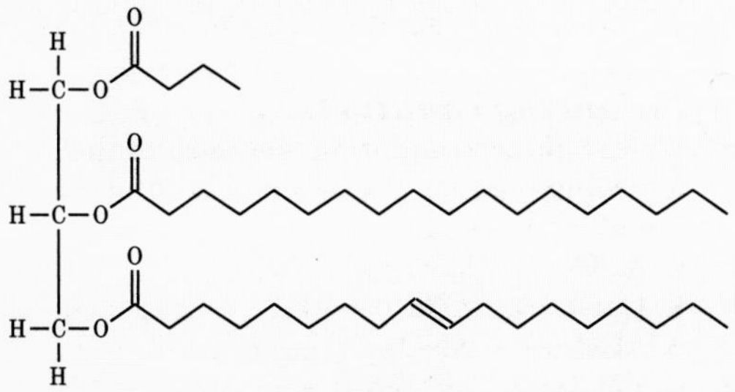

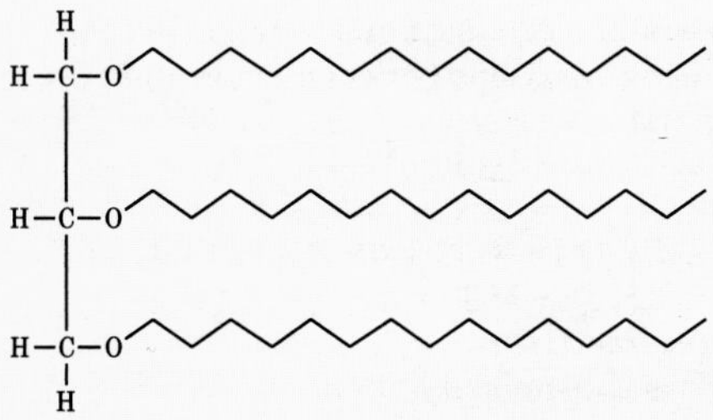

56. Classify the following compounds as organic acids or bases.

a. 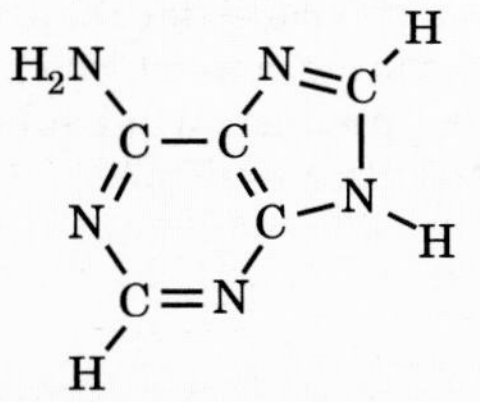 H_2N, N, C, N, H (ring structure)

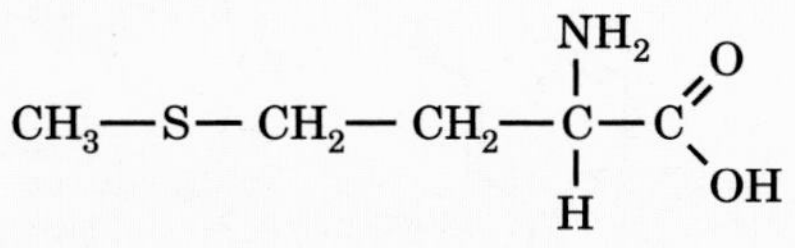

b. $CH_3-S-CH_2-CH_2-C(NH_2)(H)-C(=O)OH$

c. $H_3C-N(CH_3)-CH_3$

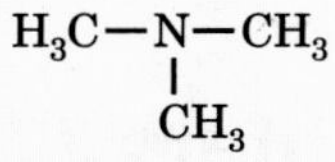

d. 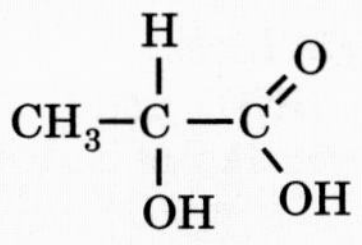 $CH_3-C(H)(OH)-C(=O)OH$

e. 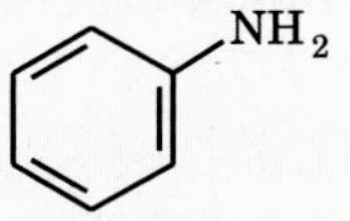 $C_6H_5NH_2$

f. 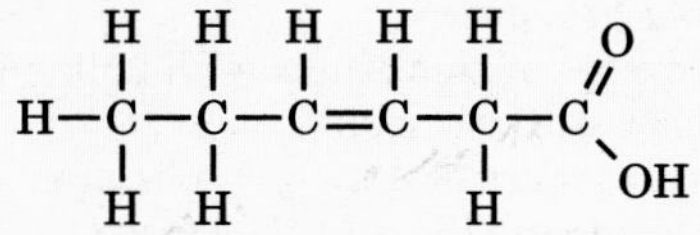 $H-CH_2-CH_2-CH=CH-CH_2-C(=O)OH$

g. 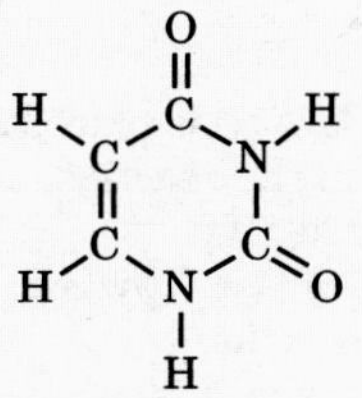 H, C, N, O (ring structure)

h. $-CH_2-C(NH_2)(H)-C(=O)OH$ (ring with N, N–H)

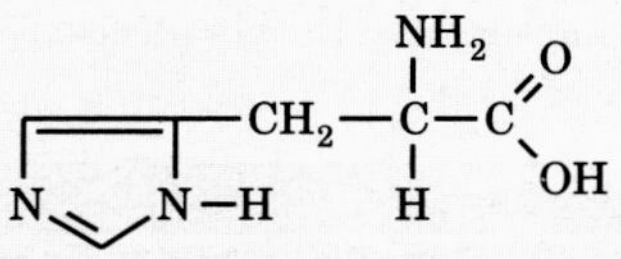

35. a. bromphenol blue
    b. The indicator should change in the acid range. Therefore, thymol blue, bromphenol blue, bromcresol green, or bromthymol blue could be used.

36. The equivalence point of a strong acid/strong base titration is pH 7. The equivalence point of a weak acid/strong base titration is in the basic range due to the formation of the conjugate base of the weak acid species. The indicator chosen should have an endpoint near the equivalence point for the titration.

37. Indicators in solution are equilibrium systems that respond to pH changes by shifting equilibrium. The equilibrium shifts brought about by the addition of acid or base cause color changes depending on the predominate form of the indicator.

38. The transition interval is the range of pH associated with the progression of color change. When the color change is complete at the end of the transition interval, the endpoint is reached.

39. The following are present at all pH levels of the titration, they just vary in concentration: Na^+, CH_3COO^-, CH_3COOH, H_3O^+, OH^-

40. The curve should be the inverse of Figure 13-16. It starts in the base range (upper left). The equivalence point should be shown at a pH of less than 7.

41. a. phenol red
    b. bromthymol blue
    c. bromphenol blue

42. $35.0 \text{ mL NaOH} \times \frac{1 \text{ L NaOH}}{1000 \text{ mL NaOH}} \times \frac{0.250 \text{ moles NaOH}}{1 \text{ L NaOH}} =$

8.75×10^{-3} mol NaOH

The mole ratio is 1 to 1.

$\frac{8.75 \times 10^{-3} \text{ mol formic acid}}{50.0 \text{ mL formic acid}} \times \frac{1000 \text{ mL formic acid}}{1 \text{ L formic acid}} =$

0.175 M formic acid

43. $125 \text{ mL NaOH} \times \frac{1 \text{ L NaOH}}{1000 \text{ mL NaOH}} \times \frac{0.75 \text{ moles NaOH}}{1 \text{ L NaOH}} = 0.0938 \text{ mol NaOH}$

The mole ratio is 1 to 1.

$\frac{0.0938 \text{ mol lactic acid}}{150.0 \text{ mL lactic acid}} \times \frac{1000 \text{ mL lactic acid}}{1 \text{L lactic acid}} =$

0.63 M lactic acid

44. $HCl(aq) + NaOH(aq) \longrightarrow NaCl(aq) + H_2O(l)$

$35.40 \text{ mL HCl} \times \frac{1 \text{ L}}{1000 \text{ mL}} \times \frac{1.000 \text{ mol HCl}}{1 \text{ L}} = 0.0354 \text{ mol HCl}$

The mole ratio is 1 to 1.

$\frac{0.0354 \text{ mol NaOH}}{67.3 \text{ mL}} \times \frac{1000 \text{ mL NaOH}}{1 \text{ L}} \times 0.526 \text{ M NaOH}$

45. $H_2SO_4(aq) + 2KOH(aq) \longrightarrow K_2SO_4(aq) + 2H_2O(l)$

$50.00 \text{ mL } H_2SO_4 \times \frac{1 \text{ L}}{1000 \text{ mL}} \times \frac{1.000 \text{ mol } H_2SO_4}{1 \text{ L}} = 0.050 \text{ mol } H_2SO_4$

The mole ratio is 1 to 2.

$0.050 \text{ mol } H_2SO_4 \times \frac{2 \text{ mol KOH}}{1 \text{ mol } H_2SO_4} =$

0.10 mol KOH

$\frac{0.10 \text{ mol KOH}}{35.4 \text{ mL KOH}} \times \frac{1000 \text{ mL}}{1 \text{ L}} =$

2.83 M KOH

46. $H_2SO_4(aq) + 2LiOH(aq) \longrightarrow Li_2SO_4(aq) + 2H_2O(l)$

$18.5 \text{ mL } H_2SO_4 \times \frac{1 \text{ L}}{1000 \text{ mL}} \times \frac{0.350 \text{ mol } H_2SO_4}{1 \text{ L}} =$

6.50×10^{-3} mol H_2SO_4

The mole ratio is 1 to 2.

$6.50 \times 10^{-3} \text{ mol } H_2SO_4 \times \frac{2 \text{ mol LiOH}}{1 \text{ mol } H_2SO_4} = 1.3 \times 10^{-2} \text{ mol LiOH}$

$1.3 \times 10^{-2} \text{ mol LiOH} \times \frac{23.95 \text{ g}}{1 \text{ mol LiOH}} =$

0.310 g LiOH

57. The structure of acetylsalicylic acid, $C_9H_8O_4$ is shown below. This chemical causes an increase in stomach acidity for some people. Aspirin manufacturers combat this problem by buffering their product. What substance could be used to buffer aspirin?

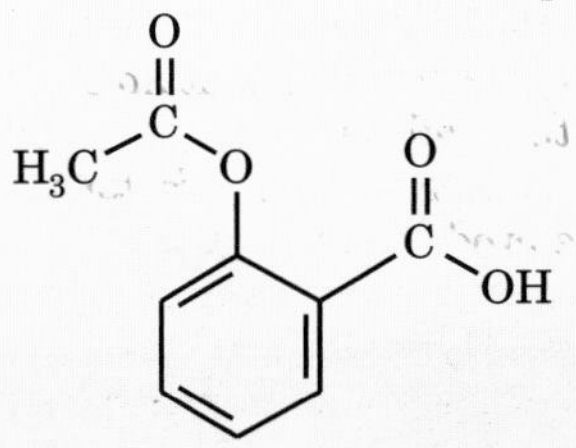

58. The amino acid tryptophan is pictured below.
a. Draw the structure of this amino acid if it is in a solution with a pH of 3.
b. Draw the structure of this amino acid if it is in a solution with a pH of 8.

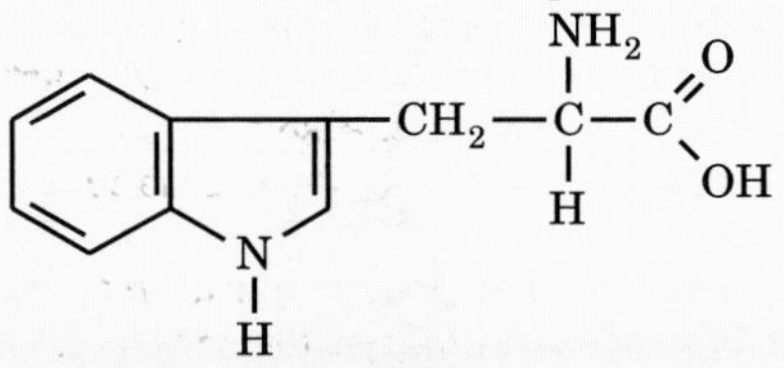

Linking chapters

1. ***Applying scientific laws***
How do nitrous acid, HNO_2, and nitric acid, HNO_3, demonstrate the law of multiple proportions?

2. **Theme: *Systems and interactions***
a. Identify the different systems that are involved in a titration.
b. Explain how each system interacts with the other systems in a titration.

3. ***Stoichiometry***
An excess of zinc reacts with 250. mL of 6.00 M sulfuric acid.
a. What mass of $ZnSO_4$ can be produced?
b. How many liters of H_2 are produced at STP?

4. ***Stoichiometry***
A seashell (mostly $CaCO_3$) is placed in a solution of HCl. From the reaction, 1.50 L of CO_2 are produced at STP.
a. How many grams of $CaCO_3$ were consumed in the reaction?
b. If the HCl solution was 2.00 M, what volume was used in this reaction?

5. ***Balancing equations***
Write balanced equations for each of the following reactions between acids and metal oxides.
a. $MgO(s) + H_2SO_4(aq) \longrightarrow$
b. $CaO(s) + HCl(aq) \longrightarrow$
c. $Al_2O_3(s) + HNO_3(aq) \longrightarrow$
d. $ZnO(s) + H_3PO_4(aq) \longrightarrow$

6. ***Balancing equations***
Write balanced equations for each of the following reactions between acids and carbonates.
a. $BaCO_3(s) + HCl(aq) \longrightarrow$
b. $MgCO_3(s) + HNO_3(aq) \longrightarrow$
c. $Na_2CO_3(s) + H_2SO_4(aq) \longrightarrow$
d. $CaCO_3(s) + H_3PO_4(aq) \longrightarrow$

7. ***Concentration***
What is the molarity of an oxalic acid dihydrate standard, $H_2C_2O_4{\cdot}2H_2O$, made from dissolving 3.000 g of the solid acid to form 150 mL of solution?

8. ***Stoichiometry***
What would the molarity of a KOH solution be if 75.00 mL of base were used to neutralize the acid in item 7? (Hint: oxalic acid is a diprotic acid.)

USING TECHNOLOGY

1. ***Computer art***
Draw the general structural formula of an amino acid. Now show that formula as a zwitterion. Show how the zwitterion reacts in both acidic and basic conditions.

2. ***Calculator***
Use your calculator to determine the following:
a. log of 1×10^{-14}
b. antilog of 4.6
c. log of 5.738×10^{-6}
d. antilog of 0

Alternative assessment

Performance assessment

1. Design an experiment to identify pH levels of four types of hair shampoo: baby shampoo, shampoo for extra body, shampoo for oily hair, and shampoo with conditioner. Chart any patterns you detect. Also compare and contrast two brands of "pH balanced" shampoo.
2. Your teacher will give you a solution to test for pH. Describe exactly how you would test the solution. If your teacher approves your plan, complete the test.
3. Design an experiment to test the neutralization effectiveness of various brands of antacid. Show your procedure, which includes all safety procedures and cautions, to your teacher for approval. Write an advertisement for the antacid you judge to be the most effective. Cite data from your experiments as part of your advertising claims.

Portfolio projects

1. ***Chemistry and you***
   Hydrochloric acid, HCl, can be found in many cleaning products. Find three products in your home that contain acids. Determine if the acids are strong or weak. Does it seem that strength is a factor in whether these products are considered harmful? Use the product warning labels as a guide.
2. ***Research and communication***
   A reaction between baking soda, $NaHCO_3$, and a baking batter that contains acidic ingredients produces carbon dioxide gas. The reaction results in a fluffier batter. Some recipes rely on baking powder instead of baking soda. Research the ingredients of regular baking powder and double-acting baking powder to discover how they differ from baking soda. Which is more likely to be used in creating a light, fluffy cake? How does the pH affect the final product?
3. ***Chemistry and you***
   Collect data on the acidity of rain in your area over the last 10 years. Graph the data and make a prediction concerning acid-rain damage in your area for the year 2010. Use evidence to support your prediction.
4. ***Cooperative activity***
   Research the Acid Rain Control Program. Debate the pros and cons of reducing oxide emission versus treating the effects of acid rain with base anhydrides. Discuss the environmental and economic impact of both plans.

47. a. $H_2C_2O_4(aq) + 2NaOH(aq) \longrightarrow Na_2C_2O_4(aq) + 2H_2O(l)$

$$1.125 \text{ g } H_2C_2O_4{\cdot}2H_2O \times \frac{1 \text{ mol } H_2C_2O_4{\cdot}2H_2O}{126.08 \text{ g } H_2C_2O_4{\cdot}2H_2O} = 8.923 \times 10^{-3} \text{ mol } H_2C_2O_4{\cdot}2H_2O$$

The mole ratio is 1 to 2.

$$8.923 \times 10^{-3} \text{ mol } H_2C_2O_4{\cdot}2H_2O \times \frac{2 \text{ mol NaOH}}{1 \text{ mol } H_2C_2O_4} = 1.785 \times 10^{-2} \text{ mol NaOH}$$

$$\frac{1.785 \times 10^{-2} \text{ mol NaOH}}{174 \text{ mL}} \times \frac{1000 \text{ mL}}{1 \text{ L}} = 0.103 \text{ M NaOH}$$

48. The conjugate base of the neutralized acid reacts with H^+ ions from water, producing OH^- ions and a basic pH.

49. If the exact amount of unknown added is recorded, then that value can be added to the initial volume of unknown, and the sum is then used in the calculations. The molarity of an unknown would not change no matter how much unknown was used.

50. Amino acids, fatty acids, and proteins buffer living systems because they exist in both the acid form and the ion form. The ions react with excess H_3O^+ ions from the solution. The acids react with any base in solution.

51. Add a small amount of an acid or base to a sample and look for a change in the pH. If the pH does not change as much as expected, the system may be buffered.

52. Even though DNA and RNA have acidic properties, they are very weak acids and are offset by many other compounds in cells including buffers.

53. The polar carboxyl group allows the short-chain molecule to dissolve in a polar solute. If a fatty acid has a long nonpolar chain, the effect of that long nonpolar segment will make the molecule insoluble in water.

Answers from page 506

15. A solution with a pH of 8 has a [OH⁻] that equals the following.

$$\frac{1 \times 10^{-14}}{1 \times 10^{-8}} = 1 \times 10^{-6}\ \text{M}$$

A solution with a pH of 11 has an [OH⁻] that equals the following.

$$\frac{1 \times 10^{-14}}{1 \times 10^{-11}} = 1 \times 10^{-3}\ \text{M}$$

The pH 11 solution has a larger [OH⁻]. It is 1000 times larger. The pH 8 solution has a larger [H_3O^+]. It is 1000 times larger.

16. pH = 3.1; [H_3O^+] = –antilog 3.1; [H_3O^+] = 7.9×10^{-4}

$$[OH^-] = \frac{1 \times 10^{-14}}{7.9 \times 10^{-4}} = 1.3 \times 10^{-11}$$

17. $HA + H_2O \longrightarrow H_3O^+ + A^-$

pH = 3.14; [H_3O^+] = –antilog 3.14; [H_3O^+] = 7.24×10^{-4}

$$K_a = \frac{[7.24 \times 10^{-4}]^2}{[1.6 \times 10^{-2}] - [7.24 \times 10^{-4}]} = \frac{[7.24 \times 10^{-4}]^2}{[1.53 \times 10^{-2}]} = 3.43 \times 10^{-5}$$

Answers from page 514

Answers to Practice

2. $2HClO_4 + Ba(OH)_2 \longrightarrow 2H_2O + Ba(ClO_4)_2$

$$46.\ \text{mL } HClO_4 \times \frac{1\ \text{L}}{1000\ \text{mL}} \times \frac{0.40\ \text{mol } HClO_4}{1\ \text{L}} = 0.0184\ \text{mol } HClO_4$$

The mole ratio is 2 to 1.

$$0.0184\ \text{mol } HClO_4 \times \frac{1\ \text{mol } Ba(OH)_2}{2\ \text{mol } HClO_4} = 0.0092\ \text{mol } Ba(OH)_2$$

$$0.0092\ \text{mol } Ba(OH)_2 \times \frac{1\ \text{L}}{0.25\ \text{mol } Ba(OH)_2} = 0.0368\ \text{L} = 37\ \text{mL}$$

3. $CH_3COOH + KOH \longrightarrow H_2O + CH_3COOK$

$$15.5\ \text{mL KOH} \times \frac{1\ \text{L KOH}}{1000\ \text{mL KOH}} \times \frac{0.215\ \text{mol KOH}}{1\ \text{L}} = 3.33 \times 10^{-3}\ \text{mol KOH}$$

The mole ratio is 1 to 1.

$$\frac{3.33 \times 10^{-3}\ \text{mol } CH_3COOH}{21.2\ \text{mL } CH_3COOH} \times \frac{1000\ \text{mL}}{1\ \text{L}} = 0.157\ \text{M } CH_3COOH$$

4. $2HCl + Ba(OH)_2 \longrightarrow 2H_2O + BaCl_2$

$$27.4\ \text{mL } Ba(OH)_2 \times \frac{1\ \text{L}}{1000\ \text{mL}} \times \frac{0.0154\ \text{mol } Ba(OH)_2}{1\ \text{L}} =$$

4.22×10^{-4} mol $Ba(OH)_2$

The mole ratio is 1 to 2.

$$4.22 \times 10^{-4}\ \text{mol } Ba(OH)_2 \times \frac{2\ \text{mol HCl}}{1\ \text{mol } Ba(OH)_2} = 8.44 \times 10^{-4}\ \text{mol HCl}$$

$$\frac{8.44 \times 10^{-4}\ \text{mol HCl}}{20.0\ \text{mL HCl}} \times \frac{1000\ \text{mL}}{1\ \text{L}} = 0.0422\ \text{M HCl}$$

Answers to Additional Problem 13H

a. $20.0\ \text{mL of HCl} \times \frac{1\ \text{L HCl}}{1000\ \text{mL HCl}} \times \frac{0.100\ \text{mol HCl}}{1\ \text{L HCl}} = 0.002\ \text{mol HCl}$

Because it is a 1 to 1 ratio, 0.002 mol of NaOH are needed.

$$\frac{0.002\ \text{mol NaOH}}{19.5\ \text{mL NaOH}} \times \frac{1000\ \text{mL NaOH}}{1\ \text{L NaOH}} = 0.103\ \text{M NaOH}$$

b. $50.5\ \text{mL HCl} \times \frac{1\ \text{L HCl}}{1000\ \text{mL HCl}} \times \frac{0.23\ \text{mol HCl}}{1\ \text{L HCl}} = 0.0116\ \text{mol HCl}$

Because it is a 1 to 1 ratio, 00116 mol of KOH are needed.

$$\frac{0.0116\ \text{mol KOH}}{39.8\ \text{mL KOH}} \times \frac{1000\ \text{mL KOH}}{1\ \text{L KOH}} = 0.292\ \text{M KOH}$$

c. $35.5\ \text{mL } H_2SO_4 \times \frac{1\ \text{L}}{1000\ \text{mL}} \times \frac{0.345\ \text{mol } H_2SO_4}{1\ \text{L } H_2SO_4} = 0.0122\ \text{mol } H_2SO_4$

Because the mole ratio is 1 to 2, 0.0244 mol KOH are necessary.

$$\frac{0.0244\ \text{mol KOH}}{42.7\ \text{mL KOH}} \times \frac{1000\ \text{mL KOH}}{1\ \text{L KOH}} = 0.571\ \text{M KOH}$$

Answers from page 526

2.

	Ion released into solution by dissociation of acid or base	Acid or base
HCl	H^+ ion	acid
$Ba(OH)_2$	OH^- ion	base

4.

Oxide	Substances formed when hydrated	Acid or base formed
SO_3	H_2SO_4	acid
BaO	$Ba(OH)_2$	base
P_2O_5	H_3PO_4	acid
SeO	H_2SeO_3	acid
K_2O	KOH	base
Li_2O	LiOH	base

Answer from page 526

14.

Acid	Formula	Triprotic	Diprotic	Monoprotic	Binary	Ternary	Strong	Weak
arsenic	H_3AsO_4	✓				✓		✓
acetic	CH_3COOH			✓		✓		✓
sulfurous	H_2SO_3		✓			✓	✓	
phosphoric	H_3PO_4	✓				✓		✓
hydrofluoric	HF			✓	✓			✓

Continued from page 531

54. a. The zwitterion forms of amino acids, the components of proteins and enzymes, change structure when they react with H_3O^+ or OH^- ions. As a result of these reactions, these ions can act as buffers.
b. Changes in pH can alter the activity of enzymes by changing the molecule to an acidic or basic form in which it may no longer fit a substrate.

55. unsaturated, unsaturated, saturated

56. a. organic base **b.** organic acid
c. organic base **d.** organic acid
e. organic base **f.** organic acid
g. organic base **h.** organic acid

57. The H^+ of the carboxyl group could be replaced with a metal ion to form a salt such as $NaC_9H_7O_4$, which would produce a buffering effect.

58. a. $[\,$indole ring (N–H)$\,–CH_2–C(H)(NH_3)–C(=O)OH\,]^+$

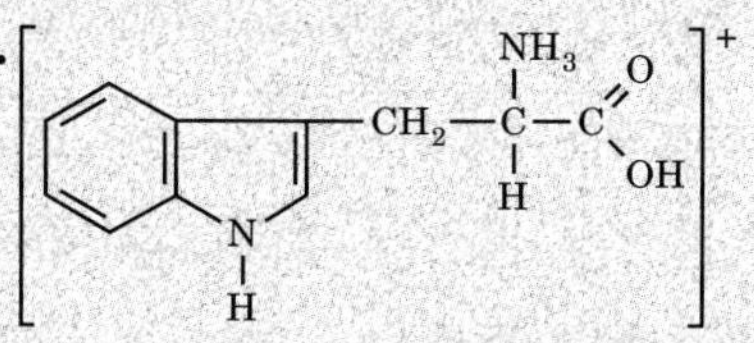

b. $[\,$indole ring (N–H)$\,–CH_2–C(H)(NH_2)–C(=O)O\,]^-$

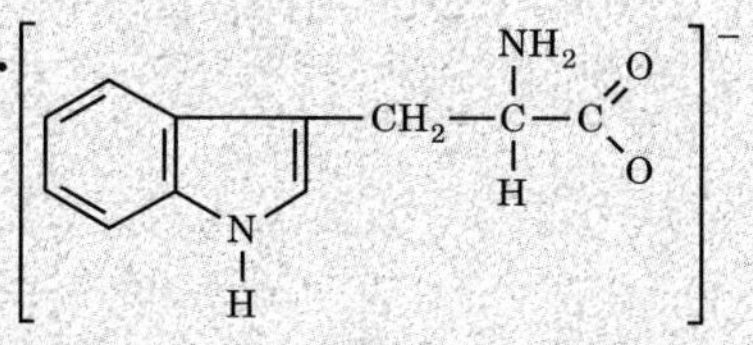

Linking chapters

1. Both HNO_2 and HNO_3 have the same types of atoms in the formula, but the atoms exist in different proportions, a 1:1:2 for HNO_2 and a 1:1:3 for HNO_3. 32 g of oxygen combine with 15.02 g of hydrogen and nitrogen in HNO_2. 48 g of oxygen combine with 15.02 g of hydrogen and nitrogen in HNO_3. The ratio of the masses of oxygen is 2 to 3.

2. a. The two systems in a titration include the solution to be titrated and the solution added as the titration occurs.
b. The acid system reacts with and neutralizes the base system.

3. a. $Zn(s) + H_2SO_4(aq) \longrightarrow ZnSO_4(aq) + H_2(g)$

$$250\text{ mL } H_2SO_4 \times \frac{6\text{ mol } H_2SO_4}{1\text{ L}} \times \frac{1\text{ L}}{1000\text{ mL}} = 1.5\text{ mol } H_2SO_4$$

The mole ratio is 1 to1.

$$1.5\text{ mol } ZnSO_4 \times \frac{161.46\text{ g } ZnSO_4}{1\text{ mol } ZnSO_4}\ 242\text{ g } ZnSO_4$$

b. $PV = nRT$

$$V = \frac{(1.5\text{ mol})\,(0.0821\text{ L·atm/mol·K})(273\text{ K})}{(1\text{ atm})} = 33.6\text{ L } H_2$$

4. a. $2HCl(aq) + CaCO_3(s) \longrightarrow CaCl_2(aq) + H_2O(l)$

$PV = nRT$

$$n = \frac{(1\text{ atm})(1.50\text{ L})}{(0.0821\text{ L·atm/mol·K})(273\text{ K})} = 0.067\text{ mol } CO_2$$

The mole ratio is 1 to 1.

$$0.067\text{ mol } CaCO_3 \times \frac{100.09\text{ g } CaCO_3}{1\text{ mol } CaCO_3} = 6.71\text{ g } CaCO_3$$

b. The mole ratio is 1 to 2.

$$0.067\text{ mol } CaCO_3 \times \frac{2\text{ mol HCl}}{1\text{ mol } CaCO_3} = 0.134\text{ mol HCl}$$

$$0.134\text{ mol HCl} \times \frac{1\text{ L}}{2.00\text{ mol HCl}} = 0.067\text{ L}$$

5. a. $MgO(s) + H_2SO_4(aq) \longrightarrow MgSO_4(aq) + H_2O(l)$
b. $CaO(s) + 2HCl(aq) \longrightarrow CaCl_2(aq) + H_2O(l)$
c. $Al_2O_3(s) + 6HNO_3(aq) \longrightarrow 2Al(NO_3)_3(aq) + 3H_2O(l)$
d. $3ZnO(s) + 2H_3PO_4(aq) \longrightarrow Zn_3(PO_4)_2(aq) + 3H_2O(l)$

6. a. $BaCO_3(s) + 2HCl(aq) \longrightarrow BaCl_2(aq) + CO_2(g) + H_2O(l)$
b. $MgCO_3(s) + 2HNO_3(aq) \longrightarrow Mg(NO_3)_2(aq) + CO_2(g) + H_2O(l)$
c. $Na_2CO_3(s) + H_2SO_4(aq) \longrightarrow Na_2SO_4(aq) + CO_2(g) + H_2O(l)$
d. $3CaCO_3(s) + 2H_3PO_4(aq) \longrightarrow Ca_3(PO_4)_2(aq) + 3CO_2 + 3H_2O(l)$

7. $$3.000\text{ g } H_2C_2O_4{\cdot}2H_2O \times \frac{1\text{ mol } H_2C_2O_4{\cdot}2H_2O}{126.08\text{ g } H_2C_2O_4{\cdot}2H_2O} = 0.0238\text{ mol } H_2C_2O_4{\cdot}2H_2O$$

$$\frac{0.0238\text{ mol } H_2C_2O_4}{150\text{ mL}} \times \frac{1000\text{ mL}}{1\text{ L}} = 0.159\text{ M } H_2C_2O_4$$

8. $2KOH(aq) + H_2C_2O_4(aq) \longrightarrow K_2C_2O_4(aq) + 2H_2O(l)$

The mole ratio is 1 to 2.

$$0.0238\text{ mol } H_2C_2O_4 \times \frac{2\text{ mol KOH}}{1\text{ mol } H_2C_2O_4} = 0.0476\text{ mol KOH}$$

$$\frac{0.0476\text{ mol KOH}}{75.00\text{ mL}} \times \frac{1000\text{ mL}}{1\text{ L}} = 0.635\text{ M KOH}$$

USING TECHNOLOGY

1. Illustrations will vary. Be sure the zwitterion can be seen in both an acid and a base system.

2. a. –14
b. 39810.72
c. –5.241
d. 1

Performance assessment

1. Answers will vary depending on the shampoos selected.

2. Answers will vary. Be sure students follow all laboratory safety precautions.

3. Answers will vary depending on the antacids selected.

Portfolio projects

1. Answers will vary. Be sure the analysis of the cleaning products includes the proper precautions.

2. Baking powder contains an acid-forming substance while baking soda does not. This allows the $NaHCO_3$ in the baking powder to cause bread to rise to a greater degree.

3. Answers will vary depending on the geographic area.

Content Background

DNA fingerprinting has been used in this country since the 1980s. This technique is most often used in establishing paternity of which there are about 100 000 profiles run per year. The O.J. Simpson murder trial in 1994-1995 brought notoriety to this technique as evidence in a criminal trial. Forensic use of DNA profiles amounts to only about 5000 cases per year.

In 1994, Timothy Spencer was the first criminal executed based upon the use of DNA profiling to establish guilt. In 1993 Walter Snyder Jr. was freed from prison when DNA profiling done on evidence from a 1985 rape case, proved that Snyder was innocent.

The Simpson trial has caused to admissibility of DNA evidence to come into question even through most courts have generally accepted such evidence up till now. Most states apply the Frye standard of admissibility which means that scientific evidence must be generally accepted in the scientific community before a court will consider it.

As of this writing, the FBI has been given the charge of setting up a national advisory board on DNA profiling quality assurance standards and regulation.

Teaching Tip

Debate the issue

In discussing the need for regulation of this industry, divide the class into two teams. Team 1 presents the side of the laboratories who feel their work is constantly scrutinized by the courts, and therefore does not need to be regulated. Team 2 plays the role of the lawyers and judges who feel that regulation would ensure that results from the labs are accurate and reliable.

Science, Technology, and Society

DNA Fingerprinting

A New Solution or a New Problem?

The molecule common to all organic life has provided forensic scientists the basis for a new and complex technique to help solve crimes. This new technique, often used in identifying victims as well as criminals, is called DNA fingerprinting, or forensic DNA profiling.

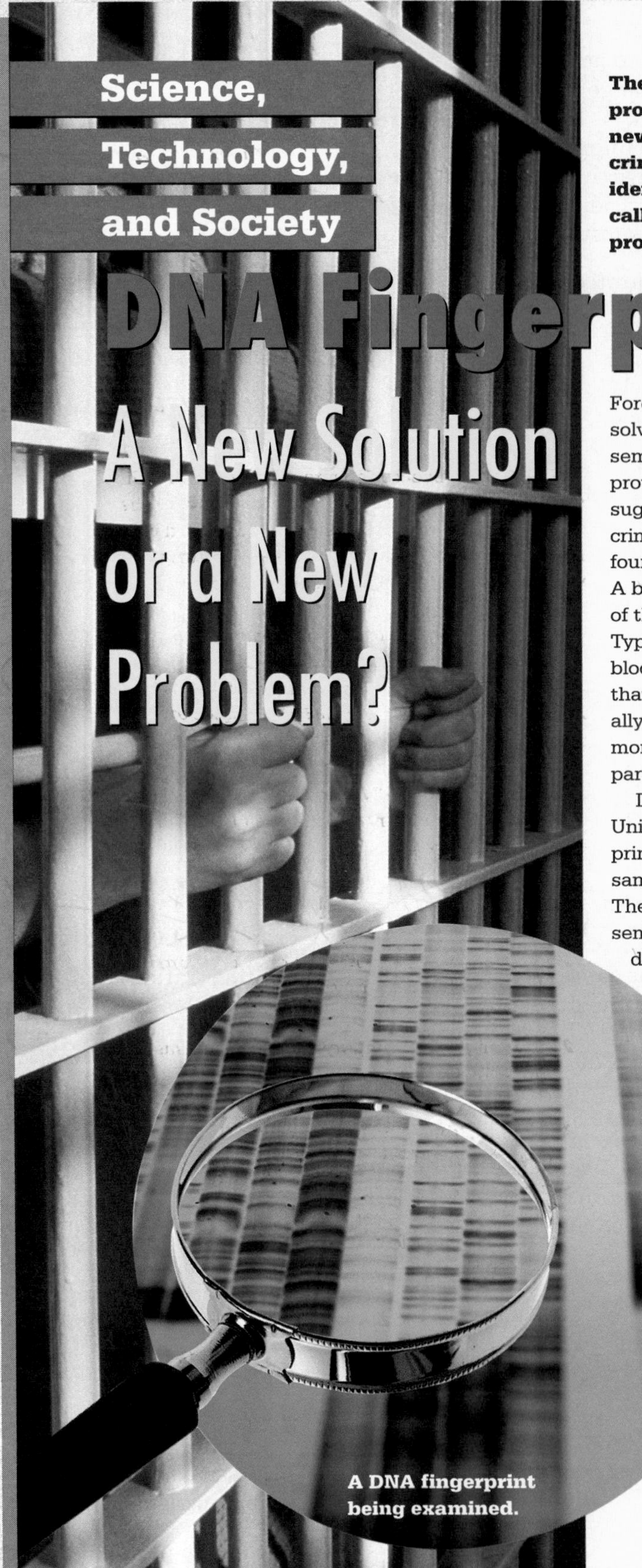

A DNA fingerprint being examined.

Forensic science uses scientific methods to help solve crimes. Fingerprinting along with blood and semen typing are widely used in forensics to help prove guilt or innocence. But these tests can only suggest that a suspect might have committed a crime. For example, if a Type A blood sample is found at the crime scene and the suspect has Type A blood, this match only suggests the possibility of the suspect's involvement. Many people have Type A blood. Although suspects with Type O blood could be ruled out, there is no guarantee that a particular suspect with Type A blood actually committed the crime. Only recently could a more specific approach be used to identify a particular individual.

Discovered in 1984 by Alec Jeffreys at the University of Leicester in England, DNA fingerprinting is used to identify individuals from samples of body tissue left at the scene of a crime. These samples might include blood, hair, or semen. Also, DNA fingerprinting is now used to determine parentage. Previously, the use of blood types to establish parentage could prove only whether someone was definitely not the parent of a child. With the introduction of DNA fingerprinting, the unique genetic profiles of possible parents can now be examined and can positively determine a parent-child relationship. Because positive DNA matches can now be made, DNA fingerprinting has been used to identify children stolen from their parents and sold on the black market during Argentina's military regime of the 1970s and 1980s.

This technique is also being used to identify what are believed to be the remains of the Russian imperial family executed in 1918 during the Bolshevik revolution.

The current Duke of Edinburgh is related to the Russian imperial family. His grandmother was the sister of Alexandra, the murdered tsarina. A conclusive DNA match is being sought in comparing DNA from the Duke of Edinburgh with that taken from the skeletal remains of the imperial family.

Another case involves a series of rape murders, the first committed just one year before Alec Jeffreys invented his DNA-fingerprinting system. In December of 1986, police in Narborough, England, were convinced that two murders had been committed by the same suspect. A man they had in custody had confessed (falsely, it soon appeared) to the second murder.

Police decided to use Jeffreys' technique to solve the matter beyond question. Jeffreys compared the suspect's DNA with the sample taken in the second murder, and determined that it did not match. Comparison with the sample from the first murder proved conclusively that the suspect had not committed that crime either. DNA fingerprinting had exonerated an innocent man, but left the Narborough police with no suspect. DNA analysis of semen samples did prove however, that the two murders had been committed by the same unknown man. If they could find the suspect whose DNA matched the samples, both murders could be solved. The Narborough police then instituted an innovative mass collection and study of DNA data. Every man in Narborough and neighboring Enderby who was a suspect in the murders was required to give a blood sample for DNA analysis. Such a genetic "police lineup" was unprecedented and not likely to be duplicated, especially in the United States, where civil rights laws would prevent it. Although the killer—an aspiring cake decorator named Colin Pitchfork—attempted to conceal his guilt by paying someone to impersonate him during the testing, the truth of the substitution emerged, and Pitchfork was convicted of both murders.

What remains to be determined is the capacity of individual forensic researchers to be accurate, reliable, and responsible in their use of the information. In the application of DNA fingerprinting to forensics, several issues must be addressed. First, the integrity of the physical evidence itself must be ensured. In theory, two samples of DNA can be placed side by side and will either match or not match. In actual practice, however, a DNA sample can become contaminated or degraded before it is studied. Also, in the matching process, a phenomenon called "band-shift" can occur spontaneously, causing differences in the pattern of DNA bands used for the comparison. In addition, the techniques used in DNA fingerprinting are the legal—and private—property of the companies holding the patents on the process, so they are not available for objective study by the scientific community at large. Thus, the body of information on this technique remains small and relatively limited. Until DNA fingerprinting information and techniques are made available for general study, the procedure cannot be used alone to prove guilt or determine family relationships. DNA fingerprinting is still subject to individual interpretation, inaccuracies, and potential abuse.

How the Duke of Edinburgh Is Related to Tsarina Alexandra

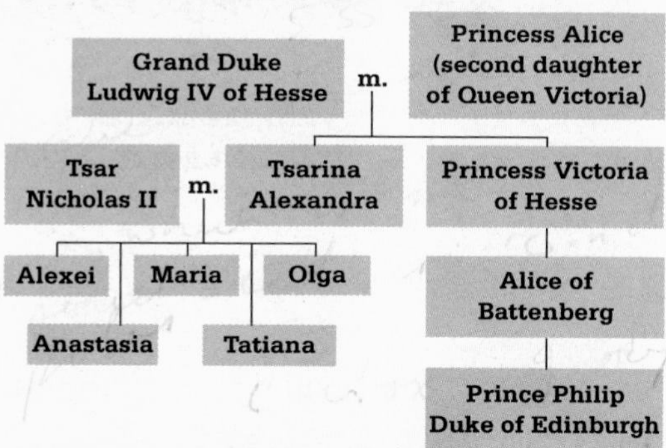

Researching the Issue

1. What standards could be set up to ensure that DNA fingerprinting is absolutely accurate?
2. How might DNA fingerprint data be misused in a courtroom? in a universal database? in employment practices?
3. Compare the advantages and disadvantages of a universal DNA database of the population.
4. Research a court case in which DNA fingerprinting was used as evidence in the United States. What role did this evidence have in determining the outcome of the case? Who did the testing, and who reported the results?
5. Find out what guidelines have been established for the use of DNA fingerprinting as evidence in the United States.
6. Find out how DNA fingerprinting is used to establish parentage.

14

PLANNING FOR

Reaction Rates

Chapter Overview

- **Section 14-1** examines the reaction that occurs when magnesium pieces are added to an aqueous solution of hydrochloric acid to develop the concept and mathematical expression for reaction rate. Students look at a graph of nitrogen(II) oxide concentration versus time to see that reaction rates generally decrease with time.
- **Section 14-2** discusses how reaction rates are affected by temperature, surface area, nature of reactant, and concentration.
- **Section 14-3** covers the rate law, reaction mechanism, and how catalysts affect reaction rate. Students look at graphs of concentration versus time to determine the order of reaction with respect to a specific reactant.

Concept Base

Students must have mastered the following concepts prior to this chapter: chemical reactions and energy of activation (Chapter 7); limiting reactant and stoichiometry (Chapter 8); kinetic molecular theory (Chapter 10); molarity (Chapter 11); concentration, dissociation, and equilibrium expressions (Chapter 12); pH (Chapter 13).

Laboratory Equipment Needs

Demonstrations

The listing of materials for Chapter 14 demonstrations begins on page T46.

Laboratory Experiments

Materials and preparation instructions for Chapter 14 are found as follows.

Text		***Page***
Exploration 14A	Reaction Rates	774
Investigation 14A	Reaction Rates—Effervescent Tablet Evaluation	778

Laboratory Experiments Manual
(refer to the table of contents in the manual)

Exploration 14-1	Catalysts
Investigation 14-1	Catalysts—Peroxide Disposal

Pacing: 10 days

Section	Inquiry Teaching Strategies
14-1 What is a reaction rate?	**Demonstrations** ■ Demo 1 *Effect of temperature on reaction rate* ■ Demo 2 *Monitoring a reaction and calculating its rate* ■ Demo 3 *Reactions slow down with time*
14-2 How can reaction rates be explained?	**Demonstrations** ■ Demo 4 *Surface area and reaction rate: dissolving Alka-Seltzer tablets* ■ Demo 5 *Reaction rates and temperature effects* ■ Demo 6 *Reaction rates for baking soda and various concentrations of HCl* **Laboratory Experiments** ■ Exp 14A *Reaction Rates* ■ Inv 14A *Reaction Rates—Effervescent Tablet Evaluation*
14-3 How can reaction rate be described?	**Demonstrations** ■ Demo 7 *Silent explosion* ■ Demo 8 *Comparison of the catalytic effect of I^- ions and catalase on the decomposition of H_2O_2* **Laboratory Experiments** ■ ■ Exp 14-1 *Catalysts* ■ ■ Inv 14-1 *Catalysts—Peroxide Disposal*

Key ■ Teaching Transparencies
■ Annotated Teacher's Edition
■ Laboratory Experiments Manual

Visual Teaching Strategies

Review and Practice Strategies

Transparencies
- F 14-2 *Noting Reaction Rate for the Reaction of Mg and HCl*

Transparency Masters
- F 14-6 *Concentration vs. Time for the Decomposition of N_2O_5 at 45°C in CCl_4 Solution*

Text Reviews
- Concept Review 1–5
- Section Review 6–9
- Chapter Review 1–9; LC 8

Study Guide Worksheets
- 14-1 Concept Review
- 14-1 Skillsheet *Using Concentration v. Time Graphs*

Transparencies
- F 14-8 *Possible Collision Orientations for the Reaction of H_2 and I_2*
- F 14-17 *Concentration vs. Possible Collisions*

Transparency Masters
- F 14-10 *Comparison of E_a for Different Reactions*
- F 14-13 *Effect of Temperature on KE of O_2*

Text Reviews
- Section Review 10–20
- Chapter Review 10–21; LC 2, 4, 5, 10

Study Guide Worksheets
- 14-2 Concept Review
- 14-2 Skillsheet *Interpreting Energy Diagrams*

Transparencies
- F 14-23 *Enzyme-Substrate Reaction Model*

Transparency Masters
- F 14-18 *Log of Concentration vs. Time for the Decompostion of N_2O_5 at 45°C in CCl_4 Solution*
- F 14-19 *Determining Reaction Order by Graphing Concentration Data*
- F 14-21 *Comparing Reaction Pathways for H_2O_2 Decompostion by Various Catalysts*
- P 566 *Determining Reaction Order*

Text Reviews
- Concept Review 21–25
- Section Review 26–32
- Chapter Review 22–44; LC 1, 3, 6, 7, 9, 10; UT 1, 2, 3, 4, 5

Study Guide Worksheets
- 14-3 Concept Review
- 14-3 Skillsheet *Using Graphs to Determine Reaction Order*

■ Pupil's Text
■ Study Guide
■ Teaching Resources

Exp *Exploration*
Inv *Investigation*

LC *Linking Chapters*
UT *Using Technology*

Assessment Options

Traditional Assessment

Test Generator
Instructional Objectives Measured:
Content mastery
Select from items 1 to 60

Performance Assessment Options

(see pages T36 and T643-15 for scoring rubrics)
Students demonstrate mastery of objectives in a hands-on environment. You may want some of these materials in the Portfolio as well.

Investigation 14A (text)
Investigation 14-1 (laboratory manual)
Instructional Objectives Measured:
Content mastery, Use of scientific methodology, Problem solving skills, Use of technology, Proficiency in written communication

Performance Assessments, page 573

1. Designing an experiment
Instructional Objectives Measured:
Content mastery, Problem solving skills, Use of scientific methodology, Scientific and chemical literacy, Proficiency in oral and written communication

2. Designing an experiment
Instructional Objectives Measured:
Content mastery, Problem solving skills, Use of scientific methodology, Scientific and chemical literacy, Proficiency in written communication

Portfolio Options

(see page T36 for scoring rubrics)
Students provide a written rationale for each selection made for the portfolio.

Concept Maps
Students use vocabulary from the Chapter Review to build a concept map.
Instructional Objectives Measured:
Content mastery

Formal Laboratory Reports
Exploration 14A (text)
Exploration 14-1 (lab manual)
Instructional Objectives Measured:
Content mastery, Use of scientific methodology, Proficiency in written communication

Portfolio Projects, page 573
Items 1 and 3. Research and communication
Instructional Objectives Measured:
Content mastery, Scientific and chemical literacy, Proficiency in oral and written communication

Item 2. Cooperative activity
Instructional Objectives Measured:
Content mastery, Scientific and chemical literacy, Proficiency in oral communication

INTRODUCING THE CHAPTER

CHAPTER 14 Reaction Rates

Story Background

Only a few strains of *E. coli* cause food poisoning, but recent outbreaks involving meat products, especially hamburger, have made *E. coli* 0157:H7 an organism of major concern. The bacteria are not in the meat because the living tissues of an animal are sterile except for the gastrointestinal system. Rather, bacteria get on the surface of the meat when it is cut. Grinding beef for hamburger increases the amount of meat surface exposed to contamination and distributes bacteria throughout the meat. Therefore, bacterial growth is of greater concern for hamburger meat than for a steak.

Because bacterial cells grow at an exponential rate, plotting the number of cells against time gives an upwardly progressive curve, but plotting the logarithm of the number of cells against time results in a straight line. Alternatively, the growth rate can be estimated from the following logarithmic equations.

$$\log_2 X_t = kt\log_2 X_O$$

or

$$\log_{10} X_t = 0.301kt\log_{10} X_O$$

X_O is the initial cell population; X_t is the cell population at a later time; t is the time between X_O and X_t expressed in hours or minutes; and k is the growth-rate constant expressed as doublings per unit of time. Growth rate is expressed as the time needed for the population to double or as the number of generations per hour.

About the Illustration

Invisible to the naked eye, *E. coli* is shown in this picture from a scanning electron microscope. Colonies of *E. coli* 0157:H7 can be present in hamburger meat that has not been thoroughly cooked.

Kinetics and Cooking Hamburgers

Alerted by Seattle area doctors on January 12, 1993, the Center for Disease Control in Washington State traced the source for the latest epidemic of food poisoning caused by *E. coli* 0157:H7 bacteria to Jack in the Box restaurants.

Samples of the hamburger meat sold to Jack in the Box were sent to the Washington State Department of Health and three independent labs for testing. *E. coli* 0157:H7 was present in 2 out of 10 hamburger patties. Acquired in the slaughterhouse through contact with the air or affected workers, the *E. coli* 0157:H7 bacteria were still present in the hamburgers served to customers because the meat had not been cooked long enough at a sufficiently high temperature to kill the bacteria.

The FDA internal temperature requirement of 140°F (60°C) must be maintained for at least 5 minutes to kill the bacteria: more than twice as long as the 2 minutes allowed by Jack in the Box. For these shorter cooking times to be effective, the internal cooking temperature must be much higher, says Michael Doyle, a food microbiology expert at the University of Georgia.

Experiments on the growth rate and survival of *E. coli* over a range of temperatures were performed by Dr. Doyle and others in the early 1980's after an *E. coli* 0157:H7 outbreak involving another fast food chain. These test results persuaded much of the fast food industry to raise their cooking temperature, because they indicate that during the short cooking times used by the fast food industry, temperatures between 150°F and 160°F are necessary to kill the *E. coli* bacteria.

Although the state of Washington had raised its requirement for the minimum internal cooking temperature to 155°F in May 1992, and the Jack in the Box equipment was designed to operate above the FDA minimum of 140°F, health department officials found internal temperatures of only 130°F when they tested patties cooked by the restaurant. In response to the outbreak of *E. coli* 0157:H7, Jack in the Box officially raised their minimum internal temperature standard from 140°F to 155°F, increased their cooking time to 2.5 minutes, and added an extra flipping of each burger.

Like the cooking process, the spoilage of meat is not immediate. It takes place over a period of time. That period of time can be lengthened by following accepted preservation techniques from the slaughterhouse to the table. The experiences of Jack in the Box and its meat supplier raise the following questions about the preparation and preservation of raw meat.

- *How does temperature affect the preparation of raw meat?*
- *What are some effective methods for preserving meat?*

Insights for answering these questions can be found in an exploration of reaction rates and the factors that affect them.

Tapping **Prior Knowledge**

Prepare students for the concepts they will need for this chapter with the following discussion topics.

1. Bread dough is set in a warm place to rise before baking. (The warm temperature favors the interaction of the yeast with the sugars present to produce the CO_2 needed to make the bread rise. At low temperatures the interaction is slow or stopped. At high temperatures the yeast is destroyed and cannot interact with the sugars.)
2. An open soda can set on a counter goes flat more quickly than a similar can left in the refrigerator. (Temperature affects the amount of CO_2 that will dissolve in water.)
3. Although both are leavening agents, baking soda cannot be substituted in a recipe that calls for baking powder. However, a combination of baking soda and cream of tartar can be substituted for the baking powder. (Baking powder is a mixture containing baking soda and an acidic component to react with a part of the baking soda. Cream of tartar is the acidic component in some baking powders.)
4. The technique of a marathon runner differs from that of a sprinter. (A marathon runner paces himself during the race, but a sprinter expends most of his energy early in the race.)

Section 14-1

FOCUS

Lesson Starter

You may wish to begin the lesson with the following demonstration.

Demonstration 1

Effect of temperature on reaction rate

Running time: 10 min

1. Place three 250 mL beakers on your desk.
2. Fill one of them with 150 mL ice water, one with 150 mL tap water, and one with 150 mL hot water.
3. Record the temperature of the water in each beaker.
4. Simultaneously add an Alka-Seltzer tablet to each beaker and begin timing with a stopwatch.
5. Record the amount of time needed for each reaction to reach completion.
6. Make a quick temperature versus time graph of the data on the board.

SAFETY

Wear safety goggles and a lab apron. Use a non-mercury thermometer.

DISPOSAL

Pour the solutions down the drain.

TEACH

Do You Know?

The study of chemical kinetics began around 1850 with Alexander William Williamson who studied the production of ether compounds.

What is a reaction rate?

Section Objectives

List examples of properties that can be monitored to determine the rate of a chemical change.

Calculate the rate of a reaction in terms of the disappearance of reactant or appearance of product.

Rate is a ratio

How could you describe the following changes to someone who has not seen them? The gas gauge went down. The bread molded. The child's hair was cut. While these statements identify the changes taking place, they would be more informative if they were expressed in terms of numbers: the gas gauge fell from 3/4 to 1/2 full; mold patches 1 cm in diameter grew on the bread; the child's hair was shortened from waist length to shoulder length. It would also be more informative to have a series of pictures showing the changes as they progress. Look at the moldy bread in **Figure 14-1**. How could you describe the changes you see? If you knew the time elapsed between each picture, you could express mold growth as a function of time according to the following equation.

$$\text{mold growth} = \frac{\text{change in diameter of mold patch}}{\text{time elapsed}}$$

This expression would be called the rate of mold growth. Any measurable change in an activity expressed as a function of time is a rate. A heart beats at the rate of 75 beats per minute. A race car travels at 78.6 kilometers per hour. Half of a sample of uranium-238 decays to lead-206 in about 4.5 billion years. When the change under consideration is part of a chemical reaction, the expression is called a **reaction rate**. **Chemical kinetics** is the study of reaction rates and the factors that affect them.

reaction rate
the change in reactant concentration per unit of time as reaction proceeds

chemical kinetics
the branch of chemistry concerned with reaction rates and reaction mechanisms

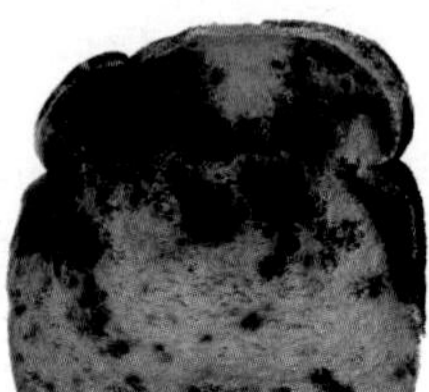

Figure 14-1
Mold growth on a bread slice can be expressed as a function of time.

Reaction rate is an experimental quantity

A reaction rate describes how rapidly a chemical change takes place. It is an experimental quantity found by measuring the disappearance of a reactant or the appearance of a product over a period of time. Expressing a rate depends on the ability to report accurately a change in some physical property such as volume, temperature, color, mass, or acidity. Study **Figure 14-2**. Magnesium metal has been added to hydrochloric acid. The equation for this reaction is written as follows.

$$Mg(s) + 2HCl(aq) \longrightarrow MgCl_2(aq) + H_2(g)$$

What observations indicate that a reaction is taking place? Which of these could be measured and used to calculate the reaction rate?

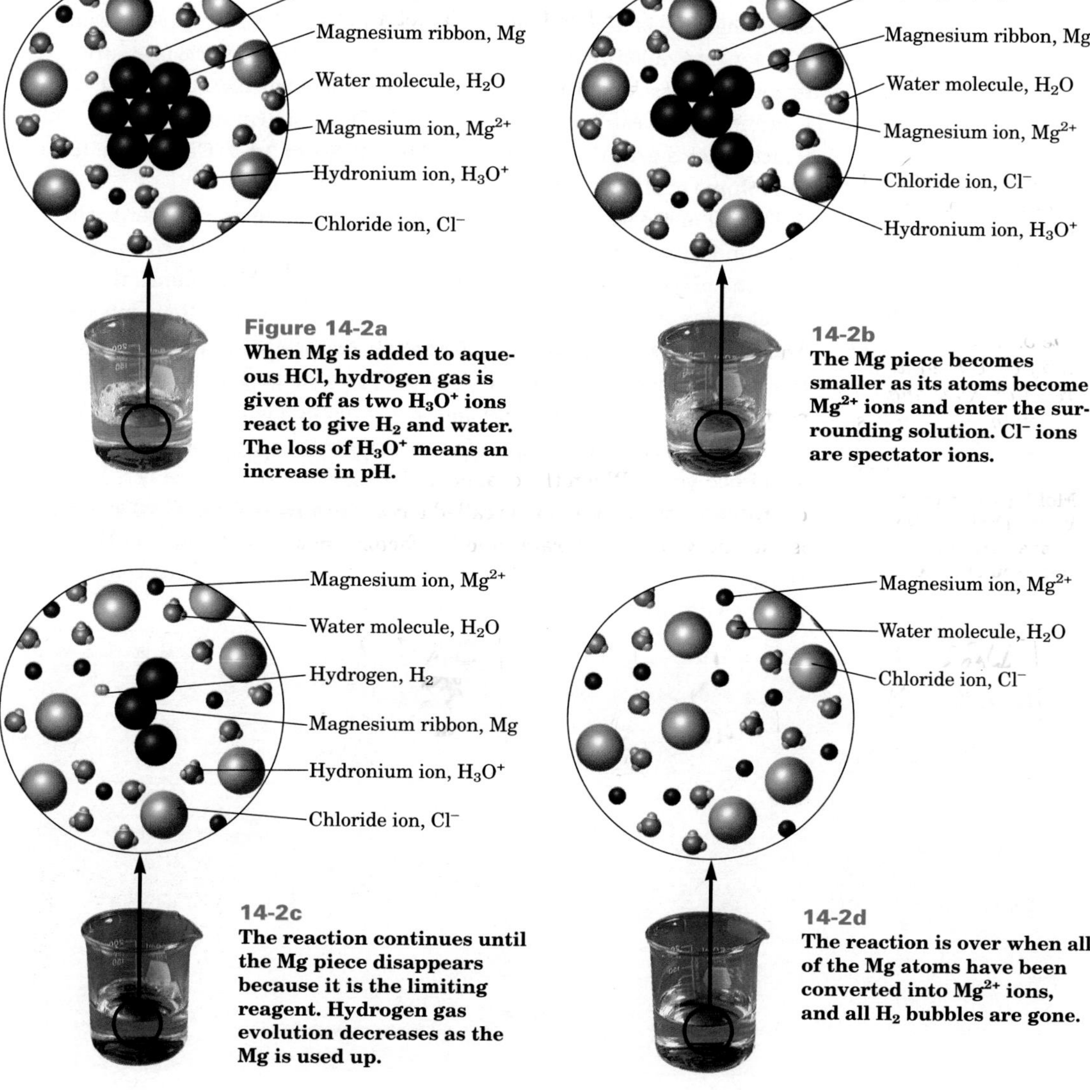

Figure 14-2a
When Mg is added to aqueous HCl, hydrogen gas is given off as two H_3O^+ ions react to give H_2 and water. The loss of H_3O^+ means an increase in pH.

14-2b
The Mg piece becomes smaller as its atoms become Mg^{2+} ions and enter the surrounding solution. Cl^- ions are spectator ions.

14-2c
The reaction continues until the Mg piece disappears because it is the limiting reagent. Hydrogen gas evolution decreases as the Mg is used up.

14-2d
The reaction is over when all of the Mg atoms have been converted into Mg^{2+} ions, and all H_2 bubbles are gone.

Visual Strategy

Figure 14-2

Review the figure insets. Point out that the number of Cl^- ions does not change because they are spectator ions and that the seven magnesium atoms in **Figure 14-2a** have all become ions in **Figure 14-2d**. Ask students to explain why beaker **d** has the highest pH, what happens to the mass of magnesium strip during the reaction, and why fewer hydrogen molecules are shown in the inset for beaker **d** than in the inset for beaker **a**.

Demonstration 2
Monitoring a reaction and calculating its rate

Approximate time: 10 min

1. Before class cut two pieces of clean magnesium ribbon 4 cm long.
2. Tie a 10 cm piece of thread around each ribbon. Tape one thread to the outside of the mouth of each eudiometer tube.
3. Pour 15 mL of 1 M HCl into each of two 50 mL eudiometer tubes. Fill one eudiometer tube with tap water and the other with ice water. Record the temperature of each tube and the atmospheric pressure.
4. Insert the ribbon into the opening of the eudiometer filled with ice water, and plug with a one-hole stopper. Invert the tube, and insert it into a 600 mL beaker filled with water.
5. Record the volume of gas produced every 15 s. Calculate the reaction rate.
6. Repeat steps **4** and **5** for the eudiometer tube filled with tap water.

SAFETY

Wear safety goggles and a lab apron.

DISPOSAL

Rinse any unreacted Mg, and put it in the trash. Combine the rinsings with the liquid from the eudiometers and beakers, neutralize with 1 M NaOH, and pour down the drain.

Teaching Tip
Gas collection methods

Not all gases can be collected by water displacement. Examples are NH_3 and HCl, which are collected by air displacement.

Your list of observations should include three distinct, measurable properties: the reduction in mass of Mg, the volume or mass of the evolved gas, and the change in acidity of the solution. Changes in the mass of Mg can be determined by running the same reaction several times and stopping each one after a specified time period. The Mg must be quickly pulled out, rinsed, dried, and its mass determined. Dividing this experimentally determined mass of Mg by its atomic mass gives moles of Mg. The initial and final masses for Mg are recorded in columns 2 and 3 of **Table 14-1**. The negative number in column 4 is the difference between these two masses, and is equal to the total mass of Mg consumed during the reaction. The mass to mole conversion is given in column 5.

The setup in **Figure 14-3** suggests one way in which the reaction between Mg and aqueous HCl can be monitored through the collection of H_2 gas. The H_2 is collected over water, so the volume indicated on the collecting tube is a combination of H_2 gas and water vapor. The pressure of the water vapor must be subtracted from the atmospheric pressure to get the pressure of the H_2 gas alone. The ideal gas law relates the volume and pressure of H_2 to the moles of H_2 as shown in the figure. Reaction data are given in columns 1 to 4 of **Table 14-1**. Conversion from the volume of H_2 to the moles of H_2 is shown in column 5.

The pH of the solution can be monitored by adding a pH meter to the setup in **Figure 14-2**. Changes in the acidity, $[H_3O^+]$, can be calculated from the pH values measured. The change in $[H_3O^+]$ determines the number of moles of H_3O^+, or of HCl, consumed in the reaction. Remember HCl is a strong acid and, as such, is completely dissociated in water. Therefore, it is present only as its ions, H_3O^+ and Cl^-; no molecular HCl is present. The following equation represents the dissociation of HCl.

$$HCl(aq) + H_2O(l) \longrightarrow H_3O^+(aq) + Cl^-(aq)$$

The number of moles of H_3O^+ and of HCl are equal because their mole ratio in the balanced equation is 1:1. Calculation of $[H_3O^+]$ from pH is shown in **Figure 14-4** on the next page. Reaction data and the conversion to moles are summarized in columns 1 to 5 of **Table 14-1**.

$$P_{H_2} = P_{atm} - P_{H_2O} = 0.947 - 0.073 = 0.874 \text{ atm}$$

$$T = 40°C + 273 = 313K$$

$$V = \frac{454 \text{ mL}}{1000} = 0.454 \text{ L}$$

$$PV = nRT$$

$$n = \frac{PV}{RT}$$

$$= \frac{(0.874)(0.454)}{(0.0821)(313)} = 0.0155 \text{ mol}$$

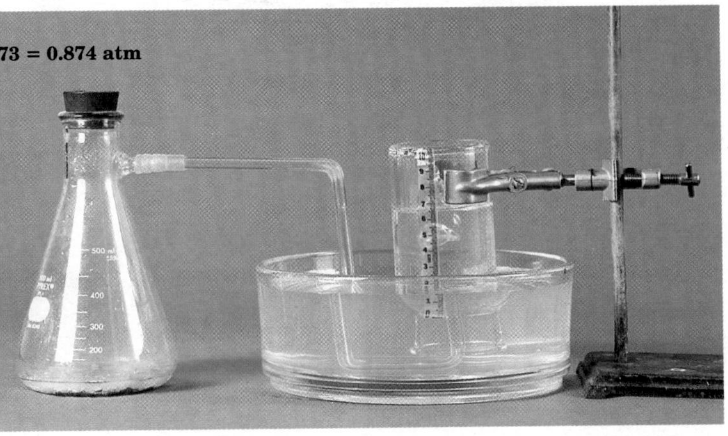

Figure 14-3
Hydrogen gas evolved when Mg reacts with aqueous HCl can be collected over H_2O and the measurement used to calculate the reaction rate.

Visual Strategy
Figure 14-3

Review the procedure for calculating pressure of a gas collected over water using the data presented here. If you ran Demonstration 2, you may wish to compare and contrast the setups.

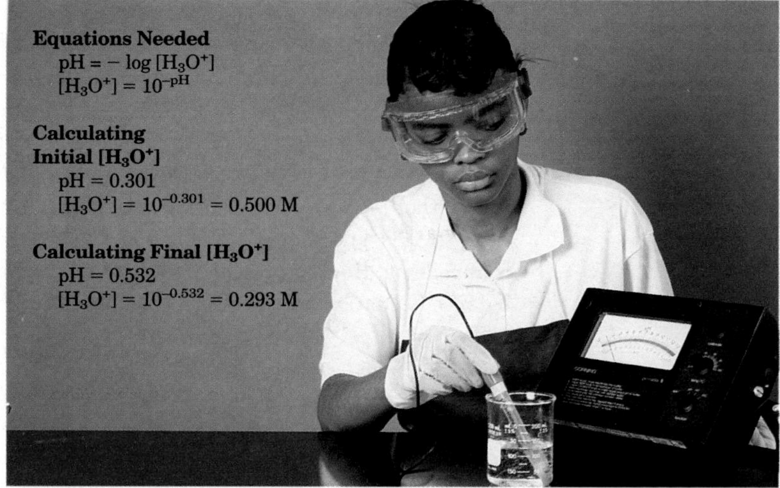

Equations Needed
$pH = -\log [H_3O^+]$
$[H_3O^+] = 10^{-pH}$
Calculating Initial $[H_3O^+]$
$pH = 0.301$
$[H_3O^+] = 10^{-0.301} = 0.500\ M$
Calculating Final $[H_3O^+]$
$pH = 0.532$
$[H_3O^+] = 10^{-0.532} = 0.293\ M$

Figure 14-4
Changes in $[H_3O^+]$ are calculated from pH readings for the Mg in aqueous HCl reaction.

These three observable properties permit you to express a change from initial to final conditions that occurs over a specified time interval. Dividing the change in the measured property by the time elapsed gives the **average rate** for the time interval as shown in column 7 of **Table 14-1**.

average rate
the change in the measured property divided by the time elapsed

Consider the entries for Mg in **Table 14-1**. At the end of the reaction you have less Mg than at the beginning; consequently, subtracting its initial mass from its final mass gives a negative number. But the change is written as a positive number in the rate column. This is because, when discussing reaction rates, chemists are more interested in how fast substances are reacting rather than whether they are formed or consumed. Thus, multiplying the negative change by a -1 makes the rate a positive number. Any time the reaction rate is calculated from changes in reactants, multiply your result by -1 to make the number positive.

Table 14-1
Reaction Rate Data and Calculations for Mg in Aqueous HCl Reaction

Col. (1) Property measured	(2) Initial quantity	(3) Final quantity	(4) Change in quantity	(5) Convert to moles so the rates will be comparable	(6) Time elapsed (s)	(7) Rate = $\frac{\text{Change in property (mol)}}{\text{elapsed time(s)}}$
Mass of Mg	0.376 g	0.0 g	−0.376 g	mol Mg = $\frac{-0.376\ g}{24.31\ g/mol}$	3.2	Rate = $\frac{0.0155\ mol\ Mg}{3.2\ s}$ = 0.0048 mol/s
Volume of H_2 collected over H_2O at 40°C 0.947 atm	0.0 mL	454 mL	454 mL	mol H_2 = $\frac{(0.874)(0.454)}{(0.0821)(313)}$	3.2	Rate = $\frac{0.0155\ mol\ H_2}{3.2\ s}$ = 0.0048 mol/s
$[H_3O^+]$ from pH of the 150 mL solution	0.500 M	0.293 M	0.207 M	mol HCl = mol H_3O^+ mol H_3O^+ = (0.207 M)(0.150 L)	3.2	Rate = $\frac{0.0310\ mol\ HCl}{3.2\ s}$ = 0.0096 mol/s

Teaching Tip
Measuring pH
You may wish to point out that quantitative measurements of pH require a pH meter or other instrumentation, but visual indicators can be used for qualitative measurements of pH.

Visual Strategy
Figure 14-4
Review the procedure for calculating concentration from pH measurements using the data presented here. Compare the values obtained with the entries for initial quantity and final quantity in **Table 14-1**. You may wish to go over all calculations in **Table 14-1**.

Teaching Tip

Stoichiometry and rate

When the rate expression has been modified to compensate for stoichiometry, the reaction has only one rate for a given set of conditions. Therefore, the rate of change in Mg, HCl, and H_2 are all equal, and the *or* could be replaced with an equals sign.

Rate can be noted from the disappearance of a reactant or appearance of a product

Look at the numbers in the rate column of **Table 14-1**. How do they compare? The reaction rate calculated from the change in the mass of Mg is equal to that calculated from the collection of H_2 gas. The reaction rate calculated from the $[H_3O^+]$, however, is twice as large as that determined from either Mg or H_2. This suggests that HCl(*aq*) is disappearing twice as fast as Mg is, and that HCl(*aq*) is also disappearing twice as fast as H_2 is appearing. How can you account for this difference? Reconsider the balanced equation for the overall reaction.

$$Mg(s) + 2HCl(aq) \longrightarrow MgCl_2(aq) + H_2(g)$$

Look at the coefficients for Mg, HCl, and H_2. Both Mg and H_2 have a coefficient of 1. So H_2 is formed as quickly as Mg disappears. The mole ratio between HCl(*aq*) and Mg, however, is 2:1. This means that 2 mol of HCl(*aq*) must be used for every 1 mol of Mg atoms consumed which produces a rate that is twice as fast. The mole ratio between HCl(*aq*) and H_2 is also 2:1. To compensate for this effect of stoichiometry on reaction rate, chemists divide the rate of change for each reactant or product by its coefficient in the balanced equation. Thus, for the reaction between Mg and aqueous HCl, rate is expressed as follows.

$$\text{rate} = \frac{1}{1}\frac{\Delta \text{ mol Mg}}{\Delta t} \text{ or } \frac{1}{2}\frac{\Delta \text{ mol HCl}}{\Delta t} \text{ or } \frac{1}{1}\frac{\Delta \text{ mol } H_2}{\Delta t}$$

The Greek letter Δ (a capital "delta") stands for "the change in." Therefore, Δ mol Mg is read as "the change in moles of magnesium." This change is equal to the number of moles of Mg at the stopping time minus the number of moles of Mg at the starting time.

The units for a reaction rate must always be expressed. The units for rate in this equation are moles per second. The most common units for a reaction rate are concentration per unit of time. When concentration units are moles of a substance per liter of solution (M), the general equation for rate is written as follows.

$$\text{rate} = \frac{-1}{a}\frac{\Delta[\text{reactant}]}{\Delta t} = \frac{1}{b}\frac{\Delta[\text{product}]}{\Delta t}$$

The minus sign makes the change in reactant concentration a positive number. The *a* is the coefficient for the reactant in the balanced chemical equation; *b* is the coefficient for the product. More specifically, when 0.1 M solutions of $BaCl_2$ and Na_2SO_4 are combined, $BaSO_4$ is precipitated according to the following equation.

$$BaCl_2(aq) + Na_2SO_4(aq) \longrightarrow BaSO_4(s) + 2NaCl(aq)$$

The rate can be expressed as follows.

$$\text{rate} = \frac{-1}{1}\frac{\Delta[BaCl_2]}{\Delta t} = \frac{-1}{1}\frac{\Delta[Na_2SO_4]}{\Delta t} = \frac{1}{1}\frac{\Delta[BaSO_4]}{\Delta t} = \frac{1}{2}\frac{\Delta[NaCl]}{\Delta t}$$

Defining rate

1. What characteristic of each of the following reactions could be used to qualitatively show the rate of reaction?
   a. $2H_2O_2(aq) \longrightarrow 2H_2O(l) + O_2(g)$
   b. $Cu(s) + 2Ag^+(aq) \longrightarrow Cu^{2+}(aq) + 2Ag(s)$
   c. $HC_2H_3O_2(aq) + NaHCO_3(s) \longrightarrow NaC_2H_3O_2(aq) + CO_2(g) + H_2O(l)$
   d. $CuCl_2(aq) + 2NaOH(aq) \longrightarrow Cu(OH)_2(s) + 2NaCl(aq)$
   e. $2C_2H_6(g) + 7O_2(g) \longrightarrow 4CO_2(g) + 6H_2O(g)$
2. For each reaction in item **1**, write one expression for the reaction rate that uses a reactant and one expression that uses a product.
3. The concentration of a substance in a reaction changes from 4.0 M to 2.0 M in 40 min. Is the substance a reactant or product? Explain how you know. Express the rate of this reaction in M/min.
4. Ammonia is formed from its elements according to the equation below.

$$N_2(g) + 3H_2(g) \rightleftharpoons 2NH_3(g)$$

   One mole of N_2 is mixed with one mole of H_2 in a one liter container. After 5 s the reaction is stopped, and the gases are remeasured. There are 0.9 mol of N_2, and 0.7 mol of H_2. Calculate the reaction rate in terms of each reactant.
5. Dinitrogen pentoxide decomposes into nitrogen dioxide and oxygen according to the following equation.

$$2N_2O_5(g) \longrightarrow 4NO_2(g) + O_2(g)$$

   If the change in O_2 concentration was found to be 2.5 M/s, what is the reaction rate in terms of $N_2O_5(g)$?

Answers for Concept Review items and Practice problems begin on page 841.

Reaction rates decrease with time

Few chemical reactions proceed at the same rate throughout the entire process. Most reactions are fast at the beginning when all of the reactant concentrations are high and slow down as the reactants are consumed, just as the amplitude of a pendulum slows over time, shown in **Figure 14-5**.

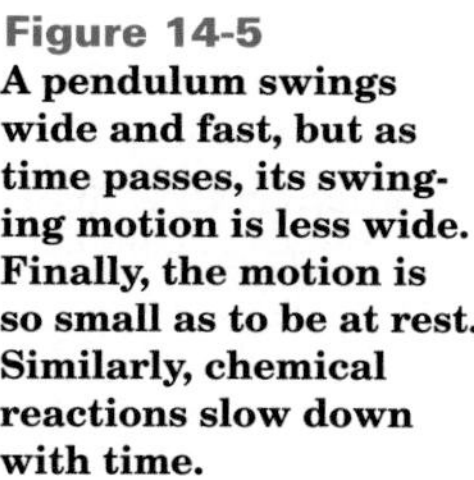
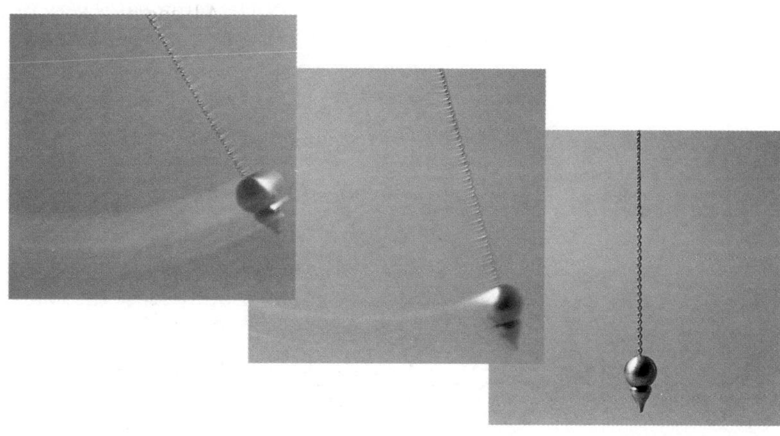

Figure 14-5
A pendulum swings wide and fast, but as time passes, its swinging motion is less wide. Finally, the motion is so small as to be at rest. Similarly, chemical reactions slow down with time.

Answers to

Concept Review

1. a. evolution of a gas
   b. change in solution's color; appearance or disappearance of metals
   c. evolution of a gas
   d. precipitation of $Cu(OH)_2$
   e. pressure changes; condensing the water out of reaction gases
2. a. $\frac{1}{2} \times \frac{-\Delta H_2O_2}{\Delta t} = \frac{\Delta O_2}{\Delta t}$
   b. $\frac{1}{2} \times \frac{-\Delta Ag^+}{\Delta t} = \frac{\Delta Cu^{2+}}{\Delta t}$
   c. $\frac{-\Delta HC_2H_3O_2}{\Delta t} = \frac{\Delta CO_2}{\Delta t}$
   d. $\frac{-\Delta CuCl_2}{\Delta t} = \frac{\Delta Cu(OH)_2}{\Delta t}$
   e. $\frac{1}{2} \times \frac{-\Delta C_2H_6}{\Delta t} = \frac{1}{4} \times \frac{\Delta CO_2}{\Delta t}$
3. reactant; concentration decreases with time; $\frac{0.050\ M}{min}$
4. N_2: $\frac{-(0.9 - 1.0)\ mol}{5\ s\ L} = 0.02\ M/s$

   H_2: $\frac{-(0.7 - 1.0)\ mol}{5\ s\ L} = 0.06\ M/s$
5. $\frac{\Delta[N_2O_5]}{\Delta t} = \frac{-2}{1} \times \frac{2.5\ M}{1\ s} = \frac{5.0\ M}{1\ s}$

   $rate = \frac{1\Delta[O_2]}{1\Delta t} = \frac{1}{2}\left(\frac{-1\Delta[N_2O_5]}{\Delta t}\right)$

   $= \frac{1}{2}\left(\frac{5.0M}{1\ s}\right) = \frac{2.5M}{1\ s}$

Teaching Tip

Necessary conditions for equilibrium

Remind the students that for an equilibrium reaction to occur, all reactants and products must be present in some amount, and the reactants must be consumed to form products as rapidly as the products are consumed to form reactants.

Demonstration 3

Reactions slow down with time

Approximate time: 15 min

1. Before class, attach a 10 cm piece of thread to each of four 20.0 cm pieces of clean magnesium ribbon.
2. Write the mass of 20.0 cm of Mg ribbon on the board to 0.001 g.
3. Tie or tape the other end of the thread to a glass stirring rod, keeping the string as short as possible.
4. Fill four 250 mL beakers with 150 mL of 0.1 M HCl.
5. Place one Mg piece into one acid solution.
6. Remove the piece of Mg after 1 min, and rinse it in a beaker of H_2O to remove any traces of acid.
7. Dry and measure the mass of the Mg to the nearest 0.001 g. Write its mass on the board.
8. Repeat steps **5–7** with the other three pieces of Mg, leaving the Mg in the acid for 2, 3, and 4 min, respectively.
9. Sketch a graph of mass versus time.

SAFETY

Wear safety goggles and a lab apron.

DISPOSAL

Put the unreacted Mg in the trash. Combine the rinsings from the Mg with the liquids from the acid solutions, neutralize with 1 M NaOH, and pour down the drain.

Rates for equilibrium reactions become constant

In Chapter 12, on page 464, you studied a reaction in which NO_2 combines with a second NO_2 molecule to make N_2O_4 according to the following equation.

$$2NO_2(g) \rightleftharpoons N_2O_4(g)$$

NO_2 is a brown gas and N_2O_4 is colorless. The brown color of the NO_2 lightened as the colorless N_2O_4 was formed. The change in color continued until a steady uniform color was achieved. The casual observer might think that this reaction stopped when the color became constant, but from your study of equilibrium constants you know otherwise. The reaction has not stopped, because the rate at which the colorless N_2O_4 is decomposing into the brownish NO_2 equals the rate at which the NO_2 is combining to form the colorless N_2O_4. You see this equivalence as constant color. When the reaction appears to stop, it has reached equilibrium, and the rate is constant.

Rates for non-equilibrium reactions decrease to zero

Carefully study the progress of the reaction between Mg and aqueous HCl presented in **Figure 14-2**. The reaction's progress can be observed by noting the amount of bubbles produced. Only a glance at **Figure 14-2** is needed to see that fewer bubbles are produced as time passes. A more detailed examination of the reaction's progress reveals that the amount of Mg is getting smaller as the amount of bubbles becomes smaller and that the H_2 gas escapes to the atmosphere because there is no lid on the beaker. The last picture in this series shows no bubbles and no Mg. From what you have learned about equilibrium in Chapter 12, you would predict that the reaction would not come to equilibrium but would continue until the magnesium was consumed. As predicted, the reaction continues until the limiting reagent, Mg, is completely used up and then stops. The rate of the reaction decreases to zero as the reaction comes to a stop.

Reaction rates can be visualized through graphs of reaction data

Dinitrogen pentoxide, N_2O_5, decomposes very rapidly. To monitor this reaction for rate data, the N_2O_5 molecules are dissolved in carbon tetrachloride, CCl_4. The CCl_4 does not react with N_2O_5, so the only reaction that occurs is the decomposition of N_2O_5 in the CCl_4 solution at 45°C. It decomposes according to the following equation.

$$2N_2O_5(soln) \xrightarrow[45°C]{\text{in } CCl_4} 4NO_2(g) + O_2(g)$$

The plot in **Figure 14-6** on the next page shows how the concentration of N_2O_5 changes with time. The reaction rate can be calculated from the slope of the curve at any point because the following relationship has been determined.

$$\text{slope} = \frac{\Delta y}{\Delta x} = \frac{\Delta [N_2O_5]}{\Delta t}$$

Figure 14-6
Rate is equal to the slope of the curve, which decreases as time passes.

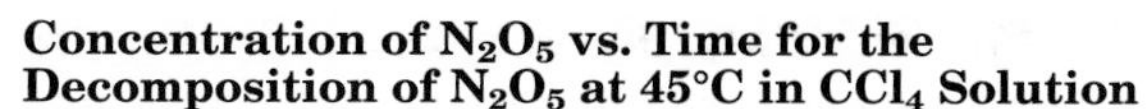

Concentration of N_2O_5 vs. Time for the Decomposition of N_2O_5 at 45°C in CCl_4 Solution

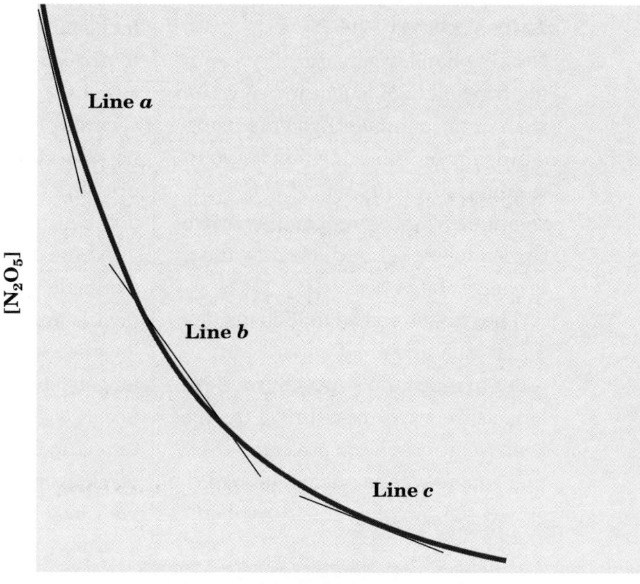

However, the plot shows a curve, and the slope for a curve is not constant like it is for a straight line, so the section of the curve you examine affects the number you get for the reaction rate. To illustrate, look at lines *a*, *b*, and *c* in **Figure 14-6**. Compare the values for their slopes. Are they identical? The line that lies closest to the curve has a slope closest to that of the curve. Which line has a slope closest to that of the curve? Line *a* is almost parallel to the beginning of the curve, whereas line *b* intersects it, and line *c* touches it at only one point. Line *a* corresponds to the first few seconds of the reaction. Its slope is very near the initial reaction rate. The **initial rate** of a reaction is the fastest rate measured very close to the start of the reaction when reactant concentrations are highest. Line *b* corresponds to an *average rate* for the time interval bounded by its points of intersection with the curve.

initial rate
the highest rate measured very close to the start of the reaction where the curve is almost linear

Rates can be manipulated

For most chemical reactions, the reaction rate is fastest at the beginning when reactant concentrations are highest and slows down as reactant concentrations decrease and products are formed. The reaction rate either becomes constant, as for an equilibrium reaction, or stops altogether because the limiting reagent has been depleted. These observations lead to the conclusion that the reaction rate can be changed by manipulating the concentration. They also bring up the question, "Other than concentration, what affects the rate of a chemical reaction?" The next section identifies four of the five factors that affect reaction rates and explains why or how the rate is affected. The fifth factor will be discussed in Section 14-3.

Teaching Tip
Comparing initial and average rates
Point out that an initial rate is a very close estimate to the actual rate because it coincides with the linear portion of the graph at the start of the reaction. An average rate, however, can be taken over any interval of the curve and its accuracy depends on how closely the chosen interval lies to the curve. Neither the initial nor the average rate tell you what is going on at any specific instant; they tell you only what happened during the entire time period chosen.

For the rate at a specific instant (provided your time scale is in seconds), you must draw a tangent line to the curve at that instant, and determine the slope of this line. Line *c* on the graph is an example of instantaneous rate.

Theme
Systems and interactions
Reaction rates can be manipulated by modifying a reaction system's environmental conditions.

Visual Strategy
Figure 14-6
Choose several points on the curve, and for every set of two points identify the slope by means of triangles. Then compare the ratios of the triangles' sides to show that the slope (rate) is changing. Point out that a triangle near the top of the curve has the highest value for its ratio, which is in keeping with the fact that reactions occur quickest when reactant concentrations are highest.

ASSESS

Alternative Assessment

Reconsider the reaction of Mg in aqueous HCl.

a. Describe the progress of this reaction if the HCl were the limiting reagent. Would you still be led to believe the reaction goes to completion rather than equilibrium? Explain.

b. Describe the progress of this reaction with generalized concentration versus time graphs for both reactants and products.

Answers to Section Review

6. reactant concentrations are highest

Answers are continued on page 573A.

Consumer Focus

Camp-out cooking takes time

Cooking hamburgers during a camping trip will take longer in the winter than in the summer if you are using a propane or isobutane fuel for your camping stove. Propane, C_3H_8, and isobutane, C_4H_{10}, are popular hydrocarbon fuels that burn cleanly and produce a lot of heat.

They are stored as liquids inside a fuel tank and must vaporize in order to reach the stove's burner. As long as the vapor pressure of the fuel is above atmospheric pressure, the fuel vapors will be pushed toward the stove's burner when it is turned on.

At 22°C the vapor pressure of either fuel is high enough for the stove to perform well. But at –12°C, a propane camping stove operates very slowly, and a butane stove may not operate at all because the vapor pressure of its fuel is no longer above atmospheric pressure.

As the amount of vapor that is available to burn decreases, less heat is produced. Less heat means lower cooking temperatures and longer cooking times. If you plan to use a cooking stove on a fall or winter camp-out, you should expect to use longer cooking times because your food heats up at a slower rate.

Section Review

6. Explain why the rate of a simple reaction is likely to be fastest at the beginning of the reaction.

7. The reaction between methanol and hydrochloric acid to form methyl chloride and water is written as follows.

$$CH_3OH(aq) + HCl(aq) \longrightarrow CH_3Cl\,(aq) + H_2O(l)$$

Time (min)	$[H_3O^+]$
0	1.83
79	1.67
158	1.52
316	1.30
632	1.00

The rate of disappearance of HCl is measured as a function of $[H_3O^+]$. The table shows the collected data. Plot the reaction data as a graph of concentration vs. time. Determine the average rate of the reaction for each time interval. Is the rate constant throughout the reaction? Explain your answer.

8. Express the reaction rate in terms of the rate of change of each reactant and each product in the following reactions.

a. $4PH_3(g) \longrightarrow P_4(s) + 6H_2(g)$

b. $S^{2-}(aq) + H_2O(l) \longrightarrow HS^-(aq) + OH^-(aq)$

c. $C_2H_4(g) + Br_2(g) \longrightarrow C_2H_4Br_2(g)$

9. Ammonia, NH_3, reacts with O_2 to form NO and H_2O as follows.

$$4NH_3(g) + 5O_2(g) \longrightarrow 4NO(g) + 6H_2O(g)$$

a. At the instant when NH_3 is reacting at a rate of 0.80 M/min, what is the rate at which O_2 is disappearing?

b. At what rate is each product being formed?

How can reaction rates be explained?

Section Objectives

Discuss how collision theory explains reaction rates.

Relate an energy distribution graph to the effect of temperature on reaction rate.

List four major factors that affect the rate of a reaction and describe how changes in each factor affect the rate.

Modeling how reactions proceed

At one time a commercial for a powdered aspirin product claimed it worked better because it got to the headache faster than any competitor's tablet. This commercial was shown to a class of 30 chemistry students. Together, they surmised that to work faster the powder would have to dissolve more quickly than the tablet. The students were then asked to use what they knew about reaction rates to evaluate the commercial's claim. **Figure 14-7** shows the initial notebook entries and equipment setup of one student, George Cynric.

George Cynric investigates rate changes in aspirin systems

George's early results appeared to justify the commercial's claim because the powder did dissolve faster in water than the tablet. But aspirin dissolves in the stomach which is an acidic environment. Maybe the results would be different in an acid. The only acid available to him was vinegar. He repeated the experiment using vinegar even though the acidic component of vinegar, acetic acid, is a weaker acid than HCl, the acid present in the stomach. When the results had been recorded, he repeated the experiment but gently warmed the vinegar before adding the aspirin tablet or powder. This way he could simulate body temperature in case temperature had an effect on how the aspirin dissolved. He then recorded his results and the conclusions derived from them as shown in **Figure 14-7** on the next page. George Cynric looked back to the kinetic-molecular theory of gases for models to help explain his observations and conclusions.

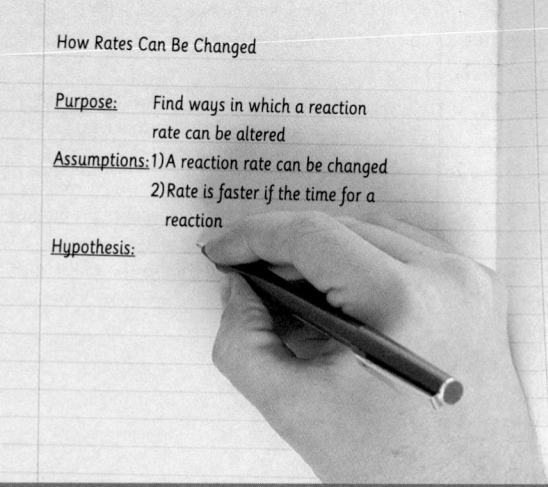

How Rates Can Be Changed

Purpose: Find ways in which a reaction rate can be altered

Assumptions: 1) A reaction rate can be changed
2) Rate is faster if the time for a reaction

Hypothesis:

Section 14-2

FOCUS

Lesson Starter

Begin the lesson with the following demonstration.

Demonstration 4

Surface area and reaction rate: dissolving Alka-Seltzer tablets

Approximate time: 10 min

1. Place three petri dishes on an overhead projector. Fill them half full with H_2O.
2. With a single-edge razor blade, score and break an Alka-Seltzer tablet into fourths. Match the pieces as evenly as possible.
3. Place the pieces on the overhead to show that they are very closely matched.
4. Using a mortar and pestle, grind up one of the Alka-Seltzer quarters into smaller chunks and a second quarter into powder. This way three different sizes can be used: full quarter, chunks, and powder.
5. Transfer each sample to weighing paper, and simultaneously add the three samples to separate petri dishes. Note the amount of time needed for each sample to quit reacting.

SAFETY
Wear safety goggles and a lab apron.

DISPOSAL
Pour solutions down the drain.

TEACH

Teaching Tip

Discussion

Direct students' attention to George Cynric's notes. Discuss how clarity in notetaking and the compilation of experimental results facilitate understanding when the experimenter later reviews what he has done.

Do You Know?

Vinegar can be produced from a variety of plants. The name of the vinegar often reflects its source. The names and sources of several types of vinegar are listed below.

malt vinegar: malt or malt and raw barley
banana vinegar: the pulp and peel of the ripe fruit
tomato vinegar: vinegar flavored with tomatoes
cider vinegar: apples
orange vinegar: orange juice and some citric acid
verjuice: crab apples, unripe grapes, or sour cider
balsamic vinegar: a strong must of the Trebbiano grape and strong ordinary vinegar stored in wooden casks for up to a century

Figure 14-7
Lab note entries for George Cynric's investigation of rate changes in aspirin systems.

How Rates Can Be Changed

Purpose: Find ways in which a reaction rate can be altered.

Assumptions: 1) A reaction rate can be changed.
2) Rate is faster if the time for a reaction decreases.

Hypothesis: Powdered aspirin will dissolve faster than tablet aspirin, so powdered aspirin works faster.

Equipment: 2 coffee cups (identical), drinking glass, colorless wooden chopsticks or other stirrer, waxed paper pieces, hammer or equivalent for crushing aspirin tablet, watch with second hand or stop watch

Chemicals: water (enough to fill cups—several times)
aspirin tablets (several)
vinegar (colorless type—small bottle)

Set up:

Procedure: 1) Fill cups almost to the brim with water from tap
2) Let it come to room temperature while crushing aspirin tablet

Setup — To crush tablet

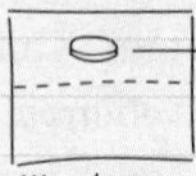

3) Put tablet into cup. Wait. Stir if necessary.
4) Record time when dissolved.
5) Put powder into cup. Wait. Stir if necessary.
6) Record time when dissolved.

Data table:

	Time
Tablet	98 min
Powder	90 min

Other things to do:
- Try again using vinegar in the cups instead of water (stomach is acidic).
- Try again using 1/2 vinegar and 1/2 water (probably stomach's not all acid).
- Try again using all warm vinegar.
- Try again using all warm water (hottest from tap).
- Try again using ice water (Just for fun and comparison to hot water).

Data Table 2

	All vinegar	1/2 vinegar 1/2 water	Warm vinegar tap	Hot water	Ice water
Tablet	20	37	12	56	> 2 1/2 hrs.
Powder	16	28	7	45	~ 2 1/2 hrs.

Results and Conclusions:

1) Powder goes faster than tablet in all trials. Particle size affects rate: if small, reaction goes faster than if the size is big.
2) Changing from water to vinegar increases reaction rate. Probably because of some natural difference between water and vinegar.
3) The hot water and warm vinegar have faster rates for both powder and tablet compared to the cooler room temperature trials. Also faster than ice water. When temperature goes up, the rate goes up. When temperature goes down, the rate goes down as well.
4) Concentrated stuff makes rate go faster because the all-vinegar trial went faster than the half and half trial.

Summary: Rate speeds up when temperature increases, the amount of stuff present increases, or the size of stuff decreases. Rate is also changed by varying reactants.

Visual Strategy

Figure 14-7

Compare and contrast your lab write-up method with what George Cynric has written. Discuss how his experiment could be redesigned for use as a science fair project.

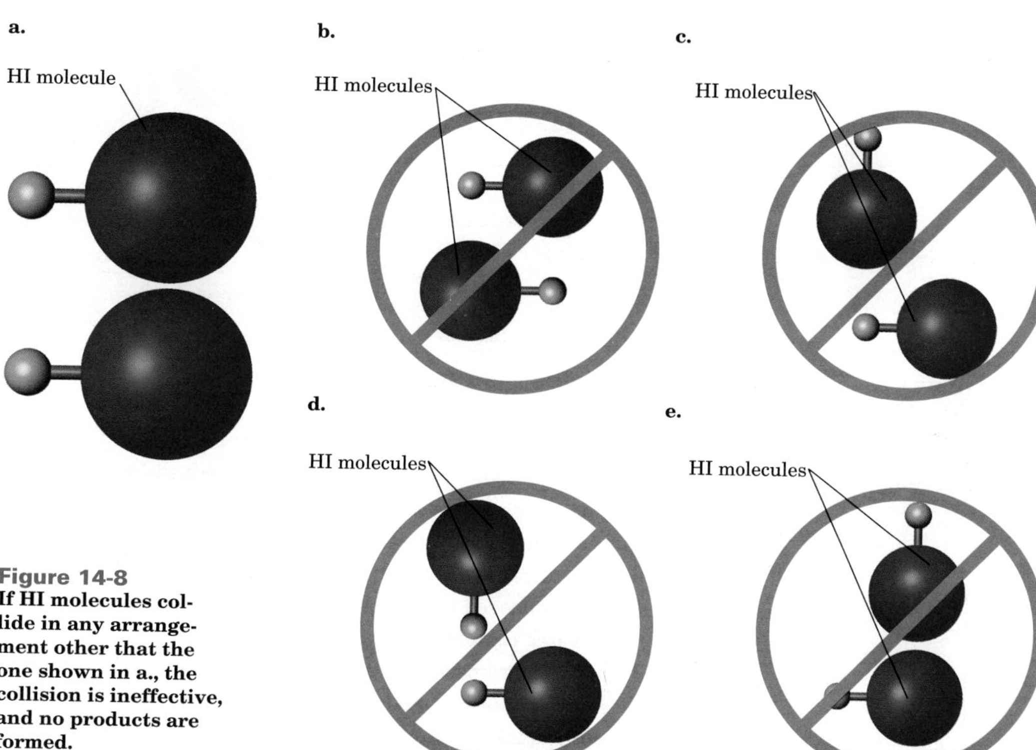

Figure 14-8
If HI molecules collide in any arrangement other that the one shown in a., the collision is ineffective, and no products are formed.

Colliding particles need the right orientation to react

A molecule moves about randomly, hitting one molecule and then another, while changing direction and speed in the process. This movement is similar to that of billiard balls on a crowded pool table. Rolling into a pool table pocket must be like forming a new molecule, George Cynric surmised. It happened sometimes but not all that often compared to the number of times the balls collided on the table top. Still, some pool players could put a particular ball into a pocket of their choosing. Success depended on having the right orientation among the cue ball, the specified playing ball, and the chosen pocket.

Molecules in motion are like the billiard balls. They must collide at just the right orientation in order to react. How specific the orientation must be depends on the type of bonding in the molecules as well as their spatial configuration. An effective collision between particles forms an activated complex. The **activated complex** is an unstable structure that represents the transition between the breaking of old bonds and the formation of new bonds. Therefore, the complexity of reacting particles, along with their geometry at the time of collision, plays an important role in collision effectiveness.

activated complex *a specific arrangement of high-energy reactant atoms or molecules, the decomposition of which leads to product or intermediate molecules; also called transition state*

Figure 14-8 shows the orientation needed for two HI molecules to form an activated complex that can split apart to form H_2 and I_2. The chances of a collision being effective are increased when the colliding particles are less complex as in **Figure 14-9** on the next page.

Groupwork Strategy

Divide the class into groups of three. Have the groups study George Cynric's notes for changes that could improve the results of his experiment. Have each group decide which of their changes is the best and present the idea to the class for discussion.

Teaching Tip

Specific orientation and effective collisions

Construct two paper cubes, one red and one green, with 8 cm sides. With a marker, put a spot on one face of each cube. Leave the other five sides blank. Roll the cubes. If both spotted sides are up, have the students shift one chair. Their movement can occur only after the cubes are rolled and stop in this orientation. Repeat with two cubes spotted on all six sides. Rolling the cubes represents collisions. Student movement represents an effective collision.

Visual Strategy

Figure 14-8

Point out that only when the two H atoms *and* the two I atoms are parallel with each other can the activated complex form. What orientation "flaw" makes the collisions in parts **b** through **e** ineffective?

Teaching Tip

Energy and effective collisions

Roll a pair of dice. As the dice roll, their different faces are revealed. The desired orientation will eventually come up. Stare at the dice, and command them to roll over. Because the dice do not roll over, the correct orientation will never be reached. The dice need the energy you supply to produce "an effective collision."

Because any collision of the Ag^+ and Cl^- ions will form solid AgCl, the E_a for this reaction is zero. The rate of reaction is governed by the ability of the ions to move through the solution.

Historical Note

The concept of minimum energy for effective collisions between particles, the energy of activation, was first suggested by Svante Arrhenius in 1889.

Theme

Systems and interactions

The energy of activation, E_a, is the minimum amount of energy needed for effective collisions, so modifying it by changing the pathway will directly alter the number of effective collisions and indirectly affect the rate.

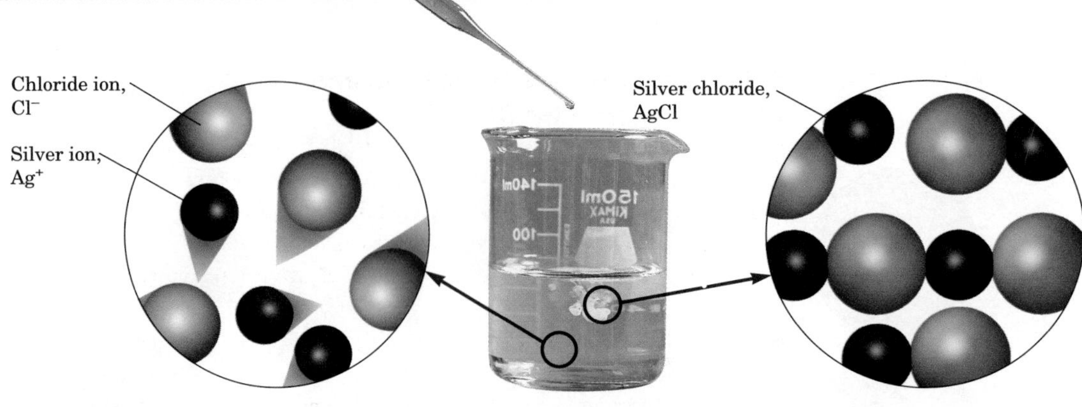

Figure 14-9
Both Ag^+ and Cl^- ions are spherically symmetrical. All orientations can produce an effective collision.

Colliding particles also need a minimum amount of energy in order to react

Even if the billiard balls and pocket were in the correct alignment, a specified ball does not always roll into the pocket. Sometimes it stops short, bounces off, or is followed into the pocket by the cue ball. It appears that a specified ball also needed a certain amount of energy as well as proper orientation. This is also true for molecules. They may collide with the proper orientation, but not have the energy needed to react, in which case the collision is not effective and no activated complex is formed.

Generalized energy diagrams for the decomposition of HI and the formation of AgCl are shown in **Figure 14-10**. Notice the shape and height of the curve in each graph. You learned in Chapter 7 that energy of activation, E_a, is the minimum amount of energy needed for colliding molecules to react. E_a is represented by the distance between the reactants' energy and the top of the energy hill. Therefore, when the curve is short, the value for E_a is small, and only a small amount of energy is required to get molecules over the hill. In **Figure 14-10b**, no energy is required to get oppositely charged ions together, so $E_a = 0$. Conversely, when the curve is tall, E_a is large, and a great deal of energy is required to get molecules over the hill.

Figure 14-10
The difference in energy between reactants and the peak of the potential energy curve is the energy of activation, E_a. The peak of the curve represents the energy of an activated complex.

Different Reactions Have Different Values for E_a

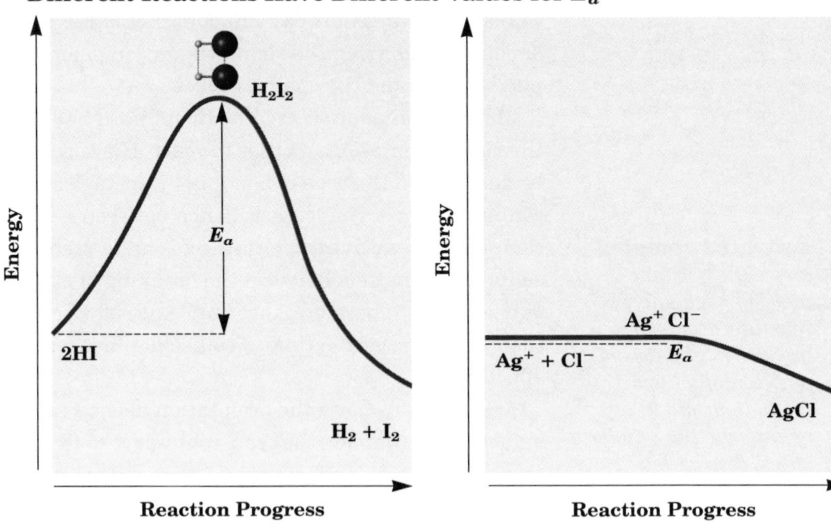

Visual Strategy

Figure 14-10

This diagram extends Chapter 9, in which the concepts of activated complex and activation energy were introduced. Associate the heights of the curves to the ease with which products are formed. Point out the pinnacle position of the activated complex, and explain that it can go back to reactants as easily as it can go forward to products.

Effective collisions must have energies equal to or greater than E_a

If two properly oriented molecules collide with energy equal to or greater than E_a for the reaction, they form an activated complex. An activated complex is represented by H_2I_2 in the HI decomposition reaction. The activated complex can then rearrange to form products.

If the energy of the collision is less than E_a, the collision will not be effective; bonds will not be broken. So how can we increase the amount of energy in a reactant sample so that collisions will have sufficient energy to break bonds? Increasing temperature adds energy to the system by increasing the random kinetic energy of the particles. This increase in energy could be modeled by the breaking of racked billiard balls. A gentle hit, as shown in **Figure 14-11** corresponds to a low temperature while the forceful hit corresponds to a higher temperature. The dashed line surrounding the outermost location of the balls after being hit by the cue ball looks somewhat like the Maxwell-Boltzman distribution curve shown in Chapter 7, where you learned that for a reaction to occur, molecules must collide with enough energy to exceed the activation energy of the reaction. But how can we tell how much kinetic energy reactants have at a given temperature?

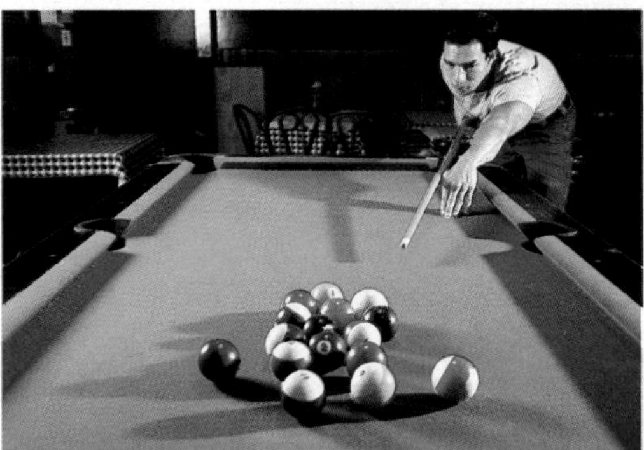

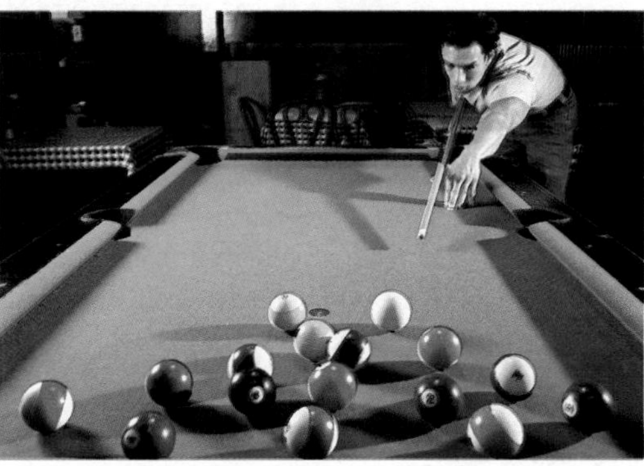

Figure 14-11a
Low temperatures, modeled by the movement of gently hit billiard balls, provide less heat energy for activation of molecules than . . .

14-11b
. . . high temperatures, which are modeled by the movement of billiard balls that have been hit forcefully.

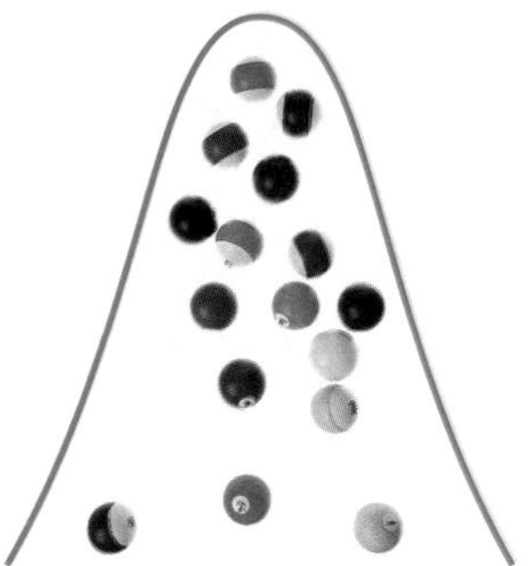

Low temperature, narrow energy distribution

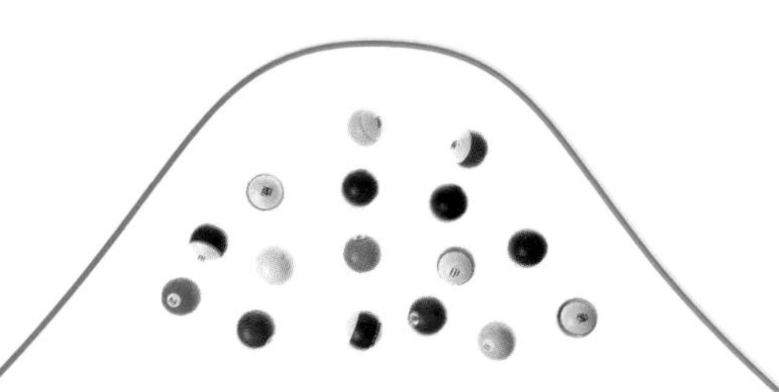

High temperature, wide energy distribution

Visual Strategy

Figure 14-11

Calculate the height-to-width ratio of each curve. Point out that the same number of balls is under each curve. The change in the position of the balls changes the shape of the curve that outlines them. Similarly, temperature influences particles (balls) to change their KE (positions) so that the curve describing their distribution changes accordingly.

Teaching Tip
Temperature and kinetic energy
Remind students that temperature is a measure of the average kinetic energy of the particles, so increased temperature relates directly to increased particle movement.

Temperature and reaction rates

Molecules at the same temperature have a range of energies

Think back to your study of gases in Chapter 10. In a sample of oxygen gas at 25°C, all the molecules have the same mass, but they do not all have the same velocity; therefore, they do not all have the same kinetic energy because of the following relationship.

$$\mathbf{KE} = \frac{1}{2}mv^2$$

Figure 14-12 represents the kinetic energy distribution of particles in a sample of oxygen gas at 25°C. The values on this graph have been verified by experiments. What does this graph mean? The area under the curve represents the range of kinetic energies for all the particles in a sample. You see from the distribution that those molecules with the highest energies (on the right side of the graph) are few in number. The molecules with the lowest energies (on the left side of the graph) are also few in number. The majority of molecules have energies close to the peak of the graph. The average kinetic energy of the gas sample is the reading that represents the largest number of molecules. This value can be used to calculate the average velocity of the molecules in the sample.

Recall from the discussion of kinetic molecular theory in Chapter 10 that the distribution of molecular velocities is sensitive to temperature. Because kinetic energy is related to velocity, you would predict that the average kinetic energy of gas molecules and the distribution of molecular energies would also vary with temperature. So how would changing the temperature affect the energy distribution? How would the number of effective collisions be changed by increasing the temperature?

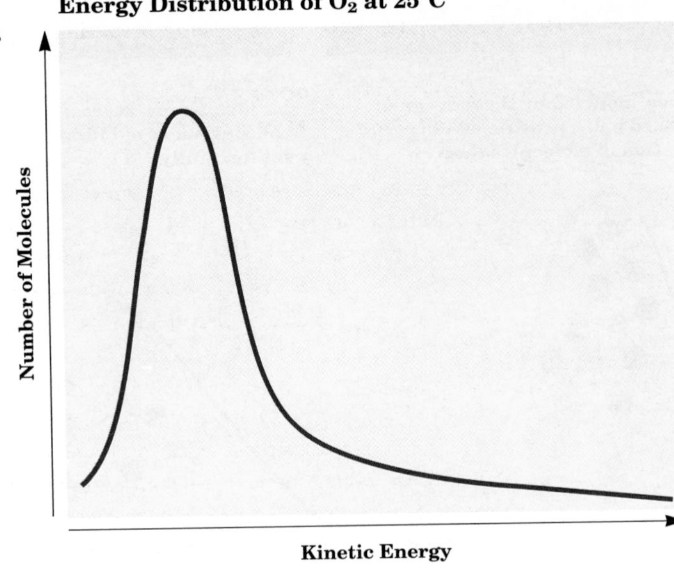

Figure 14-12
At low temperatures, most of the molecules have low KE, which accounts for the high peaked curve at the left end of the KE axis.

Visual Strategy
Figure 14-12
Pick points on the curve, and drop perpendicular lines to the axes to emphasize the relationship between the number of particles at a particular energy and the average kinetic energy of the sample.

Figure 14-13
At high temperatures, few molecules have low KE, so the height of the peak decreases. The peak occurs further down the KE axis because the average KE of the molecules is larger than it was at the lower temperatures.

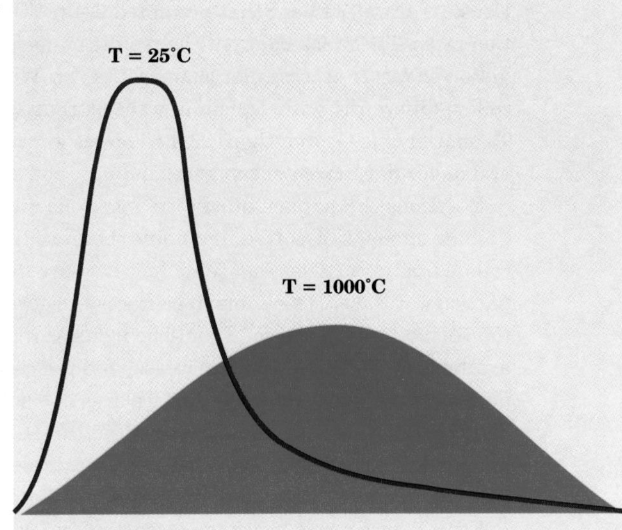

The number of effective collisions increases with increasing temperature

Figure 14-13 shows what happens to the sample of O_2 when it has been heated to 1000°C. The shape of the curve changes showing that the distribution changes. The curve is not as high and is more spread out. The molecules show a greater range of kinetic energies at the higher temperature, and the average kinetic energy of the sample is greater. An increase in temperature increases the average kinetic energy of an O_2 molecule; this can result in a greater number of effective collisions when O_2 is reacted with another substance. Cooling a reaction system has the opposite effect. Molecules move more slowly and collide with less energy, producing fewer effective collisions.

Figure 14-14
The plates on the back of a stegosaurus may have gathered heat from the sunlight to warm the dinosaur's body. As the body heated up, the dinosaur could move less sluggishly.

Rates of reaction increase (or decrease) as temperature increases (or decreases)

Look back at the results and conclusions George Cynric made concerning temperature and the reactions between vinegar and the aspirin. In both cases the time needed to dissolve decreased. Rates increased with temperature regardless of whether the aspirin was in tablet or powdered form. This observation indicates that temperature affects rate. When the reaction temperature was lowered by using an ice bath, both reactions slowed down, reinforcing the conclusion that temperature affects rate. Together these observations show that rate increases (or decreases) when temperature increases (or decreases) just as the kinetic theory suggests. For many common reactions, the reaction rate doubles for every 10°C rise in temperature.

Demonstration 5

Reaction rates and temperature effects

Approximate time: 10 min

1. Obtain three 600 mL beakers.
2. Fill one of them with 0°C H_2O, one with 20°C H_2O, and one with 60°C H_2O.
3. Break open a luminescent light stick. Note the intensity of light.
4. Darken the room, and dip the stick into the beaker of 0°C water. Allow the temperature to equilibrate. Note the change in the intensity of light.
5. Repeat step 4 with the beaker of 20°C H_2O and the beaker of 60°C H_2O.
6. Discuss the intensity of light as a function of the temperature.

SAFETY
Wear safety goggles and a lab apron.

DISPOSAL
Follow manufacturer's instructions for disposal of the light stick.

Teaching Tip

Temperature and reaction rate

You may wish to discuss the movement and function of cold-blooded animals in response to changes in their environmental temperature.

Visual Strategy

Figure 14-13

Point out that the graph is formed from ordered pairs of the form (KE, number of particles). Select several KE values. Draw a line through each value parallel to the y-axis to illustrate that for a given KE value, the number of particles at that KE can be compared for the two graphs. This comparison shows that temperature is a measure of the average KE.

Further Applications

Finely powdered forms of carbon, sometimes referred to as charcoal of various types, are made by heating such materials as coconut shells or sugar in the absence of air. The surface of this finely divided carbon is purged of impurities by heating with steam, resulting in an increased surface area ranging from 600 to 2000 m^2 per gram. The cleaned carbon powder is activated as an adsorbent. It is used to remove odors or dangerous vapors from the air and to remove impurities and bad tastes from liquids such as honey, soft drinks, and fruit juices.

Teaching Tip

Surface area and ice cubes

Discuss: Compare the melting rates of a 5 kg block of ice in a bucket of hot water and a 5 kg bag of crushed ice in a bucket of hot water. **Ask**: Which type of ice should be used in a glass of soda if you want the soda to stay cold as long as possible? **Assign**: Have students try the two situations in an at-home experiment. **Report**: Review responses from students the following day.

Do You Know?

Internal characteristics of the beef, such as fat content, will slightly alter the maximum growth and death temperatures for *E. coli*.

Theme

Systems and interactions

Individual properties of reactants determine the type of interactions that will occur between them.

Surface area and reaction rates

Consider the statement that prompted George Cynric to investigate reaction rates. Then look again at the results of his investigation. The aspirin dissolved faster as a powder than as a tablet. Why would water molecules collide differently with aspirin in a chunk than with aspirin in tiny pieces? To answer this George Cynric called on his knowledge of reaction processes and molecular movement in gases, liquids, and solids.

Reactions can happen quickly in a gaseous mixture or a liquid mixture (liquids or solids dissolved in a liquid) because the particles can mix and collide freely. Particles in a solid, however, are in fixed positions; only particles at the surface come in contact with other reactants. Therefore, if the surface area of the solid could be increased, the number of particles available for reaction would increase, and the reaction rate would increase. **Figure 14-15** shows how the surface area of a solid can be increased by dividing the solid into smaller pieces. Notice that crushing a solid reactant increases the surface area of that reactant. It also increases the rate of any reaction in which the solid participates, as George Cynric observed.

For a heterogeneous system, a reaction can occur between a gas and a liquid, a gas and a solid, or a liquid and a solid. The rate of the reaction with a solid reactant depends on the exposed area of the solid reactant. The rate usually increases if the surface area is increased but still tends to be slower overall than a reaction between two liquids or two gases.

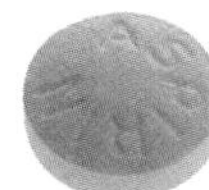

Figure 14-15a
Division of a solid . . .

14-15b
. . . makes the exposed surface of the solid larger.

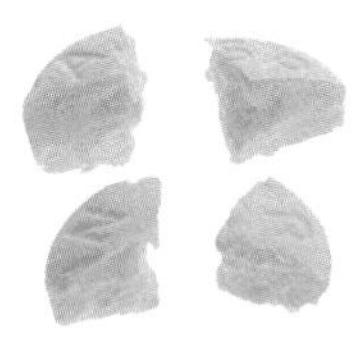

14-15c
More divisions means more exposed surface, . . .

14-15d
. . . hence more spots are available for . . .

14-15e
. . . other reactant molecules to collide into.

Nature of reactants and also concentration affect reaction rates

Regardless of what products are formed, reactions involve the making and breaking of bonds. The rate at which bonds break and re-form depends on the type of bond and the nature of the molecules in which the bonds are found. Some processes, like the Mg in HCl reaction take place very rapidly, while others, like the oxidation of lead by oxygen dissolved in water takes place very slowly, if at all. The nature of a reactant cannot be adjusted to improve reaction rate.

Figure 14-16
Pieces of magnesium ribbon react more rapidly with HCl in the beaker on the right because it is more concentrated than the HCl in the beaker on the left.

Increasing concentration increases the number of effective collisions

George Cynric concluded that increasing the concentration of a reactant increases the reaction rate because his aspirin sample dissolved faster in pure vinegar than in the vinegar and water mixture. Likewise when Mg is added to an HCl solution that is more concentrated than the one in Section 14-1, the reaction rate noticeably increases as shown in **Figure 14-16**. Remember that reaction rate represents the number of effective collisions per unit of time. Therefore, if the number of effective collisions increases, the rate increases as well. The model in **Figure 14-17** shows how an increase in concentration can increase the number of effective collisions.

The mathematical relationship between reactant concentration and reaction rate is at times very simple and at other times very complicated. This is because most reactions occur in more than one step even though only one of these steps determines the rate. The molecules participating in this **rate-determining step** govern the form of the relationship between rate and concentration. What are the steps by which a reaction occurs? How is concentration related to reaction rate and the individual steps in a reaction?

rate-determining step
the slowest step in a reaction mechanism

Figure 14-17a
One possible collision can occur between two reactants, A and B.

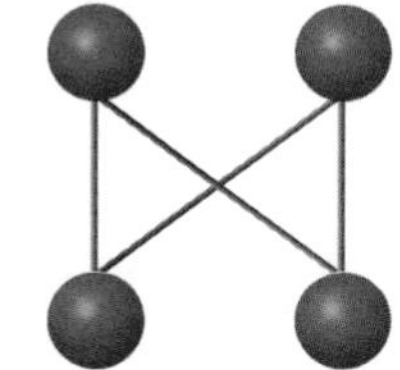

14-17b
Four possible collisions can occur in this system.

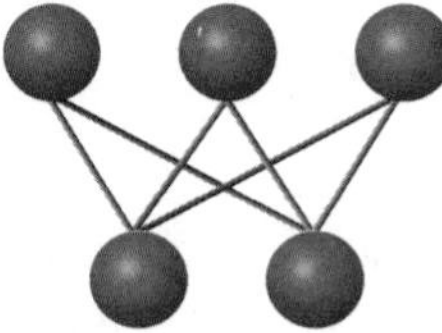

14-17c
Six possible collisions can occur in this system.

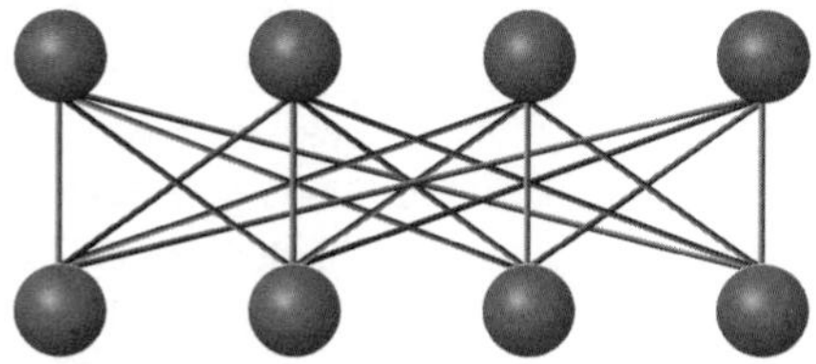

14-17d
Sixteen possible collisions can occur in this system.

Demonstration 6
Reaction rates for baking soda and various concentrations of HCl

Approximate time: 5 min

1. Fill three 600 mL beakers with 50 mL of 1 M HCl, 0.5M HCl, and 0.25M HCl, respectively, and place them on an overhead projector.
2. Simultaneously add 1 g of baking soda, $NaHCO_3$, to each beaker.
3. Note the time needed for each reaction to go to completion.

SAFETY
Wear safety goggles and a lab apron.

DISPOSAL
Combine solutions, neutralize with 1 M HCl or NaOH as appropriate, and pour down the drain.

Possible Misconception
Increasing concentration does not always produce an increase in reaction rate. Higher reactant concentrations may make side reactions or competing reactions more of a problem, which may slow down an overall reaction. This is one problem that chemical engineers face when reactions are scaled up from a flask in a laboratory to a vat in a chemical plant.

Groupwork Strategy
Divide the class into groups of three, and have each group develop a nonchemical demonstration that they can show to the class to illustrate how an increase in concentration should produce an increase in effective collisions.

Visual Strategy
Figure 14-17
Point out that the lines connecting the balls show each ball in the top row colliding with each ball in the bottom row, so as the number of balls in one or both rows increases, the number of possible collisions also increases. The ratio of effective collisions to total collisions remains the same, so if the total number of collisions increases, the number of effective collisions also increases.

ASSESS

Alternative Assessment

Discuss the industrial process for ammonia synthesis and the factors that affect its reaction rate.

Answers to Section Review

10. middle < right < left; the left

11. **a**; surface contact between a solid and a gas is less than that between the ions in **b**; precipitate formation is limited only by the ions ability to move through the solvent.

12. **a.** thick wire (conc)
b. powdered sugar (surf area)
c. at 35°C (temp)
d. with moist heat (temp)
e. in the trunk of a car (conc)

13. slower; more time is needed for the same change in conc.

14. Increasing temperature increases the total energy content of the system, so more molecules have energies at or above the energy of activation.

15. Activation energy is the minimum amount of energy needed for a collision between molecules to be effective; activated complex is the assemblage of atoms, ions, or molecules that results from an effective collision.

16. The activated complex has the highest potential energy of the three because it occurs at the peak of the energy curve; the products have less (greater) potential energy than the reactants if the reaction is exothermic (endothermic).

17. when the colliding particles have the proper orientation and the energy of the collision is greater than or equal to the activation energy

Answers are continued on page 573A.

Section Review

10. Arrange the following three energy curves in order of increasing E_a values. If each curve is a separate reaction, which reaction would you predict to be the slowest?

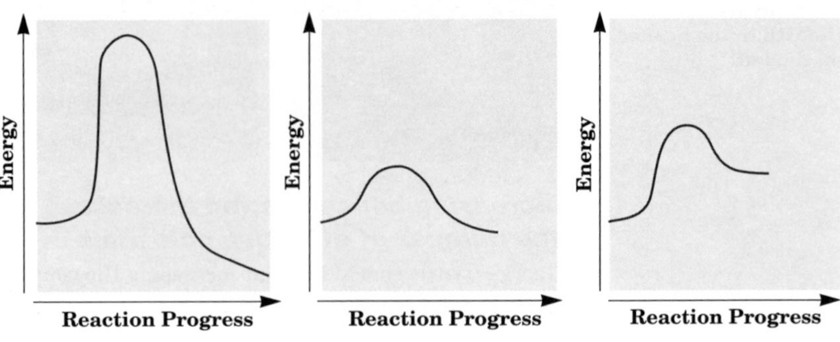

11. Look at the following two equations. Predict which would be slower and why.
a. $Cu(s) + Cl_2(g) \longrightarrow CuCl_2(s)$
b. $2KOH(aq) + Mg(NO_3)_2(aq) \longrightarrow Mg(OH)_2(s) + 2KNO_3(aq)$

12. For each of the following pairs, choose the substance or process you would expect to react more rapidly.
a. 5 cm of thick platinum wire or 5 cm of thin platinum wire
b. granulated sugar or powdered sugar
c. zinc in HCl at 20°C or zinc in HCl at 35°C
d. cooking hamburgers with moist heat or with dry heat.
e. a spark igniting a gallon of gasoline an hour after it is spilled in a warehouse or in the trunk of a car.

13. A reaction occurs in 30 s as evidenced by a color change. The same color change occurs when the reaction is run at a lower temperature, but this time it takes 60 s. Is the rate of the second reaction faster or slower than the first?

14. Using what you know about the energy of molecules, explain why reaction rates vary with temperature.

15. Compare and contrast the terms *activation energy* and *activated complex*.

16. Compare the potential energies of the following components of a reaction, all at the same temperature: reactants, activated complex, and products.

17. When will a collision lead to chemical reaction?

18. Identify the factor that affects a reaction rate but cannot be adjusted itself.

19. Story Link
Explain why Jack in the Box added an extra flipping of the burger to their cooking process.

20. Story Link
Which method is better for cooking a thick piece of food, cooking it for 10 minutes at a high temperature or cooking it for 20 minutes at a lower temperature? Explain your answer.

How can reaction rate be described?

Section Objectives

Derive a rate law from graphs relating concentration to time.

Given a rate law expression, determine the effects of changing concentrations on the rate.

Explain the relationship between the rate-determining step and the overall rate.

Explain the general effect of catalysts on a reaction mechanism and reaction rate.

The general rate law equation

Rate is directly proportional to reactant concentration, [A], raised to some power, *n*.

$$\textbf{rate} \propto [\mathbf{A}]^n$$

This proportionality is made into an equality by the inclusion of *k*, an experimentally determined constant.

$$\textbf{rate} = k[\mathbf{A}]^n$$

rate law
a mathematical expression for the rate of a reaction as a function of the concentration of one or more reactants: rate = $k[A]^n[B]^m$

order
the exponent for a specified reactant in a rate law expression

This equation is the general **rate law**. The exponent, *n*, on the reactant concentration indicates to what extent the rate for a given reaction depends on concentration. The exponent is called the **order** of reaction with respect to [A] and must be determined from experimental data. More specifically, if the concentration is doubled and the rate doubles as well, the exponent has a value of 1 and is read "first order." If the rate does not change when the concentration is doubled, the exponent has a value of 0 and is read "zero order." An exponent of 2, or second order, means the rate quadruples when the concentration is doubled. Determining the value of the exponent from graphs of experimental data for a specific reaction is part of deriving the rate law for that reaction.

Do not confuse the exponents in the rate law with those in an equilibrium equation. They are not obtained in the same way. Exponents in equilibrium equations are coefficients in the balanced chemical equation; *exponents in the rate law are derived from experimental data.* The equilibrium expression for the decomposition of N_2O_5 is derived from its balanced chemical equation on page 542 of Section 14-1.

$$K_{eq} = \frac{[NO_2]^4[O_2]}{[N_2O_5]^2}$$

N_2O_5 has an exponent of 2 because its coefficient in the balanced chemical equation is 2. However, the rate expression for this same reaction is the following.

$$\text{rate} = k[N_2O_5]^1$$

N_2O_5 has an exponent of 1. The reason the exponents differ lies within the differences between thermodynamics and kinetics, and is outside the scope of this text. The important point to remember is that exponents in rate equations must be determined experimentally.

Section 14-3

FOCUS

Lesson Starter

Begin the lesson with the following demonstration.

Demonstration 7

Silent explosion

Approximate time: 10 min

1. Before class prepare solutions **A**, **B**, and **C**.
   A: Dissolve 1.8 g KIO_3 in 1 L of water.
   B: Dissolve 0.9 g Na_2SO_3 in 1 L of water.
   C: Suspend 10 g starch in 50 mL cold H_2O. Add the suspension to 500 mL of boiling H_2O. Boil 5 min, then pour over 450 g of ice. Add 25 mL concentrated H_2SO_4 to make the volume 1 L.
2. Make solution **X** by mixing 225 mL H_2O, 50 mL solution **C**, and 25 mL solution **A**; add 50 mL of solution **B**. An intense blue color will form in 20 s.
3. Make solution **Y** by mixing 200 mL H_2O, 50 mL solution **C**, and 50 mL solution **A**; add 50 mL of solution **B**. An intense blue color will form in 10 s.

SAFETY

Wear safety goggles and a lab apron. When preparing solution **C**, also wear a face shield and gloves.

DISPOSAL

Combine all solutions. If blue, add 1 M sodium thiosulfate solution slowly, pouring and stirring alternately until the solution remains colorless for 30 min. Then neutralize with 1 M NaOH, and pour down the drain.

[Source: Humphreys. *Demonstrating Chemistry*, McMaster University, Ontario, 1983.]

Content Background

Linear explanations for data are desirable because they are direct relationships between the variables graphed on the x- and y-axes.

Teaching Tip

Logarithms and log plots
Have students calculate the rate for each concentration-time pair in the data table in the Visual Strategy box below. Then make a plot of $\log[N_2O_5]$ vs. log (rate). The result should be a straight line with a slope of 1 (the order of the reaction) and the y-intercept should equal the log k. Review the rules of logarithms with the equations below.

$\text{rate} = k[A]^n$
$\log(\text{rate}) = \log(k[A]^n)$
$\log(\text{rate}) = \log k + \log[A]^n$
$\log(\text{rate}) = n\log[A] + \log k$

From Section 14-1 you know that rate is a calculated quantity. It is a ratio found by monitoring the change in some property during a specified time interval and then dividing the change by the elapsed time. You also learned that the units for rate must always be expressed and that for chemical reactions these units are usually concentration per unit of time. Data taken from rate experiments, then, must be related to concentration and time. The conditions under which the experiment is run must also be recorded because rate is dependent on other variables, as discussed in Section 14-2. Once experimental data have been collected, they can be analyzed through graphs made from the collected data.

Rate laws can be derived from plots of concentration vs. time

Look again at the concentration vs. time plot for N_2O_5 in **Figure 14-6**. The reactant concentration decreases as time passes, so the slope of the curve decreases, changing ever so slightly from one point to the next. As mentioned in Section 14-1, rate is described by the slope of the curve, so the rate also decreases with time. This continuous change in slope is described mathematically through a process called integration. Integration is beyond the scope of this book, but its result, the *integrated rate equation*, tells chemists what they could plot to get a straight line from which the exponent in the rate law can be determined.

To illustrate, the general rate law has the following form when the exponent is 1.

$$\textbf{rate} = k[\textbf{A}]^1$$

After integration, you get the following equation.

$$\log [A]_t = k't + \log [A]_0$$

So when a plot of log [A] vs. time results in a straight line, the exponent on the concentration term of the general rate law is 1, or first order. Doubling the concentration of A doubles the rate because the order of reaction with respect to [A] is 1. Tripling the concentration of A triples the rate.

The $[N_2O_5]$ vs. time curve in **Figure 14-6** has been replotted as $\log [N_2O_5]$ vs. time in **Figure 14-18**. The straight line tells you that the exponent on $[N_2O_5]$ in the general rate law is 1, so the reaction is first order in $[N_2O_5]$.

Log of the Concentration of N_2O_5 vs. Time for the Decomposition of N_2O_5 at 45°C in CCl_4 Solution

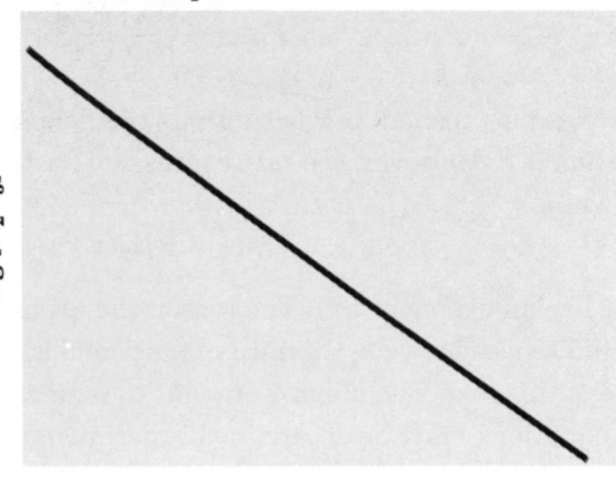

Figure 14-18
A first order relationship between concentration and reaction rate is verified when a log [A] vs. time plot gives a straight line.

Visual Strategy

Figure 14-18

Use the data table below to review how to take the logarithm of a number and how to plot log[A] vs. time.

$[N_2O_5]$	2.33	2.07	1.91	1.68	1.35	1.11	0.72	0.55
Time (s)	0	184	319	526	867	1198	1877	2315

The Order for a Given Reactant is Determined by Graphing Some Form of Concentration Data

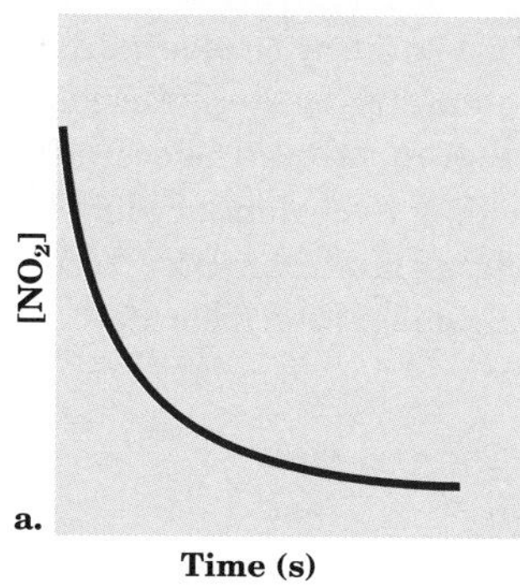

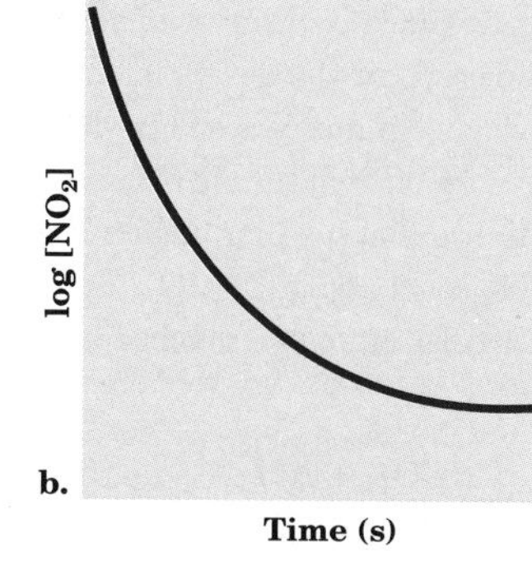

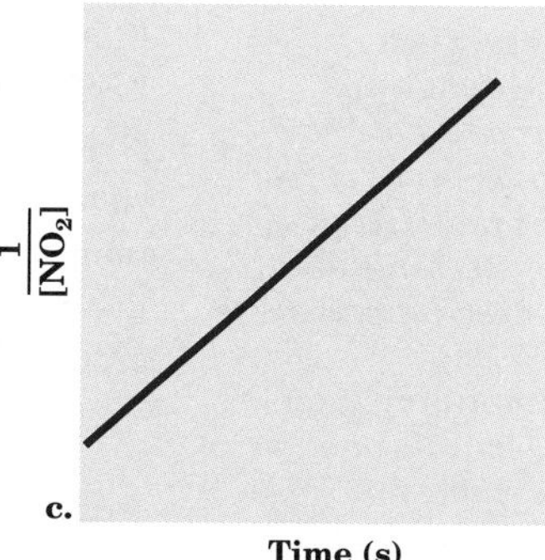

Figure 14-19
A second order relationship between reaction rate and concentration is verified when a graph of 1/[A] vs. time is a straight line.

The concentration vs. time graph for NO_2 decomposition in **Figure 14-19a** is also a curve with decreasing slope. Will the decomposition of NO_2 be first order like N_2O_5? The log $[NO_2]$ vs. time plot in **Figure 14-19b** is not a straight line. Therefore, the exponent for NO_2 in the rate equation is not 1. The experimenter must try something else. The integrated rate equation for an exponent of 2 suggests plotting 1/[A] vs. time. Such a plot for $[NO_2]$ results in a straight line, as shown in **Figure 14-19c**. The general rate law for this reaction is written as follows.

$$\text{rate} = k[NO_2]^2$$

The reaction is second order in NO_2. There is an x^2 relationship between concentration and rate; therefore, doubling the concentration of NO_2 quadruples (2^2) the rate. Tripling the concentration of NO_2 increases the rate by a factor of 9, or 3^2.

The rates calculated so far were for reactions involving only one reactant. But what about cases in which there are two or more reactants?

To handle situations in which more than one reactant is present, we revive the concept of a limiting reagent, first introduced in Chapter 8. When the limiting reagent is used up, the reaction stops. Analysis of the resulting mixture shows that in addition to the products, some of the excess reactant, is present. But if this excess reactant is grossly in excess, its concentration does not appear to change, so any change in the rate of reaction is due to the limiting reagent. This means then that for reactions in which two reactants, A and B, take part, two sets of data are recorded. In the first set, A is the limiting reagent, and B is in great excess. For the second set of data, B is the limiting reagent while A is in great excess. An appropriate concentration vs. time graph is plotted for each set of data. From them, the values for the exponents, n and m, are determined. The general rate law for the two reactants, A and B, is written as follows.

order of reaction
the sum of all the exponents on all concentration terms in a rate law expression

$$\textbf{rate} = k[A]^n[B]^m$$

[A] is the concentration of the first reactant, and n is its order. [B] is the concentration of the second reactant, and m is its order. The overall **order of reaction** is $m + n$.

Possible Misconception
When concentration is doubled, the value inside of the brackets changes, not the exponent. For example, if the reactant concentration is 0.5 M, the rate is $k[0.5]^1$, which equals $0.5k$. Doubling the concentration to 1.0 M means the rate is calculated from $k[1.0]^1$, which equals $1.0k$. Doubling the concentration doubled the rate.

Possible Misconception
For a first order reaction, you plot log[A] versus time, but the rate is equal to $k[A]^1$. Likewise, even though a straight line results from a plot of 1/[A] versus time for a second order reaction, the rate is equal to $k[A]^2$.

Teaching Tip
Analogy for reaction mechanism
The rate-determining step limits the rate of a reaction somewhat like the bagger in an assembly line making submarine sandwiches. If only one person is bagging the sandwiches, only a certain number of finished sandwiches will be done per hour no matter how many people are assembling sandwiches. In order to double the rate, you need an extra bagger. A reaction is only as fast as its slowest step.

Teaching Tip
Possible misconception
Distinguish between single-step reactions and multistep reactions. Point out why exponents in the rate law for a single-step reaction are the coefficients in the balanced chemical equation but are not exponents in the rate law for a multistep reaction.

Rate law and the reaction process

mechanism (of a reaction)
a proposed step-by-step sequence of reactions that describes how reactants are changed into products

elementary step
a chemical equation describing one stage of a reaction as it is theorized to occur

Most chemical reactions, like the cellular respiration reaction discussed in Chapter 9, take place in a series of steps called a pathway or **mechanism**. Each of these steps in the series is called an **elementary step**. An example of a process with many steps is shown in **Figure 14-20**. The overall equation for a given reaction is obtained by adding the balanced equations for individual steps just as you did for Hess's law in Chapter 9. For instance, NH_3 reacts with $Cr(CO)_6$ by replacing one carbon monoxide group (CO) according to the mechanism below.

Step 1:	$Cr(CO)_6 \longrightarrow Cr(CO)_5 + CO$	**slow**
Step 2:	$Cr(CO)_5 + NH_3 \longrightarrow Cr(CO)_5NH_3$	**fast**
Overall:	$Cr(CO)_6 + NH_3 \longrightarrow Cr(CO)_5NH_3 + CO$	

The slowest step determines the rate and defines the rate law

Although the elementary steps of a reaction occur in a particular order depending on the reacting substances, they do not all occur at the same rate. Some steps are slower than others. The slowest step determines the rate of the reaction and defines the rate law. If you know the reaction mechanism, you can write the rate law from the balanced equation for the slowest elementary step. The exponents for reactant concentrations are equal to their coefficients in the balanced equation for the slowest step. For the $Cr(CO)_6$ and NH_3 reaction, Step 1 is the slowest. It is rate determining. Because the rate law is defined by the slowest step, it can be written directly from the slowest step as follows.

$$\text{rate} = k[Cr(CO)_6]^1$$

Another example is the reaction between $CHCl_3$ and Cl_2. The following mechanism has been determined experimentally.

Step 1:	$Cl_2(g) \rightleftharpoons 2Cl\cdot(g)$	**fast and reversible**
Step 2:	$Cl\cdot(g) + CHCl_3(g) \longrightarrow HCl(g) + \cdot CCl_3(g)$	**slow**
Step 3:	$\cdot CCl_3(g) + Cl\cdot(g) \longrightarrow CCl_4(g)$	**fast**
Overall:	$CHCl_3(g) + Cl_2(g) \longrightarrow HCl(g) + CCl_4$	

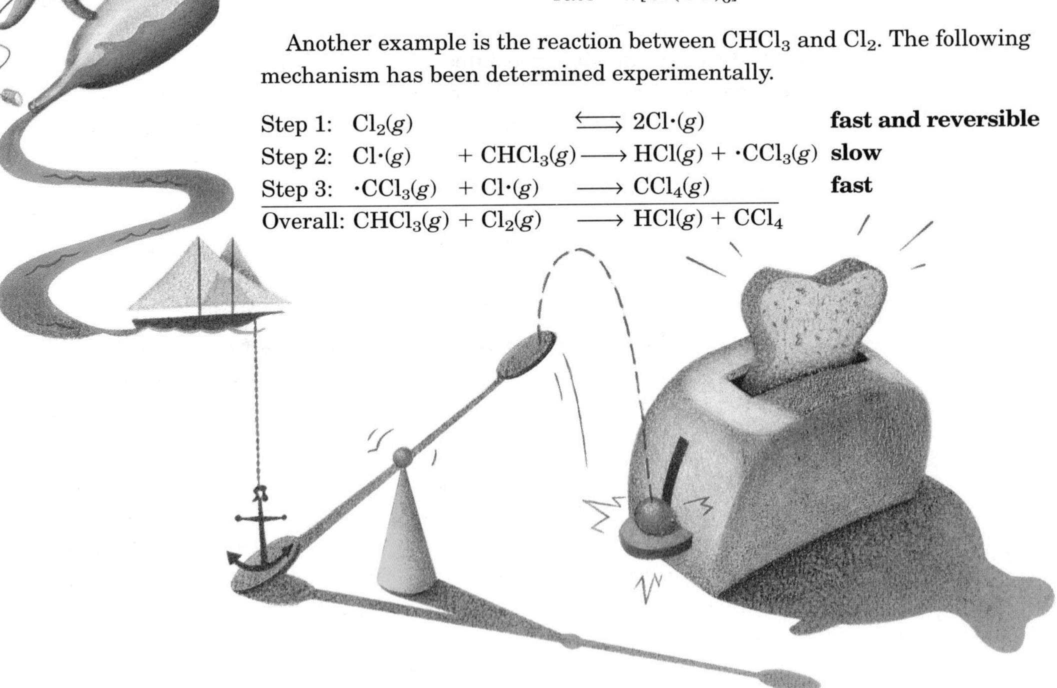

Figure 14-20
Some mechanisms are easier to figure out than others! But one step leads to the next just like a chemical reaction.

The rate law is written directly from Step 2 because it is the slowest step.

$$\text{rate} = k[\text{Cl}\cdot][\text{CHCl}_3]$$

Rates are usually written in terms of the reactants that appear in the overall balanced equation for the reaction. Cl· is not such a reactant. It is a reaction **intermediate** formed in Step 1, which is an equilibrium reaction, so [Cl·] is proportional to $[\text{Cl}_2]^{1/2}$, and the rate law can be written using only the concentrations of original reactants according to these equations.

intermediate
a structure formed in one elementary step, but consumed in a later step of a mechanism. It is neither an original reactant nor final product

$$[\text{Cl}\cdot] = (K_{eq})^{1/2}[\text{Cl}_2]^{1/2}$$

$$\text{rate} = k'[\text{Cl}_2]^{1/2}[\text{CHCl}_3]$$

From the rate law, the overall balanced equation, and experimental evidence for any reaction intermediates, we can propose pathways consistent with the data and design further experiments to test the proposals and reduce our choices. Still it is not always possible to identify just one mechanistic pathway for a reaction.

Concept Review

Reaction mechanism

21. Determine the overall balanced equation for a reaction having the following proposed mechanism.

Step 1:	$OCl^- + H_2O \rightleftharpoons HOCl + OH^-$	**rapid**
Step 2:	$HOCl + I^- \longrightarrow HOI + Cl^-$	**slow**
Step 3:	$HOI + OH^- \rightleftharpoons H_2O + OI^-$	**fast**

22. How many elementary steps are in the proposed mechanism in item **21**?
23. Identify the slowest step for the mechanism in item **21**. Write the rate law for the reaction.
24. List the reaction intermediates found in item **21**.
25. When two I *atoms* collide to form an iodine molecule, I_2, in the presence of He, the He absorbs excess heat and stabilizes the I_2 molecule while the bond is being formed. Use the following mechanism to write the rate law in terms of the original reactants.

$I\cdot + He \longrightarrow HeI\cdot$	**fast**
$HeI\cdot + I\cdot \longrightarrow He + I_2$	**slow**

Pathways are modified by catalysts

The energy diagram for a reaction indicates that if the height of the energy hill is increased, the rate slows down. If the height of the hill is reduced, the rate speeds up. Lowering the energy hill means changing the reaction mechanism and requires a catalyst. A catalyst lowers the energy barrier by forming an activated complex of lower energy than the one formed in the uncatalyzed reaction. In short, it provides the reactant molecules an energetically more favorable pathway to the products. The process by which a catalyst increases a reaction rate is **catalysis**.

catalysis
the process by which reaction rates are increased by the addition of a catalyst

Answers to

Concept Review

21. $I^- + OCl^- \rightleftharpoons Cl^- + OI^-$

22. three

23. step **2**; rate $= k[\text{HOCl}][\text{I}^-]$

24. HOCl, OH^-, HOI

25. rate $= k[\text{HeI}\bullet][\text{I}\bullet] = k[\text{He}][\text{I}\bullet]^2$

Historical Note

Joens Jakob Berzelius (1779–1848), a Swedish chemist and teacher, defined catalysts as substances which arouse the slumbering affinities of other substances but do not take part in the reaction themselves.

Demonstration 8

Comparison of the catalytic effect of I^- ions and catalase on the decomposition of H_2O_2

Approximate time: 5 min

1. Place two petri dishes on an overhead projector.
2. Pour 10 mL of 3% H_2O_2 (drugstore variety) into each dish.
3. To one dish add several drops of 1 M KI.
4. To the other dish add juice from a crushed turnip.

SAFETY
Wear safety goggles and a lab apron.

DISPOSAL
Combine solutions; when bubbling stops, pour down the drain.

Multicultural Perspective

During World War II, America used catalytic cracking of crude oil to obtain the smaller molecules that make up gasoline. Germany used catalysts to build gasoline molecules from the smaller molecules obtained by distilling coal liquids.

Hydrogen peroxide, H_2O_2, decomposes slowly over time, as represented by the following reaction.

$$2H_2O_2(aq) \longrightarrow 2H_2O(l) + O_2(g)$$

A relatively pure 30% solution of H_2O_2 will decompose at a rate of 0.5% per year at room temperature. This rate corresponds to the uncatalyzed curve with an E_a of 75 kJ/mol in the energy diagram for the decomposition of H_2O_2 shown in **Figure 14-21**. E_a is lowered to 58 kJ/mol, if iodide ions, I^-, are added to the peroxide solution. The I^- ions form a reaction intermediate, HIO^-, when they react with H_2O_2 molecules. The HIO^- ions then react with other H_2O_2 molecules to regenerate iodide ions as products are formed. The I^- ions are a **catalyst** for the decomposition of H_2O_2 because they are added into the reaction to increase the reaction rate but are not actually used up in the reaction. Because catalysts are not consumed, they can be recycled almost indefinitely. Catalysts sometimes appear in equations written above the arrow.

catalyst
a substance added to a chemical reaction to increase the rate and that can be recovered chemically unchanged after the reaction is complete

$$2H_2O_2(aq) \xrightarrow{KI} 2H_2O(l) + O_2(g)$$

Catalase and manganese dioxide also catalyze the decomposition of H_2O_2 by lowering the E_a for the reaction to 4 kJ/mol or less. Note that although a catalyst changes the value for E_a of a reaction, the value of ΔH for the reaction is left unchanged. If a reaction is exothermic without a catalyst, it is exothermic to the same extent with a catalyst.

Comparison of Pathways for the Decomposition of H_2O_2 by Various Catalysts

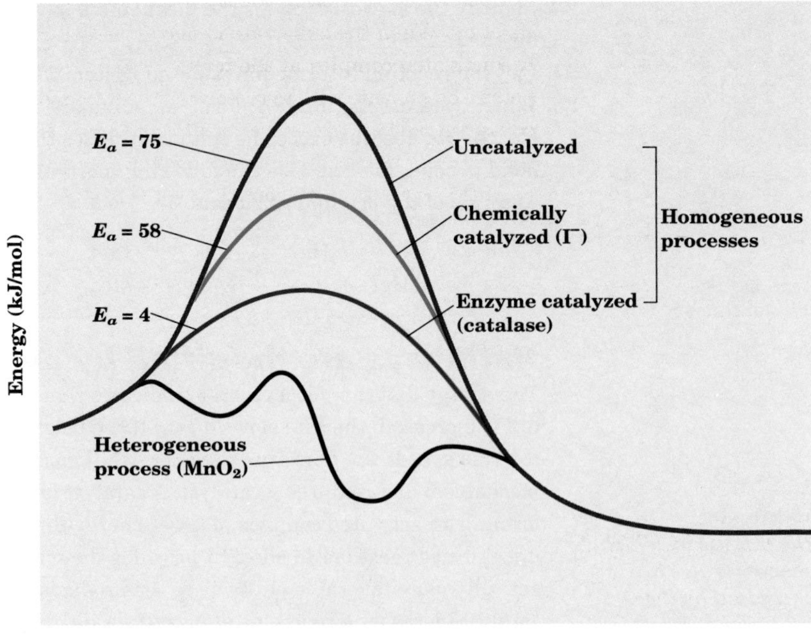

Figure 14-21 **The activation energy for a reaction can be reduced by adding a catalyst. Some catalysts work better than others.**

Visual Strategy

Figure 14-21

Identify and compare the E_a values for the four different reactions on this graph. Ask students why H_2O_2 is stored in brown plastic bottles to protect it from UV light. (UV radiation provides more energy than is needed for the uncatalyzed reaction.) You may also wish to point out that the amount of heat given off is identical regardless of the reaction pathway.

Hydrogenated vegetable oils, such as those shown in **Figure 14-22**, are made by the addition of hydrogen to the double bonds in vegetable oils in the presence of a nickel catalyst. This process is an example of a reaction that uses heterogeneous catalysis. The catalyst, nickel, is a solid, whereas the oil is a liquid. Catalysts in heterogeneous systems have the advantage of being easily separated from the reaction medium. A generalized comparison of E_a values for a homogeneously and heterogeneously catalyzed reaction appears in **Figure 14-21** on the previous page.

Figure 14-22
The naturally occurring liquid vegetable oils are converted to solid fats by catalytic addition of hydrogen.

Enzymes are nature's catalysts

Catalysts found in biological systems are called **enzymes**. Enzymes are large proteins. Like all catalysts, enzymes are not used up in a reaction, so only small amounts are needed. However, unlike most catalysts, enzymes are highly specific for a particular reaction or type of reaction. This high degree of specificity is possible because the enzyme molecule has a three-dimensional conformation. That is, the enzyme has a special location patterned to receive a particular type of reactant, much like the way that a lock is designed to receive a particular key. Look at the interaction between the enzyme and substrate shown in **Figure 14-23**. The **substrate** has to have the correct structure to bind to the enzyme and to be acted upon by the enzyme's active site. It is at the active site of the enzyme molecule that catalysis takes place. During the catalytic process, the enzyme's active site distorts the bonds in the substrate, forcing it to look more like the product. This distortion strains the substrate's bonds and weakens them. The reaction progresses more rapidly because the weaker bonds possess less energy, making the E_a value lower than that for the uncatalyzed reaction. When the substrate is most closely aligned with the active site, the activated complex at the top of the energy hill is formed. After this point, the enzyme will be restored, and the product will be formed. In some cases, there is an enzyme-product complex as an intermediate prior to the formation of the product. The overall ΔH for the reaction has not been changed, but the activation energy has been lowered.

enzyme
large protein molecule that catalyzes chemical reactions in living things

substrate
(biochemical) the molecule or molecules with which an enzyme interacts

Figure 14-23a
Energy of the system increases as the substrate nears the active site of the enzyme.

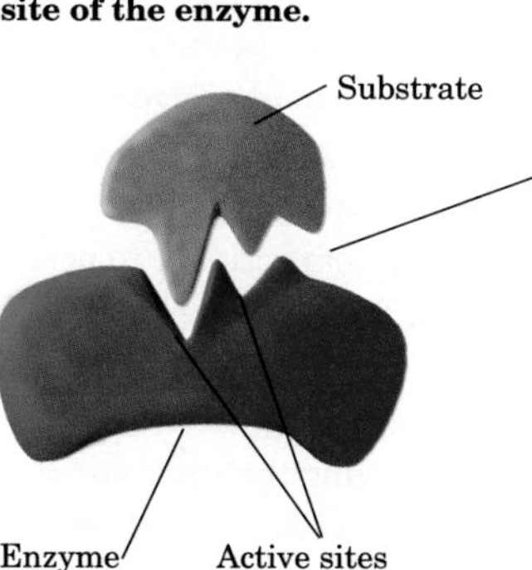

Activated complex

Figure 14-23b
The substrate-enzyme activated complex occurs at the peak of the reaction's energy curve.

Figure 14-23c
Energy of the system decreases as products form and the enzyme is restored.

Products and enzymes

Historical Note
Russian-German chemist Friedrich Wilhelm Ostwald was the first to describe the characteristics and mechanisms of catalysis using the theories of Gibbs.

Visual Strategy
Figure 14-23
Have students draw a generalized energy diagram for the enzyme-substrate reaction described here.

Do You Know?

Uncooked pineapple cannot be used in gelatin salads because the enzymes present in the pineapple keep the gelatin from setting. Cooking the pineapple destroys the enzymes and allows the gelatin to set.

Further Applications

Beeman's gum, an old brand of chewing gum still available today, contains the digestive enzyme pepsin and is still advertised as aiding digestion. Pepsin is typically derived from the glandular layer of fresh hog stomachs.

Enzymes in action

Enzymes in the body speed up the chemical processes that sustain life. Most reactions in the body are so complex that if they were not catalyzed, they would require extremely high temperatures to occur at rates fast enough for life to continue. Fortunately, enzymes alter reaction pathways so that the complex reactions of metabolism occur at 37°C (98.6°F) at a rate several billion times faster than they would without a catalyst. Life could not exist without the action of enzymes.

Enzyme supplements reduce the effects of lactose intolerance

Some people lose the ability to produce lactase, a digestive enzyme that cuts lactose, the sugar found in many dairy products. These people have what is known as lactose intolerance. Undigested lactose molecules collect in the intestine and attract water. As a result, a person with lactose intolerance can experience painful cramps or diarrhea after ingesting foods with lactose. Enzyme supplements are available that can be added to or taken with milk to hydrolyze the lactose and form its constituent sugars, glucose and galactose. The human body has hundreds of other highly specialized enzymes that control growth, reproduction, and other processes.

Enzymatic cleansers reduce protein buildup on lenses

Modern contact lenses are made of hydrophilic plastics that bind a great deal of water. Most lenses are gas permeable to enable sensitive corneas to receive enough oxygen. However, the protein matter in tears can clog the pores in contact lenses making both lenses and eyes susceptible to bacterial growth and possibly serious damage. Protein build-up is removed from the lenses by special enzymatic cleaners. Depending on the brand of cleaner, these enzymes are papain, pancreatin, or subtilisin. These enzymes are known as proteases because they cut proteins into small pieces that can be washed away, leaving the contact lenses clear and clean.

Figure 14-24
The white albedo in citrus fruits and the pectin in gelling agents can be effectively dissolved using pectinase, a safe-to-use, specific action enzyme.

Enzymes are now being used to peel fruit

Pectin, a long-chain polysaccharide found in many fruits, is the gelling agent in jams and jellies. Wine makers sometimes add an enzyme called pectinase to fruit juices to destroy the pectin and clarify the juices before fermentation. Pectinase can also digest the white inner lining of citrus fruits that makes peeling the fruit difficult. Therefore, a major citrus fruit company is experimenting with the commercial use of pectinase to peel large quantities of fruit quickly and effectively, as shown in **Figure 14-24**. The peels are scored, and the fruits are soaked in a warm solution of pectinase. After soaking, the fruit is rinsed, and the peels come off easily leaving the separate sections of fruit clean and ready to eat. If you live in southern California, you may already have eaten fruit peeled enzymatically.

An inhibitor slows down the rate of a reaction

inhibitor
a substance added to a chemical reaction to slow it down

Inhibitors are used extensively in the plastics industry to help control chain reactions from which polymers are made. Chain reactions occur in steps, so if one of the substances used to build the chain can be removed, the reaction slows down or stops. Inhibitors are added to react with substances that help build the chain.

Inhibitors and catalysts working together

If combustion in an automobile engine were 100% efficient, it would convert gasoline into carbon dioxide and water only. Unfortunately, significant amounts of carbon monoxide and unburned hydrocarbons are also produced and discharged into the air. Therefore, modern cars are equipped with a catalytic converter to reduce pollution. It converts some of the products from incomplete combustion to carbon dioxide and water. Some racing cars are supercharged by adding a nitrogen oxide compound to the fuel mixture. The reaction is complex, but the nitrogen oxide compound alters the mechanism of the combustion reaction and acts as an additional energy source. More energy is released more efficiently than with regular gasoline alone.

Gasoline is a mixture of short-chain hydrocarbons obtained from the fractional distillation of petroleum. Two components of gasoline are shown in **Figure 14-25**. Some of the hydrocarbon chains in the gasoline mixture are straight, and some are branched. The branched chains burn more smoothly and efficiently than the straight chains do. Therefore, the higher the percentage of branched chains in the gasoline mixture, the more efficiently it burns. Nevertheless, if it burns too quickly inside the cylinder, ignition occurs before the piston is in the proper position. Pressure from this early ignition pushes the piston in the opposite direction to that in which it was already traveling. This is a tremendous jolt to the crankshaft, and the subsequent shifting of the crankshaft, pistons, rods, and other parts is heard as knocking. To inhibit the fast combustion that causes knocking, tetraethyl lead was once added to gasoline. At the time, tetraethyl lead was an attractive choice because it mixed readily with gasoline, was inexpensive, and did its job well. However, scientists did not realize how much lead would be dumped into the atmosphere or how serious the effects of lead poisoning would be. Lead itself is a relatively inert metal, but its ions are toxic to most biological systems because they inhibit the actions of enzymes.

These ions also ruin the platinum catalyst in the catalytic converters of most modern cars by covering it with a layer of lead that renders the active sites useless. This is why cars with catalytic converters can use only unleaded fuels. Other antiknock agents such as ethanol and methyl-*t*-butyl ether are now used.

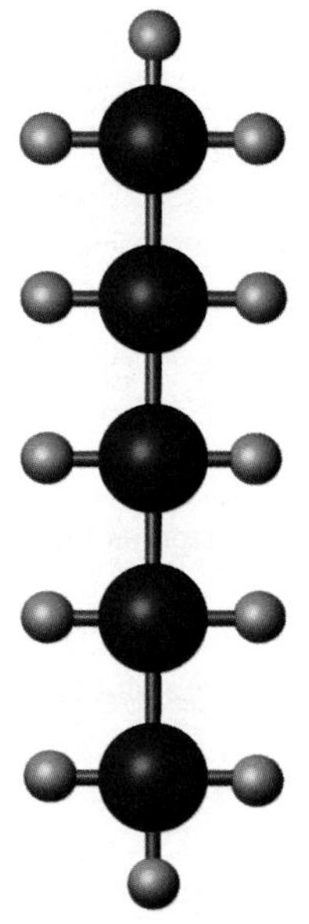

Figure 14-25a
Gasoline is a mixture of hydrocarbons. Some are straight like *n*-pentane, . . .

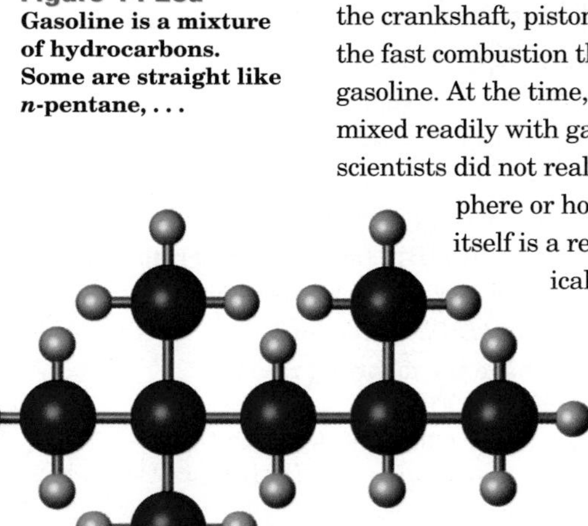

14-25b
. . . and others are branched like isooctane.

Do You Know?

N_2O is added to the engine of a race car along with extra fuel. Without the extra fuel, the fuel mixture burned in the carburetor will be too rich in oxygen and can ruin the engine.

Chapter Connections

The combustion of isooctane is discussed in Chapter 9 on page 318.

Octane ratings for gasoline are described in Chapter 6 on page 218 under Further Applications.

Visual Strategy

Figure 14-25

Review formulas for covalently bonded molecules by asking students to write the molecular formulas, structural formulas, and abbreviated organic structures for these two compounds.

ASSESS

Alternative Assessment

Use Hess's law to explain why a catalyst changes the pathway and E_a of a reaction but not the heat of reaction.

Answers to Section Review

26. To increase the rate, changes must affect either the rate-determining step or, in some cases, one of the steps preceeding it. Step **3** is rate-determining; changes to step **1** or step **2** might also affect the rate.

27. The rate is 2nd order in NO and 1st order in Cl_2. For NO and constant $[Cl_2]$, axes labels are 1/[NO] and time. For Cl_2 and constant [NO], the axes labels are $\log[Cl_2]$ and time.

28. It does not; anything raised to the zero power equals one.

29. Step **2** is rate-determining; rate $= k[N_2O_2][H_2]$ or $k'[NO]^2[H_2]$

30. increase by $(1.5)^2$

31. none

32. **a.** halved; $k[1/2A][B]^2 = 1/2k[A][B]^2$
b. increased ninefold; $k[A][3B]^2 = 9k[A][B]^2$
c. increased
d. expect increase
e. halved; $k[2A][1/2B]^2 = 2(1/4)k[A][B]^2 = 1/2k[A][B]^2$

Section Review

26. Lowering E_a for a reaction makes the reaction go faster but does not affect the overall ΔH of reaction. Consider the E_a values in the reaction diagramed below. Which, if any, would speed up the reaction if lowered? Support your answer.

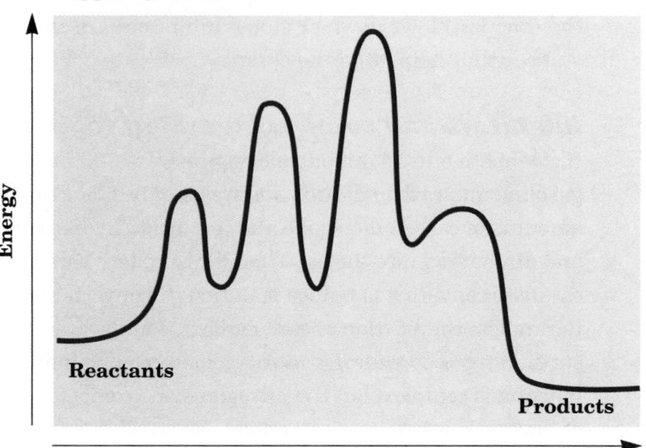

27. The observed rate law for the reaction of NO with Cl_2 is written as follows.

$$\text{rate} = k[NO]^2[Cl_2]$$

What is the order of reaction with respect to NO? with respect to Cl_2? How would the axes be labeled on a concentration versus time plot for each reactant in order to obtain a linear relationship?

28. How does a reactant with an exponent of zero in the rate law affect the reaction rate?

29. Nitric oxide can be reduced with hydrogen gas to give nitrogen and water vapor.

$$2NO(g) + 2H_2(g) \longrightarrow N_2(g) + 2H_2O(g)$$

A proposed mechanism for this reaction follows.

Step 1:	$2NO \rightleftarrows N_2O_2$	**fast**
Step 2:	$N_2O_2 + H_2 \longrightarrow N_2O + H_2O$	**slow**
Step 3:	$N_2O + H_2 \longrightarrow H_2O + N_2$	**fast**

Identify the rate-determining step. Write the rate law.

30. If the rate of a reaction is expressed by $k[B]^2$, how would the rate change if the initial concentration of B were increased by a factor of 1.5?

31. What effect does a catalyst have on ΔH for a reaction?

32. A particular reaction is found to have the following rate law.

$$\text{rate} = k[A][B]^2$$

How is the rate affected by each of the following changes?

a. the initial concentration of A is cut in half
b. the initial concentration of B is tripled
c. a catalyst is added
d. the temperature is increased
e. the concentration of A is doubled, but the concentration of B is cut in half

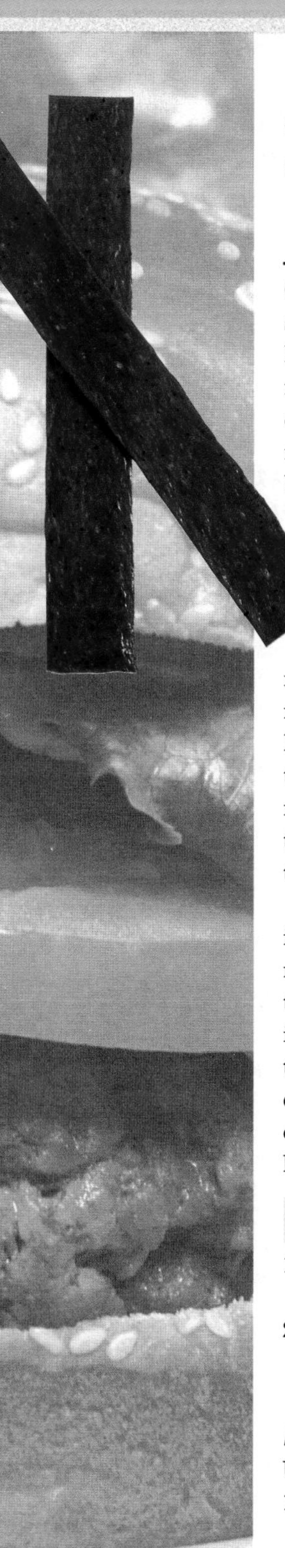

Conclusion: Kinetics and Cooking Hamburgers

At the beginning of this chapter you learned that *E. coli* 0157:H7 is responsible for meat spoilage and food poisoning. Destruction of this bacteria is related to the meat's internal cooking temperature. These points raised questions about the rate of bacterial growth and the factors that affect it. Look at these questions again.

How does temperature affect the preparation of raw meat? Chilling or freezing meat slows down decomposition reactions and bacterial growth. But warming the meat increases those reactions up to the bacteria's maximum reproduction temperature. At temperatures higher than this point, the bacteria die.

But most bacteria do not die immediately, so the meat must remain in contact with the higher temperatures long enough for its internal temperature to rise to the point that the bacteria are destroyed. Otherwise the warm center of the meat may still harbor live bacteria that multiply freely.

What are some effective methods for preserving meat? Preservation involves practices that retard bacterial growth. Bacteria are deposited on the surface of the meat through physical contact or exposure to the air. The relatively neutral pH of raw meat favors bacterial growth but lower pH values do not. In the slaughterhouse, the pH level of meat can be reduced by applying an acetic-acid-based antibacterial spray to the surface of the animal carcass after its hide is removed and before it is cut open.

The moisture in meat can be reduced by adding salt or by drying. These processes are effective preservation methods because bacteria seldom grow on foods with low water content.

Bacteria acquired during processing can be destroyed by bombarding packaged meat with gamma rays. The irradiation destroys the bacteria and the packaging prevents the surface from being recontaminated until the package is opened.

Applying Concepts

1. Water boils at a lower temperature on top of a mountain than at sea level. Explain the effect this has on the time needed to boil a hard egg.
2. Decomposition reactions involving O_2 or CO_2 present in the air cause meat to feel slimy and smell spoiled. Explain why meat spoils less rapidly when wrapped or left unsliced.

Research and Writing

Use the library to find out more about the following.

1. Write an editorial for your school or local newspaper that discusses the use of enzymes for destroying hazardous waste.
2. Why are expiration dates placed on foods and medications? Which government agencies require such dates?

Chemistry in Your Community

Investigate the packaging, handling, and storing practices of meat products at local meat markets or supermarkets. Determine how long each type of meat is allowed to be on the shelf. Find out the legal amount of time these meat products are allowed to be displayed. Construct a visual aid that compares the policies of the local stores with those of the state health department.

Answers to Applying Concepts

1. The time increases because the heat takes longer to reach the center of the egg.
2. The surface area exposed to the air is decreased.

Closure Strategy

Ask students to consider the following reaction at some temperature, T_1, and at a second, higher temperature, T_2. Then have them write for five minutes about the appearance of the concentration versus time graph for each temperature and about what information can be derived from the graphs.

$$2N_2O_5(sol) \xrightarrow[45°C]{\text{in } CCl_4} 4NO_2(g) + O_2(g)$$

CHAPTER 14

Highlights

Key terms

activated complex	**intermediate**
average rate	**mechanism**
catalysis	**order**
catalyst	**order of reaction**
chemical kinetics	**rate-determining step**
elementary step	**rate law**
enzyme	**reaction rate**
inhibitor	**substrate**
initial rate	

14-1

What is a reaction rate?

- The average reaction rate is expressed as the ratio between a measured change in a physical property or concentration and the time interval during which the change took place.
- Rates generally decrease as time passes.

14-2

How can reaction rates be explained?

- Factors affecting reaction rate include the nature of the reactants, concentration, temperature, and surface area. Increasing the concentration, temperature, or surface area increases the reaction rate because the number of collisions between reactants increases.
- Not all collisions result in a reaction. Particles must have the proper orientation and sufficient kinetic energy to break the chemical bonds in the reactants so that new bonds can form.

14-3

How can reaction rates be described?

- The reaction rate is proportional to the reactant concentration raised to some power. This exponent must be determined experimentally for each reaction.
- The mechanism of a reaction usually involves a step-by-step sequence. The slowest step determines the rate law and is called the rate-determining step.
- Catalysts increase the reaction rate by lowering the energy barrier for the reaction. Catalysts can be recovered unchanged after the reaction is complete. Inhibitors slow down the reaction.

Key problem-solving approach: Determining reaction order

Write and balance the chemical equation for the reaction; then follow the flowchart to determine the overall order of reaction.

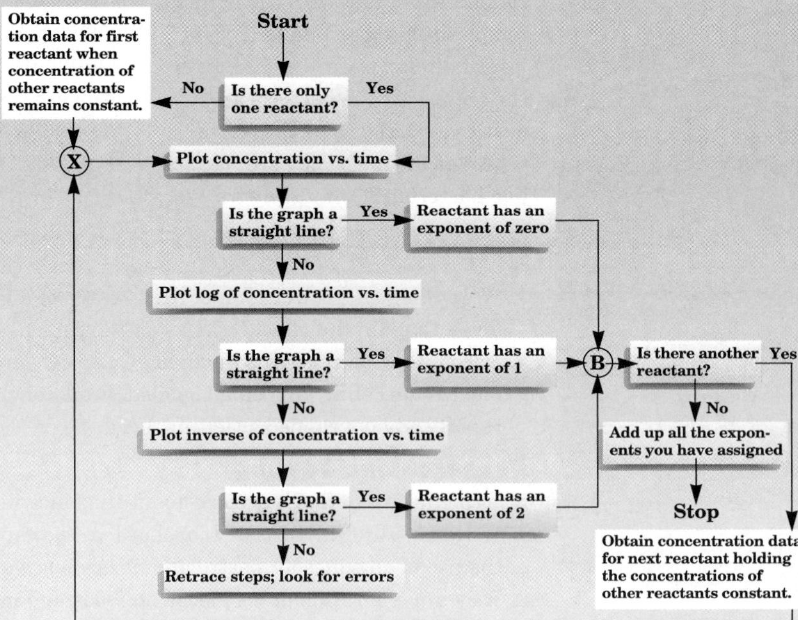

Start
Is there only one reactant?
No
Yes
Obtain concentration data for first reactant when concentration of other reactants remains constant.
X
Plot concentration vs. time
Is the graph a straight line?
Yes
Reactant has an exponent of zero
No
Plot log of concentration vs. time
Is the graph a straight line?
Yes
Reactant has an exponent of 1
B
Is there another reactant?
Yes
No
Add up all the exponents you have assigned
Stop
Obtain concentration data for next reactant holding the concentrations of other reactants constant.
No
Plot inverse of concentration vs. time
Is the graph a straight line?
Yes
Reactant has an exponent of 2
No
Retrace steps; look for errors

CHAPTER 14

Review and Assess

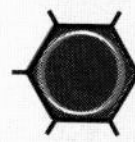

Clocking chemical reactions

REVIEW

1. Describe the rate for each of the following, using appropriate units.
   a. water dripping from a leaky faucet
   b. walking speed on an exercise path
   c. a fan blade rotating
   d. water evaporating from a 1 L container
2. Write a rate expression for the decomposition of acetoacetic acid into acetone and carbon dioxide.

$$CH_3C(O)CH_2COOH(l) \longrightarrow CH_3C(O)CH_3(l) + CO_2(g)$$

3. a. List four physical properties or observations that can be measured during the course of a reaction to determine reaction rate.
   b. For the four physical properties that you listed, what laboratory equipment would be necessary for making accurate measurements?
4. Why are some reaction rate expressions multiplied by a factor of –1?
5. Why not use a factor of –1 in the rate expression when the change in concentration of a product over time is being measured?
6. Why would measuring the rate of the following reaction be difficult?

$$H_2(g) + CO_2(g) \longrightarrow H_2O(g) + CO(g)$$

APPLY

7. Examine the following reactions. Complete the table as shown.

Reactions:
1. $CaCO_3(s) \longrightarrow CaO(s) + CO_2(g)$
2. $Br_2(l) + C_6H_6(l) \longrightarrow HBr(g) + C_6H_5Br(l)$
3. $N_2O_4(l) \longrightarrow 2NO_2(g)$
4. $NaOH(aq) + HCl(aq) \longrightarrow NaCl(aq) + H_2O(l)$

Columns:
a. Indicate whether you would measure $\frac{\Delta[\text{reactants}]}{\Delta t}$ or $\frac{\Delta[\text{products}]}{\Delta t}$
b. Indicate the substance measured.
c. Indicate how the change would be measured.

Reaction	Column A Change in reactant or product	Column B Substance measured	Column C Type of measurement
1	product	carbon dioxide	gas collection
2			
3			
4			

8. Many solutions to environmental problems involve changing a highly toxic substance into a harmless substance. If mercury(II) chloride, a highly toxic substance in aqueous solutions, is treated with sodium oxalate, mercury precipitates from solution as mercury(I) chloride, and carbon dioxide is released as a gas.

$$2HgCl_2(aq) + Na_2C_2O_4(aq) \longrightarrow 2NaCl(aq) + 2CO_2(g) + Hg_2Cl_2(s)$$

   a. Write a rate expression using Hg_2Cl_2.
   b. Why would it be easier to monitor the formation of Hg_2Cl_2 than the formation of the other two products?
   c. Examine the data in the table. Would 30 min be sufficient time to treat 0.50 M $HgCl_2$ with 0.20 M $Na_2C_2O_4$?

	Initial concentration (M)				Final (M)
Trial	$HgCl_2$	$Na_2C_2O_4$	NaCl	Δt (min)	NaCl
1	0.50	0.10	0.00	30	0.010
2	0.50	0.20	0.00	30	0.042
3	0.50	0.30	0.00	30	0.168

1. a. milliliters of water collected each hour
   b. the distance you travel measured as miles of linear displacement for each hour of exercise
   c. the number of times a point on the blade passes a fixed reference point during a minute
   d. milliliters of water lost from the beaker each hour
2. rate = k $[CH_3C(O)CH_2COOH]$
3. a. Answers will vary but may include changes in volume, temperature, color, mass or pH or the formation of a precipitate or a gas.
   b. Possible answers could include a graduated cylinder for liquid volume change, a thermometer for temperature change, a colorimeter for color change, an analytical balance for change in mass, pH paper or pH meter for pH change, an oven and analytical balance for drying and weighing a precipitate, and a calibrated gas collection device for measuring gas formation.
4. Reaction rates must be positive numbers. If the disappearance of a reactant is being measured then the subtraction of the initial concentration from the final concentration will result in a negative number. Multiplying by a factor of −1 makes the value of the rate positive.
5. The formation of products is calculated by subtracting the initial concentration from the final concentration. When a reaction starts, the initial concentration of the products is zero, so the value will always be positive.

6. All of the substances in the reaction are colorless gases. There would be problems separating, identifying, and measuring the concentrations of the individual gases as the reaction progressed.

7. *See page 573A.*

8. a. rate = $k\,[Hg_2Cl_2]^a$

b. Hg_2Cl_2 is insoluble in water and can easily be recovered by filtering the solution and drying the filtrate. NaCl is dissociated in water, and CO_2 is a gas. Both of these products would require more time and equipment to monitor.

c. No. After 30 min, only 0.042 M NaCl has been produced. For the reaction to be complete, the concentration of NaCl should be 0.500 M.

9. a. for days 1–90 from graph

$$\text{rate} = -\frac{\Delta[\text{NOCl}]}{\Delta t} = -\frac{0.58\text{ M} - 1.00\text{ M}}{90\text{ days}} = 4.7\times10^{-3}\text{ M/day}$$

for days 360–450 from graph

$$\text{rate} = -\frac{\Delta[\text{NOCl}]}{\Delta t} = -\frac{0.23\text{ M} - 0.28\text{ M}}{90\text{ days}} = 5.6\times10^{-4}\text{ M/day}$$

b. The reaction rate for days 1–90 is faster because the concentration of NOCl is higher.

10. Simple ions are spherical, and the orientation of two spheres with respect to each other is the same in all directions. See below.

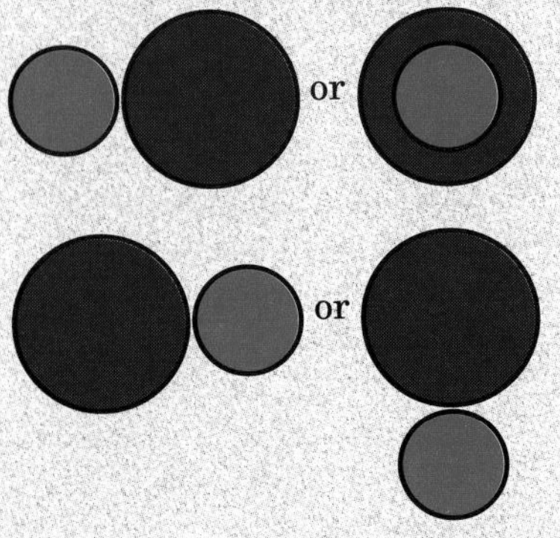

9. A mixture of nitric acid and hydrochloric acid is called aqua regia. Aqua regia will dissolve gold and platinum due to the presence of nitrosyl chloride, NOCl, and chlorine. Nitrosyl chloride slowly decomposes into nitrogen monoxide and chlorine, causing the color of the aqua regia to change from reddish brown to light green.

$$2\text{NOCl}(soln) \longrightarrow 2\text{NO}(g) + \text{Cl}_2(g)$$

a. Use the following graph to determine the rate of decrease in NOCl concentration during the first 90 days and between days 360 and 450.

b. Propose a reason why the rates for days 1–90 and days 360–450 differ.

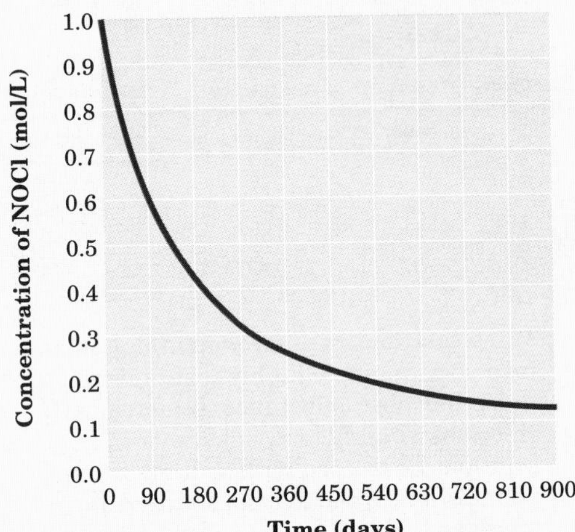

Causes for various reaction rates

REVIEW

10. Explain why many molecules must have a specific orientation to collide effectively to form a product, but simple ions such as Cl^- and Ba^{2+} are unaffected by orientation. Use the following models to illustrate your point.

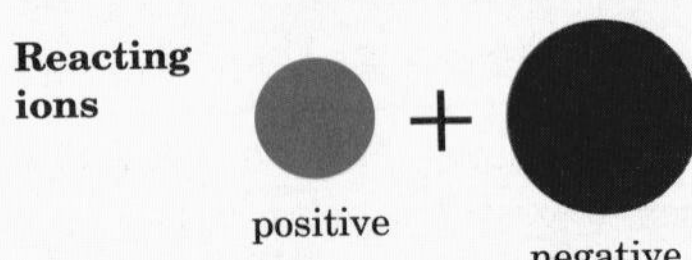

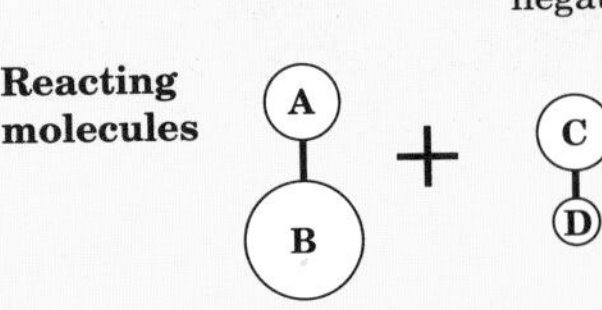

11. If we raise the energy states of reactants by increasing the temperature, are all of the reactant particles likely to have enough energy to form products? Why?
12. Why does grinding a solid cause an increase in reaction rate?
13. Iron nails will corrode slowly when placed in sulfuric acid. If iron filings are substituted, the reaction becomes instantaneous. Why?
14. Why is the basic chemical nature of a substance important when studying chemical kinetics?
15. Why does a change in concentration of reactants usually change the reaction rate?
16. Use graphs to illustrate the following situations.
    a. an exothermic reaction with a relatively large E_a value
    b. an exothermic reaction with a relatively small E_a value.

APPLY

17. At normal body temperature (37°C) the body can survive for only 5 min without oxygen. When the body temperature is lowered to a state of hypothermia (28–30°C), the body can survive almost 30 min without damage to tissues. Why?
18. The time for a reaction to come to completion is of considerable economic interest. Describe ways in which reaction rates could affect economic interests.
19. White phosphorus will ignite and burn spontaneously around 34°C.

$$4\text{P}(s) + 5\text{O}_2(g) \longrightarrow \text{P}_4\text{O}_{10}(s)$$

    a. Draw an energy diagram for this reaction.
    b. On the graph, label the following: x– and y– coordinates, reactants, products, activated complex, and E_a.
    c. Propose two ways that white phosphorus could be stored to prevent a reaction with oxygen.
20. Old newspapers that are stacked and stored will sometimes burst into flames spontaneously. Usually a high temperature is required to burn a stack of paper. How can you explain the spontaneous combustion of old newspapers when the temperature seems to be low?
21. Explosions are common in factories that grind wheat into flour. Would the heat released from the explosion of flour be greater than the heat released when whole grains of wheat burn? Why?

Representing reaction rates

REVIEW

22. How can the order of reaction with respect to reactant concentration be determined from experimental data?
23. Why aren't the coefficients in a balanced equation used to write the rate law expression?
24. Reactant A participates in two different reactions. A plot of log[A] vs. time for the first reaction shows a straight-line relationship. For the second reaction, a straight-line plot occurs when 1/[A] vs. time is plotted. What is the order of each of these reactions with respect to [A]?
25. Determine the overall order of reaction for each of the following reaction rate expressions.
    a. rate $= k[N_2O_5]$
    b. rate $= k[CH_3OH]^2\ [(C_6H_5)_3CCl]$
    c. rate $= k[O_3]\ [NO]$
26. Does the position of the slow step in the reaction mechanism have an effect on the rate law expression? Why?
27. Compare an intermediate to an activated complex.
28. In the diagram below, the energy pathway for a reaction is represented by the curved line. Draw another line representing the possible energy pathway that would result if this reaction was enzyme mediated.

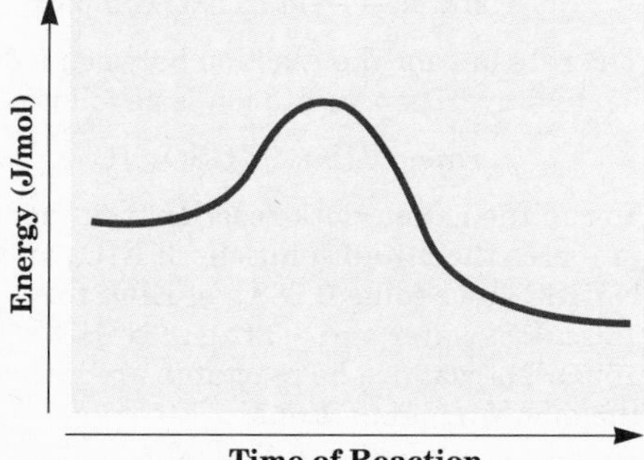

29. In terms of structure and function, compare and contrast enzymes and inorganic heterogeneous catalysts.
30. Briefly explain the process involved in the enzyme-catalyzed conversion of a substrate to a product.
31. Describe the effect of an inhibitor on the kinetics of a chemical reaction.
32. If the rate expression for a reaction is determined to be rate $= k[A]^2$, what happens to the rate if the concentration of A is tripled? What happens to the rate if the concentration of A is halved?
33. If the rate expression for a reaction is determined to be rate $= k[A][B]^2$, what happens to the rate if the concentration of A is doubled? What happens to the rate if the concentration of B is halved?

APPLY

34. What are the units for k, for each rate law in item **25**?
35. Describe how you would carry out a typical experiment with two reactants that would determine a rate law expression.
36. If you were told that the rate law for the following reaction is second order with respect to H_2O because the coefficient is 2, how would you respond?

$$CaC_2(s) + 2H_2O(l) \longrightarrow C_2H_2(g) + Ca(OH)_2(s)$$

37. At temperatures below 225°C the following reaction takes place.

$$NO_2(g) + CO(g) \longrightarrow CO_2(g) + NO(g)$$

Doubling the concentration of NO_2 quadruples the rate of CO_2 being formed if the CO concentration is held constant. However, doubling the concentration of CO has no effect on the rate of CO_2 being formed.
    a. Write a rate law expression for this reaction.
    b. Determine the order of the reaction with respect to NO_2 and with respect to CO.
    c. If NO_3 is a reaction intermediate, suggest a possible mechanism for this reaction that accounts for the experimental results.
38. Carbon monoxide gas and oxygen gas bind to the hemoglobin molecule. The rate for the release of oxygen to the tissues is measured in seconds. The release of CO from hemoglobin takes from 5 to 8 h. Both reactions are first order with respect to oxygen or carbon monoxide. Compare the kinetics of these reactions, and explain why carbon monoxide poisoning occurs.
39. The nerve gas Soman is a cholinesterase inhibitor. Antidotes to poisoning by Soman are cholinesterase reactivators. What kind of substance is cholinesterase? Suggest what may happen to it when it is inhibited by Soman.

Each molecule can have a definite spatial arrangement that is nonsymmetrical, and there are many possible combinations for the collisions. Some examples are given below.

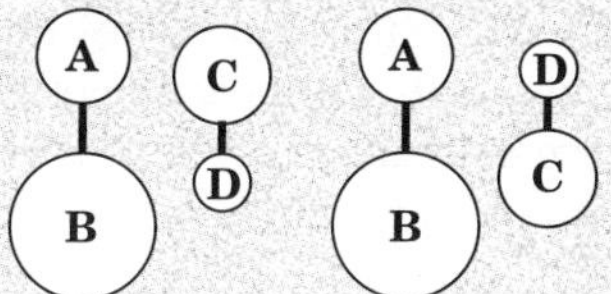

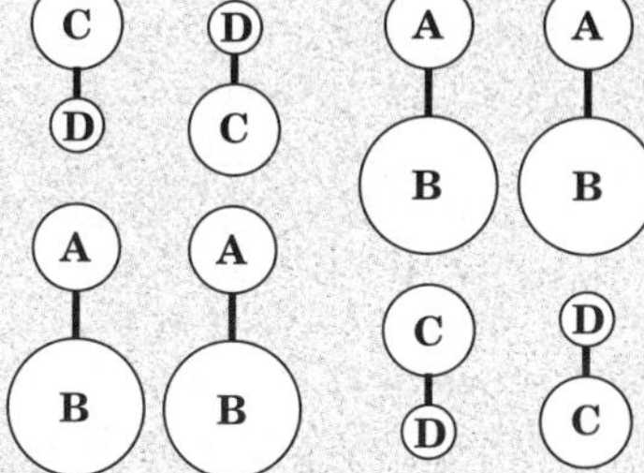

If C and A must be close to react and B and D must also be adjacent, then there is only one possible combination or orientation. All other collisions will result in no product formation.

11. Although the total energy for a system is increased when the temperature is raised, not all of the particles will have the energy required for a successful collision. There will still be a fraction of the particles with energies higher than the average and a proportionate number with energies lower than the average.
12. Grinding solids greatly increases the surface area. The smaller the particles present, the greater the total exposed surface. Rate is proportional to surface area because the particles must make physical contact to react.
13. The surface area is greatly increased when iron is ground into filings. When the total number of reacting particles is increased due to the expanded surface area, the number of particles that come into contact with the acid also increases.

14. The size, reactivity, acidic or basic character as well as other properties play a role in determining how each reactant will respond to the presence of other atoms or molecules.

15. The number of particles and the number of possible effective collisions increase when the concentration is raised.

16. a. The graph should show a haystack-like curve with steep upward and downward slopes corresponding to a large amount of activation energy. The graph should also show that the initial energy of the reactants is higher than the final energy of the products.
b. The graph should show a haystack-like curve with gradual upward and downward slopes corresponding to a small amount of activation energy. The graph should also show that the initial energy of the reactants is higher than the final energy of the products.

17. The reaction kinetics of the body are strongly temperature dependent so that a decrease in temperature drastically slows down respiration and reduces the demand for oxygen. A longer period of time is required for respiratory reactions to consume the oxygen in the tissues and for oxygen starvation or asphyxiation to take place.

18. Possible economic ramifications could include reducing production costs by increasing the reaction rate; slowing corrosion by decreasing the reaction rate; controlling the rate of reaction by altering temperature or concentration to prevent explosions. Accept all reasonable answers.

40. The following data were obtained in a bacterial growth experiment.

Time (min)	—	60.0	120	180	240	300
Concentration [(bacteria/L) × 10^{-3}]	3.50	7.00	14.0	28.0	56.0	112

a. How many minutes must pass before the population of the colony doubles?
b. When the bacteria concentration has reached 22.4×10^{-2}/L, how much time has elapsed?

41. What is the overall order of reaction for each of the following reactions if their rate constants at 773 K are $6.1 \times 10^{-4}\ s^{-1}$ and $2.88 \times 10^{-18}\ L\ mol^{-1}\ s^{-1}$, respectively?

a. 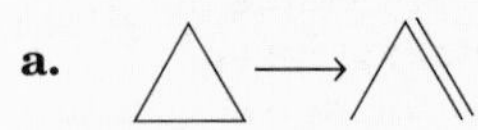

b. $Ar + O_2 \longrightarrow Ar + 2O$

42. a. Write the rate law for reaction **a** in item **41**, and determine the reaction rate if the concentration of cyclopropane is 2.0×10^{-3} M.
b. The rate constant for reaction **a** in item **41** is 5.70×10^{-4} M at 25°C. Write the rate law for this temperature.
c. Compare the rate law equations in parts **a** and **b**, and account for the difference in rate at the two temperatures.

43. The hydrolysis of thioacetamide is described by the following chemical equation.

$$CH_3CSNH_2 + H_2O \longrightarrow H_2S + CH_3CONH_2$$

The rate law for this hydrolysis is determined to be rate = $[H^+][CH_3CSNH_2]$. How does adding some solid NaOH to a solution that is 0.10 M in both hydrogen ion and thioacetamide affect the rate? the value of the rate constant?

44. Carbonic anhydrase is an enzyme that speeds the conversion of waste CO_2 to HCO_3^- in the body. The rate of the catalyzed reaction is 1×10^7 times faster than the uncatalyzed reaction. If 0.30 g of CO_2 can be converted in 1 min at 37°C using the appropriate amount of enzyme, how long would it take for the same amount of CO_2 (0.30 g) to be converted without the catalyst?

Linking chapters

1. ***Molarity***
Hydrogen peroxide solution is stored in dark bottles because light greatly increases the rate of decomposition.

$$H_2O_2(l) \longrightarrow H_2O(l) + O_2(g)$$

If a 500 mL bottle of 1.50 M H_2O_2 solution decomposes at a rate of 0.25 M/h in a clear bottle and 0.013 M/yr in a dark bottle, what is the difference in H_2O_2 concentration in the two bottles after 3 h?

2. ***Molarity and Stoichiometry***
Iodate ion converts iodide ions to iodine in acidic solution, as represented by the following reaction.

$$IO_3^-(aq) + 5I^-(aq) + 6H^+(aq) \longrightarrow 3I_2(s) + 3H_2O(l)$$

This reaction is too fast to time with a stopwatch, so sodium hydrogen sulfite is added to slow down the formation of iodine by competing with the iodide ions.

$$IO_3^-(aq) + 3HSO_3^-(aq) \longrightarrow I^-(aq) + 3SO_4^{2-}(aq) + 3H^+$$

Because HSO_3^- is a strong competitor, no I_2 can form until all the HSO_3^- ions have reacted. Still, the formation of I_2 goes undetected to the human eye because the solution is colorless. However, starch can be added to complex I_2 the instant it forms which causes the solution to turn blue-black.

$$I_2 + \text{starch} \longrightarrow \text{starch}{\cdot}I_2 \text{ complex}$$

The rate law for the reaction between iodate ion and hydrogen sulfite ion is as follows.

$$\text{rate} = k[IO_3^-]^p[HSO_3^-]^q$$

To run the iodine clock reaction, a student prepares the initial solutions of KIO_3 and $NaHSO_3$ by adding 0.788 g of KIO_3 to 150 mL of water and 0.392 g of $NaHSO_3$ to 150 mL of water. The reactants are mixed in the following proportions.

Trial	KIO_3 solution (mL)	Starch and $NaHSO_3$ solution (mL)	Water (mL)	Molarity of IO_3^- after mixing	Time (s)
1	5	8	12		115
2	7	8	10		85
3	9	8	8		67
4	11	8	6		50

a. Calculate the IO_3^- concentration for each trial.
b. What role(s) does the HSO_3^- ion play in this series of reactions? Why does its concentration remain constant?
c. What role does the starch play in this set of reactions? Why is its concentration not a factor affecting the rate?
d. Why does the amount of water added decrease as the amount of KIO_3 added increases? How would the rate be affected if the amount of water were held constant and the amount of $NaHSO_3$ were varied?
e. Explain why the reaction time decreases as the amount of KIO_3 solution increases.
f. Why do you think this series of reactions is called the "iodine clock" reaction?

3. *Weak acid equilibrium and rate*
The pollutant sulfur dioxide forms HSO_3^- ion when it dissolves in water droplets. The bisulfite ion is then oxidized to sulfate ion by oxygen which is also dissolved in the water droplets. One mechanism for this oxidation reaction is the following sequence of reactions.

$$2HSO_3^-(aq) + O_2(aq) \longrightarrow S_2O_7^{2-}(aq) + H_2O(l) \quad \textbf{fast}$$

$$S_2O_7^{2-}(aq) + H_2O(l) \longrightarrow 2SO_4^{2-}(aq) + 2H^+(aq) \quad \textbf{slow}$$

The following data were obtained in an experiment in which a 0.270 M aqueous solution of HSO_3^- ion was mixed with an aqueous solution of 0.0135 M O_2. The initial pH was 3.90.

Time (s)	$[HSO_3^-]$	$[O_2]$	$[S_2O_7^{2-}]$	$[SO_4^{2-}]$
0.000	0.270	0.0135	0.000	0.000
0.010	0.243	0.000	13.5×10^{-3}	0.000
10.0	0.243	0.000	11.8×10^{-3}	3.40×10^{-3}
45.0	0.243	0.000	7.42×10^{-3}	12.2×10^{-3}
90.0	0.243	0.000	4.08×10^{-3}	18.8×10^{-3}

a. Explain why the HSO_3^- ion concentration is constant after 0.010 s.
b. How many moles of $S_2O_7^{2-}$ have been consumed after 45 s? how many moles of SO_4^{2-} have been formed? Explain why these numbers are not the same.
c. Write a balanced equation for the conversion of the bisulfite ion to sulfate ion.
d. Write a balanced chemical equation for the formation of H_2SO_4 from SO_2 dissolved in airborne water droplets. (Hint: remember that a weak acid is in equilibrium with its ions, and that a strong acid is present as its ions.)

4. *Characteristic properties*
Sodium and silver react with water at drastically different rates. A 5 g sample of sodium completely disappears within a minute of being placed in 100 mL of water. Conversely, a 5 g sample of silver can be left in 100 mL of water for years with little detectable change. What explains this large difference in reaction rates?

5. *Particle collisions*
2-methylpropanol decomposes into 2-methyl propene and water.

$$(CH_3)_3COH(l) \longrightarrow (CH_3)_2C{=}CH_2(l) + H_2O(l)$$

For this decomposition to proceed at a perceptible rate, the temperature must be above 450°C. Use the particle collision theory to explain why 2-methylpropanol is stored at temperatures below 450°C.

6. Theme: *Equilibrium and change*
The following reaction has an equilibrium constant of 0.771 at 750°C.

$$H_2(g) + CO_2(g) \rightleftharpoons CO(g) + H_2O(g)$$

a. How do the rates of the forward and reverse reactions compare at equilibrium?
b. How do the concentrations of the products and reactants compare at equilibrium?

7. Theme: *Equilibrium and change*
a. Predict how the use of an enzyme would affect the reverse reaction of an equilibrium system.
b. What effect(s) does a catalyst have on the equilibrium constant for a given reaction? (Hint: how does a catalyst affect thermodynamic quantities such as ΔH?)

8. Theme: *Equilibrium and change*
Sketch a generalized concentration versus time graph for the formation of NO_2 as depicted in the following reaction.

$$N_2O_4 \rightleftharpoons 2NO_2$$

9. Theme: *Equilibrium and change*
The formation of nitrogen and water from nitrogen monoxide and hydrogen is believed to occur in three elementary steps. Each step is an equilibrium reaction as shown on the next page.

19. a.

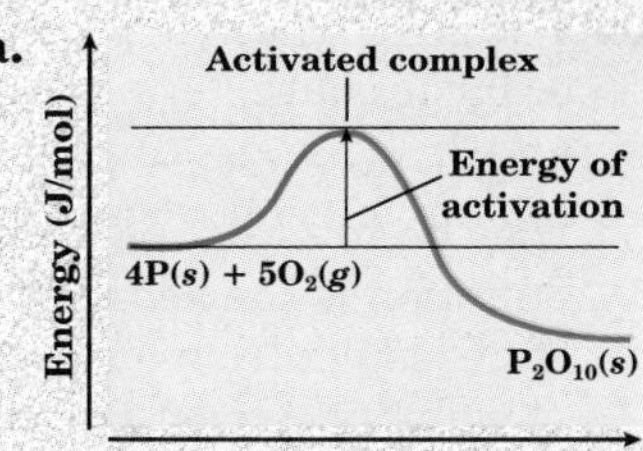

b. Also shown on the above curve

c. It could be stored under a vacuum or an inert liquid in which oxygen is insoluble, or it could be stored at a low temperature.

20. The normal oxidative processes that cause degradation of any organic product in the environment are taking place in the paper. The heat released by these processes is trapped within the stack due to poor air circulation. This heat is not rapidly dissipated from the center of a large stack, and the overall temperature in the paper stack increases. Once the combustion temperature is reached inside the stack, the paper bursts into flames.

21. No, the same amount of heat is liberated when the grain is burned whole as when it is burned as flour because the amount of heat released during a reaction is dependent on the substance burned and the amount of that substance present, not on the size of the particles that burn.

22. If one reactant's concentration is varied while all others are held constant, the effect of changes in concentration of that reactant on the reaction rate can be measured. If a plot of the log of the concentration versus time is a straight line, the rate is first order. The rate is second order if a plot of the inverse of the concentration versus time is a straight line.

23. Balanced equations do not necessarily indicate how the molecules interact, only that they *do* interact. A balanced equation is the *net* reaction that results from a series of elementary steps. The rate determining step is usually one of the elementary steps in the reaction and not the net reaction itself.

24. Reaction 1 is first order. Reaction 2 is second order.

25. **a.** first order
b. third order
c. second order

26. No. The slow step will determine the overall reaction rate regardless of position just as the progress of a group of rock climbers would be dependent on the speed of the slowest climber regardless of his/her position in the group.

27. An activated complex is a high-energy arrangement of atoms or molecules, the decomposition of which leads to a product or intermediate. An intermediate is a compound formed as the product of an elementary step. It is consumed as a reactant during a subsequent step and is not found in the net equation for the reaction. An intermediate can be a stable compound, unlike an activated complex, which is always unstable.

28.

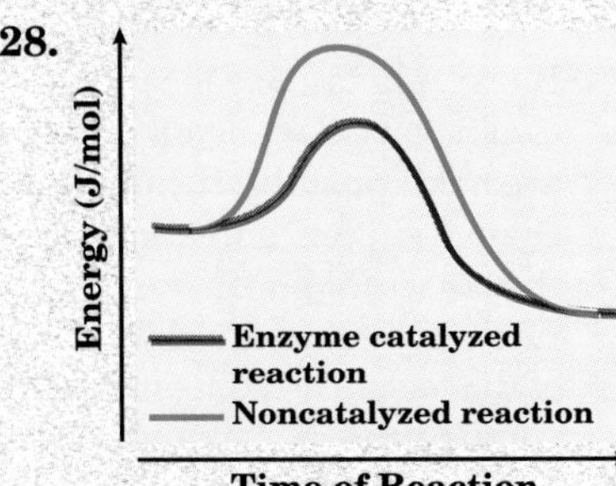

Step 1: $NO + NO \rightleftharpoons N_2O_2$

Step 2: $N_2O_2 + H_2 \rightleftharpoons N_2O + H_2O$

Step 3: $N_2O + H_2 \rightleftharpoons N_2 + H_2O$

When reactions are added, the K_{eq} of the overall reaction is equal to the product of the K_{eq} expressions for all the steps.

a. The K_{eq} expression for step 2 is as follows.

$$K_{eq} = \frac{[N_2O][H_2O]}{[N_2O_2][H_2]}$$

Explain the presence of water in the K_{eq} expression.

b. Write K_{eq} expressions for steps **1** and **3**, and determine the form of the K_{eq} expression for the overall reaction.

c. Write the balanced chemical equation for the overall reaction.

d. Why do concentration terms for the intermediates not appear in the K_{eq} expressions for the overall equation?

10. Theme: *Systems and interactions*
Iodine monochloride, ICl, a substance used to determine the iodine value of fats and oils, is a gas above 92°C. ICl reacts with hydrogen gas to form iodine gas and hydrogen chloride gas.

$$ICl(g) + H_2(g) \longrightarrow I_2(g) + 2HCl(g)$$

The rate law for this reaction is: rate = $k[ICl][H_2]$. At 230°C k has a value of 0.163 L/mol•s. Raising the temperature to 240°C increases the value of k to 0.348 L/mol•s.

a. If the initial concentration for both ICl and H_2 is 0.50 M, compare the rates at 230°C and 240°C.

b. Compare the rates at 230°C for reactant concentrations of 0.50 M and 0.25 M.

c. Which has a greater effect on the reaction rate, a 10°C increase in temperature or a doubling of the concentration?

USING TECHNOLOGY

1. ***Graphics calculator***
Verify the decomposition of NO_2 is second order with respect to $[NO_2]$ by plotting the following reaction data.

Time (s)	$[NO_2]$
0	1.00×10^{-2}
60	0.683×10^{-2}
120	0.518×10^{-2}
180	0.418×10^{-2}
240	0.350×10^{-2}
300	0.301×10^{-2}
360	0.264×10^{-2}

Begin by clearing lists L1 through L4. Press [STAT] [4] [2nd] [1] [ENTER]. Repeat this key sequence three more times except press [2], [3], or [4] instead of [1]. Now enter the data you will plot. L_1 will contain time data, L_2 concentration data, L_3 log concentration data, and L_4 the inverse concentration data. Press [STAT] [1]. Press [0], because this is the first number in the list of times, followed by [ENTER]. Continue this key sequence of number followed by [ENTER] until all data points are entered. Press [▶] and enter $[NO_2]$ data in L_2. Remember that "$\times 10^{-2}$" is entered as [2nd] [,] [(-)] [2]. Press [▶] to enter log $[NO_2]$ in L_3. Press [LOG] followed by a number in the $[NO_2]$ data list and [ENTER]. Press [▶] and enter data in L_4 by entering a number from the $[NO_2]$ data list followed by [x^{-1}] and [ENTER]. Set the limits for the range and domain, and scaling for the axes. Press [WINDOW] [▼] [0] [▼] [375] [▼] [10] [▼] [0] [▼] [0] [.] [0] [1] [▼] [0] [.] [0] [0] [0] [1]. To plot concentration vs. time press [2nd] [Y=] [1] [ENTER] [▼] [▶] [ENTER] [▼] [ENTER] [▼] [▶] [ENTER] [GRAPH]. Reset the range limits and y-axis scaling. Ymin = –3, Ymax = –2, and Yscl = 0.001. Plot log $[NO_2]$ vs time by pressing [2nd] [Y=] [2] [ENTER] [▼] [▶] [ENTER] [▼] [ENTER] [▼] [▶] [▶] [ENTER] [GRAPH]. Reset range limits and y-axis scaling to plot $1/[NO_2]$ vs. time. Select plot 3 at the STAT PLOTS screen and set parameters. Press [GRAPH].

2. ***Graphics Calculator***
A reaction involving a single reactant, A, has the rate data shown to the right. Determine the order of the reaction by graphing the data. Use the instructions in item **1** as a guide.

Time (min)	[A]
5.0	0.100
5.63	0.090
6.16	0.080
7.00	0.070

3. ***Graphics Calculator***
The decomposition of ethane, C_2H_6, is a first order reaction. Data for the reaction are to the right. From the plot of concentration versus time, estimate the rate at t = 750 s.

Time (s)	$[C_2H_6]$
0	0.01000
200	0.00916
400	0.00839
600	0.00768
800	0.00703

4. ***Graphics Calculator***
Nitrogen(II) oxide reacts with hydrogen at 800°C to form nitrogen gas and water as in the following equation.

$$2NO(g) + 2H_2(g) \longrightarrow N_2(g) + 2H_2O(g)$$

Using the data in the table to the right, graphically determine the order of reaction with respect to both H_2 and NO, and write the rate law.

Trial	[NO]	[H_2]	Rate
1	0.0010	0.0040	0.12
2	0.0020	0.0040	0.48
3	0.0030	0.0040	1.08
4	0.0040	0.0010	0.48
5	0.0040	0.0020	0.96
6	0.0040	0.0030	1.44

5. ***Computer Program***
   a. Determine the rate for the reaction in item **4** if the initial concentration of NO is 0.0024 M and the concentration of H_2 is 0.0042 M. (Hint: the value of the rate constant is constant for a given temperature.)
   b. Write a program that would allow you to determine the rate for any given set of concentrations.

Alternative assessment

Performance assessment

1. Boilers are used to heat large buildings. Deposits of $CaCO_3$, $MgCO_3$, and $FeCO_3$ can hinder the boiler operation, and aqueous solutions of HCl are commonly used to remove these deposits. The general equation for the reaction is written as follows.

$$XCO_3(s) + HCl(aq) \longrightarrow XCl_2(aq) + H_2O(l) + CO_2(g)$$

X stands for Ca, Mg, or Fe. Design an experiment to determine the effect of various HCl concentrations on the rates of this reaction. Present your design to a panel group.

2. Fats and oils decompose into fatty acids and glycerin, which cause the spoilage of many foods. Antioxidants are added to fats and oils to inhibit decomposition.
   To determine the extent of decomposition of a fat or oil, dissolve the sample in ethanol and titrate with an NaOH solution, using phenolphthalein as an indicator. The greater the volume of NaOH solution required for the solution to turn pink, the greater the concentration of free fatty acids, and the greater the extent of decomposition.
   Design an experiment to study the effects of temperature and an inhibitor on the decomposition rates of samples of fat or vegetable oil. Present your design in the form of a proposal to your teacher.

Portfolio projects

1. ***Research and communication***
Catalysts play a major role in many industrial processes. Research one particular catalyst and the industrial process that uses it. Try to determine how much of a difference the catalyst makes in the overall economics of the industrial process. Also research the environmental and occupational hazards associated with using the catalyst.
2. ***Cooperative activity***
Have a class discussion about the use of inhibitors in food products. Each student should contribute important facts to the discussion by studying content labels and researching any substance listed as a preservative, spoilage inhibitor, or antioxidant. Health-related journals in the library are a good source of information on any substance found in a commercial food product. Focus the discussion on whether sufficient research was done on the substance before it received FDA approval. Also note any instances in which hazards associated with the use of a substance are not properly emphasized on the product label.
3. ***Research and communication***
Biochemistry is a complex field of study. A high percentage of the chemical reactions that take place in living organisms are catalyzed by specific enzymes. Choose a biochemical process such as respiration, photosynthesis, vision, or muscular contraction. Prepare a diagram of this process, and detail all of the chemical reactions that take place and the enzymes associated with each reaction.

29. Both are catalysts that lower the energy of activation for a reaction without being consumed. However, enzymes are proteins used in biological systems and are organic in nature, but an inorganic heterogeneous catalyst catalyzes reactions in nonliving systems, contains no carbon, and is generally a solid.

30. The free enzyme binds to the reactant (substrate). The area to which the substrate binds is stereospecific and will allow only a select substrate(s) to "fit." An active enzyme-substrate complex forms in which the substrate bonds are broken and the bonds associated with the product are formed. When the product is formed, the enzyme releases it and is restored.

31. Inhibitors slow down the rate of a chemical reaction through competing reactions.

32. The rate increases by 3^2, or 9, times when [A] is tripled and decreases by $(1/2)^2$, or 1/4, when [A] is halved.

33. If [A] is doubled, the rate doubles. If [B] is halved, the rate decreases to one fourth of its original value.

34. a. $\frac{1}{s}$

b. $\frac{L^2}{mol^2{\cdot}s}$

c. $\frac{L}{mol{\cdot}s}$

Answers from page 544

7. a.

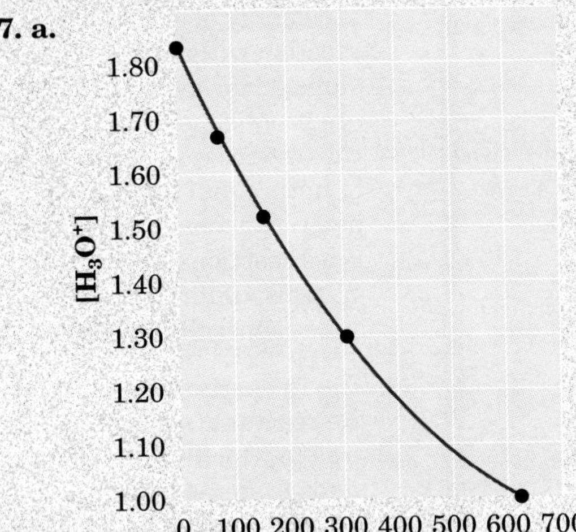

b. $\text{rate} = -\frac{(1.67 - 1.83)\ M}{(79 - 0)\ min} = 2.02 \times 10^{-3}\ M/min$

$\text{rate} = -\frac{(1.52 - 1.67)\ M}{(158 - 79)\ min} = 1.89 \times 10^{-3}\ M/min$

$\text{rate} = -\frac{(1.30 - 1.52) M}{(316 - 158)\ min} = 1.39 \times 10^{-3}\ M/min;$

$\text{rate} = -\frac{(1.00 - 1.30) M}{(632 - 316)\ min} = 9.49 \times 10^{-4}\ M/min$

c. No, less reactant is used up as time passes. The first two values are very close because they are relatively close to the beginning of the reaction.

8. a. $\text{rate} = \frac{-1}{4}\frac{\Delta PH_3}{\Delta t} = \frac{1}{6}\frac{\Delta H_2}{\Delta t} = \frac{\Delta P_4}{\Delta t}$

b. $\text{rate} = -\frac{\Delta[S^{2-}]}{\Delta t} = -\frac{\Delta[H_2O]}{\Delta t} = \frac{\Delta[HS^-]}{\Delta t} = \frac{\Delta[OH^-]}{\Delta t}$

c. $\text{rate} = -\frac{\Delta[C_2H_4]}{\Delta t} = -\frac{\Delta[Br_2]}{\Delta t} = \frac{\Delta[C_2H_4Br_2]}{\Delta t}$

9. a. $\frac{-1}{4}\frac{\Delta[NH_3]}{\Delta t} = -\frac{1}{5}\frac{\Delta[O_2]}{\Delta t}$

$\frac{\Delta[O_2]}{\Delta t} = \frac{5}{4}\left(\frac{0.80\ M}{1\ min}\right) = \frac{1.0\ M}{min}$

b. $\frac{\Delta[NO]}{\Delta t} = 4\left(\frac{1}{4}\ \frac{0.80\ M}{min}\right) = \frac{0.80\ M}{min}$

$\frac{\Delta[H_2O]}{\Delta t} = 6\left(\frac{1}{4}\ \frac{0.80\ M}{min}\right) = \frac{1.2\ M}{min}$

Answer from page 554

18. the nature of the reactants

19. to increase the exposure of the meat's surface to the grill's heat

20. If the lower temperature is adequate for preparation of the food, then 20 min at the lower temperature allows the heat more time to penetrate to the center of the food without burning the outer surface.

Answer from page 568

7.

Reaction	Column A Change in reactant or product	Column B Substance measured	Column C Type of measurement
1	product	carbon dioxide	gas collection
2	product	HBr	gas collection
3	reactant OR product	N_2O_4 NO_2	change in volume of liquid gas collection
4	reactants	H^+, OH^- ions	pH change

Continued from page 573

35. If a rate is to be measured, the type of change measured must be identified, and a method for quantifying it must be established. Once there is a clear method to determine change in rate with respect to each reactant, the concentration of one is held steady while the other is varied. The process is repeated for the second reactant. The data is examined. The reactants whose change in concentration resulted in a change in rate are used in the rate law. The value of their order is determined by the mathematical relationship of the change in concentration of the reactant and the observed change in rate.

36. Rate laws are determined experimentally and are dependent on the slowest step in the reaction mechanism. The coefficient does not correspond to the order of the reaction with respect to each reactant.

37. a. rate = $k[NO_2]^2[CO]^0$ or $k[NO_2]^2$

b. second order; zero order

c. step 1 $NO_2(g) + NO_2(g) \longrightarrow NO_3(g) + NO(g)$ slow

step 2 $NO_3(g) + CO(g) \longrightarrow NO_2(g) + CO_2(g)$ fast

net $NO_2(g) + CO(g) \longrightarrow NO(g) + CO_2(g)$

38. The rate of release of oxygen is extremely fast in comparison to the release of the carbon monoxide. Since the blood contains a finite level of hemoglobin, binding of carbon monoxide ties up the heme so that oxygen can't be carried to the cells for respiration. With enough hemoglobin tied up for hours the patient can effectively die from oxygen starvation before the carbon monoxide has dissociated from the heme.

39. Cholinesterase is an enzyme. Soman may bind at an active site on cholinesterase and prevent the normal substrate from binding. Soman might bind to cholinesterase, but not at the active site, and cause a deformation in the molecule's shape so that the active site is no longer available for binding with the normal substrate.

40. a. 60 min
b. 360 min

41. a. first order
b. second order

42. a. rate = $6.1 \times 10^{-4}\ s^{-1}$ [△]
rate = $6.1 \times 10^{-4}\ s^{-1}\ [2.0 \times 10^{-3}] = 1.22 \times 10^{-6}\ M\ s^{-1}$
b. rate = $5.70 \times 10^{-4}\ s^{-1}$ [△]
c. The rate law expressions are identical except for the value of the rate constant. Reaction rate changes with temperature because the rate constant changes and not because its dependence on concentration changes.

43. The OH^- ions will react with the H^+ ions to form water. The lowered $[H^+]$ will decrease the reaction rate. The rate constant is not affected.

44. rate of the catalyzed reaction $= \frac{0.30\ g}{1\ min} =$

(1×10^7)(rate of the uncatalyzed reaction)

rate of the uncatalyzed reaction $= \frac{1}{1 \times 10^7} \times \frac{0.30\ g}{1\ min} = \frac{3.0 \times 10^{-8}\ g}{1\ min}$

time to convert 0.30g CO_2 = $0.30\ g \times \frac{1\ min}{3.0 \times 10^{-8}\ g} = 1.0 \times 10^7\ min$

Linking chapters

1. For the clear bottle:
1.5 M − 3 hr (0.25 M/hr) = 0.75 M

For the dark bottle:
1.5 M − 3 hr (1 yr/8760 hrs) (0.013 M/yr) = 1.5 M

2. a. mol of $KIO_3 = \frac{0.788\ g\ KIO_3}{214\ g/mol} = 3.68 \times 10^{-3}$

initial $[IO_3^-] = \frac{3.68 \times 10^{-3}\ mol}{0.150\ L} = 0.0245\ M$

trial 1 $[IO_3^-] = \frac{0.0245\ M \times 0.005\ L}{0.025\ L} = 0.0049\ M$

trial 2 $[IO_3^-] = \frac{0.0245\ M \times 0.007\ L}{0.025\ L} = 0.0069M$

trial 3 $[IO_3^-] = \frac{0.0245\ M \times 0.009\ L}{0.025\ L} = 0.0088\ M$

trial 4 $[IO_3^-] = \frac{0.0245\ M \times 0.011\ L}{0.025\ L} = 0.0108\ M$

b. HSO_3^- inhibits I_2 formation and is the limiting reagent for the reaction in which SO_4^{2-} is formed. Its concentration remains constant so that the relationship between IO_3^- ion and the rate can be calculated.
c. Starch is a visual indicator for the formation of I_2. Starch does not participate in the reaction, so its concentration does not affect the rate.
d. The amount of water used is varied so that the volume of solution stays constant and the amount of increase in $[IO_3^-]$ is constant. Because the concentration of the HSO_3^- ion appears in the rate law, variations in its concentration make the rate dependent on both $[IO_3^-]$ and $[HSO_3^-]$. Therefore, the rate is affected in accordance with the value of the exponent on $[HSO_3^-]$, and the relationship between the rate and $[IO_3^-]$ cannot be determined.
e. The number of particles available for collision increases.
f. The appearance of the iodine is timed by the amount of hydrogen sulfite ion present. Accept all reasonable answers.

3. a. Oxygen, the limiting reagent, has been depleted.
b. mol of $S_2O_7^{2-}$ consumed =
$(13.5 \times 10^{-3}) - (7.42 \times 10^{-3}) = 6.08 \times 10^{-3}$
mol of SO_4^{2-} formed = 12.2×10^{-3}
The numbers are not the same because the mole ratio of $S_2O_7^{2-}$ to SO_4^{2-} in the balanced chemical equation is 1:2.
c. Add the two equations given.
$2HSO_3^-(aq) + O_2(aq) \longrightarrow 2SO_4^{2-}(aq) + 2H^+(aq)$
d. Recall that sulfuric acid is a strong acid, H_2SO_3 is a weak acid, and the addition of a nonmetal oxide to water gives an acid.
step 1 $2(SO_2 + H_2O \longrightarrow H_2SO_3)$
step 2 $2(H_2SO_3 \rightleftharpoons H^+ + HSO_3^-)$
step 3 $2HSO_3^- + O_2 \longrightarrow S_2O_7^{2-} + H_2O$
step 4 $S_2O_7^{2-} + H_2O \longrightarrow 2SO_4^{2-} + 2H^+$
net $2SO_2 + 2H_2O + O_2 \longrightarrow 2H_2SO_4$

4. Sodium and silver have different chemical natures due to the different electron configurations of the atoms.

5. At temperatures below 450°C, particles have lower kinetic energies, so fewer collisions between particles have sufficient energy to overcome the E_a barrier.

6. a. The rates of the forward and reverse reactions are the same at equilibrium.
b. The reactants are favored in this reaction because the equilibrium constant is less than 1. Mathematically, the product of the product concentrations is less than the product of the reactant concentrations.

7. a. An enzyme should affect the E_a of the reverse reaction in the same manner that it affects the E_a of the forward reaction.
b. A catalyst has no affect on the value of the equilibrium constant.

8.

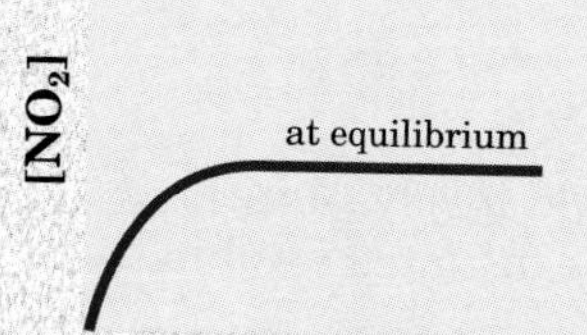

9. a. This is a gas phase reaction. The amount of water present depends on the extent of the reaction because water is only a product of the reaction rather than the solvent, as in aqueous solutions, so it is not present in great excess.

b. step 1 $K_{eq} = \frac{[N_2O_2]}{[NO]^2}$

step 3 $K_{eq} = \frac{[N_2][H_2O]}{[N_2O][H_2]}$

overall $K_{eq} = \frac{[N_2O_2]}{[NO]^2} \times \frac{[N_2O][H_2O]}{[N_2O_2][H_2]} \times \frac{[N_2][H_2O]}{[N_2O][H_2]} = \frac{[N_2][H_2O]^2}{[NO]^2[H_2]^2}$

c. $2NO + 2H_2 \longrightarrow N_2 + 2H_2O$
d. They are divided out.

10. a. $\text{rate}_1 = 0.163$ L/mol•s [0.5 mol/L][0.5 mol/L] $= 0.041$ M/s

$\text{rate}_2 = 0.348$ L/mol•s [0.5 mol/L][0.5 mol/L] $= 0.087$ M/s

b. $\text{rate}_3 = 0.163$ L/mol•s [0.50 mol/L][0.25 mol/L] $= 0.020$ M/s

c. Doubling the concentration quadruples the rate. A 10°C increase in temperature doubles the rate.

USING TECHNOLOGY

1. Students enter sets of data and plot the graphs described in **Figure 14-19** on page 557.
2. The reaction is second order. A straight line should be obtained when 1/[A] is plotted against time.
3. The rate at $t = 750$ s is approximately 9.59×10^{-6} M/s.
4. This problem compares the initial rates of several trials instead of the change in rate during one trial. Two sets of graphs are needed: log[NO] vs. log rate when $[H_2]$ is constant and log $[H_2]$ vs. log rate when [NO] is constant. The slope is the order of reaction.

   $\text{rate} = k[\text{NO}]^p[H_2]^q$

   Because $[H_2]$ is held constant for trials 1–3, the rate is expressed in terms of [NO] only.

   $\text{rate} = k'[\text{NO}]^p$

   Use logarithms to find the value of the exponent.

   $\log \text{rate} = \log k' + p(\log[\text{NO}])$

   The rate is first order in $[H_2]$ and second order in [NO]. The rate law is rate $= [H_2][\text{NO}]^2$.

5. **a.** Solve for k using data from the table in item **4** and the rate law. Then substitute k and the given concentrations into the rate law. Or use a ratio involving a set of data from the table in item **4** and the given concentrations. The example below is for trial 1.

   $$\frac{\text{rate}}{0.12} = \left(\frac{0.0024}{0.0010}\right)^2 \times \frac{0.0042}{0.0040}$$

   $\text{rate} = 0.12 \times (2.4)^2 \times (1.05) = 0.73$ M/s

   b. Answers will vary but should involve a ratio as in part **a**, or the rate law with the calculated k value.

Performance assessment

1. Generally, students will measure how fast the compounds disappear. Design criteria should include provisions for the removal of deposits that consist of mixed carbonates as well as single carbonates.
2. Students should specify sample preparation procedures, explain the significance of the titration procedure, and establish criteria for reporting and presenting their results.

Portfolio projects

1. Students may need help in finding more information on catalysts.
2. In doing their research students may require assistance finding health-related journals.
3. Look for minute details in the student diagrams.

15

PLANNING FOR

Electro-chemistry

Chapter Overview

- **Section 15-1** details the operation of a voltaic cell. Students look at a set of voltaic cell illustrations that highlight the individual cell parts or processes. The processes of oxidation and reduction are separately defined.
- **Section 15-2** focuses on the development and use of the table of standard reduction potentials, and its correlation to the activity series.
- **Section 15-3** covers oxidation numbers and their importance to redox reactions. Students look at How to boxes to determine oxidation numbers for individual atoms in compounds and ions, and to balance redox reactions by the half-reaction method.
- **Section 15-4** briefly discusses examples of rechargeable and nonrechargeable batteries as practical applications of electrochemical cells and redox reactions.

Concept Base

Students must have mastered the following concepts prior to this chapter: electronegativity and periodic trends (Chapter 4); formation and characteristics of ions (Chapter 5); conductivity in metals and salt solutions (Chapter 5); covalent bonding (Chapter 6); balancing equations, replacement reactions, and activity series (Chapter 7).

Laboratory Equipment Needs

Demonstrations

The listing of materials for Chapter 15 demonstrations begins on page T46.

Laboratory Experiments

Materials and preparation instructions for Chapter 15 are found as follows.

Text		*Page*
Exploration 15A	Redox Titration	780
Investigation 15A	Redox Titration—Mining Feasibility Study	784

Laboratory Experiments Manual
(refer to the table of contents in the manual)

Exploration 15-1	Electroplating for Corrosion Protection
Exploration 15-2	Voltaic Cells
Investigation 15-2	Voltaic Cells—Designing Batteries

Pacing: 10 days — Inquiry Teaching Strategies

15-1 How does electric current flow?

Demonstrations
- Demo 1 *Zinc in copper sulfate*

Laboratory Experiments
- Exp 15A *Redox Titration*
- Inv 15A *Redox Titration—Mining Feasibility Study*

15-2 How do you get current to flow?

Demonstrations
- Demo 2 *Predicting reactions*
- Demo 3 *Comparing voltages*

15-3 How is current flow described?

Demonstrations
- Demo 4 *Three oxidation states of manganese*

15-4 How do batteries work?

Demonstrations
- Demo 5 *Comparing flashlight batteries*
- Demo 6 *Voltage from a 6-cell fruit battery*

Laboratory Experiments
- Exp 15-1 *Electroplating for Corrosion Protection*
- Exp 15-2 *Voltaic Cells*
- Inv 15-2 *Voltaic Cells—Designing Batteries*

Key
- Teaching Transparencies
- Annotated Teacher's Edition
- Laboratory Experiments Manual

INSTRUCTION

Visual Teaching Strategies / Review and Practice Strategies

Visual Teaching Strategies	Review and Practice Strategies
Transparencies ■ F 15-1 *Particle Model for a Redox Reaction* ■ F 15-2 *Ion Movement Through a Porous Barrier* ■ F 15-3 *Electron Pathway in an Electrochemical Cell* ■ F 15-5 *Particle Models for Redox Reactions in an Electrochemical Cell*	**Text Reviews** ■ Section Review 1–5 ■ Chapter Review 1–11; UT 2 **Study Guide Worksheets** ■ ■ 15-1 Concept Review ■ ■ 15-1 Skillsheet *Determining Oxidation and Reduction* *Interpreting Half-reactions*
Transparencies ■ F 15-7 *Particle Models for the Aluminum-Copper Voltaic Cell* ■ F 15-8 *Comparing the Zn-Cu and Al-Cu Voltaic Cells* ■ F 15-9 *Comparing Reduction Potentials of Various Metals* ■ T 15-1 *Activity Series and Reduction Potentials at 25°C*	**Text Reviews** ■ Section Review 6–10 ■ Chapter Review 12–27; LC 1, 2; UT 1, 3, 4 **Study Guide Worksheets** ■ ■ 15-2 Concept Review ■ ■ 15-2 Skillsheet *Predicting* E^0 *cell* *Predicting Redox Spontaneity*
Transparencies ■ P 603 *Assigning Oxidation Numbers*	**Text Reviews** ■ Additional Examples 15A, 15B ■ Practice 1–4 ■ Concept Review 11–12 ■ Section Review 13–15 ■ Chapter Review 28–46; LC 4, 5, 6 **Study Guide Worksheets** ■ ■ 5-3 Concept Review ■ ■ 15-3 Practice *Assigning Oxidation Numbers* ■ ■ 15-3 Skillsheet *Balancing Redox Reactions*
Transparencies ■ F 15-15 *Cross Section of a Carbon-Zinc Dry Cell* ■ F 15-16 *Cross Section of an Alkaline Dry Cell* ■ F 15-17 *Cross Section of a Mercury Cell* ■ F 15-18 *Cross Section of a Lead-Acid Battery* **Transparency Masters** ■ F 15-4 *Relative Longevity of Batteries*	**Text Reviews** ■ Section Review 16–21 ■ Chapter Review 47–63; LC 3 **Study Guide Worksheets** ■ ■ 15-4 Concept Review

■ Pupil's Text
■ Study Guide
■ Teaching Resources

Exp *Exploration*
Inv *Investigation*

LC *Linking Chapters*
UT *Using Technology*

Assessment Options

Traditional Assessment

Test Generator
Instructional Objectives Measured:
Content mastery
Select from items 1 to 75

Performance Assessment Options

(see pages T36 and T643-15 for scoring rubrics)
Students demonstrate mastery of objectives in a hands-on environment. You may want some of these materials in the Portfolio as well.

Investigation 15A (text)
Investigation 15-2 (laboratory manual)
Instructional Objectives Measured:
Content mastery, Use of scientific methodology, Problem solving skills, Proficiency in written communication

Performance Assessments, page 609

1. **Designing an electrochemical cell**
   Instructional Objectives Measured:
   Content mastery, Problem solving skills, Use of scientific methodology
2. **Determining E^0 for an unknown metal**
   Instructional Objectives Measured:
   Problem solving skills, Use of scientific methodology, Proficiency in written communication
3. **Determining the usefulness of a sodium-sulfur battery**
   Instructional Objectives Measured:
   Problem solving skills, Scientific and chemical literacy, Proficiency in oral communication

Portfolio Options

(see page T36 for scoring rubrics)
Students provide a written rationale for each selection made for the portfolio.

Concept Maps
Students use vocabulary from the Chapter Review to build a concept map.
Instructional Objectives Measured:
Content mastery

Formal Laboratory Reports
Exploration 15A (text)
Explorations 15-1, 15-2 (lab manual)
Instructional Objectives Measured:
Content mastery, Use of scientific methodology, Proficiency in written communication

Portfolio Projects, page 609
Items 1 and 4. Chemistry and you
Instructional Objectives Measured:
Content mastery, Scientific and chemical literacy, Proficiency in written communication

Items 2 and 5. Research and communication
Instructional Objectives Measured:
Content mastery, Scientific and chemical literacy, Proficiency in oral and written communication

Item 3. Cooperative activity
Instructional Objectives Measured:
Content mastery, Scientific and chemical literacy, Proficiency in oral and written communication

INTRODUCING THE CHAPTER

CHAPTER 15 Electrochemistry

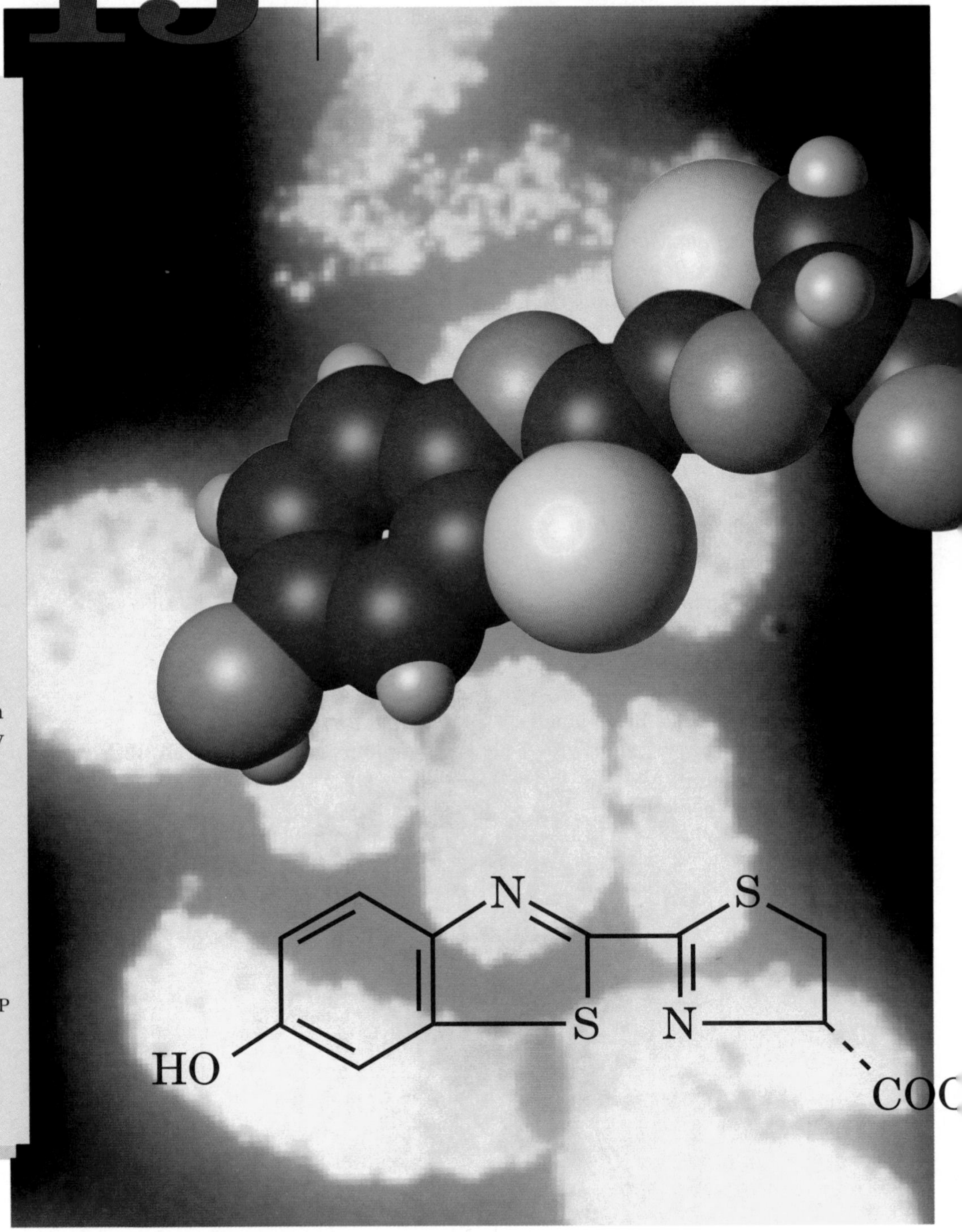

Story Background

The reaction mechanism for light production is not the same in all light-producing organisms, but in all cases it involves electron transfer and production of an excited state in a molecule that, upon returning to the ground state, emits light. In the mechanistic cycle shown below, luciferin supplies electrons to ATP as it becomes luciferyl adenylate, which then supplies electrons to molecular oxygen in the presence of the enzyme luciferase to produce CO_2, AMP, and oxyluciferin. Electrons in the oxyluciferin are in an excited state, having absorbed most of the reaction energy. Light is emitted as these electrons return to ground state. Oxyluciferin now serves as an electron acceptor and eventually is converted back to luciferin.

In the TB cultures, the TB cells themselves supply the necessary ATP, an index of bacterial liveliness. Pure luciferin and luciferase can be used in the laboratory to measure quantities of ATP as small as 10^{-12} mol by the intensity of the light flash produced.

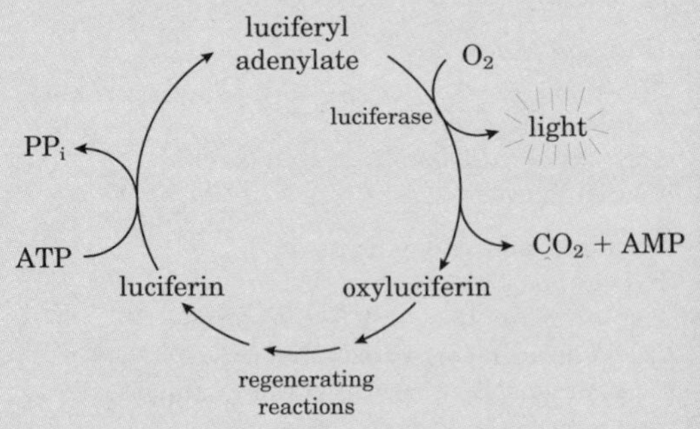

About the Illustration

Cultured tuberculosis cells glow after being treated with firefly luciferin.

Fireflies and Electrochemical Cells

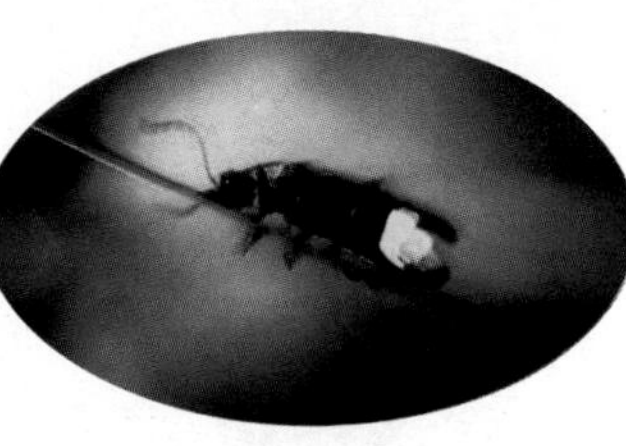

Drs. William Jacobs and Barry Bloom inject an antibiotic into a petri dish that contains glowing tuberculosis (TB) cells. They wait and watch, hoping the glow will dim and go out.

The glow is produced by the oxidation of firefly luciferin, the chemical compound responsible for light emission in the tails of fireflies. Jacobs and Bloom have placed the luciferin in a bacteria that will infect tuberculosis (TB) cells. Conditions inside active TB cells cause the luciferin to emit light or glow. The brightness of the glow is a measure of the bacteria's growth. If the glow brightens, the drug injected by Jacobs and Bloom does not kill this TB strain, and another drug must be tried. But, if the glow dims and goes out, the drug can be used to fight this TB strain.

Tuberculosis, a bacterial infection that usually affects the lungs, can be a life-threatening disease. Moreover, it is no longer easily treated because it has become more and more resistant to antibiotics. The spread of drug-resistant tuberculosis in major U.S. cities, where the death rate from this disease approaches 50%, has raised serious concerns.

One of the problems doctors face in the fight against TB is the time needed to determine if a person actually has it. Isolating and growing the bacteria responsible for tuberculosis can take from three to eight weeks. Once TB is diagnosed, another five or more weeks are needed to determine which antibiotic would be most effective. Knowing which drug is effective against a specific patient's TB strain is important to doctors because treating the disease with an ineffective drug can worsen the patient's condition. Screening possible antibiotics using this new luciferin test takes only a few days which substantially reduces the five-week waiting period before treatment can begin. In this way, the lights of fireflies, described by Smokey Mountain residents as breathtaking, may be described as breathgiving by persons with tuberculosis.

Emission of light by living organisms through the use of a chemical reaction is known as bioluminescence. Bioluminescence involves redox reactions. In this chapter, you will explore redox reactions as the basis for electrochemical cells. As a result, you will be able to answer the following questions.

- *How is the light generated by a firefly different from the light generated by a flashlight battery?*
- *How is the concentration of TB cells related to the brightness of the light emitted?*

Answering these questions requires some knowledge of how electricity is produced from chemical energy. Electrochemistry, the topic you will study in this chapter, is concerned with the interconversion of chemical and electric energy.

Tapping Prior Knowledge

Prepare students for the concepts they will need for this chapter with the following discussion topics.

1. Metals conduct electric current, as do some solutions. Do these solutions conduct electricity in the same way as metals? Why do some drivers in northern climates often take their car batteries inside on very cold nights?
2. Tarnished silver may be cleaned by lining a pan with aluminum foil, laying the tarnished piece of silver in the pan, covering with water, and heating. What reactions are taking place? Why do you think this cleaning method works?
3. How can you tell that a nail has rusted? Is rust confined to the surface of a metal? How does rust affect the performance of the nail? If rust is a metal oxide, how might rust be formed? How could you predict which of the following nails would rust first: one made of iron, one made of aluminum, or one made of stainless steel?
4. Give some examples of bonding, and describe what it means for something to be bonded. Are all bonds the same? Is a bond necessarily contributed to equally by all parties or objects involved?
5. What are batteries made of, and how do they operate? Why is one battery marked 1.5 V and another battery marked 9 V? What is the purpose of H_2SO_4 in a car battery?
6. What effects might discarded batteries have on the environment? Should you worry about throwing away one battery? If a child swallowed a mercury-zinc battery, would mercury poisoning be your first concern?

Section 15-1

FOCUS

Lesson Starter

You may wish to open this lesson with the following experiment.

Demonstration 1

Zinc in copper sulfate

Approximate time: 5 min

1. Fill a petri dish half full with 0.5 M $CuSO_4$.
2. Place it on an overhead projector.
3. To the petri dish, add several pieces of mossy Zn or Zn sheet (1 cm × 1 cm).
4. Cu should plate out as the Zn pieces disappear, and the solution should become less intensely blue, indicating the removal of Cu^{2+} ions from solution.

SAFETY

Wear safety goggles and a lab apron.

DISPOSAL

Separate the solids from the supernatant liquid, and rinse the solids with water. Combine rinsings with the supernate, and add 1 M NaOH to precipitate any dissolved copper as the hydroxide. Filter to separate $Cu(OH)_2$ from the filtrate. Put $Cu(OH)_2$ in the trash. Neutralize the filtrate with 1 M HCl, and pour down the drain.

Add approximately 500 mL of saturated NaCl solution to the rinsed solids, and slowly bubble air into this mixture for several hours until all of the Cu has dissolved. The solution will be green. Some of the Zn may not have dissolved. The now uncoated Zn may be saved for next time or discarded in the trash.

How does electric current flow?

Section Objectives

Identify the path taken by electrons in an electrochemical cell.

Describe how current flows in a voltaic cell.

Distinguish between oxidation and reduction.

Energy is released when electrons are transferred

Figure 15-1 shows a strip of zinc metal standing in an aqueous solution of copper(II) sulfate. What is the evidence that a chemical reaction has occurred? Copper metal has fallen out of solution into a pile on the bottom of the beaker. The solution has changed from a dark blue to a very light blue which suggests the loss of Cu^{2+} ions. Part of the zinc strip has been eaten away which suggests that the zinc atoms have become ions. These observations and analysis of the solution indicate a single displacement reaction with the following equation.

$$Zn(s) + CuSO_4(aq) \longrightarrow Cu(s) + ZnSO_4(aq)$$

This reaction involves the replacement of a Cu^{2+} ion by a Zn atom. If you check a periodic table, you will see that Zn has two valence electrons. For a Zn atom to replace a Cu^{2+} ion in solution, it must become an ion by transferring its two valence electrons to the Cu^{2+} ion which, in turn, becomes a solid. In this way, the system achieves stability.

Heat energy is released when redox reactions occur

Look at the solution temperature recorded in **Figure 15-1**. As electrons are exchanged between a Zn atom and a Cu^{2+} ion, energy is released in the form of heat causing the temperature to increase. As the reaction slows to a stop, less heat is produced, and the solution cools by transferring its heat to the surroundings. When the temperature of the solution equals that of the surroundings, no more heat flows. The temperature of the solution becomes constant. But what happens when the same two reactants are arranged without touching each other? Will heat still be produced?

Figure 15-1
A strip of Zn metal is suspended in an aqueous solution of $CuSO_4$. Heat energy given off when electrons are donated by Zn atoms to Cu^{2+} ions causes the solution temperature to rise.

Visual Strategy

Figure 15-1

Describe the electron transfer process taking place in the inset. Point out the increase in solution temperature. Remind students that energy must be conserved, and ask them to identify and explain what happens to the temperature of the solution if the heat energy is converted into some other form of energy.

porous barrier
any medium through which ions can slowly pass

electricity
energy associated with electrons that have moved from one place to another

Separation of reactants produces electric energy instead of heat

In **Figure 15-2**, the **porous barrier** prevents Zn atoms and Cu^{2+} ions from mixing with one another, so a Zn atom cannot directly transfer its two valence electrons to a Cu^{2+} ion. However, electrons can be transferred from Zn atoms to Cu^{2+} ions through the external connecting wire. Electrons flowing through a wire generate energy in the form of **electricity**. The temperature increases only slightly as the reaction progressed, so little energy was produced as heat. When the reactants are arranged so they do not touch each other, energy is in the form of electricity rather than heat. How can you tell that electric current is being generated in **Figure 15-2**?

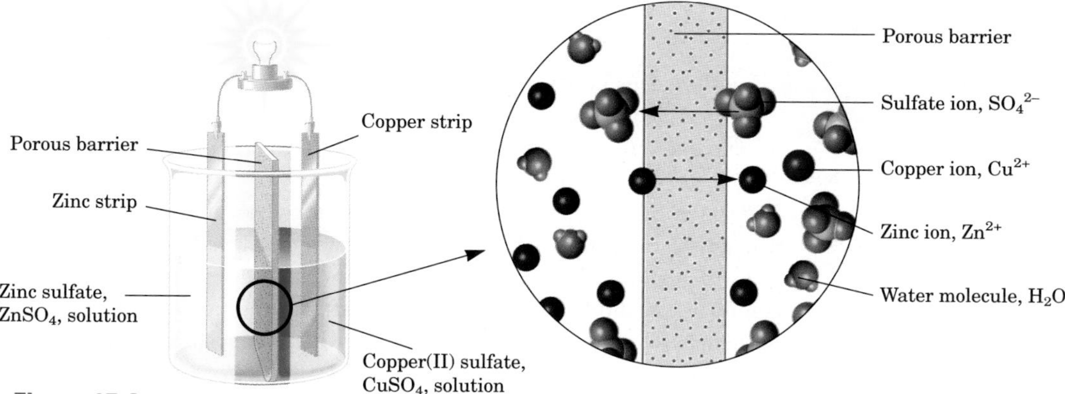

Figure 15-2
In electrochemical cells, the porous barrier keeps reagents separated. Ions flow through the barrier as the current inside the cell. Electrons flow in the external circuit.

Electricity is produced by electron transfer in voltaic cells

Electrons given up by Zn atoms pass along the wire, through the filament in the bulb, and cause it to light. The light indicates that chemical energy present in the reactants has been converted into electric energy. This conversion takes place in an *electrochemical cell.* An electrochemical cell is a container in which chemical reactions produce electricity or an electric current produces chemical change. When an electrochemical cell produces electricity, it is also known as a **voltaic cell**. The branch of chemistry concerned with the interconversion of chemical energy with electric energy is *electrochemistry*.

Separation of electrodes in a voltaic cell ensures that the major energy output is electrical. The Cu and Zn electrodes in **Figure 15-2** are immersed in sulfate solutions of their respective ions. These solutions are separated by a porous barrier that allows ions to pass through it, but prevents physical mixing of the solutions. Cu^{2+} ions gain two electrons at the surface of the Cu strip where they are deposited as Cu atoms. The removal of Cu^{2+} ions from solution reduces the positive ion concentration on the copper side of the barrier.

voltaic cell
an electrochemical cell in which a spontaneous redox reaction produces a flow of electrons through an external circuit

Add 1 M NaOH to precipitate the dissolved Cu and Zn as hydroxides. Filter to separate the hydroxides, which may be put in the trash. Neutralize the filtrate with 1 M HCl, and pour down the drain.

TEACH

Chapter Connections

Electric energy is free energy, ΔG, and because $T\Delta S$ is usually very small, electric energy is also approximately equal to the enthalpy of the system, ΔH.

Theme

Conservation
Electrons are conserved as they move from one metal to another.

Historical Note

In 1800, Alessandro Volta, an Italian physicist, discovered that two separated metals in an electrolytic solution could produce a new charge as fast as the old charge was carried off. He had invented the first modern electric battery.

Do You Know?

Voltaic cells are sometimes known as galvanic cells.

Theme

Equilibrium and change
The ions are trying to equalize both their charge concentration and their particle concentration as they migrate across the membrane.

Visual Strategy

Figure 15-2
The barrier is damp but lets only ions freely cross the barrier, making a connected yet separated container. Cations and anions are moving in their respective directions simultaneously.

Teaching Tip

Complete circuit

Flipping a wall switch interrupts the flow of electric current to a light. When the switch is in the on position, the path of the current is complete, and the light can function. When the switch is in the off position, the path of the current is incomplete, and the light can not function. Have students suggest ways in which the current flow in an electrochemical cell might be interrupted.

Content Background

Determining signs on electrodes in a voltaic cell

The sign on the electrodes in a voltaic cell can be determined from one of two vantage points: as the external circuit toward the cell or as ions within the cell. As far as the external circuit is concerned, the anode supplies electrons, so it is negatively charged. Likewise, the cathode attracts electrons, so it is positively charged. As far as the internal circuit (electrolytic solution) is concerned, the anode is losing electrons, so it is positively charged, and negative ions migrate toward it. Likewise, the cathode is gaining electrons, so it is negatively charged, and positive ions migrate toward it.

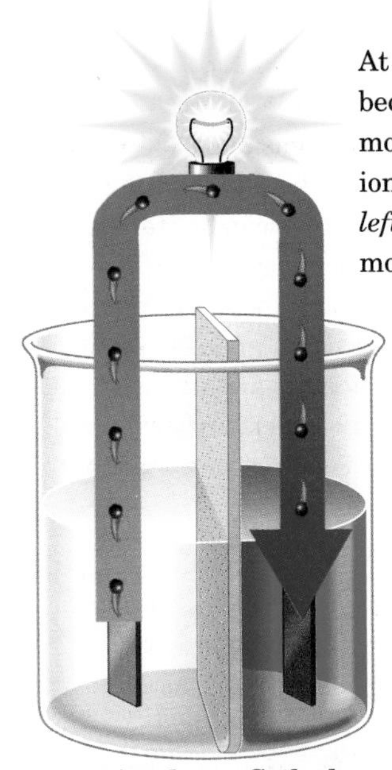

Figure 15-3
Electrons in an electrochemical cell flow from anode to cathode. Ion migration across the porous barrier closes the loop path, making a complete circuit.

At the same time, atoms in the Zn strip are losing electrons to become Zn^{2+} ions that disperse into the surrounding solution. The motion of the ions in the solution is the internal current. The SO_4^{2-} ions migrate from *right* to *left*. The Zn^{2+} and Cu^{2+} ions migrate from *left* to *right*, in the solution, in the opposite direction. The SO_4^{2-} ions move through the barrier to meet the Zn^{2+} ions, which were produced at the anode. These SO_4^{2-} ions are not needed in the cathode compartment because Cu^{2+} ions are disappearing there.

Current flows in a complete circuit

In the wire of the external **circuit** the negative electrons go from left to right (anode to cathode). There must be a corresponding motion of charges in the solution to complete the circuit. This is the motion of the sulfate ions, from right to left, or by the flow of positive ions in the opposite direction. Electrons move in the external circuit, ions move in the internal circuit. You might think of current flow as the baton in **Figure 15-4** that is being transferred between runners in a relay race.

A voltaic cell, therefore, is actually made from two separate components. Each component, a **half-cell**, consists of a metal electrode in contact with a solution of its ions. Identify the two half-cells in **Figure 15-3**. **Anode** and **cathode** are the labels used to identify the half-cells that make up a voltaic cell. These labels describe what is happening at the electrodes. *An* comes from the term *anion* which you know means a negatively charged ion. Anode then suggests a source of electrons. Similarly, *cat* from the term *cation* implies a capacity to accept electrons.

circuit
closed loop path for current to follow

half-cell
the part of a voltaic cell in which either oxidation or reduction occurs. It consists of a single electrode immersed in a solution of its ions

anode
the electrode at which oxidation occurs as electrons are lost by some substance

cathode
the electrode at which reduction occurs as electrons are gained by some substance

Figure 15-4
The baton in this relay race is transferred from one runner to the next until it has gone completely around the track. Like the movement of the baton, current flows in a complete path. It is produced by electrons moving through the external wire and by ion movement within the cell.

Visual Strategy

Figure 15-3

Emphasize that the current path has two parts: outside and inside the cell. Electrons carry the current from the anode to the cathode through a path lying outside the cell. Highlight the path of the electrons as they travel from Zn to Cu. Point out how the movement of ions in the inset for **Figure 15-2** completes the circuit by carrying the current within the cell.

Oxidation occurs when electrons are lost

You learned in Chapter 3 that individual atoms are neutral. In Chapter 5 you learned that ions have either a positive or a negative charge. Reexamine the anode for the electrochemical cell in **Figure 15-5**. The net reaction taking place is described by the following equation.

oxidation
loss of electrons by an atom or an algebraic increase in its oxidation state

$$Zn(s) \longrightarrow Zn^{2+}(aq) + 2e^-$$

Notice that Zn atoms have gone from a neutral, 0, charge in solid Zn to a 2+ charge as Zn^{2+} ion in the $ZnSO_4$ solution. Zinc has lost electrons. It is said to be *oxidized*. Any element that loses electrons is said to have undergone **oxidation**. Oxidation takes place at the anode. Zn atoms have been oxidized. But what happened to the Cu^{2+} ions?

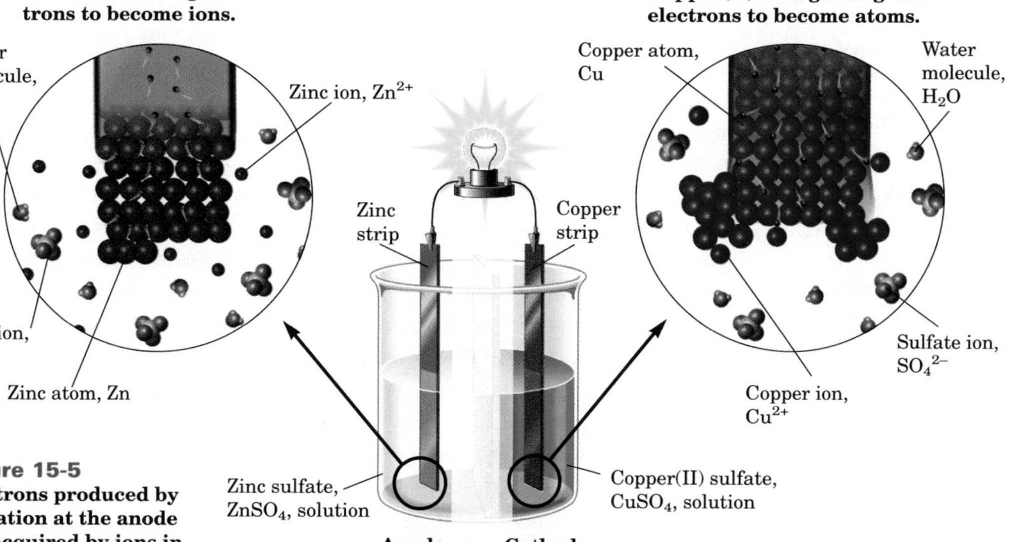

Figure 15-5
Electrons produced by oxidation at the anode are acquired by ions in solution at the cathode (reduction). The anode and cathode are half-cells.

Reduction occurs when electrons are gained

The following equation describes the net reaction occurring at the cathode in **Figure 15-5**.

$$Cu^{2+}(aq) + 2e^- \longrightarrow Cu(s)$$

reduction
the gain of electrons by an atom or the algebraic decrease in its oxidation state

Cu^{2+} ions in the $CuSO_4$ solution surrounding the Cu electrode go from a 2+ charge to a 0 charge when they are deposited as copper metal, Cu. They have gained electrons; they have been *reduced*. Any element that experiences a gain in electrons has undergone **reduction**. Since electrons must be gained to achieve a decrease in charge, reduction is defined as the gain of one or more electrons by an element. Reduction takes place at the cathode.

Teaching Tip
Vocabulary
An easy mnemonic device for remembering oxidation and reduction is **OIL RIG**: **O**xidation **I**s the **L**oss of electrons, **R**eduction **I**s the **G**ain of electrons.

Possible Misconception
Reduction is often misconstrued by students because as a cation *gains* electrons, it is *reduced*. Point out that the charge on the ion or atom, and not the ion itself, is reduced, and show by means of a number line that the charge on Cu in CuO is reduced when it decomposes to form Cu and O_2.

Content Background
Redefining oxidation and reduction
In organic chemistry it is often convenient to define oxidation as the loss of hydrogen or the gain of oxygen. Reduction is the reverse of this, namely, the gain of hydrogen or the loss of oxygen. From this vantage point, alcohols (**Table 6-10** on page 218) are oxidized when they react to form aldehydes, ketones, or carboxylic acids. Similarly, ketones are reduced when they form alkenes, which can be further reduced to form alkanes. These definitions for oxidation and reduction are of limited usefulness to inorganic chemists.

Visual Strategy
Figure 15-5
Compare the insets with those in **Figure 15-1**. As Cu^{2+} ions are reduced to Cu atoms, they adhere to the Cu strip. So the increase in mass of the Cu strip is an indirect measure of the number of e^- passed from Zn to Cu. Two moles of e^- are passed for every one mole Cu deposited. Similarly, the number of e^- passed can be determined from the Zn strip's decrease in mass and the balanced equation for the oxidation half-reaction.

Chapter Connections

Electron transfer

Electron exchange in oxidation-reduction reactions is related to ionic bonding, discussed in Chapter 5, because the formation of cations involved the loss of electrons, while the formation of anions involved the gain of electrons. Moreover, cation formation, as in Na^+, was accompanied by anion formation, as in Cl^-.

ASSESS

Alternative Assessment

Reconsider the voltaic cell in **Figure 15-2**. Suppose the Cu strip were also in $ZnSO_4$ solution. Sketch the cell and describe its operation using equations when appropriate. Would the porous barrier still be required? Could reactions of this type be commercially useful? Explain.

Answers to Section Review

1. a device in which electric energy and chemical energy are interchanged
2. Oxidation is the loss of electrons, while reduction is the gain of electrons.
3. a chemical reaction that involves both oxidation and reduction
4. **a.** from the oxidized metal, through the connecting wire and the metal at the cathode, and back to the anode through ion migration

   b.

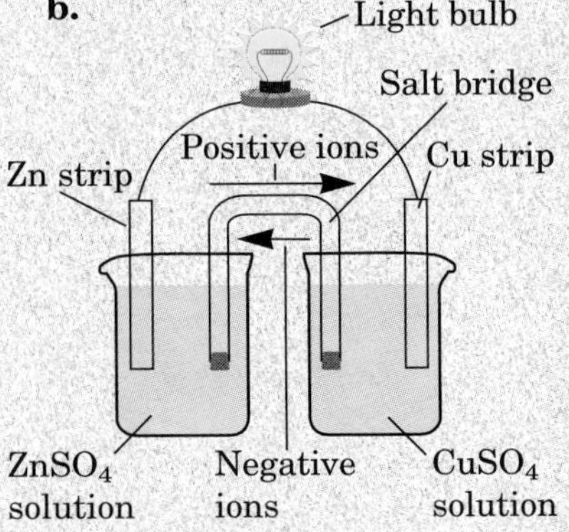

5. **a.** Zn atoms lose electrons to become Zn^{2+} ions that enter the surrounding solution.

   b. increases; decreases

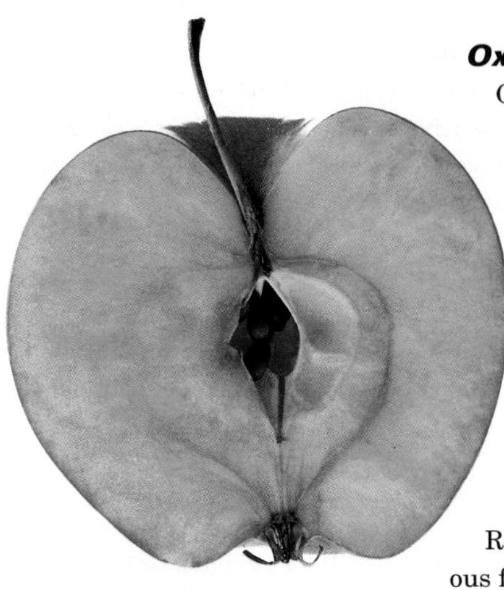

Figure 15-6
Fruit browns when it is cut open and left in contact with oxygen in the air. The discoloration is visual evidence of the redox reaction taking place.

redox reaction
a chemical reaction involving oxidation and reduction

Oxidation and reduction must occur together

Oxidation and reduction always occur simultaneously; you can't have one without the other. If one element is oxidized by losing electrons, then another element has to be reduced by taking on those electrons. Furthermore, the number of electrons lost must equal the number gained. Any reaction in which electrons are lost and gained in equal numbers is an oxidation-reduction reaction, or **redox reaction**. The net redox reaction for the electrochemical cell in **Figure 15-5** is described by the following equation.

$$Zn(s) + Cu^{2+}(aq) \longrightarrow Zn^{2+}(aq) + Cu(s)$$

Redox reactions in an electrochemical cell involve a continuous flow of electrons from anode to cathode. They are the basis of electrochemistry because their half-reactions make up voltaic cells. If voltaic cells are composed of two half-cells corresponding to the two half-reactions of a redox equation, what happens when you mix-and-match half-cells? How can you predict which direction current will flow?

Redox reactions cause the fruit in **Figure 15–6** to brown. Testing urine for sugar also involves a redox reaction. The test reagent is Benedict's solution which is an alkaline solution of copper(II) hydroxide. When glucose comes in contact with Cu^{2+} ions present in Benedict's reagent the aldehyde group of the sugar is oxidized to a carboxylic acid while the Cu^{2+} ion is reduced to Cu^{1+} ion. The Cu^{1+} ion precipitates as Cu_2O. Although Cu_2O is red, the color of the solution in which the reaction is taking place depends on the amount of sugar present. A high sugar concentration produces an orange-red solution. As the sugar concentration decreases, the color changes to yellow, to green, and, at very low sugar concentrations, to blue. Clinitest tablets that are used to test for glucose concentration in urine are a convenient, solid form of Benedict's reagent.

Section Review

1. What is an electrochemical cell?
2. How does oxidation differ from reduction?
3. What is a redox reaction?
4. **a.** Outline the path taken by current in an electrochemical cell.

   b. When electrodes and their solutions are in two separate containers, an inverted, U-shaped tube extending between both solutions replaces the porous barrier. Known as a salt bridge, this tube contains a salt solution whose anions and cations migrate at the same rate. The ends are plugged with glass wool to prevent the solutions from mixing. Using **Figure 15-5** as a guide, draw and label a diagram for this type of voltaic cell.
5. Use **Figure 15-5** to answer the following questions.

   a. What happens to the Zn atoms during the cell reaction? Where do they go? What do they become?

   b. What happens to the mass of the cathode? the anode?

How do you get current to flow?

Section Objectives

Predict the outcome of redox reactions using reduction potentials.

Explain how the activity series and reduction potentials are linked to electronegativity and to each other.

Evaluate the ability of a metal atom to reduce hydrogen in quantitative terms.

Quantifying the activity series

You learned in Chapter 7 that the activity series is a list of metals with the most active one placed at the top. Activity decreases progressively down the list. Because reactions between atoms and ions of metals involve a transfer of electrons during a redox reaction, the activity series tells you, in effect, which metals generally give up electrons and which metal ions generally accept them.

Activity series is a model that allows you to make predictions

Compare the voltaic cells shown in **Figure 15-7** with the one in **Figure 15-3**. Although the parts are similar, the composition of the electrodes differs. The anode in **Figure 15-7a** consists of a strip of aluminum metal in aqueous $AlCl_3$ instead of Zn in aqueous $ZnSO_4$. The cathode is a strip of copper metal immersed in a solution of $CuCl_2$ instead of $CuSO_4$. The solution is blue, indicating the presence of Cu^{2+} ions. The porous barrier allows ions to move freely between the anode and the cathode while keeping the electrodes separated from each other. Using **Table 15-1** on page 585, predict what will happen in this electrochemical cell.

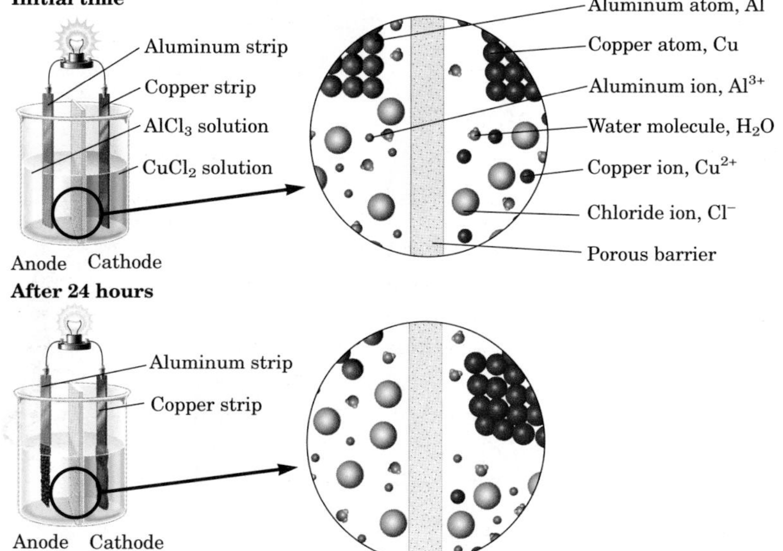

Figure 15-7a
The activity series predicts electrons will flow from Al to Cu. A flow of electricity will light the bulb.

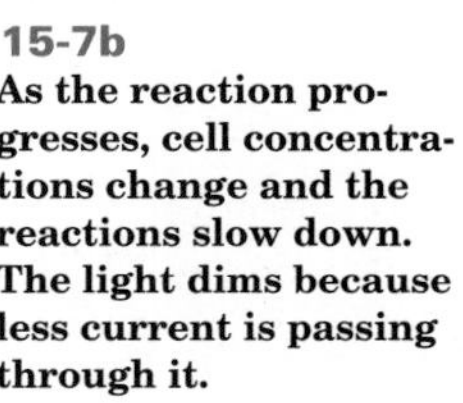

15-7b
As the reaction progresses, cell concentrations change and the reactions slow down. The light dims because less current is passing through it.

Section 15-2

FOCUS

Lesson Starter

Begin this lesson by performing the following demonstration after students have predicted the outcome of the reactions it involves. If a product is predicted for both reactions, vote on which reaction will go more quickly. Discuss the results of the demonstration in terms of periodic trends and the predictions made.

Demonstration 2

Predicting reactions

Approximate time: 10 min

1. Place two petri dishes on the overhead projector. Fill one dish half full with 0.1 M $AgNO_3$ and the other with 0.1 M $Zn(NO_3)_2$.
2. Add one polished Cu strip (0.5 cm × 2 cm) to each dish. Silver should plate out on the Cu strip in the $AgNO_3$ solution, but nothing should happen in the $Zn(NO_3)_2$ solution.

SAFETY
Wear safety goggles and a lab apron.

DISPOSAL
Remove the Cu strip coated with Ag, and scour off the Ag plate with emery paper. Put the emery paper in a labeled container for use next time; do **not** discard. Remove the other Cu strip. Save both Cu strips for future use. Save the two solutions in labeled containers for use next time.

Visual Strategy

Figure 15-7

The light bulb is a visual indicator that electrons are flowing. The light bulb is not necessary for the movement of the electrons. The bulb is actually a large resistor where electron movement slows and heat builds up, resulting in excitation of electrons in the tungsten filament and the emission of light energy. Compare the insets for evidence of a chemical change.

TEACH

Teaching Tip

The two-potato clock

Visually illustrate voltaic cell operation with the two-potato clock, an LCD digital clock that attaches to Cu and Zn electrodes embedded in two potatoes that are connected in series. It is available from science novelty suppliers. It also works with soda and fruits.

Content Background

Calculating moles of electrons passed and mass of Cu deposited

A current of 2.50 amps passing through a solution of $Cu(NO_3)_2$ for 2 h will deposit 5.93 g Cu according to the following calculations.

charge transferred:

$$2.50 \text{ amp} \times 2 \text{ h} \times \frac{3600 \text{ s}}{1 \text{ h}} = 18\,000 \text{ amp}\cdot\text{s}$$

mol e^-:

$$18\,000 \text{ amp}\cdot\text{s} \times \frac{1 \text{ mol } e^-}{9.65 \times 10^4 \text{ amp}\cdot\text{s}} =$$

0.187 mol

mass of Cu deposited:

$$0.187 \text{ mol } e^- \times \frac{1 \text{ mol Cu}}{2\,e^-} \times \frac{63.55 \text{ g Cu}}{1 \text{ mol Cu}} =$$

5.93 g

Demonstration 3

Comparing voltages

Approximate time: 15 min

1. Obtain a potato and strips of Fe, Zn, Cu, Sn, Al, Mg, and brass.
2. Polish all strips of metal with steel wool.
3. Insert the metal with the highest activity (**Table 15-1**) into the potato, and attach the negative clip of a sensitive voltmeter.

Figure 15-7b on page 581 shows the voltaic cell in **Figure 15-7a** after 24 hours. Examine it for evidence to support your prediction. Look closely at the anode. Part of the aluminum strip appears to have dissolved in the solution. If we knew the mass of the strip at the start of the reaction, we would now find that it shows a decrease in mass. Analysis of the cathode solution confirms the presence of Al^{3+} ions which were not found in this solution at the start of the reaction. Analysis of the anode solution shows an increase in Cl^- ion concentration compared to that at the start of the reaction. In spite of ion migration, Al^{3+} ions are still present in the anode solution. Thus, Al atoms have given up electrons to form Al^{3+} ions. Al atoms have been oxidized.

The strip of copper in **Figure 15-7b** has a dull, reddish brown solid deposited on it. Analysis of the electrode indicates it has increased in mass and that the solid is copper metal. The solution surrounding the copper strip is lighter blue, indicating that fewer Cu^{2+} ions are present. Thus, Cu^{2+} ions have formed Cu atoms by accepting the electrons given up by Al atoms. The Cu^{2+} ions have been reduced.

This redox reaction is exactly what should happen based on the activity series. The reaction occurred because aluminum is more reactive than copper. In other words, aluminum has a greater tendency than copper to give up its valence electrons. In fact, the entire activity series is made of such "electron-trade agreements," indicating the relative willingness of metals to give up their valence electrons. Is there a way to quantify this tendency? How much greater is the tendency for Al to give up its electrons than for Cu? or for Zn than Cu? Replacing the light bulb assembly in **Figure 15-2** and **Figure 15-7** with a voltmeter will allow the electric potential in these cells to be measured, as you can see in **Figure 15-8**. But how can this help predict the amount of electric potential in a voltaic cell made from Al and Zn? How much of the cell's electric potential is due to the anode? How much is due to the cathode?

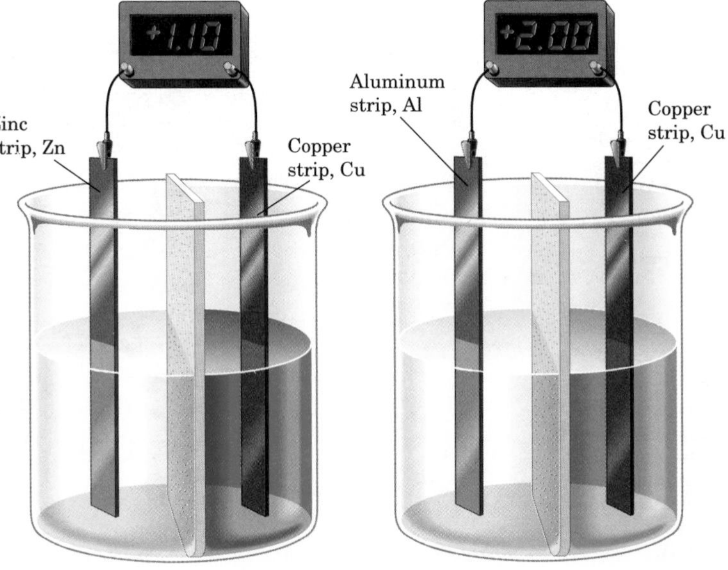

Figure 15-8
Replacing the light bulb with a voltmeter quantifies the greater tendency of Zn and Al to give up their electrons relative to Cu. What would the voltmeter read if Zn were connected to Al?

582 | Chapter 15

Visual Strategy

Figure 15-8

Point out that the difference in voltage is the result of differing relative electronegativies between the metals in the cell. Because the Zn and the Al strips are measured against the same reference metal, the voltage reading of a cell between Zn and Al would be +0.90 V, the difference between the two readings shown here.

Metals are ranked according to their ability to reduce hydrogen

Measuring the voltage generated in the separate half-cells requires the reaction that occurs at each electrode to be considered separately. The reaction occurring at an electrode is known as a **half-reaction**. Oxidation is the half-reaction occurring at the anode. Reduction is the half-reaction occurring at the cathode.

The amount of electric energy that each half-reaction can generate is determined by its **reduction potential**. Reduction potentials are measured in volts, V. The difference between the reduction potentials for the two half-reactions is the total voltage generated by the redox reaction.

As you saw in **Figure 15-8**, the voltage generated by a redox reaction is easily measurable. However, measuring the voltage generated by a half-reaction alone is impossible because there can be no transfer of electrons unless both an anode and cathode are connected to form a complete circuit. To complete the circuit, a standard half-cell is used. This standard half-cell consists of a platinum electrode that is immersed in a solution with an H^+ ion concentration of 1.00 M and surrounded by hydrogen gas at 1 atm pressure and 25°C. Metals are ranked according to their ability to reduce hydrogen under these conditions.

The anodic half-reaction occurring at the hydrogen electrode is shown by the following equation.

$$H_2(g) \longrightarrow 2H^+(aq) + 2e^-$$

The cathodic half-reaction is the reverse of this one. The voltage for both of these reactions is arbitrarily assigned a value of exactly 0 V. Consequently, any voltage measurement that is obtained can be attributed entirely to the half-cell that is connected to the hydrogen half-cell standard. Such a voltage measurement is the **standard reduction potential**, E^0. E^0 values represent the voltages generated by half-reactions.

Examine the voltaic cells in **Figure 15-9** on the next page. The standard hydrogen electrode or **SHE** is considered to be the anode in each cell. The metal electrode is the cathode. Compare the voltage readings. Some are positive values. Some are negative values. How can this observation be rationalized? Look back to **Figure 15-8**. The voltage readings are both positive values. These two cells produced electricity as drawn in the figure because you saw the lighted bulb. Positive E^0 values indicate a flow of electrons from anode to cathode as written. That is, hydrogen is more willing to give up its electrons than the metal is. A voltmeter cannot actually register a negative potential. Instead, it would show a reading of 0 V until the positions of the half-cells are reversed. Reversing the cells causes the voltage to register as a positive value, but the SHE is now the cathode instead of the anode, so electrons are flowing from the metal to the SHE. Therefore, negative E^0 values indicate that the metal is more willing to give up its electrons than hydrogen is.

half-reaction
the oxidation or reduction portion of a redox reaction

reduction potential
a measure of the tendency of a given half-reaction to occur as a reduction in an electrochemical cell

standard reduction potential
by convention, the potential, E^0, of a half-cell connected to the standard hydrogen electrode when ion concentrations in the half-cells are 1 M, gases are at a pressure of 1 atm, and the temperature is 25°C

SHE
standard hydrogen electrode has the half-reaction $2H^+(aq) + 2e^- \longrightarrow H_2(g)$; the concentration of hydrogen ion is 1 M, the temperature is 25°C, and the pressure of the hydrogen gas is 101.3 kPa (1 atm)

4. List the metals on the board. Write "Reference" next to the metal already in the potato.
5. One at a time, insert each remaining metal into the potato, attach the positive voltmeter clip, and record the voltage.

Variation: Change the reference metal, and repeat steps **4** and **5**. Compare the results with those already obtained.

SAFETY
Wear safety goggles and a lab apron.

DISPOSAL
Mark the potato "Unfit for food use," and put it in the trash. Save the strips of metal for use next time.

Do You Know?
The standard hydrogen electrode is also called the normal hydrogen electrode, NHE.

Theme
Classification and trends
The metals are classified according to the voltage produced when measured against the hydrogen electrode.

Teaching Tip
Analogy for the ordering of reduction potentials
Assign a reference height of zero to a student of average height. Measure the height of every other student against the reference student: if taller, + cm, and if shorter, − cm. The real height of the reference student is no longer important because even when two other students compare their height, they can still determine who is taller. For example, if –5 cm is compared with –3 cm, it is obvious who is the taller student. Actual height is no longer important, just the relative height.

Teaching Tip

Calculating E^0 values for half-reactions not listed in the table

The standard potential for a half-reaction not listed in the table can be found with the following two-step process.

1. Add the equations for half-reactions with known E^0 values in a Hess's law calculation to obtain the equation for the half-reaction with the unknown E^0 value. This will give you the number of electrons transferred.
2. Multiply each known E^0 value by the number of electrons in the corresponding half-reaction. Add these values together and divide by the total number of electrons transferred, as calculated in step 1.

Do You Know?

The standard potential of any half-reaction that contains OH^- ion or H^+ ion will be affected by changes in pH. For example, at a pH of 7, the value of E for the SHE decreases from 0.00 V to −0.414 V. At a pH of 14, the value of E is −0.829 V.

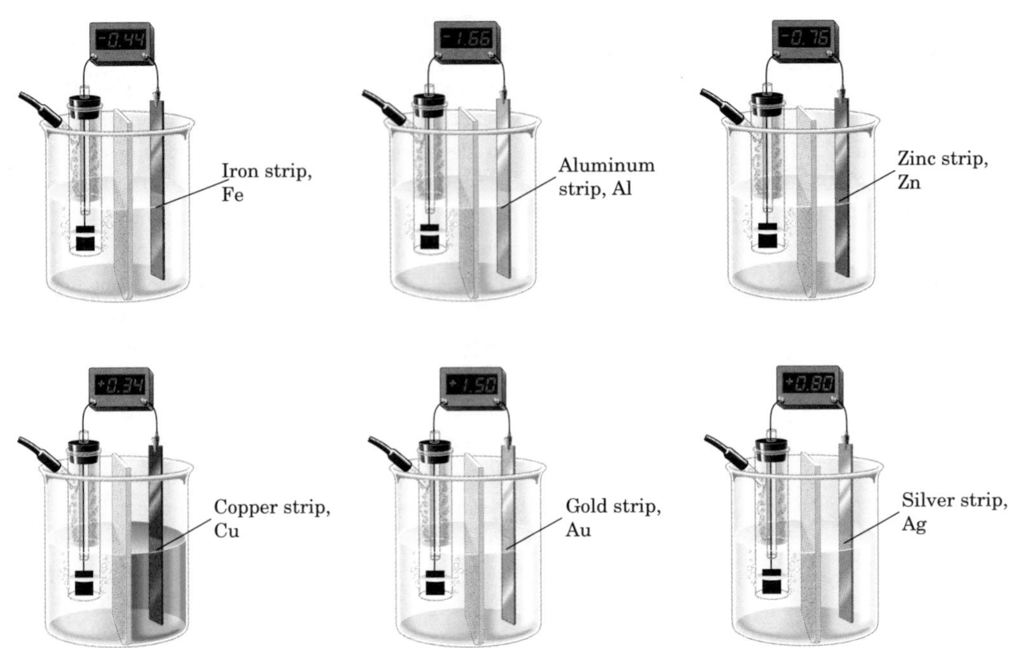

Figure 15-9
When connected to the SHE, the metal half-cell is considered the cathode. Because the SHE is arbitrarily assigned a voltage of exactly 0, the voltmeter gives the potential of the metal half-cell. Ordering the voltages produces a table of reduction potentials.

Based on these constraints, an ordering of the voltage readings in **Figure 15-9** shows Zn is less willing to give up its electrons than is Al by 0.90 V, but is 1.10 V more willing to give up electrons than Cu. Locate Cu, Al, and Zn in the activity series portion of **Table 15-1**. Zn is less reactive than Al but more reactive than Cu. E^0 values are related to the activity of metals: the more reactive the metal, the greater the negative E^0 value associated with it.

The reduction potentials shown in **Table 15-1** are an ordering of reduction half-reactions by voltage. Compare this with the activity series in the far left column of **Table 15-1**. The table of reduction potentials extends the activity series by quantifying its relativeness. How can this agreement between the two tables be explained?

Activity and reduction potential result from electronegativity differences

In your study of the periodic table, you learned that certain properties exhibit trends as you move either across a period or down a group of elements. One such property is electronegativity. You learned that electronegativity is the measure of an atom's ability to acquire electrons. The more electronegative the atom, the more readily it acquires electrons.

If the metals in the activity series were listed in order of increasing electronegativity values, the list would have a general correspondence with the activity series as well as the table of reduction potentials. You can see this correspondence by comparing the color for a metal's entry in **Table 15-1** with its color in the table for electronegativity values on page 418.

Table 15-1
Activity Series and Reduction Potentials at 25°C

Element	Symbol	Electrode	Half-reaction	E^0(V)
lithium	Li	Li^+/Li	$Li^+ + e^- \longrightarrow Li$	−3.05
rubidium	Rb	Rb^+/Rb	$Rb^+ + e^- \longrightarrow Rb$	−2.98
potassium	K	K^+/K	$K^+ + e^- \longrightarrow K$	−2.93
barium	Ba	Ba^{2+}/Ba	$Ba^{2+} + 2e^- \longrightarrow Ba$	−2.90
calcium	Ca	Ca^{2+}/Ca	$Ca^{2+} + 2e^- \longrightarrow Ca$	−2.87
sodium	Na	Na^+/Na	$Na^+ + e^- \longrightarrow Na$	−2.71
magnesium	Mg	Mg^{2+}/Mg	$Mg^{2+} + 2e^- \longrightarrow Mg$	−2.37
aluminum	Al	Al^{3+}/Al	$Al^{3+} + 3e^- \longrightarrow Al$	−1.66
manganese	Mn	Mn^{2+}/Mn	$Mn^{2+} + 2e^- \longrightarrow Mn$	−1.19
		H_2O/H_2	$2H_2O + 2e^- \longrightarrow H_2 + 2OH^-$	−0.83
zinc	Zn	Zn^{2+}/Zn	$Zn^{2+} + 2e^- \longrightarrow Zn$	−0.76
chromium	Cr	Cr^{3+}/Cr	$Cr^{3+} + 3e^- \longrightarrow Cr$	−0.74
iron	Fe	Fe^{2+}/Fe	$Fe^{2+} + 2e^- \longrightarrow Fe$	−0.44
		$PbSO_4/Pb$	$PbSO_4 + 2e^- \longrightarrow Pb + SO_4^{2-}$	−0.36
nickel	Ni	Ni^{2+}/Ni	$Ni^{2+} + 2e^- \longrightarrow Ni$	−0.25
tin	Sn	Sn^{2+}/Sn	$Sn^{2+} + 2e^- \longrightarrow Sn$	−0.14
lead	Pb	Pb^{2+}/Pb	$Pb^{2+} + 2e^- \longrightarrow Pb$	−0.13
		Fe^{3+}/Fe	$Fe^{3+} + 3e^- \longrightarrow Fe$	−0.036
(hydrogen)	(H)	H^+/H_2	$2H^+ + 2e^- \longrightarrow H_2$	0.000
		$AgCl/Ag$	$AgCl + e^- \longrightarrow Ag + Cl^-$	+0.22
copper	Cu	Cu^{2+}/Cu	$Cu^{2+} + 2e^- \longrightarrow Cu$	+0.34
iodine	I	I_2/I^-	$I_2 + 2e^- \longrightarrow 2I^-$	+0.54
		Fe^{3+}/Fe^{2+}	$Fe^{3+} + e^- \longrightarrow Fe^{2+}$	+0.77
mercury	Hg	Hg_2^{2+}/Hg	$Hg_2^{2+} + 2e^- \longrightarrow 2Hg$	+0.79
silver	Ag	Ag^+/Ag	$Ag^+ + e^- \longrightarrow Ag$	+0.80
bromine	Br	Br_2/Br^-	$Br_2 + 2e^- \longrightarrow 2Br^-$	+1.07
platinum	Pt	Pt^{2+}/Pt	$Pt^{2+} + 2e^- \longrightarrow Pt$	+1.20
		O_2/H_2O	$O_2 + 4H^+ + 4e^- \longrightarrow 2H_2O$	+1.23
		MnO_2/Mn^{2+}	$MnO_2 + 4H^+ + 2e^- \longrightarrow Mn^{2+} + 2H_2O$	+1.28
		$Cr_2O_7^{2-}/Cr^{3+}$	$Cr_2O_7^{2-} + 14H^+ + 6e^- \longrightarrow 2Cr^{3+} + 7H_2O$	+1.33
chlorine	Cl	Cl_2/Cl^-	$Cl_2 + 2e^- \longrightarrow 2Cl^-$	+1.36
		PbO_2/Pb^{2+}	$PbO_2 + 4H^+ + 2e^- \longrightarrow Pb^{2+} + 2H_2O$	+1.46
gold	Au	Au^{3+}/Au	$Au^{3+} + 3e^- \longrightarrow Au$	+1.50
		MnO_4^-/Mn^{2+}	$MnO_4^- + 8H^+ + 5e^- \longrightarrow Mn^{2+} + 4H_2O$	+1.51
		$PbO_2/PbSO_4$	$PbO_2 + 4H^+ + SO_4^{2-} + 2e^- \longrightarrow PbSO_4 + 2H_2O$	+1.69
fluorine	F	F_2/F^-	$F_2 + 2e^- \longrightarrow 2F^-$	+2.87

Electronegativity Value of the Metals

- < 1.0
- 1.0–1.9
- 2.0–2.9

Theme

Classification and trends
The similar ordering of the electronegativity values, activities, and reduction potentials allows you to predict values for elements not listed on the tables.

Groupwork Strategy

Calculating cell voltage
In groups of three, have students use the equation on page 586 to determine which half-reactions in **Table 15-1**, when combined, could produce cell voltages in the following ranges: > +2.5 V, +1.8 V to +2.0 V, +0.7 V to +0.9 V, and +0.1 V to +0.3 V.

Additional Examples

Determine E^0cell for the following sets of half-reactions.

a. $Al^{3+} + 3e^- \longrightarrow Al$
$E^0 = -1.66$ V
$Au^{3+} + 3e^- \longrightarrow Au$
$E^0 = +1.50$ V
Al is the anode.
E^0cell = (+1.50 V) − (−1.66 V) = +3.16 V

b. $Li^+ + e^- \longrightarrow Li$
$E^0 = -3.05$ V
$Ag^+ + e^- \longrightarrow Ag$
$E^0 = +0.80$ V
Li is oxidized.
E^0cell = (+0.80 V) − (−3.05 V) = +3.85 V

c. $Mg^{2+} + 2e^- \longrightarrow Mg$
$E^0 = -2.37$ V
$Fe^{3+} + 3e^- \longrightarrow Fe$
$E^0 = -0.036$ V
Mg is oxidized.
E^0cell = (−0.036 V) − (−2.37 V) = +2.334 V

Possible Misconception

When a coefficient is added to balance the number of electrons exchanged, the E^0 value is **not** multiplied by the coefficient.

ASSESS

Alternative Assessment

Find out which active metals are attached to the hulls of ships and to underground pipes to prevent corrosion. Could Na be used for this function? Explain.

Answers to Section Review

6. The activity series predicts which metals tend to accept electrons and which tend to donate them.
7. The metal's activity depends on its position in the periodic table.
8. A half-reaction is either the oxidation or reduction part of a redox reaction. It takes place at the surface of an electrode in a half-cell, which consists of an electrode immersed in a solution of its ions. A half-cell is one half of a voltaic cell.
9. The standard reduction potential represents a measure of the tendency of a given half-reaction to occur as a reduction in an electrochemical cell when ion concentrations in the half-cells are 1 M, gases are at 1 atm pressure, and the temperature is 25°C.
10.

0.84

Fe strip — Ag strip

Porous barrier

$Fe^{3+}(aq)$ — $Ag^{+}(aq)$

Anode Cathode

$Fe \rightarrow Fe^{3+} + 3e^-$ $\quad Ag^+ + e^- \rightarrow Ag$

Calculating E^0 values for electrochemical cells

Voltaic cells are made from spontaneous redox reactions. Spontaneity is represented by a positive value for E^0cell, which is calculated according to the following equation.

$$E^0 \text{ cell} = E^0 \text{ cathode} - E^0 \text{ anode}$$

When evaluating two half-reactions from which you expect to make a voltaic cell, the reaction with the lower value for E^0 in **Table 15-1** will be the anode.

Consider the zinc/copper cell discussed in Section 15-1. The two half-reactions involved and their E^0 values from the reduction potential table are as follows.

$$Zn^{2+}(aq) + 2e^- \longrightarrow Zn(s) \qquad E^0 = -0.76 \text{ V}$$

$$Cu^{2+}(aq) + 2e^- \longrightarrow Cu(s) \qquad E^0 = +0.34 \text{ V}$$

Reduction of Zn^{2+} ion has the lower value for E^0, so Zn is the anode. E^0 cell is +1.10 V, the difference between cathode and anode values for E^0. This agrees with the voltmeter reading in **Figure 15-8**.

Now consider a cell made from Al and Zn. The reduction half-reactions involved and their E^0 values from the table are as follows.

$$Zn^{2+}(aq) + 2e^- \longrightarrow Zn(s) \qquad E^0 = -0.76 \text{ V}$$

$$Al^{3+}(aq) + 3e^- \longrightarrow Al(s) \qquad E^0 = -1.66 \text{ V}$$

Al is the anode because its reaction has the lower reduction potential. E^0 cell for this reaction is + 0.90 V.

If the value calculated for E^0 cell is negative, the proposed redox reaction is not a voltaic cell because it is nonspontaneous. Voltaic cells are not limited to metal electrodes. The same approach is used for nonmetal electrodes.

Electrons move about an electrochemical cell in a circular path beginning at the anode. They are persuaded to move by differing tendencies to lose electrons at the anode and cathode. Reduction potentials assigned to half-reactions are used to predict the movement of electrons in the external circuit. But how many electrons are moving? Redox reactions require the number of electrons lost to equal the number of electrons gained. How can we be sure that electrons are conserved?

Section Review

6. What information concerning redox reactions does the activity series provide?
7. Why can the order of the metals in an activity series be considered a periodic trend?
8. Define the terms *half-cell* and *half-reaction*.
9. Explain what a standard reduction potential represents.
10. Two half-cells, one containing Fe^{3+} and Fe and the other containing Ag^+ and Ag, are connected to form a voltaic cell. Use **Table 15-1** to determine the direction of spontaneous reaction and the value for E^0 cell. Diagram the cell and label its parts. Give equations for the half-reactions.

How are electron transfers described?

Section Objectives

Assign oxidation states to the reactants and products of a redox reaction.

Identify redox reactions by changes in oxidation states.

Write balanced equations for redox reactions using the half-reaction method.

Explain how redox reactions show conservation of charge.

Oxidation states

Redox equations are the basis of electrochemistry. As the name implies, they are made of two half-reactions, namely, reduction and oxidation. Oxidation involves the loss of electrons and takes place at the anode. Reduction involves the gain of electrons and takes place at the cathode. Reduction and oxidation must occur simultaneously, but they are balanced separately and then added together to give the overall balanced redox equation. Identifying the half-reactions and balancing redox equations requires a thorough knowledge of oxidation states.

oxidation state *a numerical representation of an atom's share of the bonding electrons. In ionic compounds, it is equal to the ionic charge. In covalent compounds, it is the average charge assigned to an atom according to electronegativities*

Oxidation states represent an atom's share of the molecule's bonding electrons

Half-reactions, hence redox reactions, are characterized by the loss and gain of electrons. This trading of electrons brings about changes in the oxidation states of elements participating in the reaction. An **oxidation state**, also known as an *oxidation number*, is a value representing the apparent charge for each atom in a molecule based on the general distribution of bonding electrons among all the atoms in that molecule.

Keep in mind that an oxidation state, unlike an ionic charge, does not have an exact physical meaning. For example, an ionic charge of 1– means the gain of one electron whereas an oxidation state of –1 means a greater attraction for a bonding electron. That is, the oxidation state assigned to an atom in a molecule is related to the electronegativity value of that atom. Because electrons are negatively charged, negative numbers reflect stronger attraction for electrons than do positive numbers. The larger the negative value, the stronger the attraction. Conversely, the larger the positive value, the smaller the attraction.

The Lewis dot structures in the feature on the next page show how oxidation states are assigned to chlorine in Cl_2, NaCl, HCl, ClF_3, and ClO_4^-. As you know, NaCl is an ionic compound consisting of Na^+ and Cl^- ions. Their ionic charges are taken as a measure of their attraction for electrons. They are assigned oxidation states equal to their ionic charges.

Figure 15-10
Compounds of transition elements such as chromium produce colored solutions. The color is often an indicator of the element's oxidation state. Chromium is yellow or orange in its +6 state but green in its +3 state.

Section 15-3

FOCUS

Lesson Starter

You may wish to begin the lesson with the following demonstration.

Demonstration 4

Three oxidation states of manganese

Approximate time: 10 min

1. Fill a petri dish one-fourth full with 0.05 M $KMnO_4$ solution.
2. Place the dish on the overhead.
3. Add 1 mL of 0.1 M NaOH to the solution.
4. Gently add one drop of acetaldehyde to act as a reducing agent.
5. Let the acetaldehyde remain undisturbed on the surface. Three separate colored rings should appear, corresponding to three oxidation states of Mn: brown, +4; green, +6; purple, +7.

SAFETY

Wear safety goggles and a lab apron.

DISPOSAL

While stirring slowly, add 1 M sodium thiosulfate solution until the mixture is brown; then pour it down the drain.

TEACH

Possible Misconception

Students may try to interchange the terms *valence* and *oxidation state*.

Teaching Tip

Rules for determining oxidation states

You may wish to give specific examples for rules 1 and 2.

The following binary compounds may assist you in highlighting rule 3: SF_6, Al_2S_3, Na_2O, Na_3N, AlP, MnO_2, PbI_2.

Possible Misconception

Oxidation states of hydrogen

When hydrogen has an oxidation state of +1, it will appear as the cation in the chemical formula, as in H_2SO_4 or $NaHCO_3$. But when hydrogen has an oxidation state of –1, it will appear as the anion in the chemical formula, as in NaH or in $LiAlH_4$.

You may wish to use NH_3 as an example of the sometimes arbitrary application of rules. As written, NH_3 is a hydride, and many discussions of the descriptive chemistry for Group 15 treat it as a hydride, but from the standpoint of electronegativity alone, this should be written as a binary acid with the formula H_3N. The other members of this group have electronegativies less than or equal to that of hydrogen, so they are justifiably classified as hydrides.

How To

Assign oxidation states using electro-negativity

Step	Activity	Cl_2	NaCl
1	Identify the type of substance.	uncombined element	ionic compound
2	Draw the Lewis dot structure and determine bond polarity.	$:\ddot{\underset{..}{Cl}}:\ddot{\underset{..}{Cl}}:$	$[Na]^+ [:\ddot{\underset{..}{Cl}}:]^-$
3	For each bond, assess which of the two atoms has a greater attraction for the bonding electrons.	Both Cl atoms contribute to the bond. Both atoms have the same electonegativity, so neither atom attracts the bonding electrons more than the other one. Their share in the bond is equal.	Both atoms contribute to the bond. The Cl atom is more strongly electronegative than Na and pulls the one electron from Na's valence shell to itself. The charge on the ion formed is the oxidation state for that atom.
4	Assign each atom an oxidation state (or charge) based on the results of step 3.	Cl = 0 Cl = 0	Na = +1 Cl = −1
5	Add all of the assigned oxidation states. They should equal zero or the total charge of the ion.	0 + 0 = 0	(+1) + (−1) = 0

Rules for assigning oxidation states

From these electronegativity-based assignments for oxidation states we can derive some general rules that will allow us to assign oxidation states to atoms quickly and easily.

1. The oxidation state of any free (uncombined) element is 0.
2. The oxidation state of a monatomic ion is equal to the charge on the ion.
3. The more electronegative element in a binary compound is assigned the number equal to the charge it would have if it were an ion.
4. The oxidation state of each hydrogen atom is +1 unless it is combined with a metal, then it has a state of −1.
5. The oxidation state of fluorine is always −1 because it is the most electronegative atom.

HCl	ClF_3	ClO_4^-
diatomic covalent compound	polyatomic covalent compound	polyatomic ion
H:Cl:	:F: :Cl:F: :F:	[:O: :O:Cl:O: :O:]⁻
Both atoms contribute to the bond. The Cl atom is more electronegative than H. It attracts its own bonding electron plus that from H, creating a permanent dipole. Cl has the greater share of the bonding electrons.	For each Cl—F bond, both atoms contribute to the bond. F is more electronegative than Cl, so it creates a polar bond by attracting its own electron plus the one donated from Cl. F has the greater share of the bonding electrons. Cl has the lesser share of three separate bonds, so its oxidation state is the accumulation from all three bonds.	The overall charge on this ion indicates one valence electron is from some element which formed the cation. It is shown here as ▬ and given to one of the O atoms because O is more electronegative than Cl. The O for this Cl—O bond attracts both this electron from space and one bonding electron from Cl. The bonds between Cl and the remaining 3 O atoms are coordinate covalent bonds, i.e., both of the bonding electrons are donated by Cl toward O. Therefore, these 3 O atoms attract *both* electrons rather than one.
H = +1 Cl = −1	Cl = (+1) + (+1) + (+1)= (+3) F = −1 F = −1 F = −1	Cl = (+1) + (+2) + (+2) + (+2) = +7 O = −2 O = −2 O = −2 O = −2
(+1) + (−1) = 0	(+3) + (−1) + (−1) + (−1) = 0	(+7) + (−2) + (−2) + (−2) + (−2) = −1

6. The oxidation state of each oxygen atom in most of its compounds is −2. When combined with F, oxygen has a state of +2. In peroxides, such as H_2O_2, oxygen has an oxidation state of −1.
7. In compounds, the elements of Group 1 and Group 2, and aluminum have positive oxidation states of +1, +2, and +3, respectively.
8. The algebraic sum of the oxidation states for all the atoms in a compound is 0.
9. The algebraic sum of the oxidation states for all the atoms in a polyatomic ion is equal to the charge on that ion.

Rules 8 and 9 make it possible to assign oxidation states that are not known. The total of the known and the unknown oxidation numbers must satisfy rule 8 or rule 9, as illustrated in Sample Problems 15A and 15B.

Theme
Classification and trends
Many atoms have predictable oxidation states which allow the determination of the oxidation states for elements that have multiple oxidation states.

Additional Examples for Sample Problem 15A

a. Assign oxidation states to all atoms in potassium dichromate, $K_2Cr_2O_7$.

b. Assign oxidation states to all atoms in aluminum nitrate, $Al(NO_3)_3$.

Solutions are at the bottom of the page.

Additional Examples for Sample Problem 15B

a. Assign oxidation states to the atoms in a nitrite ion, NO_2^-.

b. Assign oxidation states to the atoms in a chromate ion, CrO_4^{2-}.

Solutions are at the bottom of the page.

Answers to Practice 15A and 15B

1. +5
2. F = −1; B = +3
3. +7
4. H = +1; N = −3

Sample Problem 15A
Assigning oxidation states to atoms in covalent compounds

Assign oxidation numbers to all atoms in sulfurous acid.

❶ *List*

- **the formula for sulfurous acid: H_2SO_3**
- **the charge on H using rule 4 from the list in the feature on page 588:** +1
- **the charge on O using rule 6 from the list in the feature on page 589:** −2

❷ *Set up*

- **You are trying to find the oxidation number for S in H_2SO_3. If more than one atom of an element is present in the formula, you must first calculate the total charge contributed by those atoms by multiplying the charge on the individual atom by its subscript in the formula.**

  H: $(+1)\,2 = +2$

  O: $(-2)\,3 = -6$

  S: $(x)\,1 = x$

- **Set up an equation using rule 8 from the feature on page 589:** −2

  $(+2) + (x) + (-6) = 0$

❸ *Calculate*

- **Solve for *x*.**

  $(x) - 4 = 0$

  $x = +4$

- **The oxidation state assigned to S in H_2SO_3 is +4; to H is +1; and to O is −2.**

Sample Problem 15B
Assigning oxidation states to atoms in polyatomic ions

Assign oxidation numbers to all atoms in $S_2O_7^{2-}$.

❶ *List*

- **the formula:** $S_2O_7^{2-}$
- **the charge on O using rule 6 from the list in the feature on page 589:** −2

❷ *Set up*

- **You are trying to find the oxidation number for S in $S_2O_7^{2-}$. If more than one atom of an element is present in the formula, you must first calculate the total charge contributed by those atoms by multiplying the charge on the individual atom by its subscript in the formula.**

  O: $(-2)\,7 = -14$

  S: $(x)\,2 = 2x$

- **Set up an equation using rule 9 from the feature on page 589.**

  $2(x) + (-14) = -2$

❸ *Calculate*

- **Solve for *x***

  $2(x) = +12$

  $x = +6$

- **The oxidation state assigned to S in $S_2O_7^{2-}$ is +6 and to O is −2.**

Practice

15A
1. **Find the oxidation state assigned to P in H_3PO_4.**
2. **Find the oxidation state for each atom in BF_3.**

15B

3. **Find the oxidation state assigned to Mn in MnO_4^-.**
4. **Find the oxidation state assigned to each atom in NH_4^+.**

Solutions for Additional Examples 15A

a. K = +1; Cr = +6; O = −2; use rules **6**, **7**, and **8**

b. Al = +3; N = +5; O = −2; use rules **6**, **7**, and **8**

Solutions for Additional Examples 15B

a. N = +3; O = −2; use rules **6** and **9**

b. Cr = +6; O = −2; use rules **6** and **9**

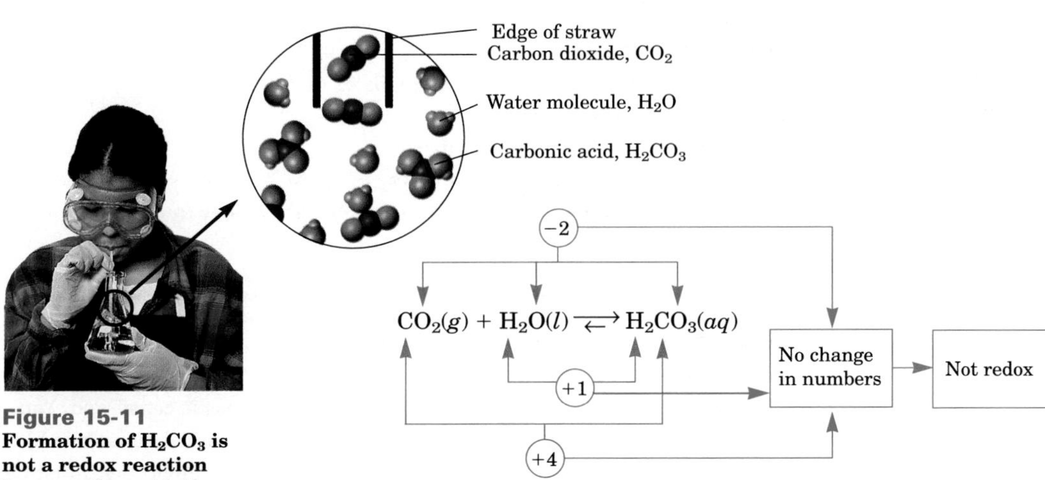

Figure 15-11
Formation of H_2CO_3 is not a redox reaction because the oxidation states for all three atoms remain the same.

Changes in oxidation states indicate a redox reaction

Figure 15-11 shows how CO_2, an acid anhydride produced by your body, reacts with water. You learned in Chapter 13 that an acid anhydride is an oxide that reacts with water to form an acid. Is this a redox reaction? To decide, first assign oxidation states to all elements, as in **Figure 15-11**. Then compare the oxidation states of C, O, and H on the left side of the arrow with those on the right side of the arrow. The oxidation states for C, O, and H did not change, as evidenced by the values on the red arrows. Nothing has been oxidized or reduced, so this is not a redox reaction.

Figure 15-12
Formation of sugar and H_2O during photosynthesis is a redox reaction because the oxidation state for C decreases from +4 to 0, and the oxidation state for some of the O increases from −2 to 0. Hydrogen's oxidation state does not change.

Now consider the reaction shown in **Figure 15-12**. Through photosynthesis, plants convert carbon dioxide and water into sugars and oxygen. The equation is not balanced as written, but oxidation states of the elements can still be assigned because they depend only on the relationship between atom and molecule.

The oxidation state for C has changed from +4 in CO_2 to 0 in $C_6H_{12}O_6$, as the arrow on the upper red line in **Figure 15-12** indicates. Carbon has been reduced because each C atom has gained four electrons. Arrows on the lower red line show that the oxidation state of O has gone from −2 in both CO_2 and H_2O to 0 in O_2. The oxidation state of O in $C_6H_{12}O_6$ is still −2; it did not change. Oxygen has been oxidized because each O atom that became part of an O_2 molecule has lost two electrons. There has been a change in oxidation states; this is a redox reaction. Any time a pure element appears as a reactant or product, the reaction is a redox reaction.

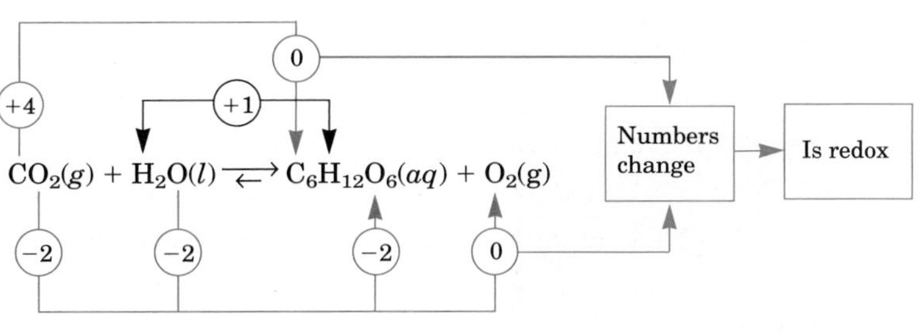

Groupwork Strategy

Assigning oxidation states

In groups of 2 or 3, have students determine the oxidation states for all of the elements in the following compounds.

a. $KMnO_4$ **b.** MnO_2
c. NH_4Cl **d.** $Sr(OH)_2$
e. $Mg(ClO_3)_2$ **f.** $MgClO$
g. $Al_2(SO_4)_3$ **h.** NaH_2PO_4
i. $Na_2CO_3 \cdot 10H_2O$

Do You Know?

Because photosynthesis converts light energy into chemical energy, it is a photochemical reaction. Photochemical reactions generally convert light energy into chemical energy very efficiently because the energy absorbed by an atom or molecule is much greater than the energy it could absorb from some other source. Production of ozone from oxygen in the upper atmosphere is also a photochemical reaction, but is it a redox reaction as well?

Chapter Connections

You may wish to point out that the hydrogen-powered car in Chapter 9 depends on a redox reaction. The hydrogen is oxidized from 0 to +1, and oxygen is reduced from 0 to −2.

Further Applications

Separating metals from their ores involves oxidation-reduction reactions.

Visual Strategy

Figure 15 -11 and **Figure 15-12**

Verify the oxidation states for each atom in the two equations. Point out that not every atom in a reaction must be oxidized or reduced.

Answers to

Concept Review

11. **a.** not redox
    b. redox
    c. not redox
12. Zn is oxidized, and Cu is reduced.

Concept Review

Redox reactions

11. Identify the following reactions as redox or nonredox.
    a. $MgCO_3 \longrightarrow MgO + CO_2$
    b. $Zn + CuSO_4 \longrightarrow ZnSO_4 + Cu$
    c. $NaCl + AgNO_3 \longrightarrow AgCl + NaNO_3$
12. For each redox reaction in item **11**, identify the element that is oxidized and the element that is reduced.

Answers for Concept Review items and Practice problems begin on page 841.

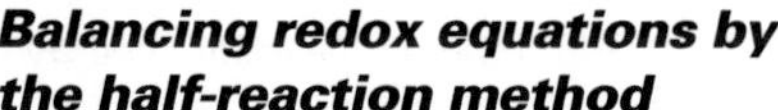

Oxidation states are used for balancing a redox equation

Chemical changes involve rearranging atoms and mass is conserved. You have seen that redox reactions are represented by chemical equations like any other chemical change. Thus, redox equations, like all others, must be balanced so that the law of conservation of mass is obeyed. In addition to mass, electrons are also transferred during redox reactions. Charge must be conserved; consequently, the sum of all the oxidation states on the left side of the equation must equal the sum of all the oxidation states on the right side of the equation. Every redox equation must be balanced in terms of mass and charge.

Balancing redox equations by the half-reaction method

The person shown in **Figure 15-13** is suspected of driving while intoxicated. A Breathalyzer is an instrument for estimating blood alcohol levels outside a laboratory. The Breathalyzer contains an orange solution of potassium dichromate, $K_2Cr_2O_7$, in aqueous acid. The $K_2Cr_2O_7$ reacts with alcohol in a person's breath to form green chromium(III) sulfate, $Cr_2(SO_4)_3$, and acetic acid. The equation for this redox reaction and instructions for how to balance it are shown on the next page.

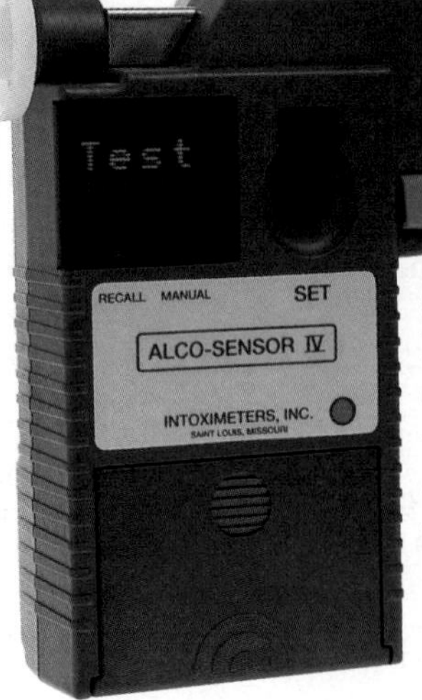

Figure 15-13
The alcohol in a person's breath is oxidized to acetic acid by dichromate ions in acidic solution. Dichromate ions are orange. The reduced Cr^{3+} ion is green. Appearance of a green color indicates an alcohol content greater than the legal limit.

One approach to balancing a redox equation involves separating it into half-reactions, balancing them separately in terms of charge and mass, and recombining them. Since each half-reaction is balanced separately, this approach is called the *half-reaction method.* The oxidation states that change in the reaction are the only ones shown. Each Cr atom in $K_2Cr_2O_7$ went from a +6 oxidation state to a +3 oxidation state in $Cr_2(SO_4)_3$. Three electrons were gained, so Cr was reduced. Each carbon atom in the alcohol has gone from a −2 oxidation state to a 0 oxidation state in the acetic acid. Because two electrons were lost, each C atom was oxidized. The electrons that a C atom loses are gained, in turn, by a Cr atom. Oxidation and reduction occur simultaneously.

Visual Strategy

Figure 15-13

People stopped by the police for driving under the influence of alcohol are often asked to blow up a balloon if a Breathalyzer is not immediately available. This is because the balloon will hold the ethanol without decomposition for a short period of time.

How To

Balance equations by the half-reaction method

Balance the following redox reaction.

$K_2Cr_2O_7(aq) + H_2SO_4(aq) + C_2H_5OH(sol) \longrightarrow$
$Cr_2(SO_4)_3(aq) + C_2H_4O_2(l) + H_2O(l) + K_2SO_4(aq)$

1. *Write the formula equation if it is not given in the problem. Then write the ionic equation.*

$$2K^+(aq) + Cr_2O_7^{2-}(aq) + 2H^+(aq) + SO_4^{2-}(aq) + C_2H_5OH(sol) \longrightarrow$$
$$2Cr^{3+}(aq) + 3SO_4^{2-}(aq) + C_2H_4O_2(sol) + H_2O(l) + 2K^+(aq) + SO_4^{2-}(aq)$$

2. *Assign oxidation numbers to each element and ion. Retain only those substances containing an element that changes oxidation state.*

$$\overset{+6}{Cr_2O_7^{2-}} + \overset{-2}{C_2H_5OH} \longrightarrow \overset{+3}{2Cr^{3+}} + \overset{0}{C_2H_4O_2}$$

3. *Write the half-reaction for oxidation.*

$$C_2H_5OH \longrightarrow C_2H_4O_2$$

When balancing redox equations, use the short-hand form of H_3O^+, which is H^+.

- ***Balance the mass.*** C atoms are balanced, but H and O atoms are not, so add 1 H_2O to the left side of the equation and 4 H^+ ions to the right. Because H_2O is the solvent and dissociates by itself into H^+ and OH^- ions, H^+ is used to balance H atoms and H_2O balances O atoms. The equation is now balanced for atoms but not for charge.

$$1H_2O + C_2H_5OH \longrightarrow C_2H_4O_2 + 4H^+$$

The left side has a net charge of 0. The right side has a net charge of +4.

- ***Balance the charge.*** Electrons are added to the side having the greater positive net charge. Add $4e^-$ to the right side to balance charge. C atoms on the left side of the equation lose $2e^-$ each.

$$1H_2O + C_2H_5OH \longrightarrow C_2H_4O_2 + 4H^+ + 4e^-$$

4. *Write the half-reaction for reduction.*

$$Cr_2O_7^{2-} \longrightarrow 2Cr^{3+}$$

The left side has a net charge of +12. The right side has a net charge of +6.

- ***Balance the mass.*** Cr atoms are balanced. Add $7H_2O$ to the right side to balance the O atoms. Then balance the H atoms by adding $14H^+$ to the left side of the equation.

$$Cr_2O_7^{2-} + 14H^+ \longrightarrow 2Cr^{3+} + 7H_2O$$

- ***Balance the charge.*** Electrons are added to the side having the greater positive net charge. Add $6e^-$ to the left side of the equation to make the charges equal. Cr atoms on the left side of the equation gain $3e^-$ each.

$$Cr_2O_7^{2-} + 14H^+ + 6e^- \longrightarrow 2Cr^{3+} + 7H_2O$$

How To continued on following page . . .

Teaching Tip

Balancing equations

If the half-reaction method proves tedious for some, equations can be balanced algebraically, and the number of electrons transferred is the product of the coefficient for the substance that was oxidized (or reduced) and the change in oxidation states for that substance.

Content Background

Balancing in basic solution

H^+ and H_2O are used to balance reactions that take place in acidic solution. For reactions that take place in basic solution, use OH^- and H_2O. Add OH^- to the side of the equation that *needs* O atoms. Add twice the number of O atoms needed. Add enough H_2O to the other side of the equation to balance the H atoms. For example, ClO^- ions react with CrO_2^- ions as follows.

$CrO_2^- + ClO^- \longrightarrow CrO_4^{2-} + Cl^-$

ox: $2(CrO_2^- + 4OH^- \longrightarrow CrO_4^{2-} + 3e^- + 2H_2O)$

red: $3(ClO^- + H_2O + 2e^- \longrightarrow Cl^- + 2OH^-)$

net: $2CrO_2^- + 3ClO^- + 2OH^- \longrightarrow 2CrO_4^{2-} + 3Cl^- + H_2O$

Additional Examples

How To Balance Redox Equations

Balance the following redox reactions using the half-reaction method.

a. $FeCl_3 + H_2S \longrightarrow FeCl_2 + HCl + S$

b. $KClO_3 \longrightarrow KCl + O_2$

c. $Al + H_2SO_4 \longrightarrow Al_2(SO_4)_3 + H_2$

d. $Cu + Br_2 + OH^- \longrightarrow Cu_2O + Br^- + H_2O$

Solutions are at the bottom of the page.

Solutions for Additional Examples

a. $2FeCl_3 + H_2S \longrightarrow 2FeCl_2 + 2HCl + S$
b. $2KClO_3 \longrightarrow 2KCl + 3O_2$
c. $2Al + 3H_2SO_4 \longrightarrow Al_2(SO_4)_3 + 3H_2$
d. $2Cu + Br_2 + 2OH^- \longrightarrow Cu_2O + 2Br^- + H_2O$

*Full solutions for items **a–d** are found on page 609A.*

ASSESS

Alternative Assessment

1. Design a flowchart for display that describes how to balance oxidation-reduction reactions. Include examples on the chart.
2. Develop a poster display that describes the black and white photographic processes in terms of the redox reactions involved.

Answers to Section Review

13. Na = +1, S = +2, O = −2; H = +1, Se = +4, O = −2; H = +1, O = −1
14. **a.** not redox
    b. not redox
    c. redox
15. **a.** red:
    $NO_3^- + 3e^- + 4H^+ \longrightarrow NO + 2H_2O$
    ox:
    $Ag \longrightarrow Ag^+ + e^-$
    net:
    $3Ag + NO_3^- + 4H^+ \longrightarrow NO + 2H_2O + 3Ag^+$
    b. red:
    $4H^+ + SO_4^{2-} + 2e^- \longrightarrow SO_2 + 2H_2O$
    ox:
    $2I^- \longrightarrow I_2 + 2e^-$
    net:
    $2H_2SO_4 + 2KI \longrightarrow K_2SO_4 + I_2 + SO_2 + 2H_2O$

How To continued from the previous page . . .

5. Conserve charge by adjusting the coefficients in front of the electrons so that the number lost in oxidation equals the number gained in reduction.

Set up a ratio of the electrons lost to the electrons gained. Reduce the ratio to its lowest terms.

The 4 comes from step 3. The 6 comes from step 4.

$$\frac{4}{6} \text{ becomes } \frac{2}{3}$$

Multiply the oxidation half-reaction by the denominator, 3.

$$3H_2O + 3C_2H_5OH \longrightarrow 3C_2H_4O_2 + 12H^+ + 12e^-$$

Multiply the reduction half-reaction by the numerator, 2.

$$2Cr_2O_7^{2-} + 28H^+ + 12e^- \longrightarrow 4Cr^{3+} + 14H_2O$$

6. Combine the half-reactions, subtracting out the electrons from both sides of the equation.

$$3H_2O + 2Cr_2O_7^{2-} + 28H^+ + 3C_2H_5OH \longrightarrow 3C_2H_4O_2 + 12H^+ + 4Cr^{3+} + 14H_2O$$

You see that H_2O and H^+ ions are on both sides of the equation. Combine H^+ ions by subtracting the smaller quantity, $12H^+$, from both sides of the equation. Do the same for H_2O.

$$2Cr_2O_7^{2-} + 16H^+ + 3C_2H_5OH \longrightarrow 3C_2H_4O_2 + 4Cr^{3+} + 11H_2O$$

7. Combine ions to form the compounds shown in the formula equation and then add the remaining ions from the ionic equation.

Ions from K_2SO_4 were eliminated as spectators in the ionic equation.

To re-form reactant compounds, combine $Cr_2O_7^{2-}$ with $2K^+$ and $2H^+$ with SO_4^{2-}. To re-form products, combine $2Cr^{3+}$ with $3SO_4^{2-}$ ions and $2K^+$ ions with a SO_4^{2-} ion.

$$2K_2Cr_2O_7 + 8H_2SO_4 + 3C_2H_5OH \longrightarrow 3C_2H_4O_2 + 2Cr_2(SO_4)_3 + 11H_2O + 2K_2SO_4$$

Section Review

13. Determine the oxidation numbers for each element in the following compounds: $Na_2S_2O_3$, H_2SeO_3, and H_2O_2.
14. Which of the following unbalanced equations represent redox reactions?
    a. $SO_3 + H_2O \longrightarrow H_2SO_4$
    b. $Pb(C_2H_3O_2)_2 + H_2S \longrightarrow PbS + HC_2H_3O_2$
    c. $I_2O_5 + CO \longrightarrow I_2 + CO_2$
15. Use the half-reaction method to balance the following equations.
    a. $Ag + NO_3^- \longrightarrow Ag^+ + NO$
    b. $H_2SO_4 + KI \longrightarrow K_2SO_4 + I_2 + SO_2 + H_2O$

How do batteries work?

Section Objectives

Identify criteria for selecting or designing a battery.

Describe the redox reactions that occur in dry-cell, alkaline, and mercury batteries.

Distinguish between an electrochemical cell and an electrolytic cell.

Describe the redox reactions that occur in rechargeable batteries.

Consumable batteries

All the batteries shown in **Figure 15-14** depend on redox reactions in which chemical energy is converted into electric energy. From what you have learned about electrochemistry, you can now examine various aspects of redox reactions that must be taken into account when designing a **battery**.

battery
single voltaic cell or group of voltaic cells that are connected together

1. Metals and their ions must be chosen so that a spontaneous redox reaction occurs.
2. The E^0 values for the half-reactions should be such that when combined, the desired voltage is obtained.
3. The redox reactions should be easily reversible if a rechargeable battery is wanted. Metals chosen for the electrodes in rechargeable batteries should be close together in the activity series.
4. The effect of reaction products on the environment must be considered, especially if they are gases or toxic materials.

Relative Longevity of Batteries

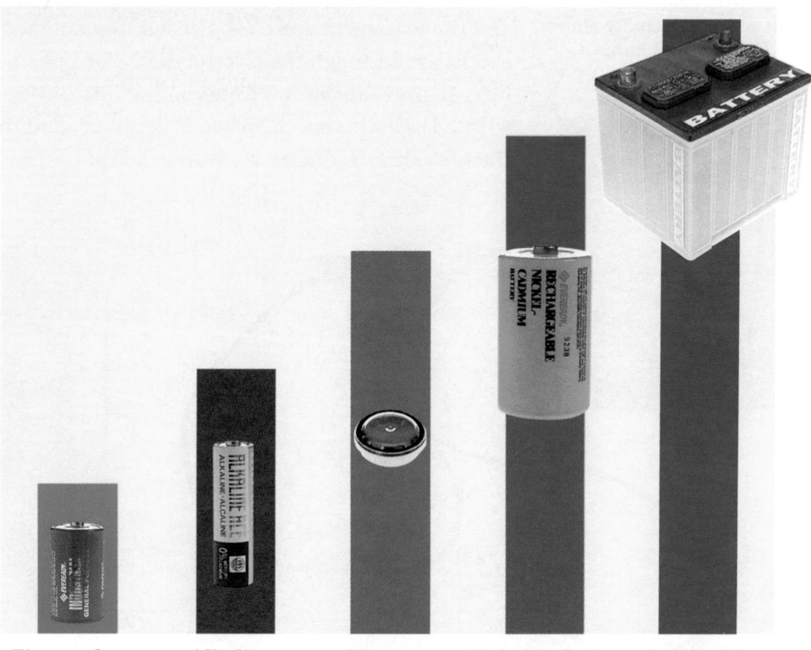

Figure 15-14
The size, voltage output, and longevity of a battery depends on the half-reactions that occur within it and the nature of the electrolyte.

Section 15-4

FOCUS

Lesson Starter

You may wish to begin this lesson with the following demonstration that compares the longevity of flashlight batteries.

Demonstration 5

Comparing flashlight batteries

Approximate time: 2 h

1. Obtain three similar flashlights.
2. Put carbon-zinc batteries in one flashlight, alkaline batteries in the second, and nickel-cadmium batteries in the third.
3. Mount the flashlights so that they shine into the classroom but not into anyone's eyes.
4. Ask the students to predict: Which flashlight will burn the longest? Which will be the brightest after 30 min? Which is cheapest to operate?
5. Turn the flashlights on. Ask the students to speak up if they see any difference in intensity of the light (this should take a while), and continue with class.

SAFETY
No special safety precautions are required.

DISPOSAL
Cover the positive terminals of the batteries with tape. Dispose of the batteries as directed by the manufacturer or the store where they were purchased.

Visual Strategy

Figure 15-14

A battery's longevity is related to its size, chemical composition, intended purpose, manufacturing cost, operating conditions, and storage temperature. Discuss the organization of the graph in terms of these criteria.

TEACH

Historical Note

The carbon-zinc dry cell we use today was invented by the French physicist Georges Leclanché.

Do You Know?

The container of a carbon-zinc dry cell is made of zinc, so if the battery could be totally depleted, the container would dissolve, being totally converted to Zn^{2+} ions.

Dry cells are so called because they are not full of a liquid electrolyte as *wet cells* are. An automobile battery is a wet cell.

Content background

Cell vs. battery

Although the terms *battery* and *cell* are used interchangeably in common speech, true batteries are 2 or more cells connected together. Cells connected in series are placed anode to cathode, causing a summation of the voltage. Cells connected in parallel have all of the anodes connected and all of the cathodes connected, causing a summation of the current. The battery in a camera is a cell, but the battery that powers an automobile is a true battery.

Demonstration 6

Voltage from a 6-cell fruit battery

Approximate time: 5 min

1. Obtain a lemon, grapefruit, or apple, and slice it into six sections.
2. Cut six Cu and six Zn strips, each 1 cm × 10 cm.
3. Clean each strip with steel wool.
4. Insert one Cu strip through the skin of each section of fruit. Also insert one zinc strip, separated from the Cu strip by at least 1 cm.

The most common battery is a zinc-carbon cell bridged by an NH_4Cl paste

The most commonly used battery, known as a dry cell, depends on the energy generated by a redox reaction involving zinc and manganese dioxide. Each year, more than 5 billion such batteries are used throughout the world, requiring over 30 metric tons of zinc per day.

These batteries are really not dry at all. Rather, they consist of a zinc container that serves as the anode and is filled with a moist paste of MnO_2, graphite, and ammonium chloride, NH_4Cl, as you can see in **Figure 15-15**. Oxidation of the zinc metal occurs at the anode.

$$Zn(s) \longrightarrow Zn^{2+}(aq) + 2e^-$$

Electrons travel through the external circuit and re-enter the battery through the carbon rod. The carbon rod functions as the cathode where MnO_2 is reduced in the presence of H_2O.

$$2MnO_2(s) + H_2O(l) + 2e^- \longrightarrow Mn_2O_3(s) + 2OH^-(aq)$$

This reaction is followed by a secondary reaction between OH^- ions and NH_4Cl. Ammonia gas is produced in this secondary reaction. Dry-cell batteries should never be recharged because the buildup of product gases can cause the sealed dry cell to explode.

This redox reaction produces a voltage of 1.5 V, so if your flashlight needs 3 V to light the bulb, you must use two of these batteries connected in series.

Unfortunately, this type of battery has two disadvantages. First, if the current is drawn too rapidly from the battery, the gaseous products cannot be removed quickly enough causing the redox reactions to slow down and the voltage to decrease. Second, the Zn and NH_4Cl react to a slight extent even though the electrolytic NH_4Cl paste is surrounded by a paper liner to prevent the two from coming into direct contact. As the Zn reacts with NH_4Cl, its concentration is depleted, and the loss of Zn decreases the life. This decreases the voltage output.

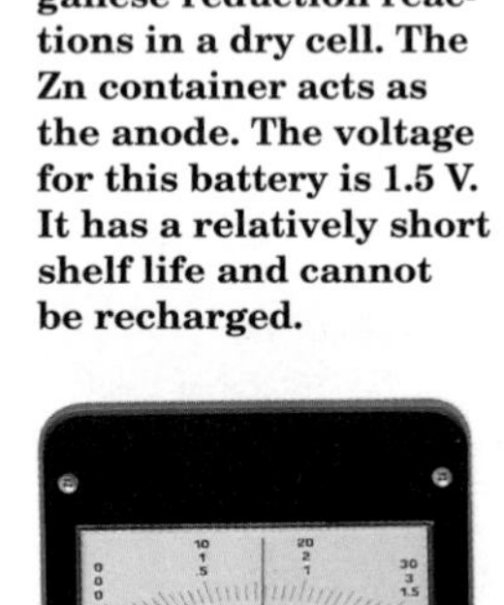

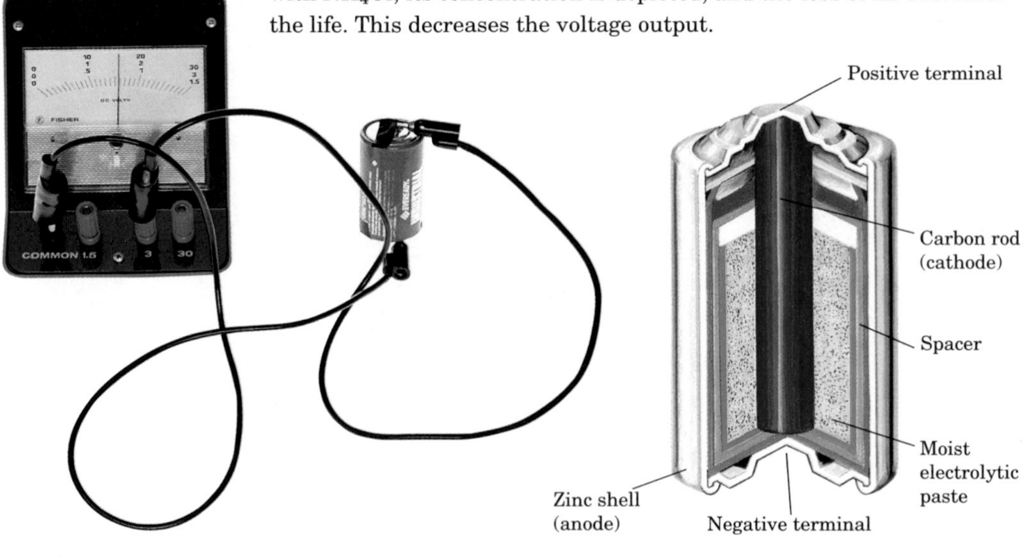

Figure 15-15
Carbon supplies the surface for the manganese reduction reactions in a dry cell. The Zn container acts as the anode. The voltage for this battery is 1.5 V. It has a relatively short shelf life and cannot be recharged.

Visual Strategy

Figure 15-15

Point out that the carbon rod serves as an inert electrode. It provides a surface on which the reduction of manganese occurs, but it does not participate in the redox reactions.

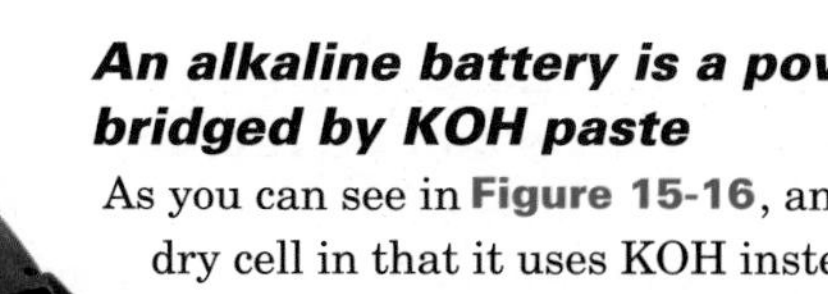

An alkaline battery is a powdered zinc-carbon cell bridged by KOH paste

As you can see in **Figure 15-16**, an alkaline battery differs from a dry cell in that it uses KOH instead of NH_4Cl. KOH is a base, so the electrolyte paste in this type of battery is alkaline. As with an ordinary dry cell, zinc is added, but this time as a powder in a gel-thickened mixture with KOH. Oxidation of Zn to either $[Zn(OH)_4]^{2-}$ or ZnO occurs at the anode while MnO_2 is reduced to $Mn(OH)_2$ at the cathode. It produces 1.5 V, the same as a dry cell battery. However, no gases are formed in an alkaline battery reaction, and no unwanted side reactions occur to reduce the voltage generated by the redox reaction. There is no acidic corrosion of the container as there is with a conventional dry cell. Elimination of the carbon post allows alkaline batteries to be made smaller than dry cells. Moreover, alkaline manganese batteries perform far better at temperatures significantly below freezing than conventional zinc-carbon batteries because their ions are more mobile.

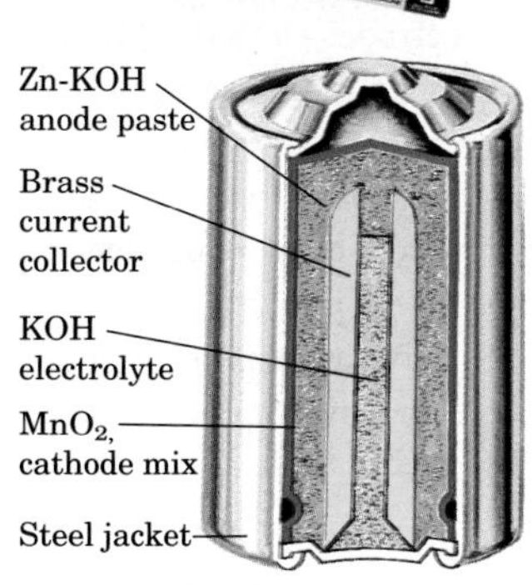

Figure 15-16
The alkaline form of the dry cell performs better in a Walkman than does the acidic form in Figure 15-15. It is also easier to miniaturize because Zn is present as a powder, and there is no graphite rod.

A mercury battery is environmentally hazardous

The mercury battery shown in **Figure 15-17** is a close relative of the alkaline battery. Again, powdered zinc is oxidized by hydroxide ion at the anode, but solid mercury(II) oxide, HgO, is reduced to elemental mercury at the cathode. The equations for the electrochemical reactions are written as follows.

$$\text{Anode: } Zn + 2OH^- \longrightarrow ZnO + H_2O + 2e^-$$
$$\text{Cathode: } HgO + H_2O + 2e^- \longrightarrow Hg + 2OH^-$$
$$\text{Net redox: } Zn(s) + HgO(s) \longrightarrow ZnO(s) + Hg(l)$$

Because the electrolyte is not used up during the redox reaction, the cell potential changes very little as the alkali concentration changes. A steady output of 1.34 V makes this redox reaction especially valuable for use in communication equipment and scientific instruments for which power fluctuations would cause problems. However, the production of liquid mercury, a poisonous metal that can cause serious health and environmental problems, is a serious disadvantage for mercury batteries. Used mercury batteries must, therefore, be recycled to recover the elemental mercury.

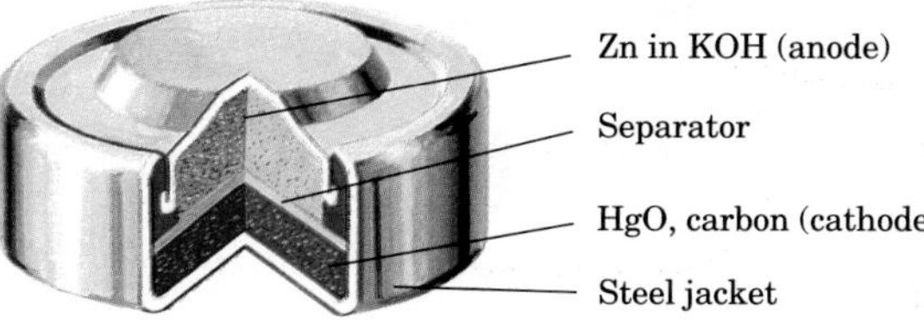

Figure 15-17
The mercury cell, used to power the flash attachment in this camera, is an alkaline form of the dry cell. The cathode is HgO mixed with graphite, and the anode is Zn mixed with KOH.

5. Lay the fruit sections side by side. Using paper clips, attach the copper strip in the first fruit section with the zinc strip in the second section. Continue this process until all six sections are connected.
6. Then connect the free copper strip to one end of a light emitting diode, or LED, and the free zinc strip to the other end of the LED which should glow brightly if the room lights are dimmed.
7. Attach a voltmeter to show the voltage that a fruit battery is capable of producing.
   Variation: Use all three fruits mentioned in step **1**, and compare the voltage output.

SAFETY
Do **not** eat the fruit. Wear goggles and a lab apron.

DISPOSAL
Mark the fruit "Unfit for food use," and throw it in the trash. Save the metal strips for next time.

Do You Know?

Most battery-powered devices are designed to operate within a range of potentials because not all batteries of the same type give exactly the same voltage or energy output. However, they are very close to one another, differing in output by no more than 5%.

Visual Strategy

Figure 15-16 and **Figure 15-17**

Correlate the parts of the mercury cell with the half-reactions in the text. Ask students what purpose the separator serves. Miniaturizing the alkaline battery in **Figure 15-16** results in a form that looks very much like that of the mercury cell. Have students try to draw the miniature form of an alkaline battery.

Do You Know?

The savings of a few pounds is very important to mileage-conscious auto makers. In the last few years, many auto makers have installed batteries that are too small for the cars they power. Because of this the battery is drained much more while starting the car, so the battery goes through greater drain-recharge cycles and wears out much faster than a larger battery would. The battery compartment, however, will hold a much larger replacement battery.

Teaching Tip

Electric discharge

Have students ascertain how an electric eel produces its charge and the effectiveness and range of the charge.

Content Background

Lead storage batteries

The sulfuric acid in a car battery is also called battery acid and is extremely corrosive. Because the electrochemical reaction consumes sulfuric acid, the degree to which the battery has been discharged can be checked by measuring the density of the electrolyte with a hydrometer.

Although thermodynamic calculations show that the voltage of electrochemical cells decreases with decreasing temperature, this decrease in voltage is only about 0.05% for a lead storage battery. Consequently, the apparent deadness of a battery in cold climates is caused by an increase in viscosity of the electrolyte as the temperature decreases. Ions move quite slowly in a viscous medium, so the ability of the electrolytic solution to conduct decreases, but its resistance increases, leading to a decrease in the battery's voltage output.

electrolysis
a process in which electric energy is used to bring about a chemical change

electrolytic cell
an electrochemical cell in which electric energy from an external source causes a nonspontaneous redox reaction to occur

Rechargeable batteries

Electrochemical cells can be recharged

A voltage will be generated by a battery only if electrons continue to be removed from one substance and transferred to another. Once this process stops, voltage is no longer produced. This process stops when an equilibrium involving both half-cells is established, and the battery is "dead."

To regenerate a dead battery, it must be charged; that is, an external voltage source must be applied to the battery's electrodes so that all half-reactions are reversed and the electrodes are returned to their original state. Thus, while the battery is being used, it operates as a voltaic cell, converting chemical energy into electric energy. But, while the same battery is being charged, it operates as an electrolytic cell, converting electric energy into chemical energy.

The process in which electric energy is used to drive a redox reaction is **electrolysis**. The container in which electric energy drives a nonspontaneous redox reaction is an **electrolytic cell**.

A car battery is a combination electrochemical-electrolytic cell

A lead-acid battery like the type used in cars is shown in **Figure 15-18**. It produces 12 V from six cells that are connected to one another in a line. The anode in each cell is lead. The electrolytic solution is sulfuric acid, H_2SO_4.

The anode reaction is described by the following equation.

$$Pb(s) + SO_4^{2-}(aq) \longrightarrow PbSO_4(s) + 2e^-$$

The electrons move through the circuit to the cathode where PbO_2 is reduced.

$$PbO_2(s) + 4H^+(aq) + SO_4^{2-}(aq) + 2e^- \longrightarrow PbSO_4(s) + 2H_2O(l)$$

The following equation describes the net redox reaction.

$$Pb(s) + PbO_2(s) + 2H_2SO_4(aq) \longrightarrow 2PbSO_4(s) + 2H_2O(l)$$

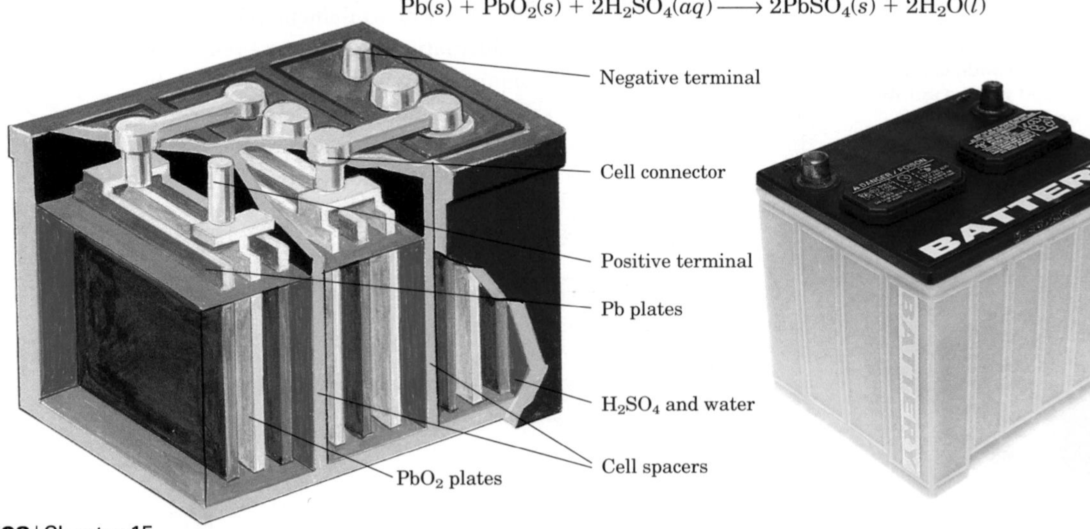

Figure 15-18
The lead-acid storage battery is a true battery because it contains six cells instead of a single cell like the nonrechargeable batteries discussed earlier.

Visual Strategy

Figure 15-18

Show how the cells are connected. Describe the function of cell parts. Caps on top of the battery lead to individual cells. The water level in each cell must be monitored so that the level of electrolyte doesn't diminish, making the cell ineffective. When a car battery wears out, often only one of the cells is truly ruined, but the lowered voltage (10 V) will not operate the starter motor at a speed adequate to start the car.

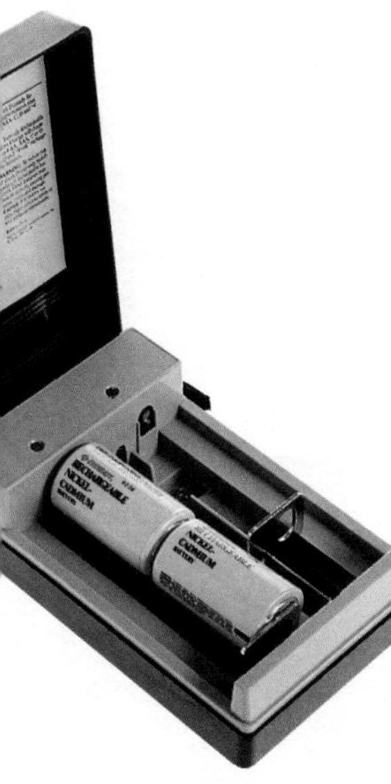

A car's battery produces the 12 V needed to start its engine. H_2SO_4, present as its ions, is consumed and $PbSO_4$ accumulates as a white powder on the electrodes. Once the car is running, the alternator produces a voltage that reverses the battery's half-reactions and regenerates Pb, PbO_2, and H_2SO_4. A lead-acid battery can undergo many thousands of discharge-recharge cycles before it finally fails. At that point, too much $PbSO_4$ has flaked off the electrodes and is no longer available as a reactant for the reverse reactions. A battery can be recharged only when all reactants necessary for the electrolytic reaction are present, and all reactions are reversible.

A ni-cad battery is also rechargeable

Rechargeable and lightweight, nickel-cadmium batteries are becoming increasingly popular. They have two advantages over other rechargeable batteries: easily regenerated electrodes and production of a nearly constant 1.4 V.

Cadmium serves as the anode.

Figure 15-19
Current from a wall outlet provides the electric energy needed for recharging these nickel-cadmium batteries.

$$Cd(s) + 2OH^-(aq) \longrightarrow Cd(OH)_2(s) + 2e^-$$

NiO(OH) is reduced at the cathode.

$$2NiO(OH)(s) + 2H_2O(l) + 2e^- \longrightarrow 2Ni(OH)_2(s) + 2OH^-(aq)$$

The net reaction occurring in a Ni-Cad battery is written as follows.

$$Cd(s) + 2NiO(OH)(s) + 2H_2O(l) \longrightarrow Cd(OH)_2(s) + 2Ni(OH)_2(s)$$

Since the late 1970s cadmium as the element or in soluble compounds has been implicated in a number of very toxic effects on humans and wildlife. Therefore, some nations control the recycling and reprocessing of ni-cad storage batteries.

Section Review

16. Which of the following pairs of electrodes would make good batteries? Explain.
a. $Cd \longrightarrow Cd^{2+} + 2e^-$; $Fe \longrightarrow Fe^{2+} + 2e^-$
b. $Ag^+ + e^- \longrightarrow Ag$; $Cu^{2+} + 2e^- \longrightarrow Cu$

17. How is a 9 V battery made?

18. What advantages does an alkaline battery have over an ordinary dry cell?

19. What disadvantage does a mercury battery have that an alkaline battery does not have?

20. Describe an electrolytic cell.

21. Explain why a rechargeable battery can be considered a combination voltaic-electrolytic cell.

Teaching Tip

Reactants in ni-cad batteries
NiO(OH) represents a mixture of nickel oxides and hydroxides.

ASSESS

Alternative Assessment

Investigate the types of batteries being considered for electric cars, and gather information on the advantages and pitfalls of each type. Develop your findings into a class presentation.

Answers to Section Review

16. a. No; Cd poses a health hazard, Fe and Cd are relatively close in activity, and the cell voltage is about +0.04 V.
b. No; silver is expensive and would have to be reclaimed. The voltage produced is only +0.46 V.

17. by connecting six 1.5 V cells in series

18. steadier voltage output, no gaseous reaction byproducts, longer service life

19. Toxic elemental Hg is a reaction byproduct.

20. a device in which electric energy is converted to chemical energy

21. When discharging, the reaction is spontaneous, and chemical energy is changed to electric energy; when charging, the reaction is nonspontaneous, and electric energy is changed to chemical energy.

Content Background

Identification of elements
The history of elemental chemistry has consisted of an understanding of the basic components of chemical substances (elements) and of the manner in which chemical substances react.

Teaching Tip

Discussion
Point out to students that in 1661, when Boyle conducted his studies, only 13 elements were known to exist, and 9 of these—carbon, iron, copper, sulfur, silver, tin, gold, mercury, and lead—had been known since antiquity. Zinc, arsenic, antimony, and bismuth were discovered in the Middle Ages. Phosphorus, the first element whose discovery was recorded, would not be isolated for another eight years. Within 300 years, 90 elements would be identified by various analytical means, electrolysis and spectroscopy being among the most powerful.

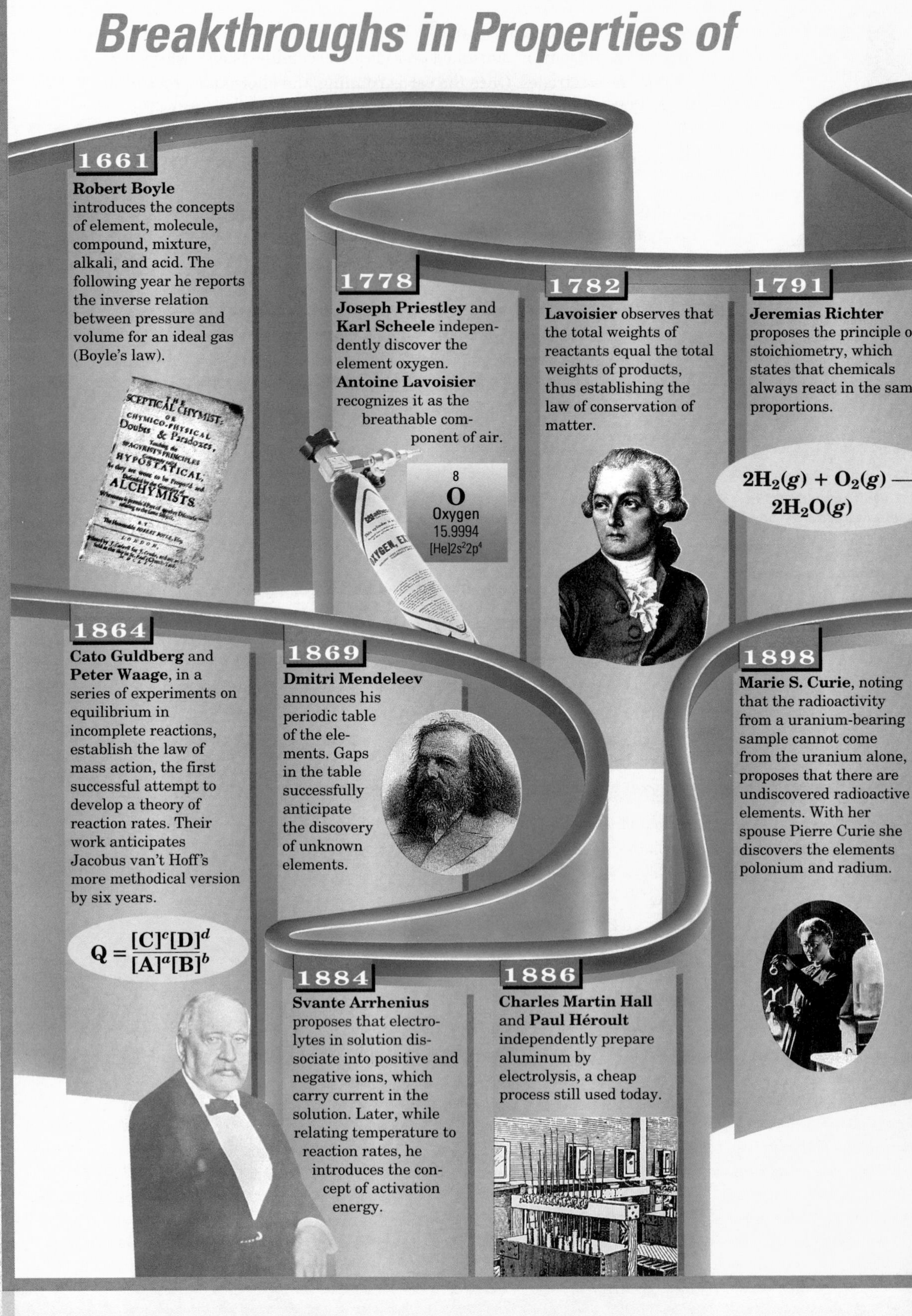

Breakthroughs in Properties of
1661
Robert Boyle introduces the concepts of element, molecule, compound, mixture, alkali, and acid. The following year he reports the inverse relation between pressure and volume for an ideal gas (Boyle's law).
1778
Joseph Priestley and Karl Scheele independently discover the element oxygen. Antoine Lavoisier recognizes it as the breathable component of air.
8
O
Oxygen
15.9994
$[He]2s^22p^4$
1782
Lavoisier observes that the total weights of reactants equal the total weights of products, thus establishing the law of conservation of matter.
1791
Jeremias Richter proposes the principle of stoichiometry, which states that chemicals always react in the same proportions.
$2H_2(g) + O_2(g)$ —
$2H_2O(g)$
1864
Cato Guldberg and Peter Waage, in a series of experiments on equilibrium in incomplete reactions, establish the law of mass action, the first successful attempt to develop a theory of reaction rates. Their work anticipates Jacobus van't Hoff's more methodical version by six years.
$Q = \frac{[C]^c[D]^d}{[A]^a[B]^b}$
1869
Dmitri Mendeleev announces his periodic table of the elements. Gaps in the table successfully anticipate the discovery of unknown elements.
1898
Marie S. Curie, noting that the radioactivity from a uranium-bearing sample cannot come from the uranium alone, proposes that there are undiscovered radioactive elements. With her spouse Pierre Curie she discovers the elements polonium and radium.
1884
Svante Arrhenius proposes that electrolytes in solution dissociate into positive and negative ions, which carry current in the solution. Later, while relating temperature to reaction rates, he introduces the concept of activation energy.
1886
Charles Martin Hall and Paul Héroult independently prepare aluminum by electrolysis, a cheap process still used today.

Inorganic Substances

The following events significantly increased our understanding of chemistry through the identification of substances and their properties and the development of theoretical explanations of those properties.

1803

John Dalton, in the process of studying the behaviors of gases, forms the idea of atomic weights and, ultimately, the first systematic atomic theory.

1807

Humphry Davy isolates the alkali metals potassium and sodium through electrolysis. The following year he repeats the procedure, isolating the alkaline-earth metals barium, strontium, and calcium.

1811

Amedeo Avogadro, in a visionary paper, postulates the idea that equal volumes of gases contain equal numbers of particles and that some elements (such as oxygen) exist as diatomic molecules. Avogadro's ideas are ignored and forgotten until Stanislao Cannizzaro revives them in 1860.

6.022×10^{23}

1833

Michael Faraday discovers that many nonconducting solids conduct electricity in a liquid state, leading to the first understanding of electrolytes. Later, Faraday notes that the amount a substance dissociates in electrolysis is proportional to the amount of current passing through the fluid.

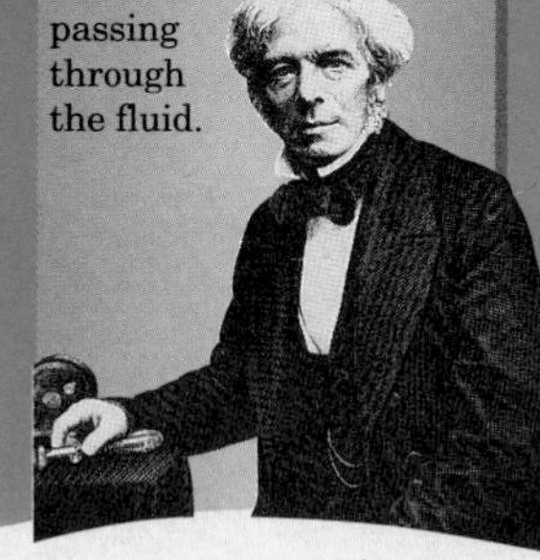

1909

Danish chemist **S.P.L. Sørensen** proposes the concept of pH as a measure of the acidity or basicity of a solution.

1923

Thomas Lowry, J. N. Brønsted, and **Niels Bjerrum** suggest that acids donate hydrogen ions and bases accept them. By contrast, **Gilbert Lewis** suggests that bases donate electron pairs and acids accept them.

$$HB(aq) + H_2O(aq) \rightleftharpoons H_3O^+(aq) + B^-(aq)$$

1934

Arnold Beckman devises the first electronic instrument to directly measure the pH of a solution.

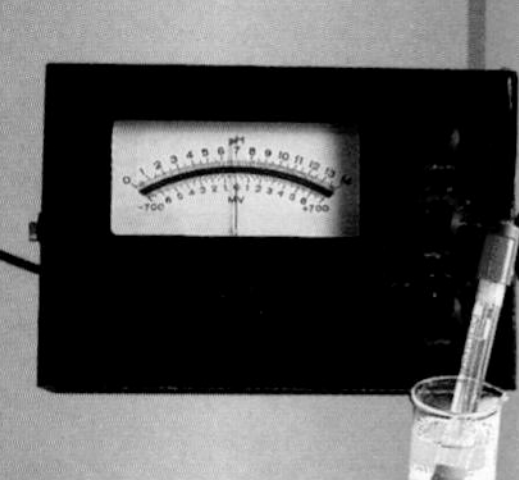

1962

Neil Bartlett demonstrates the first chemical reaction of a noble gas when he reacts xenon with platinum hexafluoride.

1967

Yuan Tseh Lee develops improved methods for observing chemical reactions. Using colliding beams of molecules he determines how energies are distributed in reactants to form products.

Content Background

Chemical processes

The identification of chemical processes can be divided into three categories: classification of chemical behavior (as was done on a primitive basis by Boyle and later to a more elaborate degree by Mendeleev); modeling of chemical mechanisms (arguably beginning with Lavoisier and continuing through the physical chemistry of Arrhenius, Brønsted, Lowry, and Lewis); and the development of the atomic model and elemental proportions (as carried out by Dalton, Richter, and Avogadro, among others).

Teaching Tip

Discussion

Stress to students the interaction of theory, experiment, and industrial application along the course of the timeline. For instance, the experimental discovery of elemental proportions in reactions gave rise to the theoretical work of Richter, Dalton, and Avogadro. This, in turn, provided a conceptual model of the atom that, combined with experimental discoveries of elements (such as those of Davy), allowed Mendeleev to make predictions about unknown elements. When these elements were isolated through experiments, the theoretical work that preceded the discoveries made the elements easier to recognize.

Chemistry in Your Community

1. A pending California law requires 2% of all cars sold in that state to be electric. Evaluate the effects this will have on consumers, and present your findings as an editorial for your school newspaper.
2. Determine what types of materials are used to make the batteries sold in local hardware, automotive, business, and camera stores. Locate the effective voltage and most suitable application for each battery type. Identify those batteries that are recyclable. Make a display chart of your findings.
3. Find out how automotive stores encourage the return and recycling of used batteries and the amount of the charge for this service in your area. Find out where distributors send the batteries to be recycled and how the process occurs. Report your findings as a large poster for display in the class.
4. Prepare a class presentation on how the electrolytic process has been used in the miniaturization and simplification of home appliances. (Hint: research circuit board production.)
5. Prepare a report on how electrochemical processes are used to remove heavy metals such as Hg from our water supply.

Answers to Applying Concepts

1. mercury battery: total mass = 281.98 amu; range = 140.99 amu/electron
   lead-acid battery: total mass = 642.58 amu; range = 321.29 amu/electron
2.

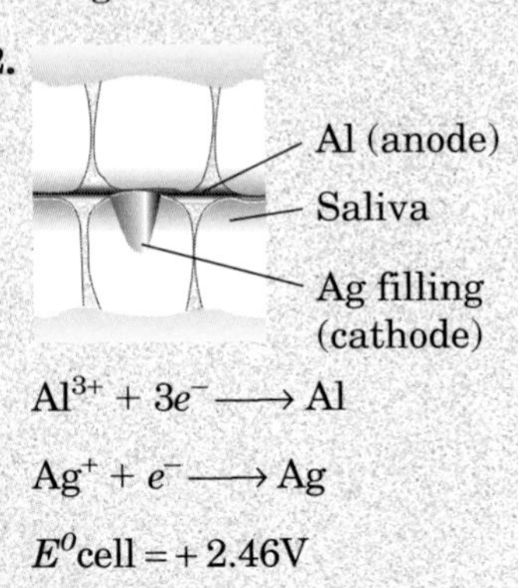

$Al^{3+} + 3e^- \longrightarrow Al$

$Ag^+ + e^- \longrightarrow Ag$

E^ocell = +2.46V

Conclusion: Fireflies and Electrochemical Cells

You have learned that luciferin is a chemical compound in the tail of a firefly. Under the proper conditions it can be made to produce light. This bioluminescence results from a series of redox reactions. You have also seen that light can be produced by redox reactions in electrochemical cells. Look again at the questions raised at the beginning of this chapter.

How is the light generated by a firefly different from the light generated by a flashlight battery? In a flashlight the electrons flow from the battery's anode to its cathode via the filament in a light bulb. Due to its resistance, the filament becomes hot and incandescent. It radiates all wavelengths in a continuous spectrum. On the other hand, special cells in the tail of a firefly oxidize luciferin, whose formula is $C_{11}H_8N_2O_3S_2$, to $C_{10}H_6N_2O_2S_2$. The electrons in the latter compound are in an excited state, and fall to the ground state, just as electrons do in excited atoms. They emit almost monochromatic light in the yellow region of the spectrum.

How is the concentration of TB cells related to the brightness of the light emitted? A firefly controls its light emission by controlling the amount of air supplied to the light-producing organs of its tail.

Adenosine triphosphate (ATP) is present in all living cells, including TB cells, and is required for the oxidation of luciferin. As long as TB cells are alive, they multiply. This makes ATP available for oxidizing luciferin so the glow continues. An antibiotic that kills the TB cells decreases the brightness of the glow by decreasing cell reproduction.

The voltage output of an electrochemical cell is related to the absolute concentration of the chemicals in the cell. Operation of the cell causes a decrease in reactants and voltage output. You see these changes as a dimmer light in a flashlight or as fainter volume in a tape player.

Applying Concepts

1. The range of a battery is estimated by adding the atomic masses of all the reactants, and dividing this sum by the number of electrons produced in the redox reaction. The answer is expressed in amu/electron. Calculate the range for both a lead-acid and a mercury battery.
2. If your gums and saliva act as a salt bridge, sketch the electrochemical cell that would produce a small jolt of pain when you bite down on an aluminum gum wrapper with a silver-filled tooth. Include half-reactions and E^o value for the cell.

Research and Writing

1. Determine how electroplating of metals is done, and cite examples of products made this way.
2. Investigate the use of cathodic protection in the construction industry.

CHAPTER 15

Highlights

Key terms

anode
battery
cathode
circuit
electricity
electrolysis
electrolytic cell
half-cell
half-reaction
oxidation
oxidation state
porous barrier
redox reaction
reduction
reduction potential
SHE
standard reduction potential
voltaic cell

Key problem-solving approach: *Assigning oxidation numbers*

When assigning oxidation numbers to molecules or polyatomic ions, first determine the oxidation numbers of the elements that are listed in the rules on pages 588–589. Then, solve for the unknown element so that the sum of the oxidation numbers equals zero or the ionic charge.

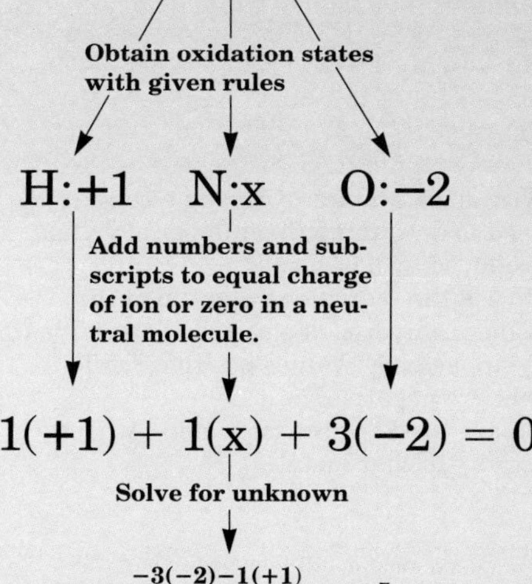
compound or ion
Ex: HNO_3
Obtain oxidation states with given rules
H:+1 N:x O:−2
Add numbers and subscripts to equal charge of ion or zero in a neutral molecule.
1(+1) + 1(x) + 3(−2) = 0
Solve for unknown
x = (−3(−2)−1(+1))/1 = +5

How does electric current flow?

- A half-cell consists of a metal electrode immersed in a solution of its ions. Two half-cells combine to form an electrochemical cell. If electric energy is produced, the cell is voltaic.
- The flow of free electrons through a circuit produces electricity in a voltaic cell.
- An element loses electrons when it is oxidized and gains electrons when it is reduced.
- Reduction and oxidation are always paired in redox reactions.

How do you get current to flow?

- Reduction potentials are linked directly to the activity series.
- The standard reduction potential, which is the voltage produced by a half-cell that is connected to the standard hydrogen electrode, can be used to predict the outcome of a redox reaction.
- Reduction potentials are assigned to the two half-reactions of an electrochemical cell. The cell potential is calculated by subtracting the lower reduction potential from the higher one.

How are electron transfers described?

- An oxidation state represents the attraction of an atom for the bonding electrons in a molecule. The oxidation state of an atom varies depending on the other atoms present in a molecule or ion.
- Redox reactions involve changes in oxidation states.
- Redox equations may be balanced by separating the reaction into half-reactions, balancing the half-reactions in terms of mass and charge, and then recombining them.
- Both mass and charge are conserved in redox reactions.

How do batteries work?

- Batteries run on redox reactions.
- A battery is "dead" when its component half-cells reach equilibrium with each other.
- Electrolytic cells are electrochemical cells in which electric energy is converted into chemical energy. They are the opposite of voltaic cells.
- Applied external current restores the half-cells of rechargeable batteries to their original, unreacted states.

Closure Strategy

An electrochemical cell with a potential of –0.25 V has the following cell reaction.

$$Ni(s) + 2H^+(aq) \longrightarrow Ni^{2+}(aq) + H_{2(g)}$$

If the concentrations of the ions are 1.0 M and the pressure of the H_2 is 1 atm, have students write a description of the cell and its operation including half-reactions and their E^0 values.

1. a. The porous barrier prevents the physical mixing of solutions.
   b. No. Without the barrier, the two electrolyte solutions would mix, and no current would be generated because electrons released at the anode would be transferred directly to cations in solution rather than migrating through the external circuit. (Compare with **Figure 15-1**.)
2. Electrons released in oxidation or gained in reduction cannot exist independently. If electrons are released or gained, the opposite reaction must take place so that electrons and charge are conserved.
3. No. Two half-cells with different potentials are required for a current to flow. Electricity is therefore not produced in a single half-cell.
4. a. It decreases.
   b. Positive ions from the anode cross the porous barrier into the cathode in an attempt to raise that compartment's cation concentration.
   c. Because a water molecule is electrically neutral, no charge would be transferred. Therefore, no current would flow, and the light bulb would not light.
5. Both the electrons in an electric circuit and the cars on a racing circuit travel in circular paths.
6. This is not true because any metal with a higher activity than the metal ion in a solution will be dissolved in that solution.
7. a. oxidation
   b. reduction
   c. reduction
   d. oxidation
8. a. The zinc ions cannot oxidize the copper metal.

CHAPTER 15

Review and Assess

The movement of electrons

REVIEW

1. a. What purpose does the porous barrier in an electrochemical cell serve?
   b. Would the cell produce electricity if the barrier were not there? Explain.
2. Why must oxidation and reduction always occur together in a reaction?
3. Can electricity be generated in a single half-cell? Explain.
4. a. Once electrons begin to flow, what happens to the cation concentration in the cathodic half-cell?
   b. How does the process in part **a** affect the cation concentration in the anodic half-cell?
   c. Would a light bulb shine if only water molecules could pass through the barrier? Explain.

APPLY

5. How is an electric circuit similar to a race track?
6. A friend tells you that only acids can dissolve metals. Is this true? Explain.
7. Identify the following half-reactions as oxidation or reduction.
   a. $Pb^{2+} \longrightarrow Pb^{4+} + 2e^-$
   b. $Pt^{2+} + 2e^- \longrightarrow Pt$
   c. $F_2 + 2e^- \longrightarrow 2F^-$
   d. $Ra \longrightarrow Ra^{2+} + 2e^-$
8. Use **Figure 15-2** to answer the following questions.
   a. Why does no reaction occur when zinc ions are in contact with the copper strip?
   b. Why do the copper ions not move towards the zinc strip?
9. Complete the following table.

	Oxidation/ reduction	Loss or gain of electrons	Change in charge
Cathode			
Anode			

10. Refer to the diagram below to answer the following questions.
    a. Why does heat production eventually stop in this reaction?
    b. What macroscopic observations suggest that a chemical reaction has taken place?
    c. What is happening at the microscopic level?

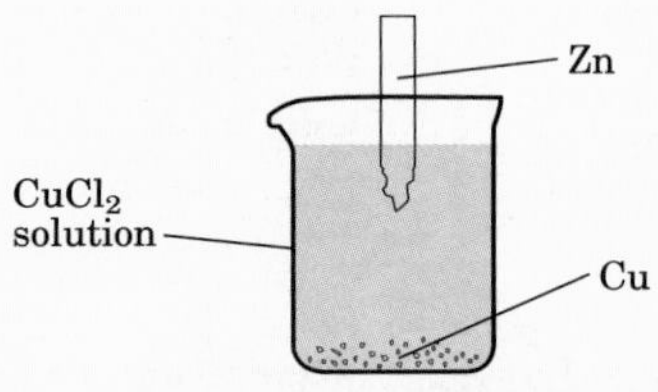

11. Draw diagrams of the following reactions at the particle level.
    a. oxidation
    b. reduction

Reduction potentials

REVIEW

12. Use **Table 15-1** to determine the E^0 cell value for the spontaneous reaction of each pair of half-cells listed below.
    a. $Ag \longrightarrow Ag^+ + 1e^-$; $Fe^{2+} + 2e^- \longrightarrow Fe$
    b. $Mg \longrightarrow Mg^{2+} + 2e^-$; $Zn^{2+} + 2e^- \longrightarrow Zn$
    c. $Li \longrightarrow Li^+ + 1e^-$; $Mn^{2+} + 2e^- \longrightarrow Mn$
    d. $Cr \longrightarrow Cr^{3+} + 3e^-$; $Pt^{2+} + 2e^- \longrightarrow Pt$
    e. $Ni \longrightarrow Ni^{2+} + 2e^-$; $Cu^{2+} + 2e^- \longrightarrow Cu$
13. a. Why is a standard hydrogen electrode used to determine the number of volts produced by a half-cell?
    b. How is the reduction potential of the standard hydrogen electrode determined?
14. Why are some E^0 values positive, while others are negative?
15. Compare the E^0 value for a metal with the reactivity of that metal.

16. How does the ranking of metals in the activity series relate to the ranking of the following?
    a. electronegativities
    b. reduction potentials
17. Explain why the reduction potential for some half-reactions in **Table 15-1** are negative.

APPLY

18. How can the fact that iron dissolves in acid help predict how the iron will react in a voltaic cell?
19. Draw a diagram of what happens at each electrode at the particle level for each half-reaction in a voltaic cell. Compare your diagram to those from item **11**.
20. Label the cathode and the anode in the diagram below. Identify the half-reactions taking place in each half-cell.

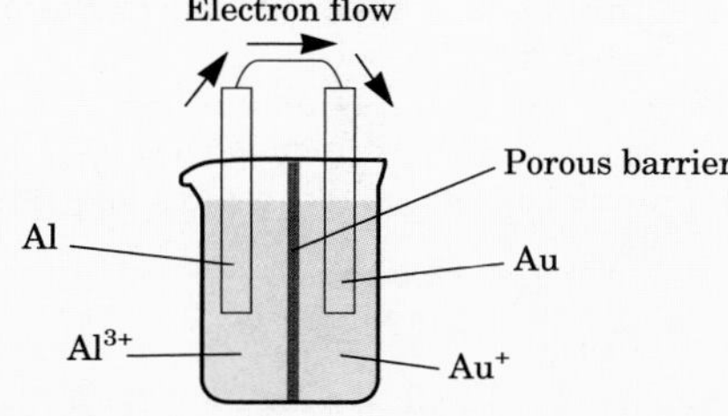

21. A friend has lost the labels for her experiment that identify the anode and cathode of the reaction below.
    a. List two observations that can determine the nature of the reaction.
    b. Label the anode and the cathode, and identify the half-reaction occurring in each of them.

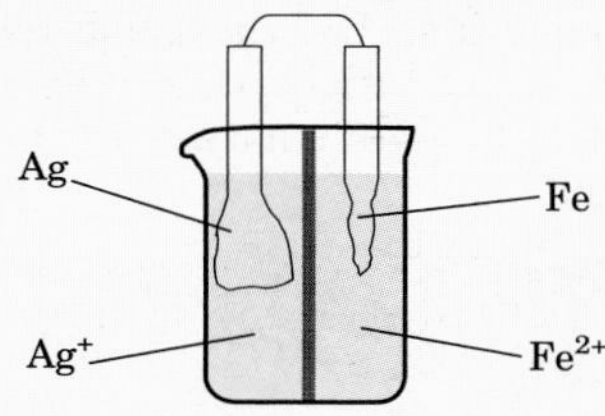

22. Evaluate a classmate's statement that because the E^0 value for H^+ is 0.00 V, hydrogen cannot be used in a voltaic cell.
23. Use **Table 15-1** to predict the voltage of the cells in the diagrams below.

a.

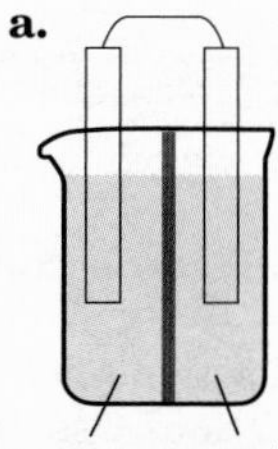

b.

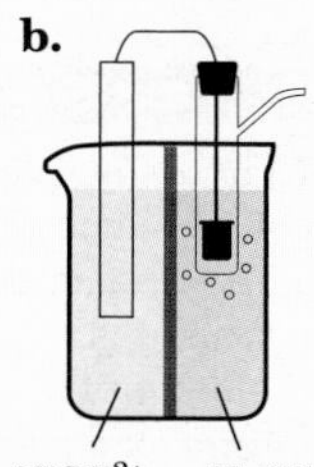

c.

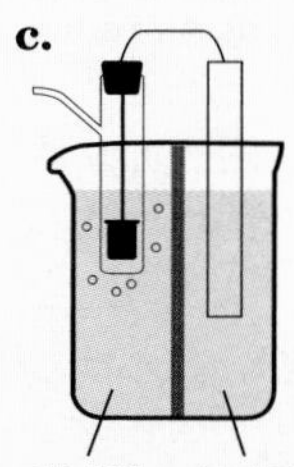

24. Identify three nonmetal electrodes listed in **Table 15-1**.
25. Why does a negative E^0 cell indicate a non-spontaneous reaction?
26. One half-cell of a voltaic cell consists of a Zn strip dipped into a 1.00 M solution of $Zn(NO_3)_2$. In the other half-cell, solid indium adsorbed on graphite is in contact with a 1.00 M solution of $In(NO_3)_2$. Indium is observed to plate out as the cell operates. The initial voltage is measured to be + 0.425 V at 25°C.
    a. Write a balanced equation for the half-reaction at the anode and for the half-reaction at the cathode.
    b. Calculate the standard reduction potential for the In^{3+}/In half-cell. (Hint: consult **Table 15-1** for the reduction potential of the Zn^{2+}/Zn electrode, and use the equation on page 586.)
27. Iron or steel is often covered with a thin layer of a second metal to prevent rusting. Tin cans consist of steel covered with tin, and galvanized iron is made by coating iron with a layer of zinc. If the protective layer is broken, iron will rust more readily in a tin can than in galvanized iron. Explain this observation by comparing the half-cell potentials for iron, tin, and zinc.

b. The solution around the zinc strip has a high concentration of positive ions because zinc ions are being formed. The copper ions are not attracted to a positive charge.

9. a. reduction, gain, from positive to zero
   b. oxidation, loss, from zero to positive

10. a. Heat production stops because the electrochemical reaction stops.
    b. The Zn strip loses mass, and the Cu metal accumulates at the bottom.
    c. Zn^{2+} ions are formed when atoms in the Zn strip lose electrons. Cu^{2+} ions acquire electrons to form Cu atoms that are deposited as metal pieces on the bottom of the beaker.

11. a. Oxidation should show electrons leaving metal atoms that become ions in solution.
    b. Reduction should show metal ions in solution gaining electrons at the surface of the metal strip (electrons do not exist free in solution) to form solid metal.

12. a. 0.80 V − (−0.44V) = 1.24 V
    b. −0.76 V − (−2.37V) = 1.61 V
    c. −1.19 V − (−3.05V) = 1.86 V
    d. 1.20 V − (−0.74V) = 1.94 V
    e. 0.34 V − (−0.25V) = 0.59 V

13. a. It is not possible to determine the potential of an isolated half-cell. The potential can be determined only through a comparison with another half-cell. Use of a standard, or reference, cell to measure reduction potentials of half-cells allows for the prediction and comparison of other cell potentials.

b. The value of the reduction potential for the standard hydrogen electrode is arbitrarily defined as 0.00 V.

14. Some elements will oxidize against a SHE. Others will reduce.

15. As the reactivity of metals increases, their E^0 values become more negative.

16. a. The activity series has a positive correlation to electronegativities.
b. The activity series has a positive correlation to reduction potentials.

17. When the tendency to acquire electrons for a specific half-reaction is less than that of the hydrogen standard electrode, the value of its reduction potential in **Table 15-1** is negative.

18. Dissolving in acid is an indication of a metal's activity. Iron is the anode when paired with a metal that does not dissolve in acid or dissolves less readily. Iron is the cathode when paired with a metal that dissolves more readily.

19. Diagrams should show electrons being added to the metal ions at the cathode and electrons being removed from a metal to produce ions at the anode. Oxidation takes place at the cathode, and reduction takes place at the anode.

20. Aluminum is the anode, and gold is the cathode.
$Al \longrightarrow Al^{3+} + 3e^-$
$Au^+ + e^- \longrightarrow Au$

21. a. Observations may include a solid depositing on the silver and the iron dissolving.
b. Fe is the anode. Ag is the cathode.
$Fe \longrightarrow Fe^{3+} + 3e^-$
$Ag^+ + e^- \longrightarrow Ag$

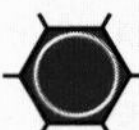

Determining oxidation states

REVIEW

28. What factors help determine the oxidation state of an element?

29. Name the distinguishing characteristic of redox reactions.

30. Do all elements in a redox reaction have to undergo oxidation or reduction? Use an example to explain.

31. Identify the following reactions as redox or nonredox.

a. $2NH_4Cl(aq) + Ca(OH)_2(aq) \longrightarrow 2NH_3(aq) + 2H_2O(l) + CaCl_2(aq)$

b. $Ca(HCO_3)_2(aq) \xrightarrow{heat} CaCO_3(aq) + CO_2(g) + H_2O(l)$

c. $2HNO_3(aq) + 3H_2S(g) \longrightarrow 2NO(g) + 4H_2O(l) + 3S(s)$

d. $[Be(H_2O)_4]^{2+}(aq) + H_2O(l) \longrightarrow H_3O^+(aq) + [Be(H_2O)_3OH]^+(aq)$

e. $Mg(s) + ZnCl_2(aq) \longrightarrow Zn(s) + MgCl_2(aq)$

32. Complete the table below.

	Electrons	Oxidation number	Charge
Oxidation			
Reduction			

PRACTICE

Answers to items in a black square begin on page 841.

33 Assign oxidation numbers to the atoms in lead chloride, $PbCl_2$, an insoluble precipitate. (Hint: see Sample Problem 15A.)

34. Assign oxidation numbers to the atoms in magnesium hydroxide, $Mg(OH)_2$, an antacid. (Hint: see Sample Problem 15A.)

35 Assign oxidation numbers to the atoms in the following compounds. PbS, MnO_2, $LiAlH_4$, Na_2O_2, HgO, NiO(OH), $PbSO_4$ (Hint: see Sample Problem 15A.)

36. Assign oxidation numbers to the atoms in the chlorate ion, ClO_3^-, used in explosives. (Hint: see Sample Problem 15B.)

37 Assign oxidation numbers to the atoms in a bicarbonate ion, HCO_3^-. (Hint: see Sample Problem 15B.)

38. Assign oxidation numbers to the atoms in the following ions. $AuCl_4^-$, $Zn(OH)_4^{2-}$, VO_2^+, $S_2O_3^{2-}$, SCN^-, $H_2BO_3^-$, BH_4^- (Hint: see Sample Problem 15B.)

APPLY

39. Why is it important to balance the electrons in a redox reaction?

40. For the following reactions, identify the elements that are oxidized and reduced.

a. $H_2(g) + CuO(s) \longrightarrow Cu(s) + H_2O(l)$

b. $4Fe(s) + 3O_2(g) \longrightarrow 2Fe_2O_3(s)$

41. The owner's manual for a car includes a warning that overcharging a battery can produce explosive hydrogen gas. The hydrogen results from the electrolysis of water within the battery. At which electrode would you expect the hydrogen to be produced? (Hint: what happens to the oxidation state of hydrogen?)

42. Use the half-reaction method to balance the reaction of nitric acid with copper metal to produce nitric oxide, a noxious pollutant.

$Cu(s) + HNO_3(aq) \longrightarrow Cu(NO_3)_2(aq) + H_2O(l) + NO(g)$

43. Arrange the following in order of increasing oxidation number of the xenon atom: $CsXeF_8$, Xe, XeF_2, $XeOF_2$, XeO_3, XeF

44. Arrange the following in order of decreasing oxidation number of the nitrogen atom: N_2, NH_3, N_2O_4, N_2O, N_2H_4, NO_3^-

45. Which of the sulfur containing species below cannot be reduced?
SO_4^{2-}, $S_2O_3^{2-}$, S^{2-}, SO_3^{2-}

46. Identify the half-reactions in the following redox equations. Use these half-reactions to balance the equations. (Hint: remember to use H^+ and H_2O when the O or H atoms do not balance.)

a. $HNO_3(g) + H_2S(g) \longrightarrow NO(g) + S(s) + H_2O(l)$

b. $Pb(s) + PbO_2(s) + 2SO_4^{2-}(aq) + 4H^+(aq) \longrightarrow 2PbSO_4(s) + 2H_2O(l)$

c. $2Na + S \longrightarrow 2Na^+ + S^{2-}$

d. $H_2(g) + OF_2(g) \longrightarrow H_2O(g) + HF(g)$

e. $Br_2(l) + SO_2(g) \longrightarrow Br^-(aq) + SO_4^{2-}(aq)$

Batteries

REVIEW

47. What factors must battery manufacturers consider when designing batteries?
48. a. What determines the voltage of a battery?
    b. How are alkaline batteries made so that they produce different voltages?
49. Compare and contrast voltaic cells and electrolytic cells.
50. How do the redox reactions for each of the following types of battery differ?
    a. dry cell b. alkaline c. mercury
51. Explain the statement "All electrolytic cells are electrochemical cells, but not all electrochemical cells are electrolytic cells."
52. What dangers exist when recharging batteries that are not designed to be recharged?
53. Although the terms *cell* and *battery* are used interchangeably by many people, the term *battery* refers to a series of cells that are connected. Which of the batteries discussed in this section are cells and which is a true battery?
54. Why do batteries "run down?"

APPLY

55. Why are dry cells referred to as dry?
56. Why is nickel not used with lithium to make a rechargeable battery?
57. Recently, battery companies have begun dating their batteries to inform the consumers of how long the battery will last in storage.
    a. Why do batteries not last indefinitely in storage?
    b. Why might batteries last longer if stored in a refrigerator?
58. The reduction potentials for the anodic and cathodic reactions in a lead-acid storage battery are, respectively, −0.356 V and +1.685 V. Calculate the cell voltage. What would be the voltage if six such cells are connected in series?
59. In the process called electroplating, the metal to be plated is connected to one terminal of an external source of electricity and is immersed in a bath of a salt of the coating metal. The other electrode is often made of the coating metal.
    a. Should the metal to be plated be the anode or the cathode? Explain.
    b. Computer floppy disks are made by plating the polymer Mylar with a metal oxide. Why must the Mylar be coated with graphite before being plated?
60. Is the statement "All chemical changes are electric in nature" true? Explain.
61. a. Use Le Châtelier's principle to explain why your tape player runs at a slower speed as the dry cell battery inside it goes dead.
    b. The carbon rod in a dry cell allows for the dissipation of H_2 gas that builds up during usage. If the gas builds up while the cell is in use, the voltage output drops. But if the load is removed and the hydrogen is allowed to escape, the cell may again produce a voltage. Explain this in terms of Le Châtelier's principle.
62. Propose a reason why rechargeable batteries must have electrodes made of metals that are close to each other in the activity series.
63. On the dry cell below, label the areas where oxidation and reduction occur. Show the flow of electrons from anode to cathode.

22. H^+ only has an E^0 of zero because it was arbitrarily chosen as the standard. As long as there is a difference in electrode potentials, hydrogen can be part of a cell with another element.

23. a. 0.80 V − (−2.93 V) = 3.73 V
    b. 0.00 V − (−0.25 V) = 0.25 V
    c. 1.20 V − 0.00 V = 1.20 V

24. Answers will vary, but should include any three of the following: H_2O/H_2, H^+/H_2, I_2/I^-, Br_2/Br^-, O_2/H_2O, Cl_2/Cl^-, F_2/F^-.

25. A negative value for E^0cell implies that electrons can flow contrary to their natural tendency, that is, from the more electronegative substance to the less electronegative substance.

26. a. anode: $Zn \longrightarrow Zn^{2+} + 2e^-$
    cathode: $In^{3+} + 3e^- \longrightarrow In$
    b. 0.425 V = E^0cath − (−0.76 V)
    E^0cath = 0.425 V + (−0.76 V) = −0.34 V

27. According to half-cell potentials, iron is more readily oxidized than tin but less readily oxidized than zinc. When iron rusts in a tin can, the iron *loses* electrons to the tin; when galvanized iron rusts, the iron *acquires* electrons from the zinc.

28. The oxidation state of an element depends on its electronegativity and the other elements with which it is combined.

29. Redox reactions involve changes in oxidation states resulting from a transfer of electrons.

30. Not all elements need to be oxidized or reduced in a redox reaction. For example, the sulfate ion in the Zn-Cu cell does not undergo either reduction or oxidation.

31. a. not a redox reaction
b. not a redox reaction
c. redox reaction
d. not a redox reaction
e. redox reaction

32. a. lose, increases, increases
b. gains, decreases, decreases

33. Pb = +2; Cl = −1
(+2) + 2(−1) = 0

34. Mg = +2; H = +1; O = −2
(+2) + 2(+1) + 2(−2) = 0

35. Pb = +2, S = −2; Mn = +4, O = −2; Li = +1, Al = +3, H = −1 (rules 4, 7, and 8 on pages 588–589); Na = +1, O = −1 (rules 5, 7, and 8 on pages 588–589); Hg = +2, O = −2; Ni = +3, O = −2, H = +1; Pb = +2, S = +6, O = −2.

36. Cl = +5; O = −2
(+5) + 3(−2) = −1

37. H = +1; C = +4 ; O = −2
(+1) + (+4) + 3(−2) = −1

38. Au = +3, Cl = −1; Zn = +2, O = −2, H = +1; V = +5, O = −2; S = +2, O = −2; S = −2, C = +4; N = −3; H = +1, B = +3, O = −2; B = +3, H = −1.

39. Electrons carry a charge, and that charge must be conserved.

40. a. H is oxidized from 0 to +1, and Cu is reduced from +2 to 0.
b. Fe is oxidized from 0 to +3, and O is reduced from 0 to −2.

41. The hydrogen gas is produced at the cathode.

42. $3Cu + 8HNO_3 \longrightarrow 3Cu(NO_3)_2 + 4H_2O + 2NO$

43. Xe, XeF, $XeOF_2$, XeF_2, XeO_3, $CsXeF_8$ (Remember that oxygen atoms are positively charged when combined with fluorine.)

44. NO_3^-, N_2O_4, N_2O, N_2, N_2H_4, NH_3

45. S^{2-}

Linking chapters

1. ***Energy and ions***
In Chapter 5, you learned that iron can form either of 2 ions, Fe^{3+} or Fe^{2+}. Use the half-reactions and E^0 values in **Table 15-1** to determine which oxidation state is preferred. Explain your reasoning.
2. ***The activity series***
Refer to **Table 15-1** to order the halogens by increasing E^0 values. Compare this ordering with their ordering in the activity series on page 259 in Chapter 7. Comment on your observations.
3. ***Free energy***
a. What would you expect the ΔG value to be for a voltaic cell?
b. What happens to the ΔG value of the original redox reaction in an electrolytic cell?
c. What is the ΔG value of a dead dry-cell?
d. Electric energy is free energy, ΔG, and, in theory, it is all available for work. Calculate the amount of electrical work generated per gram of water produced in a fuel cell that operates at 60 % efficiency using the following overall equation.

$$2H_2(g) + O_2(g) \longrightarrow 2H_2O(l)$$

(Hints: calculate ΔG for this reaction using **Table A-12** on page 799. This is the energy produced if the reaction is 100% efficient.)
4. ***Limiting reactants***
Both of the redox reactions shown below have reached completion. Identify the limiting reactant.

a.

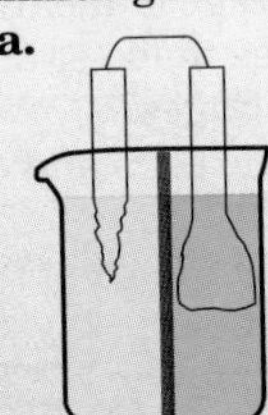

b.

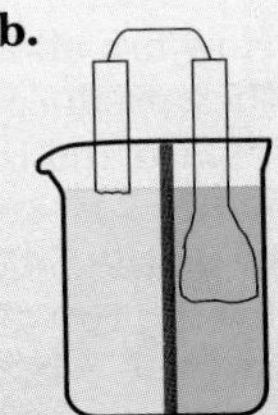

5. **Theme: *Classification and trends***
When sulfur bonds with other elements, it displays oxidation states ranging from −2 to +6. Account for this in terms of sulfur's location on the periodic table, the definition of covalent bonding, and the rules for assigning oxidation numbers.
6. **Theme: *Conservation***
What important factor is addressed when balancing redox reactions by the half-reaction method that might normally cause an error if the reaction were balanced conventionally?

USING TECHNOLOGY

1. ***Graphics calculator***
The reduction potential and electronegativity of an element are directly related. With the graphics calculator, you can estimate unknown reduction potentials. First, set the range by pressing [WINDOW] and entering the following values: Xmin = −4; Xmax = 4; Xscl = 1; Ymin = 0; Ymax = 5; Yscl = 1. Then press [STAT] [4] and clear all existing lists. Enter the list editor by pressing [STAT] [1]. Use **Table 15-1** and the electronegativity values from **Table 11-3** to enter two lists. Enter in L1, the reduction potentials of the elements in **Table 15-1**. Enter in L2, the electronegativity values of the same elements. Press [2nd] [STAT] [PLOT] and turn on Plot 1 by pressing [1] [ENTER] [▼] [ENTER] [▼] [ENTER] [▼] [▶] [ENTER] [▼] [ENTER]. Press [STAT] [▶] [5] [ENTER] to construct a line function. Enter that function into the calculator by pressing [Y=] [VARS] [5] [▶] [▶] [7]. Then press [GRAPH]. Use the [TRACE] feature to estimate the reduction potentials of the following elements listed by reading the x-value from the graph when the y-value is the electronegativity of that element.
a. cadmium electronegativity = 1.7
b. cobalt electronegativity = 1.8
c. strontium electronegativity = 1.0
Compare your estimated values with the accepted values, and comment on the accuracy of your estimate. If the percent error is greater than 10 percent, look for ways to improve the accuracy of your plot. Give reasons for your changes.
2. ***Computer art***
Use a computer to diagram a voltaic cell. Label the anode and the cathode and illustrate the movement of electrons through an external circuit. Then change the art to represent an electrolytic cell.
3. ***Computer spreadsheet***
Create a spreadsheet containing the reduction potentials of the substances listed in **Table 15-1**. Design the spreadsheet to calculate the E^0 of any cell made from the substances in the list.

Alternative assessment

Performance assessment

1. A voltaic cell produces electricity until the reactions driving it reach equilibrium or completion. Given what you know about the factors that affect chemical equilibrium, work with a partner to design an electrochemical cell that could produce electricity indefinitely. (Hint: consider Le Châtelier's principle.)
2. Your teacher will assign you a known metal with an unknown reduction potential. Devise a method to determine the E^0 value of the metal from a list of metals with known E^0 values. Present your method to your teacher in the form of a proposal.
3. Energy for electrically-driven cars comes from redox reactions that produce more energy per unit mass than a lead storage battery. Investigate the development and operation of the sodium-sulfur battery used for this purpose. Choose a stand for or against its use, and present your findings in a persuasive speech to your classmates.
4. ***Computer program***
The voltage output of a half-cell depends on the concentration of the substances present. The E^0 values in **Table 15-1** are determined when the concentrations are 1.0 M. When the concentrations are not 1.0 M, the cell potential is calculated from a form of the equilibrium expression.
Half-reactions for standard reduction potentials are represented by the following generalized equation.

$$x \text{ Ox} + ne^- \longrightarrow y \text{ Red}$$

"Ox" refers to the oxidized substance, and "Red" to the reduced substance. The half-cell potential is calculated from the following equation.

$$E = E^0 - \frac{0.0592}{n} \log \frac{[\text{Red}]^y}{[\text{Ox}]^x}$$

Write a program to calculate the half-cell potentials for half-reactions involving a metal and its ion when the ion concentration varies from 1.0 M to 0.001 M.

Portfolio projects

1. ***Chemistry and you***
For one week, keep a record of how many times you use devices powered by batteries. Record what kind of device you used and the number and type of batteries it contained. Your teacher will provide you with various batteries and a balance. Record the mass of each type of battery you used during the week. Assuming that your battery usage is typical of everyone in the country, estimate the mass of waste material produced by battery usage in one year. Write a short response offering ways to reduce the amount of waste.
2. ***Research and communication***
Research the progress in the development of electric cars. Have a class debate about the advantages and disadvantages of replacing gasoline and diesel vehicles with electric vehicles.
3. ***Cooperative activity***
Break the class into several groups. Have each group research the environmental effects of different types of batteries. Analyze both the production and waste costs. Bring the groups together to share their research with each other.
4. ***Chemistry and you***
Consumer use of rechargeable batteries is growing. Many people either own devices using rechargeable batteries or have access to them. Nickel-cadmium batteries, a common rechargeable type of battery, are used in cellular phones, shavers, and portable video-game systems. Make a list of the items that you come in contact with that use nickel-cadmium batteries or other rechargeable batteries. Write a short essay about technology that was not and could not have been available before the development of the nickel-cadmium battery.
5. ***Research and communication***
Redox reactions are not limited solely to electrochemical cells and batteries. Research common occurrences of redox reactions, and identify the chemical that is oxidized and the chemical that is reduced.

46. a. red:
$2(HNO_3 + 3H^+ + 3e^- \longrightarrow NO + 2H_2O)$

ox:
$3(H_2S \longrightarrow S + 2H^+ + 2e^-)$

net:
$2HNO_3 + 3H_2S \longrightarrow 2NO + 3S + 4H_2O$

b. red:
$PbO_2 + SO_4^{2-} + 4H^+ + 2e^- \longrightarrow PbSO_4 + 2H_2O$

ox:
$Pb + SO_4^{2-} \longrightarrow PbSO_4 + 2e^-$

net:
$PbO_2 + 2H_2SO_4 + Pb \longrightarrow 2PbSO_4 + 2H_2O$

(Sulfuric acid may be left as its ions.)

c. red:
$S + 2e^- \longrightarrow S^{2-}$

ox:
$2(Na \longrightarrow Na^+ + e^-)$

net:
$2Na + S \longrightarrow 2Na^+ + S^{2-}$

d. red:
$OF_2 + 2H^+ + 4e^- \longrightarrow H_2O + 2F^-$

ox:
$2(H_2 \longrightarrow 2H^+ + 2e^-)$

net:
$2H_2 + OF_2 \longrightarrow H_2O + 2HF$

e. red:
$Br_2 + 2e^- \longrightarrow 2Br^-$

ox:
$SO_2 + 2H_2O \longrightarrow SO_4^{2-} + 2e^- + 4H^+$

net:
$Br_2 + SO_2 + 2H_2O \longrightarrow 2Br^- + SO_4^{2-} + 4H^+$

Answers from page 593

Solution to Additional Examples

a. H^+ and Cl^- are spectators.
ox: $H_2S \longrightarrow S + 2H^+ + 2e^-$
red: $2(Fe^{3+} + e^- \longrightarrow Fe^{2+})$
net: $2FeCl_3 + H_2S \longrightarrow 2FeCl_2 + 2HCl + S$
The 2 in front of the HCl balances the H atoms and the remaining Cl atoms.

b. K^+ is a spectator.
ox: $2ClO_3^- \longrightarrow 3O_2 + 2Cl^- + 12e^-$
(6 O atoms each lose $2e^-$)
red: $2ClO_3^- + 12e^- \longrightarrow 2Cl^- + 3O_2$
(2 Cl atoms each gain 6 e^-)
net: $4ClO_3^- \longrightarrow 4Cl^- + 6O_2$
reduced to lowest terms:
$2KClO_3 \longrightarrow 2KCl + 3O_2$

c. SO_4^{2-} is a spectator.
$2Al \longrightarrow 2Al^{3+} + 6e^-$
$3(2H^+ + 2e^- \longrightarrow H_2)$
$2Al + 3H_2SO_4 \longrightarrow Al_2(SO_4)_3 + 3H_2$

d. ox: $2Cu + 2OH^- \longrightarrow Cu_2O + H_2O + 2e^-$
red: $Br_2 + 2e^- \longrightarrow 2Br^-$
net: $2Cu + Br_2 + 2OH^- \longrightarrow Cu_2O + 2Br^- + H_2O$

Continued from page 609

47. Half-cells should be chosen so that a spontaneous reaction produces the desired voltage. If a rechargeable battery is wanted, the metals chosen for the electrodes should be close together on the activity series. The effect on the environment of the materials used to make the battery should also be considered.

48. a. The voltage of a battery is equal to the difference between the reduction potentials for the cathode and the anode that make up the battery.
b. The voltages of alkaline batteries change when the cathodic reaction is changed, as in the replacement of manganese dioxide with mercuric oxide.

49. Both voltaic and electrolytic cells are electrochemical cells constructed from half-cells. A voltaic cell produces an electric current by means of a spontaneous redox reaction. In an electrolytic cell, an electric current from an external source causes a nonspontaneous redox reaction to occur.

50. a. Zinc is oxidized on a plate, and manganese dioxide is reduced at a carbon rod. The electrolyte is made of NH_4Cl, which can produce dangerous gases.
b. Zinc is oxidized, and manganese dioxide is reduced, but the KOH electrolyte does not produce dangerous gases.
c. Zinc is oxidized, and mercury(II) oxide is reduced. This gives a steady voltage but produces toxic elemental mercury.

51. Any cell in which electric and chemical energy are interchanged is an electrochemical cell. Electrolytic cells involve only the change of electric energy into chemical energy.

52. Unwanted side reactions could occur, or harmful gases could be produced.

53. The zinc-carbon battery, alkaline versions of the zinc-carbon battery, and nickel-cadmium battery, are actually cells. The lead-sulfate car battery is a true battery.

54. As the battery is used, the concentrations of reactants change, bringing the half-cells into equilibrium with each other. Electric energy is free energy, and at equilibrium, $\Delta G = 0$.

55. They do not hold a solution of ions, but rather a moist electrolyte paste.

56. A reaction involving metals close together on the activity series is easily reversible. Nickel and lithium do not fit this criterion.

57. a. Even though the components are physically separated, spontaneous reactions among the components may still occur.
b. Cold temperatures slow down many chemical reactions.

58. E^0cell $= 1.685 \text{ V} - (-0.356 \text{ V}) = 2.04 \text{ V}$
For six cells in series, the voltage is $6(2.04) = 12.2 \text{ V}$.

59. a. The object to be plated should be the cathode, a source of electrons to reduce the coating-metal ions present in the solution.
b. The Mylar, being a polymer, does not conduct electricity or ionize. The graphite serves this function.

60. No. All chemical changes involve movement of electrons, but not all changes produce electricity or electrostatic charge.

61. a. During the use of a tape player, the redox reaction in the dry cells is slowed because products of the reaction are not removed. This shifts the equilibrium of the reaction toward the reactants and lowers the voltage.
b. If the dry cell is drained too quickly, hydrogen gas builds up at the carbon rod, current flow decreases, and the voltage output drops. When the load is removed, the gas can dissipate through the carbon rod or be absorbed into the electrolytic paste. If the cell is not too far spent, current should flow when the load is reconnected.

62. While rechargeable batteries may not produce the greatest voltage, discharging them should be a gradual process that can easily be reversed.

63. Zinc is oxidized at the thin zinc casing surrounding the cell. Manganese compounds are reduced at the carbon rod in the center. Electrons flow from the bottom of the cell (negative pole) through the external circuit to the raised top of the cell (positive pole).

Linking chapters

1. Usng reduction potentials only, the preferred oxidation state for iron is +2. Fe^{2+} ion is less willing to be reduced to Fe than is Fe^{3+} ion. Fe^{3+} ion is easily reduced to Fe^{2+}.

2. The halogens increase in activity from I to F. The reduction potentials for the halogen half-cells in **Table 15-1** increase from I to F. The parallel relationship between activity and reduction potential that exists for metals also exists for nonmetals.

3. **a.** It is negative.
   b. It becomes positive.
   c. It is zero.
   d. Using **Table A-12**, $\Delta H = -571.6$ kJ, $\Delta S = -0.3265$ kJ/mol, and $T = 298$ K. Therefore, $\Delta G = -474$ kJ when the reaction is 100% efficient but -284 kJ when the reaction is 60% efficient. In terms of the 36 grams of water produced,

$$\Delta G = \frac{-284 \text{ kJ}}{36 \text{ g}} = -7.9 \text{ kJ/g}.$$

4. **a.** The solution lost all of the ions to be reduced at the cathode.
   b. The anode dissolved completely.

5. Sulfur has six electrons in its outer shell. When combined with a metal or less electronegative nonmetal, it acts as an anion having a charge of -2. When combined with oxygen or fluorine, sulfur has a positive oxidation state. The amount of its charge depends on how many oxygen or fluorine atoms are bonded to it.

6. Balancing equations by inspection may not conserve electrons.

USING TECHNOLOGY

1. **a.** estimated: -0.70 V; accepted: -0.40 V
   b. estimated: -0.45 V; accepted: -0.28 V
   c. estimated: -2.62 V; accepted: -2.89 V
   The plots may be improved by comparing only half-cells in which the ion has a +2 charge because the ions formed in **a**, **b**, and **c** are all +2. Comparing only *metals* also improves the estimated value. Removing the hydrogen half-cell because its value is an arbitrary 0 V has no affect on the estimated values.

2. Answers will vary. Be sure the half-cells are properly separated and that the electrons run from the anode to the cathode in the voltaic cell. The direction of electron travel should be reversed for an electrolytic cell.

3. Answers will vary. Be sure that the anodic reaction can be correctly identified for any selected pair of half-reactions and that E^ocell is correctly calculated.

4. Answers will vary. Be sure the full range of concentration has been accommodated and that the value for E decreases as the concentration decreases. Because only half-reactions involving metals and their ions are being used, the value of $[Red]^y$ will always be 1.

Performance assessment

1. Answers will vary. The important design points include the means both to supply fresh reactants to the cell and to remove products from it. Other points to consider are the toxicity of products and reactants and the ease with which reactants can be supplied and products can be removed.

2. Answers will vary. Be sure the oxidation and reduction electrodes are properly identified.

3. Answers will vary. Be sure the presentation includes generalized half-reactions and voltage output for the battery, that the operating temperature is around 200°C because the sodium must be in the liquid state, and why the reaction could not be run in an aqueous solution.

Portfolio projects

1. Answers will vary. This problem requires a number of factors to be considered, among them, the average lifetime of a battery and the amount of time the battery-powered devices are in use.

2. Answers will vary. Be sure to examine the environmental costs of developing and disposing of the batteries.

3. Answers will vary. Be sure to cover all common battery types.

4. Answers will vary. Many portable devices would be too expensive to run without rechargeable batteries.

5. Answers will vary. Some examples include rust forming, fruit rotting, or a newspaper turning yellow in the air.

16

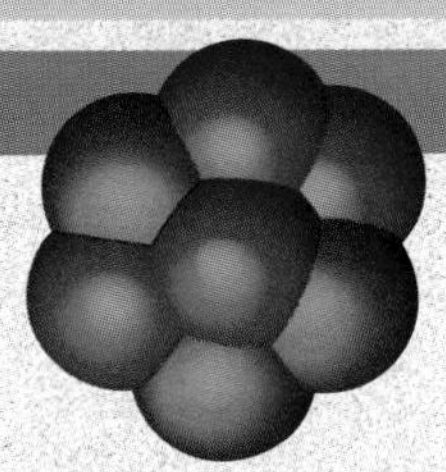

PLANNING FOR

Nuclear Chemistry

Chapter Overview

- **Section 16-1** examines nuclear stability, with attention given to the strong nuclear force, the neutron to proton ratio in stable nuclei, and binding energy.
- **Section 16-2** compares the various types of radioactive emissions, with emphasis on nuclear transmutation (both natural and artificial), radiation detection, and isotope stability. The procedure for writing equations to describe nuclear reactions is explained and compared to the procedure for balancing chemical equations. The processes that constitute nuclear fission and nuclear fusion are discussed.
- **Section 16-3** examines the applications of nuclear chemistry. Radioactive dating and the meaning and use of radioactive half-life are emphasized. Applications of radioactive isotopes to industry, including analytical and imaging techniques using artificially created and natural radioactive isotopes, are explored. Radiation effects upon health are discussed.

Concept Base

Students must have mastered the following concepts prior to this chapter: matter and energy (Chapter 2); atomic structure, atomic number, mass number, and isotopes (Chapter 3); alpha and gamma particles (Chapter 4); balancing equations and matter conservation (Chapter 7); stoichiometry (Chapter 8); and rates of reaction (Chapter 14).

Laboratory Equipment Needs

Demonstrations

The listing of materials for Chapter 16 demonstrations begins on page T46.

Laboratory Experiments

Materials and preparation instructions for Chapter 16 are found as follows.

Text		*Page*
Exploration 16A	Detecting Radioactivity	786
Investigation 16A	Detecting Radioactivity—Shielding From Alpha Particles	790

Pacing: 7 days	**Inquiry Teaching Strategies**
16-1 Why are some atomic nuclei unstable?	**Demonstrations** ■ Demo 1 *Electric forces*
16-2 What kinds of nuclear change occur?	**Demonstrations** ■ Demo 2 *Detecting and measuring radiation* **Laboratory Experiments** ■ Exp 16A *Detecting Radioactivity* ■ Inv 16A *Detecting Radioactivity—Shielding from Alpha Particles*
16-3 How is nuclear chemistry used?	**Demonstrations** ■ Demo 3 *Half-life prediction*

Key ■ Teaching Transparencies
■ Annotated Teacher's Edition
■ Laboratory Experiments Manual

Visual Teaching Strategies	Review and Practice Strategies
	Text Reviews ■ Concept Review 1–3 ■ Additional Example 16A ■ Practice 1–2 ■ Section Review 4–7 ■ Chapter Review 1–11; LC 1–4; UT 1, 2 **Study Guide Worksheets** ■ ■ 16-1 Concept Review ■ ■ 16-1 Skillsheet *Binding Energy and Stability* ■ ■ 16-1 Practice *Binding Energy*
Transparencies ■ 16-12 *Uranium-238 Decay Series*	**Text Reviews** ■ Section Review 8–10 ■ Chapter Review 12–21; LC 5, 6 **Study Guide Worksheets** ■ ■ 16-2 Concept Review ■ ■ 16-2 Skillsheet *Interpreting Nuclear Equations* *Interpreting Radioactive Decay*
Transparencies ■ F 16-20 *Potassium-40 Half-Life Comparison* **Transparency Masters** ■ P 638 *Calculating Binding Energy*	**Text Reviews** ■ Section Review 11–16 ■ Chapter Review 22–27 **Study Guide Worksheets** ■ ■ 16-3 Concept Review ■ ■ 16-3 Skillsheet *Radioactive Dating*

■ Pupil's Text
■ Study Guide
■ Teaching Resources

Exp *Exploration*
Inv *Investigation*

LC *Linking Chapters*
UT *Using Technology*

Assessment Options

Traditional Assessment

Test Generator
Instructional Objectives Measured:
Content mastery
Select from items 1 to 75

Performance Assessment Options

(see page T36 and T643-15 for scoring rubrics)
Students demonstrate mastery of objectives in a hands-on environment. You may want some of these materials in the Portfolio as well.

Investigation 16A (text)
Instructional Objectives Measured:
Content mastery, Scientific and chemical literacy, Problem solving skills, Use of scientific methodology, Proficiency in written communication.

Portfolio Options

(see page T36 for scoring rubrics)
Students provide a written rationale for each selection made for the portfolio.

Concept Maps
Students use vocabulary from the Chapter Review to build a concept map.
Instructional Objectives Measured:
Content mastery

Formal Laboratory Reports
Exploration 16A (text)
Instructional Objectives Measured:
Content mastery, Problem solving skills, Use of scientific methodology, Proficiency in written communication

Portfolio Projects, page 641
Item 1. Chemistry and you
Instructional Objectives Measured:
Content mastery, Scientific and chemical literacy, Understanding the impact of science and technology on society, Proficiency in oral communication

Items 2, 3, and 4. Research and communication
Instructional Objectives Measured:
Content mastery, Scientific and chemical literacy, Problem solving skills, Understanding the impact of science and technology on society, Proficiency in oral and written communication

INTRODUCING THE CHAPTER

CHAPTER 16 Nuclear Chemistry

Story Background

The Simon couple's discovery of the Iceman was an almost unbelievable stroke of luck. Six months before the discovery, a large dust storm in the Sahara Desert drove fine dust particles as far north as the Alps. This dust spread in a thin layer over the alpine glaciers, absorbing and reradiating sunlight until the ice began to melt, probably for the first time since its formation. The Iceman had only been exposed for three days when the Simons spotted him. Had he not been seen, snow would have covered him again within another three days.

Of the Iceman's possessions, the most significant is his copper-bladed ax. It was at first mistaken for bronze, an alloy of copper and tin that would have been common from 2200 B.C. to 1000 B.C. (the Bronze Age). The mistake resulted in the underestimation of the Iceman's age. Bronze is harder than copper, and because copper's melting point is lowered when mixed with a metal of low melting point (such as tin), bronze is easier to cast.

The Iceman lived in a time when metal was first replacing stone in tools. Copper was easy to find, but its high melting point made removing it from its ore challenging. The ore was put in a clay crucible while several workers blew through clay or wooden tubes onto the hot coals. The air from the tubes carried additional oxygen to the coals, making them heat to a temperature above 1083°C, copper's melting point.

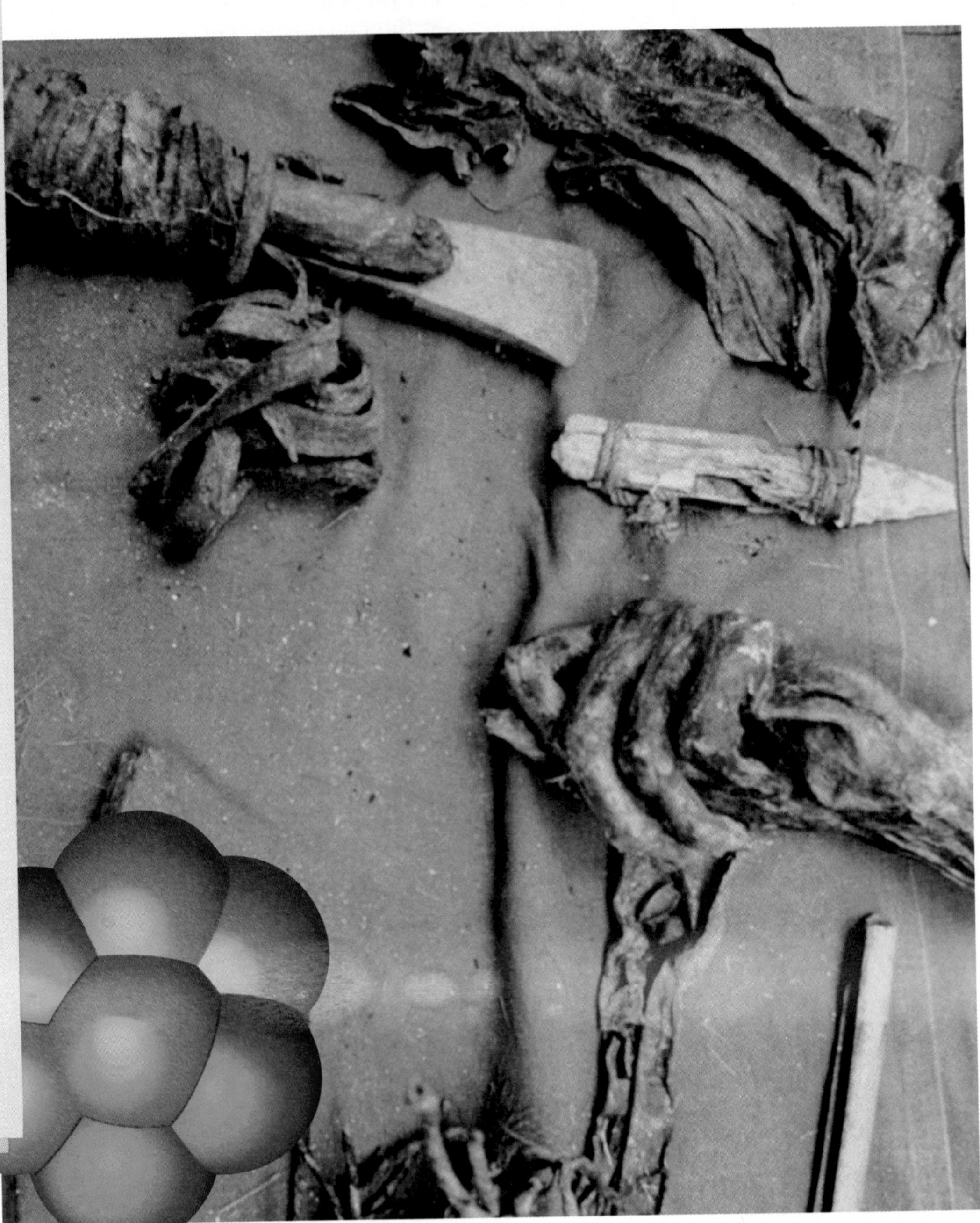

About the Illustration

The Iceman's artifacts were located 0.3–6.0 m away from his remains. Clockwise from the upper left-hand corner are the leather bindings and copper blade of the Iceman's ax; a fabric, similar to felt, that may have served as tinder; the dagger with flint blade; the Iceman's left hand, which seems to grasp at a bundle of leather strands; part of an arrow; and a bone needle.

The Iceman Meets Nuclear Chemistry

While hiking through the Alps near the Austrian border on a September day in 1991, Helmut and Erika Simon noticed something protruding from the ice. At first, the Simons thought they were looking at a doll that had been discarded and covered with ice.

But on closer inspection, the Simons discovered they had stumbled across the frozen body of a man. What they saw sticking up out of the ice were the head and shoulders of a prehistoric man, now known throughout the world as the Iceman.

For four days following the Simons' discovery, workers struggled to free the Iceman's body. They hacked away the ice using axes and ski poles, unaware of the importance of what they were uncovering. In his haste, one worker broke the Iceman's left hip with a jackhammer. Others ripped and destroyed much of the Iceman's clothing while yanking and pulling on his body. But, finally, the figure was freed. The workers were astonished by what they saw.

The body was amazingly well preserved because it had been sealed in ice. In fact, the body was in such good condition that the Iceman's eyeballs were still intact. The Iceman stood five feet two inches tall. He had wavy, medium-length, dark hair and a beard. He appeared to be relatively young, somewhere between age 25 and 40. He wore clothes made of animal skins and boots stuffed with grass to keep his feet warm. His skin bore markings in various spots, including some stripes on his right ankle and a cross behind his left knee. Perhaps these were tattoos.

The Iceman had been carrying a stone knife, a wooden backpack, a small bag containing a flint lighter and some kindling, a bow and quiver containing 14 arrows, and a copper ax. Shortly after the Iceman's body had been freed from the ice, an archaeologist examined the ax. The ax indicated that the Iceman had lived about 4000 years ago.

If the Iceman were indeed that old, then his body would be the oldest ever retrieved from an Alpine glacier. At over 10 000 feet, the site where the Iceman had been found was the highest point where any prehistoric human had ever been found in Europe. Moreover, the Iceman was one of the best-preserved early humans ever found anywhere.

Scientists immediately arranged to have the body placed in a freezer where the temperature would be maintained at a constant –6°C and the humidity would be constant at 98%. These conditions would replicate those of the ice in which the body had been preserved for so long. During the next few months, scientists closely examined the Iceman's body, with the following questions in mind.

- *Exactly when did the Iceman die?*
- *What did the Iceman look like when he was alive?*
- *How did the Iceman die?*

Answers to these questions have been obtained with the help of nuclear chemistry, the subject of this chapter.

Tapping Prior Knowledge

Use the following questions to start students thinking about and discussing the concepts to be covered.

Review questions

1. Using the periodic table, have students determine the number of each type of particle in one atom of gold-197. Tell students, as a hint, that almost all of the mass of an atom resides in the nucleus. ($79p$, $79e^-$, $118n$)
2. Define the difference between atomic number and mass number.

Experience-based questions

1. Why do X-ray technicians go behind a lead shield when taking an X ray?
2. Even though it can still be used for photography, why is it still not a good idea to send film through an X-ray detector at an airport?

Section 16-1

FOCUS

Lesson Starter

Begin the lesson by discussing the stability of the atom. Remind students what happens when an atom has fewer or more electrons than protons. (Chemically active ions try to make up for the difference in charge through chemical bonding.)

Demonstration 1

Electric forces

Approximate time: 5 min

This demonstration offers a visual analogy to the attractive and repulsive forces within the atom.

1. With two small pieces of cellophane tape, attach each pith ball (small Styrofoam ball) to a 20 cm thread.
2. Hang the balls about 1 cm apart on a ring stand.
3. Stroke a rubber (ebonite) rod with animal fur, and touch each ball. The balls should repel each other.
4. Touch both balls to remove the charge.
5. Stroke the ebonite rod with animal fur, and touch one ball. Stroke a glass rod with a piece of nylon from discarded hosiery, and touch it to the other ball. The balls should attract each other.

SAFETY

Wear safety goggles. Students must be 10 ft or more from demonstration.

DISPOSAL

Detach the balls from their strings and save for reuse. Save the fur and nylon for reuse. Put tape in the trash.

Why are some atomic nuclei unstable?

Section Objectives

Describe how the strong nuclear force contributes to the stability of an atomic nucleus.

Predict the stability of an atomic nucleus based on its neutron to proton ratio.

Interpret the band of stability on a plot of neutrons versus protons.

Define and calculate the binding energy of a nucleus.

Nuclear stability

What is the explanation for the unusual "hairstyle" shown in **Figure 16-1**? The girl's hair has become charged because she is touching a generator of static electricity. Each strand of her hair is covered with negative charge. The mutual repulsion of these negative charges forces the strands of hair to repel one another and fly apart. You might think because of this example that the protons within an atomic nucleus should also fly apart due to mutual electrical repulsion between their like, positive charges. Yet protons cluster together tightly within the nucleus along with neutrons, which are uncharged. How is this possible?

strong nuclear force *a short-range, attractive force that acts among nucleons*

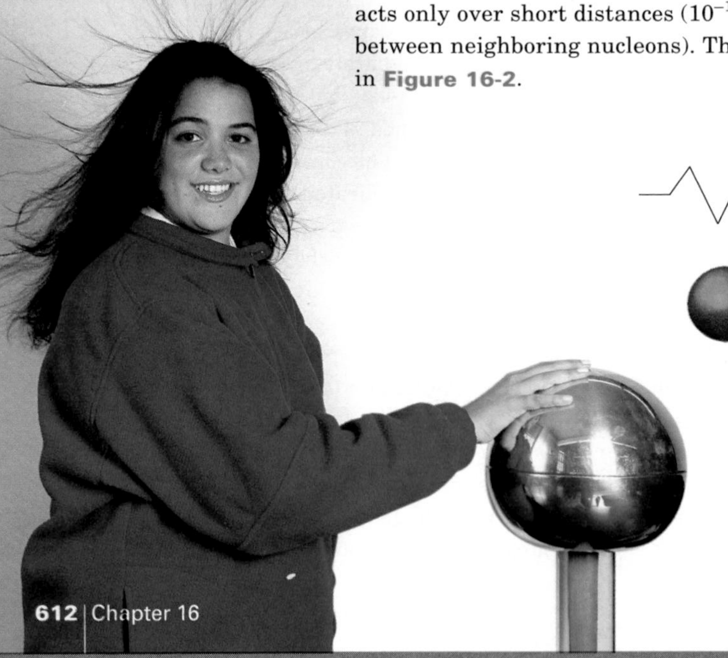

Figure 16-1
The girl in this photograph is touching a generator of static electric charge. As a result, the electric charges coat her hair and body. The negative electric charges on the strands of her hair repel one another, causing the effect you see in the photograph.

The strong nuclear force holds nuclei together

Nuclear protons *do* repel one another through the electric force.However, another, stronger force present in the nucleus overwhelms the electrical repulsion among protons. This force, appropriately called the **strong nuclear force**, is exerted by protons and neutrons (collectively called *nucleons*) on each other. Through this force, all of the nucleons of an atom attract one another. The strong nuclear force is much stronger than either the electrical or the gravitational force, even though it acts only over short distances (10^{-15} m or so, the approximate distance between neighboring nucleons). The strong nuclear force is represented in **Figure 16-2**.

Figure 16-2
The strong nuclear force attracts nucleons to one another over short distances. The force diminishes quickly with increasing distance, however.

Visual Strategy

Figure 16-1

Stress to students that the original positive and negative charges found on each hair are still present but that they are now overwhelmed by the addition of excess electrons.

Most atoms on Earth are stable

stable nucleus
a nucleus that does not spontaneously decay to become the nucleus of a different element

unstable nucleus
a nucleus that spontaneously undergoes decay to become the nucleus of a different element

transmutation
a process by which a nucleus of one element is transformed into a nucleus of a different element

radioactivity
the ability of unstable nuclei to undergo spontaneous nuclear decay

Most of the atoms found on Earth have **stable nuclei**. All of the stable nuclear isotopes have atomic numbers from 1 (hydrogen) to 83 (bismuth). For example, the common isotopes of carbon, hydrogen, oxygen, and nitrogen that make up the bulk of your body mass are stable. Stable atoms give permanence to the matter in our environment.

However, a small fraction of the atoms on Earth have **unstable nuclei**, which undergo spontaneous change. Most unstable nuclei emit electromagnetic radiation or particles of various kinds and, at the same time, increasing or decreasing protons. By increasing or decreasing one or more protons, an unstable atom changes its atomic number to become an atom of a different element. Nuclear changes of this sort are called **transmutations**. For example, the unstable carbon isotope, carbon-14, is transformed into nitrogen-14 through nuclear transmutation, as shown in **Figure 16-3**. Nuclei that undergo transmutation are said to be **radioactive**. The transmutation of radioactive isotopes is commonly called *radioactive decay* or *nuclear decay*.

Many unstable nuclei are very short lived. Some last only fractions of a second before decaying into a different element. Other unstable nuclei may exist for long periods—even billions of years—without experiencing transmutation. In all cases, a nucleus decays because it loses energy in doing so. Radioactive nuclei undergo change to increase their stability by losing energy, just as chemicals in a reaction do.

Taking into account all isotopes of the 109 known elements, there are approximately 1500 different kinds. Only about 250 of these isotopes are stable. However, stable isotopes are generally much more abundant compared to unstable isotopes.

Stability depends on the ratio of neutrons to protons

Why are some nuclei stable and others unstable? And why are *all* isotopes of *all* elements with atomic number greater than 83 unstable? The key to nuclear stability is the ratio of neutrons to protons and the resulting contest between the strong nuclear force and electric force. For a nucleus to be stable, neutrons and protons must be present in just the right ratio.

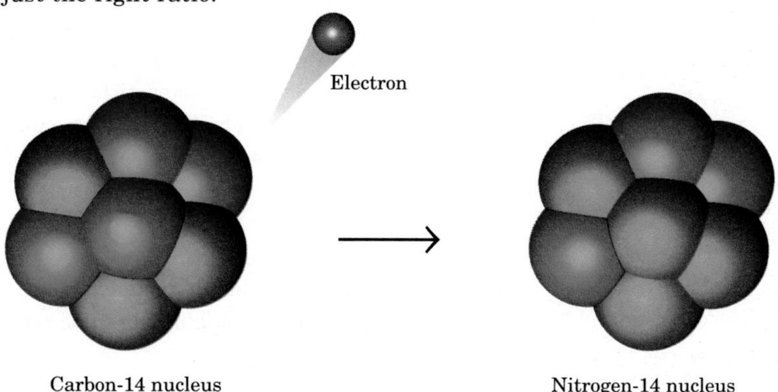

Figure 16-3 **Transmutation is the process through which a nucleus loses or gains protons to become the nucleus of another element. The unstable carbon-14 nucleus emits an electron, thereby changing a neutron into a proton, and becoming nitrogen-14.**

TEACH

Content Background
The strong nuclear force
The strong nuclear force is one of the three fundamental physical forces (the other two are the electro-weak and gravitational forces). Although it is the strongest of the three forces, it has the shortest range. By contrast, gravitation is a much weaker force but has a very long range.

Content Background
The inverse-square rule
Forces such as gravitation or electricity are found to have an inverse-square relationship with distance. For example, if the distance between two particles is doubled, the repulsive or attractive forces decrease by a factor of four. If particles are brought three times closer together, the force increases by a factor of nine.

Teaching Tip
Rule of thumb
Remind students that most processes in chemistry occur to produce the most stable atom possible.

Visual Strategy
Figure 16-3
Stress that transmutation always consists of a proton in the atomic nucleus changing into a neutron or a neutron in the nucleus changing into a proton. The total number of nucleons remains unchanged. Although particles are unstable and decay, the total mass and energy remain conserved.

Content Background
Neutrons and nuclear stability

The stability of the atom is dependent on the neutron's ability to attract protons and other neutrons by means of the strong nuclear force without introducing the repulsive electric force that occurs between two protons.

Teaching Tip
Conventions

Remind students that the number at the upper left of the chemical or particle symbol is the mass number (or the total number of nuclear particles) and the number at the lower left is the atomic number (or the total number of protons).

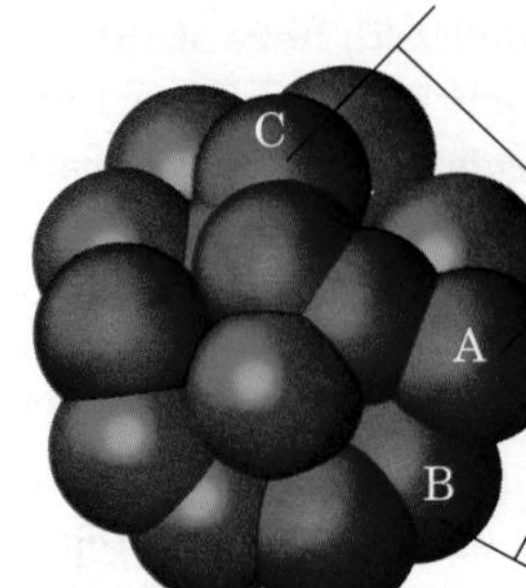

Figure 16-4
Proton A attracts proton B through the strong nuclear force but repels it through the electric force. Proton A mainly repels proton C through the electric force. The strong force applies only over the short distance of a few nucleon diameters.

Neutrons have no electric charge. Therefore they attract all nearby nucleons because they interact with them only through the strong nuclear force, which is always attractive between particles. Protons, however, repel one another through the electric force as well as attract one another through the strong nuclear force. For this reason, neutrons exert more of the force that keeps a nucleus together, as **Figure 16-4** shows. Can you see that neutrons behave as "nuclear glue"?

The more protons there are in a nucleus, the more electrical repulsion there is among them, and the more neutrons are needed to hold the nucleus together with the strong nuclear force. For example, each proton in a helium nucleus (atomic number 2) is repelled by one other proton. In a lithium nucleus (atomic number 3), each proton is pushed away by *two* other protons. But each proton in the relatively heavy gold nucleus (atomic number 79) feels the electrical repulsion of 78 other protons! This adds up to a lot of force pushing each proton out of the nucleus. Also, as **Figure 16-5** shows, in a large nucleus protons are so far apart that they fall out of the short range of the nuclear force, so they cannot attract one another through it. As you can see, the effects of electrical repulsion become an important factor for large nuclei.

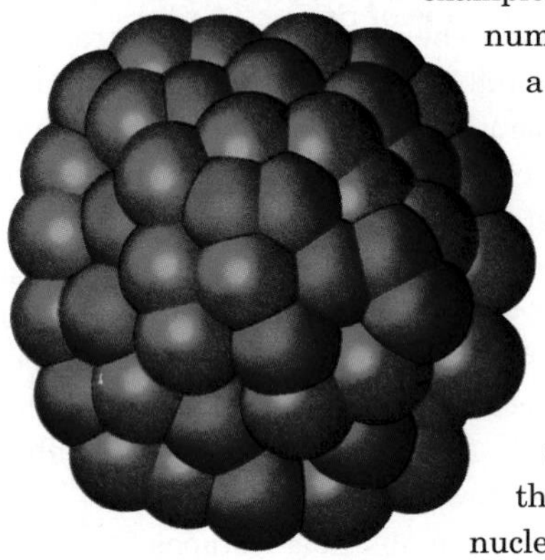

Figure 16-5
Each of the 86 protons in this radon nucleus experiences electrical repulsion from 85 other protons. Neutrons in a heavy nucleus are separated by no more than a few nucleon diameters. Electrical repulsion among protons is a significant factor for large nuclei.

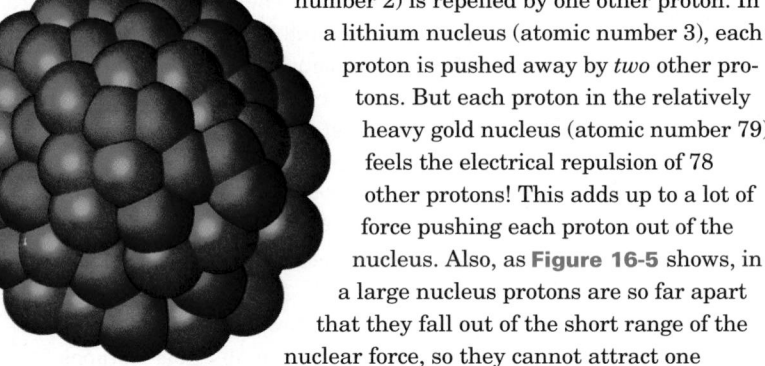

Lighter elements require only about as many neutrons as protons for stability. Carbon has six neutrons and six protons, for example, and nitrogen has seven neutrons and seven protons. Heavier elements require more neutrons to hold the nucleus together. Mercury, with 80 protons and 122 neutrons for example, has about one-and-a-half times as many neutrons as protons.

There is a limit to the stabilizing capability of neutrons, however. If isolated, neutrons themselves are not stable. Unless a neutron is near a proton, the neutron emits a beta particle and turns into a proton. This process can be represented by the following equation in which the charge of each particle is represented as a subscript and mass number is shown as a superscript.

$$^{1}_{0}n \longrightarrow {}^{1}_{1}p + {}^{0}_{-1}e$$

Although neutrons in nuclei are usually stable, some radioactive nuclei change a neutron into a proton by the same process—emitting an electron.

Visual Strategy
Figure 16-5

Stress that although the attractive nuclear force between protons is greater than the repulsive electric force, it is effective over a much smaller distance. Protons on the outer edge of a large nucleus are attracted only by nearby neutrons and protons and are repelled by all other protons in the nucleus.

As more and more neutrons and protons are added to heavier nuclei, the added protons repel all other protons. The repulsion increases steadily. However, added neutrons attract only nearby nucleons. Each added neutron gives about the same stability. Therefore, a stage is reached where additional neutrons cannot stabilize nuclei. All atoms above bismuth, atomic number 83, are unstable.

A plot of the number of neutrons versus the number of protons for stable nuclei is shown in **Figure 16-6**. Note that stable nuclei cluster together in a pattern called a *band of stability* which is highlighted in grey on the graph. As the plot shows, the ratio of neutrons to protons that leads to stability for light nuclei, such as sulfur($^{32}_{16}S$), is near 1.0. The ratio increases to about 1.5 for heavier nuclei, such as lead($^{207}_{82}Pb$). Nuclei not located on the band of stability are unstable. Unstable isotopes decay into isotopes that are located on the band of stability.

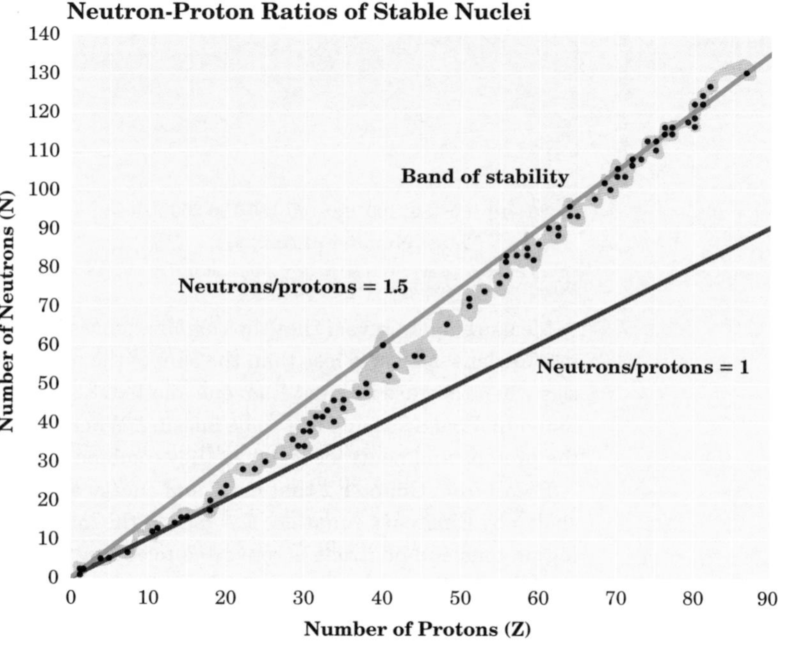

Figure 16-6
The plot shows the number of neutrons (*N*) versus the number of protons (*Z*) for the known stable nuclei. A plot of radioactive nuclei would show that they fall outside the band of stability. The maroon line shows where the data would lie for *N/Z* = 1. For stable lighter nuclei, the data falls near the maroon line. As nuclei become heavier, *N/Z* approaches 1.5 for stability and the data falls near the green line.

Concept Review

Predicting nuclear stability

1. Which is generally more stable—a large nucleus or a small nucleus? Give the reason for your answer.
2. Which is expected to be more stable: $^{6}_{3}Li$ or $^{9}_{3}Li$?
3. Use **Figure 16-6** to determine whether the following isotopes are stable or unstable: $^{12}_{6}C$, $^{3}_{1}H$, $^{206}_{82}Pb$, $^{32}_{15}P$, and $^{4}_{2}He$.

Answers for Concept Review items and Practice problems begin on page 841.

Teaching Tip

Rule of thumb

Atoms with atomic number greater than 50 are not stable if the ratio of neutrons to protons differs much from 1.5. Atoms with atomic number less than 30 are not stable if the ratio of neutrons to protons differs much from 1.0. Remind students that this rule of thumb will help them to determine the approximate number of neutrons in a stable isotope of a specified element. Stress that this rule breaks down for elements with atomic numbers from 30 to 50.

Content Background

Definition of eV and MeV

An *electron volt*, eV, is equal to 1.6022×10^{-19} J. One MeV equals 1 million eV, or 1.6022×10^{-13} J. The unit MeV is especially convenient when dealing with the energies of atomic and subatomic particles.

Answers to

Concept Review

1. A small nucleus is generally more stable because all of the nuclear particles are affected by each other's strong nuclear force. The repulsion between two protons is counteracted by their nuclear attraction.
2. $^{6}_{3}Li$ (For elements with an atomic number lower than 30, stability usually occurs with equal numbers of protons and neutrons, in this case three of each.)
3. $^{12}_{6}C$ is stable.
   $^{3}_{1}H$ is unstable.
   $^{206}_{82}Pb$ is stable.
   (The chart and the rule of thumb aren't entirely reliable in this case. $^{205}_{82}Pb$ is not stable, but should be according to both the rule and chart.)
   $^{32}_{15}P$ is unstable.
   (Again, the rule of thumb and chart are not entirely helpful. $^{30}_{15}P$ is extremely unstable, yet this is an isotope with a neutron-to-proton ratio of 1. $^{31}_{15}P$ is the stable isotope of phosphorus.)
   $^{4}_{2}He$ is stable.

Theme
Conservation
The energy needed to combine and separate protons and neutrons is conserved. Einstein's equation shows the equivalence of mass and energy. Mass can be conserved as energy (and vice versa) during nuclear processes.

Teaching Tip
Enrichment
Reinforce the concept of mass and energy equivalence by showing that the masses of the individual particles in a carbon-12 atom are greater than the atomic mass of carbon-12. A carbon-12 atom has 6 protons, 6 neutrons, and 6 electrons with the following masses.

$$(6\ \text{protons})\left(\frac{1.673 \times 10^{-27}\ \text{kg}}{1\ \text{proton}}\right)$$
$$= 1.004 \times 10^{-26}\ \text{kg}$$

$$(6\ \text{neutrons})\left(\frac{1.675 \times 10^{-27}\ \text{kg}}{1\ \text{neutron}}\right)$$
$$= 1.005 \times 10^{-26}\ \text{kg}$$

$$(6\ \text{electrons})\left(\frac{9.109 \times 10^{-31}\ \text{kg}}{1\ \text{electron}}\right)$$
$$= 5.465 \times 10^{-30}\ \text{kg}$$

$$\text{total mass} = 2.010 \times 10^{-26}\ \text{kg}$$

The mass of one atom of carbon-12 is calculated as follows.

$$12\ \text{amu} \times \frac{1.660 \times 10^{-27}\ \text{amu}}{1\ \text{kg}}$$
$$= 1.992 \times 10^{-26}\ \text{kg}$$

The difference between the two masses is accounted for by the conversion of matter into nuclear binding energy.

binding energy
the energy needed to separate nucleons within a nucleus or, equivalently, the energy released by nucleons combining to form a nucleus

Binding energy
Binding energy indicates stability

Enormous energy, called the **binding energy**, is required to tear apart a nucleus by separating its nucleons. Also, when protons and neutrons come together to form a nucleus, a large amount of energy equivalent to the binding energy is released. Binding energy indicates the stability of a nucleus. The binding energies of various isotopes are plotted in **Figure 16-7**.

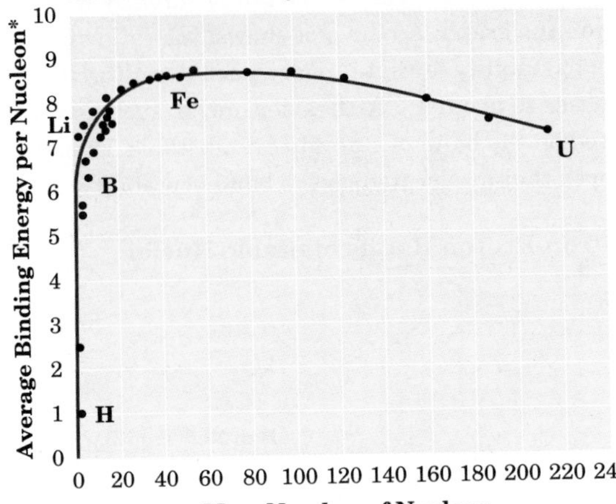

Figure 16-7
The plot of average binding energy per nucleon versus mass number indicates the relative stability of nuclei. Isotopes that have a high binding energy per nucleon are more stable. $^{56}_{26}Fe$ is the most stable atom.

* Energy is expressed in units of megaelectron-volts, an energy unit appropriate to the small scale of atomic events. One megaelectron-volt = 1.6×10^{-19} J.

Measurements reveal that the combined mass of protons and neutrons in a nucleus is always less than the sum of the masses of the same particles when they are isolated from one another. Some of the mass of the individual particles appears to be missing. Where does the excess mass of the protons and neutrons go when they combine to form a nucleus?

Recall from Chapter 2 that mass and energy are interconvertible, as shown by Einstein's equation, $E = mc^2$. If the mass that is lost by combining protons and neutrons is converted to energy, this "missing mass" is equal to the binding energy of the nucleus. Thus, the mass that neutrons and protons appear to lose when they combine is actually converted to energy that is given off as the nucleus forms. The mass of protons and neutrons that is converted to binding energy is called the **mass defect** of a nucleus and is represented by the symbol Δm.

mass defect
the difference between the mass of an atom and the sum of the masses of its individual components

Binding energy is easily calculated

Consider the stable oxygen isotope, oxygen-16, which contains the equivalent of 8 hydrogen atoms and 8 neutrons. The mass of each H atom is 1.007 825 amu. Each neutron has a mass of 1.008 665 amu. Therefore the overall mass of an oxygen-16 atom should be 16.131 920 amu. However, the measured mass of an oxygen-16 atom is 15.994 915 amu. From this information, mass defect can be easily determined as shown in Sample Problem 16A.

Visual Strategy
Figure 16-7
Stress to students that the x-axis is labeled in mass numbers, not atomic numbers. Also remind students that a MeV is a unit of energy equal to 1.6022×10^{-13} J.

Sample Problem 16A
Calculating binding energy

Calculate the binding energy of a mole of oxygen-16 atoms.

1 List

- **mass of H atom (includes 1 proton and 1 electron):** 1.007 825 amu
- **mass of a neutron:** 1.008 665 amu
- **mass of particles making up oxygen-16 nucleus:** 8(1.007 825 amu) + 8(1.008 665 amu) = 16.131 920 amu
- **mass of actual oxygen-16 atom:** 15.994 915 amu
- $c = 3.00 \times 10^8$ m/s
- **binding energy** = ? J
- **The energy unit, the *joule*, is equivalent to $kg \cdot m^2/s^2$.**

2 Set up

- **First determine the mass defect.** Δm = (total mass of particles) − (measured mass of atom)
- **Convert the mass defect to the binding energy with $E = mc^2$.** The mass defect, Δm, is the mass, m, in the equation.
- This conversion will give the binding energy for one oxygen-16 atom. For one mole of oxygen-16 atoms, the mass defect will be in grams (not amu). Because the final answer will be in joules, it is necessary to convert grams to kilograms.

3 Calculate

The mass defect can be converted directly to kg from amu by using the conversion factor 1.66054×10^{-27} kg/amu.

- **Δm = (total mass of nucleons) − (measured mass of nucleus)**

$$= 16.131\,920 \text{ amu} - 15.994\,915 \text{ amu}$$

$$= \frac{0.137\,005 \text{ amu}}{\text{atom oxygen-16}}$$

$$= \frac{0.137\,005 \text{ g}}{\text{mol oxygen-16}}$$

- $E = \Delta mc^2$

$$= 0.137\,005 \text{ g} \times \frac{1 \text{ kg}}{1000 \text{ g}} \times \frac{(3.00 \times 10^8 \text{ m})^2}{\text{s}^2} = 1.23 \times 10^{13} \text{ J}$$

Practice 16A

Answers for Concept Review items and Practice problems begin on page 841.

1. **Calculate the binding energy for one mole of deuterium atoms. Each deuterium nucleus is formed from one proton and one neutron. The measured mass of this atom is 2.0140 amu.**
2. **Calculate the binding energy of one lithium-6 atom, whose nucleus contains three protons and three neutrons. The measured atomic mass of lithium-6 is 6.0151 amu.**

Section Review

4. Define the following terms: *radioactive*, *mass defect*, and *binding energy*.
5. Explain how nuclear stability relates to the neutron-proton ratio of a nucleus.
6. Name two properties of the strong nuclear force. What is the electric force? What role do each of these two forces play in the stability of the nucleus?
7. What is the "band of stability"? How do nuclei located on the band of stability differ from those that are not on it?

Solutions for Additional Examples 16A

a. 4.66×10^{13}J
b. 2.63×10^{13}J

*Full solutions for items **a** and **b** are on page 641A.*

Additional Examples for Sample Problem 16A

a. Calculate the binding energy of a mole of manganese-55 atoms.
b. Calculate the binding energy of a mole of sulfur-32 atoms.

(Be sure to use isotopic atomic masses)

Solutions are at the bottom of the page.

Answers to Practice 16A

1. 2.25×10^{11}J
2. 5.14×10^{-12}J

*Full solutions for items **1** and **2** are on page 641A.*

ASSESS

Alternative Assessment

Have students write a short essay explaining why elements with atomic masses greater than 83 are much more likely to have unstable nuclei than elements with masses less than or equal to 83.

Answers to Section Review

4. radioactive—the ability to undergo spontaneous nuclear decay
mass defect—the difference between the mass of an atom and the sum of the masses of its individual components
binding energy—the energy needed to separate nucleons in a nucleus or, equivalently, the energy released by combining nucleons

Answers are continued on page 641A.

Section 16-2

FOCUS

Lesson Starter
Start the lesson with a discussion of element transmutation. Have students suggest what happens to the emitted particle or energy during this process. Point out that different decay processes produce different emissions, so identifying an emission permits identification of the decay process.

Demonstration 2
Detecting and measuring radiation
Approximate time: 10 min
This demonstration should give students a sense of how different types of radiation with various types of shielding register differently on a Geiger counter.

1. Obtain at least one source of each type of radiation (see Safety section for details).
2. Obtain a Geiger counter that is equipped with a loudspeaker and/or a large meter that is visible at a distance of 10 ft.
3. With the radiation sources 10 ft away from the students *and* the counter, take a reading of background radiation.
4. Measure radiation by bringing the sources up to the Geiger tube. Each source should be 5 cm away from the tube while the other sources are at least 1 m away. Note the different intensities for each sample. (Some counters are not sensitive to alpha particles.)
5. Repeat step **4**, but place a sheet of paper between the Geiger tube and sources. Note any reduction in radiation detected.
6. Repeat step **5** with a plywood board 1 cm thick.

What kinds of nuclear change occur?

Section Objectives

Describe the particles and rays that make up radioactive emissions.

Distinguish among nuclear transmutation, fission, and fusion.

Explain how transuranium elements are synthesized.

Characterize chain reactions.

Give examples of fission and fusion reactions.

You began your study of nuclear chemistry with a look at the transmutations of unstable nuclei. Transmutation is one kind of *nuclear reaction*, or nuclear change. Other nuclear reactions are *fission* and *fusion*. Nuclear fission and fusion are of tremendous practical importance because they could fulfill society's energy needs or cause catastrophic harm, depending on how they are used. In this section, you will learn more about what happens in nuclear transmutation and in the fission and fusion processes.

Radioactive emissions

Nuclear transmutation results in radioactive emissions

While studying the properties of uranium compounds around the beginning of the twentieth century, French scientist Henri Becquerel noticed a surprising phenomenon. The uranium compounds he was studying exposed photographic film that was wrapped in black paper as if the film had been exposed to light. At Becquerel's suggestion, Marie and Pierre Curie also began to investigate the curious behavior of uranium compounds, particularly those found in an ore known as *pitchblende*.

Figure 16-8
Marie Curie theorized that pitchblende, a uranium-containing material, must contain tiny amounts of an unknown, highly radioactive element. By analyzing several tons of pitchblende, she discovered the element radium.

While analyzing pitchblende, the Curies isolated and identified two new elements: *polonium* and *radium*. The Curies observed that these elements both exposed photographic film, just as the uranium compounds did. Also, polonium and radium killed bacteria and other tiny organisms that were placed near them.

The Curies also observed that the two new elements caused certain substances to glow brightly or become discolored. Furthermore, both polonium and radium raised the temperature of surrounding air by several degrees. The elements also turned air, which is normally a nonconductor, into a conductor of electricity.

Finally, Pierre Curie discovered that a sample of radium placed on his skin produced wounds that were very slow to heal.

Visual Strategy
Figure 16-8
Marie Sklodowska Curie (1867–1934) discovered with her husband Pierre the elements radium and polonium, the latter of which was named for Marie's homeland, Poland. She also determined that the only radioactive parts of uranium and thorium compounds were the uranium and thorium atoms.

It was clear to the Curies that uranium, polonium, and radium were releasing significant amounts of energy in order to bring about all these effects. Marie Curie coined the term "radioactivity" to describe the process by which such elements release energy. In the early 1900s, only uranium, polonium, radium, and thorium were known to be radioactive. Today, scientists know that all isotopes of all elements after bismuth on the periodic table are radioactive. Furthermore, all radioactive nuclei undergo transmutation and produce emissions with the properties observed by the Curies. Also, it is now known that the mysterious, energetic emissions which Becquerel and the Curies first observed consist of *alpha particles*, *beta particles*, and *gamma rays*.

Figure 16-9
This smoke detector contains a radioactive isotope. The radioactive emissions of the smoke detector ionize air to conduct a current that is sensed by the detector. When smoke interrupts current flow to the detector, an alarm sounds.

Some radioactive elements emit alpha particles

You have learned that one property of radioactive elements is that they can turn air into a conductor of electricity. This property has been put to practical use in the device shown in **Figure 16-9**. Smoke detectors use a small amount of a radioactive isotope such as americium-241. Americium-241 emits enough radiation to change the neutral molecules in air into ions. The ions can then conduct an electric current through the air inside the smoke detector. A small sensing device located inside the smoke detector monitors this current. However, when smoke particles interrupt this current, an alarm is set off, warning people of the presence of smoke and the possibility of a fire.

Recall from Chapter 4 that nuclear reactions can be represented by nuclear equations in which the element undergoing nuclear change is written on the left side of a reaction arrow and the decay products are shown on the right side. Atomic number appears as a subscript, and mass number is written as a superscript next to each isotope. The nuclear equation for the decay of americium-241 is written as follows.

$$^{241}_{95}\text{Am} \longrightarrow {}^{237}_{93}\text{Np} + {}^{4}_{2}\text{He}$$

Notice that this transmutation results in the production of a new element, neptunium, Np, and a helium nucleus, He. A helium nucleus that is emitted by a radioactive isotope is called an **alpha particle**. An alpha particle (as you may recall from studying Rutherford's experiments with the atom in Chapter 4) consists of two protons and two neutrons and has a 2+ charge. Very heavy nuclei whose neutron-proton ratio is too low for stability emit alpha particles. Emission of an alpha particle increases the neutron-proton ratio by a small amount.

alpha particle
a helium nucleus produced in nuclear decay

With its relatively large mass and charge, an alpha particle does not travel very fast or very far. Consequently, a thin sheet of aluminum foil or even a sheet of paper can stop it. Americium-241 is therefore a good choice for a radioactive isotope in a smoke detector. The emitted alpha particles ionize air that flows through the detector, but they cannot pass through the plastic covering to pose a health hazard to people.

Visual Strategy

Figure 16-9

Emphasize that ionizing smoke detectors usually carry a label explaining where and how to dispose of the unit when it no longer functions. The long half-lives of the isotopes (458 y for americium-241; 2.14 million y for neptunium-237) pose long-term environment problems if the detector is disposed of in a landfill.

7. Repeat step **6** with a piece of aluminum 1 cm thick.

SAFETY

Safety glasses and a lab apron must be worn. Students must be 10 ft or more from demonstration. Be certain that the Geiger tube has a metal or sturdy plastic mounting.

Suitable isotopes for this experiment are given in the Demonstration Materials list (page T46). Most of these may be obtained from scientific supply houses in quantities that do not require licensing. They are also sealed in plastic cases for added safety. Sources not among those listed above or not in sealed containers should not be used.

Even with small amounts of isotopes in sealed cases, gloves should be worn. Students should not handle the isotopes or the Geiger counter under any circumstances. **If the counter has a power supply that operates on AC:** Be certain that there are no bare wires, that all connections are covered with "friction" tape, that the power supply is equipped with a three-wire cord and a three-prong plug, and that the three-slot outlet is protected with a ground fault interrupter. Be sure that the equipment is off before plugging it in (and before unplugging), and that the polarity is correct. Any adjustments to the wiring must be made when the unit is unplugged. Be certain that the area under the wires is dry (as are the teacher's hands), and that wires are are kept dry and free from dangling.

DISPOSAL

None. Save all materials for future demonstrations. Unplug and disconnect all wires from the detector when finished.

TEACH

Content Background

The discovery of radioactivity

Henri Becquerel (1852–1908) originally intended to determine whether the sun emitted X rays. The uranium ore was supposed to absorb solar X rays and re-emit the energy at longer wavelengths. During an overcast day, Becquerel placed a photographic plate and the uranium in a drawer. When the plate was developed, it appeared to be no different than the plates that had been exposed to sunlight, indicating that the energy source in both cases was the uranium.

Historical Note

Ernest Rutherford named the three types of radioactive emission after the first three letters of the Greek alphabet: alpha, beta, and gamma (α, β, and γ, respectively).

Content Background

Radiation detection

The Geiger-Müller counter, commonly known as the Geiger counter, was first developed in 1908 by Hans Geiger and his teacher Ernest Rutherford as a reliable instrument for counting alpha radiation. An improved model was built in 1928 by Geiger and Walther Müller.

Content Background

The decay of $^{131}_{53}I$

The half-life of iodine-131 is 8.07 days. Its decay product, xenon-131, is chemically inert, and therefore should be harmless.

Teaching Tip

Vocabulary

Make students aware that $^{0}_{-1}e$ is the symbol for a beta particle. Beta particles, which are high-energy electrons, result from the decay of a neutron into a proton. In this decay process the total number of nucleons doesn't change, so the atomic number increases, while the mass number stays the same.

Radioactive emissions can be detected

Because of the radiation they emit, most radioactive elements can be easily detected. However, since you cannot see, hear, smell, taste, or feel radioactive emissions, you must use instruments such as the Geiger counter shown in **Figure 16-10** to detect them. The Geiger counter uses a gas-filled metal tube to detect radioactive emissions. The incoming radiation ionizes the gas in the tube, making it an electrical conductor. The current drives electronic counters or causes audible clicks from a built-in speaker. The frequency of these clicks indicates the intensity of radiation in the material or environment being tested.

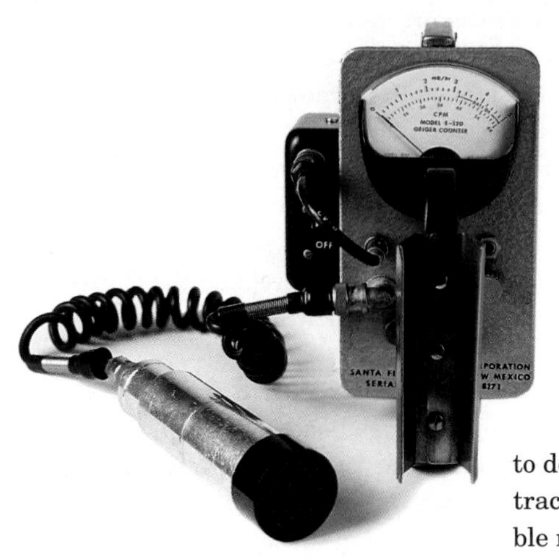

Figure 16-10
The Geiger counter detects the presence of alpha, beta, gamma, and X radiation.

Physicians depend upon the Geiger counter to detect the radiation given off by radioactive tracers—radioactive isotopes of elements whose stable nuclei are normally found in the human body.

For example, your thyroid gland needs the element iodine to synthesize the hormone thyroxin. To check for a malfunctioning thyroid, a radioactive iodine tracer can be used. The patient is given a solution of NaI containing radioactive iodine-131. The radiation emitted by the radioactive iodine can be monitored with a Geiger counter so that the physician can check the rate at which the iodine accumulates in the thyroid to determine if the thyroid is functioning properly. Radiation can also be detected by a scintillation counter, which can be used to prepare a thyroid scan or picture of the thyroid showing areas that do not absorb the isotope.

Radioactive elements can also emit beta particles

beta particle
an electron produced in nuclear decay

One kind of radiation emitted by iodine-131 consists of beta particles. A **beta particle** is a high-speed electron ejected by a radioactive nucleus. Emission of beta particles occurs in nuclei whose neutron-proton ratio is too high to be in the band of stability. Emission of a beta particle decreases the neutron-proton ratio.

The nuclear equation showing how iodine-131 emits a beta particle is written as follows.

$$^{131}_{53}\mathrm{I} \longrightarrow\ ^{131}_{54}\mathrm{Xe} + ^{0}_{-1}e$$

Notice that the element xenon, Xe, has been produced as a result of this transmutation. The beta particle, shown as $^{0}_{-1}e^{*}$, has practically no mass and half as much charge as an alpha particle. Consequently, beta particles travel much faster and have a penetrating ability about 100 times greater than that of alpha particles. They can be stopped only by several layers of aluminum foil or by thin pieces of wood, and they easily penetrate skin.

* Alternative representations for beta particles are β and $^{0}_{-1}\beta$.

Visual Strategy

Figure 16-10

The penetration of alpha, beta, and gamma emission through a given material is predictable. This makes radioactive isotopes useful in industrial and medical applications, e.g., radioactive tracers.

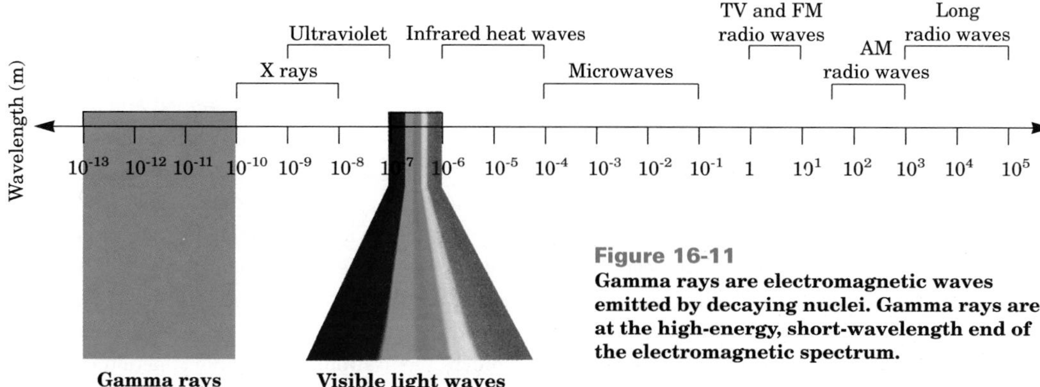

Figure 16-11
Gamma rays are electromagnetic waves emitted by decaying nuclei. Gamma rays are at the high-energy, short-wavelength end of the electromagnetic spectrum.

Gamma rays are a third type of radioactive emission

Radiation has long been used to treat cancer. For many years, the radiation given off by radium-226 was used. Today, cobalt-60, which is cheaper and emits even more radiation, is generally used for this purpose. The nuclear equations showing the radiation emitted by both of these elements are written as follows.

$$^{226}_{88}\text{Ra} \longrightarrow \,^{222}_{86}\text{Rn} + \,^{4}_{2}\text{He} + \gamma$$

$$^{60}_{27}\text{Co} \longrightarrow \,^{60}_{28}\text{Ni} + \,^{0}_{-1}e + \gamma$$

Compare these two nuclear equations. How do the emissions differ? What do both reactions have in common? The symbol γ is the Greek letter gamma. It represents the gamma rays emitted by a nuclear reaction. **Gamma rays** are high-energy electromagnetic waves. They are the same kind of radiation as visible light, but have much shorter wavelengths, as you can see in **Figure 16-11**.

gamma ray
high-energy electromagnetic radiation produced by decaying nuclei

Alpha and beta particles are never given off simultaneously from the same nucleus. Gamma rays are almost always emitted along with either alpha or beta particles. Gamma rays have no mass and no charge. Consequently, they are the most penetrating of radioactive emissions. Several feet of concrete are needed to stop gamma rays. **Table 16-1** summarizes the characteristics of the alpha, beta, and gamma emissions given off by nuclear reactions.

Table 16-1
Characteristics of Alpha, Beta, and Gamma Emissions

Particle	Symbol	Composition	Charge	Penetrating power
Alpha	$^{4}_{2}\text{He}$	2 protons 2 neutrons	2+	short range, stopped by a sheet of paper
Beta	$^{0}_{-1}e$, β	1 electron	1−	intermediate range, stopped by a few centimeters of water
Gamma	γ	electromagnetic waves	0	long range, stopped by a few centimeters of lead

Do You Know?
Particle accelerators are built underground because the gamma radiation released from the atom during a collision can penetrate solid material of considerable thickness. The earth and rock surrounding the accelerator help protect the surface region in the accelerator's vicinity.

Visual Strategy

Figure 16-11
Make students aware that visible light makes up a very small portion of the entire electromagnetic spectrum. Ask the class to imagine what our vision would be like if we were able to see all forms of electromagnetic radiation. Mention that some insects, such as bees, see primarily in the ultraviolet segment of the spectrum.

Groupwork Strategy

Divide the class into groups of three for brainstorming sessions. For the most part Americans are reluctant to buy irradiated foods, mainly because of a misunderstanding of radiation and nuclear particles. Have the students determine five methods that could be used to help educate the public about the effects, benefits, and hazards of radioactive emissions. After discussing these methods, have the groups determine five positive and five negative ways in which irradiating food might affect the United States economy.

Content Background

Positive electrons (positrons)

The positron, discovered by physicist Carl David Anderson in 1932, has properties similar to those of an electron, except that its charge is positive instead of negative. For this reason, a positron in a magnetic field is deflected in the direction opposite to that of an electron.

Consumer Focus

Irradiated foods

Gamma radiation is used to preserve food products because it can destroy the viruses, bacteria, fungi, parasites, and insects that infest and spoil food. Not only do such organisms cause food to spoil more quickly, but if consumed, they can also make people dangerously ill.

The Food and Drug Administration (FDA) regulates which foods can be irradiated and how much radiation each can receive. The first products were wheat and flour, which were approved for radiation in 1963 to destroy insects. Twenty years later, the FDA expanded the list to include spices imported from countries where health conditions and sanitary practices are not up to the standards of the United States. Meats, including both pork and poultry, fruits, and vegetables have recently been added to the list.

Nearly all food that is irradiated is exposed to the gamma rays emitted by cobalt-60. The length of exposure depends on the food. Strawberries are exposed for 8 minutes, while frozen poultry gets 20 minutes of radiation. On average, the radiation extends the shelf life of fresh fruits and vegetables from one to two weeks. Canned and frozen foods that are irradiated may have a shelf life of seven years.

While irradiated foods have been widely available for years in countries throughout the world, they are only now being introduced on a large scale in the United States food supply. Consumer reaction is mixed. Some are concerned about the chemical changes in the food that result from radiation. These changes, known as unique radiological products (URP), cannot exceed one part per million according to the FDA. Yet some suspect that even this level of exposure to URPs may be hazardous. In response to such concerns, research into the preservation of foods with gamma radiation continues. In the meantime, check your local supermarket to see if it carries irradiated food products. If so, and if you are concerned, you may want to do further research on the benefits and risks of irradiated foods.

Two other types of radioactive decay are possible

Unstable nuclei may decay by alpha, beta, or gamma emission or by two other processes. One of these is *positron emission.* A positron, generally represented as ${}^{0}_{+1}e$, is a particle identical to an electron except that it has a positive charge. For example, polonium-207 decays by positron emission to bismuth-207, as shown in the following equation.

$$^{207}_{84}\mathrm{Po} \longrightarrow {}^{0}_{+1}e + {}^{207}_{83}\mathrm{Bi}$$

Positron emission leads to a decrease in atomic number, as does another nuclear decay process called *electron capture,* EC. In electron capture, a nucleus captures an electron belonging to that atom. Beryllium-7 decays to lithium-7 via EC, as shown in the following equation.

$$^{7}_{4}\mathrm{Be} \xrightarrow{\mathrm{EC}} {}^{7}_{3}\mathrm{Li}$$

Nuclear reactions may occur in series

radioactive series
a sequence of nuclei that arise from and are transformed by radioactive decay until a stable isotope is produced

Frequently, the emission of an alpha or beta particle results in the formation of an isotope that is radioactive. The new unstable isotope may therefore initiate a series of nuclear transmutations until a stable isotope is finally produced. A sequence of such reactions is called a **radioactive series**. A radioactive series, beginning with $^{238}_{92}U$ and ending with stable $^{206}_{82}Pb$, is shown in **Figure 16-12**.

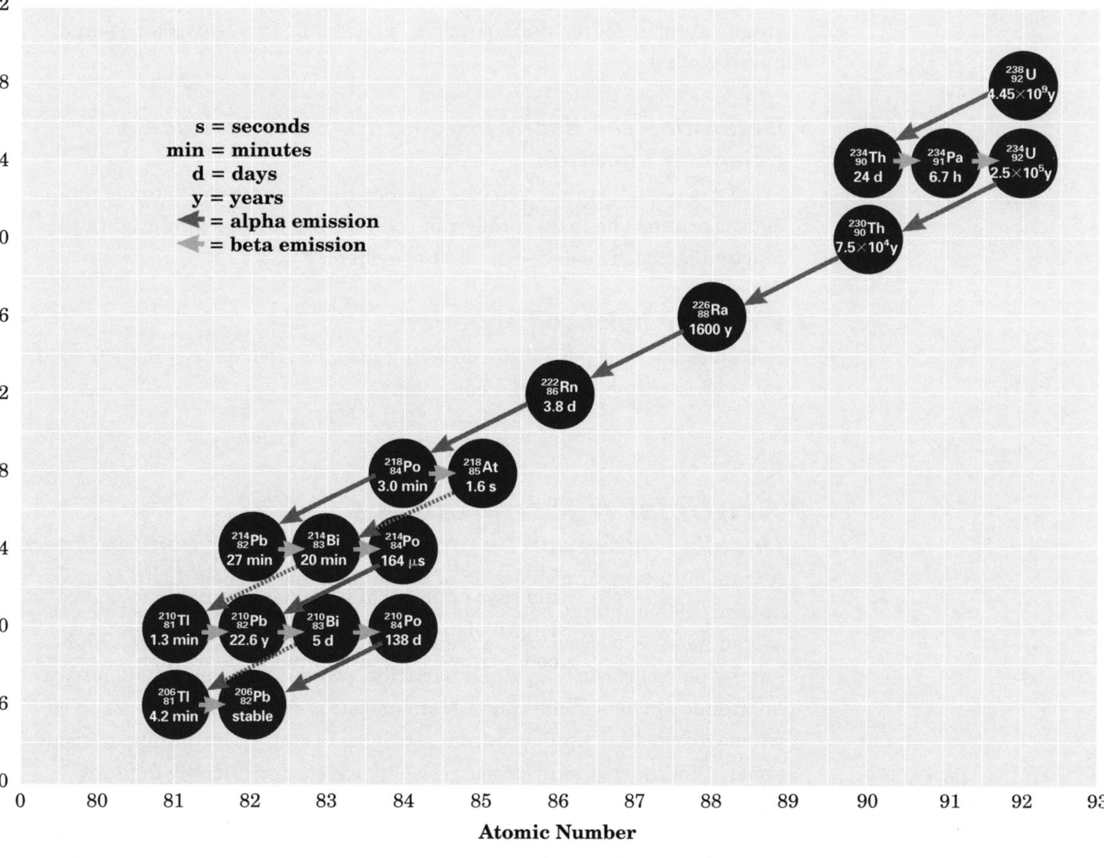

Figure 16-12
Uranium-238 decays to lead-206 in a radioactive series consisting of many steps. In the first step, uranium-238 emits an alpha particle to give thorium-234. Thorium-234 emits a beta particle to give protactinium-234, which also emits a beta particle to continue the series. The half-life of each isotope is also listed.

Nuclear equations must be balanced

Now that you know how to represent the products of nuclear decay, you can write equations to represent transmutations—almost. There is one more thing to learn. In order to write nuclear equations, keep in mind that the equations must be balanced with respect to the number of nucleons and electric charge. If you know what kind of emission is produced in decay, you can write a balanced equation by following the process shown on the next page. Emission of an alpha particle increases the neutron-proton ratio slightly. It may take several alpha emissions to increase the ratio to a degree that causes beta emission. Emission of a beta particle causes a larger decrease in the neutron-proton ratio, so long chains of beta emission reactions do not occur.

Theme

Equilibrium and change

The emission of radioactivity by an unstable nucleus leads to the formation of different nuclei, which in turn emit radioactivity until a stable atom is formed. The changes that occur in unstable nuclei are predictable events, even though the actual time at which the change occurs is not predictable.

Visual Strategy

Figure 16-12

Stress that the $^{238}_{92}U$ nucleus does not decay directly into $^{206}_{82}Pb$. Also emphasize that the $^{238}_{92}U$ nucleus, after several transmutations, turns into $^{234}_{92}U$. Two thorium isotopes, three polonium isotopes, and three lead isotopes (among many other isotopes) are created before the stable isotope $^{206}_{82}Pb$ is formed.

Teaching Tip

Modern alchemy

Students should realize at this point that the alchemists' dream can now be achieved: it is possible to produce gold from other elements in the laboratory. The problem, however, is that although it is technologically feasible, it is not economically feasible.

Theme

Systems and interactions

Atoms can be bombarded with small particles to cause a change in the original nuclear structure of the atom.

Teaching Tip

Discussion

Point out that the elements with atomic numbers 104, 105, 106, 107, 108, and 109 have been given Latin names that literally translate to the numbers 104, 105, 106, 107, 108, and 109. This was done by the International Union of Pure and Applied Chemistry (IUPAC), a group of elected scientists from all countries, to avoid disputes like those that arose when previous elements were named for scientists or geographical locations. These names are still the subject of controversy and have not been officially accepted.

How To

Write a balanced nuclear equation

1. *Balance number of nucleons using mass number.*

For example, consider the unstable $^{238}_{92}U$ nucleus that spontaneously decays by emitting an alpha particle. From the mass number of the uranium isotope, subtract the mass number of the alpha particle, which is 4. The mass number of the resulting atom is 238 − 4, or 234.

2. *Balance charge using atomic number.*

The atomic number of uranium is 92. From this number, subtract the atomic number of the alpha particle, which is 2. This leaves an atomic number of 90.

3. *Determine the decay product that will produce a balanced equation.*

From the periodic table, you can see that the element with atomic number 90 is thorium. The mass number of the thorium isotope produced in this reaction is 234. The decay product is evidently $^{234}_{90}Th$.

4. *Write the balanced equation.*

Use an arrow to separate the decaying nucleus from decay products as if you were writing a chemical equation.

$$^{238}_{92}U \longrightarrow {}^{4}_{2}He + {}^{234}_{90}Th$$

Artificial transmutations

Unstable nuclei can be produced by nuclear bombardment

A transmutation, or conversion of an atom of one element to an atom of another element, may be caused by various processes of radioactive decay, as you have seen. However, a transmutation can also occur when high-energy particles, such as alpha particles, protons, or neutrons, bombard an atomic nucleus. Sometimes, transmutations occur naturally, such as when carbon-14 is produced from nitrogen-14 in the Earth's upper atmosphere. Today, transmutations are also produced artificially in giant machines called particle accelerators or in nuclear reactors.

In 1919, Ernest Rutherford brought about the first artificial transmutation. He bombarded nitrogen gas with alpha particles emitted by radium. The nuclear equation for this reaction is as follows.

$$^{14}_{7}N + {}^{4}_{2}He \longrightarrow [{}^{18}_{9}F] \longrightarrow {}^{17}_{8}O + {}^{1}_{1}H$$

Fluorine-18 produced in this reaction is extremely unstable and immediately decomposes. Note that a proton, which is also symbolized as $^{1}_{1}H$ in nuclear reactions, has been produced. In fact, this reaction provided the first physical evidence for the existence of the proton.

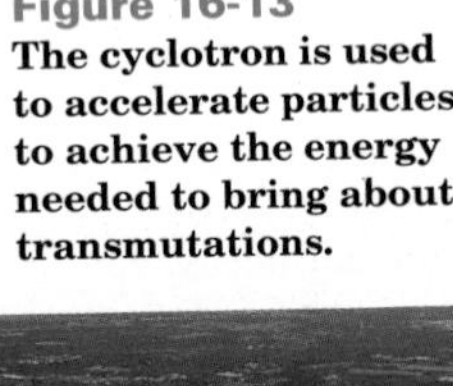

Figure 16-13 **The cyclotron is used to accelerate particles to achieve the energy needed to bring about transmutations.**

Elements with atomic numbers greater than that of uranium (92) are called transuranium elements. None of the transuranium elements are naturally occurring; they have been synthesized in particle accelerators or nuclear reactors. Each transuranium element is a product of artificial transmutations. Nuclear reactions that have been used to synthesize some of these elements are shown in **Table 16-2**.

Table 16-2
Reactions for the First Preparation of Transuranium Elements

Atomic number	Name	Symbol	Nuclear Reaction
94	plutonium	Pu	${}^{238}_{92}U + {}^{2}_{1}H \longrightarrow {}^{238}_{93}Np + 2{}^{1}_{0}n$ ${}^{238}_{93}Np \longrightarrow {}^{238}_{94}Pu + {}^{0}_{-1}e$
95	americium	Am	${}^{239}_{94}Pu + 2{}^{1}_{0}n \longrightarrow {}^{241}_{95}Am + {}^{0}_{-1}e$
96	curium	Cm	${}^{239}_{94}Pu + {}^{4}_{2}He \longrightarrow {}^{242}_{96}Cm + {}^{1}_{0}n$
99	einsteinium	Es	${}^{238}_{92}U + 15{}^{1}_{0}n \longrightarrow {}^{253}_{99}Es + 7{}^{0}_{-1}e$
101	mendelevium	Md	${}^{253}_{99}Es + {}^{4}_{2}He \longrightarrow {}^{256}_{101}Md + {}^{1}_{0}n$
102	nobelium	No	${}^{246}_{96}Cm + {}^{12}_{6}C \longrightarrow {}^{254}_{102}No + 4{}^{1}_{0}n$
103	lawrencium	Lr	${}^{252}_{98}Cf + {}^{10}_{5}B \longrightarrow {}^{258}_{103}Lr + 4{}^{1}_{0}n$
104	unnilquadium (Russia)	Unq	${}^{242}_{94}Pu + {}^{22}_{10}Ne \longrightarrow {}^{260}_{104}Unq + 4{}^{1}_{0}n$
104	unnilquadium (U.S.)	Unq	${}^{249}_{98}Cf + {}^{12}_{6}C \longrightarrow {}^{257}_{104}Unq + 4{}^{1}_{0}n$
105	unnilpentium (U.S.)	Unp	${}^{249}_{98}Cf + {}^{15}_{7}N \longrightarrow {}^{260}_{105}Unp + 4{}^{1}_{0}n$
106	unnilhexium (U.S.)	Unh	${}^{249}_{98}Cf + {}^{18}_{8}O \longrightarrow {}^{263}_{106}Unh + 4{}^{1}_{0}n$
108	unniloctium (W. Germany)	Uno	${}^{208}_{82}Pb + {}^{58}_{26}Fe \longrightarrow {}^{265}_{108}Uno + {}^{1}_{0}n$
109	unnilennium (W. Germany)	Une	${}^{209}_{83}Bi + {}^{58}_{26}Fe \longrightarrow {}^{266}_{109}Une + {}^{1}_{0}n$

nuclear fission
the process by which a nucleus splits into two smaller fragments

Nuclear fission

Some unstable nuclei can split in two

Observe what is happening to the single-celled organism shown in **Figure 16-14**. This organism is dividing in two in a process called cell division or fission. Atomic nuclei can also divide or undergo fission. **Nuclear fission** refers to a nuclear reaction in which a very heavy nucleus splits into two smaller nuclei, each with higher nuclear binding energies, as in **Figure 16-7**. Whenever fission occurs, large amounts of energy are released. For example, the fission of 1 g of uranium-235 generates as much energy as the combustion of 2 700 kg of coal.

Figure 16-14a
The single-celled paramecium is about to undergo fission, a process of reproduction. The term fission has nearly the same meaning in biology and chemistry.

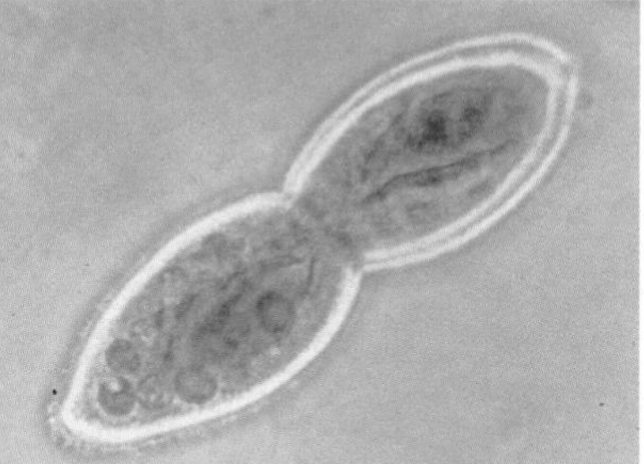

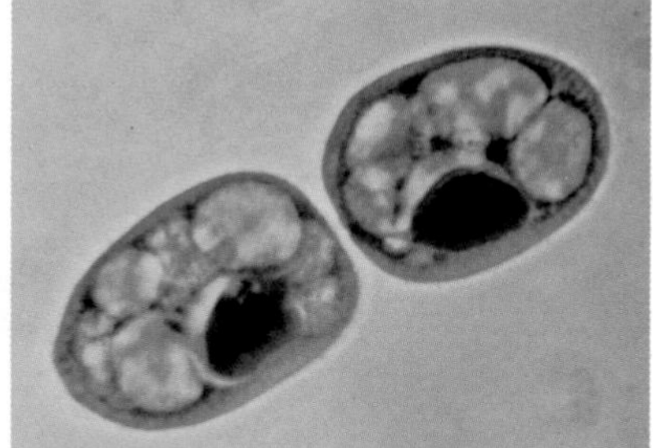

16-14b
The paramecium has fissioned, or divided, into two smaller organisms just as a nucleus divides into smaller nuclei during nuclear fission.

Teaching Tip

Chain reactions and fission

Prepare 40 envelopes with the following written on the outside of each envelope: uranium-235, $p = 92$, $n = 143$. Inside each envelope place five cards with the following words written on them: energy, krypton-94, neutron, barium-139, and neutron. The envelopes should be folded, not sealed, so that they can be reused. Give each student an envelope. Students may not open the envelope until they are given a neutron card. Neutron cards must be given away immediately to anyone with an unopened uranium-235 envelope. The teacher, who has a single neutron card, starts the chain reaction by handing the card to a student. That student will open his or her envelope, take out the five cards, and distribute the three neutron cards to three other students. Those students will then open their envelopes, take out the five cards, and distribute their three neutron cards to any students with unopened envelopes. The entire supply of uranium-235 will be quickly exhausted as the chain reaction proceeds. The number of fission reactions is represented by the number of neutron cards distributed each time, i.e., 1, 3, 9, 27, 81, 243, etc. The students should note that new fission products, the barium-139 and krypton-94 cards remaining in their envelopes, have formed as a result of the reaction. The matter and energy of the uranium-235 have not simply disappeared.

[Source: G. E. Haynes. *The Science Teacher*, Vol. 58, 1991.]

Visual Strategy

Figure 16-14

Emphasize that the biological cell analogy is not entirely satisfactory. Although the parent cell in mitosis divides to becomes two identical daughter cells, the parent nucleus in nuclear fission splits into daughter nuclei that are seldom identical, either to each other or the parent (e.g., krypton-94 and barium-139 are daughter nuclei of uranium-235).

Historical Note
Physicist Enrico Fermi and his colleagues demonstrated on December 2, 1942, that a self-sustaining nuclear fission reaction could be produced from a few lumps of natural uranium shielded by blocks of graphite. Although the reactor produced only a few watts of power, this was sufficient to show that fission could be used to produce energy.

Do You Know?
Nuclear reactors supply approximately 20% of the electricity in the United States.

Groupwork Strategy
Divide the class into groups of three, and have them devise at least five methods that could be used to dispose of nuclear waste. The students should list one advantage and one disadvantage for each method.

Fission is the source of energy generated by nuclear reactors. The main radioactive isotopes used by these reactors are uranium-235 and plutonium-239. **Figure 16-15** shows what happens to uranium-235 during fission. Notice that the nucleus of a uranium-235 atom must first be bombarded by a slow neutron. Once this is done, the nucleus splits in two, forming two new isotopes. What are these two isotopes?

In turn, each of these isotopes emits neutrons. As you can see in **Figure 16-15**, each of these neutrons when slowed down can cause the fission of another uranium-235 nucleus. Again neutrons are emitted. This process continues, forming a chain reaction. In a **chain reaction**, the material that starts the reaction is also one of the products.

chain reaction
a self-sustaining nuclear or chemical reaction in which the product from each step acts as the reactant for the next step

The amount of material used in a chain reaction is important. Sufficient radioactive material is needed to absorb the number of neutrons required to sustain the chain reaction. The amount of radioactive material needed to sustain a chain reaction is known as the **critical mass**.

critical mass
the minimum mass of fissionable material needed to produce a chain reaction

In a uranium-235 nuclear reactor, the fuel rods are surrounded by a moderator, a substance such as graphite or heavy water that slows down neutrons. The chain reaction is controlled with cadmium or boron rods. These rods absorb some of the neutrons that are produced by fission and thus control the number of neutrons that are available to bombard uranium-235 nuclei. If these control rods were not present, an uncontrolled chain reaction would occur. An uncontrolled reaction, in the worst case, could cause a nuclear meltdown. In a meltdown, dangerous radioactive material may be released. However, a nuclear meltdown does not involve a nuclear explosion, such as the one shown in **Figure 16-15**.

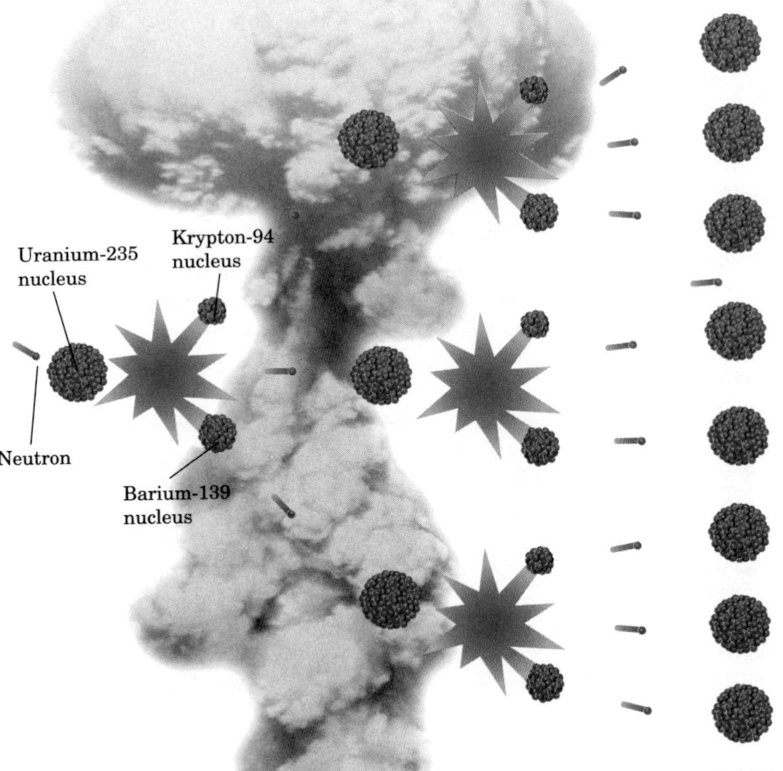

Figure 16-15
This chain reaction begins when a uranium-235 nucleus captures a stray neutron and splits into krypton and barium nuclei. Three neutrons are given off in the process. Each of these neutrons may collide with a uranium-235 nucleus, which then fissions. The fissioning uranium nuclei generate more neutrons which bombard more uranium-235 nuclei giving rise to more fission reactions. This chain reaction of uranium-235 is responsible for exploding atomic weapons.

Visual Strategy
Figure 16-15
Make sure students understand that, in this figure, the ratio of neutron release is actually 1:3:9:27:81:243, etc., and that the picture has been simplified for easier study. The escaping neutrons will continue to bombard other nearby atoms unless they encounter some type of absorbing material such as graphite, europium, or cobalt.

Nuclear fusion

Some unstable nuclei join to form heavier nuclei

nuclear fusion
the process by which two nuclei combine to form a heavier nucleus

In **Figure 16-16**, you see two examples of reactions involving nuclear fusion. **Nuclear fusion** is the process by which nuclei with low masses fuse to form heavier, more stable nuclei with higher binding energies as in **Figure 16-7**. In the process, tremendous amounts of energy are released.

In both the sun and in the explosion of a hydrogen bomb fusion reactions occur. In the case of the sun, a series of reactions cause four hydrogen nuclei to combine to form a single helium nucleus. The temperature of the Sun's core, where some of the fusion reactions occur, is about 1.5×10^{7}°C. High temperatures such as this are required to bring the nuclei together. When the hydrogen nuclei are fused to form a larger nucleus, some mass is lost and converted to energy. More energy is released per gram of reactant in fusion than in fission.

Currently, scientists are investigating ways to bring about fusion reactions at lower temperatures. Experiments using high-powered laser light to start the fusion process are being conducted. If fusion reactions are ever to become a practical source of energy, then the energy given off must be greater than the energy required to start them. At the moment, scientists have had limited success in this task.

Figure 16-16a
Nuclear fusion reactions occurring in the sun have supplied Earth with energy for billions of years.

16-16b
The hydrogen bomb uses nuclear fusion rather than fission. As a result, an exploding hydrogen bomb has even greater destructive power than a fission-based nuclear weapon.

Section Review

8. Write balanced nuclear equations for the following nuclear reactions.
   a. Uranium-233 undergoes alpha decay.
   b. Neptunium-239 undergoes beta decay.
   c. Copper-66 undergoes beta decay.
   d. Beryllium-9 and an alpha particle combine to form carbon-13. Carbon-13 then breaks apart to emit a neutron and carbon-12.
   e. Phosphorus-32 and a neutron combine to form phosphorus-33, which then decays to emit a beta particle.
9. Refer to **Figure 16-15**. Write the balanced nuclear equation for the fission reaction to show what happens to uranium-235 after it has been bombarded with a neutron.
10. One fusion reaction scientists are studying as a possible source of energy is that between a deuterium nucleus ($^{2}_{1}H$) and a tritium nucleus ($^{3}_{1}H$). One of the two products formed in this reaction is a helium nucleus. Write the balanced nuclear equation for this fusion reaction, being sure to include the other product that is formed.

Content Background

Nuclear fusion
One of the problems of developing fusion on Earth is containing the hot, ionized material (plasma). The plasma is held in a "magnetic bottle," a magnetic field that requires large amounts of energy to create and sustain. During a fusion reaction, enough energy would be created to maintain the magnetic bottle and also generate usable power.

ASSESS

Alternative Assessment

Have students identify X for each of the following reactions and name the reaction process.

a. $^{212}_{84}Po \longrightarrow ^{208}_{82}Pb + X$
$^{212}_{84}Po \longrightarrow ^{208}_{82}Pb + ^{4}_{2}He$
(alpha decay)

b. $^{239}_{93}Np \longrightarrow ^{239}_{94}Pu + X$
$^{239}_{93}Np \longrightarrow ^{239}_{94}Pu + ^{0}_{-1}e$
(beta decay)

c. $^{11}_{6}C + X \longrightarrow ^{11}_{5}B$
$^{11}_{6}C + ^{0}_{-1}e \longrightarrow ^{11}_{5}B$
(electron capture)

d. $^{13}_{7}N \longrightarrow ^{13}_{6}C + X$
$^{13}_{7}N \longrightarrow ^{13}_{6}C + ^{0}_{+1}e$
(positron emission)

Answers to Section Review

8. a. $^{233}_{92}U \longrightarrow ^{4}_{2}He + ^{229}_{90}Th$

b. $^{239}_{93}Np \longrightarrow ^{0}_{-1}e + ^{239}_{94}Pu$

c. $^{66}_{29}Cu \longrightarrow ^{0}_{-1}e + ^{66}_{30}Zn$

d. $^{9}_{4}Be + ^{4}_{2}He \longrightarrow ^{13}_{6}C$

$^{13}_{6}C \longrightarrow ^{1}_{0}n + ^{12}_{6}C$

e. $^{32}_{15}P + ^{1}_{0}n \longrightarrow ^{33}_{15}P$

$^{33}_{15}P \longrightarrow ^{0}_{-1}e + ^{33}_{16}S$

9. $^{235}_{92}U + ^{1}_{0}n \longrightarrow ^{236}_{92}U$

$^{236}_{92}U \longrightarrow ^{94}_{36}Kr + ^{139}_{56}Ba + 3^{1}_{0}n$

10. $^{2}_{1}H + ^{3}_{1}H \longrightarrow ^{4}_{2}He + ^{1}_{0}n$

Visual Strategy

Figure 16-16b
Make students aware that fusion bombs use uncontrolled fusion reactions. The goal of most scientists is to control the fusion reaction to generate electrical power.

Section 16-3

FOCUS

Lesson Starter

Start the lesson by discussing the rates of chemical reactions. Ask students whether the time it takes for a fixed number of reactions to take place will increase, decrease, or remain the same if the number of particles of the reactants is divided in half. (increase) Ask whether this principle should also apply to nuclear reactions.

Demonstration 3

Half-life prediction

Approximate time: 10 min

This demonstration shows how radioactive decay can be described by probability measurements.

1. Prior to class draw a coordinate system on the board. Label the vertical axis *Number of emissions*, and mark the range of values from 0 to 300. Number the horizontal axis from 0 to 20, and label it *Half-life*.
2. Ask four students to come forward to help. Have one student graph the data as it is collected, while the other three count the pennies.
3. Place 300 pennies in an empty box, shake the box, and pour the pennies onto the lid. Have the three student "counters" record the number of heads showing on all of the pennies. Set these coins aside in a separate pile. The student recording the data should mark on the graph the number of heads obtained after the first count (half-life = 1)

How is nuclear chemistry used?

Describe how the half-life of a radioactive isotope is used to calculate the ages of objects.

Describe how radioactive isotopes are used in the manufacture of consumer products.

Describe how the process of neutron activation is used to analyze objects.

Describe how nuclear imaging techniques are used as diagnostic tools in medicine.

Identify the possible health hazards of radiation exposure.

Half-life

At the beginning of this chapter, you read that the archaeologist who first examined the Iceman suspected that he had died some 4000 years ago. The investigator arrived at this conclusion based on his examination of the Iceman's ax. From the shape and composition of the blade, the archaeologist reasoned that the Iceman lived during the Early Bronze Age, which began around 2200 B.C.

However, the blade was found to be made of pure copper and not bronze. This indicated that the Iceman must have been older than the archaeologist first suspected because the Age of Copper began around 4000 B.C. and lasted until the advent of the Bronze Age. The Iceman could then have lived as long as 6000 years ago.

But rather than determining the Iceman's age on the basis of objects found with him, scientists have calculated his age by analyzing radioactive isotopes in the Iceman's body.

half-life
the time required for half of a sample of radioactive atoms to decay

Radioactive isotopes decay at a constant rate

The use of radioactive isotopes to determine the age of an object is called *radioactive dating*. Radioactive dating is based on the fact that the decay rates of radioactive nuclei are constant. The rate of nuclear decay is given in terms of half-lives. One **half-life** is the time that it takes for half of a sample of radioactive isotopes to decay, as you can see in Figure 16-17.

Figure 16-17
The radioactive isotope $^{131}_{53}I$ has a half-life of 8.07 days. In each successive 8.07-day period, half of the mass of $^{131}_{53}I$ in the original sample of $^{131}_{53}I$ decays to $^{131}_{54}Xe$.

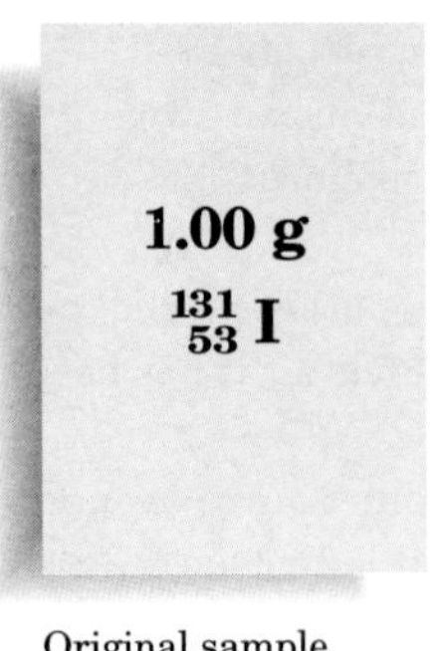

Original sample of radioactive isotope $^{131}_{53}I$

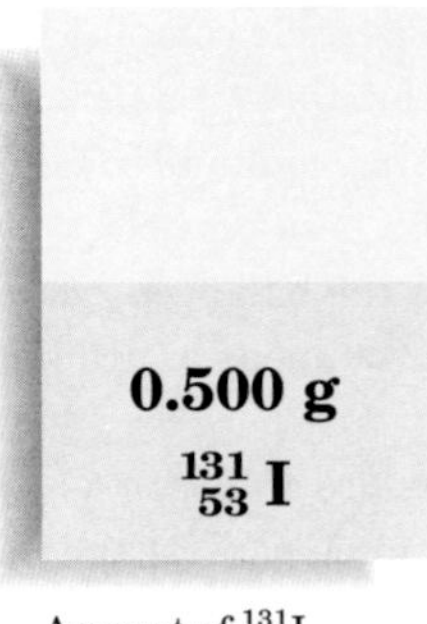

Amount of $^{131}_{53}I$ remaining after 1 half-life (8.07 days)

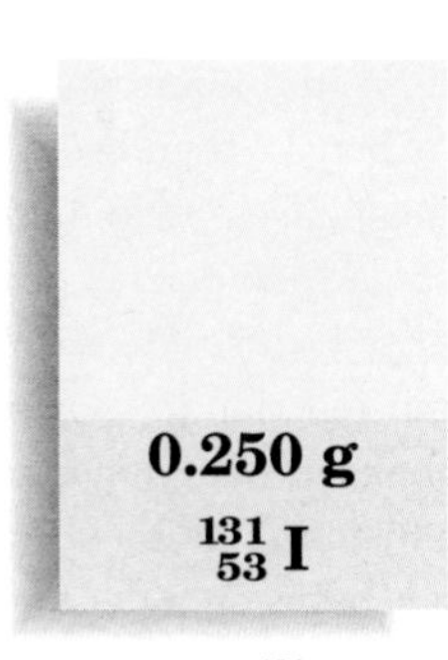

Amount of $^{131}_{53}I$ remaining after 2 half-lives (16.14 days)

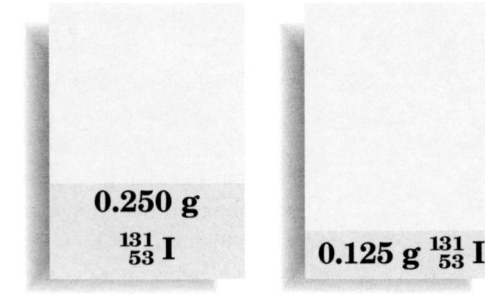

0.125 g $^{131}_{53}I$

Amount of $^{131}_{53}I$ remaining after 3 half-lives (24.21 days)

Table 16-3
Half-Lives of Some Radioactive Isotopes

Isotope	Half-life	Radiation emitted
carbon-14	5.715×10^3 y	${}^{0}_{-1}e$
potassium-40	1.3×10^9 y	${}^{0}_{-1}e, \gamma$
radon-222	3.8 days	${}^{4}_{2}He, \gamma$
radium-226	1.6×10^3 y	${}^{4}_{2}He, \gamma$
thorium-230	7.5×10^4 y	${}^{4}_{2}He, \gamma$
thorium-234	24 days	${}^{0}_{-1}e, \gamma$
uranium-235	7.0×10^8 y	${}^{4}_{2}He, \gamma$
uranium-238	4.5×10^9 y	${}^{4}_{2}He, \gamma$

Every radioactive isotope has a characteristic half-life. Like binding energy, half-life indicates the stability of radioactive isotopes. The shorter the half-life, the quicker the isotope decays, and the more unstable it is. Highly unstable nuclei have very short half-lives. For example, silicon-26 has a half-life of two seconds. On the other hand, plutonium-242 is relatively stable. Nearly 4×10^5 years are required for half of a sample of plutonium-242 to decay. **Table 16-3** lists the half-lives for various radioactive isotopes.

Radioactive dating is used to find the ages of objects

The half-life of carbon-14 provides the key to the age of the Iceman. Most of the carbon that exists is the stable isotope carbon-12; less than one-millionth of one percent of the carbon in Earth's atmosphere is the radioactive isotope carbon-14. Because both these isotopes have the same electron configuration, they react chemically in almost an identical fashion. Both carbon-12 and carbon-14 join with oxygen to become carbon dioxide, which is taken in by plants. Therefore, all plants on Earth contain a small amount of carbon-14. All animals eat plants or plant-eating animals, so animals contain some carbon-14 too. Carbon-14 undergoes radioactive decay to become nitrogen-14 in the following reaction.

$${}^{14}_{6}C \longrightarrow {}^{14}_{7}N + {}^{0}_{-1}e$$

Living plants continue to take in carbon dioxide while they are alive. In this way, they replenish the carbon-14 they lose due to decay, and the ratio of carbon-14 to carbon-12 remains constant. Plants use the carbon dioxide they consume from the atmosphere to make carbohydrates, such as glucose, which animals eat. Animals continually replenish the carbon-14 they lose through decay by consuming glucose and other carbon-containing compounds made by plants.

Figure 16-18a
This cave painting from the Cosquer cave in France dates back 16 500 years. The age was determined using carbon-14 dating.

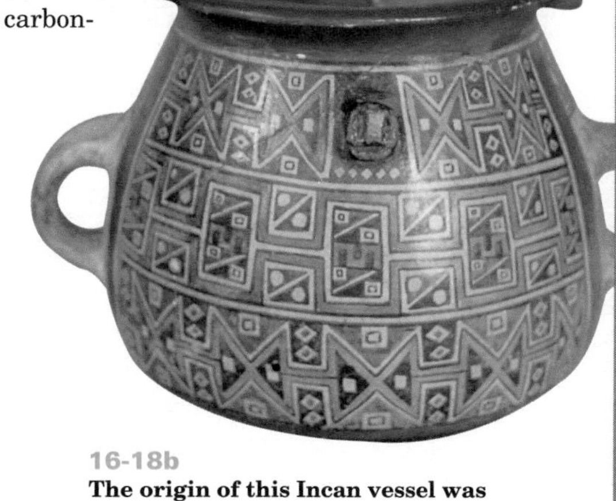

16-18b
The origin of this Incan vessel was determined to be 1450 A.D. using dating techniques.

4. Place the pennies from the "tails" pile back in the box, put the lid on the box, and shake the contents. Repeat step **3** with this set of pennies, and record the number of heads from the second count (half-life = 2).
5. Continue this process until all of the pennies have "decayed". The student should graph the data to show the rate of decay.

SAFETY
None.

DISPOSAL
None.

TEACH

Teaching Tip
Enhancement
Select one of the students in the class, and say that he or she has been given $1 000 000. The only stipulation is that every 2 s the amount of money will decrease by half. Ask the class: How long will it be until the millionaire is penniless? (In 52 s (26 half-lives) the $1 000 000 will have "decayed" to 1.5¢. The millionaire is broke before 1 min has elapsed.)

Content Background
Carbon-14 dating
Point out to students that even though the nucleus has decayed, the substance is not disappearing. Stress that decaying means that one substance, whose radioactive emissions are being measured, is changing into another substance, which may or may not be radioactive. The change in overall mass over time is very small.

Teaching Tip
Discussion
Point out to students how very difficult it must be to study an isotope that has a half-life as short as 2 s. The radioactive sample being measured would quickly disappear.

Visual Strategy
Figure 16-18
Make students realize that the amount of carbon-12 is as important to dating a sample as is the amount of carbon-14. Only by having the carbon-12 as a reference point can a sample of undetermined age be compared to samples of known age. The ratio of ${}^{14}_{6}C$ to ${}^{12}_{6}C$ is the useful measurement in carbon dating.

Groupwork Strategy

Divide the class into pairs, and have one student in each pair solve the problem for the odd-numbered isotopes, while the other student solves the problem for the even-numbered isotopes. Have team members check each other's work.
How long does it take 400 g of each carbon isotope listed in the table to decay to < 0.001 g.

Isotope	Half-life
carbon-9	0.1265 s
carbon-10	19.2 s
carbon-11	20.38 min
carbon-12	stable
carbon-13	stable
carbon-14	5715 y
carbon-15	2.449 s
carbon-16	0.75 s

(The 400 g decay to 200 g after one half-life period, to 100 g after two half-lives, and so forth until after 19 half-lives, when the amount of radioactive material remaining is 7.63×10^{-4} g < 0.001 g. Therefore, the time required for 400 g of each isotope to decay to this amount is equal to the isotope's half-life multiplied by 19.)

Time required for isotope to decay to less than 1 mg:
carbon-9; 2.404 s
carbon-10; 6.08 min
carbon-11; 6.454 h
carbon-12 and carbon-13; no decay
carbon-14; 1.086×10^5 y
carbon-15; 46.53 s
carbon-16; 14 s

Possible Misconception

Make sure that students realize that to use carbon-14 dating, the sample must have originally contained *and still contain* carbon. Petrified wood is not a candidate for carbon-14 dating because the original carbon atoms have been replaced by other atoms.

However, when a plant or animal dies, intake of carbon-containing substances ceases, so replenishment of carbon-14 comes to a halt. Then, the percentage of carbon-14 decreases—at a rate that is known and determined by its half-life. The half-life of carbon-14 is about 5715 years. So half of the carbon-14 atoms now present in a plant or animal will have decayed 5715 years after it dies; in another 5715 years half of the remaining carbon-14 atoms will have decayed; etc. The radioactive carbon-14 atoms, which can be detected with a Geiger counter, therefore continue to decrease at a steady rate after a plant or animal dies, as shown in **Figure 16-19**.

Once the relative amount of carbon-12 and carbon-14 in a fossil or other object of unknown age is measured, it is compared to the ratio of these isotopes in a sample of similar material whose age is known. For example, suppose the ratio of carbon-14 to carbon-12 in the tissues of the Iceman were found to be nearly one-half of that found in the tissues of a person who had recently died. Then, the Iceman must have lived about 5715 years ago. Using radioactive dating with carbon-14, scientists have concluded that the Iceman lived around 5000 years ago, between 3500 and 3000 B.C.

Radioactive dating is used to measure geologic time

After four half-lives, the amount of radioactive carbon-14 remaining in an object is too small to be of any use in radioactive dating. Consequently, carbon-14 is not useful for dating specimens that are more than 20 000 years old. Anything older than this must be dated with the use of a radioactive isotope that has a half-life longer than that of carbon-14.

Potassium-40 has a half-life of 1.3 billion years. Because of its long half-life, potassium-40 has been used to date ancient rocks and minerals. Potassium-40 produces two different isotopes in its radioactive decay. About 11% of the potassium-40 in a mineral decays to argon-40, which may be retained in the mineral sample. The remaining 89% of potassium-40 decays to calcium-40. The decay of potassium-40 to calcium-40 is not useful for radioactive dating because the calcium-40 cannot be distinguished from the original calcium in the rock. The argon-40, however, can be measured. From its amount, the minimum date of a rock can be determined.

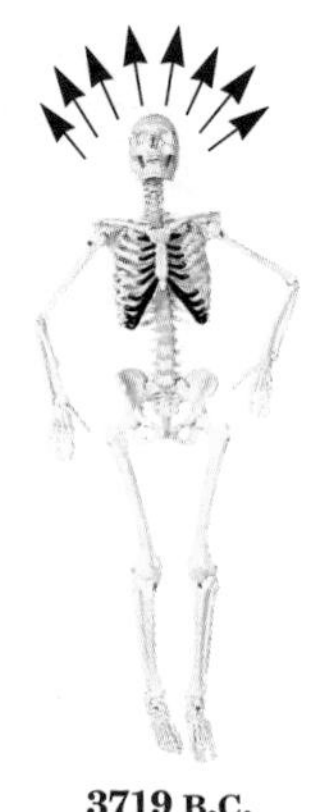

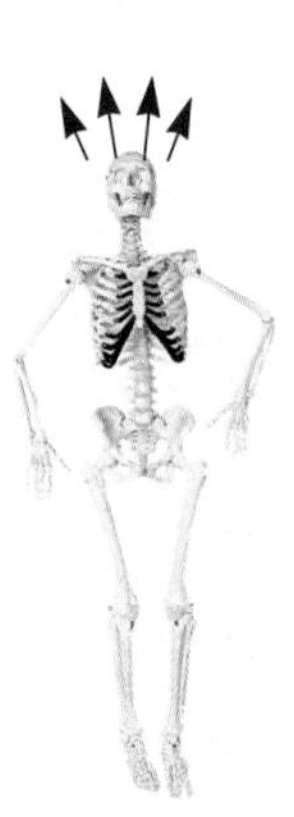

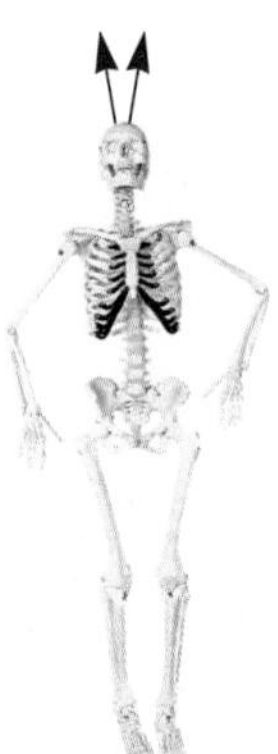

Figure 16-19
Every 5715 years, the carbon-14 isotopes in the Iceman's body decrease by one-half.

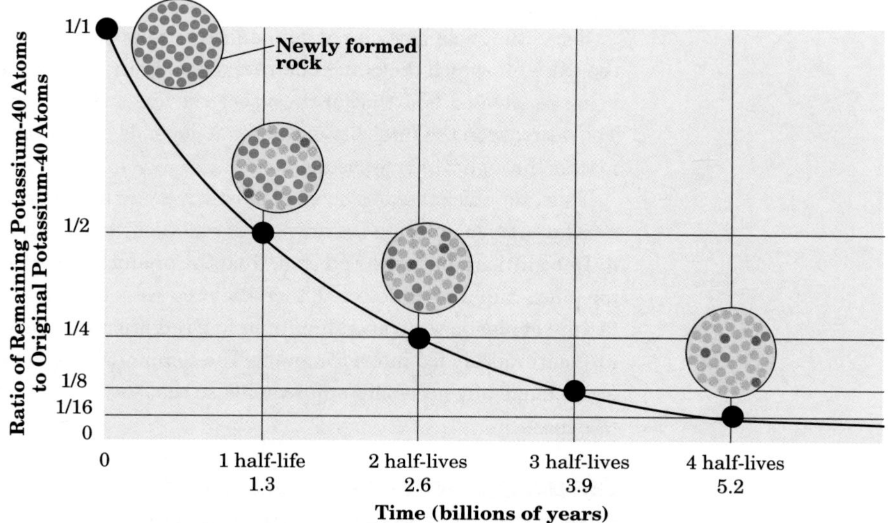

Figure 16-20 Potassium-40 decays to argon-40 and calcium-40, but scientists monitor only the ratio of potassium-40 to argon-40 to determine the age of the object.

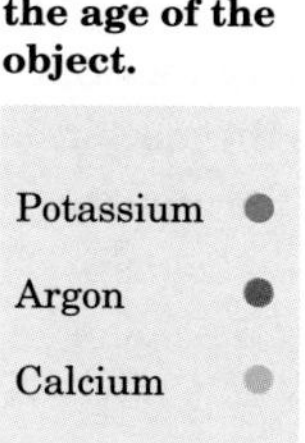

In **Figure 16-20**, you can assume that there are 40 potassium-40 atoms at zero time. After one half-life, 20 atoms remain and 20 have decayed. Of those 20 that have decayed, 11% (approximately 2) are argon-40, and 89% (approximately 18) are calcium-40. After two half-lives, another 10 have decayed, for a total of 30. Of these 30, 11% are argon-40 (approximately 3), and 89% are calcium-40 (approximately 27). After four half-lives, a total of 37.5 have decayed. Of these, 11% are argon-40 (approximately 4), and 89% are calcium-40 (approximately 33). The decay of potassium-40 is widely used in geological investigations. For example, geologists use it to study motions of the sea floor.

Other isotopes with half-lives even longer than that of potassium-40 have been used to determine the age of the Earth. From these isotopes, the Earth has been revealed to be an impressive 4.5 billion years old!

Industrial applications

Radioactive isotopes are used to make consumer products

The products you see in **Figure 16-21** have all been manufactured with the help of radioactive isotopes. The manufacturing process for each of these products depends on producing the material at the proper thickness. If plastic wrap is too thin, it may tear easily and be useless for wrapping foods. If aluminum foil is too thick, manufacturing it may not be cost-effective. Radioactive isotopes can be used to determine and control the thickness during the manufacture of each of these products.

Figure 16-21 Radiation is used to measure the thickness of metal sheets, plastic wrap, and other products. The intensity of radiation transmitted through these products depends on thickness.

Content Background

Background radiation
Stress that in potassium-40 dating the background radiation from other radioactive sources in the sample is assumed to have little or no effect on measurement of the sample's age. Background radiation becomes important when a radioactive element with a short half-life decays past the level of detection. For potassium-40, the levels become too low to measure effectively after 5 billion years, which is about the age of our solar system.

Other Applications

Potassium-40 dating was used to date moon rocks. From the results it was determined that the moon was formed at about the same time as the Earth.

Visual Strategy

Figure 16-20
Make sure students understand that the radioactive emissions from a sample never decrease to zero. The isotope whose decay is being measured will eventually exist in quantities too small to be measured, and finally will be completely transformed. However, because of the background radiation found in all materials, there will always be some measurable radioactivity.

Do You Know?
Many times the holders of precious antiquities will not allow any material to be scraped off or removed for testing, even though only a small fragment would be needed. In determining the age of the object, some damage always occurs.

Recall that beta particles or gamma rays can penetrate objects. However, the extent to which they can penetrate depends on the thickness of the object being penetrated. The thicker the object, the less radiation passes through. The decrease in the intensity of radiation depends directly on the quantity of matter through which it passes.

Thus, the thickness of a sheet of plastic wrap or aluminum foil can be measured by monitoring the amount of radiation that is able to pass through it. If too little radiation is detected, then the product is too thick. Conversely, if too much radiation is detected, then the product is too thin. During the manufacture of plastic wrap and aluminum foil, radiation intensity measurements are continuously fed into a computer. The computer then directs the machinery to make any necessary adjustments so that the product has the proper thickness.

Neutron activation analysis

Chemical composition can be determined without destroying a specimen

In order to perform a chemical analysis, part of an object being examined may have to be destroyed in the process. For example, a small sample of the specimen may have to be taken and subjected to various chemical reactions. This type of analysis may not pose any problem if the specimen is large or plentiful.

However, if the specimen is rare, valuable, or otherwise cannot be damaged, then a different approach such as *neutron activation analysis* must be used. First, the sample to be analyzed is exposed to a beam of neutrons emitted by a radioactive isotope in an artificial transmutation process. When the nuclei absorb these neutrons, some elements in the sample, in turn, become artificially produced radioactive isotopes. Each element can then be identified and its mass calculated based on the amount of radioactivity it emits.

Neutron activation analysis has been used to determine the composition of meteorites, as shown in Figure 16-22. Also, the identification of elements by neutron activation has been put to use in crime detection. Firing a gun leaves extremely small amounts of antimony, barium, and copper on a person's hand. A wax cast made of a suspect's hand picks up these elements. Subjecting the cast to neutron activation analysis will then reveal if these elements are present and thus show whether the suspect has recently fired a gun.

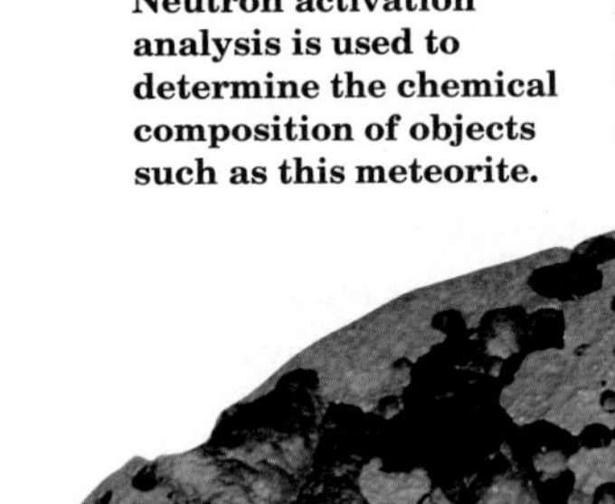
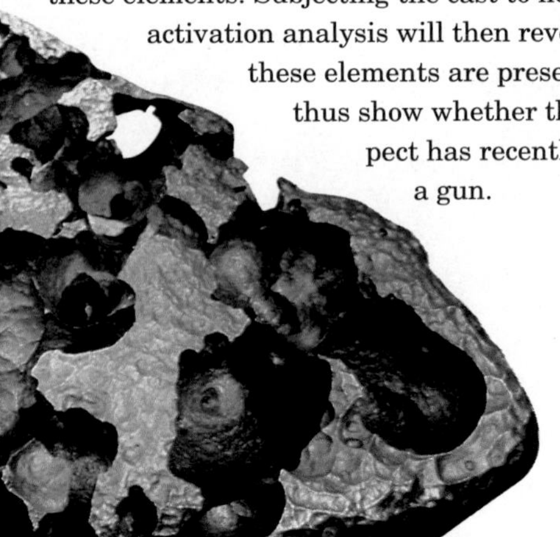

Figure 16-22
Neutron activation analysis is used to determine the chemical composition of objects such as this meteorite.

One of the most unusual applications of neutron analysis in crime detection involved the 12th president of the United States, Zachary Taylor. On July 4, 1850, President Taylor dedicated the cornerstone for the Washington Monument. Upon his return to the White House, he ate a bowl of cherries and drank a glass of milk. Soon after, he became seriously ill, suffering from diarrhea and severe vomiting. He died five days later.

Doctors attributed his death to natural causes. But some historians suspected for some time that Taylor may have been poisoned with arsenic. One hundred and fifty years after his death, Taylor's body was exhumed from his lead coffin to find out if he had been murdered. Scientists cut away some of his clothing to get samples of his hair. They also took small pieces of his nails and samples of body tissues. All of these samples were subjected to neutron activation analysis. The results revealed that arsenic was present, but only in levels that are normally present in the human body. The scientists concluded that President Taylor had not been poisoned.

Nuclear imaging

Electromagnetic and nuclear radiation are used to look inside the human body

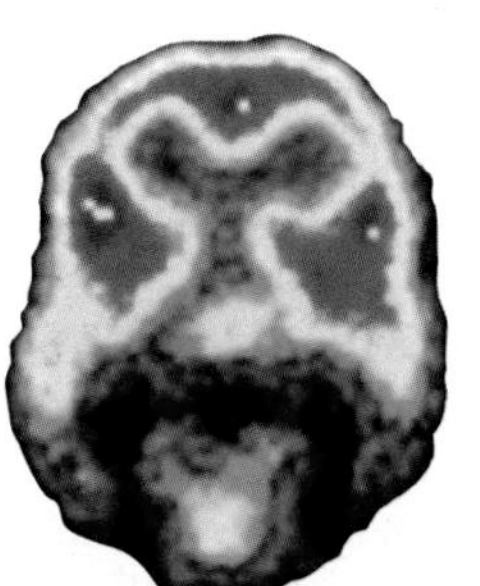

Figure 16-23a
This image of a healthy brain was obtained with NMR imaging.

X rays have been used since the late 1800s to look inside the human body to study such structures as bones and lungs. Most organs, however, do not show up well because they are soft and do not absorb the radiation given off by X rays. However, with nuclear imaging techniques, doctors can obtain a more detailed look inside a person's body.

One nuclear imaging technique is known as positron emission tomography (PET). A person is given a small dose of a radioactive isotope such as carbon-11, nitrogen-13, or oxygen-15. Each of these isotopes has a short half-life and emits a positron in the process of decay. Emitted positrons interact with electrons to produce gamma radiation, which is detected by a gamma-ray detector. A computer converts the measured gamma radiation to an image that reveals biological activity.

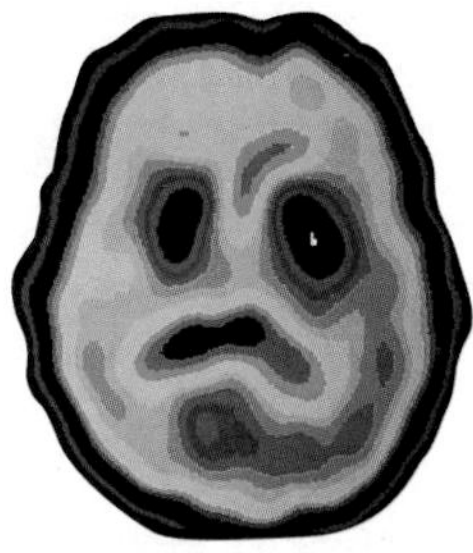

16-23b
This NMR image reveals a brain with Alzheimer's disease.

One nuclear imaging technique that does not depend on radioactivity is known as nuclear magnetic resonance (NMR). The application of NMR to produce images of human organs or tissues is called *magnetic resonance imaging* (MRI). This technique depends on how an atomic nucleus absorbs electromagnetic radiation, which, in turn, depends on the nucleus's chemical environment.

For example, H_2O, $C_6H_{12}O_6$ (a carbohydrate), $C_{57}H_{110}O_6$ (a fat), and $C_{254}H_{377}O_{75}N_{65}S_6$ (a protein) all contain hydrogen atoms. However, in each case, the hydrogen nuclei absorb electromagnetic radiation differently. Furthermore, the hydrogen nuclei in each of these compounds behave differently, depending on whether the organ containing them is normal or diseased. An MRI can check the functioning of an organ, such as the brain, by collecting data about how its hydrogen atoms absorb electromagnetic radiation. A computer converts this data into an image of the organ, such as the one shown in **Figure 16-23**.

Content Background

Conservation of matter

Even though President Taylor had been dead for a long period of time, any arsenic atoms present in the tissues of his body still would have been found.

Historical Note

Similar tests were performed on the remains of Napoleon Bonaparte. The results suggest that he, unlike Taylor, probably was poisoned.

Content Background

Radiation scanning techniques in medicine

With precision aiming and computer analysis, doctors and technicians can look at an object like the brain without ever opening the skull. These non-intrusive methods of diagnosis are welcomed by the patient, who now will not have to worry about infection.

Do You Know?

Fluoroscopy uses X rays to show moving images of the bones and organs. When fluoroscopy first came into use, it was not uncommon to find fluoroscopes in shoe stores, so that patrons could see the bones in their feet. As more was understood of the hazards of radiation, shoe-store fluoroscopes were removed.

Cultural Connection

Attempts to fix problems associated with the brain have been recorded as far back as the ancient Egyptians. Skulls have been found that had been cut open and had healed, indicating that the patient lived through the surgery.

ASSESS

Alternative Assessment

Have students determine how many half-lives are needed for 1000 kg of radioactive material to decay to 1 μg. (30 half-lives) How many years does this amount to for thorium-230? (2.25 million y)

Visual Strategy

Figure 16-23

Point out that since organs tend to function the same way in all people, a healthy organ can be readily distinguished from a diseased one. Techniques such as NMR imaging can help locate tumors and other abnormalities in the body.

Figure 16-24
These workers wear radiation detectors to monitor the level of nuclear radiation to which they are exposed.

Health concerns

Exposure to radiation must be monitored

The effects of nuclear radiation on the body are cumulative. Repeated exposure to small doses over a long period of time is dangerous if the total amount of radiation received equals the amount of a single large dose. People working with radioactivity, whether in medicine or industry, must therefore monitor the amount of radiation to which they are exposed. The people in **Figure 16-24** wear a special device to monitor their exposure to radiation. One type of device, called a film badge, contains several layers of photographic film covered with black, light-proof paper. At frequent and predetermined intervals, the film is removed and developed. The strength and type of radiation that the worker was exposed to can be determined by examining how dark the film is. The film badge doesn't protect a person from radiation; it only tells them how much radiation exposure they have had.

The biological effect of exposure to nuclear radiation is expressed in a unit known as a *rem*. **Table 16-4** lists the effects of exposure to various doses of radiation. People working with radioactive isotopes are advised to limit their exposure to 5 rems per year. This exposure is 1000 times higher than the recommended exposure level for most people, including you.

Table 16-4
Effect of Whole-Body Exposure to a Single Dose of Radiation

Dose (rems)	Probable effect
0–25	no observable effect
25–50	slight decrease in white blood cell count
50–100	marked decrease in white blood cell count
100–200	nausea, loss of hair
200–500	ulcers, internal bleeding
>500	fatal

Section Review

11. Explain how carbon-14 is used to determine the age of an object.
12. How are radioactive isotopes used in the manufacture of certain consumer goods and in the preparation of certain food products?
13. What is neutron activation analysis?
14. Describe two nuclear imaging techniques used in medicine.
15. Airline crews are exposed to about 0.5–0.7 rems of radiation per year, which is higher than that of most people. Why are airline crews exposed to a higher level of radiation?

Story Link

16. If the Iceman's body contains 0.50 times as much carbon-14 as it did when he died, how old is the Iceman? If the Iceman's body contains 0.55 times as much carbon-14 now as when he died, about how old is the Iceman?

Answers to Section Review

11. The amounts of carbon-14 and carbon-12 are measured for a sample of unknown age and are compared to those for an object whose age is known. Because the decay rate of $^{14}_{6}C$ is known, the sample's age with respect to the known object's age can be determined.
12. The thickness of materials can be determined by the amount of radiation passing through them. Isotopes in smoke detectors ionize molecules to create an electric current that is interrupted in the presence of smoke. Gamma radiation is used to kill bacteria, insects, and parasites that can cause food to spoil. (These are only a few possible answers.)
13. Neutron activation analysis uses neutrons to transmute unknown materials into radioactive isotopes. The radioactivity emitted permits identification of the material.
14. In NMR, the hydrogen atoms in different hydrocarbon molecules absorb electromagnetic radiation differently and thus indicate if the molecules, and therefore the body tissues, are normal. In PET, a positron-emitting isotope is given to the patient. The positrons combine with electrons to form gamma rays, which are measured and converted into an image that shows biological activity.
15. Radiation from space is absorbed less by the atmosphere at high altitudes than at sea level.
16. 5715 y (one half-life period); The precise answer (4929 y) requires logarithms, but a good approximation may be obtained with the following method.

$$\frac{1.00 - 0.50}{1.00 - 0.55} = \frac{5715\ \text{y}}{\text{age in y}}$$

$$\text{age} = (0.9)(5715)\ \text{y} = 5144\ \text{y}$$

Visual Strategy

Figure 16-24

Note that while considerable protection is necessary for individuals working in areas with high radiation levels, continuous exposure to background radiation is not a cause for concern. A person's average exposure to natural radiation is about 1% of a millirem per hour, or about 100 millirems per year. Most of our ingested radiation comes from radioactive potassium.

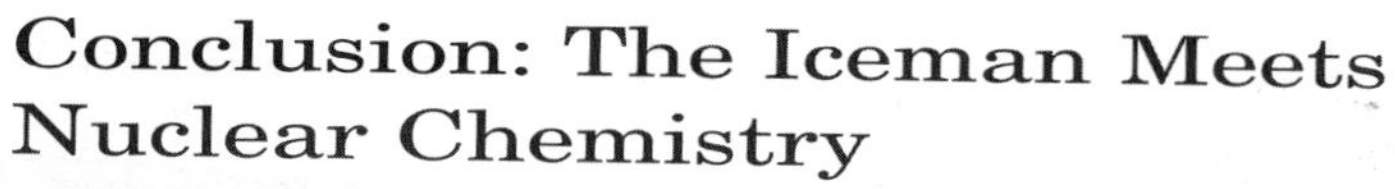

Conclusion: The Iceman Meets Nuclear Chemistry

You began your study of nuclear reactions by reading about the Iceman. Now that you have completed your study of this branch of chemistry, reconsider the questions that were asked at the beginning of this chapter.

Exactly when did the Iceman die? Radioactive dating with carbon-14 has placed the Iceman's lifetime somewhere between 3500 and 3000 B.C., during the Age of Copper. The Age of Copper was a period of global climatic warming, allowing humans to climb high up into mountains, such as the Alps where the Iceman was found.

What did the Iceman look like when he was alive? To reconstruct the Iceman's face, scientists used nuclear imaging techniques to generate three-dimensional computer images. Based on these images, scientists built a model of the Iceman's skull. They then used clay to duplicate his muscles and soft urethane to reconstruct his skin. The completed model showed that the Iceman had distinctive features—a broad nose, a protruding lower lip, and prominent chin.

How did the Iceman die? Nuclear imaging techniques provided the key to this question. Scientists believe that the Iceman died from exhaustion, perhaps from having been exposed too long to adverse weather conditions. In this state of exhaustion, the Iceman lay down on his left side in a small depression on the mountain, where he froze to death. His body was gradually covered by snow, which eventually turned into the hard ice coffin in which the Iceman lay buried for the next 5000 years.

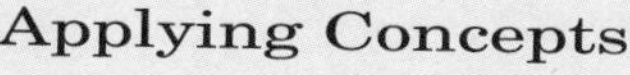

Applying Concepts

1. A fossil bone contains one-sixteenth as much carbon-14 as a living organism contains. How old is the fossil?
2. If a rock contained 1.2 mol of potassium-40 when it formed, how many grams remain after 3 billion years?

Research and Writing

Use the library to investigate the following topics.

1. Report on what procedures are proposed in the United States to bury radioactive waste. What are the problems and limitations of current disposal proposals?
2. Report on how nuclear weapons are currently tested in various countries.
3. Find out what foods the FDA has approved for irradiation. What requests are currently pending with the FDA? What pros and cons have been identified concerning the irradiation of foods? What scientific evidence supports or disproves these claims?

Chemistry in Your Community

1. Visit a local hospital, and determine what type of medical instruments that depend on nuclear radiation are being used. Find out how the workers protect themselves from excessive radiation. What is done for workers who are overexposed to radiation?
2. Visit a local dentist and investigate the use of X rays. How are the dentist and patient protected from excessive X rays? Speak with the dentist, and determine whether overexposure to X rays is a matter of concern.

Answers to Applying Concepts

1. $1/16 = (1/2)^4$, so the fossil has gone through 4 half-lives. Its age is therefore four times the half-life of carbon-14.

   $4 \times 5715 \text{ y} = 2.286 \times 10^4 \text{y}$

2. The half-life of potassium-40 is 1.3 billion y. The age of the rock, 3 billion y, is nearly 2 1/3 half-lives, so the depletion of the sample will be by a factor of $(1/2)^{2.33} = (1/2)(1/2)(1/\sqrt[3]{2}) = 0.20$, and the amount remaining of the original 1.2 mol of potassium-40 is $(0.20)(1.2 \text{ mol}) = 0.24$ mol. The atomic mass of potassium-40 is 39.1 g/mol, so the mass of the remaining potassium-40 is

   $$(0.24 \text{ mol})\left(\frac{39.1\text{g}}{1 \text{ mol}}\right) = 9.4 \text{ g}$$

Content Background

Nuclear physics/chemistry The developments that led to the current model of the atom and its nucleus were the product of both chemistry and physics. The problems of analysis and identification of elements generally have been regarded as the domain of chemistry and metallurgy, whereas the description of energy and matter was developed by physicists in the early twentieth century. Even today the distinctions are somewhat blurred—the current debate over the naming of the transfermium elements deals with elements whose positions on the periodic table suggest specific chemical properties, but which exist for such brief times and in such small quantities (sometimes only a single atom) that their chemical properties, if any, are unknown. In addition, the impact of nuclear chemistry and physics on society has brought them into the arenas of international politics, energy policy, and public safety more than perhaps any other areas of science.

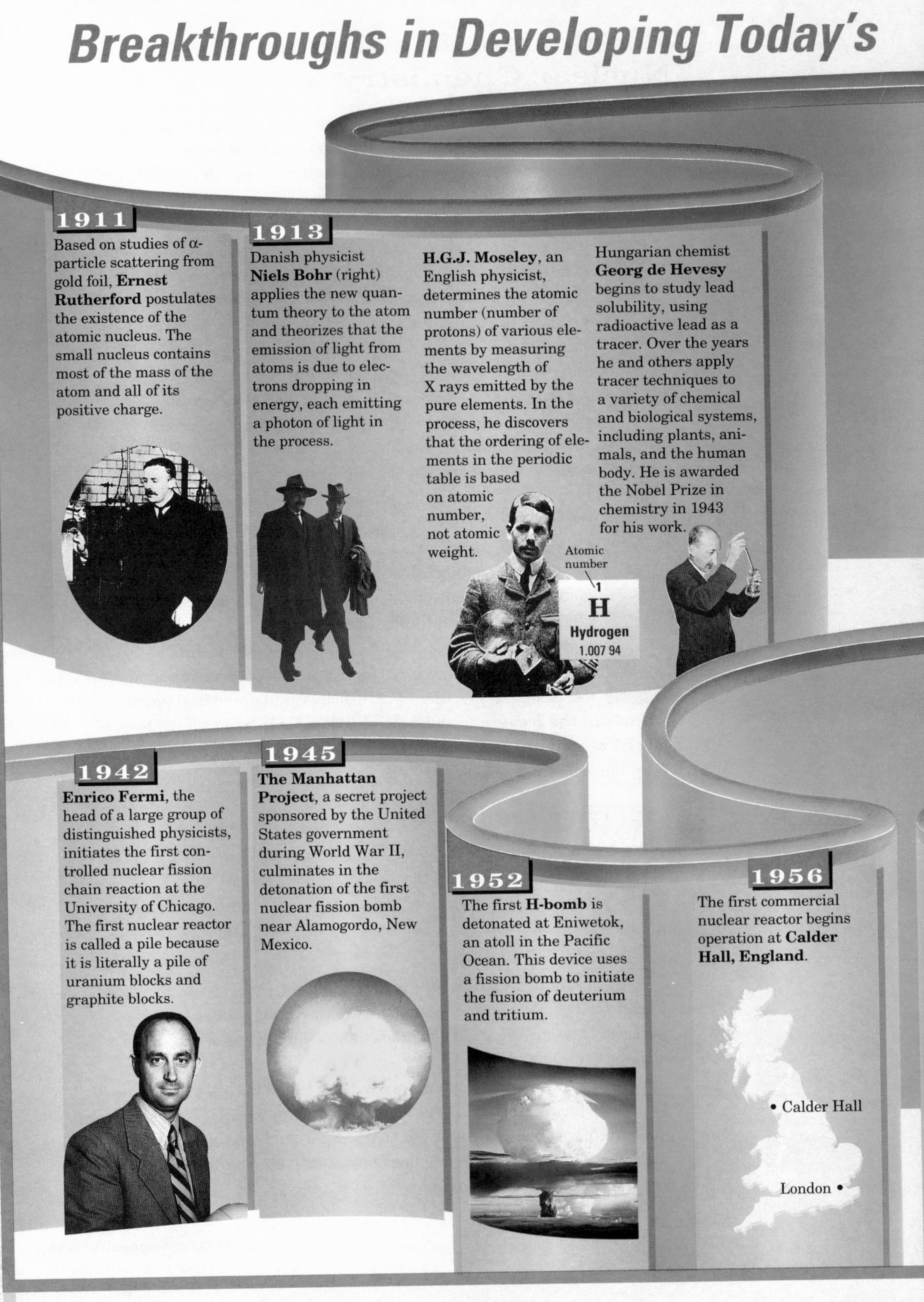
Breakthroughs in Developing Today's
1911
Based on studies of α-particle scattering from gold foil, Ernest Rutherford postulates the existence of the atomic nucleus. The small nucleus contains most of the mass of the atom and all of its positive charge.
1913
Danish physicist Niels Bohr (right) applies the new quantum theory to the atom and theorizes that the emission of light from atoms is due to electrons dropping in energy, each emitting a photon of light in the process.
H.G.J. Moseley, an English physicist, determines the atomic number (number of protons) of various elements by measuring the wavelength of X rays emitted by the pure elements. In the process, he discovers that the ordering of elements in the periodic table is based on atomic number, not atomic weight.
Atomic number
1
H
Hydrogen
1.007 94
Hungarian chemist Georg de Hevesy begins to study lead solubility, using radioactive lead as a tracer. Over the years he and others apply tracer techniques to a variety of chemical and biological systems, including plants, animals, and the human body. He is awarded the Nobel Prize in chemistry in 1943 for his work.
1942
Enrico Fermi, the head of a large group of distinguished physicists, initiates the first controlled nuclear fission chain reaction at the University of Chicago. The first nuclear reactor is called a pile because it is literally a pile of uranium blocks and graphite blocks.
1945
The Manhattan Project, a secret project sponsored by the United States government during World War II, culminates in the detonation of the first nuclear fission bomb near Alamogordo, New Mexico.
1952
The first H-bomb is detonated at Eniwetok, an atoll in the Pacific Ocean. This device uses a fission bomb to initiate the fusion of deuterium and tritium.
1956
The first commercial nuclear reactor begins operation at Calder Hall, England.
Calder Hall
London

Atomic Model

The following events significantly affected our understanding of the atomic and nuclear structure, and the uses and potentials of radioactive materials.

1931

ène and Frédéric ·liot-Curie discover a ·w form of "radiation" ·en they bombard ·ryllium metal with ·articles. Working a separate team, ·alther Bothe and ·lhelm Becker make ·imilar discovery.

1932

Sir James Chadwick, an English physicist at the Cavendish Laboratory, shows that the "radiation" discovered in 1931 is really a neutral particle with a mass close to that of the proton. The new particle, called a neutron, had been predicted by Rutherford in 1920.

E. O. Lawrence's particle accelerator, dubbed the cyclotron, produces protons with an energy of 1 million electron volts, 1 MeV. Cyclotrons and similar accelerators become indispensable to physicists in the years to come.

1937

Technetium, the first artificial element, is discovered by **Emilio Segrè** in a sample of molybdenum that was bombarded with particles accelerated in Lawrence's cyclotron. Inconclusive evidence of this element had been found as early as 1925 by Walter and Ida Noddack in samples of uranium-containing ore.

1939

The German chemists **Otto Hahn and Fritz Strassmann** discover that an isotope of barium is produced when uranium is bombarded with neutrons. Shortly after Hahn's report, two Austrian physicists, **Lise Meitner and Otto Frisch,** correctly interpret Hahn's experimental observations and conclude that the uranium nucleus was being split into two smaller pieces. They name the process nuclear fission.

963

e Limited Test Ban ·aty is signed by ·sident Kennedy, and ·ny nations agree ·alt further at·pheric testing ·uclear ·pons.

1964

Murray Gell-Mann postulates the existence of fundamental subatomic particles, which he names quarks. Protons and neutrons are theorized to consist of quarks.

mi National ·elerator ·oratory, near ·ago, begins opera- Its main accelerator is a mile in diameter produces protons an energy of 200 . In 1983, a second of superconducting ·nets is installed, ·h increases the en- to 900

1979

The Nobel Prize in physics is awarded to **Sheldon Glashow, Abdus Salam,** and **Steven Weinberg** for their theory that successfully unifies explanations of the electromagnetic and weak forces' effects on quarks, electrons, and other particles.

1994

The Superconducting Super Collider, a multibillion-dollar accelerator being built in Texas, is canceled by the United States government.

Teaching Tip

Discussion

Have students speculate on the likelihood of any more artificial elements being discovered. Have them base their arguments on issues such as nuclear stability, short lifetimes, and the almost inconceivably small amounts of synthetic nuclear material that has been produced so far. Based on their positions on the periodic table, what chemical and physical properties would be expected from these elements if they were stable? (For example, chromium, molybdenum, and tungsten (all Group 6 elements) have high melting points. Therefore it follows that unnilhexium would also have a high melting point.)

Closure Strategy

Have students describe five ways in which radiation is used to help society. Have them discuss the process that serves as the basis for the application, including the equations describing any nuclear decay that might take place. Be sure they account for the type of radiation emitted in the process and, if applicable, all radioactive materials initially required. Any hazards or risks that are involved, whether a part of preparing the application, applying it, or disposing of waste material, should also be mentioned.

CHAPTER 16 Highlights

Key terms

alpha particle	**mass defect**
beta particle	**positron emission**
binding energy	**radioactivity**
chain reaction	**radioactive tracer**
critical mass	**stable nucleus**
fission	**strong nuclear force**
fusion	**transmutation**
gamma ray	**unstable nucleus**
half-life	

Key problem-solving approach: *Calculating binding energy*

To calculate the binding energy of one mole of a substance, follow the calculation pathway below.

atomic number × mass of hydrogen (in amu) + **mass of neutrons (in amu)**

=

mass of nucleons, (in amu)

−

actual atomic mass, (in amu)

=

mass defect, (in amu)

×

conversion factor for kg $\frac{1.660\ 540 \times 10^{-27}\ \text{kg}}{1\ \text{amu}}$

×

constant c^2, $(3.00 \times 10^8\ \text{m/s})^2$

×

Avogadro's number, 6.022×10^{23}/mol

=

binding energy, (in J/mol)

Key ideas

Why are some atomic nuclei unstable?

- In spite of repulsive forces between their like charges, protons cluster tightly together within the nucleus. The strong force in the nucleus overwhelms the repulsive forces of the protons and holds protons and neutrons together.
- A small percentage of atoms have nuclei that are unstable and subject to a nuclear change called transmutation, which results in the formation of a new element. Atoms that exhibit this behavior are called radioactive.
- The stability of the nucleus is dependent on the neutron-proton ratio. Stable atoms form a pattern called the band of stability.
- The amount of energy released when a nucleus is formed is called binding energy. Binding energy can be calculated by applying Einstein's equation, $E = mc^2$.

What kinds of nuclear change occur?

- Radioactivity is composed of alpha and beta particles, gamma rays, and positron emission.
- Artificial transmutation occurs in laboratories in machines such as particle accelerators.
- Fission is a process in which a nucleus splits into two smaller nuclei, releasing large quantities of energy.
- Fusion is a process in which small nuclei combine into a larger nucleus, releasing large quantities of energy.

How is nuclear chemistry used?

- Half-life is the time required for one half of the mass of a radioactive substance to decay.
- The half-life of the carbon-14 isotope can be used to date organic material that is up to 20 000 years old. Other radioactive isotopes are used to date more ancient rock and mineral formations.
- The effects of radiation may be measured in units called rems.

CHAPTER 16

Review and Assess

Nuclear stability

REVIEW

1. Explain how the strong nuclear force overpowers the repulsive forces within the nucleus.
2. Explain why nuclear stability is a necessary condition for the stability of the environment.
3. Give at least three examples of how the instability of the nucleus affects the properties of radioactive atoms.
4. Through a nuclear reaction, an atom of a rubidium isotope becomes an atom of a krypton isotope. What has happened? Why?
5. a. What is the relationship among number of protons, number of neutrons, and the stability of the nucleus?
   b. Why is it increasingly hard to stabilize nuclei as the number of protons increases?
6. Refer to the graph at the bottom of the page to answer the following questions.
   a. Which point represents $^{102}_{45}Rh$? Is the rhodium nucleus stable? How do you know?
   b. Which point represents $^{209}_{83}Bi$? What can you predict about any isotopes heavier than this? Why?
   c. What can you say about the stability of a nucleus represented by point **d**? What about the stability of the nucleus at point **a**?
7. What is the relationship among protons, neutrons, and binding energy?
8. What is the relationship between mass defect and binding energy?

APPLY

9. Why do we compare binding energy per nuclear particle instead of the total binding energy per nucleus?

Neutron-Proton Ratios of Stable Nuclei

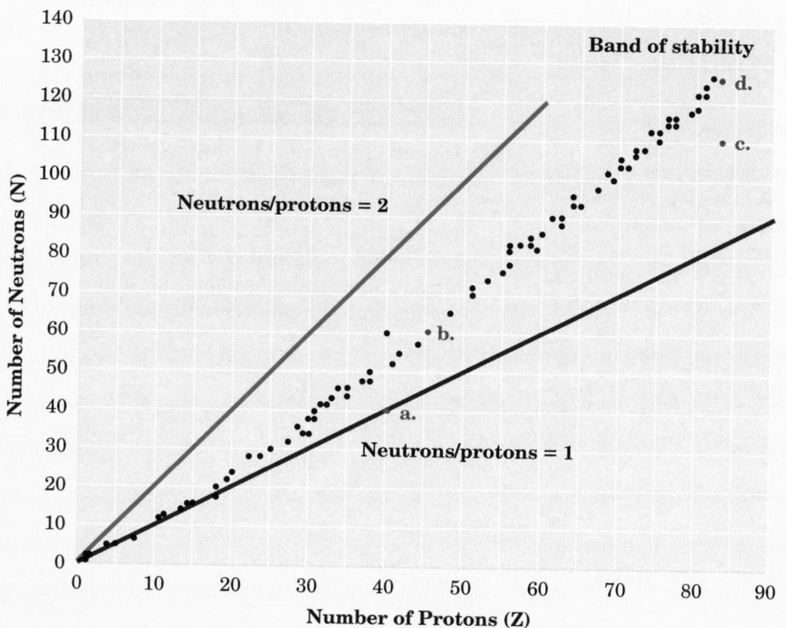

1. The strong force is attractive between all nucleons, but the electric force is repulsive only between protons. At close range the strong force also has a greater magnitude than the electric force. Neutrons act to bind to the nucleus protons that would otherwise be repelled by the non-adjacent protons in the nucleus.

2. If most nuclei were not stable, atoms would undergo transmutation, and the otherwise stable atoms and molecules that make up our bodies and our environment would be in a constant state of change.

3. Among the possible answers are the following. Because radioactive nuclei are unstable, they can be transformed into the nuclei of other elements. As they change, radioactive nuclei emit radiation or particles. The lives of many radioactive nuclei are very short because they quickly change into the nuclei of other elements.

4. Transmutation, a process of change by which one element becomes another, has taken place. It has happened because the rubidium isotope is unstable and therefore decays.

5. a. The number of protons and neutrons must be equal for a light nucleus to be stable. For heavier nuclei to be stable they must contain 1.5 times as many neutrons as protons.
   b. In larger nuclei, neutrons are needed to act as "nuclear glue" to increase the total magnitude of the strong force. However, neutrons that do not have neighboring protons are unstable. Also, the repulsion a proton in a nucleus experiences is from all other protons that are outside the range of the strong force, while the attraction it experiences is only from the nucleons that immediately surround it.

6. **a.** Point **b** represents $^{105}_{45}Rh$. The nucleus is stable; it lies along the band of stability.
b. Point **d** represents $^{209}_{83}Bi$. Isotopes heavier than this will not be stable; they do not lie along the band of stability.
c. A nucleus at point **d** is stable because it lies (barely) on the band of stability. A nucleus at point **a** would be unstable. It would decay into an isotope that lies along the band of stability.

7. When protons and neutrons combine to form a nucleus, energy is released. This energy is called binding energy.

8. The difference in the masses of individual nucleons and the mass of the complete nucleus is called the mass defect (Δm). This amount of matter is converted to binding energy through the equivalence of matter and energy, as described by Einstein's equation $E = mc^2$.

9. Although one isotope may have a greater overall binding energy than another isotope, it may not be as stable because the number of nucleons that are bound by the energy may be greater. Binding energy per nucleon is a better indicator of nuclear stability.

10. **a.** Nucleus #1: 2.69×10^{13} J/mol
Nucleus #2: 1.80×10^{13} J/mol
Nucleus #1 releases more binding energy than nucleus #2.

b. Nucleus #1: 1.28×10^{-12} J/nucleon
Nucleus #2: 1.30×10^{-12} J/nucleon
Nucleus #2 is more stable than nucleus #1.

*Full solution for item **10** is found on page 641B.*

PRACTICE

Answers to items in a black square begin on page 841.

10 **a.** Compare the binding energies of the following two nuclei, and indicate which releases more energy when formed. You will need information from the periodic table and the text.
Nucleus #1: atomic mass 34.9880 amu, atomic number 19
Nucleus #2: atomic mass 22.9898 amu, atomic number 11
b. What is the binding energy per nucleon for each nucleus? Which nucleus is more stable?

11. Using Einstein's equation, $E = mc^2$, determine how much mass (in kilograms) is lost by the formation of a nucleus of $^{56}_{26}Fe$, if the binding energy released is 7.884×10^{-11} J.

Nuclear change

REVIEW

12. **a.** What is the relationship between an alpha particle and a helium nucleus?
b. The decay of uranium-238 results in the spontaneous ejection of an alpha particle. Write the nuclear equation that summarizes this process.

13. Compare and contrast the penetrating powers of alpha particles, beta particles, and gamma rays.

14. What type of radiation is emitted in the following equation?

$$^{43}_{19}K \longrightarrow ^{43}_{20}Ca + \text{?}$$

15. Is the decay of an unstable isotope into a stable isotope always a one-step process?

16. How does artificial transmutation differ from natural radioactive decay?

17. Refer to **Table 16-2.** Describe how artificial transmutation resulted in the formation of unniloctium.

18. **a.** What role does the neutron serve in starting a nuclear chain reaction and keeping it going?
b. Why must neutrons in a chain reaction be controlled?
c. Describe the relationship between neutrons and critical mass.

19. **a.** Compare and contrast fusion and fission.
b. Explain the relationship among fusion, temperature, and energy.

APPLY

20. Are the products of nuclear transmutation higher or lower in energy than the starting material? Why?

21. The plutonium isotope $^{239}_{94}Pu$ is sometimes used in nuclear reactors. What do transmutations and the relative abundance of uranium isotopes have to do with the use of the plutonium isotope in reactors?

Using nuclear reactions

REVIEW

22. Why is the constant rate of decay of radioactive nuclei so important in radioactive dating?

23. The Environmental Protection Agency and health officials nationwide are concerned about the levels of radon gas in homes. The half-life of one radon isotope is 3.8 days. If a sample of gas taken from a basement contains 4.38 µg of radon-222, how much radon will remain in the sample after 15.2 days?

24. A pathologist working in a police laboratory wants to analyze a small sample of blood from a crime scene. He cannot damage the blood sample because it will be used in court as evidence. What kind of radioactive analysis should he use? Why?

25. Many cancer patients lose hair during radiation therapy. How is this hair loss related to rem exposure?

APPLY

26. Use an example from your everyday life to illustrate the concept of half-life.

27. Why would someone working around radioactive waste in a landfill use a radiation monitor instead of a watch to determine when the workday is over? At what point would that person decide to stop working?

Linking chapters

1. ***Structure of the nucleus***
   Compare and contrast the interaction, function, and stability of nuclear particles both inside and outside the nucleus.
2. **Theme: *Classification and Trends***
   Explain the relationship among atomic number, size of the nucleus, and stability.
3. ***Stoichiometry***
   Calculate the mass of 250 mol of plutonium-239.
4. ***Density***
   Calculate the density in g/cm^3 of uranium-235 if 8.0 m^3 has a mass of 1.5×10^5 kg.
5. ***Balancing equations***
   Balance the following.
   a. ${}^{1}_{0}n + {}^{235}_{92}U \longrightarrow {}^{89}_{37}Rb + {}^{144}_{58}Ce + {}^{0}_{-1}e + {}^{1}_{0}n$
   b. ${}^{239}_{94}Pu + {}^{1}_{0}n \longrightarrow {}^{146}_{58}Ce + {}^{90}_{38}Sr + {}^{0}_{-1}e + {}^{1}_{0}n$
   c. $Tl + O_2 + H_2O \longrightarrow Tl^+ + OH^-$
6. ***Writing equations***
   The radioactive decay series that begins with uranium-235 and ends with lead-207, shows the following sequence of emissions for the series: alpha, beta, alpha, beta, alpha, alpha, alpha, alpha, beta, beta, and alpha. Write an equation for each reaction in the series.

USING TECHNOLOGY

1. ***Computer art***
   Design a computer program that illustrates the path of particles through a magnetic field. Your program should show pathways representing movement and not simply stationary points. Use the path shown in the diagram as a model for the pathway that would be characteristic of a neutral particle.

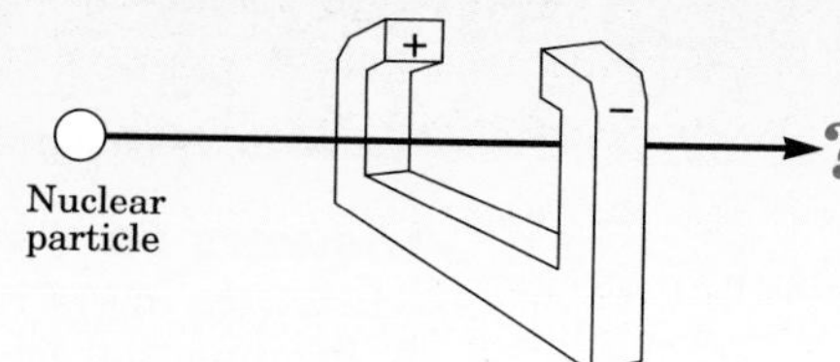

2. ***Computer program***
   In order to calculate binding energy, you need to know the actual atomic masses of the elements. Design a computer program that stores atomic masses in your computer and allows you to access them with keyboard commands. Program a spreadsheet to calculate binding energy using the process shown on page 638.

Alternative assessment

Portfolio projects

1. ***Chemistry and you***
   Your local grocery store may sell irradiated foods. Find out what stores in your area sell irradiated foods, and determine whether you have any of these foods at home. What are the shelf lives of these foods before and after irradiation? Report your findings to the class.
2. ***Research and communication***
   Compare the physiological effects of the different kinds of radiation. Find out if any adults you know are exposed to radiation at work. Ask about the kind of radiation involved and what precautions they take to avoid the harmful effects of radiation. Report your findings to the class.
3. ***Research and communication***
   a. Research some important historical findings that have been validated through radioactive dating. Report your findings to the class.
   b. You have made an important archaeological or geological find. Describe how you will use radioactive dating.
4. ***Research and communication***
   Find out about the future of nuclear power in this country. How many plants are currently supplying power? How many are in the construction stage? How does the use of nuclear power in the United States compare with that of Europe?

11. $E = mc^2$, or $E = \Delta mc^2$

$$\Delta m = \frac{E}{c^2}$$

$$\Delta m = \frac{7.884 \times 10^{-11}\ J}{9.00 \times 10^{16}\ \frac{m^2}{s^2}}$$

$$\Delta m = 8.76 \times 10^{-28}\ kg$$

12. **a.** An alpha particle is a helium-4 nucleus.
    b. ${}^{238}_{92}U \longrightarrow {}^{4}_{2}He + {}^{234}_{90}Th$

13. Alpha particles are slower, more massive, and do not penetrate as far as beta particles. Alpha particles have a positive charge, while beta particles have a negative charge. Gamma radiation has no charge and no mass. Gamma rays penetrate farther than either alpha or beta particles.

14. a beta particle (${}^{0}_{-1}e$)

15. No. In many cases the particle produced by decay is still unstable and will continue the transmutation process until a stable isotope is formed.

16. In artificial transmutation a substance is bombarded with particles that have been given high energies in a particle accelerator. The interaction produces new substances that may not be stable enough to exist in nature.

17. Lead-208 was bombarded with iron-58 nuclei to create the unstable, artificial element unniloctium. A neutron was also emitted during the reaction.

Answers from page 617

Answers to Practice 16A

1. One atom of deuterium (hydrogen-2) consists of a hydrogen atom (1 proton and 1 electron) and 1 neutron, the total mass of which is

$$\left(1\ {}^{1}_{1}\text{H atom} \times \frac{1.007825\ \text{amu}}{1\ {}^{1}_{1}\text{H atom}}\right) + \left(1\ \text{neutron} \times \frac{1.008665\ \text{amu}}{1\ \text{neutron}}\right) = 2.0165\ \text{amu}$$

mass of hydrogen-2 atom = 2.0140 amu

mass defect = 2.0165 amu − 2.0140 amu = 0.0025 amu

$$\text{binding energy} = (0.0025\ \text{amu})\left(\frac{1.66054 \times 10^{-27}\ \text{kg}}{1\ \text{amu}}\right)\left(\frac{3.00 \times 10^{8}\ \text{m}}{1\ \text{s}}\right)^2 =$$

$3.74 \times 10^{-13}\ \text{kg}\cdot\text{m}^2/\text{s}^2 = 3.74 \times 10^{-13}$ J for each deuterium nucleus

$$\text{molar binding energy} = \left(\frac{3.74 \times 10^{-13}\ \text{J}}{1\ {}^{2}_{1}\text{H atom}}\right)\left(\frac{6.022 \times 10^{23}\ \text{atoms}}{1\ \text{mol}}\right) =$$

2.25×10^{11} J/mol

For 1 mol of deuterium the binding energy is 2.25×10^{11} J.

2. One atom of lithium-6 has 3 protons, 3 electrons, and 3 neutrons, the total mass of which is

$$\left(3\ {}^{1}_{1}\text{H atoms} \times \frac{1.007825\ \text{amu}}{1\ {}^{1}_{1}\text{H atom}}\right) + \left(3\ \text{neutrons} \times \frac{1.008665\ \text{amu}}{1\ \text{neutron}}\right) = 6.0495\ \text{amu}$$

mass of lithium-6 atom = 6.0151 amu

mass defect = 6.0495 amu − 6.0151 amu = 0.0344 amu

$$\text{binding energy} = (0.0344\ \text{amu})\left(\frac{1.66054 \times 10^{-27}\ \text{kg}}{1\ \text{amu}}\right)\left(\frac{3.00 \times 10^{8}\ \text{m}}{1\ \text{s}}\right)^2 =$$

$5.14 \times 10^{-12}\ \text{kg}\cdot\text{m}^2/\text{s}^2 = 5.14 \times 10^{-12}$ J for each lithium-6 nucleus

Solutions for Additional Examples 16A

a. One atom of manganese-55 has 25 protons, 25 electrons, and 30 neutrons, the total mass of which is

$$\left(25\ {}^{1}_{1}\text{H atoms} \times \frac{1.007825\ \text{amu}}{1\ {}^{1}_{1}\text{H atom}}\right) + \left(30\ \text{neutrons} \times \frac{1.008665\ \text{amu}}{1\ \text{neutron}}\right) =$$

55.456 amu

mass of manganese-55 atom = 54.938 amu

mass defect = 55.456 amu − 54.938 amu = 0.518 amu

$$\text{binding energy} = (0.518\ \text{amu})\left(\frac{1.66054 \times 10^{-27}\ \text{kg}}{1\ \text{amu}}\right)\left(\frac{3.00 \times 10^{8}\ \text{m}}{1\ \text{s}}\right)^2 =$$

$7.74 \times 10^{-11}\ \text{kg}\cdot\text{m}^2/\text{s}^2 = 7.74 \times 10^{-11}$ J for each manganese-55 nucleus

$$\text{molar binding energy} = \left(\frac{7.74 \times 10^{-11}\ \text{J}}{1\ {}^{55}_{25}\text{Mn atom}}\right)\left(\frac{6.022 \times 10^{23}\ \text{atoms}}{1\ \text{mol}}\right) =$$

$$4.66 \times 10^{13}\ \frac{\text{J}}{\text{mol}}$$

For 1 mol of manganese-55 the binding energy is 4.66×10^{13} J.

b. One atom of sulfur-32 has 16 protons, 16 electrons, and 16 neutrons, the total mass of which is

$$\left(16\ {}^{1}_{1}\text{H atoms} \times \frac{1.007825\ \text{amu}}{1\ {}^{1}_{1}\text{H atom}}\right) + \left(16\ \text{neutrons} \times \frac{1.008665\ \text{amu}}{1\ \text{neutron}}\right) =$$

32.264 amu

mass of sulfur-32 atom = 31.972 amu

mass defect = 32.264 amu − 31.972 amu = 0.292 amu

$$\text{binding energy} = (0.292\ \text{amu})\left(\frac{1.66054 \times 10^{-27}\ \text{kg}}{1\ \text{amu}}\right)\left(\frac{3.00 \times 10^{8}\ \text{m}}{1\ \text{s}}\right)^2 =$$

$4.36 \times 10^{-11}\ \text{kg}\cdot\text{m}^2/\text{s}^2 = 4.36 \times 10^{-11}$ J for each sulfur-32 nucleus

$$\text{molar binding energy} = \left(\frac{4.36 \times 10^{-11}\ \text{J}}{1\ {}^{32}_{16}\text{S atom}}\right)\left(\frac{6.022 \times 10^{23}\ \text{atoms}}{1\ \text{mol}}\right) =$$

$$2.63 \times 10^{13}\ \frac{\text{J}}{\text{mol}}$$

For 1 mol of sulfur-32 the binding energy is 2.63×10^{13} J.

Answers to Section Review

5. Protons that lie beyond the short range of nuclear attraction of other protons will be repelled by those protons, and so will require more non-repelling neutrons to bind them in the nucleus. The larger a nucleus becomes, the more neutrons are needed to bind the protons that lie near its edge. Therefore, the ratio of neutrons to protons required for a stable nucleus increases with nuclear mass.

6. The strong nuclear force is a short-range, attractive force that acts among nucleons. The electric force is a long-range force between charged particles, the strength of which varies inversely with the square of the distance between the charges. Opposite charges attract; similar charges repel. The strong nuclear force binds nucleons together. The electric force causes a nucleus with insufficient nuclear attraction between its nucleons to transmute into a stable nucleus.

7. The band of stability is the narrow strip of neutron-to-proton ratios along which nuclei are stable. For light nuclei, the stable isotopes have a neutron-to-proton ratio of about 1:1. For heavy nuclei, the stable isotopes tend to have a neutron-to-proton ratio of 1.5:1. Nuclei not located on the band of stability tend to be unstable.

Answer from page 640

10. a. Nucleus #1: For $^{35}_{19}$K the binding energy from a 0.2993 amu mass defect is 2.69×10^{13} J/mol.

Solution: nucleus contains 19 protons, 19 electrons, and 16 neutrons
mass defect = (total mass of particles in atom) − (measured mass of atom) =

$$\left(19\ ^1_1\text{H atoms} \times \frac{1.007825 \text{ amu}}{^1_1\text{H atom}}\right) + \left(16 \text{ neutrons} \times \frac{1.008665 \text{ amu}}{1 \text{ neutron}}\right) =$$

35.2873 amu

mass of atom #1= 34.9880 amu

mass defect = 35.2873 amu − 34.9880 amu = 0.2993 amu

$$\text{binding energy} = (0.2993 \text{ amu})\left(\frac{1.66054 \times 10^{-27}\text{kg}}{1 \text{ amu}}\right)\left(\frac{3.00 \times 10^{8} \text{ m}}{1 \text{ s}}\right)^2 =$$

4.47×10^{-11} kg•m^2/s^2 = 4.47×10^{-11} J for each nucleus.

$$\text{The molar binding energy is } \left(\frac{4.47 \times 10^{-11} \text{ J}}{1 \text{ atom}}\right)\left(\frac{6.022 \times 10^{23} \text{ atoms}}{1 \text{ mol}}\right) =$$

2.69×10^{13} J/mol

Nucleus #2: For $^{23}_{11}$Na the binding energy from a 0.2003 amu mass defect is 1.80×10^{13} J/mol.

Solution: nucleus contains 11 protons, 11 electrons, and 12 neutrons

mass defect = (total mass of particles in atom) − (measured mass of atom) =

$$\left(\frac{11\ ^1_1\text{H atoms} \times 1.007825 \text{ amu}}{1\ ^1_1\text{H atom}}\right) + \left(12 \text{ neutrons} \times \frac{1.008665 \text{ amu}}{1 \text{ neutron}}\right) =$$

23.1901 amu

mass of atom #2 = 22.9898 amu

mass defect = 23.1901 amu − 22.9898 amu = 0.2003 amu

$$\text{binding energy} = (0.2003 \text{ amu})\left(\frac{1.66054 \times 10^{-27}\text{kg}}{1 \text{ amu}}\right)\left(\frac{3.00 \times 10^{8} \text{ m}}{1 \text{ s}}\right)^2 =$$

2.99×10^{-11}kg • m^2/s^2 = 2.99×10^{-11} J for each nucleus.

$$\text{The molar binding energy is } \left(\frac{2.99 \times 10^{-11} \text{ J}}{1 \text{ atom}}\right)\left(\frac{6.022 \times 10^{23}\text{atoms}}{1 \text{ mol}}\right) =$$

1.80×10^{13} J/mol

Nucleus #1 releases more binding energy than nucleus #2.

b. $\text{Binding energy per nucleon in nucleus \#1} = \frac{4.47 \times 10^{-11} \text{ J}}{35 \text{ nucleons}} =$

1.28 x 10^{-12} J/nucleon

$\text{Binding energy per nucleon in nucleus \#2} = \frac{2.99 \times 10^{-11} \text{ J}}{23 \text{ nucleons}} =$

1.30 x 10^{-12} J/nucleon

Nucleus #2 has a greater binding energy per nucleon. Because the sodium-23 nucleus has a more stable arrangement of nucleons than the potassium-35 nucleus, fewer nucleons lie beyond the range of the strong force. Therefore it is easier for the repulsion between protons to be overcome.

Continued from page 641

18. a. A chain reaction begins when a nucleus is bombarded by a neutron. It continues when the fission process releases additional neutrons, so that after each reaction the number of neutrons available for bombardment of other nuclei has been increased.

b. Fission produces many more neutrons than it absorbs, so that, uncontrolled, the reaction proceeds at a faster rate until an enormous amount of energy is liberated in an explosion.

c. Critical mass is the smallest amount of radioactive material that must be present for a self-sustaining nuclear chain reaction to occur. The reaction can continue only if all of the neutrons produced by the fission of each nucleus cause the fission of other unsplit nuclei. Therefore, the efficient use of neutrons for fission determines whether critical mass has been achieved for a particular material. Critical mass varies for different materials and for different states of those materials.

19. a. Both reactions are nuclear in nature. They also release a great deal of energy. They differ in that fusion occurs through the combination of nuclei, while fission involves the splitting of nuclei.

b. Fusion requires extremely high temperatures and a great deal of energy in order to achieve the necessary densities for the material to be fused. For each atom of fuel, fusion releases more energy than fission.

20. The products have lower energy than the starting materials. Decay processes proceed from the least stable, or highest energy states, to the most stable (lowest energy) states.

21. Plutonium-239, which is fissionable, will transmute from uranium-238. Uranium-238 is far more abundant naturally (99 %) than uranium-235 (less than 1%), but unlike uranium-235, uranium-238 is not fissionable.

22. The rate of decay must be constant for radioactive dating to be effective. For example, if certain samples of potassium-40 decayed faster than others, potassium-40 could not be used for radioactive dating.

23. Answer: 0.274 μg
Solution: For each half-life, the amount of radon-222 decreases by half. Each ⟶ represents a half-life of 3.8 days. The total number of days (15.2) divided by the half-life shows that 4 half-lives are possible.

4.38 μg ⟶ 2.19 μg ⟶ 1.095 μg ⟶ 0.548 μg ⟶ 0.274 μg

24. He can use neutron activation analysis, because it will not destroy the specimen.

25. The biological effect of a certain amount of radiation is measured in rems. If the patient were exposed to 100 or more rems of radiation per treatment, hair loss would be highly probable.

26. Answers will vary. For example, if a pelican depletes a pond of half of its fish population in a day, the number of fish it catches on any day will depend on how many fish are in the pond. For 100 fish, the pelican catches 50 the first day, 25 the next, and so forth. This model assumes no reproduction on the fishes' part.

27. The person's workday would be over when the level on the radiation monitor reached 25 rems. The worker would not need to check a watch because he or she could reach the critical exposure level at any time, depending on the type and amount of waste being handled.

Linking chapters

1. Protons are positively charged and stable outside the nucleus. Inside the nucleus, they repel one another and destabilize the nucleus. The strong nuclear force counteracts the repulsive effects of the protons' charges, but only over a short range . Neutrons are neutrally charged, and their function within the nucleus is to interact with protons to stabilize the nucleus. Neutrons are unstable, and have short lives outside the nucleus and within the nucleus in the absence of adjacent protons.

2. Elements with atomic numbers greater than 83 cannot be stabilized because of the excessive number of protons (indicated by the high atomic number) and the size of the nucleus, which reduces the number of stabilizing nucleons in contact with the outermost nucleons.

3. Taking the atomic mass of plutonium-239 to be 239 g/mol, the mass of 250 mol of the element is
$250 \text{ mol} \times 239 \text{ g/mol} = 5.98 \times 10^4 \text{ g} = 59.8 \text{ kg}$

4. mass of uranium-235 $= 1.5 \times 10^5 \text{ kg} = 1.5 \times 10^8 \text{ g}$

$$\text{volume} = 8.0 \text{ m}^3 \times \frac{10^6 \text{ cm}^3}{1 \text{ m}^3} = 8.0 \times 10^6 \text{ cm}^3$$

$$D = \frac{1.5 \times 10^8 \text{ g}}{8.0 \times 10^6 \text{ cm}^3} = 19 \text{ g/cm}^3$$

5. **a.** ${}^{1}_{0}n + {}^{235}_{92}\text{U} \longrightarrow {}^{89}_{37}\text{Rb} + {}^{144}_{58}\text{Ce} + 3{}^{0}_{-1}e + 3{}^{1}_{0}n$

   b. ${}^{239}_{94}\text{Pu} + {}^{1}_{0}n \longrightarrow {}^{146}_{58}\text{Ce} + {}^{90}_{38}\text{Sr} + 2{}^{0}_{-1}e + 4{}^{1}_{0}n$

   c. $4\text{Tl} + \text{O}_2 + 2\text{H}_2\text{O} \longrightarrow 4\text{Tl}^+ + 4\text{OH}^-$

6. The sequence of reactions by which uranium-235 decays to lead-207 is as follows.

$${}^{235}_{92}\text{U} \longrightarrow {}^{231}_{90}\text{Th} + {}^{4}_{2}\text{He}$$

$${}^{231}_{90}\text{Th} \longrightarrow {}^{231}_{91}\text{Pa} + {}^{0}_{-1}e$$

$${}^{231}_{91}\text{Pa} \longrightarrow {}^{227}_{89}\text{Ac} + {}^{4}_{2}\text{He}$$

$${}^{227}_{89}\text{Ac} \longrightarrow {}^{227}_{90}\text{Th} + {}^{0}_{-1}e$$

$${}^{227}_{90}\text{Th} \longrightarrow {}^{223}_{88}\text{Ra} + {}^{4}_{2}\text{He}$$

$${}^{223}_{88}\text{Ra} \longrightarrow {}^{219}_{86}\text{Rn} + {}^{4}_{2}\text{He}$$

$${}^{219}_{86}\text{Ra} \longrightarrow {}^{215}_{84}\text{Po} + {}^{4}_{2}\text{He}$$

$${}^{215}_{84}\text{Po} \longrightarrow {}^{211}_{82}\text{Pb} + {}^{4}_{2}\text{He}$$

$${}^{211}_{82}\text{Pb} \longrightarrow {}^{211}_{83}\text{Bi} + {}^{0}_{-1}e$$

$${}^{211}_{83}\text{Bi} \longrightarrow {}^{211}_{84}\text{Po} + {}^{0}_{-1}e$$

$${}^{211}_{84}\text{Po} \longrightarrow {}^{207}_{82}\text{Pb} + {}^{4}_{2}\text{He}$$

USING TECHNOLOGY

1. Programs may vary. The "magnet" depicted should be treated as an electric field, with the positive pole behaving as a positive charge and the negative pole as a negative charge. Positively charged particles, such as protons and alpha particles, will be deflected in the direction of the negative pole. Electrons (beta particles) will be deflected toward the positive pole. Neutrons and gamma rays undergo no deflection.

2. Programs may vary. The program should be able to access the most accurate atomic mass for a given isotope, as well as determine the number of each type of nucleon in the isotope.

Portfolio projects

1. Answers will vary. Students should show that irradiation increases the shelf life of certain foods.

2. Answers will vary. Students should learn that physiological damage depends on the penetration ability of a particle. The external and internal physiological effects of radiation should also be distinguished.

3. Answers will vary. For item **3b** students should understand that isotopes should have long half-lives to be useful in radioactive dating. Geological finds require much longer half-lives than archeological or paleontological finds.

4. Answers will vary. Economic, regulatory, and environmental aspects of plant construction, operation, and decommissioning should be included in the report. The problem of nuclear-waste storage should also be considered.

Lab Program Overview

Lab Program
Explorations
Technique Builders
Investigations
Problem Solving
CHEMYSTERY LABS, INC.
Investigations
Analysis
Material Testing
Research and Development
642 | Laboratory Program

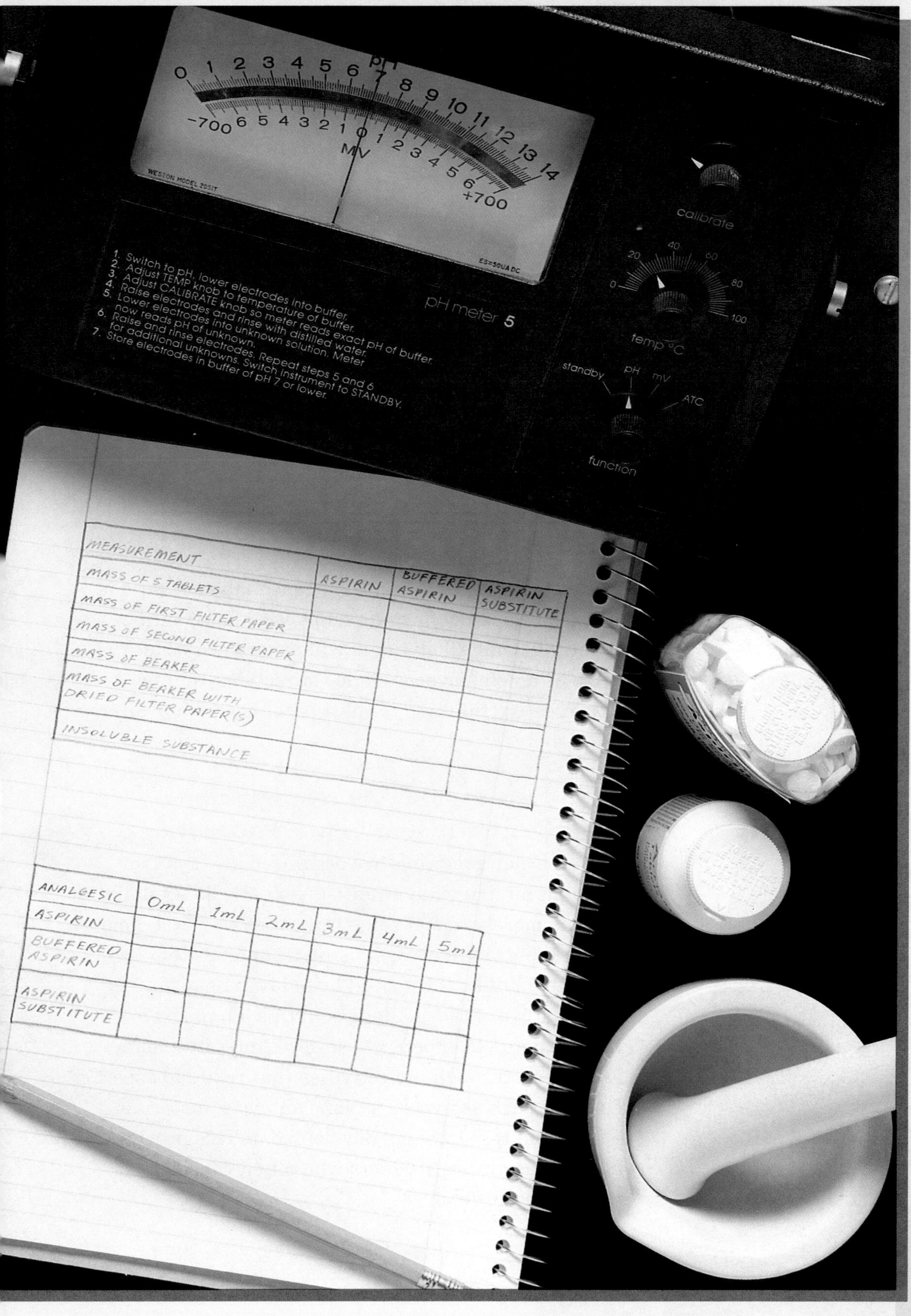
pH
MV
1. Switch to pH, lower electrodes into buffer.
2. Adjust TEMP knob to temperature of buffer.
3. Adjust CALIBRATE knob so meter reads exact pH of buffer.
4. Raise electrodes and rinse with distilled water.
5. Lower electrodes into unknown solution. Meter now reads pH of unknown.
6. Raise and rinse electrodes. Repeat steps 5 and 6 for additional unknowns. Switch instrument to STANDBY.
7. Store electrodes in buffer of pH 7 or lower.
pH meter 5
calibrate
temp °C
standby
pH
mV
ATC
function
MEASUREMENT
ASPIRIN
BUFFERED ASPIRIN
ASPIRIN SUBSTITUTE
MASS OF 5 TABLETS
MASS OF FIRST FILTER PAPER
MASS OF SECOND FILTER PAPER
MASS OF BEAKER
MASS OF BEAKER WITH DRIED FILTER PAPER(S)
INSOLUBLE SUBSTANCE
ANALGESIC
0mL
1mL
2mL
3mL
4mL
5mL
ASPIRIN
BUFFERED ASPIRIN
ASPIRIN SUBSTITUTE

Laboratory Program Overview

What's so special about this lab program?

Each lab exercise takes place in the context of a real-world scenario. As a result, students will be able to see how the techniques and concepts they are learning in the lab and classroom can relate to problems in the real world.

More than 80 all-new labs

None of the lab exercises in the textbook and the lab manual have ever been published in this form before. Even the ones that use familiar reactions and techniques have been placed in new contexts. No matter how long you've been teaching chemistry, there's bound to be something new and different you will want to add to your collection of lab ideas.

These labs take concepts into the real world

An integral part of the context of each lab's scenario is the role of the student as an employee of a scientific consulting firm. All students will experience what it is like to be an employee, to follow directions, and to apply creative thinking to solve problems with available resources.

These labs provide an opportunity to "do" science and explore careers

Most other lab programs are organized in ways that are not at all like what scientists actually do. Usually, students merely repeat some work that was first done several hundred years ago, using techniques that are completely unlike those used by today's chemists. The varied nature of the challenges in this lab program allow students to experience several "days in the life" of scientists through lab activities that relate to aspects of science such as investigations, analysis, materials testing, and research and development. This provides valuable insight for students who might consider becoming scientists.

These labs foster scientific literacy

Even those who may not become scientists will gain valuable insight into the true nature of science and how it can both create and solve problems in today's world.

Two types of labs

The lab exercises in the textbook and the lab manual are divided into two types. *Explorations* have complete procedures provided for students, along with thorough questions designed to help students make sense of what they observed in the lab. *Investigations* are complete scenarios that allow for student-designed procedures and open-ended lab work. Most of these labs are in pairs, with the *Exploration* serving as a precursor to an accompanying *Investigation*.

Classroom-tested pedagogy and procedures

The labs in this program have all been tested by our lab authors in the their chemistry labs. Some have even been used in their classes for several years. The labs have also been reviewed by teachers and a safety consultant for validity, accuracy, usefulness, and instructional value.

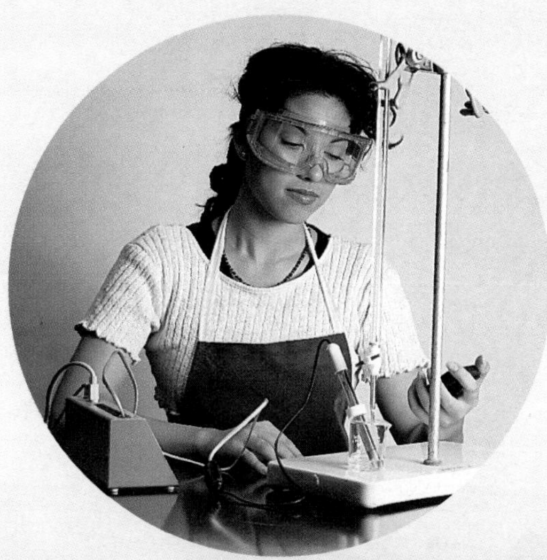

Each full-scale scientific inquiry is based on a solid foundation

- The experience with the *Exploration* helps students focus their plans for the open-ended labs on meaningful measurements and results.
- Students who know what to do in the lab based on their experiences in the *Exploration* and are much less likely to become involved in hazardous situations.

Students take responsibility for their learning

Designing their own procedures for the *Investigations*, students feel they have accomplished something. They have solved a problem using their own ingenuity. Students are given the freedom to make their own mistakes, something that most lab programs do not take into account. Students will often learn more by making a mistake than by performing a lab with skill, because they tend to be so obsessed with finding the "right way" to do things that they do not learn from their experiences. This is why the *Investigations* can include the use of budgeting and costs for equipment. If students make a mistake in the procedure but truly learn more in the process, they may lose some money on paper, but they do not need to be punished with a lower grade.

Complete safety and disposal

Complete safety information is provided for each lab. In addition, disposal instructions for all chemicals and all combinations of products are described in the teacher's notes *on the same page as the lab*. Instead of wondering, you will know you are using appropriate and environmentally sound disposal practices. For more discussion of disposal issues, see the Teacher's Notes for page 654.

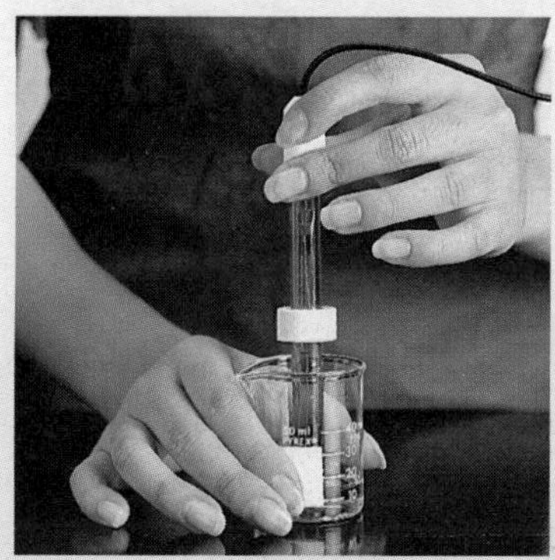

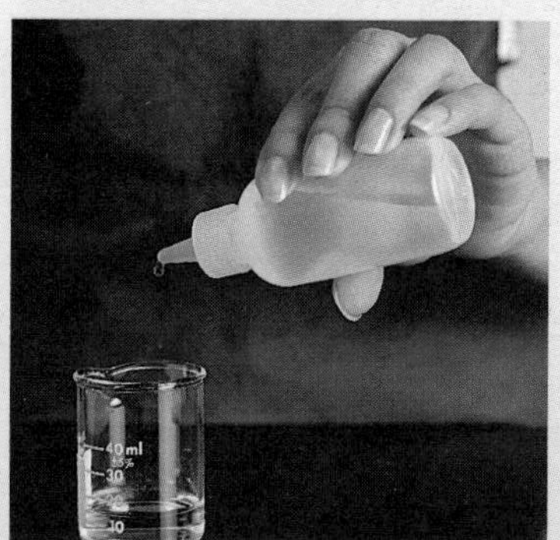

Explorations: Technique builders with procedures

Nearly all Investigations are preceded by an *Exploration*, which serves as a Technique Builder by giving students practice with lab techniques and concepts. Each *Exploration* includes a detailed, step-by-step procedure, like many traditional labs. Unlike labs from other lab programs, the *Explorations* are placed in the context of a scenario. A situation is described, and students are given a problem to solve. The procedure and the items that follow it lead students through the techniques and thought processes necessary to solve the problem.

❶ **Situation**
sets up the scenario: mysterious leaking barrels on vacant lot

❷ **Background**
links the situation to chemistry concepts

❸ **Problem**
provides a capsule overview of what will happen in the steps of the procedure

❹ **Safety**
section provides detailed warnings of potential hazards

❺ **Preparation**
section describes all steps that can be done before actual lab work begins

❻ **Technique**
section includes detailed step-by-step instructions for procedure

❼ **Cleanup and Disposal**
Specific disposal instructions match discussion in Teacher's Notes

❽ **Analysis and Interpretation** and **Conclusions**
items help students think through the lab results

❾ **Extensions**
provide more opportunities for independent work and critical thinking

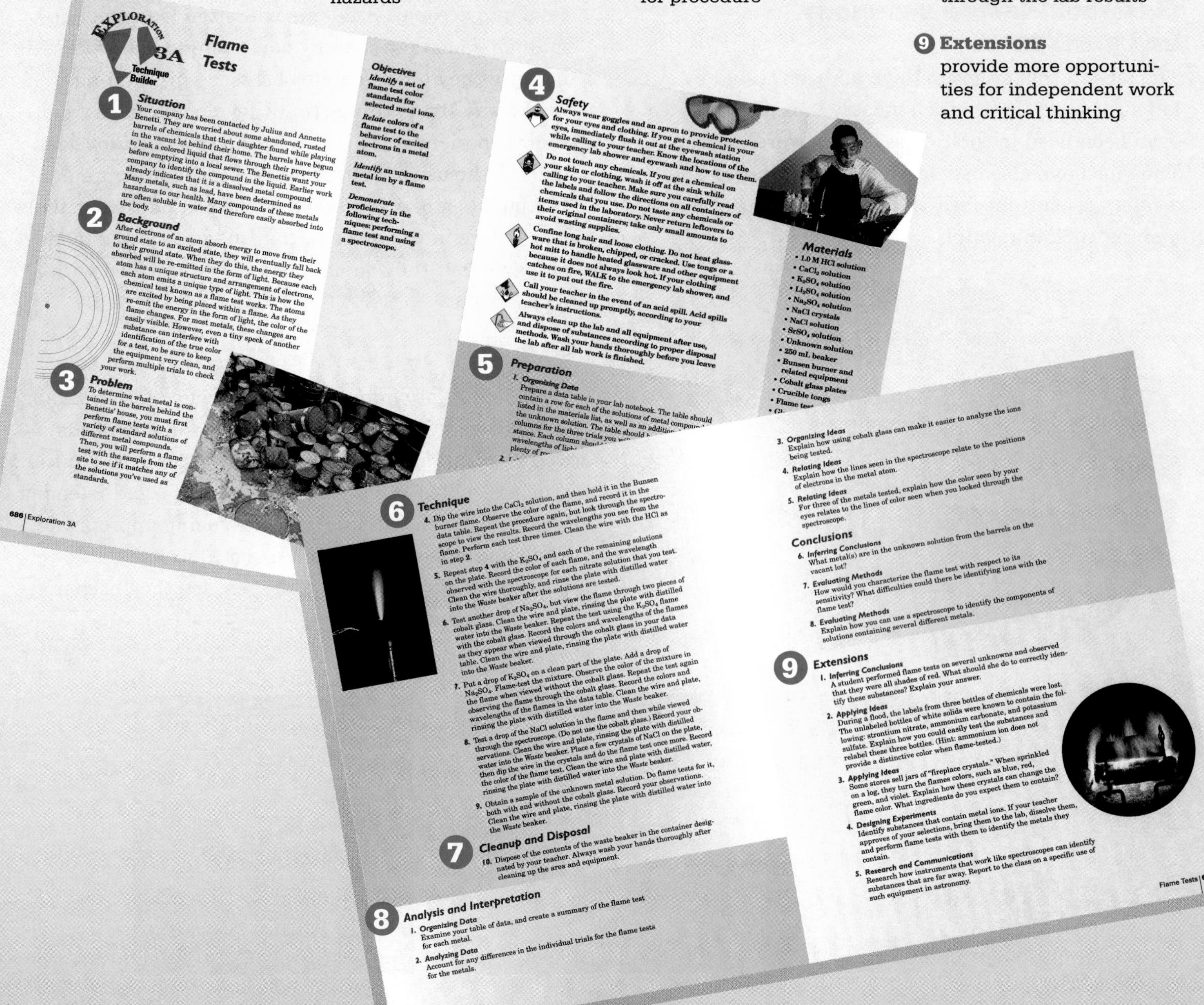

EXPLORATION 3A Technique Builder

Flame Tests

1 Situation
Your company has been contacted by Julius and Annette Benetti. They are worried about some abandoned, rusted barrels of chemicals that their daughter found while playing in the vacant lot behind their home. The barrels have begun to leak a colored liquid that flows through their property before emptying into a local sewer. The Benettis want your company to identify the compound in the liquid. Earlier work already indicates that it is a dissolved metal compound. Many metals, such as lead, have been determined as hazardous to our health. Many compounds of these metals are often soluble in water and therefore easily absorbed into the body.

2 Background
After electrons of an atom absorb energy to move from their ground state to an excited state, they will eventually fall back to their ground state. When they do this, the energy they absorbed will be re-emitted in the form of light. Because each atom has a unique structure and arrangement of electrons, each atom emits a unique type of light. This is how the chemical test known as a flame test works. The atoms are excited by being placed within a flame. As they re-emit the energy in the form of light, the color of the flame changes. For most metals, these changes are easily visible. However, even a tiny speck of another substance can interfere with identification of the true color for a test, so be sure to keep the equipment very clean, and perform multiple trials to check your work.

3 Problem
To determine what metal is contained in the barrels behind the Benettis' house, you must first perform flame tests with a variety of standard solutions of different metal compounds. Then, you will perform a flame test with the sample from the site to see if it matches any of the solutions you've used as standards.

Objectives
Identify a set of flame test color standards for selected metal ions.
Relate colors of a flame test to the behavior of excited electrons in a metal atom.
Identify an unknown metal ion by a flame test.
Demonstrate proficiency in the following techniques: performing a flame test and using a spectroscope.

4 Safety
Always wear goggles and an apron to provide protection for your eyes and clothing. If you get a chemical in your eyes, immediately flush it out at the eyewash station while calling to your teacher. Know the locations of the emergency lab shower and eyewash and how to use them.
Do not touch any chemicals. If you get a chemical on your skin or clothing, wash it off at the sink while calling to your teacher. Make sure you carefully read the labels and follow the directions on all containers of chemicals that you use. Do not taste any chemicals or items used in the laboratory. Never return leftovers to their original containers; take only small amounts to avoid wasting supplies.
Confine long hair and loose clothing. Do not heat glassware that is broken, chipped, or cracked. Use tongs or a hot mitt to handle heated glassware and other equipment because it does not always look hot. If your clothing catches on fire, WALK to the emergency lab shower, and use it to put out the fire.
Call your teacher in the event of an acid spill. Acid spills should be cleaned up promptly, according to your teacher's instructions.
Always clean up the lab and all equipment after use, and dispose of substances according to proper disposal methods. Wash your hands thoroughly before you leave the lab after all lab work is finished.

Materials
- 1.0 M HCl solution
- $CaCl_2$ solution
- K_2SO_4 solution
- Li_2SO_4 solution
- Na_2SO_4 solution
- NaCl crystals
- NaCl solution
- $SrSO_4$ solution
- Unknown solution
- 250 mL beaker
- Bunsen burner and related equipment
- Cobalt glass plates
- Crucible tongs
- Flame test ...

5 Preparation
1. **Organizing Data** Prepare a data table in your lab notebook. The table should contain a row for each of the solutions of metal compounds listed in the materials list, as well as an addition... the unknown solution. The table should ... columns for the three trials you ... stance. Each column should ... wavelengths of ligh... plenty of r...

6 Technique
4. Dip the wire into the $CaCl_2$ solution, and then hold it in the Bunsen burner flame. Observe the color of the flame, and record it in the data table. Repeat the procedure again, but look through the spectroscope to view the results. Record the wavelengths you see from the flame. Perform each test three times. Clean the wire with the HCl as in step 2.
5. Repeat step 4 with the K_2SO_4 and each of the remaining solutions on the plate. Record the color of each flame, and the wavelength observed with the spectroscope for each nitrate solution that you test. Clean the wire thoroughly, and rinse the plate with distilled water into the *Waste* beaker after the solutions are tested.
6. Test another drop of Na_2SO_4, but view the flame through two pieces of cobalt glass. Clean the wire and plate, rinsing the plate with distilled water into the *Waste* beaker. Repeat the test using the K_2SO_4 flame with the cobalt glass. Record the colors and wavelengths of the flames as they appear when viewed through the cobalt glass in your data table. Clean the wire and plate, rinsing the plate with distilled water into the *Waste* beaker.
7. Put a drop of K_2SO_4 on a clean part of the plate. Add a drop of Na_2SO_4. Flame-test the mixture. Observe the color of the mixture in the flame when viewed without the cobalt glass. Repeat the test again observing the flame through the cobalt glass. Record the colors and wavelengths of the flames in the data table. Clean the wire and plate, rinsing the plate with distilled water into the *Waste* beaker.
8. Test a drop of the NaCl solution in the flame and then while viewed through the spectroscope. (Do not use the cobalt glass.) Record your observations. Clean the wire and plate, rinsing the plate with distilled water into the *Waste* beaker. Place a few crystals of NaCl on the plate, then dip the wire in the crystals and do the flame test once more. Record the color of the flame test. Clean the wire and plate with distilled water, rinsing the plate with distilled water into the *Waste* beaker.
9. Obtain a sample of the unknown metal solution. Do flame tests for it, both with and without the cobalt glass. Record your observations. Clean the wire and plate, rinsing the plate with distilled water into the *Waste* beaker.

7 Cleanup and Disposal
10. Dispose of the contents of the waste beaker in the container designated by your teacher. Always wash your hands thoroughly after cleaning up the area and equipment.

8 Analysis and Interpretation
1. **Organizing Data** Examine your table of data, and create a summary of the flame test for each metal.
2. **Analyzing Data** Account for any differences in the individual trials for the flame tests for the metals.
3. **Organizing Ideas** Explain how using cobalt glass can make it easier to analyze the ions being tested.
4. **Relating Ideas** Explain how the lines seen in the spectroscope relate to the positions of electrons in the metal atom.
5. **Relating Ideas** For three of the metals tested, explain how the color seen by your eyes relates to the lines of color seen when you looked through the spectroscope.

Conclusions
6. **Inferring Conclusions** What metal(s) are in the unknown solution from the barrels on the vacant lot?
7. **Evaluating Methods** How would you characterize the flame test with respect to its sensitivity? What difficulties could there be identifying ions with the flame test?
8. **Evaluating Methods** Explain how you can use a spectroscope to identify the components of solutions containing several different metals.

9 Extensions
1. **Inferring Conclusions** A student performed flame tests on several unknowns and observed that they were all shades of red. What should she do to correctly identify these substances? Explain your answer.
2. **Applying Ideas** During a flood, the labels from three bottles of chemicals were lost. The unlabeled bottles of white solids were known to contain the following: strontium nitrate, ammonium carbonate, and potassium sulfate. Explain how you could easily test the substances and relabel these three bottles. (Hint: ammonium ion does not provide a distinctive color when flame-tested.)
3. **Applying Ideas** Some stores sell jars of "fireplace crystals." When sprinkled on a log, they turn the flames colors, such as blue, red, green, and violet. Explain how these crystals can change the flame color. What ingredients do you expect them to contain?
4. **Designing Experiments** Identify substances that contain metal ions. If your teacher approves of your selections, bring them to the lab, dissolve them, and perform flame tests with them to identify the metals they contain.
5. **Research and Communications** Research how instruments that work like spectroscopes can identify substances that are far away. Report to the class on a specific use of such equipment in astronomy.

686 | Exploration 3A

688 | Exploration 3A

Flame Tests | 689

Investigations: Open-ended problem solving

The *Investigation* that follows an *Exploration* gives students a chance to use their skills in *Problem Solving* as they examine another aspect of the technique, reaction, or concept that they encountered in the *Exploration*. Instead of instructions, students are given a business letter from a "client" that outlines a problem, and a memo from their supervisor that offers further suggestions. Based on the problem, they must create a plan for their procedure, select equipment from a list of available materials, and attempt to stay within a budget. Student plans are reviewed by you, the teacher, for practicality and safety before students enter the lab. On completion of the lab, instead of writing a traditional lab report, students write a business letter back to the "client" explaining their results and submitting a bill for their services.

For best results and the greatest safety, do not perform any Investigation without performing its prerequisite Exploration, if it has one. The prerequisite is identified at the beginning of the Teacher's Notes for each Investigation.

1 Business letter poses a problem for students to solve

2 Memo outlines parameters for students' budgets and reports

3 References section includes helpful hints for students planning their own procedure

4 Required Precautions provide reminders of potential hazards

5 Spill Procedures/ Waste Disposal Methods provide detailed instructions for handling and removing all products and reactants

6 Materials list includes non-essential equipment, forcing student to think critically about which and how much equipment to use

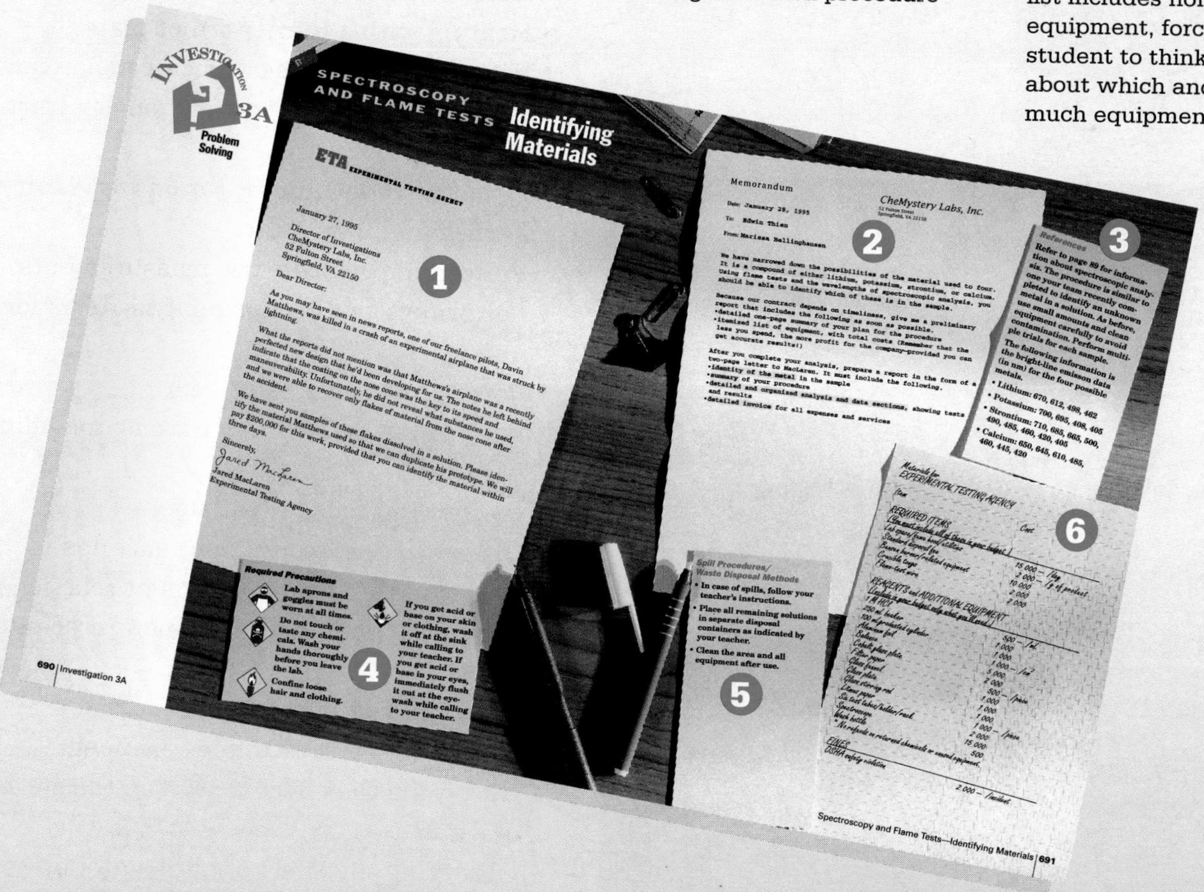

Using the tools and techniques of chemists

Like other lab programs, the labs in *Holt Chemistry: Visualizing Matter* provide ample opportunities for students to work with the same equipment and methods as practicing chemists. Unlike other lab programs, all of these tools and techniques are presented and used in the context of real-world situations.

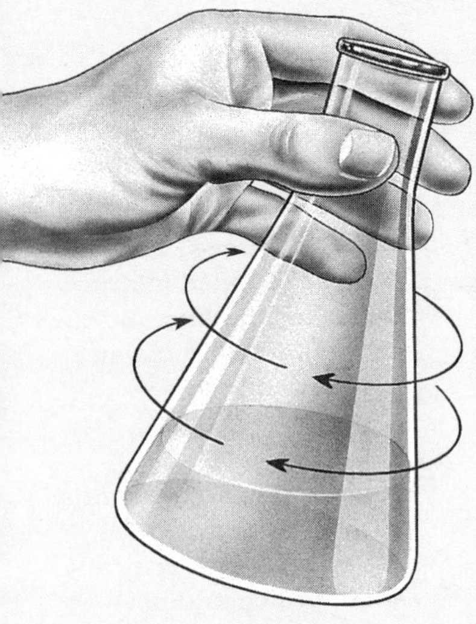

Laboratory techniques developed

- Measuring mass with a balance
- Measuring length with a ruler
- Measuring volume with a graduated cylinder, a volumetric flask, a pipet, or a buret
- Measuring temperature with a thermometer or LEAP thermistor probe
- Measuring pH with pH paper, a pH meter, or a LEAP pH probe
- Measuring wavelengths in a spectrum with a spectroscope (optional)
- Measuring time elapsed during reactions
- Measuring absorbance by eye, with a spectrophotometer, or with a LEAP colormeter
- Measuring buffering capacities
- Measuring boiling points
- Measuring radiation exposure by counting alpha-particle tracks
- Handling solids and liquids
- Heating with a burner or hot plate
- Preparing solutions of specific molar concentrations
- Performing mixture separations by fractional crystallization
- Performing vacuum filtrations or gravity filtrations
- Performing flame tests
- Performing calorimetric measurements
- Performing qualitative analysis tests for different nutrients
- Performing serial dilutions
- Performing acid-base titrations, including back-titrations
- Performing redox titrations
- Observing polymerization reactions
- Observing single-displacement reactions in solution
- Observing double-displacement precipitation reactions
- Observing decomposition reactions
- Observing neutralization reactions
- Observing the effect of soap on the surface tension of water
- Observing the effect of hard water on soap suds
- Observing the factors that affect reaction rate
- Observing redox reactions

Application of specific problem-solving skills

Multi-step determinations

- Determining empirical formulas from mass measurements
- Determining percentage composition from mass measurements
- Determining the relative activities of metals
- Determining heat of reaction from calorimetry data
- Determining concentration from absorbance data
- Determining molal concentration and molar mass from boiling-point elevation data
- Determining the concentrations of substances in an equilibrium system
- Determining the values of equilibrium constants
- Determining concentration from titration data
- Determining the relationship between rate and reactant concentration

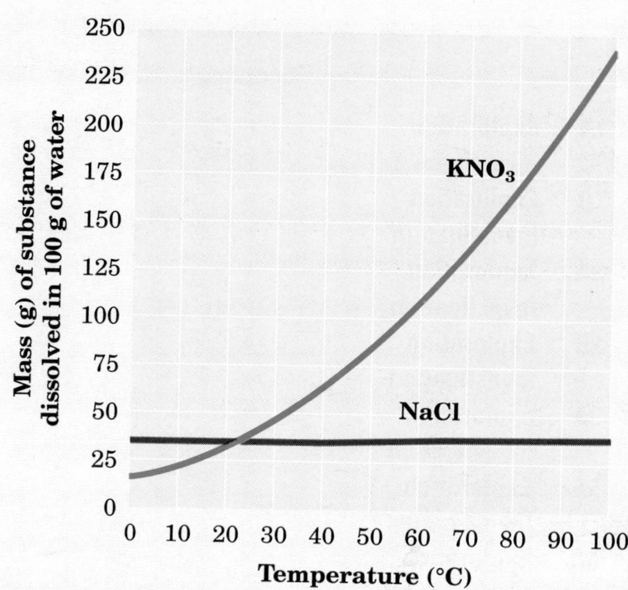

Computation

- Graphing absorbance-concentration data
- Calculating volume from measurements of length
- Calculating density from measurements of mass and volume
- Calculating percent error as a measure of accuracy
- Calculating absolute deviation, average deviation, and uncertainty as measures of precision
- Calculating molar amounts from mass measurements
- Calculating boiling-point elevations
- Calculating reaction rate from time measurements
- Calculating radon activity

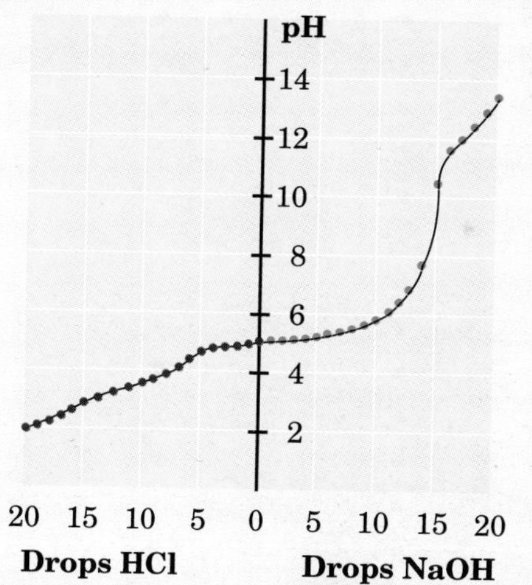

Data manipulation and interpretation

- Organizing data in tables
- Graphing data
- Graphing titration curves

Complete development of process skills

		Observing	Measuring	Organizing	Classifying	Hypothesizing	Predicting
1A	Exploration	•	•	•			
	Investigation	•	•	•			
1B	Exploration	•	•	•			
	Investigation	•	•	•	•		
2A	Exploration	•	•	•	•		
	Investigation	•	•	•	•		
2B	Exploration	•	•	•		•	•
	Investigation	•	•	•			
3A	Exploration	•	•	•	•	•	•
	Investigation	•	•	•	•		
5A	Exploration	•	•	•		•	•
	Investigation	•	•	•			
6A	Exploration	•	•	•		•	•
	Investigation	•	•	•		•	•
7A	Investigation	•		•	•	•	•
8A	Exploration	•	•	•		•	•
	Investigation	•	•	•	•		
8B	Exploration	•	•	•		•	•
	Investigation	•	•	•		•	•
9A	Exploration	•	•	•		•	•
	Investigation	•	•	•		•	•
9B	Exploration	•		•	•		
	Investigation	•		•	•		
10A	Investigation	•	•	•		•	•
10B	Investigation	•	•	•		•	•
11A	Exploration	•	•	•		•	•
	Investigation	•	•	•		•	•
11B	Investigation	•	•	•	•		
12A	Exploration	•	•	•			
	Investigation	•	•	•			
12B	Exploration	•	•	•		•	•
	Investigation	•	•	•			
13A	Exploration	•	•	•		•	•
	Investigation	•	•	•	•	•	•
13B	Exploration	•	•	•		•	•
	Investigation	•	•	•		•	•
14A	Exploration	•	•	•		•	•
	Investigation	•	•	•			
15A	Exploration	•	•	•	•	•	•
	Investigation	•	•	•			
16A	Exploration	•	•	•		•	•
	Investigation	•	•	•			

Holt Chemistry: Visualizing Matter provides for the thorough development of science process skills, as shown by the correlation below.

		Analyzing	Designing experiments	Inferring	Modeling	Communicating
1A	Exploration	•	•	•		
	Investigation	•	•	•		•
1B	Exploration	•	•	•	•	•
	Investigation	•	•	•	•	•
2A	Exploration	•	•	•		
	Investigation	•	•	•		•
2B	Exploration	•	•	•	•	•
	Investigation	•	•			•
3A	Exploration	•	•	•	•	•
	Investigation	•	•	•		•
5A	Exploration	•	•	•		
	Investigation	•	•	•		•
6A	Exploration	•	•	•	•	•
	Investigation	•	•		•	•
7A	Investigation		•	•		•
8A	Exploration	•	•	•		•
	Investigation	•	•	•		•
8B	Exploration	•	•	•		•
	Investigation	•	•	•		•
9A	Exploration	•	•	•	•	
	Investigation	•	•	•		•
9B	Exploration	•	•	•		•
	Investigation	•	•	•		•
10A	Investigation	•	•	•	•	•
10B	Investigation	•	•	•	•	•
11A	Exploration	•	•	•	•	
	Investigation	•	•	•	•	•
11B	Investigation	•	•	•		•
12A	Exploration	•	•	•	•	•
	Investigation	•	•	•		•
12B	Exploration	•	•	•	•	•
	Investigation	•	•	•	•	•
13A	Exploration	•	•	•	•	•
	Investigation	•	•	•	•	•
13B	Exploration	•	•	•	•	•
	Investigation	•	•	•	•	•
14A	Exploration	•	•	•	•	
	Investigation	•	•	•	•	•
15A	Exploration	•	•	•		•
	Investigation	•	•	•		•
16A	Exploration	•	•	•		•
	Investigation	•	•	•	•	•

Labs that meet the needs of all students

Along with a thorough development of scientific process skills, the *Investigations* provide sensible opportunities for students to sharpen the critical-thinking skills they will need whether or not they pursue science as a career. The skills used include the following.

Critical thinking skills

- identifying problems
- applying abstract concepts to specific situations
- following directions, especially those relating to safety and disposal
- finding information in a business letter, memo, or reference work
- combining information from a variety of sources to make a plan
- identifying specific measurements that must be made in order to draw conclusions
- planning a procedure to solve the problem
- deciding what equipment and materials are necessary for the plan
- developing and following a budget for the equipment, materials, and other charges
- writing a business letter that organizes the results

Practical interdisciplinary opportunities with instructional value

- **Language arts:** reading letters, memos, and other references to obtain information to identify and solve a problem; organizing and composing a business report that clearly and succinctly describes results, the procedures used to obtain them, and their validity
- **Math/economics:** keeping track of the budget for the equipment and materials necessary, and providing an invoice for the services rendered. The "profitability" of student lab groups can be used as a measure of extra credit. In addition, the Investigations lend themselves to use in portfolio assessment. See page T35 for more information on portfolio assessments.
- **Science, technology, and society:** exploring the roles of scientists in society in solving problems, helping create new products, testing existing products, testifying as expert witnesses, performing forensic analyses, etc.

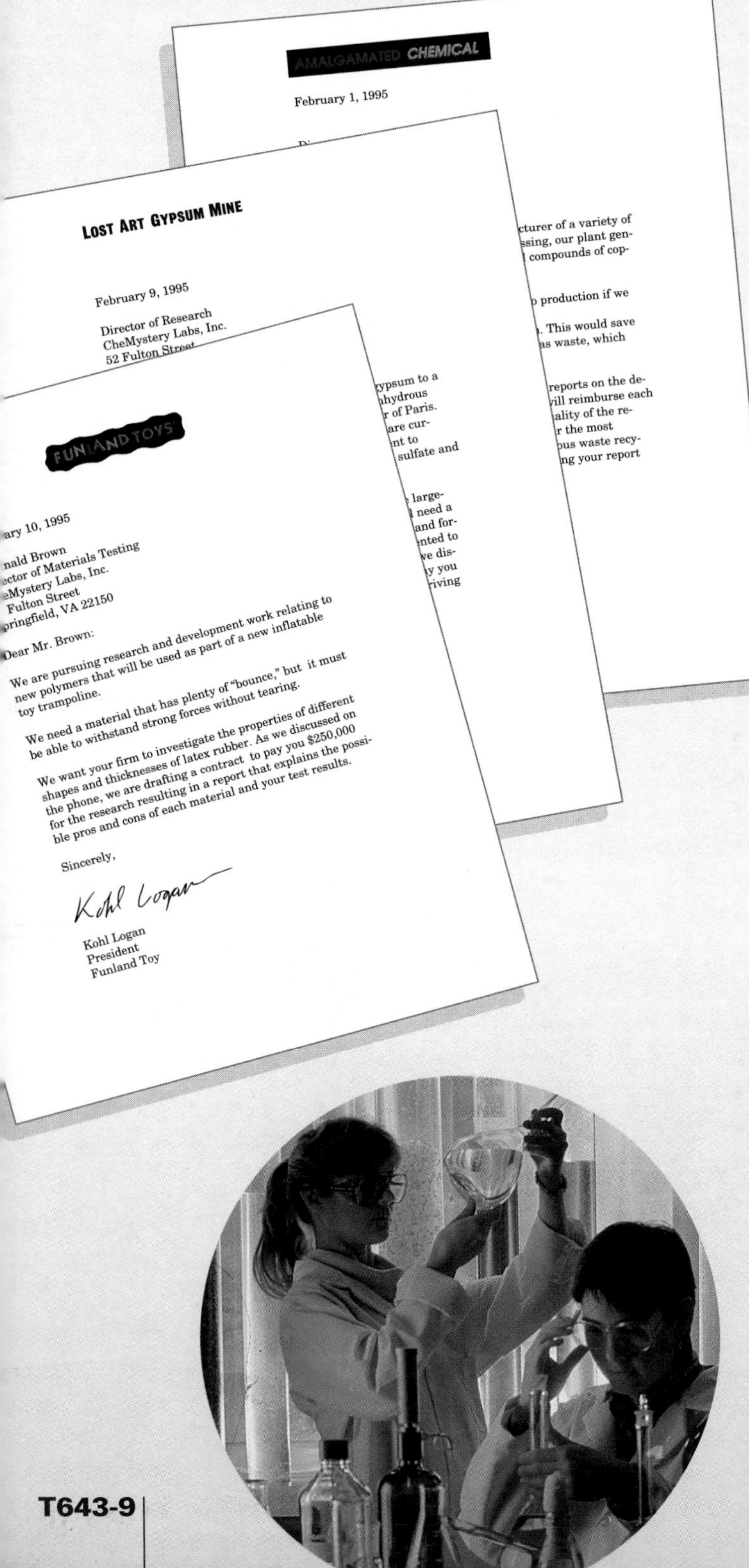

AMALGAMATED CHEMICAL

February 1, 1995

LOST ART GYPSUM MINE

February 9, 1995

Director of Research
CheMystery Labs, Inc.
52 Fulton Street

FUNLAND TOYS

ary 10, 1995

nald Brown
ector of Materials Testing
eMystery Labs, Inc.
Fulton Street
pringfield, VA 22150

Dear Mr. Brown:

We are pursuing research and development work relating to new polymers that will be used as part of a new inflatable toy trampoline.

We need a material that has plenty of "bounce," but it must be able to withstand strong forces without tearing.

We want your firm to investigate the properties of different shapes and thicknesses of latex rubber. As we discussed on the phone, we are drafting a contract to pay you $250,000 for the research resulting in a report that explains the possible pros and cons of each material and your test results.

Sincerely,

Kohl Logan

Kohl Logan
President
Funland Toy

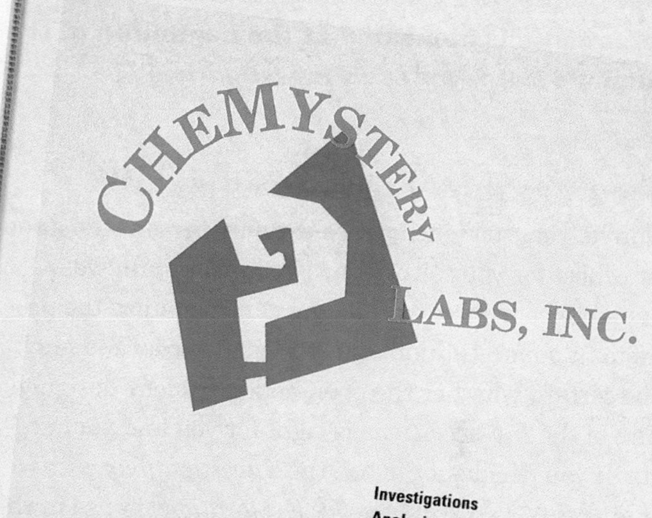

Extending the consulting firm scenario

Along with samples of their work from the laboratory, students can create other details about their business. The following optional items can help expand the scope of the year's projects, if desired.

Safety and waste management

- develop a complete safety plan for keeping the lab safe in the event of a variety of accidents, including chemical spills, fires, and natural disasters such as earthquakes, tornados, hurricanes, and floods
- develop a plan for hazardous-waste management and disposal
- create a floor plan for lab facilities that takes safety, efficiency, and cost into account

Finance and marketing

- develop a plan for creating a business, including ideas for funding and potential clients
- fill out copies of governmental forms, such as incorporation papers or corporate registration, business licenses, applications for lab permits or chemical licenses, corporate tax forms, EPA hazardous waste management forms, etc.
- fill out copies of business loan forms from local banks
- create financial statements showing income, expenses, and profit
- develop a marketing strategy for reaching new customers, including costs
- develop promotional materials describing the firm and the chemical services it provides along with fee schedules
- develop a plan for the use of any profits to reinvest in personnel, equipment, marketing, etc. (Some teachers allow lab groups that are profitable to become "venture capatalists," who loan money to lab groups that have made mistakes in planning their procedures and use of resources.)

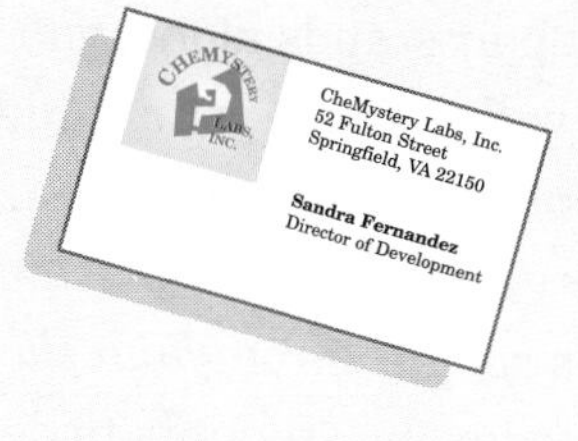

How do I get started?

For best results and the greatest safety, do not perform any *Investigations* without performing its prerequisite *Exploration*, if it has one. The prerequisite is identified at the beginning of the Teacher's Notes for each *Investigation*.

A program that maximizes flexibility

Of course, you, the teacher, are the only person who knows what is best for your situation. The lab program was designed to be flexible so that you can customize the program to fit your situation and students' needs. You may not be certain whether the open-ended student-designed nature of the *Investigations* is right for you and your students. If you decide not to use the *Investigations* with your students, you still have 19 *Explorations* to use in the textbook. Still more *Explorations* are available in the lab manual in the Teacher's Resource Organizer.

For many teachers, there will be many logistical issues to work out before they try student-designed labs. Because each *Exploration-Investigation* pair stands alone, you can choose any pair as the one you will try with your students. The authors strongly recommend that if you haven't tried this type of lab experience before, you should *start slowly, with only a few open-ended labs spread throughout the year*. Be sure to allow extra time for the students to design, refine, and perfect their procedures. After you have grown comfortable with the approach, add more in successive years.

What shortcuts can I implement to simplify things?

Even within individual *Investigations*, there are numerous adaptations that can be made to streamline the process. One possibility is to delete the budgeting and invoice tasks. Another possibility is to tell the students exactly what equipment they need instead of having them decide. A concise list is provided in the Teacher's Notes for each *Investigation*. You could even describe part of the proposed procedure provided in the Teacher's Notes, instead of requiring students to decide on a plan themselves.

Tips for carrying out Explorations

Most of the techniques you use with traditional labs will work when you assign an *Exploration*. Be sure to assign the following tasks for students to complete before entering the lab.

- Read the *Situation*, *Background*, *Problem*, *Safety*, *Preparation*, and *Technique* sections.
- Re-read the *Safety* section.
- Create fully prepared data tables according to the instructions given in the *Preparation* section.

You may even want to give students a brief quiz on the safety precautions or the procedure. Only those with a perfect score should be allowed to proceed.

Reports

After the *Exploration*, students should prepare a lab report with at least the following components.

- **title**
- **summary paragraph** describing the purpose and procedure
- **data tables and observations** that are organized and comprehensive
- **answers** to the *Analysis and Interpretation*, *Conclusions*, and any *Extensions* items you assign

Be sure to instruct students to give fully worked out answers. It may help your grading if you also tell them to circle, box, or highlight the final result of their calculations. If you will require other components for the lab reports, be certain to explain them to students.

LAB PROGRAM | Introduction

Evaluating lab work

Scoring rubrics

Traditional means of scoring labs on a percentage scale may not work as well with this type of lab program. Of course, you are the only one who can judge what is best for your students. The following scoring rubrics, which are based on a scale of 0–6, may be a useful starting point. Read the description of each level, and decide which one most accurately fits each report you grade.

Exploration lab report

Experienced level
(6 points)
- Excellent technique was used throughout the lab procedure.
- Data and observations were recorded accurately, descriptively, and completely, with no serious errors.
- *Analysis and Interpretation* was performed clearly, concisely, and accurately, with correct units and properly worked-out calculations.
- Graphs, if necessary, are drawn accurately and neatly.
- Students express their recognition of the connections between their observations and the related chemistry concepts in an exemplary manner.
- Good reasoning and logic are evident throughout the report.
- Answers to *Conclusions* items are written correctly and accurately.
- Any *Extensions* that were assigned are completed with creativity and imagination.

Competent level
(5 points)
- No errors in technique were observed during the lab procedure.
- Data and observations were recorded accurately, descriptively, and completely, with only minor errors.
- *Analysis and Interpretation* was performed accurately, with correct units and properly worked-out calculations, but it may have been slightly disorganized.
- Graphs, if necessary, are drawn accurately and neatly.
- Students effectively express their recognition of the connections between their observations and the related chemistry concepts.
- Good reasoning and logic are evident throughout the report.
- Answers to *Conclusions* items are written correctly and accurately, but there may be minor misunderstandings.
- Any *Extensions* that were assigned are completed adequately.

Intermediate level
(4 points)
- Only minor errors in technique were observed during the lab procedure.
- Data and observations were recorded accurately with only minor errors or omissions.
- *Analysis and Interpretation* was performed accurately, but some minor errors were made either in calculations or in applying correct units.
- Graphs, if necessary, are drawn accurately and neatly.
- Students satisfactorily express their recognition of the connections between their observations and the related chemistry concepts.
- Reasoning is occasionally weak in the report, but only in a few places.
- Answers to *Conclusions* items are correct, but there are some misunderstandings or minor errors.
- Students made an effort to adequately address any *Extensions* that were assigned, but their effort may not have been entirely successful.

A scoring rubric for Exploration *lab reports is given on page T643-15.*

Complete teacher support on the same page as the lab

Each lab, whether an *Exploration* or *Investigation*, has complete support for you in the Teacher's Notes in the margins of the teacher's edition.

❶ **Planning**
for the time, materials, and solutions that must be prepared before the students perform the lab activity

❷ **Pre-lab Discussion**
of concepts so that students will understand how their lab work relates to the chemistry concepts they studied

❸ **Techniques to Demonstrate**
so that students can spend their time performing the lab instead of trying to figure out how to operate the equipment

❹ **Required Precautions**
to keep you and your students safe

❺ **Sample Data and Analysis**
to help you in evaluating students' work

❻ **Disposal**
guidelines that indicate exactly what to do with each reactant and product from the lab activity

❼ **Answers**
to the items in the Analysis and Interpretation, Conclusions, and Extensions sections that are provided in full, so you can spend your grading time evaluating students' work rather than performing all the calculations yourself

Getting started, continued

Tips for carrying out Investigations

The *Investigations* require an approach different from most traditional labs. Students at this stage tend to panic when left on their own to develop a lab, particularly a procedure. Be sure to do the following to help them feel comfortable.

- **Completely discuss the *Exploration* prerequisite.**
- **Suggest that students use equipment and amounts similar to those in the *Exploration*** if they are uncertain about how to proceed.
- **Remind students to concentrate on what they need to know and how they will measure it** so that they remain focused.
- **Remind students that it is more important to understand** what is going on in the lab than to perform it with excellent technique but no understanding.
- **Provide leading questions for students to consider** as they make their plans. Suggestions are often given in the **Pre-lab Discussion** section of the Teacher's notes for the *Investigations*.
- **Hold a question-and-answer session** before students begin the lab.
- **Hint (don't tell) that some of the equipment listed may be unnecessary**.
- **Check the memo in the *Investigation* for requirements** for the preliminary report.

The Teacher's Notes for each *Investigation* contain tips that will help evaluate the pre-lab requirements. A concise list details all of the equipment and materials that each lab group should request, along with the projected total cost. A proposed procedure is also included so that you can compare it to students' suggested plans and promptly notice any discrepancies. Be sure to tell students when there are flaws in their plans. You may require them to propose solutions themselves, instead of telling them what to do.

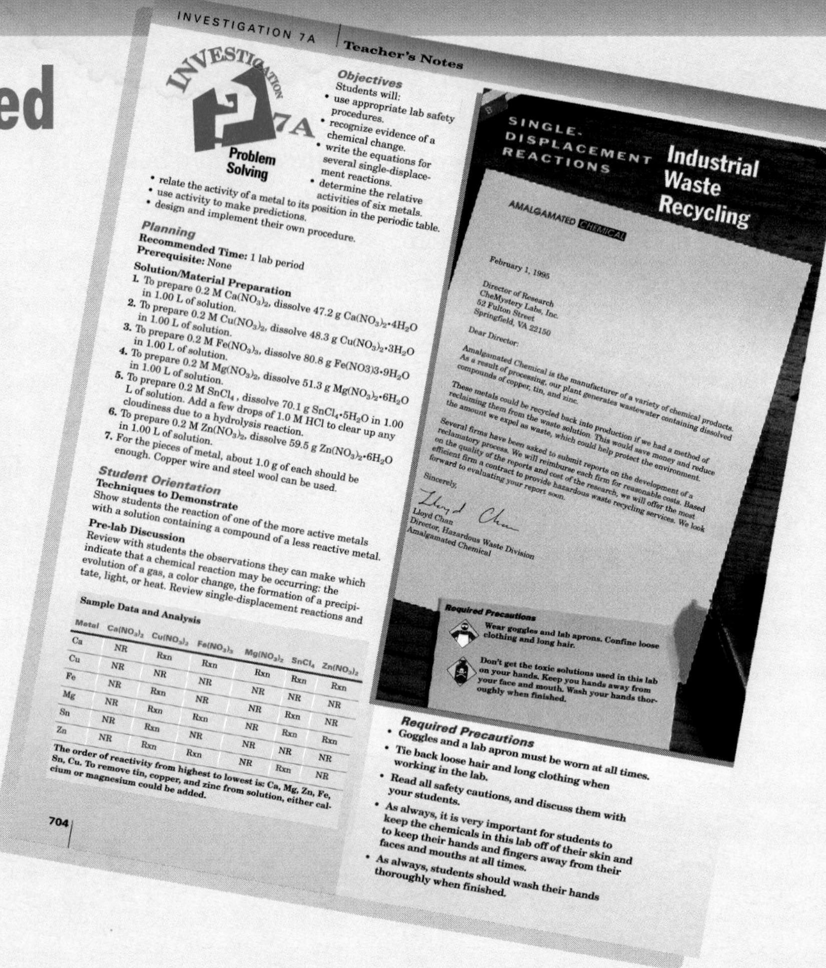

INVESTIGATION 7A | Teacher's Notes

INVESTIGATION 7A Problem Solving

Objectives

Students will:
- use appropriate lab safety procedures.
- recognize evidence of a chemical change.
- write the equations for several single-displacement reactions.
- determine the relative activities of six metals.
- relate the activity of a metal to its position in the periodic table.
- use activity to make predictions.
- design and implement their own procedure.

Planning

Recommended Time: 1 lab period

Prerequisite: None

Solution/Material Preparation

1. To prepare 0.2 M $Ca(NO_3)_2$, dissolve 47.2 g $Ca(NO_3)_2 \cdot 4H_2O$ in 1.00 L of solution.
2. To prepare 0.2 M $Cu(NO_3)_2$, dissolve 48.3 g $Cu(NO_3)_2 \cdot 3H_2O$ in 1.00 L of solution.
3. To prepare 0.2 M $Fe(NO_3)_3$, dissolve 80.8 g Fe(NO3)3$\cdot 9H_2O$ in 1.00 L of solution.
4. To prepare 0.2 M $Mg(NO_3)_2$, dissolve 51.3 g $Mg(NO_3)_2 \cdot 6H_2O$ in 1.00 L of solution.
5. To prepare 0.2 M $SnCl_4$, dissolve 70.1 g $SnCl_4 \cdot 5H_2O$ in 1.00 L of solution. Add a few drops of 1.0 M HCl to clear up any cloudiness due to a hydrolysis reaction.
6. To prepare 0.2 M $Zn(NO_3)_2$, dissolve 59.5 g $Zn(NO_3)_2 \cdot 6H_2O$ in 1.00 L of solution.
7. For the pieces of metal, about 1.0 g of each should be enough. Copper wire and steel wool can be used.

Student Orientation

Techniques to Demonstrate

Show students the reaction of one of the more active metals with a solution containing a compound of a less reactive metal.

Pre-lab Discussion

Review with students the observations they can make which indicate that a chemical reaction may be occurring: the evolution of a gas, a color change, the formation of a precipitate, light, or heat. Review single-displacement reactions and

Sample Data and Analysis

Metal	$Ca(NO_3)_2$	$Cu(NO_3)_2$	$Fe(NO_3)_3$	$Mg(NO_3)_2$	$SnCl_4$	$Zn(NO_3)_2$
Ca	NR	Rxn	Rxn	Rxn	Rxn	Rxn
Cu	NR	NR	NR	NR	NR	NR
Fe	NR	Rxn	NR	NR	NR	NR
Mg	NR	Rxn	Rxn	NR	Rxn	Rxn
Sn	NR	Rxn	NR	NR	NR	NR
Zn	NR	Rxn	Rxn	NR	Rxn	NR

The order of reactivity from highest to lowest is: Ca, Mg, Zn, Fe, Sn, Cu. To remove tin, copper, and zinc from solution, either calcium or magnesium could be added.

704

SINGLE-DISPLACEMENT REACTIONS

Industrial Waste Recycling

AMALGAMATED CHEMICAL

February 1, 1995

Director of Research
CheMystery Labs, Inc.
52 Fulton Street
Springfield, VA 22150

Dear Director:

Amalgamated Chemical is the manufacturer of a variety of chemical products. As a result of processing, our plant generates wastewater containing dissolved compounds of copper, tin, and zinc.

These metals could be recycled back into production if we had a method of reclaiming them from the waste solution. This would save money and reduce the amount we expel as waste, which could help protect the environment.

Several firms have been asked to submit reports on the development of a reclamatory process. We will reimburse each firm for reasonable costs. Based on the quality of the reports and cost of the research, we will offer the most efficient firm a contract to provide hazardous waste recycling services. We look forward to evaluating your report soon.

Sincerely,

Lloyd Chan
Director, Hazardous Waste Division
Amalgamated Chemical

Required Precautions

Wear goggles and lab aprons. Confine loose clothing and long hair.

Don't get the toxic solutions used in this lab on your hands. Keep you hands away from your face and mouth. Wash your hands thoroughly when finished.

Required Precautions
- Goggles and a lab apron must be worn at all times.
- Tie back loose hair and long clothing when working in the lab.
- Read all safety cautions, and discuss them with your students.
- As always, it is very important for students to keep the chemicals in this lab off of their skin and to keep their hands and fingers away from their faces and mouths at all times.
- As always, students should wash their hands thoroughly when finished.

Reports

The preliminary report is an essential part of these labs. If done properly, the actual procedure of the lab usually can be completed in one period. If students are disorganized, they may need to purchase more lab-space time (if you permit it) either during class or after school.

On completion of the lab, students should report their results in the form of a business letter. Specific requirements are indicated in the memo on the second page of the student's *Investigation*. Be sure to emphasize to students that the goal is to communicate their results clearly and concisely. Students should also submit an invoice that itemizes their expenses.

A scoring rubric for Investigation *preliminary reports is given on page T643-17.*

LAB PROGRAM | Introduction

Scoring rubrics, continued

Investigation final report (letter and invoice)

Experienced level (6 points)
- Excellent technique was used throughout the lab procedure.
- The report clearly expresses good reasoning, logic, and understanding of what the lab was about.
- The connections between the data and the related chemistry concepts that lead to the conclusions are clearly expressed.
- Students recognize the connection between the initial problem and the outcome of the lab work.
- Summary is clear, concise, and well organized, with few grammatical or stylistic errors.
- Data and analysis are accurate and well organized, with correct units and no errors in calculations.
- Percent error for quantitative answers is less than 15%.
- Graphs, if appropriate, are drawn and labeled

LAB PROGRAM | Introduction

Scoring rubrics, continued

Investigation preliminary report

Experienced level (6 points)
- Plan shows careful and thorough planning with good reasoning and logic.
- Plan is complete, appropriate, and safe.
- Plan is as efficient as the procedure given in the Teacher's Notes for the *Investigation*.
- Plan is expressed clearly and concisely.
- Proposed data tables are made properly and clearly indicate all measurements that must be made to solve the problem.
- Proposed list of equipment includes all equipment necessary to carry out the procedure described in the plan.
- When compared to the list given in the Teacher's Notes for the *Investigation*, the student's list of equipment includes no unnecessary pieces of equipment.
- Proposed budget is

Competent level (5 points)
- Plan shows careful and thorough planning, although the logic behind it may not be clearly expressed.
- Plan is appropriate, safe, and mostly complete, with only minor omissions.
- Plan is almost as efficient as the procedure given in the Teacher's Notes for the *Investigation*.
- Some parts of the plan could be expressed more clearly or more concisely.
- Proposed data tables indicate all measurements that must be made to solve the problem, but minor errors may have been made in preparing the data tables.
- Proposed list of equipment includes all equipment necessary for the procedure described in the plan.
- When compared to the list

Intermediate level (4 points)
- Plan shows some logic, but the reasoning could have been more careful, more thorough, or more clearly expressed.
- Plan is appropriate and safe, but there are a few omissions.
- Plan will work, but is not as efficient as the procedure given in the Teacher's Notes for the *Investigation*.
- Plan could be written more clearly or more concisely.
- Proposed data tables indicate all measurements that must be made to solve the problem, but multiple trials are not included.
- Proposed list of equipment includes almost all equipment necessary to carry out the plan described in the procedure,

A scoring rubric for the Investigation *final letter is given on page T643-19.*

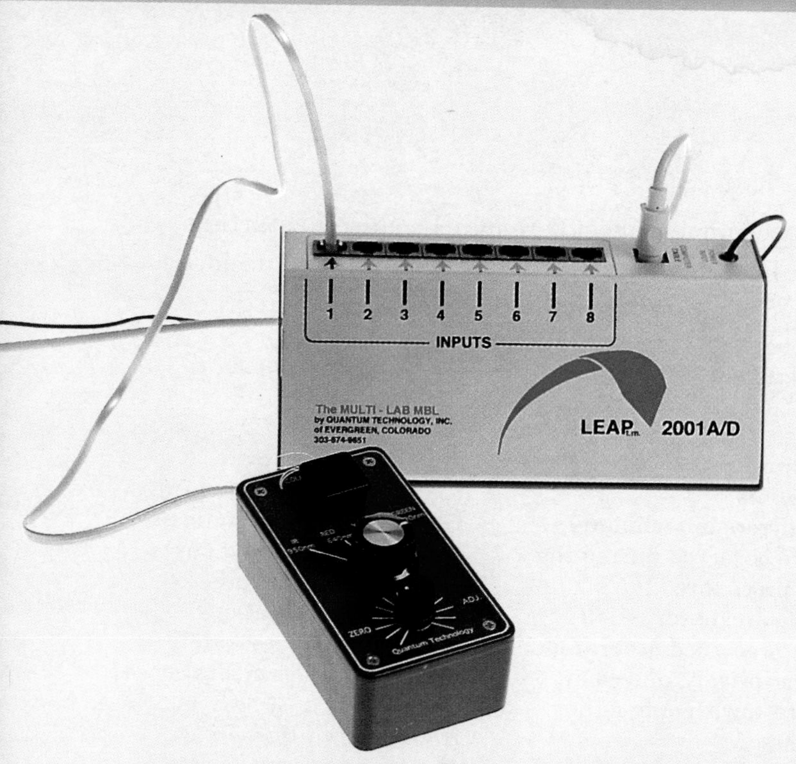

Microcomputer-based laboratory options

Some *Explorations* and *Investigations* include procedures for using equipment such as pH meters, spectrophotometers, or LEAP System microcomputer-based laboratory probes from Quantum Technology. Each time this option is presented, an alternative procedure that does not require such equipment is also presented. Thus, the lab activities remain useful and practical whether the technology is present or not.

To order the LEAP System and probes, contact Quantum Technology. Quantum has set up a variety of packages for users of ***Holt Chemistry: Visualizing Matter***.

Quantum Technology
30153 Arena Drive
Evergreen, CO 80439

Sales and Technical Support: (303) 674-9651
Fax: (303) 674-6763

Lab activity	Title	LEAP option
Exp. 1B **Inv. 1B**	*Properties of Analgesics*	pH probe
Exp. 2A	*Accuracy and Precision*	thermistor probe
Exp. 2B **Inv. 2B**	*Separation of Mixtures*	thermistor probe
Exp. 3A **Inv. 3A**	*Flame Tests*	spectroscope
Exp. 9A **Inv. 9A**	*Calorimetry*	thermistor probe
Exp. 11A **Inv. 11A**	*Colorimetry*	colormeter
Exp. 12A **Inv. 12A**	*Boiling-Point Elevation*	thermistor probe
Exp. 12B **Inv. 12B**	*Equilibrium Expressions*	colormeter
Exp. 13A **Inv. 13A**	*Acid-Base Titration*	pH probe
Exp. 13B **Inv. 13B**	*Buffering Capacity*	pH probe
Inv. 14A	*Reaction Rates*	thermistor probe

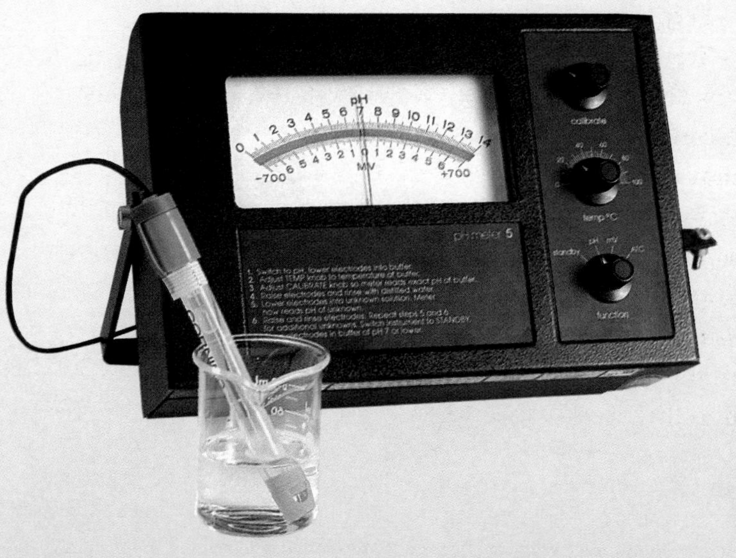

Evaluating lab work

Scoring rubrics

Traditional means of scoring labs on a percentage scale may not work as well with this type of lab program. Of course, you are the only one who can judge what is best for your students. The following scoring rubrics, which are based on a scale of 0–6, may be a useful starting point. Read the description of each level, and decide which one most accurately fits each report you grade.

Exploration lab report

Experienced level
(6 points)
- Excellent technique was used throughout the lab procedure.
- Data and observations were recorded accurately, descriptively, and completely, with no serious errors.
- *Analysis and Interpretation* was performed clearly, concisely, and accurately, with correct units and properly worked-out calculations.
- Graphs, if necessary, are drawn accurately and neatly.
- Students express their recognition of the connections between their observations and the related chemistry concepts in an exemplary manner.
- Good reasoning and logic are evident throughout the report.
- Answers to *Conclusions* items are written correctly and accurately.
- Any *Extensions* that were assigned are completed with creativity and imagination.

Competent level
(5 points)
- No errors in technique were observed during the lab procedure.
- Data and observations were recorded accurately, descriptively, and completely, with only minor errors.
- *Analysis and Interpretation* was performed accurately, with correct units and properly worked-out calculations, but it may have been slightly disorganized.
- Graphs, if necessary, are drawn accurately and neatly.
- Students effectively express their recognition of the connections between their observations and the related chemistry concepts.
- Good reasoning and logic are evident throughout the report.
- Answers to *Conclusions* items are written correctly and accurately, but there may be minor misunderstandings.
- Any *Extensions* that were assigned are completed adequately.

Intermediate level
(4 points)
- Only minor errors in technique were observed during the lab procedure.
- Data and observations were recorded accurately with only minor errors or omissions.
- *Analysis and Interpretation* was performed accurately, but some minor errors were made either in calculations or in applying correct units.
- Graphs, if necessary, are drawn accurately and neatly.
- Students satisfactorily express their recognition of the connections between their observations and the related chemistry concepts.
- Reasoning is occasionally weak in the report, but only in a few places.
- Answers to *Conclusions* items are correct, but there are some misunderstandings or minor errors.
- Students made an effort to adequately address any *Extensions* that were assigned, but their effort may not have been entirely successful.

Although the first few you grade may be time-consuming, as you grow accustomed to the rubric, it will soon become a very quick process.

A useful strategy is to keep a file of past papers that seem to exemplify each level to use as benchmarks. You may want to make this file available to students to give them examples of good work.

Transitional level
(3 points)

- Only a few errors in technique were observed during the lab procedure, but they may have been significant.
- Data and observations were recorded accurately, with only minor errors or omissions.
- *Analysis and Interpretation* was performed accurately, but some minor errors were made both in calculations and in applying correct units.
- Graphs, if necessary, are drawn accurately and neatly.
- Students recognize connections between their observations and the related chemistry concepts, but only weakly express their understanding.
- Reasoning is generally weak throughout much of the report.
- Some answers to *Conclusions* items are not correct, because of misunderstandings or minor errors.
- Students made an effort to address any *Extensions* that were assigned, but their effort was not substantial enough.

Beginning level
(2 points)

- Several serious errors in technique were observed during the lab procedure.
- Most data and observations were recorded accurately, but with several significant errors or omissions.
- *Analysis and Interpretation* was performed inaccurately, but correct units were used most of the time.
- Graphs, if necessary, are drawn adequately.
- Students may or may not recognize the connections between their observations and the related chemistry concepts, but they don't express this understanding in the report.
- Errors in logic are made in the report.
- Answers to *Conclusions* items are incorrect or poorly written.
- Students did not make a meaningful effort to address any *Extensions* that were assigned.

Inexperienced level
(1 point)

- Many serious errors in technique were observed during the lab procedure.
- Data and observations are inaccurate or incomplete.
- *Analysis and Interpretation* was performed inaccurately, with no units or incorrect ones.
- Graphs, if necessary, are drawn incorrectly.
- Students clearly do not recognize the connections between their observations and the related chemistry concepts.
- Errors in logic are made throughout the report.
- Answers to *Conclusions* items are so incorrect that it is obvious the students did not understand the lab.
- Students made no effort at all to address any *Extensions* that were assigned.

Unacceptable level
(0 point)

- All work unacceptable.
- No responses are relevant to lab.
- Major components of lab are missing.

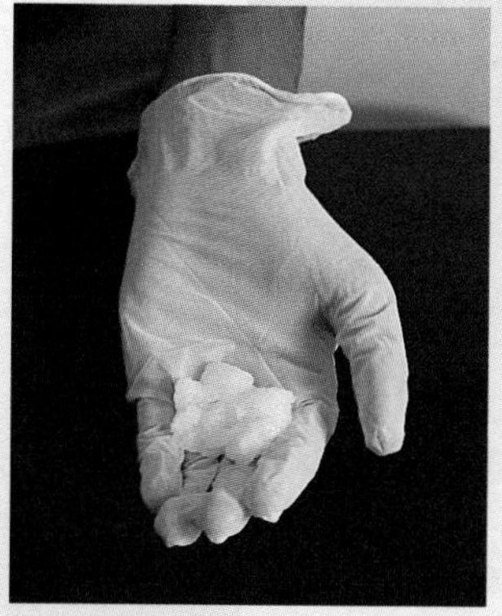

Scoring rubrics, continued

Investigation preliminary report

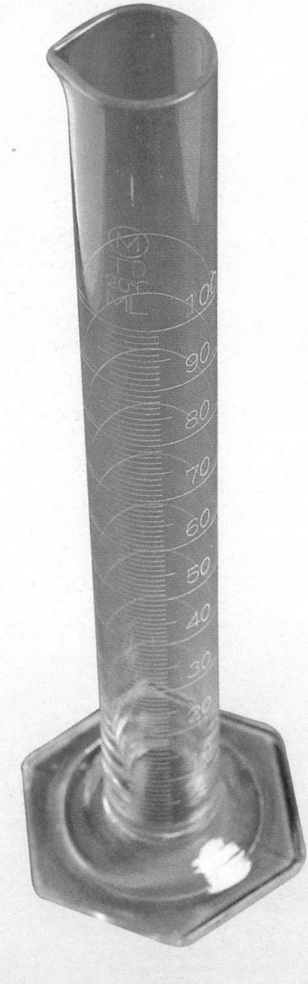

Experienced level
(6 points)
- Plan shows careful and thorough planning with good reasoning and logic.
- Plan is complete, appropriate, and safe.
- Plan is as efficient as the procedure given in the Teacher's Notes for the *Investigation*.
- Plan is expressed clearly and concisely.
- Proposed data tables are made properly and clearly indicate all measurements that must be made to solve the problem.
- Proposed list of equipment includes all equipment necessary to carry out the procedure described in the plan.
- When compared to the list given in the Teacher's Notes for the *Investigation*, the student's list of equipment includes no unnecessary pieces of equipment.
- Proposed budget is within $10 000 of the materials cost listed in the Teacher's Notes for the *Investigation*.
- Any other requirements listed in the memorandum for the *Investigation* are addressed completely, clearly, and concisely.

Competent level
(5 points)
- Plan shows careful and thorough planning, although the logic behind it may not be clearly expressed.
- Plan is appropriate, safe, and mostly complete, with only minor omissions.
- Plan is almost as efficient as the procedure given in the Teacher's Notes for the *Investigation*.
- Some parts of the plan could be expressed more clearly or more concisely.
- Proposed data tables indicate all measurements that must be made to solve the problem, but minor errors may have been made in preparing the data tables.
- Proposed list of equipment includes all equipment necessary for the procedure described in the plan.
- When compared to the list given in the Teacher's Notes for the *Investigation*, the student's list of equipment may include one or two unnecessary pieces of equipment.
- Proposed budget is within $20 000 of the materials cost listed in the Teacher's Notes for the *Investigation*.
- Any other requirements listed in the memorandum for the *Investigation* are completely addressed, but they could be more clear or more concise.

Intermediate level
(4 points)
- Plan shows some logic, but the reasoning could have been more careful, more thorough, or more clearly expressed.
- Plan is appropriate and safe, but there are a few omissions.
- Plan will work, but is not as efficient as the procedure given in the Teacher's Notes for the *Investigation*.
- Plan could be written more clearly or more concisely.
- Proposed data tables indicate all measurements that must be made to solve the problem, but multiple trials are not included.
- Proposed list of equipment includes almost all equipment necessary to carry out the plan described in the procedure, but there may be minor omissions.
- When compared to the list given in the Teacher's Notes for the *Investigation*, the student's list of equipment may include a few unnecessary pieces of equipment.
- Proposed budget is within $30 000 of the materials cost listed in the Teacher's Notes for the *Investigation*.
- Any other requirements listed in the memorandum for the *Investigation* are addressed, but not completely.

Transitional level
(3 points)
- Plan shows some logic, but not enough to completely solve the problem.
- Plan is safe, but it includes inappropriate procedures or omits necessary steps.
- Plan will probably not work and does not match the procedure given in the Teacher's Notes for the *Investigation*.
- Plan is poorly written.
- Proposed data tables include most of the necessary information, but errors have been made in preparing them.
- Proposed list of equipment omits a few pieces of equipment necessary for the procedure described in the plan.
- When compared to the list given in the Teacher's Notes for the *Investigation*, the student's list of equipment may include several unnecessary pieces of equipment.
- Proposed budget is within $50 000 of the materials cost listed in the Teacher's Notes for the *Investigation*.
- One of the other requirements listed in the memorandum for the *Investigation* is not addressed completely.

Beginning level
(2 points)
- Plan shows only a small amount of logic or understanding of what is necessary to solve the problem.
- Plan may not be completely safe.
- Plan will definitely not work and does not match the procedure given in the Teacher's Notes for the *Investigation*.
- Plan is poorly written.
- Proposed data tables include some necessary information, but there are some serious omissions.
- Proposed list of equipment has so many omissions that it would be difficult to carry out the procedure described in the plan.
- When compared to the list given in the Teacher's Notes for the *Investigation*, the student's list of equipment includes many unnecessary pieces of equipment.
- Proposed budget is within $70 000 of the materials cost listed in the Teacher's Notes for the *Investigation*.
- Few of the other requirements listed in the memorandum for the *Investigation* are addressed completely.

Inexperienced level
(1 point)
- Plan does not clearly show logic or understanding of the problem.
- Plan is unsafe.
- Plan will definitely not work and does not match the procedure given in the Teacher's Notes for the *Investigation*.
- Plan is poorly written.
- Proposed data tables include only a few of the necessary pieces of data.
- Proposed list of equipment has so many omissions that it would be impossible to carry out the procedure described in the plan.
- Not enough critical thinking has gone into the choice of equipment; most of the items on the materials list given in the *Investigation* are included, whether necessary or not.
- Proposed budget is within $100 000 of the materials cost listed in the Teacher's Notes for the *Investigation*.
- None of the other requirements listed in the memorandum for the *Investigation* are addressed completely.

Unacceptable level
(0 points)
- All work unacceptable.
- Major components of the preliminary report are missing.
- Plan is completely illogical, unsafe, or large portions are omitted.
- Plan is poorly written.
- Proposed data tables are entirely unsatisfactory.
- No thought has been given to the choice of equipment; all of the items on the materials list given in the *Investigation* are included.
- Proposed budget cost is more than $100 000 over or under the materials cost listed in the Teacher's Notes for the *Investigation*.

Scoring rubrics, continued

Investigation final report (letter and invoice)

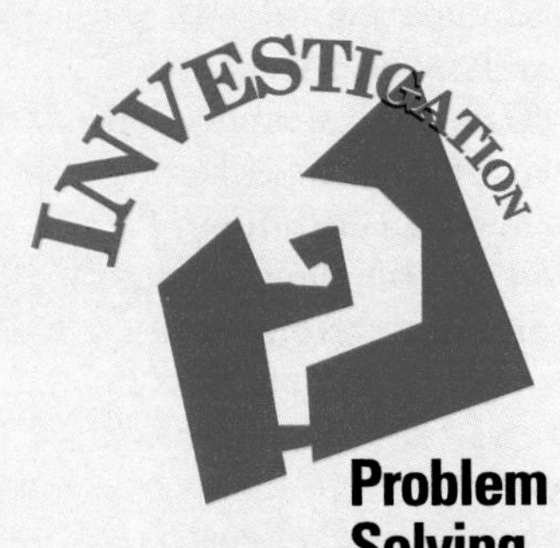

Problem Solving

Experienced level
(6 points)

- Excellent technique was used throughout the lab procedure.
- The report clearly expresses good reasoning, logic, and understanding of what the lab was about.
- The connections between the data and the related chemistry concepts that lead to the conclusions are clearly expressed.
- Students recognize the connection between the initial problem and the outcome of the lab work.
- Summary is clear, concise, and well organized, with few grammatical or stylistic errors.
- Data and analysis are accurate and well organized, with correct units and no errors in calculations.
- Percent error for quantitative answers is less than 15%.
- Graphs, if appropriate, are drawn and labeled accurately and neatly.
- Invoices are itemized to include labor costs as well as equipment prices, and total amounts will provide a profit for the firm.
- Any other requirements listed in the memorandum for the Investigation are addressed completely, clearly, and concisely.

Competent level
(5 points)

- No errors in technique were observed during the lab procedure.
- The report expresses basically sound reasoning, logic, and understanding of what the lab was about, although there may be some minor errors.
- The connections between the data and the related chemistry concepts that lead to the conclusions are expressed, but not clearly.
- Students recognize the connection between the initial problem and the outcome of the lab work.
- Summary is essentially clear, concise, and well organized, with no serious grammatical or stylistic errors.
- Data and analysis are accurate, with correct units and no errors in calculations, but they may not be well-organized.
- Percent error for quantitative answers is less than 25%.
- Graphs, if appropriate, are drawn and labeled accurately and neatly.
- Invoices are itemized, but not completely, and total amounts will provide a profit for the firm.
- Any other requirements listed in the memorandum for the *Investigation* are completely addressed, but they could be more clear or more concise.

Intermediate level
(4 points)

- Only minor errors in technique were observed during the lab procedure.
- The report expresses some logic and understanding of what the lab was about, although there are several minor errors.
- The connections between the data and the related chemistry concepts that lead to the conclusions are not clearly expressed, but students seem to understand them.
- Students recognize the connection between the initial problem and the outcome of the lab work.
- Summary is essentially correct, but it is poorly organized or contains serious grammatical and stylistic errors.
- Data and analysis are mostly accurate, but poorly organized, or have units that are incorrect, or some calculations that are not shown in full detail.
- Percent error for quantitative answers is less than 35%.
- Graphs, if appropriate, are drawn and labeled accurately but are not neat.
- Invoices are itemized, but not completely; total amounts indicate that the firm will break even.
- Any other requirements listed in the memorandum for the *Investigation* are addressed, but not completely.

Transitional level
(3 points)

- Only a few errors in technique were observed during the lab procedure, but they may have been significant.
- The report expresses some logic but little understanding of what the lab was about.
- The connections between the data and the related chemistry concepts that lead to the conclusions are not clearly expressed, and students do not seem to understand them completely.
- Students do not recognize the connection between the initial problem and the outcome of the lab work.
- Summary contains a few minor errors of fact along with grammatical and stylistic errors.
- Data and analysis are poorly organized or inaccurate because of minor errors, or calculations are not shown in full detail.
- Percent error for quantitative answers is less than 50%.
- Graphs, if appropriate, are drawn accurately but are either mislabeled or not neat.
- Invoices are not itemized; total amounts indicate that the firm will lose up to $10 000.
- One of the other requirements listed in the memorandum for the *Investigation* is not addressed completely.

Beginning level
(2 points)

- Several serious errors in technique were observed during the lab procedure.
- The report expresses little logic, and there seems to be little understanding of what the lab was about.
- The connections between the data and the related chemistry concepts that lead to the conclusions are clearly not understood.
- Students do not recognize the connection between the initial problem and the outcome of the lab work.
- Summary contains several errors of fact or is poorly written.
- Data and analysis contain a few significant errors or are poorly organized, or calculations are not shown at all.
- Percent error for quantitative answers is less than 65%.
- Graphs, if appropriate, are inaccurate or incomplete.
- Invoices are not itemized; total amounts indicate the firm will lose between $10 000 and $20 000.
- Few of the other requirements listed in the memorandum for the *Investigation* are addressed completely.

Inexperienced level
(1 point)

- Many serious errors in technique were observed during the lab procedure.
- The report expresses little logic, and there seems to be no understanding of what the lab was about.
- There is no attempt to connect the data and the related chemistry concepts to lead to a conclusion.
- Students do not recognize the connection between the initial problem and the outcome of the lab work.
- Summary contains many errors of fact or is poorly written.
- Data and analysis contain many significant errors or are poorly organized, or calculations are not shown at all.
- Percent error for quantitative answers is less than 80%.
- Graphs, if appropriate, are missing.
- Invoices are not itemized; total amounts indicate that the firm will lose between $20 000 and $30 000.
- None of the other requirements listed in the memorandum for the *Investigation* are addressed completely.

Unacceptable level
(0 points)

- Many serious errors in technique were observed during the lab procedure.
- All work is unacceptable.
- Few responses are relevant to the lab.
- Major components of the report are missing.
- The report expresses little logic, and there seems to be no understanding of what the lab was about.
- There is no attempt to connect the data and the related chemistry concepts to lead to a conclusion.
- Students do not recognize the connection between the initial problem and the outcome of the lab work.
- Summary contains many errors of fact or is poorly written.
- Data and analysis are incomplete and contain many significant errors.
- Percent error for quantitative data is more than 80%.
- Invoice is not included, or invoice indicates that the firm will lose more than $30 000.

JULIETTE BRAND FOODS
Product Development Division · 2994 Effington Lane · Thornton, NE 68181

January 9, 1995

Director of Research
CheMystery Labs, Inc.
52 Fulton Street
Springfield, VA 22150

Dear Director of Research:

Juliette Brand Foods is preparing to enter the rapidly expanding home popcorn market with a new popcorn product. As you may know, the key to making popcorn pop is the [illegible] within the kernel.

Laboratory safety

All serious accidents announce in advance that they will happen

The statistical proof of this assertion is summarized by what is called the "Heinrick Accident Triangle." The base of the triangle represents 100 000 known and unknown unsafe conditions. Statistically, these 100 000 unsafe conditions will cause 10 000 close calls, which make up the next level of the triangle. Relating this statistic to the scale of your school's labs, if you had three close calls last year in your lab, you actually had 30 unsafe conditions. Or, if you had three recordable accidents, like a burn, or a spill that caused someone to slip and fall, you probably had 30 unreported close calls and 300 unsafe conditions, some of which you did not know about and therefore did not correct.

The direct cause of most accidents is an error caused by doing something unsafe or disobeying a known safety principle. ("I knew I was supposed to read the label.") But the underlying, indirect, and fundamental cause of every accident is not recognizing the importance of the precaution that, if it had been applied, would have prevented the error of commission or omission. ("I didn't know that paying attention to the label was that important.")

Here is a list of several possible unrecognized causes of accidents and suggestions for accident prevention. Of course, the list is not complete; no such list can be. Use it to help you identify some of the known hazards in your laboratory. Once the potential causes of accidents are identified, you can then begin to work on their elimination.

Heinrick triangle	Type of occurrence	Hypothetical triangle for your lab
10	Number of fatalities	?
100	Number of permanent disablements	?
1000	Number of recordable incidents	3
10 000	Number of close calls	30
100 000	Number of unknown hazards	300

Accidents waiting to happen	Precautions for prevention
Attitudes about safety	
• There is an attitude that suggests precautions are optional if you are pressed for time.	• Practice zero-tolerance for unsafe conditions and practices. There is nothing that can be accomplished in your lab that is worth sacrificing someone's life or limb.
• There is no assessment of students' knowledge and attitudes regarding safety.	• Conduct frequent safety quizzes. Only students with perfect scores should be allowed to work in the lab.

Accidents waiting to happen

- Neither you nor your supervisors have established a comprehensive safety plan.
- The recommendations of the most recent safety inspection or audit have not been incorporated into the safety program.
- The most recent safety inspection or audit was conducted more than six months ago.
- You feel that things are safe enough already because you know that you and the other teachers will not do anything unsafe.

Precautions for prevention

- Include regular safety inspections and detailed recordkeeping in your safety plan.
- Implement improvements to the safety plan immediately. Such improvements can protect you, your life, your limb, and your liability, but only if you implement them.
- Conduct regular in-house safety and health inspections with an emphasis on improvement rather than guilt.
- Judge your safety considerations by whether they make the laboratory safe for everyone, including students, cleaning and building staff, and emergency workers.

Fires and other emergencies

Accidents waiting to happen

- Fire and other emergency drills are infrequent, and no records or measurements are made of the results of the drills.
- You have no idea what you'd do if an emergency happened in the middle of a lab exercise and the building had to be evacuated. You have never figured out how long it would take to secure the lab and any hazardous chemicals.
- Students are unsure what to do when the fire alarm or other alarm sounds.
- There are items stored in hallways along escape routes. Fire doors are propped open for ventilation.
- You know your escape route, but not any alternatives.
- The emergency phone numbers are kept only at the main switchboard.
- Fire extinguishers are placed at "dead ends" where they will not be in the way of people escaping.
- You have no idea how many fire extinguishers there are or when they were last charged. You've never used an extinguisher, and neither have the other teachers.
- Students believe that if a fire breaks out they should run for the nearest fire extinguisher and use it to put out the fire themselves, as quickly as possible.
- You and the other teachers nearby haven't had CPR or first-aid training in years.

Precautions for prevention

- Always carry out critical reviews of fire or other emergency drills. Don't wait until an emergency to find the flaws in your plans.
- Have actions preplanned in case of an emergency (e. g., establish what devices should be turned off, which escape route to use, where to meet personnel outside the building, who is designated to count people at that meeting place, who is authorized to reenter).
- Inform everyone in the lab about any alarms and what to do if they sound.
- Keep escape routes clear. Do not prop open fire doors or block them for any reason.
- Always plan alternate escape routes in case of unforeseen problems.
- Post current emergency phone numbers next to all phones.
- Place fire extinguishers near escape routes, so that they will be of use to those escaping.
- Regularly maintain fire extinguishers and train supervisory personnel in the proper use of extinguishers by extinguishing real fires.
- Instruct students (who should not use a fire extinguisher because they have not been trained in its use) that in case of a fire they should call a teacher (who should be trained).
- Get trained in CPR or first aid by your local chapter of the American Red Cross. Be sure to take frequent refresher courses.

Accidents waiting to happen	Precautions for prevention
Accident investigations	
• The investigation conducted after a serious accident was superficial and did not result in the implementation of safer procedures.	• Record who worked with what, when, and how long in order to allow meaningful retrospective studies of accidents.
• Accidents are only analyzed so that the person causing the accident can be disciplined.	• Analyze accidents to prevent repetition, not for any other reason.
Facilities and equipment	
• Eyewash fountains are present, but nobody knows anything about their specifications.	• Ensure that eyewash fountains and safety showers meet the requirements of the ANSI standard (Z358.1). See the Teacher's Notes for page 651 to be certain you know exactly how to use an eyewash fountain.
• Eyewash fountains are checked and cleaned only at the beginning of each school year.	• Flush eyewash fountains for 5 min every month to remove any bacteria or other organisms from pipes.
• No records are kept of routine checks and maintenance on the safety showers and eyewash fountains.	• Test safety showers (measure flow in gallons per minute) and eyewash fountains every six months and keep records of the test results.
• If the fan can be turned on, the lab hood is presumed to work well enough to keep everyone in the room safe.	• Ensure that laboratory ventilation and hood performance and use conforms to the requirements of the ANSI standard (Z358.1).
• Chronic contamination of breathing air is dealt with by keeping windows propped open at all times.	• Engage an industrial hygienist to conduct the appropriate measurements when contamination of breathing air is suspected. Keep records of any such measurements.
• Spills are handled on a case-by-case basis with whatever materials happen to be on hand.	• Have the appropriate equipment and materials available for spill control; replace them before their expiration dates.
• Labs are opened in the morning and locked after school is out.	• Lock all laboratory rooms whenever a teacher is not present.
• Compressed gas cylinders are carried by hand and set up in a corner of the laboratory.	• Secure all compressed gas cylinders when in use, and transport them secured on a hand truck.
• Equipment with moving parts is used without any precautions.	• All moving belts and pulleys should have safety guards.
• Electrical equipment or computers are used in the lab without special precautions.	• See the Teacher's Notes on page 652 for information on safety with electrical equipment.
Safety wear	
• Instead of goggles, you prefer to wear safety glasses with hard plastic lenses. There is no face shield present to use when you are required to prepare solutions from concentrated acids.	• Wear eye protection in the laboratory at all times; use safety goggles, ANSI Type G or H. When circumstances require, wear face, neck, and ear protection also (face shield, ANSI Type N). For details see the ANSI standard (Z87.1)
• Students or teachers who wear contact lenses use regular lab goggles.	• Require ANSI Type K cupped goggles for students or teachers who wear contact lenses.
• Gloves are used over and over again to save money. They are discarded only after they begin to fall apart.	• Check gloves routinely for pin holes, tears, or rips. Because gloves cannot resist penetration by a chemical after being handled for long periods of time, they should be replaced before that time period expires.

Accidents waiting to happen	Precautions for prevention
Safety wear	
• You wear old clothes in the lab instead of a laboratory coat or apron.	• Wear a laboratory coat or apron.
• You assume that your lab coat or apron offers the appropriate protection.	• Wear a long-sleeved shirt or blouse and slacks that extend to the ankles under the lab coat or apron.
• Your hair looks better when it is down, so you leave it that way when you work in the laboratory.	• Tie back or tuck in loose clothing, long hair, and dangling jewelry.
• You allow students to wear sandals in the laboratory.	• Do not allow any footwear in the lab that does not cover feet completely; no open-toed shoes.
• Observers are allowed into the lab but told to stay away from the lab benches.	• Keep a spare set of protective equipment on hand for visitors.
Work habits	
• New lab procedures are used with students without any planning or analysis.	• Analyze and try new lab procedures in advance to pinpoint potential hazards.
• You work alone during your preparation period to organize the day's labs.	• Never work alone in a science laboratory or storage area.
• Honor students do independent study alone in the laboratory.	• Never allow students to occupy a science laboratory unless a teacher is present.
• Although students do not eat or drink in the lab, you sometimes drink a soda or coffee in the storage area.	• Never eat, drink, smoke, or chew gum or tobacco in a science laboratory or storage area.
• You keep your lunch stored in the laboratory refrigerator because it's closer than the teacher's lounge.	• Do not store food or beverages in the laboratory environment.
• After work, you're in a hurry to get home, so you don't wash your hands.	• Always wash hands after work in a science laboratory and after spill cleanups.
• Sometimes you leave the room while you wait for a hot water bath to warm up.	• Never leave any heat sources (such as gas burners, hot plates, heating mantles, sand baths, etc.) unattended.
• The storeroom is so crowded that you decide to keep some apparatus on the lab benches.	• Do not store reagents or apparatus on lab benches, and keep lab shelves organized. Never place reactive chemicals (in bottles, beakers, flasks, wash bottles, etc.) near the edges of a lab bench.
• The windows are opened whenever volatile substances are used.	• Use a fume hood when working with volatile substances.
• When you work in the fume hood, you always put your head inside so you can see what you are doing.	• Never lean into the fume hood.
• Reagents are kept in the fume hood because you lack appropriate storage areas.	• Do not use the fume hood as a storage area.
• Students who have finished their lab work take off their goggles when they begin writing up their results.	• Make sure protection is used not only by the lab worker but also by anyone working nearby.
• You pipet by mouth because you know how it is done properly.	• Never pipet by mouth.
• No extra precautions are taken when handling liquid nitrogen.	• Tape all Dewar flasks.

Accidents waiting to happen	Precautions for prevention
Purchasing and using chemicals	
• You always prepare your solutions from concentrated stock to save money.	• Reduce risks by ordering diluted instead of concentrated substances.
• You purchase plenty of chemicals to be sure there is enough.	• Purchase chemicals in class-size quantities, if at all possible.
• You purchase several year's worth of chemicals to save money.	• Do not purchase or have on hand more than one-year's supply of each chemical. Dispose of or use up any chemicals leftover at the end of one year before the next year is over.
• When chemicals arrive, you unpack them and place them in your storage room without labeling them further.	• Label all chemicals accurately with the date of receipt, and write the initials of the person responsible for unpacking that chemical on the label.
• You are in a hurry to prepare solutions for a new lab, so you open the chemical and use it without reading the label.	• Never open a reagent package until the label has been read and completely understood.
• You never bother reading labels on chemicals because you've been using the chemicals for years and you already know all there is to know about them.	• Read each label to be sure it states the hazards and describes the precautions and first aid procedures (when appropriate) that apply to the contents in case someone else has to deal with the chemical in an emergency.
• You never bother reading the MSDS that come with your chemicals.	• Always read the Material Safety Data Sheet (MSDS) for a chemical before using it. Follow the precautions described in that Material Safety Data Sheet.
• You throw away the MSDS after you read them, because they just clutter up your storage area.	• File and organize the MSDS for each chemical in your lab where they can be found easily in case of an emergency.
• You put a copy of the warnings from the concentrated acid label on a bottle of the diluted acid.	• Label the diluted acids and bases with the hazards, precautions, and first aid procedures that apply specifically to them.
• Large reagent bottles of flammable chemicals are kept on the open shelves of the laboratory for days at a time.	• Store no more than one-day's supply of flammable liquids or solids on the open shelves of the laboratory. At the end of the day store any such unused chemical in a flammable liquid storage cabinet.
• Bottles of chemicals are kept unused on shelves in the lab during the semester. The main stockroom contains chemicals that haven't been used for many years.	• Do not leave bottles of chemicals unused on shelves in the lab for more than one week or unused in the main stockroom for more than one year.
• No extra precautions are taken when flammable liquids are dispensed from their containers.	• When transferring flammable liquids from bulk containers, ground the container, and before transferring to a smaller metal container, ground both containers.

Accidents waiting to happen	Precautions for prevention
Chemical storage	
• Students are told to put their broken glass and solid chemical wastes in the trash can.	• Have separate containers for trash, for broken glass, and for different categories of hazardous chemical wastes.
• Students retrieve chemicals and apparatus from stockrooms or storage areas.	• Lock all storage spaces; only permit teachers to enter.
• You're pretty sure you know what chemicals you have off the top of your head.	• Keep an accurate and up-to-date inventory of chemicals on hand.
• For ease of retrieval, you store your chemicals arranged in alphabetical order.	• Arrange chemicals by hazard class when storing them. (Alphabetical storage leads inevitably leads to adjacent positioning of incompatibles.)
• You're not sure what chemicals are incompatible with what other chemicals.	• Use MSDSs to determine which chemicals are incompatible.
• You keep mutually reactive chemicals stored by hazard class, but flammable liquids and solids are stored on the shelves of the storage area.	• Keep incompatible classes of mutually reactive chemicals (e.g., acids and bases, oxidizers and reducers, nitric acid and glacial acetic acid) separate from each other in the storage area. Store bulk quantities of flammable liquids and solids in a flammable materials storage cabinet or in a separate room specifically designed and designated to be used only for such storage.
• Corrosives are kept above eye level, out of reach from anyone who is not authorized to be in the storeroom.	• Always store corrosive chemicals on shelves below eye level.
• Chemicals are kept on the floor of the stockroom on the days that they will be used so that they are easy to find.	• Never store chemicals or other materials on floors or in the aisles of the laboratory or storeroom, even for a few minutes.
• Chemicals are stored permanently on laboratory shelves because the storage room is too crowded.	• Return chemicals on the laboratory shelves to storage as soon as they are no longer needed.
• Because of a budget shortfall, stacked boxes are used instead of shelves.	• Equip chemical storage shelves with lips; never use stacked boxes in lieu of shelves.
• A second-hand refrigerator is used for lab storage.	• Use only a so-called explosion-proof refrigerator for lab storage.
• Chemicals are stored without consideration of possible emergencies (fire, earthquake, flood, etc.) that could compound the hazards of the chemicals.	• Store chemicals that are incompatible with common fire-fighting media like water (such as alkali metals) or carbon dioxide (such as alkali and alkaline-earth metals) under conditions that eliminate the possibility of a reaction with water or carbon dioxide if it is necessary to fight a fire in the storage area.
• Batteries are stored without any extra precautions.	• Cover both terminals of dry cells and rechargeable batteries with insulating tape when storing them.

Prepared by Jay A. Young,
Consultant, Chemical Health and Safety,
Silver Spring, MD

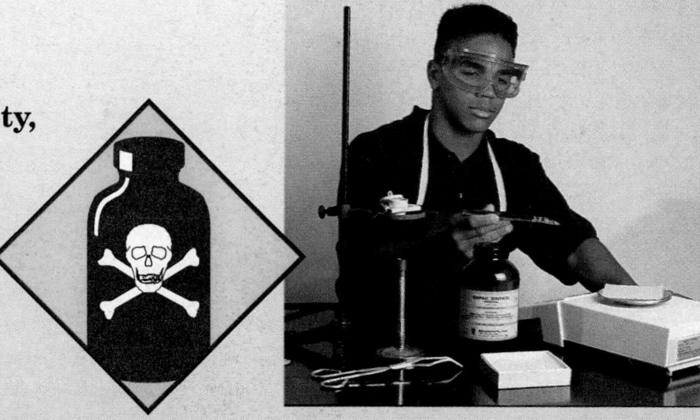

References

1. Budavari, Susan, ed.
The Merck Index, 11th ed.
Merck and Co., Rahway, NJ (1989)
Use this reference for reliable general information about a chemical instead of using a less reliable chemical dictionary or one of the many books with the words "Dangerous Properties" as part of the title.

2. Council Committee of Chemical Safety
Safety in Academic Chemistry Laboratories, 5th ed.
American Chemical Society, Washington, DC (1995).
This is the authority if you wish a brief treatment. Single copies are free. For a more extensive treatment, see Young, below.

3. "Fire Protection for Laboratories Using Chemicals" (also known as "NFPA-45")
National Fire Protection Association
Batterymarch Park, Quincy, MA (current edition).
This is the national safety code for laboratory fire protection and prevention.

4. Gerlovich, J. A., et al.
School Science Safety
Flinn Scientific, Inc., Batavia, IL (1984)
This is a practical guide in two volumes for teachers.

5. Pipitone, D. A. and D. Hedberg
"Safe Chemical Storage"
Journal of Chemical Education, 59 (1982), A159
A discussion on the proper storage of chemicals. (Also see Pipitone, D. A.
Safe Storage of Laboratory Chemicals, 2nd ed.
Wiley-Interscience, New York, NY (1991)
and the current edition of the *Flinn Chemical Catalog and Reference Manual*, Flinn Scientific, Inc., Batavia, IL.)

6. "Practice for Occupational and Educational Eye and Face Protection, Z87.1,"
"Emergency Eyewash and Shower Equipment, Z358.1,"
"American National Standard for Laboratory Ventilation, Z9.5,"
American National Standards Institute, New York, NY (current editions).
These standards are recognized worldwide as reliable. Safety goggles and face shields that meet Z87.1 requirements are marked Z87.

7. Reese, K. M.
Health and Safety Guidelines for Chemistry Teachers
American Chemical Society, Washington, DC (1980).
This source contains several useful tips on enhancing safety in your laboratory.

8. *Standard First Aid and Personal Safety*
American Red Cross
Your local Red Cross chapter probably gives a short course in first aid. This book summarizes the content of that course.

9. Young, Jay A.
"Risk Assessment and Hazard Evaluation for Undergraduate Laboratory Experiments"
Journal of Chemical Education, 59 (1982) A265.
This provides some suggestions on teaching your students how to practice safety in the lab.

10. Young, Jay A., ed.
Improving Safety in the Chemical Laboratory: A Practical Guide, 2nd ed.
Wiley-Interscience, New York, NY (1991)
The details of this topic are fully discussed.

Laboratory Program Supplies

Chemicals

This list shows the chemicals needed for 15 lab groups to perform all of the in-text laboratory exercises, both *Explorations* and *Investigations*. All these items will be consumed during the course of the laboratory, with the exception of **items indicated with asterisks, which are reusable.**

For solutions, the list includes descriptions of the concentrations and volumes required for the laboratory exercise as well as the amount of pure or concentrated reagent that should be ordered. Many chemical suppliers provide a variety of ready-made solutions. Ordering chemicals already diluted may be slightly more expensive, but it is much safer. A list of suppliers is provided on page T51.

Item	Order quantity (for 15 lab groups)	Lab(s)
Acetic acid solution, 5% (vinegar)	1.65 L	Exp. 6A, Inv. 6A
Acetic acid, 1.0 M, 750 mL; 0.1 M, 1.5 L	73.9 mL glacial acetic acid	Exp. 8B, 13B, Inv. 13B
Alpha radiation point source	15	Inv. 16A
Ammonia water, 1.0 M, 75 mL; 0.1 M, 750 mL	10.2 mL conc. $NH_3 \cdot H_2O$ (NH_4OH)	Inv. 10A, 13B
Benedict's solution	495 mL	Exp. 9B, Inv. 9B
Biuret reagent	135 mL	Exp. 9B, Inv. 9B
$CaCl_2$ crystals, 0.5 M, 1.1 L	217.8 g anhydrous $CaCl_2$	Exp. 3A, Inv. 8A, 11B, 12A
$Ca(NO_3)_2$, 0.2 M, 375 mL	17.7 g $Ca(NO_3)_2 \cdot 4H_2O$	Inv. 7A
Calcium metal, small pieces	15 g	Inv. 7A
Copper metal, small pieces	15 g	Inv. 7A
CR-39 radiation-detecting plastic†	30 pieces†	Exp. 16A, Inv. 16A
$CuCl_2 \cdot 2H_2O$	37.9 g	Inv. 11A
$Cu(NO_3)_2$, 0.2 M, 750 mL	181.2 g $Cu(NO_3)_2 \cdot 3H_2O$	Inv. 2B, 7A
$CuSO_4 \cdot 5H_2O$	75 g*	Exp. 5A
Distilled water	as needed	throughout
Ethanol solution, 50%	45 mL	Exp. 6A
$FeCl_3 \cdot 6H_2O$	535.1 g	Exp. 11A

†CR-39 plastic is available from Alpha Trak, 141 Northridge Drive, Centralia, WA 98531, (206) 736-3884

Item	Order quantity (for 15 lab groups)	Lab(s)
$Fe(NO_3)_3$, 0.2 M, 975 mL	78.8 g $Fe(NO_3)_3 \cdot 9H_2O$	Exp. 12B, Inv. 7A, 12B
$FeSO_4$, 0.15 M, 750 mL; 0.25 M, 750 mL	83.4 g $FeSO_4 \cdot 7H_2O$	Exp. 15A, Inv. 15A
Filter paper	120 pieces	Exp. 1B, 2B, 8A, Inv. 1B, 2B, 8A
Glucose	300.0 g	Inv. 12A
Gypsum	75 g $CaSO_4 \cdot 2H_2O$	Inv. 5A
Hydrochloric acid, 1.0 M, 9.7 L; 0.5 M, 1.5 L; 0.2 M, 750 mL; 0.1 M, 150 mL	898.1 mL conc. HCl	throughout
Ice	45 L	Exp. 2B, Inv. 2B
Iron metal, small pieces	15 g	Inv. 7A
Isopropanol	675 mL	Exp. 9B, Inv. 9B
KIO_3	0.32 g	Exp. 14A
$KMnO_4$ 0.0200 M, 2.25 L	7.11 g $KMnO_4$	Exp. 15A, Inv. 15A
KNO_3	480.0 g	Exp. 2B, Inv. 2B
KSCN, 0.00200 M, 750 mL	0.146 g KSCN	Exp. 12B, Inv. 12B
K_2SO_4, 0.5 M, 75 mL	6.54 g K_2SO_4	Exp. 3A
Li_2SO_4, 0.5 M, 150 mL	9.6 g $Li_2SO_4 \cdot H_2O$	Exp. 3A, Inv. 3A
Latex, liquid‡	1.65 L‡	Exp. 6A, Inv. 6A

‡Liquid latex may be purchased from Flinn Scientific, Inc., Batavia, IL.

Catalog No.	Quantity
L0004	500 mL
L0110	1 L
L0111	4 L

Item	Order quantity (for 15 lab groups)	Lab(s)
Lugol's iodine solution	135 mL	Exp. 9B, Inv. 9B
Magnesium metal, small pieces	15 g	Inv. 7A
Metal shot (Al or Cu, etc.)	300.0 g*	Exp. 2A
$MgCl_2$, 0.5 M, 750 mL	76.2 g $MgCl_2 \cdot 6H_2O$	Inv. 11B
$Mg(NO_3)_2$, 0.2 M, 375 mL	19.2 g $Mg(NO_3)_2 \cdot 6H_2O$	Inv. 7A
$MgSO_4 \cdot 7H_2O$	250.0 g	Exp. 12A
$NaC_2H_3O_2$, 0.1 M, 1.5 L	12.3 g $NaC_2H_3O_2$	Exp. 13B, Inv. 13B
NaCl	775 g	Exp. 1A, 2B, 3A, 12A, Inv. 12A
$NaHCO_3$	345 g	Exp. 8B, Inv. 10B
NaOH pellets, 150 g 6.25 M (Teacher use) 300 mL 1.0 M, 1.95 L 0.5 M,750 mL 0.1 M, 1.50 L	325 g NaOH pellets	throughout
Na_2CO_3, 0.5 M, 1.3 L	71.5 g anhydrous Na_2CO_3	Exp. 8A, Inv. 8A
$Na_2S_2O_5$	0.05 g	Exp. 14A
Na_2SO_4, 0.5 M, 75 mL	5.3 g Na_2SO_4	Exp. 3A
NH_4Cl, 0.1 M, 750 mL	4.0 g NH_4Cl	Inv. 13B
Nitric acid, 0.6 M, 750 mL	28.5 mL conc. HNO_3	Exp. 12B, Inv. 12B
pH paper, narrow-range (needed unless students will use a pH meter or LEAP System with pH probe)	2 rolls, 15 ft, pH 3.0–5.5 2 rolls, 15 ft, pH 6.0–8.0	Exp. 1B, 13B, Inv. 1B, 13A, 13B
Phenolphthalein solution, 30 mL	0.3 g phenolphthalein	Exp. 13A, Inv. 13A
Plaster of Paris	75 g $CaSO_4 \cdot \frac{1}{2}H_2O$	Inv. 5A
Rock salt	600.0 g	Exp. 2B, Inv. 2B
$SnCl_4$, 0.2 M, 375 mL	26.3 g $SnCl_4 \cdot 5H_2O$	Inv. 7A
Sodium silicate solution	180 mL water glass	Exp. 6A
$SrCl_2$, 0.5 M, 75 mL 0.3 M, 750 mL	71.0 g $SrCl_2 \cdot 6H_2O$	Exp. 3A, 8A
Starch, soluble	3.0 g	Exp. 9B, 14A, Inv. 9B
Sucrose	250.0 g	Exp. 12A
Sulfuric acid, 1.0 M, 150 mL	8.4 mL conc. H_2SO_4	Exp. 15A, Inv. 15A
Tin metal, small pieces	15 g	Inv. 7A
Weighing paper	1 package, 500 sheets	throughout
Zinc metal, mossy, small pieces	45 g	Inv. 7A, 8B
$Zn(NO_3)_2$, 0.2 M, 375 mL	22.3 g $Zn(NO_3)_2 \cdot 6H_2O$	Inv. 7A

Equipment

This list shows the equipment and laboratory apparatus needed for 15 lab groups to perform all of the in-text laboratory exercises, both *Explorations* and *Investigations*. All these items will be reusable.

In some cases, individual *Explorations* or *Investigations* can be performed more rapidly if there is more glassware available and students do not need to stop to clean it until the entire laboratory exercise is completed.

Item	Order quantity (for 15 lab groups)	Lab(s)
Balance, centigram	3 or more	throughout
Beaker, 50 mL	60	Exp. 1B, 13B, Inv. 1B, 9A, 13B
Beaker, 100 mL	90	Exp. 1B, 13A
Beaker, 150 mL	60	Exp. 2B, Inv. 2B
Beaker, 250 mL	135	throughout
Beaker, 400 mL	30	Exp. 13B, 15A, 16A, Inv. 13B, 15A, 16A
Beaker, 2 L (or plastic tub)	15	Exp. 6A, Inv. 6A
Bunsen burner (hot plate may be substituted for all uses except Exp. 3A and Inv. 3A)	15	throughout

Item	Order quantity (for 15 lab groups)	Lab(s)
Buret	30	Exp. 1B, 13A, 15A, Inv. 1B, 13A, 15A
Clamp, buret (double)	15	Exp. 1B, 13A, 15A, Inv. 1B, 13A, 15A
Clamp, thermometer	15	Exp. 12A, Inv. 12A
Cobalt glass plate	15	Exp. 3A, Inv. 3A
Crucible and cover	30	Exp. 5A, Inv. 5A
Desiccator (To make your own, see Teacher's Notes for Exp. 5A.)	15	Exp. 5A, Inv. 5A
Dropper bottle (or micropipet)	60	Exp. 5A, 8B, 14A, Inv. 9B
Drying oven	1 or more	Exp. 1B, 8A, 13A, Inv. 1B, 8A, 8B
Erlenmeyer flask, 125 mL	60	Exp. 12A, 13A, 15A, Inv. 15A
Erlenmeyer flask, 250 mL	15	Exp. 11A, Inv. 12A, 13A
Evaporating dish	15	Exp. 1A, 8B
Flame test wire	75 cm	Exp. 3A, Inv. 3A
Forceps	15	Exp. 13A
Glass plate (either 7 cm × 15 cm plate, or micro-chemistry plate with wells)	15	Exp. 3A, Inv. 3A
Glass stirring rod	15	throughout
Graduated cylinder, 10 mL	15	throughout
Graduated cylinder, 25 mL	15	Exp. 2A, 6A, Inv. 6A
Graduated cylinder, 100 mL	15	throughout
Hot plate (burner may be substituted for all uses except Exp. 9B and Inv. 9B)	15	throughout
Microscale reaction strip, 8-well	30	Exp. 14A

Item	Order quantity (for 15 lab groups)	Lab(s)
Microscope	3 or more	Exp. 16A, Inv. 16A
Mortar and pestle	15	Exp. 1B, 13A, Inv. 1B, 5A, 14A
Petri dish	15	Inv. 11B
Pipe-stem triangle	15	Exp. 5A, 8A, Inv. 5A, 8A
Pipet (see dropper bottle)		
Ring	15	throughout
Ring stand	15	throughout
Rubber tubing for gas burners	15 pieces	throughout
Rubber policeman	15	Exp. 1B, 2B, Inv. 1B, 2B
Spatula	15	throughout
Stopper, one-hole, rubber, for flask	30	Exp. 1B, 2B, 8A, Inv. 1B, 2B, 8A, 10B
Striker	15	throughout
Test-tube holder	15	Exp. 9B, Inv. 7A, 9B, 11A, 11B, 14A
Test-tube rack	15	throughout
Test tube	135	throughout
Thermometer, Celsius, nonmercury	15	throughout
Tongs, beaker	15	throughout
Tongs, crucible	15	Exp. 1A, 3A, 5A, Inv. 3A, 5A
Tray, tub, or pneumatic trough	15	Exp. 2B, 6A, Inv. 2B
Volumetric flask, 250 mL	15	Inv. 11A
Wash bottle	15	throughout
Watch glass	15	Exp. 8B, 9A, Inv. 9A
Wire gauze	15	Exp. 2B, 12A, Inv. 1A, 2B, 8B,

Miscellaneous materials

The laboratory experiences in *Holt Chemistry: Visualizing Matter* emphasize more use of consumer products than do many traditional lab manuals. This list shows the miscellaneous materials needed for 15 lab groups to perform all of the in-text laboratory exercises, both *Explorations* and *Investigations*.† All these items will be consumed during the course of the laboratory, with the exception of **items indicated with asterisks, which are reusable.**

Item	Order quantity (for 15 lab groups)	Lab(s)
Aluminum foil	7.5 ft from a roll	Inv. 1A
Aspirin substitute tablet (acetaminophen)	90	Exp. 1B
Aspirin tablet	120	Exp. 1B, Inv. 1B
Baby food		Inv. 9B
chicken	1 jar	
grape juice, white	1 jar	
fruit (apple) with yogurt	2 jars	
Bags, plastic, resealable	30	Inv. 2B
Brown wrapping paper	1 roll or several bags	Exp. 9B, Inv. 9B
Buffered aspirin tablet	90	Exp. 1B
Can with reclosable lid	15	Inv. 6A
Chemical handbook or other reference source	1 or more*	Exp. 2A
Cooking oil	100 mL	Inv. 1A, 9B
Copper wire (18 gauge recommended)	1.5 m*	Exp. 1A
Corn syrup	105 mL	Exp. 9B, Inv. 9B
Cotton swab	30	Inv. 10A
Cup, plastic foam	15*	Exp. 9A
Duct tape	1.5 m	Inv. 10B
Effervescent tablet	270 (90 each of 3 brands)	Inv. 14A
Eggshell	15	Exp. 13A
Gelatin	1 package	Exp. 9B, Inv. 9B
Glue	1 bottle	Inv. 10B
Index card, 3 in × 5 in	30	Exp. 16A, Inv. 16A
Key chain ring (for etch clamp)	15*	Exp. 16A, Inv. 16A
Paper clip	1 box of 100	Exp. 16A, Inv. 6A, 16A
Paper cup, small	45	Exp. 6A, Inv. 6A
Paper cup, medium	15	Inv. 6A

†Note: In addition to the materials listed above, students may request other items for some of the open-ended *Investigations,* especially Inv. 6A and Inv. 10B.

Item	Order quantity (for 15 lab groups)	Lab(s)
Paper cup, large	15	Inv. 6A
Paper towel	1 roll	throughout
Pennies, 150	10 each for every year from 1979 to 1993*	Inv. 2A
Pepper	1 box	Inv. 11B
Plastic cup, small (with lid), or film canister	15	Exp. 16A, Inv. 16A
Plastic soda bottle, 2 L	30	Inv. 10B
Popcorn, about 3600 kernels	1 lb bag	Inv. 1A
Pushpin	15*	Exp. 16A, Inv. 16A
Ruler, metric, clear plastic	15*	Exp. 2A, 16A, Inv. 10A, 16A
Scissors	15*	Exp. 16A, Inv. 10A, 10B, 16A
Shoe box with lid	15	Inv. 10B
Shortening (or vegetable oil)	75 mL	Exp. 9B
Soap samples, 225 mL each (use three different brands each of hand soap, dish soap, and laundry detergent)	2.025 L	Inv. 11B
Stopwatch (or clock with second hand)	15*	Exp. 13B, 14A, Inv. 10A, 13B, 14A
Straw, bendable plastic	300	Inv. 10B
Tape	roll	Exp. 16A, Inv. 16A
Tea candle	120	Inv. 10B
Toilet paper (or tissue)	30 squares	Exp. 16A, Inv. 16A
Vitamin C tablet	7	Exp. 9B, Inv. 9B
Wooden stick	15	Exp. 6A

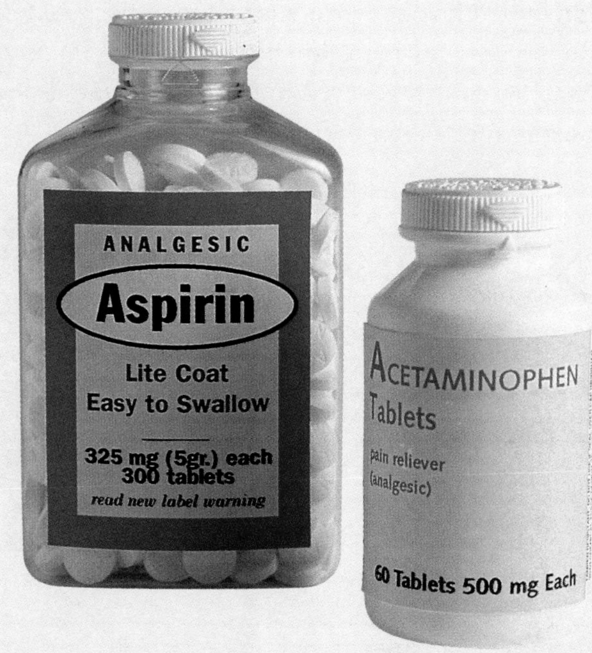

Optional materials and equipment

To help them fit your situation and resources, the laboratory experiences in *Holt Chemistry: Visualizing Matter* also contain alternative procedures for different types of equipment, including the LEAP System microcomputer-based laboratory probes.

In addition, the open-ended nature of Investigations 6A and 10B allow for students to request common items to use in testing polymers and in building a model of a CO_2 fire-extinguishing system.

Item	Order quantity (for 15 lab groups)	Lab(s)
Aspirator for spigot (vacuum filtration)	15	Exp. 1B, 2B, 8A, Inv. 1B, 2B, 8A
Büchner funnel	15	Exp. 1B, 2B, 8A, Inv. 1B, 2B, 8A
Calorimeter	15	Exp. 9A
Cuvettes	15 or more	Exp. 11A, 12B, Inv. 11A, 12B
Glass funnel	15	Exp. 1B, 2B, 8A, Inv. 1B, 2B, 8A
LEAP System with colormeter†	1	Exp. 11A, 12B, Inv. 11A, 12B
LEAP System with pH probe†	1	Exp. 1B, 13B, Inv. 1B, 11B, 13A, 13B
LEAP System with thermistor probe†	1	Exp. 2A, 2B, 9A, 12A, Inv. 2B, 9A, 12A, 14A
Marbles	1 box	Inv. 6A
pH meter	1 or more	Exp. 1B, 13B, Inv. 1B, 11B, 13A, 13B

Item	Order quantity (for 15 lab groups)	Lab(s)
Radiation source (Fiesta-ware, old glow-in-the-dark watch faces, Coleman green-label lan-tern mantles, cloud chamber needles)	15	Exp. 16A
Rubber bands	1 bag	Inv. 6A
Spectrophotometer	1 or more	Exp. 11A, 12B, Inv. 11A, 12B
Spectroscope	1 or more	Exp. 3A, Inv. 3A
String	1 roll	Inv. 6A
Test-tube clamp	15	Exp. 1B, Inv. 1B
Toothpicks	1 box	Inv. 6A
Vacuum flask (sidearm flask) and tubing	15	Exp. 1B, 2B, 8A, Inv. 1B, 2B, 8A
Wing top for burner	15	Exp. 5A
Wipes, lint-free	1 box	Exp. 11A, 12B, Inv. 11A, 12B
Wooden splints	90	Exp. 3A, Inv. 3A

†To order the LEAP System and probes, contact Quantum Technology. Quantum has set up a variety of packages for users of *Holt Chemistry: Visualizing Matter*.

Quantum Technology
30153 Arena Drive
Evergreen, CO 80439

Sales and Technical Support: (303) 674-9651
Fax: (303) 674-6763

The role of the *Explorations* in the *Holt Chemistry: Visualizing Matter* lab program is discussed further on page T643-3 of the Teacher's Edition.

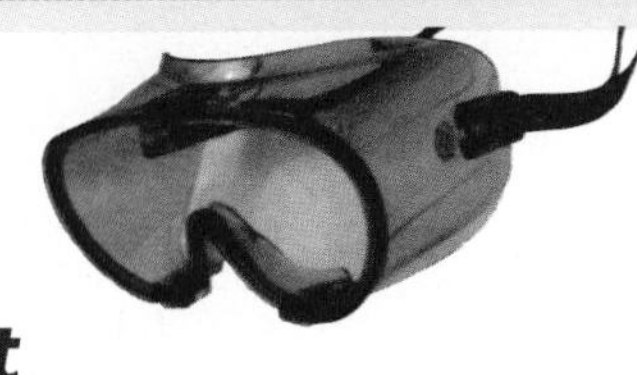

Working in the World of a Chemist

Meeting today's challenges

Even though you have already taken science classes with lab work, you will find the two types of laboratory experiments in this book organized differently from those you have done before. The first type of lab is called an *Exploration*, and it helps you gain skills in lab techniques that you will use to solve a real problem presented in the second type of lab, which is called an *Investigation*. The *Exploration* serves as a *Technique Builder*, and the *Investigation* is presented as an exercise in *Problem Solving*.

Both types of labs refer to you as an employee of a professional company, and your teacher has the role of supervisor. Lab situations are given for real-life circumstances to show how chemistry fits into the world outside of the classroom. This will give you valuable practice with real-world skills that you can use in chemistry and in other careers, such as creating a plan with available resources, developing and following a budget, and writing business letters.

As you work on these labs, you will better understand how the concepts you studied in the chapters are used by chemists to solve problems that affect life for everyone.

Explorations

The *Explorations* provide step-by-step procedures for you to follow, encouraging you to make careful observations and interpretations as you progress through the lab session. Each *Exploration* serves as a *Technique Builder* that gives you an opportunity to practice and perfect a specific lab technique or concept that will be needed later in an *Investigation*. The *Explorations* have the following sections.

- *Objectives:* what you will be expected to accomplish as you complete the Exploration.
- *Situation:* the setting in which the *Exploration* takes place
- *Background:* some of the chemistry information you will needed to work through the *Exploration*
- *Problem:* an overview of what you must do to meet the requirements of the *Situation*
- *Safety:* warnings and information on how to be safe during the *Exploration*
- *Materials:* items you will need to do the lab work
- *Preparation:* things to prepare before starting the procedure
- *Technique:* step-by-step explanations to follow for doing the lab work
- *Cleanup and Disposal:* specific guidelines for ending the lab session
- *Analysis and Interpretation:* items that will help you make sense of the data you collected and the observations you made
- *Conclusions:* items that will help you understand how your results relate to the key chemistry concepts involved in the procedure
- *Extensions:* challenging items that involve further calculations, research, or planning, many of which are good ideas for science fair projects

A scoring rubric for evaluating *Exploration* lab reports is given on page T643-15 of the Teacher's Edition.

What you should do before an Exploration

Preparation will help you work safely and efficiently. The evening before a lab, be sure to do the following.

- **Read the lab procedure** to make sure you understand what you will do.
- **Read the safety information** that begins on page 651, as well as that provided in the lab procedure.
- **Write down any questions** you have in your lab notebook so that you can ask your teacher about them before the lab begins.
- **Prepare all necessary data tables** so that you will be able to concentrate on your work when you are in the lab.

What you should do after an Exploration

Most teachers require a lab report as a way of making sure that you understood what you were doing. Your teacher will give you specific details about how to organize the lab report, but most lab reports will include the following. A sample lab report you can use as a model is shown below.

- **title** for the lab
- **summary paragraph(s)** describing the purpose and procedure
- **data tables and observations** that are organized and comprehensive
- **worked-out calculations** with proper units
- **answers**, boxed, circled, or highlighted, for items in the *Analysis and Interpretation*, *Conclusions*, and *Extensions* sections

Tamara Kieklak
Fifth Period
March 31, 1994

Lab Report

Title Exploration 12A— Boiling Point Elevation and Molar Mass

Summary *If a solute is dissolved in water, it will change the boiling point of water. First, I determined the boiling point of pure water, and then measured the boiling point with 10.0 g of each of the three solutes. The solutes used were NaCl, sucrose, and $MgSO_4$. Then I tested an unknown substance in the same manner. I compared the results to see what the unknown substance was.*

Data Table

solution	Boiling Point
H_2O	100.0°C
NaCl	103.5°C
Sucrose	100.5°C
$MgSO_4$	101.8°C
Unknown	102.0°C

Analysis and Interpretation

1. Δt_b *for* NaCl = 103.5°C − 100.0°C = 3.5°C
   Δt_b *for* Sucrose = 100.3°C − 100.0°C = 0.3°C
   Δt_b $MgSO_4$ = 101.8°C − 100.0°C = 1.8°C
   Δt_b Unknown = 101.9°C − 100.0°C = 1.9°C

2. The data for unknown most closely matches that for $MgSO_4$.

3. $m = \frac{\Delta t_b}{K_b}$

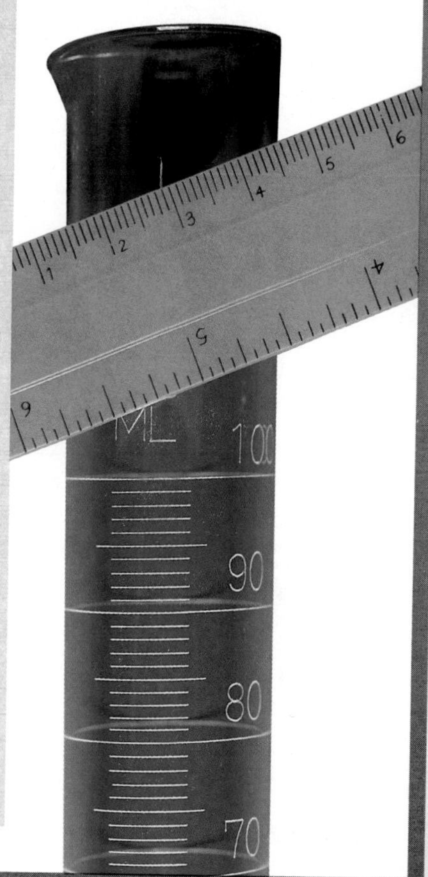

The role of the Investigations in the *Holt Chemistry: Visualizing Matter* lab program is discussed further on page T643-4 of the Teacher's Edition.

A scoring Rubric for evaluating preliminary reports for *Investigations* is given on page T643-17 of the Teacher's Edition.

Investigations

The *Investigations* may seem quite different because they do not provide step-by-step instructions. The *Investigations* require you to develop your own procedure to solve a problem presented to your company by a client. You must decide how much money to spend on the project and what equipment to use. Although this may seem very difficult, the *Investigations* contain a number of clues about what to do to be successful in *Problem Solving*.

What you should do before an Investigation

Before you will be allowed to work on the lab, you must turn in a preliminary report. Usually, you must describe in detail the procedure you plan to use, provide complete data tables for the data and observations you will collect, and list exactly what equipment you will need and the costs. Only after your teacher, acting as your supervisor, approves your plans are you allowed to proceed. A sample preliminary report is shown below. Before you begin writing a preliminary report, follow these steps.

- **Read the *Investigation* thoroughly,** searching for clues.
- **Jot down notes** in your lab notebook as you find clues.

Stacy Rivera
Third Period
March 22, 1994

Title Boiling-Point Elevation — Contamination

Summary A sample of powdered sugar contaminated by another substance will be analyzed to determine the identity of the contaminant. Because the contaminant is either glucose, salt, or calcium chloride, the procedure can be a simple boiling-point elevation test. As a control for the procedure, the boiling point of pure water is determined by heating it with a Bunsen burner and measuring the temperature at boiling with a thermometer. Then the boiling point with 10.0 g of each of the possible contaminants is measured. Once the data is obtained, the same test for the boiling-point elevation of the unknown substance can be used to identify the contaminant.

Data Table

Solution	Boiling Point
H_2O	
Glucose	
Salt	
$CaCl_2$	
Unknown	

Materials

Lab space/fume hood/utilities	\$15,000
Disposal fee for 40 g of solutes	\$80,000
Beaker tongs	\$1,000
$CaCl_2$, 10 g	\$5,000
Glucose, 10 g	\$5,000
NaCl, 10 g	\$5,000
250 mL flask	\$1,000
100 mL graduated cylinder	\$1,000
Balance	\$5,000

A scoring rubric for evaluating letters and invoices from *Investigations* is given on page T643-19 of the Teacher's Edition.

- **Consider what you must measure or observe** to solve the problem.
- **Think about *Explorations*** you have done that used a similar technique or reaction.
- **Imagine working through a procedure,** keeping track of each step and what equipment you will need.
- **Carefully consider** whether your approach is the best, most efficient one.

What you should do after an Investigation

After you finish, organize a report of your data as described in the *Memorandum*. This is usually in the form of a one- or two-page letter to the client. (Your teacher may have additional requirements for your report.) Carefully consider how to convey the information the client needs to know. In some cases, a graph or diagram can communicate information better than words can. As a part of your report, you must include an invoice for the client that explains how much they owe and how much you charged for each part of the procedure. Remember to include the cost of your work in the analysis. A sample that you can use as a model is shown below.

Kristeen McGonigal
Director of Operations
D & J Food Processing

Dear Ms. McGonigal,

The analysis has been completed on the contaminant that was accidentally included in your batch of powdered sugar. We are pleased to report that the contaminant is glucose, and you will still be able to use the processed batch.

The procedure we used tested the boiling-point elevation of 50.0 g of water with 10.0 g of each of the three possible contaminants added. In gathering our data, we found that glucose raised the boiling point of water 0.7°C, NaCl raised it 3.5°C, and $CaCl_2$ raised it 3.0°C. When we tested 10.0 g of the unknown contaminant you sent us, we found it to have a boiling point elevation of 0.6°C, indicating that the contaminant is glucose.

The following calculations were made for the molecular mass determination.

molality of particles, $m = \frac{\Delta t_b}{K_b}$

for glucose: $m = \frac{0.7°C}{0.51°C/m} = 1.4\ m$

for NaCl: $m = \frac{3.5°C}{0.51°C/m} = 6.9\ m$

for $CaCl_2$: $m = \frac{3.0°C}{0.51°C/m} = 5.9\ m$

for unknown: $m = \frac{0.6°C}{0.51°C/m} = 1.2\ m$

for glucose: molar mass $= \frac{10.0\text{ g}}{50.0\text{ g } H_2O} \times \frac{1000\text{ g}}{1\text{ kg}} \times \frac{1\text{ kg } H_2O}{1.4\text{ mol}} \times \frac{1}{1} = 140$ g/mol

for NaCl: molar mass $= \frac{10.0\text{ g}}{50.0\text{ g } H_2O} \times \frac{1000\text{ g}}{1\text{ kg}} \times \frac{1\text{ kg } H_2O}{6.9\text{ mol}} \times \frac{2}{1} = 58$ g/mol

for $CaCl_2$: molar mass $= \frac{10.0\text{ g}}{50.0\text{ g } H_2O} \times \frac{1000\text{ g}}{1\text{ kg}} \times \frac{1\text{ kg } H_2O}{5.9\text{ mol}} \times \frac{3}{1} = 100$ g/mol

for unknown: molar mass $= \frac{10.0\text{ g}}{50.0\text{ g } H_2O} \times \frac{1000\text{ g}}{1\text{ kg}} \times \frac{1\text{ kg } H_2O}{1.2\text{ mol}} \times \frac{1}{1} = 170$ g/mol

Be sure to direct students who have difficulty with graphing or with significant figures to *Appendix B*.

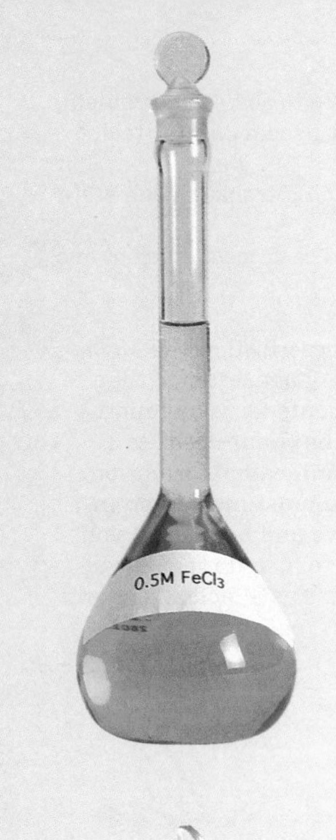

Tips for success in the lab

Whether you are performing an *Exploration* or an *Investigation*, you can do the following to help ensure success.

- **Read each lab twice** before you prepare the data tables.
- **Make sure you understand** everything that will happen in the lab.
- **Read the safety precautions** in the lab and on page 651.
- **Prepare the data tables** before you enter the lab.
- **Record all data and observations immediately** in the tables in your lab notebook.
- **Use appropriate units** whenever recording data.
- **Keep your lab desk organized** and free of clutter.

If you need help with graphing or with using significant figures, refer to *Appendix B*. The following pages contain tips to help you in lab, from preparing and filling out data tables to understanding how your teacher will grade the lab reports.

Preparing data tables

Data tables need to be prepared *before* you start any lab work because the tables must be ready for recording your data and observations *as* you do your lab work. For one example of how, you might construct a data table, refer to the instructions below and to **Figure A**.

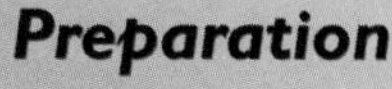

Preparation

1. Organizing Data

Prepare a data table in your lab notebook. It should have eight columns and five rows if you use the LEAP colormeter or a spectrophotometer, or four rows if this equipment will not be used. In the first row, label the boxes in the second through eighth columns *Test tube 0*, *Test tube 1*, *Test tube 2*, *Test tube 3*, *Test tube 4*, *Test tube 5*, and *Unknown*. In the first column, label the second box *mL 0.50 M* $FeCl_3$, the third box *mL* H_2O, and the fourth box *Estimates*. If you are using a colormeter or a spectrophotometer, label the fifth box in the first column *Measurements*. You will also need space for calculations.

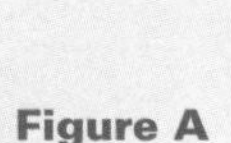

Figure A

Col.1	*Col.2* test tube 0	*Col.3* test tube 1	*Col.4* test tube 2	*Col.5* test tube 3	*Col.6* test tube 4	*Col.7* test tube 5	*Col.8* Unknown
mL 0.50 M $FeCl_3$ mL H_2O Estimates Measurements							

Completing data tables

An example of how you might fill in a data table is shown below in **Figure B**. Refer to the excerpt from an *Exploration* to understand how the procedure gives you the information you need to complete the data table.

Technique

2. Measure the mass of a clean, dry evaporating dish and watch glass to the nearest 0.01 g. Record this mass in your data table.
3. Add 2–3 g $NaHCO_3$ to your evaporating dish. Measure the mass, with the cover glass, to the nearest 0.01 g. Record this mass in your data table.
4. Slowly add 30 mL of the acetic acid solution to the $NaHCO_3$ in the evaporating dish. Add more acetic acid with a dropper or pipet until the bubbling stops.
5. If you are using a Bunsen burner, place the evaporating dish and its contents on a ceramic-centered wire gauze on an iron ring attached to the ring stand, as shown in the photograph below. Place the watch glass, concave side up, on top of the dish, making sure that there is a slight opening for steam to escape. If you are using a hot plate, place the watch glass the same way, but heat the evaporating dish directly on the hot plate.
6. Gently heat the evaporating dish until only a dry solid remains. Make sure that no water droplets remain on the underside of the watch glass. ***Do not heat too rapidly, or the material will boil, and the product will spatter out of the evaporating dish.***
7. Turn off the gas burner or hot plate. Allow the apparatus to cool at least 15 min. Determine the mass of the cooled equipment to the nearest 0.01 g. Record the mass of the dish, residue, and watch glass in your data table.
8. If time permits, reheat the evaporating dish and contents for 2 min. Let it cool and measure its mass again. You can be certain the sample is dry when there are two successive measurements within 0.02 g of each other.

Figure B

Mass of empty evaporating dish and watch glass	71.17 g *from step 2*
Mass of evaporating dish, watch glass, $NaHCO_3$	73.27 g *from step 3*
Mass of evaporating dish, watch glass, $NaC_2H_3O_2$ – First heating	73.21 g *from step 7*
Mass of evaporating dish, watch glass, $NaC_2H_3O_2$ – Second heating	73.19 g *from step 8*

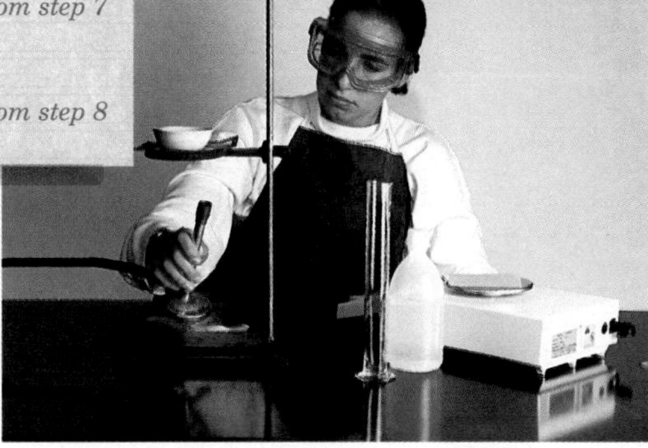

Lab report grades

Lab report requirements and grading criteria vary from teacher to teacher. A good lab report usually has the following characteristics.

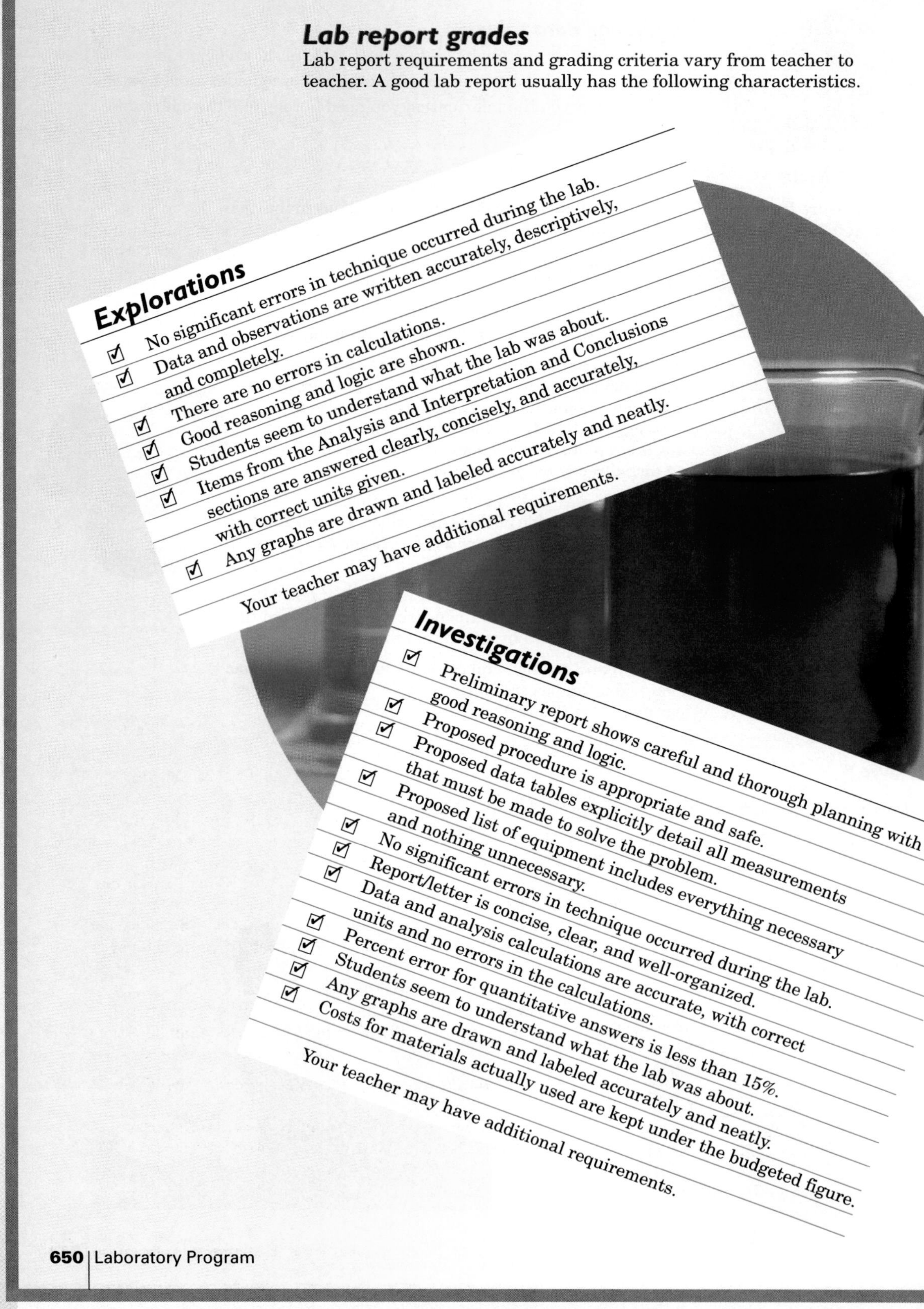

Explorations
☑ No significant errors in technique occurred during the lab.
☑ Data and observations are written accurately, descriptively, and completely.
☑ There are no errors in calculations.
☑ Good reasoning and logic are shown.
☑ Students seem to understand what the lab was about.
☑ Items from the Analysis and Interpretation and Conclusions sections are answered clearly, concisely, and accurately, with correct units given.
☑ Any graphs are drawn and labeled accurately and neatly.
Your teacher may have additional requirements.
Investigations
☑ Preliminary report shows careful and thorough planning with good reasoning and logic.
☑ Proposed procedure is appropriate and safe.
☑ Proposed data tables explicitly detail all measurements that must be made to solve the problem.
☑ Proposed list of equipment includes everything necessary and nothing unnecessary.
☑ No significant errors in technique occurred during the lab.
☑ Report/letter is concise, clear, and well-organized.
☑ Data and analysis calculations are accurate, with correct units and no errors in the calculations.
☑ Percent error for quantitative answers is less than 15%.
☑ Students seem to understand what the lab was about.
☑ Any graphs are drawn and labeled accurately and neatly.
☑ Costs for materials actually used are kept under the budgeted figure.
Your teacher may have additional requirements.

Safety in the Chemistry Laboratory

Chemicals are not toys

Any chemical can be dangerous if it is misused. Always follow the instructions for the experiment. Pay close attention to the safety notes. Do not do anything differently unless told to do so by your teacher.

Chemicals, even water, can cause harm. The challenge is to know how to use chemicals correctly so that they will not cause harm. If you follow the rules stated below, pay attention to your teacher's directions, follow cautions on chemical labels and the experiments, then you will be using chemicals correctly.

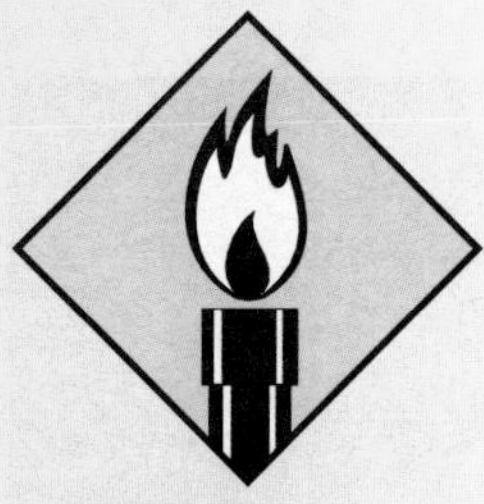

These safety rules always apply in the lab

1. ***Always wear a lab apron and safety goggles.***
   Even if you aren't working on an experiment at the time, laboratories contain chemicals that can damage your clothing. Keep the apron strings tied. More importantly, some chemicals can cause eye damage, and even blindness. If your safety goggles are uncomfortable or get clouded up, you are not the first person to have these problems. Ask your teacher for help. Try lengthening the strap a bit, washing the goggles with soap and warm water, or using an anti-fog spray.

2. ***No contact lenses in the lab.***
   Even while wearing safety goggles, chemicals could get between contact lenses and your eyes and cause irreparable eye damage. If your doctor requires that you wear contact lenses instead of glasses, then you should wear eye-cup safety goggles in the lab. Ask your doctor or your teacher how to use this very important and special eye protection.

3. ***NEVER WORK ALONE IN THE LABORATORY.***
   You should do lab work *only* under the supervision of your teacher.

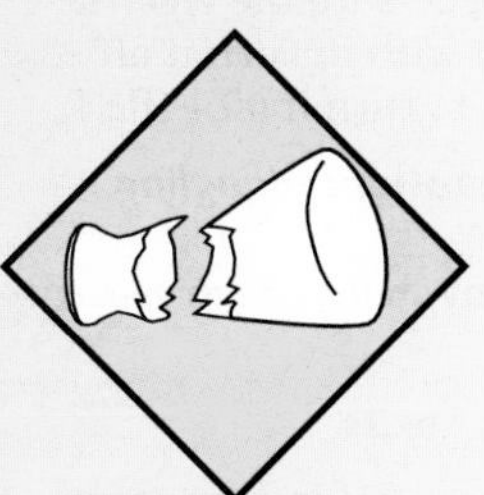

4. ***Wear the right clothing for lab work.***
   Necklaces, neckties, dangling jewelry, long hair, and loose clothing can knock things over or catch on fire. Tuck in neckties or take them off. Do not wear a necklace or other dangling jewelry, including hanging earrings. You don't have to, but it might be a good idea to remove your wristwatch so that it is not damaged by a chemical splash.

   Pull back long hair, and tie it in place. Nylon and polyester fabrics burn and melt more readily than cotton, so wear cotton clothing if you can. It's best to wear fitted garments, but if your clothing is loose or baggy, tuck it in or tie it back so that it does not get in the way or catch on fire.

   Wear shoes that will protect your feet from chemical spills—no open-toed shoes or sandals, and no shoes with woven leather straps. Shoes made of solid leather or polymer are much better than shoes made of cloth. It is also important to wear pants, not shorts or skirts.

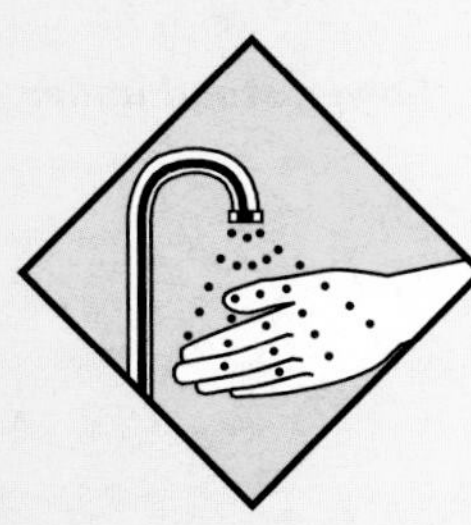

5. ***Only books and notebooks needed for the experiment should be in the lab.***
   Do not bring textbooks, purses, bookbags, backpacks, or other items into the lab; keep these things in your desk or locker.

Do you really know how to operate an eyewash fountain?

Make sure you know exactly how—it could save someone's eyesight. Be certain to instruct **anyone working in the lab** in the proper use of the eyewash fountain, as outlined below.

- If your eyes are exposed to chemicals, walk immediately to an eyewash fountain, preferably **within 15 s of exposure.**
- Immediately start flushing your eyes; **continue for at least 15 min** while someone else calls a doctor.
- **Keep your eyes wide open;** with your fingers, lift your upper and lower lids away from the eyeball so that the flowing water washes across it.
- **Move your eyes continuously** up, down, left, right, across, and around. This will help the flowing water to get inside the eyeball socket and wash out any chemicals that could attack the eyeball from the rear.

Electrical Safety

Although none of the labs in this text *require* the use of electrical equipment, several include options for the use of microcomputer-based laboratory equipment, pH meters, or spectrophotometers. The following safety precautions to avoid electric shocks must be followed any time electrical equipment is present in the lab.

- Each electrical socket in the laboratory must be a three-hole socket and must be protected with a GFI (ground-fault interrupter) type circuit.
- Check the polarity of all circuits before use with a polarity tester from an electronics supply store. Repair any incorrectly wired sockets.
- Use only electrical equipment equipped with a three-wire cord and three-prong plug.
- Be sure all electrical equipment is turned off before it is plugged into a socket. Turn off electrical equipment before it is unplugged.
- Wiring hookups should be made or altered only when apparatus is disconnected from the power source and the power switch is turned off.
- Do not let electric cords dangle from work stations; dangling cords are a tripping and shock hazard.
- Do not use electrical equipment with frayed or twisted cords.
- The area under and around electrical equipment should be dry; cords should not lie in puddles of spilled liquid.
- Hands should be dry when using electrical equipment.
- Do not use electrical equipment powered by 110–115 V alternating current for "conductivity" demonstrations or for any other use in which bare wires are exposed, even if the current is connected to a lower voltage AC or DC connection.

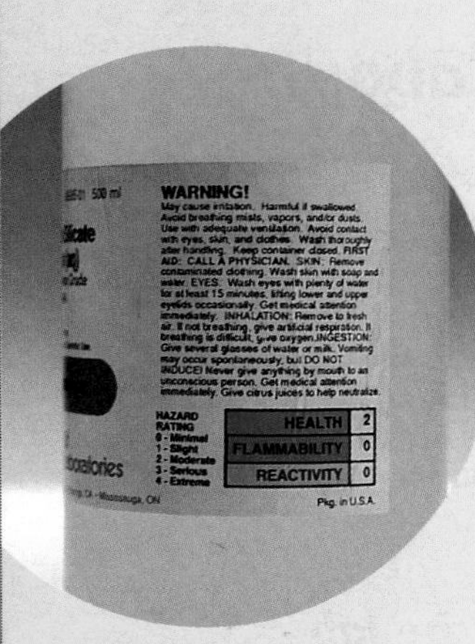

6. ***Read the entire experiment before entering the lab.***
   Memorize the safety precautions. Be familiar with the instructions for the experiment. Only materials and equipment authorized by your teacher should be used. When you do your lab work, follow the instructions and safety precautions described in the directions for the experiment.
7. ***Read chemical labels.***
   Follow the instructions and safety precautions stated on the labels.
8. ***Walk with care in the lab.***
   Sometimes you will have to carry chemicals from the supply station to your lab station. Avoid bumping into other students and spilling the chemicals. Stay at your lab station at other times.
9. ***Food, beverages, chewing gum, cosmetics, and smoking are NEVER allowed in the lab.***
   (You should already know this.)
10. ***NEVER taste chemicals or touch them with your bare hands.***
    Keep your hands away from your face and mouth while working, even if you are wearing gloves.
11. ***Use a sparker to light a Bunsen burner.***
    Do not use matches. Be sure that all gas valves are turned off and that all hot plates are turned off *and* unplugged when you leave the lab.
12. ***Be careful with hot plates, Bunsen burners, and other heat sources.***
    Keep your body and clothing away from flames. Do not touch a hot plate after it has just been turned off because it is probably hotter than you think. The same is true of glassware, crucibles, and other things after removing them from the flame of a Bunsen burner or from a drying oven.

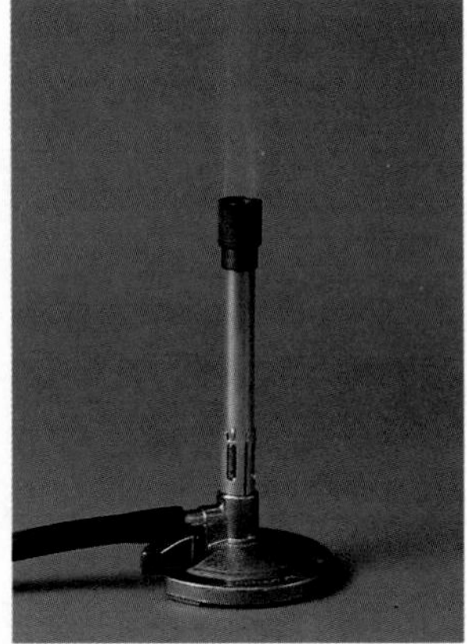

13. ***Do not use electrical equipment with frayed or twisted wires.***
14. ***Be sure your hands are dry before using electrical equipment.***
    Before plugging an electrical cord into a socket, be sure the electrical equipment is turned off. When you are finished with it, turn it off. Before you leave the lab, unplug it, but be sure to turn it off FIRST.
15. ***Do not let electrical cords dangle from work stations; dangling cords can cause tripping or electrical shocks.***
    The area under and around electrical equipment should be dry; cords should not lie in puddles of spilled liquid.
16. ***Know fire drill procedures and the locations of exits.***
17. ***Know the location and operation of safety showers and eyewash stations.***
18. ***If your clothes catch on fire, walk to the safety shower, stand under it, and turn it on.***

19. If you get a chemical in your eyes, walk immediately to the eyewash station, turn it on, and lower your head so your eyes are in the running water.
Hold your eyelids open with your thumbs and fingers, and roll your eyeballs around. You have to flush your eyes continuously for at least 15 minutes. Call your teacher while you are doing this.

20. If you have a spill on the floor or lab bench, call your teacher rather than trying to clean it up by yourself.
Your teacher will tell you if it is OK for you to do the cleanup; if not, your teacher will know how the spill should be cleaned up safely.

21. If you spill a chemical on your skin, wash it off using the sink faucet, and call your teacher.
If you spill a solid chemical on your clothing, brush it off carefully without scattering it on somebody else, and call your teacher. If you get a liquid on your clothing, wash it off right away if you can get it under the sink faucet, and call your teacher. If the spill is on your pants or somewhere else that will not fit under the sink faucet, use the safety shower. Remove the pants or other affected clothing while under the shower, and call your teacher. (It may be temporarily embarrassing to remove pants or other clothing in front of your class, but failing to flush that chemical off your skin could cause permanent damage.)

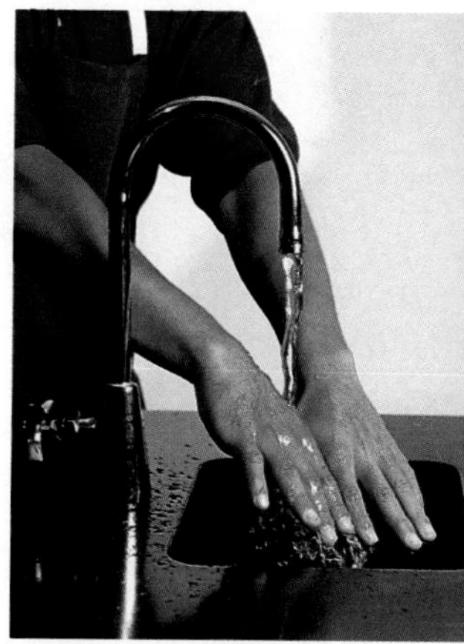

22. The best way to prevent an accident is to stop it before it happens.
If you have a close call, tell your teacher so that you and your teacher can find a way to prevent it from happening again. Otherwise, the next time, it could be a harmful accident instead of just a close call.

23. All accidents should be reported to your teacher, no matter how minor.
Also, if you get a headache, feel sick to your stomach, or feel dizzy, tell your teacher immediately.

24. For all chemicals, take only what you need.
However, if you do happen to take too much and have some left over, DO NOT put it back in the bottle. If somebody accidentally puts a chemical into the wrong bottle, the next person to use it will have a contaminated sample. Ask your teacher what to do with any leftover chemicals.

25. NEVER take any chemicals out of the lab.
(This is another one that you should already know. You probably know the remaining rules also, but read them anyway.)

26. Horseplay and fooling around in the lab are very dangerous.
NEVER be a clown in the laboratory.

27. Keep your work area clean and tidy.
After your work is done, clean your work area and all equipment.

28. Always wash your hands with soap and water before you leave the lab.

29. Whether or not the lab instructions remind you, all of these rules apply all of the time.

- Use dry cells or ni-cad rechargeable batteries as direct current sources. Do not use automobile storage batteries or AC to DC converters; these two sources of DC current can present serious shock hazards.

Disposal of Chemicals

Only a relatively small percentage of waste chemicals is classified as hazardous by EPA regulations. The EPA regulations are derived from two acts (as amended) passed by the Congress of the United States: RCRA (Resource Conservation and Recovery Act) and CERCLA (Comprehensive Environmental Response, Compensation, and Liability Act).

Some states have enacted legislation governing the disposal of hazardous wastes that differs from the federal legislation. The disposal procedures described in this book have been designed to comply with the federal legislation as described in the EPA regulations. In most cases these disposal procedures will ***probably*** comply with your state's disposal requirements. However, to be sure of this, check with your state's environmental agency. If a particular disposal procedure does not comply with your state requirements, ask that office to assist you in devising a procedure that is in compliance.

The following general practices are recommended.

- Except when otherwise specified in the disposal procedures, neutralize acidic and basic wastes with 1.0 M sodium hydroxide, NaOH, or 1.0 M sulfuric acid, H_2SO_4, added slowly while stirring.
- In dealing with a waste-disposal contractor, prepare a complete list of the chemicals you want to dispose of. Classify each chemical on your disposal list as a hazardous or non-hazardous waste chemical. Check with your local environmental agency office for the details of such classification.

Safety Symbols

To highlight specific types of precautions, the following symbols are used in the Explorations and Investigations. Remember that no matter what safety symbols are used in the lab instructions, all 29 of the safety rules previously described should be followed at all times.

- Wear laboratory aprons in the laboratory. Keep the apron strings tied so that they do not dangle.
- Wear safety goggles in the laboratory at all times. Know how to use the eyewash station. (Rules **1**, **17**, and **19** apply. Which rule says not to wear contact lenses? Do regular eyeglasses provide enough protection?)

- Never taste, eat, or swallow any chemicals in the laboratory. Do not eat or drink any food from laboratory containers. Beakers are not cups, and evaporating dishes are not bowls. (Rules **9** and **10** apply.)
- Never return unused chemicals to the original container. (Which rule applies?)
- Some chemicals are harmful to our environment. You can help protect the environment by following the instructions for proper disposal.
- It helps to label the beakers and test tubes containing chemicals. (This is not a new rule, just a good idea.)
- Never transfer substances by sucking on a pipet or straw; use a suction bulb. (This is a new rule.)
- Never place glassware, containers of chemicals, or anything else near the edges of a lab bench or table. (This is another new rule.)

Here is a question: Do any rules other than **9** and **10** apply in the lab when you see this safety symbol?

- If a chemical gets on your skin or clothing or in your eyes, rinse it immediately, and alert your teacher. (Rules **19** and **21** apply.)
- If a chemical is spilled on the floor or lab bench, tell your teacher, but do not clean it up yourself unless your teacher says it is OK to do so. (Rule **20** applies.)

Here is another question: What rules other than **19**, **20**, and **21** apply in the lab when you see this safety symbol?

- When heating a chemical in a test tube, always point the open end of the test tube away from yourself and other people. (This is another new rule.)

WARNING! From here to the end of this safety section, it is up to you to look at the list of rules and identify whether a specific rule applies, or if the rule presented is a new rule.

- Tie back long hair, and confine loose clothing. (Rule ? applies.)
- Never reach across an open flame. (Rule ? applies.)
- Use proper procedures when lighting Bunsen burners. Turn off hot plates, Bunsen burners, and other heat sources when not in use. (Rule ? applies.)
- Heat flasks or beakers on a ringstand with wire gauze between the glass and the flame. (Rule ? applies.)
- Use tongs when heating containers. Never hold or touch containers while heating them. Always allow heated materials to cool before handling them. (Rule ? applies.)
- Turn off gas valves when not in use. (Rule ? applies.)

- Use flammable liquids only in small amounts. (Rule ? applies.)
- When working with flammable liquids, be sure that no one else is using a lit Bunsen burner or plans to use one. (Rule ? applies.)

What other rules should you follow in the lab when you see this safety symbol?

- Check the condition of glassware before and after using it. Inform your teacher of any broken, chipped, or cracked glassware because it should not be used. (Rule ? applies.)
- Do not pick up broken glass with your bare hands. Place broken glass in a specially designated disposal container. (Rule ? applies.)
- Never force glass tubing into rubber tubing, rubber stoppers, or wooden corks. To protect your hands, wear heavy cloth gloves or wrap toweling around the glass and the tubing, stopper, or cork, and gently push in the glass. (Rule ? applies.)

- Do not inhale fumes directly. When instructed to smell a substance, use your hand to wave the fumes toward your nose, and inhale gently. (Some people say "waft the fumes.")

- Keep your hands away from your face and mouth. (Rule ? applies.)
- Always wash your hands before leaving the laboratory. (Rule ? applies.)

Finally, if you are wondering how to answer the questions that asked what additional rules apply to the safety symbols, here is the correct answer.
Any time you see any of the safety symbols, you should remember that all 29 of the numbered laboratory rules always apply.

- Unlabeled chemicals must be identified to the extent that they can be classified as a hazardous or nonhazardous waste. Some landfills will analyze a mystery bottle for a fee if it is shipped to the landfill in a separate package, is labeled as a sample, and includes instructions to analyze the contents sufficiently to allow proper disposal.

EXPLORATION 1A

Objectives
Students will:
- use appropriate lab safety procedures.
- use a Bunsen burner.
- determine the most efficient setting for the burner's fuel-air ratio.
- use a laboratory balance to measure mass.
- demonstrate appropriate technique in transferring solids.
- use a graduate cylinder to measure volume.
- collect data and organize it in a table.
- graph data.

Planning
Recommended Time:
1 lab period

Materials
(for each lab group)
- NaCl crystals, about 5 g
- 250 mL beakers, 2
- Balance (may be shared among lab groups)
- Bunsen burner, gas tubing, striker
- Crucible tongs (do not use beaker tongs)
- Copper wire, 18 gauge, 10 cm
- Evaporating dish
- Spatula
- Wax paper or weighing paper

Student Orientation
Pre-lab Discussion
Thoroughly discuss all safety precautions outlined in this laboratory and in the safety section. Because this laboratory will be one of the first ones of the year, be sure to make safety a priority.

Point out the location and demonstrate the operation of all of the safety equipment, especially the lab shower, eyewash station, and fire blankets. Be sure students know what steps to follow in an emergency, especially who to call. Demonstrate the proper use of safety goggles and lab apron. If students think that the lab equipment and chemicals used in this Exploration are not particularly

Laboratory Techniques

Situation
You have applied to work at a company that does research, development, and analysis work. Although the company does not require employees to have extensive chemical experience, all applicants are tested for their ability to follow directions, heed safety precautions, perform simple laboratory procedures, clearly and concisely communicate results, and make logical inferences.

The company will use their opinion of your performance on the test in considering whether to hire you and in deciding your initial salary.

Background
Pay close attention to the procedures and safety precautions because you will continue to use them throughout your work if you are hired by this company. In addition, you will need to pay attention to what is happening around you, make careful observations, and keep a clear and legible record of these observations in your lab notebook.

Problem
This laboratory orientation session will teach you some of the following techniques.
- proper use of a Bunsen burner
- how to handle solids and liquids
- how to properly use a triple-beam balance
- basic safety techniques for all lab work

Objectives
Demonstrate **proficiency in using the Bunsen burner, the triple-beam balance, and the graduated cylinder.**

Demonstrate **proficiency in properly handling solid and liquid chemicals.**

Develop **proper safety techniques for all lab work.**

Apply **data-collecting techniques, and graph data.**

Required Precautions
- Goggles and lab apron must be worn at all times.
- Loose hair and clothing must be tied back.
- If students find a gas valve that has not been turned off properly, the room must be thoroughly ventilated, using the hood and windows, in order to prevent a fire hazard.
- Read all safety cautions, and discuss them with your students.

Safety

Always wear goggles and an apron to protect your eyes and clothing. If you get a chemical in your eyes, immediately flush it out at the eyewash station while calling to your teacher. Know the locations of the emergency lab shower and eyewash and how to use them.

Do not touch any chemicals. If you get a chemical on your skin or clothing, wash it off at the sink while calling to your teacher. Make sure you carefully read the labels and follow the directions on all containers of chemicals that you use. Do not taste any chemicals or items used in the laboratory. Never return leftovers to their original containers; take only small amounts to avoid wasting supplies.

When you use a Bunsen burner, do not heat glassware that is broken, chipped, or cracked. Use tongs whenever handling glassware or other equipment that has been heated because when it is hot, it does not look hot.

Always clean up the lab and all equipment after use, and dispose of substances according to proper disposal methods. Wash your hands thoroughly before you leave the lab after all lab work is finished.

Materials

- **NaCl**
- **250 mL beakers, 2**
- **100 mL graduated cylinder**
- **Balance**
- **Bunsen burner and related equipment**
- **Copper wire**
- **Crucible tongs**
- **Evaporating dish**
- **Heat-resistant mat**
- **Spatula**
- **Test tube**
- **Wax paper or weighing paper**

Preparation

1. ***Organizing Data***
   Prepare spaces in your lab notebook for observations about the following.
   - Bunsen burner flame when the ports on the burner's base are closed and open
   - the mass of weighing paper with and without NaCl

   Prepare a data table with two columns, each having spaces to record the following.
   - the mass of an empty beaker
   - the mass of the beaker and 50 mL of water
   - the mass of 50 mL of water
   - the mass of the beaker and 100 mL of water
   - the mass of 100 mL of water
   - the mass of the beaker and 150 mL of water
   - the mass of 150 mL of water
2. Record in your lab notebook where the following items are located and how each item is used: emergency lab shower, eyewash station, and emergency telephone numbers.
3. Check to be certain that the gas valve at your lab station and at the next closest one are turned off. Notify your teacher immediately if a valve is on because the fumes must be cleared before any work continues.

Technique

4. Compare the diagram of the Bunsen burner to your burner. Construction may vary, but the air and methane gas, CH_4, always mix in the barrel, the vertical tube in the center of the burner.

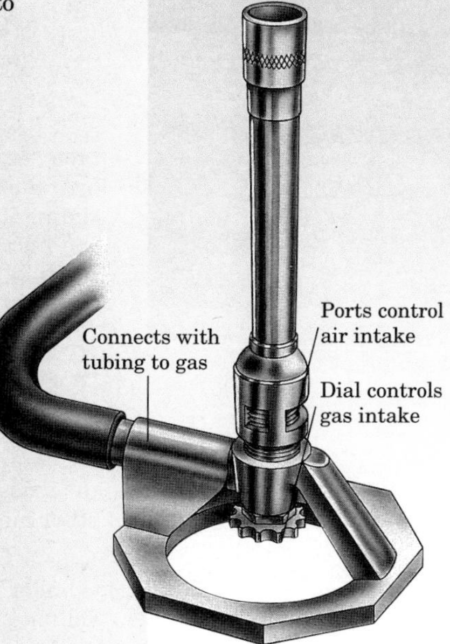

dangerous, remind them that other items in the lab may be dangerous, so safety precautions must be taken. Explain the need to tie back loose clothing and long hair when working in the lab. Emphasize the need to wear appropriate clothing, as outlined in the safety section.

Have students make the data tables before they enter the lab.

Remind students to make very careful observations, especially for the portion of the lab using the copper wire and the Bunsen burner.

Techniques to Demonstrate
Because burners vary in design, show students exactly how to light the Bunsen burners in your lab with a striker. Tell students that the burner flame is a chemical reaction in which methane and oxygen react to form carbon dioxide and water vapor.

Although lighting burners with matches is common in some places, it can be dangerous, especially if the matches are not safety matches. Rather than having to check that safety matches are used, it is best to use a striker and avoid matches entirely.

Be sure to demonstrate the proper technique for measuring mass with a balance. Now is the time to make sure students know how to help you keep your equipment in good shape. Make sure that neither chemicals nor hot objects are put directly on the balance pan. Students also have difficulty understanding how to use the vernier scale on the triple-beam balance unless they can see a demonstration. If you are using another kind of balance, you will need to explain how it works.

Explain to students how to read volumes from the meniscus. This can be demonstrated with a quick sketch on the blackboard or an overhead projector. Remind students to keep their eyes at the same level as the top of the fluid when reading volumes.

Post-Lab

Disposal

There are no special disposal guidelines. The sodium chloride can be reused by each class if you designate a special container for students to dispose of the NaCl after each part and keep it separate from the NaCl you use as a reagent. The same is true for the copper wire. Reusing these materials is more economical and environmentally sensitive than simply throwing them away.

Sample Data

When the burner's ports are closed, the flame is yellow, bright, and it makes a roaring noise. When the ports are open, the flame is blue, dim, and quiet. Answers will vary on the mass measurements involving NaCl. Be sure students measure the mass with the correct precision for your lab equipment and that they measure the mass of the weighing paper alone first.

Answers
Analysis and Interpretation

1. The hottest flame is formed when the air ports are open. The hottest part of the flame should be at the top of the light blue inner cone of the Bunsen burner flame, which corresponds to the spot labeled *c* in the photograph alongside step **6**. (Students should have observed the copper melting.)

2. Only the 3.42 g measurement could have been made by the laboratory balance. All of the other measurements indicate precision that is not possible with the laboratory balance.

3.

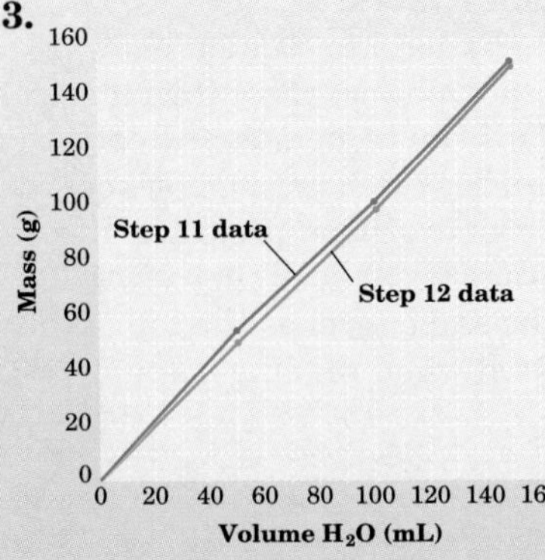

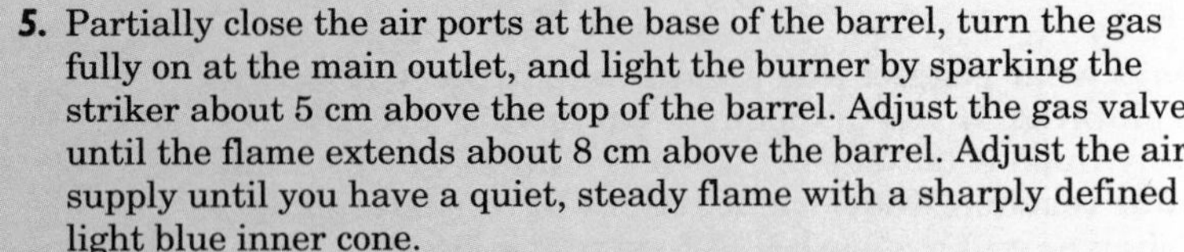

5. Partially close the air ports at the base of the barrel, turn the gas fully on at the main outlet, and light the burner by sparking the striker about 5 cm above the top of the barrel. Adjust the gas valve until the flame extends about 8 cm above the barrel. Adjust the air supply until you have a quiet, steady flame with a sharply defined light blue inner cone.
If an internal flame develops, turn off the gas valve, and let the burner cool down. Otherwise the metal of the burner can get hot enough to set fire to anything nearby that is flammable. Before you relight the burner, partially close the air ports.

6. Using crucible tongs, hold a 10 cm piece of copper wire for 2–3 s in the part of the flame labeled *a* in the photograph on the left. Repeat for the parts of the flame labeled *b* and *c*. Record your observations. Which is the hottest part of the flame?

7. Experiment with the flame by completely closing the ports at the base of the burner. Observe the color of the flame and the sounds made by the burner. Using crucible tongs, hold an evaporating dish in the tip of the flame for about 3 min. Place the dish on a heat-resistant mat, and shut off the burner. After the dish cools, examine its underside, and record your observations.

8. Before using the balance, be certain it is leveled, and the pointer rests at zero. If all slider masses are set at zero, but the pointer is not at zero, adjust the calibration knob. Use the same balance for all measurements during a lab activity. ***Never put chemicals directly on the balance pan.***

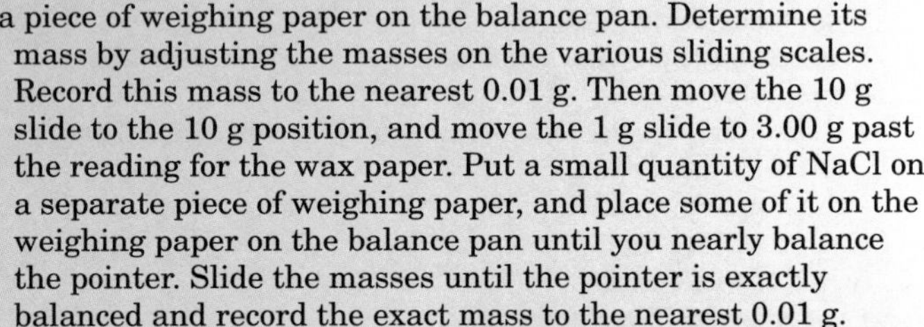

9. Place a piece of weighing paper on the balance pan. Determine its mass by adjusting the masses on the various sliding scales. Record this mass to the nearest 0.01 g. Then move the 10 g slide to the 10 g position, and move the 1 g slide to 3.00 g past the reading for the wax paper. Put a small quantity of NaCl on a separate piece of weighing paper, and place some of it on the weighing paper on the balance pan until you nearly balance the pointer. Slide the masses until the pointer is exactly balanced and record the exact mass to the nearest 0.01 g.

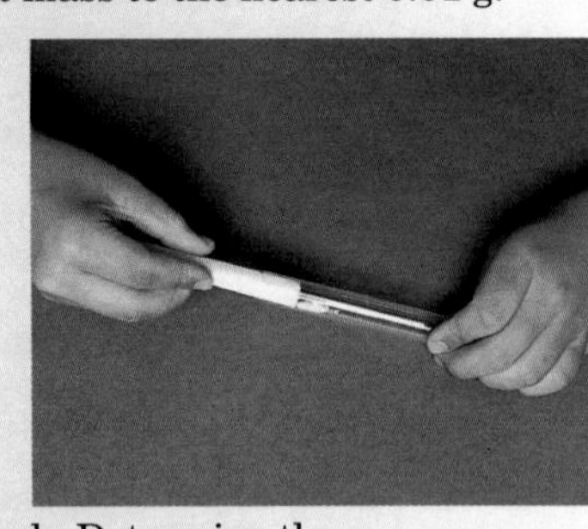

10. Remove the NaCl from the balance pan. Lay the test tube flat on the table, and transfer the NaCl into it by rolling the weighing paper and sliding it into the test tube. When you lift the test tube to a vertical position, tap the paper gently, and the solid will slip into the test tube.

11. Measure the mass of a dry 250 mL beaker, and record it in a data table. Add water up to the 50 mL mark. Determine the new mass, and record it in your lab notebook. Repeat the procedure by filling the beaker to the 100 mL mark and then to the 150 mL mark, recording the mass each time. Subtract the mass of the empty beaker from the other measurements to determine the mass of the water alone.

12. Repeat step **11** with a second dry 250 mL beaker, but use a graduated cylinder to measure the volumes of water to the nearest 0.1 mL. Read volumes from the bottom of the meniscus, the curve formed by the water's surface.

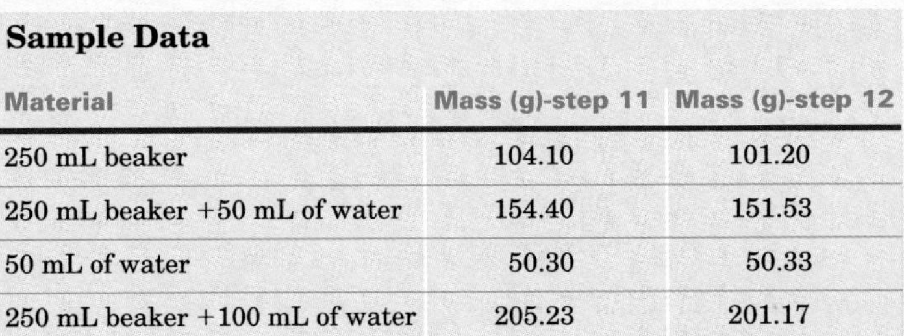

Sample Data

Material	Mass (g)-step 11	Mass (g)-step 12
250 mL beaker	104.10	101.20
250 mL beaker +50 mL of water	154.40	151.53
50 mL of water	50.30	50.33
250 mL beaker +100 mL of water	205.23	201.17

Cleanup and Disposal

13. Put the wire, NaCl, and weighing paper in the containers designated by your teacher. Pour the water from the beakers into the sink. Scrub the cooled evaporating dish with soap, water, and a scrub brush. Be certain that gas valves at your lab station and the next closest one are turned off. Be sure lab equipment is completely cool before storing it. Always wash your hands thoroughly after cleaning up.

Analysis and Interpretation

1. ***Organizing Ideas***
Based on your observations, which type of flame is hotter: the flame formed when the air ports are open or closed? What is the hottest part of the flame? (Hint: the melting point of copper is 1083°C.)

2. ***Analyzing Information***
Which of the following measurements could have been made by your balance: 3.42 g of glass, 5.666 72 g of aspirin, 0.000 017 g of paper?

3. ***Organizing Data***
Make a graph of volume vs. mass for data from steps **11** and **12**. The mass of water (g) should be graphed along the y-axis as a dependent variable, and the volume of water (mL) should be graphed along the x-axis as an independent variable.

4. ***Organizing Ideas***
For the following safety items, state the location of the item and how and when to use them: emergency lab shower, eyewash station and emergency telephone numbers.

Conclusions

5. ***Relating Ideas***
When methane is burned, it usually produces carbon dioxide and water. If there is a shortage of oxygen, the flame is not as hot, and black carbon solid is formed. Which steps demonstrated these flames?

6. ***Inferring Conclusions***
Which is the most accurate method for measuring volumes of liquids: a beaker or a graduated cylinder? Explain why.

Extensions

1. ***Resolving Discrepancies***
In Mandeville High School, Jarrold got only part way through step **7** of this experiment when he had to put everything away. Soon after Jarrold left, his lab drawer caught on fire. How did this happen?

2. ***Relating Ideas***
The density of water is equal to its mass divided by its volume. Calculate the density of water using your data from step **11**. Then calculate the density of water using data from step **12**.

3. ***Designing Experiments***
What steps would you take to make more accurate mass and volume measurements? If your teacher approves of your plan, try it. Are your new density values closer to the accepted value, 0.997 g/mL at 25°C?

Sample Data

Material	Mass (g)-step 11	Mass (g)-step 12
100 mL of water	101.13	99.97
250 mL beaker +150 mL of water	255.92	252.01
150 mL of water	151.82	150.81

***Note: this data was collected with a centigram balance. Some triple-beam balances do not have the same precision.**

4. Answers will vary depending upon the arrangement of your laboratory.

Conclusions

5. The quiet, steady light blue flame from step 5 is the one that produced carbon dioxide and water vapor. The flame with the shortage of oxygen that produced black carbon solid was the loud, yellow flame from step 7.

6. The graduated cylinder is the best method for measuring volumes of liquids because it is marked in increments of 1 mL instead of every 50 mL like the beaker.

Extensions

1. One reason for the fire could be that Jarrold did not allow the burner or the evaporating dish to cool completely. Then, when he put them away into the lab drawer, they may have been hot enough to ignite some paper.

2. Densities for step **11**
1.1 g/mL (50 mL H_2O)
1.01 g/mL (100 mL H_2O)
1.01 g/mL (150 mL H_2O)

Densities for step **12**
1.01 g/mL (50 mL H_2O)
0.997 g/mL (100 mL H_2O)
1.005 g/mL (150 mL H_2O)

3. Students suggestions for improving the procedure will vary, but could include using the average values from repeated trials, or pouring water more slowly. Be sure students get your approval before carrying out their plans.

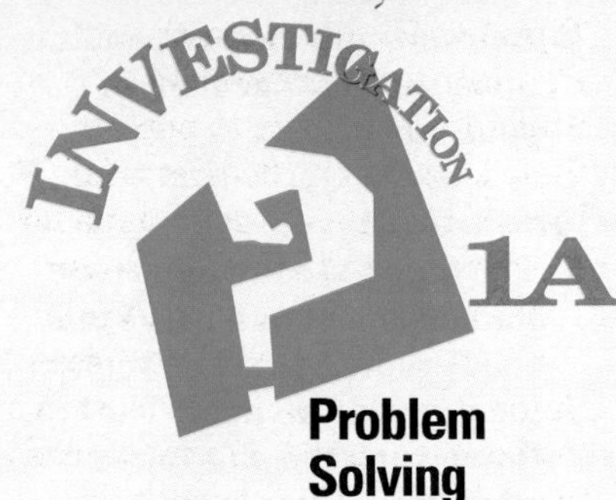

Objectives

Students will:
- use a laboratory balance.
- use a Bunsen burner.
- understand the purpose of multiple trials.
- calculate averages of data.
- design and implement their own procedure.
- relate results to the law of conservation of mass.

Planning

Recommended Time: 1–2 lab periods depending upon amount of time for students to experiment with procedure.

Solution/Material Preparation

1. Kernels for "technique beta" require no preparation.
2. Kernels for "technique gamma" should be soaked for approximately 1 hour prior to use.
3. Kernels for "technique delta" should be soaked overnight prior to use. (These pop very poorly.)
4. Be sure to pat the popcorn dry before giving the treated kernels to the students.

Student Orientation

Techniques to Demonstrate

Review the use of the balance. Students often have difficulty reading the balance to 0.01 g precision. In addition, students often forget to use weighing paper or to measure the mass of the empty glassware.

Review the use of the Bunsen burner. Remind students to close the ports at the end of the barrel on the Bunsen burner before lighting it.

Pre-lab Discussion

To help students, you may provide some of the following leading questions for the students to consider as they make their plans.
- How will you keep the popcorn from flying out of the beaker as it pops?
- Is it best to use all of the popcorn at once or to do several smaller trials?
- How often should they measure the mass of the popcorn?
- Will they need to measure the mass of any other objects or containers?

Sample Data

	Technique beta	Technique gamma	Technique delta
20 kernels	2.47 g	2.93 g	3.76 g
250 mL beaker	101.40 g	101.40 g	101.40 g
beaker + oil	101.50 g	101.50 g	101.50 g
beaker + oil + 20 kernels before	103.97 g	104.43 g	105.26 g
beaker + oil + 20 kernels after	103.50 g	103.50 g	103.80 g
mass of water in 20 kernels	0.47 g	0.93 g	1.46 g
% water in popcorn	19%	32%	39%

Popcorn prepared by technique delta pops very poorly.

B CONSERVATION OF MASS

Percentage of Water in Popcorn

JULIETTE BRAND FOODS
Product Development Division · 2994 Effington Lane · Thornton, NE 68181

January 9, 1995

Director of Research
CheMystery Labs, Inc.
52 Fulton Street
Springfield, VA 22150

Dear Director of Research:

Juliette Brand Foods is preparing to enter the rapidly expanding home popcorn market with a new popcorn product. As you may know, the key to making popcorn pop is the amount of water contained within the kernel.

Thus far, the product development division has created three different production techniques for the popcorn, each of which creates popcorn with differing amounts of water. We need an independent lab such as yours to measure the percentage of water contained in each sample and to determine which technique produces the best-popping popcorn.

I've enclosed samples from each of the three techniques, labeled "technique beta," "technique gamma," and "technique delta." Please bill us when the work is complete.

Sincerely,

Mary Biedenbecker
Director
Product Development Division

Required Precautions

Lab aprons and goggles must be worn at all times. Confine loose clothing and long hair.

Do not eat the popcorn; it is for testing purposes only.

Remember, hot glass does not look hot. Be sure to use beaker tongs.

Required Precautions

- Goggles and lab apron must be worn at all times. Loose hair and clothing must be tied back.
- Remind students that hot glass doesn't always look hot, so they need to use beaker tongs to handle any glassware.
- Remind the students that the popcorn is for testing only and must not be eaten.

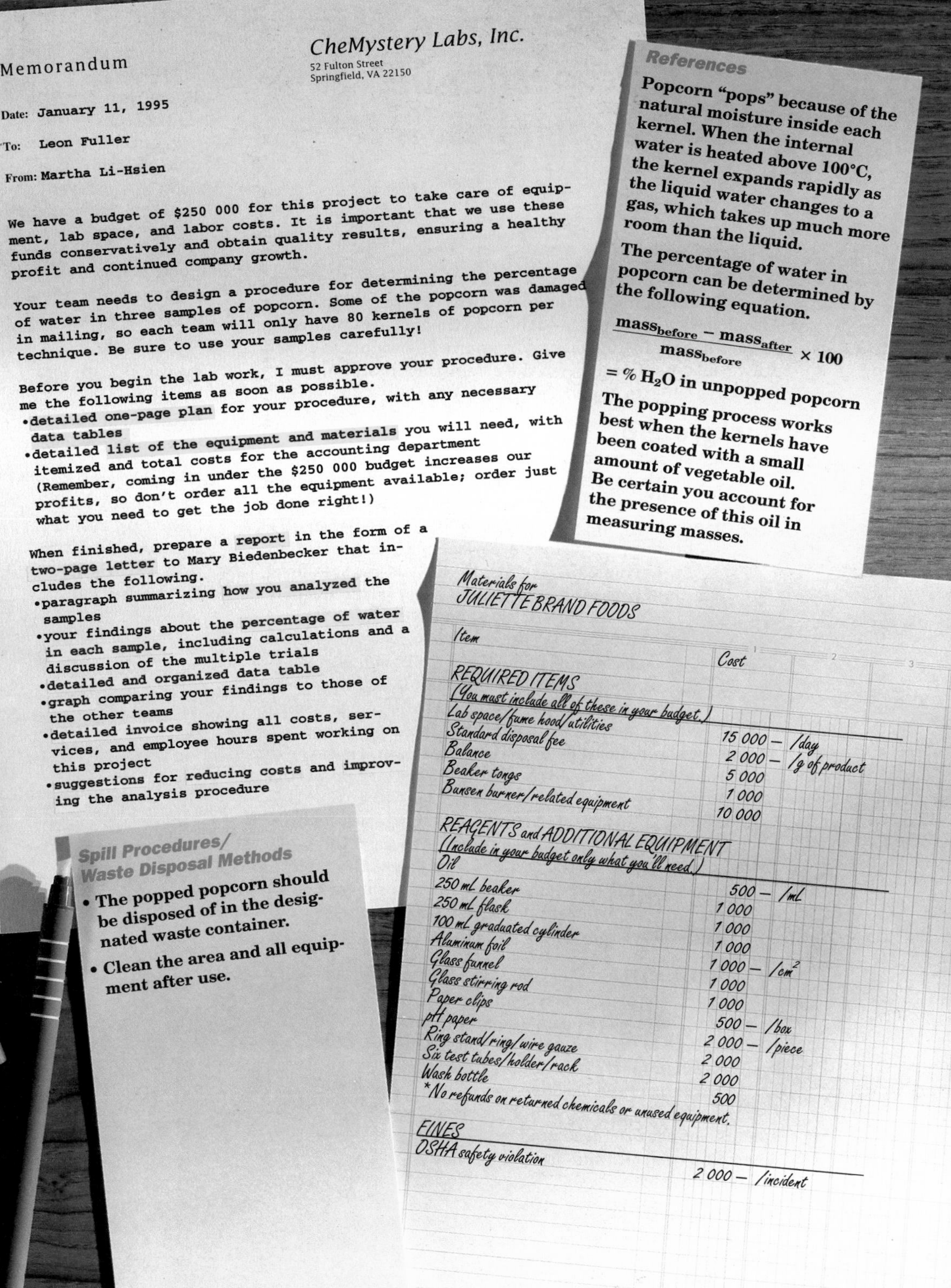

Memorandum

CheMystery Labs, Inc.
52 Fulton Street
Springfield, VA 22150

Date: January 11, 1995

To: Leon Fuller

From: Martha Li-Hsien

We have a budget of $250 000 for this project to take care of equipment, lab space, and labor costs. It is important that we use these funds conservatively and obtain quality results, ensuring a healthy profit and continued company growth.

Your team needs to design a procedure for determining the percentage of water in three samples of popcorn. Some of the popcorn was damaged in mailing, so each team will only have 80 kernels of popcorn per technique. Be sure to use your samples carefully!

Before you begin the lab work, I must approve your procedure. Give me the following items as soon as possible.
- detailed one-page plan for your procedure, with any necessary data tables
- detailed list of the equipment and materials you will need, with itemized and total costs for the accounting department (Remember, coming in under the $250 000 budget increases our profits, so don't order all the equipment available; order just what you need to get the job done right!)

When finished, prepare a report in the form of a two-page letter to Mary Biedenbecker that includes the following.
- paragraph summarizing how you analyzed the samples
- your findings about the percentage of water in each sample, including calculations and a discussion of the multiple trials
- detailed and organized data table
- graph comparing your findings to those of the other teams
- detailed invoice showing all costs, services, and employee hours spent working on this project
- suggestions for reducing costs and improving the analysis procedure

References

Popcorn "pops" because of the natural moisture inside each kernel. When the internal water is heated above 100°C, the kernel expands rapidly as the liquid water changes to a gas, which takes up much more room than the liquid.

The percentage of water in popcorn can be determined by the following equation.

$$\frac{\text{mass}_{\text{before}} - \text{mass}_{\text{after}}}{\text{mass}_{\text{before}}} \times 100 = \%\ H_2O \text{ in unpopped popcorn}$$

The popping process works best when the kernels have been coated with a small amount of vegetable oil. Be certain you account for the presence of this oil in measuring masses.

Materials for
JULIETTE BRAND FOODS

Item	Cost	1	2	3
REQUIRED ITEMS (You must include all of these in your budget.)				
Lab space/fume hood/utilities	15 000 — /day			
Standard disposal fee	2 000 — /g of product			
Balance	5 000			
Beaker tongs	1 000			
Bunsen burner/related equipment	10 000			
REAGENTS and ADDITIONAL EQUIPMENT (Include in your budget only what you'll need.)				
Oil	500 — /mL			
250 mL beaker	1 000			
250 mL flask	1 000			
100 mL graduated cylinder	1 000			
Aluminum foil	1 000 — /cm²			
Glass funnel	1 000			
Glass stirring rod	1 000			
Paper clips	500 — /box			
pH paper	2 000 — /piece			
Ring stand/ring/wire gauze	2 000			
Six test tubes/holder/rack	2 000			
Wash bottle	500			
*No refunds on returned chemicals or unused equipment.				
FINES				
OSHA safety violation	2 000 — /incident			

Spill Procedures/ Waste Disposal Methods
- **The popped popcorn should be disposed of in the designated waste container.**
- **Clean the area and all equipment after use.**

Materials
(for each pair of students)
- 80 kernels of popcorn for each of the three brands
- 250 mL beaker
- One sheet of aluminum foil
- Oil (to coat the bottom of the beaker)
- Bunsen burner
- Ring and wire gauze

Estimated cost of materials:
$65 000

Tips for Evaluating the Pre-lab Requirements

Students will want to pop all their kernels on the first trial. Encourage them to save some for a second and third trial. Other students may even plan to pop one kernel at a time. Encourage these students to use at least 10 kernels at a time. You may need to explain that, for each technique, the percentage of water in a sample should be more or less the same regardless of the size of the sample.

Be certain students' plans account for measuring the mass of the beaker after oil has been added.

Students often wonder what the foil is for. Remind them of stove-top style popcorn tins. The foil is used to keep the popcorn from being lost, and holes are made in it to allow the water vapor from the popcorn to escape.

Proposed Procedure

Cover the beaker in which the popcorn is being popped with foil. Put holes in the foil to allow the water vapor to escape.

Measure the mass of the unpopped kernels. (20 kernels work well)

Measure the mass of the beaker.

Measure the mass of the beaker with the uncooked popcorn and vegetable oil coating.

Heat the popcorn until the majority of the kernels have popped. The popcorn pops more efficiently if the beaker is held firmly with tongs and gently shaken side to side on the wire gauze.

Allow the beaker to cool for five minutes. Measure the mass of the beaker, popped corn, and vegetable oil. Subtract the before and after masses of the popcorn to determine the mass of water in the kernels.

Post-Lab

Disposal

There are no special disposal requirements, but have a container for the students to put the popcorn in to emphasize that it is not to be eaten.

EXPLORATION 1B

Objectives

Students will:

- use appropriate lab safety procedures.
- use a laboratory balance to measure mass.
- demonstrate appropriate technique in transferring solids.
- demonstrate appropriate technique in transferring liquids.
- demonstrate appropriate technique in filtering.
- use pH paper, a pH meter, or a LEAP System pH probe to measure pH.
- measure the effect on pH of adding base to solutions of analgesics in a titration.
- collect data and organize it in a table.
- graph a titration curve.
- identify physical properties that allow one to distinguish between aspirin, buffered aspirin, and aspirin substitute.

Planning

Recommended Time: 2 lab periods (Steps **1–18** one day, steps **19–23** the next; to save time, one filtration, instead of two, can be performed.)

Materials
(for each lab group)

- 0.1 M NaOH solution, 50 mL
- Aspirin tablets, 6
- Aspirin substitute (acetaminophen) tablets, 6 (Ibuprofen tablets are not recommended because they do not grind as well)
- Buffered aspirin tablets, 6
- Distilled water
- 50 mL beakers, 4
- 100 mL beakers, 6
- 250 mL beakers, 4
- 100 mL graduated cylinder
- Balance (may be shared among lab groups)
- Beaker tongs
- Drying oven (may be shared among lab groups)
- Glass stirring rod
- Mortar and pestle
- Ring and ring stand
- Rubber policeman
- Wash bottle

Properties of Analgesics

Situation

You are a research chemist at a small pharmaceutical firm that intends to break into the lucrative market of analgesics. Although large companies dominate the market, your company is determined to create an improved product that will draw consumers away from the established brand names. Your first job is to analyze the competition.

Acetylsalicylic acid

Background

An analgesic is a compound that acts as a pain reliever. There are several types of analgesics on the market today, but they can be separated into two groups: ones that contain aspirin and ones that do not. The active painkiller in all brands of aspirin is acetylsalicylic acid. Besides relieving pain, acetylsalicylic acid reduces fever and acts as an anti-inflammatory drug by increasing the flow of blood and decreasing the amount of prostaglandins released. A different type of pain reliever is called acetaminophen. Acetaminophen is a pain reliever designed for people who are allergic to aspirin or who cannot tolerate aspirin. Like aspirin, acetaminophen reduces fever and relieves pain. However this medication does not have the anti-inflammatory properties of aspirin.

Acetaminophen

In addition to the active ingredients, analgesics contains fillers. The purpose of one type of filler is to keep the tablet from falling apart. Another type of filler keeps the tablet from breaking down due to moisture in the air. Buffered aspirin contains substances that help control the pH of the medication.

Problem

Your company has asked you to analyze aspirin, buffered aspirin, and an aspirin substitute to find out what percentage of these products is insoluble filler and what percentage is active ingredient. You will also add carefully measured amounts of base to these products to determine their pH and how easily they can be neutralized.

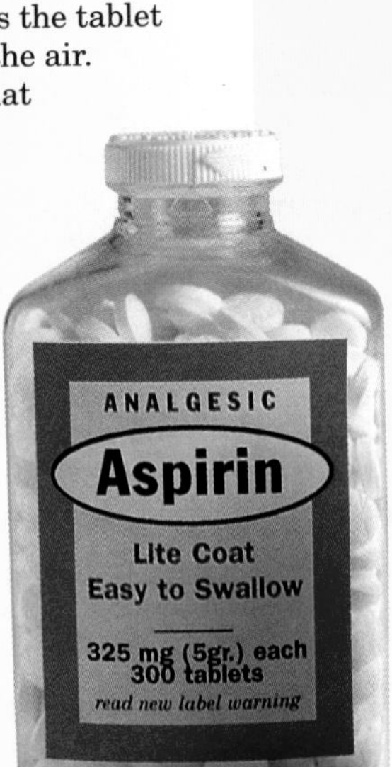
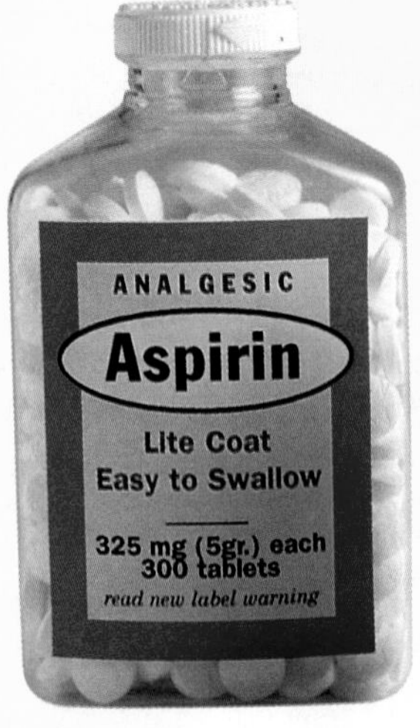

ANALGESIC
Aspirin
Lite Coat
Easy to Swallow
325 mg (5gr.) each
300 tablets
read new label warning

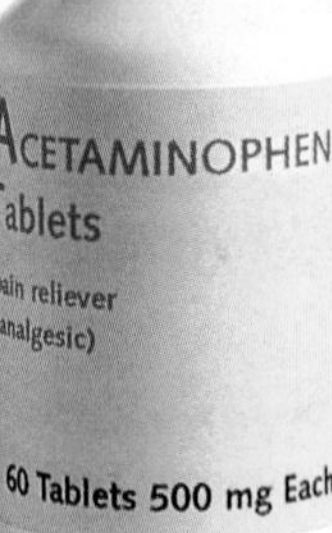

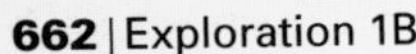

Objectives

Measure, calculate, **and *compare* the percentage of insoluble substances in aspirin, buffered aspirin, and aspirin substitutes.**

Compare **the pH of aspirin, buffered aspirin and aspirin substitute.**

Graph **the change in pH for the titration of the three analgesics.**

Demonstrate **proficiency in the following laboratory techniques: filtration, titration, and measuring pH.**

Required Precautions

- Always wear goggles and a lab apron to provide protection for your eyes and clothing.
- Read all safety cautions, and discuss them with your students.
- Students should not handle concentrated base solutions.
- Wear goggles, a face shield, impermeable gloves, and a lab apron when you prepare the NaOH. Work in a hood known to be in operating condition, with another person present nearby to call for help in case of an emergency. Be sure that you are within

Safety

Always wear goggles and an apron to provide protection for your eyes and clothing. If you get a chemical in your eyes, immediately flush it out at the eyewash station while calling to your teacher. Know the locations of the emergency lab shower and eyewash and how to use them.

Do not touch any chemicals. If you get a chemical on your skin or clothing, wash it off at the sink while calling to your teacher. Make sure you carefully read the labels and follow the directions on all containers of chemicals that you use. Do not taste any chemicals or items used in the laboratory. Never return leftovers to their original containers; take only small amounts to avoid wasting supplies.

Never put broken glass or ceramics in a regular waste container. Broken glass or ceramics should be disposed of separately.

Call your teacher in the event of an acid or base spill. Acid or base spills should be cleaned up promptly, according to your teacher's instructions.

Always clean up the lab and all equipment after use, and dispose of substances according to proper disposal methods. Wash your hands thoroughly before you leave the lab after all lab work is finished.

Preparation

1. ***Organizing Data***
   Prepare two data tables in your lab notebook. The first should contain four columns and six rows. Fill in the first row with these labels for the columns: *Measurement, Aspirin, Buffered aspirin*, and *Aspirin substitute.* Fill in the first column with these labels for the second through sixth rows: *Mass of 5 tablets, Mass of first filter paper, Mass of second filter paper, Mass of beaker,* and *Mass of beaker with dried filter paper(s) and insoluble substance.*
   Provide space below this table for recording your observations of the filtering processes.
   The second data table should contain seven columns and four rows. The columns should be labeled *Analgesic, 0 mL, 1 mL, 2 mL, 3 mL, 4 mL,* and *5 mL*. Fill in the first column with these labels for the second through fourth rows: *Aspirin, Buffered aspirin,* and *Aspirin substitute.*
   If you will be using a dropper bottle or micropipet, provide space below your table for the number of drops in a milliliter. If you will be using a buret, provide space below your table for six buret readings for each type of analgesic.
2. Label a 250 mL beaker, two 100 mL beakers, and a 50 mL beaker *Aspirin.* Label another similar set of beakers *Buffered aspirin*, and a third similar set of beakers *Aspirin substitute*. Label a 250 mL beaker *Waste*.

Materials

- **0.1 M NaOH solution, 50 mL**
- **Aspirin tablets, 6**
- **Aspirin substitute tablets, 6**
- **Buffered aspirin tablets, 6**
- **Distilled water**
- **25 mL graduated cylinder**
- **50 mL beakers, 4**
- **100 mL beakers, 6**
- **10 mL graduated cylinder**
- **100 mL graduated cylinder**
- **250 mL beakers, 4**
- **Balance**
- **Beaker tongs**
- **Ceramic spoon or rubber policeman**
- **Dropper bottle, micropipet, or buret with clamp**
- **Drying oven**
- **Filter paper**
- **Glass funnel or Büchner funnel and equipment**
- **Glass stirring rod**
- **Mortar and pestle**
- **pH paper, pH meter, or LEAP system with pH probe and test-tube clamp**
- **Ring and ring stand**
- **Wash bottle**

Gravity Filtration Option
- Glass funnel
- Filter paper

Vacuum Filtration Option
- Aspirator for spigot
- Büchner funnel (either ceramic or plastic)
- Filter paper
- One-hole rubber stopper or sleeve
- Vacuum flask (sidearm flask) and tubing

pH Paper Option
- pH paper, narrow range

pH Meter Option
- pH meter

LEAP System Option
- LEAP system with pH probe
- Test-tube clamp

Dropper Bottle or Micropipet Option
(use with pH paper or pH meter options)
- 10 mL graduated cylinder
- Dropper bottle or micropipet

Buret Option
(use with any options)
- Buret
- Buret clamp

Solution/Material Preparation
1. Dissolve 4.0 g of NaOH in a 1.00 L volumetric flask to make 0.10 M NaOH. Observe the required precautions.
2. Aspirin, buffered aspirin, and acetaminophen may be purchased over the counter at most stores.

30 s walking distance of a safety shower and eyewash station known to be in operating condition.

- In case of spills, first dilute with water. Then mop up the spill with wet cloths or a wet cloth mop designated for spill cleanup while wearing disposable gloves.
- If a pH meter or the LEAP System with pH is used, the precautions listed in the teacher's notes on page 652 must be followed to avoid electric shock.

LEAP System Option
Prior to the lab, the *Open Me First* packet and the *LEAP Chemistry Lab Manual* should be reviewed. Review the system to be sure you understand how to do the following.

- Create an experiment or use a LEAP experiment.
- Set up a probe.
- Calibrate a pH probe.
- Start and stop the system.
- Print the graph produced during the lab. (Hint: make sure the printer in *system maintenance* corresponds to the printer being used).
- Set up multiple experiments (if more than one group is using the system).
- Save data.
- Recall saved data.
- Quit the system.

Student Orientation

Pre-lab Discussion
Be sure to explain to students exactly which equipment options you will be using. For filtration (steps 1–18), either gravity filtration or vacuum filtration may be used. For measuring the pH (steps 19–20), either pH paper, a pH meter, or the LEAP System with a pH probe may be used. For the titration (steps 21–23), a dropper bottle or micropipet or buret can be used with any of the pH-measuring devices.

It is not essential that students understand everything about solubility, acids, bases, buffers, and titrations for this lab. Students should realize that the insoluble fillers will not dissolve in water and can be filtered out of the water-analgesic mixture. They should also realize that acids and bases are opposites and that the pH scale is a way to measure how acidic or basic something is.

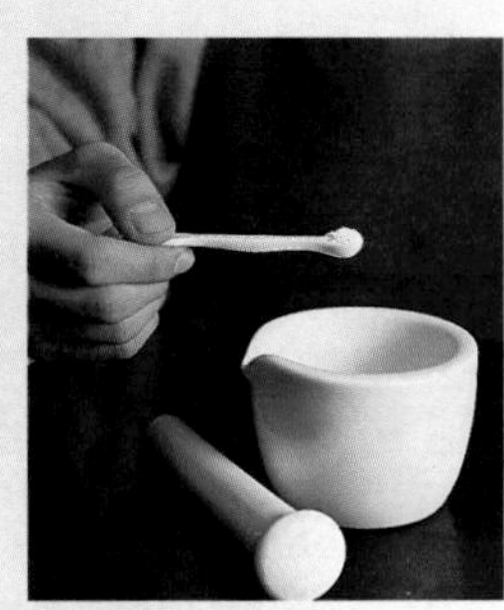

Technique

Determining Mass of Insoluble Filler

3. Determine the masses of five aspirin tablets, five buffered-aspirin tablets, and five aspirin-substitute tablets. Record the three masses in your data table.
4. Using a 100 mL graduated cylinder, pour about 85 mL of distilled water into each of the three 250 mL beakers.
5. Using a mortar and pestle, grind the five aspirin tablets, one at a time, into a fine powder. First, gently pound a tablet to pulverize it and then grind it to a powder. When removing the powdered substance from the mortar, be sure to use a nonmetal tool such as a rubber policeman or ceramic spoon, as metal can scratch the finish of the ceramic mortar. Pour the powder into the 250 mL beaker labeled *Aspirin*. Carefully clean and dry the mortar and pestle, and repeat this procedure for the buffered aspirin and aspirin substitute, pouring each into the appropriately labeled 250 mL beaker.
Do not blow into the mortar to remove any remaining powder because dust may get into your eyes and nasal passages.
6. Stir the mixture in each of the beakers for approximately 3–5 min to dissolve as much of the substances as possible.
7. Measure the mass of a piece of filter paper. If you will be using vacuum filtration and a Büchner funnel, be sure that the filter paper is big enough to cover all of the holes in the Büchner funnel. Record the mass of this first piece of filter paper in the data table.
8. Filter the mixture. If you are using vacuum filtration and a Büchner funnel, follow the instructions given step **8a**. If you are using gravity filtration and a glass funnel, follow the instructions given in step **8b**. (The vacuum filtration requires more complicated equipment, but it works faster than the gravity filtration.)

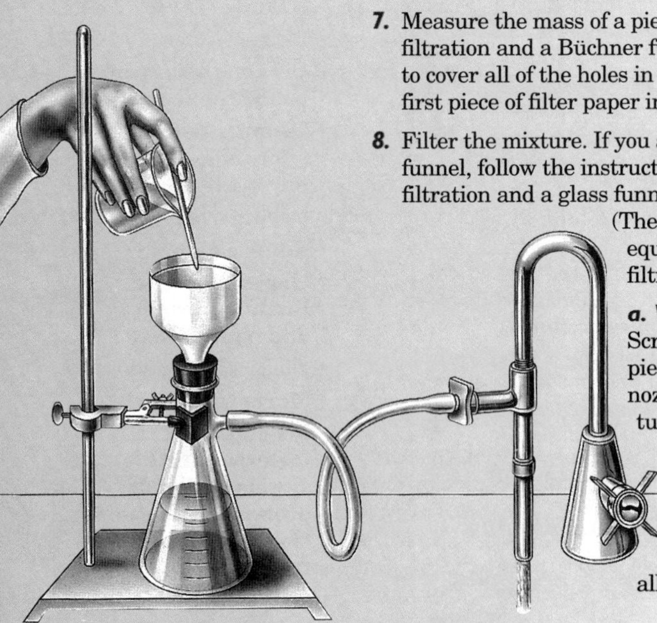

a. Vacuum Filtration with a Büchner Funnel
Screw an aspirator nozzle onto the faucet. Attach a piece of thick plastic tubing to the side arm of this nozzle. Attach the other end of the thick plastic tubing to the side arm of the filter flask. Place a one-holed rubber stopper on the stem of the funnel, and fit the stopper snugly in the neck of the filter flask as shown in the drawing to the left. Place the first filter paper on the bottom of the funnel so that it is flat and covers all of the holes in the funnel.

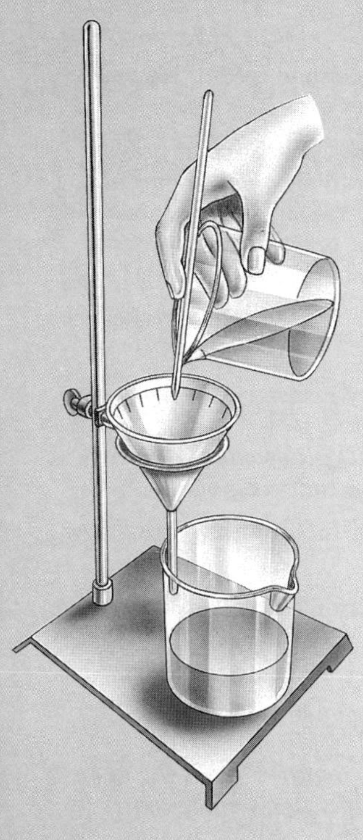

Turn on the water at the faucet. Stir the contents of the 250 mL beaker labeled *Aspirin* and pour the mixture against a glass stirring rod into the funnel. The drawing shows the proper technique for pouring. Continue pouring until the beaker is empty. Rinse the 250 mL beaker with a small amount of distilled water, and pour this into the funnel as well. When filtration is completed, pour the filtrate into one of the 100 mL beakers labeled *Aspirin*. Rinse the filter flask with a small amount of distilled water, and pour this into the 100 mL beaker as well.

b. Gravity Filtration with a Glass Funnel

Set up a glass funnel supported by a ring stand and ring as shown in the drawing to the left, with one of the 100 mL beakers labeled *Aspirin* beneath the stem of the funnel, so that it can catch the liquid filtrate after it passes through the funnel. To prepare the filter paper fold it in half along its diameter, and then fold it again to form a quadrant. Separate the folds of the filter paper so that three thicknesses are on one side and one thickness is on the other, as shown in the drawing below. Use your wash bottle to wet the funnel before the paper is placed in it. Fit the filter paper in the funnel, and wet it with a little water. Gently but firmly press the paper against the sides of the funnel so that no air can get between the funnel and the paper. The filter paper should not extend above the top of the funnel.

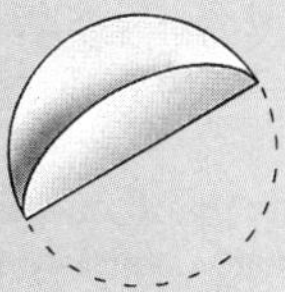

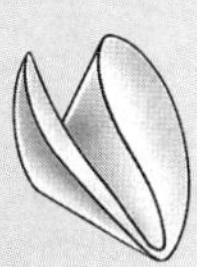

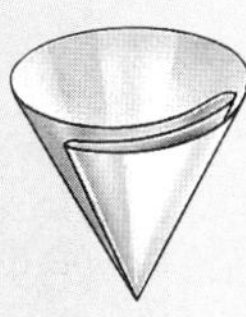

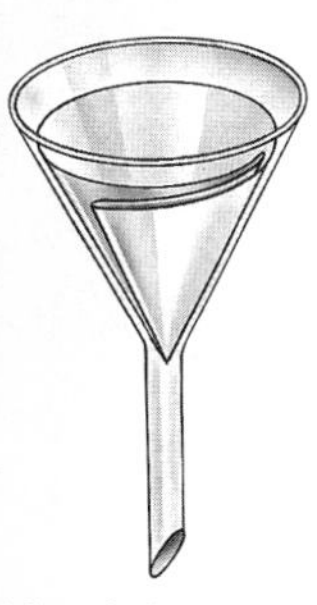

Pour the mixture from the 250 mL beaker along the side of glass stirring rod into the funnel. Be sure that you pour slowly enough so that the fluid level always remains at least 1.0 cm below the top of the filter paper. Do not fill the funnel to the top of the filter paper. Keep the funnel filled with fluid at this level so that a water column without air bubbles is established in the stem of the funnel. Rinse the 250 mL beaker with a small amount of distilled water, and pour this into the funnel as well.

9. When filtering is complete, look at the filtrate, the liquid that passed through the filter. Record your observations about the filtrate in your lab notebook. Be sure to note whether or not any undissolved particles are still visible.
10. Measure the mass of the empty 100 mL beaker labeled *Aspirin*, and record the mass in your data table.
11. Carefully remove the filter paper from the funnel, making sure you do not lose any of the insoluble substance that was left behind. Place it in the 100 mL beaker whose mass you just measured. If the filtrate in step **9** was clear, without particles, skip ahead to step **15**. If the filtrate still contained particles, consult with your teacher. If your teacher tells you to re-filter the mixture, continue with step **12**. If not, continue with step **15**.

Techniques to Demonstrate
Demonstrate proper use of the mortar and pestle. Be sure students know to gently pound the tablet first to break it up before they begin grinding it. Remind them to use a rubber policeman or other nonmetal tool to remove the powder because metal will scratch the mortar's ceramic finish.

Demonstrate the filtration technique you will use. Be sure to point out common technique errors such as pouring too much mixture into a filter or misaligning the filter paper in the funnel.

Demonstrate the pH-measuring technique you will use. Be sure to remind students to use a test buffer to calibrate the pH meter or the LEAP System pH probe. Also remind students to turn off the meter or probe and any related equipment when they are finished.

Also demonstrate the titration method you will use. If a buret will be used, remind students about washing and rinsing the buret and how to read the meniscus. Show them how to hold the stopcock to release small amounts of liquid. If a dropper or micropipet will be used, explain how adding a specified number of drops to a measured volume of fluid in a graduated cylinder can be used to determine the volume of each drop. Emphasize the need to make all drops about the same size.

12. Measure the mass of a second piece of filter paper and record this mass in your data table.

13. Filter the mixture once more, using either vacuum filtration or gravity filtration, as described earlier. For vacuum filtration, when you are finished, empty the filtrate into the now-empty 250 mL beaker labeled *Aspirin*. For gravity filtration, place this beaker underneath the stem of the glass funnel so that it can catch the filtrate.

14. Put the second filter paper into the appropriately labeled 100 mL beaker with the first filter paper.

15. Place the beaker with the filter paper in the drying oven, or leave it on the lab bench to dry overnight.
Remember to use beaker tongs to handle all glassware that may be hot, especially beakers that have been in a drying oven.

16. Repeat steps **5–15** for buffered aspirin and aspirin substitute, using new beakers and filter papers.

17. Dispose of the filtrate by pouring it down the drain with an excess of water.

18. When the insoluble substances are dry, measure the mass of each beaker containing the filter papers and the insoluble material, and record the masses in your data table.
To avoid damaging the balance, allow the beakers from the oven to cool before they are placed on the balance. Remember to use beaker tongs to handle the hot beakers.

Determining pH

19. Using a mortar and pestle, grind one aspirin tablet and transfer it to the 50 mL beaker labeled *Aspirin*. Add 20 mL of distilled water, and stir to dissolve the aspirin.

20. Measure the pH of the aspirin, and record your results in the data table in the box for *0 mL* of base added. Use the pH measurement procedure below that matches the equipment available in your lab.

    a. pH paper
    Dip a small piece of pH paper into the liquid. Remove the paper, and compare the color of the wet pH paper to the standard pH colors on a chart provided by your teacher.

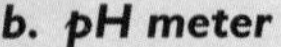

    b. pH meter
    Turn on your pH meter. Put the pH electrodes into a buffer solution provided by your teacher. Calibrate the meter to the correct pH according to your teacher's directions. Then place the pH electrodes into the solution you are testing, and read the pH on the meter. Record your measurements in your data table.

Sample Data

Measurement	Aspirin	Buffered aspirin	Aspirin substitute
Mass of 5 tablets	1.91 g	3.53 g	2.64 g
Mass of 1st filter paper	0.40 g	0.40 g	0.40 g
Mass of 2nd filter paper	0.40 g	0.40 g	0.40 g

Measurement	Aspirin	Buffered aspirin	Aspirin substitute
Mass of beaker	97.93 g	102.81 g	108.92 g
Mass of beaker with dried filter paper and substance	100.12 g	106.78 g	111.94 g

c. LEAP pH Probe

Your teacher should have the computer already set up for you. Make sure the probe is calibrated according to your teacher's directions. Hold the probe securely in the solution with a test-tube clamp attached to the ring stand. The probe should be submerged in the solution, but it should not touch the bottom of the beaker. On the computer, move the cursor to the green portion of the stoplight, and start the experiment. After the initial reading has been taken and the probe has stabilized, put the stoplight on stop, and record this initial data in your lab notebook. Then erase this initial data so that you are ready for the titration step.

Measuring pH Changes As Base Is Added

21. Using the titration method that matches your equipment, add 0.1 *M* NaOH solution to the beaker in 1.0 mL increments. After each addition is thoroughly mixed, measure the pH, recording it in your data table.

a. Dropper Bottle or Micropipet

Pour distilled water into a 10 mL graduated cylinder, until it is filled to the 5 mL mark. Remember to read the bottom of the meniscus when using a graduated cylinder. Fill your dropper or micropipet with distilled water. Count the number of drops that must be added from the dropper or micropipet before the level in the graduate cylinder reaches the 6 mL mark. Record this number in your data table. Then, add 1 mL worth of drops to the beaker labeled *Aspirin*. Swirl the beaker gently and wash down the sides with distilled water from a wash bottle. Measure and record the pH value in the data table in the box for *1 mL* of base added. Add the same number of drops again, and record the pH value in the data table in the box for *2 mL* of base added. Continue until a total of 5 mL of base have been added. Then, repeat the process for the buffered aspirin and the aspirin substitute.

b. Buret With pH Paper or pH Meter

Burets provide better precision because they are calibrated in intervals of 0.1 mL. Before using the buret, rinse the inside three times, using 5 mL of 0.1 M NaOH each time. Collect the NaOH used for rinsing in the *Waste* beaker.

Use a buret clamp to hold the buret in position on a ring stand as shown in the drawing to the left. Place the *Waste* beaker below the buret to catch any liquid that is drawn off. Into another 50 mL beaker, pour approximately 40 mL of 0.1 *M* NaOH solution from the reagent bottle. ***Carefully check the label of the reagent bottle before removing any liquid.***

Using the 50 mL beaker, fill the buret with NaOH, and then draw off enough liquid into the *Waste* beaker so that you can be sure the tip of the buret below the stopcock is filled. The level of the liquid should be on the scale. To measure the volume of the liquid, read the scale at the bottom of the meniscus.

After you have taken your first buret reading, practice using the buret and reading the volume several times. Open the stopcock and draw off approximately 1.0 mL. Read the volume of the liquid again. The exact amount drawn off is equal to the difference between your initial and final buret readings. When you can control the volume of base fairly well, record the latest buret reading in your data table.

Post-Lab

Disposal

- The soluble parts of the aspirin may be poured down the sink with running water.
- The insoluble part of the aspirin may be placed in the trash.
- Be sure to set out a single disposal container for excess liquids from the titration.
- The disposal container will contain titrated analgesic solutions, contents of the waste beaker, rinsings from the burets, and excess NaOH. This mixture should be neutralized with 0.10 M HCl until the pH is between 5 and 9. Then it can be poured down the sink. Be sure to test the pH before pouring anything down the sink.

Titration results: pH after NaOH additions

Analgesic	0 mL	1 mL	2 mL	3 mL	4 mL	5 mL
Aspirin	2.9	3.2	3.5	3.9	4.4	6.3
Buffered aspirin	6.3	7.4	7.7	7.9	8.0	8.1
Aspirin substitute	8.5	8.6	8.7	8.9	9.0	9.1

- Clear or white crystals are still visible after the first filtration. The filtrate should be clear after the second filtration.
- LEAP System graphs should show pH for buffered aspirin and aspirin substitute changed only slightly, but the pH for aspirin changed very much when NaOH was added.

Answers to Analysis and Interpretation

1. aspirin:
   100.12 g – (97.93 g + 0.80 g)
   = 1.39 g insoluble

   buffered aspirin:
   106.78 – (102.81 g + 0.80 g)
   = 3.17 g insoluble

   aspirin substitute:
   111.94 – (108.92 + 0.80 g)
   =2.22 g insoluble

2. aspirin:
   6.3 – 2.9 = 3.4 change in pH

   buffered aspirin:
   8.1 – 6.3 = 1.8 change in pH

   aspirin substitute:
   9.1 – 8.5 = 0.6 change in pH

3.

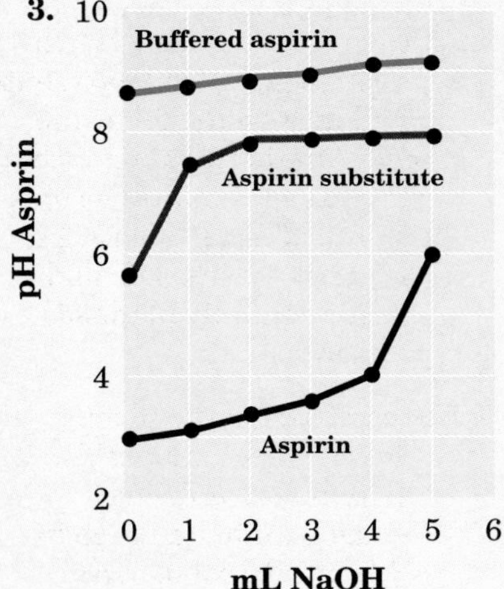

Conclusions

4. aspirin:
   $\frac{1.39 \text{ g insoluble}}{1.91 \text{ g total}} = 72.8\%$ insoluble

   buffered aspirin:
   $\frac{3.17 \text{ g insoluble}}{3.53 \text{ g total}} = 89.8\%$ insoluble

   aspirin substitute:
   $\frac{2.22 \text{ g insoluble}}{2.64 \text{ g total}} = 84.1\%$ insoluble

5. Aspirin was the most acidic and also the most affected by the addition of NaOH.

6. Students answers may vary. One point could be that if someone has a problem with ulcers or acid indigestion, buffered aspirin or aspirin substitute would be a better choice than regular aspirin. However, tablet for tablet, aspirin has the smallest amount of insoluble fillers and the greatest amount of active ingredient.

Move the *Waste* beaker aside, and use the buret to add 1.0 mL of 0.1 M NaOH to the beaker containing the mixture of aspirin and distilled water. Record the exact buret reading in your data table. Swirl gently, measure the pH, and record it in your data table. Add the NaOH 1.0 mL at a time, recording the buret reading and the pH, until a total of 5.0 mL has been added. Then, repeat the process for the buffered aspirin and the aspirin substitute.

c. Buret with LEAP pH Probe

Burets provide better precision because they are calibrated in intervals of 0.1 mL. Use a buret clamp to hold the buret in position on a ring stand, as shown in the drawing on the left. Calibrate your buret so that it will release 10 drops per second. To do this, fill your buret with water, and place the *Waste* beaker under the buret. Use a stopwatch to measure 10 second intervals. Open the stopcock just enough for the water to flow out at a rate of 10 drops per second. Practice this technique, so you know exactly how far to open the stopcock.

When you are ready to proceed, rinse the inside of the buret three times, using 5 mL of 0.1 M NaOH each time. Collect these rinses in the *Waste* beaker.

Into another 50 mL beaker pour approximately 40 mL of 0.1 *M* NaOH solution from the reagent bottle. ***Carefully check the label of the reagent bottle before removing any liquid.***

Fill the buret with NaOH, and then draw off enough liquid to fill the tip of the buret below the stopcock. The level of the liquid should be on the scale. To measure the volume of the liquid read the scale at the bottom of the meniscus. Practice this procedure several times. Place the beaker labeled *Aspirin* under the buret. Use a test-tube clamp attached to the ring stand to keep the tip of the pH probe submerged in the solution near the bottom of the beaker as shown in the photograph on the left. Start the experiment and begin the titration. Continue titrating until approximately 950 readings have been taken. Then set the stoplight to red, and print your graph. ***Only 1000 data points will be saved. If you continue taking data after this point, you will lose it.*** Record the final volume on the buret.

22. Repeat steps **19–21** for the buffered aspirin and aspirin substitute.

Cleanup and Disposal

23. The insoluble part of the aspirin can be disposed of by placing it in the trash. Soluble portions can be washed down the sink. After all titrations are complete, dispose of the contents of each of the beakers in the containers designated by your teacher.

Analysis and Interpretation

1. Organizing Data

Calculate the mass of the insoluble portion of each analgesic. This mass can be found by subtracting the mass of the beaker and each filter paper from the mass of the beaker with both filter papers and the insoluble material.

2. ***Organizing Data***
Calculate the change in pH for each substance by finding the difference between the initial pH value and the pH after adding 5.0 mL of NaOH solution.

3. ***Organizing Data***
Make a graph of your pH data. Plot the number of drops or the milliliters of NaOH solution on the x-axis and pH on the y-axis. Graph data for all three analgesics on the same grid. Label each axis, and give your graph a title.

Conclusions

4. ***Inferring Conclusions***
Which type of analgesic had the largest percentage of insoluble material? To calculate the percentage that is insoluble material, divide the mass of insoluble material by the total mass of the analgesic, and multiply by 100.

5. ***Inferring Conclusions***
Review the graph of the pH values for the three types of analgesics. Which analgesic was most acidic (lowest pH)? Which was most affected by the addition of NaOH?

6. ***Inferring Conclusions***
Using all the data you organized, write a paragraph summarizing your recommendation for the use of the three analgesics. Under what circumstances might one be better than another?

7. ***Relating Ideas***
After analyzing several analgesic products now on the market, what improvements would you suggest for the new product your company wants to design?

Extensions

1. ***Research and Communication***
Aspirin and aspirin substitutes are normally sold in a dosage of grains. Research what a grain is, and calculate the average dosage of the three analgesics in grains. Is the dosage the same for all three analgesics?

2. ***Designing Experiments***
What possible sources of error can you identify with this procedure? If you can think of ways to eliminate them, ask your teacher to approve your plan, and run the procedure again.

3. ***Designing Experiments***
When ingested, analgesics are dissolved in the acidic environment of the stomach. Design an experiment to determine the percent solubility of each analgesic in an environment similar to the stomach. If your teacher approves of your plan, try your procedure.

4. ***Designing Experiments***
Obtain several different brands of regular and extra-strength aspirin products. Run the procedures again to determine if there are differences among brands and if regular and extra-strength products differ in terms of fillers and pH.

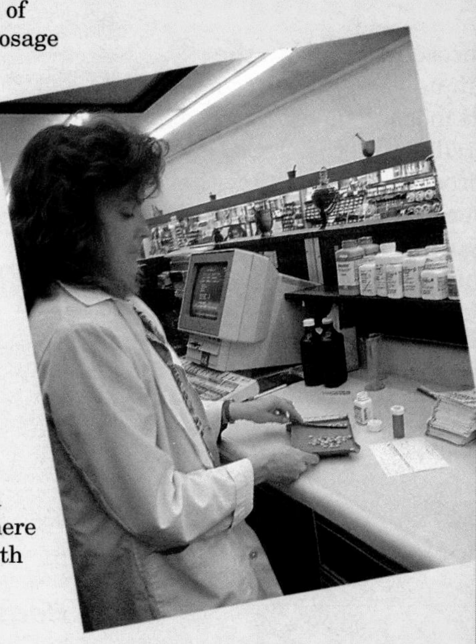

7. Students answers will vary. One possible suggestion would be to explore a better way to encapsulate the medicine so that the dose would not require insoluble fillers and would thus be smaller.

Extensions

1. A grain is the smallest unit of mass of the avoirdupois system; a pound contains 7000 grains, and an ounce contains 437.5 grains. A grain equals about 0.0648 g. For aspirin or buffered aspirin, a typical dosage is about 650 mg, or about 10 grains. A typical dosage for acetaminophen is 1.0 g, or about 15.5 grains.

2. Student suggestions for improving results will vary, but could include the following: more careful filtration techniques, use of multiple runs, and use of more tablets. Be sure answers are safe and include carefully planned procedures.

3. Student suggestions for mimicking the stomach's environment will vary. Stomach contents range in pH from 1.0 to 3.0, corresponding to a 0.1 M HCl solution and a 0.001 M HCl solution, respectively. Be sure answers are safe and include carefully planned procedures. Any waste generated must be neutralized with 0.1 M NaOH before it is poured down the drain.

4. Student answers will vary depending on the brands and products chosen.

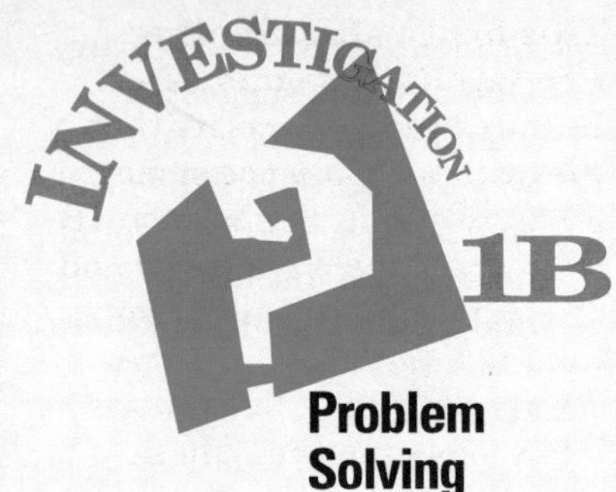

Objectives

Students will:

- use appropriate lab safety procedures.
- identify an unknown by comparing its insoluble percentage or titration curve to those of analgesics from Exploration 1B.
- design and implement their own procedure.

Planning

Recommended Time: 1 lab period (overnight drying time may be necessary)

Prerequisite: Exploration 1B

Solution/Material Preparation

1. See the Teacher's Notes for Exploration 1B for information on preparing 0.10 M NaOH.
2. Scrape off any brand name or markings on the tablets before giving them to the students.
3. You can provide one buret of NaOH per lab group, or you can set this up as a station, with three or four groups sharing one station. If you do not have burets, dropper bottles can be used, but students must calibrate them by determining the number of drops in a milliliter.

Student Orientation

Techniques to Demonstrate

Review proper techniques for filtration, pH measurement, and titration. Emphasize the need to calibrate the pH meter or the LEAP System pH probe before making any measurements.

Pre-lab Discussion

Begin by discussing the results and procedure used in the Exploration, especially any lab technique errors that occurred.

Whichever technique you use for pH measurement, discuss its advantages and disadvantages compared to the other techniques.

If students are uncertain how to proceed, suggest that they use equipment amounts and techniques similar to those in the Exploration, but scaled down for one or two tablets. Be sure they consider what is the most efficient use of the two tablets.

Sample Data

Mass of tablet	0.43 g
Mass of filter paper 1	0.40 g
Mass of filter paper 2	0.40 g
Mass of beaker	98.86 g
Mass of beaker with filter paper and insoluble substance	99.98 g
Percentage of insoluble substance:	74 %

The substance is aspirin, as confirmed by the titration curve and pH data, which show that it has a low pH, which changes very much as base is added.

B

PROPERTIES OF ANALGESICS

Forensic Analysis

Police, Vets Seek Clues in Canine Poisonings

1 Dog Dies, 2 Others Remain in Serious Condition

Veterinarians on the staff of the Animal Rescue Center are working feverishly to save the lives of two poisoned dogs. Another dog from the same neighborhood has already died from the unidentified toxin. Police are seeking clues to identify the poison.

Alma and David Brock, owners of the dog that died, discovered their pet, Shiloh, lying comatose near a bike path behind their home. Police were called to investigate after two more dogs were found unconscious in the same neighborhood. The two surviving dogs, whose owners have not been identified, are in serious condition at the Animal Rescue Center.

Police were led by David Brock to the spot where Shiloh was found. There they found the remains of some pills that could have been ingested by the dogs. Police spokesperson Deanna Jackson said that although police believe the poisoning was unintentional, they are looking for leads that will help identify the pills.

Dr. Enriqueta Sust-Newland, chief veterinarian at the clinic, suspects that they are analgesics. "Until we know what we are dealing with, we can't be certain that we are treating the affected dogs properly," she said. "Unfortunately, the clinic doesn't have the facilities or the personnel to analyze the pills."

continued on page B8

Report Discovers Shortage of ATM's in Poor Areas

Bank branches in the outer boroughs are less likely to have automatic teller machines than downtown branches, and the disparity is much greater between middle-income and poor areas within each of the boroughs, a new study has found. The probl... study released yesterday by the staff of Jerome Green, the city's Public Advocate.

And a simple, if limited, solution offered in the report is to put the machines inside or in front of police ...

Required Precautions

Goggles and lab aprons must be worn at all times.

Do not touch or taste any chemicals. Wash your hands thoroughly when finished.

If you get acid or base on your skin or clothing, wash it off at the sink while calling to your teacher. If you get acid or base in your eyes, immediately flush it out at the eyewash while calling to your teacher.

Required Precautions

- Goggles and a lab apron must be worn at all times.
- Read all safety cautions, and discuss them with your students.
- In case of base spills, dilute with water. Then mop up the spill with wet cloths or a cloth mop designated for spill cleanup while wearing disposable gloves.
- If a pH meter or the LEAP System with pH probe is used, the precautions listed on page 652 must be followed to avoid electric shock.

Memorandum

CheMystery Labs, Inc.
52 Fulton Street
Springfield, VA 22150

Date: January 4, 1995

To: Michelle Dunnigan

From: Marissa Bellinghausen

I love dogs, and when I read this news clip, I called Dr. Sust-Newland and volunteered to analyze the pills. This will be a freebie, of course, because the Animal Rescue Center is operating on a shoestring and can't pay us a dime, but I think the company will get some much needed publicity that could generate new business. The center is a nonprofit organization, so all donated materials and time are tax-deductible. But any expenses will be taken out of the company's pocket, so remember to cut costs as much as possible without compromising the results. We were only given a small sample to work with.
Before you start in the lab, I need the following items.
- detailed procedure for identifying this unknown analgesic
- detailed chart for all data to be collected
- list of materials needed, along with individual and total costs (Be careful and keep accurate records so that we can deduct these expenses from our taxes!)

When your analysis is completed, I will call Dr. Sust-Newland with preliminary results. Please also submit a report in the form of a two-page letter. It must include the following.
- your conclusion about the identity of the analgesic
- explanation of the evidence you found
- description of the procedure(s) you used, and all data, organized in a table
- graphs comparing the pH curve for the unknown analgesic to those of analgesics we've already tested
- discussion of any discrepancies in your results

It is critical that we complete this job accurately and quickly. It would be a real feather in our cap if we were able to help save these dogs' lives.

References

Refer to page 15 for more information about different types of analgesics. You must refer to Exploration 1A to compare the test results for the different types of analgesics to the results for this unknown analgesic.

Materials for
FORENSIC ANALYSIS

Item	Cost
REQUIRED ITEMS (You must include all of these in your budget.)	
Lab space/fume hood/utilities	15 000 – /day
Standard disposal fee	2 000 – /g of product
Balance	5 000
Beaker tongs	1 000
Drying oven	5 000 – /day
REAGENTS and ADDITIONAL EQUIPMENT (Include in your budget only what you'll need.)	
0.1 M NaOH	100/mL
50 mL beaker	1 000
250 mL beaker	1 000
10 mL graduated cylinder	1 000
100 mL graduated cylinder	1 000
Büchner funnel	2 000
Buret tube	5 000
Filter flask with sink attachment	2 000
Filter paper	500 – /piece
Glass stirring rod and rubber policeman	1 000
Glass funnel	1 000
Mortar and pestle	2 000
pH meter	3 000
pH paper	2 000 – /piece
pH probe—LEAP with test tube clamp	5 000
Pipet or medicine dropper	1 000
Ring stand with buret clamp	2 000
Wash bottle	500
* No refunds on returned chemicals or unused equipment.	
FINES	
OSHA safety violation	2 000 – /incident

Spill Procedures/ Waste Disposal Methods

- **In case of a spill, follow your teacher's instructions.**
- **Put excess chemicals and solutions in the containers designated by your teacher. Do not pour them down the sink or place them in the trash can.**

Materials
(for each lab group)
- 0.10 M NaOH, 50 mL
- Aspirin tablet, 2
- Distilled water
- 100 mL graduated cylinder
- 50 mL beaker, 2
- Balance
- Beaker tongs
- Drying oven
- Filter paper
- Glass stirring rod
- Mortar and pestle
- Rubber policeman
- Wash bottle

See the Teacher's Notes for Exploration 1B for information on materials for filtration and pH measurement options.

Estimated cost of materials:
$145 000–$155 000

Tips for Evaluating the Pre-lab Requirements
Be sure students specify that they will measure the same properties as in Exploration 1B for the unknown analgesic and compare to the results from the Exploration. Data tables should contain spaces for the items shown in the Sample Data section.

Students may decide to only use one test to identify the unknown. The titration test is the best one for identifying the unknown because buffered aspirin and aspirin substitute have similar percentages of insoluble fillers.

Proposed Procedure

(Students may perform either technique or both.)
Filtration: Measure the mass of a beaker and two pieces of filter paper. Grind a tablet in the mortar. Pour it into a second beaker with 20 mL of water, and stir with a glass stirring rod to dissolve as much as possible. Filter the mixture. Place the filter paper and insoluble material in the first beaker. Dry the beaker and its contents overnight and then measure the mass.
Titration: Grind a tablet in the mortar. Pour it into a beaker with 20 mL of water, and stir with a glass stirring rod to dissolve as much as possible. Measure and record the pH. Add 1 mL of 0.10 M NaOH. Measure and record the pH again. Continue adding NaOH, 1 mL at a time, until 5 mL have been added.

Post-Lab

Disposal
- The soluble parts of the aspirin may be poured down the drain.
- The filter paper with the insoluble aspirin components may be discarded in the wastebasket.
- The basic solutions from the titration should be neutralized with 0.10 M HCl until a pH between 5 and 9 is reached and then poured down the drain.

EXPLORATION 2A

Objectives

Students will:

- use appropriate lab safety procedures.
- use a laboratory balance to measure mass.
- use a graduated cylinder to measure volume.
- use a ruler to measure length.
- relate the increment on a measuring tool to the precision of measurements made with the tool.
- collect data and organize it in a table.
- calculate values from experimental measurements.
- relate accuracy to calculations of percent error.
- relate precision to calculations of absolute deviation, average deviation, and uncertainty.
- calculate an average value from class data and use it to calculate absolute deviation and average deviation.

Planning

Recommended Time: 1 lab period (2 lab periods if students set up LEAP System themselves)

Materials

(for each lab group)

- Distilled water
- Metal shot, about 20 g (Any type of metal shot will do, but aluminum and copper are recommended.)
- 250 mL beaker
- 25 mL graduated cylinder (for best results, marked in 0.2 or 0.5 mL increments)
- 100 mL graduated cylinder
- 15 cm plastic ruler
- Balance (Centigram balances will give best results, but less precise balances are acceptable.)
- Chemical handbook with density values for water and various metals (for example, *CRC Handbook of Chemistry and Physics, Lange's Handbook of Chemistry,* etc.)

Accuracy and Precision

Situation

As part of your orientation as a new employee at a chemical firm, you must participate in an assessment of your lab skills. This survey will help you evaluate the accuracy of your equipment and the precision of your lab technique.

Background

In Exploration 1A, you made measurements of volume and mass, but you did not evaluate these measurements. Measurements can be evaluated two different ways. *Accuracy* is how close to the actual value a measurement is; *precision* is how close each measurement is to others in the set.

The accuracy of a measurement can only be determined if you have some way of knowing what the measurement should be. For example, metal densities have been measured many times, so there is a consensus on the accepted value for each metal's density. One indicator of accuracy is the experimental error, which can be calculated using this equation.

$$\frac{\text{value}_{\text{accepted}} - \text{value}_{\text{measured}}}{\text{value}_{\text{accepted}}} \times 100 = \text{percent error}$$

Precision indicates whether the experimental data are consistent. Three common statistical tools used to check precision are *absolute deviation, average deviation,* and *uncertainty in measurement.*

For each experimental value, the absolute deviation is the difference between it and the average of all values. Average deviation is obtained by calculating the sum of the absolute deviations for an entire data set and dividing by the total number of measurements. To determine the uncertainty of a measurement, express one unit of the last significant digit as plus or minus to indicate that the last digit is an estimate, as in 24.65 ± 0.01 cm. The percent uncertainty is determined by dividing the uncertainty by the measurement and multiplying by 100, as shown here.

$$\frac{\pm 0.01 \text{ cm}}{24.65 \text{ cm}} \times 100 = \pm 0.04\%$$

Problem

You will use the measurement and analysis skills that you learn here to evaluate measurements of length, volume, mass, and temperature. You will also determine the identity of a metal sample by calculating its density and comparing the result to density values for various metals from chemical handbooks.

672 | Exploration 2A

Objectives

Calculate **values from experimental measurements.**

Demonstrate **proficiency in measuring masses and volumes.**

Organize **data by compiling it in tables.**

Calculate **an average value from class data.**

Relate **absolute deviation, average deviation, uncertainty, and percent error, and use them as methods for gauging accuracy and precision.**

Required Precautions

- Goggles and lab apron must be worn at all times.
- Confine loose clothing and long hair.
- Read all safety cautions, and discuss them with your students.
- If the LEAP System with thermistor probe is used, the precautions listed in the teacher's notes on page 652 must be followed to avoid electric shock.

Safety

Always wear goggles and an apron to provide protection for your eyes and clothing. If you get a chemical in your eyes, immediately flush it out at the eyewash station while calling to your teacher. Know the locations of the emergency lab shower and eyewash and how to use them.

Do not touch any chemicals. If you get a chemical on your skin or clothing, wash it off at the sink while calling to your teacher. Make sure you carefully read the labels and follow the directions on all containers of chemicals that you use. Do not taste any chemicals or items used in the laboratory. Never return leftovers to their original containers; take only small amounts to avoid wasting supplies.

Never put broken glass or ceramics in a regular waste container. Broken glass or ceramics should be disposed of separately.

Always clean up the lab and all equipment after use, and dispose of substances according to proper disposal methods. Wash your hands thoroughly before you leave the lab after all lab work is finished.

Preparation

1. ***Organizing Data***
   Prepare data tables in your lab notebook with spaces for the following.
   - smallest division on the ruler, the 100 mL and 25 mL graduated cylinders, and the balance
   - height of the 50 mL mark of the 100 mL graduated cylinder and its inside diameter
   - mass of the dry 25 mL graduated cylinder
   - temperature of water (several spaces)
   - handbook values for water density at these temperatures
   - volume of water added
   - mass of 25 mL graduated cylinder with water
   - volume of metal shot and water
   - mass of 25 mL graduated cylinder with water and metal

   Prepare another table with room for data from the rest of the class on the mass of the cylinder with water alone, the mass of the cylinder with water and metal, the volume of the water, and the volume of the water and metal.
2. Locate a handbook to use as a reference for comparing densities of metals and for determining the density of water at various temperatures.

Technique

3. Record the smallest division on your ruler and the 100 mL graduated cylinder in your data table.

Materials

- **Distilled water**
- **Metal shot**
- **250 mL beaker**
- **25 mL graduated cylinder**
- **100 mL graduated cylinder**
- **15 cm plastic ruler**
- **Balance**
- **Chemical handbook or source of density values for water and metals**
- **Non-mercury thermometer or LEAP System with thermistor probe**

- Celsius thermometer, non-mercury type, range from –10°C to 120°C (Students should use only nonmercury thermometers. If a mercury thermometer breaks, the droplets that are not cleaned up will quickly evaporate, creating toxic mercury vapors.)

Optional Equipment

- LEAP System with thermistor probe and related equipment See the LEAP System manual for instructions on setting up the thermistor probe to take temperature measurements.

Student Orientation

Pre-lab Discussion

Review the terms *accuracy, precision*, and *uncertainty*. It may help to refer students to Section 2–5, especially pages 60–61. For most students, this may be their first exposure to the concept of uncertainty in measurements. Students frequently equate science with absolute certainty, rather than with measurements of definite uncertainty.

Techniques to Demonstrate

Review the use of the analytical balance. Remind the students to check that the balance is properly calibrated before making a measurement, and be certain that students use the same balance for each measurement.

Remind students to measure volumes from the bottom of the meniscus.

Post-Lab

Disposal

There are no special disposal guidelines. The metal shot can be reused each class.

Answers to Analysis and Interpretation

1. Ruler: ± 0.5 mm (0.05 cm)
   100 mL graduated cylinder: ± 0.5 mL
   25 mL graduated cylinder: ± 0.1 mL
   Balance: ± 0.005 g

2. $V = \pi r^2 h$

   $$V = (3.14)\left(\frac{2.5\text{ cm}}{2}\right)^2(9.3\text{ cm})$$

   $V = 46\text{ cm}^3 = 46\text{ mL}$

3. Percent error:

   $$\frac{50.00\text{ mL} - 46\text{ mL}}{50\text{ mL}} \times 100 = 8\%\text{ error}$$

4. mass H_2O =
   55.98 g − 42.39 g = 13.59 g

   $m = DV$

   0.997 g/mL × 13.8 mL = 13.8 g

   Percent error:

   $$\frac{13.59\text{ g} - 13.8\text{ g}}{13.59\text{ g}} \times 100 =$$

   −2% error

5. volume of metal =
   20.2 mL − 13.8 mL = 6.4 mL

   mass of metal =
   73.98 g − 55.98 g = 18.00 g

   $D = m/V$

   $$\frac{18.00\text{ g}}{6.4\text{ mL}} = 2.8\text{ g/mL}$$

6. Student answers will vary depending on the data for the rest of the class. Be certain calculations follow rules for significant figures.

---

4. Measure the inside diameter of the 100 mL graduated cylinder with the ruler. Measure the height to the 50 mL mark. Record these measurements in your data table.
5. Examine the scale on the balance. What is the smallest division? Examine the 25 mL graduated cylinder. What is the smallest division? Record the results, with units, in your data table.
6. Using the balance, determine the mass of the dry 25 mL cylinder to the nearest 0.1 g. Record this mass in the data table.
7. Fill a beaker half full of water, and determine the temperature of the water to the nearest 0.1°C. Use step **7a** if you are using a thermometer and step **7b** if you are using the LEAP Thermistor probe. Record the temperature of the water in your lab notebook. In a handbook, look up the density of water at that temperature and record it in the data table.

   a. Thermometer
   Gently put the thermometer in the water, being careful not to break it. Wait 2 min before recording the temperature. After another 2 min, record the temperature again. Repeat this process every 2 min until two consecutive temperature measurements are the same.
   Never use a thermometer to stir anything. It will break easily. The glass wall surrounding the bulb is very thin to provide quick and accurate temperature readings.

   b. LEAP thermistor probe
   Put the LEAP thermistor probe in the water as if it were a thermometer. It should be set for your use, and all you have to do is click on the green light on the computer screen to start and click on the red light to stop. Record the temperature of the water. Keep taking measurements until the temperature remains constant for 2 min. With the thermistor, you don't have to wait as long for the probe to reach the correct temperature as you do with a conventional thermometer.
8. Put between 10 mL and 15 mL of water from the beaker in the 25 mL graduated cylinder. Read the volume to the nearest 0.1 mL, and record it. Determine the mass of the water plus the cylinder to the nearest 0.1 g. Record this value in your data table. Keep the water in the cylinder for the next step.
9. Add enough of the sample of metal shot to the cylinder to increase the volume by at least 5 mL. Determine the volume and the mass of the water and shot together in the graduated cylinder; record your measurements.

Cleanup and Disposal

10. The water can be washed down the sink after the metal shot has been removed and placed in the container designated by your teacher. Always wash your hands thoroughly after cleaning up the lab area and equipment.

Analysis and Interpretation

1. ***Organizing Data***
   What is the uncertainty in a measurement made by each of these devices: the ruler, the balance, the 100 mL graduated cylinder, and the 25 mL graduated cylinder?
2. ***Organizing Data***
   Calculate the solid volume of the 100 mL graduated cylinder up to its 50 mL mark. (Hint: $V = \pi r^2 h$ for a cylinder.)

674 | Exploration 2A

Sample Data

Measuring tool	Smallest division
Ruler	1 mm (0.1 cm)
100 mL grad. cyl.	1 mL
25 mL grad. cyl.	0.2 mL
Balance	0.01 g

Measurements of 100 mL grad. cyl.	
Inside diameter	2.5 cm
Inside height of 50 mL mark	9.3 cm

3. ***Analyzing Data***
Assuming the accepted value for the volume of a graduated cylinder at the 50 mL mark is 50.00 mL, calculate percent error in your calculations and measurements from item **2**.

4. ***Organizing Data***
Calculate the mass of water as determined by the balance and by its measured volume and known density. Using the mass obtained by the balance as the accepted value and the value calculated from the density as the experimental value, calculate percent error. Be sure to follow significant figure rules in your calculations.

5. ***Organizing Data***
Subtract the volume of the water from the combined volume of the metal and the water to calculate the volume of the metal alone. Subtract the mass of the water and cylinder from the mass of the metal, water, and cylinder to determine the mass of the metal; then calculate its density. Show all work.

6. ***Evaluating Data***
Record the metal density calculations made by the other teams, and calculate the average density of the unknown metal. Calculate the average deviation for the measurements.

Conclusions

7. ***Inferring Conclusions***
Compare your value and the class average of the unknown metal's density to the densities for metals given in handbooks. Determine the identity of the metal.

8. ***Evaluating Methods***
Calculate percent error by comparing your experimental value to the value you found in the handbook. Then, calculate percent error for the class average. Which is more accurate? The average or your value?

Extensions

1. ***Evaluating Methods***
Marie and Jason determined the density of a liquid three different times. They determined the values to be 2.84 g/mL, 2.85 g/mL, and 2.80 g/mL. The accepted value is 2.40 g/mL. Are the values determined by these two people precise? Are the values accurate? Explain your answers. Calculate the percent error and uncertainty of each measurement.

2. ***Applying Ideas***
When cutting the legs of a table to make them shorter, precision is more important than accuracy. Explain why.

3. ***Applying Ideas***
A store has a balance with a scale marked in gram units with one line halfway between each. A student working there after school measured a mass of a sample as 5.367 g using this balance. What is wrong with this measurement?

4. ***Applying Ideas***
There is a legend that the ancient Greek mathematician and scientist Archimedes used density to determine that a goldsmith had cheated when making a crown for the king. Explain what steps you would take to alter this procedure to check a metal's purity.

Measurements with 25 mL grad. cyl.	
Mass of empty cylinder	42.39 g
Volume of H_2O	13.8 mL
Mass of cylinder + H_2O	55.98 g
Volume of metal + H_2O	20.2 mL
Mass of metal + H_2O + cylinder	73.98 g

Temperature of H_2O: 25.0°C
Density of H_2O from handbook: 0.997 g/mL

Conclusions

7. Student answers will vary, depending upon which type of metal shot was used for analysis. The sample data shown so far has been for aluminum.

8. Student answers will vary. A sample calculation for aluminum is shown below. For most values, the class average will be more accurate.

Percent error:

$$\frac{2.7 \text{ g/mL} - 2.8 \text{ g/mL}}{2.7 \text{ g/mL}} \times 100 = 4\% \text{ error}$$

Extensions

1. The measurements described are precise because they are all close to each other. However, they are far from the accepted value, so they are not accurate.
The uncertainty of each measurement is ± 0.005 mL. Calculations for percent error are shown below.

$$\frac{2.40 \text{ mL} - 2.84 \text{ mL}}{2.40 \text{ mL}} \times 100 = 18\% \text{ error}$$

$$\frac{2.40 \text{ mL} - 2.85 \text{ mL}}{2.40 \text{ mL}} \times 100 = 19\% \text{ error}$$

$$\frac{2.40 \text{ mL} - 2.80 \text{ mL}}{2.40 \text{ mL}} \times 100 = 17\% \text{ error}$$

2. If the precision is poor, the legs will be of unequal length, and the table will wobble.

3. This measurement is reported with the wrong number of significant figures for the precision of the measuring tool. If the balance's uncertainty is ±0.5 g, the most precise measurement would be 5.5 g.

4. The purity of the metal could be checked by comparing volume and mass of a container of water before and after the crown was added. Then, the density could be compared to the density of an object known to be pure gold. The only change necessary would be to choose a container large enough for the crown and water.

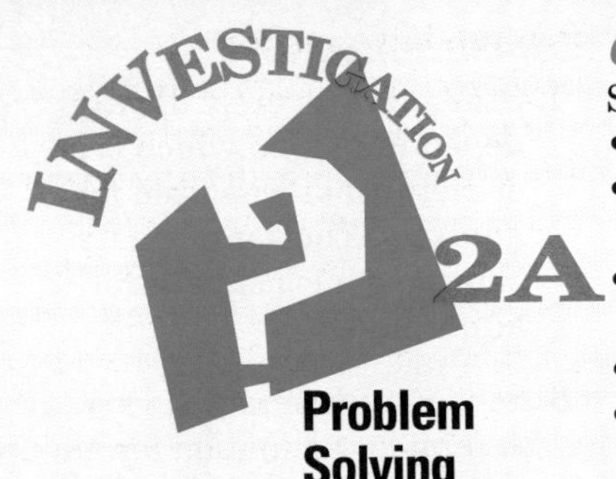

Objectives

Students will:

- use a laboratory balance.
- determine the volume of irregular objects.
- collect and analyze mass and volume data.
- calculate average densities.
- design and implement their own procedure.

Planning

Recommended Time: 1 lab period
Prerequisite: Exploration 2A

Solution/Material Preparation
Be sure that each student has at least 10 pennies for each year from 1979 to 1993. (Have students help by bringing in pennies from home that will be returned after the lab. If you cannot gather enough pennies, have lab groups take turns making measurements with different sets from different years.)

Student Orientation

Techniques to Demonstrate
Review volume measurement by water displacement with an emphasis upon reading a meniscus correctly. Also review the use of the laboratory balance.

Sample Data

Year	Average mass (g)	Average volume (mL)	Density (g/mL)
1979	3.12	0.36	8.7
1980	3.11	0.36	8.6
1981	3.11	0.37	8.4
1982	2.86	0.36	7.9
1983	2.49	0.37	6.7
1984	2.48	0.36	6.9
1985	2.50	0.36	6.9
1986	2.50	0.37	6.8
1987	2.48	0.36	6.9
1988	2.51	0.37	6.8
1989	2.50	0.36	6.9
1990	2.49	0.36	6.9
1991	2.50	0.37	6.8
1992	2.48	0.36	6.9
1993	2.49	0.36	6.9

The change occurred in 1982. According to Appendix A in this textbook, the likeliest metal is zinc, with a density of 7.14 g/mL. This what the "new" pennies contain. Also close are manganese, 7.20 g/mL, and tin, 7.28 g/mL. (Antimony is also close but should be rejected because it is listed as a metalloid.)

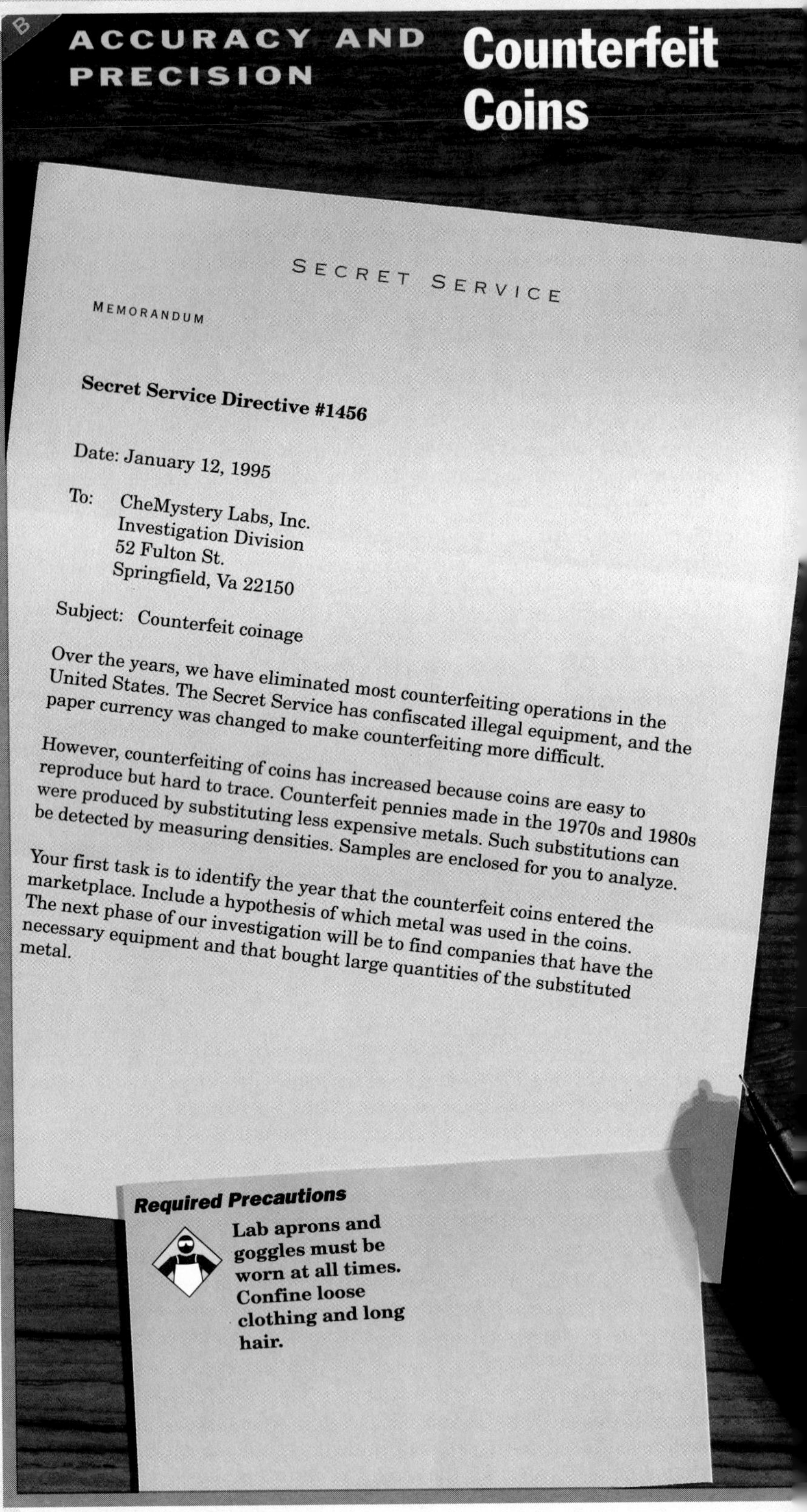

ACCURACY AND PRECISION

Counterfeit Coins

SECRET SERVICE

MEMORANDUM

Secret Service Directive #1456

Date: January 12, 1995

To: CheMystery Labs, Inc.
Investigation Division
52 Fulton St.
Springfield, Va 22150

Subject: Counterfeit coinage

Over the years, we have eliminated most counterfeiting operations in the United States. The Secret Service has confiscated illegal equipment, and the paper currency was changed to make counterfeiting more difficult.

However, counterfeiting of coins has increased because coins are easy to reproduce but hard to trace. Counterfeit pennies made in the 1970s and 1980s were produced by substituting less expensive metals. Such substitutions can be detected by measuring densities. Samples are enclosed for you to analyze.

Your first task is to identify the year that the counterfeit coins entered the marketplace. Include a hypothesis of which metal was used in the coins. The next phase of our investigation will be to find companies that have the necessary equipment and that bought large quantities of the substituted metal.

Required Precautions

Lab aprons and goggles must be worn at all times. Confine loose clothing and long hair.

Required Precautions

- Goggles and lab apron must be worn at all times.
- Read all safety cautions, and discuss them with your students.

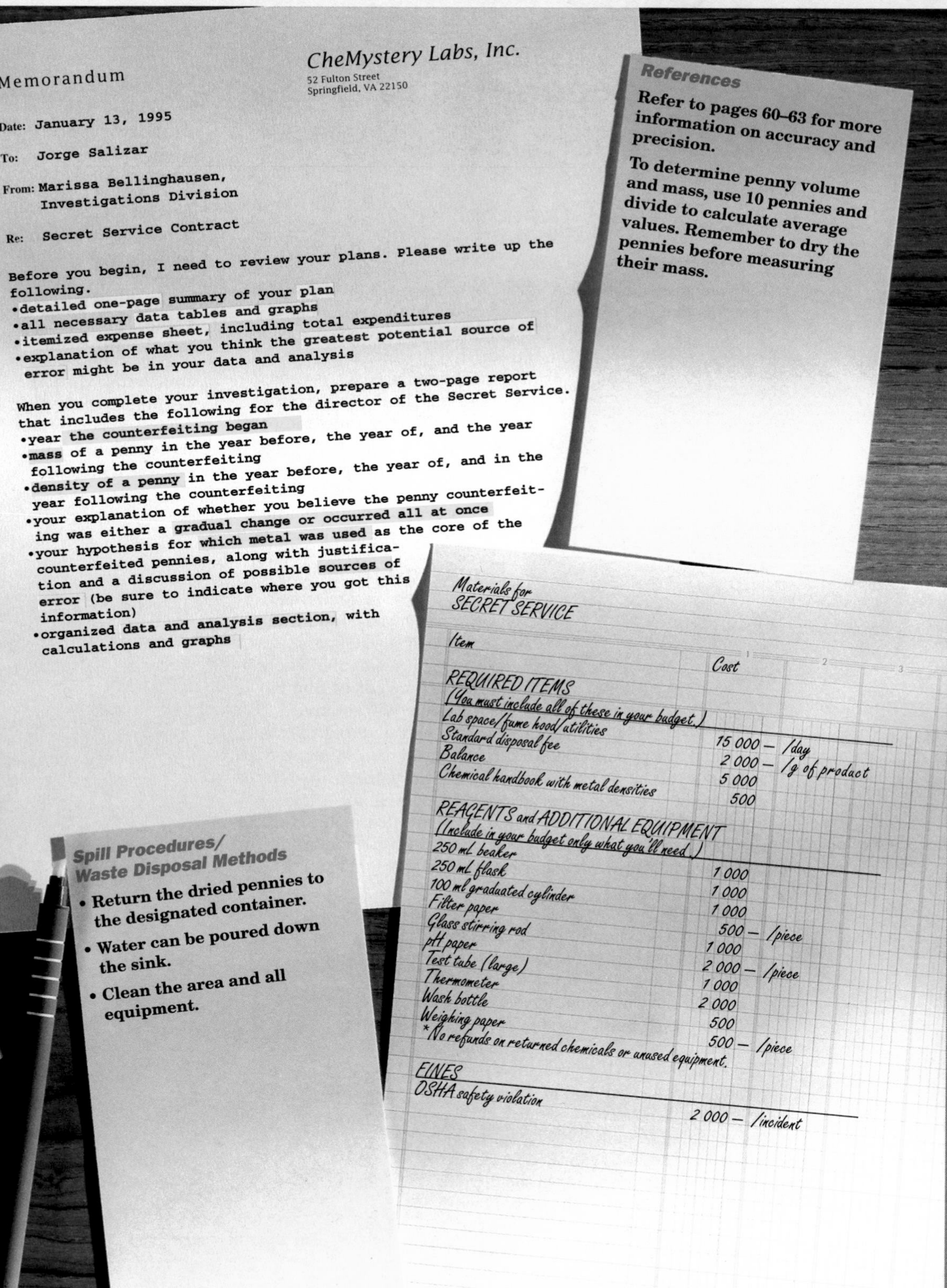

Memorandum

CheMystery Labs, Inc.
52 Fulton Street
Springfield, VA 22150

Date: January 13, 1995

To: Jorge Salizar

From: Marissa Bellinghausen, Investigations Division

Re: Secret Service Contract

Before you begin, I need to review your plans. Please write up the following.
- detailed one-page summary of your plan
- all necessary data tables and graphs
- itemized expense sheet, including total expenditures
- explanation of what you think the greatest potential source of error might be in your data and analysis

When you complete your investigation, prepare a two-page report that includes the following for the director of the Secret Service.
- year the counterfeiting began
- mass of a penny in the year before, the year of, and the year following the counterfeiting
- density of a penny in the year before, the year of, and in the year following the counterfeiting
- your explanation of whether you believe the penny counterfeiting was either a gradual change or occurred all at once
- your hypothesis for which metal was used as the core of the counterfeited pennies, along with justification and a discussion of possible sources of error (be sure to indicate where you got this information)
- organized data and analysis section, with calculations and graphs

References

Refer to pages 60–63 for more information on accuracy and precision.

To determine penny volume and mass, use 10 pennies and divide to calculate average values. Remember to dry the pennies before measuring their mass.

Spill Procedures/ Waste Disposal Methods
- Return the dried pennies to the designated container.
- Water can be poured down the sink.
- Clean the area and all equipment.

Materials for SECRET SERVICE

Item	Cost
REQUIRED ITEMS (You must include all of these in your budget.)	
Lab space/fume hood/utilities	15 000 – /day
Standard disposal fee	2 000 – /g of product
Balance	5 000
Chemical handbook with metal densities	500
REAGENTS and ADDITIONAL EQUIPMENT (Include in your budget only what you'll need.)	
250 mL beaker	1 000
250 mL flask	1 000
100 ml graduated cylinder	1 000
Filter paper	500 – /piece
Glass stirring rod	1 000
pH paper	2 000 – /piece
Test tube (large)	1 000
Thermometer	2 000
Wash bottle	500
Weighing paper	500 – /piece
* No refunds on returned chemicals or unused equipment.	
FINES	
OSHA safety violation	2 000 – /incident

Materials
(for each lab group)
- Balance (0.01 g accuracy if possible)
- Chemical handbook with metal density values (or Appendix A of this text)
- Graduated cylinder
- 10 pennies for each year from 1979 to 1993

Estimated cost of materials: $20 500

Pre-lab Discussion
Begin by discussing the results and procedure used in the Exploration, especially any errors in lab technique that occurred.

If students need help, provide some of the following leading questions for students to consider as they make their plans.
- Which measurements are necessary to determine density?
- With the measurement tools available, how many significant figures can each measurement have?
- How many significant figures will the density calculated from these measurements have?
- How would mass measurements be affected if the pennies are still wet from volume measurements?

Tips for Evaluating the Pre-lab Requirements
Students often plan to stop collecting data when they discover the year of the change. If they do this, they will not notice that although the change occurred during 1982, it was not a complete change. In 1982, both types of pennies were made, and this will make it nearly impossible to determine the identity of the new metal. Readings for later years must also be taken.

The most useful graph will be one of year vs. density.

Proposed Procedure
Measure the mass of 10 pennies at a time for each year from 1979 to 1993. Calculate the average mass of a penny for each year.

Determine the volume by pouring exactly 10.0 mL of water into a graduated cylinder, adding 10 pennies, and measuring the new volume. The change in volume will be the volume of the pennies. Then, density can be calculated for each year.

Post-Lab
Disposal
There are no special disposal guidelines. Have the students return the pennies to a designated container when they are finished.

EXPLORATION 2B

Objectives

Students will:

- use appropriate lab safety procedures.
- interpret a temperature-solubility graph.
- measure temperature.
- use a Bunsen burner and an ice bath to control temperature.
- use a laboratory balance to measure mass.
- demonstrate appropriate technique in transferring liquids.
- demonstrate appropriate technique in filtering.

Planning

Recommended Time: 2 lab periods if filtering is done once (includes overnight drying time)

Materials
(for each lab group)

- Ice (about 1.5 L)
- NaCl-KNO_3 solution, 50 mL
- Rock salt, 20 g
- 150 mL beakers, 4
- 100 mL graduated cylinder
- Balance, centigram
- Celsius thermometer, nonmercury type, range from –10°C to 120°C
- Hot plate or Bunsen burner with gas tubing and striker
- Ring stand, ring, and wire gauze (non-asbestos)
- Rubber policeman and glass stirring rod
- Spatula
- Tray, tub, or pneumatic trough for ice bath

Gravity Filtration Option

- Glass funnel
- Filter paper

Vacuum Filtration Option

- Aspirator for spigot
- Büchner funnel (either ceramic or plastic)
- Filter paper
- One-hole rubber stopper or sleeve
- Vacuum flask (sidearm flask) and tubing

Separation of Mixtures

Objectives

Recognize **how the solubility of a salt varies with temperature.**

Demonstrate **proficiency in the following techniques: fractional crystallization, and vacuum filtration or gravity filtration.**

Determine **the percentage of two salts recovered by fractional crystallization.**

Situation

Your company has been contacted by a fireworks factory. A mistake was made at the factory when a mislabeled container of sodium chloride, NaCl, was mixed with potassium nitrate, KNO_3, which is used as an oxidizer in fireworks. KNO_3 is used to be certain that the fireworks burn thoroughly. The fireworks company wants your company to investigate ways in which they could separate the two compounds. They have provided a water solution of the mixture for you to work with.

Background

The substances in a mixture can be separated by physical means. For example, if one substance dissolves in a liquid solvent but another does not, the mixture can be filtered. The one that dissolved will be carried through the filter by the solvent, but the other one will not. But both NaCl and KNO_3 dissolve in water, so this technique cannot be applied directly. However, there are differences in the way they dissolve. As you can see in the graph, about the same amount of sodium chloride will dissolve in water regardless of temperature. On the other hand, potassium nitrate is very soluble in warm water but much less soluble at 0°C.

Problem

You will make use of the differences in dissolving to separate the two salts. This technique is known as fractional crystallization. If the water solution of NaCl and KNO_3 is cooled from room temperature to a temperature near 0°C, not as much KNO_3 will be able to remain dissolved, as shown in the graph, and some of it will crystallize. This KNO_3 residue can then be separated from the NaCl solution by filtration. The NaCl can then be isolated by evaporating the water in the filtrate. To determine whether the method will work efficiently, you will measure the mass of each of the recovered substances, so that your client can decide whether this method is cost-effective.

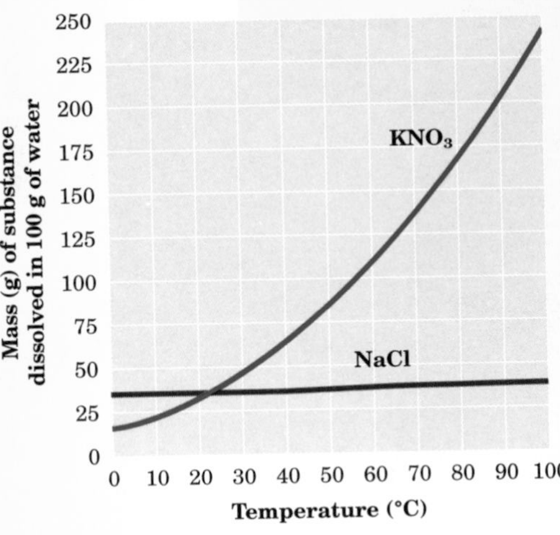

Required Precautions

- Goggles and lab apron must be worn at all times.
- Read all safety cautions, and discuss them with your students.
- Remind students that when equipment has been heated, it should be handled with tongs or a hot mitt. Hot glassware does not look hot.
- If a hot plate or LEAP System with thermistor probe is used, the precautions listed on page 652 must be followed to avoid electric shock.

Safety

Always wear goggles and an apron to provide protection for your eyes and clothing. If you get a chemical in your eyes, immediately flush it out at the eyewash station while calling to your teacher. Know the locations of the emergency lab shower and eyewash and how to use them.

Do not touch any chemicals. If you get a chemical on your skin or clothing, wash it off at the sink while calling to your teacher. Make sure you carefully read the labels and follow the directions on all containers of chemicals that you use. Do not taste any chemicals or items used in the laboratory. Never return leftovers to their original containers; take only small amounts to avoid wasting supplies.

Confine long hair and loose clothing. Do not heat glassware that is broken, chipped, or cracked. Use tongs or a hot mitt to handle heated glassware and other equipment, because it does not always look hot when it is hot. If your clothing catches on fire, WALK to the emergency lab shower and use it to put out the fire.

Always clean up the lab and all equipment after use, and dispose of substances according to proper disposal methods. Wash your hands thoroughly before you leave the lab after all lab work is finished.

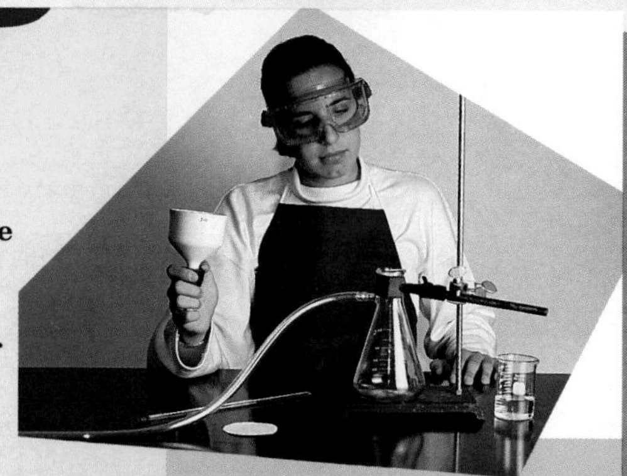

Materials

- **Ice**
- **$NaCl-KNO_3$ solution, 50 mL**
- **Rock salt**
- **100 mL graduated cylinder**
- **150 mL beakers, 4**
- **Balance**
- **Büchner funnel, one-hole rubber stopper, vacuum filtration setup with filter flask, and tubing; or glass funnel**
- **Bunsen burner and related equipment or hot plate**
- **Filter paper**
- **Glass stirring rod**
- **Nonmercury thermometer**
- **Ring and wire gauze**
- **Ring stand**
- **Rubber policeman**
- **Spatula**
- **Tray, tub, or pneumatic trough**

Optional equipment

- **LEAP System with thermistor probe**

Preparation

1. ***Organizing Data***
   Prepare a data table in your lab notebook. It should contain spaces for *Volume of salt solution added to beaker 1, Mass of beaker 1, Mass of the filter paper, Mass of beaker 1 with filter paper and KNO_3, Mass of beaker 4*, and *Mass of the beaker 4, with NaCl*. You will also need room to record the temperature of the mixture before and after cooling.
2. Obtain four clean, dry 150 mL beakers, and label them *1, 2, 3*, and *4*.

Technique

3. Measure the mass of beaker *1* to the nearest 0.01 g and record its mass in your data table.
4. Measure about 50 mL of the $NaCl-KNO_3$ solution into a graduated cylinder. Record the exact volume in your data table. Pour this mixture into beaker *1*.
5. Using a thermometer or the LEAP system with a thermistor probe, measure the temperature of the mixture. Record this temperature in your data table.
6. Measure the mass of a piece of filter paper to the nearest 0.01 g, and record the mass in your data table.

LEAP System Option

- LEAP System with thermistor probe and related equipment. See the LEAP System manual for instructions on using the thermistor probe.

Solution/Material Preparation

1. For every liter of solution, add about 140 g NaCl and 320 g of KNO_3. It is necessary to stir and gently heat the solution to completely dissolve the salts. A hot-plate stirrer is an invaluable tool.
2. Students should use only nonmercury thermometers. If a mercury thermometer breaks, the droplets that are not cleaned up will quickly evaporate, creating toxic mercury vapors.
3. Centigram balances will give best results, but less precise balances are acceptable.

Student Orientation

Pre-lab Discussion

Thoroughly discuss the procedure used in this lab. The students need to work quickly and efficiently if they are to complete this lab in a little over one lab period. Students tend to work on the lab in a step-by-step process, which can make the lab too long. Instead, encourage them to perform multiple tasks simultaneously through teamwork. While one partner is cooling the solution, the other lab partner could be cleaning, drying, and measuring the mass of the other beakers.

Students may have difficulty reading the solubility graph at first. It is not necessary for students to understand all aspects of solutions and solubility, but an understanding of this graph is important to understanding the concept of fractional crystallization. Practice taking several readings from the graph asking students how many grams of each salt would dissolve in 100 g of water at a given temperature. Relate the results to the cycles of cooling and heating in the procedure.

Results are greatly improved if a second filtration step is performed after some water has been evaporated, but this takes more time.

Students get very frustrated when their results do not match calculated ideal values. Students should realize when they answer *Conclusions* item **9** that it is impossible to achieve a perfect separation using this technique.

This exploration provides an opportunity for discussions about the uncertain nature of science and the constant need to improve on lab techniques at all levels of science.

Techniques to Demonstrate
Detailed instructions and diagrams on filtration are in Exploration 1A. Be sure to review the vocabulary associated with filtering, especially the difference between the *filtrate* and the *residue*.

If students will be using Büchner funnels, review how to set up a vacuum filtration. This method speeds up this lab considerably, and an investment in this equipment will prove valuable throughout this course. (Plastic Büchner funnels that are far less prone to breakage are available.)

If students will be using gravity filtration, review how to set up the equipment for gravity filtration, especially folding the filter paper.

Regardless of the filtering technique used, students will need to be shown how to use a rubber policeman to collect crystals.

Remind students to evaporate the water gradually to avoid violent bubbling and loss of product.

7. Set up your filtering apparatus. If you are using a Büchner funnel for vacuum filtration, follow step **7a.** If you are using a glass funnel for gravity filtration, follow step **7b**.

a. Vacuum filtration
Screw an aspirator nozzle onto the faucet. Attach a piece of plastic tubing to the sidearm of this nozzle. Attach the other end of the thick plastic tubing to the sidearm of the filter flask. Place a one-hole rubber stopper on the stem of the funnel and fit the stopper snugly in the neck of the filter flask as shown in the photograph. Place the filter paper from step **6** on the bottom of the funnel so that it is flat and covers all of the holes in the funnel. When finished, continue with step **8**.

b. Gravity filtration
Set up a ring stand with a ring. Gently rest the glass funnel inside the ring. Be sure that there is enough room between the lab bench surface and the bottom of the funnel's stem so that a 150 mL beaker can be moved under the funnel without breaking the stem. Fold the filter paper from step **6** in half along its diameter, and then fold it again to form a quadrant. Separate the folds of the filter paper so that three thicknesses are on one side and one thickness is on the other, as shown in the illustration at left. Fit the filter paper in the funnel and wet it with a little water, so that it will adhere to the sides of the funnel. Gently but firmly press the paper against the sides of the funnel so that no air is between the funnel and the filter paper. Be certain that the filter paper does not extend above the side of the funnel.

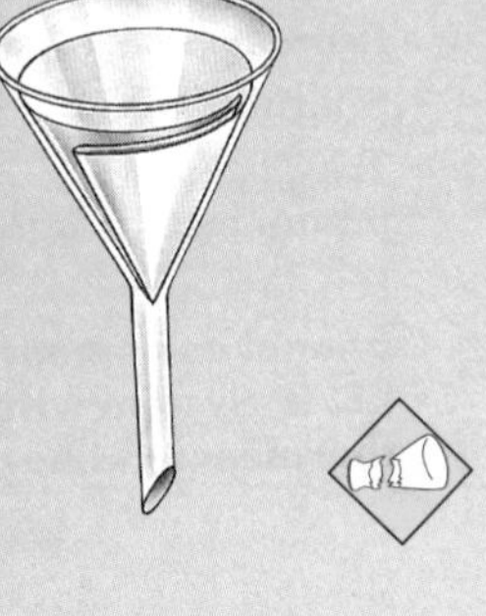

8. Make an ice bath by filling a tray, tub, or trough half full with ice. Add a handful of rock salt. The salt lowers the freezing point of water so that the ice bath can cool to a lower temperature. Fill the ice bath with water until it is three-quarters full.

9. Using a fresh supply of ice and distilled water, fill beaker *2* half full with ice and add water. Do not add rock salt to this ice-water mixture. You will use this water to wash your purified salt.

10. Put beaker *1* with your NaCl-KNO_3 solution into the ice bath. Place a thermometer or a LEAP thermistor probe in the solution to monitor the temperature. Stir the solution with a stirring rod while it cools. The lower the temperature of the mixture, the more KNO_3 will crystallize out of solution. When the temperature nears 4°C, follow step **10a** if you are using the Büchner funnel, or step **10b** if you are using a glass funnel.
Never stir a solution with a thermometer, as the bulb is very fragile.

a. Vacuum filtration
Turn on the water at the faucet that has the aspirator nozzle attached. This creates a vacuum, which helps the filtering step go much faster. If the suction is working properly, the filter paper should be pulled against the bottom of the funnel, covering all of the holes. If the filter paper appears to have bubbles of air under it, or is not centered well, turn the water off, reposition the filter paper, and begin again. Prepare the filtering apparatus by pouring approximately 50 mL of ice-cold distilled water from beaker *2* through the filter paper. After the water has gone through the funnel, empty the filter flask into the sink. Reconnect the filter flask, and pour the salt-water mixture into the funnel. Use the rubber policeman to transfer all of the cooled mixture into the funnel, especially any crystals that are visible. It may be helpful to add small amounts of ice-cold water from beaker *2* to beaker *1* to wash any crystals onto the filter paper. After all of the solution has passed through the funnel, wash the KNO_3 residue by pouring a very small amount of ice-cold water from beaker *2* over it. When this water has passed through the filter paper, turn off the faucet, and carefully remove the tubing from the aspirator. Empty the filtrate, which has passed through the filter paper and is now in the filter flask, into beaker *3*. When finished, continue with step **11**.

b. Gravity filtration
Place beaker *3* at the bottom of the glass funnel. Prepare the filtering apparatus by pouring approximately 50 mL of ice-cold water from beaker *2* through the filter paper. The water will pass through the filter paper and drip into beaker *3*. When dripping stops, empty beaker *3* into the sink. Place beaker *3* back at the bottom of the glass funnel, so that it will collect the filtrate from the funnel. Pour the salt-water mixture into the funnel. Use the rubber policeman to transfer all of the cooled mixture into the funnel, especially any crystals that are visible. It may be helpful to add small amounts of ice-cold water from beaker *2* to beaker *1* to wash any crystals onto the filter paper. After all of the solution has passed through the funnel, wash the KNO_3 by pouring a very small amount of ice-cold water from beaker *2* over it.

11. After you have finished filtering, use either a hot plate or a ring stand, ring, and wire gauze to heat beaker *3*. When the liquid in beaker *3* begins boiling, continue heating gently until enough water has vaporized to decrease the volume to approximately 25–30 mL. ***Be sure to use beaker tongs. Remember that hot glassware does not always look hot.***
12. Allow the solution in beaker *3* to cool, and then set it in the ice-bath, stirring until the temperature is approximately 4°C.
13. Measure the mass of beaker *4*. Record the mass of this beaker in your data table.
14. Repeat step **10a** or step **10b**, pouring the solution from beaker *3* onto the filter paper, and using beaker *4* to collect the filtrate that passes through the filter.

Post-Lab

Disposal
Potassium nitrate, KNO_3, cannot be disposed of in any manner because it is a strong oxidizer, but it can be reused the next time you do this lab. Allow the crystals in the disposal container you designated for the students to dry. Store the dried crystals in a sealed, labeled glass bottle completely away from bottles of reducing substances, especially those that can burn. Reuse the crystals the following year instead of ordering more. Because this is a separation exercise, it does not matter if impurities are present.

Sample Data

Volume of solution added to beaker 1	46.5 mL
Mass of beaker 1	65.80 g
Mass of filter paper	0.40 g
Mass of beaker 1 filter paper and KNO_3 (day 2)	78.22 g

Mass of beaker 4	66.25 grams
Mass of beaker 4 with NaCl	72.37 grams
Temperature before cooling	25°C
Temperature after cooling	–2°C

Answers to Analysis and Interpretation

1. 72.37 g − 66.25 g = 6.12 g NaCl

2. 78.22 g − 65.80 g = 12.42 g KNO_3

3. 12.42 g + 6.12 g = 18.54 g total

4. $1.0\ L \times \frac{6.12\ g\ NaCl}{46.5\ mL} \times \frac{1000\ mL}{1\ L} =$

130 g NaCl

$1.0\ L \times \frac{12.42\ g\ KNO_3}{46.5\ mL} \times \frac{1000\ mL}{1\ L} =$

270 g KNO_3

5. At room temperature, 25°C, about 32 g of KNO_3 and 35 g of NaCl will dissolve in 100 g of water. At −2°C, 15 g of KNO_3 and 33 g of NaCl will dissolve in 100 g of water.

Conclusions

6. $\frac{6.12\ g\ NaCl}{18.56\ g\ total} \times 100 = 33.0\%\ NaCl$

$\frac{12.42\ g\ KNO_3}{18.56\ g\ total} \times 100 = 66.92\%\ KNO_3$

7. $55\ L \times \frac{130\ g\ NaCl}{1\ L} = 7.2 \times 10^3\ g$

$55\ L \times \frac{270\ g\ KNO_3}{1\ L} = 1.5 \times 10^4\ g$

8. As much as 7 g KNO_3 could still be contaminating the NaCl if the solution was only filtered once.

9. Because the solubility of KNO_3 is never actually 0.0 g, it is always possible that some of the KNO_3 is dissolved in the solution after filtration.

10. Because the water was ice cold, the KNO_3 did not dissolve very well in it, but if there was any NaCl mixed with the KNO_3 crystals, the NaCl would have dissolved in the ice-cold water and passed through the filter paper.

11. Although the rock salt and ice mixture was cooler, it contained NaCl and would have contaminated the KNO_3 crystals if it was used in the rinsing step.

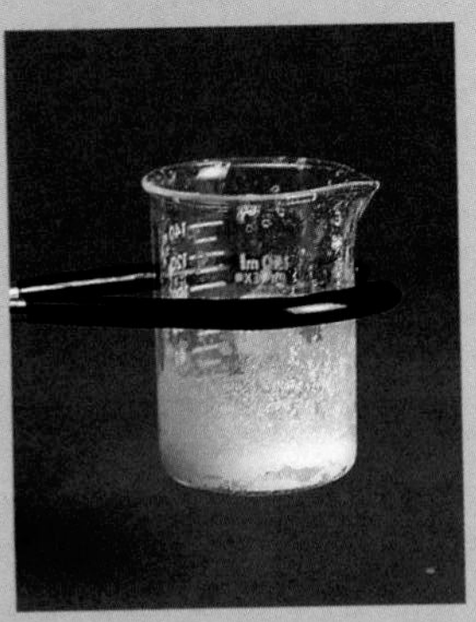

15. Wash and dry beaker *1*. Carefully remove the filter paper with the KNO_3 from the funnel and put it into the beaker. Be certain to avoid spilling the crystals. Place the beaker in a drying oven overnight.

16. Heat beaker *4* with a hot plate or Bunsen burner until it begins to boil. Keep heating gently, until all of the water has vaporized and the salt appears dry. Turn off the hot plate or burner and allow the beaker to cool. Use beaker tongs to move the beaker, even if you believe it is cool. Measure the mass of beaker *4* with the NaCl to the nearest 0.01 g and record this mass in your data table.

17. The next day, use beaker tongs to remove beaker *1* with the filter paper and KNO_3 from the drying oven. Allow the beaker to cool. Measure the mass, using the same balance you used when you measured the mass of the empty beaker. Record the new mass in your data table.

Be sure to use beaker tongs. Remember that hot glassware does not always look hot.

Cleanup and Disposal

18. Once the mass of the NaCl has been determined, add water to dissolve the NaCl and rinse the solution down the drain. Do not wash KNO_3 down the drain. Dispose of the KNO_3 in the waste container designated by your teacher.

Analysis and Interpretation

1. Organizing Data
Calculate the mass of NaCl in your 50 mL sample by subtracting the mass of beaker *4* from the mass of beaker *4* with NaCl.

2. Organizing Data
Calculate the mass of KNO_3 in your 50 mL sample by subtracting the mass of beaker *1* and the mass of the filter paper from the mass of beaker *1* with the filter paper and KNO_3.

3. Organizing Data
Calculate the total mass of the two salts.

4. Analyzing Information
How many grams of KNO_3 and NaCl would be found in a 1.0 L sample of the solution? (Hint: for each substance, make a conversion factor using the mass of the compound and the volume of the solution.)

5. Interpreting Graphics
Use the graph at the beginning of this exploration to determine how much of each compound would dissolve in 100 g of water at room temperature and at the temperature of your ice-water bath.

Conclusions

6. Inferring Conclusions
Calculate the percentage by mass of NaCl and KNO_3 in the salt mixture.

7. Applying Conclusions
The fireworks company has another 55 L of the salt mixture dissolved in water just like the sample you worked with. How many kilograms of each compound can the company expect to recover from this sample? (Hint: use your answer from item **4** to help you answer this question.)

8. ***Evaluating Methods***
Use the graph shown at the beginning of this Exploration to estimate how much KNO_3 could still be contaminating the NaCl you recovered.

9. ***Relating Ideas***
Use the graph shown at the beginning of this Exploration to explain why it is impossible to separate the two compounds completely using fractional crystallization.

10. ***Evaluating Methods***
Why was it important that you use ice-cold water to wash the KNO_3 after filtration?

11. ***Evaluating Methods***
If it was important to use very cold water to wash the KNO_3, why wasn't the salt and ice water mixture from the bath used? After all, it had a lower temperature than the ice and distilled water from beaker *2*. (Hint: consider what is contained in rock salt.)

12. ***Evaluating Methods***
Why was it important to keep the amount of cold water used to wash the KNO_3 as small as possible?

13. ***Relating Ideas***
Your lab partner tries to dissolve 95 g of KNO_3 in 100 g of water, but no matter how well he stirs the mixture, some KNO_3 remains undissolved. Using the graph, explain what your lab partner must do to make the KNO_3 dissolve in this amount of water.

Extensions

1. ***Interpreting Graphics***
Use the graph shown at the beginning of this Exploration to determine the minimum mass of water necessary to dissolve the amounts of each compound from items **1** and **2** of the Analysis and Interpretation section at room temperature and at 4°C. What volumes of water would be necessary? (Hint: the density of water is about 1.0 g/mL.)

2. ***Designing Experiments***
Describe how you could use the properties of the compounds to test the purity of your recovered samples. If your teacher approves your plan, use it to check your separation of the mixtures. (Hint: check a chemical handbook for more information about the properties of NaCl and KNO_3.)

3. ***Designing Experiments***
How could you improve the yield or the purity of the compounds you recovered? If you can think of ways to modify the procedure, ask your teacher to approve your plan, and run the procedure again.

4. ***Research and Communication***
Petroleum products are separated from crude oil through the process of fractional distillation. Contact the American Petroleum Institute, 1220 L Street, NW, Washington, D.C. 20005, for information on petroleum refining. Write a short paper on fractional distillation and note ways in which this process is similar to fractional crystallization.

5. ***Designing Experiments***
For each of the following mixtures, write a paragraph outlining the steps you would take to separate each of them.
a. water and alcohol **b.** water and food coloring **c.** sand and salt

12. By keeping the amount of water used to wash the crystals small, the amount of KNO_3 that redissolved in the water was also kept to a minimum.

13. According to the graph, the temperature must be raised to at least 60°C for that much KNO_3 to dissolve in 100 g of water.

Extensions

1. 17.5 g H_2O to dissolve 6.12 g NaCl at 25°C
38.8 g H_2O to dissolve 12.42 g KNO_3 at 25°C
18.5 g H_2O to dissolve 6.12 g NaCl at –2°C
82.8 g H_2O to dissolve 12.42 g KNO_3 at –2°C

2. Students suggestions for determining purity will vary but could include measuring the density or testing the melting point. The latter would require special equipment.

3. Students' suggestions for improving purity will vary. Be sure answers are safe and include carefully planned procedures.

4. Fractional distillation also involves a temperature-dependent property. Some components of petroleum boil at lower temperatures than other components.

5. Student answers will vary, but possible answers are shown below. Be certain student answers are a paragraph long and contain detailed instructions.
a. distillation
b. evaporation of the water
c. Add water to dissolve the salt into solution. Filter the mixture to separate the sand. Evaporate the water from the salt solution to obtain the salt.

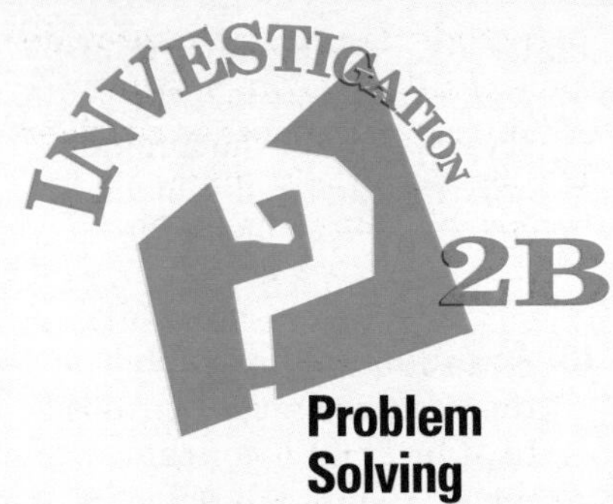

Objectives

Students will:

- use appropriate lab safety procedures.
- measure temperature.
- use a Bunsen burner and an ice bath to control temperature.
- use a laboratory balance to measure mass.
- demonstrate appropriate technique in transferring liquids.
- demonstrate appropriate technique in filtering.
- design and implement their own procedure.

Planning

Recommended Time: 2 lab periods (includes overnight drying time)

Prerequisite: Exploration 2B

Solution/Material Preparation

1. To prepare the unknown solution for 15 lab groups, add 300 g of KNO_3 and 150 g of $Cu(NO_3)_2{\bullet}3H_2O$ to 900 mL of water. Slowly heat the solution while stirring until the two salts are dissolved. Each 50 mL sample should contain approximately 15 g of KNO_3 and 7.5 g of $Cu(NO_3)_2$. There is no need to reheat the solution before lab.

Student Orientation

Techniques to Demonstrate

Review the use of the filtration apparatus.

Pre-lab Discussion

Begin by discussing the results and procedure used in the Exploration, especially any lab technique errors that occurred. Be sure students realize that KNO_3 is much less soluble than $Cu(NO_3)_2$.

Point out that repeated recrystallization can increase the yield of KNO_3. The use of plastic bags for the solution can increase yield, because the solution cools more quickly, allowing time for repeated recrystallization.

Sample Data

Volume of solution	46.5 mL
Mass of beaker 1	65.80 g
Mass of filter paper	0.40 g
Mass of beaker 1 with filter paper and KNO_3 (Day 2)	78.22 g
Mass of beaker 2	66.25 g
Mass of beaker 2 with $Cu(NO_3)_2$	72.37 g
Temperature before cooling	25°C
Temperature after cooling	–2°C

Most students will recover approximately 12 g of KNO_3 and 7 g of $Cu(NO_3)_2$ from 50 mL of solution. Given this data, the aquifer would contain 2.4×10^{14} g KNO_3 and 1.4×10^{14} g $Cu(NO_3)_2$.

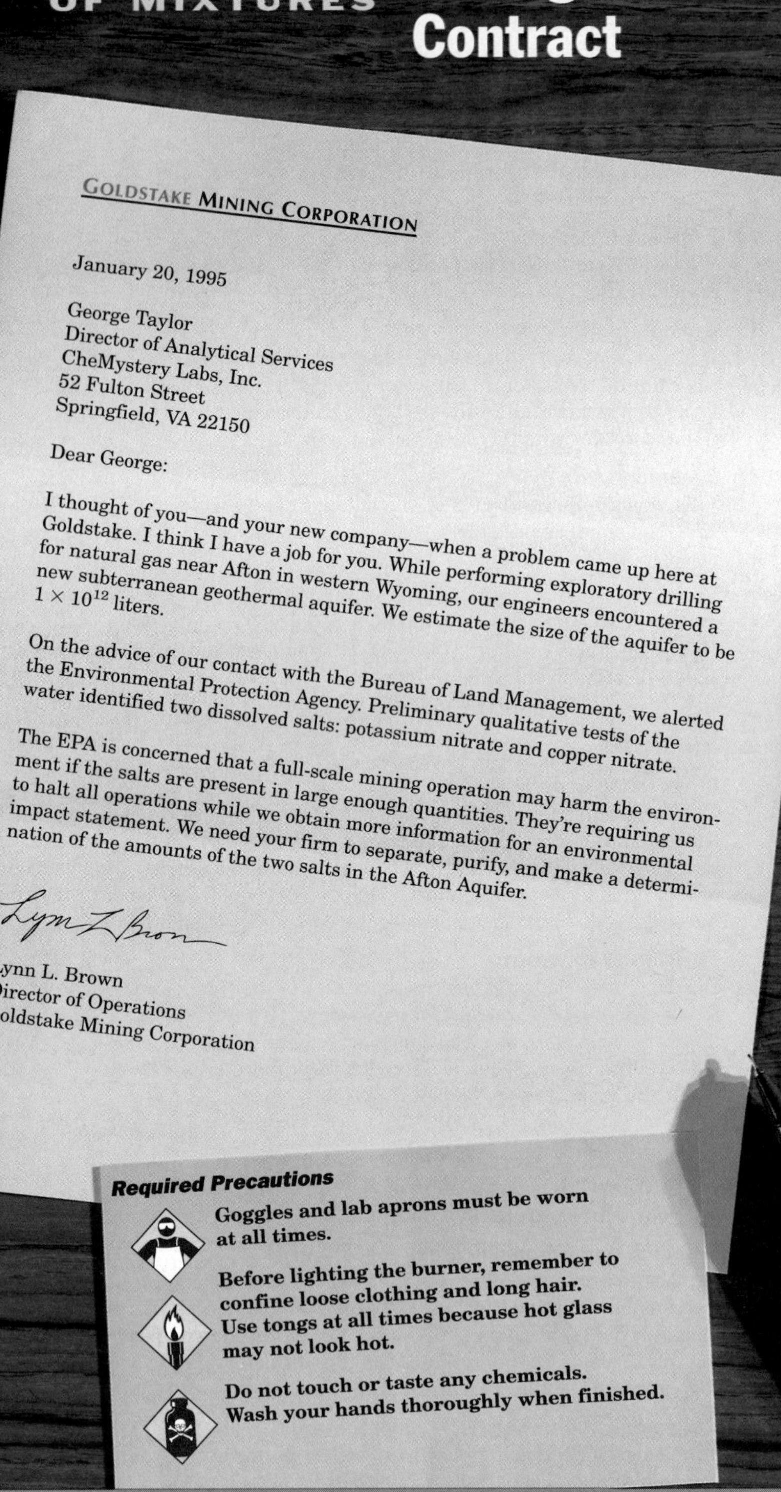

GOLDSTAKE MINING CORPORATION

January 20, 1995

George Taylor
Director of Analytical Services
CheMystery Labs, Inc.
52 Fulton Street
Springfield, VA 22150

Dear George:

I thought of you—and your new company—when a problem came up here at Goldstake. I think I have a job for you. While performing exploratory drilling for natural gas near Afton in western Wyoming, our engineers encountered a new subterranean geothermal aquifer. We estimate the size of the aquifer to be 1×10^{12} liters.

On the advice of our contact with the Bureau of Land Management, we alerted the Environmental Protection Agency. Preliminary qualitative tests of the water identified two dissolved salts: potassium nitrate and copper nitrate.

The EPA is concerned that a full-scale mining operation may harm the environment if the salts are present in large enough quantities. They're requiring us to halt all operations while we obtain more information for an environmental impact statement. We need your firm to separate, purify, and make a determination of the amounts of the two salts in the Afton Aquifer.

Lynn L. Brown
Director of Operations
Goldstake Mining Corporation

Required Precautions
Goggles and lab aprons must be worn at all times.
Before lighting the burner, remember to confine loose clothing and long hair. Use tongs at all times because hot glass may not look hot.
Do not touch or taste any chemicals. Wash your hands thoroughly when finished.

Required Precautions

- Goggles and lab apron must be worn at all times.
- Read all safety cautions, and discuss them with your students.
- Remind students to use beaker tongs or a hot mitt when handling glassware that has been heated.
- If a hot plate or the LEAP System with thermistor probe is used, the precautions listed on page 652 must be followed to avoid electric shock.

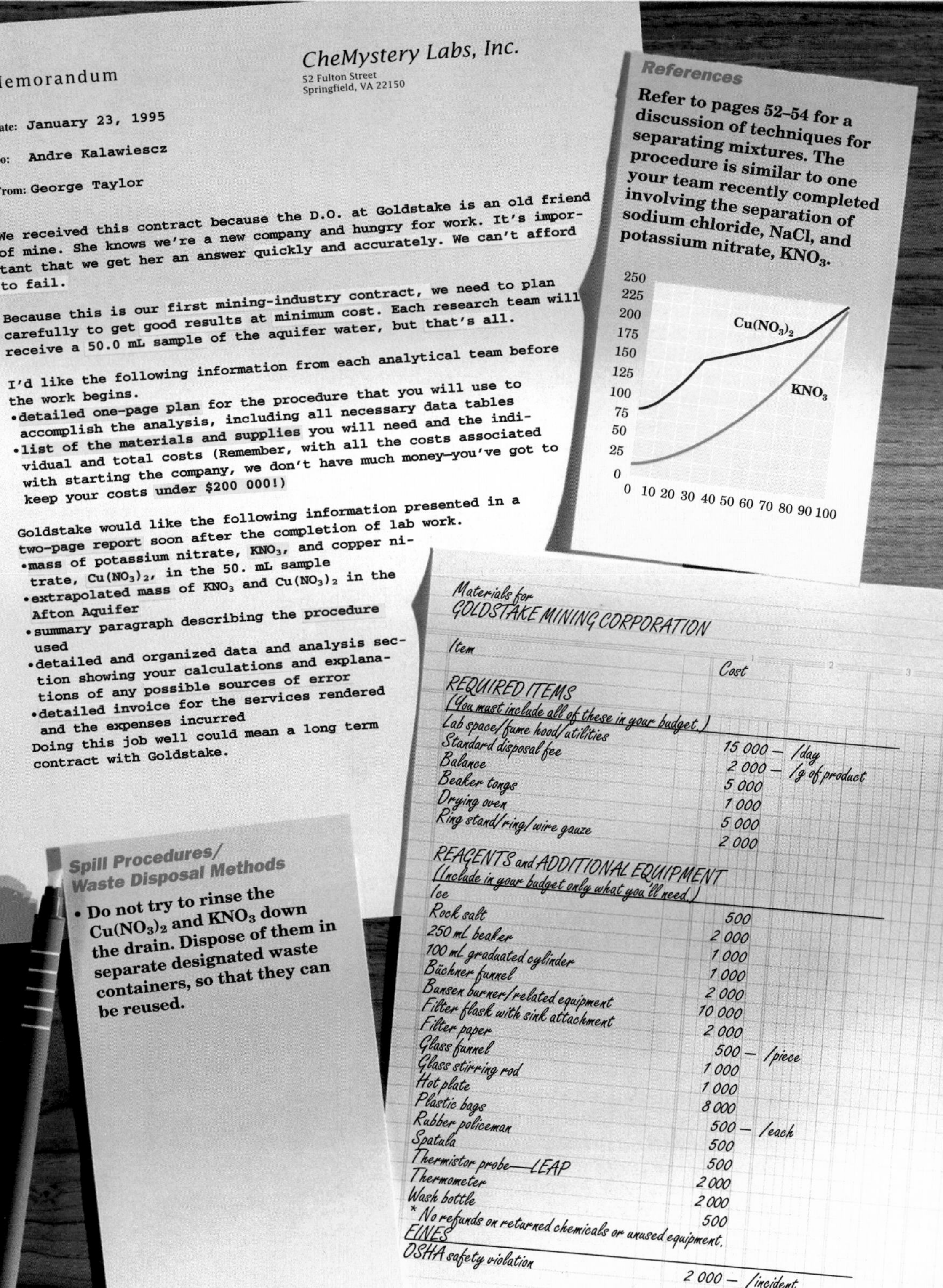

Memorandum

CheMystery Labs, Inc.
52 Fulton Street
Springfield, VA 22150

Date: January 23, 1995

To: Andre Kalawiescz

From: George Taylor

We received this contract because the D.O. at Goldstake is an old friend of mine. She knows we're a new company and hungry for work. It's important that we get her an answer quickly and accurately. We can't afford to fail.

Because this is our first mining-industry contract, we need to plan carefully to get good results at minimum cost. Each research team will receive a 50.0 mL sample of the aquifer water, but that's all.

I'd like the following information from each analytical team before the work begins.
- detailed one-page plan for the procedure that you will use to accomplish the analysis, including all necessary data tables
- list of the materials and supplies you will need and the individual and total costs (Remember, with all the costs associated with starting the company, we don't have much money—you've got to keep your costs under $200 000!)

Goldstake would like the following information presented in a two-page report soon after the completion of lab work.
- mass of potassium nitrate, KNO_3, and copper nitrate, $Cu(NO_3)_2$, in the 50. mL sample
- extrapolated mass of KNO_3 and $Cu(NO_3)_2$ in the Afton Aquifer
- summary paragraph describing the procedure used
- detailed and organized data and analysis section showing your calculations and explanations of any possible sources of error
- detailed invoice for the services rendered and the expenses incurred

Doing this job well could mean a long term contract with Goldstake.

References

Refer to pages 52–54 for a discussion of techniques for separating mixtures. The procedure is similar to one your team recently completed involving the separation of sodium chloride, NaCl, and potassium nitrate, KNO_3.

Spill Procedures/ Waste Disposal Methods

- Do not try to rinse the $Cu(NO_3)_2$ and KNO_3 down the drain. Dispose of them in separate designated waste containers, so that they can be reused.

Materials for GOLDSTAKE MINING CORPORATION

Item	Cost	1	2	3
REQUIRED ITEMS (You must include all of these in your budget.)				
Lab space/fume hood/utilities	15 000 — /day			
Standard disposal fee	2 000 — /g of product			
Balance	5 000			
Beaker tongs	1 000			
Drying oven	5 000			
Ring stand/ring/wire gauze	2 000			
REAGENTS and ADDITIONAL EQUIPMENT (Include in your budget only what you'll need.)				
Ice	500			
Rock salt	2 000			
250 mL beaker	1 000			
100 mL graduated cylinder	1 000			
Büchner funnel	2 000			
Bunsen burner/related equipment	10 000			
Filter flask with sink attachment	2 000			
Filter paper	500 — /piece			
Glass funnel	1 000			
Glass stirring rod	1 000			
Hot plate	8 000			
Plastic bags	500 — /each			
Rubber policeman	500			
Spatula	500			
Thermistor probe—LEAP	2 000			
Thermometer	2 000			
Wash bottle	500			
* No refunds on returned chemicals or unused equipment.				
FINES				
OSHA safety violation	2 000 — /incident			

Materials

(for each pair of students)
- $Cu(NO_3)_2$-KNO_3 solution, 50 mL
- Ice, 1.5 L
- Rock salt, 20 g
- 250 mL beakers, 2
- 100 mL graduated cylinder
- Balance
- Celsius thermometer, non-mercury type, range from –10 to 120°C
- Hot plate or Bunsen burner with gas tubing and striker
- Plastic bags, 2
- Ring stand, ring, and wire gauze (non-asbestos)
- Rubber policeman and glass stirring rod
- Spatula
- Tray, tub, or pneumatic trough for ice bath

See the Teacher's Notes for Exploration 1B for information on materials for filtration and pH measurement options.

Estimated cost of materials:
$100 000–$107 500

If students need help, provide some of the following leading questions for students to consider as they make their plans.
- How will your results be affected if the solution is heated quickly, and it boils and splashes out of the beaker onto the lab counter?
- Should you use warm or cold water to rinse the KNO_3 crystals?
- Should you use a small amount or a large amount of water to rinse the KNO_3 crystals?
- How will you determine whether separation is complete? (If the KNO_3 looks blue, it means there is probably $Cu(NO_3)_2$ in it.)

Tips for Evaluating the Pre-lab Requirements

Be sure students specify that they will cool to crystallize some KNO_3, heat to evaporate some of the remaining water, and then crystallize more KNO_3. Data tables should contain spaces for the items shown in the Sample Data section.

Proposed Procedure

Measure the mass of the two beakers and a piece of filter paper. Cool the solution in a plastic bag in an ice-water bath, causing the KNO_3 to crystallize. Filter the cold mixture. Rinse the crystals on the filter paper with a small amount of ice-cold distilled water. Heat the filtrate to evaporate away some of the water. Cool the solution again. Filter the cold mixture. Rinse the crystals again. Dry the filter paper in one of the beakers. Pour the filtrate into the other beaker and evaporate the water. Measure the mass of the beaker.

Post-Lab

Disposal

KNO_3 and $Cu(NO_3)_2$ cannot be disposed of because they are strong oxidizers. Allow the crystals in the separate disposal containers to dry. Store the dried crystals in a sealed, labeled glass bottle, kept away from bottles of reducing substances, especially those that can burn. Re-use the crystals the following year.

EXPLORATION 3A

Objectives

Students will:

- use appropriate lab safety procedures.
- use a Bunsen burner.
- perform flame tests with a variety of metal compounds.
- identify an unknown based on its flame test.

Planning

Recommended Time:
1–2 lab periods

Materials
(for each lab group)

- 1.0 M HCl solution, 5 mL
- $CaCl_2$ solution
- Distilled water
- K_2SO_4 solution
- Li_2SO_4 solution
- Na_2SO_4 solution
- NaCl crystals, 3 g
- NaCl solution
- $SrCl_2$ solution
- Unknown solution
- 250 mL beaker
- Bunsen burner, gas tubing, striker
- Cobalt glass plate
- Crucible tongs
- Flame test wire, 5 cm
- Glass plate (either 7 cm × 15 cm plate, or microchemistry plate with wells)

Optional Equipment

- Spectroscope
- Wooden splints

Solution/Material Preparation

1. To prepare 1.0 M HCl, observe the required precautions. Add 83 mL of concentrated HCl to enough distilled water to make 1.00 L of solution. Add the acid slowly, and stir it to avoid overheating.
2. To prepare 0.5 M $CaCl_2$, add 55 g of $CaCl_2$ to enough water to make 1.00 L of solution.
3. To prepare 0.5 M K_2SO_4, add 87 g of K_2SO_4 to enough water to make 1.00 L of solution.
4. To prepare 0.5 M Li_2SO_4, add 64 g of $Li_2SO_4 \cdot H_2O$ to enough water to make 1.00 L of solution.

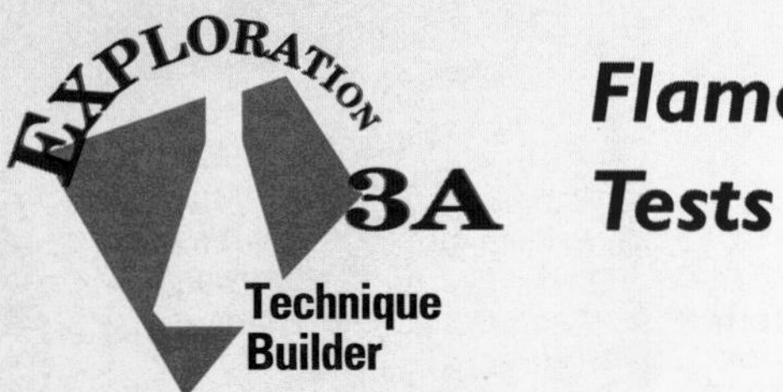

Flame Tests

Objectives

Identify **a set of flame test color standards for selected metal ions.**

Relate **colors of a flame test to the behavior of excited electrons in a metal atom.**

Identify **an unknown metal ion by a flame test.**

Demonstrate **proficiency in the following techniques: performing a flame test and using a spectroscope.**

Situation

Your company has been contacted by Julius and Annette Benetti. They are worried about some abandoned, rusted barrels of chemicals that their daughter found while playing in the vacant lot behind their home. The barrels have begun to leak a colored liquid that flows through their property before emptying into a local sewer. The Benettis want your company to identify the compound in the liquid. Earlier work already indicates that it is a dissolved metal compound. Many metals, such as lead, have been determined as hazardous to our health. Many compounds of these metals are often soluble in water and therefore easily absorbed into the body.

Background

After electrons of an atom absorb energy to move from their ground state to an excited state, they will eventually fall back to their ground state. When they do this, the energy they absorbed will be re-emitted in the form of light. Because each atom has a unique structure and arrangement of electrons, each atom emits a unique type of light. This is how the chemical test known as a flame test works. The atoms are excited by being placed within a flame. As they re-emit the energy in the form of light, the color of the flame changes. For most metals, these changes are easily visible. However, even a tiny speck of another substance can interfere with identification of the true color for a test, so be sure to keep the equipment very clean, and perform multiple trials to check your work.

Problem

To determine what metal is contained in the barrels behind the Benettis' house, you must first perform flame tests with a variety of standard solutions of different metal compounds. Then, you will perform a flame test with the sample from the site to see if it matches any of the solutions you've used as standards.

686 | Exploration 3A

Required Precautions

- Goggles and a lab apron must be worn at all times.
- Tie back loose hair and long clothing when working in the lab.
- Read all safety cautions, and discuss them with your students.
- Students should not handle concentrated acid solutions.

Safety

Always wear goggles and an apron to provide protection for your eyes and clothing. If you get a chemical in your eyes, immediately flush it out at the eyewash station while calling to your teacher. Know the locations of the emergency lab shower and eyewash and how to use them.

Do not touch any chemicals. If you get a chemical on your skin or clothing, wash it off at the sink while calling to your teacher. Make sure you carefully read the labels and follow the directions on all containers of chemicals that you use. Do not taste any chemicals or items used in the laboratory. Never return leftovers to their original containers; take only small amounts to avoid wasting supplies.

Confine long hair and loose clothing. Do not heat glassware that is broken, chipped, or cracked. Use tongs or a hot mitt to handle heated glassware and other equipment because it does not always look hot. If your clothing catches on fire, WALK to the emergency lab shower, and use it to put out the fire.

Call your teacher in the event of an acid spill. Acid spills should be cleaned up promptly, according to your teacher's instructions.

Always clean up the lab and all equipment after use, and dispose of substances according to proper disposal methods. Wash your hands thoroughly before you leave the lab after all lab work is finished.

Materials

- 1.0 M HCl solution
- $CaCl_2$ solution
- K_2SO_4 solution
- Li_2SO_4 solution
- Na_2SO_4 solution
- NaCl crystals
- NaCl solution
- $SrCl_2$ solution
- Unknown solution
- 250 mL beaker
- Bunsen burner and related equipment
- Cobalt glass plates
- Crucible tongs
- Flame test wire
- Glass test plate

Optional Equipment

- Spectroscope

Preparation

1. **Organizing Data**
   Prepare a data table in your lab notebook. The table should contain a row for each of the solutions of metal compounds listed in the materials list, as well as an additional row for the unknown solution. The table should have three wide columns for the three trials you will perform with each substance. Each column should have room to record colors and wavelengths of light. In addition, be sure that you have plenty of room for observations about each test.

2. Label a beaker *Waste*. Thoroughly clean and dry a glass plate. Put five drops of 1.0 M HCl on the plate. Clean the test wire by first dipping it in the HCl and then holding it in the colorless flame of the Bunsen burner. Repeat this procedure until there is no color from the wire in the flame. When the wire is ready, rinse the plate with distilled water, and collect the rinse water in the *Waste* beaker.

3. Put two drops of each metal ion solution except NaCl in a row on one side of the plate. Put a row of 1.0 M HCl drops on the plate across from the metal ion solutions, as shown in the illustration on the right. The wire will need to be cleaned thoroughly between each test solution, to avoid contamination from the previous test.

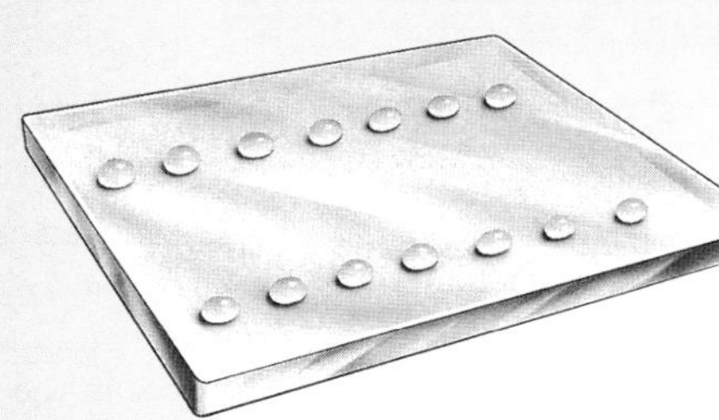

- Wear goggles, a face shield, impermeable gloves, and a lab apron when you prepare the HCl. Work in a hood known to be in operating condition with another person present nearby to call for help in case of an emergency, and within 30 s walking distance from a safety shower and eyewash station known to be in operating condition.
- In case of an acid spill, dilute first with water. Then, mop up the spill with wet cloths designated for spill cleanup while wearing disposable plastic gloves. A wet cloth mop can be rinsed out a few times and used until it falls apart.

5. To prepare 0.5 M Na_2SO_4, add 71 g of Na_2SO_4 to enough water to make 1.00 L of solution.
6. To prepare 0.5 M NaCl, add 29 g of NaCl to enough water to make 1.00 L of solution.
7. To prepare 0.5 M $SrCl_2$, add 133.3 g of $SrCl_2 \cdot 6H_2O$ to enough water to make 1.00 L of solution.
8. For the unknown solution, any of the above solutions can be used. Mixtures may prove to be too complicated for students to analyze. Excess solutions can be stored and used next year.
9. For flame-test wire, use either no. 24 platinum wire or nichrome wire. Some teachers prefer to use wooden splints for the flame tests. If they are soaked in the appropriate solutions overnight, they provide a colored flame that is longer-lasting and easier to view with the spectroscope. If this is done, however, each splint should be extinguished in the waste beaker. Be sure to label the splints with the compound they were soaked in so that they can be reused. The sodium content of the splint may interfere with some flame tests.

Student Orientation

Pre-lab Discussion

This identification technique can be used to introduce concepts that provide information about the behavior and arrangement of electrons in atoms. To emphasize this point, apply high voltage to a gas-discharge tube containing helium or neon, and place a few crystals of sodium chloride on the grating of a lit Fisher burner. Explain that the colored lights seen are actually a combination of several very specific wavelengths of light. Each wavelength of light corresponds to excited electrons moving from a different energy level to their ground state, emitting light in the process.

Techniques to Demonstrate
Demonstrate the flame test technique, including how to clean the flame-test wire. Point out that because the color lasts briefly, several trials may be necessary. If you have spectroscopes, your students can use them to identify the specific lines in the spectra of the light emitted in the flame test. One student can view through the spectroscope while the other performs the flame test. Be sure you explain how to use the spectroscope.

Post-Lab

Disposal
Set out a disposal container for the students. After all of the waste beakers have been emptied into it, neutralize the resulting solution with 0.1 M NaOH. When the solution's pH is between 5 and 9, pour it down the drain. If you use wooden splints, collect them after they have been labeled, and reuse them next year for the same compounds.

Answers to Analysis and Interpretation

(Note: assign only items 1–3 if spectroscopes are unavailable.)

1. See sample data table.
2. Student answers will vary. Some students may have had difficulty properly cleaning the wire, so the first test of a new compound may have had traces of the previous one.
3. The flame color of potassium is purple, but it is so weak that it can be overpowered by the yellow sodium light if a mixture is tested. The cobalt glass screens out the yellow sodium light.
4. Each line in the spectroscope represented the energy emitted as excited electrons moved from a specific high-energy orbital back to their ground state.
5. Answers will vary, but students should realize that the colors seen by the eye were the result of combining the colors of light seen in the line spectra.

Technique

4. Dip the wire into the $CaCl_2$ solution, and then hold it in the Bunsen burner flame. Observe the color of the flame, and record it in the data table. Repeat the procedure again, but look through the spectroscope to view the results. Record the wavelengths you see from the flame. Perform each test three times. Clean the wire with the HCl as in step **2**.
5. Repeat step **4** with the K_2SO_4 and each of the remaining solutions on the plate. Record the color of each flame, and the wavelength observed with the spectroscope for each solution that you test. Clean the wire thoroughly, and rinse the plate with distilled water into the *Waste* beaker after the solutions are tested.
6. Test another drop of Na_2SO_4, but view the flame through two pieces of cobalt glass. Clean the wire and plate, rinsing the plate with distilled water into the *Waste* beaker. Repeat the test using the K_2SO_4 flame with the cobalt glass. Record the colors and wavelengths of the flames as they appear when viewed through the cobalt glass in your data table. Clean the wire and plate, rinsing the plate with distilled water into the *Waste* beaker.
7. Put a drop of K_2SO_4 on a clean part of the plate. Add a drop of Na_2SO_4. Flame-test the mixture. Observe the color of the mixture in the flame when viewed without the cobalt glass. Repeat the test again observing the flame through the cobalt glass. Record the colors and wavelengths of the flames in the data table. Clean the wire and plate, rinsing the plate with distilled water into the *Waste* beaker.
8. Test a drop of the NaCl solution in the flame and then while viewed through the spectroscope. (Do not use the cobalt glass.) Record your observations. Clean the wire and plate, rinsing the plate with distilled water into the *Waste* beaker. Place a few crystals of NaCl on the plate, then dip the wire in the crystals and do the flame test once more. Record the color of the flame test. Clean the wire and plate with distilled water, rinsing the plate with distilled water into the *Waste* beaker.
9. Obtain a sample of the unknown metal solution. Do flame tests for it, both with and without the cobalt glass. Record your observations. Clean the wire and plate, rinsing the plate with distilled water into the *Waste* beaker.

Cleanup and Disposal

10. Dispose of the contents of the waste beaker in the container designated by your teacher. Always wash your hands thoroughly after cleaning up the area and equipment.

Analysis and Interpretation

1. ***Organizing Data***
   Examine your table of data, and create a summary of the flame test for each metal.
2. ***Analyzing Data***
   Account for any differences in the individual trials for the flame tests for the metals.

Sample Data

Metal compound	Color of flame	Wavelengths detected (nm)
$CaCl_2$	yellowish red (orange)	420, 445, 460, 485, 610, 645, 650
K_2SO_4	violet (purple)	405, 408, 695, 700
Li_2SO_4	red (carmine)	462, 498, 612, 670
Na_2SO_4	yellow	590, 595
$SrSO_4$	scarlet	405, 420, 460, 485, 490, 500, 665, 685, 710

Note: student data tables should show three trials for each compound.

3. ***Organizing Ideas***
Explain how using cobalt glass can make it easier to analyze the ions being tested.

4. ***Relating Ideas***
Explain how the lines seen in the spectroscope relate to the positions of electrons in the metal atom.

5. ***Relating Ideas***
For three of the metals tested, explain how the color seen by your eyes relates to the lines of color seen when you looked through the spectroscope.

Conclusions

6. ***Inferring Conclusions***
What metal(s) are in the unknown solution from the barrels on the vacant lot?

7. ***Evaluating Methods***
How would you characterize the flame test with respect to its sensitivity? What difficulties could there be identifying ions with the flame test?

8. ***Evaluating Methods***
Explain how you can use a spectroscope to identify the components of solutions containing several different metals.

Extensions

1. ***Inferring Conclusions***
A student performed flame tests on several unknowns and observed that they were all shades of red. What should she do to correctly identify these substances? Explain your answer.

2. ***Applying Ideas***
During a flood, the labels from three bottles of chemicals were lost. The unlabeled bottles of white solids were known to contain the following: strontium nitrate, ammonium carbonate, and potassium sulfate. Explain how you could easily test the substances and relabel these three bottles. (Hint: ammonium ion does not provide a distinctive color when flame-tested.)

3. ***Applying Ideas***
Some stores sell jars of "fireplace crystals." When sprinkled on a log, they turn the flames colors, such as blue, red, green, and violet. Explain how these crystals can change the flame color. What ingredients do you expect them to contain?

4. ***Designing Experiments***
Identify substances that contain metal ions. If your teacher approves of your selections, bring them to the lab, dissolve them, and perform flame tests with them to identify the metals they contain.

5. ***Research and Communications***
Research how instruments that work like spectroscopes can identify substances that are far away. Report to the class on a specific use of such equipment in astronomy.

Metal compound	Color of flame	Wavelengths detected (nm)
NaCl solution	yellow	590, 595
NaCl crystals	yellow	590, 595
Na only (cobalt glass)	only blue of the glass is visible	n/a
K only (cobalt glass)	violet (purple)	n/a
Na and K	yellow	n/a
Na and K (cobalt glass)	violet	n/a
Unknown	answers will vary	answers will vary

Conclusions

(Note: assign only items 6–7 if spectroscopes are unavailable.)

6. Answers will vary. Students should be able to identify the unknown by comparing it to the results for the other metal compounds tested.

7. The flame test is fairly specific because it can show an easily detectable signal with a very small amount of material. Possible difficulties include problems with contamination and the fact that some metals have similar colors when flame tested.

8. The flame test of the mixture can be examined with the spectroscope. By comparing the lines in the spectra to those for other metals, one can determine which lines are due to which metals.

Extensions

1. The student should compare the red shades to those of the known samples. If information about spectral lines is available, that would also help determine which metal is the unknown.

2. Strontium nitrate will change the color of the flame to red, potassium sulfate will change the flame color to purple, and ammonium will not change the flame color.

3. The crystals contain a mixture of metal salts. When sprinkled on a fire, they cause the flame to glow, just as if several flame tests were being performed.

4. Student answers will vary. Be sure student suggestions are safe and include carefully planned procedures. Make certain that metals chosen are not hazardous or difficult to dispose of.

5. Student answers will vary. Spectroscopes analyze the light emitted by astronomical bodies. Just as in the flame test, light is emitted by electrons returning to the ground state.

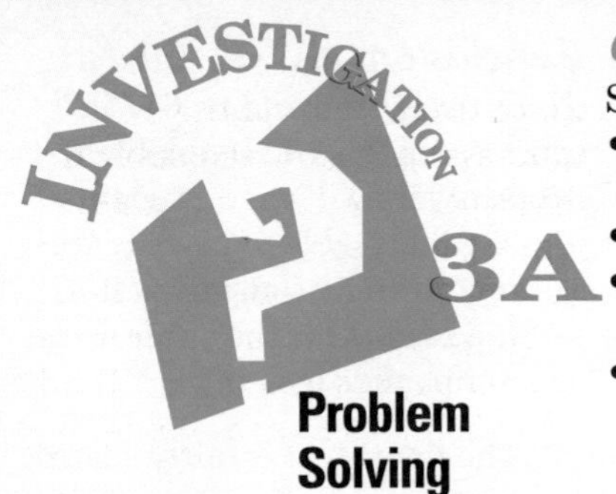

Objectives

Students will:
- use appropriate lab safety procedures.
- use a Bunsen burner.
- identify an unknown based on its flame test.
- design and implement their own procedure.

Planning

Recommended Time: 1 lab period
Prerequisite: Exploration 3A

Solution/Material Preparation

1. See the Teacher's Notes for Exploration 3A for information on preparing 1.0 M HCl.
2. To prepare 0.5 M Li_2SO_4, add 64 g of $Li_2SO_4{\bullet}H_2O$ to enough water to make 1.00 L of solution. (Other solutions from Exploration 3A can be used as the unknown, if this is preferred.)
3. If wooden splints will be used instead of flame-test wire, remember to soak them overnight in the ion solution.

Student Orientation

Techniques to Demonstrate

Review the proper way to clean the wire and glass plate. Emphasize that all rinses should be made with distilled water and collected in a waste beaker for disposal. If students will be using spectroscopes, review the techniques for viewing with a spectroscope.

Pre-lab Discussion

Begin by discussing the results and procedure used in the Exploration, especially any errors in lab technique that occurred. Discuss any disparity between the students' results and the expected flame colors and wavelengths.

If students are uncertain of how to begin planning, encourage them to start by reviewing the procedure and results from the Exploration.

Tips for Evaluating the Pre-lab Requirements

Students often want the labeled solutions from the Exploration to test again. This is not recommended. Forcing students to rely on their records of the results gained in the Exploration or on the wavelengths given in the Investigation helps students learn to relate data given in various ways and situations.

Data tables should contain spaces for the items shown in the Sample Data section.

Be certain students remember to use a waste beaker to collect all rinsings. If students suggest that they will need more than 5 mL of 1 M HCl, have them reconsider their procedure.

Sample Data

Substance	Flame color	Wavelengths
Unknown	red (carmine)	462, 498, 612, 670

Student data should have multiple trials. Data shown is for Li_2SO_4. Data will vary if a different unknown is chosen.

B

SPECTROSCOPY AND FLAME TESTS

Identifying Materials

ETA EXPERIMENTAL TESTING AGENCY

January 27, 1995

Director of Investigations
CheMystery Labs, Inc.
52 Fulton Street
Springfield, VA 22150

Dear Director:

As you may have seen in news reports, one of our freelance pilots, Davin Matthews, was killed in a crash of an experimental airplane that was struck by lightning.

What the reports did not mention was that Matthews's airplane was a recently perfected new design that he'd been developing for us. The notes he left behind indicate that the coating on the nose cone was the key to its speed and maneuverability. Unfortunately, he did not reveal what substances he used, and we were able to recover only flakes of material from the nose cone after the accident.

We have sent you samples of these flakes dissolved in a solution. Please identify the material Matthews used so that we can duplicate his prototype. We will pay $200,000 for this work, provided that you can identify the material within three days.

Sincerely,

Jared MacLaren

Jared MacLaren
Experimental Testing Agency

Required Precautions

Lab aprons and goggles must be worn at all times.

Do not touch or taste any chemicals. Wash your hands thoroughly before you leave the lab.

Confine loose hair and clothing.

If you get acid or base on your skin or clothing, wash it off at the sink while calling to your teacher. If you get acid or base in your eyes, immediately flush it out at the eyewash while calling to your teacher.

Required Precautions

- Goggles and a lab apron must be worn at all times.
- Tie back loose hair and clothing while in the lab.
- Read all safety cautions, and discuss them with your students.
- Students should not handle concentrated acid solutions.
- In case of an acid spill, dilute first with water. Then mop up the spill with wet cloths or a cloth mop designated for spill cleanup while wearing disposable gloves.

Memorandum

CheMystery Labs, Inc.
52 Fulton Street
Springfield, VA 22150

Date: January 28, 1995

To: Edwin Thien

From: Marissa Bellinghausen

We have narrowed down the possibilities of the material used to four. It is a compound of either lithium, potassium, strontium, or calcium. Using flame tests and the wavelengths of spectroscopic analysis, you should be able to identify which of these is in the sample.

Because our contract depends on timeliness, give me a preliminary report that includes the following as soon as possible.
- detailed one-page summary of your plan for the procedure
- itemized list of equipment, with total costs (Remember that the less you spend, the more profit for the company—provided you can get accurate results!)

After you complete your analysis, prepare a report in the form of a two-page letter to MacLaren. It must include the following.
- identity of the metal in the sample
- summary of your procedure
- detailed and organized analysis and data sections, showing tests and results
- detailed invoice for all expenses and services

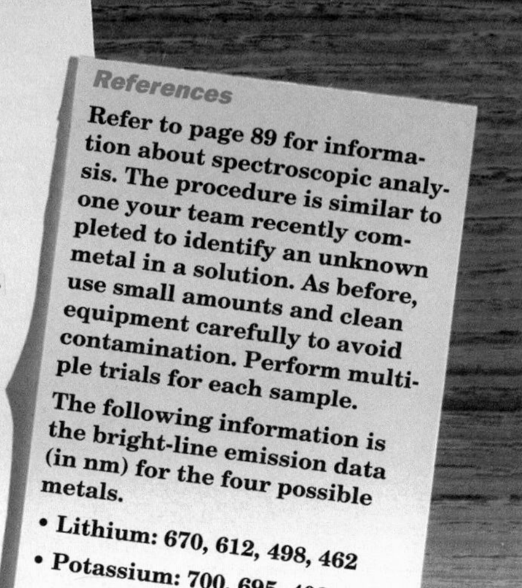

References

Refer to page 89 for information about spectroscopic analysis. The procedure is similar to one your team recently completed to identify an unknown metal in a solution. As before, use small amounts and clean equipment carefully to avoid contamination. Perform multiple trials for each sample.

The following information is the bright-line emission data (in nm) for the four possible metals.
- Lithium: 670, 612, 498, 462
- Potassium: 700, 695, 408, 405
- Strontium: 710, 685, 665, 500, 490, 485, 460, 420, 405
- Calcium: 650, 645, 610, 485, 460, 445, 420

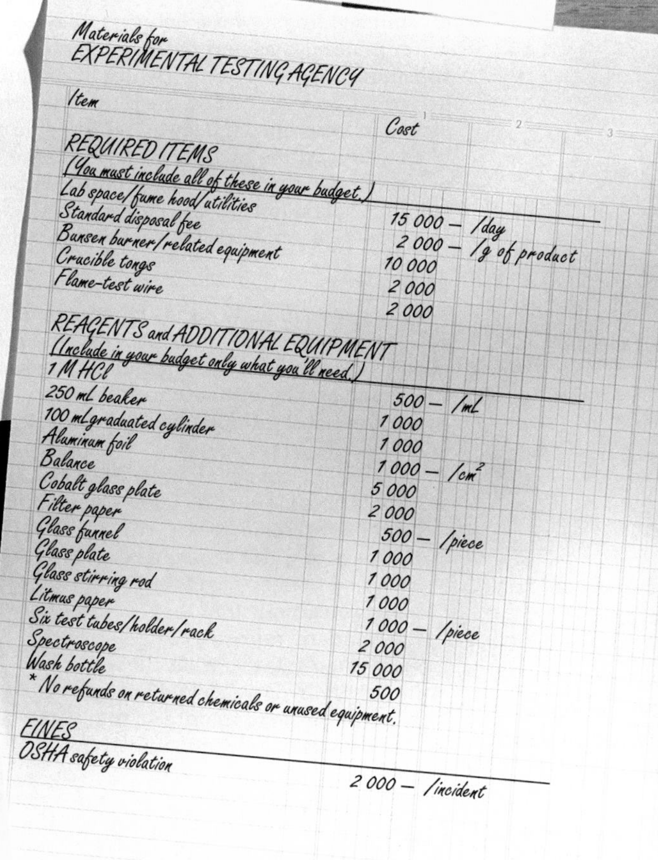

Materials for
EXPERIMENTAL TESTING AGENCY

Item	Cost
REQUIRED ITEMS (You must include all of these in your budget.)	
Lab space/fume hood/utilities	15 000 — /day
Standard disposal fee	2 000 — /g of product
Bunsen burner/related equipment	10 000
Crucible tongs	2 000
Flame-test wire	2 000
REAGENTS and ADDITIONAL EQUIPMENT (Include in your budget only what you'll need.)	
1 M HCl	500 — /mL
250 mL beaker	1 000
100 mL graduated cylinder	1 000
Aluminum foil	1 000 — /cm²
Balance	5 000
Cobalt glass plate	2 000
Filter paper	500 — /piece
Glass funnel	1 000
Glass plate	1 000
Glass stirring rod	1 000
Litmus paper	1 000 — /piece
Six test tubes/holder/rack	2 000
Spectroscope	15 000
Wash bottle	500
* No refunds on returned chemicals or unused equipment.	
FINES	
OSHA safety violation	2 000 — /incident

Spill Procedures/ Waste Disposal Methods
- In case of spills, follow your teacher's instructions.
- Place all remaining solutions in separate disposal containers as indicated by your teacher.
- Clean the area and all equipment after use.

Proposed Procedure

Place three drops of the unknown solution and several drops of 1.0 M HCl on a glass plate. Clean the flame-test wire in a drop of HCl. Dip the flame-test wire into the first drop of unknown solution. Hold it in a Bunsen burner flame, and record the color (and wavelength if a spectroscope is available). Clean the wire in another drop of HCl, and repeat the process. When finished, rinse the plate with distilled water, and collect the rinse water in a waste beaker.

Post-Lab

Disposal

Set out a disposal container for the students. After all of the waste beakers have been emptied into it, neutralize the resulting solution with 0.1 M NaOH. When the solution's pH is between 5 and 9, pour it down the drain. If you use wooden splints, collect them after they have been labeled, and reuse them next year for the same compounds.

Materials
(for each lab group)
- 1.0 M HCl solution, 5 mL
- Unknown solution (0.5 M Li_2SO_4 is recommended)
- 250 mL beaker
- Bunsen burner, gas tubing, and striker
- Cobalt glass plate
- Crucible tongs
- Flame-test wire, 5 cm
- Glass plate (either 7 cm × 15 cm plate, or microchemistry plate with wells)

Optional Equipment
- Spectroscope
- Wooden splints

Estimated cost of materials: \$86 000 (\$101 000 with spectroscope)

EXPLORATION 5A

Objectives

Students will:

- use appropriate lab safety procedures.
- use a laboratory balance to measure mass.
- use a Bunsen burner to evaporate a substance's water of crystallization.
- collect data and organize it in a table.
- calculate molar amounts from mass values.
- determine the empirical formula for the hydrated compound.
- determine the percentage of water in the hydrated compound.
- determine how much water can be absorbed by 25 g of the anhydrous compound.

Planning

Recommended Time:
1 lab period

Materials
(for each lab group)

- $CuSO_4{\cdot}5H_2O$, about 5 g
- Distilled water
- Balance (may be shared among lab groups)
- Bunsen burner, gas tubing, striker
- Crucible and cover
- Crucible tongs (do not use beaker tongs)
- Desiccator
- Dropper or micropipet
- Glass stirring rod
- Ring and pipe-stem triangle
- Ring stand
- Spatula
- Weighing paper

Optional Equipment

- Wing top for burner

Solution/Material Preparation

1. If you do not have a desiccator, you can use an old jar with a lid or a similar container. To make a resting place for a crucible within the jar, use a pipe-stem triangle with the ends bent down like legs. Fill the bottom of the jar with granular anhydrous $CaCl_2$.

Percent Composition of Hydrates

Objectives

Demonstrate **proficiency in using the balance and the Bunsen burner.**

Measure **the mass of a substance before and after its water of crystallization has been removed.**

Relate **results to the law of conservation of mass and the law of multiple proportions.**

Perform **calculations using the molar mass.**

Determine **the empirical formula and percentage by mass of water in the hydrate.**

Situation

You are a research chemist working for a company that is developing a new chemical moisture absorber and indicator. The company plans to seal the moisture absorber into a transparent porous pouch attached to a cellophane window on the inside of packages for compact disc players. This way, moisture within the packages will be absorbed, and any package that has too much moisture can be quickly detected and dried out. Your company's efforts have focused on copper(II) sulfate, $CuSO_4$, which can absorb water to become a hydrate that shows a distinctive color change.

Background

As discussed in Chapter 5, when many ionic compounds are crystallized from a water solution, they include individual water molecules as part of their crystalline structure. If the substances are heated, this water of crystallization may be driven off, leaving behind the pure anhydrous form of the compound. Because the law of multiple proportions also applies to crystalline hydrates, the number of moles of water driven off per mole of the anhydrous compound should be a simple whole number ratio. You can use this information to help you determine the formula of the hydrate.

Problem

To help your company decide whether $CuSO_4$ is the right substance for the moisture absorber and indicator, you will need to examine the hydrated and anhydrous forms of the compound and determine the following.

- the empirical formula of the hydrate, including its water of crystallization
- whether the change from hydrated to anhydrous form is obvious enough for the compound to be useful as an indicator
- the mass of water that can be absorbed by the 25 g of the anhydrous compound that the company proposes to use

Even if you can guess what the formula for the hydrate should be, carefully perform the procedure so that you know how well your company's supply of $CuSO_4$ absorbs moisture.

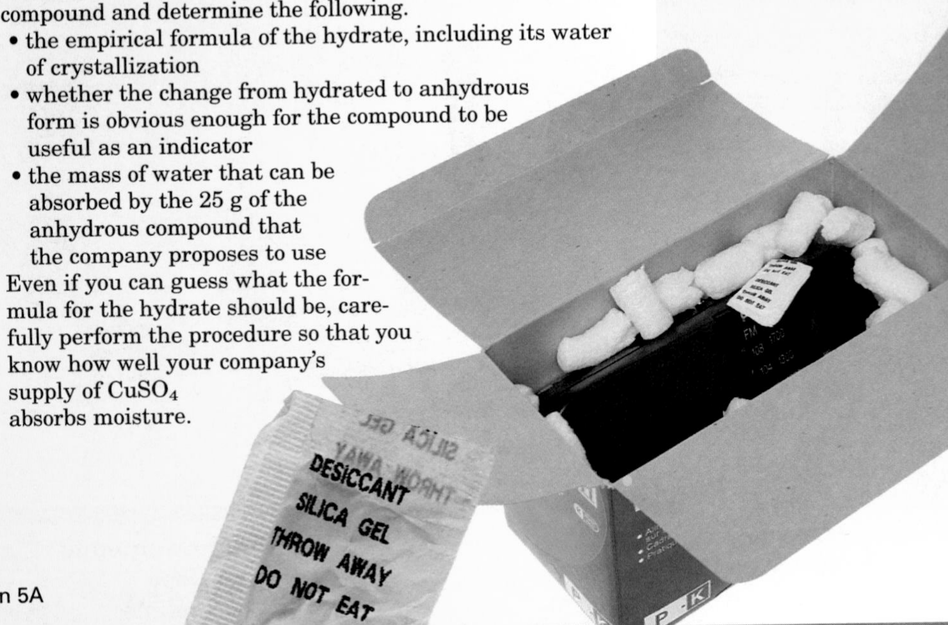

Required Precautions

- Safety goggles and a lab apron must be worn at all times.
- Read all safety cautions, and discuss them with your students.
- Remind students to confine loose clothing and long hair before lighting a burner.
- Remind students that heated objects can be hot enough to burn even if they look cool. Students should always use crucible tongs to handle crucibles and crucible covers.

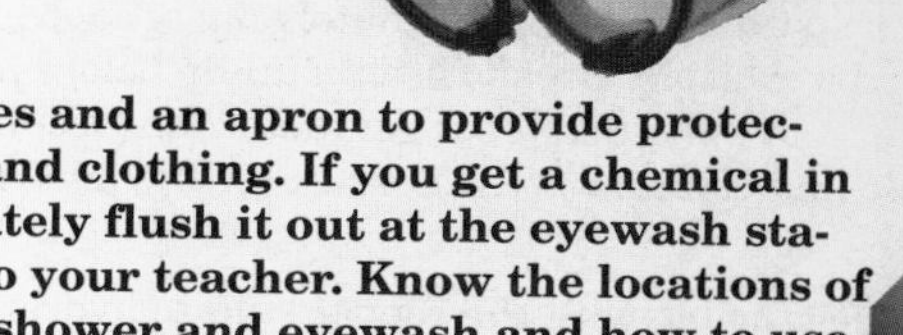

Safety

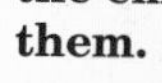

Always wear goggles and an apron to provide protection for your eyes and clothing. If you get a chemical in your eyes, immediately flush it out at the eyewash station while calling to your teacher. Know the locations of the emergency lab shower and eyewash and how to use them.

Do not touch any chemicals. If you get a chemical on your skin or clothing, wash it off at the sink while calling to your teacher. Make sure you carefully read the labels and follow the directions on all containers of chemicals that you use. Do not taste any chemicals or items used in the laboratory. Never return leftovers to their original containers; take only small amounts to avoid wasting supplies.

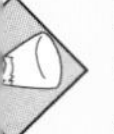

Open flames are dangerous, and you should tie back loose clothing and hair for safety reasons. Use proper methods to hold equipment being heated. The crucible and cover are very hot after each heating. Remember to handle the crucible and cover only with tongs designed to hold a crucible.

Never put broken glass or ceramics in a regular waste container. Broken glass or ceramics should be disposed of separately.

Always clean up the lab and all equipment after use, and dispose of substances according to proper disposal methods. Wash your hands thoroughly before you leave the lab after all lab work is finished.

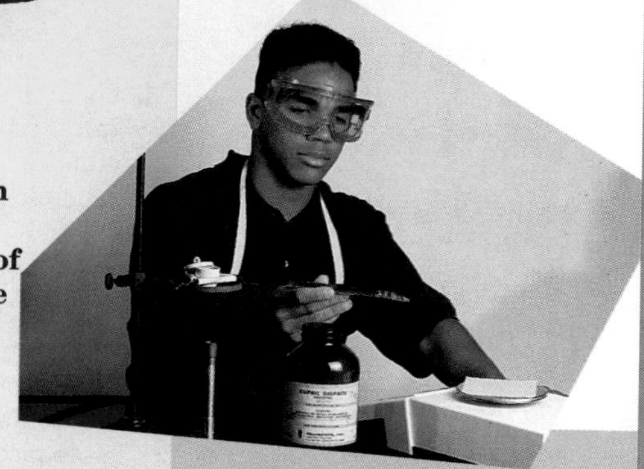

Materials

- $CuSO_4$, hydrated crystals
- Distilled water
- Balance
- Bunsen burner
- Crucible and cover
- Crucible tongs
- Desiccator
- Dropper or micropipet
- Glass stirring rod
- Ring and pipe-stem triangle
- Ring stand
- Spatula
- Weighing paper

Preparation

1. ***Organizing Data***
   Prepare a data table in your lab notebook with spaces for the mass of the empty crucible with cover and for the initial mass of the sample, crucible, and cover; and several spaces for the mass of the sample, crucible, and cover after heating. Leave room for observations about the procedure.
2. Make sure that your equipment and tongs are very clean for this work so that you will get the best possible results. Remember that you will need to cool the heated crucible in the desiccator before measuring its mass. ***Never put a hot crucible on a balance; it will damage the balance.***

Technique

3. To prepare the crucible and cover, place the crucible and cover on the triangle with the lid slightly tipped. The small opening will allow gases to escape. Heat the crucible and cover until the crucible glows slightly red. Using the tongs, transfer the crucible and cover to the desiccator, and allow them to cool for 5 min. Determine the mass of the crucible and cover to the nearest 0.01 g, and record it in your data table.

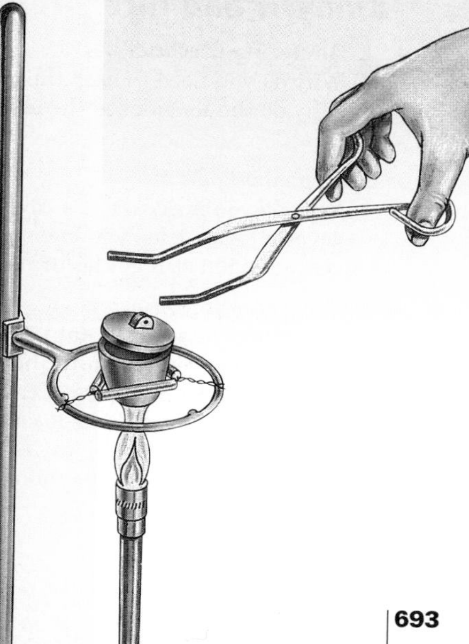

Student Orientation

Pre-lab Discussion

The discussion of hydrated crystals on pages 179–180 of Chapter 5 should be adequate preparation for this lab. Students should be familiar with the concepts of empirical formula and percentage by mass from Chapter 5, as well as the process of mass-mole conversions. Remind students that the • symbol in a hydrate is not a multiplication symbol.

To review the law of conservation of mass and the law of multiple proportions, write the following reaction on the chalkboard.

$$CuSO_4{\bullet}nH_2O \longrightarrow CuSO_4 + nH_2O$$

According to the law of conservation of mass, the mass will be the same on both sides of the equation. This allows the mass of water to be calculated by comparing the masses of the substance before and after the reaction. According to the law of multiple proportions, n, the number of water molecules per formula unit of $CuSO_4$, should be a small whole number.

Techniques to Demonstrate

Students will need to be given detailed instructions on how to use a crucible, set it on the pipestem triangle, and place the crucible lid on top of the crucible, leaving a small opening. Be certain they realize that they should always use crucible tongs to touch the crucible. Even when the crucible is cool, if they touch it, oils from their hands can contribute to errors in mass measurements. Remind students to gently break up the crystals with a stirring rod after placing them in the crucible. This will prevent cracking and breaking of the crucible as it is heated.

Students should use the same balance throughout this lab. Also remind students not to put hot crucibles on balances.

Emphasize the need to heat the copper sulfate very slowly. If it is heated too rapidly, it may splatter or decompose. As noted in the procedure, if a yellowish color persists after cooling, it is a sign of decomposition of the sulfate due to overheating.

The crystals may get slightly yellow when heated, but this color should not remain after cooling. A wing-top on the burner will produce a steady, warm flame that can help prevent decomposition.

Post-Lab

Disposal

Provide a labeled container for students to dispose of the rehydrated and anhydrous copper sulfate and any excess of the original compound. Later, redissolve the contents of the container in distilled water. Let the solution evaporate until dry, and then recover the crystals for re-use next year. Do not dispose of copper sulfate in a landfill, an incinerator, or down the drain.

Answers to Analysis and Interpretation

1. The crucible was heated to be certain it was completely dry. The tongs had to be clean so that they did not transfer any dirt or debris to the crucible. The presence of any water, dirt, or debris will cause error in the mass measurements.

2. The cover was used to keep debris out of the crystals. It also helped prevent water from condensing on the crystals as they cooled. Without it, each mass measurement would be slightly larger because the crystals would absorb water.

3. $35.42\ g - 32.18\ g = 3.24\ g\ CuSO_4$

$$\frac{3.24\ g}{159.62\ g/mol} = 2.03 \times 10^{-2}\ mol\ CuSO_4$$

4. As the sample was heated, water left the crystal to become water vapor. The mass of the water vapor and the crystals was the same as the mass of the hydrated crystals, so the law of conservation of mass was not violated.

5. $37.18\ g - 35.42\ g = 1.76\ g\ H_2O$

$$\frac{1.76\ g}{18.02\ g/mol} = 0.0977\ mol\ H_2O$$

4. Using a spatula, add approximately 5 g of copper sulfate hydrate crystals to the crucible. Break the crystal up before placing it in the crucible. Put on the crucible's cover, determine the mass of the covered crucible and crystals to the nearest 0.01 g, and record it in your data table.
5. Place the crucible with the copper sulfate hydrate on the triangle, and again position the cover so there is only a small opening. If the opening is too large, the crystals may spatter as they are heated. Heat the crucible very gently on a low flame to avoid spattering. Increase the temperature gradually for 2 or 3 min, and then heat until the crucible glows red for at least 5 min. Be very careful not to raise the temperature of the crucible and its contents too suddenly. You will observe a color change, which is normal, but if the substance remains yellow after cooling, it was overheated and has begun to decompose. Allow the crucible, cover, and contents to cool for 5 min in the desiccator, and then measure its mass. Enter the mass in your data table.
6. Heat the covered crucible and contents to redness again for 5 min. Allow the crucible, cover, and contents to cool in the desiccator, and then determine their mass, recording it in the data table. If the two mass measurements differ by no more than 0.01 g, you may assume that all of the water has been driven off. Otherwise, repeat the process until the mass no longer changes, indicating that all of the water has evaporated. Record this constant mass in your data table.
7. After recording the constant mass, set a part of your sample aside on a piece of weighing paper. Using the dropper or pipet, put a few drops of water onto this part to rehydrate the crystals. Record your observations in your data table.

Cleanup and Disposal

8. Clean all apparatus and your lab station. Make sure to completely shut off the gas valve before leaving the laboratory. Remember to wash your hands. Place the rehydrated and anhydrous chemicals in the disposal containers designated by your teacher.

Analysis and Interpretation

1. ***Analyzing Methods***
   Why do you need to heat the clean crucible before using it in this lab? Why do the tongs used throughout this lab need to be especially clean?
2. ***Organizing Data***
   Why do you need to use a cover for the crucible? Could you leave the cover off each time you measure the mass of the crucible and its contents and still get accurate results? Explain your answer.
3. ***Analyzing Information***
   Calculate the mass of anhydrous copper sulfate (the residue that remains after heating to constant mass) by subtracting the mass of the empty crucible and cover from the mass of the crucible, cover, and heated $CuSO_4$. Use the molar mass for $CuSO_4$, determined from the periodic table, to calculate the number of moles present.

Sample Data

Mass of empty crucible and cover	32.18 g
Mass of crucible, cover, and copper sulfate hydrate	37.18 g
Mass of crucible, cover, and anhydrous copper sulfate after 1st heating	36.08 g

4. ***Resolving Discrepancies***
   Explain why the mass of the sample got smaller after it was heated, despite the law of conservation of mass.
5. ***Analyzing Information***
   Calculate the mass and moles of water originally present in the hydrate using the molar mass determined from the periodic table.

Conclusions

6. ***Analyzing Information***
   Using your answers from items **3** and **5**, determine the empirical formula for the copper sulfate hydrate.
7. ***Organizing Data***
   What is the percentage by mass of water in the original hydrated compound?
8. ***Organizing Conclusions***
   How much water could 25 g of anhydrous $CuSO_4$ absorb?
9. ***Evaluating Conclusions***
   When you rehydrated the small amount of anhydrous copper sulfate, what were your observations? Explain whether this substance would make a good indicator of moisture.

Extensions

1. ***Evaluating Methods***
   What possible sources of error can you identify in your procedure? If you can think of ways to eliminate them, ask your teacher to approve your suggestions, and run the procedure again.
2. ***Applying Conclusions***
   Some cracker tins include a glass vial of drying material in the lid. This is often a mixture of magnesium sulfate and cobalt chloride. As the mixture absorbs moisture to form hydrated compounds, the cobalt chloride changes from blue-violet $CoCl_2{\cdot}2H_2O$ to pink $CoCl_2{\cdot}6H_2O$. When this hydrated mixture becomes totally pink, it can be restored to the dihydrate form by heating it in the oven. Write equations for the reactions that occur when this mixture is heated.
3. ***Applying Ideas***
   Three pairs of students obtained the following results when they heated a solid. In each case, the students observed that when they began to heat the solid, drops of a liquid formed on the sides of the test tube.

Sample number	Mass before heating (g)	Constant mass after heating (g)
1	1.92	1.26
2	2.14	1.40
3	2.68	1.78

   a. Could the solid be a hydrate? Explain how you could find out.

   b. If the solid has a molar mass of 208 g/mol after heating and a formula of XY, how many formula units of water are there in one formula unit of the unheated compound?

Mass of crucible, cover, and anhydrous copper sulfate after 2nd heating	35.43 g
Mass of crucible, cover, and anhydrous copper sulfate after 3rd heating	35.42 g

Students should reheat the crucible, cover, and anhydrous copper sulfate until successive masses are within 0.02 g.

Conclusions

6. empirical formula: $CuSO_4{\cdot}5H_2O$

$$\frac{0.0977 \text{ mol } H_2O}{2.03 \times 10^{-2} \text{ mol } CuSO_4} = \frac{4.81 \text{ mol } H_2O}{1.00 \text{ mol } CuSO_4}$$

1.00:4.81 is close to 1:5

7. $\frac{1.76 \text{ g}}{5.00 \text{ g}} \times 100 = 35.2\%\ H_2O$

8. $25.0 \text{ g } CuSO_4 \times \frac{1.76 \text{ g } H_2O}{3.24 \text{ g } CuSO_4} = 13.6 \text{ g } H_2O$

9. Anhydrous copper sulfate is a white powder. When water is added, it turns blue. This dramatic color change makes it a good indicator of moisture.

Extensions

1. Students' suggestions for improving the procedure will vary. Possible suggestions include heating larger amounts or performing multiple trials. Be sure answers are safe and include carefully planned procedures.

2. $CoCl_2{\cdot}6H_2O \longrightarrow CoCl_2{\cdot}2H_2O + 4H_2O$

3. a. The solid could be a hydrate, because it has less mass after heating. If the ratio of the mole amount of the water apparently lost to the mole amount of the apparently anhydrous compound is a small, whole-number ratio that remains constant with different samples, this is evidence of a hydrate.

b. Sample 1: $1.26 \text{ g XY} \times \frac{1 \text{ mol}}{208 \text{ g}} = 6.06 \times 10^{-3} \text{ mol XY}$

$0.66 \text{ g } H_2O \times \frac{1 \text{ mol}}{18.02 \text{ g}} = 3.7 \times 10^{-2} \text{ mol } H_2O$

$$\frac{3.7 \times 10^{-2} \text{ mol } H_2O}{6.06 \times 10^{-3} \text{ mol XY}} = \frac{6.1 \text{ mol } H_2O}{1.0 \text{ mol XY}}$$

Sample 2:

$$\frac{4.1 \times 10^{-2} \text{ mol } H_2O}{6.73 \times 10^{-3} \text{ mol XY}} = \frac{6.1 \text{ mol } H_2O}{1.0 \text{ mol XY}}$$

Sample 3

$$\frac{5.0 \times 10^{-2} \text{ mol } H_2O}{8.56 \times 10^{-3} \text{ mol XY}} = \frac{5.8 \text{ mol } H_2O}{1.0 \text{ mol XY}}$$

empirical formula: $XY{\cdot}6H_2O$

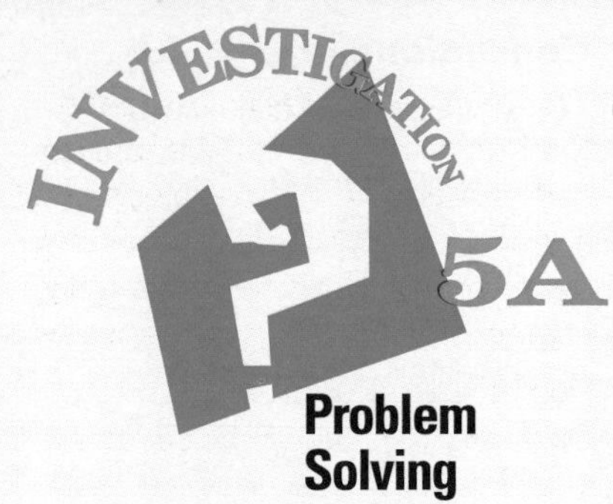

Objectives

Students will:

- use appropriate lab safety procedures.
- use a laboratory balance to measure mass.
- use a Bunsen burner to evaporate a substance's water of crystallization.
- collect data and organize it in a table.
- calculate molar amounts from mass values.
- determine the empirical formula for two hydrates of a compound.
- design and implement their own procedure.

Planning

Recommended Time: 1–2 lab periods
Prerequisite: Exploration 5A

Solution/Material Preparation

1. Gypsum and plaster of Paris are available at most hardware and art supply stores. Several days before the lab, prepare plaster of Paris according to the package directions. Allow it to dry thoroughly for several days or in a drying oven set on low temperature. To distribute the gypsum and plaster of Paris, chip off a 5 g chunk for each lab group.
2. See the Teacher's Notes for Exploration 5A for instructions on making your own desiccator.

Sample Data

	Gypsum	Plaster of Paris
Mass of empty crucible and lid	32.18 g	32.18 g
Mass of crucible, lid, and compound	37.18 g	37.18 g
Mass of crucible, lid, and compound after 1st heating	36.10 g	36.91 g
Mass of crucible, lid, and compound after 2nd heating	36.08 g	36.89 g

Analysis

$$3.90 \text{ g} \times \frac{1 \text{ mol } CaSO_4}{136.15 \text{ g}} = 2.86 \times 10^{-2} \text{ mol } CaSO_4 \text{ in gypsum sample}$$

$$1.10 \text{ g} \times \frac{1 \text{ mol } H_2O}{18.02 \text{ g}} = 6.10 \times 10^{-2} \text{ mol } H_2O \text{ in gypsum sample}$$

$$\frac{6.10 \times 10^{-2} \text{ mol } H_2O}{2.86 \times 10^{-2} \text{ mol } CaSO_4} = \frac{2.13 \text{ mol } H_2O}{1.00 \text{ mol } CaSO_4}$$

Gypsum: $CaSO_4{\cdot}2H_2O$

$$4.71 \text{ g} \times \frac{1 \text{ mol } CaSO_4}{136.15 \text{ g}} = 3.46 \times 10^{-2} \text{ mol } CaSO_4 \text{ in plaster of Paris sample}$$

$$0.29 \text{ g} \times \frac{1 \text{ mol } H_2O}{18.02 \text{ g}} = 1.6 \times 10^{-2} \text{ mol } H_2O \text{ in plaster of Paris sample}$$

$$\frac{1.6 \times 10^{-2} \text{ mol } H_2O}{3.46 \times 10^{-2} \text{ mol } CaSO_4} = \frac{0.46 \text{ mol } H_2O}{1.00 \text{ mol } CaSO_4}$$

Plaster of Paris: $CaSO_4{\cdot}\frac{1}{2}H_2O$

HYDRATES Gypsum and Plaster of Paris

LOST ART GYPSUM MINE

February 9, 1995

Director of Research
CheMystery Labs, Inc.
52 Fulton Street
Springfield, VA 22150

Dear Sir:

Lost Art Gypsum Mine previously sold its raw gypsum to a manufacturing company that used it to make anhydrous calcium sulfate, $CaSO_4$ (a desiccant), and plaster of Paris. That firm has now gone out of business, and we are currently negotiating the purchase of their equipment to process our own gypsum into anhydrous calcium sulfate and plaster of Paris.

Your company has been recommended to plan the large-scale industrial process for our new plant. We will need a detailed report on the development of the process and formulas for these products. This report will be presented to the bank handling our loan for the new plant. As we discussed on the telephone today, we are willing to pay you $250,000 for the work, with the contract papers arriving under separate cover today.

Sincerely,

Alex Farros

Alex Farros
Vice President
Lost Art Gypsum Mine

Required Precautions

Goggles and lab aprons must be worn at all times in the laboratory.

Confine loose clothing and hair.

Do not touch or taste any chemicals. Always wash your hands thoroughly when finished.

Required Precautions

- Wear safety goggles and lab apron.
- Read all safety cautions, and discuss them with your students.
- Remind students to confine loose clothing and long hair before lighting a burner.
- Remind students that heated objects can be hot enough to burn even if they look cool. Students should always use crucible tongs to handle crucibles and crucible covers.

CheMystery Labs, Inc.
52 Fulton Street
Springfield, VA 22150

Memorandum

Date: February 10, 1995

To: Kenesha Smith

From: Martha Li-Hsien

This job involves many different tasks. Your team needs to develop a procedure to experimentally determine the correct empirical formulas for both hydrates of this anhydrous compound. You will use gypsum samples from the mine and samples of the plaster of Paris product, but both are available at a cost.

Once this work is complete, our chemical engineering division will examine the most efficient way to implement the necessary procedures for manufacturing. Of the $250,000 fee we're getting from Lost Art, your end of the job must cost less than $125,000 so that the chemical engineering division has enough money to do its job.

As soon as possible, I need a preliminary report from you that includes the following.
- detailed one-page summary of your plan for the procedure, with all necessary data tables
- itemized list of equipment, with total costs for the accounting department (Remember to keep your costs under $125,000.)

After you complete the analysis, prepare a two-page report for the chemical engineering division that includes the following information.
- formulas for anhydrous calcium sulfate, plaster of Paris, and gypsum
- summary of your procedure
- detailed and organized data and analysis sections that show calculations, along with estimates and explanations of any possible sources of error
- detailed invoice of all expenses and services

This report will become Chapter 2 of our final report to Mr. Farros.

References

Refer to pages 179–180 for information about hydrates and water of crystallization. The procedure and reaction is similar to one your team recently completed involving hydrated copper(II) sulfate. Gypsum and plaster of Paris are hydrated forms of $CaSO_4$. One of the largest gypsum mines in the world is located outside of Paris, France. Plaster of Paris has less water of crystallization than gypsum. Plaster of Paris is commonly used in plaster walls and art sculptures.

Spill Procedures/ Waste Disposal Methods

- **Solids must go in the designated waste containers.**
- **Clean the lab area and all equipment after use.**

Materials for LOST ART GYPSUM MINE

Item	Cost
REQUIRED ITEMS (You must include all of these in your budget.)	
Lab space/fume hood/utilities	15 000 — /day
Standard disposal fee	2 000 — /g of product
Balance	5 000
Bunsen burner/related equipment	10 000
Crucible tongs	2 000
Ring stand/ring/pipe-stem triangle	2 000
REAGENTS and ADDITIONAL EQUIPMENT (Include in your budget only what you'll need.)	
Gypsum sample	500 — /g
Plaster of Paris sample	500 — /g
250 mL beaker	1 000
100 mL graduated cylinder	1 000
Aluminum foil	1 000 — /cm²
Crucible and cover	5 000
Desiccator	3 000
Filter paper	500 — /piece
Glass funnel	1 000
Glass stirring rod	1 000
Litmus paper	1 000 — /piece
Mortar and pestle	2 000
Six test tubes/holder/rack	2 000
Spatula	500
Wash bottle	500
Weighing paper	500 — /piece
* No refunds on returned chemicals or unused equipment.	
FINES	
OSHA safety violation	2 000 — /incident

Materials
(for each lab group)
- Gypsum sample, about 5 g
- Plaster of Paris sample, about 5 g
- Balance (0.01 g accuracy if possible)
- Bunsen burner, gas tubing, striker
- Crucibles and covers, 2
- Crucible tongs (not beaker tongs)
- Desiccator
- Mortar and pestle
- Ring stand, ring, pipe-stem triangle
- Spatula
- Weighing paper, 2

Estimated cost of materials: $75 500 (will vary depending on sample size and on disposal needs)

Student Orientation

Techniques to Demonstrate

Review proper use of the mortar and pestle to grind up the gypsum and the plaster of Paris. Review the proper technique and safety procedures for heating a crucible.

Pre-lab Discussion

Begin by discussing the results and procedure used in the Exploration, especially any errors in lab technique that occurred.

Make certain that students understand that fractional coefficients represent a fraction of a molar amount, not a fraction of a molecule.

Tips for Evaluating the Pre-lab Requirements

Students may want to heat the gypsum to create plaster of Paris, make the necessary measurements, and then heat the plaster of Paris to create anhydrous $CaSO_4$. However, it is difficult to determine when to stop heating. Instead, students should heat samples of both plaster of Paris and gypsum separately until anhydrous $CaSO_4$ is formed. You can either tell them this at the beginning or let students discover it themselves.

Many students forget to heat the empty crucible before measuring its mass at the very beginning.

Proposed Procedure

Heat a crucible and cover, cool them, and measure their mass. Add some crushed crystals of gypsum. Measure the mass, and then heat gently. Cool and measure the new mass. Keep reheating until the mass is fairly constant. Repeat for plaster of Paris.

Post-Lab

Disposal

Provide a labeled container for students to dispose of all forms of $CaSO_4$. Pour the contents of the waste container onto several layers of newspaper to absorb excess water, or wrap the waste container in newspaper. Roll up the newspaper and discard it in the trash can. Do not pour the contents of the waste container down the drain.

EXPLORATION 6A

Objectives

Students will:

- use appropriate lab safety procedures.
- use polymers to make a latex rubber ball and a ethanol-silicate polymer ball ("super ball").
- observe similarities and differences in the properties of the two balls.
- measure the bounce height of the two types of polymer balls.
- calculate volume from measurements of ball diameter.
- use a laboratory balance to measure mass.
- calculate the density of each type of ball.

Planning

Recommended Time:
1 lab period

Materials
(for each lab group)

- 5% acetic acid solution (vinegar), 10 mL
- 50% ethanol solution, 3 mL
- Distilled water
- Liquid latex, 10 mL
- Sodium silicate solution, 12 mL
- 2 L beaker or plastic bucket or tub
- 10 mL graduated cylinder
- 25 mL graduated cylinder
- 5 oz paper cup, 2
- Paper towels
- Wooden stick

Solution/Material Preparation

1. For 5% acetic acid, use white vinegar. Do not dilute glacial acetic acid.
2. Liquid latex may be purchased from Flinn Scientific, Inc., Batavia, IL. The following catalog numbers can be used: L0004—500 mL; L0110—1.00 L; L0222—4.00 L.
3. Sodium silicate solution, also known as water glass, can be purchased ready to use from many scientific supply firms.

Polymers and Toy Balls

Objectives

Synthesize **two different polymers.**

Prepare **a small toy ball from each polymer.**

Observe **the similarities and differences of the two balls.**

Measure **the density of each polymer.**

Compare **the bounce height of the two balls.**

Situation

Your company has been contacted by a toy company. They have always specialized in toy balls made from vulcanized rubber. Recent environmental legislation has increased the cost of disposing of the sulfur and other chemical byproducts of the manufacturing process for this type of rubber. The toy company wants you to research some other materials.

Background

Isoprene monomer

Rubber is a polymer of covalently bonded atoms. The simplest formula unit for the rubber monomer is called *isoprene* and is shown on the left. Three monomers bonded together are also shown. The *n* indicates a very large number, usually about 3000. The zigzag chain created when monomers bond together accounts for rubber's strength and elasticity.

When rubber is vulcanized, it is heated with sulfur, and the sulfur atoms form bonds between adjacent molecules of rubber, increasing its strength and making it more elastic.

Latex rubber is a colloidal suspension that can be made synthetically or found naturally in plants such as the para rubber tree (*Hevea brasiliensis*), in which it dries to form a waterproof layer for protection. Latex is composed of approximately 60% water, 35% hydrocarbon monomers, 2% proteins, and some sugars and inorganic salts. Commercially, latex is preserved with ammonia so that it remains a liquid until it can be molded or stretched into its final form. It is used in paints and disposable gloves.

Three isoprene monomers

Another polymer with covalent bonds can be formed from ethanol, C_2H_5OH, and a solution of sodium silicate, mostly in the form of $Na_2Si_3O_7$, which is known as water glass because it dissolves in water. When the polymer is formed, water is also a product. The polymer has the structural formula shown at left.

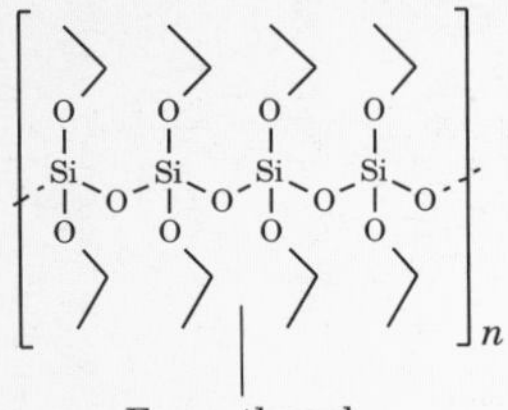

Four ethanol-silicate monomers

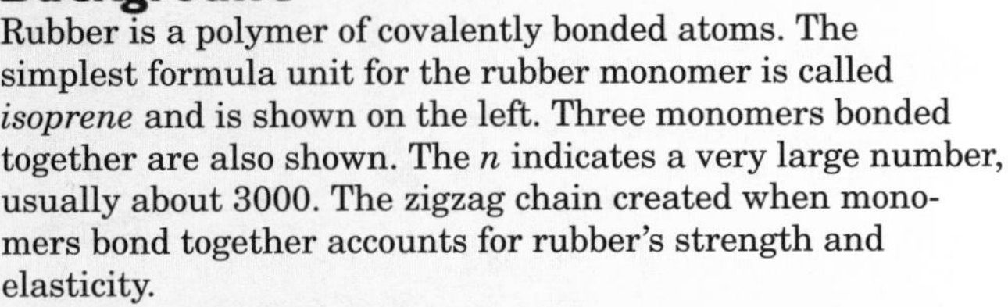

Problem

Latex and the ethanol-silicate polymer are the two materials you will investigate. You need to do the following.

- synthesize each polymer
- make a ball 2–3 cm in diameter from each polymer
- make observations about physical properties of each polymer
- measure how well each ball bounces

698 | Exploration 6A

Required Precautions

- Wear safety goggles and lab apron during the lab.
- Read all safety cautions, and discuss them with your students.
- Promptly clean up all spills with paper towels.
- Ethanol is flammable. Ensure that there are no flames anywhere in the room when an alcohol is present. Keep the alcohol in a hood, use a container with a lid, and restrict the amount kept in the hood to the minimum needed by the students.
- Do not allow students to take any ethanol-silicate polymer from the laboratory.

Safety

Always wear goggles and an apron to provide protection for your eyes and clothing. If you get a chemical in your eyes, immediately flush it out at the eyewash station while calling to your teacher. Know the locations of the emergency lab shower and eyewash and how to use them.

Do not touch any chemicals. If you get a chemical on your skin or clothing, wash it off at the sink while calling to your teacher. Make sure you carefully read the labels and follow the directions on all containers of chemicals that you use. Do not taste any chemicals or items used in the laboratory. Never return leftovers to their original containers; take only small amounts to avoid wasting supplies.

Wear disposable plastic gloves; the sodium silicate solution and the alcohol-silicate polymer are irritating to your skin.

Ethanol is flammable. Make sure there are no flames anywhere in the laboratory when you are using it. Also, keep it away from other sources of heat.

Always clean up the lab and all equipment after use, and dispose of substances according to proper disposal methods. Wash your hands thoroughly before you leave the lab when all lab work is finished.

Materials

- **5% acetic acid solution (vinegar), 10 mL**
- **50% ethanol solution, 3 mL**
- **Distilled water**
- **Liquid latex, 10 mL**
- **Sodium silicate solution, 12 mL**
- **2 L beaker or plastic bucket or tub**
- **10 mL graduated cylinder**
- **25 mL graduated cylinder**
- **Paper cup, 5 oz., 2**
- **Paper towels**
- **Wooden stick**

Preparation

1. ***Organizing Data***
   Prepare two data tables in your lab notebook, one for each polymer. Each table must have spaces for three trials of the bounce height in centimeters, the mass of each ball, and the diameter of each ball. Leave space to record observations about the balls.

Technique

2. Fill the 2 L beaker or the bucket or tub about half-full of distilled water.
3. Using a clean 25 mL graduated cylinder, measure 10 mL of liquid latex, and pour it into one of the paper cups.
4. Thoroughly clean the 25 mL graduated cylinder with soap and water. Then, rinse with distilled water.
5. Measure 10 mL of distilled water. Pour it into the paper cup with the latex.
6. Measure 10 mL of the 5% acetic acid solution, and pour it into the paper cup with the latex and water.
7. Stir the mixture with the wooden stick immediately.

4. Keep the bottles of latex and acetic acid solution in an operating hood because the vapors are annoying. Keep the bottle of ethanol in an operating fume hood because the vapors are flammable. For all three, remind students to keep the containers closed when not in use.
5. If balls need to be kept overnight, place them in plastic bags. If the ethanol-silicate polymer ball crumbles, add a few drops of water to it.

Student Orientation

Pre-lab Discussion

Review the material on polymers on page 210 of Chapter 6. Be certain students understand the differences between a monomer and a polymer. Be sure to relate the properties of these polymers, such as strength, flexibility, and elasticity, to the nature of the covalent bonds holding them together.

Techniques to Demonstrate

Show students how to roll the latex and the ethanol-silicate polymer into a ball. Students will have difficulty making a perfect sphere, but they should try to make it as regular as possible. The more irregular the shape, the more difficulty they will have in calculating volume and determining the bounce height, because the ball will not bounce straight up.

Students should use mathematical calculations and measurements of the diameters to determine the volume of the balls, not water displacement. The ethanol-silicate polymer will dissolve in water.

Remind students to be patient with the ethanol-silicate polymer, which tends to crumble. If it crumbles too much, a few drops of water will rehydrate it and allow it to be shaped into a ball.

Post-Lab

Disposal

Paper cups, paper towels, disposable gloves, latex, and ethanol-silicate polymer balls and fragments should be disposed of in the trash can. Waste liquids from this lab work can be poured down the drain.

Answers to Analysis and Interpretation

1. Isoprene monomer: C_5H_8 ethanol-silicate monomer: $SiC_4H_{10}O_3$
2. Answers will vary, but could include the following. The latex ball is more opaque, less smooth, and less crumbly than the ethanol-silicate polymer ball. The ethanol-silicate polymer ball breaks down after a period of time. Both balls bounce.
3. Ionic substances are not crumbly, nor do they tend to bounce.
4. Average bounce height for latex ball:

$$\frac{50 + 50 + 55}{3} = 52 \text{ cm}$$

Average bounce height for ethanol-silicate polymer ball:

$$\frac{58 + 60 + 50}{3} = 56 \text{ cm}$$

5. Volume for latex ball:

$$\frac{4}{3}(3.14)(3.0)^3 = 110 \text{ cm}^3$$

Volume for ethanol-silicate polymer ball:

$$\frac{4}{3}(3.14)(3.5)^3 = 180 \text{ cm}^3$$

6. Density for latex ball:

$$\frac{45.20 \text{ g}}{110 \text{ cm}^3} = 0.41 \text{ g/cm}^3$$

Density for ethanol-silicate polymer ball:

$$\frac{84.03 \text{ g}}{180 \text{ cm}^3} = 0.47 \text{ g/cm}^3$$

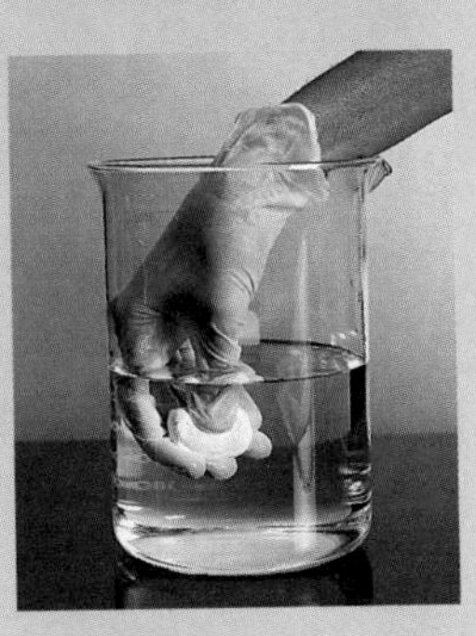

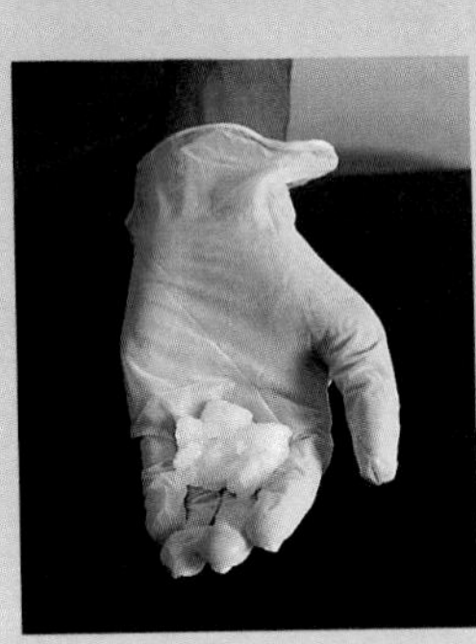

8. As you continue stirring, a polymer "lump" will form around the wooden stick. Pull the stick with the polymer lump from the paper cup and immerse it in the 2 L beaker or bucket or tub.
9. While wearing gloves, gently pull the lump from the wooden stick, keeping it immersed under the water.
10. Keeping the latex rubber under water, use your gloved hands to squeeze the lump into a ball, as shown in the figure on the left, and then squeeze several more times to remove any unused chemicals. You may remove the latex rubber from the water as you roll it in your hands to smooth the ball.
11. Set aside the latex-rubber ball to dry. While it is drying, you should begin to make a ball from the ethanol and sodium silicate solutions.
12. In a clean 25 mL graduated cylinder, measure 12 mL of sodium silicate solution, and pour it into the other paper cup.
13. In a clean 10 mL graduated cylinder, measure 3 mL of 50% ethanol. Pour the ethanol into the paper cup with the sodium silicate, and mix with the wooden stick until a solid substance is formed.
14. While wearing gloves, remove the polymer that forms, and place it in the palm of one hand. Gently press it with the palms of both your hands until a ball that does not crumble is formed. This takes a little time and patience. The liquid that comes out of the ball is a combination of ethanol and water. Occasionally moisten the ball by letting a small amount of water from a faucet run over it. When the ball no longer crumbles, you are ready to go on to the following steps.
15. Observe as many physical properties of the balls as possible, and record your observations in your lab notebook.
16. Drop each ball several times, and record your observations.
17. Drop each ball from a height of 1 m, and measure its bounce. Repeat for three trials for each ball.
18. Measure the diameter and the mass of each ball.

Cleanup and Disposal

19. Dispose of any extra solutions in the containers indicated by your teacher. Clean up your lab area. Remember to wash your hands thoroughly when your lab work is finished.

Analysis and Interpretation

1. ***Analyzing Information***
   Give the chemical formula for the latex (isoprene) monomer and the ethanol-silicate polymer.
2. ***Analyzing Information***
   List at least three observations you made of the properties of the two different balls.
3. ***Applying Models***
   Explain how your observations in item **2** indicate that the polymers in each ball are not ionically bonded.
4. ***Organizing Data***
   Calculate the average height of the bounce for each type of ball.

Sample Data

	Latex rubber	Ethanol-silicate polymer
Bounce height—trial 1	50. cm	58. cm
Bounce height—trial 2	50. cm	60. cm
Bounce height—trial 3	55. cm	50. cm

5. ***Organizing Data***
Calculate the volume for each ball. Even though they may not be perfectly spherical, assume that they are. (Hint: the volume of a sphere is equal to $\frac{4}{3} \times \pi \times r^3$, where r is the radius, which is one-half of the diameter.)

6. ***Organizing Data***
Calculate the density of each ball using your measurements for the mass and the volume from item **5**.

Conclusions

7. ***Inferring Conclusions***
Which polymer do you recommend for the toy company's new toy balls? Explain your reasoning.

8. ***Evaluating Viewpoints***
Using the table shown to the right, find the unit cost, the amount of money it costs to make a single ball. (Hint: remember to check how much of each reagent is needed to make a single ball.)

Reagent	Price (dollars per liter)
Acetic acid solution	1.50
Ethanol solution	9.00
Latex solution	20.00
Sodium silicate solution	10.00

8. ***Evaluating Viewpoints***
What are some other possible practical applications for each of the polymers you made?

Extensions

1. ***Research and Communication***
Polymers are used daily in our lives. Describe or list the polymers you come into contact with during a one-day period in your life.

2. ***Designing Experiments***
Design a mold for a polymer ball that will make it symmetrical and smooth. If your teacher approves of your design, try the procedure again with the mold.

3. ***Designing Experiments***
What possible sources of error can you identify in this procedure? If you can think of ways to eliminate them, ask your teacher to approve your plan, and run the procedure again.

4. ***Predicting Outcomes***
When a ball bounces up, kinetic energy of motion is converted into potential energy. With this in mind, explain which will bounce higher: a perfectly symmetrical, round sphere or an oblong shape that vibrates after it bounces.

5. ***Predicting Outcomes***
Explain why you didn't measure the volume of the balls by submerging them in water.

Sample Data

	Latex rubber	Ethanol-silicate polymer
Mass	45.20 g	84.03 g
Diameter	6.0 cm	7.0 cm
Other observations	Does not crumble very easily. Seems more opaque.	Tends to bounce higher than other balls. Crumbles easily. Seems translucent.

Conclusions

7. Students answers will vary, but should be based on the properties of the ball. Students may argue that because the ethanol-silicate polymer ball crumbles in time, it would be a poor candidate for a toy.

8. Unit cost for latex ball: $0.215
Unit cost for ethanol-silicate polymer ball: $0.147

9. Student answers will vary, but should be based on the properties of each substance.

Extensions

1. Student answers will vary. Some possible polymers include rubber, proteins, cellulose, nylon, Dacron, polyester, polyurethane, etc.

2. Students' suggestions for a mold will vary. Be sure that they choose a material that will not dissolve in the solvents used in this process.

3. Students' suggestions for improving the procedure will vary. Be sure answers are safe and include carefully planned procedures.

4. A ball with an oblong shape will not bounce as high because some of the kinetic energy will be transferred into the energy of vibration, leaving less to be transformed into gravitational potential energy. A sphere that bounces will transfer nearly all of its kinetic energy into gravitational potential energy.

5. The ethanol-silicate polymer ball would break down in the water, so density had to be calculated from measurements of diameter.

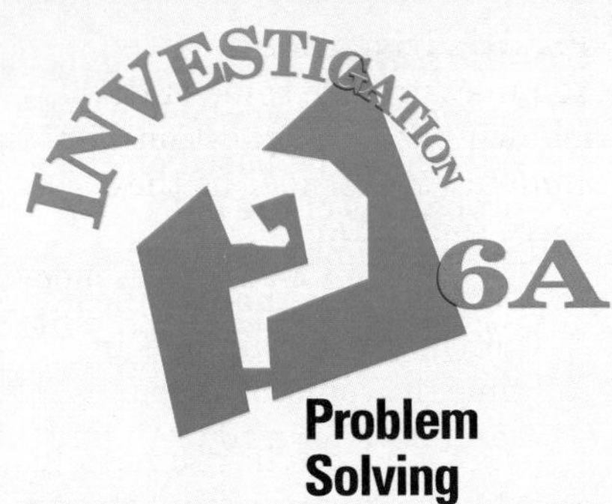

Objectives

Students will:

- use appropriate lab safety procedures.
- use latex polymers to make disks of several shapes and thicknesses.
- test the latex disks for strength and durability.
- design and implement their own procedure.

Planning

Recommended Time: 1–2 lab periods
Prerequisite: Exploration 6A

Solution/Material Preparation

1. For 5% acetic acid, use white vinegar. Do not dilute glacial acetic acid.
2. Liquid latex may be purchased from Flinn Scientific, Inc., Batavia, IL. The following catalog numbers can be used: L0004—500 mL; L0110—1.00 L; L0222—4.00 L.
3. Keep the bottles of reagents in an operating hood because the vapors are annoying. Remind students to keep containers closed when not in use.
4. Make a variety of lab supplies available to students because they may create tests that require equipment not listed here. Alternatively, students can bring their own supplies, provided that you review their choices for safety and practicality.
5. Students will need to use different sizes of molds from paper cups or foil for their latex. If foil is used, leaks can be avoided by painting the inside of the mold with paraffin or taping openings with masking tape.

Student Orientation

Techniques to Demonstrate

When the vinegar is added to the latex, the top of the latex solidifies, while the bottom remains liquid. If it is stirred too much, it forms a ball. Either layer the latex-water mixture and the vinegar, or gently lift the solidified latex, allowing the vinegar to run underneath the polymerized latex and react further.

Sample Data

Mass added (kg)	Length of Latex A (cm)	Length of Latex B (cm)	Length of Latex C (cm)
1.691	9	9	10
3.779	10	10	12
6.911	(latex broke)	(latex broke)	16
8.999			22

Three samples of latex were prepared.
Latex A used 25 mL of latex solution and measured 8 cm × 4 cm × 1 cm.
Latex B used 30 mL of latex solution and measured 8 cm × 5 cm × 1.5 cm.
Latex C used 50 mL of latex solution and measured 9 cm × 5 cm × 2 cm

B POLYMERS **Toy Trampoline**

FUNLAND TOYS®

February 10, 1995

Reginald Brown
Director of Materials Testing
CheMystery Labs, Inc.
52 Fulton Street
Springfield, VA 22150

Dear Mr. Brown:

We are pursuing research and development work relating to new polymers that will be used as part of a new inflatable toy trampoline.

We need a material that has plenty of "bounce," but it must be able to withstand strong forces without tearing.

We want your firm to investigate the properties of different shapes and thicknesses of latex rubber. As we discussed on the phone, we are drafting a contract to pay you $250,000 for the research resulting in a report that explains the possible pros and cons of each material and your test results.

Sincerely,

Kohl Logan

Kohl Logan
President
Funland Toys

Required Precautions

- Goggles and lab aprons must be worn at all times.
- Do not touch or taste any chemicals. Wash your hands thoroughly when finished.
- Disposable plastic gloves should be worn during lab work because the latex solutions can be irritating to the skin.
- Keep the bottles of reagents within a working fume hood.

Required Precautions

- Wear safety goggles and lab apron at all times.
- Read all safety cautions, and discuss them with your students.
- Promptly clean up all spills with paper towels.

Memorandum

CheMystery Labs, Inc.
52 Fulton Street
Springfield, VA 22150

Date: February 11, 1995

To: Sophia Carlucci

From: Reginald Brown

I thought of your team as soon as this research and development opportunity came in because I know you have been working on the possibility of using the same polymers in toy balls.

There are several important differences about this job. Instead of making spheres, I suggest we work on disks of the material because that will be more like a trampoline surface. You'll need to measure bounce height, and how well the material deals with strong forces. I suggest we make measurements of the size of the polymer before and after stretching, and the amount of force a sample can take before breaking. I haven't figured out exactly how we will model the repeated stress of bouncing. I'll let you figure that out. If you decide you need equipment other than what's listed below for your tests, let me know as soon as possible.

Before you get started, I need the following information.
- detailed one-page plan for the procedure and all necessary data charts, including a description of the tests you will do
- detailed list of all of the equipment and materials you will need, along with the individual and total costs (The lower you keep the costs, the more profit we can make!)

When your tests are complete, send a report to Logan in the form of a two-page letter that includes the following.
- discussion concerning the material you recommend and the reasons why you recommend it
- description of your tests and the data you collected (if at all possible, use a graph to present your data.)
- detailed invoice itemizing services rendered and expenses incurred

References

Refer to page 210 for more information on polymers. The procedure and reaction are the same as in an Exploration your team recently performed. Use toothpicks to stir the polymer mixture. Instead of taking the material out of the paper cup to form a ball, leave it in the cup. After it has polymerized and you remove any excess material, remove the solid disk, or cut away the cup.

Spill Procedures/ Waste Disposal Methods

- In case of spills, follow your teacher's instructions.
- Leftover solutions may be poured down the drain.

Materials for FUNLAND TOYS

Item	Cost
REQUIRED ITEMS (You must include all of these in your budget.)	
Lab space/fume hood/utilities	15 000 — /day
Standard disposal fee	2 000 — /g of product
REAGENTS and ADDITIONAL EQUIPMENT (Include in your budget only what you'll need.)	
5% acetic acid solution (vinegar)	500 — /mL
Liquid latex	1 000 — /mL
2 L beaker or bucket (and water)	1 000
250 mL beaker	1 000
400 mL beaker	2 000
10 mL graduated cylinder	1 000
25 mL graduated cylinder	1 000
Aluminum foil	1 000 — /cm^2
Can with reclosable lid	500
Glass stirring rod	1 000
Marbles	500 — /dozen
Paper clips	500 — /box
Paper cup (assorted sizes available)	1 000
Ring stand/ring	2 000
Rubber bands	500 — /dozen
String	500 — /10 cm
Tape	500 — /10 cm
Toothpicks	500 — /dozen
Wash bottle	500
* No refunds on returned chemicals or unused equipment.	
FINES	
OSHA safety violation	2 000 — /incident

Materials

(for each lab group)

- 5% acetic acid solution (vinegar), 100 mL (requested amounts may vary)
- Distilled water
- Liquid latex, 100 mL (requested amounts may vary)
- 10 mL graduated cylinder
- 25 mL graduated cylinder
- Bucket
- Glass stirring rod
- Paper cups, large, medium, and small
- Paper towels
- Resealable can
- Ring stand and ring

Other equipment, including marbles, paper clips, rubber bands, string, tape, toothpicks, and other items may be requested by students.

Estimated cost of materials: $179 500 (will vary depending on equipment used for testing and on disposal needs)

Pre-lab Discussion

You may provide some of the following leading questions for students to consider as they make their plans.

- How can you measure properties such as bounciness, amount of deformation, and resistance to tearing?
- Exactly what will you need to make these measurements?
- What shapes and thicknesses of latex will be easiest to test for these properties?
- How can you make a mold that will create these shapes and thicknesses of latex?

Tips for Evaluating the Pre-lab Requirements

Be ready to suggest alternate equipment if students request materials that are unsafe or unavailable. Grading should reward creativity and thoroughness.

Proposed Procedure

One way to measure strength is to insert bent paper clips through the latex. Use one clip to hang the latex from a ring on a ring stand. Use the other clip to hang a bucket from the latex, adding objects of increasing mass and measuring the length before breaking. One way to measure bounciness is to stretch a large, round, thin disk over the mouth of a can, and attach it with rubber bands. Then, drop several marbles or other objects on it, measuring bounce height each time.

Post-Lab

Disposal

Paper cups, paper towels, disposable gloves, and latex should be disposed of in the trash can. Waste liquids from this lab work can be poured down the drain.

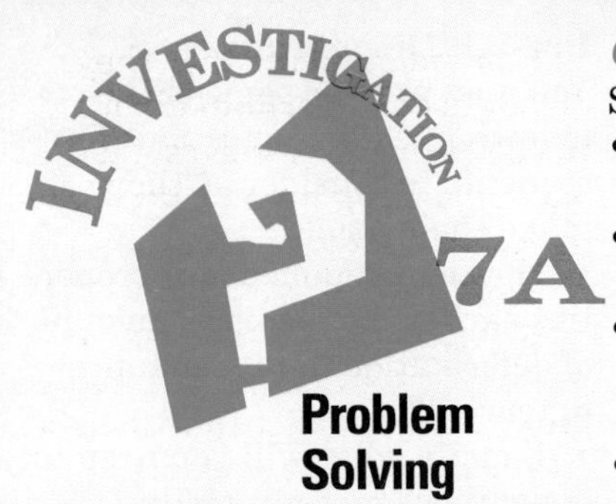

Objectives

Students will:

- use appropriate lab safety procedures.
- recognize evidence of a chemical change.
- write the equations for several single-displacement reactions.
- determine the relative activities of six metals.
- relate the activity of a metal to its position in the periodic table.
- use activity to make predictions.
- design and implement their own procedure.

Planning

Recommended Time: 1 lab period
Prerequisite: None

Solution/Material Preparation

1. To prepare 0.2 M $Ca(NO_3)_2$, dissolve 47.2 g $Ca(NO_3)_2{\cdot}4H_2O$ in 1.00 L of solution.
2. To prepare 0.2 M $Cu(NO_3)_2$, dissolve 48.3 g $Cu(NO_3)_2{\cdot}3H_2O$ in 1.00 L of solution.
3. To prepare 0.2 M $Fe(NO_3)_3$, dissolve 80.8 g $Fe(NO_3)_3{\cdot}9H_2O$ in 1.00 L of solution.
4. To prepare 0.2 M $Mg(NO_3)_2$, dissolve 51.3 g $Mg(NO_3)_2{\cdot}6H_2O$ in 1.00 L of solution.
5. To prepare 0.2 M $SnCl_4$, dissolve 70.1 g $SnCl_4{\cdot}5H_2O$ in 1.00 L of solution. Add a few drops of 1.0 M HCl to clear up any cloudiness due to a hydrolysis reaction.
6. To prepare 0.2 M $Zn(NO_3)_2$, dissolve 59.5 g $Zn(NO_3)_2{\cdot}6H_2O$ in 1.00 L of solution.
7. For the pieces of metal, about 1.0 g of each should be enough. Copper wire and steel wool can be used.

Student Orientation

Techniques to Demonstrate

Show students the reaction of one of the more active metals with a solution containing a compound of a less reactive metal.

Pre-lab Discussion

Review with students the observations they can make which indicate that a chemical reaction may be occurring: the evolution of a gas, a color change, the formation of a precipitate, light, or heat. Review single-displacement reactions and the use of the activity series. Students will need to make very careful observations about any reactions so that they will be able to determine a qualitative order of activity.

Sample Data

Metal	$Ca(NO_3)_2$	$Cu(NO_3)_2$	$Fe(NO_3)_3$	$Mg(NO_3)_2$	$SnCl_4$	$Zn(NO_3)_2$
Ca	NR	Rxn	Rxn	Rxn	Rxn	Rxn
Cu	NR	NR	NR	NR	NR	NR
Fe	NR	Rxn	NR	NR	Rxn	NR
Mg	NR	Rxn	Rxn	NR	Rxn	Rxn
Sn	NR	Rxn	NR	NR	NR	NR
Zn	NR	Rxn	Rxn	NR	Rxn	NR

The order of reactivity from highest to lowest is: Ca, Mg, Zn, Fe, Sn, Cu. To remove tin, copper, and zinc from solution, either calcium or magnesium could be added.

B

SINGLE-DISPLACEMENT REACTIONS

Industrial Waste Recycling

AMALGAMATED CHEMICAL

February 1, 1995

Director of Research
CheMystery Labs, Inc.
52 Fulton Street
Springfield, VA 22150

Dear Director:

Amalgamated Chemical is the manufacturer of a variety of chemical products. As a result of processing, our plant generates wastewater containing dissolved compounds of copper, tin, and zinc.

These metals could be recycled back into production if we had a method of reclaiming them from the waste solution. This would save money and reduce the amount we expel as waste, which could help protect the environment.

Several firms have been asked to submit reports on the development of a reclamatory process. We will reimburse each firm for reasonable costs. Based on the quality of the reports and cost of the research, we will offer the most efficient firm a contract to provide hazardous waste recycling services. We look forward to evaluating your report soon.

Sincerely,

Lloyd Chan

Lloyd Chan
Director, Hazardous Waste Division
Amalgamated Chemical

Required Precautions

Wear goggles and lab aprons. Confine loose clothing and long hair.

Don't get the toxic solutions used in this lab on your hands. Keep your hands away from your face and mouth. Wash your hands thoroughly when finished.

Required Precautions

- Goggles and a lab apron must be worn at all times.
- Read all safety cautions, and discuss them with your students.

CheMystery Labs, Inc.
52 Fulton Street
Springfield, VA 22150

Memorandum

Date: February 3, 1995

To: Sarah Muller

From: Martha Li-Hsien

This is our first chance to break into the chemical manufacturing market! We won't be able to afford any computerized analytical equipment, but I have a plan. We can add reactive metals that will undergo single-displacement reactions, causing the copper, tin, and zinc to precipitate from the solution. Then they can be separated from the wastewater. These metals are available: Mg, Ca, Cu, Fe, Sn, and Zn. We also have stock solutions of compounds of each metal.

First, predict which should work based on the activity series. Then, test each possible combination to confirm your predictions. Be sure to note whether or not the reaction occurs quickly enough to be useful. After you rank the metals by reactivity, recommend a recycling procedure. (Remember to include a suitable method for purifying the metal once it has re-formed.)

Before you begin, I will need the following items.
- detailed one-page plan for the procedure and all necessary data tables
- detailed list of all equipment and materials that you plan to use, including individual and total costs
- your prediction of which metal(s) will work best

When you've found an answer, send Mr. Chan a two-page letter that includes the following.
- detailed suggestions on how the plant could recover the tin, zinc, and copper (Remember, we don't need to perform the entire procedure at this time. We just need to show what would work.)
- metals listed in order of decreasing activity
- detailed and organized observation section
- detailed invoice for services rendered and costs incurred

References

Refer to page 258 for information on single-displacement reactions.
Refer to pages 259–260 for information on the activity series.

Materials for
AMALGAMATED CHEMICAL

Item	Cost
REQUIRED ITEMS (You must include all of these in your budget.)	
Lab space/fume hood/utilities	15 000 — /day
Standard disposal fee	2 000 — /g of product
Balance	5 000
Six test tubes/holder/rack	2 000
REAGENTS and ADDITIONAL EQUIPMENT (Include in your budget only what you'll need.)	
0.2 M $Ca(NO_3)_2$	500 — /mL
0.2 M $Cu(NO_3)_2$	500 — /mL
0.2 M $Fe(NO_3)_3$	500 — /mL
0.2 M $Mg(NO_3)_2$	500 — /mL
0.2 M $SnCl_4$	500 — /mL
0.2 M $Zn(NO_3)_2$	500 — /mL
Small pieces of metal: Ca, Cu, Fe, Mg, Zn, Sn	1 000 — /piece
250 mL beaker	1 000
Evaporating dish	1 000
Filter paper	500 — /piece
Glass stirring rod	1 000
Mortar and pestle	2 000
Ring stand with buret clamp	2 000
Rubber policeman	500
Test tube (small)	500
Wash bottle	500
Watch glass	1 000
* No refunds on returned chemicals or unused equipment.	
FINES	
OSHA safety violation	2 000 — /incident

Spill Procedures/ Waste Disposal Methods

- If any acid is spilled, use water to dilute it, and inform the teacher immediately.
- Do not rinse any solutions into the sink. Dispose of them in the designated waste containers so that they may be reused.
- Do not throw any metal in the trash can. The pieces of metal may be reused if unreacted. Rinse them with distilled water, and place them in the designated containers for recycling. Pieces of metal formed in reactions should be placed in the designated waste container.

Materials

(for each lab group)
- 0.2 M $Ca(NO_3)_2$, 25 mL
- 0.2 M $Cu(NO_3)_2$, 25 mL
- 0.2 M $Fe(NO_3)_3$, 25 mL
- 0.2 M $Mg(NO_3)_2$, 25 mL
- 0.2 M $SnCl_4$, 25 mL
- 0.2 M $Zn(NO_3)_2$, 25 mL
- Small pieces of calcium, copper, iron, magnesium, tin, and zinc metals
- Test-tube holder
- Test-tube rack
- Test tubes, 6

Estimated cost of materials: $113 000

Remind students that they do not need to perform all of the recycling themselves, just enough testing to come up with a method that will work.

Encourage students to use quantities that are as small as possible. Because this is a qualitative lab, small amounts can be used with excellent results.

Tips for Evaluating the Pre-lab Requirements

If students plan to use more than 25 mL of any solution (5 mL for each trial), have them reconsider their procedures. Be certain that students use the activity series to predict which metal will be most useful in recovering the target metals, Sn, Zn, and Cu.

Proposed Procedure

Test all possible combinations of metals and metal compound solutions, rating the strength of the reaction. Any metal that is more reactive than the target metals, Sn, Zn, and Cu, can be used in a single-displacement reaction for recovering the target metals.

Post-Lab

Disposal

The metal solutions must not be disposed of in the drain unless you have written permission on file from your local sewer authority. If permission is withheld, convert the nitrates to other salts that are acceptable for landfill disposal.

The solid metals may not be buried in a landfill. Most of the metals have value as scrap, so reuse them in future labs or recycle them instead of discarding them. You may want to consider rinsing, drying, and reusing the materials that have not gone into solution.

EXPLORATION 8A

Objectives

Students will:

- use appropriate lab safety procedures.
- observe the double-displacement reaction between solutions of strontium chloride and sodium carbonate.
- use a laboratory balance to measure mass.
- demonstrate appropriate technique in transferring liquids.
- demonstrate appropriate technique in filtering.
- perform stoichiometric calculations to determine the mass of sodium carbonate present in the solution.

Planning

Recommended Time:
2 lab periods (includes overnight drying time)

Materials
(for each lab group)

- 0.3 M $SrCl_2$ solution, 50 mL
- Na_2CO_3 solution, (unknown conc.) 15 mL
- 250 mL beakers, 2
- 100 mL graduated cylinder
- Balance
- Beaker tongs
- Distilled water
- Drying oven
- Glass stirring rod
- Paper towels
- Pipe-stem triangle
- Ring and ring stand
- Spatula
- Water bottle

Gravity Filtration Option

- Glass funnel
- Filter paper

Vacuum Filtration Option

- Aspirator for spigot
- Büchner funnel (either ceramic or plastic)
- Filter paper
- One-hole rubber stopper or sleeve
- Vacuum flask (sidearm flask) and tubing

Stoichiometry and Gravimetric Analysis

Objectives

Observe **the double-displacement reaction between solutions of strontium chloride and sodium carbonate.**

Demonstrate **proficiency with gravimetric methods.**

Measure **the mass of insoluble precipitate formed.**

Relate **the mass of precipitate formed to the mass of reactants before the reaction.**

Calculate **the mass of sodium carbonate in a solution of unknown concentration.**

Situation

You are working for a manufacturing company that makes water-softening agents for homes with hard water. Recently, there was a mix-up on the factory floor, and the sodium carbonate solution in a vat was mistakenly mixed with an unknown quantity of distilled water. You must determine the amount of Na_2CO_3 in the vat in order to properly predict the percent yield of the water-softening product. You have been given a small sample from the 575 L of new solution.

Background

When faced with problems that require them to determine the quantities of a substance by mass, chemists often turn to a technique called gravimetric analysis. In this technique, a small sample of the material undergoes a reaction with an excess of another reactant. The chosen reaction is one which almost always provides a yield near 100%. In other words, all of the reactant of unknown amount will be converted into product. If the mass of the product is carefully measured, you can use stoichiometry calculations such as those in Chapter 8 to determine how much of the reactant of unknown amount was involved in the reaction. Then, by comparing the size of the analysis sample with the size of the original material, you can determine exactly how much of the substance is present.

Professional chemists have expensive instruments available to directly measure amounts of a substance, but techniques such as gravimetric analysis are still useful when cost or equipment availability is an important consideration.

This procedure involves a double-displacement reaction between strontium chloride, $SrCl_2$, and sodium carbonate, Na_2CO_3. In general, this reaction can be used to determine the amount of any carbonate compound in a solution.

Problem

Remember that accurate results depend on precise mass measurements, so keep all glassware very clean, and do not lose any reactants or products during your lab work. You will react an unknown amount of sodium carbonate with an excess of strontium chloride. After purifying the product, you will determine the following.

- how much product is present
- how much Na_2CO_3 must have been present to produce that amount of product
- how much Na_2CO_3 is contained in the 575 L of solution

706 | Exploration 8A

Required Precautions

- Goggles and a lab apron must be worn at all times.
- Read all safety cautions, and discuss them with your students.
- Remind students that heated objects can be hot enough to burn even if they look cool. Students should always use beaker tongs to place samples in a drying oven.

Safety

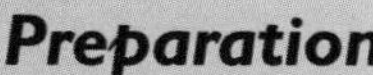

Always wear goggles and an apron to protect your eyes and clothing. If you get a chemical in your eyes, immediately flush it out at the eyewash station while calling to your teacher. Know the locations of the emergency lab shower and eyewash and how to use them.

Do not touch any chemicals. If you get a chemical on your skin or clothing, wash it off at the sink while calling to your teacher. Carefully read the labels and follow the directions on all containers of chemicals that you use. Do not taste any chemicals or items used in the laboratory. Never return leftovers to their original containers; take only small amounts to avoid wasting supplies.

Use tongs whenever handling glassware or other equipment that has been heated because when it is hot, it does not look hot.

Always clean up the lab and all equipment after use, and dispose of substances according to proper disposal methods. Wash your hands thoroughly before you leave the lab after all lab work is finished.

Materials

- **0.30 M $SrCl_2$ solution, 45 mL**
- **Na_2CO_3 solution of unknown concentration, 15 mL**
- **100 mL graduated cylinder**
- **250 mL beakers, 3**
- **Balance**
- **Beaker tongs**
- **Glass funnel or Büchner funnel with related equipment**
- **Distilled water**
- **Drying oven**
- **Filter paper**
- **Glass stirring rod**
- **Paper towels**
- **Ring and ring stand**
- **Rubber policeman**
- **Spatula**
- **Water bottle**

Preparation

1. ***Organizing Data***
   Prepare a data table in your lab notebook. It should include spaces for the following.
   - volume of Na_2CO_3 solution added
   - volume of $SrCl_2$ solution added
   - mass of dry filter paper
   - mass of beaker with paper towel
   - mass of beaker with paper towel, filter paper, and precipitate
2. Clean all of the necessary lab equipment with soap and water. Rinse each piece of equipment with distilled water.
3. Measure the mass of a piece of filter paper to the nearest 0.01 g, and record this value in your data table.
4. Set up a filtering apparatus, either a Büchner funnel or a gravity filtration, depending on what equipment is available. Instructions are given in Exploration 1B.
5. Label a paper towel with your name, your class, and the date. Place the towel in a clean, dry 250 mL beaker, and measure and record the mass of the towel and beaker to the nearest 0.01 g.

Technique

6. Measure about 15 mL of the Na_2CO_3 solution into the graduated cylinder. Record this volume to the nearest 0.5 mL in your data table. Pour the Na_2CO_3 solution into a clean, empty 250 mL beaker. Carefully wash the graduated cylinder and rinse it with distilled water.

Solution/Material Preparation

1. To prepare 0.3 M $SrCl_2$, dissolve 80.0 g of $SrCl_2 \cdot 6H_2O$ in enough H_2O to make 1.00 L of solution.
2. For the unknown, 0.5 M Na_2CO_3 is recommended. To prepare 0.5 M Na_2CO_3, dissolve 53.0 g of Na_2CO_3 in enough H_2O to make 1.00 L of solution.

Student Orientation

Pre-lab Discussion

Thoroughly discuss mass-mass stoichiometry and its application in the laboratory. It would be useful to perform calculations with sample data before performing the lab.

Techniques to Demonstrate

Review the procedures for the filtration technique your students will be using.

Remind students of the importance of using clean glassware and avoiding loss of product.

Post-Lab

Disposal

Be sure to set out two disposal containers: one for solids, and another for liquids. The solids may be disposed of in the trash. The liquids can be washed down the drain with an excess of water.

Answers to Analysis and Interpretation

1. $SrCl_2(aq) + Na_2CO_3(aq) \longrightarrow 2NaCl(aq) + SrCO_3(s)$
2. The precipitate is strontium carbonate, $SrCO_3$.
3. $1.06\ g \times \frac{1\ mol\ SrCO_3}{147.63\ g} =$ $7.18 \times 10^{-3}\ mol\ SrCO_3$
4. $7.18 \times 10^{-3}\ mol\ SrCO_3 \times \frac{1\ mol\ Na_2CO_3}{1\ mol\ SrCO_3} =$ $7.18 \times 10^{-3}\ mol\ Na_2CO_3$
5. $35\ mL\ SrCl_2 \times \frac{1\ L}{1000\ mL} \times \frac{0.30\ mol}{1\ L} = 1.05 \times 10^{-2}\ mol\ SrCl_2$ Sodium carbonate is the limiting reactant.
6. The precipitate was rinsed to remove any NaCl impurities that may have remained on the $SrCO_3$.

7. Measure about 25 mL of the 0.30 M $SrCl_2$ solution into the graduated cylinder. Record this volume to the nearest 0.5 mL in your data table. Pour the $SrCl_2$ solution into the beaker with the Na_2CO_3 solution. Gently stir the solution and precipitate with a glass stirring rod.
8. Carefully measure another 10 mL of $SrCl_2$ into the graduated cylinder. Record the volume to the nearest 0.5 mL in your data table. Slowly add it to the beaker. Repeat this step until no more precipitate forms.
9. While the precipitate settles, place the filter paper in the funnel, and wet it with a small amount of distilled water so that it will adhere to the sides of the funnel. Then, once the precipitate has settled, slowly pour the mixture into the funnel. Be careful not to overfill the funnel, as that will cause some of the precipitate to be lost between the filter paper and the funnel. Use the rubber spatula to transfer as much of the precipitate into the funnel as possible.
10. Rinse the rubber spatula into the beaker with a small amount of distilled water, and pour this solution into the funnel. Rinse the beaker several more times with small amounts of distilled water, pouring it into the funnel each time.
11. After all of the solution and rinses have drained through the funnel, slowly rinse the precipitate on the filter paper in the funnel with distilled water to remove any soluble impurities.
12. Carefully remove the filter paper from the funnel, and place it on the paper towel that you have labeled with your name. Unfold the filter paper, and place the paper towel, filter paper, and precipitate in the rinsed beaker. Then place the beaker in the drying oven. For best results, allow the precipitate to dry overnight.

Cleanup and Disposal

13. Using beaker tongs, remove your sample from the drying oven, and allow it to cool. Measure and record the mass of the beaker with paper towel, filter paper, and precipitate to the nearest 0.01 g.
14. Dispose of the precipitate in a designated waste container. Pour the filtrate in the other 250 mL beaker into the designated waste container. Clean up your equipment and lab station. Thoroughly wash your hands after completing the lab session and cleanup.

Analysis and Interpretation

1. ***Organizing Ideas***
   Write a balanced equation for the reaction. (Hint: it was a double-displacement reaction.)
2. ***Organizing Ideas***
   What is the precipitate? Write its empirical formula.
3. ***Applying Ideas***
   Calculate the number of moles of precipitate produced in the reaction. (Hint: use the results from item **2.**)
4. ***Applying Ideas***
   How many moles of Na_2CO_3 were present in the 15 mL sample? (Hint: use the mole ratio from the equation in item **1.**)

Sample Data

Volume of 0.3 M $SrCl_2$	35 mL
Volume of Na_2CO_3	15 mL
Mass of filter paper	0.30 g
Mass of filter paper + $SrCO_3$	1.36 g

5. ***Evaluating Methods***
There were 0.30 mol of $SrCl_2$ in every liter of solution. Calculate how many moles of $SrCl_2$ were added. Identify whether $SrCl_2$ or Na_2CO_3 was the limiting reactant. Would this lab have worked if the other reactant was chosen as the limiting reactant? Explain why or why not.

6. ***Evaluating Methods***
Why was the precipitate rinsed in Step **11**? What soluble impurities could have been on the filter paper along with the precipitate?

Conclusions

7. ***Inferring Conclusions***
How many grams of Na_2CO_3 were present in the 15 mL sample?

8. ***Applying Conclusions***
How many grams of Na_2CO_3 are present in the 575 L? (Hint: create a conversion factor to convert from the sample with a volume of 15 mL to the entire solution with a volume of 575 L.)

9. ***Applying Conclusions***
For every jar of water-softening product produced at the factory, 155 g of Na_2CO_3 are necessary. How many jars of water-softening product can you make from the 575 L of solution? Would the company be able to fill an order for 40 jars of water-softening product, or would they have to order more Na_2CO_3?

10. ***Evaluating Methods***
How would the calculated results vary if the precipitate wasn't completely dry? Explain your answer.

Extensions

1. ***Evaluating Methods***
Find out from your supervisor what the correct mass of Na_2CO_3 in the sample was, and calculate your percent error.

2. ***Designing Experiments***
What possible sources of error can you identify with your procedure? If you can think of ways to eliminate them, ask your teacher to approve your plan, and run the procedure again.

3. ***Designing Experiments***
In this Exploration, a chemical property of Na_2CO_3 was used to measure how much of it was dissolved in a solution. Design a technique that determines amount using other properties, such as density of solution, pH of solution, electrical resistivity, etc. Describe how to calibrate your method with solutions of known concentration. If your teacher approves your plan, carry it out, and then analyze the unknown using your new technique. Calculate percent error for your measurement.

4. ***Research and Communications***
Research some methods other than gravimetric analysis that are used to measure amounts of substances. How do they work? List advantages and disadvantages of each method. Report back to the class about the method you would select to analyze the amount of Na_2CO_3 in the solution.

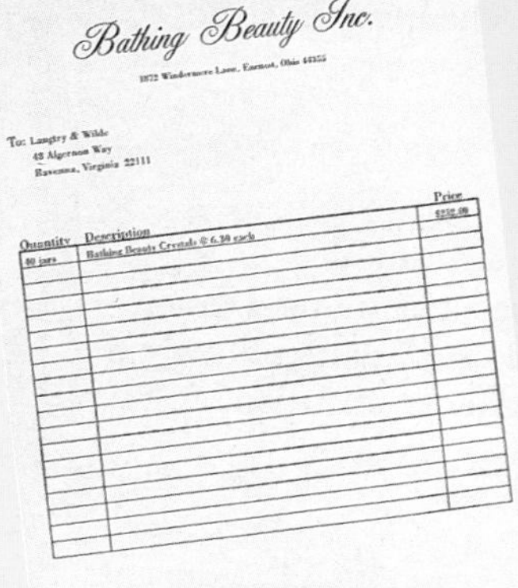

Conclusions

7. 7.18×10^{-3} mol

$$Na_2CO_3 \times \frac{105.99\ g}{1\ mol\ Na_2CO_3} =$$

$0.761 g\ Na_2CO_3/15\ mL$

8. $$575\ L \times \frac{0.761\ g\ Na_2CO_3}{15\ mL} \times \frac{1000\ mL}{1\ L} = 2.92 \times 10^4\ g\ Na_2CO_3$$

9. $$\frac{2.92 \times 10^4\ g\ Na_2CO_3}{155\ g/jar} = 188\ jars$$

There would be no need to order more Na_2CO_3.

10. If the precipitate was not dry, the excess water would cause it to have a greater mass. Then, the calculations of Na_2CO_3 present would be too large.

Extensions

1. Correct mass of Na_2CO_3: 0.795 g for every 15.0 mL (if students were given a 0.5 M solution)

Percent error =

$$\frac{0.795 - 0.761}{0.795} \times 100 = 4.3\%$$

2. Students' suggestions for improving the procedure will vary. Students may suggest using larger amounts of each reactant or running multiple trials. Be sure answers are safe and include carefully planned procedures.

3. Students' suggestions will vary. Typically, students will discover that these techniques are not as precise as gravimetric analysis. Be sure answers are safe and include carefully planned procedures.

4. Student answers will vary. Possible methods of quantitative analysis include titration, absorption spectrophotometry, and coulometry.

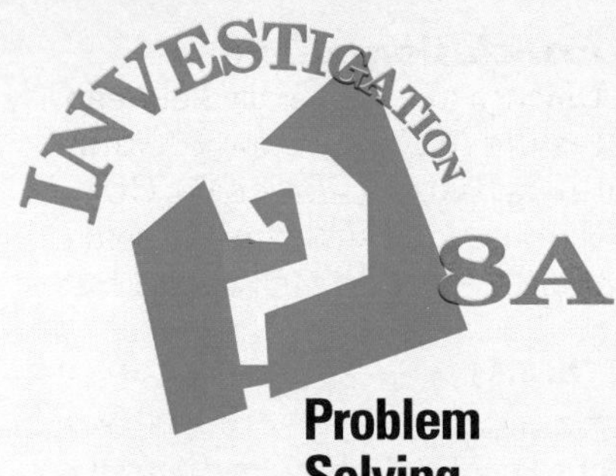

Objectives

Students will:

- use appropriate lab safety procedures.
- observe the double-displacement reaction between sodium carbonate and water containing calcium compounds.
- use a laboratory balance to measure mass.
- demonstrate appropriate technique in transferring liquids.
- demonstrate appropriate technique in filtering.
- perform stoichiometric calculations to determine the mass of calcium ions present in the solution.
- determine whether or not a water sample should be classified as hard water.
- design and implement their own procedure.

Planning

Recommended Time: 2 lab periods (includes overnight drying time)
Prerequisite: Exploration 8A

Solution/Material Preparation

1. To prepare 0.5 M Na_2CO_3, dissolve 53.0 g of Na_2CO_3 in 1.00 L of solution.
2. For the water sample, use 0.001 M $CaCl_2$ if you want students to find that the water is not hard. To prepare the 0.001 M $CaCl_2$, dissolve 0.147 g of $CaCl_2{\cdot}2H_2O$ in 1.00 L of solution. Use 0.5 M $CaCl_2$ if you want students to discover that the water is hard. To prepare the 0.5 M $CaCl_2$, dissolve 73.5 g of $CaCl_2{\cdot}2H_2O$ in 1.00 L of solution.
3. To improve the results and reduce the time needed for this lab, the calcium carbonate can be centrifuged instead of filtered and then dried with a burner. The heat must be applied slowly to avoid decomposition.

Student Orientation

Techniques to Demonstrate

Discuss any errors in lab technique that you witnessed during the Exploration lab, and demonstrate the correct technique. Demonstrate the filtering technique students will use. Vacuum filtration is recommended because it is faster than gravity filtration.

Sample Data

Volume of water sample	20.0 mL
Volume of 0.5 M Na_2CO_3	40.0 mL
Mass of 150 mL beaker	65.80 mL
Mass of filter paper	0.31 g
Mass of beaker + filter paper + $CaCO_3$	67.07 g

Analysis

$$\text{Mass of Ca} = 0.96\text{ g CaCO}_3 \times \frac{40.08\text{ g Ca}}{100.09\text{ g CaCO}_3} = 0.38\text{ g Ca} = 380\text{ mg Ca}$$

380 mg Ca/L > 120 mg/L

The sample is hard water.

GRAVIMETRIC ANALYSIS
Hard Water Testing

EDWARD F. QUIMBY, MAYOR
DANA RUBIO, CITY MANAGER

March 3, 1995

George Taylor, Director of Analysis
CheMystery Labs, Inc.
52 Fulton Street
Springfield, VA 22150

Dear Mr. Taylor:

The city's Public Works Department is investigating new sources of water. One proposal involves drilling new wells into a nearby aquifer that is protected from brackish water by a unique geological formation. Unfortunately, this formation is made of calcium minerals. If the concentration of calcium ions in the water is too high, the water will be "hard," and treating it to meet local water standards would be too expensive for us.

Water containing more than 120 mg of calcium per liter is considered hard. Enclosed is a sample that has been distilled from 1.0 L to its present volume. Please determine whether or not it is of suitable quality.

We are seeking a firm to be our consultant for the entire testing process. Interested firms will be evaluated based on this analysis. We look forward to receiving your report.

Sincerely,

Dana Rubio
City Manager

Required Precautions

Goggles and lab aprons must be worn at all times.

Use beaker tongs to remove glassware from the drying oven.

Do not touch or taste any chemicals. Wash your hands thoroughly when finished.

Required Precautions

- Goggles and a lab apron must be worn at all times.
- Read all safety cautions, and discuss them with your students.
- Remind students that heated objects can be hot enough to burn even if they look cool. Students should always use beaker tongs to place samples in a drying oven.

Memorandum

CheMystery Labs, Inc.
52 Fulton Street
Springfield, VA 22150

Date: March 4, 1995

To: Shane Thompson

From: George Taylor

We must do a very accurate and efficient job on this analysis because this contract would be valuable for us in terms of both income and prestige. On the other hand, losing the contract to some out-of-town analysis firm would be awful!

We still don't have any capital expenditure funds for elaborate equipment purchases, but we can solve this problem with some careful gravimetric analysis because calcium salts and carbonate compounds undergo double-displacement reactions to give insoluble calcium carbonate as a precipitate.

Before you begin your work, I will need the following information from you so that I can put together our bid.

- detailed one-page summary of your plan for the procedure along with all necessary data tables
- description of necessary calculations
- itemized list of equipment, with total costs (Our financial planner tells me that even though we will bill the city for this work, we can afford to spend only $200 000 on this project.)

After you complete the analysis, prepare a two-page report for Dana Rubio. Remember that this report will be seen by a variety of city officials, so be certain it projects the image we want to present. Make sure the following items are included.

- calculation of calcium concentration, in mg/L, for the aquifer water
- explanation of how you determined the amount of calcium in the sample, including measurements and calculations
- balanced chemical equation for the reaction
- explanations and estimations for any possible sources of error
- detailed invoice for services rendered and expenses incurred

References

Refer to pages 276–278 for information about mass-mass stoichiometry. The gravimetric analysis techniques are similar to those used in an Exploration you and your team recently completed. At that time, you used strontium chloride, $SrCl_2$, as a reagent to identify the amount of sodium carbonate, Na_2CO_3, present in a sample. In this investigation, you will use a similar double-displacement reaction, but Na_2CO_3 will be used as a reagent to identify how much calcium is present in a sample. Like strontium and other Group 2 metals, calcium salts react with carbonate-containing salts to give an insoluble precipitate.

Materials for
CITY OF SPRINGFIELD

Item	Cost
REQUIRED ITEMS (You must include all of these in your budget.)	
Lab space/fume hood/utilities	15 000 — /day
Standard disposal fee	2 000 — /g of product
Balance	5 000
Beaker tongs	1 000
Drying oven	5 000 — /day
REAGENTS and ADDITIONAL EQUIPMENT (Include in your budget only what you'll need.)	
0.5 M Na_2CO_3 solution	1 000 — /mL
250 mL beaker	1 000
400 mL beaker	2 000
250 mL flask	1 000
100 mL graduated cylinder	1 000
Büchner funnel	2 000
Filter flask with sink attachment	2 000
Filter paper	500 — /piece
Glass funnel	1 000
Glass stirring rod	1 000
Paper clips	500 — /box
Ring stand/ring/pipe stem triangle	2 000
Six test tubes/holder/rack	2 000
Spatula	500
Wash bottle	500
Weighing paper	500 — /piece
* No refunds on returned chemicals or unused equipment.	
Fines	
OSHA safety violation	2 000 — /incident

Spill Procedures/ Waste Disposal Methods

- **Solids must go in the trash can. Do not wash them down the sink.**
- **Liquids may be washed down the sink with an excess of water.**
- **Clean the area and all equipment after use.**

Materials
(for each lab group)

- $CaCl_2$ solution, 20 mL (see preparation notes)
- 0.5 M Na_2CO_3, 75 mL or less
- 100 mL graduated cylinder
- 250 mL beakers, 2
- Balance
- Beaker tongs
- Drying oven
- Pipe-stem triangle
- Ring and ring stand

See the Teacher's Notes for Exploration 8A for information on materials for filtration options.

Estimated cost of materials:
$78 500–$143 000, depending on amount of Na_2CO_3 and filtration technique used.

Pre-lab Discussion

Discuss any disparity between results in Exploration 8A and the actual amount of Na_2CO_3. Remind students that no reaction produces a 100% yield but that results should be close.

Encourage students to start planning by writing the balanced chemical equation for the reaction. Then provide some of the following leading questions to consider as they make their plans.

- How are you measuring the calcium content in the solution?
- Should calcium or sodium carbonate to be left over at the end?

Tips for Evaluating the Pre-lab Requirements

If students plan to use more than 75 mL of either reactant, have them reconsider their procedures. Student plans should include measuring the mass of the filter paper and beaker before work begins.

Be certain that the description of necessary calculations is accurate. Students should calculate the mass of calcium present, not the mass of $CaCO_3$. Check that student data tables contain the same headings as shown in the sample data table.

Proposed Procedure

Measure the mass of a piece of filter paper and a beaker. Add an excess of Na_2CO_3 solution of known concentration to a carefully measured volume of a solution containing calcium ions. Filter and dry the $CaCO_3$ precipitate that forms. Measure its mass.

Post-Lab

Disposal

Students may dispose of solids in the trash can. Students may wash liquids down the drain with an excess of water.

EXPLORATION 8B

Objectives
Students will:
- use appropriate lab safety procedures.
- observe the double-displacement reaction between $HC_2H_3O_2$ and $NaHCO_3$ and the subsequent decomposition of H_2CO_3.
- use a laboratory balance to measure mass.
- determine the mole ratio of reactants and products.
- calculate the number of moles of reactants and products.

Planning
Recommended Time:
1 lab period

Materials
(for each lab group)
- 1.0 M acetic acid, 50 mL
- $NaHCO_3$, about 3 g
- Balance, centigram
- Beaker tongs
- Dropper or pipet
- Evaporating dish
- Spatula
- Watch glass

Hot plate option
- Hot plate

Bunsen burner option
- Bunsen burner with gas tubing and striker
- Ring stand and ring
- Wire gauze with ceramic center

Solution/Material Preparation
1. To prepare 1.0 M $HC_2H_3O_2$, observe the required precautions. Add 87 mL of glacial $HC_2H_3O_2$ to water. Dilute to make 1.00 L of solution.
2. Better results are obtained using hot plates instead of Bunsen burners.

Stoichiometry of Reactions

Situation
Your company has a contract to determine the reaction requirements for a large-scale baking operation. The bakery purchases large quantities of ingredients and needs to know the correct proportions to avoid waste or inferior quality. They have determined that they need to produce 425 mL of carbon dioxide, CO_2, for every cake during the rising step that takes place just before baking. The bakery needs you to determine exactly what amount of ingredients are necessary to provide this amount of CO_2.

Background
Some recipes use baking soda, $NaHCO_3$, to make cakes rise. When you add a weak acid such as vinegar, which contains acetic acid, $HC_2H_3O_2$, or buttermilk, which contains lactic acid, $HC_3H_5O_3$, to baking soda, bubbles of carbon dioxide gas are produced. The word equation for the reaction with vinegar is as follows: acetic acid and sodium hydrogen carbonate yields sodium acetate, water, and carbon dioxide. But you can't merely add an excess of baking soda to be sure enough CO_2 is formed because too much baking soda can make the cakes crumbly and bitter tasting. Similarly, too much acid can cause a cake to taste sour. The amount of each reactant must be perfectly matched to produce the correct amount of CO_2, which has a density of 1.25 g/L at baking temperature.

Problem
To figure out the needs of the bakery, you will need to do the following.
- react a carefully measured mass of the reactant, $NaHCO_3$
- measure the mass of the product, $NaC_2H_3O_2$
- determine the mass and mole relationships for the other reactants and products
- calculate the number of moles and mass of each reactant required to produce 425 mL of CO_2

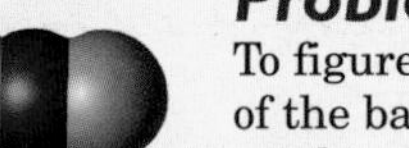
CO_2

$HC_2H_3O_2$

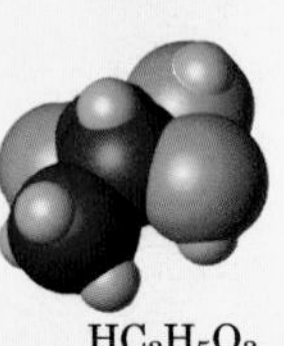
$HC_3H_5O_3$

Objectives
Demonstrate **proficiency in measuring masses.**

Determine **the number of moles of reactants and products in a reaction experimentally.**

Use **the mass and mole relationships of a chemical reaction in calculations.**

Perform **calculations that involve density and stoichiometry.**

Required Precautions
- Goggles and a lab apron must be worn at all times.
- Tie back loose hair and long clothing when working in the lab.
- Read all safety cautions, and discuss them with your students.
- Students should not handle glacial acetic acid.

Safety

Always wear goggles and an apron to provide protection for your eyes and clothing. If you get a chemical in your eyes, immediately flush it out at the eyewash station while calling to your teacher. Know the locations of the emergency lab shower and eyewash and how to use them.

Do not touch any chemicals. If you get a chemical on your skin or clothing, wash it off at the sink while calling to your teacher. Make sure you carefully read the labels and follow the directions on all containers of chemicals that you use. Do not taste any chemicals or items used in the laboratory. Never return leftovers to their original containers; take only small amounts to avoid wasting supplies.

Confine long hair and loose clothing. Do not heat glassware that is broken, chipped, or cracked. Use tongs or a hot mitt to handle heated glassware and other equipment because it does not always look hot. If your clothing catches fire, WALK to the emergency lab shower, and use it to put out the fire.

Never put broken glass or ceramics in a regular waste container. Broken glass or ceramics should be disposed of in a separate container designated by your teacher.

Call your teacher in the event of an acid or base spill. Acid or base spills should be cleaned up promptly, according to your teacher's instructions.

Always clean up the lab and all equipment after use, and dispose of substances according to proper disposal methods. Wash your hands thoroughly before you leave the lab after all lab work is finished.

Materials

- **1.0 M acetic acid**
- **2–3 g $NaHCO_3$**
- **100 mL graduated cylinder**
- **Balance**
- **Beaker tongs**
- **Bunsen burner and related equipment or hot plate**
- **Dropper or pipet**
- **Evaporating dish**
- **Ring stand and ring (for use with Bunsen burner)**
- **Spatula**
- **Watch glass**
- **Wire gauze with ceramic center (for use with Bunsen burner)**

Preparation

1. ***Organizing Data***
   Prepare a data table in your lab notebook. It should contain space to record the mass of the empty evaporating dish and watch glass, the mass of the evaporating dish with the watch glass and $NaHCO_3$, and several spaces for recording the mass of the evaporating dish with the watch glass and $NaC_2H_3O_2$ after heating.

Technique

2. Measure the mass of a clean, dry evaporating dish and watch glass to the nearest 0.01 g. Record this mass in your data table.
3. Add 2–3 g $NaHCO_3$ to your evaporating dish. Measure the mass, with the cover glass, to the nearest 0.01 g. Record this mass in your data table.
4. Slowly add 30 mL of the acetic acid solution to the $NaHCO_3$ in the evaporating dish. Add more acetic acid with a dropper or pipet until the bubbling stops.

Student Orientation

Pre-lab Discussion
This lab involves stoichiometry calculations with mass, moles, volume, and density. You may want to work through similar calculations with sample data with the class. Encourage students to use teamwork so that they will have time to dry the products twice.

Techniques to Demonstrate
Remind students to add the acetic acid very slowly. Otherwise, the reaction will be so vigorous that some of the product will be lost. Also make sure they heat the solution gently to avoid losing any of the product.

Demonstrate how to hold an evaporating dish with beaker tongs. Students should wait 15 min for the evaporating dish to cool before measuring its mass. Even then, they should continue to use beaker tongs to hold the evaporating dish.

- Wear goggles, a face shield, impermeable gloves, and a lab apron while preparing the acetic acid. Work in a hood known to be in operating condition, with another person present nearby to call for help in case of an emergency. Be sure you are within 30 s walking distance of a safety shower and eyewash station known to be in operating condition.
- In case of an acid spill, dilute first with water. Then, mop up the spill with wet cloths designated for spill cleanup while wearing disposable plastic gloves. A wet cloth mop can be rinsed out a few times and used until it falls apart.

Post-Lab

Disposal

The products may be poured down the drain.

Answers to Analysis and Interpretation

1. $NaHCO_3(s) + HC_2H_3O_2(aq) \longrightarrow CO_2(g) + H_2O(l) + NaC_2H_3O_2(aq)$

2. $NaHCO_3$ molar mass: 84.01 g/mol
$HC_2H_3O_2$ molar mass: 60.06 g/mol
CO_2 molar mass: 44.01 g/mol
H_2O molar mass: 18.02 g/mol
$NaC_2H_3O_2$ molar mass: 82.04 g/mol

3. The bubbling was caused by the formation of carbon dioxide gas.

4. The residue is entirely $NaC_2H_3O_2$ because $HC_2H_3O_2$ was added in excess to be certain that the reaction went to completion.

5. $NaHCO_3$: 2.10 g

$$2.10\text{ g} \times \frac{1\text{ mol}}{84.01\text{ g}} = 2.50 \times 10^{-2}\text{ mol}$$

$NaC_2H_3O_2$: 2.03 g

$$2.03\text{ g} \times \frac{1\text{ mol}}{82.04\text{ g}} = 2.47 \times 10^{-2}\text{ mol}$$

6. Theoretical yield of $NaC_2H_3O_2$:

$$2.50 \times 10^{-2}\text{ mol NaHCO}_3 \times \frac{1\text{ mol NaC}_2\text{H}_3\text{O}_2}{1\text{ mol NaHCO}_3} =$$

$$2.50 \times 10^{-2}\text{ mol NaC}_2\text{H}_3\text{O}_2$$

$$2.50 \times 10^{-2}\text{ mol NaC}_2\text{H}_3\text{O}_2 \times \frac{82.04\text{ g}}{1\text{ mol NaC}_2\text{H}_3\text{O}_2} = 2.05\text{ g NaC}_2\text{H}_3\text{O}_2$$

5. If you are using a Bunsen burner, place the evaporating dish and its contents on a ceramic-centered wire gauze on an iron ring attached to the ring stand, as shown in the photograph on the left. Place the watch glass, concave side up, on top of the dish, making sure that there is a slight opening for steam to escape. If you are using a hot plate, place the watch glass the same way, but heat the evaporating dish directly on the hot plate.

6. Gently heat the evaporating dish until only a dry solid remains. Make sure that no water droplets remain on the underside of the watch glass. ***Do not heat too rapidly, or the material will boil, and the product will spatter out of the evaporating dish.***

7. Turn off the gas burner or hot plate. Allow the apparatus to cool at least 15 min. Determine the mass of the cooled equipment to the nearest 0.01 g. Record the mass of the dish, residue, and watch glass in your data table.

8. If time permits, reheat the evaporating dish and contents for 2 min. Let it cool and measure its mass again. You can be certain the sample is dry when there are two successive measurements within 0.02 g of each other.

Cleanup and Disposal

9. Dispose of any unused chemicals in the containers designated by your teacher. Wash your hands thoroughly after cleaning up the area and equipment. Make sure to turn off all gas valves.

Analysis and Interpretation

1. ***Analyzing Results***
Write a balanced equation for the reaction of baking soda and acetic acid. Be sure to include states of matter for all of the reactants and products.

2. ***Organizing Data***
Use a periodic table to calculate the molar mass for each of the reactants and products.

3. ***Analyzing Results***
Explain what caused the bubbling when the reaction took place.

4. ***Analyzing Methods***
How do you know that all of the residue is actually sodium acetate rather than a mixture of sodium bicarbonate and sodium acetate?

5. ***Organizing Data***
Calculate the mass of $NaHCO_3$, the number of moles of $NaHCO_3$, the mass of $NaC_2H_3O_2$, and the number of moles of $NaC_2H_3O_2$.

6. ***Evaluating Data***
Using the balanced equation and the amount of $NaHCO_3$, determine the theoretical yield of $NaC_2H_3O_2$ in moles and grams. (Hint: see Chapter 8 for a discussion of theoretical yield, and assume the acetic acid was present in excess.)

Conclusions

7. Analyzing Conclusions
What is the percent yield for your reaction? (Hint: see Chapter 8 for a discussion of percent yield.)

8. Inferring Conclusions
What is the theoretical yield for CO_2? Using the density value given, 1.25 g/L, calculate what volume of CO_2 would be produced by this reaction in an oven. Show your calculations.

9. Applying Conclusions
How many moles of $NaHCO_3$ and $HC_2H_3O_2$ are necessary to produce 425 mL of CO_2 at baking temperature? Show your calculations. (Hint: be sure to include your percent yield for this reaction in your calculations.)

Extensions

1. Inferring Conclusions
Baking soda, $NaHCO_3$, is also known as sodium bicarbonate. Baking *powder* is a mixture of baking soda and another substance, such as cream of tartar. If you add water to baking soda, it dissolves, but water added to baking powder produces CO_2, as shown by the bubbling action. Is cream of tartar acidic or basic?

2. Designing Experiments
If your percent yield is less than 100%, explain why. If you can think of ways to eliminate any problems, ask your teacher to approve your plan, and run the procedure again.

3. Research and Communications
Many recipes for breads use yeast, instead of baking soda, as a source of CO_2. Research the use of yeast and explain what ingredients are necessary for the yeast to produce carbon dioxide. What is the balanced chemical equation for the reaction that yeast use to produce CO_2?

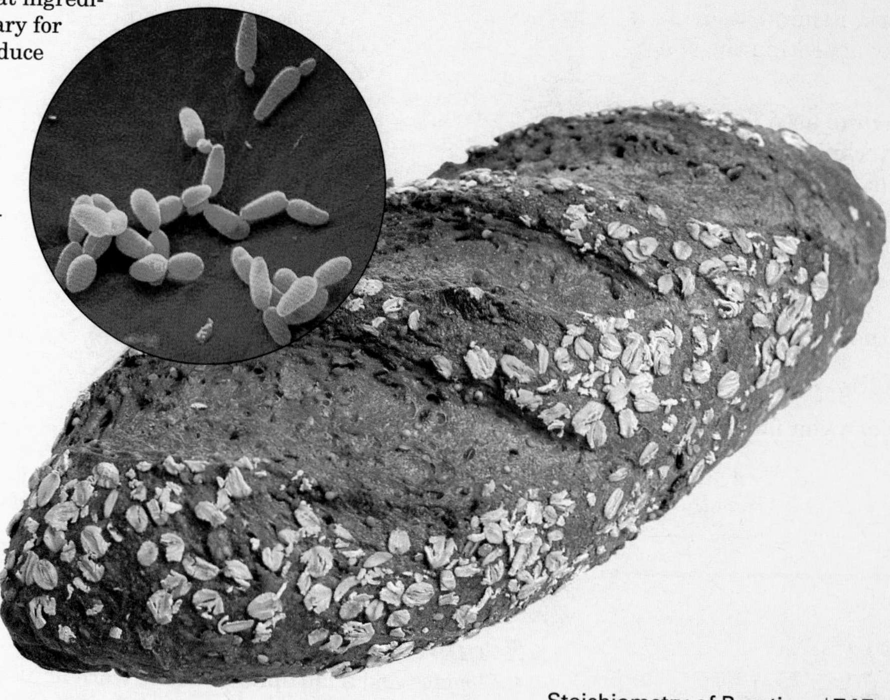

Sample Data	
Mass of dish and glass	71.17 g
Mass of dish, glass, and $NaHCO_3$	73.27 g
Mass of dish, glass, and residue ($NaC_2H_3O_2$) after 1st heating	73.22 g
Mass of dish, glass, and residue ($NaC_2H_3O_2$) after 2nd heating	73.20 g

Conclusions

7. Percent yield: $\frac{2.03 \text{ g actual}}{2.05 \text{ g theoretical}}$ = 99.0%

8. Theoretical yield of CO_2:

2.50×10^{-2} mol $NaHCO^3$ ×

$$\frac{1 \text{ mol } CO_2}{1 \text{ mol } NaHCO_3} \times \frac{44.01 \text{ g}}{1 \text{ mol } CO_2} =$$

1.10 g CO_2

$$1.10 \text{ g } CO_2 \times \frac{1 \text{ L}}{1.25 \text{ g } CO_2} = 0.880 \text{ L } CO_2$$

9. 425 mL CO_2 ×

$$\frac{2.50 \times 10^{-2} \text{ mol reactants}}{880 \text{ mL } CO_2} =$$

1.21×10^{-2} mol reactants

Extensions

1. Cream of tartar must be acidic because, like acetic and lactic acid, it reacts with $NaHCO_3$ to make CO_2 gas.

2. Students' suggestions for sources of error and plans for improving the procedure will vary. Possible suggestions could include improving technique, using larger quantities, or performing multiple trials. Be sure student suggestions for improvements are safe and include carefully planned procedures.

3. Yeast use the cellular respiration reaction to produce CO_2. This process requires oxygen and sugar, usually in the form of glucose.
$$C_6H_{12}O_6 + 6O_2 \longrightarrow 6CO_2 + 6H_2O$$

Objectives

Students will:

- use appropriate lab safety procedures.
- choose appropriate reactants to synthesize $ZnCl_2$.
- determine proper molar and mass amounts of reactants.
- use a laboratory balance to measure mass.
- design and implement their own procedure.

Planning

Recommended Time: 2 lab periods (includes overnight drying time)
Prerequisite: Exploration 8B

Solution/Material Preparation

1. To prepare 1.0 M HCl, take these precautions. Wear goggles, face shield, impermeable gloves and a lab apron. Work in a hood known to be in operating condition, with another person nearby to call for help in case of an emergency. Be sure you are within 30 s walking distance of a safety shower and eyewash station known to be in operating condition. Add 83 mL of concentrated HCl to enough distilled water to make 1.00 L of solution. Add the acid slowly, and stop to stir it in order to avoid overheating.

Student Orientation

Techniques to Demonstrate

Remind students to measure the mass of glassware that is clean and dry before they begin their work. Remind students to heat the solution gently when they are evaporating the water.

Pre-lab Discussion

Begin by discussing the results and procedure used in the Exploration, especially any errors in lab technique that occurred. To help students, you may provide some of the following leading questions for them to consider as they make their plans.

- Are you given amounts of reactants or products?
- Using what you know from Chapter 7, which reactants will react to provide $ZnCl_2$?
- How can you know what amounts of the reactants you will need?
- Which reactant should be in excess?
- When the reaction is completed, how can you purify the $ZnCl_2$ in solution?

Sample Data

Mass of 250 mL beaker	99.43 g
Mass of zinc added	2.00 g
Volume of 1.0 M HCl	25 mL
Mass of 250 mL beaker + $ZnCl_2$	101.11 g
Mass of $ZnCl_2$: 1.68 g	

STOICHIOMETRY Viral Disease Alert

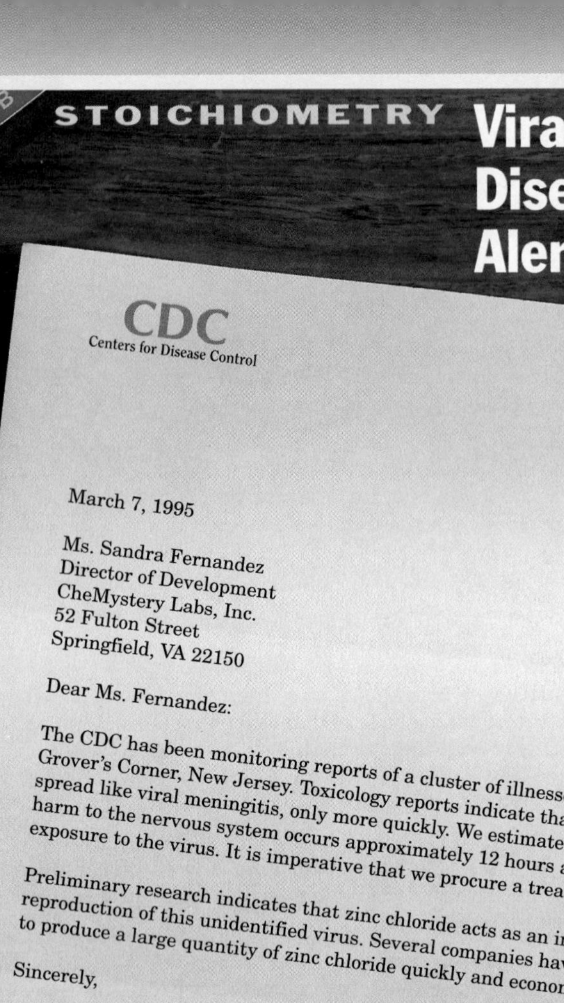

CDC
Centers for Disease Control

March 7, 1995

Ms. Sandra Fernandez
Director of Development
CheMystery Labs, Inc.
52 Fulton Street
Springfield, VA 22150

Dear Ms. Fernandez:

The CDC has been monitoring reports of a cluster of illnesses near the town of Grover's Corner, New Jersey. Toxicology reports indicate that the disease is spread like viral meningitis, only more quickly. We estimate that irreparable harm to the nervous system occurs approximately 12 hours after initial exposure to the virus. It is imperative that we procure a treatment soon.

Preliminary research indicates that zinc chloride acts as an inhibitor to the reproduction of this unidentified virus. Several companies have been contacted to produce a large quantity of zinc chloride quickly and economically.

Sincerely,

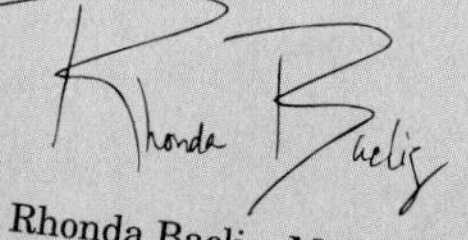

Rhonda Baclig, M.D.
Centers for Disease Control
Special Pathogens Branch

Required Precautions

Lab aprons and goggles must be worn at all times.

Do not touch or taste any chemicals. Always wash your hands thoroughly when finished.

If you get acid or base on your skin or clothing, wash it off at the sink while calling to your teacher. If you get acid or base in your eyes, immediately flush it out at the eyewash while calling to your teacher.

Required Precautions

- Goggles and a lab apron must be worn at all times.
- Read all safety cautions, and discuss them with your students.
- Students should not handle concentrated acids
- In case of an acid spill, dilute first with water. Then mop up the spill with wet cloths or a cloth mop designated for spill cleanup while wearing disposable plastic gloves.

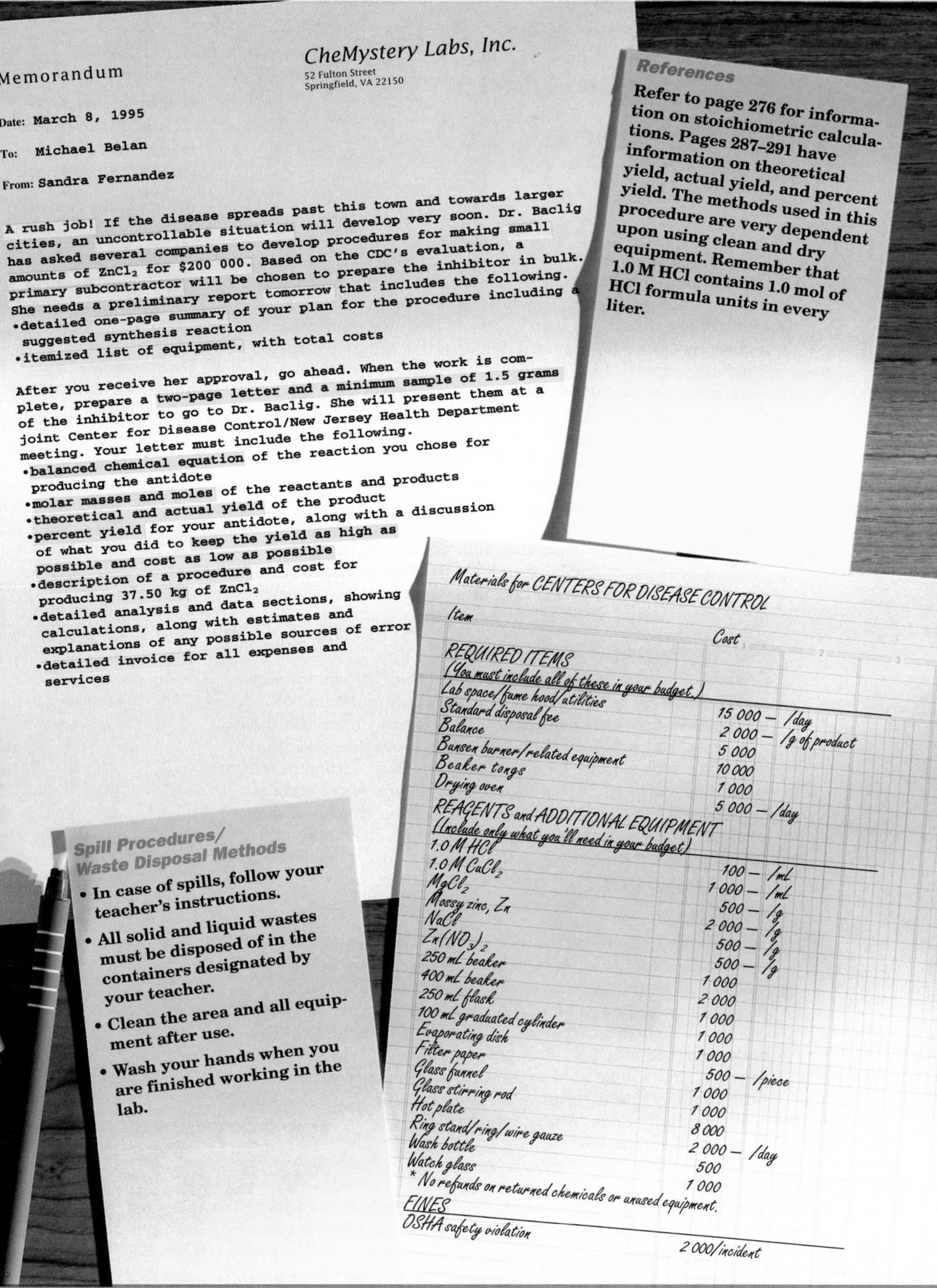

Memorandum

CheMystery Labs, Inc.
52 Fulton Street
Springfield, VA 22150

Date: March 8, 1995

To: Michael Belan

From: Sandra Fernandez

A rush job! If the disease spreads past this town and towards larger cities, an uncontrollable situation will develop very soon. Dr. Baclig has asked several companies to develop procedures for making small amounts of $ZnCl_2$ for $200 000. Based on the CDC's evaluation, a primary subcontractor will be chosen to prepare the inhibitor in bulk. She needs a preliminary report tomorrow that includes the following.

- detailed one-page summary of your plan for the procedure including a suggested synthesis reaction
- itemized list of equipment, with total costs

After you receive her approval, go ahead. When the work is complete, prepare a two-page letter and a minimum sample of 1.5 grams of the inhibitor to go to Dr. Baclig. She will present them at a joint Center for Disease Control/New Jersey Health Department meeting. Your letter must include the following.

- balanced chemical equation of the reaction you chose for producing the antidote
- molar masses and moles of the reactants and products
- theoretical and actual yield of the product
- percent yield for your antidote, along with a discussion of what you did to keep the yield as high as possible and cost as low as possible
- description of a procedure and cost for producing 37.50 kg of $ZnCl_2$
- detailed analysis and data sections, showing calculations, along with estimates and explanations of any possible sources of error
- detailed invoice for all expenses and services

References

Refer to page 276 for information on stoichiometric calculations. Pages 287–291 have information on theoretical yield, actual yield, and percent yield. The methods used in this procedure are very dependent upon using clean and dry equipment. Remember that 1.0 M HCl contains 1.0 mol of HCl formula units in every liter.

Spill Procedures/ Waste Disposal Methods

- In case of spills, follow your teacher's instructions.
- All solid and liquid wastes must be disposed of in the containers designated by your teacher.
- Clean the area and all equipment after use.
- Wash your hands when you are finished working in the lab.

Materials for CENTERS FOR DISEASE CONTROL

Item	Cost
REQUIRED ITEMS (You must include all of these in your budget.)	
Lab space/fume hood/utilities	15 000 – /day
Standard disposal fee	2 000 – /g of product
Balance	5 000
Bunsen burner/related equipment	10 000
Beaker tongs	1 000
Drying oven	5 000 – /day
REAGENTS and ADDITIONAL EQUIPMENT (Include only what you'll need in your budget.)	
1.0 M HCl	100 – /mL
1.0 M $CuCl_2$	1 000 – /mL
$MgCl_2$	500 – /g
Mossy zinc, Zn	2 000 – /g
NaCl	500 – /g
$Zn(NO_3)_2$	500 – /g
250 mL beaker	1 000
400 mL beaker	2 000
250 mL flask	1 000
100 mL graduated cylinder	1 000
Evaporating dish	1 000
Filter paper	500 – /piece
Glass funnel	1 000
Glass stirring rod	1 000
Hot plate	8 000
Ring stand/ring/wire gauze	2 000 – /day
Wash bottle	500
Watch glass	1 000
* No refunds on returned chemicals or unused equipment.	
FINES	
OSHA safety violation	2 000/incident

Materials
(for each lab group)

- 1.0 M HCl, 25 mL
- Mossy zinc, 2.0 g
- 250 mL beaker
- 100 mL graduated cylinder
- Balance, centigram
- Bunsen burner, gas tubing, and striker, or hot plate
- Drying oven
- Ring stand, ring, and wire gauze
- Weighing paper

Estimated cost of materials:
$63 000–$65 000

Tips for Evaluating the Pre-lab Requirements

Be certain students have chosen the right reactants. Although $CuCl_2$ and Zn will react to form $ZnCl_2$, the reaction is more expensive than using Zn and HCl. No other combinations of reactants will react.

Students should also designate the necessary amounts. Zn should be the excess reactant because it can be removed more easily from the $ZnCl_2$ solution than HCl. To ensure at least 1.5 g $ZnCl_2$, use about 25 mL of HCl and 2.0 g of Zn. Do not allow students to use more than 50 mL of 1.0 M HCl.

Proposed Procedure

Pour 25 mL of 1.0 M HCl into a beaker. Add an excess of Zn. When the reaction stops, remove any remaining Zn. The resulting solution contains only water and $ZnCl_2$. Evaporate the water with a Bunsen burner, and dry the crystals overnight.

Post-Lab

Disposal

Set out three disposal containers. One should be for unreacted zinc metal, which can be rinsed and reused next year. The second container is for the $ZnCl_2$ that is produced. The third is for any liquid wastes produced.

If you can use the $ZnCl_2$, evaporate the contents of the $ZnCl_2$ waste container to dryness. Save the $ZnCl_2$ crystals in a labeled sealed container.

To dispose of the contents of the liquid and $ZnCl_2$ waste containers, combine, and treat them in a working fume hood with Na_2S solution. The ZnS precipitate formed should be filtered, dried in the hood, and placed in the trash. Treat the filtrate with 1.0 M acid or base until its pH is between 5 amd 9, and pour it down the drain with an excess of water.

EXPLORATION 9A

Objectives

Students will:

- use appropriate lab safety procedures.
- use a laboratory balance to measure mass.
- use a graduated cylinder to measure volume.
- use a thermometer or the LEAP system thermistor probe to measure temperature.
- determine the maximum temperature attained by a mixture of reactants.
- use a calorimeter to determine heat of reaction for various combinations of an acid and a base.
- use measured heat of reactions in energy-stoichiometry calculations.

Planning

Recommended Time:
1 lab period

Materials
(for each lab group)

- 0.50 M HCl, 100 mL
- 1.0 M HCl, 50 mL
- 1.0 M NaOH, 50 mL
- Distilled water
- NaOH pellets, about 4 g
- 100 mL graduated cylinder
- Balance, centigram
- Glass stirring rod
- Plastic foam cups or calorimeters
- Spatula
- Celsius thermometer, nonmercury type, with a range from –10°C to 120°C
- Watch glass

LEAP System Option

- LEAP System with thermistor probe and related equipment See the LEAP System manual for instructions on setting up the thermistor probe to take temperature measurements.

Calorimetry and Hess's Law

Situation

The company you work for has been hired as an expert witness in a lawsuit. A man working for a cleaning firm was told by his employer to pour some old cleaning supplies into a glass container for disposal. Some of the supplies included muriatic or hydrochloric acid, $HCl(aq)$, and a drain cleaner containing lye, $NaOH(s)$. When the substances were mixed, the container shattered, spilling the contents onto the worker's arms and legs. The worker claims that the hot spill caused burns, and he is therefore suing his employer. The employer claims that the worker is lying because the solutions were at room temperature before they were mixed. The employer says that a chemical burn is unlikely because tests after the accident revealed that the mixture had a neutral pH, indicating that the HCl and NaOH were neutralized. The court has asked you to evaluate whether the worker's story is supported by scientific evidence.

Background

Chemicals can be dangerous because of their special storage needs. Acids cannot be stored in metal containers, and organic solvents cannot be kept in plastic ones. Chemicals that are mixed and react are even more dangerous because many reactions release large amounts of heat. Glass, although relatively nonreactive with solutions of pure substances, is heat sensitive and can shatter if there is a sudden change in temperature due to a reaction. Some glassware, such as Pyrex, is heat conditioned but can still fracture under extreme heat conditions, especially if it has been scratched.

Problem

You will carefully measure the amount of heat released by mixing the chemicals. To be sure your results are accurate, you will measure the heat of reaction in two ways. First, you will break the reaction into steps and measure the heat change of each step. Then you will measure the heat change of the reaction when it takes place all at once. When you are finished, you will be able to use the calorimetry equation from Chapter 9 to determine the following.

- the amount of heat evolved during the overall reaction
- the amount of heat for each step
- the amount of heat for the reaction in kilojoules per mole
- whether this heat could have raised the temperature of the water in the solution high enough to cause a burn

Objectives

Demonstrate **proficiency in the use of calorimeters and related equipment.**

Relate **temperature changes to enthalpy changes.**

Determine **heats of reaction for several reactions.**

Demonstrate **that heats of reaction can be additive.**

Required Precautions

- Goggles and a lab apron must be worn at all times.
- Read all safety cautions, and discuss them with your students.
- Students should not handle concentrated acid solutions.
- Wear goggles, face shield, impermeable gloves, and a lab apron while preparing the HCl. Work in a hood known to be in operating condition, with another person present nearby to call for help in case of an emergency. Be sure you are within 30 s walking

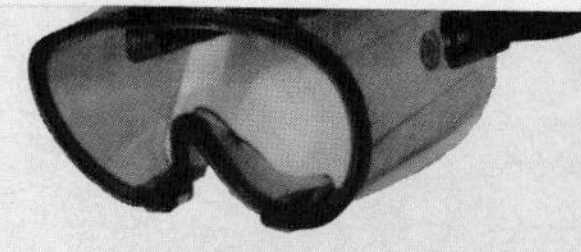

Safety

Always wear goggles and an apron to provide protection for your eyes and clothing. If you get a chemical in your eyes, immediately flush it out at the eyewash station while calling to your teacher. Know the locations of the emergency lab shower and eyewash and how to use them.

Do not touch any chemicals. If you get a chemical on your skin or clothing, wash it off at the sink while calling to your teacher. Make sure you carefully read the labels and follow the directions on all containers of chemicals that you use. Do not taste any chemicals or items used in the laboratory. Never return leftovers to their original containers; take only small amounts to avoid wasting supplies.

Never put broken glass in a regular waste container. Broken glass should be disposed of separately in the container designated by your teacher.

Call your teacher in the event of an acid or base spill. Acid or base spills should be cleaned up promptly, according to your teacher's instructions.

Always clean up the lab and all equipment after use, and dispose of substances according to proper disposal methods. Wash your hands thoroughly before you leave the lab after all lab work is finished.

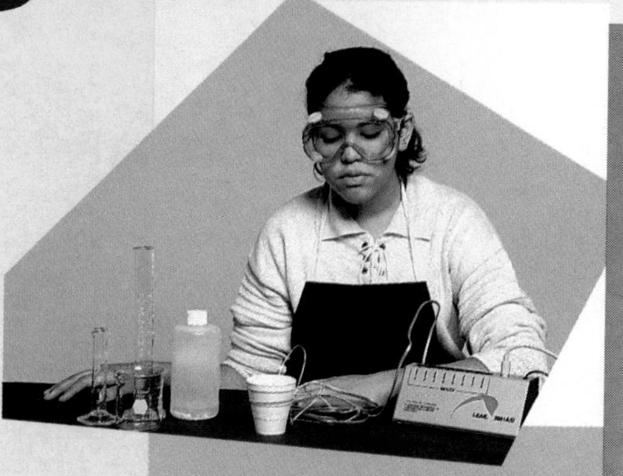

Materials

- Distilled water
- 0.50 M HCl solution, 100 mL
- 1.0 M HCl solution, 50 mL
- 1.0 M NaOH solution, 50 mL
- NaOH pellets, 4 g
- 100 mL graduated cylinder
- Balance
- Glass stirring rod
- Plastic foam cups (or calorimeters)
- Spatula
- Thermometer or LEAP System with thermistor probe
- Watch glass

Preparation

1. ***Organizing Data***
   Prepare a data table in your notebook with four columns and four rows. In the first row, label the second through fourth columns *Reaction 1*, *Reaction 2*, and *Reaction 3*. In the first column, label the second through fourth rows: *Volumes of fluid*, *Initial temperature*, and *Highest temperature*. Reactions 1 and 3 will each require two additional spaces to record the mass of the empty watch glass and the mass of the watch glass with NaOH.
2. If you are not using a plastic foam cup as a calorimeter, ask your lab supervisor for instructions on using the calorimeter. At various points in steps **3** through **13**, you will need to measure the temperature of the solution within the calorimeter. If you are using a thermometer, measure the temperature using the instructions given in step **2a**. If you are using a LEAP System with a thermistor probe, measure the temperature according to the instructions given in step **2a**.

   a. Thermometer
   Measure the temperature by gently inserting the thermometer into the hole in the calorimeter lid. The thermometer takes time to reach the same temperature as the solution inside the calorimeter, so wait to be sure you have an accurate reading.
   Thermometers break easily, so be careful with them, and do not use them to stir a solution.

distance of a safety shower and eyewash station known to be in operating condition.

- In case of an acid spill or base spill, dilute first with water. Then mop up the spill with wet cloths designated for spill cleanup while wearing disposable plastic gloves. A wet cloth mop can be rinsed out a few times and used until it falls apart.
- If the LEAP System with thermistor probe is used, the precautions listed in the teacher's notes on page 652 must be followed to avoid electric shock.

Solution/Material Preparation

1. To prepare 0.50 M HCl, observe the required precautions. Add 42 mL of concentrated HCl to enough distilled water to make 1.00 L of solution. Add the acid slowly, and stop to stir it in order to avoid overheating.
2. To prepare 1.0 M HCl, add 83 mL of concentrated HCl to enough distilled water to make 1.00 L of solution. Add the acid slowly, and stop to stir it in order to avoid overheating.
3. To prepare 1 L of 1.0 M NaOH, add 40.0 g of NaOH to enough water to make 1 L of solution.
4. To prevent the hygroscopic NaOH pellets from absorbing too much water, keep them in a reagent bottle with a stopper, and instruct students to replace the stopper after they have obtained what they need.

Student Orientation

Pre-lab Discussion
Thoroughly discuss the calorimetry equation, with reference to page 318. Consider working through some sample data with the class. At first, students may have difficulty understanding that reaction 1 and 2 are the equivalent of reaction 3. A thorough discussion of Hess's law and combining equations, as described on pages 329–330, should alleviate this problem.

Techniques to Demonstrate
Make certain that students understand how to handle NaOH pellets. They should not be picked up with fingers. The mass measurements must be taken quickly before the pellets absorb moisture from the air. Make sure that students use a watch glass instead of weighing paper for measuring the mass of the NaOH.

Post-Lab

Disposal

Set out four disposal containers. Designate one for acidic liquids, one for basic liquids, one for neutral liquids, and one for excess NaOH pellets. When students are finished, slowly combine the liquid contents of the containers, one at a time. If there are any excess NaOH pellets, add them a few at a time to the mixture, stirring constantly to be sure the pellets dissolve. Then check the pH. Add 1.0 M acid or base until the pH is within the range of 5–9, and then pour down the drain.

Answers to Analysis and Interpretation

1. $NaOH(s) \longrightarrow NaOH(aq)$
   $NaOH(aq) + HCl(aq) \longrightarrow H_2O(l) + NaCl(aq)$

   $NaOH(s) + HCl(aq) \longrightarrow H_2O(l) + NaCl(aq)$
2. equation 1 + equation 2 = equation 3
3. A good calorimeter must insulate, so that any heat created by the reaction is absorbed by the water instead of the surroundings. Plastic foam cups insulate better than paper ones, so they make better calorimeters.
4. $\Delta t_1 = 26.5°C - 21.5°C = 5.0°C$
   $\Delta t_2 = 28.1°C - 22.0°C = 6.1°C$
   $\Delta t_3 = 33.0°C - 22.0°C = 11.0°C$
5. $m = 100.0 \text{ mL } H_2O \times \frac{1.00 \text{ g}}{1 \text{ mL } H_2O} = 100.0 \text{ g } H_2O$
   for all three reactions
6. Heat for reaction 1:
   $100.0 \text{ g } H_2O \times 5.0°C \times \frac{4.180 \text{ J}}{1 \text{ g}\cdot°C} = 2100 \text{ J} = 2.1 \text{ kJ}$

   Heat for reaction 2:
   $100.0 \text{ g } H_2O \times 6.1°C \times \frac{4.180 \text{ J}}{1 \text{ g}\cdot°C} = 2500 \text{ J} = 2.5 \text{ kJ}$

   Heat for reaction 3:
   $100.0 \text{ g } H_2O \times 11.0°C \times \frac{4.180 \text{ J}}{1 \text{ g}\cdot°C} = 4600 \text{ J} \times 4.6 \text{ kJ}$

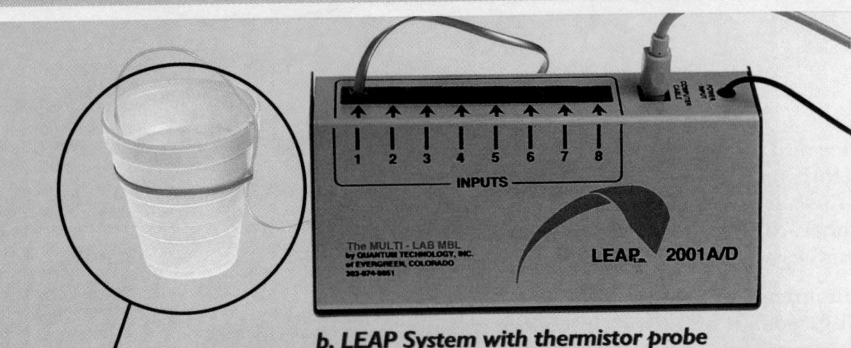

b. LEAP System with thermistor probe
Arrange the thermistor probe as shown. Lay it in the bottom of the calorimeter, and use a rubber band on the outside of the calorimeter to hold the wire for the thermistor probe in place. Then, plug the probe into the LEAP System box attached to the computer.

Technique

Reaction 1

3. Pour about 100 mL of distilled water into a graduated cylinder. Measure and record the volume of the water to the nearest 0.1 mL. Pour the water into your calorimeter. Record the water temperature to the nearest 0.1°C.
4. Determine and record the mass of a clean and dry watch glass to the nearest 0.01 g. Remove the watch glass from the balance. Obtain about 2 g of NaOH pellets, and put them on the watch glass. Measure and record the mass of the watch glass and the pellets to the nearest 0.01 g.

   It is important that this step be done quickly because NaOH is "hygroscopic." It absorbs moisture from the air, increasing its mass as long as it remains exposed to the air.
5. Immediately place the NaOH pellets in the calorimeter cup, and gently stir the solution with a stirring rod.
   Do not stir with a thermometer.

   Place the lid on the calorimeter. Record the highest temperature in the data table. When finished with this reaction, pour the solution into the container designated by your teacher for disposal of basic solutions.
6. Be sure to clean all equipment and rinse it with distilled water before continuing with the next procedure.

Reaction 2

7. Pour about 50 mL of 1.0 M HCl into a graduated cylinder. Measure and record the volume of the HCl solution to the nearest 0.1 mL. Pour the HCl solution into your calorimeter. Record the temperature of the HCl solution to the nearest 0.1°C.
8. Pour about 50 mL of 1.0 M NaOH into a graduated cylinder. Measure and record the volume of the NaOH solution to the nearest 0.1 mL. ***For this step only, rinse the thermometer or LEAP thermistor probe in distilled water, and measure the temperature of the NaOH solution in the graduated cylinder to the nearest 0.1°C. Record the temperature in your data table and then replace the thermometer or LEAP thermistor probe in the calorimeter.***
9. Pour the NaOH solution into the calorimeter cup, and stir gently. Place the lid on the calorimeter. Record the highest temperature in the data table. When finished with this reaction, pour the solution into the container designated by your teacher for disposal of mostly neutral solutions.
10. Clean and rinse all equipment before continuing with the next procedure.

Sample Data

	Reaction 1	Reaction 2	Reaction 3
Volume of fluid, mL	100.0	100.0	100.0
Initial temperature,°C	21.5	22.0	22.0
Highest temperature,°C	26.5	28.1	33.0

Reaction 3

11. Pour about 100 mL of 0.50 M HCl into a graduated cylinder. Measure and record the volume to the nearest 0.1 mL. Pour the HCl solution into your calorimeter. Record the temperature of the HCl solution to the nearest 0.1°C.

12. Measure the mass of a clean and dry watch glass, and record it in your data table. Obtain approximately 2 g of NaOH. Place them on the watch glass, and record the total mass to the nearest 0.01 g. ***It is important that this step be done quickly because NaOH is "hygroscopic." It absorbs moisture from the air, increasing its mass as long as it remains exposed to the air.***

13. Immediately place the NaOH pellets in the calorimeter, and gently stir the solution. Place the lid on the calorimeter. Record the highest temperature in the data table. When finished with this reaction, pour the solution into the container designated by your teacher for disposal of mostly neutral solutions.

Cleanup and Disposal

14. Check with your teacher for the proper disposal procedures. Any excess NaOH pellets should be disposed of in the designated container. Always wash your hands thoroughly after cleaning up the lab area and equipment.

Analysis and Interpretation

1. ***Organizing Ideas***
Write a balanced chemical equation for each of the three reactions that you performed. (Hint: be sure to include states of matter for all substances in each equation.)

2. ***Organizing Ideas***
Find a way to get the equation for the total reaction by adding two of the equations from item **1** and then canceling out substances that appear in the same form on both sides of the new equation, as was demonstrated in Chapter 9. (Hint: start with the equation that has a product which is a reactant in a second equation. Add those two equations together.)

3. ***Analyzing Methods***
Explain why a plastic foam cup makes a better calorimeter than a paper cup does.

4. ***Organizing Data***
Calculate the change in temperature (Δt) for each of the reactions.

5. ***Organizing Data***
Assuming that the density of the water and the solutions is 1.00 g/mL, calculate the mass, m, of liquid present for each of the reactions.

6. ***Analyzing Results***
Using the calorimeter equation, calculate the heat released by each reaction. (Hint: use the specific heat capacity of water in your calculations; $c_{p,H_2O} = 4.180$ J/g•°C.)

$$\text{Heat} = m \times \Delta t \times c_{p,H_2O}$$

	Reaction 1	Reaction 2
Mass of empty watch glass, g	30.15	30.15
Mass of watch glass + NaOH, g	32.15	32.16

7. Moles NaOH for reaction 1:

$$2.00 \text{ g} \times \frac{1 \text{ mol NaOH}}{40.00 \text{ g}}$$

$= 5.00 \times 10^{-2}$ mol

Moles NaOH for reaction 2:

$$50 \text{ mL} \times \frac{1 \text{ L}}{1000 \text{ mL}} \times \frac{1.00 \text{ mol NaOH}}{1 \text{ L}}$$

$= 5.00 \times 10^{-2}$ mol

Moles NaOH for reaction 3:

$$2.01 \text{ g} \times \frac{1 \text{ mol NaOH}}{40.00 \text{ g}}$$

$= 5.02 \times 10^{-2}$ mol

8. Enthalpy changes should be negative because the reactions are exothermic.

$$\Delta H_1 = \frac{-2.1 \text{ kJ}}{5.00 \times 10^{-2} \text{ mol NaOH}} =$$

$-$ 42 kJ/mol NaOH

$$\Delta H_2 = \frac{-2.5 \text{ kJ}}{5.00 \times 10^{-2} \text{ mol NaOH}} =$$

$-$ 50. kJ/mol NaOH

$$\Delta H_3 = \frac{-4.6 \text{ kJ}}{5.02 \times 10^{-2} \text{ mol NaOH}} =$$

$-$ 92 kJ/mol NaOH

9. The sum of heats for the first two reactions should equal the heat for the third reaction.

10. Reaction 1 involved heat of solution. Reaction 2 involved heat of reaction. Reaction 3 involved heat of solution and heat of reaction.

Conclusions

11. By direct measurement, ΔH should be -92 kJ/mol. By indirect calculation, it is also -92 kJ/mol ($-50.$ kJ/mol + 42 kJ/mol).

12. Molar amount of NaOH:

$$55 \text{ g} \times \frac{1 \text{ mol NaOH}}{40.00 \text{ g}} = 1.4 \text{ mol}$$

Molar amount of HCl: 1.35 mol

Heat of dissolving:

$$1.4 \text{ mol NaOH} \times \frac{-42 \text{ kJ}}{1 \text{ mol NaOH}} = -59 \text{ kJ}$$

Heat of reaction: 1.35 mol NaOH =

$$\frac{-50. \text{ kJ}}{1 \text{ mol NaOH}} = -67.5 \text{ kJ}$$

(The remaining NaOH is an excess reactant that remains unreacted.)

Total heat of reaction:

$-59\ kJ - 67.5\ kJ = -126\ kJ$

$$\Delta t_{H_2O} = 126\,000\ J \times \frac{1\ g\text{•}°C}{4.180\ J} \times \frac{1}{450.\ g\ H_2O} = 67°C$$

Final temperature = $25°C + 67°C = 92°C$

The mixture is likely to cause a burn because the temperature would be much higher than 60°C.

13. HCl is the limiting reactant. There was 0.05 mol NaOH left unreacted.

14. $\Delta t = 15°C$

$$15°C \times 450\ g\ H_2O \times \frac{4.180\ J}{1\ g\text{•}°C} \times 28\ kJ \times \frac{1\ mol\ NaOH}{92\ kJ} =$$

0.30 mol solid NaOH dissolving and reacting with a solution containing 0.30 mol HCl

Extensions

1. Students' suggestions for improving the procedure will vary. Although students may suggest using larger quantities of acid or base or more concentrated solutions, it is best to keep the concentrations of strong acids and bases given to students lower than 1.0 M. Be sure answers are safe and include carefully planned procedures.
2. Students' suggestions for improving the calorimeter will vary. Possible suggestions may include better insulation, or a bigger cup with a larger amount of water.
3. Do not allow students to try their procedures. Students' suggestions will vary but could include reacting large molar amounts of acids and bases within the glassware.
4. The chemists use an ice bath because the heat of solution for NaOH pellets is high enough to make the solution dangerously hot.

7. *Organizing Data*
Calculate the moles of NaOH used in each of the reactions. (Hint: to find the number of moles in a solution, multiply the volume in liters by the molar concentration.)

8. *Analyzing Results*
Calculate the ΔH value in terms of kilojoules per mole of NaOH for each of the three reactions.

9. *Organizing Ideas*
Using your answer to item **2** and your knowledge of Hess's law from Chapter 9, explain how the enthalpies for the three reactions should be mathematically related.

10. *Organizing Ideas*
Which of the following types of heats of reaction apply to the enthalpies calculated in item **8**: heat of combustion, heat of solution, heat of reaction, heat of fusion, heat of vaporization, and heat of formation?

Conclusions

11. *Evaluating Methods*
Use your answers from items **8** and **9** to determine the ΔH value for the reaction of solid NaOH with HCl solution by direct measurement and by indirect calculation.

12. *Inferring Conclusions*
Third-degree burns can occur if skin comes into contact for more than 4 s with water that is hotter than 60°C (140°F). Investigators believe that there were about 55 g of NaOH in the drain cleaner and 450. mL of muriatic acid solution containing a total of 1.35 mol of HCl (a 3.0 M HCl solution). If the initial temperature of the solutions was 25°C, could a mixture that is hot enough to cause burns have resulted?

13. *Applying Conclusions*
Which chemical is the limiting reactant, given the amounts in item **12**? How many moles of the other reactant remained unreacted?

14. *Applying Conclusions*
What molar amounts of each reactant would be necessary for a reaction that heats 450. mL of water from 25°C to 40°C, a much safer temperature?

Extensions

1. *Designing Experiments*
What possible sources of error can you identify with this procedure? If you can think of ways to eliminate them, ask your teacher to approve your plan, and run the procedure again.

2. *Designing Experiments*
When you work with a calorimeter, the assumptions are made that the heat energy from the reaction is used entirely to heat the water and that no heat energy escapes the calorimeter. With this in mind, design a better calorimeter. If your teacher approves your plan, build an improved calorimeter, and run the procedure again.

3. ***Applying Ideas***
How could you determine the maximum temperature that laboratory glassware can withstand? Explain your answers.

4. ***Applying Ideas***
When chemists make solutions from NaOH pellets, they often keep the solution in an ice bath. Explain why.

5. ***Applying Ideas***
When a strongly acidic or basic solution is spilled on a person, the first step is to dilute it by washing the area of the spill with a lot of water. Explain why adding an acid or base to neutralize the solution immediately is not a good idea.

6. ***Evaluating Methods***
You have worked with heats of solution for exothermic reactions. Could the same type of procedure be used to determine the temperature changes for endothermic reactions? How would the procedure stay the same? What would change about the procedure and the data?

7. ***Applying Ideas***
Thinking that two cleaners should be better than one, someone mixes a drain cleaner containing NaOH with a toilet-bowl cleaner containing HCl. Will this make a better cleanser than either of the substances alone? Explain your answer.

8. ***Applying Ideas***
A chemical supply company is going to ship NaOH pellets to a very humid place, and they've asked for your advice on packaging. Design a package for the NaOH pellets. Explain the advantages of your package's design and materials. (Hint: remember that the reaction in which NaOH absorbs moisture from the air is an exothermic one, and NaOH reacts exothermically with other compounds as well.)

9. ***Inferring Conclusions***
Which is more stable: solid NaOH or NaOH solution? Explain your answer.

10. ***Designing Experiments***
In this experiment, the calorimeter was used to measure the temperature change when two dilute solutions react. Because the solutions were dilute, it was assumed that the mixture had the same specific heat capacity as pure water. Design an experiment to determine the specific heat capacity of the solution after the reaction. If your teacher approves of your plan, test it. Using the new specific heat capacity value, recalculate the ΔH values for Reactions 2 and 3, and check to see if the adjusted results are better than your original ones.

5. If acid or base spills are neutralized instead of diluted, the heat of reaction for the neutralization could cause a heat burn, along with a chemical burn caused by the acid or base.

6. The same procedure could be used for endothermic reactions. However, the temperature of the water will decrease, and the enthalpy change for the reaction will have a positive value.

7. Sodium hydroxide reacts with hydrochloric acid to produce sodium chloride and water. This neutral salt-water solution is not a good cleaner.

8. Students' suggestions for package design will vary. Be sure each design addresses the dangers of moisture, breakage, and spills.

9. NaOH solution is more stable than solid NaOH. The products of an exothermic reaction are comparatively more stable than the reactants.

10. Students' suggestions for measuring the specific heat capacity of the solutions will vary. Be sure answers are safe and include carefully planned procedures. Possible ideas include melting an ice cube of known mass in a warm solution of NaCl. The necessary enthalpy of melting can be calculated using handbook values and the mass of the ice cube. The change in the temperature of the solution and the final mass of the solution can be used with the calorimeter equation to calculate the specific heat capacity. There should be very little difference in values for the specific heat capacity of the solutions and that of water.

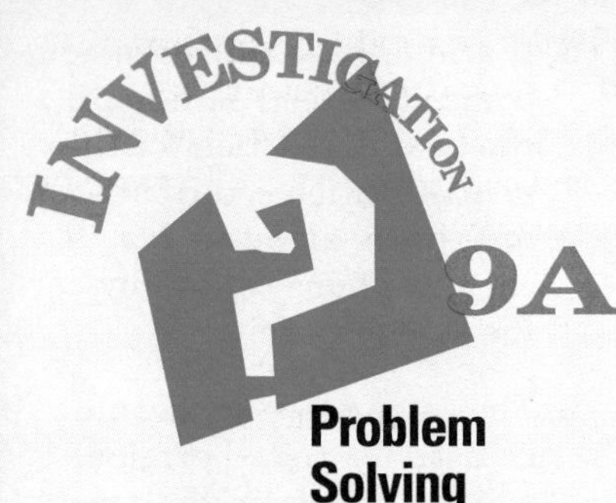

Objectives

Students will:
- use appropriate lab safety procedures.
- use a laboratory balance to measure mass.
- use a graduated cylinder to measure volume.
- use a thermometer or the LEAP System thermistor probe to measure temperature.
- design a reaction system that will keep the temperature of a nearby object between 35–40°C for 5 min.
- design and implement their own procedure.

Planning

Recommended Time: 1 lab period
Prerequisite: Exploration 9A

Solution/Material Preparation

1. See the Teacher's Notes for Exploration 9A for information on preparing 1.0 M HCl.
2. To prevent the hygroscopic NaOH pellets from absorbing too much water, keep them in a reagent bottle with a stopper and instruct students to replace the stopper after they have obtained what they need.

Student Orientation

Techniques to Demonstrate

Discuss any lab technique errors that occurred during the Exploration lab.

Pre-lab Discussion

Begin by discussing the results and procedure used in the Exploration. Remind students that the heating of the water in the 50 mL beaker will lag behind that of the actual reaction mixture. To help students, you may provide some of the following questions for students to consider as they make their plans.
- What information are you given in the lab and what information can you use from the Exploration?
- How can you determine how much of each reactant you will need?
- What will you do if your predicted method doesn't heat the 50 mL beaker enough?

Sample Data and Analysis

Δt: 40°C − 25°C = 15°C

$$\text{Heat required: } 50.\ \text{g } H_2O \times \frac{4.180\ \text{J}}{1\ \text{g}\cdot°\text{C}} \times 15°\text{C} = 3100\ \text{J}$$

$$3.1\ \text{kJ} \times \frac{1\ \text{mol NaOH}}{92\ \text{kJ}} \times \frac{40.00\ \text{g}}{1\ \text{mol NaOH}} = 1.3\ \text{g NaOH}$$

Student calculations and plans will vary. Students will discover that some heat is lost to the surroundings, so more NaOH will be required.

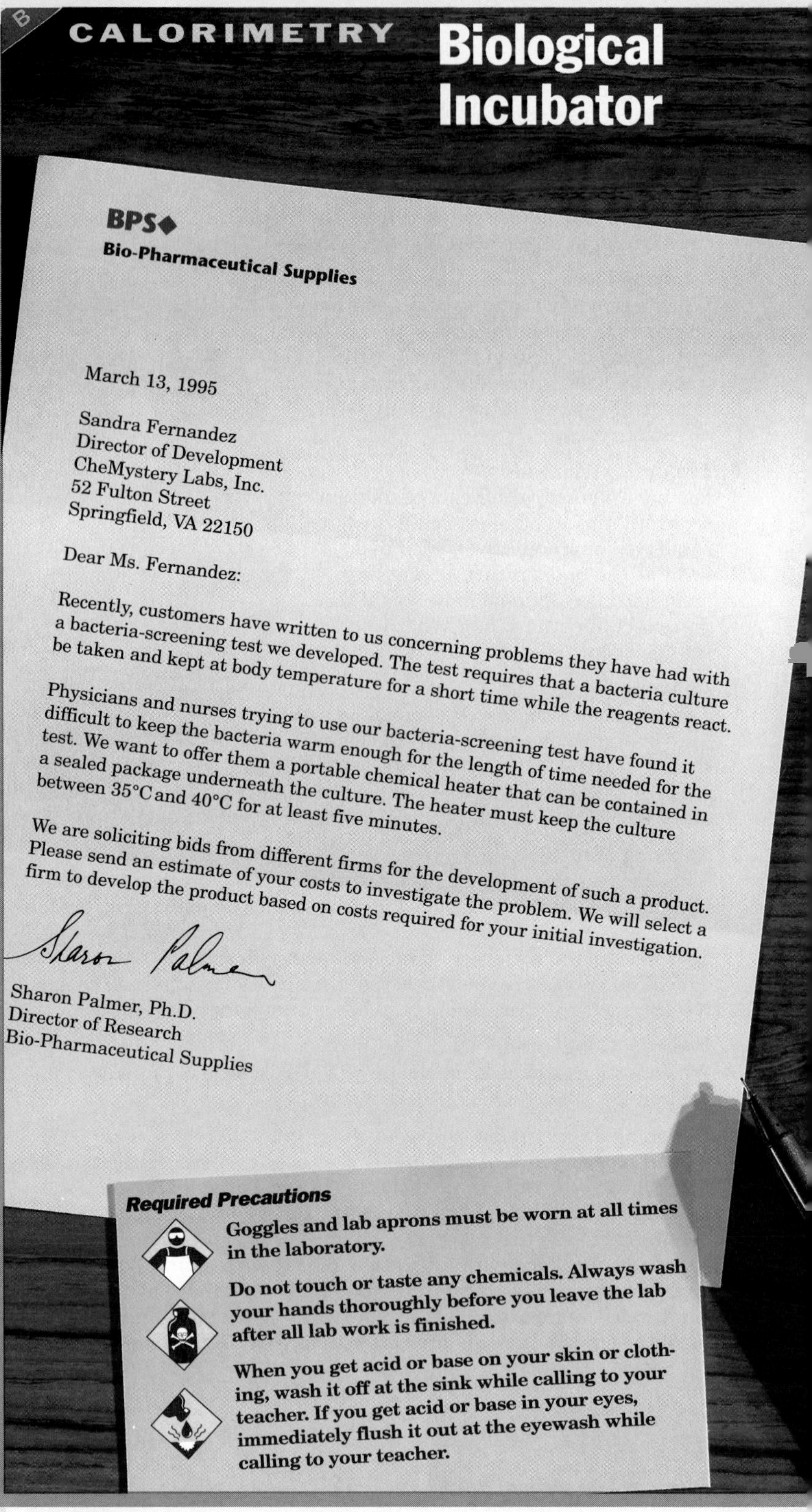

B CALORIMETRY

Biological Incubator

BPS◆
Bio-Pharmaceutical Supplies

March 13, 1995

Sandra Fernandez
Director of Development
CheMystery Labs, Inc.
52 Fulton Street
Springfield, VA 22150

Dear Ms. Fernandez:

Recently, customers have written to us concerning problems they have had with a bacteria-screening test we developed. The test requires that a bacteria culture be taken and kept at body temperature for a short time while the reagents react.

Physicians and nurses trying to use our bacteria-screening test have found it difficult to keep the bacteria warm enough for the length of time needed for the test. We want to offer them a portable chemical heater that can be contained in a sealed package underneath the culture. The heater must keep the culture between 35°C and 40°C for at least five minutes.

We are soliciting bids from different firms for the development of such a product. Please send an estimate of your costs to investigate the problem. We will select a firm to develop the product based on costs required for your initial investigation.

Sharon Palmer

Sharon Palmer, Ph.D.
Director of Research
Bio-Pharmaceutical Supplies

Required Precautions

Goggles and lab aprons must be worn at all times in the laboratory.

Do not touch or taste any chemicals. Always wash your hands thoroughly before you leave the lab after all lab work is finished.

When you get acid or base on your skin or clothing, wash it off at the sink while calling to your teacher. If you get acid or base in your eyes, immediately flush it out at the eyewash while calling to your teacher.

Required Precautions

- Goggles and a lab apron must be worn at all times.
- Read all safety cautions, and discuss them with your students.
- Students should not handle concentrated acids.
- In case of an acid spill or base spill, dilute first with water. Then, mop up the spill with wet cloths or a wet cloth mop designated for spill cleanup while wearing disposable plastic gloves.
- If the LEAP System with thermistor probe is used, the precautions listed in the teacher's notes on page 652 must be followed to avoid electric shock.

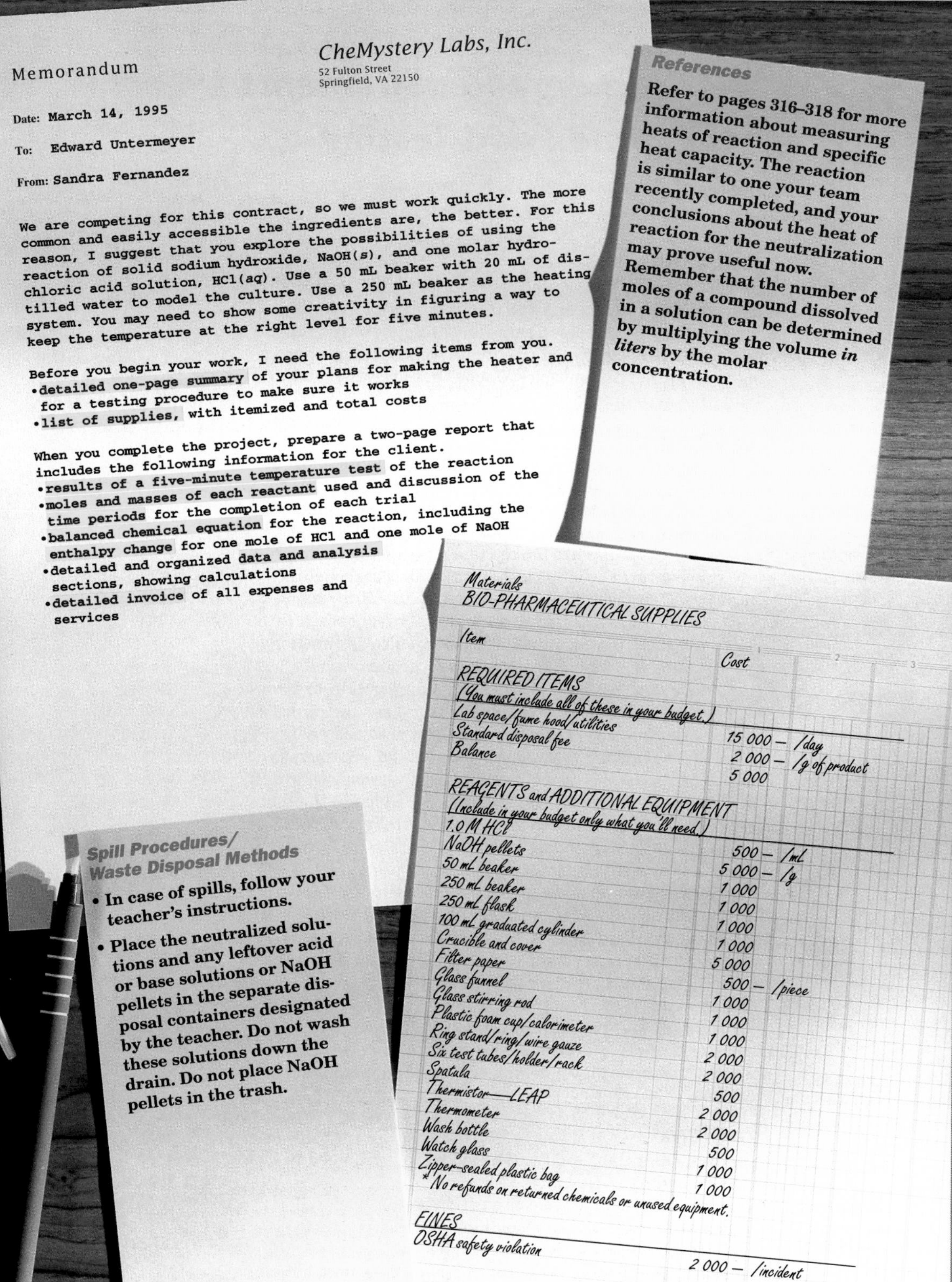

Memorandum

CheMystery Labs, Inc.
52 Fulton Street
Springfield, VA 22150

Date: March 14, 1995

To: Edward Untermeyer

From: Sandra Fernandez

We are competing for this contract, so we must work quickly. The more common and easily accessible the ingredients are, the better. For this reason, I suggest that you explore the possibilities of using the reaction of solid sodium hydroxide, NaOH(s), and one molar hydrochloric acid solution, HCl(aq). Use a 50 mL beaker with 20 mL of distilled water to model the culture. Use a 250 mL beaker as the heating system. You may need to show some creativity in figuring a way to keep the temperature at the right level for five minutes.

Before you begin your work, I need the following items from you.
- detailed one-page summary of your plans for making the heater and for a testing procedure to make sure it works
- list of supplies, with itemized and total costs

When you complete the project, prepare a two-page report that includes the following information for the client.
- results of a five-minute temperature test of the reaction
- moles and masses of each reactant used and discussion of the time periods for the completion of each trial
- balanced chemical equation for the reaction, including the enthalpy change for one mole of HCl and one mole of NaOH
- detailed and organized data and analysis sections, showing calculations
- detailed invoice of all expenses and services

References

Refer to pages 316–318 for more information about measuring heats of reaction and specific heat capacity. The reaction is similar to one your team recently completed, and your conclusions about the heat of reaction for the neutralization may prove useful now. Remember that the number of moles of a compound dissolved in a solution can be determined by multiplying the volume *in liters* by the molar concentration.

Materials
BIO-PHARMACEUTICAL SUPPLIES

Item	Cost
REQUIRED ITEMS (You must include all of these in your budget.)	
Lab space/fume hood/utilities	15 000 — /day
Standard disposal fee	2 000 — /g of product
Balance	5 000
REAGENTS and ADDITIONAL EQUIPMENT (Include in your budget only what you'll need.)	
1.0 M HCl	500 — /mL
NaOH pellets	5 000 — /g
50 mL beaker	1 000
250 mL beaker	1 000
250 mL flask	1 000
100 mL graduated cylinder	1 000
Crucible and cover	1 000
Filter paper	5 000
Glass funnel	500 — /piece
Glass stirring rod	1 000
Plastic foam cup/calorimeter	1 000
Ring stand/ring/wire gauze	1 000
Six test tubes/holder/rack	2 000
Spatula	2 000
Thermistor—LEAP	500
Thermometer	2 000
Wash bottle	2 000
Watch glass	500
Zipper-sealed plastic bag	1 000
* No refunds on returned chemicals or unused equipment.	1 000
FINES	
OSHA safety violation	2 000 — /incident

Spill Procedures/ Waste Disposal Methods

- In case of spills, follow your teacher's instructions.
- Place the neutralized solutions and any leftover acid or base solutions or NaOH pellets in the separate disposal containers designated by the teacher. Do not wash these solutions down the drain. Do not place NaOH pellets in the trash.

Materials
(for each lab group, assuming three trials)
- 1.0 M HCl, 150 mL
- Distilled water
- NaOH pellets, about 6 g
- 50 mL beaker
- 250 mL beaker
- 100 mL graduated cylinder
- Balance, centigram
- Glass stirring rod
- Spatula
- Celsius thermometer, nonmercury type, with a range from −10°C to 120°C
- Watch glass

LEAP System Option
- LEAP System with thermistor probe and related equipment

Estimated cost of materials: $143 500

Tips for Evaluating the Pre-lab Requirements
Students should use heat data from the Exploration to determine reactant amounts necessary to cause a temperature change, test their plan, and adjust it until successful. If students plan to use more than 50 mL of 1.0 M HCl or 2.0 g of NaOH in a single trial, have them reconsider their plans. To keep the NaOH reasonably dilute, students should dissolve NaOH in at least 25 mL of HCl solution per g NaOH.

Proposed Procedure

Fill a 50 mL beaker with 20 mL of water. React 50.0 mL of HCl with 1.4 g of NaOH in a 250 mL beaker. Rest the 50 mL beaker inside the 250 mL beaker, and record the temperature of the water in the 50 mL beaker every minute. Adjust the reactant amounts, and try the procedure again until temperature remains between 35°C and 40°C for 5 min.

Post-Lab

Disposal
Set out four disposal containers. Designate one for acidic liquids, one for basic liquids, one for neutral liquids and one for excess NaOH pellets. When students are finished, slowly combine the liquid contents of the containers, one at a time. If there are any excess NaOH pellets, add them a few at a time to the mixture, stirring constantly to be sure the pellets dissolve. Then, check the pH. Add 1.0 M acid or base until the pH is within the range of 5–9, and then pour down the drain.

EXPLORATION 9B

Objectives
Students will:
- use appropriate lab safety procedures.
- perform tests of several standards to determine positive test for proteins, lipids, sugars, starch, and vitamin C.
- test several foods to identify their components.
- plan a balanced menu for athletes based on the results of the tests.

Planning
Recommended Time:
1 lab period

Materials
(for each lab group, assuming four foods are tested)
- 1% starch solution, 12 mL
- Benedict's solution, 30 mL
- Biuret reagent, 6 mL
- Distilled water
- Food samples
- Gelatin, dissolved, 5 mL
- Glucose solution, 5 mL
- Isopropanol, 30 mL
- Lugol's iodine solution, 6 mL
- Shortening (or vegetable oil)
- Vitamin C solution, 5 mL
- 250 mL beaker
- 10 mL graduated cylinder
- Beaker tongs
- Brown wrapping paper
- Hot plate
- Pipet or medicine dropper
- Test-tube holder
- Test-tube rack
- Test tubes, 6

Solution/Material Preparation

Because of the small quantities of reagents used, each should be dispensed in dropper bottles.

1. To make 1% starch, dissolve 10.0 g soluble starch in 1000 mL of distilled water.
2. Benedict's solution may be purchased ready-made or as a powder which will make 1.00 L of solution.
3. Biuret solution may be purchased ready-made or as a powder which will make 1.00 L of solution.
4. Dissolve one package of gelatin in 1000 mL of distilled water.

Energy Requirements and Food Testing

Exploration 9B Technique Builder

Objectives
Test **several standards to determine positive tests for proteins, lipids, sugars, starch, and vitamin C.**

Identify **the components of several foods.**

Relate **the results of food testing to the usefulness of different foods as energy sources.**

Situation
You work for a chemical analysis firm that has been hired by a group of dietitians to analyze food in order to plan the meals of the U.S. Olympic Team. The athlete's meals must contain plenty of sources of energy, as well as the protein they need for healthy muscles. Foods that are very fatty or oily are not recommended.

Background
As discussed in Chapter 9, food contains nutrients, most of which are naturally formed organic polymers, sometimes called *macromolecules*. The three primary types of macromolecules are *carbohydrates, lipids,* and *proteins.*

Carbohydrates are broken down by organisms to provide energy. The empirical formula for a carbohydrate is CH_2O, as indicated by the two parts of the name, *carbo-* and *-hydrate*. Sugars are relatively small carbohydrates that usually have several alcohol functional groups and an aldehyde or ketone group as well. Many sugar molecules are able to link together to form polymer chains. One polymer of glucose is a molecule called *starch*, which plants use as a form of stored energy.

Glucose

Lipids provide long-term energy storage for organisms. Lipids exist in several types, but all dissolve poorly in water. Fatty acids are the structural subunits for lipids in the diet. They consist of long hydrocarbon chains with a carboxylic acid functional group on one end.

Proteins are another type of polymer. In proteins, the monomers are amino acids, molecules containing both the amine and carboxylic acid functional groups. These molecules are used by the body to make enzymes, hormones, and substances that make up the body's structure. Proteins are not usually used as a primary energy source.

In addition to carbohydrates, lipids, and proteins, people need another type of nutrient—*vitamins*, which are molecules that play a very specific role in metabolic reactions.

Problem
To help the dietitians analyze the foods, you will need to do the following.
- check samples of known composition so that you will know what positive tests look like
- perform the tests on the food samples
- based on the results, make recommendations on whether to include these foods in the athletes' meals

Required Precautions
- Goggles and a lab apron must be worn at all times.
- Tie back long hair and loose clothing when working in the lab.
- Read all safety cautions, and discuss them with your students.
- Because of the use of the hot plate, the precautions listed in the teacher's notes on page 652 must be followed to avoid electric shock.

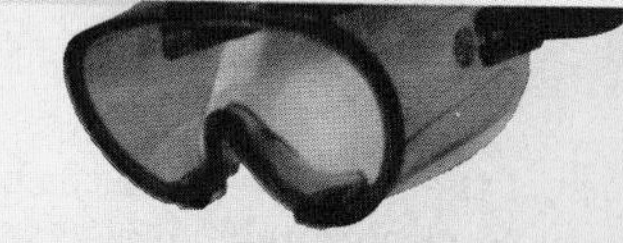

Safety

Always wear goggles and an apron to provide protection for your eyes and clothing. If you get a chemical in your eyes, immediately flush it out at the eyewash station while calling to your teacher. Know the locations of the emergency lab shower and eyewash and how to use them.

Do not touch any chemicals. If you get a chemical on your skin or clothing, wash it off at the sink while calling to your teacher. Make sure you carefully read the labels and follow the directions on all containers of chemicals that you use. Do not taste any chemicals or items used in the laboratory. Never return leftovers to their original containers; take only small amounts to avoid wasting supplies.

Use tongs whenever handling glassware or other equipment that has been heated because when it is hot, it does not look hot. Keep the brown paper away from the hot plate.

Because the isopropanol is flammable, no Bunsen burners or other open flames should be in use during this lab. Hot plates can also be a source of ignition; carry out all work with isopropanol and Lugol's solution in a hood as far away from any hot plates as possible.

Always clean up the lab and all equipment after use, and dispose of substances according to proper disposal methods. Wash your hands thoroughly before you leave the lab after all lab work is finished.

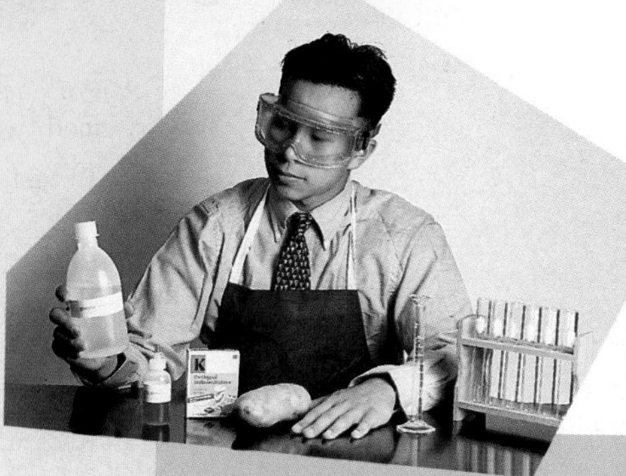

Materials

- **1% starch solution**
- **Benedict's solution**
- **Biuret reagent**
- **Distilled water**
- **Food samples**
- **Gelatin, dissolved**
- **Glucose solution**
- **Isopropanol**
- **Lugol's iodine solution in dropper bottle**
- **Shortening**
- **Vitamin C solution**
- **10 mL graduated cylinder**
- **250 mL beaker**
- **Beaker tongs**
- **Brown wrapping paper**
- **Hot plate**
- **Pipet or medicine dropper**
- **Test tubes, 6**
- **Test tube holder**
- **Test tube rack**

Preparation

1. ***Organizing Data***
   Prepare a data table in your lab notebook with six rows. Be sure there is plenty of room within the boxes to describe the results of the reactions. The boxes in the first row should be labeled *Nutrient*, *Reagent*, *Positive test*, and *Negative test (H_2O)*. Then, make enough additional columns so that there is one for each sample of food you will test. In the first column, label the second through six rows *Sugar*, *Starch*, *Protein*, *Vitamin C* and *Lipid*.
2. Label two test tubes *Positive test* and *Negative test (H_2O)*, and then label an additional test tube with the name of each food you will test.

Technique

Sugar Test

3. Using a graduated cylinder, measure about 5 mL of distilled water, and pour it into the test tube labeled *Negative test*.
4. Using a graduated cylinder, measure about 5 mL of the glucose solution, and pour it into the test tube labeled *Positive test*.

- The isopropanol alcohol is extremely flammable. It should be kept within an operating fume hood. Only place 200 mL at a time in the reagent bottle, and have a lid or stopper nearby. Students should replace the lid when they are finished. The isopropanol should not be dispensed until all of the sugar tests have been completed and all hot plates or Bunsen burners have been turned off and have cooled down.

5. To make glucose solution, add 25 g of corn syrup or glucose to 1000 mL of distilled water.
6. Lugol's iodine solution may be purchased ready-made. To make it yourself, dissolve 0.6 g of iodine crystals in 50 mL ethanol. In a separate container, dissolve 0.6 g of KI in 50 mL of distilled water. Combine the two solutions and mix.
7. To prepare the vitamin C solution, dissolve one 500 mg vitamin C tablet in 400 mL of distilled water. Filter to remove solid impurities and dilute to 500 mL. Do not heat to hasten dissolving as vitamin C decomposes with heat.
8. If hot plates are not available, a Bunsen burner with ring stand, ring, and wire gauze with ceramic center may be used to heat the Benedict's solution in a water bath.
9. Allow students to choose their own food samples.

Student Orientation

Pre-lab Discussion

Review the discussion of the different types of food and their energy content from Chapter 9. Most of these tests work because the nutrients are organic compounds that have different functional groups.

Techniques to Demonstrate

Discuss with students procedures for each indicator tested. Tests are very straightforward, and include both negative tests (distilled water) and positive tests (nutrients) for students to compare to their samples. Remind students that if the food is solid, they should add about 5 mL of water to the food and stir well. The test should be performed based on what is seen in the solution, rather than the solid food. Demonstrate for students how to set up a hot water bath for heating Benedict's solution. Hot plates are preferred over Bunsen burners because they heat the water bath more gradually and evenly.

Post-Lab

Disposal

Set out three disposal containers. Designate one for waste from the Benedict's solution test and the biuret reagent test. Another container should be for waste from the starch and vitamin C tests. The third container is for liquid wastes from the lipid test.

For the Benedict's and biuret wastes, treat the liquid and solids wastes with 1.0 M NaOH until it has a pH between 5 and 9. The solids can be put in the trash. The neutralized filtrate can be poured down the drain.

For the iodine and starch wastes, add a few drops of 1.0 M H_2SO_4 to make the mixture slightly acidic. Then add 1.0 M $Na_2S_2O_3$ slowly with stirring until the mixture is white or colorless. Then, neutralize it with 1.0 M NaOH. After checking that the pH is between 5 and 9, pour it down the drain.

For the liquid waste from the lipid test, add ten times as much water, and a little soap or detergent. Shake well to emulsify the fat, and pour it down the drain. The brown paper and any solid food samples may be put in the trash.

5. For each of the samples of food to be tested, add about 5 mL of the food to the appropriately labeled test tube.
6. To test for sugar, add 5 mL of Benedict's solution to each of the test tubes.
7. Fill a 250 mL beaker about two-thirds full with water. Place the test tubes in the beaker using the test-tube holder.
8. Place the beaker with the test tubes on a hot plate, and heat for approximately 5 min. Use beaker tongs to remove the beaker.
9. Record any color changes in the test tubes after the heating is completed.
10. Pick up the test tubes with a test-tube holder. After they have cooled, empty the test tubes into the disposal containers designated by your teacher for sugar and protein tests. Clean all of the test tubes and the graduated cylinder thoroughly, but keep the labels on the test tubes.

Starch Test

11. Using a graduated cylinder, measure about 5 mL of distilled water into the test tube labeled *Negative test*. Measure about 5 mL of the starch solution, and pour it into the test tube labeled *Positive test*.
12. For each of the samples of food to be tested, add about 5mL of the food to the appropriately labeled test tube.
13. To test for starch, add about 1 mL of Lugol's iodine solution to each of the test tubes, and record any color changes in the test tubes.
14. Empty the test tubes into the disposal containers designated by your teacher for starch and vitamin C tests. Clean all of the test tubes and the graduated cylinder thoroughly, but keep the labels on the test tubes.

Protein Test

15. Using a graduated cylinder, measure about 5 mL of distilled water into the test tube labeled *Negative test*. Measure about 5 mL of the dissolved gelatin, which contains protein, and pour it into the test tube labeled *Positive test*.
16. For each of the samples of food to be tested, add about 5 mL of the food to the appropriately labeled test tube.
17. To test for proteins, add about 1 mL of biuret reagent, and record any color changes in the test tubes.
18. Empty the test tubes into the disposal containers designated by your teacher for sugar and protein tests. Clean all of the test tubes and the graduated cylinder thoroughly, but keep the labels on the test tubes.

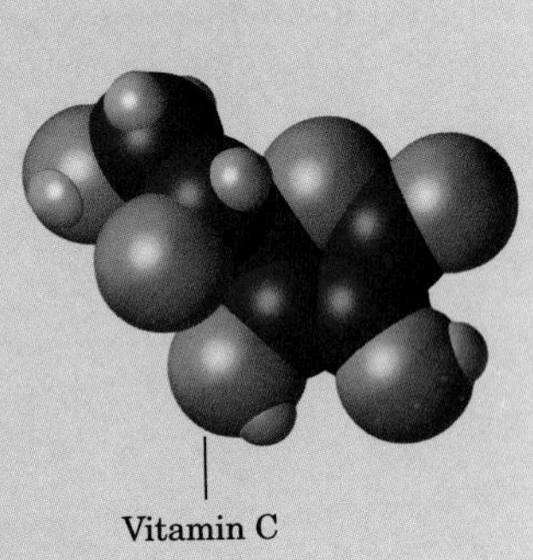

Vitamin C

Vitamin C Test

19. Place 5 drops of distilled water in the test tube labeled *Negative test*, and 5 drops of vitamin C solution in the test tube labeled *Positive test*.
20. For each of the samples of food to be tested, add about 5 drops of the liquid part of the food to the appropriately labeled test tube.
21. Put one drop of the 1% starch solution into each of the test tubes to be tested. For each test tube, add and count drops of iodine solution, one by one, until the solution in the test tube turns blue-purple and remains that color for at least 15 s. Record the number of drops in your data table in the *Vitamin C* column.

Sample Data

Nutrient	Reagent	Pos. Test	Neg. Test	Food samples
Glucose	Benedict's solution	turns orange or brick red	stays blue-green	results will vary
Starch	iodine	turns dark blue-purple	stays red-brown	results will vary
Protein	biuret reagent	turns purple or deep blue	stays lighter blue	results will vary

22. Empty the test tubes into the disposal containers designated by your teacher for starch and vitamin C tests. Clean all of the test tubes and the graduated cylinder thoroughly, but keep the labels on the test tubes.

Lipid Test

23. Measure about 2 mL of distilled water into the test tube labeled *Negative test*. Measure about 2 mL of shortening into the test tube labeled *Positive test*.
24. For each of the food samples, place about 2 mL of the food in the appropriate test tubes.
25. To test for lipids, add 5 mL of isopropanol to each sample. Mix the samples, and allow them to sit for about 10 minutes.
26. For each test tube, pour about 2–3 mL of the liquid from each test tube onto a clean part of the brown paper. Record your observations in the data table.

Cleanup and Disposal

27. The used brown paper from the lipid test may be put in the trash can. Liquids from the lipid test should be poured into their own designated disposal container. Leftover test solutions can be placed in the disposal containers designated by your teacher. Benedict's solution and biuret reagent can be disposed of together along with solutions from sugar and protein test. Starch and iodine solutions can be disposed of together along with solutions from starch and vitamin C tests.

Analysis and Interpretation

1. ***Evaluating Methods***
   Which tests were quantitative tests, ones that measured how much of a substance there was in a sample? Which were qualitative, ones that only measured whether a substance was present?

Conclusion

2. ***Applying Conclusions***
   Write a paragraph summarizing your recommendations. Which tested foods should be given to athletes? Why?

Extensions

1. ***Applying Ideas***
   Nutritional tables are available in many sources, including the *CRC Handbook of Chemistry and Physics*. Look up the foods you tested, and determine the amounts of sugar, starch, protein, and lipids in them. Using the equations given in Chapter 9, calculate the number of kilojoules of energy that would be provided by a 175 g serving of each of the foods you tested.
2. ***Applying Ideas***
   Keep a written record of the quantities of foods you consume in a single day. Look up the nutrient composition of each food and classify the foods depending on whether they contain primarily carbohydrates, proteins, or lipids.

Nutrient	Reagent	Pos. Test	Neg. Test	Food samples
Vitamin C	starch and iodine	turns dark blue-purple after a number of drops of iodine	turns dark blue-purple immediately	results will vary
Lipid	brown paper	translucent	not translucent	results will vary

Answers to Analysis and Interpretation

1. The only quantitative test was the vitamin C test. The more vitamin C present, the more drops of iodine must be added for the solution to turn blue-purple.

Conclusions

2. Student answers will vary, depending upon the foods they chose to test. Be sure that students' suggestions are well planned.

Extensions

1. Student answers will vary. Be sure that the energy relationships from page 319 are properly applied.
2. Student answers will vary. A nutritional book or the nutritional tables in the *CRC Handbook of Chemistry and Physics* can be used to check students' classifications of foods.

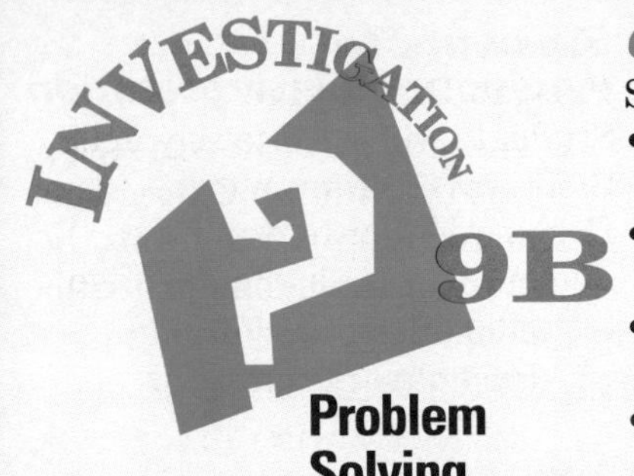

Objectives

Students will:
- use appropriate lab safety procedures.
- demonstrate proficiency in testing for nutrients.
- evaluate claims made on baby food labels
- design and implement their own procedure.

Planning

Recommended Time: 1 lab period
Prerequisite: Exploration 9B

Solution/Material Preparation

1. Because of the small quantities of reagents, dispense them in dropper bottles.
2. See the Teacher's Notes for Exploration 9B for information on preparing starch and iodine solutions.
3. Biuret and Benedict's solutions may be purchased ready-made.
4. Do not dispense isopropanol until all sugar tests have been completed and ali hot plates or Bunsen burners are turned off and cooled down.
5. Prepare as many of the following unknown baby food solutions as appropriate. For each baby food to be tested pour 30 mL into a container. Each team will receive 3 different solutions for testing. Tape labels onto each jar.
   1 Combine 900 mL distilled water, 5 oz chicken baby food, 60 mL vegetable oil, and 100 mL corn syrup. Mix in blender.
   "• 112 Democritus's All-Natural Baby Food. Contains only protein and starch. No sugar or oil."
   "• 510 Kelvin's Baby Food. Contains all the protein an infant needs, no sugar."
   2 Combine 900 mL distilled water, 4 g gelatin, 100 mL white grape juice, and four 500 mg vitamin C tablets. Mix in blender.
   "• 322 Bohr Baby Juice. Contains fruit juice only. No vitamin C for sensitive infants."
   "• 821 Nature's Finest Fruit Juice. Contains fruit juice only. Packed with natural protein and vitamin C."
   3 Combine 800 mL distilled water, 10 mL vegetable oil, 200 mL baby fruit with yogurt, 2 vitamin C tablets, 4 g gelatin. Mix in blender.
   "• 339 BabySenso for Protein Sensitive Babies. Contains no protein or sugar."
   "• 930 Johnny Appleseed Yogurt with Apples. Contains milk protein and sugar for that sweet taste a baby loves."

Sample Data

Food number	Glucose	Starch	Protein	Vitamin C	Lipids
112, 510	✔		✔		✔
322, 821	✔		✔	✔	
339, 930	✔		✔	✔	✔

Baby foods are classified by the middle digit in their code number.

FOOD TESTING — Baby-Food Label Fraud

Center for Food Safety and Applied Nutrition
FOOD AND DRUG ADMINISTRATION
6500 Fishers Lane · Rockville, MD 20857

Memorandum

Date: March 14, 1995
To: Analytical Chemist Contractors
From: R. T. Gutierrez, Infant Foods Division
Re: False advertising in "natural" baby foods

Recently, the Food and Drug Administration (FDA) has received a number of complaints about false advertising in some brands of baby foods. Spot checks by FDA's investigators have turned up a number of infractions of food labeling laws.

For example, some foods labeled as containing no starch or sugar actually contained these substances. Others advertised as "high in protein" contained no protein. It is important that these fraudulent companies are identified and prosecuted quickly.

The FDA has received authorization to request bids from private contractors to perform testing of three different brands of baby food on the market for sugar, lipids, starch, vitamin C, and protein.

Required Precautions

Goggles and lab aprons must be worn at all times.

Do not touch or taste any chemicals. Wash your hands thoroughly when finished.

Use tongs to handle equipment that has been heated.

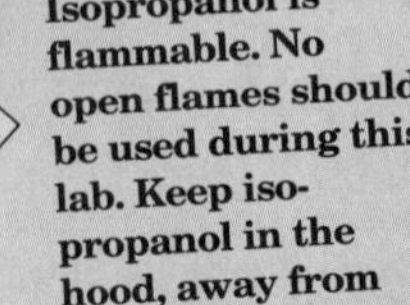
Isopropanol is flammable. No open flames should be used during this lab. Keep isopropanol in the hood, away from hot plates.

Required Precautions

- Goggles and a lab apron must be worn at all times.
- Read all safety cautions, and discuss them with your students.
- If a hot plate is used, the precautions listed in the teacher's notes on page 652 must be followed to avoid electric shock.
- Tie back loose hair and clothing while in the lab.
- The flammable isopropanol alcohol should be kept within an operating fume hood. Only make 200 mL available at a time.

Memorandum

CheMystery Labs, Inc.
52 Fulton Street
Springfield, VA 22150

Date: March 15, 1995

To: Frederica Sanchez

From: George Taylor

You know as well as I do that getting on the FDA's list of contractors would help us put the company on top. I'm counting on you to do a good job on this one.

Send Gutierrez a two-page letter containing our bid as soon as possible. She will need details on the following items.
- plan for our measurement procedure and any necessary data tables
- detailed list of the equipment and materials needed, along with the costs (Don't forget to include labor costs as a part of the bid!)

Once you get approval, begin work immediately. We want to impress the FDA with our quickness and efficiency! When you are finished, immediately send a two-page letter that includes the following.
- your recommendations of companies to be prosecuted, with explanations
- description of the tests performed, including discussion of positive and negative test standards
- complete report of your data
- detailed invoice showing all costs, services, and time for work

References

Refer to page 319 for more information on fats, carbohydrates, and proteins. The tests for the different nutrients and additives are the same as ones your team recently completed. Be very careful in observing results because the color of the food may interfere with proper reading of the indicator. It may help to dilute some of the food with water so that the results can be seen more easily.

Materials for FDA

Item	Cost	1	2	3
REQUIRED ITEMS (You must include all of these in your budget.)				
Lab space/fume hood/utilities	15 000 — /day			
Beaker tongs	1 000			
Standard disposal fee	2 000 — /g of product			
Hot plate	8 000			
REAGENTS and ADDITIONAL EQUIPMENT (Include in your budget only what you'll need.)				
1% starch solution	500 — /drop			
Benedict's solution	2 000 — /mL			
Biuret reagent	2 000 — /mL			
Brown paper	500 — /sheet			
Isopropanol	1 000 — /mL			
Lugol's iodine	2 000 — /mL			
250 mL beaker	1 000			
250 mL flask	1 000			
10 mL graduated cylinder	1 000			
Glass funnel	1 000			
Glass stirring rod	1 000			
Litmus paper	1 000			
Paper clips	1 000 — /piece			
Ring stand/ring/pipe stem triangle	500 — /box			
Ruler	2 000			
Six test tubes/holder/rack	500			
Spatula	2 000			
Weighing paper	500			
* No refunds on returned chemicals or unused equipment.	500 — /piece			
Fines				
OSHA safety violation	2 000 — /incident			

Spill Procedures/ Waste Disposal Methods

- In case of a spill, follow your teacher's instructions.
- Put excess chemicals and solutions in the containers designated by your teacher. Do not pour them down the sink or place them in the trash can unless specifically instructed to do so by your teacher.
- Pieces of brown paper may be thrown into the garbage can.

Materials

(for each lab group)
- Baby food samples, 3 (see preparation instructions)
- 1% starch solution, 3 drops
- Benedict's solution, 3 mL
- Biuret reagent, 3 mL
- Distilled water
- Isopropanol, 15 mL
- Lugol's iodine solution, 3 mL
- 250 mL beaker
- 10 mL graduated cylinder
- Beaker tongs
- Brown wrapping paper
- Dropper bottles for reagents, 4
- Hot plate
- Test-tube holder
- Test-tube rack
- Test tubes, 6

Estimated cost of materials: $73 000 (may vary due to disposal costs)

Student Orientation

Pre-lab Discussion

Students will have to rely upon their results from the Exploration for information upon the positive and negative tests for each nutrient.

Tips for Evaluating the Pre-lab Requirements

Be certain student procedures and data tables are complete.

Proposed Procedure

Test the food samples with Benedict's solution for glucose, Lugol's iodine solution for starch, biuret reagent for protein, starch and iodine for vitamin C, and on a brown paper bag for lipids.

Post-Lab

Disposal

Set out three disposal containers. One is for waste from the sugar and protein tests. Treat the liquid and solids wastes with 1.0 M NaOH until it is at a pH between 5 and 9. The solids can be put in the trash. The neutralized filtrate can be poured down the drain.

Another container should be for waste from the starch and vitamin C tests. Add a few drops of 1.0 M H_2SO_4 to make the mixture slightly acidic. Then add 1.0 M $Na_2S_2O_3$ slowly with stirring until the mixture is white or colorless. Neutralize it with 1.0 M NaOH until the pH is between 5 and 9, and pour it down the drain.

The third container is for liquid wastes from the lipid test. Add ten times as much water, and a little soap or detergent. Shake well to emulsify the oils, and pour it down the drain. The brown paper and any solid food samples may be put in the trash.

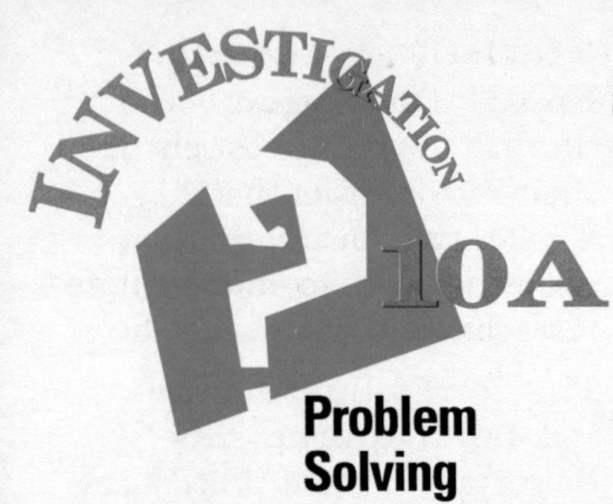

Objectives

Students will:

- use appropriate lab safety procedures.
- observe a reaction that forms a solid from two gases.
- measure the distance to the products and the time elapsed to calculate the diffusion rates of the gases.
- use Graham's law to calculate theoretical gas diffusion ratios.
- predict the location and time for chemical spill reactions, based on the data.
- design and implement their own procedure.

Planning

Recommended Time: 1 lab period
Prerequisite: None

Solution/Material Preparation

1. Wear goggles, face shield, impermeable gloves, and a lab apron when you prepare the HCl and $NH_3 \cdot H_2O$. Work in an operating fume hood, with another person nearby to call for help in case of an emergency and within 30 s walking distance of a safety shower and eyewash station known to be in operating condition.
2. To prepare 1.0 M HCl, add 83 mL of concentrated HCl to enough distilled water to make 1.00 L of solution. Add the acid slowly, and stop to stir it in order to avoid overheating.
3. To prepare the 1.0 M $NH_3 \cdot H_2O$, dissolve 68 mL of concentrated $NH_3 \cdot H_2O$ (ammonia water, also known as NH_4OH) in 1.00 L of solution.
4. Other variations of this lab use fire-polished glass tubing or glass capillary tubes. The reaction is also visible in the plastic pipet tubes suggested here. Plastic is easier and safer because there is little preparation and less chance of breakage.

Student Orientation

Techniques to Demonstrate

Show students how to clip the bulb and tip off of an extra-long pipet to make a plastic tube about 15 cm in length. Remind students that the reaction product is not brightly colored, so they will need to make very careful observations.

Sample Data and Analysis

Theoretical velocity ratio: $\sqrt{\frac{36}{17}} = \sqrt{2.1} = 1.4:1$

Experimental velocity ratio: $\sqrt{\frac{12.0 \text{ cm}}{4.2 \text{ cm}}} = 1.7:1$

Chemical reaction will occur about 1 km from the HCl plant and 1.7 km from the $NH_3 \cdot H_2O$ plant.

Time of reaction for model: 3.2 s

Predicted time of reaction for city: $1.7 \text{ km} \times \frac{1000 \text{ m}}{1 \text{ km}} \times \frac{100 \text{ cm}}{1 \text{ m}} \times \frac{3.2 \text{ s}}{12.0 \text{ cm}} \times \frac{1 \text{ min}}{60 \text{ s}} \times \frac{1 \text{ h}}{60 \text{ min}} = 13 \text{ h}$ (student answers should be between 8.0 and 15 h)

B

GAS DIFFUSION

Industrial Spill

EDWARD F. QUIMBY, M
DANA RUBIO, CITY MAN

March 17, 1995

Sandra Fernandez, Director of Development
CheMystery Labs, Inc.
52 Fulton Street
Springfield, VA 22150

Dear Ms. Fernandez:

Your company has been commissioned by the city's Hazardous Waste and Emergency Response Commission (H.W.E.R.C.) to model a chemical disaster. Recently, two new chemical plants have been built on the perimeter of our city in compliance with our new zoning laws. One plant manufactures hydrochloric acid, and the other manufactures ammonia water ($NH_3 \cdot H_2O$, or NH_4OH).

Residents are concerned about a dangerous release of chemicals from the factories. We know about the dangers of a spill at each one. Recently, however, city officials were made aware that a cloud of toxic material could form if fumes from both plants combined.

Attached is a map showing the chemical plants. Determine the likeliest location for a toxic combination, and calculate the necessary response time for local agencies. H.W.E.R.C. will evaluate reports and bids to select a company to serve as the chemical response team in the event of a local emergency.

Sincerely,

Dana Rubio
City Manager

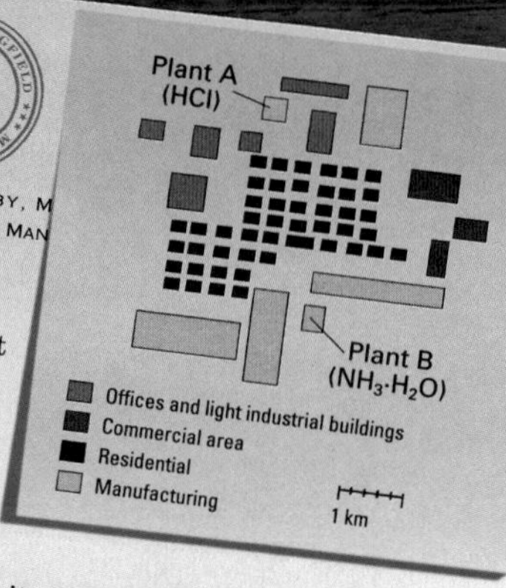

Plant A (HCl)
Plant B ($NH_3 \cdot H_2O$)
Offices and light industrial buildings
Commercial area
Residential
Manufacturing
1 km

Required Precautions

Goggles and lab aprons must be worn at all times.

Do not touch or taste any chemicals. Wash your hands thoroughly when finished.

Avoid breathing fumes from these chemicals.

If you get acid or base on your skin or clothing, wash it off at the sink while calling to your teacher. If you get acid or base in your eyes, immediately flush it out at the eyewash station while calling to your teacher.

Required Precautions

- Read all safety cautions, and discuss them with your students.
- Goggles and a lab apron must be worn at all times.
- Students should not handle concentrated acid or base solutions.
- In case of an acid or base spill, dilute first with water. Then mop up the spill with wet cloths or a wet cloth mop designated for spill cleanup while wearing disposable plastic gloves.
- Students should not breathe any fumes generated.

Memorandum

CheMystery Labs, Inc.
52 Fulton Street
Springfield, VA 22150

Date: March 18, 1995

To: Greg Buntz

From: Sandra Fernandez

The previous work we did with the city has paid off. But this contract means more than just money for us. We are in competition with several companies from nearby towns to be the local response team, and many of our employees live in Springfield. My idea for modeling the spill is to take a piece of plastic tubing, such as a piece from an extra long micropipet, and insert two cotton swabs in either end. Each swab will be soaked in solutions of one of the chemicals. Then, we can observe the fumes without making the lab into a disaster area.

I need a preliminary plan from you that includes the following.
- detailed one-page summary of the procedure, including your plans for measurements, that will enable you to calculate the necessary response time and the likeliest location of the toxic product
- balanced chemical equation for the reaction
- theoretical calculations of molecular velocity ratios for the two compounds
- itemized list of materials, with costs

When you've finished, prepare a two-page letter to Dana Rubio that includes the following.
- approximate time elapsed and location of the toxic interaction
- summary of the model and a discussion of its limitations
- theoretical and experimental molecular velocity ratios
- detailed analysis, including a balanced chemical equation, with calculations for Graham's law and experimental errors
- detailed and organized data tables
- detailed invoice for materials and services

References

Refer to page 379 for more information about the relative velocities of different gases. The longer the tube, the more accurate your results will probably be. Keep the tube horizontal at all times to maximize accuracy. The ammonia in ammonia water vaporizes easily. The likeliest reaction is a synthesis reaction with ammonia and hydrogen chloride.

Materials for CITY OF SPRINGFIELD

Item	Cost
REQUIRED ITEMS (You must include all of these in your budget.)	
Lab space/fume hood/utilities	15 000 — /day
Standard disposal fee	2 000 — /g of product
REAGENTS and ADDITIONAL EQUIPMENT (Include in your budget only what you'll need.)	
1.0 M HCl	500 — /mL
1.0 M NaOH	500 — /mL
1.0 M NH_3•H_2O	500 — /mL
250 mL beaker	1 000
100 mL graduated cylinder	1 000
Aluminum foil	1 000 — /cm^2
Balance	5 000
Cotton swab	500
Glass stirring rod	1 000
Grease pencil	500
Litmus paper	1 000 — /piece
Plastic pipet, extra long	1 000
Ring stand with buret clamp	2 000
Ruler	500
Scissors	500
Six test tubes/holder/rack	2 000
Spatula	500
Stopwatch	5 000
Wash bottle	500
* No refunds on returned chemicals or unused equipment.	
FINES	
OSHA safety violation	2 000 — /incident

Spill Procedures/ Waste Disposal Methods

- In case of spills, follow your teacher's instructions.
- Do not pour any solutions down the drain. Do not place any material in the trash can.
- Dispose of the swabs in the designated waste container.
- To clean the deposit from the middle of the pipet tube, rinse with 10 mL of 1.0 M NaOH, and pour the rinse into the disposal container designated by your teacher.
- Clean the area and all equipment.

Materials
(for each lab group)
- 1.0 M HCl, 5 mL
- 1.0 M NH_3•H_2O, 5 mL
- 1.0 M NaOH, 10 mL
- 250 mL beakers, 2
- Cotton swabs, 2
- Plastic pipet, extra long
- Ruler
- Scissors
- Stopwatch

Estimated cost of materials: $75 000

Pre-lab Discussion

You may provide some of the following leading questions for students to consider as they make their plans.
- What conversion factor relates the size of the model to the actual size of the town?
- Do you expect the molecules in your model to be traveling faster than, slower than, or as quickly as the molecules in a spill over the city? Why?
- What measurements must you make to be able to predict both the location and time of the reaction over the city?
- What limitations are there with the model?

Tips for Evaluating the Pre-lab Requirements

Students' plans should require less than 5 mL each of NH_3•H_2O and HCl. Be certain students budget for 10 mL of 1.0 M NaOH as required in the Waste Disposal section. Students must measure both the time for the reaction and the distance of the product from the reactants.

Proposed Procedure

Clip the tip and bulb off an extra-long pipet to make a plastic tube. Dip one cotton swab in HCl and another in NH_3•H_2O. Start a stopwatch, and insert one swab in each end of the plastic tube. Measure the time elapsed before a product is seen. Measure the distance from the product to the reactants.

Post-Lab

Disposal

Set out three disposal containers, one for cotton swabs, one for acid, and one for base and the rinse from the pipets. First, rinse out the pipets with 10 mL of 1.0 M NaOH to dissolve the NH_4Cl that is formed. Neutralize the contents of the acid container with 1.0 M base and the base container with 1.0 M acid, until each has a pH between 5 and 9. Place the cotton swabs in a beaker of water. After 5 min, neutralize the water with 1.0 M acid or base. All neutralized solutions may be poured down the drain.

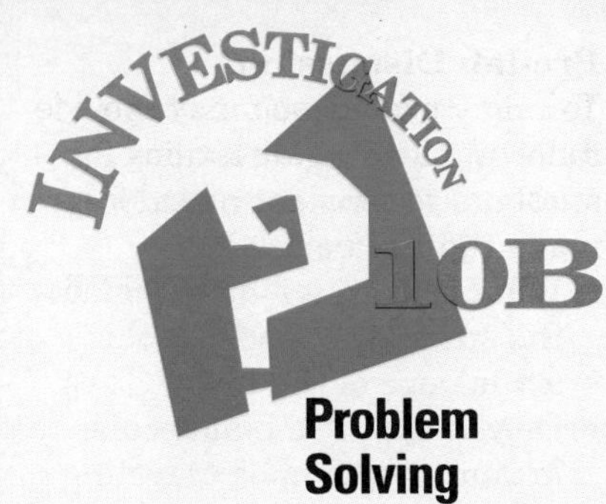

Objectives

Students will:
- use appropriate lab safety procedures.
- perform stoichiometric calculations of amounts of CO_2 gas to fill a model and a building.
- design and build a model of a fire-extinguishing system.
- test the efficiency of the fire-extinguishing system.
- design and implement their own procedure.

Planning

Recommended Time: 2 lab periods (depends upon time allotted for building model and system)
Prerequisite: None

Solution/Material Preparation

1. Wear goggles, a face shield, impermeable gloves, and a lab apron when you prepare the HCl. Work in an operating fume hood, with another person nearby to call for help in case of an emergency and within 30 s walking distance of a safety shower and eyewash station known to be in operating condition.
2. To prepare 1.0 M HCl, add 83 mL of concentrated HCl to enough distilled water to make 1.00 L of solution. Add the acid slowly, and stop to stir it in order to avoid overheating.
3. The measurements in this lab are designed so that the model can be a shoe box. Depending on the size of the boxes available, amounts may need to be adjusted. Once built, models can be reused by the next class.

Student Orientation

Techniques to Demonstrate

Show students that if they wrap the $NaHCO_3$ in a piece of tissue paper, it can be dropped into a bottle of HCl, with time to put a lid on the bottle. Be certain students use only one-hole stoppers or lids with holes drilled in them to prevent the bottle from bursting due to the pressure.

Sample Data and Analysis

Volume of model: 15 cm × 30 cm × 10 cm = 4.5 L

Amount of CO_2: $4.5\ L \times \frac{1\ mol}{22.4\ L} = 0.20\ mol\ CO_2$

Amount of $NaHCO_3$: $0.20\ mol\ CO_2 \times \frac{1\ mol\ NaHCO_3}{1\ mol\ CO_2} \times \frac{84.01\ g}{1\ mol\ NaHCO_3} =$ 17 g $NaHCO_3$

Amount of HCl: $0.20\ mol\ CO_2 \times \frac{1\ mol\ HCl}{1\ mol\ CO_2} \times \frac{1\ L}{1.0\ mol\ HCl} = 0.20\ L\ HCl$

Amounts actually used, 20 g and 250 mL, are in excess to allow for leakage.

For the building, the required amounts are not 100 times greater but 100^3 times greater, because it is a three-dimensional volume.

Amount of $NaHCO_3$ for building: 2.0×10^4 kg $NaHCO_3$

Amount of HCl for building: 2.5×10^5 L 1.0 M HCl

GAS STOICHIOMETRY — Fire Suppression System

Lyon+Abel

March 20, 1995

Director of Development
CheMystery Labs, Inc.
52 Fulton Street
Springfield, VA 22150

Dear Director:

Our small printing company needs to update our fire extinguisher system immediately to avoid losing our insurance. With our high-tech equipment and computers, a standard water sprinkler system is inappropriate. We understand that buildings with electronic equipment usually have fire suppression systems that use a nonreactive gas.

Because we are planning to move soon to a new location, we would like your firm to develop a short-term fix for our current extinguishing needs. Consider using carbon dioxide, CO_2, produced by a reaction between 1.0 M HCl solution and $NaHCO_3$ powder. The reaction vessel may be located inside or outside the building. Design requirements are enclosed.

Sincerely,

Greg Lyon

Greg Lyon
Vice-President
Lyon and Abel Printing Incorporated

Required Precautions

Goggles and lab aprons must be worn at all times in the laboratory.

If you get acid or base on your skin or clothing, wash it off at the sink while calling to your teacher. If you get acid or base in your eyes, immediately flush it out at the eyewash while calling to your teacher.

Do not touch or taste any chemicals. Always wash your hands thoroughly when finished.

Required Precautions

- Goggles and a lab apron must be worn at all times.
- Read all safety cautions, and discuss them with your students.
- Students should not handle concentrated acid
- In case of an acid spill, dilute with water. Then mop up the spill with wet cloths or a wet cloth mop designated for spill cleanup while wearing disposable gloves.
- Students should not breathe any fumes generated.

CheMystery Labs, Inc.
52 Fulton Street
Springfield, VA 22150

Memorandum

Date: March 21, 1995

To: Brad Bishop

From: Sandra Fernandez

We have been contacted about a $250 000 job. As part of the job, we will need to build a cardboard 1:100 scale model of the facility. The model should be 30 cm long, 15 cm wide, and 10 cm high, according to the plan of the facility shown at right. Tea candles will be used to represent simultaneous fires in each room. You can save money if you have the box, glue, tape, and candles that you can bring from home.

15 cm
30
model's interior walls 10 cm high

I need your design for an extinguishing system soon, along with the following.
- detailed one-page summary of your plans, including sketches if necessary
- itemized list of materials, with total costs
- stoichiometric calculations of the amount of reactants required to produce enough gas to extinguish the fires

Remember that some gas may be lost prior to its use as a fire extinguisher.

After you finish testing your design model, prepare a two-page report for Mr. Lyon that includes the following.
- description of the delivery system, including sketches if necessary
- balanced chemical equation for the CO_2-producing reaction
- results of the candle-extinguishing tests
- detailed and organized data and analysis sections
- calculations of amounts necessary for the full-scale system
- detailed invoice with costs for equipment, materials, and services

References

Refer to pages 384–386 for more information on stoichiometric calculations involving gases. Remember, 1.0 M HCl solution contains 1.0 mol/liter. Build your model at home first. Cut notches in the box for the walls and then glue or tape the walls into place. Be creative with your delivery system design, but plan it out on paper first. Get others' opinions on your system before trying it out. If you (and anyone else observing) wear goggles and lab aprons, you can try your system out at home first with $NaHCO_3$ (baking soda) and CH_3COOH (in vinegar) as your reactants for producing CO_2 gas. Base your costs for disposal on the amount of NaCl produced.

Spill Procedures/ Waste Disposal Methods

- **In case of a spill, follow your teacher's instructions. Notify your teacher immediately about any spill.**
- **Put excess chemicals and solutions in the containers designated by your teacher. Do not pour any chemicals down the drain. Do not dispose of $NaHCO_3$ in the trash can.**

Materials for LYON and ABEL PRINTING

Item	Cost
REQUIRED ITEMS (You must include all of these in your budget.)	
Lab space/fume hood/utilities	15 000 – /day
Standard disposal fee	2 000 – /g of product
Balance	5 000
REAGENTS and ADDITIONAL EQUIPMENT (Include in your budget only what you'll need.)	
1.0 M HCl	500 – /mL
$NaHCO_3$	500 – /g
250 mL beaker	1 000
2 L plastic bottle	2 000
250 mL flask	1 000
100 mL graduated cylinder	1 000
Cardboard box	500
Duct tape	500 – /cm
Glass funnel	1 000
Glue	500
Plastic straw, bendable	500
Rubber tubing	500 – /10 cm
Ring stand/ring/buret clamp	2 000
Rubber stopper, one hole	1 000 – /cm
Scissors	500
Spatula	500
Tea candle	500
Wash bottle	500
Weighing paper	500 – /piece
* No refunds on returned chemicals or unused equipment.	
FINES	
OSHA safety violation	2 000/incident

Materials
(for each lab group)
- 1.0 M HCl, 250 mL
- $NaHCO_3$, 20 g
- 100 mL graduated cylinder
- 2 L plastic soda bottles, 2
- Balance
- Bendable plastic straws, 20
- Tea candles, 8
- Duct tape, 10 cm
- Glue
- Rubber stoppers, one-hole, 2
- Scissors
- Shoe box with lid (for model)

Estimated cost for materials: $224 000

Pre-lab Discussion

To help students, you may provide some of the following leading questions for them to consider as they make their plans.
- What volume of CO_2 will fill the model?
- How much of each reactant is necessary to make enough CO_2?
- What is the simplest way to deliver the CO_2 from the reaction vessel to the rooms of the model?
- What conversion factors relate the amount of CO_2 in the model to the amount of CO_2 in the building?

Tips for Evaluating the Pre-lab Requirements

If students plan to use more than 300 mL of 1.0 M HCl, have them reconsider their procedures. Students often overdesign their delivery systems and take too long building them. Simple solutions are often the most successful. The open doors in the model allow the CO_2 to move from room to room if it is simply directed towards the middle of the model.

Proposed Procedure

Put 250 mL of 1.0 M HCl and 20 g of $NaHCO_3$ in a 2 L plastic bottle with a large straw coming out of the top. Direct the straw towards the middle of the model. Note which fires are extinguished.

Post-Lab

Disposal

Provide three labeled disposal containers for $NaHCO_3$, HCl, and the reaction mixture. Combine the HCl waste and reaction-mixture waste into a single container, pouring slowly while stirring. Dissolve $NaHCO_3$ waste in the combined liquid slowly with stirring. Neutralize with 1.0 M acid or base until the pH is between 5 and 9, and pour down the drain.

EXPLORATION 11A

Objectives

Students will:

- use appropriate lab safety procedures.
- prepare a stock solution of a specified concentration.
- measure mass using a laboratory balance.
- measure volume using volumetric flasks and graduated cylinders.
- perform dilutions to create a series of solutions of different molar concentrations.
- make colorimetric observations or measurements and relate them to molar concentration.
- plot a standard curve for the absorbance.
- determine the molar concentration of an unknown solution based on absorbance measurements and the standard curve.

Planning

Recommended Time:
1 lab period

Materials
(for each lab group)

- Distilled water
- $FeCl_3 \cdot 6H_2O$, 35 g
- 1.0 M HCl, 25 mL
- Unknown solution (0.25 M $FeCl_3$), 10.0 mL
- 10 mL graduated cylinder
- 250 mL beaker
- 250 mL flask
- Glass stirring rod
- Test tubes, 7
- Test tube rack

Optional

- LEAP System with colormeter
- Spectrophotometer
- Cuvettes
- Lint-free wipes for cuvettes

Colorimetry and Molarity

Situation

You are working in the quality control department of a pharmaceutical company. One of the company's products is a test solution for phenylketonuria. The test solution should contain iron(III) chloride, $FeCl_3$, at a concentration of 0.30 M. Lately there have been some problems in the production line in the factory. It is important that you analyze the solution to make sure it is of the proper concentration. If it is too dilute, the color change might not be noticeable enough. If it is too strong, the production process is wasting money by putting too much $FeCl_3$ in the solution.

Background

Doctors can use the $FeCl_3$ test solution to detect phenylketonuria in infants. People who have this disease are unable to break down the amino acid phenylalanine. If they eat foods containing too much phenylalanine, the toxic byproducts made when it is not completely broken down can make them sick. In the screening test, a few drops of the solution are sprinkled on a baby's wet diaper. If the solution turns a deep bluish green color, phenylpyruvic acid, a product of incomplete phenylketonuria metabolism, is present. Then a special diet with very small amounts of phenylalanine is prescribed for the child.

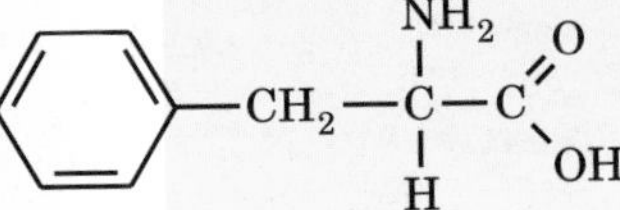

Phenylalanine

Problem

Phenyl-alanine

Like many solutions, $FeCl_3$ is colored, and in general, the more concentrated the solution, the darker its color will be. This is the basis for colorimetry, in which the color of a solution is used as a measure of its concentration. According to a relationship known as Beer's law, the amount of light of a specific wavelength that the solution absorbs, its absorbance, is proportional to its concentration. In some cases, there are slight deviations from Beer's law, so the graph of the relationship is a curve instead of a straight line. To determine the concentration of the sample pulled from the factory's production line, you must do the following.

- make several standard solutions of known concentration
- if special equipment such as a LEAP colormeter or a spectrophotometer are available, make a graph of absorbance vs. concentration
- compare the solutions of known concentration to the unknown to determine the concentration of the unknown
- compare this extrapolation to the expected value, 0.30 M

Objectives

Demonstrate **proficiency in preparing a solution and performing colorimetric measurements or observations.**

Relate **colorimetric measurements or observations to concentration.**

Determine **the molarity of a solution of unknown concentration.**

Required Precautions

- Goggles and a lab apron must be worn at all times.
- Read all safety cautions, and discuss them with your students.
- Students should not handle concentrated acid solutions.
- If the LEAP System with colormeter or a spectrophotometer are used, the precautions listed in the teacher's notes on page 652 must be followed to avoid electric shock.

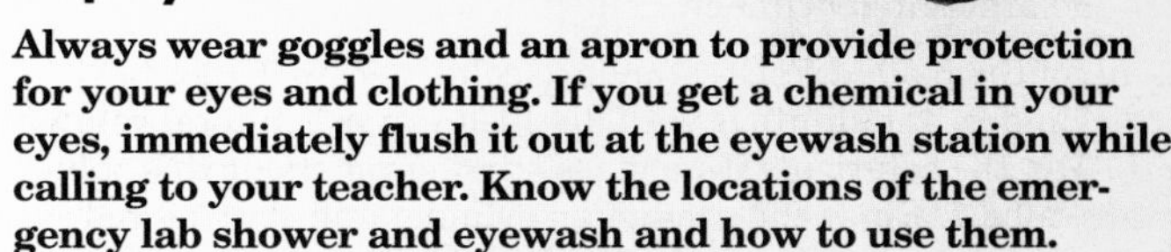

Safety

Always wear goggles and an apron to provide protection for your eyes and clothing. If you get a chemical in your eyes, immediately flush it out at the eyewash station while calling to your teacher. Know the locations of the emergency lab shower and eyewash and how to use them.

Do not touch any chemicals. If you get a chemical on your skin or clothing, wash it off at the sink while calling to your teacher. Make sure you carefully read the labels and follow the directions on all containers of chemicals that you use. Do not taste any chemicals or items used in the laboratory. Never return leftovers to their original containers; take only small amounts to avoid wasting supplies.

Always clean up the lab and all equipment after use, and dispose of substances according to proper disposal methods. Wash your hands thoroughly before you leave the lab after all lab work is finished.

Preparation

1. ***Organizing Data***
   Prepare a data table in your lab notebook. It should have eight columns and five rows if you use the LEAP colormeter or a spectrophotometer, and four rows if this equipment will not be used. In the first row, label the boxes in the second through eighth columns *Test tube 0*, *Test tube 1*, *Test tube 2*, *Test tube 3*, *Test tube 4*, *Test tube 5*, and *Unknown*. In the first column, label the second box *mL 0.50 M* $FeCl_3$, the third box *mL* H_2O, and the fourth box *Estimates*. If you are using a colormeter or a spectrophotometer, label the fifth box in the first column *Measurements*. You will also need space for calculations.
2. Label seven test tubes *0*, *1*, *2*, *3*, *4*, *5*, and *Unknown*. Label the beaker *Waste*.
3. If you will be using a spectrophotometer, turn it on now, because it must warm up for approximately 10 min.
4. Using a periodic table, determine the molar mass of $FeCl_3{\cdot}6H_2O$. Record it in your lab notebook.
5. Perform the necessary calculations to determine how many grams of $FeCl_3{\cdot}6H_2O$ would be needed to make 250 mL of a 0.50 M solution. Record this amount in your lab notebook.

Technique

Solution Preparation

6. Using the appropriate technique described in Chapter 11 and the amount calculated in step **5**, prepare 250 mL of a 0.50 M standard solution using $FeCl_3{\cdot}6H_2O$. Dilute the solution to 250 mL with 25 mL of 1.0 M HCl and more distilled water. Be sure to measure to the nearest 0.01 g. Record the amount used in your lab notebook.

Materials

- **Distilled water**
- **1.0 M HCl**
- **$FeCl_3{\cdot}6H_2O$ crystals**
- **10 mL graduated cylinder**
- **250 mL beaker**
- **250 mL volumetric flask**
- **Glass stirring rod**
- **Test tubes, 7**
- **Test tube rack**
- **Unknown solution**

Optional equipment

- **LEAP System with colormeter**
- **Spectrophotometer**
- **Cuvettes**
- **Lint-free wipes for cuvettes**

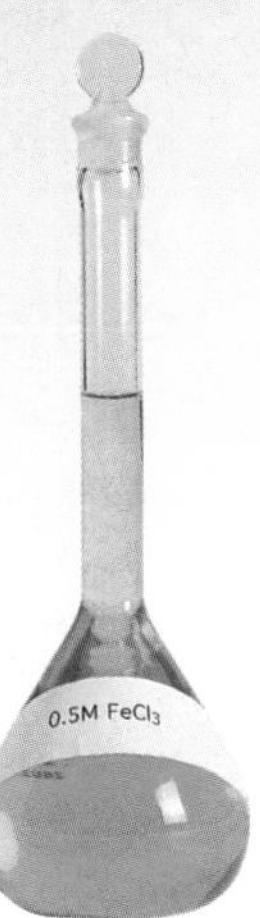

Solution/Material Preparation

1. To prepare 1.0 M HCl, observe the required precautions. Add 83 mL of concentrated HCl to enough distilled water to make 1.00 L of solution. Add the acid slowly, and stop to stir it in order to avoid overheating.
2. To prepare the unknown solution (0.25 M $FeCl_3$), add 67.58 g $FeCl_3{\cdot}6H_2O$ to 500 mL of distilled water. Add a few drops of 1.0 M HCl until cloudiness disappears. Add distilled water gradually, stopping to add HCl at the sign of any cloudiness, until the solution has been diluted to 1.00 L of solution.
3. For improved results, students can use 10 mL volumetric pipets instead of graduated cylinders in the dilution steps. Make certain to use pipets with bulbs. Never pipet with your mouth.

- Wear goggles, a face shield, impermeable gloves, and a lab apron when you prepare the HCl. Work in a hood known to be in operating condition, with another person present nearby to call for help in case of an emergency. Be sure you are within 30 s walking distance of a safety shower and eyewash station known to be in operating condition.
- In case of an acid spill, dilute first with water. Then mop up the spill with wet cloths or a cloth mop designated for spill cleanup while wearing disposable plastic gloves.

Student Orientation

Pre-lab Discussion
Discuss the concept of molarity, with an emphasis on how to calculate the concentration of a solution that is made through dilution. Relate the use of spectrophotometry and colorimetry to the discussion of atomic structure and electron excitation in Chapter 3. Be certain students understand how to make and use a standard curve.

Techniques to Demonstrate
Show students how to operate the LEAP colormeter or the spectrophotometer, if used. Emphasize the need to check the calibration of the instrument repeatedly and to wipe the cuvettes with lint-free wipes. Students should prepare all of their dilutions before using the instrument.

Make sure students add the proper amount of $FeCl_3 \cdot 6H_2O$, 33.79 g, to make their 0.5 M stock solution. If students try other amounts, it could be a sign that they have forgotten to factor in the mass of water in the hydrated compound.

7. Place the test tubes in a test tube rack. Measure 10.0 mL of distilled water into the graduated cylinder. Pour it into test tube *0*.
8. Pour 2.0 mL of the 0.50 M $FeCl_3$ standard solution into a 10 mL graduated cylinder, and dilute it with distilled water to a total volume of 10.0 mL. Mix the solution with a glass stirring rod. Pour this solution into test tube *1*. Rinse the graduated cylinder and stirring rod with distilled water, and discard the rinse in the *Waste* beaker. Record the amounts of the solution and H_2O in your data table.
9. Pour 4.0 mL of the 0.50 M $FeCl_3$ standard solution into a 10 mL graduated cylinder, and dilute it with distilled water to a total volume of 10.0 mL. Mix the solution with a glass stirring rod. Pour this solution into test tube *2*. Rinse the graduated cylinder and stirring rod with distilled water, and discard the rinse in the *Waste* beaker. Record the amounts of the solution and H_2O in your data table.
10. Pour 6.0 mL of the 0.50 M $FeCl_3$ standard solution into a 10 mL graduated cylinder, and dilute it with distilled water to a total volume of 10.0 mL. Mix the solution with a glass stirring rod. Pour this solution into test tube *3*. Rinse the graduated cylinder and stirring rod with distilled water, and discard the rinse in the *Waste* beaker. Record the amounts of the solution and H_2O in your data table.
11. Pour 8.0 mL of the 0.50 M $FeCl_3$ standard solution into a 10 mL graduated cylinder, and dilute it with distilled water to a total volume of 10.0 mL. Mix the solution with a glass stirring rod. Pour this solution into test tube *4*. Rinse the graduated cylinder and stirring rod with distilled water, and discard the rinse in the *Waste* beaker. Record the amounts of the solution and H_2O in your data table.
12. Pour 10.0 mL of undiluted 0.50 M $FeCl_3$ standard solution into test tube *5*.
13. Obtain a 10.0 mL sample of the solution of unknown concentration, and pour it into the test tube labeled *Unknown*.

Colorimetric Estimation and Measurement

14. Estimate the intensity of the color of your solutions on a scale from 0 (distilled water from test tube *0*) to 1.00 (solution from test tube *5*). To make comparisons easier, hold a piece of white paper behind the test tubes you are comparing. Record these values in your data table as *Estimates*. Based on how the unknown compares to the known solutions, estimate its concentration. If you will be using neither the LEAP colormeter nor a spectrophotometer, continue with step **19**.

15. Check with your teacher about whether you should evaluate your sample using measures of absorbancy or percent transmittance. If you are using a LEAP colormeter and measuring absorbance, set it up and calibrate it as discussed in step **15a**. If you will be using the LEAP colormeter to measure transmittance, follow the instructions in step **15b**. If you are using a spectrophotometer, set it up and calibrate it as discussed in step **15c**.

Sample Data

	Test tube 0	Test tube 1	Test tube 2	Test tube 3	Test tube 4	Test tube 4	Unknown
mL 0.5 M $FeCl_3$	0	2.0	4.0	6.0	8.0	0	n/a
mL H_2O	10.0	8.0	6.0	4.0	2.0	10.0	n/a
Estimates	0	0.2	0.4	0.6	0.8	1.0	0.6

a. *LEAP Colormeter—Absorbance*
Your teacher will have the experiment on the screen. Plug the colormeter into the input cable of your station. Set the wavelength to 590 nm, which corresponds to a yellowish green light.
Never wipe cuvettes with paper towels or scrub them with a test-tube brush. Use only lint-free tissues which will not scratch the cuvette's surface. The outside of the cuvette must be completely dry before it is placed inside an instrument, or you will get an invalid reading.
Pour some of the solution from test tube *5* into the cuvette. Be sure the outside of the cuvette is dry by wiping it clean with a lint-free tissue. Then, place the cuvette into the meter and place the cover over the sample. Adjust the ZERO ADJ knob until the reading on the LEAP graph is 1. Allow the meter a few seconds to stabilize, and readjust the ZERO ADJ knob if necessary. When the reading remains at 1, put the system on WAIT, and proceed with step **16**.

b. *LEAP Colormeter—Percent Transmittance*
Your teacher will have the experiment on the screen. Plug the colormeter into the input cable of your station. Set the wavelength to 590 nm, which corresponds to a yellowish green light.
Never wipe cuvettes with paper towels or scrub them with a test-tube brush. Use only lint-free tissues which will not scratch the cuvette's surface. The outside of the cuvette must be completely dry before it is placed inside an instrument, or you will get an invalid reading.
Fill the cuvette three-quarters full with distilled water from test tube *0*. Be sure the outside of the cuvette is dry by wiping it clean with a lint-free tissue. Because this should provide a reading of no absorbance or 100% transmittance, it is called a blank. Place this cuvette into the meter, and place the cover over the sample. Start the data collection by clicking on the green light of the traffic light on the computer screen. Adjust the ZERO ADJ on the LEAP graph on the computer screen so that the sample reads 100%. Allow the meter a few seconds to stabilize, and readjust the ZERO ADJ knob if necessary. When the reading remains at 100%, put the system on WAIT, and proceed with step **16**.

c. *Spectrophotometer—Percent Transmittance or Absorbance*
Check to be certain that the spectrophotometer has warmed up for about 10 min. Then, with the sample compartment empty and the lid closed, adjust the %T dial to 0%. Turn the right knob clockwise until you meet resistance. Insert a clean and dry cuvette. Turn the wave length knob to 625 nm (yellow light), and then turn the right front knob until the %T dial reads 100%.
Never wipe cuvettes with paper towels, or scrub them with a test-tube brush. Use only lint-free tissues which will not scratch the cuvette's surface. The outside of the cuvette must be completely dry before it is placed inside an instrument, or you will get an invalid reading.
Check with your teacher about whether you should evaluate your sample recording measures of absorbance or percent transmittance. Proceed with step **16**.

Post-Lab

Disposal
Set out a disposal container for the solutions. When all of the solutions and rinses have been added to the container, add an excess of 1.0 M NaOH to precipitate the iron as insoluble $Fe(OH)_3$. Filter and put the precipitate in the trash. Neutralize the filtrate with 1.0 M acid until its pH is between 5 and 9, and pour it down the drain.

	Test tube 0	Test tube 1	Test tube 2	Test tube 3	Test tube 4	Test tube 5	Unknown
Measurements (absorbance)	0.008	0.125	0.190	0.261	0.330	0.390	0.222
Measurements (transmittance)	100%	75%	65%	55%	47%	41%	60%

Answers to Analysis and Interpretation

(Note: omit item 4 unless percent transmittance was measured.)

1. Test tube 0: 0.0 M
   Test tube 1: 0.1 M
   Test tube 2: 0.2 M
   Test tube 3: 0.3 M
   Test tube 4: 0.4 M
   Test tube 5: 0.5 M
2. Estimated concentration is 0.25 M because the solution seems more concentrated than test tube 2, but less concentrated than test tube 3.
3. The results might be better because the volumes and amounts of solution would be more precise with a pipet than with a graduated cylinder.
4. The following is a representative sample calculation of absorbance for a solution with a 75% transmittance.

$$\text{Absorbance} = \log\left(\frac{100}{75}\right) = 0.125$$

5. 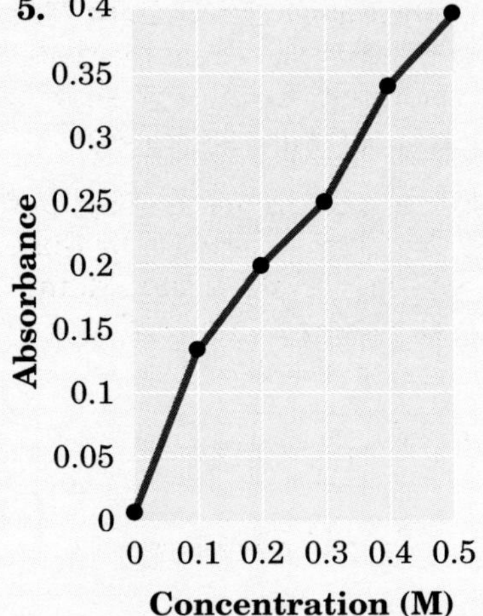

6. Answers will vary slightly. For the sample data, the graphing calculator gives the following regression equation, rounded for significant figures. absorbance = 0.74(conc.) + 0.03
7. unknown concentration =

$$\frac{0.222 - 0.03}{0.74} = 0.26 \text{ M}$$

16. Remove the cuvette, and rinse it several times with distilled water, discarding rinses in the *Waste* beaker. Fill the cuvette approximately three-quarters full with the solution from test tube *1*. Be sure the outside of the cuvette is dry by wiping it clean with a lint-free tissue. Place the cuvette in the sample compartment, and place the cover over the cuvette. If you are using the LEAP colormeter, take the system off WAIT. Allow the meter to stabilize at this reading for a few seconds, and record the percent transmittance or absorbance reading for sample *1* in your data table.
17. Remove the cuvette, and pour sample *1* back into the test tube. Rinse the cuvette several times with distilled water, discarding rinses in the *Waste* beaker. Repeat the procedure for samples from test tubes *2–5*. Be sure to dry the outside of the cuvette with a lint-free tissue after each transfer. Between samples *3* and *4* check the calibration of the meter as follows. If you are measuring percent transmittance, retest the distilled water to be certain the reading is 100%. If you are measuring absorbance, retest the solution from test tube *5* to be sure the reading is 1. If the calibration test does not give the appropriate reading, start over with step **15**. Otherwise, continue with step **18**.
18. Test the sample of unknown concentration, and record the results in your lab notebook.

Cleanup and Disposal

19. Dispose of the solutions in the container designated by your teacher. Rinse the cuvettes several times with distilled water before putting them away.

Analysis and Interpretation

1. ***Analyzing Ideas***
   What are the concentrations of $FeCl_3$ in test tubes *0*, *1*, *2*, *3*, *4*, and *5*?
2. ***Analyzing Results***
   What is your estimate for the concentration of the unknown? Explain your answer.
3. ***Analyzing Methods***
   Would you get better results if, during the dilution steps, you had used a volumetric pipette that measures exactly 2.00 mL instead of using the graduated cylinder? Explain your answer.
4. ***Organizing Data***
   If you used a colormeter or spectrophotometer and measured percent transmittance instead of absorbance, convert to absorbance using the following equation. If not, go on to item **5**.

$$\text{absorbance} = \log\left(\frac{100}{\text{percent transmittance}}\right)$$

5. ***Analyzing Data***
   Make a graph of your data with concentration of $FeCl_3$ on the x-axis and absorbance on the y-axis. (If you did not use equipment to make your measurements, use your values from the *Estimate* part of your data table as absorbance values.)

6. ***Analyzing Data***
If your graph was a straight line, give the equation for the line in the form $y = mx + b$. If it was not a straight line, explain why, and draw the straight line that comes closest to including all of your data points. Give the equation of this line. (Hint: if you have a graphics calculator, use the **STAT** mode to enter your data and make a linear regression equation using the LinReg function from the STAT menu.)

7. ***Interpreting Graphics***
Using the graph from item **5** or the equation from item **6**, determine what the concentration of the unknown must have been for it to have given the measured absorbance value.

Conclusions

8. ***Applying Conclusions***
Which of the following could be a likely source of the pharmaceutical plant's problems: too much $FeCl_3$ added to the solution, too little $FeCl_3$ added to the solution, too much water added to the solution, or too little water added to the solution?

Extensions

1. ***Evaluating Methods***
What are some advantages or disadvantages of colorimetry when compared to other methods, such as gravimetric analysis?

2. ***Relating Ideas***
Absorbance describes how much light is blocked by a sample. Transmittance describes how much light passes through a sample. As absorbance values go up, how do transmittance values change?

3. ***Designing Experiments***
What possible sources of error can you identify with this procedure? If you can think of ways to eliminate them, ask your teacher to approve your plan, and run the procedure again.

4. ***Predicting Outcomes***
How would your absorbance or percent transmittance values for the unknown solution change if someone added an additional yellow-colored compound along with the $FeCl_3$?

5. ***Applying Ideas***
Run the procedure again using one of the following wavelength settings: 950 nm or 640 nm. Can you explain why 590 nm was chosen for the lab procedure?

6. ***Evaluating Methods***
If you used a LEAP colormeter or a spectrophotometer, calculate the percent error for each of your estimates of absorbance, compared to the values given by the equipment. (Hint: divide each absorbance measurement by the largest measurement, so they will be on a scale from 0 to 1.00, just like your estimates.)

Conclusions

8. The problem is that the solution is too dilute. This could be due to too little $FeCl_3$ or too much water.

Extensions

(Note: omit items **5** and **6** if spectrophotometers and colormeters are unavailable.)

1. There is less room for human error in colorimetry because volumes and masses aren't measured, and a precipitate does not have to be purified. Colorimetry also leaves the sample intact.

2. Transmittance goes down as more light is absorbed.

3. Students' suggestions for improving the procedure will vary. Be sure answers are safe and include carefully planned procedures.

4. If an additional yellow compound was added, it would absorb even more light. The absorbance would increase, and the percent transmittance would decrease.

5. The yellowish green light of 590 nm is absorbed by the yellow solution more than the lower-energy red or infrared light.

6. $\frac{0.320 - 0.2}{0.320} = 38\%$ error

$\frac{0.487 - 0.4}{0.487} = 18\%$ error

$\frac{0.669 - 0.6}{0.669} = 10\%$ error

$\frac{0.846 - 0.8}{0.846} = 5\%$ error

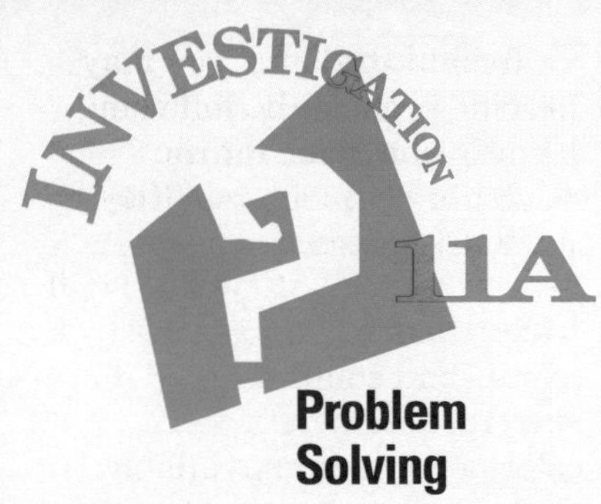

Objectives

Students will:

- use appropriate lab safety procedures.
- prepare a stock solution of a specified concentration.
- measure mass using a laboratory balance.
- measure volume using volumetric flasks and graduated cylinders.
- perform dilutions to create a series of solutions of different molar concentrations.
- make colorimetric observations or measurements and relate them to molar concentration.
- plot a standard curve for the absorbance.
- determine the molar concentration of an unknown solution based on absorbance measurements and the standard curve.
- design and implement their own procedure.

Planning

Recommended Time: 1 lab period
Prerequisite: Exploration 11A
Solution/Material Preparation
To prepare the unknown solution (0.015 M $CuCl_2$), add 2.56 g $CuCl_2{\cdot}2H_2O$ to enough distilled water to make 1.00 L of solution.

Student Orientation

Techniques to Demonstrate
Be sure students remember how to properly use the spectrophotometer or LEAP System with colormeter. Emphasize the need to check the calibration of the instrument repeatedly and to wipe the cuvettes with lint-free wipes. Students should not begin measurements until they have prepared all of their solutions.

Pre-lab Discussion
Begin by discussing the results and procedure used in the Exploration, especially any errors in lab technique that occurred. Be sure students realize that even though the absorbance at different concentrations does not give a perfect straight-line graph, it can still be used to estimate the concentration of an unknown solution.

Sample Data

	H_2O M	0.01 M	0.02 M	0.03 M	0.04 M	0.05 M	Unknown
Absorbance	0.01	0.029	0.039	0.047	0.070	0.099	0.034

Analysis

From graphing calculator:

Absorbance = 1.65 (conc.) + 0.008

$$\text{Unknown conc: } \frac{0.034 - 0.008}{1.65} = 0.016 \text{ M}$$

$$\frac{0.016 \text{ mol}}{1 \text{ L}} \times \frac{134.45 \text{ g}}{1 \text{ mol}} \times \frac{1 \text{ L solution}}{1 \text{ kg}} = \frac{2.2 \text{ g}}{10^3 \text{ g}} = 2200 \text{ ppm}$$

$$\text{Amount in reservoir: } 2.5 \times 10^9 \text{ L} \times \frac{2.2 \text{ g}}{1 \text{ L solution}} = 5.5 \times 10^9 \text{ g}$$

COLORIMETRY Reservoir Contaminant

has got a new idol PAGE B6 | police cuts PAGE B8

March 24, 1995

Contamination Closes Reservoir Indefinitely

Backup Available for Only 2 days; Firms Asked to Help

Pumps feeding water from the city's James Knox Polk Reservoir to a treatment plant were shut down suddenly last night when a water department crew discovered leaking and rusty barrels of chemicals on the reservoir's south shore.

The barrels, which were dumped illegally, appeared to be leaking directly into the reservoir. Daniel Baden, director of the Department of Health Services, ordered an immediate switch to the city's reserve water supply. Although service continued without interruption, the water department's supervisor, Jose Vaculez, warns that the city has only enough water in reserves to last for two days.

"Until we find out what chemical we're dealing with and how much is there, it's difficult to know whether the wisest course would be to try to purify the water or to find a long-term alternative," Baden told the city council, which met in emergency session this morning.

The council unanimously passed a resolution sponsored by Councilmember Joanna Wooldridge to provide a $250 000 reward to the first analytical firm that can provide an answer to the water department by 4 p.m. tomorrow.

"Our options for alternative sources are so expensive that we will be saving our taxpayers' money if we can fix what's wrong with the reservoir now," Wooldridge said. "I'm certain that the expertise to reach an answer promptly is available."

Police are still seeking leads on who dumped the barrels. "We have determined that they must have been put there since last Thursday," said Jordan Freeman, a police department spokesman. "The area was clearly marked with 'No Dumping' signs, and we intend to fine the responsible parties to the maximum amount allowable by law. We are also studying other penalties associated with creating a public health hazard."

Few Details Provided in Whitfield Proposal

Required Precautions

Goggles and lab aprons must be worn at all times. Confine loose clothing and long hair.

Do not touch or taste any chemicals. Wash your hands thoroughly when finished.

Required Precautions

- Goggles and a lab apron must be worn at all times.
- Read all safety cautions, and discuss them with your students.
- If the LEAP System with colormeter or a spectrophotometer are used, the precautions listed in the teacher's notes on page 652 must be followed to avoid electric shock.

CheMystery Labs, Inc.
52 Fulton Street
Springfield, VA 22150

Memorandum

Date: March 24, 1995

To: Antonio Gallini

From: George Taylor

One of our analysis teams got a head start on this project early this morning. They managed to identify the contaminant as $CuCl_2$, so now I need your team to finish the job by determining the concentration of $CuCl_2$ in a sample of the reservoir water. I recommend that you use colorimetric analysis techniques. We have pure crystals of $CuCl_2$ available so that you can make standard solutions.

We can't afford to waste time on mistakes, so I want to look over your plan before you start. Include the following.
- one-page procedure and necessary data tables
- detailed list of equipment and materials you will need, along with individual and total costs (I'll rush this information to the supply and accounting departments.)

As soon as you complete your work, write up a two-page report to fax to the water department. All of the following items must be covered in the report.
- chemical name and formula of the spilled compound
- concentration of the spilled compound in molarity and ppm by mass
- mass of chemical spilled (Note: the reservoir has a volume of 2.5×10^9 L.)
- summary of your procedure, including the wavelength chosen for colorimetric analysis
- detailed and organized data and analysis section that includes all data tables and calculations, as well as the graph of your standard curve for colorimetry analysis

In addition, prepare a list of your final costs for the accounting department.

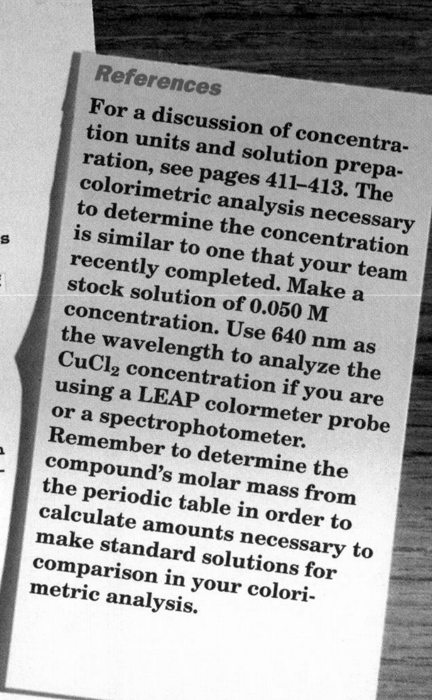

References

For a discussion of concentration units and solution preparation, see pages 411–413. The colorimetric analysis necessary to determine the concentration is similar to one that your team recently completed. Make a stock solution of 0.050 M concentration. Use 640 nm as the wavelength to analyze the $CuCl_2$ concentration if you are using a LEAP colormeter probe or a spectrophotometer. Remember to determine the compound's molar mass from the periodic table in order to calculate amounts necessary to make standard solutions for comparison in your colorimetric analysis.

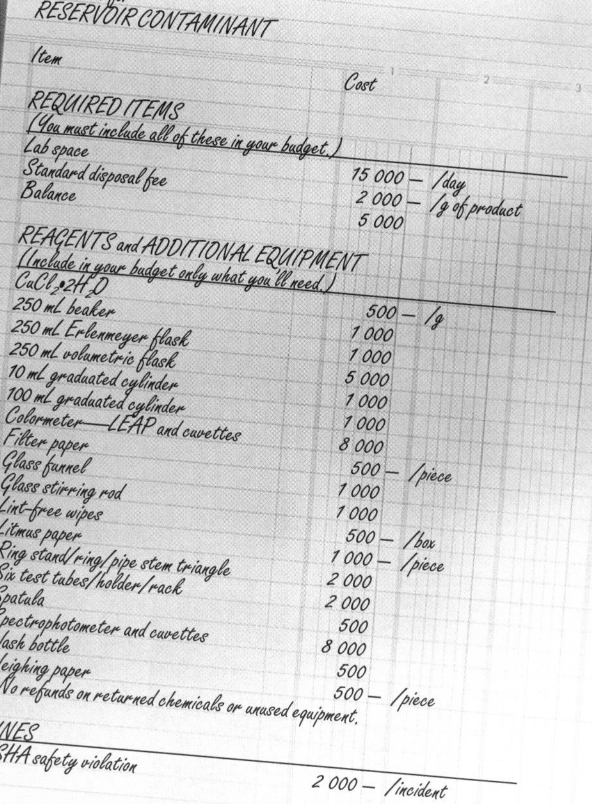

Materials for RESERVOIR CONTAMINANT

Item	Cost
REQUIRED ITEMS (You must include all of these in your budget.)	
Lab space	15 000 — /day
Standard disposal fee	2 000 — /g of product
Balance	5 000
REAGENTS and ADDITIONAL EQUIPMENT (Include in your budget only what you'll need.)	
$CuCl_2 \cdot 2H_2O$	500 — /g
250 mL beaker	1 000
250 mL Erlenmeyer flask	1 000
250 mL volumetric flask	5 000
10 mL graduated cylinder	1 000
100 mL graduated cylinder	1 000
Colormeter—LEAP and cuvettes	8 000
Filter paper	500 — /piece
Glass funnel	1 000
Glass stirring rod	1 000
Lint-free wipes	500 — /box
Litmus paper	1 000 — /piece
Ring stand/ring/pipe stem triangle	2 000
Six test tubes/holder/rack	2 000
Spatula	500
Spectrophotometer and cuvettes	8 000
Wash bottle	500
Weighing paper	500 — /piece
*No refunds on returned chemicals or unused equipment.	
FINES	
OSHA safety violation	2 000 — /incident

Spill Procedures/ Waste Disposal Methods
- In case of a spill, follow your teacher's instructions.
- Put all liquids and solids into the container designated by your teacher. Do not pour them down the sink or place them in the trash can.

Materials
(for each lab group)
- $CuCl_2 \cdot 2H_2O$, 2.5 g
- Unknown solution (0.015 M $CuCl_2$), 10 mL
- Distilled water
- 250 mL beaker
- 250 mL volumetric flask
- 10 mL graduated cylinder
- Balance
- Test-tube holder
- Test-tube rack
- Test tubes, 7
- Wash bottle
- Weighing paper

Optional Equipment
- Cuvettes
- LEAP System with colormeter
- Lint-free wipes
- Spectrophotometer

Estimated cost of materials: $202 000

To help students, you may provide some of the following leading questions for the students to consider as they make their plans.
- How much $CuCl_2 \cdot 2H_2O$ will be needed to make the stock 0.050 M solution?
- What will you need to do to make solutions of different concentrations from the stock solution?

Tips for Evaluating the Pre-lab Requirements
Make certain that proposed plans include detailed instructions on the proper technique for making solutions. If students plan to use more than 15 mL of the unknown solution, have them reconsider their plans. They should only need 2.5 g of $CuCl_2 \cdot 2H_2O$ unless they are using volumetric flasks that are larger than 250 mL.

Proposed Procedure

Place 2.13 g of $CuCl_2 \cdot 2H_2O$ in a 250 mL beaker. Dissolve it in 100 mL of distilled water. Empty the solution into the volumetric flask. Dilute to the 250 mL mark. This is a 0.050 M solution. Prepare dilutions as in the Exploration. Measure and record the absorbance at 640 nm of all solutions, of a blank with distilled water, and of the unknown solution.

Post-Lab

Disposal
Set out one disposal container for all solids and liquids. Add 1.0 M NaOH to precipitate the copper as insoluble $Cu(OH)_2$. Filter the mixture, placing the precipitate in the trash. Then neutralize the filtrate with 1.0 M acid until the pH is between 5 and 9, and pour it down the drain.

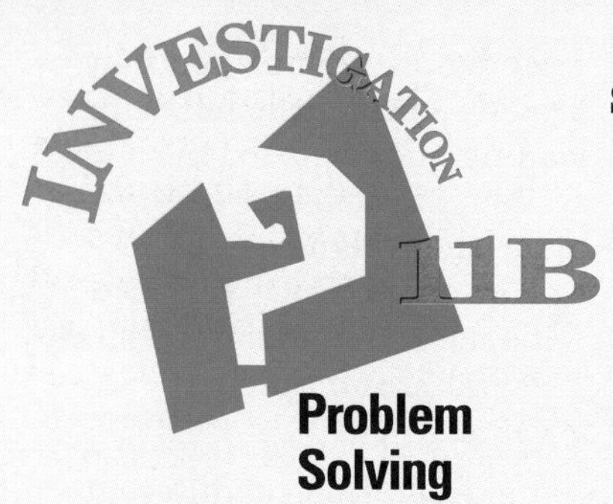

Objectives

Students will:

- use appropriate lab safety procedures.
- test several soaps for cleaning ability, amount of suds produced, and behavior in hard water.
- measure pH of several soaps.
- measure volume and mass of soaps.
- calculate density of soaps.
- observe effect of soaps on surface tension of water.
- design and implement their own procedure.

Planning

Recommended Time: 1 lab period
Prerequisite: None

Solution/Material Preparation

1. Use any commercially available soaps for this lab. If time is an important factor, or if three different soaps of each type are unavailable, use fewer.
2. To prepare the 0.5 M $MgCl_2$, add 101.6 g $MgCl_2 \cdot 6H_2O$ to enough distilled water to make 1.00 L of solution.
3. To prepare the 0.5 M $CaCl_2$, add 55.5 g of anhydrous $CaCl_2$ to enough distilled water to make 1.00 L of solution.

Student Orientation

Techniques to Demonstrate

Show students the proper procedure for shaking a test tube with a stopper in it.

Pre-lab Discussion

To help students, you may provide some of the following leading questions for them to consider as they make their plans.

- What measurements do you need to make to determine the density of the soaps?
- What equipment will you need to measure pH?
- What can you do to make sure that each soap's test of cleaning ability is comparable to the others?
- How can you quantify your measurements of cleaning ability and sudsing ability?

Sample Data

Soap	Cleaning ability	pH	Density	Suds	$MgCl_2$ and/ or $CaCl_2$	Other Observations
Hand soap F	worst	5.5	1.08 g/mL	moderate	most	pink liquid, translucent flower scent
Dish soap Q	best	5.0	1.04 g/mL	most	white ppt	yellow liquid, transparent, lemon scent
Detergent M	moderate	10.1	1.11 g/mL	least	least	green liquid, transparent but dark, perfume scent

In each of the surface tension tests, the pepper moved away from the area near where the soap was added.

EMULSIFYING AGENTS — Testing Soaps

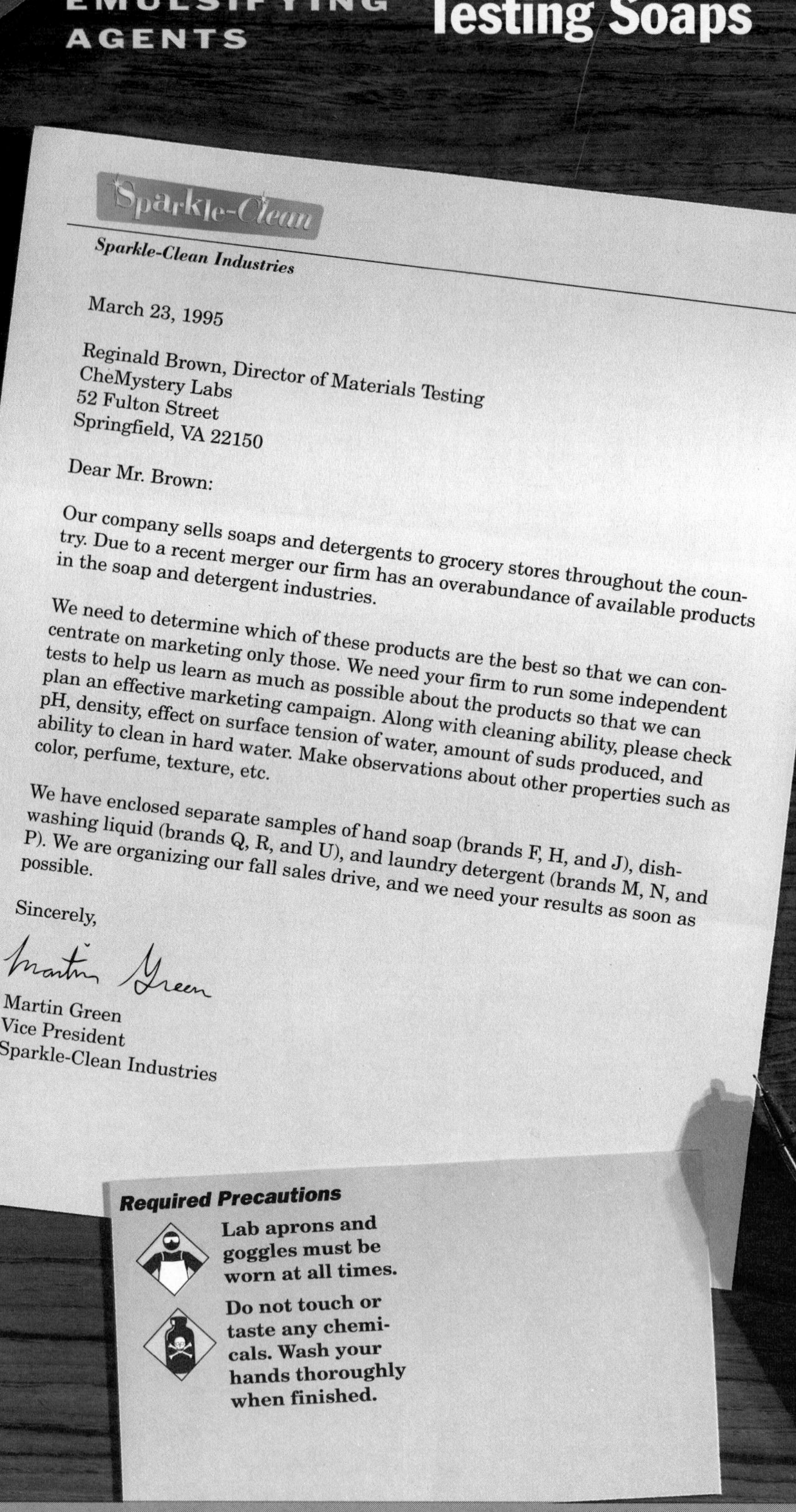

Sparkle-Clean
Sparkle-Clean Industries

March 23, 1995

Reginald Brown, Director of Materials Testing
CheMystery Labs
52 Fulton Street
Springfield, VA 22150

Dear Mr. Brown:

Our company sells soaps and detergents to grocery stores throughout the country. Due to a recent merger our firm has an overabundance of available products in the soap and detergent industries.

We need to determine which of these products are the best so that we can concentrate on marketing only those. We need your firm to run some independent tests to help us learn as much as possible about the products so that we can plan an effective marketing campaign. Along with cleaning ability, please check pH, density, effect on surface tension of water, amount of suds produced, and ability to clean in hard water. Make observations about other properties such as color, perfume, texture, etc.

We have enclosed separate samples of hand soap (brands F, H, and J), dishwashing liquid (brands Q, R, and U), and laundry detergent (brands M, N, and P). We are organizing our fall sales drive, and we need your results as soon as possible.

Sincerely,

Martin Green
Vice President
Sparkle-Clean Industries

Required Precautions

Lab aprons and goggles must be worn at all times.

Do not touch or taste any chemicals. Wash your hands thoroughly when finished.

Required Precautions

- Goggles and a lab apron must be worn at all times.
- Read all safety cautions, and discuss them with your students.
- Even though the chemicals in this lab are commercially available soaps, it is still important for students to keep the chemicals in this lab off their skin and to keep their hands and fingers away from their faces and mouths at all times.
- If the LEAP System with pH probe or a pH meter are used, the precautions listed in the teacher's notes on page 652 must be followed to avoid electric shock.

Memorandum

CheMystery Labs, Inc.
52 Fulton Street
Springfield, VA 22150

Date: March 24, 1995

To: Sabrina Moore

From: Reginald Brown

There are many tests involved with this work, and we have only $350,000 to spend, but I have some ideas. One of our references suggests that a soap's effect on surface tension can be tested by filling a petri dish about half-full with water, sprinkling pepper on it, and adding a drop of soap solution to the center. The reference didn't explain why this test works, so make careful observations.

To test sudsing ability, put similar amounts of soap detergent in test tubes with water. Stopper the tubes and shake. We can simulate the effects of hard water by performing this test with $MgCl_2$ and $CaCl_2$ solutions instead of water.

You're on your own in figuring out ways to test cleaning ability, pH, and density. Be creative, but come up with tests that are quantifiable. I recommend that you bring in some items from your home to test cleaning ability. Then rate soaps on a scale of 1 to 10.

Before you start, I will need to review the following items.
- detailed one-page summary of your plan for the procedure
- itemized list of materials, with total costs
- data tables for recording the results of your tests

After you complete the analysis, prepare a two-page report for Mr. Green that includes the following.
- your recommendation and an explanation of the reasons for your choice of the best product in each category
- summary of your procedure
- analysis, including the properties, advantages, and disadvantages of each soap or detergent
- detailed data sections, showing results of each test and observations you made
- detailed invoice for all expenses and services

References

Refer to pages 429–431 for information about soaps and detergents. It is very important to thoroughly wash and rinse glassware with distilled water between tests so that you can be sure your test results are accurate and that there is no contamination of your samples. Base your disposal costs on the amounts of $CaCl_2$ and $MgCl_2$ solution used.

Spill Procedures/ Waste Disposal Methods

- Pour any leftover soap into the disposal containers designated by your teacher.
- Clean the area and all equipment after use.

Materials for SPARKLE-CLEAN INDUSTRIES

Item	Cost
REQUIRED ITEMS (You must include all of these in your budget.)	
Lab space/fume hood/utilities	15 000 — /day
Standard disposal fee	2 000 — /g of product
REAGENTS and ADDITIONAL EQUIPMENT (Include in your budget only what you'll need.)	
$CaCl_2$ solution	500 — /mL
$MgCl_2$ solution	500 — /mL
Pepper	2 000 — /g
250 mL beaker	1 000
10 mL graduated cylinder	1 000
100 mL graduated cylinder	1 000
Balance	5 000
Filter paper	500 — /piece
Glass funnel	1 000
Glass stirring rod	1 000
Paper clips	500 — /box
Petri dish	1 000
pH meter	3 000
pH paper	2 000 — /piece
pH probe—LEAP	5 000
Pipet or medicine dropper	1 000
Rubber stopper	1 000
Six test tubes/holder/rack	2 000
Spatula	500
Wash bottle	500
*No refunds on returned chemicals or unused equipment.	
FINES	
OSHA safety violation	2 000 — /incident

Materials

(for each lab group, testing 3 brands of 3 soap types)

- 0.5 M $CaCl_2$, 50 mL
- 0.5 M $MgCl_2$, 50 mL
- 15 mL soap samples, 9
- Distilled water
- Pepper
- 250 mL beakers, 9
- 10 mL graduated cylinder
- Balance
- Petri dishes, 1
- Pipet or medicine dropper
- Test-tube holder
- Test-tube rack
- Test tubes, 9
- Wash bottle

Optional Equipment

- LEAP System with pH probe
- pH meter
- pH paper

Estimated cost of materials:
$295 000–$300 000 (depending on pH method)

Tips for Evaluating the Pre-lab Requirements

Be sure students have created a standard test of cleaning ability that is the same for all soaps. One possibility is to find several pieces of the same kind of fabric, smear the same amount of some substance on them, and soak them in the different types of soapy water for the same amount of time.

Be sure students come up with ways to quantify their data. One method is to rank the soaps from best to worst in categories such as sudsing ability or cleaning ability. Another approach is to rate them on scales from 1 to 10. The height of the suds produced when test tubes are shaken can also be recorded.

Proposed Procedure

Perform the following tests for each soap.

Measure the mass of a graduated cylinder. Pour some soap into it. Measure the volume and the mass of the graduated cylinder with soap. Calculate density.

Measure the pH with pH paper, a pH meter, or the LEAP System with pH probe.

Sprinkle pepper on a petri dish filled with water. Add a drop of soap to the water, and observe carefully.

Pour 1 mL of soap in a test tube with 5 mL of distilled water. Stopper and shake. Record the amount of suds present. Wash and rinse the test tube. Repeat the process with $MgCl_2$ and $CaCl_2$ solutions.

Place a piece of dirty fabric in a beaker with 10 mL of soap and 50 mL of water. After 5 min, remove the fabric, rinse it with 200 mL of water, and allow it to dry. Compare how clean it is to samples treated with other soaps.

Post-Lab

Disposal

Set out a single disposal container. All soap solutions may be poured down the drain.

EXPLORATION 12A

Objectives

Students will:

- use appropriate lab safety procedures.
- measure mass with a laboratory balance.
- measure volume with a graduated cylinder.
- prepare several solutions.
- measure the boiling point of water without a solute and with three non-volatile solutes (NaCl, sucrose, and $MgSO_4$).
- calculate boiling-point elevation for the solutes.
- calculate molal concentration of the solutions.
- identify an unknown using boiling-point-elevation data.
- calculate the molecular mass of three solutes using boiling-point-elevation data.

Planning

Recommended Time: 1 lab period

Materials
(for each lab group)

- Distilled water
- $MgSO_4 \cdot 7H_2O$, 10.0 g
- NaCl, 10.0 g
- Sucrose, 10.0 g
- Unknown, 10.0 g
- 125 mL Erlenmeyer flask
- Balance
- Beaker tongs
- Bunsen burner, gas tubing, striker
- Celsius thermometer, nonmercury type, with 0.1°C markings
- Glass stirring rod
- Ring stand and ring
- Thermometer clamp
- Wire gauze with ceramic center

Optional Equipment

- LEAP System with thermistor probe

Solution/Material Preparation

1. Give different groups different unknowns so that some have NaCl, some have sucrose, and others have $MgSO_4$.

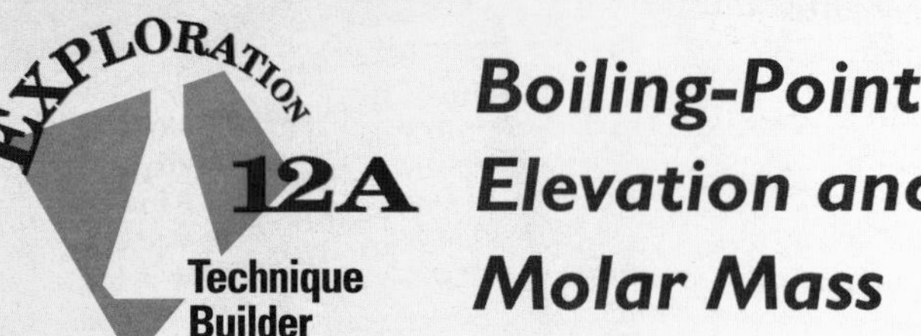

Boiling-Point Elevation and Molar Mass

Situation

Your company needs to determine the identity of a white powder. A young boy swallowed it, and doctors need to know its identity to treat him. The substance must be one of the following: sodium chloride, NaCl; sugar (sucrose), $C_{12}H_{22}O_{11}$; or hydrated magnesium sulfate (Epsom salts), $MgSO_4 \cdot 7H_2O$. Along with the unknown, you have pure samples of the three substances.

Background

The boiling point of a solution is always higher than that of a pure liquid. The reason is that the attraction of the solute for individual water molecules hinders their ability to move into the gaseous state. Properties of solutes that affect solutions such as this are known as *colligative properties.* When dealing with colligative properties, the concentration units of *molality* (m) are used instead of *molarity* (M). Molality is defined as the number of moles of solute per kilogram of the solvent.

$$m = \frac{\text{moles solute}}{\text{kilograms solvent}}$$

Experiments have determined that the boiling-point elevation of a 1.00 m solution for any molecular solute in water is 0.51°C. This value is known as the molal boiling-point constant, K_b. The constant has the units °C/m, and it can be used to predict the boiling point of any concentration of solute. For example, if 2.00 mol of sugar are dissolved in 1.00 kg of water, the boiling point will be 101.02°C (100°C + (2.00 $m \times$ 0.51°C/m).)

These properties depend upon the number of particles in solution. For example, when one mole of NaCl is dissolved in a kilogram of water, two moles of particles are formed: one mole of Na^+ ions and one mole of Cl^- ions. Thus, a 1.00 m NaCl solution will have a particle concentration of 2.00 m, and will have a boiling point like that of the 2.00 m sugar solution.

Problem

To identify the unknown, you must do the following.

- make solutions using the same solute mass for each of the standards and the unknown
- compare the boiling points of the resulting solutions to determine which matches that of the unknown

As a check on your work, you can calculate the molality of each solution based on the value of the boiling point. Then you can use the amount of solute you added and the number of moles (from the molality) to determine the molar mass.

$$\text{molar mass of solute} = \frac{\text{g of solute}}{\text{kg of solvent} \times \text{molal conc.}}$$

Objectives

***Demonstrate* proficiency in measuring masses, temperatures, and boiling points.**

***Relate* the concentration of a solution to boiling-point elevation data.**

***Determine* the molar mass of a solute from experimental data.**

Required Precautions

- Goggles and a lab apron must be worn at all times.
- Tie back loose hair and long clothing when working in the lab.
- Read all safety cautions, and discuss them with your students.
- As always, students should wash their hands thoroughly when finished.
- If the LEAP System with thermistor probe is used, the precautions listed in the teacher's notes on page 652 must be followed to avoid electric shock.

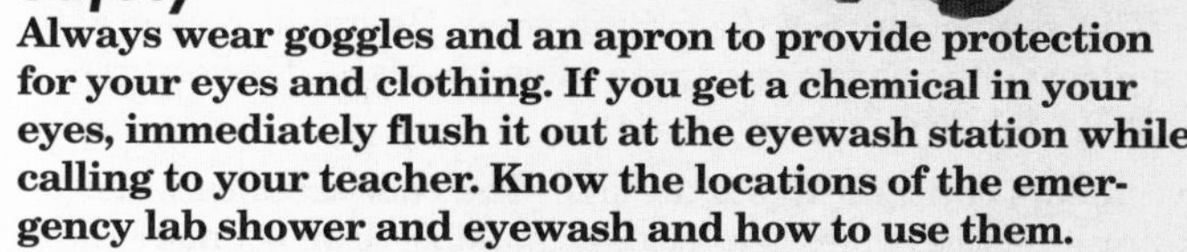

Safety

Always wear goggles and an apron to provide protection for your eyes and clothing. If you get a chemical in your eyes, immediately flush it out at the eyewash station while calling to your teacher. Know the locations of the emergency lab shower and eyewash and how to use them.

Do not touch any chemicals. If you get a chemical on your skin or clothing, wash it off at the sink while calling to your teacher. Make sure you carefully read the labels and follow the directions on all containers of chemicals that you use. Do not taste any chemicals or items used in the laboratory. Never return leftovers to their original containers; take only small amounts to avoid wasting supplies.

When you use a Bunsen burner, confine any long hair and loose clothing. Do not heat glassware that is broken, chipped, or cracked. Use tongs or a hot mitt to handle heated glassware and other equipment because hot glassware does not look hot. If your clothing catches on fire, WALK to the emergency lab shower, and use it to put out the fire.

Always clean up the lab and all equipment after use, and dispose of substances according to proper disposal methods. Wash your hands thoroughly before you leave the lab after all lab work is finished.

Materials

- **Distilled water**
- **$MgSO_4 \cdot 7H_2O$, 10.0 g**
- **NaCl, 10.0 g**
- **Sucrose, 10.0 g**
- **Unknown sample, 10.0 g**
- **Balance**
- **Beaker tongs**
- **Bunsen burner/ related equipment or hot plate**
- **125 mL Erlenmeyer flask**
- **Ring stand and ring**
- **Thermometer (nonmercury with 0.1°C markings) or LEAP System with thermistor probe**
- **Thermometer clamp**
- **Wire gauze with ceramic center**

Preparation

1. ***Organizing Data***
   Prepare a data table in your lab notebook. The data table should have four columns and six rows. The first row should contain the following headings in the first through fourth columns: *Solution, Mass of solute (g), Mass of flask + water (g)*, and *Boiling point*. In the second through sixth rows of the first column, add the following labels: *Water, NaCl, Sucrose,* $MgSO_4 \cdot 7H_2O$, and *Unknown*. Be certain there is space in your lab notebook to record the mass of the Erlenmeyer flask and perform some calculations.
2. Measure the mass of the Erlenmeyer flask. Record this mass in your lab notebook.

Technique

3. Add distilled water to the Erlenmeyer flask until the total mass is as close as possible to 50.0 g more than the amount recorded in your lab notebook for the mass of the flask. Record this mass in your data table.
4. Place the flask of water on the wire gauze on the ring stand. Using the thermometer clamp, suspend the thermistor or thermometer in the water so that it does not touch the sides or bottom of the flask. Do not use a one-hole stopper to hold the thermometer, because the solution will be boiled.

2. If you do not have access to thermometers with 0.1°C markings, increase the mass of the solutes to 15 g. (This will change the sample results and calculations.)

Student Orientation

Pre-lab Discussion

Students may initially have difficulty understanding the difference between molality and molarity and wonder why it is necessary to have different units of concentration. Point out that molarity is a measure of molar amount per liter of *solution*, while molality is a measure of molar amount per kilogram of *solvent*. Because colligative properties depend on interactions between the solute and solvent, molality is the concentration unit required.

It may help to work through the following calculations with some sample data.

Δt_b = molal concentration of particles × 0.51°C/m

Therefore, molal concentration of particles

$$= \frac{\text{mol of particles}}{\text{kg of solvent}} = \frac{\Delta t_b}{0.51°\text{C}/m}$$

Once the molal concentration has been determined, multiply by the mass of solvent to determine the number of particles. By comparing the mass of the particles to the number of particles, the molar mass can be calculated.

Techniques to Demonstrate

Demonstrate how to recognize the boiling point of a solution. Also review the use of thermometers if necessary.

Post-Lab

Disposal

Students should wash any solutions down the sink. Set out a disposal container for excess solids. Dissolve the solids in 20 times as much water, and wash them down the drain.

Answers to Analysis and Interpretation

1. Δt_b for NaCl = 3.5°C
   Δt_b for sucrose = 0.3°C
   Δt_b for $MgSO_4$ = 1.8°C
   Δt_b for unknown = 1.9°C
2. The data for the unknown most closely matches that for $MgSO_4$.

3. molality for particles, NaCl =

$$\frac{3.5°C}{0.51°C/m} = 6.9\ m$$

molality for particles, sucrose =

$$\frac{0.3°C}{0.51°C/m} = 0.6\ m$$

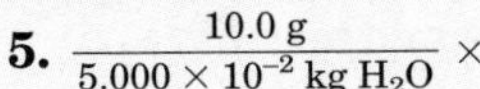

molality for particles, $MgSO_4$ =

$$\frac{0.9°C}{0.51°C/m} = 2\ m$$

molality for particles, unknown =

$$\frac{1.0°C}{0.51°C/m} = 2.0\ m$$

4. $50.0\ g \times \frac{1\ kg}{1000\ g} =$

5.00×10^{-2} kg H_2O in flask

5. $\frac{10.0\ g}{5.000 \times 10^{-2}\ kg\ H_2O} \times \frac{1\ kg\ H_2O}{6.9\ mol\ particles} \times \frac{2\ mol\ particles}{1\ mol\ NaCl} =$

58 g/mol for NaCl

$\frac{9.91\ g}{5.003 \times 10^{-2}\ kg\ H_2O} \times \frac{1\ kg\ H_2O}{0.6\ mol\ particles} \times \frac{1\ mol\ particles}{1\ mol\ sucrose} =$

300 g/mol for sucrose

$\frac{9.98\ g}{4.985 \times 10^{-2}\ kg\ H_2O} \times \frac{1\ kg\ H_2O}{2\ mol\ particles} \times \frac{2\ mol\ particles}{1\ mol\ MgSO_4} =$

200 g/mol for $MgSO_4 \cdot 7H_2O$

$\frac{10.01\ g}{4.997 \times 10^{-2}\ kg\ H_2O} \times \frac{1\ kg\ H_2O}{2.0\ mol\ particles} \times \frac{2\ mol\ particles}{1\ mol\ unknown} =$

200 g/mol for unknown

Conclusions

6. Percent error is zero, because of rounding for significant figures.

$\frac{200\ g/mol - 200\ g/mol}{200\ g/mol} \times 100 =$

0% error

7. $\frac{58.44\ g/mol - 58\ g/mol}{58.44\ g/mol} \times 100 =$

0.75% error for NaCl

$\frac{342.34\ g/mol - 300\ g/mol}{342.34\ g/mol} \times 100 =$

12% error for sucrose

$\frac{246.51\ g/mol - 200\ g/mol}{246.51\ g/mol} \times 100 =$

20% error for $MgSO_4$

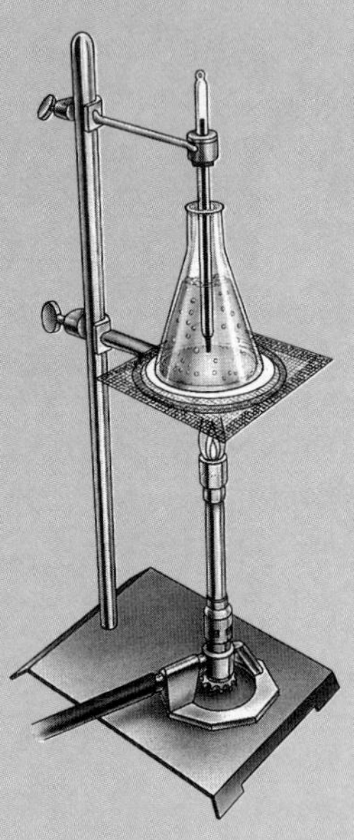

5. Heat the water until it boils vigorously and the temperature remains constant. This temperature is the boiling point. Record it in your data table. (Note: leave the box in your data table for *Mass of solute* (g) blank for this trial involving only water.)

6. Pour the contents of the flask into the sink. Then add distilled water to the flask until the total mass is as close as possible to 50.0 g more than the mass of the flask. Record this mass in your data table. Add 10.0 g of NaCl to this water, and swirl the flask gently until the salt dissolves completely. Record the mass of solute in your data table. *Do not allow the water to splash out of the flask, as it will cause inaccuracy in your results.*

7. Place your flask on the wire gauze, and suspend the thermistor or thermometer in the water as before. Heat the solution to boiling, and record the boiling point in your data table.

8. Pour the contents of the flask into the sink. Rinse it three times with distilled water. Repeat steps **6** and **7** for sucrose, $MgSO_4 \cdot 7H_2O$, and the unknown sample.

Cleanup and Disposal

9. Solutions may be rinsed down the sink. Check with your teacher for the proper disposal of other chemicals. Wash your hands thoroughly after cleaning up the area and equipment. Be certain all gas valves are turned off.

Analysis and Interpretation

1. **Organizing Data**
For the trials with solutes including the unknown, calculate the solution's change in boiling point, Δt_b. (Hint: use the boiling point you measured for the pure water as the solvent's boiling point.)

2. **Analyzing Information**
Given that similar masses of each substance were used, does the data for the unknown most closely match NaCl, sucrose, or $MgSO_4 \cdot 7H_2O$?

3. **Analyzing Data**
Calculate the approximate molalities of dissolved particles for the known and unknown solutes using the following equation and your answers to item **1**. (Hint: $K_b = 0.51°C/m$)

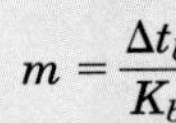

$$m = \frac{\Delta t_b}{K_b}$$

4. **Organizing Data**
For each trial, calculate the mass of water in the flask in kilograms.

5. **Analyzing Data**
Using your answers from items **3** and **4** and your data, calculate an experimental value for the molar mass of each solute. (Hint: see the Problem section of the introduction, but remember that a calculation indicating 2.00 m of Na^+ and Cl^- particles means a solution that is 1.00 m in NaCl. The water molecules that are a part of the $MgSO_4 \cdot 7H_2O$ crystal do not raise the boiling point.)

Sample Data

Solution	Mass of solute (g)	Mass of flask + water (g)	Boiling Point (°C)
H_2O	n/a	112.04	100.0
NaCl	10.0	112.12	103.5
Sucrose	9.91	112.07	100.3

Conclusions

6. ***Evaluating Conclusions***
How close is the estimated molar mass for the unknown solute to the estimated molar mass of the solute chosen in item **2**? Calculate percent error, using the unknown solute's molar mass estimate as the experimental value.

7. ***Evaluating Methods***
Determine the molar mass of NaCl, sucrose, and $MgSO_4{\cdot}7H_2O$ from a periodic table. Calculate percent error for your experimental molar mass values.

8. ***Evaluating Methods***
How accurate do you think it is to identify the unknown substance using this procedure? Can you explain any deviations that occur?

9. ***Applying Ideas***
Explain in your own words why adding a solute raises the temperature at which a substance boils. (Hint: consider what you know about bonding, phase changes, and kinetic molecular theory.)

10. ***Analyzing Methods***
Why do you need to determine the boiling point of water without a solute dissolved in it? Explain your answer.

Extensions

1. ***Designing Experiments***
What possible sources of error can you identify with this procedure? If you can think of ways to eliminate them, ask your teacher to approve your plan, and run the procedure again.

2. ***Research and Communications***
Consult chemical handbooks and other reference works to determine the possible health hazards of these solutes. For each one, what should be done if someone swallows it?

3. ***Applying Ideas***
Find out how much and what type of salt a large northern city such as New York or Chicago uses on icy roads in the winter. What problems result from use of this salt? What salt substitutes can melt ice and snow? Which would be safest for our environment?

4. ***Designing Experiments***
Obtain a water-soluble unknown from your teacher, and determine whether it is NaCl, sucrose, or $MgSO_4{\cdot}7H_2O$ using the freezing-point depression method. (As determined by experimentation, the freezing-point depression of a 1.00 *m* solution of any molecular solute in water is –1.86°C.)

5. ***Research and Communications***
Find out about the different types and ratios of ingredients in antifreeze solutions and coolants for automobile radiators, and give a report that explains how they make use of colligative properties.

6. ***Research and Communications***
What are some of the primary uses of $MgSO_4{\cdot}7H_2O$, and how is it produced for commercial use?

Sample Data

Solute	Solute Mass(g)	Mass of flask + water (g)	Boiling Point (°C)
$MgSO_4{\cdot}7H_2O$	9.98	111.89	100.9
Unknown	10.01	112.01	101.0
Mass of empty Erlenmeyer flask: 62.04 g			

8. This method is accurate if the choice is limited to several very different options. The larger the molar mass of the substance, the more possibility there was for error.

9. When a substance dissolves in water, it forms attractions with the water molecules, making it harder for individual molecules to enter the gas phase.

10. The boiling point of water is not 100°C at every location and under every condition. In order to be sure of the precision, the measurement of the pure solvent's boiling point must be made with the same instrument.

Extensions

1. Students' suggestions for improving the procedure will vary. Possibilities include using larger amounts of solute so that boiling-point elevation will be greater. Be sure answers are safe and include carefully planned procedures.

2. Sucrose and NaCl pose no health hazard. $MgSO_4$ is moderately toxic.

3. Answers may vary, but should include recognition that salt contributes to automobile rusting and the breakdown of the road surface.

4. Student answers will vary.

5. Student answers will vary, but most antifreeze solutions contain water and ethylene glycol, which are miscible, so that the molal concentration is high, raising the boiling point of the coolant.

6. Magnesium sulfate is also known as Epsom salts. It can be used in bleaching, dyeing, and fire-proofing fabrics. It can be used in making frosted papers, explosives, and matches. It is also used to treat some infections.

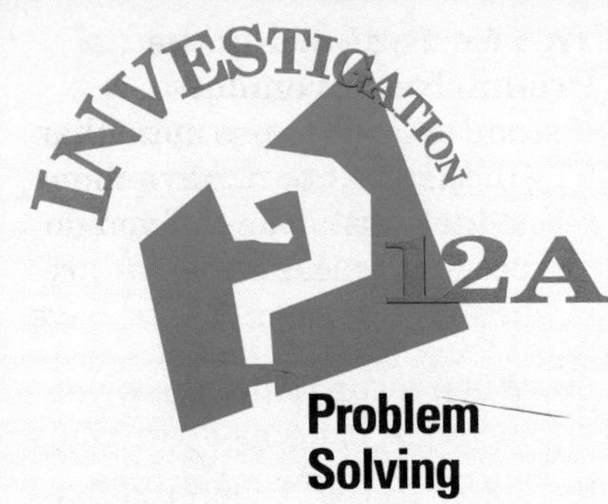

Objectives

Students will:

- use appropriate lab safety procedures.
- measure mass with a laboratory balance.
- measure volume with a graduated cylinder.
- prepare several solutions.
- measure the boiling point of water without a solute and with three non-volatile solutes (NaCl, glucose, and $CaCl_2$).
- calculate boiling-point elevation for the solutes.
- calculate molal concentration of the solutions.
- calculate the molecular mass of three solutes.
- identify an unknown using boiling-point-elevation data.
- design and implement their own procedure.

Planning

Recommended Time: 1 lab period
Prerequisite: Exploration 12A

Solution/Material Preparation

1. The unknown can be any of the substances. If you do not have thermometers marked in 0.1°C increments, make the unknown NaCl because it is easier to distinguish than the other two substances.

Student Orientation

Techniques to Demonstrate

Demonstrate how to recognize the boiling point of a solution. Also review the use of thermometers if necessary.

Pre-lab Discussion

Begin by discussing the results and procedure used in the Exploration, especially any errors in lab technique that occurred. Students should be able to recognize that the procedure will be nearly the same, but different substances are to be tested.

Sample Data

Solution	Mass of solute (g)	Mass of flask water (g)	Boiling point (°C)
H_2O	n/a	113.07	99.6
NaCl	10.00	113.02	103.1
Glucose	10.12	113.12	100.2
$CaCl_2$	9.92	112.89	102.2
Unknown	10.01	112.01	100.1

Mass of empty Erlenmeyer flask: 64.02 g

Analysis

The calculations shown are for glucose.

$$0.6^{\circ}C \times \frac{1\ m}{0.51^{\circ}C} \times 0.04910\ kg\ H_2O = 6 \times 10^{-2}\ mol$$

$$\frac{10.12\ g}{6 \times 10^{-2}\ mol} = 170\ g/mol$$

$$\frac{170 - 180.18}{180.18} \times 100 = 8\%\ error$$

B

BOILING-POINT ELEVATION

Contaminated Sugar

D&J

D & J FOOD PROCESSING

April 10, 1995

Ms. Martha Li-Hsien
Director of Research
CheMystery Labs
52 Fulton Street
Springfield, VA 22150

Dear Ms. Li-Hsien:

Each year, our company processes several thousand metric tons of sugars, seasonings, and other household products. Recently, an unknown ingredient was accidentally included in several tons of powdered sugar before we discovered the mistake. Our quality control department analyzed a sample and was able to isolate the contaminant. I am sending you a small amount of this contaminant. Please identify it as soon as possible.

Based on the production schedule, it is most likely that the contaminant is either glucose, NaCl, or $CaCl_2$. If the contaminant is glucose, we can still sell the product; otherwise, the entire batch will be a loss.

As discussed in our telephone conversation, we are prepared to pay $200 000 if this work is performed to our satisfaction. An official contract will arrive soon under separate cover.

Sincerely,

Kristeen McGonigal

Kristeen McGonigal
Director of Operations
D & J Food Processing

Required Precautions

Goggles and lab aprons must be worn at all times.

Tie back loose hair and clothing. Use tongs to handle equipment.

Do not touch or taste any chemicals. Wash your hands thoroughly when finished.

Required Precautions

- Goggles and a lab apron must be worn at all times.
- Tie back loose hair and long clothing when working in the lab.
- Read all safety cautions, and discuss them with your students.
- If the LEAP System with thermistor probe is used, the precautions listed in the teacher's notes on page 652 must be followed to avoid electric shock.

Memorandum

CheMystery Labs, Inc.
52 Fulton Street
Springfield, VA 22150

Date: April 11, 1995

To: Patricia Edgerton

From: Martha Li-Hsien

Recently, our management team has been reviewing the accuracy of the work we did in the first quarter of this year. Their reviews indicate that there is room for improvement in our cost-efficiency and accuracy. We should start with this job. To make certain that your plan is as efficient as possible, I need to review your plans before you start. Please include the following.

- detailed one-page summary of your plan for the procedure (I suggest we compare boiling-point elevations for the unknown and some standard samples.)
- any necessary data tables
- itemized list of equipment and materials, with individual and total costs

After you complete the analysis, prepare a two-page letter to McGonigal at D & J Food Processing. Make sure your letter covers the following points.

- identity of the contaminant
- brief summary of the procedure you used to identify the contaminant
- detailed analysis and data sections, showing calculations
- estimates and explanations for possible sources of error
- detailed invoice for all expenses and services

References

Refer to pages 459–460 for information about colligative properties. The procedure is similar to one your team recently completed to differentiate between NaCl, sucrose, and $MgSO_4 \cdot 7H_2O$. Remember that one mole of particles dissolved in one kilogram of water will raise the boiling point of the solution by 0.51°C. This value is called the molal boiling-point constant of water, represented by K_b. Base your disposal costs on the amounts of solute used.

Materials for
D & J FOOD PROCESSING

Item	Cost
REQUIRED ITEMS (You must include all of these in your budget.)	
Lab space/fume hood/utilities	15 000 – /day
Standard disposal fee	2 000 – /g of product
Beaker tongs	1 000
REAGENTS and ADDITIONAL EQUIPMENT (Include in your budget only what you'll need.)	
$CaCl_2$	
Glucose	500 – /g
NaCl	500 – /g
250 mL beaker	500 – /g
400 mL beaker	1 000
250 mL flask	2 000
100 mL graduated cylinder	1 000
Balance	1 000
Bunsen burner/related equipment	5 000
Drying oven	10 000
Glass stirring rod	5 000 – /day
Hot plate	1 000
Ring stand/ring/wire gauze	8 000
Spatula	2 000
Thermistor probe—LEAP	500
Thermometer	2 000
Thermometer clamp	2 000
Wash bottle	1 000
Weighing paper	500
*No refunds on returned chemicals or unused equipment.	500 – /piece
Fines	
OSHA safety violation	2 000 – /incident

Spill Procedures/ Waste Disposal Methods

- **Any solid materials must go in designated waste containers.**
- **Solutions can be washed down the drain with 20 times as much water as waste volume.**
- **Clean the area and all equipment after use.**

Tips for Evaluating the Pre-lab Requirements

If students plan to use more than 50 g of the unknown, have them reconsider their plans. If you do not have thermometers that are marked in 0.1°C increments, have students use 15 or 20 g of each unknown instead of 10 g. Be certain students plan to measure the boiling point of water again, even though it was measured previously in the Exploration.

Proposed Procedure

Measure the temperature at which 50 g of water begins to boil. Then measure the boiling points of solutions made with 10 g of each of the solutes (glucose, salt, calcium chloride, and the unknown substance). Determine which solute's data is most similar to that of the unknown substance.

Post-Lab

Disposal

Students should wash any solutions down the sink. Set out a disposal container for excess solids. Dissolve the solids in 20 times as much water, and wash them down the drain.

Materials

(for each lab group)

- $CaCl_2$, 10 g
- Distilled water
- Glucose, 10 g
- NaCl, 10 g
- Unknown, 10 g
- 250 mL Erlenmeyer flask
- 100 mL graduated cylinder
- Balance
- Beaker tongs
- Bunsen burner, gas tubing, striker
- Celsius thermometer, nonmercury type, with 0.1°C markings
- Glass stirring rod
- Ring stand and ring
- Spatula
- Weighing paper
- Wire gauze with ceramic center

Optional Equipment

- LEAP System with thermistor probe

Estimated cost of materials:
$130 000–136 500

EXPLORATION 12B

Objectives

Students will:

- use appropriate lab safety procedures.
- prepare a set of solutions by diluting a stock solution.
- perform serial dilutions to create solutions of different concentrations.
- combine two different solutions to create an equilibrium system.
- make colorimetric observations or measurements and relate them to concentration of one substance in an equilibrium system.
- plot a standard curve for the absorbance.
- estimate the concentrations of each substance at equilibrium.
- determine the equilibrium expression of a chemical reaction.

Planning

Recommended Time: 1 lab period (depends on availability of equipment)

Materials
(for each lab group)

- 0.00200 M KSCN, 10 mL
- 0.200 M $Fe(NO_3)_3$, 25 mL
- 0.6 M HNO_3, 25 mL
- Distilled water
- 10 mL graduated cylinder
- Glass stirring rod
- Test-tube rack
- Test tubes, 6

Optional Equipment

- Cuvettes
- LEAP System with colormeter
- Lint-free wipes
- Spectrophotometer

Solution/Material Preparation

1. Do not prepare the solutions too far in advance because they will oxidize and change color. The students running tests at the end of the period may have different results from those performing the same tests earlier.

Equilibrium Expressions

Objectives

Demonstrate **proficiency in preparing serial dilutions from a standard solution and comparing solutions by eye, by spectrophotometer, or with a LEAP colormeter probe.**

Relate **spectrophotometric determinations to solution concentration.**

Determine **experimentally the equilibrium expression of a chemical reaction.**

Situation

You work for a company that makes home tests for chemicals. Some homes have iron pipes, which can slowly leach iron(III) ions into the water, causing a brownish tint. Although not a health hazard, at high concentrations it can cause rust stains on laundry and dishes. Your company plans to use KSCN as a reagent to detect the iron, but an exploration of the reaction is necessary in order to calibrate the test for determining concentration. This test involves a reaction that reaches an equilibrium point, so you must determine the equilibrium constant to calibrate the test.

Background

Colorimetry can be used to measure the intensity of the red color of the $FeSCN^{2+}$ ions. Because this reaction is not always complete, it is not possible to set up a standard curve using known concentrations as in other colorimetry experiments. Consider the net ionic equation for this reaction.

$$\underset{\text{Yellow}}{Fe^{3+}(aq)} + \underset{\text{Colorless}}{SCN^-(aq)} \rightleftharpoons \underset{\text{Red}}{FeSCN^{2+}(aq)}$$

At equilibrium, $FeSCN^{2+}$ is being produced at a rate equal to the rate at which $FeSCN^{2+}$ is breaking up into Fe^{3+} and SCN^-. If the concentration of Fe^{3+} ions is increased, the equilibrium will be disturbed, and a new equilibrium will be reached to accommodate these new conditions.

As discussed in Chapter 12, the molar concentrations of substances at equilibrium can be arranged in a mathematical expression, the *equilibrium expression*, that has a constant value, the *equilibrium constant*, K_{eq}. Because this value should stay the same, you can use this method to check your determinations of $FeSCN^{2+}$ concentration.

Problem

The following must be done to adapt the test for measuring the iron(III) concentration of an unknown solution.

- prepare several solutions with different Fe^{3+} concentrations, and add similar amounts of the testing solution to each
- compare the amount of red color produced by the solutions
- relate the intensity of the red color to the concentrations of $FeSCN^{2+}$ at equilibrium
- determine the concentrations of Fe^{3+}, SCN^-, and $FeSCN^{2+}$ at equilibrium
- calculate the equilibrium constant for each trial, and evaluate your determinations of $FeSCN^{2+}$ concentration

Required Precautions

- Read all safety cautions, and discuss them with your students.
- Goggles and a lab apron must be worn at all times.
- Students should not handle concentrated acid solutions.
- In case of an acid spill, first dilute with water. Then mop up the spill with wet cloths or a wet cloth mop designated for spill cleanup while wearing disposable plastic gloves.

Safety

Always wear goggles and an apron to protect your eyes and clothing. If you get a chemical in your eyes, immediately flush it out at the eyewash station while calling to your teacher. Know the locations of the emergency lab shower and eyewash and how to use them.

Do not touch any chemicals. If you get a chemical on your skin or clothing, wash it off at the sink while calling to your teacher. Carefully read the labels and follow the directions on all containers of chemicals that you use. Do not taste any chemicals or items used in the laboratory. Never return leftovers to their original containers; take only small amounts to avoid wasting supplies.

Always clean up the lab and all equipment after use, and dispose of substances according to proper disposal methods. Wash your hands thoroughly before you leave the lab after all lab work is finished.

Materials

- **0.200 M $Fe(NO_3)_3$, 10 mL**
- **0.6 M HNO_3, 25 mL**
- **0.002 00 M KSCN, 25 mL**
- **10 mL graduated cylinder**
- **Glass stirring rod**
- **Test tube rack**
- **Test tubes, 6**

Optional equipment

- **LEAP system with colormeter probe**
- **Spectrophotometer**
- **Cuvettes**
- **Lint-free wipes for cuvettes**

Preparation

1. ***Organizing Data***
   Prepare a data table in your lab notebook. The table should contain five columns and seven rows. Label the columns of the first row: *Test tube no.*, *Fe^{3+} conc.*, *mL of Fe^{3+}*, *mL of KSCN*, and *Absorbance value*. Fill in the following test tube numbers in the second through seventh rows of the first column: *1, 2, 3, 4, 5,* and *6*. Be sure there is also room for observations.
2. If you will be using a spectrophotometer, turn it on now, as it must warm up for approximately 10 min.
3. Label six test tubes as *1, 2, 3, 4, 5,* and *6*.

Technique

4. Carefully measure 5.0 mL of 0.200 M $Fe(NO_3)_3$ into a 10 mL graduated cylinder. Pour this into test tube *1*. Record this volume and concentration in the data table.
5. Carefully measure another 5.0 mL of 0.200 M $Fe(NO_3)_3$ into the 10 mL graduated cylinder. Add 5.0 mL of 0.6 M HNO_3. Mix well with a glass stirring rod. Pour 5.0 mL of this mixture into test tube *2*. Record this volume and Fe^{3+} concentration in the data table. (Hint: the concentration of Fe^{3+} is half of what it was for test tube *1* because it has been diluted by an equal volume of HNO_3.)
6. Add 5.0 mL of 0.6 M HNO_3 to the remaining 5.0 mL of the mixture in the graduated cylinder. Mix well with a glass stirring rod. Pour 5.0 mL of this mixture into test tube *3*. Record this volume and concentration in the data table.
7. Repeat step **6** until you have filled all six test tubes.

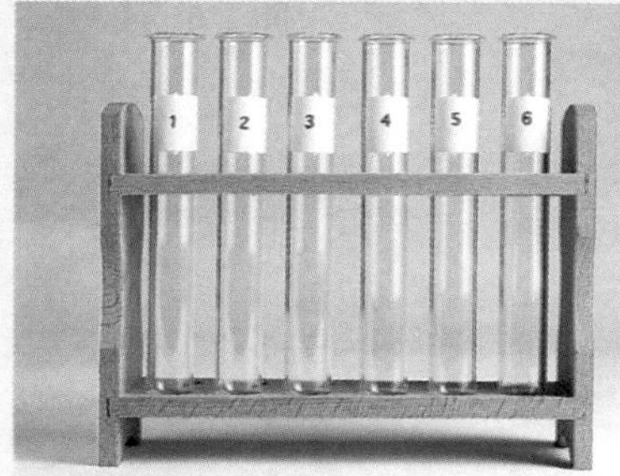

2. To prepare 0.200 M $Fe(NO_3)_3$, add 80.80 g $Fe(NO_3)_3 \cdot 9H_2O$ to enough distilled water to make 1.00 L of solution.
3. To prepare 0.6 M HNO_3, observe the required precautions. Add 38 mL of concentrated HNO_3 to enough distilled water to make 1.00 L of solution. Add the acid slowly, and stop to stir it in order to avoid overheating.
4. To prepare 0.002 00 M KSCN, add 0.194 g KSCN to enough distilled water to make 1.00 L of solution.
5. Use test tubes that are 18 mm × 150 mm or larger.

Student Orientation

Pre-lab Discussion
This experiment gives students a practical example of the concept of an equilibrium expression for a chemical reaction and provides some practice in performing equilibrium calculations. Students may be confused by the calculations and lose sight of what they are measuring and why. Remind students that if they keep track of what all of the numbers mean, it will help them avoid common errors. Because these calculations are so complicated, it is important to work through a set of sample data before the lab.

Techniques to Demonstrate
Show students how to operate the LEAP colormeter or the spectrophotometer, if used. Emphasize the need to check the calibration of the instrument repeatedly and to wipe the cuvettes with lint-free wipes. Students should prepare all of their dilutions before using the instrument. The preparation of the standard curve is similar to that in Exploration 11A.

- Wear goggles, face shield, impermeable gloves, and a lab apron when you prepare the HNO_3. Work in a hood known to be in operating condition, with another person present nearby to call for help in case of an emergency. Be sure you are within 30 s walking distance of a working safety shower and eyewash station.
- If the LEAP System with colormeter or a spectrophotometer are used, the precautions listed in the teacher's notes on page 652 must be followed to avoid electric shock.

Post-Lab

Disposal

Set out one disposal container for the solutions from this lab. Neutralize the waste with 1.0 M base until the pH is between 5 and 9, and store it.

Answers to Analysis and Interpretation

1. Student answers will vary. If the instrument was calibrated to give a reading of 1.00 for test tube 1, this step will be unnecessary, and the values in the data table will suffice.

2. Each 5.0 mL of 0.00200 M KSCN was mixed with 5.0 mL of $Fe(NO_3)_3$ solution to make a total of 10.0 mL.

$$\frac{5.0 \text{ mL } SCN^-}{10.0 \text{ mL total}} \times 0.002 \text{ M } SCN^- =$$

0.001 M SCN^- for all test tubes

3. Each 5.0 mL of $Fe(NO_3)_3$ was mixed with 5.0 mL of 0.00200 M KSCN solution to make a total of 10.0 mL.

$$\frac{5.0 \text{ mL } Fe(NO_3)_3}{10.0 \text{ mL total}} \times 0.200 \text{ M } Fe(NO_3)_3 =$$

0.100 M $Fe(NO_3)_3$ in test tube 1

$$\frac{5.0 \text{ mL } Fe(NO_3)_3}{10.0 \text{ mL total}} \times 0.0500 \text{ M } Fe(NO_3)_3 =$$

0.0250 M $Fe(NO_3)_3$ in test tube 3

$$\frac{5.0 \text{ mL } Fe(NO_3)_3}{10.0 \text{ mL total}} \times 0.0250 \text{ M } Fe(NO_3)_3 =$$

0.0125 M $Fe(NO_3)_3$ in test tube 4

$$\frac{5.0 \text{ mL } Fe(NO_3)_3}{10.0 \text{ mL total}} \times 0.0125 \text{ M } Fe(NO_3)_3 =$$

0.00625 M $Fe(NO_3)_3$ in test tube 5

$$\frac{5.0 \text{ mL } Fe(NO_3)_3}{10.0 \text{ mL total}} \times 0.00625 \text{ M } Fe(NO_3)_3 =$$

0.00312 M $Fe(NO_3)_3$ in test tube 6

---

8. Discard the contents of test tube *2*, pouring it into the waste container designated by your teacher. (This dilution is not great enough to provide a measurable difference in light absorption.)

9. Add 5.0 mL of 0.002 00 M KSCN to test tubes *1*, *3*, *4*, *5*, and *6*. Mix each solution thoroughly with a stirring rod.

10. Compare the solutions holding a piece of white paper behind the test tubes. If you are not using a spectrophotometer or LEAP colormeter, estimate the intensity of the red color on a scale from 0 (clear) to 1.00 (test tube *1*), and enter your estimate in the *Absorbance value* column of the data table.

11. If you are using a LEAP colormeter, set it up and calibrate it as discussed in step **11a**. If you are using a spectrophotometer, set it up and calibrate it as discussed in step **11b**. *For both instruments, note that you should be measuring absorbance*, NOT percent transmittance. If you are using neither, go on to step **14**.

a. LEAP Colormeter

Your teacher will have the experiment on the screen. Plug the colormeter probe into the input cable of your station. Set the wavelength to 590 nm, which corresponds to a yellowish green light. Record the wavelength in your data table. ***Never wipe cuvettes with paper towels or scrub them with a test tube brush. Use only lint-free tissues that will not scratch the cuvette's surface. The outside of the cuvette must be completely dry before it is placed inside an instrument, or else you will get an invalid reading.***

Calibrate the colormeter by pouring some solution from test tube *1* into a cuvette and inserting the cuvette in the sample compartment. Turn the ZERO ADJ knob until the absorbance value is as close to 1.00 as possible. When the reading remains at 1.00, put the system on WAIT and proceed to step **12**.

b. Spectrophotometer

If the spectrophotometer has warmed up for about 10 minutes, set the wavelength to 590 nm. Calibrate it by turning the left front knob to 0% transmittance, while using an empty sample compartment with the lid closed. Then, pour some solution from test tube 1 into the cuvette, and adjust the absorbance value with the right front knob to read as close to 1.00 as possible.

Never wipe cuvettes with paper towels or scrub them with a test tube brush. Use only lint-free tissues that will not scratch the cuvette's surface. The outside of the cuvette must be completely dry before it is placed inside an instrument, or else you will get an invalid reading.

12. With your instrument adjusted accordingly, record the absorbance value for the solution from test tube *1*. If you have only one cuvette, rinse it several times with distilled water, and then make absorbance measurements for test tube *3*. Record this value in your data table. Rinse the cuvette several times with distilled water, and measure and record absorbance values for test tubes *4*, *5*, and *6* in your data table.

Sample Data

Test tube no.	Fe^{3+} conc.	mL of Fe^{3+}	mL of KSCN	Absorbance value
1	0.200 M	5.0	5.0	1.00
2	0.100 M	5.0	5.0	n/a
3	0.0500 M	5.0	5.0	0.94

13. As a check on your measurements, retest the solution from test tube *1* at the end of the measurements. Its absorbance value should still be the same, close to 1.00. If not, repeat the procedure for all solutions.

Cleanup and Disposal

14. Dispose of all solutions in the container designated by your teacher. Wash your hands thoroughly after cleaning up the area and equipment.

Analysis and Interpretation

1. ***Analyzing Data***
   Determine how each of the absorbance values relates to that for test tube *1* by dividing each by the value obtained for test tube *1*. (Hint: after this calculation, the new value for test tube *1* should be 1.00, and the values for the other test tubes should be less than 1.00, as they were less concentrated than test tube *1*. If you were using estimates instead of colorimetry measurements, this step can be skipped.)
2. ***Analyzing Data***
   Calculate the initial concentrations of SCN^- for test tubes *1, 3, 4, 5,* and *6*. Remember that each 5.0 mL of 0.002 00 M KSCN was mixed with 5.0 mL of $Fe(NO_3)_3$ solution to give a total volume of 10.0 mL. (Hint: the value will be the same for all the test tubes.)
3. ***Analyzing Data***
   Calculate the actual initial concentration of Fe^{3+} in the test tubes in a similar way. (Hint: the values recorded in the data table show the concentrations of Fe^{3+} before being diluted by 0.002 00 M KSCN.)
4. ***Applying Ideas***
   Determine the equilibrium concentrations for test tube *1*. Because the initial concentration of Fe^{3+} was 0.100 M, much larger than the initial concentration of SCN^-, 0.001 M, assume that practically all of the SCN^- ions are consumed in the reaction. (Even though this is not necessarily true, the deviation from the true SCN^- concentration will be so much smaller than the other factors in this equation that it can be disregarded temporarily.)
5. ***Analyzing Data***
   Calculate the $FeSCN^{2+}$ equilibrium concentration for test tubes *3–6*, based on the equilibrium concentration for test tube *1* determined in item **4** and the absorbance data. (Hint: multiply the concentration for test tube *1* by the factors calculated in item **1**.)
6. ***Analyzing Data***
   Calculate the SCN^- concentration for test tubes *3–6* at equilibrium. (Hint: you know the initial SCN^- concentration from item **2**, and you know the amount of SCN^- that has formed $FeSCN^{2+}$ from item **5**.)
7. ***Analyzing Data***
   For test tubes *3–6*, calculate the Fe^{3+} concentration at equilibrium. (Hint: you know the initial Fe^{3+} concentration from item **3**, and you know the amount of Fe^{3+} that has formed $FeSCN^{2+}$ from item **5**.)
8. ***Analyzing Data***
   Graph the adjusted absorbance values from item **1** against the initial concentration values for $[Fe^{3+}]$.

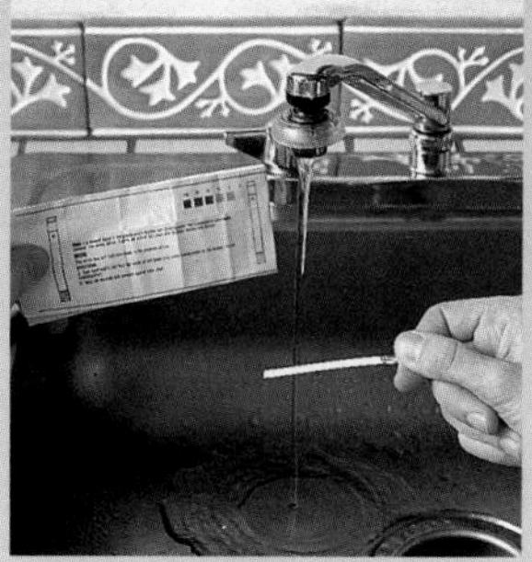

Test tube no.	Fe^{3+} conc.	mL of Fe^{3+}	mL of KSCN	Absorbance value
4	0.0250 M	5.0	5.0	0.89
5	0.0125 M	5.0	5.0	0.80
6	0.00625 M	5.0	5.0	0.72

4. Assuming that all of the SCN^- in test tube 1 is converted to $FeSCN^{2+}$, the concentration of $FeSCN^{2+}$ in the first test tube must be 0.001 M, the same as the initial SCN^- concentration. This amount must be subtracted from the initial Fe^{3+} concentration to give the final Fe^{3+} concentration: 0.100 M initial − 0.001 M used = 0.999 M final.

5. At equilibrium:
   Test tube 3:
   0.001 00 M × 0.94 = 0.000 94 M $FeSCN^{2+}$

   Test tube 4:
   0.001 00 M × 0.89 = 0.000 89 M $FeSCN^{2+}$

   Test tube 5:
   0.001 00 M × 0.80 = 0.000 80 M $FeSCN^{2+}$

   Test tube 6:
   0.001 00 M × 0.72 = 0.000 72 M $FeSCN^{2+}$

6. At equilibrium:
   Test tube 3:
   0.001 00 M − 0.000 94 M = 0.000 06 M SCN^-

   Test tube 4:
   0.001 00 M − 0.000 89 M = 0.000 11 M SCN^-

   Test tube 5:
   0.001 00 M − 0.000 80 M = 0.000 20 M SCN^-

   Test tube 6:
   0.001 00 M − 0.000 72 M = 0.000 28 M SCN^-

7. At equilibrium:
   Test tube 3:
   0.025 M − 0.000 94 M = 0.024 M Fe^{3+}

   Test tube 4:
   0.012 5 M − 0.000 89 M = 0.0116 M Fe^{3+}

   Test tube 5:
   0.006 25 M − 0.000 80 M = 0.005 45 M Fe^{3+}

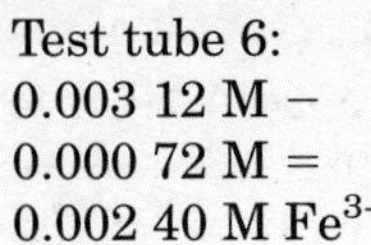
Test tube 6:
0.003 12 M −
0.000 72 M =
0.002 40 M Fe^{3+}

8.

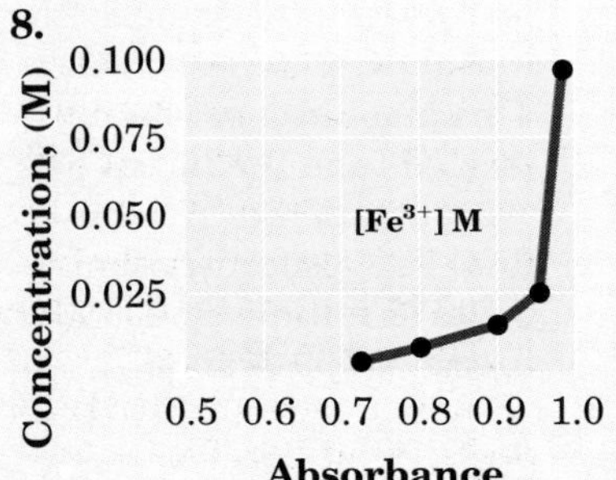

Conclusions

9. K_{eq} for test tube 3:

$$\frac{[0.000\ 94]}{[0.024]\ [0.000\ 06]} =$$

700 (rounded to 1 significant figure)

K_{eq} for test tube 4:

$$\frac{[0.000\ 89]}{[0.0116]\ [0.000\ 11]} = 7.0 \times 10^2$$

K_{eq} for test tube 5:

$$\frac{[0.000\ 80]}{[0.005\ 45]\ [0.000\ 20]} = 730$$

K_{eq} for test tube 6:

$$\frac{[0.000\ 72]}{[0.002\ 40]\ [0.000\ 28]} = 1100$$

10. average K_{eq} value = 800

absolute deviation for test tube 3: 100

absolute deviation for test tube 4: 100

absolute deviation for test tube 5: 70

absolute deviation for test tube 6: 300

average deviation:

$$\frac{100 + 100 + 70 + 300}{4} = 140$$

11. Student answers will vary. Generally, the numbers are not very precise but are within the proper order of magnitude.

12. Student answers will vary. Some possible causes for different values could be: the system did not reach equilibrium, the systems were not at equal temperatures, the measurements were inaccurate (either volume or instrument readings), or the dilutions were not exact.

13. As the concentration increased, so did the absorbance. For Fe^{3+}, this was not a linear relationship.

Conclusions

9. ***Evaluating Data***
For test tubes *3–6*, determine the value of the equilibrium constant, K_{eq}, for this reaction by using the equation given below. (Hint: use the equilibrium concentrations determined in items **5**, **6**, and **7**.)

$$K = \frac{[FeSCN^{2+}]}{[Fe^{3+}][SCN^-]}$$

10. ***Evaluating Data***
Calculate the average value for the equilibrium constant from item **9**. Calculate each value's absolute deviation from this average. Then calculate the average deviation of all of the values. (Hint: instructions on calculating absolute and average deviation can be found in Exploration 2A.)

11. ***Evaluating Data***
Based on your answers from item **10**, do you think the procedure was precise? Can you tell whether or not it was accurate? Explain your answers. (Hint: K_{eq} values that are within the same order of magnitude can be considered reasonably precise.)

12. ***Evaluating Methods***
Although the values for K_{eq} should be equal, explain several conditions that could cause these values to differ.

13. ***Interpreting Graphics***
What general statement can you make about the absorbance values compared to the concentration of Fe^{3+} and $FeSCN^{2+}$?

Extensions

1. ***Applying Conclusions***
The company developing the Fe^{3+} ion test kit needs instructions and a color key for customers that relates the different shades of red seen after adding KSCN to different concentrations of Fe^{3+} ions. Using art supplies or a computer with a color printer, create a small pamphlet or brochure that contains simplified instructions on how to perform the procedure, a color key on a strip of paper, and notations about concentration.

2. ***Designing Experiments***
How would you revise this procedure to determine the value of the equilibrium constant at different temperatures? Would you be able to maintain accurate data analysis for high or low temperature ranges? Explain your answers.

3. ***Designing Experiments***
What possible sources of error can you identify with this procedure? If you can think of ways to eliminate them, ask your teacher to approve your plan, and run the procedure again.

4. ***Applying Conclusions***
In item **4** of the Analysis and Interpretation section, the assumption was made that absolutely all of the SCN^- reacted. Now you can make

a better estimate of the actual equilibrium concentrations for the ions in test tube *1* by following these directions.

- Assign x to be the concentration of $FeSCN^{2+}$. Then the equilibrium concentrations of the other ions will be their initial concentrations subtracted by x.
- Use the equilibrium expression given in item **9**. Fill in the average value of K_{eq} calculated in item **10**. Then fill in the equilibrium concentrations of the ions from the previous step.
- Simplify the equation, and use algebra to subtract from one or both sides of the equation until there is only a zero on one side of the equation.
- Apply the quadratic formula to the equation from the previous step to determine the value of x. For any equation with the form given below, x has the values shown.

$$ax^2 + bx + c = 0$$

$$x = \frac{-b \pm \sqrt{b^2 - 4ac}}{2a}$$

- Using the value you calculate for x, determine the equilibrium concentrations of each ion.
- Assume that this newly calculated value of $FeSCN^{2+}$ concentration at equilibrium is the actual value, and calculate the percent error from assuming that the reaction went to completion and had an equilibrium $FeSCN^{2+}$ concentration of 1×10^{-3} M.

5. *Applying Ideas*
Hundreds of different equilibrium reactions are taking place constantly in your body. One very important equilibrium reaction involves oxygen, O_2, and hemoglobin, a complex protein abbreviated as Hb, to form oxyhemoglobin (HbO_2).

$$O_2(g) + Hb(aq) \rightleftharpoons HbO_2(aq)$$

In your lungs, where oxygen is in abundance, the equilibrium shifts to the right. The oxyhemoglobin then travels in your bloodstream to your oxygen-starved cells. At the cells, the equilibrium shifts to the left, releasing the oxygen. In this way, the equilibrium constant is maintained as you continue to live and breathe. Write the equilibrium expression for oxygen, hemoglobin, and oxyhemoglobin.

6. *Predicting Outcomes*
At the elevation of Mexico City, 2300 m (7500 ft), the concentration of oxygen is 75% of that at sea level. Yet, the same amount of oxygen needs to be delivered to the muscle cells. To compensate, does the body produce more or less hemoglobin? Explain your answer.

Extensions

1. Student answers will vary. Be certain students have been very explicit in their procedures.
2. Students' suggestions for measuring the constant at different temperatures will vary. One possibility would be to measure the absorbances of the same solution at a variety of temperatures. Be sure answers are safe and include carefully planned procedures.
3. Students' suggestions for improving the procedure will vary. One possible suggestion is to use more precise measures of volume, perhaps with pipets, during the dilution steps. Be sure answers are safe and include carefully planned procedures.
4. $K_{eq} = 800 = \frac{[FeSCN^{2+}]}{[Fe^{3+}][SCN^-]} = \frac{x}{(0.100 - x)(0.001 - x)}$

$800[x^2 - 0.101x + (1 \times 10^{-4})] = x$

$800x^2 - 81.8x + (8 \times 10^{-2}) = 0$

Applying the quadratic formula:

$x = \frac{81.8 \pm \sqrt{(81.8)^2 - 256}}{1600}$

$x = 0.101$ or 9.9×10^{-4} M $FeSCN^{2+}$

0.101 is too large because it is more than the initial concentration of SCN^-.

Equilibrium concentrations:

9.9×10^{-4} M $FeSCN^{2+}$
1×10^{-5} M SCN–
9.9×10^{-2} M Fe^{3+}

Percent error = $\frac{1 \times 10^{-3} - 9.9 \times 10^{-4}}{9.9 \times 10^{-4}} = 1\%$

5. $K_{eq} = \frac{[HbO_2]}{[O_2][Hb]}$

6. Although there is less oxygen at high altitudes, the muscle cells still need the same amount of oxygen in the form of HbO_2. The only way for the equilibrium constant to be maintained is for the body to produce more hemoglobin.

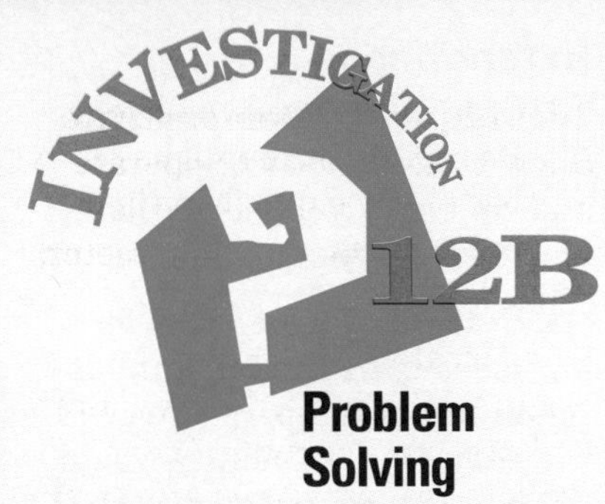

Objectives

Students will:

- use appropriate lab safety procedures.
- combine two different solutions to create an equilibrium system.
- make colorimetric observations or measurements and relate them to concentration of one substance in an equilibrium system.
- apply a standard curve to estimate concentration of one substance in an equilibrium system.
- determine the concentrations of other substances at equilibrium, using determinations of the equilibrium constant.
- design and implement their own procedure.

Planning

Recommended Time: 1–2 lab periods
Prerequisite: Exploration 12B

Solution/Material Preparation

1. Do not prepare the solutions too far in advance because they will oxidize and change color. Results may change as the period progresses.
2. See the Teacher's Notes for Exploration 12 B for information on preparing 0.200 M $Fe(NO_3)_3$, 0.6 M HNO_3, and 0.00200 M KSCN.
3. To prepare the "extract of spinach leaves," make 0.00625 M $Fe(NO_3)_3$ by adding 2.52 g of $Fe(NO_3)_3\cdot 9H_2O$ to 100 mL of 0.6 M HNO_3 and enough distilled water to make 1.00 L of solution.
4. To prepare the "extract of jasmine tea leaves," make 0.0250 M $Fe(NO_3)_3$ by adding 10.10 g of $Fe(NO_3)_3\cdot 9H_2O$ to 100 mL of 0.6 M HNO_3 and enough distilled water to make 1.00 L of solution.
5. Use test tubes that are 18 mm × 150 mm or larger.

Sample Data

Solution	Absorbance
Jasmine tea leaf extract	0.90 g
Spinach leaf extract	0.70 g

Analysis

Jasmine tea:

$$K_{eq} = 800 = \frac{[0.000\ 90]}{[x][0.000\ 10]}$$

equilibrium $[Fe^{3+}] = [x] = 1.1 \times 10^{-2}$ M

initial $[Fe^{3+}]$ before dilution $= 2(1.1 \times 10^{-2} + 9.0 \times 10^{-4}) = 2.4 \times 10^{-2}$ M

Spinach:

$$K_{eq} = 800 = \frac{[0.000\ 70]}{[x][0.000\ 30]}$$

equilibrium $[Fe^{3+}] = [x] = 2.9 \times 10^{-3}$ M

initial $[Fe^{3+}]$ before dilution $= 2(2.9 \times 10^{-3} + 7.0 \times 10^{-4}) = 7.2 \times 10^{-3}$ M

EQUILIBRIUM EXPRESSIONS — Iron Content of Tea

April 12, 1995

Reginald Brown
Director, Materials Testing Department
CheMystery Labs, Inc.
52 Fulton Street
Springfield, VA 22150

Dear Mr. Brown:

At the Far East Trading Company, we seek to dispel the myth that spinach is the best source of iron in a healthy diet. We believe that regular servings of our jasmine tea can provide just as much iron to the consumer as spinach.

Our products have never been marketed on their nutritional value, so we need to have hard data to back up these facts. Our researchers developed a procedure for extracting iron from food, and now we need an outside laboratory to perform an independent analysis of the iron content. Extracts of iron from our jasmine tea and from spinach are being sent to you.

Although we can extract the iron in our quality control lab, we do not have the analytical equipment to perform the iron content analysis. We hope your tests will verify our bold statement. We propose a $300 000 contract for your services.

Sincerely,

Richelle Pomeroy
Director of Sales and Marketing
Far East Trading Company

Required Precautions

Lab aprons and goggles must be worn at all times.

Do not touch or taste any chemicals. Wash your hands thoroughly when finished.

If you get acid or base on your skin or clothing, wash it off at the sink while calling to your teacher. If you get acid or base in your eyes, immediately flush it out at the eyewash while calling to your teacher.

Required Precautions

- Discuss all safety cautions with your students.
- Goggles and a lab apron must be worn at all times.
- Students should not handle concentrated acids.
- Dilute acid spills with water and mop them up with wet cloths or a mop designated for spill cleanup while wearing disposable plastic gloves.
- If the LEAP System with colormeter or a spectrophotometer are used, the precautions listed in the teacher's notes on page 652 must be followed to avoid electric shock.

Memorandum

CheMystery Labs, Inc.
52 Fulton Street
Springfield, VA 22150

Date: April 13, 1995

To: Alvin Horton

From: Reginald Brown

Because the solutions of iron are in the form of Fe^{3+} solutions, I believe we should use colorimetric methods to determine their concentration. However, these solutions will be somewhat dilute, so we may need to make use of the Fe^{3+}-$FeSCN^{2+}$ equilibrium to measure concentration.

We need to be efficient and very accurate because by the terms of the contract, we are liable for any false advertising claims filed against Far East Trading Company over the iron content of the tea. To help ensure accuracy, give me a preliminary report that includes the following before you begin.

- detailed one-page summary of your plan for the procedure
- all necessary data tables
- itemized list of equipment, with individual and total costs

After you complete the analysis, prepare a two-page report for Richelle Pomeroy at the Far East Trading Company. Be sure to include the following information.

- molar concentration of iron in each solution
- number of moles and grams of iron in 100.0 mL of each solution
- explanation of this analysis and its use of the concept of chemical equilibrium
- estimates and explanations of any possible sources of error
- value of the equilibrium constant
- equilibrium concentrations for each solution
- analysis and data sections, showing calculations
- detailed invoice for all expenses and services

References

Refer to pages 469–471 for information about equilibrium expressions. This analysis is similar to one you recently performed to determine the average value of the equilibrium expression for the Fe^{3+}-$FeSCN^{2+}$ equilibrium at the temperature in your lab. You can measure $FeSCN^{2+}$ concentration using the calibration data from your previous analysis. Then you can use the average value of the equilibrium expression from the previous analysis to determine the equilibrium and initial concentrations of Fe^{3+}. Or else, you can repeat the steps to be certain you have the correct values for K_{eq}.

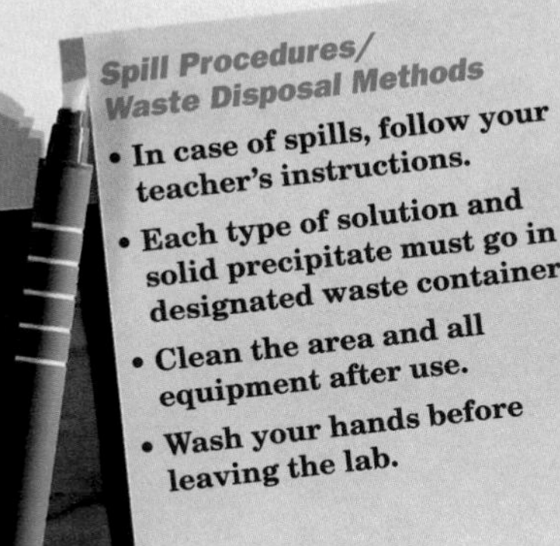
Spill Procedures/ Waste Disposal Methods

- In case of spills, follow your teacher's instructions.
- Each type of solution and solid precipitate must go in a designated waste container.
- Clean the area and all equipment after use.
- Wash your hands before leaving the lab.

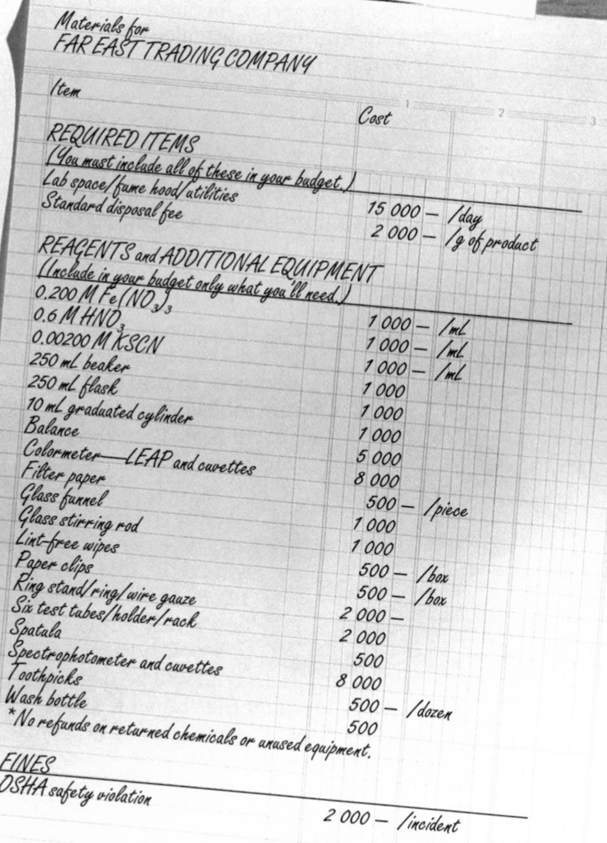
Materials for FAR EAST TRADING COMPANY

Item	Cost
REQUIRED ITEMS (You must include all of these in your budget.)	
Lab space/fume hood/utilities	15 000 – /day
Standard disposal fee	2 000 – /g of product
REAGENTS and ADDITIONAL EQUIPMENT (Include in your budget only what you'll need.)	
0.200 M $Fe(NO_3)_3$	1 000 – /mL
0.6 M HNO_3	1 000 – /mL
0.00200 M KSCN	1 000 – /mL
250 mL beaker	1 000
250 mL flask	1 000
10 mL graduated cylinder	1 000
Balance	5 000
Colormeter—LEAP and cuvettes	8 000
Filter paper	500 – /piece
Glass funnel	1 000
Glass stirring rod	1 000
Lint-free wipes	500 – /box
Paper clips	500 – /box
Ring stand/ring/wire gauze	2 000 –
Six test tubes/holder/rack	2 000
Spatula	500
Spectrophotometer and cuvettes	8 000
Toothpicks	500 – /dozen
Wash bottle	500
*No refunds on returned chemicals or unused equipment.	
FINES	
OSHA safety violation	2 000 – /incident

Materials
(for each lab group)

- "Extract of spinach leaves," 5 mL
- "Extract of jasmine tea leaves," 5 mL
- 0.200 M $Fe(NO_3)_3$, 10 mL
- 0.6 M HNO_3, 25 mL
- 0.00200 M KSCN, 35 mL
- Distilled water
- 10 mL graduated cylinder
- Glass stirring rod
- Test-tube rack
- Test tubes, 8

Optional Equipment

- Cuvettes
- LEAP System with colormeter
- Lint-free wipes
- Spectrophotometer

Estimated cost of materials:
$260 000 (with calibration)
$59 000 (without calibration)

Student Orientation

Techniques to Demonstrate
Show students how to operate and calibrate the LEAP colormeter or the spectrophotometer, if used.

Pre-lab Discussion
Begin by discussing the results and procedure used in the Exploration, especially any errors in lab technique that occurred.

As noted in the students' References section, the lab can be done very quickly if the previous data on the value of the equilibrium constant and the $FeSCN^{2+}$ calibration curve from the Exploration are used. However, this will involve some complicated calculations involving the equilibrium expression. For the most accurate results, a new calibration curve should be created. Concentration can be judged from this without the need to perform complicated calculations. Choose the option that suits your needs, and prepare students by working through similar calculations.

Tips for Evaluating the Pre-lab Requirements
Be certain students understand that they will be measuring absorbance, which is proportional to $FeSCN^{2+}$ concentration. To relate this to Fe^{3+} concentration, they will need to use the equilibrium expression.

Proposed Procedure

Students should follow the procedures used in the Exploration, preparing the dilutions and measuring the absorbance for each one or merely measuring the absorbance of the two unknown solutions and relating these values to the equilibrium constant determined in the Exploration.

Post-Lab

Disposal
Set out one disposal container for the solutions from this lab. Neutralize the waste with 1.0 M base until the pH is between 5 and 9, and store it.

EXPLORATION 13A

Objectives

Students will:

- use appropriate lab safety procedures.
- measure volume using graduated cylinders.
- measure volume with a buret or by calibrating a dropper to measure the number of drops per milliliter.
- measure mass using a laboratory balance.
- observe the reaction of $CaCO_3$ with a carefully measured excess of acid.
- perform a back-titration (titrating the excess acid with base).
- apply solution-stoichiometry concepts to determine the amounts of each reactant for the neutralization reaction and $CaCO_3$ reaction.

Planning

Recommended Time: 1–2 lab periods (shorter if eggshells are prepared in advance)

Materials

(for each lab group)

- Distilled water
- Eggshell
- Phenolphthalein solution, 1 mL
- 100 mL beaker
- 50 mL bottle, or small Erlenmeyer flask
- 10 mL graduated cylinder
- Balance
- Desiccator (optional)
- Drying oven
- Forceps
- Mortar and pestle
- Weighing paper
- White paper for background

Small-Scale Option

- 1.00 M NaOH, 15 mL
- 1.00 M HCl, 15 mL
- Medicine droppers or thin-stemmed pipets, 3

Full-Scale Option

- 1.00 M NaOH, 100 mL
- 1.00 M HCl, 100 mL
- Burets

Acid-Base Titration of an Eggshell

Objectives

Determine **the amount of calcium carbonate present in an eggshell.**

Relate **experimental titration measurements to a balanced chemical equation.**

Infer **a conclusion from experimental data.**

Apply **reaction-stoichiometry concepts.**

Situation

You are a research scientist working with the Department of Agriculture. A farmer from a nearby ranch has brought a problem to you. In the past ten years, his hens' eggs have become more and more fragile. So many of them have been breaking that he is beginning to lose money on the operation. The farmer believes his problems are linked to a landfill upstream, which is being investigated for illegal dumping of PCBs and other hazardous chemicals.

Background

Birds have evolved a chemical process which allows them to rapidly produce the calcium carbonate, $CaCO_3$, that is required for eggshell formation. The shell provides a strong protective covering for the developing embryo. Research has shown that some chemicals, like DDT and PCBs, can decrease the amount of calcium carbonate in the eggshell, resulting in shells that are thin and fragile.

PCB

Problem

To determine if the farmer's troubles are related to such weakened eggshells, you need to determine how much calcium carbonate is in sample eggshells from chickens that were not exposed to PCBs. The farmer's eggshells contain about 78% calcium carbonate. The calcium carbonate content of eggshells can easily be determined by means of an acid/base back-titration. In this back-titration, a carefully measured excess of a strong acid will react with the calcium carbonate. Since the acid is in excess, there will be some left over at the end of the reaction. The resulting solution will be titrated with a strong base to determine how much acid remained unreacted. Phenolphthalein will be used as an indicator to signal the endpoint of the titration. From this measurement, you can determine the following:

- the amount of excess acid that reacted with the eggshell
- the amount of calcium carbonate that was present to react with this acid

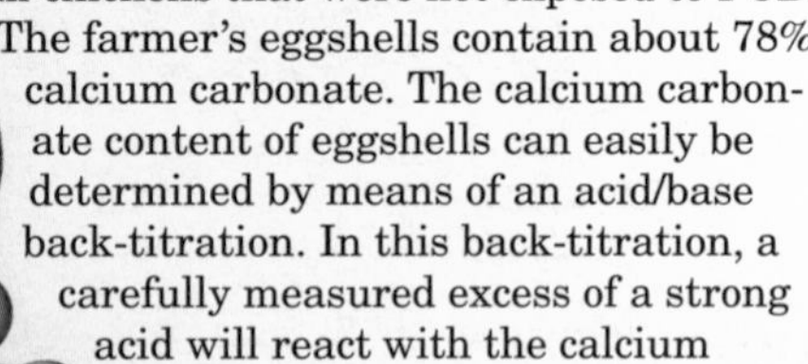

PCB

Required Precautions

- Goggles and a lab apron must be worn at all times.
- Read all safety cautions, and discuss them with your students.
- Students should not handle concentrated acid solutions.
- Wear goggles, face shield, impermeable gloves, and a lab apron when you prepare the HCl and NaOH. Work in a hood known to be in operating condition, with another person present nearby to call for help in case of an emergency. Be sure you are within

Safety

Always wear goggles and an apron to provide protection for your eyes and clothing. If you get a chemical in your eyes, immediately flush it out at the eyewash station while calling to your teacher. Know the locations of the emergency lab shower and eyewash and how to use them.

Do not touch any chemicals. If you get a chemical on your skin or clothing, wash it off at the sink while calling to your teacher. Make sure you carefully read the labels and follow the directions on all containers of chemicals that you use. Do not taste any chemicals or items used in the laboratory. Never return leftovers to their original containers; take only small amounts to avoid wasting supplies.

The oven used in this investigation is hot; use tongs to remove beakers from the oven because hot glassware does not look hot.

Call your teacher in the event of an acid or base spill. Acid or base spills should be cleaned up promptly, according to your teacher's instructions.

Always clean up the lab and all equipment after use, and dispose of substances according to proper disposal methods. Wash your hands thoroughly before you leave the lab after all lab work is finished.

Materials

- **1.00 M HCl**
- **1.00 M NaOH**
- **Distilled water**
- **Eggshell**
- **Phenolphthalein solution**
- **100 mL beaker**
- **50 mL bottle or small Erlenmeyer flask**
- **10 mL graduated cylinder**
- **Balance**
- **Desiccator (optional)**
- **Drying oven**
- **Forceps**
- **Mortar and pestle**
- **Three medicine droppers or thin-stemmed pipets**
- **Weighing paper**
- **White paper or white background**

Preparation

1. Remove the white and the yolk from the egg and dispose of them according to your teacher's directions. Wash the shell with distilled water and carefully peel all the membranes from the inside of the shell. Discard the membranes. Place ALL of the shell in a premassed beaker and dry the shell in the drying oven at 110°C for about 15 min. Continue with steps **2–5** while the eggshell is drying.
2. ***Organizing Data***
   Prepare two data tables in your lab notebook. The first should contain four columns, for initial and final volumes from the acid pipet and initial and final volumes from the base pipet. This table will require space for three trials. Also leave space to record the mass of the entire eggshell, the mass of the ground eggshell sample, the number of drops of HCl added, and the number of drops of NaOH added.
3. Put exactly 5.0 mL of water in the 10.0 mL graduated cylinder. Record this volume in the data table in your lab notebook. Fill the first dropper or pipet with water. This dropper should be labeled *Acid*. ***Do not use this dropper for the base solution***. Holding the dropper vertical, add 20 drops of water to the cylinder. ***For the best results, keep the sizes of the drops as even as possible throughout this investigation.*** Record the new volume of water in the first data table as Trial 1.

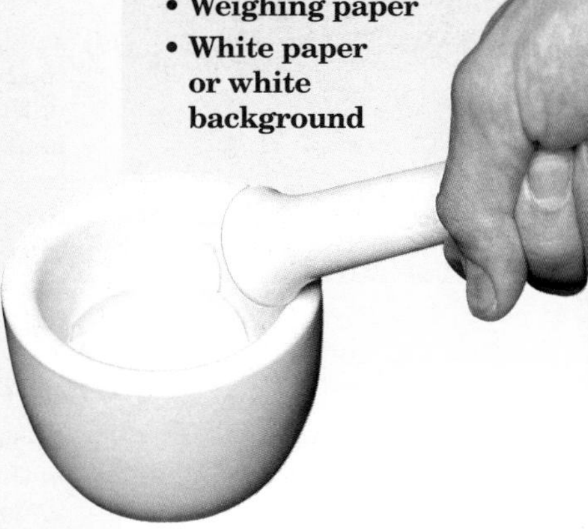

30 s walking distance of a working safety shower and eyewash station.

- In case of an acid or base spill, first dilute with water. Then, mop up the spill with wet cloths or a wet cloth mop while wearing disposable plastic gloves. Designate separate cloths or mops for acid and base spills.
- If the LEAP System with pH probe is used, the precautions listed in the teacher's notes on page 652 must be followed to avoid electric shock.
- Remind students that beakers in the oven will be hot and that they should use tongs to remove them.

Solution/Material Preparation

1. Desiccators are optional. Fewer balances, mortars, pestles, and graduated cylinders will be required if students share.
2. If droppers or pipets are labeled as *acid, base,* and *indicator,* these can be reused from class to class. For more reproducible drop sizes, place the pipet or dropper bulb in a tubing screw-clamp. Each turn of the screw will force out a reproducible drop.
3. To prepare 1.00 M HCl, follow the required precautions. Add 82.6 mL of concentrated HCl to enough distilled water to make 1.00 L of solution. Add the acid slowly, and stop to stir it in order to avoid overheating.
4. To prepare 1.00 M NaOH, add 40.0 g of NaOH pellets to enough distilled water to make 1.00 L of solution. Add a few pellets at a time, and stop to stir it in order to avoid overheating.
5. To prepare phenolphthalein solution, dissolve 1.00 g phenolphthalein in 50.0 mL of denatured alcohol, and add 50.0 mL of distilled water.
6. The LEAP system with pH probe can be used to monitor the titration and its end point. Follow the instructions given in Exploration 1B.
7. Provide miscellaneous shells (clam, oyster, snail, etc.) for Extension 3. Alternatively, students could determine the amount of $CaCO_3$ contained in pieces of classroom chalk.

Student Orientation

Pre-lab Discussion

Be sure to explain the nature of a back-titration. Contrast the steps with those in Exploration 1B. Students may not realize that the amount of $CaCO_3$ is being measured indirectly.

Procedural Changes for Full-Scale Option

- Detailed instructions on preparing burets for titrations are given in Exploration 1B.
- Step 2: Delete the first data table and "number of drops" entries in the second data table.
- Omit steps 3–6 for calibrating droppers.
- Step 8: Use a sample size near 1.00 g instead of 0.10 g.
- Step 9: Rinse burets in the same way that droppers are rinsed.
- Step 10: Instead of 100 drops, use 50.0 mL of HCl from the acid buret to react with the $CaCO_3$.
- Step 11: Titrate the mixture with NaOH from the base buret until the phenolphthalein changes color.
- Omit item 3 in the Analysis and Interpretation section.

Techniques to Demonstrate

Demonstrate titration procedures, showing the proper way to add drops from burets or droppers and to swirl the flask after adding each drop. Show the endpoint of the titration. Remind students that the pink color that first appears often disappears when the flask is swirled.

Post-Lab

Disposal

Set out three disposal containers: one for unused acid solutions, one for unused base solutions, and one for partially neutralized substances and the contents of the *Waste* beaker. One at a time, slowly combine the solutions while stirring. Adjust the pH of the final waste liquid with 1.0 M acid or base until the pH is between 5 and 9. Pour the neutralized liquid down the drain.

Answers to Analysis and Interpretation

1. $CaCO_3(s) + 2HCl(aq) \longrightarrow CO_2(g) + H_2O(l) + CaCl_2(aq)$

2. $HCl(aq) + NaOH(aq) \longrightarrow NaCl(aq) + H_2O(l)$

4. Without emptying the graduated cylinder, add an additional 20 drops from the dropper as before, and record the new volume as the final volume for Trial 2. Repeat this procedure once more for Trial 3.

5. Repeat Steps **3** and **4** for the second thin-stemmed dropper. Label this dropper *Base*. ***Do not use this dropper for the acid solution.***

6. Make sure that the three trials produce data that are similar to each other. If one is greatly different from the others, perform steps **3–5** over again. If you're still waiting for the eggshell in the drying oven, calculate and record in the first data table the total volume of the drops and the average volume per drop.

7. Remove the eggshell and beaker from the oven. Cool them in a desiccator. Record the mass of the entire eggshell in the second data table. Place half of the shell into a clean mortar and grind to a very fine powder. This will save time when dissolving the eggshell. (If time permits, dry again and cool in the desiccator.)

Technique

8. Measure the mass of a piece of weighing paper. Transfer about 0.1 g of ground eggshell to a piece of weighing paper, and measure the eggshell's mass as accurately as possible. Record the mass in the second data table. Place this eggshell sample in a clean 50 mL bottle or Erlenmeyer flask.

9. Fill the acid dropper with the 1.00 M HCl acid solution, and then empty the dropper into an extra 100 mL beaker. Label the beaker *Waste*. Fill the base dropper with the 1.00 M NaOH base solution, and then empty the dropper into the 100 mL beaker.

10. Fill the acid dropper once more with 1.00 M HCl. Using the acid dropper, add exactly 150 drops of 1.00 M HCl to the bottle (or flask) with the eggshell. Swirl gently for 3 to 4 min. Wash down the sides of the flask with about 10 mL of distilled water. Using a third dropper, add 2 drops of phenolphthalein solution. Record the number of drops of HCl used in the second data table.

11. Fill the base dropper with the 1.00 M NaOH. Slowly add NaOH from the base dropper into the bottle or flask with the eggshell mixture until a faint pink color persists, even after it is swirled gently. It may help to use a white piece of paper as a background, so you will be able to see the color as soon as possible. ***Be sure to add the base drop by drop, and be certain the drops end up in the reaction mixture and not on the sides of the bottle or flask. Keep careful count of the number of drops used.*** Record the number of drops of base used in the second data table.

Cleanup and Disposal

12. Dispose of any unused solutions, as well as the neutralized solutions, in the containers designated by your teacher. Remember to wash your hands thoroughly before leaving the laboratory.

762 | Exploration 13A

Sample Data

Mass of entire eggshell	5.27 g
Mass of ground eggshell sample	0.11 g
Number of drops of 1.00 M HCl added	150 drops
Number of drops of 1.00 M NaOH added	99 drops

Analysis and Interpretation

1. ***Organizing Ideas***
The calcium carbonate in the eggshell sample undergoes a double-replacement reaction with the hydrochloric acid in step **9**. Then, the carbonic acid that was formed decomposes. Write a balanced chemical equation for these reactions. (Hint: the gas observed was carbon dioxide.)

2. ***Organizing Ideas***
Write the balanced chemical equation for the acid/base neutralization of the excess unreacted HCl with the NaOH.

3. ***Organizing Data***
Make the necessary calculations from the first data table to find the number of milliliters in each drop. Using this mL/drop ratio, convert the number of drops of each solution in the second data table to volumes in milliliters.

4. ***Organizing Data***
Calculate what volume of the HCl solution was neutralized by the NaOH. (Hint: this relationship was discussed in Section 13-3.) Then subtract this amount from the initial volume of HCl to determine how much HCl reacted with $CaCO_3$.

Conclusions

5. ***Organizing Data***
Calculate the number of moles of $CaCO_3$ that reacted with the HCl. Determine the mass of $CaCO_3$. Then calculate the percentage of $CaCO_3$ in the eggshell sample.

6. ***Inferring Conclusions***
Compare your results to those of the farmer, given in the Problem section of this lab. Explain whether or not you think the farmer's eggs show signs of PCB-induced weaknesses. (Hint: consider how much of a difference in values is enough to indicate a link to PCBs.)

7. ***Evaluating Methods***
Other workers in a lab in another city have also tested eggs, and found that a normal eggshell is about 97% $CaCO_3$. Calculate the percent error for your measurement.

Extensions

1. ***Inferring Conclusions***
Calculate an estimate of the mass of $CaCO_3$ present in the entire eggshell, based on your results. (Hint: apply the percent composition of your sample to the mass of the entire eggshell.)

2. ***Designing Experiments***
What possible sources of error can you identify with this procedure? If you can think of ways to eliminate them, ask your teacher to approve your plan, and run the procedure again.

3. ***Applying Ideas***
Find out how much calcium carbonate is in another type of shell (clam, oyster, snail, etc.).

4. ***Research and Communications***
Investigate the procedures and costs for cleaning up a PCB-tainted site. Debate about who should pay for the cleanup operations.

3. Sample calculation for acid pipet:

60 drops is equivalent to 2.40 mL;

$$\frac{2.40 \text{ mL}}{60 \text{ drops}} = 0.040 \text{ mL/drop}$$

Sample calculation for base pipet:

60 drops is equivalent to 2.46 mL;

$$\frac{2.46 \text{ mL}}{60 \text{ drops}} = 0.041 \text{ mL/drop}$$

150 drops of acid × 0.040 mL /drop = 6.00 mL of acid

99 drops of base × 0.041 mL /drop = 4.06 mL of base

4. 4.06 mL of 1.0 M acid reacted with the 4.06 mL of 1.0 M base, so 1.94 mL of the 1.0 M acid reacted with the eggshell.

$$1.94 \text{ mL acid} \times \frac{1 \text{ L}}{1000 \text{ mL}} \times \frac{1.0 \text{ mol}}{1 \text{ L}} = 1.94 \times 10^{-3} \text{ mol HCl}$$

Conclusions

5. 1.94×10^{-3} mol HCl × $\frac{1 \text{ mol } CaCO_3}{2 \text{ mol HCl}}$ = 9.70×10^{-4} mol $CaCO_3$

9.70×10^{-4} mol $CaCO_3$ × $\frac{100.09 \text{ g}}{1 \text{ mol}}$ = 0.0971 g $CaCO_3$

$\frac{0.0971 \text{ g } CaCO_3}{0.11 \text{ g eggshell}} \times 100$ = 88% $CaCO_3$

6. Student answers will vary. Students should recognize that the difference between the values should be relatively large to indicate a link.

7. $\frac{97\% - 88\%}{97\%} = 9.3\%$ error

Extensions

1. 5.27 g total eggshell × $\frac{88\% \ CaCO_3}{100\% \text{ total eggshell}}$ = 4.6 g $CaCO_3$

2. Students' suggestions for improving the procedure will vary. Possible suggestions include using larger samples of eggshells. Be sure answers are safe and include carefully planned procedures.

3. Student answers will vary, but should be about 85–95%.

4. Student answers will vary. Be certain students recognize that the costs involved are very large.

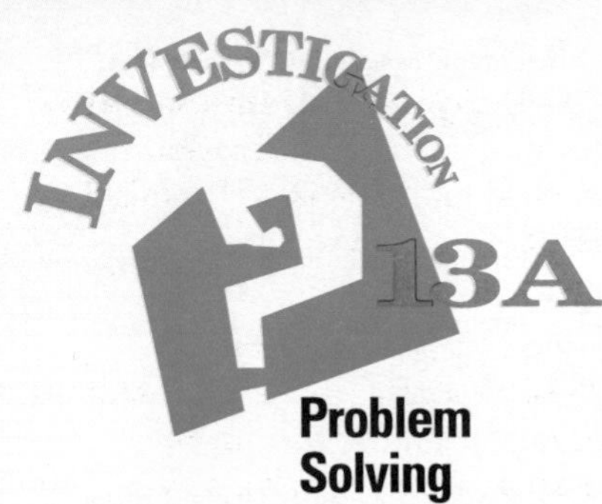

Objectives

Students will:

- use appropriate lab safety procedures.
- measure the pH of a solution using pH paper, a pH meter, or the LEAP System with pH probe.
- choose the proper reagent for titrating a solution.
- measure volume with a buret or by calibrating a dropper to measure the number of drops per milliliter.
- perform a titration and recognize the endpoint.
- use solution stoichiometry calculations to determine the concentration of an unknown solution.
- design and implement their own procedure.

Planning

Recommended Time: 1 lab period
Prerequisite: Exploration 13A

Materials/Solution Preparation

1. To prepare the unknown (0.20 M HCl is recommended) follow the precautions in the Teacher's Notes for Exploration 13A. Add 16.5 mL of concentrated HCl to enough distilled water to make 1.00 L of solution. Add the acid slowly, and stop to stir it in order to avoid overheating. (NaOH requires another indicator, so it is not recommended for the unknown.)
2. To prepare 0.50 M NaOH, add 20.0 g of NaOH pellets to enough distilled water to make 1.00 L of solution. Add a few pellets at a time, and stop to stir it in order to avoid overheating.
3. To prepare phenolphthalein solution, dissolve 1.00 g phenolphthalein in 50.0 mL denatured alcohol, and add 50.0 mL of distilled water.
4. The LEAP System with pH probe can be used to monitor the titration and its end point. Follow the instructions given in Exploration 1B.
5. The procedure can be modified to use small-scale titration techniques if the calibration procedures for drops/mL are followed as described in Exploration 13A.

Sample Data

Solution	Trial 1	Trial 2
Unknown (HCl)	25.0 mL	25.0 mL
0.5 M NaOH	9.9 mL	10.0 mL

Analysis

pH < 7, therefore unknown is HCl.

Average amount of NaOH: $10.0 \times 10^{-3}\ \text{L} \times \dfrac{0.500\ \text{M}}{1\ \text{L}} = 5.00 \times 10^{-3}\ \text{mol NaOH}$

$$5.00 \times 10^{-3}\ \text{mol NaOH} \times \frac{1\ \text{mol HCl}}{1\ \text{mol NaOH}} \times \frac{1}{25.0 \times 10^{-3}\ \text{L}} = 0.200\ \text{M HCl}$$

$$3\ \text{ponds} \times \frac{1.75 \times 10^{7}\ \text{L}}{1\ \text{pond}} \times \frac{0.200\ \text{M HCl}}{1\ \text{L HCl}} \times \frac{1\ \text{mol NaOH}}{1\ \text{mol HCl}} = 1.05 \times 10^{7}\ \text{mol NaOH}$$

(4.21×10^{5} kg NaOH for neutralization)

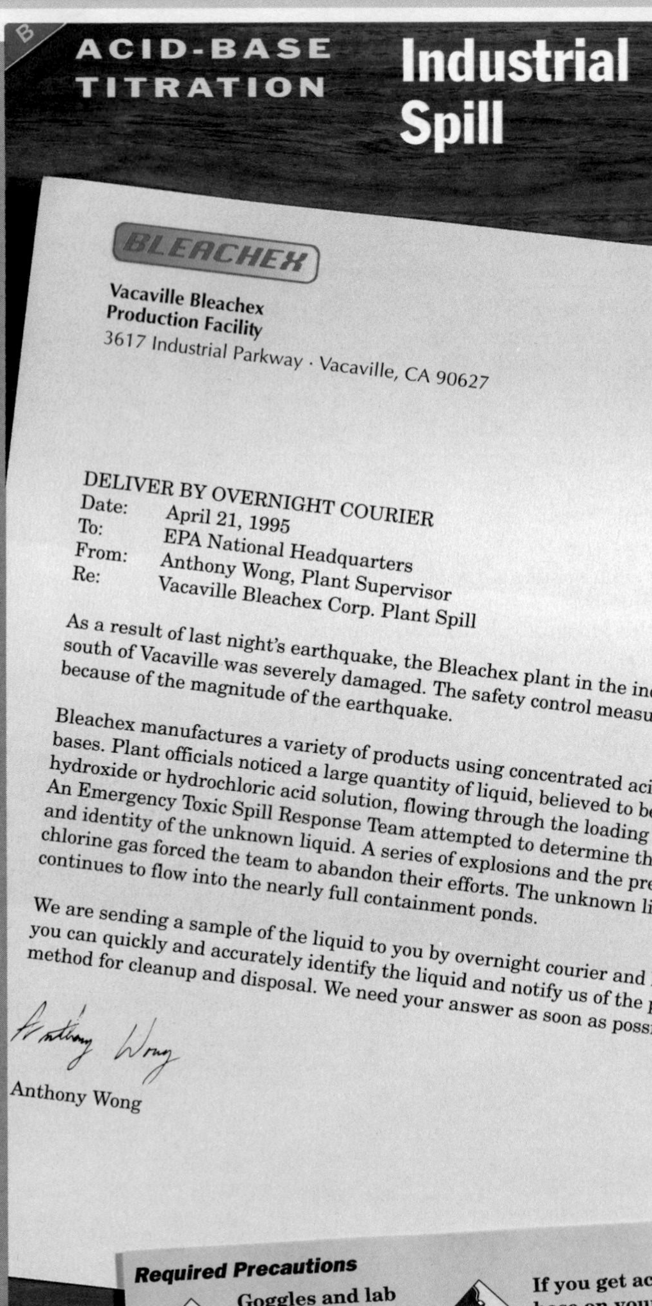

B

ACID-BASE TITRATION

Industrial Spill

BLEACHEX

Vacaville Bleachex
Production Facility
3617 Industrial Parkway · Vacaville, CA 90627

DELIVER BY OVERNIGHT COURIER
Date: April 21, 1995
To: EPA National Headquarters
From: Anthony Wong, Plant Supervisor
Re: Vacaville Bleachex Corp. Plant Spill

As a result of last night's earthquake, the Bleachex plant in the industrial park south of Vacaville was severely damaged. The safety control measures failed because of the magnitude of the earthquake.

Bleachex manufactures a variety of products using concentrated acids and bases. Plant officials noticed a large quantity of liquid, believed to be sodium hydroxide or hydrochloric acid solution, flowing through the loading bay doors. An Emergency Toxic Spill Response Team attempted to determine the source and identity of the unknown liquid. A series of explosions and the presence of chlorine gas forced the team to abandon their efforts. The unknown liquid continues to flow into the nearly full containment ponds.

We are sending a sample of the liquid to you by overnight courier and hope that you can quickly and accurately identify the liquid and notify us of the proper method for cleanup and disposal. We need your answer as soon as possible.

Anthony Wong

Anthony Wong

Required Precautions

Goggles and lab aprons must be worn at all times.

Do not touch or taste any chemicals. Always wash your hands thoroughly when finished.

If you get acid or base on your skin or clothing, wash it off at the sink while calling to your teacher. If you get acid or base in your eyes, immediately flush it out at the eyewash while calling to your teacher.

Required Precautions

- Goggles and a lab apron must be worn at all times.
- Read all safety cautions, and discuss them with your students.
- Students should not handle concentrated acids.
- In case of an acid or base spill, first dilute with water. Then, mop up the spill with wet cloths while wearing disposable plastic gloves.
- If a pH meter or the LEAP System with pH probe is used, the precautions listed in the teacher's notes on page 652 must be followed to avoid electric shock.

Memorandum

CheMystery Labs, Inc.
52 Fulton Street
Springfield, VA 22150

Date: April 21, 1995

To: Cicely Jackson

From: Marissa Bellinghausen

This is a rush job from the EPA. We have been promised \$300 000, three times their normal fee of \$100 000, so give this project top priority!

First, we must determine the pH of the unknown so that we know whether it is an acid or a base. Then titrate the unknown using a standard solution to determine its concentration so that we can advise Bleachex on the amount of neutralizing agents they will need for the three 1.75×10^7 L containment ponds.

Because we have only a limited sample, I need to approve your plans before you begin your lab work. Therefore, I need the following items.

- detailed one-page plan for your procedure with all necessary data tables (include multiple trials)
- detailed list of the equipment and materials you will need, with itemized and total costs

When finished, prepare a report in the form of a two-page letter that we can fax to Anthony Wong. The letter must include the following.

- identity of the unknown and its concentration
- pH of the unknown and how you determined it
- paragraph summarizing how you titrated the sample to determine its concentration
- detailed and organized data table
- detailed analysis section with calculations, a discussion of the multiple trials, and a statistical analysis of your precision
- your proposed method for cleanup and disposal, including amount of neutralizing agents necessary
- detailed invoice showing all costs, services, and time for this work

References

Refer to pages 508–513 for information on how to perform a titration with a buret. Before filling the buret, be sure it is clean. Then, rinse the buret three times with 5 mL of the standard solution each time before filling the buret. The equivalence point of the titration can be determined using an indicator or using the LEAP pH probe. Unlike the procedure your team recently completed using eggshells, this titration is a simple one, not a back-titration in which the unknown reacts with an excess of acid and the excess acid is titrated to determine how much must have reacted.

Spill Procedures/ Waste Disposal Methods

- In case of a spill, follow your teacher's instructions.
- Put excess chemicals and solutions in the containers designated by your teacher. Do not pour them down the sink or place them in the trash can.

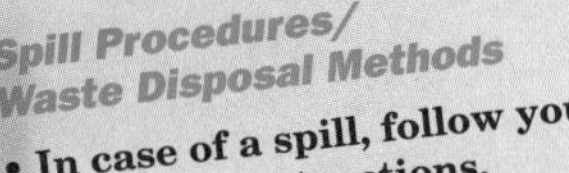

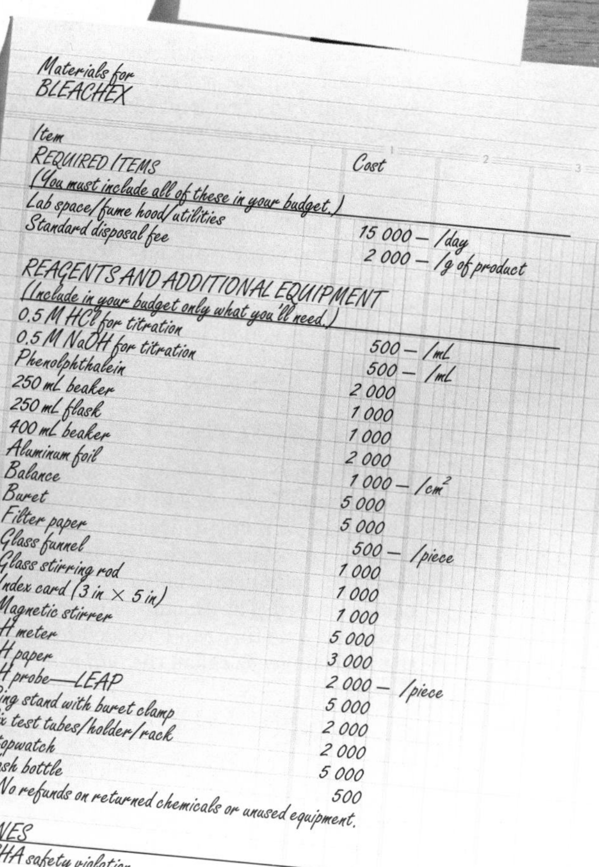

Materials for BLEACHEX

Item	Cost
REQUIRED ITEMS (You must include all of these in your budget.)	
Lab space/fume hood/utilities	15 000 – /day
Standard disposal fee	2 000 – /g of product
REAGENTS AND ADDITIONAL EQUIPMENT (Include in your budget only what you'll need.)	
0.5 M HCl for titration	500 – /mL
0.5 M NaOH for titration	500 – /mL
Phenolphthalein	2 000
250 mL beaker	1 000
250 mL flask	1 000
400 mL beaker	2 000
Aluminum foil	1 000 – /cm²
Balance	5 000
Buret	5 000
Filter paper	500 – /piece
Glass funnel	1 000
Glass stirring rod	1 000
Index card (3 in × 5 in)	1 000
Magnetic stirrer	5 000
pH meter	3 000
pH paper	2 000 – /piece
pH probe—LEAP	5 000
Ring stand with buret clamp	2 000
Six test tubes/holder/rack	2 000
Stopwatch	5 000
Wash bottle	500
* No refunds on returned chemicals or unused equipment.	
FINES	
OSHA safety violation	2 000/incident

Materials
(for each lab group)

- 0.5 M NaOH, 50 mL
- Phenolphthalein solution, 1 mL
- Unknown solution (0.20 M HCl), 50 mL
- 250 mL Erlenmeyer flask
- 250 mL beaker
- Buret tubes, 2
- Double buret clamp
- Index card
- pH paper
- Ring stand
- Wash bottle

Optional Equipment

- LEAP System with pH probe
- pH meter
- Stopwatch

Estimated cost of materials:
\$260 000–\$268 000

Student Orientation

Techniques to Demonstrate
Demonstrate the titration technique to be used. Remind students to swirl the flask after each addition.

Pre-lab Discussion
Begin by discussing the results and procedure used in the Exploration, especially any errors in lab technique that occurred. Review in detail the calculations that were necessary.

Suggest to students that they divide their sample into at least two parts and perform multiple trials. Refer them to the Exploration for ideas on how to proceed, but make it clear that procedures must be adjusted.

Tips for Evaluating the Pre-lab Requirements
If students plan to use more than 50 mL of the unknown or of the titrating solution, have them reconsider their procedure. Be sure students include a plan for identifying whether the unknown is an acid or a base before titrating.

Proposed Procedure

The unknown can be identified as an acid with phenolphthalein, pH paper, pH meter, or LEAP System with pH probe. Place 25.0 mL of unknown in a flask. After cleaning and rinsing a buret with 0.50 M NaOH, fill it with NaOH. Add small amounts of NaOH from the buret into the flask, swirling the flask after each addition, until the endpoint is reached. Repeat the procedure with another 25.0 mL of unknown.

Post-Lab

Disposal
Set out three disposal containers: one for unused acid solutions, one for unused base solutions, and one for partially neutralized substances and the contents of the *Waste* beaker. One at a time, slowly combine the solutions while stirring. Adjust the pH of the final waste liquid with 1.0 M acid or base until the pH is between 5 and 9. Pour the neutralized liquid down the drain.

EXPLORATION 13B

Objectives

Students will:

- use appropriate lab safety procedures.
- measure pH using pH paper, a pH meter, or the LEAP System with pH probe.
- measure volume using graduated cylinders.
- prepare a variety of solutions using different ratios of weak acids and weak bases.
- determine which solution is the best buffer by adding acid and base with a dropper or buret.
- measure the buffering capacity for the best buffer.

Planning

Recommended Time:

2 lab periods (Determining buffering capacity can be done in the second period.)

Materials

(for each lab group)

- 0.10 M $HC_2H_3O_2$, 50 mL
- 0.10 M $NaC_2H_3O_2$, 50 mL
- Distilled water
- Dropper bottle with 1.0 M HCl, 15 mL
- Dropper bottle with 1.0 M NaOH, 15 mL
- 10 mL graduated cylinder
- 50 mL beakers, 2
- 400 mL beaker
- pH meter or pH paper

LEAP System Option

- buret with 0.10 M HCl, 50 mL
- buret with 0.10 M NaOH, 50 mL
- LEAP System with pH probe
- Ring stand with buret clamp
- Stopwatch

Solution/Material Preparation

1. To prepare 0.10 M $HC_2H_3O_2$, follow the required precautions and slowly add 5.7 mL of glacial acetic acid to enough distilled water to make 1.00 L of solution.
2. To prepare 0.10 M $NaC_2H_3O_2$, slowly add 13.6 g of $NaC_2H_3O_2 \cdot 3H_2O$ to enough distilled water to make 1.00 L of solution.

Exploration 13B Buffering Capacity

Situation

You work in the research and development department of a company that wants to create a new hair conditioner product. Experiments have determined that the hair conditioner performs best at a pH of about 5.0. You need to determine which proportions of acetic acid and sodium acetate will work best to maintain a pH of 5.0.

Background

Acetic acid is a weak acid. When it is dissolved in water, only a small proportion of the molecules ionize. Similarly, sodium acetate is a weak base. When dissolved in water, a few acetate ions are able to bond to hydrogen to form acetic acid.

As a result, a mixture of these two compounds dissolved in water can serve as a buffer. When hydronium ions are added, they react with the acetate anions to form acetic acid, which exists mainly as non-ionized acetic acid molecules. In other words, acid is added, but the pH doesn't change.

Hydronium ion, H_3O^+

$$C_2H_3O_2^-(aq) + H_3O^+(aq) \rightleftharpoons HC_2H_3O_2(aq) + H_2O(l)$$

On the other hand, if hydroxide ions are added to this buffer solution, they react with any hydronium ions present to form non-ionized water molecules. But then, the acetic acid molecules ionize, restoring equilibrium to the buffer system. Once more, base is added, but the pH doesn't change.

$$OH^-(aq) + H_3O^+(aq) \rightleftharpoons 2H_2O(l)$$

$$HC_2H_3O_2(aq) + H_2O(l) \rightleftharpoons H_3O^+(aq) + C_2H_3O_2^-(aq)$$

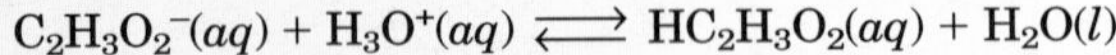

The effectiveness of any buffer system depends on the ratio of weak acid to weak base in the solution. For every ratio, there is a limit to how many hydronium ions or hydroxide ions any buffer system can absorb. Eventually, the pH rises or drops substantially because the limit has been exceeded. The amount of hydronium ions or hydroxide ions that can be added to a buffer before the pH changes is called the *buffering capacity* of the system.

Problem

To determine whether the company can use the acetic acid and sodium acetate buffer in the new hair conditioner product, you will need to do the following.

- determine the ratio of acetic acid to sodium acetate that provides the best buffer
- titrate a solution with this ratio, and determine its buffering capacity

Objectives

Demonstrate **proficiency in measuring pH and performing a titration.**

Relate **the ability of a solution to absorb acid or base and still maintain its pH to its buffering capacity.**

Determine **which ratio of acetic acid to sodium acetate provides the most efficient buffer for this system.**

Graph **the curve for the titration of the acetic acid and acetate buffer system.**

Interpret **the shape and size of the buffer titration curve.**

Calculate **the buffering capacity for the ratio you determined.**

Required Precautions

- Goggles and a lab apron must be worn at all times.
- Read all safety cautions, and discuss them with your students.
- Students should not handle concentrated acid solutions.
- In case of an acid or base spill, first dilute with water. Then, mop up the spill with wet cloths or a wet cloth mop while wearing disposable plastic gloves. Designate separate cloths or mops for acid and base spills.

Safety

Always wear goggles and an apron to provide protection for your eyes and clothing. If you get a chemical in your eyes, immediately flush it out at the eyewash station while calling to your teacher. Know the locations of the emergency lab shower and eyewash and how to use them.

Do not touch any chemicals. If you get a chemical on your skin or clothing, wash it off at the sink while calling to your teacher. Make sure you carefully read the labels and follow the directions on all containers of chemicals that you use. Do not taste any chemicals or items used in the laboratory. Never return leftovers to their original containers; take only small amounts to avoid wasting supplies.

Never put broken glass or ceramics in a regular waste container. They should be disposed of separately. Put the glass and ceramic pieces in the container designated by your teacher.

Acids and bases are corrosive. If any gets on you, wash the area immediately with running water. Call your teacher in the event of an acid or base spill. Acid or base spills should be cleaned up promptly.

Always clean up the lab and all equipment after use, and dispose of substances according to proper disposal methods. Wash your hands thoroughly before you leave the lab after all lab work is finished.

Preparation

1. Organizing Data

Prepare two data tables in your lab notebook.

- The first data table should have fifteen rows and seven columns.
- Put these labels in the columns of the first row: *Solution, Original pH, pH after 1 drop, pH after 2 drops, pH after 3 drops, pH after 4 drops,* and *pH after 5 drops.*
- Put the following labels in the first column of the second through fifteenth rows: *Solution 1/HCl, Solution 1/NaOH, Solution 2/HCl, Solution 2/NaOH, Solution 3/HCl, Solution 3/NaOH, Solution 4/HCl, Solution 4/NaOH, Solution 5/HCl, Solution 5/NaOH, Solution 6/HCl, Solution 6/NaOH, Solution 7/HCl,* and *Solution 7/NaOH.*
- The second data table is not necessary if you are working with a LEAP System with a pH probe. It should have 3 rows and 22 columns. If space is a problem, break the table, making two with 11 columns instead of one with 22.
- Put the following labels in the first row: *Solution, Original pH, pH after 1 drop, pH after 2 drops,* and so on, up to *pH after 20 drops.*
- In the first box of the second row, insert the label *HCl.* In the first box of the third row, insert the label *NaOH.*

Materials

- **Distilled water**
- **0.10 M $HC_2H_3O_2$**
- **0.10 M $NaC_2H_3O_2$**
- **1.0 M HCl in dropper bottle**
- **1.0 M NaOH in dropper bottle**
- **50 mL beakers, 2**
- **400 mL beaker**
- **10 mL graduated cylinder**
- **pH meter or pH paper**

Procedure with LEAP System and pH probe

- **0.10 M NaOH in buret**
- **0.10 M HCl in buret**
- **LEAP System with pH probe**
- **Ring stand with buret clamp**
- **Stopwatch**

- Wear goggles, a face shield, impermeable gloves, and a lab apron when you prepare the acetic acid, the HCl, and the NaOH. Work in a hood known to be in operating condition, with another person present nearby to call for help in case of an emergency. Be sure you are within 30 s walking distance of a working safety shower and eyewash station.
- If the LEAP System with pH probe is used, the precautions listed in the teacher's notes on page 652 must be followed to avoid electric shock.

3. To prepare 1.00 M HCl, follow the required precautions and add 82.6 mL of concentrated HCl to enough distilled water to make 1.00 L of solution. Add the acid slowly, and stop to stir it in order to avoid overheating. To prepare 0.10 M HCl for use with the LEAP system with pH probe, add 8.3 mL of concentrated HCl to enough distilled water to make 1.00 L of solution.
4. To prepare 1.00 M NaOH, follow the required precautions and add 40.0 g of NaOH pellets to enough distilled water to make 1.00 L of solution. Add a few pellets at a time, and stop to stir it in order to avoid overheating. To prepare 0.10 M NaOH for use with the LEAP system with pH probe, add 4.00 g of NaOH pellets to enough distilled water to make 1.00 L of solution.
5. Check the pH of your distilled water, which should be approximately 7. If it is low, boil the water shortly before the lab to drive off dissolved carbon dioxide. Store the boiled water in containers with tight-fitting lids.
6. A pH meter or LEAP System with pH probe will provide far more precise data than even narrow-range pH paper.

Student Orientation

Pre-lab Discussion

Review the discussion from the Background section that describes how buffers are able to maintain constant pH even when acid or base is added. It may help to use a model system with 10 formula units each of acetic acid and sodium acetate. Be sure to link this discussion with the concepts of equilibrium from Chapter 12.

Techniques to Demonstrate

Demonstrate the proper titration method and how to calibrate the pH meter or LEAP System with pH probe, if used.

Post-Lab

Disposal

Set out three disposal containers: one for unused acid solutions, one for unused base solutions, and one for partially neutralized substances and the contents of the *Waste* beaker. One at a time, slowly combine the solutions while stirring. Adjust the pH of the final waste liquid with 1.0 M acid or base until the pH is between 5 and 9. Pour the neutralized liquid down the drain.

Sample Data

Determining best ratio for buffer.

Solution	Original pH	pH after 1 drop	pH after 2 drops	pH after 3 drops
#1 HCl	7.2	6.4	6.2	6.1
#1 NaOH	7.2	11.2	11.5	11.4
#2 HCl	5.9	5.8	5.7	5.6
#2 NaOH	5.9	6.0	6.2	6.4
#3 HCl	5.5	5.4	5.3	5.3
#3 NaOH	5.4	5.6	5.7	5.8
#4 HCl	5.4	5.4	5.3	5.2
#4 NaOH	5.4	5.4	5.5	5.6
#5 HCl	5.2	5.2	5.1	4.9
#5 NaOH	5.2	5.3	5.4	5.5
#6 HCl	4.7	4.6	4.5	4.3
#6 NaOH	4.7	4.8	4.9	5.5
#7 HCl	4.4	4.1	3.9	3.7
#7 NaOH	4.4	4.6	4.9	5.0

2. Calibrate the pH meter or LEAP System with pH probe using a standard buffer of known pH. This step is not necessary if you are using pH paper.
3. Label the two 50 mL beakers *Solution* and *Solution + HCl*. Label the 400 mL beaker *Waste*.

Technique

Determining best ratio for buffer

4. Measure 10.0 mL of sodium acetate in a 10 mL graduated cylinder, and pour it into the 50 mL beaker labeled *Solution*. Measure 10.0 mL of distilled water, and add it to the beaker, swirling the mixture gently to mix thoroughly. This is *Solution 1*.

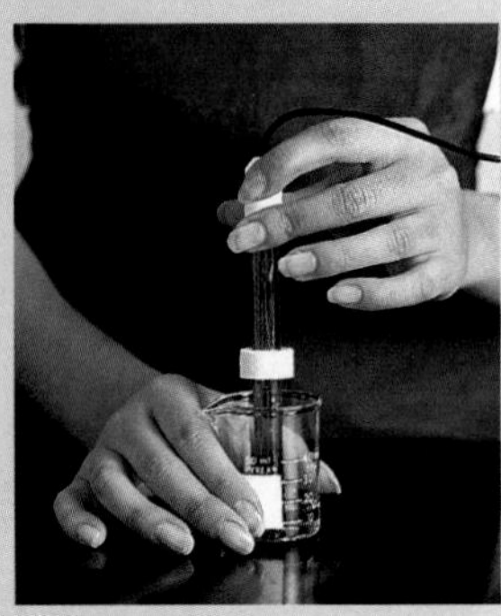

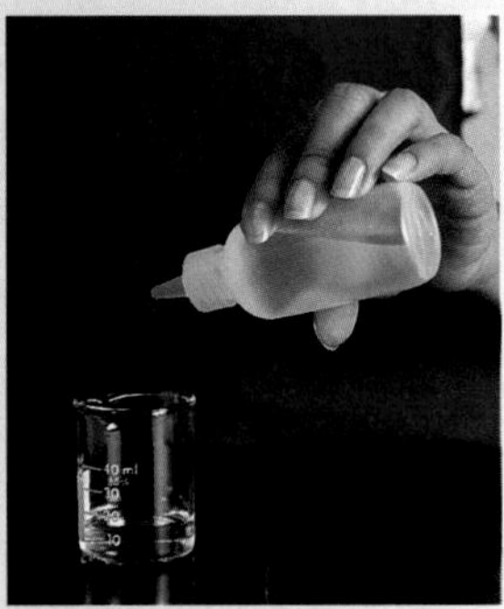

5. Measure the initial pH of *Solution 1*, and record it in the first data table under *Original pH*.
6. Pour about half of *Solution 1* into the beaker labeled *Solution + HCl*. Add one drop of 1.0 M HCl, and swirl the beaker gently. Measure and record the pH in the first data table.
7. Continue adding HCl drop by drop. After each drop, record the pH. Stop when 5 drops have been added.
8. Repeat steps **4–7** with the remainder of *Solution 1* that was left in the beaker labeled *Solution*, but this time add drops of 1.0 M NaOH. Measure and record the pH in the first data table.
9. Empty both 50 mL beakers into the *Waste* beaker. Clean both beakers and the graduated cylinder. Rinse them several times with distilled water.
10. Repeat steps **5–9** for *Solutions 2–7*. For best results, prepare solutions one at a time.
    - *Solution 2* should contain 8.0 mL $NaC_2H_3O_2$, 2.0 mL $HC_2H_3O_2$, and 10.0 mL of distilled water.
    - *Solution 3* should contain 6.0 mL $NaC_2H_3O_2$, 4.0 mL $HC_2H_3O_2$, and 10.0 mL of distilled water.
    - *Solution 4* should contain 5.0 mL $NaC_2H_3O_2$, 5.0 mL $HC_2H_3O_2$, and 10.0 mL of distilled water.
    - *Solution 5* should contain 4.0 mL $NaC_2H_3O_2$, 6.0 mL $HC_2H_3O_2$, and 10.0 mL of distilled water.
    - *Solution 6* should contain 2.0 mL $NaC_2H_3O_2$, 8.0 mL $HC_2H_3O_2$, and 10.0 mL of distilled water.
    - *Solution 7* should contain 10.0 mL $HC_2H_3O_2$ and 10.0 mL of distilled water.

Sample Data

Determining buffering capacity for solution: 5 mL $HC_2H_3O_2$, 5 mL $NaC_2H_3O_2$

Solution	Original pH	After 1 drop	After 2 drops	After 3 drops	After 4 drops	After 5 drops	After 6 drops	After 7 drops	After 8 drops	After 9 drops
HCl	4.9	4.9	4.9	4.9	4.6	4.5	4.4	4.2	3.9	3.6
NaOH	4.9	4.9	5.1	5.2	5.3	5.3	5.3	5.5	5.6	5.7

11. Clean the two beakers and the graduated cylinder, and rinse them several times with distilled water.

Determining buffering capacity

12. Examine the pH measurements in your first data table. Calculate the difference between the pH values obtained after adding 5 drops of HCl and the pH obtained after adding 5 drops of NaOH to each solution. Which ratio of weak acid to weak base shows the smallest change in pH whether acid or base is added?

13. In the beaker labeled *Solution + HCl*, prepare another batch of the solution you identified in step **12** as the best buffer. If you are working with a pH meter or pH paper, follow step **13a**. If you are using the LEAP System with a pH probe, follow step **13b**.

a. pH meter or pH paper
Add 1.0 M HCl one drop at a time. Measure and record the pH after each drop. Stop adding HCl when the pH drops suddenly, which indicates that the solution has lost its buffering capacity. Empty the solution into the *Waste* beaker, and continue with step **14**.

b. LEAP system with pH probe
Calibrate your buret to release liquid from the tip at the rate of 10 drops per second. To do this, put the *Waste* beaker under the buret, and fill the buret with water. Use a stopwatch to measure drops per second, and practice positioning the stopcock so that you have a rate of 10 drops per second.

When you are ready to proceed, empty the buret, and fill it with 0.10 M HCl. Make sure your probe is calibrated and the cap is removed. Put the probe into the solution you are testing. On the computer, move the cursor to the green portion of the stoplight and start the experiment. After you have taken the initial reading and the probe has stabilized, set the stoplight to red, and erase the data collected during calibration.

Start the experiment again, and begin the titration. Add HCl until approximately 950 readings have been taken or until the buffer solution's pH is no longer constant. (Only 1000 data points will be saved. If you continue taking data after this point, you will lose earlier data.) When finished, set the stoplight to red, and print your graph. Empty the solution into the *Waste* beaker.

14. In the beaker labeled *Solution*, prepare another batch of the solution you identified in step **12** as the best buffer. If you are working with a pH probe or pH paper, follow step **14a.** If you are using the LEAP System with a pH probe, follow step **14b**.

a. pH probe or pH paper
Add 1.0 M NaOH one drop at a time. Measure and record the pH after each drop. Stop adding NaOH when the pH rises suddenly, indicating that the solution has lost its buffering capacity. Empty the solution into the *Waste* beaker, and continue with step **15**.

Answers to Analysis and Interpretation

1. Mole ratios—mol $HC_2H_3O_2$: mol $NaC_2H_3O_2$
   solution 1—0:10
   solution 2—1:4
   solution 3—2:3
   solution 4—1:1
   solution 5—3:2
   solution 6—4:1
   solution 7—10:0
2. Solution 3 would maintain a good pH when acids are added, and solution 5 would maintain a good pH when bases are added.
3. Of all of the solutions, solution 4 kept an average pH closest to 5.0, the desired pH, whether 5 drops of acid or base were added.
4.

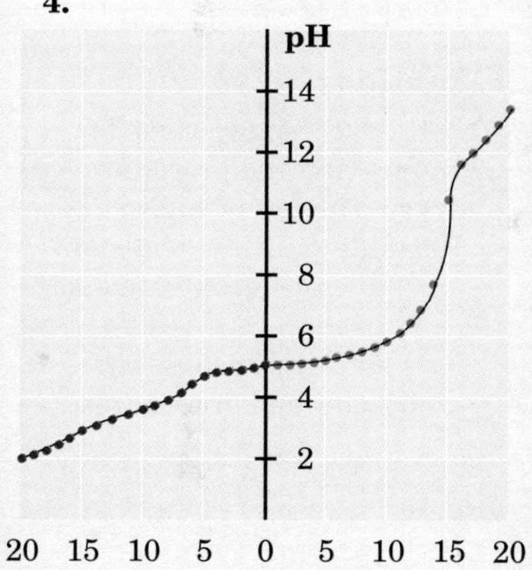

■ Drops of HCl added
■ Drops of NaOH added

After 10 drops	After 11 drops	After 12 drops	After 13 drops	After 14 drops	After 15 drops	After 16 drops	After 17 drops	After 18 drops	After 19 drops	After 20 drops
3.3	3.1	2.9	2.8	2.7	2.6	2.5	2.4	2.3	2.2	2.1
5.8	5.9	6.2	6.5	7.7	10.4	10.8	11.0	11.3	11.3	11.9

Conclusions

5. Solution 4 has a 1:1 ratio of acid to base. The change in pH for this ratio is the smallest of the seven ratios.

6. For the sample data, in the pH range from 4.0 to 6.0, the slope changes the least. This is the most useful range for this buffer. It is appropriate for maintaining a pH of approximately 5.

7. For the sample data, 7 drops of acid were added before the pH went below 4.0; 11 drops of base were added before the pH went above 6.0.

8. If a solution is exposed to the air, more CO_2 will dissolve in it, creating more H_2CO_3 and eventually more H_3O^+.

Extensions

1. Students' suggestions for improving the procedure will vary. They may suggest performing multiple trials or using less concentrated acid and base. Be sure answers are safe and include carefully planned procedures.

2. $HC_2H_3O_2(aq) + H_2O(l) \longrightarrow C_2H_3O_2^-(aq) + H_3O^+(aq)$

$$K_a = \frac{[C_2H_3O_2^-][H_3O^+]}{[HC_2H_3O_2]} = 1.75 \times 10^{-5}$$

$$\frac{(0.10)[H_3O^+]}{(0.10)}$$

$[H_3O^+] = 1.8 \times 10^{-5}$ M

pH = 4.7

This pH is in the middle of the region of the relatively constant pH from item 6.

3. Because the value of the constant is a small number, relatively few of the products are formed. Thus, the equilibrium concentration of the original reactants will be very close to the initial concentrations.

4. Student answers will vary but should reflect a procedure similar to the one in this Exploration. Be sure answers are safe and include carefully planned procedures.

b. LEAP system with pH probe
Calibrate your buret to release liquid from the tip at the rate of 10 drops per second as before. When you are ready to proceed, empty the buret, and fill it with 0.10 M NaOH. Make sure your probe is calibrated and the cap is removed. Put the probe into the solution you are testing. On the computer, move the cursor to the green portion of the stoplight, and start the experiment. After the probe has stabilized, set the stoplight to red, and erase the data collected during calibration Start the experiment again, and begin the titration. Add NaOH until approximately 950 readings have been taken or until the buffer solution's pH is no longer constant. Then set the stoplight to red and print your graph. Empty the solution into the *Waste* beaker.

Cleanup and Disposal

15. Pour the solution in the *Waste* beaker into the container designated by your teacher. Place any leftover acid solutions in a designated container. Leftover basic solutions should be placed in a different designated container. Thoroughly clean the area.

Analysis and Interpretation

1. ***Organizing Data***
Determine the seven $NaC_2H_3O_2$-$HC_2H_3O_2$ mole ratios for the seven buffer solutions you prepared in steps **4–10**.

2. ***Applying Ideas***
If you needed a solution that had to maintain the same pH with additions of acid only, which solution would you pick? If you needed a solution that had to maintain the same pH with additions of base only, which solution would you pick?

3. ***Analyzing Data***
Using your data, explain why neither of the solutions you chose in item **2** was the buffer you chose in step **12**.

4. ***Organizing Data***
Plot the pH data from the two titrations on one graph. Label the y-axis with the pH of the solution. Position a vertical line at the midpoint of the x-axis. To the left of this vertical line, plot the increasing numbers of drops of HCl. To the right of the line, plot the increasing numbers of drops of NaOH.

Conclusions

5. ***Inferring Conclusions***
Which ratio of acetic acid to sodium acetate provides the best buffering system? Use your data to support your choice.

6. ***Interpreting Graphics***
Examine the way that the slope of your titration curve changes. In what pH range does the slope change the least? In what pH range would the acetic acid and sodium acetate buffer be most suitable? Could you use this buffer system to run a reaction that requires a pH of approximately 5?

7. ***Applying Ideas***
For the buffer you chose, how many drops of acid or base was it able to absorb before the pH changed by more than 1.0 unit?

8. ***Analyzing Methods***
When water is left out in air, the equilibrium reactions shown below can take place. With this knowledge, explain why it might be important to make pH measurements with solutions prepared one at a time, so that they are not left out in the air for long periods of time.

$$CO_2(g) + H_2O(l) \rightleftharpoons H_2CO_3(aq)$$

$$H_2CO_3(aq) + H_2O(l) \rightleftharpoons H_3O^+(aq) + HCO_3^-(aq)$$

Extensions

1. ***Designing Experiments***
What possible sources of error can you identify with this procedure? If you can think of ways to eliminate them, ask your teacher to approve your plan, and run the procedure again.

2. ***Applying Ideas***
Write the balanced equilibrium equation for the dissociation of acetic acid. Also write the equilibrium expression. Determine the numerical value for the acid-dissociation constant, K_a, by consulting a chemical handbook. Substitute the numerical values for K_a and the initial concentrations of $HC_2H_3O_2$ and $C_2H_3O_2^-$ into the equilibrium expression. Solve the equation for the $[H_3O^+]$, and convert your answer to pH. Does this result support your answer for item **6** in the Conclusions section?

3. ***Applying Ideas***
Using the value of the acid-dissociation constant, K_a, explain why the assumption was made that the equilibrium concentrations were the same as the initial concentrations for $HC_2H_3O_2$ and $C_2H_3O_2^-$.

4. ***Designing Experiments***
Buffers are found in a variety of over-the-counter medicines such as antacids and analgesics. Give a detailed procedure for determining the buffering capacity of a medication and preparing a titration curve. If your teacher approves your plan, carry out the experiment.

5. ***Designing Experiments***
Many organisms and biological systems need a narrow pH range to live. Give a detailed procedure for determining the buffering capacity of some living materials such as a potato, egg white, or milk. If your teacher approves your plan, carry out the experiment.

6. ***Applying Conclusions***
Chemists officially define buffering capacity as the number of moles of H_3O^+ or OH^- needed to cause 1.00 L of the buffered solution to undergo a 1.00-unit change in pH. Calculate the buffering capacity for your solution. (Hint: in order to figure out the number of moles of H_3O^+ or OH^-, you will need to determine the volume of a drop. This can be done by measuring the volume of 50 drops and dividing to determine the volume of an individual drop.)

7. ***Research and Communication***
Find out some ways that buffering is important in chemistry and physiology. Prepare a report on your findings.

5. Student answers will vary but should reflect a procedure similar to the one in this Exploration. Be sure answers are safe and include carefully planned procedures.

6. Student results for calibrating the dropper will vary.
Sample analysis: 40 drops/mL
For 1-unit change in pH, 8 drops of HCl or 11 drops of NaOH are needed.

$$1000\text{ mL buffer} \times \frac{8\text{ drops HCl}}{20\text{ mL buffer}} \times \frac{1.0\text{ mL acid}}{40\text{ drops HCl}} \times \frac{1\text{ L}}{1000\text{ mL}} \times \frac{1.0\text{ mol HCl}}{1\text{ L}} = 0.010\text{ mol HCl}$$

$$1000\text{ mL buffer} \times \frac{11\text{ drops NaOH}}{20\text{ mL buffer}} \times \frac{1.0\text{ mL acid}}{40\text{ drops NaOH}} \times \frac{1\text{ L}}{1000\text{ mL}} \times \frac{1.0\text{ mol NaOH}}{1\text{ L}} = 0.014\text{ mol NaOH}$$

7. Student answers will vary. Some may choose to study buffers in the body, such as the H_2CO_3 buffer system in the blood. Others may choose to study other applications, such as how well dissolved minerals in lakes can act as buffers during acid precipitation.

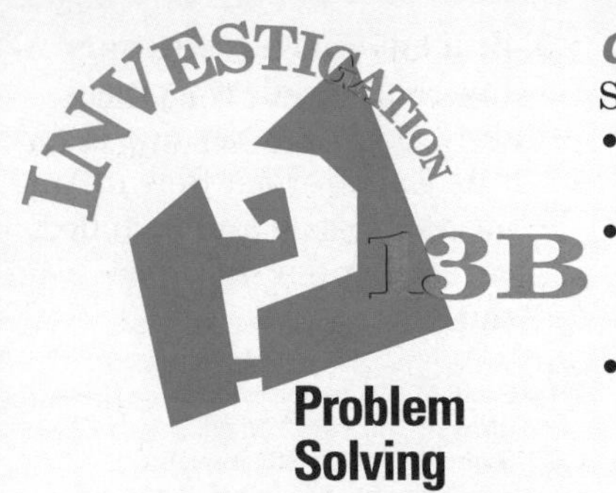

Objectives

Students will:

- use appropriate lab safety procedures.
- predict which buffer will be able to maintain a pH of 6.5–7.5.
- prepare a variety of solutions using different ratios of weak acids and weak bases.
- test which solution is the best buffer by adding acid and base with a dropper or buret.
- measure pH using pH paper, a pH meter, or the LEAP System with pH probe.
- measure the buffering capacity for the best buffer.
- design and implement their own procedure.

Planning

Recommended Time: 1 lab period
Prerequisite: Exploration 13B

Solution/Material Preparation

1. Wear goggles, a face shield, impermeable gloves, and a lab apron when you prepare acids or bases. Work in a properly functioning hood, with another person nearby to call for help in case of an emergency, and within 30 s walking distance from a properly functioning safety shower and eyewash station.
2. See the Teacher's Notes for Exploration 13B for information on preparing 1.0 M or 0.10 M HCl, 1.0 M or 0.10 M NaOH, and 0.1 M $NaC_2H_3O_2$ and $HC_2H_3O_2$ solutions.
3. To prepare 0.10 M $NH_3{\bullet}H_2O$, dissolve 6.8 mL of concentrated $NH_3{\bullet}H_2O$ (ammonia water, also known as NH_4OH) in 1.00 L of solution.
4. To prepare 0.10 M NH_4Cl, slowly add 5.3 g of NH_4Cl to enough distilled water to make 1.00 L of solution.
5. The weak acids and bases from which students will prepare the buffers can be dispensed from burets set up in strategic locations in the lab. If burets are not available, graduated pipets may be used.
6. A pH meter or LEAP System with pH probe will provide far more precise data than even narrow-range pH paper.

Student Orientation

Techniques to Demonstrate

Demonstrate the proper titration method and how to calibrate the pH meter or LEAP System with pH probe, if used. Discuss any errors in technique from the Exploration.

Sample Data and Analysis

Student answers will vary depending on measurement techniques. The data and analysis should be in a format similar to those from the Exploration. For the ammonia–ammonia chloride buffer, all pH values should be in the range of 7.5–10.5. The best buffer should be a 1:1 mixture of ammonia and ammonia chloride.

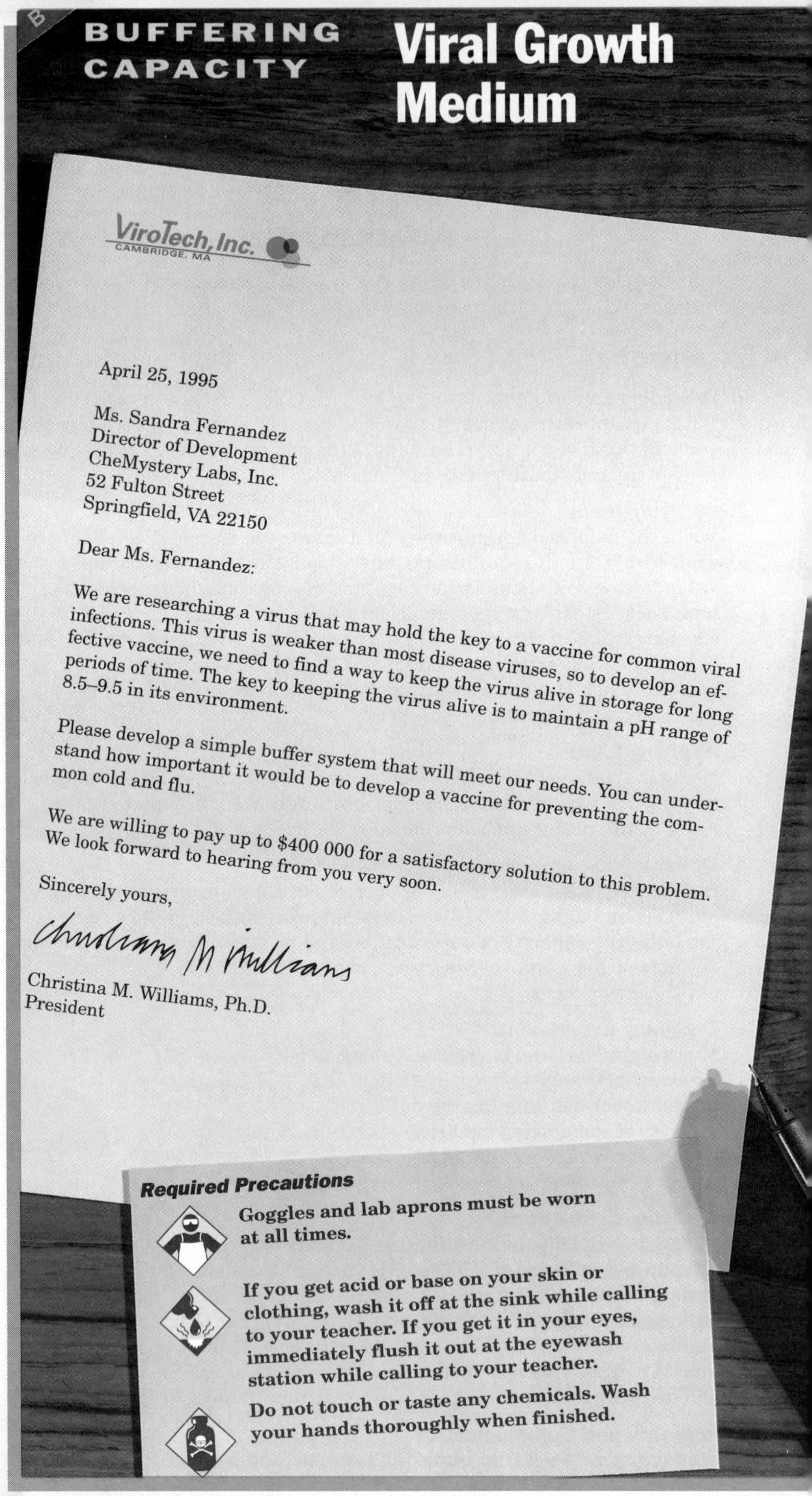

B

BUFFERING CAPACITY

Viral Growth Medium

ViroTech, Inc.
CAMBRIDGE, MA

April 25, 1995

Ms. Sandra Fernandez
Director of Development
CheMystery Labs, Inc.
52 Fulton Street
Springfield, VA 22150

Dear Ms. Fernandez:

We are researching a virus that may hold the key to a vaccine for common viral infections. This virus is weaker than most disease viruses, so to develop an effective vaccine, we need to find a way to keep the virus alive in storage for long periods of time. The key to keeping the virus alive is to maintain a pH range of 8.5–9.5 in its environment.

Please develop a simple buffer system that will meet our needs. You can understand how important it would be to develop a vaccine for preventing the common cold and flu.

We are willing to pay up to $400 000 for a satisfactory solution to this problem. We look forward to hearing from you very soon.

Sincerely yours,

Christina M. Williams

Christina M. Williams, Ph.D.
President

Required Precautions

Goggles and lab aprons must be worn at all times.

If you get acid or base on your skin or clothing, wash it off at the sink while calling to your teacher. If you get it in your eyes, immediately flush it out at the eyewash station while calling to your teacher.

Do not touch or taste any chemicals. Wash your hands thoroughly when finished.

Required Precautions

- Goggles and a lab apron must be worn at all times.
- Read all safety cautions, and discuss them with your students.
- Students should not handle concentrated acids.
- In case of an acid or base spill, first dilute with water. Then mop up the spill with wet cloths designated for that type of spill while wearing disposable plastic gloves.
- If the LEAP System with pH probe is used, the precautions listed in the teacher's notes on page 652 must be followed to avoid electric shock.

Memorandum

CheMystery Labs, Inc.
52 Fulton Street
Springfield, VA 22150

Date: April 26, 1995

To: Regina Walter

From: Sandra Fernandez

Virotech is an up-and-coming biotechnology and genetic-engineering firm. This is an excellent opportunity for our company to break into this rapidly expanding market for our services, so make sure your work is extremely efficient and accurate!

I suggest that we examine either a buffer system with ammonia and ammonium chloride or one with acetic acid and sodium acetate. Decide which buffer system you want to investigate. Determine the ratio of ingredients that provides the most efficient buffer at the pH specified in the letter. Before you begin your work, let me review the following items.

- buffer system you chose and why
- detailed one-page summary of your plan for the procedure
- any necessary data tables
- itemized list of equipment, with individual and total costs

After you complete the project, prepare a three-page report for Williams. It should include the following.

- proportions for the buffer system that works best, along with a brief description of how to prepare it
- discussion of the buffering capacity of the buffer system
- summary of your procedure
- detailed analysis and data sections showing all necessary calculations and the titration curve for the buffer
- itemized bill for materials used and services rendered

References

Refer to page 516–517 for information about buffered solutions. The procedure for determining the best proportion for a buffer and measuring buffering capacity are similar to one your team recently completed with the sodium acetate–acetic acid buffering system. Note that a different pH level is required in this Investigation.

Materials for
VIROTECH, INC.

Item	Cost
REQUIRED ITEMS (You must include all of these in your budget.)	
Lab space/fume hood/utilities	15 000 – /day
Standard disposal fee	2 000 – /g of product
REAGENTS and ADDITIONAL EQUIPMENT (Include in your budget only what you'll need.)	
0.10 M $HC_2H_3O_2$	1 000 – /mL
1.0 M HCl, in a dropper bottle or buret	8 000
0.10 M $NaC_2H_3O_2$	1 000 – /mL
0.10 M $NH_3 \cdot H_2O$	1 000 – /mL
0.10 M NH_4Cl	1 000 – /mL
1.0 M NaOH, in a dropper bottle or buret	8 000
50 mL beaker	1 000
400 mL beaker	2 000
10 mL graduated cylinder	1 000
Filter paper	500 – /piece
Glass funnel	1 000
Glass stirring rod	1 000
pH meter	3 000
pH paper	2 000 – /piece
pH probe—LEAP	5 000
Ring stand with buret clamp	2 000
Stopwatch	5 000
Wash bottle	500
Weighing paper	500 – /piece
*No refunds on returned chemicals or unused equipment.	
FINES	
OSHA safety violation	2 000 – /incident

Spill Procedures/ Waste Disposal Methods

- In case of a spill, follow your teacher's instructions.
- Place buffer solutions, leftover acid, and leftover base in separate disposal containers designated by your teacher.
- Clean the area and all equipment after use, and wash your hands before leaving the lab.

Materials

(for each lab group)

- 0.10 M $HC_2H_3O_2$, 50 mL
- 0.10 M $NaC_2H_3O_2$, 50 mL
- 0.10 M $NH_3 \cdot H_2O$, 50 mL
- 0.10 M NH_4Cl, 50 mL
- Distilled water
- Dropper bottle with 1.0 M HCl, 15 mL
- Dropper bottle with 1.0 M NaOH, 15 mL
- 10 mL graduated cylinder
- 50 mL beakers, 2
- 400 mL beaker
- pH paper or pH meter

LEAP System Option

- Buret with 0.10 M HCl, 50 mL
- Buret with 0.10 M NaOH, 50 mL
- LEAP System with pH probe
- Ring stand with buret clamp
- Stopwatch

Estimated cost of materials:
\$337 000–\$347 000

Pre-lab Discussion

Be sure to point out that the pH required in the Investigation is different from that in the Exploration, but leave it up to the students to decide which buffer to use. Some students will use their data from the Exploration to decide that the acetate buffer is unlikely to work well. Encourage students to perform multiple trials with their choice of buffers.

Tips for Evaluating the Pre-lab Requirements

Be sure that student data tables are complete and reflect the use of multiple trials and different combinations of the buffer ingredients.

Proposed Procedure

Prepare several different combinations of the buffer ingredients, as in the Exploration. Then test each by adding 1.0 M NaOH and 1.0 M HCl one drop at a time. Make another two batches of the combination that is best at maintaining a pH between 8.5 and 9.5. Measure the buffering capacity of this solution by adding drops of acid and base and measuring how many can be added before the pH changes by 1.0 unit from its original value.

Post-Lab

Disposal

Set out three disposal containers: one for unused acid solutions, one for unused base solutions, and one for partially neutralized substances and buffer solutions. Neutralize each solution with 1.0 M acid or base until the pH is between 5 and 9. Pour the neutralized liquid down the drain.

EXPLORATION 14A

Objectives
Students will:
- use appropriate lab safety procedures.
- prepare reaction mixtures of several different concentrations.
- observe chemical processes and interactions.
- measure time elapsed before the appearance of an indicator.
- relate measurements of time elapsed to reaction rates.
- define the relationship between reaction rate and concentration.

Planning
Recommended Time:
1 lab period

Materials
(for each lab group)
- Clock solution A, 5 mL ($1.0\ M\ H_2SO_4$, $Na_2S_2O_5$, and soluble starch)
- Clock solution B, 5 mL (KIO_3)
- Distilled or deionized water
- Eight-well microscale reaction strips, 2
- Fine-tipped dropper bulbs or small micro-tip pipets, 3
- Stopwatch or clock with second hand

Solution/Material Preparation:
1. To prepare Solution A, make a paste of 1 g of water-soluble starch and about 10 mL of water. Add 225 mL of boiling, distilled water. Reheat and boil for a few minutes. After solution has cooled, add 0.05 g of $Na_2S_2O_5$ (sodium metabisulfate) and 1.3 mL of $1.0\ M\ H_2SO_4$. Dilute with enough distilled water to make 250 mL of solution. (Note: this solution should be prepared within two months of use.)
2. To prepare 10 mL of 1.0 M H_2SO_4, observe the required precautions. Add 0.6 mL of concentrated H_2SO_4 to 9.4 mL of distilled water. Add the acid slowly, and stop to stir it in order to avoid overheating.

Reaction Rates

Objectives
Prepare and observe **several different reaction mixtures.**

Demonstrate **proficiency in measuring reaction rates.**

Relate **experimental results to a rate law that can predict the results of various combinations of reactants.**

Situation
Your company has been contacted by a toy company. They want technical assistance in designing a new executive desk gadget. The idea they want to investigate incorporates a reaction that turns a distinctive color in a specific amount of time. Although it will not be easy to determine the precise combination of chemicals that will work, the profit they stand to make would make it all worthwhile in the end. Executive "playtoys" are a big business, with corporate prices to match the corporate market.

Background
The reaction investigated here is a special kind of electron-transfer reaction called a *reduction-oxidation*, or *redox*, reaction. Reactions of this type will be studied further in Chapter 15. The net equation for the reaction in the gadget is written as follows.

$$3Na_2S_2O_5(aq) + 2KIO_3(aq) + 3H_2O(l) \xrightarrow{H^+} 2KI(aq) + 6NaHSO_4(aq)$$

You will measure the progress of this reaction using a starch-indicator solution. When the $Na_2S_2O_5$ solution is used up, the buildup of I_2, an intermediate in the reaction, will cause the indicator to change to a blue-black color. In the investigation, the concentrations of reactants are discussed in terms of drops of Solution A and Solution B. Solution A actually contains $Na_2S_2O_5$, the starch-indicator solution, and dilute sulfuric acid to supply the hydrogen ions needed to catalyze the reaction. Solution B contains KIO_3.

Problem
To determine the best conditions and concentrations for the reaction, you must do the following.
- prepare and observe reactions of several different concentrations of the reactants
- measure the time elapsed for each reaction
- determine the rate law for the reaction that will allow you to predict the results with other combinations
- use tools of statistical analysis to compare your data to that of the rest of the class

Required Precautions
- Goggles and a lab apron must be worn at all times.
- Read all safety cautions, and discuss them with your students.
- Students should not handle concentrated acid solutions.
- In case of an acid spill, dilute first with water. Then, mop up the spill with wet cloths designated for spill cleanup while wearing disposable plastic gloves. A wet cloth mop can be rinsed out a few times and used until it falls apart.

Safety

Always wear goggles and an apron to provide protection for your eyes and clothing. If you get a chemical in your eyes, immediately flush it out at the eyewash station while calling to your teacher. Know the locations of the emergency lab shower and eyewash and how to use them.

Do not touch any chemicals. If you get a chemical on your skin or clothing, wash it off at the sink while calling to your teacher. Make sure you carefully read the labels and follow the directions on all containers of chemicals that you use. Do not taste any chemicals or items used in the laboratory. Never return leftovers to their original containers; take only small amounts to avoid wasting supplies.

Call your teacher in the event of an acid or base spill. Acid or base spills should be cleaned up promptly according to your teacher's instructions.

Always clean up the lab and all equipment after use, and dispose of substances according to proper disposal methods. Wash your hands thoroughly before you leave the lab after all lab work is finished.

Materials

- **Distilled or deionized water**
- **Solution A**
- **Solution B**
- **Eight-well microscale reaction strips, 2**
- **Fine-tipped dropper bulbs or small micro-tip pipets, 3**
- **Stopwatch or clock with second hand**

Preparation

1. ***Organizing Data***
Prepare a data table in your lab notebook. The table should have six rows and six columns. Label the boxes in the first row of the second through sixth columns *Well 1*, *Well 2*, *Well 3*, *Well 4*, and *Well 5*. In the first column, label the boxes in the second through sixth rows *Time reaction began*, *Time reaction stopped*, *Drops of A*, *Drops of B*, and *Drops of* H_2O.
2. Obtain three dropper bulbs or small micro-tip pipets, and label them *A*, *B*, and H_2O.
3. Fill the bulb or pipet *A* with solution A, the bulb or pipet *B* with solution B, and the bulb or pipet H_2O with distilled water.

Technique

4. Using the first eight-well strip, place five drops of solution A into each of the first five wells. (Disregard the remaining three wells.) For best results, try to make all drops about the same size. Record the number of drops in the appropriate places in your data table. ***For best results, try to make all of the drops about the same size.***
5. In the second eight-well reaction strip, place one drop of solution B in the first well, two drops in the second well, three drops in the third well, four drops in the fourth well, and five drops in the fifth well. Record the number of drops in the appropriate places in your data table.

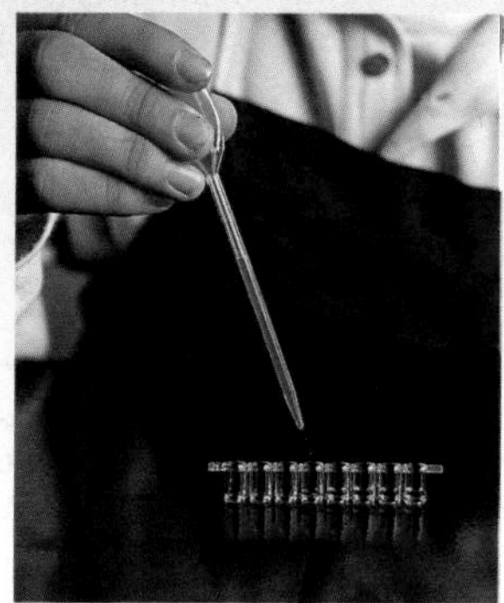

- Wear goggles, face shield, impermeable gloves, and a lab apron when you prepare the H_2SO_4. Work in a hood known to be in operating condition, with another person present nearby to call for help in case of an emergency. Be sure you are within 30 s walking distance of a working safety shower and eyewash station

3. To prepare solution B, dissolve 1.07 g KIO_3 in enough distilled water to make 250 mL of solution.
4. If the reaction suggested here is too slow, add a few milligrams of $Na_2S_2O_5$ to solution A. If it is too fast, add some distilled water to solution B to dilute the KIO_3 solution.
5. Other clock reactions may be used, but they should give distinctive end points. Be sure to test the reaction you use beforehand. Make certain you are familiar with proper disposal methods for the reactants and products.
6. Fine-tipped dropper bulbs are available commercially or may be made out of thin-stem pipets. Stretch the pipets by hand, narrowing the tube near the bulb, and then cut off the excess.

Student Orientation

Pre-lab Discussion

Thoroughly discuss the concept of rate. Work through a calculation with sample data, relating time elapsed to rate and the resulting rate data to a rate expression that involves concentration.

Techniques to Demonstrate

Point out that holding the dropper bottles vertically will help ensure consistently sized drops.

Post-Lab

Disposal

Set out one container for disposal. Treat the waste with 1.0 M $Na_2S_2O_3$ solution to be certain all iodine is reduced to iodide. Neutralize the solution with 1.0 M acid or base, and pour down the drain.

Answers to Analysis and Interpretation

1. Students will need to convert from minutes and seconds to seconds alone.
Well 1: 143 s Well 2: 67 s
Well 3: 43 s Well 4: 35
Well 5: 26 s

2.

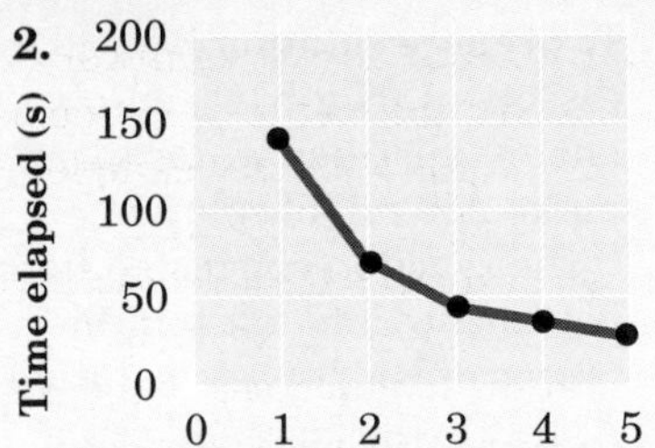

Number of Drops of Solution B

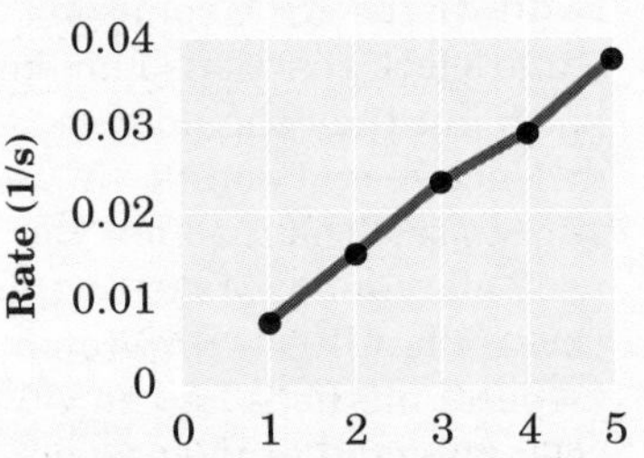

Number of Drops of Solution B

3. The most concentrated solution (5 drops of B) had the fastest reaction time. The least concentrated solution (1 drop of B, and 4 drops of water) had the slowest reaction time.

4. The total number of drops for each well should be the same to make valid comparisons of different concentrations.

Conclusions

5. The reactions began at the same time because the shake-start mixes all of the solutions at once. In this way, the reactions all take place at the same time and the same conditions.

6. The effect of reactant concentration (for solution B) on reaction rate being measured in this experiment.

7. The rate of the reaction is directly proportional to the concentration of solution B. $R = k[B]$ or $R \propto [B]$

Extensions

1. Student answers will vary, due to changes in drop size from one student group to another.

2. Students' suggestions for improving the procedure will vary. Possibilities may include minimizing error by performing multiple trials or keeping drop size standard. Be sure answers are safe and include carefully planned procedures.

6. In the second eight-well strip, which contains drops of solution B, add four drops of water to the first well, three drops to the second well, two drops to the third well, and one drop to the fourth well. Do not add any water to the fifth well.

7. Carefully invert the second strip. The surface tension should keep the solutions from falling out of the wells. Place the second strip well-to-well on top of the first strip as shown in the photograph on the left.

8. Holding the strips tightly together, record the exact time, or set the stopwatch, as you shake the solutions once, using a vigorous motion. This procedure should effectively mix the upper solutions with each of the corresponding lower ones.

9. Observe the lower wells. Note the sequence in which the solutions react, and record the number of seconds it takes for each solution to turn a deep blue color.

Cleanup and Disposal

10. Dispose of the solutions in the container designated by your teacher. Wash your hands thoroughly after cleaning up the area and equipment.

Analysis and Interpretation

1. ***Organizing Data***
Calculate the time elapsed for the complete reaction of each of the combinations of solution A and B.

2. ***Organizing Data***
Make a graph of your results. Label the x-axis *Number of drops of solution B*. Label the y-axis *Time elapsed*. Make a similar graph for drops of solution B against rate (1/time elapsed).

3. ***Analyzing Information***
Which mixture showed the fastest reaction? Which mixture showed the slowest reaction?

4. ***Evaluating Methods***
Why was it important to add the drops of water to the wells that contained fewer than five drops of Solution B? (Hint: figure out the total number of drops in each of the reaction wells.)

Conclusions

5. ***Analyzing Methods***
How can you be sure each of the chemical reactions for the gadget began at about the same time? Why is this important?

6. ***Evaluating Conclusions***
Which of the following variables that can affect rate is tested in this experiment: temperature, catalyst, concentration, surface area, or nature of reactants? Explain your answer.

7. ***Applying Ideas***
Use your data and graphs to determine the relationship between the concentration of solution B and the rate of the reaction. Describe this relationship in terms of a rate law.

Sample Data

	Well 1	Well 2	Well 3	Well 4	Well 5
Time rxn. began	00:00:00	00:00:00	00:00:00	00:00:00	00:00:00
Time rxn. stopped	00:02:23	00:01:07	00:00:43	00:00:35	00:00:26
Drops of A	5	5	5	5	5
Drops of B	1	2	3	4	5
Drops of H_2O	4	3	2	1	0

Extensions

1. ***Evaluating Data***
Share your data with other lab groups, and calculate a class average for the rate of the reaction for each concentration of B. Compare results from other groups to your results. Calculate the average deviation and the standard deviation. Explain why there are differences in the results. (Hint: instructions for these calculations are given in Exploration 2A.)

2. ***Analyzing Methods***
What are some possible sources of error in this procedure? If you can think of ways to eliminate them, ask your teacher to approve your plan, and run your procedure again.

3. ***Predicting Outcomes***
What combination of drops of solutions A and B would you use if you wanted the reaction to last exactly 2.5 min? Design an experiment to test your answer. If your teacher approves your plan, perform the experiment, and record these results. Make another graph including both the old and new data.

4. ***Predicting Outcomes***
What would your data be if the experiment was repeated, but solution A was diluted, with one part solution A for every seven parts distilled water? Design an experiment to test your answer. If your teacher approves your plan, perform the experiment. (Calculate percent error.)

5. ***Designing Experiments***
How would you determine what would be the smallest interval of time you could distinguish with the clock reaction? Design an experiment to find out. If your teacher approves your plan, perform the experiment.

6. ***Designing Experiments***
How would the results of this experiment be affected if the reaction took place in a cold environment? Design an experiment to test your answer, using materials available. If your teacher approves your plan, perform your experiment, and record the results. Make another graph, and compare it to your old data.

7. ***Designing Experiments***
How could you determine the effect of solution A on the rate law? If your teacher approves your plan, perform your experiment, and determine the rate law for this reaction.

8. ***Relating Ideas***
If solution B contains 0.02 M KIO_3, calculate the value for the constant, k, in the expression below. (Hint: remember that solution B is diluted when it is added to solution A.)
$Rate = k[KIO_3]$

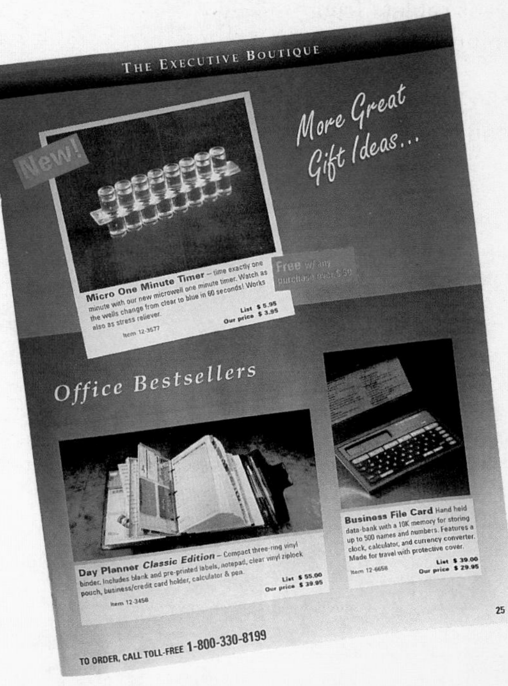

3. Students' answers will vary, depending on the conditions in your lab. For the sample data, it would require a mixture close to 4 drops A: 25 drops B: 21 drops H_2O. Students may need to use a different reaction vessel if this is too much for the micro-well strips. Be sure answers are safe and include carefully planned procedures.

4. Students' suggestions will vary. Be sure answers are safe and include carefully planned procedures. Students should find that the rate is seven times slower.

5. Student answers may vary, but the key is to determine how much of a difference in time is involved for the smallest difference in the amounts of the reactants. Be sure answers are safe and include carefully planned procedures.

6. Students' suggestions will vary. Be sure answers are safe and include carefully planned procedures. The rate is slower at lower temperatures.

7. Students may suggest using drops of water and solution A and keeping the amount of solution B constant. Another linear relationship is involved. $R = k[A][B]$.

8. Well 1: $k = \dfrac{6.99 \times 10^{-3}}{0.0020\ M} = 3.5$

Well 2: $k = \dfrac{0.015}{0.0040\ M} = 3.8$

Well 3: $k = \dfrac{0.023}{0.0060\ M} = 3.8$

Well 4: $k = \dfrac{0.029}{0.0080\ M} = 3.6$

Well 5: $k = \dfrac{0.038}{0.010\ M} = 3.8$

Average value for $k = 3.7$

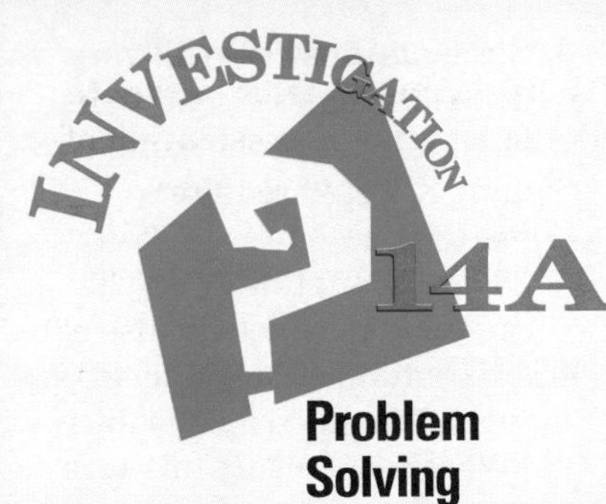

Objectives

Students will:

- use appropriate lab safety procedures.
- observe chemical processes and interactions.
- measure time elapsed for the dissolving of three brands of effervescent tablets.
- relate measurements of time elapsed to reaction rates.
- investigate different factors that affect rates of dissolving.
- suggest wording for the instructions panel on labels of effervescent tablets.
- design and implement their own procedure.

Planning

Recommended Time: 1 lab period
Prerequisite: Exploration 14A

Materials/Solution Preparation

1. Wear goggles, face shield, impermeable gloves, and a lab apron when you prepare the HCl. Work in a functioning hood within 30 s walking distance of a functioning safety shower and eyewash station with another person nearby to call for help in case of an emergency.
2. To prepare 0.10 M HCl, add 8.3 mL of concentrated HCl to enough distilled water to make 1.00 L of solution. Add the acid slowly, and stop to stir it in order to avoid overheating.
3. Obtain several different brands of effervescent tablets from a drug store. Remove them from the packaging before handing them out to students.
4. If large test tubes (18 mm × 150 mm or larger) are available, they can be used to contain the dissolving tablets.

Sample Data

Tablet A	1/4 Tablet	1/2 Tablet	3/4 Tablet	1 Tablet
Cool water	52 s	57 s	66 s	78 s
Room-temp. water	45 s	49 s	55 s	63 s
Warm water	34 s	36 s	39 s	46 s
Powdered, room temp.	30 s	33 s	38 s	44 s
Stirred, room temp.	42 s	47 s	55 s	67 s
Powdered and stirred, cool	47 s	52 s	60 s	73 s
Powdered and stirred, room temp.	22 s	24 s	27 s	31 s
Powdered and stirred, warm	15 s	17 s	21 s	26 s
Powdered and stirred in warm HCl solution	12 s	15 s	17 s	22 s

Suggested wording for label: "For best results, break up tablets, and place in a glass of warm water. Stir with a spoon until the tablet is completely dissolved."

Be certain student reports also include costs for 10 different analyses.

REACTION RATES

Effervescent Tablet Evaluation

Kravitz Morrow & Jones

May 1, 1995

Martha Li-Hsien
Director of Research
CheMystery Labs, Inc.
52 Fulton Street
Springfield, VA 22150

Dear Mr. Brown:

We are currently developing an advertising campaign for a new effervescent tablet that will relieve indigestion. Our marketing strategy will be to compare the leading brands of effervescent tablets with the new brand, now referred to as brand X. We want independent testing to be sure that our claims are valid.

We would like you to investigate how the reaction rate of brand X compares with that of two other leading brands. Enclosed are samples of these tablets, which are labeled brand A and brand B. Provide as much quantitative data as possible. Please also provide data on other factors that affect rates. This information will be used to make certain that the labeling includes instructions to help consumers make the most effective use of the product.

We have asked several analytical firms to perform this evaluation and will choose a long-term product consultant based on the reports we receive. Your bid should include the cost of completing this analysis and 10 similar analyses over the next year.

Sincerely,

Chris Strong
Director of Product Research
Kravitz, Morrow, and Jones

Required Precautions

Goggles and lab aprons must be worn.

Do not taste any of the chemicals, including the tablets. If you get any on your skin or clothing, wash it off at the sink. If it is in your eyes, immediately flush it out at the eyewash station while calling to your teacher. Wash your hands thoroughly when finished.

Confine long hair and loose clothing. Do not heat cracked glassware. If your clothing catches fire, WALK to the lab shower and put it out.

Required Precautions

- Goggles and a lab apron must be worn at all times.
- Read all safety cautions, and discuss them with your students.
- Students should not handle concentrated acids.
- In case of an acid spill, dilute first with water. Then mop up the spill with wet cloths or a mop designated for spill cleanup while wearing disposable plastic gloves.
- If the LEAP System with thermistor probe is used, the precautions listed in the teacher's notes on page 652 must be followed to avoid electric shock.

Memorandum

CheMystery Labs, Inc.
52 Fulton Street
Springfield, VA 22150

Date: May 2, 1995

To: Tom de los Santos

From: Martha Li-Hsien

Becoming the consultant for a large marketing firm could mean big profits. But we have to be certain that we provide quality work at a reasonable price. Before you begin this analysis, I'd like to review the following items.

- detailed one-page plan for the procedure and all necessary data tables
- detailed list of the equipment and materials you plan to use, along with individual and total costs
- final cost figure for our bid that reflects 10 future analyses, taking into consideration the actual costs incurred during the analysis as well as a profit margin

After you have completed your analysis prepare a report in the form of a two-page letter to Ms. Strong at Kravitz, Morrow, and Jones. The report should include the following.

- suggestions for wording to be used in advertising and on the instruction panel of the label
- rankings of the rate of reaction for each brand of tablet in a simulated stomach acid solution, as well as criteria for the rankings and examples of specific data that support the rankings
- detailed and organized data and analysis section showing any calculations you used to reach your conclusions, and graphs that show visual proof of your conclusions (Considering that this work is being done for a marketing firm, we need clean, easy-to-understand graphs to display your results.)
- detailed invoice for the services rendered and the expenses incurred for this investigation
- revised bid for the other 10 proposed analyses

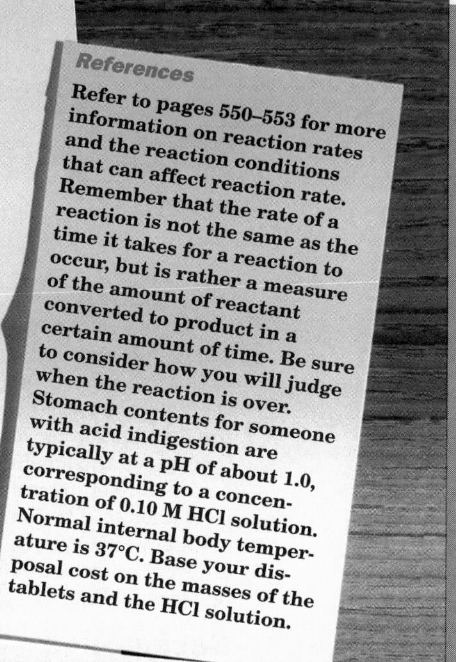

References

Refer to pages 550–553 for more information on reaction rates and the reaction conditions that can affect reaction rate. Remember that the rate of a reaction is not the same as the time it takes for a reaction to occur, but is rather a measure of the amount of reactant converted to product in a certain amount of time. Be sure to consider how you will judge when the reaction is over. Stomach contents for someone with acid indigestion are typically at a pH of about 1.0, corresponding to a concentration of 0.10 M HCl solution. Normal internal body temperature is 37°C. Base your disposal cost on the masses of the tablets and the HCl solution.

Materials for KRAVITZ, MORROW, and JONES

Item	Cost
REQUIRED ITEMS (You must include all of these in your budget.)	
Lab space/fume hood/utilities	15 000 – /day
Standard disposal fee	2 000 – /g of product
Beaker tongs	1 000
REAGENTS and ADDITIONAL EQUIPMENT (Include only what you'll need in your budget.)	
0.10 M HCl	100 – /mL
Ice	500
250 mL beaker	1 000
250 mL flask	1 000
100 mL graduated cylinder	1 000
Balance	5 000
Bunsen burner	10 000
Glass stirring rod	1 000
Hot plate	8 000
Mortar and pestle	2 000
Ring stand/ring/wire gauze	2 000
Rubber policeman	500
Six test tubes (large)/holder/rack	2 000
Stopwatch	5 000
Test tube (small)	500
Thermistor—LEAP	2 000
Thermometer	2 000
Wash bottle	500
Weighing paper	500
* No refunds on returned chemicals or unused equipment.	
Fines	
OSHA safety violation	2 000 – /incident

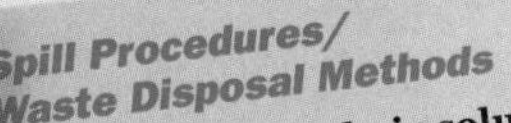

Spill Procedures/ Waste Disposal Methods

- All tablets and their solutions may be washed down the sink after they have been diluted with 20 times as much water.
- Any leftover acid should be disposed of in the container designated by your teacher.

Student Orientation

Techniques to Demonstrate

Demonstrate the dissolution of a single effervescent tablet. Discuss with students how to determine the end of this reaction. Determination of when the reaction stops is very subjective, and answers between groups will vary a great deal.

Pre-lab Discussion

Students may be confused at first because this lab appears very different from the Exploration. This Investigation is designed to reinforce the same concepts of elapsed time, rate, and the factors affecting rate in an entirely different context.

Emphasize the need to quantify data as much as possible. Students should be discouraged from using qualitative labels such as "very fast" or "not so that fast."

Suggest that students divide tablets into equal masses so that they can perform multiple trials.

Tips for Evaluating the Pre-lab Requirements

If student plans suggest the use of more than six tablets per brand, have them reconsider their procedures. Be certain students plan to test dissolution under a variety of conditions, such as temperature, pH, stirring, and size of particle (broken tablets compared to whole ones).

Proposed Procedure

Divide a tablet into several pieces so that its rate of dissolving can be tested under a variety of conditions. For each trial, measure the time elapsed during dissolving with a stopwatch. Record the time when dissolving is finished.

Post-Lab

Disposal

All solutions may be flushed down the drain.

Materials
(for each lab group)

- Effervescent tablet A, 6
- Effervescent tablet B, 6
- Effervescent tablet X, 6
- HCl, 0.10 M, 150 mL
- Ice, 250 mL
- 100 mL graduated cylinder
- 250 mL beaker
- Balance and weighing paper
- Beaker tongs
- Glass stirring rod
- Mortar and pestle
- Hot plate, or Bunsen burner with gas tubing and striker
- Ring stand/ring/wire gauze
- Spatula
- Stopwatch
- Test-tube holder and rack
- Test-tubes, 6
- Thermometer, nonmercury

Optional Equipment

- LEAP System with thermistor probe

Estimated cost of materials: $226 500

EXPLORATION 15A

Objectives

Students will:

- use appropriate lab safety procedures.
- measure volume using burets.
- perform a redox titration.
- recognize the end point of a redox reaction.
- write a balanced oxidation-reduction reaction.
- determine the concentration of a solution using stoichiometry and titration data.

Planning

Recommended Time:
1 lab period

Materials
(for each lab group)

- 0.0200 M $KMnO_4$, 50 mL
- 1.0 M H_2SO_4, 5 mL
- Distilled water
- Unknown $FeSO_4$ solution (0.15 M), 50 mL
- 250 mL beakers, 2
- 400 mL beaker
- 125 mL Erlenmeyer flasks, 4
- 100 mL graduated cylinder
- Burets, 2
- Double buret clamp
- Ring stand
- Wash bottle

Solution/Material Preparation:

1. To prepare 0.0200 M $KMnO_4$, add 3.16 g of $KMnO_4$ to enough distilled water to make 1.00 L of solution. For best results, this solution must be prepared shortly before the lab.
2. To prepare 0.15 M $FeSO_4$, dissolve 41.70 g of $FeSO_4•7H_2O$ to enough distilled water to make 1.00 L of solution.
3. To prepare 1.00 L of 1.0 M H_2SO_4, observe the required precautions. Slowly add 56 mL of concentrated H_2SO_4 to enough distilled water to make 1.00 L of solution. Add the acid slowly, and stop to stir it in order to avoid overheating.

Redox Titration

Objectives

Demonstrate **proficiency in performing redox titrations and recognizing endpoints of a redox reaction.**

Write **a balanced oxidation-reduction equation for a redox reaction.**

Determine **the concentration of a solution using stoichiometry and volume data from a titration.**

Situation

You are a chemist working for a chemical analysis firm. A large pharmaceutical company has hired you to help salvage some of their products after a small fire broke out in their warehouse. Although there was only minimal smoke and fire damage to the warehouse and products, the sprinkler system ruined the labeling on many of the pharmaceuticals. The firm's best-selling products are iron tonics, used to treat low-level anemia. The tonics are produced from hydrated iron(II) sulfate, $FeSO_4•7H_2O$. The different types of tonics contain different concentrations of $FeSO_4$. You need to help them figure out what labels to put on the bottles of tonic.

Background

Hydrated iron(II) sulfate, $FeSO_4•7H_2O$, is a useful compound of iron in the +2 oxidation state, that is used as a reducing agent and in medicines. Because this substance is a reducing agent, it can undergo a redox reaction with an oxidizing agent. Redox reactions involve the processes of oxidation (the loss of electrons by a reactant, Fe^{2+}) and reduction (the gain of electrons by a reactant, in this case MnO_4^-).

You have already studied acid-base titrations in Chapter 13, in which an unknown amount of acid is titrated with a carefully measured amount of base. In this procedure, a similar approach called a *redox titration* is used. In a redox titration, the reducing agent, Fe^{2+}, is oxidized to Fe^{3+} by the oxidizing agent, MnO_4^-. When this process occurs, the Mn in MnO_4^- changes from a +7 to a +2 oxidation state and has a noticeably different color. You can use this color change in the same way that you used the color change of phenolphthalein in acid-base titrations—to signify a redox reaction "endpoint." When the reaction is complete, the addition of excess MnO_4^- will give the solution a pink or purple color.

Problem

To determine how to label the bottles, you need to determine the concentration of iron(II) ions in the sample from an unlabeled bottle from the warehouse.

- titrate the Fe^{2+} sample with a $KMnO_4$ solution of known concentration
- use mole ratios from the balanced redox reaction and the volume data obtained from the titration to determine the concentration of the sample
- identify which tonic the sample is, given information about the concentration of each tonic

Required Precautions

- Goggles and a lab apron must be worn at all times.
- Read all safety cautions, and discuss them with your students.
- Students should not handle concentrated acid solutions.

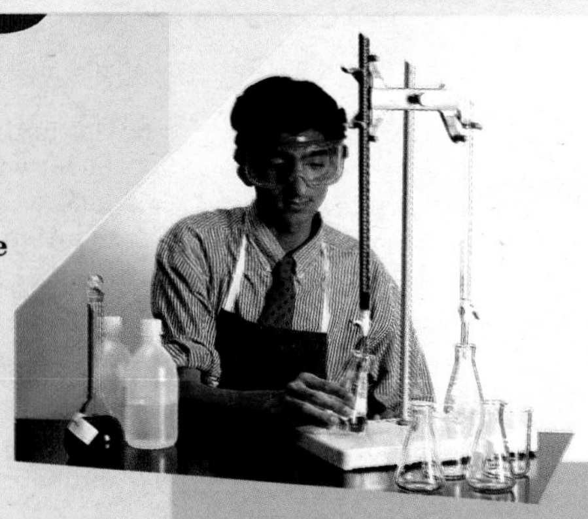

Safety

Always wear goggles and an apron to provide protection for your eyes and clothing. If you get a chemical in your eyes, immediately flush it out at the eyewash station while calling to your teacher. Know the locations of the emergency lab shower and eyewash and how to use them.

Do not touch any chemicals. If you get a chemical on your skin or clothing, wash it off at the sink while calling to your teacher. Make sure you carefully read the labels and follow the directions on all containers of chemicals that you use. Do not taste any chemicals or items used in the laboratory. Never return leftovers to their original containers; take only small amounts to avoid wasting supplies.

Never put broken glass in a regular waste container. Broken glass should be disposed of separately according to your teacher's instructions.

Call your teacher in the event of an acid, base, or potassium permanganate spill. Such spills should be cleaned up promptly. Acids and bases are corrosive; avoid breathing fumes. $KMnO_4$ is a strong oxidizer. If any of the oxidizer should spill on you, immediately flush the area with water and notify your teacher.

Always clean up the lab and all equipment after use, and dispose of substances according to proper disposal methods. Wash your hands thoroughly before you leave the lab after all lab work is finished.

Materials

- **0.0200 M $KMnO_4$**
- **1.0 M H_2SO_4**
- **Distilled water**
- **$FeSO_4$ solution**
- **250 mL beaker, 2**
- **400 mL beaker**
- **125 mL Erlenmeyer flask, 4**
- **100 mL graduated cylinder**
- **Burets, 2**
- **Double buret clamp**
- **Ring stand**
- **Wash bottle**

Preparation

1. Prepare a data table in your lab notebook. The table should have four rows and five columns. Label the boxes in the top row *Trial, Initial $KMnO_4$ volume, Final $KMnO_4$ volume, Initial $FeSO_4$ volume*, and *Final $FeSO_4$ volume*. In the first column, number the second through fourth rows *1, 2*, and *3*.
2. Clean two 50 mL burets with a buret brush and distilled water. Rinse each buret at least three times with distilled water to remove any contaminants.
3. Label two 250 mL beakers *0.0200 M $KMnO_4$*, and *$FeSO_4$ solution*. Label three of the flasks *1, 2*, and *3*. Label the 400 mL beaker *Waste*. Label one buret *$KMnO_4$* and the other *$FeSO_4$*.
4. Measure approximately 75 mL of 0.0200 M $KMnO_4$ and pour it into the appropriately labeled beaker. Obtain approximately 75 mL of $FeSO_4$ solution and pour it into the appropriately labeled beaker.
5. Rinse one buret three times with a few milliliters of 0.0200 M $KMnO_4$ from the appropriately labeled beaker. Collect these rinses in the *Waste* beaker. Rinse the other buret three times with small amounts of $FeSO_4$ solution from the appropriately labeled beaker. Collect these rinses in the *Waste* beaker.

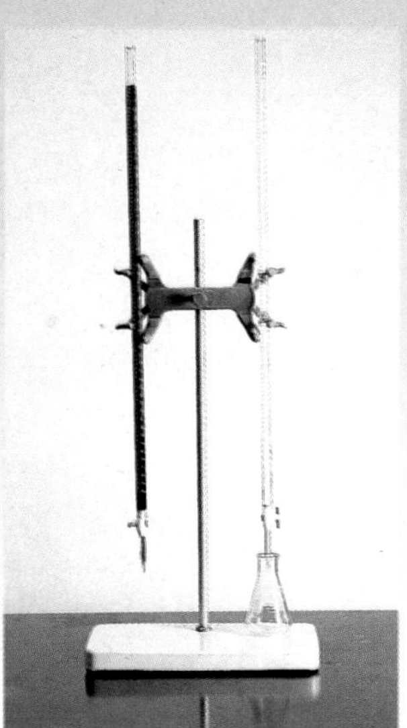

Student Orientation

Pre-lab Discussion

Thoroughly discuss the concept of oxidation-reduction and how this type of reaction can result in a color change. Review the concept of molar ratios and balancing redox equations.

Techniques to Demonstrate

At this point, students should have had plenty of practice in operating burets with acid-base titrations. Remind students to create the end-point standard in step 7.

Post-Lab

Disposal

Set out one container for disposal. The mixture that results should be acidic. If the mixture is purple, add $FeSO_4$ solution slowly, while stirring until the color is gone. Precipitate the iron by adding 1.0 M NaOH. Filter and put the precipitate in the trash. Neutralize the filtrate with 1.0 M acid or base until its pH is between 5 and 9, and pour it down the drain.

- Wear goggles, face shield, impermeable gloves, and a lab apron when you prepare the H_2SO_4. Work in a functioning hood within 30 s walking distance of a working safety shower and eyewash with another person present nearby to call for help in case of an emergency.
- In case of an acid spill, dilute first with water. Then mop up the spill with wet cloths or a cloth mop designated for spill cleanup while wearing disposable plastic gloves.

Answers to Analysis and Interpretation

1. $MnO_4^-(aq) + 8H^+(aq) + 5Fe^{2+}(aq) \longrightarrow 5Fe^{3+}(aq) + Mn^{2+}(aq) + 4H_2O(l)$

2. Trial 1: 15.0 mL $KMnO_4$ ×

$\frac{1\ L}{1000\ mL} \times \frac{0.020\ mol\ KMnO_4}{1\ L} =$

3.0×10^{-4} mol $KMnO_4$

Trial 2: 14.5 mL $KMnO_4$ ×

$\frac{1\ L}{1000\ mL} \times \frac{0.020\ mol\ KMnO_4}{1\ L} =$

2.9×10^{-4} mol $KMnO_4$

Trial 3: 15.5 mL $KMnO_4$ ×

$\frac{1\ L}{1000\ mL} \times \frac{0.020\ mol\ KMnO_4}{1\ L} =$

3.1×10^{-4} mol $KMnO_4$

3. The ratio of Fe^{2+} to MnO_4^- is 5:1.

Trial 1: 3.0×10^{-4} mol MnO_4^- ×

$\frac{5\ mol\ Fe^{2+}}{1\ mol\ KMnO_4^-} =$

1.5×10^{-3} mol Fe^{2+}

Trial 2: 2.9×10^{-4} mol MnO_4^- ×

$\frac{5\ mol\ Fe^{2+}}{1\ mol\ KMnO_4^-} =$

1.4×10^{-3} mol Fe^{2+}

Trial 3: 3.1×10^{-4} mol MnO_4^- ×

$\frac{5\ mol\ Fe^{2+}}{1\ mol\ KMnO_4^-} =$

1.6×10^{-3} mol Fe^{2+}

4. average molarity =

$\frac{1.5 \times 10^{-3}\ mol\ Fe^{2+}}{10.0\ mL} \times \frac{1000\ mL}{1\ L} =$

0.15 M Fe^{2+}

5. The concentrations of each solution is very important, and if you only rinse the buret with distilled water, the solutions might be diluted if there was any water left over in the buret. By using the solution being measured to rinse the buret, you can be sure any solution left behind will not dilute the solution added.

6. Set up the burets as shown in the photograph on page 781. Fill one buret with approximately 50 mL of the 0.0200 M $KMnO_4$ from the beaker and the other buret with approximately 50 mL of the $FeSO_4$ solution from the other beaker.
7. With the *Waste* beaker underneath its tip, open the $KMnO_4$ buret long enough to be sure the buret tip is filled. Repeat for the $FeSO_4$ buret.
8. Add 50 mL of distilled water to one of the 125 mL Erlenmeyer flasks, and add one drop of 0.0200 M $KMnO_4$ to the flask. Set this aside to use as a color standard, to compare to the titration and determine the endpoint.

Technique

9. Record the initial buret readings for both solutions in your data table. Add 10.0 mL of the hydrated iron(II) sulfate, $FeSO_4 \cdot 7H_2O$, solution to flask *1*. Add 5 mL of 1.0 M H_2SO_4 to the $FeSO_4$ solution in this flask. The acid helps keep the Fe^{2+} ions in the reduced state to allow you time to titrate.
10. Slowly add $KMnO_4$ from the buret to the $FeSO_4$ in the flask, while swirling the flask. When the color of the solution matches the color standard you prepared in step **7**, record the final readings of the burets in your data table.
11. Empty the titration flask into the *Waste* beaker. Repeat the titration procedure in steps **9** and **10** with flasks *2* and *3*.

Cleanup and Disposal

12. Dispose of the contents of the *Waste* beaker in the container designated by your teacher. Also pour the color-standard flask into this container. Always wash your hands thoroughly after cleaning up the area and equipment.

Analysis and Interpretation

1. **Organizing Ideas**
Write the balanced equation for the redox reaction of $FeSO_4$ and $KMnO_4$.
2. **Evaluating Data**
Calculate the number of moles of MnO_4^- reduced in each trial.
3. **Analyzing Information**
Calculate the number of moles of Fe^{2+} oxidized in each trial.
4. **Applying Conclusions**
Calculate the average concentration (molarity) of the iron tonic.
5. **Analyzing Methods**
Explain why it was important to rinse the burets with $KMnO_4$ or $FeSO_4$ before adding the solutions. (Hint: consider what would happen to the concentration of each solution if it was added to a buret that had been rinsed only with distilled water.)

Conclusions

6. **Inferring Conclusions**
The company makes three different types of iron tonics: *Feravide A* with a concentration of 0.145 M $FeSO_4$, *Feravide Extra-Strength* with 0.225 M $FeSO_4$, and *Feravide Jr.* with 0.120 M $FeSO_4$. Which tonic is your sample?

782 | Exploration 15A

Sample Data

Trial	Initial $KMnO_4$ volume (mL)	Final $KMnO_4$ volume (mL)	Initial $FeSO_4$ volume (mL)	Final $FeSO_4$ volume (mL)
1	50.0	35.0	50.0	40.0
2	35.0	20.5	40.0	30.0
3	20.5	5.0	30.0	20.0

Extensions

1. ***Evaluating Methods***
Calculate the absolute and average deviation for your measurements of the concentration of the $FeSO_4$. What reasons can you give for any differences in the values? (Hint: see Exploration 2A for information on absolute and average deviations.)

2. ***Designing Experiments***
What possible sources of error can you identify with this procedure? If you can think of ways to eliminate them, ask your teacher to approve your plan, and run the procedure again.

3. ***Applying Ideas***
Hydrogen peroxide, H_2O_2, was once widely used as an antiseptic. It decomposes by the oxidation and reduction of its oxygen atoms. The products are water and molecular oxygen. Write the balanced redox equation for this reaction.

4. ***Applying Ideas***
When gaseous hydrogen sulfide burns in air to form sulfur dioxide and water, the oxidation number of hydrogen does not change, but that of sulfur changes from a –2 state to a +4 state, and that of oxygen changes from a zero state to a –2 state. Which substances are oxidized, and which are reduced? Write the balanced chemical equation for the combustion reaction.

5. ***Research and Communication***
Blueprints are based on a photochemical reaction. The paper is treated with a solution of iron(III) ammonium citrate and potassium hexacyanoferrate(III) and dried in the dark. When a tracing-paper drawing is placed on the blueprint paper and exposed to light, Fe^{3+} ions are reduced to Fe^{2+} ions, which react with hexacyanoferrate(III) ions in the moist paper to form the blue color on the paper. The lines of the drawing block the light and prevent the reduction of Fe^{3+} ions, resulting in white lines. Find out how sepia prints are made, and report on this information.

6. ***Relating Ideas***
Electrochemical cells are based on the process of electron flow in a system with varying potential differences. Batteries are composed of such systems and contain different chemicals for different purposes and price ranges. You can make simple experimental batteries using metal wires and items such as lemons, apples, and potatoes. What are some other "homemade" battery sources, and what is the role of these food items in producing electrical energy that can be measured as battery power? Explain your answers.

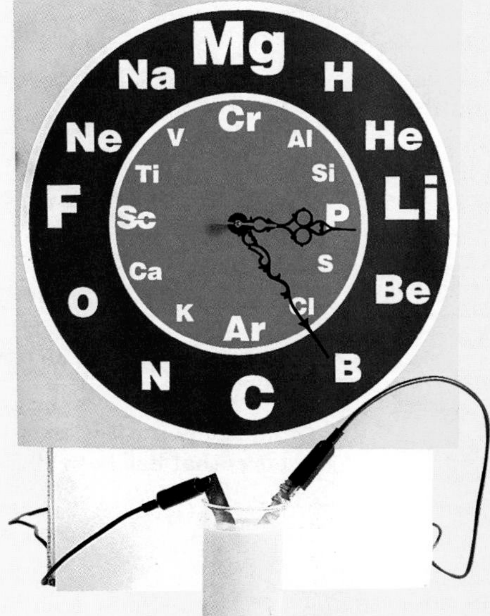

Conclusions

6. The sample tonic is most likely Feravide A, 0.145 M $FeSO_4$.

Extensions

1. Trial 1: Absolute deviation: 0.005 M

   Trial 2: Absolute deviation: 0.000 M

   Trial 3: Absolute deviation: 0.010 M

   Average deviation:

$$\frac{0.005\text{ M} + 0\text{ M} + 0.010\text{ M}}{3} = 0.005\text{ M}$$

2. Students' suggestions for improving the procedure will vary. Possible suggestions include performing repeated trials or using larger volumes for the titration. Be sure answers are safe and include carefully planned procedures.

3. $2\underset{+1\,-1}{H_2O_2} \longrightarrow 2\underset{+1\,-2}{H_2O} + \underset{0}{O_2}$

4. $\underset{+1\,-2}{H_2S} + \underset{0}{O_2} \longrightarrow \underset{+4\,-2}{SO_2} + \underset{+1\,-2}{H_2O}$
oxidized: sulfur (loss of electrons)
reduced: oxygen (gain of electrons)

5. Sepia printing is similar to blue-print printing. In sepia printing, the potassium hexacyanoferrate(III) oxidizes the crystals of silver in the print so that they become silver ions again. Then the print is redeveloped in sodium sulfide.

6. Student answers will vary. The food is the conductor that completes the circuit between the two metals, each of which has a different reduction potential. In terms of a traditional cell, the food acts as both the porous barrier that separates the metal and the solution of aqueous ions that completes the circuit.

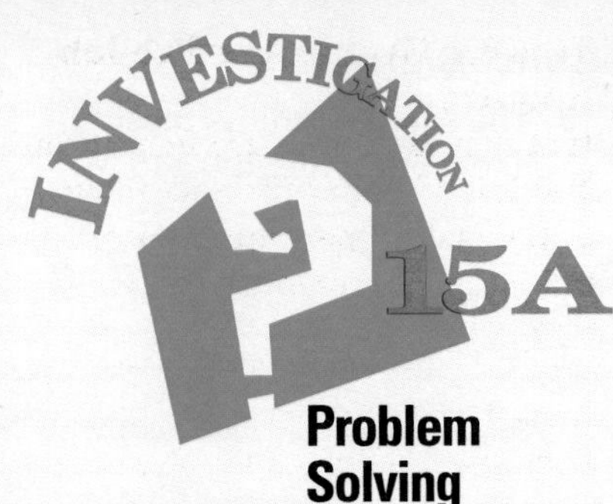

Problem Solving

Objectives

Students will:

- use appropriate lab safety procedures.
- measure volume using burets.
- perform a redox titration.
- recognize the end point of a redox reaction.
- write a balanced oxidation-reduction reaction.
- determine the concentration of a solution using stoichiometry and titration data.
- design and implement their own procedure.

Planning

Recommended Time: 1 lab period
Prerequisite: Exploration 15A

Solution/Material Preparation

1. See the Teacher's Notes for Exploration 15A for information on preparing 0.0200 M $KMnO_4$ and.1.0 M H_2SO_4.
2. To prepare 0.25 M $FeSO_4$, dissolve 69.51 g of $FeSO_4 \cdot 7H_2O$ to enough distilled water to make 1.00 L of solution.

Student Orientation

Techniques to Demonstrate

Correct any technique errors that occurred in the Exploration. Remind students to create an end-point standard to compare with as they titrate.

Pre-lab Discussion

The process is very similar to that of the Exploration, but because a different solution is being titrated, reaching the end-point will require a different volume of $KMnO_4$. Encourage students to use 10 mL samples of the solution, so the amounts of $KMnO_4$ necessary are not excessive.

Sample Data

Trial	Initial $FeSO_4$ volume (mL)	Final $FeSO_4$ volume (mL)	Initial $KMnO_4$ volume (mL)	Final $KMnO_4$ volume (mL)
1	50.0	40.0	50.0	25.0
2	40.0	30.0	50.0	25.2
3	30.0	20.0	50.0	24.8

Analysis

$$25.0\text{ mL} \times \frac{1\text{ L}}{1000\text{ mL}} \times \frac{0.020\text{ mol }KMnO_4}{1\text{ L}} = 5 \times 10^{-4}\text{ mol }KMnO_4$$

$$5\times10^{-4}\text{ mol }KMnO_4 \times \frac{5\text{ mol }FeSO_4}{1\text{ mol }KMnO_4} = 2.5\times10^{-3}\text{ mol }FeSO_4 \text{ (also the amount of Fe)}$$

mass $FeSO_4$ = 0.38 g $FeSO_4$; mass Fe = 0.14 g Fe

$$\text{percentage Fe} = \frac{55.85\text{ g Fe}}{151.91\text{ g }FeSO_4} \times 100 = 36.77\%$$

$$\text{annual yield: } \frac{0.15\text{ g}}{10\text{ mL}} \times \frac{1000\text{ mL}}{\text{L}} \times \frac{1\text{ kg}}{1000\text{ g}} \times \frac{1 \times 10^5\text{ L}}{\text{day}} \times \frac{365\text{ days}}{\text{year}} =$$

5.5×10^5 kg/year

REDOX TITRATION

Mining Feasibility Study

GOLDSTAKE MINING CORPORATION

May 11, 1995

George Taylor
Director of Analytical Services
CheMystery Labs, Inc.
52 Fulton Street
Springfield, VA 22150

Dear George:

Because of the high quality of your firm's work in the past, Goldstake is again asking that you submit a bid for a mining feasibility study. A study site in New Mexico has yielded some promising iron ore deposits, and we are evaluating the potential yield.

Your bid should include the costs of evaluating the sample we're sending with this letter and the fees for 20 additional analyses to be completed over the next year. The sample is a slurry extracted from the mine using a special process that converts the iron ore into $FeSO_4$ dissolved in water. The mine could produce up to 1.0×10^5 L of this slurry daily, but we need to know how much iron is in that amount of slurry before we proceed.

The contract for the other analyses will be awarded based on the accuracy of this analysis and the quality of the accompanying report. Your report will be used for two purposes: to evaluate the site for quantity of iron and to determine who our analytical consultant will be if the site is developed into a mining operation. I look forward to reviewing your bid proposal.

Sincerely,

Lynn L. Brown
Director of Operations
Goldstake Mining Corporation

Required Precautions

Goggles and lab aprons must be worn at all times.

If you get acid, base, or permanganate on your skin or clothing, wash it off at the sink while calling to your teacher. If you get acid, base, or permanganate in your eyes, immediately flush it out at the eyewash station while calling to your teacher.

Do not touch or taste any chemicals. Wash your hands thoroughly when finished.

Required Precautions

- Goggles and a lab apron must be worn at all times.
- Read all safety cautions, and discuss them with your students.
- Students should not handle concentrated acids.
- In case of an acid spill, dilute first with water. Then mop up the spill with wet cloths or a cloth mop designated for spill cleanup while wearing disposable plastic gloves.

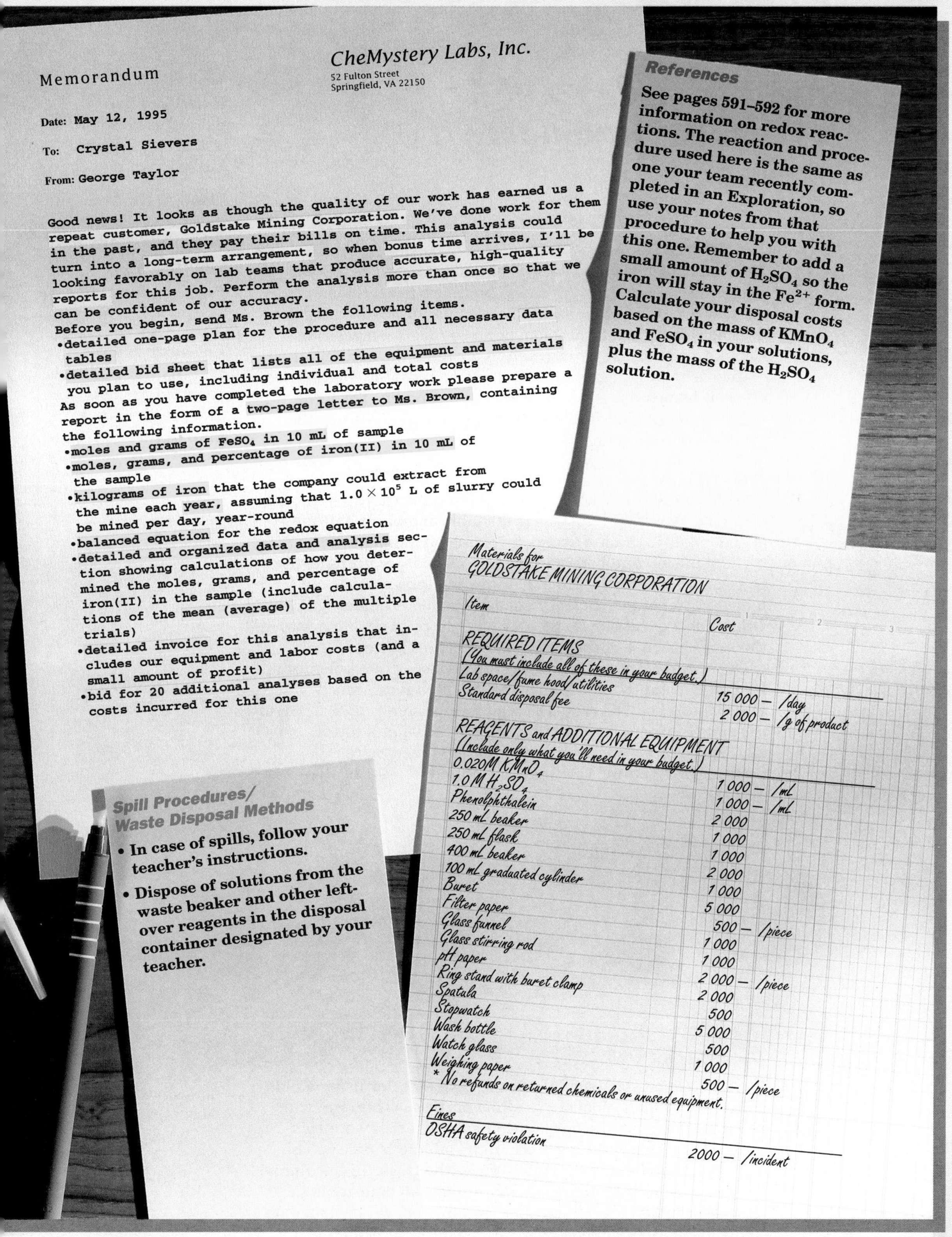

Memorandum

CheMystery Labs, Inc.
52 Fulton Street
Springfield, VA 22150

Date: May 12, 1995

To: Crystal Sievers

From: George Taylor

Good news! It looks as though the quality of our work has earned us a repeat customer, Goldstake Mining Corporation. We've done work for them in the past, and they pay their bills on time. This analysis could turn into a long-term arrangement, so when bonus time arrives, I'll be looking favorably on lab teams that produce accurate, high-quality reports for this job. Perform the analysis more than once so that we can be confident of our accuracy.

Before you begin, send Ms. Brown the following items.
- detailed one-page plan for the procedure and all necessary data tables
- detailed bid sheet that lists all of the equipment and materials you plan to use, including individual and total costs

As soon as you have completed the laboratory work please prepare a report in the form of a two-page letter to Ms. Brown, containing the following information.
- moles and grams of $FeSO_4$ in 10 mL of sample
- moles, grams, and percentage of iron(II) in 10 mL of the sample
- kilograms of iron that the company could extract from the mine each year, assuming that 1.0×10^5 L of slurry could be mined per day, year-round
- balanced equation for the redox equation
- detailed and organized data and analysis section showing calculations of how you determined the moles, grams, and percentage of iron(II) in the sample (include calculations of the mean (average) of the multiple trials)
- detailed invoice for this analysis that includes our equipment and labor costs (and a small amount of profit)
- bid for 20 additional analyses based on the costs incurred for this one

References

See pages 591–592 for more information on redox reactions. The reaction and procedure used here is the same as one your team recently completed in an Exploration, so use your notes from that procedure to help you with this one. Remember to add a small amount of H_2SO_4 so the iron will stay in the Fe^{2+} form. Calculate your disposal costs based on the mass of $KMnO_4$ and $FeSO_4$ in your solutions, plus the mass of the H_2SO_4 solution.

Materials for
GOLDSTAKE MINING CORPORATION

Item	Cost
REQUIRED ITEMS (You must include all of these in your budget.)	
Lab space/fume hood/utilities	15 000 – /day
Standard disposal fee	2 000 – /g of product
REAGENTS and ADDITIONAL EQUIPMENT (Include only what you'll need in your budget.)	
0.020M $KMnO_4$	
1.0 M H_2SO_4	1 000 – /mL
Phenolphthalein	1 000 – /mL
250 mL beaker	2 000
250 mL flask	1 000
400 mL beaker	1 000
100 mL graduated cylinder	2 000
Buret	1 000
Filter paper	5 000
Glass funnel	500 – /piece
Glass stirring rod	1 000
pH paper	1 000
Ring stand with buret clamp	2 000 – /piece
Spatula	2 000
Stopwatch	500
Wash bottle	5 000
Watch glass	500
Weighing paper	1 000
	500 – /piece
* No refunds on returned chemicals or unused equipment.	
Fines	
OSHA safety violation	2000 – /incident

Spill Procedures/ Waste Disposal Methods
- In case of spills, follow your teacher's instructions.
- Dispose of solutions from the waste beaker and other left-over reagents in the disposal container designated by your teacher.

Materials
(for each lab group)
- 0.0200 M $KMnO_4$, 100 mL
- 1.0 M H_2SO_4, 5 mL
- Distilled water
- Unknown $FeSO_4$ solution (0.25 M), 50 mL
- 250 mL beakers, 2
- 400 mL beaker
- 125 mL Erlenmeyer flasks, 4
- 100 mL graduated cylinder
- Burets, 2
- Double buret clamp
- Ring stand
- Wash bottle

Estimated cost of materials:
$158 000

Tips for Evaluating Pre-lab Requirements
If students plan to use more than 100 mL of either the unknown or potassium permanganate, have them reconsider their procedures.

Proposed Procedure
Prepare the burets for the titration as in the Exploration, using $KMnO_4$ in one buret and $FeSO_4$ in the other. Put a drop of $KMnO_4$ in one Erlenmeyer flask to serve as a color standard to recognize the end point.

Put 10.0 mL of the $FeSO_4$ solution in an Erlenmeyer flask. Add 5 mL of 1.0 M H_2SO_4. Slowly add $KMnO_4$ from the buret. Swirl after each drop added. When the color matches the standard, record the final volumes of the burets. Repeat the titration procedure twice.

Post-Lab
Disposal
Set out one container for disposal. The mixture that results should be acidic. If the mixture is purple, add $FeSO_4$ solution slowly while stirring until the color is gone. Precipitate the iron by adding 1.0 M NaOH. Filter and put the precipitate in the trash. Neutralize the filtrate with 1.0 M acid or base until its pH is between 5 and 9, and pour it down the drain.

EXPLORATION 16A

Objectives

Students will:

- build a radiation detector using CR-39 plastic.
- use the detector to measure radon exposure.
- observe and count the tracks of alpha particles with a microscope.
- calculate the activity of radon.
- convert measurements from units of picocuries to becquerels.
- evaluate radon activity over a large area by using class data to draw a map of the activity.

Planning

Recommended Time: 2 lab periods (with exposure time of three weeks—can be shortened with a source of alpha particles)

Materials

(for each lab group)

- CR-39 plastic
- 3 in × 5 in index card
- Clear plastic ruler or stage micrometer
- Etch clamp (the ring from a key chain)
- Microscope
- Paper clips
- Push pin
- Scissors
- Small plastic cup with lid
- Tape
- Toilet paper or other tissue

Optional

- Source of radiation (Fiestaware, Coleman green-label lantern mantles, old glow-in-the-dark watch faces, cloud chamber needles)

For Teacher Use Only

- 6.25 M NaOH, 10 mL per lab group
- Small test tube, 1 per lab group
- 1 L beaker or 2 400 mL beakers
- Bunsen burner or hot plate

Detecting Radioactivity

Situation

Local officials have received calls from concerned citizens regarding the possibility of radon in their homes. Radon is a major source of natural background radiation. Ordinarily, it does not pose a health problem because radon is a gaseous element and normal air circulation prevents the accumulation of the gas in buildings. However, in some air-tight, energy-efficient homes, there could be cause for concern. The local health commissioner has asked your company to survey the area to determine the level of radon emissions.

Background

The element radon is the product of the radioactive decay of uranium. The $^{222}_{86}$Rn nucleus is unstable and has a half-life of about 4 days. Radon decays by giving off alpha particles (helium nuclei) and beta particles (electrons) according to the equations below with $^{4}_{2}$He indicating an alpha particle, and $^{0}_{-1}\beta$ indicating a beta particle. Chemically, radon is a noble gas. Other noble gases can be inhaled without causing damage to the lungs because of the gases' nonreactive nature. When radon is inhaled, however, it can rapidly decay into polonium, lead, and bismuth, all of which are solids that can react and lodge in body tissues. The lead isotope shown at the end of the chain of reactions, $^{210}_{82}$Pb, has a half life of 22.6 years, and it will eventually undergo even more decay before creating the final stable product, $^{206}_{82}$Pb.

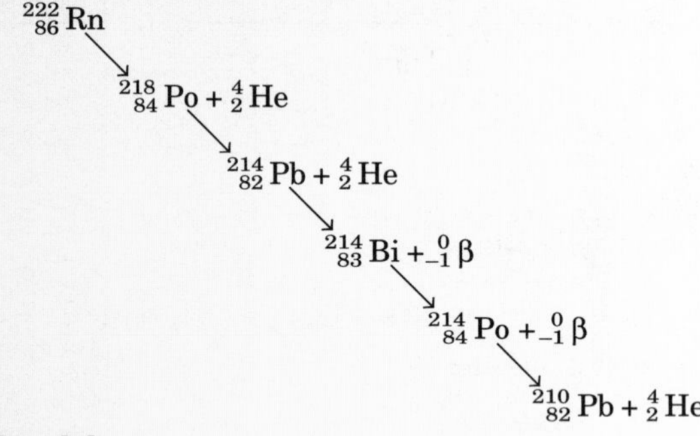

Problem

First, you must construct an inexpensive detector using a plastic (CR-39) that is sensitive to alpha particles. Then you will place the detectors in your home or somewhere in your community for a three-week period. In the lab the plastic will be etched with sodium hydroxide to make the tracks of the alpha particles visible with a microscope, so you can determine the activity of radon at each location. Finally, you will pool all data to make a map of radioactivity throughout your community.

Objectives

Build **a radon detector, and use it to detect radon emissions.**

Observe **the tracks of alpha particles microscopically and count them.**

Calculate **the activity of radon.**

Evaluate **radon activity over a large area using class data, and draw a map of its activity.**

Required Precautions

- Wear safety goggles and lab apron.
- Read all safety cautions, and discuss them with your students.
- Students should not handle NaOH solutions with concentrations greater than 1.0 M.
- If a hot plate is used, the precautions in the teacher's notes for page 652 should be followed to avoid hazard from an electric shock.

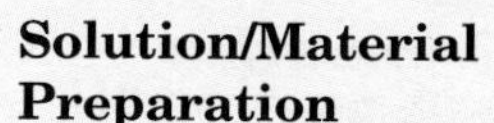

Safety

Although you will not use any chemicals in this investigation, there are hazardous chemicals in the lab where you will work, so wear goggles and an apron to provide protection for your eyes and clothing. You may remove your goggles only while you are using the microscope. Wash your hands thoroughly before you leave the lab after all lab work is finished.

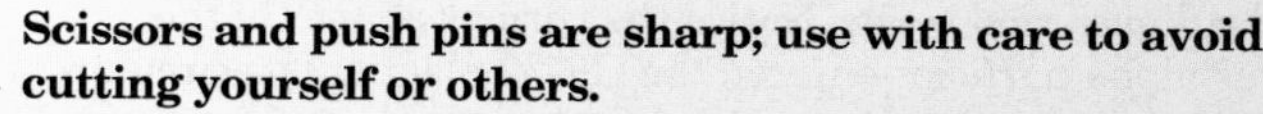

Always clean up the lab and all equipment after use, and dispose of substances according to proper disposal methods. Wash your hands thoroughly before you leave the lab.

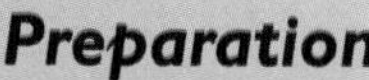

Scissors and push pins are sharp; use with care to avoid cutting yourself or others.

Materials

- **CR-39 plastic (with polyethylene film covering on both sides)**
- **3 in × 5 in index card**
- **Clear plastic metric ruler or stage micrometer**
- **Etch clamp**
- **Microscope**
- **Paper clip**
- **Push pin**
- **Scissors**
- **Small plastic cup with lid**
- **Tape**
- **Toilet paper or other tissue**

Preparation

1. ***Organizing Data***
Read each section of the laboratory guidelines and prepare tables for data. Your first data table should provide space to record the date and the time of day that the detector was put in place, the location of the detector in your home, the floor level of the location, and the date and the time of day the detector was removed. Your second data table should provide space for recording the number of tracks that you count in each of 10 fields, as well as the diameter of the microscope's field of view. You will also need room for other observations.

Technique

Detector Construction

2. Cut a rectangle, 2 cm × 4 cm, from the 3 in × 5 in index card.
3. Locate the side of the CR-39 plastic that has the felt-marker lines on it. Peel off the polyethylene film, and use the push pin to inscribe an ID number or other identification near the edge of the piece. With a short piece of transparent tape, form a loop with the sticky side out. Place the tape on the back of the piece of CR-39 plastic (the side that is still covered with polyethylene), and firmly attach it to the index card rectangle.
4. With a permanent marker, write the ID number or other identification on the outside of the plastic cup. Place the paper rectangle, with the CR-39 plastic on top, in the cup.
5. Cut a hole in the center of the plastic lid. Place a small piece of tissue over the top of the cup to serve as a dust filter. Then snap the lid onto the cup.
6. Place the completed detector in the location of your choice. (Check with your instructor first.) The detector must remain undisturbed at that location for at least three weeks.
7. At the end of the three weeks, return the entire detector to school for the chemical-etching process and the counting of the radiation tracks.

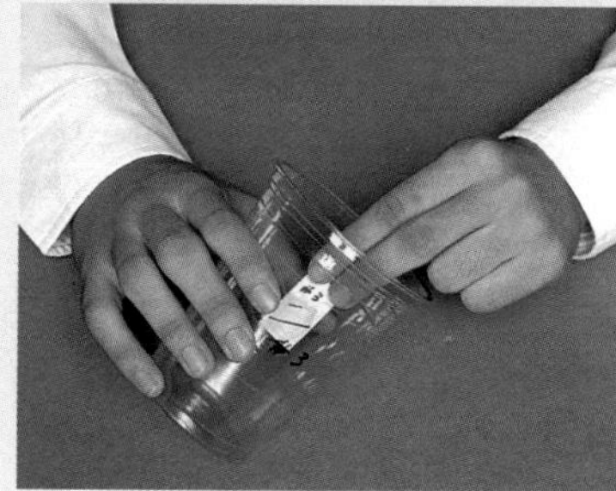

- If students use radioactive sources, they must wear impermeable disposable gloves when handling these objects. "One-size-fits-all" polyethylene gloves are available at most grocery stores. When finished, the gloves should be removed by turning them inside out as they are taken off the hand and fingers. Keep them inside out when disposing of them in the trash can.

Solution/Material Preparation

1. Wear goggles, face shield, impermeable gloves, and a lab apron when you prepare and use the 6.25 M NaOH. Work in a hood known to be in operating condition with another person present nearby to call for help in case of an emergency. Be sure you are within 30 s walking distance of a working safety shower and eyewash station.
2. To prepare 1 L of 6.25 M NaOH, dissolve 250. g of NaOH in enough distilled water to make 1.00 L of solution. The NaOH solution will get hot as the pellets dissolve. To prevent boiling and splattering, add the pellets a few at a time while stirring, instead of adding all of the pellets at once.
3. An empty film canister can be used in place of the small plastic cup with lid.
4. CR-39 plastic is available from Alpha Trak, 141 Northridge Drive, Centralia, WA 98531, (360) 736-3884.

Etching Instructions

1. If students observe the etching process, they should wear goggles, a face shield, and a lab apron, and they should stand at least five feet from the caustic solution.
2. Pour enough 6.25 M NaOH into each test tube so that the CR-39 plastic will be completely immersed when the paper clip is hooked over the test tube edge.
3. Check the paper clip and etch clamp to be certain the students have attached them properly to the CR-39 plastic.
4. Hook the paper clip over the test tube so that the CR-39 plastic is completely immersed in the NaOH.
5. Add water to the beaker(s). Place the test tubes in it.
6. Heat the water bath to boiling for about 30 min. The water may have to be replenished periodically.
7. Carefully remove the CR-39 from the hot water bath, and rinse it thoroughly. Dry the plastic with a soft tissue.

Student Orientation

Pre-lab Discussion

Thoroughly discuss all safety precautions outlined in this laboratory and in the safety section. Radioactivity is a poorly understood topic, so take this opportunity to discuss practical ways to reduce exposure. Discuss how the properties of radon as a gas cause problems in airtight buildings.

Encourage teams/individuals to test a wide geographical area. Be sure students have organized data and calculation tables in their lab notebooks before beginning this lab.

Discuss the calculations for Bq/L prior to the students doing it themselves. They often do not realize that *activity* = 2373 tracks/cm^2/day was determined as a standard from 370 pCi/L of air (13.69 Bq/L), which was used to calculate unknown concentrations of radon.

Explain that the CR-39 plastic is covered with the polyethylene film to protect the CR-39 from radiation until it is ready for use. The NaOH bath strips away a thin layer of the plastic to expose the pathway of the alpha particles.

Techniques to Demonstrate

Either demonstrate how to build the detector, or have one on hand for students to examine. You should not need to explain the use of the microscope, but students will need help measuring and calculating the field of view. Be certain that they have focused on the proper level of the plastic.

Post-Lab

Disposal

Neutralize waste NaOH solution by adding 1.0 M acid slowly, while stirring, until the pH is between 5 and 9. After checking that the pH is in this neutral range, the solution may be poured down the drain. Save the etched plastic for examination by students next year.

Paper clip

Etch clamp

CR-39 plastic

Etching

8. Remove the CR-39 plastic from the detector and the index card, and peel the polyethylene film from the back. Slip the ring of an etch clamp over the top of the CR-39 plastic, and hook it onto a large paper clip that has been reshaped to have "hooks" at each end, as shown in the diagram.
9. When you have completed this work, give it to your teacher for the etching step. During this step, NaOH solution will be used to remove the outer layer of the plastic so that the tracks of the alpha particles become visible.

Counting the Tracks

10. Examine the illustration below. Notice the various shapes of the tracks left as alpha particles entered the CR-39 plastic. The circular-shaped tracks are due to alpha particles that entered straight on, and the teardrop shaped tracks are due to alpha particles that entered at an angle.
11. Use a clear plastic metric ruler or a stage micrometer to measure the diameter of the microscope's field of view. Make this measurement for low power (10×). Record this diameter in the data table in your lab notebook.
12. The tracks are on the top surface of the CR-39 plastic. Make certain that you focus on that surface and that the tracks look like those in the illustration. These tracks were produced by placing the CR-39 plastic within a radium-coated clay urn. Your radon detector should not have nearly as many tracks. If there are too many tracks, switch to a high power (40×) objective, after recording the diameter of the field of view. Place your piece of CR-39 plastic on a microscope slide. Count and record the number of tracks in 10 different fields. Record these numbers in the data table.

Analysis and Interpretation

1. ***Organizing Data***
   In order to increase the accuracy of your data, you counted tracks in 10 different areas. Find the average number of tracks in a single area for your piece of CR-39 plastic.
2. ***Evaluating Data***
   Find the absolute deviation and average deviation for each of the 10 measurements. This calculation was described in Exploration 2A.
3. ***Organizing Data***
   You counted the number of tracks within the field of view of your microscope. In order to calculate the number of tracks per cm^2, you need to know the area of the field. Use the diameter of the field to calculate the area. (Hint: the field is circular; $d = 2r$; $A = \pi r^2$.)
4. ***Organizing Data***
   Using your answers to items **1** and **3**, calculate the average number of tracks per cm^2.
5. ***Organizing Data***
   In item **4** you calculated the number of tracks that accumulated per cm^2 over the total period the detector was in place. Divide this number by the number of days the detector was in place to calculate tracks/cm^2/day.

Alpha particle tracks

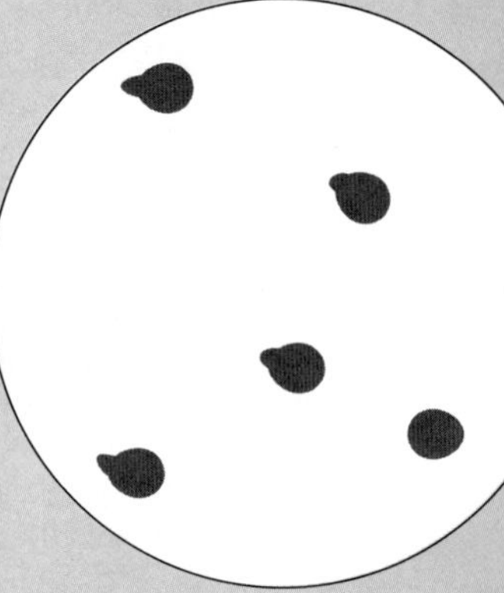

Sample Data

Date detector was put in place	12/3/94
Time detector was put in place	3:45 p.m.
Location of detector (e.g., bedroom, kitchen, dining room, furnace room, under the stairwell, etc.)	kitchen

Floor level of location	1st
Date detector was removed	12/27/94
Time detector was removed	8:15 p.m.

Conclusions

6. *Organizing Data*
Several pieces of CR-39 plastic were sent to a facility that had a radon chamber in which the activity of the radon was known to be 13.69 Bq/L (becquerels per liter of air), as measured by a different technique. A becquerel is the name for the SI unit of activity for a radioactive substance and is equal to 1 decay/s. The exposed pieces of CR-39 plastic were etched, counted, and found to have an activity of 2370 tracks/cm^2/day. Calculate the radon activity measured by your detector in becquerels/liter. (Hint: use the proportion 13.69 Bq/L : 2370 tracks/cm^2/day as a conversion factor to convert your data.)

7. *Relating Ideas*
The Environmental Protection Agency and other government agencies use non-SI units of picocuries per liter (pCi/L) to measure radiation. One curie is 3.7×10^{10} Bq, and one picocurie is 10^{-12} curies. Convert your data into units of pCi/L.

8. *Organizing Data*
Combine your data with that of other teams in your class, and jointly construct a map of your region that shows the levels of radon activity in both Bq/L and pCi/L at various locations throughout your community.

Extensions

1. *Applying Ideas*
The half-life of $^{222}_{86}$Rn is 3.823 days. After what time will only one-fourth of a given amount of radon remain?

2. *Designing Experiments*
Factors other than geographic location can have an effect on radon emissions. For example, readings taken in basements are likely to be higher than those in attics. Design an experiment to test different aspects of the three highest and three lowest regions of radiation on the map to examine the influence of these other factors. If your teacher approves your suggestion, try it.

3. *Designing Experiments*
CR-39 plastic could be used for further investigations of naturally occurring radiation. Design an experiment to explore one of the following areas. If your teacher approves your plan, carry out the experiment.
- the range of alpha particles
- the radon activity in soil
- radioactivity of common items such as lantern mantles (green Coleman label), glow-in-the-dark clock faces, and pieces of Fiestaware

4. *Research and Communication*
Investigate the ways in which people are exposed to nuclear radiation. Is there an acceptable level of radiation? Is there a place where you can go to avoid all radiation? What can you do to avoid excessive radiation? Present your findings in the form of a chart or a poster to share with your class.

Field	1	2	3	4	5	6	7	8	9	10
Number of tracks	3	4	4	5	9	7	4	3	8	6

Answers to Analysis and Interpretation

1. 5.3 tracks/field on average

2. Field 1 $5.3 - 3 = 2.3$
Field 2 $5.3 - 4 = 1.3$
Field 3 $5.3 - 4 = 1.3$
Field 4 $5.3 - 5 = 0.3$
Field 5 $5.3 - 9 = -3.7$
Field 6 $5.3 - 7 = -1.7$
Field 7 $5.3 - 4 = 1.3$
Field 8 $5.3 - 3 = 2.3$
Field 9 $5.3 - 8 = -2.7$
Field 10 $5.3 - 6 = -0.7$
average deviation = 1.76 tracks

3. $A = \pi r^2 = (3.14)\left(\frac{0.17 \text{ cm}}{2}\right)^2 = 2.3 \times 10^{-2} \text{ cm}^2$

4. $\frac{5.3 \text{ tracks}}{0.023 \text{ cm}^2} = 230 \text{ tracks/cm}^2$

5. $\frac{230 \text{ tracks/cm}^2}{24.19 \text{ days}} = 9.5 \text{ tracks/cm}^2\text{/day}$

Conclusions

6. $9.5 \text{ tracks/cm}^2\text{/day} \times \frac{13.69 \text{ Bq/L}}{2370 \text{ tracks/cm}^2\text{/day}} = 5.5 \times 10^{-2} \text{ Bq/L}$

7. $5.5 \times 10^{-2} \text{ Bq/L} \times \frac{1 \text{ Ci}}{3.7 \times 10^{10} \text{ Bq}} \times \frac{10^{12} \text{ pCi}}{1 \text{ Ci}} = 1.5 \text{ pCi/L}$

8. Maps should correctly reflect the data gathered and should have appropriate amounts in units of Bq/L and pCi/L.

Extensions

1. One-fourth will remain after 2 half-lives, or 7.646 days.

2. Students' suggestions will vary. Be sure that students' plans meet all necessary safety guidelines before allowing students to try them.

3. Students' suggestions will vary. Be sure that students' plans meet all necessary safety guidelines before allowing students to try them.

4. Student answers will vary, but students should reach the conclusion that there are low levels of radiation everywhere.

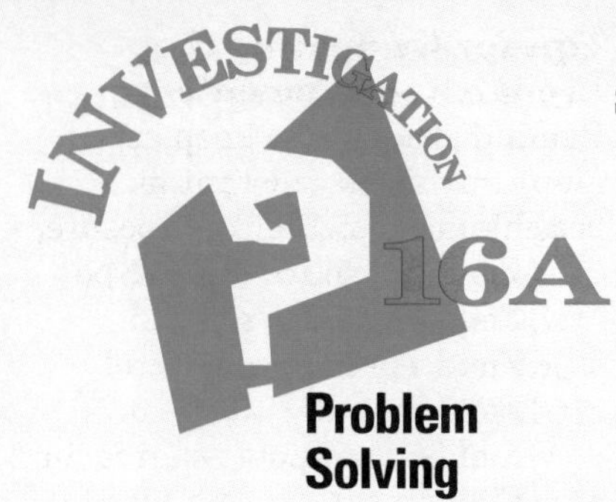

Objectives

Students will:

- build a radiation detector using CR-39 plastic.
- expose various shielding materials to alpha particles.
- observe and count the tracks of alpha particles with a microscope.
- convert measurements from units of picocuries to becquerels.
- determine which material is the most effective shield for alpha particles.

Planning

Recommended Time: 1–2 lab periods
Prerequisite: Exploration 16A

Solution/Materials Preparation

1. See the Teacher's Notes for Exploration 16A for information on preparing 6.25 M NaOH.
2. CR-39 plastic is available from Alpha Trak, 141 Northridge Drive, Centralia, WA 98531, (206) 736-3884.
3. Any of the following items could be used as a point source of alpha particles: pieces of Fiestaware, Coleman green-label (not gold) lantern mantles, old glow-in-the-dark clock faces, cloud chamber radiation source, smoke detector apparatus (contains americium).
4. Groups may coordinate their testing and combine results to decide on the best shielding material for the price. The CR-39 may be cut into smaller pieces to increase the number of trials and shielding materials that the teams test.
5. Follow the etching instructions given in the Teacher's Notes for Exploration 16A.

Student Orientation

Techniques to Demonstrate

Review the calculations for field of view and for converting Bq/L to pCi/L.

Pre-lab Discussion

Begin by discussing the results and procedure used in the Exploration, especially any lab technique errors that occurred. Make sure students understand how this lab differs. Instead of measuring how many particles cause tracks, the students are determining how well different materials can shield the detector. If students are not certain how to proceed, advise them to purchase the same quantities of materials as used in the Exploration.

Sample Data and Analysis

Typically, a maximum of 20 particles/field are found. Depending on the shielding ability of the materials tested, fewer may be found.

B DETECTING RADIOACTIVITY

Shielding From Alpha Particles

SOUTH WEST VIRGINIA
STATE UNIVERSITY

May 19, 1995

Mr. Reginald Brown
Director of Materials Testing
CheMystery Labs, Inc.
52 Fulton Street
Springfield, VA 22150

Dear Mr. Brown:

Recently, Professor Madeline Hoffmann joined our physics department faculty. She is eager to begin work on alpha-radiation, but we have told her this work cannot continue until we create a lab site that will not endanger anyone with radiation.

Professor Hoffmann assures me that alpha particles are relatively easy to block, but for legal reasons, I must make sure that we have independent confirmation for this claim.

Please test a variety of materials for their ability to block alpha radiation, so we can work with our maintenance and engineering services team to create a lab that will be safe yet inexpensive.

The university's president has authorized me to offer you $200 000 for this work, as outlined in the contract you will receive from the university's general counsel.

Sincerely,

John C. Hill

John C. Hill
South West Virginia State University

Required Precautions

Goggles and lab aprons must be worn at all times in the laboratory.

Wear disposable polyethylene gloves when you handle point source radioactive material. When finished, remove the gloves by turning them inside out as they are taken off. Keep them turned inside out, and dispose of them in the trash.

Do not touch or taste any chemicals. Wash your hands thoroughly when finished.

Required Precautions

- Goggles and lab aprons must be worn at all times.
- Read all safety cautions, and discuss them with your students. Be sure all safety precautions from the Exploration lab are followed.
- Students should not handle NaOH solutions with concentrations greater than 1.0 M.

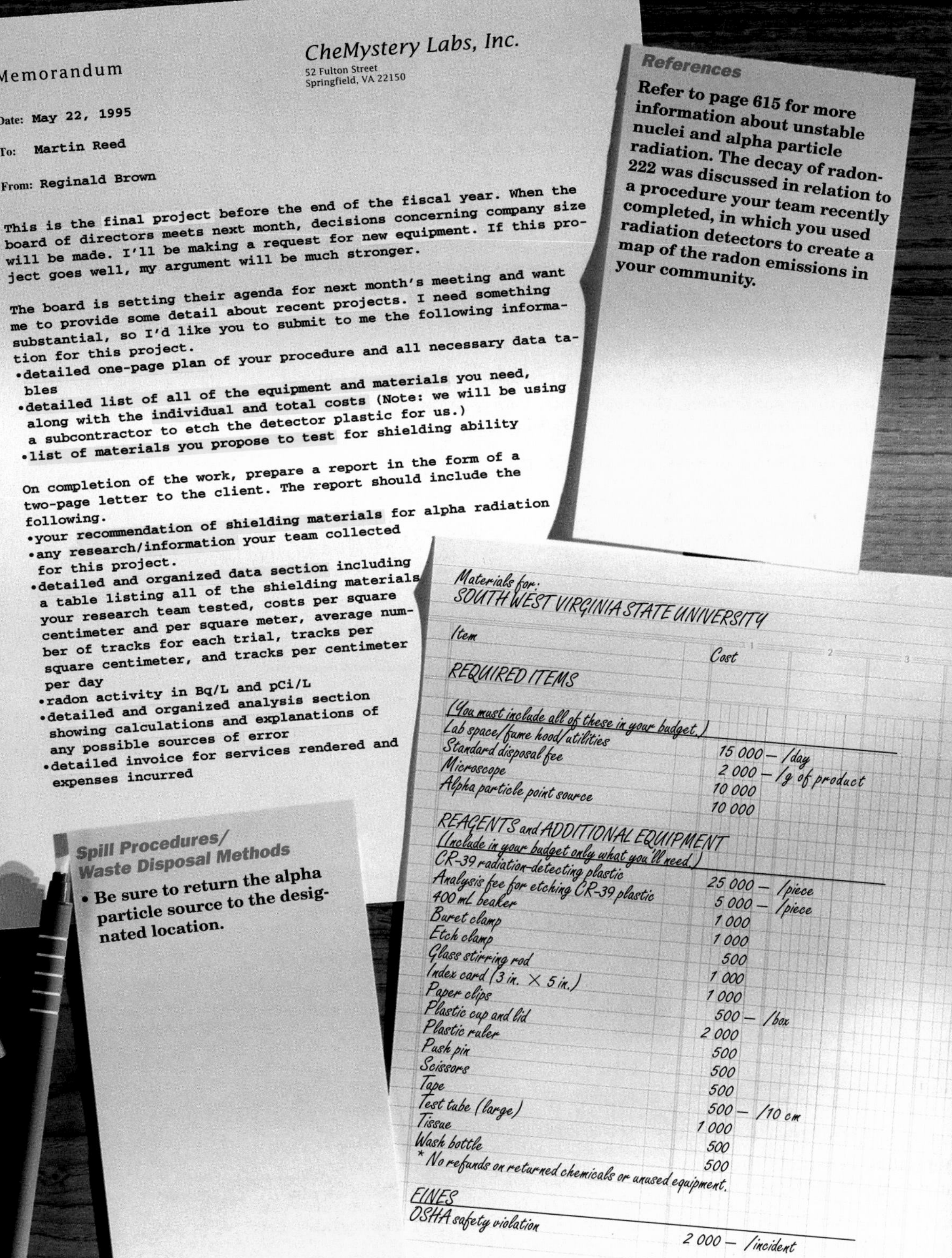

CheMystery Labs, Inc.
52 Fulton Street
Springfield, VA 22150

Memorandum

Date: May 22, 1995

To: Martin Reed

From: Reginald Brown

This is the final project before the end of the fiscal year. When the board of directors meets next month, decisions concerning company size will be made. I'll be making a request for new equipment. If this project goes well, my argument will be much stronger.

The board is setting their agenda for next month's meeting and want me to provide some detail about recent projects. I need something substantial, so I'd like you to submit to me the following information for this project.

- detailed one-page plan of your procedure and all necessary data tables
- detailed list of all of the equipment and materials you need, along with the individual and total costs (Note: we will be using a subcontractor to etch the detector plastic for us.)
- list of materials you propose to test for shielding ability

On completion of the work, prepare a report in the form of a two-page letter to the client. The report should include the following.

- your recommendation of shielding materials for alpha radiation
- any research/information your team collected for this project.
- detailed and organized data section including a table listing all of the shielding materials your research team tested, costs per square centimeter and per square meter, average number of tracks for each trial, tracks per square centimeter, and tracks per centimeter per day
- radon activity in Bq/L and pCi/L
- detailed and organized analysis section showing calculations and explanations of any possible sources of error
- detailed invoice for services rendered and expenses incurred

References

Refer to page 615 for more information about unstable nuclei and alpha particle radiation. The decay of radon-222 was discussed in relation to a procedure your team recently completed, in which you used radiation detectors to create a map of the radon emissions in your community.

Spill Procedures/ Waste Disposal Methods

- Be sure to return the alpha particle source to the designated location.

Materials for:
SOUTH WEST VIRGINIA STATE UNIVERSITY

Item	Cost
REQUIRED ITEMS	
(You must include all of these in your budget.)	
Lab space/fume hood/utilities	15 000 — /day
Standard disposal fee	2 000 — /g of product
Microscope	10 000
Alpha particle point source	10 000
REAGENTS and ADDITIONAL EQUIPMENT	
(Include in your budget only what you'll need.)	
CR-39 radiation-detecting plastic	25 000 — /piece
Analysis fee for etching CR-39 plastic	5 000 — /piece
400 mL beaker	1 000
Buret clamp	1 000
Etch clamp	500
Glass stirring rod	1 000
Index card (3 in. × 5 in.)	1 000
Paper clips	500 — /box
Plastic cup and lid	2 000
Plastic ruler	500
Push pin	500
Scissors	500
Tape	500 — /10 cm
Test tube (large)	1 000
Tissue	500
Wash bottle	500
* No refunds on returned chemicals or unused equipment.	
FINES	
OSHA safety violation	2 000 — /incident

Materials
(for each lab group)

- Alpha radiation point source
- CR-39 plastic
- 3 in × 5 in index card
- Clear plastic ruler or stage micrometer
- Etch clamp (the ring from a key chain)
- Microscope
- Paper clips
- Push pin
- Scissors
- Small plastic cup with lid
- Tape
- Toilet paper or other tissue

For Teacher Use Only

- 6.25 M NaOH, 10 mL per lab group
- Small test tube, 1 per lab group
- 1 L beaker or 400 mL beakers, 2
- Bunsen burner or hot plate

Estimated cost of materials: $187 000

Tips for Evaluating the Pre-lab Requirements

Remind students to keep as many variables constant as possible, such as time of exposure, distance from source, and type of source used. Also remind students to plan for a control method.

When using a point source for alpha particles, the amount of exposure time should be no longer than 5 min.

Proposed Procedure

Prepare several CR-39 pieces in detectors. Surround them with materials to test. After each has been exposed for about 5 min, prepare the CR-39 for etching. Use a microscope to count the alpha tracks. Make a comparison of each material tested.

Post-Lab

Disposal

Neutralize waste NaOH solution by adding 1.0 M acid slowly, with stirring, until the pH is between 5 and 9. After checking that the pH is in this neutral range, the solution may be poured down the drain. Save the etched plastic for examination by students next year.

Appendix A

Table A-1
SI Measurement

Most measurements in this book are expressed in SI units. Scientists throughout the world use SI and the metric system, and you will always use these units when you make measurements in the laboratory. The official name of the measurement system is the Système International d'Unités, or International System of Units.

SI is a decimal system—that is, all relationships between units of measurement are based on powers of 10. Most units have a prefix that indicates the relationship of that unit to the seven base units presented in Chapter 2.

Metric Prefixes

Prefix	Symbol	Factor of Base Unit
giga	G	1 000 000 000
mega	M	1 000 000
kilo	k	1000
hecto	h	100
deka	da	10
deci	d	0.1
centi	c	0.01
milli	m	0.001
micro	µ	0.000 001
nano	n	0.000 000 001
pico	p	0.000 000 000 001

Mass

1 kilogram (kg)	= SI base unit of mass
1 gram (g)	= 0.001 kg
1 milligram (mg)	= 0.000 001 kg
1 microgram (µg)	= 0.000 000 001 kg

Length

1 kilometer (km)	= 1 000 m
1 meter (m)	= SI base unit of length
1 centimeter (cm)	= 0.01 m
1 millimeter (mm)	= 0.001 m
1 micrometer (µm)	= 0.000 001 m
1 nanometer (nm)	= 0.000 000 001 m
1 picometer (pm)	= 0.000 000 000 001 m

Area

1 square kilometer (km^2)	= 100 hectares (ha)
1 hectare (ha)	= 10 000 square meters (m^2)
1 square meter (m^2)	= 10 000 square centimeters (cm^2)
1 square centimeter (cm^2)	= 100 square millimeters (mm^2)

Volume

1 liter (L)	= common unit for liquid volume (not SI)
1 cubic meter (m^3)	= 1000 L
1 kiloliter (kL)	= 1000 L
1 milliliter (mL)	= 0.001 L
1 milliliter (mL)	= 1 cubic centimeter (cm^3)

Teaching Tip

Chapter 10

Students will need to use the data in **Table A-3** to work some of the problems in Chapter 10.

Table A-2
Symbols and Abbreviations

Symbol		Meaning
α	=	helium nucleus (also ${}^4_2He$) emission from radioactive materials
β	=	electron (also ${}^{0}_{-1}e$) emission from radioactive materials
γ	=	high-energy photon emission from radioactive materials
Δ	=	change in a given quantity (e.g., ΔH for change in enthalpy)
c	=	speed of light in vacuum
c_p	=	specific heat capacity (at constant pressure)
D	=	density
E_a	=	activation energy
E^0	=	standard electrode potential
E^0cell	=	standard potential of an electrochemical cell
G	=	Gibbs free energy
ΔG^0	=	standard free energy of reaction
$\Delta G^0{}_f$	=	standard molar free energy of formation
H	=	enthalpy
ΔH^0	=	standard enthalpy of reaction

Symbol		Meaning
$\Delta H^0{}_f$	=	standard molar enthalpy of formation
K_a	=	dissociation constant (acid)
K_b	=	dissociation constant (base)
K_{eq}	=	equilibrium constant
K_{sp}	=	solubility-product constant
KE	=	kinetic energy
m	=	mass
N_A	=	Avogadro's number
n	=	number of moles
P	=	pressure
pH	=	measure of acidity ($-\log [H_3O^+]$)
R	=	ideal gas law constant
S	=	entropy
S^0	=	standard molar entropy
T	=	temperature (thermodynamic, in kelvins)
t	=	temperature (± degrees Celsius)
V	=	volume
v	=	velocity

Abbreviations		
amu	=	atomic mass unit (mass)
atm	=	atmosphere (pressure, non-SI)
Bq	=	becquerel (nuclear activity)
°C	=	degree Celsius (temperature)
J	=	joule (energy)
K	=	kelvin (temperature, thermodynamic)

Abbreviations		
mol	=	mole (quantity)
M	=	molarity (concentration)
N	=	newton (force)
Pa	=	pascal (pressure)
s	=	second (time)
V	=	volt (electric potential difference)

Table A-3
Vapor Pressure of Water at Selected Temperatures

Temperature °C	Pressure (kPa)	Temperature °C	Pressure (kPa)	Temperature °C	Pressure (kPa)	Temperature °C	Pressure (kPa)
0	0.61	**18**	2.06	**23**	2.81	**35**	5.63
5	0.87	**18.5**	2.13	**23.5**	2.90	**40**	7.38
10	1.23	**19**	2.19	**24**	2.98	**50**	12.34
12.5	1.45	**19.5**	2.27	**24.5**	3.10	**60**	19.93
15	1.71	**20**	2.34	**25**	3.17	**70**	31.18
15.5	1.76	**20.5**	2.41	**26**	3.36	**80**	47.37
16	1.82	**21**	2.49	**27**	3.57	**90**	70.12
16.5	1.88	**21.5**	2.57	**28**	3.78	**95**	84.53
17	1.94	**22**	2.64	**29**	4.01	**100**	101.32
17.5	2.00	**22.5**	2.72	**30**	4.25		

Table A-4
Physical Constants

Quantity	Symbol	Value
Atomic mass unit	amu	$1.660\,5402 \times 10^{-27}$ kg
Avogadro's number	N_A	$6.022\,137 \times 10^{23}$/mol
Electron rest mass	m_e	$9.109\,3897 \times 10^{-31}$ kg
Ideal gas law constant	R	8.314 L•kPa/mol•K = 0.0821 L•atm/mol•K
Molar volume of ideal gas at STP	V_M	22.414 L/mol
Neutron rest mass	m_n	$1.674\,9286 \times 10^{-27}$ kg
Normal boiling point of water	T_b	373.15 K = 100.00°C
Normal freezing point of water	T_f	273.15 K = 0.00°C
Proton rest mass	m_p	$1.672\,6231 \times 10^{-27}$ kg
Speed of light in a vacuum	c	$2.997\,924\,58 \times 10^{8}$ m/s
Temperature of triple point of water		273.16 K = 0.01°C

Table A-5
Common Ions

Cation name	Symbol	Anion name	Symbol
aluminum	Al^{3+}	**acetate**	CH_3COO^-
ammonium	NH_4^+	**bromide**	Br^-
arsenic(III)	As^{3+}	**carbonate**	CO_3^{2-}
barium	Ba^{2+}	**chlorate**	ClO_3^-
calcium	Ca^{2+}	**chloride**	Cl^-
chromium(II)	Cr^{2+}	**chlorite**	ClO_2^-
chromium(III)	Cr^{3+}	**chromate**	CrO_4^{2-}
cobalt(II)	Co^{2+}	**cyanide**	CN^-
cobalt(III)	Co^{3+}	**dichromate**	$Cr_2O_7^{2-}$
copper(I)	Cu^+	**fluoride**	F^-
copper(II)	Cu^{2+}	**hexacyanoferrate(II)**	$Fe(CN)_6^{4-}$
hydronium	H_3O^+	**hexacyanoferrate(III)**	$Fe(CN)_6^{3-}$
iron(II)	Fe^{2+}	**hydride**	H^-
iron(III)	Fe^{3+}	**hydrogen carbonate**	HCO_3^-
lead(II)	Pb^{2+}	**hydrogen sulfate**	HSO_4^-
magnesium	Mg^{2+}	**hydroxide**	OH^-
mercury(I)	Hg_2^{2+}	**hypochlorite**	ClO^-
mercury(II)	Hg^{2+}	**iodide**	I^-
nickel(II)	Ni^{2+}	**nitrate**	NO_3^-
potassium	K^+	**nitrite**	NO_2^-
silver	Ag^+	**oxide**	O^{2-}
sodium	Na^+	**perchlorate**	ClO_4^-
strontium	Sr^{2+}	**permanganate**	MnO_4^-
tin(II)	Sn^{2+}	**peroxide**	O_2^{2-}
tin(IV)	Sn^{4+}	**phosphate**	PO_4^{3-}
titanium(III)	Ti^{3+}	**sulfate**	SO_4^{2-}
titanium(IV)	Ti^{4+}	**sulfide**	S^{2-}
zinc	Zn^{2+}	**sulfite**	SO_3^{2-}

Table A-6
The Elements—Symbols, Atomic Numbers, and Atomic Masses

Name of element	Symbol	Atomic number	Atomic mass
actinium	Ac	89	[227.0278]
aluminum	Al	13	26.981539
americium	Am	95	[243.0614]
antimony	Sb	51	121.757
argon	Ar	18	39.948
arsenic	As	33	74.92159
astatine	At	85	[209.9871]
barium	Ba	56	137.327
berkelium	Bk	97	[247.0703]
beryllium	Be	4	9.012182
bismuth	Bi	83	208.98037
boron	B	5	10.811
bromine	Br	35	79.904
cadmium	Cd	48	112.411
calcium	Ca	20	40.078
californium	Cf	98	[251.0796]
carbon	C	6	12.011
cerium	Ce	58	140.115
cesium	Cs	55	132.90543
chlorine	Cl	17	35.4527
chromium	Cr	24	51.9961
cobalt	Co	27	58.93320
copper	Cu	29	63.546
curium	Cm	96	[247.0703]
dysprosium	Dy	66	162.50
einsteinium	Es	99	[252.083]
erbium	Er	68	167.26
europium	Eu	63	151.965
fermium	Fm	100	[257.0951]
fluorine	F	9	18.9984032
francium	Fr	87	[223.0197]
gadolinium	Gd	64	157.25
gallium	Ga	31	69.723
germanium	Ge	32	72.61
gold	Au	79	196.96654
hafnium	Hf	72	178.49
helium	He	2	4.002602
holmium	Ho	67	164.93032
hydrogen	H	1	1.00794
indium	In	49	114.818
iodine	I	53	126.90447
iridium	Ir	77	192.22
iron	Fe	26	55.847
krypton	Kr	36	83.80
lanthanum	La	57	138.9055
lawrencium	Lr	103	[262.11]
lead	Pb	82	207.2
lithium	Li	3	6.941
lutetium	Lu	71	174.967
magnesium	Mg	12	24.3050
manganese	Mn	25	54.93805
mendelevium	Md	101	[258.10]
mercury	Hg	80	200.59
molybdenum	Mo	42	95.94
neodymium	Nd	60	144.24
neon	Ne	10	20.1797
neptunium	Np	93	[237.0482]
nickel	Ni	28	58.6934
niobium	Nb	41	92.90638
nitrogen	N	7	14.00674
nobelium	No	102	[259.1009]
osmium	Os	76	190.23
oxygen	O	8	15.9994
palladium	Pd	46	106.42
phosphorus	P	15	30.973762
platinum	Pt	78	195.08
plutonium	Pu	94	[244.0642]
polonium	Po	84	[208.9824]
potassium	K	19	39.0983
praseodymium	Pr	59	140.90765
promethium	Pm	61	[144.9127]
protactinium	Pa	91	231.03588
radium	Ra	88	[226.0254]
radon	Rn	86	[222.0176]
rhenium	Re	75	186.207
rhodium	Rh	45	102.90550
rubidium	Rb	37	85.4678
ruthenium	Ru	44	101.07
samarium	Sm	62	150.36
scandium	Sc	21	44.955910
selenium	Se	34	78.96
silicon	Si	14	28.0855
silver	Ag	47	107.8682
sodium	Na	11	22.989768
strontium	Sr	38	87.62
sulfur	S	16	32.066
tantalum	Ta	73	180.9479
technetium	Tc	43	[97.9072]
tellurium	Te	52	127.60
terbium	Tb	65	158.92534
thallium	Tl	81	204.3833
thorium	Th	90	232.0381
thulium	Tm	69	168.93421
tin	Sn	50	118.710
titanium	Ti	22	47.88
tungsten	W	74	183.84
unnilennium	Une	109	[266]
unnilhexium	Unh	106	[263.118]
unniloctium	Uno	108	[265]
unnilpentium	Unp	105	[262.114]
unnilquadium	Unq	104	[261.11]
unnilseptium	Uns	107	[262.12]
uranium	U	92	238.0289
vanadium	V	23	50.9415
xenon	Xe	54	131.29
ytterbium	Yb	70	173.04
yttrium	Y	39	88.90585
zinc	Zn	30	65.39
zirconium	Zr	40	91.224

Table A-7
Properties of Common Elements

Name	Form/color at room temperature	Density† (g/cm^3)	Melting point (°C)	Boiling point (°C)	Common oxidation states
aluminum	silver metal	2.702	660.37	2467	3+
antimony	blue-white metalloid	6.684^{25}	630.5	1750	3+, 5+
argon	colorless gas	1.784*	–189.2	–185.7	0
arsenic	gray metalloid	5.727^{14}	817 (28 atm)	613 (*sublimes*)	3–, 3+, 5+
barium	bluish-white metal	3.51	725	1640	2+
beryllium	gray metal	1.85	1278 ± 5	2970 (0.0066 atm)	2+
bismuth	white metal	9.80	271.3	1560 ± 5	3+
boron	black metalloid	2.34	2300	2550	3+
bromine	red-brown liquid	3.119	–7.2	58.78	1–, 1+, 5+
calcium	silver metal	1.54	839 ± 2	1484	2+
carbon	diamond graphite	3.51 2.25	3500 (63.5 atm) 3652 (*sublimes*)	3930 —	2+, 4+
chlorine	green-yellow gas	3.214*	–100.98	–34.6	1–, 1+, 5+, 7+
chromium	gray metal	7.20^{28}	1857 ± 20	2672	2+, 3+, 6+
cobalt	gray metal	8.9	1495	2870	2+, 3+
copper	red metal	8.92	1083.4 ± 0.2	2567	1+, 2+
fluorine	yellow gas	1.69‡	–219.62	–188.14	1–
germanium	gray metalloid	5.323^{25}	937.4	2830	4+
gold	yellow metal	19.31	1064.43	2808 ± 2	1+, 3+
helium	colorless gas	0.1785*	–272.2 (26 atm)	–268.9	0
hydrogen	colorless gas	0.0899*	–259.34	–252.8	1–, 1+
iodine	blue-black solid	4.93	113.5	184.35	1–, 1+, 5+, 7+
iron	silver metal	7.86	1535	2750	2+, 3+
lead	bluish-white metal	11.3437^{16}	327.502	1740	2+, 4+
lithium	silver metal	0.534	180.54	1342	1+
magnesium	silver metal	1.74^{5}	648.8	1107	2+
manganese	gray-white metal	7.20	1244 ± 3	1962	2+, 3+, 4+, 7+
mercury	silver liquid metal	13.5462	–38.87	356.58	1+, 2+
neon	colorless gas	0.9002*	–248.67	–245.9	0
nickel	silver metal	8.90	1455	2730	2+, 3+
nitrogen	colorless gas	1.2506*	–209.86	–195.8	3–, 3+, 5+
oxygen	colorless gas	1.429*	–218.4	–182.962	2–
phosphorus	yellow solid	1.82	44.1	280	3–, 3+, 5+
platinum	silver metal	21.45	1772	3827 ± 100	2+, 4+
plutonium	silver metal	19.84	641	3232	3+, 4+, 5+, 6+
potassium	silver metal	0.86	63.25	760	1+
radium	white metal	5(?)	700	< 1140	2+
radon	colorless gas	9.73*	–71	–61.8	0
silicon	gray metalloid	2.33 ± 0.01	1410	2355	2+, 4+
silver	white metal	10.5	961.93	2212	1+
sodium	silver metal	0.97	97.8	882.9	1+
strontium	silver metal	2.6	769	1384	2+
sulfur	yellow solid	1.96	119.0	444.674	2–, 4+, 6+
tin	white metal	7.28	231.88	2260	2+, 4+
titanium	white metal	4.5	1660 ± 10	3287	3+, 4+
tungsten	gray metal	19.35	3410 ± 20	5660	6+
uranium	silver metal	19.05 ± 0.02^{25}	1132.3 ± 0.8	3818	3+, 4+, 6+
xenon	colorless gas	5.887 ± 0.009*	–111.9	–107 ± 3	0
zinc	blue-white metal	7.14	419.58	907	2+

† Densities obtained at 20°C unless noted (superscripted). *Densities of gases are given in g/L at STP.
‡ Density of fluorine is given in g/L at 1 atm and 15°C.

Table A-8
Densities of Gases at STP

Gas	Density (g/L)
air, dry	1.293
ammonia	0.771
carbon dioxide	1.997
carbon monoxide	1.250
chlorine	3.214
dinitrogen monoxide	1.977
ethyne (acetylene)	1.165
helium	0.1785
hydrogen	0.0899
hydrogen chloride	1.639
hydrogen sulfide	1.539
methane	0.7168
nitrogen	1.2506
nitrogen monoxide (at 10°C)	1.340
oxygen	1.429
sulfur dioxide	2.927

Table A-9
Density of Water

Temperature (°C)	Density (g/cm^3)
0	0.999 84
2	0.999 94
3.98 (maximum)	0.999 973
4	0.999 97
6	0.999 94
8	0.999 85
10	0.999 70
14	0.999 24
16	0.998 94
20	0.998 20
25	0.997 05
30	0.995 65
40	0.992 22
50	0.988 04
60	0.983 20
70	0.977 77
80	0.971 79
90	0.965 31
100	0.958 36

Table A-10
Solubilities of Gases in Water
Volume of gas (in liters) at STP that can be dissolved in 1 L of water at the temperature (°C) indicated.

Gas	0°C	10°C	20°C	60°C
air	0.029 18	0.022 84	0.018 68	0.012 16
ammonia	1130	870	680	200
carbon dioxide	1.713	1.194	0.878	0.359
carbon monoxide	0.035 37	0.028 16	0.023 19	0.014 88
chlorine	—	3.148	2.299	1.023
hydrogen	0.021 48	0.019 55	0.018 19	0.016 00
hydrogen chloride	512	475	442	339
hydrogen sulfide	4.670	3.399	2.582	1.190
methane	0.055 63	0.041 77	0.033 08	0.019 54
nitrogen*	0.023 54	0.018 61	0.015 45	0.010 23
nitrogen monoxide	0.073 81	0.057 09	0.047 06	0.029 54
oxygen	0.048 89	0.038 02	0.031 02	0.019 46
sulfur dioxide	79.789	56.647	39.374	—

*Atmospheric nitrogen—98.815% N_2, 1.185% inert gases

Table A-11
Solubilities of Compounds
Solubilities are given in grams of solute that can be dissolved in 100 g of water at the temperature (°C) indicated.

Compound	Formula	0°C	20°C	60°C	100°C
aluminum sulfate	$Al_2(SO_4)_3$	31.2	36.4	59.2	89.0
ammonium chloride	NH_4Cl	29.4	37.2	55.3	77.3
ammonium nitrate	NH_4NO_3	118	192	421	871
ammonium sulfate	$(NH_4)_2SO_4$	70.6	75.4	88	103
barium carbonate	$BaCO_3$	—	$0.0022^{18°}$	—	0.0065
barium chloride dihydrate	$BaCl_2{\cdot}2H_2O$	31.2	35.8	46.2	59.4
barium hydroxide	$Ba(OH)_2$	1.67	3.89	20.94	$101.40^{80°}$
barium nitrate	$Ba(NO_3)_2$	4.95	9.02	20.4	34.4
barium sulfate	$BaSO_4$	—	$0.000\,246^{25°}$	—	0.000 413
cadmium sulfate	$CdSO_4$	75.4	76.6	81.8	60.8
calcium acetate dihydrate	$Ca(C_2H_3O_2)_2{\cdot}2H_2O$	37.4	34.7	32.7	29.7
calcium carbonate	$CaCO_3$	—	$0.0014^{25°}$	—	$0.0018^{75°}$
calcium fluoride	CaF_2	$0.0016^{18°}$	$0.0017^{26°}$	—	—
calcium hydrogen carbonate	$Ca(HCO_3)_2$	16.15	16.60	17.50	18.40
calcium hydroxide	$Ca(OH)_2$	0.189	0.173	0.121	0.076
calcium sulfate	$CaSO_4$	—	$0.209^{30°}$	—	0.1619
cerium(III) sulfate nonahydrate	$Ce_2(SO_4)_3{\cdot}9H_2O$	21.4	9.84	3.87	—
cesium nitrate	$CsNO_3$	9.33	23.0	83.8	197
copper(II) chloride	$CuCl_2$	68.6	73.0	96.5	120
copper(II) sulfate pentahydrate	$CuSO_4{\cdot}5H_2O$	23.1	32.0	61.8	114
lead(II) chloride	$PbCl_2$	0.67	1.00	1.94	3.20
lead(II) nitrate	$Pb(NO_3)_2$	37.5	54.3	91.6	133
lithium chloride	$LiCl$	69.2	83.5	98.4	128
lithium sulfate	Li_2SO_4	36.1	34.8	32.6	$30.9^{90°}$
magnesium hydroxide	$Mg(OH)_2$	—	$0.0009^{18°}$	—	0.004
mercury(I) chloride	Hg_2Cl_2	—	$0.000\,20^{25°}$	$0.001^{43°}$	—
mercury(II) chloride	$HgCl_2$	3.63	6.57	16.3	61.3
potassium aluminum sulfate	$KAl(SO_4)_2$	3.00	5.90	24.8	$109^{90°}$
potassium bromide	KBr	53.6	65.3	85.5	104
potassium chlorate	$KClO_3$	3.3	7.3	23.8	56.3
potassium chloride	KCl	28.0	34.2	45.8	56.3
potassium chromate	K_2CrO_4	56.3	63.7	70.1	$74.5^{90°}$
potassium iodide	KI	128	144	176	206
potassium nitrate	KNO_3	13.9	31.6	106	245
potassium permanganate	$KMnO_4$	2.83	6.34	22.1	—
potassium sulfate	K_2SO_4	7.4	11.1	18.2	24.1
silver acetate	$AgC_2H_3O_2$	0.73	1.05	1.93	$2.59^{80°}$
silver chloride	$AgCl$	$0.000\,089^{10°}$	—	—	0.0021
silver nitrate	$AgNO_3$	122	216	440	733
sodium acetate	$NaC_2H_3O_2$	36.2	46.4	139	170
sodium chlorate	$NaClO_3$	79.6	95.9	137	204
sodium chloride	$NaCl$	35.7	35.9	37.1	39.2
sodium nitrate	$NaNO_3$	73.0	87.6	122	180
sucrose	$C_{12}H_{22}O_{11}$	179.2	203.9	287.3	487.2
ytterbium(III) sulfate	$Yb_2(SO_4)_3$	44.2	$22.2^{30°}$	10.4	4.7

Dashes indicate values not available.

Table A-12
Standard Thermodynamic Properties (P = 100 kPa = 0.987 atm; T = 298.15 K = 25°C)

Substance	Standard enthalpy of formation ΔH^o_f (kJ/mol)	Standard entropy S^o (J/mol·K)
Aluminum		
$Al(s)$	0.0	28.3
$AlCl_3(s)$	–705.6	110.7
$Al_2O_3(s$, corundum)	–1676.0	51.0
Bromine		
$Br_2(l)$	0.0	152.2
$Br_2(g)$	30.9	245.5
$HBr(g)$	–36.4	198.6
Calcium		
$Ca(s)$	0.0	41.6
$CaCO_3(s$, calcite)	–1206.9	92.9
$CaCl_2(s)$	–795.8	104.6
$CaO(s)$	–634.9	38.2
$Ca(OH)_2(s)$	–986.1	83.4
Carbon		
$C(s$, graphite)	0.0	5.7
$C(s$, diamond)	1.9	2.4
$CCl_4(l)$	–135.4	216.2
$CCl_4(g)$	–95.8	309.9
$CH_4(g)$	–74.9	186.3
$CH_3OH(l)$	–239.1	127.2
$C_2H_2(g)$	226.7	201.0
$C_2H_4(g)$	52.5	219.3
$C_2H_6(g)$	–83.8	229.1
$C_2H_5OH(l)$	–277.7	161.0
$C_3H_8(g)$	–104.7	270.2
$C_4H_{10}(g$, n–butane)	–125.6	310.1
$C_4H_{10}(g$, isobutane)	–134.2	294.6
$C_6H_6(l)$	49.0	173.4
$C_6H_{12}O_6(s)$	–1273.3	212.1
$C_6H_{14}(g$, n–hexane)	–167.1	388.4
$C_7H_{16}(g$, n–heptane)	–187.7	427.9
$C_8H_{18}(g$, n–octane)	–208.6	466.7
$C_8H_{18}(g$, isooctane)	–224.0	423.2
$CO(g)$	–110.5	197.6
$CO_2(g)$	–393.5	213.8
$CS_2(g)$	117.1	237.8
$HCOOH(l)$	–425.1	129.0
Chlorine		
$Cl_2(g)$	0.0	223.1
$HCl(g)$	–92.3	186.8
Copper		
$Cu(s)$	0.0	33.2
$CuCl_2(s)$	–220.1	108.1
$CuSO_4(s)$	–770.0	109.3
Fluorine		
$F_2(g)$	0.0	202.8
$HF(g)$	–272.5	173.8
Hydrogen		
$H_2(g)$	0.0	130.7
$H_2O(l)$	–285.8	70.0
$H_2O(g)$	–241.8	188.7
$H_2O_2(l)$	–187.8	109.6
$HCN(g)$	135.1	201.7
Iron		
$Fe(s)$	0.0	27.3
$FeCl_3(s)$	–399.4	142.3
$Fe_2O_3(s$, hematite)	–825.5	87.4
$Fe_3O_4(s$, magnetite)	–1120.9	145.3
Lead		
$Pb(s)$	0.0	64.8
$PbCl_2(s)$	–359.4	136.2
$PbO(s)$	–219.4	66.3
Lithium		
$Li(s)$	0.0	29.1
$LiOH(s)$	–484.9	42.8
$LiCl(s)$	–408.8	59.3
Magnesium		
$Mg(s)$	0.0	32.7
$MgCl_2(s)$	–641.6	89.6
Mercury		
$Hg(l)$	0.0	76.0
$Hg_2Cl_2(s)$	–265.2	192.5
$HgO(s)$	–90.8	70.3
Nitrogen		
$N_2(g)$	0.0	191.6
$NH_3(g)$	–45.9	192.8
$NH_4Cl(s)$	–314.5	94.6
$NO(g)$	90.3	210.8
$NO_2(g)$	33.1	240.0
$N_2O(g)$	82.4	220.0
$N_2O_4(g)$	9.1	304.4
$HNO_3(g)$	–134.3	266.4
Oxygen		
$O_2(g)$	0.0	205.1
$O_3(g)$	142.7	238.9
Potassium		
$K(s)$	0.0	64.7
$KCl(s)$	–436.7	82.6
$KNO_3(s)$	–494.6	133.1
$KOH(s)$	–424.7	78.9
Silicon		
$Si(s)$	0.0	18.8
$SiCl_4(g)$	–657.0	330.9
$SiO_2(s$, quartz)	–910.9	41.5
Silver		
$Ag(s)$	0.0	42.7
$AgCl(s)$	–127.1	96.2
$AgNO_3(s)$	–124.4	140.9
Sodium		
$Na(s)$	0.0	51.5
$NaCl(s)$	–411.2	72.1
$NaOH(s)$	–425.9	64.4
Sulfur		
$S(s)$	0.0	32.1
$SO_2(g)$	–296.8	248.1
$SO_3(g)$	–395.8	256.8
$H_2S(g)$	–20.5	205.7
$H_2SO_4(l)$	–814.0	156.9
Tin		
$Sn(s$, white)	0.0	51.6
$Sn(s$, gray)	–2.1	44.1
$SnCl_4(l)$	–511.3	258.6
Zinc		
$Zn(s)$	0.0	41.6
$ZnCl_2(s)$	–415.1	111.5
$ZnO(s)$	–348.3	43.6

+ = endothermic
- = exothermic
exo - Products side
A+B → C+D+heat

Teaching Tip

Graphing

Review this material with students when you cover density in Chapter 2.

Appendix B

How To

Graph scientific data

Graphs are a useful tool for displaying scientific data because they show relationships among variables in a compact, visual form. You probably know how to make and interpret several types of graphs such as pie charts and bar graphs (or *histograms*). You may have also used *x-y* graphs (or *Cartesian* graphs) in your math classes. However, you may not know how to use *x-y* graphs to display experimental data in chemistry laboratory work. The following guidelines will help.

1. Determine the independent variable

- Determine which of the quantities that you will be graphing is the *independent variable* and which is the *dependent variable*. The independent variable, denoted as x, is the variable whose values are chosen by the experimenter. The independent variable is plotted on the horizontal axis. The dependent variable, denoted y, is plotted on the vertical axis. Values of the dependent variable are determined by the independent variable.
- For example, the data shown in the table to the left was gathered in an experiment in which the temperature of a gas was increased and the resulting volume increase was measured. In this case, temperature was the independent variable and volume was the dependent variable. In the graph for this experiment, temperature is plotted on the horizontal axis and volume is plotted on the vertical axis.

Temperature (K)*	Volume (L)*
120	1
240	2
360	3
480	4

*Values specified at standard pressure.

2. Scale the axes

- Each axis must have a scale with equal divisions.
- Allow as much room as possible between divisions.
- Each division must represent a whole number of units of the variable being plotted, such as 1, 2, 5, 10 or some multiple of these. To decide which multiple to use for the horizontal axis, divide the maximum value of the independent variable by the number of major divisions on your graph paper. For example, **Graph A**, on the next page, shows 10 divisions along the grid on the horizontal axis. The data used to plot the curve for **Graph A** is shown to the left. The maximum value of T is 480 K. Divide the number of divisions into the maximum value of the variable to get 480 K/10 divisions or 48 K per division. To simplify, round up to allow 50 K per division on the horizontal axis.
- To scale the vertical axis follow the same procedure. The maximum value of the dependent variable, V, is 4 L. The grid allows for 6 divisions in the vertical direction. Divide 4 L/6 divisions to obtain 0.66 L per division. Round up to allow 1.0 L per division. Then there will be 2 divisions left over on the top of the vertical axis. Check **Graph A** to see how this looks.

- Label each axis with the quantity to be plotted and the units used to express each measurement. For example, the axes of **Graph A** are *Volume* (L) and *Temperature* (K).

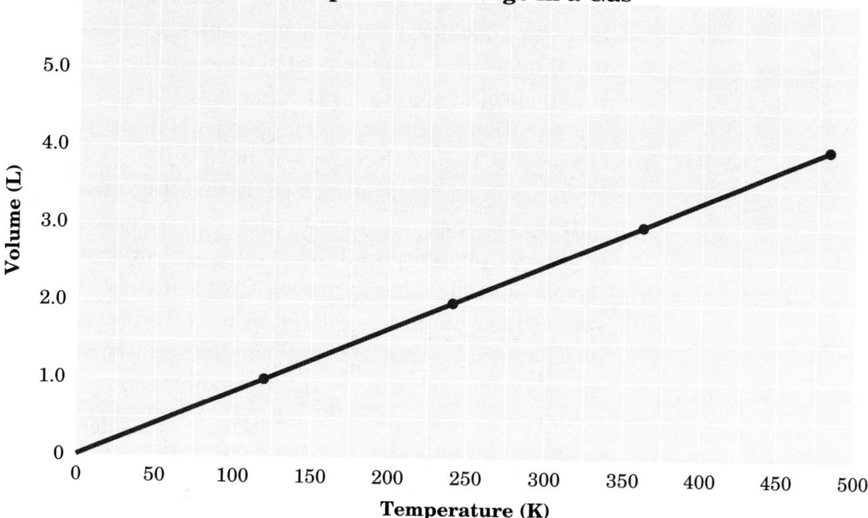

3. Plot the data

- Plot each data point by locating the proper coordinates for the ordered pair on the graph grid. If the data points look like they fall roughly on a straight line, use a transparent ruler to find the line of best fit for the data points. Draw the best-fit line through or between the points.
- If the data points clearly do not fall along a straight line, but appear to fit another smooth curve, lightly sketch in the smooth curve that connects the points.
- Once you have sketched a smooth curve, draw over it in ink.

4. Title your graph

- Title your graph to indicate the x and y variables. If you can also tell how the variables relate to one another without making the title too long, include this information. For example, "Volume Versus Temperature Change in a Gas" is a suitable title for **Graph A**. Write the title at the top of the graph.

How To continued on following page . . .

Teaching Tip

Interpreting graphs

Review this material with your students when you cover the gas laws in Chapter 10.

How To continued from the previous page . . .

Pressure (atm)*	Volume (L)*
0.100	224
0.200	112
0.400	56.0
0.600	37.3
0.800	28.0
1.00	22.4

*Values specified at constant temperature.

5. Interpret your graph

- If your data points lie roughly along a straight line, the x and y variables have a **linear relationship** or are **directly proportional**. This means that as one variable increases, the other does too, in a constant proportion—as x doubles, y doubles; as x triples, y triples; etc. Directly proportional quantities, x and y, relate to one another through mathematical equations of the form $y = mx + b$, where m is a constant and b is zero. The equation for the directly proportional linear relationship shown in **Graph A** is $V = kT$. Here, $m = k$ and $b = 0$.
- If your data points lie along a curve that drops from left to right as shown in **Graph B**, then the quantities have an **inverse relationship** or are **inversely proportional**. In an inverse relationship, one quantity increases as the other decreases. **Graph B** shows that gas pressure and volume have an inverse relationship; as the pressure of a gas increases, its volume decreases. The mathematical relationship that expresses an inverse relationship is $y = 1/x$. The expression relating gas pressure and volume follows the form $PV = k$. Note that inverse relationships are nonlinear because the increase of one variable is not accompanied by a constant rate of decrease in the other variable.

Graph B

Volume Versus Pressure Changes in a Gas at Constant Temperature

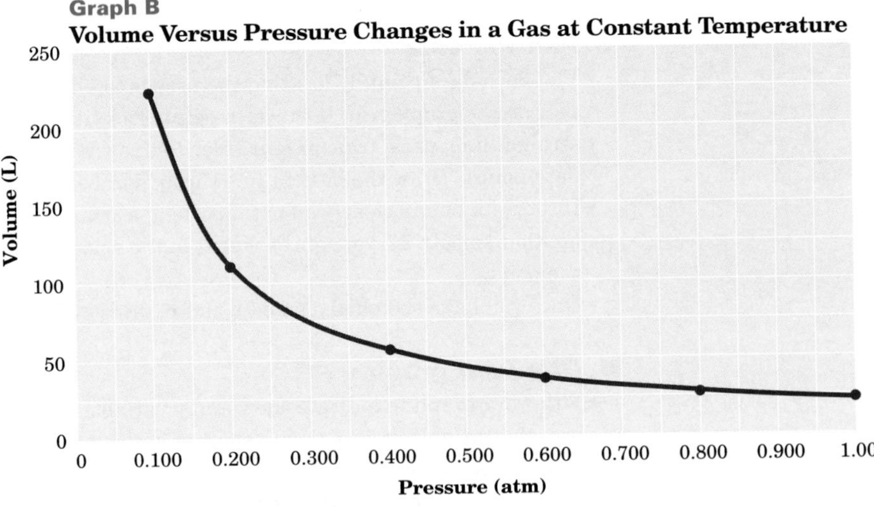

Pressure (atm)*	1/P (1/atm)	Volume (L)*
0.100	10.0	224
0.200	5.00	112
0.400	2.50	56.0
0.600	1.67	37.3
0.800	1.25	28.0
1.00	1.00	22.4

*Values specified at constant temperature.

6. Use your graph

- Straight-line graphs are the easiest graphs to analyze and to express as equations. More complex graphs illustrate inversely proportional, exponential, or logarithmic relationships. It is often useful to replot a nonlinear graph to obtain a straight-line graph.

  Graph C shows the inverse relationship $PV = k$ replotted as a straight line. To obtain this graph, both sides of the equation $PV = k$ were divided by P.

$$\left[\frac{PV}{P} = \frac{k}{P}\right] = \left[V = k \times \frac{1}{P}\right]$$

The resulting equation, $V = k \times 1/P$, has the same form as $y = mx$, which if plotted would produce a straight line that passes through the origin. To plot the actual data, the pressure values in the table must be converted to 1/P values. The first pressure conversion is as follows.

$$\frac{1}{0.100} = 10.0$$

V is plotted on the y axis and $1/P$ is plotted on the x axis.

Graph C
Volume Versus the Reciprocal of Pressure for a Gas at Constant Temperature

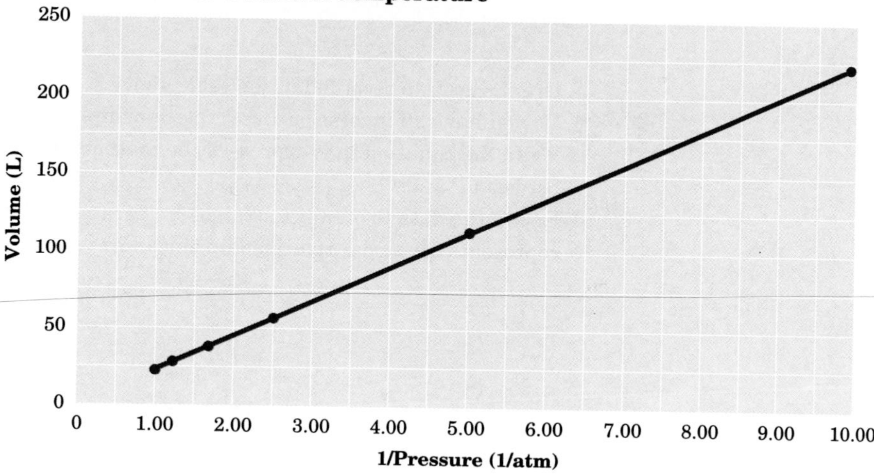

Answers to

Concept Review

1. The axis should extend from 0 to 120 grams. Check that the scale is consistent between divisions. Students should not scale their axes by the points listed.

2. The x axis should extend from 0°C to about 110°C as the minimum. The y axis should extend from 0 g to about 250 g as the minimum. Check that both axes are correctly labeled using the labels given in the data table. The solubility of KNO_3 at 65°C is about 120 g. The solubility of KNO_3 at 105°C is about 265 g.

3. a. The x axis should extend from 0 atm to about 10 atm as the minimum. The y axis should extend from 0 L to about 250 L as the minimum. Check that both axes are correctly labeled using the labels given in the data table.
b. inverse relationship
c. $PV = k$
d. $V = k \times \frac{1}{P}$

Graphing

1. Draw a horizontal axis on graph paper that is suitable for plotting the following mass measurements: 20.5 g, 39.7 g, 61.0 g, 92.8 g, and 116.9 g.
2. Prepare a graph representing the solubilities of potassium nitrate, KNO_3, in water at the following temperatures. From your graph, estimate the solubility of KNO_3 at 65°C and at 105° C.

Temperature (°C)	Solubility (g solute/100 g water)
0	13.9
20	31.6
40	61.3
60	106
80	167
100	245

3. The pressure of 1 mol of ammonia gas was varied and the following volume measurements were made. Temperature was kept constant at 25° C.

Pressure (atm)	Volume (L)
0.100	245
0.200	122
0.400	61.0
0.800	30.4
2.00	12.2
4.00	5.98
8.00	2.92

a. Graph the data recorded in the table above.
b. Do you have an inverse or direct relationship?
c. Write the equation that expresses the mathematical relationship between P and V.
d. Replot the data to obtain a straight-line graph. Write the equation that represents the linear relationship shown by your graph.

Answers for Concept Review items and Practice problems begin on page 841.

How To

Use significant figures and scientific notation

Scientists use significant figures to report information about the certainty of measurements and calculations. With this method, a measurement of 2.25 m means that while the 2 in the ones place and the 2 in the tenths place are certain, the 5 in the hundredths place is an estimate. If this measurement is combined with several other measurements in a formula, there must be some way of tracking the amount of uncertainty in each measurement and in the final result. For example, using a calculator to find the volume of a cube that measures 2.25 m on a side, you get 11.390625 m^3. This answer indicates far greater precision in the volume measurement than is realistic. Remember that the 5 in 2.25 is an estimated digit.

The rules and examples that follow will show you how to work with the uncertainty in measurements to express your results with an appropriate level of precision.

1. Determining the number of significant figures

The first set of rules shows you how to look at a measurement to determine the number of significant figures. A measurement expressed to the appropriate number of significant figures includes all digits that are certain and one digit in the measurement that is uncertain.

Rules for Determining the Number of Significant Figures

Rule	Example
The following are always significant	
• All nonzero digits	673 has three, 2.8 has two
• All zeros between nonzero digits	506 has three, 1.009 has four
• Zeros to the right of a non-zero digit and left of a *written* decimal point	34 800. mL has five, 200. cm has three
• Zeros to the right of a non-zero digit and right of a *written* decimal point	4.0 kg has two, 57.50 K has four, 2.90×10^3 has three
The following are never significant	
• Zeros to the left of the decimal point in numbers less than one	0.984 kg has three, 0.6 has one
• Zeros to the right of a decimal point, but to the left of the first non-zero digit	0.067 has two, 0.004 has one
Exceptions to the rules	
• Exact conversion factors are understood to have an unlimited number of significant figures	By definition there are exactly 100 cm in 1 m so the conversion factor 100 cm/1 m is understood to have an unlimited number of significant figures.
• Counting numbers are understood to have an unlimited number of significant figures	There are exactly 30 days in June, not 30.1 or 29.005, so an unlimited number of significant figures is understood in the expression "30 days".

2. Calculating with significant figures

When measurements are used in calculations, you must apply the rules regarding significant figures so that your results reflect the number of significant figures of the measurements.

How To continued on following page . . .

Teaching Tip

Significant figures

Use this material to reinforce the discussion from Section 2-5.

How To continued from the previous page . . .

Rules for Making Calculations with Significant Figures

Rule	Example
Addition and subtraction The answer must be rounded so that it contains the same number of digits to the right of the decimal point as there are in the measurement with the smallest number of digits to the right of the decimal point.	2.89 m + 0.00043 m = 2.89043 m = 2.89 m
Multiplication The product or quotient should be rounded off to the same number of significant figures as in the measurement with the fewest significant figures.	3.5293 mol × 34.2 g/mol = 120.70206 g = 121 g

3. Rounding answers to get the correct number of significant figures

To obtain the correct number of significant figures in a measurement or calculation, numbers must often be rounded. To round numbers correctly, observe the following rules.

Rules for Rounding

Rule	Example
If the digit immediately to the right of the last significant figure you want to retain is	
• Greater than 5, increase the last digit by 1.	56.87 g ⟶ 56.9 g
• Less than 5, do not change the last digit.	12.02 L ⟶ 12.0 L
• 5, followed by nonzero digit(s), increase the last digit by 1.	3.7851 ⟶ 3.79
• 5, not followed by a nonzero digit and preceded by odd digit(s), increase the last digit by 1.	2.835 s ⟶ 2.84 s
• 5, not followed by nonzero digit(s), and the preceding significant digit is even, do not change the last digit.	2.65 mL ⟶ 2.6 mL

4. Expressing numbers in scientific notation

Measurements made in chemistry often involve very large or small numbers. To express these numbers conveniently, *scientific notation* is used. In scientific notation, numbers are expressed in terms of their order of magnitude. For example, 54 000 can be expressed as 5.4×10^4 in scientific notation, and the number 0.000 008 765 can be expressed as 8.765×10^{-6}.

As the preceding examples show, each value expressed in scientific notation has two parts. The first factor is always between 1 and 10, but it may have any number of digits. To write the first factor of the number, move the decimal point to the right or left so that there is only one nonzero digit to the left of it. The second factor of the number is written raised to an exponent of 10 that is determined by counting the number of places the decimal point must be moved. If the decimal point is moved to the left, the exponent is positive. If the decimal point is moved to the right, the exponent is negative.

For calculations involving scientific notation, the rules are the same as for exponents in algebra.

Rules for Calculations with Numbers in Scientific Notation

Rule	Example
Addition and Subtraction All values must have the same exponent before they can be added or subtracted. The result is the sum or difference of the first factors all with the same exponent of 10.	$4.5 \times 10^6 - 2.3 \times 10^5 =$ $45 \times 10^5 - 2.3 \times 10^5$ $= 42.7 \times 10^5$ $= 4.3 \times 10^6$
Multiplication The first factors of the numbers are multiplied and the exponents of 10 are added.	$(3.1 \times 10^3)(5.01 \times 10^4) =$ $(3.1 \times 5.01) \times 10^{4+3}$ $= 16 \times 10^7 = 1.6 \times 10^8$
Division The first factors of the number are divided, and the exponent of 10 in the denominator is subtracted from the exponent of 10 in the numerator.	$\frac{7.63 \times 10^3}{8.6203 \times 10^4} = \frac{7.63 \times 10^{3-4}}{8.6203}$ $= 0.885 \times 10^{-1}$ $= 8.85 \times 10^{-2}$

5. Expressing significant figures using scientific notation

Using scientific notation along with significant figures is especially useful for measurements such as 200 L, 2560 m, or 10 000 kg, because it is unclear which zeros are significant. In such cases follow the procedure described below.

Rules for Expressing Scientific Notation with Significant Figures

Rule	Example
Use scientific notation to eliminate all placeholding zeros. Convert the number to scientific notation and eliminate zeros before an unwritten decimal point that are not significant figures.	$2400 \longrightarrow 2.4 \times 10^4$ (if both zeros are not significant) $600 \longrightarrow 6.0 \times 10^2$ (if only one zero is significant) $750\ 000 \longrightarrow 7.5000 \times 10^5$ (if all zeros are significant)

Significant figures and scientific notation

1. How many significant figures are there in these expressions?
   a. 470. km **b.** 0.0980 m **c.** 30.8900 g
   d. 0.09709 kg **e.** 1000 g/1 kg **f.** 4.870×10^5 s
2. Perform the following calculations and express the answers in significant figures.
   a. 32.89 g + 14.21 g **b.** 34.09 L − 1.230 L **c.** 100 m + 0.7 m
   d. 1.8940 cm × 0.0651 cm **e.** 24.897 mi / 0.8700 h **f.** 111.0 in × 1.020 in
3. Perform the following calculations.
   a. $\frac{8.369 \times 10^3 + 4.58 \times 10^2 - 6.30 \times 10^3}{4.156 \times 10^7}$
   b. $(6.499 \times 10^2)(5.915 \times 10^4 + 3.4733 \times 10^5)$
   c. $(7.23780 \times 10^{-3} - 3.65 \times 10^{-5})(3.6792 \times 10^2 + 2.67)$
   d. $\frac{(2.1267 \times 10^{-5})(3.3456 \times 10^{-2} - 0.012)}{(2.6 \times 10^{-2} - 3.23 \times 10^{-2})}$

Answers to Concept Review

1. **a.** 3
   b. 3
   c. 6
   d. 4
   e. unlimited number
   f. 4
2. **a.** 47.10 g
   b. 32.86 L
   c. 100 m
   d. 0.123 cm^2
   e. 28.62 mi/h
   f. 113.2 in^2
3. **a.** 6.08×10^{-5}
   b. 2.642×10^8
   c. 2.67
   d. -7.2×10^{-5}

Teaching Tip

Concept mapping

Review this material when covering Chapter 1.

Additional information along with tips for evaluating students' maps can be found on page T44.

Appendix C

How To

Make concept maps

A concept map presents key ideas, meanings, and relationships for the major concepts being studied. A concept map for a chapter can be thought of as a visual road map for learning the material in the chapter. Using concept maps, this learning happens efficiently because you work with only the key ideas and how they fit together.

The concept map shown as **Map A** was made from most of the vocabulary terms in Chapter 2. Vocabulary terms are generally labels for concepts, and concepts are generally nouns. Concepts are linked using linking words to form propositions. A proposition is a phrase that gives meaning to the concept. For example, in the map shown "matter is changed by energy" is a proposition.

Studies show that people are better able to remember materials presented visually. The concept map is better than an outline because you can see relationships among many ideas. Because outlines are linear there is no way of linking the ideas from various sections of the outline. Read through the map, to become familiar with the information presented. Look at the map and in relation to all of the text pages in Chapter 2, which would be more useful to study before an exam?

Map A

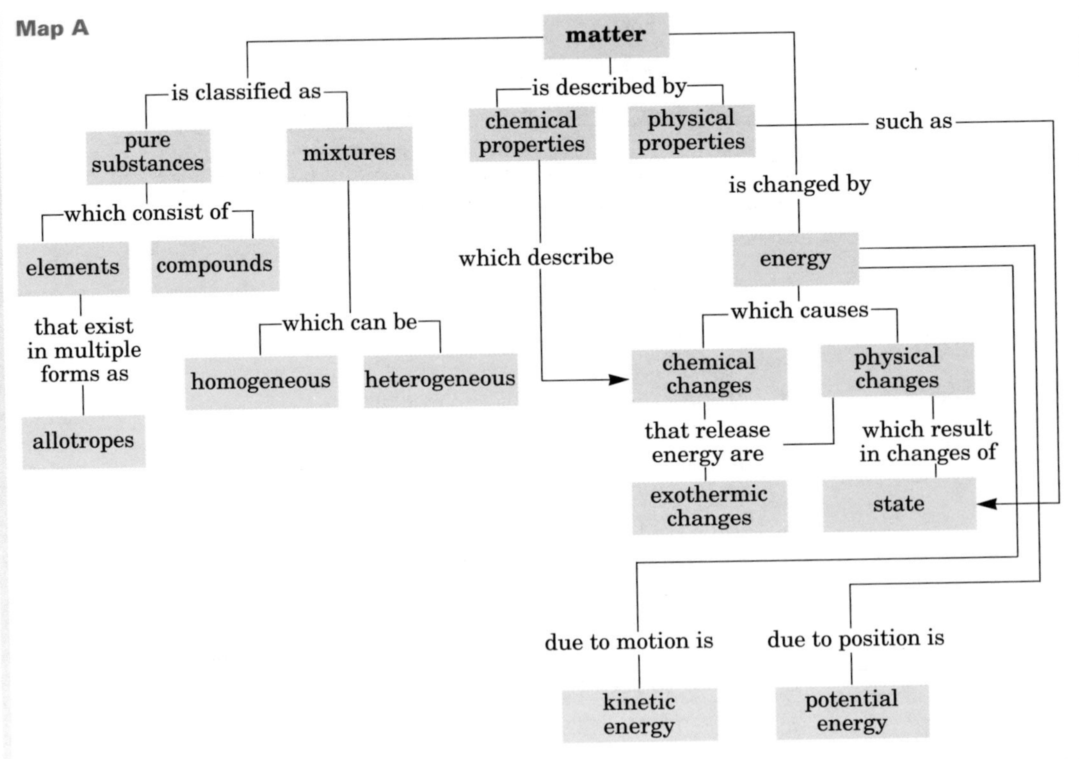

1. Start by taking a text section and listing all the important concepts.

We'll use the boldface terms in Section 2–3.

pure substance
mixture
homogeneous mixture
heterogeneous mixture
phase
element
compound
allotrope

- From this list, group similar concepts together. For example, one way to group these concepts would be into two groups—one that is related to mixtures, and the other that is related to pure substances.

mixture	*pure substance*
homogeneous mixture	element
heterogeneous mixture	compound
phase	allotrope

2. Select a main concept for the map.

We'll use matter as the main concept for this map.

3. Start building the map by placing the concepts according to their importance under the main concept, matter.

One way of arranging the concepts is shown in **Map B**.

Map B

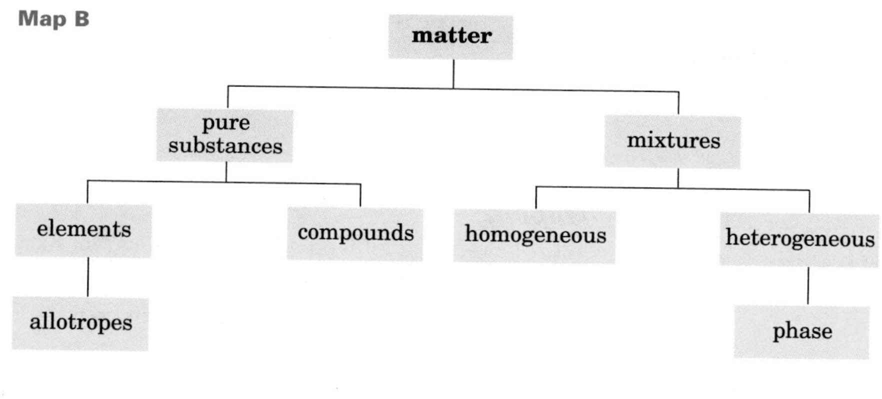

4. Now add the linking words that give meaning to the arrangement of concepts.

When adding the links, be sure that each proposition makes sense. To distinguish concepts from links, place your concepts in circles, ovals, or rectangles—like the ones in the map shown. Now look for cross links. Cross links are made of linking words and lines that connect information from various parts of the map. Cross links can be shown as links made with an arrow head. Map C is the finished map covering the main ideas of Section 2-3.

At first, making maps might seem difficult. However, the process of making maps forces you to think about the meanings and relationships among the concepts. Therefore, if you don't understand those relationships, you can get help.

One strategy to try when practicing mapping is to make concept maps about topics you know. For example, if you know a lot about a particular sport (such as basketball) or you have a particular hobby (such as music), you can use those topics to make practice maps. You'll perfect your skills with information that you know very well and you'll then feel more confident about your skills when you make maps from the information in a chapter.

Remember, the time you devote to mapping will pay off when it is time to review for an exam.

Map C

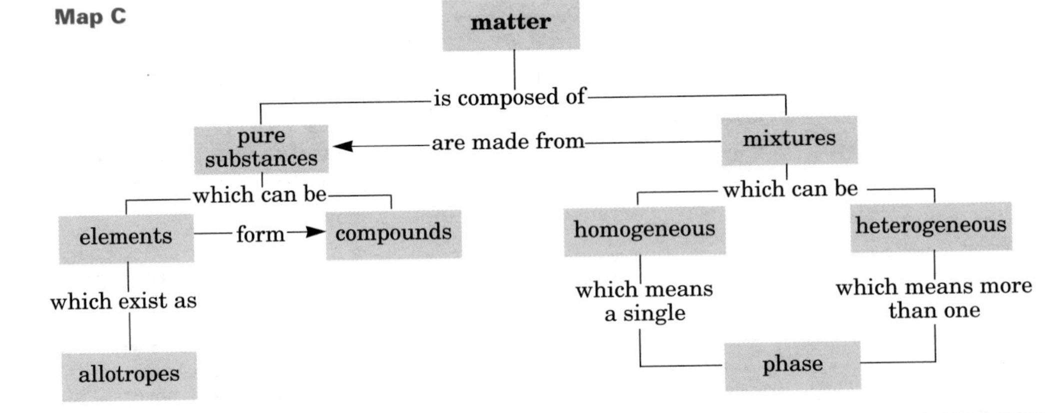

Concept mapping

1. Classify each of the following as either a concept or linking word(s).

a. classification	**b.** is classified as	**c.** forms
d. is described by	**e.** reaction	**f.** reacts with
g. metal	**h.** defines	

2. Write three propositions from the information in **Map A**.
3. List two cross links shown on **Map A**.

Answers for Concept Review items and Practice problems begin on page 841.

Answers to

Concept Review

1. **a** concept
   b. linking words
   c. linking word
   d. linking words
   e. concept
   f. linking words
   g. concept
   h. linking word
2. Answers will vary. Check student responses against Map A. Some possible responses are the following. *matter is described by chemical properties, matter is classified as mixtures, chemical changes that release energy are exothermic changes, physical properties such as state, and so on.*
3. Cross links are denoted by the arrowhead. The two cross links in Map A are: *chemical properties which describe chemical changes, physical properties such as state*

Glossary

absolute temperature scale: a temperature measurement made relative to absolute zero—the lowest possible temperature (369)

accuracy: the extent to which a measurement approaches the true value of a quantity (60)

acid: a class of compounds whose water solutions taste sour, turn blue litmus to red, and react with bases to form salts (5)

acid anhydride: an oxide that forms an acid when reacted with water (491)

acid-dissociation constant: a quantity derived from the ratio of the concentrations of the products and reactants at equilibrium for a weak acid equilibrium system (498)

actinides: metallic elements with atomic numbers 90 through 103 that fill the 5*f* orbitals (117)

activated complex: a specific arrangement of high-energy reactant atoms or molecules, the decomposition of which leads to product or intermediate molecules; also called transition state (547)

activation energy: the minimum amount of energy that must be supplied to a system to start a chemical change (326)

activity series: arrangement of elements in the order of their tendency to react with water and acids (259)

actual yield: measured amount of product actually produced from a given amount of reactant (287)

alkali metals: highly reactive metallic elements which form alkaline solutions in water, burn in air, and belong to Group 1 of the periodic table (113)

alkaline-earth metals: reactive, metallic elements which belong to Group 2 of the periodic table (114)

allotropes: different molecular forms of an element in the same physical state (49)

alloy: a solid or liquid mixture of two or more metals (406)

alpha particle: a helium nucleus produced in nuclear decay (619)

amino acid: a carboxylic acid that is the structural subunit of a protein (518)

amphiprotic: having the property of behaving as an acid and base (496)

anion: ion with a negative charge (156)

anode: the electrode at which oxidation occurs as electrons are lost by some substance (578)

atom: the basic unit of matter (35)

atomic mass: the mass of an atom in atomic mass units (133)

atomic mass unit: one-twelfth the mass of the carbon-12 isotope (133)

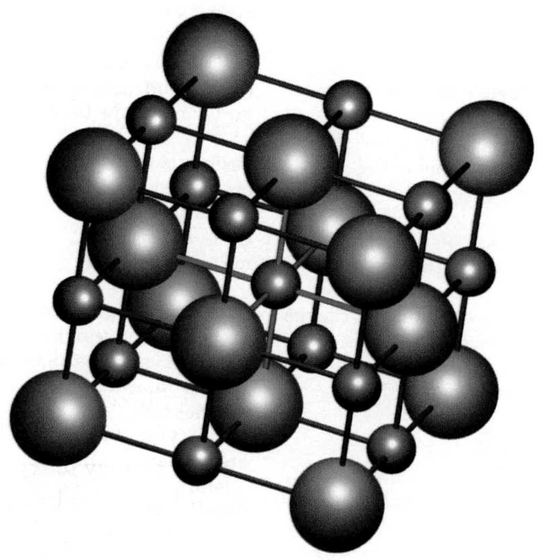

atomic number: the number of protons in the nucleus of an atom (93)

atomic radius: one-half of the distance from center to center of two like atoms (121)

average rate: the change in the measured property divided by the time elapsed (539)

Avogadro's number: 6.022×10^{23}, the number of particles in a mole (134)

Avogadro's principle: equal volumes of different gases under the same conditions have the same number of molecules (366)

basal metabolic rate: one's resting energy expenditure measured in the morning, at least 12 hours after the last meal (344)

base: a class of compounds that taste bitter, feel slippery in water solution, turn red litmus to blue, and react with acids to form salts (5)

base anhydride: an oxide that forms a base when reacted with water (493)

base-dissociation constant: a quantity derived from the ratio of the concentrations of the products and reactants at equilibrium for a weak base equilibrium system (499)

battery: single voltaic cell or group of voltaic cells that are connected together (595)

beta particle: an electron produced in nuclear decay (620)

binary compound: compound composed of two elements (166)

binding energy: the energy needed to separate nucleons within a nucleus or, equivalently, the energy released by nucleons combining to form a nucleus (616)

bond energy: the energy required to change the atoms in one mole of bonds from a bonded state to an unbonded state (153)

bond length: average distance between the nuclei of two bonded atoms (193)

Boyle's law: the volume of a gas at constant temperature is inversely proportional to the pressure (372)

buffer: a system that is able to withstand small additions of acid or base without significant change in pH; the system is composed of a conjugate acid-base pair (516)

cathode: the electrode at which reduction occurs as electrons are gained by some substance (578)

cation: ion with a positive charge (156)

catalyst: a substance added to a chemical reaction to increase the rate and that can be recovered chemically unchanged after the reaction is complete (560)

chain reaction: a self-sustaining nuclear or chemical reaction in which the product from each step acts as the reactant for the next step (626)

Charles's law: the volume of a gas at constant pressure is directly proportional to the absolute temperature (370)

chemical: any substance formed by or used in a chemical reaction (4)

chemical change: a change that produces one or more new substances (45)

chemical equation: symbols that describe a chemical reaction and indicate identities and relative amounts of reactants and products (238)

chemical kinetics: the branch of chemistry concerned with reaction rates and reaction mechanisms (536)

chemical properties: properties that can be observed only when substances interact with one another (38)

chlorofluorocarbons: a family of organic compounds in which the hydrogen atoms have been replaced by fluorine and chlorine (359)

circuit: closed loop path for current to follow (578)

coefficient: numeral used in a chemical equation to indicate relative amounts of reactants or products (239)

colligative property: a physical property that is dependent on the number of particles present rather than on the size, mass, or characteristics of those particles (459)

combustion: exothermic reaction that usually involves oxygen to form the oxides of the elements in a reactant (252)

common ion: an ion that comes from two or more substances making up a chemical solution (477)

common ion effect: a process in which an ionic compound becomes less soluble upon the addition of one of its ions by adding another compound (478)

complex ions: an ion having a structure in which a central atom or ion is bonded by coordinate bonds to other ions or molecules (called ligands) (466)

compounds: pure substances composed of two or more different elements (50)

concentration: ratio of solute to solvent or solution (411)

conductance: the measurement of a solution's ability to conduct electrical energy (452)

conjugate acid: the particle formed when a base has accepted a H^+ ion (497)

conjugate base: the particle formed when an acid has donated a H^+ ion (497)

covalent bond: bond formed when atoms share pairs of electrons (191)

critical mass: the minimum mass of fissionable material needed to produce a chain reaction (626)

crystal lattice: three-dimensional arrangement of atoms or ions in a crystal (158)

Dalton's law of partial pressures: the total pressure in a gas mixture is the sum of the partial pressures of the individual components (375)

decomposition: chemical reaction in which a single compound is broken down to produce two or more simpler substances (257)

diffusion: the process by which particles disperse from regions of higher concentration to regions of lower concentration (378)

dissociation: a process using energy to separate a compound into ions in water (456)

dissociation constant for water: the ion product constant for water that is equal to 1.00×10^{-14} (505)

double bond: covalent bond formed by the sharing of two pairs of electrons between two atoms (201)

double-displacement: chemical reaction in which two elements in different compounds exchange places (261)

effusion: the motion of a gas through an opening into an evacuated chamber (379)

electricity: energy associated with electrons that have moved from one place to another (577)

electrolysis: a process in which electric energy is used to bring about a chemical change (598)

electrolyte: any substance that, when it is dissolved in a solution, will conduct an electric current by means of movement of ions (452)

electrolytic cell: an electrochemical cell in which electric energy from an external source causes a nonspontaneous redox reaction to occur (598)

electromagnetic spectrum: the total range of electromagnetic radiation, ranging from the longest radio waves to the shortest gamma waves (90)

electron affinity: the energy change that accompanies the addition of an electron to an atom in the gas phase (124)

electronegativity: the tendency for an atom to attract electrons to itself when it is combined with another atom (125)

elementary step: a chemical equation describing one stage of a reaction as it is theorized to occur (558)

elements: the 109 simplest substances from which more complex materials are made (48)

empirical formula: simplest whole-number ratio of atoms that matches the relative ratio found in a chemical compound (169)

emulsifying agent: stabilizes an emulsion that would otherwise separate into different phases (428)

emulsion: colloidal-sized droplets (about 100 nm wide) of one liquid suspended in another liquid (427)

end point: a point in a titration indicated by the color change of an indicator (515)

endothermic change: a physical or chemical change in which a system absorbs energy from its surroundings (44)

energy: the capacity to do work (40)

energy level: a specific energy or group of energies that may be possessed by an electron in an atom (87)

enthalpy: total energy content of a system (324)

enthalpy change: heat energy released or absorbed when a physical or chemical change occurs at constant pressure (324)

entropy: a measure of the randomness or disorder of a system (333)

enzyme: large protein molecule that catalyzes chemical reactions in living things (561)

equilibrium: a reversible reaction in which the rates of the forward and reverse reactions are equal (22)

equilibrium constant (K_{eq}): for a reversible reaction, a number expressing the relationship between the mathematical product of the molar concentrations of the products divided by the mathematical product of the molar concentrations of the reactants, each raised to the power of its coefficient in the balanced equation (470)

equilibrium vapor pressure: a measure of the tendency of the particles of a liquid substance to enter the gas phase at a given temperature (389)

equivalence point: the point in a titration process where the moles of standard are stoichiometrically equivalent to the moles of substance titrated (510)

excess reactant: reactant that will not be used up in a reaction that goes to completion (285)

excited state: the condition of an atom in a state higher than the ground state (89)

exothermic change: a physical or chemical change in which energy is released by a system to its surroundings (44)

fatty acid: the long chain carboxylic acid subunit of a fat and oil (521)

formula unit: simplest collection of atoms from which a compound's formula can be established (169)

free energy: a quantity of energy related to the capacity of a system to do work, which can be used to predict spontaneity (339)

functional group: group of atoms that determines an organic molecule's chemical properties (219)

gamma ray: high-energy electromagnetic radiation produced by decaying nuclei (621)

Graham's law of effusion: the rates of effusion for two gases are inversely proportional to the square roots of their molar masses at the same temperature and pressure (379)

greenhouse effect: an increase in the warming effects of infrared radiation absorption brought about by an increase in levels of carbon dioxide and other greenhouse gases in the atmosphere (357)

ground state: the lowest energy state of a quantized system (89)

group: a series of elements that form a column in the periodic table (75)

half-cell: the part of a voltaic cell in which either oxidation or reduction occurs. It consists of a single electrode immersed in a solution of its ions (578)

half-life: the time required for half of a sample of radioactive atoms to decay (628)

half-reaction: the oxidation or reduction portion of a redox reaction (583)

halogens: elements that combine with most metals to form salts and that belong to Group 17 of the periodic table (119)

heat: the total of kinetic energy of random motion of molecules, atoms, or ions in a substance (315)

heat of reaction: the amount of heat energy absorbed or released during a chemical change (320)

Hess's law: the total enthalpy change for a chemical or physical change is the same whether it takes place in one or several steps (329)

heterogeneous mixture: a mixture containing substances that are not evenly distributed (47)

homogeneous mixture: a mixture containing substances that are uniformly distributed (47)

Hund's rule: the most stable arrangement of electrons is that with the maximum number of unpaired electrons, all with the same spin direction (97)

hydration: the process by which water molecules surround each ion as it moves into solution (456)

hydrogen bond: attraction occurring when a hydrogen atom bonded to a strongly electronegative atom is also attracted to another electronegative atom, often of a different molecule (197)

hydronium ion: a hydrogen ion covalently bonded to a water molecule, written as H_3O^+ (454)

hydroxide ion: the OH^- anion (490)

hypothesis: a proposition based on certain assumptions that can be evaluated scientifically (18)

ideal gas: a model that effectively describes the behavior of real gases at conditions close to standard temperature and pressure; a gas for which the product of the pressure and volume is proportional to the absolute temperature (360)

ideal gas law: the equation of state for an ideal gas in which the product of the pressure and volume is proportional to the product of the absolute temperature and the amount of gas expressed in moles (382)

immiscible: indicates liquids or gases that will not dissolve in each other (405)

indicator: (as applied to acid-base chemistry) a particle that reversibly changes color depending on the pH (508)

inhibitor: a substance added to a chemical reaction to slow it down (563)

initial rate: the highest rate measured very close to the start of the reaction where the curve is almost linear (543)

inorganic compound: all compounds outside the organic family of compounds (5)

insoluble: does not dissolve appreciably in a particular solvent (406)

intermediate: a structure formed in one elementary step, but consumed in a later step of a mechanism. It is neither an original reactant nor final product (559)

intermolecular forces: attraction resulting from forces between molecules (196)

ion: an atom or group of atoms that has gained or lost one or more electrons to acquire a net electric charge (123)

ionic bond: bond formed by the attraction of oppositely charged ions (157)

ionic compound: chemical compound composed of cations and anions combined so that the total positive and negative charges are equal (157)

ionization energy: the amount of energy needed to remove an electron from a specific atom or ion in its ground state in the gas phase (123)

isotope: one of two or more atoms having the same number protons but different numbers of neutrons (93)

kinetic energy: energy that moving objects possess by virtue of their motion (40)

kinetic molecular theory: the theory that explains the behavior of gases at the molecular level (360)

lanthanides: shiny, metallic elements with atomic numbers 58 through 71 that fill the *4f* orbitals (117)

lattice energy: energy released when a crystal containing one mole of an ionic compound is formed from gaseous ions (162)

law of conservation of energy: the observed fact that in any chemical or physical process, energy is neither created nor destroyed (43)

Le Châtelier's principle: a principle stating that if a system at equilibrium is disturbed by applying stress, the system will adjust in such a way as to counter the stress (464)

Lewis structure: diagram showing the arrangement of valence electrons among the atoms in a molecule (199)

ligand: a functional group, atom, or molecule that is attached to the central atom of a complex ion (467)

limiting reactant: reactant that is consumed first in a reaction that goes to completion (285)

main-block elements: elements that represent the entire range of chemical properties and belong to Groups 1, 2, and 13 through 18 in the periodic table (118)

mass: the quantity of matter in an object (34)

mass defect: the difference between the mass of an atom and the sum of the masses of its individual components (616)

mass number: the total number of protons and neutrons in the nucleus of an atom (94)

matter: anything that has mass and volume (34)

mechanism (of a reaction): a proposed step-by-step sequence of reactions that describes how reactants are changed into products (558)

metal: any of a class of elements that generally are solid at room temperature, have a grayish color and shiny surface, and conduct electricity (78)

metalloid: an element having properties of metals as well as nonmetals (79)

miscible: indicates liquids or gases that will dissolve in each other (405)

mixture: a collection of two or more pure substances physically mixed together (46)

molar mass: sum of the molar masses of all atoms represented by 1 mol of formula units (176)

molar volume: the volume of one mole of a substance at STP (366)

molarity: concentration unit, expressed as moles of solute per liter of solution (411)

mole: the fundamental SI unit used to measure the amount of a substance (134)

mole fraction: the number of moles of an individual substance compared with the total number of moles in the mixture expressed as a ratio (377)

molecular compound: substance consisting of atoms that are covalently bonded (191)

molecular formula: gives type and actual number of atoms in a chemical compound (209)

molecule: a neutral group of atoms held together by chemical bonds (35)

monatomic ion: cation or anion formed from a single atom (164)

mutagen: any substance or agent that causes a noticeable increase in the frequency of mutations (358)

neutralization reaction: a reaction between an acid and a hydroxide base in which H^+ and OH^- react to form H_2O (507)

noble gas: an element that exists in the gaseous state at normal temperatures and is nonreactive with other elements (79)

nonelectrolyte: a substance that, when dissolved in an aqueous solution, will not conduct an electric current (453)

nonmetal: any chemical element that is neither a metal, metalloid, or a noble gas (78)

nonpolar covalent bond: covalent bond in which the bonding electrons are shared equally between the two bonding atoms (196)

normal boiling point: the temperature at which a substance boils at 1.0000 atm of pressure (391)

normal freezing/melting point: the temperature at which a substance melts and freezes at 1.0000 atm of pressure (392)

nuclear fission: the process by which a nucleus splits into two smaller fragments (625)

nuclear fusion: the process by which two nuclei combine to form a heavier nucleus (627)

nuclear reaction: a reaction that involves a change in the nucleus of an atom, as opposed to a chemical reaction which involves changes to the arrangement of electrons that surround the nucleus (128)

nucleic acid: a biological polymer consisting of phosphoric acid, a 5-carbon sugar, and four nitrogen bases (520)

nucleus: the central region of an atom made up of protons and neutrons (86)

octet rule: main-block elements form bonds by rearranging electrons so that each atom has a stable octet in its outermost energy level (151)

orbital: a region of an atom in which there is a high probability of finding electrons (91)

order: the exponent for a specified reactant in a rate law expression (555)

order of reaction: the sum of all the exponents on all concentration terms in a rate law expression (557)

organic compound: any covalently bonded compound containing carbon (except carbonates and oxides) (5)

oxidation: loss of electrons by an atom or an algebraic increase in its oxidation state (579)

oxidation number: apparent charge assigned to an atom based on the assumption of complete transfer of electrons (205)

oxidation state: a numerical representation of an atom's share of the bonding electrons. In ionic compounds, it is equal to the ionic charge. In covalent compounds, it is the average charge assigned to an atom according to electronegativities (587)

partial pressures: the pressure of an individual gas in a gas mixture that contributes to the total pressure of the mixture (375)

pascal: a unit of pressure equal to the force of 1 N/m^2 (365)

percent yield: ratio of actual yield to theoretical yield, multiplied by 100 (289)

period: a series of elements that form a horizontal row in the periodic table (77)

periodic law: properties of elements tend to change with increasing atomic number in a periodic way (109)

pH: the negative logarithm of the hydronium ion concentration of an aqueous solution; used to express acidity (499)

phase: any part of a system that has uniform composition and properties (47)

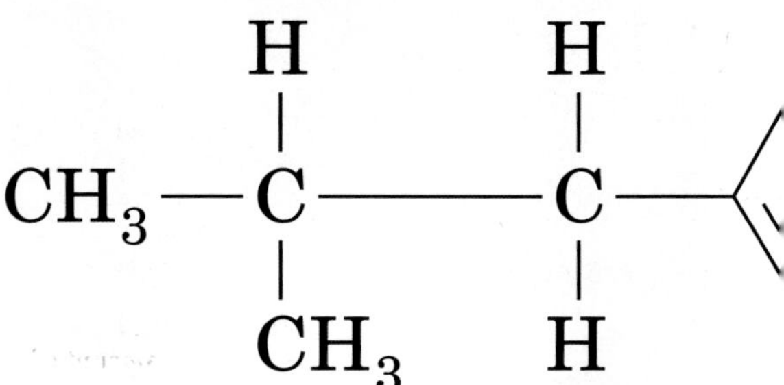

phase diagram: a graphic representation of the relationships between the physical state of a substance and its pressure and temperature (391)

physical change: a change that affects only physical properties (44)

physical properties: properties that can be observed or measure without changing the composition of matter (37)

polar covalent bond: covalent bond in which the bonding electrons are more strongly attracted by one of the bonding atoms (196)

polyatomic ion: ion made of two or more atoms bonded together that function as a single ion (170)

polymer: large molecule made of many repeated small subunits, each of which is a small molecule or group of atoms (210)

polymerization: chemical reaction in which many simple molecules combine in chains to form a very large molecule (256)

porous barrier: any medium through which ions can slowly pass (577)

potential energy: energy an object possesses because of its position (41)

pressure: the force exerted per unit area (363)

precision: the degree of exactness or refinement of a measurement (60)

principal energy level: a specific energy or group of energies that may be possessed by an electron in an atom (87)

pure substance: matter composed of only one kind of atom or molecule (46)

quantum theory: the field of physics based on the idea that energy is quantized and that this has significant effects on the atomic level (91)

radioactivity: the ability of unstable nuclei to undergo spontaneous nuclear decay (613)

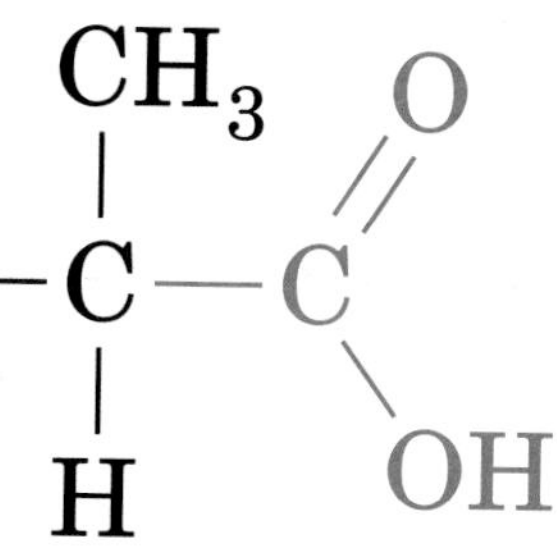

radioactive series: a sequence of nuclei that arise from and are transformed by radioactive decay until a stable isotope is produced (623)

rate law: a mathematical expression for the rate of a reaction as a function of the concentration of one or more reactants: rate = $k[A]^n [B]^m$ (555)

rate-determining step: the slowest step in a reaction mechanism (553)

reaction rate: the change in reactant concentration per unit of time as reaction proceeds (536)

reagent: a chemical used to convert one substance into another substance in a chemical reaction (7)

redox reaction: a chemical reaction involving oxidation and reduction (580)

reduction: the gain of electrons by an atom or the algebraic decrease in its oxidation state (579)

reduction potential: a measure of the tendency of a given half-reaction to occur as a reduction in an electrochemical cell (583)

resting metabolic rate: energy expended by a person at rest in a thermally neutral environment (344)

salt: a compound with an ionic lattice that is formed from an acid when H^+ is replaced by a metal ion or cation (157, 508)

saturated: containing the standard amount of solute specified by the solubility at a given temperature (408)

saturated fatty acid: the long chain carboxylic acid subunit of a fat and oil with no double bonds (521)

SHE: standard hydrogen electrode has the half-reaction $2H^+(aq) + 2e^- \rightarrow H_2(g)$. The concentration of hydrogen ion is 1 M, the temperature is 25°C, and the pressure of the hydrogen gas is 101.3 kPa (1 atm) (583)

shielding effect: the reduction of the attractive force between a nucleus and its outer electrons due to the blocking effect of inner electrons (122)

single bond: sharing of one pair of electrons between two atoms (199)

single-displacement: chemical reaction in which one element replaces another element in a compound (258)

solubility: the maximum amount of a chemical that will dissolve in a given amount of a solvent at a specified temperature while the solution is in contact with some undissolved solute (407)

solubility-product constant (K_{sp}): the equilibrium constant for a solid in equilibrium with its ions in a saturated solution; used for substances that are described as "insoluble" because they are only very slightly soluble (475)

soluble: can be dissolved in a particular solvent (406)

solute: the material dissolved in a solution (405)

solvent: the material dissolving the solute to make the solution (405)

specific heat capacity: the amount of heat energy required to increase the temperature by one gram of a substance by one degree Celsius (315)

spontaneous change: a change that will occur because of the nature of the system, once it is initiated (323)

stable nucleus: a nucleus that does not spontaneously decay to become the nucleus of a different element (613)

standard reduction potential: by convention, the potential, E^0, of a half-cell connected to the standard hydrogen electrode when ion concentrations in the half-cells are 1 M, gases are at a pressure of 1 atm, and the temperature is 25°C (583)

standard temperature and pressure (STP): standard conditions for a gas of 0°C and 1.0000 atm (365)

state: the condition of being a gas, liquid, solid, plasma, or neutron star (37)

stoichiometry: mass and quantity relationships among reactants and products in a chemical reaction (272)

strong electrolyte: a substance that is completely or largely dissociated in an aqueous solution (453)

strong nuclear force: a short-range, attractive force that acts among nucleons (612)

structural formula: indicates the spatial arrangement of atoms and bonds within a molecule (211)

sublevel: one orbital or a group of orbitals within an energy level which have the same value of ℓ (95)

sublimation: a change of state in which a solid is transformed directly to a gas without going through the liquid state (392)

substrate: (biochemical) the molecule or molecules with which an enzyme interacts (561)

supersaturated: containing more than the standard amount of solute specified by the solubility at a given temperature (408)

surface tension: the measure of a liquid's tendency to decrease its surface area to a minimum (393)

surfactant: a compound that stabilizes an emulsion by acting at the surface between two immiscible substances (431)

suspension: mixture that appears uniform while being stirred, but separates into different phases when agitation ceases (405)

synthesis: the process of building compounds from elementary substances through one or more chemical reactions (6)

system: all the parts that form a unified whole (22)

technology: the application of scientific knowledge for practical purposes (15)

temperature: a measure of the average kinetic energy of random motion of the particles in a sample of matter (315)

theoretical yield: calculated maximum amount of product possible from a given amount of reactant (287)

theory: an explanation of an observation that is based on experimentation and reasoning (19)

theory of ionization: the explanation of the process by which electrolytes break apart in solution in the form of freely moving ions that can conduct an electrical current (456)

titration: an analytical procedure used to determine the concentration of a sample by reacting it with a standard solution (508)

titration standard: a solution of precisely known concentration; also called a titrant (508)

transition elements: metallic elements that have varying properties and belong to Groups 3 through 12 of the periodic table (116)

transition interval: the pH range over which an indicator exhibits different colors for its acidic and alkaline forms (508)

transmutation: a process by which a nucleus of one element is transformed into a nucleus of a different element (613)

triple bond: covalent bond formed by the sharing of three pairs of electrons between two atoms (201)

triple point: the temperature and pressure at which all three states of a substance exist in equilibrium (392)

unit cell: the simplest portion of a crystal lattice that portrays the three-dimensional structure of the entire lattice (159)

unsaturated: containing less than the standard amount of solute specified by the solubility at a given temperature (408)

unsaturated fatty acid: a fatty acid with one or more carbon-carbon double bonds (521)

unshared pair: pair of electrons that is not involved in covalent bonding, but instead belongs exclusively to one atom (198)

unstable nucleus: a nucleus that spontaneously undergoes decay to become the nucleus of a different element (613)

Valence Shell Electron Pair Repulsion (VSEPR): system for predicting molecular shape based on the idea that pairs of electrons orient themselves as far apart as possible (213)

valence electron: electron present in the outermost energy level of an atom (198)

volatile: a term used to describe a substance that is readily vaporized at low temperature (390)

voltaic cell: an electrochemical cell in which a spontaneous redox reaction produces a flow of electrons through an external circuit (577)

volume: the amount of space an object occupies (34)

weak electrolyte: a compound that experiences only a small degree of dissociation in an aqueous solution (453)

zwitterion: the reactive dipolar form of an amino acid (519)

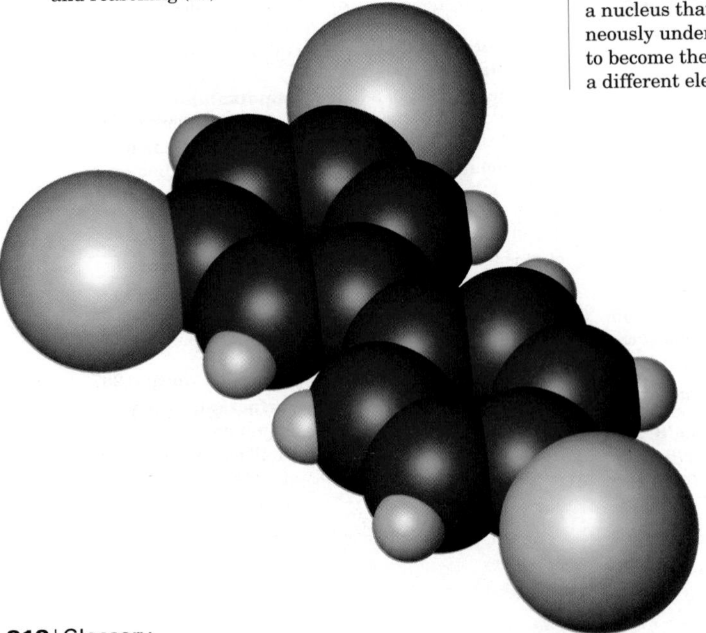

Index

Page numbers in boldface type refer to figures. Page numbers preceded by the letter *t* refer to tables.

819

825

827

200ml
200ml

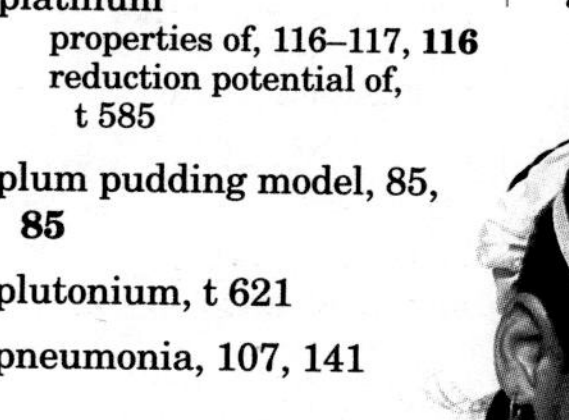

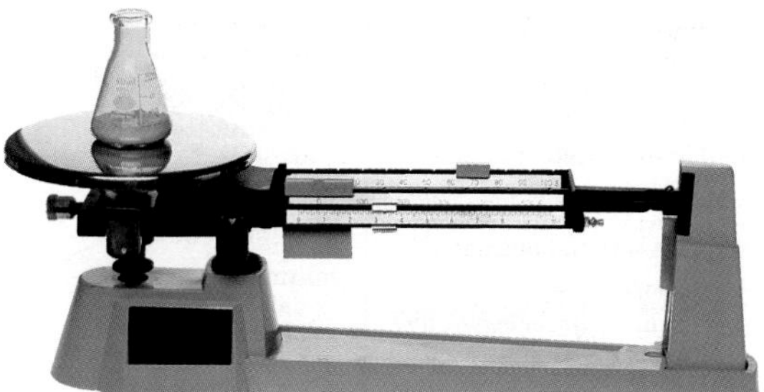

835

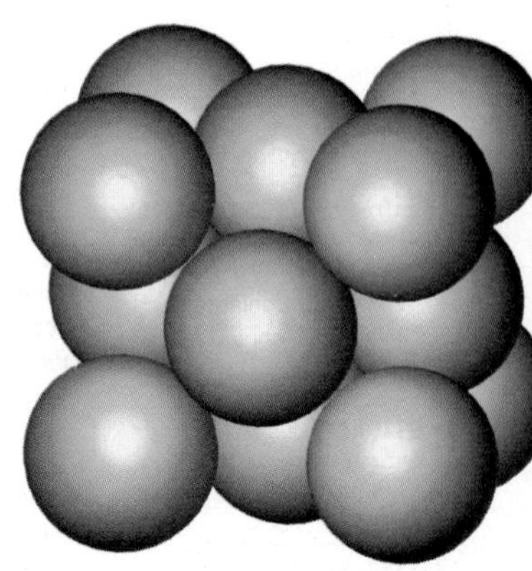

Credits

List of Abbreviations
All safety symbols were designed by Michael Helfenbein/Evolution Design.

Special thanks to the students, faculty, and staff of Poly Prep Country Day School, Brooklyn, NY for the use of their chemistry laboratory and other facilities. Many of the student models in this book are students at the school.

Cover (coal) © Telegraph Colour Library/FPG International Corp., (diamond structure, orbital) Foca, (diamonds) © 1991 Martin Rogers/FPG International Corp., (full cover wrap) © Goavec Photography/The Image Bank; **I** (all) Foca; **II** (l) Foca, (b) Sergio Purtell/Foca; **III** (bc) Foca, (br) Sergio Purtell/Foca; **IV** (tl, b) Sergio Purtell/Foca, (bl) Foca; **IV–V** (hot air balloon) © Jenny Hager/ Adventure Photo; **V** (tr) Sergio Purtell/Foca, (tc, c, br) Foca; **VI** (all) © Pat Crowe/Courtesy of Rohm and Haas Research Lab-oratories, (fructose, water) Foca; **VII** (all) Foca; **VIII** (all) Foca; **IX** (all) Sergio Purtell/ Foca; **X** (tr) Greg LaFever/Scott Hull Associates, (tl) Sergio Purtell/Foca, (bl) Steven Nau/Deborah Wolfe Ltd, (br) Foca; **XI** (tl) Foca, (b) Sergio Purtell/Foca; **XII** (tl, tr) Sergio Purtell/Foca, (b) Foca; **XIII** (tl) Lucas Deaver/Creative Freelancers Management, Inc., (tr, b) Foca; **XIV** (tl, bl) Sergio Purtell/ Foca; **XV** (tl, tr) Sergio Purtell/ Foca, (l) Greg LaFever/Scott Hull Associates; **XVI** (all) Sergio Purtell/Foca; **1** (tl, magnifying glass) Sergio Purtell/Foca, (DNA fingerprint, enlargement) © Sinclair Stammers/Science Photo Library/Photo Researchers, (Curie) © The Bettmann Archive, (Bohr/ Einstein) © AIP Niels Bohr Library, (Hodgkin) © UPI/ Bettmann, (Lee) © AP/Wide World Photos; **2** (bkgrd) Sergio Purtell/Foca, (c) Foca; **3** (tr) © John Lemker/Earth Scenes; **4** (b) Sergio Purtell/Foca; **5** (b) Foca, (bl, bottle, hand) Sergio Purtell/ Foca; **6** (tl) Sergio Purtell/Foca, (b) Foca; **7** (t, b) Foca, (l) © 1989 Robert J. Erwin/Photo Researchers, (br) © Rod Planck/Photo Researchers; **8** (l) © Will & Deni McIntyre/Photo Researchers; **8** (r) © 1993 Charles D. Winters; **9** (all) © Pat Crowe/Courtesy of Rohm and Haas Research Laboratories; **10** (all) Foca; **11** (t) Foca; **12** (brochure, brochure cover, lotion, warning label) Sergio Purtell/Foca, (container) © Mike English/MediChrome, **12** (OSHA label) © Mike English/ MediChrome; **13** (bottle, box, facts, warning label) Sergio Purtell/Foca, **13** (hazard sign, tr) © Richard Laird/The Stock Shop; **14** (tl, bl) Sergio Purtell/ Foca, **14** (tr) Sergio Purtell/Foca/ Copyright © 1994 MDL Information Systems, Inc., San Leandro, CA; **15** (tl, b) Foca; **17** (l) Foca, (br) © Science Photo Library/ Photo Researchers; **18** (tl, bl) Sergio Purtell/Foca; **19** (tl) © Comstock Inc., (b) Sergio Purtell/ Foca/Copyright © 1994 by The New York Times Company. Reprinted by permission; **19** (bl) Foca; **20** (blueprint, chemist, mouse) Sergio Purtell/ Foca, (c) Foca; **21** (bl) Sergio Purtell/Foca, (journal) Sergio Purtell/Foca/ Excerpted with permission from Science Magazine, Vol. 262, pp.416-17, "Structural Features of Polysaccharides That Induce Intra-Abdominal Abcessses." by Arthur O. Tzianabos. Copyright 1993 American Association for the Advancement of Science; **22** (t, b) Sergio Purtell/Foca, (r) Foca; **23** (c) Foca, (bl) Sergio Purtell/ Foca; **24** (bl) Sergio Purtell/Foca; **25** (l) Stacy Danon/ Foca; **26** (bkgrd) Sergio Purtell/ Foca, (inset) © Robert J. Erwin/ Photo Researchers; **28** (all) Foca; **30** (bkgrd) Roger Ressmeyer/© 1994 Corbis (tl) Foca; **31** (tr) © 1992 Mason Morfit/FPG International Corp.; **32** (bkgrd) © Ray Pfortner/Peter Arnold, Inc., (tl) Foca, **33** (tr) Sergio Purtell/Foca; **34** (bl) Sergio Purtell/Foca; **35** (t, bl) Sergio Purtell/Foca, (cl, c, br) Foca; **36** (all) Sergio Purtell/ Foca/ The Merck Index: An Encyclopedia of Chemicals, Drugs, and Biologicals, Eleventh Edition, Susan Budavari et al., Eds. (Merck & Co., Inc., Rahway, N.J., U.S.A., 1989; **37** (tl, br) Sergio Purtell/ Foca, (c,bl, bc) Foca; **38** (bottle, label, tablets) Sergio Purtell/ Foca, (br) © Obremski/The Image Bank; **39** (apple, pencil) Sergio Purtell/ Foca; **39** (bc) Foca; **39** (bicycle) © Chuck Kuhn/The Image Bank; **40** (all) © Erwin & Peggy Bauer; **41** (all) Sergio Purtell/Foca; **42** (l) © Don Spiro/The Stock Shop; **43** (l) © 1986 Howard Sochurek/The Stock Shop; **44** (tl) Jim McNee/FPG International Corp., (b) © Joanna McCarthy/The Image Bank; **45** (all) Sergio Purtell/Foca; **46** (cl, br) Foca, (gold ring) Sergio Purtell/Foca, (nugget) Neal Mishler/FPG International Corp.; **47** (bl) Sergio Purtell/Foca; **47** (inset) © Bio-Photo Associates/ Science Source/Photo Researchers; **48–49** (hot air balloon) © Jenny Hager/Adventure Photo; **49** (balloon) Sergio Purtell/Foca, (tc, bc) Foca; **50** (all) Foca; **51** (all) Foca; **52** (tl) Sergio Purtell/Foca, (bl) © Peter Pearson/Tony Stone Images, (br) © C. Bruce Forster/AllStock, Inc.; **53** (tl) © Jaime Villaseca/ The Image Bank, (tr) © Jeff Smith/The Image Bank, (bl, bc, br) Sergio Purtell/Foca; **54** (b) George Koizumi; **55** (all) Sergio Purtell/Foca; **56** (bl) Sergio Purtell/Foca; **57** (all) Foca; **59** (all) Foca; **60** (all) Sergio Purtell/Foca; **61** (all) Sergio Purtell/Foca; **63** (tl) Sergio Purtell/Foca; **64** (l) © Ray Pfortner/Peter Arnold, Inc.; **68** (br) Foca; **70** (bkgrd, bottle) Sergio Purtell/Foca; **71** (r) © COMSTOCK INC./Michael Geissinger; **72** (bkgrd) © 1991 Scott Suchman/Uniphoto, (br) Foca; **73** (tr) Sergio Purtell/Foca; **74** (atoms) © IBM Research, (bl) Science Photo Library/Photo Researchers, (br) Sergio Purtell/Foca; **75** (tl) Sergio Purtell/Foca, (Curie) © The Bettmann Archive; **77** (all) Foca; **78** (periodic table blocks) Foca, (spike) © Jon Riley/The Stock Shop, (carbon) © Leonard Lee Rue III/Bruce Coleman, Inc., (knee, mercury) Sergio Purtell/ Foca, (phosphorous, selenium) © Tom Pantages, (silver, titanium) Paul Silverman/Fundamental Photographs, NYC; **79** (arsenic) Richard Megna/ Fundamental Photographs, NYC (periodic table blocks) Foca, (detergent) Sergio Purtell/Foca/ Reproduced with permission from U. S. Borax, Inc., (krypton light) Sergio Purtell/Foca/OCS Associates/Art Service, (germanium) Paul Silverman/ Fundamental Photographs, NYC (helium, neon) Sergio Purtell/ Foca, (tellurium) © Tom Pantages; **80** (tl, c) Sergio Purtell/Foca, (bl) Foca; **81** (l) Foca, (r) Ron Thomas/ FPG International Corp., (bc) © D. Raines/Uniphoto; **82** (all) Foca; **83** (l, r) Sergio Purtell/Foca, (tr) Foca; **84** (t) Peter Bollinger, (br) Sergio Purtell/Foca; **85** (tl) Richard Megna/Fundamental Photographs, NYC (bl) Foca, (br) Sergio Purtell/Foca; **86** (t) Peter Bollinger, (marble demonstration) Sergio Purtell/Foca; **87** (tl) © John Zoiner/Uniphoto, (tc, bl) Sergio Purtell/Foca; **88** (tl) Foca; **89** (b) Foca, (calcium, mercury, prism) Sergio Purtell/Foca, (calcium spectrum, mercury spectrum) Wabash Instrument Corporation/Fundamental Photographs, NYC; **90** (c) Foca, (bl) © Dr. Jeremy Burgess/Science Photo Library/Photo Researchers; **91** (all) Foca; **92** (tl) Science Photo Library/ Photo Researchers, (tc) Foca; **93** (l) Foca, (r) Foca; **94** (l) Foca; **95** (all) Foca; **98** (t) Foca; **100** (bkgrd) © 1991 Scott Suchman/ Uniphoto, (tl, black powder, yellow powder) Sergio Purtell/ Foca; **103** (all) Foca; **104** (all) Foca; **106** (bkgrd) © Jim Pickerell; (tr) Foca; **108** (tl) Sergio Purtell/Foca, (bl) Foca; **110** (t) Foca; **112** (cl) Foca, (bl) ©

Comstock Inc./Adam Tanner, (r) Foca; **110–111** (b, c) Foca; **113** (tl, r) Foca, (explosion, jar) Sergio Purtell/Foca; **114** (cl, r) Foca, (bc) © Werhner Krutein/Gamma Liaison, (bl) © Comstock Inc./Russ Kine; **115** (bkgrd) Sergio Purtell/Foca; **116** (tl) Foca, (buckets) Sergio Purtell/Foca; **116–117** (t) Foca; **117** (t) Sergio Purtell/Foca/Ira Moskovitz Jewelry, N.Y.C.; **118** (tr) © Christian Vioujard/Gamma Liaison, (l, c) Foca, (lizard) Sergio Purtell/ Foca; **119** (tl, tr, cl) Foca, (b) © Telegraph Colour Library/FPG International Corp.; **120** (all) Sergio Purtell/Foca; **121** (all) Foca; **122** (t) Foca; **123** (t) Foca; **124** (all) Foca; **125** (t) Foca; **126** (all) Foca; **127** (bl) © Stanford Linear Accelerator Center/US Department of Energy, (br) © Stanford Linear Accelerator Center/US Department of Energy; **128** (b) Foca, (bl) Roger Ressmeyer/© 1994 Corbis; **129** (t) Foca, (b) © Telegraph Colour Library/FPG International Corp.; **130** (all) © Lawrence Berkeley Laboratory, University of California; **131** (c) Foca; **132** (all) Foca; **134** (bl) © The Bettmann Archive, (eggs, paper) Sergio Purtell/Foca; **135** (t) Sergio Purtell/Foca; **141** (all) © Jim Pickerell; **143** (all) Foca; **144** (all) Foca; **145** (all) Foca; **148** (all) Foca; **149** (tr) © Gary Milburn/Tom Stack & Associates; **150** (b) Sergio Purtell/Foca; **152** (all) Foca; **153** (all) Foca; **154** (bl) © Yoav Levy/Phototake, (br) Foca, (br) Sergio Purtell/Foca; **155** (tl) © Dr. Dennis Kunkel/Phototake, (bl, br, r) Foca; **156** (r) Foca; **157** (c) Foca; **158** (t, bkgrd, b, c) Sergio Purtell/Foca, (others) Foca; **159** (all) Foca; **161** (all) Foca; **162** (c) Foca, (bl) © Lionel Atwill/Peter Arnold, Inc., (bl) Sergio Purtell/Foca; **163** (candle, charcoal) Sergio Purtell/Foca, (oil) Sergio Purtell/Foca/Reckitt & Colman Inc., (others) Foca; **165** (all) Foca; **167** (all) Sergio Purtell/ Foca; **168** (c) Foca, (bl) Sergio Purtell/Foca; **170** (all) Sergio Purtell/Foca; **177** (br) Sergio Purtell/Foca; **179** (all) Sergio Purtell/Foca; **181** (all) Foca; **183** (all) Foca; **184** (all) Foca; **188** (bkgrd) Landsat imagery courtesy of EOSAT., (br) Foca; **189** (tr) Sergio Purtell/Foca; **190** (l, r) Triune Pictures, L.P.; **191** (bl, br) Sergio Purtell/Foca; **191** (iodine, salt) Foca; **192** (all) Foca; **193** (all) Foca; **194** (all) Foca; **195** (all) Foca; **196** (all) Foca; **197** (tc) Foca, (tr) Sergio Purtell/Foca; **198** (all) Foca; **199** (all) Foca; **200** (bl) © Photri, (others) Foca; **201** (tl, tr) Sergio Purtell/Foca, (tc, br) Foca, (bl) © Phil Degginger/Picture Perfect USA TM; **202** (all) Foca; **203** (tl) Sergio Purtell/Foca, (all others) Foca; **204** (tl) Sergio Purtell/Foca, (c) Foca; **206** (all) Foca; **207** (tl) Sergio Purtell/Foca, (c, glucose) Foca; **209** (all) Foca; **210** (bl) Sergio Purtell/Foca, (chain) Foca; **211** (all) Foca; **212** (bkgrd) © Professors P.M. Motta & S. Correr/Science Photo Library/Photo Researchers, (g. acid, valine) Foca, (healthy cells, sickle cells) © Science Source, (normal hemoglobin, sickle cell hemoglobin) © Kenneth Eward/ BioGrafx; **213** (all) Foca; **214** (all) Foca; **215** (all) Foca; **216** (tl, molecules) Foca, (b) Sergio Purtell/ Foca; **218** (all) Foca; **219** (all) Foca; **220** (bl) Sergio Purtell/ Foca, (all others) Foca; **221** (all) Foca; **222** (all) Foca; **223** (bkgrd) Reproduced by permission of Earth Observation Satellite Company, Lanham, Maryland, U.S.A.; **224** (trash bag) Sergio Purtell/ Foca, (Carothers) © E.I. Du Pont Co., (diamond structure, polyethylene, stereoisomers, tetrafluoroethylene) Foca, (Lavoisier) © The Bettman Archive, (Pluncket, Staudinger, Woodward) © AP/Wide World Photos, (Scheele) © The Bettman Archive, (urea) Smith/Custom Medical Stock Photo; **224–225** (green ribbons) Greg LaFever/Scott Hull Associates; **225** (Barton) © AP/Wide World Photo, (Beilstein, Couper) © Chemical Heritage Foundation, (benzene, buckyball, carbon tetrahedral, cyclobutane, penicillin) Foca, (Hodgkins) © UPI/Bettmann, (Kekule) © The Bettman Archive, (mauve cloth) Sergio Purtell/Foca, (Sheehan) © UPI/Bettman; **227** (all) Foca; **228** (all) Foca; **229** (all) Foca; **230** (all) Foca; **231** (all) Foca; **232** (bkgrd) © James Rudnick/The Stock Market, (all others) Foca; **233** (tr) © Ewing Galloway; **234** (bkgrd) Sergio Purtell/Foca; **235** (tl) Sergio Purtell/Foca; **236** (tl) © Ken Kay/Fundamental Photographs, NYC; (all others) Foca; **237** (tr) Sergio Purtell/Foca/Diamond Brands Inc., Minneapolis, MN; **238** (all) Foca; **239** (all) Foca; **240** (b) Blazer and Blazer Body designs are trademarks of Chevrolet Motor Division, General Motors Corp., (all others) Foca; **245** (b) Sergio Purtell/Foca; **246** (tl, tr) Sergio Purtell/Foca, (bl) © J. Zimmerman/FPG International Corp.; **248** (tl, b) Sergio Purtell/Foca, (cl, br) Claire Booth Medical Illustration; **249** (all) Foca; **250** (all) Sergio Purtell/ Foca; **253** (tl) Sergio Purtell/Foca; **254** (tl) Sergio Purtell/Foca/H.J. Murray & Co., (CO_2 extinguisher) Sergio Purtell/Foca/Walter Kidde, The Fire Extinguisher Co., (CO_2 label, sodium label) Sergio Purtell/Foca, (sodium extinguisher) Sergio Purtell/Foca/ Amerex Corp., Model 453; **256** (tl, b) Sergio Purtell/Foca, (all others) Foca; **257** (all) Foca; **258** (t, bkgrd, b) Sergio Purtell/Foca, (all others) Foca; **259** (all) Sergio Purtell/Foca; **260** (all) Sergio Purtell/Foca; **261** (bl) Sergio Purtell/Foca; **262** (all) Sergio Purtell/Foca; **263** (penny) Sergio Purtell/Foca; **264** (bkgrd) © James Rudnick/The Stock Market; **265** (all) Foca; **266** (all) Foca; **267** (all) Foca; **268** (br) Sergio Purtell/ Foca, (others) Foca; **269** (all) Foca; **270** (bkgrd) Courtesy of U.S. Army; **271** (tr) Sergio Purtell/ Foca; **272** (banana, bubble gum) Sergio Purtell/Foca, (all others) Foca; **273** (tl) Sergio Purtell/Foca, (all others) Foca; **274** (tr, br) Foca, (l) Sergio Purtell/Foca; **275** (tl, br) Foca, (bl) Sergio Purtell/Foca; **278** (all) Foca; **279** (all) Foca; **281** (all) Foca; **284** (l) Sergio Purtell/ Foca, (all others) Foca; **285** (b) Sergio Purtell/Foca; **287** (l) Sergio Purtell/Foca, (c, r) Foca; **288** (all) Foca; **289** (baseballcard) Jackie Robinson's name and likeness are used with permission of Rachel Robinson/ National Baseball Library and Archives, Cooperstown, New York/The Topps Co., Inc.; **290** (all) Foca; **291** (all) Foca; **292** (all) Foca; **293** (all) Foca; **294** (bl) Courtesy of Ford Motor Company; **295** (all) Peter Bollinger; **297** (tl, br) Foca, (b) Sergio Purtell/Foca; **298** (all) Peter Bollinger; **300** (b) © J. Blakeway/Panoramic Images, Chicago 1995; **301** (tl) Courtesy of AC Delco Systems Division GMC, (b) © J. Sohm/ Panoramic Images, Chicago 1995; **303** (bkgrd) Courtesy of U.S. Army, (FRH) Sergio Purtell/Foca/Zesto Therm Inc., (r) Foca; **305** (all) Foca; **306** (all) Foca; **307** (all) Foca; **308** (all) Foca; **310** (bkgrd, inset) © Mazda Motor of America, Inc., (all others) Foca; **311** (tr) Sergio Purtell/Foca; **312** (all) Sergio Purtell/Foca; **314** (b) Sergio Purtell/Foca, (all others) Foca; **316** (tl) Sergio Purtell/Foca; **317** (all) Foca; **318** (l) Sergio Purtell/ Foca, (calorimeter) Greg LaFever/Scott Hull Associates; **319** (bl) Sergio Purtell/Foca; **323** (bl) Science Photo Library/Photo Researchers; **324** (all) Sergio Purtell/Foca; **325** (cr) Foca, (cl) Sergio Purtell/Foca; **327** (all molecules) Foca, (chloroplast, mitochondrion) Claire Booth Medical Illustration, (bkgrd) Sergio Purtell/Foca; **330** (bc) © Daniel Grogan/Uniphoto; **333** (all) Sergio Purtell/Foca; **334** (t, b, bkgrd) Sergio Purtell/Foca, (all others) Foca; **335** (t) Foca, (bl, br) Triune Pictures, L.P.; **336** (l) Sergio Purtell/Foca; **338** (all) Foca; **341** (all) Foca; **342** (bl) Science Photo Library/Photo Researchers, (br) © NASA/Science Source/Photo Researchers; **345** (l) Sergio Purtell/Foca; **346** (tl) Sergio Purtell/Foca; **347** (bkgrd) © Mazda Motor of America, Inc., (inset) © Telegraph Colour Library/FPG International Corp.; **354** (bl) Foca, (bkgrd) Photograph provided by Raven Ind. of Sioux Falls, SD; **355** (tr) © NASA/Sipa Press; **356** (b) Foca; **357** (c) Foca; **358** (bl) Stephen Bauer/Deborah Wolfe Ltd; **359** (tl, tc, tr) © NASA/Science Source, (all others) Foca; **360** (tl, b) Foca, (l) Sergio Purtell/Foca; **363** (b) © NASA/Sipa Press, (others) Foca; **364** (bl, br) Sergio Purtell/Foca, (manometers) Steven Nau/Deborah Wolfe Ltd., (all others) Foca; **366** (tl, tc, tr) Foca, (all others) Sergio Purtell/Foca; **368** (all) Sergio Purtell/Foca; **371** (t, bkgrd, b) Sergio Purtell/Foca, (others)

Foca; **375** (tr, cr, br) Foca, (manometers) Steven Nau/ Deborah Wolfe Ltd.; **376** (t) Sergio Purtell/Foca, (tr) Foca; **378** (tr, c, cr) Foca, (all others) Sergio Purtell/Foca; **379** (all) Foca; **389** (flasks) Steven Nau/Deborah Wolfe Ltd; **390** (tr, bl) Sergio Purtell/Foca, (tr) Foca; **391** (t) Foca; **392** (b) Foca; **393** (l) Sergio Purtell/Foca, (c) Foca; **394** (tl, bkgrd) Photograph provided by Raven Ind. of Sioux Falls, SD; **397** (cl) Foca; **401** (cl) Foca; **402** (c, bkgrd) Foca, (bkgrd) © Jim Brandenburg/ AllStock, Inc.; **403** (tr) © Alec Pytlowany/Masterfile; **404** (t, b, bkgrd) Sergio Purtell/Foca, (tr, br) Foca; **405** (t, b, bkgrd) Sergio Purtell/Foca, (tr, br) Foca; **406** (tl, b) Sergio Purtell/Foca, (tl) Foca; **408** (tl, tr) Sergio Purtell/Foca, (bl, bc, br) Sergio Purtell/Foca/ Prism Technologies, Inc.; **409** (t) Peter Cross/© 1995 Light Sources Stock, (b) © Will & Deni McIntyre/AllStock, Inc., (tr, br) Foca; **412** (tr, c, bl) Sergio Purtell/ Foca; **413** (all) Sergio Purtell/ Foca; **417** (tr, r, br) Foca, (all others) Sergio Purtell/Foca; **419** (tr) Foca, (bl) Sergio Purtell/Foca; **420** (all) Foca; **421** (all) Foca; **422** (all) Foca; **423** (all) Foca; **424** (all) Foca; **425** (all) Foca; **426** (all) Foca; **427** (bl) Sergio Purtell/ Foca, (others) Foca; **428** (bl) Sergio Purtell/Foca, (br) Foca; **430** (tl) Foca, (bl) © The Bettmann Archive, (tr) Sergio Purtell/Foca; **431** (all) Foca; **432** (b) Foca, (others) Sergio Purtell/ Foca; **433** (l, c) Foca, (bl) Sergio Purtell/Foca; **434** (tl, bl) Sergio Purtell/Foca/The contour bottle shape is a Registered Trademark of The Coca Cola Company., (tr, br) Foca; **435** (tl) © Leo de Wys/ Sipa Press/Gamba; **436** (tl) Sergio Purtell/Foca, (bl, br) Foca; **437** (all) Foca; **438** (all) Sergio Purtell/Foca; **439** (l) Courtesy Johns Hopkins School of Public Health, (others) Foca; **440** (l) © Jim Brandenburg/AllStock, Inc., (inset) © Tom Van Sant/ Geosphere Project, Santa Monica/Science Photo Library/Photo Researchers; **442** (all) Foca; **444** (all) Foca; **445** (all) Foca; **448** (mammogram) Steinmark/Custom Medical Stock Photo (plane) © Andy Sacks/Tony Stone Images; **449** (tl) © XX NCI/CNRI/Phototake; **450** (bkgrd) © Arnulf Husmo/ Tony Stone Images, (bl) Foca; **451** (tr) © J. Zimmerman/FPG International Corp.; **452** (br) Sergio Purtell/ Foca; **453** (tl, tr, bl, br) Sergio Purtell/Foca, (microviews) Foca; **454** (bl) Sergio Purtell/Foca; **455** (b) Sergio Purtell/Foca, (br) Claire Booth Medical Illustration; **456** (tl) Sergio Purtell/Foca, (c) Foca; **457** (bl, br) Sergio Purtell/Foca; **457** (others) Foca; **458** (equation) Foca; **459** (tl) Sergio Purtell/Foca; **460** (microviews) Foca, (all others) Sergio Purtell/Foca; **461** (microviews) Foca; **461** (l) Sergio Purtell/Foca; **462** (bl) Sergio Purtell/Foca, (bc) Foca; **464** (tl, tc, tr) Sergio Purtell/Foca, (bl) © Chemical Heritage Foundation; **466** (tl, tr) Sergio Purtell/Foca/ Selima Optique, N.Y.C., (bl) Foca; **467** (all) Sergio Purtell/Foca; **468** (tl) Sergio Purtell/Foca; **469** (l) Stacy Danon/Foca; **471** (bl) © 1992 James Kay/PhotoBank, Inc.; **472** (all) Claire Booth Medical Illustration; **473** (c) Sergio Purtell/Foca; **477** (tl) © 1991 Science Photo Library/Custom Medical Stock Photo, (b) Claire Booth Medical Illustration; **478** (c) Sergio Purtell/Foca, (microviews) Foca; **479** (l) © Arnulf Husmo/Tony Stone Images, (inset) © 1994 Tom Prettyman/PhotoEdit; **481** (all) Foca; **482** (all) Foca; **483** (all) Foca; **486** (bkgrd) Excavations of The Metropolitan Museum of Art, 1929; Rogers Fund, 1930. (30.3.31) © 1978 by The Metropolitan Museum of Art, (molecule) © Science Vu-NIH, R. Feldman/Visuals Unlimited; **487** (tr) © Jim Coxe/ American Museum of Natural History; **488** (all) Sergio Purtell/ Foca; **489** (c) Sergio Purtell/Foca; **490** (bl, br) Foca, (tl) © E. Scott Bar Collection/AIP Emilio Segrè Visual Archives; **491** (all) Foca; **492** (l) Foca, (bl) Sergio Purtell/ Foca, (factory) © Comstock Inc., (lake) Jeremy Barnard/© 1995 Light Sources Stock; **493** (tl) © Ray Pfortner/Peter Arnold, Inc., (tr) © Ray Pfortner/Peter Arnold, Inc., (tc) © NYS Dept. of Environmental Conservation; **494** (all) Foca; **495** (all) Foca; **497** (all) Foca; **500** (all) Foca; **501** (b) Foca, (all others) Sergio Purtell/Foca; **504** (tl) Foca; **507** (all) Sergio Purtell/Foca; **508** (all) Sergio Purtell/Foca; **510** (all) Sergio Purtell/Foca; **511** (all) Sergio Purtell/Foca; **512** (all) Sergio Purtell/Foca; **513** (all) Sergio Purtell/Foca; **515** (t, b) Sergio Purtell/Foca, (tr, br) Foca; **516** (l) Sergio Purtell/Foca; **518** (all) Foca; **519** (all) Foca; **520** (DNA, RNA) Claire Booth Medical Illustration, (all others) Foca; **521** (all) Foca; **522** (all) Claire Booth Medical Illustration; **523** (tl) Sergio Purtell/Foca, (b, c) Foca; **524** (bkgrd) Excavations of The Metropolitan Museum of Art, 1929; Rogers Fund, 1930. (30.3.31) © 1978 by The Metropolitan Museum of Art,, (inset) M.C.R Communications/ Gamma Liaison; **529** (all) Foca; **530** (all) Foca; **532** (bkgrd) © Comstock Inc./Russ Kinne, (DNA fingerprint, magnified DNA) Science Photo Library/Photo Researchers, (magnifying glass) Sergio Purtell/Foca; **534** (bkgrd) Sergio Purtell/Foca, (inset) © David M. Phillips/Visuals Unlimited; **535** (tr) Sergio Purtell/Foca/ Atkins Technical, Inc.; **536** (bread) Sergio Purtell/Foca; **537** (microviews) Foca, (beakers) Sergio Purtell/Foca; **538** (b) Sergio Purtell/Foca; **539** (t) Sergio Purtell/Foca; **541** (all) Sergio Purtell/Foca; **543** (all) Foca; **545** (all) Sergio Purtell/ Foca; **546** (all) Sergio Purtell/ Foca; **547** (all) Foca; **548** (tc) Sergio Purtell/Foca, (molecules) Foca; **549** (all) Sergio Purtell/ Foca; **550** (all) Foca; **551** (t) Foca, (bl) © C.R. Knight/American Museum of Natural History; **552** (all) Sergio Purtell/Foca; **553** (tl, tr) Sergio Purtell/Foca, (all others) Foca; **554** (all) Foca; **556** (all) Foca; **557** (all) Foca; **558** (b) Wallace Keller; **560** (all) Foca; **561** (tl) Sergio Purtell/Foca, (all others) Foca; **562** (bl) Sergio Purtell/Foca; **563** (all) Foca; **564** (c) Foca; **565** (all) Sergio Purtell/ Foca; **574** (b) Foca; **574** (t) Foca, (bkgrd) © Dr. William R. Jacobs, Jr./Howard Hughes Medical Institute; **575** (tr) © Gregory K. Scott/Photo Researchers; **576** (b) Sergio Purtell/Foca, (all others) Foca; **577** (l) Lucas Deaver/Creative Freelancers Management, Inc., (r) Foca; **578** (tl) Lucas Deaver/Creative Freelancers Management, Inc, (bkgrd) © Petit-Vandy-Loubat/Agence Vandystadt/Photo Researchers, (all others) Sergio Purtell/Foca; **579** (microviews) Foca; **579** (c) Lucas Deaver/Creative Freelancers Management, Inc.; **580** (tl) Sergio Purtell/Foca; **581** (microviews) Foca, (others) Lucas Deaver/Creative Freelancers Management, Inc.; **582** (all) Lucas Deaver/Creative Freelancers Management, Inc.; **584** (all) Lucas Deaver/Creative Freelancers Management, Inc.; **587** (all) Sergio Purtell/Foca; **588** (all) Foca; **589** (all) Foca; **591** (tl, bl) Sergio Purtell/Foca, (tc) Foca; **592** (c) © Jim Pickerell/ The Image Works, (bl) Sergio Purtell/Foca/Alco-Sensor IV, manufactured by Intoximeters, Inc.; **595** (all) Sergio Purtell/Foca; **596** (bl) Sergio Purtell/Foca, (br) Ed Lipinski/Creative Freelancers Management, Inc.; **597** (tl, bl) Sergio Purtell/Foca, (others) Ed Lipinski/Creative Freelancers Management, Inc.; **598** (bl) Ed Lipinski/Creative Freelancers Management, Inc., (br) Sergio Purtell/Foca; **599** (tl) Sergio Purtell/Foca; **600** (Arrhenius) © E. Scott Barr Collection/AIP Emilio Segrè Visual Archives, (Boyle) © Chemical Heritage Foundation (C. M. Hall) © Alcoa, (Curie, Lavoisier) © The Bettman Archive, (Mendeleev) © Culver Pictures, (oxygen) Sergio Purtell/Foca/Courtesy Matheson Gas Products, Secaucus, NJ; **600–601** (green ribbons) Greg LaFever/Scott Hull Associates; **601** (pH meter, pH paper) Sergio Purtell/Foca, (chemical structure) Foca, (Faraday, Lee) © AP/Wide World Photos, (H. Davy) © Michael Holford Library; **602** (bkgrd) © Dr. William R. Jacobs, Jr./Howard Hughes Medical Institute, (inset) © Comstock Inc./Townsend P. Dickinson; **604** (r) Foca; **605** (all) Foca; **607** (bl) Ed Lipinski/Creative Freelancers Management, Inc.; **608** (all) Foca; **610** (bkgrd) © SNS/Sipa Press, (bl) Foca; **611** (tr) © Hanny/Gamma Liaison; **612** (bl) Sergio Purtell/Foca, (br) Foca; **613** (all) Foca; **614** (all) Foca; **618** (bl) © The Bettmann

Archive; **619** (detector) Sergio Purtell/Foca, (smoke) © Comstock Inc./Georg Gerster; **620** (tl) Sergio Purtell/Foca; **621** (t) Foca; **622** (all) Sergio Purtell/Foca; **623** (c) Foca; **624** (bl) © Fermilab Visual Media Services; **625** (all) © A.M.Siegelman/Visuals Unlimited; **626** (b) Dept. of Defense/ FPG International Corp.; **627** (l) Roger Ressmeyer/© 1994 Corbis, (c) © The Bettmann Archive; **628** (b) Foca; **629** (bl) © Gaillarde Raphael/Gamma Liaison, (br) © American Museum of Natural History/Courtesy Department Library Services/K4676; **630** (all) Sergio Purtell/Foca; **631** (bl) Sergio Purtell/Foca; **632** (b) © American Museum of Natural History/P. Hollembeak/Courtesy Department of Library Services/3316; **633** (t) SPL/Custom Medical Stock Photo, (b) © 1990 Science Photo Library/Custom Medical Stock Photo; **634** (tl) Sergio Purtell/ Foca, (badge) © Southern Illinois University/Photo Researchers, (bunny suit) © Nuclear Energy Institute; **635** (l) © SNS/Sipa Press, (iceman head and skull) © Kenneth Garrett/National Geographic Imaging Collection; **636** (Bohr/Einstein) © AIP Emilio Segrè Visual Archives, (Fermi) © The Bettman Archive, (H-bomb) © Los Alamos National Laboratory, (Hevesy) © UPI/Bettmann, (hydrogen block, map) Foca, (Manhattan Project) © Los Alamos National Laboratory/ Mark Marten/Photo Researchers, (Moseley) Public Domain/Sergio Purtell/Foca, (Rutherford) © Stock Montage; **636–637** (green ribbons) Greg LaFever/Scott Hull Associates; **637** (cyclotrons, technetium block, neutron) Foca, (Chadwick) © Nobel Foundation/Courtesy AIP Emilio Segre Visual Archives, (Curies) © The Bettman Archive, (Fermi accelerator) © Fermilab Visual Media Services, (Gell-Mann) © Archive Photos/Express Newspapers, (Glashow,Salam, Weinberg) © UPI/Bettman, (Hahn) © 1967 Archive Photos/DPA, (Kennedy) © UPI/Bettmann, (Lawrence) © Lawrence Berkeley Laboratory, University of California, (Meitner) © UPI/Bettmann, (Super Collider) © Bruno Torres/UPI/ Bettmann; **641** (r) Foca; **642–643** (spread) Sergio Purtell/Foca; **644** (tl) Sergio Purtell/Foca; **645** (all) Sergio Purtell/Foca; **646** (all) Sergio Purtell/Foca; **647** (all) Sergio Purtell/Foca; **648** (all) Sergio Purtell/Foca; **649** (tl, bl, cr) Foca; **649** (tr, l, bl, r) Sergio Purtell/Foca; **650** (bkgrd) Sergio Purtell/Foca; **656** (br) Sergio Purtell/Foca; **657** (tr) Sergio Purtell/Foca, (br) Greg LaFever/ Scott Hull Associates; **658** (tl) Sergio Purtell/Foca, (br) Sergio Purtell/Foca, (c) Greg LaFever/ Scott Hull Associates; **659** (r) Sergio Purtell/Foca; **662** (cl, bl) Foca, (br) Sergio Purtell/Foca; **663** (tr) Sergio Purtell/Foca; **664** (tl, tc, tr) Sergio Purtell/Foca, (bl) Greg LaFever/Scott Hull Associates; **665** (all) Greg LaFever/ Scott Hull Associates; **666** (bl) Sergio Purtell/Foca; **667** (all) Greg LaFever/Scott Hull Associates; **668** (l) Sergio Purtell/Foca; **669** (br) © Comstock Inc./ Russ Kine; **672** (br) Sergio Purtell/ Foca; **673** (all) Sergio Purtell/ Foca; **675** (tr) Sergio Purtell/ Foca; **678** (tl) © Telegraph Colour Library/FPG International Corp.; **679** (tr) Sergio Purtell/ Foca; **680** (tl) Sergio Purtell/Foca, (bl) Greg LaFever/Scott Hull Associates; **681** (l) Sergio Purtell/ Foca; **682** (tl) Sergio Purtell/Foca; **683** (br) David Bartruff/ FPG International Corp.; **686** (l) Foca, (br) © Gary Milburn/Tom Stack & Associates; **687** (tr) Sergio Purtell/Foca, (br) Greg LaFever/Scott Hull Associates; **688** (tl) Sergio Purtell/ Foca; **689** (br) © R. Carlton/Photo Researchers; **692** (all) Sergio Purtell/Foca; **693** (tr) Sergio Purtell/Foca, (br) Greg LaFever/ Scott Hull Associates; **694** (br) Sergio Purtell/Foca; **695** (r) Sergio Purtell/Foca; **698** (br) Josef Beck/FPG International Corp., (others) Foca; **699** (all) Sergio Purtell/Foca; **700** (all) Sergio Purtell/Foca; **701** (br) © 1991 Eugen Gebhardt/FPG International Corp.; **706** (all) Sergio Purtell/Foca; **707** (all) Sergio Purtell/Foca; **708** (tl) Sergio Purtell/Foca; **709** (br) Sergio Purtell/Foca; **712** (br) © Tom Tracy/FPG International Corp., (all others) Foca; **713** (all) Sergio Purtell/Foca; **714** (tl) Sergio Purtell/Foca; **715** (br, bread) Sergio Purtell/Foca, (inset) © Science Photo Library/Photo Researchers; **718** (br) © Harvey Lloyd/Peter Arnold, Inc.; **719** (all) Sergio Purtell/Foca; **720** (t) Sergio Purtell/Foca; **721** (tl) Sergio Purtell/Foca; **722** (r) Sergio Purtell/Foca; **723** (tr) Sergio Purtell/Foca; **726** (br) © Benelux Press/Photo Researchers, (c) Foca; **727** (tr) Sergio Purtell/ Foca; **728** (tl) Sergio Purtell/Foca, (bl) Foca; **729** (br) Sergio Purtell/ Foca; **736** (bl, r) Foca, (br) Sergio Purtell/Foca/Provided courtesy Mead Johnson Nutritional Group, **737** (all) Sergio Purtell/ Foca; **739** (all) Sergio Purtell/ Foca; **741** (tr) Sergio Purtell/ Foca/Reprinted by Permission of Texas Instruments.; **746** (br) Sergio Purtell/Foca; **747** (tr) Sergio Purtell/Foca; **748** (tl) Greg LaFever/Scott Hull Associates, (all others) Sergio Purtell/Foca; **749** (br) Ron Rovtar/FPG International Corp.; **752** (br) Sergio Purtell/Foca; **753** (tr) Sergio Purtell/Foca, (br) Sergio Purtell/Foca; **754** (tl) Sergio Purtell/Foca, (bl) Sergio Purtell/ Foca; **756** (br) Sergio Purtell/ Foca; **757** (br) © AP/Wide World Photos; **760** (br) © M. Bertinetti/ Photo Researchers, (others) Foca; **761** (tr, br) Sergio Purtell/Foca, (c) Greg LaFever/ Scott Hull Associates; **762** (c) Greg LaFever/Scott Hull Associates; **763** (br) Sergio Purtell/ Foca; **766** (c) Foca, (br) Sergio Purtell/Foca; **767** (tr) Sergio Purtell/Foca; **768** (all) Sergio Purtell/Foca; **769** (l) Sergio Purtell/Foca; **771** (br) Sergio Purtell/Foca; **774** (br) Sergio Purtell/Foca; **775** (tr) Sergio Purtell/Foca, (br) Sergio Purtell/Foca; **776** (tr) Sergio Purtell/Foca; **777** (br) Sergio Purtell/Foca; **780** (br) © Ewing Galloway; **781** (all) Sergio Purtell/Foca; **782** (tl) Sergio Purtell/Foca; **783** (br) Sergio Purtell/Foca; **786** (br) ©Comstock Inc./ Van Garmon; **787** (all) Sergio Purtell/Foca; **788** (tl) Sergio Purtell/Foca, (br) Foca; **789** (br) Sergio Purtell/Foca; **811** (tr) Foca; **813** (c) Sergio Purtell/ Foca; **814** (bl) Foca; **815** (tl) Sergio Purtell/Foca; **816–817** (t) Foca; **818** (b) Foca; **820** (tl) Sergio Purtell/Foca; **824** (bl) Sergio Purtell/Foca; **825** (tr) Foca; **828** (tl, cl) Sergio Purtell/Foca; **831** (br) Sergio Purtell/Foca; **834** (bl) Sergio Purtell/Foca; **835** (tr) Sergio Purtell/Foca; **836** (all) Foca.

Answers to Selected Problems

Chapter 2

Concept Review

page 43

6. The skier has both kinetic and potential energy. The potential energy arises from the skier's elevation. The kinetic energy comes from the skier's motion.
7. 40 J
8. Answers may vary but should include chemical energy (or internal energy) transformed to thermal energy transformed to electrical energy.

Concept Review

page 48

12. c, d, e, h, i
13. a. true
b. false
c. true
d. true
14. a. Pictures should show NO particles only within the field of view. Spaces between particles should reflect those of the model for a gas.
b. Pictures should show diatomic oxygen and diatomic nitrogen molecules within the field of view. Spaces between particles should reflect those of the model for a gas. The picture should show an even mixing of the gas molecules.

Concept Review

page 62

24. a. 4.42
b. 0.684
c. 2.2×10^{-3}
d. 4.4
e. 24.78
f. 3.5
25. 2.0×10^4 kg•m^2/s^2
26. $V = (5 \text{ cm})^3 = 125 \text{ cm}^3$
$D = 8.56 \times 10^{-2} \text{ g/cm}^3$

Chapter 3

Concept Review

page 77

1. a. Li
b. Ba
c. Cr
d. P
e. Pu
f. S
2. a. hydrogen
b. chlorine
c. iron
d. californium

Chapter 4

Practice

page 133

1. 28.09 amu
2. 15.99 amu
(Note: Due to rounding of isotope masses, this value varies slightly from that stated in the periodic table.)

Practice

page 137

1. 2.2×10^{24} atoms Na
2. 9.33×10^{25} atoms As
3. 9.40×10^{2} mol Xe
4. 4.796×10^{-9} mol Ag

Practice

page 139

1. 690 g Br
2. 212 g Si
3. 3.2 mol C
4. 2 mol H

Practice

page 140

1. 1.68×10^{-24} g/atom H
2. 2.52×10^{-22} g/atom Eu

Chapter Review and Assess

pages 145, 146

40. 6.94 amu
47. 3.01×10^{22} O atoms
49. 2.49×10^{-7} mol Au
51. 0.908 g Ne
53. 1.52×10^{-3} mol Na
55. 4.37×10^{-22} g/atom

Chapter 5

Concept Review

page 165

18. O^{2-}
19. P^{3-}
20. Sn^{2+}, tin(II) ion, and Sn^{4+}, tin(IV) ion
21. a. 5+
b. 4+, 7+
c. 3+
d. 1+
e. 2+
(from Figure 5-12)

841

Practice
page 167
1. a. $CaCl_2$
b. FeO
c. MgO

Practice
page 173
1. a. $Al_2(SO_4)_3$
b. $Mg(OH)_2$
c. $Cu(C_2H_3O_2)_2$
d. H_2O_2
e. Fe_2S_3
f. $Pb_3(PO_4)_2$

Practice
page 176
1. a. 122.55 g/mol
b. 234.06 g/mol
c. 132.17 g/mol
d. $NaHCO_3$; 84.01 g/mol
e. $K_2Cr_2O_7$; 294.20 g/mol
f. $Mg(ClO_4)_2$; 223.21 g/mol

Practice
page 177
1. 34.6% Cu, 30.4% Fe, 34.9% S
2. pure chalcocite because the percentage of copper is greater

Practice
page 178
1. Mn_2O_3
2. $Cd(OH)_2$

Practice
page 180
1. 233 g hydrate
2. 34.4 g $CaCl_2$

Chapter Review and Assess
pages 185, 186
31. a. AlF_3
b. MgO
c. CaS
d. $SrBr_2$
33. a. K_2HPO_4
b. $Sr(NO_3)_2$
c. Li_2SO_4
d. $Mg(H_2PO_4)_2$
45. 197.34 g/mol
47. 119.99 g
49. 40.04% Ca, 12.00% C, 47.96% O
51. 1.87 mol N, 7.43 mol H, 1.87 mol Cl

$$\frac{7.43}{1.87} = 3.97$$

$$\frac{1.87}{1.87} = 1.00$$

1.00 mol N: 3.97 mol H: 1.00 mol Cl
NH_4Cl

53. First compound:
1.00 mol P: 1.50 mol O
2.00 mol P: 3.00 mol O
P_2O_3
Second compound:
1.00 mol P: 2.50 mol O
2.00 mol P: 5.00 mol O
P_2O_5
55. 54.73% $Cr_2(SO_4)_3$

Chapter 6

Practice
page 202
1. a. $\ddot{O}=C=\ddot{O}$

b. $(:\ddot{Cl})_2C=C(\ddot{Cl}:)_2$

c. $:N\equiv N:$

d. $H:C\equiv N:$

2. (benzene ring: six C atoms in a ring with alternating double bonds, each C bonded to one H)

3. (benzene ring with five H atoms and one $-C(=\ddot{O})-\ddot{O}-H$ group)

Concept Review
page 204
7. $\ddot{S}$ bonded to $:\ddot{O}$ (single) and $\ddot{O}:$ (double) $\longleftrightarrow$ $\ddot{S}$ bonded to $\ddot{O}$ (double) and $:\ddot{O}:$ (single)

8. $[:\ddot{O}-\ddot{Cl}(-\ddot{O}:)-\ddot{O}:]^-$ $[:\ddot{O}-S(-\ddot{O}:)(-\ddot{O}:)-\ddot{O}:]^{2-}$

9. $[C(-\ddot{O}:)(=\ddot{O})(-\ddot{O}:)]^{2-} \longleftrightarrow [C(=\ddot{O})(-\ddot{O}:)(-\ddot{O}:)]^{2-} \longleftrightarrow [C(-\ddot{O}:)(-\ddot{O}:)(=\ddot{O})]^{2-}$

Practice
page 208
1. $K_2Cr_2O_7$; moles 0.6793 K, 0.6810 Cr, 2.377 O; mole ratio 1:1:3.5; whole-number ratio 2:2:7

2. P_2O_5; moles 0.1431 P, 0.3573 O; mole ratio 1:2.5; whole-number ratio 2:5
3. Na_2SO_4; moles 1.409 Na, 0.7063 S, 2.812 O; mole ratio 2:1:4

Practice
page 210
1. C_6H_6
2. $C_{18}H_{34}O_2$

Chapter Review and Assess
pages 228, 229, 230
15. a. 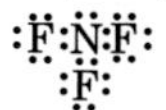 :F:N:F: / :F: **b.** 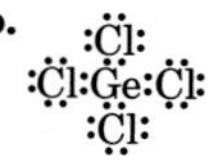 :Cl: / :Cl:Ge:Cl: / :Cl: **c.** 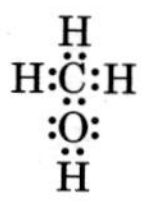 H / H:C:H / :O: / H

17. a. 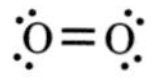 $\ddot{O}=\ddot{O}$

b. 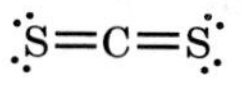 $\ddot{S}=C=\ddot{S}$

c. H–C(H)(H)–C(H)(H)–C(=Ö:)–Ö–H

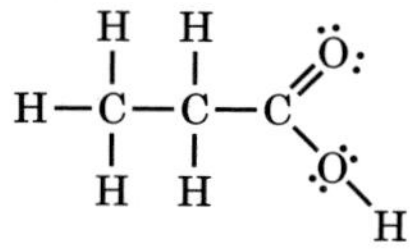

31. $C_6H_8O_7$
33. The molecular formula is $4 \times CH_2O$, or $C_4H_8O_4$.
39. a. 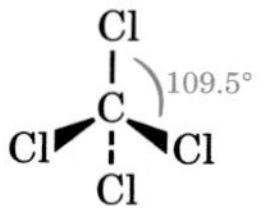 CCl_4 (tetrahedral, 109.5°)

b. 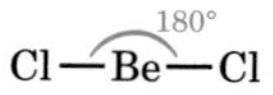 Cl—Be—Cl (180°)

c. 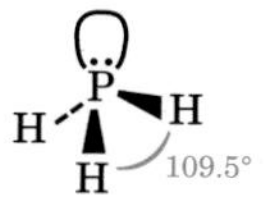 PH_3 (lone pair on P, 109.5°)

48. 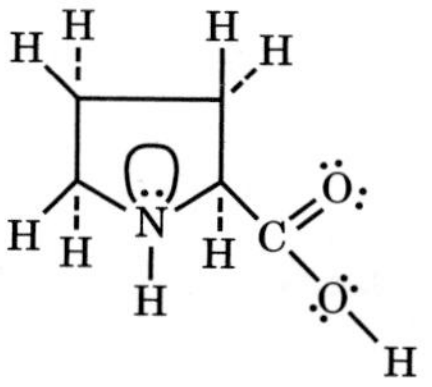 (ring structure with N, lone pair, C=Ö:, Ö–H)

Chapter 7
Practice
page 244
1. a. $3CaSi_2 + 2SbCl_3 \longrightarrow 6Si + 2Sb + 3CaCl_2$
b. $2C_2H_2 + 5O_2 \longrightarrow 4CO_2 + 2H_2O$
c. $2Al + 6CH_3OH \longrightarrow 2Al(CH_3O)_3 + 3H_2$
2. $Ca + 2H_2O \longrightarrow Ca(OH)_2 + H_2$
$Ca(OH)_2 \xrightarrow{\Delta} CaO + H_2O$

Concept Review
page 253
19. $C_3H_8 + 5O_2 \xrightarrow{\Delta} 3CO_2 + 4H_2O$

Practice
page 255
1. $4C + S_8 \longrightarrow 4CS_2$
2. $4Fe + 3O_2 \longrightarrow 2Fe_2O_3$

Concept Review
page 260
20. a. $Ba(s) + 2H_2O(l) \longrightarrow Ba(OH)_2(s) + H_2(g)$
b. $2Rb(s) + 2H_2O(l) \longrightarrow 2RbOH(s) + H_2(g)$
c. $Zn(s) + 2H_2O(l) \longrightarrow Zn(OH)_2(s) + H_2(g)$
21. a. $4Au(s) + 3O_2(g) \longrightarrow 2Au_2O_3(s)$
b. $2Mn(s) + O_2(g) \longrightarrow 2MnO(s)$
(Other reactions are possible.)
c. $2Pb(s) + O_2(g) \longrightarrow 2PbO(s)$
(Other reactions are possible.)

Practice
page 261
Reactions are listed from most to least likely.
2. $Mg(s) + Pb(NO_3)_2(aq) \longrightarrow Pb(s) + Mg(NO_3)_2(aq)$
1. $Zn(s) + CuCl_2(aq) \longrightarrow Cu(s) + ZnCl_2(aq)$
4. $Cu(s) + 2AgNO_3(aq) \longrightarrow 2Ag(s) + Cu(NO_3)_2(aq)$
3. $Ni(s) + Al_2(SO_4)_3(aq) \longrightarrow$ no reaction

Chapter Review and Assess
pages 267, 268
12. formula equation: $CH_4 + Cl_2 \longrightarrow CCl_4 + HCl$
balanced equation: $CH_4 + 4Cl_2 \longrightarrow CCl_4 + 4HCl$
30. $2Mg(s) + O_2(g) \longrightarrow 2MgO(s)$
32. Zinc is higher than lead on the activity series, so zinc will displace lead from compounds.
$Zn(s) + PbCl_2(aq) \longrightarrow ZnCl_2(aq) + Pb(s)$

Chapter 8
Practice
page 279
1. 215 g SnF_2
2. 37.72 kg $C_{20}H_{12}O_5$

Practice
page 280
1. 0.453 g CO_2
2. 0.186 g H_2O

Practice
page 281
1. 3.57 L CO_2
2. 648 g $NH_2C_6H_4CO_2H$

Practice
page 286
1. 0.933 mol HCl
0.154 mol $Al(OH)_3$
0.311 mol $Al(OH)_3$ needed
0.154 mol $Al(OH)_3$ less than 0.311 mol $Al(OH)_3$; there is not enough $Al(OH)_3$

2. $N_2 + 3H_2 \longrightarrow 2NH_3$
   1.32×10^5 mol H_2 present
   9.93×10^3 mol H_2 required
   9.93×10^3 mol $H_2 < 1.32 \times 10^5$ mol H_2, therefore N_2 is the limiting reactant.

Practice
page 291

1. 37.2%
2. Because the formula mass and mole ratios are the same as in item **1**, the theoretical yield for *o*-nitrotoluene is also 819 g.
   57.1%

Practice
page 296

1. $2NaN_3 \longrightarrow 2Na + 3N_2$
   $6Na + Fe_2O_3 \longrightarrow 3Na_2O + 2Fe$
   37.7 g Fe_2O_3
2. $6Na + Fe_2O_3 \longrightarrow 3Na_2O + 2Fe$
   $Na_2O + 2CO_2 + H_2O \longrightarrow 2NaHCO_3$
   119 g $NaHCO_3$
3. 54.1 mL $NaHCO_3$
4. 128 g CH_3COOH, 179 g $NaHCO_3$
5. 175 g CH_3COONa, 38.4g H_2O

Practice
page 299

1. **a.** yes, 35 000 L >26 110 L
   b. no, 2.00×10^5 L $< 2.15 \times 10^5$ L
   c. no, 400. L < 499 L
2. **a.** Note that this is a mixture and two equations are needed.
   $2C_8H_{18}(g) + 25O_2(g) \longrightarrow 16CO_2(g) + 18H_2O(g)$
   $2C_7H_{14}(g) + 21O_2(g) \longrightarrow 14CO_2(g) + 14H_2O(g)$
   b. 7824 L + 850. L = 8674 L of air

Chapter Review and Assess
pages 305, 306, 307, 308

6. $H_2CO_3 \longrightarrow H_2O + CO_2$
   0.177 g CO_2
8. 1.53 mol O_2
10. 112 L O_2
17. $Cu(s) + 2AgNO_3(aq) \longrightarrow 2Ag(s) + Cu(NO_3)_2(aq)$
    Because there is less $AgNO_3$ than is needed to react with the Cu, $AgNO_3$ is the limiting reactant.
    63.5 g Ag
23. 72.46% yield
25. 710. g CH_4
31. 10.7 g baking powder
33. 3130 L H_2O
35. 12.5 mol *p–tert*-butylphenol

Chapter 9

Practice
page 316

1. –23.4 kJ
2. 66.7 kJ

Practice
page 320

1. heat gained by H_2O = heat lost by oatmeal
   ΔH = 22 990 J = 23 kJ
2. 26.8°C

Practice
page 322

1. mass of C_8H_{18} = 420 g
   mass of O_2 = 1500 g
2. 1750 kJ

Concept Review
page 326

9. No; the amount of heat in the initial and final states has not changed, so ΔH_{rxn} has not changed.
10. E_a for the reverse reaction is greater than that for the forward reaction, so more heat is required for the reverse reaction. The reactants are at a higher enthalpy than the products, so they are less thermodynamically stable.

Practice
page 331

1. 1.9 kJ/mol
2. −592.7 kJ

Practice
page 337

1. − 3.3 J/K; no
2. + 5.3 J/K

Practice
page 340

1. 638.3°C
2. −394.4 kJ/mol
   It is spontaneous.
3. raise the temperature
4. −869 kJ/mol

Chapter Review and Assess
pages 349, 350, 351, 352

7. -1.9×10^7 J
9. 50. kJ
   12 Cal
11. 49.4 g $C_6H_{12}O_6$
23. −114 kJ
31. −550 J/K
42. 155 kJ
    The reaction is not spontaneous.

844

Chapter 10

Practice
page 371
1. 9.0×10^2 K
2. 0.67 L

Practice
page 373
1. 5.0×10^2 kPa
2. 0.50 m^3

Practice
page 374
1. 7.4 L
2. 2 L
   The balloon compresses to a 2 L volume.
3. 38.8 kPa
4. 200. mL
5. 10. L

Practice
page 377
1. 82.66 kPa
2. 98.51 kPa

Practice
page 380
1. 236 m/s
2. v_1 = 450 m/s
   v_2 = 213 m/s
   v_3 = 255 m/s
   Hydrogen sulfide will be smelled first, followed by benzaldehyde, and then methyl salicylate.

Practice
page 383
1. 29.0 L of CO_2
2. 22.4 L
3. 0.17 mol
4. 3.9×10^3 kPa
5. 5.2×10^4 K

Practice
page 384
1. 4.0 m^3
2. 6.02×10^3 kPa
3. 165 K
4. 81 m^3

Practice
page 386
1. Assume that the pressure and temperature for each gas is the same.
   6.0 L
2. **a.** 19 L O_2
   b. 5.25 mol H_2O
   c. 14 g
3. 8.1×10^3 g

Chapter Review and Assess
pages 397, 398
14. 2.18 L
16. 10.0 L
18. 4.00×10^8 L
30. Partial pressure of water at 20.0°C = 2.34 kPa, so
    P_{oxygen} = 98.0 kPa − 2.34 kPa = 95.7 kPa
32. 1.32×10^3 m/s
34. 70.8 g/mol
40. 248 kPa
42. 482 mL
44. 10.4 L H_2
46. 0.484 g Mg

Chapter 11

Concept Review
page 407
1. 98 g
2. RbCl at 0°C, $NaNO_3$ at 70°C
3. $NaNO_3$

Practice
page 416
1. 0.840 M
2. 486 mL
3. 1.70 L
4. 4.68 g $CuSO_4 \cdot 5H_2O$
5. 1.38 kg $C_{12}H_{22}O_{11}$
6. 676 g H_2O, 4390 g $Ca(H_2PO_4)_2$
7. 9.45×10^{-5} M

Practice
page 425
1. BCl_3 is a nonpolar molecule with polar bonds.
2. SO_2 is a polar molecule with polar bonds. It can dissolve in water.

Concept Review
page 435
23. Because CO_2 is only moderately soluble in water at normal air pressure, higher CO_2 pressures are necessary for more CO_2 to dissolve in the water.
24. Keep the soda covered tightly, and keep it cold. Less CO_2 remains in solution as the temperature increases, and the uncapped bottle allows more CO_2 to escape.

Chapter Review and Assess
pages 443, 444, 445
19. **a.** 3.5 M $NaNO_3$ **b.** 2.5 M KOH **c.** 0.80 M Na_2S
21. **a.** 4.0 mol $AgNO_3$ **b.** 1.25 mol HCl **c.** 0.100 mol HNO_3
23. **a.** 0.250 M NaOH **b.** 0.953 M NH_4Br **c.** 1.61 M $CuCl_2$
25. **a.** 87.7 g NaCl **b.** 315 g NaOH **c.** 327 g K_2SO_4
27. 698 g $Ca_3(PO_4)_2$, 243 g H_2O
36. C_4H_{10} (2.5 − 2.1 = 0.4); will dissolve in CS_2 (2.5 − 2.5 = 0) because both are nonpolar substances.
50. 2.8×10^{-5} M $C_6H_8O_6$

Chapter 12

Concept Review

page 458

1. a. $CaCl_2(s) \xrightleftharpoons{H_2O} Ca^{2+}(aq) + 2Cl^-(aq)$

   b. $Al_2(SO_4)_3(s) \xrightleftharpoons{H_2O} 2Al^{3+}(aq) + 3SO_4^{2-}(aq)$

2. Because more energy is usually released in the hydration of ions that are smaller in size or higher in charge, smaller $AlCl_3$, with a cation charge of 3+, should release more energy during hydration than $MgCl_2$ with a cation charge of 2+.

3. a. $HCl(aq) + H_2O(l) \rightleftharpoons H_3O^+(aq) + Cl^-(aq)$

   b. $H_2CO_3(aq) + H_2O(l) \rightleftharpoons H_3O^+(aq) + HCO_3^-(aq)$

   c. $Na_2CrO_4(s) \xrightleftharpoons{H_2O} 2Na^+(aq) + CrO_4^{2-}(aq)$

   d. $NH_4NO_3(aq) + H_2O(l) \rightleftharpoons NH_3(aq) + H_3O^+(aq) + NO_3^-(aq)$

Concept Review

page 471

15. a. $K_{eq} = \frac{[ClO^-][H^+]}{[HClO]}$

    b. $K_{eq} = \frac{[[Cu(NH_3)_4]^{2+}]}{[[Cu(H_2O)_4]^{2+}][NH_3]^4}$

16. a. $K_{eq} = \frac{[NO_2]^2}{[N_2O_4]}$

    b. $K_{eq} = \frac{[CO][H_2O]}{[CO_2][H_2]}$

    c. $K_{eq} = \frac{[NO]^2}{[N_2][O_2]}$

    d. $K_{eq} = \frac{[CH_3COO^-][H_3O^+]}{[CH_3COOH]}$

    e. $K_{eq} = \frac{[H_2][I_2]}{[HI]^2}$

Practice

page 474

1. 0.720
2. 3.6×10^{-1}
3. 0.046 mol/L
4. 1.1 mol/L

Practice

page 476

1. 2.06×10^{-35}
2. 3.62×10^{-5} M
3. 1.69×10^{-3} M

Chapter Review and Assess

pages 482, 483

28. 1.74×10^{-5}
30. 1.12×10^{-2} M
32. 7.91×10^{-9} M

Chapter 13

Concept Review

page 494

1. $Na_2O + H_2O \longrightarrow 2NaOH$
   $Cl_2O_7 + H_2O \longrightarrow 2HClO_4$
2. chloric acid, nitric acid, perchloric acid, periodic acid, permanganic acid, tetrafluoroboric acid, acetic acid, hydrocyanic acid, hypochlorous acid, nitrous acid
3. CsO, base
   TeO, acid
   I_2O_5, acid
   SrO, base

Concept Review

page 496

4. a. base, CH_3NH_2; acid, H_2O
   b. acid, HSO_4^-; base, H_2O
5. a. base
   b. acid

Practice

page 503

1. a. 5
   b. 2
2. a. 5.60
   b. 3.058
3. a. 1.0×10^{-13}
   b. 1.0×10^{-11}
4. a. 4.0×10^{-8}
   b. 3.2×10^{-4}
5. a. 10.7
   b. 10.4

Practice

page 505

1. a. 1×10^{-11}
   b. 5.0×10^{-8}
   c. 1.5×10^{-4}
   d. 6.3×10^{-4}
   e. 3.2×10^{-11}
   f. 2.0×10^{-12}

Practice

page 514

1. $HCl + NaOH \longrightarrow NaCl + H_2O$
   0.57 M NaOH
2. $2HClO_4 + Ca(OH)_2 \longrightarrow 2H_2O + Ca(ClO_4)_2$
   37 mL
3. $CH_3COOH + KOH \longrightarrow H_2O + CH_3COOK$
   0.157 M CH_3COOH
4. $2HCl + Ba(OH)_2 \longrightarrow 2H_2O + BaCl_2$
   0.0422 M HCl

Concept Review

page 517

25. $H_2PO_4^- + OH^- \longrightarrow HPO_4^{2-} + H_2O$
    Added OH^- accepts a proton from $H_2PO_4^-$ to form H_2O.

846

Chapter Review and Assess
pages 527, 528
21. 2.59
23. 7.87
25. antilog (–3.45) = 3.55×10^{-4}
27. pH = 9.4
29. pH = 2.4
42. 0.175 M HCOOH
44. 0.526 M NaOH
46. 0.310 g LiOH

Chapter 14

Concept Review
page 541

1. **a.** evolution of a gas
   b. change in solution's color; appearance or disappearance of metals
   c. evolution of a gas
   d. precipitation of $Cu(OH)_2$
   e. pressure changes; condensing the water out of reaction gases

2. **a.** $\frac{1}{2} \times \frac{-\Delta H_2O_2}{\Delta t} = \frac{\Delta O_2}{\Delta t}$

   b. $\frac{1}{2} \times \frac{-\Delta Ag^+}{\Delta t} = \frac{\Delta Cu^{2+}}{\Delta t}$

   c. $\frac{-\Delta HC_2H_3O_2}{\Delta t} = \frac{\Delta CO_2}{\Delta t}$

   d. $\frac{-\Delta CuCl_2}{\Delta t} = \frac{\Delta Cu(OH)_2}{\Delta t}$

   e. $\frac{1}{2} \times \frac{-\Delta C_2H_6}{\Delta t} = \frac{1}{4} \times \frac{\Delta CO_2}{\Delta t}$

3. Reactant; concentration *decreases* with time; 0.050M/min
4. N_2: 0.02 M/s
   H_2: 0.02 M/s
5. 5.0 M/s

Concept Review
page 559
21. $I^- + OCl^- + H_2O \rightleftharpoons Cl^- + OI^- + H_2O$
22. three
23. step 2; rate = $k[HOCl][I^-]$
24. HOCl, OH^-, HOI
25. rate = $k[HeI\bullet][I\bullet] = k[He][I\bullet]^2$

Chapter 15

Practice
page 590

1. +5
2. F = −1; B = +3
3. +7
4. H = +1; N = −3

Concept Review
page 592
11. **a.** not redox
b. redox
c. not redox
12. Zn is oxidized, and Cu is reduced.

Chapter Review and Assess
page 606
33. Pb = +2; Cl = −1
(+2) + 2(−1) = 0
35. Pb = +2, S = −2; Mn = +4, O = −2; Li = +1, Al = +3, H = −1 (rules 4, 7, and 8 on pages 588–589); Na = +1, O = −1 (rules 5, 7, and 8 on pages 588–589); Hg = +2, O = −2; Ni = +3, O = −2, H = +1; Pb = +2, S = +6, O = −2.
37. H = +1; C = +4 ; O = −2
(+1) + (+4) + 3(−2) = −1

Chapter 16

Concept Review
page 615

1. A small nucleus is generally more stable because all of the nuclear particles are affected by each other's strong nuclear force. The repulsion between two protons is counteracted by their nuclear attraction.
2. $^{6}_{3}Li$ (For elements with an atomic number lower than 30, stability usually occurs with equal numbers of protons and neutrons, in this case three of each.)
3. $^{12}_{6}C$ is stable.
   $^{3}_{1}H$ is unstable.
   $^{206}_{82}Pb$ is stable.
   (The chart and the rule of thumb aren't entirely reliable in this case. $^{205}_{82}Pb$ is not stable but should be, according to both the rule and chart.)
   $^{32}_{15}P$ is unstable.
   (Again, the rule of thumb and chart are not entirely helpful. $^{30}_{15}P$ is extremely unstable, yet this is an isotope with a neutron-to-proton ratio of 1. $^{31}_{15}P$ is the stable isotope of phosphorus.)
   $^{4}_{2}He$ is stable

Practice
page 617

1. 2.25×10^{11} J for 1 mol of deuterium
2. 5.14×10^{-12} J for each lithium-6 nucleus

Chapter Review and Assess
page 640
10. **a.** Nucleus #1: 2.69×10^{13} J/mol
Nucleus #2: 1.80×10^{13} J/mol
Nucleus #1 releases more binding energy than nucleus #2.

847

b. Binding energy per nucleon in nucleus #1 =

$$\frac{4.47 \times 10^{-11}\ \text{J}}{35\ \text{nucleons}} = 1.28 \times 10^{-12}\ \text{J/nucleon}$$

Binding energy per nucleon in nucleus #2 =

$$\frac{2.99 \times 10^{-11}\ \text{J}}{23\ \text{nucleons}} = 1.30 \times 10^{-12}\ \text{J/nucleon}$$

Nucleus #2 has a greater binding energy per nucleon. Because the sodium-23 nucleus has fewer nucleons than the potassium-35 nucleus, fewer nucleons lie beyond the range of the strong force. Therefore it is easier for the repulsion between protons to be overcome.

Appendix B

Concept Review

page 804

1. The axis should extend from 0 to 120 grams. Check that the scale is consistent between divisions. Students should not scale their axes by the points listed.

2. The x-axis should extend from 0°C to about 110°C as the minimum. The y-axis should extend from 0 g to about 250 g as the minimum. Check that both axes are correctly labeled using the labels given in the data table. The solubility of KNO_3 at 65°C is about 120 g. The solubility of KNO_3 at 105°C is about 265 g.

3. a. The x-axis should extend from 0 atm to about 10 atm as the minimum. The y-axis should extend from 0 L to about 250 L as the minimum. Check that both axes are correctly labeled using the labels given in the data table.

b. inverse relationship

c. $PV = k$

d. $V = k \times \frac{1}{P}$

Concept Review

page 807

1. a. 3
b. 3
c. 6
d. 4
e. unlimited number
f. 4

2. a. 47.10 g
b. 32.86 L
c. 101 m
d. 0.123 cm^2
e. 28.62 mi/h
f. 113.2 in^2

3. a. 6.08×10^{-5}
b. 2.642×10^{8}
c. 2.67
d. -7.2×10^{-5}

Appendix C

Concept Review

page 810

1. a. concept
b. linking words
c. linking word
d. linking words
e. concept
f. linking words
g. concept
h. linking word

2. Answers will vary. Check against Map A. Some possible responses are the following. *matter is described by chemical properties, matter is classified as mixtures, chemical changes that release energy are exothermic changes, physical properties such as state, and so on.*

3. Cross links are denoted by the arrowhead. The two cross links in Map A are: *chemical properties which describe chemical changes, physical properties such as state.*

848

Periodic Table of the Elements

6	Atomic number
C	Symbol
Carbon	Name
12.011	Average atomic mass
$[He]2s^2 2p^2$	Electron configuration

Period	Group 1	Group 2	Group 3	Group 4	Group 5	Group 6	Group 7	Group 8	Group 9
1	1 **H** Hydrogen 1.00794 $1s^1$								
2	3 **Li** Lithium 6.941 $[He]2s^1$	4 **Be** Beryllium 9.012182 $[He]2s^2$							
3	11 **Na** Sodium 22.989768 $[Ne]3s^1$	12 **Mg** Magnesium 24.3050 $[Ne]3s^2$							
4	19 **K** Potassium 39.0983 $[Ar]4s^1$	20 **Ca** Calcium 40.078 $[Ar]4s^2$	21 **Sc** Scandium 44.955910 $[Ar]3d^1 4s^2$	22 **Ti** Titanium 47.88 $[Ar]3d^2 4s^2$	23 **V** Vanadium 50.9415 $[Ar]3d^3 4s^2$	24 **Cr** Chromium 51.9961 $[Ar]3d^5 4s^1$	25 **Mn** Manganese 54.93805 $[Ar]3d^5 4s^2$	26 **Fe** Iron 55.847 $[Ar]3d^6 4s^2$	27 **Co** Cobalt 58.93320 $[Ar]3d^7 4s^2$
5	37 **Rb** Rubidium 85.4678 $[Kr]5s^1$	38 **Sr** Strontium 87.62 $[Kr]5s^2$	39 **Y** Yttrium 88.90585 $[Kr]4d^1 5s^2$	40 **Zr** Zirconium 91.224 $[Kr]4d^2 5s^2$	41 **Nb** Niobium 92.90638 $[Kr]4d^4 5s^1$	42 **Mo** Molybdenum 95.94 $[Kr]4d^5 5s^1$	43 **Tc** Technetium (97.9072) $[Kr]4d^6 5s^1$	44 **Ru** Ruthenium 101.07 $[Kr]4d^7 5s^1$	45 **Rh** Rhodium 102.90550 $[Kr]4d^8 5s^1$
6	55 **Cs** Cesium 132.90543 $[Xe]6s^1$	56 **Ba** Barium 137.327 $[Xe]6s^2$	57 **La** Lanthanum 138.9055 $[Xe]5d^1 6s^2$	72 **Hf** Hafnium 178.49 $[Xe]4f^{14} 5d^2 6s^2$	73 **Ta** Tantalum 180.9479 $[Xe]4f^{14} 5d^3 6s^2$	74 **W** Tungsten 183.84 $[Xe]4f^{14} 5d^4 6s^2$	75 **Re** Rhenium 186.207 $[Xe]4f^{14} 5d^5 6s^2$	76 **Os** Osmium 190.23 $[Xe]4f^{14} 5d^6 6s^2$	77 **Ir** Iridium 192.22 $[Xe]4f^{14} 5d^7 6s^2$
7	87 **Fr** Francium (223.0197) $[Rn]7s^1$	88 **Ra** Radium (226.0254) $[Rn]7s^2$	89 **Ac** Actinium (227.0278) $[Rn]6d^1 7s^2$	104 **Unq*** — (261.11) $[Rn]5f^{14} 6d^2 7s^2$	105 **Unp*** — (262.114) $[Rn]5f^{14} 6d^3 7s^2$	106 **Unh*** — (263.118) $[Rn]5f^{14} 6d^4 7s^2$	107 **Uns*** — (262.12) $[Rn]5f^{14} 6d^5 7s^2$	108 **Uno*** — (265)† $[Rn]5f^{14} 6d^6 7s^2$	109 **Une*** — (266)† $[Rn]5f^{14} 6d^7 7s^2$

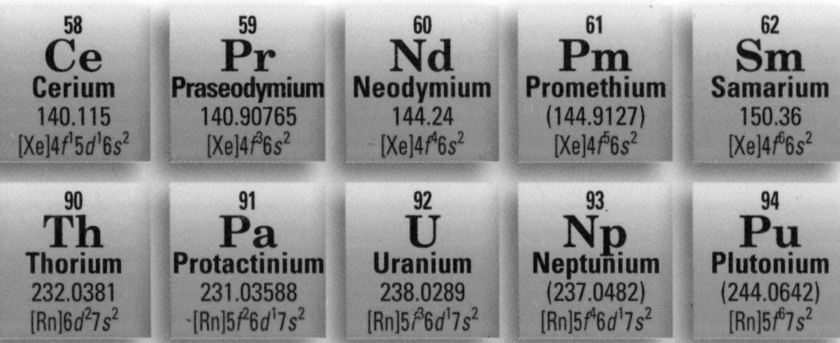

58 **Ce** Cerium 140.115 $[Xe]4f^1 5d^1 6s^2$	59 **Pr** Praseodymium 140.90765 $[Xe]4f^3 6s^2$	60 **Nd** Neodymium 144.24 $[Xe]4f^4 6s^2$	61 **Pm** Promethium (144.9127) $[Xe]4f^5 6s^2$	62 **Sm** Samarium 150.36 $[Xe]4f^6 6s^2$
90 **Th** Thorium 232.0381 $[Rn]6d^2 7s^2$	91 **Pa** Protactinium 231.03588 $[Rn]5f^2 6d^1 7s^2$	92 **U** Uranium 238.0289 $[Rn]5f^3 6d^1 7s^2$	93 **Np** Neptunium (237.0482) $[Rn]5f^4 6d^1 7s^2$	94 **Pu** Plutonium (244.0642) $[Rn]5f^6 7s^2$

* The systematic names and symbols for elements greater than 103 will be used until the approval of trivial names by IUPAC.

† Estimated from currently available IUPAC data.

Atomic masses listed in this table reflect the precision of current measurements. However, atomic mass measurements throughout the text have been rounded to two places to the right of the decimal.